Industrial Minerals and Rocks

6th Edition

Senior Editor

Donald D. Carr
Indiana Geological Survey
Bloomington, Indiana

Associate Editors

A. Frank Alsobrook
Alsobrook and Company, Inc.
Medford, New Jersey

George S. Austin
New Mexico Bureau of Mines
Socorro, New Mexico

James M. Barker
New Mexico Bureau of Mines
Socorro, New Mexico

Robert L. Bates
The Ohio State University
Columbus, Ohio

James W. Baxter
Illinois State Geological Survey
Champaign, Illinois

H. Lyn Bourne
Consulting Geologist
Northville, Michigan

James R. Dunn
Rust Environment and Infrastructure
Albany, New York

Bruce H. Mason
Indiana Mineral Aggregates Association
Indianapolis, Indiana

Henry N. McCarl
University of Alabama at Birmingham
Birmingham, Alabama

Haydn H. Murray
Indiana University
Bloomington, Indiana

James C. Norman
Simon Hydro-Search
Huntington Beach, California

Keith G. Papke
Geologic Consultant
Sparks, Nevada

Lawrence F. Rooney
US Geological Survey
Denver, Colorado

Nelson Severinghaus, Jr.
Franklin Industries, Inc.
Nashville, Tennessee

Published by

Society for Mining, Metallurgy, and Exploration, Inc.
Littleton, Colorado • 1994

Printed in the United States of America
by Braun-Brumfield, Inc., Ann Arbor, Michigan

Library of Congress Catalog Card Number 93-84488
ISBN 0-87335-103-7

PREFACE

Industrial Minerals and Rocks, Sixth Edition (IMAR6), is a comprehensive reference and source of information for scientists, engineers, and administrators in the field of industrial minerals, as well as for a wider circle of potential users. It consists of 94 chapters produced by more than 130 authors, 14 Associate Editors, the Senior Editor, and a Manager of Book Publishing for SME. Such a work would not have been possible without much interaction of the participants and the encouragement and financial support of various governmental agencies, universities, individuals, industries, and the Society for Mining, Metallurgy, and Exploration.

From its inception, with the first edition in 1937, *Industrial Minerals and Rocks* has been a handbook, its chapters written by the specialist or specialists most appropriate for the subject covered. This approach has carried over through ensuing editions—second in 1949, third in 1960, fourth in 1975, and fifth in 1983—with many chapters building upon those of previous editions.

The concept of IMAR6 was approved by the membership of the Industrial Minerals Division at the SME annual meeting in Las Vegas, February 1989. Subsequently, it received approval of the SME Book Publishing Committee and the SME Board of Directors. An Editorial Board for IMAR6 was organized in February 1990. In 1989, I volunteered to work as Senior Editor because I thought I would learn something and that it would be fun. I learned that the work was more than the fun.

Early in the planning of IMAR6, we decided to query the membership to determine how much revision was necessary. The response was decidedly in favor of a major change. Consequently many new chapters were proposed, and many old chapters were either omitted or combined with others to achieve a more logical presentation. We enlarged the numbers of introductory chapters dealing with the broad and diverse field of industrial minerals to reflect the common thread of concerns that tie the field together. Most industrial minerals do share economic, regulatory, legal, and environmental concerns, especially in the industrialized countries. Industrial minerals production and consumption now must be balanced against energy supplies and the health, safety, and environmental concerns of the citizenry.

The handbook format is useful, but it lacks a unifying theme, style, and length of chapters. To help remedy these shortcomings, we prepared an Author's Guide. We asked the authors to consider the following basic outline when writing chapters:

- Introduction
- Geology
  - Mineralogy
  - Chemical properties
  - Physical properties
  - Origin
  - Distribution of major deposits
    - United States
    - World
- Industry structures (if for uses)
- Technology
  - Exploration techniques
  - Processing (if for uses)
  - Mining and processing
  - Specifications
  - Testing
  - Byproducts and coproducts
- Uses
- Economic factors
  - Costs
  - Markets
  - Transportation
  - Alternative materials
- Governmental considerations
  - Taxes, subsidies, depletion provisions
  - Zoning and land use
  - Environmental regulations
  - Safety regulations
- Problems and future trends
- Bibliography or Reference

Although the outline served a useful purpose, the chapters are too diverse in content and scope to conveniently fit a single mold. The authors too come from different backgrounds, and their writing is refracted through their expertise, whether it be geology, mineral processing, engineering, or economics. Hence, we gave authors much freedom. The volume is not intended to be a teaching text, lacking as it does in treatment of basic principles, but it makes an excellent reference and has been used extensively in the past for that purpose.

Also early in our planning the Editorial Board decided to take a bold step in metrication. This volume uses the International System of units (Systeme International d'Unites) or "SI." The field of industrial minerals has become increasingly directed by international trade; world markets require SI units. The United States has been painfully slow in moving to SI units, and it now finds itself with only two other countries (Myanmar and Liberia) that primarily use inch/pound units. However, the amended Metric Conversion Act of 1975 and the 1991 Executive Order 12770 of the President of the United States provide both the rationale and the mandate for a transition to the use of metric units. We hope that by using SI units in this volume, we are, in a small way, nudging the industrial minerals industry along the evolutionary process to metrication. Most disconcerting for those of us who used a McGuffey reader is the substitution of spaces for commas in numbers, such as 1 000 000.

Most disconcerting too is the way the names of countries seem to change on an almost daily basis. It's been hard to keep up. When an author refers to the USSR, we have tried to insert "former," and some of the authors have actually used Commonwealth of Independent States. A similar problem exists with Germany and the Balkan states. Attempts have been made to pinpoint the actual locations under discussion, but the reader is asked to bear with us in this matter.

I thank the Associate Editors for suggestions, some voluminous but always thoughtful, that helped mold this volume, and we, together, value written comments of those individual SME members who took time to respond in depth to our questionnaire: John F. Burst, David H. Hebb, Duane B. Jorgensen, Robert B. Kistler, Dave E. Leppert, John D. Macfadyen, Gordon C. Presley, Henry E. Reed, and Mark J. Zdunczyk. I give special thanks to Lawrence F. Rooney, Associate Editor for the fourth, fifth, and sixth editions, and my former supervisor at the Indiana Geological Survey, whom I turned to for advice time and again. His high standards of accuracy, timeliness, and editing were an inspiration to aim for, although not possible to achieve.

I am grateful to Aldo F. Barsotti, Chief, Branch of Industrial Minerals, US Bureau of Mines, for his support and for the support of his staff, many of whom supplied statistical data and some of whom contributed to chapters of this volume. I give thanks to Lee R. Rice, President, Data Technology Services and SME member, who generously donated his time and expertise to convert author disks, written on word processing programs long antique, to a format that we could use for this volume.

I would like to acknowledge the great contribution of Marianne Snedeker, retired Manager of Book Publishing, SME. Ms. Snedeker was involved in *Industrial Minerals and Rocks* from the first edition when Joe Gillson would visit the AIME offices at 39th Street in New York City with a pile of manuscripts. She would sit down with him and work out problems. This scenario would be repeated every couple of weeks until the volume was completed. Ms. Snedeker sat down with me during the organizational phases of the sixth edition and guided me in handbook publishing.

I am grateful to Barbara Dygert who took over responsibilities for the volume when Ms. Snedeker retired. Ms. Dygert has a sunny, good humor and the uncanny sense of knowing when clouds are blackest and then calling with words of encouragement. I give special thanks to Marilyn DeWees, my secretary, who effectively assisted and stuck with me through the entire process. She managed the flow of manuscripts and correspondence and sought out ways to make my job easier.

I am indebted to the Indiana Geological Survey at Indiana University for its support.

Donald D. Carr, Senior Editor
Industrial Minerals and Rocks, Sixth Edition

Bloomington, Indiana
January 1994

GUIDE TO USE OF THE SIXTH EDITION

Industrial Minerals and Rocks, Sixth Edition (IMAR6), is organized in two parts: first, an introduction, consisting of chapters describing the broad field of industrial minerals and second, the chapters on commodities and uses arranged in alphabetical order.

As Robert L. Bates explains in his Overview of the Industrial Minerals, industrial minerals are diverse and defy simple classification. For the purposes of organizing this volume, and to identify chapters on specific subjects, we classified the field of industrial minerals under the broad terms of "commodities" and of "uses." The latter may be thought of as the markets in which industrial minerals are used. The outline below provides a schematic overview of the chapters in Part 2 and may be useful for finding information under a general topic.

COMMODITIES

- Asbestos
- Barium minerals
- Bauxite
- Beryllium minerals
- Boron and borates
- Bromine resources
- Chromite
- Clays
  - Introduction
  - Bentonite
  - Common clay
  - Hormites: palygorskite (attapulgite and sepiolite)
  - Kaolin
- Diamond industrial
- Diatomite
- Emery
- Feldspars
- Fluorspar
- Garnet
- Gilsonite
- Glauconite
- Graphite
- Gypsum and anhydrite
- Iodine
- Limestone and dolomite
- Lithium resources
- Magnesite and magnesia
- Manganese minerals
- Mica
- Nepheline syenite
- Olivine and dunite
- Perlite
- Phosphate rock
- Potash resources
- Pumice and volcanic cinder
- Pyrophyllite
- Rare earth minerals
- Salt
- Sand and gravel
- Silica
  - Industrial sand and sandstone
  - Specialty materials
- Sillimanite group
  - Andalusite
  - Kyanite and related minerals
- Soda ash
- Sodium sulfate resources
- Staurolite
- Stone, crushed
- Stone, dimension
- Strontium minerals
- Sulfur resources
- Talc
- Titanium minerals
- Tripoli
- Vermiculite
- Wollastonite
- Zeolites
- Zirconium and hafnium minerals

USES

- Abrasives
- Agricultural materials
  - Fertilizers
  - Soil additives
- Biotechnical materials
- Catalysts
- Chemicals
- Construction uses
  - Adobe and similar materials
  - Aggregates: markets and uses
  - Cement and cement raw materials
  - Clay-based ceramic raw materials
  - Gypsum plasters and wallboards
  - Insulation
  - Lightweight aggregates
  - Roofing granules
  - Stone, conservation
  - Stone, decorative
- Electronic and optical materials
- Environmental uses
  - Asbestos substitutes
  - Clay liners and barriers
  - Flue gas desulfurization
  - Neutralization
  - Water and sewage treatment
- Fillers and coatings
- Filters and absorbents
- Glass raw materials
- Lime
- Metallurgical uses
  - Fluxes for metallurgy
  - Fluxes for soldering, brazing, and welding
  - Foundry sands
- Pigments, iron oxide
- Refractories
- Well drilling materials

The field of industrial minerals is so extensive and so large that it is impossible for each chapter to be inclusive. We have, therefore, encouraged authors to include a comprehensive bibliography so that readers may pursue a topic further if they need greater detail. The experienced user of industrial minerals information is likely to know which chapter to consult for specific information, but the inexperienced reader may be overwhelmed by the vast amount of information, most of which is "buried" within a chapter and whose location is not readily apparent. The latter should find the extensive index at the back of IMAR6 a starting point to locate information.

Unless stated otherwise, reference to "reserves" and "resources" in this volume follows terminology adopted for joint use by the US Geological Survey and the US Bureau of Mines (1980, Principles of a reserve/resource classification for minerals: US Geological Survey Circular 831, 5 p.). This terminology is shown and described as follows:

INDUSTRIAL MINERALS AND ROCKS

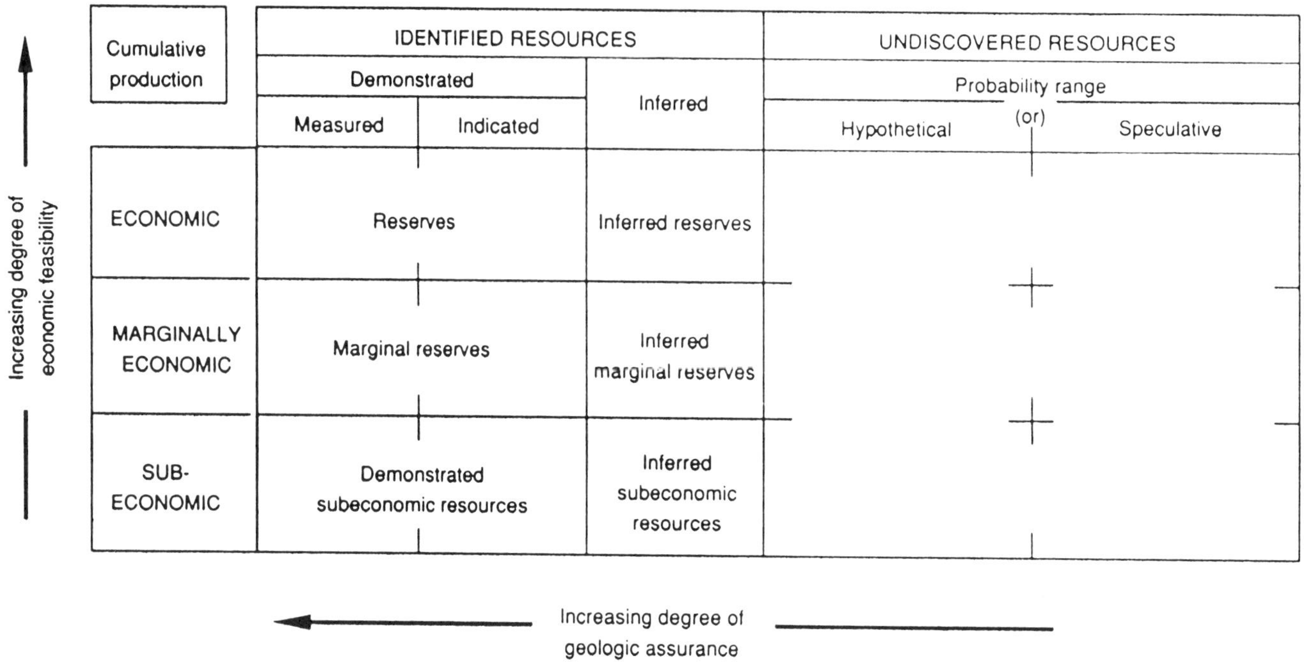

Resource. A concentration of naturally occurring solid, liquid, or gaseous materials in or on the Earth's crust in such form that economic extraction of a commodity is regarded as feasible, either currently or at some future time.

Identified resource. A resource the location, grade, quality, and quantity of which are known or can be estimated from specific geologic evidence. Identified resources include economic, marginally economic, and subeconomic resources.

Undiscovered resources. Undiscovered bodies of mineral-bearing material the existence of which is surmised from broad (regional) knowledge and theory.

Reserve. That portion of an identified resource from which a usable mineral or energy commodity can be economically and legally extracted at the time of determination. The term "ore" applies to reserves of some kinds of mineral commodities, generally metallic, but for want of another term it is sometimes applied to industrial minerals commodities.

Measured. Materials the quality and quantity of which have been determined, within a margin of error of less than 20% by quantitative data, including appropriate analyses, from closely spaced and geologically well-known sample sites.

Indicated. Materials the quality and quantity of which have been estimated partly from analyses and measurements and partly from reasonable geologic inferences.

Demonstrated. A collective term for the sum of materials in both measured and indicated resources.

Inferred. Materials in identified but unexplored deposits the quality and quantity of which have been estimated from geologic projections.

Hypothetical resources. Undiscovered materials that may reasonably be expected to exist in known mining districts under known geologic conditions.

Speculative resources. Undiscovered materials that may occur in known types of deposits in geologic settings where no previous discoveries have been made or in as-yet-unknown types of deposits that remain to be recognized.

Industry sometimes uses the terms *proved*, *probable*, and *possible* to describe the economic evaluation of a mineral deposit. *Proved* closely approximates *measured*, as here defined, and *probable* and *possible* both fall within the meaning of *indicated*.

As mentioned in the Preface, IMAR6 uses SI units almost entirely. Where conversion to SI seemed improper or ambiguous, we placed the inch/pound units in quotations or in some way tried to indicate that the use was nonstandard. An extensive table giving conversions of inch/pound and SI units is in the appendix; a classified list of conversion factors follows:

A. Factors for converting SI metric units to inch/pound units

To convert from	To	Multiply by
	Length	
millimeter (mm)	inch (in)	0.03937
meter (m)	foot (ft)	3.281
	yard (yd)	1.094
kilometer (km)	mile (mi)	0.6214
	mile, nautical (nmi)	0.5400
	Area	
meter^2 (m^2)	foot^2 (ft^2)	10.76
	yard^2 (yd^2)	1.196
	acre	0.0002471
hectometer^2 (km^2)	acre	2.471
kilometer^2 (km^2)	mile^2 (mi^2)	0.3861
	Volume	
centimeter^3 (cm^3)	inch^3 (in^3)	0.06102
decimeter^3 (dm^3)	inch^3 (in^3)	61.02
	pint (pt)	2.113
	quart (qt)	1.057
	gallon (gal)	0.2642
	foot^3 (ft^3)	0.03531
meter^3 (m^3)	foot^3 (ft^3)	35.31
	yard^3 (yd^3)	1.308
	gallon (gal)	264.2
	barrel (bbl), (petroleum, 1 bbl=42 gal)	6.290
	acre-foot (acre-ft)	0.0008107
hectometer^3 (hm^3)	acre-foot (acre-ft)	810.7
kilometer^3 (km^3)	mile^3 (mi^3)	0.2399
	Volume per unit time (includes flow)	
decimeter^3 per second (dm^3/s)	foot^3 per second (ft^3/s)	0.03531
	gallon per minute (gal/min)	15.85
	barrel per day (bbl/d), (petroleum)	543.4
meter^3 per second (m^3/s)	foot^3 per second (ft^3/s)	35.31
	gallon per minute (gal/min)	15 850
	Mass	
gram	ounce avoirdupois (oz avdp)	0.03527
kilogram (kg)	pound avoirdupois (lb avdp)	2.205
megagram (Mg)	ton, short (2 000 lb)	1.102
	ton, long (2 240 lb)	0.9842
	Pressure	
kilopascal (kPa)	pound-force per inch^2 (lbf/in^2)	0.1450
	atmosphere, standard (atm)	0.009869
	bar	0.01*
	inch of mercury at 60°F (in Hg)	0.2961
	Temperature	
kelvin (K)	degree Fahrenheit (°F)	([1])
degree Celsius (°C)	degree Fahrenheit (°F)	([2])

[1]Temp °F=1.8 temp K-459.67 [2]Temp °F=1.8 temp °C+32

B. Factors for converting inch/pound units to SI metric units

To convert from	To	Multiply by
	Length	
inch (in)	millimeter (mm)	25.4*
foot (ft)	meter (m)	0.3048
yard (yd)	meter (m)	0.9144*
mile (mi)	kilometer (km)	1.609
mile, nautical (nmi)	kilometer (km)	1.852*
	Area	
foot^2 (ft^2)	meter^2 (m^2)	0.09290
yard^2 (yd^2)	meter^2 (m^2)	0.8361
acre	meter^2 (m^2)	4 047
	hectometer^2 (hm^2)	0.4047
mile^2(mi^2)	kilometer^2 (km^2)	2.590
	Volume	
inch^3(in^3)	centimeter^3 (cm^3)	16.39
	decimeter^3 (dm^3)	0.01639
foot^3 (ft^3)	decimeter^3 (dm^3)	28.32
	meter^3 (m^3)	0.02832
yard^3 (yd^3)	meter^3 (m^3)	0.7646
pint (pt)	decimeter^3 (dm^3)	0.4732
quart (qt)	decimeter^3 (dm^3)	0.9464
gallon (gal)	decimeter^3 (dm^3)	3.785
	meter^3 (m^3)	0.003785
barrel (bbl), (petroleum, 1 bbl=42 gal)	meter^3 (m^3)	0.1590
acre-foot (acre-ft)	meter^3 (m^3)	1 233
	hectometer^3 (hm^3)	0.001233
mile^3 (mi^3)	kilometer^3 (km^3)	4.168
	Volume per unit time (includes flow)	
foot^3 per second (ft^3/s)	decimeter^3 per second (dm^3/s)	28.32
	meter^3 per second (m^3/s)	0.02832
gallon per minute (gal/min)	decimeter^3 per second (dm^3/s)	0.06309
	meter^3 per second (m^3/s)	0.00006309
barrel per day (bbl/d), (petroleum)	decimeter^3 per second (dm^3/s)	0.001840
	Mass	
ounce avoirdupois (oz avdp)	gram (g)	28.35
pound avoirdupois (lb avdp)	kilogram (kg)	0.4536
ton, short (2 000 lb)	megagram (Mg)	0.9072
ton, long (2 240 lb)	megagram (Mg)	1.016
	Pressure	
pound-force per inch^2 (lbf/in^2)	kilopascal (kPa)	6.895
atmosphere, standard (atm)	kilopascal (kPa)	101.3
bar	kilopascal (kPa)	100.*
inch of mercury at 60°F (in Hg)	kilopascal (kPa)	3.37
	Temperature	
degree Fahrenheit (°F)	kelvin (K)	([3])
degree Fahrenheit (°F)	degree Celsius (°C)	([4])

[3]Temp K=(temp °F+459.67)/1.8 [4]Temp °C=(temp °F-32)/1.8

SI PREFIXES

MULTIPLICATION FACTORS	PREFIX	SI SYMBOL
1 000 000 000 000 000 000 = 10^{18}	EXA (QUINTILLION)	E
1 000 000 000 000 000 = 10^{15}	PETA (QUADRILLION)	P
1 000 000 000 000 = 10^{12}	TERA (TRILLION)	T
1 000 000 000 = 10^{9}	GIGA (BILLION)	G
1 000 000 = 10^{6}	MEGA (MILLION)	M
1 000 = 10^{3}	KILO (THOUSAND)	K
100 = 10^{2}	HECTO* (HUNDRED)	H
10 = 10^{1}	DEKA* (TEN)	DA
0.1 = 10^{-1}	DECI*	D
0.01 = 10^{-2}	CENTI*	C
0.001 = 10^{-3}	MILLI	M
0.000 001 = 10^{-6}	MICRO	M
0.000 000 001 = 10^{-9}	NANO	N
0.000 000 000 001 = 10^{-12}	PICO	P
0.000 000 000 000 001 = 10^{-15}	FEMTO	F
0.000 000 000 000 000 001 = 10^{-18}	ATTO	A

*Not recommended but occasionally used.

SI (International System of Units) a modernized metric system of measurement. An asterisk after the last digit of the factor indicates that the conversion factor is exact and that all subsequent digits are zero; all other conversion factors have been rounded to four significant digits. Use of hectare (ha) as an alternative name for square hectometer (hm^2) is restricted to the measurement of small land or water areas. Use of liter (L) as a special name for cubic decimeter (dm^3) is restricted to the measurement of liquids and gases. No prefix other than milli should be used with liter. Metric ton (t) as a name for megagram (Mg) should be resirtcted to commercial usage.
Source: Modified from Suggestions to Authors of the Reports of the United States Geological Survey, Seventh Edition, 1991.

Editorial Board

Contributing Authors

A. Frank Alsobrook
Silica: Specialty Materials
President
Alsobrook and Company, Inc.
Medford, New Jersey

John A. Ames
Cement and Cement Raw Materials
Senior Consultant
Sewickley, Pennsylvania

R. Lee Aston
Mining and Eviromining Law for Industrial Minerals and Rocks
Attorney-at-Law, Mining Engineer, Geologist
Aston Mineral Law and Engineering
Elberton, Georgia

George S. Austin
Soil Additives; Adobe and Similar Materials; Stone, Decorative
Senior Industrial Minerals Geologist
New Mexico Bureau of Mines
Socorro, New Mexico

Gordon T. Austin
Garnet
Commodity Specialist
Branch of Industrial Minerals
US Bureau of Mines
Washington, DC

James M. Barker
Stone, Decorative; Perlite
Senior Industrial Minerals Geologist
New Mexico Bureau of Mines
Socorro, New Mexico

James J. Bartels
Phosphate Rock
Senior Project Geologist
Gurr and Associates, Inc.
Lakeland, Florida

Robert L. Bates
Overview of lhe tndustrial Minerals
Professor of Geology, Emeritus
The Ohio State University
Columbus, Ohio

Ronald R. Bauer
Wollastonite
Vice President, Exploration
NYCO Minerals, Inc.
Willsboro, New York

James Watson Baxter
Fluxes for Soldering, Brazing, and Welding
Senior Geologist, Emeritus
Industrial Minerals Section
Illinois State Geological Survey
Champaign, Illinois

Richard B. Berg
Tripoli
Industrial Minerals Geologist
Montana Bureau of Mines and Geology
Montanta Tech
Butte, Montana

Subhash B. Bhagwat
The Role of Industrial Minerals in the US Economy
Senior Mineral Economist and Section Head
Industrial Minerals and Mineral Economics
Illinois State Geological Survey
Champaign, Illinois

Frederick M. Bodycomb
Insulation
Consultant-Industrial Minerals
Springfield, Colorado

H. Lyn Bourne
Glass Raw Materials
President
Consulting Geologist
Northville, Michigan

Richard O.Y. Breese
Diatomite; Perlite
Lyons, Colorado
prepared under the auspices of Celite Corporation, Lompoc, California

Donald A. Brobst
Barium Minerals
Consulting Geologist
Reston, Virginia

John F. Burst
Clay-based Ceramic Raw Materials
Principal
IMMI Consulting Group
Rolla, Missouri

Donald D. Carr
Limestone and Dolomite
Principal Geologist, Industrial Minerals
Indiana Geological Survey
Bloomington, Indiana

August V. Castelli
Well Drilling Materials
Consultant
Houston, Texas

Stephen B. Castor
Rare Earth Minerals
Industrial Minerals Geologist
Nevada Bureau of Mines and Geology
Reno, Nevada

Robert L. Cheek, Jr.
Sodium Sulfate Resources
Vice President
Ozark-Mahoning Company
Elf Atochem North America, Inc.
Brownfield, Texas

Peter A. Ciullo
Pyrophyllite
Manager-Minerals Development
Research and Development Div.
R.T. Vanderbilt Company
Norwalk, Connecticut

Joseph R. Copeland
Wollastonite
Vice President-Marketing
NYCO Minerals, Inc.
Willsboro, New York

William E. Cutcliffe
Cement and Cement Raw Materials
East Regional Manager
Rust Environment and Infrastructure
Albany, New York

Lorne R. Duncan
Magnesite and Magnesia
President
International Refractories Engineering
Coraopolis, Pennsylvania

Jessica Elzea
Bentonite
Thiele Kaolin Co.
Sandersville, Georgia

Harry L. Francis
Water and Sewage Treatment
Technical Manager
National Lime Association
Arlington, Virginia

Robert C. Freas
Flue Gas Desulfurization; Lime; Limestone and Dolomite
Vice President, Marketing
Franklin Industrial Minerals
Nashville, Tennessee

Derek G. Fullerton
Diamonds, Industrial
President
Exmin Corporation
Bloomington, Indiana

Robert B. Fulton, III
Chemicals; Fluorspar; Staurolite
Technical Advisor
Kennett Square, Pennsylvania

Thomas E. Garnar
Titanium Minerals; Zirconium and Hafnium Minerals
Titanium Consultants, Inc.
Keystone Heights, Florida

Ronald P. Geitgey
Pumice and Volcanic Cinder
Industrial Minerals Geologist
Oregon Department of Geology and Mineral Industries
Portland, Oregon

Noel A. Genco
Talc
Senior Business Analyst
Specialty Minerals Inc.
Easton, Pennsylvania

Harold B. Goldman
Sand and Gravel
Geological Consultants, Inc.
San Francisco, California

A.P. Grobbelaar
Andalusite
Technical Manager
Cullinan Refractories
Olifantsfontein, Republic of South Africa

G. Robert Guillet
Nepheline Syenite
Consulting Geologist
Industrial Minerals Specialist
Woodview, Ontario

Theodore M. Gurr
Phosphate Rock
President
Gurr and Associates, Inc.
Lakeland, Florida

Kenneth A. Gutschick
Water and Sewage Treatment
Technical Consultant
National Lime Association
Arlington, Virginia

Robert W. Hagemeyer
Fillers and Coatings
Consultant
Atlanta, Georgia

Peter W. Harben
World Distribution of Industrial Mineral Deposits
Peter W. Harben, Inc.
Morris, New York

Edwin A. Harre
Fertilizers
Agricultural Economist
National Fertilizer and Environmental Research Center
Tennessee Valley Authority
Muscle Shoals, Alabama

Fred G. Heivilin
Hormites: Palygorskite (Attapulgite) and Sepiolite
Corporate Manager
Oil-Dri Corp. of America
Ochlocknee, Georgia

Cahit Helvaci
Boron and Borates
Professor
Bornova-Izmir, Turkey

Robert J. Henning
Olivine and Dunite
Vice President, Technical Marketing/Product Development-Minerals
Metals/Minerals Group
AIMCOR
Pittsburgh, Pennsylvania

Dean H. Herrick
Stone, Crushed
Senior Geologist
Continental Placer, Inc.
Albany, New York

V.G. Hill
Bauxite
Consultant, Bauxite and Industrial Minerals
Kingston, Jamaica

James R. Hindman
Vermiculite
Publisher
Vermiculite Newsletter
Salt Lake City, Utah

Robert J. Hite
Potash Resources
Emeritus Geologist
US Geological Survey
Denver, Colorado

Gretchen K. Hoffman
Soil Additives
Senior Coal Geologist
New Mexico Bureau of Mines
Socorro, New Mexico

David A. Holmes
Zeolites
Industrial Minerals Consultant
Lakewood, Colorado

R.A. Holmes
Manganese Minerals
Consultant
Linthicum, Maryland

William G. Holroyd
Emery
Vice President Minerals and Ores
F & S Alloys and Minerals Corporation
New York, New York

R.B. Hoy
Diamonds, Industrial
Retired
Golden, Colorado

Randall E. Hughes
Clay-based Ceramic Raw Materials
Senior Geologist
Industrial Minerals and Minerals Economics Section
Illinois State Geological Survey
Champaign, Illinois

Kenneth S. Johnson
Iodine Resources
Oklahoma Geological Survey
Norman, Oklahoma

Duane B. Jorgensen
Gypsum and Anhydrite
Consultant
formerly Manager, Geological Services
US Gypsum
Ludington, Michigan

Rustu S. Kalyoncu
Roofing Granules
US Bureau of Mines
Washington, DC

Roger A. Kauffman
Feldspars
Vice President, Industrial Minerals
Hecla Mining Company
Coeur d'Alene, Idaho

Harlan L. Kebel
Gypsum Plasters and Wallboards
Vice President, Manufacturing
Tibbals Flooring Company
Oneida, Tennessee

Karan S. Keith
Clay Liners and Barriers
Doctoral Student
Department of Geological Sciences
Indiana University
Bloomington, Indiana

David P. Keller
Pigments, Iron Oxide
Harcros Retiree (formerly Marketing Manager)
Harcros Pigments, Inc.
Easton, Pennsylvania

Robert B. Kistler
Boron and Borates
Chief Geologist
U.S. Borax, Inc.
Valencia, California

Harold R. Kokal
Fluxes for Metallurgy
Senior Staff Engineer
Research and Development
Magnetics International, Inc.
Burns Harbor, Indiana

Dennis S. Kostick
Salt; Soda Ash
Mineral Commodity Specialist
Branch of Industrial Minerals
US Bureau of Mines
Washington, DC

Ezra L. Kotzin
Foundry Sands
Vice President Engineering Services
American Foundrymen's Society, Inc.
Des Plaines, Illinois

Deborah A. Kramer
Beryllium Minerals
Physical Scientist
US Bureau of Mines
Washington, DC

Ihor A. Kunasz
Lithium Resources
Cyprus Foote Minerals
Malvern, Pennsylvania

S.J. Lefond (*deceased*)
Diamonds, Industrial
Consultant, Industrial Minerals
Evergreen, Colorado

Harry D. Lewis
Gilsonite
American Gilsonite Co.
Salt Lake City, Utah

Michael A. Linkous
Silica: Industrial Sand and Sandstone
U.S. Borax & Chemical Corp.
Raleigh, North Carolina

John D. Macfadyen
Cement and Cement Raw Materials
Senior Consultant
Penta Engineering Corp.
St. Louis, Missouri

E.L. Mann
Asbestos
Consultant
Denver, Colorado

Bruce H. Mason
Lightweight Aggregates
Executive Director
Indiana Mineral Aggregates Association
Indianapolis, Indiana

Henry N. McCarl
International Trade in Industrial Minerals; Aggregates: Markets and Uses
Professor of Economics & Geology
School of Business
The University of Alabama at Birmingham
Birmingham, Alabama

Edward F. McCarthy
Talc
Luzenac America
Englewood, Colorado

Derek J. McCracken
Emery
Manager Industrial Minerals and Glass Raw Materials
F & S Alloys and Minerals Corporation
New York, New York

W.H. McCracken
Magnesite and Magnesia
International Refractories Engineers Ltd.
Coraopolis, Pennsylvania

Sid McIlveen, Jr.
Sodium Sulfate Resources
Geologist/Environmentalist
Ozark-Mahoning Company
Elf Atochem North America, Inc.
Brownfield, Texas

Gill Montgomery
Fluorspar
Consulting Geologist
Eldorado, Illinois

John J. Mortvedt
Fertilizers
Senior Scientist
National Fertilizer and Environmental Research Center
Tennessee Valley Authority
Muscle Shoals, Alabama

Haydn H. Murray
Clays: Catalysts; Clay Introduction; Bentonite; Common Clay; Hormites: Palygorskite (Attapulgite) and Sepiolite; Kaolin; Clay Liners and Barriers
Professor
Department of Geological Sciences
Indiana University
Bloomington, Indiana

Joyce A. Ober
Electronic and Optical Materials; Strontium Minerals
Physical Scientist
Division of Mineral Commodities
Branch of Industrial Minerals
US Bureau of Mines
Washington, DC

John F. Papp
Chromite
Chromium Specialist
Division of Mineral Commodities, Branch of Metals
US Bureau of Mines
Washington, DC

Sam M. Pickering, Jr.
Kaolin
Owner and Exploration Geologist
Industrial Mineral, Land, & Geological Services
Macon, Georgia

Arthur C. Pincomb
Neutralization
Rust Environment and Infrastructure
Naperville, Illinois

Robert J. Piniazkiewicz,
Talc
Project Manager
Luzenac America
Englewood, Colorado

George Podolsky
Pigments, Iron Oxide
Technical Director
Harcros Pigments Inc.
Easton, Pennsylvania

W. Robert Power
Stone, Dimension
Consultant
Frisco, Colorado

Madhu G. Ranade
Fluxes for Metallurgy
Manager
Ironmaking Operating Technology/Process Research
Inland Steel Flat Products Co.
East Chicago, Indiana

K. Reckling
Diamonds, Industrial
Diamond Abrasives Corp.
New York, New York

Lawrence F. Rooney
Limestone and Dolomite
Supervisory Geologist
US Geological Survey
Denver, Colorado

Unni H. Rowell
Diamonds, Industrial
Senior Research Geologist
Exmin Corporation
Bloomington, Indiana

Ken Santini
Wollastonite
Principal Geologist
Santini and Associates
Lakewood, Colorado

Alan R. Shapiro
Neutralization
Vulcan Materials Company
Countryside, Illinois

James P. Searls
Potash Resources
Physical Scientist
Division of Mineral Commodities
US Bureau of Mines
Washington, DC

Nelson R. Shaffer
Biotechnical Materials
Geologist
Indiana Geological Survey
Bloomington, Indiana

Nenad Spoljaric
Glauconite
Senior Scientist
Delaware Geological Survey
University of Delaware
Newark, Delaware

Staff
United States Industrial Mineral Statistics and Prices
US Bureau of Mines
Washington, DC

Kerry J. Stanaway
Titanium Minerals
Geologist
BHP International Division
Herndon, Virginia

Charles T. Steuart
Tripoli
Mine Superintendent
Malvern Minerals Company
Hot Springs, Arkansas

Steven J. Stokowski, Jr.
Insulation
Stone Products Consultants, Inc.
Ashland, Massachusetts

Palmer C. Sweet
Kyanite and Related Minerals
Geologist
Department of Mines, Minerals, and Energy
Division of Mineral Resources
Charlottesville, Virginia

James T. Tanner, Jr.
Mica
Professor
North Carolina State University
Minerals Research Laboratory
Asheville, North Carolina

Harold A. Taylor, Jr.
Graphite
Physical Scientist
Branch of Industrial Minerals
US Bureau of Mines
Washington, DC

C.S. Thompson
Pyrophyllite
Consulting Mineralogist
Norwalk, Connecticut

Nikhil C. Trivedi
Fillers and Coatings
Vice President
Research Development & Technology
Specialty Minerals Inc.
Bethelem, Pennsylvania

Louis J. Trostel, Jr.
Refractories
Independent Technical Consultant
Princeton, Massachusetts

Dean Van Dyk (*deceased*)
Feldspars
Chief Metallurgical Engineer
Technical Services Department
Hecla Mining Company
Coeur d'Alene, Idaho

Steve Van Kouteren
Filters and Absorbents
Project Manager
Kline & Company
Fairfield, New Jersey

Robert L. Virta
Asbestos; Asbestos Substitutes
Physical Scientist
Division of Mineral Commodities
US Bureau of Mines
Washington, DC

William W. Wellborn
Abrasives
Vice President Technology Corporate
Washington Mills Company
North Grafton, Massachusetts

Barbara B. Wessel
Index
Lakewood, Colorado

Gregory R. Wessel
Sulfur Resources; Index
Consulting Geologist
Lakewood, Colorado

M.J. Wilhelm
Bromine Resources
Marketing Manager, Bromine & Derivatives
Ethyl Corporation
Baton Rouge, Louisiana

K.C. Williams
Bromine Resources
Manager, Planning & Development
Ethyl Corporation
Baton Rouge, Louisiana

Sherilyn Williams-Stroud
Potash Resources
Geologist
Branch of Sedimentary Processes
Geologic Division/Office of Energy and Marine Geology
US Geological Survey
Denver, Colorado

Erhard M. Winkler
Stone, Conservation
Consultant
South Bend, Indiana

Mark J. Zdunczyk
Silica: Industrial and and Sandstone
Rust Environment and Infrastructure
Albany, New York

Contents

1

INTRODUCTION

OVERVIEW OF THE INDUSTRIAL MINERALS

ROBERT L. BATES

An industrial mineral, says the *Glossary of Geology*, is "any rock, mineral, or other naturally occurring substance of economic value, exclusive of metallic ores, mineral fuels, and gemstones; one of the nonmetallics." Probably most geologists and engineers would accept this simple and straightforward definition, at least as a general statement.

On analysis, however, certain anomalies crop up. For example, what is metallic and what is nonmetallic? Bauxite and ilmenite certainly qualify as metallic ores, but both are also raw materials of important nonmetallic substances; hence they are included in this book. Another anomaly is that, by common consent, the term industrial minerals may include not only rocks and minerals but also certain manufactured products, such as cement and lime, and even mixed groups of natural and artificial products, such as abrasives and refractories. Thus the central realm of the industrial minerals is clear enough, but its boundaries tend to be indefinable. Since a definition that allowed for every irregularity in the outer fringes would be excessively unwieldy, the one quoted above may be considered satisfactory.

The expression *industrial minerals* is commonly equated with *nonmetallics*, and with the somewhat more precise phrase *industrial minerals and rocks* as in the title of this book.

CHARACTERISTICS OF THE FIELD

The industrial minerals as broadly interpreted are a highly diverse group of materials; indeed, we may say that diversity is about all they have in common. At one end of the spectrum are ordinary earth materials such as gravel and crushed stone, with a value of a few dollars per ton; at the opposite end are industrial diamonds, which sell (by the carat) at the rate of roughly $20 million per ton. Neither the metals nor the fuels, and few if any categories of manufactured products, include items that differ in value by more than six orders of magnitude. Most gravel and stone are of no value unless close to a market; diamonds and other exotic substances, on the other hand, are so unique and desirable that their occurrence in remote corners of the earth is no deterrent to their use. The broad range of industrial minerals falls between these extremes of unit value and place value.

The industrial minerals penetrate the entire fabric of modern industrialized society, from such fundamental operations as drilling for oil and smelting iron ore to the manufacture of virtually the whole array of consumer products. There is no industrial-minerals industry; there are many industries. The simple term *field* is convenient as an all-inclusive expression. An industry is considered to belong in the industrial-minerals field if its operations consist of extraction and processing, but not manufacturing. Thus the mining and beneficiation of feldspar is a mineral industry, but the manufacture of ceramic products is not. In the cement and lime industries, processing merges into manufacture, and the line is less distinct. It disappears altogether in a plant that extracts salt as artificial brine, and makes sodium and chlorine chemicals, all under the same roof.

An aspect of many industrial minerals that differentiates them from the metals is that their physical properties carry over into ultimate use of the product. Such minerals as mica, asbestos, diatomite, kaolin, and graphite—not to mention architectural stone and crushed stone—owe their value to physical properties that persist from quarry face to final application. Geologists and engineers who work with metallic ores lose interest by the time the bars of metal are produced; ultimate use of the copper or aluminum is not their concern. Those whose field is the industrial minerals cannot afford indifference to final use of the product.

Some industries in the field are faced by problems of substitution and synthesis. Natural abrasives have been essentially replaced by artificial ones; dimension stone is hard-pressed by brick, tile, metal, and concrete. Replacement of a natural mineral by its artificial analog, made synthetically, faces producers of natural industrial diamond and quartz crystal. Some industrial minerals, on the other hand, are essentially immune from attacks of this kind. For example, because the humble diatom did things with silica—by the billions of billions, and down to the micron level—that man cannot do, incidentally producing a material of great industrial versatility, diatomite is not likely to suffer eclipse through replacement by other materials or by a synthetic equivalent.

The great diversity of the industrial minerals gives the field as a whole a rather acute problem of identity. The Perlite Institute speaks for perlite, and the National Sand and Gravel Association for the producers of sand and gravel; but no agency can speak for the entire field. This is probably not a very serious matter, except in the departments of geology and mineral engineering in our universities. There the mineral industries have traditionally meant metal mining, and economic geology has been equated with metallic ore deposits. Some schools are still dominated by what has been called the "metallic mentality," and the image of the immense field of the industrial minerals seems destined to continue somewhat vague and unfocused. It sometimes helps to remind the uninformed that the value of industrial minerals consumed annually in the United States is twice that of the metals.

CLASSIFICATION

The wide range and numerous peculiarities of the industrial minerals make their classification both challenging and difficult.

To serve as a basis for a book on geology of the industrial minerals, Bates (1960) devised a combined end-use and genetic classification. A division was first made into industrial rocks and industrial minerals, on the basis of economic factors; each group was then subdivided on geologic criteria. Wright and Burnett (1962) utilized unit price and bulk produced to develop a threefold classification, in order to introduce the prospector and small-mine operator to the field of industrial minerals. A more elaborate classification was made by Fisher (1969). He used unit value, place value, and representative volume of production as criteria for distinguishing six groups. This is the most detailed classification that has appeared. Kline (1970) simply divided the subject into two groups, chemical minerals and physical minerals, each with different characteristics and uses. A matrix classification in chart form, presented by Dunn (1973), is an ingenious attempt to relate the facts of geologic occurrence to those of end use.

In a comprehensive book on the industrial minerals and rocks, Kužvart (1984) included a two-page table summarizing several of the classifications mentioned above, plus a number that have been used by Russian and eastern European authors. For his book, Kužvart divided the subject on the basis of economics into industrial minerals and industrial rocks; within these groups the commodities are arranged alphabetically. In a book on geology of the nonmetallics, Harben and Bates (1984) used a subdivision that was solely geological—this is, by mode of origin. This turned out to have certain disadvantages, and in a successor volume (1990), the authors adopted a straight alphabetical arrangement. As they remark, this system is designed for ready reference; can accommodate all occurrences of a given rock or mineral under one head even though such occurrences may be of more than one geological type; and, best of all, avoids the problem that a cynic has referred to as "classifying the unclassifiable." The present volume, like its predecessors, retains the alphabetical arrangement.

ENVIRONMENTAL CONCERNS

The inevitable friction between a growing population and an expanding resource base is sharply defined in the field of industrial minerals. Every stage of a company's operation—from mining or quarrying to disposal of wastes and potential pollution—is subject to regulation by federal, state, or local government. A proposal to open a new mine may also face a meeting of concerned citizens. No matter how carefully the company has prepared its proposals for mining, pollution control, and reclamation, objections at such a "hearing" are likely to ignore technical matters and simply to be based on opposition to any mining activity whatsoever. Special problems of land use and zoning face producers of aggregate, in or near areas of high population density. In the years since 1960, governments have become progressively more involved, some would say more restrictive and high-handed, owing in the last analysis to the pressure of a constantly growing population on a planet with finite resources of air, water, land, and minerals.

A mineral industry may encounter problems connected with the fabrication and use of its products, as has happened with asbestos. Citing health hazards from breathing asbestos dust, in 1989 the US Environmental Protection Agency announced a ban on almost all asbestos-containing products, to be accomplished in three stages ending in August 1996. The result has been the effective dismantling of the asbestos industry in North America. Over-regulation of allegedly hazardous products by zealous but ill-informed agencies may be hazardous to industry.

CHALLENGES AND CHANGES

There may be sectors of the industrial minerals field that are essentially static, but the field as a whole conveys an impression of energy, activity, and growth. The situation is not unlike that in manufacturing—with the notable difference that the product is based on a naturally occurring substance, not on material that can be ordered at will from a supplier. The producers of industrial minerals and rocks—-ranging in properties from talc to diamond, kaolin to black granite, pumice to barite—have an enviable capacity for creating products of use to man.

Shifts in structure or emphasis take place from time to time. The asbestos industry, with its North American and western European markets shriveling, is looking toward eastern Europe, the Middle East, southeast Asia, and South America for new opportunities, especially for such infrastructure requirements as asbestos-cement pipe. At the end of the 1980s, the sulfur industry threatened to become even more unsettled than usual as a method was announced for converting sulfur dioxide of stack gases into elemental sulfur. Producers of Wyoming trona, having conquered North American markets for soda ash, continue to assault European strongholds of the Solvay process. In May 1991, fluorspar producers met at an international conference to assess the effects of decreasing use of chlorofluorocarbons. A major cement producer started using waste organic solvents as a fuel. Increased interest in historic preservation, creating a demand for stone to match older structures, together with an architectural trend toward the use of stone in large buildings, have produced a resurgence in the Indiana limestone and other dimension-stone districts; a marble quarry in western Colorado was re-opened after being idle for 50 years.

Among commodities, the industrial minerals and rocks are members of the international set. High-quality pumice, imported from Turkey, gives blue jeans that disreputable stone-washed look so desired by younger customers. Dimension stone for construction in the United States is imported from Italy, France, Spain, Canada, and other countries. Three-fourths of the country's barite comes from China, India, and Morocco, and most of the graphite from Mexico, China, and Brazil. An enterprising Scot has even shipped crushed stone from his country to the US Gulf Coast as a back-haul cargo. On the other hand, soapstone from Virginia may form the handsome stoves—the traditional European space heaters—in Swiss homes. Special-purpose materials such as Wyoming bentonite, Georgia kaolin, California borax, and Montana talc help to balance the books in international trade. Today, the whole fraternity of producers, processors, and consumers of industrial minerals operates on a worldwide scene.

COMMUNICATION

Until the 1960s, producers of industrial minerals and rocks had little of the feeling of community that existed among metal miners or oil producers. Communication among the diverse segments of the field was confined pretty much to previous editions of this book, to the Industrial Minerals Division of the Society of Mining Engineers (as it then was), and to occasional limited-appeal conferences such as the First Salt Symposium in 1963. To help stimulate common interests, in 1965 the Forum on Geology of Industrial Minerals was organized. It has focused on practical problems of geological occurrence and on how these problems affect extraction and use. The Forum has met in a different state or province each year since its founding; its proceedings have been indexed (Bates, 1991). An event of major significance was the appearance in 1967 of a monthly magazine, *Industrial Minerals*. Published in London, it is dominantly but not exclusively a trade journal. Its coverage is worldwide, and the magazine has become essential reading for those interested in the field. The annual review of industrial minerals in *Mining Engineering* is another source of authoritative information.

About 400 geologists, engineers, chemists, and economists, mostly from the United Kingdom and western Europe, met in

London in July 1974, at the first Industrial Minerals International Congress. They heard papers on a wide range of topics, from geologic aspects of an undeveloped deposit to the testing of final products, and also had opportunity for informal exchange of views and information. Several field trips followed the meeting. Similar congresses have been held, at two-year intervals, in Munich, Paris, Atlanta, Toronto, Monte Carlo, Boston, Sydney, and San Francisco. Attendance at these meetings has exceeded 500 persons, representing as many as 40 countries. Proceedings are published by the sponsoring organization, the publisher of *Industrial Minerals* magazine. The same sponsor has organized occasional conferences on raw materials for such products as drilling muds, glass and ceramics, and refractories.

The fourth (1975) and fifth (1983) editions of the present volume, by industry-wide authorities under the editorship of Stanley J. Lefond, have documented the ever-expanding field of the industrial minerals and rocks.

SELECTED REFERENCES

Bates, R.L., 1960, "Classification," *Geology of the Industrial Rocks and Minerals*, Harper, New York, pp. 15–19; reprinted in Dover edition, 1969.

Bates, R.L., 1991, "Index to Proceedings of the Forum on Geology of Industrial Minerals," Special Paper 24, Oregon Dept. of Geology and Mineral Industries, 43 pp.

Carr, D.D., and Herz, N., 1989, *Concise Encyclopedia of Mineral Resources*, Massachusetts Institute of Technology Press, Cambridge, 426 pp.

Dunn, J.R., 1973, "A Matrix Classification for Industrial Minerals and Rocks," *Proceedings,* 8th Forum on Geology of Industrial Minerals, Public Information Circular 5, Iowa Geological Survey, pp. 185–189, chart.

Fisher, W.L., 1969, "The Nonmetallic Industrial Minerals: Examples of Diversity and Quantity," *Mining Congress Journal*, Vol. 55, No. 2, Feb., pp. 120–126.

Harben, P.W., and Bates, R.L., 1984, *Geology of the Nonmetallics*, Metal Bulletin, Inc., New York, 392 pp.

Harben, P.W., and Bates, R.L., 1990, *Industrial Minerals: Geology and World Deposits*, Metal Bulletin Plc, London, 312 pp.

Kline, C.H., 1970, "Industrial Minerals Are Big Business," *Mining Engineering*, Vol. 22, No. 12, Dec., pp. 46–48.

Kužvart, M., 1984, *Industrial Minerals and Rocks*, Developments in Economic Geology 18, Elsevier, Amsterdam, 454 pp.

Prentice, J.E., 1990, *Geology of Construction Materials,* Chapman and Hall, London, 202 pp.

Wright, L.A., and Burnett, J. L., 1962, "The Search for Industrial Minerals," *Mineral Information Service*, California Div. of Mines and Geology, Vol. 15, No. 1, Jan., pp. 1–8.

UNITED STATES INDUSTRIAL MINERAL STATISTICS AND PRICES

STAFF, BRANCH OF INDUSTRIAL MINERALS, US BUREAU OF MINES

Assembling consistent data for a 40-year time series is a challenge because of the many changes that have occurred in this period. In many cases, because of process improvements and consumer requirements, grades of materials change in a 40-year span. This can lead to a wide fluctuation in price data simply because of increased costs for premium grades. Sources for obtaining price data can also change during a 40-year period. Changing the data source also introduces anomalies in a time series. In some cases, published price data are not representative of the real value of the material and provide little basis for further analysis. For some industrial mineral commodities, there are only a few producers, and value and production data are considered proprietary.

In the data presented in this chapter, every attempt was made to ensure that the data were as consistent and representative of the industry as possible. However, the factors mentioned above may have introduced inconsistencies in some of the data series.

Many different factors influence trends for industrial minerals because of the diversity of end uses for these commodities. Trends in production and prices for most of the industrial minerals can be related to one or more of the following: changing demand or technologies in end-use industries, development of new end uses, the effect of environmental regulations, production and processing technology changes, energy costs, foreign competition and substitution by other materials, and production as a byproduct of other minerals. World economic conditions, such as recessions and inflation rates, also play a part in some production and price trends.

Data for 40 years of production for industrial minerals in the United States are presented in Table 1 in 10-year increments. Unit values for the industrial minerals, in actual and constant 1990 dollars, are shown in Tables 2 and 3.

CHANGING END-USE PATTERNS AND TECHNOLOGY

Construction Industry

Construction industry variations were reflected in production of crushed stone, construction sand and gravel, gypsum, mica, perlite, pumice, asbestos, and talc. Because of the construction of the Interstate Highway System between 1950 and 1973, the growth in the production of crushed stone and construction sand and gravel was almost continuous, paralleling the increased demand for construction aggregates. Following the reduction in the volume of work in the Interstate Highway Program in the late 1960s, the crushed stone and construction sand and gravel industries, while still growing, became more sensitive to the ups and downs of the economy. The 1974 to 1975 and 1982 recessions were reflected by low levels of production of crushed stone and construction sand and gravel in those years.

Gypsum and mica are used mainly in the construction industry—gypsum in wallboard, building plaster, and as an ingredient in portland cement; and mica as a filler in wallboard joint compound, in paints, and in roofing. From 1950 until the 1970s, demand for housing was strong. Domestic mica production more than doubled, and gypsum production increased, although imports from Canada supplanted some US production. The oil embargo in the mid-1970s depressed construction activity and the building industries, and demand for gypsum and mica declined.

Steel Industry

Changes in the steel industry, in both production and technology, significantly influenced production and prices for several industrial minerals. Magnesia, graphite, zircon, clays, and pyrophyllite are used as refractory materials in the iron and steel industry. Lime and fluorspar are used as fluxes in steelmaking.

In the mid-1950s, the steel industry began switching from open hearth furnace technology to basic oxygen furnace (BOF) technology, and later to electric arc furnace technology. These changes created a demand for different types and higher quality refractory linings. In the 1960s, US production of steel expanded, and the use of refractories by this industry also expanded. Contraction of the domestic steel industry in the 1980s led to a corresponding shrinkage in refractory end uses for most minerals.

Changes in refractory consumption were mirrored in production rates for magnesia, dead-burned dolomite, graphite, pyrophyllite, fire clay, and zircon. By 1990, domestic dead-burned magnesia production declined by more than 40% from a high of about 800 kt in the mid-1960s. Dead-burned dolomite production declined by more than 80% from its high in the mid-1950s. Pyrophyllite and fire clay production also decreased. High zircon prices led to the substitution of spinel refractories for zircon and zirconia refractories. Conversely, world natural graphite production has multiplied almost four times since 1950 because of growth in demand for specialized graphite-based refractories in the steel industry.

For lime, the switch from open hearth to BOF meant consumption rates of about 90 kg/t of steel produced in BOF, compared with 14 kg/t in open hearth furnaces. This, coupled with increased US production, resulted in lime consumption by the steel industry in excess of 8 Mt annually. By the end of the 1980s, lime consumption by the steel industry had dropped to about 4.6 Mt annually. The overcapacity left by the lost steel market helped cause stagnation of prices during the 1980s and a steady decline of constant 1990 dollar prices.

Other Industries

The automotive industry is the single largest user of abrasive materials. Changes in the automotive manufacturing industry affected the types and quantities of abrasives used. Automotive parts made from plastics require less grinding and finishing than parts

Table 1. US Industrial Mineral Production (thousand t unless noted otherwise)

Commodity	1950	1960	1970	1980	1990
Abrasives:					
Tripoli (crude) (t)	39 662	52 356	61 784	109 981	94 390
Special silica stone (t)	7 179	2 303	2 843	572	454
Garnet (t)	8 440	9 545	17 089	24 411	47 009
Emery (t)	5 397	7 411	W	W	W
Silicon carbide (t)	58 971	120 854	151 500	154 241	105 251
Aluminum oxide (t)	127 325	177 723	176 901	175 121	185 479
Metallic abrasives (t)	130 937	101 952	189 085	216 098	224 139
Industrial diamond[a]	—	2 000	13 000	50 000	90 000
Asbestos	37	41	113	80	W
Barite	630	648	775	2 037	447[b]
Beryllium concentrate (t)	507	462	W	6 758	4 548
Boron oxide	173	294	510	710	565
Bromine	45	80	159	172	175[b]
Cement:					
Portland	39 614	56 036	66 225	66 820	68 399
Masonry	[c]	[c]	2 722	2 768	2 911
Clays:					
Ball	294	403	644	811	985[b]
Bentonite	883	1 151	2 298	3 796	3 381[b]
Common	23 950	31 116	35 991	29 479	28 130[b]
Fire	8 651	8 995	5 859	1 901	703[b]
Fuller's earth	359	370	891	1 391	1 820[b]
Kaolin	1 588	2 477	4 468	7 148	9 021[b]
Construction sand and gravel	320 433	621 914	760 332	692 273	826 083
Diatomite	210	437	542	625	631[b]
Feldspar	415	510	659	644	630
Fluorspar	274	208	244	84	64[b]
Graphite:					
Low-modulus fiber	NA	NA	NA	153	380
High-modulus fiber	NA	NA	NA	278	3 394
Gravel industrial	NA	NA	NA	767	1 315
Gypsum:					
Calcined	6 660	7 794	7 665	10 748	15 924
Mined	7 433	8 913	8 560	11 227	14 883
Lime:					
Quicklime	3 478	7 503	13 833	14 490	13 392
Hydrated	1 710	2 463	2 836	2 308	2 098
Dead-burned dolomite	1 596	1 768	1 246	448	342
Magnesium compounds:					
Refractory	304	459	727	663	335
Caustic-calcined	38	78	111	143	135
Mica (scrap and flake)	63	89	108	105	109
Perlite:					
Expanded	79	221	377	487	478
Processed	92	283	414	579	565
Phosphate rock	10 948	17 796	35 144	54 415	46 343
Potash (K_2O equivalent)	1 168	2 393	2 476	2 239	1 713
Pumice	652	545	445	493	443
Pyrophyllite	106	113	109	103	82
Rare earths (kg)	W	W	W	15 986	22 713
Salt:					
Rock salt	3 563	5 866	12 855	10 710	12 772
Salt in brine	8 503	13 527	23 794	20 168	17 374
Solar salt	719	1 298	1 598	2 207	2 985
Vacuum pan and open pan	2 301	2 425	3 391	3 521	3 662
Sand, industrial	15 639	16 629	26 435	26 046	24 454
Soda ash:					
Natural	307	716	2 430	7 507[d]	9 156
Synthetic	3 621	4 135	4 005	W	—
Sodium sulfate:					
Natural	169	W	542	529	345
Synthetic	676	601	703	504	313
Stone (dimension):					
Limestone	733	780	437	268	NA
Marble	81	117	57	55	NA
Granite	483	544	523	601	526[b]
Sandstone	241	398	264	154	NA
Slate	153	108	90	45	NA

(cont.)

Table 1 (cont.)

Commodity	1950	1960	1970	1980	1990
Stone (crushed):					
Limestone	163 394	408 590	567 275	656 045	731 794[3]
Dolomite	4 451	[f]	[f]	[f]	44 728[e]
Marble	161	1 374	1 562	1 223	11 088[e]
Calcereous marl	NA	1 164	1 578	3 374	4 643[e]
Shell	NA	7 177	19 698	9 901	6 280[e]
Granite	19 956	38 236	78 139	107 002	152 800[e]
Traprock	20 736	79 228	70 050	73 841	89 957[e]
Sandstone and quartzite	8 015	18 664	21 562	26 194	28 887[e]
Slate	538	267	1 036	958	1 894[e]
Miscellaneous stone	14 764	19 411	31 027	NA	25 460[e]
Total	228 713	557 490	787 100	892 217[b]	1 100 779[e]
Volcanic cinder	NA	1 460	2 310	2 914	NA
Sulfur:					
Frasch	5 276	5 118	7 196	6 390	3 726
Recovered	144	779	1 480	4 073	6 536
In all forms	6 082	6 768	9 710	11 866	11 560
Talc	454	552	824	1 022	1 185
Titanium:					
Ilmenite	410	716	835	539	W
Rutile	6	8	W	W	W
Titanium dioxide pigment	271	413	594	660	979
Vermiculite:					
Concentrate	189	181	259	306	209[b]
Exfoliated	NA	137	200	255	142

NA Not available. W Withheld to avoid disclosing company proprietary data.
[a]Thousand carats, [b]estimated, [c]included with portland cement, [d]includes synthetic soda ash in order to avoid disclosing company proprietary data, [e]1989 production data, [f]Dolomite included under limestone.

made from metal, thus less abrasives are required. Extruded cast parts require less grinding than normal cast parts, and stamped parts require less work than cast parts. In the past 20 years, industrial diamond has replaced silicon carbide in many finishing operations, and in the past 10 years, cubic boron nitride has been replacing fused aluminum oxide.

The erosion of the glass container market by polyethylene terephthalate (PET) bottles and the growing use of cullet by the glass industry affected soda ash consumption for most of the 1980s. Excess soda ash production capacity and stagnant domestic growth contributed to the decline in average soda ash value. Part of the stagnant growth is because of two economic recessions that affected the domestic building construction and automotive industries, both of which are major consumers of flat glass.

Changes in the aircraft industry had a significant effect on the US production of graphite fiber. Since 1972, domestic production of high-modulus graphite fiber has multiplied more than 200 times as military and commercial aerospace demand has grown. This fiber, as a component of composite airframes, is used in stealth aircraft. US production of low-modulus graphite fiber and cloth has multiplied over seven times since 1972 with the increasing demand for aircraft brakes and fireproof materials, particularly for heat shields.

As much as 95% of barite consumption is used as a weighting agent in petroleum drilling mud. More barite is required for drilling deeper wells, which are likely to run into higher pressure fluids or gases than shallow wells. Because of this, the demand for barite has generally trended upwards as the shallower, cheaper pools were drilled out, and exploratory efforts shifted to deeper wells. Opening of offshore drilling acreages with deep high-pressure pools in 1955 and increased drilling between 1973 and 1982 as an attempt to reduce dependence on imported oil led to increased barite consumption during these periods.

Bromine production and consumption was affected primarily by two end uses—as a scavenger for lead in gasoline antiknock additives and as a fire retardant in polymers. Environmental concern about lead during the 1970s banned leaded gasoline and decreased the use of bromine after 1979. But as the use of polymers grew during the 1980s as a replacement for metals in structural parts of electronic equipment, bromine additives began to be used as fire retardants. The net result is that bromine consumption decreased about 20% between 1979 and 1990, but the value of the products increased.

In the 1960s and 1970s, sodium sulfate's primary end use was in the Kraft paper pulping process. Then fillers for powdered soap and detergents became the primary end use for sodium sulfate. In 1986 consumers began using liquid detergents and superconcentrates which do not use sodium sulfate in their formulation. Consequently, by 1990 the US sodium sulfate industry had excess capacity to supply a declining market.

DEVELOPMENT OF NEW END USES

The development of color television had a major impact on demand for rare earths and for strontium. World production of celestite (strontium ore) has increased by a factor of almost 20 since 1950. Up until 1950, most strontium compounds were used in pyrotechnics and military applications. Since that time pyrotechnic applications have remained important, but the quantity of strontium compounds used in these applications has been dwarfed by the quantity used in color television faceplate glass. For rare earths, demand for europium and yttrium as components of a newly developed color television phosphor increased dramatically in 1964.

EFFECTS OF ENVIRONMENTAL REGULATIONS

Changing environmental regulations and environmental concerns have had mixed impacts on industrial minerals. In cases such as asbestos, dimension stone, salt, soda ash, bromine, thorium oxide

Table 2. Industrial Mineral Unit Values Actual Dollars ($/t unless noted otherwise)

Commodity	1950	1960	1970	1980	1990
Abrasives:					
Emery	13.96	19.16	NA	NA	NA
Garnet	94.02	103.30	113.30	78.16	147.61
Special silica stone	48.49	104.63	233.90	3,376.81	506.61
Tripoli	29.59	4.72	8.42	6.15	33.65
Silicon carbide	123.85	170.75	158.67	417.18	487.75
Aluminum oxide	93.92	152.55	154.90	364.78	419.23
Metallic abrasives	89.35	110.23	181.57	371.51	397.58
Industrial diamonds ($/carat)	3.24	3.94	3.67	5.06	1.43
Asbestos	76.00	104.00	94.00	382.00	NA
Barite	9.82	13.24	16.52	32.39	41.89[a]
Beryl ore ($/ton unit BeO)	27.56	31.97	38.58	76.06	111.33
Borates:					
Borax pentahydrate	33.58	63.93	82.67	186.29	272.00[a]
Boric acid	NA	123.46	152.12	557.77	NA
Boron oxide contentrate	91.70	161.80	170.31	516.32	NA
Bromine (¢/kg)	55.10	70.50	38.10	54.80	73.40[a]
Cement:					
Portland	13.95	19.89	19.50	56.09	54.53
Masonry	[b]	[b]	20.57	68.46	75.92
Clays:					
Ball	13.51	14.83	15.76	33.11	42.46
Bentonite	9.69	13.04	11.10	30.36	35.40
Common	.97	1.33	1.62	3.89	6.29
Fire	3.35	5.03	5.89	18.95	25.76
Fuller's earth	18.10	24.74	26.84	57.30	88.04
Kaolin	15.08	18.44	25.98	73.74	109.85
Construction sand and gravel	.82	1.05	1.27	2.89	4.19[a]
Dead-burned dolomite	13.61	18.36	20.77	63.15	78.94
Diatomite	NA	55.12	60.63	160.94	218.56
Feldspar	6.17	9.36	14.64	36.05	43.49
Fluorspar:					
Acid-grade	51.26	52.36	75.51	176.92	192.50
Metallurgical	40.79	41.89	52.91	121.25	92.50
Graphite (Index of unit value):					
High-modulus fiber	NA	NA	NA	50.00	50.00
Low-modulus fiber	NA	NA	NA	125.00	166.00
Gravel, industrial	NA	NA	NA	8.23	10.54
Gypsum:					
Calcined	9.08	15.52	17.23	25.15	17.49
Mined	3.06	4.00	4.10	9.18	6.69
Iodine ($/kg)	3.21	1.66	5.59	10.41	13.80
Kyanite:					
Andalusite	NA	NA	NA	191.80	209.00
35 mesh	31.97	47.40	58.42	77.16	77.16
Sillimanite	NA	NA	NA	407.85	361.00
Lime	12.27	14.72	16.02	49.05	57.07
Lithium:					
Lithium carbonate	NA	1.48	1.15	2.67	4.03
Lithium hydroxide monohydrate	NA	1.59	1.30	3.53	5.11
Magnesium compounds	NA	NA	58.00	231.00	292.00
Mica	28.00	30.00	23.00	60.00	54.00
Perlite:					
Expanded	60.63	58.42	66.14	142.20	189.60
Processed	7.05	9.41	11.85	28.51	33.80
Phosphate rock	5.67	6.57	5.80	22.78	23.20
Potash	38.82	37.46	38.15	160.00	176.80
Pumice	4.08	5.07	2.77	8.66	24.13
Pyrophyllite	NA	15.40	15.17	8.17	NA
Rare earths ($/kg REO):					
Bastnasite concentrate (60% REO)	NA	NA	.66	2.03	2.87
Monazite concentrate (55% REO)	.37	.33	.37	.81	1.19

(cont.)

Table 2 (cont.)

Commodity	1950	1960	1970	1980	1990
Salt:					
Rock	5.45	7.67	7.28	16.15	17.76
Salt in brine	.96	3.43	4.05	7.17	5.39
Solar	4.97	7.73	7.30	17.25	34.61
Vacuum pan and open pan	12.51	24.65	28.12	84.26	110.58
Sand, industrial	2.06	3.55	3.83	11.00	15.61
Soda ash	23.69	28.43	23.18	99.04	91.33
Sodium sulfate	13.00	21.32	20.15	68.81	96.63
Stone (dimension) ($/m^3):					
Georgia rough monumental granite	NA	86.00	139.00	186.00	NA
Indiana rough block limestone	NA	37.00	58.00	112.00	NA
Stone (crushed):					
Volcanic cinder	NA	8.16	4.66	5.90	NA
Total, average	1.71	1.55	1.74	3.66[a]	NA
Strontium minerals (celestite)	17.64	26.46	24.25	59.52	80.00
Sulfur:					
Frasch	18.59	22.73	23.28	97.36	91.17
Recovered	NA	23.06	20.56	74.13	73.89
All elemental	18.59	22.76	22.77	89.06	80.14
Talc	NA	6.38	7.84	18.19	26.21
Thorium oxide ($/kg)	16.53	15.43	19.84	16.00	55.00
Titanium:					
Ilmenite	15.00	24.00	36.00	54.00	72.00
Rutile	88.00	88.00	204.00	482.00	565.00
80% titanium dioxide slag	NA	NA	44.00	113.00	280.00
85% titanium dioxide slag	NA	NA	NA	135.00	298.00
Titanium dioxide pigment ($/kg)	.51	.62	.60	1.40	2.23
Vermiculite:					
Concentrate	11.24	17.22	25.14	76.86	91.53[a]
Exfoliated	NA	75.23	93.82	213.79	276.56[a]
Wollastonite:					
325 mesh	NA	29.76	35.27	99.21	200.00
400 mesh	NA	43.54	47.18	116.84	210.00
Zircon	NA	220.00	197.00	279.00	368.00

NA Not available.
[a]Estimated, [b]included with portland cement.

and pyrophyllite, environmental regulations have had a negative impact on consumption and production. For lime, however, some environmental regulations have been beneficial to demand. Passage of the Clean Air Act in 1970 created a new market for lime and soda ash—flue gas desulfurization. This market has grown steadily since 1970, and by 1990, it accounted for 10% of total US lime consumption and 3% of soda ash demand. Implementation of the Clean Air Act amendments of 1990 will expand the flue gas desulfurization market further.

In the late 1960s and early 1970s the general public became concerned about the risks posed by environmental exposure to asbestos. This concern became a national issue, involving environmental groups, Congress, national research groups, regulatory agencies, etc. As a result efforts have been made to ban the use of asbestos, and consumption declined from 1974 to 1990. Asbestos in the ore of the largest US vermiculite producer was responsible for closure of the mine in 1990, and this has had a substantial effect on US output of vermiculite.

In the 1980s, the use of salt was criticized by the environmental and medical communities. Studies indicated that long-term and excessive salt use for highway deicing caused damage to roadside vegetation and contributed to high levels of sodium in ground water supplies. Corrosion to automobiles and metal bridge surfaces and road reinforcement rods were attributed to the use of salt. Salt also was reported to be the cause of hypertension in humans, and efforts were made to reduce the quantity of salt added to food by the food processing industry. Although salt for human consumption represents only about 3% of total domestic demand, about 200 kt of the 2.1 Mt used annually by the food processing industry were lost due to changes made in this industry. The use of chlorine-based compounds, produced from salt, became an environmental issue in 1990. Chlorine used in pulp bleaching and in treating municipal water supplies reportedly produced carcinogenic compounds. Its use in chlorofluorocarbons contributed to ozone depletion in the upper atmosphere. Because chlorine is produced from salt, these problems will reduce the quantities of salt consumed.

Synthetic soda ash production decreased significantly in the 1970s because the growing national awareness regarding clear air and clean water prompted environmental legislation. Because the Solvay process discharged byproduct calcium chloride and sodium chloride in wastewater effluent, synthetic soda ash plants in Ohio, Texas, and Virginia were shut down rather than retrofitted with the antipollution equipment required to meet Federal and State regulations. By 1990, environmental quality became an international topic. Several small synthetic soda plants around the world closed because of pollution problems, and plans to construct others were suspended.

Thorium oxide price increases since 1980 were directly related to the increased costs of environmental compliance. Thorium's natural radioactivity represents a significant problem to the indus-

Table 3. Industrial Mineral Unit Values 1990 Constant Dollars ($/t unless noted otherwise)

Commodity	1950	1960	1970	1980	1990
Abrasives:					
Emery	76.81	81.54	NA	NA	NA
Garnet	517.28	439.63	354.73	119.90	147.61
Special silica stone	266.80	445.31	732.34	5,180.02	506.61
Tripoli	162.78	20.08	26.37	9.44	33.65
Silicon carbide	681.42	726.71	496.80	639.95	487.75
Aluminum oxide	516.75	649.25	484.99	559.57	419.23
Metallic abrasives	491.60	469.14	568.50	569.90	397.58
Industrial diamonds ($/carat)	17.83	16.77	11.49	7.76	1.43
Asbestos	418.15	442.62	294.31	585.99	NA
Barite	54.04	56.34	51.72	49.68	41.89[a]
Beryl ore ($/ton unit BeO)	151.62	136.05	120.80	116.68	111.33
Borates:					
Borax pentahydrate	184.76	272.10	258.85	285.77	272.00[a]
Boric acid	NA	525.44	476.28	855.62	NA
Boron oxide content	504.54	688.61	533.23	792.04	NA
Bromine (¢/kg)	303.00	300.00	119.50	84.20	73.40[a]
Cement:					
Portland	76.75	84.65	61.05	86.04	54.53
Masonry	[b]	[b]	64.40	105.01	75.92
Clays:					
Ball	74.33	63.10	49.35	50.79	42.46
Bentonite	53.31	55.50	34.75	46.57	35.40
Common	5.34	5.68	5.07	5.97	6.29
Fire	18.44	21.39	18.43	29.07	25.76
Fuller's earth	99.59	105.28	84.04	87.90	88.04
Kaolin	82.97	78.49	81.35	113.12	109.85
Construction sand and gravel	4.51	4.49	3.99	4.43	4.19
Dead-burned dolomite	74.90	78.16	65.02	96.87	78.94
Diatomite	NA	234.57	189.82	246.88	218.56
Feldspar	33.95	39.84	45.83	55.30	43.49
Fluorspar:					
Acid-grade	282.02	222.84	236.42	271.40	192.50
Metallurgical	224.40	178.28	165.66	186.00	92.50
Graphite (Index of unit value):					
High-modulus fiber	NA	NA	NA	76.70	50.00
Low-modulus fiber	NA	NA	NA	191.75	166.00
Gravel, industrial	NA	NA	NA	12.62	10.54
Gypsum:					
Calcined	49.97	66.06	53.94	38.58	17.49
Mined	16.86	17.03	12.84	14.08	6.69
Iodine ($/kg)	17.66	7.06	17.00	15.97	13.80
Kyanite:					
Andalusite	NA	NA	NA	294.22	209.00
35 mesh	175.88	201.73	182.91	118.37	77.16
Sillimanite	NA	NA	NA	625.64	361.00
Lime	67.50	62.63	50.15	75.25	57.07
Lithium:					
Lithium carbonate	NA	6.29	3.59	4.09	4.03
Lithium hydroxide monohydrate	NA	6.76	4.07	5.41	5.11
Magnesium compounds	NA	NA	181.60	354.35	292.00
Mica	154.06	127.68	72.01	92.04	54.00
Perlite:					
Expanded	333.57	248.65	207.08	218.13	189.60
Processed	38.79	40.06	37.10	43.73	33.80
Phosphate rock	31.20	27.96	18.16	34.94	23.20
Potash	213.59	159.43	119.45	245.44	176.80
Pumice	22.44	21.58	8.66	13.29	24.13
Pyrophyllite	NA	65.54	47.49	12.53	NA
Rare earths ($/kg REO):					
Bastnasite concentrate (60% REO)	NA	NA	2.07	3.11	2.87
Monazite concentrate (55% REO)	2.04	1.40	1.16	1.24	1.19
Salt:					
Rock	29.96	32.65	22.78	24.77	17.76
Salt in brine	5.28	14.59	12.67	11.00	5.39
Solar	27.34	32.89	22.85	26.46	34.61
Vacuum pan and open pan	68.84	104.90	88.04	129.26	110.58
Sand, industrial	11.34	15.11	11.98	16.88	15.61

(cont.)

Table 3 (cont.)

Commodity	1950	1960	1970	1980	1990
Soda ash	130.33	120.99	72.58	151.93	91.33
Sodium sulfate	71.51	90.73	63.09	105.55	96.63
Stone (dimension) ($/m^3):					
Georgia rough monumental granite	NA	366.02	435.21	285.32	NA
Indiana rough block limestone	NA	157.47	181.60	171.81	NA
Stone (crushed):					
Volcanic cinder	NA	34.73	14.59	9.05	NA
Total, average	9.41	6.60	5.45	5.61[a]	NA
Strontium minerals (celestite)	97.04	112.61	75.93	91.31	80.00
Sulfur:					
Frasch	102.28	96.74	72.89	149.35	91.17
Recovered	NA	98.14	64.37	113.72	73.89
All elemental	102.28	96.87	71.29	136.62	80.14
Talc	NA	27.16	24.54	27.90	26.21
Thorium ($/kg)	90.95	65.67	62.12	24.54	55.00
Titanium:					
Ilmenite	82.53	102.14	112.72	82.84	72.00
Rutile	484.18	374.53	638.72	739.39	565.00
80% titanium dioxide slag	NA	NA	137.76	173.34	280.00
85% titanium dioxide slag	NA	NA	NA	207.09	298.00
Titanium dioxide pigment ($/kg)	2.81	2.64	1.88	2.15	2.23
Vermiculite:					
Concentrate	61.86	73.28	78.72	117.91	91.53[a]
Exfoliated	NA	320.19	293.74	327.96	276.56[a]
Wollastonite:					
325 mesh	NA	126.66	110.43	152.19	200.00
400 mesh	NA	185.31	147.72	179.23	210.00
Zircon	NA	936.32	616.81	427.99	368.00

NA Not available.
[a]Estimated, [b]included with portland cement.

try, and costs for storing, monitoring, and disposing of waste thorium continue to escalate.

Changes in environmental regulations covering insecticides and competition from mineral substitutes affected pyrophyllite consumption by the insecticide industry. Mine health and safety regulations affected the prices of dimension granite and limestone. Boron production was affected by a moratorium between 1976 and 1980 that limited surface disturbances in Death Valley National Monument. Consequently, extraordinary means were required to develop a significant underground deposit. Bromine consumption was affected by the ban on leaded gasoline, as mentioned previously.

PROCESSING TECHNOLOGY CHANGES

Changes or improvements in process technology can have a significant impact on prices for industrial minerals. The use of more efficient mining equipment and procedures were principally responsible for the overall decline in value, in constant-dollar terms, of mica, gypsum, crushed stone, and construction sand and gravel. By achieving economies of scale for high-modulus graphite fiber production, the unit value of high-modulus graphite fiber dropped to one-half of its 1973 value by 1980. The change from underground to open pit mining in 1957 was reflected in significant increases of production and decrease in prices for borax pentahydrate.

The demise of the Solvay process, mentioned earlier, impacted not only soda ash production, but lime production as well. During the 1950s, the major end use of lime was in Solvay process alkali plants that produced synthetic soda ash. As domestic synthetic soda ash production was replaced by natural soda ash production from 1971 until the last Solvay plant was closed in 1986, the captive lime industry lost more than 3 Mt of production capacity.

ENERGY COSTS

Many industrial minerals experienced a sharp increase in prices in response to the energy crises in the 1970s. In some cases, high energy costs led to closures in several US industries. In addition to environmental problems, the oil embargo in 1973 adversely affected US synthetic soda ash producers. The synthetic soda ash process required more energy to produce one ton of product, compared with one ton produced from trona. Higher energy costs contributed to a higher selling price for synthetic soda ash, thereby establishing a two-tier price structure with natural soda ash. By 1979, only one synthetic soda ash plant remained in operation.

Several closures and consolidations of salt operations occurred in the 1970s. Small rock salt and solar salt producers were acquired by medium-sized companies in order to be competitive with the major salt companies, all of which had foreign subsidies on which to rely to supply less-expensive imported salt. In 1977, the US Federal Energy Administration purchased a large Louisiana rock salt mine to establish the nation's first Strategic Petroleum Reserve, an oil storage facility to use in national emergencies.

The domestic fluorspar industry shrank between 1975 and 1990 from 16 mines to 3 mines, which account for nearly all of US production. In general, the industry was beset by problems of increased operating, equipment, and material costs. The substantial increase in labor and fuel costs in the 1970s forced closure of many marginal operations, especially in the western states, where freight rates to the Midwest and East became prohibitively high.

Because fuel is the largest cost of lime operations, price increases in natural gas and bituminous coal of 41 and 82%, respectively, caused operating costs to soar in 1984. This increased the rate at which marginal lime operations went out of business. Dramatic price increases reported in 1974 to 1975 were the direct result of higher fuel costs.

FOREIGN COMPETITION AND SUBSTITUTION

Materials produced in the United States are subject to substitution or competition from other materials when production costs are high or when the value of the US dollar is high relative to other currencies. Production of dimension stone decreased dramatically in a 40-year period. Limestone production dropped because of competition from alternate, cheaper materials, particularly concrete. Marble production dropped drastically because of competition from similar Italian marble available at lower prices. Slate production dropped with the substitution of asphalt shingles for slate roofing and the loss of the billiard table market to Italian slate. Sandstone production dropped with the decline in use of load-bearing dimension stone in favor of alternate materials or thin dimension stone products; sandstone is difficult to make into thin products.

The entrance of China into world markets in the 1980s affected fluorspar and magnesia. During the 1980s, China's share of the world fluorspar market rose dramatically until it became the world's largest producer. Its production capacity, plus a pricing policy that is 20 to 30% less than the competition, has discouraged any expansion of domestic production. Dead-burned magnesia imports from China also affected US magnesium compound production in the 1980s; import reliance for magnesium compounds rose from less than 10 to 25%, primarily because of increases in imports of cheaper magnesia from China.

Between 1970 and 1990, import reliance for bromine from Israel increased significantly from 9 to a high of 96% and was 78% in 1990. Despite the US-Israel Free Trade Agreement, tariffs on some bromine compounds that were determined to be sensitive were retained until 1995.

Production of domestic gypsum, feldspar, and potash has been affected by imports from Canada. From 1955 to 1970, crude gypsum production in the United States remained constant even though demand continued to increase throughout the period. Imports, primarily from Canada, increased from about one-fourth of US consumption in 1955 to about one-third of US consumption by 1970. Growth of domestic feldspar production also has been affected for many years by competition from imported nepheline syenite from Canada. For potash, Canadian beneficiating plants are closer to the large market in the upper midwest than the US plants, and potash production costs are lower in Canada than those in the United States.

PRODUCTION AS A BYPRODUCT OF OTHER MINERALS

Rare earth minerals, such as monazite and zircon are normally recovered as byproducts of titanium minerals production. Production of these minerals is dependent upon the demand for titanium minerals. Consequently, achieving a supply-demand balance between minerals produced as byproducts is difficult.

Sodium sulfate is also produced as a byproduct, associated with production of ascorbic acid, cellulose, flue gas desulfurization, hydrochloric acid, lithium carbonate, rayon, resorcinol, silica, and sodium dichromate. Because it is produced as a manufacturing byproduct, sodium sulfate is priced as a waste product to facilitate disposal.

OTHER FACTORS AFFECTING PRODUCTION AND PRICES

Production and consumption of some minerals used in agriculture, including phosphate and potash, were generally affected by increases in world population and food demand. Fertilizer production and consumption change from year to year due to inclement weather, recessions in the US and world economies, and US Government programs designed to control farm production.

Some industrial mineral prices were affected primarily by external factors such as inflation rates, recessions, etc. Talc, perlite, and lithium prices increased primarily in response to inflation rates.

Other industrial minerals production and price changes were prompted by internal factors, principally individual mineral supply-demand balances. This was the case for sulfur and titanium minerals (ilmenite and rutile).

In general over the past 40 years, production of most industrial minerals is directly related to demand in their principal end uses, although other factors mentioned in this chapter can influence production levels. Prices tended to be high in times of short supply and great demand, and depressed in times of abundant supplies and weaker demand.

SOURCES OF PRODUCTION AND PRICE DATA

The following sources were used to generate the information presented in Tables 1, 2, and 3:

US Bureau of Mines, *Minerals Yearbook*, Vol. 1, 1950–90 editions.

US Bureau of Mines, *Mineral Facts and Problems*, 1975, 1980, and 1985 editions.

US Bureau of the Census, 1950–90.

Chemical Marketing Reporter, 1972–90.

Oil, Paint, and Drug Reporter, 1950–71.

Engineering and Mining Journal, 1950–87.

Industrial Minerals, 1971–90.

Metal Bulletin, 1965–90.

American Metal Market, 1963–74.

Molycorp Inc., 1967–90 (Bastnasite).

Metals Week, 1967–90.

Rhône-Poulenc Inc., 1950–90 (Thorium).

E.I. Dupont de Nemours & Co. Inc., 1971–90.

US Lithium Producers, 1986–90 (Lithium).

WORLD DISTRIBUTION OF INDUSTRIAL MINERAL DEPOSITS

PETER W. HARBEN

Once considered the dowdy country cousins of the glamorous metallics, industrial minerals are shedding their old image. They are neither common nor easy and their time has come in an age of increasing specialization. Consider that of the 45 or so industrial minerals and rocks listed in Table 1, almost three-quarters have ten or fewer significant suppliers (i.e., those individual countries contributing 2% or more of total world production); in more than half the cases, 85% of world production is accounted for by only five countries or fewer. Curiously, even materials that appear to be virtually ubiquitous like crushed rock and common salt have significant production vacuums. For example, there is a severe lack of sound aggregates along the US Gulf Coast and common salt production on a large scale is virtually absent in central Africa. These examples all underline the obvious, but acute truism that "The single most important fact about mineral resources is that they are not distributed equally over the world" (Flawn, 1966).

Just as significant today is that human resources are not distributed equally over the world, and so arise some interesting commercial incongruities. In certain regions raw materials are plentiful but consumers are not, and without a market a mineral deposit is merely a geological curiosity. Elsewhere, there may be a market but no local raw material supply. For example, despite a huge market there is no or virtually no commercial production of chromite, diamonds, and manganese in the United States and Canada, nor phosphate rock, diamonds, rutile, and zirconium minerals in western Europe. In contrast, Australia with its small domestic market is the world's largest supplier of bauxite, diamonds, ilmenite, natural and synthetic rutile, and zircon; the same is true for South Africa which is a leading producer of chromite, manganese, diamonds, andalusite, ilmenite, rutile, and zircon (Table 2). In the commercial world, Nature's uneven distribution is counterbalanced by deep sea international trade.

GEOLOGICAL DISTRIBUTION

Continuing with the obvious theme, Stanton (1972) concluded that most ore deposits appear to be closely related to their geological environments, and since these geological environments vary over time, it follows that particular ores should have been conspicuously concentrated in certain places at certain times. In the metallic world, these places are known as metallogenic provinces (Govett and Govett, 1976), a concept that can be applied to a more limited degree to nonmetallics. Consequently, the basic distribution pattern of mineral resources is obviously determined by geology. However, the distribution of commercial production, that is reserves as opposed to resources, is influenced by a diverse host of factors. These include mineral grade and consistency, amenability to mineral beneficiation, geographic location, demographics, labor rates, tax and investment incentives, political stability, entrepreneurial skills, transportation, market demands, research and development, price competitiveness, economic climate, environmental regulations, government intervention, and timing. Natural and man-made factors have combined to produce the distribution pattern described below.

PRODUCTION PATTERNS

The industrial minerals may be broadly placed into geological pigeon holes (Table 3), although several find themselves residing comfortably in several slots.

IGNEOUS INTRUSIVE

Olivine, Chromite, Nepheline Syenite, Diamonds

Both olivine and chromite deposits are closely associated with ultramafic plutonic rocks. The bulk of chromite reserves occur in the large laterally extensive stratiform or Bushveld-type form occurring in stable shield areas as exemplified by the Bushveld Igneous Complex in South Africa, the Great Dyke of Zimbabwe, northern Finland, and Bahia State, Brazil. In contrast, the smaller podiform or Alpine-type deposits occur in mobile belts such as the Urals of the USSR, the Tethyian mountain chain of the Balkans, Turkey, and Iran, and in the Circum-Pacific belt. Overall, significant chromite reserves and production are restricted to less than ten countries, and nonmetallurgical grades to still fewer, namely South Africa, the Philippines, Turkey, Greece, Finland, Albania, and India.

Commercial olivine and dunite deposits are common in the Alpine-type ultrabasic terrains. A limited market restricts production to Norway with 75% of world production; smaller producers are Spain, Italy, Japan, and the United States. The modest US production is from North Carolina and Washington state.

Nepheline syenite is a relatively common silica-deficient magmatic intrusive rock. However, commercial production is limited to large operations in Canada, Norway, and the USSR due to the limited market size, competition from feldspar, and the requirement for a consistently low iron content. Production from Canada and Norway is virtually all exported; this accounts for 70 and 30% respectively of world production, excluding the USSR.

The primary geological habitat for natural diamond is kimberlite, an ultrabasic intrusive rock associated with stable shield regions. Diamondiferous kimberlites are concentrated in southern Africa, the Siberian Platform, Brazil, and Western Australia. Ages range from Precambrian in South Africa to Recent in Tanzania. Diamonds are also produced commercially from placer deposits such as in Namibia (see the sedimentary section that follows). Overall, Africa is a prime region for diamond production, in particular South Africa, Botswana, Namibia, Lesotho, Swaziland, and

Table 1. Distribution of World Industrial Minerals Production (tons)

	Production	% world production	% world production cumulative
Asbestos	4 325 487		
USSR	2 600 000	60	60
Canada	709 800	16	76
Brazil	230 000	5	81
Zimbabwe	190 000	4	85
China	160 000	4	89
South Africa	150 000	3	92
Greece	80 000	2	94
Barite	5 712 050		
China	1 750 000	31	31
Mexico	566 000	10	41
USSR	540 000	10	51
Turkey	460 000	8	59
India	395 000	7	66
Morocco	320 000	6	72
United States	290 000	5	77
Germany	190 000	3	80
Bauxite	105 029 000		
Australia	38 583 000	37	37
Guinea	16 523 000	16	53
Jamaica	9 395 000	10	63
Brazil	8 500 000	8	71
India	4 768 000	5	76
USSR	4 600 000	4	80
China	4 000 000	4	84
Suriname	3 530 000	3	87
Yugoslavia	3 252 000	3	90
Hungary	2 900 000	3	93
Greece	2 400 000	2	95
Bentonite	9 794 127		
United States	3 112 365	32	32
USSR	2 900 000	30	62
Greece	1 200 000	12	74
Japan	526 131	5	79
Italy	220 000	2	81
Brazil	210 000	2	83
Germany	200 000	2	85
Mexico	188 500	2	87
Romania	180 000	2	89
Beryl & bertrandite	7 586		
United States	4 592	60	60
USSR	2 000	26	86
Brazil	900	12	98
Boron minerals	2 989 000		
Turkey	1 400 000	47	47
United States	1 114 000	37	84
Argentina	200 000	7	91
USSR	200 000	7	98
Bromine	422 900		
United States	174 600	41	41
Israel	120 000	28	69
USSR	65 000	15	84
United Kingdom	25 000	6	90
France	18 000	4	94
Japan	15 000	3	97
Chromite	11 901 300		
South Africa	3 800 000	32	32
USSR	3 800 000	32	64
India	800 000	8	72
Finland	750 000	6	78
Albania	700 000	6	84
Turkey	650 000	5	89

Table 1. Distribution of World Industrial Minerals Production (tons) (cont.)

	Production	% world production	% world production cumulative
Zimbabwe	570 000	5	94
Brazil	225 000	2	96
Diamond (carats)			
Industrial, natural	51 821 000		
Australia	17 540 000	34	34
Zaire	16 150 000	31	65
USSR	6 500 000	13	78
South Africa	5 106 000	10	88
Botswana	4 576 000	9	97
Gem, natural	43 016 000		
Australia	17 540 000	41	41
Botswana	10 676 000	25	66
USSR	4 500 000	10	76
South Africa	4 010 000	9	85
Zaire	2 850 000	7	92
Namibia	970 000	2	94
Angola	950 000	2	96
Diatomite	1 838 000		
United States	631 000	34	34
France	250 000	14	48
Romania	250 000	14	62
USSR	250 000	14	76
Spain	100 000	5	81
Korea	75 000	4	85
Denmark	60 000	3	88
Mexico	55 000	3	91
Germany	47 000	2	93
Feldspar	4 728 000		
Italy	1 320 000	28	28
United States	715 000	15	43
USSR	370 000	8	51
Germany	330 000	7	58
France	230 000	5	63
Korea	198 000	4	67
Thailand	198 000	4	71
Spain	154 000	3	74
Brazil	138 000	3	77
Mexico	110 000	2	79
South Africa	93 000	2	81
Romania	90 000	2	83
Fluorspar	5 731 340		
China	1 700 000	30	30
Mexico	861 000	15	45
Mongolia	800 000	14	59
USSR	410 000	7	66
South Africa	368 340	6	72
France	183 000	3	75
Spain	180 000	3	78
Italy	140 000	2	80
United Kingdom	110 000	2	82
Kenya	100 000	2	84
Morocco	100 000	2	86
Thailand	98 000	2	88
Fuller's earth	3 079 308		
United States	1 881 511	61	61
Senegal	98 882	3	64
Spain	45 000	2	66
Garnet*	133 175		
United States	63 503	48	48
India	27 216	20	68
China	18 144	14	82

Table 1. Distribution of World Industrial Minerals Production (tons) (cont.)

	Production	% world production	% world production cumulative
Australia	14 515	11	93
Norway	7 257	5	98
* capacity			
Graphite	631 213		
China	200 000	32	32
Korea	98 000	16	48
USSR	84 000	13	61
India	55 000	9	70
Mexico	45 000	7	77
Brazil	30 000	5	82
Czechoslovakia	25 000	4	86
N. Korea	25 000	4	90
Madagascar	14 000	2	92
Turkey	13 000	2	94
Zimbabwe	11 000	2	96
Gypsum	108 624 000		
United States	17 624 000	16	16
Canada	9 590 000	9	25
Iran	9 300 000	9	34
China	8 900 000	8	42
Japan	6 900 000	6	48
France	6 266 000	6	54
Spain	6 060 000	6	60
Thailand	6 038 000	6	66
Mexico	5 305 000	5	71
Germany	2 382 000	2	73
Australia	1 980 000	2	75
Iodine	15 622		
Japan	7 600	49	49
Chile	4 000	26	75
USSR	2 000	13	88
United States	1 508	9	97
China	500	3	100
Natural iron oxide	282 684*		
India	145 000	51	51
United States	42 034	15	66
Spain	31 000	12	78
France	15 000	5	83
Cyprus	10 000	3	86
Austria	9 000	3	89
Chile	8 300	3	92
Brazil	6 000	2	94
Germany	5 000	2	96
* reported producers only			
Kaolin	25 669 083		
United States	8 973 668	35	35
United Kingdom	3 140 000	12	47
USSR	2 000 000	8	55
Colombia	1 800 000	7	62
France	1 400 000	5	67
Korea	1 219 174	5	72
Brazil	846 200	3	75
Germany	825 000	3	78
India	755 000	3	81
Czechoslovakia	680 000	3	84
Kyanite and related minerals	477 000		
South Africa	270 700	57	57
United States	90 000	19	76
India	53 000	11	87
France	50 000	11	98

Table 1. Distribution of World Industrial Minerals Production (tons) (cont.)

	Production	% world production	% world production cumulative
Lithium*	12 000		
United States	5 600	47	47
Chile	1 700	14	61
Australia	1 300	11	72
USSR	1 089	9	81
China	730	6	87
Zimbabwe	730	6	93
Canada	500	4	97
Brazil	290	2	99
* capacity of lithium content			
Magnesite & magnesia*	10 401 000		
USSR	2 200 000	21	21
N. Korea	1 250 000	12	33
United States	1 013 000	10	43
China	1 010 000	10	53
Czechoslovakia	700 000	7	60
Greece	612 000	6	66
Austria	576 000	5	71
Japan	485 000	4	75
Brazil	345 000	3	78
Turkey	285 000	3	81
Yugoslavia	240 000	2	83
India	229 000	2	85
Spain	205 000	2	87
* capacity in MgO equivalent			
Manganese*	8 443 000		
USSR	2 630 000	31	31
South Africa	1 520 000	18	49
Gabon	1 215 000	14	63
Australia	1 008 000	12	75
Brazil	605 000	7	82
China	544 000	6	88
India	508 000	6	94
* metal content			
Mica	265 387		
United States	119 142	45	45
USSR	50 000	19	64
Korea	30 000	11	75
India	26 200	10	85
Canada	12 000	5	90
France	11 000	4	94
Nepheline syenite	3 250 000		
USSR	2 500 000	77	77
Canada	510 000	16	93
Norway	240 000	7	100
Nitrogen (ammonia)	99 516 000		
USSR	20 100 000	20	20
China	16 960 000	17	37
United States	12 544 000	13	50
India	6 712 000	7	57
Canada	3 331 000	3	60
Netherlands	3 000 000	3	63
Romania	2 603 000	3	66
Indonesia	2 525 000	3	69
Poland	2 200 000	2	71
Mexico	2 095 000	2	73
Olivine & dunite*	7 098 000		
Norway	5 300 000	75	75
Spain	1 000 000	14	89

Table 1. Distribution of World Industrial Minerals Production (tons) (cont.)

	Production	% world production	% world production cumulative
Japan	350 000	5	94
Italy	200 000	3	97
* estimated capacity			
Perlite	1 777 000		
United States	580 000	33	33
USSR	544 000	31	64
Greece	160 000	9	73
Turkey	145 000	8	81
Hungary	110 000	6	87
Japan	77 000	4	91
Italy	71 000	4	95
Czechoslovakia	43 500	2	97
Mexico	36 000	2	99
Phosphate rock*	50 488 000		
United States	14 893 000	29	29
USSR	10 850 000	21	50
Morocco	8 000 000	16	66
China	4 100 000	8	74
Jordan	1 800 000	4	78
Tunisia	1 800 000	4	84
Brazil	1 290 000	3	87
Israel	1 231 000	3	90
Togo	1 208 000	2	92
South Africa	1 040 000	2	94
* P_2O_5 content			
Potash*	29 789 000		
USSR	10 500 000	35	35
Canada	7 458 000	25	60
Germany	5 440 000	18	78
United States	1 595 000	5	83
Israel	1 271 000	4	87
France	1 200 000	4	91
Jordan	790 000	3	94
Spain	745 000	3	97
* K_2O equivalent			
Pumice/pozzolan	10 964 348		
Italy	5 225 000	48	48
Greece	1 600 000	15	63
Spain	900 000	8	71
Turkey	700 000	6	77
United States	443 000	4	81
France	400 000	4	85
Yugoslavia	400 000	4	89
Chile	290 000	3	92
Guadeloupe	220 000	2	94
Pyrophyllite	2 205 000		
Japan	1 231 300	56	56
S. Korea	640 000	29	85
United States	83 301	4	89
Brazil	75 000	3	92
N. Korea	70 000	3	95
India	60 000	3	98
Rare earth minerals*	70 000		
United States	26 000	37	37
China	20 350	29	66
Australia	11 020	16	82
Malaysia	3 900	6	88
Brazil	2 200	3	91
India	2 200	3	94
USSR	1 500	2	96

*REO; excludes production from South Africa and monazite from the United States

Table 1. Distribution of World Industrial Minerals Production (tons) (cont.)

	Production	% world production	% world production cumulative
Salt	209 988 000		
United States	38 856 000	19	19
China	30 850 000	15	34
Germany	17 804 000	8	42
USSR	16 300 000	8	50
Canada	12 279 000	6	56
India	9 900 000	5	61
France	8 860 000	4	65
Mexico	8 435 000	4	69
Australia	8 100 000	4	73
Poland	6 280 000	3	76
Romania	5 500 000	3	79
United Kingdom	6 390 000	3	82
Brazil	4 840 000	2	84
Italy	4 834 000	2	86
Netherlands	4 080 000	2	88
Soda ash	31 663 638		
United States	9 065 434	29	29
USSR	4 898 000	15	44
China	2 979 000	9	53
Germany	2 400 000	8	61
France	1 297 000	4	65
Japan	1 120 000	4	69
Bulgaria	1 097 000	3	72
India	998 000	3	75
United Kingdom	998 000	3	78
Poland	910 000	3	81
Romania	798 000	3	84
Italy	617 000	2	86
Spain	562 000	2	88
Sodium sulfate	4 732 000		
United States	705 000	15	15
Spain	615 000	13	28
USSR	608 000	13	41
Mexico	454 000	10	51
Germany	345 000	7	58
Canada	317 000	8	66
Japan	268 000	6	72
Belgium	255 000	5	77
Turkey	177 000	4	81
France	155 000	3	84
Italy	131 000	3	87
Austria	120 000	3	90
Sweden	100 000	2	92
United Kingdom	90 000	2	94
Silica sand	124 713 000		
United States	26 489 000	21	21
Netherlands	25 640 000	21	42
Argentina	9 439 000	8	50
France	7 483 000	6	56
Germany	5 986 000	5	61
Peru	4 998 000	4	65
Italy	4 499 000	4	69
Japan	4 377 000	4	73
United Kingdom	4 199 000	4	77
Venezuela	3 809 000	3	80
India	3 102 000	3	83
Canada	2 721 000	2	85
Brazil	2 703 000	2	87
Belgium	2 630 000	2	89
South Africa	2 181 000	2	91
Yugoslavia	2 104 000	2	93
Australia	2 004 000	2	95
Spain	2 000 000	2	97

Table 1. Distribution of World Industrial Minerals Production (tons) (cont.)

	Production	% world production	% world production cumulative
Strontium minerals	244 300		
Mexico	91 000	37	37
Turkey	59 000	24	61
Spain	40 000	16	77
United Kingdom	25 000	10	87
Iran	22 000	9	96
Algeria	5 400	2	98
Sulfur	58 348 000		
United States	11 592 000	20	20
USSR	9 350 000	16	36
Canada	6 625 000	11	47
Poland	5 090 000	9	56
China	4 900 000	8	64
Japan	2 523 000	4	68
Germany	2 175 000	4	72
Mexico	2 367 000	4	76
Saudi Arabia	1 400 000	2	78
Iraq	1 270 000	2	80
Spain	1 235 000	2	82
France	1 051 000	2	84
Talc & steatite	6 497 000		
United States	1 171 871	18	18
China	750 000	12	30
USSR	530 000	8	38
Brazil	425 000	7	45
India	385 000	6	51
Finland	380 000	6	57
France	340 000	5	62
Australia	195 000	3	65
S. Korea	160 000	3	68
Italy	155 000	3	71
Austria	135 000	2	73
Canada	110 000	2	75
N. Korea	110 000	2	77
Norway	100 000	2	79
Titanium minerals*			
Ilmenite	4 260 000		
Australia	1 100 000	26	26
Canada	830 000	19	45
South Africa	640 000	15	60
Norway	450 000	11	71
Malaysia	275 000	6	77
USSR	250 000	6	83
India	200 000	5	88
United States	210 000	5	93
China	90 000	2	95
Brazil	83 000	2	97
Sri Lanka	80 000	2	99
Rutile	520 000		
Australia	260 000	50	50
Sierra Leone	120 000	23	73
South Africa	56 000	11	84
United States	26 000	5	89
India	19 000	4	93
Brazil	15 000	3	96
Sri Lanka	13 000	2	98
USSR	10 000	2	100
Synthetic rutile	580 000		
Australia	250 000	43	43
India	138 000	24	67
United States	100 000	17	84
Japan	46 000	8	92
Malaysia	46 000	8	100
* capacity in TiO_2 content			
Vermiculite	595 705		
United States	304 000	51	51
South Africa	227 791	38	89
Argentina	22 267	4	93
Brazil	17 000	3	96
Japan	17 000	3	99
Zirconium minerals	993 088		
Australia	546 000	55	55
South Africa	160 000	16	71
United States	118 388	12	83
USSR	90 000	9	92
Malaysia	19 700	2	94
Brazil	19 000	2	96
India	16 000	1	97
China	15 000	1	98
Thailand	5 000		
Sri Lanka	4 000		

Source: *Industrial Minerals HandyBook,* 1992.

Some pegmatites are rich in lithium minerals such as spodumene, petalite, and lepidolite which are separated to form a lithium concentrate plus byproduct feldspar, quartz, and/or mica. The main production area is around Kings Mountain, NC, Bikita, Zimbabwe, western Canada, the USSR, and Western Australia. These pegmatitic sources of lithium, which account for over 70% of the supplies, are being strongly challenged by lithium extracted from brines (see following). Beryl is also associated with pegmatites as in Brazil, the USSR, and western Canada. Production of bertrandite in Utah has increased to account for 60% of the world's beryllium supply.

Fluorspar is a ''persistent'' mineral occurring in various ore deposits including Mississippi Valley (lead/zinc) type deposits, hydrothermal veins, stratabound or Manto deposits, contact metamorphic terrains, and alkali rock complexes. Over 70% of world production comes from China, Mexico, Mongolia, and South Africa. Much of the production is exported. Mexico dominates the North American industry since mines in the midwestern United States and eastern Canada have closed. In Europe, significant production comes from Spain, Italy, France, and the United Kingdom whose combined total is 10% of world supplies. Other producers, largely for export, include Thailand, Brazil, Morocco, and Kenya.

SURFICIALLY ALTERED

Sands, Kaolin, Bauxite, Manganese, Vermiculite

Feldspar-rich deposits subject to weathering break down to form feldspathic sand deposits such as those exploited in the western United States and in Spain. Further weathering forms a mixture of feldspar, silica, and kaolin such as that mined in Bavaria. Still further decomposition through weathering and/or hydrothermal activity eliminates much of the mica and silica and yields premier quality deposits of kaolin such as those mined in Cornwall, Georgia and the Carolinas in the southeastern United States, and in the Amazon Basin of Brazil. The United States, plus the United Kingdom, produce almost 50% of the world's kaolin, and more significantly, virtually all the coating-grade material. Other suppliers of quality kaolin include France, Germany, Czechoslovakia, Brazil, Malaysia, and Australia.

Vermiculite is a supergene alteration product formed by the combined effects of weathering and circulating groundwater. Large-scale commercial production is confined to South Carolina and Virginia in the United States and the Palabora Complex in South Africa which contribute 90% of world supplies. Minor quantities come from Japan, Brazil, Argentina, and Kenya.

Since residual bauxite deposits result from the tropical weathering of a variety of source rocks, their distribution is based on climatology rather than lithology. Formation is encouraged by long periods of tectonic stability permitting deep and thorough weathering. Most bauxite deposits are post-Cretaceous in age and many occur in modern tropical regions. Bauxite provinces have been defined as follows:

- Guiana Shield of South America (Venezuela, Guyana, Surinam, Guiana, and parts of Brazil and Colombia)
- Northern Brazilian Shield Province
- Caribbean Shield Province (Costa Rica, Jamaica, Dominican Republic, Haiti, and Puerto Rico)
- Guinea Shield Province (Guinea-Bissau to Togo)
- Cameroon Province (Cameroon, Zaire)
- Australian Province
- European Province (France, Greece, Hungary, and Yugoslavia)
- Others (United States, China, USSR, India, Malaysia).

A dozen countries contribute 95% of world production with 70% plus coming from Australia, Guinea, Jamaica, and Brazil. Nonmetallurgical grades, that is for refractories, abrasives, chemicals, and aluminum cement, come from China, Australia, Guinea, Brazil, Guyana, and Suriname.

Manganese is found in most geological environments; the more important commercially being sedimentary and residual. Large sedimentary marine deposits of manganese are exploited in the Ukraine, the USSR, the Kalahari Basin of South Africa, Groote Eylandt, Australia, and in Mexico. Residual deposits are important in Ghana and Gabon in west Africa and in Amapá, Brazil. Over 90% of world production comes from seven countries—the USSR, South Africa, Gabon, Australia, Brazil, China, and India. Nonmetallurgical grades constitute a relatively small percentage of output from these major suppliers, plus smaller tonnage producers such as Ghana, Morocco, and Greece.

Iron oxides are generally associated with volcanic activity and sulfide deposits combined with subsequent leaching and diagenic alteration. The world's largest supplier with over 50% of production is India, followed by the United States with an additional 15%. US production is concentrated in Georgia, Virginia, and Missouri. European production, accounting for over a fifth of world production, is centered on Spain, United Kingdom, France, Italy, and Austria (the latter being the main source of natural micaceous oxide). Cyprus is noted for its variety of iron oxides including ocher and umber.

Tripoli is a microcrystalline and friable high-silica (98 to 99%) material formed from the weathering of siliceous limestone. The only large-scale commercial producers are in the United States, specifically the southwestern Missouri-northeastern Oklahoma region, southern Illinois, and the Ouachita Mountain region of Arkansas. Deposits in the latter region are associated with novaculite.

Natural zeolites are formed through the reaction of pore water with volcanic glass, clay, feldspar, and a variety of other rocks and minerals. Although zeolites have been recognized in virtually all parts of the world, large-scale commercial production is restricted to the western United States, Cuba, Japan, and several eastern European countries including Bulgaria.

VOLCANIC EXTRUSIVE

Pumice and Perlite

Because of their susceptibility to devitrification and alteration, commercial deposits of extrusive volcanic rocks like perlite and pumice are generally confined to younger geological terrains. Perlite deposits, which are rarely older than Oligocene, are exploited in the western United States, Mexico, Greece, Turkey, Italy, western USSR, Hungary, Czechoslovakia, and Japan. Between them, the United States and the USSR account for over two-thirds of world production. Pumice is common on volcanic islands such as Lipari near Sicily, Yali and Nisisros in Greece, and the Canary Islands of Spain. These three countries control over 70% of the world's pumice supply. In addition, large deposits are worked in the western United States, particularly California, Arizona, New Mexico, and Nevada. Smaller tonnages come from France, Yugoslavia, Chile, and Guadeloupe.

SEDIMENTARY

Silica, Ball Clays, Titanium & Zirconium Minerals, Rare Earths, Diamonds

Sedimentary deposits are formed through the erosion, transportation, and redeposition of minerals that can survive the rigors of

Table 2. Regional Distribution of Industrial Minerals Production (% World Production)

AFRICA	South Africa	Algeria	Angola	Botswana	Gabon	Guinea	Kenya	Madagascar	Morocco	Namibia	Senegal	Sierra Leone	Togo	Tunisia	Zaire	Zimbabwe	Total
Asbestos	3															4	7
Barite									6								6
Bauxite						16											16
Bentonite																	—
Beryl																	—
Boron minerals																	—
Bromine																	—
Chromite	32															5	35
Diamond																	
industrial	10			7											31		50
gem	9		2	25						2					7		45
Diatomite																	—
Feldspar	2																2
Fluorspar	6						2		2								10
Fuller's earth											3						3
Garnet																	—
Graphite								2								2	4
Gypsum																	—
Iodine																	—
Iron oxide																	—
Kaolin																	—
Kyanite etc.	57																57
Lithium																6	6
Magnesite																	—
Manganese	18				14												32
Mica																	—
Nepheline syenite																	—
Nitrogen																	—
Olivine & dunite																	—
Perlite																	—
Phosphate rock	2								16				2	4			24
Potash																	—
Pumice																	—
Pyrophyllite																	—
Rare earths																	—
Salt																	—
Soda ash																	
Sodium sulfate																	—
Silica sand	2	2															4
Strontium minerals		2															2
Sulfur																	—
Talc & steatite																	—
Titanium minerals																	
Ilmenite	15																15
Rutile	11											23					34
Syn. rutile																	—
Vermiculite	38																38
Zirconium minerals	16																16

(cont.)

Table 2. Regional Distribution of Industrial Minerals Production (% World Production) (cont.)

EASTERN EUROPE	Albania	Bulgaria	Czechoslovakia	Hungary	Poland	Romania	USSR	Yugoslavia	Total
Asbestos							60		60
Barite							10		10
Bauxite				3			4	3	10
Bentonite						2	30		32
Beryl							26		26
Boron minerals							7		7
Bromine							15		15
Chromite	6						32		38
Diamond									
industrial							13		13
gem							10		10
Diatomite						14			14
Feldspar						2	8		10
Fluorspar							7		7
Fuller's earth									—
Garnet									—
Graphite			4				13		17
Gypsum									—
Iodine							13		13
Iron oxide									—
Kaolin			3				8		11
Kyanite etc.									—
Lithium							9		9
Magnesite			7				21	2	30
Manganese							31		31
Mica							19		19
Nepheline syenite							77		77
Nitrogen							20		20
Olivine & dunite									—
Perlite			2	6			31		39
Phosphate rock							21		21
Potash							31		31
Pumice								4	4
Pyrophyllite									—
Rare earths									—
Salt					3	3	8		14
Soda ash		3			3	3	15		24
Sodium sulfate							13		13
Silica sand								2	2
Strontium minerals									—
Sulfur					9		16		25
Talc & steatite							8		8
Titanium minerals									
Ilmenite							6		6
Rutile							2		2
Syn. rutile									—
Vermiculite									—
Zirconium minerals							9		9

Table 2. Regional Distribution of Industrial Minerals Production (% World Production) (cont.)

WESTERN EUROPE

	Austria	Denmark	Finland	France	Germany	Greece	Italy	Netherlands	Norway	Spain	Sweden	United Kingdom	Total
Asbestos						2							2
Barite					3								3
Bauxite						2							2
Bentonite					2	12	2						16
Beryl													—
Boron minerals													—
Bromine				4								6	10
Chromite			6										6
Diamond													
industrial													—
gem													—
Diatomite		3		14	2					5			24
Feldspar				5	7		28			3			43
Fluorspar				3			2			3		2	10
Fuller's earth					22								22
Garnet									5				5
Graphite													—
Gypsum				6	2					6			14
Iodine													—
Iron oxide	3			5	2					12			22
Kaolin				5	3							12	20
Kyanite etc.				11									11
Lithium													—
Magnesite	5					6				2			13
Manganese													—
Mica				4									4
Nepheline syenite									7				7
Nitrogen								3					3
Olivine & dunite							3		75	14			92
Perlite						9	4						13
Phosphate rock													—
Potash				4	18					3			25
Pumice				4		15	48			8			75
Pyrophyllite													—
Rare earths													—
Salt				4			2	2				3	11
Soda ash				4	8		2			2		3	19
Sodium sulfate				3	7		3			12	2	2	29
Silica sand				6	5		4	21		2		4	42
Strontium minerals										16		10	26
Sulfur				2	4					2			8
Talc & steatite	2		6	5			3		2				18
Titanium minerals													
Ilmenite									11				11
Rutile													—
Syn. rutile													—
Vermiculite													—
Zirconium minerals													—

(cont.)

Table 2. Regional Distribution of Industrial Minerals Production (% World Production) (cont.)

NORTH AMERICA	Canada	Mexico	United States	Total
Asbestos	16			16
Barite		10	5	15
Bauxite				—
Bentonite		2	32	34
Beryl			60	60
Boron minerals			37	37
Bromine			41	41
Chromite				—
Diamond				
industrial				—
gem				—
Diatomite		3	34	37
Feldspar		2	15	17
Fluorspar		15		15
Fuller's earth			61	61
Garnet			48	48
Graphite		7		7
Gypsum	9	5	16	30
Iodine			9	9
Iron oxide			15	15
Kaolin			35	35
Kyanite etc.			19	19
Lithium	4		47	51
Magnesite/magnesia			10	10
Manganese				—
Mica	5		45	50
Nepheline syenite	16			16
Nitrogen		2	13	15
Olivine & dunite				—
Perlite		2	33	35
Phosphate rock			29	29
Potash	25		5	30
Pumice			4	4
Pyrophyllite			4	4
Rare earths			37	37
Salt	6	4	19	29
Soda ash			29	29
Sodium sulfate	8	10	15	33
Silica sand			21	21
Strontium minerals		37		37
Sulfur	11	4	20	35
Talc & steatite	2		18	20
Titanium minerals				
Ilmenite	19		5	24
Rutile			5	5
Syn. rutile			17	17
Vermiculite			51	51
Zirconium minerals			12	12

Table 2. Regional Distribution of Industrial Minerals Production (% World Production) (cont.)

SOUTH AMERICA & CARIBBEAN	Argentina	Brazil	Chile	Colombia	Guadeloupe	Jamaica	Peru	Suriname	Venezuela	Total
Asbestos		5								5
Barite										—
Bauxite		8				10		3		21
Bentonite		2								2
Beryl		12								12
Boron minerals	7									7
Bromine										—
Chromite		2								2
Diamond										
industrial										—
gem										—
Diatomite										—
Feldspar		3								3
Fluorspar										—
Fuller's earth										—
Garnet										—
Graphite		5								—
Gypsum										—
Iodine			26							26
Iron oxide			2	3						5
Kaolin			3	7						10
Kyanite etc.										—
Lithium		2	14							16
Magnesite		3								3
Manganese		7								7
Mica										—
Nepheline syenite										—
Nitrogen										—
Olivine & dunite										—
Perlite										—
Phosphate rock		4								4
Potash										—
Pumice			3		2					5
Pyrophyllite		3								3
Rare earths		3								3
Salt		2								2
Soda ash										—
Sodium sulfate										—
Silica sand	8	2					4		3	17
Strontium minerals										—
Sulfur										—
Talc & steatite		7								7
Titanium minerals										
Ilmenite		2								2
Rutile		3								3
Syn. rutile										—
Vermiculite	4	3								7
Zirconium minerals		2								2

(cont.)

Table 2. Regional Distribution of Industrial Minerals Production (% World Production) (cont.)

AUSTRALASIA

	Australia	China	India	Indonesia	Japan	North Korea	South Korea	Malaysia	Mongolia	Sri Lanka	Thailand	Total
Asbestos		4										4
Barite		31	7									38
Bauxite	37	4										41
Bentonite												—
Beryl												—
Boron minerals												—
Bromine					3							3
Chromite												—
Diamond												
industrial	34											34
gem	41											41
Diatomite							4					4
Feldspar							4				4	8
Fluorspar		30							14		2	36
Fuller's earth												—
Garnet	11	14	20									45
Graphite		32	9			4	16					71
Gypsum	2	8			6						6	22
Iodine		3			49							51
Iron oxide			51									51
Kaolin			3				5					8
Kyanite etc.			11									11
Lithium	11	6										17
Magnesite		10	2		4	12						28
Manganese	12	6	6									24
Mica			10				11					21
Nepheline syenite												—
Nitrogen		17	7	3								27
Olivine & dunite					5							5
Perlite					4							4
Phosphate rock		8										8
Potash												—
Pumice												—
Pyrophyllite					56	3	29					88
Rare earths	16	29	3					6				54
Salt	4	15	5									24
Soda ash		9	3		4							16
Sodium sulfate					6							6
Silica sand	2		3		4							9
Strontium minerals												—
Sulfur		8			4							12
Talc & steatite	3	12	6			2	3					26
Titanium minerals												
Ilmenite	26	2	5					6		2		41
Rutile	50		4							2		56
Syn. rutile	43		24		8			8				83
Vermiculite					3							3
Zirconium minerals	55		1					2				59

Table 2. Regional Distribution of Industrial Minerals Production (% World Production) (cont.)

MIDDLE EAST	Cyprus	Iran	Iraq	Israel	Jordan	Saudi Arabia	Turkey	Total
Asbestos								—
Barite							8	8
Bauxite								—
Bentonite								—
Beryl								—
Boron minerals							47	47
Bromine				28				28
Chromite							5	5
Diamond								
industrial								—
gem								—
Diatomite								—
Feldspar								—
Fluorspar								—
Fuller's earth								—
Garnet								—
Graphite							2	2
Gypsum								—
Iodine								—
Iron oxide	3							3
Kaolin								—
Kyanite etc.								—
Lithium								—
Magnesite							3	3
Manganese								—
Mica								—
Nepheline syenite								—
Nitrogen								—
Olivine & dunite								—
Perlite							8	8
Phosphate rock				3	4			7
Potash				4	3			7
Pumice							6	6
Pyrophyllite								—
Rare earths								—
Salt								—
Soda ash								—
Sodium sulfate							4	4
Silica sand								—
Strontium minerals		9					24	33
Sulfur			2					2
Talc & steatite								—
Titanium minerals								
Ilmenite								—
Rutile								—
Syn. rutile								—
Vermiculite								—
Zirconium minerals								—

Source: Industrial Minerals HandyBook, 1992.

Table 3. Geological Classification of Industrial Minerals (Adapted From Harben and Bates, 1984)

Primary commercial environment	Alternative sources
IGNEOUS	
Intrusive	
Olivine	Sedimentary (placer)
Chromite	Sedimentary (placer)
Nepheline syenite	
Granite	
Pegmatitic & hydrothermal	
Feldspar	Alteration/sedimentary (sand)
Mica	
Quartz crystal	Synthetic
Lithium minerals	Evaporate
Beryllium minerals	
Fluorspar	
Extrusive	
Basalt & related rocks	
Pumice, pumicite, & scoria	
Perlite	
SEDIMENTARY	
Clastic	
Sand & gravel	
Sandstone	
Clays	
Titanium & zirconium minerals	Intrusive
Rare-earth minerals	Intrusive
Diamonds	Intrusive/synthetic
Biogenetic	
Limestone & dolomite	
Diatomite	
Phosphate	
Sulfur	Hydrothermal/byproduct
Chemical	
Barite	Alteration/sedimentary
Salt	
Sodium carbonate	Synthetic
Sodium sulfate	Byproduct
Nacholite & dawsonite	
Gypsum	Byproduct
Potassium minerals	
Borates	
Celestite	
Nitrates	Synthetic
Bromine	
Iodine	
SURFICIALLY ALTERED	
Vermiculite	
Manganese minerals	Sedimentary
Bauxite	
Iron oxide	
Tripoli & novaculite	
Zeolites	
METAMORPHIC	
Marble	
Slate	
Asbestos	
Magnesite & magnesia	Igneous/sedimentary/synthetic
Graphite	Synthetic
Corundum & emery	Synthetic
Garnet	Sedimentary (placer)
Wollastonite	Synthetic
Sillimanite minerals	
Pyrophyllite	

transportation. The most common is silica which forms a number of materials including silica sand, sand and gravel, and flint. The precursor is igneous quartz (for example, in granite), and then the sedimentary deposit may undergo metamorphism and recementing to produce quartzite. Sand and gravel for construction use is extremely common, and production is more dependent on local markets than availability. However, certain areas are noted for the production of industrial sand sufficiently pure to be used in the manufacture of glass, ceramics, sodium silicate and the like. Examples include the midwest region of the United States, Badgeley Island, Ontario, Canada, Cheshire in northwest England, the Loch Aline area of Scotland, areas of Belgium and the Netherlands, Cape Flattery Island, Queensland, Australia, and Sarawak, Malaysia. In many cases, the use of local sand is based on price rather than quality. The United States and the Netherlands are the largest producers of industrial sand, each accounting for over one-fifth of world production. Production of flint is much more restricted, based largely on the chalk deposits of southern England and northern France.

Several clays composed mainly of kaolinite are of sedimentary origin. Premier deposits of ball clay, the carbon content of which indicates that it was deposited in swampy conditions, occur in the Kentucky-Tennessee area of the United States, Devon in southwest England, and in Czechoslovakia. Flint clay, as produced commercially in the United States, China, Israel, Australia, and Argentina, is generally derived from the weathering of soil and deposition in shallow basins. Fire clay or refractory kaolin is a kaolinite material common in many parts of the world, particularly in association with coal deposits. A 400-km belt of kaolinite-rich rocks extends from Aiken, South Carolina, to Eufaula, Alabama, and includes areas supplying high- and medium-quality kaolin and refractory kaolin. Another belt of kaolin, bauxite, and bauxitic and kaolinitic clays extends from western Tennessee into northeastern Mississippi. Other areas include southwest England and over the Channel into France (kaolin and ball clay), various parts of Czechoslovakia (kaolin and ball clay), Spain, the Amazon Basin in Brazil (bauxite, kaolin), Japan (kaolin, refractory clay, roseki, and toseki), and eastern Australia (bauxite, kaolin).

Volcanic ash deposited as part of a sedimentary sequence eventually forms sodium or calcium bentonite. Important bentonite deposits occur in the United States in the Wyoming and Montana region (sodium-based bentonite) and in the Mississippi-Texas region (calcium-based). Over 30% of the world's bentonite production is from these and some smaller deposits in the United States. More modest tonnages are produced over the borders in Mexico and Canada. In Europe, bentonite is mined on Milos Island, Greece, Cyprus, Turkey, Sardinia, Bavaria, southern England, and Spain. In Asia production is centered on Japan, India, and China. Attapulgite and sepiolite (fuller's earth) are more restricted, being produced in Georgia and Florida in the United States, Germany, the United Kingdom, Senegal, and Spain.

Placer and palaeo-placer mineral deposits are important sources of heavy minerals such as ilmenite, rutile, and zircon. Monazite and xenotime, rare earth sources, are invariably associated with the mineral sands deposits. Many titanium/zirconium/rare earth mineral deposits are Tertiary and Quaternary in age since this was a period of geological uplift which provided the correct conditions for accumulation plus the fact that older examples have been destroyed. Important areas include the east and west coasts of Australia, parts of Florida and Georgia in southeast United States, around Richards Bay, South Africa, and Sierra Leone in Africa, the coastal areas of Tamil Nadu and Kerala states in southern India extending into eastern Sri Lanka, and the coastal areas of Brazil. Consequently, supplies are dominated by Australia, South Africa, United States, Sierra Leone, India, and Brazil. Hard-rock ilmenite deposits are exploited in Quebec, Canada, and in Norway. The sole commercial source of baddeleyite (ZrO_2) is as a byproduct of phosphate and copper production at Palabora, Transvaal, South Africa. Except for the United States, most of the production is exported.

Major diamondiferous beach placers extend along the southwest coast of Africa and are exploited in South Africa and Namibia.

BIOGENIC

Limestone/Dolomite, Diatomite, Phosphate Rock, Sulfur

Limestone is an extremely common rock formed as shell beds on a shallow sea floor. Purity depends on the environment of deposition and the subsequent mineralogical and tectonic history which may include metamorphism to marble. Limestone is exploited for uses ranging from construction aggregates and railroad ballast through cement and lime manufacturing, glassmaking to functional fillers in paper, plastics, and paint. The relatively modest price even for the high-calcium and high-brightness grades (less than $200/ton) means that consumption is generally close to the point of production, i.e., a local or regional market. In the United States, for example, crushed limestone is produced in all states except for Louisiana (which does produce shell), includes over 2 500 quarries, and accounts for some 66% of the nation's crushed stone output. High-quality filler-grade calcium carbonate produced in Vermont, Massachusetts, and Maryland serves the northeast, Georgia and Alabama the southeast, Illinois the midwest, Texas the south and south-central region, and California the west. In Canada production is in Ontario and British Columbia. In western Europe chalk is important in the United Kingdom, France, and Belgium, whereas crushed marble is often used in Italy and Greece.

Dolomite has many of the uses outlined for limestone, plus several others including refractories, seawater magnesia and magnesium metal production, and as a dimension stone. Although less common than limestone, dolomite production, particularly for aggregates, is extremely widespread. In the United States some 136 dolomite quarries operating in 25 states contribute 4% of the country's crushed rock output. Nonaggregate production is concentrated in California, Ohio, Michigan, Alabama, Texas, Connecticut, and Pennsylvania. Europe has an active dolomite industry where it is used extensively as a raw material for refractories and seawater magnesia production. The main producers are Spain, the United Kingdom, Belgium, France, Germany, Norway, Sweden, and Finland.

Diatomite deposits are formed through the accumulation of the frustules of diatoms, small animals that thrived after the Cretaceous period. Diatoms require marine or freshwater rich in nutrients like phosphates, nitrates, and silica, but relatively free of sediment. In many cases the diatomite is associated with volcanic activity which may be the source of silica. Important diatomite production sites include the western United States—especially California, Nevada, Washington, and Oregon. More than three-quarters of the world's production comes from the United States, France, Romania, and the USSR, with lesser amounts from Denmark (mainly moler, an impure diatomite product), Spain, Germany, Italy, Iceland, and Korea.

The bulk of commercial phosphorus-based compounds is derived from marine sedimentary phosphate rock deposits with much of the rest derived from igneous deposits. Sedimentary phosphate deposits are concentrated in two main belts, the trade-wind belt aligned north-south from the Equator to 50° and the Equatorial belt oriented west to east in low latitudes. Examples include:

- Trade-wind belt
  - Southeast and northwest United States
    - Florida, North Carolina, Idaho
  - Baja California, Mexico
  - Sechura Desert, Peru
  - Caribbean Sea
- Equatorial belt
  - North Africa
    - Morocco, Tunisia, Algeria, Togo
  - Middle East
    - Jordan, Israel, Syria, Senegal, Iran
  - USSR
  - Venezuela
  - Colombia

Some two-thirds of the world's phosphate rock production comes from the US, the USSR, and Morocco, the latter being the leading exporter. There are a series of middle-range producers including China, Jordan, Tunisia, Brazil, Israel, Togo, and South Africa. The production in the USSR, Brazil, and South Africa is largely derived from igneous deposits which may also yield rare earths, vermiculite, and/or copper.

EVAPORITE MINERALS

Barite, Salt, Sodium Carbonate, Sodium Sulfate, Gypsum, Potassium Minerals, Borates, Celestite, Nitrates, Bromine, Iodine, Lithium, Magnesia

Certain minerals are found dissolved in seawater and various other brines; evaporite deposits form when the concentration reaches saturation and precipitation occurs. This process requires favorable conditions such as a barred basin or broad shelf environment, plus a hot dry climate that encourages evaporation. Once formed, the fragile deposit needs to be preserved through subsequent geological events such as burial. Halite, gypsum, and anhydrite, often interbedded with limestone and dolomite, are the most common minerals present in marine evaporites which often extend over hundreds of square kilometers and attain a thickness in excess of several thousand meters. In rarer cases other evaporite minerals are present such as potassium minerals, borates, and strontium minerals.

Common salt lives up to its name by being ubiquitous in most brines and many evaporate deposits. It is exploited commercially from seawater where the modern climate allows evaporation, for example in Western Australia, Mexico, Bahamas, Netherlands Antilles, Sri Lanka, India, Brazil, Italy, and Spain. Large buried rock salt and salt dome deposits are important in the northeastern, southern, and the midwestern areas of North America, central and northern Europe, parts of the USSR, and the Middle East. Saline lakes are worked in the western United States, for example the Great Salt Lake in Utah (which yields or has yielded sodium sulfate, potassium minerals, magnesium chloride, bromine, and lithium as well as salt) and Searles Lake in California (sodium sulfate and borates). Despite its widespread production, three-quarters of the world production is accounted for by just ten producers.

In some cases, potassium minerals are found associated with salt. During the Permian period in Europe, for example, the Zechstein Basin extended from northern Britain through the Netherlands, Denmark, Germany, to Poland. This was a shallow stable sea that allowed thick evaporite sequences to accumulate resulting in the large potash and salt deposits exploited today. To the east in the USSR, three basins contain vast reserves of potash—Stebnik/Kalush in the Ukraine near the border with Poland, Soligorsk near Minsk in Byelorussia, and Solikamsk/Berezniki west of the Urals. In western Canada during the Silurian to mid-Devonian period, the land subsided to form a large basin stretching southward into the United States. This allowed vast thicknesses of evaporites to be formed including the potash resources of Saskatchewan, Canada. A large potash/salt rich basin identified in Thailand has potential for commercial production. Despite its strategic importance as a fertilizer, large-scale potash production is fairly restricted with just eight countries—the USSR, Canada, Germany, the United States, Israel, France, Jordan, and Spain—accounting for 97% of world production.

In addition to salt, other sodium-rich minerals are concentrated in modern brines or evaporate deposits. Vast deposits of natural sodium carbonate or trona around Green River, Wyoming, yield over 10 Mtpy of refined soda ash and form a soda feedstock for a variety of chemical products. Smaller deposits are known in China and Turkey. Sodium carbonate is also extracted from saline lakes in the western United States, Mexico, China, the USSR, and in Africa. Lake Magadi, Kenya, which has been producing sodium carbonate for most of this century, is one of numerous sodium-rich lakes in the 6 030-km rift valley stretching from Turkey through Arabia and East Africa to Tanzania. This valley also includes the Dead Sea (dividing Israel and Jordan) which currently produces salt and potash and has the potential to produce sodium carbonate, magnesium oxide, and bromine. Natural soda ash production constitutes almost 30% of total world production (virtually all from the United States) with the balance produced in Solvay plants using a salt and limestone feedstock.

Sodium sulfate is a common coproduct in brine-based operations. In addition, natural sodium sulfate is exploited on a large scale from lakes in Saskatchewan and Alberta in western Canada, California and Texas in the US, Mexico, the USSR, and Turkey. Buried deposits are mined in Spain. Sodium sulfate is also a byproduct of several industrial processes including rayon manufacture. Although production is broadbased, over 50% is in the US, Spain, the USSR, and Mexico.

There are three important commercial sulfate minerals—gypsum (calcium), celestite (strontium), and barite (barium). In North America, large deposits of gypsum occur in the northeast (New York in the United States and Ontario and the Maritime Provinces in Canada), midwest (Michigan, Iowa, and Indiana), the southwest and west (Oklahoma, Texas, Kansas, New Mexico, and California). In the US Gulf Coast region accumulations of salt, gypsum, and sulfur stretch into Mexico. Overall, the three countries in North America account for 30% of world gypsum production. In Australasia, China, Japan, Thailand, and Australia contribute over 22% of world production, and in Europe the large evaporate deposits outlined previously also contain gypsum with the main suppliers being France, Spain, Germany, and the United Kingdom. Several countries such as Germany produce gypsum as a byproduct from FGD or phosphoric acid plants competing with the natural product in the marketplace. Despite its low unit value, gypsum is often exported on a large scale from coastal locations in Mexico, Canada, and Spain to the United States.

Some 98% of the world's celestite is mined in just six countries. In northern Mexico the output from several mines accounts for more than one-third of world production; Turkey contributes a further quarter, with Cyprus, Iran, and Pakistan being smaller producers in the Middle East-Asian region. Production in Europe, particularly southwest Spain (associated with barite and gypsum), southwest England (correlated with gypsum), and smaller deposits in Italy, contribute another quarter.

As outlined above, barite may be associated with evaporate minerals such as celestite and gypsum. In addition, barite is also found as a hydrothermal vein filling associated with stratiform massive sulfide deposits and as a residual deposit. China has emerged as the world leader in barite production, accounting for over 30% of the world total. In contrast, barite production in the United States has declined to 5% of world production because of the availability of imports from China, Mexico (10% of world production), and even Morocco (6%). Production is largely centered in Nevada and to a lesser extent Missouri and Georgia. In Europe the main suppliers are Germany, France, Italy, the United Kingdom, Eire, Romania, and the USSR.

About 85% of the world's production of borates is controlled by the United States and Turkey, with Argentina and the USSR supplying most of the balance. Part of the production is based on brines and encrustations in Searles Lake, California, as well from small concentrations along a stretch of the Andes Mountains encompassing Argentina, Bolivia, Chile, and Peru. The major source of borate minerals is a buried mass of sodium borate at Boron in the Mojave Desert of California. Other buried and surface borate deposits occur in Death Valley, including the Billie colemanite deposit which was mined until the mid-1980s. In Turkey borates are concentrated in six areas—the most important being the Emet and Kirka areas (both sodium borate) and the Bigadiç area (calcium borates).

In addition to hard-rock deposits, lithium is found concentrated in brines. Large-scale production occurs at Silver Peak, Nevada, and there is potential from several of the western United States saline lakes already described. This output combined with the pegmatitic deposits described above contributes 47% of world production. Commercial production is being developed in the salt-encrusted playas or "salars" in South America, in particular Bolivia, Argentina, and Chile. This last country is the most important with 14% of world production.

In addition to magnesite, dolomite, and olivine, magnesia is concentrated in subterranean and seawater brines. Magnesium-rich brines are exploited on a commercial scale in Michigan in the United States, Mexico, the Netherlands, and Israel. Magnesia is also extracted from seawater in Japan, South Korea, the United States, Mexico, the United Kingdom, France, Italy, Eire, Norway, and the USSR.

Natural nitrate production is now confined to an area of northern Chile which has an annual rainfall of less than 1 cm. In addition, iodine is coproduced from the caliche ore. Elsewhere, commercial iodine production is based on brines largely in Japan, the United States (Oklahoma), China, and the USSR. Bromine is also extracted from well brines in Arkansas and Michigan in the United States, the Dead Sea in Israel, potash brines in Germany and France, and from seawater in the United Kingdom, France, Spain, and Japan.

METAMORPHIC

Asbestos, Talc, Graphite, Corundum and Emery, Garnet, Wollastonite, Sillimanite Minerals, Pyrophyllite

Metamorphism produces a range of minerals, many of which have unique properties that are utilized commercially. Asbestos is found in several metamorphic environments usually associated with ultramafic rocks and serpentinization. Large deposits are exploited in Quebec and Newfoundland, eastern Canada, and over the border in Vermont, United States, in the Transvaal and Cape Province of South Africa, as well as Swaziland and Zimbabwe in Africa, the USSR, Italy and Greece in Europe, New South Wales in Australia, Brazil, and India.

The United States and Canada account for around 20% of world talc production with output from the asbestos areas of Vermont, upstate New York, Montana, Texas, and California in the United States, and Quebec and Ontario in Canada. In Australasia, China, India, North and South Korea, Japan, and Australia are the major suppliers, with a combined output representing some 26% of world production. In Europe significant talc producers include France, Italy, Austria, Finland, and Norway, along with the USSR. In South

America, Brazil is the largest producer and the fourth largest producer worldwide after the US, China, and the USSR.

Wollastonite is formed through the metamorphism of rocks containing silica and calcium. Major producing areas of high-quality wollastonite include the Adirondack Mountains of upstate New York in the United States, southeastern Finland, several provinces in China, India, and the USSR. Garnet is associated with some wollastonite deposits including one at Willsboro, New York. In the same area of the state a large-scale hard-rock garnet mine is in operation. Elsewhere in the United States, although there are hard-rock garnet deposits in Maine and Nevada, the most important commercially are placer deposits in Idaho. Garnet is also produced as a byproduct of mineral sand operations in Western Australia, India, and Sri Lanka.

The kyanite group of minerals occurs in aluminous metamorphic rocks and their weathered derivatives. Production is restricted to a handful of countries including South Africa, the USSR, the United States, France, India, Sweden, Spain, China, and Zimbabwe. Like garnet, some are found associated with placer mineral sand deposits—particularly in India. Weathering may form sericite or pyrophyllite, the hydrous aluminum silicate. The main pyrophyllite deposits, however, are formed through the hydrothermal alteration of acidic volcanic rocks. This is particularly well developed in areas of Japan and Korea—between them accounting for 85% of world production. Smaller producers include Canada, the United States, India, China, Thailand, Australia, Brazil, and Argentina. Natural corundum is another alumina-rich mineral formed through metamorphism. The main producers are Zimbabwe and South Africa, the USSR, and India. Production of the impure form, emery, is restricted to Turkey and Greece.

When certain organic matter is metamorphosed, deposits of graphite may form. However, world production is concentrated in fewer than 20 countries, with over 60% in Asia—China, Korea, Sri Lanka, and India. In the Americas, Mexico and Brazil are well established producers, and Canada is emerging as a major supplier. In Europe, the main producers are Germany, Austria, Czechoslovakia, Norway, Romania, Turkey, and the USSR, and in Africa, Zimbabwe and Madagascar.

THE EMERGING PATTERN

The uneven distribution of industrial rocks and minerals production counters the concept of a group of common, low priced commodities destined for local markets. Granted local markets are the most important for some industrial minerals, especially for developing countries which should adopt simpler approaches to exploiting their domestic resources. Nevertheless, the overall view of industrial minerals is an international one of intriguing complexity.

For instance, production of borates, beryl, iodine, nepheline syenite, celestite, and vermiculite is restricted to just a handful of countries, and essential materials such as titanium, zirconium, and rare earth minerals, sulfur, graphite, phosphates, and potash are extremely active in deep sea trade. Even relatively common materials enter international trade as consumers demand higher and more consistent quality (feldspar and silica sand) or find it more cost effective to import (soda ash, salt, barite, gypsum, and fluorspar). The international supply and demand pattern is dynamic as new producers and markets come and go. A decade ago Australia's diamond industry was a kimberlite pipe dream, today it accounts for over one-third of world production. Twenty years ago celestite was used solely for pyrotechnics, today the much larger market is dominated by color TV screen glass. The pattern can be influenced by political events such as the opening up of trade with China, which promptly became a dominant force in world markets for magnesite, talc, barite, bauxite, rare earths, and graphite. This shifting pattern of supply and demand will continue to offer future opportunities and challenges to the increasingly sophisticated industrial minerals industry. The streets of this century and the next are paved with industrial minerals not gold. There will be more of them, and the traffic promises to be heavy.

BIBLIOGRAPHY AND REFERENCES

Anon., 1990, US Bureau of Mines, *Minerals Yearbook Preprints*, various chapters.

Brobst, D.A., and Pratt, W.P., eds., 1973, *United States Mineral Resources*, US Geological Survey Professional Paper 820, 722 pp (various chapters).

Flawn, P.T., 1966, *Mineral Resources*, Rand McNally, Chicago, IL, 406 pp.

Govett, G.J.S., and Govett, M.H., eds., 1976, *World Mineral Supplies: Assessment and Perspective*, Elsevier Scientific Publishing Company, Amsterdam, 472 pp.

Harben, P.W., and Bates, R.L., 1984, *Geology of the Nonmetallics*, Metal Bulletin Inc., New York, 392 pp.

Harben, P.W., and Bates, R.L., 1990, *Industrial Minerals Geology and World Deposits*, Industrial Minerals Division, Metal Bulletin Plc, London, 312 pp.

Harben, P.W., 1992, *Industrial Minerals HandyBook—A Guide to Markets, Specifications, and Prices*, Industrial Minerals Division, Metal Bulletin Plc, London, 160 pp.

Stanton, R.L., 1972, *Ore Petrology*, McGraw-Hill, New York, 713 pp.

INTERNATIONAL TRADE IN INDUSTRIAL MINERALS

HENRY N. McCARL

In the period 1982–1992, world trade has increased in real terms about 65% (Anon., 1992b). The nominal value of this trade in the early 1990s is approximately $5 trillion. In 1970, mineral commodities constituted approximately 20% of the value of world trade. By 1980, despite rising petroleum prices, mineral commodities made up about 18% of the total value of world trade, with crude oil amounting to 16% and other minerals the remaining 2%. By 1987, the major worldwide economic slowdown and falling oil prices meant that minerals constituted only about 10% of the total value of world trade, or about $500 billion in 1987 (Kimbell and Zajac, 1988). The importance of minerals in world trade since 1987 has remained at about these levels. Approximately 40% of world mineral commodity production is traded internationally. Some economists predict that minerals and other commodities are likely to continue to decline in relative importance in terms of the value of world economic output, as well as world trade, into the foreseeable future (Thurow, 1992). This projected decline in commodity demand may also be related to much slower global economic growth over the next decade (Alperovitz, 1992).

Industrial minerals and rocks constituted roughly 2.5% of total mineral commodity exports in the late 1980s, but their relative importance as a portion of mineral commodity exports nearly doubled in the decade 1977–1987 (Kimbell and Zajac, 1988). Industrial minerals constitute the most rapidly growing share of total mineral commodity exports. In addition, price declines among industrial minerals were uncommon in the decade 1977–1987 (Kimbell and Zajac, 1988). While prices of sulfur and most of the fertilizer minerals have seen some erosion and slow growth in markets in the period 1987–1992, other industrial minerals have increased in both value and share of the export market (USBM, 1992). The place of industrial minerals in world trade becomes even more significant when one looks at shipments of chemical fertilizers and basic chemicals that are counted as manufactured products rather than industrial mineral commodities.

The United States is the major world producer of industrial mineral commodities and products made from these nonmetallic minerals and rocks. The former Soviet Union, now The Commonwealth of Independent States (CIS), and China have very significant deposits of industrial minerals. Japan exports significant quantities of cement. The US, CIS, and Morocco export large amounts of phosphate rock. Canada and Germany export substantial amounts of potash. Sulfur from Poland, industrial diamonds from South Africa, Zaire, and Brazil, and fluorspar from Mexico are all important exports.

MAJOR INDUSTRIAL MINERALS IN INTERNATIONAL TRADE

The following review of industrial mineral commodities is intended to focus on some of the most significant nonmetallic minerals and rocks entering international trade.

Cement

Cement is produced largely for domestic consumption, but significant quantities do cross international borders to serve regional markets. Some countries do not have sufficient cement production capacity to serve their domestic needs and must import substantial quantities from countries with excess capacity. World cement production is normally in excess of 1 Gty, and the heavily populated countries of China, India, Japan, the United States, and the European Community are the major producers and exporters of cement (Table 1). Between 10 and 20% of world cement production enters international trade. In 1990, this amount would have been worth over $9.5 billion. Japan and the European Community were the largest exporters with significant amounts shipped to Africa and South America. Many smaller countries in Asia, South America, and Africa, with limited reserves of limestone, are net importers of cement (McCarl and Waters, 1985).

The United States imports between 10 and 20% of its cement needs primarily from Canada, Mexico, the European Community (EC), and the Bahamas. Other countries that have been historically important in supplying US cement needs include Venezuela, Japan, Greece, and Spain (the latter two are members of the EC). Major anti-dumping actions were taken against Mexico, Japan, and Venezuela in the late 1980s and early 1990s, resulting in reductions in US imports due to the imposition of duties on shipments from these nations (*Mining Engineering,* 1992). Cement production demonstrates significant economies of scale. Marginal cost pricing by cement exporters can yield prices that are significantly below their average costs of production. This commonly leads to charges of dumping (when export price is actually lower than average sales price in the country of origin).

Phosphate Rock

In the past 30 years, world consumption of chemical fertilizers has more than tripled (USBM, historical data). With between 80 and 90% of phosphate rock production consumed in chemical fertilizers (McCarl and Waters, 1985; *Mining Engineering,* 1992), the consumption of phosphate rock has paralleled this growth. Roughly 50% of phosphate rock production enters international commerce, and to this can be added significant amounts of phosphate rock that has been chemically converted into internationally traded fertilizer materials. World production of phosphate rock is shown in Table 2.

The United States is the world's largest exporter of phosphate rock, shipping large quantities to Canada, the European Community, Asia, and South America (Fig. 1). Morocco may have the largest reserves of phosphate raw materials and, with Tunisia, supplies major imports to the European Community. The CIS contributes over 20% of world production and is a major exporter to Europe and Asia. Important reserves are found in Israel/Jordan, Tunisia/Algeria, Senegal, and Togo. These countries also supply phosphate to European, Asian, and South American markets. Various Pacific islands have been historically important sources of phosphate from guano, but these reserves are rapidly depleting.

Table 1. World Production of Hydraulic Cement

Country	Proportion of World Output (circa 1990)
China	18 %
European Community	14
Commonwealth of Independent States (formerly USSR)	12
Japan	7
United States	6.3
India	3.7
Others (Eastern Europe, Korea, Brazil, Mexico, Taiwan, Canada, Australia, etc.)	39
	100 %
Approximate World Production	1.14 Gt
Approximate value at cement mills in United States	$55 per ton

[e] estimated
Source: US Bureau of Mines *Mineral Commodity Summaries* and *World Economy Minerals Yearbook.*

Table 2. World Production of Phosphate Rock[e]

Country	Proportion of World Output (circa 1990)
United States	29 %
Commonwealth of Independent States (formerly USSR)	23
Morocco	13.5
China	11
Tunisia	4.2
Jordan	3.8
Others (Brazil, Israel, Togo, Senegal, South Africa, etc.)	15.5
	100 %
Approximate World Production	143 Mt
Approximate value at US mines	$23 to $24 per ton

[e] estimated
Source: US Bureau of Mines *Mineral Commodity Summaries* and *World Economy Minerals Yearbook.*

Important reserves are known in China, South Africa, Russia (CIS), and Vietnam, and are likely to be important in future world consumption. Most of China's current production is consumed domestically (McCarl and Waters, 1985).

The breakup of the USSR and the changes in Eastern Europe could dramatically alter the picture in international phosphate trade as Russia attempts to increase international trade in an effort to earn hard currencies. The depressed state of the global economy has contributed to a decline in world phosphate production in the period 1988–1992 (*Mining Engineering,* 1992).

Nitrogen Compounds

As with phosphate rock, the major markets for nitrogen compounds such as ammonia are as components of chemical fertilizers. Most nitrogen is derived from the fractional distillation of liquid air. Significant nitrogen production capacity is located in areas of abundant natural gas resources used to fuel the cryogenic production process. The CIS, various states of the Middle East, and densely populated developing nations in Asia, have all added important ammonia production capacity in recent years (*Mining Engineering,* 1992). Somewhere between 30 and 50% of nitrogen compound production enters world trade as nitrogen or as components of chemical fertilizers. Nations with production capacity in excess of domestic needs are normally regional international exporters of either nitrogen compounds or chemical fertilizers.

World production of nitrogen in ammonia in 1990 is shown in Table 3. In 1991, 118 Mt of ammonia were produced by 66 countries. With nitrogen comprising just over 82% of ammonia by weight, this was the equivalent of about 97 Mt of nitrogen. Roughly 85% of global ammonia production was consumed in chemical fertilizers; the remaining 15% is used in industrial chemicals and as nitrogen gas. Economically developed nations in North America, Western Europe, and Oceania (primarily Japan) account for about 25% of world ammonia capacity. Developing nations in Asia, South America, and Africa, and the countries of Eastern Europe and the CIS control 75% of world ammonia production capacity (*Mining Engineering,* 1992).

Sulfur

While only about 10% of sulfur production is shipped to international markets as elemental sulfur (*Mining Engineering,* 1992), most is recovered as or converted to sulfuric acid, an important commodity in international trade. Consumers commonly purchase sulfuric acid rather than taking crude sulfur and converting it to sulfuric acid. Sulfuric acid is not only one of the most important basic industrial chemicals, but the largest use for sulfuric acid is in the chemical process of converting phosphate rock into components of chemical fertilizers. Significant amounts of sulfur production therefore enter international trade as sulfuric acid, industrial chemicals, and chemical fertilizer. Sulfur is an important byproduct in the production of (sour) natural gas, and sulfuric acid is commonly reclaimed as a byproduct in the smelting of sulfide ores. Elemental sulfur comprises over two-thirds of western-world sulfur output (*Mining Engineering,* 1992). This does not include the important output of elemental sulfur from Poland (Table 4), nor does it consider the significant byproduct sulfuric acid production of the CIS. Table 4 indicates the estimated world production of elemental sulfur (including output from sour natural gas and other sources). It is difficult to say just how much of this output eventually enters world trade in one form or another. Sulfur exports have been historically important from Canada and Poland.

Salt (Sodium Chloride)

Salt is an abundant raw material found in seawater, natural brines, and in beds of rock salt that precipitated from seawater in the geologic past. Salt domes are derived from the flow of rock salt into dome-shaped structures that intrude bedded deposits and create an environment associated with the accumulation of gypsum, salt, oil, and natural gas. World salt production in 1990 is detailed in Table 5. Salt production is commonly associated with the production of industrial chemicals and muriatic (hydrochloric) acid. Virtually all sodium and chlorine compounds are derived from common salt. The amount of salt consumed as sodium chloride is relatively small compared to the amount used to produce chemicals such as sodium hydroxide, chlorine, and chlorides (e.g., chlorinated hydrocarbons). Most countries utilize their domestic salt output in the production of chemicals derived from sodium chloride. As in the case of sulfur and nitrogen, most salt enters international trade as a chemical derivative.

Potash

Potash is a generic term that covers several potassium compounds found in nature. These include muriate of potash (potassium

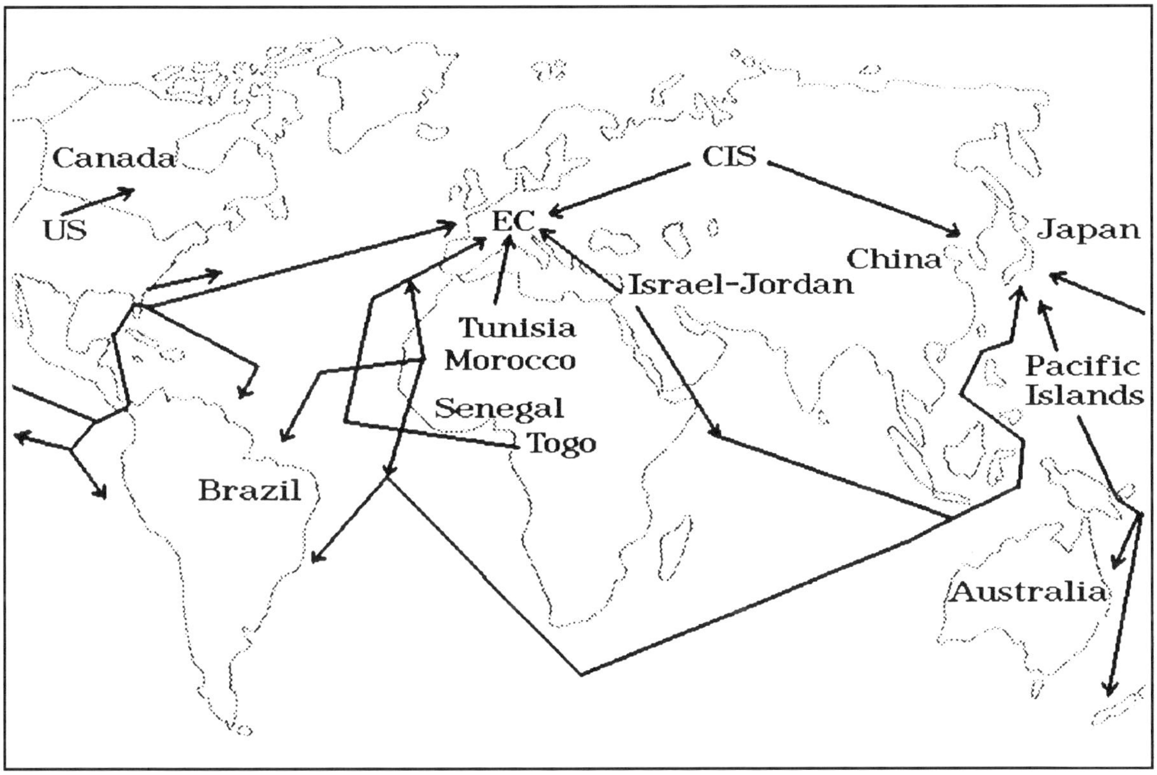

Fig. 1. Trade patterns in phosphate rock.

Table 3. World Production of Nitrogen in Ammonia[e]

Country	Proportion of World Output (circa 1990)
Commonwealth of Independent States (formerly USSR)	21 %
China	16.4
United States	12.8
European Community	11
India	7
Eastern Europe	7
Canada	3
Others (Indonesia, Mexico, Japan, Trinidad & Tobago, Pakistan, etc.)	21.8
	100 %
Approximate World Production	99 Mt
Approximate value at US ports Gulf Coast	$106 per ton

[e] estimated
Source: US Bureau of Mines *Mineral Commodity Summaries* and *World Economy Minerals Yearbook.*

Table 4. World Production of Elemental Sulfur[e]

Country	Proportion of World Output (circa 1990)
United States	20 %
Commonwealth of Independent States (formerly USSR)	16
Canada	12
European Community	11
Poland	8.7
China	8.5
Japan	4.7
Mexico	4.1
Saudia Arabia	2.6
Others (Iraq, South Africa, Finland, Yugoslavia, Sweden, Kuwait, Brazil, Iran, etc.)	12.4
	100 %
Approximate World Production	58 Mt
Approximate value at US mines Gulf Coast	70 to $80 per ton

[e] estimated
Source: US Bureau of Mines *Mineral Commodity Summaries* and *World Economy Minerals Yearbook.*

chloride), sulfate of potash (potassium sulfate), sulfate of potash-magnesia (potassium magnesium sulfate), niter (potassium nitrate), and, at times, potassium carbonate (*Mining Engineering,* 1992). Potassium forms the third component of most chemical fertilizers (NPK: nitrogen, phosphorous, and potassium). The two most important deposits of potash are in Saskatchewan, Canada, and just west of the Ural Mountains in Russia (CIS). Germany (formerly DFR and DDR), France, and Israel/Jordan have also been important historical sources of potash. Table 6 and Fig. 2 provide production and generalized trade patterns for potash.

OTHER INTERNATIONALLY TRADED INDUSTRIAL MINERALS

Other industrial minerals and rocks that find their way into international trade include barite (used in drilling muds and as a source of barium compounds), bauxite and alumina (used for abrasives, but mostly an ore of aluminum), borates, bromine, chromite

Table 5. World Production of Salt (Sodium Chloride)[e]

Country	Proportion of World Output (circa 1990)
European Community	23 %
United States	20
China	12
Commonwealth of Independent States (formerly USSR)	8.4
Canada	5.8
India	4.6
Australia	4
Others (Mexico, Eastern Europe, Brazil, Japan, Turkey, etc.)	22.2
	100 %
Approximate World Production	184 Mt
Approximate value at US mines	
Rock Salt	$16.00 per ton
Salt from brines	$ 5.00 per ton

[e] estimated
Source: US Bureau of Mines *Mineral Commodity Summaries* and *World Economy Minerals Yearbook.*

Table 6. World Production of Marketable Potash[e]

Country	Proportion of World Output (circa 1990)
Commonwealth of Independent States (formerly USSR)	34 %
European Community	27
Canada	26
United States	6
Israel	5
Others (China, Jordan, Brazil,etc.)	2
	100 %
Approximate World Production	28 to 30 Mt
Approximate value at US mines	
K_2O equiv.	$130 per ton

[e] estimated
Source: US Bureau of Mines *Mineral Commodity Summaries* and *World Economy Minerals Yearbook.*

(used for refractories, but mostly an ore of chrome), various clays (kaolin, bentonite, attapulgite or fuller's earth, flint clay, ball clay, etc.), diatomite (as filter media, filler, and source of high-purity silica), dimension stone, fluorspar, garnet, gemstones, gypsum, industrial diamonds, kyanite, lime, lithium minerals and compounds, magnesite, perlite, pumice, rare earths, silica sand, soda ash, sodium sulfate, strontium minerals, talc, ilmenite, rutile and other titanium minerals (as a source of titanium dioxide in paint), vermiculite, zeolites, and zircon. Some portion of virtually all industrial minerals and rocks enter international trade. In cases such as gemstones and industrial diamonds, nearly all output enters international trade in one form or another. In other cases, such as borates, kaolin, fluorspar, chromite, ilmenite, lithium compounds, and rutile, very significant portions of world production enter world trade. More detailed discussion of international trade in many of these commodities may be found in the relevant commodity chapters of this current volume.

MARINE TRANSPORTATION

Over 85% of all industrial minerals and rocks entering international trade are moved by water. Ocean shipping adds approximately 15% to the value of dry bulk commodities (Kimbell and Zajac, 1988). Transportation cost can be the factor that makes one deposit more suitable than another. A quarry on the Mexican coast can serve US Gulf Coast ports, but that same quarry 160 km inland is unlikely to be an economic success.

Marine transportation rates vary considerably based on:

1) value of the cargo being shipped
2) size of the ship and the degree of specialization for a specific cargo
3) availability of different size ships and/or the use of containers
4) cost of fuel used to propel the vessel
5) general economic conditions in the *importing* countries
6) availability of back-haul cargo.

Source: Kimbell and Zajac, 1988.

Shipping rates are often scaled to reflect the value, bulk, or fragility of the cargo. Cement would be more expensive to ship than limestone due to the possibility of moisture ruining the cargo. Chemical fertilizers cost more to ship than phosphate rock or other components of the fertilizers. Nitrates are often prorated due to the potential danger of explosion. Ammonium nitrate is a high explosive that has been responsible for port disasters in places such as Houston, TX. Smaller loads of more valuable minerals often bear higher freight rates. Containerization helps to simplify the handling of many industrial mineral materials, but it is also normally more expensive per ton than the same commodity shipped in bulk.

Large bulk carriers are generally less expensive to employ than smaller vessels per ton shipped. Specialization of a vessel, such as the marine carriers designed by the Vulcan Materials Company to self load and unload crushed limestone from their quarry on the Yucatan to ports on the US Gulf Coast, makes it cheaper to use if they are continuously engaged and loaded to capacity. Specialization, however, does limit the potential for back hauls and the overall versatility of the ship for different kinds of cargo.

If a large number of different size ships are available for marine transport, this permits the selection of the proper size vessel to match the size of the shipment. It is not unusual to find ships with partial loads due to the unavailability of the proper sized vessel. This often increases the cost of shipment.

While capital costs take most of the cash flow of marine transport, fuel costs will constitute the major share of the variable costs. Significant swings in the price of fuel, most often fuel oil, do have an impact on marine transportation rates. In times of high petroleum prices, it is not uncommon to see the upward adjustment of marine transportation rates.

General economic conditions in the *importing* country can influence shipping costs due to the availability or absence of modern dock and unloading facilities. Port conditions can be responsible for significant time delays in unloading marine cargo. While the producer of an industrial mineral or rock can install the most suitable loading facilities at the port of origin, there is rarely the same opportunity for a producer to assure the availability of appropriate unloading facilities at the port of destination. The existence or absence of modern unloading equipment at the receiving port can have a significant impact on total handling costs. Primitive methods of bulk cargo handling can play havoc with the delivered cost of industrial mineral commodities. Kaolin is often shipped as a slurry, and specialized loading and unloading facilities make this method of shipment quite cost effective. The absence of such specialized facilities will normally require the use of individual bags, the most expensive method of shipping most bulk cargo.

Since a ship, like a railroad car or truck, is less expensive per kilometer when it is used to capacity in continuous service, the

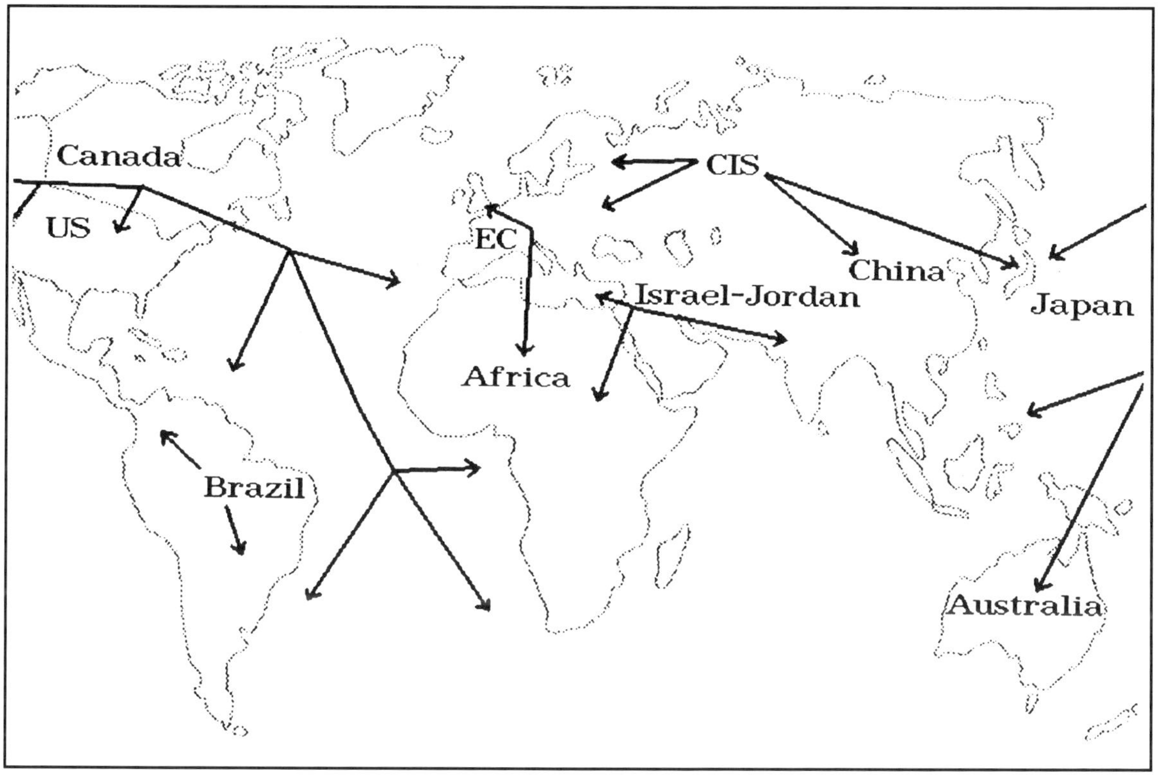

Fig. 2. Trade patterns in potash.

ability to carry cargo in both directions can lower shipping costs. While it is more difficult to find suitable bulk cargoes for a specialized marine carrier, when this is possible a lower shipping cost can normally be negotiated. Phosphate rock shipped to the Mediterranean may be shipped at lower cost if pumice or marble can be hauled back to the US Gulf Coast. Limestone can be shipped more economically if coal can be hauled back to Newfoundland.

OTHER FACTORS AFFECTING INTERNATIONAL MINERAL TRADE

Trade in mineral commodities is influenced by most of the things that impact on international trade in other goods and services. These factors include monetary exchange rates, political and economic relations between the countries with potential products and services to exchange, membership in economic and political alliances or trade agreements, government policies of assistance or discouragement of production or trade, business relationships between governments and multinational corporations, freedom of currency movements between countries, and military or defense considerations. This section will attempt to briefly examine the more important factors affecting the international trade in nonmetallic mineral raw materials and basic products with high industrial mineral content.

BARTER AND COMMODITY EXCHANGE AGREEMENTS

The exchange of one commodity for another is known as barter or countertrade. Countries without freely convertible currencies have used this form of trade because of the problems in determining the true market value of their currencies. Countertrade agreements are normally based on a predetermined ratio of quantities to be exchanged that, in turn, is based on observed free market price ratios for the same commodities. This occurs when countries do not have supplies of internationally acceptable (hard) currency, but do have commodities they wish to exchange (McCarl and Waters, 1985). In the mid-1970s, countertrade amounted to less than 2% of international commerce. By the 1980s, countertrade accounted for more than 20% of trade between countries outside the Soviet Bloc and closer to 40% if all economic systems are included (Truell, 1983; Welt, 1982). Countertrade in mineral commodities is becoming more the rule than the exception. The exchanges may be two-way, such as Romanian petrochemicals for phosphate rock, or they may involve a third country such as an American company sending phosphate rock or fertilizer to Poland, Polish sulfur sent to Spain, and the Spanish chemical producer could choose to pay the American company in hard currency or some other commodity. Exchanges might also involve a country lending capital for the development of a mine or chemical plant and taking payment in the commodity or commodities produced by that plant. This is quite common with Japanese business organizations and has led to significant capital investment in natural resource development in both developed economies, such as the United States, and in developing nations, such as Mexico and Brazil.

GOVERNMENT PROGRAMS

Purchases of strategic minerals, such as quartz crystal, sheet mica for electronic use, and ilmenite and rutile as a source of titanium metal, have had historic impact on industrial mineral markets, but they are rapidly becoming less important with the advent of new technology and the apparent easing of tensions between East and West. Some price support programs, such as those for fluorspar, have also been affected by pressures to cut government spending. Military needs are changing with technology, and it is becoming less and less important to maintain defense stockpiles of strategic raw materials. The impact of such defense stockpiles has been controversial for many years, with many industrialists holding the opinion that defense stockpiles tend to destabilize mineral commodity markets and restrict international trade (McCarl and Waters, 1985).

Government programs of price supports and subsidies are being subjected to closer scrutiny as interference with fair trade. Fines and penalties such as those recently imposed on various foreign cement producers for dumping are becoming more common. Questions of this nature continue to surface in the debate over European economic unity and in trade negotiations with the various countries of the former Soviet Bloc.

OTHER FACTORS

International cartels such as OPEC could be formed in fluorspar, phosphate rock, potash, industrial diamonds, and other nonmetallic minerals with major concentrations of reserves in relatively few geographic locations. This is unlikely due to the domestic antitrust policies of the United States and the questionable history of attempts to do this in copper, iron ore, mercury, tungsten, and bauxite (McCarl and Waters, 1985).

Capital and technical assistance programs of the United Nations (UNDP, UNCTAD, etc.) can have an important impact on individual mineral development projects, but funding is often limited. The privatization in former communist economies of Europe and Asia could lead to increased private capital flows and investment in mineral developments in those countries. However, it will be many years before they have stabilized sufficiently and developed the legal and economic framework that will make such investment more common.

While there are still tariffs and quotas on many mineral commodities, these are subject to continuing negotiations to dissolve such barriers to international trade. The recent agreements between the United States and Canada and the United States and Mexico have the potential to expand trade in commodities such as fluorspar and potash. The specific details of these and future agreements with other nations are still unclear in terms of their impact on industrial minerals trade and domestic US production (Anon., 1992b).

GENERAL OBSERVATIONS ON THE FUTURE OF TRADE

Mineral commodities have been a significant part of international trade between nations for most of recorded history. Despite their shrinking role in recent decades and the skepticism of some economists about their future importance (Thurow, 1992), the historic growth trend of nonmetallics as compared to fuels and metals gives us hope that industrial mineral commodities may provide a ray of light in an otherwise dim future for the mining industry. Privately owned multinational corporations are likely to take an increasingly important role in industrial mineral production as government ownership and operation declines throughout much of the former Soviet Bloc and attitudes toward private investment moderate in the developing world as well.

Potential new discoveries may be found as more knowledge is gained about the world's ocean basins, Central Asia, Arctic Canada and Siberia, Interior Australia, the Amazon Basin, and Central Africa. Such new discoveries will undoubtedly impact on international mineral trade.

While technology may have the potential to diminish the importance of many metals and fuels, this same technology seems to create growing international markets for ceramic, plastic, and other industrial mineral consuming products.

REFERENCES

Anon., 1980, *World Mineral Trends and US Supply Problems*, Resources for the Future (RFF), Washington, DC.

Anon., 1992a, *Mineral Commodity Summaries 1992*, US Department of the Interior, US Bureau of Mines, Washington, DC, 204 pp. (annual publication).

Anon., 1992b, *World Development Report*, The World Bank, Washington, DC (biannual report).

Alperovitz, G., 1992, "Facing Up to Economic Stagnation," *Technology Review*, Vol. 95, No. 2, Feb./Mar., pp.72–73.

Bates, R.L., and Jackson, J.A., 1982, *Our Modern Stone Age*, William Kauffman, Los Altos, CA, 136 pp.

Eulenstein, K.H., 1979, *United States Dependence on Imports of Nonfuel Minerals*, US Navy Research and Anaylsis Office, Washington, DC.

Harben, P.W., and Bates, R.L., 1990, *Industrial Minerals Geology and World Deposits*, Metal Bulletin PLC, London, 312 pp.

Jensen, R.G., Shabad, T., and Wright, eds., 1984, *Soviet Natural Resources in the World Economy*, University of Chicago Press, Chicago.

Kimbell, C.L., and Zajac, W.L., 1988, "Minerals in the World Economy," *Minerals Yearbook, 1988, Vol. III, Area Reports: International*, US Dept. of the Interior, Bureau of Mines, Washington, DC.

McCarl, H.N., and Waters, G., 1985, "International Trade in Mineral Commodities," *Economics of the Mineral Industries*, 4th ed., W.A. Vogely, ed., AIME, New York, pp. 51–95.

Mining Engineering, June 1992, "Annual Review of Industrial Minerals." Commodities are reviewed by different members of the Industrial Minerals Division of the Society for Mining, Metallurgy, and Exploration, Inc. each year in the June issue.

Nankani, G.T., 1980, "Development Problems of Nonfuel Mineral Exporting Countries," *Finance and Development*, Vol. 1, No. 3, Mar., pp. 6–10.

Papp, J.S., 1982, "Soviet Non-Fuel Mineral Resources: Surplus or Scarcity," *Resources Policy*, Vol. 8, No. 3, Sept., pp. 155–176.

Radetzki, M., 1977, "Where Should Developing Countries' Minerals Be Processed? The Country View Versus the Multinational Company View," *World Development*, Vol. 5, No. 4, Apr., pp. 325–334.

Radetzki, M., 1985, *State Mineral Enterprises: An Investigation Into Their Impact on International Mineral Markets*, Resources for the Future, Washington, DC.

Risch, B.W.K., 1978, "The Raw Material Supply of the European Community: The Importance of Secondary Raw Materials," *Resources Policy*, Vol. 4, No. 3, Sept., pp. 181–188.

Rosson, R., and Varon, B., 1977, *The Mining Industry and Developing Countries*, Oxford University Press, New York.

Stone, P., 1992, "Forecast Cloudy: The Limits of Global Warming Models," *Technology Review*, Vol. 95, No. 2., Feb./Mar., pp. 32–40.

Strauss, S.D., 1983, "Role of Third World Countries in Mineral Supply," *Mining Congress Journal*, Vol. 68, No. 2, Feb., pp. 33–37.

Thurow, L., 1992, *Head To Head*, William Morrow, New York, 320 pp.

Truell, P., 1983, "Barter Accounts for a Growing Portion of World Trade Despite Its Inefficiency," *Wall Street Journal*, Aug. 15, p. 21.

Van Rensburg, W.C., Jr., 1981, "Global Competition for Strategic Mineral Supplies," *Resources Policy*, Vol. 7, No. 1, Mar., pp. 4–13.

Welt, L.G.B., 1982, *Countertrade—Business Practices for Today's World Market*, American Management Association, New York.

THE ROLE OF INDUSTRIAL MINERALS IN THE US ECONOMY

SUBHASH B. BHAGWAT

No economic activity would take place without human ingenuity, driven by our needs and wants. Economic activity is made possible by the energy from fossil and other energy sources, our technological hardware, the buildings and infrastructure made from materials such as metals, plastics, wood, water, and numerous industrial minerals. Industrial minerals are difficult to define; however, they generally include utilitarian minerals other than metals and fuels. Some of the industrial minerals form an exception to this categorization and, as explained elsewhere in this book, belong to more than one category because of the manner in which they are used.

Industrial minerals are so numerous, and their applications so ubiquitous that they often do not receive the attention given to metals or fuels. Nevertheless, without industrial minerals, many of our most fundamental economic activities such as construction of our cities and towns, control of floods, transportation of our people, the search for minerals, the smelting of metals, and the delivery of supplies of clean water would be impossible. How then do we assess the economic importance of such a variety of materials lumped together under the general designation of industrial minerals? The price per ton or the total amount consumed do not satisfactorily explain the relative value of these materials to the economy. For example, the current price of limestone is about \$5/t and the price of fluorspar is about \$190/t. However, nearly 1.1 Gt of limestone, valued at about \$5.5 billion, are consumed in the United States annually, compared with 590 kt of fluorspar, valued at about \$112 million. It would not, however, be accurate to conclude that the mineral with the higher total value is always more economically important. Many other questions must be answered before making judgments of this type. These questions include how many people are employed in the production, processing, and distribution of the mineral? How is the mineral used? Is it partly or entirely imported? Where do the imports come from? How is the mineral transported? Is the mineral considered critical to national security? Is there a substitute material available in case the mineral supply is interrupted? Where would the substitute be produced? What would be the basic material from which the substitute will be produced?

Even when these questions have been answered satisfactorily, the economic importance of the mineral is not completely known until the consuming sectors are studied as well. For example, even though the manufacture of fighter aircraft is vital to the national security, it employs only a fraction of the number of people employed in the housing and construction industry or the automobile industry or the appliance industry. Mineral commodities consumed in these latter sectors of the economy are more significant than those consumed in the production of fighter aircraft, because nearly two-thirds of the Gross Domestic Product (GDP) depends on the purchasing power of individual consumers. In comparison, the entire defense budget accounts for only 6% of the GDP.

There is no easy way to present the importance of industrial minerals to the US economy because most minerals have diverse applications and products made from industrial minerals often contribute to the production of other goods and services. One way of highlighting the value of a mineral would be to relate it to the GDP reported by categories such as construction, machinery, stone, clay and glass products, electrical and electronic equipment, paper and allied products, and chemicals and allied products. These GDP categories represent industries fundamental to the economy totaling \$644 billion, or about 12% of the GDP in 1990. However, these categories are not convenient for discussion of the many minerals that are used in our complex economy. In a general way, minerals are the 5% of our economy that make the other 95% possible. This chapter, therefore, discusses a selected few industrial minerals to illustrate their importance to various sectors of the economy.

INDUSTRIAL MINERALS IN MAJOR ECONOMIC SECTORS

Construction Materials

The construction industry constitutes a major sector of the US economy and is one of the motors of domestic economic activity. In 1990, the construction industry contributed about \$260 billion or 4.8% to the GDP of the United States (Mohr, 1991). Many of the materials used in the construction industry are derived from industrial minerals. The consumption, value, and employment in the industrial minerals commonly used in construction—sand and gravel, crushed stone, cement, gypsum, and clays—are summarized in Table 1.

The total value of these five mineral commodities exceeds \$15 billion, and their production provides employment to a large number of people in counties where the minerals are produced. Fluctuations in the construction business can, therefore, affect these often rural counties directly.

There are other industrial minerals used in the construction industry, especially as glass and insulation materials. Although the use of glass is not limited to the construction industry, construction represents a major glass market and, consequently, a major portion of sales of silica sand. The industrial sand consumption in the United States in 1990 was about 25.4 Mt and worth about \$400 million. In 1990, about 43% of industrial sand production was used in glass manufacture. Although industrial sand extraction provided a modest 1 600 jobs in 1990, the significance of these jobs should not be underestimated at the county level. More important than the jobs associated with mining is employment in sand-processing and in the glass manufacturing and glass products industries. All are dependent on the mining of industrial sands.

Other industrial minerals are used in glass production, either to enhance specific characteristics of glass or to facilitate the production process. Soda ash and lime are two of the most important other materials in glass with nepheline syenite, boron, feldspar, and

Table 1. Industrial Minerals and Mineral Products Used in Construction Industries

	Consumption (10^6 tpy)	Value ($ billion)	Employment
Sand and Gravel	837.6	3.50	33 000
Crushed Stone	1 106.5	5.60	65 000
Cement	82.1	4.50	19 900
Gypsum	24.3	0.33	6 800
Clays	40.0	1.60	11 100
Total		15.53	136 000

Source: *Mineral Commodity Summaries 1991*, US Bureau of Mines.

kaolin as minor ingredients (O'Driscoll, 1990; Harben, 1991). Soda ash constitutes about 20% of glass wool, but only about 1% of fiberglass. On the other hand fiberglass production requires 26 to 28% kaolin or ball clay. About 51% of US soda ash consumption (3.4 Mt worth $425 million) went into glass production in 1990. The soda ash industry also provided employment to 2800 persons. Similarly, the glass industry is the major consumer of boron compounds. The value of boron compounds consumed by the glass industry in 1990 was about $210 million, and the boron industry employed about 1 100 persons.

Another major component of the construction industry is the insulating material made from fiberglass and referred to as glass wool to distinguish it both from glass fiber used as reinforcement fiber in roofing and as electrical insulation or E-glass fiber (Russell, 1991). Silica sand of 95% purity constitutes between 53 and 57% of glass wool. Glass fiber production requires 99% pure sand. Boric oxide, an expensive material averaging about $275/t in 1990, is added in small quantities of up to 5% to the glass wool production batch as an energy-saving flux. About 54% of the 565 kt of boric oxides produced in the United States in 1990 were consumed in the glass industry. Many other industrial minerals such as magnesia, calcium oxide, and sodium sulfate are used in glass manufacture. The GDP reporting category "Stone, Clay, Glass and Concrete Products" includes flat glass, glass containers, structural clay products, pottery, concrete, gypsum, cut stones, as well as abrasives and many others. Many of these product groups have been discussed already as they primarily consume industrial minerals. Similar products also are included in the construction materials category. Thus a considerable overlap exists between GDP categories.

With the exception of gypsum, the United States does not import significant quantities of the industrial minerals discussed here. Although the total dollar value of all the minerals does not exceed 5% of the GDP, the minerals are fundamental to the construction industry as there are few substitutes for them. The near complete self-sufficiency of the nation in these materials makes the construction industry an important motor of the US economy because every dollar spent on construction has multiplier effects in the supplier industries.

Abrasives and Cutting Materials

The machinery sector does not seem to be an obvious user of industrial minerals. Although commonly associated with metals and alloys, industrial minerals also play an important part in the production of machinery. The cutting and polishing of metal parts for precision fit and finish would not be possible without industrial minerals, such as diamonds and sands. Wear- and heat-resistant coatings, belt drives, and lubricants also require industrial minerals. Hight (1983) describes various types of abrasives products. A wide variety of industrial minerals are used in the production of abrasives including silica sand, corundum, diamond, feldspar, pumice, diatomite, tripoli, garnet, flint, chert, fused alumina, silicon carbide, and many others. Loose abrasive grains are produced in many sizes. The finer grain sizes are used in metal-polishing in the machinery industry while the coarser ones are mostly used in the stone-cutting, wood, and construction industries. Bonded abrasives are used in grinding wheels, but the trend here is towards using more manufactured abrasives than natural abrasives. Coated abrasives made from silicon carbide and aluminum oxide are commonly used in the metal-working industries.

The versatility of abrasives makes it impossible to account for their use by specific industries. According to the US Bureau of Mines (USBM) (Austin, 1991), the total value of the principal abrasives produced in the United States in 1990 exceeded $230 million. An additional $470 million worth of abrasives were imported for consumption, but about $295 million worth were also exported. Grinding wheels and millstones accounted for about 8.7 Mt of abrasives imports worth over $80 million. About 96 million carats of diamonds in various forms, worth over $136 million, also were imported in 1990. Despite their relatively low volumes and dollar value, abrasives play a fundamental economic role in the machinery industry because they are critical in the precision work vital for excellent technical performance and product finish, which can win a marketing edge for products over their competitors.

Graphite is extensively used in the machinery industry as a component in nonautomobile brake linings and as a lubricant, although its major use is in refractory brick linings in steel production. Comprehensive statistics on graphite were compiled by Taylor (1991). Graphite is typical of industrial minerals in that its consumption volume and value are low. In 1990, the total consumption of graphite in the United States was 36 334 t, and its total value was about $38.4 million. Graphite consumed for brake linings and as lubricant totaled 12 869 t and was valued at about $14.25 million. Most of the graphite needs of the United States are met by imports from Sri Lanka and Madagascar. Although graphite prices in 1990 ranged from 968 to $1,251/t, substitutes cost even more and have thus helped maintain the market for graphite in industry. Despite its critical uses and a total import dependence, graphite cannot be considered as strategic or critical because the countries from which graphite is imported are not politically unstable and domestic stockpiles are adequate.

Materials for Electronic and Electrical Equipment

About 80% of all the elements in the Periodic Table are classified as metals. Chynoweth (1976) lists 25 principal elements and minerals used in the electrical and electronics industries, and Lawrence (1986) and Murray (1989) describe the uses of ceramics in the electrical and electronics industries. Most of the 25 materials, except garnet, mica, and quartz, are metals. However, due to the small amounts used for industrial purposes, a number of these metallic elements are conventionally considered as industrial minerals. Table 2 indicates that of the 25 mineral commodities, copper, aluminum, lead, and zinc were consumed in quantities of 1.2 to 4.8 Mt in 1990. The remaining materials were consumed in much smaller quantities.

The total value of commodities consumed in the electrical and electronic sector in 1990 was about $145 billion. The largest single commodity consumed was copper with the total value for electrical and electronic consumption of $141 billion. Excluding the common metals such as copper, the platinum group, gold, silver, aluminum, nickel, lead, and zinc, the value of the rest of the mineral commodities consumed in the manufacture of electrical and electronic equipment was about $157 million. These dollar figures exclude those minerals for which the fraction consumed in the electrical/electronic sector was not available.

Table 2. Materials Consumed in the Electrical/electronic Industries in the United States

	Total US Consumption, all uses (tons)	% used in electrical/electronic industries	Value of consumption in electrical/electronic industries ($ millions) except when otherwise stated	Import Reliance(%)
Yttrium	50	NA	1.0–5.75	100
Indium	28	15	6.5	NA
Gallium	8	86	4.2	NA
Beryllium	142	66	76.5	Net exporter
Germanium	34	10	36.0	47
Tantalum	391	60	27.0	86
Copper	2 200 000	24	590,000.0	5
Platinum Group	99	32	$488/oz Pt $125/oz Pd $2658/oz Rh	88
Mercury	1 200	33	8.7	W
Cobalt	7 850	NA	173.0	85
Gold	230	38	2810.0	6
Silver	4 000[e]	25	643.0	7
Tin	48 900	22	296.0	76
Aluminum	4 800 000	10	8026.0	Net exporter
Nickel	154 000	10	1355.0	83
Tungsten	8 500	22	46.0	73
Antimony	41 300	10	75.4	64
Lead	1 220 000	75[a]	1235.0	4
Rare Earths	17 500	NA	$650–2650/t	9
Niobium	NA	NA	NA	NA
Manganese	680 000	NA	450.0	100
Zinc	1 270 000	10	2096.0	37
Mica (sheet)	1 150	100	2.3	100
Quartz	1 000	100	58.0	17
Garnet	36 803	7	$60–$2000/t	Net exporter

Source: *Mineral Commodity Summaries, 1991*, US Bureau of Mines.
[a]mostly in automotive batteries, [e]estimated, NA not available, W withheld, conversion factor: 1 oz = 31.104 g.

As the last column of Table 2 indicates, United States reliance on imports of many of the minerals listed is high and, in some cases, total. Substitutes, when available, are either technologically inferior or more costly. As a result, these minerals and metals are critical to the electrical and electronic industries that form the backbone of the high technology sector of the US economy. With the intense competition in basic and low technology industries, the high technology areas represent the future of US competitiveness, and many industrial minerals play a crucial role in this respect.

Industrial Minerals In Paper and Allied Products

The paper and allied products group includes pulp, paper, paperboard, and their many applications. This group of products accounts for only about 1% of the US gross domestic product, but pervades every other industry and all our lives. The paper industry uses minerals as fillers and coatings to reduce acidity, increase strength, increase whiteness and gloss, increase opacity, improve print quality, and replace more expensive pulp. Kaolin, soda ash (sodium carbonate), talc, precipitated calcium carbonate and sodium silicate are the primary minerals consumed in the paper and allied products industries. Smaller quantities of other minerals or mineral-based products such as alum and sulfuric acid also are used. As shown in Table 3, the total value of the major minerals exceeds $4 billion, of which about $1.5 billion worth of minerals are consumed in the paper industries. Kaolin, used as a filler/extender and as paper coating, accounts for about 75% of the mineral consumption, by weight, of the paper industry (Van Kouteren, 1989). Almost 90% of all minerals consumed in the paper and pulp industry are used as extenders, fillers, and coating agents (Seddon, 1991). Their importance has grown in the late 1980s due to rising pulp prices. Over $1 billion worth of kaolin was produced and consumed in the United States in 1990, about half of it in the paper industry. The economic importance of kaolin production was most notable for Georgia, which accounted for 84% of kaolin production, followed by 7% in South Carolina. Kaolin, valued at about $125/t, is facing strong competition as a filler material from precipitated calcium carbonate (PCC), in spite of its higher price, because of PCC's compatibility with alkaline paper. A derivate of limestone, PCC is less expensive at 200 to $500/t than wood fiber pulp at 600 to $800/t and titanium dioxide at about $2000/t (Reisch, 1991). The PCC market shows excellent growth prospects because of its use in increasingly popular alkaline paper. Alkaline paper lasts 200 to 300 years compared with conventional paper, which begins to yellow within a few years and to crumble after 30 to 40 years. Mineral fillers can constitute 15 to 25% of the paper's weight without an excessive loss of strength. The potential for PCC in the market is best illustrated by noting that about 97% of the books in the United States Library of Congress are printed on acidic paper, a quarter of which are already too fragile for circulation. Paper producers are rapidly converting their operations to produce acid-free paper in the United States. It is estimated that over half of all paper production has already been converted to acid-free products. Another growth market for PCC is paper recycling. PCC contributes to the bright-

Table 3. Mineral Commodities Consumed in the US Paper and Allied Industries

	Total US Consumption, all uses (tons)	Value of total US Consumption, all uses ($ million)	Percent of US total consumed by paper industry	US import dependence in percent
Kaolin	9 021 000	1100	46	Net exports
Soda Ash (sodium carbonate)	6 560 000	608	2	Net exports
Talc	1 021 004	111	18	Net exports
Titanium Dioxide (non-metal)	924 000	2200	26	Net exports
Precipitated Calcium Carbonate	680 000	250	100	NA
Sodium Silicate	25 000[a]	NA	100	NA

Sources: *Mineral Commodity Summaries 1991*, US Bureau of Mines, 1990 Annual Commodity Reports by USBM; "Precipitated Calcium Carbonate Finds Growing Niche in Paper Market," *Chemical and Engineering News*, Apr. 29, 1991, M.S. Reisch, pp. 9–11; "Paper industry changes shake supplier lineup," *Chemical and Engineering News*, Nov. 11, 1991, E. Anderson, pp. 15–16.
[a]North American consumption, NA Not available.

ness and printability of recycled paper and, therefore, is preferred by industry.

Soda ash contributes to the reduction of acidity in paper and enjoys a growing demand from the pulp and paper industry. There are some 104 pulp mills in the United States. They are rapidly reducing their consumption of chlorine for bleaching and increasing the use of caustic soda, a co-product of soda ash. Soda ash could take the place of caustic soda in the process, but soda ash requires 30% larger quantities to accomplish the function fulfilled by caustic soda (Kostick, 1990). With the pulp industry growing at an annual rate of 2.75%, the prospects for soda ash look bright. The United States remains a major exporter of soda ash. However, the structure of the industry has changed rapidly and five out of six companies producing soda ash are now owned by foreign investors from Australia, France, Japan, and Korea.

Research continues in developing new, better, and/or cheaper alternative materials to substitute, at least in part, for traditional materials. One of the more difficult substitutions is for the fibers used in paper production. The current technology is primarily based on organic fibers, mostly derived from trees. With environmental and economic pressures on the industry mounting, the search for inorganic fibers for paper production has intensified. Mineral wool is one such man-made material with substantial fibrous and non-fibrous components. New methods have been developed in the past ten years that allow the separation of fibrous mineral wool from the rest (Koenig, 1984). Among the advantages of mineral wool fiber are its finer fiber diameter, refractory nature, and lower costs. The most significant advantages, however, are that it is produced from minerals available domestically in large quantities and that it helps reduce the consumption of trees.

Minerals In the Chemical and Allied Industries

The use of minerals by the chemical and allied industries is perhaps the broadest in terms of number of applications and products. Minerals are consumed in the production of industrial inorganic chemicals, soaps and detergents, fertilizers, and other chemicals used in industrial processes. Minerals also are used to assist in the production of other materials, such as the barite in drilling mud used in oil and gas production, or in the production of commodities such as iron, with fluorspar as a flux.

Minerals are in increasing demand for their environmental applications such as the use of limestone for the removal of sulfur and nitrogen oxides from the combustion gases of fossil fuels. Limestone also is used in treating acidic lakes, and quicklime has been found to assist in the neutralization of toxic polychlorinated biphenyls (PCB)(Anon., 1991). Quantitatively, the use of minerals in chemical and allied products is the most difficult to measure. Table 4 indicates, however, that industrial minerals provide a wide range of inorganic chemicals in a variety of industries. These minerals have few affordable substitutes in the chemical industry, a vital segment of the US economy.

CONCLUDING OBSERVATIONS

This chapter summarizes how industrial minerals are used in several sectors of the US economy, from the most common construction sand to high-technology electronics materials. In dollar amounts, the industrial minerals are dwarfed by many other sectors such as housing and automobiles. However, without industrial minerals, nearly all economic activity would come to a stop. In this regard industrial minerals can be compared with food in a middle class American family's budget. The percentage of family disposable income spent on food ranks lower than housing and transportation, and perhaps even lower than clothing. Yet without food no one would survive. The relatively low contribution of industrial minerals to the GDP also indicates that the industry does not provide large numbers of jobs by itself. However, jobs directly involved in producing industrial minerals are significant at local and county levels, and, in some cases, at state levels because of the predominantly rural base of the mineral industry. Industrial minerals are

Table 4. Mineral Commodities and Their Uses in Chemical Products

Mineral	Type of chemical use
Antimony	flame retardants, plastics, solvents
Barite	oil and gas drilling mud
Bismuth	pharmaceuticals
Boron	soaps and detergents
Diatomite	filter aid, chemicals
Fluorspar	water fluoridation, CFC propellants, fluxes
Garnet	rubber products
Gypsum	agricultural acid neutralizer
Lime	agriculture, sulfur removal from flue gases
Magnesium	pharmaceuticals, chemicals
Mica	paints
Phosphate rock	fertilizers, detergents
Potash	fertilizers, chemicals
Sulfur	paints, paper, pharmaceuticals, preservatives
Rare Earths	oil refining
Vermiculite	paints, agriculture

much more important for the large number of jobs they generate in major sectors of the economy that require these minerals.

Another aspect of the importance of industrial minerals is the fact that the United States is largely self-sufficient in the production and processing of most industrial minerals. Industrial minerals generally are not considered strategic or critical materials in the conventional sense of the definition, because most of these minerals are not imported in large quantities and will not cause unmanageable domestic emergencies. Viewed in another way, however, many industrial minerals are strategic and critical because they are so basic to all the rest of our economic activity.

Over decades, the demand for most industrial minerals has continued to grow because alternative materials are either unavailable, inferior, more expensive, or some combination of the three. Where substitutes are available, they often are made from other industrial minerals. New materials are created to meet expectations for higher quality and to meet new demands created by technological advances. The number of industrial minerals and their role in the US economy is, therefore, expected to continue to grow in the future.

REFERENCES

Anon., 1991, "Findings Show Lime Can Destroy Toxic PCBs," *Rock Products*, May, p. 28.

Austin, G.T., 1991, "Abrasive Materials," *US Bureau of Mines Annual Report*, 22 pp.

Chynoweth, A.G., 1976, "Electronic Materials: Functional Substitutes," *Materials: Renewable and Nonrenewable Resources*, P.H. Abelson and A.L. Hammond, eds., Special Science Compendium No. 4, American Association for the Advancement of Science, pp. 123–130.

Harben, P., 1991, "Glass raw materials—Aspects of quality, quantity, and prices," *Industrial Minerals*, July, pp. 31–43.

Hight, R.P., 1983, "Abrasives," *Industrial Minerals and Rocks*, 5th ed., Vol.1, S.J. Lefond, ed., AIME, New York, pp. 11–32.

Koenig, A.R., 1984, "Processed mineral fiber in mats and paper," *Tappi Journal*, Vol. 67, No. 10, pp. 52–54.

Kostick, D.S., 1990, "Soda Ash," *US Bureau of Mines Minerals Yearbook, Vol. 1, Metals and Minerals*, pp. 1039–1059, Washington, DC.

Lawrence, R.D., 1986, "High Technology and Industrial Minerals: A Contradiction in Terms?" *Proceedings of the 7th Industrial Minerals International Congress*, Monte Carlo, pp. 23–31.

Mohr, M.F., 1991, "Gross National Product by Industry 1987–89," *Survey of Current Business*, Apr., pp. 25–27.

Murray, H.H., 1989, "Clay Minerals for Advanced Ceramics," *Mining Engineering*, Vol. 41, No. 11, pp. 1123–1125.

O'Driscoll, M., 1990, "U.S. Glass Industry—Overcapacity shadows strong demand," *Industrial Minerals*, No. 269, Feb., pp. 34–53.

Reisch, M.S., 1991, "Precipitated Calcium Carbonate Finds Growing Niche in Paper Market," *Chemical and Engineering News*, Apr. 29, pp. 9–11.

Russell, A., 1991, "Minerals in Fibre Glass—Current malaise in a growth industry," *Industrial Minerals*, Nov., pp. 27–52.

Seddon, M., 1991, "Kaolin Consumption Increase Could Mean Tight Supply in Early 1990s," *Pulp and Paper*, May, pp. 88–90.

Taylor, H.A., Jr., 1990, "Graphite," *US Bureau of Mines Minerals Yearbook, Vol. 1, Metals and Minerals*, pp. 523–532, Washington, DC.

Van Kouteren, S., 1989, "Inorganic Mineral, Chemical Market Expected to Remain Strong in 1989," *Pulp and Paper*, May, pp. 123–125.

MINING AND ENVIROMINING LAW FOR INDUSTRIAL MINERALS AND ROCKS

R. LEE ASTON

With the advent of the environmental "green" movement of the 1960s, eventually leading to Earth Day 1970, Congress responded by passage of the National Environmental Policy Act (NEPA), 1970. NEPA gave impetus to a flood of federal acts and laws in the subsequent three decades. Many of those protecting man's environment, health, and well-being also affect mining.

For purposes of this chapter, a current mineral law summarization lends itself to a division into two parts: viz., Part I, General Mining Law, and Part II, Enviromining Law. The word enviromining has been coined by the author and is defined and used herein as those federal and state statutes and regulations that were created to provide protection and preservation of man's environment, health, safety, and welfare, and they have particular application to and affect on mining activities. They are listed and treated under Enviromining Laws in Part II.

Mining, as a basic industry to provide the essential minerals for mankind's necessities and well-being, truly reflects the mining industry's slogan, "If it can't be grown, it must be mined." Without the basic minerals and rocks won from the earth by mining and mineral processing for their valuable physical properties or chemical constituents, man would soon be forced to revert to the cave-dwelling age. Unfortunately, for man to survive in the "manner of life to which he has become accustomed," he is caught between the demands of cleaning up his deteriorating environment and a continued supply of his basic requirements and amenities for daily living supplied by mining of the earth's minerals. He must have "clean living" at the same time with "good living;" neither is expendable. A very high pricetag accompanies this essential and critical choice which must be paid by society.

The United States is internationally recognized as the world leader in environmental matters. Many countries are following suit, principally the other Anglo nations. Because US law is founded and based on English law, on occasion, Anglo-American law is referred to and cited as authority in litigation anywhere in the Anglo-American world of jurisprudence. For this reason, and because mining and the industry is international in nature and operations, some British Commonwealth cases will appear in various topics. If the case is of interest to the reader, further inquiry at a law library may be made through use of the case's style and citation given. "Style" refers to the name of the litigating parties. "Citation" refers to the law publication in which the case and decision may be found, as well as the jurisdiction and court of decision. Citation of cases from the United States and British Commonwealth nations is explained under Case Citation Explanation.

Space and time limit exhaustive treatment of subjects and covering all geographic and jurisdictional areas. This chapter is intended as a broad outline and is based on the author's master's thesis at the University of Missouri-Rolla, 1992. The following Case Citation Explanation provides guidance on formatting of citations.

CASE CITATION EXPLANATION

Citations are the references to case decisions used in legal research and writings.

United States state cases, are generally from case reporters published by the West Publishing Company, St. Paul, MN. A list of their regional state case reporters follows. The number preceding the reporter symbol, e.g., P. = Pacific Reporter, refers to the volume number, while the number following the reporter symbol refers to the page number on which the case begins. At times where a quotation is given from the case, a second page number refers to the page where the quotation or specific reference may be found, or the quotation or reference page may be given as "at 405." The jurisdiction, as Montana, will follow the page number, and the court rendering the decision will follow, and lastly the date of the decision; Ex. W.R. Grace & Co. (plaintiff) v. Montana Dept. of Revenue (defendant), 779 (vol. no.) P. (Pacific Reporter) 2d (series) 470 (page number) (Mont. [Montana Supreme Court] 1989).

State Reporter series are: Atlantic (A), Northeastern, (N.E.), Northwestern (N.W.), Southern (So.), Southeastern (S.E.), Southwestern (S.W.), and Pacific (P.). There are also two states that have individual state reporters for cases; California (Cal. Rptr) and New York (NY Rprtr). If "2d" appears after the Reporter symbol, it indicates the second, or current series.

US Federal Court citations generally occur at three levels: District Court (F. Supp.), Court of Appeals (F.), and Supreme Court (US). The reporters are the Federal Supplement (F.Supp.) for the US District Courts, Federal (F.) for the US Courts of Appeals, and US Supreme Court Reports (US).

Where a federal case is from a US District Court (F.Supp.), the jurisdiction will follow in the citation, as F. Supp. (M.D. Ga.), meaning the Middle District of Georgia; or another example, E.D. Mo., meaning the Eastern District of Missouri.

Where the federal case decision is from a US Court of Appeal (F.), the jurisdiction follows in the citation as 815 F.2d 1356 (10th Cir. 1987), meaning the 10th Circuit Court of Appeals which is located in Denver, CO, and 1987 is the year of the decision. There are 13 federal judicial circuits. Another federal court cited herein is the US Claims Court (U.S Cl.Ct.).

British Commonwealth case citations are generally taken from provincial law reviews. Examples are:

Australia: Federal Commissioner of Taxation v. Reynolds Australia Alumina Ltd, ALR (Australian Law Review) [1987] (year of decision);

Australia: Byrnes v. Groot Eylandt Mining Co Pty Ltd. 93 ALR (Australian Law Review) (1990) (CA/NSW) (Court of Appeals/New South Wales)

Canada: Procan Explora.Co. v. Cadillac Explora. [1985] NWTR (NorthWest Territories Review) 153

Canada: Corona v. Lac Minerals, [1987] 62 O.R. (2d) (Ontario Reports, 2d series) 1 Ont.C.A. (Ontario Court of Appeals)

New Zealand: Police v. Purser Asphalts & Contractors, 1 NZLR (New Zealand Law Review) 693 [1988].

PART I: GENERAL MINING LAW AND LITIGATION

1(a). Mineral Deeds, Estates, and Reservations

A mineral estate is created by reservation in a deed granting or retaining an interest or ownership in the subsurface estate. It may be for certain specified minerals or for the total mineral interest in the subsurface estate when properly described by the deed. Mineral deeds may attempt to cover all minerals in the subsurface by adding the phrase ". . . and all other minerals." A mineral interest and grant in the subsurface estate should be specific and name the minerals of interest, e.g., "reserving an interest in all coal, oil and gas," or ". . . all metallic ore deposits as copper, zinc, lead, gold, silver, et al.," or "all phosphate, gypsum, and fossil fuels," or ". . . all marble deposits."

The mineral reservation cannot be expected to name all known minerals as a safeguard to future legal questions. If a general grant of all minerals is intended, then it can and should be so worded as to convey that intent without question. The current trend in case law is to question whether certain unspecified common rocks and construction materials are minerals and were intended to be included in the mineral reservation. Common materials such as sand, gravel, limestone, clays, and most other stone (rock) materials are being questioned in many states as to whether they fall within the ". . . and all other minerals" category. Where the questioned materials are found not to be included within the mineral grant, they belong to the surface estate owner. Courts are tending to classify common rocks, sand, and gravel as part of the surface; to place them as part of the mineral kingdom in a mineral reservation would often be to deny the surface owner his estate.

An illustrative Missouri case, showing the pitfalls of questionable deed construction of wording in a mineral interest grant, is that of McDonald v. Snyder Construction Co. [744 S.W. 2d 550(Mo.App.1988)]. The case also illustrates the frequently litigated status of sand and gravel, stone and clay, as to whether they are "minerals" or not within the meaning of mineral reservations.

The Missouri court's decision was influenced by similar holdings (case law) in 14 other states (Alabama, California, Colorado, Illinois, Kansas, Kentucky, Michigan, Mississippi, New Mexico, North Dakota, Oklahoma, South Dakota, Texas, and West Virginia) whose courts had refused to give the term "minerals" such "broad meaning as to include gravel, sand, clay," and even limestone in some instances. In McDonald, it was held that the construction materials did not belong to the mineral estate but to the surface owner.

Other recent cases illustrating various court decisions concerned with the ambiguity caused by the phrase ". . . and other minerals," or ". . . all other minerals" are:

i) Coastal Petroleum Co. v Intn'l. Minerals & Chemicals Corp., 709 F.Supp. 1093 (N.D. Fla. 1988): phosphate included in IMC's lease, but not in Coastal's.

ii) Kunkel v. Meridian Oil, Inc., 755 P.2d 470 (Wash.App. 1989): re: sand, gravel, and stone, not minerals.

iii) Need to Record Mineral Rights: The need for recording a deed registering mineral rights is stressed. The failure to record it at the local county courthouse where the land is located is emphasized by a recent West Virginia case, Belcher v. Greer, 382 S.E. 2d 33 (W.Va. 1989).

1(b). Are Sand, Gravel, Limestone, and Other Rocks "Minerals" for Reservation Purposes?

Limestone:

i) Griffith v. Cloud, 764 P. 2d 163 (Okla. 1988): The Oklahoma courts have held that the (mineral) reservation ". . . oil, gas and other minerals" did not include limestone.

ii) Canada Cement LaFarge Ltd. v. Manitoba, 65 Man. R.(2d) 28 [1990]: The issues were whether limestone was a mineral and whether it belonged to the surface owner or to the Crown where minerals had been reserved to the Crown. The Manitoba Court found that when used as a "building stone," or for "building purposes," it was not classified as a "mineral." When used for the manufacture of cement, it was a mineral and not for "building purposes." A similar finding was made by the British Columbia Supreme Court for dolomite in International Marble & Stone Co., v. Adams Estate, [1991] B.C.J. No. 156 Vancouver, B.C.

iii) West v. Godair, 538 So. 2d 322 La. App., 3 Cir. (1989): The excavation of sand and gravel has a detrimental effect on the use and development of the surface and was not intended to be conveyed with other minerals.

iv) Miller Land & Mineral Co. v. Wyo. Hwy. Comm., 757 P.2d 1001 (Wyo. 1988): The Wyoming Supreme Court relied on a Texas case "substances such as sand and gravel, and limestone are not minerals unless they are rare and exceptional in character, or possess a peculiar property" citing sand for glassmaking and limestone for manufacturing cement as "peculiar" qualities. Miller's claim for payment for sand and gravel taken from his mineral estate was denied.

v) Cochran v. US, 19 Cl. Ct. 455 (1990): Sand and gravel were found to be minerals on the Belknap Indian Reservation in Montana and included in the mineral interest owners' estate rather than in the surface estate.

vi) Watt v. Western Nuclear Inc., 462 U.S. 36, 103 S.Ct. 2218 (1983): The US Supreme Court held that gravel was included as a mineral under the government's reservation of "all minerals" under the Stock Raising Homestead Act of 1916.

vii) Millsap v. Andrus, 717 F.2d 1326 (10th Cir. 1983): The court found the mining of dolomite fell within the meaning of the mineral reservation on an indian reservation.

Common Sand v. Uncommon Sand:

viii) In 1989 in the case of Duval, the Bureau of Land Management (BLM) issued a patent for Duval's 1,000-acre claim for a sand deposit located in the Oregon Dunes National Recreation Area. Duval was successful because he was able to establish that the sand he claimed was "valuable" and an "uncommon" variety, and not to be mined for sand, per se, but for its constituent minerals, viz., the quartz and feldspar. In addition, the deposit had a lack of impurities and was of uniform grain size.

ix) Comm. Department of Mines v. May Bros., Inc., 396 S.E. 2d 695 (Va.App. 1990): The excavation of dirt, as borrow material, for use at another site was found not to be "mining" within the meaning of the Virginia law to require a mining permit.

2(a). Surface Rights for Mineral Estates

i) Where the subsurface/mineral estate has not been severed from the surface estate, the mineral rights belong to the surface owner. Where there has been a severance of the minerals from the surface estate by a mineral reservation, the granting deed will sometimes include a description of the extent to which mining rights may use the surface estate for its mining purposes. The rights conveyed to use the surface, when and if mining occurs are in many cases inadequately described, if at all. When mining does occur, the foreseeable conflict arises between the mining operator and the surface owner over how much of the surface may legitimately and reasonably be used for mining purposes.

Although coal was the mineral at issue, the Indiana case of Consolidated Coal Co. v Mutchman, 565 N.E. 2d 1074 (Ind.App. 1 Dist. 1990), serves as a good illustration of the problems and conflicts to be resolved over the subsurface owners' right to use the surface for mining. The legal principals involved are the same for any minerals to be mined by either surface or underground mining methods.

Mutchman was a class action by surface landowners for 116 deeds under which Consolidated owned the coal mining rights. At issue were the mining company's rights to use the surface, to what extent, including strip/open pit mining, the validity of the mineral deeds severing the "minerals" from the surface estates, and loss of mineral ownership under Indiana's Dormant Mineral Act.

The 116 deeds were divided by the court into five categories granting various amounts of surface uses to the mineral owner. They ranged from necessary to limitations, with and without damage liability to the surface, to payments for surface use, and options to buy the surface necessary for mining. Two categories of deeds denied surface-use rights. It may be seen from Mutchman that there are extreme variations in granted rights to use the surface for mining.

2(b). Subsurface Rights for Surface Owners

i) In a New York case, subsurface rights of the surface estate arose. The litigated question was to settle ownership of the void space in the ground after the minerals had been removed. In International Salt Co. v. Geostow et al., 697 F.Supp. 1258 (W.D.N.Y. 1988), Geostow planned to store incinerator ash in certain portions of the salt mine that they contended were mined out. Geostow sought to acquire agreements from the surface owners to convey the rights to the empty chambers from the room and pillar mining process which were below their respective tracts. International Salt brought an action against Geostow seeking a declaration of title to the "empty chambers."

The New York court noted in the Westerman decision, 260 Pa. 140, 103 A. 539 (1918), that "... neither the coal, nor the space, revert to the (surface owner) for temporary suspension of mining. The deed gives the (mineral owner) the coal, without any restriction as to time when it shall be mined; so, there is nothing upon which to base any claim of reversion" (to the surface owner). The New York court concluded that since "... some salt remained in all chambers, mined out or not, beneath the surface owners' respective tracts of land, and the lack of any language in the deeds limiting the quality of their estate or duration of the mineral interest therein, or of its right to mine and remove the salt ... the salt company has the present and exclusive right to the use of the mining chambers beneath the surface owners' properties."

ii) In Morgan v. Ga. Vitrified Brick & Clay Co., 397 S.E. 2d 49 (Ga.App. 1990), a surface landowner sued the mineral rights owner for damage to the surface from clay mining and failure to restore the surface. The Georgia Court of Appeals found that the mining company had left the surface in such a "wrongfully altered state ... and rough terrain," which was not reasonably necessary to conduct mining operations. Damages were awarded.

3. Dormant and Mineral Lapse Statutes

Where there has been a separation of the mineral estate ownership (and rights to mine) from the surface estate ownership, several of the States have devised mineral statutes to ensure that reserved and deeded mineral rights are either used or forfeited to the surface estate. These statutes are usually called dormant mineral, mineral lapse, or marketable title statutes, but occasionally tax statutes affect mineral estates. The purpose of the statutes is to encourage the development of mining and to remove encumbrances to titles so that surface land can be put to productive use without fear of interference from future mining activity.

Indiana's dormant mineral statute is one of the best known having made its way to the US Supreme Court for constitutional challenge and been upheld in Texaco v. Short, 45 U.S. 516 (1982). About 14 states have varying degrees of dormant mineral or mineral lapse statutes: Florida, Georgia, Illinois, Indiana, Kansas, Kentucky, Michigan, Nebraska, North Carolina, North Dakota, Oregon, South Dakota, Virginia, and Wisconsin. The statutes generally specify that if the reserved minerals are not actively pursued by some manner of evaluation, exploration, mining activity, or payment of taxes within a specified time limit, the two estates may be reunited by legal action of the surface owner into one fee simple estate.

Georgia's mineral lapse statute, similar in nature and requirements to the Indiana statute, with only work and time variations, underwent three trials involving industrial minerals from 1989 to 1990. The cases are illustrative of the dormant and lapse type statutes. See: Mixon v. One Newco & General Chemical Co, 863 F.2d. 846, (11 Cir. 1989); Parker v. Reynolds Metals Co., 747 F.Supp.711 (M.D.Ga. 1990); and Georgia Marble Company v. Whitlock, 392 S.E. 2d 881 (Ga. 1990).

In Reynolds the decision was in favor of the mining company retaining its mineral interest. In One Newco and Georgia Marble, the mining companies lost their mineral estates by failure to comply with the statutorily required duties, and the subsurface estates were returned to the surface estate owners.

4. Mineral Contracts, Leases, Royalties, Operating Agreements, Joint Ventures, and Ethics

Recent case law decisions serve well to update this area.

i) Lease agreement alteration: In Oswell et al. v. Tennessee Land & Exploration Co., 774 F.2d 1556 (N.D.Ala.1985), a printed lease form altered while negotiating with the landowner by adding a typewritten clause later gave rise to a dispute over a conflict of terms with the printed form terms. Tennessee Land lost its lease because of the addition of the typewritten clause.

A general principle of contract law as stated by the federal court was that: Where terms are inconsistent, typewritten terms added control over printed terms; and, handwritten terms added prevail over printed terms, typed or otherwise, "as it is presumed to have commanded stricter attention of the parties."

ii) Lease (sand and gravel) Is there A Duty to Mine? Dvorken v. Lone Star Inds. Inc., 740 S.W. 2565 (Tex.App.1987): The lessor sought to cancel the lease and deny renewal because of a lack of development during the initial term. Lessor's claim exceeded a limitation of four years which applied.

iii) Lease; No Impled Duty to Mine: Davidson Minerals v. Baird & Vulcan Materials Co. v. Baird, 390 S.E. 2d 33 (Ga. 1990): The Georgia Supreme Court found that an aggregate company had no implied duty to mine under the terms of its lease. The Supreme Court also referred to a precedent kaolin clay case , Higginbotham v. Thiele Kaolin Co., 304 S.E. 2d 365 (1983).

iv) Lease; Royalties (kaolin): Smith V. Freeport Kaolin Co., 687 F.Supp. 1550 (M.D. Ga. 1988); re: a dispute over royalty payment for processed kaolin instead of crude ore.

v) Mining Rights—Profit a Prendre: In Stevens Mineral Co.

v. State of Michigan, 418 N.W. 2d 130 (Mich. App. 1987), a 30-year limit on a profit was upheld for mineral extraction.

vi) Mining Rights—Profit a Prendre: In Willamette Quarries v. Wodtli, Compton Constr'n. & Cascade Crushing Inc., 761 P.2d 1356, (Or. App. 1988), 781 P.2d 1196 (Or.S.Ct. 1989), Willamette's exclusive right to quarry stone was limited to rip rap and revetment stone. Compton and Cascade were quarrying stone on the same owner's land. During a period of Willamette's inactvity, Willamette's stockpiles of stone were taken. The Supreme Court found for Willamette on the claim of conversion under a profit a prendre and not a lease. Under a profit, once the materials are severed from the earth, title to the stone is in the quarrier and not the landowner.

vii) Royalties; mispayment not grounds for cancellation of lease (brick clay): In Rogers v. Salisbury Brick Corp., 382 S.E. 2d 915 (S.C. 1989), the Supreme Court of South Carolina held that: (1) lessee's unintentional errors in royalty payments were not substantial enough to constitute a breach to warrant recision of lease; (2) lessor was entitled to an accounting where there was an alleged discrepancy in amount of royalties due.

viii) Lease: lost for lack of miner-like actvity: In Fry v. Dayon and Sack, 56 B.C.L.R. 123 [1983]: the Yukon Territory Court of Appeal dismissed a miner's claim for an interest in a mineral deposit based on a contract to win the ore for a percentage of the profits. Fry had breached the contract by failing to work the deposit continuously in a miner-like fashion.

ix) Sand and Gravel Contract Mining and Royalty: In T. L. James & Co. v. Kenner Landing, Inc. 550 So.2d 1378 (La.App. 5 cir. 1989), the dredging contractor on the Mississippi River for sand and gravel brought an action against landowner and corporate officer to recover sums owed under the contract. The Louisiana Department of Wildlife and Fisheries intervened to recover unpaid royalties for minerals in state waters. The Court held: the officer of Kenner was not held personally liable for the dredging contract debt; (2) the Department of Wildlife and Fisheries was not due to recover any royalties as the contract was completed before the state's royalties act was passed.

x) Contract Mining and Judgement Interest: In Mesa Sand & Gravel Co. v. Landfill, Inc. 776 P.2d 362 (Colo. 1989), Mesa had a contract whereby Landfill would excavate the gravel, but Landfill was required to mine down to bedrock to remove the "maximum quantity of gravel." Landfill was filling the disposal site without mining to bedrock. Mesa sued Landfill for breach of contract and was awarded damages. Mesa was denied prejudgment interest. On appeal, the Supreme Court of Colorado held Mesa was entitled to prejudgment interest at 8% from the date the case was filed.

xi) Overburden and royalties: In Grove v. Winter, 554 N.E. 2d 722 (Ill. App. 5 Dist. 1990) the owner of a limestone property brought an action against the lessee aggregate quarrier for breach of contract, seeking damages for conversion of the overburden and failure to pay all royalties for the stone quarried. The court upheld a lease where the owners had insisted on enforcing a clause in the lease that stated in pertinent part ". . . the overburden or soil resting on top of said rock shall be replaced upon the area from which the same is removed after the operation is finished . . ."

xii) Mining Contract—Due Diligence: Teton Exploration Drilling v. Borkum Resources, 818 F.2d 1521 (10th Cir. 1987) is a contract mineshaft drilling case where Teton sued Bokum to recover for services under the contract. Bokum counterclaimed against Teton asserting breach of contract and negligence.

Teton had contracted to drill five large diameter shafts with steel cased linings adequate to withstand hydrostatic pressures exerted by underground water in the formations being drilled. When Teton was lowering a section of casing into the shaft, Teton dropped the entire string about 30 m. to the bottom. Teton tried to repair the casing but was unable to prevent water from flooding the shaft through a rupture at the bottom. The court found that the collapse was due "to a combination of inadequate design and improper fabrication of the casing, including welding."

The chief point of the dispute was the meaning of a clause concerning workmanship to be performed in "a good workmanlike manner . . . with due diligence and care," and also liability for losses. These phrases counteracted with an exception clause charging risk and liability for losses to Teton where there was gross negligence by Teton. Teton's negligence was found to be the direct cause of a failure and had materially breached the contract.

xiii) Contracts—Drilling: Employer liabilities for torts of contractor, independent or agent? Professional conduct: see Bradbury v. Phillips Uranium Corp., 815 F.2d 1356 (10th Cir. 1987).

xiv) Leases' surrender is effective: In Kennecott Corp. v. Union Oil of California, 242 Cal Rptr. 403 (Cal. App. 4 Dist. 1987), an Imperial Valley, California, property underlain by mineral deposits and geothermal formations, produced steam for electrical energy generation. The property was leased to Imperial Thermal Products (ITP) as operator by several corporate owners. Kennecott became ITP's successor. Liability and payment for the high costs for disposing of waste products under water quality control agencies was the issue. The other owners had surrendered their interests to ITP. Effectiveness of interest surrender to avoid costs-liability was questioned by Kennecott.

xv) Lease—termination, breach of contract, reformation: In Monte Vista v. Anaconda Co., 755 P.2d 1358 (Mont. 1988), Anaconda notified Monte in 1980 of termination of their chrome leases. Monte questioned Anaconda's right to terminate after Monte had paid minimum rentals for three years without production from the leases. The leases provided for termination after "... the end of the third full successive year of the accrual of minimum rentals."

Monte's interpretation of the "accrual" was that it referred to an accrual of unpaid rentals and having paid the required minimum rentals, that their leases were maintained in good order for extension. The Montana Court of Appeals interpreted the phrase "accrual of minimum rentals" as meaning "accrual of paid or unpaid minimum rentals." Since Monte had held the leases for three years without producing, paying only minimum rentals, Anaconda had rightfully terminated. The Supreme Court of Montana affirmed the decision.

xvi) Joint venture: Partnership liability for unpaid wages: In Procan Exploration Co. v. Cadillac Exploration, [1985] NWTR 153, the question before the Supreme Court of the Northwest Territories was one of liability for unpaid wages to employees in a joint venture when the "operator" has filed individually for bankruptcy. The court held that the joint venture was still liable for payment of wages. Whether the solvent venturer is liable was left undetermined. Under US partnership law, bankruptcy of any partnership dissolves the partnership, and the solvent partner must pay all outstanding obligations and creditors, including unpaid employees' wages.

xvii) Joint venture ethics: Obligation of confidence and fiduciary trust in negotiations—Ethics International Corona

Resources Ltd. v. Lac Minerals Ltd., [1987] 62 O.R.(2d) 1 Ont.C.A.; affirmed by the Supreme Court of Canada, this case is a classic for Anglo-American jurisprudence with regard to professional ethics, fiduciary trust, and responsibility as required in negotiating for mining properties.

Corona held valuable, proven gold claims in Ontario. Lac approached Corona with an expressed interest. Corona revealed its mineral values and explorational results to interest Lac. Lac proposed a joint venture taking 60%. Corona's exploration continued with increased valuable results during talks with Lac. Corona revealed an offer made to acquire a valuable adjoining property. During the continuing negotiations, Lac bought the adjoining property and proceeded with a new mine and mill of its own. Negotiations were dropped and Corona sued Lac for breach of duty of confidence for misuse of confidential information and for breach of a fiduciary trust.

The court found that Lac was unjustly enriched to the detriment of Corona and had breached an obligation of confidence. The remedy was a constructive trust of the Lac property (with estimated reserves of 9 M oz of gold), the mine and mill held for Corona. The total damages awarded to Corona were Cdn$700 million.

xviii) Joint venture: Breach of fiduciary trust and fraud—Ethics: Ontex Resources Ltd. v. Metalore Resources Ltd., 75 O.R. 514 [1990], was similar to Corona, except the parties were operating under a written contractual arrangement. Ontex held gold prospects and offered Metalore a share, with option for 100% ownership, for doing the exploration and valuation. Metalore breached its fiduciary trust and committed fraudulent acts to take over 100% ownership. The Ontario Court found for a constructive trust for Ontex. Ontex received its properties back, as well as 459 additional adjoining claims that Metalore had subsequently obtained as a result of its fraud, and their contract was rescinded.

xix) Misuse of valuable mineral prospect information—Ethics: In Kent v. Graham Construction and Engineering Ltd., 84 Sask. R. 152 [1990], prospector Kent revealed to Graham information about a valuable stone property he had assessed. No agreements were made between them. Subsequently Graham opened a dimension stone quarry on the site Kent had revealed. Kent filed a builder's lien on Graham. The Saskatchewan Court found a lien was an improper claim and dismissed, but suggested that a different claim for remedy might be proper.

xx) Minerals investments—mine promotion—joint venture—fraud—sampling—"Open Fields" Exception—Ethics: In US v. Jos. P. Fahey Co., 769 F.2d 829 (1985), Fahey was convicted of 14 counts of mail and wire fraud in connection with precious metals market sales and mine promotional efforts. Fahey misappropriated investors' funds in a commodity pool for purchasing silver and gold from four unpatented, nonproducing mining claims near Death Valley in Nevada. Fahey's promotion included incorporation of a prospective ore report for an adjacent property. The Massachusetts Securities Division served a cease and desist order. The Federal Bureau of Investigation investigated Fahey's promotion and sampled his prospects. Fahey objected to the sampling in that his 4th Amendment rights had been violated by a warrantless, illegal search, and moved that the evidence be suppressed at trial.

The Massachusetts federal court held that legal title to the unpatented claims was in the United States. As holder of an unpatented claim, Fahey had a nonexclusive possessory interest in the land for purposes of mining only; the United States retained the right to manage and dispose of the other resources on the land, with right of entry, and the general public's use for recreational or other purposes as long as they did not interfere with mining. Fahey lacked any reasonable expectation of privacy from inspection and observation of mining activity. The government's taking of samples from the site was permissable under the "open fields" exception to the warrant requirement for search and seizure. Open fields do not provide the setting for intimate activities intended to be sheltered. There was no reasonable distinction between growing of crops and the separating of minerals from the surface of the land.

xxi) In Cusac Inds. v. Erickson Gold Mining Corp. [1990] 45 B.C.L.R. (2d) 347(C.A.), C.C.L.6811, an exercised option for division of profits when commercial production of a mine was reached was upheld by a British Columbia court. Optionee had choice of "commercial production" method or the "payment/work expenditure" method subject to 40/60% division of net profits from commercial production after the first "12,000 s.t had been mined."

xxii) Mining agreement-escrowed shares of stock-conditions subsequent: In Amalgamated Mining Western Ltd. v. Stikine Resources Ltd. (June 18, 1990), Doc. No. CA012557 (B.C.C.A.), C.C.L.8097, shares of corporate stock were traded for a mining claim and placed in escrow with provisos for cancellation unless conditions subsequent were performed by the purchaser. The British Columbia Court dismissed the claim holding that the purchaser was never entitled to the value of the escrowed shares.

5. Taxation of Minerals and Mineral Lands

i) Loss of mineral interest by failure to pay taxes: In Georgia Marble Co. v. Whitlock, 329 S.E.2d 881 (Ga.1990), the surface owners sued the mining company owning the mineral estate to gain title under the Georgia mineral lapse statute. The statute in pertinent part states "the owner of the real property in fee simple may gain title to such mineral rights by adverse possession if the owner of the mineral rights has neither worked nor attempted to work the mineral rights, nor paid any taxes due on them for the seven years immediately preceding the filing of a petition to gain title by adverse possession." The trial court found for the surface owners, with Georgia Marble losing its mineral estate.

Georgia Marble admitted that it had neither worked, nor attempted to work the mineral rights. It alleged that it had paid taxes on those rights. The court found that the company had paid a lump sum tax on mineral properties, but a specific tax amount had never been paid on the Whitlock's property. The loss of the mineral estate was upheld.

ii) Mineral lands: Reassessment of tax on change of ownership: In Howard v. Amador County (California), 269 Cal. Rptr. 807 (1990), under California's Proposition 13, a tax reassessment is called for upon change of ownership in mineral properties and mineral leases where the period is 35 years or more, inclusive of renewal options. Howard alleged an improper mineral property tax assessment and revaluation for the entire mineral property when only one of two of the mineral leases on his property changed ownership. There had been no change in ownership of the leased sand mining operation, but the lignite mine on the same property had a change in leased ownership. Only the lignite mining acreage was subject to reassessment. Howard was due a refund by Amador on taxes paid.

iii) Mining-reassessment-income computation: In Mattabi Mines Ltd. v. Ontario (Mine Assessor), (Feb 7, 1990) Doc. No. CA 650/87 (Ont.C.A.); C.C.L. 4442, deductions for processing costs and regulation allowances could only be taken for production from its own mine and not for custom processing of minerals from other mines through its mill.

iv) Mining-Mine net proceeds tax: In W.R. Grace & Co. Montana Dept. of Revenue, 779 P.2d 470 (Mont. 1989), the operator of a vermiculite open pit sought judicial review of a Montana State Tax Appeal Board decision denying certain deductions for computing the net proceeds tax. The Supreme Court of Montana held that: (1) general overhead, office, and administrative expenses could not be deducted from gross value to arrive at tax on mine's net proceeds; (2) costs of transporting and marketing mine product after point of beneficiation, premanufacturing phase, were not deductible to arrive at tax on mine's net proceeds; and (3) limitations period for assessment revisions of property taxes, rather than two-year statute of limitations for liability created by statute other than penalty or forfeiture, applied to deficiency assessments arising out of improper deductions.

v) Mining-Allowable Sales Tax Deductions for Mine Equipment: In Federal Commissioner of Taxation v. Reynolds Australia Alumina Ltd., Austl. L.R. 543 [1987], the issue before the Australian Federal Court concerned whether a new 51-km conveyor from mine to plant site qualified for the sales tax exemption for equipment as part of the ''alumina mining operation.'' The court had to determine the extent of: (1) the mining operations; (2) the mineral processing covered by the mining operations, and (3) the ''premises'' for a mining operation.

A former case of FCT v. Northwest Iron Co. Ltd (1986) was found to be similar where ''mining operations extend until the completion of the pellet-producing process . . . A slurry pipeline there is essential to the end product . . . and is not different in essence from a necessary conveyor line conveying alumina ore from one section to another within the complex.'' The entire process was considered a single set of mining premises over which Reynolds had exclusive control.

vi) Exploration costs—Tax Exemption-Australia -1991: The Australian Government introduced tax deductions, effective July 1, 1991, for exploration and related expenditures made before a mining lease is acquired.

vii) Exploration/Development Costs-Deductible: In 1990, the Internal Revenue Service (IRS) issued a public ruling, Rev. Rul.90–20, IRB 1990–9, 8 which allows a carrying party to deduct 100% of its exploration and mine developments costs in the hard mineral context. A ''carrying party'' is one who has been assigned an interest in a mining property in return for bearing all the costs of exploration, development, and operation. The ''carrying party'' is entitled to all income until the payout date when it recovers its investment. After the payout date, the ''carrying party'' assumes an alienable ownership of its agreed-upon share, but it may not transfer any of its interest until the payout date.

6. Zoning and Permitting of Mining Operations

Mine permitting in many localities, other than some federal/public lands, is dependent upon local zoning. Environmental concerns are treated in the agency permitting process which will alleviate many of the concerns of the local citizens at the public zoning hearing. The ubiquitous Not In My Backyard (NIMBY) syndrome, will be encountered at the zoning hearing. Well-prepared plans to cope with environmental issues are essential in assuring the citizenry that their backyard will not deteriorate because of the presence of a new mining operation. On the contrary, benefits will be reaped. (See Part II, 16, Eminent Domain/ Wetlands/Regulatory Takings, (ii) Florida Rock case, infra, for support.)

i) Denial of nonconforming use permit to recover stone: In Appeal of Eureka Stone Quarry, Inc, 539 A.2d 1375 (Pa.Cmwlth. 1988), a Pennsylvania aggregate stone quarrier challenged a local zoning ordinance that prohibited expansion into a 133-m setback area where the quarrier had mined prior to the ordinance. Eureka was denied a variance to a nonconforming use permit. The court upheld the denial unless the quarrier could show that (1) there were unique physical circumstances peculiar to that property; (2) the property could not be developed in strict conformity with the provisions of the ordinance; (3) the ordinance created unnecessary hardship; and (4) that any variance would not be detrimental to the public.

ii) Revocation of zoning permit: In Re: Crushed Rock, Inc. 557 A.2d 84 (Vt. 1988), the Town of Clarendon, VT, had approved quarrying of sand, gravel and stone. Quarrier sought amendment to operate an asphalt plant. Opposition to the asphalt plant brought complaints on blasting violations and action to revoke the permit. [See Part II, 15(ii)(a), infra, In Re Crushed Rock, Inc., for blasting violations.]

iii) Zoning regulations for quarrying found unconstitutional: In Davidson Mineral Properties v. Monroe County, 257 Ga. 215 (1987), an aggregate stone quarrier was denied zoning on the basis of the County Commissioner Board's enactment of a ''Moratorium on Commercial Development.'' The moratorium, as found by the Georgia Supreme Court, lacked sufficient standards to meet due process requirements, lacked objective criteria for permit approval, and gave the Board uncontrolled discretion. Zoning was approved for the quarry.

iv) Denial of nonconforming use to quarrier: In Meridian Minerals Co. v. King Co. (Wash), 810 P.2d 31 (Wash. App. 1991), an aggregate stone quarrier produced ballast only for a parent railroad company. Quarry operation preceded zoning. Application for expansion for commercial production for public use was denied on the basis that enlargement of a nonconforming use was prohibited.

v) Zoning regulations-vagueness: In Lithonia Asphalt Co. v. Hall County (Ga.) Planning Comm., 364 S.E. 2d 860 (1988), a building permit denied by County Planning Commission for objectionable conditions, e.g., ''too much truck traffic generated.'' The local code was found to be arbitrary and vague; ''traffic'' was not to be a usable criteria for denial.

vi) Zoning—local ordinances pre-empted when conflict with US Mining Laws: See Duval, 1 b) viii, Common Sand v. Uncommon Sand, supra. Duval was confronted by opponents' claims that local zoning would prevent mining after patenting was approved for the sand claim. The US Bureau of Land Management and the US Forest Service presented a precedent case decision, U.S. v. The Dredge Corp. (1981), that prepatent discovery and dates are taken into consideration; that prior to patent, a local zoning ordinance would be preempted to the extent to which it conflicted with the federal mining laws [Ventura County v. Gulf Oil Corp. (1979)].

vii) Zoning-government attempted recission of quarrying permit: In England, in Kirklees Metropolitan Borough Council v. Calder Gravel Ltd., (U.K.) Town and Country Planning Act [1990] C.O.D. 149–219, the local government attempted recission of permission for a 40-year old sand and gravel operation. No record of original approval for mining was found. The Chancery Court granted continuation of mining on the basis of longevity and a ''presumption of regularity'' as being a case where there was original permission.

viii) Zoning ordinances-land use restrictions eased in Oregon: A 1989 revision of Or.Rev. Stat. Sect. 517.960–517.976 makes mineral exploration permits easier to obtain by leaving the expertise to state agencies instead of to the counties. County zoning jurisdiction had become extremely limited under the federal doctrine established in California Coastal Commission v. Granite Rock Co., 107 S.Ct. 1419 (1987).

ix) Zoning-attempted revocation of special use (mining) permit-impaired rights:

a) In Bing Construction Co. of Nevada v. County of Douglas, 810 P.2d 768 (Nev. 1991), a gravel pit operator who received a special use permit allowing him use of county roads sued for injunctive relief after county resolution impaired some of his transportation rights under the special use permit. The Supreme Court of Nevada reversed and remanded the case to the County to afford Bing a hearing and proper notice on the resolution revoking his rights.

b) In Re John Meaker, 588 A.2d 1362 (Vt.1991), the Zoning Board of Adjustment (ZBA) for the town of Waterbury, VT, issued a conditional use permit for operation of a sand and gravel pit. Citizens objected and sued to revoke on basis of adverse affect on "community facilities," specifically road use and traffic. The trial court found no adverse effects and approved the permit. The Supreme Court of Vermont reversed and remanded for rehearing by ZBA. Although roads were not listed in the zoning bylaws as a community facility, evidence showed that an adverse impact would be made on the roads of the community and should be considered before permitting the pit operation.

x) Mining permit required for removal of 24-year old stockpiles: In Crowell Constructors v. State, 393 S.E. 2d 312 (N.C. App. 1990), the North Carolina Department of Environmental Health and Natural Resources (DEHNR) required an asphalt manufacturer to apply for a mining permit to remove 24-year old stockpiles of sand from a property. The former operation had ceased in 1960 and the sandpiles were left in place. Crowell contended that such short term removal was not mining. Crowell was required to submit a soil erosion and sedimentation control plan. The plan approved, Crowell continued removal of the sand. DEHNR fined Crowell $10,000 stating that off-site sedimentation had occurred and restoration efforts were ineffective. The Mining Commission upheld the violation and fine. The North Carolina Court of Appeals held that "mining" had occurred and a mining permit was required.

xi) Mining permit-attempted revocation by citizens group: In Concerned Citizens v. Rhodes, 394 S.E.2d 462 (N.CApp. 1990), a citizens' action group sued to have a mining permit for a crushed stone quarry revoked. The Superior Court declared the mining permit void. Appeal was made. Anyone challenging decisions of DEHNR must comply with the Administrative Procedure Act and exhaust administrative remedies before seeking judicial review. The citizens' action group had failed to make an administrative appeal. The Superior Court had no jurisdiction to void the mining permit. The Court of Appeals reversed the permit's voidance and remanded the claim for proper procedure.

xii) Permitting for continued mining and expansion-"Grandfathering": In Atlantic Cement Co. v. N.Y., 516 NYS. 2d 523(AD3 DEPT. 1987), Atlantic had operated a limestone pit and cement manufacturing facility since 1961. New York's Mined Land Reclamation Law became effective in 1975. Atlantic was required to obtain a mining permit from the New York Department of Environmental Conservation (DEC). In order to continue mining, Atlantic had to submit, inter alia, property maps, mining and reclamation plans, and a reclamation bond. Atlantic received permit renewals in 1978 and 1981 with only minimal paperwork. In 1984, DEC rejected Atlantic's normal application and required a State Environmental Quality Review (SEQR) and additional information for DEC's Life of the Mine Review Policy. Atlantic refused to provide the information. DEC issued notice in April 1985 for a draft EIS, and if not received by May 1, 1985, the permit renewal would be denied. Atlantic began Article 78 proceedings challenging the authority of DEC for its added requirements for permit renewal. DEC denied the permit renewal. The court held: (1) DEC had taken longer than the required 15 days to notify Atlantic of insufficiency of application information; (2) Atlantic's mining activities were grandfathered under the SEQR Act and Atlantic was not required to file an EIS; and (3) DEC had already determined in previous years Atlantic's operation would have an environmental finding of no significant impact (FONSI). The Court of Appeals affirmed the finding that DEC's permit denial was illegal, arbitrary, and capricious.

xiii) Grandfather Act—One court's test for permitting of expansion of operations: In Town of Wolfeboro (Planning Bd.) v. Smith , 556 A.2d 755 (N.H. 1989), Smith owned a sand and gravel pit on 35-acres that had been in continuous operation since 1950. By 1989 only ten acres remained to be mined. The Town of Wolfeboro filed a petition for a cease and desist order to prevent further pit expansion on Smith's land without a permit. Smith claimed he was not required to obtain a mining permit as he was grandfathered in when New Hampshire's law (RSA Chapter 155-E:2 Laws 1979, 481:3,) took effect on August 24, 1979. The statute allows the owner of an existing excavation "to continue such existing excavation without a permit." At the time of effectiveness of the law, only 8 of the 35 acres had been mined. Additional excavation continued from 1979 to 1989 without a mine permit until only ten acres remained to be mined There was no dispute between the parties that Smith was entitled to continue mining the existing pit. The sole issue was whether the statute's clause entitled Smith to excavate the remaining ten acres without a mining permit. The Town argued "that by its very language, the provision exempts only the 'land area' which is being used, as of August 24, 1979, and, therefore, exempts only the land already excavated as of the effective date of the chapter." The Town argued the operator could only continue mining in depth and not laterally. The trial court disagreed and ruled that Smith was entitled under the grandfather clause to mine his entire parcel without a permit. On appeal, the Supreme Court of New Hampshire also found difficulty in accepting the Town's interpretation, stating "If the phrase 'continue such existing excavation' as found in the grandfather clause were understood to allow only vertical and not lateral expansion, such an interpretation would lead many owners to find that they could not continue their existing operations very long, if at all. Only by allowing the continued excavation of land previously appropriated for that use would the owner truly be able to continue an excavation which he had begun." The Court further stated that the lateral expansion of an existing pit ". . . will be considered a continuation of a previous excavation" if the land was appropriated for excavation prior to the effective date of the law.

A decision seemingly in favor of grandfathered surface mining operations was quickly reversed by the Court's subsequent condition that the operator had to "objectively" show appropriation of the land for mining before the law's passage. "Intent alone is not enough." The Court believed that the phrase "continue such existing excavations" contained some limitation on increases in area or intensity of mining. It added that "An increase in the intensity which serves to change the character of purpose of the non-conforming use will be considered to have changed the use." This could end its exemption from the law as a grandfathered use.

In its decision, the New Hampshire Supreme Court established a three-pronged test for operators wishing to continue mining without a permit in New Hampshire: The operator must prove: (1) the activities were actively being pursued when the law became effective; (2) the area the operator wishes to excavate was clearly intended to be excavated, as measured by objective actions, not by subjective intent; and (3) continued operations do not, and will not, have a substantially different and adverse impact on the neighborhood.

QUESTION AND COMMENT: Under the rationale and final test of the New Hampshire Supreme Court, how was the operator able to expand mining from the eight acres already mined at the date of the effectiveness of the law for 10 more years without a mining permit to the date that Wolfeboro demanded no further mining of the last 10 acres without a mining permit ? It would appear that the mining of the 17 acres in between the two dates would imply that it was more than "subjective intent" to mine the whole 35 acres and the act of having mined 17 acres after August 24, 1979, shows "by objective actions" that the whole 35 acres was "clearly intended to be excavated." Hopefully, the test will not spread to other states.

7. Public Lands—National Forest Mining—General Mining Law of 1872—Crown Lands

Nearly ⅓ of the United States land area is publicly owned and roughly 55% of that is open to mining. The BLM manages the largest part, approximately 1.9 Tm^2, lying in 11 western states, and 2/3 of the total is in Alaska. The National Forest Service (NFS) manages about 768 Gm^2 in 40 states. NFS claims the presence of more than 40 industrial minerals on NFS lands. Aggregate stone production from NFS lands has made it the one of the larger crushed stone producers in the United States.

There are three different federal policy systems governing the exploration, development, and production of minerals on US unappropriated public lands: (1) hardrock (generally, metallic) claims under the General Mining Law of 1872 [30 U.S. Code Sec. 22 (1982)] ; (2) mineral leasing under the Mineral Leasing Act of 1920; and (3) the minerals disposal system under the Materials Act of 1947. The latter two plans generally deal with industrial minerals.

Under the Leasing Act of 1920, specific minerals, such as phosphate, trona, potash, et al., were removed from the General Mining Law's coverage and placed under the Mineral Leasing Act. Zeolites, as double salts, become problematic as to whether they are locatable or leasable minerals. A two-pronged test to determine whether a zeolite is locatable or leasable was developed in a 1977 case [see GFS(MIN) 34 (1977)]. Under the test, if a zeolite is classified as a silicate of sodium, it is locatable provided the sodium present is in sufficient quantity to be commercially valuable, or if the presence of sodium or any other element listed in the Mineral Leasing Act is essential to the existence of the mineral.

Also, under the Mineral Leasing Act, deposits of borates, carbonates, chlorides, nitrates of potassium, and sulfates are authorized to be leased. In addition, a potassium lease may include magnesium, aluminum, or calcium that is associated with the potassium. However, in the case of Foote Mineral Co. v. U.S., 654 F.2d 81 (Cl.Ct. 1981), it was determined that lithium is not a deposit associated with potassium. The leasing method requires annual rentals until production, with royalties paid thereafter. The responsible agency has complete discretion to accept or reject offers.

The Materials Act makes available common minerals as sand, gravel, pumice, volcanic cinder, stone, and clay at a market price usually determined by competitive bidding (contract sale). The nonmetallic minerals are acquired by mining claim or by a sale contract which is based on their chemical composition, physical properties, and prospective use. Also, uncommon varieties of sand, gravel, stone, pumicite, clay, etc., are locatable under the Act of 1955. Common quartz varieties such as jasper, obsidian, and opal are not locatable unless they exhibit uncommon qualities. Semiprecious minerals such as amethyst, garnets, topaz, and geodes, and precious stones such as emeralds, ruby, and sapphire are locatable. Also, geothermal steam and associated geothermal resources on BLM lands are leasable under the Geothermal Steam Act of 1970 [30 U.S.C. 1001–1025, 1976; Suppl.V, 1981]. Under the Mineral Leasing Act and the Materials Act agency permits are required for explorational activity.

i) Unpatented mining claims-sampling: In U.S. v. Michael R. Ware, 113 IBLA 1, GFS (MIN) 14 (1990), a miner's claim samples were challenged by BLM as a valid claim. BLM showed that mining costs would be 15 to 20 times higher than costs from the miner's sample values. Claim would be uneconomical and not a valid claim. The Department of the Interior's Board of Appeals held that: (1) BLM had proven a case of nondiscovery; (2) claimant's assay evidence alone was inadequate to refute BLM's assays; and (3) it was claimant's responsibility to refute the BLM's evidence.

Also, See Part I, 4 (xx), supra, re: U.S. v. Jos. Fahey Co., 769 F.2d 829 (1985); "Open Fields" exception for governmental sampling unpatented claims.

National Forests and Public Lands—Mining Regulations:

As national forests, parks, and wildlife preserves become increasingly protected in the Anglo nations, mining is increasingly restricted or prohibited.

In Australia, the New South Wales National Parks and Wildlife (Mining Prohibition) Amendment Act 1990 specifically prohibits mineral prospecting in national parks, nature reserves, state game reserves, and aboriginal areas. Existing mining interests are allowed to continue.

i) Anticipated haze control: Mining companies operating on public lands may expect the EPA to exercise some control over "regional haze" on the public lands in the near future. As an amendment to the Clean Air Act, signed into law, the EPA was directed to study the haze affect on wilderness and national park areas.

ii) Forest Service public notice rule change: Effective April 1990, implementation of an approval of a mining project on forest service lands cannot start until seven calendar days after notice of the decision is announced in a newspaper of general circulation. This will allow interested parties time for an appeal of the decision.

iii) Alaska National Parks-unpatented mining claims: In Northern Alaska Environmental Center v. Lujan, 872 F.2d 901 (9th Cir. 1989), environmental organizations challenged and appealed the federal district court's decision upholding the Secretary of Interior's (SI) approval of operation plans for unpatented mining claims within Alaskan national parks. The Court affirmed that the Alaska National Interest Lands Conservation Act did not require on-site inspection and mineral examination to verify discovery of valuable minerals before the SI could determine the validity of an unpatented mining claim and approve mining operations within a national park.

iv) Unpatented mining claims-validity: In Kile v. Belisle, 759 F.2d 1292 (Alaska 1988), a federal order withdrawing lands from mineral location was found to divest mining claims on which there had not been prior substantial compliance with annual assessment requirements.

v) National Forest—unpatented mining claims-jail penalty: In U.S. v. Smith , 866 F.2d 1092 (9th Cir. 1989), holders of unpatented mining claims were convicted and sentenced to confinement by US District Court for Alaska for residing and working on claims without having sought, filed, or obtained a permit (Plan of Operations) from NFS. On appeal, convictions were reversed as they were precluded by the Paper Reduction Act where the government failed to use a current control number on its information request to the parties.

vi) National Forest-quarry permit violation-imprisonment penalty: In U.S. v. Hatch, 919 F.2d 1394 (9th Cir. 1990), Hatch had operated an aggregate stone quarry in Nevada's Humboldt National Forest since 1953. The quarrier was convicted for constructing a road and disturbing the surface of the land without authorization or an approved operations plan from the NFS. Before sentencing and on appeal, the court reversed and held that a miner could not be subject to a criminal penalty for constructing a road on NFS land without authorization or an approved plan where the NFS failed to comply with the Paper Reduction Act. Government information collection requests that do not display a current control, or explanation, are considered bootleg requests and may be ignored by the public.

vii) National Forest land-criminal citation for residential occupation violation by miner: In U.S. v. Hall, 751 F.Supp.1380 (E.D. Cal. 1990), a miner of precious metals in the Tahoe National Forest of California was criminally cited in 1987 for occupying NFS lands for residential purposes. The miner's operating plan, approved by the Forest Service, had authorized his residential usage. His plans had been subsequently amended and approved several more times, each including residential usage. A federal court found that the NFS failed to give notice to the miner before attempting to revoke his operational plan to exclude residential usage. The miner had a vested property interest requiring due process of law before revocation could proceed. Criminal penalties could not attach since he had not been given fair notice of violation of the law.

viii) National Forest land—unpatented claim—nonpossessory interest—removal of structures: In U.S. v. Burnett, 750 F.Supp 1029 (D. Idaho, 1990), in the Nez Perce National Forest of Idaho, the NFS sued to enjoin the purchaser of unpatented mining claims from trespassing on the claims and to have mining structures removed from the claims. Where prior claim holders had validly relocated their claims, the subsequent purchaser of interest in unpatented mining claims succeeded only to the mining claim with no possessory interest in the structures on the claims. The successor in interest was required to remove the structures from the claim since his operating plan had been rejected by the NFS. The structures were a trespass on government property. The purchaser of the claims bought only the right to remove the structures and not the right to possess them on the claims. If the purchaser failed to remove the structures before a court-set deadline, NFS had the right to remove them.

ix) National Forest land—denial of mineral patent: In Shiny Rock Mining Corp. v. U.S., 906 F.2d 1362 (9th Cir. 1990), Shiny Rock was denied a mineral patent for its claim in Oregon's Willamette National Forest. BLM had withdrawn part of the area for mineral entry by a Public Land Order (PLO) in 1964. Shiny Rock challenged the PLO as being in error and depriving them of property without due process of law. The court applied a six-year statute of limitation dating from the 1964 PLO. Shiny Rock's 1983 date for accrual was defeated, with the mineral patent denial confirmed.

x) National Forest—privately owned mineral rights—strip mining prohibited: In U.S. v. Stearns Coal & Lumber Co., 816 F.2d 279 (6th Cir. 1987), where there was a reservation of mining rights in a 1937 deed of 64,000 acres to the United States for the Daniel Boone National Forest, the court held no right reserved to surface mine. See U.S. v. Stearns Co. 873 F.2d 134 (6th Cir. 1989); re-affmd. Nov. 18, 1991,WL 236899, 6th Cir.

xi) National Forest-permit denial for strip mining: In Kerr McGee v. Hodel and Global Exploration & Development Corp. v. Hodel, (combined), 630 F. Supp. 621 (D.D.C. 1986), the mining companies had discovered valuable phosphate deposits in the Osceola National Forest within the meaning of Sect. 211(b) of the Mineral Leasing Act. Leases were denied. The companies filed suit to compel issuance of leases by the Department of the Interior. An Environmental Impact Statement (EIS) concerning the impact of phosphate mining on the hydrology and endangered species was required. The creation of national forests was to protect and improve the forest for securing favorable conditions of water flows and to furnish a continuous supply of timber for the public. Nowhere in the legislative history of the Osceola NF was there a reference to mineral development or phosphate mining. The court found the mining incompatible with the primary purposes of the forest. The court distinguished between public domain lands and acquired lands. The former were those that never left the control of the United States, while acquired lands in the NF may be leased for mineral development only upon conditions imposed by the Mineral Leasing Act for Acquired Lands, 30 U.S.C. Sect. 351 et seq.

The fact that the United States Geological Survey (USGS) had determined valuable deposits discovered and mining companies' discoveries under their prospecting leases did not vest any right to receive mining leases. It was within the discretion of the Secretary of the Interior to find that the requisite discovery had not been shown. Denial of the leases was upheld.

xii) Public Lands under General Mining Law of 1872-pre-1920 claims: In Marathon Oil Co. v. Lujan,751 F.Supp. 1454 (D.Colo. 1990), valid claims in existence before Feb. 25, 1920 are protected by 30 U.S.C. Sect. 193, (the savings clause) which allows such claims to qualify for patents so long as they are "thereafter maintained in compliance with the laws under which they were initiated, which claims may be perfected under such laws, including discovery."

xiii) Public Lands-mineral reservation of patented lands under 1914 Act includes sodium minerals: In a 1991 decision of the Interior Board of Land Appeals (IBLA), the Board affirmed the decision of BLM (Shell Western E & P, Inc. v. BLM), that sodium minerals included in oil shale reservations for patented mineral lands granted under the Act of 1914 are reserved to the United Strates. IBLA stated that the government reservation of shales for oil could not be limited to only one component of a deposit; therefore, the sodium minerals commingled with oil shale deposits are included in the reservation to the United States.

xiv) Public Lands-BLM bonding policy: As of Aug. 14, 1990, the BLM required reclamation bonds for exploration and mining operations on public lands. The regulations require those operators with a good compliance record with BLM, but without a state bond, to obtain a bond. Bonding is not to exceed \$1,000/acre for exploration, or \$2,000/acre for mining. Those with a history of noncompliance will require BLM bonding for 100% of the estimated reclamation costs, and state bonds will be inapplicable to them. The policy remains in effect until rulemaking is completed by BLM.

8. Workman's Compensation—Wrongful Death Claims

i) Wrongful Death-Underage Employee-Negligence: In S.V. Blancato Sr. v. The Feldspar Corp., 522 A.2d 123 (1987), a 17 year old was hired at a feldspar mining pit. Connecticut law prohibits minors working in "hazardous" industry. The minor was buried alive and died. The deceased's father recovered under workman's compensation and sued for wrongful death in addition. In Connecticut, the Worker's Compensation Act was the "exclusive remedy" for wrongful death. Additional compensation was denied. The Supreme Court held: (1) the exclusive remedy of the Workman's Compensation Act did not bar a wrongful death action because of illegal employment, and (2) facts in evidence sustained company's negligence.

ii) Wrongful Death—Charge of Quarrier's/Owner's Negligence: In Turnbaugh v. Anderson, 793 P.2d 939 (Utah App. 1990), the surviving spouse of a front-end loader operator brought a wrongful death claim against the owner of the loader and the nonoperating owner of mineral interest of a cinder (scoriaceous lava) pit when the loader rolled into the open pit killing the operator. Judgment was in favor of the owners. The Utah Court held: (1) could not recover on theory of pit as public or private nuisance; (2) "fencing" statute did not apply to an active mine; and (3) owner of loader had no duty to warn operator or operator's employer.

iii) Wrongful Death-Manslaughter: In Police v. Purser Asphalts & Contractors Ltd., 1 NZLR 693 [1988], the liability of the employer for an act of the employee that resulted in a wrongful death was the issue in a New Zealand case. The company was charged under the Transport Act 1962 of causing the death of a pedestrian by an employee's careless use of a company-owned motor vehicle. A proper locking pin in towing a trailer and lack of a safety chain was the cause of the accident. The corporation was charged with and convicted of manslaughter. The company argued that a corporation could not be convicted of manslaughter. The High Court of New Zealand found that the Act applied to incorporated bodies, and, as an alternative to a jail sentence, a fine could be applied.

iv) Wrongful Death-Abandoned Mine Shaft-Discharge of Duty to Warn: In Davis v. Esparado Mining Co., 750 S.W.2d 887 (Tex. 1988), Esparado Mining Co. was a prior owner of the mineral estate and not the surface estate. The inactive mine shafts existed prior to Esparado's ownership. Esparado sold its mineral rights, retaining only a 1/16th royalty interest, and disclosed to the new mineral owner existence of the abandoned shafts. Esparado was charged with a wrongful death action for a minor who fell into a shaft.

The Texas court found that Esparado owed no duty to the decedent as it had no possessory interest in the minerals, only a royalty interest. Esparado had discharged its duty as a seller and had no liability by disclosing (in the deed) to the new owner at the time of sale that a dangerous condition existed.

v) Workman's Compensation-Strike Unemployment Compensation: In Brannan Sand and Gravel v. Industrial Claims Appeals, 762 P.2d (Colo.App 1988), the Colorado court rejected the employer's claim that strikers were voluntarily unemployed and held that: (1) the employer-employee relationship is not terminated, but merely suspended during a strike; and (2) the relationship is terminated if the employer permanently replaces the striking employee.

vi) Worker's Injuries-Liability-Parent Company not Liable: In Fiscus v. Atlantic Richfield, 770 P.2d 158 (Wyo. 1989), an injured open pit motor grader operator filed a claim against the parent company for injuries incurred on the job. The Supreme Court of Wyoming found that for a parent company to be liable for the conduct of a subsidiary operation resulting in injuries to an employee of the subsidiary operation, it must have assumed some independent legal control over some aspect of the operation of the subsidiary which was involved in the accident resulting in the claimant's injuries.

vii) Workman's Compensation-Injury-Jurisdiction of Claim: In Byrnes v. Groot Eylandt Mining Co Pty Ltd, 93 ALR 131 [1990](CA/NSW), a manganese miner in Australia's Northern Territory (NT) was injured and awarded compensation limited to three years under the Workmen's Compensation Act. Subsequently he moved to New South Wales and filed a claim there for the same injury. The Supreme Court of New South Wales ruled that the law of the jurisdiction in which the injury occurred is to be applied in whatever state the action might may be sought, and that Sect. 23 of the NT Workmen's Compensation Act extinguished the common law rights to recover greater damages.

PART II: ENVIROMINING LAWS

9. Air Quality—Clean Air Act (CAA) and as Amended [Nov. 15, 1990 PUB. L.No. 101–549, 104 STAT. 2399 (1990), U.S.C. Sect. 7401–7642]

Sweeping amendments of 1990 revamped the CAA (1977). The Amended 1990 CAA has not been in effect long enough to have developed many case illustrations as rules are not yet fully promulgated. Mining operations, including aggregate operations, are required to obtain air permits.

Parts of the 1990 CAA that are expected to affect mining are:

Title I, Section 107, Nonattainment: states must designate regions for nonattainment for national ambient air quality standards (NAAQS) and Section 172(b) sets designations for ozone, carbon monoxide, and fine particulate matter (PM10= less than 10 microns, aerodynamic diameter). Daily emissions standards have been set, viz., primary for 24 hr. avg. @ 150 micrograms/m^3 and half that (75 micrograms/m^3) for secondary; annual limit for primary and secondary is 50 micrograms/m^3.

Source performance standard for nonmetallic mines and mineral processing plants are: (1) no greater than 0.05 g/dscf from any source and (2) no greater than 7% opacity from any source.

Title II, Mobile Sources: more stringent standards are required for tailpipe standards to control exhaust pollutants from cars and trucks, improvements in evaporative emissions, required use of reformulated gasoline and diesel to reduce volatile organic compounds (VOC), and toxic air pollutants.

Title III, Hazardous Air Pollutants: Under Section 112(b) a list of hazardous air pollutants must be created for organic chemicals and metals common to industrial processes. A source categories list must be published by the Environmental Protection Agency (EPA) in twelve months from November 1990. A major source is any stationary source emitting 10 tons per year or more of any listed hazardous pollutant. Existing sources are exempt for six years otherwise applicable control technology-based standard through "early reductions."

Title IV, Acid Deposition Control: Mineral producing plants using fossil fuels in kilns, roasters, dryers, etc., must be aware of this Title and reduction standards for sulfur dioxide emissions. Excess emission penalties are provided for at $2,000 per ton for SO_2.

Title V, Permits: The permitting program is to be eventually turned over to the states after approval of their programs by the EPA, but the EPA will still retain overview powers. The states have three years from November 1990 to develop and submit to the EPA for approval state permitting programs.

An Administrative Penalty Scheme [Section 113 (d)] provides that the EPA administrator may assess penalties for violations not exceeding $200,000 and that the proceeding must be brought within one year of the first alleged date of violation. Field inspectors may assess violations at $5,000/day for appropriate minor violations. Section 113(b)(2) of the 1990 Amendment allows for actions for violations of the new sections where an operator has violated the Act with civil penalties up to $25,000 per day for each violation, not ''per day of violation.'' Citizens awards of $10,000 may be made for information and services in reporting violators that lead to a conviction. Increases in larger assessments for violations are being evidenced, e.g., a phosphate mining operator in North Carolina was assessed a $5.7 million civil penalty for air pollution violations in 1987. (See Criminal Penalties for Violations, Part II, Miscellaneous 19(d) , infra.)

Compliance for open pit air permits will call for keeping pit, crusher, plant, stockpiles, and road areas watered down and dust free. Rule of thumb estimates for quarry dust emissions are 30 to 50% due to crushing, screening, and processing, 20 to 40% for roadways, and lesser amounts from stockpile-wind erosion. Adding and replacing equipment will likely require permitting permission. Letters of notification to the state EPA are usually required before any new construction starts. Plant equipment under federal New Source Performance Standards (NSPS) must meet strict standards for emissions. Equipment manufactured after Aug. 31, 1983, is subject to controlled specifications, e.g., screens, conveyors, bins—10% opacity limit, crushers—15% opacity limit. Equipment manufactured before Aug. 31, 1983, is exempt from the stricter opacity limits, but limited to a range of 20 to 40%.

EPA mine permitting requires: a list of emission sources, calculation of annual uncontrolled emissions, specification of controlled apparatus and estimate of control efficiency, and calculation of annual controlled emissions. Simply stated, if dust can be seen, there is a violation.

i) Permit Violations-Denial of Renewal Application: In Joliet Sand & Gravel Co. v. Pollution Control Board, 516 N.E.2d 955 (Ill.App. 3 Dist. 1987), denial by the Illinois Pollution Control Board for Joliet's permit renewal was for violation of excessive particulate emissions at various points in its plant. Joliet planned corrective measures with new equipment. The agency refused permit renewal in spite of assurance by Joliet. Joliet brought action for denial of due process at the hearing. Denial for permit was affirmed.

ii) Air Inspections-Monitoring-Premises Entry: In Re: Bunker Hill Mining & Smelting Complex, 728 F.Supp. 626 (D. Idaho 1990), the court held that Section 114 of the CAA grants authority to EPA to inspect all records operator is required to maintain, but the search must be related to the purpose of the inspection. The court quashed a portion of EPA's administrative warrant as being overly broad. The company has the right to search its own records and provide those it found that are responsive to the EPA warrant.

In Koppers Inds., Inc. v. US EPA, 902 F.2d 756 (9th Cir. 1990), the Court found it premature to suppress the use of air emissions data gathered under a warrant when the administrative process has not been completed.

10. Water Quality—Clean Water Act (CWA)

Under Section 402, the National Pollutant Discharge Elimination Systems (NPDES), August 1990, more commonly known as ''stormwater runoff,'' NPDES permits are required for mining operations. State laws controlling mine and plant waste waters may require a closed-loop, no-discharge system. Compliance with the water permit regulations for mine and plant is critical for uninterrupted operation. Self-monitoring is essential in maintaining discharge limits into public streams and waters. Sedimentation and turbidity limits for discharges are being closely monitored by government inspectors.

EPA's actions with regard to CWA permits are not subject to agency-hearing review within the time frames of Section 307(a)(2) CWA where EPA promulgates technology-based standards pursuant to Sections 301, 304, and 306, CWA (see Rybachek v. EPA, infra). The requirements of Section 307 apply only when EPA promulgates health-based standards for specific toxic pollutants. Also, judicial review of EPA's NPDES permitting action is unavailable by a federal court where an Individual Controlled Strategy (ICS) is involved under Section 304(1). (See Glatfelter, Westvaco, and Natural Resources Defense Council example cases, 10(i) (a), (b), (c) infra.)

i) State-Issued NPDES permits-EPA approval:

a) In P.H. Glatfelter Co. v. EPA, No. 90–1488, slip op. (4th Cir., Dec. 10, 1990), the 4th Circuit US Court of Appeals held that judicial review was unavailable for EPA's conditional approval of a state-issued NPDES permit as an Individual Control Strategy (ICS) under CWA Section 304(1). The Court noted that it had jurisdiction only if the EPA had itself promulgated an ICS under Section 509(b)(1)(G) of the CWA.

b) In Westvaco Corp. v. EPA, 899 F.2d 1383 (4th Cir. 1990), the 4th Circuit US Court of Appeals held that judicial review was unavailable for EPA's preliminary disapproval of a 304(1) list of ''impaired waters'' submitted by the states of Virginia and Maryland. [Also, see American Paper Inst., Inc. v. EPA, 726 F.2d 1256 (S.D. Ala. 1989) where a Federal District Court of Alabama held that an EPA policy statement giving guidance to states within Region IV regarding discharge of toxic pollutants into rivers and streams was not a final action under Section 304(1).]

c) In Natural Resources Defense Council v. EPA, 915 F.2d 1314 (9th Cir. 1990), NRDC sought review of a final rule issued by EPA for some of the ''polluted'' waters listed under Section 304(1) that states must identify responsible point sources and develop ICS's for such sources.

The 9th Circuit Court of Appeals held that: (1) the EPA lacked the authority to restrict the statutory scope of Section 304(1), (2) states must identify point sources for all listed waters under Sections 304(1)(A) and (B), and (3) on remand, the EPA must reconsider whether ICSs are required for each point source pursuant to Section 304(1)(D).

ii) Waste Water-Pretreatment: In Internat'l Union (UAW) v. Amerace Corp., Inc., 740 F.Supp. 1072 (D.N.J. 1990), a citizen's suit alleged violations of pretreatment and reporting standards. The court held that: (1) federal pretreatment, discharge, and reporting requirements apply to an indirect discharger even if they have not been included in a permit issued by a public owned water treatment facility, (2) discharger must report all monitoring results even if monitoring was not required, and (3) a NPDES permit violation is not required to hold an industrial user liable for violating categorical standards.

iii) Permit Violations: In Thompson & Phillips Clay Co. v. Dept. of Environmental Resources (DER), 582 A.2d 1162 (Pa. Cmwlth. 1990), T&P's clay pit was downhill from an inactive coal mine. DER denied the operator's application for a second stage release of its reclamation bond of its mine site for acid mine drainage being discharged from their pit into a nearby stream. The acid water contained excessively high concentrations of metals and sulfates not allowable under Pennsylvania's Clean Streams Law, Act of 1937, P.L. 1987.

T&P argued that it did not cause the pollution. It maintained that the acid drainage was flowing through forces of gravity from the uphill abandoned mine to its mine; therefore, it was not responsible for treating the drainage.

The Court responded that ''the statutory prohibition against unauthorized discharges into waters of the Commonwealth did not require a causal link between the clay mine's activities and acid mine drainage allegedly originating from another's mine.'' The clay mine operator was held liable and denial of release for the reclamation bond was upheld until corrective measures were made for the contaminated discharge.

iv) Violation-Dumping Minerals-Penalty: In Regina v. Gulf Canada Corp., [1987] N.W.T.R. 277, dumping of excess amounts of barite and cement into the Beaufort Sea by Gulf Canada on eight consecutive days cost the company a total of C$180,000 in fines for violations of Canada's Ocean Dumping Control Act.

v) Stream Pollution-Violation: In Regina v. St. Lawrence Fluorspar Ltd., [1989] 80 Nfld. & P.E.I.R. 171, 249 Apr. 171 (Nfld. Prov. Ct.), a fluorspar producer was found guilty of severe water pollution under Canada's Fisheries Act and fined C$15,000 for extensive damage to the environmental life of fish. The fine followed repeated warnings to keep effluents to a minimum. The Newfoundland Court noted that ''pollution was a serious crime.''

vi) Navigable Waters Pollution-Mining Discharge Violation: In Rybachek v. U.S. EPA, 904 F.2d 1276 (9th Cir. 1990), an Alaskan mine operator and Miner's Association challenged EPA's regulations dealing with treatment required for discharges of untreated dredged soil and rock directly into navigable streams. The EPA interprets dredged soil and rock as pollutants and requires settling pond treatment before discharge.

vii) Mine permit—Denial—Potential Water Contamination: In Merman Pty Ltd v. Parker, Minister for Minerals & Energy [1986] WAR 159, an aggregate producer in Western Australia applied for a mining license for a new site, and encountered the Metropolitan Water Supply & Drainage Act 1909. The location was in a water reserve and catchment area. Approval was allowed by the Minister for Minerals and Energy. Since the Mining Act of 1978 was subject to the Environmental Protection Act of 1971, an environmental impact report (EIR) was required before approval. The aggregate producer was required to file an EIR before its license to mine could be reviewed.

11. Wetlands

Section 404 of the CWA prohibits the discharge of dredged or fill materials (overburden) onto wetlands without a permit. The US Congress gave authority to the US Army Corps of Engineers (Corps) to administer Section 404 permits by jurisdictional reasoning that wetlands were part of the ''navigable waters'' of the United States. The Corps and EPA share ultimate responsibility for 404 permitting. Prior to the Corps' permit issuance, a certification from the state or the EPA is required assuring the Corps that any discharge will comply with provisions of the CWA.

Wetlands include marshes, swamps, bogs, and similarly ''water-logged'' areas that ''are inundated or saturated by surface or groundwater at a frequency and duration sufficient to support, and under normal circumstances do support, a prevalence of vegetation typically adapted for life in saturated soil conditions.''

Where a mining activity in the past has created a ''man-made'' wetland on its own property that has become a part of the environment such that it supports plant and animal life, it can be susceptible to wetland regulations. The three wetland parameters that must be present for classification are: (1) wetland hydrology, (2) wetland hydric soil, and (3) wetland hydrophytic vegetation. The presence of all three are required to come under wetland regulations.

Several challenges have been made to the Corps' permitting activities:

i) Sect. 404 Permit Denial: In Borough of Ridgefield v. U.S. Army C.O.E., No. 89–3180, slip op. (D.N.J. July 2, 1990), the court held that the Corps may properly consider local zoning in determining whether practicable nonwetland alternatives were available.

ii) Zoning Permittal Denial-Alleged Taking: In Cioffoletti v. Planning & Zoning Commission, 552 A.2d 796 (Ct. 1989), Cioffoletti owned and operated a 23-acre sand and gravel pit in Ridgefield, CT and applied for pit expansion within and adjacent to existing wetlands. The operator's application was referred to the local soil and water conservation district, the US Department of Agriculture's Soil Conservation Service, and the state EPA, all of which approved it. At public hearings, an opponent expert witness testified to adverse impact on the wetlands, and the government agencies had failed to address the possibility of damage to the groundwater supply in event of a fuel spill.

Operator was granted a special use permit with restrictions and a $100,000 bond. Cioffoletti sued the City of Ridgefield attempting to show that the detrimental economic effect of the restrictions effected a ''taking'' of his property without just compensation. The trial court rejected his arguments. The Supreme Court of Connecticut found: (1) the exclusion of the operator's economic effect evidence limited the court's ability to rule on the property-taking issue and ordered that the taking-claim be heard, (2) the Commission's right to regulate mining and excavation in areas adjacent to wetlands under the Inland Wetlands Act was proper, (3) due process was met as claimant was given a fair hearing before the Commission, and (4) the Commission's right to require a performance bond.

iii) Permit Denial: In Florida Dept. of Environmental Regulation v. Goldring, 477 So.2d 532 (1985), Goldring appealed an order of DER denying his application for a permit to mine limestone on a site in Dade County. The Florida Supreme Court upheld that DER's dredge and fill jurisdiction depended on the predominance of listed aquatic vegetation on the property, along with an exchange of waters, either one-way or two-way.

iv) Section 404 Permit Review-Jurisdiction: In Leslie Salt Co. v. U.S., 896 F.2d 354 (9th Cir. 1990), the court found that the Corps' jurisdiction extended over isolated wetlands where presence of water was seasonal and where activities of the state and federal agencies created the wetlands.

v) Section 404 Permit- a wetland Regulatory ''taking'' of Mineral Land: In Florida Rock Inds. v. U.S., 21 Cl.Ct. 161 (1990), Florida Rock, in 1972, purchased 1,560 acres solely for limestone mining. In early 1978 dragline operations were started. In December 1977, prior to mining operations, CWA was amended to include wetlands. The Corps ordered the mining stopped. Florida Rock complied and restored the surface. The Corps advised Florida Rock that only 98 acres could be permitted at a time. Florida Rock applied for a 404 permit. After two years, the Corps denied the permit. The Corps' denial was based on the allegation that Florida Rock's limestone mining posed a risk of contamination of the sole aquifer in the area. Florida Rock filed suit against the United States seeking compensation for a regulatory ''taking'' of its property. In 1990, the US Claims Court entered judgment in favor of Florida Rock, and awarded the mining company $1,029,000 in damages for the taking of its 98 acres. For other issues of the case, see Eminent Domain-Regulatory Takings of Mineral Properties. For more details, see Part II, 16 (ii), infra.

vi) CWA-claims-statute of limitations: In Public Interest Research Group v. Powell Duffryn Terminals,Inc., 21 Cl.Ct. 153 (1990), the Court found that a state may not impose a more strict statute of limitations than the five-year federal statute, finding that states may not be allowed to set their own limitation periods on a cause of action created by the US Congress. The Court also held that the limitations period began running when the Discharge Monitoring Reports listing the violations were filed, but are extended 60 days by the required 60-day notice.

Also see five-year federal limitation (28 U.S.C. sect. 2462) holding for citizens suits: Public Interest Research Group, Inc. v. Witco Chemical Corp., 31 E.R.C. 1571 (D.N.J. 1990).

vii) CWA-claims-federal immunity waived: In Ohio v. US Dept. of Energy (DOE), 31 Env't. Rep.Cas.(BNA) 1448 (6th Cir. 1990), the 6th Circuit, US Court of Appeals, found that Sect. 313 of the CWA waived sovereign immunity for CWA claims.

12 (a). Mine Waste Cleanup—Comprehensive Environmental Compensation and Liability Act (CERCLA)—Solid and Hazardous Waste Disposal, Mine and Mill Waste Disposal

CERCLA was reauthorized for three more years to Sept. 30, 1994, under the Omnibus Budget Reconciliation Act of 1990. In 1990 several administrative developments concerning mine wastes were promulgated by the EPA.

i) Mine wastes. In May 1990 EPA released Strawman II Guidance for disposing of mine wastes under subtitle D of the Resources Control and Recovery Act (RCRA).

ii) Mineral processing wastes. The EPA's report on wastes from mineral processing listed twenty wastes. Public hearing was completed in September 1990.

iii) Mine wastes. In January 1990, five of the twenty mineral processing wastes were eliminated under a Final Rule: (1) furnace off-gas solids from elemental phosphorus production, (2) air pollution control dust/sludge from lightweight aggregate production, (3) sulfate process waste acids from titanium dioxide production, (4) sulfate process waste solids from titanium dioxide production, and (5) process waste water from primary lead processing.

iv) Hazardous mineral waste disposal-discarded materials: The elements for liability for hazardous waste under Section 107(a) are: (1) the defendant falls within one of the classes of potentially responsible parties (PRP) in Section 107(a), (2) the site is a facility, (3) a release or threatened release of hazardous substance has occurred, and (4) the United States has incurred response costs as a result.

(a) In American Mining Congress v. U.S. EPA, 907 F.2d 1179, @ 1192 (D.C. Cir. 1990) the court defined "discarded materials" to include materials extracted from the manufacturing process and held that sludges formed in the wastewater treatment impoundments, which in the future may be discarded, are subject to regulation as hazardous waste.

(b) In Louisiana-Pacific Corp. v. Asarco, 909 F.2d 1260 (9th Cir. 1990), the court held that the term "release" does not include any quantitative requirement. And in U.S. v. Western Processing Co., 734 F.Supp. 930 (W.D. Wash. 1990), the court stated that a "threatened release" is sufficient for liability.

(c) In Louisiana-Pacific v. Asarco, supra, the court further held that the use of a waste slag byproduct from the production of mineral wool in road building constituted a disposal.

v) Hazardous mineral waste cleanup-costs-liability-state remedial plans:

a) Several cases have upheld that CERCLA liability is strict liability, i.e., there can be no defense against it. (See U.S. v. Shaner, Civ. No. 85–1372 (E.D. Pa. June 25, 1990) and U.S. v. Consolidated Rail Corp.,729 F.Supp. 1461 (D. Del. 1990.) Shaner also held that unless the harm is divisible, the liability imposed is joint and several.

b) A classic case of hazardous waste contamination from mine dumps that resulted in a state CERCLA (cleanup) claim was that of State of Colorado v. Idarado Mining Co., 707 F.Supp. 1227 (D. Colo. 1989) and its appeal, 916 F.2d 1486 (10th Cir. 1990). In Idarado, Colorado instituted an action for a declaratory judgment that the mine owners and operators were liable for the cleanup of hazardous wastes and for natural resource damage at and near their facilities. Colorado sought enforcement of its own remedial plan. The US District Court of Colorado allowed injunctive relief for the State of Colorado to enforce the State's plan and held that: (1) the mine owners and operators were liable for cleanup of tailings piles, (2) defendants were liable for the cost of a 5-year blood lead screening program for residents of the town, and (3) the town was not allowed to intervene.

On appeal, the court found that CERCLA, as amended by the 1986 Superfund Amendments and Reauthorization Act (SARA), does not authorize a state to seek injunctive relief compelling responsible parties to perform a cleanup. The section of the Act providing that a state may enforce any federal or state standard, requirement, criteria, or limitation to which remedial action is required to conform, envisions remedial action selected by the federal government or its delegates. Under CERCLA, a state must comply with otherwise legally applicable or relevant and appropriate federal, state and local requirements, including permit requirements, not as a condition to selecting its remedial plan, but as a condition of recovering its costs. The 10th Circuit Court found that a federal District Court did not have inherent equitable authority to issue an injunction imposing state-selected remedial plans under CERCLA.

In addition, where the District Court had ordered relocation of residents of a trailer camp operated by the responsible parties, the 10th Circuit Court held that a court cannot, under Sect. 121(e)(2) of CERCLA, require a defendant to pay for permanent relocation costs of residents involved in a cleanup area. Only the President of the United States can order a relocation. The Appeals Court also upheld that pre-CERCLA costs that meet the definition of response costs are recoverable.

vi) Hazardous waste-successor liability:

(a) In Louisiana-Pacific Corp. v. Asarco, Inc., supra, iv(b), the court questioned whether a corporation should be held liable as a successor. Applying traditional rules of successor liability, the court held the successor was liable.

(b) In Idarado (see 12 (a) (v.) (b) supra), despite the current owner's contention that it was responsible for only approximately 3% of all metal contaminants loaded into the stream and water system and that neither they, nor their affiliates, disposed of the tailings on the property. The State had incurred cleanup response costs for which the current owners were liable as they had assumed all obligations, debts and liabilities of the predecessor owners. [CERCLA, (1980), Sect. 107(a), as amended 42 U.S.C.A. Sect. 9607(a)].

vii) Hazardous waste-cleanup liability for parent company: In U.S. v. Kayser-Roth Corp, 910 F.2d 24 (1st Cir. 1990), the court held a parent corporation of a wholly-owned, dissolved subsidiary directly liable for cleanup costs as an operator under CERCLA based on its active involvement in and control over the activities of the subsidiary. However, in Joslyn Mf'g. Co. v. T.L. James & Co., 893 F.2d 80 (5th Cir. 1990), the court found that the definition of owner or operator in CERCLA did not include a parent company where the facts did not justify piercing the corporate veil. And, in Columbia River Service Corp.v. Gillman, 32 Env't. Rep. Cas. (BNA) 1375 (W.D. Wash. 1990), the court held that a company that dissolved pursuant to state law could not be sued under CERCLA, but that shareholders might be held liable.

viii) Waste liability-even where not major contributor: See Louisiana-Pacific Corp.v. Asarco 12.(a) vi (a) supra, and also Thompson & Phillips Clay Co. v. Dept. of Environmental Resources (DER), supra at 10 (iii).

In Asarco, the court held that the defense that the release and damages were caused solely by a third party is not available, even where a generator's disposal of waste may have been only minimal. Also, see Idarado, supra, 12 (a)(v.)(b), contributing only 3% of contaminants. ix) Waste-definition of a release: In The Fertilizer Institute v. EPA, 1991 WL 96474 (D.C. Cir., June 11, 1991), EPA's definition of a ''release'' was rejected by the DC Court of Appeals. Under CERCLA, parties must report to the National Response Center when any hazardous substance of reportable quantity is released from a facility into the environment.

The EPA interpreted the placement of a hazardous substance in an ''unenclosed containment structure'' as constituting a release. The Court concluded that under CERCLA ''nothing less than the actual release of a hazardous material into the environment triggers its reporting requirements.''

It appears that only the actual escape of hazardous materials into the air, the soil, or surrounding waters, or into groundwaters through the soil, must take place before a reportable release has occurred. This limitation may disqualify storage of nonvolatile materials in open containers and piles, thus avoiding Superfund liability.

12(b). Former Mill Site Hazardous Waste Liability

i) Continental Concrete Pipe Corp. v. K&K Sand & Gravel, Inc., No. 88-C-9375 (N.D. Ill. Jan.22, 1990), where the seller, K&K, contended that a CERCLA action for hazardous waste on the property was barred by releases in settlement of prior litigation; the Court refused to dismiss the action on that basis because of other disputed facts.

ii) Former Mill Site Liability Under Uranium Mill Tailings Radiation Control Act (1978) UMTRCA: In 1990, several producers and former producers of thorium, uranium, and other radioactive rare earth minerals litigated the issue of former government contracts seeking to make the United Sates a responsible party and reduce their cleanup cost liability of former mill sites containing hazardous radioactive wastes.

UMTRCA provides for payment of waste control costs based on a time division. For pre-1978 tailings piles accumulated from government supply contracts, the federal government would pay 90% of the cleanup or stabilization costs, with the state government paying the balance, and the producers paying nothing.

Title I of UMTRCA provides that the United States will pay for the stabilization and decommissioning of all inactive mill sites which were not licensed prior to Jan. 1, 1978.

Title II of UMTRCA authorized the EPA to develop regulations for stabilization and decommissioning of mill tailings piles of producers which remained active after Jan. 1, 1978. Under Title II, licensees are responsible for complying with, and costs of, the EPA regulations. UMTRCA makes no provision to pro-rate cleanup costs for sites whose waste piles contained both pre- and post-1978 radioactive tailings.

a) In Atlas Corp. v. U.S., 895 F.2d 745 (Fed. Cir. 1990), a group of thorium, radioactive rare earth, and uranium producers (Kerr McGee Chemical,Western Nuclear, Union Carbide, Homestake, Quivira, et al.) brought an action against the United States to recover costs of stabilization of radioactive mill tailings. All companies to the action were licensed under Title II and that, according to the Nuclear Regulatory Commission (NRC), made them liable for their own cleanup costs. As to the producers' allegations of mutual mistake about the hazardous nature of the tailings and seeking reformation of former contracts with the government, the court held that the hazardous nature of the tailings in the future was ''not knowable'' at the time of the government contracts; therefore, there could be no ''mutual mistake'' for reforming the contracts. In the absence of fraud, mistake, or illegality, the court could not change the terms of the contract to include cleanup costs.

The Sovereign Acts Doctrine barred the producers' claim that increased obligations were imposed on them by Congressional post-contract passage of UMTRCA and that reimbursement was implied. The court replied that the government's requirement of tailings stabilization and mill decommissioning was not an exercise of its contractual powers, but of its powers as the sovereign to act for the public good. As to allegations of a ''taking'' of the producers' properties, the court found none since its actions were done under the power of eminent domain and were for the public's health, good, and welfare. The cleanup requirement did not make the production of minerals unprofitable, thus, no ''taking'' had occurred.

b) In Kerr-McGee v. U.S. Nuclear Regulatory Comm., 903 F.2d 1 (D.C. Cir. 1990), the mill owner sought review of the NRC's decisions defining ''by-product'' material in the mill tailings which differed for ''on-site'' and ''off-site'' use. A provision of UMTRCA defines ''by-product material'' as tailings or waste produced by extraction or concentration of thorium or uranium from ore processed primarily for its source material content.

Part of Kerr-McGee's mill tailings were primarily from the processing for rare earths and partly for thorium, but all were eventually processed for thorium until the mill closed in 1973. Those tailings sold for land fills were labelled ''off-site,'' while those retained at the mill site were ''on-site.'' The NRC attempted to delegate treatment of the ''off-site'' tailings to the State of Illinois, to which Kerr-McGee objected. The court found the NRC's interpretations under UMTRCA not permissable and granted review of the claims affecting treatment routes.

c) In Hecla Mining Co. v. U.S., 909 F.2d 1371 (10th Cir. 1990), Hecla sued the US Department of Energy (DOE) seeking to require the government to cleanup the mill tailings site under UMTRCA as a former buyer of radioactive minerals from Hecla. The DOE refused to classify Hecla's site as a former processing site since the tailings had been moved and reprocessed to recover uranium and other metals. The reprocessing had been for commercial sales and not for a government contract. The appeals court affirmed DOE's and the Colorado District Court's decisions that the United States was not required to pay for the cleanup.

d) In FMC Corp. v. U.S. Dept. of Commerce, 31 Env't. Rep. Cas.(BNA)1959 (E.D. Pa. July 18, 1990), the court held that the United States might be liable for hazardous wastes as a past operator of a plant during World War II based upon its decisions that regulated the entire market even if it was not involved in internal plant management. This recent decision is contra to Title II UMTRCA

13. Mined Land Reclamation-Surface Mining Control and Reclamation Act (SMCRA)—Office of Surface Mining Reclamation and Enforcement (OSMRE)

The federal SMCRA 1977 regulates only coal mining. There is no comparable federal act that regulates hardrock surface mining operations. However, a majority of states have enacted their own version of state SMCRAs for non-coal surface mining and the regulations apply to all mining operations within a state's boundaries, whether federal or state lands. Federal pre-emption may apply to BLM lands within a state.

i) Reclamation-Environmental Impact Statement (EIS) not Required: In the California case of City of Ukia v. County of Mendocino and Ford Gravel Co., 238 Cal.. Rptr. 139 (Cal.App.1987), Ukia brought an action to compel the county to set aside approval of a gravel operator's reclamation plan and compel an EIS. The court ruled that the operator's reclamation plan provided sufficient substantive description of natural streambed restoration and did not require an EIS.

ii) Surface Reclamation-SMCRA Versus Leasing Agreements:

(a) Quality Ready Mix v. Mamone (DOR), 520 N.E2d 192 (OH 1988), an Ohio sand and gravel operator appealed an enforcement order of the Division of Reclamation that the operator was delinquent in reclaiming mined land. The trial court voided DOR's reclamation order because the operator had paid $20,000 in a damages action to the landowner for failure to restore the land's surface as called for in the lease. The Ohio Supreme Court held that the duty to reclaim is not contractual, but arises from state SMCRA regulations and reclamation responsibility was not excused by the prior judgment. In a 1992 decision, Quality Ready Mix won reimbursement of the $20,000 paid in damages.

(b) Grove v. Winter, 554 N.E. 2d 722 (Ill.App. 5 Dist. 1990) [see Part I, 4 (xii) supra] an Illinois court held that a lessor of a limestone quarrying property could assert a breach of contract against the quarrier for failure to comply with Illinois SMCRA, which states in part "All grading shall proceed in conjunction with surface mining and shall be carried to completion by the operator prior to the expiration of 11 months after June 30 of the fiscal year in which the mining occurred." [Ill. Rev. Stat. 1985, ch. 96–1/2, para. 4501, et seq., 4507(k)].

(c) Barton v. Gifford Hill & Co., Inc., 760 F.Supp. 98 (W.D. La. 1991), lessor of a sand and gravel property sued the operator for breach of contract for failure to restore the surface. The Louisiana court held that the miner had a duty under the Louisiana Mineral Code to partially restore the land after its operations were finished by sloping the sides of the gravel pit, even if the leasing agreement did not contain any provision for surface restoration.

iii) SMCRA-New Performance Standards:

Several federal SMCRA reclamation procedures and performance standards for postmining lands were litigated in the 1990 case of National Wildlife Federation v. Lujan, 733 F.Supp.419 (D.D.C. 1990), which has been appealed. These procedural changes in federal law will likely be reflected in state regulations for non-coal mining.

Revegetation [30 CFR 816.116 (c)(4)/817.116(c)(4)]: The court upheld the rule that rill and gully repair can be considered normal practice that will not restart extended responsibility period for operator, 31 ERC 1617, 1621–15 (June 8, 1990).

Reforestation [30 CFR 816.116(b)(3)/817.116(b)(3)]: The court upheld the requirement that both state and federal regulatory authority require approval for reforestry and wildlife programs. The success standard requires that 80% of trees be in place for 60% of the applicable period of responsibility (i.e., five years in the East and ten years in the West), 31 ERC at 1625–31.

Reclamation Period of Success for Postmining Land: The court upheld the precipitation requirements, viz., for areas (grazing, pasture, and crop lands) with more than 26 inches of precipitation, two years will meet the success standard of responsibility; for areas with less than 26 inches of precipitation the success standard period shall be met in the last two successive years of responsibility. OSM shall have discretion in accounting for weather variability during the periods. ERC at 1631–33.

Water Impoundments: [30 CFR 816.79, 817.49]: The court upheld rules for class (b) and (c) dams (those that failure would cause loss of life or serious property damage). Temporary impoundments and spillways were not ruled on while reevaluation by the Secretary is being made 31 ERC 2034, 2059–61 (August 1990).

Recharge Capacity of Water for Underground Mining 30 CFR 817.41(b)(2): The court upheld deletion of the requirement that underground mines be required to recharge capacity of the disturbed area and that enforcement was legally "debatable," ERC 1617, 1633–36 (June 30, 1990).

Reclamation Roads: The court upheld the rule requiring any road used to meet reclamation performance standards be retained as a primary road regardless of frequency and nature of traffic. Operators are also responsible for prevention of mitigation of lands within the boundaries of reclamation as listed in Sect. 522(e)(1) that may result from roads into the area.

iv) SMCRA-Valid Existing Rights: In Belville Mining Co., Inc. v. U.S., 763 F.Supp.1411 (S.D. Ohio 1991), the court ruled that OSMRE cannot reconsider a SMCRA determination that a lessee has valid existing rights to strip mine after denial of a mining permit. It also found that deprivation of that right by OSMRE might be found to be a property "taking" under the 5th Amendment. The court ordered OSM to issue the necessary permits for strip mining.

v) SMCRA Rule Challenges and Reviews: In Amerikohl Mining Inc. v. U.S., 899 F.2d 1210 (Fed. Cir. 1990), challenges made to SMCRA rules were found to be exclusively reviewable in the DC Circuit Court of Appeals.

14. Mine Safety and Health (MSHA):

i) Authority to Inspect Non-mining, Mineral Processing/ "Milling" Plants: In Donovan/Sec. Labor) v. Carolina Stalite Co., 2 MSHC 1665 (1985), Stalite was a lightweight aggregate manufacturer who bought slate gravel from a slate quarry adjoining its operation. It did not consider its operation to be a mine and subject to MSHA inspections or authority. It had no interest, stock or corporate relationship with the quarrying company. Stalite denied admission to an MSHA inspector. MSHA threatened closure. An Administrative Law Judge (ALJ) ruled that Stalite was a "mine" for purposes of MSHA. The court ruled that a determination must be made between milling and manufacturing. Milling of minerals comes under MSHA, while manufacturing does not.

ii) MSHA Shutdown- Operator Liability for Miners' Wages: In Clinchfield Coal Co. v. Federal MSH Rev. Comm., 895 F.2d 773 (D.C. Cir. 1990), the court determined where miners are idled by a Sect. 103(k) control order and an MSHA safety order is issued shortly before a withdrawal order, or the fact that the withdrawal order does not specify any violations on the mine operator's part, a compensation award is not precluded, but in fact is extended with prejudgment interest to the miners.

iii) MSHA-Rule challenges:

(a) International Union, UMWA v. MSHA, 900 F.2d 384 (D.C. Cr. 1990): Sect. 101(d) of MSHA provides for judicial review of mandatory health or safety standards promulgated by the SI. In the case at bar, a mining company petitioned on the 60th day, and it was rejected as untimely. The court stated that the 59-day statutory time limit seeking review under 101(k) is "jurisdictional and unalterable."

b) In Sec. of Labor v. Western Fuels-Utah, Inc., 900 F.2d 318, (D.C. Cir. 1990), the court found that supervisory personnel performing miner's duties are subject to the Mine Safety and Health Act's requirements for task training.

c) In Davidson Mining, Inc. v. MSHA, No. 99–1136, slip op. (4th Cur. Aug. 28, 1989), a Review Commission upheld a mine ventilation violation significant and substantial with civil penalty and closure of the mine, the judgment was affirmed. In spite of conflicting evidence of proper ventilation by the operator, the court had discretion to choose the more believable as conclusive.

d) In UMWA v. Dole, 870 F.2d 662 (D.C. Cir. 1989), the court found that the SI's promulgation of new regulations for roof bolt support and removal failed to make the necessary finding and was remanded to the Secretary.

15. Noise Abatement-Noise Control Act of 1972-Blasting

Noise Exposure: MSHA regulation 30CFR, Part 56, provides for the hearing protection of surface mining employees. The limitation for an 8-hour working day is 90 dBA. For every 5 dBA increase in the sound level, the allowable time of exposure is reduced by one-half, e.g., 95 dBA exposure for 4 hrs, 100 dBA for 2 hrs, etc.

i) In Sec. Labor v. Landwehr Materials, Inc., 3 MSHC 2128 (1986), the limestone quarrier was charged and assessed a penalty for excessive noise level violation of 30CFR 56.5–50(b).An MSHA inspector found the shovel operator was exposed to 96 dBA for an 8-hour day in excess of the 90 dBA limit. The shovel operator had been wearing ear plugs/muffs, but the citation was issued because better "engineering controls" were not being used to lessen the noise level.

ii) Noise-Blasting Violations-Permit Revocation-Damage Claims:

(a) In Re: Crushed Rock, Inc. 557 A.2d 84 (Vt. 1988), the Environmental Board revoked a stone quarrier's mining permit finding that he had exceeded the permit's limit on explosive per shot, explosive per delay, and blasted outside the permitted days and hours on several occasions. The court upheld the Board's right to revoke for cause, but remanded for rehearing because the quarrier's right to offer evidence at the revocation hearing had been denied.

b) In Monk v. Peabody Coal Co., 530 N.E. 2d 76 (Ind. App. 4 Dist. 1988), the legal principles apply to any open pit mining. Claimant Monk sued for alleged damages to his home resulting from pit blasting by the mine operator. The court found that Monk had failed to prove any causal connection between the blasting and the alleged damages to his residence. The claim was dismissed.

iii) In Sellersburg Stone Co. v. FMSHRC & MSHA, 3 MSHC 1385, Sect. 56.5–106 (MSHA), requires that at mining and quarrying operations, the faces and muckpiles be examined for undetonated explosives. An openpit limestone quarrier was fined $9,500 in penalties after a driller was injured by secondary drilling into oversize shot-rock which contained undetonated explosives.

iv) Challenge to Blasting Regulations-Differences Between Non-coal and Coal Regulations: In Croner, Inc. v. Department of Environmental Resources, 589 A.2d 1183 (Pa. Cmwlth. 1991), Croner, an open pit coal operator, challenged Pennsylvania's DER restrictions on his overburden and blast limitations. The Pennsylvania Environmental Hearing Board (EHB) dismissed Croner's claim on grounds that his constitutional challenge to the State's Code constituted a challenge to federal regulations under the federal Surface Mining Control and Reclamation Act (SMCRA) and that its jurisdiction was preempted by federal code [30 U.S.C. Sect. 1275 (a)(1)], and Croner's claim could only be brought in the US District Court for the District of Columbia. Croner appealed. The Court found that Croner's challenge to the state's air blast regulations were correctly lodged for a Pennsylvania statute in a Pennsylvania Court and did not have to be filed in a federal court.

Blasting limitations were placed for a dwelling closer than 300 feet in distance. Croner was limited to a peak particle velocity of one (1) inch per second and maintenance of an air over pressure of 132 dBL. The residents of a nearby dwelling were specifically named (Hartman) as not permitted to give a release.

Croner contended that: (1) the limitations were more stringent than those imposed on non-coal blasting operations, (2) a coal landowner is precluded from having blasting limitations waived while a non-coal miner is not, (3) that there was no reasonable basis for distinction between blasting limitations for coal and non-coal mining operators, and (4) the disparate treatment in blasting regulations violated his equal protection rights under the law.

The Commonwealth Court reviewed and found that: (1) the EHB had jurisdiction to hear Croner's claim and should not have dismissed it on grounds of preemption of federal law, and (2) Croner's notice of appeal sufficiently raised an issue of whether a regulation imposing air blast and peak particle limitations violated the statute which permitted a dwelling owner to waive prohibition against surface mining activities within 300 feet of a dwelling.

Since equal protection constitutional rights are challenged, the decision will likely be appealed.

16. Eminent Domain—Regulatory Takings of Mineral Properties[1]

As stated by Thomas J. Schoenbaum, "Environmental land use regulations are particularly subject to attack as a taking."[2] Environmental regulatory preclusion of private mineral land uses may serve as grounds for the landowner to allege a taking of his property.

"Property is deemed taken . . . when it is totally destroyed or rendered valueless, or when it is damaged by a public use in connection with an actual taking by the exercise of eminent domain, or when there is interference with the use of the property to the owner's prejudice with resulting diminution in value thereof."[3] The crucial issue for the courts is whether the government action is a "taking," requiring just compensation, or merely a regulation under the police power, not requiring compensation. As Chief Justice Holmes said in Pennsylvania Coal v. Mahon, "if a regulation goes too far it will be recognized as a taking."[4] The judicial system may review challenged agency final actions to assure that the claimed

---

[1]R. L. Aston, A Review of Recent Litigation of Environmental Regulatory 'Takings' of Mineral Resources, Graduate School, University of Georgia, 1991.

[2]T. J. Schoenbaum, Environmental Policy Law, 1985 ed Note on the Taking Issue, p. 438.

[3]Black's Law Dictionary, 5th ed., p.1 303 (take).

[4]260 U.S. 393, 415–416. This case has not been overruled.

compliance with NEPA is authorized and reasonable and also determine whether a taking has occurred.

i) Keystone Bituminous Coal Ass'n. v. De Benedictis, 107 S.Ct. 1232 (1987)[5]: Although a coal case, the principles involved are germane to nonmetallic mining as well, particularly where the room and pillar system of mining is employed in underground limestone or salt mining, or in the case of longwall mining as used in underground trona mining operation in Wyoming.

In Keystone, the coal companies alleged a taking of coal under the Pennsylvania's Subsidence Act[6] that required the operators to leave in place some 27 million tons of minable coal to prevent subsidence of the surface. The coal companies argued that they were relieved of a duty to support the surface because they had obtained waivers for damage to the surface estates' structures and land.

The courts have determined that it is insufficient for the claimant to show that there has been merely a diminution in value of the property by reason of the alleged taking.[7] The claimant bears the burden of proving that he has been denied of all viable use of the property and that it is virtually worthless.

The Keystone Court held that: (1) there was public purpose for the Subsidence Act, (2) public interests in the legislation were adequate to justify the impact of the Act on the coal companies' contracts with the surface owners, and (3) the Act did not work an unconstitutional taking on its face.

ii) Florida Rock Inds. v. U.S., 21 Cl. Ct. 161 (1990)[8]:

In Florida Rock, the US Army Corps of Engineers (Corps) denied a Section 404 permit to Florida Rock to operate a limestone quarry for crushed stone in Dade County, Florida.

The Sect. 404 permit regulates discharged dredge and fill material into waters controlled under the CWA. After two years of waiting, the Corps denied the application. Florida Rock filed a suit against the United States seeking compensation for a regulatory taking of their land.

The government argued that the denial could not be viewed as a ''taking'' because the Plaintiff was not legitimately entitled to use the land for quarrying. Former court decisions held that quarrying was a ''noxious use'' and a public nuisance.[9] and that ''. . . the government need not compensate 'individual owners for pecuniary losses they may sustain, by reason of their not being permitted by a noxious use of their property, to inflict injury upon the community.' ''[10] The government contended that Florida Rock posed a risk of contamination of the sole aquifer furnishing drinking water for Dade County/Miami. It did not contend that the limestone mining would actually contaminate the aquifer, and it had produced no evidence to quantify that risk. The Court had observed that there were many operational limestone quarries and pits in the area close to the Plaintiff's property. ''Rock mining of the type planned for the property never has been considered a nuisance;'' and such a fact ''belies any claim that a nuisance is involved here.''[11]

The court observed that desirable reclamation uses had been made of the mined-out land in that area. Some former pits were turned into small lakes and overburden that had been removed had been placed to raise the ground level above the highwater level affording building sites for residential development. It was further noted that a zone of development followed depleted mining operations of former years. The Court further stated that, ''here, it is clear that the nuisance exception to the Fifth Amendment's requirement of just compensation is inappropriate.''[12]

The measure of economic impact on the claimant was determined by the ''before and after permit denial'' fair market values of the property, and whether there existed any other uses for the property.

The Court accepted a valuation of $500/acre after denial, and a predenial fair market value of $10,500/acre. Limestone mining was found to be the highest and best use for the property. In this case, the diminution of value was not slight, but substantial, being diminished by 95% of its prepermit denial value. The fact that Florida Rock had to purchase other property to meet its mining needs illustrated the extent of the economic impact on the company.

In summation, the Court quoted the US Supreme Court ''When a regulation goes too far, the Fifth Amendment requires compensation.'' (Pennsylvania Coal, 260 U.S. at 413.) The Court's decision in Florida Rock was: (1) the proposed use of the property for limestone mining would not have constituted a nuisance and (2) denial of permit constituted a taking for which landowner was entitled to damages in amount of full fair market value of property at the time of taking. The amount of damages was fixed at $1,029,000, plus interest, from the date of the taking. The permit denial date was the date of taking.

iii) Whitney Benefits, Inc.& P. Kiewit Sons' Co. v. United States, 18 Cl.Ct. 394 (1989) in another post-Keystone coal land taking case, Whitney Benefits, a regulatory taking was found by virtue of a subsequent Congressional enactment and denial of a surface mining permit. Environmental restrictions in the legislative act were the basis for the prohibition of mining.

Part of the SMCRA's guidelines provided that no permit or application shall be approved if it should ''interrupt, discontinue or preclude farming on alluvial valley floors that are irrigated or subirrigated. . . .'' Thus, Whitney's right to mine the coal on its property, located on an alluvial valley floor, was invalidated by the enacted legislation of SMCRA and was the basis for the alleged taking in 1983.

The government argued that an exercise of police power in carrying out government regulations for the health and welfare of the public can never result in a taking. The Court responded by quoting from Florida Rock that ''it is no longer asserted that a regulation, just by its nature of being a regulation, cannot be an exercise of eminent domain.''

After extensive market and mining cost investigations,

[5]This case has not been overruled.

[6]Subsidence Act, Pa. Stat. Ann., Tit. 52 Sect. 1406.1 et seq. (Purdon Supp . . . 1986).

[7]Keystone, id. p.1246, quoting Agins v. City of Tiburon, 447 U.S. 255. Agins has not been overruled.

[8]This case has not been overruled.

[9]58 Am.Jur.2d, ''Nuisances'' Sect.7: ''public nuisance'' has been defined as the doing or failure to do something that injuriously affects the safety, health, or morals of the public . . .'' Whatever tends to endanger life or generate disease, and effects the health of the community, etc. . . . is a public nuisance.'' State v. Turner, 198 S. C. 487, 18 S.E.2d 372. Turner has not been overruled.

[10]Florida Rock, id. at 166, quoting Keystone, id. at 1244, relying on Mugler v. Kansas, 123 U.S. 623, at 668–669.

[11]id. at p. 167.

[12]id. at p. 167.

the Court established a final sum of $60,296,000 for the total 1977 value of recoverable Whitney Coal for an assumed annual production rate of 2.5 million tons and a cost of $2 million for backfilling. Interest was payable to Whitney from Aug. 3, 1977, to date of payment.

The US Claims Court held that: (1) the enactment of SMCRA totally eliminated economic value of plaintiffs' coal and constituted a taking under the Fifth Amendment, (2) the taking occurred at the time SMCRA became effective, (3) the valuation method incorporating discounted cash flow approach offered reliable method for determining the fair market value of the coal on the date of the taking, and (4) the plaintiffs were entitled to prejudgment interest.

iv) United Nuclear Corp. v. United States, 912 F.2d 1432 (Fed. Cir. 1990) another recent mineral land taking claim concerned a uranium property. United Nuclear Corp. (United) had been awarded two mining leases on reservation land by the Navajo Tribal Council. United lost its leases by undue delay of the Secretary of Interior to approve the United's mining plans. United had spent over $5 million in exploration costs and had found valuable uranium deposits. The Court found that there had been a taking of United's property interests in the mining leases. The case was remanded to the Claims Court to determine the amount of just compensation.

v) Yuba Natural Resources, Inc. v. United States, 904 F.2d 1577 (1990) in Yuba, a temporary taking of a mining property for six years was found. The mineral title of the property the company had operated and dredge-mined for gold for sixty years came into question by the United States and was finally taken in 1976 . The government notified Yuba that it was the fee simple owner and further dredging was prohibited until the Corps of Engineers had completed impact studies. Yuba began a quiet title action to establish its title, finally gaining title to its property after six years. Thus, the "taking" was considered "temporary," even though the government's intent at the time of "taking" was meant to be permanent.

The basis for just compensation was determined to be the fair rental value for use of the property for the period of the temporary taking. The Court ruled that Yuba was not entitled to recover the difference between the value of gold during the period of taking, and the value of gold at the end of taking, i.e., lost profits.[13] A value of $580,555.40 was fixed by the court for six years. Yuba contended that they should be awarded additional royalty amounts for lost profits had the property been operated as was their intention. Yuba's intended lessee was to pay them 4 and 6% royalties on net sales of gold. The court denied Yuba's argument stating that lost profits were consequential damages and not compensable in a temporary taking.[14]

COMMENT ON YUBA- It should be noted that Yuba had a sixty-year production record which was a sound basis for extrapolating production and royalties for the six years of erroneous taking. The denial of royalties meant a loss of between 35 and $690 million to Yuba. There is seeming inconsistency in the Court's argument to establish the fair value during the six-year period. The court stated "just compensation" is the value of the use of the property during the temporary taking, i.e., the amount which the owner lost as a result of the taking. Yet the court denied the additional royalties to Yuba which Yuba would have realized by continued operation if not interrupted by a wrongful taking by the government, and which was not meant to be temporary, but permanent.

COMMENT ON TAKINGS—Keystone was a fair decision. No mining entity should be allowed noxious-use activity of their property that harms another's property (except by waiver), or the public health and welfare. As to alleged values for compensation of losses for less than full use of its properties and recovery of the coal reserves required to be left in place to support the surface, the Court's analogy was right on point that a statutory "requirement that a building occupy no more than a specified percentage of the lot on which it is located could be characterized as a taking of the vacant area as readily as the requirement that coal pillars be left in place. Similarly, "one could always argue that a set-back ordinance requiring that no structure be built within a certain distance from the property line constitutes a taking because the footage represents a distinct segment of property for takings law purposes."[15] Obviously, as Kusler and Meyers state, "Not every square foot of private property can be used for its 'highest and best use'."[16]

vi) Crown Lands-Town and Country Planning Act [C.O.D. 149–219]-British Commonwealth, the Town and Country Planning Act in the United Kingdom, as enacted by several of the British Commonwealth nations, bears a slight resemblance to the US power of eminent domain when a taking has occurred by its provision for a compensatory award to a property owner where permission is denied for a mining use (et al.).

In Lopinot Limestone Ltd v. Atty-Gen of Trinidad and Tobago, (P.C.) [1987] WLR 797, Lopinot Limestone applied for a permit to develop 22 acres of its 200-acre tract as a limestone quarry under the Town and Country Planning Act. The Minister refused.

Section 26(1) of the Act provides for payment of compensation if planning permission is refused. Upon refusal, Lopinot sought the compensation from the Minister of Finance, but was denied. Lopinot brought an action against the Attorney-General in the High Court of Trinidad and Tobago. On final appeal, the Privy Council held that compensation would be payable to Lopinot in the amount equivalent to the difference in the land's value with permission for quarrying limestone (subject to limitations) and its value without permission. Lopinot's application for quarrying permission was remanded to the Minister, and if refused again, Lopinot would be entitled to compensation subject to the provisions.

17. Miscellaneous

a) Coastal Zone Management Act (CZMA): The Act [(16 U.S.C. 1451–1464 (1988)] requires all coastal states to develop a management plan. There are approximately 30 states under the Act, viz., states bordering both Pacific and Atlantic Oceans, states bordering the Great Lakes, and Nevada bordering Lake Tahoe.

A state CZM plan, approved by the US Secretary of Commerce, gives a state great power over activities, including permitting. A federal agency cannot issue a permit if the proposed mining activity is inconsistent with the state's plan. However, the states must operate under regulations of the US Department of Commerce (15 CFR 930). Those regulations require a public-interest review pro-

[13] Yuba Natural Resources, Inc. v. U.S., 904 F.2d 1577, at p. 1581.

[14] id. at p. 1583.

[15] Keystone, id. at p. 1249.

[16] Kustler & Meyers, National Wetlands Newsletter, Nov./Dec., "Litigation: Is the Claims Court All Wet?" p. 7.

cess The applicant for mining must certify to the state that the project is consistent with the state's CZM plan. For a CMZA case, see California Coastal Commission v. Granite Rock Co., 107 S.Ct. 1419 (1987).

As part of the Omnibus Budget Reconciliation Act of 1990, Congress enacted provisions which reauthorize and fund the CZMA until 1995.

b) The Endangered Species Act (ESA)(16 U.S.C. 1531): Permits will be denied for mining projects that will harm any species or habitat. The applicant must provide project data and information for a ''jeopardy opinion'' from the US Fish & Wildlife Service (USFWLS).

In April 1990, the USFWLS designated the desert tortoise a threatened species. Mining activities on public lands in the southwest desert areas of California, Nevada, Arizona, and Utah are subject to the consultation requirements of Sect. 7 of the ESA.

Other wildlife acts that could affect a mining project are: Fish & Wildlife Coordination Acts, Migratory Birds Treaty Act, Bald Eagle Protection Act, Wild Free-Roaming Horses and Burros Act, Marine Mammal Protection Act, Fisheries Conservation & Management Act, etc.

c) The National Historic Preservation Act (NHPA)(16 U.S.C. 470): The permitting agency is required to give the Advisory Council on Historic Preservation an opportunity to comment on any proposal that may affect properties listed, or eligible for listing, in the National Register of Historic Places.

d) The Native America Grave Protection and Repatriation Act [(Pub. L. No. 1011–106, 104 Stat. 3048 (1990)]: A provision of this Act concerns any development activity on public lands that any person who discovers, or has reason to know of, Native American cultural items in the course of activities such as mining, construction, etc., must stop all activity in the area of discovery and make a reasonable effort to protect the items, notifying the secretary of the agency with primary management authority over the land. Mining and construction activities may resume following notification and after a reasonable period of time.

National Indian Youth Council v. Andrus, 501 F.Supp. 649 (D.N.M. 1980), aff'd. 664 F.2d 220 (10th Cir. 1981), is a leading minerals case to date. In National Indian Youth, a federal mining lease was involved in an area of New Mexico where the Indian Council felt that Native American archeological and paleontological culture was in jeopardy from mining. The Indians argued that studies and surveys should be completed before a lease was given. The court determined that the award of a mining lease did not constitute approval of a mining plan. The court further noted that to complete a cultural study on a 40,000-acre tract before mine planning for a mere 8-acre tract bordered on the ''absurd.''

e) Criminal Penalties for Environmental Violations

It is apparent from the 1990 reports of the US Department of Justice (DOJ) that criminal sentences for environmental offenses are on the increase. A record number of successful convictions was reported as 95% of the indictments. Jail sentences were given to more than half of those convicted and large fines were levied on corporations.

Title VII of the 1990 Amended CAA made criminal the release of hazardous pollutants into the air by a Knowing Endangerment clause. An individual violator is subject to 15 years imprisonment, and a corporate violator fined up to $1 million. Lessor, misdemeanor imprisonment up to one year and fines may be made for negligent pollution releases. In addition, the amended Act authorized a greater number of environmental inspectors (''police'') for stepped up enforcement. Additional provisions have also been made to coordinate the CWA with violations other environmental statutes. Knowing violations of reporting and monitoring requirements are punishable as two-year felonies. Five-year imprisonment and fines may be imposed for knowing violations of implementation plans, new source performance standards, etc. Primacy states may also impose their own punishments. The case of People v. Martin, 211 Cal. App. 3d 699 (1990), decided that criminal sanctions were also applicable to persons who ''should have known'' that their act of unpermitted hazardous waste disposal would cause injury and harm. The California Code provides for imprisonment on conviction for 16 to 36 months in state prisons.

A policy known as The Responsible Officer Doctrine was approved in the Case of U.S. v. Dee 912 F.2d 741 (4th Cir. 1990). The case serves as precedent in case law and empowers a federal judge to instruct a jury to convict a manager of an operation for a felony based on failure to prevent or detect an environmental violation. Numerous cases in 1990 illustrate the serious nature of court-imposed jail sentences and fines for managers and corporate officers who were responsible for environmental violations.

FINAL COMMENT: At present, environmental concerns and regulations for mineral resources activities are suffering ''growing pains.'' A resolution for compatibility for the essential mineral needs of man and safeguarding the environment simultaneously will be forthcoming, given time, truth, and less hysteria. Environmentalists are prevailing with the everpresent NIMBY syndrome in public zoning and mine permitting hearings. It is encouraging to note that Vermont Supreme Court Justice Dooley (joined by Justice Morse) stated in his dissent in the case In re: John Meaker for zoning permitting [see Part I (6)(x)(b)], ''Under the standards imposed, we are in serious danger of expanding not-in-my backyard into not anywhere.''

It behooves the minerals industry as a whole, to join the environmental movement as prime supporters for a cleaner environment. Improving the bad image that mining has had for decades in the eyes of the lay environmentalist, can only result in large benefits to the mining industry. Mining can no longer resort to its age-old claim that it is a basic industry necessary for man's beneficial living and must be tolerated for better or worse. When left to a choice in zone permitting for a new mining operation that creates new jobs and taxes, the lay citizen will choose clean air and water first.

2

COMMODITIES AND USES

ABRASIVES

WILLIAM W. WELLBORN

Abrasives are substances both natural and synthetic that are used to grind, polish, abrade, scour, clean, or otherwise remove solid material usually by rubbing action (as in a grinding wheel), but also by impact (pressure blasting). Facts about the discovery, early use, adaptation, and development of abrasives prior to the 20th century are few. From the sparse, earliest information that is available, it would appear that the science of abrasives began when, to fashion his tools, man began to select certain kinds of rocks in preference to others because of their peculiar properties. There is evidence that stones were sawed by some sort of grinding machine as early as 4000 BC in Egypt. The real expansion of abrasives use coincided with the beginning of metallurgy in the Middle East around 2000 BC. The earliest confirmed use of abrasives for grinding metal comes from a steel dagger found along with a sharpening stone which was dated at about 1500 BC.

Natural and manufactured abrasive substances play an important role in fashioning and finishing numerous products with a wide variety of end-use applications. The mining and preparation of crude, natural abrasives for their markets and the manufacture of synthetic and other abrasive materials contribute significantly to the US economy (Table 1). In addition, just as sawing to specific dimensions enhanced the value of the Egyptian stone, the use of abrasives in fabrication of modern products, and the addition of abrasives to others, add value.

GENERAL CONSIDERATIONS

The most important physical properties of materials that qualify as abrasives are hardness, toughness (or rigidity), grain shape and size, character of fracture (or cleavage), and purity (or uniformity). To fabricate bonded abrasive products such as grinding wheels, additional considerations include stability under high heat and bonding characteristics of grain surfaces. The economic factors of cost and availability are always important.

No one singular property is paramount for any use. For some applications extreme hardness and toughness are needed, as in diamonds for drill bits. For other uses, factors of greatest importance are hardness and ability to break down slowly and, in the process, develop fresh cutting edges when grains become worn. In garnet sandpaper, for example, highly cleavable, friable, or extremely tough grains are not desirable. For other uses, extreme hardness may be objectionable. Abrasives for dentifrices and for glass-cleaning soaps are cases in point. For the most efficient uses in more critical applications, different types of abrasives are rarely completely interchangeable. While crushed quartz and garnet are both used in sandpaper, the two abrasives are not at all interchangeable in their applications.

In the final analysis, the choice of a high-grade abrasive depends upon the quality and quantity of work performed by the abrasive per unit of cost. Initial cost of an artificial abrasive may be much greater than that of a natural abrasive, but the artificial mineral may do so much better work and do it so much faster that the ultimate cost is less. It is for this reason that artificial abrasives have largely replaced natural abrasives.

Abrasive Value

Mineralogical hardness or "scratch" hardness as expressed in Mohs scale is an important property in evaluating abrasive materials, but it is only one of several essential properties. The mineral hardness of pure crystal almandite garnet is about 7.5. However, if the crystal is crossed by incipient fracture planes, or if it contains inclusions of other minerals, the useful hardness may be much lower. While the quartz grains in a sandstone have a hardness of 7, the bond holding the grains together may be so weak that the stone is valueless as a commercial abrasive. In artificially bonded wheels and stones; the hardness, strength, and character of the bond are fully as important as the hardness of the abrasive grains. Thus, in an overall consideration of abrasive hardness of loose abrasive grains, both "scratch" hardness and toughness must be considered. In natural or artificially bonded abrasive stones, a third factor, the character of the bond, is of equal if not greater importance.

The problem of abrasive hardness is further complicated by the shortcomings of methods for testing hardness and for expressing relative values. The Mohs scale is inadequate because the methods of testing are very crude and the intervals between steps in the scale are not uniform. There is far less difference between a hardness of 6 and 7 on the scale than between 9 and 10. Numerous attempts have been made to remedy these deficiencies. Ridgeway, Ballard, and Bailey (1933) proposed an extension of Mohs scale to include artificially prepared substances. Knoop, Peters, and Emerson (1939) devised a diamond indentation method of measuring hardness which gives reproducible results on a wide variety of materials and provides definite numerical values over a wide range, up to diamond hardness. Knoop hardness of various abrasive materials, given in Table 2, are mostly from Thibault and Nyquist (1947), as reported by Hight (1983). Hardnesses of corundum and aluminum oxide plus titania are as reported by Coes (1971).

Another consideration, perhaps, even more important than true hardness in appraising abrasive value, is toughness or friability. Very hard materials are neither necessarily tough nor resistant to fracture. Diamond is an excellent example. It is the hardest material known to man, yet it can fracture easily under impact. Several techniques for testing friability have been developed and are in widespread use in the quality control of abrasive grain. The basic operating principle common to all strength-of-grain tests is to impact a given sample of abrasive grain under a set of standardized conditions followed by measurement of the amount of breakdown. Naturally, the higher the survival rate—the tougher the grain, other factors being equal.

Table 1. Salient US Abrasives Statistics

		1986	1987	1988	1989	1990
Natural abrasives production by producers:						
Tripoli (crude)	tons	106 298	104 259	99 928	105 230	94 390
Value	thousands	$918	$975	$864	$2 537	$3 194
Special silica stone[a]	tons	973	1 378	1 892	898	454
Value	thousands	$501	$489	$566	$147	$230
Garnet[b]	tons	24 298	38 353	42 506	42 605	47 009
Value	thousands	$2 603	$4 350	$4 707	$4 408	$6 939
Emery	tons	2 611	1 764	869	W	W
Value	thousands	W	W	W	W	W
Staurolite	tons	W	W	W	W	W
Value	thousands	W	W	W	W	W
Manufactured abrasives[c d]	tons	438 043	[r]462 773	[r]544 899	[r]535 213	514 869
Value[d]	thousands	$173 858	[r]$167 593	[r]$209 998	$227 761	$218 207
Foreign trade (natural and artificial abrasives):						
Exports (value)[e]	do.	$207 624	$238 522	$281 633	$260 363	$270 928
Reexports (value)[e]	do.	$27 011	$21 192	$19 302	$33 771	$24 545
Imports for consumption (value)[e]	do.	$406 572	$424 640	$501 707	[r]$419 084	$469 781

Source: Adapted from Austin, 1991, Table 1.
[a]Includes grindstones, oilstones, whetstones, and deburring media. Excludes grinding pebbles and tube-mill liners.
[b]Primary garnet; denotes first marketable product. Includes crude concentrates.
[c]Includes Canadian production of crude silicon carbide and fused aluminum oxide and shipments of metallic abrasives by producers.
[d]Excludes United States and Canadian production and value of aluminum-zirconium oxide.
[e]Bureau of the Census.
[r]Revised. W withheld to avoid disclosing company proprietary data.

Table 2. Relative Hardness of Abrasive Materials on the Knoop Scale

Substance	Knoop Hardness (K100)
Quartz	820
Spinel	1 270
Garnet	1 360
Topaz	1 340
Alumina/25% Zirconia	1 450
Alumina/40% Zirconia	1 600
65% Alumina/30% Zirconia/5% Titania	1 700
Sintered Bauxite	1 900
Corundum (parallel to C axis)	1 950
Al_2O_3 + 3% TiO_2	1 950
Corundum	2 050
Heat treated fused aluminum oxide	2 090
Corundum (perpendicular to C axis)	2 150
Al_2O_3 + .3% Cr_2O_3	2 150
Silicon Carbide	2 480
Boron Carbide	2 760
Cubic Boron Nitride	7 800
Diamond	8 000

Source: Adapted from Hight, 1983, Table 1.

A simple and inexpensive device for performing this test is an ordinary ball mill. A detailed description of this test is given in the "American Standard Recommended Practice for Ball Mill Test for Friability of Abrasive Grain" (Anon., 1987).

Grain shape is another fundamental physical property which must be considered in selecting the best abrasive for a particular application. Blocky and nearly equidimensional grains are identified as strong-shaped and are less friable than slivery or flaky grains which are considered weak-shaped. In grinding wheels strong-shaped grains are preferred in extra heavy duty applications involving high heat and pressure, while weaker-shaped grains are selected for precision grinding to close tolerances and where high surface finishes are required. Strong- and weak-shaped grains, respectively, are illustrated in Figs. 1 and 2.

In modern abrasive technology, close control of abrasive shape is necessary. The determination of the bulk density of an abrasive is a dependable indicator of shape, inasmuch as bulk density is a function of shape. A sample of weak-shaped grain will not pack as efficiently and will have a lower bulk density than the same abrasive with a strong shape.

Various techniques are commonly used for the measurement of the bulk density (shape) of abrasive grain. The "loose-pack density test" is described in "American National Standard Test for Bulk Density of Abrasive Grain" (Anon., 1988). In performing this test, care must be taken to avoid vibration. The container is filled to overflowing and then carefully leveled with a straight edge. The weight of the grain filling the container is obtained and the bulk density calculated using the traditional formula $D=W/V$ in which D is the bulk density of the sample in grams per cubic centimeter; W is the weight of the grain in grams; and V is the volume of the cylinder in cubic centimeters.

Characteristics other than shape also influence the strength of abrasive grains. The manufacture and processing of aluminous abrasives permit control of some of these important characteristics. For example, it has been clearly established by laboratory and field tests that the finer the crystal size of an abrasive, the stronger that abrasive will be. The cooling history of a fused alumina melt is the dominant factor influencing the crystal size of the finished product. Perhaps at the cost of over-simplification, it may be said that rapid cooling will result in fine crystal size and slow cooling will produce coarse crystals. As a frame of reference, a slow-cooled product might typically have a crystal size on the order of 3 000 μm and a rapidly cooled product, 400 μm.

A rapid and inexpensive quality control check for crystal size is the "monocrystalline count." Any convenient grit size can be used, but 24-grit has been found by experience to be ideal. Each grain is examined on a petrographic microscope to determine whether it consists of a single crystal fragment or, perhaps, of two or more crystals. The former grain is monocrystalline, while the latter is polycrystalline. A count is made of, say, 100 grains and the

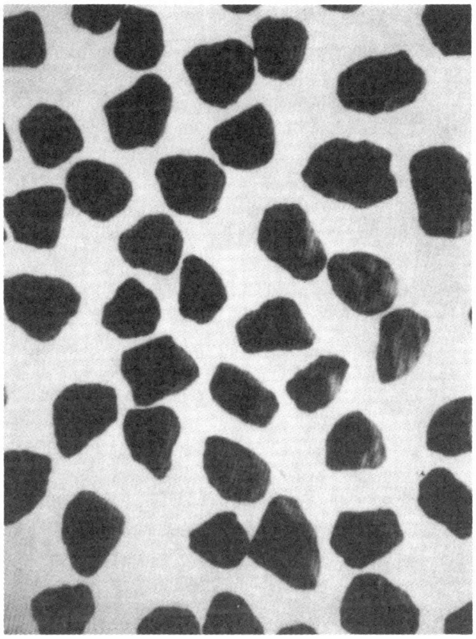

Fig. 1. Strong-shaped grain.

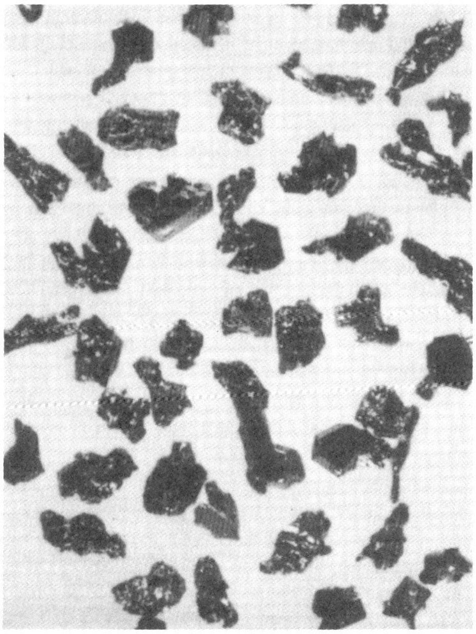

Fig. 2. Weak-shaped grain.

percentage which is found to be monocrystalline is by definition "the monocrystalline count." Generally speaking the higher the monocrystalline count the weaker the grain. It is theorized that a polycrystalline grain is less apt to disintegrate since a fracture starting through the grain might pass through one crystalline unit making up the grain, but then be deflected upon encountering another crystalline unit not in crystallographic continuity with the first.

A secondary treatment which has been found to increase the strength of fused aluminum oxide abrasive grain is heat treatment or roasting. Here it is concluded that the heated slag (impurities between crystals) softens and flows out onto the surface of the grain, cementing imperfections and healing incipient fractures.

TYPES OF ABRASIVE PRODUCTS

Loose Abrasive Grains

Abrasive grains are produced from a wide variety of materials, both for use as grains and for incorporation into other products. Thus abrasive grains are not only important products themselves, but are also the starting point for making bonded shapes, coated abrasives, abrasive tools, polishes, cleaners, grinding pastes, and other compounds. For pressure blasting, in addition to silica sand, other natural mineral grains such as corundum, garnet, flint, and chert are used. Manufactured products such as fused alumina, silicon carbide, and steel shot are also employed. All except the steel shot have a hardness of 7 or more. Besides hardness, the physical properties of toughness, grain shape, grain size, uniformity, and specific gravity are important. Grains should be tough so that they will not disintegrate readily under impact. For some uses, rounded grains are considered desirable. For others, sharp cutting edges are favored. Uniformity of grain size and other physical properties are always desirable. The higher the specific gravity, the greater is the force of impact for grains of equal size. Each use has its own special requirements. Common uses for pressure blasting are: cleaning stone and concrete, cleaning metal castings, preparing surfaces for painting, and etching glass and plastics. Loose grains of relatively coarse sizes are used for sawing stone, rough grinding plate glass, and surfacing stone. Materials applied include quartz sand, garnet, corundum, emery, aluminum oxide, and silicon carbide.

Grains of somewhat smaller sizes of the same materials find applications for grinding lenses, rough polishing gems and ornamental stones, and dressing and polishing wood surfaces. Fine-grained powders are specified for polishing and lapping glass, tile and artificial stone flooring, gems, semi-precious stones, and wood surfaces. The abrasive materials used are rouge, crocus, tin oxide, aluminum oxide, chromium oxide, cerium oxide, diamond dust, feldspar, garnet, rottenstone, pumice, diatomite, tripoli, ground silica, clay, whiting, and zirconium oxide. Most automobile body cleaners and polishes contain diatomite. A familiar bathroom and kitchen scouring powder for glass and porcelain items contains feldspar or volcanic pumice. In one method of drilling small holes in glass, porcelain, gem stones, and similar materials, a brass tube charged with one of several abrasives (diamond, corundum, emery, aluminum oxide, silicon carbide, or boron carbide dust) is used in a vehicle of water, oil, or grease. Diamond-tipped drills are also applied for this purpose.

Emery, magnetite, silicon carbide, and aluminum oxide are cast into the treads of steps and ramps to increase traction.

Bonded Abrasives

Abrasive grains, closely sized, are bonded and pressed or molded into a wide variety of bonded abrasives such as grinding wheels, honing stones, and tumbling media. Originally the abrasives in such products were corundum or emery. With the introduction of the electric-furnace abrasives, aluminum oxide and silicon carbide, the use of the natural abrasives has declined greatly.

There are five main types of bonded abrasives depending on the type of bond and method of manufacture:

1) Vitrified wheels with a clay-feldspar bond which is vitrified in ceramic kilns. Because of their high rigidity and dimensional stability, vitrified bonded wheels are preferred for precision grind-

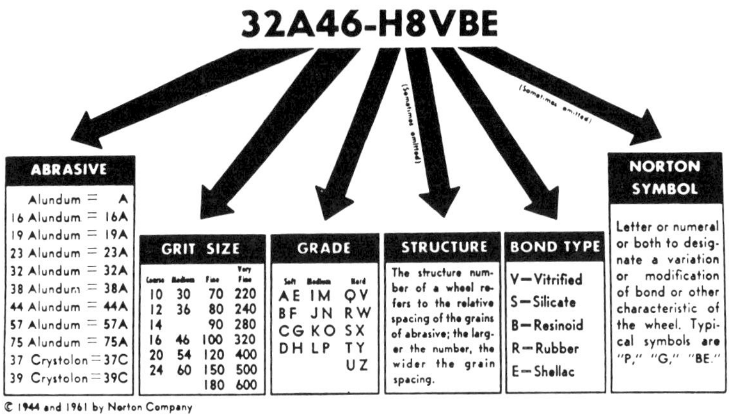

Structure numbers are not always shown in the marking.
However, such products may nevertheless be made to a definite structure.

Fig. 3. Grinding wheel marking system.

ing operations. They are unaffected by water, acids, oils, and ordinary temperature variations.

2) Resinoid wheels with a hard synthetic resin bond. These are high speed wheels used in foundries, in welding shops, and for billet conditioning. They are also used in cutoff and thread grinding operations.

3) Rubber wheels, bonded with natural or synthetic rubber, are somewhat elastic in nature. They are used for grinding ball races, most centerless feed wheels, and for portable snagging operations where finish is an important consideration. Rubber-bonded cutoff wheels can be made very thin.

4) Shellac wheels used for producing high finishes on such items as camshafts and paper mill rolls.

5) Silicate bonds with application in operations where heat generated in grinding must be kept to a minimum. Silicate bonded wheels are mild acting and are used in grinding edge tools of all kinds.

Wheels are made that vary in at least five physical properties aside from size and shape:

1) type of abrasive grain (aluminum oxide, silicon carbide, diamond, etc.);

2) grit (size of abrasive grains ranging from 8- to 1500-mesh);

3) grade (strength of bond encompassing 18 to 20 grades);

4) structure (grain spacing with 10 to 12 spacings);

5) bond (vitrified, etc., five bond types).

Fig. 3 shows a wheel marking system that is widely used in the abrasive industry (Anon., 1970).

When all these variables are combined with an almost infinite number of sizes, shapes, and spindle diameters that are either standard or available on special order, the complexity of the abrasive wheel business is apparent.

In addition to grinding wheels, bonded abrasives are made into blocks, bricks, and sticks, and used as sharpening and polishing stones: oil stones, scythestones, and as razor and cylinder hones. Curved blocks and segments can be combined into large wheels, such as pulpstones, or used mounted or unmounted for grinding or polishing the interior or exterior of curved surfaces.

Rubber pencil and ink erasers contain abrasive grains and similar soft rubber wheels, sticks, and other forms are made for finishing soft metals.

Coated Abrasives

Coated abrasives consist of sized abrasive grains cemented to a paper or cloth backing. Originally ''sandpaper'' was perhaps coated with silica sand, but today crushed quartz is used on silica-type paper because its grains have much sharper cutting edges. The principal abrasives used today for this purpose are aluminum oxide, silicon carbide, alumina-zirconia, sol-gel alumina, crushed quartz, and garnet. Most ''emery'' cloth today is coated with silicon carbide or aluminum oxide. ''Flint'' paper in the United States is coated with crushed quartz, but true flint is used in Europe. With the exception of sintered abrasives, all abrasive grains, natural or artificial, are made by graded crushing and close sizing. Crushing usually is done in rolls to avoid producing extreme fines for which there is little demand. Close screening is situated between the sets of two rolls. Products are passed through magnetic separators to remove iron-bearing material, abraded from the rolls, and again precisely sized on screens. Careful sizing is essential to prevent contamination of grain grades. The finest powders are sized by air or water flotation and sedimentation.

In earlier practice, grade numbers were more or less arbitrarily assigned to grain sizes and garnet, flint, and emery scales were all different. Artificial abrasives were given numbers based on screen mesh designations now specified by the American National Standards Institute (Anon., 1977). Gradually garnet and ''flint'' paper grade numbers have come to include both the old and the new numbers. Table 3 shows the relationship between the different scale sizes.

The bond for coated abrasives may be either glue or a synthetic resin adhesive. The backing is either paper or cloth. Garnet and ''flint'' papers and cloths are most widely used for wood, leather, hard rubber, plastics, felt, and rubbed paint and varnish finishes on metals. Cloth, coated with silicon carbide, is used chiefly in the metal working industries. Alumina-zirconia and Sol-Gel abrasives are finding increasing use in heavy duty cloth belts for industrial use.

Grains and Powders for Soap, Cleaners, and Polishes

Many different materials, mostly natural, but some manufactured, are used in making soaps, cleaners, and polishes. Feldspar, pumice and pumicite, sand, ground quartz, tripoli, diatomaceous earth, clay, and wood flour are ingredients in hand and scouring soaps. Often low price is the primary factor in the selection of the abrasive. Occasionally, the abrasive is an important factor in the formulation and use of a product. A household cleanser should not contain quartz or any mineral of hardness equal to or greater than that of glass or that of enamels, nearly all about 6 or lower. Otherwise, the glass or enamel surface would be readily scratched. Such

Table 3. Size Grades for Abrasives

Modern Mesh Designation	Old Garnet Scale	Flint Scale	Emery Scale
400			
320			
280	8/0		
240	7/0[a]	5/0	
220	6/0	4/0	
		3/0	
180	5/0		3/0
150	4/0	2/0	2/0
120	3/0		1/0
		1/0	
100	2/0		
		1/2	1/2
80	1/0		1
		1	
60	1/2		1 1/2
50	1	1 1/2	2
		2	2 1/2
40	1 1/2		
36	2		3
30	2 1/2	3	
24	3	3 1/2	
	3 1/3[b]		

Source: Adapted from Hight, 1983, Table 2.
[a]Approximately 220-mesh Tyler.
[b]Approximately 20-mesh Tyler.

cleansers should not contain calcium carbonate, calcium sulfate, or other easily reactive lime compounds since lime reacts with most soaps to form insoluble substances that are difficult to remove from glass or enamel surfaces.

An ideal mineral abrasive for soaps, cleaners, and polishes would be a mineral or rock between 3 and 5 in hardness, occurring in abundance in easily mined deposits, close to rail transportation, and within a reasonable freight-haul distance of important markets. The material should be uniform in texture and physical properties. It should not contain any impurities harder than 6 and no reactive impurities such as lime compounds.

The ultimate grain size of abrasives in this group depends upon the finished product. Sizes range from extremely fine air-separated and water-floated particles with sizes expressed in micrometers (much finer than 325-mesh), up to 100-mesh or even coarser for heavy-duty scouring soaps.

CLASSIFICATION

Abrasives may be divided into two general classes, natural and manufactured. The former includes all rocks and minerals used for abrasive purposes without chemical or physical change other than crushing, shaping, or bonding into suitable forms. Manufactured or artificial abrasives are made either by heat or chemical action from metals or mineral raw materials. Table 4 lists most of the important abrasives, classified as to inherent types and the forms in which they are used industrially.

For most types of applications, there are manufactured products that can be substituted for natural products, usually at higher initial cost, but with greater efficiency. This is not always the case. For example, there is no satisfactorily manufactured garnet substitute for making coated abrasive paper and cloth. For some abrasives, such as chaser stones and others with declining use, the production of manufactured substitutes has not been economically attractive. For a low-priced commodity such as pressure blasting sand; steel shot, fused aluminum oxide, and silicon carbide grains may be substituted.

The use of many natural abrasives such as sand and quartz has seriously declined because of government regulations which impose tight restrictions on their use. Free crystalline silica has been determined to be a carcinogen. This reduction in usage of siliceous abrasives has been accompanied by replacement with manufactured abrasives. The transition has not been a net loss to the mineral industry since virtually all manufactured abrasives are made from mineral raw materials.

Natural Abrasives

In this book separate chapters are devoted to such commodities as diamonds, diatomite, tripoli, pumice and pumicite, silica sand, and quartz. Thus, in this chapter reference to these rocks and minerals is restricted to their use as abrasives.

Corundum and Emery: Corundum and emery have become relatively unimportant on a world scale and from the standpoint of domestic production and consumption. No corundum is mined today in the United States or Canada, and, in 1990, there was only one North American producer of emery, Oregon Emery, Albany, OR. Imports from Greece and Turkey accounted for most of the domestic consumption in 1990. Most of the emery is used as hardener for heavy duty concrete floors, as nonslip material on bridges and at toll booths. There is some limited use of the material as tumbling media.

Production. It is estimated that the United States consumed approximately 9 kt of emery in 1990 (Austin, 1991). Production of processed emery in recent years has come from three producers and has averaged about 9 ktpy.

Diamonds—Industrial: The United States has no production of natural diamonds and is therefore dependent on other countries for its natural diamond supply.

Natural Industrial Diamonds. There are four major types of natural industrial diamonds: 1) bort which includes off-color, flawed or broken fragments of diamonds unsuitable for gems; 2) carbonado or black diamond which is a very hard and an extremely tough aggregate of very small diamond crystals; 3) ballas, a very hard, tough, globular mass of diamond crystals radiating from a common center; and 4) diamond crystals which are used for grinding wheel dressing. Carbonadoes come only from Bahia, Brazil; ballas, chiefly from Brazil, but some from South Africa. Bort comes from all diamond-producing centers. In addition, there is a considerable production of diamond dust and powder waste from cutting gem diamonds.

Uses. The industrial diamond has become one of the most important and essential materials in modern industry. Diamond drilling, once used only for locating metallic ores, is now widely used for exploring geologic structures for oil and gas, testing foundations of dams, buildings and heavy machinery, exploring the internal condition in heavy concrete structures such as dams, step mining, explosive demolitions under special conditions, and for miscellaneous other purposes.

Some of the most important uses are: diamond drill bits for drilling rock and concrete; diamond dies for wire drawings; diamond tipped tools for truing abrasive wheels and for turning and boring hard rubber, fiber, vulcanite, hard plastics, etc; diamond-toothed (segmented) saws and rim-impregnated (continuous rim) saws for sawing stone, glass, quartz, and metals and for slicing expansion joints in concrete highways, etc.; abrasive wheels, both for grinding and cutoff work in which the working face consists of diamond grit bonded with a resinoid, metal or ceramic; diamond-tipped tools for cutting glass and for engraving gems; and diamond powder for cutting gems. Cemented carbides and other exceedingly

Table 4. Classification of Abrasives

Natural Abrasives						
Superior Hardness (above 7 in Mohs scale)	Intermediate Hardness (H-5.5 to 7) Silica Abrasives	Intermediate Hardness (H-5.5 to 7) Other Rocks and Minerals	Inferior Hardness (H- under 5.5)	Manufactured Abrasives		Types of Abrasive Products
Diamond H-10	Buhrstone	Argillaceous limestone	Apatite	Boron carbide	Lampblack	Abrasive grains and powders, loose
Corundum H-9	Chalcedony	Basalt	Calcite	Boron nitride	Lime	Abrasive grains bonded into wheels, blocks and special shapes
Emery H-7 to 9	Chert	Feldspar	Chalk	Calcium carbonate (pptd.)	Magnesia (pptd.)	Coated abrasives; grains bonded to paper and cloth
Garnet H-6.5 to 7.5	Flint	Granite	Clay	Calcium phosphate	Manganese dioxide	Abrasive grains and powders; paste form; oil or water vehicles
Staurolite H-7.0 to 7.5	Novaculite	Mica schist	Diatomite	Cerium oxide	Periclase (artif.)	Abrasive grains and powders; brick and stick form; grease, glue and wax binders
	Quartz	Perlite	Dolomite	Chromium oxide	Silicon carbide	Natural rocks shaped into grindstones, pulpstones, chaser stones, millstones, etc.
	Quartzite	Pumice and pumicite	Iron oxides	Clay (hard burned)	Tantalum carbide	Natural rocks shaped into sharpening stones, such as oil stones, whetstones, scythe stones, razor hones, etc.
	Sandstone	Quartz conglomerate	Limestone	Diamond	Tin oxide	Natural stones shaped into rubbing and polishing stones such as holystones and pumice scouring blocks
	Silica sand		Rottenstone	Fused alumina	Titanium carbide	Natural stones shaped into blocks for tube-mill and pebble-mill liners
			Siliceous shale	Glass	Tungsten carbide	Pebbles, natural and manufactured, for grinding mills
			Silt	Iron oxides	Zirconium oxide	
			Talc		Zirconium silicate	
			Tripoli	Metallic abrasives, including steel wool, steel shot, angular steel grit, brass wool, and copper wool		
			Whiting	Porcelain blocks for mill liners and grinding pebbles		

Source: Adapted from Hight, 1983, Table 3.

hard, tough alloys can be cut and shaped efficiently with diamond tools. Diamond tipped tools are essential for the rapid and accurate shaping, truing, and dressing of abrasive wheels.

Quality Control and Testing. A diamond wheel marking system (Fig. 4), analogous to that previously shown for ordinary grinding wheels, is used by the abrasives industry (Anon., 1970).

With diamond, as with ordinary abrasives such as aluminum oxide and silicon carbide, the previously mentioned properties of size, shape, and friability must be closely controlled so that diamonds with the proper physical properties may be selected for a particular application.

Resinoid or vitrified bonded diamond wheels require a fairly friable, weak-shaped, diamond abrasive grain. Metal bonded wheels require a more durable, strong-shaped grain.

Akin to the bulk density apparatus used for ordinary abrasives, a similar device is applied to determine the bulk density of diamond abrasive samples (Hight, 1971). The cardinal rule in any determination of this type is close reproducibility of conditions from one test to another. Another shape test is a microscopic technique, developed at the Diamond Research Laboratory, Johannesburg. It is fast and inexpensive, but is quite subjective. This method was further refined by Custers and Raal (1959). In this method four shape categories are recognized. Shape No. 1 is a near perfect cube or sphere while shape No. 4 is a thin flake or a needle-like grain. Shapes No. 2 and No. 3 are intermediate; No. 2 is nearer to No. 1 than it is to No. 3 and No. 3 is similarly nearer to No. 4. The arithmetic mean of 100 grains is determined and this number becomes the "shape count" for that particular sample.

A friability test was devised by Belling and Dyer (1965) at the Diamond Research Laboratory, Johannesburg.

US Imports for Consumption. The United States is the largest single consumer of industrial diamonds. It is estimated that apparent domestic consumption of both natural and synthetic industrial diamonds in 1990 was approximately 86 million carats. US imports

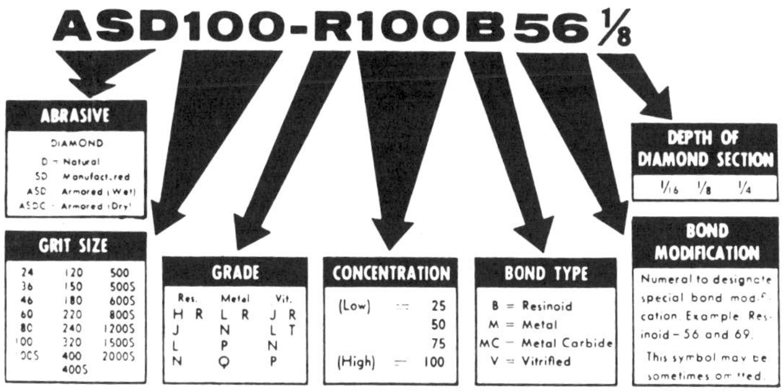

Fig. 4. Diamond wheel marking system.

of natural diamond stones in 1990 were 6.2 million carats with a value of $34.6 million; imports of natural diamond powder and dust, 12.3 million carats with a value of $13.1 million; and 2.5 million carats of imported ''miners' diamond,'' of which a part was natural, had a value of $32 million (Austin, 1991, Table 12). The United States was a net exporter of diamond grit and powder, a net importer of industrial stones, but enjoyed an excess trade balance of $14.4 million (Austin, 1991).

***Garnet:** The name ''garnet'' is given to a group of iron-aluminum silicate minerals having similar physical properties, crystal forms, and general chemical formulae.

Composition. In garnet, general formula $3RO\text{-}R_2O_3\text{-}3SiO_2$, the bivalent element may be calcium, magnesium, ferrous iron, or manganese. The trivalent element includes aluminum, ferric iron, or chromium, rarely titanium. Further, silicon is also replaced by titanium in some examples. There are three prominent groups and various subdivisions under each, many of which blend into each other. These are as follows:

1) Aluminum Garnet
   a. Grossularite, calcium-aluminum garnet ($3CaO{\cdot}Al_2O_3{\cdot}3SiO_2$)
   b. Pyrope, magnesium-aluminum garnet ($3MgO{\cdot}Al_2O_3{\cdot}3SiO_2$)
   c. Almandite, iron aluminum garnet ($3FeO{\cdot}Al_2O_3{\cdot}3SiO_2$)
   d. Spessartite, manganese-aluminum garnet ($3MnO{\cdot}Al_2O_3{\cdot}3SiO_2$)
2) Iron Garnet
   Andradite, calcium-iron garnet ($3CaO{\cdot}Fe_2O_3{\cdot}3SiO_2$)
3) Chromium Garnet
   Uvarovite, calcium chromium garnet ($3CaO{\cdot}Cr_2O_3{\cdot}3SiO_2$)

Properties. COLOR—colors in garnet vary greatly but generally are as follows:

Grossularite—white, pale green, or yellow
Pyrope—deep red to black
Almandite—deep red, brownish red to black
Spessartite—brown to red
Andradite—black, green, and yellow green
Uvarovite—emerald green.

HARDNESS—Hardness ranges from 6 (grossularite) to 7.5 (almandite); some almandite has a hardness of between 8 and 9.

*This section on garnet, originally prepared for the 3rd edition by H.H. Vogel, of Barton Mines Corp and updated by Barton Mines Corp. staff for both the 4th and 5th editions, has been further updated and edited. Some parts, less directly related to abrasives, are omitted.

CRYSTAL SYSTEM—Cubic, commonly as rhombic dodecahedrons or tetragonal trisoctahedrons, or in combination of the two.

FRACTURE—Garnets have a glassy structure and usually have a marked conchoidal fracture. However, sometimes the mineral tends to break into thin flakes. In some garnets the fracture is sharp and uneven.

INDEX OF REFRACTION—The index of refraction of the garnet group ranges from 1.735 to 1.94.

TENACITY—Aggregates composed of individual crystals of various size are brittle and shatter readily. Massive garnet and well-formed crystals are remarkably tough and shatter with difficulty.

CLEAVAGE—Occasionally an indistinct dodecahedral cleavage is observed. Some species of almandite possess a pronounced laminated structure; the laminae are planes of weakness along which the mineral separates. This parting has no relation to the crystal form and is not a true cleavage.

FUSIBILITY—Garnets having a high iron content, such as almandite, fuse at a temperature of about 1200°C. White garnets containing a considerable percentage of chromium are infusible.

OTHER PROPERTIES—Minerals of the garnet group have specific gravities that range from 3.5 to 4.2, luster that is vitreous, resinous, or dull, and are transparent to opaque.

Occurrence. Garnet commonly occurs as an accessory mineral in a large variety of rocks, more particularly in gneisses and schists, but also in contact metamorphic deposits, crystalline limestones, pegmatites, and serpentines. It occurs as gangue in ore veins formed at high temperatures. As most varieties of garnet are resistant to chemical and mechanical erosion, they tend to be concentrated in the sands of present day or preexisting beaches, streams, or other alluvial deposits.

Deposits of garnet are found in many foreign countries and in nearly every state in the United States. However, New York and Maine have the only known high-quality garnet reserves in the nation. No production of garnet from Maine, where an estimate of reserves near Rangely is more than 1.8 Mt, was reported in 1990 (Austin, 1991). In addition to material mined and processed by four firms, International Garnet Abrasives Inc. operated a garnet reclaiming plant in Jefferson Parish, LA.

NEW YORK—Based on data from Austin (1991), Barton Mines Corp. produces almandite garnet from an igneous-metamorphic rock of uncertain origin in Warren County, NY, for use in coated abrasives, glass grinding and polishing, and metal lapping; the NYCO Div. of Processed Minerals Inc., Essex County, NY, recovers and sells crude garnet concentrate, a byproduct of their wollastonite operation, for additional refining and marketed refined garnet for use as blasting and filtration media; and International

Abrasives Inc., Clinton County, NY, produces garnet, also for pressure blasting and filtration. Reserves in New York are high; for example a rough estimate suggests that more than 600,000 kt of high quality garnet can be recovered from the Barton Mines' Ruby Mountain deposit and other similar deposits occur in the area (Austin, 1991).

IDAHO—Emerald Creek Garnet Milling Co. has mining operations, two jigging plants, and a mill in Benewah County, ID, where garnet produced, recovered from garnet-bearing gravels, is primarily for blasting and filtration media (Austin, 1991).

FOREIGN COUNTRIES—Garnet is or has been commercially mined in Madagascar, Japan, Argentina, India, and Tanzania. Accurate data on world occurrence is not available, but Austin (1991) states that reserves are significant in Australia, India, and the former USSR, and that huge inferred and undiscovered resources are likely to exist throughout the world.

Preparation for Market. GRADING AND SIZING—Practically 100% of the garnet sold today is a sized and graded product. Sizings are governed by mining techniques and end use. They generally conform to specific standards: e.g., "Standards for Quality Criteria" and "Standard Methods of Test, Rating, Certification" established in cooperation with the National Bureau of Standards to provide a uniform basis for fair competition. Grading standards for coated abrasive use and micrometer sizing for grades used in glass grinding and metal lapping have exceedingly close tolerances. Wheel grades, polishing grades, and sand blast grades each have different standards and much greater tolerances. Specifications can be found in "Specifications For Grading of Certain Abrasive Grain on Coated Abrasive Products," (Anon., 1977) and in "Specifications for the Size of Abrasive Grains for Grinding Wheels, Polishing and General Industrial Uses (Anon., 1976).

HEAT TREATMENT—Practically all garnet grain used today is heat treated as a processing function rather than for the purpose of improving inherent abrasiveness. Impurities picked up during processing adhere to garnet particle surfaces and destroy cleanliness and capillarity needed in later processing to give adhesion for bonding. Early in the 1930s, it was thought that "roasting" temperatures improved the hardness, toughness, and fracture of the garnet. In the late 1940s, it was thought higher temperatures would further increase the abrasive quality. Temperatures approaching 1 000°C were used for some products. By the early 1960s, scientific investigation had determined that excessively high temperatures could destroy the abrasive quality of garnet. It was concluded that heat treatment solely contributed to cleansing of particle surfaces or, at most, it produced superficial healing of microscopic fractures on the grain surface. In the mid-1960s, the coated abrasive industry accepted uniform color standards for cleanliness of the grain surface and presented a color standard most closely approaching the natural red color of abrasive garnet. This degree of heat exposure is known as RT treatment.

TEST FOR ABRASIVE QUALITY OF GARNET—There is no exact method for testing the abrasive quality of garnet or any loose abrasive except by practical application. However, several rough tests and examinations serve to indicate their abrasive possibilities. Fracture, sharpness and shape of the grain, character of the grain structure, and the presence of inclusions of other minerals which weaken the grain structure can be studied under the microscope. Hardness and toughness of grains can be determined roughly by placing a definite grade, such as #60, between two glass microscopic slides and rubbing them together. The relative scratch-hardness of the grains and the amount of breakdown can be assessed in this manner.

The US Bureau of Standards has developed an abrasive tester designed primarily for testing abrasive quality of corundum. This can be adapted for testing the abrasive quality of any loose, abrasive grain. Loose, abrasive grain is also tested on various production or laboratory machines that evaluate glass (stock) removal on flat plates. This data can be correlated with the anticipated performance of the material to be abraded as an evaluation of actual abrasive performance. In all these methods, actual stock removal is determined either by weighing or measuring. In the coated abrasive industry, test belts are made and stock removal is measured on test blocks. These tests measure stock removal only and are a valuable indicator when other important variables such as bond, backing, coating flex, grit size, etc. are closely controlled.

In judging the abrasive quality of garnet, as well as all other abrasives, stock removal is important, but is not the only factor to be considered. Surface finish is perhaps, in most cases, a more important factor. An abrasive used on some materials can be too hard and cause scratches in the substructure that are so deep that they cannot be removed in subsequent operations. Also, if the abrasive does not have the proper grain shape or, while in use for grinding or surfacing, becomes dull and does not break down to yield new sharp cutting edges, it will tend to burn or gouge the material being worked, rendering it unusable.

Uses and Markets. ABRASIVE PAPERS AND CLOTH—Garnet coated papers and cloths are used primarily for wood sanding, but also for finishing of leather, hard rubber, plastics, glass, and softer metals. Coated abrasives have been improved in recent years, mainly by using resin bonds in place of glue bonds for certain applications. Also, electrostatic coating is now in general use. In this process, an electrostatic charge causes the grains to imbed upright in the wet bond on the backing. In effect, the sharp cutting edges of the grain are bonded perpendicular to the backing and, due to individual grain repulsion, individual grain particles are more evenly spaced.

Garnet paper and cloth products comprise approximately 10 to 11% of the total volume of coated abrasives produced for the past several years. This percentage has remained fairly constant notwithstanding the general growth of the economy and substitutions of abrasive papers and cloths for other grinding methods.

GLASS SURFACING—During World War II, when imported corundum became scarce for grinding optical lenses, the War Production Board searched for substitute abrasives. Garnet, being readily available and possessing desirable properties was the logical mineral. However, it had never been processed for this application. After development, garnet fulfilled the critical need, and, by the time the War ended, the desirable qualities of garnet were widely known. Its use decreased lens scratching and produced a superior finish, and as a result, improved the quality of the product. Use of garnet reduced rejections and decreased polishing time, both important factors in development of more economical lens grinding. The properties that made garnet a competitive abrasive for this application continue to produce new markets in the glass, plastics, ceramics, and softer metals industries.

LOOSE GRAINS—A major nontechnical use of garnet in loose form is for sandblasting. Garnet has the advantage of having no free silica and cannot cause silicosis. Being heavier than quartz, garnet delivers harder blows than quartz grains of equal size. Practically all production of Idaho garnet is sold to US West Coast aircraft industries as a sandblast agent because of its price and location.

MISCELLANEOUS USES—Garnet, in different grain sizes, is also used for the manufacture of grinding wheels, as tumbling media, for nonskid applications, and as filter media. In powder sizes, it is made into lapping and buffing compounds.

Markets and Prices. In general, every abrasive has an application for which its qualification surpasses others, but there is keen competition, based on relative cost and grinding and finishing abilities, in applications where qualifications overlap. Garnet has substantially kept its relative position in the expanding abrasives market through exploration and application of new technology. Market prices of garnet grains and powders vary widely, depending

Table 5. Garnet Sold or Used by Producers in the United States

Year	Quantity (tons)	Value (thousands)
1986	28 899	$ 6 748
1987	35 812	7 744
1988	41 896	11 144
1989	41 320	9 768
1990	45 935	11 759

Source: Adapted from Austin, 1991, Table 9.

on garnet type (almandite, pyrope, etc.), purity, particle size, grade control, shipping quantities, and packaging.

Production. In 1990, the United States was the largest producer and consumer, producing about 45% of the world's garnet. Reported production of crude garnet was 47 009 t with a value of $8.3 million. Abrasive garnet sold or used by producers in the United States from 1986–1990, as published by the US Bureau of Mines, is shown in Table 5. Figures for 1990 were 45 935 t valued at $11.76 million (Austin, 1991).

Staurolite: Staurolite, a complex iron aluminum silicate, has been commercially recovered from various river placer deposits in the southeastern United States, one is a DuPont operation in Clay County, FL. Hardness, 7 to 7½ on Mohs scale, is equal or slightly higher than that of quartz. Staurolite is used mostly for pressure blasting. Production data is not disclosed, but it is reported that 1990 industrial grade staurolite production decreased 8% in quantity and 10% in value compared to 1989 (Austin, 1991).

Silica: *Silica Sand.* Silica sand is extensively used for pressure blasting, for the initial grinding or surfacing of plate glass, and as a cutting medium for gang saws on dimension stone. For pressure blasting the main centers of production in the United States are Ottawa, IL, and Cape May, NJ. The Ottawa material occurs as a friable sandstone that is mined hydraulically, broken down to its natural grain, and washed and screened. The grains are spherical in shape. The Cape May sand is sub-angular and uneven in shape and yields larger grain sizes than those obtainable from the Ottawa sand.

In Canada, pressure blast sand has been obtained from the decomposed rock containing friable quartz and china clay in Lac Remi, north of Montreal, Quebec; from a similar type of material near Smoky Falls, north of Lake Nipissing, 40 km west of North Bay, Ontario; and from a friable quartzite at East Templeton, Quebec, a few kilometers northeast of Ottawa, Ontario. The grain of the Canadian sands is sharp. The approximate range of grain size is: No. 1, (20- to 35-mesh) for light work; No. 2 (10- to 28-mesh); No. 3, (6- to 10-mesh); and No. 4 (4- to 8-mesh). The last two sizes are used for heavy, cast-iron and steel work.

Pure, clean, beach and river sands and Illinois sand are used for the preliminary or coarse surfacing of plate glass. The crushed sand is water-graded into a number of grades at glass plants and fed to surfacing machines. Approximately three tons of sand are required to surface one ton of plate glass. Cutting sand, composed of sharp, solid quartz grains, is used as an abrasive for sawing stone. It is usually ungraded and about equivalent to a No. 1 pressure blasting sand.

Burnishing sand is a fine, rounded-grain silica sand of uniform size between 65- and 100-mesh used in rolling down and burnishing gold decorations on porcelain.

SILICA SAND PRODUCTION—Abrasive sand accounted for 8% of the total amount of industrial sand and gravel sold or used in the United States in 1990 (Bolen, 1991). US Bureau of Mines statistics indicate that, converted to metric, the amount used for sand blasting was 2 052 kt valued at $45.248 million with an average value of $22.05/t. In addition, an undisclosed amount of ground sand, used in scouring powders, was sold at an average price of $31.70. It should be noted that blast furnace slags have been processed and used as substitutes for pressure blasting sand. Increased use of such materials may, in the future, be expected because of the possible health and environmental disadvantages of silica products.

Quartz. Crushed and graded quartz are used for the abrasive backing of "flint" sandpapers. Almost any deposit of massive white quartz is suitable. Being the cheapest of all the abrasive coated paper, this product is still sold in fair amounts, mainly in hardware stores and by small jobbers. It is made only in the form of paper, not as cloth. True chalk flint from England and France is extensively employed for this purpose in Europe and has better cutting qualities and longer life than ordinary quartz.

Crushed and ground sand and sandstone and powdered quartz and silt are sometimes used in handsoaps, scouring compounds, and harsher metal polishes.

Soft Siliceous Powder Abrasives. Some natural, highly siliceous materials, which either occur as a powder or find abrasive use only in a powdered form, are used for mild abrasives. For the majority of these materials, use as an abrasive is of minor importance as compared with their principal applications. Among these, the abrasive uses of diatomite, pumice, and tripoli are reviewed here.

DIATOMITE—Diatomite production in the United States comes primarily from California and Nevada, with small outputs from Oregon and Arizona. Abrasive uses include metal (silver) polishes, dental powders and pastes and, occasionally, as a friction agent in the manufacture of matches. More significant markets, in nonabrasive applications, are filtration media, mineral fillers, and insulation.

PUMICE—Included within the category of pumices are: lump pumice and pumicite, or volcanic dust, a natural powder. Lump pumice is used by manufacturers of furniture and musical instruments for dressing wood and metal surfaces, by silver platers for preparing metal surfaces, by lithographers for cleaning stone surfaces, for rubbing down and polishing fine tools and instruments, by restaurants for scouring grills and cooking utensils, and in household cleansers and hand cleaners. Pumicite or ground pumice is mainly used as a cleanser where thin, sharp, and striated grains are particularly suitable. Recently, use of pumice as an abrasive has increased markedly because of the application of pumice lumps in "stone-washing" blue jeans. Minor uses for abrasive pumice entail applications in polishing compounds, in polishing powders for bone, celluloid, and hard rubber, dental use, and in the manufacture of some rubber erasers.

TRIPOLI, MICROCRYSTALLINE SILICA, AND ROTTENSTONE—The fine-grained, porous materials, tripoli, microcrystalline silica, and rottenstone are known to the trade as "soft silicas." In the United States, white tripoli or "microcrystalline silica" comes from southern Illinois. Buff to rose colored tripoli is mined and processed in Garland County, AR, and similar material, mined in northeast Oklahoma, is also milled in southwest Missouri. The colors are imparted by minor iron content. Tripoli from the Oklahoma-Missouri district is mainly used in the form of madeup tripoli grease bricks or tripoli compositions for buffing and polishing. The compound is applied to a rapidly revolving belt or to canvas wheels and used for finishing or buffing metals, plated products, etc. Tripoli is composed of microcrystalline quartz grains that do not have distinct sharp edges and therefore perform a mild abrasive function in toothpaste, industrial soaps, and metal polishing compounds. It is employed to a small extent in the manufacture of some scouring and cleaning powders and for rubbing down painted surfaces such as automobile bodies.

Microcrystalline, sometimes erroneously termed "amorphous," silica of southwestern Illinois is differentiated from tripoli on the basis of its low iron content, high reflectance properties and,

Table 6. Processed Tripoli Sold or Used by Producers in the United States, by use

Use		1986	1987	1988	1989	1990
Abrasives	tons	33 188	26 637	26 360	26 192	24 090
Value	thousands	$3,590	$3,089	$3,151	$3,172	$3,083
Filler	tons	67 048	71 160	68 618	63 080	56 468
Value	thousands	$8,588	$9,855	$9,876	$9,185	$11,041
Other	tons	W	W	W	W	W
Value	thousands	W	W	W	W	W
Total	**tons**	100 237	97 796	94 978	89 272	80 558
Total value	**thousands**	$12,178	$12,944	$13,027	$12,357	$14,124

Source: Adapted from Austin, 1991, Table 5.

Table 7. Special Silica Stone Finished Products Sold or Used by Producers in the United States[a]

Year	Quantity (tons)	Value (thousands)
1986	463	$6,520
1987	598	7,367
1988	371	5,415
1989	377	5,459
1990	450	6,328

Source: Adapted from Austin, 1991, Table 6.
[a]Includes grindstone, oilstones, and whetstones. Excludes grinding pebbles, tube-mill liners, and deburring media.

due to its white color, acceptance for use in wide range of filler and extender uses. It is also employed in buffing and polishing compounds and is in high demand for use in white "coloring" operations on high class work. Chemically precipitated amorphous silicas are also used in polishing and buffing compositions.

Rottenstone, a fine-grained, gray-buff siliceous-argillaceous limestone, comes from Antes Forte, Lycoming County, PA, and is used as a base for automobile polishes.

The amount of processed tripoli, including amorphous silica and Pennsylvanian rottenstone, used or produced in the United States in the 5-year period 1986–1990, is shown by use in Table 6. In 1990, 24 090 t, valued at $3.083 million are credited to use in abrasive application (Austin, 1991).

Other Soft Abrasives: *Ground Feldspar.* Ground feldspar is extensively used in scouring and cleaning compounds and for a window cleaner.

Chalk. Chalk (calcium carbonate) is a soft, compact, fine-grained, white limestone composed of the calcareous remains of microscopic, marine shells. A small amount of this chalk, mainly from England and France, known in the trade as "whiting," is used as a very mild abrasive for hand polishing nickel, gold, silver, or plated ware, buttons, and similar materials.

China Clay. China clay (kaolin) and some pipe clays have been used successfully in polishing powders. Pipe clay at one time was the standard polish for naval and military tunic buttons.

Bath Brick. Bath brick, used for scouring steel utensils, is made from a very fine-grained, quartzose clay found along the banks of the Parrot River in England.

Sharpening Stones: Hand-used, sharpening stones include scythestones, whetstones, oilstones, water stones, razor hones, holystones, and rubbing stones. While in general, stones made from bonded artificial abrasive grains have largely replaced natural sharpening stones, there is still some small production of oilstones and whetstones in Arkansas, whetstones in Indiana, and scythestones in New Hampshire. Arkansas novaculite stones are still preferred for sharpening fine edged tools for surgeons, carvers, and engravers. It is claimed that these stones give a smoother and longer-lasting edge.

Natural abrasive stones are made from a wide variety of materials including sandstone, novaculite, mica schist, siliceous argillite, shale, slate, and pumice. The superior cutting quality of some of these stones is due to the inclusion of well disseminated, fine-grained inclusions of garnet or other minerals of superior hardness.

The quantity and value of finished grindstone, oilstone, and whetstone products sold or used by producers in the United States during the 5-year period 1986–1990, as reported by the US Bureau of Mines under the category "special silica stone products," is given in Table 7. Production in 1990 was 450 t valued at $6.328 million. Arkansas accounted for 99% of the value and 97% of the total quantity. The estimated value of US apparent consumption was $1.3 million, and the value of exported special silica stone products was $6.2 million (Austin, 1991).

Grinding Pebbles and Tube-Mill Linings: Tube-mills, conical mills, and cylindrical batch mills are used more often than others for the fine grinding of hard ores, minerals, paints, chemicals, ceramic bodies, glazes, enamels, portland cement clinker, and similar materials. Most efficient grinding can be done in mills with iron or steel liners and iron or steel balls, slugs, or rods. Where contamination of the color or chemical purity of the product by metals and metal oxides must be avoided, the grinding surfaces consist of blocks or bricks for the mill linings and natural pebbles or artificially prepared balls are specified for grinding media. These materials should (1) have high abrasive hardness, (2) have great toughness and freedom from flaws which will result in breaking and chipping, (3) produce a white powder when abraded, and (4) contain no dark-colored metallic impurities such as iron or manganese. While most of the natural and artificial mill linings and pebbles have a specific gravity close to that of quartz (2.6), pebbles with greater density have obvious advantages. The heavier the pebbles per unit of size, the greater are the impact and abrasive grinding forces. Thus, with heavy pebbles less mill volume is needed for a given mill capacity (throughput). With equal pebble volume, faster grinding can be done. Small balls, having more surface area per kilo than large balls, grind faster, but with small pebbles the unit weight is low unless high density pebbles are used.

The most favored natural mill-lining material for most purposes is Belgian silex which is a very hard, tough, more or less cellular quartzite resembling French buhrstone. This is imported in rectangular blocks essentially shaped to fit the curve of a mill. During World Wars I and II, when imports were cut off, and to a much lesser extent at other times, domestic substitutes have been used. These have consisted chiefly of quartzite from near Jasper, MN, Iron City, TN, and Baraboo, WI, and granites from Salisbury, Lilesville, and Faith, NC. Some of these materials are reported to give service equal to that of Belgian silex. Special hard, dense, porcelain blocks are used in some mills for grinding paint, ceramic, and chemical materials.

For grinding media, Danish flint pebbles, when available, have long been standard because of their superior hardness, toughness, and uniformity. These pebbles are found on the shores of Greenland, but are marketed through Denmark. Other foreign sources of similar pebbles exist along the seacoasts of Belgium, France (between Le Havre and St. Valery-sur-Somme), Norway, and England. Seven sizes of Danish pebbles are marketed, ranging from 25.4 to 203 mm.

Domestic substitutes for grinding media are natural pebbles of flint, quartz, and quartzite as well as pebbles artificially rounded by tumbling small blocks of quartzite, granite, chalcedonized rhyolite, or other siliceous materials in rotating cylinders. During World War II, natural pebbles were shipped from Encinitas Beach, CA (true flint), Jasper, MN (quartzite), the Austin chalk beds in south-central Texas (true flint), Salisbury and Faith, NC (granite), Baraboo, WI

Table 8. Crude Manufactured Abrasives Produced In The United States and Canada, by kind

Kind		1986	1987	1988	1989	1990
Silicon carbide[a]	kt	112	113	118	117	105
Value	thousands	$48,064	$48,790	$50,559	$56,430[r]	$51,336
Aluminium oxide (abrasive grade)[a]	kt	137	151[r]	205	195	185
Value	thousands	$50,584	$56,393[r]	$71,325	$81,785	$77,758
Aluminum-zirconium oxide	kt	W	W	W	W	W
Value	thousands	W	W	W	W	W
Metallic abrasives[b]	kt	189	198	222	224	224
Value	thousands	$75,210	$62,410	$88,114	$89,546	$89,113
Total[c]	kt	438	462[r]	545[r]	536	515
Total value[c]	thousands	$173,858	$167,593[r]	$209,998	$227,761	$218,207

Source: Adapted from Austin, 1991, Table 15.
[a]Figures include material used for refractories and other nonabrasive purposes.
[b]Shipments for US plants only.
[c]Excludes US and Canadian production and value of aluminum-zirconium oxide.
[r]Revised. W withheld to avoid disclosing company proprietary data.

(quartzite), and several other points in the United States. Extensive deposits of quartzite pebbles in southwestern Saskatchewan have been reported. Beach pebbles were reported to be nonuniform in hardness and to require careful sorting.

Ceramic pebbles are made from very dense, tough porcelain, alumina, and sillimanite (mullite). A dense, heavy pebble having a specific gravity of 3.7 is made from zircon. For certain purposes, "blocks" of ore are used as grinding media. This practice has been followed to some extent in the "autogenous grinding" of some nepheline syenite.

Manufactured (Artificial) Abrasives

The discovery of the electric-furnace method for making silicon carbide by Edward G. Acheson in 1891 began the revolution in the high-grade abrasives industry. Prior to that time, all abrasives were natural minerals and rocks except for minor chemically prepared materials such as rouge and Vienna lime. Today, artificial abrasives, with a few notable exceptions such as diamond and garnet, dominate the field of high-grade abrasives. Manufactured abrasives are essential to modern industry because they are not only superior to natural abrasives but they are also much more uniform in quality and their properties can be varied to meet differing needs. Modern, high-speed, streamlined, production methods require very accurate, specialized tools with a degree of uniformity and dependability that cannot be met with natural abrasives.

Today, manufactured abrasives may be divided into three main groups: 1) electric-furnace products, 2) chemical precipitates, and 3) miscellaneous additions.

Electric-Furnace Abrasives: Electric-Furnace abrasives include silicon carbide, fused aluminum oxide (alumina-titania, alumina-chromium oxide, alumina-zirconia, alumina-zirconia-titania), and boron carbide.

Silicon Carbide. Silicon carbide, with the chemical formula SiC, is commonly known by the trade names Carborundum, Crystolon, and Carbolon. It is made by reacting a mixture of high-grade silica sand and carbon in an electric furnace. The preferable form of carbon used today is petroleum coke, but anthracite and coke made from coal low in ash have been used. High iron and alumina in the ash are objectionable because they tend to assist in the formation of elemental graphite and metallic silicon during the reaction. High-ash carbons make impure grades of silicon carbides.

Silicon carbide is made by charging a resistance type electric furnace with pure glass sand and finely ground petroleum coke. The silicon of the sand combines with the coke which is carbon to form silicon carbide. The temperature at the core of the furnace is about 2 200°C. If the temperature is too high, silicon carbide decomposes. The silica is volatilized and the carbon becomes graphitized. The time of a run is about 60 hrs (36 hrs heating and 24 hrs cooling). At the end of the run, the core consists of loosely knitted silicon carbide crystals surrounded by unreacted or partially reacted raw materials. The quality of the product depends largely on the raw materials used and the accuracy of control and operation of the furnace. Several types of products are made in which the toughness of the grains varies. A special surface treatment is given to grains for coated abrasives to increase adhesion of the glue.

The lump material from the furnace is first crushed in jaw crushers and then sequentially processed in rolls, hammer mills, or ball mills. It is passed through a magnetic separator to remove magnetic impurities. The grains are screened mechanically into 20 or more standard sizes from 8- to 240-mesh. The finer flour sizes are prepared by air or water flotation and sedimentation.

PRODUCTION OF SILICON CARBIDE—Austin (1991, Table 15) shows crude manufactured abrasives produced in the United States and Canada during the years 1986–1990, by kind. Total production of silicon carbide in 1990 was 105 kt compared to 117 kt in 1989 and a high of 118 t in 1988 (Table 8). Included in the 1990 total is 54 339 t of abrasive-grade product. The average value of abrasive-grade silicon carbide in 1990, as reported by producers, was $514/t (Austin, 1991).

Fused Aluminum Oxide. Aluminum oxide abrasives are made in electric-arc furnaces. Bauxite, the crude raw material, should have the following typical chemical properties:

Al_2O_3	85.0 to 87.0%
TiO_2	3.0 to 4.5%
SiO_2	3.0 to 5.5%
Fe_2O_3	6.0 to 10.3%
LOI (plus combined H_2O)	1.0 to 2.0%

The ore as mined is crushed, then calcined to remove both free and combined water. The calcined ore is mixed with iron borings and ground coke and charged into the furnace. The coke reduces the impurities which combine with the iron and sink to the bottom of the furnace.

At the end of the melting and reacting period, the charge is poured into cooled molds and allowed to cool under conditions that are controlled to obtain desired crystal size. The finished, crystallized alumina typically consists of about 96% Al_2O_3, 2.7% TiO_2, 0.67% SiO_2, 0.11% Fe_2O_3, and 0.40% other oxides. For certain purposes, Bayer-process alumina is used instead of calcined bauxite making a product with above 99% Al_2O_3.

The cooled mass from the furnace is crushed, screened, and cleaned in about the same way that silicon carbide is treated. The several types of grain made vary in toughness, type of fracture, glue adhesion, and other properties. Another form of fused Bayer process alumina is achieved with the addition of chromium oxide in percentages varying from 0.15 to 3.0%. The addition of chromium oxide increases the hardness of these abrasives without substantially increasing the toughness. These properties are useful in preparing grinding wheels which produce a cool grind without resulting damage to the metal being ground.

Other heavy-duty fused abrasives have been developed. One is an electric furnace fusion product made up primarily of intimately crystallized mixtures of aluminum oxide and zirconia. Although other proportions have been successful, two mixtures, one containing approximately 25% and the other about 40% zirconia, have found the most widespread commercial applications. Another fused abrasive is one that is composed of a mixture of alumina, zirconia, and titania which is fast cooled or quenched. A favored mixture contains 34% zirconia, 3% titania, and the balance alumina. Fused alumina-zirconia abrasives are noted for their extreme durability and are particularly well suited for foundry snagging and steel conditioning operations.

PRODUCTION—Austin (1991) provides data on the production of fused alumina for the 5-year period, 1986–1990 (Table 8). Data on the use of alumina-zirconia abrasives are withheld. Production in 1990 of fused aluminum oxide in the United States and Canada was 185 kt with a value of $77.758 million. Of this, 164 kt was regular grade and 21 kt was high-purity (Austin, 1991).

Boron Carbide. Boron Carbide is an artificial abrasive introduced in 1934 by the Norton Company under the name Norbide. Its chemical formula is B_4C. Although considerably harder than silicon carbide, it is far from the hardness of the diamond (see Table 2). It is made from boric oxide (B_2O_3) and carbon in the form of petroleum coke in a carbon resistance furnace or arc furnace at a temperature of about 2 600°C. The finished product is crushed and ground to make a range of sizes of grains and powders. In grain form, it is used for grinding and lapping operations previously possible only with diamond dust. Pulverized boron carbide may be molded under very high heat and pressure to form extremely wear-resistant products such as pressure blast nozzle liners, thread guides extrusion dies, and all types of extremely accurate plug, snap, and ring gages. It is not manufactured into bonded wheels or sharpening stones.

Sintered Abrasive: A different family of abrasives is the sintered product. This group differs drastically in method of manufacture from the more traditional fused product in that the raw materials are heated in kilns only to the sintering temperature (1 200 to 1 400°C) rather than being carried all the way to the fusion point. Although other basic starting materials have been used, bauxite, with or without additives and still in the green (unfired) state, is formed into predetermined sized and shaped granules by certain granulation techniques or by extrusion through an appropriately shaped orifice. The resultant presized and preshaped granules are then sintered to make the finished abrasive grain. Process steps such as crushing, shaping, and sizing (common to manufacturing fused products) are eliminated in making a sintered product.

Sol-Gel Abrasives: Sol-Gel abrasives are prepared by jelling a form of aluminum oxide known as Boehmite, adding various precursors to enhance the properties along with a seed or nucleation agent (various ones have been used), evaporating the water from the gel, and producing a hard, solid material. The resulting product is then crushed, graded into sizes and fired at temperatures of 1 200 to 1 500°C. Such abrasives are described in US Patent #4 744 802, Inventor: Mark G. Schwabel (assigned to Minnesota Mining and Manufacturing Company); US Patent #4 623 364, Inventors: Cottringer, van de Merwe, and Bauerr (assigned to Norton Company); and US Patent #4 799 938, Inventor: Janz and Gottschamel (assigned to Treibacher Chemische Werke AG).

Although most of the raw materials used in the Sol-Gel process are considerably more expensive than those going into regular aluminum oxide abrasive, the performance of the Sol-Gel abrasives, particularly in paper, cloth, and grinding wheel applications, seems to justify the additional cost in a number of uses.

Manufactured Diamond: In February 1955, General Electric Co. announced that its laboratories had succeeded in producing diamonds from carbonaceous material subjected to pressures of around 5.5 GPa at very high temperatures for prolonged periods. Crystals up to 1.6 mm long were made in 16 hrs, but smaller ones were produced in much shorter times. The press is capable of maintaining temperatures above 2 760°C at pressures up to about 11 GPa. Costs were reported to be about twice that of comparable natural diamonds and the sizes were small. However, today economical commercial production has been achieved. Diamonds can now be made in a relatively few minutes, and their performance in specific end-use applications, previously discussed under "Industrial Diamond (Natural)," equals that of natural diamond.

There are several types of manufactured diamond, both domestic and foreign, now on the market, and each has distinct characteristics. In resinoid bonded grinding wheels, the adherence of the bond to the abrasive seems to be improved if the diamond is metal-coated. Metal-coated (nickel, copper) diamond represents still another commercially available diamond type.

Production. The use of manufactured diamond has steadily increased. In 1969, domestic production of manufactured industrial diamonds amounted to approximately 13 million carats. By 1981, productivity had soared to approximately 57.0 million carats. The United States, with a rated annual capacity of 100 million carats, was the largest producer of synthetic diamonds in 1990 (Austin, 1991). Production data is not disclosed, but during the last ten years production is reported to have increased at an average rate of 9%, compounded annually.

Boron Nitride: Boron nitride in cubic form is a manufactured abrasive discovered by General Electric Co. laboratories in 1957 and trademarked BORAZON®. Unlike manufactured diamond, it has no known counterpart in nature. It is made at temperatures and pressures comparable to those required for diamond manufacture. The thermal stability of cubic boron nitride exceeds that of diamond. It is stable at temperatures above 1 371°C, while diamond reverts to graphite at above 816°C. The Knoop hardness (k100) is 7 800, well above ordinary abrasives, but inferior to diamond.

In resinoid bonded grinding wheels, BORAZON® has been found to be effective in grinding hard and tough tool steels.

Metallic Abrasives: Metallic abrasives include crushed steel, steel shot, angular steel grit, steel wool, brass wool, and copper wool. The following description of the first three of these products is by Johnson and Schauble (1939):

"Crushed steel is made from high-carbon and crucible sheet steel specially treated to impart brittleness. It is then crushed to sizes ranging from 2- to 200-mesh. After screening, each batch is heat-treated and separated into 25 sizes ranging from 20- to 200-mesh. Sizes from 70-mesh upward are screened on silk bolting cloth, and the finer sizes in powder form are used in steel cement, various chemical compounds, and fireworks sparklers.

"Steel shot is merely chilled cast iron. Only raw materials of the highest grade, including selected scrap and charcoal iron are used it its manufacture. These are melted in a cupola. During the casting period, the molten metal is separated into small spherical globules by directing high-pressure steam or heated compressed air against the stream of metal. The globules are blown into water and cooled. The shot, made brittle by the rapid cooling, is heat-treated to impart a temper of hardness and graded by mechanical means into 15 sizes that range from 4 to 90-mesh. Coarse material left on the 4-mesh screen is granulated by specially designed crushing equipment and in this form is known as angular steel grit. The crushed material is heat-treated to impart toughness and durability. It is then graded into 15 sizes ranging from 7 to 100-mesh. Steel shot and angular steel grit are used for many abrasive purposes.''

Steel wool, made from both ordinary and stainless steel in various degrees of fineness, is used for finishing wood and soft metals such as aluminum and for scouring and cleaning. It is made in several types of machines by shaving or scraping continuously moving wire with a fixed serrated cutting tool. Brass and copper wool are used chiefly for household cleaning.

Production. Production, shipments and annual capacities of metallic abrasives in the United States, by product, for the years 1989–1990 have been reported by Austin (1991). Total production of 224 139 t valued at $89.113 million are reported for 1990 (Table 8).

Chemical Precipitates: Chemical precipitates, mainly oxides, all have a very fine grain size and are used as final polishing agents.

Iron Oxides. ROUGE—Rouge, a ferric oxide, has its largest use in the polishing of plate glass, but is also widely used in powder, paste, coated cloth, and stick form for polishing precious metals, stone, and other materials to give a final high luster.

CROCUS—Crocus is similar to rouge in composition, but purplish red in color. Crocus is used in finishing cutlery and some brass work.

Other Precipitates. CHROMIUM OXIDE—Chromium oxide or ''Green Rouge'' is used chiefly for platinum and stainless steels.

TIN OXIDE—Tin Oxide is used largely for polishing glass and precious stones.

PUTTY POWDER—Putty Powder is a mixture of tin oxide and oxalic acid used for about the same purposes as tin oxide.

CERIUM OXIDE—Cerium Oxide is a polishing agent for glass and may be substituted for tin oxide in other applications.

MAGNESIA—Magnesia in the soft precipitated form has been noted as a mild polishing powder. The hard-burned variety, artificial periclase, has also been used as an abrasive.

PRECIPITATED CALCIUM CARBONATE—Precipitated calcium carbonate is used as a mild abrasive in some dentifrices.

Calcium and Magnesium Oxides (Lime): Lime (CaO), manufactured by calcination of limestone, and dolomitic lime (CaO·MgO), derived by calcination of dolomite, have important abrasive uses. This has been especially true of those marketed under the names of Vienna lime or Sheffield lime. The so-called Vienna lime (originally from Vienna, Austria), which is used on the American continent in a grease-brick buffing composition, is obtained from certain beds of dolomite at Francis Creek and Manitowoc, WI. Vienna lime is made by calcining the dolomite and cleaning and grinding to a certain fineness. It is packed in carefully sealed containers. As soon as the lime becomes hydrated, it ceases to function and also attacks the grease composition. This product is used for buffing brass, copper, bronze, steel, pearl, celluloid, and similar materials. Its principal use is for coloring nickel after plating. Vienna lime gives nickel a deep undersurface blue peculiar to the metal. Lime attacks aluminum and is not applied with this metal.

Miscellaneous Manufactured Abrasives: *Porcelain.* The use of porcelain blocks and pebbles for grinding mills has been noted. This ''porcelain'' may consist largely of zirconium oxides or silicate, or of mullite converted from andalusite, kyanite or dumortierite.

Glass. Crushed and screened glass has been used for coated abrasives to make glass paper.

Lampblack. ''Lampblack'' finds some use for polishing celluloid and bone.

Clay. Clay, burned to a very hard state and finely pulverized, is reported to have been used as an abrasive for polishing metal.

ACKNOWLEDGMENT

The author wishes to acknowledge the original work done by Richard P. Hight in the 5th edition of *Industrial Minerals and Rocks.*

BIBLIOGRAPHY AND REFERENCES

Anon., 1965, ''American Standard Test for Ball Mill Test for Friability of Abrasive Grain,'' American Standards Association, B74.8.

Anon., 1970, *Abrasives and Grinding Wheels*, Norton Co., Worcester, MA, 36th printing, Mar., pp. 22–62.

Anon., 1970a, Technical Publication SMD-B-9–500, General Electric Co., Dec.

Anon., 1976, ''Specifications for the Size of Abrasive Grains for Grinding Wheels, Polishing and General Industrial Uses,'' B74.12–1976, American National Standards Institute, New York.

Anon., 1977, ''Specifications for Grading of Certain Abrasive Grain on Coated Abrasive Products,'' B74.18–1977, American National Standards Institute, New York.

Anon., 1987, ''American Standard Practice for Ball Mill Test for Friability of Abrasive Grain,'' B74.8–1987, American National Standards Institute, New York.

Anon., 1988, ''American National Standard Test for Bulk Density of Abrasive Grain,'' B74.4–1988, American National Standards Institute, New York.

Austin, G.T., 1991, ''Abrasive Materials,'' *Annual Report for 1990,* US Bureau of Mines, 21 pp.

Belling, N.G., and Dyer, H.B., 1965, *Impact Strength Testing of Diamond Abrasive Grit*, Diamond Research Laboratories, Johannesburg, Republic of South Africa.

Bolen, W.P., 1991, ''Industrial Sand and Gravel,'' *Annual Report for 1990,* US Bureau of Mines, 15 pp.

Coes, L., Jr., *Applied Mineralogy-Abrasives*, 1971, Springer-Verlag, Vienna/New York.

Custers, J.F.H., and Raal, F.A., 1959, ''A New Method of Determining the Average Shape of Diamond and Other Particles,'' *Industrial Diamond Review*, Vol 19, p. 72.

Harben, P., 1978, ''Abrasives—Taking the Rough with the Smooth,'' *Industrial Minerals*, No. 134, Nov., pp. 49–73.

Heywood, J., 1942, *Grinding Wheels and Their Uses*, Penton Publ. Co., New York, 436 pp.

Hight, R.P., 1971, ''Diamonds, Industrial,'' *Encyclopedia of Industrial Chemical Analysis*, Vol. 11, John Wiley, New York, pp. 462–484.

Hight, R.P., 1983, ''Abrasives,'' *Industrial Minerals and Rocks,* 5th ed., S.J. Lefond, ed., AIME, New York, pp. 11–32.

Johnson, B.L., and Schauble M., 1939, ''Abrasive Materials,'' *Minerals Yearbook 1939*, US Bureau of Mines, pp. 1225–1240.

Knoop, F., et al., 1939, National Bureau of Standards Research Paper 1220, *Journal of Research*, 23 pp.

Oskam, J., 1980, ''Garnets, Abrasive and Gem,'' Preprint 80–382, SME-AIME Fall Meeting, Minneapolis, Oct., 5 pp.

Ridgeway, R.R., Ballard, A.H., and Bailey, B.L., 1933, ''Hardness Values for Electrochemical Products,'' *Journal of the Electrochemical Society*, May.

Thibault, N.W., and Nyquest, H.L., 1947, ''The Measured Knoop Hardness of Hard Substances and Factors Affecting its Determination,'' *Transactions*, American Society for Metals, Vol. 38, pp. 271–330.

AGRICULTURAL MATERIALS

FERTILIZERS

EDWIN A. HARRE AND JOHN J. MORTVEDT

Plants obtain nutrients from both air and soil. They absorb carbon dioxide mainly from the air, although some carbon dioxide may be dissolved in soil water which is absorbed by plant roots. Carbon dioxide, a gaseous form of carbon and oxygen, supplies the carbon which usually makes up 50% or more of plant structure.

Growing plants use water more than any other biological substance. Plant nutrients dissolve in soil water and enter the plant via root absorption. Water is then given off into the air as the plant breathes or transpires. Water also supplies hydrogen and oxygen to plants. Plant roots absorb nutrient cations or anions by contact exchange with soil clay particles.

Inorganic or mineral nutrients essential to plants are supplied through the soil. Inorganic or mineral composition of plants usually is less than 5% of their weight. In most crop-growing areas, much of this nutrition is inadequately supplied from soil reserves and must be provided by fertilizers. The role of the fertilizer industry is to provide economical, efficient, good quality fertilizers to improve or maintain soil productivity for optimum crop production.

The essential plant nutrients other than carbon, hydrogen, and oxygen which often must be added to soils or plants are categorized as primary nutrients (nitrogen, phosphorus, and potassium), secondary nutrients (calcium, magnesium, and sulfur), and micronutrients (boron, chlorine, copper, iron, manganese, molybdenum, and zinc). These classifications are based on quantity of nutrients utilized by plants and do not imply their function or degree of necessity.

One point of major significance to the fertilizer industry is that not all fertilizer minerals contain the plant nutrients in a chemical form available to the plant. A product which may be considered a fertilizer mineral may require substantial processing or treatment before it becomes useful as a nutrient source or fertilizer.

AGRICULTURAL PRODUCTION AND PLANT NUTRIENT USE

Until about 1900, demands for higher crop yields were met simply by bringing new agricultural lands into cultivation. Since that time great progress has been made in increasing yields through fertilization and improved crop varieties and cultural practices. At the same time, land under cultivation throughout the world has remained relatively constant or, as in the United States in recent years, has been taken out of production. Fig. 1 illustrates the relationship of agricultural production to crop acreage and consumption of the primary plant nutrients in the United States over the past 25 years.

In the past 25 years, particularly during the 1960s and early 1970s, the growth of the US fertilizer industry has been phenomenal. In 1962, US farmers used 24.1 Mt of fertilizer materials at a total expenditure of $1.5 billion. In 1990, total fertilizer consumption exceeded 43.2 Mt valued at over $7 billion. Table 1 indicates farm expenditures and plant nutrient consumption in the United States from 1965 to 1990. Phosphate and potash are expressed as P_2O_5 and K_2O, respectively. To convert P_2O_5 to P, multiply P_2O_5 by 0.44 and multiply K_2O by 0.83 to convert to K.

Food demand for a growing world population stimulated crop production in the United States during the 1960s and 1970s. Farmers rapidly expanded plant nutrient application rates in an attempt to increase crop production and farm income. Then, as farm commodity prices fell in the face of over-production, farmers began to reduce application rates. However, crop yields have continued to set new records as farmers have increased the efficiency of fertilizer use through improved soil fertility and management practices. Fig. 2 illustrates the pattern of US plant nutrient application rates for corn over the past three decades.

As application rates of P and K have declined in recent years, crop removal of these nutrients is depleting the nutrient supply in some soils. With the current emphasis on sustainability in agriculture, nutrient removal will become a critical factor in future crop production. Fig. 3 illustrates the change in application/removal ratios for the United States from 1965 through 1990.

World population is forecast to exceed 6 billion people early in the next century. Forecasts of widespread food shortages and famine date back to Malthus in the 1700s when he prophesied that population growth would far exceed the world's ability to feed itself. As indicated in Fig. 4, world per capita food production has kept pace with population growth and diets have steadily improved. There may always be pockets of limited food supplies, but with the proper use of fertilizers and other agricultural inputs, the success of world agriculture will continue. Fertilizer has been and will remain the keystone to adequate food and fiber production in the coming years.

PRIMARY PLANT NUTRIENTS IN FERTILIZERS

Primary nutrients (nitrogen, phosphorus, and potassium) are designated as such because of the amounts required and their significance to plant growth. Practically all soils are deficient in nitrogen. Most soils are also deficient to varying degrees in available phosphorus and potassium. Consequently, these nutrients comprise the vast majority of fertilizers applied to soils.

Nitrogen

Although nitrogen is not a fertilizer mineral since it is not a mined product, it is the most important fertilizer nutrient. Nitrogen

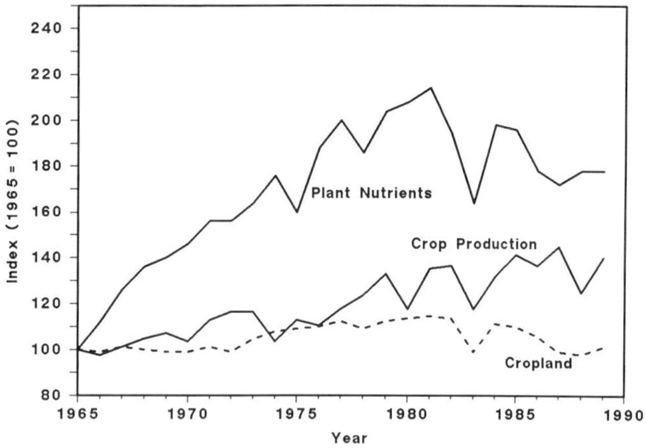

Fig. 1. Index of crop production, cropland, and plant nutrient consumption in the United States (Anon., 1991, Hargett et al., 1986–1991).

Table 1. Estimated Farm Expenditures in Billions of Dollars for Fertilizers, and Plant Nutrient Consumption in the United States (Modified from Hargett et al., 1986–1991)

Year†	Total Fertilizer Expenditure	Primary Plant Nutrients (thousand tons)			
		Nitrogen	Phosphate	Potash	Total
1965	1.921	4 209	3 187	2 572	9 967
1970	2.224	6 767	4 150	3 661	14 577
1975	6.447	7 809	4 092	4 040	15 941
1980	8.269	10 348	4 928	5 665	20 942
1985	8.114	10 426	4 226	5 037	19 689
1990	7.036	10 048	3 942	4 720	18 710

†Fertilizer year ending June 30

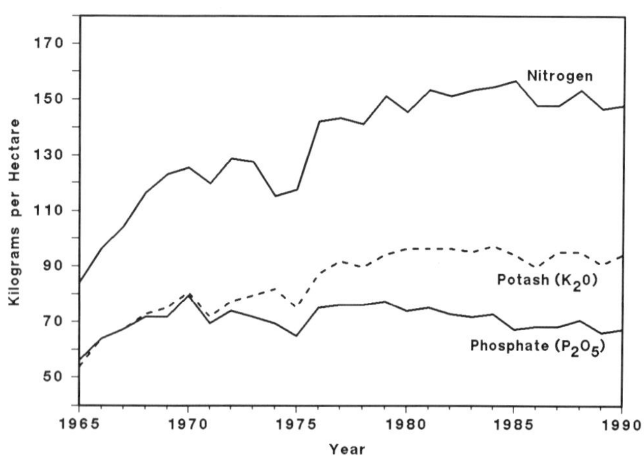

Fig. 2. Average rates of plant nutrient application on corn in the United States (Anon., 1990–1991, Vroomen, 1989).

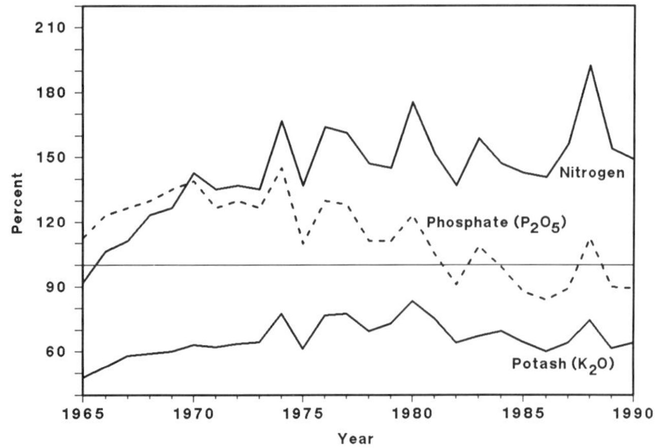

Fig. 3. Ratio of plant nutrient use to removal by crops in the United States (Wallingford, 1991).

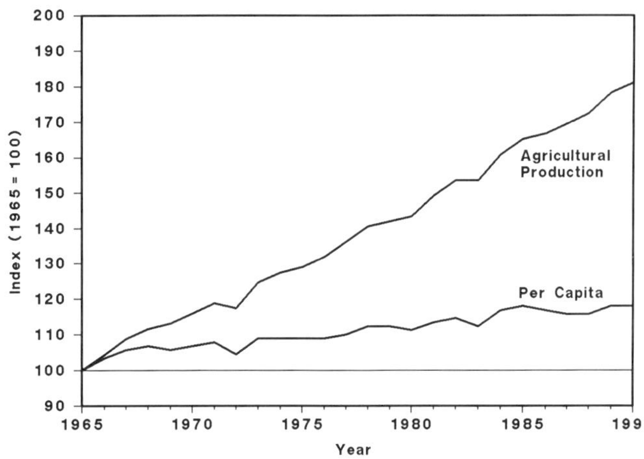

Fig. 4. Index of world agricultural production and agricultural production per capita (Anon., 1965–1990).

accounts for about half of the total nutrients applied as fertilizers in the United States. Table 2 shows consumption of nitrogen in North America and other world markets in recent years. North America and Western Europe have decreased in importance in the world market as fertilizer use has increased dramatically in Eastern Europe and Asia. In 1990, the People's Republic of China accounted for 24% of world nitrogen use, followed by the United States and the USSR, with 12.7% each.

The manufacturing process for nitrogen fertilizers involves the fixation of atmospheric nitrogen. Commercially available sources of nitrogen include both organic and inorganic forms. Regardless of the original source, however, all nitrogen must be converted to the nitrate or ammonium form before it can be absorbed by the plant. Nitrogen is mainly absorbed from the soil by plants as nitrate (NO_3^-), but may also be taken up as ammonium (NH_4^+) ions. After entering the plant, nitrogen is converted into complex compounds and ultimately into proteins.

In the past, virtually all nitrogen was applied to soil in the form of animal manures. Certain soils in Europe have a "man-made" surface layer more than 0.9 m thick due to centuries of manuring. One of the earliest commercial sources of nitrogen was guano (excreta and remains of sea fowl found in deposits on certain arid islands). This material can contain as much as 13% nitrogen and about 9% phosphorus. Neither manure nor guano is an important commercial source of fertilizer today. Animal manures are applied to soils primarily because of the need for disposal, but manures may supply significant amounts of nitrogen to fields near their source of origin. The major organic sources of nitrogen in use today are urea (NH_2CONH_2) and urea formaldehyde ($NH_2CONH_2 \cdot CH_2OH$). Both are synthetic products.

Commercially important inorganic sources of nitrogen include

Table 2. World Nitrogen Fertilizer Consumption (Anon., 1965–1990a)

	Developed Regions				Developing Regions				
Year†	North America	Europe and USSR	Oceania	Other	Africa	Latin America	Near East	Far East	Total
	(million tons, N)								
1965	4.38	7.59	0.09	0.84	0.13	0.69	0.39	1.21	15.32
1970	7.04	12.84	0.18	1.07	0.24	1.18	0.71	5.43	28.70
1975	8.34	17.93	0.20	0.95	0.41	1.88	1.03	7.83	38.58
1980	11.18	22.03	0.26	1.21	0.54	2.66	1.78	17.55	57.24
1985	11.68	25.52	0.37	1.16	0.69	3.20	2.64	25.36	70.64
1990	11.24	25.41	0.45	1.07	0.90	3.88	3.35	32.76	79.08

†Fertilizer year ending June 30.

ammonium nitrate (NH_4NO_3) available in both solid (33 to 34% N) and liquid (19 to 21% N) forms, ammonium sulfate [$(NH_4)_2SO_4$—21% N], calcium nitrate [$Ca(NO_3)_2$—15% N], potassium nitrate [KNO_3—13 to 15% N), urea (NH_2CONH_2—45 to 46% N), anhydrous ammonia (NH_3—82% N), and nitrogen solutions (28 to 32% N)—a combination of ammonium nitrate and urea solutions. The relative importance of these products in the United States is shown in Fig. 5. Worldwide, urea has become the leading nitrogen fertilizer because of its high analysis and applicability to most crop management practices.

Nitrogen is applied to soils either prior to planting, at planting, or after planting as a sidedress application. All inorganic sources of nitrogen convert to the usable nitrate form rather rapidly in warm, moist soils, but nitrates may leach rapidly under certain conditions. The crop, soil, and weather conditions dictate which materials and application methods are best.

Phosphorus

Of the three primary nutrients in fertilizers, phosphate is the least used. In 1990, total phosphate used in the United States was 3.94 Mt. Although its growth has not paralleled that of nitrogen, world consumption of phosphate has increased from 13.39 to 31.21 Mt from 1965 to 1985 and to 37.35 Mt in 1990 (Table 3).

Phosphorus is absorbed from soils by plants principally as the $H_2PO_4^-$ ion. Once inside the plant it enters into the composition of phospholipids and nucleic acids. A major metabolic role of phosphorus in the cells of both plants and animals is the providing of energy which results from the high energy phosphate bonds. In the growing plant, phosphorus is most abundant in growing tissues

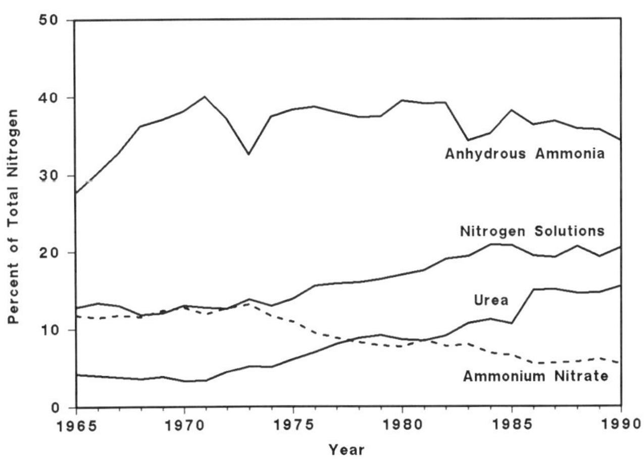

Fig. 5. US consumption of major nitrogen fertilizer materials (Anon., 1965–1985, Hargett et al., 1986–1991).

where it is utilized in the synthesis of nucleoproteins. As the plant matures, phosphorus accumulates in the seed and fruit.

Fertilizer phosphorus is highly reactive with clays and organic constituents in soils and forms relatively insoluble compounds. As a result, it does not move downward with percolating ground waters, but tends to accumulate in the surface layer of soil. Thus, after several years of high phosphorus fertilization, farmers are able to reduce their phosphorus fertilization rates without loss in crop yields. However, high crop yields usually require a maintenance application of phosphate fertilizer.

The development of the modern phosphate fertilizer industry dates back to 1840 when Justus von Liebig, a German scientist, demonstrated that the fertilizer value of bones was significantly increased by treatment with sulfuric acid. Two years later, the process by which phosphate rock (apatite) was treated with sulfuric acid to produce ordinary superphosphate (20% P_2O_5) was patented in England by John Bennett Lawes. Commercial production began in the United States in 1852.

In today's multimillion-dollar fertilizer phosphate industry, nearly all products originate from phosphate rock. It is still acidulated with sulfuric acid using essentially the same principle employed in England more than 100 years ago. Ordinary superphosphate dominated the phosphate market until recent years when new materials began to replace it, primarily because of their higher analysis and lower transportation costs. The three products currently dominating the world phosphate market include normal superphosphate [$Ca(H_2PO_4)_2 \cdot H_2O + CaSO_4 \cdot 2H_2O$], concentrated superphosphate [$Ca(H_2PO_4)_2 \cdot H_2O$], and complex fertilizers including monoammonium phosphate ($NH_4H_2PO_4$), diammonium phosphate [$(NH_4)_2HPO_4$], and nitric phosphates of varying analyses. Fig. 6 illustrates how the production of ammonium phosphates has grown since its introduction in the US fertilizer market in the 1950s.

Nearly all commercially available sources of phosphorus are highly water-soluble. However, when the calcium phosphates, such as normal superphosphate, are ammoniated to make other complex fertilizer products, the high pH of the manufacturing process often causes further reaction of calcium with the phosphorus, resulting in precipitation of insoluble calcium phosphate compounds. These products are more effective on acid soils than on alkaline soils.

A major consideration in phosphorus application to soil and its utilization by plants is soil pH. The soil pH range for optimum phosphorus availability is 6.0 to 7.2. In more acid soils, phosphorus can be complexed by iron and aluminum compounds. At alkaline pH levels, phosphorus reacts with calcium, forming insoluble calcium phosphates. These problems are partially overcome by liming acid soils and applying only highly water-soluble phosphate sources on alkaline soils.

Because of its relative immobility in soils, phosphorus must be incorporated into the soil for optimum effectiveness. Broadcasting

Table 3. World Phosphate Fertilizer Consumption (Anon., 1965–1990a)

	Developed Regions				Developing Regions				
Year†	North America	Europe and USSR	Oceania	Other	Africa	Latin America	Near East	Far East	Total
	(million tons, P_2O_5)								
1965	3.44	6.95	1.21	0.70	0.10	0.45	0.13	0.41	13.39
1970	4.43	9.22	1.20	0.98	0.20	0.77	0.31	1.69	18.80
1975	4.59	11.13	0.92	1.05	0.33	1.48	0.49	2.68	22.69
1980	5.56	14.80	1.27	1.29	0.39	2.48	1.03	4.38	31.21
1985	4.95	14.84	1.10	1.21	0.55	2.42	1.51	7.43	34.02
1990	4.56	15.58	0.75	1.01	0.65	2.46	1.75	10.60	37.35

†Fertilizer year ending June 30.

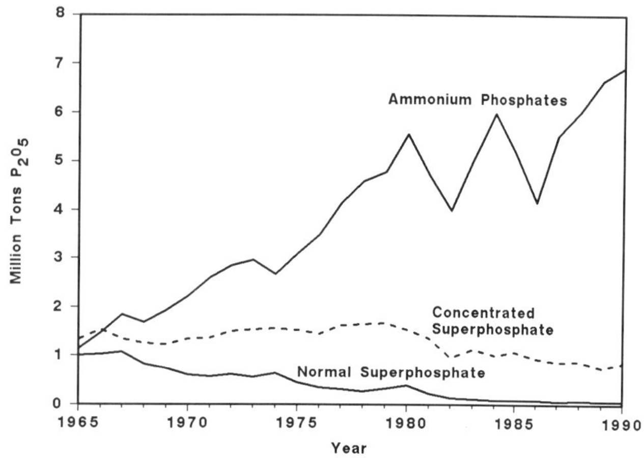

Fig. 6. US production of phosphate fertilizers (Anon., 1965–1990b).

phosphate fertilizers with subsequent incorporation is an effective method of application. However, recovery of applied phosphorus is usually higher when it is placed in a band beside the seed row, particularly when available soil phosphorus levels are quite low. At low rates, some phosphate fertilizers have been effectively applied in direct contact with the seed. Broadcasting still dominates phosphorus fertilizer application in the US. Band application at seeding continues to grow in importance, however, especially with the use of liquid phosphate materials.

Potassium

From 1965 to 1975, world potash consumption nearly doubled. Table 4 shows that it increased from 10.94 to 19.82 Mt in that period. By 1990, consumption increased to 26.85 Mt. Muriate of potash (KCl) is the dominant source of fertilizer potassium throughout the world. Other important, but much less significant, sources are sulfate of potash (K_2SO_4) and sulfate of potash-magnesia ($K_2SO_4 \cdot MgSO_4$).

Growing plants absorb potassium as the K^+ ion. Potassium is unlike the other mineral constituents in that its exact metabolic role in the plant is unknown. It occurs in the ionic state and not as a component of compounds within the plant. However, potassium is indispensable for plant growth and is required in large quantities by most plants. Although not fully understood, the fundamental roles of potassium in the plant are undoubtedly regulatory or catalytic.

Potassium and phosphorus application methods are very similar. Most potassium is broadcast with a lesser amount band applied near the seed row at planting time. Muriate of potash is usually mixed with phosphorus fertilizers and the two nutrients are applied simultaneously to soils. Almost two-thirds of the potassium used in the United States is applied to soils of the Midwestern corn belt.

SECONDARY PLANT NUTRIENTS IN FERTILIZERS

Calcium, magnesium, and sulfur—the secondary micronutrients—are required by plants in smaller amounts than the primary nutrients. In many areas, sulfur and magnesium are becoming increasingly important as fertilizer nutrients. Increasing crop yields, sandy soils, and leaching due to high rainfall or irrigation are factors influencing the use of magnesium and sulfur in fertilizers. The use of fertilizer calcium strictly as a nutrient is limited and dictated by special crop needs and unusual soil conditions.

Common fertilizer materials containing these secondary nutrients are potassium sulfate, sulfate of potash-magnesia, gypsum, normal superphosphate, magnesium sulfate, elemental sulfur, and calcium nitrate.

Calcium

Calcium is important not only as a plant nutrient, but also because of its very favorable influence on maintaining desired physical and biological conditions in the soil. Of the exchangeable bases found in soil (potassium, magnesium, calcium, and sodium), calcium is dominant, normally accounting for 50 to 80% of the total exchangeable bases.

Calcium is absorbed by plants as the Ca^{2+} ion. Inside the plant, calcium tends to accumulate in older tissues. Much of the calcium in plants is permanently fixed in the cell walls as a calcium salt of the pectic compounds in the middle lamella, giving cell walls their strength. Calcium is also important in the reduction of nitrates in the formation of plant proteins.

Most soils contain sufficient calcium for plant growth, and calcium-deficient plants are seldom observed. However, many soils are too acid for optimum plant growth. Soil pH can be increased to a desirable level most economically with broadcast applications of finely ground agricultural liming materials. Lime supplies are virtually unlimited and commercial deposits have been found in most agricultural areas. Common lime materials are calcite ($CaCO_3$), dolomite ($CaCO_3 \cdot MgCO_3$), and to a lesser extent, basic slag which is a calcium silicate material containing some phosphorus and magnesium. These materials are sometimes used as fillers in mixed fertilizers, although more than 95% of agricultural lime is applied directly to soils. Table 5 shows agricultural lime consumption in the United States.

Table 4. World Potash Fertilizer Consumption (Anon., 1965–1990a)

	Developed Regions				Developing Regions				
Year†	North America	Europe and USSR	Oceania	Other	Africa	Latin America	Near East	Far East	Total
	(million tons, K_2O)								
1965	2.69	6.85	0.16	0.65	0.09	0.27	0.02	0.21	10.94
1970	3.84	9.26	0.19	0.77	0.12	0.59	0.02	0.80	15.59
1975	4.25	11.88	0.23	0.86	0.21	0.93	0.04	1.41	19.82
1980	6.01	13.29	0.22	0.88	0.22	1.60	0.05	1.77	24.06
1985	5.43	14.78	0.28	0.79	0.27	1.67	0.09	2.57	25.89
1990	5.08	14.53	0.25	0.72	0.34	2.08	0.16	3.68	26.85

†Fertilizer year ending June 30.

Table 5. Use of Agricultural Liming Materials in the United States (Anon., 1991)

Year†	Liming materials (million tons)
1949	27.9
1954	19.0
1959	22.7
1964	27.0
1969	28.4
1974	27.8
1979	31.0
1984	26.6
1989‡	20.7

† Fertilizer year ending June 30.
‡Tepordei (1991).

Soils of the humid regions east of the Mississippi River generally require higher rates and more frequent applications of lime. Soil calcium and magnesium may be leached from those soils by relatively high rainfall, resulting in a decrease in the soil pH. In addition, application of nitrogen fertilizers indirectly increases soil acidity. As a general rule, 3 kg/ha of $CaCO_3$ are required to neutralize the soil acidity resulting from every kg/ha of nitrogen applied to soils.

Besides creating a desirable soil pH level for optimum plant growth and resulting in a more desirable soil physical condition, a secondary effect of lime is that soil pH influences the availability of certain other nutrients. While phosphorus availability is greatest at pH 6.0 to 7.2, the micronutrients copper, iron, manganese, and zinc generally are made less available as the soil pH approaches neutrality. These micronutrients must often be applied to the soil or crop as fertilizers.

Crops such as peanuts and tomatoes require a relatively large amount of soluble calcium. Gypsum ($CaSO_4 \cdot 2H_2O$) is often used where these crops are grown and is usually applied directly to soil rather than as a component of mixed fertilizers. Sources of much less significance for their calcium content are normal and concentrated superphosphates and calcium nitrate.

Magnesium

Magnesium is absorbed by plants as the Mg^{2+} ion. Its role in the plant as a constituent of the chlorophyll molecule is of major significance. It is also important in phosphorus metabolism and sugar and starch production. The desired exchangeable or available level of magnesium in soils is usually about one-tenth that of calcium. The use of high-calcium lime materials and the natural tendency of many soils to be low in available magnesium has stimulated the practice of including soluble magnesium sources in some premium fertilizer products. These soluble magnesium materials may increase the crop production efficiency of the primary nutrients.

Magnesium sources either added to other fertilizers or applied directly to soil can be classified by their chemical nature or by their the degree of water-solubility. A source must either be water-soluble or convertible to a water-soluble form before the magnesium can be absorbed by the plant.

Magnesium sources which are soluble or readily available are hydrated magnesium sulfates and sulfate of potash-magnesia (langbeinite: $K_2SO_4 \cdot 2MgSO_4$). Included in the hydrated magnesium sulfate group are kieserite ($MgSO_4 \cdot H_2O$) and epsomite ($MgSO_4 \cdot 7H_2O$). Sources which are only slightly water-soluble but which become soluble over a period of 2 to 4 years are: 1) dolomite ($CaCO_3 \cdot MgCO_3$), 2) brucite ($MgO \cdot H_2O$), 3) magnesite ($MgCO_3$), 4) basic slag (which contains various nutrients and usually from 2 to 8% Mg), and 5) periclase (MgO) produced in fertilizer manufacture. Although most of the soluble magnesium materials are mixed with other fertilizers, nearly all other magnesium sources are directly applied to soils. Dolomite, by far the most significant source, is applied to soils at rates of 1 to 3 t/ha.

Sulfur

Sulfur has been recognized as an essential plant nutrient for more than 140 years. Further, it has been basic to the fertilizer industry since its inception due to the treatment of phosphate rock with sulfuric acid in the manufacture of superphosphate. In that process the sulfuric acid reacts with calcium in phosphate rock, resulting in the formation of gypsum ($CaSO_4 \cdot 2H_2O$). Gypsum is not removed from normal superphosphate so that the product contains 12% sulfur. At one time, normal superphosphate was the chief source of sulfur in fertilizers and nearly all mixed fertilizers contained some normal superphosphate. In recent years, however, increasing importance has been placed on high analysis phosphates. In these products sulfuric acid is used to treat phosphate rock, but the sulfur is removed with further processing to higher analysis phosphate materials.

Some plants require almost as much sulfur as they do phosphorus. Sulfur enters the plants as the SO_4^{2-} ion and becomes a necessary component of protein. It is also a constituent of the mustard oil glycoside which imparts characteristic odors and flavors to such species as mustard, onions, and garlic.

The major sources of sulfur in today's fertilizers are ammonium sulfate, normal superphosphate, potassium sulfate, langbeinite, and gypsum. In chemically manufactured mixed fertilizers, sulfuric acid is still in wide use for neutralization of the ammonia used in their formulation. In certain areas where response to sulfur is obtained, langbeinite ($K_2SO_4 \cdot 2MgSO_4$) is increasing in importance as a sulfur source. Elemental sulfur also is commonly used, but its oxidation to the available sulfate form requires time and reduces soil pH levels.

Ammonium thiosulfate [$(NH_4)_2S_2O_3$] is a widely used sulfur source in the fluid fertilizer industry. It is made by reacting SO_2 and aqueous ammonia, with further reaction with elemental sulfur. It contains about 26% sulfur which soon oxidizes in soil to the sulfate (SO_4^{2-}) form and is completely soluble in liquid N-P-K mixtures. It is important to the fluid fertilizer industry because fluids are normally sulfur-free and are not compatible with most other sulfur sources.

MICRONUTRIENTS IN FERTILIZERS

Micronutrients are essential to plants, but are required only in minute quantities. They are measurable in plant tissues only in parts per million (ppm) rather than percentages. Most occur in soils in very small quantities. Micronutrients which must be added to certain soils as fertilizers include boron, copper, iron, manganese, molybdenum, and zinc. Chlorine is also a micronutrient, but there are no known areas of chloride-deficient soils.

Increased crop production, high yield goals, and diminished soil reserves of available micronutrients have combined to increase the consumption of micronutrient fertilizers, particularly in recent years. Data on US micronutrient consumption was not collected until 1967 when the USDA obtained information from known producers of copper, iron, manganese, molybdenum, and zinc. Boron use data were not collected because there were so few producers. Use of manganese and zinc is much higher than that of the other micronutrients (Table 6). This is because manganese and zinc are mainly applied to corn, wheat, soybean, cotton, and rice, which are grown over the largest acreage in the United States. The dramatic increase in use of manganese and zinc consumption from 1976 to 1980 resulted from a change in reporting procedures to include processors of industrial byproducts containing manganese and/or zinc for fertilizers.

Micronutrient sources vary considerably in physical state, chemical reactivity, cost, and availability to plants. They are classified as inorganic, synthetic chelates, and natural organic complexes. Fritted glass products (frits) constitute a much smaller fraction of micronutrient products. Most micronutrient sources are applied with multinutrient mixed fertilizers. They may be incorporated in chemically manufactured fertilizers, in blends, or in fluids. Lesser amounts of micronutrients are applied directly to soils or used as foliar sprays.

Micronutrients, just as the primary and secondary nutrients, must be water-soluble before they can be absorbed by plants. The solubility or availability of copper, iron, manganese, and zinc to plants decreases with increasing soil pH, while that of boron and molybdenum generally increases with increasing soil pH.

Table 6. Amounts of Micronutrients (on Elemental Basis) Sold in the United States (Anon., 1965–1985)

Year†	Cu	Fe	Mn	Mo	Zn
	(thousand tons)				
1968	2.2	3.1	10.5	0.07	13.1
1972	0.6	1.2	11.2	0.09	14.4
1976	0.5	2.3	8.1	0.10	14.7
1980	1.5	5.1	13.2	0.12	40.2
1984	1.1	5.9	15.1	‡	37.3

†Fertilizer year ending June 30.
‡Not reported.

Boron

Boron is added to more fertilizers than any of the other micronutrients. Boron fertilizers usually are required for optimum growth of alfalfa, cotton, and clovers. A large number of plant species will not tolerate more than traces of soluble boron. Conversely, many plant species will not produce seed without adequate boron. Boron is added to mixed fertilizers as water-soluble borax ($Na_2B_4O_7 \cdot 10H_2O$), colemanite ($Ca_2B_6O_{11} \cdot 5H_2O$), and frits. Availability of boron in frits is related to particle size and composition of these glass-like silicate-based products.

Copper

Although copper stimulates crop growth, the functions of this nutrient in plant growth are not fully understood. Copper is known to stimulate seed production and probably functions as enzyme cofactor and regulator. Since plants absorb copper readily, its application with mixed fertilizers is one of the important ways of ensuring that this essential nutrient in animal nutrition is in sufficient concentrations in animal feeds.

Many soils—especially the leached, sandy, and excessively cropped soils—need copper applications for optimum crop production. Highly organic soils are also often copper-deficient. Copper is usually added to mixed fertilizers as sulfates, oxides, and chelates. Chalcocite (Cu_2S), chalcopyrite ($CuFeS_2$), and cuprite (Cu_2O) are the most important minerals from which copper fertilizers originate.

Iron

Iron is an important nutrient closely associated with chlorophyll formation and as a catalyst in plants. Lack of sufficient iron in plants results in chlorosis or yellowing of the plant. Although most soils contain sufficient iron, it may not be available to the plant, particularly in calcareous, high pH soils. Therefore, supplemental iron is applied as fertilizers or foliar sprays. Soil applications of inorganic iron sources generally are not effective in supplying iron for crops growing on high pH soils. Inorganic sources such as ferrous sulfate ($FeSO_4$) or ferric sulfate [$Fe_2(SO_4)_3$] may be used as foliar sprays, but they are rapidly converted to forms which are not as available to plants when they are applied to soils alone or with other fertilizers. Some iron chelates are more effective for soil application, but their high cost restricts their use to high cash value crops.

Manganese

Another essential micronutrient closely associated with iron and having somewhat similar functions in the plant is manganese. Manganese will not take the place of iron, but either its absence or excess reduces the effectiveness of iron. Manganese, copper, and iron probably are the main oxidation-reduction regulators in plants. Available manganese is often deficient in agricultural soils, particularly those that have been recently limed. Manganese is supplied in the form of sulfates, oxides, oxysulfates, and chelates. Usually the less soluble oxides are less effective, although this depends mainly on granule size. The chief minerals which supply these manganese materials for fertilizer are pyrolusite (MnO_2) and psilomelane (mixed Mn-oxides), but the manganese in these sources must be reduced to the divalent form to be available to plants. Byproduct manganous oxide (MnO) is also used as a source of manganese.

Molybdenum

Of all the essential elements, molybdenum is required by plants in the lowest amounts. Little is known regarding the exact functions of molybdenum, although molybdenum is known to be associated with the reduction of nitrates in plants. There are some soils on which little plant growth can be sustained without molybdenum applications. Molybdenum very often stimulates the growth of legumes because of its role in nitrogen metabolism. Molybdenum is closely associated with copper nutrition in animals. Excesses of molybdenum can be just as injurious as inadequate amounts. Molybdenum is incorporated, alone or as part of a general micronutrient mixture, in many mixed fertilizers in very low concentrations. The main molybdenum sources are sodium molybdate and ammonium molybdate, which are produced from the mineral molybdenite (MoS_2).

Zinc

Functions of zinc in plant nutrition are closely associated with those of copper, mainly as a cofactor in enzyme systems. Most zinc-deficient plants are stunted or grow improperly, and maturity is often delayed. While zinc deficiency occurs in wide areas of the United States, zinc fertilizers are mainly applied to corn and rice. Zinc is usually incorporated in mixed fertilizers as sulfates, oxysulfates, oxides, and chelates. Most of these sources have been prepared from the mineral sphalerite (ZnS) or from byproduct oxides and sulfates from zinc-processing operations and other industries.

ACKNOWLEDGMENT

The authors wish to acknowledge the original chapters written by Bryant (1975, 4th edition) and Kincheloe (1983, 5th edition). This 6th edition chapter has been extensively revised and updated, but these previous two editions provided a welcome base.

BIBLIOGRAPHY AND REFERENCES

Anon., 1965–1990, "Production," Annual Reports, Food and Agriculture Organization of the United Nations, Rome.

Anon., 1965–1990a, "Fertilizer," Annual Reports, Food and Agriculture Organization of the United Nations, Rome.

Anon., 1965–1990b, "Fertilizer Materials," Annual Reports, US Department of Commerce.

Anon., 1965–1985, "Commercial Fertilizers," Annual Reports, US Department of Agriculture, Statistical Reporting Service.

Anon., 1990–1991, "Agricultural Resources, Inputs, Situation, and Outlook Report," Annual Reports, US Department of Agriculture, Economic Research Service.

Anon., 1991, "Production and Efficiency Statistics, 1989," US Department of Agriculture, Economic Research Service, 54 pp.

Bryant, J.P., 1975, "Fertilizer Minerals," *Industrial Minerals and Rocks*, 4th ed., S.J. Lefond, ed., AIME, New York, pp. 225–233.

Engelstad, O.P., 1985, *Fertilizer Technology and Use*, 3rd ed., Soil Science Society of America, Madison, WI, 656 pp.

Hargett, N.L., et al., 1986–1991, *Commercial Fertilizers*, Annual Reports, Tennessee Valley Authority, Muscle Shoals, AL.

Hauck, R.D., 1984, *Nitrogen in Crop Production*, American Society of Agronomy, Madison, WI, 804 pp.

Khasawneh, F.E., et al., 1980, *The Role of Phosphorus in Agriculture*, American Society of Agronomy, Madison, WI, 910 pp.

Kincheloe, S., 1983, "Fertilizer Minerals," *Industrial Minerals and Rocks*, 5th ed., S.J. Lefond, ed., AIME, New York, pp. 233–241.

Mengel, K., and Kirkby, E.A. 1982, *Principles of Plant Nutrition*, International Potash Institute, Bern, 655 pp.

Mortvedt, J.J., et al., 1991, *Micronutrients in Agriculture*, 2nd ed., Soil Science Society of America, Madison, WI, 760 pp.

Munson, R.D., 1985, *Potassium in Agriculture*, American Society of Agronomy, Madison, WI, 1223 pp.

Nelson, L.B., 1990, *History of the US Fertilizer Industry*, Tennessee Valley Authority, Muscle Shoals, AL, 522 pp.

Tabatabai, M.A., 1986, *Sulfur in Agriculture*, American Society of Agronomy, Madison, WI, 688 pp.

Tepordei, V.V., 1991, "Crushed Stone—1989," US Department of Interior, Bureau of Mines, p. 15.

Vroomen, H., 1989, "Fertilizer Use and Price Statistics, 1960–1988," US Department of Agriculture, Economic Research Service, 56 pp.

Wallingford, G.W., 1991, "The US Nutrient Need is in the Red," *Better Crops with Plant Food* (Fall), Potash and Phosphate Institute, Atlanta, pp. 16–18.

AGRICULTURAL MATERIALS

SOIL ADDITIVES

GRETCHEN K. HOFFMAN AND GEORGE S. AUSTIN

Soil amendments or soil conditioners generally carry the connotation of being useful for problem soils or soils needing special treatments to produce maximum yields. Many different materials have been utilized, both inorganic and organic. Only the most commonly used and those that appear to have potential to become commonly used will be discussed.

Soil productivity depends on many factors. Soil amendments mainly modify physical characteristics and microbial action, and/or change the pH of a soil, rather than directly supplying plant nutrients. They give soil the needed physical properties to allow plant growth, fight erosion, or save water. Some soil amendments provide nutrients in addition to altering the physical, biological, or chemical characteristics of soil. The term "soil amendment," as it is used by Soil Science Society of America, allows soil amendments to be materials other than those used primarily as fertilizers (Anon., 1987). Soil amendments as fertilizers of any type will not be considered here because this aspect is more relevant to the previous chapter entitled "Agricultural Materials: Fertilizers" in this volume.

Several inorganic soil amendments have been used since the early 20th century to alter the character of soil: limestone, lime, and similar calcium-rich carbonates and oxides for acid soils, and calcium sulfate (gypsum) and native or uncombined sulfur for alkaline soils. The use of these amendments (limestone, lime, gypsum, and sulfur) or their equivalents has been studied and documented in many basic soils tests (McCaslin and Boyle, 1980). Perlite and vermiculite are considered in separate chapters and only a brief statement of their use as soil conditioners will be made in this chapter.

Peat is the principal natural organic soil amendment, but manure, both as a fertilizer and soil conditioner, is used in every part of the world. In less-developed countries, manure is the primary soil fertilizer/conditioner. An organic-rich material designated as "humate" has been marketed as a soil amendment for many years and will be addressed in this paper.

Other materials are discussed by the Soil Science Society of America in its Special Publication 7, "Soil Conditioners" (Stewart, 1975). They include bitumen emulsion, polyacrylamide, polyvinyl acetate, polyvinyl alcohol, bentonite, nonionic surfactants, and natural mulches. Zeolites can be added to the list of miscellaneous soil amendments (McCaslin and Boyle, 1980), as can calcium chloride (Pressler and Pelham, 1985), sulfuric acid, ferric and ferrous sulfate, aluminum sulfate, ammonium polysulfide, calcium polysulfide, and ammonium bisulfite (Stromberg and Tisdale, 1979). Sludge will not be considered as a soil amendment because it has properties more consistent with fertilizers. All these "other materials" account for a very small amount of the volume and value of soil amendments used and will be only briefly discussed.

Problem Soils

Soils amenable to treatment with soil amendments are in five general categories: acid, saline, alkaline, saline-alkaline, and calcareous. Optimum plant growth occurs for a great majority of crops and plants at a soil pH of 6–6.5 to 7 (neutral). In acid soils (pH 3.5 to 6), calcium is leached by relatively high rainfall, lowering the soil pH. Acid soils also are more readily formed from the weathering of acid igneous rocks, such as granites, and siliceous sedimentary rocks, such as sandstones, than from the weathering of basic igneous rocks, such as basalts. Aluminum also comes into play to the extent that highly acid soils can result from aluminum saturation. Soils become acid because the cations of soil colloids, primarily calcium, are replaced by hydrogen ions. This involves ion exchange and other adsorption reactions that are associated with the colloids, such as the clay mineralogy, the amount of organic matter, and similar attributes. Another and much larger part of total acidity is usually referred to as the potential acidity, representing the hydrogen ions held on the colloidal surface. Since clay minerals and organic particles have large surface areas, they usually have much higher total acidity than sandy soils, and it takes a larger volume neutralizing soil amendment to raise the pH of these soils.

A saline soil is one having an electrical conductivity of the saturated extract (EC_e) of more than 4 millisiemens[1] per centimeter (mS/cm) at 25°C and an exchangeable sodium percentage (ESP)[2] of less than 15 (Stromberg and Tisdale, 1979). Ordinarily the pH of these soils is less than 8.2. A saline soil can be recognized by the accumulation of white salts on the surface. The chief cations are Na^+, Ca^{2+}, and Mg^{2+} and the chief anions are Cl^- and SO_4^{2-}, with lesser amounts of NO_3^- and HCO_3^-. Soluble carbonate is rarely present in saline soils. An alkali or sodic soil is one which has an ESP of greater than 15 and an EC_e of less than 4 mS/cm. The soil pH is usually greater than 8.5 and may reach 10 or more. The high ESP of alkali soils causes them to lose their granular structure. Because of the destruction of soil aggregates responsible for the presence of large pores, the soils become almost impervious to water and air. Root penetration is impeded, clods are hard, and seedbeds are difficult to prepare. Saline-alkaline soils are both saline and alkaline, i.e., they have an EC_e of greater than 4 and an ESP greater than 15. Their physical properties are similar to saline soils as long as excess salts are present. As soon as they are irrigated and the soluble salt content of the surface horizon is reduced to a low level, they act as alkaline soils. Calcareous soils contain free calcium carbonate throughout the profile. They are nonsaline and have a pH ranging from about 7.5 to a maximum of 8.2. The EC_e is below 4 mS/cm and the ESP is usually less than 10.

---

[1]A millisiemens is equivalent to a millimho.

[2]The total amount of exchangeable cations a soil can retain is called the cation-exchange capacity (CEC) and is usually expressed in milliequivalents per 100 g of soil. The ESP is the ratio of the exchangeable sodium to the CEC.

On some alkaline and calcareous soils many crops exhibit symptoms of nutritional deficiencies. This is often attributed to the failure of the plant to assimilate some of the nutrient elements, such as phosphorus, iron, copper, manganese, and zinc; all of which are micronutrients except phosphorus. Deficiencies of these elements are sometimes exhibited by a yellowing of the leaves, often termed "lime-induced chlorosis." This an oversimplification of a more complex problem. It must be recognized that crops grown on many calcareous soils with a pH of 8.0 do not show these symptoms. However, the solubility increases by a hundred- to a thousand-fold from a pH of 8.0 to 7.0. Micronutrient deficiencies may also exist in acidic soils. In such cases, they may be true deficiencies and not a case of the plants being unable to absorb the nutrients because of a high soil pH (Stromberg and Tisdale, 1979).

RAW MATERIALS

Inorganic and organic soil amendments (Table 1) will be considered separately because of the dissimilarities in materials, functions, and/or marketing.

Inorganics

Limestone and dolostone: The most common carbonate rocks used as soil amendments to treat acidic soils are limestone and dolostone (or dolomite). Raising the pH in acid soils (pH 3.5 to 6) is accomplished most economically in a process called liming. This is commonly done with pulverized limestone (termed "agricultural limestone"). The neutralization is by base exchange with calcium and magnesium cations displacing acid hydrogen ions in the soil. Numerous agricultural experiments and actual farming have shown increased yields of 15 to 35% by liming to optimum conditions (Boynton, 1966). In the United States, soils in the humid states east of the Mississippi River and moist-soil areas of Washington, Oregon, and California generally require higher rates and more frequent applications of agricultural lime.

Besides raising the soil pH to a desirable level for optimum plant growth, soil pH adjusted by liming influences availability of certain other nutrients. Liming does not necessarily supply these nutrients, but merely makes them more accessible to plants. In particular, the availability of soil phosphorous and molybdenum is greatly enhanced. Conversely, the micronutrients boron, manganese, iron, copper, and zinc are made less available by increasing pH and must be added back into the soil in fertilizers.

Table 1. Agricultural Use of Inorganic and Organic Soil Amendments

Agricultural Use	Soil Amendment
Increase pH	Agricultural limestone (and dolostone), lime, hydrated lime, quicklime
Reduce pH	Peat, humate, gypsum, sulfur
Increase percolation	Gypsum, perlite
Increase availability of fertilizer	Agricultural limestone (and dolostone), lime, hydrated lime, quicklime, humate, peat, sulfur
Increase organic acid content	Humate, peat
Improve workability of soil	Agricultural limestone (and dolostone), lime, hydrated lime, quicklime, gypsum, peat, humate, vermiculite, perlite
Increase soil stability	Agricultural limestone (and dolostone), lime, hydrated lime, quicklime
Retain moisture	Humate, peat, perlite, vermiculite

Liming soil increases organic matter by fostering larger and more prolific plant growth. The earthworm population generally increases as the pH is elevated to neutral. When organic matter, the food of earthworms, is increased, microorganisms increase, further contributing to soil fertility. Liming improves soil tilth by rendering the soils more workable and friable, facilitating cultivation, and it also increases the efficiency of fertilizers (Boynton, 1966).

Mineralogy. Limestone is composed dominantly of the mineral calcite ($CaCO_3$). Aragonite ($CaCO_3$) has the same chemical composition as calcite but a different crystal structure and is common only in modern deposits, such as coquina (composed of sea shells) and oolitic limestone. With time, aragonite alters to calcite. Dolostone is composed of the mineral dolomite ($CaMg(CO_3)_2$) and most commonly is formed from limy sediments that have reacted with magnesium to produce the double salt. Even so, many if not most limestones contain minor amounts of magnesium and most dolostones can have more than 54.3% $CaCO_3$, which is the theoretical percent for pure dolomite.

Chemical Properties. The chemical analysis of carbonate rocks is essential for estimating the neutralizing value of agricultural carbonate rocks. This value commonly is expressed in terms of calcium carbonate equivalents (CCE). Pure calcite ($CaCO_3$) is assigned a CCE of 100, although pure dolomite ($CaMg(CO_3)_2$) has a theoretical value of 108.6. Dolomite is 8.6% more effective than calcite as a neutralizer because the molecular weight of $MgCO_3$ is 84.32, as compared to $CaCO_3$ with a molecular weight of 100.9. However, because of the difference in solubility of dolomite in contrast to limestone, it takes longer to neutralize with pure dolomite than with pure limestone. The composition of limestone and dolomite varies greatly from pure materials and includes impurities such as clay minerals, quartz, and organic matter. It is also very important to establish the exact chemical composition of the soil before a decision is made about the liming material.

Gypsum: Gypsum is the most common of the naturally occurring sulfate minerals. Crude gypsum is marketed for use in cement, agriculture, and fillers. Since the 18th century, gypsum has been used as a soil conditioner in Western Europe. Much of the western United States has saline, alkaline, saline-alkaline, or calcareous soils. As a soil amendment, gypsum neutralizes alkaline, saline, and alkaline-saline soils, improves the permeability of argillaceous materials, and provides sulfur and catalytic support for maximum fertilizer utilization and leguminous productivity (Pressler, 1985).

Although the amount of agricultural gypsum used is relatively small compared to other uses, it still is about 1 Mt each year. Pressler (1985) quoted a projection that 1.5 Mt of gypsum would be used each year for agriculture in the United States by the year 2000. He added that this figure was probably high.

Mineralogy. Evaporation of seawater produces the deposition of anhydrite ($CaSO_4$). Seawater contains only 0.13% $CaSO_4$ and pure bedded sulfate deposits originate either on a sabka supratidal shelf or with associated chloride accumulations in a restricted, brine-filled basin. Later hydration by meteoric waters of the anhydrite to gypsum ($CaSO_4 \cdot H_2O$) occurs at depths ranging from 0 to 325 m below the topography. The depth of hydration generally is related to topography, structure, and climate, because these factors affect the depth of groundwater and surface-water penetration.

Chemical and Physical Properties. Gypsum has a composition of 79% calcium sulfate and 21% water, $CaSO_4 \cdot 2H_20$, and is used as a commercial and generic term for all calcium sulfate minerals. Microbial action on gypsum releases hydrogen ions, lowering the pH. In addition, permeability is increased.

Sulfur: A small amount of elemental sulfur is used as a soil amendment and as a nutrient. At times it is difficult to distinguish between sulfur as a nutrient and sulfur as a soil conditioner. As a conditioner, sulfur lowers the pH in alkali soils. However, this action also improves the uptake of sulfur by the soil as a fertilizer.

The market for agricultural sulfur is considered to be expanding because sulfur can be added in relatively impure forms which are less expensive than other products (Ober, 1992, oral communication). The sulfur is oxidized by soil bacteria, releasing hydrogen ions (Messick, 1992, oral communication). It also acts as protection against mildew, powdery mildew, and other damaging fungi.

Vermiculite: The finer size grades of vermiculite are used for soil conditioning as a plant-growth medium, chemical fertilizer carrier, extender, anticaking agent, and carrier for pesticides and herbicides (Meisinger, 1985). The high cation exchange capacity (CEC) of vermiculite facilitates many of these applications (Strand and Stewart, 1983). When added to soil, vermiculite helps to aerate and improve the workability of the soil. An estimated 31% of the total 1990 vermiculite production in the United States (240.6 kt) was used for agricultural purposes (Potter, 1991). Meisinger (1985) predicted 108.9 kt of vermiculite would be used in agriculture by the year 2000.

Perlite: Perlite is often used as a substitute for vermiculite or peat or in conjunction with these materials as a soil amendment. The important properties of perlite for this use are (1) being chemically inert, therefore long lasting, and (2) retention of several times its own weight in water, providing moisture to the plant root system. Perlite is also used to carry insecticides, and pesticides, as well as fertilizers (Meisinger, 1985a). In addition, perlite aerates the soil which is essential for optimum plant development. Perlite as a soil additive is a minor part of its total consumption; Meisinger (1985a) indicated that 8% (35.4 kt) of the total US perlite production in 1983 was used for agricultural purposes. He predicted an average of 3.5% annual growth to the year 2000.

Organics

Humate: Organic soil amendments are widely regarded as containing humic acids that are beneficial to plants. Humic material is not a pure substance and much confusion exists regarding its terminology in the scientific and business communities. ''Humic acid'' is defined as base-soluble humic, fulvic, and ulmic acids and their salts, formed during partial or complete decay of organic matter. This decay releases a high-molecular-weight material that is darkly colored, partly colloidal, and weakly acidic (Roybal and Barker, 1986).

Three different categories of humic materials are used here to denote the different origins, but some generalization must be made initially to discuss the origins. Humate commonly is associated with both coal and lignite (Fig. 1). Humate associated with oxidized lignite in the Williston Basin in North Dakota and in the Canadian provinces of Saskatchewan and Alberta has been called leonardite, after A.G. Leonard of the North Dakota Geological Survey (Kohanowski, 1970). The humic acid content of lignite decreases rapidly as coal rank increases and oxidation decreases (Abbott, 1963). The third category of humates is organic-rich mudstones which are within the coal-bearing sequences. These mudstones generally have a lower humic acid content than either the oxidized lignites or coals.

Chemical Properties. The definition of the term ''humic acid'' varies between geology and chemistry, with geologic humic acid including additional smaller molecules and having greater acidity than chemical humic acid as derived in the laboratory (Krauskopf, 1967). Humic acids have an average molecular weight between 5 000 and 50 000, are amorphous with poorly defined X-ray patterns, and have a large CEC ranging from 200 to 500 meq/100 gm at pH 7 (Senn and Kingman, 1973). This high CEC may explain the acidity variation between humic acid and ''humate'' (chemical) because a molecule with hydrogen-filled exchange sites yields

Fig. 1. Humate (organic-rich mudstone) interbedded with coal (black) in deformed Menefee Formation (Upper Cretaceous) in northwest New Mexico.

humic acid with a lower, but still modest, pH than exchange sites filled with cations, such as calcium, sodium, aluminum, and/or iron, that yield chemical humate (Senn and Kingman, 1973). Humates possess a high ion exchange capacity that results in retention of nutrients and better utilization of fertilizer (Siemers and Waddell, 1977).

Physical Properties. ''Humate'' used as a lithologic term includes oxidized coals and lignites, organic-rich mudstones and claystones, and concentrations of humic substances, often found in sandstones. Because they are organic colloids, humic acids behave somewhat like clays, increasing the water-retention capacity of the soil (Gosz et al., 1978). Humic substances promote the decomposition of rocks and minerals to form soil and increase the workability of the soil. They increase the acidity of alkali soil, thus making several elements more available to plants (Siemers et al., 1975) and stimulate growth of microorganisms that process plant nutrients.

Distribution of Major Deposits. Economic deposits of humate-rich material occur in Arkansas, Florida, Louisiana, New York, North Dakota, Michigan, Minnesota, Texas, and Wyoming (Burdick, 1965). Oxidized lignite is present in southeastern Saskatchewan, Canada (Broughton, 1972). Appreciable amounts of humate (leonardite) are present in the Estevan and Willow Bunch coalfields, Saskatchewan (Broughton, 1973; Guliov, 1991). Deposits are also found in the Battle River area of central Alberta, Canada (Hamilton, 1991). In the past some production of humic material for drilling mud conditioners came from this area in Alberta, but no production has been reported since 1980. No rigorous evaluation of humate resources is available. Kohanowski (1970) estimated the reserves in western North Dakota to be 7.38 Mt in 1954. Current production in North Dakota is 28.1 ktpy, primarily sold as a drilling mud additive. Siemers and Wadell (1977) estimated 4.5 to 6.8 Gt of humate in the San Juan Basin, New Mexico. Production in New Mexico is about 18.1 ktpy (Roybal and Barker, 1986).

Peat: Peat is formed in swamps, bogs, and other wetlands in favorable environments for profuse plant growth. Peat consists of the partial decomposition of mosses, sedges, and trees preserved under anaerobic conditions (Davis, 1985). The types of biologic material and the level of decomposition determine the end use of the peat.

Chemical Properties. Peat is defined (ASTM D2607–69) as organic matter of geologic origin with less than 25% ash, excluding coal, formed from dead plant remains in water and in the absence of air. Water level is a major factor in peat development, covering the dead growth so that complete decomposition by aerobic organisms cannot take place. Through the loss of H_2O, CO_2, and CH_4, plant cellulose partially decomposes to peat ($C_{62}H_{72}O_{28}$) (Davis, 1985). The surrounding rock, soil, and water dictate the acidity of the peat deposit. The pH of peat varies from 2.8 to 4.0 (Harben, 1992). As a soil additive, peat can lower the pH of alkali soil. With peat's high ionic exchange capacity, nutrients first are absorbed and then slowly released, which increases fertilizer effectiveness. Peat is generally lower in sulfur and ash content and higher in volatile matter, oxygen, and nitrogen than lignite or bituminous coals (Davis, 1985).

Physical Properties. Peat has several physical properties that make it a beneficial soil additive: high water retaining capacity, low density, low heat conductivity, high porosity, and high resistance to decomposition (Prud'homme, 1989). Sphagnum moss peat in particular has an anticompaction effect on soils (Davis, 1985). These physical properties of peat vary with the level of decomposition and the type of peat.

Distribution of Major Deposits. Peat occurs in all 50 states, with known resources of 281 Gt on an air-dried basis (Davis, 1985). The majority (95%) of the peat resources are in Alaska, Minnesota, Michigan, Florida, and Wisconsin. The resource in the lower 48 states is only about 8 to 14% of the total, principally from deposits in Minnesota, Michigan, Florida, Wisconsin, Louisiana, North Carolina, and Maine (Davis, 1985). Large resources of peat are found in Canada (339 Gt) and Europe, particularly in Russia, which has an estimated 698 Gt peat (Cantrell, 1988). Estimates of the world peat resources are 1.7 Tt.

Miscellaneous Amendments

Several miscellaneous soil amendments are listed at the beginning of this chapter. Some of these are polymers or organic compounds; others are naturally occurring substances such as bentonite, zeolite, bitumen, and tree bark; still others, such as calcium chloride, sulfuric acid, ferric and ferrous sulfate, aluminum sulfate, ammonium polysulfide, calcium polysulfide, and ammonium bisulfite, are inorganic chemicals. Information on some of these soil amendments and their applications is limited.

Mineralogy. Of the miscellaneous soil amendments, only bentonite and zeolite are composed of or are minerals. Bentonite is a clay formed from the decomposition of volcanic ash or tuff and consists of clay minerals in the smectite group, complex hydrous aluminosilicates with substitution of various ions including Na^+, Ca^{2+}, Mg^{2+}, Fe^{2+} and Fe^{3+} for Al^{3+} and Si^{4+}. Bentonites tend to be either sodium- or calcium-rich, most bentonites used as soil additives are sodium-rich. Zeolites are hydrated aluminosilicates of the alkali and alkaline earth metals. Commercial occurrences are bedded deposits formed from volcanic ash deposited in alkaline lakes.

Chemical Properties. Many of the miscellaneous soil additives identified in Table 2 are either natural or manufactured organic compounds. Natural zeolites, particularly clinoptilolite, have extremely large ion-exchange capacities. Asphalt or bitumen occur naturally but are also derived from natural gas and liquid petroleum feed stocks. An emulsified asphalt is a stable suspension of minute globules of asphalt in water, or a suspension of minute globules of water in a liquid asphalt (Bodzer, 1980). Natural calcium chloride is produced from brines by five companies in California and Michigan. Synthetic calcium chloride is produced by four companies in New York, Louisiana, and Washington (Pressler and Pelham, 1985). Although the principal use of calcium chloride is as a paving deicing agent, about 20% is used in dust control and roadway stabilization.

Physical Properties. Bitumen emulsions tend to be hydrophobic and will flocculate around fertilizers when they are concentrated salts, such as KCl (De Boodt, 1975). The polyacrylamide, polyvinyl alcohol, and polyvinyl acetate are also hydrophobic, which promotes aggregation and can help to control erosion. Nonionic surfactants lower the surface tension of water and are used to increase infiltration rates in water-repellent soils (Letey, 1975). The added surface area of shredded tree bark when mixed with soil increases aggregate stability and improves percolation. Zeolites have the ability to absorb excess moisture, which is particularly advantageous for clay-poor sandy soil (Mumpton, 1977). Bentonite is known for its ability to absorb moisture and swell. When combined with NaCl or Na_2CO_3, bentonite can reduce infiltration, water movement, and percolation losses in soil (Das and Dakshinamurti, 1975).

Table 2. Agricultural Use of Miscellaneous Soil Amendments

Agricultural Use	Soil Amendment
Improved aggregation	Bitumen emulsion[1], polyacrylamide (PAM) with kaolinite, polyvinyl acetate, polyvinyl alcohol (PVAs)[2], shredded tree bark
Reduce infiltration	Bentonite with sodium chloride and sodium carbonate, bitumen emulsion[1], zeolites (particularly clinoptilolite)
Increase infiltration	Nonionic surfactants, bitumen emulsion, sulfuric acid
Control erosion	Polyvinyl acetate, polyvinyl alcohol[3], bitumen emulsion
Reduce percolation	Bentonite with sodium chloride and sodium carbonate
Prevent compaction	Bitumen emulsion[1]
Reduce leaching of fertilizer	Bitumen emulsion, zeolites[4]
Stabilize soil	Calcium chloride
Reduce pH	Ferric and ferrous sulfate, aluminum sulfate, ammonium polysulfide, calcium polysulfide, ammonium bisulfite

[1]Bitumen emulsion has greater effect on soils rich in clay fraction than on soils with higher silt content (Pla, 1975).
[2]PVAs in emulsion form increase the stability in sodic soils (Carr and Greenland, 1975).
[3]PVA and PAMs are most effective on soils with 30% clay (Mausbach and Shrader, 1975).
[4]Most effective in sandy soils (Mumpton, 1977).

INDUSTRY STRUCTURE

Inorganics

Specifications for crushed stone are numerous. Various national specifications and tests exist and the most active organizations in preparing and reviewing them are the American Society for Testing and Materials (ASTM) and the American Association for State Highway and Transportation Officials (AASHTO). However, requirements for agricultural limestone and gypsum are not very rigid. The $CaCO_3$, $MgCO_3$, and $CaSO_4 \cdot 2H_2O$ equivalent or content is of major significance. In 1989, 14.8 Mt of agricultural limestone was produced (Tepordei, V.V., 1992, oral communication). Of this total, 2 700 t was either shells (coquina) or calcareous marl. In the same year 2.4 Mt of dolostone was used for this purpose.

Organics

The advantage of peat as an agricultural amendment was recognized in the early 1900s. The agricultural peat industry has grown, not only in the United States, but worldwide. World production of peat in 1990 is estimated at 196 Mt, most of which is from the former USSR (93%), followed by Ireland, Finland, the Federal Republic of Germany, Canada, and the United States (Cantrell, 1991). The former USSR used 83% of its peat for agricultural purposes, and the remainder was used for energy production. Ireland and Finland use most of their peat production as a source of fuel. The majority of Germany, Canada, and US peat production is used as a soil amendment (Davis, 1985).

In the United States, peat is produced in 22 of the 48 contiguous states and in Alaska near Anchorage and Fairbanks; 65% of the total US peat production comes from Florida, Michigan, and Minnesota. Reed-sedge peat is 58% and humus is 25% of the total production, followed by 8% sphagnum moss, 3% hypnum moss, and 6% unclassified forms of peat. Bulk and packaged peat control equal shares of the market (Cantrell, 1991). Although reed-sedge moss accounts for more than half of domestic production, 50% of the peat consumption in the United States is sphagnum moss.

Canada's primary peat production is from sphagnum moss. The majority of this production is from Quebec (48%) and New Brunswick (31%). Alberta, Manitoba, and Saskatchewan also produce peat. Minor production comes from Nova Scotia, Prince Edward Island, Newfoundland, and Ontario. Domestic consumption of peat in 1989 was about 12% of the total production; the rest is exported, principally to the United States and Japan (Prud'homme, 1989).

TECHNOLOGY

Inorganics

Limestone and Dolostone: At a typical quarry in relatively flat-lying strata, a limestone or dolomite bed near the surface is stripped of overburden and quarried in a conventional manner, then crushed and screened to appropriate sizes at a nearby plant. Most liming of soils is with crushed and sized limestone. The maximum size as regulated by the state varies, but is generally between 8 and 20 mesh. The smallest size fraction is <120 mesh and the optimum size fraction is 60 to 40% <60 mesh (Remick, oral communication, 1992). In 1990, several thousand operations across the United States produced 1.103 Gt of all types of crushed stone; 71% was limestone and dolostone (Tepordei, 1991). It is estimated that 2% of the total crushed stone is used in agriculture.

Lime and Quicklime: Although most agricultural liming today is achieved with pulverized limestone, many farmers still use ground quicklime (CaO) and hydrated lime ($Ca(OH)_2$), particularly in the middle Atlantic states. Lime costs more than limestone on an equivalent calcium oxide basis, but it reacts faster, neutralizing soil acidity rapidly so crop yields do not suffer. In some sandy soils, dolomitic lime is used due to the persistent magnesium deficiencies.

Lime is, generally speaking, a calcined or burned form of limestone, commonly known as quicklime, calcium oxide, or calcia. When water is added, it converts to calcium hydroxide and is called hydrated lime or slaked lime. Calcination refers to a broad class of reactions of which limestone to lime is only one. In calcination a substance is heated to just below the melting point, resulting in a product that weighs less than it did originally. In calcining limestone to produce lime, the basic chemical reaction is as follows:

$$CaCO_3 \text{ (limestone)} + \text{heat } (100° \text{ to } 1\,300°C) = CaO \text{ (quicklime)} + CO_2 \uparrow$$

or

$$CaCO_3 \cdot MgCO_3 \text{ (dolostone)} + \text{heat } (900° \text{ to } 1\,200°C) = CaO \cdot MgO \text{ (quicklime)} + 2CO_2 \uparrow$$

In 1990, the principal lime-producing states (limestone and dolostone) in order of output were Missouri, Ohio, Pennsylvania, Kentucky, Alabama, Texas, and Illinois (Miller, 1991). These seven states produced 64% of the total lime output of 10.2 Mt. Total use for quicklime (CaO) and hydrated lime ($Ca(OH)_2$) in the United States was 8.669 Mt and 1.565 Mt for Jan. 1 to Sept. 30, 1991. Agricultural use of quicklime (CaO) and hydrated lime ($Ca(OH)_2$) in the same period was 0.012% and 1.6% of the totals respectively.

Gypsum: Agricultural gypsum requires little processing and is used as mined. For agricultural use, the raw material is pulverized to 100 mesh or finer. Some low-grade ore is upgraded by log-washing, sink-float, or selective crushing and screening, both wet and dry. These methods produce a commercial gypsum product of 80% $CaSO_4 \cdot 2H_2O$ or higher (Pressler, 1985).

Sulfur: In agriculture, refined sulfur, with a purity of 99.5%, is ground and applied directly to the soil. Crude sulfur, formed into granules or pellets prior to grinding, or as fines separated from bulk sulfur, is used. In minor cases, sulfur in the form of sulfuric acid or as gypsum is used as a soil amendment.

Organics

Humate: *Processing.* Humate generally is mined by front-end loaders and stockpiled to dry before processing at the mill, where the humate is crushed and screened (Fig. 2). For some uses, the material is custom blended to obtain a specific humic acid content. After the material is crushed, it is shipped in bulk, bagged, or run through an extraction process to produce a liquid. Most humate is sold to commercial farms, but some of the bagged or liquid form is sold at nurseries.

Testing. Humates are tested for the humic acid content that in part determines the end use of the product. Most testing procedures include extraction of humic acid using varying concentrations of NaOH and leach time, in nitrogen or air atmosphere (Fig. 3). Studies by Verploegh and Brandvold (1990) show leach solutions of 0.1M NaOH and leach times of one hour are sufficient to determine the humic acid content. Reporting on humic and fulvic acids on an ash-free basis eliminates metals inherent in the humate structure (Stevenson, 1985) and contaminants such as Na^+ and Cl^- as humate. This is especially true for fulvic acids. The addition of an ashing step to the procedure and reporting on an ash-free basis would make the analyses of humate consistent by eliminating differences in results due to varying leach times and/or NaOH concentrations (Verploegh and Brandvold, 1990).

Fig. 2. Plant facilities at the Agronics Clod Buster No. 1 humate operation, New Mexico.

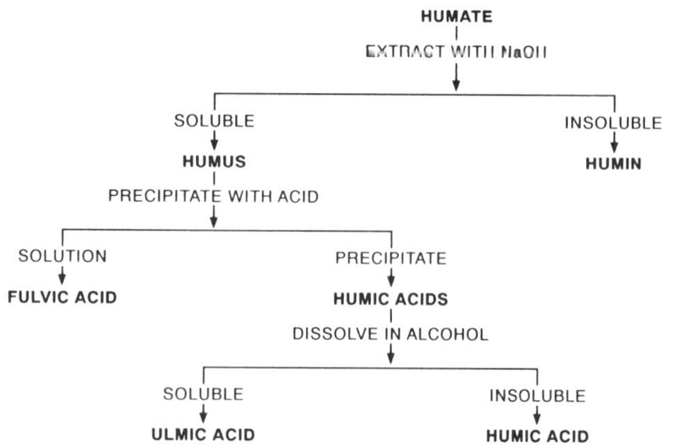

Fig. 3. General analytical flow chart for determining the various components of humate (after Siemers and Wadell, 1977).

Byproducts and Coproducts. Humate could be considered a byproduct of coal and lignite, particularly in surface mining operations where the coal outcrops are weathered. Weathering lowers the calorific value below acceptable levels for energy use, but increases the humic acid content.

Peat: *Processing.* Peat is removed from peat bogs by several methods. The peat bogs have to be drained and large plant debris such as trunks and limbs cleared before peat harvesting can begin. Most methods involve loosening the top layer of material with harrows and then scraping or vacuuming the loosened material from the bog. The peat is transported to the processing plant for drying, shredding, screening, grinding, and packaging. The drying process is by natural and/or artificial means. After drying the peat is shredded and screened. The larger-size material is ground or put through a hammer mill and screened again. Some material is left in bulk form and blended for large contracts, such as golf courses. The processed peat is blended with vermiculite, perlite, or sand for potting soil. Large bales (1.1 m^3 or 1.7 m^3) of peat are compressed to a 2:1 ratio.

Testing. ASTM has established requirements for different types of peat. The accepted value for air-dried peat is 55% moisture and <25% ash on a dry-weight basis. Minimum particle size is 0.15 mm and maximum is 12.5 mm. The ASTM D2607 classification of peat is based on the genetic origin and fiber content on an oven-dried basis. The resulting categories are moss peat, which includes sphagnum, hypnum, and other mosses; reed-sedge peat, whose main organic constituents are derived from reeds or sedges, or other swamp material; and humus, which is made up of decomposed material whose biological origin cannot be determined. Sphagnum peats are light and fibrous and low in moisture content; therefore they can be compressed into bales and sold by volume. Other peats tend to have a greater degree of decomposition, are higher in moisture content, and are sold by weight (Davis, 1985).

The USDA developed another classification of peat based on the level of decomposition and resulting fiber content. Fibric peat has undergone little decomposition and contains >66 2/3% plant fibers. Hemic peat shows moderate decomposition and contains 33 1/3% to 66 2/3% plant fibers. Sapric peat has undergone extensive decomposition and contains <33 1/3% fibers (Davis, 1985).

Still another method of classifying peat used in Canada is the von Post system, which assigns grades according to the degree of decomposition or the level of humification (Troyer, 1985). There are ten grades in this system from H1 through H10. Horticultural peat, which has little or no decomposition of plant material, is in the H1 to H4 range on the von Post scale. With greater degrees of decomposition or humification, peat loses its absorptive properties and has increased density. This peat has a higher von Post grade (H5 to H10) and is used for fuel.

Byproducts and Coproducts. Future byproducts may include synthetic natural gas, ammonia, and sulfur. Currently, there are no byproducts or coproducts from peat harvesting (Davis, 1985).

USES

Inorganics—Limestone and Lime

Limestone: Agricultural limestone is dispersed onto the surface to raise the pH in pasture soil (Fig. 4). For deep-root plants, it is disked into the soil to a maximum depth of about 15 cm. From 1.1 to 1.5 t/ha (0.11 to 0.15 kg/m^2) is considered the maximum amount of agricultural lime that should be applied to soil during a year (Remick, 1992, oral communication).

Lime: The total amount of lime for soil stabilization for highways and off-highway uses exceeds total structural lime consumption (Boynton and Gutschick, 1983). Clay minerals and lime react to form complex cementing compounds which bind soil into a hard, stable mass that is not sensitive to water hydration. The plasticity and shrink and swell of the soil are markedly reduced. The amount of the lime used generally ranges between 3 to 5% of the weight of the dry soil. Over 80% is hydrated lime with the remainder being pebble or ground quicklime. Lime is also used with pozzolan fly ash and aggregate in the preparation of highway base material at highway mix plants. Additionally, contractors use lime to speed up construction by drying up muddy subgrades and to construct temporary all-weather haul roads.

Organics

Humate: Application of humate to fields in the dry form is done at 448 to 672 kg/ha, depending on the pH of the alkali soil (Shomaker and Hiss, 1974). Liquid humate may be used alone or added to other nutrients and sprayed on to the soil (Spence, 1991, oral communication).

Peat: Peat is tilled into clay soils to loosen the soil and added to sandy soils to retain moisture. The organic matter in peat helps to enrich depleted soils. Peat is mixed with fertilizers, vermiculite, and perlite to form a substrate. It is also mixed with limestone, soil,

Fig. 4. Spreading crushed limestone on a field before spring planting (*courtesy:* B.W. Remick, *National Stone Association*).

and fertilizers for potting soil (Prud'homme, 1989). The principal uses of peat are as a base for building lawns, golf courses, and for mulching (Davis, 1985). Peat also is used in mushroom beds and as an earthworm-culture medium. Common uses are as a seed carrier and in the production of peat pots for sprouting plants (Prud'homme, 1989).

Miscellaneous

These materials have a variety of agricultural uses as soil amendments (Table 2). This outline is gleaned predominantly from articles in Stewart (1975). Uses for the others listed (calcium chloride, sulfuric acid, and zeolites) are from Pressler and Pelham (1985), Stromberg and Tisdale (1979), and Mumpton (1977), respectively.

ECONOMIC FACTORS

Inorganics

Cost: In 1989, crushed limestone for application as a soil amendment sold for \$4.50/t FOB plant (Tepordei, 1992, oral communication), the cost of gypsum was \$8.26/t, and sulfur was \$89.26/t. Lime was priced at \$55.10/t.

Organics

Cost: Humate in dry form sells for 55 to \$165/t depending on the humic acid content, type of packaging, and size of the purchase. Liquid humate sells for 1.30 to \$5.30/l, depending on unit volume.

The price of peat is dependent on the type of peat, such as sphagnum, hypnum, reed-sedge, humus, and other, and whether it is in bag, bale, or bulk form. In 1990, the average price for all types of baled or bagged peat was \$33.55/t, 42% higher than the average bulk price, \$19.52/t (Cantrell, 1991). Domestic sphagnum moss, baled or packaged peat, cost \$114.03/t in 1990 and imported sphagnum moss bulk or bale was \$160.64/t (Cantrell, 1991). The actual price of peat has fluctuated in the 1980s from a low of \$22.57/t in 1980 to a high of \$28.27/t in 1983. The actual price in 1990 (preliminary) was \$25.10/t (Cantrell, 1991).

Markets-Transportation: Transportation of humate is predominantly by truck, although rail is used for some shipments. Overseas shipments are generally of the liquid or bagged product. Markets for humate from New Mexico sold as a soil additive include all the contiguous states, Hawaii, Central America, the Caribbean, and Taiwan (Taylor, 1991, oral communication, Spence, 1991, oral communication).

Most peat produced in the United States is sold within a 240-km radius of the source because of the high moisture content and the resulting density (approximately 370 kg/m^3), making it economically prohibitive to transport great distances. Sphagnum moss, because of its lower density (171 kg/m^3), is baled and shipped greater distances, generally by truck (Cantrell, 1991, Cantrell, 1991, oral communication).

Canadian peat producers have formed a national peat association to increase sales in the United States (Prud'homme, 1989). The former USSR controls a large portion of the European peat market because of its large reserves and production. The majority of Ireland's and Finland's peat exports are to England, with smaller shipments to the Netherlands and France.

Alternative Materials: For moisture retention, peat substitutes are straw and tree bark. Sand may be a substitute for peat as a filler in fertilizer. Humate could be considered a peat substitute in some of its applications as a soil conditioner.

GOVERNMENT CONSIDERATIONS

Organics

In New Mexico most humate operations are on private land and face little or no regulation from the state (Russell, 1991, oral communication). The Bureau of Land Management (BLM) views humate as a fill material, similar to sand and gravel. Permitting on federal land requires biological, archeological, plant surveys, and bonds for reclamation. The BLM also charges a royalty of \$0.55/t (Yost, C., 1991, oral communication).

Very few federal regulations govern the peat industry. All peat entering the United States is duty free (Cantrell, 1991). The US government does not promote the production of peat through incentives and peat is not stockpiled by the US government. Federal tax laws allow peat producers a depletion allowance of 5% (Davis, 1985).

Prior to harvesting, peat bogs must be zoned for commercial operations. Environmental and reclamation procedures must be addressed in the plan of operation to obtain local and state permits. Peat harvesting requires the bog to be cleared of all surface vegetation and drained. Often water drained from the bog has a very different chemical composition than that from normal runoff. Therefore, wetlands legislation often requires bog water be treated before release into nearby streams. This, and other environmental restrictions, may place an economic hardship on small producers.

PROBLEMS AND FUTURE TRENDS

Inorganics

Regulations governing quarry operations for limestone, dolostone, and gypsum continue to increase in number and complexity while industry standards are becoming more stringent. The suburbanization of what used to be farmland near large cities has meant that quarries increasingly are becoming surrounded by a populace that looks more for peace and quiet than for jobs. As a consequence, quarries either must locate farther from their principal markets or devote far more resources to placating neighborhood unrest. The escalating cost, time, and effort make quarry development economically less attractive. Smaller operations become less economic more rapidly than larger ones, but larger companies need larger markets, resulting in considerable flux in the industry. There appears to be little likelihood that this trend will be reversed.

Organics

As with all extractive endeavors, the humate industry will have to comply with more environmental regulations. Demand for humate is dependent on economic factors and has been affected by the 1991–1992 recession. However, in New Mexico production increased during the 1980s from approximately 6.0 ktpy to about 18.0 kt/yr in 1991 (Taylor, 1991, oral communication, Spence, 1991, oral communication). This may be in part due to the increased interest in using ''natural'' products instead of chemicals for organic farming.

Production of peat in the United States was 0.7 Mt for 1990, although consumption was 1.3 Mt. Since 1985, both domestic production (approximately 0.9 to 0.7 Mt) and the number of producers (94 to 82) have declined in the United States. This is due to several factors: the economic recession, environmental pressures, resurgence of composting of materials, and an increased market share in the United States from Canada (Cantrell, 1991).

Recently, Canadian producers have consolidated operations to increase their profitability. Consolidation has occurred through mergers and joint ventures in research and development. However, development of new peat bogs is expected to decline with more efficient processing centers (Prud'homme, 1989).

ACKNOWLEDGMENTS

The manuscript was critically reviewed by C.R. Keizer, M.H. Austin, and our colleague, J.M. Barker. All made many comments that greatly improved both the content and syntax of the finished product.

BIBLIOGRAPHY

Anon., 1987, *Glossary of Soil Science Terms*, Soil Science Society of America, Madison, WI, 44 pp.

Anon., 1990, "Standard Classification of Peats, Mosses, Humus, and Related Products, D2607–69," *Annual Book of ASTM Standards*, Vol. 4.08, American Society of Testing Materials, p. 318.

Abbott, G.A., 1963, "Leonardite: A Material of Industrial Promise," US Bureau of Mines Information Circular 8164, pp. 80–86.

Bodzer, D., 1980, "Sales of Asphalt in 1980," *Energy Data Reports*, Department of Energy, Energy Information Administration DOE/EIA 0112(80), p. 2.

Boynton, R.S., 1966, *Chemistry and Technology of Lime and Limestone*, John Wiley & Sons, New York, 520 pp.

Boynton, R.S., and Gutschick, K.A., 1983, revised by R.C. Freas, and J.L. Thompson, "Lime," *Industrial Minerals and Rocks*, 5th ed., Vol. 2, S.J. Lefond, ed., AIME, New York, pp. 809–831.

Broughton, P.L., 1972, "Identification of Leonardite, a Naturally Oxidized Lignite, by Low-angle X-ray Scattering Method," *Journal of Sedimentary Petrology*, Vol. 42, pp. 356–358.

Broughton, P.L., 1973, "Humic Acid Complexes from Naturally Oxidized Lignite: their Genesis, Chemistry, and Utilization," *Proceedings*, Eighth Forum on the Geology of Industrial Minerals, Public Information Circular 5, Iowa Geological Survey, pp. 135–157.

Burdick, E.M., 1965, "Commercial Humates for Agriculture and the Fertilizer Industry," *Economic Botany*, Vol. 19, pp. 152–159.

Cantrell, R.L., 1991, "Peat in 1990," *Mineral Industry Surveys, 1990*, US Bureau of Mines, 10 pp.

Cantrell, R.L., 1988, "Peat," *Minerals Yearbook, 1987*, US Bureau of Mines, pp. 715–723.

Carr, C.E., and Greenland, D.J., 1975, "Potential Application of Polyvinyl Acetate and Polyvinyl Alcohol in the Structural Improvement of Sodic Soils," *Soil Conditioners*, B.A. Stewart, ed., Special Publication 7, Soil Science Society of America, pp. 47–63.

Das, D.K., and Dakshinamurti, C., 1975, "Bentonite as a Soil Conditioner," *Soil Conditioners*, B.A. Stewart, ed., Special Publication 7, Soil Science Society of America, pp. 65–75.

Davis, C.L., 1985, "Peat," *Mineral Facts and Problems, 1985*, Bulletin 675, US Bureau of Mines, pp. 563–571.

Davis, L.L., 1991, "Gypsum," *Mineral Commodity Summaries, 1991*, US Bureau of Mines, pp. 68–69.

De Boodt, M., 1975, "Use of Soil Conditioners Around the World," *Soil Conditioners*, B.A. Stewart, ed., Special Publication 7, Soil Science Society of America, pp. 1–12.

Gosz, J.R., et al., 1978, "An Evaluation of New Mexico Humate Deposits for Restoration of Mine Spoils," *The Reclamation of Disturbed Arid Lands*, R.A. Wright, ed., University of New Mexico Press, Albuquerque, pp. 181–188.

Guliov, P., 1991, "Overview of Industrial Minerals in Saskatchewan," *Proceedings*, 27th Forum on the Geology of Industrial Minerals, Industrial Minerals of Alberta and British Columbia, Canada, Z.D. Hora, et al., eds., Information Series 115, Alberta Geological Survey, pp. 19–36.

Harben, P.W., 1992, *The Industrial Minerals Handybook: A Guide to Markets, Specifications & Prices*, Metal Bulletin PLC, London, p. 148.

Hamilton, W.N., 1991, "Industrial Minerals in Alberta," *Proceedings*, 27th Forum on the Geology of Industrial Minerals, Industrial Minerals of Alberta and British Columbia, Canada, Z.D. Hora, et al., eds., Information Series 115, Alberta Geological Survey, pp. 9–18.

Kohanowski, N.N., 1970, "Leonardite in North Dakota," *North Dakota Quarterly*, Vol. 38, No. 2, pp. 38–42.

Krauskopf, K.B., 1967, *Introduction to Geochemistry*, McGraw-Hill, New York, 721 pp.

Letey, J., 1975, "The Use of Nonionic Surfactants on Soils," *Soil Conditioners*, B.A. Stewart, ed., Special Publication 7, Soil Science Society of America, pp. 145–154.

Lofty, G.J., et al., compilers, 1990, *World Mineral Statistics, 1984–88*, British Geological Survey, Keyworth, Nottingham, 373 pp.

McCaslin, B.D., and Boyle, F.W., 1980, "Report of Research on Soil Conditioners," Research Report 411, New Mexico State University, Agricultural Experiment Station, Las Cruces, New Mexico, 39 pp.

Mausbach, M.J., and Shrader, W.D., 1975, "Influence of Surface Treatment of Selected Subsoil Materials on Infiltration and Erosion," *Soil Conditioners*, B.A. Stewart, ed., Special Publication 7, Soil Science Society of America, pp. 83–93.

Meisinger, A.J., 1985, "Vermiculite," *Mineral Facts and Problems, 1985*, Bulletin 675, US Bureau of Mines, pp. 917–922.

Meisinger, A.J., 1985a, "Perlite," *Mineral Facts and Problems, 1985*, Bulletin 675, US Bureau of Mines, pp. 571–577.

Miller, M.M., 1991, "Lime," *Mineral Commodity Summaries, 1990*, US Bureau of Mines, pp. 92–93.

Mumpton, F.A., 1977, "Utilization of Natural Zeolites," *Mineralogy and Geology of Natural Zeolites*, F.A. Mumpton, ed., Short Course Notes, Mineralogical Society of America, Vol. 4, pp. 177–198.

Pla, I., 1975, "Effect of Bitumen Emulsion and Polyacrylamide on some Physical Properties of Venezuelan soils," *Soil Conditioners*, B.A. Stewart, ed., Special Publication 7, Soil Science Society of America, pp. 35–46.

Potter, M.J., 1991, "Vermiculite," *Mineral Commodity Summaries, 1990*, US Bureau of Mines, pp. 180–181.

Pressler, J.W., 1985, "Gypsum," *Mineral Facts and Problems, 1985*, Bulletin 675, US Bureau of Mines, pp. 349–356.

Pressler, J.W., and Pelham, L., 1985, "Lime, Calcium, and Calcium Compounds," *Mineral Facts and Problems, 1985*, Bulletin 675, US Bureau of Mines, pp. 453–460.

Prud'homme, M., 1989, "Peat," *1989 Canadian Minerals Yearbook*, Mineral Report No. 38, Energy and Mines and Resources Canada, pp. 45.1–45.11.

Roybal, G.H., and Barker, J.M., 1986, "Geology and Production of Humate and Weathered Coal in New Mexico," *Transactions SME*, Vol. 280, pp. 2105–2111.

Senn, T.L., and Kingman, A.R., 1973, "A Review of Humus and Humic Acids," Research Series, No. 145, South Carolina Agricultural Experiment Station, 18 pp.

Shomaker, J.W., and Hiss, W.L., 1974, "Humate Mining in Northwestern New Mexico," New Mexico Geological Society, 25th Field Conference, pp. 333–336.

Siemers, C.T., et al., 1975, "IV. Humate and Associated Deposits of the Lower Part of the Menefee Formation (Stop 2)," *Field Trip to Central New Mexico*, AAPG-SEPM Rocky Mountain Section Meeting, Albuquerque, June 1–4, pp. 39–46.

Siemers, C.T., and Wadell, J.S., 1977, "Humate Deposits of the Menefee Formation (Upper Cretaceous), Northwestern New Mexico," New Mexico Geological Society, Supplement to the 28th Field Conference, pp. 1–22.

Stewart, B.A., 1975, *Soil Conditioners*, Special Publication 7, Soil Science Society of America, 186 pp.

Stevenson, F.J., 1985, "An Introduction to Humic Substances in Soil, Sediment and Water," *Humic Substances in Soil, Sediment and Water*, G.R. Aikin, ed., John Wiley & Sons, Canada, pp. 1–52.

Strand P.R., and Stewart, O.F., 1983, "Vermiculite," *Industrial Minerals and Rocks*, 5th ed., Vol. 2, S.J. LeFond, ed., AIME, New York, pp. 1375–1381.

Stromberg, L.K., and Tisdale, S.L., 1979, "Treating Irrigated Arid-land Soils with Acid-forming Sulphur Compounds," Technical Bulletin 24, The Sulphur Institute, Washington, DC, 26 pp.

Tepordei, V.V., 1985, "Crushed Stone," *Minerals Facts and Problems*, Bulletin 675, US Bureau of Mines, pp. 757–768.

Troyer, D.R., 1985, "Peat Resources in Saskatchewan," Report 218, Saskatchewan Energy and Mines, p. 10.

Verploegh, J.M., and Brandvold, L.A., 1990, "Optimization of the Analyses of Humic and Fulvic Acid in Soil Humates," Unpublished Report, New Mexico Bureau of Mines and Mineral Resources.

ASBESTOS

ROBERT L. VIRTA AND E.L. MANN

Asbestos is a generic name given to six fibrous minerals that have been used widely in commercial products. The six types of asbestos are chrysotile, crocidolite, amosite, anthophyllite asbestos, tremolite asbestos, and actinolite asbestos (Table 1). Chrysotile is the asbestiform variety of serpentine. Crocidolite, amosite, anthophyllite asbestos, tremolite asbestos, and actinolite asbestos are the asbestiform analogues of riebeckite, cummingtonite-grunerite, anthophyllite, tremolite, and actinolite, respectively. Although amosite is an acronym for the Asbestos Mines of South Africa (AMOSITE) and does not specify a mineral type, it generally is composed of cummingtonite-grunerite asbestos so the terms are assumed to be synonymous.

PROPERTIES

Physical

Asbestos fibers are characterized by high tensile strength, flexibility, resistance to chemical attack and thermal degradation, large surface area, and ability to be woven. Each type of asbestos has different physical characteristics.

Chrysotile is a white, fibrous material. The fibers are extremely thin, and most are soft and flexible enough to be woven. It has a tensile strength between 550 and 700 MPa, making it one of the stronger asbestos types. It is extremely heat resistent and is used routinely in commercial products that are exposed to temperatures exceeding 700°C. The fusion temperature for chrysotile is 1 521°C. Chrysotile fibers are readily degraded by exposure to acids and bases. Because of the fibrillar structure, chrysotile has an extremely large surface area, on the order of 13 to 18 m^2/g (Hodgson and White, 1967).

Table 1. Types of Asbestos

Type	Theoretical Formulae
Chrysotile Hydrated magnesium silicate	$Mg_6[(OH)_4Si_2O_5]_2$
Crocidolite Complex sodium iron silicate (riebeckite) commonly called Blue Asbestos	$Na_2Fe_5[(OH)Si_4O_{11}]_2$
Amosite Iron magnesium silicate (grunerite)—same as anthophyllite except for high iron content	$MgFe_6[(OH)Si_4O_{11}]_2$
Anthophyllite asbestos Magnesium silicate with varying amounts of iron in the lattice	$(Mg,Fe)_7[(OH)Si_4O_{11}]_2$
Tremolite asbestos Calcium magnesium silicate	$Ca_2(Mg,Fe)_5[(OH)Si_4O_{11}]_2$
Actinolite asbestos Calcium magnesium silicate with varying amounts of iron	$Ca_2(Mg,Fe)_5[(OH)Si_4O_{11}]_2$

Amphibole asbestos fibers generally are harsher, more springy and more brittle than those of chrysotile. They also are more resistant to chemical attack, have very good filtration rates, have a greater hardness (4 to 6 on the Mohs scale), and are comparatively long, ranging up to several centimeters. Their color ranges from white for tremolite through yellowish-brown for amosite to lavender or blue for crocidolite. Tensile strengths range from 28 MPa for anthophyllite asbestos to a maximum of 2 100 MPa for crocidolite. All of the forms of amphibole asbestos withstand temperatures exceeding several hundred degrees without degradation. The fusion temperature for all of the asbestiform amphiboles exceeds 1 093°C. The resistance to attack by acids and bases ranges from fair for actinolite asbestos to very good for anthophyllite asbestos (Table 2). The surface area of amphiboles ranges from 2 to 8 m^2/g (Addison et al., 1966).

Chrysotile fibers are hexagonally close packed and parallel to each other, having cross-sectional diameters ranging from 180 to 300 Å. They are tubular in structure (Figs. 1 to 3). The curvature is the result of the mismatch in the dimensions of the silica $[Si_2O_5]$ and brucite ($Mg(OH)_2$) layers, which are on the inside and outside, respectively, of a double sheet.

Amphibole fibers (Figs. 4 to 6) are many times larger in cross section, have considerably smaller surface areas, and are less symmetrical than those of chrysotile. The structure of the amphiboles consists of two ribbon chains based on Si_4O_{11} units separated by a band of seven cations with two hydroxyl groups attached to the central cation of each unit cell. These stack together to form a sandwich of ribbons in the final structure, with a marked cleavage developed between the ribbons. The fiber bundles of both chrysotile and amphibole asbestos can be physically separated into fibers of various lengths and cross sections.

Chemical

The theoretical compositions of the six varieties of asbestos are given in Table 1. Crocidolite, amosite, anthophyllite asbestos, tremolite asbestos, and actinolite asbestos are amphiboles and thus have the same basic composition. Chrysotile, a serpentine mineral, has a composition that is substantially different.

Theoretical compositions are different from those observed in most deposits. The difference for most chrysotile can be attributed to impurities in the samples. Magnetite is one of the common impurities and accounts for higher than normal iron concentrations in some chrysotile samples. The difference between the theoretical and observed compositions for asbestiform amphiboles is attributed to differences in the chemistry of the deposits. Cations can be easily substituted for one another in the amphibole crystal structure. Consequently, variations in the chemistry of the deposit during meta-

Table 2. Properties of Asbestos Fibers (Badollet, 1951)

Property	Actinolite asbestos	Amosite	Anthophyllite asbestos	Chrysotile	Crocidolite	Tremolite asbestos
Structure	Reticulated long prismatic crystals and fibers	Lamellar. Coarse to fine fibrous and asbestiform	Lamellar. Fibrous asbestiform	Usually highly fibrous fibers fine and easily separable	Fibrous in ironstones	Long. Prismatic and fibrous aggregates
Mineral association	In limestone and in crystalline schists	In crystalline schists, etc.	In crystalline schists and gneisses	In altered periodotite adjacent to serpentine and limestone near contact with basic igneous rocks	Iron rich silicious agrillite in quartzose schists	In Mg limestones as alteration products of highly magnesian rocks. Metamorphic and igneous rocks
Origin	Metamorphic	Metamorphic	Metamorphic	Metamorphism of basic igneous rocks	Regional metamorphism	Metamorphic
Veining	Slip or mass fiber	Cross fiber	Slip or mass fiber	Cross and slip fibers	Cross fiber	Slip or mass fiber
Essential composition	Ca,Mg,Fe silicate with some water	Fe,Mg silicate with some water	Mg silicate with some iron	Mg silicate with some water	Na,Fe silicate with some water	Ca,Mg silicate with some water
Crystal system	Monoclinic	Monoclinic	Orthorhombic	Monoclinic	Monoclinic	Monoclinic
Color	Greenish	Ash gray or brown	Grayish white, brown-gray or green	White, gray, green	Lavender blue, metallic blue	Gray-white, greenish, yellowish, bluish
Luster	Silky	Vitreous. Somewhat pearly	Vitreous to pearly	Silky	Silky to dull	Silky
Hardness	6±	5.5–6.0	5.5–6.0	2.5–4.0	4	5.5
Specific gravity	3.0–3.2	3.1–3.25	2.85–3.1	2.4–2.6	3.2–3.3	2.9–3.2
Cleavage	110 perfect	110 perfect	110 perfect	010 perfect	110 perfect	110 perfect
Optical properties	Biaxial negative, extinction inclined	Biaxial positive, extinction parallel	Biaxial positive, extinction parallel	Biaxial positive, extinction parallel	Biaxial ±, extinction parallel	Biaxial negative, extinction inclined
Refractive index	1.63 ± weakly pleochroic	1.64 ±	1.61 ±	1.51–1.55	1.7 pleochroic	1.61 ±
Length	Short to long	2 to 11 in., varies	Short	Short to long	Short to long	Short to long
Texture	Harsh	Coarse but somewhat pliable	Harsh	Soft to harsh. Also silky	Soft to harsh	Generally harsh, sometimes soft
Specific heat, J per kg per °K	505	449	488	619	468	493
Tensile strength, kPa	6 895 and less	110 316 to 620 528	27 579 and less	551 581 to 689 476	689 476 to 2 068 427	6 895 to 55 158
Temperature at maximum ignition loss, °C	—	871 to 982	982	982	648	982
Filtration properties	Medium	Fast	Medium	Slow	Fast	Medium
Electric charge	Negative	Negative	Negative	Positive	Negative	Negative
Fusion point, °C	1 393	1 399	1 468	1 521	1 229	1 316
Spinnability	Poor	Fair	Poor	Very good	Fair	Poor
Resistance to acids and alkalies	Fair	Good	Very good	Poor	Good	Good
Magnetite content	—	0	0	0–5.2	3.0–5.9	0
Mineral impurities present	Lime, iron	Iron	Iron	Iron, chrome, nickel, lime	Iron	Lime
Flexibility	Poor	Good	Poor	High	Good	Poor
Resistance to heat	—	Good. Brittle at high	Very good	Good. Brittle at high	Poor, fuses	Fair to good
Ionizable salts, μmhos (relative electrical conductance)	—	1.34	0.58	1.82	0.84	—

Fig. 1. Chrysotile. X150 000. Reduced approximately 33% in reproduction.

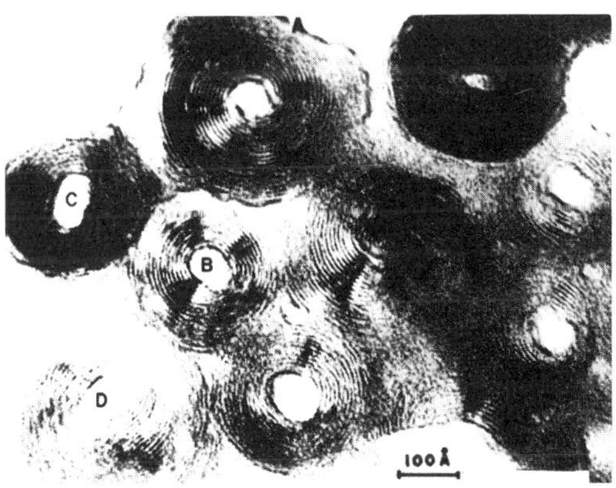

Fig. 2. High resolution electron micrograph of chrysotile. Reprinted with permission from *Acta Chrystalogica* (Yada, 1967). Reduced approximately 38% in reproduction.

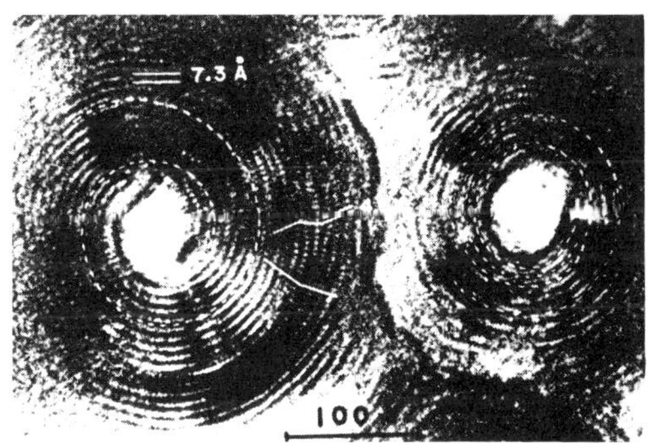

Fig. 3. High resolution electron micrograph of chrysotile. Reprinted with permission from *Acta Chrystalogica* (Yada, 1967). Reduced approximately 38% in reproduction.

morphism are reflected in the compositions of the amphiboles that form. Table 3 gives the range of compositions of commercial asbestos. Many trace elements are also present, but rarely in amounts greater than 0.20%.

Chrysotile has a surface charge that can be positive to negative depending on its source. Most chrysotile has a positive charge, reflecting the net positive charge on the surface layer of the fiber. Fibers from which weathering has removed MgOH layers, exposing the siliceous layer on the fiber surface, have negative charges. The size and sign of the charge also change when the fiber is suspended in solutions of different pH and composition. The ions in the solution are adsorbed onto the fiber's surface, changing the surface charge (Hodgson, 1986).

The surface charge for asbestiform amphiboles is negative. The negative charge is attributed to the siliceous layers exposed on the fiber surface.

The isoelectric point or pH for zero potential for chrysotile ranges from 10.5 to 11.3, depending on the source of the fiber and the method of measurement. Chrysotile flocculates from solution when the isoelectric point is reached.

Effect of Heat on Asbestos

The weight loss for different types of asbestos when heated for 2 hr at successive temperatures from 204 to 982°C is shown in Table 4 (Badollet, 1951, Badollet and Streib, 1955). The weight loss for

Fig. 4. Electron micrograph of crocidolite. X8 000. Reduced approximately 33% in reproduction.

Fig. 5. Amosite, Penge, South Africa. X5 000. Reduced approximately 33% in reproduction.

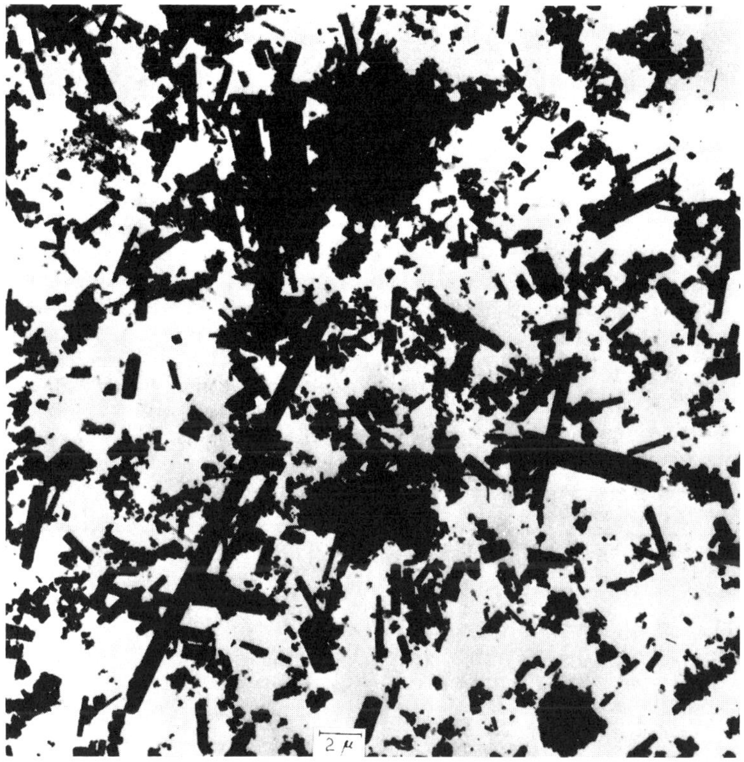

Fig. 6. Anthophyllite, Georgia. X5 000. Reduced approximately 33% in reproduction.

Table 3. Approximate Chemical Analysis of Asbestos in Percent

	Chrysotile	Crocidolite	Amosite	Anthophyllite asbestos	Tremolite asbestos
SiO_2	37–44	49–53	49–53	56–58	53–62
MgO	39–44	0–3	1–7	28–34	0–30
FeO	0–6	13–20	34–44	3–12	1.5–5
Fe_2O_3	0.1–5	17–20	—	—	—
Al_2O_3	0.2–1.5	—	2–9	0.5–1.5	1–4
H_2O	12–15	2.5–4.5	2–5	1–6	0–5
CaO	tr-5	—	—	—	0–18
Na_2O	—	4.0–6.5	—	—	0–9
$CaO+Na_2O$	—	—	0.5–2.5	—	—

the amphiboles is considerably less than that for chrysotile because much less -OH is available to be driven off as water.

Table 5 shows the effect of heating chrysotile on the tensile strength of the fiber. The fiber bundles tested for tensile strength were very small so that heat transfer would be completed in the 3-min heating time. Even after heating at 648°C, chrysotile retains 32% of its strength.

Structural changes in chrysotile can occur under conditions of intense grinding. Localized heating within the fibrils produces dehydroxylation as the tremendous impact energies are absorbed. The heating apparently causes the structure to become amorphous.

Table 4. Effect of Temperature on Loss in Weight of Asbestos Fibers

°C	Time, hr	% Loss in Wt: Amosite	Anthophyllite asbestos	Chrysotile	Crocidolite	Tremolite asbestos
204	2	0.23	0.05	0.30	0.08	0.04
316	2	0.57	0.24	0.85	0.25	0.08
371	2	0.80	0.30	1.78	0.49	0.13
427	2	0.98	0.38	2.17	0.73	0.22
482	2	1.07	0.41	2.83	0.83	0.26
538	2	1.16	0.44	3.99	0.86	0.29
593	2	1.36	0.52	10.38	1.00	0.37
649	2	1.39	0.54	12.75	1.04	0.37
760	2	1.43	0.54	13.43	1.03	0.47
816	2	—	0.64	—	—	0.56
871	2	1.52	1.12	13.62	0.93*	0.67
927	2	—	1.73	—	—	1.40
982	2	1.53	2.39	13.77	0.77*	2.18

*Iron changing in weight due to oxidation.

Table 5. Effect of Heat on Tensile Strength of Canadian Chrysotile Crude

	Tensile strength, kPa	Percent of original tensile strength
Original crude, no heat	903 000	—
Heated 3 min at 316°C	827 000	91.6
Heated 3 min at 427°C	662 000	73.3
Heated 3 min at 538°C	538 000	59.5
Heated 3 min at 649°C	290 000	32.0

Table 6. Solubility of Asbestos

	Percent Loss in Weight, Refluxing 2 hrs, 25% acid or caustic				
	HCl	CH_3COOH	H_3PO_4	H_2SO_4	NaOH
Actinolite	20.31	12.28	20.19	20.38	9.25
Amosite	12.84	2.63	11.67	11.35	6.97
Anthophyllite	2.66	0.60	3.16	2.73	1.22
Chrysotile	55.69	23.42	55.18	55.75	0.99
Crocidolite	4.38	0.91	4.37	3.69	1.35
Tremolite	4.77	1.99	4.99	4.58	1.80
	Percent Loss in Weight, Room Temperature 26°C for 528 hrs, 25% acid or caustic				
	HCl	CH_3COOH	H_3PO_4	H_2SO_4	NaOH
Actinolite	22.55	12.14	20.10	20.60	9.43
Amosite	12.00	3.08	11.83	11.71	6.82
Anthophyllite	2.13	1.04	3.29	2.90	1.77
Chrysotile	56.00	24.04	56.45	56.00	1.03
Crocidolite	3.14	1.02	3.81	3.48	1.20
Tremolite	4.22	1.41	4.89	4.74	1.65

Chemical Resistance

Asbestos often is used in chemical applications where resistance to chemicals is of paramount importance. Table 6 gives data on the effect of various acids and caustic soda on asbestos after long exposure at room temperature and for short exposure at boiling temperatures. Chrysotile is aggressively attacked by strong acids. It is resistant to 25% caustic, but is attacked by strong caustic (70%) at its boiling temperature.

Most amphibole fiber varieties are more acid resistant than those of chrysotile. However, they are not entirely acid resistant and can exhibit weight losses of 2 to 20% when subjected to boiling concentrated acids.

GEOLOGY

General

Chrysotile constitutes about 98% of the current world production of asbestos. The host rock for most of the world's chrysotile production is ultrabasic in composition (Keith and Bain, 1932, Berger, 1963, Bates, 1969). The remaining chrysotile production is derived from serpentinized dolomitic limestone. Amosite and crocidolite are found in metamorphosed ferruginous sedimentary formations (banded ironstones, ferruginous quartzites, and iron-rich silicified argillite) and, together, account for 2% of world production. Tremolite and anthophyllite make up the balance of the production and are associated with highly metamorphosed ultrabasic rocks.

The age of asbestos deposits ranges greatly from earliest Precambrian in Zimbabwe and Swaziland to Upper Jurassic in California. The Ontario, Ungava, and Brazil deposits are Precambrian. Deposits of the Eastern Townships, Vermont, and Newfoundland, are Mid-Paleozoic in age and are associated with early folding in the Appalachian mountain belt. The deposits of western Canada are associated with mountain building during the Late Paleozoic. The deposits in the former USSR range from Early Paleozoic to Late Paleozoic or Triassic in age. For location of major deposits see Fig. 7.

Structures

Faulting, folding, or shearing evidently played a major role in the formation of asbestos deposits (Cooke, 1937). The introduction of swarms of small intrusive dikes of acidic to intermediate composition also contributed to the fracturing and opening up of the rock. The fracturing, in turn, led to serpentinization and the formation of chrysotile fiber.

Most chrysotile ore deposits are found in serpentinized peridotite rather than in dunitic host material. It is believed that the serpentinized dunite bodies tend to flow when subjected to stress, thereby aiding in the development of fracturing in the surrounding, more brittle peridotite.

Asbestos Veining

Chrysotile asbestos is referred to as cross fiber when the fibers lie transverse to the vein and as slip fiber when the fiber lies in the plane of the vein (Cooke, 1937, Gold, 1967). The orientation of cross fiber chrysotile ranges from normal to sharply inclined to the vein walls. When sharply inclined, it is referred to as oblique fiber. The fiber in cross fiber and slip fiber veins may be straight, gently curved, or contorted. Fiber containing sharp flexures is apt to break easily.

Partings are present in many of the veins, either as microscopic discontinuities or as irregularly shaped inclusions composed of picrolite, serpentine, brucite, and magnetite, or combinations of these. The veins may be relatively persistent or short and lenticular, in the form of stockworks or in parallel systems known as ribbon fiber. They may be fissure fillings, replacements, or stress-relief features, and many are the result of a combination of these processes. Some very short fiber is derived from a variety referred to as "mass" fiber. Mass fiber consists of either the complete replacement of most of the original mineral constituents by fiber, each grain of which has a different fiber orientation, or by a dense network of fine, reticulating replacement veins. Both varieties result in a very high content of short fiber, and the rock, which appears to be largely fibrous, is sometimes referred to as "fur" rock.

Other varieties of mass fiber include "platy" fiber, of which the Coalinga deposit is unique. A similar, but less important, variety is known as "mountain leather."

CHRYSOTILE DEPOSITS IN ULTRABASIC ROCKS

North America

Canada: *Eastern Townships, Quebec.* Chrysotile asbestos was discovered in two localities in the Eastern Townships in 1878 and has been actively mined at Thetford Mines and Asbestos for more than 100 years. The chrysotile deposits of Quebec's Eastern Townships are found along a major serpentine belt that arcs northeastward into the Gaspé Peninsula in one direction and southward into the Appalachian mountain belt of Vermont in the other.

Most of the major deposits are located along a region 90 km

long that lies parallel to and 80 km southeast of the St. Lawrence River. The region extends from a point near East Broughton, Quebec, southwestward past Thetford Mines and Black Lake to Asbestos and Danville.

This serpentine belt is a typical ophiolitic complex of gabbroic and dioritic rocks, and of serpentinized pyroxenite, peridotite and dunite. The complex is associated with granitic, rodingitic, and talc-carbonate rocks and was described as a partially serpentinized ultrabasic intrusive emplaced in the crust along faulted zones of weakness. It may have been extruded onto, or just under, unconsolidated aluminous and siliceous sediments in a eugeosyncline ocean basin. Magmatic differentiation within the ultramafic magma is thought to have been responsible for the formation of the various pyroxenite, peridotite, and dunite phases found in the complex.

The small granitic masses found within the ultramafic complex have been considered intrusive but may be the result of the ingestion and digestion of fragments of underlying sea floor sediments. Subsequent faulting and shearing caused numerous dislocations within and adjacent to the ultramafic complex.

Consideration of this complex as a submarine extrusive simplifies the explanations of the source of water for serpentinization, the allowance for large volume change, the presence of fiber as open fracture fillings in stockworks, and the banded fiber in ribbon zones. It also explains the presence of "pillows," most of which are gradationally serpentinized from the surface inward.

The fiber formed mainly in highly serpentinized peridotite (harzburgite containing 10 to 15% enstatite) and is found in numerous places along the strike of the complex.

Five deposits presently are mined in the Eastern Townships and numerous others are either dormant or completely mined out. For purposes of description, the deposits are grouped on the basis of their distribution within the belt, i.e., the Pennington dike deposits, those of Thetford Mines area, the Black Lake deposits, and those of Asbestos.

Pennington Dike, Thetford Mines-East Broughton Area (Merrill, 1957, Riordon, 1957, Rowbotham, 1970, Anon., 1972). The Pennington "dike" is separated from the main Thetford ultrabasic intrusive. It extends northeastward for approximately 30 km and has been mined over much of its length in the past. The serpentinite body actually is sill-like and closely conformable with the country rocks, which dip as much as 60°SE. Shears are conspicuous throughout its length. The cause is interpreted as a major thrust fault, the plane of which lies at the base of the sill. Where the sill is narrow, shears extend from wall to wall, and much of the rock has been steatized. The wider portions, which are as thick as 240 m, are less affected. They contain most of the ore. Slip fiber predominates in all the deposits, though some cross fiber is found in the wider, less sheared sections, particularly toward the southwest. The narrowness of the ore bodies and the attitude of the sill are obstacles to mining at any appreciable depth.

Thetford Mines Area (Cooke, 1937, Riordon, 1957, 1957a,b, Riordon and Laliberté, 1957). A large mass of peridotite is host to the Thetford Mines and Black Lake deposits. The Thetford group lies at the northeast end of this mass and contains four closely connected deposits in a well-defined zone along the hanging-wall side of a prominent fault structure. There has been extensive talc-carbonate alteration along the fault. The zone is 1 830 m long and 460 m wide. The largest and deepest deposits extend to a depth of 460 m or more.

A number of subsidiary shears, most of which dip in the opposite direction to the main fault and strike roughly parallel to it, break the ore zone into segments. Most of these subsidiary shears are accompanied by bodies of syenite, and evidently developed along the original zones of weakness into which these acid rocks were introduced. Some large bodies of serpentinized dunite within the ore zone contributed to the fracturing in the surrounding peridotite by behaving as relatively incompetent masses that tended to flow under stress.

Black Lake Area (Riordon, 1957a,c,d, Riordon and Laliberté, 1957, Anon., 1972). The three major ore bodies in the general vicinity of Black Lake are the British Canadian, the Lake Asbestos, and Normandie deposits. The British Canadian mine is more than 1.6 km long and 0.8 km wide, and occupies a position adjacent to a pronounced flexure in the peridotite contact immediately east of the town of Black Lake. This ore body is cut by a series of roughly parallel shear zones and syenite dikes striking tangentially to the folded contact. Large bodies of serpentinized dunite are also found within the ore body and are low grade to barren.

Southwest of the town of Black Lake are the ore bodies of the Lake Asbestos mine, the lake having been drained to permit mining of the deposit beneath. The easternmost of these ore zones borders the transverse fault that dips to the northeast. Like the major fault in Thetford, this zone has also undergone intensive steatization. Another ore body to the west lies beneath a large, flat-lying acid dike that evidently localized fracturing and was probably responsible for the formation of fiber in the vicinity.

The Normandie deposit, about 1.6 km to the west of Lake Asbestos and 3.2 km southwest of the town of Black Lake, consists of at least three separate ore bodies along a northeasterly plunging fold axis. The old Vimy Ridge deposit, which is now completely worked out, is the most southerly of these and close to surface; much of the upper portions of the ore body have been truncated and removed by erosion.

The central zone lay completely buried beneath a capping of drift and overburden, and was only discovered by regular grid pattern diamond drilling in the peripheral areas surrounding the old Vimy Ridge mine.

Further drilling on the downdip extension of the Normandie ore body led to the discovery of a deep ore zone, the Penhale ore body. Mining of the Normandie ore body ended several years ago.

The origin of these three ore bodies is closely related to the folding but may also be dependent on a fault zone trending southwest that parallels the fold axis to the east of the ore bodies.

Whereas the other deposits in the Eastern Townships are made up of a stockwork of asbestos veins, the Normandie, Vimy Ridge, and Penhale ore bodies are composed predominantly of a prominent system of parallel veins of ribbon fiber. These veins form a conjugate pattern but dip predominantly to the northeast with the rake of the ore zone.

The Vimy Ridge and Normandie ore bodies are separated by numerous large syenite dikes and some shears, whereas the Normandie and Penhale deposits are separated by a zone of low-grade to barren material.

Asbestos-Shipton Area (Allen et al., 1957, Bourassa, 1957, Riordon and Laliberté, 1957, Rowbotham, 1970). The Jeffrey and the Nicolet ore bodies lie along the north and footwall side of a peridotite-dunite intrusive. The Jeffrey deposit, discovered more than 100 years ago, has been producing since the early 1880s and is now the largest asbestos mine outside the former USSR. The ore body is ellipsoidal to cylindrical or roughly pipe-like and about 610 m in average diameter, and its major axis plunges about 55° to the southwest. The ore zone is bounded on both the footwall and hanging-wall contacts by major faults or zones of shearing that dip south-southeast at about 65 to 70°. The ore body also is cut by several irregular shears that trend roughly parallel to the footwall contact and divide the deposit into five major ore zones. Numerous acid dikes of dioritic to syenitic composition follow the shear zones, but a few cut across. Most of the asbestos is good- to medium-grade cross fiber, but some is slip fiber and mass fiber.

The Nicolet ore body occupies a narrow zone along the footwall contact, bounded on its south side by a prominent zone of shearing.

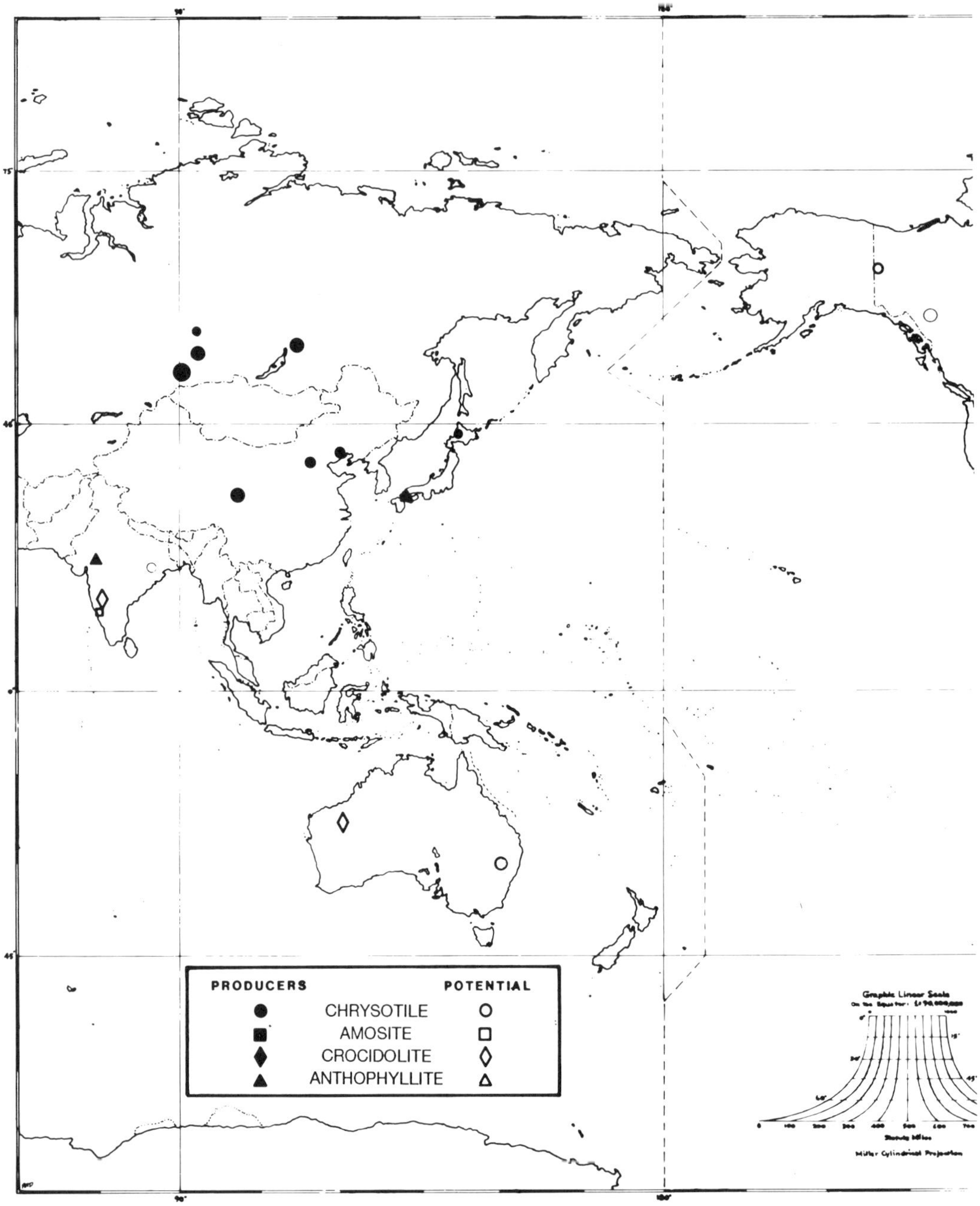

Fig. 7. World asbestos deposits.

Baie Verte, Newfoundland. The Advocate ore body near Baie Verte forms a continuous fiber zone within the highly serpentinized periphery of a massive elliptical pyroxenite body, 1.6 km long by 0.4 km wide. The inward-dipping fiber zone surrounds the massive pyroxenite core and is in turn surrounded by highly sheared serpentinite. The serpentinite is part of the 90-km belt of ultrabasic rocks that separate the schists of the Ordovician Fleur de Lys group from the volcanics to the east.

The strong, parallel, nearly vertical bands of pyroxene-rich and pyroxene-poor phases of the massive pyroxenite core suggest that it is a fragment of differentiated oceanic crust caught up in a subsequent ultramafic flow. The fiber probably formed in fractures on the highly stressed, highly serpentinized periphery of this peridotite core. The fiber is present as a stockwork of good quality fiber veins averaging 1/4 to 5/16 in. in length.[1] Mining of the Baie Verte deposit ceased in 1990.

[1]Because asbestos is graded by the Quebec Standard Test (see section below on evaluation), which uses English units, fiber length in this chapter is given in English units.

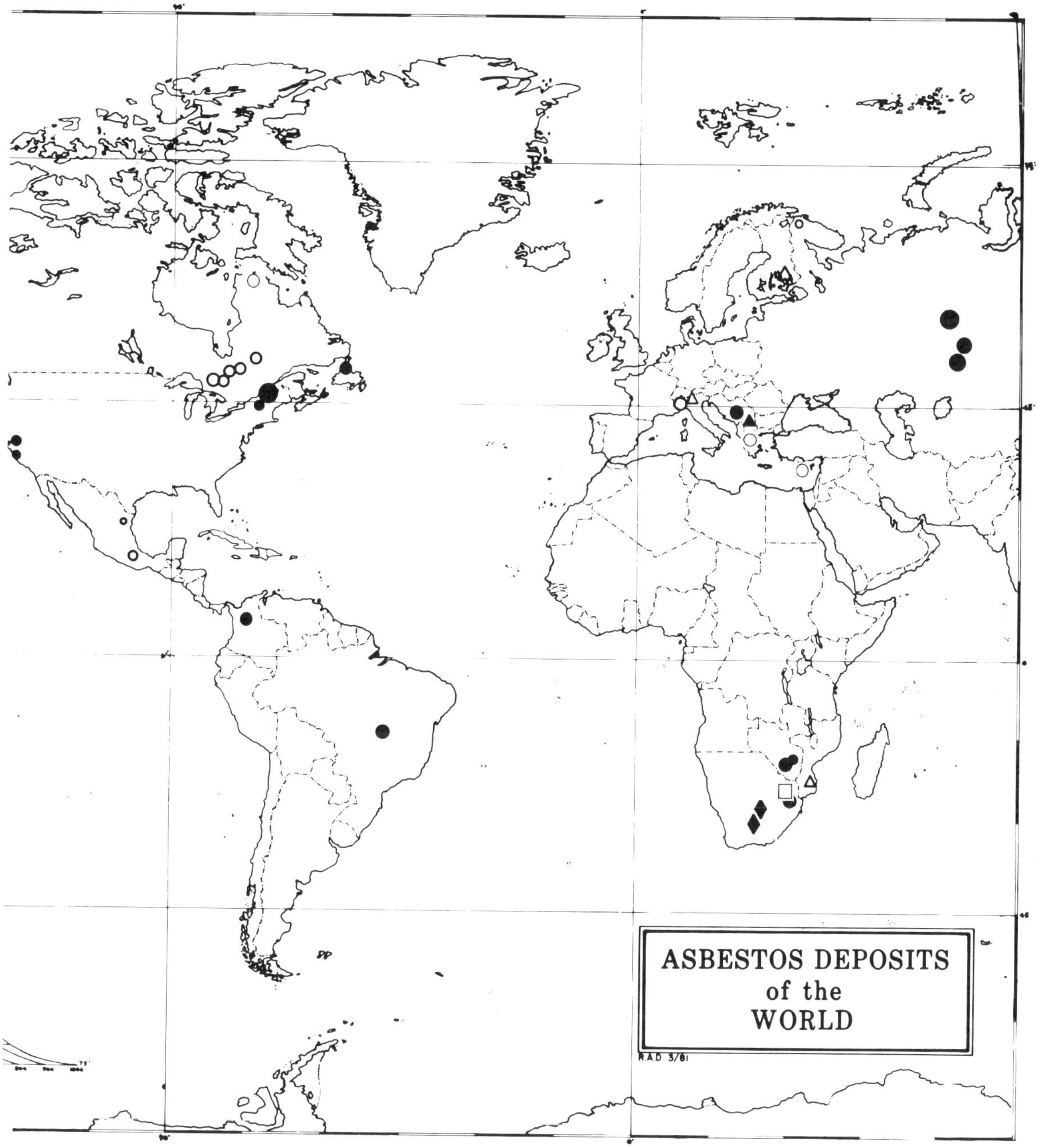

Fig. 7. cont.

Reeves Township, Timmins, Ontario. The chrysotile deposit at Reeves, 60 km southwest of Timmins (Hendry and Conn, 1957, Douglas, 1970, Rowbotham, 1970) lies near the northern tip of a large serpentine mass in contact with Precambrian metavolcanics. The south, or footwall, side of the ore body is bordered by a major east-trending fault that dips 45°N. Extensive talc-carbonate alteration is common along this fault. Two north-trending diabase dikes cut the ore zone. Fibers along the contacts with these dikes have been strongly altered.

Mining ceased in 1974 because the grade was marginal and alteration was pervasive.

The deposit in Munro Township in eastern Ontario lies within a narrow ultrabasic dike that is bounded by faults containing pronounced talc-carbonate alteration. The ore zone is offset by cross faulting, but mining of the deposit ended several years ago.

Cassiar, Northern British Columbia. The Cassiar asbestos deposit is at an elevation of more than 1 830 m near the peak of Mount McDame in the Cassiar Range (Smitheringale, 1957; Douglas, 1970). Fiber in the Cassiar deposit is long and of exceptional grade and quality. It is associated with an ultrabasic sill that intruded into a folded succession of volcanic flows and sedimentary rock of Devonian age. The asbestos is present as cross fiber veins in dark-green serpentine commonly 13 to 26 mm wide and as much as 5 m long. Mining ceased in 1992.

Clinton Creek, Yukon Territory. The deposit lies south of the Yukon River and within 390 km of the Arctic Circle. Cross fiber averaging close to 1/4 in. in length is present in a small ultrabasic body enclosed by phyllites, argillites, and quartzites of Paleozoic age. A portion of the deposit has undergone strong talc-carbonate alteration. Mining of asbestos at Clinton Creek near the Canada-Alaska boundary was short-lived and the relatively small deposit is now closed (Douglas, 1970).

Asbestos Hill, Putunig, Ungava, Quebec. The Asbestos Hill deposit, about 60 km from tidewater on Quebec's Hudson Strait (Stewart, 1978), is on the western margin of the large, ultrabasic to basic intrusive. The deposit is about 760 m long and as much as 60 m wide. It lies where the dip of the serpentine changes from gently inward dipping through vertical to steeply outward dipping from the intrusive complex. A small drag fold at the most abrupt change in dip appears to form the locus of the ore zone. Talc alteration is present in some zones. The ore is exceptionally high grade; most of the fiber ranges from 3/16 to 7/16 in. in length and the amount of short fiber is minimal. The Asbestos Hill deposit no longer is mined.

Other Potential Deposits. Several other deposits have been considered at various times for possible production: the Abitibi deposit 80 km north of Amos, Quebec; the McAdam deposit 30 km east of Chibougamau, Quebec; the Lloyd Lake Midlothian Township deposit 60 km south of Timmins, Ontario; and the Garrison deposit 110 km east of Timmins.

United States of America: Chrysotile has been mined for more than 100 years in the United States. Most of the production has come from ultrabasic deposits in California and Vermont. Two of the deposits produced cross and slip fiber and the third produced mass fiber.

Pacific Asbestos, Copperopolis, CA. The Pacific Asbestos deposit, in the foothills of the Sierras about 200 km east of San Francisco (Leney and Loeb, 1972), lies within a belt of ultrabasic rocks that trend northwest for 30 km and reach a maximum width of 6 km. Strike faulting is common, and the peridotites and dunites have been altered to serpentine. The ore body occupies a zone about 630 m wide between two bands of antigorite schist. The schist is believed to represent zones of shearing within the intrusive. Cross fiber as much as 25 mm in length is present, but most of the fiber is 1/4 in. or less. The ore body is cut by narrow dioritic dikes. Talc-carbonate alteration is prevalent in some portions of the ore body. Mining of this deposit has now ceased.

New Idria-Coalinga, CA. The New Idria deposit, northwest of Coalinga, is in an ultrabasic intrusive (Munro and Reim, 1962). The ultrabasic body trends northwestward and covers an area about 20 km in length and 5 to 8 km wide. It was intruded into Jurassic sediments in Late Jurassic or Early Cretaceous times. The intrusive is a highly sheared serpentinite characteristically extremely platy and slickensided. Boulders of massive serpentinized material are scattered throughout the loose platy serpentine. Landslides, for which there is evidence, may have contributed to the extreme deformation of the serpentine. The ore contains abundant short chrysotile fiber similar to the Canadian variety.

Eden, VT. The southern extension of the Eastern Townships serpentine belt of Quebec continues intermittently southward down the Appalachian chain into northern Vermont (Rowbotham, 1970). Fiber is found over a considerable area in Lamoille and Orleans Counties. The host rock is a highly serpentinized peridotite in which the fiber is found along a contact between gneiss and amphibolite schist. A deposit near Eden was mined for years and consisted of a substantial ore body, about 910 m wide. Slip fiber was the more predominant variety mined, but some good quality cross fiber was also recovered.

Mexico: *Cuicatlan, Oaxaca*. A large deposit of slip fiber was discovered at Concepcion Papalo, east of Cuicatlan and about 80 km north-northwest of Oaxaca. The Pegaso deposit is exposed on two sides of a ridge in highly dissected terrain. It appears to be part of a tabular serpentinite body that forms the core of a highly folded and sheared sequence of rocks. The low tensile strength of the fiber and its presence as a slip fiber render the deposit very marginal.

Ciudad Victoria, Tamaulipas. Chrysotile fiber is present in the Sierra Madre Oriental Mountains west of Ciudad Victoria in the State of Tamaulipas (Rowbotham, 1970) as generally small, discontinuous zones of ribbon fiber, and is of little economic significance.

South America

Cana Brava, Goiás, Brazil: The Cana Brava asbestos deposit, about 190 km north of Brasilia and 100 km northeast of Uruaçá in the State of Goiás, is in a large ultrabasic body that intrudes gneisses of the Brazilian Precambrian Shield. The ore is a good-quality cross fiber of medium length in two separate zones 550 m long and roughly 60 m wide.

Las Brisas, Antioquia, Colombia: A deposit of chrysotile asbestos was mined at Las Brisas, Antioquia, in central Colombia. The deposit is about 130 km north of Medellin or roughly midway between Bogota and the Caribbean port of Cartagena. It is reported to comprise at least two separate zones containing fiber of distinctly different grades.

Europe

Balangero, Northern Italy: Most of the Italian chrysotile was produced from a large, low-grade slip fiber deposit about 30 km north-northwest of Turin (Rowbotham, 1970). The ore zone is in a highly sheared serpentine body about 460 m wide that trends north. It is intrusive into paragneisses.

Zidani, Kozani, Northern Greece: About 120 km southwest of Thessaloniki and 30 km due south of Kozani is a large deposit of low-grade slip fiber, which is not in production. The deposit is in a highly sheared and fractured serpentinite, probably of dunitic composition, which is flanked by crystalline marble to the west.

Troodos, Cyprus: Cyprus was an ancient source of asbestos. It was known and exploited by the ancient Greeks and Romans (Bear, 1963, Vokes, 1964). It contains some of the largest deposits of chrysotile in the European area. The chrysotile is in a body of bastite-serpentinite covering an area some 6 km long and 4 km wide, being roughly oval in plan. The best fiber is confined to the southern part of the area near the faulted eastern margin of the peridotite, where deformation was most intense. The fiber is present in small discontinuous lenticles or in larger composite veins of cross fiber and picrolite as much as 13 mm wide and 2 m long. The fiber seldom exceeds 1/4 in. in length. These deposits are probably Upper Cretaceous in age.

Yugoslavia: Short-fiber chrysotile has been produced from a number of mines in Yugoslavia for many years (Rowbotham, 1970). There are commercial deposits at Korlace in Serbia and near Tuzla, Bosnia. Details of the geology of these deposits are not available.

Union of Soviet Socialist Republics (former USSR): The USSR (Nalivkin, 1960, Rowbotham, 1970, Smirnov, 1971, Stenho and Fronek, 1977, Petrov and Znamensky, 1978) is now the largest producer of asbestos in the world. Its known reserves are vast and are considered to surpass those of any other country.

Four major areas of asbestos production in the former USSR are: the Bazhenovo deposit of the central and northern Urals, the Dzhetygara and Kiembay districts in the southern Urals, and the Ak-Dovurak deposits in Tuva, west of Lake Baikal.

Bazhenovo (Bajenovsk) District. Mines in the Bazhenovo district are the leading producers in the USSR and possibly the world. These deposits are in the Bazhenovo-Alapajevsk or eastern serpentine belt, which extends about 180 km in the central Urals. Most of the deposits are near Asbest, which is 80 km northeast of Sverdlovsk and 1 440 km east of Moscow.

The deposits are associated with fault zones in serpentinites that formed predominantly as the result of alteration of peridotites and, to a lesser extent, of pyroxenites. The veins of chrysotile are thought to have been formed by the filling of fissures in a massif of ultrabasic rocks.

Dzhetygara District, Northwest Kazakhstan. This is the second largest producing district in the USSR. The geology of the Dzhetygara district is similar to that of the Bazhenovo district. The district lies about 220 km southwest of Kustenay. The deposits are in serpentinized peridotites of Early Hercynian (Middle Devonian) age.

Kiembay (Kijembai), Southern Urals. This deposit, the third largest in the USSR, is 140 km east of Orsk in the Orenburg Administrative Region. The deposit is in a completely serpentinized ultrabasic intrusive. Severe weathering 50 to 60 m deep has caused the surface fiber to become weaker and relatively brittle. The deposit has been developed recently by a collective combine of the USSR, the Czech Republic, Poland, Hungary, the former East Germany, Romania, and Bulgaria.

Ak-Dovurak (Aktovrak), Tuva. Important deposits of asbestos near Ak-Dovurak, Tuva, are associated with a belt of serpentinized ultrabasic rocks of the Salair complex, which forms part of the Lower Hercynian of the eastern Altai-Sayan Province. The ultrabasic intrusions are in belts that extend along major faults. At least six significant deposits have been discovered, and the district is reported to contain vast reserves near the village of Kyzyl-Mazalyk and in the Kedrovsk deposit 50 km to the south.

Other Deposits. Chrysotile asbestos deposits are known at Molodeznoje (Molodjezhnoje) about 210 km east of Lake Baikal. The deposit is of excellent grade. The Ilcirsk and Sajan deposits in the same general area also have good potential for production. A peculiar fiber deposit is located near Pechenge, close to the Norwegian border. The deposit is large but has a high iron content, and it is doubtful that the deposit will ever prove to be economically feasible.

Africa

Shabani, Zimbabwe: The Shabani mine, situated about 160 km east of Bulawayo, has the largest reserves of fiber in southern Africa (Laubscher, 1964, 1968, Oldham, 1968, Wilson, 1968). The deposit lies near the base of a lenticular ultrabasic sill. The sill is 14 km long and 2 to 5 km wide and intrudes Early Precambrian gneisses. The sill strikes northwest and dips about 40°SW. It is composed of dunite at the base, overlain by peridotite, pyroxenite, and gabbro. A swarm of younger diabase dikes cut the ore zone. The dikes are truncated by strong zones of shearing, both parallel and oblique to the basal contact. The intrusion and shearing have caused strong serpentinization of the dunite and heavy talc-carbonate alteration, particularly along the footwall shear zone. The fiber is present as a stockwork of cross fiber veins described as "stress-controlled dilation seams." The quality is good and the length of the fiber as much as 1 1/2 in. in places.

Fiber is also mined at other localities in Zimbabwe, notably in the Mashaba, Belingwe, and Filabusi districts.

Msauli, South Africa: The Msauli ore body is located 30 km south-southeast of Barberton in the eastern Transvaal and only 10 km southwest of the Havelock mine across the Swaziland border (Hall, 1930, Pelletier, 1964, Van Biljon, 1964). The serpentine in which the ore formed is very similar to that of the Havelock deposit. It appears to be the faulted continuation of the same sill-like ultrabasic intrusive. The serpentine body averages 140 m in width and dips about 65° to the east. The best fiber is reported to be to the north in association with major faulting. The fiber is present as a stockwork of veins from 1/16 to 3/4 in. wide. It is of the same general high quality as that of Havelock, although some sections contain a higher percentage of magnetite.

Havelock, Swaziland: The Havelock deposit is located within 1 km of Swaziland's western border with the Republic of South Africa, and lies about 20 km south-southeast of Barberton in the eastern Transvaal (Pelletier, 1964, Van Biljon, 1964). The fiber is in a sill-like body of serpentine that is intercalated in the Early Precambrian Fig Tree sedimentary sequence. All of the beds dip south at 55°. The ore zone has a strike length of 1 520 m and widths of 20 to 110 m. The average width is about 30 m. The ore body is cut by a longitudinal fault zone with associated transverse faulting. Talc-carbonate alteration is common in parts of the deposit. The deposit is a stockwork of high-quality cross fiber seams, in which the fibers average 1/2 in. or more in length and compose 3 to 4% of the rock.

Other Deposits: Many small deposits of chrysotile asbestos are found at many locations in northern and eastern Transvaal, but their combined production has never been appreciable. Most of the deposits are elongated ribbon fiber zones in ultrabasic rocks or associated with dolomitic limestones.

Chrysotile also is reported to be present in the Ingessana Hills and Qala en Nahl districts of eastern Sudan. Both areas have been explored extensively, and several small deposits have been located, but the total potential remains small.

Australia

Woodsreef (Transpacific Asbestos), New South Wales: The Woodsreef asbestos deposit, east of Barraba in northeastern New South Wales (Straw, 1955, Butt, 1978), is part of the Great Serpentine Belt of New South Wales. The ore body is in a layered peridotite-dunite complex that has intruded sediments of Mid-Paleozoic age. It is approximately 700 m long and 80 m wide and is bounded on the western margin by prominent faults and shears. Strong shear zones are common within the fiber zone. The fiber is present mainly as a stockwork of cross fiber seams and is generally short to medium in length.

New Zealand

Pyke Asbestos, South Island: A small deposit of chrysotile fiber is reported near the headwaters of Pyke River (Babcock, 1978). The ore body is in the Southern Alps, about 80 km northeast of Milford Sound. Because of its isolated location and relatively low grade, it cannot be considered an economic deposit.

Asia

China: The Lai-Yuan district of Hopeh, approximately 160 km southwest of Peking, was for many years the only source of chrysotile in China. Now, the major producing deposits are in the Shihmien area of southwest Szechwan (the word *Shihmien* is translated as *rock cotton*). Chrysotile production also has been reported in Qinghai, Liaoning, Shanxi, and Shensi provinces.

India: Indian asbestos production is small and includes chrysotile of both ultrabasic and limestone origin, together with some amphibole varieties.

Chrysotile has been produced from the Roro deposit, located in southern Bihar State, about 280 km west of Calcutta. The ribbon fiber is in five separate discontinuous zones along highly serpentinized peridotite bands and massive-to-sheared clot peridotite. One of the five zones is economic and contains from 7 to 15% fiber.

Japan: Chrysotile asbestos is mined at two localities in the Yamabe district of Hokkaido, Japan (Rowbotham, 1970). Most fiber is short.

CHRYSOTILE DEPOSITS IN DOLOMITE

In addition to the large deposits of chrysotile asbestos found in serpentinized peridotite, comparatively small tonnages are mined from serpentinized dolomitic limestones (Hall, 1930, duToit, 1946, Pelletier, 1964, Van Biljon, 1964, Rowbotham, 1970). Much of such fiber is of high quality and free of magnetite commonly associated with most deposits of ultrabasic origin.

Chrysotile of this type has been mined in the Carolina district in the Transvaal and in the Salt River and Sierra Ancha regions in Arizona. The Arizona deposit, approximately 60 km northeast of Globe, is tabular in shape and is in a serpentinized dolomitic limestone formation. Serpentinization occurred during the intrusion of diabase sills. Chrysotile is found in thin discontinuous veins. The fiber ranges from harsh to soft and from a few millimeters to several centimeters long. Unlike the ultrabasic deposits, the Globe deposit is relatively free of magnetite. The only chrysotile of this variety known in the USSR is present near Aspagash just south of Krasnovask on the Yenesei River.

CROCIDOLITE AND AMOSITE DEPOSITS

In South Africa, crocidolite and amosite are found in deposits known as banded ironstones, ferruginous quartzites, or iron-rich silicified argillites (Hall, 1930, duToit, 1946, Frankel, 1953, Cilliers et al., 1961, Keep, 1961, Cilliers and Genis, 1964, Pelletier, 1964). Cross fiber veins are present as closely spaced ribbons roughly conformable with the bedding, which in some localities is distorted and steeply dipping. Most crocidolite veins are less than 50 to 80 mm thick, and most of the fiber is between ¼ to ¾ in. in length. Amosite veins commonly are as much as 250 to 300 mm thick.

Crocidolite is found over a large area of Cape Province in a belt of the Lower Griquatown series of the Transvaal system. One of the oldest producing areas is in the vicinity of Prieska on the Orange River, with notable production coming from Westerberg and Koegas. Another producing area of blue asbestos in the Cape Province is farther north in the Kuruman district, and at Pomfret, just south of the Botswana border.

Crocidolite and amosite are found in similar formations in the vicinity of Pietersburg in northern Transvaal. In some places, the two varieties are side by side in the same vein. A vein composed of crocidolite at one end and amosite at the other is called a doublet.

In the Pietersburg area, a silky, commonly dusty blue variety of amosite called *montasite* has been found. This variety of amosite is sometimes confused with crocidolite.

Amosite and crocidolite are found in the Bababudan Hills north of Chikmagalur in Mysore, India. Both varieties of fiber are present in an assemblage of banded ironstones and shales similar to those of northeastern Transvaal, although the amosite is noticeably shorter.

Deposits of crocidolite are found in the Hammersley Range of Western Australia and in Bolivia.

An amosite-bearing banded ironstone formation crops out for a distance of more than 30 km near Penge in the Lydenburg district of the Transvaal. The fiber-bearing banded ironstones and associated sedimentary members dip southwestward at about 20°. This sequence has been intruded by thin, persistent sills of dolerite that are conformable with the bedding.

A detailed cross section of the country rock was developed through a diamond drilling program. The following column is from a 370-m diamond drillhole that intersects the fiber-bearing horizon at a depth of 300 m:

Shales
Dolerite sills
Shales
Bevetts conglomerate
Banded ironstone (main fiber zone)
Dolerite sill
Banded ironstone (sometimes fiber-bearing)
Dolomite.

The Bevetts conglomerate is considered an important marker for the fiber-bearing formation. Diabase cuts vertically across the whole formation. Evidence as to possible relationship of these dikes to the genesis of the amosite mineralization is inconclusive.

With the steady depletion of known amosite ore reserves, extensive exploration for additional ore was begun more than a decade ago. Considerable success was achieved by using a novel structural approach developed by the Cape Asbestos Group.

Careful mapping of the main marker horizons with respect to detailed topographic contours disclosed a system of broad, gentle north-trending folds having amplitudes of about 15 m and wavelengths of 1 220 to 1 520 m. A second, superimposed pattern of east-trending cross folds having similar amplitude and a wavelength approaching 2 130 m also was detected. Careful structural mapping further indicated that each of the known or producing deposits was located on the loci of intersections of either anticlinal or synclinal structures. By drilling similar loci, five new deposits were discovered.

This record of successful drilling gives considerable weight to the hypothesis of a structural control for the genesis of amosite fiber. Similar patterns of cross-folding have been noted in many crocidolite deposits in the Northwest Cape and elsewhere.

It is interesting to consider this structural origin for the genesis of chrysotile fiber as well because the effects of faulting, folding, and dilation by shearing or intrusion almost invariably occur in close proximity to chrysotile-bearing formations.

ANTHOPHYLLITE, TREMOLITE, AND ACTINOLITE DEPOSITS

Production of amphibole asbestos, other than crocidolite, includes small amounts of anthophyllite and tremolite and token amounts of fibrous actinolite (Rowbotham, 1970).

Of the countries in which anthophyllite deposits are known, Finland was the most important producer. The famous deposits at Paakkila in the parish of Tuusniemi in eastern Finland and those of Maljasalmi in Kuusjarvi Parish have now been mined out.

The Finnish anthophyllite deposits were a series of lenses of amphibolitized and serpentinized ultrabasic material. The lenses appeared to have become detached from larger bodies during an early period of severe deformation. The serpentinized amphibolite lenses were found in a biotite gneiss that had been intruded by Late Karelian granites and pegmatite. The size of these bodies was small, ranging from 40 to 60 m in length and 9 to 20 m in width.

A small quantity of anthophyllite has been mined near Greenmountain, Yancey County, in North Carolina. The deposits are associated with altered peridotites and pyroxenites. Most of the deposits mined consisted of mass fiber, although cross and slip fiber

Table 7. World Asbestos Production (tons)

	1981	1982	1983	1984	1985	1986	1987	1988	1989	1990
North America:										
Canada	1 122 000	834 000	858 000	836 654	750 190	662 381	664 546	710 357	701 227	682 200
United States	75 618	63 515	69 906	57 422	57 457	51 437	50 600	18 233	17 427	W
Total	1 197 618	897 515	927 906	894 076	807 647	713 818	715 146	728 590	718 654	682 200
South America:										
Argentina	1 280	1 218	1 240	1 093	1 244	1 697	332	2 328	225	300
Brazil	138 417	145 998	158 885	134 788	165 446	204 460	212 807	227 653	206 195	210 000
Total	139 697	147 216	160 125	135 881	166 690	206 157	213 139	229 981	206 420	210 300
Europe:										
Bulgaria	400	600	700	500	400	300	400	400	400	400
Greece	457	17 016	31 811	45 376	46 811	51 355	60 134	71 000	72 500	72 500
Italy	137 086	116 410	139 054	147 272	136 006	115 208	118 352	94 549	54 500	20 000
USSR	2 105 000	2 700 000	2 500 000	2 500 000	2 500 000	2 400 000	2 554 600	2 600 000	2 600 000	2 400 000
Yugoslavia	13 591	11 657	10 502	8 556	6 916	8 596	10 964	17 030	7 105	7 000
Total	2 256 534	2 845 683	2 682 067	2 701 704	2 690 133	2 575 459	2 744 450	2 782 979	2 734 505	2 499 900
Africa:										
Mozambique	1 425	852	600	400	55	—	—	—	—	—
Zimbabwe	247 600	197 682	153 221	165 385	173 580	163 984	193 295	186 581	187 066	190 000
South Africa	235 943	211 860	221 111	167 389	164 247	138 862	135 074	145 786	155 462	147 500
Swaziland	35 264	30 145	26 287	25 832	25 130	24 475	25 925	22 804	27 291	35 000
Total	520 232	440 539	401 219	359 006	363 012	327 321	354 294	355 063	369 819	372 500
Asia:										
China	106 000	110 000	160 000	135 000	150 000	150 712	144 673	157 478	160 000	160 000
Cyprus	24 440	18 952	17 288	11 298	16 360	13 011	18 070	14 585	—	—
India	24 515	26 761	24 873	25 240	30 183	25 236	29 110	31 123	36 502	37 000
Iran	—	—	—	—	2 500	3 000	3 300	3 400	3 500	3 500
Japan	3 950	4 135	4 000	3 140	2 971	3 593	3 143	3 000	3 000	3 500
Korea, Republic of	14 084	15 933	12 506	8 062	4 703	2 983	2 518	2 428	2 361	2 300
Taiwan	2 317	2 392	2 819	1 355	625	—	—	—	—	—
Turkey	3 860	958	1 510	1 499	1 500	1 098	806	50	—	—
Total	179 166	179 131	222 996	185 594	208 842	199 633	201 620	212 064	205 363	206 300
Oceania:										
Australia	45 494	18 587	1 240	—	—	—	—	—	—	—
Total World	4 349 466	4 559 495	4 428 867	4 311 842	4 248 988	4 029 364	4 235 458	4 316 443	4 242 973	3 979 500

W Withheld to avoid disclosing company proprietary data.
Source: US Bureau of Mines Minerals Yearbooks, 1985–1990.

are more common in other parts of the state. Similar material is known in the nearby state of Georgia.

Italy has produced some long-fiber tremolite from small deposits at Val Malenco, in the Sondrio district 100 km north of Milan. Tremolite fiber has been found in the Aosta district north of Turin in the Italian Alps.

Amphibole asbestos also has been found in Bulgaria, India, Romania, Taiwan, Turkey, and Yugoslavia. Most of the Indian production is from Rajasthan. Little is known about the geology of deposits or amphibole production in the other countries. Although some deposits of fibrous actinolite have been reported, production is extremely small.

PRINCIPAL DEPOSITS

In the previous sections, the major asbestos deposits of the world have been briefly described. Global distribution of these deposits appears to be rather haphazard, but closer examination reveals that the deposits generally coincide with the earth's major chains of mountain building of all ages. This is particularly true of the Urals, the Appalachians, and the Rocky Mountains. The relationship is more difficult to recognize in Precambrian fold belts.

PRODUCTION

Approximately 4 Mt of asbestos fiber was produced worldwide in 1990. Slightly more than 98% of the asbestos produced was chrysotile. The remainder was amosite and crocidolite. Canada and the former USSR accounted for 17 and 60%, respectively, of production. They were followed by: Brazil, 5%, Zimbabwe, 5%, the Republic of South Africa, 4%, and China, 4%. Seventeen other countries also produced asbestos (Table 7).

Production of fiber increased steadily to a peak of 4.9 Mt in 1979 and slowly declined during the 1980s (Fig. 8). Only Brazil, India, Iran, Greece, and the former USSR increased production between 1981 and 1990. Production in Canada and most other producing countries remained constant or declined during this same period (Table 7).

Canada, while still the leading market economy producer of asbestos, reduced its production by 39% between 1981 and 1990 in response to shrinking markets. Many Canadian mines and mills were closed as a result of restructuring of the Canadian asbestos industry in the 1980s. The Eastern Townships of Quebec (which had been producing asbestos for the past 100 years) account for most of the fiber produced. The Jeffrey mine at Asbestos remains the largest single producer.

Production in the former USSR has steadily increased since

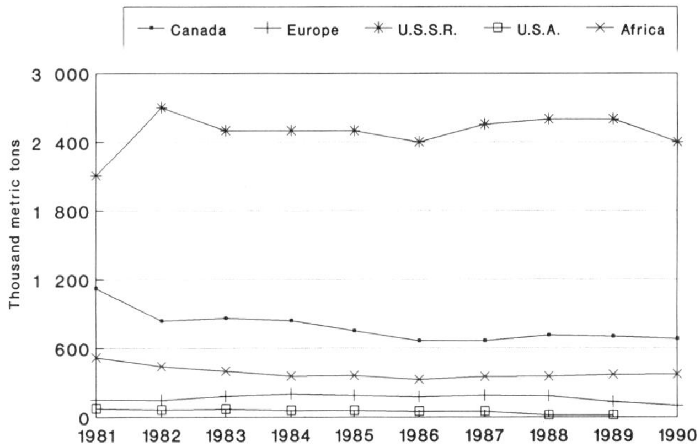

Fig. 8. Asbestos production, major areas.

Table 8. Production of Amphibole Asbestos by the Republic of South Africa (tons)

	1987	1988	1989	1990
Amosite	26 026	20 325	26 124	26 566
Crocidolite	7 326	11 885	13 918	31 521

Minor amounts of tremolite, actinolite, and anthophyllite may be produced in Bulgaria, India, Italy, Romania, Taiwan, Turkey, and/or Yugoslavia although general information is inadequate for the formulation of reliable estimates of output levels.

production began in the early 1900s. Russian production of medium- and short-fiber chrysotile is concentrated in the central and southern Urals where the Bazhenovo, Dzhetygara, and Kiembay combines are located.

Production in the Republic of South Africa and Zimbabwe has declined over the past 10 years, dropping from 483 543 t in 1981 to 337 500 t in 1991. European production has slowly increased over the past 10 years. Most of the increase can be attributed to the USSR. Italy led all other European countries in production until 1988, when production from Greece reached 72 500 t. Bulgaria, and the former USSR are the only European countries for which production currently is reported, although the Czech Republic and Romania also may produce asbestos.

The United States ranked eleventh in world production in 1990. Asbestos was produced in California and Vermont. Production peaked in 1973 at 136 000 t, but has decreased significantly in recent years. The decline was in response to environmental issues and maturing markets.

Crocidolite and amosite were produced only in the Northern Cape and northeastern Transvaal regions of the Republic of South Africa. Production of amosite has remained constant, and the production of crocidolite has increased significantly since 1987.

Production of the other varieties of amphiboles asbestos is comparatively small. Bulgaria, India, Italy, Romania, Taiwan, Turkey, and Yugoslavia may produce some anthophyllite and tremolite asbestos, but production is not officially reported and reliable production estimates cannot be made from available information.

Table 8 gives the production of crocidolite and amosite in the Republic of South Africa.

EXPLORATION AND EVALUATION

The following is a review of exploration and evaluation methods employed in Canada for chrysotile deposits that are present in ultrabasic rock.

Geophysical

Aeromagnetic and ground magnetic surveys often are employed in the early stage of exploration where there are few outcrops of asbestos-bearing ultrabasic rocks (Low, 1951, Conn, 1967). Ground magnetic surveys may be used to check and define in more detail anomalies obtained by an airborne survey of a large area. The ground survey may be used by itself if the area for exploration is small.

Magnetic surveys are used to locate ultrabasic bodies and define potential ore deposits because of their association with secondary magnetite formed during extensive serpentinization. Many asbestos deposits in ultrabasic rocks contain more magnetite than does barren serpentine. Fig. 9 shows the results of ground magnetometer surveys carried out over two known ore bodies.

Diamond Drilling

Diamond drilling is used to assess and define the limits of an asbestos deposit. Because most asbestos ore bodies are large in volume, it is customary to drill vertical holes on a grid pattern. In the first stages of exploration, a wide interval is used. A narrower interval is used when an asbestos-bearing zone is encountered. When a deposit is elongated in one direction, holes are spaced at closer intervals across the strike. Narrow, tabular targets are best explored by angle holes planned to give the attitude and true thickness of the body.

One or two preliminary drillholes should be used to determine whether the fiber has a preferential vein angle or not. If the angles of the vein intersections appear random, no changes need be made to the drilling program. If the deposit exhibits prominent vein angles that diverge from the average 45°, then the attitude of all later drillholes should be changed to correct for this variation.

In the past, drill-evaluation programs frequently used small-diameter core sizes for evaluating deposits where shearing was minimal and where holes did not exceed 91 m. Large core sizes are preferable for better depth penetration, more geologic data, and to aid in fiber logging and dry milling of the core.

In regions where permafrost conditions are expected, precautions should be taken to avoid freezing the string of drill rods down the hole. A suitable low-freezing drill-lubricant media, such as a concentrated brine solution, generally is used to allow drilling in most areas of permafrost. However, there often is a high percentage of fiber cuttings in the water return, so the brine solution has to be renewed at frequent intervals. This is a factor that considerably increases the cost of drilling asbestos prospects in permafrost areas. Some drill ventures, completed successfully under permafrost conditions, showed the frozen ground to persist to at least 305 m below surface.

In areas where drilling is impractical, exposure by trenching or exploration beneath the surface by adit or shaft and lateral workings may offer the only means of assessing a deposit.

Evaluation

An asbestos deposit is evaluated on size, grade, and quality of fiber (Dean and Mann, 1968, Conn and Mann, 1971). Dimensions of a mineral zone are established by conventional methods, such as mapping, magnetic surveys, trenching, and diamond drilling. The value of the contained asbestos fiber is dependent on physical properties such as fiber length, strength, flexibility, harshness, and color.

The determination of grade cannot be based on a simple chemical analysis because both the fiber and the wall rock have essentially the same chemical composition. To avoid the complete crush-

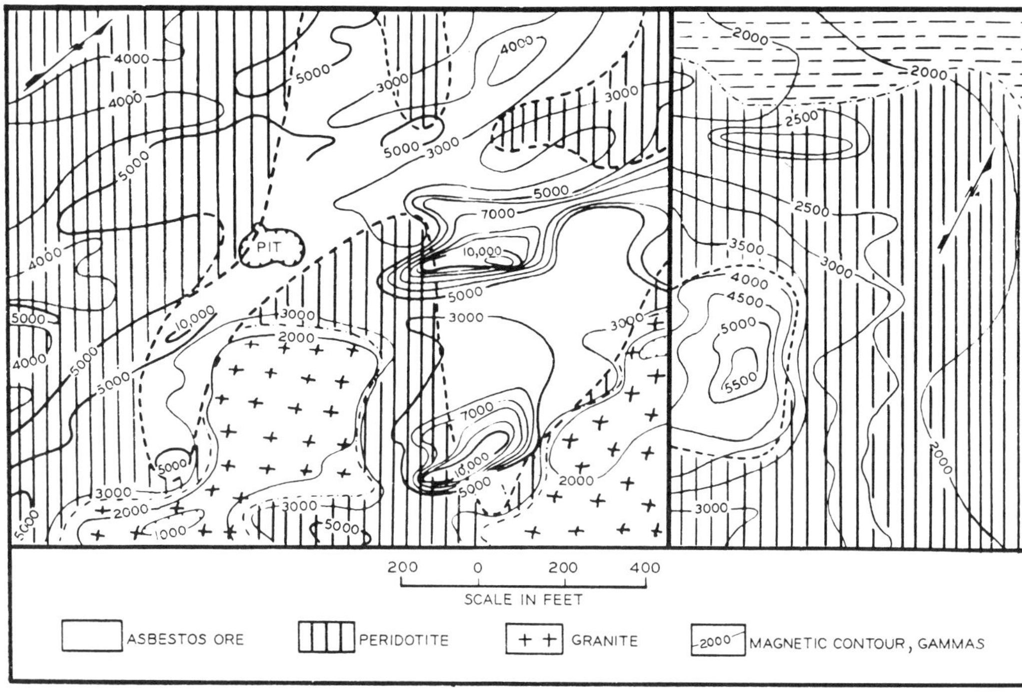

Fig. 9. Results of ground magnetometer surveys of two known ore bodies.

ing of the rock and physical separation of the fiber into different lengths, a method of visual evaluation has been developed. The number of fiber veins and the average length of fiber in each vein is determined for the core section. The fiber lengths, if expressed in sixteenths of an inch give an approximation of the grade.[2] Each 1/16 in. vein of fiber in a 5-ft section corresponds to approximately 0.1% fiber.

Because the price of fiber varies according to length, the use of grade based solely on its percentage fiber content is of little significance. Instead, the product of these two variables, expressed in dollars per ton of rock, offers a far more meaningful value, which can be used for direct comparison in the final evaluation.

Drill Core: The visual method is ideal for evaluating drill-core sections. Each vein of cross fiber is recorded and the length of fiber in the vein is carefully measured and recorded in multiples of 1/16 in. Because many veins are of irregular width, the average width of each vein is estimated. Some veins contain partings, or the fiber may have kinks, which cause it to break into shorter lengths. The fiber in some veins may be at right angles to the vein walls, whereas, in others, the fiber is sharply inclined. Allowance must be made for these conditions to determine the true fiber lengths.

If more precise determinations of fiber content are required, the angle of each vein in the core is measured to permit individual volume corrections. In practice, it is usually assumed that the angle of each vein is 45° and a factor of 1.414 (inverse of the sine of 45°) is used. If the average angle is small, the logging results should be checked against laboratory recovery and bulk sampling.

Slip fiber is present as a minor constituent in many deposits made up mainly of cross fiber, and in others the slip fiber may be the predominant type. It is not easy to determine the slip fiber length by the normal visual methods of logging core. However, its presence may be recorded and the percentage of slip fiber determined in the same manner as used for cross fiber. Laboratory assistance is invariably required to evaluate slip fiber deposits.

Regardless of the method used to evaluate a deposit, the value of the fiber, the yield, and the value of the ore in dollars per ton should be determined. Laboratory extraction determines only the yield of an ungraded fiber whose value is unknown. To place a value on the ungraded fiber resulting from laboratory extraction, a graph is used (Fig. 10). The curve on this graph is derived experimentally from the Quebec Standard Test. For the Quebec Standard Test, a 16 oz sample is separated into +1/2 in. fiber, −1/2 in. to +4 mesh-per-inch fiber, −4 to +10 mesh-per-inch fiber, and −10 mesh-per-inch fiber using sieves. The weight fraction for each fiber size is determined. The weights are multiplied by a factor to give each grade a point rating. The factors for the size fractions are 100 for +1/2 in., 40 for −1/2 in. to +4 mesh per inch, 10 for −4 to +10 mesh per inch, and 1 for −10 mesh per inch. The weighting factors have no real significance and have been chosen only to control the shape of the curve.

As an example, a sample with 5.7 oz of fiber on the 1/2 in. sieve, 7.5 oz on the 4 mesh-per-inch sieve, 2.3 oz on the 10 mesh per inch sieve, and 0.5 oz in the pan would have a point rating of:

$$\begin{array}{rcr} 5.7 \times 100 & = & 570 \\ 7.5 \times 40 & = & 300 \\ 2.3 \times 10 & = & 23 \\ 0.5 \times 1 & = & \text{—} \\ \hline & & 893 \end{array}$$

and from the curve the fiber value would be about \$455/t (1973 prices).

The visual method determines only the yield. Further calculations are required to estimate the fiber and ore value. It is necessary to correlate the observed lengths with the corresponding Quebec Standard Test. In order to do so, it is assumed that fibers with a

[2] English units are used for grading asbestos according to the Quebec Standard Test. Metric units are used in all other situations.

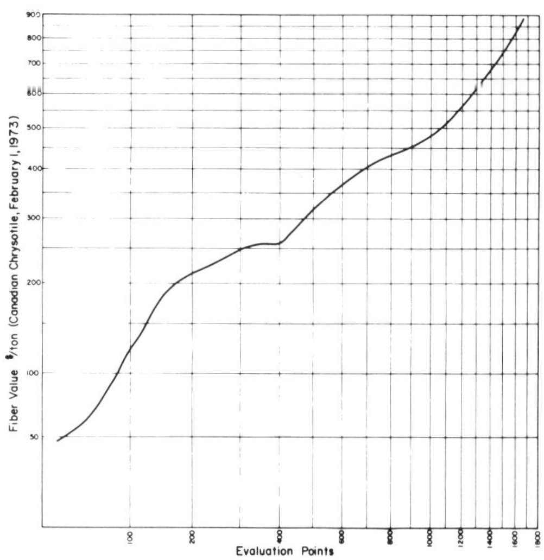

Fig. 10. Graph for evaluating ungraded fiber based on 1973 prices. (Reproduced by permission of Asbestos Corp. Ltd.)

length of 1/2 in. or more are collected on the 1/2 in. sieve, 1/4 to 7/16 in. fibers are collected on the 4 mesh-per-inch sieve, 1/8 to 3/16 in. fibers are collected on the 10 mesh-per-inch sieve, and the −1/16 in. fiber are collected in the bottom pan.

To illustrate the procedure, a core length of 85 ft is used. Assume that the core contains 4 veins of 1/2 in. length, 2 of 9/16 in., 1 of 5/8 in., 2 of 7/8 in., and 1 of 1 in.; 16 of 1/4 in., 3 of 5/16 in., 2 of 3/8 in., and 2 of 7/16 in.; 23 of 1/8 in. and 10 of 3/16 in.; and 74 of 1/16 in. These fiber lengths are then expressed in sixteenths of an inch. There are 104 sixteenths of +1/2 in. fiber, 105 of 1/4 to 7/16 in. fiber, 76 of 1/8 to 3/16 in. fiber, and 74 of 1/16 in. fiber for a total of 359 sixteenths.

This information is correlated to the Quebec Standard Test as follows:

1/2 in.	4 Mesh/in.	10 Mesh/in.	Pan	Total
104 sixteenths	105	76	74	359
104/359x100	105/359x100	76/359x100	74/359x100	100%
4.6	4.7	3.4	3.3	16.0 oz

Then to find the fiber value from the graph:

$$\begin{aligned} 4.6 \times 100 &= 460 \\ 4.7 \times 40 &= 188 \\ 3.4 \times 10 &= 34 \\ 3.3 \times 1 &= 3 \\ \hline & 685 \text{ points} \end{aligned}$$

From the graph the fiber value is found to be $400/t. To convert the dollar amount into a dollar per ton value, the yield must be calculated. In 85 ft of core section there are 85 × 12 × 16 sixteenths; therefore, the uncorrected yield is

$$\frac{359 \text{ sixteenths}}{85 \times 12 \times 16} \times 100 = 2.2\%$$

A correction must be made next for the average vein angle, which here will be assumed to be 45°:

$$\text{Corrected yield} = \frac{2.2}{\sin 45^\circ} = 3.1\%$$

From the corrected yield, the indicated ore value is calculated to be 3.1% of $400 or $12.40/t (1973 prices).

Face Readings

On the surface or in underground workings, channel sampling may be employed. It also is possible to log these surfaces in a similar manner to that used for drill core. One method is to take a linear reading along either wall of a drift or crosscut and another is to take cross sections at intervals of 1.5 m on the back and both walls to ensure that veins running parallel to the drive are not excluded.

An alternative method is to record all the veins in the face, walls, and back after each round. A factor based on the area involved is applied to the reading for each surface, and an average percentage for the round is determined. Because the rock tends to break along one or more of the numerous fault planes present rather than across the fault-bounded blocks to expose the veins within them, it is difficult to obtain representative results.

Bulk Sampling

The various methods of logging and sampling outlined seldom give entirely dependable results. Bulk sampling is the most accurate means of determining yield, fiber value, and ore value. The visual methods of logging usually will produce dependable results only where there is a relatively low content of short fibers. Slip fiber, as already noted, presents a problem. Even laboratory results, which are dependent on complete extraction, usually give an appreciably higher fiber value than that obtained in a conventional mill. It is not easy to simulate the conditions of an operating mine and mill in the laboratory. From the time of blasting to the final product, the fiber is subjected to a good deal of handling. Some of the handling is rather severe and breakage of the fibers can occur.

Bulk sampling often is used as a means to check and to arrive at a suitable factor to be applied to drill-core data. The sampling may be done by diamond drilling a block of ground at close intervals prior to mining and milling. The cores are read visually and then milled. If precautions are taken to avoid contamination and the sample is sufficiently large, results should be reasonably reliable despite the erratic distribution of fiber in the rock.

To compare the results of bulk sampling to the results of the visual method, a yield factor must be used. A yield factor of two was necessary to correlate the visual ore value with that found by milling of a 12 000-t bulk sample from one deposit. There were approximately equal amounts of slip and cross fiber in the deposit, and the visually determined yield was low and the fiber value was high.

	Fiber Value per ton, $	%	Ore Value per ton, $	Yield Correction Factor
Mill recovery	111	6.80	7.53	1.00
Laboratory recovery	164	4.59	7.70	0.98
Visual analysis	202	1.87*	3.78	1.99

*No correction for vein angles.

A 50 000-t sample assessed in the same manner also required a factor of two although the fiber value determined visually was fairly close to that found by milling. This agreement is attributed to the absence of slip fiber that, when present, makes the estimation of fiber value more difficult.

	Fiber Value per ton, $	%	Ore Value per ton, $	Yield Correction Factor
Mill recovery	153	2.75	4.20	1.00
Laboratory recovery	160	2.52	4.03	1.04
Visual analysis	148	1.27*	2.03	2.07

*No correction for vein angles.

Tensile Strength

In the preceding paragraphs, it is important to point out that the measurements and evaluations are based entirely on the length of the fiber. For many applications, the fiber is valued for the strength it lends to the product. Strength evaluations should include tests on the product made with the subject fiber. Once the inherent strength of the fiber from a particular ore body has been determined, it is usually possible to equate strength to length-distribution and dust-content measurements.

Tonnage and Grade

To estimate the tonnage and grade of a deposit from diamond drill-core data, individual drillholes may be weighted according to their interval by using the polygonal method. An alternative procedure is to use cross sections, or groups of cross sections and, by weighting the individual holes in each section, determine the average grade for each section.

A third method employs contoured cross sections. The contouring is based on a reasonable interpolation of the intervening area between drillholes. This method permits the estimator to make use of all available geological information in his interpolation. In open pit operations, contoured horizontal sections that correspond to expected mining level intervals may be prepared. These serve as a useful guide to mining. Separate horizontal sections contoured for rock value and fiber value per ton permit the mine operator to produce a more balanced mill feed with respect to both fiber content and grades.

EXPLOITATION

Mining

The development of asbestos mining in Canada was accompanied by a succession of mining practices and equipment. The simplest hand methods were employed in shallow open pit workings. When pits reached considerable depth, elaborate overhead cableway derricks were popular. Power shovels and heavy duty trucks have now replaced all other loading and transporting equipment where quarry methods are employed.

During the late 1940s and early 1950s, most asbestos was mined in Canada by underground methods. Underground methods that have been used in the past include glory holes, shrinkage and sublevel stoping, and block caving. This trend has been reversed and now open pit mining prevails. Surface mining offers advantages in recovery, grade control, economy, and safety. The shift to open pit mining is mostly due to the introduction of several technological innovations. Large capacity power shovels and trucks with 200-t payload limits are now available, improved rock drilling equipment has been introduced, and new blasting agents and techniques have been used. These innovations permit mining when the waste to ore ratio reaches 3:1 or more, thus lengthening the economic life of open pits.

In Africa, most chrysotile is mined by underground methods. Most ore bodies are tabular in shape with a pronounced dip so that the economic limit for quarry mining was reached at a comparatively early stage. Ore widths in the larger mines commonly range from 20 to 60 m and as much as 120 m. Some ore bodies, notably in the Shabani district of Zimbabwe, are long. In one case, development extended for 5 km along the strike, and was being developed or diamond drilled to more than 300 m in depth.

Several underground methods have been used. Sublevel stoping and caving may be initiated by blasting holes drilled upward from sublevel crosscuts, starting first on the hanging-wall side and retreating over a considerable width toward the footwall. Development and retreat also may be along the strike of the ore. In some cases, high pressures tend to develop from an arching effect. These pressures can be largely released by cutting a vertical slot that may extend to the surface. In the sublevel stoping method, a slot also may be opened across the center of the ore body. The long holes that are fanned out from the sublevel drifts are then blasted toward the slot, and mining proceeds as a systematic retreat in two directions away from the opening.

Much of the amosite and crocidolite in Africa has been mined from small, narrow open cuts or adits following the fiber-bearing band. In the Penge area the operations have been developed for larger scale and deeper mining. The main ironstone reef is approximately 1.4 m thick and dips about 20°. A sloping face is advanced along the strike by drilling rounds with jackhammers. The broken material is shoveled back from the face into temporary pillars for roof support. The remaining material is handloaded into small cars, hoisted to a haulage level, and transported to a sorting plant for further removal of waste rock before being sent to the cobbing plant and mill. Formerly, the fiber-bearing portion was hand-sorted underground, and excessive values were left behind in the pillars. The main fiber-bearing reef commonly contains as much as 30% asbestos, so that the proportion of waste to barren works out well for the system used. The method is locally referred to as shrinkage stopage, although it is not the same as the shrinkage stopage employed in other areas.

Milling

Mill flow lines are designed to suit the characteristics of a given ore body. Most raw fiber is recovered using the dry process that has been used almost exclusively since the inception of asbestos milling in Canada some 90 years ago.

There have been few major changes in the basic process. The major improvements have been in material handling, packaging, and shipping methods. The improved methods developed by the asbestos mining industry have made it possible for the producer and customer to store, ship, receive, and use fiber in a relatively dust-free environment.

The major portion of the production is packaged in woven, polylaminate, 50-kg pressure-packed bags for the domestic and export markets. Fiber in water-soluble, unsized, kraft paper bags is provided to some customers who consume the bag, along with the fiber, in wet, paper-making production processes. Unitized bags are supplied by some producers. The present trend is toward 100% mechanical handling of the asbestos at both the mines and point of use.

The description that follows of the essential steps of milling operations is based on Figs. 11–14.

Mill feed is derived from the underground or pit operation. Primary crushing may be done in underground stations, in the case of an underground mine, or in a surface plant, in the case of an open pit mine.

Large jaw crushers (the 1.2 × 1.5-m size is the most popular) are used in many Canadian plants either underground or on the

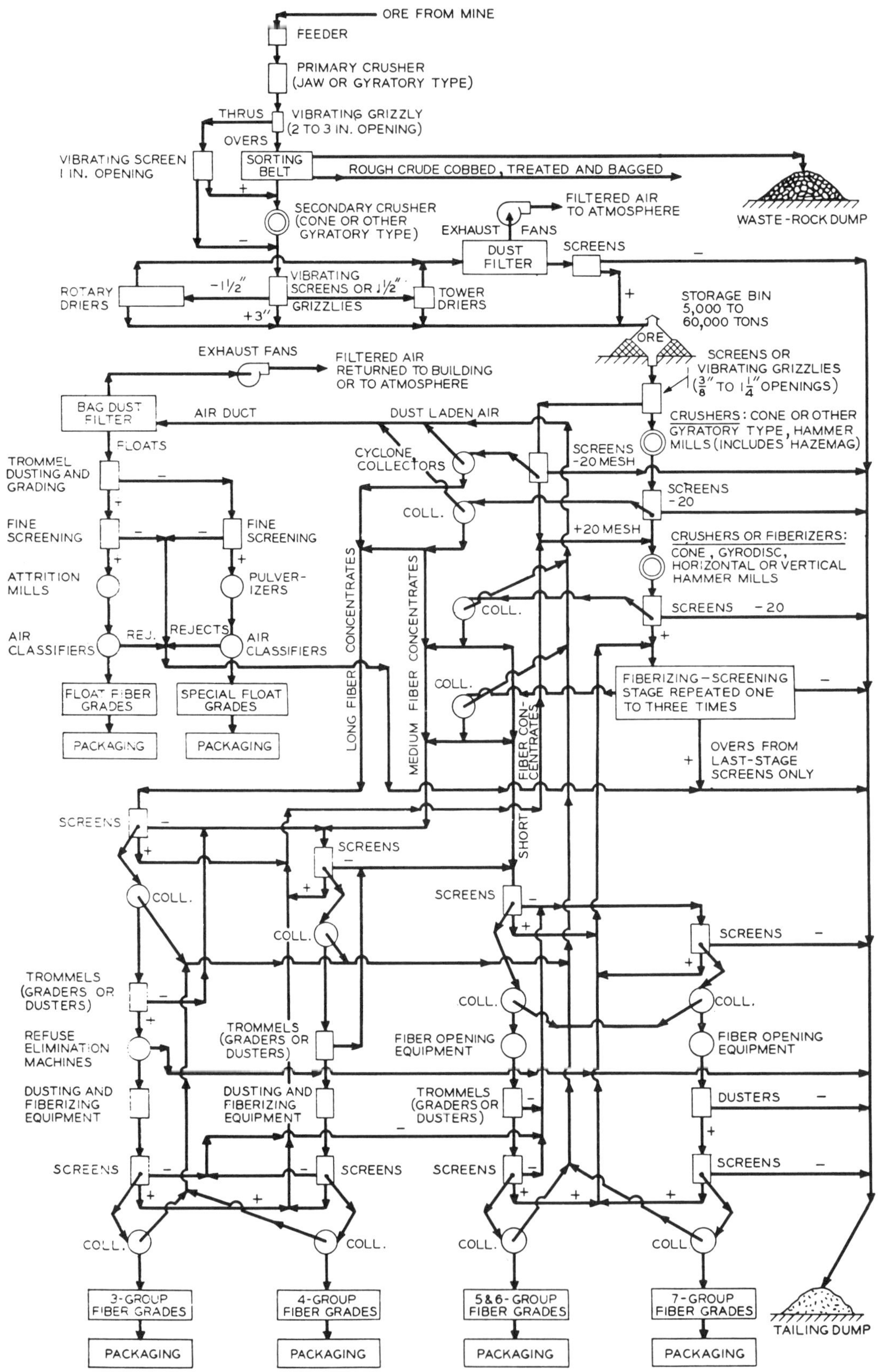

Fig. 11. Flowsheet of a Canadian asbestos mill.

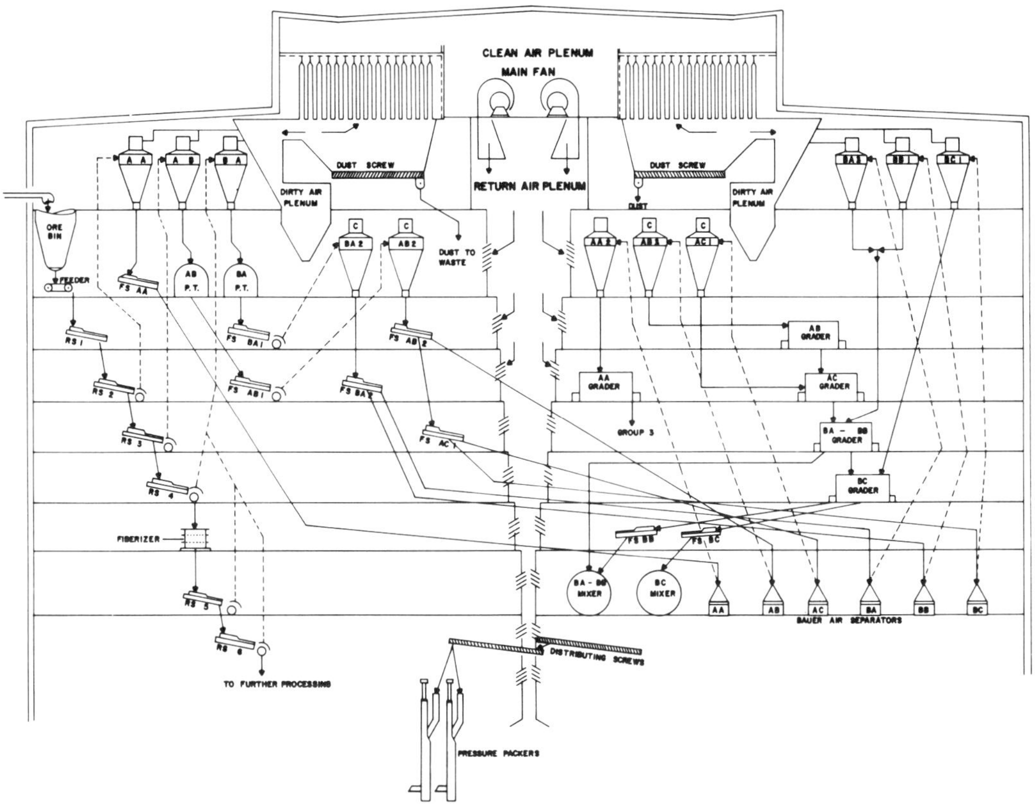

Fig. 12. Section of a Canadian asbestos mill showing the primary rock line and fiber cleaning circuit.

surface. Heavy duty pan feeders also proved practical and comparatively trouble-free, particularly for handling open pit ore containing large unbroken pieces, which are likely to cause blockages. The trend has been toward the use of large capacity gyratory crushers, which eliminate the need for feeders and bins.

In countries where cheap labor is still available, some hand sorting continues before and/or after primary crushing in order to remove barren rock and also to recover pieces of the larger asbestos veins to be used in the production of No. 1 or No. 2 crudes. Reduction in primary crushers of the size mentioned is normally to about 0.2 m, and is followed quite commonly by 2.1-m or smaller cone crushers with settings ranging from 19 to 64 mm.

Ore concentration is an important step in the milling of asbestos ore and is particularly important for lower grade ore bodies. It is not uncommon to discard as much as 40% of the mine ore through selective impacting and screening in the primary and secondary crushing circuits. Some producers use magnetic pulleys for upgrading the mine ore, although not all asbestos ore bodies are amenable to this type of separation.

The next step is to dry the ore. It is considered good practice to provide a storage large enough for one shift's production of mine ore to assist in regulating the flow rate to the dryers. Most of the moisture is contained in the -38 mm fraction. At some plants, this fraction is screened and dried separately from the oversize (38 to 178 mm) material.

The selection of a dryer is based largely on local preference. The two most commonly used dryers are the rotary and vertical dryers. Fluid-bed dryers also have been used. Generally, there is less mechanical damage to the fiber when vertical-tower and fluid-bed dryers are used than when rotary kiln-type dryers are used. However, rotary dryers are preferable and are more effective for open pit ores that might contain snow and ice.

Where conditions permit, oversize material (approximately +38 to 76 mm) may be separated before drying and bypassed to reduce the pebble-milling action of rotary dryers. When vertical dryers are used, the feed should not be much coarser than a 38-mm grizzly opening to avoid blockages in the gridwork of arresting bars in the dryers.

There is a limit to the air velocity that may be used in the counterflow vertical-type dryer where the products of combustion from an oil- or coal-fired furnace rise through the falling ore. Near the bottom of the tower, where the hot gases enter, fluffed particles of fiber may be temporarily held in suspension in a damaging-temperature environment.

To prevent fiber damage, some of the hot gases may be introduced at the top of the dryer to be drawn downward with the falling material. In dryers using this principle, the discharge-gas takeoff is midway up the height of the dryer, and thus air velocities and temperature remain within permissible limits toward the bottom of the dryer.

Cool air may be blown through the hot ore as it is discharged from the dryer. The cool air removes additional moisture from the ore. The air from this operation may be used as preheated makeup air in the dryers.

The air discharged from the dryers carries a considerable volume of dust. Bag filters have been used successfully and are a positive means of removing the dust and keeping the plant area clean.

It is desirable to have substantial storage capacity between drying and the main stage of milling. Besides assuring an adequate and more uniform supply of ore to the mill, further drying (up to 1% moisture loss) takes place during the storage. The size of storage facilities varies with mill capacity.

Fiber is released and separated from gangue by successive stages of crushing or comminution by impact (Fig. 13). Impactors

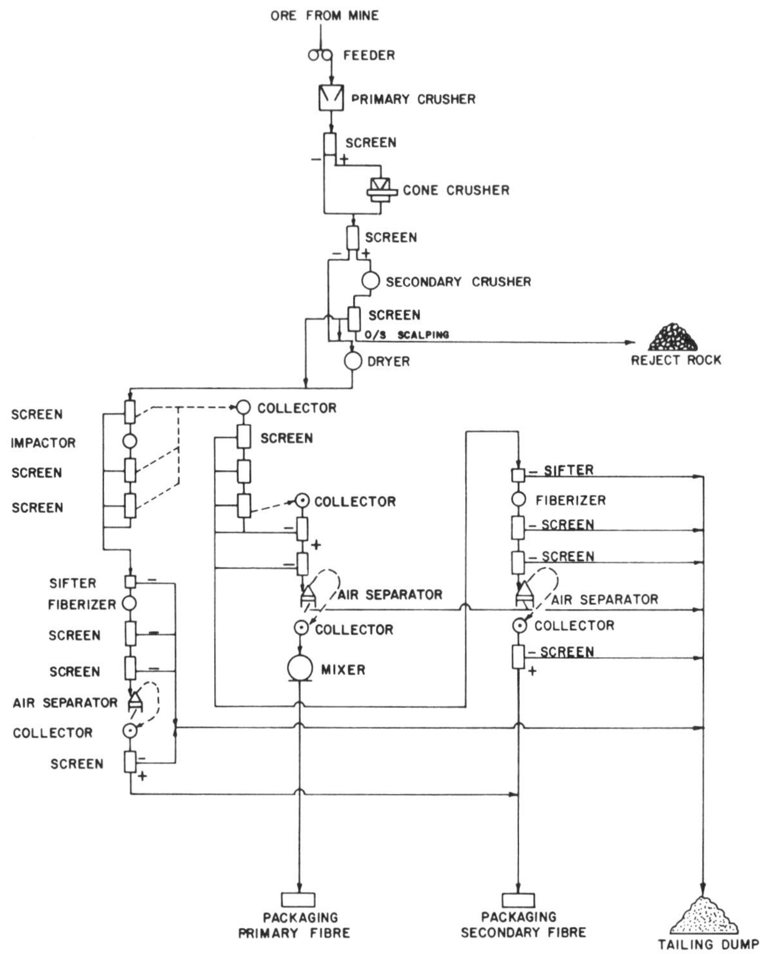

Fig. 13. Schematic flow of a two-fiber-product Canadian asbestos mill.

are designed to release the fiber from the host rock and at the same time produce a minimum of fines.

Fiber released by crushing is lifted by air suction, leaving most of the rock as a reject to go to the next stage of impacting and eventually to tailings.

Finer fractions generally are screened out prior to air separation to avoid aspirating a large portion of rock fines along with the fiber. Rotary aspirators designed to present a uniform layer of material to the aspirating hood are commonly used to improve the aspirating efficiency. The products of these first stages of separation may be considered as concentrates. They contain a large percentage of rock particles and must be further treated.

The concentrates are further graded and undergo a series of cleaning operations for the purpose of removing sand and dust. Screens, trommels, specific gravity air separators, and modified forms of these, such as graders and dusters, further clean the fiber and separate it into standard grade lengths.

In the grading mill, the fiber within each grade is further subdivided according to fiber quality. It is then subjected to several stages of screenings by means of shaking screens, rotating screens, conventional trommels, trommel-like graders, and rotary dusters.

When well opened or fluffed out grades are required, the fiber is specially processed in one or more of a variety of machines. These range from graders or Willows (fixed shell trommels having a rotating center shaft to which beater arms are attached) to one of several types of high speed hammer mills, disk grinders, or pulverizers. The type of machine or machines used depends upon the length and type of fiber to be processed and the degree of opening or fluffing required. This additional treatment is generally given to the shorter fibers.

A large volume of air is used in an asbestos mill, most of which is used for separating fiber from rock and dust by air suction and classification. An additional volume of air is required to reduce dust levels in the mill building. Mills having a capacity of approximately 5 000 tpd of ore, use 236 to over 378 m^3/s of free air. The average consumption in Canadian mills is approximately 7 to 14 t of air per ton of ore treated (Rozovsky, 1957, Kula and Wiser, 1970). Modern practice requires that the air be filtered before being discharged from the mill into the open. Extensive bag filter units have been installed in nearly all mills.

In Zimbabwe, South Africa, and Swaziland, beneficiation of chrysotile does not differ greatly from Canadian practice. One feature of African milling is the need to control moisture content carefully. It may be too low, particularly during the dry season. Fibers may break or be damaged more easily under extremely dry conditions.

Crocidolite and amosite are prepared largely by hand sorting, crushing, and screening, as with the crude grades in Canada. In the Penge area of the Transvaal, milling practice is somewhat more

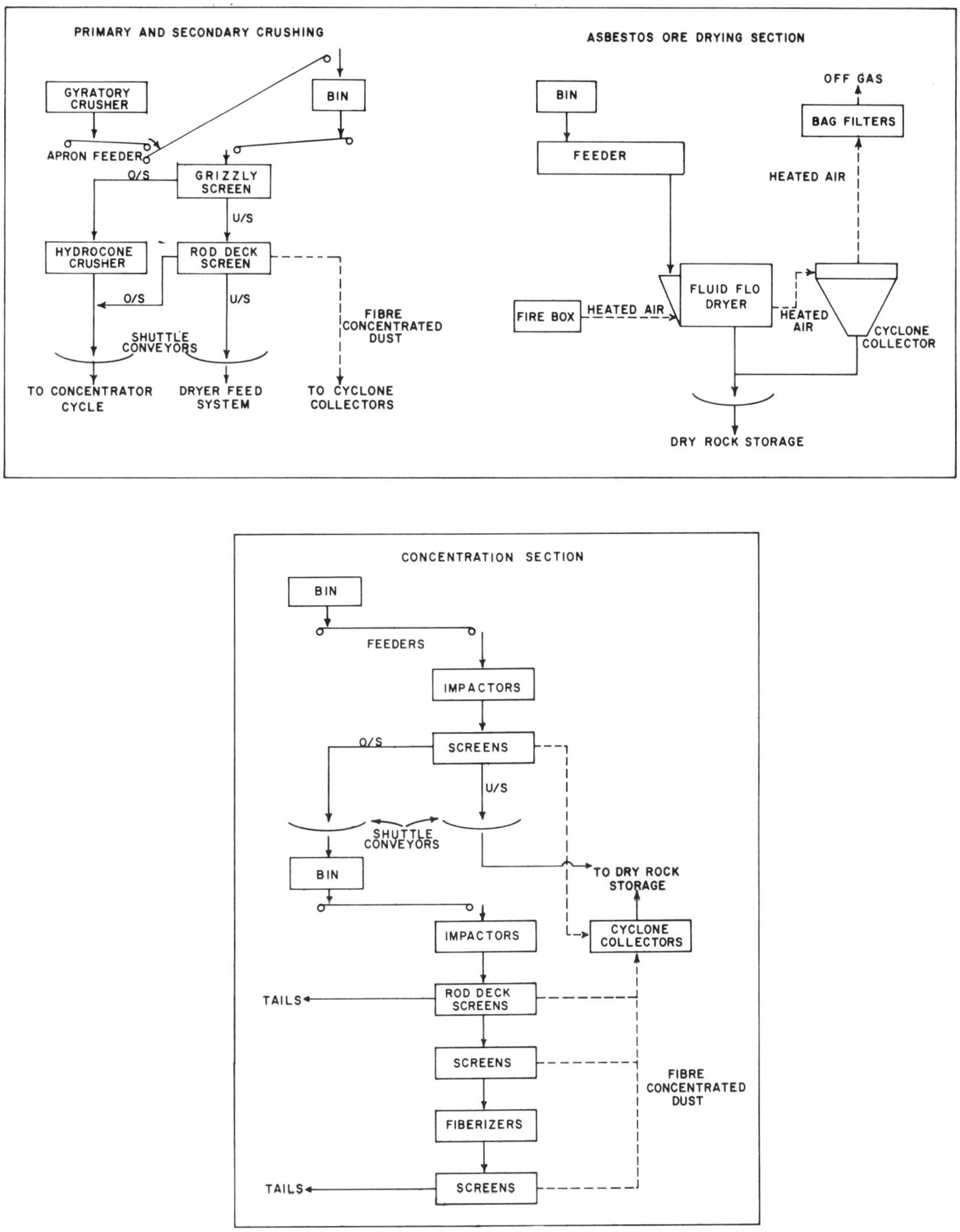

Fig. 14. Canadian installation of a primary and secondary crushing, drying, and ore concentration system.

elaborate, and several milled grades of amosite are produced. In addition to the conventional sorting and cobbing, the fiber is treated in mill circuits equipped with jaw crushers, cone crushers, rolls, hammer mills, and trommels.

As an example the longest grade, D_3 is prepared as follows:

Grizzly 76 × 356 mm (3 × 14 in.)

↓

Reduction to 38 mm (1 1/2 in.) in small jaw crusher

↓

Crushing rolls 0.6 × 1.2 m (24 × 48 in.) to 19 mm (3/4 in.)

↓

Trommel — 38 mm (1 1/2 in.) — undersize to next shorter grade circuit

↓

Crushing rolls 0.3 × 0.9 m (12 × 36 in.) to 12.7 mm (1/2 in.)

↓

Trommel — 19 mm (3/4 in.) undersize to next short grade circuit

↓

Table 9. Quebec Grading as Set Forth by the Quebec Asbestos Mining Association*

SPECIFICATION

Group No. 1
No. 1 Crude. Cross fiber veins having 3/4-in. staple and longer.

Group No. 2
No. 2 Crude. Cross fiber veins having 3/8-in. staple up to 3/4-in. Run-of-mine Crude consists of unsorted crudes. Sundry Crudes consist of crudes other than above specified.

	Guaranteed Minimum Shipping Test			
	1/2 In., oz	4 Mesh, oz	10 Mesh, oz	Pan, oz
Group No. 3 (Commonly referred to as textile or shipping fibers)				
3F	10.5	3.9	1.3	0.3
3K	7	7	1.5	0.5
3R	4	7	4	1
3T	2	8	4	2
3Z	1	9	4	2
Group No. 4 (Commonly referred to as asbestos cement fiber)				
4A	0	8	6	2
4D	0	7	6	3
4H	0	5	8	3
4H	0	5	7	4
4K	0	4	9	3
4M	0	4	8	4
4R	0	3	9	4
4T	0	2	10	4
4Z	0	1.5	9.5	5
Group No. 5 (Often referred to as paper stock grades)				
5D	0	0.5	10.5	5
5K	0	0	12	4
5M	0	0	11	5
5R	0	0	10	6
5Z	0	0	8.6	7.4
Group No. 6 (Paper and shingle fibers)				
6D	0	0	7	9
6F	0	0	6	10
Group No. 7 (Shorts and floats)				
7D	0	0	5	11
7F	0	0	4	12
7H	0	0	3	13
7K	0	0	2	14
7M	0	0	1	15
7R	0	0	0	16
7T	0	0	0	16
7RF and 7TF Floats	0	0	0	16
7W	0	0	0	16
Group No. 8 & 9 (Sands and gravels)				
8S	0	0	0	16
		Minimum 50 lb per cu ft		
8T	0	0	0	16
		Minimum 75 lb per cu ft		
9T	0	0	0	16
		More than 75 lb per cu ft		

*As of 1993, the grading standards have not been converted to the metric system.

Table 10. Cassiar Asbestos Grades, Northern British Columbia

C-1	Crude $\frac{3}{4}$ - in. staple and longer.	
AAA	Extra long spinning fiber	Canadian Group 3
AA	Long spinning fiber	Canadian Group 3
A	Spinning fiber	Canadian Group 3
AC	Spinning fiber	Canadian Group 3
CC	Spinning fiberd	Canadian Group 3
AK	Asbestos-cement fiber	Canadian Group 4
CP	Asbestos-cement fiber	Canadian Group 4
AS	Asbestos-cement fiber	Canadian Group 4
CT	Asbestos-cement fiber	Canadian Group 4
AX	Asbestos-cement fiber	Canadian Group 5
AY	Asbestos-cement fiber	Canadian Group 5
CY	Asbestos-cement fiber	Canadian Group 5
AZ	Asbestos-cement fiber	Canadian Group 6
CZ	Asbestos-cement fiber	Canadian Group 6

Degritting trommels (2 in parallel) — 6.4 mm (1/4 in.)
↓
Bagging

In the circuits for shorter grades, cone crushers, hammer mills, and other fiberizing and cleaning equipment are used. Trommels have proved effective as a cleaning device for medium, as well as for long grades.

For the very short fibers, high speed vibrating screens are found to be more efficient. Air separation is practiced in various forms, such as air traps, suction hoods, and screens, depending on the fiber grade being treated.

TESTS AND SPECIFICATIONS

The only test that is generally recognized by the government in Canada is the Quebec Asbestos Mining Association (QAMA) classification known as the Quebec Standard (QS) Test. This test is performed in the Quebec Standard Testing Machine, which consists of a stack of three sieve boxes with 1/2 in., 4 mesh-per-inch, and 10 mesh-per-inch screens, from top to bottom, and a bottom box serving as a pan. Sixteen ounces of asbestos is placed in the top sieve, and the stack of sieves is mechanically shaken on a rotary shaker for 600 revolutions at a speed of 328 rpm. The Quebec Asbestos Mining Association has set specifications for the various fiber grades based on the quantities of fiber that remain on each screen. The test is used for production control and determines the specification of the grade of fiber for sale. This test has been partially adopted by other countries, but in some cases the results are expressed in percentages instead of ounces.

Wet classification and screening also have been used to monitor fiber lengths, nonfibrous fractions, and dust content during production. The McNett or the Clark wet testing machine may be used for this purpose.

A Ro-Tap machine is sometimes used to evaluate further the Quebec Standard sieve fractions. Many asbestos mill operators believe that the Ro-Tap gives a better distribution of the fiber lengths than the Quebec testing machine. The Ro-Tap test is not accepted by the industry because the type and size of screens, lengths of time of Ro-Tapping, and the quantity of fiber used in the test are not standardized.

Many customers use density measurements as a guide in producing specific products. The density of the asbestos fiber is determined by filling a cubic-foot box or a fraction of a cubic-foot box with fiber. The box is weighed and the density is expressed as pounds per cubic foot.

Many customers have developed their own tests for evaluating

Table 11. Classification of African Chrysotile*

Grade	Source
C&G1 Long, crudy textile fiber C&G2 Textile fiber C&G3 Long shingle fiber C&G4 Shingle fiber C&G5 Short shingle fiber or paper stock	From mines in the Shabani District of Zimbabwe
VRA-2 Textile fiber VRA-3 Long shingle fiber VRA-4 Shingle fiber	From the Mashaba District of Zimbabwe
HVL2 Textile fiber HVL3 Long shingle fiber HVL4 Shingle fiber HVL5 Short shingle or paper stock	From the Havelock mine in Swaziland
Msauli—Grade 4 Msauli—Grade 5	From the Msauli mine near Swaziland border in the Transvaal
Amianthus 1 and 2—Textile fibers F—Long shingle fiber AA—Shingle fiber	From the Barberton District of the Transvaal
Munnik-Myburgh M1—Textile fiber M3—Shingle fiber M4—Short shingle fiber	From the Barberton District near Nelspruit of the Transvaal

*In the absence of a current listing of Zimbabwean, South African, and Swaziland grades, this list from the 1975 edition is included to indicate type and source.

Table 12. Classification of Chrysotile in the Former USSR

Grade	Type	Texture	Mark according to USSR standards, 1972
0	Spinning fiber	Harsh, crudy	DV-0 80, DV-0 55
1	Spinning fiber	Harsh, crudy Semi-crudy	J-1–50, J-1–38 PRJ-1–75, PRJ-1–50
2	Spinning fiber	Harsh, crudy Semi-crudy Semi-open	J-2–20 PRJ-2–30, PRJ-2–15 P-2–30, P-2–15
3	Asbestos-cement fiber	Harsh, crudy Semi-open	J-3–40 P-3–70, P-3–60 P-3–50, M-3–70 M-3–60
4	Asbestos-cement fiber	Semi-open Open,soft	P-4–40, P-4–30 P-4–20, P-4–5 M-4–40, M-4–30 M-4–20, M-4–5
5	Paper fiber	Semi-open Open, soft	P-5–67, P-5–65 P-5–52, P-5–50 M-5–65, M-5–50
6	Paper and shingle Fiber and shorts	Semi-open Open, soft	P-6–45, P-6–30 M-6–40, M-6–30 K-6–45, K-6–30 K-6–20, K-6–5
7	Unguaranteed		7–300, 7–370 7–450, 7–520

the performance of fiber in their finished products. The tests cover properties such as viscosity, absorption, grit, troweling, bulk, penetration, soluble salts, magnetite content, compressibility, flexing, tensile strength, color, strength units, and surface area (Cossette and Delvaux, 1979).

Fiber producers in other countries such as Zimbabwe, USSR, Australia, Cape Province, and Transvaal, South Africa, have developed their own testing methods for grading and classifying fibers, but do use the Quebec Testing Machine to a certain degree.

Table 13. Classification of African Amosite

Grade	Approximate range average fiber length, in.
S 11	$1 - 1\frac{1}{2}$
W 3	$\frac{1}{2} - 1\frac{1}{2}$
K3	$\frac{1}{2} - 1\frac{1}{2}$
SK	$\frac{3}{16} - \frac{3}{4}$
S 33	$\frac{1}{8} - \frac{1}{2}$
S 33/65	$\frac{1}{8} - \frac{1}{2}$
GW	$\frac{1}{8} - \frac{1}{4}$
GK	$\frac{1}{8} - \frac{1}{4}$
S 44	$\frac{1}{8} - \frac{1}{4}$
RK	$\frac{1}{8} - \frac{1}{4}$
6605	$\frac{1}{16} - \frac{1}{4}$

The grades listed above are produced in the Penge area from the Penge mine and the Weltevreden and Kromellenboog mines. The fiber lengths shown are not necessarily exact, but have been included in order to give some indication of the relative lengths of fibers of the different grades.

Table 14. Cape Blue (Crocidolite)

Grade	Approximate range average fiber length, in.	Typical values surface area (Rigden), cm^2/g
C	$1\frac{1}{4} - 1\frac{3}{4}$	1 500
S	$\frac{1}{4} - \frac{3}{4}$	5 500
S 80	$\frac{1}{4} - \frac{3}{4}$	9 500
P 25	$\frac{1}{4} - \frac{3}{4}$	8 300
H	$\frac{1}{8} - \frac{1}{2}$	7 500
H 80	$\frac{1}{8} - \frac{1}{2}$	10 000
713	$\frac{1}{8} - \frac{1}{2}$	13 000
WDS	$\frac{1}{8} - \frac{3}{8}$	9 000

The classifications for various grades from the more important producing areas are given in Tables 9–14.

USES

Several thousand products are manufactured that take advantage of the unique properties of asbestos. Some of the major ones are listed, along with a brief discussion of the products involved (Badollet, 1948). Although crocidolite and amosite are not graded according to the QAMA specifications, they are discussed below in the equivalent QAMA categories.

Crudes No. 1 and No. 2—Most chrysotile crudes are processed by the customer to produce a long, spinning fiber for use in textiles. A desirable fiber for textiles is one that is flexible, soft, low in soluble salts and magnetite, and easily carded without an excess drop in shorts. Fiber of this quality is desirable for all textile uses, including those for the electrical industry. In some cases, this grade of fiber is used as felts in laminates, along with resins, to form a strong molded sheet for use in airplanes, boats, etc.

Crocidolite crudes that are processed to produce long fibers can be used in textiles, gaskets, ropes, or in laminates with resins.

Amosite crudes, after reprocessing, will give a long, bulky fiber that is used in blanket insulation or in products requiring a low density and good insulation value. Spinning of amosite can be accomplished, but is difficult.

Group 3 Milled Fibers—Chrysotile fibers that meet this classification generally are used in textiles. Some are used in long-fiber asbestos papers, packings, gaskets, brake linings, clutch facings, electrolytic diaphragms, pipe coverings, and insulating blocks. Some of the fibers of this group are now being used in laminates with resins.

Crocidolite may be used for most of the purposes listed above, providing color is not an objection. It generally should be prepared in a well opened condition.

Group 4 Milled Fibers—Many grades of fibers in this classification are used in asbestos-cement processes to produce pipe, jackets, boards, sheets, and a variety of hand-molded articles. Other uses are for papers, pipe coverings, packings, gaskets, millboards, and plastics.

Crocidolite, similar to Group 4 chrysotile, is used in asbestos-cement pipes, in some packings and in some gaskets.

Amosite is used in magnesia blocks, pipe coverings, and other insulation compositions when a light density is desired. It is also used in acetylene cylinders to give strength to the calcium silicate mix during curing.

Group 5 Milled Fibers—Fibers of this group sometimes are used as replacements for Group 4 fibers. Products made with these fibers include asbestos-cement sheets, corrugated or flat boards, pipe, electrical panels, papers, millboards, pipe coverings, gaskets, packing, brake linings, and plastics.

Crocidolite of this classification is used in asbestos-cement pipes.

Group 6 Milled Fibers—Chrysotile fibers of Group 6 are used in asbestos-cement shingles, flat sheets, corrugated sheets, boards, brake lining, papers, millboards, and plastics.

Crocidolite of this grading is used in the production of some asbestos-cement products. The balance is considered filler and used wherever its physical properties could be employed to advantage.

Amosite of this classification is considered too short for most applications and probably would be used as an inexpensive filler to free up asbestos-cement slurries prior to pressing.

Group 7 Milled Fibers—This grade of chrysotile is used in certain papers, cements, asphalt roof coatings, paints, welding rods, floor tile, and plastics.

Crocidolite and amosite fibers of a length equivalent to Group 7 chrysotile are not known to be used in commercial products unless as a cheap filler where color is not objectionable.

Floats—Chrysotile floats (Badollet, 1952, 1956) are used extensively in plastics and coatings.

The selection of asbestos fiber for a particular application will depend upon the processing method, as well as the desired properties of the end product. The following examples outline some of the major uses for asbestos fiber and illustrate pertinent factors that influence the choice of fiber grade and type.

Asbestos-Cement Products—Pressure pipe that must conform to hydrostatic test specifications is produced from high quality Group 4 fiber. Most of it is a blend of chrysotile and crocidolite to ensure a good modulus of rupture. Flexural and impact strength requirements for asbestos cement sheets usually can be met by using a Group 6 fiber. Formulations for corrugated sheets generally include some Group 5 material to improve adhesion of the wet sheets during the forming process.

In all asbestos-cement products made by the wet machine process, drainage is an important fiber characteristic because it has direct bearing on the production rate. For this reason, preference may be given to fast-filtering fibers, and amosite may be included in the formulation as a filter aid.

The strength or drainage requirements for shingles made by a dry process are not so exacting and can be satisfied with a lower quality 6 or 6–7 blend.

Asbestos Paper—Traditionally, asbestos fiber has been used with an organic binder to manufacture a variety of paper and millboard products, such as roofing felt, pipe coverings, and electrical insulations. Various blends of Groups 4, 5, and 6 fibers are used for this application, depending upon the desired strength and porosity of the paper.

A latex-asbestos process is used to produce some paper products. A long clean Group 7 fiber having a high surface area is coated with latex rubber by a chemical precipitation method. The coated fiber then is formed into a continuous sheet on a paper machine. A large proportion of this latex asbestos paper is used as an underlayment for vinyl rolled floor covering. Less is used in the manufacture of gaskets.

Friction Materials—A wide variety of products are classified under the friction materials category. As a consequence, the products uses fiber ranging from Group 3 spinning grades to the shorter Group 7.

Clutch plates are made from an asbestos open-weave cloth impregnated with resin and bonded to a steel disk. A similar product can be manufactured by molding a dry resin-fiber blend under conditions of high temperature and pressure onto a packing plate. For the first method, a Group 3 fiber is required, whereas the molding process uses a Group 5.

Most automobile brake linings bonded to a steel shoe are made from Group 7 fiber in a semi-wet extrusion process, whereas heavy blocks for railcars and large vehicles use Group 5 or 6 fiber dry-molded and machined to finished dimensions.

Group 5 fiber is also used extensively in disk brake pad formulations.

Sheet Packing—Latex asbestos paper made from Group 7 fiber can be densified and used for gasketing, but most sheet packing material is formed on a sheeter machine by a calendering process. This latter method requires a longer fiber in the Group 4 to 5 range that has been cleaned and opened. The fiber is blended with natural or synthetic rubber, plasticizers, and other ingredients in a high-shear mixer to form a dough that is later calendered into sheets of various thicknesses.

Asphalt Products—Group 7 asbestos fiber, in combination with asphalt and various solvents, form the basis of a wide variety of products often classified under the heading of "Blackline." These

Table 15. Asbestos Prices—Canada, United States

Standard Grade Designation	Per ton, FOB mine Canadian dollars
Grade 3	$729
Grade 4	554
Grade 5	474
Grade 6	307
Grade 7	180

Source: Canadian Minerals Yearbook, 1991.

Table 16. Annual Exports of Asbestos from Canada by Areas

	1988		1989	
	Quantity (t)	Value (1 000$C)	Quantity (t)	Value (1 000$C)
United States	79 029	27 348	43 206	17 177
Europe[1]	158 169	110 813	126 566	95 267
Asia[2]	223 312	103 544	197 959	102 386
Others	224 575	148 075	164 720	118 851
Total	685 085	389 780	532 451	333 681

[1]Includes Belgium, Denmark, France, former West Germany, Greece, Ireland, Italy, Luxembourg, Netherlands, Portugal, Spain, and the United Kingdom.
[2]Includes Indonesia, Japan, Republic of South Korea, Malaysia, Taiwan, and Thailand.

Table 17. Exports from the Former USSR of Asbestos to Selected Countries 1987, 1988, 1989 (tons)

	1987	1988	1989
Austria	3 699	216	1 633
Belgium-Luxembourg	5 013	1 199	2 022
Czechoslovakia	35 327	35 330	30 670
France	12 149	15 885	16 675
West Germany	3 304	3 410	1 824
Italy	3 069	230	7 702
Japan	44 593	47 844	49 016
Poland	70 106	60 924	54 832
Yugoslavia	28 356	29 810	19 245
Other	65 784	55 152	42 981
Total	271 400	250 000	226 600

include spray or brush-on roof coatings, sound deadeners for automobile body panels, and caulking components.

Group 7 fibers are used in automobile underbody protective coatings applied by airless spray equipment. Because the finished compound must pass through an orifice 0.53 to 0.71 mm in diameter under high pressure, the fiber used must meet critical specifications on fineness and viscosity-building properties.

Short Group 7 chrysotile asbestos added to hot asphalt paving mix helps to improve characteristics of toughness, flexibility, and water impermeability. Asbestos-modified pavements have proved successful in high traffic areas, such a busy street intersections, bus stops, and bridge decks.

Caulking Compounds—Combinations of long asbestos with cement and other ingredients, along with waterproofing resins, are used to produce special types of caulking compounds. In many cases, short asbestos and floats are also combined with various types of resins and other materials to produce a soft plastic caulking compound, which can be controlled either to remain soft or to set up as a hard mass.

Plastics—Asbestos, or a combination of asbestos and glass, is used to reinforce structural plastics. The asbestos may be in the form of a mat, or as paper or cloth, to form laminates with resins such as polyesters, phenolics, thermosetting silicones, melamines, and furanes.

The use of long-fiber chrysotile, crocidolite, or in some cases amosite, in the form of felts or papers and impregnated with resins, produces a tough product of high strength and good heat resistance. These products have been used in aeroplane wings in England and in small sailing boats, radar scanner aerials, aircraft tanks, automobile bodies, and other products, including rocket tubes, missile nose cones, and other parts.

Short Group 7 fiber and floats also are used as a fibrous filler for the production of molded phenolic resin and polyester parts, such as automobile heater and air conditioner housings, electric kettle bases, and other appliance parts. In such applications, freedom from abrasive particles is especially important to hold die wear at a minimum level.

Joint Filler—Short Group 7 fiber and floats are used to manufacture joint-filler cements and texture paints. Fineness and whiteness are critical properties of the fiber. Because the viscosity of the mix after the addition of the prescribed amount of fluid must be consistent from batch to batch, the absorptive capacity of the asbestos must be controlled within very narrow limits.

Prices for Canadian fiber are listed in Table 15.

MARKETS

In the late 1960s and early 1970s, the consumption of asbestos fiber increased at the rate of 3 to 4% yearly. During the 1980s, consumption declined. The largest decreases in consumption occurred in North America and Western Europe where markets were maturing and environmental and liability issues became increasingly important. These losses have been partially offset by increased sales in Asia and in some developing countries because of the high demand for proven, low cost construction materials.

On a worldwide basis, consumption is estimated at 4 Mt. Canada supplies about 17% of world requirements. As indicated in Table 16, Western Europe and Asia account for 24 and 38% of total Canadian exports, respectively.

The largest producer is the USSR with approximately 2.4 Mt or 60% of the world's production. The USSR exports approximately 250 000 tpy (Table 17). More than half of the exports are shipped to Europe.

Major end uses for asbestos fiber are asbestos-cement sheet, asbestos-cement pipe, friction materials, roofing, and coatings. In the United States, 37% of the fiber consumed was used in roofing products, 26% in friction materials, and 20% in asbestos-cement products in 1990 (Table 18).

ENVIRONMENTAL CONSIDERATIONS

Environmental and liability issues continue to affect the asbestos industry. The trend in the United States and other countries is to enact stricter exposure regulations as the concern over the health risks posed by low level exposures to asbestos increases. More emphasis is being placed on environmental exposures than in the past.

The risk posed by low level exposure to asbestos continues to be debated. Over the past five years, there has been a gradual swing to the feeling that some of the risk has been grossly exaggerated. There has been increasingly stronger criticism by some scientists (Ross, 1987), reflected lately also by editorials in newspapers and on television.

Table 18. US Asbestos Consumption by End Use, Grade, and Type (thousand tons)

	Chrysotile[1]									
End use	Grades 1–3	Grade 4	Grade 5	Grade 6	Grade 7	Total[2]	Amosite	Crocid-dolite	Other[3]	Total Asbestos[4]
1989	0.2	5.7	3.8	3.4	38.3	51.5	[5]	0.6	3.2	52.1
1990:										
Asbestos-cement pipe	—	3.2	0.5	0.9	—	4.6	—	0.8	—	5.4
Asbestos-cement sheet	—	—	—	0.9	0.9	1.8	—	—	—	1.8
Coatings and compounds	—	—	—	—	1.7	1.7	—	—	—	1.7
Friction products	—	—	1.0	0.5	7.9	9.3	—	—	—	9.3
Packing and gaskets	—	[5]	0.7	0.1	1.8	2.7	—	—	—	2.7
Paper	—	—	—	—	0.2	0.2	—	—	—	0.2
Plastics	—	—	—	—	0.3	0.3	—	—	—	0.3
Roofing products	—	—	—	—	12.9	12.9	—	—	—	12.9
Other	0.1	0.4	[5]	—	0.4	0.9	—	—	—	0.9
Total	0.1	3.6	2.1	2.4	26.0	34.3	—	0.8	6.2	35.1

[1]Estimated distribution based upon data provided by the Asbestos Institute, Montreal, Canada, and the US Bureau of Mines asbestos producer survey.
[2]Data may not add to totals shown because of independent rounding.
[3]*Source*: Bureau of the Census. "Other" category contains unspecified fiber type and end use.
[4]Does not include "Amosite" and "Other" category in total.
[5]Less than $\frac{1}{10}$ unit.

ASBESTOS SUBSTITUTES

The use of asbestos substitutes and alternative products will increase in response to environmental and liability issues. Product manufacturers gradually have been replacing asbestos with substitute materials, redesigning old products to eliminate the need for asbestos, or designing new products which require neither asbestos nor asbestos substitutes. Some of the factors considered in developing the substitutes include substitute cost, additional manufacturing costs, product design costs, product performance, and possible health risks.

Examples of materials substituted for asbestos include aramid fiber, cellulose fibers, ceramic fiber, fibrous glass, graphite flake and fiber, mica, polyethylene fiber, polypropylene fiber, polytetrafluoroethylene fiber, steel fibers, and wollastonite. Examples of alternative products include aluminum, vinyl, and wood siding, aluminum pipe and sheet, asphalt coatings, ductile iron pipe, fiberglass sheet, polyvinylchloride pipe, prestressed concrete and reinforced concrete pipe, semimetallic brakes, urethane coatings, and vinyl composition floors. No single substitute has proved to be as versatile and as cost effective as asbestos.

BIBLIOGRAPHY AND REFERENCES

Anon., *Minerals Yearbook*, US Bureau of Mines, annually.

Anon., 1966, *Testing Procedures for Chrysotile Asbestos Fiber*, 2nd ed., Quebec Asbestos Mining Association.

Anon., 1972, *Geological Guide to the Asbestos-Mining Region of Southeast Quebec*, Quebec Asbestos Mining Association, p. 32.

Anon., 1984, "Asbestiform Fibers—Nonoccupational Health Risks," National Research Council, National Academy Press, Washington, DC, 334 pp.

Anon., 1989, "Asbestos," *Mining Magazine*, Vol. 41, No. 11, pp. 411–414.

Addison, W.E., Neal, G.H., and White, A.D., 1966, "Amphiboles-Part IV, Surface Properties of Amosite and Crocidolite," *Journal of the Chemical Society*, Sect. A, pp. 79–81.

Allen, C.C., Gill, J.C., and Koski, J.S., 1957, "The Jeffrey Mine of Canadian Johns-Manville Company, Limited," *The Geology of Canadian Industrial Mineral Deposits*, 6th Commonwealth Mining and Metallurgical Congress, Toronto, pp. 27–36.

Anderson, H.V., and Clark, G.L., 1929, "Application of X-Rays in the Classification of Fibrous Silicate Minerals Commonly Termed Asbestos," *Industrial & Engineering Chemistry*, No. 10, pp. 924–933.

Aruja, E., 1944, "Displacement of X-Ray Reflections," *Nature*, Vol. 154, p. 53.

Aruja, E., 1944a, "An X-Ray Study of the Crystal Structure of Antigorite," *Mineralogical Magazine*, Vol. 27, pp. 65–74.

Avery, R.B., Conant, M.L., and Weissenborn, H.F., 1958, "Selected Annotated Bibliography of Asbestos Resources in the United States and Canada," Bulletin 1019-L, US Geological Survey, pp. 817–865.

Babcock, R.C., 1978, "The Pyke Asbestos Deposits, New Zealand," SME Preprint No. 78-H-84, AIME Annual Meeting, Denver.

Badollet, M.S., 1937, US Patent 2,068,219, Jan. 19.

Badollet, M.S., 1948, "Asbestos," *Encyclopedia of Chemical Technology*, Vol. 2, R. Kirk and D. Othmer, eds., Interscience, New York, pp. 134–142.

Badollet, M.S., 1951, "Asbestos, A Mineral of Unparalleled Properties," *Transactions*, Canadian Institute of Mining & Metallurgy, Vol. 54, pp. 151–160.

Badollet, M.S., 1952, "Asbestos Floats," *Transactions*, Canadian Institute of Mining & Metallurgy, Vol. 55, May, pp. 185–189.

Badollet, M.S. 1956, "The Role of Asbestos in Plastics," *Transactions*, Canadian Institute of Mining & Metallurgy, Vol. 59, pp. 283–288.

Badollet, M.S., 1963, "Asbestos," *Encyclopedia of Chemical Technology*, Vol. 2, Kirk-Othmer, ed., Interscience, New York, pp. 734–747.

Badollet, M.S., and Streib, W.C., 1947, US Patent 2,616,801, Sep. 20.

Badollet, M.S., and Streib, W.C., 1955, "The Heat Treatment of Chrysotile Asbestos Fibers," *Transactions*, Canadian Institute of Mining & Metallurgy, Vol. 58, pp. 33–37.

Bates, R.L., 1969, "Metamorphic Minerals—Asbestos," *The Geology of Industrial Rocks and Minerals*, Dover Publications, New York, pp. 317–328.

Bates, T.F., et al., 1950, "Tubular Crystals of Chrysotile Asbestos," *Science*, Vol. 3, pp. 512–513.

Bear, L.M., 1963, "The Mineral Resources and Mining Industry of Cyprus," Bulletin No. 1, Geological Survey, Cyprus, 208 pp.

Berger, H., 1963, *Asbestos Fundamentals*, Chemical Publishing Co., New York, 171 pp.

Bourassa, P.J., 1957, "The Asbestos Mine of Nicolet Asbestos Mines Limited," *The Geology of Canadian Industrial Mineral Deposits*, 6th Commonwealth Mining and Metallurgical Congress, Toronto, pp. 26–27.

Bowles, O., 1955, "The Asbestos Industry," Bulletin 552, US Bureau of Mines, 122 pp.

Bragg, W.L., 1937, "The Pyroxene and Amphibole Groups," Atomic Structure of Minerals, University of Cornell Press, Ithaca, NY, p. 184.

Brindley, G.W., and Zussman, J., 1957, "A Structural Study of the Thermal

Transformation of Serpentine Minerals to Forsterite,'' *American Mineralogist*, Vol. 42, No. 7–8, pp. 461–474.

Butt, B.C., 1978, ''Exploration Forecasts and Exploitation Realities of the Woodsreef Mine, New South Wales, Austrialia,'' SME Preprint No. 78-H-71, AIME Annual Meeting, Denver.

Carroll-Porczynski, C.Z., 1956, *Asbestos*, The Textile Institute, Manchester, England.

Chidester, A.H., and Shride, A.F., 1962, ''Asbestos in the United States, Exclusive of Alaska and Hawaii,'' Mineral Investigation Research Map MR 17, US Geological Survey.

Cilliers, J.J. le R., 1964, ''Amosite at the Penge Asbestos Mine,'' Vol. 2, *The Geology of Some Ore Deposits of Southern Africa*, Geological Society of South Africa, pp. 579–591.

Cilliers, J.J. le R., and Genis, J.H., 1964, ''Crocidolite Asbestos in the Cape Province,'' *The Geology of Some Ore Deposits of Southern Africa*, Vol. 2, Geological Society of South Africa, pp. 543–570.

Cilliers, J.J. le R., et al., 1961, ''Crocidolite from the Koegas-Westerberg Area, South Africa,'' *Economic Geology*, Vol. 56, pp. 1421–1437.

Conn, H.K., 1967, ''Geophysics and Asbestos Exploration,'' *Mining and Groundwater Geophysics*, Economic Geology Report No. 26, Geological Survey of Canada, pp. 485–491.

Conn, H.K., and Mann, E.L., 1971, ''Evaluation of Asbestos Deposits,'' SME Preprint No. 71-H-27, AIME Annual Meeting, New York, 9 pp.

Cooke, H.C., 1937, ''Thetford, Disraeli and Eastern Half of Warwick Map Areas, Quebec,'' Memoir 211, Geological Survey of Canada, pp. 86–140.

Cossette, M., and Delvaux, P., 1979, ''Technical Evaluation of Chrysotile Asbestos Ore Bodies,'' Short Course in Mineralogical Techniques of Asbestos Determination, Mineralogical Association of Canada, University of Laval, Quebec, pp. 79–110.

Dean, A.W., and Mann, E.L., 1968, ''The Evaluation of Chrysotile Asbestos Deposits,'' *Ore Reserve Estimation and Grade Control*, Special Vol. 9, Canadian Institute of Mining & Metallurgy, pp. 281–286.

Douglas, R.J.W., 1970, *Geology and Economic Minerals of Canada*, Economic Geology Report No. 1, Geological Survey of Canada, Dept. of Energy, Mines and Resources, 838 pp.

duToit, A.L., 1946, ''The Origin of the Amphibole Asbestos Deposits of South Africa,'' *Transactions*, Geological Society of South Africa, Vol. 48, pp. 161–206.

Fankuchen, I., and Schneider, M., 1944, ''Low Angle X-Ray Scattering from Chrysotiles,'' *Journal of American Chemical Society*, Vol. 66, No. 3, Mar., pp. 500–501.

Frankel, J.J., 1953, ''South African Asbestos Fibres,'' *Mining Magazine*, London, Nos. 2 and 3, pp. 73–83, 89, 142–149.

Genis, J.H., 1964, ''The Formation of Crocidolite Asbestos,'' *The Geology of Some Ore Deposits in Southern Africa*, Vol. 2, Geological Society of South Africa, pp. 571–578.

Gold, D.P., 1967, ''Local Deformation Structures in a Serpentinite,'' *Ultramafic and Related Rocks*, P.J. Wyllie, ed., John Wiley, New York, pp. 200–202.

Graham, R.P.D., 1944, ''Serpentine Belt, Eastern Townships,'' *Geology of Quebec,* Geology Report 20, Dept. of Mines, Quebec, Vol. 2, pp. 439–443.

Griffin, J., 1986, ''Asbestos—Rising in East but Setting in West,'' *Industrial Minerals*, No. 223, London, pp. 21–37.

Gross, P., and Braun, D., 1984, *Toxic and Biomedical Effects of Fibers*, Noyes Publications, Park Ridge, 257 pp.

Hall, A.L., 1930, *Asbestos in the Union of South Africa*, Memoir 12, 2nd ed., Geological Survey of South Africa, p. 324.

Harris, H.I., 1973, ''How Nicolet Proved and Evaluated Columbian Asbestos Deposit,'' *World Mining*, Vol. 26, No. 13, pp. 43–46.

Hendry, N.W., 1972, ''The Outook for Asbestos in Canada,'' *Bulletin*, Canadian Institute of Mining & Metallurgy, Vol. 65, No. 724, Aug., pp. 40–44.

Hendry, N.W., and Conn, H.K., 1957, ''The Ontario Asbestos Properties of Canadian Johns-Manville Company, Limited,'' *The Geology of Canadian Industrial Mineral Deposits*, 6th Commonwealth Mining and Metallurgical Congress, Toronto, pp. 36–44.

Hillier, J., and Turkevich, J., 1949, ''Electron Microscopy of Colloidal Systems,'' *Analytical Chemistry*, Vol. 21, No. 4, Apr., pp. 475–485.

Hodgson, A.A., and White, A., 1967, ''Measurement of the Surface Area of Asbestos Fibers,'' 1st International Asbestos Conference, Oxford.

Hodgson, A.A., 1985, *Alternatives to Asbestos and Asbestos Products*, Anjalena Publications Ltd., Crowthorne, 230 pp.

Hodgson, A.A., 1986, *Scientific Advances in Asbestos, 1967 to 1985*, Anjalena Publications, London, 186 pp.

Keep, F.E., 1961, ''Amphibole Asbestos in the Union of South Africa,'' *Transactions*, Vol. 1, 7th Commonwealth Mining and Metallurgical Congress, Johannesburg, pp. 90–120.

Keith, S.B., and Bain, G.W., 1932, ''Chrysotile Asbestos: 1. Chrysotile Veins,'' *Economic Geology*, Vol. 27, pp. 169–188.

Kula, J., and Wiser, J.P., 1970, ''Msauli Asbestos Mill,'' *World Mining*, Sep., pp. 26–29.

Lamarche, R.Y., 1972, ''Ophiolites of Southern Quebec,'' *Canadian Contributions 1–11 to the Geodynamics Project—A Symposium*, Earth Physical Branch, Dept. of Energy, Mines and Resources, Ottawa.

Laubscher, D.H., 1964, ''The Occurrence and Origin of Chrysotile Asbestos and Associated Rocks, Shabani, Southern Rhodesia,'' *The Geology of Some Ore Deposits of Southern Africa*, Vol. 2, The Geological Society of South Africa, pp. 593–624.

Laubscher, D.H., 1968, ''The Origin and Occurrence of Chrysotile Asbestos in the Shabani and Mashaba Areas, Rhodesia,'' Symposium on Rhodesian Basement Complex, *Transactions*, Geological Society of South Africa Annexure, Vol. 71, pp. 195–204.

Leney, G.W., and Loeb, E.E., 1972, ''The Geology and Mining Operations at Pacific Asbestos Corporation,'' *Asbestos*, Vol. 54, No. 4, pp. 4–14.

Levadie, B., ed., 1984, ''Definitions for Asbestos and Other Health-Related Silicates,'' ASTM Special Publication 834, 213 pp.

Low, J.H., 1951, ''Magnetic Prospecting Methods in Asbestos Exploration,'' *Transactions*, Canadian Institute of Mining & Metallurgy, Vol. 54, pp. 388–395.

Mamen, C., ed., 1973, ''China's Mineral Industry,'' *Canadian Mining Journal*, Vol. 94, No. 1, pp. 21–31.

May, T.C., and Lewis, R.W., 1970, ''Asbestos,'' *Mineral Facts and Problems*, Bulletin 650, US Bureau of Mines, pp. 851–863.

Merrill, R.J., 1957, ''The Carey-Canadian Asbestos Deposit,'' *The Geology of Canadian Industrial Mineral Deposits*, 6th Commonwealth Mining and Metallurgical Congress, Toronto, pp. 45–49.

Miles, K.R., 1942, ''The Blue Asbestos-Bearing Banded Iron Formations of the Hammersley Range, Western Australia,'' Bulletin No. 100, Geological Survey of Western Australia, Pt. 1, pp. 5–37.

Mossman, B.T., et al., 1990, ''Asbestos: Scientific Developments and Implications for Public Policy,'' *Science*, Vol. 247, Jan. 19, pp. 294–301.

Munro, R.C., and Reim, K.M., 1962, ''Coalinga Asbestos Fiber—A Newcomer to the Asbestos Industry,'' *Canadian Mining Journal*, Vol. 83, No. 8, Aug.; *Mining Engineering*, Vol. 14, No. 9, pp. 60–62.

Nalivkin, D.V., 1960, *The Geology of the USSR—A Short Outline*, S.I. Tomkeiff and J.E. Richey, trans. eds., Pergamon Press, New York, 170 pp.

Oldham, J.W., 1968, ''A Short Note on the Recent Geological Mapping of the Shabani Area,'' Symposium on Rhodesian Basement Complex, *Transactions*, Geological Society of South Africa Annexure, Vol. 71, pp. 189–194.

Parry, J., 1980, ''Zimbabwe Opens Its Doors,'' *Industrial Minerals*, No. 158, p. 55.

Pauling, L., 1930, ''The Structure of the Chlorites,'' *Proceedings*, National Academy of Science, Vol. 16, p. 578.

Pelletier, R.A., 1964, *Mineral Resources of South-Central Africa*, Oxford University Press, 277 pp.

Petrov, V.P., and Znamensky, V.S., 1978, ''Asbestos Deposits of the USSR,'' SME Preprint No. 78-H-106, AIME Annual Meeting, Denver.

Pundsack, F.L., 1955, ''The Properties of Asbestos. I. The Colloidal and Surface Chemistry of Chrysotile,'' *Journal of Physical Chemistry*, Vol. 59, No. 9, Sep., pp. 892–895.

Pundsack, F.L., 1956, ''The Properties of Asbestos. II. The Density and Structure of Chrysotile,'' *Journal of Physical Chemistry*, Vol. 60, No. 3, Mar., pp. 361–364.

Pundsack, F.L., and Reimschussel, G., 1956, ''The Properties of Asbestos. III. Basicity of Chrysotile Suspensions,'' *Journal of Physical Chemistry*, Vol. 60, Sep., pp. 1218–1222.

Rabbit, J.C., 1948, ''A New Study of the Anthophyllite Series,'' *American Mineralogist*, Vol. 33, May-June, pp. 263–323.

Ralston, J., and Kitchener, J.A., 1975, ''The Surface Chemistry of Amosite Asbestos, an Amphibole Silicate,'' *Journal of Colloid and Interface Science*, Vol. 50, No. 2, pp. 242–249.

Rice, S.J., 1963, "California Asbestos Industry," Mineral Information Service, California Div. of Mines, Vol. 16, No. 9, pp. 4–6.

Riordon, P.H., 1952, "Geology of the Thetford Black Lake District of Quebec with Particular Reference to the Asbestos Deposits," PhD thesis, McGill University, unpublished.

Riordon, P.H., 1955, "The Genesis of Asbestos in Ultrabasic Rocks," *Economic Geology*, Vol. 50, No. 1, pp. 67–81.

Riordon, P.H., 1957, "The Structural Environment of the Thetford-Black Lake Asbestos Deposits," *Proceedings*, Geological Association of Canada, Vol. 9, pp. 83–93.

Riordon, P.H., 1957a, "The Asbestos Belt of Southeastern Quebec," *The Geology of Canadian Industrial Mineral Deposits*, 6th Commonwealth Mining and Metallurgical Congress, Toronto, pp. 3–8.

Riordon, P.H., 1957b, "The Asbestos Deposits of Thetford Mines, Quebec," *The Geology of Canadian Industrial Mineral Deposits*, 6th Commonwealth Mining and Metallurgical Congress, Toronto, pp. 9–17.

Riordon, P.H., 1957c, "The British Canadian Mine," *The Geology of Canadian Industrial Mineral Deposits*, 6th Commonwealth Mining and Metallurgical Congress, Toronto, pp. 17–21.

Riordon, P.H., 1957d, "Normandie and Vimy Ridge Mines," *The Geology of Canadian Industrial Mineral Deposits*, 6th Commonwealth Mining and Metallurgical Congress, Toronto, pp. 21–26.

Riordon, P.H., and Laliberté, R., 1957, "Asbestos Deposits of Southern Quebec," *Excursion B-08 Guidebook*, 24th International Geological Congress, Canada, pp. 1–21.

Robinson, K., and Shaw, E.R.S., 1952, "Summarized Proceeding of a Conference on Structures of Silicate Minerals (November 1951)," *British Journal of Applied Physics*, Vol. 3, Sep., pp. 277–282.

Rosato, D.V., 1959, *Asbestos, Its Industrial Applications*, Reinhold, New York, pp. 198–199.

Ross, M., 1987, "Minerals and Health: The Asbestos Problem," Special Paper, *Aggregates to Zeolites in Arizona and the Southwest*, H.W. Peirce, ed., State of Arizona, Bureau of Geology and Mineral Technology, pp. 83–89.

Rowbotham, P.I., ed., 1970, "World Asbestos Industry," *Industrial Minerals*, No. 28, Jan., pp. 17–29.

Rozovsky, H., 1957, "Air in Asbestos Milling," Ventilating Conference, Keller Center for Continuing Education, Michigan State University, East Lansing, MI.

Russell, A., 1990, "Asbestos—Cement Markets," *Industrial Minerals*, No. 278, London, pp. 39–57.

Sinclair, W.E., 1959, *Asbestos-Its Origin, Production, and Utilization*, Mining Publications, Ltd., London, 512 pp.

Skinner, H.C.W., Ross, M., and Frondel, C., 1988, "Asbestos and Other Fibrous Materials," Oxford University Press, New York, 204 pp.

Smirnov, V.I., 1971, *Essays on Metallogeny*, E.A. Alexandrov, trans., Queens College Press, Flushing, NY, 96 pp.

Smitheringale, W.V., 1957, "The Mine of Cassiar Asbestos Corporation Limited, Cassiar, B.C.," *The Geology of Canadian Industrial Mineral Deposits*, 6th Commonwealth Mining and Metallurgical Congress, Toronto, pp. 49–53.

Speil, S.S., and Leineweber, J.P., 1969, "Asbestos Minerals in Modern Technology," *Environmental Research*, Vol. 2, No. 3, Apr., pp. 166–208.

Stehno, J., and Fronek, R., 1977, "The Soviet Asbestos Deposits. A Guarantee for the Continuing Development of the Czechoslovakian Asbestos Cement Industry," *Stavivo* (Building Materials), No. 11, pp. 430–432. (Translation from Czechoslovakian technical journal.)

Stewart, P., French, G., and Anthony, D., 1990, "Baie Verte Wet Process Plant," *Industrial Minerals*, No. 273, London, pp. 51–55.

Stewart, R.V., 1978, "Geology and Evaluation of the Asbestos Hill Orebody," SME Preprint No. 78-H-69, AIME Annual Meeting, Denver.

St. Julien, P., 1967, "Tectonics of Part of the Appalachian Region of Southeastern Quebec," Special Publication 10, Royal Society of Canada, pp. 41–47.

Straw, D.J., 1955, "A World Survey of the Main Chrysotile Asbestos Deposits," *Canadian Mining & Metallurgical Bulletin*, Vol. 48, pp. 610–630.

Streib, W.C., 1978, "Asbestos," *Kirk-Othmer: Encyclopedia of Chemical Technology*, Vol. 3, 3rd ed., R. Kirk and D. Othmer, eds., John Wiley & Sons, Inc., New York, pp. 267–283.

Tallman, D.N., and Pahlman, J.E., 1987, "MgO Filtration Research," US Bureau of Mines, Information Circular 9138, pp. 32.

Van Biljon, W.J., 1964, "The Chrysotile Deposits of the Eastern Transvaal and Swaziland," *Geology of Some Ore Deposits in Southern Africa*, Vol. 2, Geological Society of South Africa, pp. 625–669.

Vokes, F.M., 1964, "Asbestos Bearing Claims on Troodos," US Special Fund Project, Cyprus, United Nations unpublished report.

Warren, B.E., 1932, "Structure of Asbestos—An X-Ray Study," *Industrial & Engineering Chemistry*, Vol. 24, No. 4, pp. 419–422.

Warren, B.E., 1942, "X-Ray Study of Chrysotile Asbestos," *American Mineralogist*, No. 27, p. 235.

Warren, B.E., and Bragg, W.L., 1928, "The Structure of Diopside," *Zeitschrift fuer Kristallographie*, Vol. 69, pp. 168–193.

Warren, B.E., and Hering, K.W., 1941, "The Random Structure of Chrysotile Asbestos," *Physical Reviews*, No. 59, p. 925.

Whittaker, E.J.W., 1952, "The Unit Cell of Chrysotile," *Acta Chrystalogica*, Vol. 5, pp. 143–144.

Whittaker, E.J.W., and Zussman, J., 1956, "The Characterization of Serpentine Minerals by X-Ray Diffraction," *Mineralogical Magazine*, No. 31, pp. 107–126.

Wilson, J.F., 1968, "The Mashaba Igneous Complex and Its Subsequent Deformation," *Symposium on Rhodesian Basement Complex*, Geological Society of South Africa Annexure, Vol. 71, pp. 175–188.

Yada, K., 1967, "Study of Chrysotile Asbestos by a High Resolution Microscope," *Acta Chrystalogica*, Vol. 23, pp. 704–710.

Yada, K., 1971, "Study of Microstructure of Chrysotile Asbestos by High Resolution Electron Microscopy," *Acta Chrystalogica*, Vol. A27, pp. 659–664.

Zussman, J., Brindley, G.W., and Comer, J.J., 1957, "Electron Diffraction Studies of Serpentine Minerals," *American Mineralogist*, No. 42, pp. 133–153.

BARIUM MINERALS

DONALD A. BROBST

The minerals barite ($BaSO_4$ — barium sulfate) and witherite ($BaCO_3$ — barium carbonate) are the chief commercial sources of the element barium and its compounds, whose many uses are nearly hidden among the technical complexities of modern industrial processes and products.

Barite, the major ore mineral, is extremely vital to the petroleum industry, which in 1990 consumed about 90% of the world's production of 4.9 Mt as a major ingredient of the heavy fluid called mud that is circulated in rotary drilling of oil and gas wells. The remaining 10% of the barite production was consumed chiefly in the manufacture of barium chemicals and glass and as a pigment, filler, and extender. Barite is found throughout much of the world and is available from three major geologic types of deposits—vein and cavity-filling, residual, and bedded—in sufficient quantity at competitive prices to meet current demands. The world's demand for barite is expected to increase because of the world's need for oil and gas. Geologic circumstances are favorable for the discovery of new deposits of commercial value.

Witherite, desirable as a raw material for the production of barium chemicals, is much less common and abundant than barite.

GEOLOGY

Mineralogy

Barite ($BaSO_4$ — barium sulfate), also called barytes, tiff, cawk, or heavy spar, is the most common and abundant ore mineral of barium. Pure barite contains 58.8% barium (or 65.7% BaO and 34.3% SO_3) and has a calculated specific gravity of 4.5, but inclusions in natural barite may reduce this value considerably. Barite crystallizes in the dipyramidal class of the orthorhombic system; well formed crystals are generally tabular and have three cleavages, the most perfect of which is parallel to {001}. The barite of many deposits, however, has an uneven fracture and an apparent cleavage caused by separation along planes between successively deposited layers. The barite in most commercial deposits occurs as irregular masses, concretions, nodules, rosette-like aggregates, and in laminated to massive beds of fine crystallinity. The mineral occurs in many colors, although shades of white to dark gray and black are most common in commercial deposits. The hardness is 2.5 to 3.5 on Mohs scale (in which the hardness of a fingernail is about 2). Commercially, the terms "hard" and "soft" refer to the ease with which barite can be ground. Barite is relatively insoluble in water and acid and thus can be used as a chemically inert material.

Barite is most commonly associated with the following minerals: quartz, chert, jasperoid, calcite, dolomite, siderite, rhodochrosite, celestite, fluorite, and various sulfide minerals such as pyrite, chalcopyrite, galena, sphalerite, and their oxidation products. Residual deposits of barite are included in residual ferruginous clays that make up much of the deposit. Barite is a common gangue mineral in many types of ore deposits, especially veins that are mined principally for other commodities, especially lead, zinc, gold, silver, fluorite, and rare earth minerals. Barite is a common constituent of some large siderite-rich iron deposits in Europe, but deposits of this type have not been found in the United States.

Barite in vein and residual deposits may contain as much as several percent strontium, which substitutes for barium in the crystal structure because of the close similarity in the ionic radius of the two metals in their bivalent state. Other metals also occur in trace amounts in barite, but the suites of elements and the amounts of each generally are smaller in the bedded deposits, particularly those in the barite of bedded deposits not associated with metal sulfides.

Witherite ($BaCO_3$ — barium carbonate) contains 70% barium (or 77.7% BaO and 22.3% CO_2), and it has a calculated specific gravity of 4.2 and a hardness of 3 to 3.5 on the Mohs scale. Like barite, this mineral crystallizes in the dipyramidal class of the orthorhombic system. The cleavage parallel to {010} is distinct, to {110} is imperfect. The fracture is uneven and the luster is vitreous to resinous. Crystals or complex masses are colorless to milky white or gray or have tints of yellow, brown, or green. Witherite is a minor accessory mineral in some barite deposits. Witherite is highly desirable as a source for the production of barium chemicals because of its solubility in acid. Only the mine at El Portal, Mariposa County, CA, has produced witherite in commercial amounts domestically, but operations ended about 1950. The world's most productive deposits of witherite in Northumberland, England, went out of production in the late 1980s. Discoveries of large deposits of witherite have been reported in China (Wang and Li, 1991). It seems that commercial deposits of witherite constitute only a small part of the world's resources of barium.

Sanbornite ($BaSi_2O_5$ — barium silicate) contains 50% barium and is of potential interest as a source of barium for chemicals. Unlike many silicate minerals, sanbornite is soluble in acid, a fact that permits the recovery of barium by further chemical processing. The mineral, formerly considered rare, has been found in abundance with other barium silicate minerals in contact zones with metamorphic rocks, mostly quartzite, in roof pendants of the Sierra Nevada batholith in Fresno County, California (Matthews and Alfors, 1962). The resource potential of sanbornite is virtually unstudied.

Barium occurs in other minerals, but in concentrations so low that none of these minerals is likely to be a potential source of the element soon.

Types of Deposits

Barite occurs in many geological environments in sedimentary, igneous and metamorphic rocks. The commercial deposits of barite are classified geologically into three types—vein and cavity filling deposits, residual deposits, and bedded deposits—that may be further subdivided into eight categories (Clark, Gallagher, and Poole, 1990).

Vein and Cavity-filling Deposits: The vein and cavity-filling deposits are those in which barite and associated minerals occur along faults, gashes, joints, bedding planes, breccia zones, and solution channels and in various sink structures. The deposits in solution channels and sink structures are most common in limestone. The host rocks are Precambrian to Tertiary in age. Many deposits in the western states are associated with igneous rocks of Tertiary age, although notable exceptions are vein deposits of barite and rare earth minerals associated with igneous rocks of Precambrian age at Mountain Pass, San Bernardino County, California, and in the Wet Mountains, Custer County, Colorado. In most of the eastern and midwestern states, an association of these deposits with igneous activity is not obvious.

Most of the barite in the vein and cavity-fillings is dense, white to gray, and is associated with other minerals including those of the metallic elements already mentioned. The grade of the barite ore differs by deposit and within deposits. The deposits commonly have sharp contacts with wall rocks; large-scale replacement of the host rocks beyond ore-controlling structures is rare. Within individual districts, the deposits commonly are scattered and irregular and range in thickness from a few centimeters to a few tens of centimeters and in length from a few to hundreds of meters. Barite cementation of breccia zones forms thin deposits of small extent in the eastern United States.

Rich but small barite deposits in collapse and sink structures are common in central Missouri and in the Appalachian states.

On weathering, vein and cavity-filling deposits may form valuable bodies of residual ore because of the insolubility of barite in the weathering regime.

The barite and other minerals of the vein and cavity-filling deposits are typical of an epithermal suite precipitated from low-temperature hydrothermal solutions. Such an origin is postulated for most vein deposits containing barite in the western states. Evidence has accumulated to suggest that vein deposits of barite and base metal sulfides in the midwestern and eastern states might have formed from circulating connate brines as well as near-surface waters (Leach, 1980, Cathels and Smith, 1983, Kesler, Gesink, and Haynes, 1989).

Residual Deposits: Residual barite deposits are found in unconsolidated material and are formed by weathering of preexisting deposits. Many residual deposits of commercial value in the United States lie within the clayey residuum derived from Cambrian and Ordovician limestone and dolomite. Classic examples of these deposits are found in southeastern Missouri (Kaiser et al., 1987) and in the Appalachian region from Pennsylvania to Alabama (Clark, 1987), but especially in the Sweetwater District, Tennessee (Maher, 1970), and in the Cartersville District, Georgia (Kesler, 1950).

Most of the residual barite is white and translucent to opaque; it occurs in mammillary, fibrous, or dense fine-grained masses, but subhedral to euhedral crystals also occur. The barite in these deposits ranges in size from microscopic particles to irregular boulders weighing hundreds of pounds. Most of the masses, however, are 2 to 15 cm in diameter. Chert, jasperoid, and drusy quartz are common in many deposits. Small amounts of pyrite, galena, and sphalerite occur in or on some of the barite, and locally the lead and zinc minerals have been recovered as a byproduct. Incompletely weathered rock fragments and yellow, red, or brown illitic clays typically make up the rest of most deposits.

The grade of these residual barite ores varies greatly. Much residuum in commercial deposits contains less than 10 to about 20% barite.

The size and shape of the residual barite deposits also vary greatly. Some of the larger deposits in Georgia, Missouri, and Tennessee extend over many hectares. The shape of residual deposits generally reflects that of the original deposit. Deposits derived from solution channels and vein systems tend to be elongate and those derived from sink structures tend to be circular.

The depth of residual deposits differs according to local conditions. Some residual deposits in Georgia terminate at the contact with bedrock as much as 45 m beneath the surface. The deposits in Missouri terminate at bedrock that in many places is only 3 to 5 m beneath the surface. The barite may be distributed randomly through the residuum or concentrated near the contact with bedrock. The contact of the residuum and the bedrock is irregular. Pillars or ribs of bedrock that protrude into, or even through, the residuum are common in many deposits.

The bedrock in many areas of residual barite deposits contains some veins filled with barite, fluorite, calcite, and quartz and locally some sulfide minerals including pyrite, chalcopyrite, galena, and sphalerite. In other deposits, only thin veinlets or disseminated lumps and pods of barite occur in the carbonate bedrock.

Bedded Deposits: Bedded deposits include those in which barite occurs as a principal mineral or cementing agent in stratiform bodies in layered sequences of rock. The term also includes the deposits other workers have called *stratiform* (to describe the geometry of the body) or *stratabound* (to connote either confinement to a specific stratigraphic unit or, in some cases, a specific genesis). In the older literature, the term *stratabound* implied epigenetic replacement, but now it has come to mean a syngenetic origin that covers all processes from early diagenesis through epigenetic remobilization (de Brodtkorb, 1989).

The bedded deposits of barite are the most commercially valuable and sought of all types because they tend to be high grade and individual deposits may contain several million tons of ore. In the past few decades many new deposits have been found all over the world. Most of these deposits occur in sequences of sedimentary rocks characterized by abundant chert and black siliceous shale and siltstone ranging from Precambrian to Tertiary age but especially in rocks of early to mid-Paleozoic age (Ramos and de Brodtkorb, 1989). The barite occurs in beds or zones from a few centimeters to 30 m thick. Exposures of these zones or beds may be traced in some deposits for more than 1 km. The deposits may extend over many hectares. The individual beds are massive to laminated and fine grained (0.05 to 1.5 mm in diameter). Individual beds may contain as much as 50 to 95% barite that can be used with little or no beneficiation other than grinding. Rosettes or nodules of barite with concentric or radiating bladed structures may make up substantial parts of some beds. Conglomeratic beds consisting of nodules or fragments of barite, phosphate (apatite), chert, and rock fragments in a finer-grained matrix of barite occur in some deposits. Organic carbonaceous material is common in many beds: the odor of H_2S is commonly detected when the barite is struck by a hammer. Carbonate minerals are uncommon in many deposits.

These bedded barite deposits occur in association with large concentrations of base metal sulfides, commonly known as "massive sulfide" deposits, or peripherally to them, as in the Red Dog deposit in the Brooks Range, northern Alaska (Moore et al., 1986), at Meggen, Germany (Krebs, 1981), and in the Selwyn Basin, Yukon Territory, Canada (Morin, 1986). Other barite deposits in the black siliceous shale environment such as those in Arkansas (Zimmermann and Amstutz, 1989), Nevada (Papke, 1984, Poole, 1988, Jewell and Stallard, 1991), and the Qinling District, China (Wang and Li, 1991), are unrelated to massive sulfide deposits.

Maynard and Okita (1991) divided the black bedded barite deposits in siliceous rocks into two types based on their stratigraphic succession, tectonic setting, and geochemistry. They proposed the name "cratonic-rift type" for those hosted by coarse clastic rocks associated with turbidity currents and deposited in intercratonic rifts with strong local relief. Volcanic rocks occuring with these deposits generally are alkalic basalts. Deposits of this group are commonly associated with economically valuable concentrations of base metal sulfides formed from submarine volcanic exhalations. The second group of deposits, the ''continental-margin type,'' consists of barite interbedded with chert and siliclastic rocks formed in small ocean basins lying between a passive continental margin and a subduction zone. This group does not have associated sulfides of lead and zinc. Geochemical data suggest considerable influence of oceanic crust. The cerium geochemistry and the distribution of organic carbon and phosphorous, more abundant in the barite-rich than in the chert- and shale-rich parts of the rock sequence, suggest to Jewell and Stallard (1991) that coastal upwelling in a Tethyan-type ocean is associated with the formation of these barite beds and that such an upwelling favors a biogenic origin for the barite.

Other important bedded barite deposits occur in volcanic piles, such as those of the Kuroko-type in Japan (Marumo, 1989) and at Buchans, Newfoundland, Canada (Dawson, 1985).

Distribution of Deposits

United States: *Nevada.* According to Papke (1984), 120 deposits of bedded barite and 61 deposits of barite in veins are widely distributed in 11 counties. Most of the estimated more than 8 Mt of barite produced through 1990 has come from bedded deposits in the Tuscarora and Independence Ranges, Elko County, the Shoshone Range, Lander County, and the Toquima Range, Nye County. Many large, high grade bedded deposits occur in sequences of sedimentary rocks composed chiefly of black chert and carbonaceous shale and siltstones in the Slaven Chert of Devonian age, but other valuable deposits occur in the Vinini Formation, part of a similar sequence of rocks of Ordovician age. Barite has been extracted from 88 bedded deposits, and 10 have yielded ¼ to 2 Mt of product. Twenty-seven vein deposits have yielded barite, but not nearly in the quantity as that obtained from the bedded deposits.

Missouri. The major deposits of Missouri are clustered in the Washington County area, about 80 km southwest of St. Louis (Kaiser et al., 1987). This area, commonly considered part of the Southeast Missouri Lead District, had yielded about 12 Mt of barite through 1987. The barite is concentrated in residue overlying Potosi and Eminence dolomites of Cambrian age. Single areas of many hectares may be ore-bearing, but the richest parts of the deposits form ''runs'' or ''leads'' 3 to 6 m wide and 100 m long that generally lie within 5 m of the surface. Some deposits contain as much as 25% barite, but much residuum contains less than 10%.

The several hundred small barite deposits of the Central District, chiefly in Cole, Miller, Moniteau, and Morgan Counties, are mostly in Gasconade and Jefferson City dolomites of Ordovician age and Burlington limestone of Mississippian Age. These so-called circle deposits are bell-, flask-, or cone-shaped in the vertical dimension and are approximately circular in plan.

Arkansas. The Chamberlain Creek syncline near Malvern, Hot Spring County, contained the largest single deposit of bedded barite known in the United States. An estimated 9 Mt of barite was produced from 1942 to 1982 from a zone of black siliceous rocks about 18 m thick in the lower part of the Stanley Shale of Mississippian age. The ore grade varied, but the richest parts of the body contained more than 70% barite. This and other bedded deposits, such as Fancy Hill and Gap Mountain, and some vein deposits in the Stanley Shale and the underlying Arkansas Novaculite (Devonian and Mississippian age) in Montgomery, Pike, and Polk Counties have been discussed in detail by Scull (1958), Howard and Hanor (1987) and Zimmermann and Amstutz (1989). Barite that locally cements parts of the Pike Gravel, the basal formation of the Trinity Group of Cretaceous age, was mined in Howard and Sevier Counties.

Georgia. For many years, barite has been mined from the residual deposits of the Carterville District, Bartow County (Kesler, 1950, Reade et al., 1980). The deposits are associated with the residuum of the Weisner, Shady, and Rome Formations of Cambrian age. The general features of the deposits are similar to those of the other residual deposits already described, except that some barite has been recovered from pits as much as 45 m deep. The Cartersville District has produced most of the more than 4 Mt of barite mined through 1987 in Georgia. Residual deposits in Polk, Cherokee, Gordon, Murray, and Whitfield Counties also have been worked from time to time.

Tennessee. An estimated 1.5 Mt of barite has been mined through 1987 in Tennessee, mostly from the residual deposits of the Sweetwater District that lies in parts of Loudon, McMinn, and Monroe Counties (Maher, 1970). The barite is associated with the residuum overlying the upper part of the Knox Group of Ordovician age. The bedrock contains thin veins and shatter zones filled with barite, fluorite, pyrite, quartz, and calcite. The large deposits cover several hectares and are as much as 18 to 25 m thick. Other residual deposits are in the Fall Branch-Greenville District in northeast Tennessee. In the Del Rio District, Cocke County, barite occurs with pyrite, fluorite, chert, quartz, hematite, ankerite, and dolomite in veins and replacements of fault breccia associated with thrust faults involving sedimentary rocks of Precambrian and Cambrian age.

Veins of barite, fluorite, and calcite with smaller amounts of sphalerite, galena, pyrite, and local traces of chalcopyrite in host rocks of Ordovician age in the Central Tennessee District now worked for base metals might offer a possible barite byproduct. The nature and origin of these veins are discussed by Gratz and Misra (1987).

California. Barite occurs in more than 150 localities in 22 counties of California, but more than 90% of the estimated nearly 1 Mt of barite produced in the state came from five deposits in the Sierra Nevada region (Pemberton, 1983). Some of the state's production recently has come from the bedded deposits in rocks of Devonian age on Girard Ridge in the Klamath Mountains, Shasta County (Weber and Matthews, 1967). Barite also occurs in the silver-bearing veins of the Calico District, San Bernardino County. In the Mountain Pass District, San Bernardino County, rare earth minerals are mined from large reserves in carbonate bodies that also contain 20 to 25% barite that is recoverable as a byproduct (Olson et al., 1954).

Alaska. The widespread barite deposits of Alaska are massive and nodular replacement bodies and major or gangue minerals in both hydrothermal veins and bedded deposits of various ages (Clark and Poole, 1989). Most of Alaska's barite production of about 1 Mt came from the Lime Point deposit on Prince of Wales Island and the Castle Island deposit, at and below tide level, near Petersburg. Of considerable interest is the barite associated with large bedded deposits of base metal sulfide in sedimentary rocks of Mississippian and Pennsylvanian age on Red Dog and Drenchwater Creeks in the Brooks Range (Lange et al., 1985, Moore et al., 1986). Barite masses as thick as 30 m occur in the Red Dog deposit.

Kelley et al. (1993) described seven recently discovered stratiform deposits of barite interbedded with chert, limestone, and black shale of Mississippian age in the remote Howard Pass region of the northwestern Brooks Range, AK. The barite is high grade (about 95% barium sulfate) and contains little iron and titanium and neither visibly associated metallic sulfides nor precious metals. Barite of this quality is especially suited to the manufacture of barium chemicals or use in glass, fillers, extenders, and pigments. The indicated resources of these deposits alone are estimated to contain about 50 Mt of barite that currently are about one-third of the identified resources in the United States.

Other States. Barite deposits are scattered in many other states.

Information on deposits in Paleozoic sedimentary rocks in 15 east-central states from Georgia to Wisconsin has been compiled (Clark, 1987). The barite deposits of Kentucky are described by Anderson, Trace, and McGrain (1982). Horton (1990) considered significant the barite resources of the Kings Mountain area near Charlotte, NC. Barite veins are associated with some early Mesozoic basins of the eastern United States (Robinson and Woodruff, 1988). Many barite deposits of all types occur in the western states, Arizona, Colorado, Idaho, Montana (Berg, 1988), New Mexico, and Utah. Barite has been produced in recent years from vein and bedded deposits in Stevens and Pend Orielle Counties, WA (Moen, 1964).

Foreign Deposits: About 150 deposits and occurrences of barite are found in almost all provinces of Canada, but those of recent economic interest are mostly in Newfoundland, Nova Scotia, Ontario, and British Columbia (Andrews and Collings, 1990). Additional significant deposits of bedded barite, some associated with base metal sulfides, occur in clastic rocks of Cambrian to Mississippian age in the upper Proterozoic and Paleozoic cordilleran miogeocline in northern British Columbia and the Selwyn Basin of the Yukon and Northwest Territories (Morin, 1986).

In Mexico, large vein deposits in the Muzquiz District, Coahuila, and the Galeana District, Nuevo Leon, have yielded much barite. Bedded barite in siliceous marine rocks of mid-Paleozoic age is produced at the Barita de Sonora mine, near Mazatan, Sonora (Clark and Poole, 1989).

In South America, the vein deposits of Peru and the bedded deposits of Camamu Bay, Brazil (Cassedanne, 1989) have been the recent major producers. The 16 districts containing bedded deposits of barite and celestite in Mesozoic rocks of the Neuquen-Mendoza Basin, Argentina, are of economic interest (de Brodtkorb, Shalamuk, and Ametrano, 1989). Bedded deposits also have been described in early Cretaceous rocks in the Atacama region of northern Chile (Diaz, 1986).

In Europe, barite has been produced from all types of deposits, but especially from bedded deposits in recent years (Fuchs, 1989). Major producers include Germany, (Robbins, 1985), France, Ireland, Sardinia (Padalino, Tocco, and Violo, 1989), Greece, Turkey, the former USSR, and the UK.

Vein deposits in Morocco and Algeria have been the largest sources of barite in Africa. Vein deposits also occur in many other countries. Bedded deposits of Archean age have been described in South Africa (Reimer, 1980).

In Asia, barite has been mined principally in China, India, Iran, Japan, Pakistan, the Philippines, and Thailand.

China became the world's largest producer of barite during the 1980s. Old mining areas are in Guangxi, Fujian, Hubei, and Shandong (Gloria, Harnisch, and Braumann, 1985). The Xiangzhow mine is the largest producer in Guangxi where pods in limestone yield white barite. In south China, large stratiform deposits of barite and witherite have been found in Lower Cambrian black shales and chert in the Qinling geosyncline and the Yantgtze platform to the south (Wang and Li, 1991). These are the largest concentrations of witherite to be discovered in many decades.

The large reserves and resources of the bedded deposits in Proterozoic rocks in the Mangampeta area, Cuddapah District, Andhra Pradesh, dwarf most of the other barite deposits of India (Bose and Vaidyanathan, 1979).

Many stratabound sulfide deposits containing barite and fluorite occur in carbonate rocks of Mesozoic age in a large, complexly evolved terrane that extends from Pakistan across Iran and lies between the Arabian, Indian, and Eurasian plates (Jankovic, 1984).

Barite occurs with bedded sulfides in the Kuroko-type areas of Japan. The deposits of the Hokuroku District and those in southeastern Hokkaido are major domestic suppliers of barite (Marumo, 1989).

In Oceania, barite deposits occur in Australia, New Zealand, and the Fiji Islands. Major production has been from vein deposits of Precambrian and Cambrian age in the Flinders and Mount Lofty Ranges, South Australia. A large deposit in the North Pole mining center, Pilbara District, Western Australia, consists of barite originally deposited in sedimentary rocks of Archean age that has been mobilized and injected into fractures in an overlying series of volcanic rocks domed up by a granitic pluton (Hickman, 1983).

The US Bureau of Mines reported that 45 nations produced barite in 1989.

Reserves and Resources

In 1993, the United States reported barite reserves of 30 Mt in a reserve base of 60 Mt. World reserves were reported at 170 Mt in a reserve base of 500 Mt. The identified resources of barite in the United States were estimated at 170 Mt, of which more than 50 Mt are in Alaska. The identified barite resources of the world were an estimated 500 Mt. The distribution of the world's major resources of barite were reviewed by Clark, Gallagher, and Poole (1990).

The current status of barite reserves, reserve base, and resources and the meaning of the terms is reviewed in the barite chapter of the *Mineral Commodity Summaries*, an annual publication of the US Bureau of Mines.

TECHNOLOGY

Exploration

The success of future exploration will be increased if prospecting is guided by a knowledge of the geologic association and the role of barium in various geologic environments, especially in the sedimentary environment (Large, 1980, Cathels and Smith, 1983, Lange et al., 1983, Brobst, 1984, Lange et al., 1985, de Brodtkorb, 1989, Kesler, Gesink, and Haynes, 1989). Many deposits of dark-gray bedded barite are large and worthy targets for exploration. The beds of dark barite may be overlooked in the field because they are easily confused with dark impure limestone, except for the noticeably greater weight of the barite-rich rocks. No longer should the search for barite deposits of commercial value be guided only by a close association with intrusive igneous activity. With the apparent positive correlation between bedded barite deposits and organic materials, the possibility of finding barite deposits at the sites where oxygenated and anoxic waters mixed certainly should be considered (Miller, Brobst, and Beck, 1977, Jewell and Stallard, 1991). Careful mapping of barite occurrences, phosphate concentrations, and the cerium geochemistry may be effective tools in the successful search for bedded barite deposits in siliceous rocks not associated with base metal sulfides. With so much information available, many explorationists are turning to the assistance of computer models as tools in the search (Orris, 1986).

Geophysical techniques to explore for barite deposits have not been used widely. Gravity surveys for residual barite deposits were undertaken in Missouri (Uhley and Scharon, 1954), where they indicated that large concentrations of barite could be outlined and the tonnage computed with an expected error of 35% or less. In eastern Germany, the use of geophysical techniques to search for hidden deposits of barite and fluorspar was described as less than ideal (Klaus and Schroder, 1967).

Useful geochemical techniques are available for exploration. Trace element analysis of rocks and minerals can be carried out by inductively coupled plasma spectrometer, atomic absorption, emission and X-ray fluorescence spectrography, and various wet chemical tests. A relatively simple and inexpensive turbidimetric test to detect small amounts of barium in rocks and soils was applied in

Arkansas with satisfactory results (Brobst and Ward, 1965). The use of a portable radioisotope fluorescence analyzer in geochemical prospecting for barite and celestite in western England has been described by Ball et al. (1979).

There is an excellent chance for the discovery of new, valuable deposits of barite in the coming years by the application of the geologic research reported in recent years. Bedded deposits might well be found in places where they are now unknown or virtually unsought, as in the Appalachian region (Neulle and Shelton, 1986).

Evaluation of Deposits

Evaluating the commercial potential of a specific barite prospect is a complex process that requires careful study of the relation of pertinent geologic, technologic, and economic factors. Geologic factors to consider are the type of deposit and its structure, size, grade, and mineralogy. Knowledge of zoning, texture, grain size, and chemical composition of the ore minerals and the associated gangue is also critical to the geologic evaluation. Geologic information puts limits on technical factors such as the choice of mining and beneficiation methods to be used. Economic factors of major importance include: location of the deposit with respect to current and future markets and transportation to them, costs of land acquisition, costs of labor and equipment to mine, beneficiate, process the ore, and produce a marketable product, and problems of royalties, financing, and taxes.

Barite prospects generally are evaluated geologically by detailed geologic mapping and surveying and by digging test pits and trenches for further examination and sampling of the rock body. Drilling is commonly used for evaluation of vein and bedded deposits and also has been used successfully in evaluating residual deposits (Kesler, 1949).

Beneficiation tests must be made to assure that the anticipated mineral separation and recovery are attainable and that barite products can be made profitably to meet the specifications of the purchaser. In some cases, barite is only one of the salable products recoverable from a deposit. Some barite users do not like to buy byproduct barite. They fear that the availability of the barite is unsteady because its production is tied to the demand for the mine's primary product. Byproduct barite probably will find greater acceptance in some future markets. Its potential value should not be ignored when estimating the economic value of an ore deposit.

Mining and Beneficiation

Residual barite deposits generally are mined in open pits. After the overburden is stripped away, the barite-bearing residuum is dug with 1 to 1.5 m^3 power shovels, although draglines and front-end loaders are used in some mines. The ore is loaded into 15- to 30-ton off-the-road dump trucks for hauling to the washer plant, rarely more than 5 km from the operating pit.

The ore-bearing residuum is dumped into a hopper from which it is fed to rotary breakers using water at high pressure. The softer barite breaks up more readily than the gangue materials; the barite moves on through the circuit, but the coarse gangue is diverted to the dump. The barite and remaining gangue move on to the log washers, which separate the clay from the gravel and the barite. The clay particles flow away, by gravity, to the tailings pond. A secondary rotary breaker or trommel screen receives the product of the log washer. Oversized material from the trommel is byproduct used for road building. The baritic material that passes the trommel screen is moved to jigs, which separate, by gravity, barite from gravel. The gravel-sized waste from this step is used as road building material. The barite product is then ready for grinding or other processing. A typical flowsheet for this process is shown in Fig. 1.

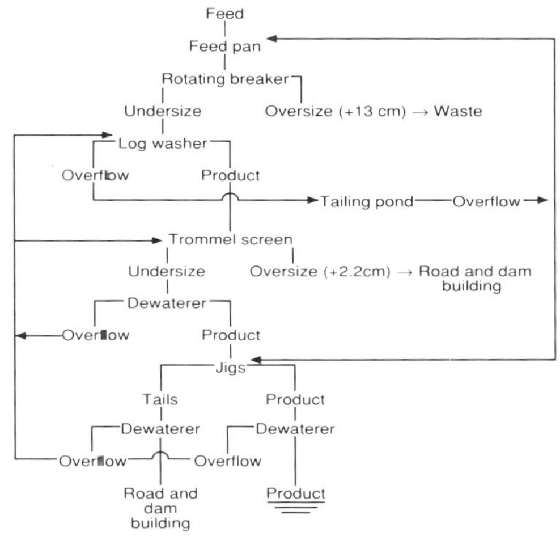

Fig. 1. A flowsheet for the washing and concentration of barite from residual deposits in Missouri (Sackett, 1962).

Washer plants vary in size, but the capacity generally ranges from about 50 to 90 m^3/h of residuum. As much as 0.3 m^3/s of water is used in large washers.

It is said that 60 to 112 kg of barite is recovered per cubic meter of residuum measured in the ground, but the efficiency of the plants differs greatly. The recovery of the barite in the jig product ranges from 65 to 85% (Wharton, 1972).

Conventional crushing equipment is not currently used in the beneficiation of residual barite in Missouri because the beneficiation process itself accomplishes this mission. The major component of these ores is clay, but the solid fragments other than the barite and dolomite are hard abrasive forms of silica: chert, chalcedony, and drusy quartz. In the normal agitation of the beneficiation process, the soft and brittle barite is dislodged from the other materials. Considerable amounts of barite are lost with the coarse gangue, but about two-thirds of the total losses are believed to travel with the slimes to the tailings ponds. Perhaps 20 to 30% of barite in the original ore is lost and comes to rest in the tailings ponds. Wharton (1972) estimated that 1.7 Mt of barite might be available in 67 tailings ponds in the Washington County District, Missouri, an amount equal to about 14% of the recorded production of the district through 1987.

Bedded and vein deposits may be mined by either open pit or underground methods, depending on the characteristics of each deposit. Open pit methods are easier and less costly than methods generally employed underground. At Chamberlain Creek, near Malvern, AR, the large bedded deposit has been mined by both open pit and underground methods. The bedded deposits of the Shoshone Range, Nevada, are mined in open pits where blasting is not always required because the rocks in many deposits have been tectonically deformed and broken.

Barite from many vein deposits requires concentration to meet the buyer's specifications. It depends upon the suite of minerals in the deposit, but beneficiation by simple, inexpensive gravity separation is generally ineffective, and more complicated and expensive flotation techniques are needed. The long troublesome history of the attempt to separate intergrown barite and fluorite has been solved by flotation methods developed by Eddy and Browning (1964). Many vein deposits containing sulfide minerals could yield barite to meet some commercial specifications as a coproduct.

The ore from many bedded deposits requires no beneficiation other than grinding to meet some user's specifications, especially

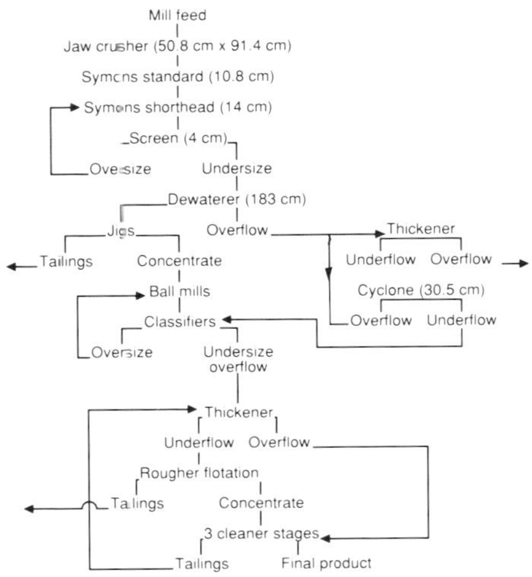

Fig. 2. A flowsheet for the concentration of bedded barite by flotation methods. The diagram is modified from Sackett (1962).

the requirements for drilling mud, but ore from many deposits needs some concentration. The bedded ore from the deposit near Malvern, AR, was beneficiated by double crushing, jigging, and wet grinding in ball mills to pass a 44 micron screen; flotation of barite with anionic reagents followed according to the general scheme shown in Fig. 2 (Sackett, 1962). In 1991, however, barite for drilling mud was no longer produced by flotation processes in the United States.

Barite is available from producers as crude lumps or a powder. For many uses the lump or jigged concentrate must be ground to meet the user's specifications. Grinding is done either wet or dry, but much barite is ground wet because it is still wet from prior steps of beneficiation, or will be wet again in further steps of processing, as in bleaching with sulfuric acid to remove objectionable iron stains. Barite is ground in ball mills if iron contamination is unimportant. Raymond mills are now in wide use for grinding barite. Most ground barite is packed in paper bags holding about 45 kg, except bleached barite which may be packed in bags of about 25 kg.

Preparation of Chemicals

The first step in the chemical preparation of barium compounds is the roasting of hard lump barite with carbon in a rotary kiln at a temperature of about 1 200°C. This process reduces the barium sulfate, which is insoluble in water, to barium sulfide (black ash), which is soluble in water. The barium sulfide product is leached to yield a solution that, because of hydrolysis, contains $Ba(OH)_2$ and $Ba(SH)_2$. The solution is filtered to remove excess carbon and impurities. High-purity barite is needed for the roasting process because iron oxide, silica, and alumina can react with barium to form insoluble compounds. One percent iron oxide can convert as much as 4% barium sulfide to insoluble compounds.

The solution leached from the black ash is the basic raw material for making lithopone, blanc fixe, and other barium chemicals. If a solution of zinc sulfate is added, the resulting precipitate is a mixture containing about 30% ZnS and 70% $BaSO_4$. This precipitate is filtered, washed, dried, and calcined under carefully controlled conditions, quenched in water, wet ground, and then dried to yield the white pigment known as lithopone. Approximately one ton of barite is required to manufacture each ton of lithopone, even though the product is only 70% $BaSO_4$.

If sodium sulfate is added to the black ash solution, the chemically pure precipitate of barium sulfate, called blanc fixe, is formed.

If sodium carbonate is added to the black ash solution, the precipitate is barium carbonate. The precipitate may be filtered, washed, dried, and ground for use as the carbonate or for conversion to other barium compounds. Heating the carbonate in the presence of carbon yields barium oxide, which then can be oxidized to barium peroxide or hydrated to barium hydroxide. The nitrate and chloride of barium generally are prepared by treating barium carbonate with the corresponding acids.

Barium metal may be produced by electrolyzing barium chloride in the fused state or by the Guntz process (Johnstone and Johnstone, 1964), in which barium oxide is reduced by metallic aluminum or silicon at about 1 200°C under reduced pressure in an electric furnace.

As noted earlier, barium sulfate is virtually inert, but the water-soluble compounds are toxic if ingested.

Barium chemicals were produced domestically in 1991 by the Chemical Products Corp., Cartersville, GA.

Specifications

The barite product for use in drilling mud should have a specific gravity of at least 4.2 (about 90% $BaSO_4$) and a maximum of 250 mg/kg of water-soluble alkaline earth metals such as calcium, according the American Petroleum Institute (API) standard 13A of 1985. A few percent iron oxide is permitted. At least 90 to 95% of the ground barite should pass a 44 micron screen. Lundie (1986) and Garrett (1987) discussed the needs and reasons for meeting or exceeding API standard 13A. It is evident that special circumstances may require special care and testing to ensure the efficient and safe use of the mud. For these reasons, about half of all mud users allow only a maximum of 3 000 mg/l carbonate and 50 mg/l sulfide in caustic soluble form in the barite-mud product.

Glassmakers have their own rigid specifications for the barite used in their products: 96 to 98% $BaSO_4$; no more than 0.1 to 0.2% Fe_2O_3; and only a trace of TiO_2. Some specifications allow no more than 1.5% SiO_2 and 0.15% Al_2O_3. Sizing is one of the stringent requirements set by individual users. Most manufacturers want a product that will pass a 1.19 mm screen, but with as little as 5 to as much as 40% passing a 149 micron screen. Material of finer size is undesirable because it tends to ball up in the melt.

Specifications for barite to be used as filler, extender, and weighting agent are set by various users, but most manufacturers want a fine-grained product that is virtually all -44 microns. Color is important to many users.

The American Society for Testing and Materials publishes specifications for barite used in pigments (D602–42) that require a minimum of 94% $BaSO_4$ and no more than 0.05% Fe_2O_3, 0.2% water-soluble materials, 0.5% moisture, and 2% foreign matter. Specifications for blanc fixe are similar, except that a minimum of 97% $BaSO_4$ is required and a maximum of only 0.02% combined iron oxides is permitted. The manufacture of lithopone requires 95 to 98% $BaSO_4$ and no more than 1% combined iron oxides.

Manufacturers of barium chemicals want a raw material containing at least 95% $BaSO_4$, less than 1% combined iron oxides and 1% $SrSO_4$, and only traces of fluorine. The barite for use in chemical plants generally is sized between 0.84 and 16 mm, although some producers buy crude lump ore and grind it to their own specifications. Barite that is too finely ground is lost as flue dust and barite that is too coarse mixes poorly with carbon and yields poor results on roasting.

Most barite for heavy aggregate in concrete is gravel-sized material.

USES

Most of the world's barite production since 1926 has been used as a weighting agent for the muds circulated in rotary drilling of oil and gas wells. The muds fundamentally are mixtures of water, clay, and barite in different proportions that vary according to local reservoir conditions. Muds with a specific gravity as great as 2.5 are used. The mud is pumped down the hollow drill stem, passes through the bit at the bottom of the hole, and rises to the surface in the space between the drill stem and the wall of the hole. In the course of this circulation the drill bit is cooled, cuttings are removed, the drill stem is lubricated, the walls of the hole are sealed, and the hydrostatic head of the column of weighted fluid helps to confine high oil and gas pressures. The latter feature aids in prevention of gushers, thus reducing both environmental pollution and waste of oil and gas resources as well as conserving the natural rate of recovery of the products contained in the rocks. Barite is particularly well suited for drilling mud because it is clean, easy to handle, soft (nonabrasive), heavy, and virtually inert compared to many other available heavy materials. (See chapter on Well Drilling Materials.)

Barite is used in the manufacture of glass in continuous tanks. The addition of barite homogenizes the melt and gives greater brilliance and clarity to the finished glass.

Ground barite, both unbleached and bleached by sulfuric acid, is a common industrial filler, extender, and weighting agent. The rubber industry is a major consumer of barite as a filler. Barite also is added to bristolboard, heavy printing paper, playing cards, rope finishes, brake linings, clutch facings, plastics, and linoleum, to name but a few uses. Bleached barite has long been an extender in white lead paint because of its weight. The low index of refraction of barite makes its ability to cover marks poorer than some other substances, but its low capacity for absorption of oil is a good feature. Off-color or unbleached barite can be used as an extender in colored paints.

In the construction industry, some lump barite is used in concrete aggregate to weigh down pipelines buried in marshy areas and to shield nuclear reactors. Because barite absorbs gamma radiation well, its use reduces the amount of expensive lead shielding otherwise necessary in nuclear installations. The cavities of tires of heavy road construction equipment may be partly filled with a ground barite mixture to add weight as an aid in packing areas of earth fill.

Barium ceramics have some important industrial uses. Unusual electromechanical properties make barium titanate ceramics useful in the electronics industry. Barium ferrate ceramics are used in permanent magnets.

As a pigment, barite is used as the natural or bleached white sulfate, as blanc fixe, and as lithopone. Blanc fixe is used where higher purity is required than found in natural barite. Blanc fixe is probably best known as an indicator in medical X-ray photography.

A survey of the uses of barium chemicals listed more than 2 000 specific industrial applications in 17 major classifications (Anon., 1961). Various manufactured barium compounds, including carbonate, chloride, nitrate, oxide, peroxide, hydroxide, sulfide, sulfate, and some organic varieties, are widely used as reagents and catalysts: as in sugar refining; sizing, waterproofing, and fireproofing for textiles; drugs: in the manufacture of vitamins, hormones, and blood coagulants; various pyrotechnics, especially green flares; stabilizers for glue and starch; coagulation of synthetic rubber; plastics; insecticides, herbicides, and germicides; case hardening of steel; phosphors for fluorescent lamps; bonding phosphors to glass for black and white TV tubes; reagents for beneficiation of ores and paper manufacture; welding fluxes; additives for oil and grease; the melting and refining of magnesium; the recovery of indium; the recovery of zinc from dross; and as a scum preventative in the ceramic and brick industries.

Barium metal is used as a "getter" (deoxidizer) in degasifying TV and other vacuum tubes. Barium is alloyed with other metals such as aluminum, magnesium, lead, and calcium. An alloy of barium, lead, and calcium is used for bearings. Barium in an oxide system with yttrium and copper is of interest in research on superconductivity.

ECONOMIC FACTORS

Prices

Current prices for some barite products are available regularly in the monthly issues of *Industrial Minerals* (London). In May 1991, crude lump barite, specific gravity 4.25 FOB Morocco sold for $40/t and specific gravity 4.10 for $34/t. Barite ground to API standards sold for $75 to $80/t FOB the Gulf Coast.

The published prices in many cases are flexibly structured, depending on the desired specifications and the quantity needed by the purchaser. Most of the finished mud-grade product is sold with engineering services by the major companies, the largest of which are the Baroid Corp., M-I Drilling Fluids, and Milpark. Small producers commonly sell their barite to major users according to specially negotiated terms that generally vary somewhat from published prices.

Markets

The principal marketing areas for barite in the United States are the eastern, midwestern, and Gulf Coast states, and California. All of these areas use crude, crushed, or ground barite. Most of the barite for markets in the eastern states is mined in Georgia and Tennessee, and additional supplies from abroad arrive at East Coast ports. Most of the barite consumed in this region is used for chemicals and as a raw material in various manufacturing processes and products.

Domestic barite for drilling mud needed in the the Gulf Coast area has been mined for decades in Arkansas, Missouri, and Nevada. Domestic supplies in this major consuming region were augmented in the last decade by imports of crude barite, chiefly from China (86% in 1990), India, and Morocco, with lesser amounts from Canada and Mexico. In 1991, about a dozen grinding plants were operating along the Gulf Coast. For many years, mines in Missouri and Illinois have supplied the manufacturing industries of the Midwest. Mud-grade barite for the Rocky Mountain and Northern Plains regions in recent years has been mined in Idaho, Montana, Nevada, and Washington. Domestic barite for the West Coast markets is supplied by the mines of California, Nevada, and Washington. Some mud-grade barite has been produced in Alaska for use by the petroleum industry north of the Brooks Range.

The total domestic consumption of barite fluctuates annually with the rise and fall of drilling in the petroleum industry. Statistics in the *Minerals Yearbook*, compiled annually by the US Bureau of Mines, recount the history in detail. The high year of domestic production and import was 1981; when US production reached 2.5 Mt and imports 0.8 Mt and world production was 7.2 Mt. Since then, domestic production generally has been under 0.5 Mt, but imports have averaged about 1 Mt annually. About 90% of the barite has been used in the drilling industry, and the remainder goes into glass, chemicals, and many products as a filler and extender.

Transportation

Barite is a commodity of small unit value, and therefore, the cost of transportation to market is an important economic consideration. Long truck and rail hauls to market currently are prohibitively expensive in the United States. Marine transport, however, has been relatively cheap in recent years, which makes much foreign barite available in US markets, especially on the Gulf Coast, at less cost than the domestic product.

Substitutes

Substitutes for barite have been tried in drilling mud, but all are generally no better. Celestite ($SrSO_4$) has been rejected because of its higher cost and lower specific gravity (3.95 to 3.97). Iron ores are more abrasive and, because of the color, unpleasant to handle. Galena (PbS) is too expensive.

The use of lithopone as a pigment has nearly ended because of greater use of titanium dioxide pigments that have higher birefringence.

Barium compounds will continue to be used for most of their current industrial uses because they are cheaper than satisfactory substitutes, and they will be used increasingly in new, highly technical applications.

GOVERNMENTAL CONSIDERATIONS

Taxes and Tariffs

Barite producers have a depletion allowance of 14% on both foreign and domestic production. Some states have severance taxes.

On Jan. 1, 1990, the tariffs were \$1.25/t on crude barite and \$3.20/t on ground barite for Most Favored Nations (MFN) and \$3.94/t on crude barite and \$7.38/t on ground barite for non-MFN.

Barite and barium compounds are not included in the strategic materials stockpile.

Environmental Considerations

The barite industry has no especially difficult ecological problems to solve other than those attendant upon the mining and beneficiation of ores. Barite flotation plants and washers need tailings ponds to store their generally inert wastes. Open pit mines can be rehabilitated for other subsequent uses. Some toxic gases are generated in chemical plants, especially hydrogen sulfide, in the manufacture of barium chloride. The toxic gases can be withheld from discharge and the hydrogen sulfide is converted to other sulfide products.

For some years, Alaska has required that mud-grade barite contain no more than 3 ppm cadmium and 1 ppm mercury. This standard may soon be applied to California and the Gulf of Mexico states. Such potential regulations have been discussed in the *Federal Register* (Mar. 13, 1991, p. 10693, and Apr. 16, 1991, pp. 15349–15368). These standards can be met by barite from many sources, but especially from bedded barite deposits not associated with sulfide deposits, such as those in Nevada. This standard, however, precludes the use of barite from many vein and bedded deposits that contain associated metal sulfides, especially those with zinc, which commonly also contain higher than usual amounts of cadmium. Further proposed rules would not permit any discharge from drilling platforms within 6.4 km of the shore, but beyond that limit discharges of fluids containing no more than 1 ppm each of cadmium and mercury would be allowed.

Almost none of the barite mined is recycled. The residuum of the drilling mud is generally abandoned in covered waste ponds after the well is completed. The barite or barium compounds used in various manufacturing processes and products are lost to further use because the use of the products is widely distributed and because it would be virtually impossible to reclaim the small amounts in any one product.

FUTURE PROBLEMS AND TRENDS

The fortunes of the barite industry rise and fall with those of the drilling of oil and gas wells. The growing world population and its insatiable need for energy in liquid and gas form will stimulate demand for barite in world markets. The pace of drilling will likely increase by early in the 21st century as older wells and oilfields pass their peak years of production. The amount of barite needed to drill each well also is likely to increase as wells go deeper and into higher pressure areas. Enforcement of the standard limiting cadmium to 3 ppm and mercury to 1 ppm in barite for drilling mud mentioned in the previous section might cause some supply problems in a high-intensity world search for oil and gas in the years ahead.

Consider also that the world has produced about 160 Mt of barite in the past 45 years and that half of that production occurred in the past 15 years. These factors imply that to continue barite production at current or increased rates will require a considerable effort in research and exploration. Finding and developing new deposits for production takes time, generally measured in years. Geologic factors suggest that world barite production can be increased and that new deposits will be found.

Vein deposits mined chiefly for barite are yielding major amounts of ore in many countries of the world, and substantial amounts will continue to be obtained from vein deposits of all sizes. Smaller vein deposits, however, probably will lose their economic significance in the more advanced stages of industrialization as has already happened in the United States.

An increasingly important source of barite for some uses may well be as a coproduct from the mining of iron, base metal, fluorspar, and rare earth ores in deposits already known in many parts of the world. Larger volumes of barite also may be recovered from mine dumps and tailings ponds, where the barite already is partly ground and concentrated.

BIBLIOGRAPHY

Anon., 1961, *Barium Bibliography*, Mineral Products Div., Food Machinery and Chemical Corp., New York, 433 pp.

Anderson, W.H., Trace, R.D., and McGrain, P., 1982, "Barite Deposits of Kentucky," Bulletin 1, Series 11, Kentucky Geological Survey, 56 pp.

Andrews, P.R.A., and Collings, R.K., 1990, "Barytes in Canada," *Industrial Minerals*, No. 268, Jan., pp 46–57.

Ball, T.K., et al., 1979, "Geochemical Prospecting for Barite and Celestite Using a Portable Radioisotope Fluorescence Analyser," *Journal of Geochemical Exploration*, Vol. 11, pp. 277–284.

Berg, R.B., 1988, "Barite in Montana," Memoir 61, Montana Bureau, Mines and Geology, 100 pp.

Bose, R.N., and Vaidyanathan, N.C., 1979, "Gravity Surveys Around Mangampeta, Cuddapah District, Andhra Pradesh," *Journal Geological Survey of India*, Vol. 20, No. 11, pp. 540–547.

Brobst, D.A., 1984, "The Geological Framework of Barite Resources," *Transactions*, Institution of Mining and Metallurgy, Sec. A, Mining Industry, Vol. 93, July, London, pp. A123-A130.

Brobst, D.A., and Ward, F.N., 1965, "A Turbidimetric Test for Barium and Its Geological Application in Arkansas," *Economic Geology*, Vol. 60, No. 5, pp. 1020–1040.

Cassedanne, J., 1989, "Brazilian Stratabound Barite Ore Fields," *Nonmetalliferous Stratabound Ore Fields*, M.K. de Brodtkorb, ed., Van Nostrand Reinhold, New York, pp. 69–92.

Cathels, L.M., and Smith, A.T., 1983, "Thermal Constraints On the Formation of Mississippi Valley-type Lead-Zinc Deposits and Their Implications for Episodic Dewatering and Deposit Genesis," *Economic Geology*, Vol. 78, No. 5, pp. 983–1002.

Clark, S.H.B., 1987, "Metallogenic Map of Zinc, Lead, and Barite Deposits and Occurrences in Paleozoic Sedimentary Rocks, East-Central United States," Miscellaneous Investigations Series Map MI-1773, US Geological Survey.

Clark, S.H.B., Gallagher, M.J., and Poole, F.G., 1990, "World Barite Resources: a Review of Recent Production Patterns and a Genetic Classification," *Transactions*, Institution of Mining and Metallurgy, Sect. B: Appl. Earth Science, Vol. 99, London, pp. B125-B132.

Clark, S.H.B., and Poole, F.G., 1989, "Barite in North America (Excluding Akansas)," *Nonmetalliferous Stratabound Ore Fields*, M.K. de Brodtkorb, ed., Van Nostrand Reinhold, New York, pp. 93–117.

Dawson, R.K., 1985, "Geology of Barium, Strontium, and Fluorine Deposits in Canada," Economic Geology Report 34, Geological Survey of Canada, 136 pp.

de Brodtkorb, M.K., ed., 1989, *Nonmetalliferous Stratabound Ore Fields*, Van Nostrand Reinhold, New York, 332 pp.

de Brodtkorb, M.K., Schalamuk, I.B.A., and Ametrano, S., 1989, "Barite and Celestite Stratabound Ore Fields in Argentina," *Nonmetalliferous Stratabound Ore Fields*, M.K. de Brodtkorb, ed., Van Nostrand Reinhold, New York, pp. 41–68.

Diaz, L., 1986, *Stratabound Zinc, Barium, (Silver) Ore Deposits of the Lower Cretaceous in the Atacama Region, Northern Chile*, Heidelberger Geowissenschaftliche Abhandlungen, Bd. 3, Heidelberg, Germany, 186 pp.

Eddy, W.H., and Browning, J.S., 1964, "Selective Flotation of a Barite-Fluorspar Ore from Tennessee," Report of Investigations RI-6491, US Bureau of Mines, 8 pp.

Fuchs, Y.A., 1989, "Stratiform Ore Fields in Western Europe," *Nonmetalliferous Stratabound Ore Fields*, M.K. de Brodtkorb, ed., Van Nostrand Reinhold, New York, pp. 149–164.

Garrett, R.L., 1987, "Quality Requirements for Industrial Minerals Used in Drilling Fluids," *Mining Engineering*, Vol. 39, No. 11, pp. 1011–1015.

Gloria, H.G., Harnisch, D.H., and Braumann, F., 1985, *Die Energie, Rohswirtschaft der Volksrepublik China*, Verlag Gluckauf, Essen, Germany, 490 pp.

Gratz, J.F., and Misra, K. C., 1987, "Fluid Inclusion Study of the Gordonsville Zinc Deposit, Central Tennessee," *Economic Geology*, Vol. 82, No. 7, pp. 1790–1804.

Hickman, A.H., 1983, "Geology of the Pilbara Block and its Environs," Bulletin 127, Geological Survey of Western Australia, 267 pp.

Horton, J.W., Jr., 1990, "Barite in Quartz, Sericite Schist and Schistose Pyroclastic Rocks of the Battleground Formation, Kings Mountain," "Mineral Resources of the Charlotte 1° × 2° Quadrangle North Carolina and South Carolina," Professional Paper 1462, J.E. Gair, ed., US Geological Survey, pp. 115–118.

Howard, K.W., and Hanor, J.S., 1987, "Compositional Zoning in the Fancy Hill Stratiform Barite Deposit, Ouachita Mountains, Arkansas, and Evidence for the Lack of Associated Massive Sulfides," *Economic Geology*, Vol. 82, No. 5, (Aug.), pp. 1377–1385.

Jankovic, S., 1984, "Strata-bound Low Temperature Lead, Zinc, Barite, and Fluorite Deposits in Carbonate Rocks of Western Asia: Geotectonics Setting and Main Metallogenic Features," *Syngenesis and Epigenesis in the Formation of Mineral Deposits*, A. Wauschkuhn et al., eds., Springer Verlag, Berlin, pp. 373–390.

Jewell, P.W., and Stallard, R.E., 1991, "Geochemistry and Paleoceanographic Setting of Central Nevada Bedded Barites," *Journal of Geology*, Vol. 99, pp. 151–170.

Johnstone, S.J., and Johnstone, M.C., 1964, "Barium," *Minerals for the Chemical and Allied Industries*, 2nd. ed., John Wiley, New York, pp. 59–69.

Kaiser, C.J., et al., 1987, "Geologic and Geochemical Controls of Mineralization in the Southeast Missouri Barite District," *Economic Geology*, Vol. 82, No. 3, May, pp. 719–734.

Kelley, J.S., et al., 1993, "Barite Deposits in the Howard Pass Quadrangle and Possible Relations to Barite Elsewhere in the Northwestern Brooks Range, Alaska," Open File Report 93–215, *US Geological Survey*, 13 pp.

Kesler, S.E., Gesink, J.A., and Haynes, F.M., 1989, "Evolution of Mineralizing Brines in the East Tennessee Mississippi Valley-type Ore Field," *Geology*, Vol. 17, No. 5, pp. 466–469.

Kesler, T.L., 1949, "Occurrence and Exploration of Barite Deposits at Cartersville, Georgia," *Transactions*, AIME, Vol. 184, New York, pp. 371–375.

Kesler, T.L., 1950, "Geology and Mineral Deposits of the Cartersville District, Georgia," Professional Paper 224, US Geological Survey, 97 pp.

Klaus, D., and Schroder, N., 1967, "Erfahrungen beim Einsatz Geophysikalischer Messverfahren in der Flusspat—und Schwerspaterkundung," *Zeitschrift fur Angerwandte Geologie*, Vol. 13, No. 11–12, pp. 610–618.

Krebs, W., 1981, "The Geology of the Meggen Ore Deposit," *Handbook of Strata-bound and Strati-form Ore Deposits*, K.H. Wolf, ed., Vol. 9, Elsevier Scientific Publishing Co., Amsterdam, pp. 509–549.

Lange, S., Chandur, S., and Clauer, N., 1983, "Strontium Isotopic Evidence for the Origin of Barite and Sulfides from the Mississippi Valley-type Ore Deposits in South East Missouri," *Economic Geology*, Vol. 78, No. 6, pp. 1255–1261.

Lange, I.M., et al., 1985, "Geologic Setting, Petrology, and Geochemistry of Stratiform Sphalerite-Galena-Barite Deposits, Red Dog Creek and Drenchwater Creek Areas, Northwestern Brooks Range, Alaska," *Economic Geology*, Vol. 80, No. 7, pp. 1896–1926.

Large, D.E., 1980, "Geological Parameters Associated with Sediment-hosted Submarine Exhalative Lead-Zinc Deposits, an Empirical Model for Mineral Exploration," *Geologishes Jahrbuch*, Series D, Vol. 40, pp. 59–129.

Leach, D.L., 1980, "Nature of Mineralizing Fluids in the Barite Deposits of Central and Southeastern Missouri," *Economic Geology*, Vol. 75, pp. 1168–1180.

Lundie, P., 1986, "Standardisation of Minerals for Drilling Fluids," *Industrial Minerals*, No. 222, Mar., pp. 113–117.

Maher, S.W., 1970, "Barite Resources of Tennessee," Report of Investigations 28, Tennessee Division of Geology, 40 pp.

Marumo, K., 1989, "The Barite Ore Fields of Kuroko-type of Japan," *Nonmetalliferous Stratabound Ore Fields*, M.K. de Brodtkorb, ed., Van Nostrand Reinhold, New York, pp. 201–232.

Matthews, R.A., and Alfors, J.T., 1962, "Sanbornite from Rush Creek, Fresno County," *Mineral Information Service*, California Division of Mines and Geology, Vol. 15, No. 6, pp. 1–3.

Maynard, J.B., and Okita, P.M., 1991, "Bedded Barite Deposits in the United States, Canada, Germany, and China: Two Major Types Based on Tectonic Setting," *Economic Geology*, Vol. 86, No. 2, pp. 334–376.

Miller, R.E., Brobst, D.A., and Beck, P.C., 1977, "Fatty Acids As a Key to the Genesis and Economic Potential of Black Bedded Barites in Arkansas and Nevada," *Organic Geochemistry*, Vol. 1, No. 1, pp. 11–26.

Moen, W.S., 1954, "Barite in Washington," Bulletin 51, Washington Division of Mines and Geology, 112 pp.

Moore, D.W., et al., 1986, "Geologic Setting and Genesis of the Red Dog Zinc-Lead-Silver Deposit, Western Brooks Range, Alaska," *Economic Geology*, Vol. 81, No. 7, pp. 1696–1727.

Morin, J.A., ed. 1986, "Mineral Deposits of the Northern Cordillera," *Proceedings*, The Mineral Deposits of Northern Cordillera Symposium, Dec. 1983, Special Volume 37, Canadian Institute of Mining and Metallurgy, 378 pp.

Nuelle, L.M., and Shelton, K.L., 1986, "Geologic and Geochemical Evidence of Possible Bedded Barite Deposits in the Devonian Rocks of the Valley and Ridge Province, Appalachian Mountains," *Economic Geology*, Vol. 81, No. 6, pp. 1408–1430.

Olson, J.C., et al., 1954, "Geology of the Rare-Earth Deposits of the Mountain Pass District, San Bernardino County, California," Professional Paper 261, US Geological Survey, 75 pp.

Orris, G.J., 1986, "Descriptive Models of Bedded Barite," D.P. Cox, and D.A. Singer, eds., "Mineral Deposit Models," Bulletin 1693, US Geological Survey, pp. 216–218.

Padalino, G., Tocco, S., and Violo, M., 1989, "Stratabound and Karstic Deposits of Barite in Sardinia," *Nonmetalliferous Stratabound Ore*

Fields, M.K. de Brodtkorb, ed., Van Nostrand Reinhold, New York, pp. 165–200.

Papke, K., 1984, ''Barite in Nevada,'' Bulletin 98, Nevada Bureau of Mines and Geology, 125 pp.

Pemberton, H.E., 1983, *Minerals of California*, Van Nostrand Reinhold, New York, 591 pp.

Poole, F.G., 1988, ''Stratiform Barite in Paleozoic Rocks of the Western United States,'' *Proceedings*, 7th Quadrennial IAGOD Symposium, E. Zachrisson, ed., pp. 309–319.

Ramos, V.A., and de Brodtkorb, M.K., 1989, ''Celestite, Barite, Magnesite, and Fluorspar: Stratabound Setting Through Time and Space,'' *Nonmetalliferous Stratabound Ore Fields*, M.K. de Brodtkorb, ed., Van Nostrand Reinhold, New York, pp. 297–321.

Reade, E.H., et al., 1980, ''Barite Deposits in the Cartersville District, Bartow County, Georgia,'' *Excursions in Southeastern Geology*, R.W. Frey, and T.L. Neathery, eds., Vol. 2, American Geological Institute Guidebook for the Geological Society of America, 1980 Annual Meeting, pp. 379–384.

Reimer, T.O., 1980, ''Archean Sedimentary Baryte Deposits of the Swaziland Supergroup (Barberton Mountain Land, South Africa),'' *Precambrian Research*, Vol. 12, pp. 393–410.

Robbins, J., 1985, ''The Industrial Minerals of West Germany,'' *Industrial Minerals*, No. 219, pp. 15–47.

Robinson, G.R., Jr., and Woodruff, L.G., 1988, ''Characteristics of Base-Metal and Barite Vein Deposits Associated with Rift Basins, with Examples from Some Early Mesozoic Basins of the Eastern North America,'' *Studies of Early Mesozoic Basins of the Eastern US*, Bulletin 1776, A.J Frolich and G.R. Robinson, Jr., eds., US Geological Survey, pp. 377–390.

Sackett, E.L.H., 1962, ''Barite—Little-Known Industry That Means Mud to Oil Men,'' *Mining Engineering*, Vol. 14, No. 5, May, pp. 46–49.

Scull, B.J., 1958, ''Origin and Occurrence of Barite in Arkansas,'' Information Circular 18, Arkansas Geology and Conservation Commission, 101 pp.

Uhley, R.P., and Scharon, L., 1954, ''Gravity Surveys for Residual Barite Deposits in Missouri,'' *Mining Engineering*, Vol. 6, No. 1, pp. 52–56.

Wang, Z., and Li, G., 1991, ''Barite and Witherite in Lower Cambrian Shales of South China: Stratigraphic Distribution and Geochemical Characterization,'' *Economic Geology*, Vol. 86, No. 2, pp. 354–363.

Weber, F.H., and Matthews, R.A., 1967, ''Prospecting For Barite in Northern Shasta County,'' *Mineral Information Service*, Vol. 20, No. 9, California Division of Mines and Geology, pp.107–114.

Wharton, H.M., 1972, ''Barite Ore Potential of Four Tailings Ponds, Washington County Barite District, Missouri,'' Report of Investigations 53, Missouri Geological Survey and Water Resources, 91 pp.

Zimmermann, R.A., and Amstutz, G.C., 1989, ''Barite Deposits in Arkansas,'' *Nonmetalliferous Stratabound Ore Fields*, M.K. de Brodtkorb, ed., Van Nostrand Reinhold, New York, pp. 119–148.

BAUXITE

V.G. HILL

The term *bauxite* was first used by O.P.A.P. Dufrenòy in 1845 for an ore which was discovered by Pierre Berthier in 1821 near the village of Les Beaux de Provence near Arles in southern France. His data suggested that this mineral was an alumina hydrate with two molecules of water. However, it was subsequently shown by thermal analysis that the mineral was a fortuitous mixture of the monohydrate and the tri-hydrate. Diaspore had been previously discovered in 1801 and gibbsite in 1822. The term was however retained as a rock name. The name of the village was changed to Les Baux around 1885; Sainte Claire Deville suggested that the mineral name should reflect this, and it was changed to bauxite (Bracewell, 1962, Régnier, 1988).

Mining of bauxite began around 1860 in the Auriol region in the Bouches du Rhône area in the south of France. This ore was transported to Marseilles by carts and was processed in the Salindres plant, which at that time was the only plant producing alumina using the Deville process and aluminum by the chemical process. During that period white bauxite (low iron content) was particularly sought after for the production of aluminum sulfate, refractory bricks, and quick setting cement. Export of bauxite from France began in 1880 and was initially delivered to Chemische Fabrik, Goldschmischen near Breslau, and to Guilini's Ludwigshafen plant in Germany in 1885. These companies produced alum and aluminum sulfate (Régnier, 1988). The next bauxite mining operation was in Antrim County in Northern Ireland in 1873. This was first used for the manufacture of alumina refractory bricks and quick setting cement (Bracewell, 1962).

It is important to emphasize that initially almost all of the bauxite which was produced came from France and was mainly used for non-metallurgical purposes. At that time even the production of alumina was mainly for use as a mordant in textile dying. However, with the development of the Hall-Héroult process, the alumina which was available was increasingly used for smelting to aluminum. In spite of this the demand for non-metallurgical bauxite continued to increase and a set of applications were developed. These include refractories, abrasives (including the recently developed proppants), high alumina cements, aluminum chemicals, activated bauxite, and alumina. It became increasingly evident that the high alumina-low alkali raw materials have special advantages. The question however was one of cost benefit.

Chin (1989) has pointed out that the non-metallurgical operations have been traditionally grafted on a bauxite/alumina/aluminum industry tree. In this context the non-metallurgical alumina produced was, with few exceptions, regarded as either coproduct or byproduct i.e.:

Bauxite Trunk Sector

- Chemical-grade bauxite for the production of aluminum chemicals
- Refractory-grade bauxite to produce high alumina refractories
- Abrasive-grade bauxite for the production of abrasives.

Alumina Trunk Sector

- Bayer aluminum tri-hydroxide, Bayer ATH and white ATH to produce different categories of:
  - fillers and pigments for plastics and elastomers
  - aluminum chemicals (e.g., aluminum fluoride aluminum sulfate and sodium aluminate)
  - activated aluminas for desicants, catalysts, and catalyst supports
  - special calcined aluminas for ceramics, refractories, abrasives, soft burned calcined aluminas
  - hard burned calcined aluminas to produce tabular alumina and calcium aluminate cements.

Aluminum Trunk Sector

Aluminum used to produce high purity aluminum chemicals.

This review is mainly concerned with non-metallurgical products from the bauxite trunk sector. Today the non-metallurgical aluminas play an increasing role as the demand for superior products grows, and the marginal cost of producing additional alumina tri-hydrate in some new plants approaches that of the equivalent non-metallurgical bauxite type. The difference between the alumina and aluminum sectors as sources of non-metallurgical products is simply one of product purity and cost. Emphasis here is on the non-metallurgical bauxites, and every effort is made to isolate the data for non-metallurgical grades from that for the metallurgical grade. Differences between some of the non-metallurgical grades make it desirable to disaggregate the data, because this information permits improved understanding of the time trends, developments, and prospects for the industry.

The database used in this study was initially collected over several years because of interest in the aluminum industry. The initial data was collected mainly from US Bureau of Mines publications, particularly the *Minerals Yearbook,* and from *Metallgesellschaft.* This data was subsequently refined and that for non-metallurgical bauxites separated out. Company reports, mine visits, and personal contacts in the industry were used in this refinement and upgrading of the database. It must be pointed out that the statistics for the different products (i.e., refractory grade, RGB, chemical grade, CGB, and aluminous cement, ACGB) have different conversion factors for crude bauxite mined to product and so are not equivalent. However, because of the need to determine time trends, we have as a first approximation added some of these numbers after correction for loss of combined water due to calcination where necessary.

GEOLOGY

The geology of bauxite deposits has been treated at length by three excellent works to which, readers are referred (Bárdossy,

1982, Bárdossy and Aleva, 1990, and Patterson et al., 1986), so this aspect of the industry will not be discussed here.

Mineralogy

Although bauxite deposits contain a wide variety of minerals, most of the non-metallurgical bauxites usually consist almost entirely of the aluminum oxyhydroxide minerals. In this regard some refractory bauxites consist of gibbsite with minor amounts of kaolin. The most important of the minerals found in non-metallurgical bauxites are the aluminum oxyhydroxides: gibbsite, boehmite, and diaspore. Kaolinite, quartz, hematite, goethite, rutile, and anatase are impurities. Kaolinite may be considered as a dilutant in some cases.

Hill and Ostojic (1984) proposed a bauxite classification scheme to reflect the impact of different bauxites on Bayer technology. In this scheme most of the non-metallurgical bauxites are either the purest members of the Guyana and Weipa types or belong to the Mediterranean-2 and -3 types. The exceptions are the high iron bauxites which are used in the production of some cements.

Major Bauxite Provinces of the World

The major bauxite deposits of the world may be grouped into provinces which are identified below and are shown in Fig. 1. These bauxites are also grouped into karstic (deposits associated with limestone) and lateritic (associated with other rock types). The first four provinces which are listed below are karstic and the others lateritic. The extent and description of these bauxite provinces are:

The Caribbean Province: extends from Jamaica through Hispaniola and terminates in Puerto Rico. The known deposits occur mainly as infilling in steep-walled depressions formed by the coalescence of usually uniform cylindrical vertical pipes. Essentially only metallurgical bauxite has been produced from bauxite from this province, but Kaiser exported some high silica cement grade bauxite to the Bahamas Cement Company from a deposit in Comfort, St. Elizabeth (24.1 and 48.3 kt in 1964 and 1965 respectively).

The Mediterranean Province: extends from Turkey through Greece, Yugoslavia, Hungary, and France into Italy, with smaller associated deposits in Spain and Austria. The principal ore mineral is usually boehmite, but in a few deposits gibbsite is predominant, while in others significant amounts of diaspore are present. This was the first bauxite province developed.

The Central Urals-Kazakhstan Province: has subsidiary occurrences in the Tikhvin area near Boksitogorsk, east of St. Petersburg, the Ukraine, and the Moscow Basin. These deposits vary from bauxite horizon grading into bauxitic clay and finally clay into a series of bauxite horizons in clay. The mineralogy varies from predominantly gibbsitic in many areas of Kazakhstan to predominantly boehmitic in the Urals. The deposits in the Tikhvin district, the Urals, and the Ukraine are lower in silica than the others.

The China Province: extends from Chagpu in Fukien Province southwesterly to Kwangi Province in southern China, and into Vietnam. Much of the bauxite is boehmitic, with significant amounts of diaspore, particularly in Kweichow Province and Vietnam. This province is developing into the second major non-metallurgical grade bauxite producing province of the world.

African Province: comprises the Guinea and Cameroon Shields. The Guinea Shield area extends from Guinea-Bissau to Togo, with branches projecting north-northeasterly into Upper Volta and Hogger in the central Sahara. The deposits are part of an extensive area of laterization which is associated with Precambrian gneisses, and in many deposits the boehmite content increases to as much as 10 and even 15%. The Cameroon area contains deposits in the large laterized area in the shield and extends from the west coast and across Central Africa to Mozambique.

South Asia—Australian Province: comprises the following sub-provinces:

(i) *The Australian Area:* includes the major bauxite deposits of Gove, Weipa, the Mitchell Plateau, and the Darling Ranges. These contain both in situ and transported materials which were deeply weathered during the Cretaceous and Miocene ages. The mineralogy is dominated by gibbsite, but a significant amount of boehmite is present.

(ii) *Indian Sub-Province:* includes the deposits on the Deccan Plateau of India such as the large finds in Orissa and the deposits in Gujarat, Maharashtra, and Madhya Pradesh. The deposits are usually of the blanket type and are part of a large laterized area. Gibbsite is the principal aluminum mineral.

(iii) *Southeast Asian Sub-Province:* includes the bauxite deposits in Malaysia, Ria Archipelago, and elsewhere in Indonesia. The mineraology of these deposits is gibbsitic, with goethite and some clay as impurities minerals. They appear to result from both in situ and transported bauxites. The latter are associated with the sediments fringing nepheline syenite. Gibbsite is the predominant aluminum ore mineral, with minor amounts of quartz.

The North American Province: comprises the Arkansas and southeastern coastal plain and the Appalachian Valley. These deposits are gibbsitic.

The South America Province: includes the bauxite deposits associated with the Guiana Shield, the Central Brazilian Shield, the Atlantic Shield, and the recent sediments of the Amazon syncline, as well as those fringing the Precambrian Shield areas. The province extends from Venezuela through Guyana, Surinam, and French Guiana into Colombia and Brazil. There is extensive laterization on the erosion surface overlaying younger basic dolerites and gabbros. This province contains the major reserves of non-metallurgical bauxites of the world and is the principal producing area.

The bauxite occurrences described here contain most of the known bauxite reserves of the world, but there are other minor areas outside these which contain less significant bauxite deposits.

COMMERCIAL APPLICATIONS

Approximately 90% of the bauxite which is produced today is used for the production of alumina, and about 85% of this is consumed for metallurgical applications. Alumina tri-hydrate is used for the production of the pure aluminum chemicals, high quality refractories, and other high aluminum products, while aluminum is used for the preparation of the purest aluminum chemicals.

Table 1 gives typical analysis of the requirements for raw bauxites for the production of metallurgical, abrasive, chemical, and refractory grades. The selection of a particular bauxite for any use other than metallurgical purposes depends not only on the composition of the run of the mine bauxite, but also on the ease of benefication of selectively mined bauxite to meet the requirements of the particular product. Sometimes a mixture of high quality bauxite and alumina are used as the starting materials in order to further lower the iron content. A generalized flow chart from bauxite through various processing stages to a variety of end uses is given in Fig. 2.

The principal non-metallurgical applications are as follows.

Aluminum Chemicals

Bauxite and alumina are the principal raw materials for the production of aluminum chemicals such as aluminum sulfate, alu-

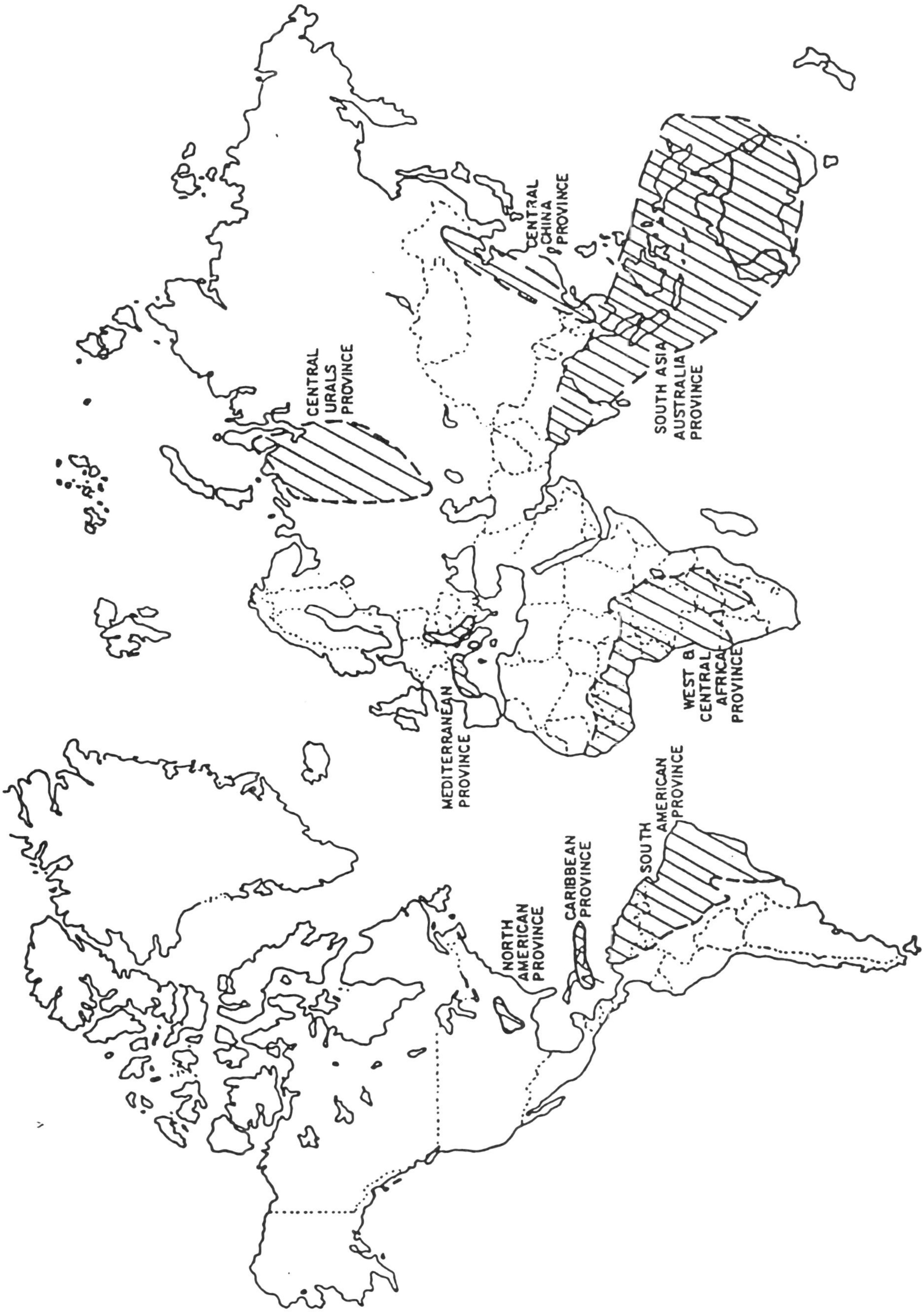

Fig. 1. Major bauxite provinces.

Table 1. Chemical Analysis of Typical Raw Bauxite of Grades (%)

Grade	Al_2O_3	SiO_2	Fe_2O_3	TiO_2
Metallurgical	50–55	0–15	5–30	
Abrasive	Min. 55	Max. 5	Max. 6	Min. 2.5
Chemical	Min. 55–58	Max. 5–12[2]	Max. 2	
Refractory	Min. 59–61	Max. 1.5–5.5	Max. 2	Max. 2.5

1. Most of the remainder represents water of crystallization.
2. A major portion of the bauxite used by the US chemical industry exceeds 15%.
Source: Everts, 1984.

minum chloride, aluminum fluoride, sodium aluminate, and aluminum acetate. The selection of bauxite for the direct production of aluminum sulfate depends on the relative solubility of the phases present in sulfuric acid. Firstly, gibbsite is the aluminum oxyhydroxide which is most readily soluble, boehmite less so, and diaspore the least. Secondly, the ratio of alumina/iron needs to be considered, but chemical grade bauxite should have an Al_2O_3/Fe_2O_3 ratio of 100 to 1. However the normal specification requires Al_2O_3/Fe_2O_3 ratio of 23 to 1. This ratio gives an acceptably pale straw-yellow colour solution, which is preferred by most consumers. This ratio is based on historical preferences rather than on strictly scientific considerations, since ferric sulfate is also a flocculent in its own right. The AWWA standard permits the use of aluminum sulfate with higher iron content provided the iron content of the treated water is not increased. Silica levels of 10% in the aluminum sulfate can be tolerated. However, toxic ions must be avoided and high levels of organic compounds adversely affects water quality (Andrews, 1984, Benbow, 1988).

High Alumina Refractories Grade

This is the highest priced group of non-metallurgical grade bauxite. Bauxites for this application should have a high alumina content, low TiO_2 content (generally < 2.5% after calcination), and only traces of alkalis and alkaline earths. High alumina refractories are a relatively recent development in refractories which date back to the 17th century, while the high alumina type dates back to 1906 when bauxite from Missouri was tried for the production of liners for the hot zones in Portland Cement kilns (Anon., 1964). However, it was not until 1920 that commercially acceptable high alumina bricks were produced from the Missouri diaspore clays. Development continued and other low iron bauxites were tried, but it was not until after World War II that production of refractories from bauxite from the Guianas began. The characteristics of bricks made from Guyana bauxite were superior to those from the US diaspore clays. The kaolinitic-bauxite deposits in Georgia were also developed to produce a range of refractories. These mixtures sometimes include low alkali refractory aluminas. Table 2 gives some properties of some alumina containing refractories.

Abrasives

Abrasive grade bauxite is used for the production of fused alumina for the manufacture of grinding and polishing equipment. The highest quality fused alumina is made from alumina, while bauxites of different grades are used for producing brown fused alumina. The usual criteria for the assessment of bauxite for producing brown alumina are both chemical and physical. The important combinations are Al_2O_3, SiO_2, Fe_2O_3, TiO_2, alkali, and water. The alumina content should be as high as possible to maximize productivity and reduce impurities.

The normal requirements are:

Al_2O_3	>80%
Fe_2O_3	<10%
Si	<10%

Activated Grade

This bauxite product is used as a drying agent for gases and organic liquids. The material has a high surface area which enables it to absorb sulfur and chlorine based species. It also has some applications as a catalyst mount and filtering agent particularly in the petrochemical industry. However it is not as efficient as material which is produced from alumina (Benbow, 1988).

Aluminous Cements

Two types of aluminous cements are produced (low iron and high iron varieties). The low iron variety, which is also known as calcium aluminate cement, is used as the refractory cement for bonding the high alumina refractories. The advantages are greater density, lower porosity, and less shrinkage of the castable body. These conditions reduce penetration by molten metals, and the strength of the cured and fired product is high.

The high alumina high iron cement is sometimes referred to as cement fondu. Some low grade bauxite with high iron and silica is also used as an additive in the production of Portland cement.

TECHNOLOGIES

The production of non-metallurgical grade bauxite involves the same exploration techniques as metallurgical grade bauxites. However development drilling is usually more closely spaced in order to facilitate selective mining. Separate stockpiles are maintained if more than one type of bauxite is being produced. Each grade is then crushed to −5 cm, the silica and iron selectively removed by washing, and the product stockpiled. Chemical grade bauxite is then dried and shipped. The refractory and the abrasive grades are calcined prior to shipment and are both reported as calcined bauxite, but the type should be specified.

Aluminum Chemicals

Chemical grade bauxite is reacted with sulfuric acid diluted to 45% in lead lined steel tanks. The reaction is allowed to proceed to completion, and water is then added to dilute to the desired strength of about 17% Al_2O_3 for commercial and municipal grades prior to shipment.

Calcined Bauxite

Bauxite required for most non-metallurgical applications, except the production of aluminum chemicals, is usually calcined prior to shipment. The specific calcination conditions depend on the mineralogy of the bauxite and the use to be made of it. Fig. 3 illustrates the options in thermal transformation among the aluminum oxyhydroxide phases. The series of structural forms which develop are determined by the structure of the starting material, its particle size, and the conditions of calcination. Perhaps the most significant feature is that diaspore converts directly to α-Al_2O_3 starting below 500°C. This is in contrast to boehmite and gibbsite which decompose to a series of meta-stable phases on calcination. Boehmite first converts to Γ-Al_2O_3 series and then to δ-Al_2O_3, and

```
BAUXITE                                                           ELECTRIC
 MINES                                    :--> ABRASIVE GRADE    --> ARC       -----> CRUDE BROWN FUSED
   :                                      :       BAUXITE            FURNACE          ALUMINUM OXIDE ABRASIVE
   v                                      :
CRUSHER     :------>   CALCINING  ------:--> CHEMICAL GRADE
   :        :            KILN             :       BAUXITE
   v        :                             :
WASHER      :                             :--> REFRACTORY GRADE                  :--  DETERGENTS
   v        :                                     BAUXITE                        :--  MORDANTS,DYES,ORGANIC LAKES
DRYER  ----:                                                                      :--  PAPER MANUFACTURERS
   v                                                                              :--  SEWAGE TREATMENT
METALLURGICAL        :-------------------> CHEMICAL INDUSTRY  -------------->:--  CEMENT WATERPROOFING
GRADE                :                                                            :--  DEGASSING FLUID
BAUXITE              :                                                            :--  LEATHER TANNING
   :                 :                                                            :--  WATER TREATMENT
   :                 :                                                            :--  GLASS & CERAMICS
   v                 :
ROD MILL             :                                                            :--  TOOTH PASTE
   :                 :                    :--> PHARMACEUTICALS    -------------->:--  DEODORANTS
   v                 :                    :                                       :--  ANTACIDS
BAYER ----> SCALE    :                    :                                       :--  COSMETICS
ALUMINA        :     :                    :
PLANT          v     :                    :                                       :--  CARPET BACKING
   :        :--------->:BALL MILL  ------:--> FLAME RETARDANTS   -------------->:--  FOAM & SOLID PLASTICS
   :        :              :              :                                       :--  PLASTIC WIRE INSULATION
   :        :              :              :                                       :--  WALLBOARD
   :        :              :              :
   :        :              :              :--> FILLERS & COATINGS -------------->:--  PAINTS,PAPER
   v        :              :              :                                       :--  ADHESIVES
ALUMINA  --:               v              :
TRIHYDRATE             PELLETISER         :                                       :--  WALLPAPER,BOXES
   :                   (SPHERES)          :--> PIGMENTS ------------------------>:--  INK,PAINT,PAPER
   :                       :
   :                       v                                                      :--  CLADS PROCESS SULFUR
   :                   CONTROLLED                                                 :--  ADSORBANTS FOR GASSES,LIQUIDS
   :                   DEHYDRATION        ---> ACTIVATED ALUMINA  -------------->:--  DESICANTS
   :                                                                              :--  PETROLEUM REFINING
   v                                                                              :--  CATALYST SUPPORTS,AUTO EXHAUST
STATIONARY :------>    BALL MILL  ------:
CALCINER   :               :              :                                       :--  WHITEWARE,FINE CHINA
   :       :               :              :--> LOW SODA ALUMINA   -------------->:--  PORCELAIN INSULATORS,SPARK PLUGS
   :       :               v              :                                       :--  KILN FURNITURE
   :       :           PELLETISER         :
   :       :           (SPHERES)          :                                       :--  SYNTHETIC RUBIES & SAPHIRES
   v       :               v              :                                       :--  CUTTING TOOLS
CALCINED   :           DRAFT  KILN        :--> HIGH PURITY ALUMINA    ----------->:--  POLISHING COMPOUNDS
ALUMINA    :           (SINTERING)                                                :--  SUBSTRATES,MICROCHIPS,CATALYSTS
   :       :               :                                                      :--  TRANSLUCENT ALUMINA PRODUCTS
   :       :               v
   :       :           TABULAR            :----------------------------------->:--  GLASS,CERAMICS
   :       :           ALUMINA            :                                       :--  REFRACTORIES
   :       :               :              :
   :       :               :              :                                       :--  WELDING ELECTRODES COATING
   :       :               :              :--> FILLERS & COATINGS -------------->:--  EPOXY,SILICONE,POLYESTER
   :       :               v              :
   :       :           CRUSHER ---------:----------------------------------->:--  CATALTST SUPPORT BEDS
   :       :                                                                      :--  ABRASIVES
   v       :
PRIMARY    :           ELECTRIC                CRUDE,WHITE,FINE &    GRINDING    :--  WHITE FUSED ABRASIVE GRAINS
ALUMINUM   :-------->    ARC    ------------>                    -->    &     -->:--  FINE FUSED ABRASIVE GRAINS
SMELTER                FURNACE                 RUST FUSED ALUMINA     SIZING     :--  RUST FUSED ABRASIVE GRAINS
```

Fig. 2. Specially alumina production flowchart (Baumgardner, 1990).

θ-Al_2O_3, before transforming to α-Al_2O_3 at about 1 100°C. In the case of gibbsite, if it is heated slowly to about 100°C, or if either coarse grained or the water vapour pressure in the surrounding atmosphere is high, then it is transformed firstly to boehmite and then behaves like boehmite on further heating. On the other hand, if gibbsite is heated rapidly or is very fine grained, then it transforms to χ-Al_2O_3 and then to κ-Al_2O_3 at higher temperatures. This explains the behavior of gibbsite as against diaspore bauxites available from China and Greece (Wefers and Misra, 1987).

Proppants are a new development to enhance hydrocarbon recovery from reservoir rocks by preventing fracture closure. They are manufactured by the comminution of calcined bauxite to less

Table 2. Properties of Alumina Containing Refractories

Type refractory	% Al_2O_3	Typical ↑ cone numbers	Pyrometric equiv. ↑ softening °C	Cold crushing strength MPa	Flexture strength MPa	t 25 psi °C	Load test > linear subsidence %
Fireclay							
low-duty	22–33	25–27	2 795/535	13.8–41.4	6.9–17.2		
medium-duty	25–38	29–31	3 040/671	11.7–41.4	5.5–17.2	13 649	3–10
high-duty	35–41	31.5–33	3 105/707	10.3–48.3	3.4–20.7	13 649	0.5–0.7
super-duty	40–44	33–34	3 185/752	10.3–41.4	4.1–11	14 649	2–8
High-alumina							
45–48%	45–48						
50%	50	35	3 245/785	17.2–41.3	6.9–11	14 649	4–6
60%	60	36–37	3 295/813	12.4–48.2	4.1–12.4	14 649	2.5
70%	70	37–38	3 310/821	27.6–62	8.3–15.2	14 649	2.5
Mullite	71.8	38	3 360/849	24.1–62	6.9–24.1	16 093	1.5
80%	80	39	3 390/866	27.6–62	7.6–20.7	14 649	1.5
85%	85	39–41	3 410/877				
90%	90	40–41	3 425/885	27.6–62	8.3–24.1	16 093	0.3
Corundum	99+	42–	3 630/999	34.5–62	12.4–20.7	16 649	0.5–1.5

Source: Alcoa, 1964.

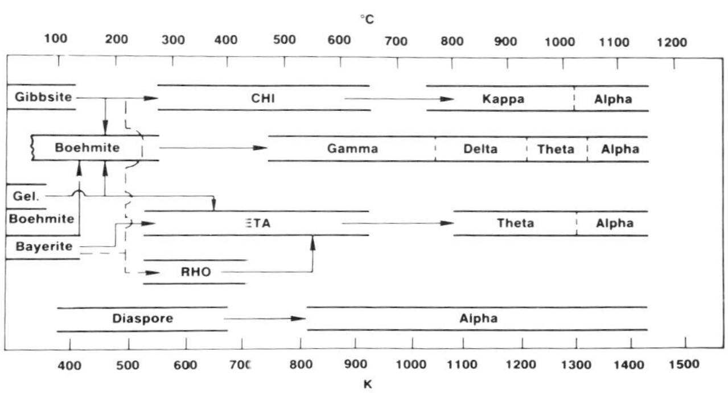

Fig. 3. Transformation sequence $Al(OH)_3$—Al_2O_3 (Wefers and Misra, 1987).

than 12 μ particle size. These are then pelletized and processed to a high degree of roundness prior to sintering at 2 000°C. This results in the crystallization of minute randomly oriented α-Al_2O_3 crystals. In order to maximize permeability the material is screened into fractions between 6 and 80 mesh range, of which 20 to 40 mesh is the most popular. Comalco simply screen the washed bauxite to separate out the required size fractions which are then calcined.

Physical Beneficiation of Bauxite for Refractory Purposes

Since 1983 almost all of the refractory grade bauxite consumed or stockpiled in the United States is imported from Guyana, Surinam, and China. However, approximately 10 to 15 Mt in the Arkansas bauxite deposits could meet US National Stockpile specifications for refractory grade bauxite but for the iron contents. Iannicelli (1984) reported on the application of high intensity wet magnetic separation of iron from bauxite. This technique has been used to beneficiate Georgia kaolin. The unit operations which are involved are identified here (Iannicelli, 1984):

1. Blunge (makedown bauxite in water slurry to approximately 25% solids)
2. Screen slurry on 200 mesh (to remove oversize particles which could plug matrix in magnet)
3. Magnetic separation (to remove most of Fe_2O_3 and TiO_3)
4. Filtration (increase solids from 25 to 60%)
5. Spray drying (dry to less than 5% moisture)
6. Pellitized spray dried bauxite to round pellets (to compact and deaerate product)
7. Extrude and calcine beneficiated bauxite (pressure extrusion to achieve high density after calcining)
8. Screen and package calcined bauxite for shipping.

The question is the cost effectiveness of this technology, as against imported refractory grade bauxite taking the US national security aspect into consideration.

Environmental Impact

The principal areas of possible adverse environmental impact of non-metallurgical bauxite operations are in the mines, bauxite processing plants including kilns, and in the alumina facilities.

The higher value of the non-metallurgical bauxites, as compared to the metallurgical grade, usually make it economical for the open cast operations to be deeper, so it is often necessary to dispose of large quantities of overburden. Where there is abundant water, artificial lakes are sometimes constructed, and the overburden removed by suction dredges. This is so in the case of the Onverdacht mine in Surinam. These lakes need to be designed to avoid wide scale destruction of the forest. Additionally there is the need for a well designed reclamation program to ensure that the reclamation meets acceptable standards.

In the producing areas where bauxite is beneficiated by crushing and washing, proper sedimentation facilities are necessary so that the water can be re-used and the tailings properly disposed. The drying and calcining facilities also need to be fitted with dust precipitators to prevent high dust levels leaving the stacks. In the production of aluminum chemicals from bauxite, the disposal of the acidic residue poses a problem at some locations.

PROFILES OF PRINCIPAL OPERATIONS

Table 3 identifies the principal companies involved in the production of non-metallurgical bauxites worldwide in 1991. France and the United States were the major producing countries until 1938, when major production began in Guyana, which with

Table 3. Non-Metallurgical Grade Bauxite Mines of the World: 1991

COUNTRY	COMPANY	LOCATION	STARTUP	MINE TYPE	FACILITIES: MINE-ktpy	FACILITIES: BENEFIC.*	FACILITIES: KILNS	PROD. CAP. ktpy: CBG	PROD. CAP. ktpy: RGB	PROD. CAP. ktpy: AGB	PROD. CAP. ktpy: ACGB	REMARKS
USA	ALCOA	BRYANT, AR		S	0							CLOSED MAY 1990
	AMERICAN CYNAMID CO	BRYANT,AR		S	100						(100)	CLOSED 1988
	DIDIER TAYLOR REFRACTORY	EUFAULA,AL		S	68							Bx/KAOL
	HARBISON-WALKER REFRAC.	EUFAULA,AL		S	60		2R		300			Bx/KAOL
	MULLITE CO. OF AMERICA	EUFAULA,AL		S	60				300			Bx/KAOL
	MULLITE CO. OF AMERICA	ANDERSONVILLE,GA		S			7R		400			Bx/KAOL
BRAZIL	CIA BRASILEIRA DE BAUXITA	PARAGOMINAS, PARA	1989	S	100	C,W	R		100			
	MAGNESITA,SA.	BELO HORIZONTE					R					
	MINERACAO SANTA LUCRECIA,SA	ALMEIRIM,PARA	1985	S	200	C,W,S	R		60		*	EXP.TO 100 kty PL.
	MINERACAO CURIMBABO LTD.	P.CALDAS,MG	1989	S	400		R	300	*	75	*	
GUYANA	GUYMINE—LINDEN OPERATION	E.MONTGOMERY	1979	S	1500	C,W,S	4R		1000	50		RGB
	GUYMINE—LINDEN OPERATION	KARA KARA	1962	S	1000							RGB
	GUYMINE—LINDEN OPERATION	N.DORABECA	1981	S	500							RGB
	BERBICE MINING CO	KWAKWANI		S	500	C,W,S		100	100	100	100	RGB,CGB
	BERBICE MINING CO	GREEN CREEK		S								
	BERBICE MINING CO	MANAKA		S								
SURINAM	SURALCO	MOENGO	1939	S	0							CL 1984/87
	BMS	ONVERDACHT	1941	S	0		(5R)					CLOSED 1982
GREECE	BAUXITES PARNESSE MINING CO	MT.PARNASSE	1932	S,U	300			*	*	300	*	
	BAUXITES PARNESSE MINING CO	MT.GHIONA	1952	S,U	60					50		
	ELEUSIS BAUXITE CO	ELEUSIS	1952	S,U	50			50		50	50	
FRANCE	SABAP	BEDARIEU	1985	S	60					60		
YUGOSLAVIA	ENERGOINVEST	B.KRUPA,W.BOSNIA	1950	S/U	50							
GUINEA	CIE DES BAUXITES DE GUINEA	SANGAREDI,BOKE	1973	S	150	C	1R			150		
MOZAMBIQUE ZIMBABWE	MEIKLES CO.	PENHALONGA UMTALI	1935	S	6			6	*		*	
AUSTRALIA	COMALCO LTD	WEIPA	1972	S	300		2R			300		
	GOVE ALUMINIUM LTD.	GOVE	1971	S	250						250	
CHINA		HENAN					S		450	250		
		SHANXI					R		*	*		
		GUIZHAU			700		S,R		*	*		
		GUANDONG					S		*	*		
		HEBEI					S		*	*		
MALAYSIA	JAHORE MINING & SMELTING LTD	PENERANG	1956	S	200	W,S		*	*		*	
TURKEY	ETIBANK	MILOS REGION		S	150						100	
INDIA	GUJARAT MIN.DEV.CORP.	JAMANGAR		S	50	C,W,S	R		50			
	BOMBAY MINERAL SUPPLY CO.LTD.	HAPA,JAMNAGAR		S	150	C,W,S	R		133			
	SAURASHTRA CALCINED BAUXITE LTD	PORBANDAR, JUNAGADH		S	180	C,W,S	*		180			
	CARBORANDUM UNIVERSAL LTD.	OKHA,JAMNAGAR		S	30	C,W,S	R			27		
	MAHARSHTRA MINERALS CORP.LTD	KURWADE,MAHARASTRA		S	5	C,W,S	*	*		*	*	

* BENEFICIATION: C= CRUSHING, W= WASHING, S= SCREENING, H= HYDROCYCLONE
MINES: S= SURFACE, U= UNDERGROUND
Sources: Company reports, USBM Minerals Yearbooks, Bhatt 1991, Benbow, 1988.

Surinam subsequently dominated the worldwide production. Other centers subsequently started bauxite mining, but these have been mainly concerned with the production of metallurgical grade bauxite.

France

France, the birthplace of the industry, began importing refractory grade bauxite after World War II. This importing continued to increase until 1991, when bauxite mining ceased in France due to a combination of stricter environmental impact requirements and decreasing competitiveness with imported bauxite. However SABP now has permission to rework some of the dumps to provide about 59 ktpy bauxite for cement production.

United States

The non-metallurgical bauxite mining operations in the United States faced serious competition following World War II because the levels of mining which were necessary to meet the War requirements necessitated some high grading. The consequence was increasing reliance on imported bauxite, particularly the low silica refractory grade. In this new situation all of the underground operations became uncompetitive and were closed. In addition, both of the alumina plants in Arkansas progressively switched to the production of non-metallurgical alumina. The Reynolds Hurricane Creek Alumina Plant was closed in 1984, and the Alcoa Bauxite, plant in May 1990. In addition American Cyanamid Company closed most of their mining operations in Arkansas, but until May 1990 continued to process a small amount of chemical grade bauxite for the production of aluminum sulfate at the company's facilities in Michigan and Illinois. Consequently only a few small mining operations presently remain. These also supply the requirements of Porocel Corporation at Bryant, AR, for the production of activated bauxite for the petroleum industry.

Today, the only other active bauxite mining operations in the United States are at Eufaula, AL, and Andersonville, GA. These continue to operate successfully because they were based on low iron bauxitic kaolin which is used for the production of some high alumina refractories. Presently Harbison-Walker Refractories, a Division of Dresser Industries Inc., mines the material at Eufaula for calcination at its local plant and sells some to Carbo Ceramic Company (formerly Standard Oil Proppants Co.) for the manufacture of proppants. Similarly Mullite Company of America (Mulcoa) produce bauxite from their mines in Henry County, Alabama, and Summer County, Georgia, for calcination at their plant near Andersonville, GA. They have also purchased the mines and plant facilities of A.P. Green Refractories Company in Baker Hill, AL. These latter facilities were closed in 1987. However the demand for these US domestic ores was relatively weak during 1990 so mine output was intermittent (Plunkert and Sehnke, 1992).

Guyana

Bauxite production began in Guyana (then British Guiana) by Alcoa through their local subsidiary the Demerara Bauxite Company in 1917. However the operations were transferred to Alcan, Alcoa's Canadian associate in 1929. Formal production of non-metallurgical bauxite started in 1938 as part of an overall expansion of the operations to meet the demands of the war by the Allies. The government of Guyana acquired the operations from Alcan in 1974.

The highest grade bauxite deposits occur on the right bank of the Demerara River within a 3 km radius of Linden. These deposits are mainly mined for refractory grade bauxite, the production of which is the deciding factor in the mining program. The Berbice area is the other important mining area, and both metallurgical and chemical bauxite grades are produced. Table 3 summarizes basic information on these two mining areas.

The ore is transported by rail from the mines in the Linden area to the processing plant, while in the Berbice area it is barged down the Berbice River to the processing plant at Everton. The crude ore from the mines is stockpiled at the processing plants by grades. Each grade is then crushed to −8.9 cm and transferred to vibrating screens on which it is washed by jets of high pressure water. The exposed clay is washed through the screens along with some loose gibbsite. The material remaining on the vibrating screens is further crushed to −3.8 cm and again re-washed to remove any exposed clay. The RASC grade bauxite is then calcined at 1 550°C in rotary kilns which are up to 91 m long and 3.5 m in diameter. The Everton operations are similar, but only drying of the product is done at that facility.

Surinam

Alcoa started production of metallurgical grade bauxite in 1922 and production of chemical grade bauxite started at the Moengo mine in 1939 in order to support the Allied War effort. Production of metallurgical grade bauxite ceased at that location in 1966 in favor of production of calcined grade bauxite, which had commenced there in 1953, but was terminated in 1981. Billiton also started production of abrasive, chemical, and refractory grades of bauxite at their Onverdacht mine. The facilities in the Moengo and Onverdacht areas include crushing equipment, washing equipment, and rotary kilns. Suralco (the local Alcoa subsidiary) ceased production of non-metallurgical bauxites at the Moengo Mine in 1981 because it was not competitive. However, production of metallurgical grade bauxite continued. Billiton (now a Shell subsidiary) also ceased production in 1989. The Billiton facilities, which include five rotary kilns with an aggregate capacity of 250 ktpy, have been mothballed.

Guinea

The Sangaredi bauxite deposit in the Boké region has pockets of very high grade bauxite which have been selectively mined since 1973 and stockpiled separately prior to shipment to Kamsar by rail.

Crushing and calcination facilities are located at Port Kamsar. These include storage stockpiles, crushing and 150 ktpy rotary kilns, and storage at the port for export.

Australia

Prior to 1970 only a few thousand tons per year of bauxite were produced from deposits in Victoria, Queensland, and New South Wales and were mainly for use in their domestic, chemical, and steel industries (Bracewell, 1962).

However, Australia's first bauxite calcination plant was commissioned at Weipa in 1970. The facility initially had a capacity of 100 ktpy of abrasive grade calcined bauxite and cost $3.2 million. A second kiln with an annual capacity of 100 ktpy was completed in 1975. The facilities include 65 kt product storage and are equipped with quality control facilities.

In addition, Gove Aluminium Ltd. is a leading supplier of bauxite and laterite to Portland cement producers.

China

The major bauxite mining operations in China are identified in Table 3. Henan and Shanxi are the two most important centers for

Table 4. Organized Sector Bauxite Base Units in Gujarat, India

Unit	Location	Product	Plant Capacity
Carborandum Universal Ltd.	Okha Jamnagar District	Calcined Bauxite	75 tpd
Natraj Ceramic & Chemicals	Khambhalaya Jamnagar District	High Alumina Refractory	3 000 tpd
Orient Abrasive	Porbandar Junagadh District	Calcined Bauxite	7 000 tpd
Interkilan Ceramic Pvt. Ltd.	Kadi Ahmedabad District	High Alumina Refractory	1 500 tpd
Saurashtra Calcined Bauxite Ltd.	Porbandar Junagadh Ltd.	Calcined Bauxite	50 tpd
Bombay Mineral P. Ltd.	Hapa Jamnagar District	Calcination	10 tpd
Special Refractory	Vatva Ahmedabad District	High Alumina Refractory	2 000 tpd

the production of non-metallurgical grade bauxite. Shaft, round, and rotary kilns are operated in both these provinces. Most of the shaft kilns in China are reported to have no temperature control and depend on visual inspection for quality control. Most of their high quality bauxite is diasporic, so chemical grade bauxite is not produced in China (Benbow, 1988). The operating unit is at the county level with capacity from 10 ktpy. The output from several of these groups is combined for transport to the mineral trading houses for export.

India

India has one of the oldest bauxite mining industries in the world outside Europe and the United States. Production started in 1908 in the Katni District of Madhya Pradesh, increasing from 6.7 kt in 1923 to 24 kt in 1924, but ceased operations between 1931 and 1932. This production supplied small orders from oil companies for the purification of kerosene and the production of aluminum sulfate. Efforts were made to utilize Indian bauxite for the manufacture of refractories, abrasives, and aluminous cement within the country. However, production did not increase significantly until after aluminum smelting started. From then until 1988 national emphasis was on the aluminum industry, and the export of calcined bauxite was prohibited. A change in policy occurred in 1988 and was due, firstly, to the availability of excess bauxite, alumina, and aluminum from the Orissa project of the National Aluminum Company. Secondly, there was the availability of export markets, principally in the Middle East, particularly for high silica bauxite for cement manufacturers. Thirdly, the scattered nature of the deposits in Gujarat and the size of the reserves were the reasons for the policy change.

Gujarat Mineral Development Corporation (GMDC) was given the exclusive right to develop bauxite reserves for a proposed alumina plant and so several areas were reserved for exclusive exploration by them. However, in the new circumstances private sector non-metallurgical bauxite projects are being authorized. Some of these are identified in Table 4. These are in addition to current facilities such as those of the Maharashtra Minerals Crop. Ltd., producing 5 kty of +48% Al_2O_3 from its mine at Kurwade, Raigad District. This bauxite is used in the production of cement, chemicals, petrochemicals, and abrasives. This company also produces 3 kty of black emery for the production of abrasives (Russell, 1991, Bhatt, 1991, Anon., 1990).

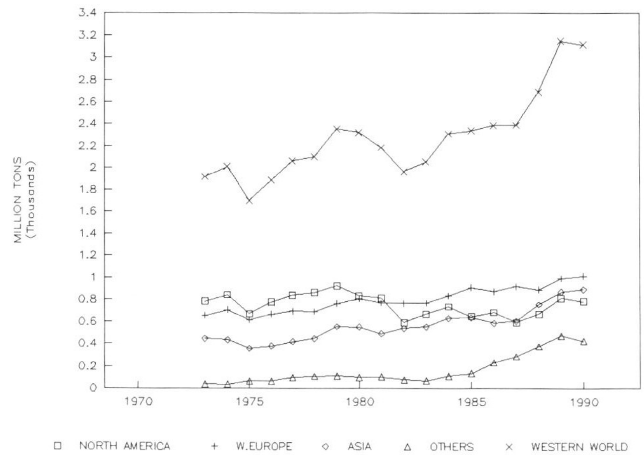

Fig. 4. Western world non-metallurgical alumina production 1973–90 (*Data Source:* IPAT).

ECONOMIC FACTORS

Markets

The major consumers of non-metallurgical bauxites are the United States, Western Europe, Japan, and the Commonwealth of Independent States (CIS), the successors to the USSR. Of these only Japan initially lacked a domestic bauxite source. However the others also had to increasingly import some of their requirements after 1950. In this evaluation, the United States is used as the typical example of these major markets.

A number of limitations were observed in the data collected. The first of these is that some metallurgical grade bauxite has been used for the production of aluminum chemicals. This is suggested by a report that about 48% of the bauxite which was used in the production of aluminum chemicals in the United States was imported and came mostly from Guyana, (then British Guiana). Secondly, abrasive grade bauxite exported to Canada for fusion is again re-exported to the United States as alumina. Thirdly, alumina is added in the production of some refractories, abrasives, and aluminum chemicals and it is unclear whether this is included in the estimation of bauxite consumed. Fig. 4 gives data on the western world production of non-metallurgical alumina from 1973 to 1990. This plot shows that Western Europe and Asia show a continuous growth in production over the period. This is in contrast to North America, which shows a discontinuity in growth in 1982. It is significant that an increase in production started in Latin America, while Oceania began production of this material at the Weipa Mine at about this time.

Figs. 5A–D illustrate the time trend plots of US non-metallurgical bauxite consumption by groups for 1950 to 1990. These show that the abrasive grade bauxite (AGB) consumption trend is relatively constant over the period, but the annual variations are usually high. The same is true of the ''other'' that is the residual group. The abrasive grade bauxite has been largely supplied by imports after 1956. In contrast, the chemical grade bauxite (CGB) consumption increased between 1953 and 1968, and since that time there has been an overall declining trend. On the source side, the trend of the domestic supply of this material showed increased growth between 1953 and 1975. This was followed by a discontinuity, and then a gradual decline starting in 1975, and continuing until the data was withheld in 1988. In the case of refractory grade bauxite (RGB), there was initially slow growth. The rate of growth then began to

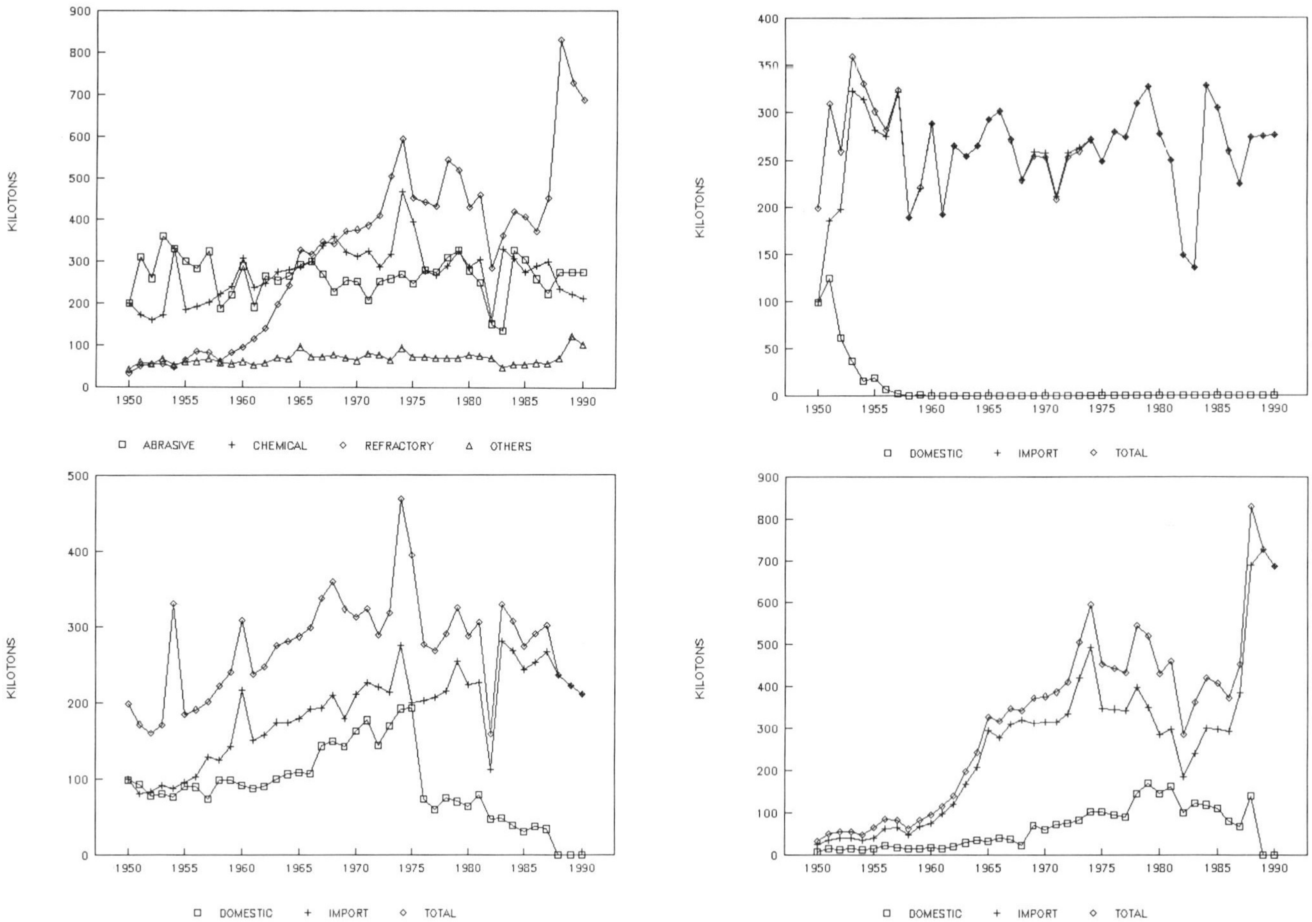

Fig. 5. US non-metallurgical bauxite consumption: 1950–90. (*Data Source: USBM Minerals Yearbooks.* A. Total by grades, B. abrasive grade, C. chemical grade, D. refractory grade.)

increase rapidly between 1960 and 1965, and then slowed down until about 1974, when the rate of growth began to decline. In 1989, the data for domestic production, which had always been the minor one, was withheld because there were too few suppliers. This withholding of data explains the apparent high increase after 1989.

Supply Systems

Bauxite mining first commenced in France and satisfied the worldwide requirement until 1887, when US production began. Other producers started operations later. By the early 1960s an average of 12% of the bauxite production of France was used for non-metallurgical applications (6% for cements, 4% for abrasives, and 2% for refractories) (Bracewell, 1962). Since that time imports have become increasingly important to the non-metallurgical bauxite supply in France, and since 1991 only the requirements for cement production have been supplied locally.

The industry in the United States developed slightly differently from that in France, but the abnormally high levels of US domestic bauxite production during World War II greatly depleted the already limited bauxite reserves. In addition the availability of large reserves of very high quality bauxites in Guyana and Surinam, with their much lower mining costs, left US bauxite increasingly uncompetitive. The consequence has been a switch in production from France and the United States to Guyana and Surinam. Fig. 6 illustrates the time trend non-metallurgical bauxite production from

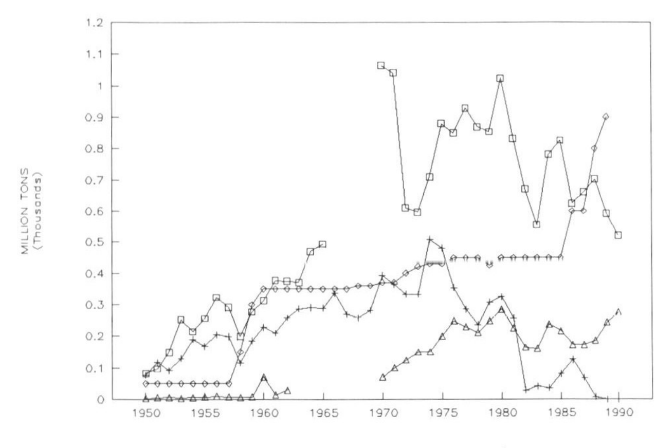

Fig. 6. Non-metallurgical bauxite supply from major exporting countries 1950–90. (*Data Source:* BIDCO, Database.)

1950 to 1990 for the major exporting countries (Guyana, Surinam, China, and Australia). Guyana dominated the worldwide refractory grade bauxite market and Surinam the chemical grade market until early 1970, when China successfully entered the international mar-

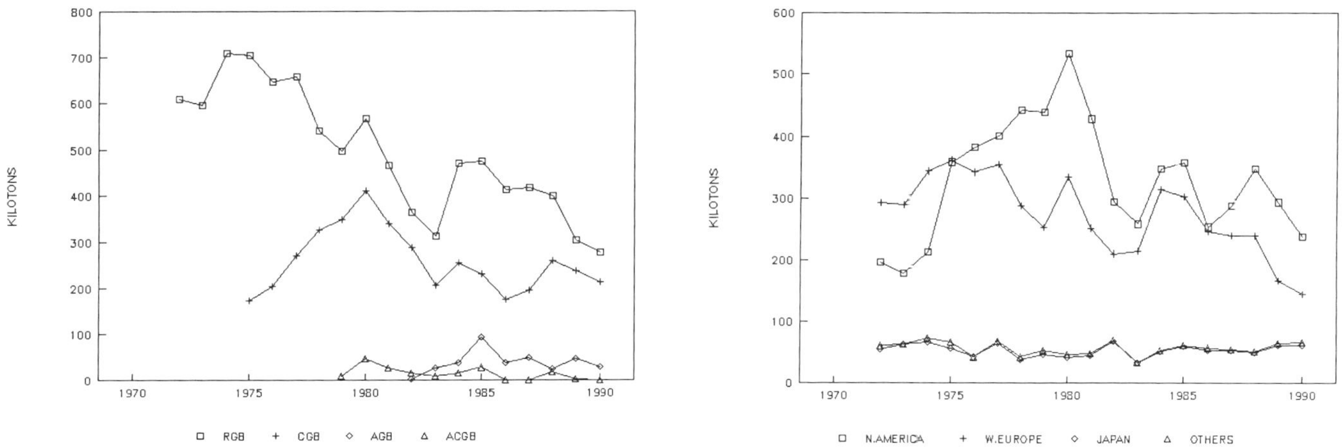

Fig. 7. Non-metallurgical grade bauxite exports from Guyana 1972–90. (*Data Source:* BIDCO. A. by Products, B. by destinations.)

ket for calcined bauxite for refractory and abrasive applications. This was followed by the commencement of production of abrasive grade bauxite by Comalco in 1970 at their Weipa mine in Australia. Sangaredi then started production of abrasive grade bauxite in 1976. Comalco's production initially went mainly to Europe and Japan, while production from Sangaredi mainly to Europe. Today Guyana and China are the major competitors supplying refractory grade bauxite to the United States, Western Europe and Japan. In the United States domestic producers such as Combustion Engineering, Harbisson-Walker Refractories, and Norton dominate the US domestic refractories market from fireclay to mullite. In addition Alcoa supplies tabular alumina which is required for the production of 90 and plus 99% grades alumina refractories. It is the intermediate high alumina refractories for which the overseas bauxite producers supply raw materials (see Table 2).

Fig. 7 illustrates the annual non-metallurgical grade bauxite exports from Guyana from 1972 to 1992 by products and destinations. Fig. 7A highlights the overall decline in export of refractory grade bauxite over the period, while chemical grade bauxite began its decline about 1980. Neither the abrasive nor the aluminous cement grades have been major products from Guyana. Fig. 7B demonstrates that North America and Western Europe have been the principal markets for non-metallurgical bauxite from Guyana, with the former region being slightly more important. These graphs suggest a decline in both of these markets. In contrast the Japanese market has been a minor but relatively steady one. Fig. 8 shows the time trends in calcined grade bauxite exports from China by destinations. It demonstrates that Western Europe has been the principal market for China since 1970. Also, there has been marked growth in export from China to Western Europe and North America since 1979. Japan has been the minor market, but it has developed steadily.

Fig. 9A illustrates the time trend plots of calcined refractory grade bauxite imports into the United States from China, Guyana, and Surinam over the period 1981 to 1990. This disaggregation of the data is necessary as the demand for these commodities reflects demand by specific industries. It shows that imports from Guyana have been gradually declining since 1984. In contrast, annual imports from China, although more erratic, suggest an overall slight increasing trend. Surinam, as a minor source, ceased production during this period. It is interesting to compare the time trend in the annual average prices for this commodity from Guyana as compared to China (Fig. 9B). This plot demonstrates that the average annual price of the product from China has been consistently lower than

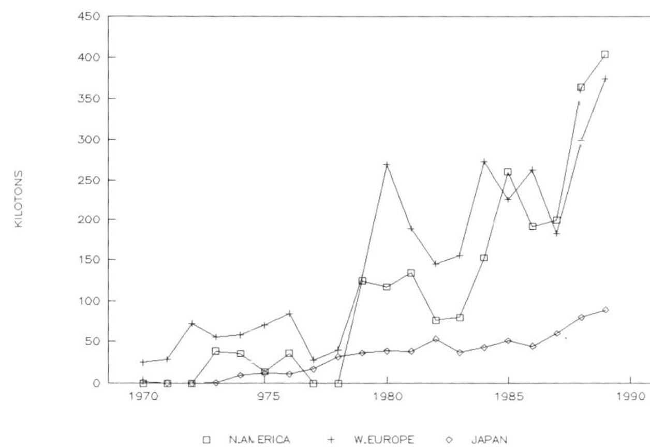

Fig. 8. Calcined grade bauxite exports from China by destinations. (*Data Source: USBM Minerals Yearbooks.*)

that from Guyana over the period. This is so although the product from Guyana is of a much higher and more uniform quality. This suggests that it is possible to produce a more competitive product by blending calcined refractory grade bauxite from China.

In contrast to the refractory grade bauxite, suppliers of bauxite for the production of brown abrasive market are Australia, Guinea, China, Guyana, and Greece. Surinam ceased production of both calcined bauxite for refractories and abrasives in 1981. Alumina is the raw material for the production of white fused alumina. This corundum is a superior product and requires alumina which contains less than 0.4% Na_2O and less than 50 ppm sulfur.

Alumina and chemical grade bauxite are the starting materials for the production of aluminum chemicals. Gibbsite, because of its greater ease of solubility must be the aluminum oxyhydroxide phase in the starting material. The important factor here is that the cost of disposing of the residue after extraction of the alumina is now resulting in the preference for alumina hydrate. Suppliers of chemical grade bauxite include Comalco in Australia, Bauxite Industry Development Company (BIDCO) in Guyana, Mineracào Curimbaba in Brazil, and Jahore Mining and Stevedoring Company in Malaysia. On the alumina side, the principal producers are B.A.

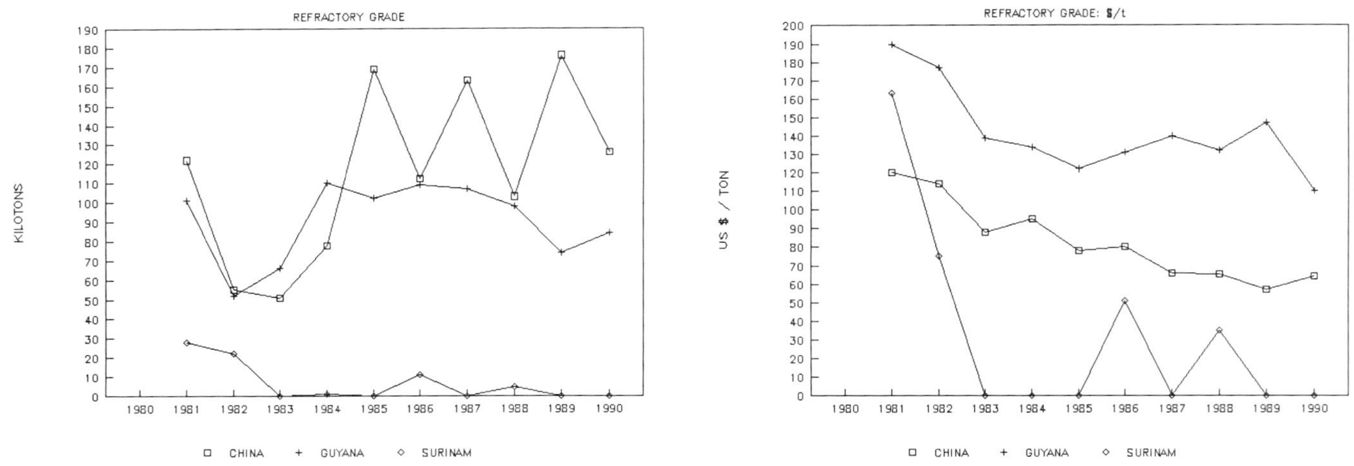

Fig. 9. Refractory grade bauxite imports into the United States. (*Data Source: USBM Minerals Yearbooks.* A. Sources B. Price.)

Chemicals, a subsidiary of British Alcan, Pechiney, VAW, Alusuisse Martinswerk GMbH in Europe, Alcoa in the US, and Alcan in Canada.

Raw materials for the production of high alumina cements are: a high iron type for the production of cement fondu and a low iron type for the production of the low iron type. Cement fondu can contain up to 25% iron oxide. In Europe Bauxite Parnasse and Eleusis Bauxite Mines in Greece, Etibank in Turkey, Jahore Mining and Stevedoring Company, Gove Aluminum Limited, and recently, Comalco in Australia are the major suppliers.

Capacities and Production

Table 3 identifies the principal non-metallurgical grade bauxite mines and processing facilities worldwide. The available non-metallurgical grade bauxite resources in Guyana, and production levels in Guyana, Surinam, China, and Australia suggest that there is no worldwide shortage, since adequate resources exist. Further the refining of all types of bauxite to give alumina hydrate for the production of hydrate and calcined low alkali alumina provides the capability to produce alumina chemicals and high quality alumina for the production of refractories and abrasives.

PROBLEMS AND FUTURE TRENDS

The non-metallurgical grade bauxites (except for the high iron cement grades) must be high in alumina and low in iron, and for the refractory grades, also low in alkalis. These requirements have traditionally placed a preminum on the bauxite from the Guiana Coastal belt and some diaspore bauxite which can be readily beneficiated. However the present need to increase the productivity of Bayer plants in order to decrease alumina production cost has resulted in some additional alumina at relatively lower cost, and the subsequent closure of some older Bayer plants which were not cost effective. The result is that alumina hydrate is now increasingly produced in plants with high liquor productivity. In addition there is increasing pressure to reduce the adverse environmental impact associated with the disposal of residues from the acid extraction of bauxite. The combination of these factors favors the use of alumina hydrate in the production of high alumina refractories, abrasives, and aluminum chemicals. These trends are demonstrated by Fig. 6 which shows that of the four major non-metallurgical bauxite exporting countries, only China presently shows an overall increase in exports and this started in 1986. In contrast, Fig. 4 shows, firstly, increasing production of non-metallurgical aluminas in the western world. Secondly, although production of non-metallurgical alumina decreased significantly in North America in 1982, it then continued to increase. In contrast, production in Western Europe and Asia has increased steadily. It is also significant that the production of non-metallurgical alumina in Latin America and Oceania have also been increasing steadily.

The question which must be asked is whether the observed decrease in non-metallurgical grade alumina in North America represents a new trend and one which will probably be followed by Western Europe and Asia. In other words, will there be a new trend for increased markets for non-metallurgical alumina in Western Europe and Japan, and who are the possible suppliers. In this regard it is noteworthy that the Friguia plant, which has been the only Bayer plant in Africa, has to date not produced non-metallurgical alumina. Also, the Sangaredi mine is nearing exhaustion, and the present surplus of mined refractory grade bauxite is currently stockpiled. GBG therefore has a limited supply from the stockpile after Sangaredi is mined out. There is presently no other deposit like Sangaredi in Guinea as far as the author is aware.

The contributions of the major non-metallurgical grade bauxite producers (i.e., Guyana and China for refractory grade, Australia and Guinea, China, and Guyana for the abrasive grade) need detailed analysis against the background of the inroads made by non-metallurgical alumina into the aluminum chemicals market. This trend is seen as an opportunity for alumina producers in bauxite producing countries to increase their production of aluminum chemicals. This is critical when it is seen that metallurgical grade alumina is increasingly produced near bauxite mines. More data is required to produce a definitive answer so these should be collected and analysed.

In addition the potential for other non-metallurgical grade bauxite producers to enter the international market place in a similar manner to China needs evaluation. The recent development of the aluminum industry in Orissa, India, and the consequent policy change regarding the utilization of bauxite from Gujarat make India a potential new entrant on the export market for refractory, abrasive, and chemical grades bauxite. The depth of overburden over most of the present non-metallurgical grade bauxite reserves in Guyana, and subsequent high mining cost, makes this a distinct possibility. This situation needs full evaluation particularly in Guyana whose only Bayer plant is presently closed and yet who needs to have all options open in order to retain the competitiveness of its industry.

The overall observation is that bauxite will undoubtedly continue to have a great future as an industrial mineral contributing to the production of refractories, abrasives, aluminous chemical, and other products. However non-metallurgical alumina will also contribute to this, so Bayer plants in bauxite producing countries should make every effort to participate in these markets as these industries continue to grow.

ACKNOWLEDGMENTS

Many persons contributed greatly to the completion of this task by providing information which has been included in the database. Among these are Errol D. Sehnke of the US Bureau of Mines who readily forwarded copies of publications which were requested. In addition the Hon. Dr. Harold Pollack, Minister of Natural Resources, Surinam, and S.L. Carmichael of the Bauxite Industry Development Company, Guyana, provided data on Surinam and Guyana respectively, while Dr. I.N. Char and Dr. M.M. Deshmukh greatly assisted my obtaining data on the industry in India. The International Bauxite Association permitted the full use of their library.

REFERENCES

Anon., 1964, "Alcoa Aluminas in Refractories," Alcoa, Pittsburgh, PA, 11 pp.

Anon., 1990, "Bauxite," Indian Minerals Yearbook, Indian Bureau of Mines, pp. 308–329.

Anon., 1991, Member Country Profile: Guyana, IBA Quarterly Review, Vol. 17, No. 2, pp. 7–10.

Anon., quarterly and annual reports, "Statistical Summaries," International Primary Aluminium Institute, London.

Anon., various annuals, Metallstatistik, Metallgesellschaft, Frankfurt, Germany.

Anon., annual, Minerals Yearbook, US Bureau of Mines, Washington, DC.

Andrews, W.H., 1984, "Uses and Specifications of Bauxite," Chap. 3, *Bauxite*, L. Jacob, Jr., ed., AIME, New York, pp. 49–66.

Bárdossy, G., 1982, *Karst Bauxite: Bauxite Deposits on Carbonate Rocks*, Elsevier, Amsterdam, 441 pp.

Bárdossy, G., and Aleva, G.J.J., 1990, *Lateritic Bauxites*, Elsevier, Amsterdam, 624 pp.

Baumgardner, L.H., 1990, "World Production and Economics of Alumina Chemicals," *Alumina Chemicals, Science and Technology Handbook*, L.D. Hart, ed., American Ceramics Society, pp. 7–11.

Benbow, J., 1988, "Bauxite Aluminating Non-metallurgical Sectors," *Industrial Minerals*, No. 252, pp. 67–83.

Bhatt, J.V., 1991, *Scope for Bauxite Based Projects in Gujarat State*, Industrial Extension Bureau, Ahmedabad, Gujarat, India, 201 pp.

Bracewell, S., 1962, "Bauxite, Alumina and Aluminium," Overseas Geological Survey, HM Stationary Office, London, 233 pp.

Chin, L.A.D., 1989, "Co-products and By-products of Alumina Plants," *Travaux*, Vol 19, No. 22, Zagreb, Yugoslavia, pp. 351–355.

Everts, J.A., 1984, "Calcined Bauxite," Chap. 5, *Bauxite*, L. Jacob, Jr., ed., AIME, New York, pp. 84–93.

Hill, V.G., and Ostojic, S., 1984, "The Characteristics and Classification of Bauxites," Chap. 2, *Bauxite*, L. Jacob, Jr., ed., AIME, New York, pp. 31–48.

Hoppe, R.W., 1977, "Sangaredi—an African Plateau of Bauxite," *Engineering and Mining Journal*, Vol. 178, No. 8.

Iannicelli, J., 1984, "Beneficiation of Bauxite to Refractory Grade Quality by High Intensity Magnetic Separation," Chap. 32, *Bauxite*, L. Jacob, Jr., ed., AIME, New York, pp. 681–689.

Patterson, S.H., Kurtz, H.F., Olson, J.C., and Neeley, C.L., 1986, "World Bauxite Resources," US Geological Survey, Professional Paper No. 1076B, US Government Printing Office, Washington, DC, 151 pp.

Plunkert, P., and Sehnke, E.D., 1992, *Aluminium, Bauxite and Alumina: 1990*, US Bureau of Mines, Washington, DC, 31 pp.

Règnier, J., 1988, "Bauxite—Its Technical and Economical History During the Last Hundred Years," *Light Metals*, TMS, Warrendale, PA, pp. 13–30.

Russell, A., 1991, "India's Industrial Minerals: Completing the Picture," *Industrial Minerals Supplement*, pp. 45–48.

Wefers, K., and Misra, C., 1987, "Oxides and Hydroxides of Aluminum," Technical Paper No. 19, revised, Alcoa, Pittsburgh, PA, 92 pp.

BERYLLIUM MINERALS

DEBORAH A. KRAMER

Although beryllium was initially discovered in 1797, it was not until 1916 that the first significant quantity of beryllium metal ingot was produced in the United States. Before 1940, most of the small quantity of beryllium metal used in the United States was supplied by Germany. Beryllium was not used in its metallic form; it was used as an alloying addition to other metals, mainly copper and nickel. World consumption of beryllium at that time was estimated to be a few hundred kilograms annually.

Because of increased tensions in Europe before World War II, imports of metal from Germany ceased in 1940, and domestic production was required to meet the needs of the military. As a result, by 1946, the United States became the principal supplier of beryllium products for the world, a condition that prevails today.

Because of beryllium's high strength, light weight, and high thermal conductivity, it has been used in a large number of industrial and defense applications as a metal, alloy, and oxide. The United States is the only country other than China and the former USSR currently producing beryllium products from beryllium ores and concentrates. Bertrandite, mined in Utah, is the principal ore source of domestically produced products. Imported ore, mainly from Brazil, supplements domestic ore supplies. Production and consumption data for the United States is shown in Figs. 1 and 2. Production data for 1969–1979 is withheld to avoid disclosing company proprietary data.

GEOLOGY OF BERYLLIUM DEPOSITS

Because of its strategic importance, beryllium mineral occurrences in the United States were studied extensively in the years after World War II, particularly by the US Geological Survey and the US Bureau of Mines. During this period, the United States did not have a reliable domestic supply of beryllium minerals and was reliant on imports to supply most of its requirements. Most domestic beryllium was recovered from beryl produced as a byproduct of mining other pegmatite minerals. After production of bertrandite began in Utah in 1969, providing the United States with a secure beryllium source, little efforts were made to evaluate additional beryllium mineral deposits. Some beryllium mineral compositions are detailed in Table 1.

Beryllium in Pegmatites

Granitic pegmatites are the world's principal source of beryl. Beryl has been mined from zones, fracture-filling units, and replacement bodies in heterogeneous pegmatites. The principal beryl deposits are zones. Most beryl-rich zones contain only a few thousand tons of rock, but a few may contain as much as 1 Mt. Beryl occurs in concentrations of approximately 2% in zones containing plagioclase, quartz, and muscovite, from which mica is the principal product. It is also as abundant in zones from which spodumene, amblygonite, and feldspar are recovered.

Beryl tends to be finer-grained in outer zones than in inner zones of pegmatites. Much of the fine- and medium-grained beryl of pegmatites in the Black Hills, SD, contains abundant inclusions of other minerals (Tullis, 1952). Large masses of beryl containing almost no inclusions have been found in inner zones and may extend across several zones. Beryl is evenly distributed in some deposits, but in others it is irregularly distributed in rich aggregates in various parts of a zone.

None of the other beryllium minerals in pegmatites are of commercial importance. Chrysoberyl is concentrated in parts of some pegmatites but is much less abundant than beryl. Other beryllium minerals are exceedingly rare.

Beryllium in Quartz Veins

Beryl has been found in quartz veins containing cassiterite, wolframite, molybdenite, and other minerals. The concentration of beryl in the outer parts of these veins, just as beryl is concentrated in the outer part of quartz zones in some pegmatites, suggests that these quartz veins are genetically similar to pegmatites. Beryl crystals in quartz veins are normally less than 2.5 cm long. Beryl has been studied in tin veins in South Dakota, England, and elsewhere. Beryl also has been reported in a tungsten vein in New Mexico (Jahns, 1944) and in quartz veins in Colorado (Adams, 1953); however, the mineral has not been reported in the more important tin and tungsten deposits of the world.

Beryllium in Replacement Deposits

Replacement deposits that contain beryllium are primarily tactites and emerald-bearing schists. Helvite is the chief beryllium mineral of tactites and related rocks, but idocrase, grossularite, chlorite, and possibly other minerals also contain beryllium. The most thoroughly documented beryllium deposits in tactites in the United States are in New Mexico (Jahns, 1944) and Alaska (Sainsbury, 1963), although the Alaskan deposits are of minor significance.

Beryllium in Other Deposits

Nepheline syenite and other nepheline rocks have been reported to contain as much as 0.1% beryllium oxide. Deposits of this type have not been studied thoroughly enough to determine the commercial feasibility of their development, although small quantities of beryllium-rich deposits have been found in Arkansas and Colorado (Warner et al., 1959).

Beryllium also has been found in coals. Sampling of a variety of US coal ashes indicated a significant distribution pattern for beryllium, with an average of 46 ppm for all of the samples (Stad-

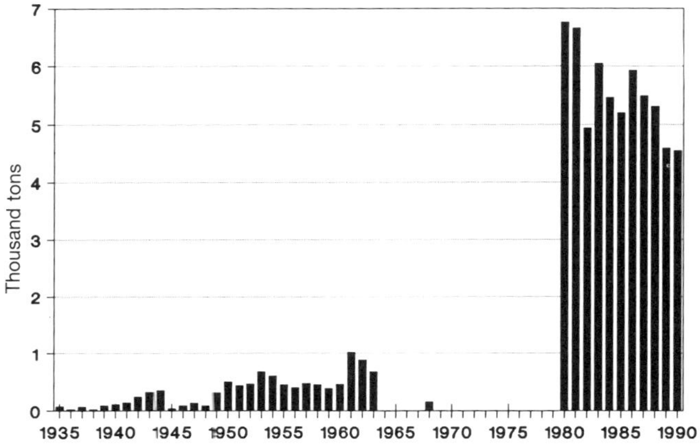

Fig. 1. US beryllium mineral production.

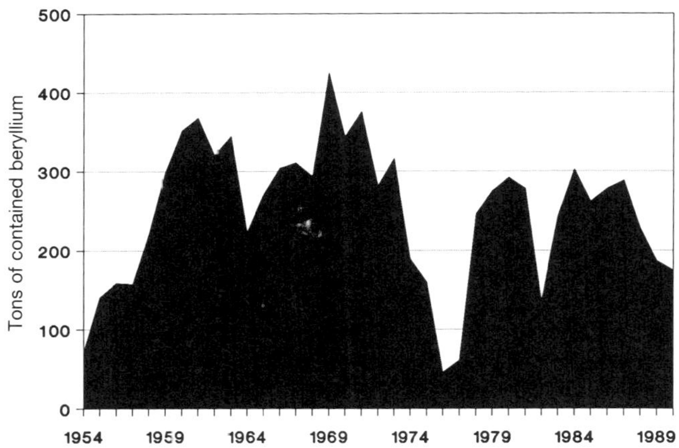

Fig. 2. US beryllium demand.

Table 1. Composition of Some Beryllium Minerals

Mineral	Chemical composition	% BeO
Barylite	$BaBe_2Si_2O_7$	15.4–15.8
Bertrandite	$Be_4Si_2O_7(OH)_2$	39.6–42.6
Beryl	$Be_3Al_2(SiO_3)_6$	10.0–14.0
Beryllonite	$NaBePO_4$	19.8
Chrysoberyl	$BeAlO_4$	16.9–19.7
Danalite	$Fe_4Be_3Si_3O_{12}\cdot S$	12.7–13.8
Eudidymite	$HNaBeSi_3O_8$	10.6–11.1
Helvite	$Mn_4Be_3Si_3O_{12}\cdot S$	10.5–15.0
Herderite	$CaBePO_4(OH,F)$	15.0–15.8
Phenacite	Be_2SiO_4	44.0–45.6

Adapted from Warner, 1959.

nichenko, Zubovic, and Sheffey, 1961). Accumulation of beryllium in coal was determined to be a syngenetic process, and the beryllium content of the coal primarily is dependent on the rocks that contributed material to the coal-depositional sites.

DISTRIBUTION OF MAJOR DEPOSITS

United States

One method of evaluating the importance of individual deposits is examining the production record of each deposit. Since 1969, nearly all of the US beryllium mineral production has been from one mine in Utah, and production of beryl has nearly ceased. The few beryl-producing mines that operate are almost all family owned, and production is sporadic. In the late 1950s, however, beryl was recovered as a byproduct at as many as 200 operations in 12 different states. Quantities produced during this period ranged from a few kilograms to hundreds of tons. Because of the wide variation in beryl concentration within an ore body, an individual mine may produce byproduct beryl for a few years, and then cease beryl recovery entirely.

New England: In general, New England beryllium occurrences are either very small or low grade and as such are not economically exploitable. Pegmatitic deposits in New England are grouped into five major districts, with other scattered occurrences. The major districts are Middletown, CT, Keene and Grafton, NH, Topsham and Paris-Rumford, ME. Additional scattered pegmatites containing beryl occur in western Connecticut, western and central Massachusetts, eastern New Hampshire, and Maine. New Hampshire pegmatites were estimated to contain 1 630 t of beryl (Page and Larrabee, 1962). Nonpegmatitic deposits in New England include beryl-bearing granite in western Connecticut, central Massachusetts, and eastern Maine; phenacite-danalite-helvite bodies in granite in central New Hampshire; and beryl-bearing quartz veins in northern Rhode Island.

The principal beryllium mineral in all the deposits except one is beryl, sometimes with associated small quantities of herderite, bertrandite, beryllonite, or chrysoberyl. Because of ionic substitution of alkali metals in the crystal lattice, beryl in New England generally contains about 12.0 to 12.5% BeO and may contain as little as 9%, significantly less than the theoretical BeO content (Barton and Goldsmith, 1968).

Beryllium mineral production in New England was limited to the recovery of beryl as a byproduct of pegmatite mining. Mica and feldspar mining in New Hampshire, Connecticut, and Maine began in the 1800s, and beryl was recovered sporadically from some of these operations until 1963. Most of the beryllium-bearing mines in New England were small open pits, and mining equipment was minimal.

Colorado: Colorado was one of the largest producers of beryllium minerals in the late 1950s. Much of the production was from a nonpegmatite deposit in Park County. Beryllium was found in dumps of the Boomer mine, a lead-silver prospect dating back to the 1890s, and this led to the discovery of high-grade, vein-type beryllium ore in the old workings. Production began in 1956 and continued for a few years. This was most likely the first nonpegmatite ore mined in commercial quantities. The most important minerals present in the Boomer deposits are beryl and bertrandite (Meeves, 1966).

South Dakota: During the late 1950s and early 1960s, South Dakota was the largest beryllium-mineral producing state in the United States. Beryl was recognized in the pegmatites of the Black Hills in the late 1800s, but substantial quantities were not produced until 1914. Most beryl was recovered as a byproduct of feldspar and mica mining. Minor quantities of beryl continue to be recovered from deposits in Custer County.

The principal minerals in the South Dakota pegmatites are feldspar, quartz, and muscovite. Beryl was reported in all pegmatite zones; commercial production, however, was predominantly from intermediate zones. Beryl was reported in fair quantities in the wall zones of certain pegmatites, and production was reported from the core. Both fracture fillings and replacement bodies may contain beryl.

Although some deposits in the Black Hills have been investigated, reserve estimates have varied widely, because many deposits were not investigated fully or developed. Runke, Mullen, and Cunningham (1952) published reserve estimates of 13 300 t, with

2 400 t in dump material at 75 properties and 10 900 t in unmined portions of the deposits examined.

Texas: The Sierra Blanca region in the far western part of Texas was investigated in the 1980s by Cabot Corp. as part of a US-wide beryllium geochemical reconnaissance survey. This region is characterized by five microgranite bodies that intrude sedimentary rocks. Mineralization in the microgranite bodies includes bertrandite, behoite, a beryllium hydroxide mineral, phenacite, and minor quantities of chrysoberyl. Reserve estimates of these deposits were projected to be 11 300 t of BeO contained in deposits grading more than 2% BeO (Anon, 1986). A test mine was erected and ore samples were evaluated to determine optimum processing methods, but because of a downturn in the beryllium market, a commercial operation was not started.

Utah: The principal beryllium mineralization in Utah is bertrandite, which occurs in the Spor Mountain area. In 1959, beryllium occurrences were discovered in a rhyolitic tuff on the lower slopes of Spor Mountain in Juab County. The deposit is associated with fluorspar and uranium. The beryllium mineralization occurs principally in a Tertiary rhyolitic tuff that is overlaid by hard, abrasive rhyolite. The tuff rests on a surface of Devonian dolomite and limestone. Drilling in the area indicated that the tuff increases in thickness from a few meters to more than 80 m.

Montoya, Havens, and Bridges (1962) studied the chemical, mineralogical, and physical properties of the tuff. Two beryllium-bearing minerals were found to be present—a hydrated form of bertrandite and a berylliferous saponite. Each of the mineral forms are predominant in certain areas of the district, but the beryllium is distributed erratically throughout the tuff in individual mineralized bodies. Samples of the tuff that were examined were found to contain from 0.2 to 1.8% BeO.

Brush Wellman Inc. began mining this deposit in 1969 by open pit methods. As the company continues to develop new open pits, resource information on this area is updated. In its 1990 annual report, Brush Wellman estimated proven reserves to be 6 131 000 t of bertrandite, grading 0.251% BeO. Probable bertrandite reserves were estimated to be 6 624 000 t, grading 0.281% BeO.

Other Countries

Brazil: Brazil is traditionally the largest beryl supplier in the international market. The principal beryl-producing areas are the Rio Doce Valley and adjacent areas in the State of Minas Gerais, the Campina Grande District in the State of Rio Grande do Norte, the Critais-Berilandia District in the State of Ceara, and the Jequitinhonha River Province in the State of Bahia. Soja and Sabin (1986) estimated that more than 106 Mt of ore containing more than 15 000 t of beryllium was present in some properties evaluated in these areas, and almost 99% of the contained beryllium is available from the State of Minas Gerais.

All commercial-grade pegmatites in Brazil are complex-zoned or heterogeneous pegmatites. Heterogeneous pegmatites are normally composed of three zones, which are progressively coarser grained toward the quartz nucleus. The intermediate zone contains beryl as well as spodumene, columbite-tantalite, cassiterite, and semiprecious gemstones.

Canada: Beryllium occurrences were first noted in the Thor Lake area, near Yellowknife, Northwest Territories, in 1978. Further geological investigations resulted in the definition of a number of unusual mineral assemblages that appear to be products of the somatic replacement of syenitic and granitic types. Beryllium mineralization, primarily as phenacite, is widespread in all the rock types in the area except the syenite (Trueman, Pedersen, and de St. Jorre, 1984).

In the mid-1980s Highwood Resources Ltd. began drilling in the area and identified four different zones—the R zone, the S zone, the T zone, and the Lake zone. Much of the beryllium mineralization occurred in the T zone, and reserves were estimated to be 450 000 t of ore grading 1.4% BeO in the T zone and 1.3 Mt of ore grading 0.66% BeO south of the T zone (Anon, 1985). Preliminary studies indicated that recoveries of 80 to 90% could be obtained, producing a concentrate containing 10 to 13% BeO. Because of difficulties in marketing the products, a mine was never opened at this location.

Large concentrations of barylite and eudidymite have been reported near Seal Lake in Labrador. The only other significant occurrence of beryllium in Canada is the Hellroaring Creek prospect west of Kimberley, British Columbia. The prospect is owned 80% by Bearcat Exploration Ltd. and 20% by Colt Exploration Ltd. The property is underlain by a beryl-bearing pegmatitic body, with indicated ore resources of 500 000 t, grading 0.1% BeO (Schiller, 1985).

Other occurrences: Beryllium has been recovered from small deposits in many countries throughout the world, but production has been sporadic. Over the past 40 years, the countries with the largest contribution to world production, outside Brazil, the US and the USSR, are Argentina, India, Mozambique, and Zimbabwe. Because of the nature of beryl mining, production is subject to internal political and social pressures in individual countries, particularly those in Africa. As a result of limited production, deposits in most other countries are not well documented.

TECHNOLOGY

Exploration

In many cases, the presence of beryl is visible by the naked eye. It appears as greenish-white crystals, and because it is usually associated with colorless quartz, large crystals are easily distinguished. For other beryllium minerals and finely disseminated beryl crystals, a beryllometer is normally used in the field for detecting beryllium mineralization. The beryllometer uses a gamma radiation supplied by an antimony 124 radioisotope to generate neutrons when the radiation field comes into contact with a beryllium-bearing mineral. This equipment is similar to a Geiger counter used to detect uranium. Beryllometers can be made with a sensitivity of 0.01% beryllium or less. Detection of beryllium in ore is usually followed by diamond drilling and assaying to delineate beryllium deposits further.

Mining

Most of the beryl produced outside the United States is from Brazil, where labor-intensive, pick-and-shovel mining of weathered pegmatites is the primary method used. Garimpeiros (laborers) perform all of the mining and cobbing of the ore. Because the garimpeiros mine on private land and sell directly to ore buyers, data concerning mining methods and production is scarce.

Pegmatite mining in Brazil essentially consists of stoping into exposed, weathered pegmatites with picks and shovels. Mines are established wherever beryl-rich pegmatites exist. Each stope is worked by three to five men who pick beryl crystals out of the host rock and muck out the waste. Other garimpeiros (usually women and children) sort through the mucked-out material for beryl crystals that were overlooked. Stopes are approximately 1.5 m high and 1.0 m wide. A mine can contain as many as 40 stopes (Soja and Sabin, 1986).

In Utah, Brush Wellman mines bertrandite by open pit methods. Unlike beryl, in which the mineral can be identified by color and crystal structure, bertrandite mineralization cannot be recognized by the naked eye. Consequently, Brush Wellman conducts geologic and geochemical evaluations on a specific area, followed by a drilling program to determine if an economic ore body exists.

After delineating an ore body, overburden is removed to within 2 m of the ore by a local earth-moving contractor during the winter and spring. This stripping operation usually requires about 9 months for completion. In the 2-m cover remaining, drill benches are constructed on 7.5-m centers to take samples of the ore body at 0.6-m intervals. Information obtained from analyzing the samples allows cross-sectional and contour maps to be developed. These maps are used to plan the mining and processing operations.

After the maps are prepared, the remainder of the overburden is removed, and the ore is mined, typically with a self-loading scraper. Because of the irregular ore-grade distribution in the ground, the ore is mined from predetermined areas and placed in a stockpile in layers to obtain a homogeneous blend. Drilling, sampling, and assaying the stockpiled ore generate a map that identifies ore-grade distribution throughout the stockpile. Stockpiled ore is trucked to the mill, which is about 80 km away, for further processing.

Beryllium Hydroxide Production

Beryllium is recovered commercially from two ores—beryl and bertrandite. Beryllium hydroxide, the intermediate product recovered from these ores, is subsequently converted to beryllium metal, alloys, or oxide. Brush Wellman is the only company with facilities to extract the beryllium values from both ores. Because of the difference in beryllium content of beryl and bertrandite, the ores are treated separately until the beryllium is dissolved, then the two solutions are combined. A basic flowsheet for Brush Wellman's process is shown in Fig. 3.

Bertrandite is crushed and wet-milled to yield a slurry at about -20 mesh. The slurry then is leached with a sulfuric acid solution at about 95°C to dissolve the beryllium, forming a beryllium sulfate solution. In contrast, after crushing, beryl ore is heated to 1 700°C and quenched rapidly in water to form a frit. Heat treating the frit at 1 000°C, grinding to about -200 mesh, and then leaching with a concentrated sulfuric acid solution at 250 to 300°C dissolves the contained beryllium, forming a beryllium sulfate solution.

After separating the solids from both beryllium sulfate solutions, the two solutions are combined. Solvent extraction is used to remove additional impurity elements that were extracted with the beryllium. In solvent extraction, the beryllium sulfate solution contacts an organic solution of di-2-ethylhexylphosphoric acid in kerosene. Beryllium is selectively dissolved in the organic solution. The slow rate of extraction at room temperature is accelerated by warming the extractant and leach solution. The loaded organic phase is treated with an aqueous ammonium carbonate solution, and beryllium is stripped from the organic phase forming $(NH_4)Be(CO_3)_3$.

In addition to beryllium, the solvent extraction step dissolves iron, aluminum, and fluoride. Heating the strip solution to about 70°C separates the iron and aluminum as hydroxide or carbonate precipitates. Two hydrolysis steps at 95 and 165°C, respectively, remove the ammonia and carbon dioxide, resulting in the formation of beryllium hydroxide, which is filtered and then drummed for shipment. Approximately 80% of the beryllium content of both the beryl and bertrandite is recovered by this process (Walsh, 1979).

Beryllium-Copper Alloy Production

To produce beryllium-copper master alloy, beryllium hydroxide, electrolytic copper, and carbon are combined in an electric-arc furnace. The resultant melt, containing about 4% beryllium, is cast

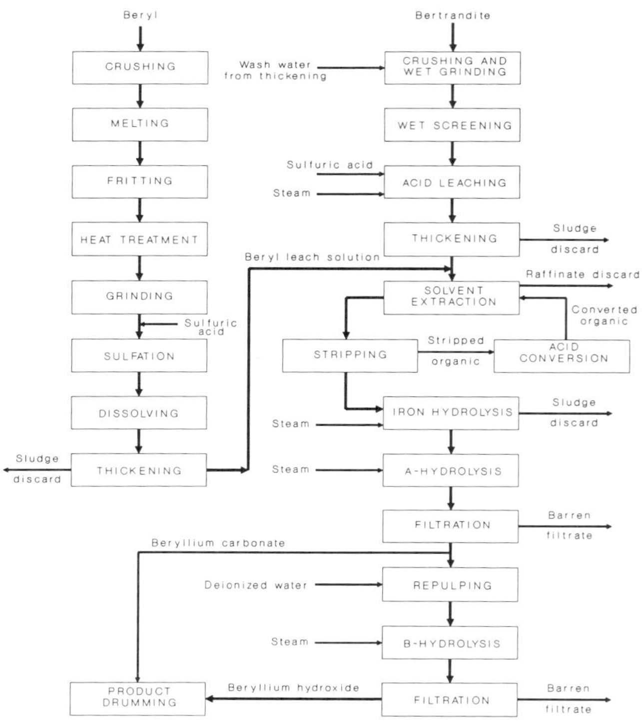

Fig. 3. Beryllium hydroxide production from bertrandite and beryl. (*Source*: Brush Wellman Inc.)

into ingots. Re-melting master alloy ingots with additional copper and other alloying elements yields the desired beryllium-copper alloy, which is then cast into slabs or billets. Slabs of beryllium-copper alloys are processed further into strip or plate, and billets are extruded into tube, rod, bar, and wire products. Specifications for some beryllium-copper alloys are shown in Table 2.

Beryllium Metal Production

To produce beryllium metal, beryllium hydroxide is dissolved in an ammonium bifluoride solution. Filtering and treating the solution with ammonium sulfide removes impurities such as iron, lead, manganese, and zinc. The purified solution in concentrated in an evaporator to yield an ammonium beryllium fluoride salt. After the salt is removed with a centrifuge and dried, it is fed to a continuous induction furnace. Decomposing the salt at 700 to 900°C in the induction furnace produces molten anhydrous beryl-

Table 2. Beryllium-copper Alloy Specifications (weight %)

	Brush alloy no.			
Constituent	25	165	10	50
Be	1.80–2.00	1.60–1.79	0.40–0.7	0.25–0.50
Co	—	—	2.4–2.7	1.4–1.7
Co + Ni, min	.20	.20	—	—
Co + Ni + Fe, max	.60	.60	—	—
Ag	—	—	—	0.90–1
Cu	Bal.	Bal.	Bal.	Bal.
Density, g/cm^3	8.36	8.41	8.82	8.82

Source: Brush Wellman Inc.

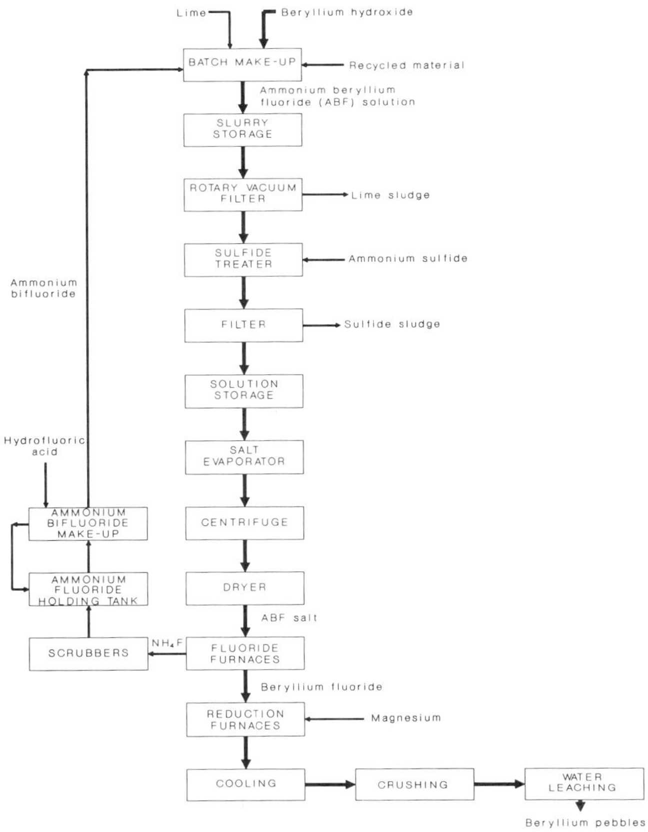

Fig. 4. Beryllium pebble production. (*Source*: Brush Wellman Inc.)

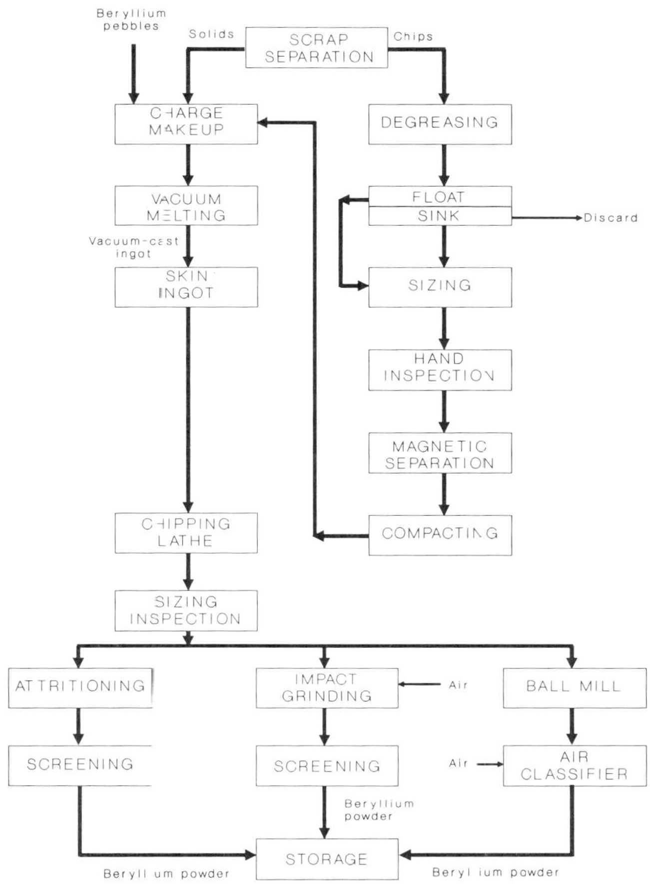

Fig. 5. Beryllium powder production. (*Source*: Brush Wellman Inc.)

lium fluoride and ammonium fluoride gas, which is captured in scrubbers for reuse. Cooled beryllium fluoride is reacted with magnesium in batch induction furnaces to produce metallic beryllium and magnesium fluoride. Cooling the mixture produces a solid cake that contains beryllium pebbles, magnesium fluoride, and unreacted beryllium fluoride. Crushing this mixture followed by water leaching yields beryllium metal and magnesium fluoride. After separating the magnesium fluoride, the pebbles are vacuum melted to remove any slag trapped in them and cast into ingots.

Vacuum-cast ingots are normally further processed into a powder produced by machining the ingot into chips. Chips cut by the lathe are removed by a vacuum system, and after passing through a cyclone, are automatically loaded into drums. Grinding in a ball mill, impact grinding, or attrition grinding reduces the size of the chips. Each of these methods produces a different particle size powder. In attrition grinding, chips are ground into a powder between a fixed, grooved beryllium plate and a rotating, grooved beryllium plate. In impact grinding, a high velocity stream of air blasts a mixture of chips and oversize powder onto a solid beryllium target. Upon impact, the chips break into small particles that are removed from the chamber by vacuum conveying. A ball mill is used when extremely fine powder is required. After grinding, the beryllium powder is sized, either with screens or an air classifier. Beryllium powder is formed into billets by vacuum hot-pressing, hot-isostatic pressing, or cold-isostatic pressing, depending on the end use of the material. Beryllium metal production is detailed in Figs. 4 and 5, and specifications are shown in Table 3.

Table 3. Beryllium Specifications (weight %)

Constituent	Commercial beryllium powder SP-200F	National Defense Stockpile hot pressed powder billet Grade A	Grade B	Grade C
Be, min	98.5	98.0	98.0	94.0
BeO, max	1.5	1.5	2.2	9.0
Al, max	.1	.07	.10	.16
C, max	.15	.10	.15	.25
Fe, max	.13	.12	.15	.25
Mg, max	.08	.08	.08	.08
Si, max	.06	.08	.08	.08
Other, each, max	.04	.04	.04	.10

Sources: Brush Wellman Inc., National Stockpile Purchase Specifications P-110a-R1, Mar. 25, 1983 and P-110B, Oct. 18, 1985.

Beryllium Oxide Production

Beryllium oxide powder is produced by dissolving beryllium hydroxide in sulfuric acid. The resulting beryllium sulfate solution is concentrated by evaporation and cooled until the solution is super saturated. During cooling, beryllium sulfate tetrahydrate crystals are produced. Wet salt crystals are removed from the mother liquor in a centrifuge. Calcining these crystals in a car-hearth furnace at

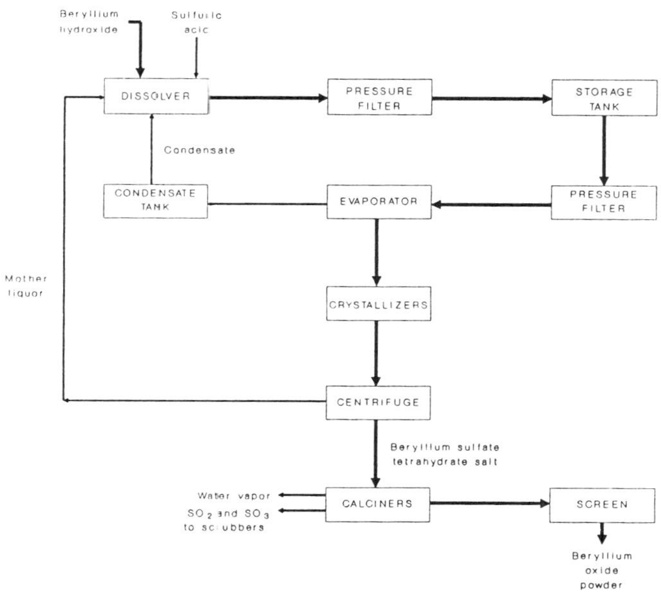

Fig. 6. Beryllium oxide production. (*Source*: Brush Wellman Inc.)

temperatures up to 1 430°C decomposes the salt to beryllium oxide, water vapor, and SO_2 and SO_3 gases. The gases are fed through scrubbers, and the beryllium oxide is cooled, screened, and bagged for shipment. Beryllium oxide production is shown in Fig. 6.

INDUSTRY STRUCTURE

The United States is the only country that can process beryllium ore and concentrates into beryllium products and supplies most of the rest of the world with these products. Brush Wellman mines bertrandite and converts this ore, along with beryl, to beryllium hydroxide at its facility in Delta, UT. Beryllium hydroxide is shipped to the company's plant in Elmore, OH, where it is converted into beryllium alloys and metal. Brush Wellman also produces beryllium oxide at a plant in Tucson, AZ, from the beryllium hydroxide recovered at Delta.

One other company in the United States has the capability to produce beryllium alloys. NGK Metals Corp., a subsidiary of NGK Insulators of Japan, produces beryllium alloys at a plant in Reading, PA. Because NGK Metals does not have facilities to process the raw materials, the company purchases imported beryl, and Brush Wellman converts the beryl into beryllium hydroxide under a tolling arrangement.

Ore production from labor-intensive operations outside the United States is sporadic, except for the largest producers—Brazil, China, and the former USSR. Most of the beryl recovered in Brazil is exported to the United States, and that country is the largest US beryl import source. Some beryl from China has been imported into the United States in the past, but it is assumed that China has some capability to process beryl because products including beryllium-copper master alloy originating there have surfaced in the world market. Although little is known about the USSR, it is assumed that it has an integrated beryllium industry because little, if any, beryl produced in that country has been traded.

PROPERTIES AND USES

Beryllium-copper Alloys

Beryllium-copper alloys are used in a wide variety of applications and average about 65% of annual US consumption on a beryllium metal equivalent basis. These alloys, most of which contain approximately 2% beryllium, are used because of their high electrical and thermal conductivity, high strength and hardness, good corrosion and fatigue resistance, and nonmagnetic properties. Beryllium-copper strip is manufactured into springs, connectors, and switches for use in applications in automobiles, aerospace, radar and telecommunications, factory automation, computers, home appliances, and instrumentation and control systems. The principal use of large diameter beryllium-copper tubing is in oil and gas drilling equipment and in bushings and bearings in aircraft landing gear and heavy machinery. Connectors in fiber optic telecommunications systems are the main application for beryllium-copper rods. Small, pluggable sockets for joining integrated circuits to printed circuit boards are the main application for beryllium-copper wire. Beryllium-copper bar and plate are used in resistance-welding parts, components for machinery and materials-handling systems, and for molds to make metal, glass, and plastic components.

Beryllium also is used in small quantities in nickel- and aluminum-base alloys. Miniature electronic connector components that operate at high temperatures are the main use for beryllium-nickel alloys. Beryllium-aluminum alloys are used as castings in the aerospace industry. Addition of small quantities of beryllium to magnesium alloys inhibits oxidation.

Beryllium Metal

Beryllium metal, which averages about 20% of annual US demand, is used principally in aerospace and defense applications. Its high stiffness, light weight, and dimensional stability over a wide temperature range make it useful in satellite and space vehicle structures, inertial guidance systems, military aircraft brakes, and space optical system components. Because beryllium is transparent to X-rays, it is used in X-ray windows. In nuclear reactors, beryllium also serves as a canning material, as a neutron moderator, in control rods, and as a reflector. In the past, the metal has been used as a triggering device in nuclear warheads. Other applications for metallic beryllium include high-speed computer components, audio components, and mirrors. In the US space shuttles, several structural parts and brake components use beryllium.

Beryllium Oxide

Beryllium oxide (beryllia) is an excellent heat conductor, with high hardness and strength. This material also acts as an electrical insulator in some applications. Beryllium oxide, averaging about 15% of domestic demand, serves mainly as a substrate for high-density electronic circuits for high-speed computers, automotive ignition systems, lasers, and radar electronic countermeasure systems. Because it is transparent to microwaves, microwave communications systems and microwave ovens also use beryllium oxide.

Table 4. Yearend Beryllium Prices

Year	Beryllium metal ($/lb)	Beryllium-copper strip (Alloy 25) ($/lb)	Beryllium oxide ($/lb)	Beryl ore[a] ($/short ton unit of BeO content)
1970	$ 60.00	$2.93	$13.50	$ 35.00
1971	60.00	3.05	13.50	33.00
1972	61.00	3.14	18.00	30.00
1973	61.00	3.14	18.00	30.00
1974	61.90	3.78	26.00	30.00
1975	61.90	3.96	26.00	30.00
1976	61.90	4.00	26.00	36.00
1977	96.00	4.31	26.00	40.00
1978	103.00	4.77	26.00	40.00
1979	120.00	5.54	35.38	47.00
1980	148.00	6.36	37.50	69.00
1981	166.00	6.61	40.00	94.00
1982	166.00	7.10	49.60	121.00
1983	178.00	7.10	49.60	126.00
1984	178.00	7.10	52.50	88.00
1985	196.00	7.25	55.70	87.00
1986	204.00	7.65	59.30	88.00
1987	229.00	8.00	61.35	84.00
1988	244.00	8.00	65.65	93.00
1989	261.00	8.90	65.65	90.00
1990	268.83	9.25	72.50	101.00

Source: American Metal Market.
[a]Average yearly value of US imports.
Conversion factor: 1 lb = 0.45 kg; 1 short ton **unit** = 20 lb.

ECONOMIC FACTORS

Prices

Prices for imported beryl and domestically produced beryllium products are shown in Table 4. After Brush Wellman opened its bertrandite mining operation in 1969, applications for beryllium metal were expanded in the aerospace and nuclear industries. As these new uses were developed, prices remained relatively steady to encourage continued expansion. In the late 1970s, prices escalated rapidly in response to high inflation rates and increased energy costs. At the same time, the Occupational Safety and Health Administration (OSHA) and the Environmental Protection Agency (EPA) promulgated new regulations concerning beryllium air emissions. In order to comply with these regulations, plants were required to install additional pollution control equipment. The additional costs were ultimately passed on to the consumer. In the 1980s, greater regulation of waste products from beryllium mining and processing increased disposal costs. Again these cost increases resulted in higher prices for the consumer.

Tariffs and Depletion Allowances

Tariffs for beryl and beryllium products are shown in Table 5. Beryllium ores carry a 22% domestic depletion allowance and a 14% foreign depletion allowance.

Alternative Materials

Because of its high cost compared with those of other materials, beryllium is used in applications in which its properties are crucial. Steel, titanium, or graphite composites substitute for beryllium metal in some applications, and phosphor bronze substitutes for beryllium-copper alloys, but these substitutions result in substantial loss in performance. In some cases, aluminum nitride may be substituted for beryllium oxide.

SAFETY AND ENVIRONMENTAL REGULATIONS

Beryllium dust and fumes have been recognized as the cause of beryllosis, a serious chronic lung disease. In the 1940s the disease was diagnosed among industry employees and their relatives, who had handled dusty workclothes prior to the establishment of suitable hygienic procedures. Cases were also reported among residents of communities surrounding beryllium-processing plants. Although uncertainties related to the cause of the disease still exist, the problem appears to be controlled when established preventative measures are exercised. In beryllium-processing plants, harmful effects are prevented by maintaining clean workplaces; requiring the use of safety equipment such as personal respirators; collection of dust, fumes, and mists at the source of deposition in dust collectors; medical programs; and other procedures to provide safe working conditions. Control of potential health hazards adds significantly to the final cost of beryllium products.

Under the Clean Air Act, EPA has issued standards for certain hazardous air pollutants, including beryllium. Beryllium emissions for plants that process beryllium materials are limited to 10 g of beryllium over a 24-hour period. EPA also established water quality criteria for beryllium under the Clean Water Act. The Clean Water Act is intended to ensure a water quality sufficient to protect a use for which the water body has been designated. Typical designations include protection of aquatic life and wildlife habitat and protection of human health. A concentration level of 130 μg/l of beryllium was established to protect aquatic organisms from acute health effects, and a level of 5.3 μg/l was established for protection of chronic health effects. For human health, EPA established a concentration of 3.7 ng/l to protect human health during the ingestion of water and fish, and a concentration of 64.1 ng/l was established to protect human health during the ingestion of fish only. Beryllium currently is not regulated by EPA under the Safe Drinking Water Act.

OSHA also issues standards for airborne beryllium particles.

Table 5. US Import Duties

Tariff item	HTS No.	Most favored nation (MFN) Jan. 1, 1991	Non-MFN Jan. 1, 1991
Ore and concentrate	2617.90.0030	Free	Free
Unwrought beryllium	8112.11.6000	8.5% ad valorem	25.0% ad valorem
Beryllium, wrought	8112.19.0000	5.5% ad valorem	45.0% ad valorem
Beryllium waste and scrap	8112.11.3000	Free	Free
Beryllium-copper master alloy	7405.00.6030	6.0% ad valorem	28.0% ad valorem
Beryllium oxide or hydroxide	2825.90.1000	3.7% ad valorem	25.0% ad valorem

Current OSHA standards limit the 8-hour exposure level to an average of 2 $\mu g/m^3$, with a peak of 25 $\mu g/m^3$ not to exceed 30 minutes, and a ceiling concentration of 5 $\mu g/m^3$. Mine Safety and Health Administration regulations affect only Brush Wellman's operation in Utah and are the same standards as those of OSHA.

PROBLEMS AND FUTURE TRENDS

The beryllium industry is unique because essentially one producer outside China and the USSR controls raw material processing for the rest of the world. Although several other companies have investigated beryllium as a potential business to develop, it is unlikely that any new plants will be constructed in the United States, primarily because of the associated costs of construction of a plant that meets environmental regulations. Offshore construction of beryllium processing plants would be more economically feasible, if a fairly large ore source was developed.

Because of its high cost and potential health hazards, designers and consumers specify beryllium only when there is a clear technical and economic advantage over other materials. In 1985, Brush Wellman introduced a series of patented beryllium-copper alloys. Alloy 174, created primarily to compete with other copper-based alloys, such as phosphor bronze, is less expensive than traditional beryllium-copper alloys. Alloy 174 is used in applications that require physical properties between those of traditional copper alloys and those of other beryllium-copper alloys. In particular, Alloy 174 is replacing brass and phosphor bronze in automotive relays, switches, and other electrical devices. Continued development of beryllium materials such as Alloy 174 could increase the use of beryllium by lessening some of the cost barriers. Brush Wellman also is developing alternative processes for producing beryllium metal designed to reduce costs and maintain quality. Until technologies such as this are fully developed to reduce the cost of the finished products, beryllium demand is unlikely to increase significantly.

REFERENCES

Anon., 1985, "Highwood Bulk Sampling," *Industrial Minerals*, No. 212, May, pp. 9–10.

Anon., 1986, "Cyprus and Cabot Forming Joint Venture," *Metals Week*, Vol. 57, No. 42, Oct. 20, p. 7.

Adams, J.W., 1953, "Beryllium Deposits of the Mount Antero Region, Chaffee County, Colorado," Bulletin 982-D, US Geological Survey, 24 pp.

Barton, W.R., and Goldsmith, C.E., 1968, "New England Beryllium Investigation," Report of Investigations 7070, US Bureau of Mines, 177 pp.

Jahns, R.H., 1944, "Beryllium and Tungsten Deposits of the Iron Mountain District, Sierra and Socorro Counties, New Mexico," Bulletin 945-C, US Geological Survey, 37 pp.

Meeves, H.C., 1966, "Nonpegmatitic Beryllium Occurrences in Arizona, Colorado, New Mexico, Utah, and Four Adjacent States," Report of Investigations 6828, US Bureau of Mines, 68 pp.

Montoya, J.W., Havens, R., and Bridges, D.W., 1962, "Beryllium-Bearing Tuff From Spor Mountain Utah: Its Chemical, Mineralogical, and Physical Properties," Report of Investigations 6084, US Bureau of Mines, 15 pp.

Page, J.J., and Larrabee, D.M., 1962, "The Beryl Resources of New Hampshire," Professional Paper 353, US Geological Survey, 49 pp.

Runke, S.M., Mullen, D.H., and Cunningham, J.B., 1952, "Progress Report on Pegmatite Investigations in South Dakota for Fiscal Year Ended June 30, 1951," Report of Investigations 4928, US Bureau of Mines, 46 pp.

Sainsbury, C.L., 1963, "Beryllium Deposits of the Western Seward Peninsula, Alaska," Circular 479, US Geological Survey, 18 pp.

Schiller, E.A., 1985, "Beryllium—Geology, Production and Uses," *Mining Magazine*, Vol. 152, No. 4, Apr., pp. 317–322.

Soja, A.A., and Sabin, A.E., 1986, "Beryllium Availability—Market Economy Countries," Information Circular 9100, US Bureau of Mines, 19 pp.

Stadnichenko, T., Zubovic, P., and Sheffey, N.B., 1961, "Beryllium Content of American Coals," Bulletin 1084-K, US Geological Survey, 43 pp.

Trueman, D.L., Pedersen, J.C., and de St. Jorre, L., 1984, "Geology of the Thor Lake Beryllium Deposits: An Update," *Contributions to the Geology of the Northwest Territories*, Vol. 1, I.N.A.C., pp. 115–120.

Tullis, E.L., 1952, "Beryl Resources of the Black Hills South Dakota," Report of Investigations 4855, US Bureau of Mines, 19 pp.

Walsh, K.A., 1979, "Extraction," *Beryllium Science and Technology*, Vol. 2, D.R. Floyd and J.N. Lowe, eds., Plenum Press, New York, pp. 1–11.

Warner, L.A., et al., 1959, "Occurrence of Nonpegmatite Beryllium in the United States," Professional Paper 318, US Geological Survey, 198 pp.

BIOTECHNICAL MATERIALS

NELSON R. SHAFFER

Biotechnology has become a household word of the nineties, and it is expected to become as important in the next century as the computer is in the present. Numerous books and articles portray biotechnology developments as nothing less than a scientific revolution. Almost everywhere one looks new biotechnical breakthroughs are being reported that offer almost limitless opportunities to harness the force of living things to produce materials and manipulate their properties. Biotechnology has been broadly defined as any applications of biological organisms, systems, or processes to manufacturing and service industries. This seemingly new technology is, in fact, one of mankind's oldest scientific activities (Table 1), which has been recently revolutionized by techniques of genetic engineering that arose out of basic research in biology, biochemistry, genetics, and information sciences. From fields as old as agriculture and medicine to those as new as monoclonal antibodies, transgenic plants, or biocomputers are encompassed by biotechnology. Like most human endeavors, industrial minerals play critical roles in biotechnology. In addition biotechnology holds real potential to improve extraction and beneficiation of certain industrial minerals themselves.

BIOTECHNOLOGY OVERVIEW

Companies using established biotechnical techniques make up large and diverse groups such as agriculture, chemicals, and pharmaceuticals. The massive US Pharmacopia (Anon., 1990a) provides detailed specifications for minerals used in medicines. Alumina, zirconia, apatite, and bioactive glass have seen service as implant materials (Williams, 1990) and new uses for minerals in health sciences are being actively researched. Agriculture produced \$361 billion worth of food and drink during 1991 in the United States; organic chemicals, pharmaceuticals, and enzymes accounted for \$68, \$59, and \$42 billion, respectively (Anon., 1992a). It is not possible to separate the contributions of industrial minerals to biotechnical products, but they represent a very large and rapidly growing new field of uses.

The new biotechnology has nearly 300 small companies, plus 15 established companies with 742 biotechnology-related firms (Dibner, 1991b). Revenues exceeded \$2 billion in 1990 and are expected to grow to \$50 billion by 2000 (Anon., 1992c), with worldwide sales exceeding \$100 billion (Burrill and Roberts, 1992). Federal research amounted to \$3.4 billion in 1990 (Anon., 1992b). The United States is the world leader in biotechnology, but other countries have large, well-funded programs. Despite debate about safety, obstacles to new biotechnology products are declining (Embers, 1992, Gibbons, 1991). Fifteen biotechnology drugs valued at \$1.2 billion (Thayer, 1991a) are on the market, and more than 100 are in various stages of testing (Edington, 1992). Many diagnostic tests are also in use or development (Demain, 1983, Anon., 1992c). Large scale efforts to produce or transform important chemicals are also underway (Ng et al., 1983, Hinman, 1991), and research into geologic uses of biotechnology has begun.

Much has been published about microbial mining, oil recovery, desulfurization, bioremediation, and other geologic aspects of biotechnology, but this chapter is the first attempt to explore interactions of biotechnology and industrial minerals. This chapter examines uses of minerals in biotechnology; how biotechnology can be used to discover, recover, and beneficiate industrial minerals; and speculates on some potential, but as yet untried uses.

Definitions

What exactly does the word biotechnology mean? Bud (1989) states that the first use of the term was by Karl Ereky in 1919 to cover the interaction of biology and technology, and in 1933 the term was used in *Nature*. After citing seven different definitions, Smith (1988) concludes that biotechnology is a series of enabling technologies involving practical applications of organisms or their subcellular components to manufacturing and service industries or to environmental management. Walker and Cox (1988) suggest a definition of "the practical applications of biological systems to the manufacturing and service industries and to the management of the environment." Primrose (1991) says that it is "the commercial exploitation of living organisms or their components."

There is essentially an older broad use of the term and a new use. The US Office of Technology Assessment (Anon., 1984) uses a broad definition that includes any technique that uses living organisms (or parts of organisms) to make or modify products, to improve plants or animals, or to develop micro-organisms for specific uses. Definitions are different, but they all have several fundamental elements that include the control, management, or manipulation of living things for commercial, industrial, or useful ends. While such a definition encompasses all of agriculture in practice the "new" biotechnology is restricted to processes involving microorganisms—plant and animal cells, or enzymes.

Many consider biotechnology to be recent, but it is one of our oldest technologies as evidenced by the prehistoric origin of brewing, cheese-making, and other techniques. Table 1 gives some of the important developments in the history of biotechnology. Smith (1988) breaks down historical developments of biotechnology into four phases: 1) prehistoric with no understanding of underlying processes; 2) nonsterile processes; 3) sterile processes after 1940; and 4) genetic and recombinant DNA technology—deliberate design of special organisms or processes.

Information Sources

Biotechnology has a long history, during most of which it was known as industrial microbiology, and which covers such essential

Table 1. Historical Development of Biotechnology (from Anon., 1984, Smith, 1988, Ehrlich, 1990, Scragg, 1991, Prentis, 1984, Ralph, 1985a, 1985b)

before 3000 BC	Proving bread with leaven
before 3000 BC	Fermentation of juices to alcoholic beverages
before 3000 BC	Knowledge of vinegar formation from fermented juices
3rd century BC	Manufacture of beer in Sumaria, Babylonia, and Egypt
about 150 BC	Copper recovery from China
3rd century AD	Wine making promoted in Germany
from 1150 AD	Production of spirits of wine (ethanol)
14th century	Vinegar manufacturing industry near Orleans
after about 1650	Visualization of yeast cells by Leeuwenhoek
before 1670	Copper mined at Rio Tinto, Spain
1857	Description of lactic acid fermentation by Pasteur
1881	Microbiological production of lactic acid
1888	Winogradsky observed bioaccumulation of iron
1890	Bacterial role in rock weathering by Muentz
1897	Detection of fermentation enzymes in yeast by Buchner
from 19th century	First communal sewage plants in Berlin, Hamburg, Munich, Paris, and other cities
1915	German process for the manufacture of bakers' yeast
1914–1916	Process for the large-scale production of food and fodder yeast by Delbruck, Hayduck, and Henneberg
1915/1916	Weizmann process for the manufacture of butanol and acetone
from about 1920	Manufacture of citric acid in a surface process
1928/1929	Discovery of penicillin by Fleming
1937	Microbiological transformations discovered by Mamoli and Vercellone
1941/1944	Beginning of penicillin manufacture
1944	Discovery of streptomycin by Schatz and Waksman
1947	Colmer and Hinkle show microbial oxidation
1949	Submerged process for the production of acetic acid
from about 1949	Manufacture of vitamin B_{12}
from about 1949	Technical microbiological transformations
from 1950	Discovery of many other antibiotics
1950s	ZoBell's work with sulfate reducing bacteria and petroleum microbiology
1953	Discovery of the structure of deoxyribonucleic acid by Watson and Crick
after 1957	Manufacture of glutanic acid by Kinoshita et al.
about 1955/1960	Manufacture of citric acid in the submerged process
1962	Microbial Uranium begins in Canada
Modern Era	
1973	First cloning of a gene
1973	Brazilian alcohol fuel program
1975	First hybridoma created
1976	First US firm to exploit rDNA technology
1980	Supreme court rules that microorganisms can be patented
1981	First monoclonal antibody diagnostic kits
	First automated gene synthesizer marketed
1982	First rDNA pharmaceutical product (human insulin) approved
1983	First expression of a plant gene in a plant of a different species
1985	Permit issued by EPA for release of a genetically altered organism
1986	Technology Transfer Act of 1986 rights to commercialize government research
1987	US Patent and Trademark Office announces animals are patentable
1988	NIH extablishes program to map the human genome
	First US patent on an animal—transpenic mouse
1989	Bioremediation gains attention
1990	FDA approves recombinant renin, an enzyme used to produce cheese
	First approval of human gene therapy clinical trial
1991	Biotechnology companies sell $17.7 billion in new stock
	EPA approves the first genetically engineered biopesticide for sale
1992	Transgenic food approved
	Gene therapy trials

topics as waste water treatment, brewing, food making, and production of bulk chemicals, amino acids, and pharmaceuticals. Accounts of industrial microbiology appeared in the Sept. 1981 issue of *Scientific American*. The industry is very large, diverse, and important to modern life. Reed (1982), or Miller and Litsky (1984) provide details of the field and the series, *Developments in Industrial Microbiology,* gives annual updates.

Because of its multidisciplinary nature, biotechnology has a literature that is scattered through many fields. Crafts-Lighty (1986) made a good, but now dated, guide to the biotechnology literature, and Woodman (1989) produced a more recent guide. Johnson and Rader (1990), Bruce et al. (1989), and Crafts-Lighty (1991) profile databases available (Table 2). Journals that include biotechnology research or review articles are shown in Table 2. Dictionaries by Coombs (1986) and Walker and Cox (1988) help clarify complex terms. Numerous business sources provide current data through newsletters, clipping services, and other functions.

Great public interest in biotechnology has resulted in many popular books (Davis, 1991, Hodgson, 1989, Antebi and Fishlock, 1986, Sylvester and Klotz, 1983). Prentis (1984), Teitelman (1989),

Table 2. Sources with Information About Biotechnology

Journals	
Advances in Biochemical Engineering/Biotechnology	Current Microbiology
Advances in Biotechnology Processes	Developments in Industrial Microbiology
Annual Review of Microbiology	Enzyme and Microbial Technology
Applied and Environmental Microbiology	European Journal of App. Micro. and Biotechnology
Applied Microbiology and Biotechnology	FEMS Microbiology Letters
Archives of Microbiology	Folia Microbiologic
ASM News Biomaterials	Geomicrobiology Journal
Biorecovery	International Biodeterioration
Bioresearch Today	Journal of Applied Bacteriology
Bioseparation	Journal of Bacteriology
Bio/Technology	Journal of Biotechnology
Biotechnology Advances	Journal of Chemical Technology and Biotechnology
Biotechnology and Applied Biochemistry	Journal of Fermentation Technology
Biotechnology and Bioengineering	Journal of General and Applied Microbiology
Biotechnology and Genetic Engineering	Journal of General Microbiology
Bioscience, Biotechnology, and Biochemistry	Letters in Applied Microbiology
Biotechnology Letters	Microbiology (Russian Translation)
Biotechnology Progress	Microbiological Reviews
Biotechnology Techniques	Process Biochemistry
Canadian Journal of Microbiology	Progress in Industrial Microbiology
Chemical & Engineering News	Research in Microbiology
Chemical Market Reporter	Systematic and Applied Microbiology
Chemical Week	Trends in Biotechnology
Critical Reviews in Biotechnology	World Journal of Microbiology and Biotechnology
Critical Reviews in Microbiology	

Computer Bibliographies		
Agricola	Biosis	Georef
Biobusiness	Biotechnology Citation Index	Life Sciences Abstracts
Biocommerce Abstracts	Chemical Abstracts	Medline

Directories	
Biotechnology Directory (Coombs and Alston, 1991)	International Biotechnology Handbook (Hack et al., 1988)
Biotechnology Guide for Japan (Yuan and Dibner, 1991)	State by State Biotechnology Directory
Federal Biotechnology Information Resources	Biotechnology Guide USA (Dibner, 1991a)

and Yoxen (1983) look at the business and social aspects. General microbiology books, such as those by Postgate (1992), and Gest (1987), commonly contain sections on biotechnology.

Brock and Madigan (1991) and Prescott, Harley, and Klein (1990) are good microbiology texts with special sections on biotechnology. Laskin and Lechevalier (1977) and Williams, Sharpe, and Holt (1989) give comprehensive data on microbiology. O'Leary (1989) and Gerhardt (1981) present practical aspects of growing microbes. Many other texts devote varying amounts of attention to biotechnology.

The new era of biotechnology began in the early 1970s when genetic engineering allowed manipulation of DNA. By inserting selected genetic material into plasmids then placing plasmids into functioning cells, those new cells could be induced to produce desired products, commonly in large amounts. Details of genetic engineering can be found in Primrose (1991), Watson et al. (1992), or in simplified expositions, such as Drlica (1992) or Witt (1990). Techniques continue to improve so that virtually any cell can now be domesticated.

A number of specialized information sources have grown to serve the burgeoning industry. The US Office of Technology Assessment (OTA) published a review of the impacts of genetic engineering (Anon., 1981). Moo-Young (1985) and Rehm and Reed (1981) try to cover the entire field. Handbooks by Atkinson and Mavituna (1983) or Bailey and Ollis (1986) and books by Wang et al. (1979), Peppler and Perlman (1979), Reed (1982), Vogel (1983), Sikyta et al. (1983), Stanbury and Whitaker (1984), and Smith, Berry, and Kristiansen (1980) cover industrial microbiology practices. Multi-author volumes address details of biotechnology (Wiseman, 1983, Higgins, Best, and Jones, 1985, Ouellette and Cheremisinoff, 1985a, 1985b, Demain and Solomon, 1986, Bu'lock and Kristiansen, 1987, Marx, 1989, and Moses and Cape, 1991). Crueger and Crueger (1990) published a comprehensive textbook about biotechnology.

Several culture collections nurture and distribute known strains of microorganisms. In the United States the preeminent one is the American Type Culture Collection (ATCC) at Rockville, MD, which can provide more than 50 000 well characterized bacteria strains (Gherna and Pienta, 1992) plus fungi, yeasts, and other items (Edwards, 1988). Numerous countries have similar collections and Crafts-Lighty (1991) and Kirsop (1988) present guides to services by various collections. These collections provide organisms for biotechnology along with considerable literature and practical experience.

BIOTECHNOLOGY INDUSTRY

The biotechnology industry contributes greatly to the United States and world economics. Table 3 shows a select few of the products of biotechnical processes. In addition to staples, such as bread, cheese, and wine, other fermented food products are found around the world (Odunfa, 1988, Hesseltine, 1981, Batra and Millner, 1976). Substantial amounts of organic acids (Mattey, 1991), enzymes, amino acids (Aida et al., 1986), and specialty chemicals

Table 3. A Few Selected Products of Biotechnology

	Food/Beverages		
Bread Single cell protein Beer	Cheese Yeast Wine	Yogurt Natto Whiskey	Vinegar Soy Sauce Sake
	Fuels		
Ethanol	Butanol	Methane	Hydrogen
	Bulk Chemicals		
Organic Acids		Amino Acids	
Lactic acid Citric acid Itaconic acid Tartric acid Lipase	Acetic acid Glutamic acid 2-Ketogluconic acid Erythorbic acid	Lysine Tryptophan Valine Glycine Methionisene	Aspartic acid Arylglycines Leucine Cysteine
	Fine/Special Chemicals		
Amalyse Glucose isomerase	Protease Pectinase	Invertase Rennin	Glucose Oxidase Amyloglucosidase
	Antibiotics		
Penicillin Chloramphenicol Griseofulvin Polymyxin B	Bacitracin Cycloheximide KanamycinLincomycin Streptomycin	Cephalosporin Cycloserine Neomycin	Tetracycline Erthromycin Nystatin
	Engineered Products		
Blood Proteins		Hormones	
Tissue plasminogen activator Erythropo etin	Factors VII, VIII, IX Urokinase	Insulin Epidermal growth factor β-Endorphin Atriopeptin	Human growth hormone Parathyroid hormone Bone growth factor
Immune Modulators		Vaccines	
Interferons Tumor necrosing factor Lysozyme	Interleukin-2 Colony stimulating factor Allogenic factor	Hepatitis B Measles AIDS	Cytomegalovirus Cholera Rabies

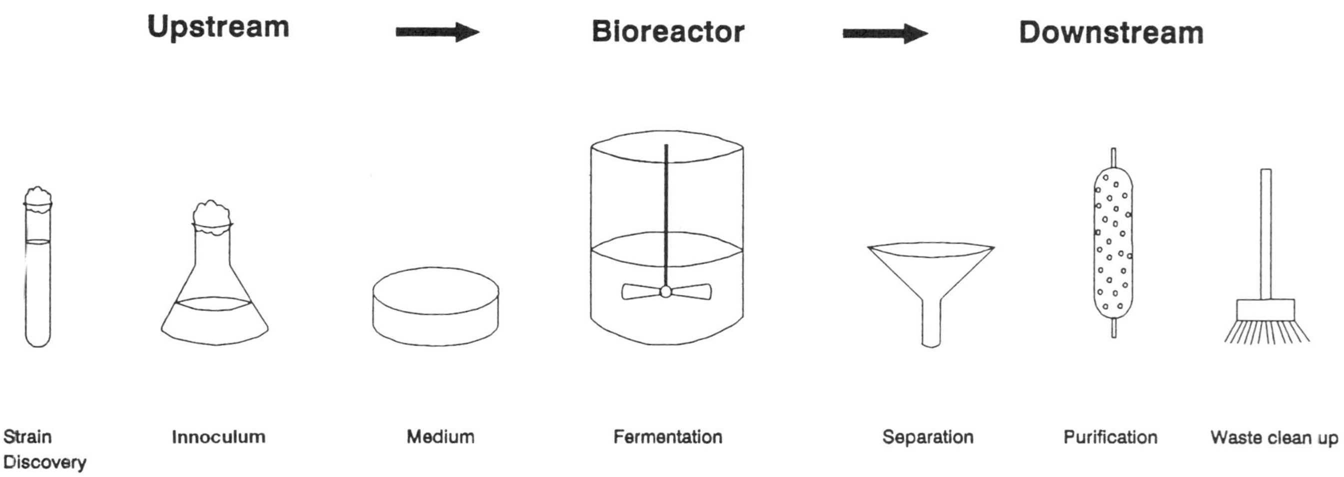

Fig. 1. Generalized flow diagram of fermentation processes. Industrial minerals find use in many facets of biotechnology.

are produced by fermentation. Fuels, especially ethanol, are very important to some economies, such as Brazil. The most visible products of biotechnology are new drugs, vaccines, and medical products or techniques. Many of these depend on fermentation, which can be defined as any process that produces a useful product by mass culture of microorganisms. A flow diagram of fermentation processes is given in Fig. 1. Fermentation is serviced upstream by provision and improvement of bacteria or inoculum, raw material acquisition, preparation of growth media, and sterilization. Very well known microbes are added to the media and grown under carefully controlled conditions in bioreactors where products are formed by very specific biological reaction. After microbial growth, products are harvested by various downstream separation processes then purified to final products.

Bioreactors

Fermentation is carried out in fermenters or bioreactors (Fig. 2) which serve a central role in biotechnological processing. These are fully sterilizable vessels in which an organism can be grown in a controlled manner to 1) produce cell mass, 2) produce a product, such as enzyme or metabolite, or 3) transform an existing compound added to the reactor (Scragg, 1991). These tanks are usually cylindrical ranging up to about 400 000 l and operate in batch, fed batch, or continuous modes. Batch fermenters operate by loading, sealing, sterilizing, inoculating, running reactions, then removing and cleaning reactors analogous to brickmaking in periodic kilns. In the fed-batch mode, nutrients are added incrementally to produce metabolites that could be interfered with by rich media. In continuous fermentation, media and microorganisms are added continuously, while media with product are removed analogous to a tunnel kiln operation. Some processes work best with fixed- or fluidized-bed reactors in which microbes are immobilized on solid substrates.

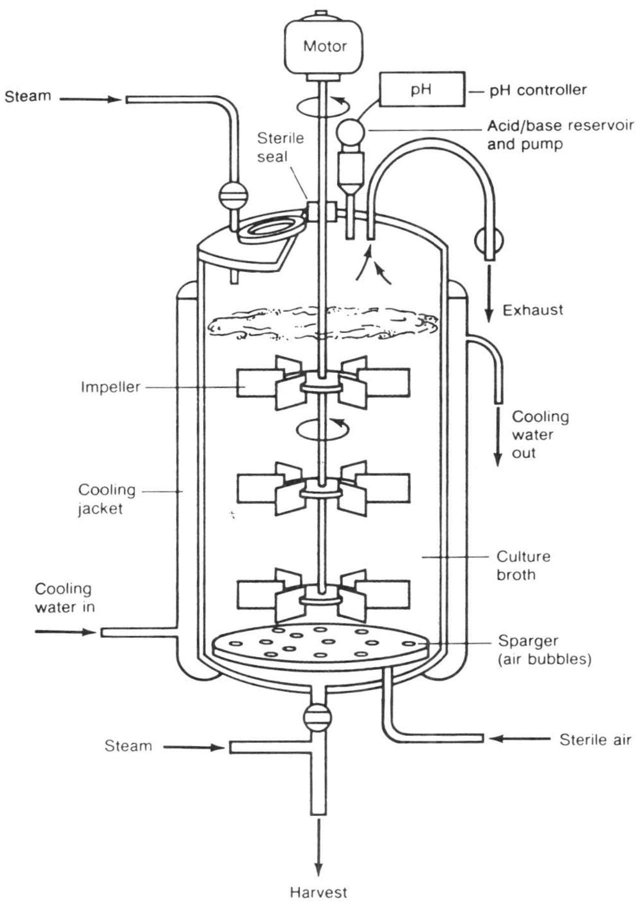

Fig. 2. Schematic diagram of generic large-scale, stirred bioreactor or fermenter (from Brock and Madigan, 1991). Industrial minerals are used in pH control, in formulating culture broth, and in sterilizing water, media, and air.

Microbial growth produces heat that is removed by cooling jackets. Reactors are closely monitored so that growth can be optimized. Most commercial microbes need air, and so most reactors are stirred to provide equal access to nutrients and air. A different type, the air-lift or loop reactor, bubbles air through the media. Many other designs are available, but most current industrial batch processes use stirred tank-aerated bioreactors. Large scale animal cell culturing presents special problems (Ho and Wang, 1991). Air and media can be sterilized by heating, irradiation, chemicals, scrubbing, or filtration, but only heat and filtration are practical at industrial scales. At one time activated carbon, mineral-wool fibers, or asbestos were common filter material, but now glass-fiber filters and cartridges are the most used sterilization items (Crueger and Crueger, 1990).

Culture Media

Each organism is grown in a specifically formulated culture medium to optimize cell growth or production of desired product. This aqueous (usually) medium must provide all essential nutrients for life, plus raw material to build new cells and produce product. The mineral component can account for 4 to 14% of cost in some cases. One of the major uses for industrial minerals is in making growth medium (Table 4). All organisms require eleven essential elements to live and grow, and only four of these (H, O, C, and N) make up 99% of most cells. Other essential elements are Na, K, Ca, Mg, P, S, and Cl. Seventeen additional trace elements V, B, Cr, Mn, Fe, Ca, Ni, Cu, Zn, Mo, Si, Se, F, I, Ar, Cr, and Sn occur sporadically and in very low amounts (Frausto-DeSilva and Williams, 1991). Considerable amounts of industrial minerals are consumed as animal and plant food (Allen, 1992, Kilmer, 1979). An additional, but unknown amount is needed to grow microorganisms for biotechnology.

Industrial minerals also play a role in control of fermentation processes (Table 5). Calcium carbonate can be used for pH control, as can phosphoric and other mineral acids. Silicones, derived indirectly from silica (Buchner et al., 1989), can be used to control foam. Some materials, such as calcium carbonate or clay, aid cell growth and are used even though their exact functions are unknown. Sikyta et al. (1983) and Smith, Berry, and Kristiansen (1980) noted that carbonates from different sources showed very different effects on growth. Calcium carbonate or $CaCl_2$ improved ethanol production, but NaCl decreased output (Sreekumar and Basappa, 1992). Xu and Hang (1989) inhibited one troublesome strain of microbe by using sodium or potassium bicarbonate. Park et al. (1985) showed suppression of yeast growth with increased salt. Water for growth must be low in trace elements because certain elements, such as iron, can cause problems with excess cell growth. Jones and Greenfield (1983), Jayme and Blackman (1985), and Calam (1967) reviewed media for industrial fermentation. Brune, Reach, and O'Connor (1981) investigated petroleum-related saltwater as media for biomass production with mixed results. Klausmeier et al. (1963) demonstrated ten-fold changes in fungus development by varying mineral salts.

Clay minerals affect microbial growth in soils so it is not surprising that clays show considerable effects in industrial fermentation. Weaver and Dugan (1972) reported increased bacterial methane oxidation when clay particles were present. Polysaccharide production is enhanced by clay additives (Lanza and Watson, 1984) but continuous culture growth of one strain was decreased by addition of kaolinite (Maigetter and Pfister, 1975). Perez-Rodriguez, Carretero, and Maqueda (1989) noted that treated sepiolite enhanced good methanogenic bacteria while repressing sulfate reducers in anaerobic bioreactors. Large populations of bacteria in kaolinite slurries showed little effect on rheology or clay products (Eagon et al., 1987).

Table 4. Inorganic Compounds and Minerals Used in Making Culture Media for Biotechnology

Phosphates	Nitrates	Sulfates		Chlorides		Carbonates	Acids/Bases	Others
KH_2PO_4	NH_4NC_3	$CaSO_4$	$MgSO_4$*	NaCL	KCL	$CaCo_3$	HCl	KI
K_2HPO_4	KNO_3	$FeSO_4$*	$ZnSO_4$*	$CaCl_2$*	$FeCl_3$	$NaHCO_3$	H_2SO_4	NaBr
Na_2HPO_4	$NaNO_3$	$MnSO_4$*	$CuSO_4$*	$MgCl_2$	$MnCl_2$	Na_2CO_3	HBO_3	NaF
Na_3PO_4	$Ca(NO_3)_2$	K_2SO_4	$CoSO_4$	NH_4Cl	$ZnCl_2$	$NaKCO_3$	H_3PO_4	$Na_2S \cdot 9H_2O$
$CaHPO_4$	$Na(NO_3)_2 \cdot 6H_2O$	VSO_4*	Al_2SO_4*	$SrCl_2 \cdot 6H_2O$	$CuCl_2 \cdot 2H_2O$	K_2CO_3	KOH	$Na_2B_4O_7 \cdot 10H_2O$
		Na_2SO_4*	NH_4SO_4	$NiCl_2 \cdot 6H_2O$	$CoCl_2 \cdot 6H_2O$	$AlCO_3 \cdot 6H_2O$	NaOH	$NaM_0O_4 \cdot 7H_2O$
		$Fe(NH_4)_2(SO_4)_3$	$AlK(SO_4)_2$	LiCl	$SnCl_2$			$NaHSeO_3$
		$NiSO_4 \cdot 6H_2O$	$CaSO_4 \cdot 2H_2O$					

*Plus variously hydrated compounds

Sources: Demain and Solomon, 1986, Crueger and Crueger, 1990, Gherna and Pienta, 1992, Stanbury and Whitaker, 1984, Atkinson and Mavituna, 1983.

Table 5. Inorganic Compounds and Minerals Used as Process Aids

pH Control	Sterilization	Growth Aids
Phosphoric acid	Glass Fiber	Clays
Sulfuric Acid	Mineral Wool	Vermiculite
Boric Acid	Asbestos	Limestone
Calcite	Norite	Gypsum
Caustic Soda		
Sodium Hydroxide		
Potassium Hydroxide		

Sources: Demain and Soloman, 1986, Crueger and Crueger, 1990, Atkinson and Mavituna, 1983.

Clays are known to interact readily with organic compounds and so to make good carriers for pesticides, and numerous studies of clay/pesticide/organisms interactions appear in the literature (Boudot, 1992, Oya et al., 1991, Stockmeyer, 1991). Bacteria and viruses are known to adhere to clay particles (Lipson and Stotzky, 1984). One theory, championed by Cairns-Smith (1982) and Cairns-Smith and Hartman (1986), even suggests that life arose by reaction of clays and prebiotic organic compounds. Further study on clay mineral reactions with bacteria under industrial production conditions would be beneficial.

Immobilization

Daniels (1972), Kolot (1981a, 1981b), and Kovalenko and Sokolovskii (1992) reviewed inorganic materials used for immobilization (Table 6). Adsorption onto surfaces is used in industrial processes by immobilizing (attaching) cells or enzymes to inorganic substrates, thus enhancing activity or preventing bacteria from being washed out of the bioreactor. Immobilization is an important topic (Tampion and Tampion, 1987, Hartmeier, 1988) because it improves production by increasing cell density, prolonging life, protecting cells from shear stresses or high pressures, providing mechanical stability, and allowing recovery of cells or enzymes.

Immobilized materials act as biocatalysts that can increase reaction rates by 10^3 to 10^{11} times over non-catalyzed rates, and they constitute a multi-billion-dollar industry (Klibanov, 1983). Cells or enzymes can be immobilized, chemically or physically, by entrapment in hollow fibers, ceramic monoliths, and spun fibers or in polymer matrices, such as agar, carragennan, alginites, cellulose, or other organic materials (Wu and Wisecarver, 1992). This latter microencapsulation requires gellification of the polymer/cell matrix by reaction with $CaCl_2$, NaCl, or KCl (Rosevear, 1984, Fukushima et al., 1988).

Treated glass beads, porous ceramics, diatomite, or even sand (Anderson et al., 1990, Bon et al., 1984) physically entrap cells into microcarriers that are used in submerged, fixed bed, or fluidized bed reactors. Magnetite has been used as a microcarrier (Garcia, Oh, and Engler, 1989) because it allows recovery of bioactive materials by magnetic extraction (Setchell, 1985). A different use of ceramics is as large, monolithic supports for continuous culture (Shiraishi et al., 1989, Lydersen et al., 1985). Much remains to be learned about ceramic and mineral immobilization techniques.

Chemical immobilization occurs both by adsorption to surfaces, ionic or covalent bonding, or by cross linking; most inorganic substrates function by adsorption. Much of industry relies on organic materials, but inorganic materials have definite advantages under rigorous conditions and show promise for future use because of lower costs. Attachment to minerals by adsorption can be affected by such properties as temperature and pH; mineral attachment is useful for affinity chromatography, a technique for purifying products. A large number of inorganic materials have the ability to immobilize cells or enzymes (Table 6); doubtless other material could be used.

Separation and Purification

After fermentation is completed, the broth with cells, spent media, and product must be separated through downstream processes to yield a pure product. Efficient handling is critical because 20 to 60% of total manufacturing cost can come from product recovery (Stanbury and Whitaker, 1984). Several industrial minerals play a role in various techniques of separation and purification (Table 7). Most notable are clay and diatomite, used extensively as filter aids, and numerous chlorides and sulfates, used as flocculants or crystallization aids. Zeolites, clays, silica sands, and other materials are also used in purification steps. In addition to general reviews by Asenjo (1990), Verrall and Hudson (1987), and Dechow (1989), two journals, *Separation and Purification Methods* and *Separation Science and Technology*, are devoted to purification studies.

If desired products are held inside cells, the cells must be disrupted. Heating, detergents, osmotic shock, or grinding with abrasives are used to help disintegrate cells. Solids are separated from the broth by filtration or by centrifuge. Considerable quantities of diatomite are used as a filter aid in certain kinds of filtration. Flocculants, such as alum, calcium or iron salts, bentonite, or organic material, as well as temperature and pH changes, are useful in enhancing efficiency of centrifugation (Nakamura, 1961, Purchas, 1968, Gasner and Wang, 1970). Some products that are concentrated by precipitation can be enhanced by inorganic materials, such as NH_4SO_4, NaCl, $NaSO_4$, NaF, $CaSO_4$, $BaCO_3$, or $Ca(OH)_2$, in a process known as salting out (Donaldson, 1984). After initial separation, purification of the finished product is usually necessary. Several methods are used including four varieties of chromatography: 1) adsorption, 2) ion-exchange, 3) gel filtration, and 4) affinity for purification. Inorganic absorbants, such as alu-

Table 6. Inorganic Supports for Immobilizing Cells or Enzymes

Alumina	Clay	Montmorillonite
Alumina (soil)	Coal	Nickel Oxide
Aluminum Hydroxide	Coke	Pumice
Aluminum Phosphate	Diatomite=cellite	Pyrite
Anthracite	Diatomaceous Earth=Kieselguhr	Pyroxene+
Antimony Pentasulfate	Emery	Quartz+
Asbestos	Ferric Hydroxide	Sand
Ash	Ferric Oxide	Seipiolite
Attapulgite	Fullers Earth	Silica
Ball Clay+	Glass	Silica Gel
Barium Sulfate	Graphite+	Stainless Steel
Bentonite	Halloysite	Sodium Aluminum Silicate
Bentonite Acid	Hornblende	Sulfur
Bentonite Amine	Hydroxyapatite	Talc
Brick	Illite	Titania
Calcium Carbonate	Kaolinite	Vermiculite
Calcium Oxalate	Lava	Volcanic Rock (Kisiris)
Calcium Phosphate	Magnesium Hydroxide	Zeolites
Ceramics	Magnesium Oxide	Zinc Hydroxide
Charcoal	Magnesium Pyrophosphate	Zirconia
Cinnabar	Manganese Oxide+	Zirconium-Earth
Cordierite	Meerschaum	Various rocks, ores, etc.

+Used in enzyme immobilization only
Sources: Daniels, 1972, Kolot, 1981a, 1981b, Chibata and Wingard, Jr., 1983, Bon et al., 1984, Lydersen et al., 1985, Setchell, 1985, Omar and Rehm, 1988, Perez-Rodriguez, Carretero, and Maqueda, 1989, Tampion and Tampion, 1987, Shiraishi et al., 1989, Kovalenko and Sokolovskii, 1992, Hartmeier, 1988.

Table 7. Techniques of Downstream Processing and Some Minerals Used for Separation and Purification

Techniques			
Disruption	**Separation**	**Concentration**	**Purification**
Osmotic Shock	Flocculation	Adsorption	Crystallization
Heat/Cold	Floatation	Membrane Filtration	Ultrafiltration
Detergents	Filtration	Precipitation	Chromatography
Chemical Lysis	Precipitation	Solubilization	adsorption, gel
Abrasion	Centrifugation	Solvent Extraction	affinity, ion exchange
Minerals Used			
Salt	Iron Salts	Alumina	Apatite
Glass	Calcium Salts	Clays	Coal
Alkalis	Sodium Silicate	Zeolites	Clays
	Alum	Silica	Silica
	Diatomite	Greensands	Zeolites
	Clays	Activated Charcoal	Diatomite
			Zr Compounds

Sources: Asenjo, 1990, Crueger and Crueger, 1990, Atkinson and Mav tuna, 1983, Stanbury and Whitaker, 1984, Dechow, 1989.

minum oxide, aluminum hydroxide, magnesium oxide, silica gel, carbon, diatomite, barite, bentonite, glass, alumina and salts, calcium phosphate, barium sulfate, rutile (Miles and Thomson, 1974), and organic resins, are used in adsorption chromatography (Table 7).

Burton (1983) reviewed purification techniques for enzymes with special attention to affinity chromatography whereby products are selectively attached to chromatographic matrices, then flushed by a different eluent to yield purified product. Many of the comments made about immobilized materials applies to this type of separation. Bentonite is especially well known for purifying alcoholic beverages. Activated carbon, clays, silica gel, and zeolites are also useful in biochemical processing (Gonzalez-Pradas et al., 1990, Blum and Eriksson, 1991, Cheng and Lee, 1992).

Many industrial minerals, rocks, and manufactured products, such as glass or ceramics, play large roles in biotechnological production. Although it is impossible to determine tonnages, biotechnology usages of certain materials are substantial and will increase as the industry grows and as economics dictates use of the best, most efficient raw materials. Research into details of mineral properties and their relationships to biotechnologic requirements is slight and scattered. Scientific efforts to better understand mineral-bacteria interactions will doubtless provide new avenues of production and improved economics. In addition, application of biotechnology to geology shows considerable potential.

BIOTECHNOLOGY AND GEOLOGY

The relatively new field of geomicrobiology investigates interactions of microbial life and geologic materials and processes.

Microorganisms have great geologic effects, destructing or forming rocks and economic mineral deposits, recycling of elements, and modifying the atmosphere and hydrosphere. Microorganisms are the earliest recognized life (Knoll, 1977), and they live virtually everywhere on earth. They have the ability to efficiently perform every conceivable, and probably many not yet thought of, low-temperature chemical reactions. "Microbes can and will do anything: microbes are smarter, wiser, and more energetic than microbiologists, chemists, and engineers" (Perlman, 1980). An excellent overview of geomicrobiology is provided by Ehrlich (1990). Earlier treatments by Zajic (1969), Krumbein (1983), Fenchel and Blackburn (1979), Trudinger and Swaine (1979), and Trudinger, Walter, and Ralph, (1980) are generally arranged around mineral cycles of certain elements. Ralph (1985a, 1985b) relates geomicrobiology to the new biotechnology with special emphasis on microbial leaching.

Microbial Mining

The potential for microorganisms to aid in mining or recovering sulfide minerals and uranium was recognized early (Brierley, 1978), and the process has been studied extensively, resulting in a vast literature about microbial leaching or mining (Ehrlich and Brierley, 1990). Numerous modern examples of microbial leaching are published (Ehrlich and Holmes, 1986, Scheiner, Doyle, and Kawatra, 1989, Murr and Brierley, 1978, Schwartz, 1977) and others have speculated on new uses for biotechnology in mining or processing (Nicolaidis, 1987, Smith, Misra, and Dubel, 1991, Borman, 1991). *Proceedings of the International Biohydrometallurgy Symposium* (Salley, McCready, and Wichlacz, 1989) also cover microbiological leaching. Commercial leaching of copper, uranium, and gold ores are in operation; however, in gold operations the leaching is usually a pretreatment phase (Hutchins, Brierley, and Brierley, 1988).

Coal Desulfurization

Many of the organisms and techniques used in metal-leaching have been used to remove sulfur from coal as reported by papers in International Conferences on Utilization of High-Sulfur Coals, in proceedings of the Biohydrometallurgy Symposia, and in other reviews of microbial leaching (Attia, 1990a, Andrews and Maczuga, 1984, Kargi, 1986, Kilbane, 1989). Certain organisms help to solubilize coal (Faison, 1991).

Enhanced Oil Recovery

A great body of research exists on using microorganisms to enhance oil recovery (MEOR) (Bryant, 1989, Updegraff, 1990, Yen, 1990, Donaldson, Chilinganian, and Yen, 1989, Zajic et al., 1983, Burchfield and Bryant, 1988). Bacteria can cause production problems mainly by biofouling of porosity and metal corrosion, so bactericides are injected into wells to help forestall such problems (Atlas, 1984). Thus, oil producers are cautious about introducing bacteria into wells; most microbially enhanced oil recovery is used on largely spent wells. Microorganisms, primarily bacteria, can improve oil recovery by producing gases, such as CH_4, CO_2, or H_2, that increase gas pressure in the reservoir, or surfactants that reduce oil viscosity and interfacial tension, which improves mobility. Certain bacteria can promote porosity enlargement by acid production and subsequent dissolution of carbonate rock. Some strains produce polymers or even biomass that can selectively plug depleted areas.

Large deposits of oil shale contain immense reserves of organic matter that is not currently extractable by conventional techniques. One of the most probable uses for the new biotechnology is development of organisms that can economically modify kerogen or asphalt into a fluid that can be produced, transported, and processed like common petroleum. Numerous papers have dealt with microbial oil recovery from oil shales or the interaction of microorganisms and oil shale (Yen, 1976, 1977, Scott and Irish, 1988, Jones, 1990).

Miscellaneous Uses

A large number of minor but interesting uses of biotechnology applied to mining exist and more will certainly appear. One method uses organisms to modify methane to less explosive gases (Apel, Dugan, and Wiebe, 1991). Sensitive biosensors, combinations of electronics and biochemicals, can be made to monitor trace amounts of gas or solutes like microbial canaries. Exploration for metal deposits using microbes has been reported by Carlisle et al. (1986) and by Parduhn (1991). Various microbial transformations of mineral species can be harnessed. One example is production of sulfur from gypsum (Cork, 1985). Many other transformations relate directly to mineral matters and many suitable bacterial processes have been applied to industrial minerals processing.

BIOTECHNOLOGY AND INDUSTRIAL MINERALS

Although most research efforts have gone into biological leaching of sulfide or uranium ores, several papers have shown the efficacy of microbial techniques to mining or beneficiation of industrial minerals. Ralph (1985a) and Groudev (1987) summarize research on microbes and industrial minerals. Three major microbial processes — solubilization, accumulation or precipitation, and biotransformations — have been applied and other many untried possibilities exist. No genetic engineering experiments to improve microbe performance in industrial minerals applications have been reported.

Solubilization

Major work on solubilizing phosphate from ore has been reported (Rogers et al., 1989) and a research program is underway at the Idaho Falls National Laboratories where a substantial number of bacteria and fungi samples have shown the ability to solubilize phosphate. Groudev and Groudeva (1986a) reported that aluminum can be leached by microbial action. Rossi (1978) extracted potassium from leucite using bioleaching. Glombitza et al. (1988) succeeded in microbiologically leaching rare earth elements from zircon. Rossi and Ehrlich (1990) and Ralph (1985a) review microbial leaching of nonmetallic materials. Many other minerals could be solubilized using biotechnological techniques thus allowing inexpensive extraction of many industrial minerals.

A related approach, sometimes called reverse solution, solubilizes unwanted contaminants to leave an improved mineral. Groudev and Groudeva (1985, 1986b) demonstrated that iron could be removed from quartz sand using microbial leaching to remove ferrugineous compounds. This group has also demonstrated microbial removal of iron from clays and improvement of kaolin and ceramic properties via microbial action (Groudev et al., 1989, 1991). Groudev and Groudeva (1988a, 1988b) improved five different bauxite ores by reducing their silica content in as little as five days of microbial leaching. Karavaiko, Rosse, and Avakyan (1989) investigated bauxite dressing by several strains of bacteria. McFarlane (1987) reported the importance of microorganisms to bauxite formation. Mohanty and Mishra (1989) reported experiments that optimized silica removal from magnesite ore.

Accumulation

Another currently used microbial technique is absorption of elements from waste water or other fluids. Elements, such as strontium (Faison et al., 1990), germanium (Chmielowski and Klapcinska, 1986), and even radioactive elements and actinides, have been absorbed by microbial means (Macaskie, 1991). The US Bureau of Mines at Salt Lake City, UT, has an active program of biosorption research. Many aspects of microbial absorption are covered by Ehrlich and Brierley (1990). As was the case with microbiological leaching, most efforts have been directed toward metal absorption (Volesky, 1990, Hughes and Poole, 1989, Beveridge and Doyle, 1989).

Surface Changes

In addition to leaching activities, microbes can also alter mineral surfaces and so lead to different or enhanced properties. Groudev et al. (1989, 1991) demonstrated improved ceramic properties such as bending strength, plasticity, shrinkage, and absorption of kaolin after microbial treatment. It has also been shown that bacteria affect surfaces of biotite, pyrite, and many other minerals (Boyle, Voight, and Sawhney, 1967, Attia, 1990b). Leaching of carbonate minerals could increase surface area and so improve brightness for paper or reactivity during SO_2 scrubbing. Modification of floatation properties via bacterial action have also been demonstrated, and it may be possible to enhance or suppress floatation of a particular mineral by first treating it microbially, probably by inexpensive heap leaching.

Biodeterioration

Microorganisms have long been known to be important in decay of natural stone and in soil formation (Berthelin, 1983, Krumbein and Dyer, 1985), whereby they produce the marvelous mantle of soil that makes possible all life. Microorganisms can cause unwanted decay in man-made materials, however, so an entire field, known as biodeterioration, has developed to study and ameliorate effects of biological degradation of materials. A recent bibliography lists more than 600 articles on stone biodeterioration (Koestler and Vedral, 1991). A journal on *International Biodeterioration* and a series of international symposia also cover aspects of stone decay and attempts to limit such decay.

Various techniques have been advanced to determine the kind and quantity of microorganisms in stone. Higher numbers of organisms commonly occur in more degraded stones (Caneva et al., 1989). Special attention has been paid to biodeterioration of ancient stone statues and buildings made of limestone or marble and to ways of combating decay (Felix, 1985). Krumbein (1987), Jones and Wilson (1985), and Strzelczyk (1981) give reviews of the overall problem.

Though much research has involved biodeterioration of marble, other building materials, such as sandstone, concrete, brick, and plaster are also affected by bacteria. Hueck-van der Plas (1968) tabulated microorganisms known to degrade porous building materials and compounds used to control them. Even granite, brick, ceramics, and glass are subject to biodeterioration (Palmer and Hirsch, 1991, Krumbein, Urzi, and Gehrmann, 1991). Most exposed materials are subject to microbial attack, but surprisingly little research is reported about suppressing microbial growth and deterioration.

Grant and Bravery (1981) established test procedures and evaluated several materials for destruction of algae on construction materials. Griffin et al. (1991), Schnabel (1991), and Richardson (1987), have reported techniques for control of microorganisms on stone. Physical methods of stone cleaning as well as biocide treatment must be repeated and so are no panacea. There remains to be found a long-term solution to the problems of biodeterioration by microorganisms. The tendency of bacteria to degrade materials can, however, be turned to good use.

Bioremediation

Bioremediation is the general term used to denote any of numerous modern microbial methods of removing harmful pollutants or modifying them to be less harmful. Biological systems commonly take in or sequester metals. The US Bureau of Mines has developed several methods of biosorption to clean water and recover metals. A variety of plants are also known to sequester metals as do several special strains of microbes. Details of metal uptake and storage are covered in much literature.

Microbial processes have shown promise in removing clays from phosphate mining wastes (Brierley, Lanza, Scheiner, 1981, Leslie, Greasham, Hulbert, 1984) or other suspensions (Avnimelech, Troeger, and Reed, 1982), and it has been suggested that red mud wastes from barite production might also be cleaned by microorganisms. Bacteria have long been known to degrade cyanide. Carta, Ghiami, and Rossi (1980) pointed out the potential for biochemical beneficiation of effluents from nonmetallic flotation plants. Omar and Rehm (1988) found that organisms immobilized on granular clay pellets degraded oily sludge twice as fast as free cells. In situ bioremediation can be affected by soil clays (Miller and Alexander, 1991, Boudat, 1992), and considerable work on clay-organism interactions remains to be done. Diatomaceous earth modified by precipitation of metal hydroxides proved useful in removing viruses from water (Farrah et al., 1991). There are undoubtedly many areas where biotechnology can be used to improve mining, beneficiation, and reclamation.

A growing portion of the $4 billion annual expense of remediating waste sites is being used by bioremediation, and as many as 250 future projects are being evaluated (Swett, 1992). Most often bioremediation is used to transform organic contaminate into less noxious substances (Mitchell, 1992, Thayer, 1991b), but several biotechnical techniques show promise in cleaning mining wastes even though the discipline is still in its infancy.

SUMMARY

Biotechnology and its more established progenitor, industrial microbiology, have a lot in common with industrial minerals in that all represent a somewhat ill-defined, very diverse group of procedures and products. Like industrial minerals, biotechnologic products range from very inexpensive, such as alcoholic drinks at pennies per dose, to very costly, such as pharmaceuticals at thousands of dollars. Many diverse geologic processes operate to produce industrial mineral deposits, just as biotechnology encompasses very diverse techniques. Industrial minerals commonly form the basis for large industries, as does biotechnology, and our modern world could not function without either.

Many industrial minerals are used in biotechnological processes, such as media, process aids, separation or purification materials, or in waste cleanup. It is not possible to provide tonnages of minerals used, but amounts are substantial and biotechnology industries, along with their mineral demands, are growing rapidly.

Biotechnology applied to extraction of industrial minerals is in its infancy, but already several promising lines of research have emerged. Production of phosphate, aluminum, and other materials through microbiological solubilization show distinct promise, even though far behind metal leaching research. Beneficiating bulk commodities by leaching away contaminants can also become an important tool for the industrial minerals industry. Concentration of elements from dilute sources also holds considerable promise for certain important elements. Complete transformations of mineral species or modification of surface properties through microbial action have been demonstrated, as has improvement in ceramic properties. Arresting biodegradation of natural stone or man-made construction materials will improve durability and prolong usability of materials. Application of biotechnology to clean up mine or process wastes, as well as in bioremediation of otherwise contaminated sites, also holds great promise. Thus far biotechnology research has been carried out using only existing organisms. When the full arsenal of modern molecular biology abilities are brought to bear upon problems of industrial mineral production, it is most likely that great progress will be made.

ACKNOWLEDGMENTS

In any new venture it is necessary to seek help, and I have been fortunate to have had considerable assistance for which I wish to thank the following people. Conversations were held with T.D. Brock, Howard Gest, and Floyd Huber, who kindly provided a plant tour, and considerable personal experience about fermentation. Norman Herz provided data vital to the biodeterioration section. Visits with Ollie Touovinen and Y.S. Attia were most useful in gaining information about mineral leaching and coal desulfurization. Corale Brierley provided considerable insight into leaching and searched her references for industrial minerals items. Marilyn DeWees deserves great thanks for efficiently typing the manuscript, and Kathryn Shaffer for contributing to figures and other invaluable assistance. This manuscript was helped considerably by comments of several reviewers, whom I also wish to thank.

BIBLIOGRAPHY

Anon., 1981, *Impacts of Applied Genetics Micro-Organisms, Plants, and Animals*, US Office of Technology Assessment, Washington, DC, 321 pp.

Anon., 1984, *Commercial Biotechnology an International Analysis*, US Office of Technology Assessment, Washington, DC, 612 pp.

Anon., 1987, *Federal Biotechnology Information Resources Directory*, OMEC Internatiaonal, Washington, DC, 151 pp.

Anon., 1990a, *The United States Pharmacopeia XXII*, US Pharmacopial Convention, Rockville, MD.

Anon., 1990b, *State-By-State Biotechnology Directory Centers, Companies, and Contacts*, Bureau of National Affairs, Wahington, DC, 163 pp.

Anon., 1992a, *U.S. Industrial Outlook '92*, US Department of Commerce, Washington, DC.

Anon., 1992b, *Biotechnology for the 21st Century*, Federal Coordinating Council Science Engineering Technology Committee on Life Sciences and Health, Washington, DC, 51 pp.

Anon., 1992c, *Biotechnology in a Global Economy*, US Office of Technology Assessment, Washington, DC, 283 pp.

Aida, K., et al., 1986, "Biotechnology of Amino Acid Production," *Progress in Industrial Microbiology*, Vol. 24, Elsevier, New York.

Allen, J.D., 1992, "Minerals in Animal Feed, Future Use Assured in Shrinking Market," *Industrial Minerals*, Jan., pp. 35–39.

Anderson, W.A., et al., 1990, "Adsorption of Streptococcus faecalis on Diatomite Carriers for Use in Biotransformations," *Journal of Chemical Technology and Biotechnology*, Vol. 47, pp. 93–100.

Andrews, G.F., and Maczuga, J., 1984, "Bacterial Removal of Pyrite from Coal," *Fuel*, Vol. 63, Mar., pp. 297–302.

Antebi, E., and Fishlock, D., 1986, *Biotechnology Strategies for Life*, MIT Press, Cambridge, MA, 239 pp.

Apel, W.A., Dugan, P.R., and Wiebe, M.R., 1991, "Use of Methanotrophic Bacteria in Gas Phase Bioreactors to Abate Methane in Coal Mine Atmospheres," *Fuel*, Vol. 70, pp. 1001–1003.

Asenjo, J.A., 1990, *Separation Processes in Biotechnology*, Marcel Dekker, New York, 801 pp.

Atkinson, B., and Mavituna, F., 1983, *Biochemical Engineering and Biotechnology Handbook*, Nature Press, New York, Vol. 1, 119 pp.

Atlas, R.M., 1984, *Petroleum Microbiology*, Macmillan, New York, 692 pp.

Attia, Y.A., 1990a, "Biotechnology in Mineral Processing," *Mining Engineering*, Vol. 42, No. 5, pp. 497.

Attia, Y.A., 1990b, "Feasibility of Selective Biomodification of Pyrite Floatability in Coal Desulfurization by Froth Floatation," *Resources, Conservation and Recycling*, Vol. 3, pp. 169–175.

Avnimelech, Y., Troeger, B.W., and Reed, L.W., 1982, "Mutual Flocculation of Algae and Clay: Evidence and Implications," *Science*, Vol. 216, Apr. 2, pp. 63–65.

Bailey, J.D., and Ollis, D.F., 1986, *Biochemical Engineering Fundamentals*, 2nd ed., McGraw-Hill, New York, 984 pp.

Batra, L.R., and Millner, P.D., 1976, "Asian Fermented Foods and Beverages," *Developments in Industrial Microbiology*, Vol. 17, pp. 117–127.

Berthelin, J., 1983, "Microbial Weathering Processes," *Microbial Geochemistry*, W.E. Krumbein, ed., Oxford, UK, pp. 223–262.

Beveridge, T.J., and Doyle, R.J., eds., 1989, *Metal Ions and Bacteria*, John Wiley & Sons, New York, 461 pp.

Blum, Z., and Eriksson, H., 1991, "Utilization of Zeolite Y in the Removal of Anionic, Cationic and Nonionic Detergents During Purificaiton of Proteins," *Biotechnology Techniques*, Vol. 5, No., pp. 49–54.

Bon, E., et al., 1984, "Immobilization of Glucoamylase on Chitin," *Biotechnology and Bioengineering Symposium*, No. 14, pp. 485–492.

Borman, S., 1991, "Bacteria that Flourish above 100°C Could Benefit Industrial Processing," *Chemical and Engineering News*, Nov. 4, pp. 31–35.

Boudot, J.P., 1992, "Relative Efficiency of Complexed Aluminum, Noncrystalline Al Hydroxide, Allophane and Imogolite in Retarding the Biodegradation of Citric Acid," *Geoderma*, Vol. 52, pp. 29–39.

Boyle, J.R., Voight, G.K., and Sawhney, B.L., 1967, "Biotite Flakes: Alteration by Chemical and Biological Treatment," *Science*, Vol. 155, pp. 193–195.

Brierley, C.L., 1978, "Bacterial Leaching," *CRC Critical Reviews in Microbiology*, Vol. 6, Nov., pp. 207–262.

Brierley, C.L., Lanza, G.R., and Scheiner, B., 1981, "Microbial Processes for Removal of Suspended Clays from Selected Industrial Wastewaters," *Biotechnology and Bioengineering Symposium*, No. 11, pp. 507–520.

Brock, T.D., and Madigan, M.T., 1991, *Biology of Microorganisms*, 6th ed., Prentice Hall, Englewood Cliffs, NJ, 974 pp.

Bruce, N.G., et al., 1989, "Finding Biotech Information: Databases and Vendors," *Bio/technology*, Vol. 7, No. 5, pp. 455–458.

Brune, D.E., Reach, D.C., and O'Connor, J.T., 1981, "Inland Saltwater as a Medium for the Production of Biomass," *Biotechnology and Bioengineering Symposium*, No. 11, pp. 79–93.

Bryant, R.S., 1989, "Laboratory and Field Studies on Microbial Enhanced Oil Recovery," *Developments in Industrial Microbiology*, Vol. 30, pp. 255–267.

Burchfield, T.E., and Bryant, R.S., 1988, *Proceedings of the Symposium on Applications of Microorganisms to Petroleum Technology*, US Dept. of Energy Conf-870858.

Buchner, W., et al., 1989, *Industrial Inorganic Chemistry*, VCH, New York, 614 pp.

Bud, R., 1989, "Janus-Faced Biotechnology: An Historical Perspective," *TIBTECH*, Sep., Vol. 7, pp. 230–233.

Bu'lock, J., and Kristiansen, B., 1987, *Basic Biotechnology*, Academic Press, Boston, MA, 561 pp.

Burrill, G.S., and Roberts, W.J., 1992, "Biotechnology and Economic Development: The Winning Formula," *Bio/technology*, Vol. 10, June, pp. 647–653.

Burton, C.J., 1983, "Large-scale Purification of Enzymes," *Philosophical Transactions of Royal Society of London Series B*, Vol. B 300, pp. 249–261.

Cairns-Smith, A.G., 1982, *Genetic Takeover and the Mineral Origins of Life*, Cambridge University Press, New York, 477 pp.

Cairns-Smith, A.G., and Hartman, H., 1986, *Clay Minerals and the Origin of Life*, Cambridge University Press, New York, 193 pp.

Calam, C.T., 1967, "Media for Industrial Fermentations," *Process Biochemistry*, June, pp. 19–22.

Caneva, G., et al., 1989, "Correlation Analysis in the Biodeterioration of Stone Artworks," *International Biodeterioration*, Vol. 25, pp. 161–167.

Carlisle, D., et al., 1986, *Mineral Exploration: Biological Systems and Organic Matter*, Vol. 5, Prentice-Hall, Engelwood Cliffs, NJ, 465 pp.

Carta, M., Ghiami, M., and Rossi, G., 1980, "Biochemical Beneficiation of Mining Industry Effluents," *Biogeochemistry of Ancient and Modern Environments*, pp. 669–688.

Cheng, Y.L., and Lee, T.Y., 1992, "Separation of Fructose and Glucose Mixture by Zeolite Y," *Biotechnology and Bioengineering*, Vol. 40, pp. 498–504.

Chibata, I., and Wingard, L.B., Jr., 1983, "Immobilized Microbial Cells," *Applied Biochemistry and Bioengineering*, Vol. 4, Academic Press, New York, 355 pp.

Chmielowski, J., and Klapcinska, B., 1986, "Bioaccumulation of Germanium by *Pseudomonas Putida* in the Presence of Two Selected Substrates," *Applied and Environmental Microbiology*, Vol. 51, No. 5, pp. 1099–1103.

Coombs, J., 1986, *Dictionary of Biotechnology*, Macmillan, London, 330 pp.

Coombs, J., and Alston, Y.R., 1991, *Biotechnology Directory*, Stockton Press, New York, 673 pp.

Cork, D.J., 1985, "Microbial Conversion of Sulfate to Sulfur—An Alternative to Gypsum Synthesis," *Advances in Biotechnological Processes*, Vol. 4, Alan R. Liss Co., pp. 183–209.

Crafts-Lighty, A., 1986, *Information Sources in Biotechnology*, 2nd ed., Stockton Press, New York, 403 pp.

Crafts-Lighty, A., 1991, "Communication," *Biotechnology The Science and the Business*, pp. 117–127.

Crueger, W., and Crueger, A., 1990, *Biotechnology A Textbook of Industrial Microbiology*, Sinauer Associates, Sunderland, MA, 357 pp.

Daniels, S.L., 1972, "The Adsorption of Microorganisms onto Solid Surfaces: A Review," *Developments in Industrial Microbiology*, Vol. 13, pp. 211–253.

Davis, B.D., 1991, *Genetic Revolution*, Johns Hopkins University Press, Baltimore, MD, 295 pp.

Dechow, F.J., 1989, *Separation and Purification Techniques in Biotechnology*, Noyes Publication, Park Ridge, NJ, 490 pp.

Demain, A.L., 1983, "New Applications of Microbial Products," *Science*, Vol. 219, pp. 709–714.

Demain, A.L., and Solomon, N.A., 1986, *Manual of Industrial Microbiology and Biotechnology*, American Society for Microbiology, Washington, DC, 466 pp.

Dibner, M.D., 1991a, *Biotechnology Guide USA*, 2nd ed., Sep., Stockton Press, New York, 650 pp.

Dibner, M.D., 1991b, "Tracking Trends in US Biotechnology," *Bio/technology*, Vol. 9, pp. 1334–1337.

Donaldson, E.C., Chilinganian, G.V., and Yen, T-F, 1989, "Microbial Enhanced Oil Recovery," *Developments in Petroleum Science*, Vol. 22, Elsevier, New York, pp. 227.

Donaldson, T.L., 1984, "Phase Separation of Organics/Water Mixtures Using Salts," *Biotechnology and Bioengineering Symposium*, No. 14, pp. 573–579.

Drlica, K., 1992, *Understanding DNA and Gene Cloning*, Wiley, New York, 240 pp.

Eagon, R.G., et al., 1987, "The Effect of High Microbial Populations on the Physical Properties of Kaolin Slurries," *International Biodeterioration*, Vol. 23, pp. 61–68.

Edington, S.M., 1992, "Biotech Vaccines' Problematic Promise," *Bio/Technology*, Vol. 10, pp. 763–766.

Edwards, M.J., ed., 1988, *ATCC Minerals & Cells at Work*, American Type Culture Collection, Rockville, MD, 232 pp.

Ehrlich, H.L., 1990, *Geomicrobiology*, 2nd ed., Marcel Dekker, New York, 643 pp.

Ehrlich, H.L., and Brierley, C.L., 1990, *Microbial Mineral Recovery*, McGraw-Hill, New York, 454 pp.

Ehrlich, H.L., and Holmes, D.S., 1986, *Workshop on Biotechnology for the Mining, Metal-Refining and Fossil-Fuel Processing Industries*, John Wiley & Sons, New York, 386 pp.

Embers, L., 1992, "No Special Rules Needed for Biotech Food Products," *Chemical and Engineering News*, June 1, pp. 5–6.

Faison, B.D., 1991, "Biological Coal Conversions," *Critical Reviews in Biotechnology*, Vol. 11, No. 4, pp. 347–366.

Faison, B.D., et al., 1990, "Binding of Dissolved Strontium by *Micrococcus luteus*," *Applied and Environmental Microbiology*, Vol. 56, pp. 3649–3656.

Farrah, S.R., et al., 1991, "Use of Modified Diatomaceous Earth for Removal and Recovery of Viruses in Water," *Applied and Environmental Microbiology*, Vol. 57, pp. 2502–2506.

Felix, G., ed., 1985, *5th International Congress on Deterioration and Conservation of Stone*, 2 Vols., Presses Polytechniques Romandes, Lausanne, Switzerland, 1124 pp.

Fenchel, T., and Blackburn, T.H., 1979, *Bacteria and Mineral Cycling*, Academic Press, New York, 225 pp.

Frausto-de Silva, J.J.R., and Williams, R.J.P., 1991, *The Biological Chemistry of the Elements*, Clarendon Press, Oxford, UK, 561 pp.

Fukushima, Y., et al., 1988, "A New Immobilization Technique of Whole Cells and Enzymes with Colloidal Silica and Alginate," *Biotechnology and Bioengineering*, Vol. 32, pp. 584–594.

Garcia, A., III, Oh, S., and Engler, C.R., 1989, "Cellulase Immobilization on Fe_3O_4 and Characterization," *Biotechnology and Bioengineering*, Vol. 33, pp. 321–326.

Gasner, L.L., and Wang, D.I.C., 1970, "Microbial Cell Recovery Enhancement through Flocculation," *Biotechnology and Bioengineering*, Vol. 12, pp. 873–887.

Gerhardt, P., ed., 1981, *Manual of Methods for General Bacteriology*, American Society of Microbiology, Washington, DC, 524 pp.

Gest, H., 1987, *World of Microbes*, Science Tech, Madison, WI, 249 pp.

Gherna, R., and Pienta, P., 1992, *A.T.C.C. Bacteria and Phages*, 18th ed., 1992, American Type Culture Collection, Rockville, MD, 694 pp.

Gibbons, A., 1991, "Biotech Pipeline: Bottleneck Ahead," *Science*, Vol. 254, Oct., pp. 369–370.

Glombitza, F., et al., 1988, "Bacterial Leaching of Zircon Mineral for Obtaining Trace and Rare Earth Elements (REE)," *Proceedings, International Symposium on Biohydrometallurgy*, P.R. Norris and D.P. Kelly, eds., University of Warwick, UK, pp. 407–418.

Gonzales-Pradas, E., et al., 1990, "Removal of Aromatic Amines from Aqueous Solution by Activated Sepiolite," *Journal of Chemical Technology and Biotechnology*, Vol. 47, pp. 15–22.

Grant, C., and Bravery, A.F., 1981, "Laboratory Evaluation of Algicidal Biocides for Use on Constructional Materials, 1. An Assessment of Some Current Test Methods," *International Biodeterioration Bulletin*, Vol. 17, No. 4, Winter, pp. 113–123.

Griffin, P.S., et al., 1991, "The Biodeterioration of Stone: A Review of Deterioration Mechanisms, Conservation Case Histories, and Treatment," *International Biodeterioration*, Vol. 28, pp. 187–207.

Groudev, S.N., 1987, "Use of Heterotrophic Microorganisms in Mineral Biotechnology," *Acta Biotechnology*, Vol. 7, pp. 299–306.

Groudev, S.N., et al., 1989, "Improvement of the Quality of Kaolins by Means of Microbial Treatment," *Second World Congress on Non-Metallic Minerals*.

Groudev, S.N., et al., 1991, "Improvement of the Ceramic Properties of Kaolin," *7th Euroclay Conference*, Dresden.

Groudev, S.N., and Groudeva, V.I., 1985, "Biological Removal of Iron from Quartz Sands, Kaolins and Clay, *15th International Mineral Processing Congress*, Cannes, France, Vol. 2, pp. 378–387.

Groudev, S.N., and Groudeva, V.I., 1986a, "Biological Leaching of Aluminum from Clays," *Biotechnology and Bioengineering Symposium*, No. 16, pp. 91–99.

Groudev, S.N., and Groudeva, V.I., 1986b, "Iron from Quartz Sands—A Microbial Approach," *Industrial Minerals*, Mar., pp. 81–84.

Groudev, S.N., and Groudeva, V.I., 1988a, "Microbial Removal of Silicon from Mineral Raw Materials," *Proceedings, International Symposium on Biohydrometallurgy*, University of Warwick, UK, pp. 397–406.

Groudev, V.I., and Groudev, S.N., 1988b, "Dressing of Bauxite Ores by Means of Microorganisms," *16th International Mineral Processing Congress*, Stockholm, Sweden.

Hack, R., et al., 1988, *International Biotechnology Handbook*, Facts on File Publications, New York, 244 pp.

Hartmeier, W., 1988, *Immobilized Biocatalysts*, Springer-Verlag, New York, 212 pp.

Hesseltine, C.W., 1981, "Future of Fermented Foods," *Process Biochemistry*, Apr./May, pp. 2–6, 13.

Higgins, I.J., Best, D.J., and Jones, J., 1985, *Biotechnology Principles and Applications*, Blackwell Scientific Publications, Boston, MA, 224 pp.

Hinman, R.L., 1991, "A Role for Biotech in Producing Chemicals?" *Bio/Technology*, Vol. 9, June, pp. 533–534.

Ho, C.S., and Wang, D.I.C., 1991, *Animal Cell Bioreactors*, Butterworth-Heinemann, Boston, MA, 494 pp.

Hodgson, J., 1989, *Bio-Technology Changing the Way Nature Works*, Cassell, London, 130 pp.

Hueck-van der Plas, E.H., 1968, "The Micro-Biological Deterioration of Porous Building Materials," *International Biodeterioration Bulletin*, Vol. 4, No. 1, pp. 11–28.

Hughes, M.N., and Poole, R.K., 1989, *Metals and Micro-organisms*, Chapman and Hall, New York, 412 pp.

Hutchins, S.R., Brierley, J.A., and Brierley, C.L., 1988, "Microbial Pretreatment of Refractory Sulfide and Carbonaceous Ores Improves the Economics of Gold Recovery," *Mining Engineering*, Vol. 40, No. 4, pp. 249–254.

Jayme, D.W., and Blackman, K.E., 1985, "Culture Media for Propagation of Mammalian Cells, Viruses, and Other Biologicals," *Advances in Biotechnological Processes*, Vol. 5, pp. 1–30.

Johnson, L.M., and Rader, R.A., 1990, "An Overview of Biotechnology Information Resources," *TIBTECH*, Vol. 8, Nov., pp. 318–323.

Jones, D.R., 1990, "Batch Leaching Studies of Rundle Oil Shale," *Journal of Environmental Quality*, Vol. 19, July-Sep., pp. 408–413.

Jones, D., and Wilson, M. J., 1985, "Chemical Activity of Lichens on Mineral Surfaces—A Review," *International Biodeterioration*, Vol. 21, No. 2, pp. 99–104.

Jones, R.P., and Greenfield, P.F., 1983, "A Review of Yeast Ionic Nutrition," *Process Biochemistry*, Vol. 21, No. 6, pp. 183–194.

Karavaiko, G.I., Rosse, G., Avakyan, Z.A., 1989, "Microbiological Processing of Bauxite," *Biohydro-metallurgy*, Canmet SP89–10, pp. 93–102.

Kargi, F., 1986, "Microbial Methods for Desulfurization of Coal," *TIBTECH*, Vol. 4, pp. 293–297.

Kilbane, J.J., 1989, "Desulfurization of Coal: The Microbial Solution," *TIBTECH*, Vol. 7, pp. 97–101.

Kilmer, V.J., 1979, "Minerals and Agriculture," *Biogeochemical Cycling of Mineral-Forming Elements*, P.A. Trudinger and D.J. Swaine, eds., Elsevier Science Publishers, New York, pp. 515–558.

Kirsop, B., 1988 "Computerised Data Bases for Locating Microorganisms: Functions and Differences of Major Information Resources," *MIRCEN*, Vol. 4, pp. 419–424.

Klausmeier, R.E., et al., 1963, "Mineral-Salts Media for Fungus Resistance Testing," *Developments in Industrial Microbiology*, Vol. 4, pp. 306–311.

Klibanov, A.M., 1983. "Immobilized Enzymes and Cells as Practical Catalysts," *Science*, Vol. 219, Feb., pp. 722–727.

Knoll, A.H., 1977, "Paleomicrobiology," *CRC Handbook of Microbiology*, 2nd ed., Vol. 1, CRC Press, Cleveland, OH, pp. 1–21.

Koestler, R.J., and Vedral, J., 1991, "Biodeterioration of Cultural Property: A Bibliography," *International Biodeterioration*, Vol. 28, pp. 229–340.

Kolot, F.B., 1981a, "Microbial Carriers—Strategy for Selection," *Process Biochemistry*, Aug./Sep., pp. 2–9.

Kolot, F.B., 1981b, "Microbial Carriers—Strategy for Selection—Part 2," *Process Biochemistry*, Oct./Nov., pp. 30–33, 46.

Kovalenko, G.A., and Sokolovskii, V.D., 1992, "Epoxidation of Propene by Microbial Cells Immobilized on Inorganic Supports," *Biotechnology and Bioengineering*, Vol. 39, pp. 522–528.

Krumbein, W.E., 1983, *Microbial Geochemistry*, Blackwell Scientific, Boston, MA, 330 pp.

Krumbein, W.E., 1987, "Concrete and Stone Microbial Interations with Mineral Materials," *Biodeterioration 7*, pp. 78–100.

Krumbein, W.E., and Dyer, B.D., 1985, "This Planet Is Alive," *The Chemistry of Weathering*, J.I. Drever, ed., pp. 143–160.

Krumbein, W.E., Urzi, C.E., and Gehrmann, C., 1991, "Biocorrosion and Biodeterioration of Antique and Medieval Glass," *Geomicrobiology Journal*, Vol. 9, pp. 139–160.

Lanza, G.R., and Watson, J.B., 1984, "Enhancement of Polysaccharide Production with Clay Additives," *Developments in Industrial Microbiology*, Vol. 25. pp. 657–668.

Laskin, A.I., and Lechevalier, H.A., 1977, *CRC Handbook of Microbiology*, CRC Press, Cleveland, OH, 9 vols.

Leslie, J.F., Greasham, R.L., and Hulbert, M.H., 1984, "A Cyanobacterial System for Consolidation of Phosphate Slimes," *Geomicrobiology Journal*, Vol. 3, No. 4, pp. 343–358.

Lipson, S.M., and Stotzky, G., 1984, "Effect of Proteins on Reovirus Adsorption to Clay Minerals," *Applied and Environmental Microbiology*, Vol. 48, No. 3, pp. 525–530.

Lydersen, B.K., et al., 1985, "Ceramic Matrix for Large Scale Animal Call Culture," *Bio/Technology*, Jan., pp. 63–67.

Macaskie, L.E., 1991, "The Application of Biotechnology to the Treatment of Wastes Produced by the Nuclear Fuel Cycle," *Critical Reviews in Biotechnology*, Vol. 11, pp. 41–112.

Maigetter, R.Z., and Pfister, R.M., 1975, "A Mixed Bacterial Population in a Continuous Culture With and Without Kaolinite," *Canadian Journal of Microbiology*, Vol. 21, pp. 173–180.

Marx, J.L., 1989, *Revolution in Biotechnology*, Cambridge University Press, New York, 227 pp.

Mattey, M., 1991, "The Production of Organic Acids," *Critical Reviews in Biotechnology*, Vol. 12, pp. 87–132.

McFarlane, M.J., 1987, "The Key Role of Micro-Organisms in the Process of Bauxitisation," *Modern Geology*, Vol. 11, pp. 325–344.

Miles, B.J., and Thomson, A.R., 1974, "New Inorganic Materials for Protein Separation and Recovery," *Process Biochemistry*, May, pp. 11–16.

Miller, B.M., and Litsky, W., 1984, *Industrial Microbiology*, McGraw-Hill Book Co., New York, 465 pp.

Miller, M.E., and Alexander, M., 1991, "Kinetics of Bacterial Degradation of Benzylamine in a Montmorillonite Suspension," *Environmental Science and Technolology*, Vol. 25, No. 2, pp. 240–245.

Mitchell, R., ed., 1992, *Environmental Microbiology*, John Wiley & Sons, New York, 411 pp.

Mohanty, B.K., and Mishra, A.K., 1989, "Optimisation of Silica Release from Magnesite Ore by Bacillus Sp.," *Indian Journal of Microbiology*, Vol. 29 , No. 4, pp. 361–370.

Moo-Young, M., ed., 1985, *Advances in Biotechnology*, Pergamon Press, New York, 4 vols.

Moses, V., and Cape, R.E., 1991, *Biotechnology the Science and the Business*, Harwood Academic Publishers, New York, 596 pp.

Murr, L.E., and Brierley, J.A., eds., 1978, *Metallurgical Applications of Bacterial Leaching and Related Microbiological Phenomena*, Academic Press, New York, NY, 526 pp.

Nakamura, H., 1961, "Chemical Separation Methods for Common Microbes," *Journal of Biochemical and Microbiological Tehcnology and Engineering*, Vol. 3, No, 4, pp. 395–403.

Ng, T.K., et al., 1983, "Production of Feedstock Chemicals," *Science*, Vol. 219, pp. 733–739.

Nicolaidis, A.A., 1987, "Microbial Mineral Processing: the Opportunities for Genetic Manipulation," *Journal of Chemical Technology and Biotechnology*, Vol. 38, pp. 167–185.

Odunfa, S.A., 1988, "Review: African Fermented Foods: From Art to Science," *MIRCEN*, Vol. 4, pp. 259–273.

O'Leary, W.M., 1989, *Practical Handbook of Microbiology*, CRC Press, Boca Raton, FL, 681 pp.

Omar, S.H., and Rehm, H-J., 1988, "Degradation of n-alkanes by Candida Parapsilosis and Penicillium Frequentans Immoblized on Granular Clay and Aquifer Sand," *Applied Microbiology and Biotechnology*, Vol. 28, pp. 103–108.

Ouellette, R.P., and Cheremisinoff, P.N., 1985a, *Essentials of Biotechnology*, Technomic, Lancaster, PA, 226 pp.

Ouellette, R.P., and Cheremisinoff, P.N., 1985b, *Applications of Biotechnology*, Technomic, Lancaster, PA, 247 pp.

Ōya, A., et al., 1991, "An Antimicrobial and Antifungal Agent Derived From Montmorillonite," *Applied Clay Science*, Vol. 6, pp. 135–142.

Palmer, R.J., Jr., and Hirsch, P., 1991, "Photosynthesis-Based Microbial Communities on Two Churches in Northern Germany: Weathering of Granite and Glazed Brick," *Geomicrobiology Journal*, Vol. 9, pp. 103–118.

Parduhn, N.L., 1991, "A Microbial Method of Mineral Exploration: A Case History at the Mesquite Deposit," *Journal of Geochemical Exploration*, Vol. 41, pp. 137–149.

Park, D.H., et al., 1985, "The Effect of Carbon Dioxide, Aeration Rate and Sodium Chloride on the Secretion of Proteins from Growing Bakers' Yeast (Saccharomyces Cerevisiae)," *Journal of Biotechnology*, Vol. 2, pp. 337–346.

Peppler, H.J., and Perlman, D., 1979, *Microbial Technology*, Academic Press, New York, 536 pp.

Perez-Rodriguez, J.L., Carretero, M.I., and Maqueda, C., 1989, "Behavior of Sepiolite, Vermiculite and Montmorillonite as Supports in Anaerobic Digesters," *Applied Clay Science*, Vol. 4, pp. 69–82.

Perlman, D., 1980, "Some Problems on the New Horizons of Applied Microbiology," *Developments in Industrial Microbiology*, Vol. 31, p. XV-XXIII.

Postgate, J., 1992, *Microbes and Man*, 3rd ed., Cambridge University, New York, 297 pp.

Prentis, S., 1984, *Biotechnology A New Industrial Revolution*, George Braziller, New York, 192 pp.

Prescott, L.M., Harley, J.P., and Klein, D.A., 1990, *Microbiology*, Wm. C. Brown, Dubuque, IA, 883 pp.

Primrose, S.B., 1991, *Molecular Biotechnology*, 2nd ed., Blackwell Scientific Publications, Boston, MA, 196 pp.

Purchas, D.B., 1968, "Flocculation and Coagulation," *Process Biochemistry*, Oct., pp. 17–27.

Ralph, B.J., 1985a, "Biotechnology Applied to Raw Minerals Processing," *Comprehensive Biotechnology*, Vol. 4, Pergamon Press, New York, pp. 201–234.

Ralph, B.J., 1985b, "Geomicrobiology and the New Biotechnology," *Developments in Industrial Microbiology*, Vol. 26, pp. 23–59.

Reed, G., 1982, *Prescott & Dunn's Industrial Microbiology*, 4th ed., AVI Publishing, Westport, CT, 883 pp.

Rehm, H.J., and Reed, G., 1981, *Biotechnology*, Verlag Chemie, Deerfield Beach, FL, 8 vols.

Richardson, B.A., 1987, "Control of Microbial Growths on Stone and Concrete," *Biodeterioration 7*, pp. 101–106.

Rogers, R.D., et al., 1989, "Biosolubilization of Phosphate Ore," *Biotechnology in Minerals and Metal Processing*, B.J. Scheiner, F.M. Doyle, and S.K. Kawatra, eds., Society of Mining Engineers, Littleton, CO, pp. 161–168.

Rosevear, A., 1984, "Immobilised Biocatalysts—a Critical Review," *Journal of Chemical Technology and Biotechnology*, Vol. 34B, pp. 127–150.

Rossi, G., 1978, "Potassium Recovery Through Leicite Bioleaching: Possibilities and Limitation," *Metallurgical Applications of Bacterial Leaching and Related Microbiological Phenomena*, pp. 297–319.

Rossi, G. and Ehrlich, H.L., 1990, "Other Bioleaching Processes," *Microbial Mineral Recovery*, pp. 149–170.

Salley, J., McCready, R.G.L., and Wichlacz, P.L., 1989, "Biohydrometallurgy," *Proceedings*, International Symposium, Jackson Hole, WY, Canmet SP89–10, 771 pp.

Scheiner, B.J., Doyle, F.M., and Kawatra, S.K., eds., 1989, *Biotechnology in Minerals and Metal Processing*, Society of Mining Engineers, Littleton, CO, 209 pp.

Schnabel, L., 1991, "The Treatment of Biological Growths on Stone: A Conservator's Viewpoint," *International Biodeterioration*, Vol. 28, pp. 125–131.

Schwartz, W., 1977, *Gesellschaft fur Biotechnologische Forschung mbH Braunschweig-Stockheim*, Verlag Chemie, New York, 270 pp.

Scott, C.B., and Irish, G.E., 1988, "Scottish Shale Oil—Production, Occupational Exposure and Health Effects," *Journal of Chemical Technology and Biotechnology*, Vol. 41, pp. 1–12.

Scragg, A.H., 1991, *Bioreactors in Biotechnology a Practical Approach*, Ellis Horwood Co., New York, 330 pp.

Setchell, C.H., 1985, "Magnetic Separations in Biotechnology—a Review," *Journal of Chemical Technology and Biotechnology*, Vol. 35B, pp. 175–182.

Shiraishi, F., et al., 1989, "Characterization of Production of Free Gluconic Acid by Gluconobacter Suboxydans Adsorbed on Ceramic Honeycomb Monolith," *Biotechnology and Bioengineering*, Vol. 33, pp. 1413–1418.

Sikyta, B., et al., 1983, *Methods in Industrial Microbiology*, John Wiley & Sons, New York, 349 pp.

Smith, J.E., 1988, *New Studies in Biotechnology*, 2nd ed., Edward Arnold Co., New York, 130 pp.

Smith, J.E., Berry, D.R., and Kristiansen, B., 1980, *Fungal Biotechnology*, Academic Press, New York, 308 pp.

Smith, R.W., Misra, M., and Dubel, J., 1991, "Mineral Bioprocessing and the Future," *Minerals Engineering*, Vol. 4, pp. 1127–1141.

Sreekumar, O., and Basappa, S.C., 1992, "Effect of Calcium and Sodium Salts on Ethanol Production in High Sugar Fermentation by Free Cells of *Zymomonas Mobilis*," *Biotechnology Letters*, Vol. 14, No. 6, pp. 511–514.

Stanbury, P.F., and Whitaker, A., 1984, *Principles of Fermentation Technology*, Pergamon Press, New York, 255 pp.

Stockmeyer, M.R., 1991, "Adsorption of Organic Compounds on Organophilic Bentonites," *Applied Clay Science*, Vol. 6, pp. 39–57.

Strzelczyk, A.B., 1981, "Stone," *Economic Microbiology*, Vol. 6, A.H. Rose, ed., Academic Press, New York, pp. 61–80.

Swett, G.H., 1992, "Bioremediation: Myths Vs. Realities," *Environmental Protection*, May, pp. 23–26.

Sylvester, E.J., and Klotz, L.C., 1983, *Gene Age*, Charles Scribner's Sons, New York, 203 pp.

Tampion, J., and Tampion, M.D., 1987, *Immoblized Cells: Principles and Applications*, Cambridge University Press, New York, 257 pp.

Teitelman, R., 1989, *Gene Dreams*, BasicBooks, New York, 237 pp.

Thayer, A.M., 1991a, "Biopharmaceuticals Overcoming Market Hurdles," *Chemical and Engineering News*, Feb. 25, pp. 27–48.

Thayer, A.M., 1991b, "Bioremediation: Innovative Technology for Cleaning Up Hazardous Waste," *Chemical and Engineering News*, Aug. 26, pp. 23–44.

Trudinger, P.A., and Swaine, D.J., 1979, *Biogeochemical Cycling of Mineral-Forming Elements*, Elsevier Scientific Publishing, New York, 612 pp.

Trudinger, P.A., Walter, M.R., and Ralph, B.J., 1980, "Biogeochemistry of Ancient and Modern Environments," *Proceedings, 4th International Symposium on Environmental Biogeochemistry*, Australian Academy of Science, Canberra, ACT, 723 pp.

Updegraff, D.M., 1990, "Early Research on Microbial Enhanced Oil Recovery," *Developments in Industrial Microbiology*, Vol. 31, pp. 135–148.

Verrall, M.S., and Hudson, M.J., 1987, *Separations for Biotechnology*, Society of Chemical Industry, Ellis Horwood Limited, Chichester, UK, 502 pp.

Vogel, H.C., 1983, *Fermentation and Biochemical Engineering Handbook*, Noyes, Park Ridge, NJ, 440 pp.

Volesky, B., ed., 1990, *Biosorption of Heavy Metals*, CRC Press, Boca Raton, FL, 396 pp.

Walker, J.M., and Cox, M., 1988, *Language of Biotechnology, A Dictionary of Terms*, American Chemical Society, Washington, DC, 254 pp.

Wang, D.I.C., et al., 1979, *Fermentation and Enzyme Technology*, John Wiley & Sons, New York, 374 pp.

Watson, J.D., et al., 1992, *Recombinant DNA*, W.H. Freeman Co., New York, 626 pp.

Weaver, T.L., and Dugan P.R., 1972, "Enhancement of Bacterial Methane Oxidation by Clay Minerals," *Nature*, Vol. 237, June 30, pp. 518.

Williams, D., 1990, ed., *Concise Encyclopedia of Medical & Dental Materials*, Pergamon Press, New York, 379 pp.

Williams, S.T., Sharpe, M.E., and Holt, J.G., eds., 1989, *Bergey's Manual of Systematic Bacteriology*, 1989, Williams & Wilkins, Baltimore, MD, 4 vols., 2,648 pp.

Wiseman, A., 1983, *Principles of Biotechnology*, Chapman & Hall, New York, 217 pp.

Witt, S.C., 1990, *Biotechnology Microbes and the Environment*, Center for Science Information, San Francisco, 219 pp.

Woodman, W.F., 1989, *Biotechnology and the Research Enterprise: A Guide to the Literature*, Iowa State University Press, Ames, IA, 358 pp.

Wu, K-Y.A., and Wisecarver, K.D., 1992, "Cell Immobilizaton Using PVA Crosslinked with Boric Acid, *Biotechnology and Bioengineering*, Vol. 39, pp. 447–449.

Xu, W., and Hang, Y.D., 1989, "Inhibition of Geotrichum Candidum by Bicarbonate," *MIRCEN*, Vol. 5, pp. 109–113.

Yen, T.F., 1976, *Recovery of Hydrocarbons from Microbial Attack on Oil-bearing Shales*, Institute of Petroleum, London, pp. 22–46.

Yen, T.F., 1977, "Microbial Oil Shale Extraction," *Microbial Energy Conversion*, H.G. Schlegel and J. Barnea, eds., Pergamon Press, New York, pp. 309–323.

Yen, T.F., 1990, *Microbial Enhanced Oil Recovery: Principle and Practice*, CRC Press, Boca Raton, FL, 257 pp.

Yoxen, E., 1983, *Gene Business*, Oxford University Press, New York, 230 pp.

Yuan, R.T., and Dibner, M., 1991, *Japanese Biotechnology*, Stockton Press, New York, 200 pp.

Zajic, J.E., 1969, *Microbial Biogeochemistry*, Academic Press, New York, 345 pp.

Zajic, J.E., et al., 1983, *Microbial Enhanced Oil Recovery*, PennWell Publishing, Tulsa, OK, 174 pp.

BORON AND BORATES

ROBERT B. KISTLER AND CAHIT HELVACI

Borates are defined as "salts or esters of boric acid; a compound containing the radical B_2O_3" (Bates and Jackson, 1987). Borates are defined by industry as any compound that contains or supplies boric oxide. A large number of minerals contain boric oxide, but the three that are the most important from a worldwide commercial standpoint are: borax, ulexite, and colemanite. These are produced in a limited number of countries (Fig. 1), dominated by the United States and Turkey, which together furnish about 90% of the world's borate supplies (Lyday, 1991). Production in the United States originates in the Mojave Desert of California; borax and kernite are mined by US Borax from its large deposit at Boron. Borate containing brines are pumped from Searles Lake by North American Chemical Co., and a limited amount of colemanite is mined by Newport Mineral Ventures from Death Valley. Turkish production is controlled by Etibank, the national mining enterprise, which supplies most of the commercially traded ulexite and colemanite from mines in the Bigadic and Emet Districts, plus borax from the huge deposit at Kirka.

Borate minerals have been employed in a wide range of uses for many centuries, dating from at least the 8th century when they were used primarily as a flux for assaying and refining gold and silver. Their valuable properties and relative rarity soon stimulated international trade in borates; Marco Polo claimed to have transported Chinese borate minerals from Tibet to Europe, and his home port of Venice was the center for borate imports and use throughout the Middle Ages (Travis and Cocks, 1984).

Borates were traded at relatively high prices for highly specialized applications into the late years of the 19th century. At that time they were being used for medicines, food preservatives, ceramic glazes, and in expanded applications as metal fluxes. It was in that era that borax was discovered in Clear Lake, CA, and more significantly, the playa deposits of western Nevada were found and exploited by F.M. "Borax" Smith and others. Research and development by Smith's company began the evolution of borates from a high priced commodity with a few specialized uses to the relatively low priced chemical raw material used in large tonnages on a worldwide scale today.

Borates are often defined and sold by their boric oxide or B_2O_3 content, and most statistical data are listed in tons of B_2O_3. Of the products sold by the industry, borax pentahydrate or 5 mol and boric acid are the most commonly traded commodities. Boric acid plants are operated by all of the major borate producers. Glass fiber insulation is the major end use in the United States followed by textile glass fiber and borosilicate glass, detergents, and ceramics. Detergent usage continues to be a major end use in Europe.

GEOLOGY

Mineralogy

Over 150 minerals are known to contain the element boron (B). These minerals may be divided for convenience into three broad groups according to their origin and geological environments: A) skarn minerals related to intrusives, mainly silicates and iron oxides; B) magnesium oxides related to marine sediments; and C) hydrated sodium and calcium borates related to continental sediments and volcanic activity.

In groups A and B are the source minerals for the Russian and some of the Chinese production, the major ones being datolite and szaibelyite. In group C are borax, kernite, colemanite, and ulexite which provide the source for most of the world's production from Turkey, South America, and the United States. Table 1 provides a more complete list of the common borates.

For description of minerals in detail, readers are referred to mineralogy texts. Here, mineral names and compositions are taken from Dana's System of Mineralogy (Palache et al., 1951). Some names of long standing are persistently used in the borate industry, such as *tincal,* an old name for borax which is often used to distinguish the naturally occurring mineral from the identical manufactured compound; others are indicated by parentheses on Table 1.

Borax is by far the most important mineral for the borate industry. It crushes freely and dissolves readily in water; its solubility and rate of solution increases with water temperature. Borax in large tonnages is present in the deposits at Boron, CA, Kirka, Turkey, and Tincalayu, Argentina. Kernite, the metamorphic phase of borax, is present in minor amounts at Kirka and Tincalayu, but it makes up about a third of the total reserve at Boron. Kernite has a higher B_2O_3 content than borax, but its excellent cleavage causes it to form into fibers that mat and clog handling equipment, and being slowly soluble in water, it requires autoclaves or pre-refinery hydration for efficient conversion into refined products. It is currently used primarily as feed for the boric acid plant at Boron.

Colemanite is the preferred calcium-bearing borate used by the non-sodium fiberglass industry. It has low solubility in water, although it dissolves readily in acid. Some colemanite is used in Europe in chemical plants to produce boric acid because the supply from Turkey provides B_2O_3 at lowest cost. Turkey is the world's major source of high grade colemanite. The United States has important reserves in the Death Valley area, but only limited amounts are produced there at this time. Colemanite is not known to occur in major deposits outside Turkey and North America, although the higher hydrate, inyoite, is mined on a limited scale in Argentina and in eastern Europe.

Ulexite is the usual borate found on or near the surface, in playa-type lakes and marshes of Recent to Quaternary Age throughout the world, where it occurs as soft, often damp, masses of fibrous crystals. These "cotton balls" or "papas" are collected in major amounts in South America and China. Ulexite of Neogene Age, which is mined and produced in Turkey and which occurs at Boron and Death Valley in the United States, is well lithified and is, therefore, harder, more dense, and commonly well-bedded.

Szaibelyite (ascharite) is a major source of both Chinese and Russian borate material. It is a magnesium borate and, like colemanite, has low solubility in water. Although it is less satisfactory,

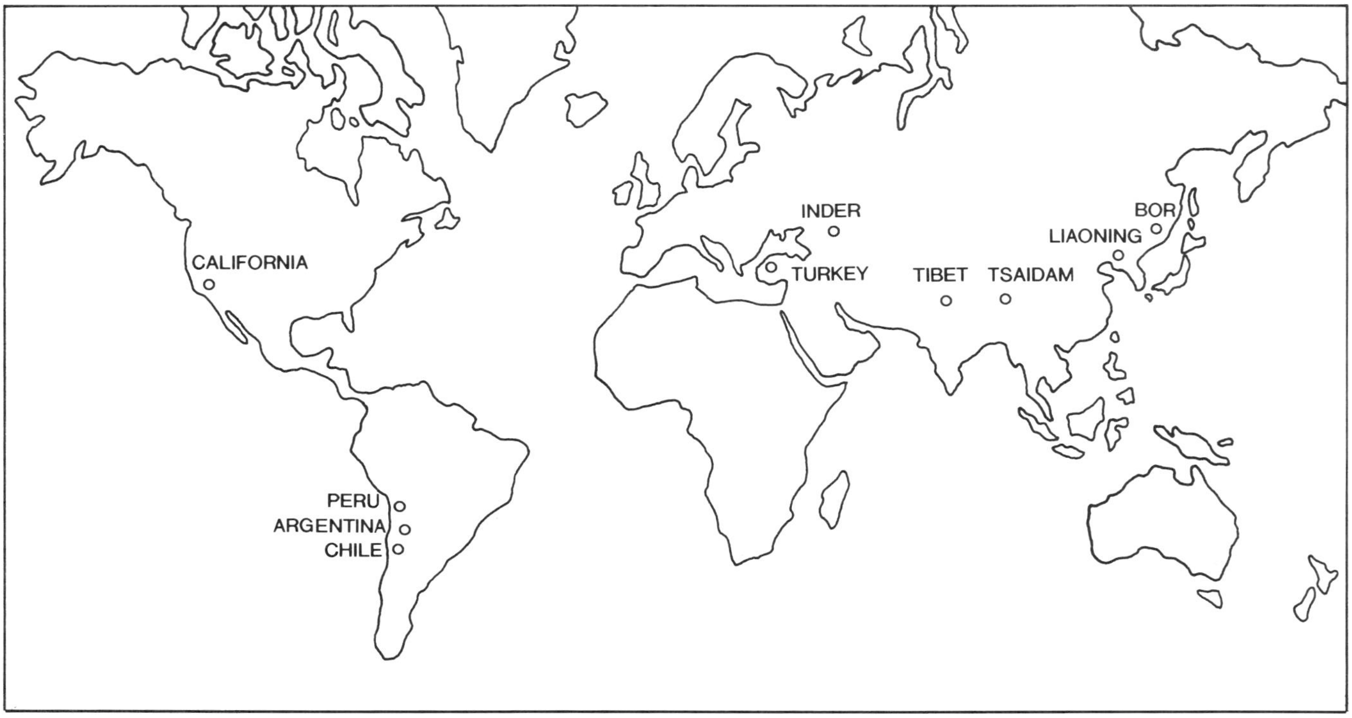

Fig. 1. Current world sources of borates.

Table 1. Common Borate Minerals

Mineral	Empirical Formula	B_2O_3 Content, Wt %
Sassolite	$B(OH)_3$ or $B_2O_3{\cdot}3H_2O$	56.4
Borax (Tincal)	$Na_2B_4O_7{\cdot}10H_2O$	36.5
Tincalconite	$Na_2B_4O_7{\cdot}5H_2O$	48.8
Kernite	$Na_2B_4O_7{\cdot}4H_2O$	51.0
Ulexite	$NaCaB_5O_9{\cdot}8H_2O$	43.0
Probertite	$NaCaB_5O_9{\cdot}5H_2O$	49.6
Priceite (Pandermite)	$Ca_4B_{10}O_{19}{\cdot}7H_2O$	49.8
Inyoite	$Ca_2B_6O_{11}{\cdot}13H_2O$	37.6
Meyerhofferite	$Ca_2B_6O_{11}{\cdot}7H_2O$	46.7
Colemanite	$Ca_2B_6O_{11}{\cdot}5H_2O$	50.8
Hydroboracite	$CaMgB_6O_{11}{\cdot}6H_2O$	50.5
Inderborite	$CaMgB_6O_{11}{\cdot}11H_2O$	41.5
Kurnakovite	$Mg_2B_6O_{11}{\cdot}15H_2O$	37.3
Inderite	$Mg_2B_6O_{11}{\cdot}15H_2O$	37.3
Szaibelyite (Ascharite)	$Mg_2B_2O_5{\cdot}H_2O$	41.4
Suanite	$Mg_2B_2O_5$	46.3
Kotoite	$Mg_3B_2O_6$	36.5
Pinnoite	$MgB_2O_4{\cdot}3H_2O$	42.5
Boracite (Strassfurite)	$Mg_3B_7O_{13}Cl$	62.2
Datolite	$Ca_2B_2Si_2O_9{\cdot}H_2O$	21.8
Cahnite	$Ca_2AsBO_6{\cdot}2H_2O$	11.7
Danburite	$CaB_2Si_2O_8$	28.3
Howlite	$Ca_4Si_2B_{10}O_{23}5H_2O$	44.5
Vonsenite (Paigeite)	$(Fe,Mg)_2FeBO_5$	10.3
Ludwigite	$(Fe,Mg)_4Fe_2B_2O_7$	17.8
Tunellite	$SrB_6O_{10}{\cdot}4H_2O$	52.9

due to its magnesium content, for most uses than either borax or colemanite, substantial tonnages are utilized in eastern Europe, Russia, and Asia; it is not traded internationally as a mineral concentrate on a major scale.

The Russians also produce substantial amounts of borates utilizing the skarn borosilicates, mainly datolite, with some reports of minor amounts of danburite, ludwigite, and tourmaline. These minerals must first be liberated, concentrated, and then dissolved in acid to make a usable product because their natural melting points exceed those of the other minerals used in common glass furnaces. They also usually contain iron and manganese, which must be removed because they color glass.

The list of boron minerals in Table 1 includes some that were utilized in the past even though they are difficult to treat. Priceite (pandermite) was mined in Turkey and hydroboracite was mined in Russia and Argentina. Other minerals such as inyoite, howlite, meyerhofferite, and kurnakovite occur intimately associated with the major ores. Boracite was used in Germany prior to 1945 where it and minor magnesium borates were recovered as a byproduct of potash mining.

A large number of silicates contain boron in their lattices in varying amounts. Axinite, suanite, kotoite, and others are listed in the literature as occurring with the Russian borosilicate ores. Sassolite has only mineralogic interest at most occurrences, for the quantity found is generally very small. In the Lardarello region of Italy, however, natural steam carries boric acid recoverable as sassolite, and for a long period prior to 1965, several thousand tons per year were produced.

Chemistry

Boron is the fifth element of the periodic table and is the only electron-deficient nonmetallic element. Thus, boron has a high affinity for oxygen, forming strong covalent boron-oxygen bonds in compounds known as borates. Boron is also the only light element with two abundant isotopes, B^{10} and B^{11}; the former has a large capture cross-section that makes it an excellent neutron absorber.

The chemistry of borates is somewhat analogous to, but more

complex than, silicates. The boron atom in borates can be either three or four coordinate-bonded to oxygen, forming either planar trigonal BO_3 units or negatively charged tetrahedral BO_4^{-1} units; crystalline sassolite or boric acid, $B(OH)_3$, has only BO_3 units. Other borate minerals are typically salts containing a mixture of BO_3 and BO_4^{-1} units.

The nomenclature for borates can be confusing because three different formula systems are in common use. For example, ulexite can be described by an empirical formula, $NaCaB_5O_9{\cdot}8H_2O$, an oxide formula, $Na_2O{\cdot}2CaO{\cdot}5B_2O_3{\cdot}16H_2O$, and a crystal structure formula, $NaCa[B_5O_6(OH)]{\cdot}5H_2O$. Only the latter formula conveys information about how atoms are organized in the crystal. Empirical formulae are used to describe the borate minerals listed in Table 1.

Many borate minerals contain waters of hydration and, with the exception of borax, are stable at ambient conditions. Borax dehydrates to tincalconite under ambient conditions unless the relative humidity is high. Upon heating, borates lose their attached water molecules before fusing to a melt. The common borate salts melt at low temperatures; the melts are excellent fluxes or solvents for dissolving other more refractory oxides.

Only alkali-metal borates, ammonium borates, and boric acid have appreciable solubility in water. In dilute aqueous solutions, borates exist as an equilibrium mixture of two species, un-ionized molecular boric acid molecules, $B(OH)_3$, and metaborate ions, $B(OH)_4^{-1}$. For example, borax dissolves in water giving a 1:1 mixture of $B(OH)_3$ and $B(OH)_4^{-1}$ and is a good buffer system at pH 9.4.

$$B(OH)_3 + 2H_2O \leftrightharpoons B(OH)_4^{-1} + H_3O^{+1} + Na_2B_4O_7{\cdot}10H_2O \text{ (borax)}$$

Most borate minerals dissolve in strong mineral acids to liberate the weaker boric acid; for example, boric acid can be produced by mixing ulexite with sulphuric acid. Boric acid is a weak Lewis acid, electron pair acceptor, with a solution pH of about 5.

Crystalline borate minerals are made up of both trigonal and tetrahedral borate units. These borate units can link through B-O-B bonds to form chain polyborates or they may join into rings which can contain three or four boron atoms, that is, tri- and tetraborates. These rings can in turn link together to form penta-, hexa-, and higher borates.

Non-borate boron-containing compounds (those containing no boron-oxygen bond) are not found in nature. These include elemental boron, borides, boron halides, boranes, and organoboron compounds. In general, these manmade materials are either stable inert solids, or they hydrolyze in the presence of moisture or oxidize in the presence of oxygen to form the more stable borates (Smith, 1992).

Borates are widely distributed in nature in low concentrations as alkali-metal and alkaline-earth borate and borosilicate minerals, and less commonly as boric acid. They are typically found in soil and rock in concentrations of up to about 450 ppm total boron, distributed in over 150 minerals primarily as salts of sodium, calcium, and magnesium. The average concentration of boron dissolved in land surface water is about 0.1 ppm and, in sea water, about 4.6 ppm (Sprague, 1972).

At low levels, water soluble boron is an essential micronutrient for the growth and viability of plants; the range between insufficient and excess boron is narrow (0.25 to 15.0 ppm boron) and most soils are usually within this range (Mass, 1986, Sprague, 1972). Borate has been found in animal tissues at about 1 ppm, resulting from ingestion of fruits and vegetables; it is not known to have an essential biochemical function, although it may play a role in the body's ability to use calcium. Borate is transported in plants and animals usually complexed with polyalcohols in the aqueous phase (Smith, 1992).

Origin

Major borate deposits throughout the world are found in tectonically active extensional terrain associated with plate boundaries, (Ozol, 1977, Alonso et al., 1988). Most of the commercial borate deposits in the United States, South America, and Turkey are thought to be associated with continental sediments and volcanism of Neogene Age. Many of the older skarn deposits also appear to be related to continental volcanic sources. Marine borate deposits are apparently the product of evaporation of seawater in a restricted basin probably associated with a sea floor borate source and/or progressive decanting that preferentially concentrated the borates many times above natural seawater values. Borates associated with igneous and some metamorphic rocks are thought to be an end phase of specialized magmatic segregation, or leached from the intruded rocks by associated hydrothermal fluids. Boron isotope studies indicate that most borates are derived from crustal sources (Swihart et al., 1986, Ol et al., 1989).

Borate Deposits in Nonmarine Basins: The largest known borate deposits originated as chemical precipitates and are found interbedded with clays, mudstone, tuffs, limestones, and similar lacustrine sediments. There is evidence that most of these deposits were closely related in time to active volcanism. Thermal springs and hydrothermal solutions associated with this volcanic activity are therefore regarded as the most likely source of the boron (Waring, 1965, Alonso et al., 1988). As examples, several South American springs in volcanically active areas are depositing borates (Muessig, 1966), and the first borax discovered in the United States was found in the muds associated with warm springs at Clear Lake, CA, a volcanically active area (Hanks, 1883). The Italian (Tuscany) steam vents from which sassolite was recovered also represent an active volcanic source.

In addition to the concentrated source of the borates and a "basin" in which they can collect, an arid to semiarid climate also seems to be an essential requirement during deposition and concentration of economic amounts of the soluble borates. These soluble borates can, in the long run, only be preserved by burial; however, the lack of deposits of soluble borates older than mid-Tertiary may indicate that even burial is not adequate for their protection over long periods of geologic time.

Hydrated borates may accumulate in several ways within a nonmarine basin. They may be deposited in layers in a spring apron around a borate spring, with ulexite, borax, or inyoite as the primary borate mineral (Muessig, 1966). Borates may also form in a pool dominantly fed by a borate spring, with borax crystals formed in bottom muds or at the intermittently dried margins as at Clear Lake, CA (Hanks, 1883, White 1957), and at Salar de Surire, Chile (Fig. 2). If the spring flow is low or intermittent, evaporation develops a surface efflorescence or precipitate, or an accumulation of crystals just below the surface; these are the *marsh* or playa deposits of early mining in California and Nevada and some of the salar deposits of

Fig. 2. Warm spring borate source, Salar de Surire, Chile (photograph: R. Kistler, May 1987).

South America (Papke, 1976). Finally, there are lake deposits, whose occurrence required much more than seasonal flooding; these are the borax deposits such as Boron and Kirka, formed by chemical precipitation in a closed basin. There is some disagreement as to whether the thicker ulexite deposits, such as those in Turkey and Death Valley, for example, are spring apron or lake deposits. In fact, there is probably a complete gradation between spring apron and small lake deposits. These borate lakes are essentially *mono mineralic* in the sense that no other salts occur in major quantities. There is, however, another type of borate lake deposit, with mixed salts and/or brine containing borates in sufficient quantity to justify recovery. These are becoming increasingly important on a worldwide scale. Searles Lake, CA, has been cited as a type example of a multicomponent lake formed by evaporation of lake waters (Smith, 1966, 1979). Numerous studies of Searles Lake have concluded that boron and the other dissolved constituents originated along the eastern front of the Sierra Nevada, were concentrated and decanted in a series of updrainage lakes, and finally were precipitated and preserved in Searles Lake itself. This scenario seems to fit several of the Chinese and Tibetan lakes as well, although they are far less studied (Lock, 1991). The borates found in the large South American salars, such as Uyuni and Atacama, may also have formed by leaching of surrounding rocks and subsequent evaporation, although the role of local mineralized spring waters containing boron has not been fully evaluated at this time.

Marine Evaporites: Borates of marine origin have been found in commercial quantities only in eastern Europe. These are magnesium borates associated with Permian salt deposits. They were produced in Germany as a byproduct of potash mining, and in the Inder region of Kazakhstan. The Inder deposits, where the borates occur as veins in the cap of a very large salt dome complex, are reported to be remobilized and concentrated from the salt during the intrusion of the salt dome itself. The source of the unusual amount of marine boron, however, remains a problem. Some of the Chinese deposits of the Liaoning Peninsula may be of similar origin although they occur as veins in metamorphosed limestone and magnesite of Precambrian Age.

The Inder Lake brines which are also a source of Kazakhstan production appear to be simply a sump accumulation of borate values leached from the huge Inder salt dome complex. These would be similar in occurrence to the complex lake brines, referenced above, although leached from marine salts rather than continental rocks. The Kara-Bagaz-Gol Lagoon borates on the east shore of the Caspian Sea would appear to be leached from marine brines as well (Lyday, 1991).

Magmatic Sources: Pegmatites and contact metamorphic rocks contain assemblages of various boron-containing minerals such as datolite, ludwigite, paigeite, and tourmaline. These represent concentrations of boron that relate more or less directly to the crystallization of intrusive granitic magma. Analyses show that granites average about 10 ppm boron with a few exceptions ranging up to 300 ppm. However, boron does not enter readily into the crystal structure of the common rock-forming minerals; hence, when magma crystallizes, the boron is thought to leave with the released water. The conditions of high temperature, 300 to 400°C, and fluids under high pressure at an intrusive contact also provide the possibility that some of the boron may be extracted from the adjacent country rocks. In either case, borate skarn deposits may be developed. These deposits, some associated with iron ores and magnesium deposits of commercial grade, are mined in both eastern Russia and China (Zhang and Jin, 1991, Lisitsyn and Pastushenko, 1983).

Major Deposits

United States: In 1856, J.A. Veach discovered borax in the muds associated with a mineral spring at what is now Clear Lake,

Fig. 3. Location of major US borate deposits.

in Lake County, CA. Small-scale production began there in 1864. Prospecting soon led to the discovery of larger deposits in the desert areas of western Nevada and southeastern California. The first 63 years of domestic production saw several changes in both the type and location of the mineral deposits mined, but in general, production increased and the price of borax decreased as larger and better deposits were found and refinery techniques improved. The year 1927 was a benchmark in US borate history as both the large borax deposit at Boron and the Searles Lake brines with their recoverable borax content were brought into large scale production and soon replaced most of the other domestic sources. Borates are presently produced at three locations in Southern California: Boron, Searles Lake, and, to a lesser extent, the Death Valley area (Fig. 3).

Boron, CA (Kramer)—The Boron (Kramer) deposit is located between the towns of Mojave and Barstow in the northwestern Mojave Desert, about 145 km northeast of Los Angeles. Geologically, the deposit lies near the edge of a large Tertiary basin. The "basement" rocks consist of Jurassic and Cretaceous granitics and metavolcanics. These are overlain by a lower Tertiary sequence consisting of continental sediments, primarily arkosic sands and silts apparently derived from the earlier granites, with occasional freshwater limestones and volcanics. The borates are immediately underlain by a Tertiary basalt flow and overlain by a series of continental arkosic sands, Fig. 4 (Siefke, 1991). The Boron deposit has been dated as upper Middle Miocene on the basis of mammalian fossils found above the borates (Whistler, 1984) and from tuffs within the ore.

The Boron deposit consists of a lenticular mass of borax, kernite, and interbedded montmorillonite-illite clays, dipping about 10°S. It is approximately 1.6 km long, 0.8 km wide, and up to 100 m thick. Outward from this central borax core, there is a facies change to ulexite and clay, which also exists above and below the borax. The deposit is thought to have formed in a shallow "permanent" lake fed by sodium- and boron-rich thermal (volcanic) springs which accompanied late stage volcanic activity. Several tuff and tuffaceous clay beds, some of which exhibit ripple marks and other evidence of shallow water deposition, occur interbedded with the borax.

The original lake basin was controlled by faulting contemporaneous with the deposit. Continued fault movement tilted and dropped the lake sediments; they were then buried under at least 762 m of late Miocene and Pliocene arkosic sediments (Christ and Garrels, 1959). The more deeply buried portions of the borax deposit were metamorphosed to kernite at this time. Renewed movement in late Pliocene-Pleistocene uplifted the deposit, removed a portion of it by erosion, and caused the outer portions of

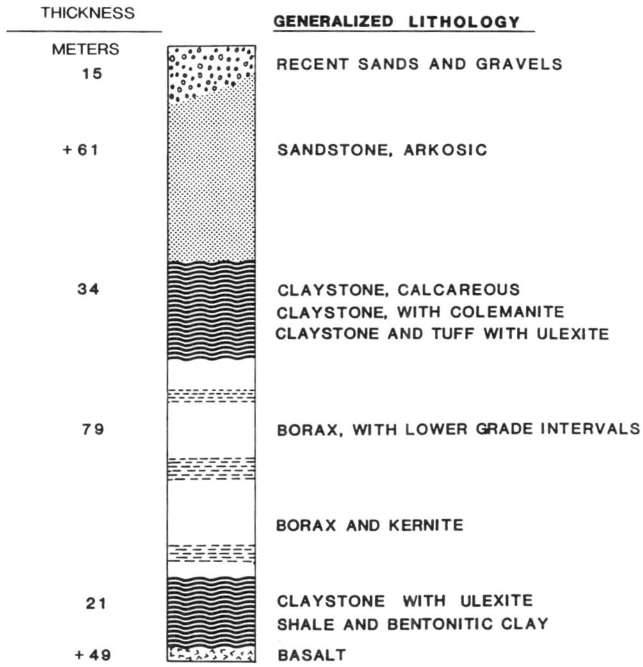

Fig. 4. Stratigraphic section, Boron, CA.

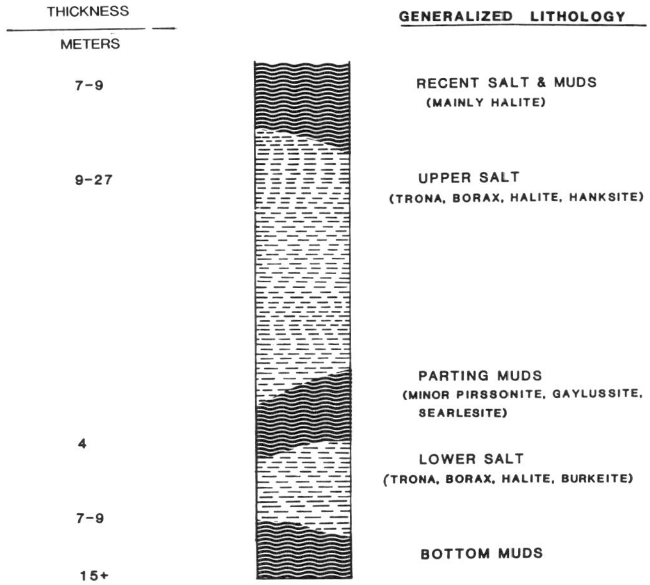

Fig. 5. Stratigraphic section, Searles Lake, CA.

the kernite facies to rehydrate to borax. The deposit was then covered by 9 to 21 m of Pleistocene and Recent sands and gravels (Barnard and Kistler, 1966, Siefke, 1991).

Over 80 minerals have been reported as occurring in the Boron deposit (Morgan and Erd, 1969) including at least 15 borates (Puffer, 1975). Of these, borax comprises about two-thirds of the ore body with kernite and clays, mainly montmorillonite, comprising most of the remainder. The ulexite and claystone facies which envelops the ore body contains most of the 80 minerals mentioned in the literature, including minor amounts of colemanite which occurs at the upper ulexite-clay contact.

Five faults cut the borax body in a NNW direction. Most of these exhibit both late Tertiary strike-slip movement and evidence of movement contemporaneous with the deposition of borax. Borax crystals in the main portion of the ore body are commonly euhedral to sub-euhedral, and range in length from less than 0.5 to over 2.5 cm (Barnard and Kistler, 1966).

The Boron deposit did not crop out; it was discovered in 1913 by Suckow, a homesteader, who struck colemanite accidentally while drilling a water well on the northwest fringe of the deposit. The sodium borates were first detected by exploration drilling in 1925, and production of kernite and borax began in 1927. Mining, soon entirely in borax, continued by large scale underground methods until 1957, when production began from the present open pit. US Borax presently mines both borax and kernite by truck/shovel methods with the ore being removed from the pit by conveyor and processed in the adjacent sodium borate refinery and boric acid plant. Reserves of the Boron deposit were estimated by the US Bureau of Mines at 140 Mt (Norman, 1991).

Searles Lake, CA—The Searles Lake playa occupies the central portion of a small desert valley midway between Death Valley and Boron, in southeastern California. The "lake" consists of a central salt flat, approximately 103 km^2 in extent, which overlies a crystalline mass of admixed salts, mainly sodium carbonates, with borax, clays, and interstitial brines (Fig. 5). The salts and brines are thought to be the end results of concentration, desiccation, and decantation, in a series of late Quaternary lakes which extended along the east front of the Sierra Nevada. The boron contained within the salt bodies probably originated in the hot springs along the Sierra Nevada frontal faults in Long and Owens Valleys (Smith, 1979, Mumford, 1954, Gale, 1914). There are two main salt producing horizons: the "upper salt," which ranges from 9 to 27 m thick, averaging 15 m, with a brine content averaging 1.0% B_2O_3; and the "lower salt," which is 8 to 14 m thick, averaging 12 m, with a brine content averaging 1.2% B_2O_3. The upper salt is overlain by a recent surficial mud and salt crust about 7 to 9 m thick and is separated from the lower salt by 3 to 6 m, averaging 4 m, of relatively impervious clays called the "parting muds." The interstitial multicomponent brine, which makes up almost 45% by volume of these salt horizons, is pumped from a series of wells and treated in two adjacent refineries. A relatively constant brine content is maintained by natural recharge and dilute refinery effluent which dissolves additional salts. The two producing salt horizons are pumped separately as they differ somewhat in chemical makeup and therefore require different plant treatment.

Playa-type borax was discovered on the shores of Searles Lake by D. Searles and others perhaps as early as 1863. Small amounts of borax were produced from the playa surface until 1919 when, after extensive potash exploration drilling during World War I, borax along with trona and various other salts began to be recovered from brine wells sunk into the central portion of the lake, with large scale production beginning in 1927. At the present time, North American Chemicals Co. is the only company producing borates from the lake. Reserves in Searles Lake have been estimated to contain 40 Mt of sodium borates (Norman, 1991).

Death Valley, CA—Death Valley is a long, north-trending valley near the Nevada border with Southern California. It is fault-bounded on both the east and west sides by recently active faults which have lowered the area near Badwater to 86 m below sea level while raising the adjacent Panamint Range to 3 367 m. The valley contained several lakes during the Pleistocene and over 300 m of salts, mainly halite and gypsum, accumulated in the Badwater area at that time. Minor amounts of the playa borates, borax and ulexite, which occur with these salts, accumulated in the surface muds and seeps and were gathered in the early 1880s. The more productive playa area near Harmony, 32 km north of Badwater, near the present National Park headquarters, was the source for the famous 20 Mule Team borates (1883–1897). This area of the Valley floor appears to be enriched by fluids derived from the leaching of the adjacent borate-containing Tertiary beds.

In the late Neogene, over 2 130 m of lake sediments were deposited in what is now Furnace Creek Wash, in and adjacent to the east side of Death Valley. It is these lake sediments of the Furnace Creek Formation which contain the bedded borates that were the main source of domestic borates from 1907 to 1927 and that are currently being mined.

The area has undergone several periods of faulting since the Tertiary borates were deposited. Many deposits are fault-bounded and bedding dips of up to 45° are common. Geologic mapping indicates that the deposits are elongate parallel to the old shore lines of the Tertiary lake. Laterally, the deposits grade into shallow water limestones or limey mudstones and tuffaceous clays.

"The main deposits (Tertiary colemanite-ulexite deposits) are low in the Furnace Creek Formation, starting at the top of the lowest conglomerate and continuing to about 152 m higher. The deposits form elongate lenses as much as 61 m thick in which the borates are interlayered with mudstone, limestone, conglomerate, and gradational variations of these lithologies. The three minable borate minerals, where found in the same deposit, are distributed in three zones in the following order of increasing abundance: probertite in the core or deepest zone, ulexite next, and colemanite in the surrounding outermost zone. The boundaries of the zones are generally transitional, but colemanite veins sharply transect some of the ulexite. Probertite is a significantly large part of only one of the deposits, whereas ulexite is a large part of many, and colemanite is predominant in most" (McAllister, 1970). A generalized stratigraphic section based on more recent mapping and drill data is shown on Fig. 6.

The 16 deposits that were commercially worked in the Death

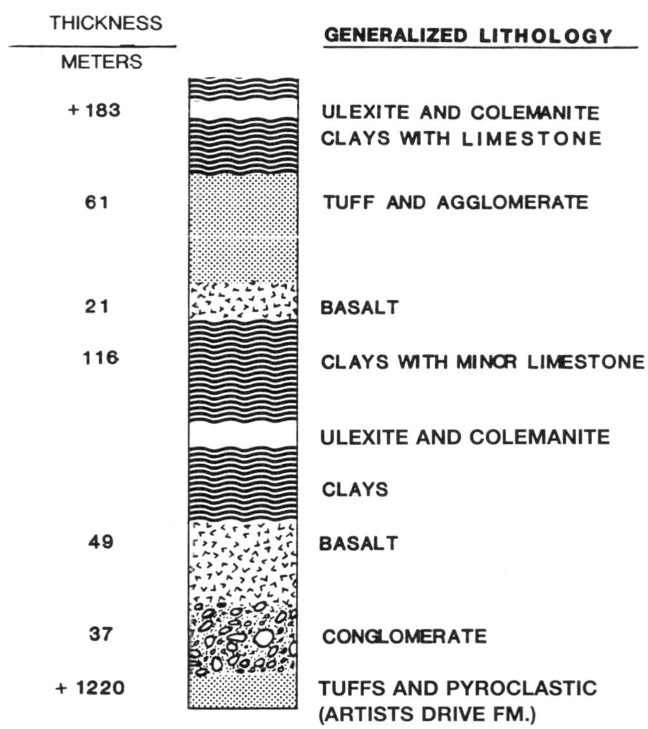

Fig. 6. Stratigraphic section, Death Valley, CA.

Valley area range in size from 181 kt to over 13.6 Mt, averaging 18 to 24% B_2O_3; all but four were underground operations. Production is presently from the Billie underground mine, operated by Newport Mineral Ventures (formerly American Borate Company) and from the Gerstley underground mine of US Borax located near Shoshone, CA, about 50 km to the southeast of the Billie.

Turkey: The borate deposits of Turkey occur in Western Anatolia, south of the Marmara Sea within an area roughly 300 km east-west by 150 km north-south. The main borate districts are Bigadic, Kestelek, Sultancayiri, Emet, and Kirka (Fig. 7).

These borate deposits were formed in lacustrine environments during periods of volcanic activity which commenced in early Tertiary time and continued at least to the beginning of the Quaternary. Although the lithologies of the borate deposits show some differences from one to another, they are generally interbedded with conglomerate, sandstone, tuff, tuffite, claystone, marl, and limestone, and are usually enveloped by, or grade into, limestones or claystones (Fig. 8). Sediments in the borate basins often show clear evidence of cyclicity and exhibit both lateral and vertical facies changes.

Volcanic rocks in the vicinity of the playa-lakes in which the borate deposits were formed are extensive. The volcanic rocks are generally represented by a calc-alkaline series of flows ranging from acidic to basic and by pyroclastic rocks which are interbedded with the sediments.

Bigadic—The Bigadic borate deposits are located in two different zones of Neogene playa-lake sediments within a northeast trending basin. Volcano-sedimentary sequences in the Bigadic district consist of, from bottom to top: basement volcanics, lower limestone, lower tuff, lower borate zone, upper tuff, upper borate zone and olivine basalt. The Neogene sequence in the region rests unconformably on a Paleozoic and Mesozoic basement complex (Helvaci, 1983, Meixner, 1965, Özpeker 1969). The Bigadic borate deposits were formed in two separate but interconnected lacustrine playa-lakes, fed by thermal springs associated with local volcanic activity and under arid climatic conditions. The borates are interbedded with tuffs, tuffites, clays, and limestones (Helvaci and Alaca, 1984, Helvaci, 1989).

The borate minerals formed in two distinct zones, lower and upper, separated by thick tuff beds transformed during diagenesis to montmorillonite and chlorite and to zeolites with heulandite being most common. Colemanite and ulexite predominate in both borate zones, but other borates include howlite, probertite, and hydroboracite in the lower borate zone; inyoite, meyerhofferite, priceite, terschite, hydroboracite, howlite, tunellite, and rivadavite are found in the upper borate zone. Calcite, anhydrite, gypsum, celestite, potassium feldspars, analcime, heulandite, clinoptilolite, quartz, opal-CT, montmorillonite, chlorite, and illite are the associated minerals. These are the largest colemanite and ulexite deposits known in the world (Helvaci, 1989).

Sultancayiri—In the Sultancayiri borate deposit, the sequence consists of, in ascending order: basement conglomerate, tuff, limestone, borate zone, gypsum zone, and limestone units. The borates are interbedded with gypsum, claystone, limestone, and tuff. Priceite is abundant, but other borates include colemanite and howlite. Gypsum exists abundantly and calcite, zeolite, smectite, illite, and chlorite are the other associated minerals in this deposit. Howlite, which has apparently grown in the clays alternating with thin priceite and colemanite bands, indicates a period of increasing silica concentration. As a result of diagenetic events, some small howlite nodules are also embedded in the priceite and colemanite nodules. In the Sultancayiri deposit, the initial solutions must have had abundant boron, calcium, and sulfate (Helvaci, 1990).

Kestelek—The Neogene sediments which contain borates in the Kestelek region rest unconformably on the Paleozoic and Mesozoic basement complex. The sequence consists of, from bottom to top:

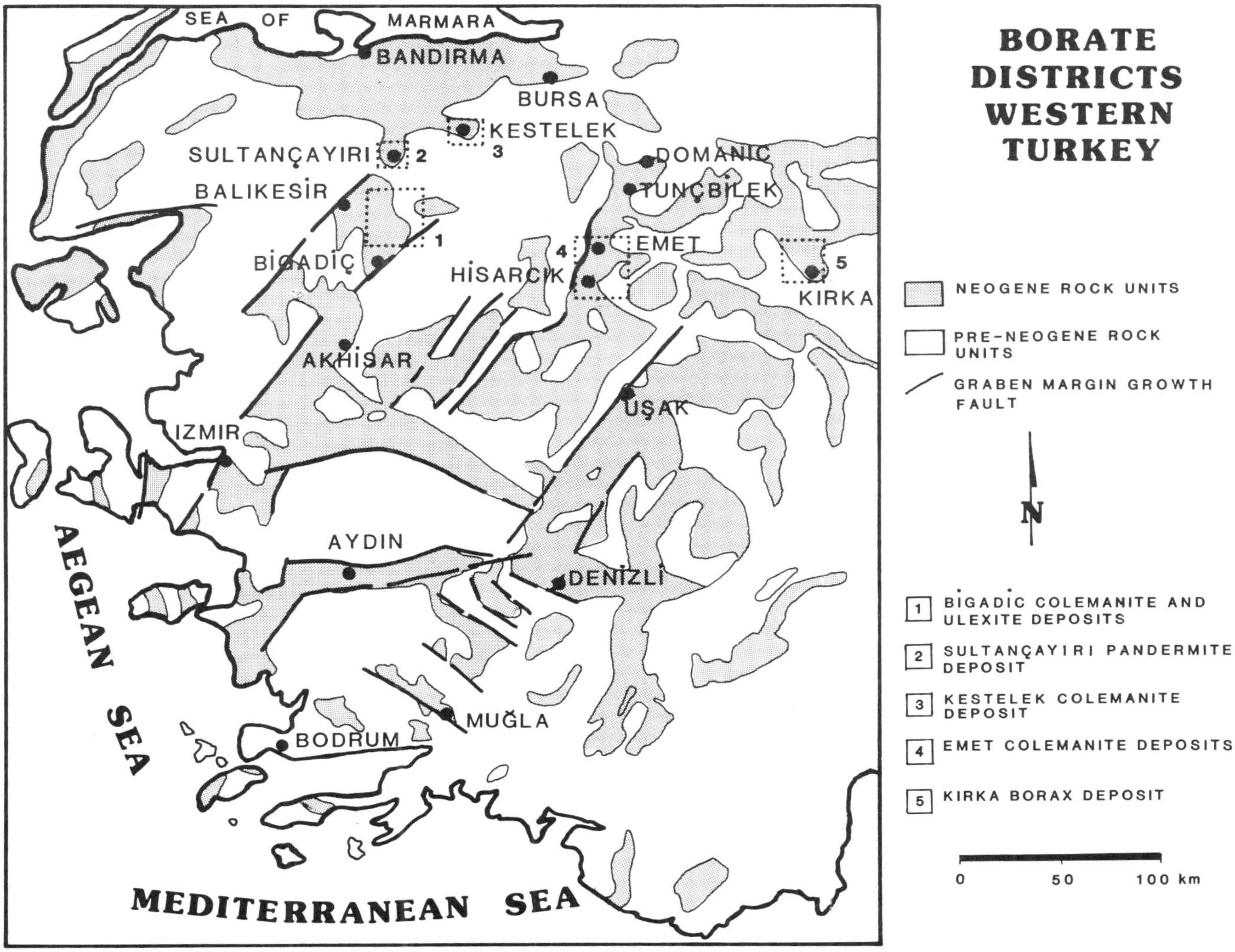

Fig. 7. Borate districts of western Turkey.

basement conglomerate and sandstone, claystone with lignite seams, marl, limestone, tuff, and agglomerate; the borate zone consists of clay, marl, limestone, tuffaceous limestone, tuff, and borate. These sediments were deposited during a tectonically stable period with extensive volcanic activity. During this period, the volcanic activity gradually increased and produced tuff, tuffite and agglomerate, and andesitic and rhyolitic volcanics that are interbedded with the sediments. This sequence is capped by loosely cemented conglomerate, sandstone, and limestone (Helvaci, 1990, Özpeker, 1969).

The borate minerals occur interbedded with clay as nodules or masses and as thin layers of fibrous and euhedral crystals. Colemanite, ulexite, and probertite predominate, with hydroboracite occurring rarely. Calcite, quartz, zeolite, smectite, illite, and chlorite are the associated minerals. Secondary colemanite occurs as transparent and euhedral crystals in the cavities of nodules, in cracks and in vugs. Probertite, which forms in the same chemical environment with ulexite in the Kestelek deposit, indicates a period of higher temperature within the playa lake. The initial solutions crystallizing the borates in the Kestelek deposit are deduced to have had a very low concentration of chloride, low concentrations of sulphate and high concentrations of boron and calcium, with some sodium. Reserves here are about 10 Mt (Helvaci, 1990).

Emet—The Neogene sequence in the Emet region rests unconformably on Paleozoic metamorphic rocks that consist of marble, mica-schist, calc schist, and chlorite schist. The Neogene consists of the following sequence in ascending order: conglomerate and sandstone; thin bedded lower limestone with lenses of marl and tuff; intermediate and acid volcanics, tuff, and agglomerate; a red bed containing conglomerate, sandstone, clay, marl, limestone with coal, and gypsum bands; clay, tuff, tuffite, and marl containing the borate deposits; an upper limestone containing clay, marl, and chert layers; and a capping basalt (Helvaci, 1977). The unit of clay, tuff, tuffite, and marl containing the borate deposits has abundant realgar and orpiment at some horizons indicating that arsenic and boron have a genetic relationship and volcanic origin at Emet (Helvaci, 1984).

The borates are interlayered with tuff, clay, and marl with limestone occurring above and below the borate lenses. The principal borate mineral is colemanite, with minor ulexite, hydroboracite, and meyerhofferite. The Emet borate deposits are the only Turkish deposits known to contain any of the minerals veatchite-A,

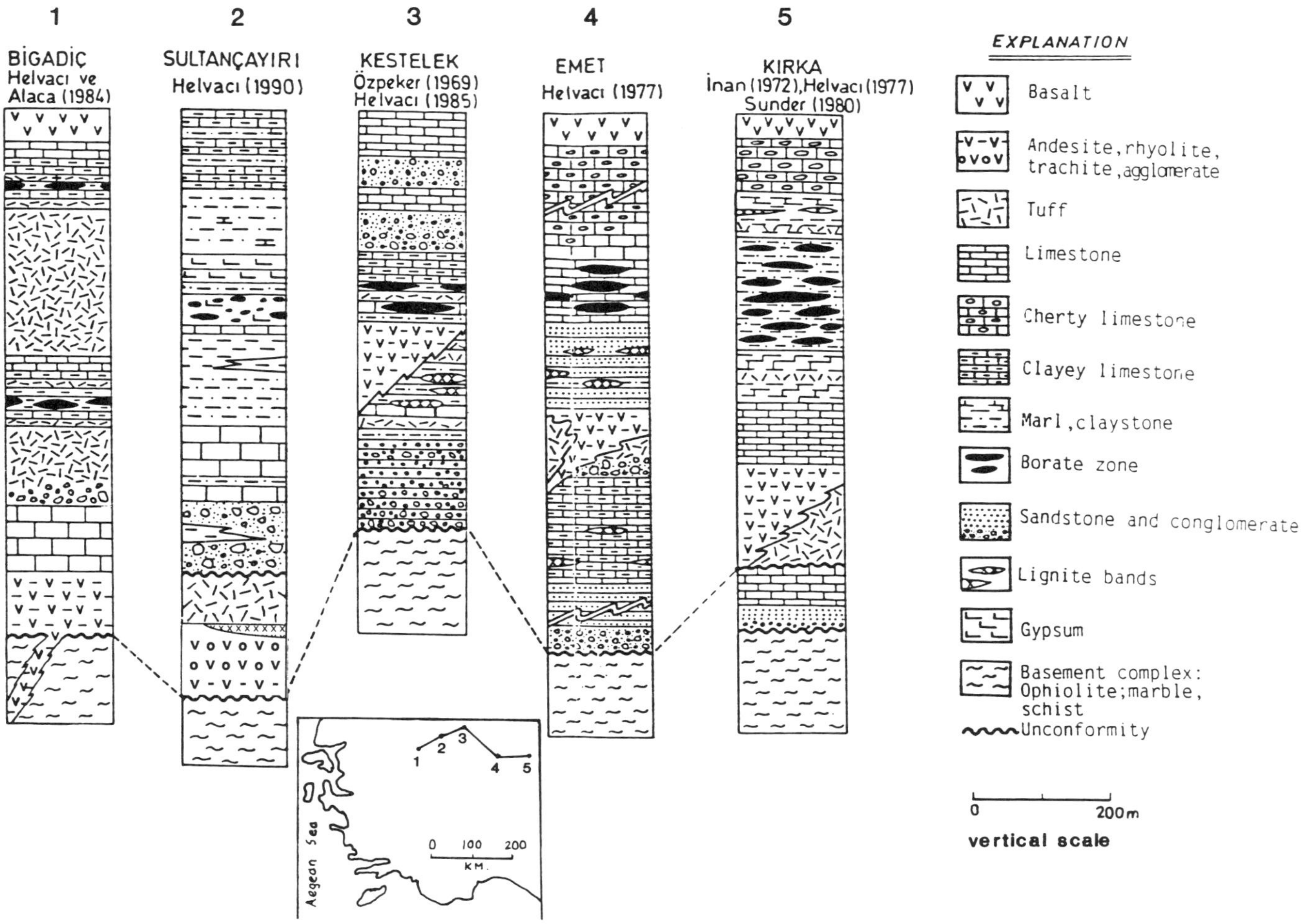

Fig. 8. Stratigraphic sections of Turkish borate deposits.

tunellite, teruggite, and cahnite. Montmorillonite, illite, and chlorite are the only clay minerals that have been identified; montmorillonite is the dominant clay mineral in all the samples and occurs as Al-, Mg-, or Al-Mg-Fe montmorillonite. Illite is only a minor component and is distributed randomly. Chlorite is widely distributed within the deposits and is relatively abundant near or within the horizon of tuffs and tuffites. Zeolites are abundant along the tuff and tuffite horizons. Native sulfur, realgar, orpiment, and celestite occur in the borate zone throughout the area. Gypsum associated with borate minerals has been observed in the southern deposits. Calcite is found in outcrops and adjacent to faults as a result of modern weathering (Helvaci, 1984). Reserves are extensive, estimated at 110 Mt of 45% B_2O_3 (Helvaci and Firman, 1976).

Kirka—The Neogene volcano-sedimentary sequence in the Kirka Basin rests unconformably partly on Paleozoic metamorphics, a Mesozoic ophiolite complex, and Eocene fossiliferous limestone. The Neogene sequence consists of, from bottom to top: volcanic rocks and tuffs; lower limestone with marl and tuff interbeds; borate zone; upper claystone; upper limestone containing tuff and marl with chert bands; and basalt (Helvaci, 1977, Inan et al., 1973, Sunder, 1980).

The principal mineral in the Kirka borate deposit is borax with lesser amounts of colemanite and ulexite. In addition, inyoite, meyerhofferite, tincalconite, kernite, hydroboracite, inderborite, inderite, kurnakovite, and tunnelite are found (Inan et al., 1973, Helvaci, 1977, 1978, 1983). This is the only deposit in Turkey that contains the sodium borates (borax, tincalconite, and kernite), together with inderborite, inderite, and kurnakovite. The borax body is enveloped by a thin ulexite facies, followed outward by a colemanite facies. The whole is enclosed by calcite and limestone.

The borate layers contain minor amounts of realgar, orpiment, gypsum, celestite, calcite, and dolomite, and the clay partings contain some tuff layers, quartz, biotite, and feldspar. The clay is made up of smectite-group minerals and, less frequently, illite and chlorite minerals. Zeolites occur within the tuff horizons (Helvaci, 1983).

This deposit is distinct from similar borax deposits at Boron and Tincalayu in having very little intercrystalline clay; the clay at Kirka is very pale green to white and is high in carbonate. The borax crystals are fine, 10 to 20 mm, and quite uniform in size. The ore body has an indicated thickness of up to 145 m, averaging 20 to 25% B_2O_3 with reserves of plus 100 Mt (Wendel, 1978).

South America: There are over 40 borate deposits located along an 885-km trend in the high Andes near the common borders of Argentina, Bolivia, Chile, and Peru, of which at least 14 are currently in production (Fig. 9). This is an arid segment of the Andean tectonic-volcanic belt, characterized by compressional tectonics and many closed basins with playas or salt flats, called salars. Reserves are said to total millions of tons of B_2O_3, but the figures lack supporting data. For many years (1852–1967), Chile was a

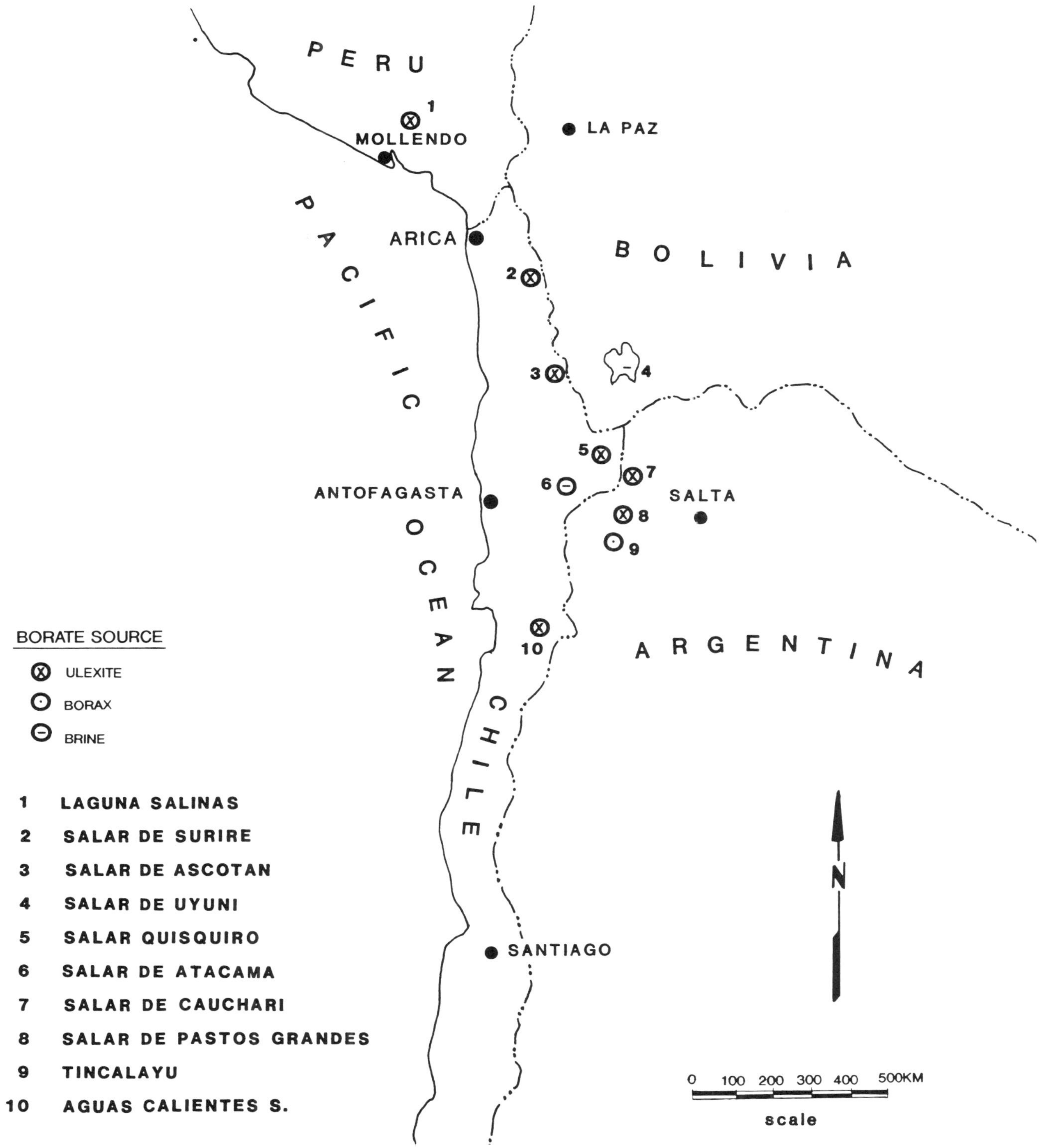

Fig. 9. Location of major South American borate deposits.

major producer of borates, but the industry gradually shifted to adjacent areas of Argentina. Recently, Chile has again become a major borate producer.

Borate spring deposits are better developed in South America than anywhere else in the world. These deposits consist of cones and aprons of ulexite and in one case borax, built up around vents from which warm to cool waters and gas are still issuing in some cases (Fig. 10).

Fig. 10. Arethuzar Borate Spring Mounds, Argentina (photograph: R. Kistler, Mar. 1991).

Fig. 11. Tincalayu mine, Argentina, at elevation 4 118 m (photograph: R. Kistler, Sep. 1990).

Most deposits are associated with calcareous tufa which occurs as a late-stage capping over the borates, and sometimes with halite and gypsum as well. Recent volcanic activity is indicated by basaltic to rhyolitic flows in adjacent areas, and a volcanic source for the borates is presumed. Older rocks in the vicinity of the deposits range from Precambrian to Tertiary and appear to have no genetic relationship to the deposits. The deposits currently forming are quite small, ranging from a few hundred to approximately 9 kt of borates averaging 20% B_2O_3 (Muessig, 1966).

The salar deposits of South America consist of beds and nodules of ulexite with some borax or inyoite, associated with Recent playa sediments, primarily mud, silt, halite, and gypsum. The borate-bearing playas occupy individual basins which range in size up to several hundred km^2. Springs and seeps appear to be the source of the borates. The brines of Atacama, Chile, and Uyuni, Bolivia, contain borates of possible commercial interest.

Argentina—Argentina produces about 200 ktpy of boron minerals and compounds, mostly from the Tincalayu borax deposit (Lyday, 1991). The borax there occurs as a lens of very fine crystalline borax about 100 m in diameter and more than 30 m thick, which was apparently pushed upward along with associated playa sediments by a plug of massive halite (Alonso, 1986). The borax rests on gypsum and halite, and it is partially overlain by about 60 m of reddish, salty, detrital sediments that include thin layers containing, separately, ulexite, inyoite, kurnakovite, and some kernite. The deposit and the surrounding sediments are complexly folded into an anticlinorium and truncated by a fault. The deposit is mined by open pit methods with relatively modern equipment by Borax Argentina S.A. (Fig. 11), and the ore is transported by truck and rail some 400 km to the refinery near Salta. Reserves are about 10 Mt (Alonso and Helvaci, 1988).

The Pastos Grandes (Sijes) area north of Tincalayu contains a number of borate deposits that are located east and south of the Salar de Pastos Grandes. The borates are found in the Sijes Formation of upper Miocene Age. There are at least six small open cut mines in the area which produce colemanite, inyoite, and hydroboracite. Reserves are on the order of 20 Mt (Sureda et al., 1986). In addition, at least six Argentinian salars currently produce ulexite and occasionally borax crystals. The ulexite occurs as "papas" or large cotton balls and as irregular bedded masses, some with a high sulfate and chloride content. The crude ulexite is dried and bagged for shipment to local and Brazilian boric acid plants.

Chile—Ulexite from Chile currently enters the world market from at least five salars. About half of the total production, estimated at 31 ktpy, comes from the Salar de Surire, a relatively small salar situated near the border with Bolivia. The ulexite occurs 6 to 14 cm below the surface as irregular masses and in beds that attain up to a meter in thickness and averaging perhaps 30 cm. The surface muds and salts are removed by a small front end loader, but the ore is mined by hand; it is then stacked to drain and dry, and finally trucked to a boric acid plant near Arica. Other salars currently in production are Ascotan, Aguas Calientes Sur, Atacama, and Quisquiro (Diabolique).

Peru—Peru exported ulexite from about 1883 until the 1920s; production of 4 to 6 ktpy is again reported (Lyday, 1991). The source is Laguna de Salinas, which occupies a large mountain basin east of Arequipa, reached by 56 km of mountain road that crosses a mountain pass at nearly 5 200 m. The basin, at an altitude of 4 328 m, has a seasonally flooded floor, and the ulexite is recovered from shallow pits along the lake margins in the dry season. Ulexite occurs as irregular masses and lenses 25 to 130 cm thick in fine, sandy detritus and green muds, 0.6 to 2.9 m thick, which lie below a surface layer of black to green muds with disseminated ulexite, sulfate, and halite. Reserves are estimated at 9 Mt of ulexite (Norman and Santini, 1985, Muessig, 1958).

Bolivia—Very limited production of borates is reported from Bolivia. Reserves of playa borates and brines containing borates in the giant Salar de Uyuni have been known for many years, but supply and transport difficulties have hindered production (Anon, 1991c).

China: China produces about 27 ktpy of borate minerals and compounds (Lyday, 1991) from deposits in Liaoning, Xizang (Tibet), and Qinghai Provinces. There are few published data on the sources of this production, but the output appears to be rising rapidly. The Liaoning production comes from a number of relatively small, open pit and underground mines on the Liaodong Peninsula of northeast China. There are reported to be more than 55 operations on some 112 separate deposits of which 12 contain 98% of the reserves. The ores are magnesium borates, szaibelyite (ascharite), with ludwigite and suanite associated with magnesite, magnetite, and rare earths (Dongming, 1988). The ores occur as veins in early Proterozoic (2.3 Ga), magnesian marbles of the middle Liaojitite suite. The ore bodies are closely related to, and conformable with, the Liaoji granites (Zhang, 1988). Average grades are low by western standards, 5 to 18% B_2O_3, and many of the smaller deposits are mined by labor intensive methods.

The Qinghai production comes from three playa lakes in the Tsaidam Basin of south-central China: Da Qaidam, Xiao Qaidam, and Mahai. The main borates there are ulexite, pinnoite, hydroboracite, and borax. Recent reports indicate that the surface crusts of ulexite and pinnoite are mainly worked out and that most production is now from the brines, which are associated with mirabilite ($Na_2SO_4 \cdot 10H_2O$) that is also produced. In addition, China has substantial reserves on the Tibetan Plateau where 57 lakes have been identified as containing borates. These are not believed to be in large scale production at this time due to transport and supply difficulties (Lock, 1991).

Other: There are few data in English on the sources for borate production from the former USSR, which is estimated to be about 200 000 tpy (Norman, 1991). For many years, the Inder (Guryev) District of Kazakstan was considered the principal source (Anon, 1989). Borates, mainly szalbelyite and hydroboracite, occur there as veins and fracture fillings in a large (250 km^2) Permian marine salt dome associated with gypsum, carbonate, and clay, and in the nearby brine lakes. Reserves of up to 7 Mt of 20% B_2O_3 are reported. However, recent (1990) exchanges with Russian geoscientists indicate that 95% of the Russian supply now comes from the Bor deposit at Dalnegorsk in Primorsk near the Sea of Japan. The ore is a datolite skarn assemblage formed in Triassic limestones and siltstones intruded by a large porphyry dike swarm. The borosilicate skarns occur over a length of 2 km and are up to 500 m wide; they have drilled ore grade material to 1 000 m. Ore grade is estimated at 8 to 10% B_2O_3 (30 to 45% datolite). The ore is mined by open pit methods and trucked to an adjacent refinery.

Minor production of ludwigite has been reported from small pipe-like bodies in North Korea at Khol-don (Hol-Kol) and Raitakuri, southwest of Pyongyang during the 1950s (Kurmen, 1958). There have been no recent reports that these deposits are currently in operation.

The colemanite body at Magdalena, Sonora, owned jointly by US Borax and Vitro (Mexico's major glass producer) remains on standby status pending an upturn in the market. The development of this deposit has been hampered by processing problems and the relatively high arsenic content of the ore.

The occurrence of low grade borates, predominantly howlite and colemanite, has been reported from the Jarandol Basin of Yugoslavia. The borates here apparently lie above the commercial coal seams and are associated with zeolites, tuff, gypsum, calcite, and bitumen (Obradovic, et al., 1991).

A small colemanite occurrence has been noted on the Greek island of Samos just off the Turkish coast (Stamatakis and Economou, 1991), the first borates reported from the Greek Isles.

TECHNOLOGY

Exploration

Borate exploration consists of detailed prospecting of favorable areas followed by drilling and utilizes all the tools available to the exploration geologist. The recognition of trends of favorable host rocks and structures is an important guide to areas that are of possible interest. Satellite imagery, both real and false color, and standard photo interpretation can be successfully used under certain conditions.

In most parts of the world, the identification of a Cenozoic suite of non-marine fine-grained sediments and tuffs is the usual starting point for the field geologist, because most commercial borates are found associated with these rocks. Any accumulation of salts in these sediments is examined closely. Because many borates are associated with volcanic rocks, volcanic centers, flows, ash deposits, and tuffs, particularly if zeolite-bearing, may also be favorable guides to borate prospecting (Barker and Lefond, 1979).

The skarn borates of eastern Europe and Asia were found by careful prospecting in geologically preserved fold belts where limy sediments are in contact with potassic to alkaline volcanics. Datolite and danburite occur in skarns where the limestone was originally rich in calcium and silica; axinite, kotoite, and ludwigite skarns are hosted by dolomitic limestones. The magnesium borates and those associated with iron ores are generally of low grade. Marine borates are sought in tectonically stable areas with shallow to outcropping salt structures, with gypsum caps, where near surface borate deposits are identified by whitish soil cover and scanty vegetation (Lisitsyn and Pastushenko, 1983).

In arid regions, ulexite often accumulates at, or just beneath, the current surface of salt flats and playas, indicating that boron is moving in the system. These recent crusts may also indicate brine deposits containing boron values of interest. In either case, additional prospecting is usually justified; playas are usually prospected by pitting on a 100 to 500 m grid. Springs and recent spring deposits containing anomalous borates may also be used as a guide to ore in certain areas.

Geochemical surveys are useful methods of narrowing down prospect areas to a drill target. Both soil and rock chip sampling techniques are utilized in exploration programs with boron, strontium, arsenic, and lithium as a common suite of elements. Beryllium is also used in the search for skarn borates in Russia, as are complex B-Mg-Ca-Cl ratios. Water sampling, both surface and well (subsurface), may be useful. Certain plants are boron sensitive and vegetation surveys might prove interesting, but only the Russians have done much work in this area.

Geophysical surveys, particularly gravity and magnetics, are used to outline target basins or structures beneath sedimentary basin fill. Resistivity and seismic surveys have been used to define basin structures and formations which may be associated with the borates in that area. Various downhole well logging techniques, including natural gamma and neutron probes will indicate the approximate percentage of borates and clay in zones of special interest.

Geologic mapping followed by drilling is still the definitive test in most areas of the world. While rotary drill methods may be used, cores are generally taken of the most prospective zones. Assays of B_2O_3 and other associated elements (arsenic, lithium, strontium) are then run on the horizons that appear favorable for borates. Because the saline borates are water soluble, short core runs are used, but the common borates generally core well with recoveries above 90%.

In areas of doubt, two easy field tests are available to determine the presence of borates. The original flame test, where the mineral is soaked with sulfuric acid and alcohol and ignited (''she burns green, Rosie. . . .''), is still one of the most diagnostic. The tumeric test where hydrochloric acid and tumeric solution turns the specimen red-brown is also used. As chemical tests performed in the laboratory are more diagnostic and accurate, most field samples are sent there for analysis.

Under current economic conditions, bedded deposits of borax, colemanite, and ulexite are not generally sought at depths greater than 500 m. Brines with a high borate content, particularly those associated with other salts of value, might be extracted from greater depths under certain circumstances. The skarn and magnesium borates are economical only from surface and near surface excavations at this time.

Mining/Processing

Most commercial borate deposits in the world are mined by open pit methods (Fig. 12). The world's major borate operations, the Boron mine of US Borax at Boron (Kramer), CA, and the Kirka mine of Etibank in Turkey are huge open pit mines utilizing large trucks and shovels and front end loader methods for ore mining and overburden removal. Ores and overburden are drilled and blasted for easier handling. The Boron operation uses a belt conveyor to move ore from the in pit crusher to a coarse ore stockpile from which

Fig. 12. Boron open pit mine, California (photograph: R. Kistler, Oct. 1990).

Fig. 13. Mining at Salar de Surire, Chile (photograph: R. Kistler, May 1987).

it is reclaimed by a bucketwheel that blends the ore before it is fed to the refinery. Kirka utilizes trucks which haul to a crusher near the refinery which is about 0.5 km from the current ore faces.

Smaller operations in Argentina, Chile, China, Turkey, and Russia use similar methods, but on a scale down-sized to the scale of the operation. Some of the South American and Chinese salar operations utilize hand labor to mine the thin salar borates, generally after stripping of overburden with a small dozer or front end loader (Fig. 13).

Borates are mined by underground methods in the Liaoning area of northeast China, at the Billie and Gerstley mines in Death Valley, CA, and in Turkey at Emet and the Simav mine at Bigadic. Borate brines are recovered at Searles Lake, CA, and in the Qinghai Basin of China; brines may also be utilized in the Inder region of Kazakhstan. Borate containing brines are being considered for production from several salars in South America, but there have been no borates produced there as of late 1991. Also under study is the recovery of low grade borates by inplace acid leaching, producing a crude boric acid for subsequent refining. This latter methodology is being pilot-tested by Fort Cady Minerals just east of Barstow, CA.

Processing techniques are related to both the scale of the operation and the ore type, with either the upgraded or refined mineral (borax, colemanite, ulexite) or boric acid as the final product for most operations (Fig. 14).

Borax-kernite ores (Boron, Kirka, Tincalayu) are crushed to −2.5 cm and then dissolved in hot water/recycled borate liquor. The resultant strong liquor is clarified and concentrated in large counter-current thickeners, filtered, fed to vacuum crystallizers, centrifuged, and then dried. The final product is refined borax decahydrate or pentahydrate, or fused anhydrous borax, or is used as feed for boric acid production.

The ulexite from most of the South American salars is air dried, screened, and bagged. It is then combined with locally available sulfuric acid to produce a relatively low grade boric acid or exported as feed for boric acid plants elsewhere.

Colemanite concentrates are used directly in specific glass melts or used as a feed for boric acid plants. The magnesium borates are generally concentrated, dissolved in acid to remove the magnesium, and then converted to boric acid or sodium borates. The borosilicates of the Bor deposit in eastern Russia with their relatively low B_2O_3 grades, are crushed, and then run through a complex plant, which includes magnetic separators, heavy media separators, and flotation cells. The concentrates are then dried, leached, and calcined before being converted to boric acid or to a sodium borate.

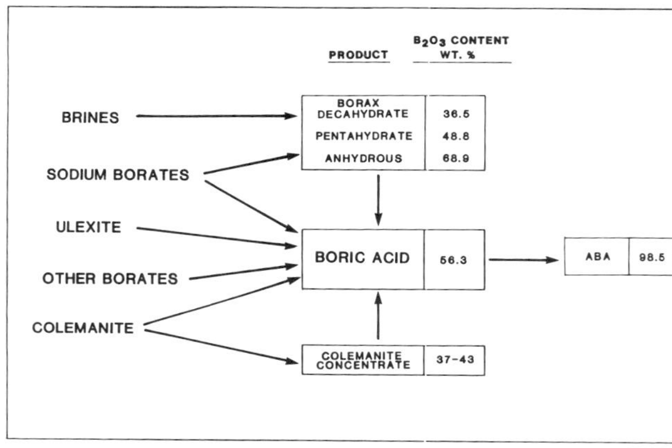

Fig. 14. Borate sources and their primary products.

Brines from Searles Lake, and presumably the Chinese sources, are recovered by either controlled evaporation or carbonation. In the latter process, carbon dioxide produced from lime kilns or flue gas is bubbled through the brine to crystallize sodium bicarbonate; borax is then crystallized in vacuum crystallizers. In the "evaporation" process, a rapid, controlled cooling selectively crystallizes the various salts. The remaining borate liquor is fed to tanks containing borax seed crystals which aid in the recovery of borates from the liquor. The resultant slurry is filtered, washed, redissolved, and fed to vacuum crystallizers that produce dehydrated borax products or boric acid.

Boric acid is one of the final products produced from most of the processes. The world's largest boric acid facility is located adjacent to the Boron pit. It utilizes either refined borates or hydrated kernite as feed, with sulfuric acid as the reacting agent. Other smaller facilities around the world make use of smelter acid or other locally available acid feed-stock to produce products acceptable to their local markets.

USES

The principal uses of borates have not changed much in the past decade, although sales have declined slightly since the peak year of 1987. The current market breakdown is shown in Fig. 15. Major domestic markets include fiberglass insulation, textile or continuous-filament glass fibers, glass, detergents and bleaches, enamels and frits, fertilizers, and fire retardants. In Europe, bleaches and detergents are the major end use; however, sales for glass and glass fibers, including fiberglass, are increasing.

Boron fiber-reinforced plastics continue to be utilized in quantity for aerospace frame sheathing where they combine flexibility and light weight with strength and ease of fabrication. Relatively minor uses that are expected to increase in the near future include those in fertilizers, wood preservatives, alloys and amorphous metals, fire and flame retardants, and insecticides. However, the promising field of boron-iron-silicon electrical transformers has not developed as rapidly as predicted due to various cost factors.

As noted in Table 2, "miscellaneous" is a sizable portion of total US sales. It includes pharmaceuticals, cosmetics, anti-corrosion compounds, adhesives, abrasives, insecticides, metallurgical processes, and nuclear shielding, to name but a few. Research is continuing in many areas. One of the more publicized is in supermagnets, where borates, combined with rare earths, nickel and iron, produce an alloy that can be used to make electromagnets for computer drives, high fidelity speakers, automobile starter motors, and various household appliances. Another new field is biological growth control agents.

Borates can be utilized to protect the environment by aiding in converting heavy metals in industrial waste streams into recoverable free metals and by removing impurities from polymers used in bleaching wood pulp for paper production. Borates also aid in the control of the refractive index in optical fibers for medical research where precise control is needed. Other medical applications include cancer research where the B^{10} isotope reacts with low energy neutrons to give off short range alpha particles that can be used for microsurgery in previously inoperable areas of the brain (Barth et al., 1990). Current tests on boron analogues indicate they may be effective in reducing serum cholesterol and other disease-causing proteins.

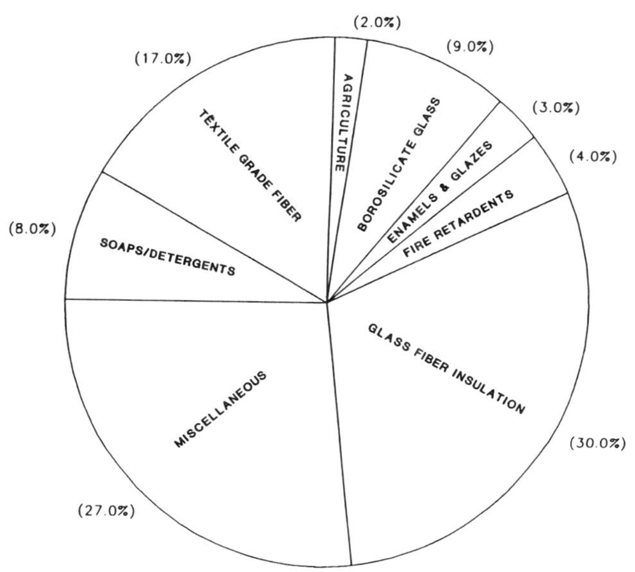

Fig. 15. Borate end uses 1990 (United States only).

Table 2. US Borate End Uses

Borate End Use	Percent of Total
Glass Fiber Insulation	30
Textile-grade fiber	17
Enamels Borosilicate Glass	9
Soaps and detergents	8
Fire and flame retardants	4
Enamels and glazes	3
Agriculture	2
Miscellaneous	27

ECONOMIC FACTORS

Borates have become a relatively modestly priced industrial mineral commodity in recent years following the development of the large deposits at Boron, and more recently, Kirka. Prices are directly related to the costs of production, of which the major cost is fuel for drying, dehydrating, and melting the refined ore into the products desired by industry. Industry prices for most products have held steady with the rate of inflation despite added costs imposed by government regulations.

Both Western Europe and Japan, neither of which have local borate sources, are major markets for US and Turkish production. South America is largely self sufficient with an increasing amount in excess of their needs which is exported mainly to Europe and Japan. Russia and China both appear to be self sufficient in borates at this time, although their costs of production per ton of B_2O_3 are thought to be relatively high.

Borates are a lightweight commodity and are generally sold in bulk by carload lots, in IBC's (Intermediate Bulk Containers) or "super-sacks," and by palletized bags. In the United States, about a third of the production now moves to market by truck, a major change from the recent past when most moved by rail. Overseas shipments are made mostly in bulk from special terminals at Los Angeles, CA, and from Bandirma on the Bosporus Strait in Turkey, to similar terminals in the Netherlands, Belgium, and the United Kingdom, from where it is moved by barge, as well as rail and truck. Other imports to Europe arrive in Italy and Spain. Imports to the Far East are generally sold in small bag lots. Bulk imports to the United States (mainly colemanite) usually land in Charleston, SC, where there are grinding facilities; this colemanite is then shipped to eastern fiberglass manufacturers. There is no import duty on borates brought into the United States.

GOVERNMENTAL CONSIDERATIONS

Newly discovered deposits of borax and kernite, as well as brines that contain sodium borates in the United States can be exploited on federal lands only under the provisions of the Mineral Leasing Acts of 1917 and 1920 applicable to sodium minerals. The federal government's position is that these borates are "sodium minerals." Other borates can be claimed under the mining law. Borate minerals carry a 14% depletion allowance.

Most of the known colemanite reserves in the United States are within Death Valley National Monument with stringent controls on development. Most of the favorable prospects for borates are located in the adjoining California Desert Conservation Area, portions of which are being considered for permanent withdrawal under bills currently in Congress. In addition, laws and regulations protecting various endangered species, such as the Mojave ground squirrel and desert tortoise, also affect where and when exploration and development can take place (Anon, 1991b).

In Turkey, only Etibank, the Government mining company, can sell borates, effectively limiting participation by others. South American governments are generally favorable in their treatment of foreign mining enterprises. The situation in both China and Russia is in a state of flux at this time. It is believed that, while both countries will eventually permit exploration by foreign concerns, development and subsequent export of products and profits may be somewhat difficult.

In all countries, pollution control has become a major concern. With regard to borates, this concern relates mainly to fugitive dust control and disposal of trace elements contained in the tailings.

In the United States, the National Institute of Occupation Safety and Health and a research team sponsored by US Borax conducted studies of employees at Boron who were exposed to borate dust in the work place. Both studies, comprised of almost 800 employees of US Borax, showed slightly healthier lung functions than predicted by the national averages. These results confirm that there are no respiratory or lung effects from even long term exposure to borate dust. In general, naturally occurring boron compounds have relatively low toxicity for man and other mammals. Boron in medium to large doses can be toxic to plants and is used in nonselective herbicides. It is also toxic to certain insects and prevents bacteria from forming.

FUTURE TRENDS

Boron consumption is directly related to the usage of glass, glass fibers, and ceramics. These materials, along with certain plastics that contain borate products, are seen as having a steady consumer demand in the construction and housewares markets well into the next century. Other major uses, detergents, plant foods, wood preservation, etc., are expected to show a slowly rising demand. Specialty uses are expected to increase, but these will probably have only a minor affect on total borate sales. Total world borate demand is expected to grow at about 3% in the near future, based on industry forecasts.

Known reserves of borate minerals are large, particularly in Turkey, South America, and the United States, and production from Turkey and the United States will continue to dominate the world market. However, borates from other areas will probably take up an increasing share of the world market. This trend is already evident with boric acid from Chile reaching the Far East and Europe, and both Russia and China beginning to export. The latter two countries may or may not become major players depending upon what role ''real world economics'' play there, because it appears that their sources are inferior in grade and recovery to those now in production elsewhere.

There are few substitutes for borates. In most applications, they provide unique chemical properties at a reasonable price; this is particularly true in glass fibers and in the field of heat and impact resistant glass. Borates are an essential part of certain plant foods. Their use in nuclear reactor shielding and control is well documented.

As with many products in the mineral industry, there are two somewhat opposing borate trends: 1) toward less refined and therefore less expensive borates for some users, 2) toward super refined, very pure, and more expensive products. Both trends are at the expense of those borates currently produced.

Future markets are difficult to predict. Based upon recent history, the major world consumers of borates will continue to be the developed countries of North America, Europe, and Japan. The large populations of China, India, and Russia would appear to want an ever-increasing amount of borates. The markets for borates in the less developed countries are predicted to remain relatively minor at least for the near term.

REFERENCES

Anon., 1987–1991, *Borax Review*, Nos. 1–10, Borax Holdings, Ltd., London, 16 pp. each.

Anon., 1989, *The Economics of Boron*, 6th ed., Roskill, London, 145 pp.

Anon., 1991a, *Chemical Week*, Vol. 149, No. 22, pp. 18–34.

Anon., 1991b, *Federal Register*, Vol. 56, No. 234, Dec. 5, pp. 63, 741–63, 742.

Anon., 1991c, ''Mineral Deposits and Occurrences of the Bolivian Altiplano and Cordillera Occidental,'' Open File Report 91–0286, US Geological Survey, 365 pp.

Alonso, R.N., et al., 1988, ''A New Tertiary Borax Deposit in the Andes,'' *Mineralium Deposita*, Vol. 23, pp. 299–305.

Alonso, R.N., and Helvaci, C., 1988, ''Mining and Concentration of Borates in Argentina,'' *Proceedings*, 2nd International Mineral Processing Symposium, Y. Aytekin, ed., Izmir, Turkey, pp. 301–305.

Alonso, R.M., 1986, ''Occurrences, Stratigraphic Position and Genesis of Borate Deposits in the Puna, Argentina,'' PhD thesis, Salta University, Argentina (in Spanish), 196 pp.

Ataman, G., and Baysal, O., 1978, ''Clay Mineralogy of Turkish Borate Deposits,'' *Chemical Geology*, Vol. 22, pp. 233–247.

Barker, J.M., and Lefond, S.J., 1979, ''Some Additional Borate and Zeolites from the Mesa Del Alamo Borate District, North Central Sonora, Mexico,'' Preprint 79–367, Society of Mining Engineers, Littleton, CO, 12 pp.

Barnard, R.M., and Kistler, R.B., 1966, ''Stratigraphic and Structural Evolution of the Kramer Sodium Borate Body, Boron, California,'' *Second Symposium on Salt*, Vol. 1, Northern Ohio Geological Society, Cleveland, OH, pp. 133–150.

Barth, R.F., Soloway, A.H., and Fairchild, R.G., 1990, ''Boron Neutron Capture Theory for Cancer,'' *Scientific American*, Vol. 263, No. 4, Oct., pp. 100–107.

Bates, R.L., and Jackson, J.A., eds., 1987, *Glossary of Geology*, 3rd ed., American Geological Institute, Alexandria, VA, 778 pp.

Baysal, O., 1972, ''Mineralogic and Genetic Studies of the Sarikaya (Kirka) Borate Deposits,'' PhD thesis, Hacettepe University, Turkey, 157 pp.

Christ, C.L., and Garrels, R.M., 1959, ''Relations Among Sodium Borate Hydrates at the Kramer Deposit, Boron, California,'' *American Journal of Science*, Vol. 257, pp. 516–528.

Countryman, R.L., 1979, ''The Subsurface Geology, Structure and Mineralogy of the Billie Borate Deposit, Death Valley, California,'' MS thesis, University of California at Los Angeles, 131 pp.

Dongming, L., 1988, ''Boron Mineral Resources in the North Eastern Region of China—A Technical and Economic Evaluation,'' Bureau of Geology and Mineral Resources, Beijing, 5 pp. (translated by D. Lock).

Evans, J.R., et al., 1976, ''Mines and Mineral Deposits in Death Valley National Monument, California,'' Special Report 125, California Division of Mines and Geology, 60 pp.

Gale, H.S., 1914, ''Salines in the Owens, Searles, and Panamint Basins, Southeastern California,'' Bulletin 580-L, US Geological Survey, pp. 251–323.

Hanks, H.G., 1883, ''Report on the Borax Deposits of California and Nevada,'' *California State Mining Bulletin*, 3rd Annual Report, Part 2, 111 pp.

Heller, G., 1986, ''A Survey of Structural Types of Borates and Polyborates,'' *Topics Current Chemistry*, No. 131, pp. 39–98.

Helvaci, C., 1977, ''Geology, Mineralogy and Geochemistry of the Borate Deposits and Associated Rocks at the Emet Valley, Turkey,'' PhD Thesis, University of Nottingham, UK, 338 pp.

Helvaci, C., 1978, ''A Review of the Mineralogy of the Turkish Borate Deposits,'' *Mercian Geology*, Vol. 6, pp. 257–270.

Helvaci, C., 1983 ''Mineralogy of the Turkish Borate Deposits,'' *Jeologi Muhendisligi*, Vol. 17, pp. 37–54.

Helvaci, C., 1984, ''Occurrence of Rare Borate Minerals Veatchite-A, Tunnelite, Terrugite and Cahnite in the Emet Borate deposits,'' *Mineralium Deposita*, Vol. 19, pp. 217–226.

Helvaci, C., 1986, "Stratigraphic and Structural Evolution of the Emet Borate Deposits, Western Anatolia, Turkey," Research Paper MM/JEO-86, AR 008, Dokuz Eylul Univ., Izmir, Turkey, 28 pp.

Helvaci, C., 1989, "Mineralogy and Genesis of Bigadiç Borate Deposits, Western Turkey," *Abstracts*, 28th International Geological Congress, Washington, DC, Vol. 2, pp. 49–50.

Helvaci, C., 1990 "Mineral Assemblages and Formation of the Kestelek and Sultancayiri Borate Deposits," *Extended Abstracts*, International Earth Sciences Congress on Aegean Regions, Izmir, Turkey, pp. 11–12.

Helvaci, C., and Alaca, O., 1984, "Geology and Mineralogy of the Bigadiç Borate Deposits," *Book of Abstracts*, 38th Scientific and Technical Congress of the Geological Society of Turkey, pp. 110–111.

Helvaci, C., and Firman, R.J., 1976, "Geological Setting and Mineralogy of Emet Borate Deposits, Turkey," *Transactions*, Institute of Mining and Metallurgy, Vol. 85, Sec. B., pp. 142–152.

Helvaci, C., and Inci, U., 1989, "Geological Framework of Beypazari District and Neogene Trona Deposit of the Region, Turkey," *Book of Abstracts*, Vol. 2, 28th International Geological Congress, Washington, DC, p. 49.

Helvaci, C., et al., 1985, "Neogene Stratigraphy and Economic Potential of Western Turkey," *Abstracts*, 8th Congress on Mediterranean Neogene Stratigraphy, Budapest, Hungary, pp. 263–265.

Helvaci, C., et al., 1987, "Neogene Stratigraphy and Economic Potential of Turkey," *Journal of Isparta Engineering*, Faculty of Akdeniz University, No. 3, pp. 31–45.

Inan, K., Dunham, A.C., and Esson, J., 1973, "Mineralogy, Chemistry and Origin of Kirka Borate Deposit, Eskisehir Province, Turkey," *Transactions*, Institute of Mining and Metallurgy, Sec. B, Vol. 82B, pp. 114–123.

Kistler, R.B., and Smith, W.C., 1983, "Boron and Borates," *Industrial Minerals and Rocks*, 5th ed., S.J. Lefond, ed., Society of Mining Engineers, New York, pp. 533–560.

Kurmen, I.M., 1958, "The Pacific Ocean and Mid-Asiatic Boron Belts," *Odtel Geol.—Geograf. Nauk.*, No. 1m, USSR (in Russian), pp. 470–486.

Kunh, R., 1962, "Geochemistry of the German Potash Deposits," Special Paper 88, Geological Society of America, pp. 460–466.

Lefond, S.J., and Barker, J.M., 1979, "A Borate and Zeolite Occurrence near Magdalena, Sonora, Mexico," *Economic Geology*, Vol. 74, pp. 1883–1889.

Lisitsyn, A.E., and Pastushenko, I.I., 1983, "Prospecting, Exploration and Evaluation of Boron Occurrences," All-Union Scientific Research Institute of Geology, Moscow, 173 pp.

Lock, D., 1991, "Notes on the Borate Resources of the People's Republic of China," M.R.I. Report, Australian National University, 53 pp.

Lyday, P.A., 1982, "Boron," *Minerals Yearbook 1980*, Vol. 1, US Bureau of Mines, pp. 133–142.

Lyday, P.A., 1991, "Boron—1990," *Mineral Commodity Summaries*, US Bureau of Mines, 9 pp.

Mass, E.V., 1986, *Applied Agricultural Research,* Vol. 1, No. 1, pp. 12–26.

McAllister, J.F., 1970, "Geology of the Furnace Creek Borate Area, Death Valley, Inyo County, California," Map Sheet 14, California Division of Mines and Geology, text 9 pp., map scale 1:24,000.

Meixner, H., 1965, "Borate Deposits of Turkey," *MTA Institute Bulletin*, No. 125, Ankara, pp. 1–12.

Morgan, V., and Erd, R.C., 1969, "Minerals of the Kramer Borate District, California," *Minerals Information Service*, Vol. 22, Nos. 9,10, California Division of Mines and Geology, pp. 143–154, 165–172.

Muessig, S., 1958, "First Known Occurrence of Inyoite in a Playa, at Laguna Salinas, Peru," *American Mineralogist*, Vol. 43, Nov-Dec., pp. 1144–1147.

Muessig, S., 1966, "Recent South American Borate Deposits," *2nd Symposium on Salt*, J.L. Rau, ed., Vol. 1, Northern Ohio Geological Society, Cleveland, OH, pp. 151–159.

Mumford, R.W., 1954, "Deposits of Saline Minerals," Chap. 8, *Geology of Southern California*, Bulletin 170, Division of Mines, California Department of Natural Resources, pp. 15–22.

Norman, J.C., 1991, "Boron: A Review of 1990 Activities," *Mining Engineering*, Vol. 43, No. 7, July, pp. 740–741.

Norman, J.C., and Santini, K.N., 1985, "South American Borate Deposits," *Borates: Economic Geology and Production*, J.M. Barker and S.J. Lefond, eds., Society of Mining Engineers, New York, pp. 55–69.

Obradovic, J., Djurdjevic, J., and Vasic, N., 1991, "Magnesite and Borates or Boron-Bearing Minerals from Neogene Lacustrine Basins in Yugoslavia," *Comparative Sedimentology of Lacustrine Sequences in Neogene Strike-Slip Basins*, H.E. Kerey, ed., IGCP/GLOPALS field seminar, Turkey, Abstract, pp. 13.

Ol, T., et al., 1989. "Isotopic Compositions of Some Boron Minerals," *Geochimica Cosmochimica Acta*, Vol. 53, No. 12, pp. 3189–3195.

Ozol, A.A., 1977, "Plate Tectonics and the Process of Volcanogenic-Sedimentary Formation of Boron," *International Geology Review*, Vol. 20, No. 6, pp. 692–698.

Özpeker, I., 1969, "Western Anatolian Borate Deposits and Their Genetic Studies," PhD thesis, Technical University of Istanbul, Turkey, 116 pp.

Palache, C., Berman, H., and Frondel, C., 1951, *Dana's System of Mineralogy*, 7th ed., Vol. 2, John Wiley, New York, pp. 320–389.

Papke, K.G., 1976, "Evaporites and Brines in Nevada Playas," Bulletin 87, Nevada Bureau of Mines and Geology, 33 pp.

Puffer, J.H., 1975, "The Kramer Borate Assemblage," *Mineralogical Record*, Vol.10, No. 10, Mar.-Apr., pp. 84–91.

Russell, A., 1991, "Minerals in Fibre Glass," *Industrial Minerals*, Nov., pp. 27–41.

Schaller, W.T., 1930, "Borate Mines from the Kramer District, Mojave Desert, California," Professional Paper 1581, US Geological Survey, pp. 137–170.

Siefke, J.W., 1991, "The Boron Open Pit Mine at the Kramer Borate Deposit," *The Diversity of Mineral and Energy Resources of Southern California*, M.A. McKibben, ed., SEG Guidebook Series, Vol. 12, pp. 4–15.

Sheppard, R., and Gude, A., 1968, "Distribution and Genesis of Authigenic Silicate Minerals in Tuffs of Pleistocene Lake, Tecopa, Inyo County, California," Professional Paper 587, US Geological Survey, 38 pp.

Smith, G.I., 1966, "Geology of Searles Lake—A Guide to Prospecting for Buried Continental Salines," *2nd Symposium on Salt*, J.L. Rau, ed., Northern Ohio Geological Society, Cleveland, pp. 167–180.

Smith, G.I., 1979 "Surface Stratigraphy and Geochemistry of Later Quaternary Evaporites, Searles Lake, California," Professional Paper 1047, US Geological Survey, 140 pp.

Smith, G.I., et.al., 1973, "Evaporites and Brines," Professional Paper 820, US Geological Survey, pp. 207–208.

Smith, G., et al., 1983, "Core-KM-3, A Surface to Bedrock Record of Late Cenozoic Sedimentation in Searles Valley, California," Professional Paper 1256, US Geological Survey, 24 pp.

Smith, R.A., 1992, "US Borax Research," Report, US Borax and Chemical Co., 7 pp.

Sprague, R.W., 1991, "Boron, Metals and Minerals, Annual Review—1991," *Mining Magazine*, p. 108.

Sprague, R.W., 1972, *The Ecological Significance of Boron*, Ward Ritchie Press, New York, 58 pp.

Stamatakis, M.G., 1986, "Boron Distribution in Hot Springs, Volcanic Emanations, Marine Evaporites and Volcanic Sedimentary Rocks of Cenozoic Age in Greece," PhD thesis, University of Athens, 495 pp.

Stamatakis, M.G., and Economou, G., 1991, "A Colemanite and Ulexite Occurrence from Lake Miocene, Karolovassi Basin, Samos Island, Greece," *Economic Geology*, Vol. 86, pp. 187–194.

Sunder, M.S., 1980, "Geochemistry of the Sarikaya Borate Deposits (Kirka-Eskisehir)," *Bulletin of the Geological Congress of Turkey*, Vol. 2, pp. 19–34.

Sureda, R.J., et al., 1986, "Aspectos Metalogenicos del Noroeste Argentino," *Capricornio*, J.A. Salfity, ed., National University of Salta, Republic of Argentina, Vol. 1, No. 1–96, pp. 39–96.

Swihart, G.H., Moore, P.B., and Callis, E.L., 1986, "Boron Isotopic Composition of Marine and Non-Marine Evaporite Borates," *Geochimica Cosmochimica Acta*, Vol. 50, pp. 1297–1301.

Travis, N.S., and Cocks, E.S., 1984, *The Tincal Trail*, Chap. 1, Harrap, London, 311 pp.

Utekhin, G.M., 1965, "The Problem of Distribution Features and Conditions of Boron Concentration in Skarns," E.A. Alexandrov, reviewer, *Economic Geology*, Vol. 60, No. 8, Dec., pp. 1750.

Valyashko, M.G., 1970, "Genesis and Exploration of Borate Deposits Related to Marine Salt Deposits," *International Geology Review*, Vol. 12, No. 6, pp. 711–719.

Waring, G.A. (rev. by Blankenship and Bentall), 1965, "Thermal Springs of the United States and Other Countries of the World—A Summary," Professional Paper 492, US Geological Survey, 383 pp.

Wendel, C., 1978, "Special Report of Borate Resources," US Department of Interior, National Park Service, Mining and Minerals Division, Apr., 139 pp.

White, D.E., 1957, ''Magmatic Connate and Metamorphic Waters,'' *Geological Society of America Bulletin*, Vol. 68, pp. 1659–1657.

Whistler, D.P., 1984, ''An Early Hemingfordian (Early Miocene) Fossil Vertibrate Fauna from Boron, Western Mojave Desert, California,'' *Contributions in Science*, No. 355, Allen Press, Lawrence, KS, 36 pp.

Wood, W.W., and Sanford, W.G., 1990, ''Groundwater Control of Evaporite Deposits,'' *Economic Geology*, Vol. 85, pp. 1226–1235.

Zhang, Q., 1988, ''Early Proterozoic Tectonic Styles and Associated Mineral Deposits of the North China Platform,'' *Precambrian Research,* Vol. 39, pp. 1–20.

Zhang, G., and Jin, S., 1991, ''Geological Characteristics and Genesis of the Heping Boron Deposit, Changking Co., Zhejiang Province (China),'' *Kuangchuang Dishi*, Vol. 10, No. 2, p. 192.

BROMINE RESOURCES

M.J. WILHELM AND K.C. WILLIAMS

Bromine is the intermediate member of the halogen family of elements between iodine, a solid, and chlorine, a gas. The name is derived from the Greek *bromos,* meaning stench. Bromine is the only nonmetallic element that is liquid at room temperatures, with a melting point of −7.2°C and a boiling point of 58.78°C. Reddish brown in color, it dissolves easily in water to yield a red solution. Bromine volatilizes at ordinary temperatures to a red vapor that is very irritating to the eyes, nose, and throat. Density of the liquid is 3.12, and gas weighs 7.59 gpl in standard conditions. Because the element is highly active, it exists in nature only as bromide compounds. Bromine was discovered by Antoine Jerome Balard in seawater bitterns in 1825 and began to find commercial applications shortly thereafter in photography, medicine, and dyes. The first commercial production of bromine was in 1846 in Freeport, PA, followed by production in Germany in 1865.

OCCURRENCE

Bromine is widely distributed in nature, and compounds occur in solution in natural brines, in seawater, saline lakes, oil and gas well brines, and in evaporite chloride minerals. The concentration of bromine salts ranges from 60 to 65 ppm in seawater to several thousand ppm in oil well brines and saline lakes. The distribution of bromine in marine evaporite minerals and in the solutions from which they crystallized was determined many years ago by H.E. Boeke (1908). Theoretical aspects regarding the geochemistry of bromine have been published in numerous papers (Raup and Mite, 1969). Bromine occurs in solid solution in carnallite and sylvite associated in potash deposits, in halite in salt beds, salt in oil shales, and in trona deposits. Its presence is useful as a guide in geochemical exploration.

The analysis of small amounts of bromides in the presence of large amounts of chlorides is discussed by W. Schwerdtner (1963). Manganese, iron, and iodine may interfere with the determination. Other analytical methods are discussed at length in literature (Stenger and Atchison, 1964).

COMMERCIAL SOURCES AND RESERVES

The supply of bromine is virtually unlimited because it is found in recoverable quantities in the oceans (Table 1). It is estimated that 99% of all commercially recoverable bromine is in the ocean, and, as a result, there is the potential for bromine production in most areas of the world. However, subterranean and surface brines, having a higher bromine content than seawater, continue to be the most economic and preferred sources for production. The two largest commercial production sources today are the Arkansas subterranean brines and the Dead Sea brines. In Arkansas, bromine is extracted from brines containing 5 000 ppm bromides. The brine is contained in limestone of the Smackover Formation at depths as great as 2 500 m, from which it is pumped to the surface for bromine recovery. The Dead Sea contains 4 000–5 000 ppm bromides. However, potash is also produced from the Dead Sea brines, and the bromine is extracted from a concentrated (12 000 ppm) effluent stream from that production.

PRODUCTION

Processing Techniques

Bromine is produced currently by two processes known as *steaming out* and *blowing out.* The steaming out process is used for brines with bromide concentrations greater than 1 000 ppm, such as the Dead Sea brines and Arkansas brines. The blowing out process is used for brines with concentrations under 1 000 ppm, such as seawater. In both processes, the essential steps are (Lyday, 1981):

1. treatment with chlorine to oxidize the bromide to bromine
2. removal of bromine vapor from the solution
3. condensation of bromine vapor or reaction to form a salt or acid
4. purification of bromine.

Product Specification

Merchant bromine produced in the United States contains less than 200 ppm of impurities and is therefore at least 99.98% pure. Impurities are mainly chlorine, water, and traces of organic material. Consumers require strict control of organic impurities, which may be introduced into the product along with chlorine.

Producers

Dead Sea Bromine, Great Lakes Chemical, and Ethyl Corporation are the world's largest bromine producers with 75% of the world's bromine capacity (Table 2). In addition, Great Lakes owns 88% of the Associated Octel group. Ethyl and Great Lakes produce from subterranean brines in Arkansas, Dead Sea from potash bitterns in Israel, and Associated Octel (UK), Atochem (France), and Tosoh (Japan) from seawater plants. Dead Sea, Great Lakes, and Ethyl use the steaming out process for the concentrated bromide brines, whereas Tosoh, Atochem, and Associated Octel use the blowing out process for the less concentrated seawater.

Markets

About 85% of all bromine produced is consumed at the producing site by the bromine producers. Major products include flame

Table 1. Sources and Reserves of Bromine

Country	Source	Bromine concentration, ppm
United States	Arkansas—subterranean brines	5 000
	California—Searles Lake	85
	Michigan—subterranean brines	2 500
France	Alsatian potash fields	—
	Seawater	60–65
Japan	Seawater	60–65
United Kingdom	Seawater	60–65
Germany	Stassfurt potash	1 500–2 500
USSR	Subterranean brines	—
Israel	Dead Sea	5 000

Table 2. Bromine Production Capacity

Company	Location	Capacity (thousand tons/yr)
Dead Sea Bromine	Israel	150
Ethyl Corporation	United States (2 plants)	120
Great Lakes Chemical	United States (4 plants*)	115
USSR	USSR (6 plants)	68
Associated Octel	United Kingdom	30
Atochem	France	13
Tosoh	Japan	20
Others	Various	24
TOTAL		570

*Includes Arkansas Chemical, jointly owned by Great Lakes and PPG Industries.

retardants, soil and space fumigants, oilfield drilling fluids, and antiknock compounds for leaded gasoline (Table 3). The balance of the elemental bromine is shipped to a large variety of chemical processors, who convert it into such products as agricultural chemicals, dyes, pharmaceuticals, biocides, halons, and photographic chemicals.

The largest single end use for bromine is in flame retardants, which consumed more than 25% of the bromine in 1990. Most of these retardants are ring-brominated aromatic compounds that impart flame retardancy to a variety of polymer systems, such as epoxies, high impact polystyrene, acrylonitrile-butadiene-styrene (ABS), and polyurethanes.

The second largest end use for bromine is agriculture, consuming more than 20% of production. The principal product is methyl bromide, which functions as a complete soil sterilant with very effective herbicidal and nematocidal properties. Methyl bromide also is used as a space fumigant, insecticide, and rodenticide in food processing plants, grain silos, warehouses, and houses.

The third major use of bromine is as ethylene dibromide, an additive in leaded gasoline. Because its function is to help remove lead from the engine, this use is declining with the phaseout of leaded gasoline. In 1972, ethylene dibromide consumed nearly 70% of all the bromine produced. Today, it consumes less than 13%. Phase down of leaded gasoline continues on a worldwide basis with a corresponding reduction expected in ethylene dibromide production. The use of ethylene dibromide as a soil fumigant has also been banned in the United States, further contributing to the decline in production.

Solutions of inorganic bromide salts have found use in oil and gas production: in well completions, work-overs, and packer fluids in high pressure geological formations. Clear brine fluids have the advantage over traditional mud systems, such as barite, which tend to clog the pores of the formation and thus greatly reduce the flow

Table 3. World Usage of Bromine, 1990

	Thousands of tons of contained bromine	%
Flame retardants	110	26
Agricultural	85	20
Fuel additives	55	13
Oil field chemicals	35	8
Other uses	135	33
	420	100

of oil and gas fluids to the well bore. This application consumed slightly more than 8% of worldwide bromine production in 1990.

Bromine is used in various other miscellaneous applications, such as halon fire extinguishing agents, pharmaceuticals, dyes, biocides, photographic chemicals, and rubber compounds.

Prices

Approximately 85% of all bromine manufactured is converted to bromine derivatives at the producing site. The balance constitutes the merchant market for elemental bromine and is transported to a large variety of small and large chemical processors. Elemental bromine is contained and transported in special glass, lead-lined steel, and nickel-lined steel containers. Because of weight limitations, most tank trailers are lined with nickel, whereas rail cars are lined with lead. Bromine is classified by the US Department of Transportation as a corrosive liquid.

Most large volume (greater than 500 000 kg/yr) purchasers of bromine enter into supply contracts with the producers, and sale prices are not available. According to published information, bromine prices at the end of 1990 were as follows:

Drums, truckload, works: $2.27/kg

Bulk, tank trailers or tank cars, works: $0.95–1.02/kg.

FUTURE CONSIDERATIONS AND TRENDS

The bromine industry has undergone major changes in the past ten years and will face continued challenges in the future. New markets for bromine chemicals have emerged to replace declining markets. Government legislation, which has greatly reduced the largest end use market for bromine (ethylene dibromide for gasoline), has created new markets by making fire retardancy requirements more stringent. This flame retardant sector should continue to grow at a 5 to 6% annual rate. In addition, new markets, such as the use of bromide salts in completion, work-over, and packer fluids, should continue to grow as the technology is accepted throughout the world. The use of bromine in biocidal applications is rapidly expanding. Bromine compounds are replacing chlorine in water sanitation where its ability to maintain its biocidal activity at wider pH ranges and temperatures is an important parameter (hot tubs, spas, cooling towers). In addition, the superior biodegradability of bromine residues over chlorine residues has sparked interest for their use in municipal and utility water treatment.

Another area that has the potential for growth is the use of zinc bromide in rechargeable batteries. Applications include electric vehicles and load leveling in power plants. It is expected that this market will begin to develop in the later 1990s.

Interest in the use of bromine for recovery of precious metals has also been growing. Ores from which the noble metals are most easily extracted by cyanide are beginning to play out, and, as the ores for which the cyanide process is not practical are mined, bromine may play an increasing role.

Somewhat offsetting this growth is recent government regulation of ozone depleting agents, such as the bromine-containing halons. Although the market for halons is small in terms of bromine usage, environmental and legislative activity is expected to continue to significantly affect the overall bromine market over the next ten years.

HEALTH AND SAFETY FACTORS

Skin contact with bromine liquid and inhalation of the vapor present serious hazards. The liquid rapidly attacks the skin, and comparatively low concentrations of the vapor are highly irritating to the eyes and the entire respiratory tract (Stenger and Atchison, 1964). Vapor concentrations of 500 to 1 000 ppm (by volume) are fatal on short exposure; concentrations of 40 to 60 ppm are dangerous to life on exposure for 0.5 to 1 hr. The maximum concentration considered safe for 8-hr exposure is less than 1 ppm. The American Conference of Governmental Industrial Hygienists' (accepted) threshold limit is 0.1 ppm in air. At this level bromine can be detected by its odor.

Plants using bromine should provide for the rapid disposal of spilled liquid and for adequate venting of vapors.

ACKNOWLEDGMENT

The authors wish to acknowledge the original work done by J.H. Jensen many years ago and upon which this chapter is based.

BIBLIOGRAPHY AND REFERENCES

Anon., 1968, "Bromine, Its Properties and Uses," Michigan Chemical Corp., Chicago, IL, 62 pp.

Anon, 1968a, "How to Handle Bromine," Michigan Chemical Corp., Chicago, IL, 36 pp.

Anon, 1968b, "The Dow Chemical Co., 80 Years of Leadership . . . and Still Pioneering, Bromine and Brominated Compounds," Midland, MI, 89 pp.

Anon, 1972, "Extraction of Chemicals from Seawater, Inland Brines and Rock Salt Deposits," United Nations, New York, 59 pp.

Anon., 1988, *The Economics of Bromine*, 5th ed., Roskill Information Services, Ltd.. 93 pp.

Absalom, S.T., 1981, "Bromine," Preprint, 1978–1979, US Bureau of Mines, 7 pp.

Boeke, H.E., 1908 "Uber das Kristallisationsschema der Chloride, Bromide, Jodide von Natrium, Kalium und Magnesium, sowie uber das Vorkommen das Broms und das Fehlen von Jod in der Kalisalzlagerstatten," *Zeitschrift fuer Kristallographie*, Vol. 45, pp. 346–391.

Jolles, Z.E., 1966, *Bromine and Its Compounds*, Academic Press, Inc., New York, 940 pp.

Keiser, H.D., 1960, "Bromine" in "Minor Industrial Minerals," *Industrial Minerals and Rocks*, J.L. Gillson, ed., AIME, New York, pp. 607–609.

Llewellyn, T.O., and Browning, J.S., 1971, "Continuous Heavy Liquid Concentration of Kyanite," Report of Investigations No. 7481, US Bureau of Mines, Mar., 11 pp.

Lyday, P.A., 1981, "Bromine," *Mineral Commodity Summaries 1982*, US Bureau of Mines, pp. 24–25.

MacMillan, R.T., 1970, "Bromine," *Mineral Facts and Problems*, Bulletin 650, US Bureau of Mines, pp. 893–901.

Price, D., Iddon, B., and Wakefield, B.J., 1988, *Bromine Compounds Chemistry and Applications*, Elsevier Science Publishing, Amsterdam, 422 pp.

Raup, O.B., and Mite, R.J., 1969, "Bromine Distribution and Paleosalinities from Wall Cuttings, Paradox Basin, Utah and Colorado," *3rd Symposium on Salt*, Northern Ohio Geological Society, pp. 40–47.

Reineke, C.E., and Stenger, V.A., 1979, "Bromine," *Encyclopedia of Chemical Technology*, Vol. 4, 3rd ed., R.F. Kirk and D.F. Othmer, eds., Wiley-Interscience Publishers, New York, pp. 226–263.

Schwerdtner, W., 1963, "Analysis of Small Amounts of Bromides in Presence of Large Amounts of Chlorides," *1st Symposium on Salt*, Cleveland, OH, pp. 247–248.

Stenger, V.A., and Atchison, G.J., 1964, "Bromine," *Encyclopedia of Chemical Technology*, Vol. 3, 2nd ed., R.F. Kirk and D.F. Othmer, eds., Interscience Publishers, New York, pp. 750–783.

Tallmage, J.A., et al., 1964, "Minerals From Sea Salt," *Industrial and Engineering Chemistry*, Vol. 56, No. 7, pp. 44–65.

Tippin, R.B., et al., 1970, "Continuous Heavy Liquid Concentration of Spodumene," Report of Investigations 7379, US Bureau of Mines, May, 24 pp.

CATALYSTS

HAYDN H. MURRAY

The three major catalyst markets that utilize industrial minerals as raw materials are petroleum refining, chemical processing, and emission control. Because of economic, competitive, and regulatory pressures, catalyst producers are pushed to develop products with improved performance to meet changing environmental standards. Also, refiners are demanding new families of cracking catalysts that will crack residual fuels. In 1987, worldwide cracking catalyst sales amounted to $457 million (Occelli, 1988a). Recently enacted federal and state laws to improve air quality have prompted strong growth for environmental catalysts, which in 1989 accounted for more than a third of the $1.9 billion catalyst market in the United States (Farrauto, Heck, and Speronello, 1992). A catalyst is a material that promotes or enhances a chemical reaction but does not itself enter into the reaction.

One of the more important mineral groups used as catalysts is clays. The first recorded use of clay as a catalyst was its use in the dehydration of alcohol (Bondt et al., 1797). Clay minerals that are used as catalysts are kaolinite, halloysite, smectite, palygorskite, and sepiolite. Minerals used as catalysts, catalyst supports, or as a raw material to make zeolites are kaolinite, halloysite, sodium montmorillonite, calcium montmorillonite, hectorite, palygorskite, sepiolite, talc, calcined kaolinite, gibbsite, chlorite, and titanium dioxide.

CATALYSTS FOR PETROLEUM REFINING

Although clays are used as a support for many different kinds of catalysts, by far the largest use of clay as a catalytic material is in catalytic cracking (Hettinger, Jr., 1991). With the advancement of engine technology high octane gasoline was needed, and Eugene Houdry, a French mechanical engineer, began experimenting in 1927 with cracking Venezuelan gas oil. He used kaolin-supported nickel (Oblad, 1983). By 1942 Houdry had developed his process and licensed it to Socony Vacuum Oil Co. (Mobil), Sun Oil Company, and Esso (Exxon) so that three processes were available to the US Government, which enabled the production of high octane gasoline for aviation use in World War II. The composition of the cracking catalysts were clay based and synthetic silica-alumina based. After World War II the Houdry Process Corporation focused intensive work on clay based catalysts, including kaolin, montmorillonite, and halloysite. This research work was conducted jointly with the Filtrol Corporation. During this period Nalco Corporation in Chicago discovered that a sodium zeolite, when exchanged with ammonia, was an excellent cracking catalyst, but it was not until many years later, in the early 1960s, that zeolite catalysts were introduced by Mobil Oil Co. As research and development work continued, kaolin based zeolite catalysts proved to be very successful. Today there are many variations and modifications to cracking catalysts, but a common key ingredient in all of them is kaolin (Fig. 1).

At the present time about 400 kt of cracking-catalyst manufacturing capacity is in the United States, Europe, and Japan, and the selling price per ton ranges between 1,200 to $2,000. The cracking-catalyst industry is now dependent on kaolin clay as a necessary ingredient. It is estimated that approximately 200 kt of kaolin is used annually to produce petroleum cracking catalysts (Hettinger, Jr., 1991). Occelli (1988a) estimated that the worldwide sales of catalysts to the petroleum, petrochemical, and chemical industry exceeded $2.4 billion, and by the year 2000 catalysts would become a $5 billion per year global business.

Each succeeding year brings about new and improved catalysts that contain calcined rare earth exchanged zeolites that offer improved chemical as well as thermal and hydrothermal stability. Metals such as nickel, vanadium, iron, and copper contaminate and poison cracking catalysts. Much research and development is currently underway to modify the catalyst composition and microporous structure to reduce the effect of the metal contaminants. New large pore, shape selective materials are being developed, such as pillared clays and delaminated clays (see chapter on clay). Pillared clays use smectites as the starting material. Work on the chemical modification of natural and synthetic clays, zeolites, and pillared clays is continuing in both academic and industrial research laboratories. Occelli (1988a) reported that catalyst manufacturers offered 45 new catalysts in 1987, which indicates the interest, necessity, and activity in the refining industry to improve the production of the lighter hydrocarbon fractions from crude petroleum feedstocks.

Occelli (1988b) reported that the addition of sepiolite, a hydrated magnesium silicate, to a high activity commercial fluid cracking catalyst enhances vanadium resistance and gives improved coke and hydrogen selectivity. Fuller's earth from near Wrens, GA, and chlorite also enhance metal resistance and give improved gasoline fraction results (Albers, personal communication, 1992). In recent years catalysts have been developed to better crack the heavy residuum in the crude oil (Hettinger, Jr., 1988). The Reduced Crude Cracking Catalyst (RCC) developed by Ashland Petroleum Co. contains an ultrastable Y zeolite, treated with rare earth to enhance cracking activity and resist vanadium poisoning, and kaolin. Other residual cracking catalysts have been developed and are now produced in large commercial quantity, and these significantly increase the recovery of gasoline and other lighter fractions from the crude oil feedstock.

ENVIRONMENTAL CATALYSTS

Federal and state laws to improve air quality have defined limits for potentially harmful emissions from industry transportation, power generation, and other sources. The emissions that are targeted by the federal and state laws are such air pollutants as carbon monoxide, volatile organic compounds, nitrogen oxide, sulfur ox-

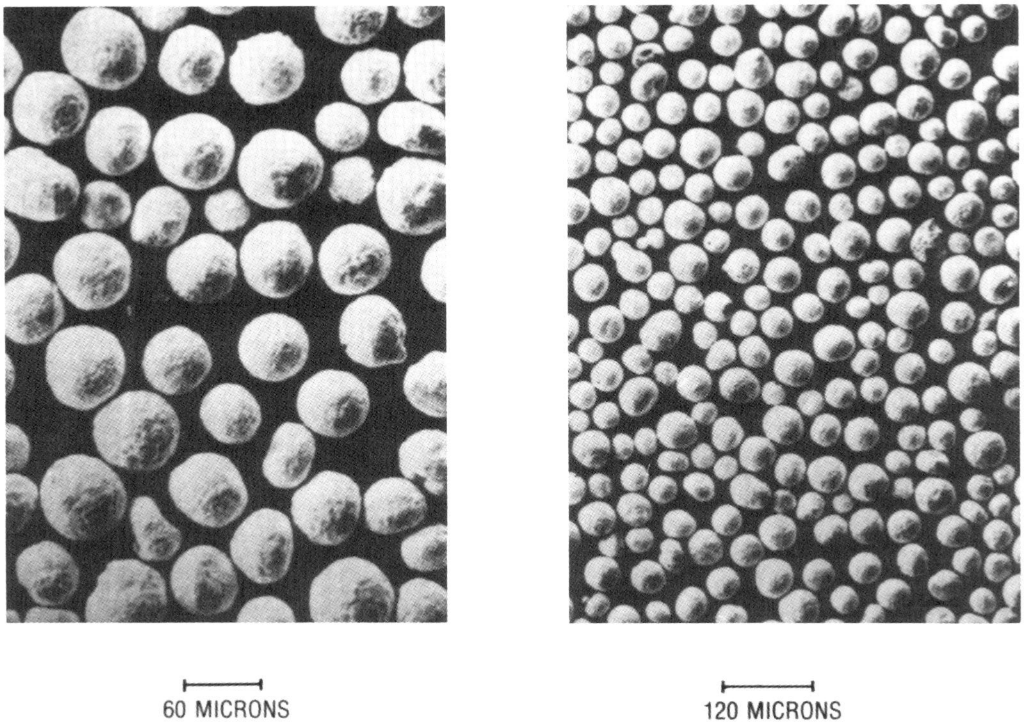

Fig. 1. Electron micrograph of spray-dried beads of a kaolin-based molecular sieve cracking catalyst.

ides, and particulates. The 1990 amendments to the Clean Air Act set limits for 189 toxic air pollutants.

Carbon monoxide is a partial combustion product of many fuels. Volatile organic compounds (VOCs) are emitted in many industrial processes. Nitrogen oxides form when nitrogen and oxygen react in power plants or combustion engines at temperatures above 1 500°C, in burning nitrogen-containing fuels, or in certain chemical operations. Sulfur oxides form when sulfur compounds in fuels burn, especially in boilers, power plants, and diesel trucks. Particulates cover a wide range of materials derived from diesel engines, coal fired boilers, coal fired power plants, and other industrial dryers.

Catalysts have been used in the United States to control automobile emissions since the early 1970s and to control gaseous emissions from industrial facilities since the mid-1940s. The new regulations are forcing new and better catalytic performance and will require: automobile catalytic converters to last 161 000 km vs. the current 80 000 km and also reduce pollutant emissions from the present standards; many diesel trucks and buses to be equipped with ''aftertreatment'' devices to control particulates and nitrogen oxides; catalytic oxidation of VOCs from industrial exhausts containing certain constituents such as halogen containing compounds; and use of reformulated gasolines in heavily polluted urban areas.

After 1996 all new automobiles must have a 161 000 km catalytic converter, nonmethane hydrocarbon emissions must be reduced to 0.078 g/km by 2004, carbon monoxide to 1.1 g/km, and nitrogen oxides to 0.1 g/km. At present allowable nonmethane hydrocarbon emissions are 0.25 g/km, carbon monoxide 2.1 g/km, and nitrogen oxides are 0.6 g/km. All are significant reductions in the allowable emissions.

Automotive exhaust emissions are controlled by catalytic converters located in the exhaust system so that all exhaust gases must pass through them. Oxidation catalysts in the converter convert the carbon monoxide and hydrocarbons produced by incomplete combustion into carbon dioxide and water. Catalytic materials in the automobile converters are supported on a ceramic honeycomb monolith (Fig. 2). This honeycomb contains 46 to 62 square channels per sq cm and each channel is coated with an activated alumina layer called the washcoat. Platinum, palladium, and rhodium catalysts are dispersed in the washcoat. The ceramic monolith is cordierite which has a very low coefficient of expansion and thus can withstand continued heating and cooling cycles. The raw materials used to make this cordierite monolith are calcined kaolin, calcined talc, alumina, and kaolinite which must be very plastic and have a high green and dry strength (Murray, 1989).

Nitrogen oxides reduction systems must be integrated with the engine so the exhaust stream will have the type and amount of hydrocarbons needed to reduce these oxides at the optimum temperature for the particular hydrocarbon. At present a mordenite zeolite Y containing cerium and Al_2O_3-supported cobalt are promising nitrogen oxides reduction catalysts (Farrauto et al., 1992).

Emissions and particulates from truck diesel engines must be significantly reduced by 1994. Exhaust aftertreatments will be needed to reduce soot, soluble organic fraction, and sulfates. Porous ceramic filters will trap 60 to 90% of particulates. As soot builds up in the filter the resistance to flow increases and when this resistance becomes too great the filter must be regenerated by heating to burn off the soot. Oxidation catalysts will be used to remove most of the soluble organic fraction and nitrogen oxides will be reduced by using a catalytic lean nitrogen-oxides aftertreatment and perhaps alternate fuels.

The major generator of nitrogen oxides are stationary sources including boilers, incinerators, process heaters, combustion turbines, and reciprocating piston engines that burn fuels ranging from coal, oil, and natural gas to wood and solid waste. Nitrogen oxides are removed most efficiently by selective catalytic reduction. Am-

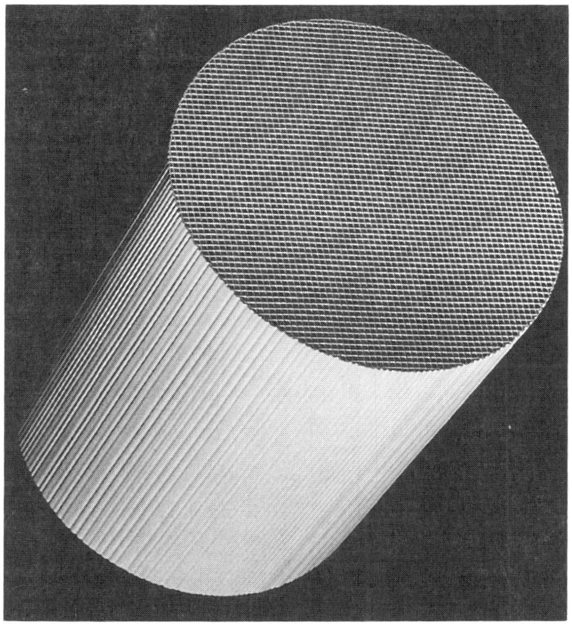

Fig. 2. Ceramic monolith (cordierite) used in automobile converters.

monia is added to an oxygen containing exhaust stream that passes over selective catalytic reduction catalysts supported on a ceramic or metallic honeycomb.

Carbon monoxide is also emitted from stationary sources, but it can be controlled by a precious metal oxidation catalyst on a ceramic or metal honeycomb. As described previously there will be an increasing need for ceramic raw materials to make the mono-filters that will withstand the shock of expansion and contraction caused by heating and cooling. High strength plastic kaolin, calcined kaolin, calcined talc, and alumina will be in increasing demand. In 1990 more than 150 kt of high strength plastic kaolin were used in making the ceramic monoliths, and it is estimated that this will increase to more than 300 kt in 1994.

CATALYSTS FOR CHEMICAL PROCESSING

There are many special catalysts used in the manufacture and processing of industrial chemicals. These catalysts generally are clay and zeolite based and have a known exchange capacity. Rare earth and precious metal ions are exchanged onto the clay and/or zeolite to achieve the special catalytic reaction that is desired. Again there is extensive ongoing research and development to produce better and more effective catalysts and catalyst supports.

SUMMARY

Catalysts play an important and necessary role particularly in the refining of petroleum; controlling potentially harmful emissions from industry, transportation, power generation, and other sources and in the large and diverse chemical manufacturing and processing industries. Industrial minerals are an important raw material in the catalyst industry, and with the growth of the use of catalysts in environmentally related applications they will become increasingly important. Those that are now used in significant quantity are kaolin, smectite, palygorskite, sepiolite, talc, titanium dioxide, rare earth minerals, silica, and gibbsite (bauxite). With additional research and development, natural zeolites may become an important raw material and could replace some of the synthetic zeolite that is now used in large quantities.

REFERENCES

Bondt, N., et al., 1797, *Anm. Chim. Phyc.*, Vol. 21, pp. 48.

Farrauto, R.J., Heck, R.M., and Speronello, D.K., 1992, "Environmental Catalysts," *Chemical and Engineering News*, Vol. 70, No. 36, pp. 34–44.

Hettinger, W.P., Jr., 1988, "Development of a Reduced Crude Cracking Catalyst," *Fluid Catalytic Cracking*, ACS Symposium Series No 375, Chap. 19, M.L Occelli, ed., Washington, DC, pp 308–340.

Hettinger, W.P., Jr, 1991, "Contribution to Catalytic Cracking in the Petroleum Industry," *Applied Clay Science*, Vol. 5, pp. 445–468.

Murray, H.H., 1989, "Clay Minerals for Advanced Ceramics," *Mining Engineering*, Vol. 41, No. 11, pp. 1123–1126.

Oblad, A.C., 1983, "The Contributions of Eugene J. Houdry to the Development of Catalytic Cracking," *Heterogeneous Catalysis—Selected American Histories*, B. David and W.P. Hettinger, Jr., eds., ACS Symposium Series No. 222, Washington, DC, p. 61.

Occelli, M.L., 1988a, "Recent Trends in Fluid Catalytic Cracking Technology," *Fluid Catalytic Cracking*, ASC Symposium Series No. 375, M.L. Occelli, ed., Chap. 1, Washington, DC, pp. 1–16.

Occelli, M.L., 1988b, "Cracking Metal-Contaminated Oils with Catalysts Containing Metal Scavengers: Effects of Sepiolite Addition on Vanadium Passivation," *Fluid Catalytic Cracking*, ACS Symposium Series No. 375, M.L. Occelli, ed., Chap. 10, Washington, DC, pp. 162–181.

CHEMICALS

ROBERT B. FULTON, III

The objective of this chapter is to discuss the interrelationship between industrial minerals and chemical manufacturing. It is intended to supplement rather than duplicate the commodity chapters. Particular emphasis is given to the pertinent chemical element and to market factors.

Condensing this broad subject into a few pages of this handbook permits treating only the most important elements derived from industrial minerals. Hydrocarbons, which quantitatively dominate as raw materials for the chemical industry, are omitted, as are the metallic elements and the minerals covered in other "use" chapters such as phosphorous, potassium, and nitrogen for fertilizers, and titanium dioxide for pigments. The remaining six elements of major importance are: boron, bromine, chlorine, fluorine, sodium, and sulfur. These elements are treated individually under separate headings.

Table 1 affords an overview of the main industrial minerals, the chemical products derived from them, and end uses of the products.

Salt brines have particular importance as raw material sources for the chemical industry. Table 2 is a chart of the chemical compounds derived from four types of brines: (1) Owens Lake-type brines, which are sources of boron and sodium compounds; (2) Midland-type brines, from which bromine, iodine, and chlorides of calcium, magnesium, potassium, and sodium are derived; (3) Searles Lake-type brines, yielding boron, bromine, lithium, magnesium, potassium, and sodium compounds; and (4) Silver Peak-type brines, produced mainly for lithium.

MARKET ATTRIBUTES

Some of the important market traits common to industrial minerals used by the chemical industry are:

1. They are international commodities, such as fluorspar and sulfur, which largely move to foreign consumers.
2. Grade, and freedom from deleterious elements are important factors affecting their usability in chemical processes. An example is salt (NaCl) used in electrolysis where ultrapure evaporated salt is required to meet rigid specifications.
3. Purified products take on the characteristics of specialty items and command a distinctly higher price than the basic commodity from which they are derived.
4. In practically all cases, chemical users require some sort of cleaning or beneficiation of the naturally-occurring mineral to bring it to specification, and individual specifications may vary from user to user for essentially the same use.
5. In some instances it is necessary to strike a balance between what the vendor can supply and what the buyer requires, with the result that specifications have to be eased to afford the needed materials in marginal cases.
6. Because they tend to be bulk commodities, low cost for handling and transportation are important and such costs may limit the area from which a chemical user can draw his supply.
7. Shipments are usually in bulk and frequently in multiple-car, full-trainload or full-shipload lots to reduce transport costs, which in turn may require large terminal investment facilities.
8. Purchases are generally by contract of one year or longer term, with spot buying playing only a minor role.
9. Contract prices are usually fixed in short term commitments, but may vary according to assay, with premiums and penalties for content above or below the norm; however, general practice is for specifications to be fixed in the contract with minimums being set for the desired material and maximums for undesired elements. In longer term contracts, prices are often escalated on labor, fuel, and other vendor processing costs.
10. Suppliers of individual commodities to the chemical industry tend to be limited in number and are generally medium- to large-size producers that supply a few major consumers.
11. The bulk of the mineral volume is for basic chemical uses, sulfur suppliers to sulfuric acid producers and fluorspar for hydrofluoric acid producers being typical examples. These basic chemical products then are used for the production of other products.
12. Shortage of a supply of adequate quality leads consumers to seek substitutes. In the case of fluorspar, much work is being done on recovery of fluorine from phosphate rock. Success in the form of fluorosilicic acid and/or hydrofluoric acid production could, in time, affect the hydrofluoric acid chemical industry.
13. Markets tend to be characterized by cycles of shortage followed by oversupply, with attendant wide price fluctuations.
14. Barriers to trade can have an adverse effect on the necessary movement of industrial minerals used by the chemical industry in international trade. Antidumping laws, quotas, and tariffs can disrupt or dislocate normal markets.
15. Chemical industry consumers may back-integrate for security of supply or for favorable economics, sometimes by joint ownership and often with experienced mining partners.

ELEMENTS OF MAJOR IMPORTANCE

Boron

Boron is used by the chemical industry mainly as borax, (disodium tetraborate decahydrate). It is derived principally from desert evaporite deposits. Major commercial boron minerals include: borax ($Na_2B_4O_7{\cdot}10H_2O$), coleminate ($Ca_2B_6O_{11}{\cdot}5H_2O$), tincalconite ($Na_2B_4O_7{\cdot}5H_2O$), ulexite ($NaCaB_5O_9{\cdot}8H_2O$) and kernite ($Na_2B_4O_7{\cdot}4H_2O$). Commercially, sodium borates predominate and are supplied by refineries in a variety of grain sizes in

Table 1. From Industrial Mineral to Inorganic Chemical

Commodity	Chemical Formula	Reaction Type	Reagents	Products: Formula	Products: Name	Uses for End Products
Spodumene	$LiAl(SO_4)_3$	Acid decomposition	Sulfuric acid	Li_2SO_4	Lithium sulfate	Ceramics, glass, fluxes, flame colorant
		Alkali decomposition	Limestone	$LiOH$	Lithium hydroxide	Absorption refrigeration, greases, high strength glass
Rock salt	$NaCl$	Electrolytic	Electricity, water	$NaOH$	Caustic soda	*Miscellaneous chemicals (30%), rayon (17%), pulp and paper (8%), metals (8%)
				Cl_2	Chlorine	Miscellaneous chemicals including organics (83%), pulp and paper (13%)
				H_2	Hydrogen	Ammonia synthesis, miscellaneous chemicals, metallurgy
		Ammoniation	Ammonia, lime, carbon dioxide	Na_2CO_3	Soda ash	Miscellaneous chemicals (37%), glass (30%) metallurgical (10%), soaps (16%)
				$CaCl_2$	Calcium chloride	Road treatment, dust suppression
Sylvite	KCl	Electrolytic	Electricity, water	KOH	Potassium hydroxide	General chemicals, fertilizers, detergents
				Cl_2	Chlorine	See above
				H_2	Hydrogen	See above
Beryl	$3BeO \cdot Al_2O_3 \cdot 6SiO_8$	High temperature fluorination	Sodium fluoroferrate	$Be(OH)_2$	Beryllium hydroxide	Fluorescent tubes, TV tubes, glass, ceramics
Limestone	$CaCO_3$	Calcination	—	CaO	Quicklime	Caustic soda manufacture, effluent treatment, water treatment
				CO_2	Carbon dioxide	General chemicals, soda ash
Anhydrite	$CaSO_4$	Carbothermic reduction	Natural gas, coal, air	H_2SO_4	Sulfuric acid	Fertilizers (36%), chemicals (26%), petroleum refining (12%), pigments (9%), metallurgical
				CaO	Quicklime	See above
Seawater	$MgCl_2$	Precipitation, calcination	Dolime	MgO		
Magnesite	$MgCO_3$	Calcination	—	MgO		
Brine	$MgCl_2$	Steam decomposition		MgO	Magnesium oxide	Refractories, general chemicals, fertilizers, cements
				HCl	Hydrochloric acid	Miscellaneous chemicals, metal pickling, organic chemicals
Celestite	$SrSO_4$	Carbothermic reduction	Coal	$SrCO_3$	Strontium carbonate	Glass, ceramics, pyrotechnics
Barytes	$BaSO_4$	Carbothermic reduction	Coal	$BaCO_3$	Barium carbonate	Glass, ceramics, TV tubes, pyrotechnics, oxygen source, ferrite magnets, medicine, pigments, bricks, metal soaps
				$BaCl_2$	Barium chloride	

Bastnaesite Monazite Xenotime	Ce, La(PO_4)	Acid decomposition	Sulfuric acid	Ce, La O	Rare earth oxides	Glass colorants, polishing powders, TV tubes, lighter flints, arc carbons, catalysts, phosphors, ceramic stains, camera lenses
Zircon	$ZrSiO_4$	Chlorination	Chlorine, coke	ZrOCl	Zirconium oxychloride	Metal production, special chemicals
Rutile (natural & synthetic)	TiO_2	Reductive chlorination	Chlorine, coke	$TiCl_4$	Titanium tetrachloride	Pigments, titanium metal, waterproofing agents
Chromite	Fe_3CrO_4	Alkali decomposition	Soda ash	Na_2CrO_4	Sodium chromate	Chromium chemicals, electroplating baths, oxidants
Pyrolusite	MnO_2	Acid decomposition Carbothermic reduction	Hydrochloric acid Coal, nitric acid	$MnCl_2$ $Mn(NO_3)_2$	Manganous chloride Manganous nitrate	Dry batteries, soaps (dryers), fertilizers, glass, enamels, catalysts
Colemanite	$Ca_2B_4O_{11} \cdot 5H_2O$	Alkali decomposition	Soda ash	$NaBO_2$	Sodium borate	Detergents, fluxes
Bauxite	$Al_2O_3 \cdot 2H_2O$	Alkali solution Acid decomposition	 Sulfuric acid	$Al(OH)_3$ $Al_2(SO_4)_3$	Aluminum hydroxide Aluminum sulfate	Water treatment chemicals, flocculants, dye mordants, refractories, catalysts
Quartz, silica	SiO_2	Alkali reaction	Soda ash	Na_2SiO_3	Sodium silicate	Adhesives, cements, deflocculants, soaps
		Reductive chlorination	Chlorine, coke	$SiCl_4$	Silicon tetrachloride	Organic chemicals (silicone fluids and rubbers)
Caliche	$NaNO_3$	Acid decomposition	Sulfuric acid	HNO_3	Nitric acid	Fertilizers, general chemicals
Phosphate rock	$Ca_5F(PO_4)_3$	Acid decomposition	Sulfuric acid	H_3PO_4	Phosphoric acid	Fertilizers, detergents, foodstuffs, organic chemicals, water treatment
		Electric furance reduction	Silica, coke	P	Phosphorus	See phosphoric acid
Brimstone	S_6	Burning	Air	SO_2	Sulfur dioxide	Fungicide, insecticide, reducing agent, solvent
		Catalytic oxidation	Air	H_2SO_4	Sulfuric acid	See above
Iron pyrite	FeS_2	Roasting, catalytic oxidation	Air	H_2SO_4	Sulfuric acid	See above
Fluorspar	CaF_2	Acid decomposition	Sulfuric acid	HF	Hydrofluoric acid	Organic chemicals, fluxes, etchants, metallurgy, isotope separation
Seawater, brines	NaBr	Chlorination	Chlorine	Br_2	Bromine	Organic synthesis, pharmaceuticals, petroleum additives
Seawater, brines, caliche	NaI	Chlorination Chemical reduction	Chlorine Sodium bisulfite	I_2	Iodine	Foodstuffs, sanitizers, organic chemicals

*US consumption percentages in parentheses.
Source: Jones, 1973.

Table 2. Salt Brine Derivatives

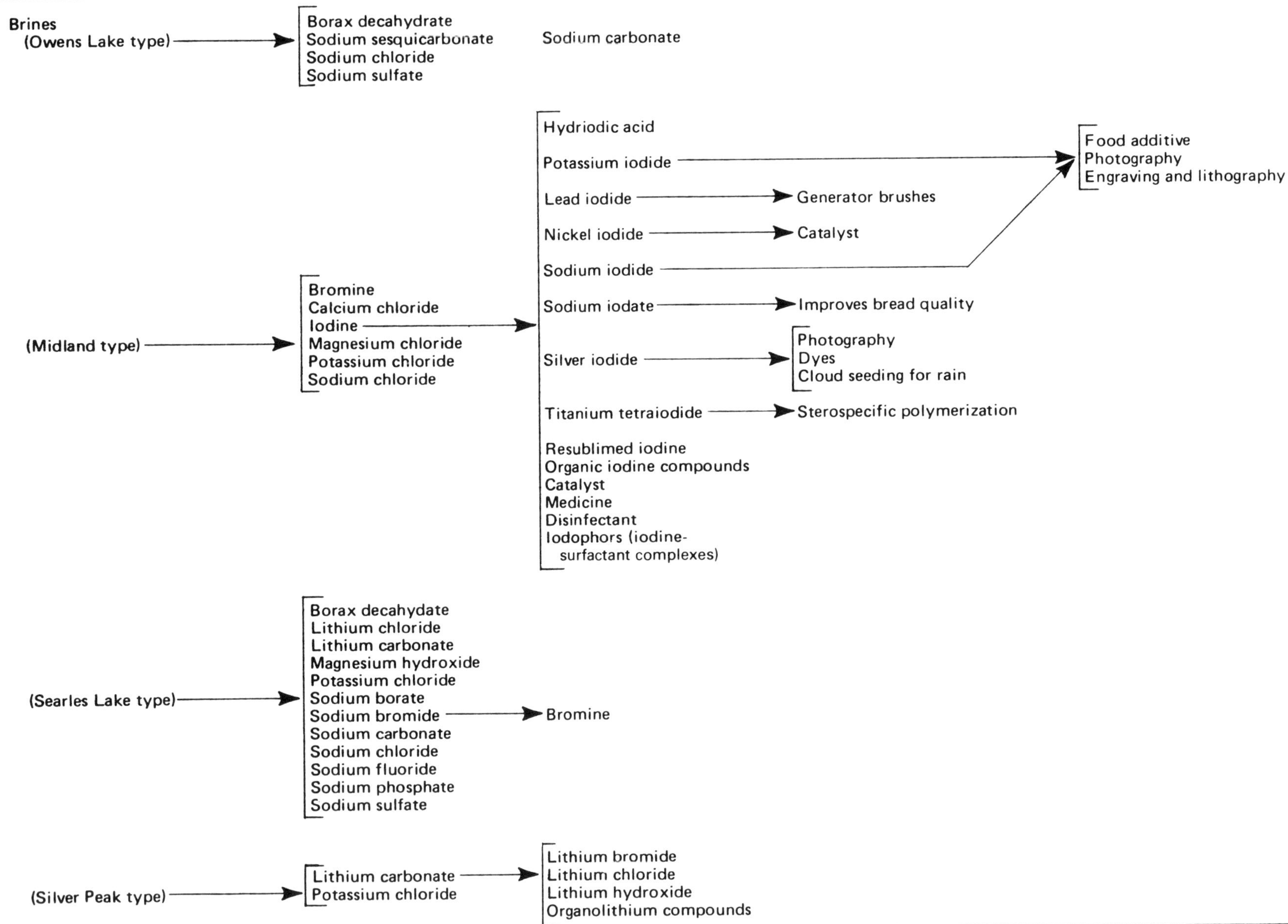

Source: Anon., 1967 (courtesy Stanford Research Institute).

either the purified borax form or as fused, crushed, anhydrous borax. Boric acid and boric oxide are also available. The industry defines boron content in terms of B_2O_3; e.g., granular borax decahydrate contains 36.5% B_2O_3, anhydrous borax, 69.1% and boric acid, 56.3%. These products are sold at 99.5% purity. Raw and semi-processed ores are also sold.

Borates are used as fluxes in metal processing, in ceramics, as a constituent of heat resistant glass, as an aid in glass fiber manufacture, as cleaning agents, and in textile manufacture. Borax and boric acid are used as extenders in dye making. In elemental form, boron acts as a neutron absorber in shielding of radioactivity. Organic boron compounds are used as gasoline additives and in insecticides and pharmaceuticals. The tri-fluoride etherate is used for manufacture of fluorinated hydrocarbons and in production of antiknock compounds for gasoline. Cubic boron nitride has extreme hardness and stability at high temperature. Fuels containing boron and hydrogen, outstanding for high energy yield, have not been widely used, mainly because of pollution effects and high cost. Boron starting-materials and end products are depicted in Table 3.

According to the US Bureau of Mines (Lyday, 1991), US production of boron minerals and compounds in 1990 was 1 094 kt, compared to 1975 at 1 060 kt and 1981 at 680 kt, illustrating wide fluctuation possible in production of industrial minerals. Approximate year end 1990 FOB prices quoted per ton in bulk were:

Boric acid (99.9%, Tech)	$775–784
Tech. pentahydrate borax	$321

Opportunities for wider applications of boron compounds, in addition to the growth in established fields of use, lie in such uses as glass fibers for reinforced plastics used for weight reduction in automobiles, fuels, and as a possible replacement of fluorspar as a fluxing agent in steel making. Boron's capability of forming a large number of coordination compounds offers avenues for investigation which are being pursued to yield a variety of diverse products. The reader is referred to Lyday's comprehensive 1990 Annual Report (Lyday, 1991a, 1991b) for coverage of new uses and other significant information.

Bromine

Bromine production in the United States is in the range of 140 to 180 ktpy, derived mainly from deep well brines in Arkansas. Worldwide, Israel is second as a bromine producer, at 105 to 135 ktpy. Other producers in order of rank are Republics of the former USSR (60 ktpy), the United Kingdom (28 ktpy), France (18 ktpy) and Japan (15 ktpy). The reader is referred to Lyday (1991b) for comprehensive coverage of bromine.

Bromine is used, mainly in the form of ethylene dibromide (EDB), by the chemical industry in making tetraethyl lead mixture, an antiknock additive for gasoline. The bromine compound inhibits deposition of lead when the gasoline burns. Methyl bromide has similar application. A variety of organobromides are used in agriculture as fumigants for controlling nematodes in soil and insects in storage of grain and seed. Elemental bromine is consumed in disinfectants and bleaching agents. Sodium, potassium, and ammonium bromides have minor uses in textile fiber manufacturing, photography, and medicines. Bromine compounds are used in small amounts in making dyes, resins, leather, and rubber products and in production of textile intermediates and finishing agents. Newer uses are in the field of flame retardants and extinquishants and for gas- and oil-well completion fluids, the latter as calcium bromide.

According to the US Bureau of Mines (Lyday, 1991b), US production in 1990 was about 177 kt, down substantially from highs of about 230 ktpy in the 1970s, the drop being mainly in the use of EDB for antiknock. Current US production compares to production capacity of about 260 ktpy.

Bromine prices at year end 1990 were 43 to 46 cents per 0.5 kg, having increased from 22.5 cents in the mid 1960s through about 30 cents in the early 1980s. The price outlook is for further increases due mainly to increased cost of process energy.

Chlorine

Chlorine production in the United States in 1990 was about 11 Mt, representing about 85% of installed capacity. Large chlorine producers, with plants based on the electrolysis of natural salt brines, are located in the New York-Ohio-Michigan region and on the Gulf Coast.

A number of principal consumers in the chemical industry are situated in the producing regions and in many cases are back-integrated with obvious advantages. Chlorine consumers who are not back-integrated either purchase chlorine or buy chlorine compounds such as hydrochloric acid, chloroform, ethyl chloride, ethylene dichloride, carbon tetrachloride, perchlorethylene and the like, depending on process requirements and economics. Choice of purchase mix is determined by relative costs; for example, the choice between either buying carbon tetrachloride, or buying chlorine and hydrocarbons for in house production of the chlorocarbon. This choice has ramifications, such as the internal generation of hydrochloric acid, produced as a byproduct of hydrocarbon chlorination and of fluorination. The internally produced hydrochloric acid can then be oxidized back to chlorine in a closed circuit, reducing the intake of purchased chlorine and thereby reducing costs.

Similarly, manufacturers of polyvinylchloride can choose to buy ethylene dichloride or if they have chlorine and ethylene available, they can elect to make it themselves. Recycling byproduct HCl from vinyl chloride monomer plants, is widely practiced with use of the HCl in oxychlorination to make ethylene dichloride.

Principal chemical industry uses for chlorine are for intermediates and for synthesis of organic and inorganic chlorides used in plastic solvents, agricultural chemicals, automotive fluids, sanitation products, etc. Among the fastest growing is the use of ethylene dichloride intermediate for vinyl chloride. Other uses of ethylene dichloride are for making solvents, such as trichloro- and perchloroethylene, and as cheaper, though less effective, lead scavenger than ethylene dibromide.

An important inorganic intermediate is titanium tetrachloride; found in TiO_2 pigment manufacture by the so called chloride process, now supplanting the older sulfate process.

Chlorine is also required for the manufacture of organic intermediates such as adiponitrile from butadiene. In this instance both chlorine and caustic are consumed, fortunately, in the same ratio as they are produced by electrolysis of brine. This ratio, known industrially as one electrochemical unit (ECU), consists of one ton of chlorine to 1.1 t of caustic soda. Neither the chlorine nor the caustic appears in the end product, but other routes to adiponitrile exist, affording the manufacturer an option to use or not use chlorine.

Growth of elastomeric products, made either by using chlorine, butadiene, and caustic, or by using hydrochloric acid and acetylene, has contributed significantly to world demand for chlorine.

Product obsolescence can play an important role in the chlorine market, as in the case of byproduct chlorine from metallic sodium manufacture. Sodium production is expected to decrease if demand for use in gasoline declines, reducing the amount of chlorine available from this source.

Phosgene is another chlorine chemical important to chemical processing, e.g., in the production of agricultural chemicals, isocyanates, and other fiber intermediates and elastomers.

Table 3. Boron Minerals and Chemicals

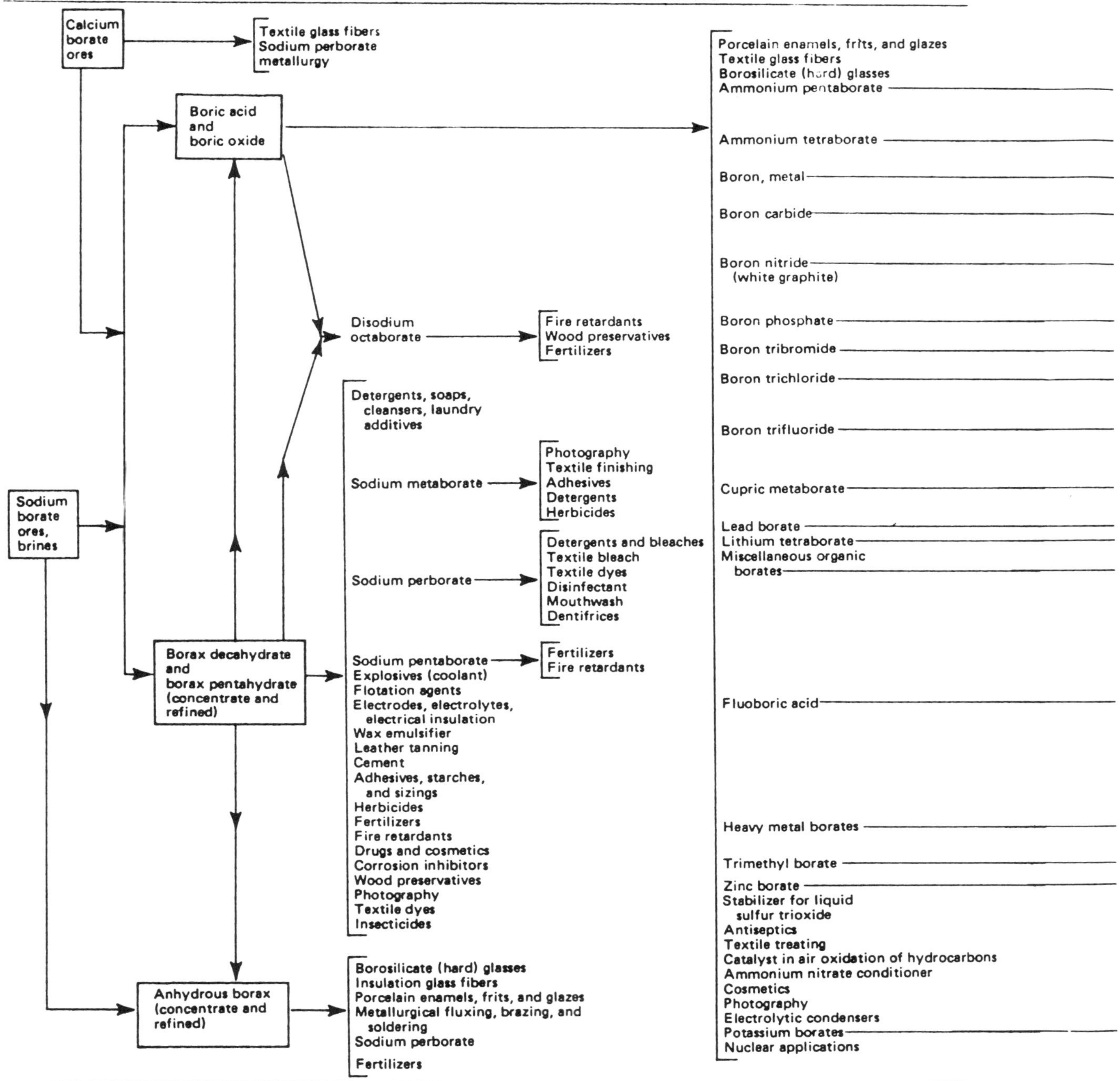

Source: Anon., 1970d (*courtesy:* Stanford Research Institute).

Table 3. Boron Minerals and Chemicals (cont.)

→ Electrical condensers; Fireproofing; Paper coating

→ Neutralizing agent; Fireproofing

→ Delay fuses; Flare igniters; Boron phosphide → High temperature transistors; Hexaboron silicide, Tetraboron silicide → Refractory material; Getter in manufacture of vacuum tubes; Abrasives; Grain refiner in aluminum manufacture; Nuclear shielding; Coating material for solar batteries; Semiconductors

→ Abrasives; Control rods for nuclear reactors

→ High-temperature lubricant; Refractory; Fibers; Thermionic devices

→ Ultra low loss dielectric

→ Diborane

→ Boron trifluoride; Diborane → (Pentaborane → Ethyl pentaborane; High energy fuels; Higher boron hydrides → Carborane polyesters, Rayon delustrant, Rubber vulcanization, Fluxing agent; Polymerization catalyst for ethylene; Copolymerization of styrene and butadiene); Organo-boron compounds; Trichloroborazole → Gelling agent, Catalyst; Depolymerization of paraldehyde; Flux; Polymerizing agent for driers; Stabilizer for liquid sulfur trioxide

→ Neutron analysis; Catalyst

→ Fireproofing agent for wood; Fungicides → Wheat rust compounds; Oil pigments; Pigments for ceramics

→ Galvanoplastic work; Paint and varnish driers; Lead glass

→ Ceramics

→ Glycol borates, Alcohol borates → Lubricant and hydraulic fluid, Gasoline antiknock agent; Jet fuel deicer; Epoxy resin curing agents; Nuclear shield via neutron-detecting; Bactericide in fuels; Trimethoxyboroxine → Extinguishing metal fires, Epoxy foams

→ Ammonium fluoborate → Molding sand additive in magnesium casting; Cadmium fluoborate, Cupric fluoborate, Lead fluoborate, Nickel fluoborate → Electroplating baths; Potassium fluoborate → Binder and cooling agent in abrasive wheels, Electrolytic brightening of aluminum; Sodium fluoroborate → Electrochemistry, Heat treating magnesium alloys, Nonferrous flux; Stannous fluoborate, Zinc fluoborate → Electroplating baths

→ Fireproof and mold resistant paints

→ Sodium borohydride → Lithium borohydride; Potassium borohydride; Pharmaceuticals; Reduction of aldehydes and ketones; Vat dyes; Blowing rubber and epoxy resin foams; Paper bleach and pulp brightener; Hair waving formulations

→ Ceramics; Medicines; Fire-retardant; Fungicides

→ Inorganic boron compounds; Photography; Casein solvent; Metallurgical welding and brazing; Diazotype developer solutions

Demand of chlorine in the United States grew at a rate of about 8.5% per year during the 1960s, spurred by increased production of vinyl chloride and other chloro-organics. The rise was followed by a modest decline to about 5% from 1970 to 1971, mainly due to the curtailed use of chlorinated pesticides. In 1972–1973 demand recovered, exceeding the 1970 record and causing a tight supply situation which was alleviated by adding to capacity in the late 1970s. Profitability has since suffered because rising costs have not been adequately compensated for by price increases. In the 1980s, chlorine was long in the market. In some cases spot is sold at nearly zero dollars with incremental sales of caustic providing enough income on a marginal ECU basis. Most spot business in 1990 ranged from $50 to over $60, with older contracts being about double those amounts. In 1991, spot business deteriorated with prices ranging from −$20 to +$10. By mid-1993, extreme volatility was again demonstrated as another wide fluctuation occurred with list prices of chlorine reaching as high as $195/t and market prices near that figure, while 1993 caustic list price was $300/t, its market price was far below at about $150/t and dropping as low as $40/t for spot export prices. As one commentator put it "prices are all over the map."

Domestic growth is flat and will continue a low or negative rate due to: (1) nonchlorine based processes supplementing some important chlorine required, (2) slackened growth in cyclic organic detergents and chlorinated solvents which have been attacked on environmental grounds, (3) leveling off in growth of other major products containing chlorine such as vinyl chloride, (4) the problem with dioxin poisoning as a negative influence in pulp and paper bleaching and 5) curtailment of chlorofluorocarbon manufacture and use.

Environmental issues are profoundly affecting the chlor-alkali business. Discontinuing use of lead additives and impending curtailment of carbon tetrachloride production are prime examples. The generally accepted thesis that chlorine in the upper atmosphere is responsible for a chain reaction destroying the earth's protective ozone layer will doubtless lead to further significant reductions in manufacture and use of chlorine compounds into the next century.

The economics of the chlor-alkali industry was characterized in the 1960s by the ability to compensate for sharply rising costs by means of technical improvements and economies-of-scale in the manufacturing process, but the industry has now achieved whatever could be gained through these routes. Major consumers react now by seeking opportunities for back-integration or by making long term contracts for purchases at the lowest negotiable price. The choice between these alternatives is determined by capital costs and contract terms and prices.

Chlorine-caustic market pricing in the 1960s and 1970s provides insight to the pricing mechanism at work. Chlorine pricing in the early 1960s was characterized by vendor willingness to make long term contracts at a set price because, while raw material costs were slightly rising, technical improvements were offsetting costs. Such contracts, with terms of five to seven years, provided the seller with an adequate, though modest, return on investment. ECU prices, equally divided between chlorine and caustic, were about $70. With the decline in chlorine demand along with the general economy in 1968–1971, chlorine prices could not be increased. The manufacturers took advantage of the tightness in the caustic market by raising caustic prices, thus compensating for the weakness in chlorine and giving a slight increase in overall profitability. Industry return on investment is not so great as to give strong incentive to accelerated expansion. As the economy recovered after 1971, the chlorine market improved and the industry was operating at capacity by the beginning of 1973, giving the opportunity for further price increases. These increases started in 1972 when an ECU delivered Gulf Coast was in the low $80s and rose to the level of about $100 early in 1974. Prices rose to the $100+ range in the late 1970s and into 1980, although spot prices weakened sharply late in 1980 due to the recession. The same scenario was played again in the 1980s when chlorine demand slackened in the late 1980s recession.

Since most of the chlorine produced in the United States is for captive use (about 60%), merchant sales have been used by the manufacturers to achieve high rates of operation, thereby minimizing unit costs of their captive requirements. These merchant sales are largely influenced by proximity of vendor to purchaser. The overall market demand must be viewed in terms of chlorine chemicals; some of the ostensibly captive chlorine requirement of manufacturers is truly just one move downstream along the route to a final product. In the case of chlorinated hydrocarbons, the chlorine manufacturer takes the next step of treating hydrocarbons and selling the chlorinated compound.

About three-quarters of US salt production is used in the chemical industry where industrial capacity for salt brine electrolysis has barely kept ahead of rising demand. Individual new plants and additions to existing ones are now being sized at 700 tpd compared to about 200 tpd in the mid-1960s. Each major addition to capacity is thus 2 or 3% of the industry capacity. Despite the forecast of flat to mildly negative domestic chlor-alkali growth, a major new ion-exchange diaphragm chlorine/caustic manufacturing plant of Formosa Plastics Texas, begun in 1989, is due on stream in late 1993 at point Comfort, TX, with a chlorine output capacity of 567 ktpy and 620 ktpy caustic. Chlorine will feed Formosa's new EDC unit. Caustic will be sold. EDC capacity of 544 ktpy will be used to produce VC monomer and PVC resins, as well as EDC for sale.

The following table lists the top ten US chlorine producers' rated capacities as of 1990:

Rated Capacity of US Chlorine Producers (ktpy)

Producer	Capacity
Dow Chemical USA	3 490
Occidental Chemical	3 170
PPG Industries	1 590
Olin Chemicals	730
Vulcan Chemicals	590
Georgia-Gulf	420
LCP Chemicals	330
Pioneer	290
Atochem-Basic	280
Niachlor	240

An interesting change of technology in chlorine manufacture is the decline of mercury cells in favor of diaphragm cells due to environmental control measures, even though diaphragm cells have higher operating costs, albeit lower initial investment. Prior to this restructuring, the trend was toward larger and larger mercury cells, up to cells producing 20 t of chlorine per day. Through the impact of pollution control, the emphasis has changed so that more than 70% of the chlorine produced by brine-electrolysis comes from diaphragm cells, in part also stimulated by development of the dimensionally stable electrode. This electrode eliminates the problem of gradual chlorination of the organic binders used in graphite anodes which resulted in inefficient cell operation as the anodes disintegrated. The new electrode made of titanium metal protected by a film of inert metal oxide permits construction of larger cells with higher current densities and results in a three-to-five year electrode life compared to three-to-six months for graphite electrodes. Diaphragm cells with a productive capacity of 3 t of chlorine per day have a life of nine months to a year, being limited by diaphragm problems. Solution of these diaphragm difficulties is centered on new materials and techniques of construction.

A further cell modification is a bipolar cell, in which anodes are directly connected to adjacent cathodes to reduce the voltage drops now experienced in bussbar connections between cells. Development of bipolar membrane cells will permit higher concentrations of caustic in the cell, eliminating subsequent costly evaporation.

The long range market outlook is for continued modest, steady growth in use of chlorine by the chemical industry, despite adverse factors cited. A trend can be expected toward a greater percentage of foreign manufacturing based on chlorine and petrochemicals.

Fluorine

Fluorine's role in the chemical industry is mainly as a constituent of fluorocarbons and of inorganic fluorides. Hydrofluoric acid is the intermediate for practically all fluorine chemicals. It is produced by reacting fluorspar with sulfuric acid, yielding hydrogen fluoride gas and calcium sulfate. The gas is usually liquefied as anhydrous hydrogen fluoride, but may be adsorbed in water to make aqueous hydrofluoric acid, usually 70% HF. In commercial practice about 2.2 t of acid grade fluorspar (97% CaF_2) are required to make 1 t of hydrofluoric acid. Contaminants found in commercial, acid grade fluorspar are calcium carbonate, silica, and sulfur; maxima are usually set. Other objectionable constituents, for which maxima may be set, include arsenic, chloride, and phosphate. Grain size and moisture content are also specified.

The fluorocarbons comprise a large family of useful compounds; the first to obtain commercial acceptance were used as refrigerants and still constitute a major part of the market. Because of environmental concern their use for aerosol propellants has been discontinued in the United States. They are used for the manufacture of plastics characterized by chemical inertness, toughness, and electrical resistivity. Other fluorocarbons are used as fire extinguishing agents, dielectrics, surfactants, anesthetics, and medicinals.

Increasing concern for the ozone depleting potential of chlorofluorocarbons has led to serious effort to develop non-ozone depleting alternatives which contain less or no chlorine in the molecule. In other work, substitutes are sought, so called ''not in kind'' substitutues, for some applications. Early indications are that the alternative fluorocarbon molecules will contain a higher percentage of fluorine atoms than the chlorofluorocarbons they will replace. Depending on the amount of the market captured by not-in-kind substitutes and the amount gained by the new higher fluorine-containing alternatives, the usage of fluorspar by the chemical industry may stay about level or even increase moderately.

Inorganic fluorides are produced by reacting hydrofluoric acid with the appropriate base or basic salt. Principal uses of inorganic fluorides are outside of the chemical industry, such as: 1) the very large use of synthetic cryolite (Na_3AlF_6) used in the production of aluminum metal, 2) the relatively minor uses of sodium and potassium silicofluorides for fluxes and as constituents of enamels and glass, 3) aluminum fluoride as brazing flux, 4) sodium-aluminum-fluoride in resin-bonded grinding wheels, and 5) the use of sodium silicofluoride in water supply systems to control human tooth decay.

Hydrofluoric acid is used as a source of fluorine in reactions with uranium to produce gaseous uranium hexafluoride from which U^{235} is diffusion-separated from the more abundant U^{238}. Hydrofluoric acid is also widely used as a catalyst in the alkylation of petroleum during high-octane gasoline production. A minor use of hydrofluoric acid is for etching glass, such as electric light bulbs.

The reader is referred to the chapter on fluorspar for further specific information.

Sodium

Caustic soda, soda ash (sodium carbonate), and salt cake (sodium sulfate) are useful to the chemical industry mainly as sources of alkalinity and of the sodium ion. Supply of caustic is entirely dependent on chlorine production.

Demand for caustic is in part related to chlorine, particularly with respect to adiponitrile production from butadiene, but is unrelated to chlorine in many of its applications. In cases where chlorine and caustic are related in direct proportion to the electrochemical unit (ECU) (described earlier in this chapter under *Chlorine*) a price advantage for the total ECU is possible because the vendor has no surplus of either commodity. The general case, however, is that pricing of the two are independent.

For the manufacture of certain sodium compounds, such as bicarbonate, cyanide, perborate, nitrate, and silicate, various sodium raw materials may be used. In making sodium bicarbonate and sodium silicate, soda ash is the usual starting materials, but in the case of the silicate, caustic is sometimes used, while for nitrate, caustic is preferred.

Caustic is used in making such diverse products as cellophane, nylon intermediates, and pigments. It has important applications in waste treatment processes, and pulp, paper, and alumina manufacture. Increased demand for these uses has periodically led to tight supply and rising prices, partly because chlorine demand failed to keep pace with demand for caustic. This was again true in 1990 and 1991, when environmental issues adversely impacted chlorine use in its principal markets, including pulp and paper, chlorofluorocarbons, chlorinated solvents, and PVC, while caustic soda demand continued its steady, though moderate, growth. As a result, the price of chlorine dropped and caustic rose. The discrepancy between weak caustic supply, dependent on chlorine production, and steadily growing demand has created a gap which is likely to increase, causing instability in the market on an ECU basis. In 1991, caustic consumption in the United States was 7% below 1990 and prices declined for the first time since 1987.

Shortages or non-competitive prices for caustic push users to buy soda ash for alkalinity requirements. A case in point is the use of trona (sodium sesquicarbonate), which can be used on its own or can be converted to caustic. Two plants for caustic production from trona are operational by FMC and Tenneco, totalling 120 ktpy, while another by Texas Gulf is due in 1992 at 150 ktpy.

The principal use of soda ash is in the glass industry (54%). Chemicals use 20%, pulp and paper 4%, soaps and detergents 7%, and water treatment 4%. Other domestic uses account for 11% of consumption. In 1990, US soda ash production was about 9 160 kt and domestic consumption was about 6 840 kt, the balance being exported.

Caustic is expected to grow in international trade, where US exports have more than doubled since the mid-1960s, mainly due to the growing demand for use in alumina production.

Sulfur

Mined sulfur is normally converted to sulfuric acid for use in fertilizer manufacture and chemical processing. About 50% of sulfur production is used for fertilizer manufacture and 20% for chemicals, followed by less than 5% each for pigments, textiles, steel and petroleum. Nonacid uses of sulfur are in pulp and paper manufacture ($<5\%$), carbon disulfide ($<5\%$) and insecticides ($<5\%$).

Sulfur's usefulness is so widespread and pervasive in the chemical industry that its consumption closely follows the pattern of industrial activity. Tables 4 and 5 provide a remarkably clear condensation, succinctly portraying the role of sulfur and sulfuric acid derivatives in organic and inorganic reactions and process applications. The history of sulfur in its relation to the chemical industry is interestingly told by Haynes (1959).

The following publications are recommended for current and comprehensive coverage of sulfur matters: 1) the magazine

Table 4. Sulfur Derivatives

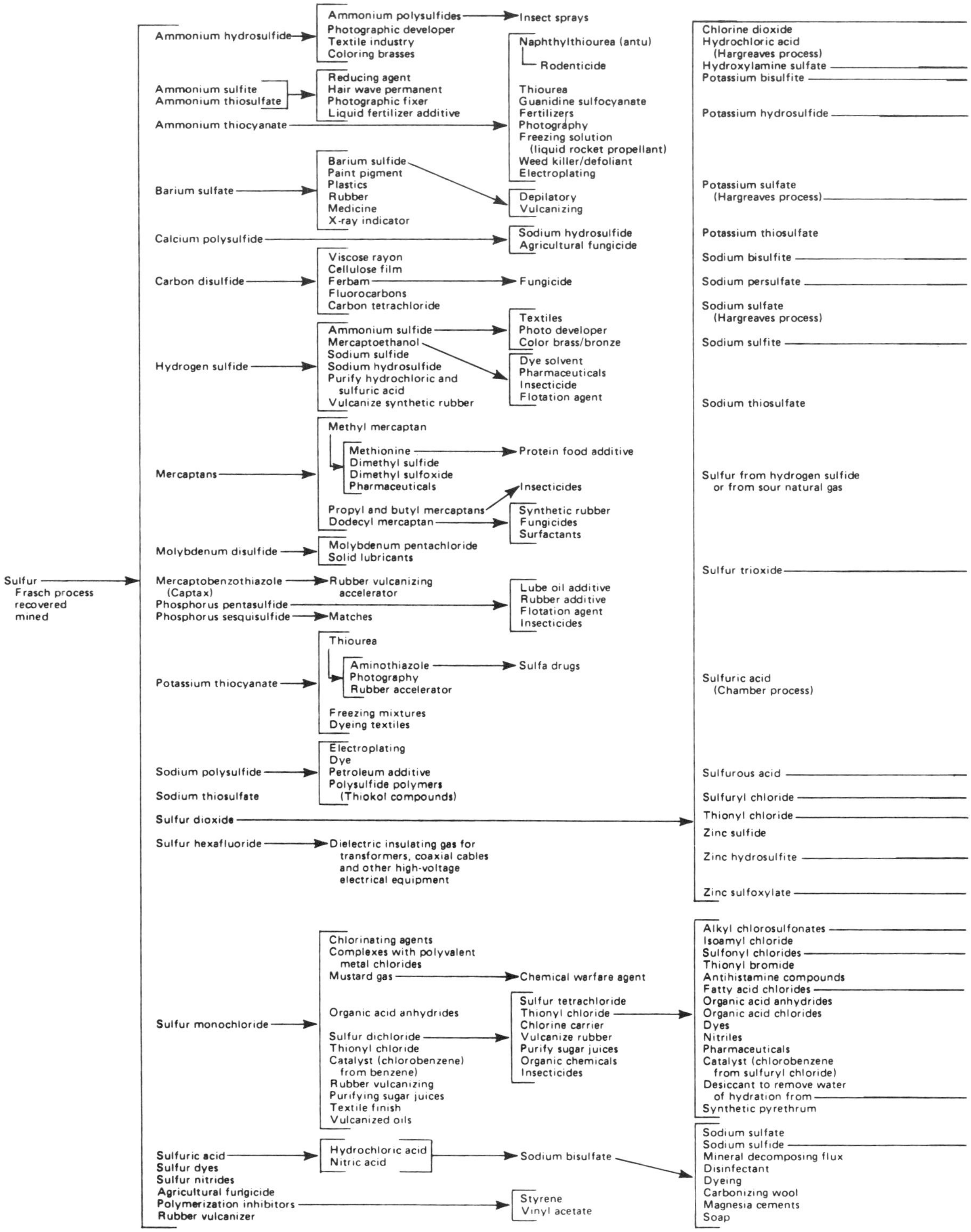

Source: Anon., 1967 (*courtesy:* Stanford Research Institute).

Table 4. Sulfur Derivatives (cont.)

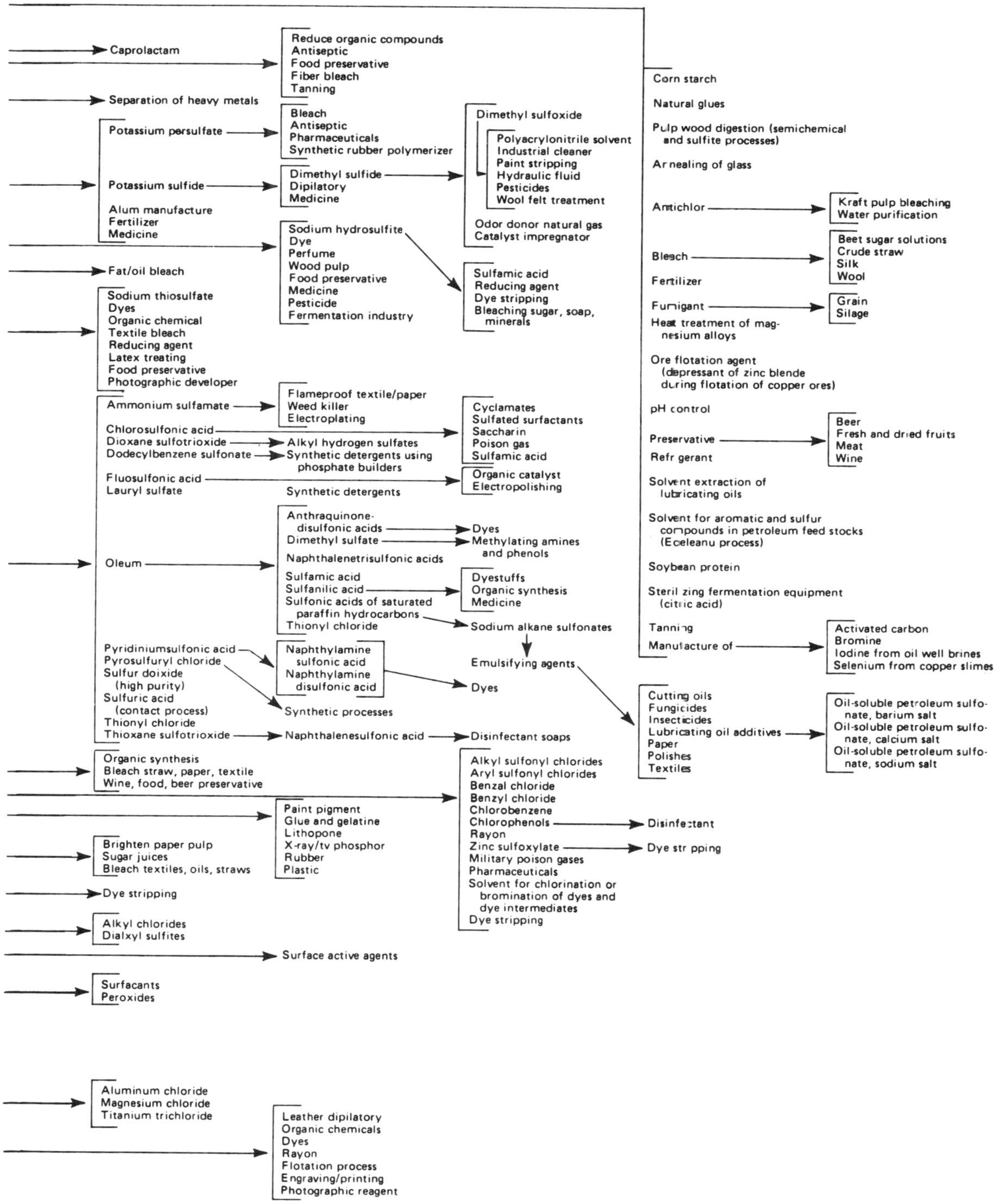

Table 5. Sulfuric Acid Derivatives

Sulfuric acid* from elemental sulfur, from pyrites, from smelter gases, from sludge acid, refortified acid, from H_2S → Inorganic reactions:

- Aluminum sulfate; Iron free aluminum sulfate → Sizing paper; Water purification; Dye mordant; Sewage treatment; Fire fighting; Fat/oil clarifier; Aluminum sodium sulfate → (Textile waterproofing; Tanning; Paper precipitant; Sugar refining)
- Ammonium bisulfate → Ammonium persulfate → Hydrogen peroxide
- Ammonium sulfate (from coke-oven gas) → Fertilizer
- Ammonium sulfate (from synthetic ammonia) → Ammonium chloride → Dry cells; Cobaltous ammonium sulfate; Nickel ammonium sulfate → Electroplating; Fertilizer; Flame retardant in fabrics; Industrial ethyl alcohol from molasses (nutrient for yeast culture); Viscose rayon
- Antimony sulfate → Explosives
- Barium sulfate (synthetic blanc fixe) → Barium oxide; White pigment for paints
- Beryllium sulfate → Beryllium oxide
- Boric acid
- Cadmium sulfate → Cadmium electric cells; Medicine; Vacuum tube manufacture
- Chlorosulfonic acid → Sulfamic acid; Sulfonated oils → Textile finishing agents; Sulfuryl chloride; o-Toluene sulfonyl chloride → (Saccharin; Dyestuffs); Sulfated sufractants
- Chromic acid → Zinc chromate; Andizing aluminum and aluminum alloys; Ceramic glaze; Chrome plating baths; Copper stripping; Dyestuffs; Metal surface treatment; Pharmaceuticals; Rubber pigment
- Chromic sulfate → Green paint/varnish; Leather tanning; Textiles
- Cobaltous sulfate → Ceramics; Pigments; Glazes; Plating
- Copper sulfate → Fungicide
- Ferric sulfate; Ferrous sulfate → Aluminum etching; Pigment; Soil amendment; Coagulant in water and sewage treatment
- Fluoboric acid → Electroplating
- Fluosulfonic acid → Aryl sulfonyl fluorides → Lubricating oil additives; Catalyst (alkylation of isoparaffins); Electropolishing; Wood preservative
- Hydrochloric acid (from sodium chloride)
- Hydrofluoric acid (from fluorspar)
- Hydrocyanic acid (from sodium cyanide) → Acrylonitrile; Acrylates; Adiponitrile; Dyes; Fumigants; Rubbers; Plastics; Chelating agents
- Hydrogen sulfide
- Lead sulfate, basic → Ceramics; Leaded zinc oxide; Pigment; Stabilizer in vinyl plastics
- Lithium sulfate → Pharmaceuticals
- Magnesium sulfate → Mineral water; Magnesium tungstate → Luminescent paint; Fireproofing formulations; Leather tanning; Medicine; Paper sizing
- Mercuric sulfate → Medicine; Gold/silver extractant; Galvanic batteries
- Mercurous sulfate → Laboratory batteries
- Nickel sulfate → Nickel-ammonium sulfate → Electroplating; Nickel catalyst; Nickel plating; Printing textiles
- Oleum
- Phosphoric acid (wet process) → Monoammonium phosphate → Diammonium phosphate → Fertilizer; (Enriched superphosphate; Triple superphosphate) → Fertilizers; Rust proofing metals; Soft drink; Sugar refining; Water treatment
- Potassium bisulfate → Food preservative; Tartrates
- Potassium dichromate
- Potassium sulfate (from potassium chloride) → Potassium aluminum sulfate (potash alum) → (Medicine; Dye mordant; Baking powder); Potassium bisulfate; Potassium persulfate; Potassium sulfide → (Medicine; Depilatory); Antiflash agent in smokeless powders; Fertilizers; Manufacture of gypsum wall board
- Silica gel
- Silicon tetrafluoride (by product of phosphate rock acidulation) → Fluosilicic acid
- Sodium bisulfate; Sodium dichromate → Metallurgical flux; Pickling metal; Leather bleach; Cleaning agents
- Sodium sulfate
- Sulfamic acid
- Sulfuryl chloride
- Titanium dioxide → Pigment for → (Lacquers; Paint; Printing inks; Varnish); Delustering synthetic fibers; Single crystals for transducers; Filler for → (Coated fabrics and textiles; Floor coverings; Paper; Rubber); Welding rod
- Tungstic acid → Tungsten, metal; Textiles; Plastics
- Vanadyl sulfate; Zinc sulfate → Mordant; Aniline black; Ceramic coloring

* Varying strengths of sulfuric acid are used as processes require (sulfuric acids =100% H_2SO_4, or oleum 100% H_2SO_4)

Source: Anon., 1970d (*courtesy:* Stanford Research Institute).

Table 5. Sulfuric Acid Derivatives (cont.)

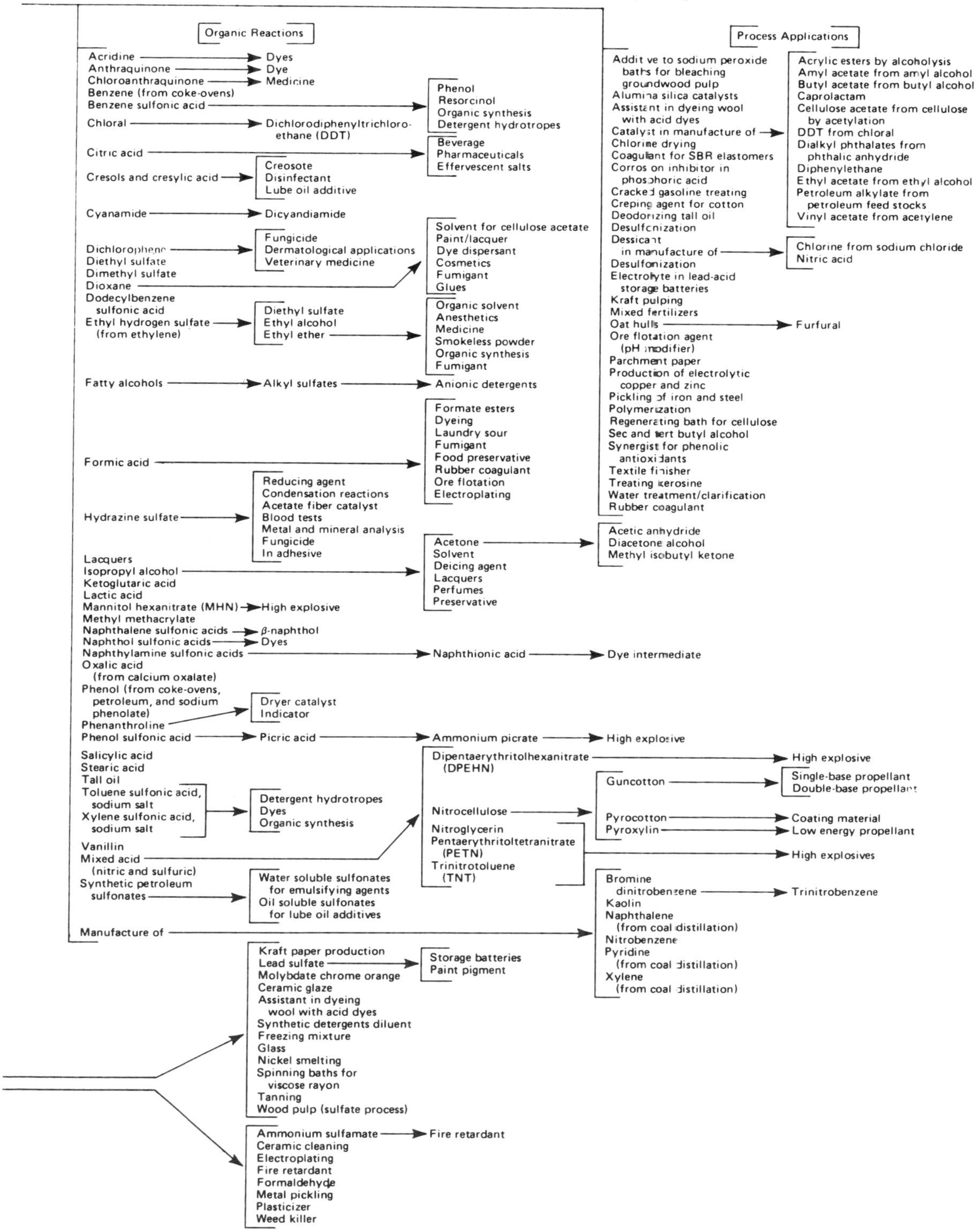

Sulphur, published by the British Sulphur Corp. Ltd., 2) the half-yearly *Statistical Supplement,* published by the British Sulphur Corp. Ltd. in conjunction with *Sulphur,* and 3) publications of The Sulphur Institute, Washington, DC.

Frasch process sulfur mining has traditionally satisfied most sulfur requirements. Other important supplies of sulfur are coproduct sulfur from natural gas and petroleum, and sulfur from sulfide minerals, such as iron pyrite. A new supply of sulfur may be the byproducts of pollution control. The amount of this new supply is difficult to quantify but may be large enough to give incentive for new uses such as road paving, and to have an impact on sulfur and sulfuric acid prices.

Marketing of sulfur reflects the well established preference of bulk consumers for deliveries in liquid form. High labor and energy costs for remelting, together with environmental aspects, continue to stimulate the preference for transport and use of liquid, yet distribution equipment is in short supply. Both producers and consumers are reluctant to make the financial commitments required for the new transport equipment and terminals required to increase shipping capacity.

US sulfur production in 1990 was about 11 560 kt, divided among Frasch 32%, recovered 57% and other sources 11%. Import was about 2 570 kt. Comparative figures for 1986 were 11 090 kt production, divided 36, 52, and 11%, respectively, with import much lower at 1 230 kt. Average prices in 1990 were about $90 for Frasch, $74 for recovered, and $90 for other, compared to 1986 figures of $125, $92, and $86, respectively.

Future pricing, as in the past, will depend on relatively small excess or shortfall in supply being reflected by exaggerated price movements.

ACKNOWLEDGMENT

The assistance of Lynn S. Jacoby of DuPont in updating the sections on chlorine and sodium is gratefully acknowledged.

BIBLIOGRAPHY AND REFERENCES

Anon., 1967, *Chemical Origins and Markets,* Stanford Research Institute, Menlo Park, CA, 99 pp.

Anon., 1970a, "The Adequacy of Acid-Grade Fluorspar Supplies Questioned," *Industrial Minerals,* No. 34, July, pp. 11–19.

Anon., 1970b, "Trends in the Usage of Fluorspar," PB 198 339, National Technical Information Service, National Materials Advisory Board, US Dept. of Commerce, 54 pp.

Anon., 1970c, et seq., *Chemical Economics Handbook,* Stanford Research Institute, Menlo Park, CA.

Anon., 1970d, *Chemical Economics Handbook,* Stanford Research Institute, Menlo Park, CA.

Anon., 1971, "Fluorspar, Cryolite and Major Fluorine Chemicals," Roskill Information Services, London, 140 pp.

Anon., 1972, "Du Pont Halon 1301 Fire Extinguishant," Organic Chemicals Dept., E.I. duPont de Nemours & Co., Wilmington, DE, 14 pp.

Farmer, M.H., and Bertrand, R.R., 1971, "Long Range Sulfur Supply and Demand Model," PB 208 993, Esso Research and Engineering Co., National Technical Information Service, US Dept. of Commerce, 69 pp.

Fulton, R.B., 1971, "World Economics of Selected Industrial Minerals," *AIME Centennial Volume,* AIME, New York, pp. 74–87.

Gould, R.F., ed., 1961, *Borax to Boranes,* Advances in Chemistry Series No. 32, American Chemical Society, Washington, DC, 244 pp.

Haynes, W., 1959, *Brimstone, the Stone that Burns,* Van Nostrand, New York, 308 pp.

Johnstone, S.J., and Johnstone, M.G., 1961, *Minerals for the Chemical and Applied Industries,* 2nd ed., John Wiley, New York, 788 pp.

Jones, G.K., 1973, "Industrial Minerals and the Chemical Industry," *Industrial Minerals,* No. 73, Oct., pp. 9–29.

Kliff, E.J., et al., 1971, *World Supply and Demand for Fluorspar and Fluorine Products,* Charles H. Kline & Co., Fairfield, NJ, 466 pp.

Lyday, P.A., 1991a, "Boron," Annual Report, 1990, US Bureau of Mines, 9 pp.

Lyday, P.A., 1991b, "Bromine," Annual Report, 1990, US Bureau of Mines, 8 pp.

Raymont, M.E.D., 1972, "Sulphur Sources and Uses—Past, Present and Future," *Canadian Mining & Metallurgical Bulletin,* Nov., pp. 49–54.

Voskuil, W.H., 1955, *Minerals in World Industry,* McGraw-Hill, New York, 324 pp.

The US Bureau of Mines annual *Minerals Yearbook* and the *Mineral Industry Surveys* are also very helpful.

CHROMITE

JOHN F. PAPP

Chromite is used both as a metallic and nonmetallic mineral. Chromite ore is used as the source of chromium by the metals industry and is used, without chemical alteration, in the refractory and foundry industries. Chromite ore is also the source of chromium for the chemical industry. On average worldwide far more chromite ore is used by the metallurgical industry than by any other industry. Thus, although the primary purpose of this chapter is to describe the nonmetallic uses of chromite, the metallurgical uses must also be discussed in order to provide a complete description of this commodity.

GEOLOGY

Mineralogy

Chromium is found in a wide variety of minerals including both oxide and silicate minerals. Only two chromium minerals are of significance: chromite and crocoite. Chromite is of commercial importance because it is the only mineral mined for chromium and for the mineral itself. Chromium metal does not occur naturally on earth. Crocoite is of historical importance because it was the mineral from which chromium was first isolated. Other minerals associated with chromite include the elements Si, Ni, Ca, and Ti. Chemistry and physical conditions determine which minerals form.

It was about 1760 when crocoite found its way to Europe from deposits in the Ural Mountains. However, it was not until 1797–1798 that chromium was isolated by Nicholas Vauquelin. The wide variety of colorful compounds derivable from crocoite led Vauquelin to name the newly discovered element chromium, a name derived from chroma, the Greek name for color. Crocoite, also called Siberian red lead, was found to produce a yellow pigment which became popular. Thus paint became the first commercial application of chromium. Chromium was soon discovered in chromite, a much more common mineral, also from the Ural Mountains.

Minerals are classified by their crystal structure and chemical composition. Chromite is an oxide mineral of spinel structure. The general chemical formula for spinels is XY_2O_4. This general structural type includes magnetite and spinel. There are four oxygen atoms (anions denoted by O), associated with cations denoted by X and Y. In the spinel structure, the elements represented by X are Fe, Mg, Mn, Ni, and Zn; Y are Al, Cr, and Fe. The Y cations are the basis of subdividing the minerals of spinel structure into the spinel series as shown in Table 1. The spinel structure series includes the spinel series (based on Al), the chromite series (based on Cr), and the magnetite series (based on Fe). The latter two series are named for the major mineral constituent of that series. Note that the term ''spinel'' is used in reference to the generic structure, to the aluminum series, and to a specific mineral (i.e., $MgAl_2O_4$). The chromite series forms a solid solution including $FeCr_2O_4$ (chromite) and $MgCr_2O_4$ (magnesiochromite). In its pure form, $FeCr_2O_4$ is 32.0% FeO and 68.0% Cr_2O_3, a value never reached in naturally occurring ores.

Spinel minerals form a close packed, face-centered cubic lattice with X ions occupying tetrahedral sites and Y ions occupying octahedral sites. Physically, chromite is black in color and is opaque except in thin sections. Its index of refraction is reported in the range of 1.2 to 2.16. According to Thayer, it has a specific gravity of 5.09 and hardness of 5.5 mhos. However, for naturally occurring ores a range of specific gravity values are reported in the literature ranging from 3.8 to 5.09 with values of about 4 to 4.5 appearing most commonly. A range of hardness values are reported ranging from 5.5 to 8 mhos. Chromite is weakly magnetic, and typical grain sizes range from 1 to 5 mm.

Origin

Chromite deposits result from the crystallization of chromite from cooling magma. The origin of chromite is discussed in general by Stowe (1987) and by Thayer (1956). The origin of many significant deposits are discussed by Stowe (1987) and by Petrascheck (1988). Two types of hard rock chromite deposits are known: stratiform and podiform. Stratiform deposits, as the name implies, are tabular, like coal seams, and occur in layered igneous intrusions. They typically have large lateral extent. These intrusions are formed in ancient continental shield areas and those that contain economic chromite deposits are all older than 1.9 billion years. Podiform deposits, named for the pod-like shape many have, may take on many shapes, from highly irregular to nearly tabular. Podiform deposits occur in geologic settings very different from those of stratiform deposits. Podiform deposits are found in ophiolite complexes that originated at spreading centers in the oceans and were subsequently thrust up onto folded mountain belts of continents or island arcs by tectonic forces. Stratiform deposits tend to be larger than podiform and easier to mine. Extensions of stratiform deposits are also easier because of their tabular nature and conformity to layering in the host rock. Podiform deposits were probably being deformed while they were crystallizing which accounts for their highly irregular shapes.

Distribution of Major Deposits

There are many attributes, other than geometry, associated with podiform and stratiform deposits. These include: method and condition of emplacement, mineral characteristics, and host rock characteristics. Podiform deposits outnumber stratiform deposits, but the podiform deposits are typically smaller. Chromite deposits also form as the result of weathering and erosion of chromite bearing rocks. These are placer deposits.

Table 1. Minerals of Spinel Structure

Series	Elemental basis	Mineral name	Chemical composition
Chromite	Chromium	Chromite	$FeCr_2O_4$
		Magnesiochromite[1]	$MgCr_2O_4$
Spinel	Aluminum	Spinel	$MgAl_2O_4$
		Hercynite	$FeAl_2O_4$
		Gahnite	$ZnAl_2O_4$
		Galaxite	$MnAl_2O_4$
Magnetite	Iron	Magnetite	$FeFe_2O_4$
		Magnesioferrite	$MgFe_2O_4$
		Franklinite	$ZnFe_2O_4$
		Jacobsite	$MnFe_2O_4$
		Trevorite	$NiFe_2O_4$

[1]Also called picrochromite.

The major stratiform chromite deposits include the Bushveld Complex (Republic of South Africa), Campo Formoso (Brazil), Great Dyke (Zimbabwe), Kemi (Finland), Orissa Complex (India), and the Tsaratanana District (Madagascar). Non-producing stratiform deposits include Bird River (Canada), Coobina (Australia), Fiskenaesset (Greenland), and the Stillwater Complex (United States). Major podiform deposits are found in Albania, Cuba, Greece, Iran, Pakistan, the Philippines, Sudan, Turkey, USSR (Kazakh Republic), and Yugoslavia. Non-producing podiform deposits occur in Cyprus and New Caledonia. Placer chromite deposits are found in Indonesia, Papua New Guinea, the Philippines, the United States, Vietnam, and Zimbabwe.

TECHNOLOGY

Exploration Techniques

Exploration has non-technologic aspects. Non-technologic aspects of exploration include general and commodity specific economic factors and politics. Exploration requires investment, which in turn, requires economic decisions. The response of chromium commodity prices to market conditions, and supply relative to current and projected demand, may indicate the need to increase supply. General economic conditions may suggest the need to increase or reduce exploration investment. Politics may encourage or discourage exploration within specific geographic areas by certain exploration teams, or may proscribe or prescribe commodities sought by these teams.

Having identified the mineral in which chromium appears as an ore, the rock types that contain chromite, and the tectonic setting and conditions under which the ore deposits formed, geologists can combine this information with knowledge of the geologic history of the earth and deposit formation processes to enhance the likelihood of further discoveries. The first step in this exploration process is deposit description, followed by cross correlation of deposit information. Deposit descriptions covering many important and less well known deposits have been published and analyzed. Government and academic geologists have correlated geologic aspects of chromite deposits, defining the general geologic conditions that are consistent with known chromite deposits.

Both stratiform and podiform deposits are associated with ultramafic rocks even though the origins of these two types of chromite deposits differ. For stratiform deposits, the regular layering can be used to locate chromite deposits concealed by faulting or segmentation. Podiform deposits cannot be reliably inferred. So far, no consistently reliable geophysical or geochemical exploration technique has been found for podiform deposits. Without chromite-specific physical indicators, the traditional methods of ore body location, outcrop analysis, trenching, and drilling, remain the most reliable way to locate chromite deposits. Drilling and drifting are used to locate or extend underground deposits. When an ore body has been located, structural analysis may be used to locate deposit extensions if they exist.

Short supply of ferrochromium in 1988–1990 resulted in increased ferrochromium prices followed by ferrochromium production capacity expansions. Mine production capacity was increased to meet added smelter capacity. Some South African mines produce chemical, refractory, and/or foundry sand grade chromite in addition to metallurgical grade chromite, so these expansions also increased availability of certain nonmetallurgical grades of chromite. The large reserve base of chromite in the Bushveld Complex is adequate to meet world demand for a long time. Additional resources from the Bushveld Complex can easily be upgraded from inferred to demonstrated or from indicated to measured by exploration.

Mining, Beneficiation, and Processing

A wide variety of techniques are applied to the recovery of chromite. Recovery includes surface and underground mining using unmechanized to mechanized methods. Beneficiation to marketable chromite products varies from hand sorting to gravimetric and electromagnetic separation methods. Processing of chromite to produce chromium products for the refractory, chemical, and metallurgical markets includes crushing and grinding and size sorting by pneumatic and hydraulic methods, kiln roasting, and electric furnace smelting (see Table 2 and Fig. 2). Labeling of material as it moves from the earth to the consumer is not uniform. The terms chromite or chromite ore are used here to refer to material in the ground, run-of-mine ore (i.e. material removed from the ground), or material supplied to the marketplace. For the purpose of trade, imports are called chromite ore and concentrate made therefrom. This description is frequently abbreviated to chromite ore and concentrate, chromite ore, or simply chromite. Some sources use chromite ore to refer to material in the ground and material removed from the ground before processing. The term chromite products is then used to refer to material supplied to the marketplace. Historically, mining operations supplied minimally processed material. Beneficiation and processing shown in Fig. 2 may be carried out at the mine site, at a plant which serves several mines in one geographic area, or at a plant associated with end users. This variety in processing further complicates labeling of material. Today, quality control leads consumers to seek chromite supplies that do not vary significantly in physical or chemical properties over time. As a result, chromite ore is typically beneficiated to produce a physically and chemically uniform product before it reaches the marketplace.

Mining methods are carefully chosen to meet the characteristics of a deposit, including the ore and its environment. Since both small and large, podiform and stratiform, high grade and low grade, subsurface and near surface, massive and disseminated chromite deposits are exploited, a variety of mining methods are used. Since, typically, surface mining is less expensive than underground mining and ore bodies are found by their outcrops, surface mining at an outcrop precedes underground mining.

The amount and kind of beneficiation required depends on the ore deposit characteristics and mining technique. An operation where miners can extract only high grade ore and avoid host rock (a massive deposit) may require only hand sorting and screening. Beneficiation is necessitated by geologic conditions that result in intermixing of chromite with other minerals or the use of mechanized mining methods that are nonselective.

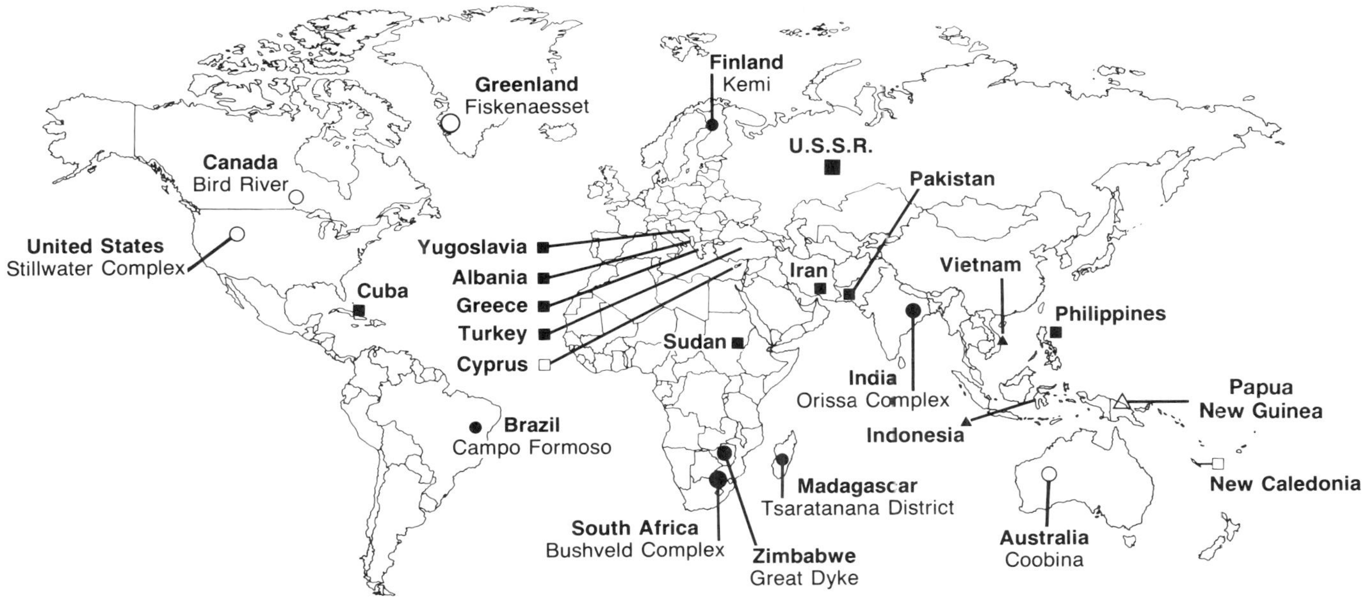

Source: Bureau of Mines and US Geological Survey.

Explanation: Geologic Deposit Type

Stratiform			Podiform			Laterite		
Symbol		Resources	Symbol		Resources	Symbol		Resources
Nonproducing	Producing	(tons)	Nonproducing	Producing	(tons)	Nonproducing	Producing	(tons)
○	●	> 10^9	None	None	>10^9	△	None	>10^9
○	●	10^6-10^9	□	■	10^6-10^9	△	▲	10^6-10^9
○	●	< 10^6	□	■	< 10^6	△	▲	< 10^6

Fig. 1. Location, geologic type, and size of major chromite deposits.

Table 2. Factors Used to Characterize Chromite Deposits and Mining, Beneficiation, and Processing Methods Applied to Chromite Deposits

Mineral	Mineral appearance in ore	Deposit types	Host rock	Mining method	Beneficiation method	Processing method
Chromite (a spinel structure oxide mineral)	Massive (spinel mineral of varying chemistry) Disseminated (spinel mineral intermixed with other minerals)	Stratiform (bedded) Podiform (lenticular)	Mafic Ultramafic	Surface Subsurface	Hand sorting Screening Gravity concentration Electromagnetic separation Heavy media separation	Sizing Roasting and dissolution Smelting

The purpose of beneficiation is to increase desirable ore attributes and decrease undesirable ones. For example, depending on end use, increasing chromic oxide content, chromium-to-iron ratio, or alumina content is desirable. Reducing silica or other host rock associated with chromite is desirable. Depending on end use, certain sizes may be selected or rejected. The techniques used to accomplish these tasks depend on the physical properties and sizes of the minerals present. Beneficiation cannot change the chemical characteristics of the chromite mineral. However, since chromite ore is a mixture of minerals, the characteristics of the ore can be changed by altering its mineral mix. A deposit producing lumpy ore in which the chromite is easily distinguished may require only hand sorting and screening. When the chromite cannot easily be distinguished visually from associated minerals and the ore is lumpy, then heavy media separation can be used. A deposit that yields an ore of chromite thoroughly intermixed with other minerals, however, may

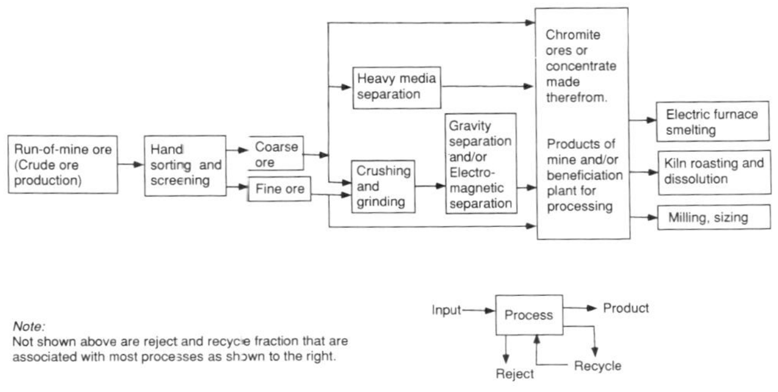

Fig. 2. A simplified composite representation of chromite mining, beneficiation, and processing.

require milling and sizing followed by gravimetric and/or electromagnetic separation methods to produce marketable chromite products.

Beneficiation may also be selected to process tailings once sufficient quantities have been stockpiled and the technology of beneficiation and processing has been established (see Byproducts and Coproducts section).

Specifications

Domestic and foreign companies supply imported chromite ore to US consumers. The chemical and physical specification of chromite ore varies among consumers and producers. Typically, consumers do not reveal detailed ore specification. However, chromite producers do make typical specifications available to prospective customers. Typical chemical and physical specifications for a variety of chromite ores available to US consumers have been assembled here. The chemical specifications of several chromite ores are shown in Table 3 by end use. The physical specifications associated with those ores are shown in Table 4. The suppliers of the chromite ore listed in Tables 3 and 4 are shown in Table 5.

Byproducts and Coproducts

Chromite ore is not a coproduct or byproduct of any other mining operation, nor are there any coproducts or byproducts associated with chromite mining operations. Here, byproduct or coproduct is assumed to mean a mineral product that is different from the primary product and not different grades of the primary mineral product. A single mining operation is likely to produce more than one grade of its product. Grades of chromite products are distinguished by ore size and chemistry.

It is possible that chromite may become a byproduct of platinum mining in South Africa. Mintek, South Africa's mining research organization, has demonstrated smelting of beneficiated chromite-containing waste material from certain platinum mines. Platinum has been mined from the Merensky Reef, a chromite-free seam of the Bushveld Complex. As those platinum mines deplete their reserves, platinum mining is expected to move to the chromite-containing UG-2 seam, generating more chromite-containing tailings. UG-2 seam platinum mining has started. Mintek has also demonstrated the feasibility of recovering platinum from tailings resulting from chromite mining of the LG-6 chromitite layer. It remains for the economics of these processes to be favorably demonstrated and commercially initiated.

USES

Chromium and Chromite Uses

Chromite is used in the metallurgical, chemical, and refractory industries. In the metallurgical industry, chromite is processed into ferrochromium or chromium metal, then it is used as an alloying metal to make a variety of ferrous and nonferrous alloys. The major end use is in stainless steel, a ferrous alloy made resistant to oxidation and corrosion by the addition of chromium. Chromite is used in the chemical industry to make sodium dichromate which is both a chemical industry product and an intermediate product used to make other chromium chemicals. Chromium chemicals find a wide variety of end uses including pigments, and plating and surface finishing chemicals. Chromite is used in the refractory industry to produce refractory materials including shapes, plastics, and foundry sands. These refractory materials are then used in the production of ferrous and nonferrous alloys, glass, and cement. Chromite is useful in the refractory industry because it retains its physical properties at high temperatures and is chemically inert.

Based on the chromium contained in ore, chromium ferroalloy, metal, and chemicals, it is estimated that US chromium demand averages about 410 000 tpy (contained chromium). On average, about 70% of US chromium consumption goes into the metallurgical industry and about 70% of this goes into stainless steel. The remaining chromium consumed in the United States goes into the chemical industry (about 20%) and the refractory industry (about 10%). The allocation of chromium consumption in other market economies is similar to that in the United States. The material flow associated with the use of chromite is shown in Fig. 3. This figure demonstrates the interrelationship between industrial and metallurgical uses of chromium materials. (Chromium consumption differs from chromite consumption because chromium consumption includes chromite, alloys, and chemical while chromite consumption includes only chromite ore.)

Chromite Consumption

Reported chromite consumption in the United States averaged (from 1986 through 1990) about 481 ktpy, a decline from the 1940–1990 average of about 938 ktpy. Virtually all of this chromite was imported. The chromite was used to make chromium ferroalloys and chemicals, and chromite refractory materials including casting sand. The major reason for declining domestic chromite use is the shift from domestic to foreign ferrochromium supply as the source of chromium units for the metallurgical industry. US chromite consumption and trade is detailed in Table 6 and shown graphically in Fig. 4, quantitatively, and Fig. 5, proportionately.

Fig. 4 shows an increase in US chromite use from 1940 to the mid-1950s, and declining use since then. The largest changes have been in the metallurgical industry. Chemical industry use increased slowly from 1940 to 1985. Refractory industry use appears to have reached a plateau from about 1950 to 1965, after which it slowly declined. On average from 1940 through 1990, reported consumption of chromite ore was 99.9% of net trade. (Net trade is imports for consumption minus exports.) Fig. 4 shows that metallurgical industry chromite use increased its market share from 1940 through about 1980, when its share dropped quickly. As a result of the drop in metallurgical use, chemical use took a larger share of the US market. Declining chromite use in the metallurgical industry resulted from declining domestic ferrochromium production. Over the time period in which the metallurgical, chemical, and refractory industry data were reported separately (i.e., 1940–1985), the metallurgical industry accounted for 56% of reported chromite consumption, refractory industry, 27%, and chemical industry, 18%.

Table 3. Chemical Specifications for Chemical, Foundry, and Refractory Grade Chromite Ore [Representative % Content of Specified Chemicals and Chromium-to-iron Ratio (Cr:Fe)]

Source	Cr_2O_3	FeO	Fe	SiO_2	Al_2O_3	Cr_2O_3+ Al_2O_3	MgO	TiO_2	V_2O_5	CaO	Ni	P	S	Cr:Fe
						Chemical								
Brazil:														
Vila Nova	≥49	—	17–19	≤3	11–13	—	5–7.5	—	—	0.1	—	—	—	≥1.7
South Africa:														
Elandsdrift	46.3	25.4	—	1.2	15.4	—	9.8	0.55	0.36	0.33	0.08	0.005	0.000	—
Hernic	46.0	—	26.0	1.0	15.20	—	8.70	0.54	0.43	0.26	—	0.002	0.003	1.55
Lebowa Development Corp:														
Dilokong	46.92	26.14	—	0.85	14.59	—	9.80	0.55	—	0.21	—	—	—	1.58
Rand Mines:														
Henry Gould	46.00	26.50	—	1.00	15.50	—	9.80	0.50	0.40	0.25	—	0.003	0.002	1.53
Winterveld	45.00	24.70	—	2.50	15.80	—	11.40	0.40	0.30	0.20	—	0.005	0.002	1.60
Samancor:														
Tweefontein[1]	46.80	27.03	—	0.50	15.17	—	9.23	—	—	0.16	—	—	—	1.52
						Foundry								
South Africa:														
Elandsdrift	45.9	25.4	—	1.3	15.4	—	9.8	0.55	0.36	0.33	0.08	0.005	0.000	—
Rand Mines:														
Henry Gould	46.20	26.50	—	1.00	15.50	—	9.80	0.50	0.40	0.25	—	0.003	0.002	—
Millsell	46.00	25.20	—	1.00	14.70	—	10.20	0.60	0.40	0.20	—	0.003	0.002	—
Winterveld 1	46.60	25.00	—	0.80	15.20	—	11.20	0.50	0.30	0.23	—	0.003	0.002	—
Winterveld 2	46.30	25.00	—	0.90	15.20	—	11.20	0.50	0.30	0.23	—	0.003	0.002	—
Samancor[2]	47.05	25.77	—	0.70	14.84	—	10.75	0.69	—	0.05	—	—	—	1.60
						Refractory								
South Africa:														
Anglo American:														
Marico	48.3	20.7	—	0.94	17.2	—	12.4	0.39	—	0.01	—	—	—	2.05
Rand Mines:														
Winterveld 1	46.60	24.90	—	0.70	15.40	—	11.50	0.50	0.30	0.25	—	0.003	0.002	—
Winterveld 2	46.50	24.50	—	0.75	15.60	—	11.50	0.50	0.30	0.25	—	0.003	0.002	—
Winterveld 3	46.94	24.50	—	0.30	15.60	—	11.50	0.50	0.30	0.25	—	0.003	0.002	—
Samancor[3]	47.07	25.77	—	0.56	14.84	—	10.75	0.69	—	0.05	—	—	—	1.60
Philippines:														
Benguet Corp.:														
Masinloc 1	31.55	—	11.34	5.84	—	58.79	—	—	—	—	—	—	—	—
Masinloc 2	32.18	—	10.83	5.42	—	60.02	—	—	—	—	—	—	—	—
Masinloc 3	33.88	—	10.79	3.30	—	62.21	—	—	—	—	—	—	—	—
Masinloc 4	35.73	—	12.45	1.69	—	63.50	—	—	—	—	—	—	—	—

≥ Greater than or equal to. ≤ Less than or equal to. Numbers separated by "-" indicate a range of values.
[1] Samancor also produces chemical grade chromite ore at Mooinooi Mine.
[2] Samancor produces foundry grade chromite ore at Waterkloof and Montrose Mine.
[3] Samancor produces refractory grade chromite ore at Groothoek and Jagdlust Mine.

More recently (from 1976–1985) metallurgical industry chromite use has decreased to 51%, while refractory use declined to 18% and chemical use increased to 31%. Over the most recent five year period (1986–1990), refractory use has declined further to 11% of chromite use, while chemical and metallurgical use was 89% of reported consumption.

Metallurgical

Chromite ore consumption by the metallurgical industry is shown in Table 6 and in Fig. 4 and metallurgical industry chromite market share is shown in Fig. 5. Fig. 4 shows that the metallurgical industry dominated domestic chromite consumption from 1940 to 1980. The patterns of total chromite consumption and net imports are similar to those for metallurgical consumption over that time period. More recently, the number of metallurgical chromite consumers has fallen below the minimum three companies required by the Bureau of Mines to report collected data. As a result, in 1986, metallurgical and chemical industry chromite consumers were combined into a single group for the purpose of reporting chromite consumption. The dramatic drop in metallurgical consumption of chromite ore in 1980–1983 resulted in chemical industry consumption surpassing that of the metallurgical industry for the first time since 1940. The decline of metallurgical market share is shown in Fig. 5.

Chemical

Chromium is used in the chemical industry to make a variety of products including biocides, catalysts, corrosion inhibitors, metal plating and finishing chemicals, oil field chemicals, pigments, printing chemicals, and tanning compounds. Chromium was first used in pigments and tanning compounds. Chromium plating, the electrodeposition of chromium from a solution of chromic acid, started in the early 1900s. A more recent use for chromium is in wood preservation. Chromium - copper - arsenate (CCA) impregnated

Table 4. Size Specifications for Chemical, Foundry, and Refractory Grade Chromite Ore (Representative % that is of Specified Size)

Metric (mm):[1]	4.75	3.35	2.36	2.00	1.70	1.18	.850	.710	.600	.500	.425	.300	.212	.180	.150	.106	.075	.053	.045
US Standard:	4	6	8	10	12	16	20	25	30	35	40	50	70	80	100	140	200	270	325
Tyler Sieve:	4	6	8	9	10	14	20	24	28	32	35	48	65	80	100	150	200	270	325
Source																			
CHEMICAL																			
Brazil:																			
Vila Nova	(maximum of 10% > 0.50 mm; minimum of 80%, 0.50 to 0.125 mm; maximum of 10% < 0.125 mm)																		
South Africa:																			
Elandsdrift	—	—	—	—	—	—	—	—	—	—	1.6	18.8	30.0	26.1	17.1	4.6	1.4	0.4	—
Hernic[2]	—	—	—	—	—	0.02	0.06	0.12	0.124	—	2.29	10.78	20.36	—	30.96	17.32	13.80	3.39	—
Lebowa Development Corp:																			
Dilokong	(14.5% between 0.25 and 0.50 mm; 85% less than 0.25 mm)																		
Rand Mines:																			
Henry Gould	—	—	—	—	—	—	1.2	—	7.3	—	25.4	33.9	23.5	6.1	1.8	0.5	0.3	—	—
Winterveld	—	—	—	—	—	9.7	11.3	—	16.1	—	17.4	16.0	11.9	—	6.8	4.7	2.8	1.5	1.8
Samancor:																			
Tweefontein[3]	—	0.0	—	—	0.0	—	0.3	—	4.0	—	20.3	27.5	24.4	—	14.2	7.3	2.0	0.2	—
FOUNDRY																			
South Africa:																			
Elandsdrift	(Not available)																		
Rand Mines:																			
Henry Gould	—	—	—	—	—	0.4	2.1	—	11.4	—	29.6	31.4	—	17.6	5.5	1.6	0.3	0.1	—
Millsell	—	—	—	—	—	0.1	1.0	—	11.0	—	34.1	34.0	—	14.4	4.2	0.7	0.3	0.1	—
Winterveld 1	—	—	—	—	—	0.1	0.4	—	1.9	—	8.5	32.5	—	34.6	15.6	5.1	1.1	0.2	—
Winterveld 2	—	—	—	—	—	—	0.1	—	0.8	—	4.7	14.8	—	24.0	26.6	18.6	8.4	1.6	0.4
Samancor[4]	—	—	—	0.1	—	—	0.1	—	2.2	—	15.4	33.7	27.1	—	12.8	6.0	1.6	0.9	—
REFRACTORY																			
South Africa:																			
Anglo American:																			
Marico chrome	(2.0mm > 0.2% > 1.4mm, 17.1% > 1.0mm, 48.2% > 0.500mm, 29.5% > 0.250mm, 4.8% > 0.125mm, 0.2% < 0.125mm)																		
Rand Mines:																			
Winterveld 1	—	—	—	—	—	6.2	14.8	13.3	13.4	—	29.2	16.3	5.0	1.1	0.3	0.3	0.1	—	—
Winterveld 2[5]	—	—	0.0	—	0.1	1.3	4.8	8.6	8.3	—	31.6	24.8	12.7	—	5.2	1.9	0.5	0.1	—
Winterveld 3[6]	—	—	1.2	—	14.6	30.0	21.7	19.3	4.8	—	5.8	1.3	0.5	—	0.3	0.2	0.1	0.1	—
Samancor	—	—	—	—	—	—	8.20	—	21.10	18.90	15.30	25.20	9.40	—	1.60	0.30	0.10	—	—
Philippines:																			
Benguet Corp.:																			
Masinloc 1[7]	26.5	19.93	15.93	—	6.7	3.71	—	—	—	—	—	5.50	0.93	—	—	—	—	—	—
Masinloc 2[8]	26.2	22.81	9.19	—	12.37	2.18	—	—	—	—	—	2.09	0.67	—	—	—	—	—	—
Masinloc 3[9]	—	—	1.3	—	7.1	7.8	12.3	—	13.3	—	5.8	23.5	12.9	6.3	3.4	4.3	1.3	—	—
Masinloc 4[10]	—	—	—	—	—	—	—	—	—	—	—	2.75	5.13	—	18.50	38.10	24.69	8.68	1.12

mm Millimeter. % Percent. < Less than. > Greater than.

[1] Metric equivalent to US Standard from ASTM standard E11–87. Tyler equivalent to US Standard from *Chemical Engineering Handbook, 5th Edition*. ASTM Sieve designation is the same as US Standard Series sieve number.

[2] 0.67% < 0.053 mm.

[3] 0.1% < 0.053 mm.

[4] 0.1% < 0.053 mm.

[5] 0.1% < 0.053 mm.

[6] 0.1% < 0.053 mm.

[7] 17.57% > 4.75 mm and 3.25% < 0.212 mm.

[8] 22.86% > 4.75 mm and 1.63% < 0.212 mm.

[9] 0.7% < 0.075 mm.

[10] 1.03% < 0.044 mm.

wood can be protected from weathering, insects, and rotting for 30 years.

The terms chromium and chrome are used in the chemical industry and mean the same, the element chromium. Similarly, the terms bichromate and dichromate are used interchangeably in the chemical industry. Historically, the term chrome was used more commonly than chromium. Thus many chemical products, namely pigments, have chrome in their name. To be consistent with the chemists' convention for naming compounds, dichromate would be used instead of bichromate. However, in trade and commerce, the term bichromate is used.

Chromite is used in the chemical industry to produce sodium dichromate from which other chromium chemicals are manufactured. Chromite ore is pulverized and mixed with soda ash (sodium carbonate) and a diluent. The diluent could be lime (calcium oxide) or recycled material from the production processes. The mixture is

Table 5. Selected Suppliers of Chromite to the US Market

Company	Source	Cre		
		Grade		
		Chemical	Foundry	Refractory
Caemi International, Inc. Stamford, CT 06902–6732	Brazil	X		
F & S Alloys and Minerals Corp. Pittsburgh, PA 15276	South Africa	X	X	X
Larson Sales Company Pittsburgh, PA 15275	South Africa	X		
Lebowa Development Corp. Ltd Pietersburg 0700, Republic of South Africa	South Africa	X		
M & M Ferrous America, Ltd. New York, NY 10017	Philippines			X
Minerais U.S. Inc. Exalmet Division Bridgewater, NJ 08807	South Africa			X
North American Minerals Corp. Ann Arbor, MI 48108	South Africa	X	X	
Samancor Chrome Limited Johannesburg 2000, Republic of South Africa	South Africa	X	X	X

roasted in a rotary kiln to produce a compound containing sodium chromate which is leached out and treated with acid to produce sodium dichromate and then purified. Sodium chromate and dichromate are produced by this process in the United States. The major commercial product is sodium dichromate dihydrate. Table 7 lists several primary chromium chemicals produced by US chemical companies that produce chromium chemicals from chromite. Many other chromium chemicals are manufactured from sodium dichromate. The primary chromium chemicals and their end uses are shown in Fig. 6.

A chromium chemical end use with which many people are familiar is pigments. Medium yellow pigment, a lead-chromate base pigment, is used in traffic marking yellow paint found on all major streets and highways. Copper-chrome pigment is used in the black coating found on outdoor grills and wood-burning stoves. An important use of chromium pigments is in anticorrosion coatings. The military and the federal government are major users of chromate metal primers. Chromium oxide green pigment is used in camouflage because it has desirable infrared reflectance properties.

Chromite ore consumption by the chemical industry is shown in Table 6 and in Fig. 4 and chemical industry chromite market share is shown in Fig. 5. Fig. 4 shows that chemical industry chromite consumption has been increasing slowly since 1940. Annual consumption at the end of the 1940–1985 time period was nearly double that of the beginning of the period yielding an average annual growth rate of about 1.4% over the period. The radical increase of chemical industry market share in 1982 shown in Fig. 5 resulted from the decrease in metallurgical industry chromite consumption and not from a major increase in consumption by the chemical industry.

Russia has the world's largest chromium chemicals production capacity followed closely by the United States and the United Kingdom. The United States is a major consumer, producer, and world supplier of chromium chemicals. Table 8 shows world chromium chemical production capacity by country. World production capacity in 1992 was about 329 kt, contained chromium, or about 944 kt sodium dichromate dihydrate equivalent. Many small chromium chemical plants are located in chromite producing countries. However, the major chromium chemical producing countries include Germany, Japan, the former USSR, the UK, and the US. Of these, only the USSR is a significant chromite ore producer with about one-third of world chromite production and a significant chromium chemical producer with the world's estimated second largest production capacity. Japan produces chromite ore, but the quantities are insignificant compared to its industrial demand. Therefore, all of the major chromium chemical producers, except the USSR, are located in industrialized countries that do not produce chromite. Table 9 shows recent US chromium chemical trade. The table shows that, on average, over the five year period, US chromium chemical exports have exceeded imports by about 41% in terms of weight and by about 7% in dollar value. Sodium dichromate and chromic acid accounted for the largest volume of exports and of imports.

Fig. 7 shows production of major domestically produced chromium chemicals: sodium dichromate, chromic acid, and chromium pigments. Sodium dichromate production was reported until 1985, when production was about 120 kt. Chromium pigment production has declined, apparently starting in 1977, from about 50 ktpy to the current level of about 15 ktpy. Chromic acid production was reported until 1973, when production peaked at about 27 kt and reporting was discontinued. The sodium dichromate and chromic acid production data shown in Fig. 7 are reported, whereas the chromium pigments production shown is a composite of the reported production of several chromium pigments. The production of those chromium pigments, including chrome yellow and orange, chrome molybdate orange, chrome oxide green, zinc yellow, and chrome green, are shown in Fig. 8. Chromium pigment production is by far the greatest for chrome yellow and orange followed by chrome molybdate orange. In both cases, production appears to have peaked in the 1973–1977 time period and declined since. Chrome yellow and orange annual production has declined from about 34 000 t to the current level of about 15 000 t, while chrome molybdate orange has declined from about 12 000 t in the 1974–1977 time period to about 6 000 t in 1987 when the production series was discontinued. Zinc yellow annual production declined from 7 000 t in the 1965–1967 time period to about 1 000 t in 1987 when the series was discontinued. The zinc yellow production curve was generated using straight line interpolation over intervening years when production data was withheld. Chrome oxide green annual production appears to have remained between about 4 000 and

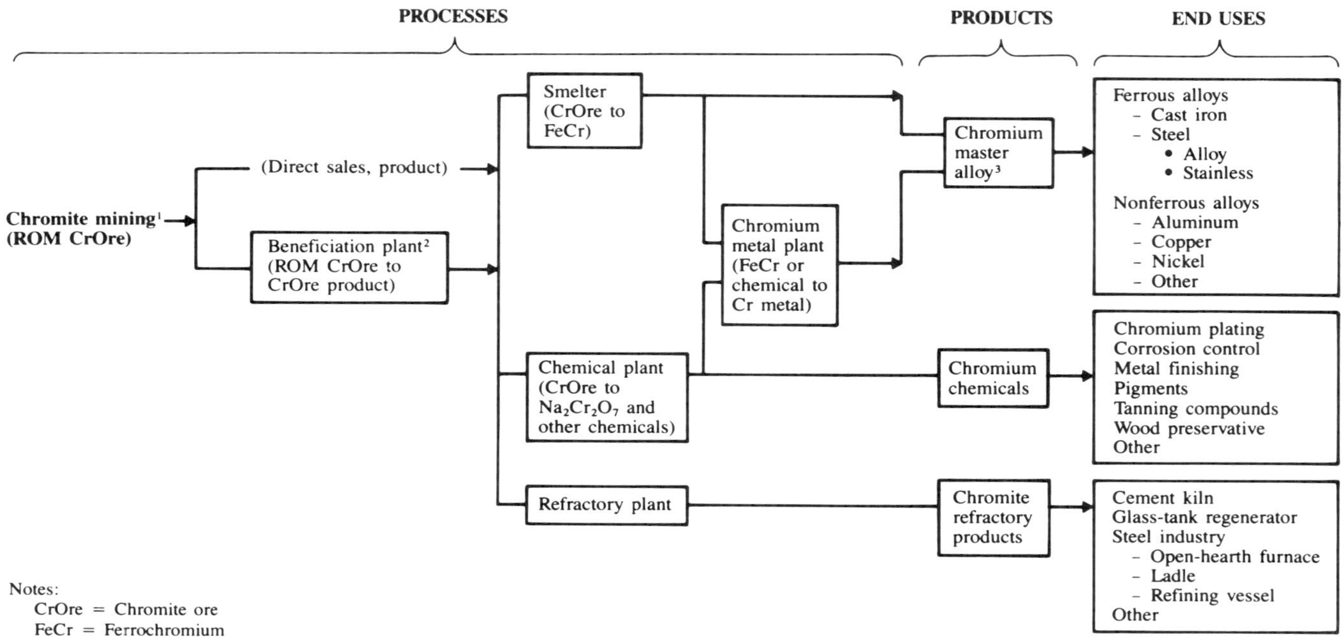

Notes:
CrOre = Chromite ore
FeCr = Ferrochromium
$Na_2Cr_2O_7$ = Sodium bichromate
ROM = Run-of-Mine

[1]Mining includes screening and hand sorting.
[2]Beneficiation includes crushing, grinding, and separation techniques including gravimetric, heavy media, magnetic, and spiral.
[3]Master alloy = an alloy used as a feed stock to produce other alloys.

Fig. 3. Chromium material flow from chromite mining to end use.

Table 6. Reported US Consumption and Stocks of Chromite by End Use and Trade, and Chromite Imports, Exports, and Net Trade (t)

	1981	1982	1983	1984	1985	1986	1987	1988	1989	1990
Reported consumption:										
Chemical	216 332	176 842	166 569	171 422	201 659	([1])	([1])	([1])	([1])	([1])
Metallurgical	456 360	257 170	58 341	204 776	247 557	342 281	458 536	494 713	516 931	358 620
Refractory	134 130	72 357	65 363	88 422	59 189	45 303	45 737	88 570	43 780	43 670
Total	806 822	506 369	290 273	464 620	508 406	387 584	504 272	550 636	560 711	402 290
Reported stocks:										
Chemical	370 463	312 808	239 420	233 260	207 191	([1])	([1])	([1])	([1])	([1])
Metallurgical	229 800	119 540	140 324	24 442	44 361	249 291	308 870	366 268	368 482	333 224
Refractory	128 210	113 233	75 832	69 619	48 635	35 379	23 144	23 353	23 791	21 437
Total	728 473	545 581	455 576	327 321	300 187	284 670	332 014	387 621	392 273	354 661
Trade:										
Imports	814 652	459 943	172 748	276 310	375 911	442 890	489 901	615 332	525 044	305 507
Exports	64 113	7 407	10 008	49 830	91 453	83 555	1 145	3 931	40 445	6 321
Net imports[2]	750 539	452 536	162 740	226 480	284 458	359 331	488 756	611 401	484 599	299 186

[1]Included with metallurgical.
[2]Net imports is imports (for consumption) minus exports.

8 kt until reporting was discontinued in 1985. Chrome green annual production varied between about 2 500 and 4 000 t until reporting was discontinued in 1972.

The number of producers of selected chromium chemicals (sodium dichromate and chromate, chrome yellow and orange, and chrome molybdate orange) are shown in Fig. 9. Figs. 8 and 9 show that not only has production of chromium pigments declined, the number of producers has also declined. The number of sodium dichromate and chromate producers has also declined. However, production has not declined, suggesting that production has consolidated with the remaining two producers. Reported sodium dichromate and chromate production was more erratic than reported chemical industry chromite ore consumption. Nevertheless, the general trend of reported production appears to be consistent with that of chromite ore consumption. Since chemical industry chromite ore consumption and sodium dichromate and chromate production

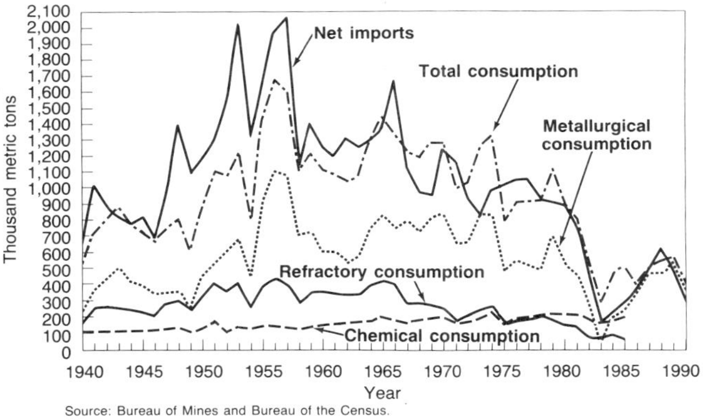

Fig. 4. US chromite consumption by end use and net imports.

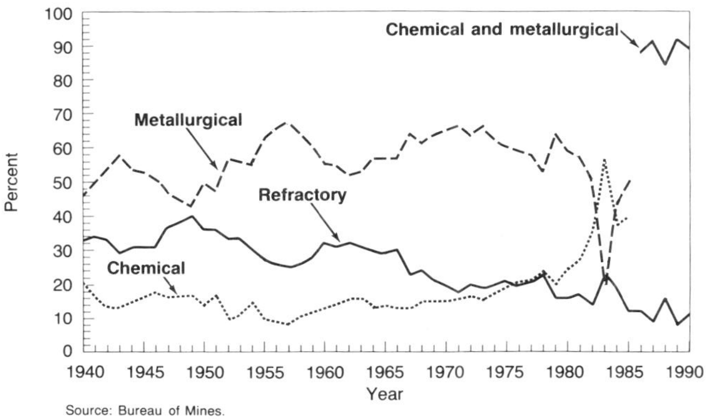

Fig. 5. US chromite percent consumption by end use.

have been increasing and reported production of specific chromium chemicals produced from sodium dichromate and chromate has been declining, exports of sodium dichromate and chromate or the production of unreported chromium chemicals, or both, have been increasing.

Refractory

Refractory materials resist degradation when exposed to heat. Chromite is a refractory material. Unlike the chemical and metallurgical industries, where chromite is processed to extract its chromium content, chromite is used chemically unmodified in the refractory industry. Refractories are broadly categorized according to their material composition into clay and nonclay refractories. The predominant nonclay refractory material is silica. Nonclay refractories materials also include alumina (bauxite), carbon (graphite), chromite, dolomite, forsterite, magnesia (magnesite), mullite, pyrophilite, silicon carbide, and zirconia (zircon). Basic refractories are a type of nonclay refractory, so called because they behave chemically as bases. Basic refractories are made of chromite, dolomite, magnesite, or various combinations of magnesite and chromite. In the refractory industry, chromite-containing refractories are called chrome refractories. Chrome-magnesite refractories are those in which more chromite than magnesite is used. Magnesite-chrome refractories are those in which more magnesite than chromite is used.

Refractories are further categorized by the form in which they are supplied as shaped or unshaped. Shaped refractories are manufactured to fit together to form a desired geometric structure, like building blocks. Unshaped refractories include mortars (materials used to hold shaped refractories together), plastics (materials that may be formed into whatever shape is desired), and gunning (material that may be sprayed onto a surface). In the refractory industry, the term monolithics is commonly used to describe refractories that are not shaped. The units used to report shipments of shaped and unshaped refractories differ. Shaped refractories have been reported in thousand "9-inch brick" equivalents; unshaped refractories, in tons. A 9-inch brick equivalent is a solid volume of 0.165919 m^3 and used in the United States as a "standard unit" for refractory bricks.

Chromite-containing refractory producers are shown in Table 10 along with refractory industry products and the end users and uses of those products. The major end users for chromite refractories are in the cement, copper, glass, nickel, and steel industries. Basic refractories are used in copper and nickel furnaces. In the glass industry, chromite refractories are used in glass tank regenerators. In the cement industry, chromite refractories are used primarily in the transition zones of cement kilns. Basic refractories are typically used in open hearth and electric arc steel-making furnaces.

Chromite refractories were used heavily in steel production using the open hearth furnace method. Contemporary steel-making processes which use the basic oxygen furnace or the electric arc furnace use much less chromite-containing refractories. Whereas open hearth furnaces used about 30 kg of refractories per ton of steel, the basic oxygen furnace uses about 1 kg/t and the electric arc furnace uses in the range of from 1 to 2 kg/t. As a result, the steel industry demand for basic refractories has declined dramatically as open hearth furnace steel making has been phased out.

The historical trends of chromite-containing refractory shipments are shown in Figs. 10 through 12. These figures show shipments of chromite-containing refractories since 1960. Basic refractory shipments trends shown in these figures indicate trends for chromite refractory shipments as well. Since the chrome and chrome-magnesite refractories are predominantly chromite, at least half of their content represents chromite consumed in refractories. The American Society for Testing and Materials sets specifications for the identification of chrome, chrome-magnesite, and magnesite-chrome brick. Chrome brick is identified as a refractory brick manufactured substantially or entirely of chrome ore. Chrome-magnesite and magnesite-chrome brick are classified by nominal and minimum magnesia (MgO) content. Nominal MgO content ranges from 30 to 80%. So the chromite content of magnesite-chrome refractories is over 20%. However, the distribution between magnesite and magnesite-chrome within the magnesite and magnesite-chrome category is unknown.

Basic unshaped refractories may include chromite, dolomite, forsterite, magnesite, or zircon. Thus, these shipments data should be viewed as indicative of the performance of a segment of the refractory industry of which chromite is a part.

Fig. 10 shows shipments of chromite-containing shaped refractories. The figure clearly shows a downward trend in chrome and chrome-magnesite refractory shipments from 1960 through 1981 when the data series was discontinued. Over the same time period the number of chrome and chrome-magnesite refractory producers fell from nine to seven. Since 1981, the number of chromite refractory producers has fallen from seven to four, a clear trend of declining production and use. Magnesite and magnesite-chrome refractories showed growth from 1960 through 1965, after which large variations in shipments occurred. The magnesite-chrome refractories shipments trend from 1982 through 1990 declined from 28 to 18 million "9-inch brick" equivalents. The chrome and chrome-magnesite curves are based on reported shipments. The magnesite and magnesite-chrome and the magnesite-

Table 7. Primary Chromium Chemical Producers and Chromium Chemical Products

Producer	Product: Formula	Product: Name
American Chrome & Chemicals Inc.	$Na_2Cr_2O_7 \cdot 2H_2O$	Sodium bichromate dihydrate
Buddy Lawrence Drive	$Na_2Cr_2O_7$	Sodium bichromate anhydrous
P.O. Box 9912	$Na_2CrO_4 \cdot 4H_2O$	Sodium chromate tetrahydrate
Corpus Christi, TX 78469	Na_2CrO_4	Sodium chromate anhydrous
	Cr_2O_3	Chromic oxide
	CrO_3	Chromic acid
Occidental Chemical Corporation	$Na_2Cr_2O_7 \cdot 2H_2O$	Sodium bichromate dihydrate
Occidental Tower, 5005 LBJ Freeway	Na_2CrO_4	Sodium chromate anhydrous
P.O. Box 809050	$Na_2CrO_4 \cdot 4H_2O$	Sodium chromate tetrahydrate
Dallas, TX 75380–9050	CrO_3	Chromic acid
	$K_2Cr_2O_7$	Potassium bichromate

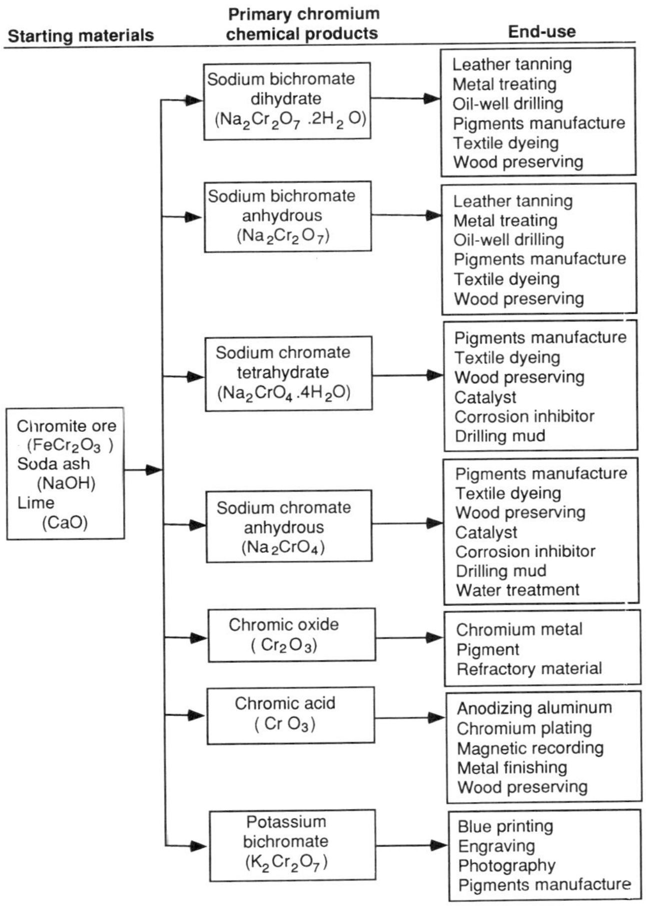

Fig. 6. Primary chromium chemicals and their end uses.

chrome curves are composites of reported shipments data (see following detailed discussion and Fig. 12).

The trend in shipments of basic unshaped refractories is shown in Fig. 11. Mortar shipments show a rapid decline from 1964 to 1970, followed by a very slow decline until 1986 when the series was discontinued. Plastic shipments show a slow decline as does gunning. The data are not sufficiently discriminating to show trends among different materials used to make basic unshaped refractories.

Table 8. World Chromium Chemical Production Capacity (ktpy, Contained Chromium)

Country	December 1992 Capacity
Argentina	6
Brazil	12
China	16
Germany	20
India	5
Iran	2
Japan	21
Kazakhstan	42
Macedonia	5
Mexico	5
Pakistan	3
Poland	7
Romania	5
Russia	63
Turkey	10
United Kingdom	52
United States	55
Total	329

Note: To get sodium dichromate dihydrate equivalent, multiply capacity in contained chromium by 2.9.

The components of the composite magnesite and magnesite-chrome (1971–1981) and the magnesite-chrome (1982–1990) refractory shipments curves are shown in Fig. 12. Magnesite and magnesite-chrome shaped refractory shipments (1960–1970) were reported and are shown over the 1960–1970 time period in Fig. 10. Magnesite and magnesite-chrome shaped refractory shipments (1971–1981) were reported as pitch bonded, pitch impregnated, or other. Shipments from these three categories were combined to form the magnesite and magnesite-chrome shaped refractory shipments curve over the 1971–1981 time period shown in Figs. 10 and 12. Magnesite and magnesite-chrome refractory shipments (1982–1990) were reported as high fired magnesite-chrome, except high fired magnesite-chrome, and magnesite and magnesite-carbon refractories. Thus, the magnesite-chrome were reported separately from other magnesite refractory shipments starting in 1982. Magnesite-chrome shaped refractory shipments (1982–1990) were reported as high-fired and except high-fired. These two categories were combined to form the magnesite-chrome shaped refractory shipments curve over the 1982–1990 time period shown in Figs. 10 and 12.

Since the chromite-containing magnesite refractories were separated from the chromite-free magnesite refractories since 1982, the

Table 9. US Chromium Chemical Trade, Quantity and Value, 1986–1990

Material	1986 Quantity (t)	1986 Value ($1,000)	1987 Quantity (t)	1987 Value ($1,000)	1988 Quantity (t)	1988 Value ($1,000)	1989 Quantity (t)	1989 Value ($1,000)	1990 Quantity (t)	1990 Value ($1,000)
Exports:										
Oxides of chromium:										
Chromic acid	5 077	$8,944	4 086	$8,361	4 174	$9,436	3 929	$8,120	4 082	$9,811
Other	NA	NA	NA	NA	NA	NA	448	2,822	1 226	4,590
Chromium sulfates	NA	NA	NA	NA	NA	NA	1	39	44	68
Chromates and bichromates of:										
Zinc and lead	NA	NA	NA	NA	NA	NA	607	1,255	425	859
Sodium	14 367	9,943	15 019	12,063	19 977	17,404	18 556	10,397	16 663	10,611
Potassium	19	21	9	9	27	47	46	58	48	95
Other	NA	NA	NA	NA	NA	NA	171	416	228	707
Pigments	2 260	7,618	3 216	9,530	3 239	11,066	2 264	7,654	2 643	9,252
Total	21 723	26,526	22 330	29,963	27 417	37,953	25 567	31,280	25 143	35 701
Imports:										
Oxides of chromium:										
Chromic acid	4 197	7,520	2 558	4,039	1 542	3,155	661	2,727	1 677	3,502
Other	NA	NA	NA	NA	NA	NA	3 880	10,624	3 512	11,018
Chromium sulfates	NA	NA	NA	NA	NA	NA	313	312	192	176
Chromates and bichromates of:										
Zinc and lead	NA	NA	NA	NA	NA	NA	607	1,255	425	859
Sodium	6 946	4,103	4 755	3,173	3 756	4,960	5 489	3,777	1 661	1,548
Potassium	750	1,038	1 006	1,387	1 327	1,714	827	1,483	680	1,264
Other	NA	NA	NA	NA	NA	NA	1 010	2,296	1 860	4,491
Pigments:										
Chrome green	24	43	94	97	160	336	NA	NA	NA	NA
Chrome oxide green	2 566	5,481	2 411	5,540	4 034	10,073	NA	NA	NA	NA
Chrome oxide green-hydrated	—	—	15	41	—	13	NA	NA	NA	NA
Chrome yellow	1 933	3,061	3 355	5,573	3 999	7,720	3 688	7,352	3 643	7,636
Molybdenum orange	749	1,754	1 106	2,461	1 131	3,128	1 046	2,858	916	2,579
Strontium chromate	119	267	119	308	122	361	NA	NA	NA	NA
Zinc yellow	1 288	1,751	1 207	1,787	1 098	1,279	542	1,166	436	880
Other	NA	NA	NA	NA	NA	NA	687	1,979	657	2,613
Total	18 572	25,018	16 626	24,406	17 169	32,769	18 750	35,829	15 659	36,566

NA Not available.
Note: In 1989, the United States adopted the Harmonized Tariff code. As a result several categories changed from 1988 to 1989.

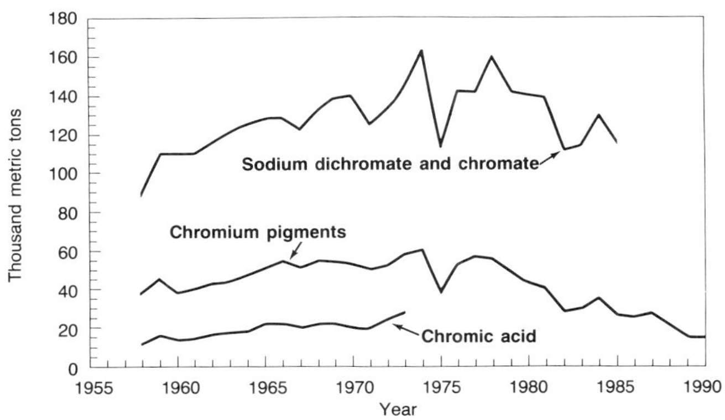

Source: Bureau of the Census.

Fig. 7. US chromium chemical production.

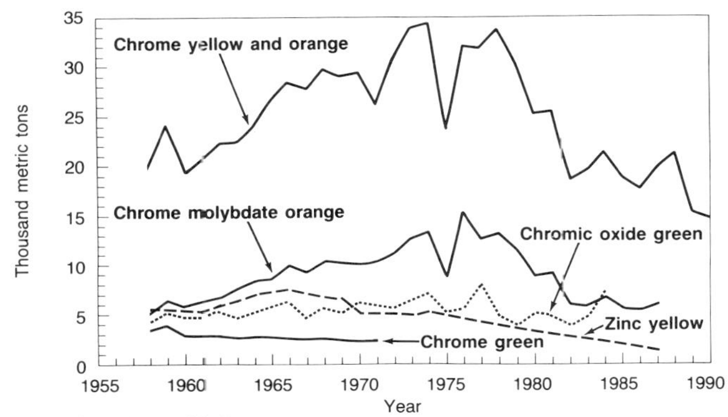

Source: Bureau of the Census.

Fig. 8. US chromium pigment production.

fraction of magnesite and magnesite-chrome refractory shipments that contain chromite can be calculated over the 1982–1990 time period. By volume, magnesite-chrome refractory shipments averaged 50% of magnesite and magnesite-chrome refractory shipments from 1982–1990 with specific values ranging from 41 to 57%.

In 1985, shapes started to be reported in tons also. No specific conversion factor is applicable because the weight equivalent of the volume of shipments is dependent upon the distribution of minerals used to make the bricks. From 1985–1990, the weight per volume of magnesite-chrome refractories averaged 4.7 kt per million 9-inch brick equivalents (kt/M9be), with a range of from 4.3 to 5.0 kt/M9be.

The general decline in refractory use results, at least in part, from the more cost efficient use of refractories. Longer lasting

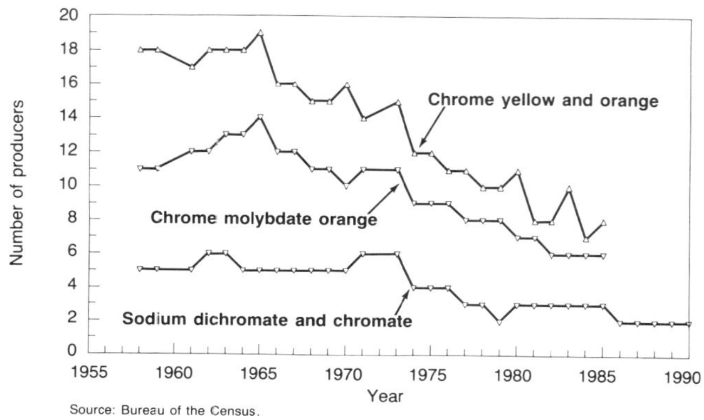

Fig. 9. Number of US chromium chemical and pigment producers by product.

refractories result in lower labor cost to change the refractories and higher production equipment availability because of less down time for relining. A specific reason for the decline in chromite-containing refractory use results from changes in steel industry production practice. The major end use for basic chrome refractories was in the production of steel in open hearth furnaces. As steel production technology has shifted away from open hearth furnace steel making, chromite refractory use has declined.

Chromic oxide, a chemical industry product, is also used to make refractories for the glass industry. Chromic oxide refractories are used in glass contact areas of glass melting furnaces to achieve long furnace life.

Foundry Sand

Foundry sand use of chromite is a modern application. Sand is used to contain molten metal in a desired shape until the metal has solidified. Sand used in the foundry industry is washed, graded, and dried. Since silica sand is common and inexpensive, it is the most commonly used mineral. However, when physical or chemical conditions dictate, other sands are chosen, such as zircon, olivine, or chromite. Chromite foundry sand is used in the ferrous and copper casting industries.

Casting sands are defined by function and by processing. Mold and core sands are for the exterior and interior of a casting, respectively. Facing sand is used on the surface of a core or mold. Flour or paint may be applied to the facing sand. As indicated by its name, flour is finer in size than sand. Before casting, sand is naturally or chemically bonded. There are a variety of methods for bonding sand before casting. Chromite sand is compatible with the commonly used methods. After casting, foundry sand is reclaimed.

Chromite sand is compatible with steel castings. It is typically used as facing sand in heavy section (greater than 4 t) casting and enjoys a technical advantage over silica sand in casting austenitic manganese steel. Chromite sand does not react with the manganese in the steel. Chromite and zircon, having a higher melting point than silica, are chosen when casting temperatures exceed those acceptable for silica sand. US foundry sand producers, products, and end use industries are shown in Table 11. Chromite sand is also used in copper-base nonferrous casting.

Chromite sand casting was developed in South Africa where chromite fines are readily available as an inexpensive grade of chromite associated with chemical, refractory, and metallurgical chromite production. After satisfactory results in South Africa in the late 1950s, use expanded in the 1960s to include the United Kingdom followed closely by the United States. Use of chromite sand was facilitated first by a shortage of zircon sand supply then by the higher price of zircon sand. The foundry characteristics that make chromite sand desirable include: good thermal stability, good chill properties, not easily wetted, resistant to metal penetration, highly refractory, and chemically unreactive. Its disadvantages, compared with the zircon sand it replaces, include: higher thermal expansion, occasional presence of hydrous mineral impurities, and different bonding practice with some binders.

Chromite sand for US foundry use was estimated to have increased from about 20 kt in 1965 to about 36 kt in 1971. Industry sources estimate US chromite foundry sand use in 1989 and 1990 to be about 40 ktpy.

Reclamation is an integral part of the foundry industry. It includes mechanical, pneumatic, wet, and thermal processes, and combinations thereof. Using these processes, as much as 90% of chemically bonded foundry sand (average over all minerals used) can be reclaimed (Heine, 1989). Chromite sand is adaptable to these processes. After casting, chromite sand, typically used as facing sand, becomes mixed with the bulk sand (silica). Since chromite sand has a size distribution similar to that of silica sand, mechanical separation is not applicable. Hydraulic spiral separation and magnetic separation are effective at separating chromite sand from silica and zircon sand. Silica and zircon sands are nonmagnetic. Some chromite sand was found to degrade during use. However, degraded sand tends to adhere to the castings, so it does not become part of the reclaimed sand. Reclaimed chromite sand was found to be interchangeable with new chromite sand. The actual amount of chromite sand reclaimed, like the amount used, is unknown. However, Sontz (1972) estimated that about half of the foundry industries chromite demand could be met by reclaimed chromite sand.

ECONOMIC FACTORS

Production, Consumption, and Trade

Chromite is not currently produced in the United States. The USSR was the original source of chromite in the 1700s and was the major world supplier into the 1800s. During the 1800s, chromite was discovered and mined in the United States, in Maryland and Pennsylvania and later in California. As a result, the United States became the world's leading chromite producer until about 1900. Chromite was then discovered in Turkey, which soon surpassed the United States as the leading world supplier. United States production soon declined to virtually nothing and commercial production ended. Some noncommercial production continued sporadically into the 1960s (see Fig. 13). Today, South Africa and the USSR are the leading world chromite producers. World chromite production by producer country is shown in Table 12. World chromite production and US chromite consumption are shown in Fig. 14.

Fig. 13 shows that US chromite production was very small except during times of national emergency. Fig. 14 shows US chromite ore consumption and world chromite production. Major declines in world chromite ore production occurred during the 1940–45 and 1978–83 time periods. As a result of industrial mobilization preceding, high production rates during, and post war recovery programs after World War II, the US dominated world chromite ore consumption. Since then, world chromite ore production has grown while US chromite consumption was steady during the 1960s time period followed by slow average decline since then. Reduced US chromite ore consumption since about 1970 resulted from the displacement of chromite by ferrochromium as a source of chromium units for the domestic metallurgical industry and, secondarily, from declining use of chromite refractories.

Table 12 shows world production by country. The two major

Table 10. US Chromite Refractory Producers, Products, and End Use Markets

Producers	Products	End users and uses
General Refractories Company	Bricks and shapes	Steel industry
US Refractories Division	Mortar	AOD vessels
Pittsburgh, PA 15219	Plastic	Barrel
	Gunning	Trunnion and Tuyere area
		Bottom
Harbison-Walker Refractories,		Electric arc furnaces
Division of Dresser Industries Inc.		Sidewall
Pittsburgh, PA 15222		Slagline
		Steel ladles
		Slagline
National Refractories and Minerals		Ladle metallurgical furnaces
Livermore, CA 94550		Slagline
		Sidewall and bottom
		Vacuum degassers
North American Refractories		Sidewall
Cleveland, OH 44115		Snorkel
		Open hearth furnace
		Backwalls
		Endwalls
		Lower walls
		Roofs
		Furnace ports
		Copper and nickel industry
		Electric furnaces
		Slagline
		Bottom
		Flash furnaces
		Roof and sidewall
		Bottom
		Pierce Smith converters
		Tuyere areas
		Barrel and endwall
		Top blown rotary converters
		Upper cone
		Bottom and barrel
		Anode furnace
		Barrel and endwall
		Fire refining and secondary furnaces
		Bottom
		Sidewall and endwall
		Bottom
		Cement and lime industry
		Rotary kilns
		Burning zone
		Upper transition zone
		Lower transition zone
		Glass industry
		Glass furnace regenerators
		Checker
		Wall
		Crown

chromite producers, South Africa and Kazakhstan, consistently produce more than 1 Mtpy each. Together they account for about two-thirds of world annual production. There are seven significant chromite producers, Albania, Brazil, Finland, India, the Philippines, Turkey, and Zimbabwe, with annual production consistently in the range of from 100 kt to 1 Mt. Together they account for about one-third of world annual production. The remaining producers account for about 3% of world annual production.

Chromite is typically consumed by industries primarily located in the developed countries: steel, chemicals, and refractories. Historically, chromite has been mined, exported, processed, and then consumed. However, chromite producing countries have been moving to more vertical integration in their production processes. Turkey started chromium chemical production in 1986 to take more complete economic advantage of its chromium resources. Since about 1970, ferrochromium production has shifted from areas that historically have produced stainless steel (i.e., Japan, United States, and Western Europe) to chromite producing countries. South Africa and the USSR are the only major industrialized countries independent of external chromite or chromium supply. The development of ferrochromium production in chromite producing countries has reduced international trade of chromite. Japan continues to produce substantial, but declining quantities of ferrochromium. The United States and Western Europe retain marginal production capacity. US chromite trade is shown in Table 6 and Fig. 4.

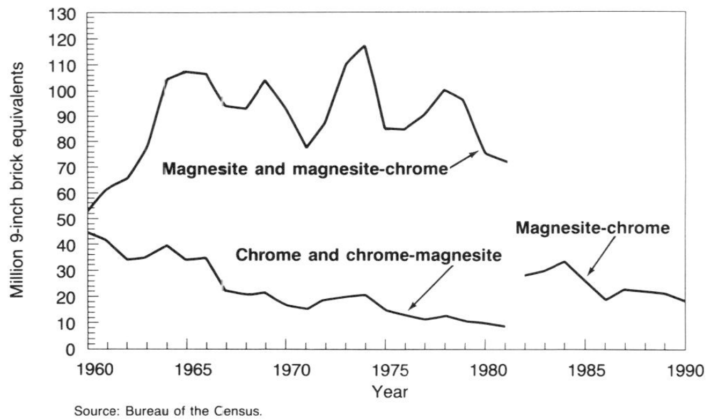

Fig. 10. US chromite-containing shaped refractory shipments.

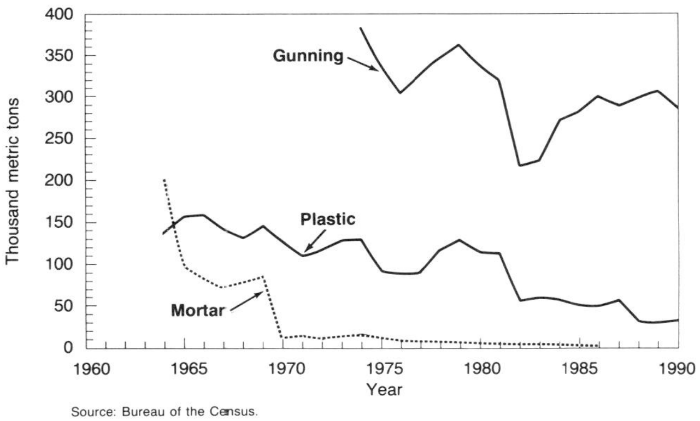

Fig. 11. US chromite-containing unshaped refractory shipments.

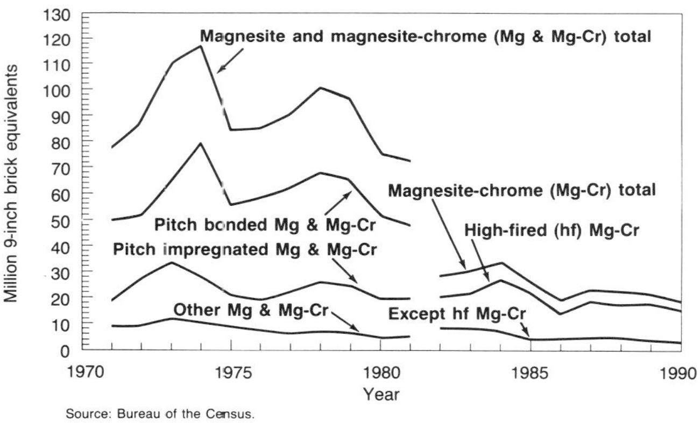

Fig. 12. US chromite-containing shaped refractory shipments by type.

Prices

Chromite is not traded in open market exchanges like gold, silver, nickel, and other metals. (No chromium products are traded on market exchanges.) As a result, chromite does not have an easily identifiable price. The price of chromite is usually negotiated between buyer and seller and is known only to them. Price speculation is, of course, a very popular activity. As a service to their readers, some publications survey chromite sellers and buyers, and calculate a composite price. Included among these are *Metals Week*, *Metals Bulletin*, *Industrial Minerals*, and *Metals Price Report*. Unfortunately, the volume of trade at the reported price is unknown. Upon being imported to the United States, the value of imported material at the port of export is declared. This is called the FOB (free on board) value. Using this value, a value history for chromite by grade has been generated. Fig. 15 shows the actual and the adjusted value of chromite. Here actual value is that reported in the specified year. Adjusted value is actual value adjusted for inflation to 1982 dollars. Fig. 16 shows actual value of chromite by grade over the same time period. Values are given in dollars per ton, gross weight.

Fig. 16 shows that the value of chromite was relatively stable through about 1970, when the value started to rise. The value of chromite declined from 1982 to 1988, the time period during which steel production in general and stainless steel production in particular was weak. Strong stainless steel recovery in the 1988–1990 time period resulted in short supply of ferrochromium. The shortage nearly doubled the price of ferrochromium and stimulated capacity expansion in that industry, primarily in South Africa. The price of chromite ore rose following that of ferrochromium. Additional chromite production capacity was added to meet the anticipated additional demand from added ferrochromium production capacity, primarily in South Africa and India.

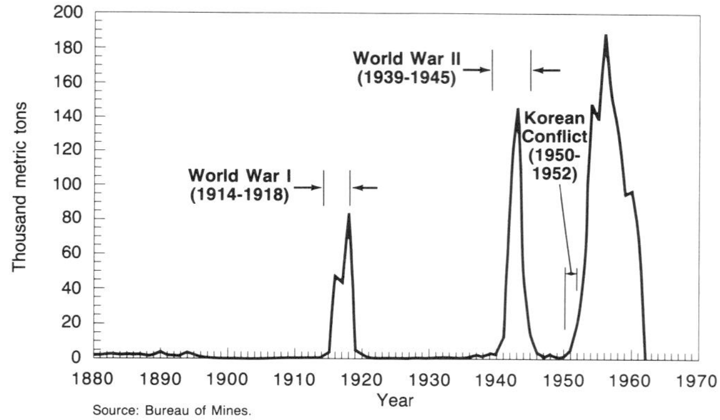

Fig. 13. US chromite production.

Table 11. US Foundry Sand Producers, Products, and End Uses

Producers	Products	End uses
American Colloid Company Arlington Heights, IL	Chromite sand	Casting facing sand Mold coating
	Chromite flour	Colorant Architectural brick
American Minerals Wayne, PA 19087		Glass Ceramic Brake shoes

Availability

The US Bureau of Mines has studied the availability of chromium. The Bureau analyzed for the simultaneous availability of chromium contained in chromium ferroalloy products and in exportable chromite products (metallurgical, chemical, refractory, and foundry sands) in ten Market Economy Countries (MEC).

Table 12. Chromite production by country (thousand t)

Country	1983	1984	1985	1986	1987	1988	1989	1990	1991	1992
Albania	685	720	825	850	1 075	1 109	1 200	910	800	150
Brazil	161	260	190	353	338	410	476	267	340	340
China	—	—	—	50	32	26	25	25	25	25
Cuba	34	38	38	50	52	52	51	50	50	50
Egypt	—	—	—	—	—	1	2	([1])	1	1
Finland	246	446	506	678	543	700	498	489	458	480
Greece	27	61	59	60	64	50	47	22	32	—
India	360	423	569	630	624	821	1 003	939	995	1 000
Indonesia	—	—	—	—	—	8	8	8	2	2
Iran	48	59	56	54	92	60	73	77	90	100
Japan	8	7	12	11	12	10	12	8	8	8
Kazakhstan[2]	—	—	—	—	—	—	—	—	—	3 600
Macedonia[3, 4]	—	—	—	—	—	—	—	—	—	9
Madagascar	60	127	83	107	107	64	63	73	73	63
Morocco	—	—	—	—	—	1	1	([1])	1	1
New Caledonia	92	84	79	72	62	70	60	6	—	—
Oman	24	7	—	5	—	—	13	—	—	—
Pakistan	6	3	5	8	10	3	27	18	31	30
Philippines	267	261	272	174	188	124	217	186	184	132
Russia[2]	—	—	—	—	—	—	—	—	—	121
South Africa	2 466	3 407	3 699	3 907	3 789	4 245	4 951	4 618	5 110	3 361
Sudan	20	20	9	9	13	8	25	13	10	10
Thailand	—	—	([1])	([1])	([1])	1	([1])	—	—	—
Turkey	346	487	589	618	762	851	1 100	800	870	850
USSR[5]	2 939	2 940	3 360	3 640	3 570	3 700	3 800	3 800	3 800	—
Vietnam	16	16	4	4	4	4	4	4	4	4
Yugoslavia[4, 6]	—	—	10	9	13	12	13	11	9	—
Zimbabwe	420	477	536	533	570	561	627	643	564	560
Total	8 210	9 776	10 945	11 797	11 919	12 896	14 294	12 968	13 445	10 896

[1]Less than 1/2 unit.
[2]Part of USSR before 1992.
[3]Part of Yugoslavia before 1992.
[4]All of 1988–91 Yugoslavian production came from Macedonia.
[5]Dissolved in Dec. 1991.
[6]Dissolved in Apr. 1991.
Source: US Bureau of Mines, *Chromium Minerals Yearbook,* chromite chapters, various editions.

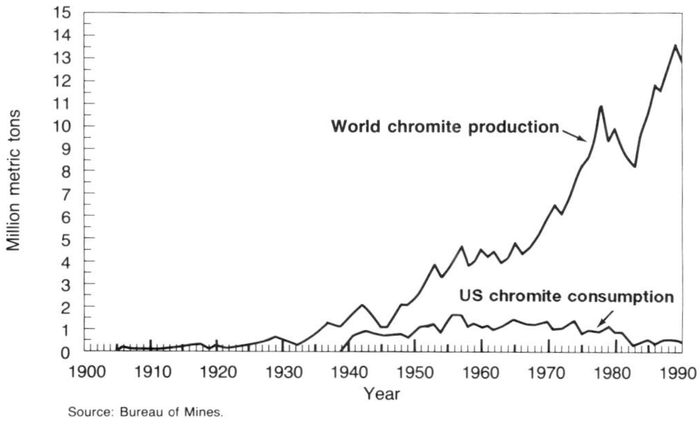

Fig. 14. World chromite production and US chromite consumption.

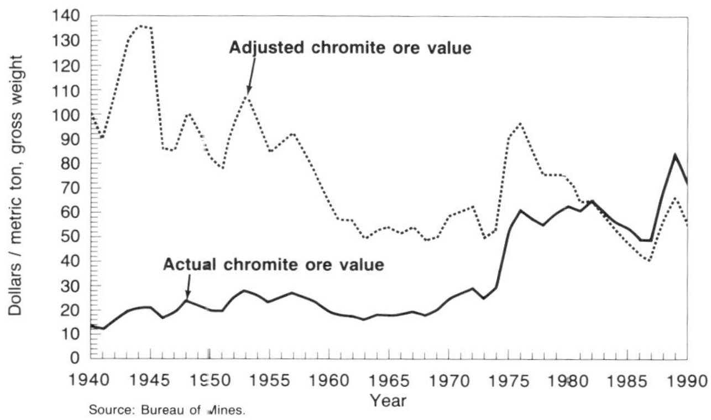

Fig. 15. Average actual and adjusted chromite ore value by year.

A total of about 874 Mt gross weight of in situ material containing about 203 Mt of chromium was analyzed. Extraction and beneficiation of this material was estimated to result in about 475 Mt of chromite products, of which 289 Mt would be available for export and the remaining 187 Mt would be smelted in the country in which it was mined to produce about 80 Mt of chromium ferroalloys. The chromium ferroalloys would then be available for use in the country of production or for export. The 80 Mt of chromium ferroalloy included about 74 Mt of high carbon ferrochromium, 4 Mt of low carbon ferrochromium, and 2 Mt of ferrochromium-silicon.

The countries of South Africa and Zimbabwe contained about 80% of the in situ contained chromium. India and Finland accounted

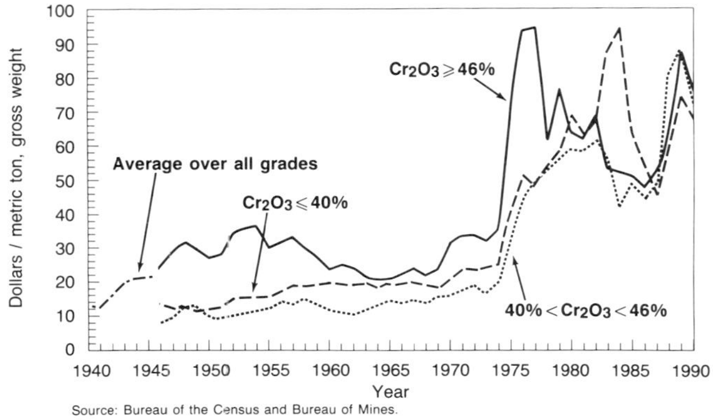

Fig. 16. Value of chromite ore by grade and by year.

for an additional 11% of the contained chromium; another 8% of the contained chromium was fairly evenly divided among Brazil, the Philippines, Turkey, and the United States; the remainder was in Greece and Madagascar.

Based on Cr_2O_3 content of in situ chromite ore, the ten MECs split into two groups, a high grade group and a low grade group. The high grade group, those with ore grades ranging from 33.96 to 43.01% Cr_2O_3, included the countries of India, Madagascar, South Africa, Turkey, and Zimbabwe. The low grade group, those with ore grades ranging from 9.16 to 26.65% Cr_2O_3, included the countries Brazil, Finland, Greece, the Philippines, and the United States. The grade differences between the two groups resulted in a wide disparity in the respective weighted average Cr_2O_3 contents. The high grade group averaged 38.76% Cr_2O_3 and the low grade group, only 15.90% Cr_2O_3.

Table 13 shows the results of the Bureau's analysis. Chromium material costs were calculated on a weighted average basis, free on board ship at the port of export. The chromite cost shown in Table 13 includes mining and beneficiating the ore (including mine capital and operating costs and taxes) and transportation of ore and products to port facilities. Chromium ferroalloy cost includes chromite ore but excludes smelter capital cost. On a weight basis, the Bureau's analysis shows that South Africa and Zimbabwe could produce about 78% of metallurgical chromite ore, 93% of chemical chromite ore, 85% of refractory chromite ore, and 93% of foundry chromite ore that could be produced at their respective estimated break even cost. South Africa and Zimbabwe could also produce about 69% of the high carbon ferrochromium, 89% of the low carbon ferrochromium, and 100% of ferrochromium-silicon that could be produced at their respective estimated break even cost. The product breakdown between chromium ferroalloy and chromite and among the grades within those product categories was based on mine and smelter production capacities and known operating relationships circa the 1987–1988 time period.

Table 13. Availability of Chromite and Chromium Ferroalloys from Ten Market Economy Countries

Chromium materials	Quality available (million t gross weight)	Cost[1] ($ per t) Weighted average	Range
Chromite:			
Chemical grade	64.3	53	35–174
Foundry sand grade	16.4	49	39–83
Metallurgical grade:			
Primary product	145.4	101	42–705
Secondary product	35.6	54	33–117
Subtotal	181.0	92	33–705
Refractory grade	26.8	87	54–180
Total	288.5		
Chromium ferroalloys:			
Ferrochromium:			
High carbon ferrochromium	74.3	473	417–1,286
Low carbon ferrochromium	3.9	937	635–1,309
Ferrochromium-silicon	2.0	737	578–814
Total	80.2		

[1]Cost of production for 0% discounted cash flow rate of return in January 1989 dollars per metric ton, gross weight, of product.

GOVERNMENT CONSIDERATIONS

Strategic Factors, Stockpile

It is the policy of the Federal Government of the United States to maintain a National Defense Stockpile for use in the event of a national defense emergency. The US Government has maintained a stockpile in the period since World War I. The Defense Logistics Agency, Department of Defense, is currently responsible for the National Defense Stockpile. Chromium materials included in the National Defense Stockpile are chromite ore (metallurgical, chemical, and refractory grades), chromium ferroalloys (high and low carbon ferrochromium and ferrochromium-silicon), and chromium metal. Table 14 shows the goals and current inventories for chromium materials in the National Defense Stockpile.

The National Defense Logistics Agency is currently upgrading chromium materials in the National Defense Stockpile. In recognition of the United States' declining capacity to promptly convert chromite into ferrochromium for use by the nation's steel industry, the President started a program, which was subsequently legislatively mandated, to convert chromite into ferrochromium. The historical and projected conversions under that program are shown in Table 15. In addition to the chromite-to-high-carbon ferrochromium program, the Defense Logistics Agency is converting nonspecification grade low-carbon ferrochromium to chromium metal. The historical and projected conversions under that program are shown in Table 16.

Health Factors

Chromium is a trace mineral required by the human body. The National Research Council recommends a daily intake in the range of 50 to 200 μg of chromium. Chromium is a cofactor for insulin, a hormone that participates in carbohydrate and fat metabolism. A cofactor is a material that acts with another material. The dietary chemical form of chromium is as trivalent compounds. Because the human body cannot convert trivalent chromium to hexavalent chromium, the carcinogenicity of hexavalent chromium compounds bears no relevance to the nutritional role of trivalent chromium.

The toxic effect of an element depends on its chemical form and concentration and on the conditions and duration of exposure. At about 185 ppm, chromium is the thirteenth most common element in the Earth's crust; at about 0.2 ppb, the twenty-sixth most common element in seawater. Chromium generally forms chemical compounds in which chromium has either the hexavalent or trivalent oxidation state. Hexavalent chromium compounds are generally recognized as toxic. Chronic occupational exposure to hexavalent chromium has been associated with an increased incidence of bron-

Table 14. Goals and Inventory for Chromium Materials in the National Defense Stockpile in December 1992 (thousand t, gross weight)

Material	Stockpile goals	Physical inventory: Stockpile-grade	Nonstock-pile-grade	Total[1]
Chromite ore:				
Metallurgical	34	1 208	71	1 278
Chemical		220	—	220
Refractory	159	355	—	355
Chromium ferroalloys:				
High carbon ferrochromium	621	707	42	749
Low carbon ferrochromium		274	14	288
Ferrochromium-silicon		52	1	53
Chromium metal	27	4	2	6

[1]Data may not add to total shown owing to independent rounding.
Source: Defense Logistics Agency.

Table 15. Historical and Projected Performance of the National Defense Stockpile Chromite Ore to Ferrochromium Conversion Program

Contract year	Ore converted (t)	High carbon ferrochromium produced (t): Gross	Content	Cost (millions)
HISTORICAL PERFORMANCE				
1984	113 968	45 590	30 180	$22.3
1985	124 298	44 872	29 630	22.5
1986	85 301	31 944	20 898	17.6
1987	125 739	52 414	34 415	28.8
1988	111 105	41 511	26 727	26.7
1989	120 651	46 940	30 392	31.5
1990	113 453	42 881	28 082	30.9
1991	111 531	42 577	28 181	31.0
1992	113 944	41 120	27 511	32.8
PROJECTED PERFORMANCE[e]				
1993	179 988	73 255	NA	46.0

[e]Estimated. NA Not available.
Source: Defense Logistics Agency.

Table 16. Historical and Projected Performance of the National Defense Stockpile Low Carbon Ferrochromium to Chromium Metal Conversion Program

Time period	Low carbon ferro-chromium[1] (t)	Chromium metal (t)	Cost (millions)
HISTORICAL PERFORMANCE			
October 1989–September 1990	2 129	1 046	$ 7.3
October 1990–September 1991	1 090	780	7.0
October 1991–September 1992	2 049	1 022	10.1
PROJECTED PERFORMANCE[e]			
October 1992–September 1993	2 894	1 452	13.5

[e]Estimated.
[1]Nonspecification grade.
Source: Defense Logistics Agency.

chial cancer. The toxic status of trivalent chromium compounds is not clear. However, trivalent chromium compounds are less toxic than hexavalent chromium compounds. Chemical compounds containing chromium in lower valence states (i.e., divalent or less) are generally recognized as benign.

Environmental Regulations

Effluents: Chromium in water effluents is manageable. The solubility of trivalent chromium compounds in neutral water usually results in a chromium concentration below that required by the Environmental Protection Agency (EPA) for drinking water (0.1 ppm). Thus, when water is neutralized, chromium can be removed by filtration. If hexavalent chromium compounds are present, they must first be reduced to trivalent, a technically manageable operation.

Emissions: Congress enacted the Clean Air Act Amendments Law of 1990 (Public Law 101–549), completely revising the Air Toxics Program. Congress identified 189 hazardous air pollutants to be regulated. Chromium compounds—defined as any chemical substances that contain chromium as part of their structure—were included among those hazardous air pollutants. Under the revised Air Toxics Program, Congress instructed EPA to regulate hazardous air pollutants by regulating the source of those pollutants. Congress required EPA to identify pollution sources by November 1991, then to set emission standards for those sources. EPA eliminated the use of chromium chemicals in comfort cooling towers. This was one source of chromium emissions.

Solid Waste: EPA regulates solid waste generated by the chemical industry in the production of sodium chromate and dichromate. Chromium-containing treated residues from roasting and/or leaching of chrome ore is regulated under subtitle D of the Resource Conservation and Recovery Act. EPA found no significant danger associated with treated residue from roasting and/or leaching of chrome ore based on waste characteristics, management practices, and damage case investigations.

EPA regulates solid waste containing chromium. EPA determined that chromium-containing wastes exhibit toxicity. Therefore, they have established a policy that, if the extract from a representative waste sample contains chromium at a concentration greater than or equal to 5.0 mg/l (total chromium) as measured by a specified toxicity characteristics leaching procedure, it is hazardous. EPA has promulgated a treatment standard for chromium-containing refractory brick wastes based on chemical stabilization. (Stabilization is a process that keeps a compound, mixture, or solution from changing its form or chemical nature.) EPA has also determined that some chromium-containing refractory brick wastes can be recycled as feedstock in the manufacture of refractory bricks or metal alloys. EPA recognizes that there is insufficient capacity to process inorganic solid debris (i.e., treatable material that is greater than 9.5 mm in size and requires cutting or crushing and grinding in mechanical sizing equipment) prior to chemical stabilization. As such a material, chromite-containing refractory bricks were granted a 2-year capacity variance (i.e., a variance until May 1992).

Safety Regulations

The Occupational Safety and Health Administration (OSHA) reviewed health, risk, and feasibility evidence for chromium metal, chromic acid and chromates, and chromyl chloride. Table 17 summarizes OSHA permissible exposure limits.

Table 17. Occupational Health and Safety Permissible Exposure Limits for Chromium Compounds

Material	Permissible exposure limit
Chromic acid and chromates	0.1 mg of CrO_3 m^3, ceiling
Chromium metal and insoluble salts	1.0 mg of Cr m^3, 8 hr time weighted average
Soluble chromic and chromous salts	0.5 mg of Cr m^3, 8 hr time weighted average

PROBLEMS AND TRENDS

In the United States, chromite use for the production of chromium chemicals has been slowly increasing. Chromite use for the production of chromite-containing refractories has been declining, as has use in the metallurgical sector. However, chromium use is strong and increasing.

Environmental regulations are impacting chromite use, primarily through chemical and refractory industries. Regulations limiting worker exposure and limiting chromium-containing emissions, effluents, and solid wastes increase operating costs and, for new operations, capital costs. Which chromium- or chromite-related end uses remain economic under new and emerging regulations is yet to be demonstrated.

ACKNOWLEDGMENTS

The author wishes to acknowledge Harry Mikami for his authorship of interesting and informative chromite chapters for previous editions of *Industrial Minerals and Rocks*. Much of the information provided in those chapters remains useful. The author also wishes to thank Bruce Lipin of the US Geological Survey and Edward H. Boyle, Jr. of the US Bureau of Mines for their helpful comments and recommendations. Several industry sources were very helpful and generous in providing information about their respective industries and products.

BIBLIOGRAPHY AND REFERENCES

Mineralogy

Duke, V.W.A., 1983, "Chromium, The Shining Metal," *Chromium Review*, No. 1, International Chromium Development Association, Paris, France, pp. 2–5.

Ford, W.E., 1957, *Dana's Textbook of Mineralogy*, John Wiley and Sons, 4th ed., John Wiley and Sons, New York, 851 pp.

Frye, K., 1974, *Modern Mineralogy*, Prentice-Hall, Englewood Cliffs, NJ, 325 pp.

Hurlbut, C.S., and Klein, C., 1977, *Manual of Mineralogy (after James D. Dana)*, 19th ed., John Wiley and Sons, New York, 532 pp.

Lipin, B.R., 1983, "Low-Grade Chromium Resources," *Cameron Volume on Unconventional Mineral Deposits*, W.C. Shanks, ed., AIME, New York, pp. 47–60.

Mikami, H.M., 1983, "Chromite," *Industrial Minerals and Rocks*, 5th ed., Vol. 1, S.J. Lefond, ed., AIME, New York, pp. 567–584.

Stowe, C.W., 1987, "The Mineral Chromite," *Evolution of Chromium Ore Fields*, C.W. Stowe, ed., Van Nostrand Reinhold Co., New York, pp. 1–22.

Thayer, T.P., 1956, "Mineralogy and Geology of Chromium," *Chromium*, Chap. 2, M.J. Udy, ed., American Chemical Society Monograph 132, Reinhold Publishing Corp., New York, Vol. 1, pp. 14–52.

Thayer, T.P., 1973, "Chromium," *United States Mineral Resources*, US Geological Survey Professional Paper 820, pp. 111–121.

Origin

Mikami, H.M., 1983, "Chromite," *Industrial Minerals and Rocks*, 5th ed., Vol. 1, S.J. Lefond, ed., AIME, New York, pp. 567–584.

Petrascheck, W., et al., eds., 1988, *Chromites*, Theophrastus Publications, Athens, Greece, 339 pp.

Stowe, C.W., ed., 1987, *Evolution of Chromium Ore Fields*, Van Nostrand Reinhold Co., New York, 340 pp.

Thayer, T.P., 1956, "Mineralogy and Geology of Chromium," *Chromium*, Chap. 2, M.J. Udy, ed., American Chemical Society Monograph 132, Reinhold Publishing Corp., New York, Vol. 1, pp. 14–52.

Thayer, T.P., 1973, "Chromium," *United States Mineral Resources*, US Geological Survey Professional Paper 820, pp. 111–121.

Distribution of Major Deposits

Anon., 1985, "The Chromitites of the Stillwater Complex," *Chromium Review*, No. 4, International Chromium Development Association, Paris, France, pp. 5–13.

Boyle, E.H., Shields, D.J., and Wagner, L.A., 1993, *Chromium Availability in Market Economy Countries and Network Flow Model Analysis of World Chromium Supply*, Bureau of Mines Information Circular 9337, 131 pp.

DeYoung, J.H., Lee, M.P., and Lipin, B.R., 1984, *International Strategic Minerals Inventory Summary Report—Chromium*, US Geological Survey Circular 930-B, 41 pp.

Duke, V.W.A, 1982, *Chromium—A Mineral Commodity Review*, Internal Report No. 86, Council for Mineral Technology, Republic of South Africa, 317 pp.

Gruenewaldt, G., 1983, "Chromium Deposits of the Bushveld Complex," *Chromium Review*, No. 1, International Chromium Development Association, Paris, France, pp. 8–11.

Kujanpää, J., 1989, "Finland's Kemi Chromite Deposit," *Chromium Review*, No. 10, International Chromium Development Association, Paris, France, pp. 5–8.

Prendergast, M.D., 1984, "Chromium Deposits of Zimbabwe," *Chromium Review*, No. 2, International Chromium Development Association, Paris, France, pp. 5–9.

Rabchevsky, G.A., 1985, "Chromium Deposits of Albania," *Chromium Review*, No. 3, International Chromium Development Association, Paris, France, pp. 14–17.

Silk, M.S., 1988, *World Chromite and Ferrochromium Production*, Special Publication No. 11, Council for Mineral Technology, Republic of South Africa, 149 pp.

Stowe, C.W., 1987, "Extended Tabulation of Chromite Ore Fields," *Evolution of Chromium Ore Fields*, C.W. Stowe, ed., Van Nostrand Reinhold Co., New York, pp. 297–332.

Thomas, P.R., and Boyle, E.H., 1984, *Chromium Availability—Market Economy Countries*, US Bureau of Mines Information Circular 8977, 86 pp.

Vermaak, C.F., 1985, "The UG2 Layer—South Africa's Slumbering Chromitite Giant," *Chromium Review*, No. 5, International Chromium Development Association, Paris, France, pp. 9–22.

Exploration Technique

Albers, J.P., 1986, "Descriptive Model of Podiform Chromite," *Mineral Deposit Models*, US Geological Survey Bulletin 1693, S.A. Singer and D.P. Cox, eds., pp. 34.

Page, N.J., 1986, "Descriptive Model of Bushveld Cr," *Mineral Deposit Models*, US Geological Survey Bulletin 1693, S.A. Singer and D. Cox, eds., pp. 12–13.

Petrascheck, W., et al., eds., 1988, *Chromites*, Theophrastus Publications, Athens, Greece, 339 pp.

Singer, D.A., and Page, N.J., 1986, "Grade Tonnage Model of Minor Podiform Chromite," *Mineral Deposit Models*, US Geological Survey Bulletin 1693, S.A. Singer and D.P. Cox, eds., pp. 34–38.

Singer, D.A., Page, N.J., and Lipin, B.R., 1986, "Grade Tonnage Model of Major Podiform Chromite," *Mineral Deposit Models*, US Geological Survey Bulletin 1693, S.A. Singer and D.P. Cox, eds., pp. 38–44.

Stowe, C.W., ed., 1987, *Evolution of Chromium Ore Fields*, Van Nostrand Reinhold Co., New York, 340 pp.

Thayer, T.P., 1973, "Chromium," *United States Mineral Resources*, US Geological Survey Professional Paper 820, pp. 111–121.

Tilton, J.E., Eggert, R.G., and Landsberg, H.H., eds., 1988, *World Mineral Exploration: Trends and Economic Issues*, Resources for the Future, Washington, DC, 464 pp.

Wynn, J.C., 1983, "Strategic Minerals Geophysical Research: The Chromite Example," *Mining Engineering*, Vol. 35, No. 3, pp. 246–251.

Mining, beneficiation, and processing

Boyle, E.H., Shields, D.J., and Wagner, L.A., 1993, *Chromium Availability in Market Economy Countries and Network Flow Model Analysis of World Chromium Supply*, US Bureau of Mines Information Circular 9337, 131 pp. Company brochures.

Lipin, B.R., and Thayer, T.P., 1980, "The Outlook for Small Chromite Mines," *The Future of Small Scale Mining*, R.F. Meyer and J.S. Carman, eds., United Nations Institute for Training and Research, New York, pp. 191–196.

Mikami, H.M., 1983, "Chromite," *Industrial Minerals and Rocks*, 5th ed., Vol. 1, S.J. Lefond, ed., AIME, New York, pp. 567–584.

Thomas, P.R., and Boyle, E.H., 1984, *Chromium Availability—Market Economy Countries*, US Bureau of Mines Information Circular 8977, 86 pp.

Byproducts and coproducts

Overbeek, P.J., Loo, J.P., and Dunne, R.C., 1984, "The Development of a Concentration Procedure for Platinum-Group Metals and Chromite from the UG-2 Reef of the Bushveld Complex," presented at 1st International Symposium on Precious Metals Recovery, June 10–14, Reno, NV, 26 pp.

Von Gruenewaldt, G., and Hatton, C.J., 1987, "Platinum-Group Metals—A Resource in the Tailings of Chromium Mines in South Africa," *Journal of The South African Institute of Mining and Metallurgy*, Vol. 87, No. 9, pp. 265–268.

Uses

Granville, A., and Statham, E.F., 1989, *Profits From Processing*, Mintek Special Publication No. 13, Mintek, Randburg, Republic of South Africa, pp. 10–11.

Papp, J.F., 1985, "Chromium," *Minerals Facts and Problems*, US Bureau of Mines Bulletin 675, pp. 139–156.

Papp, J.F., 1986, "Chromium contained in U.S.-Stainless and Heat-Resisting Steels," *Chromium-Chromite: Bureau of Mines Assessment and Research. Proceedings of Bureau of Mines Briefing Held at Oregon State University, Corvallis, OR, June 4–5, 1985*, C.B. Daellenbach, comp., US Bureau of Mines Information Circular 9087, p. 5–11.

Chromite Consumption

Various authors, various years, "Chromium," chapters in *Minerals Yearbook*, US Bureau of Mines.

Metallurgical

Boyle, E.H., Shields, D.J., and Wagner, L.A., 1993, *Chromium Availability in Market Economy Countries and Network Flow Model Analysis of World Chromium Supply*, US Bureau of Mines Information Circular 9337, 131 pp.

Elyutin, V.P., et al., 1957, *Production of Ferroalloys Electrometallurgy*, Office of Technical Services, US Department of Commerce, 2nd ed., 446 pp.

Papp, J.F., 1991, *Chromium, Nickel, and Other Alloying Elements in U.S. Produced Stainless and Heat-Resisting Steel*, US Bureau of Mines Information Circular 9275, 41 pp.

Sully, A.H., and Brandes, E.A., 1967, *Metallurgy of the Rarer Metals—1. Chromium*, 2nd ed., Plenum Press, New York, 373 pp.

Udy, M.J., ed., 1956, *Chromium, Vol. 2, Metallurgy of Chromium and Its Alloys* American Chemical Society Monograph 132, Reinhold Publishing Corp., New York, 433 pp.

Westbrook, J.H., 1979, "Chromium and Chromium Alloys," *Kirk-Othmer Encyclopedia of Chemical Technology*, 3rd ed., Vol. 6, Wiley-Interscience, New York, pp. 54–82.

Chemical

Anon., various years, *Current Industrial Reports: Inorganic Chemicals*, Series MA28A, US Department of Commerce, US Bureau of the Census.

Anon., 1985, "The Use of Chromium in Timber Preservation—An Editorial Review," *Chromium Review*, No. 3, International Chromium Development Association, Paris, France, pp. 3–10.

Anon., 1985, "Problems of Traditional Chromium-based Electroplating, and a Novel Method to Overcome Them—An Editorial Review," *Chromium Review*, No. 4, International Chromium Development Association, Paris, France, pp. 15–16.

Anon., 1985, "Notes on Pigment Technology (and especially lead chromate pigments)," *Chromium Review*, No. 4, International Chromium Development Association, Paris, France, pp. 20–21.

Chemical company literature.

Covington, A.D., 1985, "Chromium in the Leather Industry," *Chromium Review*, No. 5, International Chromium Development Association, Paris, France, pp. 2–9.

Duffield, P. A., 1987, "Chromium in Wool Dyeing," *Chromium Review*, No. 8, International Chromium Development Association, Paris, France, pp. 12–18.

Hartford, W.H., 1979, "Chromium Compounds," *Kirk-Othmer Encyclopedia of Chemical Technology*, 3rd ed., Vol. 6, Wiley-Interscience, New York, pp. 82–120.

Hartford, W.H., 1981, "Trends and Technology in Industrial Chromium Chemicals," *Specialty Inorganic Chemicals*, Royal Society of Chemistry Special Publication No. 40, London, pp. 311–345.

Lewis, P.A., ed., 1988, *Pigment Handbook, Vol. I Properties and Economics*, Wiley-Interscience, New York, 930 pp.

Ohtani, H., 1987, "Chromium Trioxide Enhances the Durability of Wooden Doors," *Chromium Review*, No. 6, International Chromium Development Association, Paris, France, pp. 4–7.

Patton, T.C., ed., 1973, *Pigment Handbook, Vol. I Properties and Economics*, Wiley-Interscience, New York, 989 pp.

Udy, M.J., ed., 1956, *Chromium, Vol. 1, Chemistry of Chromium and its Compounds*, American Chemical Society Monograph 132, Reinhold Publishing Corp., New York, 433 pp.

Refractory

Anon., various years, *Current Industrial Reports: Refractories, Summary*. Series MQ32C, US Department of Commerce, US Bureau of the Census.

Anon., 1961, *Modern Refractory Practice*, Harbison-Walker Refractories Co., Pittsburgh, PA, 607 pp.

Anon., 1987, *Refractories*, The Refractories Institute, Pittsburgh, PA, 48 pp.

Anon., 1991, "Standard Classification of Chrome Brick, Chrome-Magnesite Brick, Magnesite-Chrome Brick, and Magnesite Brick, Designation: C455–84 (Reapproved 1988)," *1991 Annual book of ASTM Standards*, Sec. 15 General Products, Chemical Specialties, and End Use Products, Vol. 15.01 Refractories; Carbon and Graphite Products; Activated Carbon, American Society for Testing and Materials, Philadelphia, pp. 110–111.

Benbow, J., 1989, "Steel Industry Minerals," *Industrial Minerals*, No. 265, Oct., pp. 27–43.

Benbow, J., 1990, 'Cement Kiln Refractories," *Industrial Minerals*, No. 268, Jan., pp. 37–42.

Crookston, J.A., and Fitzpatrick, W.D., 1985, "Refractories," *Industrial Minerals and Rocks*, 5th ed., Vol. 1, S.J. Lefond, ed., AIME, New York, pp. 373–385.

Geiger, G., 1991, "Trends in Refractory Technology," *The American Ceramic Society Bulletin*, Vol. 70, No. 6, pp. 994–1003.

Hayhurst, A., 1988. "Chromite in Modern Refractories," *Chromium Review*, No. 9, International Chromium Development Association, Paris, France, pp. 13–19.

Udy, M.J., ed., 1956, *Chromium, Vol. 1, Chemistry of Chromium and its Compounds*, American Chemical Society Monograph 132, Reinhold Publishing Corp., New York, 433 pp.

Foundry

Garnar, T.E., 1977, "Mineralogy of Foundry Sands and Its Effect on Performance and Properties," *AFS Transactions*, pp. 399–416.

Heine, H.J., 1989, "Key Fundamentals for Sand Preparation and Reclamation," *Foundry Management and Technology*, Apr., pp. 34–41.

Middleton, J.M., and Brownes, F.F., 1970, "The Properties of chromite and its application in the Foundry," *The British Foundryman*, Aug., pp. 248–260.

Murphy, M.T., 197[illegible], "How to Use Chromite Sand," *Foundry*, Mar., pp. 58–61.

Sanders, C.A., 1983, "The History of Copper-Base Non-Ferrous Molding Sands Part II," *The Crucible*, Sep./Oct., pp. 6–9.

Sontz, A., 1972, "Recovery of Chromite Sand," *AFS Transactions*, pp. 1–12.

Wilborg, H.E., and Henderson, G.V., 1983, "Foundry Sand," *Industrial Minerals and Rocks*, 5th ed., Vol. 1, S.J. Lefond, ed., AIME, New York, pp. 271–278.

Production, Consumption, and Trade

Anon., 1991, *The Economics of Chromium*, 7th ed., Roskill Information Services Ltd., London, 774 pp.

Silk, M.H., 1988, *World Chromite Resources and Ferrochromium Produc-*

tion, Special Publication No. 11, Council for Mineral Technology, Randburg, Republic of South Africa, 149 pp.

Various authors, various years, "Chromium," *Minerals Yearbook*, US Bureau of Mines.

Prices

Various authors, various years, "Chromium," *Minerals Yearbook*, US Bureau of Mines.

Availability

Boyle, E.H., Shields, D.J., and Wagner, L.A., 1993, *Chromium Availability in Market Economy Countries and Network Flow Model Analysis of World Chromium Supply*, US Bureau of Mines Information Circular 9337, 131 pp.

Thomas, P.R., and Boyle, E.H., 1984, *Chromium Availability—Market Economy Countries*, US Bureau of Mines Information Circular 8977, pp. 86.

Strategic Factors, Stockpile

Anon., various issues, *Strategic and Critical Materials Report to the Congress, Operations under the Strategic and Critical Materials Stockpiling Act* US Department of Defense, published biannually.

Health Factors

Anon., 1984, *Health Effects Assessment for Trivalent Chromium*, Environmental Protection Agency, Report EPA/540/1–86–035, 32 pp.

Anon., 1984, *Health Effects Assessment for Hexavalent Chromium*, Environmental Protection Agency, Report EPA/540/1–86–019, 49 pp.

Anon., 1988, *Chromium*, Environmental Health Criteria 61, World Health Organization, Geneva, 197 pp.

Anon., 1989, *Recommended Dietary Allowances*, National Research Council, National Academy of Science, Washington, DC, pp. 241–243.

Burrows, D., 1983, *Chromium: Metabolism and Toxicity*, CRC Press, Boca Raton, FL, 184 pp.

Langård, S., ed., 1982, *Biological and Environmental Aspects of Chromium*, Topics in Environmental Health, Vol. 5, Elsevier Biomedical Press, New York, 285 pp.

Nriagu, J.O., and Nieboer, E., eds., 1988, *Chromium in the Natural and Human Environments*, Advances in Environmental Science and Technology, Vol. 20, Wiley-Interscience, New York, 571 pp.

Taylor, S.R., and McLennan, S.M., 1985, *The Continental Crust: Its Composition and Evolution*, Blackwell Scientific Publications, Oxford, pp. 15–16, 67.

Effluents

Barnhart, R.J., 1991, "The Current Environmental Issues Concerning Chrome from a Chromium Compound Supplier's Viewpoint," presented at 93rd Annual Meeting of the American Ceramic Society, Cincinnati, OH, Apr. 29, 1991 (Refractory Ceramic Division, Paper No. 2-R-91).

Emissions

Anon., 1991, "The Clean Air Act of 1990—A Detailed Analysis," *The Hazardous Waste Consultant*, Vol. 9, Issue 1, Jan.-Feb., pp. 4.1–4.30.

Public Law 101–549, Nov. 15, 1990.

Solid Waste

Anon., 1990, *Report to Congress on Special Wastes From Mineral Processing. Volume I: Summary and Recommendations*, Environmental Protection Agency, pp. 2, 11–12.

Barnhart, R.J., 1991, "The Current Environmental Issues Concerning Chrome from a Chromium Compound Supplier's Viewpoint," presented at 93rd Annual Meeting of the American Ceramic Society, Cincinnati, OH, Apr. 29, 1991 (Refractory Ceramic Division, Paper No. 2-R-91).

Bray, D.J., 1985, "Toxicity of Chromium Compounds Formed in Refractories," *The American Ceramic Society Bulletin*, Vol. 64, No. 7, pp. 1012–16.

Federal Register, 1990, Vol. 55, No. 15, Jan. 23, p. 2322.

Federal Register, 1990, Vol. 55, No. 106, June 1, pp. 22563–22565.

Marvin, C.G., 1991, "Chrome-Bearing Hazardous Waste," presented at 93rd Annual Meeting of the American Ceramic Society, Cincinnati, OH, Apr. 29, 1991 (Refractory Ceramic Division, Paper No. 3-R-91).

Safety Regulations

Anon., 1988, *Issues Document*, Industrial Health Foundation, Pittsburgh, PA, 50 pp.

Federal Register, 1989, Occupational Safety and Health Administration (Dept. of Labor), Air Contaminants, Vol. 54, No. 12, Jan 19, pp. 2332–2983.

CLAYS

INTRODUCTION

HAYDN H. MURRAY

The term clay is somewhat ambiguous unless specifically defined, because it is used in three ways: 1) as a diverse group of fine-grained minerals, 2) as a rock term, and 3) as a particle size term. As a rock term, clay is difficult to define because of the wide variety of materials that comprise it; therefore, the definition must be general. Clay is a natural, earthy, fine grained material comprised largely of a group of crystalline minerals known as the clay minerals. These minerals are hydrous silicates comprised mainly of silica, alumina, and water. Several of these minerals also contain appreciable quantities of iron, alkalies, and alkaline earths. Many definitions state that a clay is plastic when wet. Most clay materials do have this property, but some clays are not plastic, e.g., halloysite and flint clay.

As a particle size term, clay is used for the category that includes the smallest particles. The maximum size particles in the clay size grade are defined differently on various grade scales. Soil investigators and mineralogists generally use 2 µm. as the maximum size, whereas the widely used scale by Wentworth (1922) defines clay as material finer than approximately 4 µm.

Some authorities find it convenient to use the term clay for any fine-grained, natural, earthy, argillaceous material (Grim, 1968). When used this way, the term includes clay, shale, or argillite, and some argillaceous soils.

Even though no standard definition of the term clay is accepted by geologists, agronomists, engineers, and others, the term is generally understood by those who use it. Clay is an abundant natural raw material, and it has an amazing variety of uses and properties which will be discussed in this chapter.

Clays rank among the leading industrial minerals in both tonnages produced and total value. Their importance is indicated in Table 1, which lists the 1988, 1989, and 1990 productions according to the US Bureau of Mines (Ampian, 1992).

As industrial minerals, clays are a complex group that consists of several mineral commodities, each having somewhat different mineralogy, geologic occurrence, technology, and uses. In this section these commodities are organized as follows: 1) bentonite, 2) common clay, 3) hormites, and 4) kaolin. Fuller's earth is discussed in both the bentonite and hormite chapters. Fuller's earth is a term based on use, and both bentonites and hormites are sold as fuller's earth. The overlapping is particularly evident where both bentonite and non-bentonite fuller's earth are used for the same purposes or products, such as in drilling mud, bleaching or clarifying fats and oils, and carriers for insecticides and fertilizers. Kaolin, ball clay, halloysite, and refractory clays are grouped together because they consist mainly of minerals of the kaolin group. Common clay is a grouping of several fine-grained materials, including shale, sometimes referred to as miscellaneous clay.

The scientific and technological publications pertaining to clays are so numerous that to cover all the material in the format used for other chapters would require a book. Therefore, the author has attempted to summarize the subject matter and wherever possible has indicated by bibliographic reference where more detailed information is available. For example, the reference to Bicker (1970) is to a bulletin containing only brief summary information on the economic geology of bentonite in Mississippi, but that bulletin contains references to virtually all published reports on bentonite in that state.

The remaining parts of this introduction will be mainly an outline of some of the major advancements in the broad field of clays since the 5th edition of *Industrial Minerals and Rocks* (Patterson and Murray, 1983). The intent is to note where some of the contributions, mainly in clay mineralogy, have been made, so that those who are interested can find detailed information not included in this report.

A vast amount of new information on the mineralogy, geology, and technology of clays has become available in recent years. One of the principal sources is The Clay Minerals Society. This professional and scientific society was organized in 1963 from an informal group sponsored and partially supported for the preceding years by the National Academy of Sciences-National Research Council. For five years after that date, the papers presented at the Society's annual meetings were published in volumes by Pergamon Press. Since 1968, The Clay Minerals Society has published its own journal under the title *Clays and Clay Minerals*. A second major source of information on clays is the publications of the Association International pour l'Étude des Argiles (AIPEA); these publications are principally volumes containing the papers presented at the meetings of the association. A third excellent source of information on clays is *Clay Minerals*, the journal of The Clay Minerals Group of The Mineralogical Society (London). In addition, clay mineral groups or societies in France, Spain, Italy, Japan, Argentina, Germany, and other countries have periodic publications. Considerable information on kaolin resources in several countries is available in three volumes resulting from a symposium held at the International Geological Congress in Czechoslovakia in 1968 (Malkovský and Vachtl, 1969a, 1969b) and in the report of another symposium on kaolin held at the 1972 International Clay Conference (AIPEA) in Spain in 1972 (Serratosa, 1972). In 1968 the International Geological Correlation Program (IGCP) sponsored by UNESCO organized project 23 on the genesis of kaolin. Ten kaolin symposia were held in various parts of the world and proceedings were published on each.

Several major textbooks and reference volumes on clays have been published in the last four decades. They include one concerned mainly with mineralogy and another on applied mineralogy of clays by Grim (1962, 1968). The Clay Minerals Group of the Mineral-

Table 1. Clay Tonnages (thousand tons) 1988, 1989, and 1990

	1988	1989	1990
Kaolin	8 080	8 080	8 790
Ball Clay	890	810	710
Fire Clay*	720	740	560
Bentonite	2 580	2 800	3 130
Fuller's Earth	1 610	1 690	2 080
Common Clay	26 190	23 900	23 350
Total	40 100	38 000	38 600

*Refractory uses only

ogical Society (London) has a classic series on clay minerals, including a volume concerned with differential thermal analysis, edited by Mackenzie (1957), one on the X-ray identification and crystal structures of clay minerals edited by Brindley and Brown (1980), and a third on the electron-optical investigation of clays edited by Gard (1971). A volume on the electron microscopy of clay minerals has been prepared by Beutelspacher and Van Der Marel (1968), and in 1967, Zvyagin published a book concerned with the electron diffraction analysis of clays. A Special Paper of The Geological Society of America by Carroll (1970) outlines guides to the X-ray identification of clays. An excellent volume on colloid chemistry of clays has been authored by Van Olphen (1977). Gillott's 1968 volume outlines the role of clays in engineering geology. Volumes concerning clays used in ceramic materials include books by Grimshaw (1972) and Clews (1969). An excellent volume on the geology and origin of clays is one by Millot (1964), which has been translated from French into English. A Special Paper of The Geological Society of America by Weaver et al. (1972) contains considerable information on the geochemistry of clays related to diagenesis and metamorphism. Another volume by Weaver and Pollard (1973) contains information on the chemistry of clays, including oxide data and structural formulae, as does a volume by Newman (1987). Other volumes of note are the *SEPM Petrology Atlas* (Welton, 1984) with excellent electron micrographs of clays; the Mineralogical Society of America volume on *Hydrous Phyllosilicates* (Bailey, 1988); Velde's (1985) volume on clay minerals explaining their occurrence from a physical-chemical standpoint; a book on the formation and properties of clay-polymer complexes by Theng (1979); an understandable and well written book on the identification and analysis of clay minerals by X-ray diffraction by Moore and Reynolds, Jr. (1989); a book on clay sedimentology by Chamley (1989); and a history of fuller's earth by Robertson (1986). There are many other excellent books on clays that could be mentioned. A great deal of information on the geology, technology, and uses of clay has also been published by various provincial, state, and federal geological surveys, bureaus of mines, and other agencies.

Many changes have taken place in the various clay industries in recent years. Several of these have come about because of technological advancements, changing economic conditions, new uses, shifts in demand, and increases in both domestic and export markets. The changing conditions, particularly increasing freight rates, the energy crisis, other production and marketing costs, and the pressures brought about by governmental controls and mining laws resulting from requirements and restrictions related to environmental considerations, etc., have motivated the formation of organizations to look after the interests of some of the clay producing industries. One of these organizations is the Sorptive Minerals Institute. This institute was formed mainly to represent the producers of absorbent granules and other fuller's earth products. A similar organization, the Bentonite Producers Association was organized in 1971 in Casper, WY, and represents the interests of the producers of bentonite. The kaolin industry organized the US Clay Producers Traffic Association which was incorporated in 1953 in New Jersey. In addition to these organizations dealing specifically with clays, the clay industry and other mining industries formed mining associations in many states, including Georgia and Texas in 1972, in order to protect their interests against restrictive state and federal legislation.

GOVERNMENTAL CONSIDERATIONS

MINERAL OWNERSHIP

In the United States and the United Kingdom, minerals are owned by the owner of the land, who must be dealt with by mineral lease, inground mineral purchase, or land and mineral purchase. Almost all of the kaolin producing areas of the United States are in private ownership, so the government is not a prominent owner of kaolin mineral rights. In the Western United States the government owns the land so claims must be staked and assessment work done to keep the claims in force. However, most other nations of the world have very different mineral law, and the government is the owner of many or most types of minerals regardless of surface land ownership. It can be very time consuming to acquire mining rights in many nations. It is usually also necessary to pay surface damages, or even to buy the land where kaolins, bentonites, or hormites are intended to be mined.

ENVIRONMENTAL REGULATIONS

LAND RECLAMATION

In the United States reclamation of land mined for coal is regulated by the federal government, but other mining reclamation is determined by laws of the individual states. In Georgia, where most US kaolin is mined, the 1968 Surface Mining Act requires that all land disturbed by mining be reclaimed to a completely vegetated, non-eroding condition. In order to obtain a permit to open a surface mine, it is necessary to:

1. Identify the mining company, the mineral to be mined, mining methods intended, present land use, boundaries of the pit, location of the overburden dump, haul road, and other areas to be disturbed.
2. Delineate erosion and sedimentation control to be used, the catchment basin where runoff water is to be discharged from the mine, and a settlement basin to impound and clarify runoff from a 50-year storm. Show the size and type of impound dam planned for review of dam safety.
3. Describe any pit backfilling intended, highwall reduction, grading and sloping, lake development, site cleanup, cover crops planned, intended land use after completion of reclamation, etc.
4. Provide approval from county and/or city authorities for the new mine.
5. Have public notice of intent to mine sent out to interested organizations throughout the state to enable public comment.
6. Post bond to assure money is available to complete reclamation in case of default, or apply for bonding exemption.
7. Provide assurance of the right to enter leased property for a minimum of three years after completion of mining in order to complete reclamation.

Within 18 months after completion of mining each individual pit, reclamation work must be under way. Release from bond and/or obligation will not be made until the disturbed area is leveled to blend with the surrounding topography, with a total, permanent cover crop of grasses or trees, and in a no-erosion condition. Many tens of thousands of hectares disturbed since 1969, when the law took effect, have been successfully reclaimed. Most of this reclamation has used grasses which are good wildlife foods and fix nitrogen naturally into the soil. Much of this acreage was planted in select pine seedlings after the grass had become established. Hundreds of lakes and ponds have been made from deep kaolin mines and stocked with sport and food fish. In addition, most of the large kaolin mining companies have voluntarily completed reclamation not required by the state of all their land which had been mined before the law was passed. Current cost for reclaiming land disturbed by kaolin mining in Georgia is estimated to be from $1500 to $3500 per acre. Bentonite, commom clay, and hormite mines are reclaimed according to the state regulations in which the particular mine is located.

WETLAND ACCESS

The type of terrain and topography on which major kaolin deposits are found in the Georgia-South Carolina mining district is such that many millions of tons of kaolin reserves are in areas that could be considered wetlands, swamps, etc. Wetlands (even though private property) are now protected from most types of development by the federal government in the United States, by Section 404 of the 1972 Clean Water Act. This legislation requires that anyone interested in depositing dredge or fill material must apply for and receive a permit from the US Army Corps of Engineers. Activities covered include, but are not limited to, placement of fill material, ditching by sidecast, levee and dike construction, landclearing involving relocation of soil, land leveling, most road construction, and dam construction. This regulation has been interpreted with increasing rigidity recently. There have been many definitions proposed for what wetlands covered by Section 404 may be. The US Army Corps of Engineers and the Environmental Protection Agency have jointly defined wetlands as: *"Those areas that are inundated or saturated by surface or ground water at a frequency and duration sufficient to support, and that under normal circumstances do support, a prevalence if vegetation typically adapted for life in saturated soil conditions. Wetlands generally include swamps. marshes, bogs, and similar areas."* (Anon., 1987). However, since 1987 many new criteria for wetland recognition have been proposed, and the regulatory status of wetland permitting is most unclear.

Several wetland brick clay, peat, and sand and gravel mines near the Georgia-South Carolina kaolin district have received 404 exemptions from the Corps to mine in river swamps, but the writer knows of no permitted kaolin mines in wetlands since section 404 passage in 1972. Perhaps the effect most felt in the kaolin industry so far has been the difficulty of obtaining permits to construct water treatment impounds in stream bottoms. These impounds are required by the federal government, under NPDES point and non-point source regulations, to avoid water pollution. Bentonite and hormite mining permits are also affected by the wetlands issues.

PROCESSING AND MINE WASTEWATER CONTROL

Kaolin refining plants are required to flocculate suspended clay from their wastewater and neutralize to an acceptable pH level. Plant discharge is regulated under the federal National Pollutant Discharge System (NPDES) and Georgia's 1964 Water Quality Control Act. Muddy water runoff from the entire area disturbed by mining is also regulated under provisions of the NPDES. Such water is caught in impounds before leaving the property mined and clarified to meet standards.

AIR QUALITY AND DUST CONTROL

By their nature, kaolin, bentonite, and hormite plants which produce dry products in powder form are prone to emit dust in the atmosphere. Although clay dust is not toxic and is abundantly present in the air along dirt roads and fields being plowed during dry weather, it is regulated under federal and state law. The Georgia Air Quality act of 1978 requires an extensive system of dust collectors, which have greatly reduced unsightly white dust around kaolin plants.

ZONING AND LAND USE LIMITATIONS

Zoning of land use is not applied to the rural areas where kaolin and hormite clays are mined in the Georgia-South Carolina district. However, recently regulations were put into effect which will require the approval of local officials (county commissioners and/or city government) in order for a Georgia mine permit to be issued. It is not known at this time (1993) what effect this will have on new mining.

SAFETY AND HEALTH REGULATIONS

The Mine Safety and Health Administration (MSHA) of the US Department of Labor has responsibility for regulating the clay mining industry throughout the United States. MSHA programs consist of setting standards for safe practices in mines, processing plants, laboratories, and for exploration crews. Recent emphasis in the kaolin industry has been on hazardous materials in the workplace (chemicals in clay refining and laboratory testing) and on training miners in safe operation of heavy earthmoving equipment.

DEPLETION ALLOWANCE AND TARIFFS

Kaolin has a depletion allowance of 14% domestic or foreign. This is the same allowance as for other commercial clays, such as ball clay, bentonite, fuller's earth, and common clay.

Tariffs are set in the United States on the basis of whether a country is classed as a Most Favored Nation or not. At the beginning of 1991, kaolin tariff for Most Favored Nations was $0.325/t, and for non-Most Favored Nations it was $2.46/t.

REFERENCES

Anon., 1987, *Recognizing Wetlands*, US Army Corps of Engineers.

Ampian, S.G., 1992. "Clays," *Mineral Commodity Summaries, 1992*, US Bureau of Mines, pp. 50–51.

Bailey, S.W., ed., 1988, *Hydrous Phyllosilicates*, Mineralogical Society of America, Reviews in Mineralogy, Vol. 19, 725 pp.

Beutelspacher, H., and Van Der Marel, H.W., 1968, *Atlas of Electron Microscopy of Clay Minerals and Their Admixtures*, Elsevier, Amsterdam, 333 pp.

Bicker, A.R., Jr., 1970, "Economic Minerals of Mississippi," Bulletin 112, Mississippi Geological, Economic and Topographical Survey, 80 pp.

Brindley, G.W., and Brown, G., eds., 1980, *Crystal Structures of Clay Minerals and Their X-Ray Identification*, Mineralogical Society Monograph No. 5, London, 995 pp.

Carroll, D., 1970, *Clay Minerals: A Guide to Their X-Ray Identification*, Special Paper 126, Geological Society of America, 80 pp.

Chamley, H., 1989, *Clay Sedimentology*, Springer-Verlay, Berlin, Germany, 623 pp.

Clews, F.H., 1969, *Heavy Clay Technology*, 2nd ed., Academic Press, New York, 481 pp.

Gard, J.A., ed., 1971, *The Electron-Optical Investigations of Clays*, Mineralogical Society (Clay Minerals Group), London, 383 pp.

Gillott, J.E., 1968, *Clay in Engineering Geology*, Elsevier, New York, 296 pp.

Grim, P.E., 1962, *Applied Clay Mineralogy*, McGraw-Hill, New York, 422 pp.

Grim, R.E., 1968, *Clay Mineralogy*, 2nd ed., McGraw-Hill, New York, 596 pp.

Grimshaw, R.W., 1972, *The Chemistry and Physics of Clays and Other Ceramic Raw Materials*, 4th ed., rev., Wiley-Interscience, New York, 1024 pp.

Mackenzie, R.C., ed., 1957, *The Differential Thermal Investigation of Clays*, Mineralogical Society (Clay Minerals Group), London, 456 pp.

Malkovský, M., and Vachtl, J., eds., 1969a, *Proceedings, Symposium I. Genesis of Kaolin Deposits,* 23rd International Geological Congress, Prague, 1968 Report, Vol. 14, 135 pp.

Malkovský, M., and Vachtl, J., eds., 1969b, *Proceedings, Symposium I. Kaolin Deposits of the World, A. Europe,* 23rd International Geological Congress, Prague, 1968 Report Vol. 15, 322 pp.

Malkovský, M., and Vachtl, J., eds., 1969c, *Proceedings of Symposium I. Kaolin Deposits of the World, B. Overseas Countries,* 23rd International Geological Congress, Prague, 1968 Report, Vol. 16, 157 pp.

Millot, G., 1964, *Geologic des Argiles*, Masson, Paris 320 pp. (Trades by W.R. Farrand and H. Paquet, published as *Geology of Clays*, Springer-Verlag, New York, 1970, 425 pp.)

Moore, D.M., and Reynolds, R.C., Jr., 1989, *X-Ray Diffraction and the Identification and Analysis of Clay Minerals*, Oxford University Press, Oxford, UK, 332 pp.

Newman, A.C.D., ed., 1987, *Chemistry of Clays and Clay Minerals*, Mineralogical Society Monograph No. 6, London, 480 pp.

Patterson, S.H., and Murray, H.H., 1983, "Clays," *Industrial Minerals and Rocks*, 5th ed., S.J. Lefond, ed., AIME, New York, pp. 585–651.

Robertson, R.H.S., 1986, *Fullers Earth—A History of Calcium Montmorillonite*, Volturna Press, Hythe, Kent, UK, 421 pp.

Serratosa, J.M., ed., 1972, *Kaolin Symposium*, International Clay Conference, Madrid, 195 pp.

Theng, B.K.G., 1979, *Formation and Properties of Clay-Polymer Complexes*, Developments in Soil Science 9, Elsevier, Amsterdam, the Netherlands, 362 pp.

Van Olphen, H., 1977, *An Introduction to Clay Colloid Chemistry*, 2nd ed., John Wiley and Sons, New York, 318 pp.

Velde, B., 1985, *Clay Minerals—a Physico-chemical Explanation of Their Occurrence*, Elsevier, Amsterdam, the Netherlands, 427 pp.

Weaver, C.E., et al., 1972, *Clay Water Diagenesis During Burial: How Mud Becomes Gneiss*, Special Paper 132, Geological Society of America, 99 pp.

Weaver, C.E., and Pollard, L.D., 1973, *The Chemistry of Clay Minerals*, Elsevier, New York, 224 pp.

Welton, J.E., 1984, *SEM Petrology Atlas*, American Association of Petroleum Geologists, Tulsa, OK, 237 pp.

Wentworth, C.K., 1922, "A Scale of Grade and Class terms for Elastic Sediments," *Journal of Geology,* Vol. 30, pp. 377–392.

Zvyagin, B.B., 1967, *Electron-Diffraction Analysis of Clay Mineral Structures*, Plenum Press, New York, 364 pp.

CLAYS

BENTONITE

JESSICA ELZEA AND HAYDN H. MURRAY

The term *bentonite* was first proposed in 1898 by Knight, a year after he had named this clay taylorite, because taylorite had been previously used for another mineral. The name *taylorite* was after the Taylor ranch, the site of the first mine, near Rock River, WY, and the name *bentonite* is from the Benton Shale in which the clay was thought at that time to occur. The Benton Shale was, in turn, named after Fort Benton, MT, located more than 640 km miles north of Rock River.

Early in the 20th century, several geologists recognized that bentonite, mainly in beds in Cretaceous and Tertiary rocks, originated from transported volcanic materials. This recognition led to definitions based on origin, the one most widely quoted and generally accepted by geologists is the following definition by Ross and Shannon (1926):

"Bentonite is a *rock* composed essentially of a crystalline clay-like mineral formed by devitrification and the accompanying chemical alteration of a glassy igneous material, usually a tuff or volcanic ash; and it often contains variable proportions of accessory crystal grains that were originally phenocrysts in the volcanic glass. These are feldspar (commonly orthoclase and oligoclase), biotite, quartz, pyroxenes, zircon, and various other minerals typical of volcanic rocks. The characteristic clay-like mineral has a micaceous habit and facile cleavage, high birefringence and a texture inherited from volcanic tuff or ash, and it is usually the mineral montmorillonite, but less often beidellite."

The difficulty in applying the foregoing definition to bentonite as an industrial mineral commodity is that it is based on origin and is restricted to an ash, tuff, or volcanic glass parent material. Therefore, deposits consisting of the clay minerals required by this definition, but having uncertain origin or parent materials, cannot properly be called bentonite. Furthermore, many deposits in the western United States and in other countries that have formed from rocks other than the types required by the definition are being mined and sold as bentonite.

Perhaps the best definition of bentonite as an industrial mineral is one given by R.E. Grim in a plenary lecture at the International Clay Conference (AIPEA) at Madrid, Spain, June 27, 1972. According to this redefinition, which will be used in this chapter, bentonite is a clay consisting essentially of smectite minerals (montmorillonite group of some usages), regardless of origin or occurrence. This definition solves the problem of the difference between the geologic and industrial usages of the term and overcomes the difficulty in assigning a name to smectite clay that formed from igneous rock other than ash, tuff, or glass, or those of sedimentary or uncertain origin. However, bentonite, when used with this meaning is still a rock term (consisting of more than one mineral), and it will not be possible to distinguish it from fuller's earth in many instances.

One way of classifying bentonite is based on its swelling capacities when wet or added to water. Bentonite having sodium (Na+) as either the dominant or as an abundant exchangeable ion typically has very high swelling capacities and forms gel-like masses when added to water. Bentonite in which exchangeable calcium (Ca++) is more abundant than other ions has much lower swelling capacities than sodium varieties. Some calcium types swell little more than common clay, and most crumble into granular masses in water. Intermediate calcium-sodium bentonites, the so called mixed types, tend to swell moderately and to form gels of lesser volumes than equal masses of the sodium type bentonite. Because of the general relationship of swelling and exchangeable ion characteristics bentonite is commonly divided into the high swelling or sodium, low swelling or calcium, and moderate swelling or intermediate types. The term *sub-bentonite* (Davis et al., 1940) is used inconsistently in industry for the low or moderate swelling varieties. The authors believe the use of this term should be discouraged because of its implication of low quality or low value and the lack of a mineralogical or use basis for it.

In the United States bentonite is also classified by geographic location and the uses for which it is sold. Inasmuch as most of the low swelling calcium type occurs in states bordering the Gulf of Mexico, this variety is commonly called *Southern bentonite*. The largest high swelling sodium bentonite deposits and the major producing districts are in Wyoming and adjacent states. Therefore, this bentonite is commonly called the *Wyoming* or *Western* type. Such terms as drilling mud bentonite, foundry bond bentonite, and taconite bond bentonite relating directly to use are applied in marketing. Other terms including high and low yield bentonite, high and low gel bentonite, and high and low strength bentonite are also used to distinguish different grades.

The varied classifications notwithstanding, bentonite occurs in so many different varieties that some cannot logically be classified according to any of the foregoing groupings. One of these types is hectorite, which is a high swelling lithium-bearing variety of smectite occurring mainly in California and adjacent states. It is, therefore, not a Wyoming type, and it occurs even farther west than the so called western bentonites. Others are bentonites having magnesium (Mg++) or hydrogen (H+) as the most abundant or dominant exchangeable ion. These types are neither sodium nor calcium varieties, and apparently some are low swelling, whereas others have rather high swelling capacities.

Still another type of bentonite, the so-called potassium type, K-bentonite, or metabentonite, occurs in Ordovician and other Paleozoic rocks at many places in Appalachian and Mississippi Valley regions and elsewhere. This bentonite, which is generally thought to have formed from volcanic ash, consists mainly of illite and mixed layer minerals. Smectite minerals are ordinarily present in only minor quantities. Because of its mineral composition, this bentonite contains appreciably more potassium than most other types. The meta prefix originates from the idea that this bentonite was altered by low grade metamorphism or diagenesis. Apparently, the only attempt to use K-bentonite in the United States was when five or six carloads were mined near Dayton, Rhea County, TN, about 1928; this bentonite was used for purifying lard (Gildersleeve, 1946), and no further mention of this type will be made.

The term *fuller's earth* is more or less a catchall for clay or other fine-grained earthy material suitable for bleaching and absorbent and certain other uses. It has no compositional or mineralogical meaning. The origin of the term dates back into antiquity (Robertson, 1986); it was first applied to material used in cleansing and fulling wool, thereby removing the lanolin and dirt from it. When in the latter half of the last century it was found that some earths used for fulling would also serve in decolorizing and purifying mineral, vegetable, and animal oils, the term *fuller's earth* was modified to include this usage.

Extensive use of fuller's earth in processing mineral oils in the first half of this century and the virtual end of its use in fulling fiber led to the general application of the term, fuller's earth, as meaning primarily earth used in petroleum processing. Further modification of the meaning came about as other uses developed for the earth and replaced oil processing as the major use. The term fuller's earth is retained, however, for fine-grained materials used for many purposes. Most of these uses require absorbent properties in one form or another, but the properties required for some uses, such as certain drilling mud and fillers, are other than absorbency and, therefore, result in further modification of the term.

The classification and understanding of fuller's earth are also complicated by several terms having more or less duplicate or overlapping meanings. When applied in the oil processing sense, fuller's earth has the same meaning as *naturally active clay*. Fuller's earth and other clays which are treated with acid or otherwise altered to improve their desirable properties are called *activated clay* (Torok and Thompson, 1972). The terms *bleaching clay* and *bleaching earth* are applied mainly to both naturally active and activated clay, but they also include activated bauxite (Rich, 1960). Both naturally active and activated clays are included under the term *absorbent clays* (Nutting, 1943); therefore, this term has nearly the same meaning as bleaching clay. The term *absorbent clay* is applied to fuller's earth used for a wide variety of absorbing purposes which are different from that of processing oils. Many bentonites are fuller's earth if they are suitable for bleaching and absorbent uses.

HISTORY AND USE

Bentonite

Bentonite mining began on the Taylor ranch near Rock River, WY, in 1888. This bentonite reportedly was shipped crude to Philadelphia where it sold for $25 a ton and was used in making cosmetics. A mine was opened in the Upton, WY, district on the western side of the Black Hills in 1903. The early method of drying was to rake the bentonite by hand over steel plates heated by wood or coal fires.

The date of the first mining of southern bentonite is somewhat conjectural because of the confusion between the terms bentonite and fuller's earth. One fuller's earth deposit near Gonzales, TX, that is almost certainly bentonite was mined as early as 1906, and another deposit in that same state mined during the Civil War (Lang et al., 1940) may also have been bentonite. Large scale mining of calcium-type bentonite began in Mississippi in the early 1930s (Bicker, Jr., 1970), and bentonite deposits were used as fuller's earth in several states about this time.

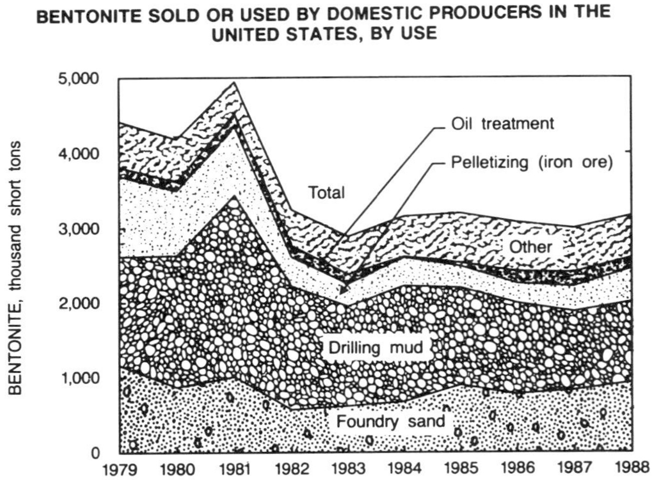

Fig. 1. Bentonite uses and volumes.

The value of bentonite as foundry sand bond was recognized in the 1920s, and the iron and steel foundries have since been major consumers of bentonite. Bentonite was first used as drilling mud in the late 1920s or early 1930s, and bentonite is still one of the most efficient materials for drilling mud where the rocks penetrated contain only freshwater. The high swelling Wyoming bentonite is the most efficient type for drilling mud, but the high viscosity properties of hectorite make it useful in drilling mud (Larsen, 1955). Some Texas bentonite is also used for drilling mud, particularly in shallow wells, and commonly it is treated with soda ash, polymers, or other chemicals to make it suitable for this use. Low quality bentonite when sold for use in drilling mud is commonly classified merely as clay rather than bentonite (Anon., 1972).

Bentonite production remained small during the first few decades of mining, and the first yearly total sufficiently large to be separated from miscellaneous clays in the US Bureau of Mines statistics was in 1930 when production was only a few thousand tons. Demands for bentonite increased during World War II, but a yearly total of 1 t was not reached until 1950. More than 2 Mt was produced in 1966, and the peak production year was 1978, when 4.5 Mt were sold or used by producers. Bleaching clay (oil refining, filtering, clarifying, and decolorizing) drilling mud, iron ore pelletizing, and foundry sand bonding have been the major uses of bentonite since World War II (Fig. 1). Beginning about 1950, steel companies used Wyoming-type bentonite as a bond for pelletized taconite ore. This use grew rapidly, and in the last three decades approximately one-third of the national production has been used for this purpose.

In addition to the major uses, bentonite is used in many miscellaneous products, and hundreds of patents for specialty uses have been issued or applications filed. The specialty uses include filtering agents (one of which is a high value product for clarifying wine, and another is a less costly one for treating waste water), water impedance (preventing seepage loss from reservoirs, irrigation ditches, and waste disposal ponds, and seepage through basement walls, tunnel walls, and other structures), ingredients in cosmetics, animal feed, pharmaceuticals, colloidal fillers for certain types of paints, an additive to ceramic raw materials to increase plasticity, fire retarding materials, and for many other purposes (Papin, 1964). The white bentonites occurring in Texas, Oklahoma, and Nevada and imported from Italy and Mexico are particularly suitable for many of the specialty products, and a great deal of research evaluating them for many uses has been done. Also a new application begun in 1991 is the use of high swelling bentonite in ''scoopable'' cat litter. One of the unique uses of high swelling bentonite that is likely to increase is in the ''slurry trench'' or ''diaphragm wall'' method of excavation in construction in areas of unconsolidated rock or soil (Blackman, 1969, Lang, 1971). In this method, the trench or hole being excavated is filled with bentonite slurry and the earth being excavated is removed through it. A thin filter cake on the walls of the excavation prevents loss of fluid, and the hydrostatic head of the slurry prevents caving and running of loose soils, which makes costly shoring unnecessary. Another use for bentonite is as a coating on some types of computer paper and on no-carbon required copy paper. Considerable quantities of bentonite were once used in making catalysts for petroleum refining, but the market for these types of catalysts in the United States declined after World War II. However, sizable quantities of catalysts have been produced from bentonite and similar clays in Japan, the United Kingdom, and elsewhere. Acid-activated bentonite is used for bleaching oils and in making multiple-copy paper requiring no carbon paper.

Fuller's Earth

Published reports give rather confusing accounts of the discovery of fuller's earth in the United States. Several state that fuller's earth was discovered in this country in 1893 near Quincy, FL, by the Owl Commercial Co. (predecessor of the present makers of White Owl cigars) during an attempt to burn brick from clays on tobacco property. The clay was not suitable for making brick, but an Alsatian immigrant employed as a farm worker recognized that it was similar to fuller's earth mined in Germany. His observation led to development of the first mine near Quincy two years later and to the use of this clay in processing mineral oils. Fuller's earth had been mined on a small scale, however, in Arkansas in 1891 (Miser, 1913), and tested for use in the refining of cottonseed oil.

Both of these reported discoveries are related to the use of fuller's earth in processing oils and fail to note its earlier use for other purposes. Clays and other earthy materials were undoubtedly used by the early settlers from Europe in cleansing wool and other materials. Little effort was made to search the historical records and to document this, but it was found that soldiers stationed near Perth Amboy, NJ, used nearby Woodbridge fire clay for cleansing buckskins during the Revolutionary War (Cook and Smock, 1878). Fuller's earth associated with an iron ore bed near Kent, CT, had been mined in the early 1800s (Silliman, 1820), and fuller's earth near Falls City, TX, had been used to bleach sugar during the Civil War (Lang et al., 1940). Also, American Indians were reported to have used bentonite in cleaning blankets, and they probably dug it for other cleansing purposes before the Columbian period.

The use of fuller's earth in the refining of oils continued to be the major one for many years. The demand for fuller's earth for processing mineral oil increased rapidly in the first part of this century and reached a peak of approximately 317 kt in 1930; this was 91% of the total US production that year. The production of fuller's earth for this purpose began to decrease thereafter. The market for this use was also depressed by improvements in refining methods, which produce oils requiring little purification or bleaching. In the last several decades, the production of fuller's earth for use in refining mineral oils has maintained a rather uniform rate of 35 to 40 ktpy. The use of fuller's earth in processing animal and vegetable oils has never been a major one and had decreased to the point where it is included as a part of the miscellaneous uses in the US Bureau of Mines *Minerals Yearbooks*.

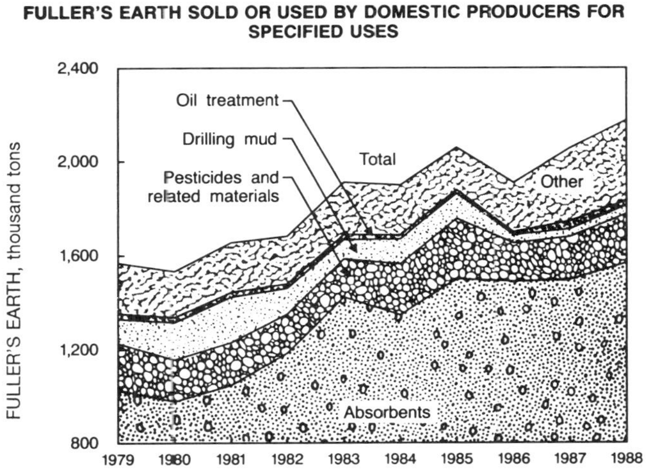

Fig. 2. Fuller's earth uses and volumes.

Fuller's earth was used in significant quantities as a carrier for insecticides and fungicides by 1950, and the market for this use has grown at a rather uniform rate since that year. In 1990, about 7% of the fuller's earth produced was used for this purpose.

The use of fuller's earth granules for absorbent purposes began during the 1930s, but this use did not expand significantly until the World War II period when fuller's earth was used as an absorbent for greases, oil, water, chemicals, and other undesirable substances on the floors of factories, filling stations, canning plants, aircraft hangers, decks and engine rooms of ships, and other installations. The absorbent granules are porous and ordinarily weigh less than 18 kg per m^3. Because of their light weight, size, porosity, and absorbent properties, the granules are suitable for many uses, and since World War II many different markets for them have developed (Fig. 2). Among other uses, they are now sold for litter and bedding for poultry, pets, and other animals, and as a soil conditioner in greenhouses and for golf courses.

The foregoing discussions apply only to those earths that are classified separately by the US Bureau of Mines in the *Minerals Yearbooks*. In addition, smaller quantities of fuller's earth are or have been produced for many miscellaneous uses.

PRODUCT SPECIFICATIONS

Product specifications have been tentatively standardized for bentonite and/or fuller's earth sold for drilling mud, foundry sand bond, absorbent granules, and oil bleaching and will be outlined briefly. Specifications for most of the many other products made from these clays are prepared to meet the requirements of the individual customer, and they commonly vary with each.

Drilling Mud

The most critical specifications for bentonite for drilling mud are the suspension properties, wet screen analysis, and moisture as shipped.

Suspension Properties: The test for suspension properties involves the preparation of a suspension consisting of 22.5 g of bentonite in 350 cc of distilled water. The suspension is aged and the viscosity determined and yield point calculated from dial readings at 300 and 600 rpm with a direct indicating viscometer, according to procedure and equipment requirements outlined in

American Petroleum Institute specification API STD 13A, 5th edition, 1969. Another important suspension requirement is the filtrate test, which is a measure of the volume of water lost from the prepared suspension when tested in a pressurized filter press according to procedure outlined in American Petroleum Institute specification API RP 13. According to these specifications, to be drilling-mud quality, a bentonite must have a dial reading at 600 rpm of 30 minimum, a yield point kilograms per 9.3 m^2 of 3 times plastic viscosity maximum, and a filtrate volume of 13.5 maximum.

Many companies still use a yield specification for bentonite. Yield is a term used in an earlier American Petroleum Institute specification for the number of barrels of 15 centipoise viscosity mud that can be made from a ton of bentonite. The yield requirement is ordinarily 90 bbl per ton minimum.

Wet Screen Analysis: Wet-screen analysis (grit test) is a measure of the material in bentonite mud that is coarser than 200-mesh US Series sieve. The specified test is made by mixing 10 g bentonite in 350 cc water containing 0.2 g dispersing agent, stirring, aging, stirring, and washing through a sieve with a specified spray system (API STD 13A). The residue on the sieve is then dried, weighed, and the percentage of the original bentonite is determined. Bentonite to fulfill the specifications for drilling mud must have no more than 2.5% residue (grit) and palygorskite no more than 8% residue.

Bentonite and fuller's earth are tested for use in drilling mud by adding water and mixing slurries of specific concentrations and measuring viscosity, yield, yield point, and filtrate (a measure of wall building properties), and making a wet screen analysis, according to standard procedures recommended by the American Petroleum Institute (Anon., 1969, 1969a). The Oil Companies Materials Association (Anon., 1969b) in the United Kingdom also has specifications for testing drilling mud. Tests for foundry sand-bonding properties require preparation of standard mixtures of bentonite and sand and measurements of compression strength of carefully prepared test pieces in a manner similar to the methods outlined by Fisk (1946), the American Foundrymen's Society (Anon., 1962, 1963), and the Steel Founder's Society of America (Anon., 1965).

Tests for taconite pelletizing bonds vary considerably (Davison, 1969), but some require certain strengths in pellets measured by the number of times they can be dropped in specified intervals onto a hard surface without breaking (Stone et al., 1971, Sastry and Fuerstenau, 1971, Wakeman, 1972).

Oil-bleaching tests of fuller's earth and bentonite are commonly made according to procedures described by the American Oil Chemists' Society (Anon., 1958). Ross (1964) and Nutting (1943) also have given procedures for evaluating fuller's earth for use in bleaching oils.

Tests for a variety of different insecticide, dust diluents, and carrier uses have been described by Weidhaas and Brann, Jr. (1955). Most tests for absorbent granules are made according to procedures in Federal Specifications P-A-1056A, Absorbent Material, Oil and Water (for floors and decks) available from the US General Services Administration.

No standard procedures have been developed for testing bentonite for use in water impedance, but rather detailed procedure for testing bentonites for canal and pond sealants have been described by the Colorado State University Experiment Station (Dirmeyer and Skinner, 1968).

Moisture: The maximum moisture content of bentonite when shipped from the plant where it is processed is 10% and that for palygorskite is 16%.

Foundry Sand Bond

Tentative specifications for Western bentonite for bonding foundry sand are outlined in Steel Founders Society of America SFSA Designation 13T-65 issued in 1965. This specification requires the following characteristics and properties for Western bentonite, as tested by methods described in the specification: 1) water content shall not exceed 12% or be less than 6%, 2) pH value shall be equal to or greater than 8.2, 3) calcium oxide content shall not exceed 0.70%, and 4) the liquid limit shall not be less than 600 or greater than 850. Each foundry has its own green, dry, and hot strength specifications for bentonite, which vary with the type of metal, size of castings, and foundry production procedures. Methods of testing strengths of bentonite-bonded sands are outlined in detail in publications of the American Foundrymen's Society (Anon., 1962, 1963). The tests require the preparation of a specified mixture of bentonite, standard testing sand, and water, the preparation of test cylinders, and testing strengths with approved apparatus, according to controlled procedures.

Iron Ore Pelletizing

Specifications for bentonite used for pelletizing taconite-type iron ore have not been standardized, and several tests are used (Sastry and Fuerstenau, 1971, Wakeman, 1972). As green pellets must be capable of withstanding handling, compaction, and drying and the dry pellets must be even stronger, mixtures of bentonite, iron ore, and water are commonly tested for wet-drop strength, wet compression strength, plastic deformation, and dry compression strength.

Absorbent Granules

Most absorbent granules marketed are prepared to fulfill the requirements outlined in Federal Specification P-A-1056A, Absorbent Material, Oil and Water (for floors and decks), outlined for purchases by the US General Services Administration. A similar set of specifications is outlined in the American Society for Testing and Materials (ASTM) Standard: C 431–65 Standard Methods for Sampling and Evaluation of Sorptive Mineral Products Used as Floor Absorbents. The size and absorbent characteristics required of granules as outlined in the federal specification are listed in Table 1.

Table 1. Characteristics of Absorbent Granules

Characteristic	Max.	Min.
Matter retained on a No. 6 US standard sieve	1.0%	0.0%
Matter retained on a No. 30 US standard sieve	99.0%	52.0%
Matter retained on a No. 40 US standard sieve	99.8%	73.0%
Matter retained on a No. 60 US standard sieve	100.0%	90.0%
Ability to absorb lubricating oil (per gram of sample)	—	0.8 ml
Ability to absorb distilled water (per gram of sample)	—	0.9 ml
Solubility in distilled water	1.5%	—

The specifications require that absorbent granules consist of a uniform mixture of minerals of the silicate type. They also must be clean, uniform, and free of lumps or foreign matter, and no more than 10% of them can pass through an 80-mesh sieve in the attrition resistance test. The attrition resistance test is made by shaking with steel balls on a screen, according to a specified procedure.

Bleaching Oils

Test methods for evaluating bentonite and other types of fuller's earths for bleaching soybean and cottonseed oils are outlined in the

Fig. 3. (a) SEM of Wyoming bentonite, (b) SEM of Texas bentonite.

American Oil Chemists Society AOCS Official Method Cc 8b-52, revised April 1952, and AOCS Official Method Cc 8a-52, corrected 1958. These specifications contain instructions on bench-type tests, including stirring time, heating rates and temperatures; approved equipment; quantities of raw oil and clay required; methods of color determinations; and so forth. They also require the comparison of the material tested with an official natural bleaching earth approved by the American Oil Chemists Society. The color determination of the bleached oils and their comparisons with the oil bleached by the official earth in the same test are the basis for purchase specifications.

GEOLOGY

Mineralogy

Smectite: The major mineral constituent of bentonite, and the mineral that gives bentonite its unique physicochemical properties, is smectite (Fig. 3). Smectite is the general term given to a group of clay minerals that includes beidellite, montmorillonite, nontronite, saponite, and hectorite (Table 2). Each of these minerals shares a similar structure but each is chemically distinct. Nontronite, for instance, is an iron-rich smectite and hectorite is a lithium-rich smectite.

Table 2. Smectite Group Minerals

End-Member Mineral	Structural Subgroup	Chemical Composition
Montmorillonite	Dioctahedral	u, v, y = 0
Beidellite	Dioctahedral	v, y, z = 0
Nontronite	Dioctahedral	v, x, z = 0
Hectorite	Trioctahedral	z > 0
Saponite	Trioctahedral	z = 0

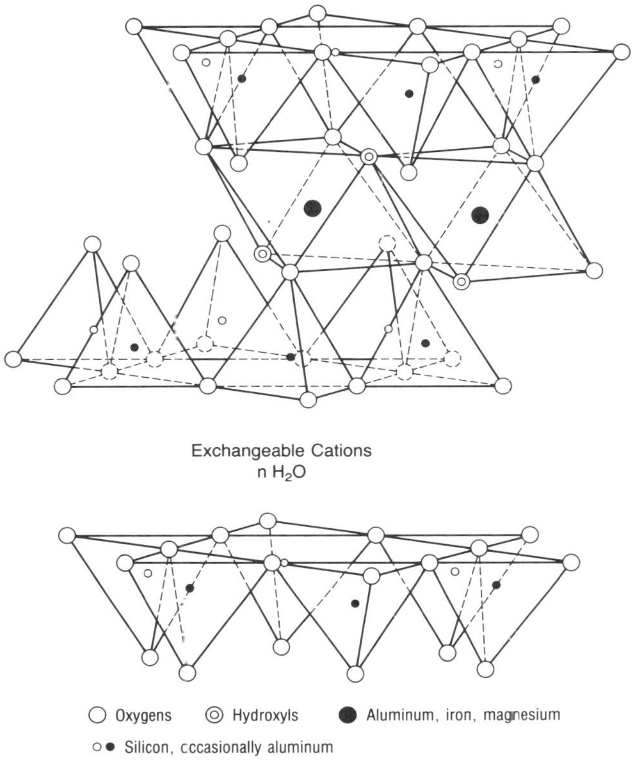

Fig. 4. Structure of smectite.

Structure: The most common smectite group mineral found in economic bentonite deposits is montmorillonite. The calcium and sodium varieties, named based on the predominant exchangeable cation, are the most abundant. Hectorite is also a major constituent of important commercial deposits.

Structurally these minerals are made up of stacked units comprised of a layer of octahedrally coordinated ions sandwiched between two layers of tetrahedrally coordinated ions (Fig. 4). These units or layers are often referred to as TOT units.

The smectite minerals are divided into two structural groups depending on whether all three octahedral sites are occupied (trioctahedral) or two out of three are occupied (dioctahedral) (Güven, 1988). The theoretical structural formula for the dioctahedral smectite minerals montmorillonite, beidellite, and nontronite is

$$[Al_xFe_yMg_z)_{2.00}(Si_{4.00-(u+v)}Fe_vAl_u)O_{10}(OH)_2]M^+_{u+v+z}$$

where M^+ represents the exchangeable cations. The trioctahedral smectites, hectorite and saponite, have the following theoretical formula:

$$[(Mg_{3.00-z}Li_z)(Si_{4.00-u}Al_u)O_{10}(OH_2]M^+_{z+u}.$$

These minerals exhibit substitution of various ions in both the tetrahedral and octahedral sheets. It is this substitution that gives smectite its surface properties which, in large part, control the physical properties of the bentonite. Smectite surface chemistry is determined by the amount and type of ionic substitution that takes place in the clay structure. The substitution of lower valence ions for higher valence ions, for example Al^{3+} for Si^{4+} in the tetrahedral sheet, results in a negative surface charge. This negative surface charge is balanced by cations which are loosely held on the surface and are therefore exchangeable.

Exchangeable Cation Chemistry: The most common exchangeable cations in smectite are sodium, calcium, magnesium, hydrogen, and potassium. These cations are found between neighboring TOT units. Water also enters between these layers. The amount of water is controlled by the type of cation present. Calcium, for example, allows 2 layers of water to enter between the sheets and sodium allows 1 or more layers of water to occupy this interlayer space. Organic polar molecules are also found in this position.

The ability that smectite has for exchanging cations, expressed as the cation exchange capacity (CEC), is measured in milliequivalents per 100 g of clay (meq/100 g). Most bentonites have CECs that range between 60 and 170 meq/100 g. As mentioned previously, the CEC is largely controlled by the substitution of lower valence for higher valence ions. It has also been found that the oxidation state of iron in the octahedral sheet controls the CEC (Stucki et al., 1984).

An excellent example of this phenomenon is found in the sodium bentonites mined in Wyoming. These bentonites exhibit blue (reduced) or yellow (oxidized) coloration depending on the oxidation state of the octahedrally coordinated iron. For many decades it has been known that, in general, the yellow or weathered bentonite possesses superior colloidal properties. Speculations were made as to why certain bentonite properties improved with weathering. It was originally proposed that weathering caused an increase in the concentration of exchangeable calcium and that this somehow enhanced colloidal properties (Knechtel and Patterson, 1962). Recently it was found that with oxidation there is a decrease in the negative surface charge which is accompanied by a decrease in the concentration of exchangeable calcium resulting in a high exchangeable Na/Ca (Elzea, 1990).

The amount and type of exchangeable cation influences the clay surface chemistry and charge which determines how the clay particles will interact in an aqueous suspension (Wolfbauer 1977, Foster 1953, Alther 1986, Lagaly 1989, Van Olphen 1977). It has been observed that sodium smectites compared to calcium smectites have higher swelling capacities and superior colloidal properties. This behavior is due to the presence of sodium as the major exchangeable cation and the hydration and charge properties of the sodium ion.

Under some conditions, and for certain applications, calcium smectite exhibits superior properties. For example, calcium smectite is more suitable than the sodium variety for drilling mud that will come in contact with saltwater. This is because sodium smectite will more easily flocculate in solutions with high ionic strengths or salinities. In addition, calcium bentonite, in general, has higher green strengths when used as a foundry bond clay.

It is standard industry practice to chemically treat calcium bentonite with soda ash or other additives to enhance its colloidal properties. This is done especially where natural sodium bentonite is not available; however, the quality of sodium exchanged calcium bentonite rarely matches that of natural sodium bentonite.

Other Minerals: Other minerals commonly found in bentonites fall into three categories: 1) those that are volcanic in origin, 2) those that form in situ as secondary minerals in response to diagenesis and weathering, and 3) those that are detrital contaminants. Typical volcanic minerals include feldspars, biotite, quartz, cristobalite, apatite, zircon, magnetite, amphiboles, and sometimes unaltered volcanic glass.

Fig. 5. Exposed surface of Wyoming bentonite showing a "popcorn" texture.

Minerals of low temperature origin include various forms of poorly ordered opaline silica (Elzea et al., 1991), zeolites, calcite, selenite, other clay minerals including kaolinite and illite, and various iron sulfate minerals. Detrital minerals are the least abundant and include all of the major rock forming minerals found in most sedimentary environments. These minerals rarely constitute more than 10% of a commercially viable deposit.

OCCURRENCE

Bentonite is typically found as beds in marine and nonmarine strata ranging in age from Jurassic to Pleistocene. These beds which tend to be parallel to the overlying and underlying strata, range in thickness from several centimeters to tens of meters and can be geographically very extensive. Bedded bentonite can also occur as small lens shaped bodies with a limited lateral extent. A less common occurrence of bentonite is as irregular shaped bodies that grade into unaltered host rock.

Bentonite ranges in color from white to yellow to olive green to brown to blue. It has a characteristic soapy texture and a waxy appearance. Weathered sodium bentonite has a distinctive "popcorn" texture (Fig. 5). Calcium bentonite weathers to an "aligator skin" texture.

ORIGIN

Bentonite most commonly forms as a result of the in situ alteration of volcanic ash or tuff. Another less common origin is as

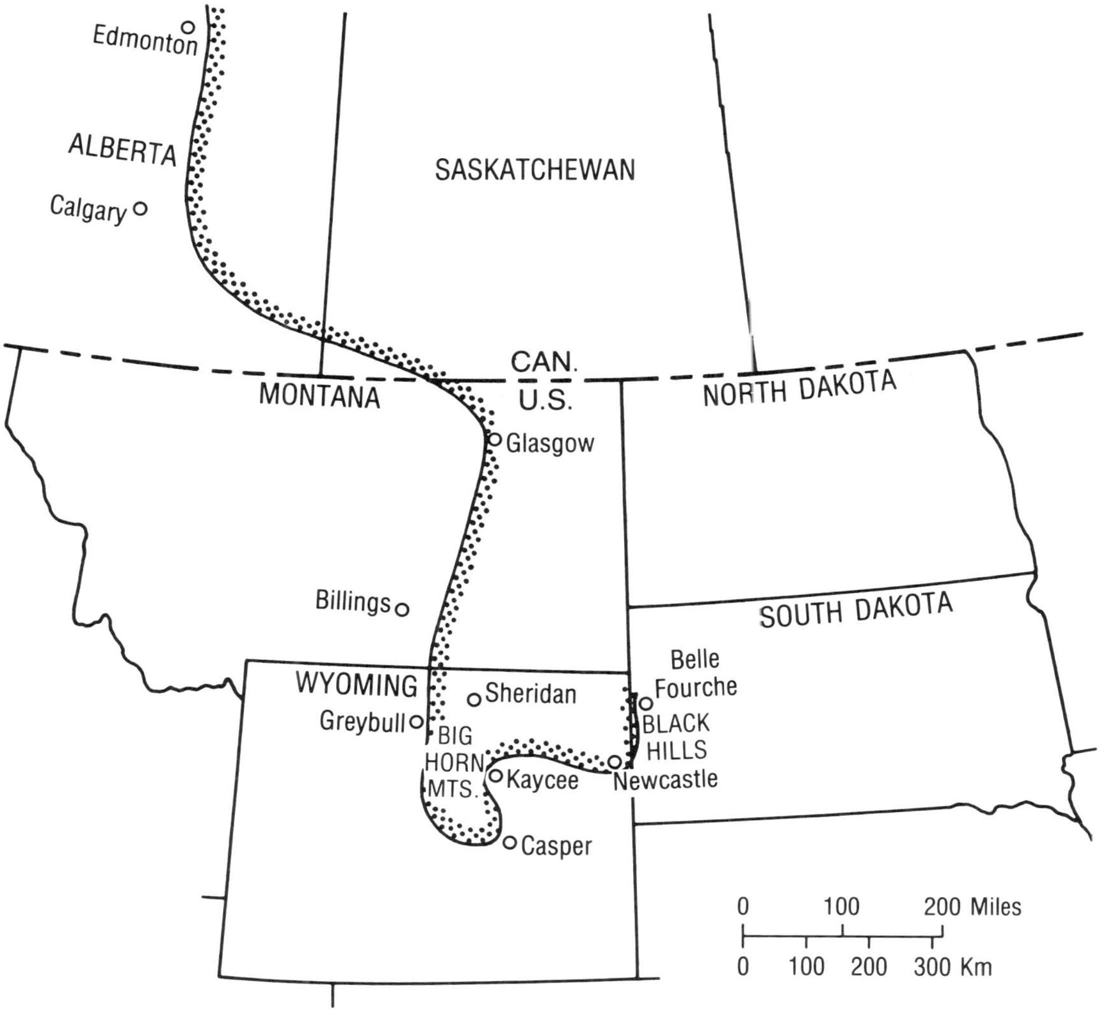

Fig. 6. Map of Western bentonite outcrop area.

the hydrothermal alteration of volcanic rocks. The well known hectorite deposits in California appear to have formed under hydrothermal conditions. The parent rock for the majority of bentonite deposits was most likely siliceous and probably ranged from dacitic to rhyolitic in composition.

The alteration of ash to smectite is a hydration reaction that requires magnesium and produces excess silica as well as alkali ions. Because of these requirements bentonite is most commonly found in marine strata where the ash apparently altered in contact with seawater. In some instances it also formed in contact with alkaline lake water and groundwater.

The evidence for bentonite having altered from volcanic ash is: 1) mineralogy—bentonite beds often contain euhedral minerals characteristic of volcaniclastic rocks, 2) chemistry—the major element and the rare earth element chemistry (Elzea and Murray, 1990) of the smectite is consistent with a volcanic precursor, 3) texture—relict glass shards pseudomorphically replaced by smectite have been observed both in thin section and under the scanning electron microscope. The most compelling evidence for a volcanic parent, however, is the presence of unaltered glass shards in some deposits.

The timing of alteration has been disputed. Some researchers believe that alteration begins as soon as the ash contacts water and others believe that alteration occurrs soon after the ash reaches the seafloor or lake bottom. Still others have proposed that alteration takes place after burial. It has been shown that the Wyoming bentonites altered after burial within the top several hundred meters of subseafloor sediment. Alteration occurred in contact with diagenetic pore fluids that evolved from Mowry seawater (Elzea, 1990).

DISTRIBUTION OF DEPOSITS

North America

Western Bentonite: Western or Wyoming bentonite is mined in three major districts across Wyoming and into Montana and South Dakota (Fig. 6). These districts supplied nearly 70% of the total 1990 US production of bentonite.

The easternmost mining district is centered near Colony, WY, and extends south to Upton, WY, and east to Belle Fourche, SD. This district, which is commonly referred to as the Black Hills District, produces high quality Na-bentonite primarily for the

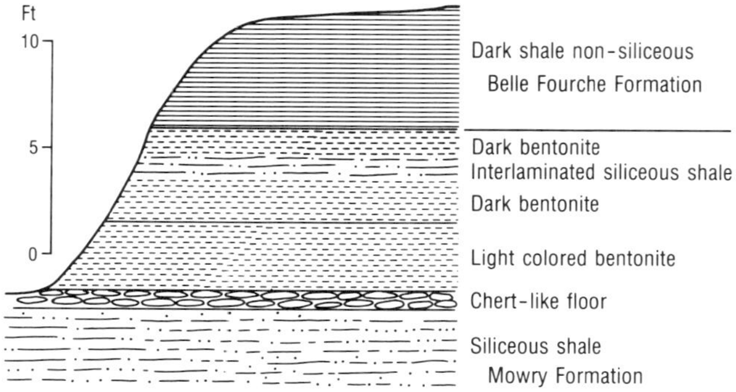

Fig. 7. Stratigraphic section of the Clay Spur bentonite.

foundry and drilling mud industries although new markets such as civil engineering applications and cat litter are becoming increasingly important. Currently there are 5 plants producing Na-bentonite in this area which is the oldest Na-bentonite mining district in the world.

Bentonite is also mined in central Wyoming along the eastern flank of the Big Horn Mountains near the towns of Casper and Kaycee (Fig. 6). One plant, near Casper, processes most of the clay mined in this area.

Along the western flank of the Big Horn Mountains is the third major area for Na-bentonite mining in Wyoming. Mines are located near Tensleep, Lovell, Greybull, Worland, and Thermopolis. Plants are located near Lovell, Greybull, Worland, and Thermopolis.

Approximately 8 beds are mined in Wyoming and the majority of these occur in the Cretaceous Mowry Shale (Fig. 7), which is an organic rich marine shale that was deposited in the Western Interior Seaway. Several commercial bentonite beds also occur in the Belle Fourche Shale and the Frontier Formation which overlie the Mowry Shale. In the Big Horn Basin, several beds are mined from older formations.

All of the Wyoming bentonite beds, except for the Newcastle Bed, which is mined in the Black Hills District, are intercalated with marine sediments. The Newcastle bed has a fairly limited geographic extent and appears to have been deposited in fresh water.

Southern Bentonite: Southern, or calcium bentonite, is mined in Texas, Mississippi, and Alabama primarily for use in foundry molding sands, as a refining agent in oils and fats, as a rheological additive, and in a variety of specialty applications. A limited market for oil well drilling mud also exists. Next to Wyoming this is the second most productive bentonite mining area in the United States.

Bentonite deposits are found throughout the Gulf Coast region and occur in Upper Cretaceous to Middle Tertiary age strata. In Mississippi commercially developed bentonite is produced from beds in the Eutaw Formation (Upper Cretaceous), Ripley Formation (Upper Cretaceous), and the Vicksburg Formation (Tertiary). Bentonite deposits in Alabama are also mined from the Ripley Formation.

In Texas, extensive bentonite deposits have been developed in the Jackson and Gueydan Formations both of which are Tertiary in age. Some of these clays are treated to produce organophilic clays for use as rheological additives in liquid organic systems such as paint, ink, grease, and cosmetics. A major market for untreated calcium bentonite mined in Texas is as a rheological additive for ceramic glazes, as a plasticizer in whiteware, and as a plasticizer and binder in plastic refractory materials. Texas bentonite is also finding a major use in water treatment.

Fuller's Earth: The principle bentonite fuller's earth producing district in the United States is the paleocene Porters Creek

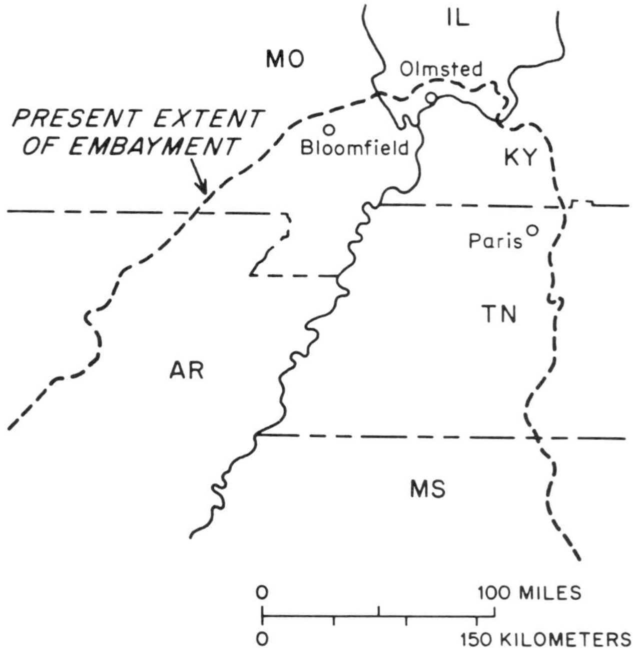

Fig. 8. Outcrop area of the Porter's Creek Formation (dotted line).

Formation with producing plants located on the Mississippi embayment outcrop area from Bloomfield, MO, to the west up through the southern tip of Illinois and then south through Tennessee and into Mississippi (Fig. 8). The total thickness of the Porters' Creek Formation ranges from 3 to 40 m in the north (Thomas, 1981) to up to 150 m in Mississippi (Hughes, Jr., 1958). The portion that is most heavily mined is the upper 3 to 12 m of weathered tan clay which contains less pyrite and is easier to process. The minerals contained in the Porters' Creek Clay are calcium-magnesium smectite, opal C-T, kaolinite, illite, and trace amounts of halloysite and clinoplitolite. Pyrite is concentrated in the top portion of the underlying black to dark gray clay (Reisch, 1992).

Fuller's earth is also provided from the Miocene Monterrey Formation near Taft, CA, in the southwestern corner of the San Joaquin Valley. The major use of this material is cat litter and some oil absorbents. The Monterrey Formation is several hundred meters thick, but only those areas where the bulk density is less than 19 kg per m^3 is mined.

Other areas where fuller's earth is mined and processed are from the Hawthorne Formation near Ocala FL, from the Twiggs Clay Member of Barnwell Formation near Wrens, GA, and from a Miocene diatomaceous earth near Christmas Valley, OR.

Hectorite: Hectorite, a magnesium-lithium smectite, has unique properties that allow it to impart certain thixotropic and rheological properties in aqueous systems. Most hectorite is mined from Tertiary age deposits near Hector, CA, which is the type locality for this mineral. These deposits are altered alkaline lake sediments.

Hectorite is also mined from a deposit in the Amargosa Valley, Nevada. This deposit consists of approximately 25% hectorite and 75% calcium carbonate. It averages approximately 4.6 m thick and extends over a distance of several kilometers. West of McDermitt, NV, relatively pure to impure hectorite beds and lenses, associated with volcanic sediments, zeolites, dolomites, and limestones have been in commercial production since 1986 (Odom, 1992).

Other Deposits

Most of the major bentonite production in the United States is centered around Wyoming and the Gulf Coast. In 1990 Wyoming and Mississippi accounted for 90% of the total US bentonite production. A number of small, but significant, deposits are mined in other areas.

Southwest of Boise, ID, sodium bentonite is mined from the Miocene Chalk Hills Formation. This clay is used primarily for the preparation of livestock feed, and it is also used in cat litter. Elsewhere in the western United States, bentonite is mined from deposits in the Death Valley region of California, western Nevada, and Arizona. The clay mined in Arizona comes from near Chambers and has been exploited for a number of years for use as a desiccant and as a bleaching earth. Sodium bentonite is also mined near Adrian, OR. This bentonite is used in the civil engineering and drilling mud industries. Several small deposits of sodium and calcium bentonite located in San Pele County, Utah, are mined primarily for civil engineering applications. Another use for these clays is as a binder in animal feed.

Canada: Bentonite deposits occur throughout western Canada in Alberta, Manitoba, British Columbia, Saskatchewan, and the Northwest Territories. These deposits are distributed throughout Cretaceous and Tertiary age sediments and are most extensively developed in Upper Cretaceous marine sedimentary horizons (Grim and Guven, 1978). They are most commonly associated with shales and limestones, but also occur as partings in coal beds. These bentonites are the same age as those in Wyoming, Montana, and the Dakotas, but differ in that calcium is most commonly the major exchangeable cation. Sodium bentonites are much less common.

Commercially developed sodium bentonite deposits are found in Saskatchewan and are mined for use in drilling mud, foundry, civil engineering, and animal feed applications. Another major mining district is the Pembina District in Manitoba. Low swelling calcium bentonite is extracted from beds in the Upper Cretaceous Vermillion River Formation and is processed in plants located in Morden, Miami, and Winnipeg. The plant in Winnipeg produces acid-activated clay which is used in the refinement and decolorization of a variety of oils and waxes.

Mexico: Bentonite and fuller's earth deposits are distributed throughout central Mexico where volcanoclastic rocks are pervasive. The majority of these bentonites contain calcium as the dominant exchangeable cation and many of them occur in marine rocks of Tertiary age.

Both calcium and sodium bentonites are mined in Mexico which produced 136 kt of bentonite in 1986 (O'Driscoll, 1989). The largest calcium bentonite deposits are located in Tlaxcala and Morelos in Hidalgo. The principal sodium bentonite producing region is in Durango. The sodium bentonite is marketed to the foundry industry and is also used in detergents, pencil manufacture, and cosmetics. Acid activated and sodium exchanged calcium bentonites are also produced in plants in Durango and Puebla.

South America

Brazil: Bentonite deposits have been reported in the states of Minas Gerais, Rio Grande Sul, Santa Catarin, Pariba, Parana, and Sao Paulo. Most of these clays are of little commercial value because they contain high percentages of minerals other than smectite. Relatively pure bentonite deposits exist in the states of Pariba, Sao Paulo, and Bahia and these supplied approximately 99% of the 206 kt of bentonite produced in Brazil in 1986 (O'Driscoll, 1989).

Peru: Calcium and acid activated bentonites are mined and processed in Piury, Vichayal. Occurrences of bentonite and fuller's earth have been reported in the provinces of Pisco, Paita, Canete, and Contralmirante Villar (Cabrera, 1963). These deposits are found in Tertiary strata located in the coastal ranges of northern and south central Peru.

Chile: Currently extensive deposits of calcium bentonite are being mined in the Arica, I Region. This clay is sold as a fish meal pelletizing agent and will probably be sold as an acid-activated clay for bleaching in the very near future (Donoso and Theume, 1990).

Europe

United Kingdom: Calcium bentonite (known as fuller's earth in British terminology) deposits are well known and commercially exploited in Upper Cretaceous and Jurassic age sedimentary rocks. The major mining areas are in Surrey, Oxfordshire, and Bedfordshire. Beds in this region supply bentonite to most of the major markets including drilling mud, foundry, cat litter, civil engineering, animal feed, and other miscellaneous uses (O'Driscoll, 1989).

Germany: Germany is one of the three largest producers of bentonite in the world. The bentonite, which is the calcium variety, is extracted from deposits in Bavaria where it occurs in marine marls and tuffaceous sands in the Upper Miocene Molasse (Grim and Güven, 1978). These bentonites probably formed as the result of the in situ laterization of acidic tuff. Many of the bentonite beds in this region have no commercial value because they contain large amounts of illite and kaolinite.

Several grades of bentonite are produced in Germany. Calcium, acid-activated, and sodium exchanged bentonites are supplied to the European and world markets. The acid-activated clay makes a high quality decolorizing product and the sodium exchanged bentonite is sold primarily for uses in oil well drilling and foundry. Expanding markets include cat litter and soil sealants.

Sweden: There is only one bentonite deposit known in Sweden, but it is too small and too low in quality to be a commercially viable deposit. It is located in Kinnekulle.

France: Currently there is one bentonite mine in France at La Buisson in the Dordogne region. This mine produces a small tonnage (approximately 6 to 7 ktpy) for animal feed. Other bentonite beds have been discovered in France, but there has been very little commercial development of these deposits because they are small and generally poor in quality.

Greece: Extensive and very thick bentonite deposits, up to 30.5 m at some localities, occur on the island of Milos. Some of these deposits formed due to the alteration of volcanic ash deposited in a marine environment during the Pliocene (Wetzenstein, 1972). Others may have formed as the result of the hydrothermal alteration of dacitic tuffs (Grim and Güven, 1978).

Greece is second in world production after the United States. Both natural calcium and sodium exchanged calcium bentonite are produced for a variety of uses including drilling mud and foundry. Most of the clay is exported to western Europe and some is exported to Canada for use in iron ore pelletizing.

Italy: Extensive calcium bentonite deposits occur on the island of Sardinia and are mined by a number of companies that produce calcium and sodium exchanged bentonite. These deposits formed by the hydrothermal alteration of trachytes and trachtic tuffs (Annedda, 1956). Major uses for these bentonites include drilling mud, foundry, civil engineering, and animal feed pelletization.

Spain: Calcium bentonite is mined in Madrid and Almeria. The deposits in Almeria formed by the hydrothermal alteration of rhyolite and andesites (Martin-Vivaldi and Linares, 1968). Natural, acid-activated and Na exchanged grades are available for a variety of applications. At present these clays are used in drilling mud, foundry, civil engineering, animal feed, wine clarification, oil bleaching, ceramics, and other specialty applications such as carbonless copy paper.

Cyprus: Calcium bentonite is mined near Limassol. Oil well drilling, animal feed, civil engineering and absorbent grade bentonite is exported to Europe and the Middle East.

Turkey: Deposits of sodium and mixed Na/Ca bentonite are known in many parts of Turkey, but only a few are in production.

Africa

Morocco: Large reserves of various types of bentonite occur in Cretaceous and Tertiary sedimentary rocks that contain interbedded volcaniclastics. Evidence for extensive volcanic activity in this region includes lava flows and intrusions of various compositions. Conditions and mode of alteration vary from deposit to deposit. For example, some deposits formed as a result of hydrothermal or deuteric alteration while others underwent alteration in saline lakes (Grim and Güven, 1978).

In the past many of these deposits have been commercially exploited. Currently calcium bentonite with a high magnesium content is mined in Morocco by two producers. Approximately 4 kt are mined for local consumption in the drilling industry. The deposits are located at Trebia, Bouhoua, and Providencia.

South Africa: Sodium and calcium bentonite are mined from deposits at Koppies, Orange Free State, and Heidelberg, Cape Province. A number of different grades are produced for a variety of applications. Sodium bentonite and Na/Ca blends are sold for use in the foundry (40%), pelletizing (40%), drilling (10%), and civil engineering (10%) industries. Acid-activated bentonite for bleaching and polymer treated bentonite for toxic waste containment are also produced (O'Driscoll, 1989).

Australia

Bentonite deposits are distributed throughout Australia and are found associated with a number of different rock types varying in age from Permian to Pliocene. For example the Permian—Triassic coal measures in New South Wales contain several bentonite beds that appear to have formed from the alteration of some sort of igneous material (Grim and Güven, 1978). Another interesting occurrence is reported in Upper Jurassic sandstones also in New South Wales.

Calcium bentonite is produced in Cressfield, NSW, for foundry use. Sodium bentonite is mined at Gurulmundi in Queensland, which is located approximately 350 km west of Brisbane. Markets which this clay supplies include animal feed, drilling, foundry, civil engineering, and iron ore pelletizing. In the past fuller's earth has been mined in the Dubbo District of New South Wales.

Asia

India: Commercial bentonite and fuller's earth deposits are located in the states of Bihar, Gujarat, Jammu, Kashmir, and Kutch where production totals approximately 400 ktpy (Russell, 1991). Many grades of bentonite are produced in India including foundry, drilling mud, bleaching, iron ore pelletizing, civil engineering, and pharmaceutical grades. Calcium, sodium, sodium exchanged calcium, acid-activated and organo clays are the primary products available from India. The organo clays are used as a rheological additive in paints, in tire manufacture, and in resins.

Pakistan: Calcium bentonite and fuller's earth deposits are mined in Pakistan. The calcium bentonite is exchanged with sodium so that it can be used as an additive to drilling mud. The fuller's earth is acid activated for use in the refinement and decolorization of oils.

Japan: Bentonite has been produced commercially in Japan since 1938 (Grim and Guven, 1978). Sodium bentonite and a more unusual hydrogen, or acid, clay are the two major types of economically valuable deposits in Japan. Japan is the fourth largest producer of bentonite with an annual production of 400 ktpy, and it produces all major grades of bentonite.

These clays altered from a variety of volcanic rock including pumice, tuff, and ash. Modes of alteration appear to have encompassed a range of processes including deuteric alteration, hydrothermal alteration, and subaqueous devitrification. The formation of the acid clays was probably greatly influenced by post-alteration weathering processes. Most of the deposits occur in Miocene and Pliocene age strata including marine mudstones, siltstones, sandstones, and coal (Sudo and Shimoda, 1978).

EVALUATION OF DEPOSITS

Exploration

Exploration for bentonite and fuller's earth deposits consists of several steps. The first step in exploration is to identify areas that are likely to contain bentonite deposits. This step requires a complete understanding of the genesis of bentonite and the types of geologic environments likely to have favored bentonite formation. For example, bentonite is known to most commonly form as the result of the subaqueous alteration of volcanoclastic material. It is also known that bentonite deposits are commonly found associated with Cretaceous and Tertiary marine strata. Therefore, a likely target for a bentonite deposit would be an area where volcanism is known to have taken place and where water, either ocean or lake, was available at the time of alteration. This is a simplified example but provides some insight into how experience and knowledge of bentonite occurrence and geology is essential to a successful exploration program.

Once target areas have been identified, a field reconnaisance is usually conducted to establish whether or not clay deposits are present. If clay is found then the next step is to map the extent of the deposits. In some areas, such as in Wyoming, the bentonite is easily recognized on the surface and the deposits can be mapped with relative ease. In other areas drilling must be conducted to accurately map the deposits. Usually both surface mapping and drilling are utilized to determine the extent of available reserves.

After a deposit has been located and mapped other factors must be considered before capital is invested in evaluation and extensive testing of the deposit. Factors such as amount of overburden, distance to markets, available transportation, and amount of reserves must be considered.

Mineralogical Investigations

The mineralogy, and the chemistry of a clay are important properties that must be considered when evaluating the economic use of a deposit. A detailed understanding of the particle size and shape of the clay fraction can also be essential to making the appropriate determination for the end-use of a clay.

Simple physical testing is no longer sufficient for a number of reasons. One reason is that these clays are being used in an increasing number of specialty products that require very stringent specifications, many of which are related to chemistry, particle size, and shape. Another is that reserves of the high quality clays, for which the standard testing procedures and specifications were written, are becoming depleted and different quality clays are being mined. These clays often must be beneficiated to meet the traditional specifications and this requires a detailed understanding of mineralogy and chemistry. A third reason for characterizing mineralogy and chemistry is that there is an increasing interest in the research and development of these clays for use in more highly technical industrial applications such as catalysts, non-linear optics, advanced ceramics, and polymer composites.

Analytical methods for studying clay minerals include X-ray diffraction, scanning electron microscopy, transmission electron microscopy, infrared spectroscopy, differential thermal analysis, and optical microscopy. These methods are used to study the structure and chemistry of raw clays and clay based products. Other methods used to investigate clays include various methods for determining structural chemistry, surface chemistry, surface area, and particle size distribution to name a few. Many companies that mine and produce these clays have in house research and development programs. Companies that do not have these resources can obtain analytical data from consultants experienced in the use and interpretation of these analytical methods.

Appraisal of Field and Laboratory Results

Testing: The testing of bentonite and fuller's earth is handicapped by the general absence of universally accepted procedures and specifications. As a result, these clays are tested and their valuable properties are expressed in many different ways. Some of the reasons for this are the widely varying properties of the clays and the many different specifications outlined by purchasers for many uses. The purchaser of a single grade of bentonite may require tests of a slurry of different concentrations, preparation procedures, aging, temperature specifications, and even measurements of liquid and plastic limits. Clearly, the absence of standardized test procedures indicates the need for an effort on the part of the American Society for Testing and Materials or another organization, with the cooperation of industry, to establish useful procedures having wide application. Rather than go into details on the many empirical and tentative test procedures that are now in use in evaluating bentonite and fuller's earth and that are likely to be superseded by more scientific and technically accurate methods, only a few generalized comments on tests will be made.

As swelling bentonites are generally more suitable for uses requiring colloidal properties and low-swelling ones are used in other ways, a preliminary indication of the potential value of bentonite can be determined by a swelling test. Perhaps the most common test is the simple procedure of adding 2 g of dried and ground bentonite to a 100-ml graduate filled with distilled water and determining the volume of the settled bentonite in milliliters (Ross, 1964).

Mineralogy: In addition to the testing of the type outlined in the foregoing paragraph, information on the mineralogical and chemical composition of bentonite and fuller's earth is also required. Mineralogical composition, including the purity and characterization of undesirable contaminants, is determined using X-ray diffraction, electron microscopy, differential thermal analysis, petrographic microscopy, and other methods of investigating clays.

Appraisal of Field and Laboratory Results: The final step in the evaluation of bentonite and fuller's earth deposits involves the synthesis of data collected during the field investigations, sampling and testing, and other laboratory programs. Maps are prepared locating drill holes and showing the thicknesses and quality of clay and thicknesses of overburden. Reserves of various grades of clay are calculated, and the mining cost is determined from such maps and from geologic and engineering notes taken during the fieldwork. Factors such as amount of overburden, distance to the plant, environmental permitting, and accessibility to roads must be considered when estimating the cost of mining.

Preparation for Markets

Mining Methods: In the United States, virtually all bentonite is mined by stripping methods. Bulldozers and scrapers or pans are most commonly used in removing overburden. In the typical pit, the overburden is removed in panels, the clay mined by loading trucks with dragline or endloaders, and the overburden from an adjacent panel shifted to the mined out area. One company hauling on privately owned roads operates trucks having 100-t capacity, but most hauling requires use of public highway and lighter trucks. Thicknesses of overburden removed vary considerably. Most Wyoming-type bentonite mined is under less than 9 m of overburden, but in a few places overburden is as much as 12 m thick. The thickness of overburden removed from some southern bentonite deposits is as much as 30 m (Teague, 1972), and beds of the same thickness have been stripped from low swelling bentonite in one district in Arizona. The overburden stripped from fuller's earth deposits ranges from a few meters to more than 23 m.

The rocks overlying most bentonite deposits are soft and can be loaded directly by self-propelled or pushed scrapers, but in places they must be loosened by bulldozers and rippers before loading. Volcanic rocks above hectorite deposits in California also must be blasted where stripping is done. Bentonite is mined underground by the square-set method in the New Discovery mine near Beatty, NV (Papke, 1969), and insofar as the authors are aware, this is the only bentonite now mined underground in North America. Hectorite in California and bentonite in the Ash Meadows District, Nevada (Papke, 1969), and the Chambers [Cheto], Arizona, district were formerly mined underground.

Most bentonite produced in other countries is also strip mined, but underground methods are used in a few places, including the Combe Hay District in the United Kingdom (Anon., 1969d). Deposits in this district are selectively mined by hand, using compressed air picks. The clay is extracted in lumps 7 to 45 cm across, and it is hauled to underground storage facilities in mine cars drawn by electric locomotives. It is loaded as needed from the storage dumps onto trucks and hauled to the plant.

Because of variable physical properties, most bentonite deposits are selectively mined. Bentonite from a single pit or bed may be separated into as many as three stockpiles at the plant, each with different physical properties. Various grades of bentonite suitable for different uses are then prepared from the separate piles and by blending clay from more than one pile. Blending bentonite as it is dumped on stockpiles using earth-moving or cultivating equipment to obtain a uniform clay, is also common practice.

Beneficiation and Processing Techniques: Bentonite is processed by simple milling techniques that involve removal of water and in some instances, other volatile matter, and grinding to suitable sizes (Fig. 9). The Wyoming or high swelling bentonite, when delivered to the plant, contains approximately 30% moisture; the free water in the southern or calcium-type bentonite is about 25%. Some fuller's earth contains as much as 50% volatile matter and 10% undesirable impurities. The processed bentonite ordinarily contains only 7 or 8% water, but because it is hygroscopic, it may contain considerably more when used. Moisture content of the fuller's earth is approximately the same as that of bentonite. Most absorbent granules, however, are heated to a temperature high enough to drive off interlayer and zeolitic water in the minerals present. Any moisture in the granules has been absorbed after processing.

In most plants, the raw clay is passed through some sort of a clay "slicer" to break up the large chunks before drying. A fluid-bed dryer is in operation in one bentonite plant at Colony, WY. The temperature in dryers varies with the intended use of the clay. The desirable properties of bentonite is reduced greatly if the clay is

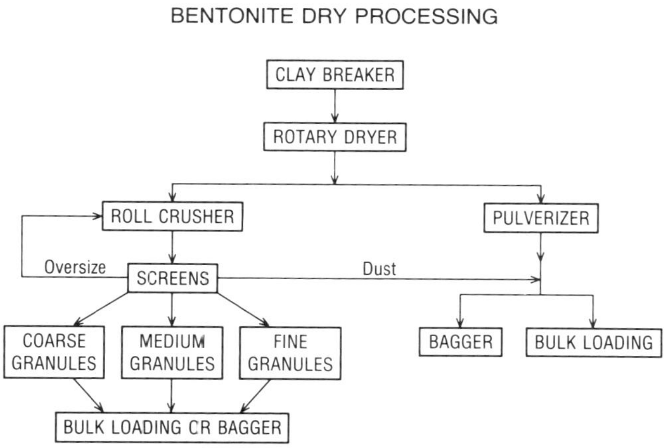

Fig. 9. Flow sheet of bentonite processing.

heated too much. The temperature in dryers processing bentonite is likely to be approximately 800°C at the inlet, 100 to 200°C at the outlet, and 400 to 500°C in the main drying zone. The bentonite itself is kept at temperatures of less than 150°C (Ross, 1964).

The dried clay is ground and sized in several ways. In some plants, rods in rotary dryers do much of the grinding, but most of the powdered product is ground with roll and hammer mills or other pulverizers, and screened. Most bentonite is ground to approximately 90% finer than 200 mesh (US Standard sieve size). Granular grades of bentonite are also sold, and the wide use of absorbent granule-type fuller's earth requires the preparation of a coarse product. The preparation of coarse absorbent granules is, in effect, a beneficiation process, because much of the fine and medium-grained sand impurities in the fuller's earth are in the size fraction discarded.

Hectorite and some Southern bentonite, in addition to being processed like other bentonite, are beneficiated by a hydroclassification process. In this process, the clay is dispersed in water and pumped into a multistage concentrating circuit where nonclay and other undesirable materials are separated and removed. The slurry is then passed through drum or spray-type dryers, and the concentrate is pulverized. A high value material is prepared in one plant by centrifuging a dilute slurry.

Transportation: Several types of packaging are used for both bentonite and fuller's earth. For other than bulk shipments, multiwalled paper bags, some of which have plastic liners, are most common. Fiber drums are used for specialty products sold in rather small quantities. Metal drums of 45 kg capacity are required for some uses, particularly absorbent granules used aboard ships and where lengthy periods of storage or repeated handling are necessary. One French company uses the modern containerized method for export shipments (Anon., 1969d). Most of the granules sold for animal litter are packaged in 4.5 to 13.6 kg paper bags, and packaging is a major item in the cost of the product.

The method of transporting bentonite depends on use and location of markets. Most of the processed bentonite is shipped by rail in bulk hopper cars, but boxcars and bulk trucks are also used. Bagged bentonite is commonly palletized for shipment by either rail or truck. Trains carrying only bentonite are used for large shipments to distant points and for overseas consignments. Ore carrier type ships having bentonite as the only cargo are commonly used for Great Lakes and overseas shipments. The bentonite transported this way is only partly dried at the point of origin and is shipped semicrude. This method makes it possible to load and unload it by bulk handling equipment, avoiding the costs of bagging, but it requires further drying and grinding at or near the place used. Much of the bentonite used in pelletizing taconite ore, as well as most of that sold in the European markets is handled in this way.

Synthetic and Organic-clad "Bentonite"

Synthetic hectorite is made in the United Kingdom by Laporte Industries, Ltd., and sold under the brand name *Laponite CP* (Green et al., 1970). This material is purer than naturally occurring varieties, as it contains no fluoride. It presumably is suitable for some of the same uses that synthetic bentonite was developed for in the United States. The synthetic hectorite and montmorillonite plant in Houston is no longer in production.

For several decades an organic treated bentonite has been on the market under the registered trade name *Bentone* (Grim, 1962). This product is processed in such a way that the original inorganic exchangeable ion on the bentonite is replaced by an alkyl amine organic cation. This reaction produces a hydrophobic clay because the inorganic ions that can be hydrated are removed, and a large part of the clay mineral surface formerly capable of absorbing water is coated by hydrocarbon chains. Organic-clad bentonites are used in making paint, to gel organic liquids, to produce greases having superior properties for adherence to metal, ability to repel water, and resistance to high temperatures, and in other products.

Organic-clad bentonite is also made in the United Kingdom, France, and Germany (Anon., 1970) and in Japan (Takeshi and Kato, 1969). Some of the Japanese product is sold under the brand name *Nikkagel*. It is prepared by treating bentonite with alcohol under pressure and heat followed by ion exchange with an organic ammonium salt. A product called *Organite* is prepared by further exchanging *Nikkagel* with an amine having a benzene nucleus.

Future Considerations and Trends

The overall demands for both bentonite and fuller's earth seem destined to increase at a rate approximately equal to the growth in Gross National Product. According to Cooper (1970), in the year 2000, production of bentonite in the United States is likely to be three to five times the 1968 amount. The growth rate in fuller's earth requirements will probably exceed the rate forecast for bentonite.

Both bentonite and fuller's earth are used in so many different products and in so many ways that there are many possible substitutes for specific uses and both have lost markets to other materials in the past. One of the most obvious losses of fuller's earth to other materials occurred when the use of synthetic materials in petroleum refining increased before World War II, and more efficient methods of refining were introduced. Bentonite sales have also suffered mildly from the use of oil-based and specialty mud; however, bentonite is likely to remain the cheapest and most efficient material for mud in certain types of drilling for years to come. Bentonite is in some danger of a market loss which may be caused by a breakthrough in the use of another material for bonding taconite pellets. Though the proportion of bentonite added in taconite bonding is small, it is still a contaminant, and some type of bonding material that will burn off completely or even add heat in the steel furnace will continue to be sought.

The manufacture of synthetic hectorite and other montmorillonite-like mineral products formerly in the United States and now in the United Kingdom marks a technological breakthrough. Worldwide markets for organic-clad bentonite and washed or otherwise purified white bentonite for specialty products are also likely to increase with general economic growth and the development of new uses and products.

Rising costs of transportation and fuel are two of the major pressures on the bentonite and fuller's earth industries that are likely to have a significant influence on future trends. The cost spread between rail and ocean transportation is the principal reason for the loss to Greek producers of some of the Wyoming-type bentonite market for bonding taconite ore in eastern Canada. The comparatively low cost of ocean transportation is also the primary motivation behind an overseas search for bentonite reserves near deep water. Progressive increases in fuel costs, particularly for the low sulfur coal type required for environmental reasons, seem to be almost a certain forecast.

REFERENCES

Anon., 1958, ''Bleaching Tests,'' A.O.C.S. Official Method, Cc 8a 52, American Oil Chemists Society, 2 pp.

Anon., 1962, *Moulding Methods and Materials*, American Foundrymen's Society, Chicago, IL, 619 pp.

Anon., 1963, *Foundry Sand Handbook*, 7th ed., American Foundrymen's Society, Chicago, IL, 200 pp.

Anon., 1965, ''Tentative Specifications for Western Bentonite,'' Designation BT-65, Steel Founder's Society of America, 5 pp.

Anon., 1969, ''API Recommended Practice Standard Procedure for Testing Drilling Fluids,'' API RP 13B, 2nd ed., American Petroleum Institute, 19 pp.

Anon., 1969a, ''API Specifications for Oil Well Drilling Fluid Materials,'' API STD 13a, 5th ed., American Petroleum Institute, 11 pp.

Anon., 1969b, ''Bentonite,'' Specifications No. DFCP.4, Oil Companies Materials Associations, London.

Anon., 1969c, ''Laporte's Lead in Fuller's Earth Continued Expansion at Redhill,'' *Industrial Minerals*, No. 24, Sep., pp. 15–16.

Anon., 1969d, ''Ste. Francaise des Bentonites, A Range of Quality Clays,'' *Industrial Minerals*, No. 24, Sep., p. 45.

Anon., 1970, ''An Emphasis on Minerals Marketing, Berk Ltd.'s Mineral Products Division,'' *Industrial Minerals*, No. 36, Sep., pp. 40–41.

Anon., 1972, ''Industrial Minerals in Oil Well Drilling,'' *Industrial Minerals*, No. 60, Sep., pp. 9–31.

Alther, G.R., 1986, ''The Effect of Exchangeable Cations on the Physico-Chemical Properties of Wyoming Bentonites,'' *Applied Clay Science*, Vol. 1, pp. 273–284.

Annedda, V., 1956, ''Deposits of Bentonite in Sadali and Vallanova Tulo Territories, Sardinia,'' Resuconti Association Min. Sarda, 60, pp. 50–59.

Bicker, A.R., Jr., 1970, ''Economic Minerals of Mississippi,'' Bulletin 112, Mississippi Geological, Economic and Topographical Survey, 80 pp.

Blackman, A.G., 1969, ''Bentonites Major Market: Two, Civil Engineering,'' *Industrial Minerals*, No. 25, pp. 23–25.

Cabrera, A. LaR., 1963, ''I. Bentonitas, II. Diatomitas,'' Peru, Ministerio de Fomento y Obras Publicas, Minerales No-metalicos Serie, Memoir No. 7, 107 pp.

Cook, G.H., and Smock, J.E., 1878, ''Report on the Clay Deposits of the Woodbridge, South Amboy and Other Places,'' New Jersey Geological Survey, Trenton, 381 pp.

Cooper, J.D., 1970, ''Clay,'' *Mineral Facts and Problems* 1970, US Bureau of Mines, Bulletin 650, pp. 923–938.

Davis, C.W., et al., 1940, ''Bentonite, Its Properties, Mining, Preparation and Utilization,'' Technical Paper 609, US Bureau of Mines, 83 pp.

Davison, J., 1969, ''Bentonites Major Markets: Three, Iron Ore Pelletising,'' *Industrial Minerals*, No. 25, Oct., pp. 27–31.

Dirmeyer, R.D., and Skinner, M.M., 1968, ''Evaluation of Colorado Clays for Sealing Purposes,'' Technical Bulletin 83, Colorado State University Experiment Station, 36 pp.

Donoso, R., and Theume, C., 1990, ''Non-Metallic Mining on Chile,'' *Industrial Minerals*, No. 277, pp. 63–71.

Elzea, J.M., 1990, ''Geology, Geochemistry, Selected Physical Properties and Genesis of the Cretaceous Clay Spur Bentonite in Wyoming and Montana,'' PhD dissertation, Indiana University, Bloomington, IN, 201 pp.

Elzea, J.M., et al., 1991, ''Characterization of Low Temperature Silica Polymorphs in Calcium Bentonites, Sodium Bentonites, and Fuller's Earths by XRD, SEM/EDS and TEM,'' 28th Annual Meeting of the Clay Minerals Society, Houston, TX, Oct. 5–10.

Elzea, J.M., and Murray, H.H., 1990, ''Variation in the Mineralogical, Chemical and Physical Properties of the Cretaceous Clay Spur Bentonite in Wyoming and Montana (USA),'' *Applied Clay Science*, Vol. 5, pp. 229–248.

Fisk, H.G., 1946, ''Bentonite with Test Methods and Results of Tests of Wyoming Bentonites,'' Bulletin 2, Wyoming, University Natural Research Institute, 39 pp.

Foster, M.D., 1953, ''Geochemical Studies of Clay Minerals: II. Relation Between Ionic Substitution and Swelling in Montmorillonites,'' *American Mineralogist*, Vol. 38, Nos. 11,12, pp. 994–1006.

Gildersleeve, B., 1946, ''Minerals and Structural Materials of East Tennessee,'' Report B, Tennessee Valley Authority Regional Products Research Division, 26 pp.

Green, J.M., et al., 1970, ''Thermal Reactions of Synthetic Hectorite,'' *Clays and Clay Minerals*, Vol. 18, pp. 339–346.

Grim, R.E., 1962, *Clay Mineralogy*, 2nd ed., McGraw-Hill, New York, 596 pp.

Grim, R.E., 1973, ''Technical Properties and Application of Clays and Clay Minerals,'' *Proceedings, International Clay Conference 1972* (AIPEA), Madrid, pp. 719–721.

Grim, R.E., and Guven, N., 1978, *Bentonites—Geology, Mineralogy, Properties and Uses*, Developments in Sedimentology, 24, Elsvier, Amsterdam, the Netherlands, 256 pp.

Güven, N., 1988, ''Smectites,'' *Reviews in Mineralogy*, Vol. 19, S.W. Bailey, ed., pp. 497–559.

Hughes, R.V., Jr., 1958, ''Kemper County Geology, Mississippi Geological Survey,'' Bulletin 84, 274 pp.

Knechtel, M.M., and Patterson, S.H., 1962, ''Bentonite Deposits of the Northern Black Hills District, Wyoming, Montana, and South Dakota,'' Bulletin 1082-M, US Geological Survey, pp. 893–1030.

Knight, W.C., 1898, ''Bentonite,'' *Engineering & Mining Journal*, Vol. 66, No. 17, 491 pp.

Lagaly, G., 1989, ''Principles of Flow of Kaolin and Bentonite Dispersions,'' *Applied Clay Science*, Vol. 4, pp. 105–123.

Lang, W.B., et al., 1940, ''Clay Investigations in the Scuthern States, 1934–35,'' Bulletin 901, US Geological Survey, 346 pp.

Lang, W.J., 1971, ''Bentonite: The Demand and Markets of the future,'' SME Preprint No. 71H29, AIME Centennial Annual Meeting, New York, Mar., 9 pp.

Larsen, D.H., 1955, ''Use of Clay in Drilling Fluids,'' *Clays and Clay Technology*, J.A. Pask and M.D. Turner, eds., California Division of Mines Bulletin 169, pp. 269–281.

Martin-Vivaldi, J.L., and Linares, G.J., 1968, ''Las Bentonitas de Cabo de Gata: II Yacimiento de Palma del Muerto,'' *Span. Institute Geologico y Minero Boletin*, Vol. 79, No. 6, pp. 65–71.

Miser, H.D., 1913, ''Developed Deposits of Fuller's Earth in Arkansas,'' Bulletin 530, US Geological Survey, pp. 207–220.

Nutting, P.D., 1943, ''Adsorbent Clays, Their Distribution, Properties, Production and Uses,'' Bulletin 928-C, US Geological Survey, pp. 127–221.

O'Driscoll, M., 1989, ''Bentonite: Overcapacity in Need of Markets.'' *Industrial Clays—A Special Review*, G. Clarke, ed., Industrial Minerals Division of Metal Bulletin, p. 10.

Odom, I.E., 1992, ''Hectorite Deposits in the McDermitt Caldera of Nevada,'' Preprint no. 92–155, Annual Meeting, Society for Mining, Metallurgy, and Exploration, Littleton, CO, 12 pps.

Papin, R., 1964, ''Bentonite,'' *Encyclopedia of Chemical Technology*, Kirk-Othmer, ed., Vol. 3, pp. 339–360.

Papke, K.G., 1969, ''Montmorillonite Deposits in Nevada,'' *Clays and Clay Minerals*, Vol. 17, No. 4, pp. 211–222.

Reisch, F.J., 1992, ''Investigation of the Geology, Mineralogy and Economic Utilization of the Porters Creek Formation in Tippah County, Mississippi,'' MS thesis, Indiana University, Bloomington, IN, 71 pp.

Rich, A.D., 1960, ''Bleaching Clay,'' *Industrial Minerals and Rocks*, 3rd ed., J.L. Gillson, ed., AIME, New York, pp. 93–101.

Robertson, R.H.S., 1986, *Fuller's Earth—A History of Calcium Montmorillonite*, Volturna Press, Hythe, Kent, UK, 421 pp.

Ross C.S., and Shannon, E.V., 1926, ''The Minerals of Bentonite and Related Clays and their Physical Properties,'' *American Ceramic Society Journal*, Vol. 9, No. 2, pp. 77–96.

Ross, J.S., 1964, "*Bentonite in Canada*," Mines Branch Monograph 873, Canada Dept. of Mines and Technical Surveys, 61 pp.

Russell, A., 1991, "India—A Treasure Trove of Minerals. Lifting the Lid on Exports," *Industrial Minerals*, No. 287, pp. 17–33.

Sastry, K.V.S., and Fuerstenau, D.W., 1971, "A Laboratory Method for Determining the Balling Behavior of Taconite Concentrates," *Trans. SME-AIME*, Vol. 250, No. 1, pp. 64–67.

Silliman, B., 1820, "Sketches of a Tour in the Counties of New Haven and Litchfield in Connecticut, with Notices of the Geology, Mineralogy and Scenery, etc.," *American Journal of Science*, Vol. 2, pp. 201–235.

Stone, R.L., et al., 1971, "Muller Mixing of Bentonite and Iron Ore Concentrate at Cornwall," *Trans. SME-AIME*, Vol. 250, pp. 236–239.

Stucki, J.W., et al., 1984, "Effects of Reduction and Reoxidation of Structural Iron on the Surface Change and Dissolution of Dioctahedral Smectites," *Clays and Clay Minerals*, Vol. 32, pp. 350–356.

Sudo, T., and Shimoda, S., 1978, *Clays and Clay Minerals of Japan*, Elsevier, Developments in Sedimentology No. 26, Amsterdam, the Netherlands, 326 pp.

Takeshi, H., and Kato, C., 1969, "Minerals," *Clays of Japan*, ed. Editorial Subcommittee, International Clay Conference, Geological Survey of Japan, pp. 103–120.

Teague, K.H., 1972, "Southern Bentonite," SME Preprint No. 72H328, SME Fall Meeting, Birmingham, AL, Oct., 7 pp.

Thomas, A.R., 1981, "The Porters Creek Formation (Paleocene) of the Northern Mississippi Embayment: Clay Mineral Variations and Depositional Interpretation," MA thesis, Indiana University, Bloomington, IN, 100 pp.

Torok, A., and Thompson, T.D., 1972, "Activated Bleaching Clay for the Future," *Trans. SME-AIME*, Vol. 252, pp. 15–17.

Van Olphen, H., 1977, *An Introduction to Clay Colloid Chemistry*, 2nd ed., John Wiley and Sons, New York, 318 pp.

Wakeman, J.S., 1972, "Method for Evaluation of Data from the Batch Testing of Green Pellets," *Trans. SME-AIME*, Vol. 252, pp. 83–86.

Weidhaas, D.E., and Brann, J.L., Jr., 1955, *Handbook of Insecticide Dust Diluents and Carriers*, 2nd ed., Dorland Books, Cladwell, NJ, 233 pp.

Wetzenstein, W., 1972, "Die Bentonitlagerstatten im Ostteil der Insel Milos und Ihre Mineralogische zusammensetung," *Geological Society of Greece Bulletin*, Vol. 9, No. 1, pp. 144–171.

Wolfbauer, C.A., 1977, "Exchangeable Cations in Cretaceous Bentonites from Wyoming and Montana," US Geological Survey, Open-File Report 77–158, 38 pp.

CLAYS

COMMON CLAY

HAYDN H. MURRAY

Common clay includes a wide variety of clay and other fine-grained rocks that are used in many ways. These clays are sometimes classified as miscellaneous clay and shale. Most products made from them are fired and include such things as structural and face brick, drain tile, vitrified pipe, quarry tile, flue tile, conduit, pottery, stoneware, and roofing tile. Very large tonnages of these materials are bloated by firing to form lightweight aggregate, and large quantities are used in making portland cement (see the specific chapters in this volume). Uses not requiring firing of a finished product include filler for paint and other products and shale and clay used for packing dynamite in blastholes and for plugging oil and gas wells that are no longer in use.

PROPERTIES

Clay and shale are used in so many different structural clay products that they necessarily have a wide range in physical properties. The properties are plasticity, green strength, dry strength, drying and firing shrinkage, vitrification range, and fired color. The properties desired vary with the structural clay product made. For example, a clay used in making conduit tile must be very plastic and have high green and dry strengths and uniform shrinkage, but for drain tile or common brick these properties do not have to be controlled so closely.

Most clays are plastic when naturally wet or mixed with water. Plasticity can be defined as the property of a material to undergo permanent deformation in any direction without rupture under stress beyond that of elastic yielding. Some clay materials are very plastic and are called *fat* clay; others having little plasticity are called *lean* clay. The type of clay mineral, particle size, particle shape, organic matter, soluble salts, adsorbed ions, and the amount and type of nonclay minerals influence the plastic properties of clay.

Green-strength and dry-strength properties of clay and shale are important because most structural clay products are handled at least once and must be strong enough to maintain shape. Green strength is the strength of the clay material in the wet or plastic state. Dry strength is the strength of the clay after it is dried. Plasticity and green strength are closely related and are influenced mainly by the same variables. Dry strength is dependent on the proportion of fine particles present, the shape of the individual particles, the degree of hydration of the clay fraction, the method of forming the ware, and the rate and thoroughness of the drying. The presence of a small amount of smectite, which occurs in very fine hydrated particles, generally increases dry strength.

Both drying and firing shrinkages are critical properties of clay used for structural clay products. Shrinkage is the loss in volume of a clay as it is dried or fired. Drying shrinkage is dependent on the water content, the character of the clay minerals, and their particle size. Most plastic clays shrink appreciably with drying, which tends to produce cracking and warping. Sandy clay or clay having low plasticity has low shrinkage properties but tends to produce a weak porous body. Smectite minerals in rather large amounts (10 to 25%) commonly cause excessive shrinkage, cracking, and slow drying. Firing shrinkage depends on the density of the clay, volatile materials present, the types of crystalline phase changes that take place during firing, and the dehydration characteristics of the clay minerals.

The temperature range of vitrification or glass formation during firing is a very important property of clay and shale used in structural clay products. Vitrification is a process of gradual fusion in which some of the more easily melted constituents produce increasing amounts of liquid as the temperature is increased. This liquid produces glass as the material is cooled, and the glass is the bonding material in the final fired product. Some clay has a short vitrification range, and when such clay is fired the temperature of the kiln must be very closely regulated. Clay consisting chiefly of illite, smectite, or chlorite has lower vitrification temperature than kaolinite rich clay. Some mineral impurities in clay, such as calcite and feldspar, lower vitrification temperatures by acting as fluxes. The degree of vitrification attained also depends on the duration of firing and the temperature. Commonly the degree of vitrification is regulated by the amount of shrinkage and porosity needed in the final product.

Uniform color is an essential property of many structural clay products. The color of a product is influenced by the state of oxidation and particle size of the iron minerals, the firing temperature and degree of vitrification; the proportion of alumina, lime, and magnesia in the clay material; and the composition of the gases in the kiln during the burning operation. High grade white burning clay contains less than 1% Fe_2O_3, buff burning clay ordinarily contains 1 to 5% Fe_2O_3, and red burning clay contains 5% or more Fe_2O_3. Other constituents also affect the color, but finely divided iron bearing minerals generally are the principal materials causing color in fired products.

OCCURRENCE

Common clay occurs in many types of rocks ranging in age from Precambrian to Holocene. They include glacial clay, soils, alluvium, loess, shale, weathered and fresh schist, slate, and argillite. Some fire clay and kaolin are also included in the miscellaneous clay group, particularly when used in the manufacture of structural-clay products. This group consists of so many different types of rocks that only a few broad generalizations about their mineral makeup are possible. The most common mineral in many of them is one of the members of the mica group, and the dominant one in a given deposit may be illite, sericite, or one of the micas normally occurring in coarser grain sizes, such as muscovite and biotite. In addition to illite, the clay minerals present commonly are kaolinite, smectite, mixed layer varieties, and chlorite. Some materials in this group actually contain more quartz and other detrital minerals than clay minerals.

Common clay is now mined in every state in the United States, except Alaska and Rhode Island, according to US Bureau of Mines *Minerals Yearbooks*. In 1990, 25.9 Mt of these materials were produced (see Table 1 in the Introduction to this section). The states producing more than 1 Mt were Alabama, California, Georgia, Illinois, Indiana, Iowa, Louisiana, Maryland, Michigan, Missouri, New York, North Carolina, Ohio, Pennsylvania, South Carolina, Texas, and Virginia. Texas was the leading state in production, and North Carolina was second.

TESTING, PROCESSING, AND MARKETS

The testing of common clay for most products requires the preparation of test pieces of specific dimension, drying, and firing (Clews, 1969, Grimshaw, 1972, Klinefetter and Hamlin, 1957). Fired test pieces are tested for such properties as shrinkage, warping, cracking, permeability, modulus of rupture, and crushing or compression strengths. Color and soluble salt content are also important for some products.

Common clay is dug from open pits. Pits must be near plants and the digging must be cheap, because both the raw material and finished products are heavy, and the margin of profit is ordinarily low. Most products made from these materials are processed and marketed much in the manner described for refractory clays.

FUTURE CONSIDERATIONS AND TRENDS

The long range trends in demands for some products made from common clay can be expected to keep pace with the growth in the gross national product, but demands for others are likely to fall short of this growth rate. Demands for clay and shale required for portland cement and lightweight aggregate are increasing, and this trend is likely to continue. Though demands for heavy structural-clay products will continue, these materials are facing strong competition from cement, wood, glass, plastics, aluminum, and other metals. Growth in demands for structural-clay products is also hampered by the heavy weight of these products, which limits their market range, and by increasing production costs. The rise in fuel costs, which has been forecast, seems likely to be a major problem faced by the industry.

Another problem for the industry using common clay has been the shift to rural areas. For a long time, plants were near population centers because of transportation costs. Suburban growth in many populations centers has forced plants to close or to relocate because of environmental pressures and raw material supply problems. This trend seems certain to continue until virtually all active plants are in rural areas, which, in turn, tends to favor a substitute material because of increased transportation costs and other factors.

REFERENCES

Clews, F.H., 1969, *Heavy Clay Technology*, 2nd ed., Academic Press, New York, 481 pp.

Grimshaw, R.W., 1972, *Chemistry and Physics of Clays and Other Ceramic Raw Materials*, 4th ed., rev., Wiley-Interscience, New York, 1024 pp.

Klinefetter, T.A., and Hamlin, H.P., 1957, "Syllabus of Clay Testing," Bulletin 565, US Bureau of Mines, 67 pp.

CLAYS

HORMITES: PALYGORSKITE (ATTAPULGITE) AND SEPIOLITE

FRED G. HEIVILIN AND HAYDN H. MURRAY

The ''Hormite Group'' was proposed for palygorskite (attapulgite) and sepiolite for their complex magnesium silicate composition and elongate crystals (Martin-Vivaldi and Robertson, 1971). These minerals occur in close association with each other and more complex structural variations may exist (Bailey, 1972).

In 1862 Savchenkov used the name palygorskite to describe a mineral from the Palygorsk locality (Hay, 1975), near the Ural Mountains. Ovecharenko and Kukovsky (1984) mention that when mountain leather deposits were prospected in the Palygorsk Division mine it was assumed this unusual mineral was a variety of asbestos. Early mineralogists used the terms ''mountain cork'' or ''mountain leather'' when referring to palygorskite. Robertson (1986) mentions that it appears palygorskite was known since Theophrastus' time, ca. 314 BC. J. de Lapparent used ''attapulgite'' for clays from Attapulgus, GA, and Mormoiron, France, because he thought them different from palygorskite, but the two types were proved to be the same (Bailey et al., 1971). The name attapulgite is still used for the Florida and Georgia deposits when the crystal length to diameter ratio does not exceed 10:1 (Merkl, 1989). Georgia palygorskite clays are of much shorter length compared to classic palygorskite.

In 1847 Glocker first used the name sepiolite which was called ''Meerschaum'' by Werner (1788) and Hauy (1801) named it ''Ecume de Mer.'' Brochant (1802) described low density and white magnesium silicates adding the name Talcum Plasticum and Ecume de Mer.

In the Meigs-Attapulgus-Quincy district palygorskite (attapulgite) commonly occurs in two distinct forms referred to as short length palygorskite (Meigs Member) and long length palygorskite (Dogtown Member) (Merkl, 1989). Long length palygorskite crystals (> 10 μm) are rarely observed in the Meigs and Dogtown Members, but when present are in association with dolomite crystals. The short length form is usually less than 2 μm in length and has a low magnesium content whereas the long length form has a high magnesium content and a length greater than 2 μm.

The distinctions in morphology are not only important because of the relationship to the origin of the deposits, but also in relation to activity in causing membranolytic activity related to data on palygorskite samples from 9 locations ranging from relatively inert to active in work reported by Nolan et al. (1989). The > 10 μm lengths amounted to only 51 of 17,401 fibers sized. The shortest lengths (< 0.5 μm) were relatively inert. This study pointed out that surface activity, morphology, and chemical differences may be distinctly different within the definition of palygorskite, or for that matter for any individual mineral so that health and other properties must be measured because the name alone does not necessarily indicate uniformity.

Palygorskite (attapulgite) fuller's earth was first sold for drilling mud in 1941. The market for this use expanded slowly and has maintained a level of 7 to 10% of the total US production during the last few years. Most of the fuller's earth sold for drilling mud comes from the southern part of the Meigs-Attapulgus-Quincy district of Georgia and Florida. Palygorskite clays produced in this area are superior to most other fuller's earth for mud used in drilling salt formations, but because of high water loss, they are inferior to bentonite where the rocks drilled contain no saltwater.

According to Oulton (1965), more than 90 different grades of fuller's earth are produced. Some of these grades are used for pharmaceuticals designed to absorb toxins, bacteria, and alkaloids; for treatment of dysentery; for purifying water and dry cleaning fluids, dry cleaning powders and granules; for the manufacture of NCR (no carbon required) multiple copy paper; for the manufacture of wallpaper; and as extenders or fillers for plastic, paint, and putty. Fuller's earth mined near Ellenton, FL, was used for making lightweight aggregates for the construction of concrete barges during World War II (Calver, 1957). Still other uses of fuller's earth and its suitability for uses in new products are outlined by Haden, Jr., and Schwint (1967), Haden, Jr., (1972), and Haas (1970). One special use of fuller's earth is as a carrier of platinum catalysts that are made in the United Kingdom from sepiolite clays mined in Spain. Other uses of sepiolite fuller's earth (Chambers, 1959) are similar to those of the palygorskite (attapulgite) type mined in the United States.

OCCURRENCE

Mineralogy

The hormite clays, palygorskite (attapulgite) and sepiolite, are the principal fuller's earth deposits along with some montmorillonite-based bentonite clays. Palygorskite is a hydrated alumino-magnesium silicate with an elongate morphology and an ideal formula with Mg partially replaced by Al or even Fe (Bradley, 1940). The structure (Fig. 1) consists of two silica chains linked in amphibole chain-like structures which has both monoclinic and orthorhombic symmetry (Christ et al., 1969). The structural formula for sepiolite is based on the model of Brauner and Preisinger (1956). The sum of the octahedral ions (Mg, Al, Fe^{+2}, Fe^{+3}) is between 3.76 and 4.64 (Newman and Brown, 1987) with a mean value of 4.0 indicating the mineral is dioctahedral.

One orthorhombic and three different monoclinic geometric unit cells identified give palygorskite much more diversity than sepiolite. This accounts for the long fiber forms in Russia which were once mistaken for asbestos, the shorter fiber gelling clay in southern part of Meigs-Attapulgus-Quincy district and the non-

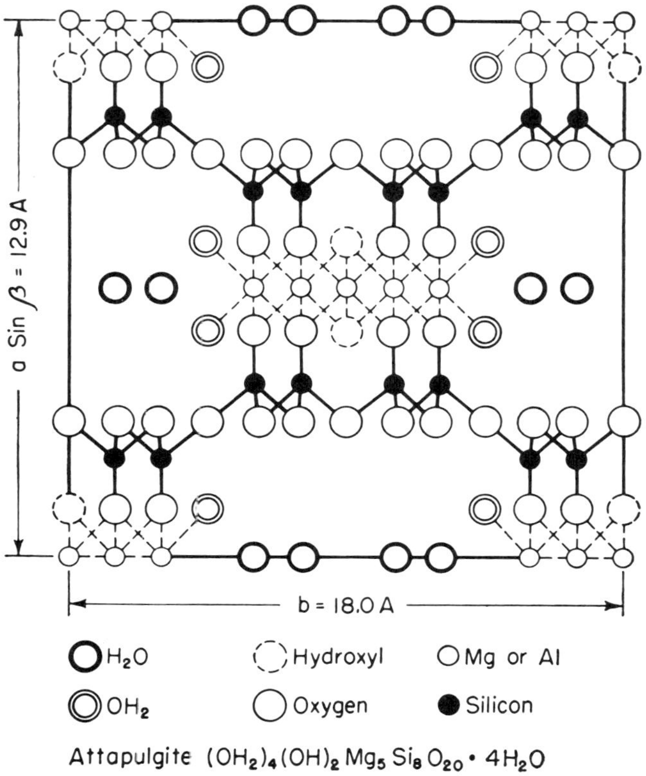

Fig. 1. Structure of palygorskite (attapulgite).

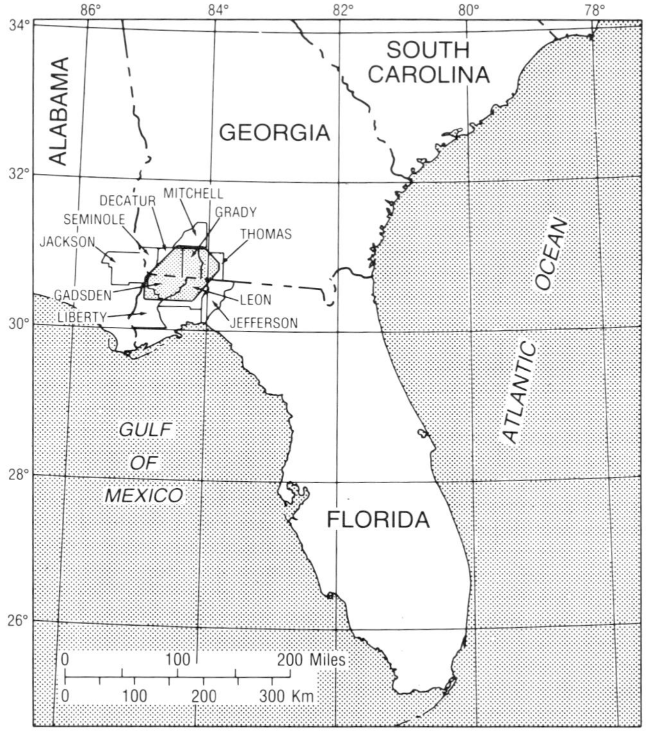

Fig. 2. Location map showing the fuller's earth district in Georgia and Florida.

gelling type of even shorter fibers in northern part of the district. Jeffers and Reynolds (1987) suggest a mechanism by which a chain structure may be rearranged into a sheet phyllosilicate such as a smectite.

The palygorskite crystals can retain their morphology even when altered to kaolin (Keller, personal communication). Until the mid-1970s the blue unaltered part of the Meigs Member was not mined because it was more expensive to process than the denser gray and tan clays which had been exposed to weathering or surface water alteration. These alterations, poor development of crystals, and possible oxidation of samples while waiting for testing may have caused a reduction in the amount of palygorskite indicated in samples analyzed using X-ray diffraction. There are also indications that Al replacement of Mg occurs in the Meigs Member. Weaver and Beck (1977) give ample evidence of the complexity of the mineralogy and the need for much additional research.

Associated minerals with the palygorskite in the South Georgia-North Florida deposits are smectite, quartz, dolomite and calcite, along with very small amounts of kaolinite and clinoptilolite in isolated locations.

ORIGIN

The Georgia-Florida deposits are of early to middle Miocene age. Their origin has been attributed to alteration of volcanic detritus, diagenetic alteration of montmorillonite derived from a crystalline upland, and direct crystallization (Merkl, 1989). The deposition of these palygorskite rich deposits took place in a shallow water trough which connected the Apalachicola Embayment on the Gulf of Mexico with the Southeast Georgia Embayment on the Atlantic Ocean. The water in this trough was characterized by fluctuating salinities and at times lagoons were closed off from normal circulation and sufficient Mg was present to form palygorskite. The South Georgia-North Florida District extends from Quincy, FL, on the south to the Meigs, GA, area on the north (Fig. 2), a distance of about 80 km. The mineralogical content changes from dominantly palygorskite (attapulgite) in the Quincy, FL-Attapulgus, GA area to a mixture of smectite and palygorskite in the northern part of the district around Ochlocknee and Meigs, GA. Millot (1970) believes that smectite is transformed to palygorskite and sepiolite in a saline alkaline water environment and that small amounts of sepiolite are precipitated. Patterson (1974) believes the deposits formed in very shallow marine lagoons or tidal flats and that the magnesium needed to form the palygorskite and sepiolite was introduced by reaction of diluted material with magnesium rich brines and salt resulting from the evaporation of entrapped seawater.

PHYSICAL PROPERTIES

Palygorskite and sepiolite are both elongate minerals that are comprised of magnesium aluminum silicate. The structure (Fig. 1) contains open channels which give these minerals some unique properties particularly in the area of sorption of various materials. Because of substitutions of magnesium and iron for aluminum in the octahedral layer of the clay there is an excess negative charge. The surface area is on the order of 75 to 150 m^2/g (Van Olphen and Fripiat, 1979). This large surface area that may have extensive negatively charged sites then becomes an effective sorbent for any polar molecules or positive ions.

The cation exchange capacity of palygorskite and sepiolite is in the range of 20 to 50 milliequivalents per 100 g which is not as high as for smectite but is considerably higher than kaolinite. Because of the relatively high surface area and the surface charge, the hormite minerals are excellent sorbents for both water and oil.

The elongate particle shape of the hormite clays is another important physical property that makes palygorskite and sepiolite useful as a viscosifier and suspending agent. Because of the particle shape these hormite minerals are used extensively in drilling fluids. Drilling muds containing these clays show very small variations in viscosity and gel strength with large variations in electrolyte content. This is particularly useful when drilling in areas where salt water brines are encountered and in systems where electrolytes and soluble salts are present. The yield value of attapulgite clays from the South Georgia-North Florida area is about 100 bbl/t. The explanation of the resistance to salt flocculation lies in the fact that the hormite clays when dispersed in water consist of a bulk of randomly intermeshed elongate particles. The elongate shape prevents flocculation, as happens in the case of smectites, so that the hormites remain in suspension because of the physical interference of the individual elongate particles.

Another physical property unique to the hormite clays is the holes or channels which are present in the structure as noted previously. These channels or holes are about 6 Å in diameter and in nature are filled with water molecules or hydrated cations. This water or hydrated ions can be removed by heating to a temperature of about 500°C. Rehydration is very slow so that when dried at a temperature high enough to evaporate this water in the channels, which is sometimes referred to as zeolitic water, these open channels can then readily absorb polar organic or inorganic molecules. In industrial practice these hormite clays are designated as LVM clay which means low volatile material. Those hormite clays that are dried at a lower heat of about 200°C are designated as RVM clays which means regular volatile material.

WORLD PRODUCTION AND MAJOR PRODUCING AREAS

The world production figures are difficult to ascertain because the figures for hormite clay production are combined with smectites that are used as sorbents and called fuller's earth. However, an educated estimate for world production is about 2.5 Mt. The Attapulgus, GA-Quincy, FL district in the United States dominates the world's production of palygorskite and Spain dominates the world's sepiolite production. Sepiolite is also produced in Turkey. Palygorskite is also produced in Senegal, Turkey, India, Russia, China, France and Australia. Some of this production is a mixture of palygorskite and smectite.

United States

The Meigs-Attapulgus-Quincy palygorskite type fuller's earth district is in southern Georgia and extends a short distance into northern Florida. This palygorskite type fuller's earth is found in two distinct stratigraphic intervals (Fig. 3). The younger Meigs Member of early middle Miocene age unconformably overlies the Dogtown Clay Member of late lower Miocene age. Both of these members are part of the Hawthorne Group. The Meigs Member contains smectite, short length palygorskite (Fig. 4), a small amount of sepiolite, Opal-A, and minor and sporadic amounts of clinoptilolite along with diatoms which in some instances comprise up to 60% of the fabric in SEM photos (Fig. 5). The Dogtown Member is comprised of longer length palygorskite (Fig. 6) and minor amounts of quartz, smectite, and dolomite with trace amounts of pyrite. At present five producers operate in this district. The deposits

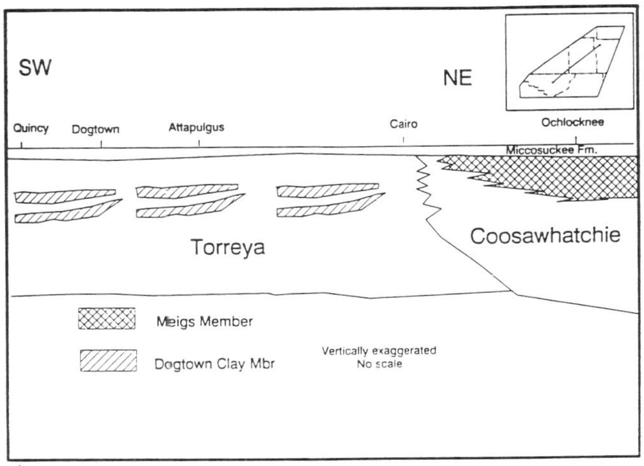

Fig. 3. Schematic stratigraphic cross-section showing the relationship of the Meigs and Dogtown Members of the Hawthorne Formation.

Fig. 4. Scanning electron micrograph of short length palygorskite (Meigs Member).

in the Dogtown Clay Member in the southern part of the district range from 2 m to slightly more than 3 m in thickness and the deposits in the Meigs Member, in the northern part of the district, range from 9 to 14 m in thickness. The production from this district is over 1.5 Mt and reserves are adequate to last for at least 30 years.

Sepiolite is mined in small quantities near Lathrop Wells, NV, 160 km northwest of Las Vegas in the Amargosa Valley. The sepiolite occurs in thin beds in saponite deposits and is selectively mined for use in suspension fertilizers, ceramics, and other uses. The deposit is Pleistocene in age (Papke, 1969).

Spain

The production of sepiolite clays in 1988 amounted to 508 kt and for attapulgite 43 kt in 1988 (Griffiths, 1991). The major uses for this material were pet litter absorbents, agricultural carriers and drilling mud. The sepiolite formed in Tertiary age lakes in central Spain in an evaporite sequence (Huertas et al., 1971). The largest producing operation is at Vicalvaro in Madrid where approximately 360 kt were produced in 1988. The clay is sun dried at the mine

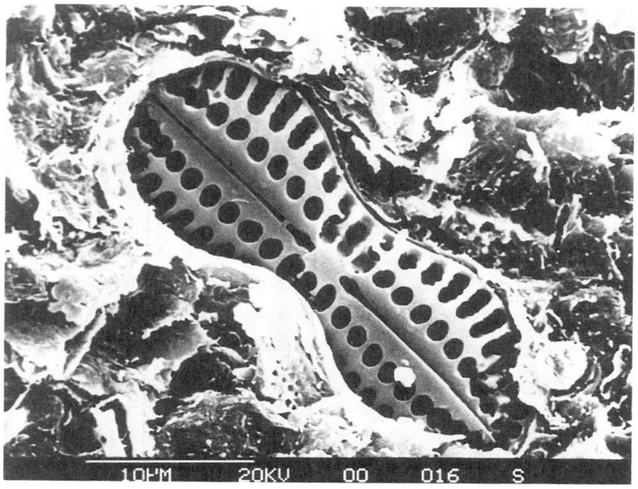

Fig. 5. Scanning electron micrograph of a diatom frustule surrounded by attapulgite and smectite.

Fig. 6. Scanning electron micrograph of long length palygorskite (Dogtown Member).

which reduces the moisture from 45 to about 32%. The clay is dried at the plant from 32 to 12% and is then classified into various granular and pulverized grades before packaging. Other uses for this material are in fertilizers, plastics, mastics, catalysts, bleaching earth and ceramics. Other locations where hormite clays are produced in Spain are Toledo, Torrejon el Rubo, Lebija, Almeria, and Maderuelo (Griffiths, 1991).

Senegal

Attapulgite is produced from mines near Thies which is about 100 km east of Dakar. The latest production figures indicated that 111 kt were produced in 1987 (Clarke, 1989). The production capacity of the processing plant in Senegal is 300 kt. This attapulgite clay is used primarily for industrial absorbents and pet litter, although about 15 kt are used in drilling mud applications. The attapulgite deposit is Miocene in age and is overburden on the phosphate deposits which are mined at Thies.

India

Attapulgite is mined and processed in the Bhaumagar District in the State of Gujurat. The material is extruded and dried and pulverized or screened for use as drilling mud and pet litter and industrial absorbents. The estimated production is 30 kt.

Australia

About 30 kt of attapulgite are produced from a surface deposit at Lake Narromyne, some 160 km northeast of the port of Geraldton in Western Australia. The material is dried, crushed, and sized for use as industrial absorbents, pet litter and agricultural carriers.

Ukraine

Large deposits of palygorskite of lower Miocene Age have been described in the Ukraine (Ovcharenko et al., 1967). In the Cherbassey District near Kiev, about 10 kt are mined and processed for use as a drilling mud. Reserves of palygorskite are estimated at 10 Mt.

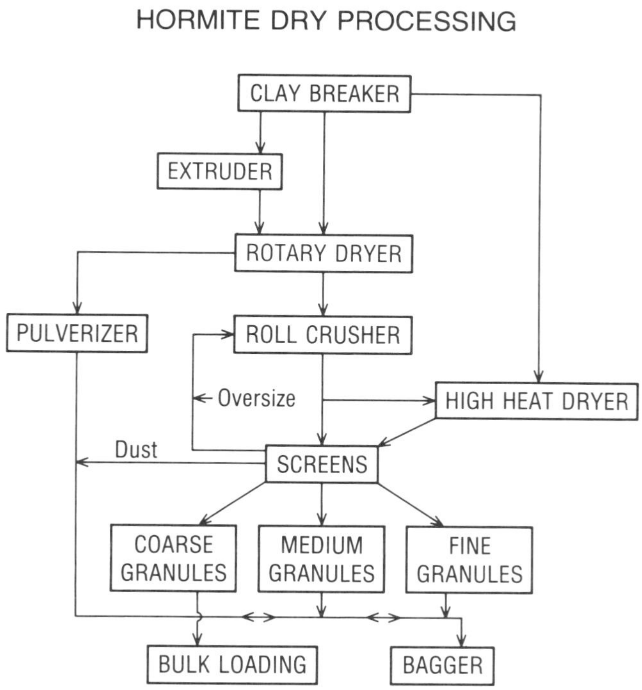

Fig. 7. Typical process flow sheet for hormites.

China

Attapulgite is mined from Miocene age clays near the town of Xuyi which is about 100 km north of Nanjing. The latest production figure available was 1987 when 5 kt were produced for drilling muds. The clay is extruded and dried at a small plant in Xuyi.

There is small production of palygorskite and sepiolite in Turkey, South Africa, and France, but the tonnage produced is not known.

MINING AND PROCESSING

Palygorskite and attapulgite around the world are mined from open pits. The stripping is done with scrapers, draglines or bulldozers and the clay is mined with shovels, backhoes, small draglines or front end loaders. The clay is transported by trucks to the processing plant. A general flow sheet is shown in Fig. 7. The process in general is quite simple involving crushing, drying, classification, and pulverizing. Processes used for special purposes include extrusion to separate the elongate particles of the hormite clay along with the addition of 1 or 2% of MgO to improve the viscosity characteristics when used as a drilling mud; high heat drying to drive the water out of the structural channels or holes in the hormite clay which produces an LVM clay to improve its sorbent properties; and ultra fine pulverization to improve suspension properties for certain pharmaceutical applications.

In the United States all the mining operations are reclaimed as specified by the federal or state laws that are in place. All reclamation must meet the federal requirements and the state requirements if they are more stringent than the federal.

PRODUCT SPECIFICATIONS

Specifications for palygorskite and sepiolite for use in drilling must fulfill the same viscosity specifications as bentonite. Tests are made by adding water and mixing slurries of specific concentrations. Measurements are then made for the viscosity, yield, yield point, filtrate (a measure of wall building properties), and a wet screen analysis according to standard procedures recommended by the American Petroleum Institute (Anon., 1969, 1969a). The Oil Companies Materials Association (Anon., 1969b) in the United Kingdom also has specifications for testing drilling mud. The only difference between the test methods for palygorskite (attapulgite) and bentonite is that the suspension when made up with palygorskite uses water containing 40 g of salt (NaCl) per 100 cc of water.

Oil bleaching tests of palygorskite-attapulgite fuller's earth are made using procedures described by the American Oil Chemists Society (Anon., 1958). Ross (1964) and Nutting (1943) also have described procedures for evaluating fuller's earth for use in bleaching oils.

Tests for a variety of different insecticide, dust diluents, and carrier uses have been described by Weidhaus and Brann, Jr., (1955). Most tests for absorbent granules are made according to procedures in Federal Specifications P-A-1056A, Absorbent Material, Oil and Water (for floors and decks), available from the US General Services Administration.

Many customers also have specifications regarding granule sizes, dust content, specific sorption requirements, and others that the producers must meet. In general, the moisture requirements for specific end uses are set by the customers. For example, the maximum moisture content for palygorskite used as a drilling mud is 16%, whereas the maximum moisture content for bentonite used as a drilling mud is 10%.

EVALUATION OF DEPOSITS

Deposits of hormite clays are evaluated from samples taken from core drilling. The cores are taken on a grid pattern normally on drill cores spaced 50 m down to 15 m apart, dependent upon the size and uniformity of the deposit. Usually, mine plans and processing requirements require a minimum of 20 years of known reserves and quality. The cores are sampled at 1 m intervals and are tested for drilling muds or gelling clays and for absorbent use. The gelling clays must meet standard viscosity and grit requirements and the absorbent tests are conducted with water and oil and must meet the standards of the Federal Specifications P-A-1056A as noted in the section on Product Specifications. Some specialty uses have their own set of tests that are required by the end user which may require chemical analyses, brightness values, color, and special absorption or adsorption tests. In Senegal and China, test pits are dug down through clay rather than using a core drill because it is less expensive to dig them by hand.

Based on the drilling and testing, maps are compiled which show the overburden thickness, clay thickness, and quality. After the quantity of clay is calculated and the deposit is proved to have adequate quality and size, then it is advisable to open a small test pit so that a relatively large quantity can be run through a typical processing plant. This would include crushing the clay, drying the product to RVM and LVM moisture, screening the dried product to estimate the granule size distribution, pulverizing the material to various fine grinds, i.e., –200 mesh, –325 mesh, and even finer sizes to determine the quality and recovery. In addition, some material should be extruded with the addition of MgO in order to see if the properties of the drilling mud can be enhanced. Also most granular clay is now dedusted so that some of the material should be tested to measure the amount of tactile dust on the granules and whether or not the granules can be dedusted using standard dedusting techniques.

If the testing of samples from drill cores or test pits show a good quality and sizable deposit that can be economically mined and that the processing of a sizable quantity proved successful, then a mine plan can be established with the probability that the clay can be successfully and profitably processed.

FUTURE CONSIDERATIONS AND TRENDS

Both palygorskite and sepiolite have particular attributes that provide a distinct set of end uses although there are several common markets with calcium montmorillonite. Both of these clays, i.e., hormites and calcium montmorillonite, are classed as fuller's earth. The markets for palygorskite and sepiolite fuller's earth continue to grow, and it is estimated that the market will grow at a rate of 3 to 5% per year for the next decade. New markets for specialty applications will develop and new processes will be developed which will enhance certain desirable properties. It is possible that new deposits will be discovered particularly in China, India, and Africa, which will certainly affect certain markets.

The most significant considerations for the future are the effect of environment and health and safety regulations on the clay industry particularly in the United States and Europe. Reclamation requirements, wetland regulations that affect mining permits, clean air and water regulations, crystalline silica regulations, and potential threat of legislation against elongate minerals all must be carefully monitored because any of these factors could seriously affect the cost and subsequent profitability of the industry. Disposal of absorbent clays on which oil or other compounds are sorbed may become a problem, although research has shown that the oil or other organic compounds are bound tightly on the surface of the clay particles.

A potential new use for hormite clays may be in clay liners and barriers where the sorptive capacity and stability against flocculation can be a very definite advantage (*see* chapter on Environmental Uses: Clay Liners and Barriers in this volume).

REFERENCES

Anon., 1858, "Bleaching Tests," A.O.C.S. Official Method Cc 8a 52, American Oil Chemists Society, 2 pp.

Anon., 1969, "API Recommended Practice Standard Procedure for Testing Drilling Fluids," APIRP 13B, 2nd ed., American Petroleum Institute, 19 pp.

Anon., 1969a, "API Specifications for Oil Well Drilling Fluid Materials," API STD 13a, 5th ed., American Petroleum Institute, 11 pp.

Anon., 1969b, "Bentonite," Specifications No. DFCP.4, Oil Companies Materials Associations, London.

Bailey, S.W., 1972, "Recent Advances in the Crystal Structure and Crystal Chemistry of Layer Silicates," *Proceedings,* International Clay Conference, Madrid, Spain, Div de Ciecians, CSIC, pp. 3–25.

Bailey, S.W., et al., 1971, "Report of Nomenclature Committee, 1969–1979," *Clays and Clay Minerals,* Vol. 19, Clay Mining Society, pp. 123–133.

Bradley, W.F., 1940, "The Structural Scheme of Attapulgite," *American Mineralogist*, Vol. 25, No. 6, pp. 405–410.

Brauner, K., and Preisinger, A., 1956, "Structure and Enstehung des Sepioliths," *Tschermaks*, Minerology Petrography Mit., Vol. 6, pp. 120–140.

Brochant, A.S.M., 1802, *Traite de Mineralogie*, Viller, Paris, 451 pp.

Calver, J.H., 1957, "Mining and Mineral Resources," Bulletin 39, Florida Geological Survey, 132 pp.

Chambers, G.P.C., 1959, "Some Industrial Applications of the Clay Mineral Sepiolite," *Silicates Industrials*, Vol. 24, pp. 181–189.

Christ, C.L., et al., 1969, "Palygorskite: New X-ray Data," *American Mineralogist*, Vol. 54, Nos. 1,2, pp. 198–205.

Clarke, G., 1989, "Industrial Clays: A Special Review," Industrial Minerals Division, Metal Bulletin plc, London, 89 pp.

Glocker, E.F., 1847, "Synopsis, Halle," *A System of Mineralogy*, 5th ed., J.D. Dana, ed., 1868, Trubner and Co, London, p. 456.

Griffiths, J., 1991, "Spains Minerals: Mixed Fortunes," *Industrial Minerals*, No. 285, London, pp. 23–47.

Haas, C.Y., 1970, "Attapulgite Clays for Future Industrial Mineral Markets," *Mining Engineering*, Vol. 22, No. 12, p. 63.

Haden, L.W., Jr., and Schwint, I.A., 1967, "Attapulgite, TS Properties and Applications," *Industrial and Engineering Chemistry*, Vol. 59, No. 9, pp. 8–69.5

Haden, L.W., Jr., 1972, "Attapulgite," SME Preprint No. 72H-327, AIME New York, 8 pp.

Hauy, R.J., 1801, "*Traite de Mineralogie*," Paris, Vol. 4, p. 443 (new ed., 1968, Culture et Civilization, Bruxelles).

Hay, M.H., 1975, *Chemical Index of Minerals*, 2nd ed., British Museum of London, 728 pp.

Huertas, F., et al., 1971, "Minerals Fibrosis de la Arcilla en Cuenca Sedimentarias Espanolas, I Cuenca de Tajo," *Span. Institute Geology Minero Boletin*, Vol. 82, No. 6, pp. 28–36.

Jeffers, J.D., and Reynolds, R.C., 1987, "Expandable palygorskite," *Clays and Minerals*, Vol. 35, pp. 473–476.

Martin-Vivaldi, J.I., and Robertson, R.H.S., 1971, "Palygorskite and Sepiolite (The Hormites)," J.A. Gard, ed., *Electron optical investigation of clays*, Mineralogical Society, Monograph 31, London, pp. 255–275.

Merkl, R.S., 1989, "A Sedimentological, Mineralogical, and Geochemical Study of the Fuller's Earth Deposits of the Miocene Hawthorne Group of South Georgia-North Florida," PhD dissertation, Indiana University, Bloomington, IN, 182 pp.

Millot, G., 1970, *Geologic des Argiles*, Masson, Paris, 320 pp. (translated by W.R. Farrand and H. Paquet, published as *Geology of Clays*, Springer-Verlag, New York, 1970, 424 pp.

Newman, H., and Brown, G., 1987, "The Chemical Constitution of Clays," *Chemistry of Clays and Clay Minerals*, A. Newman, ed., Monography No. 6, Mineralogical Society, 128 pp.

Nolan, R.P., et al., 1989, "Membranolytic Activity and Morphological Characterization of Palygorskite and Sepiolite," *Effect of Mineral Dusts on Cells*, B. Mossman, and R. Begin, eds., MATO ASI Series Vol. 30, Springer-Verlag, Berlin, pp. 37–48.

Nutting, P.D., 1943, "Absorbent Clays; Their Distribution, Properties, Production, and Uses," Bulletin 928-C, US Geological Survey, pp. 127–221.

Oulton, T.D., 1965, "Mining, Production, and Uses of Attapulgite Clay Products," SME preprint 65H 39, AIME, New York, 10 pp.

Ovcharenko, F.D., et al., 1964, *Colloid Chemistry of Palygorskite*, Israel Program for Scientific Translation, Jerusalem, 101 pp.

Ovcharenko, F.D., et al., 1967, *Investigations of The Physicochemical Mechanics of Clay-Mineral Dispersions* (trans. from Russian by Z. Lerman), Israel Program Science Translations, Jerusalem, 140 pp.

Ovcharenko, F.D., and Kukovsky, G., 1984, "Palygorskite and Sepiolite deposits in the USSR," *Palygorskite-Sepiolite Occurrences, Genesis, and Uses*, A. Singer and E. Galan, eds., Developments in Sedimentology Series 37, Elsevier, Amsterdam, pp. 233–241.

Papke, K.G., 1969, "Montmorillonite Deposits in Nevada," *Clays and Clay Minerals*, Vol. 17, No. 4, pp. 211–222.

Patterson, S.H., 1974, "Fuller's earth and other industrial mineral resources of the Meigs-Attapulgus-Quincy district, Georgia and Florida," Professional Paper 828, US Geological Survey, 45 pp.

Robertson, R.H.S., 1986, *Fullers earth: A history of calcium Montmorillonite*, Volturna Press, Hythe, Kent, UK, 421 pp.

Ross, J.S, 1964, *Bentonite in Canada*, Mines Branch Monograph 873, Canada Dept. of Mines and Technology Surveys, 61 pp.

Savchenkov, T.V., 1862, "Definition," Verlag Russ-Kaiser, Gessellschaft Mineralogy, St. Petersburg, USSR, p. 102.

Van Olphen, H., and Fripiat, J.J., 1979, *Data Handbook for Clay Materials and Other Non-metallic Minerals*, Pergamon Press, Oxford, 346 pp.

Weaver, C.E., and Beck, K.C., 1977, "Miocene of the S.E. United States: A Model for Chemical Sedimentation in a Perimarine Environment," *Sedimentary Geology*, Vol. 17, pp. 1–236.

Weidhaus, D.E., and Brann, J.L., Jr., 1955, *Handbook of Insecticide Dust Diluents and Carriers*, 2nd ed., Dorland Books, Caldwell, NJ, 233 pp.

Werner, C.E., 1868, "Definition of The Species," *A System of Mineralogy*, 5th ed., J.D. Dana, ed., Trubner and Co., London, 456 pp.

CLAYS

KAOLIN

SAM M. PICKERING, JR. AND HAYDN H. MURRAY

The term kaolin is used here, in a commercial sense, for near-white clay deposits which are dominantly comprised of the mineral kaolinite, or for products refined from such deposits. The term ''china clay'' has been used in the past as a synonym for kaolin, a reference to its first being mined some 3000 years ago at a hill near Jauchau Fu, Jiangxi Province, China as a porcelain raw material. The name kaolin was derived from the Chinese *Kauling*, which means high ridge.

For most modern industrial applications kaolin must be extensively refined and processed from the crude state to enhance whiteness, purity, and other important commercial characteristics. After such processing, kaolin ranks as one of the most valuable of industrial clays, and it is used to some extent in most manufactured consumer products. In 1991, the United States alone produced kaolin valued at $1,072,366,000. It is applied very widely as a pigment, filler, coater, extender, ceramic raw material, catalyst base, electrical insulator, and pharmaceutical. Some more prominent uses include paper filling and coating; paint, plastic, adhesive, and ink pigment; rubber reinforcing agent; ceramic raw material for porcelain, dinnerware, tile, and enamels; catalyst base for petroleum cracking and auto exhaust emission catalytic control devices; cosmetics base; and digestive coating remedy. Those properties of kaolin clay which make it suited to these applications are related to its physical, optical, mineralogical, and chemical nature.

OCCURRENCE

Mineralogy

Kaolin group minerals include kaolinite, by far the most common, along with dickite, nacrite, and halloysite. All are hydrous alumino-silicates, and are all in the monoclinic or triclinic crystal classes.

One of the important mineralogical characteristics of kaolin clay in industrial applications is its softness or lack of abrasivity. Hardness is measured on the Mohs scale, which compares ten minerals on a 1 to 10 scale from the softest, talc, to the hardest, diamond. Kaolin ranks at a Mohs hardness of 2 to 2.5 (Dana, 1949), which means that it is softer than most common white minerals except talc and gypsum. This softness is an advantage in many industrial uses, as it reduces abrasion of process and application equipment.

Clays are very small crystalline materials, smaller than can be clearly resolved with the optical microscope. The size, habit, and surface charge of clay crystals determine many of the physical characteristics which allow the industrial uses of clays. Kaolinite occurs as micron size (one micron is one-thousandth of a millimeter) pseudo-hexagonal crystal plates. Kaolinite also may occur as thick stack-like crystals which may be many times longer than their diameter, at times even in strange vermicular shapes. The stack-like crystals have excellent basal cleavage parallel to the [001] crystal axis. The scanning or transmission electron microscope shows kaolin group minerals to be made up of all degrees of crystal perfection, from well-formed euhedral pseudo-hexagonal plates and stacks, to anhedral, featureless shapes (Fig. 1).

The crystalline nature of clays is also revealed by their X-ray diffraction patterns, which offer the best means of identification of most clay minerals. A typical X-ray pattern for well crystallized kaolinite is shown in Fig. 2. Natural kaolins vary greatly in their perfection of crystallization, or how precisely X-rays are diffracted by the atomic lattice. The sharpness and perfection of the X-ray diffraction pattern is thus thought to be an approximation of the lattice regularity of the crystals, and a ''Crystallinity Index'' (Hinckley, 1963) is used to rank this degree of atomic lattice ordering. However, a fine kaolin which the electron microscope shows to be made up of sharp, perfectly euhedral small crystal plates will often have an indistinct, poorly organized X-ray pattern (Fig. 3) compared with a coarse kaolin with stack-like crystals. This is perhaps because the thin, small plates of the fine kaolin do not diffract the X-ray beam well, rather than any lack of order in the crystals' atomic lattice.

Origin

Kaolin clays are most often formed by the alteration of aluminum silicate minerals in a warm, humid environment. Feldspar is the most common source mineral. Plagioclase (calcium and/or sodium based) feldspars are usually the first to become kaolinized; potash feldspars are slower to alter and produce kaolins mixed with fine-grained sericite, illite, or hydro-muscovite (Bristow, 1990). Alteration may be weathering or hydrothermal action, as explained by the following general equation for potash feldspar (Murray, 1989):

$$2KAlSi_3O_8 + 3H_2O = Al_2Si_2O_5(OH)_4 + 4SiO_2 + 2KOH$$

All potassium must be lost by solution from this alteration system from K-feldspars, or authigenic illite will also form.

Kaolins which have altered in situ generally retain the texture and form of the parent rock and are referred to as *primary* occurrences. Those kaolins which have been deposited by sedimentation in fresh or brackish water environments are called *secondary*. Both types of kaolins are mined and used commercially. The origin of these two types of kaolin deposits are discussed briefly below and in more detail in the descriptions of individual worldwide occurrences.

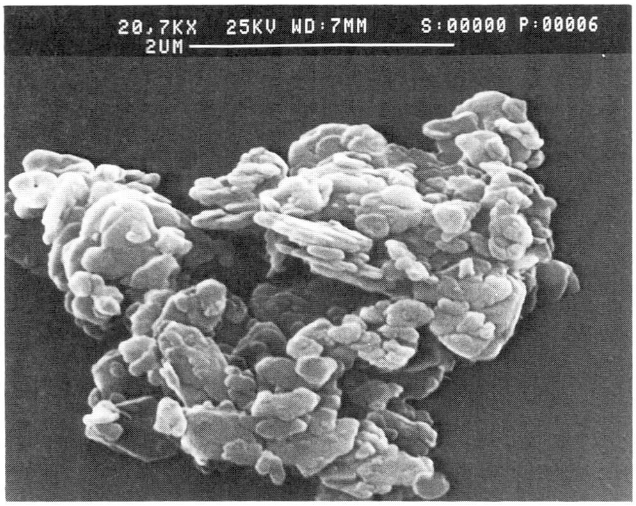

Fig. 1. Scanning electron micrograph of a coarse cretaceous kaolin, Wilkinson County, Georgia.

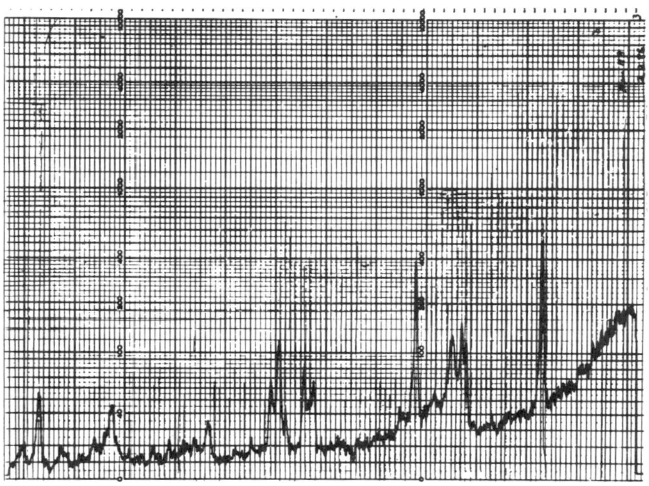

Fig. 2. X-ray diffraction pattern of a well crystallized coarse central Georgia kaolin.

Primary Kaolins

The most common type of primary kaolin deposit was formed by the oxidation weathering of feldspathic source rocks in moist, warm climatic conditions. The parent rock is generally sialic, such as granite, gneiss, syenite, or feldspar-rich volcanic ash. Much of the world's soil cover was formed by weathering in humid, warm climates, and is often rich in kaolin and associated bauxite. Contaminating iron oxides and other mineral pigments make such red and brown soils unsuitable for refined industrial kaolin products. However, the sub-soils of areas where weathering has penetrated to depths of tens of meters in tropical or sub-tropical areas often include a pallid zone containing very pure white kaolin. Such soils usually retain much of the texture of the parent rock. These primary weathering-type kaolin deposits may be much thicker along shear and fracture zones which have allowed meteoric ground water to penetrate deep into the parent rock. Below the depth of water saturation, the kaolin often is mixed with elongate tubular crystals of halloysite, which under special conditions may be the first mineral formed in the alteration process. When halloysite is exposed to seasonal wetting and drying in the soil-forming process, it loses some of its formula water and is altered to kaolinite. The tubular halloysite crystals may unroll to form plates of kaolinite, or they may retain their elongate shape and be referred to as dehydrated or 7Å.

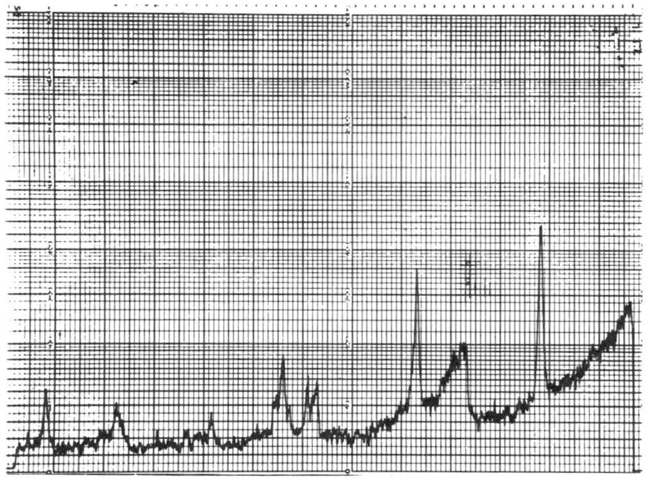

Fig. 3. X-ray diffraction pattern of a poorly crystallized fine East Georgia kaolin.

Primary kaolins may also form by hydrothermal alteration of feldspathic rock, either by heated magmatic solutions, volcanic exhalations (solfatara), or circulating ground water warmed by radiogenic heat from uranium or thorium minerals in the parent rocks. Hydrothermal kaolinizing fluids also commonly follow the higher porosity of fracture and shear zones, causing confusion with deposits formed by meteoric ground water weathering along such permeable fissures. The two types of kaolin deposits can usually be distinguished by their associated minerals; hydrothermal clays usually have at least traces of metallic sulfides, tourmaline, fluorite, or alunite, etc. Additional insight on the magmatic/meteoric nature of the alteration fluids is provided in the deuterium/hydrogen and oxygen isotype work of Sheppard (1977) and Jackson et al. (1982) on the Cornwall and Devon District kaolins of England.

Secondary Kaolins

Secondary deposits of kaolin have formed by sedimentary processes. The kaolin may have altered elsewhere and was then deposited in the basin of deposition as a clay. This may occur when suspended clay carried by a stream or river is flocculated by a change of pH when it encounters brackish water in estuarine conditions. Such sedimentary kaolins may, if deposited in low energy environments, contain little or no sand, mica, and other contaminating minerals. More commonly, such sedimentary kaolins are rather sandy. Typical kaolin concentrations in sandy sediments are at the tops of fining-upward sand sequences capped by kaolin clay beds. These sequences may be superimposed on one another in cyclical deposition (Pickering and Hurst, 1989).

Sedimentary sand kaolin deposits may also result from post-depositional alteration of feldspar clasts in arkosic sand. Other sedimentary kaolin deposits are formed by the solution of carbonate from argillaceous limestone or dolomite, leaving an insoluble residue comprised mostly of clay.

Post-kaolinization Alteration

Most kaolin deposits were far from commercial quality when first crystallized or sedimented. The role of post-depositional

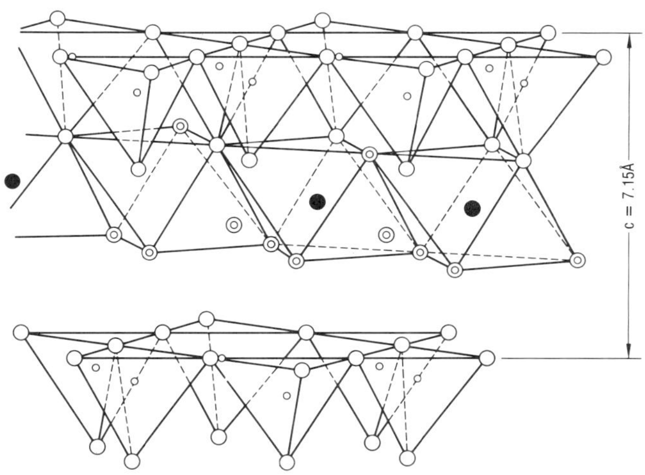

Fig. 4. Structure of kaolinite.

weathering and diagenesis in upgrading the quality of kaolin is just beginning to be understood (Pickering and Hurst, 1989). Many very white commercial kaolins were deposited in a much reduced state. Subsequent lateritic or semi-lateritic oxidation then leached out dark organic matter, altered pyrite and iron silicate and carbonate minerals to easily removed iron oxides, rinsed soluble salts away, and resulted in considerable recrystallization of the kaolinite itself. The final result is a deposit of extremely insoluble residues from weathering. All these changes may be quite beneficial to the quality of the resulting oxidized clay deposit, and may accomplish naturally what is very costly to do in industrial processing.

CHEMICAL PROPERTIES

Composition

The first step in testing most types of mineral prospects is a careful chemical analysis. Most kaolin users, however, are remarkably uninterested in a report of the chemical composition of a kaolin deposit; their main interest is in the optical and physical characteristics of the clay. However, a knowledge of the chemistry of kaolin is key to understanding some ceramic applications.

Kaolin minerals belong to the 1:1 group of clay minerals. This designation refers to their atomic lattice structure being made up of one sheet of silica tetrahedra (four oxygen atoms bonded about one silicon atom) linked to one sheet of alumina octahedra (eight oxygen atoms bonded to one aluminum atom) (Fig. 4). The sheets are linked together to make a layer silicate crystal structure. The excellent [001] cleavage occurs as the layers are sheared parallel to the weak hydrogen bonds. The kaolin minerals form one of the two subgroups of the 1:1 group of clay minerals.

The kaolin sub-group minerals kaolinite, dickite, nacrite, and halloysite are polymorphous, and have the approximate formula $2H_2O\ Al_2O_3\ 2SiO_2$. Above 400°C kaolin sub-group minerals dehydroxylate, losing water of crystallization and their distinctive crystal structure. Pure kaolin has the following ideal composition:

Alumina	39.5%
Silica	46.5%
Formula Water	14.0%

In nature, however, small variable percentages of iron and perhaps titanium substitute for aluminum in the crystal lattice, somewhat distorting the atomic structure and reducing the perfection of crystallinity.

Reactivity

One of the more important industrial characteristics of kaolin is its inert chemical nature. As a last remaining non-reactive residual mineral in some of the most intense weathering environments on earth, kaolin is stable under most industrial chemical conditions, over a pH range of at least 3 to 9. Thus it can be used as a pigment, coater, or filler without concern about decomposition, discoloration, or reaction with other components.

Typical Impurities

Natural deposits of kaolin contain a wide variety of impurities, most of which must be removed in order to make a commercially useful clay product. The coarser impurities, generally quartz, muscovite, and heavy minerals, are rather easily separated by settling or screening. However, there usually are micron-size impurities such as anatase, limonite, hematite, pyrite, organic matter, illite (hydromuscovite), montmorillonite (smectite), etc. The partial or complete removal of these impurities in an economical manner has been the subject of much research and process equipment development in the industry over the past thirty years (Prasad et al., 1991), and has enabled the production of kaolin products of remarkable purity from deposits formerly regarded as too impure to mine.

Table 1 shows typical weight-percent chemical composition of degritted, refined paper-coating clay products of similar brightness, from several major commercial kaolin mining districts.

PHYSICAL PROPERTIES

Optical Characteristics

One of the most important industrial attributes of kaolin is its highly reflective, quite opaque white color. The measurement and control of optical properties is central to the production of kaolin products for almost all uses. The most used determination is of brightness. Brightness is a measure of reflectivity and is tested by two methods: the GE (TAPPI) system, used in the United States and with modifications in Japan, and the ISO system, used in Europe.

Table 1. Chemical Composition of Refined Paper Coating Clays

	Georgia coarse secondary	Georgia fine secondary	Brazil fine secondary	Cornwall, UK coarse primary
SiO_2	45.2	45.0	46.0	47.2
Al_2O_3	39.2	38.0	37.0	37.6
TiO_2	0.53	1.6	0.98	0.04
Fe_2O_3	0.58	1.0	1.8	0.68
MgO	0.08	0.09	0.07	0.20
CaO	0.06	0.06	0.02	0.08
Na_2O	0.03	0.29	0.08	0.08
K_2O	0.02	0.13	0.0	1.39
H_2O	13.3	14.0	14.3	12.7

Color is measured by a variety of methods, but perhaps the most used in the kaolin industry is the Hunter System. This system expresses all shades of colors and grayness in three numerical values, also discussed in the section on specifications.

Opacity is a very important characteristic of a mineral pigment and is based on its ability to hide from view any pattern or boundary of reflectivity behind it. Opacity is indicated on a 0 to 100% scale, and measures the difference in show-through on a pure white surface coating of a pure black pattern on the back of the sheet. Opacity depends on such factors as index of refraction, particle size and aspect ratio, nature of particle interfaces, and brightness. Higher brightness pigments, all other factors being equal, are of lower opacity due to their higher contrast with dark ink show-through.

Other optical factors such as purity, saturation, luminosity, and surface gloss are sometimes specified for industrial pigments, but for most purposes kaolin optical factors can be compared by measurement of brightness, color, and opacity. Modern instruments can make a variety of measurements from a single pigment sample at the same time, and record the values on paper tape and/or digital memory. This greatly simplifies the evaluation of kaolins.

Particle Size and Geometry

Another of the main industrial cost advantages of kaolin is its naturally fine particle size. Most other minerals used as pigments, fillers, coaters, and extenders, must be reduced to fine powder by very costly grinding and milling, resulting in rather random size, irregularly fractured fragments. Kaolin is made up of loosely agglomerated minute crystals of rather consistent size and shape, a major advantage in producing functional powdered products. Both the natural particle size of many kaolins, and the needs of the commercial user, average rather close to two microns in diameter. Thus, it has become customary to discuss kaolin particle size as the weight percent finer than two microns. Measurement systems used assume that kaolin crystals are spherical in shape, which of course is not the case. Systems of particle size measurement used in the kaolin industry, and their inconsistencies, are discussed in the testing section.

Particle shape is important in kaolins as a determinant of both opacity and viscosity. Scanning electron microscopy is used to image the basic shape of the crude clay crystals. Thin crystal plates provide good opaque covering power when they are bonded with a binder of differing index of refraction, and provide a smooth coating surface for papers (Fig. 5) and paints. Large stack-like crystals (Fig. 6) which in past decades were discarded as waste are now cleaved by attrition delamination to form thin plate-shaped particles of very high aspect ratio (Fig. 7). High speed centrifugation can separate particles into narrow size fractions to suit the precise industrial application.

Viscosity

Because kaolin is often applied in industrial processes as either a water or polymer liquid suspension, the viscosity of these slurries is of key importance. Viscosity is a measure of the fluidity of liquids. Those which flow easily, like water, are referred to as low viscosity fluids. Those which flow sluggishly, like syrup or heavy oil, are designated high viscosity fluids.

Kaolins are valued principally for their very low, fluid viscosity at high solids. This is the opposite of bentonite clays, which are prized in industry generally for their very high viscosity at low solids percentages. Kaolins used for paper coating, molding in polymer suspension, or for casting at high solids must meet strict specifications for low viscosity.

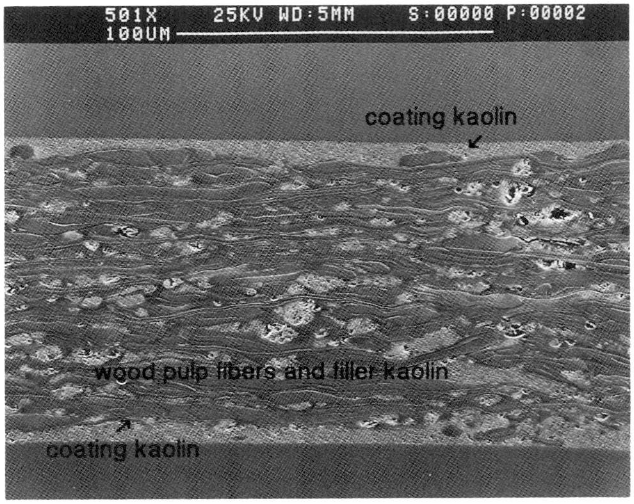

Fig. 5. Microtome section of a sheet of coated paper (*courtesy*: ECC/America, Larry Little, photographer).

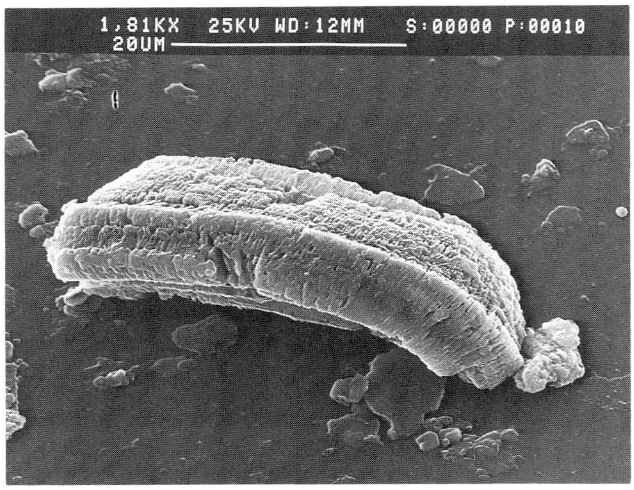

Fig. 6. Electron micrograph of a large crystal book or stack typical of coarse cretaceous central Georgia kaolin.

The kaolin industry has made remarkable technical advances in recent decades in improving the brightness, color, particle size, and abrasivity of its products (Prasad et al., 1991). Viscosity, however, has proven more difficult to improve when it is unacceptable in the crude state, by any practical process other than blending. Much research has been, and will continue to be, devoted to processes for viscosity reduction in kaolins occurring throughout the world.

Kaolins may have high viscosity in water slurry for at least one or more of the following reasons:

1. The presence of smectite, fine hydro-muscovite, or illite mica, as impurities.
2. A very uniform, well sorted size distribution of kaolinite crystals is usually an indication of high viscosity. Low viscosity kaolins tend to have bimodal or broad unsorted particle size distribution, with at least a scattering of fine crystals to ease interaction of coarse crystals between each other almost like ball bearings (Fig. 8). Thus, when coarse crude kaolin lacks

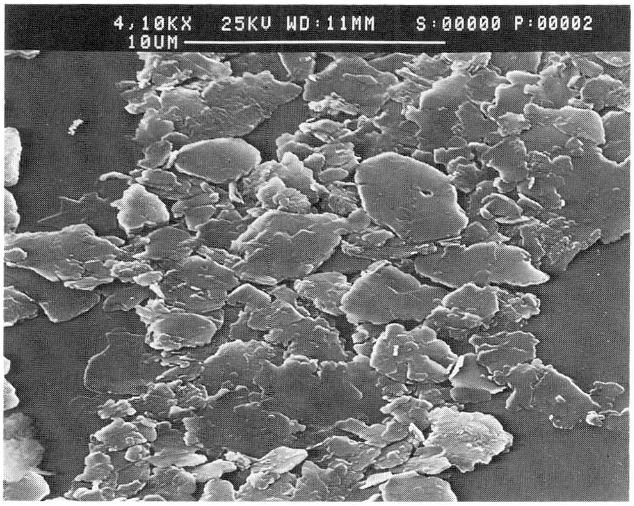

Fig. 7. Electron micrograph of delaminated kaolinite plates.

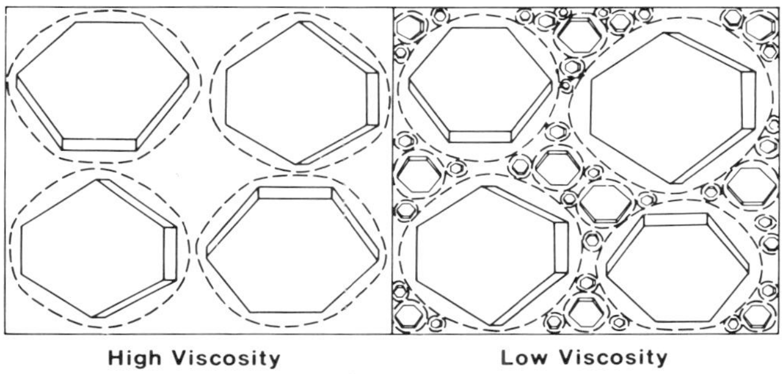

Fig. 8. Representation of the relationship between particle packing and viscosity.

bimodal or broad size distribution, viscosity is often improved by blending small percentages of fine kaolin during processing. Thus a mining district which has distinct deposits of both coarser and finer crude kaolins is in a commanding position in terms of viscosity.

3. The presence of chemical salts in the slurry, which inhibit the electrical charge on the kaolinite crystals so they fail to repel each other. Thorough rinsing with pure water during filtration may remedy this type of high viscosity.

There are two systems in common use for measuring the viscosity of liquid kaolin slurries: the Brookfield or low-shear method and the Hercules or high-shear method. The two measure very different fluid properties and it is quite common for a particular kaolin to have low Brookfield and high Hercules, or vice versa. Many kaolin customers require their products to meet demanding specifications for both. More complete descriptions of Brookfield and Hercules measurement apparatus and procedures will be found in the specifications section.

The demand in recent years for kaolins of very low Hercules, or high shear viscosity, in the paper coating industry has been rather intense. This has been driven by the following factors:

1. The higher running speeds of the most modern paper coating machines. The clay slurry is commonly spread onto the paper by a stationary blade as the paper moves beneath at speeds as high as 100 kph. High viscosity, unacceptable Hercules kaolin products cause the blade to drag across the sheet, causing streaking or even ripping of the paper.
2. As energy costs rise, papermakers try to use higher solids coating formulations in order to minimize drying time and cost. Viscosity of a kaolin suspension increases rapidly as it is thickened to higher solids percentage.

Table 2. Kaolin Production by Country 1988

Nation	t
United States	8 992 000
United Kingdom	3 149 000
Soviet Union* (former)	2 000 000
Brazil	795 000
Czechoslovakia*	697 000
India*	695 000
West Germany (former)	553 000

*may represent crude t production

An exception to the general commercial demand for kaolins of very low viscosity is in the ceramic dinnerware and table china industry. Here the ware (plates, bowls, cups, etc.) is slurry injected into porous molds which draw water away, leaving "greenware" which must be handled considerably before firing. If typical low viscosity kaolin is used to make such pieces, they have little strength and much costly breakage results. A very rare type of kaolin with a natural white magnesium-calcium smectite content of 10 to 30% is mined in a limited small area of Twiggs County, GA. This high smectite white-firing kaolin, completely unsuited to other uses, is sufficiently viscous and sticky to have excellent "green strength" and to greatly reduce breakage before firing. The air floated product from this unusual crude clay is in demand for restaurant white dinnerware or bone china. The ceramic catalyst support matrix used in the automotive industry to reduce exhaust pollution are also made from viscous kaolin. In this case, however, the higher viscosity is due to lack of bimodal particle size distribution.

World Production

Kaolin is mined and processed in many nations. It is one of the few industrial minerals of sufficient value, in its highly refined state, to be widely exported into international trade throughout the industrialized world. The United States, England, Brazil, and Australia are the major exporting nations. Most export kaolin is of paper coating grade.

The latest US Bureau of Mines production statistics (*Minerals Yearbook 1988*, Vol. III, Area Reports, International) report 1988 total world kaolin production to be 25.7 Mt. However, this total probably includes estimates reported for some eastern European and Asian countries which commonly are listed in crude tons; actual purified kaolin product recovered by processing in these nations may be considerably less. Kaolin production in 1988, as reported by the US Bureau of Mines, is shown in Table 2.

It should also be noted that commercial kaolin production covers a wide range of value from unprocessed crude clay to various degrees of refinement, depending upon grade, specifications, and market suitability. Kaolin ranges in approximate value per ton from 10 to $US500, without including transportation to the user. The quality of kaolin in any particular nation's production, and therefore its value, thus may be only generally proportionate to the tonnage reported.

NATIONAL OCCURRENCES

Major Kaolin Producing Countries

United States: Georgia-South Carolina District.

Geology. The world's major kaolin production district is located on the upper Coastal Plain of Georgia and South Carolina in Late Cretaceous and Early Tertiary age sediments. The kaolin deposits occur in lens-shaped beds at the tops of interlayered, cross-bedded, fining upward sand sequences. The commercial kaolin bodies, which are essentially monomineralic kaolinite, may be as thick as 12 m, but average closer to 3 m in thickness. The clay mined from these deposits is extremely pure compared with most kaolin mined elsewhere; it averages less than 5% "grit" coarser than 45 μ. The strata in the district dip seaward at 3 to 6 m/km and are very little folded or faulted.

There are two kaolin-bearing stratigraphic units of commercial interest in the district (Buie and Fountain, 1967, and Austin, 1973): the Late Cretaceous Buffalo Creek Formation of the Oconee Group (Pickering and Hurst, 1989), and the Middle Eocene Jeffersonville Member of the Huber Formation of the Oconee Group (Huddlestun and Hetrick, 1991). The Buffalo Creek Formation kaolins are rather soft, coarse in particle size (averaging 65% finer than 2 μm), with a smooth, conchoidal fracture and abundant stack-like crystals packed face-to-edge. The Jeffersonville Member kaolins are harder and tougher, considerably finer in particle size (averaging 85% finer than 2 μ), contain more hematite and anatase, and are dominantly fine plate-shaped crystals which are packed face-to-face. The blending combinations possible from these two differing sources of crude clay have enabled this district to be particularly versatile, producing almost every type of kaolin product needed by any sort of consuming industry.

Origin. As early as 1898, Ladd concluded that weathered metamorphic and igneous gneisses and granites to the northwest were the source of kaolinitic detritus that, through sorting and modification, became the kaolins of the district today. This general view remains unchallenged, but many differing interpretations of the composition of the detritus, depositional conditions, and post-depositional alteration have since been advanced.

Veatch (1909) speculated that the kaolins were laid down "in the deeper and quieter waters of offshore lakes and sounds" as "non-persistent and lenticular beds of pure white clay, while in the shallower water under conditions of shifting currents, the cross-bedded sands and, occasionally, beds of pebbles were laid down." He recognized that some of the sediments were re-subjected to the action of currents, partly eroded, and redeposited. These early observations of Veatch have formed the basis of all more recent thought on the origin of the Georgia-South Carolina commercial kaolins.

More recently, Kesler (1963) thought that he found evidence that feldspathic sands from the Piedmont were rapidly deposited in a littoral or deltaic environment, and that the kaolin deposits then originated by chemical weathering of these sediments, followed sometimes by erosion and redeposition of the winnowed clay as kaolin lenses and beds. Most subsequent observers have agreed with Kesler's proposed littoral or deltaic depositional environment, but they can find little evidence to support the deposition as feldspathic sands. Most, like Murray (1976), view the kaolins as having formed from already weathered clay minerals, with the Tertiary clays perhaps eroded and redeposited from the Cretaceous kaolins.

Pickering and Hurst (1989) suggested that the Lower Tertiary hard, fine kaolins could not have been eroded from source beds in the Cretaceous soft, coarse kaolins, because of their consistently differing crystallinity values, crystal habit, and physical and chemical characteristics. Hassanipak and Eslinger (1985) showed by oxygen isotope analysis that the hard, Tertiary kaolins of Georgia could not have been derived from the Cretaceous kaolins by winnowing unless post-sedimentation recrystallization of one or both had occurred. Sprague et al. (1987) used carbon isotope ratios to show that the hard kaolins were deposited in significantly more saline brackish water than the soft kaolins.

Fig. 9. Overburden removal using shovels and trucks.

Dombrowski (1992) correlated the soft kaolins with granite and gneiss source rocks, and the hard kaolins with phyllite metavolcanic source rocks by trace element analysis. Pickering and Hurst (1989) presented evidence that the soft kaolins have been rather completely recrystallized by lateritic alteration, while the younger hard kaolins have not. Webb and Sprague (1991) showed that the originally deposited sediment which is now the hard kaolins was deposited in intensely reducing conditions, as it had a high organic content. Their evidence is that it has been very extensively bioturbated by burrowing worms and crustaceans which could not have survived if organic debris were not abundant in the original clay. Thus, the kaolin must have been much changed since deposition, as the organic matter and pyrite have since been removed by intense oxidation weathering.

Mining. In the past five years or so, overburden stripping in this area has largely changed to using hydraulic hoes and shovels, which load large, often all wheel drive trucks (Fig. 9). This equipment has proven to be more cost effective, able to function in wetter conditions, and quicker. Also used are self loading pan type scrapers. Hydraulic and scraper equipment are used by a number of companies that contract for overburden earthmoving for many kaolin producers. Large draglines were much used in the past, but have lost favor due to their high initial capital cost, need to be disassembled to move to new mining areas, low clay recovery and difficult quality control, inability to work in severe artesian water conditions, and difficulty of reclaiming the rough terrain they leave after mining is completed.

Once overburden is removed, clay is mined by hydraulic hoe or loader, or by small dragline. It may be mined directly into highway-capable trucks to be hauled to the production plant (Fig. 10), or blunged in the mine and transported by slurry pipeline (Fig. 11). Some advantages of blunging in the mine area are:

1. Lower cost than truck hauling, often by a factor of 5 or 10, once the pipeline construction cost has been amortized.
2. Much larger, more statistically useful kaolin quality control units. Each tank of kaolin slurry, holding 800 t of clay on a dry basis, is intimately mixed and is thoroughly tested by a single easily collected sample.

   Tanks of clay can then be blended easily to obtain the exact quality required. Each 20-ton truckload, however, is too small

Fig. 10. Loading kaolin into trucks.

Fig. 11. Dragline feeding blunger for slurry transport by pipeline.

to be tested economically, and too large for a single grab sample off the top to be representative.

3. Pipeline transport of kaolin slurry, often for distances of 30 to 80 km, keeps large, heavy trucks from congesting and wearing the light duty roads of the area, and removes clay transport from government regulation. Slurry transport also saves fuel.
4. When clay is slurried in the mine area, nearby old mine pits usually are available for disposal of sand and other coarse impurities removed by degritting. This is generally simpler than building large disposal impounds in the congested area of the production plant.

Production. After transport to the production plant, the clay is refined by either water washing or air flotation. In the Georgia-South Carolina District, the water wash plants have a finished capacity from 300 to 1 200 kty. Air float plants range from 50 to 300 kty. At present, there are some 10 to 15 water wash kaolin plants in operation in the district, and 20 to 30 airfloat plants.

As has been mentioned, the availability of both coarse and fine crude clay in this district has allowed a very versatile range of kaolin products. These products are also among the highest value added

Table 3. Tonnage and Value of Various Kaolins Produced in United States

Processing Type	t	% t	$US Value	% Value
Unprocessed	1 245 800	13	17,337,400	1
Air float	1 450 700	15	84,576,500	8
Water washed	4 231 900	43	469,040,400	44
Delaminated	1 328 400	14	160,519,700	15
Calcined**	1 505 000	15	340,892,000	32
TOTAL	9 761 800		1,072,366,000	

* Data adapted from US Bureau of Mines.
**Includes both pigment grades produced in the Georgia-South Carolina District and Texas and refractory grades produced from kaolin-bauxite mixtures elsewhere.

clay products in the world. In 1988 Georgia and South Carolina produced 96% of the United States' total value of kaolin.

The paper industry uses more than three-quarters of the nearly 9 Mt of kaolin produced annually in Georgia and South Carolina. Paper kaolin products produced include number 1, 2, and 3 coating grades, ultrafine glossing clay, delaminated clays, de-fined light-scattering opaque grades, coarse and fine calcined clays, and filler clays. GE brightness of paper coating kaolins produced ranges from 86 to 90; calcined kaolins range from 89 to 94, and filler kaolins range from 78 to 90.

Plastic and paint pigments are produced in a wide range of particle size and brightness, both as is and coated with titanite and silane functional surface modifiers.

Ceramic kaolins produced in the Georgia-South Carolina District vary widely in casting and dewatering characteristics, chemistry, green strength, and fired color, depending on the needs of the market. Table 3 details the major tonnage and value produced in the United States in 1988 (almost all from Georgia and South Carolina).

The diverse range of kaolin products produced in the Georgia-South Carolina District sell for a wide range of prices. Unprocessed, crude clay may sell for less than $10/t; highly refined kaolin for a very specialized market may bring $500 or more per ton. Table 4 shows a recent compilation for some of the more commonly traded grades.

In 1990 the US Bureau of Mines reported that the United States was a strong net exporter of kaolin, almost exclusively from Georgia-South Carolina production. Total US exports for that year increased 21% to 2.83 Mt, valued at $428 million, or $151.24/t. The major receiving nations were Japan, Canada, Italy, the Netherlands, Finland, and Mexico. Tonnage exported for various markets are shown in Table 5.

Due also largely to production in the Georgia-South Carolina district, the United States is quite self-sufficient for kaolin to meet domestic needs. In 1990 only 3 067 t of kaolin was imported, valued at less than $1,500,000.

California. Kaolin is mined and processed as a byproduct from a sand operation near Ione in the Central Valley of California, from the Ione Formation of Eocene age (Bates, 1945). The Ione Formation is a cross-bedded and much channeled sand with variable but small percentages of kaolin. The basal Ione contains much lignite, pyrite, and siderite. The upper zone, where kaolin is recovered, has been weathered to a well oxidized state where iron is present as more easily removed limonite and hematite.

The Ione Formation is a marine to brackish water sediment deposited in fluvial, deltaic, and lagoonal conditions. Originally it was deposited in a highly reducing environment, as shown by the lignite, pyrite, and siderite content present in the unoxidized lower zone.

The main mining operation is for sand, and kaolin is washed

Table 4. 1990 Price Quotes for Kaolin Products from Georgia*

Dry ground, air floated, soft, FOB Georgia, per short ton	$ 52.00
Filler, general purpose, same basis, per short ton	$ 64.00
No. 3 coating, water washed, uncalcined, in bulk carload lots, same basis, per short ton	$ 85.00
No. 2 coating, water washed, same basis as for No. 3, per short ton	$ 87.00
No. 1 coating, water washed, same basis as for No. 3, per short ton	$110.00
National formulary, powder, colloidal, bacteria controlled, 50 lb. bags, 5,000 lb. lots, per short ton	$273.00
Delaminated, water washed, uncalcined, paint grade, 1 micron average, FOB Georgia, per short ton	$284.00
Calcined, water washed, in bags, carload lots, same basis, per short ton	$323.00
Calcined, water washed, paper grade, same basis, per short ton	$470.00

*Adapted from US Bureau of Mines, 1990, as quoted from *Chemical Marketing Reporter*, 1990. Note that kaolin prices are often significantly discounted, particularly during times of oversupply.

Table 5. 1990 Market Percentage of Total Kaolin Exported

Paper coating	69%
Paper filling	18%
Rubber filling	5%
Paint	2%
Other	6%

out and recovered as a byproduct. Clay recovery is 20 to 25% and products are made which range in GE brightness from 76 to 83. The kaolin is marketed to the ceramic, paper filling, and fiberglass industries. Magnetic separation and reduction leaching are necessary to attain 80 to 83 brightness. Production in the Ione area is less than 100 kty.

Florida. Near Edgar, FL, kaolin is recovered as a coproduct from a sand mine. Kaolinitic sand in the Fort Preston Formation of Middle Miocene age (Calver, 1949) is mined to a depth of 6 to 17 m. An average of 15% plastic kaolin is recovered from the sand. The kaolin is sold to the ceramic industry for whiteware, tile, glaze, and refractories.

Texas. Kaolin is mined from Wilcox Group (Late Paleocene) sands near Kosse, TX (Fisher et al., 1965). It is recovered as a byproduct from a glass sand operation and sold as a coarse calcined product of 90 to 93 GE brightness.

Minnesota. There has been exploration interest in primary kaolin deposits altered from granite and gneiss in the Minnesota River valley from Redwood Falls to Fairfax, in south-central Minnesota (Parham, 1970). The alteration seems to have occurred during the Cretaceous, and may be somewhat localized beneath black organic, pyritic Cretaceous shales (due perhaps to enhanced kaolinization by oxidizing organic matter forming carbonic acid, and pyrite oxidizing to produce sulfuric acid). Kaolin content of the granite residuum is 30 to 60%, but the clay is mixed with greenish chlorite mica. There are several small mines where kaolin is produced as an additive in Portland cement manufacture to increase alumina, and also as a soybean meal filler for animal feeds.

Considerable leasing and exploration has been done by several companies for paper grade kaolin deposits in Minnesota, due no doubt to the transportation advantage over the Georgia-South Carolina District in delivery of clay to the western Great Lakes paper manufacturing market. The state legislature recently provided a $480,000 grant to the Minnesota Mineral Resources Research Center, the Department of Natural Resources, and the Minnesota Geological Survey to study the kaolin deposits. To date, however, there has been no announcement of discovery of a paper quality kaolin deposit in Minnesota.

United Kingdom

Cornwall-Devon Kaolin District

Geology. The Cornwall-Devon area of southwestern England is the second largest kaolin producing area in the world. Production is from primary deposits, which are developed on different areas of the Cornubian granite batholith of Variscan age, especially the St. Austell and Tregonning low iron lithium mica granite plutons.

The kaolinite is quite coarse in crude crystal size, averaging perhaps 25 to 35% finer than 2μ. Most of the kaolin crystals are stack-like in habit. Small amounts of smectite, derived from alteration of biotite, is found near alteration fronts (Exley, 1976). Secondary mica (illite or hydro-muscovite) is fairly common, although distribution is patchy. Minor accessory minerals which are easily removed in the degritting process include tourmaline, topaz, apatite, fluorite, zircon, cassiterite, wolframite, rutile, hematite, and turquoise. Kaolinization has followed fractures, joints, and fissures in the parent rock, generally on north-south and northwest-southeast trends. The shape of the kaolinized bodies is generally similar to a funnel with the narrowed end extending downward as much as several hundreds of meters in some cases. Depending on rock fracture orientation, however, kaolinized bodies may be trough, slab, or irregular in shape. The most extensive kaolinized ore bodies are often found where several fracture zones intersect or overlap (Exley, 1959, 1976, Bristow, 1977, Bristow and Exley, 1989).

Origin. The main kaolinization of the Cornwall-Devon district took place from Late Triassic to the present, during Bristow and Exley's Stage V alteration of the Variscan granite batholiths of the area. The altering fluids were meteoric in origin (Sheppard, 1977, Jackson et al., 1982). These altering fluids moved as convection cells (Fehn, 1985, Durrance et al., 1982). Kaolinization occurred in the downflow areas of the convection cells, while upflow was in granite and plutonic margins. The convection cells are apparently still active; kaolinization is perhaps still in progress in the district. Heat to drive the convective process and activate the alteration fluids was derived from radioactivity of apatite, zircon, monazite, xenotime, and especially uraninite. Chief among the parent minerals altered to kaolinite was sodic plagioclase (Bristow, 1990).

Mining. Due to the loose, friable texture of the high grit Cornwall and Devon kaolin ores, they can be mined very economically by hydraulic monitor jet. The clay slurry from the hydraulic mining face is collected in low elevation sumps. This allows much of the grit to settle out and the slurry is then pumped to concentrators for further degritting.

Production. The hydraulically mined kaolin slurry is pumped from concentration sumps to spiral or wheel type degritting separators to remove quartz and mica. Degritting is then completed by cyclones to remove material coarser than 50 μm. The clay slurry is thickened to about 30% solids by settling. Centrifuging then separates the kaolin into a 75% finer than 2 μm coating clay fraction, and a coarse crystal residue. The coarse crystals are refined by ultraflotation, and then delaminated to thin cleavage plates by sand attrition. Further screening, filtration, and drying and/or calcination provides the finished kaolin products. These include paper coating clay at 80% finer than 2 μm and 85.5 to 87.5 ISO brightness; paper filling clay at 45% finer than 2 μm and 79.5 to 82.5 ISO brightness; and ceramic kaolins at 57 to 70% finer than 2 μ which fire relatively white.

Commonwealth of Independent States (formerly USSR)

Ukrainian Republic. In the middle to late 1980s several Western kaolin producing companies were approached by the then USSR about the possibility of a joint venture to produce high quality paper coating kaolin products for the Scandinavian fine paper industry, and eventually for domestic Soviet consumption. A number of Western kaolin specialists visited the clay mines and production plants of the southern Ukraine, where kaolin has been mined for about 100 years.

Geology. Most production of refined kaolin comes from primary deposits developed on Ukrainian Shield granites and gneisses (Petrov, 1969). The main production areas are at Prosyanovski, north of the Sea of Azov and southeast of Dnepropetrovsk, and at Glhovetski 200 km southwest of Kiev and 70 km north of Vinitsa. The kaolinized zone is covered with Neogene sediments deposited when the level of the Black Sea was considerably higher than at present; at Prosyanovski these younger sediments contain extensive smectite fuller's earth and are saturated with saline ground water.

Kaolinization of the Ukrainian Shield primary commercial clay deposits thickens along fracture zones in the parent granite and gneiss, and extends to a depth of from 5 to at least 50 m. At Gluhovetski the kaolinized zone reaches a thickness of 65 m along a prominent west-north-west tending fracture zone in the parent rock.

The Ukrainian kaolin ore retains the texture and lithic features of the parent rock, and is rather soft and friable. Beneath the ore zone there is a hydromica zone of incompletely altered material some 5 to 20 m thick above unaltered rock. The unaltered parent rock consists of coarse post-metamorphic biotite granite, very coarse injection migmatite, and finer grained, much older contorted gneiss and interlayered amphibolite.

Origin. The primary kaolins of the Ukraine are said to be of hydrothermal origin, but the shape of the deposits, apparent absence of veins of hydrothermal type, and the lack of prominent minerals of hydrothermal derivation, suggest that alteration was at least in part from weathering by meteoric waters. These kaolins differ from the primary kaolins of the Cornwall-Devon District by their common occurrence with halloysite in the deeper zones.

The best kaolins of the district were derived from post-metamorphic granites and pegmatites, which may provide ore which averages from 30 to 50% kaolin of 80 to 90 ISO brightness after processing. The migmatite-derived kaolins are usually of lower quality, and those altered from the gneiss-amphibolite generally of much lower quality.

Also in the Ukraine in the Transcarpathian region Rusko (1973) mentions deposits of nearly pure dickite, and mixed deposits of kaolinite, dickite, and halloysite.

Mining. At Prosyanovski kaolin is mined from the older Versinski mine and the newer West Dibrovski mine. The kaolinized zone is more consistent at the newer Dibrovski mine. There is 0 to 10% montmorillonite (smectite) at each mine; perhaps a benefit to the ceramic uses of the kaolin, but detrimental to paper applications.

At Gluhovetsky the crude kaolin seems to contain less montmorillonite and less saline pore water. Mining is done from the Turbov pit. Crude kaolin quality here seems better and more consistent, however permission to mine may be more difficult because of conflict with local agriculture.

Overburden removal and mining in both areas are generally accomplished with large bucket wheel excavators, which load overburden or clay ore onto conveyor belts for transport to the dump or to production plants.

Production. After transport to the processing area, the crude clay from Prosyanovski is blunged at a ceramic plant in washing drums and degritted with sand rakes. No centrifuging or leaching is done at this ceramic plant, which produces a product approximately 52% finer than 2 µm at 72 to 73 ISO brightness. The same slurry is also fed from the ceramic plant to the paper coating plant, where it is centrifuged to approximately 75% finer than 2 µm and leached to a 84+ and a 86+ ISO brightness product. Some of the clay is processed by a small magnetic separator. The 30 to 40% oversize reject kaolin from the centrifuging process is returned to the ceramic plant. Due perhaps to brackish mine and process water, the kaolin product from Prosyanovski is of marginal Brookfield and Hercules viscosity.

Kaolin production is less complex at Gluhovetsky than at Prosyanovski, as there is no centrifuging or leaching done. The crude clay is similarly degritted, filtered, dried, and shipped without further treatment.

Kaolin products from the Ukraine are used domestically or exported to other republics and to eastern Europe, for applications in the paper, ceramic, rubber, and chemical industries.

Other Republics. Other kaolin deposits have been described by Petrov (1969) in Kazakhstan, near Pologa Station on the Konok River, and near Vladivostok in the Maritime Territory.

Brazil

Amazon Region

Geology. A major secondary or sedimentary kaolin deposit was discovered in 1967 on the east bank of the Jari River about 100 km north of its junction with the Amazon River (Moeri et al., (1989) and about 350 km west of Belem, the large city near the mouth of the Amazon. The kaolin of commercial quality is 25 to 32 m thick, very fine in particle size (at least 75% finer than 2 µm), and extends over an area of several tens of square kilometers. The age of this kaolin is apparently Pliocene, in the Belterra Formation of the Barreiras Group (Klammer, 1971). The clay is very similar to the hard, fine Tertiary kaolins of the Jeffersonville Member of the Huber Formation in the Georgia-South Carolina District of the United States.

The Jari kaolin lies just beneath the flat tops of a very irregular, extensive series of steep-sided plateaus (Klammer, 1984) covered with thick tropical rain forest. These plateaus are largely unexplored for kaolin or other minerals; a plateau area explored has been shown to contain more than 250 Mt of good, commercial grade kaolin (Moeri et al., 1989). Mining began on the Jari deposits at the Felipe mine in 1976 and continues to the present.

Another large secondary kaolin prospect, which has not yet been mined commercially, is located on both the east and west sides of the Capim River about 150 km southwest of Belem. Also occurring in the Barreiras Formation (Hurst and Bosio, 1975), this sedimentary kaolin is very different from the Jari deposits. The Capim River kaolins are soft, rather coarse in crude particle size (about 60% finer than 2 µm), with a smooth, conchoidal fracture, low iron and titania content, a high percentage of stack-like crystals packed face-to-edge, and a high Hinckley crystallinity index. In almost all respects, therefore, the Capim River kaolins closely resemble the Late Cretaceous Buffalo Creek Formation kaolins of central Georgia in the United States. Analyses by Hurst and Bosio (1975) report leached GE brightness as high as 89, with low Brookfield and Hercules viscosity. They state that this clay is a good prospect for wet processed paper-grade kaolin production. Although there are very few good kaolin outcrops along the river, test pitting and core drilling which has been done indicates at least tens of millions of tons of paper-coating-quality kaolin are present in the subsurface.

Origin. The Amazon area kaolins were derived from weathered granitic rocks of the Guyana Shield and were deposited in a very large lacustrine basin in a generally deltaic environment. Rain forest growth is abundant, resulting in plentiful organic acids. This acidic soil water, combined with the very heavy rainfall and hot climate, have resulted in intense lateritic leaching which have removed all reactive mineral and organic matter from the kaolin strata. It is very likely that the Capim River kaolins have undergone diagenetic recrystallization similar to that of the Cretaceous kaolins of central Georgia in the United States.

Kaolin is also produced at Mogi das Cruzes in Sao Paulo State in southeastern Brazil for the domestic paper industry (Anon., 1989).

Mining and Processing. The Jari River kaolin deposits are mined from the Felipe pits on the east side of the river. Stripping of overburden, and clay mining, are done with self-loading scraper pans. The pit is operated by benches approximately 6 m high. Six benches are now in operation to work the 32 m of kaolin under about 11 m of overburden. Clay is blunged and degritted at a stockyard in the mining area and transported by pipeline under the river to the production plant at Muguba on the west bank.

Processing at Muguba includes centrifugation to 97% finer than 2 μm, reduction leaching, filtration, and extrusion and apron drying. Finished clay is loaded in bulk directly from the plant into large ocean going ships which easily navigate up the Jari River at all seasons of the year. The product is 88 GE brightness, or 86.5 ISO brightness, and has very good Hercules and Brookfield viscosity. It is used for paper coating, principally to enhance gloss and to control viscosity. Most of production is exported to Europe and Japan.

Selected Lesser Kaolin Producing Nations

Czechoslovakia. The main producer in eastern Europe is Czechoslovakia where primary kaolin is mined from altered granites in the Karlovy-Vary area for paper filler, rubber filler, fine porcelain, and refractories (Murray, 1989). Kaolinization alteration depth is more than 50 m (Kuzvart, 1969), and the district extends over about 85 km^2. In 1987 Czech kaolin production was listed at 697 kt (Anon., 1989); it is not known how much of this is crude, unprocessed clay and how much is refined product.

Germany. Secondary kaolin is mined from Triassic age slightly clayey feldspathic sand in the Hirshau and Schnaitenbach areas of Bavaria, in southwest Germany. The sediments were derived from granitic source rocks in the Naab basin mountains and are preserved in an east-west trending half-graben. The deposit averages about 76% quartz, 12% feldspar, and 12% kaolin. The quartz has an average grain size of 1 mm, the feldspar about 1/10 mm, and the kaolin about 2 μm; economic separation depends upon efficient size fractionation.

Overburden is removed and raw ore mined by bucketwheel excavator and transported by conveyor belt to the adjacent production plant. The ore is blunged by chain stirrers in stone-lined vats. The sand is removed by screening, and the slurry is extensively cycloned to remove fine quartz and feldspar. Additional very fine feldspar is removed by flotation. The feldspar and quartz sand concentrates are cleaned and sold to local glass industries. Much more quartz sand results than can be sold; at one operation it is dewatered and stacked into a very large hill known as *Montekaolino*. The quartz sand hill is much used for summertime skiing.

Kaolin recovery is 12 to 16% by weight from the raw ore. The kaolin concentrate is reduction leached, filtered, and dried. The product is used as paper filler and coater, as paint, plastic, and rubber filler and pigment, and for ceramics.

In Tirschenreuth a kaolinized pegmatite is mined for ceramic and pigment grade clay products. In northwest Saxony in southeastern Germany, between Dresden and Leipzig, primary kaolins derived from the alteration of volcanic ignimbrites and porphyrytic andesites of the Bohemian Massif are extensively mined. These kaolins are used as pigments in paper, paint, and plastic; as filler in rubber, and as raw material in fine art ceramics and porcelains. Kaolin altered from pitchstone, felsite, and quartz porphyry has been mined at Seilitz, near Meissen, continuously since 1764 (Murray, 1989).

Spain. In central Spain there is widespread occurrence of a slightly kaolinitic sand in the region from Guadalajara to Valencia (Martin-Vivaldi, 1969). The sand is used in glass manufacture, and the kaolin is washed and concentrated for ceramic and pigment applications. The sands are thought to have been originally arkosic, the feldspar originating from granitic and feldspathic rocks of the *Meseta Uberica* during the early Cretaceous. Alteration of the feldspar in the sands is greatest at the top, indicating alteration by percolating ground water (Murray, 1989).

France. Most kaolin production in France is from Cornwall-type primary deposits in Brittany, which are used for ceramics, fillers, and pigments. Both primary and secondary deposits are mined around the *Massif Central* (Damiani and Trautman, 1969).

India. Secondary kaolins are mined from the narrow southwestern Coastal Plain from Cochin to Quilon to Trivandrum. The kaolin is in strata in sedimentary sands, and was deposited in deltaic, fluvial conditions. It was derived from the alteration and erosion of nearby charnokite, gneiss, and granite. Production is by both air classification and water washing; the product is used domestically for refractories, rubber filler, and miscellaneous pigments. A production plant at Veli, Trivandrum, refines kaolin from Thonnakkal, 30 km to the north, by fractionation, reduction leaching, filtering, and drying. The product is equivalent to a US made No. 2 coating clay, and it is used by the domestic paper industry.

In the Jaipur District of Rajasthan a primary kaolin derived from a pegmatitic granite is produced. This area is unusual in that much of the clay mining is underground. The northwestern area near Kutch also produces some kaolin.

Indonesia. Much of the porphyritic biotite granite areas of the Indonesian islands of Belitung and Bangka, east of Palembang in Sumatra, are altered to very coarse kaolin with halloysite and smectite (Murray et al., 1978). This area is a good example of weathering alteration in a moist, tropical climate. The kaolin is separated for use as paper filler, dinnerware, and sanitaryware.

People's Republic of China. Kaolin was first discovered, named, and put to use in China (Johnson and Blake, 1867) for ceramics, porcelain, and paper filling. Still mined in many small scale operations throughout the nation for domestic ceramic and filler applications, it is ironic that China depends on imports to meet its refined kaolin needs for paper coating.

Primary kaolins are mined and used for fine ceramic ware in many areas of China. In Kiangsi Province in the Ching-Tae-Chen District near Fuliang altered, kaolinized phyllite is mined and refined for art ceramics (Murray, 1989). Weathered granite is kaolinized near Hong Kong in Kwangtung Province. A particularly low iron Jurassic rhyolitic ash tuff has been weathered and altered to an abundant very white kaolin-illite clay 30 km west of the port city of Raoping in Guangdong Province, about 300 km northeast of Hong Kong. This primary clay may have good potential for high brightness delaminated paper coating and filling applications.

Secondary kaolin occurs as lenses of sandy clay of Late Tertiary age near the city of Maoming in Guangdong Province 350 km southwest of Hong Kong. This kaolin may also be suitable for refinement to paper coating or filling clay products (Yuan and Murray, 1990).

Australia. Secondary kaolin is mined for paper coating clay at Weipa, on the west coast of the Cape York Peninsula in northeast Queensland. The sandy kaolin stratum is beneath an extensive lateritic bauxite. The sandy kaolin is left exposed by the bauxite mining. The clay zone is 8 to 12 m thick, and averages 40 to 70%

kaolin (Murray, 1989). It is processed to make a fine No. 1 coating clay product of excellent high shear viscosity.

There are many primary kaolin deposits in Australia which formed as a result of deep weathering of silicic and mafic parent rocks (Gaskin, 1969). Secondary kaolin is also found in association with Western Australia coal deposits. Other than the sandy kaolins of the Cape York Peninsula, Australian kaolins are known to be abundant and often of good brightness and color, but quite viscous and abrasive (Patterson and Murray, 1983).

New Zealand. A very white halloysite-kaolinite mixture is mined from the Maungaparerua deposit on North Island, about 210 km north of Auckland. The deposit was formed by the combination of hydrothermal and weathering alteration of volcanic rhyolite flow rocks (Murray et al., 1977). Altered clay averages 50% quartz, fine amorphous silica, and cristobalite, and 50% halloysite, kaolinite, and allophane (Murray, 1989). The refined clay is prized as a whiteware and dinnerware raw material because of its fired whiteness and translucency. The clay is also used in paper, paint, plastic, and rubber manufacture.

TECHNOLOGY EXPLORATION

Strategy

New kaolin districts and deposits are located by assessing factors which are favorable to kaolin formation. Important determinants include: rocks with a source of feldspar or other aluminous minerals, moist lateritic and/or hydrothermal conditions, and shallow burial in areas of little overburden.

In the Georgia-South Carolina District of the southeastern United States, properties of high commercial kaolin potential are selected by:

1. Location in the correct delta-front paleoenvironmental position on the prograding Coastal Plain;
2. Optimum elevation, based on structure-contour mapping, between the unconformity at the top of the Lower Tertiary Huber Formation, and the unconformity at the base of the Late Cretaceous Buffalo Creek Member; and
3. Proximity to present or former kaolin mines.

Sampling

Accurate and representative samples are crucial in kaolin exploration, because quality testing is so complex, and decisions involving large capital investments hinge on proven reserves. In major kaolin mining areas deposits are located, evaluated for quality and tonnage, and mine plans developed, by drilling on a surveyed grid pattern. In less developed areas where drilling machines are not readily available and hand labor is more economical, test pits are commonly dug by hand.

Drilling

In the Georgia-South Carolina kaolin mining district, truck mounted core drills are used. Capable of penetration to depths of 30 to 90 m, they commonly use double tube core barrels of NX rating, recovering 5.5 cm diameter core, and recover nearly 100% of all clayey sediment encountered. Wireline core-pulling capability is useful for deeper drilling. A portable aluminum sump is used, kept filled by a 4 000 to 6 000 l water truck with a built-in rapid filling pump replenished from nearby streams and lakes. A small bulldozer is used to push access to the grid drill sites. Grid locations are rapidly staked by laser ranging surveys. Drill core is packed into wooden cases on site and delivered each evening to the testing laboratory. Particular care is taken to avoid core freezing in the field, which destroys important physical properties of the crude kaolin.

Table 6. Purpose of Kaolin Drilling Program

Drill Spacing (m)	
120	Property exploration
60	Proving assured kaolin reserve
30	Mine planning and development
15	Kaolin quality control and mine inventory volume and tonnage measurement

An experienced three-person drill crew with this type of equipment can complete two to five core holes each day. Initial drilling of a new property may spot-locate a few holes to establish the key elevation range for the kaolin strata. The area of higher promise is then surveyed and staked on a grid. The spacing of drill holes on this grid depends upon the purpose for which the drill information is intended. Table 5 shows the grid spacings that are in general use in the southeastern United States.

Drilling usually is started at a spacing of 120 m between holes. Those holes which encounter substantial kaolin are resurveyed, and drilled at 60 m spacing. Kaolin deposits proven and tested at this spacing are generally regarded as authentic mineral reserves, of known quality and size. From this information, volume is calculated in cubic yards. The conversion is then made from volume to tonnage by the commonly accepted approximation:

1 in-ground cubic yard = 1.62 wet short clay tons, or
1 in-ground cubic yard = 1.25 dry short clay tons.

These conversion factors are very important, as kaolin is commonly inventoried in the mines, and royalties paid to landowners by surveyed volume, and the clay then hauled, processed, and sold by the ton. A 20% safety factor is often used for kaolin reserves proven by 60 m spaced drilling, in order to account for erosional losses from irregularities on the clay surface.

When a proven kaolin deposit is needed for mining in the next one to two years, it is usually redrilled and tested on a spacing of 30 m. The kaolin deposit is then regarded as sufficiently well defined to justify the cost of overburden removal, using a 10% safety factor to account for inevitable mine losses. Precisely scaled drill maps, posted with laboratory testing results, allow detailed mine planning, overburden removal contracting, and designing of product quality forecasts. The availability of special beneficiation equipment in the processing plant is also allocated at this time.

After the overburden is removed and the upper surface of the kaolin is excavated, most companies redrill the new mine on 15 m spacing. This exposed clay drilling provides exact volume and tonnage inventory, and allows quality testing for selective mining and blending for statistical product quality control. Because there is no loose overburden to contaminate the drill samples, this exposed drilling is often done by very rapid pneumatic air-sampling, which allows 15 to 20 holes per day to be completed.

In other countries exploratory drilling is done by the military (as in the former Soviet Union), the State Department of Mining and Geology (India), or by People's Geological Teams (China).

Test Pitting

In less developed countries where mechanized drilling is not possible or practical (Brazil, India, etc.), hand-dug test pits are used for kaolin exploration. Typically, such pits are at least 1 m in diameter, and may be dug to depths of as much as 15 to 30 m. These pits, where carefully managed, do provide excellent representative samples. They have the advantages of relatively low cost, very large sample size, and availability for rapid re-sampling years later. A sufficiently slim geologist can climb down into these pits and study the kaolin strata in detail, if not in comfort. However, test pitting has the disadvantages of being slow, rather dangerous to the diggers, and limited in depth. A perched ground water table above many kaolin beds can stop the digging of a pit well before the total thickness of the kaolin stratum is sampled.

Geophysics

Geophysical prospecting methods, so successful in exploration for other minerals, have not proven helpful to date in the location of high quality kaolin deposits. Persistent attempts have been made by seismic, EM, electric self-potential, and ground penetrating radar techniques. The most hopeful results have been obtained from ground-penetrating radar; however, to date insufficient signal power has resulted in penetration of only 5 to 7 m. Probably the lack of success of geophysics is due to the common presence of a masking layer of dense smectite fuller's earth clay overburden, and to the fact that kaolin beds of inferior quality are encountered in virtually every deep drill hole. A geophysical method that could reliably locate clay strata would thus predict success on virtually every drill hole in the Georgia-South Carolina kaolin district.

Kaolin Mine Operations
Overburden Stripping

Open pit methods are used in all major kaolin mines today. Depending on the size and scope of the operation, availability of heavy equipment in the area, depth of overburden, overburden-to-clay ratio, water conditions, climate, etc., a variety of stripping methods are used. Hydraulic hoes and shovels which load large all wheel drive offroad trucks are commonly used. Pan-type self loading scrapers have often been used, but are difficult in soft or wet dirt. Large draglines have been used in the past, but have lost favor due to their large initial capital investment, need to be disassembled to move to new mine sites, lack of precision and low clay recovery in mining, difficulty of handling artesian water, and inability to leave a smooth, gently sloped surface for reclamation. Bucket wheel excavators loading conveyor belts are used in many European kaolin mines.

In some less industrialized countries where heavy equipment is difficult to obtain and operate, hand removal of earth is still utilized. Picks and shovels are used, and both overburden and clay are carried out in baskets or pans often on the heads of bearers, or by wheelbarrow.

Mining the Clay

Although the kaolin may be removed by much the same methods used to strip overburden, it must be handled with much more care to assure quality and to recover all possible worthwhile ore-grade material. As many as four or five separate grades of crude kaolin may be separated and produced from a single mine. Thus the actual mining of kaolin generally uses smaller equipment capable of more exact control. While the stripping of overburden is often done by contractors, clay mining is generally done by company operators using precise maps and plans to follow statistical quality control guidelines.

The crude kaolin may be mined by hydraulic hoe, shovel, or loader, small dragline, self loading pan, or by hand. In secondary kaolins where the clay is stratified horizontally by sedimentary deposition, quality differences are generally strongly layered. Thus more effective blending can be done by mining vertically by hydraulic hoe or shovel, resulting in better mixed crude ore, more uniform quality in the plant, and less wastage of crude kaolin. The clay may be hauled by truck directly from the mine to the processing plant, piled into stockpiles by quality, moved by conveyor belt, or blunged in the mine and transported by slurry pipeline.

Potential Future Mining Methods

In the future, kaolin in areas of high artesian pressure, and in wet areas, might perhaps be stripped and/or mined by large cutterhead dredges. This could open up kaolin deposits which currently cannot be economically stripped.

Research has also been done by the US Bureau of Mines on possibly mining kaolin through cased boreholes without the necessity or cost of overburden removal. The clay would be cut, slurried, and removed by hydraulic jet at the well head. Unfortunately, trials in a mine in Georgia showed that the very low grit kaolins of the area are too tough for this type of jet cutting. The system would also require a strong cap rock over the kaolin stratum to avoid caving and contamination of the clay by overburden.

PROCESSING AND BENEFICIATION

Two basically different processes are used to produce a salable kaolin product—dry process and a wet process. The dry process is simple and yields a lower cost and lower quality product than the wet process. In the dry process, the properties of the finished kaolin reflect generally the properties originally found in the crude kaolin. Therefore, deposits containing the desired qualities of brightness, low grit content, and particle size distribution must be located by drilling and testing. In the dry process, the kaolin is crushed to approximately egg size, dried in rotary dryers, pulverized, and air floated. The air floating process removes most of the grit.

The wet process (Fig. 12) has been highly developed in recent years to produce kaolin products of uniform and predetermined physical and chemical properties. Crude kaolin of variable quality can be wet processed to produce a uniform high quality product.

This slurry is pumped from the blunger to a degritting station, where, by the use of settling boxes, screens, and/or hydrocyclones, the coarse foreign particles called grit are removed (Fig. 13). Grit is defined as that material coarser than about 44 μm. After degritting, the kaolin slurry is collected in large storage tanks (Fig. 13) and then pumped through pipelines to the beneficiation plant where it is stored in large tanks. Kaolins of various qualities are blended at the mine tanks or in the terminal tanks at the plant to provide a feed for the beneficiation plant that will produce products to meet specifications (Fig. 14). The first step in the wet process is to fractionate the kaolin slurry into coarse and fine fractions by using continuous centrifuges, hydrocyclones, or hydroseparators (Fig. 12). After fractionation, the kaolin is leached to remove ferruginous coloring compounds. In this step the kaolin is acidified with sulfuric acid to a pH of about 2.5 to dissolve the iron. A strong reducing agent such as hydrosulfite is then added, and the iron reacts and forms a soluble sulfate, which is removed during the dewatering step. The dewatering of the flocculated kaolin slip is accomplished by using evaporators, rotary vacuum filters, or filter presses. Sometimes the flocculated slip is thickened by using thickeners or centrifuges before the filtration. After filtration, the kaolin filter cake can be extruded to noodle form and dried in apron dryers or rotary dryers or it can be redispersed to the fluid state and spray dried, drum dried, or shipped in slurry form as 70% solids in tank cars.

The foregoing paragraph is a simplified description of the wet

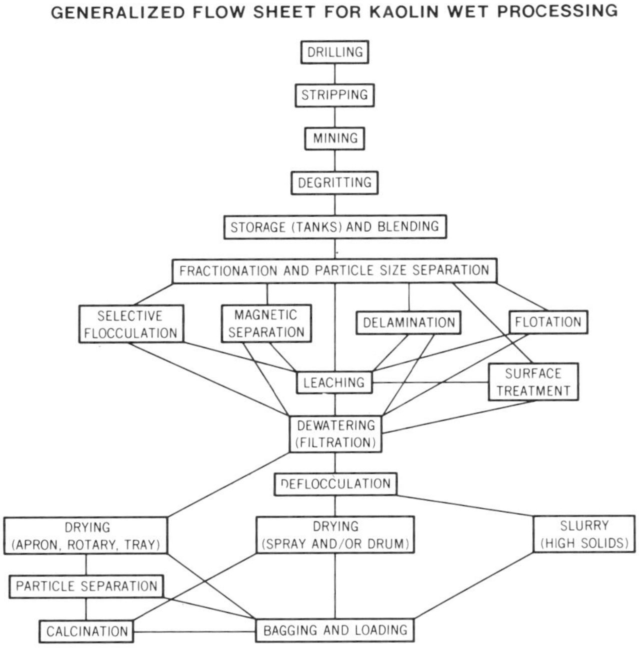

Fig. 12. Wet process flowsheet for kaolin coating and filler production.

Fig. 13. Photo of degritting station and mine storage tanks.

Fig. 14. Photo of kaolin wet processing plant.

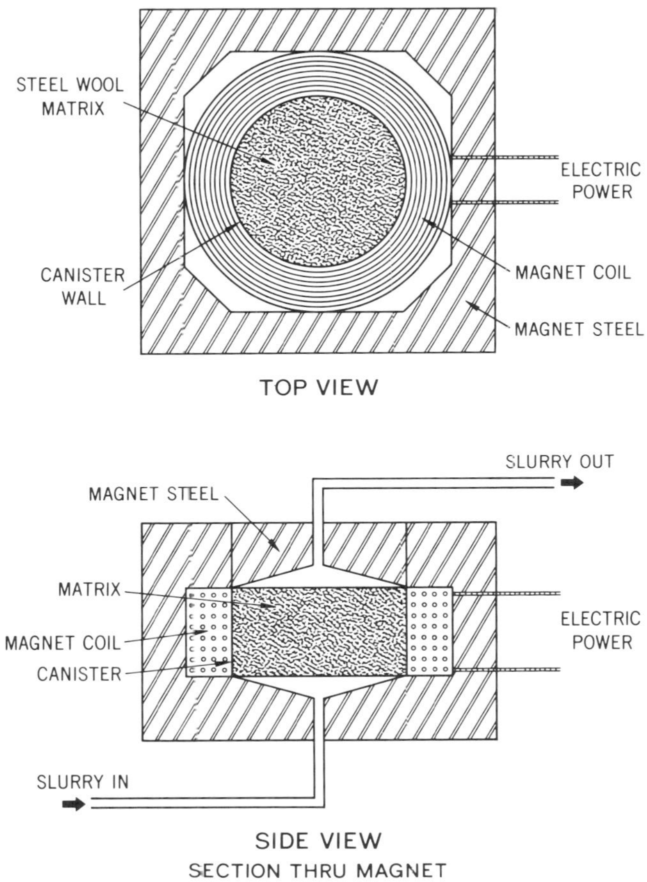

Fig. 15. Diagrammatic sketch of high gradient magnetic separator.

process. Other processes are used to produce brighter and whiter kaolins and to reduce viscosity. Flotation (Greene and Duke, 1962) and selective flocculation (Bundy and Berberich, 1969) are used to remove iron and titanium impurities to produce kaolin in the range of 90% brightness.

A standard method for improving brightness and whiteness is a high intensity magnetic separator which removes substantial quantities of iron and titanium mineral impurities. The intensity of these magnets range from 18 to 50 kilogauss (1.8 to 5.0 Tesla). The main feature of the separator system (Fig. 15) is a canister filled with a special type of fine steel wool which provides a uniform magnetic field through which the kaolin slurry is pumped and the magnetic particles retained in the matrix. Periodically, the magnet is de-energized, and a high volume of water under pressure is back flushed through the matrix to flush out the impurities (Oder and Price, 1973).

Delaminated kaolin is produced by splitting apart the larger books or stacks of kaolin (Fig. 16); this process yields a product in which the individual plates tend to be wider, thinner, and whiter. The process used for delamination is wet attrition grinding (Gunn and Morris, 1965). Delaminated kaolin has special application in lightweight paper coatings, in paint films, and in special rubber applications.

Calcination is another widely used process to produce special

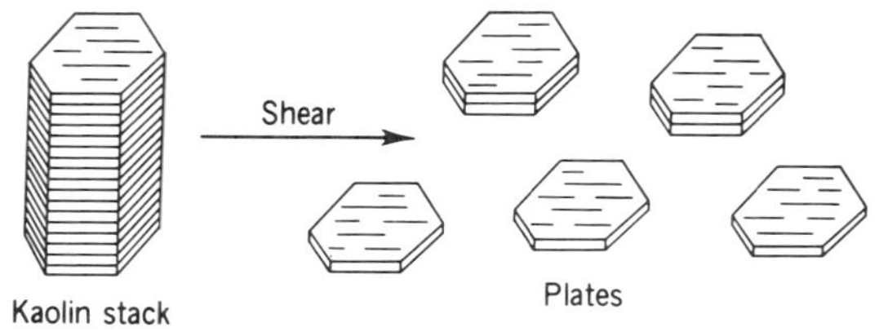

Fig. 16. Diagrammatic representation of delamination of a kaolin stack into large thin plates.

products from kaolin. When kaolin is heated to approximately 1 050°C, it is converted to mullite, silica-alumina spinel, and cristobalite. The product is whiter, brighter, has better hiding properties when used in thin film applications, but is more viscous and abrasive.

Other special processes are used to modify the surface of the kaolin chemically to improve its function in many systems, such as ink, rubber, paint, and plastics. These surface-modified kaolins can be hydrophilic, hydrophobic, or organophilic.

PRODUCT SPECIFICATIONS

Paper Filler and Coater

The most important specifications for kaolin used in paper applications are brightness, particle size and distribution, viscosity, and screen residue. Some purchasers also require tests of abrasiveness. The Technical Association of the Pulp and Paper Industry [TAPPI] (Lyons, 1966) gives detailed information regarding the properties and test procedures for kaolin used in paper; only brief descriptions of the more important tests will be outlined below.

Brightness

In the United States this test consists of drying, pulverizing, forming a plaque, comparing with a standard of known brightness, and measuring the brightness at 457 mμ with a reflectance meter constructed, calibrated, and operated according to the requirements of TAPPI Standard T646-om-86. This test must be very carefully done by an experienced and competent technician in order to obtain accurate and reproducible results (Fig. 17). In general, a minimum brightness of 80% is required for filler clay and 85% for coating clays. In Europe, the ISO system (TAPPI 534-om-86) with different optical geometry is used.

Particle Size

Many methods of determining particle size and particle-size distribution are used, but the most rapid and commonly used is based on the Stokes principle. Size determinations can be made by the application of this principle, because settling rates in fluids are controlled by particle diameters when the attraction of gravity, densities of particles and fluids, shape of particles, and the viscosity of the fluids are identical. Determinations are approximate, as kaolin particles are shaped quite differently from the sphere assumed by Stokes law. Coarse particles, therefore, settle more rapidly than finer ones, and determinations of a given particle size and distribution can be made by the time-rate principle. The time-rate principle is equally valid for settling by gravity or when settling is augmented by centrifugation. A rapid test for plant control consists of the preparation of a standard, dilute deflocculated slip of the clay

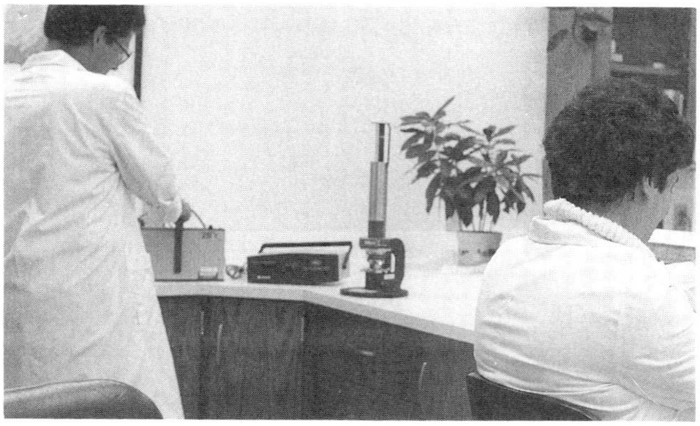

Fig. 17. Photo of a brightness meter.

in question, settling by controlled centrifugal force for predetermined short time intervals under such conditions as to effect reproducible packing of the sediment, and measuring the apparent volume of sediment at each time interval. Typical particle-size analyses of filler and coating grades are shown on Fig. 18.

Viscosity

Flow properties of kaolin are very important to the paper coater (TAPPI standard 648-om-88) because they affect the functioning of the coating operation as well as the final quality of the paper product. Two viscosity tests are used in the kaolin industry, one to measure high shear viscosity and the other to measure low shear viscosity. The measurements are generally made on slurry suspensions at 71% solids. A Brookfield four speed viscometer is used to measure low shear viscosity. A Hercules viscometer is used to measure high shear viscosity. Standard procedures and conditions are rigorously followed in making these tests.

Screen Residue

Material held on a 325-mesh screen is regarded (TAPPI 681-cm-83) as grit or residue. A 100-g sample is thoroughly mixed and dispersed with a dispersing chemical. After carefully determining the weight and percent solids of the dispersed slip, the amount of residue retained on the screens can be accurately weighed and percentages calculated.

Filler for Paint and Plastics

In addition to several of the properties required for use in paper, the kaolin used in paint and plastics must meet certain electrical resistivity specifications. The resistivity test is designed to give an indication of the amount of residual salts that might be contained in the clay. The higher resistivity values reflect a low soluble salt content. A certain weight of kaolin is added to a specified weight of distilled water. The suspension is boiled for 5 min and reweighed. The loss of distilled water from boiling is replaced, and the conductivity is determined using an industrial conductivity bridge.

For certain paint applications, the Hegman fineness-of-grind test specification is important. This test is intended to measure the degree of dispersion (Anon., 1970) and consists of a fineness gauge and a scraper. The gage is calibrated from 0 to 8, and the fineness is reported by numbers. The higher the number, the finer the particles and the better the dispersion. Some paint customers specify

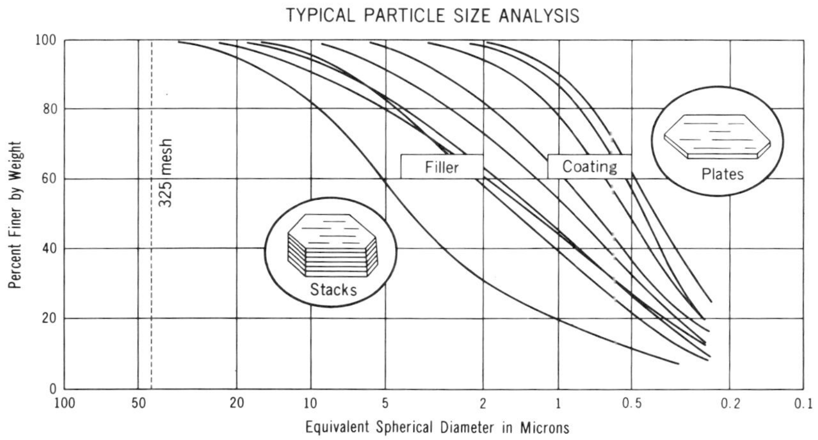

Fig. 18. Particle-size distribution of some typical kaolin coating and filler grades.

that a pigment or extender must be finer than a particular Hegman number.

Ceramics

The most important tests for kaolin sold for ceramic products are modulus of rupture, casting rate, pyrometric cone equivalent (PCE), fired color, and shrinkage. The measurement of strength of clay bodies is a basic requirement by the ceramic industry. The modulus of rupture (MOR) is the fracture strength of a material under a bending load. The MOR measurement is made on a long bar of either rectangular or circular cross section, supported near its ends; a load is applied to the central part of the supported span. ASTM has standard methods of test for MOR of ceramics.

Casting rate is important, and fine-grained bodies cast more slowly than coarse ones. Viscosity of a slip must be carefully controlled, for if it is too viscous the slip will not properly fill the mold or drain cleanly. Therefore, viscosity is measured and controlled on shipments of kaolin used in the casting process.

For certain applications, particularly for refractory purposes, the pyrometric cone measures the combined effects of temperature and time (Anon., 1952, Klinefelter and Hamlin, 1957, Norton, 1968). The cones consist of a series of standardized unfired ceramic compositions molded into the shape of triangular pyramids. The sample of kaolin is molded into the standard cone shape and is heated along with several standard cones so that its end point can be determined in terms of an equivalent cone number (Table 7). ASTM test method C24–56 gives the procedure of making this PCE test.

The fired color of kaolin is also important because some burn white and some burn buff or slightly pink. This is a visual test, and the test specimen, fired to a predetermined temperature, is compared with a standard.

Shrinkage is an important property. Ceramic articles undergo shrinkage at several different points in the manufacturing sequence. During drying, the article will shrink in varying amounts depending upon its composition and the amount of water present. The article shrinks further during firing. Therefore, it is important to know both the drying shrinkage and the firing shrinkage. Linear shrinkage and volume shrinkage can both be measured, although linear shrinkage is more commonly reported (Jones and Bernard, 1972).

Table 7. End Points of Small Orton Pyrometric Cones

Cone No.	End Point °C	Cone No.	End Point °C
07	1 008	15	1 430
06	1 023	16	1 491
05	1 062	17	1 512
04	1 098	18	1 522
03	1 131	19	1 541
02	1 148	20	1 564
01	1 178	23	1 605
1	1 179	26	1 621
2	1 179	27	1 640
3	1 196	28	1 646
4	1 209	29	1 659
5	1 221	30	1 665
6	1 255	31	1 683
7	1 264	31½	1 699
8	1 300	32	1 717
9	1 317	32½	1 724
10	1 330	33	1 743
11	1 336	34	1 763
12	1 355	35	1 785
13	1 349	36	1 804
14	1 398		

Table 8. Kaolin Quality Factors

• Brightness (TAPPI, ISO, etc.)	• Color (X,Y,Z; L, a, b)
• Reduction leach brightness	• Oxidation brightness
• Magnet brightness	• Flotation brightness
• Brookfield viscosity	• Hercules viscosity
• Crystal particle size	• Crystal aspect ratio
• Delamination response	• Calcination response
• Dispersant demand	• Fired brightness
• Cation exchange capacity	• Einlehner abrasivity
• Pyrometric cone temperature	• Ink receptivity
• Sheet opacity, brightness	• Water release rate
• Differential thermal analysis	• Water retention rate
• Fe, Ti, K, Ca, Mg, Na content	• Oil absorption
• Bulk density, packing	• Filterability
• Bacteria and fungus content	• Resistivity

Miscellaneous Uses and Products

Because of the many different uses of kaolin, the industry is plagued by many different purchase specifications and product requirements. Virtually none of these, except those outlined in foregoing discussions, have been standardized, and purchase specifications are frequently changed. However, the grades of kaolin now marketed by the major producers, those grades described in sales and advertising brochures, can, with or without blending, fulfill a wide variety of applications.

PRODUCTS AND GRADES

Mineral professionals new to the kaolin industry are frequently surprised by the variety of complex products and specifications required by manufacturers who purchase refined kaolin from producers. A large kaolin producer may mine as many as fifteen different types of crude clay, and produce up to 100 distinct products selling for a very wide range of prices. Table 8 shows some of the testing that may be needed to meet customer specifications for the different refined kaolin products. A few of the most common kaolin product types are listed in Table 9 with some ranges of specified quality factors.

KAOLIN TESTING

The number and type of laboratories large kaolin producers find necessary is often remarkable to newcomers to the industry. It is important to understand the need for each type of testing, and why the work often cannot be done in a single central laboratory.

Table 9. Paper Coating Kaolin Specifications

Product type	GE brightness	Non-stokes particle size %—2 μ	Brookfield Viscosity @ 10 RPM, #3 disc
No. 1 coating			
standard	87–88	90 to 94%	500 cpe. @ 70% sol.
premium	89–91	90 to 94%	500 cpe. @ 70% sol.
No. 2 coating			
standard	86–87	80 to 84%	400 cpe. @ 70% sol.
premium	88–90	80 to 84%	400 cpe. @ 70% sol.
High glossing			
standard	86–88	95 to 98%	700 cpe. @ 70% sol.
premium	89–90	98 to 100%	700 cpe. @ 70% sol.
Delaminated			
standard	87–89	78 to 82%*	300 cpe. @ 67% sol.
premium	89–90	78 to 82%*	300 cpe. @ 67% sol.
Calcined**			
hi opaque	80–85	78 to 80%	500 cpe. @ 50% sol.
standard	90–92	84 to 86%	500 cpe. @ 50% sol.
premium	92–94	90 to 94%	500 cpe. @ 50% sol.

*Delaminated particle size inexact due to aspect ratio.
**Also sold for paper filling, paint, and plastic pigment.

Paper Filling Kaolin Specifications (may also use coating grades):

Waterwash filler			
standard	82–84	60 to 70%	400 cpe. @ 70% sol.
premium	84–85	60 to 65%	400 cpe. @ 70% sol.
Air float filler			
standard	76–79	50 to 60%	400 cpe. @ 70% sol.
premium	79–83	50 to 60%	400 cpe. @ 70% sol.

Ceramic Kaolin Specifications

	Unfired Brightness	Non-stokes Particle size %—2 μ	Methylene blue C.E.C. or % iron oxide
Hi strength	78–83	55 to 65%	9 to 12 meq.
Sanitaryware	75–80	less than 60%	less than .5%
Fiber glass	none	less than 70%	less than .5%

Note: This is a very general listing, individual kaolin company specifications for particular products may vary somewhat.

Crude-Exploration Laboratory

The Crude-Exploration Laboratory is used to evaluate exploration core drilling samples. Thus it must be equipped to work with very small samples which are costly to obtain, and difficult to replace if consumed. Also, this lab must operate as a miniature version of the production plant, actually degritting, fractionating, beneficiating, filtering, and drying the crude clay to make small samples of the real refined product. The information reported by the crude lab is used for:

1. Proving of assured kaolin reserves, in order to provide confidence that sufficient clay exists to justify the cost of production plant construction or expansion.
2. Selection of the kaolin deposits which must be opened each year to meet the quantity and quality required to meet the sales forecast.
3. Mine planning, in order to assure that kaolin is exposed ready for use when and where it is required.
4. Justification for the planned purchase of needed plant beneficiation equipment, to be certain that it is available to beneficiate the exact type of crude clay which is obtainable from the mines.

Quality-Control Laboratory

The Quality-Control Laboratory is used to test each batch of product as it is produced by the production plant. It has large size samples available of the actual product for testing and, thus, does not need to go through the process of refining the clay. However, the quality control lab is generally always rushed to complete analyses in time for production to be shipped. The Quality-Control Laboratory serves the following functions:

1. To monitor the hour-to-hour quality of the products produced to assure that they are neither unacceptably below or above customers' specifications. If quality is below contract specifics then the customer will be lost; if consistently above specifications then the customer will require future shipments to meet unrealistically high standards.
2. To retain accurate samples from all shipments in order to provide objective resolution of customer quality problems.
3. To maintain a system of statistical reporting and control of all quality factors, in order to assure customers of long term consistency of all product specifications.
4. To serve as authority for plant operation decisions affecting product quality.

Technical Service Laboratory

When customers have a problem applying a kaolin product in a manufacturing process, the Technical Service Laboratory works to resolve the difficulty. Its function is to:

1. Process retained samples to identify the physical or chemical cause of the problem and communicate possible remedies to the customer.
2. Often operate as an open lab, where technicians from producer and customer work together on equipment which duplicates the buyer's manufacturing process.
3. Provide periodic lab instrument calibration throughout the company, originate interlab blind testing samples, develop and

authorize testing procedures, and do limited training of company lab technicians.

Research and New Product Development Laboratory

This laboratory assures that a kaolin production company operates for the future, instead of repeating the past. It is responsible for:

1. Using advanced instruments to determine the physical and chemical characteristics of the available variety of kaolin crudes and finished products.
2. Understanding and explaining the basic principles behind present kaolin processing equipment, and customers' manufacturing equipment which applies kaolin to make their finished product.
3. Keeping in contact with advances in technology that could prove useful in processing kaolin to meet customers' present or future needs.
4. Applying the above knowledge and experience, in close communication with the customer, to develop innovative new products to serve existing and developing markets.

Of necessity, smaller kaolin producing companies must combine the functions, instrumentation, and technicians of these laboratories into a few or even a single facility. However, regardless of the size of the kaolin producer, it is important to remember the differing necessary functions of each type of lab to be certain that each can and will be performed.

Key Laboratory Equipment

The scale of equipment should be sized to fit the size of sample available for testing. Basic models of many instruments will be satisfactory for the lab where rapidity of testing is not required; however, the higher cost of more automated units will be quickly repaid when technicians are overloaded with work which must be completed on a fixed schedule.

Essential instruments for a paper coating and filling kaolin laboratory include: brightness reflectometer suitable for measuring compressed pigment powder, Brookfield viscometer, Hercules viscometer, a system of particle size measurement, a variety of graduated standard mesh size screens, vacuum filters, large 100°C drying ovens, accurate weighing scales, graduated cylinders, and a range of specific gravity/liquid density tubes, and perhaps a 1 500°C electric furnace with accurate controls for temperature and rate of temperature change for calcination. Where appropriate for the production processes used, lab-size magnetic separator, ozone generator, conditioning and/or flotation cells may be needed. The paper-coating kaolin Technical Service Laboratory will need blade coating, drying, calendaring, and printing apparatus. A kaolin research lab generally finds scanning and transmission electron microscopes, microprobe, X-ray diffractometer, and advanced chemical analysis systems necessary.

Information from a paper-grade kaolin laboratory will have considerable value regarding paint, plastic, rubber, and ceramic applications. However, each use is specialized, and additional apparatus and expertise is needed for each.

ECONOMIC FACTORS OF THE KAOLIN INDUSTRY

Kaolin Markets

Kaolin Production Costs

Production of refined kaolin to meet the complex specifications of the modern marketplace, especially by waterwash technology, is a very capital-intensive business. A new wet-processing kaolin plant was recently estimated (Sukel, 1987) to require an initial grass roots investment of approximately 250 to $300 per annual finished ton. With consideration of the rather flat price increases for kaolin products in recent years, some overcapacity in the world kaolin industry, and historical increases in power, fuel, chemicals, labor, and capital investments, Sukel concluded that new hydrous kaolin pigment plants are very difficult to justify. A general look at some of the more readily available costs to produce kaolin in the Georgia-South Carolina District may be useful in updating Sukel's conclusions.

Exploration and Mine Development Costs

It is usually necessary to drill many hundreds of core holes, and to test thousands of drill samples, in order to discover and develop enough crude kaolin ore to sustain a substantial-size plant operation for one year. One core drill hole in the Georgia-South Carolina District probably costs between 700 to $US1000 to complete, plug, test, record, store data, and interpret. A complete permit from the State of Georgia Environmental Protection Division to open a single new mine (if no wetland is involved) may take three to six months to prepare and obtain, and it may involve several thousand dollars in company field and office work. Some large kaolin companies have 10 to 25 mines in one stage of completion or another in any given year.

Stripping, Mining, and Crude Clay Transport

Overburden removal may be arranged with any of several experienced contractors for a price of 0.50 to $0.90 per m^3 depending on length of haul to the overburden dump site, type and ratio of overburden, and extent of artesian and perched mine water.

Kaolin royalty payments on existing leases probably range from about 0.50 to $4.00 per cubic yard (0.46 to $2.47 per inground wet short ton) of consumed clay, depending on clay quality, minimum annual lease payment, and required distance to transport the clay. Crude kaolin transported by truck may cost 2.50 to $5.00 per wet ton to transport 16 to 40 km on reasonably good roads. Pipeline transport from an in mine blunger location probably averages one-tenth the cost of truck haul once the cost of pipe, land right-of-way, and construction are amortized. Kaolin mining, blunger loading and operation, primary dispersant chemical, and mine-area degritting may cost 7 to $16, depending on productivity of the blunger, particle size and hardness of the clay, local power costs, etc. Thus, it can cost anywhere from 15 to $30 per ton to deliver crude kaolin to the processing plant.

Kaolin Processing and Drying

A typical waterwash kaolin plant varies widely in operating cost, depending on special beneficiation required, labor rates and productivity, range and types of products made and economy of scale, whether the product is shipped as slurry, dry bulk, or dry bagged, cost of power and fuel, age and condition of the plant, and percentage of capacity utilized. No realistic attempt can be made to generalize plant operation cost per ton of product, cost of sales, etc.

Kaolin Product Transport

Unfortunately, the Georgia-South Carolina kaolin district is located almost across the continent from the center of the fine coated-paper manufacturing industry (the western Great Lakes area), a railroad distance of approximately 2 400 km. In 1989 the cost of shipping kaolin from Georgia to Grand Rapids, MN, was about $53.50 a net ton in bulk, and about $65.30 per dry ton in slurry (Anon., 1989). Bulk kaolin is shipped as liquid slurry at 70% solids, in lined tank cars, and as dry powder in covered hopper cars. The demand for efficient, rapid transport of kaolin to the northern midwest was so great that in 1988 one of the major rail lines began a six-day-per-week unit train service, The Kaolin Express, which has since been increased to two trains per day (Ampian, 1988). This has had the effect of considerably improving shipments and reducing rail car turn-around time.

Overseas export cost includes rail transport to the ports of Savannah or Charleston, SC, and shipping to the destination seaport, plus tariffs and fees. Overseas transport, when compared with slurry cost, must also include the cost of slurry make-down and redispersal at the receiving port.

Ball Clay

History and Uses: Probably the first deposits of the material now known as ball clay mined in the United States were those worked near Paris, TN, in 1860 for use in a local pottery (Hosterman, 1973). The first clay shipped to out of state consumers was that produced in west Tennessee in 1894 by I. Mandle. Later the industry expanded into Kentucky, and the ball clay districts in the western parts of these two states competed with districts in the United Kingdom (Anon., 1969) for the world leadership in production. In 1990, more than 788 kt of ball clay were produced in the United States (Ampian, 1991). More than 80% of this tonnage was produced by companies active in Tennessee and Kentucky. The other states producing ball clay are Mississippi, Texas, and California. Ball clay is also produced intermittently in Maryland.

Ball clay is used principally in the manufacture of vitreous china, sanitary ware, electrical porcelain, floor and wall tile, dinnerware, and artware (Phelps, 1972). It is also used in refractory products, ceramic glazes, and porcelain enamel slips.

Geology: *Mineralogy.* Kaolinite is the principal mineral in the ball clay deposits in Tennessee and Kentucky, but other clay minerals are present and nonclay minerals vary appreciably in abundance. The kaolinite in these deposits is fine-grained, and much of it is poorly ordered. The main clay minerals other than kaolin are illite, smectite, chlorite, and mixed-layer clay. Quartz is by far the most abundant nonclay mineral; in some deposits it makes up as little as 5% of the clay, but in others, as much as 30% (Olive and Finch, 1969). The other nonclay minerals ordinarily present occur in very minor or trace amounts and include plagioclase, potassium feldspar, and calcite. Organic matter in the form of black leaf imprints and disseminated lignitic and peaty materials are common in most ball clays.

Occurrence and Origin. Ball clay is extremely plastic. Deposits range from very light gray to nearly black. Most deposits are lenticular bodies varying considerably in size and shape. Some are more than 9 m thick and extend over areas 30 m wide and 760 m long, but most deposits are smaller. The physical properties of a typical deposit vary considerably, and layers and pockets mined from a given deposit are commonly separated in different stockpiles suitable for a particular use.

Most deposits in Tennessee and Kentucky are in beds of Claiborne age (middle Eocene), but a few, including one mined, are in Wilcox beds (lower Eocene) (Olive and Finch, 1969). The deposits mined in Texas near Henry's Chapel, northeastern Cherokee County (Fisher et al., 1965) are also in Wilcox beds. The ball clay mined in Panola County, MS (Bicker, Jr., 1970), is in a formation of Claiborne age, and the deposits are approximately the same age as younger ones in Tennessee and Kentucky. The deposits mined in Stanislaus County, California, are in beds of Paleocene or Eocene age (Cleveland, 1957). Those mined on a small scale in the Hart District in eastern San Bernardino County, California, are in hydrothermally altered Tertiary rhyolites (Kelley, 1966).

Most ball clay occurs in transported sedimentary deposits, but one minor producing district in California contains hydrothermally altered clay. The deposits in the Tennessee, Kentucky, and Mississippi districts are in sedimentary formations cropping out along the eastern edge of the Mississippi embayment. According to one theory of origin (Olive and Finch, 1969), they accumulated in an elongate broad valley. The abundance of plant remains and organic matter associated with this clay indicates that the depositional basins probably were swampy.

Distribution of Deposits: *United States.* Any attempt to estimate the resources of ball clay is based on piecemeal information. The reserves, primarily in Tennessee and Kentucky, total probably as much as 100 Mt, and are owned by several different companies; they are thought adequate to sustain the industry for 100 years (Phelps, 1972). Additional reserves are in producing districts in Mississippi, Texas, and California. Potential resources of low grade ball clay, which probably would require beneficiation for market, occur in the vicinities of all of these districts. The prospects are also favorable for the discovery of more deposits of kaolinitic clay having the physical properties of ball clay in several southern states. Furthermore, if the term ball clay is applied to such deposits as the hydrothermally formed clay at Hart, CA, many areas in the western states are likely to contain large deposits.

Other Countries. The principal ball clay deposits in the United Kingdom occur in the Bovey Basin southeast of Dartmoor in Devonshire, the vicinity of Petrockstow in north Devonshire, and in the Wareham and Poole regions of Dorsetshire. These ball clays occur in sedimentary rocks of Tertiary age, consisting mainly of lenticular units of sand, clay, and lignite of probable lacustrine origin (Scott, 1929). The clay is very fine-grained plastic kaolin ranging in color from off-white to dark brown and gray. The darker colors, as in the ball clay in Kentucky and Tennessee, apparently are caused by carbonaceous impurities, and the clay burns white. The deposits are believed to have been transported from altered granite and other igneous rocks.

In Germany, the largest ball clay producing district is the Westerwald, located north and east of Roblenz on the Rhine River. The fine-grained, plastic clays are Tertiary in age and were deposited in subsidence depressions in Devonian age rocks. Most of the clay mined is supplied to the ceramic industry and includes clay exported to Italy for use in ceramic tile. Approximately 1 Mt of plastic clays are produced annually from the Westerwald area of which a large portion is for the manufacture of heavy clay products, bricks, and pipes.

Ball clay occurs at several localities in India (Arogyaswamy, 1968) but production is limited, and in recent years India has been an importer of ball clay. The best quality ball clay produced is in the Than District of Gujarat. Other deposits are mined in two districts in Kerala.

Evaluation of Deposits: The prospecting techniques for ball clay are very similar to those outlined for kaolin. The most commonly used tests are very similar to those used for evaluating the ceramic properties of kaolin and heat resistant properties of refractory clays. Chemical analysis for soluble sulfate in ball clays is commonly required because this material can be tolerated in only very minor amounts in most ceramic products.

Preparation for Markets: Ball clay is mined by stripping methods and trucked to plants. Most plant processing involves primarily drying and grinding to a suitable size. A few plants air-float part of their output. Most ball clay is packaged in paper bags, but some is shipped in bulk carload lots in both crude and processed form. In recent years, considerable ball clay is shipped in tank cars. This clay is prepared for shipment by deflocculation in water, and it is shipped as 60 to 62% solids (Phelps, 1972). Such slurries are used by plants manufacturing ceramics by slip-casting methods.

Future Considerations and Trends: The production of ball clay in the United States has grown rather consistently for the last several decades, and continuation of this trend is forecast. The major growth is predicted to be in exports to Central and South America. In general, the ball clay industry faces little competition from foreign sources and exports are appreciably greater than imports. However, the major districts in Tennessee and Kentucky are in the central part of the country and as rail transportation costs increase, a greater proportion of the demands in the populous coastal region could be satisfied by imports if new foreign sources are discovered.

Halloysite

History and Uses: Apparently the first halloysite used in the United States was that mined on a small scale near Rising Fawn, GA, in the late 1800s. This clay is reported to have been sold as a filler or extender in food (Veatch, 1909). Deposits near Gore, GA, were mined and used for making alum about 1912 (Broadhurst and Teague, 1954). A few hundred tons of halloysite was mined from deposits in Indiana and used for pottery and alum during and after World War I (Callaghan, 1948). The major source of halloysite in this country has been the approximately 1 Mt mined from the Dragon mine, Tintic District, Utah, by the Anaconda Co., but the source is now depleted and the mine is no longer operating. It was processed into petroleum-cracking catalyst at the Filtrol Corp. plant in Salt Lake City, UT. Mining to supply this plant began in 1949. Minor tonnages of halloysite were produced from this same mine before 1949 for other uses. Halloysite occurring at several other places in Utah has also been used in making light-colored brick, firebrick, tile, and as a paper filler (Van Sant, 1964). Halloysite in Nevada has been used in portland cement in recent years (Papke, 1971). Deposits in Idaho consisting of mixed kaolinite and halloysite are used in refractories (Hosterman et al., 1960), and deposits of similar clay-mineral composition in North Carolina (Parker, 1946) are mined for use in ceramics and other products.

Geology: *Mineralogy.* Halloysite is so similar to kaolinite that the two cannot be distinguished without careful mineralogical work (Brindley et al., 1963). Very fine tubular structures commonly observable in electron micrographs of dehydrated halloysite are ordinarily diagnostic of this mineral. The shortcoming of this criterion is that not all dehydrated halloysite occurs in the tubular form. A reliable criterion is that the basal (001) spacing of halloysite is at approximately 10A; as halloysite converts irreversibly to dehydrated halloysite, the spacing collapses to about 7.2A, which is essentially the same spacing as for kaolinite. Halloysite, therefore, can be identified by the greater spacing in X-ray diffractograms of samples containing natural moisture.

Nonclay minerals and other impurities in halloysite vary considerably in different deposits. Some deposits, such as those in the Dragon mine, Utah, are exceptionally pure and contain only minor quantities of contaminants, consisting chiefly of pyrite, silica, dolomite, iron-bearing minerals, noncrystalline materials, and manganese oxides (Morris, 1968). In most deposits, silica is the principal impurity; it occurs in the form of quartz, cristobalite, chert, and poorly crystalline forms. Many halloysite deposits are intermixed with kaolinite, and several in western United States are closely associated with alunite. Gibbsite and manganese minerals also occur in some halloysite deposits.

Occurrence and Origin. Small deposits of halloysite occur at many places in the United States, but known deposits large enough to mine profitably are rare. The deposits in the Dragon mine, Utah, were two large pipelike bodies replacing lower Paleozoic limestone near the contact with monzonite porphyry (Kildale and Thomas, 1957). The two bodies are separated by a zone of iron-oxide deposits extending along the Dragon fissure (Morris, 1968). The occurrence of these deposits and the minerals associated with them clearly indicate that they formed by hydrothermal activity. Other deposits thought by Papke (1971) to have formed hydrothermally occur in the Terraced Hills, Washoe County, Nevada. These deposits are bedded and apparently altered mainly from tuff. The halloysite in the Gore District, Georgia, is bedded and occurs in the Armuchee Chert of Devonian age (Broadhurst and Teague, 1954). An adequate explanation of these deposits has never been advanced, but the extent of the beds suggests that they are of sedimentary origin. Halloysite mixed with kaolinite in the Spruce Pine District, North Carolina (Hunter and Hash, 1953, Parker, 1946), and the Bovill District, Idaho (Hosterman et al., 1960), are mainly residual and formed by the weathering of igneous rocks. Several deposits in foreign countries are thought to have formed hydrothermally, and some are the result of surficial weathering.

Distribution of Deposits: *United States.* The major productive halloysite mine in the United States has now been shut down as reported in the foregoing discussion. Most of the many other known scattered deposits are either too small or too impure to be considered as significant sources of halloysite. Possible exceptions include the following: 1) the halloysite mixed with kaolin in weathered nepheline syenite masses in the Arkansas bauxite region (Gordon et al., 1958), 2) the body of halloysite reported to have been exposed for 107 m in the adit of a mine in the Bullion (Railroad) District, Nevada (Olson, 1964), and 3) rather extensive deposits reported near Bartow, GA (Anon., 1959). Possibilities for the discovery of new large high grade deposits are not easy to evaluate. The likelihood of major discoveries does not seem promising, but rocks in many large areas in western United States have been altered hydrothermally, and several of these could contain large deposits of halloysite. Also, sources of halloysite in the old Gore mining district in Georgia are estimated at more than a million tons (Broadhurst and Teague, 1954), and some use may again be found for these deposits.

Other Countries. Halloysite occurs in many countries, but, as in the United States, most deposits are small; the authors are aware of small scale mining only in Morocco, Japan, Korea, Czechoslovakia, and New Zealand. The deposits mined in Morocco are in the Maaza District about 6 km west of Melilla (Hilali et al., 1969, Martin-Vivaldi and Girela-Vilchez, 1959). These deposits occur above travertine and are overlain by pyroclastic materials. They are in a bed of Pliocene age, ranging in thickness from 1.5 to 2.5 m. Halloysite mined in Japan includes deposits that formed by hydrothermal alteration of granite and quartz porphyry and others that were transported and deposited in sedimentary basins (Takeshi and Kato, 1969). Hydrothermal deposits occur at the Taishu mine, and white sedimentary halloysite is in the Tajimi and Arikabe deposits. The halloysite mined in Korea is presumably from the San Chong and Tan Song deposits described by Kim and Kim (1964). The halloysite deposits mined in Czechoslovakia are in the Michalovce district in the eastern part of the country (Kuzvart, 1969). These deposits are thought to have formed by the hydrothermal alteration of dacite. They extend over an area 500 m^2 and are as much as 36 m thick. They occur below overburden 30 to 40 m thick. The halloysite mined is reported to be suitable for use in ceramics, refractories, paper and rubber fillers, and in refining sugar. The deposits in New Zealand occur in the Maungaerua District on North Island. This clay is mined by New Zealand China Clays, Ltd. and sold for use in paper filler, ceramics, paint extenders, and filler for adhesives and cosmetics. The halloysite is in massive deposits formed by hydrothermal alteration of andesite of middle Cenozoic age (Bowen, 1969, Murray et al., 1977).

Several of the kaolin deposits in other countries consist of mixtures of halloysite and kaolinite. One such mixed deposit occurs in altered felsitic volcanic rocks in the Burela District, Spain (Anon., 1972), the principal kaolin-producing center in that country. The kaolin produced contains appreciably more halloysite than kaolinite, but the clay must be considered to be a mixed type. Another mixed halloysite-kaolinite deposit is in the Bukidon District in the Philippines (Comsti, 1969).

Future Considerations and Trends: Halloysite will probably be in demand in the United States for use in petroleum catalysts and other products as long as suitable high quality deposits can be mined at a reasonable cost. As mentioned previously, known high grade deposits are limited, and though additional deposits may be found, the point may be reached when it will be cheaper to use substitute materials than to prospect for new deposits. Catalysts made from halloysite are in competition with those manufactured from kaolin and synthetic materials containing alumina and silica. The products other than catalysts made from halloysite in recent years are all low value or high weight items lacking the broad market base of catalysts. All these other products can be made from other clays or other materials, but halloysite will continue to be used for them until reserves are exhausted, cheaper substitutes found, or local markets disappear.

Refractory Clays

History and Uses: The first plant in the United States for making refractory products from clays began operation at Salamander, NJ, in 1825. The refractory industry grew rapidly and expanded from New Jersey to centers in Pennsylvania, the Ohio River valley, Missouri, and to western states.

Refractory clays are used mainly in making firebricks and blocks of many shapes, insulating bricks, saggers, refractory mortars and mixes, monolithic and castable materials, ramming and air gun mixes, and other products. A product called *chamotte* in Europe (Anon., 1972a, 1972b) and elsewhere is made by calcining high grade fire clays and other kaolinitic clays. A similar material produced in the United States is called merely calcined clay or calcined kaolin. A related material called mullite refractory is made by calcining bauxitic clay or clayey bauxite. Further fabrication into finished refractory products is required for chamotte and calcined clay. Considerable tonnages of fire clay also have been used in the past in the United States in the manufacture of light-colored face brick, tile, stoneware, and other products.

Product Specifications: The specifications for refractory clays are as many as the different uses. As resistance to heat is the most essential property, many specifications include reference to pyrometric cones (Table 7), indicating the heat duty required. Resistance to shrinkage, warping, cracking, and abrasion is also very important in many products, and expressions of the requirements for these properties appear in some specifications. Many of the end products must fulfill rigid specifications on dimensions. Restrictions on raw clay commonly include cutoffs on plasticity, impurities, and content of alkalies, alkaline earths, and other constituents that reduce fusion points.

Geology: The mineralogy, occurrence, and origin of refractory kaolin and refractory ball clay have been discussed in the foregoing sections. Therefore, the following discussions will be concerned with these subjects only as they apply to fire clay.

Mineralogy. The degree of ordering of the kaolinite, the principal mineral in fire clay, varies considerably. Most of the kaolinite in the harder and purer varieties of fire clay (flint clay) is very well ordered, and it is even more perfectly crystalline than the kaolinite in many kaolins. The kaolinite in the hard fire clay also tends to occur in coarser particles, and in the harder varieties the grains are interlocking. The kaolinite in the plastic varieties of fire clay, however, is ordinarily characterized by imperfectly crystalline structures. The more disordered form of this kaolinite has been widely referred to as fire clay mineral in an effort to distinguish it from the more perfectly crystallized forms.

Some of the better grades of refractory clays contain minerals that are richer in aluminum than kaolinite, and, therefore, higher alumina products can be made from them. Some refractory kaolin is a mixture of kaolinite and gibbsite, $Al_2O_3{\cdot}3H_2O$. The better grades of fire clay, that are exceptionally rich in aluminum, consists of kaolinite and diaspore, $Al_2O_3{\cdot}H_2O$, and minor quantities of boehmite, $Al_2O_3{\cdot}H_2O$.

Occurrence and Origin. The physical characteristics of fire clay vary considerably; the clays range from soft and plastic to flintlike. Fire clay, therefore, is divisible into plastic, semiplastic, semiflint, and flint types. Flint clay has unique properties for a clay; it lacks plasticity, breaks with a conchoidal parting into shardlike particles, and most of it is as brittle as limestone. Typical diaspore-bearing flint clay is brittle and has an oolitic, pisolitic, or nodular texture. Such clay is referred to as burley, birdseye, or nodular clay.

Most fire clay occurs in sedimentary rocks, and deposits range in age from Pennsylvanian to Tertiary. Fire clay is particularly common in rocks of Pennsylvanian age, and deposits of this age occur as underclays (seat earths) immediately below or closely associated with coalbeds. Fire clay of Cretaceous and Tertiary age occurs mainly in lenticular bodies. Some deposits of these ages are associated with lignite and are probably underclays, and others have apparently been transported and deposited in local basins in a nearshore, swamp, or flood plain environment.

Opinions differ considerably on the origin of the extensive fire clays occurring below coalbeds. Some geologists (Keller, 1970, Patterson and Hosterman, 1963), including the authors, believe that most underclay deposits formed mainly by the alteration of aluminous sediments in a swampy environment. Some mineralogists, however, believe that these clays were transported and some sort of a sedimentary winnowing process caused the rather pure accumulations of kaolinite in fire clay. Some clays of Cretaceous and Tertiary age have been transported and formed by a process similar to those thought to have formed the transported kaolin deposits.

Distribution of Deposits: *United States.* Fire clay of Pennsylvanian age is widely scattered throughout the Appalachian region and parts of the Mississippi Valley (Anon., 1967). The very high quality diaspore-bearing clay, as well as other grades of fire clay, occur in the Clearfield District, Pennsylvania (Anon., 1964, Bolger and Weitz, 1952), and the Ozark region in Missouri (Keller, 1952, Keller et al., 1954). High grade kaolinitic flint clay and semiflint clays are mined in the Olive Hill District, Kentucky (Patterson and Hosterman, 1963); Oak Hill District, Ohio (Stout et al., 1923); and Somerset District, Pennsylvania (Hosterman et al., 1968). Major districts producing fire clay suitable for low and moderate-heat-duty refractory products include the Allegheny Valley and Beaver Valley Districts, Pennsylvania, Cordova District, Alabama, East Liverpool District, Ohio and West Virginia, and the Tuscarawas Valley and Hocking Valley Districts, Ohio (Hosterman et al., 1968), and a region including parts of Monroe, Audrain, Callaway, and Montgomery Counties, Missouri (McQueen, 1943).

The largest and best quality fire clay deposits in the Rocky Mountain region are in Fremont, Pueblo, Custer, Huerfano, Jefferson, and Las Animas Counties, Colorado. These deposits are in the Purgatoire Formation and Dakota Sandstone of Cretaceous age (Waagé, 1953). Fire clay in these deposits occurs as isolated tabular lenses ranging in thickness from 1 to 6 m.

In addition to the fire clay mentioned in the foregoing discussions, other deposits are scattered throughout the western states (Anon., 1967, Mark, 1963, Van Sant, 1959, 1964). These deposits are of several types, and they include transported, residual, and hydrothermally formed clay. Districts in which these scattered deposits occur include several areas in King County and the Castle

Rock area in Cowlitz County, Washington, the Molalla and Hobart Butte areas, Oregon, and the Alberhill area, California. Most of these deposits are of Cenozoic age.

Refractory kaolin is produced primarily in the following districts: 1) Georgia-South Carolina kaolin belt, 2) Andersonville, Georgia (Anon., 1972e), 3) Eufaula, Alabama, 4) Arkansas bauxite region, 5) scattered districts in Texas, 6) Latah County, Idaho, and 7) Ione, California (Anon., 1972a). Refractory ball clay is produced primarily in the western parts of Kentucky and Tennessee, but some of the ball clay produced in Texas may be used in refractory products.

The total refractory clay resources in the United States suitable for low-heat-duty refractory products may be as much as 7 Gt (Hosterman, 1973). However, the reserves—those clays that can be mined and processed now at competitive costs and are suitable for low-heat-duty products or better—are probably no more than 1 Gt. The reserves are primarily in the fire clay and refractory kaolin and ball clay districts outlined previously. The remaining resources of high grade diaspore clay are mainly under considerable overburden in Pennsylvania, and only small resources are present in Missouri. The bauxitic kaolin resources are primarily in the Arkansas bauxite region, and the Eufaula, AL, and Andersonville, GA, districts. Some deposits also occur in other parts of Alabama and Georgia.

Other Countries. Refractory clays are produced in many countries; however, the information available to the authors on worldwide production is sketchy and incomplete. This is partly because the distinction between fire clay and miscellaneous clay is not made in some countries, and clay used for refractory products is lumped with kaolin in others.

Those countries that produce more than 1 Mt of refractory (fire) clay annually include the United Kingdom, the former West Germany, and Japan. The production in France exceeded 1 Mt a few years ago, but has apparently dropped below this figure in recent years. Other countries producing 100 kt to 1 ktpy include Argentina, Australia, India, Italy, Mexico, New Zealand, Sweden, the United Arab Republic, Uruguay, and Yugoslavia. Refractory clay is exported from the People's Republic of China, which is apparently a major producer. Fire clay is also mined in the former USSR, Czechoslovakia, Hungary, Poland, and other East European countries.

As in the United States, fire clay in other countries is in several types of deposits. Kaolinitic fire clay associated with coalbeds of Carboniferous age is mined in the United Kingdom, France, Germany, and elsewhere. Very high alumina diaspore bearing flint clays are sources of refractory materials in Scotland, South Africa, Israel, and the People's Republic of China. Refractory kaolin is mined in Japan, India, Hungary, Italy, Czechoslovakia, Poland, Sweden, Yugoslavia, Mexico, Argentina, Brazil, Chile, Peru, the Republic of South Africa, the United Arab Republic, Australia, New Zealand, Iran, and other countries; most such deposits are no older than Cretaceous.

Exploration and Evaluation: The methods used in exploring and evaluating fire clay deposits are in general similar to those used for bentonite and kaolin, except that different drilling and testing procedures are required. Drilling fire clay deposits, particularly those of Pennsylvanian age, ordinarily requires diamond bits and core barrels of the type used for minerals in hard rock. This is because the clay itself and the rocks overlying the deposits are hard. The most common test in the evaluation of fire clay is the determination of PCE.

In addition to the PCE requirement, fired test pieces also may be tested for such properties as high-temperature stability, hot strength, porosity, and spalling (Anon., 1972h, Norton, 1968). Chemical analyses are also required for some evaluations, and the contents of alkalies, alkaline earths, iron oxides, and a few less common elements are critical, because they act as fluxes when clay refractories are heated.

Preparation for Markets: Fire clay and other refractory clays are mined by methods similar to those used for other types of clay. However, a much higher percentage of fire clay mining in the past has been underground. The underground mining has been required because much of the higher quality clay occurs at considerable depths, and the overlying rocks are too hard to strip profitably.

Firebrick and related products are prepared for markets by grinding the clay, shaping or forming the firebrick or other products, and firing. Commonly several varieties of fire clay and grog (crushed previously fired clay) are blended to make a mix desired for a certain product, and blends are varied considerably to prepare different grades of firebrick. The harder clays are ordinarily the most refractory, and as they lack plasticity, small proportions of kaolin or plastic fire clay or both are added to aid in forming and bonding the brick until firing. Most firebrick is formed by hydraulic or other mechanical presses, but several types of products are intricately shaped and must be formed by hand.

Most firebrick is fired in tunnel kilns heated by gas or oil. A few plants, particularly those making specially shaped products or having a limited market, use downdraft kilns. Finished products are palletized and shipped by rail or truck. Heavy crating with padded packaging is required for overseas shipments. Calcined clay is ordinarily fired in rotary kilns and is shipped in both paper bags and in bulk carloads.

Future Considerations and Trends: The peak output of refractory (fire) clay in the United States was 11.8 Mt in 1957. Production has decreased, with only minor yearly reversals of the trend, since that date. In 1990, the total refractory clay produced was 625 kt. This trend in decreasing production of fireclay has been due mainly to the following changes: 1) the development of more durable aluminous refractory materials, 2) the shift to the basic oxygen process in steelmaking, which requires basic rather than alumina-type firebrick in furnaces, 3) changes in statistical reporting, whereby fire clay used in light-colored facebrick, stoneware, and ceramic products is now classified as miscellaneous clay rather than fire clay.

The decrease in demands for refractory clays brought about by the factors outlined in the previous paragraph has probably bottomed out, and markets for refractory clay and products made from it can be expected to increase consistently.

The availability of high grade and very high grade refractory clays at competitive costs is one of the long range problems faced by the refractories industry. Deposits of very high grade diaspore clay in Pennsylvania and Missouri that can be mined by stripping methods are nearing exhaustion, and underground mining increases costs considerably. The cost of underground mining is one of the reasons why the production of refractory bauxitic and kaolinitic clay has increased in the Andersonville, GA, and Eufaula, AL, districts in recent years. The trend for the increasing use of these bauxitic clays is expected to continue. Another trend may be an increase in the beneficiation of kaolin both to improve purity and to recover a coproduct, such as glass sand, to make the cost competitive. Kaolin and glass sand are now produced as coproducts from impure deposits in Texas, Idaho, and California.

REFERENCES

Anon., 1952, *Manual of ASTM Standards on Refractory Materials*, Committee C-8, American Society for Testing and Materials, 294 pp.

Anon., 1959, "Halloysite in the Cartersville District," *Georgia Mineral Newsletter*, Georgia Geological Survey, Vol. 12, No. 2, pp. 42–43.

Anon., 1964, "Map of the Mercer Clay and Adjacent Units in Clearfield, Centre, and Clinton Counties, Pennsylvania," 4th Series, No. 12, Pennsylvania Geological Survey.

Anon., 1967, "Potential Sources of Aluminum," Information Circular 833S, US Bureau of Mines, 148 pp.

Anon., 1967a, "Area Reports: International," *Minerals Yearbook, 1965*, Vol. 4, US Bureau of Mines, p. 343.

Anon., 1969, "Buoyant Market for Ball Clays," *Industrial Minerals*, No. 17, Feb., pp. 9–15.

Anon., 1970, "Fineness of Dispersion of Pigment-Vehicle Systems," Annual Book of ASTM Standards, 1970, American Society for Testing and Materials, D1210–64, pp. 210–215.

Anon., 1972, "An Introduction to Refractories," *Industrial Minerals*, No. 58, July, pp. 9–23.

Anon., 1972a, "CE Minerals: Georgia Bauxite and Kaolin Calcined for Refractory Grog," *Industrial Minerals*, No. 56, May, pp. 17–22.

Anon., 1972b, "Field Trip Guides," *Proceedings*, International Clay Conference, Madrid, Assoc. International pour L'Etude des Argiles (AIPEA), pp. 11–1 to 11–10, III-599.

Anon., 1972c, "Ione Calcined Kaolin, Interpace's Refractory Aggregate," *Industrial Minerals*, No. 59, Aug. pp. 37–41.

Anon., 1972d, "Refractory Raw Materials, The Producers Reviewed," *Industrial Minerals*, No. 59, Aug. pp. 9–30.

Anon., 1989, "Industrial Clays," *An Industrial Minerals Special Review*, G. Clarke, ed., Industrial Minerals Division of Metal Bulletin Plc., pp. 8–47.

Ampian, S.G., 1988, "Clays," *Minerals Yearbook*, US Bureau of Mines.

Ampian, S.G., 1991, "Clays," *Mineral Commodity Summaries*, US Bureau of Mines.

Arogyaswamy, R.N.P., 1968, "Clays," *Economic Geology Bulletin 29*, India Geological Survey, Series A, 220 pp.

Austin, R.S., 1973, "Distinction Between Cretaceous and Tertiary Kaolins of the Irwinton District, Georgia," Abstract, Geol. Soc. of America, 22nd Annual Meeting, Southeastern Section, Knoxville, TN, p. 10.

Bates, T.F., 1945, "Origin of the Edwin Clay, Ione, California," *Geological Society of America Bulletin*, Vol. 56, No. 1, pp. 1–38.

Bicker, A.R., Jr., 1970, "Economic Minerals of Mississippi," Bulletin 112, Mississippi Geological, Economic, and Topographical Survey, 80 pp.

Bolger, R.C., and Weitz, J.H., 1952, "Mineralogy and Origin of the Mercer Fire Clay of North-Central Pennsylvania," *Problems of Clay and Laterite Genesis*, AIME, New York, pp. 81–93.

Bowen, F.E., 1969, "Kaolin Deposits of New Zealand," *Proceedings of Symposium I. Kaolin Deposits of the World, B. Overseas Countries*, 23rd International Geological Congress, Prague, 1968, M. Malkovsky and J. Vachtl, eds., Vol. 16, pp. 151–157.

Brindley, G.W., et al., 1963, "Mineralogical Studies of Kaolinite—Halloysite Clays: Part I Identification Problems," *American Mineralogist*, Vol. 48, Nos. 7,8, pp. 897–910.

Bristow, C.M., 1977, "A Review of the Evidence for the Origin of the Kaolin Deposits in S.W. England," *Proceedings*, 8th International Kaolin Symposium, Madrid-Rome, Sept. 7–16, 1977, 19 pp.

Bristow, C.M., 1990, "The Genesis of the China Clays of South-West England—A Multistage Story," Abstract, Clay Minerals Society 27th Annual Meeting, Columbia, MO, p. 27.

Broadhurst, S.D., and Teague, K.H., 1954, "Halloysite in Chattooga County, Georgia," *Georgia Mineral Newsletter*, Vol. 7, No. 2, pp. 56–61.

Buie, B.F., and Fountain, R.C., 1967, "Tertiary and Cretaceous Age of Kaolin Deposits in Georgia and South Carolina," Geological Society of America, Southeastern Section, 1967 Annual Meeting Program, p. 19 (Special Paper No. 115, p. 465).

Bundy, W.M., and Berberich, V.P., 1969, "Kaolin Treatment," US Patent 3477809, Nov. 11.

Callaghan, E., 1948, "Endellite Deposits in Gardner Mine Ridge, Lawrence County, Indiana," Bulletin 1, Indiana Division of Geology, 47 pp.

Calver, J.L., 1949, "Florida Kaolins and Clays," Information Circular 2, Florida Geological Survey, 59 pp.

Cleveland, G.B., 1957, "Clay," *Mineral Commodities in California*, L.A. Wright, ed., California Division of Mines, Bulletin 176, pp. 131–152.

Comsti, F.A., 1969, "Kaolin Deposits of the Philippines," *Proceedings of Symposium I. Kaolin Deposits of the World, B. Overseas Countries*, 23rd International Geological Congress, Prague, 1968, Report, M. Malkovsky and J. Vachtl, eds., Vol. 16, pp. 39–42.

Damiani, L., and Trautman, F., 1969, "Les Depots de Kaolins Francais," *Proceedings of Symposium I. Kaolin Deposits of the World A. Europe*, 23rd International Geological Congress, Prague, 1968, Report, M. Malkovsky and J. Vachtl, eds., Vol. 15, pp. 141–178.

Dana, E.S., 1949, *A Textbook of Mineralogy*, 4th ed., rev. W.E. Ford, John Wiley and Sons, New York.

Dombrowski, T., 1992, "The Use of Trace Elements to Determine Provenance Relations Among Different Types of Georgia Kaolins," PhD dissertation, Indiana University, Bloomington, IN, 132 pp.

Durrance, E.M., et al., 1982, "Hydrothermal Circulation and Post-Magmatic Changes in Granites of Southwest England," *Proceedings*, Ussher Society, Vol. 5, pp. 304–320.

Exley, C.S., 1959, "Magmatic Differentiation and Alteration in the St. Austell Granite," *Quarterly Journal*, Geological Society of London, Vol. 114, pp. 197–230.

Exley, C.S., 1976, "Observations on the Formation of Kaolinite in the St. Austell Granite," *Clay Minerals*, Vol. 11, pp. 51–63.

Fehn, U., 1985, "Post-Magmatic Convection Related to High Heat Production Granites of Southwest England; A Theoretical Study," *High Heat Production (HHP) Granites, Hydrothermal Circulation and Ore Genesis*, Institution of Mining and Metallurgy, London.

Fisher, W.L., et al., 1965, "Rock and Mineral Resources of East Texas," Report of Investigations 54, Texas Bureau of Economic Geology, 439 pp.

Gaskin, A.J., 1969, "Kaolin Deposits of Australia," *Proceedings of Symposium I. Kaolin Deposits of the World, B, Overseas Countries*, 23rd International Geological Congress, Prague, 1968, Report, M. Malkovsky and J. Vachtl, eds., Vol. 16, pp. 115–150.

Gordon, M., Jr., et al., 1958, "Geology of the Arkansas Bauxite Region," Professional Paper 299, US Geological Survey, 268 pp.

Greene, E.W., and Duke, J.B., 1962, "Selective Froth Flotation of Ultrafine Minerals or Slimes," *Trans. SME/AIME*, Vol. 223, pp. 389–395.

Gunn, F.A., and Morris, H.H., 1965, "Delaminated Domestic Sedimentary Clay Products and Method of Preparation Thereof," US Patent 3171718, Mar 2.

Hassanipak, A.A., and Eslinger, E., 1985, "Mineralogy, Crystallinity, O^{18}/O^{16}, and D/H of Georgia Kaolins," *Clays and Clay Minerals*, Vol. 33, No. 2, pp. 99–106.

Hilali, E.A., et al., 1969, "Les Gites de Kaolin du Maroc," *Proceedings of Symposium I. Kaolin Deposits of the World, B. Overseas Countries*, 23rd International Geological Congress, Prague, 1968, Report, M. Malkovsky and J. Vachtl, eds., Vol. 16, pp. 55–59.

Hinckley, D.N., 1963, "Variability in Crystallinity Values Among the Kaolin Deposits of the Coastal Plain of Georgia and South Carolina," *Clays and Clay Minerals*, Vol. 11, pp. 229–235.

Hosterman, J.W., 1973, "Clays," Professional Paper 820, US Geological Survey, pp. 123–131.

Hosterman, J.W., et al., 1960, "Investigations of Some Clay Deposits in Washington and Idaho," Bulletin 1091, US Geological Survey, 147 pp.

Hosterman, J.W., et al., 1968, "Clay," Professional Paper 580, US Geological Survey, pp. 167–188.

Huddlestun, P.F., and Hetrick, J.H., 1991, "The Stratigraphic Framework of the Fort Valley Plateau and the Central Georgia Kaolin District," Georgia Geological Society Guidebooks, Vol. 11, No. 1.

Hunter, C.E., and Hash, L.J., 1953, "Halloysite Deposits of Western North Carolina," Bulletin 58, North Carolina Division Mineral Resources, 33 pp.

Hunter, D.R., and Urie, J.G., 1969, "Kaolin Deposits of Swaziland," *Proceedings of Symposium I. Kaolin Deposits of the World, B. Overseas Countries*, 23rd International Geological Congress, Prague, Report, M. Malkovsky and J. Vachtl, eds., Vol. 16, pp. 67–74.

Hurst, V.J., and Bosio, N.J., 1975, "Rio Capim Kaolin Deposits, Brazil," *Economic Geology*, Vol. 70, pp. 990–992.

Jackson, N.J., et al., 1982, "Hydrothermal Activity in the St. Just Mining District, Cornwall, England," A.M. Evans, ed., *Metallization Associated with Acid Magmatism*, John Wiley and Sons, Chichester, UK, pp. 137–179.

Johnson, S.W., and Blake, J.M., 1867, "On Kaolinite and Pholerite," *American Journal of Science*, 2nd ser., Vol. 43, pp. 531–561.

Jones, J.T., and Bernard, M.F., 1972, *Ceramics: Industrial Processing and Testing*, Iowa State University Press, Ames, IA, 213 pp.

Keller, W.D., 1952, "Observations on the Origin of Missouri High-Alumina Clays," *Problems of Clay and Laterite Genesis*, AIME, New York, pp. 115–134.

Keller, W.D., 1970, "Environmental Aspects of Clay Minerals," *Journal of Sedimentary Petrology*, Vol. 40, pp. 788–813.

Keller, W.D., et al., 1954, "The Origin of Missouri Fire Clays," *Clays and*

Clay Minerals—Proceedings of the Second National Conference on Clay Minerals—1953, A. Swineford, ed., US National Research Council Publication 327, pp. 7–46.

Kelley, F.R., 1966, ''Clay,'' Mineral and Water Resources of California, p. 1, Mineral Resources, US 89th Congress, 2nd Session, Senate Committee on Interior and Insular Affairs, Committee Print, pp. 126–134 (reprinted in Bulletin 191, California Division of Mines and Geology, pp. 126–134).

Kesler, T.L., 1963, ''Environment and Origin of the Cretaceous Kaolin Deposits of Georgia and South Carolina,'' *Georgia Mineral Newsletter*, Vol. 16, Nos. 1–2, pp. 2–11.

Kildale, M.B., and Thomas, R.C., 1957, ''Geology of the Halloysite Deposits at the Dragon Mine,'' Utah Geological Society Guidebook to the Geology of Utah, No. 12, pp. 94–96.

Kim, J.H., and Kim, D.H., 1964, ''Investigation Report of the San Chong and Tan Song Halloysite Deposits, San Chong Gun, Kyongsany Nambo, Korea,'' Bulletin 7, Korea Geological Survey, pp. 67–88.

Klammer, G., 1971, ''Uber Plio-pleistozame und ihre Sedimente in unteren Amazonasgebiet,'' *Zeitschrift fur Geomorphologie*, N.F., Vol. 15, No. 1, pp. 62–106.

Klammer, G., 1984, ''The Amazon—Limnology and landscape ecology of a mighty tropical river and its basin,'' H.E. Dumont, ed., Dr. W. Junk Publishers, Dordrecht, Germany, Vol. 56.

Klinefelter, T.A., and Hamlin, H.P., 1957, ''Syllabus of Clay Testing,'' Bulletin 565, US Bureau of Mines, 67 pp.

Kuzvart, M., 1969, ''Kaolin Deposits of Czechoslovakia,'' *Proceedings of Symposium I. Kaolin Deposits of the World, A. Europe*, 23rd International Geological Congress, Prague, 1968, Report, M. Malkovsky and J. Vachtl, eds., Vol. 15, pp. 47–73.

Ladd, G.E., 1898, ''A Preliminary Report on a Part of the Clays of Georgia,'' Bulletin No. 6, Georgia Geological Survey, pp. 81–91.

Lyons, S.C., 1966, ''Clays,'' *Paper Coating Pigments*, Technical Association Pulp and Paper Industry Monograph 30, H.H. Murray, ed., pp. 57–124.

Mark, H., 1963, ''High-Alumina Kaolinite Clay in the United States,'' Mineral Resource Investigations Map MR-37, US Geological Survey.

Martin-Vivaldi, J.L., 1969, ''Kaolin Deposits of Spain,'' *Proceedings of Symposium I. Kaolin Deposits of the World, A. Europe*, 23rd International Geological Congress, Prague, 1968 Report, M. Malkovsky and J. Vachtl, eds., Vol. 15, pp. 225–261.

Martin-Vivaldi, J.L., and Girela-Vilchez, F., 1959, ''A Study of the Halloysite from Maazza (North Morocco),'' *Silicates Industriels*, Vol. 24, Nos. 7–8, pp. 380–385.

McQueen, H.S., 1943, ''Geology of the Fire Clay Districts of East-central Missouri,'' Missouri Geological Survey and Water Resources, Vol. 28, 2nd Ser., 250 pp.

Moeri, E.N., et al., 1989, ''Geological Model and Computer Control for Planning and Operating of the Jari Kaolin Mine, Morro do Felipe, Brazil,'' *Proceedings*, 13th World Mining Congress, p. 129.

Morris, H.T., 1968, ''The Main Tintic Mining District, Utah,'' *Ore Deposits of the United States*, 1933–1967, Vol. 2, J.D. Ridge, ed., AIME, New York, pp. 1043–1073.

Murray, H.H., 1963, ''Mining and Processing Industrial Kaolins,'' *Georgia Mineral Newsletter*, No. 16, pp. 3–11.

Murray, H.H., 1976, ''The Georgia Sedimentary Kaolins,'' International Geological Correlation Program Project 23, 7th Symposium on Genesis of Kaolin, Tokyo, Japan, pp 114–125.

Murray, H.H., 1989, ''Kaolin Minerals: Their Genesis and Occurrences,'' *Hydrous Phyllosilcates*, S.W. Bailey, ed., *Reviews in Mineralogy*, Vol. 19, Mineralogical Society of America.

Murray, H.H., et al, 1977, ''Mineralogy and Geology of the Maungaparerua Halloysite Deposit in New Zealand,'' *Clays and Clay Minerals*, Vol. 25, pp. 1–5.

Murray, H.H., et al., 1978, ''Alteration of a Granite to Kaolin— Mineralogy and Geochemistry,'' Schriftenr, Geol. Wiss, No. 11, Berlin, pp. 197–208.

Norton, F.H., 1968, *Refractories*, McGraw-Hill, New York, 228 pp.

Oder, R.R., and Price, C.R., 1973, ''A New Clay Technology—Its Implication for the Paper Industry,'' TAPPI Coating Conference 1973, Technical Association of the Paper Industry, Atlanta, GA, pp. 23–27.

Olive, W.W., and Finch, W.I., 1969, ''Stratigraphic and Mineralogic Relations and Ceramic Properties of Clay Deposits of Eocene Age in the Jackson Purchase Region, Kentucky, and Adjacent Parts of Tennessee,'' Bulletin 1282, US Geological Survey, 33 pp.

Olson, R.H., 1964, ''Clays,'' Mineral and Water Resources of Nevada, US Geological Survey and Nevada Bureau of Mines, US Congress, 88th, 2nd Session, Senate Document 87, pp. 185–189.

Papke, K.G., 1971, ''Halloysite Deposits in the Terraced Hills, Washoe County, Nevada,'' *Clays and Clay Minerals*, Vol. 19, No. 2, pp. 71–74.

Parham, W.E., 1970 ''Clay Mineralogy and Geology of Minnesota's Kaolin Clays,'' SP-10, Minnesota Geological Survey, 142 pp.

Parker, J.M., 1946, ''Residual Kaolin Deposits of the Spruce Pine District, North Carolina,'' Bulletin 48, North Carolina Division Mineral Resources, 45 pp.

Patterson, S.H., and Hosterman, J.W., 1963, ''Geology and Refractory Clay Deposits of the Haldeman and Wrigley Quadrangles, Kentucky,'' Bulletin 1122-F, US Geological Survey, 113 pp.

Patterson, S.H., and Murray, H.H., 1983, ''Clays,'' *Industrial Minerals and Rocks*, 5th ed., S.J. Lefond, ed., AIME, New York, pp. 585–651.

Petrov, V.P., 1969, ''Kaolin Deposits of the USSR,'' *Proceedings of Symposium I. Kaolin Deposits of the World, A. Europe*, 23rd International Geological Congress, Prague, 1968, Report, M. Malkovsky and J. Vachtl, eds., Vol. 15, pp. 289–319.

Phelps, G.W., 1972, ''The Ball Clays of Tennessee and Kentucky,'' SME Preprint No. 72H305, SME Fall Meeting, Birmingham, AL, Oct., 11 pp.

Pickering, S.M., Jr. 1971, ''Lithostratigraphy and Biostratigraphy of the North Central Georgia Coastal Plain,'' 6th Annual Georgia Geological Society Field Trip Guidebook, Georgia Geological Survey, 15 pp.

Pickering, S.M., and Hurst, V.J., 1989, ''Commercial Kaolins in Georgia; Occurrence, Mineralogy, Origin, Use,'' *Excursions in Georgia Geology*, W.J. Fritz, ed., Georgia Geological Society Guidebook No. 9, pp. 29–75.

Pickering, S.M., Jr., et al., 1988, ''Industrial Implications of Kaolin Clay Stratigraphy in Georgia,'' *Proceedings*, 24th Forum on the Geology of Industrial Minerals (Abstract), South Carolina Geological Survey (1990), p. 153.

Prasad, M.S., et al., 1991, ''Kaolin: Processing, Properties, and Applications,'' *Applied Clay Science*, Vol. 6, pp. 87–119.

Rusko, Y.A., 1973, ''Dickite Kaolin of the Beregovian Deposit,'' (Inst. Gookim. Fiz. Miner. Kiev USSR), Akademiya Nauk, Ukrainskoy RSR Dopovidi, Seriya B, Vol. 35, No. 1, pp. 43–46 (in Russian).

Scott, A., 1929, ''Ball Clays,'' Special Report of the Mineral Resources of Great Britain, Memoir, Great Britain Geological Survey, Vol. 31, 73 pp.

Sheppard, S.M.F., 1977, ''The Cornubian Batholith, Southwest England: D/H and O^{18}/O^{16} Studies of Kaolinite and Other Alteration Minerals,'' *Journal*, Geological Society of London, Vol. 133, pp. 573–591.

Stout, W.G., et al., 1923, ''Coal-Formation Clays of Ohio,'' Bulletin 26, Ohio Geological Survey, 4th Series, 588 pp.

Sprague, E.K., et al., 1987, ''Apparent Salinity of Depositional Environments of Commercial Sedimentary Kaolinites, Georgia, by Carbon Isotope Analysis (Delta 13c/12c),'' GSA Southeastern Section.

Sukel, T.M., 1987, ''Grass Roots Kaolin Plants—Can They be Justified?'' SME Preprint 87–103, SME Annual Meeting, Denver, CO, 14 pp.

Takeshi, H., and Kato, C., 1969, ''Minerals,'' *The Clays of Japan*, ed. by Editorial Subcommittee for 1969 International Clay Conference, Geological Survey of Japan, pp. 103–120.

Van Sant, J.N., 1959, ''Refractory-Clay Deposits of Colorado,'' Report of Investigations 5553, US Bureau of Mines, 156 pp.

Van Sant, J.N., 1964, ''Refractory-Clay Deposits of Utah,'' Information Circular 8213, US Bureau of Mines, 176 pp.

Veatch, O., 1909, ''Second Report on the Clay Deposits of Georgia,'' Bulletin 18, Georgia Geological Survey, 53 pp.

Waagé, K.M., 1953, ''Refractory Clay Deposits of South-Central Colorado,'' Bulletin 993, US Geological Survey, 104 pp.

Webb, H., and Sprague, E.K., 1991, ''Timing of Kaolinitization in Hard Kaolins of the Huber Formation in Central and Eastern Georgia: Trace Fossil Microtextures Documented by Scanning Electron Microscopy,'' (Abstract) Clay Minerals Society 28th Annual Meeting, Houston, TX, p. 171.

Yuan, J., and Murray, H.H., 1990, ''Industrial Properties of Maoming Kaolin from Guangdong Province, South China,'' Abstract, Clay Minerals Society 27th Annual Meeting, Columbia, MO, Oct. 6–11, 1990, p. 131.

CONSTRUCTION USES

ADOBE AND SIMILAR MATERIALS

GEORGE S. AUSTIN

Mud is one of the oldest building materials used by man, but it is not only of historic interest. Even today in many parts of the world, particularly lightly forested areas, it is the chief building material. Many authorities believe that at least 30% and perhaps 50% of the world's present population lives in earthen dwellings (Dethier, 1985, Coffman et al., 1990). Not all of these people are members of primitive societies or live in third world countries. European, African, Asian, and both North and South American countries contain a large number of such structures. In the United States, the American West, and particularly the Southwest, is known for this type of construction.

Some of the earliest remains of adobe structures are those discovered in the ruins of Neolithic farming villages in Mesopotamia dating as far back as 7000 BC (Steen, 1972). The word *adobe* has its roots in Egyptian hieroglyphs denoting brick and evolved to its present form through Arabic and Spanish (Lumpkins, 1977). The Spanish conquest of the New World spread the use of wooden molds to produce a standard adobe brick. Today, the word adobe is used to describe various earth building materials and techniques, usually referring to sun-dried adobe brick now used in the United States, but is also applied to puddled adobe structures, mud-plastered logs or branches (Jacal or waddle-and-daub), pressed-earth blocks, and rammed-earth walls or pisé (Smith and Austin, 1989, Ferm, 1985).

Modern mud construction is used in many countries in various parts of the world. In the United States, the states from Texas to California are perhaps best known for this type of construction. Of these states, New Mexico has the dominant reputation for adobe use. Indeed, in New Mexico the Santa Fe style (Fig. 1) has made adobe not only acceptable, but preferred.

RAW MATERIALS

Adobe soil used by present day adobe producers, and probably past adobe producers as well, is principally sandy loam (50% clay and silt), although clayey silts are used in some areas (Coffman et al., 1990). In New Mexico, the best adobe soils are those developed on stream deposits, particularly Holocene (Recent) terrace deposits and older, loosely compacted geologic formations, such as the Santa Fe Group (Tertiary) located in the Rio Grande Valley. Some modern adobe producers use a mixture of materials from the screened fines of aggregate operations and mud from irrigation ditches in the river valleys combined with varying amounts of sand to produce the proper blend.

Mineralogy

Bulk Mineralogy: X-ray diffraction analyses of whole rock samples from many parts of the world where adobe is the dominant construction material show the major constituents of adobe are quartz and feldspar, with lesser amounts (in order of abundance) of calcite, clay minerals, and gypsum (Coffman et al., 1990). Adobes from arid climates contain considerable calcite; in some cases calcite is second only to quartz in volume. Quartz, feldspar, most of the clay minerals, and some calcite commonly are derived from the mechanical/chemical breakdown of older rocks units. Some of the clay minerals, much of the calcite and perhaps all of the gypsum are precipitated from evaporating water.

Clay Mineralogy: Clay-size minerals consist dominantly of clay minerals, but nearly clay-size fractions contain minor amounts of quartz and calcite, and occasionally other nonclay minerals. In earth construction materials from New Mexico, clay-size particles are the most compositionally variable in commercial adobe soils. However, the clay mineral groups in this size fraction consist of about equal parts of expandable clay minerals (smectite and mixed-layer illite/smectite or I/S) and non-expandable clay minerals (kaolinite, illite, and chlorite), with minor quartz, calcite, and feldspar (Smith and Austin, 1989, Austin, 1990). In the arid American Southwest, smectite is commonly calcium-rich and the I/S is disorganized and randomly interstratified. In a study of 42 New Mexican commercial adobe soils, only two contained chlorite, and vermiculite, sepiolite, and palygorskite were not found.

Soils in temperate climates, as are found in the midwestern and northeastern United States, contain an abundance of illite (Potter et al., 1975). The clay minerals in earthen structures of that region and in Europe commonly contain illite with less amounts of kaolinite, smectite, I/S, chlorite, and vermiculite. In contrast, the clay-size fractions of soils of humid tropical areas are typically acid, with kaolinite as the dominant clay mineral and lesser amounts of I/S, illite, smectite, and others (Chamley, 1989).

Chemical Properties

Soils in the arid New Mexican climate are typically alkaline. Ground water near the Rio Grande Valley is generally hard to extremely hard, containing total dissolved solids (TDS) ranging from about a hundred to several thousand parts per million (Wilkens, 1986, Anderholm, 1987). Soluble salts, notably calcium carbonate and calcium sulfate, precipitate as this water evaporates. White crusts of these salts at the surface are common in marshy areas of the state during drier times of the year. The stability of the desert surfaces causes precipitation of calcium carbonate from ground water, initially producing calcium-rich soils, followed by impermeable pedogenic calcite layers called caliche just above the ground water table (Gile et al., 1981).

Adobe walls in New Mexico are remarkably durable in this climate. With proper care, walls hundreds of years old are durable.

Fig. 1. Modern adobe home in Santa Fe, NM.

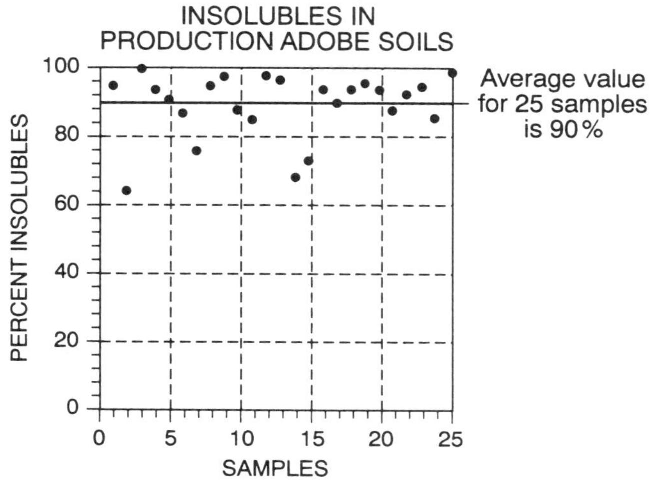

Fig. 2. Summary of leaching analysis of 25 commercial New Mexico adobe soils. The average insoluble fraction (insoluble after boiling for 4 hr in EDTA) is 90% (Austin, 1990).

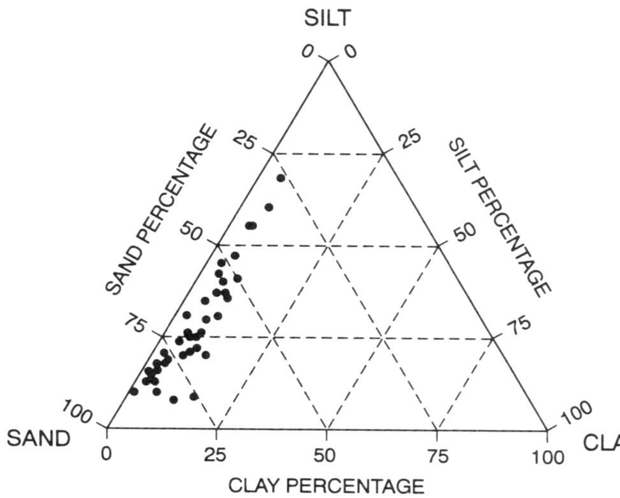

Fig. 3. Plot of sand-and-larger-, silt-, and clay-size fractions of 42 production adobe soils used in New Mexico. The diagram shows the dominance of larger particles as compared to clay-size material (Austin, 1990).

Great care is taken to keep the wall interior dry. Walls in some Indian Pueblos are re-covered with a natural mud plaster on a yearly basis as part of community service by the inhabitants. The natural plaster has about the same mineral and chemical composition as the walls themselves, but is slightly finer grained. The plaster is expected to slowly wash off during the year and to be replaced during the next replastering. The slow removal of the plaster apparently is due to calcite and gypsum precipitation from the mixing water that forms a caliche-like bond between grains as the mud slowly dries (Austin, 1990). The clay component on the surface of most vertical adobe surfaces limits the penetration of precipitation; it runs off before it can soak into the wall and erosion is limited in New Mexico to "1 inch in 20 years" (McHenry, Jr., 1992, written communication). This assumes a reasonable amount of clay because an adobe soil with very little clay will erode very rapidly in a driving rainstorm. On a horizontal surface, however, moisture tends to puddle and soak in; then subsequent raindrops physically damage the softened surface.

Leaching tests with EDTA (ethylenedinitrilotetraacetic acid) on 25 commercial New Mexican adobe soils (Fig. 2) suggest that commercial soils contain an average of about 90 wt% insoluble and 10 wt% soluble material; the latter is dominantly calcite and some gypsum (Austin, 1990). In that study, the soluble material ranged from 36 wt% to essentially zero. Adobe soils with the smallest amount of soluble material were also the highest in sand and larger-size particles. A study of a limited number of adobe from other parts of the world found a similar range of soluble material (Coffman et al., 1990).

Physical Properties

Particle-size Distribution: Adobes from some historic structures from the American Southwest, as well as other parts of the world, have clay contents from 1 to 38 wt%, while sand-and-larger particles range from 4 to 82 wt% on a limited number of samples (Coffman et al., 1990). The common statement by modern New Mexican adobe producers is that their soil mix is usually one-half sand and one-half clay or fines (silt and clay), but commercial adobe soils in New Mexico range from 85 to 99 wt% nonclay-size particles (Smith and Austin, 1989). Tests of commercially produced New Mexican adobes show that the soil contains 27 to 89 wt% sand-and-larger-grain size, 8 to 68 wt% silt-size, and 1 to 15 wt% clay-size grains (Fig. 3). The average grain-size composition was 67 wt% sand-and-larger, 27 wt% silt, and 6 wt% clay.

Large scale commercial adobe producers in New Mexico use adobe soils with less clay-size material than do small scale commercial and non-commercial adobe producers. Some of the former are as low as about 1 wt% clay, whereas many of the latter are between 8 and 15 wt% (Smith and Austin, 1989). In part, this is because large scale commercial adobe producers use stabilizers that not only protect blocks from rain damage, but aid in consolidation of the drying soil mix as well. An abundance of clay-size particles in adobe soils causes excessive cracks as blocks dry in an adobe yard. To combat cracking, producers add straw and/or additional sand to the mud mixture.

Thermal Properties: Traditionally, materials are evaluated for thermal properties based on measurements known as R- and U-values. The R-value is an indicator of the ability of a wall to insulate effectively. The R-value is calculated by dividing the thickness of the wall by the wall's thermal conductivity, the amount of heat flowing from the hotter side to the cooler side of the wall. The

U-value is the reciprocal of the R-value and reflects the rate at which heat is conducted through a material. The total R-value may be calculated for a given wall by adding up the values of the individual components of a wall structure, including all insulation, interior sheathing, framing, air-space resistance, and masonry. Adobe walls have very low R-value because they commonly consist of 26 or 36 cm blocks covered with a thin stucco on the outside and thin gypsum plaster on the inside.

R- and U-values do not tell the full story in determining what is a high quality, thermally efficient wall (Fine, 1976). Both these values reflect the rate heat passes through a wall only after the steady state of heat flow (the state of heat energy is passing uninterrupted from one side of the wall to the other at a constant rate) has been achieved. What is not considered, and is of critical importance in masonry-mass walls such as adobe, is the heat storage capacity of the wall, which determines the length of time that passes before a steady state of heat flow is achieved. The higher the heat storage capacity of the wall, the longer period of time it will take for heat flow to reach a steady state. In real situations, external temperatures, in particular, are changing constantly, so that a true steady-state condition is rarely achieved. Because diurnal changes in the arid Southwest are typically 15 to 25°C, the fly-wheel effect thus will keep adobe buildings' daytime temperatures cool in the summer and warm in the winter.

Resistance of Sound: Thick massive walls of a typical adobe are well known sound deadeners, making these homes remarkably quiet. Windows in older adobe building are normally small, further adding to the quietness. Newer solar adobe homes take advantage of the many sunny days in arid climates with large windows but use well-insulated glass to retain much of the sound deadening characteristics of adobe dwellings.

Other Physical Properties: There are many other physical properties that make adobe construction appealing. Homes so constructed are water resistant, flame retardant, unaffected by termites, and energy efficient. In addition to the preservation of nighttime cool temperatures in the summer and daytime heat in the winter, the sun-dried method of production, rather than using high heat to produce masonry brick or cement, increases the energy efficiency. Wright (1978) stated that it takes over 300 times more commercial energy to produce a concrete block than a sun-dried adobe block (equal in volume).

An unusual test of a physical property on adobe in Nicaragua determined that adobes were resistant to penetration of bullets (Lola, 1981). Five shots from a machine gun at a 20-m distance penetrated an adobe wall only 4 to 5 cm.

INDUSTRY STRUCTURE

Little formal adobe industry structure exists in the United States. Most noncommercial adobe makers come from generations of adoberos (adobe makers) and learned the techniques at home. Some educational institutions in the Southwest are offering formal or semi-formal courses in adobe construction to those without that background, but with a desire to know. Most commercial adobe producers know others nearby, but few structured contacts, such as during industry meetings, are made. The Getty Conservation Institute of Los Angeles, CA, directs national attention to the preservation of adobe construction, particularly in the American Southwest. International structures for preservation are headed by the International Centre for the Preservation and the Restoration of Cultural Property (ICCROM) and the International Centre for Earth Construction (CRATerre). ICCROM focuses international attention on adobe by holding international meetings on the preservation of earthen construction. Its US arm is US/ICOMOS (the acronym ICOMOS stands for the International Council for Monuments and Sites). CRATerre, affiliated with the Ecole d'Architecture in Grenoble, France, trains students in earthen architecture and construction. CRAterre also contracts with government agencies, particularly in emerging nations to train students and build large housing projects on site in those countries. The Department of Architecture at the University of New Mexico has offered courses in earth architecture for more than ten years, in both Spanish and English.

TECHNOLOGY

Expandable clay minerals tend to be more sticky than nonexpandable varieties and thus are more effective in binding silt and sand particles together. Expandable clay minerals also form colloidal suspensions with water and therefore moisture, whether as rainfall or ground water, has the greatest effect on adobe soils with the largest proportion of smectite and I/S.

For past and present adobe producers in New Mexico, expandable clay minerals were and are sometimes a problem. Cracking of drying adobe brick is due most probably to the somewhat large proportion of smectite and I/S in adobe soil; soils with higher clay content, but lower smectite and I/S content, will have less tendency to crack. Cracking is extreme on windy days when the shrinking clay structure is changing rapidly. Drying slowly over many calm days allows multiple layers of finely crystalline calcite (and some gypsum) to form on a clay-size scale, strengthening the bricks and preventing cracks. The resulting adobe wall can resist torrential late summer rains for long periods if the adobe bricks do not contain too much clay material and are properly cured.

Several varieties and sizes of earthen brick are produced throughout the American Southwest; these include traditional adobe, semistabilized and stabilized adobe, New Mexican terrónes (cut-sod brick), quemados (burnt adobe), and machine-pressed-earth block. In addition rammed-earth walls are constructed without brick (McHenry, Jr., 1984, Smith and Austin, 1989). The two major types of adobe brick currently produced in New Mexico are the traditional adobe brick and the semistabilized adobe brick. Stabilized adobes are available on special order. In Arizona, the most common adobes are stabilized, and traditional or unstabilized adobes are all but impossible to find (McHenry, Jr., 1992, written communication).

Production Techniques

Traditional (untreated) Adobe Bricks: Often called untreated or sun-dried adobe brick, traditional adobe is made with soil composed of sand with some larger particles, and of silt and clay. Straw is sometimes added for strength and to prevent excessive cracking during drying. The moistened soil mixture commonly is packed into a brick-like mold, released, and allowed to dry and cure for several weeks before use (Fig. 4).

Pressed-earth Blocks: Pressed-earth blocks presently make up a small portion of earth brick used in New Mexico (Smith and Austin, 1989). The CINVA-Ram hand operated press was developed by a Chilean engineer in the 1950s and has been used in New Mexico, but most pressed-earth blocks in the state are made by gasoline- or diesel-powered machines (Fig. 5). Several have been designed and used in the past in New Mexico to press the adobe soil mixture into a form, minimizing the amount of time required between forming the block and placing it into the wall. Portland cement or asphalt emulsion has been used to partly or fully stabilize pressed-earth blocks. Most pressed-earth blocks come from small volume and/or part time, or non-commercial producers.

Rammed-earth Walls: Rammed-earth homes commonly have

Fig. 4. Adobe being packed into a 10-brick ladder form at a commercial adobe yard in New Mexico. After 2 to 3 hours of drying time, the form is removed (see blocks on ground to left). After 2 tc 3 days, the adobe is trimmed by hand, turned on edge, and dried for an additional two weeks. Blocks then are stacked in the yard (see extreme left) or on pallets ready for sale.

Fig. 5. Large-size commercial pressed-earth-block machine operation at the job site in northern New Mexico.

much thicker walls than most other earthen dwellings, up to 0.9 m thick. Wooden or metal concrete-type forms are put in place on stone or concrete footings and 15- to 20-cm-thick layers of moistened soil are put between the walls of the forms. Hand or hydraulic tampers are used to pound the soil into the shape of the form, compacting and reducing the volume of the mixture by 25 to 30% (McHenry, Jr., 1984, Middleton, 1987). Once the layers of tamped soil reach the desired height, the forms are removed and the wall is allowed to dry (Fig. 6). Portland cement is the common stabilizer used. Producers say rammed-earth walls continue to harden, or cure, during the first year after construction. During 1987, New Mexico's two rammed-earth construction firms built three homes (Smith and Austin, 1989).

Fig. 6. Exterior 2-m-thick rammed earth wall of a home under construction. Note the layering produced by tamping and the concrete bond beam topped by the wooden frame at the top of the wall.

Stabilized Adobe Bricks: Fully stabilized adobe brick is defined by the New Mexico Building Code as water-resistant adobe made of soil with certain admixtures that limit the brick's seven day water absorption to less than 4 wt%. A fully stabilized adobe brick usually is made with 6 to 12 wt% of asphalt emulsion (Anon., 1963, Scheuch and Busch, 1988). Exterior walls constructed with stabilized mud mortar and brick require no additional protection and can be left exposed without stucco. The production of fully stabilized adobe brick is very low because most walls are stuccoed with water-resistant plaster, and the additional waterproofing agent adds extra cost.

Semistabilized Adobe Bricks: Semistabilized adobe brick was developed by major adobe producers in New Mexico and is classified as a water-resistant brick because of the addition of 3 to 5 wt% of a stabilizer or waterproofing agent (Anon., 1963, Scheuch and Busch, 1988). The stabilizer protects the brick from rainstorm damage during the curing process. Asphalt emulsion is the primary stabilizer because of the ease of use and the low cost, but 5 to 10 wt% portland cement produces the same result. Semistabilized adobe is made the same way as traditional adobe, except for mixing the stabilizer into the adobe soil prior to packing it into a form.

Specifications

Government specifications for adobe or mud construction are not widely used in the United States, because it is uncommon in many states and there is no universal demand. The state with the most adobe construction, New Mexico, probably has the most complete approach. The New Mexico code for unburned clay masonry is Section 2412 of the *New Mexico Building Code* (Anon., 1988). A summary of the most significant rules regarding adobe blocks is given here:

1. *Compressive Strength:* The units shall have an average compressive strength of 300 lbs/in^2 when tested.
2. *Modulus of Rupture:* This unit shall average 50 lbs/in^2 in modulus of rupture when tested according to the specified procedures.
3. *Moisture Contents:* The moisture content of untreated units shall not be more than 4%.
4. *Adsorption:* A dried cube cut from a sample shall not adsorb more than 4% moisture by weight when placed upon a con-

Fig. 7. The 1000-year-old Taos Pueblo originally built of puddled adobe. Mud plastering is carried out each year to maintain the structure.

stantly water-saturated porous surface for seven days. An adobe unit that meets this specification shall be considered stabilized.

5. *Shrinkage Cracks:* No unit shall contain more than three shrinkage cracks, and no shrinkage crack shall exceed 2 in. in length or 1/8 in. width.

USE AND DISTRIBUTION

United States

Examples of early mud construction are present in many parts of the United States. Perhaps the most famous is the still occupied, 1,000 year-old Taos Pueblo in New Mexico (Fig. 7). Large earthen mounds built by North American aboriginal people called mound builders are present in many parts of the eastern United States as well (Emerson and Woods, 1990). The common building method in these pre-Columbian structures was puddled-clay or clay-lump construction without the adobe mold introduced by the Spanish in the 16th century. Similar construction was used in several sections of the country into the 19th century. Western New York state (Dassler, 1990) has a number of good examples. Sod was the principal building material of many immigrants on the treeless Great Plains during the same period. As modes of transportation broadened and the cost of moving building material fell, earth construction declined in most areas of the county.

An earlier improvement in adobe construction occurred in the 1880s, when the railroads reached New Mexico. Local inhabitants soon learned how sloped roofs, covered by corrugated iron sheets from the east, were far more effective than the flat, dirt covered roofs used prior to that time. Modern improvements, such as the use of soil stabilizers, further improved earth blocks and their ability to withstand moisture damage.

The American Southwest has long had a love affair with adobe, and the landscapes of New Mexico, Arizona, Texas, and California contain many examples of enduring adobe homes. Old military forts, churches, and commercial buildings also attest to its popularity. New Mexico, both historically and today, is the largest domestic producer and user of adobes. During the past decade, 3 to 4 million adobe bricks and pressed-earth blocks have been produced in New Mexico each year by 50 commercial manufacturers (Smith and Austin, 1989).

More than 59,000 adobe buildings, representing one-third of the adobe dwellings in the United States, are in use in New Mexico (Gerbrandt and May, 1986). Although today less than 3% of new homes built each year in New Mexico are constructed with adobe, the number usually averages between 500 and 600 new dwellings. From examination of building permits in New Mexico and discussions with local architects, realtors, and planners throughout the Southwest, it appears that most new adobe dwellings are built for high income customers. The houses usually contain above average square footage with the addition of extensive southwestern architectural details and style. In some areas, particularly in parts of rural northern New Mexico, many backyard adobe producers build their own homes (McHenry, Jr., 1985). This is often done on a part time basis, while working full time for a local business or government agency. Often the house is completed without an extensive debt or long term mortgage payments.

A breakdown of New Mexican 1987 adobe-brick production shows that 27% were traditional (untreated) bricks, 69% were semistabilized bricks, and 5% were stabilized bricks (Smith and Austin, 1989). These percentages appear to be typical of the 1980s, as semistabilized adobe is generally the adobe brick of choice. Before 1970, most adobe buildings were built with traditional adobes.

Today, most builders purchase the adobe bricks from commercial yards located throughout New Mexico. The adobe-block operation is a labor intensive, but fuel efficient seasonal industry with the production of blocks usually limited by the number of frost free days. The principal standard-size adobe brick produced and used in New Mexico measures 10 x 26 x 36 cm and weighs approximately 17 kg (Smith and Austin, 1989).

World

Many parts of the world contain evidences of past earth construction. Recently constructed earthen buildings are widely known from Mexico to the tip of South America and from Egypt across northern Africa (McHenry, Jr., 1984). Less well known are examples of earthen structures in Germany, Italy, and Belgium. France has both a history of mud construction and CraTERRE, a modern school teaching the method (Doat et al., 1979). In India, 100 million people live in earthen buildings (Krishna and Chandra, 1981). Other widely divergent countries containing a significant number of earthen dwellings include: the United Kingdom (Bouwens, 1990), China (Jiyao and Weitung, 1990), central Asia (Chayet et al., 1990), Sri Lanka (Nandadeva, 1990), Australia (Middleton, 1987), Nicaragua (Lola, 1981), Zimbabwe, (Ndoro, 1990), and Saudi Arabia (Emrick and Meinhardt, 1990).

ECONOMIC FACTORS

Cost

In New Mexico, adobe brick production varies from a labor intensive traditional hoe-shovel-and-wheelbarrow technique to a mechanical large scale operation producing 5,000 to 20,000 bricks per day. The production of adobe bricks is seasonal and is usually limited by the number of frost free days for a particular adobe yard. In New Mexico, the production season lasts from five to nine months, depending on climate and weather conditions.

Smith and Austin (1989) report that producers using the handcraft technique for traditional (untreated) adobe marketed their products at 21 to 40¢ per brick at their yards. One-person commercial yards produced 100 to 300 bricks per day. Semimechanized adobe yards with 8 to 10 employees can produce 5,000 to 6,000 semistabilized bricks per day that are sold for 40 to 45¢ per brick at the adobe yard (Smith, 1992, written communication). The largest mechanized adobe yard with 3 to 4 employees can produce 5,000 to 7,000 semistabilized adobes per day that are sold for 40 to 45¢ at the adobe yard. Most producers say that the stabilizers (asphalt emulsion and portland cement add between 5 and 10 cents to the price of an adobe. Adobe bricks can be delivered to the job site via flatbed trucks that carry up to 1,000 palletized or stacked adobes and off-loaded. A 1,000-brick load would weigh 16 to 18 t; typical adobe haulers use 2-t trucks. Pressed-earth blocks in a production yard are sold for comparable prices, but some producers take their machine to the job site and use local soil for their blocks at a slightly reduced price. These producers are the contractors as well and build the desired walls or a complete house.

Markets

Tradition is the most important factor in determining markets for adobe materials. In areas that have a strong tradition of mud construction, adobe is appealing, even preferred. In other areas where the population is not familiar with adobe, or worse, considers it beneath them to live in such buildings, new adobe buildings will not be built and old ones will disappear. An example of the former area is Santa Fe, NM. The Santa Fe style of construction is adobe pueblo and territorial style (Smith and Austin, 1989). Adobe buildings are preferred by many wealthy landowners and even contractors who use wood or concrete block of construction mimic the adobe styles.

Mud construction is less popular in nearly all non-southwestern parts of the United States and is rarely seen. Hydraulically pressed mud blocks appear to have some appeal in some parts of the country, but no large scale construction efforts using this method are known. Mud construction is no longer used in most of Great Britain, France, and Germany, countries with long histories of earthen construction and with excellent examples of multi-storied rammed-earth buildings existing even today. Manufacturers of hydraulically pressed block machines, both in New Mexico and elsewhere, appear to believe that markets for such machines exist in eastern Europe and Central and South America where the population is without the capital or resources to use conventional wood construction. However, labor is cheap enough in many of these countries to prohibit use of expensive hydraulic pressed-earth-block machines. In other parts of the world, those opposed to widespread lumbering practices see mud construction as a viable alternative for shelter.

Transportation

Adobe may be made on the construction site, but in recent years it is more likely to be made in adobe yards and transported to building sites on flatbed trucks (Fig. 8). Although transport to distant construction sites is uncommon, some producers ship adobe several hundred kilometers. The problem is not normally in the availability of the raw material. Acceptable, if not superior, adobe can be made with most native raw materials, providing a qualified adobe maker is involved. Long distance shipping occurs when the landowner wants an adobe home and contractors in the immediate area have neither the training nor talent for this type of construction.

Rammed-earth construction methods require that the walls be made at the building site. Consequently, the normal method of construction is to use local materials or materials that have not been transported far.

Alternate Materials

The most commonly used alternative material in the past was concrete block. Many adobe-style homes even in the American Southwest were constructed of stucco-clad concrete block. Although smaller and lighter than adobe and easy to use, concrete blocks do not have the physical properties of adobe and must be carefully insulated to achieve insulating and sound-deadening properties similar to those of adobe. In recent years, the cost of cement and labor has made this type of construction less appealing. Wooden frame structures resembling adobe-style homes are also common, but they have similar deficiencies and increasing expense, as do concrete block homes.

Fig. 8. Semi-truck load of adobe bricks on pallets ready for delivery in northern New Mexico.

In underdeveloped countries something as basic as the availability and the cost of cement can make adobe the preferred type of construction. The use only of available native materials can be a strong incentive in countries too short of funds to buy cement or to build a cement plant. Many local officials have been educated in Western counties and teach Western standards and the result is that adobe is considered to be for the homes of the poor and many governments have little sympathy for its use.

GOVERNMENTAL CONSIDERATIONS

In most regions of the United States where adobe is widely used in construction, locally adopted amendments to the Uniform Building Code establish building specifications. In New Mexico, specifications for adobe were last modified in 1988. The New Mexican code defines the acceptable earthen construction materials and products and the necessary tests to be used are listed and explained. Similar codification of earthen construction specifications has been produced in other states.

Health and Safety Regulations

Radon and Seismic Activity: The identification of radon gas as a health hazard in homes and the low strength materials used in adobe homes in seismically-active areas have caused owners to be concerned about the safety of their adobe structures. Modest changes in construction and care commonly will prevent problems in all but the most severe cases (Smith and Austin, 1989).

Radon is a colorless, odorless, heavier-than-air radioactive gas derived from the breakdown of some radioactive elements in soil and rock. The most stable radon isotope, ^{222}Rn, has a half-life of 3.8 days (Wilkening, 1980). Radon enters buildings through cracks, particularly when the buildings are closed and have a negative air pressure as is commonly true during the heating months in winter, and it accumulates in low spots. Although adobe buildings have not been shown to contain significantly more radon than other types of construction, good ventilation and positive interior air pressure are the easiest ways to prevent a buildup of the gas.

Seismic activity is very destructive in many parts of the world that use low strength masonry walls. Adobe buildings in the United States are commonly constructed with one and two stories and on concrete slab foundations. Designing the slab to resist cracking both during normal life of the structure and possible earthquakes is prudent in seismic areas. Recent work in California suggests that a combination of proper slab construction, reinforcing walls with rebar, the use of wire mesh both inside and outside the building beneath the plaster and stucco, interconnected bond beams and roof beams at the top of walls, and buttresses can serve to reduce hazards relating to earthquakes (Tibbets, 1986).

The building code for earthen construction is concerned with safety for the inhabitants. In New Mexico and California, two states with sizable adobe structures and seismic activity, the building code specifically addresses the need of adobe structures to endure considerable earth movement during seismic events (Smith and Austin, 1989). The New Mexico code forbids the construction of adobe structures exceeding two stories in height. All bearing walls must be topped with a continuous belt course or tie beam. The belt course is commonly reinforced concrete and tie beams are solid wood at least 15 cm thick. The California Building Code requires that walls have internal reinforcement with rebar as well. This results in adobe structures faring about as well during seismic events as buildings constructed with more conventional materials. Some California Building Code regulators also now allow the flexibility of substitution of some horizontal for some vertical reinforcement in earth construction (McHenry, Jr., 1992, written communication).

PROBLEMS AND FUTURE TRENDS

Adobe and similar earth construction is gaining respectability, particularly in the American Southwest, as the Santa Fe style of earth construction continues to be popular. It will continue to account for 3 to 5% of the homes constructed in this region for the foreseeable future. It also will be the preferred construction of both the low income and high income groups. Low income groups use this construction because of the availability of local materials, and they can build the structures themselves, often in their spare time. Adobe is also the preferred type of construction for the wealthy because a superior home can be constructed that is in harmony with nature and currently chic.

Although adjustments must be made for such problems as insulating properties, possible damage during earthquakes, and radon gas accumulation, these adjustments are neither very costly nor impossible to make. In addition, on a worldwide basis, adobe construction offers a considerable advantage of housing a rapidly expanding population with a maximum amount of local building material with a minimal use of expensive energy resources.

ACKNOWLEDGMENTS

Edward W. Smith, Tesuque, NM, and Paul G. McHenry Jr., Albuquerque, NM, reviewed the manuscript and made excellent comments on its content. Marjorie H. Austin was of significant help on grammar. Ed Smith introduced me to the fascinating world of earth construction materials and still talks to me even though I do live in a frame house.

BIBLIOGRAPHY

Anon., 1963, "The Manufacture and Use of Asphalt Emulsion Stabilized Adobe Bricks " California Research Corporation, Richmond, CA, 17 pp.

Anon., 1988, "New Mexico Building Code," Construction Industries Division, General Construction Bureau, Santa Fe, NM, pp. 21–26.

Anderholm, S.K., 1987, "Hydrogeology of the Socorro and La Jencia Basins, Socorro County, New Mexico," US Geological Survey, Water Resources Investigation Report 84-4224, 62 pp.

Austin, G.S., 1990, 'Adobe and Related Building Materials In New Mexico, USA," *Adobe 90 Preprints*, 6th International Conference on the Conservation of Earthen Architecture, The Getty Conservation Institute, Los Angeles, pp. 417–423.

Bouwens, D., 1990. "Clay Lump—the English Adobe," *Adobe 90 Preprints*, 6th International Conference on the Conservation of Earthen Architecture, The Getty Conservation Institute, Los Angeles, pp. 14–19.

Chamley, H., 1989, *Clay Sedimentology*, Springer-Verlag, Berlin, 623 pp.

Chayet, A., et al., 1990, "Earth Used for Building in the Himalayas, the Karakoram, and Central Asia," *Adobe 90 Preprints*, 6th International Conference on the Conservation of Earthen Architecture, The Getty Conservation Institute, Los Angeles, pp. 29–34.

Coffman, R., et al., 1990, "Adobe Mineralogy: Characterization of Adobes from Around the World," *Adobe 90 Preprints*, 6th International Conference on the Conservation of Earthen Architecture, The Getty Conservation Institute, Los Angeles, pp. 424–429.

Dassler, L., 1990, "Nineteenth Century New York State Earthen Homes: An Investigation of Their Material Composition," *Adobe 90 Preprints*, 6th International Conference on the Conservation of Earthen Architecture, The Getty Conservation Institute, Los Angeles, pp. 430–437.

Dethier, J., 1985, "A Back to Earth Approach to Housing," *UNESCO Courier*, Mar., pp. 31–33.

Doat, P., et al., 1979, "Construire en Terre," CRATerre, Centre de Recherche et d'Applications—Terre, Haut-Brié, France, 272 pp.

Emerson, T.E., and Woods, W.I., 1990, "The Slumping of the Great Knob: An Archaeological and Geotechnic Case Study of the Stability of a Great Earthen Mound," *Adobe 90 Preprints*, 6th International Conference on the Conservation of Earthen Architecture, The Getty Conservation Institute Los Angeles, pp. 219–224.

Emrick, M., and Meinhardt, C., 1990, "Reconstruction of Traditional Structures in the Al-Turaif Quarter, Dir'lyah, Kingdom of Saudi Arabia," *Adobe 90 Preprints*, 6th International Conference on the Conservation of Earthen Architecture, The Getty Conservation Institute, Los Angeles, pp. 153–158.

Ferm, R.L., 1985, "Stabilized Earth Construction: An Instructional Manual," The International Foundation for Earth Construction, Washington, DC, 74 pp.

Fine, A.H., 1976, "The Thermal Properties of Building Materials," *Adobe News*, Albuquerque, NM, No. 11, pp. 14, 26.

Gerbrandt, H., and May, G.W., 1986, "The Extent of Adobe Use in the United States," *Solar Earthbuilder International*, Issue 47, Las Cruces, NM, pp. 12–15, 56–59.

Gile, L.H., et al.,1981, "Soils and Geomorphology in the Basin and Range Area of Southern New Mexico—Guidebook to the Desert Project," New Mexico Bureau of Mines and Mineral Resources, Memoir 39, 222 pp.

Jiyao, H., and Weitung, J., 1990, "Earth, Culture, Architecture—The Protection and Development of Rammed Earth and Adobe Construction in China," *Adobe 90 Preprints*, 6th International Conference on the Conservation of Earthen Architecture, The Getty Conservation Institute, Los Angeles, pp. 72–76.

Krishna, J., and Chandra, B., 1981, "Earthen Buildings in Seismic Areas of India," *Proceedings*, International Workshop Earthen Buildings in Seismic Areas, G.W. May, ed., Vol. 1, May 24–28, 1981, University of New Mexico, Albuquerque, NM, pp. 227–234.

Lola, C., 1981, "Research Efforts on Soil Stabilizers for Low-Cost Housing in Nicaragua," A.T. International, Washington, DC, p. 10.

Lumpkins, W., 1977, "Adobe (from the Arabic 'Atobe')," unpublished paper, Museum of New Mexico, Santa Fe, NM, 25 pp.

McHenry, P.G., Jr., 1984, *Adobe and Rammed Earth Building Design and Construction*, John Wiley & Sons, New York, 217 pp.

McHenry, P.G., Jr., 1985, *Adobe, Build It Yourself*, University of Arizona Press, Tucson, AZ, rev. ed., 158 pp.

Middleton, G.F., 1987, *Earth-Wall Construction*, Australian Government Publishing Service, Bulletin 5, 4th ed., (rev. by L.M. Schneider), 65 pp.

Nandadeva, B.D., 1990, "Traditions and Techniques of Earthen Architecture of Sri Lanka," *Adobe 90 Preprints*, 6th International Conference on the Conservation of Earthen Architecture, The Getty Conservation Institute, Los Angeles, pp. 105–110.

Ndoro, W., 1990, "An Investigation into the Pattern of Deterioration of Daga (earth) Structures at Zimbabwe Type Monuments," *Adobe 90 Preprints*, 6th International Conference on the Conservation of Earthen Architecture, The Getty Conservation Institute, Los Angeles, pp. 377–382.

Potter, E., et al., 1975, "Clay Mineralogy of Modern Alluvial Muds of the Mississippi River Basin," Bulletin du Centre de la Recherche, Pau., Vol. 9, pp. 353–389.

Scheuch, K.E., and Busch, R.D., 1988, *New Mexico Conservation Code Application Manual, Residential Buildings*, Energy, Minerals, and Natural Resource Department, Santa Fe, NM, 108 pp.

Smith, E.W., and Austin, G.S., 1989, "Adobe, Pressed-Earth, and Rammed-Earth Industries in New Mexico," New Mexico Bureau of Mines and Mineral Resources, Bulletin 127, 60 pp.

Steen, C.R., 1972, "An Archaeologist's Summary of Adobe," Vol. 77, No. 4, Museum of New Mexico Press, Santa Fe, NM, pp. 29–39.

Tibbets, J.M., 1986, "Of Major Importance to Every Earthbuilder," *Solar Earthbuilder International*, Issue 47, Las Cruces, NM, pp. 26–29, 64.

Wilkening, M., 1980, "Radon Transport Processes Below the Earth's Surface," *Natural Radiation Environment III*, CONF-780422, Vol. 1, Technical Information Center/US Department of Energy, Washington, DC, pp. 90–104.

Wilkens, D.W., 1986, "Geohydrology of the Southwest Alluvial Basins Regional Aquifer-Systems Analysis, Parts of Colorado, New Mexico, and Texas," US Geological Survey, Water Resources Investigation Report 84-4224, 61 pp.

Wright, D., 1978, *Natural Solar Architecture*, Van Nostrand Reinhold, New York, 245 pp.

CONSTRUCTION USES

AGGREGATES: MARKETS AND USES

HENRY N. McCARL

From the earliest use of crushed and broken stone as landfill for eroded gullies in trails and roads, to the mixing of sand and gravel with cement to make concrete, men and women have found mineral aggregates to be useful materials in satisfying human wants. Crushed stone, sand and gravel, volcanic cinders, pumice, manufactured lightweight aggregates, slag, cinders, expanded perlite and vermiculite, and a great array of recycled materials such as grog (broken or crushed brick) and fly ash have served as mineral aggregates. These materials have provided bulk and strength in portland cement concrete, bituminous concrete, road base, fill, concrete block, and plaster and stucco finishes.

Mineral aggregates provide many special characteristics such as compressive and tensile strength, surface textures, weight and density, thermal and acoustical insulation, abrasion resistance, and impermeability to various concrete products and mixes. While most materials used as mineral aggregates are consumed on the basis of their low unit values, the large volumes extracted in the United States each year make them one of the most valuable mineral resources in current production. In 1960, US production of construction aggregate was just over a billion tons per year (Meyer and Zelnick, 1991). Two decades of accelerated interstate highway, residential, and commercial construction brought the total annual production of mineral aggregates to roughly 1.4 Gtpy by 1970 and just over 1.6 Gtpy by the late 1970s. General economic and related construction slowdowns in the early 1980s saw annual US production of aggregates fall to below 1.1 Gt in 1982, and then rise to just over 1.7 Gtpy in 1990, a new peak in mineral aggregate production and use.

Per capita annual consumption of mineral aggregates was approximately 5.68 t/capita in 1960, 6.89 t/capita in 1970, rose to a high of 7.25 t/capita in 1979, and has since settled to just under 7 tpy for every man, woman, and child counted in the United States Census of 1990. The value of these materials came to approximately \$9 billion in 1990. During the period 1960–1990, roughly half of all mineral aggregates were used in the construction of residential, commercial, and industrial structures, and the remainder was consumed in highways, bridges, and other transportation facilities such as railway roadbeds, airports, and water-related projects (Meyer and Zelnick, 1991). During the period 1988–1992, public works (highways, waterway projects, road maintenance, etc.) represented closer to 60% of annual mineral aggregate demand with residential and non-residential construction approximately 40%. The growing market share of public works projects has been mostly due to the dramatic slowdown in residential, commercial, and industrial construction during the 1988–1992 period.

WHAT GIVES MINERAL AGGREGATES VALUE?

There are various characteristics of materials that make them useful as mineral aggregates. Bulk, weight, durability and compressive strength, inert chemistry or lack of reactivity, uniformity of composition, and special features such as color, insulating value, and texture, when employed in special uses.

BULK, WEIGHT, DURABILITY, AND COMPRESSIVE STRENGTH

Mineral aggregates are mostly used in portland cement concrete, concrete products, bituminous mixes, or as base or fill materials in highway construction, railroad ballast, waterway construction (dams and riprap to prevent erosion), or other construction uses. They are used as a relatively cheap component in concrete, mixes, base, and fill where they basically take up space (bulk) while providing compressive strength necessary in the final product. Aggregates used in portland cement or bituminous mixes must have surface properties that permit the chemical and physical bonding that holds these products together. The stronger these bonds, the higher the compressive strength and durability of the concrete or bituminous mix. It is highly desirable that concrete is both strong and durable for many years. This makes the construction both long lasting and relatively inexpensive to maintain.

Mineral aggregates may be used in many gradations and combinations of particle sizes. In base, fill, railroad ballast, and other uses, aggregates may contain a range of particle sizes or gradations in particle size in order to provide stability and high compressive strength when compacted. Aggregates may also be mostly large size pieces in uses such as riprap and jetty stone where they provide protection for river banks, harbors, and shorelines from the ravages of water erosion and storm damage. Regardless of particle size or gradation, aggregates should withstand weathering and physical compression to provide stability and protection for underlying materials and foundation support for overlying structures such as highway surfaces, bridges, and office, industrial, or residential buildings.

INERT CHEMISTRY OR LACK OF REACTIVITY

In concrete as well as in other uses, mineral aggregates should not deteriorate or cause the products containing them to deteriorate, discolor, or degrade. Components such as raw shale, clays, iron or other metallic sulfides or oxides are often responsible for reactivity and chemical changes that impact on the durability of concrete and other products containing mineral aggregates. The ideal mineral aggregate does not contain any deleterious substance. This is not always possible due to natural variations in the mineral and chemical composition of earth materials used as mineral aggregates. Specifications are generally set by state highway departments, the American Association of State Highway and Transportation Officials (AASHTO), American Society for Testing and Materials

(ASTM), American Railway Engineering Association (AREA), and other agencies such as the Federal Government, to limit the permissible content of deleterious materials in various mineral aggregate uses. Naturally the content of reactive or deleterious substances may vary with the intended use. Reactivity of some impurities may not be as deleterious in bituminous mixes as in portland cement concrete and concrete products. Discoloration may not be a problem when concrete is not visually exposed, and some substances may not react if they are protected from water and/or exposure to the atmosphere. The purity and necessity of strictly limiting the content of deleterious contaminants will vary depending on the use and combination with other components.

UNIFORMITY OF COMPOSITION

Specifications for mineral aggregates written by the ASTM, AASHTO, and other agencies, generally call for some type of uniform size or mineral composition. The size gradation and uniformity of particle size or mineral composition varies depending on the specific end-use of the aggregate. Portland cement and bituminous mixes are normally tailored for specific size gradations and aggregate compositions. Finer size gradations require more portland cement or asphalt, and this can dramatically change the final cost of the product when it is put in place. Uniformity in physical and chemical composition permits the design mix to maintain uniformity in the final product. Concrete and bituminous mixes are designed for certain size and compositions of mineral aggregate components. Silica sand, granite aggregate, limestone aggregate, slag, or trap rock aggregate are each likely to yield better final products with different optimal amounts of portland cement or asphalt. Mixtures of sand and coarse aggregate of different composition are often used with cement or asphalt content tailored to those specific mixtures. Repetitive procedures are simplified and costs are minimized if the aggregate components in concrete products are consistently uniform in composition.

SPECIAL FEATURES

In some uses such as decorative surfaces, exposed or exterior walls, or floors, features such as color, texture, or insulating value of the mineral aggregate are important. Architectural design generally specifies what features are desirable. Terrazzo not only requires certain color characteristics, but the final surface should also hold a polish. Marble or limestone do not wear as well in floors as granite or trap rock, but they do polish more easily and are often the materials of choice in terrazzo. Uniform texture is often important in walls, ceilings, and floors that will have a visual impact on the people who see them. Insulating values are often important in concrete block and lightweight concrete used in walls and floors. These special features normally carry a premium price, and mineral aggregates meeting these specifications are often more costly to produce.

SPECIFICATIONS

Some of the most important considerations in producing and selling mineral aggregates are the specifications that the products must meet to be acceptable for various uses. These are summarized in Table 1. Buyers of mineral aggregates want a dependable supplier who can produce a consistent product at the lowest delivered cost to the geographic point of use.

Table 1. Functional Uses and Related Specifications for Mineral Aggregates as Designated by the United States Bureau of Mines

Code	Use[1]	Specification
304	Concrete aggregate—nominal gradations from 88.9 mm (3.5 in.) maximum to 12.7 mm (0.5 in.) maximum	American Society for Testing and Materials (ASTM) specification C33
305	Bituminous aggregate—nominal gradations from 88.9 mm maximum to 12.7 mm maximum	ASTM specification D602
306	Macadam aggregate—nominal gradations from 88.9 mm maximum to 12.7 mm maximum	ASTM specification D693
307	Sense graded roadbase aggregate—nominal gradations from 38.1 mm (1.5 in.) maximum	ASTM specification D2940
308	Surface treatment aggregate—nominal gradations from 25.4 mm (1 in.) maximum	ASTM specification D1139
309	Unspecified construction aggregate and roadstone	n.a.
310	Riprap and jetty stone—heavy, irregular rock chunks used for riverbank, harbor and shoreline stabilization	Specifications of Corps of Engineers (formerly US Army Corps of Engineers) as appropriate
311	Railroad ballast—nominal gradations from 63.5 mm (2.5 in.) to No. 4 sieve (4.75 mm)	Specifications of American Railway Engineering Association (AREA) or equivalent ASTM specifications
312	Filter aggregate—any porous aggregate through which water is filtered	Federal (GSA) specification SS744a
313	Manufactured fine aggregate—(stone sand or crushed gravel) 100% passing 9.5 mm (3/8 in.)	ASTM specification C33
314	Terazzo and exposed aggregate—small chips or pieces of crushed stone or other aggregate 12.7 mm (0.5 in.) to 19.1 mm (0.75 in.)	n.a.

Sources: Schenck, G.H.K., and Torries, T.F., 1983, "Construction Materials: Aggregates—Crushed Stone," *Industrial Minerals and Rocks*, 5th ed., S.J. Lefond, ed., AIME, New York, pp. 60–61. Anon., 1991, *Annual Book of ASTM Standards*, American Society for Testing and Materials, Philadelphia, PA. Anon., 1990, *AASHTO Standard Methods of Sampling and Testing*, 15th Ed., American Association of State Highway and Transportation Officials, Washington, DC.

[1]Since specifications are written based on English units of measurement, the English units are presented in parentheses to clarify the origin of the metric measurements given in the table. While metric specifications are more common throughout most of the world, the United States still uses English units of measurement to formulate size specifications for mineral aggregates.

RELATIVE IMPORTANCE OF MARKET

The two major markets served by producers of mineral aggregates are public works (highways, dams, airports, waterway construction and stabilization, and other publicly funded construction) and private construction (residential and non-residential construction). Railroad maintenance and new railroad construction are nor-

Table 2. Mineral Aggregate Use Factors

Use	Estimated Aggregate Consumption (all types)
Interstate Highway	
Rural	
6-lane	60 000 – 72 000 tons per km
4-lane	39 000 – 45 000 tons per km
Urban	
8-lane	96 000 –105 000 tons per km
6-lane	75 000 – 81 000 tons per km
4-lane	51 000 – 57 000 tons per km
Primary 2-lane	
with bridges & interchanges	13 500 – 16 200 tons per km
w/o bridges & interchanges	72 000 – 9 000 tons per km
Secondary 2-lane	
Urban Roads	72 000 – 9 000 tons per km
Rural Roads	6 000 – 7 500 tons per km
Highway Maintenance (resurfacing & shoulders)	1 000 – 1 800 tons/lane km
Residential Housing	
Single-family Dwelling	230–250 tons per unit
Multiple Family Dwelling (High rise)	120–200 tons per unit

mally lumped with public works due to the similarity of product specifications. Table 2 gives some general guidelines on the quantities of mineral aggregates consumed in some of the major market segments. Quantities in other uses such as industrial and commercial construction vary greatly due to variations in methods of construction and requirements for parking. Quantities of aggregates used in the construction of parking lots parallel the guidelines indicated in Table 2 for different types of highway construction and maintenance.

Guidelines for mineral aggregate use cover aggregates used in ready mixed concrete, concrete blocks, concrete pipe, prestressed or preformed concrete construction units, road and foundation base, and other intermediate products containing mineral aggregate raw materials. Aggregate producers sell their output to the producers of concrete or concrete products, who in turn sell their output to the final contractors, builders, or agencies doing the actual construction work. The mineral aggregate producer is often vertically integrated into the ready mixed concrete or precast concrete products business, and under such circumstances, aggregate producers may be their own customers for intermediate products such as concrete block or concrete pipe. Competition in many aggregate markets often makes this vertical integration attractive in order to assure profitable markets for the mineral aggregate producer. In markets with limited competition or even a single producer, mineral aggregate prices are normally higher and construction costs and profitability commonly reflect this lack of competition.

Markets for mineral aggregates are heavily dependent on population density and the resulting demand for building and highway construction. Tables 3, 4, and 5, and related Figs. 1 and 2, present a recent picture of geographic mineral aggregate demand as reflected in the two major components: 1) sand and gravel and 2) crushed stone. Note that the production of mineral aggregates is heavily concentrated in those areas of the United States with the highest population densities. With the exception of unique situations, prices and output of mineral aggregate materials tend to be dependent on availability of deposits of raw materials and population-based demand. The availability of different types of transportation (especially waterways) also has a major impact on the location of production. Mineral aggregate producers located on major waterways and railroads, or with relatively easy access to

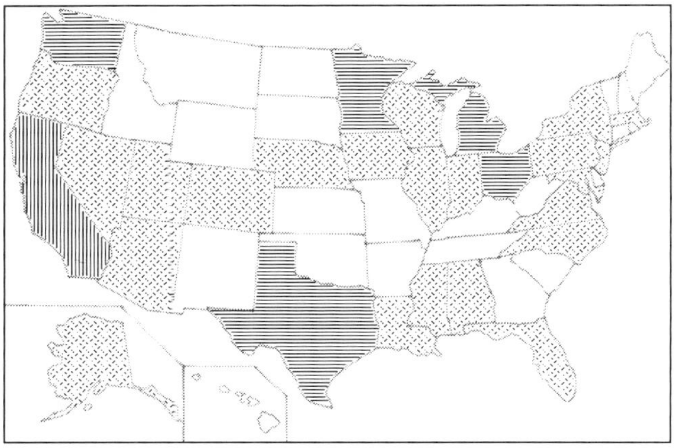

Legend
- Over 100 million tons
- 30 to 50 million tons
- 10 to 30 million tons
- Less than 10 million tons

Fig. 1. US sand and gravel production 1990.

Source: USBM, 1992, *State Mineral Summaries 1992*, US Department of the Interior, Bureau of Mines, Washington, DC.

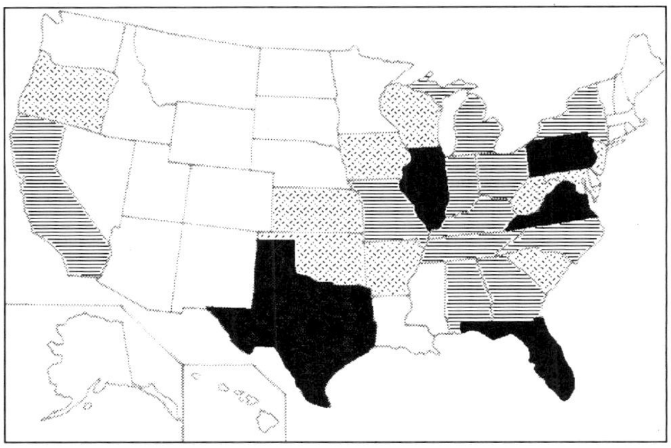

Legend
- 50 to 90 million tons
- 30 to 50 million tons
- 10 to 30 million tons
- Less than 10 million tons

Fig. 2. US crushed stone production 1990.

Source: USBM, 1992, *State Mineral Summaries 1992*, US Department of the Interior, Bureau of Mines, Washington, DC.

interstate highways are often able to save enough in terms of transportation costs to still be competitive in many markets. Such savings in transportation costs, combined with the lack or depletion of locally available materials, make it possible for crushed stone

Table 3a. US Sand and Gravel Production in 1990 (by State)

State	Tonnage (000)	Value $ (000)	Unit Value
Alabama	12 821	50 243	$3.92
Alaska	13 727	41 800	$3.05
Arizona	25 377	92 166	$3.63
Arkansas	8 785	35 475	$4.04
California	120 195	626 000	$5.21
Colorado	22 671	86 541	$3.82
Connecticut	7 765	37 943	$4.89
Delaware	1 985	6 967	$3.51
Florida	16 793	59 123	$3.52
Georgia	4 689	16 644	$3.55
Hawaii	398	2 459	$6.18
Idaho	8 384	25 590	$3.05
Illinois	29 436	104 728	$3.56
Indiana	21 708	76 886	$3.54
Iowa	13 594	46 432	$3.42
Kansas	9 875	24 170	$2.45
Kentucky	8 002	29 581	$3.70
Louisiana	13 263	55 902	$4.21
Maine	7 150	29 349	$4.10
Maryland	16 610	104 023	$6.26
Massachusetts	11 613	51 466	$4.43
Michigan	48 845	153 057	$3.13
Minnesota	30 790	77 502	$2.52
Mississippi	11 847	45 817	$3.87
Missouri	8 403	25 097	$2.99
Montana	4 649	14 319	$3.08
Nebraska	10 412	30 056	$2.89
Nevada	16 706	59 008	$3.53
New Hampshire	7 183	26 599	$3.70
New Jersey	12 602	64 245	$5.10
New Mexico	9 420	39 708	$4.22
New York	27 045	121 525	$4.49
North Carolina	10 666	44 872	$4.21
North Dakota	6 949	17 219	$2.48
Ohio	40 502	165 394	$4.08
Oklahoma	8 395	21 993	$2.62
Oregon	14 350	60 928	$4.25
Pennsylvania	18 985	97 348	$5.13
Rhode Island	1 790	9 042	$5.05
South Carolina	7 843	24 941	$3.18
South Dakota	8 808	23 689	$2.69
Tennessee	6 926	23 474	$3.39
Texas	41 894	158 080	$3.77
Utah	12 365	44 881	$3.63
Vermont	3 341	11 948	$3.58
Virginia	11 905	48 950	$4.11
Washington	36 592	133 067	$3.64
West Virginia	2 916	14 950	$5.13
Wisconsin	26 884	73 750	$2.74
Wyoming	3 935	14 446	$3.67
US Total 1990	827 789	3 249 393	$3.93

Source: United States Bureau of Mines, 1992.

Table 3b. US Crushed Stone Production in 1990 (by State)

State	Tonnage (000)	Value $ (000)	Unit Value
Alabama	32 818	202 400	$6.17
Alaska	2 455	19 800	$8.07
Arizona	4 818	13 500	$2.80
Arkansas	16 182	76 900	$4.75
California	38 636	200 600	$5.19
Colorado	6 909	36 100	$5.23
Connecticut	9 273	70 600	$7.61
Delaware	n.a.	n.a.	n.a.
Florida	67 273	317 400	$4.72
Georgia	48 182	317 300	$6.59
Hawaii	6 364	55 400	$8.71
Idaho	3 909	12 900	$3.30
Illinois	57 000	283 100	$4.97
Indiana	33 364	147 700	$4.43
Iowa	26 364	118 600	$4.50
Kansas	18 909	79 200	$4.19
Kentucky	45 545	182 900	$4.02
Louisiana	1 909	16 800	$8.80
Maine	1 545	8 700	$5.63
Maryland	27 727	163 900	$5.91
Massachusetts	8 364	54 500	$6.52
Michigan	39 182	129 000	$3.29
Minnesota	8 273	31 900	$3.86
Mississippi	1 273	5 500	$4.32
Missouri	48 273	190 900	$3.95
Montana	3 636	15 300	$4.21
Nebraska	3 636	21 200	$5.83
Nevada	1 455	5 000	$3.44
New Hampshire	545	2 500	$4.58
New Jersey	19 273	131 700	$6.83
New Mexico	2 182	12 800	$5.87
New York	36 273	207 600	$5.72
North Carolina	48 091	276 200	$5.74
North Dakota	909	4 600	$5.06
Ohio	44 000	190 900	$4.34
Oklahoma	23 000	89 500	$3.89
Oregon	16 364	86 600	$5.29
Pennsylvania	87 091	502 700	$5.77
Puerto Rico	7 545	46 500	$6.16
Rhode Island	1 455	8 800	$6.05
South Carolina	23 818	135 400	$5.68
South Dakota	4 364	16 800	$3.85
Tennessee	49 636	268 600	$5.41
Texas	74 364	285 700	$3.84
Utah	4 182	20 200	$4.83
Vermont	3 364	35 000	$10.41
Virginia	54 000	320 000	$5.93
Washington	n.a.	n.a.	n.a.
West Virginia	10 909	45 200	$4.14
Wisconsin	24 182	91 000	$3.76
Wyoming	2 000	14 000	$7.00
US Total 1990	1 100 818	5 569 400	$5.06

Source: United States Bureau of Mines, 1992.

producers on the Yucatan Peninsula in southern Mexico or in Newfoundland to be competitive as suppliers of mineral aggregates along the US Gulf Coast and in the major metropolitan coastal markets in the northeastern US from Norfolk to Boston.

SELECTION OF TRANSPORTATION METHOD

The sale of most mineral aggregates is based on the cheapest delivered price for material meeting the desired specifications. This may be modified by considerations of reliability of meeting delivery schedules, but normally the least expensive supplier wins the sale. Since transportation or haulage cost is a significant portion of delivered price, the cost of haulage to the point or area of use is normally the determining factor in the selection of a mineral aggregate supplier.

Truck haulage is the most common mode of transporting mineral aggregate. Contract trucking is the most flexible method of delivery to the customer. For distances up to 50 km between producers and consumers, trucks are the predominant choice. Trucks

Table 4a. US Sand and Gravel Production in 1990 (by Tonnage)

State	Tonnage (000)	Value $ (000)	Unit Value
California	120 195	626 000	$5.21
Michigan	48 845	153 057	$3.13
Texas	41 894	158 080	$3.77
Ohio	40 502	165 394	$4.08
Washington	36 592	133 067	$3.64
Minnesota	30 790	77 502	$2.52
Illinois	29 436	104 728	$3.56
New York	27 045	121 525	$4.49
Wisconsin	26 884	73 750	$2.74
Arizona	25 377	92 166	$3.63
Colorado	22 671	86 541	$3.82
Indiana	21 708	76 886	$3.54
Pennsylvania	18 985	97 348	$5.13
Florida	16 793	59 123	$3.52
Nevada	16 706	59 008	$3.53
Maryland	16 610	104 023	$6.26
Oregon	14 350	60 928	$4.25
Alaska	13 727	41 800	$3.05
Iowa	13 594	46 432	$3.42
Louisiana	13 263	55 902	$4.21
Alabama	12 821	50 243	$3.92
New Jersey	12 602	64 245	$5.10
Utah	12 365	44 881	$3.63
Virginia	11 905	48 950	$4.11
Mississippi	11 847	45 817	$3.87
Massachusetts	11 613	51 466	$4.43
North Carolina	10 666	44 872	$4.21
Nebraska	10 412	30 056	$2.89
Kansas	9 875	24 170	$2.45
New Mexico	9 420	39 708	$4.22
South Dakota	8 808	23 689	$2.69
Arkansas	8 785	35 475	$4.04
Missouri	8 403	25 097	$2.99
Oklahoma	8 395	21 993	$2.62
Idaho	8 384	25 590	$3.05
Kentucky	8 002	29 581	$3.70
South Carolina	7 843	24 941	$3.18
Connecticut	7 765	37 943	$4.89
New Hampshire	7 183	26 599	$3.70
Maine	7 150	29 349	$4.10
North Dakota	6 949	17 219	$2.48
Tennessee	6 926	23 474	$3.39
Georgia	4 689	16 644	$3.55
Montana	4 649	14 319	$3.08
Wyoming	3 935	14 446	$3.67
Vermont	3 341	11 948	$3.58
West Virginia	2 916	14 950	$5.13
Delaware	1 985	6 967	$3.51
Rhode Island	1 790	9 042	$5.05
Hawaii	398	2 459	$6.18
US Total 1990	827 789	3 249 393	$3.93

Source: United States Bureau of Mines, 1992.

Table 4b. US Crushed Stone Production in 1990 (by Tonnage)

State	Tonnage (000)	Value $ (000)	Unit Value
Pennsylvania	87 091	502 700	$5.77
Texas	74 364	285 700	$3.84
Florida	67 273	317 400	$4.72
Illinois	57 000	283 100	$4.97
Virginia	54 000	320 000	$5.93
Tennessee	49 636	268 600	$5.41
Missouri	48 273	190 900	$3.95
Georgia	48 182	317 300	$6.59
North Carolina	48 091	276 200	$5.74
Kentucky	45 545	182 900	$4.02
Ohio	44 000	190 900	$4.34
Michigan	39 182	129 000	$3.29
California	38 636	200 600	$5.19
New York	36 273	207 600	$5.72
Indiana	33 364	147 700	$4.43
Alabama	32 818	202 400	$6.17
Maryland	27 727	163 900	$5.91
Iowa	26 364	118 600	$4.50
Wisconsin	24 182	91 000	$3.76
South Carolina	23 818	135 400	$5.68
Oklahoma	23 000	89 500	$3.89
New Jersey	19 273	131 700	$6.83
Kansas	18 909	79 200	$4.19
Oregon	16 364	86 600	$5.29
Arkansas	16 182	76 900	$4.75
West Virginia	10 909	45 200	$4.14
Connecticut	9 273	70 600	$7.61
Massachusetts	8 364	54 500	$6.52
Minnesota	8 273	31 900	$3.86
Puerto Rico	7 545	46 500	$6.16
Colorado	6 909	36 100	$5.23
Hawaii	6 364	55 400	$8.71
Arizona	4 818	13 500	$2.80
South Dakota	4 364	16 800	$3.85
Utah	4 182	20 200	$4.83
Idaho	3 909	12 900	$3.30
Nebraska	3 636	21 200	$5.83
Montana	3 636	15 300	$4.21
Vermont	3 364	35 000	$10.41
Alaska	2 455	19 800	$8.07
New Mexico	2 182	12 800	$5.87
Wyoming	2 000	14 000	$7.00
Louisiana	1 909	16 800	$8.80
Maine	1 545	8 700	$5.63
Rhode Island	1 455	8 800	$6.05
Nevada	1 455	5 000	$3.44
Mississippi	1 273	5 500	$4.32
North Dakota	909	4 600	$5.06
New Hampshire	545	2 500	$4.58
Washington	n.a.	n.a.	n.a.
Delaware	n.a.	n.a.	n.a.
US Total 1990	1 100 818	5 569 400	$5.06

Source: United States Bureau of Mines, 1992.

offer the distinct advantage of delivery on relatively short notice and backhaul arrangements may reduce haulage charges in some cases. Materials can normally be delivered on an as needed basis from producer stockpiles at the pit, quarry or production site.

Rail haulage is often used for intermediate distances of 50 to 100 km when producers have access to rail connections and stockpiles can be maintained in close proximity to mineral aggregate consumers. Rail transport requires a siding at both the producing end and the consuming end of the haul as well as loading and unloading equipment and facilities. Delivery schedules are not as flexible as with trucking, and trucks are often required to make final delivery to the customer from the stockpile in or near the market.

Water transport by barge or special marine vessel is becoming more common for very long haulage from producer to consumer. This is generally the case in distances over 100 km where water access is available. The US Gulf Coast is now being served by a major crushed stone producer located on the Yucatan Peninsula in Mexico, with delivery to major US markets by a specially designed self-unloading marine vessel. Stockpiles are maintained at dockside in central locations from which delivery can be made by truck. Similar plans have been made for quarries in Newfoundland to supply US coastal markets in the Northeast. Weather can play an

Table 5a. US Sand and Gravel Production in 1990 (by Unit Value)

State	Tonnage (000)	Value $ (000)	Unit Value
Maryland	16 610	104 023	$6.26
Hawaii	398	2 459	$6.18
California	120 195	626 000	$5.21
Pennsylvania	18 985	97 348	$5.13
West Virginia	2 916	14 950	$5.13
New Jersey	12 602	64 245	$5.10
Rhode Island	1 790	9 042	$5.05
Connecticut	7 765	37 943	$4.89
New York	27 045	121 525	$4.49
Massachusetts	11 613	51 466	$4.43
Oregon	14 350	60 928	$4.25
New Mexico	9 420	39 708	$4.22
Louisiana	13 263	55 902	$4.21
North Carolina	10 666	44 872	$4.21
Virginia	11 905	48 950	$4.11
Maine	7 150	29 349	$4.10
Ohio	40 502	165 394	$4.08
Arkansas	8 785	35 475	$4.04
Alabama	12 821	50 243	$3.92
Mississippi	11 847	45 817	$3.87
Colorado	22 671	86 541	$3.82
Texas	41 894	158 080	$3.77
New Hampshire	7 183	26 599	$3.70
Kentucky	8 002	29 581	$3.70
Wyoming	3 935	14 446	$3.67
Washington	36 592	133 067	$3.64
Arizona	25 377	92 166	$3.63
Utah	12 365	44 881	$3.63
Vermont	3 341	11 948	$3.58
Illinois	29 436	104 728	$3.56
Georgia	4 689	16 644	$3.55
Indiana	21 708	76 886	$3.54
Nevada	16 706	59 008	$3.53
Florida	16 793	59 123	$3.52
Delaware	1 985	6 967	$3.51
Iowa	13 594	46 432	$3.42
Tennessee	6 926	23 474	$3.39
South Carolina	7 843	24 941	$3.18
Michigan	48 845	153 057	$3.13
Montana	4 649	14 319	$3.08
Idaho	8 384	25 590	$3.05
Alaska	13 727	41 800	$3.05
Missouri	8 403	25 097	$2.99
Nebraska	10 412	30 056	$2.89
Wisconsin	26 884	73 750	$2.74
South Dakota	8 808	23 689	$2.69
Oklahoma	8 395	21 993	$2.62
Minnesota	30 790	77 502	$2.52
North Dakota	6 949	17 219	$2.48
Kansas	9 875	24 170	$2.45
US Total 1990	827 789	3 249 393	$3.93

Source: United States Bureau of Mines, 1992.

Table 5b. US Crushed Stone Production in 1990 (by Unit Value)

State	Quantity	Value	Unit Value
Vermont	3 364	35 000	$10.41
Louisiana	1 909	16 800	$8.80
Hawaii	6 364	55 400	$8.71
Alaska	2 455	19 800	$8.07
Connecticut	9 273	70 600	$7.61
Wyoming	2 000	14 000	$7.00
New Jersey	19 273	131 700	$6.83
Georgia	48 182	317 300	$6.59
Massachusetts	8 364	54 500	$6.52
Alabama	32 818	202 400	$6.17
Puerto Rico	7 545	46 500	$6.16
Rhode Island	1 455	8 800	$6.05
Virginia	54 000	320 000	$5.93
Maryland	27 727	163 900	$5.91
New Mexico	2 182	12 800	$5.87
Nebraska	3 636	21 200	$5.83
Pennsylvania	87 091	502 700	$5.77
North Carolina	48 091	276 200	$5.74
New York	36 273	207 600	$5.72
South Carolina	23 818	135 400	$5.68
Maine	1 545	8 700	$5.63
Tennessee	49 636	268 600	$5.41
Oregon	16 364	86 600	$5.29
Colorado	6 909	36 100	$5.23
California	38 636	200 600	$5.19
North Dakota	909	4 600	$5.06
Illinois	57 000	283 100	$4.97
Utah	4 182	20 200	$4.83
Arkansas	16 182	76 900	$4.75
Florida	67 273	317 400	$4.72
New Hampshire	545	2 500	$4.58
Iowa	26 364	118 600	$4.50
Indiana	33 364	147 700	$4.43
Ohio	44 000	190 900	$4.34
Mississippi	1 273	5 500	$4.32
Montana	3 636	15 300	$4.21
Kansas	18 909	79 200	$4.19
West Virginia	10 909	45 200	$4.14
Kentucky	45 545	182 900	$4.02
Missouri	48 273	190 900	$3.95
Oklahoma	23 000	89 500	$3.89
Minnesota	8 273	31 900	$3.86
South Dakota	4 364	16 800	$3.85
Texas	74 364	285 700	$3.84
Wisconsin	24 182	91 000	$3.76
Nevada	1 455	5 000	$3.44
Idaho	3 909	12 900	$3.30
Michigan	39 182	129 000	$3.29
Arizona	4 818	13 500	$2.80
Washington	n.a.	n.a.	n.a.
Delaware	n.a.	n.a.	n.a.
US Total 1990	1 100 818	5 569 400	$5.06

Source: United States Bureau of Mines, 1992.

important role in water delivery reliability depending on the season and location of both producer and consumer. Newfoundland quarries would necessarily have a limited production season, and water delivery would be unlikely during winter months.

COMPETITION

Every market for mineral aggregates has at least one optimally located producer. When justified by market size or growth, and if suitable deposits of raw materials are available, there may be more than one producer. Most metropolitan areas have one or more producers. All major population centers have as many producers as available deposits and competition will allow. Prices normally reflect the circumstances and intensity of competition. Markets with one producer generally have higher prices.

Pricing is generally more competitive with larger numbers of producers. Producers constantly search for available deposits of raw materials and attempt to purchase or lease those deposits that are optimally located to serve existing or projected market demand.

Producers routinely consider the pricing strategies and policies of existing or potential competitors when determining their own prices. Geographic location of deposits and their proximity to the location of mineral aggregate consumers gives a competitive advantage to the nearest producer. Transportation costs are such an important part of the delivered price of mineral aggregates that the nearest producer generally makes the sale. There are instances where high production costs or lack of availability of local deposits make long distance haulage competitively advantageous, especially if the long distance source has low production costs or enjoys significant economies of scale. Crushed stone output is particularly sensitive to volume, and often demonstrates significant economies of scale for very large producers. It is simply cheaper to run a crushed stone or sand and gravel facility near its production capacity. This may lead to the accumulation of large seasonal stockpiles and marginal cost pricing (often called *dumping* by disadvantaged competitors) by producers of mineral aggregates in some markets.

Competition for major highway contracts is often intense, even in markets with a few producers. Producers will often establish temporary production facilities in order to have a competitive locational advantage in bidding on some highway construction projects. This may lead to price cuts by existing producers in order to prevent the incursion of new competition in traditionally served markets. Pricing, even below apparent production costs, has been known to occur to protect market territories and preserve market share. While this could be construed as cut-throat price competition and a possible violation of antitrust legislation, it is very difficult to prove due to the relatively low marginal costs of mineral aggregate production when operating near full capacity, as compared to operating at significantly less than full capacity. Existing large producers of mineral aggregates have even been know to provide financial assistance to medium or smaller sized producers to prevent incursion into their traditional geographic market territory by nearby larger competitors. Mineral aggregate markets display most of the competitive characteristics of the classic oligopoly market model.

FUTURE TRENDS

While US demand for residential and nonresidential construction may be reduced in the early 1990s from recent historical levels, there is likely to be increased activity in highway and other transportation related construction. This will shift markets more heavily to public works projects in the near term. The intermediate term for residential and nonresidential construction is clouded by the slow recovery of the US economy and the questions about sustained US economic growth. However, the long run outlook is bright for all types of construction, and a return to historical trends is likely by the turn of the century. This sets the picture for markets for mineral aggregate materials—slow but steady growth in demand, with increasing potential in those areas of more rapid economic growth.

There are likely to be continuing problems of urban encroachment on existing deposits of mineral aggregate raw materials, forcing longer distance hauling and increasing delivered costs in many urban markets. This will make foreign producers even more competitive in areas with potential marine delivery such as the US Gulf Coast and the Northeastern Coastal United States.

The overall economic outlook for mineral aggregate producers may be offset to some degree by the innovative use of worked out stone quarries as water reservoirs, storage sites, recreation facilities, or conversion to sanitary landfills. Reclaimed sand and gravel operations will become increasingly valuable as sites for residential and commercial development after extraction of the mineral values. Innovative thinking and land use planning can and will turn the potential sow's ear into the proverbial silk purse.

REFERENCES

Anon., 1990, *AASHTO Standard Methods of Sampling and Testing*, 15th ed., American Association of State Highway and Transportation Officials, Washington, DC.

Anon., 1991, *Annual Book of ASTM Standards*, American Society for Testing and Materials, Philadelphia, PA.

Anon., 1992, *State Mineral Summaries 1992*, US Department of the Interior, US Bureau of Mines, Washington, DC.

Barksdale, R.D, ed., 1991, *The Aggregate Handbook*, National Stone Association, Washington, DC.

Langer, W.H., 1988, *Natural Aggregates of the Conterminous United States*, US Geological Survey Bulletin 1594, US Geological Survey, US Government Printing Office, Washington, DC.

Meyer, D.A., and Zelnick, Jr., S.P., 1991, "Structure and Economics," *The Aggregate Handbook*, R.D. Barksdale, ed., National Stone Association, Washington, DC, pp. 2–1 to 2–17.

Schenck, G.H.K., and Torries, T.F., 1983, "Construction Materials: Aggregates—Crushed Stone," *Industrial Minerals and Rocks*, 5th ed., S.J. Lefond, ed., AIME, New York, pp. 60–61.

Tepordei, V.V., 1988–92. *Crushed Stone and Sand and Gravel*, Mineral Surveys, US Department of the Interior, US Bureau of Mines, Branch of Industrial Minerals, Washington, DC.

CONSTRUCTION USES

CEMENT AND CEMENT RAW MATERIALS

JOHN A. AMES, WILLIAM E. CUTCLIFFE, AND JOHN D. MACFADYEN

Webster's dictionary nearly equates portland cement with its current primary definition of cement. Although such equation may be a triumph of common usage, the confusion between the terms cement and concrete unfortunately is very widespread. Full definition of cement properly would extend to many substances, but for purposes of this chapter cement is restricted to varieties of portland cement including portland-pozzolan cement.

Among the usually recited historical notes concerning cement is that it was developed by the Romans. Their use of cement in the great structures of Rome, and even in the far corners of their Empire, such as Hadrian's Wall in the north of England, testifies to the antiquity of a major cement industry. From English sources we know of John Smeaton's carefully proportioned hydraulic cement mixed in 1756 and of Joseph Aspdin's patented cement to which, in 1824, he gave the name portland cement. (He claimed that his cement mortar looked like the famous dimension limestone quarried from the Isle of Portland on England's south coast.) The year 1971 marked the centennial of portland cement manufacture in the United States. A fair historical account of the many real contributors to the cement industry since 1756 unfortunately would be too extensive for this chapter. Development of rotary kiln practice and burning at higher temperatures clearly was the major move into today's technology.

Cement manufacture is the processing of selected and prepared mineral raw materials to produce the synthetic, mineral mixture (clinker) that can be ground to a powder having the specified chemical composition and physical properties of cement. Cement making incorporates a number of distinct steps, but it is characterized by the key pyroprocessing (or burning) step, which brings about the necessary changes in the raw materials. This pyroprocessing is a chemical process. Conversion of large volumes of mineral raw materials into specification-controlled cement at relatively low cost is the objective.

Statistics on the cement industry are of uneven quality. The US Bureau of Mines (USBM) is the principal collector of available data. These statistics can be only as good as the quality of information supplied by the reporting companies. The data perhaps most often questioned relates to productive capacity of the industry. The questions submitted by USBM are valid, nevertheless, and the information gained from them, as good as is available, is what is generally used. The Portland Cement Association also is a source of much quality information.

In analyzing cement industry production rates, it is well to keep in mind that measurement of raw materials requires the reader to recognize that approximately 1.6 t of dry raw materials are required to make 1 t of clinker. This number can vary depending on the amount of kiln dust wasted. Also that to produce 1 t of portland cement required .94 to .96 t of clinker, the remainder being gypsum or anhydrite. A ton of masonry cement or pozzolanic cement requires considerably less clinker.

RAW MATERIALS

Concrete is so widely used as a construction material that cement plants are spread geographically to serve most areas. The major parameters that orient plant locations within the cement industry are markets, transportation, raw materials, available fuel, and power. Standards are determined by each company for quality and quantities of raw materials it deems requisite for its needs; 40 to 100 years' supply is the usual amount sought for a new plant. Sites are dictated primarily by market and/or transportation considerations. Long distance transport of raw materials is avoided when possible. A number of exceptions to each such rule can be cited. Selection among raw materials that potentially could be moved to or obtained at each given site is determined by the economic and chemical advantages of each.

The trend to site new or enlarged plants near major waterways and easily accessible to interstate highways to reduce transportation costs continues. This has accentuated the installation of terminals in market areas served by a single large plant.

Objectives and Needs

The primary requirement for cement manufacture is a source of lime (CaO) generally available as calcium carbonate ($CaCO_3$). Calcite, a crystalline form of $CaCO_3$, is the mineral raw material most often used and is found naturally as limestone in many varieties or in limestone's close relatives, marble and marl. Other sources include shell (oyster or coquina) in which the organically secreted calcium carbonate is present mainly as aragonite. Recently, use of oolitic aragonite sands (Bahamian) as a cement raw material added another source material of aragonite mineralogy. Lime-rich slag and other industrial waste products occasionally have been adapted as primary sources of lime for cement in the United States. Anhydrite (and/or gypsum) has long been used at a few plants in the world to produce both cement clinker and sulfuric acid from calcium sulfate. In Russia, feldspars (calcium alumino-silicates) have been the primary source of lime in cement raw material for at least one plant.

Secondary raw materials are required to supply the balance of silica and alumina when the lime content of the primary material is higher than needed. Except in the case of white cement manufacture, iron in the raw materials also is sought or is added. Although the sources of these chemical components are diverse, both

Table 1. Raw Material Used in the Manufacture of Portland Cement Clinker in United States[1]

Types of Raw Materials	
Calcium Carbonate Sources	Alumina Sources
Limestones, including:	Shale
Lithified limestone	Clay and mud
Chalks	Loess
Marbles	Slag
Marls	Fly ash
Oyster shell (reef)	Bauxite
Coquina shell	Alumina process waste (red mud)
Aragonite sands	Staurolite
Slag	Mill fines (granite)
Silica Sources	Iron Sources
Sand, sandstone, or quartzite	Iron ores
Clay and claystone	Blast furnace flue dusts
Shale	Pyrite cinder
Loess	Mill scale
Slag	Fly ash
Mill fines	
Fly ash	

	Volumes Reported t*		
	1988	1989	1990
Calcium carbonates (primary sources)	94 081	94 924	97 506
Primary alumina materials	8 364	8 642	8 775
Primary silica materials	2 725	2 355	2 367
Primary iron materials	940	931	935
Blast furnace slag	0	0	0
Miscellaneous	0	0	0

Source: 1991 US Bureau of Mines Minerals Yearbooks, Cement Chapter, Table 5.
*Amounts in thousands of tons.
[1]Includes Puerto Rico.

from natural deposits and waste products, the ratio of silica/alumina (often also evaluated as silica/R_2O_3 with R_2O_3 including alumina plus the ferrous oxide value of the iron source) has to be closely controlled. Among the naturally occurring sources of silica are the sediments, sand, silt, clay, and loess, or their corresponding rocks, sandstone, siltstone, shale, or mudstone. Alumina sources generally are mud, clay, loess, and related rocks. Wastes that supply secondary components include slag, fly ash, red mud (from bauxite processing), and wash plant or mill tailings. Ash from the fuel (especially coal) used for processing the cement raw materials adds secondary components, sometimes in significant amounts. Table 1 lists types of raw materials used in the United States and provides a generalized statement of volumes used.

The considerable flexibility in selection of raw materials for a blended kiln feed lies mainly in the fact that cement making is a chemical process. Nevertheless, mineral and physical characteristics of the selected raw materials have a great deal to do with the costs and control of recovery, preparation, and processing of those materials. Powerful arguments can be made that impure limestone called cement rock, which requires virtually no addition of secondary raw materials, is preferable to the artificially blended materials, each of which in its own right may be relatively pure. However, the proverbial cement rock or limestone that contains the correct balance of lime, silica, alumina, and iron for even Type I cement is rare.

In addition to those raw materials that are blended to provide an acceptable kiln feed, a source of SO_3 is required. The SO_3 material is ground with clinker to provide predicable control of the setting times for concrete made with the cement. Although the clinker itself normally contains SO_3 and can be produced near levels required by specification, natural gypsum or a gypsum-anhydrite source of SO_3 is preferred, and, in some quantity, gypsum is mandatory. The need to meet environmental regulatory constraints on SO_2 emissions and the adoption of pre-heater technology in conjunction with roller mills for drying and grinding the raw materials, has raised inherent SO_3 levels in clinker and has constrained the addition of gypsum. It is anticipated that synthetic gypsum from SO_2 scrubbers on power plant stacks will replace some natural gypsum in portland cement.

Portland cement masonry mortar includes additional materials interground with clinker, generally dolomite or limestone, more rarely lime or other materials. Portland pozzolan cements and slag cement, of course, contain interground pozzolanic and slag materials.

Raw materials and fuels introduced into the cement kiln contain components other than those sought. These components are permitted, within specified limits, and certain of them, in some ways, are beneficial. If present above defined levels of concentration, these unsought components may constitute intolerable impurities. Magnesium compounds perhaps are the most familiar and ubiquitous of these. At low levels of concentration magnesia's (MgO) role as a fluxing agent is beneficial. As long as it is present in low concentrations, usually less than 5% ignited basis, magnesia is considered innocuous. As its concentration in the cement increases, however, it becomes an intolerable impurity because certain magnesium compounds, such as periclase, cause expansion and result in disruption of concrete. Other uninvited components are comparable in cement in that they are acceptable at lower but intolerable at higher concentrations.

Evaluation

In the areas of the United States where new cement plant sites are considered, the general geology of surface deposits is well enough known so that the process of exploration perhaps would be better characterized as evaluation. To be sure, the job of location still must be accomplished, but prudent selection among alternatives is a better description of the task and the objective than the term exploration implies. The complex set of factors to be evaluated importantly includes plant design requirements necessary for the recovery, handling, and processing of the specific raw materials available to each site. Zoning and environmental considerations must be understood and their requirements satisfied at every operating location. Potential profitability still must stand as the final arbiter. Decisions concerning opening new plants or modernizing and expanding old ones obviously must be based on hard, reliable data indicating adequate raw materials available at affordable costs. Cement plants are closed for many reasons, including the depletion of local raw materials, but that factor perhaps is the one most controllable by prudent selection of the plant site.

Recovery Steps

The winning or recovery of the mineral raw materials for cement manufacture adds up to a surprisingly large volume of mining activity. Roughly 100 Mt of raw feed (expressed in dry terms) went into US kilns to produce cement in 1990. This feed weight does not include fuel, water, or air. In the processes of recovery and reclamation of cement raw materials, the tonnage of overburden handled each year may well equal the amounts of raw materials reported used. Open quarry methods typically are employed to produce the required cement raw materials. Usual steps in quarry operation include:

Clearing and stripping
Drilling
Blasting and breaking
Loading
Transportation (hauling) to the crushing department
Reclamation

Exceptions to this sequence can be cited for various operations, and the relative cost of each step may differ greatly from one operation to another. Some quarries have so little overburden that stripping is not necessary. In other quarries as much as 6 t of overburden must be stripped for every ton of limestone recovered. Drilling, blasting, and breaking may be required to win hard rock in conventional limestone quarries or, at some plants, these methods may be eliminated because softer materials can be recovered by ripping and/or scooping operations. Loading is accomplished by a variety of equipment ranging from electric or diesel shovels or draglines to the numerous types of end loaders and scooploaders' Truck hauling is the most common means of transporting stone out of or within the quarry, although belts, skip hoists, rail cars, and wheeled loaders (pick and carry use) are important among the wide variety of equipment used for transportation to the crusher.

The relatively few plants that depend on underground mining have dwindled in number. Compared to most open quarries, underground mining predictably is more costly. Plants whose lime requirements are met by materials other than limestone (limestone defined broadly to include rock types from marble through sedimentary limestone to marl) continue to decline. Shell beds (mostly oyster reefs up to many thousands of years old) were sources of cement raw material. Although some of the small reefs are buried in mud and support no live oysters, some reefs do, and recent pressure by environmentalists to force dredgers to stop disturbing living oysters has made shell no longer an acceptable cement raw material. The only cement plant using aragonite (dredged off the Bimini Islands) has been shut down. Operation methods range from dredging to modified quarrying to recover shell or other poorly or nonconsolidated carbonate materials.

PROCESSING

Fig. 1 illustrates the classic steps in the manufacture of portland cement. Fig. 2 illustrates the modern dry-process pre-heater calciner technology that has replaced the long dry or wet process kiln. The process steps in manufacturing generally are visualized as being those which physically or chemically change the materials as they progress from their raw condition to the finished product. From the first step of recovery to the finished cement, much of the processing involves the physical changes of size reduction and proper blending and mixing. The chemical changes brought about in the kiln most importantly include calcining of calcium carbonate to calcium oxide (lime) in an environment that promotes the lime's reaction with other components to form the synthetic silicate minerals and glasses that comprise clinker. While size and shape of clinker particles are of great range, a typical range is buckshot to golf ball size; shapes of larger particles are rounded by rolling in the kiln and tend to be roughly spherical. The processes of clinker grinding primarily are physical although chemical changes are involved.

The whole scope of processing is difficult to describe in generalities. Each plant is uniquely designed in an effort to apply the best process art and equipment available at the time of that plant's construction to make, economically, products of competitive quality from its particular raw materials. Cost, type of fuels available, power and opinions of what is best, as well as environmental considerations, obviously also influence which process flow and equipment are designed into a given plant.

Processing During Recovery

Recovery of the raw materials is a first step in processing. In contrast to preparing various size grades of crushed stone aggregates, the objective in processing cement raw materials is to arrive at the fine kiln feed sizes as quickly and economically as possible. Drilling, blasting, and breaking are done so as to achieve size-reduction results consistent with the objectives of fines production and economy. In many quarries, care is taken to recover the raw materials in distinct chemical grades to maximize the recovery and control undesirable impurities, such as alkalis and sulfur. Beyond these objectives, quarry practice is limited mainly by considerations of the community environment. "Let the explosives do the work" is an often cited rule; however, use of explosives is modified by safety, noise, and vibration standards. Work is the term descriptive of the energy applied to do the job of size reduction, and the processes through which work is applied again are a matter of plant design and practice. Quarrying, in the cement industry, most often is described to extend from breaking ground to the delivery of quarried materials into the crushing department. In many quarries blending for quality control begins with selection of working faces because the chemistry of the stone in each face is to some extent known. A variety of methods of quality monitoring is applied. Practice at some quarries includes close diamond-drilling surveillance as quarry faces progress; other deposits are more predictable.

Depending upon the raw material and the design of the milling system the crushing system might be elaborate or it may not be present at all. The crushing process is expected to produce material of a given maximum size. Screens generally determine this crusher-run size, which typically is 10 mm up to 65 mm. Primary and secondary crushing stages, sometimes followed by a tertiary crushing stage, produce the crushed product for the mills. Virtually all varieties of crushing machines found in crushed stone plants are also found in the crushing departments of cement plants. Size grading is accomplished mainly to recycle coarse fractions for further crushing, to remove properly sized material from the stream and move it on to the next process, or to facilitate transportation. In a few

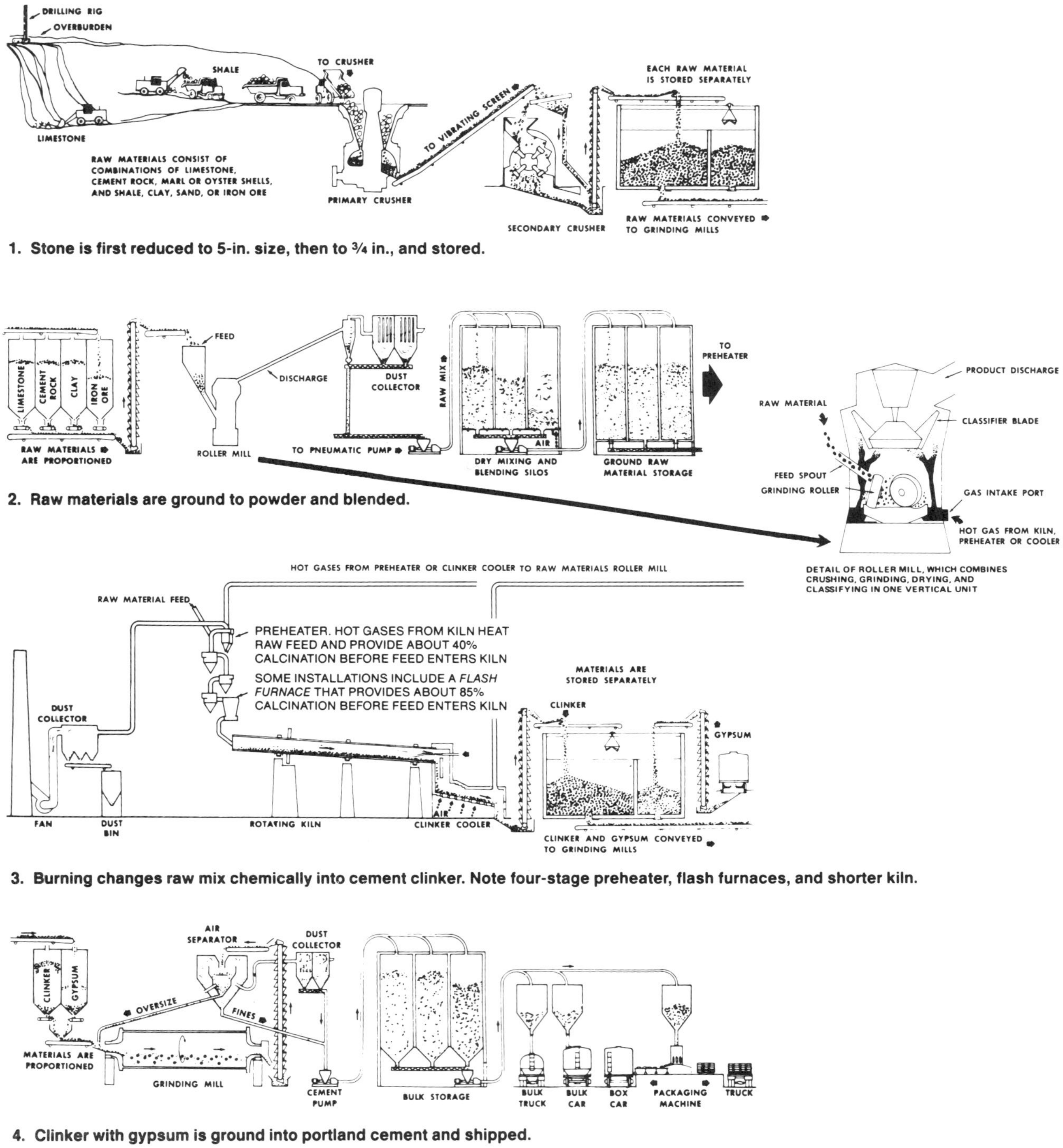

Fig. 1. Steps in the manufacture of portland cement (*source:* Portland Cement Association).

instances discard of some or all of certain screened fractions is indicated for control of chemistry. Crushed stone may move directly to bins from which mills are fed, but more often it moves to storage buildings and pre-blending piles then to the bins. The stone storage and the bins provide additional opportunities for blending.

Beneficiation is employed at some plants. Beneficiation may be defined as a process step by which an undesirable component of the raw material is taken out through flotation or by an equally sophisticated technique. That type of beneficiation is employed at very few cement plants. However, the elimination of undesirable components from the recovered materials may be as simple as drying to eliminate water or as complex as roasting to drive off

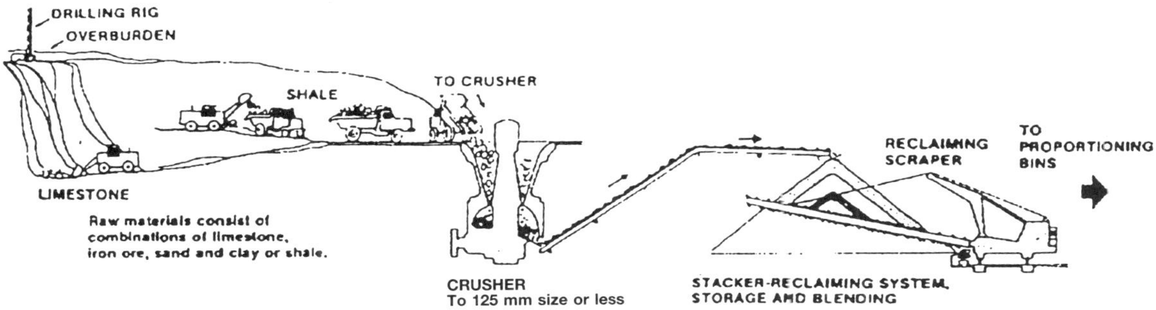

1. Quarrying and blending of raw materials.

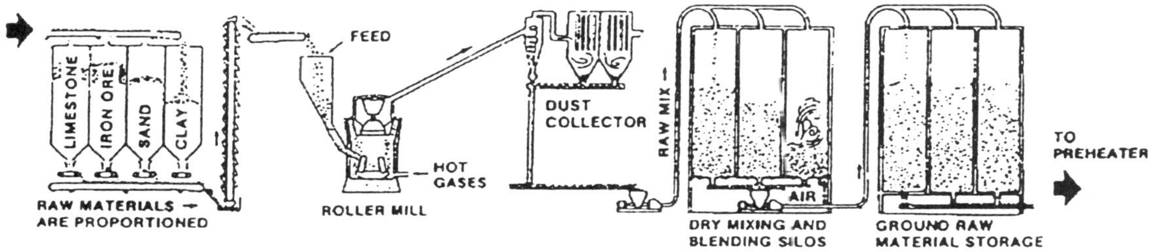

2. Proportioning and fine grinding of raw materials.

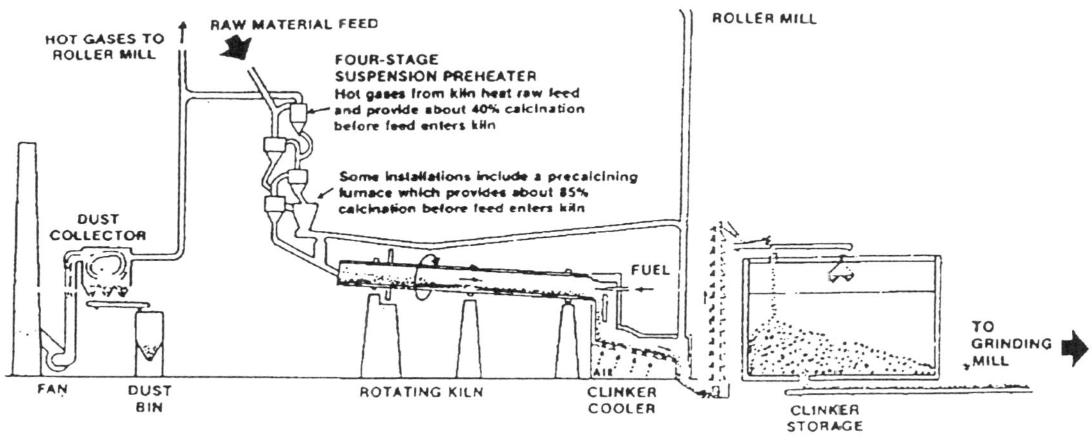

3. Kiln system. Preheating, burning, cooling and clinker storage.

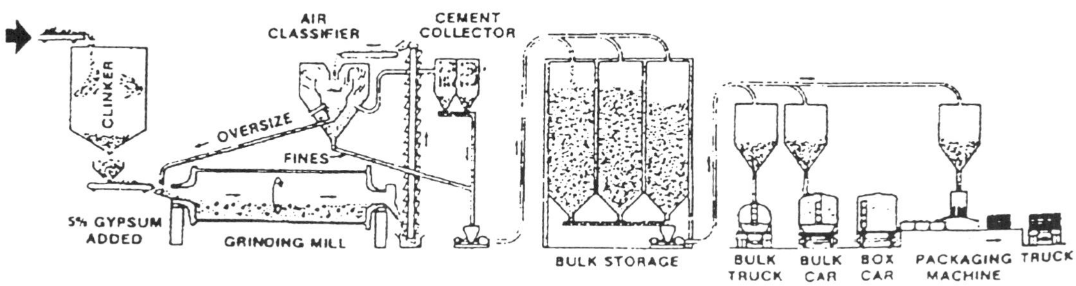

4. Finish grinding and shipping.

Fig. 2. Steps in the manufacture of portland cement by the dry process using a preheater.

kerogens, and this later case has to be regarded as a type of beneficiation. Screening, washing, or other methods of segregating unwanted materials for discard also is benefication. The term beneficiation may be expanded from the general upgrading of raw material to include the addition of relatively pure materials needed to bring the mix to the desired quality. A large number of cement plants employ this type of beneficiation.

Raw Milling

The primary objective of milling is to prepare sizes and mixtures of raw material for kiln feed that effectively and economically can be processed in the kiln to produce the desired cement. Standards of maximum size and percent through a given screen (say 90% through a 200 mesh screen with none coarser than 50 mesh) are set for the finished raw mill product. Determination of these standards usually is empirical. A sizable fraction of the mill-feed stream may occur in nature in qualifying finenesses, but no plants eliminate milling as a process. The mill acts as a superior mixing device in addition to accomplishing size reduction.

Historically, the early plants all were dry process. As the technology progressed a number of factors led to the development of wet-process plants, prime among them being wet raw materials, the efficiency of wet blending, and the then low cost of fuel. Wet marl, shell and other raw materials can be handled and processed more conveniently without a drying state. For many years the wet-process dominated the design of new plants. Recent upward surges of fuel costs provide strong incentive for the designing of new plants as dry process. Fairly recent developments in drying-grinding mills and heat recuperating preheaters, added to available, efficient dry-blending systems, go far toward removing historical objections to dry-process installations. Conversion of existing wet plants to dry plants poses more difficult problems, and economic justification of this move is not automatic.

Raw grinding practice varies rather widely, with the biggest difference between wet and dry milling. Nearly all wet plants employ mills ranging from rod or rod-peb to ball and tube mills. Water is added directly to the mills or may be introduced as a clay-wash mill product (clay slip). Wet milling produces a slurry of the ground kiln feed in which the water content is kept as close as possible to the minimum (usually 30 to 45%) that can be pumped and handled successfully. Nearly all wet milling systems employ positive sizing equipment from conventional vibrating screens to DSM screen installations, hydroclassifiers, hydroseparators, and thickeners. Coarse fractions are returned for further grinding and finished fractions are moved out of the milling system. Circulating loads are large. The slurry is pumped to storage tanks or basins, where blending is achieved. Mechanical and/or air agitation are main methods of keeping the slurry mixed. Closely controlled and variable kiln feeding is very important. The preferred method uses a magnetic flow meter and V-Ball valve in closed loop with a set point controller and slaved to kiln speed. The ferris wheel feeder is obsolete.

In traditional dry plants where raw material moisture exceeds 5%, one or more of the materials are dried between the stone storage area and the raw mill-feed bins. Rotary dryers are the most widely used. They are fossil fuel fired, but may also use recuperated kiln heat. During grinding the material moisture content is usually reduced to 0.5% by the introduction of hot air either from a fossil fuel fired hot air generator or recuperated kiln heat. In modern dry process plants the entire drying requirements are accomplished by combined drying-grinding mills. The variety of mills is greater in dry process plants, where vertical, roller, and ball race mills are employed as well as ball mills and tube mills. The preferred modern milling system is the air swept vertical roller mill with its integral classifier. Its drying capacity is highest, uses kiln and/or clinker cooler gases for drying and it tolerates feed up to 100 mm in size. In ball mill circuits air separators classify the milled product and return the coarse fraction to the milling system. The circulating loads typically are large.

Dry blending is accomplished in a number of types of systems. A plant designer now has several options other than the ubiquitous and trade marked Airmerge silo which uses large quantities of compressed air to fluidize, agitate and mix the incoming feed. These alternatives include the huge storage silo with computer controlled withdrawal from multiple ports to the inverted cone discharge systems with computer controlled radial withdrawal to multiple small capacity initial bins which feed a large secondary silo. These alternatives replace the high operating cost of the compressed air based blending system with high capital cost systems which may fail rigorous economic evaluation but nonetheless provide other attractive advantages. Positive feeding of the prepared blended dry kiln feed with weigh feeders or flowmeters with calibration bins completes the system.

The preheaters now used reclaim heat recuperated from the kiln gases by directly bringing these hot gases into contact with the prepared raw feed in a series of 2 to 6 stacked cyclones. The combination of counter and concurrent flow of the raw materials within these cyclones as the gases move up results in heating these raw materials up to 800°C before they enter the kiln. This European innovation matured in the 1970s. The 1980s has seen the introduction of 5 and 6 stages of cyclones and the adoption of low pressure drop cyclone design.

Although preheating generally results in partial calcining of carbonates, the 1970s also witnessed the development of true precalcining systems. Enough heat is applied in the precalcining equipment to drive off nearly all of the CO_2 before the materials are introduced into the rotary kiln. A number of precalcining systems are now in operation in US, Canadian, and Mexican cement plants. These precalcining units are appreciably more complex than traditional clinker-producing equipment, but they achieve higher rates of production for the same invested capital (also less expense for refractories). However, preheating and precalcining systems may cause plugging problems because of chlorine, sulfur, and alkalis. Compounds are formed by these impurities which become liquid at the temperatures in the lower preheater vessels, causing build ups which plug the ducts connecting the vessels. Understanding the alkali processing relationships is critical to the evaluation and selection of raw materials and fuels for successful precalcining operations. A primary objective in these new designs is the more efficient use of fuel.

Another preheater technology requires that the kiln feed be pelletized (requiring addition of some water). The pellets are preheated on a traveling grate prior to being introduced into the kiln. This preheater technique has also been applied where wet raw grinding is used. The blended slurry is dewatered using filter presses, the filter cake then fed to a twin roll pelletizer to produce small cylindrical pellets which are fed to the traveling grate.

In many plants, collected kiln dust is part of the raw material. In the preheater process where the hot gases are used to sweep the roller mill or ball mill (for drying the raw materials), kiln dust is intimately mixed with fresh raw feed and eventually is returned to the kiln. In older, long dry or wet-process kilns, some or all of the dust captured from the kiln gases is returned to the kiln with the fuel stream (insufflation) or through controllable feeding devices into the feed end of the kiln or through dust scoops downstream of the kiln chains. Where an electrostatic precipitator is used to clean the gases, the dust from the downstream sections typically contain higher levels of alkalis. This permits alkali reduction by wasting the alkali-rich dust. Although there are no US plants now doing so, dust leaching has been developed to remove these undesirable alkalis from the dust permitting the return of all the dust.

A recent development combines the leaching of alkalis from kiln dust with removal of sulfur dioxide from power plant gases. The system uses kiln dust as the sulfur-absorption medium and processes the reacted dust to produce sodium sulfate, potassium sulfate, and calcium sulfate. The cleaned power plant gases are discharged to atmosphere and the kiln dust returned to the cement plant as kiln feed or re-used by the power plant scrubber.

Burning

Pyroprocessing (burning) is the key process in the manufacture of cement. Burning at high temperatures causes properly constituted and prepared raw materials to react and combine to produce clinker containing a balance of synthetic calcium silicate compounds that are ground into the desired cement. Rotary kilns produce all domestic clinker. Worldwide practice includes vertical kiln installations of many varieties and a number of other types of plants. Raw material variations range from use of gypsum (or anhydrite) in plants that make coproduct cement and sulfuric acid to use of calcitic feldspar or lime, rather than calcium carbonate, as a primary raw material. The kiln and its clinker cooler are the heart of the cement plant. Basically everything else is sized to serve it, and it should be run as near as possible to capacity. Plant capacity expansion or reduction implies changes in kiln capacity.

Rotary kilns essentially are steel cylinders, ranging up to 7.6 m in diameter and 230 m in length and lined with fairly thick refractories (mostly brick). One function of the refractories is to develop a coating of adhering raw materials, especially in the hot zone, so that finished materials are processed over similar materials. Kilns are inclined at 4 to 5° from the horizontal so that their rotation (usually at 70 to 90 rph in long wet or dry kilns and 120 to 180 rph in preheater kilns) moves the materials being processed from the feed end to the discharge end of the kiln at the desired rate of speed. Retention time of the processed material in the kiln is is 2 to 3 hours in a long wet or dry kiln and 15 minutes in a preheater kiln. The burning zone in which the calcium silicates are formed occupies only about the last 20 to 30 m of the kiln length. Variable factors that influence kiln performance are legion, and their interaction perhaps is becoming better understood and more predictable. The effects of each variable and of changes in variables often cause kiln performance to be cyclic. A balanced, smooth operation is an objective in kiln operation. The calciner and preheater kiln system, which is much more stable, has solved many of these problems, and results in much improved refractory life.

Fuel is introduced into the rotary kiln under slight pressure through a burner pipe. The burner pipe is positioned in the product discharge end of the kiln, and the fuel blown through the pipe is ignited into a flame, which extends well up into the kiln. The thermal inertia in the burning zone is enormous and results in a stable flame even when burning low-volatility fuels. Powdered coal and coke is the preferred kiln fuel for economic reasons. No.6 fuel oil or natural gas are used depending on cost and availability. Several kilns use wastes as fuel, particularly wet process kilns. These wastes include used lubricating oil, spent solvents, waste paints, waste printing inks, chlorinated hydrocarbons and other organic compounds classified as hazardous by the US Environmental Protection Agency. Other ignitable wastes such as whole or shredded tires, rice hulls, and nut shells are used. The driving force of these alternate fuels is economics.

As the materials being processed move from the feed end of the kiln toward and into the zone of the flame, they pass through a series of stages. These stages include drying to lose water, calcination of the carbonates (i.e., the driving off of CO_2), and clinkering or burning. In the drying zone the remaining heat in the kiln gases heat incoming raw material. In both long wet and dry kilns chain sections are used as heat exchangers. The chain links are alternately heated by the hot gases and then heat the raw materials as the kiln rotates. In the calcining zone, the raw material is heated to about 1 000°C at which temperature the carbon dioxide is driven off from the limestone forming lime (CaO). This is necessary in order to provide the main chemical for the subsequent formation of cement clinker.

The calcined raw material then moves into the clinkering zone. This zone is the heart of the pyroprocess in which the incipient fusion, partial melting, and reaction to form clinker compounds takes place. Reaction (burning) temperatures begin at about 1 190°C with the formation of calcium alumino ferrite, proceed to 1 300°C with the formation of tricalcium aluminate and reach 1 500°C with the formation of dicalcium silicate and tricalcium silicate. The clinker then exits the kiln. The clinker compounds formed in cement kilns have been intensively investigated and described. The Onoda Cement Co. (Japan) was a leader in developing a unique method of analyzing clinker microscopically to assist in kiln operation and improvement in clinker quality. Although the Onoda method is approximate, it is quite practical and rapid. The brief statement descriptive of these compounds by Clausen (1960) is repeated here as Table 2.

Gas exit temperatures from long dry process kilns of 850 to 900°C are much higher than exit temperatures from wet kilns (200 to 300°C). The thermal economy of dry kilns is thus improved markedly by the addition of a preheater. Exit gas temperatures from preheater kilns varies from 500°C (2 stage preheater) to 275°C (6 stage preheater). Gases above 275°C are required for effective drying of the raw materials in combination drying-grinding mills such as the vertical roller mill.

Successful kiln burning produces clinker that may form the classic rounded clinkers or that may be sand- or dust-sized. The clinker leaving the kiln is typically at or above 1 100°C and in US and Canadian plants is cooled in either a reciprocating grate cooler or a planetary cooler. On a worldwide basis there are a wide variety of ways to cool clinker—from primitive pits to rotary units to the sophisticated systems used in North America. The preferred technique is the reciprocating grate cooler in which moderate pressure air (up tp 1 100 mm water gauge) is forced up through the deep (900 mm) bed of hot clinker as it slowly moves over the perforated plates. An objective of clinker-cooler design is the efficient reuse of the escaping heat and the proper cooling of the clinker to achieve good quality. The cooling and tempering of clinker is critical to the achievement of subsequent cement strengths. Improper cooling can result in the formation of undesirable clinker compounds with little if any hydraulic properties. The preheating of combustion air is achieved in the hot end of the grate clinker cooler. Modern designs have greatly improved fuel efficiency. A development of the 1980s has been the cooling of the waste air from the clinker cooler in air-to-air heat exchangers before passage to dust removal systems. Several plants have also adopted recirculation of this cooled and cleaned air to the undergrate fans, thereby eliminating an emission point and achieving a small improvement in heat recuperation.

Finish Grinding

Beyond the cooler, clinker generally is moved by conveyors into storage, where it may be segregated, tested, blended, and moved into bins for feeding to the cement (finish) mills. The type of clinker storage available and the timing of clinker grinding both have quality implications. Clinker that was ground hot and old clinker that has become hydrated each produce poor quality cement.

The SO_3 content of finished cement is critical in providing the desired setting time for concrete made with that cement. Gypsum (3 to 6%) traditionally has been interground with clinker to provide this SO_3. Anhydrite and other sources of SO_3 have been found satisfactory substitutes for part, but not all, of the gypsum.

Table 2. Probable Cement Compounds

Compound	Abbreviation	Common Abbreviation	Oxide Composition	Stoichiometric Composition	Approx. Content in Type I Portland Cement, %†
Tricalcium silicate (Alite)	3CS	C_3S	$(CaO)_3SiO_2$	Ca_3SiO_5	45
Dicalcium silicate (Belite)	2CS	C_2S	$(CaO)_2SiO_2$	Ca_2SiO_4	27
Tricalcium aluminate	3CA	C_3A	$(CaO)_3Al_2O_3$	$Ca_3Al_2O_6$	11
Tetracalcium-aluminoferrite*	4CAF	C_4AF	$(CaO)_4(Al_2O_3)(Fe_2O_3)$	$Ca_4Al_2Fe_2O_{10}$	8

Source: Adapted from Clausen, 1960.
*This composition of the iron-containing phase is an approximation and may range from $(CaO)_2Fe_2O_3$ to $(CaO)_6(Al_2O_3)_2(Fe_2O_3)$.
†Commercial cements contain 4–6% gypsum or anhydrite (for regulation of the "setting time" of the concrete), approximately 0.5% each of alkali oxides (Na_2O and K_2O) and uncombined CaO, plus a few percent impurities, largely MgO.

Clinker is ground into cement in ball mills in North American plants. In other parts of the world clinker grinding occurs in rod, ball, vertical roller, and ball race mills. In the US and Canada, all cement grinding systems are closed circuit systems in which air separators provide classification. This is necessary because of strength requirements dictated by the marketplace requiring cement typically ground to not less than 3 300 Blaine and more often 4 000 Blaine. High early strength and specialty cements are ground to 5 000 to 7 000 Blaine. A major development in the 1980s has been the adoption of high efficiency classifiers pioneered by the Onoda Cement Co. (Japan) and the development of a pregrinding clinker crusher. This device, known as a roll press, places the clinker under intense pressure between two counter rotating rolls causing fractures along grain boundaries. In the subsequent grinding in a ball mill, much higher throughputs are achieved for a given power input. The benefit of both of these developments is a reduction in total grinding power of 30 to 35% at the typical Blaine fineness of Type I cements. The reduction is twice as much on high fineness cements such as Type III.

Careful proportioning of clinker and gypsum feed to the mills is a requisite, and the additions of air-entraining agents or other small amounts of ingredients must be accurately metered. Good practice includes cooling of cement enroute to storage. The product is sampled and its quality is verified before it is loaded for shipment. Concrete storage silos generally are the major structures of the storage and shipping department, and that department is a significantly large capital cost factor in the whole plant. A feature of US and Canadian cement plants is the very small proportion of cement shipped in bags. Several plants have no provision to ship cement in bags. Several plants ship a majority of their output as either cement or clinker by barge or ship to terminals (silo type storage warehouses) and satellite grinding plants.

Dust

Cement processing requires the creation and handling of fine mineral materials. Problems arise through the escape of these materials as dust. Dust collection, for many years, has required elaborate and costly equipment. Within the plant most dust-generating points can be equipped with adequate dust collectors; however, the big problem is the cleaning of kiln exit gases. Mechanical collectors (including glass bag installations), electrostatic precipitators, and wet scrubbers all have their own advantages and disadvantages. With legal requirements for collection becoming more demanding, technology gives little assurance that even new plants will operate fully in compliance with the standards. Particulate emissions standards have become so stringent that kiln and mill particulate control systems must typically achieve 21 mg per dry standard cu m in order to comply with Best Available Control Technology (BACT).

MANUFACTURING CONTROL

In the broad sense, manufacturing control might be related to nearly every calibration, screen size, speed of drives, and so on, but the term now has come to imply the systems of electric devices that have been developed to control cement processing. Control capabilities have greatly improved in the last 20 years, and controls perhaps more refined than are justified may be imposed on some process steps. Safety of men and equipment has high priority in the establishing of safeguards of a control nature.

Historically, cement manufacturing control developed through the use of instruments to help guide the operators. Methods of installing the instruments and watching, recording, and reacting to what they sensed were improved, along with the understanding of the information that they provided. From these bases came improvements in instruments, their housing, and finally, in centralized control systems and practices. Experience within and outside the cement industry prompted the development of automation and the imposition of computer control.

Objectives

The manufacturer must justify expenditures for manufacturing control on the basis that it improves his costs and his products. Process control objectives relate to specification limits, which define chemical and physical product characteristics, and to the processes used to achieve them. Plant standards further narrow the limits for each specific operation. These parameters provide the reference framework for each process step and define the limits that the control sensors bring under surveillance and within which they call for action. Along with quality improvement, savings on labor, fuel, and power cost, as well as on equipment maintenance, are prime objectives. Effective controls are expected to accomplish smoother and more continuous operations to produce more cement as well as improve brick and chain life.

Modern laboratories can be equipped and staffed to provide quick information on sampled materials. Samples may be taken by continuous, online devices or be collected by more traditional hand methods. Computer evaluation of sample analyses (commonly determined by X-ray spectrophotometer) at many plants provides information in unprecedented depth. Many plants now also have atomic adsorption flame photometers for quick determination of metals.

Devices and Systems

Control devices and sensors usually function to determine temperatures, pressures, speeds, rates of flow, weight, fullness (or emptiness), presence or absence of a component, and so on. A great variety of instruments is used. Closed circuit television in some plants provides better than direct views of processes. Analog and digital computers are part of some systems. Computerized evaluation of sensing and printouts of results can be achieved, although computer determinations more frequently signal desired action or reaction through other controls. Mill loads and efficiencies can be deduced by sonic sensing and feed added or withheld accordingly. Flow of electricity, radioactive beam interruption, viscosity, and many other conditions can be sensed; action or reaction can be initiated as desired. Stopping or starting equipment commonly is automated, sequenced off or on reactions are routine, and variable feeding or withdrawal are part of the control provided for the operation of most plants.

Individual circuits are fitted together in systems that commonly become elaborate and complex. Servicing of the systems generally can be boiled down to routine practice although troubleshooting may be demanding. Reliability and miniaturization of equipment continue to improve rapidly, and design of unitized components makes servicing less complicated at the plant control room level.

PRODUCTS

Cement has been selected by some to exemplify a one product industry, but the product and the market are more complex than a superficial view reveals. Within the primary product line, pertinent specifications describe the characteristics of cement.

Five basic types of cement have been developed:

Type I—The general-type cement, qualified for a wide range of uses, mainly construction.
Type II—Cement qualified for general concrete construction that requires moderate resistance to sulfates and moderate heats of hydration.
Type III—Cement developed for high early strength uses.
Type IV—Cement developed for uses in which low heats of hydration particularly are desired.
Type V—Cement developed for uses requiring high sulfate resistance.

These five types are modified further or combined with other materials to qualify better for various uses. Cement into which an air-entraining agent has been introduced will produce concrete containing dispersed, tiny air bubbles, which act as buffers against certain types of deterioration (mainly those induced by freeze-thaw). Specification controlled amounts of air in concrete, due to these bubbles, is sought mainly as modification of cement Types I, II, and III; these variations are designated as Types IA, IIA, and IIIA. The air-entraining agent is interground with the particular type of clinker to produce the A varieties of cements. The agents are numerous and many are proprietary chemicals offered under a market name rather than a compositional formula.

A number of cements sold under specifications are known by names descriptive of their uses of compositions. Some of these are masonry, oil well, portland-pozzolan, portland blast furnace slag, waterproofed, and white cements. Natural cements, cement-containing and cementlike products also are prepared and sold under the manufacturer's guarantee whether they conform to accepted specifications or not. Premixed dry concrete or mortar is widely offered to the public; nearly all of it is packaged. Calcium aluminate cement is being produced in larger quantities than ever before.

Portland blast furnace slag cement usually contains, by weight, 35 to 45% granulated blast furnace slag, although current specifications permit a range as great as from 25 to 65% of slag. Granulated blast-furnace slag of suitable quality is produced from a hot stream (approximately 1 370°C) of slag by rapidly quenching it in water or steam and air. Quenching essentially produces a glass that chemically resembles a low-lime clinker. This granulated slag is somewhat more difficult to grind than clinker. When finely interground with Type I clinker, the slag imparts certain hydraulic qualities to the finished Type I-S cement. Granulated slag is generally cheaper than clinker.

Portland-pozzolan cement has proved itself in large western US dams. An important advantage is cost savings. Portland-pozzolan cement is between 15 and 40% by weight pozzolan. Pozzolans include some diatomaceous earths, opaline cherts and shales, tuffs, volcanic ashes or pumicites, clays, and fly ashes. The American Society for Testing and Materials (ASTM) definition of pozzolan is, "A siliceous or siliceous and aluminous material, which in itself possesses little or no cementitious value but will, in finely divided form and in the presence of moisture, chemically react with calcium hydroxide at ordinary temperatures to form compounds possessing cementitious properties." Pozzolans may be natural, prepared, or artificial so long as they have the desired pozzolanic reaction in concrete as it sets. Portland cement hydration, in the concrete environment, liberates calcium hydroxide, which then reacts with the pozzolan, when it is available, to form compounds that are cementitious. Tested alone, pozzolans have little or no cementitious properties. Pozzolans in concrete counteract or prereact with aggregates, which may later cause alkali-aggregates problems if soda or potash are present above specified levels. Noncrystalline or cryptocrystalline substances qualify as pozzolans and include hydrous or opaline silicas, natural glasses such as pumicites, and synthetic glasses such as fly ashes. Clays, shales, diatomites, and many other materials seem to have pozzolanic properties enhanced by calcining.

Specifications

As technology replaces art in cement and concrete manufacture, use, and evaluation, specifications become more discriminating. Product quality requirements are formalized. Definitions of current practice or state-of-the-art standards generally are condensed into specifications. The importance of construction and its materials dictates procedures for arriving at meaningful specifications that serve to establish product quality. Quality consistency and reliability generally are predictable over large geographic areas because of specifications control.

The primary framework of specification guidance is that evolved by the ASTM. This society annually publishes formal specifications that result from considerable work and debate of the various committees of specialists, who are the key members of the ASTM. Membership of the committees is balanced with respect to producer and nonproducer interests, and the resulting specifications might be viewed as compromises. However, the standards are better characterized as representing what is possible and acceptable. Consumer interests generally are effectively protected. Other groups importantly involved in formulating specifications or standards (mainly for concrete) include the American Association of State Highway and Transportation Officials (AASHTO), the American Concrete Institute (ACI), the Army US Corps of Engineers, the US Bureau of Reclamation, and other government agencies and authorities. A notable trend is that private customers continue to become more knowledgeable and demanding buyers, and they dictate specific requirements. Table 3 is a compact statement summarizing cement specifications.

Table 3. Comparison of Requirements as Determined by the American Society for Testing and Materials (ASTM) and the American Association of State Highway and Transportation Officials (AASHTO)[a]

	Type of Cement					
	Type I		Type IA		Type II	
Item	ASTM	AASHTO	ASTM	AASHTO	ASTM	AASHTO
Standard Chemical Requirements						
Silicon dioxide (SiO_2), min. %	—	—	—	—	20.0	21.0
Aluminum oxide (Al_2O_3), max. %	—	—	—	—	6.0	6.0
Ferric oxide (Fe_2O_3), max. %	—	—	—	—	6.0	6.0
Magnesium oxide (MgO), max %	6.0	6.0	6.0	6.0	6.0	6.0
Sulfur trioxide (SO_3)[b], when:						
$3CaO \cdot Al_2O_3$[c] is 8% or less, max. %	3.0	3.0	3.0	3.0	3.0	3.0
$3CaO \cdot Al_2O_3$ is 8% or more, max. %	3.5	3.5	3.5	3.5	NA	NA
Loss on ignition, max. %	3.0	3.0	3.0	3.0	3.0	3.0
Insoluble residue, max. %	0.75	0.75	0.75	0.75	0.75	0.75
Tricalcium silicate ($3CaO \cdot SiO_2$), max. %	—	—	—	—	—	55
Dicalcium silicate ($2CaO \cdot SiO_2$), min. %	—	—	—	—	—	—
Tricalcium aluminate ($3CaO \cdot Al_2O_3$), max. %	—	15	—	15	8	8
Tetracalcium aluminoferrite + twice the tricalcium aluminate ($4CaO \cdot Al_2O_3 \cdot Fe_2O_3$) + $2(3CaO \cdot Al_2O_3)$ or solid solution ($4CaO \cdot Al_2O_3Fe_2O_3$ + $2CaO\ Fe_2O_3$) applicable, max. %	—	—	—	—	—	—
Optional Chemical Requirements (Note: These optional requirements apply only when specifically requested.)						
Tricalcium aluminate ($3CaO \cdot Al_2O_3$)						
for moderate sulfate resistance	—	—	—	—	—	—
for high sulfate resistance	—	—	—	—	—	—
Sum of tricalcium silicate and tricalcium aluminate[e], max. %	—	—	—	—	58	58
Alkalies[f] (Na_2O + 0.658 K_2O), max %	0.60	0.60	0.60	0.60	0.60	0.60
Standard Physical Requirements						
Air content of mortar[g], volume						
max. %	12	12	22	22	12	12
min. %	—	—	16	16	—	—
Fineness[h], specific surface, m^2/kg[i]						
Turbidimeter test:						
Average value, min.	160	160	160	160	160	160
Min. value, any one sample	150	150	150	150	150	150
Average value, max.	—	220	—	220	—	220
Max. value, any one sample	—	230	—	230	—	230
Air permeability test						
Average value, min.	280	280	280	280	280	280
Min. value, any one sample	260	260	260	260	260	260
Average value, max.	—	400	—	400	—	400
Max. value, any one sample	—	420	—	420	—	420
Soundness, autoclave expansion, Max. %	0.8	0.8	0.8	0.8	0.8	0.8
Time of setting[j], alternate methods						
Gillmore test						
Initial set, minutes, not less than	60	60	60	60	60	60
Final set, hr, not more than	10	10	10	—	10	—
Vicat test						
Initial set, minutes, not less than	45	45	45	45	45	45
Final set, hours, not more than	8	8	8	8	8	8
Strength, not less than the values shown for the ages indicated below[k]:						
Compressive strength, psi (MPa)						
1 day	—	—	—	—	—	—
3 days	1800 (12.4)	1800 (12.4)	1450 (10)	1450 (10)	1500 (10.3) 1000[l] (6.9)[l]	1500 (10.3) 1000[l] (6.9)[l]
7 days	2800 (19.3)	2800 (19.3)	2250 (15.5)	2250 (15.5)	2500 (17.2) 1700 (11.7)	2500 (17.2) 1700 (11.7)
28 days	—	—	—	—	—	—

Table 3. Comparison of Requirements as Determined by the American Society for Testing and Materials (ASTM) and the American Association of State Highway and Transportation Officials (AASHTO)[a] (cont.)

Type of Cement

Type IIA		Type III		Type IIIA		Type IV		Type V	
ASTM	AASHTO	ASTM	AASHTO	ASTM	AASHTO	ASTM	AASHTO	ASTM	AASHTO
20.0	21.0	—	—	—	—	—	—	—	—
6.0	6.0	—	—	—	—	—	—	—	—
6.0	6.0	—	—	—	—	6.5	6.5	—	—
6.0	6.0	6.0	6.0	6.0	6.0	6.0	6.0	6.0	6.0
3.0	3.0	3.0	3.0	3.5	3.5	3.5	3.5	3.5	3.5
NA	NA	4.5	4.5	4.5	4.5	NA	NA	NA	NA
3.0	3.0	3.5	3.5	3.5	3.5	2.5	2.5	3.0	3.0
0.75	0.75	0.75	0.75	0.75	0.75	0.75	0.75	0.75	0.75
—	55	—	—	—	—	35	35	—	—
—	—	—	—	—	—	40	40	—	—
8	8	15	15	15	15	7	7	5	5
—	—	—	—	—	—	—	—	20[d]	20[d]
—	—	8	8	8	8	—	—	—	—
—	—	5	5	5	5	—	—	—	—
58	58	—	—	—	—	—	—	—	—
0.60	0.60	0.60	0.60	0.60	0.60	0.60	0.60	0.60	0.60
22	22	12	12	22	22	12	12	12	12
16	16	—	—	16	16	—	—	—	—
160	160	—	—	—	—	160	160	160	160
150	150	—	—	—	—	150	150	150	150
—	220	—	—	—	—	—	220	—	220
—	230	—	—	—	—	—	230	—	230
280	280	—	—	—	—	280	280	280	280
260	260	—	—	—	—	260	260	260	260
—	400	—	—	—	—	—	400	—	400
—	420	—	—	—	—	—	420	—	420
0.8	0.8	0.8	0.8	0.8	0.8	0.8	0.8	0.8	0.8
60	60	60	60	60	60	60	60	60	60
10	10	10	10	10	10	10	10	10	10
45	45	45	45	45	45	45	45	45	45
8	8	8	8	8	8	8	8	8	8
—	—	1800	1800	1450	1450	—	—	—	—
		(12.4)	(12.4)	(10)	(10)				
1200	1200	3500	3500	2800	2800	—	—	1200	1200
(8.3)	(8.3)	(24.1)	(24.1)	(19.3)	(19.3)	—	—	(8.3)	(8.3)
800[f]	800[f]								
(5.5)[f]	(5.5)[f]								
2000	2000	—	—	—	—	1000	1000	2200	2200
(13.8)	(13.8)					(16.9)	(16.9)	(15.2)	(15.2)
1350[f]	1350[f]								
(9.3)	(9.3)								
—	—	—	—	—	—	2500	2500	3000	3000
						(17.2)	(17.2)	(20.7)	(20.7)

(cont.)

Table 3. Comparison of Requirements as Determined by the American Society for Testing and Materials (ASTM) and the American Association of State Highway and Transportation Officials (AASHTO)[a] (cont.)

	Type of Cement					
	Type I		Type IA		Type II	
Item	ASTM	AASHTO	ASTM	AASHTO	ASTM	AASHTO
Optional Physical Requirements (Note: These optional requirements apply only when specifically requested.)						
Heat of hydration, cal/g (kJ/kg)						
7 days, maximum	—	—	—	—	70 (290)[m]	70 (290)[m]
28 days, maximum	—	—	—	—	80 (335)[m]	80 (335)[m]
False set, final penetration, min. %	50	50	50	50	50	50
Strength, not less than the values shown						
Compressive strength, psi (MPa)						
28 days	4000 (27.6)	3500 (24.3)	3200 (22.1)	2800 (19.3)	4000 (27.6) 3200[m] (22.1)[m]	3500 (24.1) 2800[m] (19.3)[m]
Sulfate expansion[n], 14 days, max. %	—	—	—	—	—	—

[a] ASTM C-150; AASHTO M85-79I.
[b] Sulfur trioxide. ASTM specifications state that there are cases where optimum SO_3 for a particular cement exceeds the limit in this specification. Where it has been demonstrated that by Method C563 that this condition exists, an additional amount of SO_3, in no case more than 0.5 by weight percent of cement, is permissible provided that, when the cement with the additional calcium sulfate is tested by Method C265, the calcium sulfate in the hydrated mortar at 24± ¼ hr expressed as SO_3 does not exceed 0.5 g/L. When the manufacturer supplies cement under this provision he will, upon request, supply supporting data to the purchaser.
[c] The expressing of chemical limitations by means of calculated assumed compounds does not necessarily mean that the oxides are actually or entirely present as such compounds.
[d] Does not apply when the sulfate expansion limit under "Optional Physical Requirements" is specified.
[e] This limit applies when moderate heat of hydration is required and tests for heat of hydration are not requested.
[f] Alkalies. This limit may be specified when the cement is to be used in concrete with aggregates that may be deleteriously reactive. Reference should be made to ASTM Specification C33 for suitable criteria of deleterious reactivity.
[g] Air content. Compliance with the requirement of this specification does not necessarily ensure that the desired air content will be obtained in concrete.

(cont.)

As more complete standards and specifications are accepted for control of a product, regardless of its place of manufacture, a curious set of problems arises out of this uniformity. The cement industry wrestles with problems of a single product image, wherein quality becomes less effective as a selling device. Practices of cement marketing resulting from this situation invite charges of oligopoly against the industry because of the single, uniform product image. Informed observers realize that the imposition of uniformity of product by outside agencies bears little relationship to whether industry members act in concert in the market.

CONSUMPTION

End Uses—Quantities and Trends

The total production of cement is significant (Table 4). Moreover, cement is peculiar in that nearly all of it actually finds its use as a minority component of concrete. Uses of concrete are diverse, and many do not lend themselves to being as readily defined as do the markets for cement. The status of construction activity largely determines how much concrete is used and how much cement is produced and sold. Identifying the general type of buyer provides one classification useful in thinking about consumption.

Substitutes—Alternative Materials

In the context of applying high technology to developing alternate construction materials, cement seems vulnerable to substitutes. Capital investment for a new plant is high enough to prompt the investor to evaluate the long term threat of substitutes prior to commitment. The current situation appears to be that those materials that compete on the basis of performance cost too much, whereas those that compete economically do not perform as well in most applications. For the foreseeable future there is no satisfactory substitute.

Basic and product research by individual cement companies and by the Portland Cement Association for years have suffered substantial cutbacks due to low profits. This situation stimulated predictions that substitute materials were poised to take large segments of the market. Time will tell, but threatened long range shortages of some of the basic substitute materials weigh in favor of cement. A number of countries place considerable emphasis on substituting concrete products for many of the materials that the US industry is concerned will competitively replace concrete.

ECONOMIC FACTORS

The cement industry statistic most meaningful to the customer is the price one pays for delivered cement. The components of that price tend to be lumped into the categories of cost, transportation, discounts and allowances, and profit. Costs may be accounted in detail or they may be so generalized as to be viewed as mill (plant) and as overhead costs. Obviously the more costs are generalized for marketing purposes (to arrive at a price), the less the interrelationships of those costs may be understood. The earlier practice of cement companies' paying the transportation bill, and then including it in the price to the customer, has given way to customer pickup at the plant or terminal. Discounts and/or allowances, which amount to price cutting devices, continue in widespread use. Although discounts are justified for a variety of reasons including a means to compete for the large volume jobs or customers, they result in lower prices and have a negative overall effect on profits. This is particularly noticeable when supplies are tight (i.e. plants are sold out) and prices cannot be adjusted to reflect this condition. For the cement industry, profit has been an elusive goal in recent years in many market areas. Permanent closing of cement plants, although a very costly move, has been the obvious answer.

Table 3. Comparison of Requirements as Determined by the American Society for Testing and Materials (ASTM) and the American Association of State Highway and Transportation Officials (AASHTO)[a] (cont.)

Type of Cement									
Type IIA		Type III		Type IIIA		Type IV		Type V	
ASTM	AASHTO	ASTM	AASHTO	ASTM	AASHTO	ASTM	AASHTO	ASTM	AASHTO
70 (290)[m]	70 (290)[m]	—	—	—	—	60 (250)	60 (250)	—	—
—	80 (335)[m]	—	—	—	—	70 (290)	70 (290)	—	—
50	50	50	50	50	50	50	50	50	50
3200	2800	—	—	—	—	—	—	—	—
(22.1)	(19.3)								
2560[m]	2240[m]								
(17.7)[m]	(15.4)[m]								
—	—	—	—	—	—	—	—	0.045	0.045

[h] Fineness. Either of two alternative fineness methods may be used at the option of the testing laboratory. However, when the sample fails to meet the requirements of the air permeability test, the turbidimeter test shall be used and the requirements for the turbidimetric method shall govern.
[i] These values were revised in ASTM specifications because of the change in unit from cm^2/g to m^2/kg.
[j] Time of setting. The purchaser should specify the type of setting time test required. In case he does not so specify the requirements of the Vicat test only shall govern.
[k] Strength. The strength at any specified age shall not be less than that attained at any previous testage.
[l] When the optional heat of hydration or the chemical limit on the sum of the tricalcium silicate and tricalcium aluminate is specified.
[m] When the heat of hydration requirements are specified, the sum of the tricalcium silicate and tricalcium aluminate shall not be specified. These strength requirements apply when either heat of hydration requirements or the sum of tricalcium silicate are specified.
[n] When the sulfate expansion is specified, it shall be instead of the limits of C_3A and $C_4AF + 2C_3A$ listed under "Standard Chemical Requirements."

Table 4. Annual Production, Shipments, and Capacity of the Cement Industry of the United States and Puerto Rico, and Average Prices for Cement Sold in the United States (thousands of t)

						Capacity		Prices[4]	
Year	Shipments[1,2]	Production[1]	Imports[1]	Exports	No. of active portland cement plants	Finish grinding capacity	Clinker capacity	Average mill ($/st)[5]	ENR* delivered ($/st)
1981	66 201	66 163	3 626	275	147	91 052	81 221	52.46	60.24
1982	59 015	58 369	2 658	184	143	93 467	78 569	51.43	62.07
1983	66 102	64 725	3 872	107	143	96 735	80 105	50.45	61.80
1984	73 629	71 395	8 025	73	141	97 465	79 388	52.24	63.75
1985	76 207	71 540	13 142	89	136	94 156	77 056	51.87	64.91
1986	80 488	72 499	14 804	54	133	93 573	76 251	50.73	63.35
1987	82 034	72 122	16 081	47	132	98 265	75 661	49.70	65.33
1988	82 020	70 989	15 865	92	127	93 205	75 982	49.36	65.48
1989	79 480	71 268	14 280	464	123	90 730	73 969	49.71	63.34
1990	79 547	71 310	12 041	503	119	88 211	73 533	50.20	NA

NA = Not available.
*Engineering News Record.
[1]Portland and masonry cement, and cement produced from imported clinker, for the 50 States and Puerto Rico.
[2]Includes portland cement shipped by domestic producers and distributors.
[3]Includes clinker.
[4]Annual prices are based on an average per ton value of cement sales, FOB plant, reported by producers. Monthly prices are based on an average per ton value of cement delivered to 20 cities as published monthly by Engineering News Record.
Source: US Bureau of Mines, unless noted otherwise, converted to SI.

Costs

Elements of the cost of cement manufacture may be organized and accounted for in many ways. Changes in systems and practices of accounting are encouraged by new office equipment, tax laws, and evolving methods. Classifying expenditures, selecting categories into which costs fall, and relating them to each other in meaningful terms are formidable tasks. Operating personnel tend to prefer unit-cost accounting so that their surveillance of jobs being done is measured in terms that permit comparisons and reflect changes. Unit cost examples would be: cost per ton, cost per meter drilled, cost per MJ, etc. General management may prefer having costs lumped into categories, such as costs of labor, power, fuel, maintenance supplies, and so on. Discrimination between fixed and variable costs becomes most useful in understanding what happens to costs as a result of plant shutdowns or as a result of sales over a break even level of operation. Many heroic efforts to lower manufacturing costs are overshadowed by whatever can be saved by operating the plant nearer to its rated capacity. Compared to the effect of raising prices or lowering transportation costs, the savings from lowering manufacturing costs are small.

Each plant has its own unique combination of initial capital, operating, and transportation costs; those costs partly are predetermined by the design, location, and manageability of the operation. In this capital intensive industry the initial investment is so large that the cost of money is very significant. Particularly notable in the 1970s was the soaring cost of labor. Approximate doubling of the cost of a worker brought the pay of cement industry workers to a new, relatively high level compared to that of workers in other industries.

The 1970s were also marked by the explosive increase in energy costs. General inflation of capital costs made questionable the wisdom of investment in sorely needed new plants in the United States, whereas larger and more numerous plants characterized the cement industry's moves in other countries.

The 1980s saw a dramatic change in the cost structure. Fuel costs stabilized or dropped and labor cost increases slowed as many plants became non-union or abandoned prior union associations. Electrical power costs continued to escalate, particularly in those regions in which nuclear plants came on line. Profitability of most cement plants declined sharply as low cost imports held down or decreased prices of cement. Investment in new plants declined drastically and those few planned experienced considerable problems obtaining environmental permits.

Transportation

As is often pointed out, essentially every other step in the manufacture of cement is transportation. Transportation of fuels and purchased raw materials is a sizable cost to most plants, as is transportation of raw materials to many plants. Clinker-transportation costs are important in some plants. But, unless otherwise qualified, the industry's transportation costs almost automatically are interpreted to be the cost of transporting finished cement to the consumer.

Cement is heavy in relation to its value so that the cost of moving it is proportionally high. Methods of transport are critical determinants of costs. For cement, they include various means of shipment by truck, rail, and barge or other waterborne vessel. Direct delivery to the customer by a single type of carrier predominates although cement that is moved through distribution terminals often moves the first leg of its journey via water or rail and then out of the terminal by truck. The 1960s witnessed schemes of cement transport more complicated than before, and the product often was moved greater distances. Handling of cement through distribution terminals reached it peak in the 1960s, but the trend is now back to direct shipment.

Marketing

The direct efforts of selling cement probably do virtually nothing toward increasing the total amount sold within a defined geographic area and within a limited period of time, such as a year. Achievement in the selling of cement long has been measured by a company's success in capturing or improving its share of the cement sold in any given market. This type of market change takes place much more dramatically upon the closing of a producing plant than upon the building of a new plant. Important as it is to push volume of sales to or beyond a given percentage of market goal, it is far more significant to sell the product for prices that will produce a reasonable return on the investment. During the 1960s failure to be reasonably profitable severely eroded the management personnel of many companies. The soaring 1970s predicted by enthusiastic prognosticators turned out to be very spotty. The 1980s were a repeat of the 1960s aggravated by unprecedented levels of imports in the last half of the decade. Lack of the promise of reasonable returns on investment restrains new plant construction in the United States. Imported cement and clinker by both producers and marketing companies that built terminals at dockside or used floating terminals (ships configured as storage and shipping facilities) took 20% of the US market. In 1991, the first successful antidumping ruling was obtained and imports dropped. Prices, however, have not recovered to allow adequate profitability and renewed plant modernization. The industry still has considerable unused capacity. The reader is referred to the names and locations of US cement plants given in Tables 5 and 6 and plant cement grinding and clinker capacity by state given in Tables 7 and 8.

The percentage of capacity at which a cement plant must be operated to break even, say in a given year, is high and orients the marketing of cement toward high volumes of sales. During the 1960s the effects of industry overcapacity were disastrous. In some years, prices dropped to levels at which relatively few plants were profitable. While the economics of each plant and each market area is unique, still those areas overlap, and the problems generated by the condition of overcapacity spread throughout the whole country during the decade of the 1960s. During the 1970s, several new plants were built or existing plants modernized using the new preheater and calciner technology. However, in the early 1980s, such programs declined markedly in response to the deep recession of 1981 to 84. Recovery in the 1985 to 91 period never occurred due to pressure from low cost imports. The lack of adequate return on investment continues to plague the industry.

Selling tools have included the touting of color differences, offering technical assistance, custom-manufacturing of cement, providing facilities or special services, and selling aggregates as well as cement. Personal relationships of sales people, extension of credit terms, encouragement of customer pickup (the customer sends their own trucks into the cement plant or distribution terminal to pick up the cement), and price concessions (discounts) also are among devices used to win (or buy) customers. In retrospect, many of the competition generated schemes to establish better service for the customer were countered by competitors' lowering the price. Cement company profits sank from the best profit performance of all major industries in 1955 to the poorest in 1967 and has continued at an unsatisfactory low level since then. This experience was extremely damaging and has resulted in the demise of several companies.

The cement industry continues to labor under government constraints. Although free from Federal Trade Commission and Justice Department actions in recent years, it now faces restrictive if not hostile environmental regulations.

The question often is asked as to what government subsidies, direct or indirect, particularly favor the cement industry. Tax provisions of such wide effect as investment credit are held by some to favor particularly capital intensive industries, and the cement industry certainly is in that group. However, special benefits for the cement industry as a whole due to this tax provision were not large because the industry hardly was in financial condition to make the investments from which these tax advantages might flow. Of course, some companies were engaged in capital programs that happened to be well timed, by coincidence, with the credit provisions. Pollution control equipment, which required substantial investment, was being installed when the credits were offered. Local bond financing for law-imposed pollution control devices has helped make possible prompt compliance for installations that otherwise may have imposed a difficult financial burden on the companies involved.

Perhaps the subsidy granted the mineral industries that is most often pilloried is the familiar percentage depletion allowance granted by the US Congress to compensate for the physical using-up of nonrenewable mineral resources. The cement industry scarcely is a front running beneficiary of this eminently sound tax relief measure. Because profitability is necessary to the realization of this

Table 5. United States Gray Cement Plant Capacities

Rank	Clinker (000 t)	Name and location
1	1 773	Lafarge Corporation, Alpena, MI
2	1 514	Mitsubishi Cement Corp., Lucerne Valley, CA
3	1 452	Kaiser Cement Corporation, Permanente, CA
4	1 406	Southdown Inc., Victorville, CA
5	1 390	Blue Circle Inc., Ravena, NY
6	1 308	Holnam, Inc., Theodore, AL
7	1 237	Medusa Cement Company, Charlevoix, MI
8	1 190	Holnam Inc., Clarksville, MO
9	1 139	Texas Industries, Midlothian, TX
10	1 089	Southdown Inc., Brooksville, FL
11	1 076	Lafarge Corporation, Grand Chain, IL
12	1 070	River Cement Company, Festus, MO
13	1 041	Gifford-Hill & Company, Inc., Oro Grande, CA
14	1 013	Roanoke Cement Company, Cloverdale, VA
15	1 002	Lone Star Industries, Cape Girardeau, MO
16	991	Holnam Inc., Holly Hill, SC
17	966	California Portland Cement, Rillito, AZ
18	943	California Portland Cement, Mojave, CA
19	933	Tarmac Florida, Inc., Medley, FL
20	907	Boxcrow Cement, Midlothian, TX
21	900	Lehigh Portland Cement, Union Bridge, MD
22	895	Texas-Lehigh Cement Company, Buda, TX
23	880	Holnam Inc., Dundee, MI
24	874	Essroc Materials, Nazareth, PA
25	872	Ash Grove Cement Co., Louisville, NE
26	865	Lafarge Corporation, New Brunfels, TX
27	863	Essroc Materials, Speed, IN
28	857	Ash Grove Cement Co., Foreman, AR
29	844	Allentown Cement Company Inc., Blandon, PA
30	816	North Texas Cement, Midlothian, TX
31	806	Holnam Inc., Mason City, IA
32	789	Giant Cement Company, Harleyville, SC
33	780	Holnam Inc., Florence, CO
34	778	Lafarge Corporation, Buffalo, IA
35	776	Capitol Aggregates, Inc., San Antonio, TX
36	767	Natl. Cement Company, Ragland, AL
37	746	Capitol Cement Corporation, Martinsburg, WV
38	726	RMC Lonestar, Davenport, CA
39	695	Medusa Cement Company, Clinchfield, GA
40	695	South Dakota Cement, Rapid City, SD
41	689	Lafarge Corporation, Whitehall, PA
42	689	Lehigh Portland Cement, Mason City, IA
43	689	Lehigh Portland Cement, Mitchell, IN
44	689	Texas Industries, New Braunfels, TX
45	680	Alamo Cement Co., San Antonio, TX
46	680	California Portland Cement, Colton, CA
47	657	Kosmos Cement Co., Louisville, KY
48	656	Hercules Cement Company, Stockertown, PA
49	655	Lafarge Corporation, Demopolis, AL
50	649	Lone Star Industries, Greencastle, IN
51	640	Phoenix Cement Company, Clarkdale, AZ
52	638	Medusa Cement Company, Wampum, PA
53	623	Lone Star Industries, Pryor, OK
54	611	Monarch Cement Company, Humboldt, KS
55	591	Calaveras Cement Company, Redding, CA
56	591	Lehigh Portland Cement, Leeds, AL
57	590	Natl. Cement Co. of California, Lebec, CA
58	565	Lone Star Industries, Nazareth, PA
59	560	Gifford-Hill & Company, Inc., Harleyville, SC
60	555	Blue Circle Inc., Atlanta, GA
61	553	St Marys Peerless Cement Corp., Detroit, MI
62	553	Southdown Inc., Fairborn, OH
63	547	Carlow Group, Zanesville, OH
64	546	Keystone Cement Company, Bath, PA
65	544	Blue Circle Inc., Calera, AL
66	544	Holnam Inc., Ada, OK
67	544	Continental Cement Co., Inc., Hannibal, MO
68	544	Ash Grove Cement Co., Nephi, UT
69	544	Southdown Inc., Knoxville, TN
70	544	Blue Circle Inc., Tulsa, OK
71	518	Florida Crushed Stone, Brooksville, FL
72	512	Rinker Portland Cement Corp., Miami, FL
73	506	Lehigh Portland Cement, Cementon, NY
74	499	Southdown Inc., Odessa, TX
75	486	Essroc Materials, Bessemer, PA
76	475	Dixon-Marquette, Dixon, IL
77	464	Independent Cement Corporation, Catskill, NY
78	457	Holnam Inc., Artesia, MS
79	454	Ash Grove Cement Co., Durkee, OR
80	452	Independent Cement Corp., Hagerstown, MD
81	450	Ash Grove Cement Co., Chanute, KS
82	449	Lone Star Industries, Sweetwater, TX
83	449	Glens Falls Cement Co., Inc., Glens Falls, NY
84	448	Holnam Inc., Fort Collins, CO
85	448	Holnam Inc., Tijeras, NM
86	445	Lafarge Corporation, Paulding, OH
87	437	Lafarge Corporation, Sugar Creek, MO
88	429	Holnam Inc., Seattle, WA
89	422	Lone Star Industries, Oglesby, IL
90	418	Centex, Laramie, WY
91	413	Dragon Products Company, Thomaston, ME
92	408	Signal Mountain Cement Co., Chattanooga, TN
93	408	Southdown Inc., Lyons, CO
94	386	Calaveras Cement Company, Tehachapi, CA
95	376	Centex, Fernley, NV
96	372	Centex, La Salle, IL
97	367	Essroc Materials, Logansport, IN
98	357	Kosmos Cement Co., Pittsburgh, PA
99	347	Lafarge Corporation, Fredonia, KS
100	336	Essroc Materials, Frederick, MD
101	335	Holnam Inc., Saratoga, AR
102	305	Heartland Cement Co., Independence, KS
103	298	Holnam Inc., Morgan, UT
104	283	Holnam Inc., Three Forks, MT
105	281	Armstrong Cement & Sup. Corp., Cabot, PA
106	272	Monarch Cement Company, Des Moines, IA
107	254	Ash Grove Cement Co., Montana City, MT
108	239	Hawaiian Cement Company, Ewa Beach, HI
109	191	Ash Grove Cement Co., Inkom, ID
Total USA:	73 271	

Source: Portland Cement Association, 1991.

source of tax relief, many plants have not benefited significantly from this source through many years. The biggest potential source of this benefit is the calcium carbonate raw materials valued at the kiln feed state; 14% of this value is allowable. Import tariffs are regarded to be an item of government subsidy by some. Except for the anti-dumping tariffs levied on some imports, there are no tariffs on portland cement.

Price Trends and Profits

The price of finished cement, a relatively sophisticated product, has had a rather remarkable recent history. Prices of cement decreased from the 1960 high to lows in the middle 1960s, and recovery was delayed until the 1970s, peaking in 1980. Since then prices have declined and shown little inclination of recovering. On a constant dollar basis, prices are at or below the low in the 1960s. Average prices per t sold are indicated in Table 4.

Contract selling of cement generally offers little advantage to

Table 6. United States Gray Cement Plant Capacities

Rank	Finish grinding (000 t)	Name and location
1	1 814	Lafarge Corporation, Alpena, MI
2	1 587	Holnam Inc., Theodore, AL
3	1 578	Holnam Inc., Holly Hill, SC
4	1 542	Mitsubishi Cement Corp., Lucerne Valley, CA
5	1 497	Southdown Inc., Victorville, CA
6	1 479	Blue Circle Inc., Ravena, NY
7	1 451	Kaiser Cement Corporation, Permanente, CA
8	1 451	South Dakota Cement, Rapid City, SD
9	1 430	California Portland Cement, Mojave,CA
10	1 270	Holnam Inc., Clarksville, MO
11	1 257	California Portland Cement, Rillito, AZ
12	1 243	Medusa Cement Company, Charlevoix, MI
13	1 197	Southdown Inc., Brooksville, FL
14	1 179	Alamo Cement Co., San Antonio, TX
15	1 179	Texas Industries, Midlothian, TX
16	1 179	LaFarge Corporation, Grand Chain, IL
17	1 179	River Cement Company, Festus, MO
18	1 090	Ash Grove Cement Co., Louisville, NE
19	1 088	Lafarge Corporation, Buffalo, IA
20	1 088	Lone Star Industries, Cape Girardeau, MO
21	1 088	Roanoke Cement Company, Cloverdale, VA
22	1 088	Texas-Lehigh Cement Company, Buda, TX
23	1 034	Essroc Materials, Nazareth, PA
24	998	Allentown Cement Company Inc., Blandon, PA
25	998	Holnam Inc., Dundee, MI
26	989	Tarmac Florida, Inc., Medley, FL
27	980	St Marys Peerless Cement Corp., Detroit, MI
28	975	Essroc Materials, Speed, IN
29	970	Holnam Inc., Florence, CO
30	934	Lehigh Portland Cement, Union Bridge, MD
31	916	Ash Grove Cement Co., Foreman, AR
32	907	Boxcrow Cement, Midlothian, TX
33	907	Natl. Cement Company, Ragland, AL
34	907	Lafarge Corporation, New Braunfels, TX
35	907	Monarch Cement Company, Humboldt, KS
36	868	Gifford-Hill & Company, Inc., Oro Grande, CA
37	862	California Portland Cement, Colton, CA
38	862	Giant Cement Company, Harleyville, SC
39	816	Capitol Cement Corporation, Martinsburg, WV
40	816	North Texas Cement, Midlothian, TX
41	816	Gifford-Hill & Company, Inc., Riverside, CA
42	816	Holnam Inc., Mason City, IA
43	794	Gulf Coast Portland Cement Co., Houston, TX
44	794	RMC Lonestar, Davenport, CA
45	794	Lehigh Portland Cement, Mason City, IA
46	771	Capitol Aggregates, Inc., San Antonio, TX
47	748	Lafarge Corporation, Demopolis, AL
48	734	Medusa Cement Company, Clinchfield, GA
49	726	Essroc Materials, Bessemer, PA
50	726	Lafarge Corporation, Whitehall, PA
51	726	Hercules Cement Company, Stockertown, PA
52	726	Texas Industries, New Braunfels, TX
53	703	Phoenix Cement Company, Clarkdale, AZ
54	689	Blue Circle Inc., Atlanta, GA
55	689	Lehigh Portland Cement, Mitchell, IN
56	685	Holnam Inc., Ada, OK
57	682	Lone Star Industries, Greencastle, IN
58	667	Kosmos Cement Co., Louisville, KY
59	662	Lehigh Portland Cement, Leeds, AL
60	649	Medusa Cement Company, Wampum, PA
61	635	Calaveras Cement Company, Redding, CA
62	635	Independent Cement Corporation, Catskill, NY
63	635	Dragon Products Company, Thomaston, ME
64	635	Holnam Inc., Seattle, WA
65	635	Natl. Cement Co. of California, Lebec, CA
66	635	Gifford-Hill & Company, Inc., Harleyville, SC
67	635	Lone Star Industries, Pryor, OK
68	617	Blue Circle Inc., Calera, AL
69	617	Southdown Inc., Fairborn, OH
70	612	National Portland Cement Co., Palmetto, FL
71	608	Centex, Laramie, WY
72	604	Rinker Portland Cement Corp., Miami, FL
73	600	Dixon-Marquette, Dixon, IL
74	599	Holnam Inc., Tijeras, NM
75	597	Lone Star Industries, Nazareth, PA
76	590	Keystone Cement Company, Bath, PA
77	590	Southdown Inc., Knoxville, TN
78	590	Lafarge Corporation, Tampa, FL
79	571	Continental Cement Co., Inc., Hannibal, MO
80	571	Ash Grove Cement Co., Nephi, UT
81	567	Lehigh Portland Cement, Cementon, NY
82	544	Blue Circle Inc., Tulsa, OK
83	544	Hawaiian Cement Company, Ewa Beach, HI
84	544	Southdown Inc., Odessa, TX
85	544	Holnam Inc., Artesia, MS
86	544	Florida Crushed Stone, Brooksville, FL
87	522	Holnam Inc., Saratoga, AR
88	503	Lafarge Corporation, Paulding, OH
89	501	Ash Grove Cement Co., Chanute, KS
90	499	Carlow Group, Zanesville, OH
91	499	Centex, Fernley, NV
92	499	Lone Star Industries, Sweetwater, TX
93	499	Independent Cement Corp., Hagerstown, MD
94	481	Holnam Inc., Fort Collins, CO
95	476	Ash Grove Cement Co., Durkee, OR
96	457	Lafarge Corporation, Sugar Creek, MO
97	454	Essroc Materials, Essexville, MI
98	454	Glens Falls Cement Co., Inc., Glens Falls, NY
99	454	Centex, La Salle, IL
100	449	Essroc Materials, Frederick, MD
101	447	Lone Star Industries, Oglesby, IL
102	433	Signal Mountain Cement Co., Chattanooga, TN
103	431	Southdown Inc., Lyons, CO
104	408	Calaveras Cement Company, Tehachapi, CA
105	408	Heartland Cement Company, Independence, KS
106	390	Essroc Materials, Logansport, IN
107	369	Lafarge Corporation, Fredonia, KS
108	363	Ash Grove Cement Co., Inkom, ID
109	363	Kosmos Cement Co., Pittsburgh, PA
110	336	Ash Grove Cement Co., Montana City, MT
111	331	Holnam Inc., Morgan, UT
112	327	Holnam Inc., Three Forks, MT
113	302	Armstrong Cement & Sup. Corp., Cabot, PA
114	272	Monarch Cement Company, Des Moines, IA
115	181	Essroc Materials, Middlebranch, OH
116	63	Alaska Basic Industries, Anchorage, AK
Total USA:	87 862	

Source: Portland Cement Association, 1991.

the cement companies. Customers rather freely break contracts, if it is convenient to them, and the producers usually regard themselves bound to honor such agreements. Cost escalation provisions in such sales contracts are rare. Requirements of the environmental protective laws also force up prices because of the cost of compliance. How rapidly and how far cement prices will move up is an unanswerable question.

Listening to similar arguments for years, and viewing precious little by way of results, cement company shareholders and boards of directors understandably have become arch skeptics. When will the industry become attractively profitable? Perhaps that day is still

Table 7. United States Clinker Capacities by State (includes gray and white plants)

Rank	Clinker (000 t)	States
1	9 426	California
2	7 788	Texas
3	6 025	Pennsylvania
4	4 442	Michigan
5	4 242	Missouri
6	3 864	Alabama
7	3 050	Florida
8	2 809	New York
9	2 567	Indiana
10	2 545	Iowa
11	2 345	Illinois
12	2 339	South Carolina
13	1 712	Kansas
14	1 712	Oklahoma
15	1 687	Maryland
16	1 636	Colorado
17	1 605	Arizona
18	1 545	Ohio
19	1 250	Georgia
20	1 192	Arkansas
21	1 013	Virginia
22	952	Tennessee
23	872	Nebraska
24	842	Utah
25	746	West Virginia
26	695	South Dakota
27	657	Kentucky
28	537	Montana
29	457	Mississippi
30	454	Oregon
31	448	New Mexico
32	429	Washington
33	418	Wyoming
34	413	Maine
35	376	Nevada
36	239	Hawaii
37	190	Idaho
Total:	73 518	

There are no clinker producing plants in the following states: Alaska, Connecticut, Delaware, Louisiana, Massachusetts, Minnesota, New Hampshire, New Jersey, North Carolina, North Dakota, Rhode Island, Vermont, Wisconsin or in the District of Columbia.

Table 8. United States Grinding Capacities by State (includes gray, white and grinding plants)

Rank	Finish grinding (000 t)	States
1	11 046	California
2	9 499	Texas
3	6 848	Pennsylvania
4	5 487	Michigan
5	4 566	Missouri
6	4 536	Florida
7	4 521	Alabama
8	3 135	New York
9	3 075	South Carolina
10	2 970	Iowa
11	2 736	Indiana
12	2 680	Illinois
13	2 185	Kansas
14	1 960	Arizona
15	1 882	Colorado
16	1 882	Maryland
17	1 858	Oklahoma
18	1 800	Ohio
19	1 451	South Dakota
20	1 438	Arkansas
21	1 423	Georgia
22	1 090	Nebraska
23	1 088	Virginia
24	1 022	Tennessee
25	902	Utah
26	816	West Virginia
27	667	Kentucky
28	662	Montana
29	635	Maine
30	635	Washington
31	608	Wyoming
32	599	New Mexico
33	544	Hawaii
34	544	Mississippi
35	499	Nevada
36	476	Oregon
37	363	Idaho
38	63	Alaska
Total:	88 193	

There are no cement producing plants in the following states: Connecticut, Delaware, Louisiana, Massachusetts, Minnesota, New Hampshire, New Jersey, North Carolina, North Dakota, Rhode Island, Vermont, Wisconsin or in the District of Columbia.

remote and unpredictable for the industry as a whole, but certain companies are perceived to be attractive both to acquisitors and to investors. The Red Queen's sage advice to Alice, "it takes all the running you can do to keep in the same place," unfortunately is all too appropriate to describe the experience of many segments of the cement industry. The entry of industrial conglomerate companies into the industry in the 1960s and the diversification of many companies that have their roots in the industry now has a significantly long history. Some criticism has been leveled at US cement producers for devoting some of their resources to diversification, rather than plowing those resources all back into new or modernized cement production facilities. Surely a careful study of actual investments in bigger and better cement plants since the 1960s would confirm that the capital would have been more profitably invested elsewhere. Financial results generally have been disastrous.

The 1980s saw a major and dramatic change in corporate structure. The ownership of plants by conglomerates disappeared. Massive corporate consolidations occurred through acquisitions and mergers reducing the number of companies and plants. Foreign ownership became dominant with the foreign giants Lafarge Coppee Group and Holderbank Financier Glaris Ltd. taking major positions. Approximately 65% of industry capacity is now in foreign hands.

ENVIRONMENTAL CONSIDERATIONS

The Regulatory Environment

The US Environmental Protection Agency (EPA) recognizes the cement industry as a major source of air pollutants. The industry therefore comes under the Major Source group of regulations. The

Air Toxics provisions of the Clean Air Act Amendments of 1990 impose potential additional regulations. However the regulations pertaining to air toxics from cement plants will probably not be promulgated until 1997 and not become effective until 1999. Plants that commenced operations prior to the Clean Air Act rules that became effective in 1971 are regulated solely by State standards in effect at the time. New plants and modifications of existing plants since then must comply with the Clean Air Act regulations.

The EPA also has set standards for water and solid wastes. Waste water emissions are regulated under the National Pollutant Discharge Elimination System (NPDES) for both ordinary waste water and for storm water discharges. Solid wastes are regulated under the Resource Conservation and Reclamation Act (RCRA) for ordinary non-toxic wastes (e.g., trash) under Sub-title C, for toxic wastes (e.g., solvents, greases, and lubricating oils) under Sub-title D and for Cement Kiln Dust (CKD) as a Special Waste under the Bevill Amendment. The EPA is mandated to study the CKD matter and recommend whether and how it should be regulated. The study will occur in 1992 and 1993 and the report is to be issued in 1993. Regulations are anticipated shortly thereafter.

Enforcement of regulations became a major EPA objective in the late 1980s. As a result, each state has become much more vigilant in monitoring emissions from major sources. Tolerance for emissions exceeding permitted levels from cement plants has reached unprecedented levels. Additionally, the Clean Air Act Amendments of 1990 have imposed expensive fines for failure to remain within permitted air emission limits and jail sentences for willful violation of regulations. These amendments also enable citizens to bring action in the event state environmental agencies or the EPA fail to take action in the event of a violation of regulations. There has been a requirement at plants permitted since 1977 for continuous emissions monitors (CEMs) for particulates on points of major emissions, such as stacks discharging kiln gases. Exceedences of permitted opacity levels have required prompt reporting to the state environmental agency. A pattern of violations prompts a visual inspection by compliance personnel from the agency and often result in Notice(s) of Violation (NOVs). NOVs often result in fines or other regulatory enforcement action.

The enforcement of emission and permit limits is anticipated to become stricter and more pervasive. CEMs are expected to become the norm to monitor not only particulates, but sulfur dioxide, oxides of nitrogen, volatile organic compounds (VOC), and carbon monoxide. The public is less tolerant than at any time in recent history of pollution and generally is not persuaded that restrictive emission standards result in more expensive products, lower levels of employment, lower levels of domestic production, a lower standard of living and greater imports of those products subject to our restrictive environmental standards. The realization appears to have escaped the public, the EPA and Congress that greater global pollution occurs when these products are made in third world nations with minimal or non-existent emission standards. This is very apparent in the case of the US where such a significant amount of cement is imported from sources with minimal or no standards.

Cement Plant Emission Standards

The Federal Air Pollution Regulations impose a maximum emission standard for particulates of "0.3 lbs/st" kiln feed from the kiln gases and "0.1 lbs/st" kiln feed from the clinker cooler gases. Additionally, the regulations impose an opacity limit of 20% on these sources as well as emissions from all other plant sources. However, the mechanism for obtaining a permit for significant modifications to an existing process or the addition of capacity to an existing plant or the construction of a new plant is far more complex than these standards imply. These arcane procedures and standards require many months of permitting work and many thousands of dollars of expense.

To obtain a permit requires a Prevention of Significant Deterioration of Ambient Air Quality Review (PSD Review). This is a complex process requiring extensive modeling of emissions from the cement plant and all other sources within its area of influence. It must be demonstrated that the proposed changes or new plant will not cause or contribute to a violation of ambient air quality standards nor consume more than the allowable increment in emissions. The proposed plant or modifications must demonstrate that Best Available Control Technology (BACT) is being used for all criteria pollutants (particulates, sulfur dioxide, oxides of nitrogen, volatile organic compounds, and carbon monoxide). If the plant is located in a non-attainment area, i.e., an area defined by the state in which the state or federal ambient air quality standards are not met for a particular pollutant, then the permitting process becomes even more complex and requires the use of Lowest Achievable Emissions Reduction (LAER) technology and the application of offset regulations. This requires the plant to show that emissions of a particular pollutant in the area are being reduced by at least as much as the proposed project will add. The construction of new plants and the modification of existing plants has been seriously restrained by these arcane procedures and standards, the many months added to the project due to permitting work, and the thousands of dollars of added expense. The BACT requirements have resulted in a moving target that now requires the cement industry to demonstrate particulate emission levels one-third of the statutory "0.3 lbs/st" kiln feed even when the modeling work demonstrates that such low levels of emissions are not required to comply with ambient air quality standards.

Cement Plant Emissions

Although present BACT indicates that particulate emissions are limited to 21 mg per standard dry m^3 of gas, as a general guide the following table shows the expected controlled emissions from various sources in a cement plant when fabric filters are used as the emissions control devices.

Sources	Particulates (mg/Nm^3)	Sulfur dioxide* (mg/Nm^3)	Oxides of nitrogen* mg/kg clinker
PreHeater calciner kiln			
with roller mill	50	130	1 100
without roller mill	50	330	1 100
Clinker cooler	50	-0-	-0-
Cement mill	50	-0-	-0-
Materials Handling, and Storage	35	-0-	-0-

*Coal Fired, 0.5% sulfur limit in fuel.

Emissions of carbon monoxide and carbon dioxide are of increasing concern to the EPA and receiving public attention from some environmental activists. Carbon monoxide emissions from kiln systems vary significantly, typically in the range of 150 ppm to 1 000 ppm. Raw materials are known to provide a major amount at some plants and the calciner is also a major source. Long wet process kilns typically generate the lowest levels from combustion. The major source of carbon dioxide is generated by the calcining of the limestone. For each ton of clinker produced approximately 0.56 t of carbon dioxide are evolved. Combustion of the fuel generates the remainder of the carbon dioxide.

A SUMMARY VIEW

The 1980s have seen a marked slowdown in the modernization of the US cement industry. The financial performance of most of

Table 9. Modernization and Expansion Programs

Ash Grove Cement, Seattle, WA
Capital Aggregates, San Antonio, TX
Centex Mountain Cement, Laramie, WY
CBR-Monolith, Tehachapi, CA
Puerto Rican Cement, Ponce, PR
Southdown, Victorville, TX
Southdown, Brooksville, FL
Texas-Lehigh Cement, Buda, TX

the cement industry has been dismal. The two largest companies, Ideal Basic Industries (Ideal Cement) and Lone Star Industries (Lone Star Cement), have undergone actual or de facto bankruptcy. The former becoming part of the Holderbank group of cement companies. Another company, Penn-Dixie, disappeared from the scene, also in a cloud of bankruptcy. Columbia Cement simply closed down and Martin-Marietta sold off its properties. General Portland Cement was acquired by Lafarge-Coppee. More than 65% of the US cement industry became foreign owned. Approximately 40 plants have been closed since the 5th edition of *Industrial Minerals and Rocks* was published in 1983, although a few continue to grind clinker and ship cement. Only two new plants (Box-Crow Cement, Midlothian, TX, and Florida Materials, Brooksville, FL) have been built and seven plants have been expanded and modernized (Table 9).

Table 10. Hydraulic Cement Production, World and by Country (top 20 countries as of 1990—1000 t)

	1986	1987	1988	1989	1990
China	161 479	179 623	203 209	206 838	224 000
USSR (former)	135 119	137 404	139 499	140 436	151 000
Japan	71 264	71 551	77 554	79 717	90 400
USA/incl. Puerto Rico	72 499	72 122	70 989	71 268	78 606
India	36 400	36 980	40 733	42 100	48 500
Germany	38 567	37 698	38 725	41 018	44 100
Italy	35 938	37 257	37 013	36 469	40 200
Republic of Korea	23 403	25 662	28 995	30 474	37 038
Brazil	25 297	25 470	25 328	25 883	27 550
Spain	24 040	23 405	24 040	24 494	27 600
Turkey	20 003	21 979	22 675	23 800	26 600
France	23 496	23 560	25 300	24 040	26 500
Mexico	19 751	22 347	22 513	22 766	25 091
Taiwan	14 806	15 627	17 281	18 043	20 350
Thailand	7 913	9 850	11 514	15 024	19 900
Poland	15 800	16 100	17 001	17 100	18 700
North Korea	7 983	8 981	11 793	16 329	18 000
United Kingdom	13 413	14 311	16 506	15 966	17 600
Indonesia	10 941	11 843	12 242	14 099	15 170
Romania	14 216	14 297	13 971	13 971	14 300
Total	772 327	806 066	856 880	879 835	971 205

Table 11. Some Notable Developments in the US Cement Industry

1960s	1970s to 1990s
Manufacturing	
Wet process installations reached their peak with giant kiln plants of the 1960s, which will be active for years. Manpower reduction primary cost reduction strategy.	New plants are dry process preheater calciner systems with roller mills due to rapidly increasing fuel cost. Large single production lines replace multiple old small lines, reducing manpower. Dry process dominant reduce to fuel costs.
Quarry Crushing	
Elevating pan scrapers were adapted for stripping where they can handle the material and the volume is larger.	Wheeled loaders increasingly used as "pick and carry" units to serve mobile primary crushers located in the quarries with the crushed product being conveyed to secondary crushing plants generally located out of active quarry areas.
Smaller drill holes, closer patterns, and lower quarry faces became common practice and probably improved safety and decreased nuisance. Dropball breakage mostly replaced secondary blasting of large pieces.	Impact crushers selected in place of gyratory and jaw types in cement-rock quarries.
Most blasting converted to use of ANFO, but success with other slurries and water gels indicate blasting agents and methods probably will continue to change.	In larger capacity quarries, some return to larger diameter blast practice has occurred.
Wheeled loaders and other end loaders were displacing shovels for the loading of trucks, both with shot rock and overburden.	Shovels disappear from scene. Tracked backhoe type excavators applied in quarries with difficult conditions.
Raw Milling-Drying-Blending	
Increase in size of wet grinding mills in the 1960s was spectacular, and overall efficiencies generally were achieved; nearly all were wet process, and some plants included rod or rod-peb mill.	Roller-dryer milling systems more commonly employed where the raw materials permit.
	Dry process blending systems dramatically improved, characterized by large capacity silos and computer directed random withdrawal schemes.
DSM screen positive classifying systems in wet process mills became widely accepted reducing cost and increasing capacity.	Particulate emission control from roller or ball mill dry grind circuits improved. Sulfur dioxide removal becomes a mandatory requirement driving selection of fabric filters for gas cleaning in place of electrostatic precipitators.
Pipeline transportation of slurry over relatively long distances was adapted to cement raw materials.	

Table 11. Some Notable Developments in the US Cement Industry (cont.)

1960s	1970s to 1990s
Preheating-Burning-Clinker Cooling	
The ever-increasing size of wet kilns reached its maximum. Both wet and dry kiln design effectively were committed to computer control with generally good results.	The dry counterflow preheaters, although not new, are improved and applied in a majority of new or replacement installations. The calciner is developed and adopted in most installations since 1975. Existing pre-heater systems adopt various schemes of adding fuel to the riser duct, including shredded or whole tires.
Chain use and effect generate almost as much theory as new results in the 1960s, with many variations being tried. Fuel efficiency in new wet process kilns significantly improved.	Conversion to the use of coal as the primary fuel and use of waste products as fuels widespread. Technology for destroying organic hazardous wastes by incineration in kilns developed. Improved planetary coolers are introduced. Large units prove to be unreliable. Grate coolers are improved using high bed-depth and high pressure fans.
Finish Cement Milling	
The trend to giant mills with ever higher horsepower characterized finish mill installations in the 1960s. Air operated clutches developed to permit starting large mill motors. Large gear reducers developed for center drive mills. Experimentation with dual drive motors and gear reducers for center drive mills. Sonic mill controls and cement cooler installations became common.	Awareness of mill system inefficiencies accelerate innovations in configuration, grinding media, and grinding environment. Classifying mill liners adopted. Widespread replacement occurs of older air-separator type classifiers with high efficiency separators. High pressure roll press type precrushers increasingly applied. Mill productivity improved. First wrap-around mill motor developed.
Storage Distribution	
Sales of cement in bags dwindled to still lower percentage. Long distance water transport increased and trucking made further drastic inroads on rail haulage. Nevertheless, railroads' pressure-flow cars did much to hasten the demise of many cement distribution terminals. Managements more intimately involved themselves in trucking problems than they did in rail carrier problems in prior years.	Conveyor belt transport of cement (vs. pneumatic pumps) increased for short-distance steady moves. Inflation in costs and local shortages of cement permitted some cement plants to choose closer, presumably more profitable markets in which to sell. Paradoxically, the same situation created the willingness of more distant customers to pay more for the transportation of their cement and encouraged imports to markets served by port facilities. The contraction in demand due to lower economic activity in the 1980s in conjunction with truck deregulation, excess overseas cement production and ocean shipping capacity, dramatically increases imports, lowers prices and causes economic hardships in all sectors of the industry.
Controls	
Centralized controls were an objective during the 1960s with development of far more discriminating sensors, better input intelligence, and long steps made toward automating responses to some of the input.	Controls rapidly continued to be refined and miniaturized, and whole groups of responses were automated. Motor control by programmable logic controllers replaced relays, mimic and control panels replaced by CRTs and sophisticated process system displays. Computer control strategies and artificial intelligence schemes extensively applied. Control-system costs reduced with better performance.
Pollution Control	
Sizing of dust collectors advanced rapidly to keep up with rapidly increasing plant sizes. Efficiency demands were higher than ever. Glass bag fabric filter installations came into prominence and electrostatic precipitators were improved to meet stringent requirements.	Costly efforts requiring an impressive array of equipment continued to be made to solve the very difficult technical problem of cleaning stack gases, and other sources of air pollution. Heat exchangers installed on clinker cooler exhaust gases to protect fabric filters from high gas temperatures. PSD regulations and permit conditions force much lower emissions than statutory limits.
Product	
Shrinkage compensated and rapid setting concretes introduced due to developments in cements. The strong push for the use of fly ash in concrete spread the use of that type of pozzolan. Higher fineness and high early strengths of Type I became the norm driven by market strategies for product differentiation.	Organic compound additions to cement along with curing of concrete by radiation continued to be developed as fast as needs and costs justified, because of remarkable product properties. Plastics and cement were joined to produce better materials instead of competing separately.

Table 11. Some Notable Developments in the US Cement Industry (cont.)

1960s	1970s to 1990s
Product *continued*	
Ferrocement received perhaps more press than use.	Colored cements and special architectural (i.e., buff) cements developed but have short term market appeal. Considerable efforts made to improve on fast setting cements used in concrete for highway repair.
	The adoption of high efficiency classifiers results in higher cement strengths at the same Blaine fineness or the same strength at lower fineness.
Use Applications	
The 1960s was the second decade and a far more successful one, in the remarkable growth of the pre-stressed field.	Modular systems construction with concrete, so widely used in Europe, found a more effective place in the United States in the 1970s.
The established technology of shotcrete was improved, and applications, particularly underground in rock, matured.	Metal fiber reinforcement of concrete found some specialized uses.
Continuous slip forming of concrete in roads and many varieties of structures moved into new prominence.	Intense effort in 1970s by government agencies and others on the potential use of fly ash and slag in finished cement results in the major commitment to develop granulated slag facility at an East Coast location.
Lighweight concrete generally matured in its many tested and accepted uses.	
Organizational	
Cement companies entered the 1960s as a diversified group but commonly characterized by operations engineering personnel dominating, or having a strong voice, in management. As sales situations became desperate, a swing toward more sales-oriented managements took place; however, the move toward financially expert managers was the indicated direction by the end of the decade.	With the vast array of available equipment to provide quick, accurate accounting, and computers adding decision making, it seems almost paradoxical that decentralization of management was the established trend of the 1970s. Widespread profit centers with considerable autonomy were either in effect or under formation, in many companies, as the decade ended.
With growing diversification, the professional manager, who might not have had an intimate cement background, more commonly was in charge.	Severe profitability problems of the 1980s lead to extensive restructuring, changes in ownership, bankruptcies, closing of plants, and decline of the unionization of the workforce.
	Although management's clear charge is to achieve an adequate return on investment, the inability to generate profits, the burden of societal and environmental standards, and the constraints caused by imports results in the failure to achieve this goal.

Considerable capacity remains in need of modernization. Much of the wet process capacity continues to produce cement only because a significant portion of the fuel to fire the kilns are ignitable wastes, classified as hazardous by the EPA, and therefore available on favorable economic terms. Modernization of older less efficient capacity will be restrained in the balance of this decade by the lack of sufficient cash flow, the imports of cement and clinker, and environmental permit constraints.

The Canadian cement industry has established itself in the US market and since it has the capacity to produce about 5 Mt more than its domestic market requires, the excess capacity is available to flow to the United States. The preponderance of Canadian plant capacity is owned by Lafarge, Holderbank, St. Marys, Cimenteries CBR, S.A., and Ciments Française. These companies have significant US capacity or market positions.

The Mexican cement industry is poised to take a major role in the US. Although currently constrained by anti-dumping charges, the Mexican cement industry is expected to be a major supplier in the southern United States after the US-Mexico-Canada free trade agreement comes into effect. Other Latin American cement companies are also poised to participate in the US market provided ocean freight rates remain favorable. Table 10 is a summary of cement production by country. The US ranked fourth in 1990.

This summary would not be complete without a brief review of the technology advances of the industry. The 1970s witnessed the acceptance of more economical pyroprocessing system, principally the adoption of the calciner technology in conjunction with 4- stage preheaters. The 1980s saw further refinement with the introduction of 5- and even 6-stage preheaters and low pressure drop preheater vessels. The major development of the 1980s was the reduction of power consumption in grinding by the adoption of high efficiency classifiers (air separators) in cement grinding circuits and the introduction of the roll press to prefracture cement clinker. Together these developments have reduced power consumption by about 30% at 3 500 to 3 600 Blaine fineness and even greater reduction at higher Blaine numbers. Manpower productivity gains have been achieved by even larger kiln systems made possible by the calciner. Kilns with a capacity of 5 ktpd are becoming common, and those with 10 ktpd are expected. The manpower level required to operate and maintain these single large systems is the same as a single kiln system of one fifth the capacity. Table 11 summarizes these and other technical improvements over the past 25 years. The decline of union representation at US cement plants in the 1980s has also contributed to improvement in productivity, but this aspect can hardly be called a technical improvement.

ACKNOWLEDGMENT

This chapter in the 5th edition of *Industrial Minerals and Rocks* was written by John A. Ames and William E. Cutcliffe, and this version in the 6th edition is by John D. Macfadyen.

BIBLIOGRAPHY AND REFERENCES

Literature and information sources dealing with the subject matter of this chapter are vast beyond assessment here, even if they were limited to writings in English and mainly generated in the United States. The magnitude of sources may be hinted by the fact that more than 60 national associations exist that have something to do with cement and concrete. Government publications related to the subject fill a small library. This bibliography includes representative and major sources pertinent to the subject. Bibliographies of the literature cited here contain references that would comprise an adequate library on the subject. Publications of associations, societies, and institutes, along with other trade journals, are important current sources of information, as are knowledgeable people in cement-related companies, design engineers, materials and equipment suppliers, and product users.

This bibliography is organized under headings intended to be descriptive.

Standard References

Blanks, R.F., and Kennedy, H.L., 1955, *The Technology of Cement and Concrete*, John Wiley & Sons, New York, 422 pp.

Bogue, R.H., 1955, *The Chemistry of Portland Cement*, 2nd ed., Reinhold Publishing Corp., New York, 793 pp.

Bransome, M., 1985, Draft Summary Report on Hazardous Waste Combustion in Calcining Kilns, US Environmental Protection Agency, Cincinnati, 67 pp.

Clausen, C.F., 1960, "Cement Materials," *Industrial Minerals and Rocks*, 3rd ed., J.L. Billson, ed., AIME, New York, pp. 203–231.

Eckel, E.C., 1928, *Cements, Limes and Plasters*, 3rd ed., John Wiley & Sons, New York, 401 pp.

Hansen, W.C., and Offutt, J.S., 1969, *Gypsum and Anhydrite in Portland Cement*, 2nd ed., United States Gypsum Co., 81 pp.

Harben, P., 1979, "The Clamor for Cement," *Industrial Minerals* No. 143, Aug., pp. 52–53.

Lea, F.M., 1971, *The Chemistry of Cement and Concrete*, 3rd ed. (1st American ed.), Chemical Publishing Co., New York, 760 pp.

LeLonde, W.S., and Janes, M., 1961, *Concrete Engineering Handbook*, McGraw-Hill, New York, 1216 pp.

Lesley, R.W., 1924, *History of the Portland Cement Industry in the United States*, Chicago International Trade Press, 330 pp.

MacDonald, L.P., et al., 1977, "Burning Waste Chlorinated Hydrocarbons in a Cement Kiln," Environmental Protection Service Report No. EPS-4-EP-77–2, 221 pp.

Meade, R.K., 1926, *Portland Cement*, Chemical Publishing Co., New York, 707 pp.

Peray, K.E., and Waddell, J.J., 1972, *The Rotary Cement Kiln*, Chemical Publishing Co., Inc., New York, 194 pp.

Queen, D.M., 1971, *The US Cement Industry (A Memorandum)*, Evans & Co., Inc., (investment brokers), New York, 50 pp.

Randall, R., 1961, "A Reference for Manufacturing Process Department Literature," MP-100, Portland Cement Association, Dec., 172 pp.

Taylor, H.F.W., et al., 1964, *The Chemistry of Cements*, 2 vols., Academic Press, London and New York, 442 pp.

Troxell, G.E., Davis, H.E., and Kelly, J.W., 1968, *Composition and Properties of Concrete*, 2nd ed., University of California, Berkeley, Civil Engineering Series, 513 pp.

Witt, J.C., 1947, *Portland Cement Technology*, Chemical Publishing Co., New York, 346 pp.

Some Trade Journals Include

Cement and Concrete Research, published bi-monthly, Pergamon Press, New York.

Concrete Construction, published monthly.

Concrete Products, published monthly.

Engineering News Record, published weekly.

International Cement Review, published monthly.

Zement Kalk Gips, published monthly.

Modern Concrete, published monthly.

Pit & Quarry, published monthly.

Rock Products, published monthly.

Some Organizations and Some of Their Publications

American Association of State Highway and Transportation Officials, Washington DC.

American Ceramic Society, *Journal of the American Ceramic Society*.

American Concrete Institute, Detroit, MI, *Journal of the American Concrete Institute*, published monthly, *ACI Manual of Concrete Practice* (3 parts), current edition, ACI Special Publications.

Society for Mining, Metallurgy, and Exploration, Littleton, CO, *Mining Engineering*, published monthly.

American Society for Testing and Materials, Philadelphia, PA, *1980 Annual Book of ASTM Standards*, "Part 14, Concrete & Mineral Aggregates."

Portland Cement Association, Skokie, IL, various publications.

US Dept. of Defense, Army Corps of Engineers, various publications; e.g., *Handbook for Concrete and Cement*, 1949, with loose leaf revisions, 928 pp.

US Dept. of the Interior, US Bureau of Mines; *Minerals Industry Survey* (Cement Monthly), *Minerals Yearbook* (Annual Cement Chapter).

US Dept. of the Interior, Bureau of Reclamation various publications; e.g., *Concrete Manual* (current edition), usually 600 to 700 pp.

CONSTRUCTION USES

CLAY-BASED CERAMIC RAW MATERIALS

JOHN F. BURST AND RANDALL E. HUGHES

This chapter describes the nature and occurrence of the principal clay minerals that are used as ceramic raw materials, explains how these raw materials are used to produce finished goods, and summarizes positive and negative trends that will affect future demand for raw materials and products in the ceramic industry.

The term ceramics was derived from the ancient Greek *keramikos* (a clay jar). A ceramic product is a material that has been fired to at least 468.3°C (Mitchell, 1983). In addition to being one of our earliest manufacturing and construction materials, ceramic products are among the earliest records of civilization. The enduring significance of ceramics is shown by the fact that, even today, the manufacture of whitewares, brick and related red-fired wares, refractories, and engineered ceramics continues to be a major sector of the US economy.

A wide range of materials can be made into ceramic products, although relatively few earth materials make up most of the ceramics in today's markets (Table 1). Most ceramic products are made from clay or a related material. The long history of clay use in the ceramics industry stems from its variable and sometimes unique physical and chemical properties, as well as its ubiquitous occurrence and relatively low cost. Clay is important to ceramics because of its plasticity, moldability and filterability, chemical content, color, refractoriness, solubility, corrosion resistance, electrical properties, and other properties that result from the essentially two dimensional morphology of the individual clay mineral platelets.

Because of their basic relationships to the chemical and physical properties of clays, the chemical and mineralogical content have been the subject of continuous studies over the past few decades. Advances in knowledge about the chemical, mineralogical, and physical properties of clays are responsible for most of the advances in processing. The need to remove off-colored contaminants, minerals with low melting points, and soluble salts has resulted in beneficiation by wet processing methods such as water washing. Water washing is restricted to kaolins, although low cost methods of water washing have been extended more recently to ball clays in Europe. High intensity magnetic filtration and flotation technologies have noticeably improved the quality of kaolins and broadly expanded their use in the ceramics industry.

In addition to products for construction (e.g., brick, tile, sewer pipe, field drain pipe, cement, fiberglass, roof granules, and paint) and manufacturing (e.g., refractories, kiln furniture, sanitary ware, pottery, and dinnerware), clays are used in calcined form as ceramic fillers, carriers, extenders, and grogs. Ceramics suppliers are under economic pressure in several of these product areas. Manufacture of refractories in America has declined as steel production has decreased. Similarly, dinnerware and tile production have declined because of foreign competition. Fortunately, new uses for clays, such as the ceramic monolith for catalytic converters, offer promise of another millennium of clay-based ceramics.

RAW MATERIALS

Mitchell (1983) separated the components of ceramic raw materials into seven groups: 1) skeleton formers, 2) glass formers and bonding agents, 3) fluxes, 4) special property producers or modifiers, 5) processing or fabrication aids, 6) environmental or fabrication conditioners, and 7) secondary and overlapping constituents. Refractory components constitute the first group. Nonrefractory minerals and soluble salts make up most of groups 2 and 3. Group 4 materials give rise to color and special properties such as texture, insulating capacity, shielding properties, machinability, and related properties. The processing or fabrication aids in group 5 include water and oils, various organic materials, and some clays, especially those with high plasticity. Group 6 is made up of various gases that can be introduced into the kiln. Finally, several materials provide two or more functions in the ceramic body (Mitchell, 1983).

Common Clays, Shales, and Fireclays

Common clays and shales are the most widespread ceramic materials in terms of both occurrence and numbers of mines and plants. Because transportation is a relatively large factor in the final cost of both the raw materials and finished products, mines and plants often are located near one another and near large metropolitan markets. Economic forces of scale, automated plants, and artificially low shipping rates to the upper Midwest on freight cars returning from southern ports have worked against small, geographically widespread brick and tile plants (Hughes, White, and Warren, 1989). Competition from several products made from wood, plastic, cement, and metal has significantly reduced the demand for similar products made from clays and shales.

Occurrence, Mineralogy, and Geology: Most clayey and silty materials near the earth's surface can be used to make brick, structural tile, drain pipes, flower pots and vases, roof tile, or similar products (O'Connor, 1989, Hughes, White, and Warren, 1989). Generally, materials that offer a range of color and firing characteristics and that can be extracted from a single location or several nearby locations are most likely to be profitably used. Most raw materials are either illitic or kaolinitic in composition. Typical red-firing shale contains abundant illite, with lesser amounts of

Table 1. Ceramic Raw Materials (after Burst, 1991)

Soils, clays, and shales	
Fireclays and flint clays	
Plastic	pce 15–26
Flint clay	pce 26–33
Burley flint clay	pce 33–35
Ball Clay	
Kaolin	
Bauxite and bauxitic clay	
Talc	
Pyrophyllite	
Andalusite	
Kyanite	
Sillimanite	
Quartz	
Feldspar	

chlorite, kaolinite, and quartz, and small amounts of mixed-layered illite/smectite (I/S) and feldspars. Producers generally avoid materials containing expandable clay minerals such as smectite, vermiculite, and I/S because their presence in significant quantities leads to excessive shrinkage during drying as well as other problems.

Because the terminology for kaolinitic raw materials is often confusing, several terms are defined in Table 2. Fireclays are clays associated with coals, commonly as underclays, and contain sufficient kaolinite and low enough amounts of mineral fluxes to meet the requirements of a low to medium grade refractory clay. Most of the fireclays associated with coals are paleosols (ancient soils) in which weathering of detrital clay and nonclay minerals by silica-extracting plants and/or percolating water produced a deposit rich in kaolinite and mixed-layered kaolinite/expandables (K/E) or kaolinite/smectite (Hughes et al., 1987, Hughes, Moore, and Reynolds, 1993). The plasticity and firing properties that distinguish fireclays from kaolins with a similar alumina content result both from I/S and relatively abundant K/E. In fireclays, K/E forms as the major weathering product of illite, illite/smectite, smectite, and vermiculite.

Table 2. Definitions of Kaolinitic Raw Materials (partly after Bates and Jackson, 1980)

Ball clay	A plastic, kaolinitic clay with minor to abundant organic matter producing high green strength and fired shrinkage in ceramic bodies and usually firing to white. The term comes from the old English practice of rolling clays in ~25 cm balls.
Fireclay	A plastic, kaolinitic clay with sufficient Al_2O_3 to be refractory. The clay usually occurs as an underclay. Ampian's (1991) inclusion of flint clays within the fireclay category is not supported by the differences between the two clays.
Flint clay	A hard, nonplastic, kaolinitic claystone that breaks with conchoidal fracture and does not disperse in water. "Burley flint clay" is a claystone in which aluminum-rich minerals such as diaspore raise the Al_2O_3 above that of kaolinite (~40%).
Kaolin	A soft, white, relatively nonplastic but dispersible, kaolinitic clay. In primary and sandy kaolins that are less than one-half kaolinite, the term kaolin reflects the purity and properties of the fine-fraction separate.
Underclay	A stratum of soft, dispersible clay or claystone that typically underlies coal. Hughes et al. (1987) distinguished three facies: (1) shale-type or illitic, (2) soil-type or fireclay-type (kaolinitic), and (3) gley-type (rich in mixed-layered illite/smectite). They regarded some claystones without an overlying coal to be underclays, if those clays showed alteration patterns typical of underclays.

Table 3. Product Grades of Refractory Raw Materials (modified from Burst, 1991)

% Al_2O_3	Source clay
15–50	Fireclay—Flint Clay
~60	Alabama bauxite
70—85	Flint clay/bauxite blends
85–90	Bauxite
90–99	Synthetic bauxite

Flint Clays

Missouri flint clays contain about 40% Al_2O_3 and are extensively used for refractories. Clays for refractories must be low in iron and alkalis, and they normally fire to a white color. Table 3 lists the grades and sources of materials that are standard in refractories markets.

Occurrence, Mineralogy, and Geology: Missouri fireclays and flint clays usually occur as sinkhole deposits. The Missouri deposits vary from plastic fireclays with less than 35% Al_2O_3 to flint clays with an alumina content of kaolinite of about 40% (Keller, Wescott, and Bledsoe, 1954). Above 40% Al_2O_3, diaspore becomes more abundant and the clays are given the name burley flint clay. Flint clay, referred to as tonstein in Europe, derives its name from its hardness and conchoidal fracture, which result from the fact that the crystallites of kaolinite are intergrown to such a degree that they appear to have been derived from a solution or *melt* much as would be observed for an igneous rock (Keller, 1981). Fireclays are plastic and the individual crystallites can be dispersed in water; flint clays are nonplastic and cannot be dispersed.

Flint clays are thought to form by the alteration of detritus in depositional environments associated with coal formation (Keller, Wescott, and Bledsoe, 1954, Hughes et al., 1987, Gardner, Williams, and Holbrook, 1988). The Al_2O_3 content of these deposits has been increased by leaching of alkalies, alkaline earths, and SiO_2 from detritus. The parent detritus contained illite with lesser chlorite and I/S, quartz, and minor K-feldspar and plagioclase. Leaching is assumed to have occurred in sinkholes or swampy lowlands, and the distribution of alumina in the deposits suggests that the leaching was more intense toward the center of the deposit. As a result, the typical Missouri deposit is characterized by a core with a high alumina content of up to 70% Al_2O_3, surrounded by clay with a lower alumina content (Keller, Wescott, and Bledsoe, 1954). This outer zone varies considerably in alumina content from place to place in the deposit. Therefore, the outer zones of a deposit often are mined first, and alumina content is controlled by blending core material to produce a product with the desired Al_2O_3 content. Clay producers and consumers often blend varieties of clay for additional control of specific properties.

Ball Clays

Ball clays are used in a wide variety of ceramic products. Because most ball clays burn to a light color and readily accept glaze, sanitary ware manufacturers are large users of ball clay. Other products made from ball clay include wall tile and floor tile, granules for roofing, dinnerware and earthenware, and glass fiber insulation. Ball clays contain more kaolinite than most common clays and shales, and they are more plastic than kaolins and flint clays. Some fireclays have properties that are within the range of colloidal properties generally associated with ball clays. One difference between ball clays and fireclays is abundant organic matter of colloidal size in ball clays. The plasticity and related properties of ball clays probably result both from the universal presence of expandable clay minerals (K/E, I/S, and smectite) and the common presence of fine-grained organic matter.

Ball clays are often classified according to particle size. This parameter is thought to be the single most useful indicator of forming properties (Anon., 1972). Clays are classified by their content of less than 0.5 μm particles. The 0.5 μm content of a coarse clay is 0 to 39%, medium is 40 to 49%, and fine is above 50%.

Occurrence, Mineralogy, and Geology: Major ball clay deposits in the United States lie in a generally north-south trending line east of and roughly parallel to the Mississippi River in Kentucky, Tennessee, and Mississippi (Patterson and Murray, 1984). Ampian (1991) also reports production of ball clays in California and Texas. The deposits within the Mississippi Embayment sedimentary sequence are found in the Claiborne Formation (Eocene). Many of the deposits contain lignite, and the clays seem to have been deposited in swampy peat-forming environments (Olive and Finch, 1969). Mineralogical differences between typical ball clay deposits and those developed in situ as paleosols on the smectitic Porters Creek Formation (Kirk and Bougher, 1986) suggest that most of the deposits were formed by erosion of thick soils from paleogeographic highs and that only minor alteration or weathering occurred at the site of deposition.

Large deposits of ball clays are mined in Germany, Thailand, and southern Britain. The recent political changes in the former Soviet bloc and China will probably result in increased production and export of ball clays from these areas.

Ball clays are composed of variable amounts of kaolinite, smectite, illite, and quartz with minor amounts of K-feldspar, plagioclase, muscovite, I/S, and organic matter. Little is known about the organic matter, but it probably has a major effect on the properties of clays from some deposits.

A recent, unpublished survey (Hughes and Leach, Kentucky-Tennessee Clay Company) of representative materials from the Mississippi Embayment district showed that all samples contained kaolinite (30 to 90% by weight based on estimates from X-ray diffraction, XRD). Illite was present universally and varied from about 1 to 30% of the sample. About half the samples contained smectite (0 to 10%). Few samples showed I/S; however, the XRD traces of most samples contained a shoulder on the illite peak that is typical of I/S with a low proportion of smectite interlayering. Quartz in these ball clays varied from about 10 to 50%. K-feldspar was present in about two of three samples and plagioclase was detectable in one in eight. Neither feldspar exceeded 5% of any sample.

Kaolins

Kaolin is used in products that require its low content of iron and other contaminants, particle size and shape, and white color. Kaolins from the southeastern United States are the material of choice in many ceramics applications because of their purity. The distinction between kaolin, ball clay, and fireclay is somewhat arbitrary. However, kaolins generally contain more than 90% kaolinite as mined, and most are white, light yellow, or light gray. Part of the distinction is based historically on the low ion-exchange capacity and plasticity of Cretaceous kaolins in the Southeast. Tertiary kaolins in the Southeast and kaolins from other localities have properties intermediate between Cretaceous kaolins, ball clays, and fireclays.

Occurrence, Mineralogy, and Geology: Major deposits of sedimentary kaolin occur along the Fall Line (the location of a structural monocline and the boundary between the Piedmont and the Coastal plain physiographic provinces) in Georgia and to a lesser extent in Alabama and South Carolina. Murray (1988) proposed that the Cretaceous kaolins in this area are derived from the granite and gneiss of the Piedmont to the north and northwest of the deposits. He also proposed that Tertiary kaolins in the same area are reworked Cretaceous deposits. However, analyses of the trace elements show that Tertiary kaolins in east Georgia and South Carolina were derived from felsic metavolcanics, and those of Tertiary age in mid-Georgia were derived from both granite-gneiss and metavolcanic sources (Dombrowski, Murray, and Gilstrap, 1989). The principal contaminants in kaolins of both ages are quartz, muscovite, and anatase; smectite is much less common in these deposits than it is in ball clays. Some gray deposits contain pyrite and/or marcasite. Iron content varies from about 0.3 to 1%. Only 0.2 to 0.5% Fe_2O_3 is substituted within the structure of the kaolinite found in Cretaceous deposits. Deposits of Tertiary age, however, contain about 1% Fe_2O_3 within the kaolinite structure (Murray, 1988). Most of the Georgia deposits are soft and easily dispersed. In the Tertiary clays, bonding between fine particles increases the hardness and difficulty of dispersion. Kaolin deposits in California, Florida, Minnesota, North Carolina, and Texas are generally sandy; however, primary clays also contain abundant nonclay fragments (Patterson and Murray, 1984).

Cretaceous clays from the Georgia district are relatively coarse grained and have a bimodal particle-size distribution. The coarser mode, about 2μm, consists primarily of vermicular kaolinite particles; the finer mode, about 0.5μm, contains mainly plate-like crystallites of kaolinite (Hughes and Smith, 1972). Cretaceous clays are dominated by vermicular particles, although particle-size curves of fine-grained clays of Tertiary age have only a shoulder for the coarser mode and generally contain mostly finer, platy particles. The differences in particle size and shape for the two types of kaolins give rise to significant differences in properties. The typically low iron content of Cretaceous clays makes them especially suitable for markets where color or electrical resistance is important. Cretaceous clays also are used extensively if higher rates of filtration and/or relatively nonplastic clays are required. Delaminated clays made by milling the vermicular particles provide clays with a high aspect ratio (width to thickness). These clays are particularly suited to dinnerware and similar uses because their thin, platy particles give rise to special properties in the fired ware. Tertiary clays are favored for their higher plasticity and in processes where their platyness is advantageous and higher iron content poses no problems.

Raw Materials Other Than Clays

Nonclay raw materials include silica, magnesite-chromes, graphite, zirconia, forsterite, mullite, kyanite-group materials, bauxites and related materials with more than 60% Al_2O_3, silicon carbide, dead-burned magnesia or magnesite, and dead-burned dolomite (Ampian, 1991). The geology and properties of the naturally occurring materials in this group are covered in the commodities chapters of this book and will not be reviewed here.

BENEFICIATION

Clays, Shales, Fireclays, and Flint Clays

Common clays, shales, and fireclays require and receive little beneficiation. Generally, these materials are only crushed, shredded, or ground and shipped. Occasionally, concretions and other deleterious matter are removed by dry screening. A few clays within this category are beneficiated by drying and air flotation in order to meet ball clay or kaolin product specifications. Clays used in refractories are often blended to meet product standards.

Ball Clays

Ball clays are usually mined in several grades so that they can be blended to maintain standards. Air classification of dried and ground material is still the standard preparation in the ball clay industry, but slurry production is becoming increasingly common. Other wet cleaning methods borrowed from the kaolin industry, including particle-size classification, leaching and filtration, magnetic separation, froth flotation, differential flocculation, and spray drying, have been evaluated and are being employed in Europe. Various surface treatments of both dry and slurry clays are possible. These treatments improve forming of the ceramic body in water-based or organic-based systems.

Beneficiation of any clay by any type of process is driven by cost and performance. For example, optimum products can be produced by processing ball clays as slurries. Slurries also eliminate the cost and difficulty of producing slurry at the site of the ceramic consumer. Most slurries save energy at the site of product manufacture, although the amount of these savings must be balanced against the cost of shipping water to the user location. At present, the most promising beneficiation processes include improved slurry production, size classification, leaching, filtration, and magnetic separation.

Kaolins

All of the beneficiation steps cited above for ball clays were developed originally for kaolins. In addition, the process of delaminating vermicular particles was discovered about 30 years ago (Billue, 1962). This process produces clays that are particularly suited to production of dinnerware and pottery. Because kaolins usually command a higher price, more expensive processing often can be justified.

USE IN CERAMICS

Forming

Most clays processed for common brick and similar products are formed by creating a plastic mass with clay, dispersants, and water, and then extruding the clay mass through a die. Various shapes can be achieved by changing the die and cutting or drilling the extruded product. The product is then stacked, dried, and fired.

Pottery and related products are often cast from a slurry. This process works by depositing the clays and other additives onto the wall of a mold. The cast product is then dried and often coated with glaze before firing.

Many new products are formed with powder technology. Wet or dry materials can be sprayed on, dip coated, or electrocoated. Some finished products can be applied by spraying or painting, or they can be extruded at the site of use, for example, into a furnace.

Small variations in chemical and physical characteristics of clay materials can have a significant effect on their forming properties. Soluble salts in clays affect casting properties; greater amounts of soluble salts increase the effective range of deflocculants. Colloidal organic matter affects slip properties as well as fired color. Even small amounts of smectite-group clay minerals and the smectitic

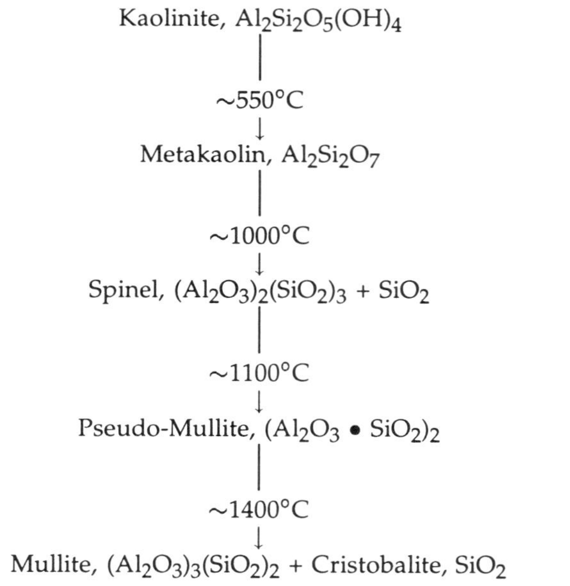

Fig. 1. Schematic diagram of the firing of kaolinitic raw materials (after Burst, 1991).

layers in K/E and I/S have dramatic effects on the rheological properties of casting slurries. Viscosity and surface area are important properties. For kaolins, particle shape is important to the forming process and the rate at which the green product can be dried.

Fine-grained ball clays add plasticity for ease of forming. Coarse-grained materials increase drying rate, decrease excessive plasticity, and reduce shrinkage and cracking. Clays with natural or manmade (delaminated) platy particles give dinnerware a desirable translucence and can create increased strength and other useful properties in many products. Many other guides to the selection of raw materials can be used to maximize efficiency during forming and firing, and thus consistently produce products at minimum cost. Because many of these guides are based on indirect measures of the properties required, these guides change as technology advances and are beyond the scope of this chapter.

Firing of Ceramic Materials

The essence of ceramics is the conversion of raw materials into a durable object through the application of heat. Both the chemical content and crystallite size of constituent minerals affect the response of materials to the firing process. Materials with more refractory minerals require a higher temperature to achieve adequate firing. For example, a typical red-firing clay for brick manufacture will contain enough fluxes to fire to full hardness by 900 to 1 200°C. However, most kaolinitic raw materials require higher temperatures to create a hard, durable product. Many problems can occur during firing that will lead to prohibitive warping or breaking in the kiln or to other deleterious characteristics. For example, underfiring can result in moisture expansion of bricks due to hydroxyl absorption. This expansion causes damage months or years after the brick is incorporated into a structure (Hughes, Bargh, and White, 1983).

Kaolinite, illite, K/E, I/S, montmorillonite, and muscovite react similarly during firing. The first reaction is dehydroxylation to an anhydride such as metakaolin (Fig. 1). This reaction generally takes

Table 4. Three Ceramic Product Groups (modified from Burst, 1991)

Whitewares	Structural and refractory products	Advanced technology ceramics
Sanitary ware	Brick and structural tile	Engineered ceramics
Vitreous china	Floor and wall tile	Abrasives
Earthenware	Refractories	Vitreous enamel
Electrical porcelain	Structural clay products	Glass
Pottery		

place between 400 and 750°C, depending on the rate of heating, the mineralogical composition, crystallite size, and degree of structural order within the individual clay mineral. Between about 650 and 1 000°C, the anhydride is converted to phases such as silicon spinel, γ-alumina, and silicon-aluminum spinel (Fig. 1). Mullite is usually formed concurrently and becomes more abundant at higher temperatures. The exothermic differential thermal analysis (DTA) peak at about 980°C for kaolinite-rich materials is thought to result from the crystallization of γ-Al_2O_3 from kaolinite and K/E in fireclays, ball clays, and halloysites, and of mullite and γ-Al_2O_3 from well-crystallized kaolinite in kaolins (Tsuzuki, 1961, Watanabe, et al., 1987).

As shown in Fig. 1, mullite and cristobalite are the dominant phases formed above about 1 100°C; by about 1 400°C, the kaolinitic raw material has been completely converted to mullite plus cristobalite (Brindley and Nakahira, 1959, Brindley, 1976). Mullite persists to 1 850°C, but melting of cristobalite begins at about 1 600°C; therefore, mullite refractories are normally not fired above about 1 650°C.

The high melting temperature of mullite and its interlocking acicular crystals produce a product with several valuable properties, especially strength and low specific gravity. Mullite forms with a range of compositions between the end-members $(Al_2O_3)_3(SiO_2)_2$, $(Al_2O_3)_2(SiO_2)$, and $(Al_{0.9}Fe_{0.1})_2O_3)_3(SiO_2)_2$ (Burst, 1991). These compositions are similar to those of many kaolins, fireclays, and flint clays, and these clays are the primary raw materials for the production of mullite refractories.

Ceramic Products

Burst (1991) separated the range of ceramic products into three main groups: whitewares, structural clay products and refractories, and advanced technology or engineered ceramics (Table 4). These groups are somewhat arbitrary, but the raw materials for each generally have similarities that distinguish them from the other groups. The third group differs from the others because raw materials other than clays are used to make most of these products. Raw materials for the first group are generally kaolinitic, and the products are all produced by molding. Products within the second group are largely made from clays, and most are extruded during forming.

Whitewares: As the name implies, whitewares include products that fire white or near-white. Sanitary ware, earthenware, wall tile, stoneware, porcelain, and electrical porcelain are the principal products in this group (Robbins, 1984). Kaolin, ball clay, quartz, and fluxes make up most whitewares; they are normally fired between 1 150 and 1 250°C (Burst, 1991). Electrical porcelain must be fired to about 1 400°C to produce an acceptable product. Two types of electrical porcelains are produced: 1) a steatite-type that contains a small amount of bonding clay, and 2) a cordierite-type that contains about 70% talc and 30% clay (Burst, 1991).

Most whitewares are composed primarily of kaolinite-rich raw materials-ball clay, kaolin, fireclay, and flint clay. Preferred materials contain mostly kaolinite and little iron, typically much less than 1% as Fe_2O_3. Kaolinite aids in the forming of whitewares and increases the strength of unfired and fired products (Robbins, 1984). Use of kaolinitic clays ranges from near 100% of porcelains to 20 to 30% of sanitary ware products (Burst, 1991).

Ball clays are used in whitewares to improve the plasticity and dry strength and to control the rheological properties of the mixture during forming or slip-casting (Robbins, 1984). Ball clays also contain minerals that increase the vitrification temperature range. Wall tile, earthenware, stoneware, sanitary ware, and electrical porcelain insulators each require special ball clays to impart distinctive properties (Robbins, 1984).

Brick and Structural Clay Products: The high costs of breakage and warping, energy, and beneficiation have forced the structural clay products industry to use higher quality raw materials. Although some field drain tiles are still manufactured from glacial lake clay, most manufacturers of bricks and related products have survived because they obtain raw materials that minimize production costs and provide the variety of products required in the market place.

An ideal raw material for bricks and related fired-clay products contains minerals with a range of melting points. This range makes it possible to fire the ceramic body to a steel-hard state without extensive melting and concomitant distortion of the body.

Noncalcareous illitic to kaolinitic shales are often optimal, especially when the ratio of clay minerals to nonclay minerals and the ratio of illite to kaolinite can be controlled by selective mining. Variation in mineral content often can be obtained at one site by mining the fireclay facies of an underclay and an upward coarsening roof shale of the same coal. This control of raw materials facilitates complete firing and production of ceramics with a range of colors and strengths (Hughes, White, and Warren, 1989).

Some clays and shales contain concretions or pebbles that must be removed by screening. On occasion, carbonate pebbles are so abundant that the fired ware must be treated with steam during the cool down in the kiln to inhibit pops that ruin the products (Hughes, White, and Warren, 1989).

For efficient extrusion, clays must have sufficient plasticity; however, excessive plasticity generally is associated with firing problems. Some clays and shales are too fine for optimal firing. A shale or clay deposit with variation in particle sizes allows the producer to maintain enough plasticity (from fine particles) and obtain rapid and complete firing (from coarse particles).

Many clays contain contaminants, such as pyrite, that cause firing and pollution problems. Several minerals cause serious forming and firing problems not reflected in their melting points. Dissolved salts, expandable clay minerals, iron-rich minerals, carbonates, and sulfides and sulfates are undesirable minerals in raw materials.

Careful selection of raw materials can raise the efficiency in other aspects of manufacture. For example, materials can be chosen to provide a natural range of fired colors at minimum cost. Broad firing range may be the most important raw material property. The challenge for the producer is to generate a product with minimal product losses, deviations from specifications, energy input, and processing costs. Given the variety of properties required for products and the natural variations in clays and shales, optimization will remain a problem.

Refractory Products: Clay-based refractory producers have suffered in North America because consumers have changed their production processes (Burst, 1991). The new processes require materials that can withstand higher temperatures and resist corrosion and other types of degradation caused by new metallurgical techniques. Burst (1991) points out, however, that ball clay consumption has increased, due primarily to the increased production of gunning materials (plastic mixes that are extruded or sprayed).

Materials for making refractories should contain high levels of

Al_2O_3 and minimal amounts of low melting point minerals and soluble fluxes. In metals applications, refractories in contact with molten metal should resist erosion and incorporation into the melt. Generally this resistance is obtained by using raw materials with a high melting temperature or pyrometric cone equivalent (PCE). However, some low to moderate PCE ball clays perform well in contact with molten metal because they expand slightly during use and seal the spaces between bricks and similar elements. This expansion is caused by illite and/or similar micaceous minerals that expand significantly at high temperatures. Like the manufacturers of whitewares and structural clay products, refractory producers often use the inherent variations in natural raw materials to meet a large number of product requirements.

Advanced Technology Ceramics: Several of the products manufactured for markets in this area are made primarily from non-clay raw materials and are not discussed further in this chapter. Kiln furniture and related refractories have been made for many years from raw materials that fire to cordierite. Burst (1991) states that cordierite's low specific gravity, low coefficient of thermal expansion (CTE), and long service life have led to important new uses for this material. New uses include monolithic catalyst carriers, ceramic membranes, permanent foundry molds, heat exchanger parts, and investment casting molds (Burst, 1991).

Catalyst carriers. Burst (1991) states that the construction of the monolithic catalytic carrier is an engineering marvel. It has been standard emission control equipment for the past 15 years in the United States. Cordierite is normally used for automotive applications, but mullite is used in some applications. Cordierite is preferred because it has a lower CTE than mullite. Other more exotic formulations are used for special applications in certain hostile environments.

Ceramic membranes. Ceramic membranes or filters are a new group of ceramic materials that are closely related to automotive catalyst supports. They can be used in a wide variety of applications, from beer filtration to hot metal separation. Ceramic filters need to withstand processing temperatures as high as 1 000°C and the effects of highly corrosive cleaners and live steam, which is also used as a sanitizer and sterilizer. Burst (1991) also points out that ceramic membranes that are cleaned automatically during operation are being developed for diesel engines and power generation facilities. These devices should make it possible to keep diesel vehicles and stationary installations in service, drastically reduce emissions, and obtain long service life for the membrane.

Permanent foundry molds. The normal foundry system of constructing a mold of sand, clay, and other binders and additives for a single casting is extremely inefficient. Permanent ceramic molds probably could replace this inefficient system, although maintenance of precise dimensions may be a problem as the mold is used.

Heat exchangers. Burst (1991) argues that low-CTE materials such as cordierite should be excellent ceramic heat exchangers. He states, however, that silicon carbide may dominate the field because it was the first material to be successfully used. "Perhaps cordierite will not find use as an exchange tube component but the lower cost and ease of manufacture of cordierite shapes still argue for their application somewhere in the overall process" (Burst, 1991).

Investment molds. These molds are another possible market for cordierite. The process involves coating a binder-saturated, granular cordierite around a pattern, melting the pattern, casting the metal object within the cordierite mold, and removing the cordierite mold after the metal has cooled (Burst, 1991).

Present Production and Future Market Trends

Production statistics for clays in the United States and rest of the world are given by Ampian (1991) and summarized in Table 5 and Fig. 2. A total of 14.1 Mt of common clay and shale were mined in 47 of the 48 contiguous states. Ball clays were produced from California, Kentucky, Mississippi, Tennessee, and Texas, and total production for ceramics equaled 664.8 kt (Ampian, 1991). More than two-thirds of the production and just less than two-thirds of the value of ball clays came from Tennessee. Fireclay was mined in 11 states, although Missouri had about 60% of all the mines and produced over one-half of the tonnage and value. Fireclay and flint clay production for ceramics equaled 698.9 kt. Kaolin was produced in 13 states; Georgia produced about 84% (7.5 Mt) of the tonnage and over 92% ($945 million) of the value (Ampian, 1991). Much of the value of kaolin production was for paper coatings and fillers; ceramics markets for kaolin accounted for 1.35 Mt (about 15%) of the total production (Ampian, 1991). The amount of bentonite and fuller's earth used in ceramics was 74.2 and 28.4 kt, respectively.

Table 5. Raw Material Clays Used or Produced for Various Ceramic Markets in 1989 (in thousands of tons; after Ampian, 1991)

Product	Clay and shale	Fireclay[1]	Ball clay	Kaolin	Bentonite[2]
		Whitewares			
China, etc.			3.4	15.2	
Crockery		0.2		18	
Electrical porcelain			21.5	15.3	
Pottery	18.8	20.9	125.5	30.2	24.4
Sanitaryware			194.1	35.1	
Other	92.4+	1.6	73.9	89.2	W
Total	111.2+	22.7	418.4	203	24.4+
		Clay products			
Brick	12 433.4	19.8	W	303.2	W
Drain tile	10.3				
Floor and wall tile	24	4.2	176.7+	46	4.5*
Flower pots	45.1				
Flue linings	10.1	7.3		6.4	
Roofing tile	3.9				0.1*+
Sewer pipe	366.9	0.1		8.6	
Structural tile	49				
Terra Cotta	18.8	W			
Other	243	3.7	W	19.3	3
Total	13 204.5	35.0+	176.7+	383.3	7.7†
		Refractories			
Firebricks and blocks	100.2	546.8	W	115.3	W
High-Al brick		35.6	21.4	6.6	
Kiln furniture	2.8		1.5	5.9	
Mortar and cement	414.3			1.3	
Other	16.6		46.7	9.3	70.5†
Total	533.9	582.4	69.6	138.4	70.5†
		Engineered ceramics			
Grogs and calcines[3]		58.8		629	

W withheld.
[1]Fireclay and flint clay together.
[2]Bentonite includes bentonite and fuller's earth (*); †denotes both clays.
[3]These are mostly used for monoliths.

Table 5 gives production values for whitewares, clay products, refractories, and engineered ceramics. If clay products and refractories were combined, the resulting three categories would correspond directly with the product types outlined in Table 4. Fig. 2 shows pie diagrams of the total tonnage produced or used in 1989 by product market and type of clay resource. Although the amount

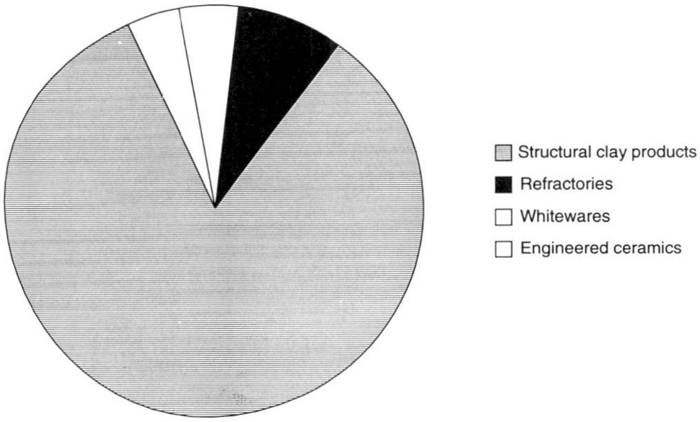

Fig. 2a. Schematic diagram of clay consumption for ceramic products in 1989.

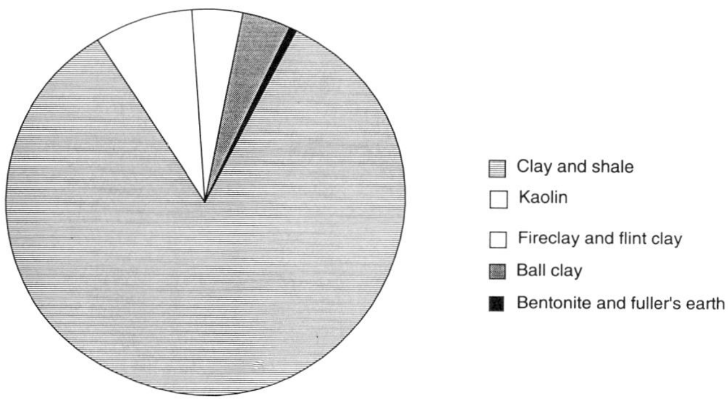

Fig. 2b. Schematic diagram of clay resources produced or used in 1989 by type of clay (after Ampian, 1991).

of common clay and shale produced is 10 to 15 times greater than the amounts for the other categories, the monetary values of the production for each category are much closer to being equal. The total value of refractory products made from clays and nonclays, respectively, was $617.5 and $737.8 million (Ampian, 1991).

Predictions are difficult to make in any market area, and raw materials markets for ceramics in the United States are no exception. However, trends during the last few decades suggest that loss of market share for conventional products such as clay-based refractories will be at least partly replaced by new products, such as engineered ceramic monoliths. Also, part of the loss of ceramic production of dinnerware and tile in the United States also has been offset by increased export of clays and other raw materials. As socio-political factors in foreign locations reduce their economic advantages, technological advantages will exert greater control on where new production facilities are established. Furthermore, most ceramics have high production costs and many still have excessive levels of property variation that reduce their value. These problems are opportunities for innovative approaches. Opportunities in both established industries and high technology products seem more than sufficient to provide for long term economic health for the ceramic raw materials industry.

SUMMARY

The mineral content of ceramic raw materials controls the way ceramic products are formed and fired. The minerals in raw materials can be described as skeleton formers, glass formers and bonding agents, fluxes, special property producers or modifiers, processing or fabrication aids, environmental or fabrication conditioners, and secondary and overlapping constituents. The elemental composition and the distribution of these elements within natural raw materials affect the nature of fired products. For example, the segregation of refractory and nonrefractory elements in different minerals causes a wide range of firing temperature and makes it possible to produce a durable product at temperatures well below those at which the whole mass deforms by melting. The nature of the product is strongly influenced by the phase changes, hydroxyl loss, and selective melting of individual minerals during firing.

Markets for ceramic raw materials have been reasonably stable. However, part of this stability has come from replacement of production lost to foreign competition by new products such as engineered ceramic monoliths. Between the opportunities for improvements in a product as old as common brick and the need for new products as new as catalyst supports, the future should bring many useful and profitable advances.

ACKNOWLEDGMENTS

The authors are indebted to J.H. Goodwin, B.E. Leach, and D.M. Moore for their helpful reviews. We also appreciate permission granted by Kentucky-Tennessee Clay Company to summarize the results of an unpublished study of the composition of representative ball clays. Discussions between the junior author and W.D. Keller, H.H. Murray, and W.A. White were invaluable to the preparation of this chapter.

BIBLIOGRAPHY

Anon., 1972, "Ball Clay Classification," *Ceramic Age*, July, pp. 45–55.

Ampian, S.G., 1991, "Clays," *US Bureau of Mines Minerals Yearbook 1989*, US Dept. of the Interior, 34 pp.

Bates, R.L., and Jackson, J.A., eds., 1980, *Glossary of Geology*, 2nd ed., American Geological Institute, Falls Church, VA, 751 pp.

Billue, R.F., 1962, "Fractured Clay," US Patent, 3,058,671, 9 pp.

Brindley, G.W., 1976, "Thermal Transformations of Clays and Layer Silicates," S.W. Bailey, ed., *Proceedings, International Clay Conference*, Mexico City, 1975, Applied Pub. Ltd., Wilmette, IL, pp. 119–129.

Brindley, G.W., and Nakahira, M., 1959, "The Kaolinite-Mullite Reaction Series," *Journal American Ceramic Society*, Vol. 42, pp. 311–324.

Burst, J.F., 1991, "The Application of Clay Minerals in Ceramics," *Journal Applied Clay Science*, Vol. 5, pp. 421–443.

Dombrowski, T., Murray, H.H., and Gilstrap, M., 1989, "A Geologic Model to Explain Chemical Differences Between Kaolins in the Georgia-South Carolina Kaolin District," *Program Abstracts*, 26th Annual Meeting, The Clay Minerals Society, Sacramento, CA, p. 26.

Gardner, T.W., Williams, E.G., and Holbrook, P.W., 1988, "Pedogenesis of Some Pennsylvanian Underclays; Ground-Water, Topographic, and Tectonic Controls," Geological Society America Special Paper 216, pp. 81–101.

Hughes, R.E., Bargh, B.L., and White, W.A., 1983, "Expansion and Weathering of Brick," C.H. Ault and G.S. Woodard, eds., *Proceedings, 18th Forum on the Geology of Industrial Minerals*, Indiana Geological Survey Occasional Paper 37, pp.183–198.

Hughes, R.E., Moore, D.M., and Reynolds, R.C., Jr., 1993, "The Nature, Detection, Occurrence, and Origin of Kaolinite/Smectite," H.H. Murray, W.M. Bundy, and C.C. Harvey, eds., Special Publication No. 1, *Kaolin Genesis and Utilization*, Clay Minerals Society, Boulder, CO, pp. 291–324.

Hughes, R.E., et al., 1987, "Origin of Clay Minerals in Pennsylvanian Strata of the Illinois Basin," L.G. Schultz, H. van Olphen, and F.A. Mumpton, eds., *Proceedings, International Clay Conference*, The Clay Mineral Society, Boulder, CO, pp. 97–104.

Hughes, R.E., and Smith, S.J., 1972, ''Testing Kaolin Clays for the Paper Industry,'' *TAPPI*, Vol. 55, pp. 904–908.

Hughes, R.E., White, W.A., and Warren, R.L., 1989, ''The History, Geology, and Future of Industrial Clays in Illinois,'' R.E. Hughes and J.C. Bradbury, eds., *Proceedings, 23rd Forum on the Geology of Industrial Minerals*, Illinois State Geological Survey, Industrial Minerals Notes 102, pp. 59–70.

Keller, W.D., 1981, ''The Sedimentology of Flint Clay,'' *Journal of Sedimentary Petrology*, Vol. 51, pp. 233–244.

Keller, W.D., Wescott, J.F., and Bledsoe, A.O., 1954, ''The Origin of Missouri Fire Clays,'' A. Swineford, ed., *Clays and Clay Minerals, Proceedings 2nd National Conference on Clays and Clay Minerals*, National Academy of Sciences-National Research Council Publication 327, pp. 7–46.

Kirk, L.G., and Bougher, A.K., 1986, ''Geology, Mining, and Marketing of Mississippi Embayment Ball Clays,'' *Program Abstracts*, Fall Meeting, SME, St. Louis, MO, p. 11.

Mitchell, L., 1983, ''Ceramic Raw Materials,'' *Industrial Minerals and Rocks*, 5th ed., Vol. 1, S.J. Lefond, ed., AIME, New York, pp. 33–39.

Murray, H.H., 1988, ''Kaolin Minerals: Their Genesis and Occurrences,'' *Hydrous Phyllosilicates (Exclusive of Micas), Reviews in Mineralogy*, Vol. 19, The Mineralogical Society of America, pp. 67–89.

O'Connor, B.J., 1989, ''An Overview of the Geology of Clays, Shales, and Slates Utilized in Ceramic and Structural Clay Products in Georgia,'' R.E. Hughes and J.C. Bradbury, eds., *Proceedings of the 23rd Forum on the Geology of Industrial Minerals*, Illinois State Geological Survey, Industrial Minerals Notes 102, pp. 91–95.

Olive, W.W., and Finch, W.I., 1969, ''Stratigraphic and Mineralogic Relations and Ceramic Properties of Clay Deposits of Eocene Age in Tennessee,'' Bulletin 1282, US Geological Survey, 35 pp.

Patterson, S.H., and Murray, H.H., 1984, ''Kaolin, Refractory Clay, Ball Clay, and Halloysite in North America, Hawaii, and the Caribbean Region,'' Professional Paper 1306, US Geological Survey, 56 pp.

Robbins, J., 1984, ''Ceramic Whiteware—An Overview of Raw Materials Supply,'' *Industrial Minerals*, Vol. 204, pp. 31–63.

Tsuzuki, Y., 1961, ''Mechanism of the 980°C Exotherm of Kaolin Minerals,'' *Journal Earth Science*, Vol. 9, pp. 305–344.

Watanabe, T., et al., 1987, ''^{29}Si- and ^{27}Al-MAS/NMR Study of the Thermal Transformations of Kaolinite,'' *Clay Minerals*, Vol. 22, pp. 37–48.

CONSTRUCTION USES

GYPSUM PLASTERS AND WALLBOARDS

HARLAN L. KEBEL

Gypsum, the dihydrate form of calcium sulfate, has a history of usefulness to man dating back several thousand years, and a worldwide industry has been built on the mining and processing of this versatile industrial mineral. However, anhydrite, the other principal calcium sulfate mineral, has very limited use, so much so that separate statistics for it are not always kept, but are usually included with those for gypsum.

Although often found in close association, the two minerals have important chemical differences in that anhydrite is the anhydrous form of calcium sulfate ($CaSO_4$) whereas gypsum is the dihydrate ($CaSo_4 \cdot 2H_2O$). These two molecules of combined water result in significant differences in hardness, density, and solubility of the two minerals. Of most importance, however, is the difference in thermal properties wherein gypsum, upon the application of a modest amount of heat, can be converted to the hemihydrate of calcium sulfate ($CaSO_4 \cdot \frac{1}{2}H_20$) or plaster of paris, an intermediate product upon which is based over 90% of the value of all calcium sulfate end products. Anhydrite, on the other hand, is inert at these temperatures and its uses have been limited to only those few cases where calcium sulfate in the anhydrous form may have an advantage.

The earliest known use of gypsum plaster dates back some 5000 years to the time of the Pharaohs of Egypt who had discovered that gypsum rock, when exposed to fire, broke down into a powder which, when mixed with water, would form a putty that could be plastered on a rough mud brick or stone wall to make a smooth finish. It was used down through the centuries in a limited way for ornamental purposes, but did not achieve wide acceptance because of the quick (25 to 30 min) setting or hardening time which made it difficult to use. The first real understanding of gypsum chemistry was developed in France about 1755, and the gradual growth of its present utilization dates from that time.

In the United States, the first utilization of gypsum was to grind it to a fine material for use as a soil conditioner at about the time of the Revolutionary War, and the term *land plaster* for ground gypsum was coined and has remained in use. Another cementitious material—lime or calcium oxide—had become the usual plastering material, but by about 1870 a method for retarding the set of gypsum plaster was discovered; today it has almost completely replaced lime as a plastering material.

In recent years, gypsum products sold in the United States have reached a total valuation of approximately $2 Gpy as illustrated in Table 1.

STRUCTURE OF THE INDUSTRY

In the technologically more advanced countries such as the United States, Canada, Western Europe, and Japan where the industry is well developed, the structure of the industry tends to be organized into large, well financed, vertically integrated companies. On the other extreme, in those countries where the usage of gypsum is not extensive, its production is usually accomplished by small operators with limited objectives and capabilities. Several North American and western European companies operate on an international basis through subsidiaries or other arrangements (equity or otherwise) but, in general, the percentage of total business done outside their home country is small.

There are a total of 29 companies (1990) who mine and/or process gypsum in the United States, ranging in size from large, multiplant, multiproduct organizations to small, one-plant, one-product companies who serve a local market for portland cement retarder or for agricultural gypsum. Of these, 13 companies have calcining and wallboard manufacturing facilities, and the remaining 16 ship only uncalcined products. Of the 13 companies with calcining and manufacturing facilities, eight are multiplant operations with anywhere from two to 23 different locations, with the remainder being single-plant operations. In 1990 there were 73 active calcining and manufacturing plants.

The larger companies are completely integrated from mining to the marketing and distribution of finished products. As the importance of wallboard has grown, some of the companies have also built paper mills to supply part or all of their needs for the face and back paper used on the board. This trend towards vertical integration has resulted in strong competition, and in a continuing need to control all phases of their business. As a consequence, there is practically no market for crude gypsum rock to the calcining-manufacturing companies.

WORLD TRADE

World trade in gypsum consists primarily in the movement of crude gypsum rock, although, to a limited extent, certain manufactured products do move across international boundaries. The United States is the world's largest producer and the largest importer of gypsum rock. Canada is the largest exporter of gypsum, with almost all of it being in the form of crude rock to the United States.

In Europe, France is both the largest producer and the largest exporter of gypsum, with its exports being in the form of crude rock. Spain also exports substantial quantities of gypsum rock.

The primary reason for the limited amount of world trade in gypsum or its products is found in the widespread occurrence of deposits, and in the relatively low ratio of manufacturing costs to product weight. This results in freight being a major segment of delivered cost, a fact that discourages long, international movements. For further discussion of occurrence and the concept of place value, and for statistics regarding US imports, the reader is referred to the chapter on Gypsum and Anhydrite.

Table 1. Gypsum Products (Made from Domestic, Imported, and Byproduct Gypsum) Sold or Used in the United States, By Use

Use	1988 Quantity, 1 000 t	1988 Value, $1,000	1989 Quantity, 1 000 t	1989 Value, $1,000	1990 Quantity, 1 000 t	1990 Value, $1,000
Uncalcined:						
Portland cement	3 617	47,622	3 104	41,834	3 950	46,660
Agriculture & Other	1 259	27,851	1 899	32,310	1 904	39,143
Total*	4 876	75,473	5 003	74,145	5 854	85,803
Calcined:						
Plasters	718	98,411	W	W	W	W
Prefabricated products+	17 376	1,416,901	19 550	1,852,531	17 450	1,627,045
Total calcined (1)	18 094	2,015,313	19 550	1,852,531	17 450	1,627,045
Grand total (1)	22 970	2,090,786	24 553	1,926,676	23 304	1,712,848

Source: Davis, 1989, 1990.
*Data may not add to totals shown because of independent rounding.
+Includes weight of paper, metal, or other materials.
(1) Does not include plasters data in 1989 and 1990.
W Withheld to avoid disclosing proprietary data.

BASIC PROCESSING METHODS

The processing of gypsum can be divided into three basic steps: 1) rock preparation; 2) calcining; and 3) formulating and manufacturing. This is illustrated graphically in Fig. 1. The specific means used to accomplish each of these steps will vary with the quality of the gypsum being processed and with the type of product to be made.

Rock Processing

Primary crushing is accomplished by gyratory, jaw, or impact crushers, with the type selected depending upon the size of the mine run rock, the desired throughput, and the type of subsequent processing. Secondary crushing is done with a variety of standard units, but hammer mills and cone-type crushers are most commonly utilized. Fine grinding of both uncalcined and calcined gypsum is generally accomplished by air-swept roller mills fitted with integral air separators for better control of particle size, although high energy impact mills plus air classifiers also have been used.

Both primary and secondary crushing steps are usually conducted with vibrating screens in the circuit, in part to maximize crushing efficiency and to reduce the production of ultrafines, but also to recover portland cement rock, this being the first marketable product of gypsum processing. The particle size of this product will vary with the requirements of the cement plant using it, but in the United States, it most often is controlled within a range of 38 to 51 mm top size by 6 to 13 mm bottom size.

Drying (the removal of free moisture) may or may not be used either before or after the secondary crushing stages, with its need being a function of the amount of free moisture in the mine run rock. Crushed gypsum is difficult to handle;—5 mm material is not freeflowing, particularly if wet, so a drying step is often added to rock preparation to assure the free flow of material in subsequent steps. This most often is accomplished in rotary dryers, and must be carefully controlled so that the temperature of the rock does not exceed 49°C, this being the point at which dissociation of the combined water begins to take place.

In many cases, the purity or quality of gypsum as mined is sufficiently high that it can be used without upgrading. However, the use of beneficiation techniques is becoming more common as the quality of final products is improved to meet the higher requirements of the marketplace. The most common form of beneficiation is simple classification of particle size by dry screening, air separation, or other means wherein a size fraction containing a significant portion of impurity is discarded. In general, this method is used to reduce clay or sand (dirt) impurities. In some cases, impurities which are harder than gypsum, and hence tend to concentrate in the coarser size fractions after crushing, can also be reduced by screening.

Washing and/or wet screening is used in a few cases, particularly where the need for a white color exists. The use of heavy media sink-float separation for the removal of impurities is currently employed at two operations in Canada and one in the United States. In many cases, gypsum is amenable to other gravity methods of beneficiation, froth flotation, or to color sorting. Some laboratory work has been reported on these approaches (French, 1967, MacPherson, 1950). However, the relatively high cost of these methods and the availability of deposits which do not require this type of beneficiation usually eliminate commercial application.

Synthetic Gypsum

In recent years, synthetic gypsum has to some degree begun to replace natural gypsum. This byproduct of sulfuric acid from industrial acid neutralization or flue gas desulfurization in fossil fuel burning power plants can contain purities of up to 98% $CaSO_4 \cdot 2H_20$. Synthetic gypsum is usually made available in its natural granular form (30 to 60 μm) with a moisture content varying from 7 to 15%. The product is normally dried by the end user to less than 1% moisture prior to calcining, then handled in the same manner as natural gypsum throughout the manufacturing process. There are no significant advantages or disadvantages to using synthetic gypsum versus natural gypsum other than delivered cost. The decision is one of economic consideration with the cost of transportation usually being the determining factor. The availability and use of synthetic gypsum is likely to increase over the next fifteen years as requirements for reducing sulfur dioxide emissions are implemented throughout the power generating industry.

Calcining

As applied to the gypsum industry, calcining is the step of reducing the dihydrate of calcium sulfate ($CaSO_4 \cdot 2H_2O$) to the

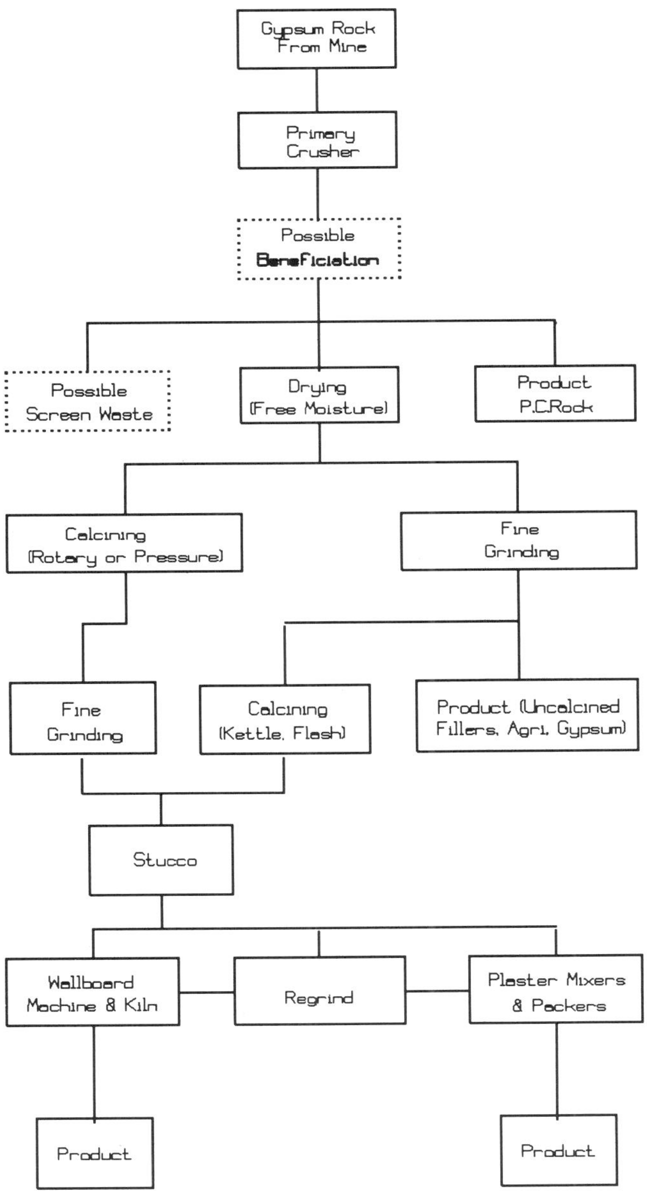

Fig. 1. Generalized flow diagram of gypsum processing.

hemihydrate ($CaSO_4 \cdot \frac{1}{2}H_2O$) or anhydrous ($CaSO_4$) forms. In the very early history of the making of gypsum plaster, pieces of rock were simply heated in an open wood fire to bring about the dehydration process. Then crude ovens were built in which fist or head size pieces of gypsum were placed, and in a variation of this, vertical kilns were used. These latter may still be found in less technically developed countries, operating with crude burners, and with little or no prior preparation of the rock. However, where the utilization of calcined gypsum has developed on a broad scale, more sophisticated methods of calcination have evolved which require prior preparation of the gypsum as described in the preceding paragraphs.

Considerable theoretical work has been performed as regards the thermodynamic properties of the system $CaSO_4 \cdot H_2O$ (Kelley et al., 1941, Riddell, 1950) and the literature is quite complete in this regard. In its practical application, four calcined products are made on a commercial basis: the two forms (alpha and beta) of hemihydrate; a soluble anhydrite; and an insoluble or dead-burned anhydrite. All hemihydrates, no matter how made, are usually called *stucco* within the industry, but this term is not to be confused with the mixture of portland cement and sand stucco sometimes used for exterior application on buildings. The term plaster of Paris is also often used for hemihydrate, and had its origin from the large deposits of gypsum under and near the city of Paris, France.

It is the consensus that hemihydrate is the only lower hydrate of calcium sulfate whose identity has been established with any degree of certainty (Riddell, 1950). Two forms of hemihydrate—alpha and beta—have been identified; however, the beta form has a higher energy content and a higher solubility rate. The two forms can only be distinguished by highly sophisticated analytical methods. The alpha form of hemihydrate is less reactive than the beta form, and has a slower rate of strength development. This is a disadvantage in many uses for stucco, nevertheless rehydrated alpha hemihydrate makes a denser, stronger plaster which is advantageous in other uses. The principal characteristics of these two forms of hemihydrate are given in Table 2.

Alpha hemihydrate can be prepared by dissociation of gypsum in a water saturated environment above a temperature of 97°C, and in practice this is usually accomplished at elevated pressure in an autoclave in the presence of steam. Beta hemihydrate can be prepared by the dissociation of gypsum in a vacuum at 100°C, but in practice is accomplished in a less than saturated environment at atmospheric pressure.

The greatest utilization of calcined gypsum is in the manufacture of wallboard and the formulation of plasters for building markets, uses for which early strength development is necessary; the more reactive beta hemihydrate is most suitable for this purpose. Alpha hemihydrate makes a denser, higher strength cast, and may be preferred for industrial uses where these characteristics are important. Other factors to be considered in the utilization of these two forms of hemihydrate are cost (beta is considerably less expensive to make) and water demand (alpha requires less excess water, over and above that necessary for rehydration, to make a slurry of equivalent consistency or viscosity).

By far the most commonly used vessel to calcine gypsum is the kettle, the basic design of which originated about 1870. Its general features are illustrated in Fig. 2. In its simplest form, the kettle is a cylindrical steel vessel with a height greater that its diameter enclosed in a refractory shell (Fig. 3). Heat is introduced in a firebox below the vessel and flows upward around it. To improve heat transfer to its pulverized gypsum contents, four or more horizontal cross-flues are installed, and agitation is added by means of a vertical, rotating shaft with sweeps or rabble arms which stir the contents. The kettle is filled from the top, and an opening at the bottom with a quick operating gate provides for fast dumping at the desired moment.

For the production of stucco with strength development and set characteristics desirable for wallboard production, the kettle is operated in a continuous mode in which pulverized gypsum (landplaster) of a fineness of approximately 90% −100 mesh is metered into the kettle at a constant rate by a screw or rotary feeder supplied by a large bin. Continuous kettles discharge finished stucco through an overflow arrangement connected to the hot pit or product bin. As fresh, cold landplaster enters the kettle top, it sinks through the hot, lighter, calcining mass forcing calcined stucco out the overflow pipe by gravity. Stucco is typically discharged from the continuous kettle at from 138 to 154°C. Various systems of compressed air addition to the overflow are used to fluidize the stucco to assist in steady overflow and discharge. A variant on the continuous kettle is the use of submerged combustion in which the combustion of the fuel takes place inside the kettle shell; this results in better energy efficiency, but produces a more violent calcination.

Table 2. Characteristics of Alpha and Beta Hemihydrates

	Normal Consistency*	Setting Range In Min, Vicat	Typical Set Expansion, m per m +	Avg. Compressive Strength, kPa, Dry+	Reaction
Alpha	40–43	20–30	0.003	38 000	Neutral
Beta	64–66	10–20	0.002	14 000	Neutral

*Parts water to 100 parts stucco by weight to make a pourable slurry.

A kettle is run in a batch mode to produce stucco with high plasticity, high strength, and density. These characteristics are desirable for construction and industrial plaster. The full cycle from filling to dumping the batch requires 2 to 3 hours. Kettles have been built with 1.8 to 27 t capacity of finished product. However, the usual size is 13.6 to 18 t. The normal cycle requires from 20 to 30 min to fill the kettle and 90 to 120 min to bring the mass up to the desired temperature after filling is complete. Dumping is usually accomplished in 5 min or less, with the finished stucco falling by gravity through the dump gate into the vented hot pit where it is held up to one hour to allow as much of the entrapped steam as possible to escape before it is transferred to a storage bin at the point of use.

In the calcining process, gypsum begins to lose its water of crystallization when it reaches a temperature of 43 to 49°C. During the filling part of the cycle, the firing rate is controlled to keep the temperature of the kettle contents about 104°C. When filling has been completed, the firing rate is increased, with different rates producing different properties in the finished stucco. As the temperature of the mass rises, water of crystallization is released as steam, and at 116 to 121°C the mass boils vigorously as the vapor pressure of the released steam equals atmospheric pressure. After boiling ceases and little or no more steam is given off, the contents of the kettle settle to 12 to 15% less than the original volume. This quiet period indicates that most of the gypsum particles are dehydrated to the hemihydrate. Heating is continued to dry the mass until its temperature reaches 149 to 166°C, at which point it is dumped into the hot pit and the cycle is complete.

The product of the foregoing operation is known as first settle stucco, and it contains from 5 to 6% combined water, as against 20.9% contained in 100% pure gypsum, i.e., approximately 75% of the water of crystallization has been driven off. Because this reaction takes place at atmospheric pressure in a less than saturated environment, most of the resulting hemihydrate is in the beta form, although varying amounts of alpha hemihydrate may also be present depending upon the design of the kettle as regards the removal of steam vapor and the temperatures used.

If desired, a second settle stucco can be made after reaching the first boil and first settle stages. This is done by increasing the temperature to about 177°C, at which point the gypsum again starts to boil and give off steam. This second boil is of shorter duration and less vigorous than the first because there is a smaller percent of water of crystallization to be removed. At about 204°C, the material is dumped, at which point almost all of the water of crystallization has been removed, and the material is usually referred to as soluble anhydrite. It differs from first settle stucco in that it has less plasticity, but after rehydration will make a product with greater density and strength. Also, it has a high affinity for moisture and is primarily useful for its desiccant properties.

Aridized stucco refers to first-settle kettle stucco to which a small percentage of a soluble salt has been added during the filling part of the cycle. In general practice, approximately 0.68 kg of calcium chloride is used per ton of stucco, with the exact amount depending upon the purity of the gypsum rock and the kind and amount of impurities present. This treatment increases the vapor pressure within the kettle so that calcining takes place at a somewhat lower temperature, and the resulting stucco is used in the formulation of many industrial plasters.

Low water demand stucco can also be produced by adding water to the kettle after the batch calcination cycle is complete. The stucco is quenched to about 93°C and then reheated to drive off excess water before dumping to the hot pit. Alternatively, low water demand stucco can be made by adding a measured amount of water to the stucco after it has left the kettle.

Other methods have been utilized in the production of beta hemihydrate. Hollow flight screw conveyors with heat applied by circulating hot liquid or gas through the hollow flight and/or by heated jackets around the conveyor box have seen limited use. Flash calcining in a hot air stream with subsequent separation in a cyclone or bag house dust collector can be done either after or in conjunction with final grinding. High temperature hammer mills or roller mills have seen increased use as hot air swept grinder—calciners for efficient stucco production. These mills are typically fed by –25 mm gypsum rock. Flash calcination is not as controllable as kettle calcination and is not used where high quality stuccos are required. The rotary kiln is still used for calcination in some parts of the world but has been almost totally replaced in the United States by kettle or flash calciners.

Pressure calcining in steam autoclaves, with or without the addition of a crystal modifier, is used to make superstrength plasters for certain industrial uses. The gypsum may be introduced into the vessel either as a crushed, sized rock or as a slurry, with the stucco being dried as soon as calcination is completed. This method produces all alpha hemihydrate, is relatively high cost, and its use is limited to just those products requiring this type of specialized plaster. Alpha hemihydrate may also be produced by dehydration of gypsum in hot, salty water.

Dead-burned gypsum or insoluble anhydrite is made in beehive kilns, rotary calciners, or flash calciners using much higher temperatures (up to 482°C) than are required for other calcining methods. The resulting product is an anhydrous calcium sulfate that does not take up water (rehydrate) at any appreciable rate; it is the least soluble of any of the forms of calcium sulfate (Riddell, 1950). The primary uses of this material are to make Keenes Cement by the addition of chemical accelerators to restore its setting properties, and to produce a mineral filler by fine grinding.

Hydraulic gypsum refers to lump (head size) gypsum which has been heated to about 899°C at which temperature the material is at a red heat and calcium sulfate begins to break down to release sulfur as sulfur dioxide gas. This has the effect of increasing the calcium oxide content, and results in a material which, when mixed with water, is slow setting as compared to other calcined gypsum. However, it provides a very hard, dense product that is quite durable, and is sometimes referred to as *Estrich gypsum* (a German term). Although it has been used in Europe as a flooring material, it has not found application in the United States.

FORMULATION AND MANUFACTURING

The direct product of gypsum calcining is almost never used without some further processing such as grinding, mixing with

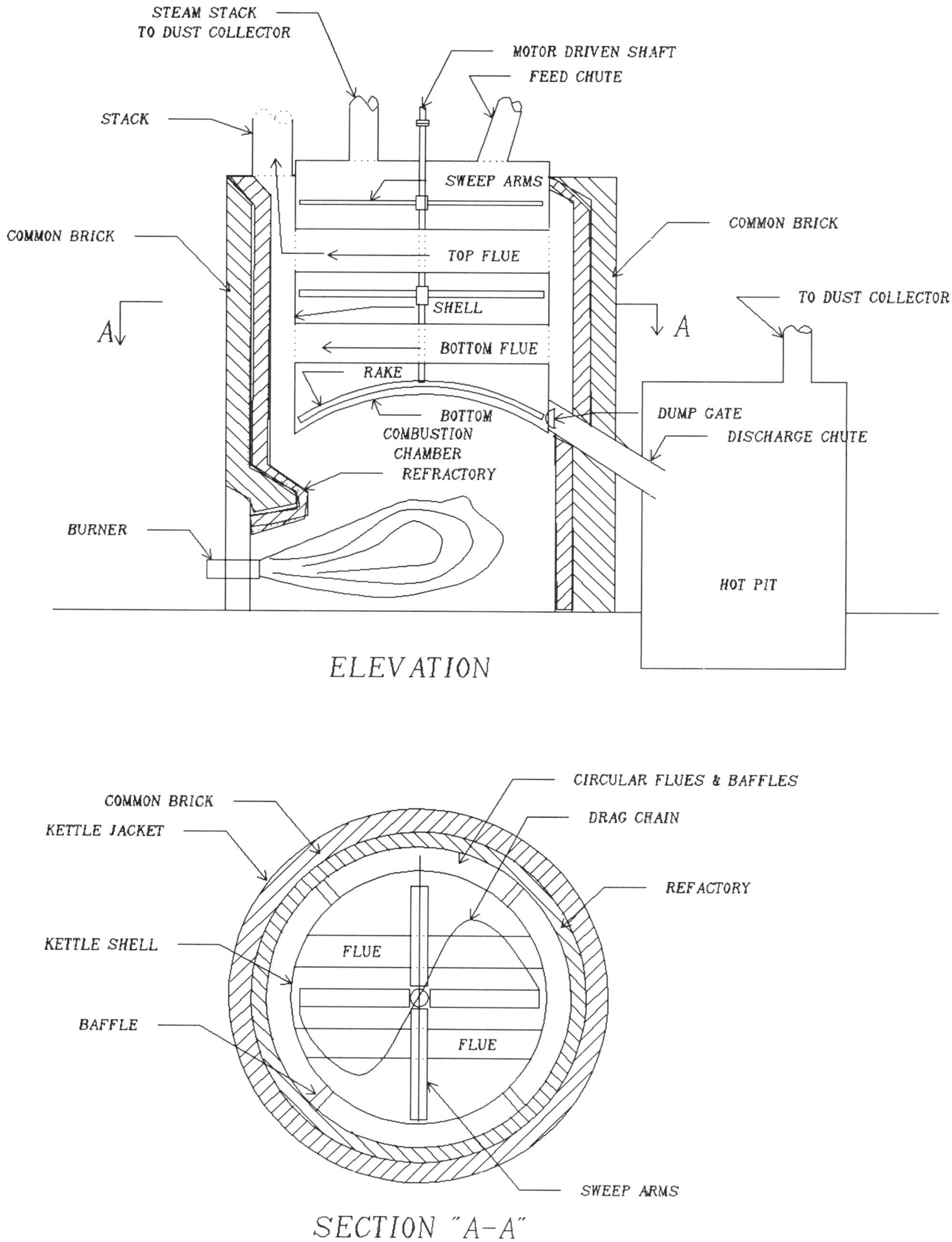

Fig. 2. General features of kettle used to calcine gypsum.

Fig. 3. Exterior view of gypsum kettle.

additives, or rehydrating and casting into block or wallboard. Kettle stucco, the most common calcined material, is not just a single form of calcium sulfate, but is a mixture of particles in varying states of dehydration from the dihydrate to anhydrous forms, a factor which varies with the method of calcination. The amount of variation can be minimized by careful control of the calcining process; this may be a deciding factor in considering the type of subsequent treatment to be given the stucco.

Probably the earliest attempts at formulating calcined gypsum to improve its utility were the addition of fibers and/or aggregates. However, the singlemost important discovery was that setting time could be either retarded or accelerated to very exact time limits by mixing certain materials with stucco (plaster of paris). Properly calcined gypsum when mixed with water will have a natural setting (hardening) time of 15 to 25 min, which is usually too fast for satisfactory application on walls or ceilings. However, in the middle 1870s, a method was discovered to retard the set to 2 or 3 hr (or longer). This led to the manufacture of "slow set plaster," which was the key to the rapid growth in the use of gypsum in the building industry. Retarders are made from organic compounds using several different materials. For the most part, these are formulated and manufactured by the major gypsum companies for their own use. Accelerators can be natural or synthetic salts, such as potassium sulfate, or gypsum itself. Finely ground raw gypsum, or rehydrated stucco cast into a block and then finely ground, are both commonly used to accelerate the set of stucco for the manufacture of wallboard.

By the proper use of these materials, gypsum plaster can be accelerated to set within 3 to 4 min, or be retarded so that its set does not occur for several hours. However, adjustment of the set can cause variations in other properties, especially strength, so that care is advised in the use of these materials.

The type of stucco and the characteristics of the plaster being formulated can be controlled by other additives mixed in precise proportions prior to bagging, and much research has been conducted to determine the formulas used. Among other products, plasters premixed with sand, expanded perlite or exfoliated vermiculite aggregates are offered for the building trades, and a wide range of formulated plasters for industrial use are also available.

Gypsum is often re-ground after calcining to impart special qualities to the stucco. The classic buhrstone mill once used for this purpose has been replaced by roller mills operating in closed circuit with air separators. In recent years, high energy impact mills have successfully been employed. When stucco is to be used with aggregate (as for sanded or perlited wall plaster), a tube mill is used for regrinding.

One of the early uses of gypsum as a construction material was to cast it into blocks or tile. Continuous casting machines were developed for this operation along with kilns to dry the product by hastening the removal of excess moisture. Blocks are cast either as solid masses or as hollow tiles, and can be made in any desired dimension. Metal reinforcing is used for some products, and some are cast in a metal form to produce a metal edged tile or plank. The addition of fiber for reinforcing is a common practice and use of an accelerator hastens the initial set of the stucco in the mold.

However, the largest end use for gypsum today is in the manufacture of wallboard; approximately 65 to 70% of the gypsum consumed in the United States is used for this purpose. In its simplest form, wallboard is a thin, flat slab of gypsum made by casting a water slurry of stucco in a paper envelope, i.e., a gypsum core with specially prepared paper bonded to both sides and the edges. This is a sophisticated product, made on high speed, highly automated machines to a wide range of precise specifications.

The bond between the paper and the gypsum core is produced by the growth of gypsum crystals locking themselves into the fibers of the paper as the stucco is rehydrated, rather than by the use of adhesives. Extensive research has gone into the formulation of the gypsum slurry used in wallboard, but essentially it consists of beta hemihydrate, an accelerator, fibers, starch, and a foaming agent, with stucco making up at least 95% of the materials used prior to mixing with water. Fig. 4 shows a general process flow diagram of a wallboard plant.

In the manufacturing process, the dry components are brought together with water in a high energy mixer designed for a continuous flow-through of material, and the resulting slurry is dropped on a moving strip of paper, the edges of which have been turned up to form a wide, shallow trough. This mass then moves under a second strip of paper, and the roughly formed board moves between edge guides (Fig. 5) and under a 610 mm diameter roller called a master roll or a stationary steel forming plate, which imparts the exact desired width and thickness of the board. The edges of the bottom paper are turned over to form square or rounded corners and an adhesive is added so that the top paper is bonded along its edges to the lower. The thickness of the board can be varied by raising or lowering the master roll or forming plate. The now fully formed board moves along on a flat conveyor belt several hundred meters in length.

The speed of this board line varies with the design of the machine and the product being made, but usually ranges between 0.5 and 1.5 m/s. The length of the board line is designed to provide sufficient time (5 to 6 min) for the stucco to set, at which point the moving strip is cut at desired intervals by a rotating knife to produce individual boards ready for drying to remove excess free moisture. This is accomplished by moving them through a continuous, multideck kiln heated directly by hot air or indirectly by steam coils.

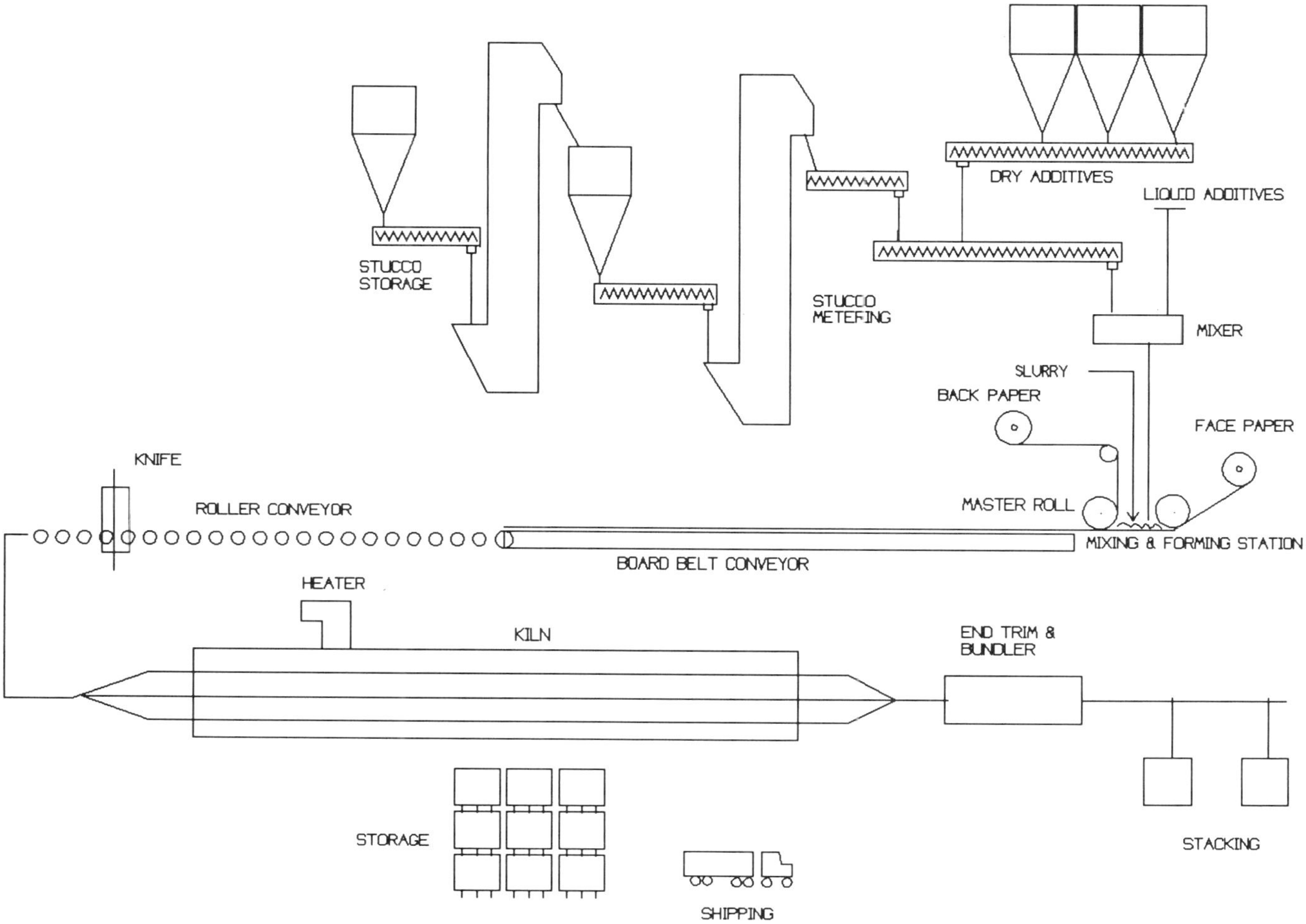

Fig. 4. Simplified flow diagram for complete small gypsum plant.

Fig. 5. Mixing and forming section of wallboard machine showing stucco slurry feeding onto bottom paper at master roll.

Close control is required to remove the free moisture without starting to re-calcine the gypsum core.

The most popular size of wallboard is 1.2-m × 2.4-m × 12.7-mm thick, but a wide variety of sizes is made and offered to the market. After emerging from the kiln, the board moves automatically through devices that smooth the ends and cut to the precise length desired. Usually, one face is made with light colored, smooth texture paper which forms the finished wall surface. Two boards are placed with these faces together to form a bundle for shipping. If desired, the board can move on to a finishing machine where a special facing is added, such as vinyl or imitation wood veneer paper to make a pre-decorated panel.

Gypsum sheathing is made in the same general manner except that asphalt-impregnated paper is used. Asphalt may also be added to the core to improve its water resistant qualities. Plasterboard is a 406 mm wide board used as a base for plastered walls instead of wood or metal lath, and is made on the same machine as wallboard. Shipment of finished material is by truck or rail, and most wallboard plants have large warehousing space adjacent to the manufacturing area to store adequate stocks of various types and sizes of products.

ENERGY CONSIDERATIONS

The gypsum industry is energy intense, particularly as regards fuel which is used in the rock drying and calcining operations, as well as in the manufacture of paper and the drying of excess

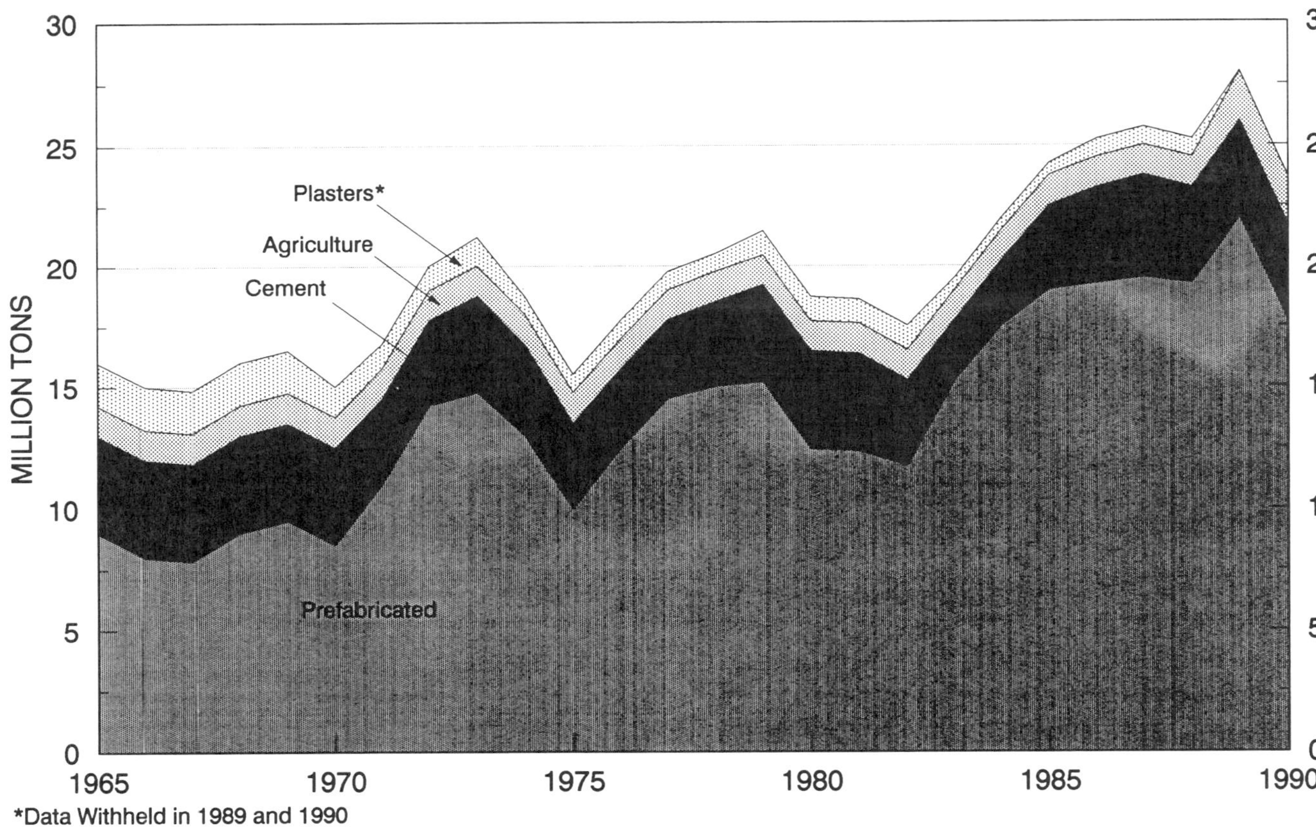

Fig. 6. Sales of gypsum products, by use.

moisture from wallboard. In total, an average 2.8 GJ are required per 100 m^2 of 12.7-mm-thick wallboard for these operations. The drying of both rock and wallboard is usually accomplished in direct fired kilns (although steam coils can be used for wallboard) in which a clean-burning fuel is required to prevent contamination of the product, and natural gas is the preferred fuel for this purpose. The calcination step, particularly when carried out in a kettle, requires the ability to change the intensity of the heat quickly, and either natural gas or fuel oil is preferred for this property, although coal can also be used.

The rapid rise of energy cost in recent years has encouraged extensive development and use of conservation methods, as well as investigation of process technology to reduce the energy requirement. Fuel substitution possibilities may be required as a result of changing availability and pricing policies.

MAJOR CATEGORIES OF END USES

Gypsum product lines are divided into several categories based upon end uses and/or processing methods. Three major groups are generally recognized: 1) construction, i.e., all those products used directly as a construction or building material; 2) industrial—all those products used in the manufacture or processing of other materials; and 3) agricultural. As indicated in Table 1 and Fig. 6, the construction group utilizes about 75% of the gypsum rock used in the United States and produces approximately 95% of total product valuation. The industrial group uses about 17% of the rock but yields only 3% of the product value; the balance is accounted for by agricultural gypsum.

The proportions produced for each use will vary considerably from country to country, and in many instances the only use for gypsum will be in the industrial category as portland cement retarder. In the following discussion, a few of the more important end uses will be reviewed.

CONSTRUCTION GROUP

Gypsum products are used in the building and construction industry as covering and finishing materials. They are not structural components in the sense that they provide the frame or structure of a building; instead, this requirement is met with wood, steel, concrete, concrete block, and fired or adobe brick. It is in the covering of a structural frame (Fig. 7), or in the finishing of a masonry wall, that gypsum products find their greatest utilization, and they compete in these applications with other covering or finishing materials, such as wood products, plastics, or other cementitious products (cement, lime, or in primitive regions, even mud).

By far the largest market for gypsum is in wood frame buildings, and because wood has historically been the basic structural material in the United States, most of the development for this usage has taken place in this country. Gypsum construction products are also compatible with steel or aluminum framing, and may be used in conjunction with masonry construction, although their economics in this latter application are more difficult to justify.

The basic characteristics of gypsum which give it value in the manufacture of construction materials are its fireproofing qualities, the versatility which comes from being able to use it in many forms, and its economy of application. It is used entirely in its calcined

Fig. 7. Installing gypsum wallboard partition using steel studs.

form, with stucco being formulated and/or manufactured to make plasters, block or tile, and wallboards. In recent years, approximately 90% of the gypsum used in the United States for construction materials is in the manufacture of various sizes and types of wallboard with a total surface area of some 1 500 km^2 per year.

Gypsum is noncombustible and is one of the most effective fireproofing materials known, a property which has its basis in the combined water content of the product. When the hemihydrate form of gypsum (stucco) is rehydrated to make a covering material it reverts back to the dihydrate, so that when heat from fire hits a plaster wall (or gypsum wallboard) it again begins to lose its combined water in the form of steam which inhibits the transmission of heat. Extensive research into the utilization of this property has resulted in a number of products with a high fire rating, and this is an important reason for the wide acceptance of gypsum products by the construction industry, particularly when wood framing is used.

The versatility of gypsum is also a major advantage. Plaster can be used in many forms, as can wallboard and cast block or tile, and by the use of different calcining methods and/or additives, physical properties can be varied to meet a wide range of requirements. These characteristics give the architect and builder several choices of design, and enable maximum flexibility in producing a structure to meet the desired specifications.

The economics of application of building materials is usually measured in terms of cost per square meter of finished wall, or floor space, or of cubic meter of content. This in turn, is a function of the cost of material delivered to the job plus the cost of installation.

In the final analysis, the wide use of gypsum construction products is based upon the industry's ability to keep its manufacturing costs low and to minimize the use of labor in their installation. The relatively low processing cost of gypsum is due to its nonabrasiveness and low calcining temperature. The wide geographical distribution of deposits and the resulting low cost of transportation contribute to economical delivery.

Plastered walls and ceilings, which in the early 1900s were widely used in the United States, have given way to dry wall construction employing gypsum wallboard. However, one coat veneer-type plasters have been developed for application over 1.2 m wide plaster bases, and many plasters are now formulated for machine mixing on the job and pumping to the point of use. Plasters continue to be used where special effects are desired, and their flexibility enables wide latitude in finished design.

Gypsum wallboard was originally developed in the United States, and the first product, made about 1890, consisted of at least four sheets of felt or paper with thin layers of gypsum stucco between them. By 1914, the concept of just a top and bottom sheet of paper with a single layer of gypsum had been perfected, and the use of this product received a big boost in the construction of buildings during World War I. Following World War II, as the increase in hourly wage rates raised the in place cost of plastered walls, the use of gypsum wallboard expanded rapidly because it provided fireproof construction, with a lower labor requirement than the traditional two coat plaster wall.

The present dominant position of wallboard stems not only from its economy of application, but also from the variety of installation systems which the industry has developed. For example, one such installation is the movable partition; another is a method for enclosing elevator shafts in high rise buildings; and still other types produce optimum acoustical properties in multiple unit housing.

The use of gypsum partition and roof tile has declined in recent years, although they are still preferred for some applications. Poured gypsum roof decks are often specified for industrial buildings; a specially formulated plaster slurry is pumped in place over a wallboard base to provide a completely fireproof roof. Other construction uses include water-resistant sheathing, plasterbase board, and fireproofing building steel.

INDUSTRIAL GROUP

Gypsum finds a wide variety of markets in industrial applications which bear little or no relationship to the construction products discussed previously. Industrial gypsum can be divided into three broad categories: calcined, anhydrous, and uncalcined, each of which utilizes distinct properties of the mineral.

The industrial segment of the gypsum industry is well developed in the United States; however, markets are widely scattered geographically. Many product lines tend to be specialized, and with the exception of portland cement retarder rock, individual customers are likely to want tens or hundreds of tons of a product, rather than thousands of tons.

Calcined Industrial Gypsum: Perhaps the largest use of calcined gypsum by industry is for mold making in the manufacture of sanitary ware, pottery, metal casting, and decorative objects.

Molding plaster is made from high purity gypsum (95%) at select locations and is marketed on the basis of its water demand, strength, white color, setting time, and closely controlled expansion/shrinkage qualities. It may be made from either beta or alpha hemihydrate, or a combination of the two.

The qualities which make gypsum an excellent material for mold making also contribute to its use as a casting plaster; gypsum casts are widely used by many industries. Included in this category is dental plaster, a highly specialized white material made for dental and orthopedic work which is accelerated to produce a setting time of only 3 to 4 min.

Another important industrial use is as a cementing agent in well drilling, particularly in the oil and gas industries. This usually is formulated from low consistency alpha hemihydrate (pressure calcined) and is effective for sealing off porous or cavernous rock stratas often encountered in the drilling of oil, gas, or water wells.

A growing use of high density, high strength plasters is in the maintenance of floors, and for quick patching high use surfaces (such as highways) that cannot be closed down for more conventional repairs. In both this and the well drilling application, gypsum plaster may be mixed with other cementitious materials to obtain optimum results.

Anhydrous Gypsum

In addition to the hemihydrate industrial plasters described previously, there are several products made from calcium sulfate with no water of crystallization, i.e., from second settle stucco made when the calcining temperature has been carried up to about 200°C. This material, often referred to as soluble anhydrite, has a high affinity for water, making it an efficient drying agent. It is sold in various particle size gradations for use as a desiccant in laboratory and commercial applications. A finely ground soluble anhydrite is used as a carrier for certain types of insecticides which must remain absolutely dry to maintain their toxicity.

When the calcination process is carried to a temperature of 480°C, a dead-burned or insoluble anhydrite product is formed which is used for industrial filler applications or for the manufacture of Keenes Cement. Dead-burned gypsum is preferred for some filler uses over uncalcined gypsum, particularly where the temperature of the product in which the filler is used may go over 50°C, the point at which natural gypsum begins to release water of crystallization. Dead-burning usually produces a whiter product, and is preferred where color (i.e., lack of color) is important. Dead-burned fillers are used as a source of calcium in food products, and for yeast and beer processing. They also serve as diluents or extenders in such compositions as pills, rubber, artificial wood, plastics, paper, and pigments.

Keenes Cement is a generic name for dead-burned gypsum which by the use of additives can be made to set and harden after being mixed with water. The usual set specifications fall in the range of 4 to 12 hr, and its major use is as a wall plaster where extra density, strength, and hardness are desired. It is made in only a few locations, and only in small quantities.

Uncalcined Industrial Gypsum

This category, sometimes referred to as raw gypsum, includes the largest volume use for the mineral outside that for wallboard, i.e., as a retarder for portland cement. In 1990 (Table 1) 3.9 t or 17% of all the gypsum used in the United States was consumed by this market.

It has long been known that calcium sulfate compounds will control the setting time of portland cement. These compounds control both the rate at which the cement paste develops strength, and the shrinkage of cement products during drying. The use of calcium sulfate compounds in cement has been the subject of extensive research, and is closely regulated by standards of the American Society for Testing and Materials. The amount of gypsum (or gypsum/anhydrite blend) used depends upon: 1) the SO_3 content of the gypsum, 2) the type of cement, and 3) the mix of raw materials used to make the cement clinker. In practice, the amount of gypsum used will vary from 4 to 6%, averaging approximately 5% by weight of finished cement. It is mixed with the clinker as it is fed into the grinding mill, and the two materials are then ground together.

The setting and strength development of portland cement is based upon several complex chemical reactions. As these are better understood, raw material specifications are being tightened, and particular emphasis is being placed upon the uniformity of all constituents. There is a trend towards the use of more SO_3 per unit of cement, with the result that the gypsum industry is being asked to provide increasing quantities of uniform gypsum and anhydrite products for portland cement manufacture.

The foregoing statistics on portland cement retarder usage include anhydrite, and in fact, this is the largest market in the United States for this mineral. However, it is a highly fragmented market because not all cement plants want anhydrite in their retarder rock. Where used, it is generally blended with gypsum, the usual blend containing 40% of the mineral anhydrite. This means that any one cement plant might use only 4.5 to 9.0 ktpy of anhydrite, and because many plants prefer straight gypsum to a gypsum/anhydrite (GA) blend, it is estimated that probably no more than 5% of the 3.9 Mt of gypsum reported used for portland cement retarder in 1990 was the mineral anhydrite.

Terra alba is a term applied to high purity, finely ground raw gypsum. Usual specifications call for a 95%-44 µm, very white color material whose exact specifications as to color and particle size vary with the end use. It has the same general utilization as dead-burned (anhydrous) gypsum, that is, as an inert filler or diluent, and/or as a source of soluble and biologically available calcium. However, as noted under that discussion, the two products do have different characteristics, and are not completely interchangeable. When used in bread and bakery products, or in pharmaceuticals, terra alba must meet Food Chemicals Codex specifications, and generally is made from +97% pure gypsum. In one of its more interesting applications, the presence of calcium sulfate fillers in water for the brewing of beer has been found to be desirable because it assists in the development of proper yeast fermentation, helps the settling out of the yeast, and aids in the clarification of the beer.

Glass-batch gypsum is the uncalcined mineral ground and sized to be comparable to the fineness of sand, and is used in the manufacture of container glass. The gypsum, which must be fairly dust free, is thoroughly mixed with the other ingredients of a batch prior to charging the furnace, and performs three functions: as an oxidizing agent, as a fining agent, and as a scum removing agent.

AGRICULTURAL GROUP

The oldest use of gypsum in the United States, dating back to the 1700s, is as a soil conditioner. Generally called land plaster this material can be produced from either gypsum or anhydrite, and is usually ground in an air-swept mill to a fineness ranging between 75 and 90%-150 µm. A screened and air separated granular-type product is available in certain locations in limited quantities which will average only 25% through 150 µm. In a few western states, particularly California, gypsite (an impure, unconsolidated form of gypsum) is sold in large quantities, and generally requires no processing except a hammer mill to break down the lumps.

The principal purposes for applying gypsum on land are to:

1. Improve the structure or physical conditions of soil by breaking up compacted clays, which, in turn, increases porosity and aids drainage.
2. Supply neutral, soluble calcium for peanuts and other crops.
3. Supply quickly available sulfate sulfur.
4. Neutralize sodium compounds in alkali soils.
5. Correct high sodium irrigation waters.
6. Improve availability and utilization of nitrogen (S-N relationship).
7. Clear up muddy farm ponds.
8. Stimulate soil microorganisms.

In other agricultural related uses, ground calcium sulfate, either in the dihydrate or anhydrous form, is often added as an ingredient in formulating feeds and feed premixes for beef cattle, dairy cows, and sheep. In this use, it supplies total sulfur requirements in safe, easy to mix form, increases the efficiency of nonprotein nitrogen in urea feeds, is an ideal supplement for ensilage enhancers, and is an effective regulator for self-feeding on the range. Terra alba gypsum is also used as a carrier for insecticides and for micro-

Table 3. Average Gypsum Prices (1990)*

Portland cement retarder	$ 11.82 per t
Agricultural gypsum	16.17 per t
Plasters	141.57 per t
Wallboard 12.7 mm	87.59 per t

**Source:* Based upon total quantities and value of these products as shown in US Bureau of Mines *Minerals Year Book* (Davis, 1990).

Table 4. Price Index for Gypsum Products (1982 = 100)*

1984	135.4
1985	132.3
1986	137.0
1987	125.2
1988	112.9
1989	110.0
1990	105.2

**Source: Construction Review,* Vol. 37, No. 2 Mar.–Apr., 1991, US Dept. of Commerce.

nutrients, with granulated and pelletized gypsum being produced in limited quantities for the same purpose.

MINOR USES FOR CALCIUM SULFATE

The use of anhydrite to construct load-bearing packwalls and to provide support for longwall gateroad steel arches in underground coal mines is practiced extensively in Germany and to a lesser degree in Great Britain and France. Experimental work is in progress toward application of this support system at the Lingan Colliery in Nova Scotia, operated by Cape Breton Development Corporation.

Anhydrite is crushed and sized, transported underground and then placed pneumatically at the application site. An accelerator consisting of potassium sulfate and ferrous sulfate is mixed into the crushed anhydrite and the material is wetted during placement. This process produces a quick setting, high compressive strength, monolithic material.

Gypsum and anhydrite constitute the world's largest reserve of sulfur, and considerable research has gone into the technology of the use of this sulfur. However, the economics of utilizing either gypsum or anhydrite for this purpose are not competitive with other sources of sulfur.

Although the process was offered to and considered in the United States during the sulfur shortage of the late 1960s, no plants were built. A plant of sufficient size to be economical in cement production would produce an excess of sulfuric acid. Capital requirements would be correspondingly high, and the risk of balancing two markets—cement and sulfur—which historically have wide price swings was considered to be great.

A variation from portland cement-sulfuric acid route is production of high quality lime and sulfuric acid (Wheelock and Boylan, 1968). In this process, gypsum or anhydrite is heated in a reducing atmosphere (without the addition of alumina, silica, and iron required for portland cement) to produce a relatively pure calcium oxide. Although considerable experimentation has been performed, no commercial operation has yet been developed. In both the cement acid and lime acid processes, anhydrite is the preferred mineral because of the higher yield of sulfur per ton, and elimination of the energy required to drive off the water of crystallization in gypsum.

Another use of gypsum for its sulfur value is in the production of ammonium sulfate. Several plants have been built in sulfur-poor countries, and the technology is well developed (Habashi, 1969). In 1968 to 1969, a plant for this purpose was built in California; however, it was closed down about a year later.

PRICES

In Table 3, average prices of five representative gypsum products are shown for 1990, and are typical of the spread in value between different products. These figures represent mill net prices and are averages across the United States. Prices at any given location will vary widely from these averages and freight and other distribution costs must be added. Delivered costs, based upon list prices, are quoted for several cities in *Engineering News Record.*

The recent trend of gypsum prices is shown in Table 4, and reflects the drop in building and construction activity in 1988 to 1991 when the industry was operating below capacity.

ACKNOWLEDGMENT

The author wishes to acknowledge his debt to Frank C. Appleyard, the author of this chapter in the 5th edition of *Industrial Minerals and Rocks*. A former Vice President with US Gypsum in Chicago, Mr. Appleyard is now retired and living in Arizona. His excellent work provided a base for this author's contribution to the 6th edition.

BIBLIOGRAPHY AND REFERENCES

Anon., 1970, "World Trade in Gypsum," *Industrial Minerals,* Oct., pp. 9–25.

Anon., 1978, *The Economics of Gypsum*, 2nd ed., Roskill Information Services Ltd.

Ashizawa, R.Y., 1969, "Gypsum," *Minerals Yearbook,* US Bureau of Mines, pp. 547–553.

Bedwell, W.L., 1952 "The Production of Sulphuric Acid From Calcium Sulphate," The Royal Institute of Chemistry, Lectures Monographs, No. 3, 21 pp.

Davis, L.L., 1990, "Gypsum," *Minerals Yearbook* 1990, US Bureau of Mines.

Dawes, R.E., 1970, "Gypsum," *Minerals Yearbook*, US Bureau of Mines, pp. 559–566.

Duda, W.H., 1966, "Simultaneous Production of Cement Clinker and Sulfuric Acid," *Minerals Processing,* Aug., pp. 10–13, 26.

Fisher, H.C., 1953, "The Setting of Gypsum Plaster," *ASTM Bulletin,* Sep., pp. 43–47.

French, R.R., 1967, "Beneficiation of Low Grade Gypsum by Electronic Color Sorting," Preprint 67–H–318, SME Fall Meeting, 15 pp.

Habashi, F., 1969, "Processes For Sulfur Recovery From Ores," *Mining Congress Journal,* Vol. 55, No. 6, June, pp. 38–46.

Hull, W.Q., et al., 1957, "Sulfuric Acid From Anhydrite," *Industrial & Engineering Chemistry,* Vol. 49, No. 8, pp. 1204–14.

Kelley, K.K., et al., 1941, "Thermodynamic Properties of Gypsum and Its Dehydration Products," Technical Paper 625, US Bureau of Mines, 73 pp.

Lane, M.K., 1968, "Disintegration of Plaster Particles in Water," *Rock Products,* Mar., pp. 60–63, 108; Apr., pp. 73–75, 116–17.

MacPherson, A.R., 1950, "Recent Investigations Into The Beneficiation of Canadian Gypsum," Canada Department of Mines & Technical Surveys, Memorandum Series No. 111, Nov., pp. 1–28.

Moyer, F.T., 1939, ''Gypsum and Anhydrite,'' Information Circular 7049, US Bureau of Mines, Feb., 45 pp.

Riddell, W.C., 1950, ''Physical Properties of Gypsum,'' *Rock Products,* May, pp. 68–71, 102.

Riddell, W.C., 1954, ''Effect of Some Inorganic and Organic Compounds on the Solubility, Setting Time and Tensile Strength of Calcined Gypsum,'' *Rock Products,* Oct., pp. 109, 113, 117.

Robinson, V., and McFarlane, J.D., 1969, ''Recovery of Sulphur Values From Gypsum and Anhydrite,'' *CIM Bulletin,* Vol. 62, No. 689, Sep., pp. 967–971.

Smith, F.H., 1963, ''Aging of Caclium Sulfate Hemihydrate,'' *Nature,* Vol. 198, pp. 1055–1056.

Wheelock, T.D., and Boylan, D.R., 1968, ''Production of Sulfuric Acid From Calcium Sulfate By Reductive Reaction,'' American Institute of Chemical Engineers, Preprint 6B, Annual Meeting, Dec., 23 pp.

Wheelock, T.D., 1980, ''Desulfurization of Phosphogypsum,'' International Symposium on Phosphogypsum, Nov. 5–7.

Wilder, F.A., 1923, *Gypsum: Its Occurrence, Origin, Technology and Uses,* Iowa Geological Survey, Vol. 28, 1917–18, pp. 47–537.

Wyman, R.A., 1969, ''The Floatability of Twenty One Non-Metallic Ores,'' Canada Department of Energy, Mines & Resources, TB 108, Jan., 51 pp.

CONSTRUCTION USES

INSULATION

FREDERICK M. BODYCOMB AND STEVEN J. STOKOWSKI, JR.

This chapter deals with the use of industrial minerals in insulating materials used to control our thermal and acoustical environment. There are numerous products on the market used for thermal and acoustical insulation; they consume substantial quantities of a variety of industrial minerals. Following is a brief review of the theory and function of insulating materials.

Acoustics is the science of sound, and architectural acoustics deals with sound in buildings. If this definition is broadened to include transportation vehicles, it is this end use for which acoustical insulations are primarily manufactured. Sound is defined as the auditory sensation in the ear, or the disturbance in a medium which causes this sensation. Acoustical insulation is used to reduce or eliminate noise, defined as unwanted sound. Noise control may be accomplished by the reduction of sound radiation at the source, by utilizing sound reducing construction methods such as staggered studs, etc., to control sound transmission, or by using sound absorbing material on walls, ceilings, and floors (Rossing, 1985).

Acoustical insulation functions by absorbing the energy in sound waves in a porous material. As the sound waves penetrate the pores, they are dissipated by reducing the amplitude of vibration of the air molecules by contact with the pore walls, with subsequent conversion of the sound energy to heat energy. Acoustical insulation is generally highly porous, 1.3 cm or greater in thickness, and has interconnected pores. Acoustical insulating materials include fibrous vegetable or mineral matter and cellular organic and inorganic material, such as natural or manufactured foams (Sabine and Moulder, 1979). Acoustical materials are most commonly porous boards and tiles, which are often composite materials with perforated facing. Roof insulation board and lightweight concrete can also contribute to the absorption of noise.

Thermal insulation is defined as material used to control the flow of heat. It may be used for heat control from low cryogenic temperatures (below -100°C) to in excess of 2 200°C for applications such as ablative shields for spacecraft.

Heat transfer occurs by conduction, convection, or radiation. Conduction is the transfer of heat without the movement of the conducting medium, as through a metal. Heat convection through a liquid or gas takes place by movement of the medium. Forced convection is the basis for heating and cooling equipment, utilizing pumps or blowers to transfer the thermally conditioned fluid. Free or natural convection occurs due to density differences in the fluid. Radiation is the transfer of energy in space (Eckert, 1985).

Conduction of heat through a solid is reduced by insulating the body with foams, or fibrous or small celled granular material. Heat transfer through a gas (convection) is controlled by creating many small pores which inhibit the movement of the fluid, substituting a gas of low thermal conductivity, or evacuating the space. Heat radiation can be reduced by the use of materials that reflect or absorb the energy (Glaser, 1967).

Thermal insulating materials may be vegetable or mineral fibers, organic or inorganic particles, or foams with small pores, generally used to form a low density thermal barrier. They may be blocks, poured lightweight cementitious materials, bricks, etc.

RAW MATERIALS AND PRODUCTION

Industrial minerals and rocks used in acoustical and thermal insulations include: perlite, vermiculite, limestone, clays, gypsum, raw materials for the production of portland cement, glass, and mineral or rock wool, etc. (Coombs, 1989).

Calcium Silicate Insulation

Calcium silicate insulations are primarily composed of industrial minerals, hydrated lime, and a silica source such as silica flour, diatomaceous earth, or perlite ore. Small quantities of organic or inorganic fibers are added as reinforcement. Lime is produced by many companies at approximately 102 locations in the United States, and it is discussed in detail in another chapter in this book. The silica source for the production of calcium silicate thermal insulations may be finely divided silica from quartzite, etc. (silica flour), perlite or diatomaceous earth. The use of a natural grade of diatomite is common.

Due to its high surface area, diatomite may be more reactive in the hydrothermal chemical reaction with lime than silica flour. Details on diatomite are provided elsewhere in this book; however, a brief review of its origin and production is included here.

Diatomite is a rock composed of siliceous skeletal remains of diatoms, an algae which lived in salt or fresh water. It is usually composed of 85% or more amorphous silica, with bound water, alumina, iron, and other minor constituents. It is generally mined by open pit methods, and the production of natural (uncalcined) grades consists of crushing, drying, and classification by cyclones and separators. Diatomite production in the United States is located in Arizona, California, Oregon, Washington, and Nevada. Production abroad occurs in many countries including Mexico, Chile, Spain, France, Germany, the former USSR, Japan, Iceland, and elsewhere.

Silica flour is essentially all crystalline silica, and natural diatomite frequently contains 3% or more crystalline silica. Crystalline silica was classified as ''probably carcinogenic to humans'' by the International Agency for Research on Cancer in March 1988.

To reduce or eliminate the amount of unreacted crystalline silica in the end product, finely divided (baghouse) perlite ore is sometimes used as the silica source in the production of calcium silicate insulations, as mentioned previously.

The production of calcium silicate insulations is relatively straightforward. Hydrated lime, the silica source material, reinforcing fibers, and water are mixed and reacted at an elevated temperature. The slurry is then pumped to molds and pressed to dewater and form the finished part. The product may then be autoclaved for completion of the chemical reaction and oven dried.

Thermal insulating products made by this process include pipe covering, blocks for the insulation of pressure vessels, etc., and low density structural sheets used for paneling on ships, chutes for the conveyance of molten aluminum, and other specialty items. Calcium silicate insulations are classified as Type I, service temperature to 649°C, and Type II, for service to 871°C (American Society for Testing and Materials C533).

Glass Fiber Insulation

Fibrous glass, or glass wools, are used in a wide variety of thermal and acoustical insulating applications. The major industrial minerals used in the manufacture of fibrous glasses are silica sand, soda ash, and limestone, the major constituents. Other raw minerals include dolomite, feldspar, nepheline syenite, fluorspar, and various borates (borax, colemanite, ulexite, and szaibelyite). These materials are discussed in detail elsewhere in this book.

Glass wools may be produced by two methods. In the lesser used pot-and-marble system, the constituents are mixed, melted, and formed into marbles. These preforms are then remelted over platinum nozzles. The melt, passing through the nozzle, may be drawn into continuous filaments or formed into discontinuous fibers by steam or, more commonly, by flame attenuation. Generally, however, insulating glass wools are produced by melt-spinning or centrifugal spinning. In one method, the glass melt is introduced into a bowl, or inner centrifuge. The melt exits through apertures in the bowl and enters a second stage centrifuge. Fiberization takes place when the material exits through holes in the second stage centrifuge (Jarvela et al., 1990). The glass fibers are bonded with organic resin binders. Glass wool products may be rated for use at temperatures as high as 538°C, using phenolic binders, but most products are designed for use at lower temperatures.

Glass fiber products are exhaustively tested not only for thermal and acoustical insulating performance, but also for fire resistance, flame spread, smoke generation, corrosivity, water absorption, resistance to fungus, physical strengths, etc. Some of the applicable test methods, designated by The American Society for Testing and Materials, are ASTM E-96, D-781, D-828, E-84, C-335, C-356, C-411, C-547, C-585, C-795, C-518, D-792, D-638, D-790, D-257, D-3679, C-612, E-136, C-553, and numerous other government, Underwriters Laboratory, and other specifications.

Fiberized glass is used in many forms for both thermal and acoustical insulation. Typical products and applications include: wall, ceiling and attic or roof insulation, both with and without vapor barriers, vessel and pipe insulation, formed boards for plenums, air ducting, and other compressed board products. An important market for glass fiber is in transportation vehicles, where it is used for both thermal and acoustical insulation. Ease of installation by semiskilled labor contributes to the use of glass fiber in new and retrofit residential construction.

One potential problem faced by glass fiber, as with other fibrous mineral insulation materials, is that of health considerations in its use. The industry is diligently pursuing the investigation of this aspect of glass fiber production and usage, and it must be recognized as a possible factor in the future of the industry.

Table 1. Chemical Composition of Materials for Making Mineral or Rock Wool

Compound	Wool rock%	Slag%
SiO_2	21–41	30–40
Fe_2O_3	1–4	0.5–3.0
Al_2O_3	0–12	10–25
CaO	17–35	15–45
MgO	1–18	2–20
Volatile matter	20–30	0

Mineral or Rock Wool Insulation

Mineral, rock, and slag wools are manmade inorganic insulating materials composed of vitreous glass fibers. For the last 10 years there have been just under 15 producers of raw fiber, along with several additional manufacturers of finished products.

Mineral wools range in color from white or green to brown or black. In present terminology, a mineral wool has slag, generally an iron blast furnace slag, as the primary raw material. More generally, mineral wool designates a silicate-glass-wool insulating material. Rock wools utilize naturally-occurring rock as the raw material. Slag wool, currently little used, designates a product made from slag, and is synonymous with the present usage of mineral wool.

Rock wools are of great historic importance, with significant production from Illinois and Ohio (Lamar et al., 1934). Production also occurred in California, Indiana, Kentucky, Minnesota, Pennsylvania, Washington, and Wisconsin (Thoenen, 1938). The units mined generally were argillaceous limestones and dolomites, sometimes along with quartzites, shales, granites, and other rock types used as additives. Woolrocks also occur in Alabama, the Carolinas, Kansas, Mississippi, West Virginia, Virginia, and possibly other states (Gonge, 1931, Logan, 1932, Morse and McCutcheon, 1945, Plummer, 1937).

Sub-woolrocks are even more common than woolrocks. They are generally considered to be the carbonate rocks that are less argillaceous than the woolrocks, but can even be more shaley or sandy. They must be mixed with a flux rock, such as shale, sandstone, limestone, or dolomite, to produce a mixture with the proper chemical composition for wool production. Granites, wollastonite, basalt, and peridotites have also been used as sub-woolrocks or flux rocks.

Slags are also used as the raw material. Iron blast furnace slags are often used and are typically available in areas near the market for the product. Copper, phosphate, and other slags are also suitable.

The raw materials used in wool production must be within the compositional limits given by Lamar et al. (1934), as excerpted into Table 1, to melt at the lowest possible temperature that will yield a liquid with the proper viscosity for forming fibers. The fiber diameter is directly related to the acid/base ratio (on a molecular basis) in the molten glass, with values less than 1.8 yielding satisfactory diameters. An empirical rule-of-thumb regarding woolrocks is that natural rocks have the proper composition if they have 20 to 30% CO_2 (Lamar et al., 1934).

The raw material must have a 4 to 25 cm size to forestall material from being blown out of the melting furnace and maintain the proper draft. This is very significant to the raw materials producer, largely because of the possibility of having to waste the smaller material.

The production of both the mineral and rock insulating wools starts first with melting, generally in an unlined, water-jacketed steel kiln. For woolrocks, melting generally occurs around 1 400 to 2 100°C. Slags melt at a lower temperature, 1 100°C; this energy savings is the primary reason that many slags are used at current plants. Slags also have transportation and yield advantages because they are already essentially free of volatiles, such as CO_2.

Two processes can be used to produce the wool fibers. In one process, steam at 5.5 to 6.9 x 10^{+5} Pa is blown through a molten stream of glassy raw material. This produces a product that forms as the fibrous tail of the globules made by the steam. In another process, the molten stream of glass is poured onto a spinning, cast iron disc, which centrifugally forms the material into thin fibers.

The wool product consists of fibers and shot. The fibers are generally 2 to 10 µm in diameter and up to several centimeters in length. The shot, small spherules or teardrops of glass, are about 50 µm in diameter. Producers limit the content of shot in the product during the processing operation.

The collected mineral wool fibers are either used loose, or shaped into batts, blankets, sheets, or blocks with a resin, generally phenolic, binder. This material is most often used for heat insulation, where it has a better R-value than its main competitor, glass fiber. It also has value as sound insulation and may be more vermin resistant than other materials. The Thermal Insulation Manufacturers Association reports over 35 000 applications. Major applications are in industry, appliances, and automobiles.

Perlite Insulation

Perlite is discussed in a separate chapter in this book, however a brief overview of the material is included here.

Perlite is a naturally occurring volcanic rhyolite glass. It is most commonly found in domes and flows of Tertiary or Quaternary age. It is composed primarily of aluminum silicate, is formed from the hydration of obsidian, and typically contains 2 to 5% water (Breese and Piper, 1985). Perlite deposits are widespread in the western United States. New Mexico is the leading producing state, and there is also commercial production in Arizona, California, Colorado, Idaho, Nevada, and Oregon. Major foreign producing countries include the former USSR, Greece, Turkey, and Hungary. Although there are many known deposits of perlite, it is the efficiency and degree to which the ore expands, and the characteristics of the expanded material, which determine whether the perlite is of commercial quality and for which markets it may be used.

Mining and milling methods are very straightforward. The ore from open pit mines is ripped or blasted. The milling procedure is basically one of crushing, drying, and screening to produce closely-sized ore grades which are thermally expanded for various end product uses.

The commercial value of perlite derives from the fact that when flash heated, the contained water is converted to steam which expands the viscous glass particle. This pyroprocessing results in particles with a bulk density as low as 32 kg/m^3 (Jackson, 1986). Expansion is accomplished in vertical or horizontal furnaces at approximately 870°C. The cost of shipping the low density expanded perlite mandates that it be used in close proximity to the expansion facility. Therefore, the crushed and graded ore is normally shipped from the mining/milling location to the expansion plant, which is located near the market or manufacturing location. Most perlite mining companies own expansion plants remote from the mining location, but near the consumers of expanded perlite. In addition, there are numerous independent companies which expand ore purchased from the mining companies.

Nearly all perlite is used in its expanded form, and finds use in a variety of thermal and acoustical insulation applications (Anon., 1991). Small particle size perlite expanded to a very low density is used as cryogenic insulation in storage tanks and vessels containing industrial gases such as liquid natural gas, hydrogen, oxygen, helium, or nitrogen, etc., which are stored and transported at very low temperature. The annular space in a double walled vessel is pneumatically filled with perlite, then air may be evacuated from the cavity for thermal insulation. Portable expanders are frequently used at the jobsite to expand the perlite for cryogenic use.

A substantial market for expanded perlite is block fill. After a cement block or other cavity type masonry wall is constructed, loose perlite is simply poured into the block cavities for thermal insulation. Perlite fills the irregular void spaces without bridging. The material is usually silicone treated to reduce water absorption (ASTM C-549). Perlite used in this manner can increase the fire rating of a block wall from two to four hours. Competitive materials for this application include vermiculite and expanded polystyrene beads and inserts.

Perlite is used as a lightweight aggregate in portland cement and gypsum plasters for exterior and interior applications, for encasement of structural steel members and placement on floor and roof assemblies, as a base for ceramic and masonry veneers and tiles, and in stucco formulations. These plasters serve as fire protection and contribute to thermal and acoustical insulation, depending on usage. As an aggregate for lightweight concrete, perlite contributes sound and heat insulating properties when the concrete is used as a substrate for builtup roofing, floor fill, structural decks, etc. Perlite concrete is also used as a thermal insulating base for swimming pools and as a permanent insulation and structural support for underground piping. Another use for this material is to reline existing chimney flues to prevent damage to the mortar in the structure.

A major market for perlite is in the production of acoustical tile for residential and commercial buildings. Perlite may comprise approximately 15% of the formulation. Formed cellulosic boards that may contain as much as 70% perlite are used for a thermal insulating substrate for builtup roofs in commercial and industrial buildings.

Miscellaneous uses for perlite as a thermal insulating material include loose fill between the flue and body of chimneys. Thermal-insulating block and pipe covering is made from expanded perlite and sodium silicate or other suitable binder (ASTM C-610). One of the few uses of perlite ore is as ladle topping in the foundry industry. Added to pouring ladles, the perlite acts as a slag coagulant and forms a crust on the surface of the molten metal to maintain the melt temperature. One additional use of perlite in graded ore form is as a silica source in the production of calcium silicate insulations, for which baghouse material is sometimes used. Additional information on this use is included in the section on calcium silicate insulation.

Refractory Insulation

Refractories are insulating materials used at temperatures above 538°C, and may be divided into two broad categories: brick and fired shapes and specialties or monolithic items. Approximately 70% of refractories are aluminosilicates, 15% are silica, and 15% are basic refractories (magnesite, dolomite, chromite, etc.) (McMahon, 1982). Refractory materials must withstand not only high temperature environments, but also damage from physical and chemical abuse. Although certain refractories are designated as insulating products, i.e., low density insulating brick, etc., all refractory products may be considered to be thermal insulations in the sense that they serve to contain and control heat. This is a large industry with annual shipments in the United States of approximately $1.5 billion (Anon., 1987).

Various raw materials are used in refractory products (Norton, 1967). Fireclay refractories including firebrick are made from plastic and flint kaolin clays (hydrous aluminum silicates) formed and fired. High alumina refractories, originally made from diaspore, are now primarily manufactured from bauxite and have higher service temperatures than fireclay products. Silica brick is manufactured from quartzite. Basic refractory materials include magnesite brick (magnesite), chrome brick (chromite), etc. Insulating (fireclay) brick has low density, achieved by including an organic burnout material, perlite, vermiculite or expanded clay, etc., to achieve high porosity and reduce thermal conductivity. Zircon, andalusite, sillimanite, and kyanite are also used in the production of refractories (Crookston and Fitzpatrick, 1983).

Refractories are used to contain and control heat in a myriad of end-use applications, ranging from residential fireplaces to linings for high temperature furnaces, kilns, ovens, boilers, etc., in a wide variety of industries.

A separate chapter in this book deals with refractories in detail.

Vermiculite Insulation

Vermiculite, a magnesium-iron-aluminum sheet silicate, is a hydrated alteration product of biotite or phlogopite which results in a micaceous morphology. A comprehensive discussion of vermiculite is presented elsewhere in this book.

Vermiculite deposits are quite widespread in the United States, but commercial production is limited to South Carolina and Virginia, with very minor production reported from Montana. The Libby mine in Montana has ceased operations, although stockpiled concentrate from that location continued to be sold until 1992. Foreign production is dominated by the Palabora Mining Co. in the Republic of South Africa, which exports major quantities of concentrate to the United States. Some Brazilian vermiculite is also being shipped to this country. Other producing countries include China, India, the former USSR, and Zimbabwe.

Vermiculite is generally mined by open pit methods, and milling/concentration consists of wet processing including flotation using a variety of surfactants. The mine product is a vermiculite concentrate, which is shipped to exfoliation plants near the end markets, due to the higher transportation costs involved in shipping the low density exfoliated product.

At the exfoliation plants, the concentrate is flash heated at a temperature of approximately 750 to 800°C. Bound water between the micaceous sheets explosively turns to steam and is responsible for the expansion of the particles. Additional water may be liberated above approximately 870°C, resulting in a weak powdery product. Exfoliated vermiculite can have a bulk density as low as 50 kg/m^3.

W.R. Grace Co., long a dominant factor in the vermiculite industry in the United States, mines vermiculite and has an extensive organization of exfoliating plants. Other companies also mine and process the concentrate. In addition, there are a number of non-mining firms which operate exfoliation plants and buy their concentrate from outside sources.

As with perlite, vermiculite is used in a wide variety of thermal and acoustical insulations due to its low bulk density (Yang, private communication, 1991). Indeed, both materials compete for use in many end products. Vermiculite, treated with either asphalt emulsion or a silicone for water resistance, is used as cavity fill for thermal insulation in cement block and other masonry structures exactly as is perlite (ASTM C-516). On the other hand, coarser grades of vermiculite are used as loose fill in residential attics and walls, an inappropriate use for perlite due to its dustiness. Vermiculite used in these applications also contributes to the acoustical insulation of the buildings.

Vermiculite is used as an aggregate in lightweight concrete in direct competition with perlite. Reported drawbacks to the use of vermiculite for this application are a greater susceptibility to water and lower strengths, as compared to perlite (Power, 1986). Vermiculite is used in portland cement and gypsum plasters, enhancing both thermal and acoustical insulating properties. It is also employed in acoustical panels and ceiling tiles, however, its dark color is a limiting factor in this market. The concentrate is used as ladle topping, as is perlite.

Large quantities of unexfoliated vermiculite concentrate are used in fire rated wallboard. In this application, as heat from a dwelling fire increases, the vermiculite particles expand, compensating for the shrinkage of the gypsum matrix and maintaining the structural integrity of the board. An important market for vermiculite is in formulations for sprayable fireproofing for structural steel members. Another use is in intumescent coatings. Upon heating, the material expands, forming a low density thermal barrier. Vermiculite is used in the manufacture of refractory brick, castables, ramming mixes, and other articles for service temperatures up to 1 250°C.

Other Insulation Materials

There are many insulating products containing industrial minerals as a relatively minor constituent or ancillary use, that have insulating properties as a secondary attribute of their overall performance, or that represent a relatively small market or market segment. Several of these insulation products are briefly discussed here.

Foamed Glass: Powdered glass and a blowing agent are heated to form a plastic mass. Gas expansion from the blowing agent forms a closed cell structure. Foamed glass insulation is used for pipe and vessel insulation and also in walls and roofs. Service temperatures for some of these products are as high as 538°C.

Foamed Metals: Aluminum and zinc are heated with blowing agents, similar to foamed glass, to form insulating materials.

Foamed Portland Cement: Aluminum powder and flake, added to concrete, react with the high pH matrix to form hydrogen gas, resulting in bubbles or cells in the material. The low density product is useful for its insulating properties.

Diatomaceous Earth: The use of this material as a silica source in the production of calcium silicates is discussed above. Other uses of diatomite as insulation include insulation in safes, fireproof filing cabinets, etc. It has been used as a loose fill insulation, but this is now limited due to its dustiness.

Refractory or Ceramic Fibers: A wide variety of these materials are being produced, including those made from kaolin/alumina, silica, zirconia, lanthanum chromite, and others. One rather exotic use is in ablative materials, which operate in extremely high temperature environments such as spacecraft heat shields. They act as heat sinks, and heat dissipation is accomplished by mass removal. Refractory fibers are used as reinforcement in some of these products. Other uses are in the reinforcement or various resin systems for many applications.

Lightweight Concrete Aggregates: Lightweight concretes are widely used for their contribution to the thermal and acoustical insulation of a structure. Their low density derives from the lightweight aggregates used in the formulation. Perlite and vermiculite usage was discussed previously. In addition to those materials, clays, shale, and slate are heated until they bloat. Pumice, cinders, and other naturally occurring volcanic materials are also used as lightweight aggregate, as well as processed slags.

Asbestos: Due to health and environmental reasons, asbestos usage for insulation in the United States is virtually nil. In some countries, however, asbestos fiber is still used as a reinforcement in calcium silicate insulation. In such products as insulating board, a sandwich construction of two sheets of asbestos cement with a core of insulating material is produced for both interior and exterior use (ASTM C-551).

Ceiling Tile: Acoustical ceiling tile is mentioned here as an example of an insulation product that consumes large quantities of industrial minerals and rocks in the United States. The product may contain 15% clay, 55% mineral wool, plus the 15% perlite mentioned previously.

INDUSTRY TRENDS

The oil shortage in the 1970s focused attention on energy conservation, and interest in and demand for thermal insulating products consequently increased. Abundant, although relatively expensive supplies of oil in the 1980s and early 1990s maintained the demand for insulation materials. It is possible that the environmental revolution of the late 1980s, which promises to continue through the decade of the 1990s, could lead to increased demand for thermal insulation as a means of reducing energy consumption and thus pollution. Reduced energy usage equates with lower carbon dioxide and sulfur dioxide emissions, in turn reducing the effect of these pollutants on global warming and acid rain. A trend toward higher living standards in many countries of the world should result in an increase in residential construction, with a concurrent increase in demand for insulations.

Encouragement by the federal government for the construction of more energy efficient residential and commercial buildings is expected to continue. The *1989 Model Energy Code*, published by the Council of American Building Officials, recommends higher insulation standards which are predicted to increase residential insulation usage by perhaps 10% as the Code is adopted by more state and local governments (Anon., 1991a). Regulations of the US Department of Housing and Urban Development and standards developed by the American Society of Heating, Refrigerating, and Air Conditioning Engineers will certainly not be relaxed with regard to insulation requirements for residential and commercial buildings, and in all probability will become more stringent over the long term. Also, programs are in place which provide for the upgrading of insulation in homes occupied by, but not owned by, low income families.

Insulation is one of the few items in new or retrofit residential construction that can be shown to have a finite payback period, actually saving the homeowner money. As energy costs inevitably increase over the long term, insulation usage should increase based solely on economics.

A factor to be addressed by the producers of most mineral insulation materials is that their manufacturing processes are energy intensive. To remain competitive with other insulation materials, ongoing research should be conducted to reduce the energy required in their processes, lest the very factors that encourage the increased use of insulation prohibitively increase their production costs.

The use of mineral and mineral-containing insulation will depend on the development and marketing of cost effective products of high quality, stressing permanence, ease of installation, safety, and performance vs. competitive products.

BIBLIOGRAPHY

Anon., 1987, Refractories, The Refractories Institute, Pittsburgh, PA.

Anon., 1991, The Perlite Institute, Inc., Chicago, IL.

Anon., 1991a, *Manville Today*, June-July newsletter, Denver, CO.

ASTM C-516, Standard Specification for Vermiculite Loose Fill Thermal Insulation.

ASTM C-533, Standard Specification for Calcium Silicate Block and Pipe Thermal Insulation.

ASTM C-549, Standard Specification for Perlite Loose Fill Insulation.

ASTM C-551, Standard Specification for Asbestos-Cement-Fiber-Board Insulating Panels.

ASTM C-610, Standard Specification for Expanded Perlite Block and Pipe Thermal Insulation.

Breese, R.O.Y., and Piper, J.R., 1985, "Occurrence and Origin of Perlite," Preprint 85-359, SME Fall Meeting, Oct. 16–18, Albuquerque, NM, Society of Mining Engineers, Littleton, CO.

Coombs, G., 1989, "Insulation Raw Materials," *Concise Encyclopedia of Mineral Resources*, D.D. Cahn and N. Herz, eds., Pergamon Press, Oxford, pp. 175–183.

Crookston, J.A. and Fitzpatrick, W.D., 1983, "Refractories," *Industrial Minerals and Rocks*, 5th ed., S.J. LeFond, ed., Vol. 1, Society of Mining Engineers, pp. 373–386.

Eckert, E.R.G., 1985, "Heat Transfer," *The Encyclopedia of Physics*, 3rd ed., R.M. Besancon, ed., Van Nostrand Reinhold, New York, pp. 556–561.

Glaser, P.E., 1967, "Thermal Insulations," *Standard Handbook for Mechanical Engineers*, T. Baumeister, ed., McGraw-Hill, New York, pp. 6–203 to 6–205.

Gonge, M.F., 1931, "Raw materials for the manufacturing of rock wool in the Niagara penninsula of Ontario," Memorial Series No. 50, Mines Branch, Dept. of Mines, Ottawa, Canada.

Jackson, F.L., 1986, "Processing Perlite for Use in Insulation Applications," *Mining Engineering*, Vol. 38, No. 1, pp. 40–46.

Jarvela, P., et al., 1990, "Modeling Porous Glass Wool Composite Strength," *Handbook of Ceramics and Composites*, Vol 1, Synthesis and Properties, Marcel Dekker Inc., pp. 207–228.

Lamar, J.E., et al., 1934, Rock Wool from Illinois Mineral Resources, Bulletin No. 61, Illinois State Geological Survey, 201 pp.

Logan, W.N., 1932, "The Mineral Wool Industry in Indiana," AIME Preprint.

McMahon, J.F., 1982, "Refractory," *Encyclopedia of Science and Technology*, Vol. 15, McGraw-Hill, New York, pp. 254–256.

Morse, W.C. and McCutcheon, T.E., 1945, "Rock Wool," Bulletin No. 62, Mississippi State Geological Survey, 19 pp.

Norton, F.H., 1967, "Refractories," *Standard Handbook for Mechanical Engineers*, T. Baumeister, ed., McGraw-Hill, New York, pp. 6–206 to 6–215.

Plummer, N., 1937, "Rock Wool Resources of Kansas," Mines Resource Circular 5, Vol. 38, Kansas State Geological Survey.

Power, T., 1986, "Perlite and Vermiculite—The Market Overlap," *Industrial Minerals*, Nov., No. 230, pp. 39, 40, 43, 45, 47–49.

Rossing, T.D., 1985, "Acoustics," *The Encyclopedia of Physics*, 3rd ed., R.M. Besancon, ed., Van Nostrand Reinhold, New York, pp. 29–30.

Sabine, H.J., and Moulder, R., 1979, "Sound Absorptive Materials," *Handbook of Noise Control*, C.M. Harris, ed., McGraw-Hill, New York, pp. 21–1 to 21–11.

Shreve, R.N., 1956, *The Chemical Process Industries*, 2nd ed., McGraw-Hill, New York, pp. 250–251.

Thoenen, J.R., 1938, "Mineral Wool," Information Circular 6984, US Bureau of Mines.

CONSTRUCTION USES

LIGHTWEIGHT AGGREGATES

BRUCE H. MASON

Lightweight aggregates are classified as those minerals, natural rock materials, rock-like products, and byproducts of manufacturing processes that are used as bulk fillers in lightweight structural concrete, concrete building blocks, precast structural units, road surfacing materials, plaster aggregates, and insulating fill. Other uses include architectural wall covers, suspended ceilings, soil conditioners, and other agricultural uses (Mason, 1989). One unusual agricultural use recently reported was for bedding on a crocodile farm on the south coast of Natal. It seems that female crocodiles became upset because there was no vermiculite in which to lay their eggs due to the late arrival of a consignment of vermiculite. The vermiculite was being used as a substitute for river sand to nest crocodile eggs (Griffiths, 1989).

The various lightweight aggregate materials may be classified into four groups as follows:

1. *Natural Lightweight Aggregate Materials*—prepared by crushing and sizing natural rock materials, such as pumice, scoria, tuff, breccia, and volcanic cinders.
2. *Manufactured Structural Lightweight Aggregates*—prepared by pyroprocessing shale, clay, or slate in rotary kilns or on traveling grate sintering machines.
3. *Byproduct Lightweight Aggregates*—prepared by crushing and sizing foamed and granulated slag, cinders, and coke breeze. This group of lightweight materials, with the exception of slag, has become less significant and will not be discussed in this chapter. Other lightweight aggregates are beginning to appear in the marketplace. Some are the result of manufacturing processes on industrial byproducts and/or waste products, such as fly ash and incinerator residue (Bigner, 1992).
4. *Manufactured Insulating Ultralightweight Aggregates*—prepared by pyroprocessing ground vermiculite, perlite, and diatomite. The application of diatomite in brick and in extruded and aggregate forms has declined in recent years (Kadey, 1983) and will not be considered in this chapter.

Lightweight aggregates for structural concrete and concrete masonry units, according to the American Society for Testing and Materials (1986) consist of "aggregates of low density used to produce lightweight concrete, including: pumice, scoria, volcanic cinders, tuff and diatomite; expanded shale, clay, slate, perlite, vermiculite or slag; and end products of coal or coal combustion" (Mason, 1989).

Lightweight aggregates are distinguished from other mineral aggregate materials by their lighter unit weights—1 281 to 1 602 kg/m^3 or less for loosely packed lightweight aggregates—as compared to 2 002 kg/m^3 or more for crushed stone, sand, gravel, air-cooled slag, and other common mineral aggregates. The specific gravities of these lightweight aggregates are also used to classify them for use in concrete; they range from the extremely light materials, used for insulative and non-structural concrete to expanded shale, clay, slate, and slags used for structural concrete. Vermiculite and perlite are sometimes referred to as super lightweights. Insulating concrete (non-load bearing) weighing as little as 240 kg/m^3 can be made with these aggregates.

Natural aggregates such as pumice, pumicite, and scoria are normally made into concrete weighing about 400 to 480 kg/m^3. Expanded shale, clay, and slate produced in a rotary kiln make structural concrete ranging from 1 201 to 1 842 kg/m^3. Expanded shale and clay produced by sintering make concrete that ranges from 1 442 to 1 922 kg/m^3. Air-cooled slag produces concrete ranging from 2 162 to 2 403 kg/m^3.

The low density lightweight aggregates are used primarily for insulating purposes. They have relatively low compressive strength. The middle density aggregates are used for insulation and for fill. Those at the upper end develop excellent compressive strength and are used in many structural applications (Anon., 1971).

In recent years, the production of lightweight aggregates in the United States for all uses (including insulation and construction materials) has averaged about 19 Mtpy (Table 1). The production and consumption of lightweight aggregate is closely tied to the fortunes of the construction industry and the need to construct energy efficient buildings. Recent years have seen the production of lightweight aggregate approach the levels of the early 1970s. The period of 1979 to 1981 was a low point in production levels. The low levels of production were caused by significantly higher plant energy costs due to the large amounts of fuel and energy used in the production of most types of lightweight aggregate. Lower levels of construction activity also affect production levels.

This chapter focuses on the use of natural lightweight aggregates and manufactured insulating ultralightweight aggregates, and on both the production and use of manufactured structural lightweight aggregates. Some additional comments are made on byproduct materials. Separate commodity chapters on vermiculite, perlite, pumice, and volcanic cinders are elsewhere in the volume. Discussion of the commodities covered elsewhere in the book will be limited to observations on their utilization and relation to other lightweight aggregate materials.

RAW MATERIALS

Most materials considered as lightweight aggregates are derived from some initial natural raw material. While historical definitions have included such organic materials as straw and peat as

Table 1. Domestic Production of Lightweight Aggregate, 1983 to 1989 (thousand tons)

Type of Aggregate	1983	1984	1985	1986	1987	1988	1989
Clay	3 988	3 971	4 274	3 344	4 140	3 809	3 382
Slag							
Air-cooled	11 231	13 903	12 123	12 248	12 200	12 920	12 504
Expanded	1 066	1 317	1 580	1 705	1 609	1 504	1 548
Perlite	430	452	470	460	484	522	545
Pumice & Pumicite	407	455	461	503	356	352	424
Vermiculite	256	286	285	288	275	276	250
Totals	17 278	20 384	19 193	18 548	19 064	19 384	18 653

Sources: Minerals Yearbook, 1983 to 1989, U.S. Bureau of Mines *Mining Engineering* Annual Review, 1983 to 1989

lightweight aggregates, the discussion in this chapter will, in the most part, discuss lightweight aggregates derived from mineral or rock materials. Some lightweight materials produced from waste materials, which have recently appeared in the marketplace, are briefly discussed.

Actual lightweight material in use may be a manufactured product derived from an ore or raw material that is obviously a mineral resource. In this context, then, raw materials for lightweight aggregates may include iron ore and flux stone (the components of blast furnace slag), coal (the raw material from which cinders are derived), and shale, clay, and slate that display the characteristics of bloating or controlled expansion when subjected to elevated temperatures (871 to 1 316°C).

A new potential source of lightweight aggregate is the residue from incineration of municipal refuse (Bigner, 1992). If the lightweight aggregate produced from industrial and municipal refuse is accepted in the marketplace, then it could be argued that in certain metropolitan areas municipal refuse should be considered as a reserve. At the present time, 137 waste-to-energy plants in 36 states handle nearly 16% of the 185 Mt of solid waste generated in the United States. The US Environmental Protection Agengy estimates that by the year 2000, there will be more than 300 US waste-to-energy plants handling about 25% of the nation's solid waste. Not all of these facilities will be sources of bottom ash suitable for lightweight aggregate, but some will. Land disposal of solid waste remains the most common method of handling municipal solid waste. Even after the solid waste has been processed (converted to energy), the residual material (bottom ash) remains (Hammon, 1992, Rogoff, 1992).

The principal properties desired in materials used as lightweight aggregates vary according to end use, but normally include:

1. *Lightweight.* Reduces dead load (weight in place) and facilitates the physical handling of the materials. This lighter weight increases productivity in construction where lightweight aggregates are used in concrete structural products.
2. *Thermal and Acoustical Insulation Properties.* Normally the result of airspaces, voids, or pores in the lightweight aggregate materials.
3. *High Fire Resistance.* Low probability of physical breakdown at temperatures below the melting point of the aggregate.
4. *Toughness.* Minimum tendency to crack or break if nailed or stressed in construction use.

Other properties that are normally considered desirable include:

Substantial compressive strength
Low water absorption
Good resistance to freezing and thawing
Low shrinkage on drying and minimum thermal expansion
Good bonding with cement
Chemical inertness
Good elastic properties
Abrasion resistance

The actual specifications and tolerances vary according to the end use or final product in which the lightweight aggregate is consumed. Specifications for various lightweight aggregate materials by use have been compiled by the American Society for Testing & Materials (ASTM) and standards may be consulted as required. ASTM standards C330 (lightweight aggregate for structural concrete), C331 (lightweight aggregate for concrete masonry units), and C332 (lightweight aggregate for insulating concrete) are the most relevant for most lightweight aggregate materials.

The variety and use of materials as lightweight aggregate is extensive. They are mostly bulky and of low unit value, and normally, serve local and regional markets. However, some structural lightweight aggregates are economically shipped long distances because the density of these aggregates are less than half of the density of locally available crushed stone, sand, and gravel. Any comparison of lightwieght aggregate and locally available aggregate such as crushed stone, sand, and gravel should be made on a volume basis. For example, lightweight aggregate may be four times as expensive per ton as local aggregate, but may yield two to three times the volume needed. One ton of lightweight aggregate will normally make two to three times as much end product as one ton of normal weight aggregate (Ries, 1992).

Products such as pumice or expanded slag may be used in substantial quantities in some localities and may not even be available (at competitive prices) in others. Perlite and vermiculite deposits are limited in location, but serve national markets due to relatively high market value and because their ores are shipped prior to pyroprocessing and physical expansion. Most of these pyroprocessing and expansion plants reflect speciality uses in regional and national markets. The more specialized uses command higher market prices that lead to extensive geographic distribution and use.

PRODUCTION

Naturally occurring lightweight aggregate raw materials, such as pumice and volcanic cinders, are normally mined by open pit or quarry methods, depending on the degree of consolidation of the raw materials. Mining costs are similar to those for sand and gravel, except where substantial blasting and crushing are required. In the latter case, production costs parallel those for crushed stone. Selling prices for pumice and pumicite are approximately four times those of sand, gravel, and crushed stone and have ranged from $10.39 (1986) to $25.92 (1991) per ton. A twenty city average price for gravel and crushed stone (19 × 9.5 mm) is $9.41 and $7.39 respectively (Anon., 1992). Domestic output of pumice and pumicite

Table 2. Value of Domestic Lightweight Production, 1983–1989 (thousands $)

Value of Production	1983	1984	1985	1986	1987	1988	1989
Clay	$ 33,532	36,692	31,746	47,020	38,098	38,650	28,345
Slag							
Air-cooled	46,265	73,071	68,991	64,925	67,342	76,517	73,385
Expanded	12,461	21,100	26,777	42,833	37,203	35,727	32,125
Perlite	15,665	16,646	17,537	15,648	16,512	17,814	16,322
Pumice	4,482	4,923	4,543	5,761	4,497	4,128	8,217
Vermiculite	27,217	31,526	32,415	34,451	33,114	33,925	30,427
Totals	139,622	183,958	182,009	210,638	196,766	206,306	188,821

Source: *Minerals Yearbook*, 1983 to 1989, U.S. Bureau of Mines

has increased in recent years, from a low of 356 kt in 1987 to a high of 443 kt in 1990 (Presley, 1992).

The number of domestic pumice producers increased by three in 1991 as compared to 1990. The principal producers are located in Oregon, New Mexico, Idaho, and California. Price increases have been attributed to the continued large demand for stone for washing designer jeans and for landscaping uses. The decrease in volume in the last year is attributed to the decreased demand in the construction industries.

Shale, clay, and slate are used as raw materials in the manufacture of building brick as well as lightweight aggregate. Most production plants for brick, however, do not seek the heat expanding or pyroplastic properties desired by producers of lightweight aggregate. The shale, clay, and slate used as lightweight aggregate raw materials should reach a pyroplastic state at elevated temperatures and then generate a gas that will expand or bloat the resulting product to a porous substance that will retain its physical strength despite its lighter unit weight when cooled.

Shale, clay, and slate are normally mined by open pit and quarry methods, then dried in large sheds or open stockpiles to control water content in the raw feed prior to high temperature pyroprocessing in either rotary kilns or sintering machines. The resulting clinker may then be screened or crushed before screening to yield proper gradation mixes for final use as an aggregate in concrete masonry units, structural concrete, or other applications. The specific gravity of the expanded shale, clay, and slate can be controlled to some degree by the type of kiln and/or sinterling grate used to process the raw material.

Table 1 summarizes the annual output of lightweight aggregate in the United States in the period 1983 to 1989 by quantity and estimated value. In recent years, significant price increases for expanded perlite and exfoliated vermiculite give them approximately 24% of the market in terms of value despite contributing only about 5% of the tonnage. Expanded shale, clay, and slate are widely used. In 1989 they accounted for about 18% of quantity and 15% of value. Table 2 summarizes the stated values of the production for the same period.

While raw material availability is important, most lightweight manufacturing plants require a proximity to the concrete masonary, structural concrete, and asphalt pavement markets to be economically viable. Western Aggregates Inc., a subsidiary of Inco United States, recently purchased and refurbished the Ideal Basic plant, 8.5 km south of Boulder, CO, which services the large Denver Metropolian area market (Carter, 1992).

Most producers of expanded shale, clay, and slate will find long run economic survival rather precarious if they do not engage in substantial raw material testing, geologic and mining studies, and marketing research investigations prior to the location of a new plant (Conley et al., 1948, Shildeler, 1961).

Production cost for expanded shale, clay, and slate have increased in general along with the cost of energy, raw materials and supplies. During the period 1980 to 1990 cost associated with the installation of air pollution control equipment (baghouses) and their added cost of operation have had an effect on operational costs. The cost of labor is considerably higher today in proportion to other cost centers. More direct cost increases were felt as more power was required per unit of production than in the past. Conservation efforts continue to lower the costs of energy-related items, such as refractories and electric power. Most plants continue to use coal as a primary source of fuel.

A number of producers are now using alternate fuels to fire kilns. Some of these alternate fuels are better known by as waste derived fuels and solvents from various industrial processes. These waste derived fuels can provide some producers with an excellent source of low cost fuels, provided air emmissions parameters can be met. In some cases, the producer using these fuels actually receives an income stream from the generator of the waste derived fuel. This has caused some market distortions, but has served to keep the production costs of some producers stable, while at the same time providing a reasonable method of disposing of large quantities of non-health threatening waste (Singletary, 1991, Kelly, 1992).

Expanded slag is a byproduct of iron and steel production. It is manufactured by spraying a stream of water through molten blast furnace slag as it is drawn from the furnace. The resulting foamed slag is then crushed and screened for use in concrete block or structural concrete. Air-cooled slag is also a byproduct of iron and steel production. It is manufactured by pouring the molten blast furnace slag into pits where it is cooled by water. It is then excavated, crushed, and screened into various sizes.

Cinders, a byproduct of the burning of lump coal, are still a small component in the marketplace. Cinders are more of a factor in the southwestern United States than in other parts of the country. Ash is the solid residue that remains after combustion. Bottom ash includes all solid materials that pass through the furnace unburned, such as tramp metal, glass, rocks, and dirt. Most coal-consuming processes utilize pulverized coal in combustion and produce fly ash rather than cinders. Fly ash production continues to increase with the increased use of pulverized coal in electric power generation. Roughly 54.4 Mtpy of fly ash are produced, but utilization is only about 25% (Huhta and Stearn, 1980). The production of lightweight aggregate from fly ash was once heralded as a major potential market, but very little is currently being utilized as a lightweight aggregate raw material because the technology is not fully developed (Ries, 1992).

Recently many state legislatures, in an effort to reduce the volume of fly ash being disposed of in landfills, charged many utilities to investigate alternate uses for fly ash. In one location, fly ash is being mixed with Milk-of-Lime to produce a lightweight aggregate. In another location, fly ash is being mixed with portland cement. Acceptance of these products and their preformance in the marketplace has been mixed to poor.

In recent years, 42 states have adopted legislated recycling goals. The reauthorization of the Resource Conservation Recovery Act (RCRA) in 1993 could present the EPA with mandates to issue regulations that will severely limit the disposal of fly ash and bottom ash in landfills. If this happens, then additional pressures will be applied to those generators of such waste materials to find alternate means of disposal. This could lead to the appearance in the marketplace of many new lightweight aggregate sources. Not all of these will be based on pure market economics. It has be suggested that in some instances, solid waste will contribute an income stream to the processing facility in the form of tipping fees. In the future, a great acceptance of these and other waste product lightweight aggregates could have a profound impact on the lightweight aggregate market as we know it today (Gershman and Miller, 1992).

The construction of the first of a series of plants to process municipal solid waste and sewage into a lightweight aggregate has been announced. This plant is scheduled to break ground on an 8 hectare site in northwestern Indiana in 1993. The reality and long term viability of this type of process is still unproven. This proposed plant will process a combination of pulverized municipal solid wastes (which has had the recyclables and oversized material removed) with a mixture of milled clay and liquid non-hazardous waste (sewage sludge) to form a pellet. The pellet will be processed through a three stage kiln. In the first stage, the pellet will be pyrolyzed, or chemically altered by heat, which causes combustible gases to escape. These gases will not only bloat the pellet into a lightweight aggregate, but will also provide some of the required fuel for the kiln. The second stage will remove most of the pellet's carbon through oxidation. This will add to the lightweight character of the aggregate and will supply some of the energy requirements of the process. The third and final stage will vitrify the pellet.

This new process has been in development for approximately 12 years. The process is projected to be self-sustaining with respect to its energy requirements. Income is expected from solid waste disposal fees, use of non-hazardous liquids, and the sale of recovered recyclables. The company expects to market the aggregate for approximately $33 per ton.

Questions abound as to whether this process will pass present air emissions standards and if the lightweight product produced will be accepted in the marketplace. A pilot plant operated as a test facility near Brisbane, Australia, in 1988 and many technical and process problems were encountered. The process may have great promise if all the technical and processing problems can be addressed. It may be the late 1990s or early 2000s before this process has much impact on the marketplace. This process has also been tested in Germany (Bigner, 1992; Ries, 1992).

The majority of perlite applications require that the mineral be used in the expanded form. Run-of-mine perlite is crushed and ground before any thermal treatment can take place. Run-of-mine material is normally reduced in size by using a jaw crusher. Depending upon the proposed end use, it may then be futher reduced by cone crushers and air classified (McMichael, 1989). Crude perlite production is primarily located in the western United States. New Mexico accounts for 85% of the domestic production. The remaining production comes from Arizona, California, Colorado, Nevada, and Idaho (O'Driscoll, 1989).

CONSUMPTION

Domestic pumice production was estimated to be approximately 439 kt in 1991 and apparent domestic consumption was 559 kt. Domestic production supplied approximately 78% of the demand in 1991 with net imports of 135 kt supplying the remaining 22%. In 1991, concrete admixtures, decorative building block, concrete aggregate, ballast, and road surfacing were the major uses of pumice and pumicite, accounting for 69% of the annual consumption. Landscaping uses, large demand in the washing of designer jeans, insulation, and other end-uses constituted the balance of the production (Bolen, 1992b, O'Driscoll, 1989, 1990, Presley, 1992).

Table 3. Total Shipments of Expanded Shale, Clay, and Slate in United States (thousand tons)

Year	Tons
1986	4 390
1987	4 400
1988	4 310
1989	4 173
1990	4 010

Source: Expanded Shale, Clay, and Slate Insititute.

Consumption of expanded clay, shale, and slate has declined from about 10.9 Mtpy in the early 1970s to 3.4 Mty in 1989. During this time, increasing production costs have caused producers to significantly increase prices, which may have contributed to a decrease in the amounts consumed. From 1986 to 1989 sales of expanded shale declined approximate 9% due to lower levels of construction activity, higher interest rates, and the general state of the nation's economy. Total shipment of expanded shale, clay and slate during the period from 1986 to 1990 are shown in Table 3.

Prices have more than doubled in the past ten years from $11 to $20 per ton in the early 1980s (McCarl, 1983) to $33 to $57 per ton (depending upon gradation and aggregate type) in 1990 to 1991 (Konczak, 1992). Raw materials for these structural lightweight aggregates have also increased in cost during this period. Typical values for lightweight aggregate raw materials (ores) and the expanded products at the production plant are shown in Table 4.

Expanded slag consumption has varied from 1.01 Mtpy in 1983 to 1.7 Mty in 1986 to 1.3 Mtpy in 1990. The lowest production year, 1983, was due largely to low levels of construction activity. The tonnages reported for 1989 and 1990 illustrate that production and sales of expanded slag are closely related to the strength of the construction industry. The tonnages consumed by users in the midwestern United States are approximately half of those reported for 1990 for the same area (Konczak, 1992). More than 95% of all expanded slag is used in lightweight concrete and concrete products. The consumption of expanded slag in the market place by end use, varies from location to location. Some area markets seem to show a preference (by specification) towards one lightweight aggregate rather than another. This trend is probably related to a preference by some architectural/engineering firms toward specifying lightweight aggregates that they have used successfully. The value of expanded slag has increased from about $7.71 per ton in 1980 to an average of about $20.75 per ton in 1989. Downturns in the construction industry in 1992 have dropped those prices back to the 8 to $13 per ton range.

Expanded perlite typically makes up 80% of domestic lightweight aggregate consumption (Larsen, 1990). Domestic expanded perlite consumption in the period 1981 to 1990 ranged from 429 to 452 kty. The value of perlite ore (FOB mine) during the period has ranged from 32.85 to $33.80 per ton. The value of the expanded perlite is estimated to be approximately $209 per ton. The higher unit values were for expanded perlite treated with silicone. Much of the growth in perlite has occurred in building materials, particularly for ceiling tiles and loose fill insulation. Increased use has been reported as a concrete filler to achieve stronger lightweight

Table 4. Average Unit Value of Lightweight Aggregate Materials at Production Plants

Type of Material	1988	1990
Exfoliated vermiculite	$221.29	$230.00
Crude vermiculite (concentrate)	111.51	142.00
Expanded perlite	186.00	209.00
Crude perlite (ore)	30.65	34.24
Expanded clay, shale, and slate	33.00	40.00
Clay and shale (raw)	10.15	11.50
Expanded slag	13.45	13.00
Pumice and volcanic cinders	11.70	19.83

applications (O'Driscoll, 1989). A comparison of perlite end uses in 1987 and 1988 is presented in Table 5 to indicate levels of consumption in the various segments of the market (Bolen, 1992a).

Production levels of domestic vermiculite have declined from a high of 290 kt in 1981 to a low of 209 kt in 1990 (estimated). Apparent domestic consumption of the concentrate has averaged 285 kty. Imports of concentrate seem to be having an impact on domestic production. Imports have increased every year since 1982, from a low of 19 kt in 1982 to a high of 50 kt in 1989. Producers in Brazil have recently expanded their production capabilities (Kinch, 1989). The amount of vermiculite concentrate exported (principally to Canada) is reported to have been 18 kt valued at $5.2 million. Unit values for vermiculite concentrate in 1991 were reported to be $74.05 per ton. The actual prices of vermiculite used in North America vary. Vermiculite is imported into the United States for $142 per ton, and exported for $179 per ton as a major substitute material for many insulating uses. In 1990, 31% of all exfoliated vermiculite was used in agriculture, 31% in plaster and cement premixes, 19% in insulation products, 15% in lightweight concrete aggregate, and 4% in miscellaneous uses (Potter, 1991).

SPECIFICATIONS

The consumption of lightweight aggregate materials is closely tied to construction codes and other specifications dealing with unit weight, size gradation, deleterious substances, and concrete-making properties. More detailed specifications may be obtained from ASTM, Perlite Institute, Vermiculite Institute, National Slag Association, Expanded Shale, Clay and Slate Institute, Lightweight Aggregate Producers Association, American Concrete Institute, National Concrete Masonry Association, National Ready Mixed Concrete Association, and various federal, state, county, and local government building codes.

Unit Weight

Dry loose weight may vary from 881 to 1 121 kg/m^3 for structural lightweight aggregates to 96 to 192 kg/m^3 for exfoliated vermiculite and expanded perlite. Specific dry weights are given in Table 6.

Size Gradation

The size of particles relate to various other characteristics, such as compressive strength, thermal and acoustical insulation values, heat resistance, and cement content in concrete mixes. It is generally expressed as ranges of percent by weight passing certain sieve or screen size openings. This gives an idea of the gradation of particle size in the raw material being used as aggregate. Although similar to grading in other aggregates, there are some variations. Size

Table 5. Perlite Uses in the United States, 1979 and 1988

Use	1979 %	1988 %
Filter aid	18	13
Plaster aggregate	4	2
Concrete aggregate	8	2
Horticultural products	7	10
Low temperature insulation	1	1
Formed products	14	59
Masonry and cavity-fill insulation	4	2
Fillers	2	5
Other uses	42	4

Table 6. Dry Weights for Various Lightweight Aggregates

Aggregate Type and Size Designation	Dry Loose Weight, Max., kg/m^3
Structural lightweight aggregate	
Fine Aggregate	1 121
Coarse aggregate	880
Combined fine and coarse aggregate	1 041
Vermiculite	160 (96 min.)
Perlite	190 (96 min.)

gradation specifications vary according to use as shown in Tables 7 and 8.

Deleterious Substances

The amount of contaminants in lightweight aggregates must be limited by weight so that the quality of the final product (normally concrete or concrete block) will not be adversely affected. Organic impurities and iron oxides must be limited to avoid discoloration and staining. Clay lumps are limited to less than 2% by dry weight. Loss on ignition should be less than 5%, except in the case of cinders where ignition loss should be less than 35%.

Concrete-Making Properties

A number of tests must normally be made with the lightweight aggregate in various concrete mixes. Specific quantities of fine and coarse aggregate are mixed with cement and water. This mixture is cast into cylinders and bars for a series of tests. The tests include compressive strength and unit weight (of concrete), drying shrinkage, visual inspection for popouts, and various durability tests, such as freezing and thawing or sulfate reactivity or soundness tests.

Specifications vary depending on the final product to be made. ASTM or other standards should be consulted for specific details on testing and product requirements.

Other Specifications

In specialized uses, such as insulating fill or an insulating-concrete aggregate, other standards such as thermal conductivity, fire rating, and physical flow characteristics, become important in judging the suitability of various lightweight aggregates. The actual requirements for any specialized application should be reviewed before any material is used for that purpose.

Virtually any naturally occurring lightweight material with the desired physical and chemical characteristics may be considered a

Table 7. Grading Requirements for Lightweight Aggregates Percentages (by Weight) Passing Sieves Having Square Openings

Size Designation	25.0 mm	19.0 mm	12.5 mm	9.5 mm	4.75 mm	2.36 mm	1.18 mm	0.29 mm	0.149 mm
Lightweight Aggregates for Structural Concrete (ASTM C330)									
Fine Aggregate:									
4.75 to 0 mm	. . .	. . .	. . .	100	85–100	. . .	40–80	10–35	5–25
Coarse aggregate:									
25.0 to 4.75 mm	95–100	. . .	25–60	. . .	0–10	. . .	. . .	. . .	. . .
19.0 to 4.75 mm	100	90–100	. . .	10–50	0–15	. . .	. . .	. . .	. . .
12.5 to 4.75 mm	. . .	100	90–100	40–80	0–20	0–10	. . .	. . .	. . .
9.5 to 2.36 mm	. . .	. . .	100	80–100	5–40	0–20	0–10	. . .	. . .
Combined fine and coarse aggregate:									
12.5 to 0 mm	. . .	100	95–100	. . .	50–80	. . .	. . .	5–20	2–15
9.5 to 0 mm	. . .	. . .	100	90–100	65–90	35–65	. . .	10–25	5–15
Lightweight Aggregates for Concrete Masonry Units (ASTM C331)									
Fine aggregate:									
4.75 to 0 mm	. . .	. . .	. . .	100	85–100	. . .	40–80	10–35	5–25
Coarse aggregate:									
12.5 to 4.75 mm	. . .	100	90–100	40–80	0–20	0–10	. . .	. . .	. . .
9.5 to 2.36 mm	. . .	. . .	100	80–100	5–40	0–20	0–10	. . .	. . .
Combined fine and coarse aggregate:									
12.5 to 0 mm	. . .	100	95–100	. . .	50–80	. . .	. . .	5–20	2–15
9.5 to 0 mm	. . .	. . .	100	90–100	65–90	35–65	. . .	10–25	5–15

Table 8. Grading Requirements for Lightweight Aggregates for Insulating Concrete (ASTM C332) Percentages (by weight) Passing Sieves Having Square Openings

Size Designation	19.0 mm	12.5 mm	9.5 mm	4.75 mm	2.36 mm	1.18 mm	600-µm	300-µm	150-µm
Group I (ultra-lightweights)									
Perlite	. . .	. . .	. . .	100	85–100	40–85	20–60	5–25	0–10
Vermiculite—Coarse	. . .	. . .	100	98–100	60–100	30–85	2–45	1–20	0–10
Vermiculite—Fine	. . .	. . .	. . .	. . .	100	85–100	35–85	2–40	0–10
Group II									
Fine Aggregate:									
4.75 to 0 mm	. . .	. . .	100	85–100	. . .	40–80	. . .	10–35	5–25
Coarse Aggregate:									
12.5 to 4.75 mm	100	90–100	40–80	0–20	0–10	. . .	. . .	. . .	. . .
9.5 to 2.36 mm	. . .	100	80–100	5–40	0–20	. . .	. . .	. . .	. . .
4.75 to 2.36 mm	. . .	. . .	100	90–100	0–20	. . .	. . .	. . .	. . .
Combined Fine Aggregate and Coarse Aggregate:									
12.5 to 0 mm	100	95–100	. . .	50–80	. . .	. . .	. . .	5–20	2–15
9.5 to 0 mm	100	90–100	65–90	35–65	. . .	. . .	10–25	5–15	. . .

lightweight aggregate. Small quantities of diatomite have been used for such purposes, and even meerschaum (sepiolite), has found use as a lightweight building material. Perhaps the only significant commercial substitutes are chemical foams and plastics, such as styrofoam or polyurethane, that compare in strength to exfoliated vermiculite. Mineral wool and other insulating materials are unlikely to make a significant impact on the growth in use of lightweight aggregates.

Substitution of manufactured lightweights for byproduct materials such as cinders and slag is likely to continue. Since lightweight aggregates commonly substitute for materials, such as stone, sand, and gravel in concrete, about the only competitive threat from these natural materials could occur if fuel prices and availability of fuel supplies severely limit the ability to produce lightweight aggregates or render them uneconomical.

ECONOMIC FACTORS

Price Trends

Prices for manufactured lightweight aggregates have risen dramatically in the past ten years due to general inflation and to higher energy costs. Fuel and transportation costs continue to rise. Other energy costs, such as electric power, are continuing to increase more rapidly than inflation. Cost of refractories is a significant cost item as prices reflect higher energy costs in manufacture. Similar increases have been influencing the costs of expanded perlite and exfoliated vermiculite. The prices of lightweight aggregates will continue to reflect the rising cost of energy, and prices are likely to parallel energy costs.

Transportation

All forms of surface transport are used in moving lightweight aggregates. Truck movement on highways and streets is especially important in the delivery of expanded aggregates either in bulk form or in containers. Economics of lightweight aggregate haulage are very similar to transportation of crushed stone and other aggregates. Due to the higher value of lightweights, however, some additional haulage can be tolerated with proportionately less delivered price impact. Transportation costs are likely to amount to from 15 to 35% of the delivered price for most structural lightweights. The impact of transportation costs on delivered prices of vermiculite and perlite is naturally less due to higher unit value.

A good rule of thumb is that distance of movement of aggregate materials is likely to be directly proportional to unit price and inversely proportional to unit weight and bulk. Perlite and vermiculite ores are transported by rail from mines to delivery points throughout the United States and by ocean freighter throughout the world. Some vermiculite ores come to the United States by boat from South Africa, and rail haulage from the western United States to the East Coast is common. The exfoliated or expanded products, however, are not likely to be distributed outside local and regional markets due to great bulk.

Rail transportation of other lightweight aggregates is largely regional in nature; it is normally limited by the proximity of competing producers. Trucking may be local or regional depending on the availability of rail transport and relative costs. Trucks are normally most economical for hauls of 80 km or less, with railroads best suited for hauls in the 160 to 480 km range. The transportation method used is dictated by the access to other competitive modes, as well as haulage costs. The object is to keep delivered price down to maximize returns to the producer.

Marketing

Most producers of natural and manufactured lightweight aggregates maintain their own sales staff and technical service representatives, some producers market products through brokers. Products are sold on the basis of specifications and cost (in place) of the final product. In many cases the ability to reduce dead loads (the weight of concrete in structures) through the use of lighter aggregates (in the concrete), will permit reductions in the cost of structural steel. Insulation properties of ultralightweight aggregates in concrete and masonry units may enable the construction of thinner walls and/or cut down on the need for other insulation. These factors often permit cost saving in the construction of buildings despite higher unit prices for lightweight aggregate. Product yield commonly can be improved by the use of manufactured lightweights. Surface textures and uniformity are commonly superior to naturally occurring or byproduct aggregates.

Byproduct materials are normally handled by wholesale and retail dealers. The actual producers often handle slag sales, but cinders, fly ash, coke breeze, and other byproducts are usually sold through brokers.

Producers of vermiculite and perlite ores generally license or franchise brokers for the exfoliated or expanded product. Trademarks, promotional materials, specifications, and other sales aids are normally provided to the licensed or franchised dealer by the raw material supplier. The dealer pyroprocesses the ore and packages the product in bags or ships in bulk. The dealer acts as wholesaler in the distribution to large users and retail building material outlets.

Marketing research studies are usually important in the location of a new lightweight aggregate plant. The actual use and potential use of lightweight aggregate materials in the intended market must be measured or estimated from planned and prospective construction trends. The proximity of competitors and transportation costs in the market area must be carefully studied to determine the economic feasibility of a new production plant.

Lightweight aggregate producers that do not maintain a marketing research department are advised not to initiate a new plant location without some market analysis. The choice of a consultant or specialized marketing research firm should be based on both the cost of the service and past experience in similar or related evaluations.

Government Considerations

Taxes, depreciation, and depletion allowances for lightweight aggregate producers are similar to those for the producers of other construction materials. Zoning and land use regulations are restrictive in urban and other heavily populated areas, and they are closely related to the situations encountered by producers of crushed stone or sand and gravel. However, the more limited occurrence of lightweight aggregate raw material deposits and the smaller number of producers generally leads to the location of plants at greater distance from the more heavily populated market centers. Thus, lightweight aggregate producers tend to have less problem with zoning and land use regulation. In cases where plants are simply pyroprocessing raw materials mined at distant locations or handling byproducts, they are in the same category as heavy industry and are located in industrial areas.

To date there have been no significant reductions in government regulations on air and water pollution. Environmental standards continue to present challenges to the industry. If some relaxation in environmental standards is achieved there will be some cost saving, but this is unlikely to create any significant reduction in production costs. The same observation can be made for safety and health regulations. Once implemented, the additional costs must be included in operating costs. Any relaxation in government regulations could, of course, hold down future cost increases.

OUTLOOK

The drop in the output of structural lightweights is related to construction activity. Some of the decreased use is undoubtedly due to loss of market share to materials, such as crushed stone and gravel. The costs of producing structural lightweight aggregate are related to the cost of energy and raw materials. Transportation costs of the finished aggregate have always had an effect on the price of the various lightweight materials available in the marketplace. Prices for crushed stone and sand and gravel have also risen, but the differences in unit cost between natural aggregates and manufactured lightweight aggregates are much greater today than they were in the early 1980s (Konczak, 1992).

In the future, total consumption of lightweight aggregates will be tied to the construction industry. Presently the glut in commercial buildings in most regional markets combined with a decline in construction activity means that future opportunities for the industry are limited. The reduction of interest rates to historic lows, however, provides some hope that construction activities will return to the higher levels of pervious years.

Some increases in demands have been noted in markets for pumice and volcanic cinders, as these materials become relatively more attractive in terms of prices when compared to manufactured lightweights. Markets for insulating materials have increased as energy costs have increased, leading to higher production and use levels for expanded perlite and exfoliated vermiculite. Perlite use has increased roughly 3% per year while vermiculite consumption has grown at about 2% per year over the last five years (Meisinger,

1988). Continued growth in markets for these materials is likely to continue into the foreseeable future.

The decline in markets for expanded clay, shale, and slate is related to construction activity. Some increase in construction activity would cause related growth in demand for these structural lightweight aggregates.

In the long term, modest growth in demand for lightweight aggregates is unlikely to exceed 2% per year, and this demand will come mostly from natural lightweights and insulating materials due to favorable price considerations and continued high energy costs.

The final products of the lightweight aggregate industry are energy-efficient buildings. This means that the necessity to conserve energy resources will continue to create demand for lightweight aggregates as well as contribute to cost of production. The insulation qualities of lightweight aggregates will continue to provide markets for these materials, even though the relative shares of the market will continue to shift in future years based on the costs and values of the final products compared to the costs and qualities of the raw materials.

ACKNOWLEDGMENTS

This chapter has benefited from the excellent work in the fifth edition, authored by Henry N. McCarl, Professor of Economics and Geology, School of Business, The University of Alabama at Birmingham, Birmingham, AL.

BIBLIOGRAPHY AND REFERENCES

Anon., 1971, ''Lightweight Concrete, History, Applications, Economics,'' Expanded Shale Clay and Slate Institute, Washington, DC, Oct. 1971, pp. 14–15.

Anon., 1992, ''Material Prices,'' *Engineering News Record*, June 1, 1992, p. 41.

Bigner, M., 1992, ''Lightweight Aggregate Plant Will Utilize Garbage,'' *Rock Products*, Vol. 95, No. 6, pp. 23, 25–26.

Bolen, W.P., 1992a, ''Perlite,'' *Mineral Commodity Summaries—1991,* US Bureau of Mines, Department of Interior, Washington, DC.

Bolen, W.P., 1992b, ''Pumice and Pumicite,'' *Mineral Commodity Summaries—1991,* US Bureau of Mines, Department of Interior, Washington, DC.

Bush, A.L., 1973, ''Lightweight Aggregates,'' United States Mineral Resources, D.A. Brobst, ed., Professional Paper No. 820, US Geological Survey, pp. 333–355.

Carter, R.A., 1992, ''Rocky Mountain Light,'' *Rock Products*, Vol. 94, No. 8, pp. 50–56.

Conley, J.E., et al., 1948, ''Production of Lightweight Concrete Aggregates from Clays, Shales, Slates, and Other Materials,'' Report of Investigations 4401, US Bureau of Mines, Nov., 121 pp.

Coxon, P., and Moss, D.W., 1975, ''Sintering Machine Converts Fly Ash to Lightweight Aggregate,'' *Rock Products*, Vol. 78, No. 6, pp. 68, 70, 75.

Gershman, H.W., and Miller, D.R., 1992, ''Recycling: Running Up Against The Hard Realities,'' *Engineering News Record*, Vol. 228, No. 25, pp. E-52–55.

Griffiths, J.B., 1989, ''South Africa's Minerals,'' *Industrial Minerals*, No. 263, p. 42.

Hammon, M.D., 1992, ''Landfills; The Sophisticates We Can't Live Without,'' *Enginering News Record*, Vol. 228, No. 25, p. E-57.

Horler, D.B., 1977, ''Some Advantages of Lightweight Aggregates,'' *Precast Concrete*, Vol. 8, No. 3, pp. 137–140.

Huhta, R.S., and Stearn, E., 1980, ''The Aggregate Scene Around the Country'' and ''Lightweight Aggregates-New Markets Help Fill Construction Void,'' *Rock Products Forecast*, Vol 83, No. 12, pp. 46–52, 60–61.

Kadey, F.L., 1983, ''Diatomite,'' *Industrial Minerals and Rocks*, 5th ed., Vol. 1, S.J. Lefond, ed., AIME, New York, p. 694.

Kelly, K.E., 1992, ''Is Burning Waste Hazardous to Health,'' *Rock Products*, Vol. 95, No. 4, pp. 32, 68.

Kinch, D., 1989, ''Brazil Set to Boost Vermiculite Market,'' *Industrial Minerals*, No. 262, pp. 40–41.

Konczak, J., 1992, personal communication, technical sales representative, The Levy Co., Portage, IN.

Larsen, J.R., 1990, ''Industrial Minerals Review—1989'', *Mining Engineering*, Vol. 42, No. 6, p. 570.

Mason, B.H., 1989, ''Construction Materials: Lightweight Aggregate,'' *Concise Encyclopedia of Mineral Resources*, 1st eds., D.D. Carr & N. Herz, eds., Pergamon Press Inc., New York, pp. 75–78.

McCarl, H.N., 1983, ''Lightweight Aggregates,'' *Industrial Minerals and Rocks,* 5th ed., S.J. Lefond, ed., AIME, New York, pp. 81–95.

McMichael, B., 1989, ''Pyroprocessing Techniques—The Fiery World of Industrial Minerals,'' *Industrial Minerals*, No. 263, p.66

Meisinger, A.C., 1980, ''Vermiculite,'' *Mining Engineering*, Vol. 32, No. 5, pp. 588–589.

Meisinger, A.C., 1981, ''Perlite,'' *Minerals Yearbook, 1978–1979,* US Bureau of Mines, pp. 671–675.

Meisinger, A.C., 1981a, ''Vermiculite,'' *Mining Engineering*, Vol. 33, No. 5, p. 592.

Meisinger, A.C., 1981b, ''Vermiculite,'' *Minerals Yearbook 1978–1979,* US Bureau of Mines, pp. 977–980.

Meisinger, A.C., 1982, ''Pumice and Volcanic Cinder,'' *Mineral Commodity Summaries 1982,* US Bureau of Mines, pp. 118–119.

Meisinger, A.C., 1988a, ''Perlite,'' *Mineral Yearbook 1987–1988,* US Bureau of Mines.

Meisinger, A.C., 1988b, ''Vermiculite,'' *Minerals Yearbook 1987–1989,* US Bureau of Mines.

Milanese, R.S., 1980, ''Perlite,'' *Mining Engineering*, Vol. 32, No. 5, p. 579.

Milanese, R.S., 1981, ''Perlite,'' *Mining Engineering*, Vol. 33, No. 5, p. 583.

O'Driscoll, M., 1989, ''US Pacific Northwest,'' *Industrial Minerals*, No. 259, pp. 36–37.

O'Driscoll, M., 1990, ''Minerals in the US Southwest,'' *Industrial Minerals*, No. 272, pp. 77–79.

Potter, M.J., 1991, ''Vermiculite,'', *Mineral Commodity Summaries 1991,* US Bureau of Mines, pp. 180–181.

Presley, G.C., 1992, ''Industrial Minerals—1991,'' *Mining Engineering*, Vol. 44, No. 6, pp. 570–571.

Rogoff, M.J., 1992, ''Trends In The Waste-To-Energy Industry,'' *Engineering News Record*, Vol. 228, No. 25, pp. E-61–63.

Shildeler, J.J., 1961, ''Manufacture and Uses of Lightweight Aggregates for Structural Concrete,'' Bulletin D40, Research and Development Laboratories, Portland Cement Association, Jan., 17 pp.

Singletary, D.L., 1991, ''Cement Industry Repsonds to Safety Charges—Kilns Are A Safe Recycling Option,'' *Rock Products*, Vol. 94, No.10, pp. 58–64.

Additional Sources of Information

American Society for Testing and Materials (ASTM), ''Lightweight Aggregate for Structural Concrete,'' Specification C330, revised periodically; ''Lightweight Aggregate for Insulating Lightweight Aggregate for Insulating Concrete,'' Specification C332, revised periodically.

Cahners Publications, *Brick and Clay Record,* published monthly.

Expanded Shale, Clay and Slate Institute.

''Lightweight Aggregate Outlook and Forecast,'' Annual reviews in *Pit and Quarry* and *Rock Products*, December issues.

Pit and Quarry Publications, *Lightweight Aggregate Producing Plants in the US and Canada* (Map), Chicago, IL, 1963 ed., 1 p.; 1965 ed., 1p.: 1969 ed., 1p.

CONSTRUCTION USES

ROOFING GRANULES

RUSTU S. KALYONCU

Rainbow colors of roofing granules, adorning the roofs of buildings from single family units to largest office buildings, have their origin in the roofing shingles as burlap soaked with tar appearing as early as the 1700s.

Dusting of talc and mica on the coated surfaces, to keep roofing from sticking to itself in storage, represents an early improvement to the original construction product. Sometimes, colored, natural mineral dusts were added as color pigments within the dusting material, for initial color. Prolonged exposure to sun caused the asphalt to become brittle, and it eventually cracked and leaked. Search for solutions to cracking and subsequent leaking problems eventually led to the use of crushed mineral granules on asphalt coated shingles. This provided the impetus for further growth in the asphalt roofing industry. Application of a layer of crushed mineral on the asphalt coating minimized the coating deterioration from exposure to ultraviolet light, thus significantly improving the shingle life.

In the 1900s, screened waste from slate quarries was the most commonly used granule in asphalt roofing products. Records indicate the use of coarse slate screenings for asphalt roofing as early as 1906. As the asphalt roofing industry further developed, other naturally colored minerals, such as silica sand, quartz, mica, feldspar, talc, slag, granite, rhyolite, and oyster shells were used on roofing surfaces. Later on, a number of manufactured granular materials, such as crushed brick and tile, fire clay, and white porcelain were used.

The first artificially colored roofing granule was produced in 1914 by running molten slag into a solution of alkaline metal silicate, which was absorbed by the porous slag. Various colors were obtained by adding pigments to the silicate coating and by subsequently heat treating the granules. In 1922, another technique was developed in which green slate granules were saturated with copper sulfate solution and heated to 760°C to form copper oxide. Since then, many other granule coloring processes have been patented, one of the most significant being the sodium silicate coatings introduced around 1975.

Granules as used in construction (specifically roofing materials) are defined as natural or synthetic-mineral grains smaller than gravel and larger than sand in particle size. They are used primarily for protective and decorative purposes on the weather-exposed surface of asphalt composition roofing. Development of artificial coloring techniques has made it possible for the roofing industry to provide asphalt roofing shingles in a wide variety of colors complementing the color and style of buildings. Roofing granules, as we see today, are the products of long technological evolution.

RAW MATERIALS

Rock types used as a natural granule base (Table 1) can be broadly classified into igneous, sedimentary, and metamorphic in terms of their geologic origin. Igneous rocks are the result of solidification from molten and partially molten material (magma). Sedimentary rocks, as the term indicates, are formed by consolidation of mechanically or chemically deposited particles, whereas metamorphic rocks are pre-existing igneous or sedimentary rocks that have undergone mineralogical and structural alterations as a result of pressure, temperature, and chemical changes within their environment.

In view of the properties demanded of roofing granules most metamorphic rock types are not very suitable for granule production, because they often exhibit a pronounced plane of weakness (cleavage or foliation). The exception is thermally altered rock types, occurring in small localized areas, or very weakly metamorphosed greenstones and orthoquartzites. Most sedimentary rocks, e.g., clays, shales, and limestones, are highly prone to weathering, and thus are not suitable for use as roofing granules with the exception of arkose and arkosic quartzite.

Igneous rocks usually meet most or all of the property requirements as granule sources, except those with coarse grain structure that exceed the size range of roofing granules. Therefore, medium to fine grained types are used as reflected in Table 2. Igneous rocks are classified according to grain size (depth of origin) and chemical composition (mineralogy).

Intrusive rocks, formed deep within the earth's crust, exhibit coarse microstructure due to a slow cooling process, which promotes formation of large mineral grains within the rock. This coarse grained texture makes them unsuitable as a source of granules. Examples of such rock types include gabbro, diorite, dunite, and coarse granite. Shallow intrusive rocks are medium to fine grained and are appropriate sources for granule production. Among such types are diabase, granite, granodiorite porphyry, and dacite porphyry.

Extrusive igneous rocks formed by volcanic activity on the surface of the earth exhibit fine grain microstructure as a consequence of rapid cooling. Rocks, such as basalt, rhyolite, and andesite are the most desired rock types for granule production. Greenstones are slightly metamorphosed basalt which retain a fine grained texture and can be used as granule sources. Similarly, argillites (graystones) are slightly altered, highly indurated claystones and siltstones that do not exhibit lamination characteristics and, therefore, lend themselves to granule production. They are, however, less

Table 1. Rock Types Used as Granule Base

West Coast
Basaltic river gravels
Dacite porphyry
Midwest
Andesite
Argillite (graystone)
Granite
Nepheline syenite
East Coast
Rhyolite (partially metamorphosed)
Diabase
Greenstone
Arkosic quartzite
Canada
Basalt

Table 2. Classification of Igneous Rocks

	Texture		
	Extrusive	Intrusive	
Color and Composition	Fine Crystalline Volcanic	Medium or Fine Crystalline	Coarse Crystalline
Light colored (felsic) High silica and potassium Low iron and magnesium	Dacite	Rhyolite porphyry to quartz dacite	Granite to quartz diorite
Light colored (felsic) Quartz poor	Trachite	Syenite porphyry	Syenite
Dark color Low silica, sodium and calcium	Basalt	Diabase	Gabbro, peridotite, dunite, serpentinite

desirable because they are less resistant to weathering than other rock types used for granules.

Deposits suitable for producing granules, in addition to several necessary favorable operating conditions, rarely occur within reasonable distances of most large market areas. Areas that have suitable deposits must also have space for quarry and mill sites which include large areas for the disposal of overburden and wastes. Waste fractions can go as high as 50% and usually have no market value.

Fig. 1 depicts locations of ceramic coated colored roofing plants in the United States, and also illustrates the extent of vertical integration in the roofing industry. Several plants, located in California, Wisconsin, Arkansas, Delaware, and New York, produce granules for general use only. Several plants (South Carolina, California, Oregon and Massachusetts, and some others) show complete integration, producing granules entirely for their own use in manufactured roofing products.

TECHNOLOGY

Manufacturing technology for roofing granules calls for exploration, plant site location, quarrying, crushing, screening, and coloring phases. Detailed geological exploration, laboratory evaluations, and marketing considerations determine the plant location. Accessibility to mass transportation routes, water, power, and labor sources are important economic factors to be considered.

Quarrying operations are similar to those of construction aggregates. Roofing granules are produced principally in open pit operations. After removal of overburden, drilling, and blasting are followed by hauling the rock to the primary crusher. Crushing is carried out by either jaw or gyratory crushers. Depending on the quarry and weather conditions, a washing operation might be necessary. Secondary crushing is usually performed by cones, rolls or hammer mills in circuit with vibrating screens.

In the modern granule plant, as illustrated in Figs. 2 and 3, elaborate dust control systems are necessary, not only to protect employees' health, but also to comply with local, state and federal air pollution control laws in the United States. The crushed and screened granules are transported to the coloring section of the plant, where they are mixed with pigments and other coating materials in rotary mixers.

The coated granules are then fed into a rotary kiln for the final heat treatment to mature the bond. Finally, the granules are discharged from the kiln into a rotary cooler or similar device. Rescreening on vibratory screens is usually necessary for quality control. All modern granule plants are supported by elaborate quality control departments in order to meet the stringent requirements imposed by rigid product specifications.

The manufacturing process just described is not much different from what it was in the early years of roofing technology and has not very significantly changed in the past three decades, with the exception of the addition of pollution prevention equipment in the form of dust collectors and the introduction of computerized process control systems. Also, quality control has increased significantly in recent years.

In addition to natural material for roofing granules, recent advances have been made in synthetic granule technology. Methods have been developed for the production of granules, organic and inorganic, for use in chemical and pollution control technologies. The methods yield granules with controlled morphology, particle size, particle shape, and flow characteristics, plus extreme inertness to various liquid and gaseous media in purification applications.

Specifications

During the past few decades, specifications for roofing granules have taken a new direction and as a result, no single rock type can be classified as the only suitable material for granules. Today, a mineral deposit is carefully scrutinized for its physical and chemical properties in view of the formal requirements imposed upon it as a potential granule source. Quality control steps need to be implemented in every stage of operations from the quarry to the finished roofing shingle.

Several criteria are used to evaluate a potential deposit.

Weathering: The weathering properties of a roofing granule are of primary importance. If it does not adequately withstand exposure to temperature and humidity extremes, the granule will not provide the necessary protection for the asphalt matrix, nor will it give color permanence to the roofing shingle.

Coloring: If coloring is to be done after crushing, many mineral granules do not have the surface characteristics necessary to accept a color coating. For example, quartz with its conchoidal fracture is very difficult to coat. Also, the mineral constituents must be chemically inert to coating chemicals. Many rocks spall during the heating process because of their thermal expansion characteristics and/or the presence of water within the crystal structure. This condition alters the grade and leaves uncolored surfaces on the granule. To maintain consistent products, it is of the utmost importance that the rock possess uniform quality and homogeneous color as well as in other physical characteristics.

Porosity: High porosity in the rock allows the penetration of moisture which in turn allows the seepage of water into the asphalt

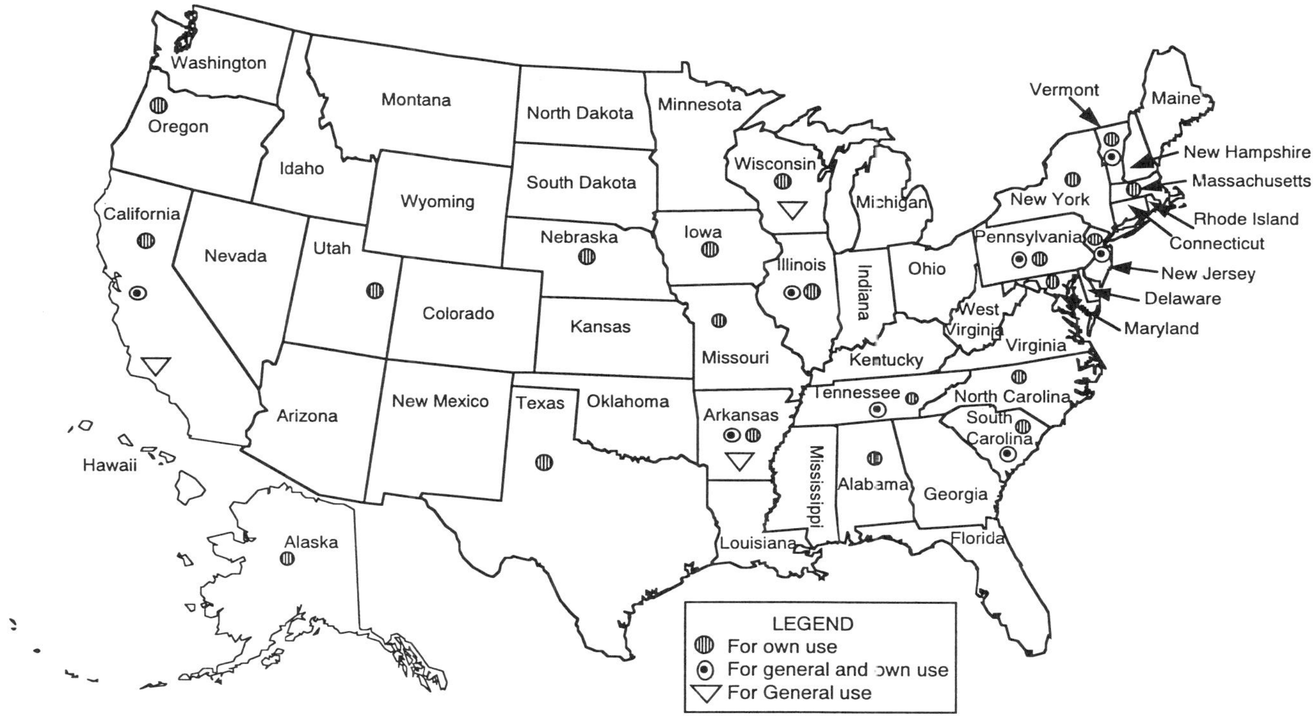

Fig. 1. Major roofing granule plants in the United States.

shingle, causing it to blister and/or crack. Therefore, rocks with low porosity are preferred raw materials.

Opacity: The roofing granule must be opaque to ultraviolet light. Sun radiation also causes photochemical activity (actinism) which can rapidly deteriorate asphalt roofing. Opacity is important to shield the asphalt both from ultraviolet radiation and actinic radiation.

Rock Uniformity: To assure quality, and acceptably reproducible properties, the deposit must be homogeneous and uniform in appearance as well as physical characteristics. By the same token, the deposit must be sufficiently large in tonnage to make it an economically viable operation. A given quarry must be large enough to serve its market for 30 to 40 years.

Toughness and Fracture Behavior: The quality of toughness is especially important during the manufacturing cycle where unavoidable attrition by conveyors, chutes, screens, and elevators could severely reduce the size of the original grains. Granules must have sufficient hardness and mechanical strength to withstand attrition forces during handling, installation, and use. To provide isotropic non-directional light reflection characteristics, the bedrock should exhibit a conchoidal fracture pattern, that is, it should not have any particular cleavage behavior. The granules should display no illusionary color pattern on the roof due to directional reflection of light. A deposit with a high quartz content is desirable for it fracture behavior. For instance, white roofing shingles manufactured from quartz granules exhibit the most consistent and uniform color.

Testing

Quality control procedures require certain testing procedures for grading size, shape, color, adhesion, opacity, and resistance to blistering.

The particle size distribution and shape of the granules influence the color and overall appearance of the roofing shingle and affect the amount of granules needed to cover it. A reasonably narrow size distribution is desirable for a given grade of roofing granules. Table 3 depicts some typical gradation requirements for several grades of granules.

Granules are also tested for color acceptance, color performance, opacity, adhesion to the asphalt base, staining, and blistering. Blistering is the formation of bubbles in the asphalt coating due to expansion of entrapped water. Accelerated weathering tests and freeze-thaw tests are commonly carried out as a part of quality control practices in the plants.

The crushing and grinding operations in the production of roofing granules result in materials that range in particle size from almost gravel to powder. Only certain size ranges qualify as roofing granules and the rest make up the byproducts. These byproducts are still used by the roofing industry principally as headlap (unexposed portion of the shingle) granules. The finer sized ground powder portion in the asphalt mix act as a stabilizer and as a backing material. Currently, up to 60% of the asphalt composition is stabilizers. Particle size of stabilizers ranges from 0.044 to 0.25 mm. Most common mineral types used for asphalt stabilization are rhyolite, granite, basaltic rocks, slate, limestone, and silica.

USES

The principal use of granules is in asphalt roofing production. With the increasing importance of the quality and color of roofing materials, ceramic coated and/or artificially colored granules have become the dominant surfacing materials in asphalt roofing products. Many ceramic processes employ granulated raw materials and raw material mixes in the production of the final ceramic product. Such is the case in the refractories industry. Sintered granular

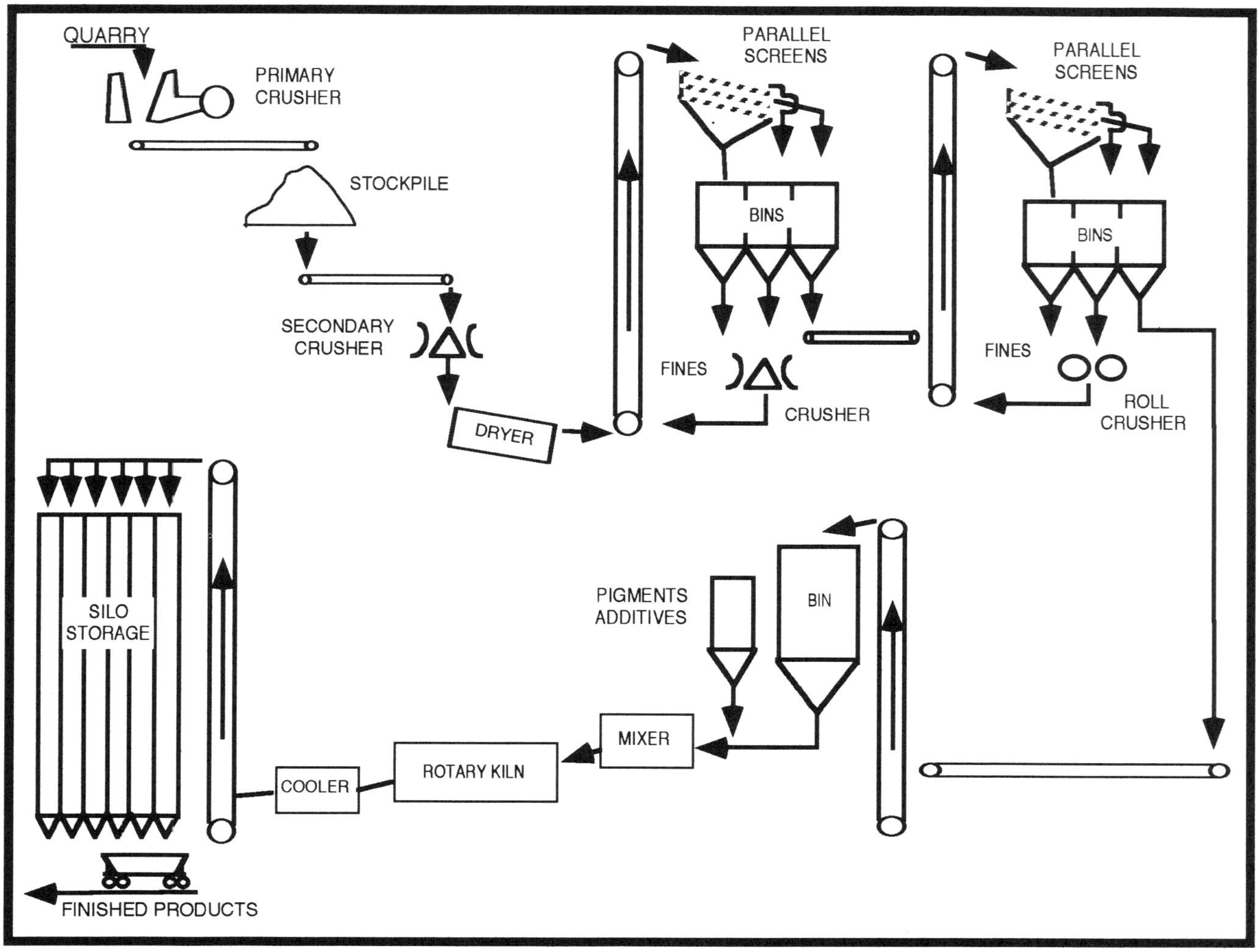

Fig. 2. Flowsheet for a typical roofing granule plant (after Jewett et al., 1975).

Table 3. Typical Grading Specifications for Roofing Granules

Sieve Number		Percentage Retained					
		No. 8 Grade*		No. 9 Grade*		No. 11 Grade*	
US Standard	Tyler	Min.	Max.	Min.	Max.	Min.	Max.
6	6	0	2	0	0	0	0
8	8	15	30	0	8	0	0
12	10	30	50	35	50	0	2
16	14	18	37	25	45	30	40
20	20	2	12	10	20	30	45
30	28	0	5	0	7	15	30
40	35	0	0	0	0	2	9
50	48	0	0	0	0	0	2

*Standard Specification Sieve Analysis of Granular Mineral Surfacing for Asphalt Roofingand Singles, ASTM Designation: 451.

materials enable the ceramic technologist to produce the refractory wares of high density necessary for improved high temperature mechanical properties and resistance to slag attack.

Today the range of uses for synthetic granules, organic and inorganic, goes beyond roofing shingles. Uses for synthetic granules include pollution control, industrial filtration, and the production of high technology ceramics. An example of the use of granules in pollution control technology involves the various filtration devices such as granular bed filters used in industrial cleaning processes. Mathematical models have been developed describing the filtration efficiency and pressure drop across granular bed filters. Clarification of mine waters by filtration through granular charges is well known.

ECONOMIC FACTORS

The United States began a recession in 1989 and leading economic indicators pointed to continued uncertainty in 1991 and the first part of 1992. New construction represents 20% of the US roofing market; the other 80% of the market relies on the reroofing business. The roofing market, therefore, remains relatively dependent on new construction activity.

The roofing market is divided into new and reroofing, as well as residential and commercial. Tables 4 and 5 depict the 1990 (actual) and 1991 (projected) sales for residential and commercial

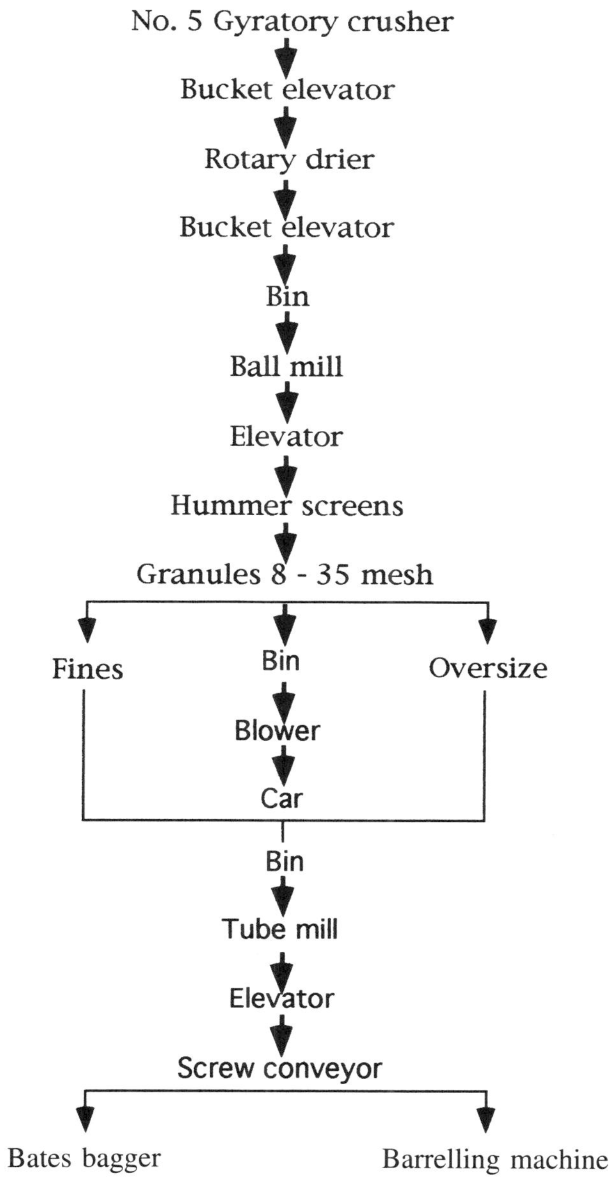

Fig. 3. Flowsheet of a mill for manufacturing slate granules.

markets respectively, for four geographic regions of the country. The 1991 projected values are based on 850 responses to a questionnaire sent out to commercial and residential roofing contractors by the National Roofing Contractors Association (NRCA). According to the survey, the total roofing market eased to $16.7 million, $1 million down from the projected 1990 level, but still a strong figure despite a worsening economy. A decline in construction activity caused a decrease in volumes in the Midwest (9.3%) and in the South (6.5%). Of the $16.7 million, the commercial market represented $11.6 million, 70% of 1990 market. The remaining $4.9 million (approximately 30%) went to the residential sector. Reroofing, as always, dominated both markets accounting for $8.3 million (71.3%) of commercial and $4 million (83%) of the residential business.

The 1991 roofing market is forecast at $16.9 million, much less than the 8 to 12% annual increases the industry enjoyed during the mid to late 1980s. The commercial section is projected to be $11.8 million, relatively unchanged from 1990, accounting for approximately 70% of all roofing business, the remaining going to the residential market.

The construction aggregates industry supplies some of the raw materials used to make roofing granules. The tonnage and dollar value of granules shipments to the US roofing industry in three categories are given in Table 6.

GOVERNMENT CONSIDERATIONS

Almost every state and county in the United States has zoning laws that regulate land use within its jurisdiction. A mining permit is invariably required from the local government to start a quarry operation for natural granules. If the land is under both city and country jurisdiction, permits usually must be obtained from each governmental unit. An acceptable reclamation plan that meets the guidelines and regulations of the governmental agency must also be approved. In certain states, an Environmental Impact Report (EIR) or Environmental Impact Statement (EIS) may be required for a new quarry operation. Environmental requirements usually arise because a great number of operations are located in highly populated areas. Respirable and fugitive dust problems must then be effectively addressed. Moreover, discharge of processing water, noise abatement, and employee and public health and safety concerns must be dealt with.

In our society, the search for new mineral resources and efforts to find additional uses for them is a continuous process. In the development of new industrial processes and new products, health considerations have become the principal public concern. Along these lines, the carcinogenic nature of a number of natural and manmade fibrous materials have been well established.

Asbestos is undoubtedly the most well publicized of such materials. Health effects of asbestos and efforts to get rid of asbestos products in public school buildings are well documented. All natural minerals and synthetic materials used in the manufacture of roofing granules contain from 10 to 100% crystalline silica.

Occurrence of pulmonary diseases due to the inhalation of silica and silicate containing mineral dust among workers in a variety of industries, such as coal mines, sand, stone, clay, shale, and many non-metallic mineral mills, have also been studied. However, a 1987 study on the effects of exposure to crystalline silica by the International Agency for Research on Cancer (IARC) has identified crystalline silica as a potential human carcinogen. According to the language of IARC, ''In the absence of adequate data on humans, it is biologically plausible and prudent to regard agents for which there is sufficient evidence of carcinogenicity in experimental animals as if they presented a carcinogenic risk to humans'' (IARC Monograph 42).

Most countries recognize that some forms of silica are potentially hazardous and many countries mandate or recommend permissible exposure limits (PEL). In this respect, the United States imposes stricter rules regarding free silica. Exposure determinations may be based on measurements of airborne dust, respirable dust, or analyzed concentrations of silica in the air. For the purpose of establishing exposure limits, the duration of work is standardized for an 8-hr day, 40-hr week, at an activity level corresponding with average physical work. The control level for free silica in the United States, as promulgated by the Mine Safety and Health Administration (MSHA) and the Occupational Safety and Health Administration (OSHA) approaches 0.1 mg/m^3 of respirable quartz.

As a result of the IARC study, crystalline silica has been classified as a probable carcinogen, regulated by the Hazard Communication Standard. The revised Hazard Communication Standard

Table 4. Residential Roofing Market ($ millions)

	1990				1991[P]			
Region	New	Reroofing	Fiberglass	Organic	New	Reroofing	Glassfiber	Organic
Northeast	200	1 140	850	140	200	1 240	900	140
Midwest	190	990	690	230	170	1 000	1 420	250
South	220	1 140	840	50	190	1 070	810	40
West	290	800	420	40	770	850	410	50

P-Projected survey dollar figures are for the value of contracts including labor, materials, profit, etc.
Source: National Roofing Contractors Association, Rosemont, IL/US Bureau of Mines, Washington, DC

Table 5. Commercial Roofing Market ($ billions)

	1990				1991			
Region	New construction	Reroofing	BUR	EPDM	New construction	Reroofing	BUR	EPDM
Northeast	880	2 560	550	1 480	780	2 560	560	1 410
Midwest	690	2 010	690	970	640	2 180	730	960
South	1 000	2 440	1 420	650	960	2 580	1 470	650
West	720	1 200	980	140	690	1 120	860	130

P-Projected survey dollar figures are for the value of contracts including labor, materials, profit, etc., BUR-Built Up Roofing, EPDM-Ethylene, Propylene, Diene Monomer.
Source: National Roofing Constractors Association, Rosemont, IL/US Bureau of Mines, Washington, DC.

Table 6. Value of Industrial Sand and Gravel Used for Roofing Granules and Filler

1990	$7,985,000	1984	$3,881,000
1989	$13,453,000	1983	$3,778,000
1988	$12,452,000	1982	$4,886,000
1987	$7,782,000	1981	$2,448,000
1986	$5,056,000	1980	$2,390,000 (quantity 1980— 171 818 tons)
1985	$5,056,000		
Quantity and Value of Construction Sand and Gravel Used in Roofing Granules			
1990	Not yet available	1985	NA
1989	NA	1984	NA
1988	566 354 tons $2,517,000	1983	NA
		1982	NA
1987	NA		
		1981	NA
1986	512 727 tons $3,102,000	1980	NA
Quantity and Value of Crushed Stone Used in Roofing Granules			
1990	NA	1985	2 311 818 tons $14,349,000
1989	2 804 tons $14,063,000	1984	NA
1988	NA	1983	1 324 545 tons $7,973,000
1987	1 520 000 tons $8,717,000	1982	NA
		1981	4 077 273 tons $18,094,000
1986	NA		
		1980	4 080 000 tons $17,556,000

contains certain requirements for employers in industries that expose the workers to silica in respirable dust environments. Among the requirements, employers must develop and implement programs addressing dust exposure problems. Up-to-date Material Safety Data Sheets (MSDS) for hazardous products must be maintained. Employers must provide the workers information and training on hazardous materials.

The US industry has responded to these regulations by implementing methods and installing equipment to reduce the respirable silica levels to below the permissible exposure limits (PEL). There is an increased emphasis from mining and milling operations on dust reduction programs, worker protection, and respirable dust surveys. Several trade associations have prepared training programs and training literature for workers in their industries. The Chemical Manufacturers Association, the International Diatomite Producers Association, and the Refractories Institute are examples of such organizations.

Several producers have also developed analytical methods for determining free silica in their mineral products. Among the government agencies associated with the respirable dust issues, the Mine Safety and Health Administration (MSHA) and the Occupational Safety and Health Administration (OSHA) have the responsibility of promulgating and enforcing the regulations. The National Institute of Occupational Safety and Health (NIOSH) is concerned with the medical issues. The Bureau of Mines (BOM) conducts research and development activities in respirable dust control to help the industry meet MSHA and OSHA requirements.

FUTURE TRENDS

The current roofing materials industry has reached a level of maturity that provides practical, economical, and aesthetically attractive roofing materials for both residential and commercial uses with improved performance and reliability. Therefore, no significant developments are predicted to take place in the roofing industry for the foreseeable future. In the long term, however, it is reasonable to expect that innovations in materials for building, industry, and changing architectural styles will dictate the major changes in the roofing industry. Anticipated increases in plastics use in housing construction may lead to use of plastic roofing materials that may simulate natural slate and ceramic tile roofs in the future.

In the early days of roofing shingle production, colored granules were very commonly used in roofing shingle production because natural slates of green, black, and red were available. Increased demands combined with the dictates of architectural designs, plus homeowner preference for variety, led to the development of artificially colored ceramic-coated roofing granules.

Today additional colors ranging from light pastels to dark rust and earth tones adorn the roofs of many family homes. Homeowner preferences have also dictated the development of thicker and more rugged looking roofing shingles, with the use of coarser grade granules or two layer laminated shingles to project the desired appearance. Developments and innovations in the roofing granules field take place slowly. Patent literature points out the fact that attention has been paid to more innovative color development, as well as methods of algicidal treatment.

BIBLIOGRAPHY

Bowles, O., 1922, *The Technology of Slate*, USBM B 218, Government Printing Office, Washington, DC, pp. 87–103.

Bowles, O., 1934, *The Stone Industries*, 1st ed., McGraw-Hill, New York, pp. 461–64.

Brown, J.A., 1960, "Granules," *Industrial Minerals and Rocks*, 3rd ed., J.L. Gillson, ed., AIME, New York, pp. 443–454.

Fehner, R.H., 1963, "Soot-Resistant Roofing Granules," US Patent 3,397,073, Dec. 11.

Henry, E.C., 1948, *Utilization of Slate Waste*, *Journal of Canadian Ceramics Society*, Vol. 17, pp. 27–31.

Jewett, C.L., 1975, "Construction Materials," *Industrial Minerals and Rocks*, 4th ed., S.J. Lefond ed., AIME, New York, pp. 175–184.

Jewett, C.L., et al., 1983, "Roofing Granules," *Industrial Minerals and Rocks*, 5th ed., S.J. Lefond, ed., AIME, New York, pp. 203–12.

Joedicke, I.B., 1981, "Light Colored Roofing Granules," US Patent 4,359,505, Feb 11.

Joedicke, I.B., 1981, "Silicate Coated Roofing Granules," US Patent 4,378,408, Sept. 4.

Johnstone, S.J., 1954, *Minerals for the Chemical and Allied Industries*, Wiley, New York, pp. 461–64.

Josephson, G.W., 1949, "Granules," *Industrial Minerals and Rocks*, 2nd ed., S.H. Dolbear, ed., AIME, New York, pp. 404–416.

Kalyoncu, R.S., 1986, *Construction Materials: Roofing Granules in Encyclopedia of Materials Science and Engineering*, Vol. 2, M.B. Bever, ed., Pergamon, Oxford, pp. 73–75.

Larabee, D.M., 1968, "Roofing Granules," *Mineral Resources of the Appalachian Region*, Professional Paper 580, US Geological Survey, Washington, DC, pp. 252–54.

McMahon, W.A., 1966, "Zinc Containing Algicidal Surfacing Method and Granules," US Patent 3,507,676, Dec. 15.

Meyer, J.E., 1974, "Roofing Granule Treatment by Coating with a Metallic Algicidal," US Patent 4,092,441, June 30.

Nelson, A.S., 1972, "Roofing Granules with Enhanced Algicided Post Treatment," US Patent 3,094,877, Oct. 25.

Remayne, M.B., 1967, "Colored Roofing Granules," US Patent 3,752,696, Feb. 17.

Saxena, S.C., 1980, *Mechanistic Modeling of Granular Bed Filters for the Removal of Particulate Matter from High-Temperature High-Pressure Gas Streams: A Critical Assessment*, Special Publication DOE/METC/SO-80/24, US Department of Energy, Washington, DC.

Sherrington, P.J., and Oliver, R., 1981, "Charnockites and the Granulites of Southern India: A Review," *Dansk Geologie Forening*, Vol. 17, pp. 473–93.

Sloan, R.J., 1966, "Colored-Coated Roofing Granules," US Patent 3,479,201, Jan. 18.

Tohata, H., 1979, "Development of Granulation Technology," *Kemikaro Enjiniyaringu*, Vol. 24, No. 1, pp. 76–81.

CONSTRUCTION USES

STONE, CONSERVATION

ERHARD M. WINKLER

The rapid decay and disfiguring of stone monuments in urban and desert rural areas has challenged conservators to protect stone surfaces from premature decay. They attempt to halt the natural process of stone decay and possibly to restore the original strength lost mostly by chemical weathering and the loss of binding cement. A general solution is not possible because the physical and chemical characteristics must be considered for different stone types. The failures of stone preservation and restoration are greater in number than the cures. The need for repair of stone decay goes back to evidence of Roman replacement of decaying stone.

The presence of excess water in buildings has long been recognized. Moisture tends to enter masonry from air in humid climates, a most important but often underrated factor (Fig. 1) suggesting that sealing should be the answer. Undesirable staining and efflorescence result in accelerated scaling. Today, the great variety of chemicals available to the modern conservator for sealing, consolidating, or hardening stone fall into two very different categories: surface sealers and penetrating stone consolidants, or a combination of both.

SEALERS

Sealers develop a tight, impervious skin which prevents access of moisture. Surface sealing has saved monuments from decay by eliminating the access of atmospheric humidity. Pressure tends to develop behind the stone surface by moisture escape. Efflorescence, crystal growth action, and freezing can cause considerable spalling (Anderegg, 1949). Flaking results when moisture is trapped behind the sealed surface. Yellowing and blotchiness are also frequently observed.

The following sealants are in common use today: linseed oil, paraffin, silicone, urethane, acrylate, and animal blood on stone and adobe. Extensive cracking and yellowing has resulted soon after application. In the past many such treatments have created more problems than cures:

1. Linseed oil and paraffin have been in use for centuries. Embrittlement and yellowing occur rapidly because these are readily attacked by solar ultraviolet radiation.
2. Animal blood as paint has temporarily waterproofed adobe mud and stone masonry. The origin of blood paint has a religious background rooted in the Phoenician and Hebrew cultures. Instant water soluble dried blood can substitute for fresh blood. Winkler (1956) described the history and technique of the use of blood.
3. Silicones have proven very effective and are long lasting. In contrast, acrylates, urethane, and styrene are generally rapidly attacked by UV radiation (Clark et al., 1975).

Sealing of Different Rock Types

Granitic rocks have a natural porosity traced to 4.5% contraction of quartz, during cooling of the parent magma, compared with only 2% contraction of all other minerals; protection against the hygric forces may require waterproofing of granite in some instances. The Egyptian granite obelisk in London is an example. Soon after its relocation from Egypt to London, Cleopatra's Needle was treated, in 1879, with a mixture of Damar resin and wax dissolved in clear petroleum spirit; surface scaling became evident after half a year of exposure to the humid London atmosphere. The treatment of the ancient granite monument from Egypt has denied access of high relative humidity (RH) in London to the trapped salts inherited from the Egyptian desert and has protected the monument from decay (Burgess and Schaffer, 1952). The sister obelisk set up in Central Park, New York City, has fared less favorably because similar treatment was done too late, only after the salts hydrated and hundreds of kilograms of scalings disfigured the obelisk surface (Winkler, 1980). Surface coating of other common stones may be needed.

Crystalline marble absorbs moisture from high RH atmospheres; dilation may ensue when curtain panels bow as the moisture starts to expand during daily heating-cooling cycles. A good sealer may prevent the moisture influx provided that no moisture can enter from the inside of the building.

Limestones, dolomites and all carbonate rocks are subject to dissolution attack by rainwater, especially in areas where acid rain prevails (Fig. 2). The interaction of sulfates in the atmosphere with the stone can be halted by waterproofing to avoid the formation of soft and more soluble gypsum. The stone surface attack can be diminished if nearly insoluble Ca-sulfite crusts can form, instead of Ca-sulfate. Replacement of fluorite or barium compounds at the stone surface acts as a hardener, rather than a sealant.

Sandstones have generally high porosity and rapid water travel can occur along unexpected routes and from any direction. Any surface sealing may do more damage by scaling and bursting than if the stone is left without treatment. Sealing of sandstones is therefore not advised at any time.

Testing the efficiency of sealants: Several authors discuss waterproofing materials, silicones, urethanes, acrylates and stearates, as to their water absorption, spreading rates of water on the treated surface, water vapor transmission, resistance to efflorescence, and general appearance (Clark et al., 1975). De Castro (1983) measured the angle of contact of a microdrop (0.004 cm^3) on a stone surface as characteristic of the wettability. Laboratory tests and limited field performance are described by Heiman (1981). The crest of a Gothic sandstone arch, which was sealed with silicone,

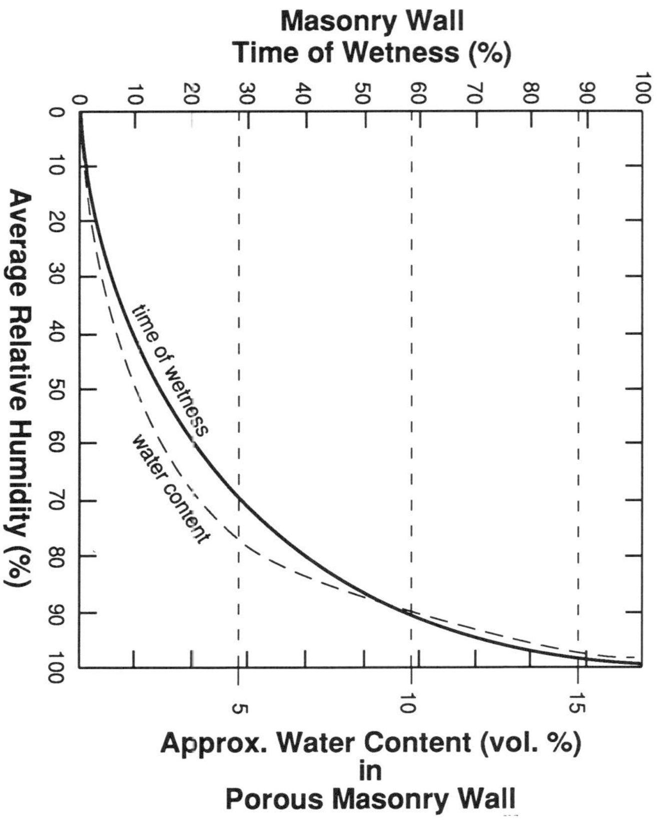

Fig. 1. Approximate water content by moisture absorbed as a function of the relative humidity and a similar curve of the time of wetness in a porous masonry wall. Data adapted from Voss and Tammes (1968).

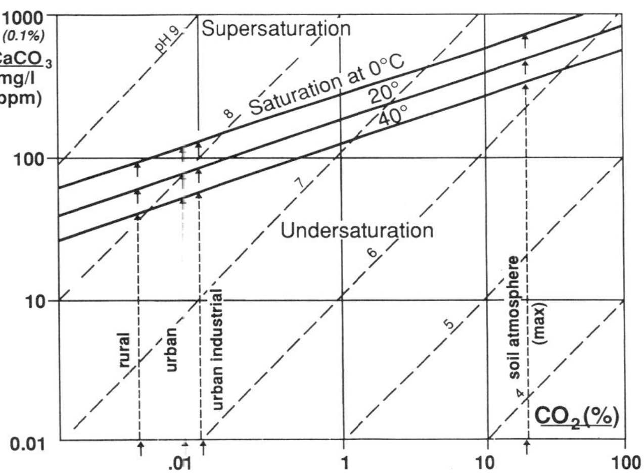

Fig. 2. Saturation diagram for calcite in water of different CO_2 content and temperature. Saturation equilibria are drawn for rural, urban, urban industrial and soil atmospheres. Adapted from Winkler (1987).

illustrates spalling due to trapped salts. Sealing should be only advised in special cases where no moisture can enter from anywhere behind the treated surface or surfaces.

CONSOLIDANTS

Stone Consolidation (Table 1), a method to replace natural grain cement lost by weathering from the stone, has been practiced since Roman times. Treatment with waxes, linseed oil, and limewater were in use for many centuries, but lasting cures were not obtained. The chemical industry has developed a large variety of a few inorganic and many organic chemical compounds to strengthen stone with little or no visual disturbance and high effectiveness. It should be clearly distinguished between a mere filler of stone pores and filler-strengtheners. Inorganic and organic strengtheners are summarized by Clifton (1980), Snethlage (1984) and Koblischek (1985). Wihr (1980) compiled practical methods of treatment with chemicals in his field handbook for conservators and stone masons.

Inorganic Consolidants

The introduction of inorganic solutions into crumbling stone, mostly sandstones and limestones, to replace substances lost by weathering is a natural and logical method. The following chemicals are in use today, often with limited success: water of lime, pozzolan-cements, silica as waterglasses, fluates, and baryte water.

Lime: The limewater technique is the oldest and most natural method to restore calcitic stone. Caroe (1987) describes the history of limewater treatment on the sculptures of 13th century Wells Cathedral carved on porous Doulting limestone. The soft stone soon required renovation on which the burnt lime technique was applied, and which has been improved through the centuries. Price (1984) experimented with limewater by applying freshly burnt lime as paste held firmly against the stone surface with sacking, a technique commonly used during the Middle Ages. The decaying statues above the main church entrance of Wells Cathedral in England were the focus of this method. Price found an increase in strength despite the repetition of the process while the surface remained moist during the time of treatment. The failure of the limewater method, explains Gutschick (1988) is the slow rate of calcification. For instance, lime mortars applied during the Roman period have not all reverted to calcite. The limited solubility of burnt lime in water (18 mg/l $Ca(OH)_2$) is another inefficiency. Calcite has a similar solubility which can be increased by a factor of 10 by the introduction of CO_2 in the presence of 15% CO_2 (Fig. 2). A concentration of 100 to 200 mg/l of $CaCO_3$ makes calcite precipitation possible, when CO_2 is lost from the solution and the pH is controlled (Fig. 2), and is similar to natural diagenetic processes.

Fluates: The replacement of the carbonates, such as limestone, dolostone, and crystalline marble, by colorless fluorite (CaF_2), is reviewed by Kessler (1983). The process was developed about 90 years ago. The newly formed mineral fluorite is virtually insoluble in rainwater through the entire pH range, in contrast to the limited water solubility of calcite. Volume by volume replacement of calcite by fluorite, a well documented natural geological process, may take thousands of years or longer and may form a gray surface.

Barytization: Replacement of soluble carbonates with nearly insoluble $BaCO_3$ in the presence of urea as a catalyst, has been developed and field tested by Bear and Lewin (1970). It yields decreased solubility and increased surface and near-surface hardening of the exchanged mineral material. The slow reaction rate and change of color make this process impractical.

Silica as Waterglass: The dissolution of silica in pure water depends on the pH, the temperature of the solvent, and on the degree of crystallinity (Fig. 3). Yet, pure SiO_2 introduction has not been successful although silica can be readily introduced as waterglasses (Table 2).

The introduction of silica into crumbling quartz sandstones to replace the grain cement was first attempted 150 years ago with

Table 1. Summary of Consolidants

Chemical	Binds to	Penetration	Solvent	Discoloring	Crusts
A) Inorganic:					
1) fluoric acid	fluates fluorite	limited	water	graying	—
$CaCO_3 + MgSiF_6 \rightarrow 6\ CaF_2 + MgF_2 + SiO_2$					
2) barytization +urea (catalyst)	baryte $CaCO_3$ to $BaCO_3$	limited	water	graying	—
3) limewater $Ca(OH)_2 + CO_2 \rightarrow CaCO_3$	calcite	limited	water	graying	—
5) waterglass Na_2SiO_3 K_2SiO_3	silica gel	limited	water	initial wetness	Na-crust
B) Organic:					
1. oil, waxes	organics	surface	org. solvents	strong yellowing	crusts
2. acrylic resins	polymers	limited	org. solvents	some seals	gloss
3. epoxy resins	polymers	limited	org. solvents	yellowing seals	
4. ethyl silicates	silicone	deep	alcohol, ketone	none	
5. silanes=siloxenes 2 w. SiO_2	silanol($SiOH_3$) ester	deep	alcohols	none	reacts with quartz
6. silicic acid $Si(OC_2H_5) + 4H_2O = SiO_2 + alcohol$					

Source: Snethlage (1984), Koblischek (1985), Weber and Zinsmeister (1991).

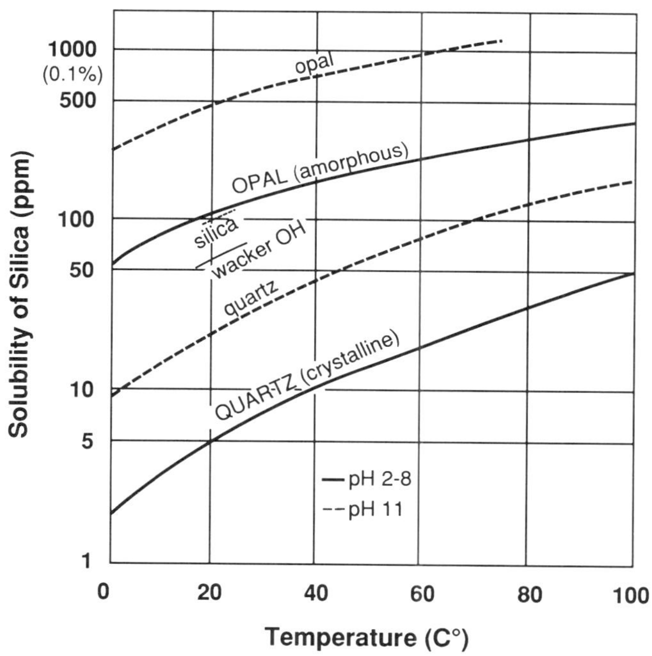

Fig. 3. Solubility of silica (marked with dashed lines) as a function of the degree of crystallinity, temperature and the pH of the solvent. Data from Kastner (1981); amorphous silica precipitated from Wacker, OH, from Snethlage (1984).

water soluble sodium or potassium silicates. Today, commercial aqueous solutions of silica as waterglasses of sodium or potassium contain 10 to 30% solids whereby the viscosity of the fluid increases rapidly with the increase of the dissolved solids. Once the solution contacts the mineral or matter of the stone, the waterglass solidifies to Na_2SiO_3 or K_2SiO_3 and binding starts in less than 1 hr. The subsequent interaction of the hardened waterglass with the CO_2 of the atmosphere separates the remaining amorphous SiO_2 from Na_2CO_3, or K_2CO_3. Both compounds have a very different solubility in rainwater. The solubility of sodium-carbonate is limited to about 150 g/l in water; the potassium-carbonate dissolves almost 10 times more at 1 100 gm/l. The carbonate of sodium tends to form a white efflorescence in contrast, the carbonate of potassium, remains in solution permanently above 40% RH, forming temporary wet areas on the masonry until rain has washed away the dissolved potassium-compound. Sodium, rejected by all plants, circulates from the masonry into the ground and back into the masonry with increasing concentration. In contrast, potassium is eagerly absorbed as an important plant food and disappears from the environment. The more expensive potassium-silicate is therefore preferred to sodium-silicate. Waterglasses, however, have caused many problems in the past when applied to stone.

Laboratory experiments have recorded better strength and binding with quartz sandstones than with crystalline marble. Residual sodium and potassium can form compounds with salts in solution in the masonry which may combine to form insoluble efflorescent compounds leading to spalling and flaking. The amorphous silica residue cements the grains of quartz or calcite. Extensive experiments with crumbling Carrara marble and quartzose sandstone from Berea, OH, show a substantial increase of strength from the treatment with potassium-waterglass, more for the sandstone than for marble (Fig. 4). High ambient RH reduces the strength of the silica residue, more at the beginning of the application than after half a year of progressive recrystallization. The initial great difference between oven dry and exposure to 100% RH atmospheres is ascribed to the great hygroscopicity of the potassium-silicate above 40% RH. Removal of the potassium-carbonate leaves behind a pure amorphous silica binder which appears to increase its strength, probably by gradual crystallization from the metastable amorphous silica on the way to the microcrystalline form. The present development of new waterglass-type compounds without the presence of alkali metals will eliminate the adverse effects of the carbonates of sodium or potassium. The silica binder reduces the pore spaces by about one-third of the original volume. Micro-cracking by shrinking of silica gel could open up new pores for increased sensitivity to frost action near the surface. There is no danger of color or other alteration by ultraviolet (UV) radiation. Polluted urban rainwater is not likely to attack stone reinforced with silica, but a sun heated alkalic solution will dissolve silica gel. The complexity of the process of curing is reflected in the time-strength graph (Fig. 4).

The geologic record shows evidence that a similar process

Table 2. Waterglasses: Advantages and Disadvantages

Chemical Reactions	Advantages	Disadvantages
Na_2SiO_3 $Na_2SiO_3 + H_2O = Na_2SiO_3$ (liquid)	good; very hard	strong white efflorescence
Na_2SiO_3(solid)+$CO_2 = SiO_2 \cdot H_2O$ (amorphous)+ Na_2CO_3 (150gm/l)	hard	hard white crusts
$K_2SiO_3 + H_2O = K_2SiO_3$ (liquid) =	similar to Na compound	occasional white efflorescence;
K_2SiO_3 (solid)+$CO_2 = SiO_2$ H_2O (amorphous) + K_2CO_3 (1100 gm/l)		temporary wet blotches by hygroscopic attraction

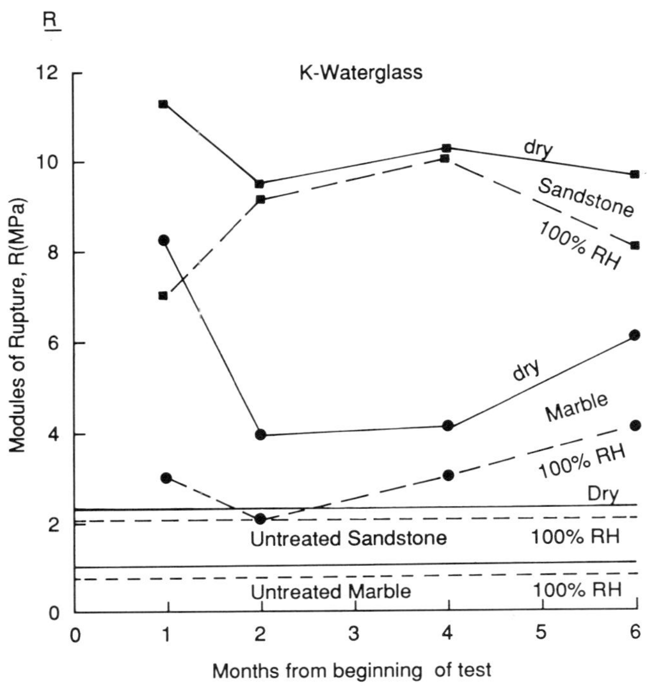

Fig. 4. Effectiveness and weatherability of potassium-waterglass impregnated stone discs 35 mm in diameter, 2½ mm thick; samples of Hindostan siltstone, Bloomington, IN, and deeply weathered Carrara marble. All discs were impregnated 3 times and leached 1 hr in artificial "acid rain" of pH=4.2 once a month and the modulus of rupture, R, tested every other month (Winkler, 1986). Gradual change and release of K_2CO_3 shows in the data.

involving natural waterglass has probably led to the formation of feldspathic grain cement. The brownstone building fronts of the northeastern United States have occasional microcrystalline feldspar cement (Hubert and Reed, 1978) which weathers rapidly to clay while losing strength and durability. This Triassic-Jurassic sandstone was formed in a desert valley near the present east coast of the United States in the presence of alkalic waters. A pH of 9 or greater could readily dissolve silica and alumina, the basis for the formation of feldspars in the presence of additional potassium, sodium, or calcium. Natural mineral materials and processes are the most logical consolidant, although proper catalysts are still lacking.

Organic Consolidants

The use of epoxies and plastics has quickly become a very profitable field in stone consolidation. The proliferation of proprietary new organic consolidants, often with minor variations from one compound to another, makes a discussion of specific consolidants impossible. Some organic compounds, such as the silicic esters, cure to inorganic, hydrous, amorphous silica gel when exposed to moisture in stone pores. Many stone conservators do not consider a stone infiltrated with plastic a "stone" any longer.

The most important basic types of organic consolidants are: ethyl silicates, organic resins, and silicones. Good summaries were compiled by Snethlage (1984), Koblischek (1985), and Weber and Zinsmeister (1991), and the chemistry of consolidants and consolidation was given by Amoroso and Fassina (1983). The variation of chemicals is great and changes continuously with an ever increasing number of patented products. Only the major groups of organic consolidants are here discussed.

Ethyl silicates: Silicic acid ester is the most used consolidant for sandstones. The silicic acid ester reacts with moisture as follows:

$$Si(OC_2H_5)_4 + 4\ H_2O \rightarrow SiO_2 + 4\ C_2H_5OH$$

$$\text{ethyl silicate + pore water} \rightarrow \text{silica gel + ethanol}$$

The reaction takes several weeks to complete which permits good penetration into the weathered stone surface. This technique has been used successfully for many years with good durability and no discoloring of the resulting inorganic amorphous silica. The chemical cannot be used in air lacking sufficient relative humidity to assure capillary moisture for polymerization through hydrolysis. Residual moisture, on the other hand, may remain trapped behind the solidified surface zone often leaving pockets of untreated stone. Shrinkage of the silica gel by gradual dehydration can develop undesirable cracks in the grain cement. The treated surface is usually pervious enough to permit the escape of moisture before flaking can take place. The pure residual silica gel derived from Wacker, OH, is of similar solubility to silica deposited from waterglasses (Fig. 5). Several commercial products of silicic esters are presently produced. Zinsmeister et al. (1988) give an overview of physical properties and testing of Ohio sandstone treated with silicic esters.

Silanes, siloxenes, siliconates: Silanes, siloxenes, and siliconates are all characterized by hydrolysis forming silicon resins with capillary moisture leading from silanes, possibly to siloxenes by cross-linking with a silicon atom of the stone. It appears that this group of silanes requires a quartz rock host to become effective.

Resins: Resins comprise a large variety of compounds and processes. The most important are epoxy resins, methacrylic resins, unsaturated polyester resins, and polyurethane resins. Snethlage (1984) discusses the chemistry in more detail.

Epoxy resins harden by the addition of accelerators with active hydrogen atoms in accurately prescribed proportion and temperature. Epoxies lack penetration, seal out moisture, and yellow readily in exterior exposure to UV light.

Methacrylic resins are readily polymerizable to polymethyl methacrylate (PMMA), generally known as plexiglass, which is

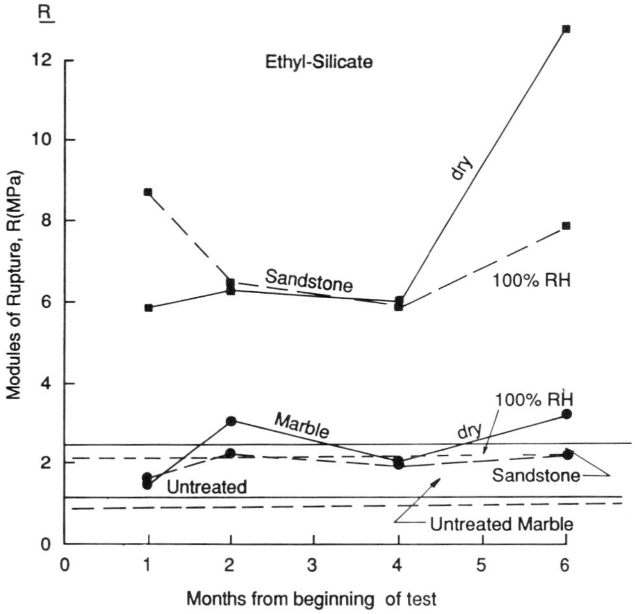

Fig. 5. Effectiveness and weatherability of ethyl silicate impregnated stone discs as in Fig. 4, also reflects the gradual conversion to amorphous silica. The change is also reflected in the graph. Data show comparison of untreated with treated stone. Marble shows minor improvement when treated.

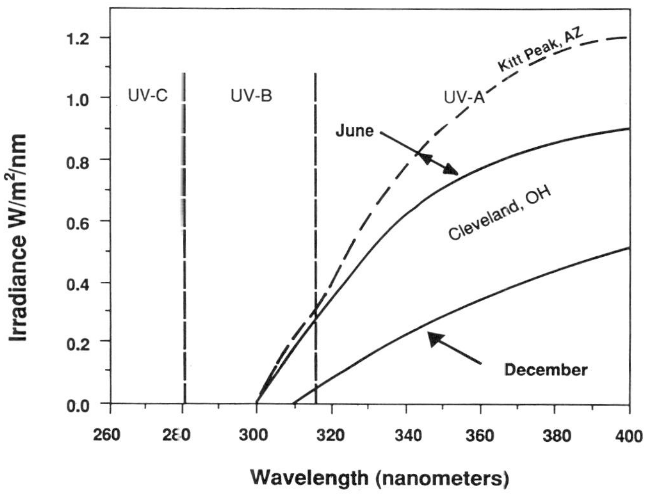

Fig. 6. Seasona variation and range of solar radiation in the UV portion of the spectrum. Little difference between the 42nd and 33rd parallel, but much higher at 2 300 m elevation. UV-A (400–315 nm): damage to polymers, tanning of human skin. UV-B (315–280 nm): shortest wave length on the earth surface; severe damage to polymers, bactericide, strongly carcinogenic. Data adapted from Brennan and Fedor (1987).

resistant to weathering, especially to UV radiation. Accurate dosing of the ingredients is necessary to obtain favorable strength and resistance to weathering. PMMA treated stone hardens and prevents travel of moisture. The low viscosity of water permits deep penetration and is therefore used for vacuum infiltration of small crumbling art objects. Total impregnation should only be performed on small objects in a vacuum vat if a statue cannot be saved in any other way. A fully impregnated stone, however, should not be called a natural stone any longer.

Polyester resins are neither resistant to alkalies nor to UV radiation. Such resins should not be used any longer in the presence of much better chemicals on the market.

Polyurethane resins tend to form elastic polymers which are not practical as consolidants, but are favorable waterproofers.

UV Radiation Attack

Most organic compounds and plastics tend to react with the UV part of the light spectrum. Fig. 6 shows the solar distribution of UV radiation at summer and winter solstice for Cleveland, OH, on the 42nd parallel, for Phoenix, AZ, on the 33rd parallel, and for Kitt Peak (west of Tucson, AZ), on the 32nd parallel and 2 300 m above sea level with a considerable summer UV increase. The solar irradiance maxima lie in the mid-latitudes. The UV ranges (adapted from Brennen and Fedor, 1987) react differently:

UV A (400–315 nm): Damage to polymers, tanning of human skin, formation of eye cataracts.

UV B (315–280 nm): Shortest wave length in nature; severe damage to polymer, bactericide, strongly carcinogenic, and absorbed by window glass.

UV-C (280–100 nm): Found only in outer space

Yellowing and cracking of most organic sealers and consolidants can lead to eventual crumbling after only a few years of exposure. Consolidant fillings primarily in sandstones may become quite sensitive to frost and other hygric actions; the degree of damage varies with the compound. Inorganic silica, even if derived from organic, silicic-acid esters, is not affected by UV. Some polyurethane compounds are very vulnerable, but PMMA reacts little and may be considered safe.

Effectiveness of Consolidation

The formation of crusts is the major concern in the use of most chemicals. The critical problem is to achieve a uniform continuous hardness profile from the surface inward to the unweathered stone substrate. The penetration of a consolidant through a case hardened surface can be a major problem in many sandstones and pervious limestones. The data on duration of consolidation effectiveness are still scarce.

Clarke and Ashurst (1972) studied 24 different public buildings after exposure for 20 years to natural weathering. The results of these early organic treatments were not satisfactory, and the great cost did not yield improved durability. More recently, Charola et al. (1985) infiltrated three different porous, fossiliferous limestones (Indiana limestone, Vicenza limestone, and Lecce limestone) with Acryloid B 72 (supplied by Rohm and Haas). Depending on the solvent exposure, rain has gradually restored the stones to their original pretreated properties and porosity. The author believes that the solvent in the consolidants is lost gradually as a function of the exposure to the number of natural wet/dry cycles.

Predictions of durability should be based on absolute terms per 1 000 mm of rain at a pH of 4.0. Durability tests for consolidants should be kept simple and quantitative. Field exposures on rooftops under urban or rural conditions can approximate the expected life. Tucci et al. (1985) discuss the reduced effectiveness of acrylic and silicone resins separately exposed to UV and acid rain in a chamber for accelerated decay (CAD). The amount of decay of the organic filling and bonding was observed under the scanning electron microscope (SEM). The authors observed depolymerization of polymers in 3 000 hr of exposure to UV while acid fog ate away the calcite crystals in limestone leaving behind a residual acrylic skeletal network. Success with silicic acid ester was described by Zinsmeister et al. (1988) for quartz sandstones which were carefully solidified resulting in a pure inorganic silica gel residue and tested for the apparent porosity (ASTM C20), density and water absorption (ASTM C97), modulus of rupture (ASTM C99), compressive strength (ASTM C170), and abrasion resistance (ASTM C241). DeWitte et al. (1984) tested 12 inorganic and organic consolidants, ethyl silicate, siloxenes, silicic acid, potassium-silicate, fluosilicate, polyurethane, epoxy, and PMMA for the following easily obtainable values for comparison: consumption of the consolidant, impregnation depth after drying, decrease of capillary water absorption, scratch hardness, micro-sand blasting, and SEM photos. Pure organic polymers discolor readily. The ethyl silicates are considered most promising whereas the inorganic silicates showed poor performance. The literature for the evaluation of the efficiency of consolidants and testing techniques is plentiful, yet there is no quick and efficient evaluation method available.

Suggested Testing of Consolidants

A recent unpublished series of simple tests sheds some light on the durability of consolidants. Thin discs, approximately 34 x 3 mm, are cored from brick-sized stone blocks and sliced. They are subsequently immersed for 3 sec into a consolidant three times. Testing is at intervals of 2 months and at least six discs should be made for each consolidant and test. The thin slices permit entry of consolidants from all sides. All specimens are dipped in artificial acid rain at a pH of 4.2 for 1 hr every month before being tested dried at 60°C or after exposure to 100% RH for 3 days. Progressive weight loss and the modulus of rupture are then determined.

Testing using simple techniques is summarized:

1. **Water absorption:** Specimens, submersed 48 hr in the laboratory reflect the available total pore space. Water absorption on a stone surface can be closely estimated with a small glass tube, the Kersten permeameter. It is a reliable, consistent, easily read, and a simple field instrument to measure permeability when attached to a horizontal or vertical stone surface before and after treatment. The degree of absorption per minute or second read in milliliters through the stone substance at or near the stone surface measures the degree of solidification, the effectiveness of a sealant, or the case hardening of a surface of limited permeability. Gale (1989) describes the Kersten tube and its operation, long used in Europe, for the American conservator. Large capillaries move the water downward by gravity, capillaries narrowed by consolidant filling can change the permeability, and the water from the tube often travels upward leaving visible wet areas around the point of attachment to the stone with putty (Winkler, 1988). Determine water absorption from the atmosphere in stone slices after exposure to 3 days of 100% RH and the strength after such exposure.

2. **Modules of Rupture:** Determine the modulus of rupture, R, following each leaching, both oven dried and at 100% RH (Winkler, 1986).

3. **Light Reflection:** Determine light reflection with a light spot meter, and the change of color shades compared using the US Soil Color Chart.

4. **Porosity:** Prepare a porosity profile by grain size, for each depth from the surface inward, as a three-dimensional presentation for progressive change of critical pore sizes.

Tests 1, 2, and 3 reflect the progressive loss of the consolidant during weathering experiments, especially expressed in the pore size distribution. All figures should be compared with those from the untreated original stone. The literature pays much attention to comparative SEM tests which explain the kind and degree of filling within the stone fabric; quantitative measurements, however, are not possible with SEM pictures.

The great variation of the stone substance from bed to bed in a natural outcrop should be considered when a testing program is planned. Successful consolidants should be applied at dilutions in which the strength and porosity does not exceed those of the unweathered stone thus avoiding the formation of hard crusts.

Cleaning of Stone Masonry

Cleaning of buildings in polluted urban areas has become not only a practice of building maintenance for esthetic reasons, but also for the prolonged life of the buildings and monuments. A multitude of techniques using numerous modern chemicals have entered the market and have been generously applied, often without proper understanding and testing of the possible interaction between the chemical compound and the stone substance. The excessive use of corrodents has frequently caused much damage, such as accelerated scaling and discoloring. Many special reports are available published by producers of chemicals. Unbiased reports on the evaluation of cleaners were published by the National Bureau of Standards (Godette et al., 1975), Anderegg (1949), ASTM (Clifton, 1987), the Australian Mineral Development Laboratories (AMDEL) (Spry, 1982; Spry and West, 1985), and by many in short articles devoted to special problems. Boyer (1984) and Jones (1984) discuss the state of the art of modern masonry cleaning. The techniques range from pure washing with water to high pressure steam and water jet cleaning, to the use of chemicals, and to abrasive methods, often with adverse results. Carbonate rocks respond differently to cleaning chemicals than quartz sandstones or granitic rocks. Grimmer's (1988) practical field handbook includes the problem of graffiti removal and a summary chart that is the best guide for masonry cleaning to date, starting with the most gentle methods and avoiding the harshest whenever possible. Such labor intensive and time consuming techniques are practiced by the cautious US National Park Service and are now accepted by most conservators. The basic methods are described in the following:

Water Washing: Washing starts with gentle irrigation with enough water running down the wall to keep it wet while it softens the crusts of soot, often intermixed with gypsum and dust, followed by a mild water spray. Three hours are often sufficient to remove most of the deposits from the stone surface when aided by gentle hand operation with soft nylon brushes.

Washing with elevated pressure and water volume can range from mild to high pressure jets, with operating pressures ranging from near 13.78×10^5 Pa to a still safe 68.9×10^5 Pa. Boyer (1984), as well as Jones (1984), give nozzle pressures and water consumption at different pressures. Pressure washing can cause dissolution of carbonate rocks in areas with soft washing water. Forcing water under pressure into the capillary system may cause undesirable travel routes. High pressure jets may also cause mechanical erosion of delicate features. Hot water is occasionally helpful in removing grease and paint.

Steam cleaning with nozzle pressures of 0.69×10^5 to 5.51×10^5 Pa has been successful for many years when cooler water jets cannot remove the crusts.

Abrasive Cleaning, Dry or Wet: A great number of abrasives and equipment have been developed to remove crusts and stains. Quartz sand blown through a nozzle at high pressure against a stone surface has become restricted by laws against both flying fragments and dust in the atmosphere which can cause silicosis. Upon elimination of silica sand, milder and safer abrasives were introduced,

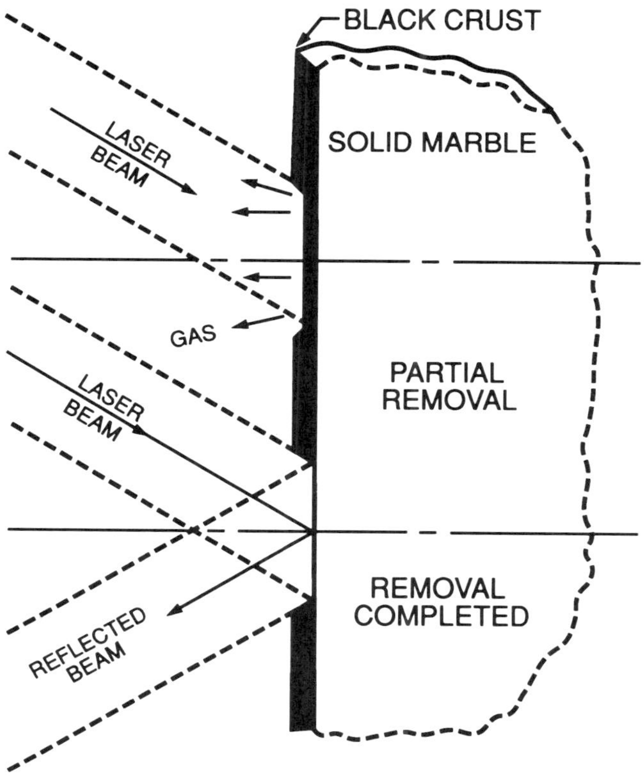

Fig. 7. Cleaning of marble surface by laser beam; black crust of soot (and gypsum). Laser gasifies the organic matter until the marble surface is reached; beam reflects off white marble then. Adapted from Asmus, Murphy, and Munk (1973).

like soft glass beads, which do not cause silicosis. Corn husk grit and other mild abrasives are often included in a water jet. Mechanical abrasion tends to model the stone surface, often smooths edges of fossil shells, but accentuates relief between the shells and generally softer matrix in fossiliferous limestones. Relief can be readily measured across the surface, to an accuracy of 0.2 mm, with macro stereophotogrammetry (Winkler, 1983). Such relief tends to trap soot and dust where it may adhere to the stone substance more strongly than before the treatment.

The removal of secondary lime crusts has become a serious problem. Structural concrete or lime mortar leak from joints and tend to dissolve readily and redeposit as gray crusts of lime and occasionally as dripstone formations. The strong adherence of such crusts to any natural stone surface makes their removal very difficult. Matero (1984) describes field tests with a variety of commercial wet abrasives on the Commodore Perry Memorial column on Bass Island, OH, with granite facing a concrete structure. Multiple roof leaks supplied the water for leaching and deposition of the lime crusts. Mechanical removal appears to be the only solution.

Laser Cleaning: Asmus et al. (1973) evaporated thin layers of black soot on marble statues, where no other means of cleaning could have reached, using long pulse ruby laser beams to burn away the black crusts in several short bursts while the white marble substrate remains unaffected. The soot burns off at 4 000 to 5 000° C, but very little heat is conducted into the marble substrate. Fig. 7 shows a section across an encrusted marble. The laser technique has recently evolved to be a useful, but limited and expensive tool.

Chemical Cleaning: The chemicals fall into three categories: acidic cleaners, alkaline cleaners, and organic solvents.

Acidic cleaners range from harsh hydrofluoric acid (HF) to other acid precleaning solutions. Hydrofluoric acid is generally only used on silicate rocks. HF attacks silicate minerals, etching and corroding them and often leaving non-shiny, non-reflecting feldspar surfaces regardless of the strength of the acid. Eisenberg and Milton (1972) report dulling of originally sparkling feldspars on a granite facade of the Old Cleveland Trust Building, Cleveland, OH. Porous sandstones and limestones readily retain acid in the pore systems where it remains active. Carbonate rocks appear to be endangered most by acidic cleaners.

Sodium hydroxide (NaOH) or ammonium hydroxide (NH_4OH) are the most common alkaline cleaning agents for use in removing surface stains. Careful application and subsequent rinsing are very important. Iron or silica are known to mobilize, during alkaline treatment, from inside the stone to its surface forming rusty blotches or gray coatings.

Organic solvents, aromatic hydrocarbons, and chlorinated hydrocarbons are often used for the removal of oil, grease, and other bituminous materials. Flammability and toxicity of many organic solvents restrict their use.

Several authors have accumulated and summarized chemical cleaning data in charts useful to the masonry cleaner (Spry, 1982, Boyer, 1984), but the most complete and practical work is by Grimmer (1988).

BIBLIOGRAPHY

Amoroso, G.G., and Fassina, V., 1983, "Stone Decay and Conservation," Materials Science Monographs, Vol. 11, Elsevier, New York, 453 pp.

Anderegg, F.O., 1949, "Testing Surface Water Proofers," ASTM Bulletin, TP33, pp. 71–77.

Asmus, J.F., Murphy, C.G., and Munk, W.H., 1973, "Studies on the Interaction of Laser Radiation with Art Artifacts, Development in Laser Technology, II," *Proceedings, Society of Photo-Optical Instrumentation Engineers*, Vol. 41, pp. 19–27.

Bear, N.S., and Lewin, S.Z., 1970, "The Replacement of Calcite by Fluorite: A Kinetic Study," *The American Mineralogist*, Vol. 55, Mar.-Apr., pp. 466–476.

Boyer, D.W., 1984, "Masonry Cleaning—the State of the Art," Cleaning Stone and Masonry, J.R. Clifton, ed., ASTM Special Technical Publication 935, pp. 25–51.

Boyer, D.W., 1987, "A Field and Laboratory Program—Determining the Suitability of Deteriorated Masonries for Chemical Consolidation," *Association Preservation Technology Bulletin*, Vol. 19, No. 4, pp. 45–52.

Brennan, P., and Fedor, C., 1987, "Sunlight, UV and Accelerated Weathering," *Plastics Compounding*, Vol. 12, No. 1, pp. 44–49.

Burgess, G., and Schaffer, R.J., 1952, "Cleopatra's Needle," *Chemistry and Industry*, pp. 1026–1029.

Caroe, M., 1987, "Conservation of Figure Sculptures, Wells Cathedral 1975–86," *Association Preservation Technology Bulletin*, Vol. 19, No. 4, pp. 11–15.

Charola, A.E., Laurenzi Tabasso, M., and Santamaria, U., 1985, "The Effect of Water on the Hydrophobic Properties of an Acrylic Resin," *Proceedings*, 5th International Congress on Deterioration and Conservation of Stone, Lausanne, pp. 739–747.

Clark, E.J., Campbell, P.G., and Frohnsdorff, G., 1975, "Waterproofing Materials for Masonry," US Department of Commerce, National Bureau of Standards, NBS Technical Note 883, 76 pp.

Clarke, B.L., and Ashurst, J., 1972, "Stone Preservation Experiments," Building Research Establishment, Department of the Environment, 78 pp.

Clifton, J.R., 1980, "Stone Consolidating Materials—A Status Report," US Department of Commerce, National Bureau of Standards, NBS Technical Note 1118, 46 pp.

Clifton, J.R., 1987, "Preliminary Performance Criteria for Stone Treatments for the United States Capitol,'' US Department of Commerce, National Bureau of Standards, NBS IR 87–3542, 12 pp.

De Castro, E., 1983, ''Studies on Stone Treatments,'' Laboratorio Nacional de Engenharia Civil, Memoria No. 584, Lisbon, 6 pp.

DeWitte, E., Charola. A.E., and Cherryl, R.P., 1984, ''Preliminary Tests on Commercial Stone Consolidants,'' *Proceedings*, 5th International Congress on Deterioration and Conservation of Stone, Lausanne, pp. 709–718

Eisenberg, K.S., and Milton, C., 1972, ''Case History of a Structural Granite Cleaning Problem,'' *Proceedings*, 1st International Symposium on the Deterioration of Building Stones, LaRochelle, France, pp. 129.

Gale, F., 1989, ''Measurement of Water Absorption,'' *Association Preservation Technology Bulletin*, Vol. 21, No. 3/4, pp. 8–9.

Godette, M., Post, M.. and Campbess, P.G., 1975, ''Graffiti-resistant Coatings: Methods of Test and Preliminary Selection Criteria,'' US Department of Commerce, National Bureau of Standards, NBS IR 75–798, 35 pp.

Grimmer, A.E., 1984, ''A Glossary of Historic Masonry Deterioration and Preservation Treatments,'' US Department of the Interior, National Park Service, Preservation Assistance Division, TH 5321.G56, 65 pp.

Grimmer, A.E., 1988, ''Keeping It Clean,'' US Department of the Interior, National Park Service, Preservation Assistance Division, 33 pp.

Gutschick, K.A., 1988, ''Lime, Important Construction Material,'' *ASTM Standardization News*, July, pp. 32–35.

Hawkins, W.L., 1973, ''Environmental Deterioration of Polymers,'' *Polymer Stabilization*, W.L. Hawkins, ed., Wiley, New York, pp. 1–28.

Heiman, J.L., 1981, ''The Preservation of Sydney Sandstone by Chemical Impregnation,'' Experimental Building Station, Technical Record 469, Chatswood, New South Wales, 22 pp.

Hubert, J.F., and Reed, A.A., 1978, ''Red-bed Diagenesis in the East Berlin Formation, Newark Group, Connecticut Valley,'' *Journal of Sedimentary Petrology*, Vol. 48, pp. 175–184.

Jones, L.D., 1984, ''Criteria for Selection of a Most Appropriate Cleaning Method,'' Cleaning Stone and Masonry, J.R. Clifton, ed., ASTM Special Technical Publication 935, pp. 52–70.

Kastner, M., 1981, ''Authigenic Silicates in Deep Sea Sediments, Formation and Diagenesis,'' *The Sea, The Oceanic Lithosphere*, Vol. 7, C. Emiliani, ed., Wiley Interscience, New York, pp. 915–940.

Kessler, L., 1983, ''Sur un Procédé de Durcissement des Pierres Calcaires Tendre au Moyen des Fluosilicates à Base d'loxydes Insolubles, Comptes-Rendus,'' *Académie des Sciences, Institut de France*, Vol. 96, pp. 1317–1319.

Koblischek, P., 1985, ''Polymers in the Renovation of Buildings Constructed of Natural Stone,'' *Proceedings*, 5th International Congress on Deterioration and Conservation of Stone, Lausanne, pp. 749–758.

Lewin, S.Z., 1972, ''Recent Experience with Chemical Techniques of Stone Preservation,'' *Proceedings*, Meeting for the Conservation of Stone, Bologna, ct. 1–3, 1971, pp. 139–144

Matero, F.G., 1984, ''A Diagnostic Study and Treatment Evaluation for the Cleaning of Perry's Victory and International Peace Memorial,'' *Association Preservation Technology Bulletin*, Vol. 16, No. 3/4, pp. 34–51.

Price, C.A., 1984, ''The Consolidation of Limestone Using a Lime Poultice and Limewater,'' *Internatl. Coll. Natursteinkonservierung*, Munich, May 21–22, Arbeitsheft, pp. 31.

Riederer, J., 1989, *Bibliography on the Cleaning of Stone*, Rathgen-Laborberichte, pp. 1–5.

Sleater, G.A., 1983, ''Stone Preservatives: Methods of Laboratory Testing and Preliminary Performance Criteria,'' US Department of Commerce, National Bureau of Standards, NBS Technical Note 941, 72 pp.

Snethlage, R., 1984, ''Steinkonservierung (Stone Conservation),'' Forschungs Programm des Zentrallabors f. Denkmalpflege 1979–1983, Bayer. *Landesamt f. Denkmalpflege*, Arbeitsheft 22, Munich, 203 pp.

Spry, A.H., 1982, ''Principles of Masonry Cleaning Buildings,'' Australian Council of National Trust, Technical Bulletin 3.1, 30 pp.

Spry, A.H., and West, D.G., 1985, ''The Defense Against Graffiti,'' Australian Mineral Development Laboratories, Report 1571, 196 pp.

Torraca, G., 1976, ''Treatment of Stone in Monuments, A Review of Principles and Processes,'' *The Conservation of Stone*, R. Rossi-Manaresi, ed., Bologna, June 19–21, 1973, pp. 297–315.

Torraca, G., 1978, ''Solubility and Solvents for Conservation Problems,'' International Centre for the Study of the Preservation and the Restoration of Cultural Property, 60 pp.

Tucci, A., et al., 1985, ''The Influence of Acid Rain and UV Radiation on the Aging of Acrylic and Silicone Resins,'' *Proceedings*, 5th International Congress on Deterioration and Conservation of Stone, Vol. 2, Lausanne, Sept. 25–27, pp. 891–898.

Voss, B.H., and Tammes, E., 1968, ''Flow of Water in the Liquid Phase,'' Institute TNO for Building Materials and Building Structures, Report No-B-1-68-38, Delft, Holland, 45 pp.

Weber, H., and Zinsmeister, K.J.H., 1991, *Conservation of Natural Stone: Guidelines to Consolidation, Restoration and Preservation*, Expert Verlag, Germany, 163 pp.

Wihr, R., 1980, *Restauricrauz von Steindeukurölern*, VerCamp Calliog Munich, 230 pp.

Winkler, E.M., 1956, ''The Influence of Blood on Clays,'' *Soil Science*, Vol. 82, No. 3, pp. 193–200.

Winkler, E.M., 1980, ''Historical Implications in the Complexity of Destructive Salt Weathering—Cleopatra's Needle, New York,'' *Association Preservation Technology Bulletin*, Vol. 12, No. 2, pp. 94–102.

Winkler, E.M., 1983, ''A Macro Stereogrammetric Technique for Measuring Surface Erosion Losses on Stone,'' Cleaning Stone Masonry, J.R. Clifton, ed., ASTM Special Technical Publication 935, pp. 153–161.

Winkler, E.M., 1986, ''A Durability Index for Stone,'' *Bulletin Association of Engineering Geologists*, Vol. 23, pp. 344–347.

Winkler, E.M., 1987, ''Weathering and Weathering Rates of Natural Stone,'' *Environmental Geology and Water Science*, Vol. 9, No. 2, pp. 85–92.

Winkler, E.M., 1988, ''Response to a Field and Laboratory Program,'' *Association Preservation Technology Bulletin*, Vol. 20, No. 3, pp. 28.

Zinsmeister, K.J.H., Weiss, N.R., and Gale, F.R., 1988, ''Laboratory Evaluation of Consolidation Treatment of Masillon (Ohio) Sandstone,'' *Association Preservation Technology Bulletin*, Vol. 20, No. 3, pp. 35–39.

CONSTRUCTION USES

STONE, DECORATIVE

JAMES M. BARKER AND GEORGE S. AUSTIN

Stone, one of the oldest building materials, today remains a well-established material throughout the construction industry. The use of natural stone is much less prevalent now than in the past. It is still widely considered to be the most aesthetically pleasing, prestigious, and durable building material. New and re-opened quarries are coming onstream to meet increased demand related to new building technology and increased residential use of stone.

CLASSIFICATION

No classification can completely eliminate overlap between dimension stone, aggregate, and decorative stone because most stone is multi-purpose. Many used for decorative purposes are not produced specifically for that end use. Rock otherwise considered waste in dimension stone or aggregate quarries can be decorative stone coproducts (Fig. 1). Many uses require a compromise between decorative and structural qualities (Bowles, 1992, written communication).

Shipley (1945) used decorative stone interchangeably with ornamental stone. Gary et al. (1972) defined decorative stone as that used for architectural decoration, such as mantels, columns, and store fronts, but added that it is sometimes set with silver or gold in jewelry as curio stones. Bates and Jackson (1987) also restricted decorative stone to that used for architectural decoration. Meanings of otherwise identical terms used in the stone industry differ between geologists, engineers, and quarriers. They often carry a much broader meaning for quarriers and engineers compared to their very specific use by geologists (Makens et al., 1972).

Decorative stone, including ornamental stone, is more broadly defined by geologists as any stone used primarily for its color, texture, and general appearance. It is not used primarily for its strength or durability, such as construction stone, or in specific sizes, such as dimension stone. The decorative stone industry uses a much wider range of stone types compared to stone that is dimensioned. Decorative stone usually serves some structural purpose, but is not load-bearing to any great extent. Weak or costly stones serve in decorative, not structural, applications.

STATISTICS AND END USES

Decorative and dimension stone data are difficult to separate because the US Bureau of Mines keeps statistics only on dimension stone and crushed stone. The value of domestic dimension stone production in 1990, which includes some decorative stone, was about $210 million compared to imports of about $524 million and exports of about $35 million. Production was 1 080 t of which at least one-third was for decorative uses (Taylor, 1992). The principal uses are rough blocks in building construction (23%) and monuments (18%); the remainder is used as ashlar (18%), curbing (12%), and miscellaneous (29%). Major rock types are granite (50%), limestone (30%), sandstone (10%), slate (3%), marble (2%), and other (5%) (Harben, 1990).

Crushed stone valued at $5.6 billion was produced in the United States in 1990 by 1700 companies operating 3400 active quarries in 48 states (Tepordei, 1991). About 52% is used in construction, 9% in cement and lime manufacturing, 2% in agriculture, 2% in industrial uses, and 35% for unspecified uses including decorative aggregate. Limestone and dolomite comprise about 71%, granite 14%, and traprock 8% of the stone crushed in the United States. The remaining 7% are, in descending quantity, sandstone, quartzite, miscellaneous rock, marble, shell, calcareous marl, volcanic cinder and scoria, and slate. The basic types of decorative stone are: rough stone, aggregate, cut or dressed stone, and manmade stone (Table 1).

Rough Stone

Rough stone is used as it is found in nature with very limited processing such as minor hand shaping, edge fitting, and size or quality sorting (Perath, 1992, written communication). This stone type is often marketed locally in relatively small tonnages and includes fieldstone and flagstone. The primary end uses of rough stone are landscaping, edging, paving, or large individual stone landscape or interior accents (Fig. 2).

Fieldstone: Fieldstone is picked up or pried out of the ground (gleaned) without extensive quarrying and includes garden or large landscaping boulders (Austin et al., 1990, Hansen, 1969). Boulders and cobbles may be split or roughly trimmed for use in rubble walls and veneers, both interior and exterior. Popular fieldstone rock types include sandstone, basalt, limestone, gneiss, schist, quartzite, and granite, but many others are suitable. Much fieldstone is collected by individuals or small companies because the industry is labor intensive and markets are small. The stone may be sold locally in small quantities from the back of vehicles (Austin et al., 1990). Fieldstone includes many rock types, sizes, and shapes with the only common denominator that it must be set by hand and be durable (Power, 1992, written communication).

Moss Rock. Moss rock is fieldstone partially covered by algae, mosses, lichens, and fungi that give the rock an aged and variegated patina (Austin et al., 1990). The plants are supported by moisture and nutrients in the stone. Moss rock is used for landscaping, walls, and fireplaces. Although almost any durable rock can be a moss rock, most are slabby or rounded sandstone and limestone (Fig. 3).

Flagstone: Flagstone or flagging consist of thin irregular slabs used for paving, walkways, and wall veneers. Random-shaped flagging is produced widely in the United States. Suitable stone

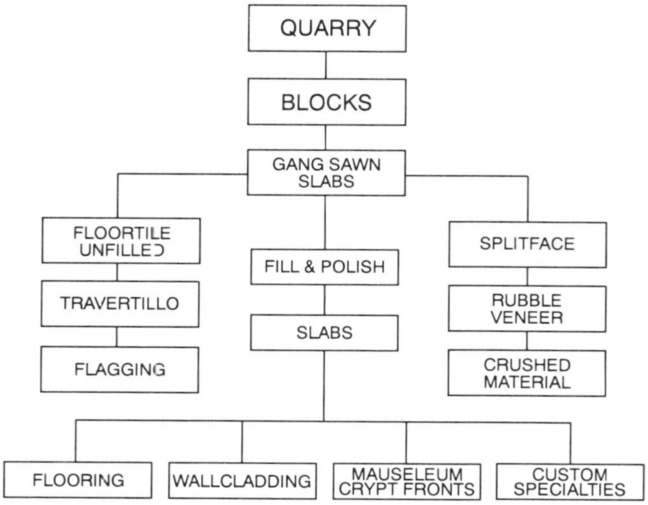

Fig. 1. Materials flow from quarry (top) to numerous products, produced by New Mexico Travertine west of Belen, NM. Travertillo is travertine tile with rounded corners similar to ceramic saltillos produced in Mexico.

Fig. 2. Stone accents, flagging, and walls at the College of Santa Fe, Santa Fe, NM.

Table 1. The Classification of Decorative and Ornamental Stone Used in this Chapter

- ROUGH STONE (unprocessed or slightly processed)
  - Fieldstone
    - Moss Rock
  - Flagstone
- AGGREGATE (lightly processed; screened or not)
  - Uncrushed Stone
    - River Rock
    - Scoria and Cinder
    - Fused Argillaceous Rock
  - Crushed Stone
    - Rubble
    - Exposed Aggregate
    - Dash
    - Terrazzo
- CUT or DRESSED STONE (moderately to highly processed; no set size)
  - Statuary and Objet d'Art
  - Fireplace Rocks and Hearthstones
  - Ashlar
  - Monuments and Memorials
  - Tile and Paving Blocks
  - Veneer and Wall Cladding
  - Miscellaneous
- MANMADE STONE (made from various raw materials)

Fig. 3. Red sandstone moss rock used in a retaining wall. The blocks are mottled due to attached lichens.

breaks very easily in one direction producing flags. Any fissile stone can be used, but sandstone (bedding planes) and slate (cleavage surfaces) are best. Limestone and dolostone are quarried as flagstone in the Great Lakes area of the United States (Johnson, 1992, written communication). The Coconino sandstone is quarried extensively in Arizona and produces very high quality, red to pink to white flagstone (Townsend, 1962). Sandstone flags up to 0.5 m^2 can be split to a thickness of 3 cm or less. Flagstone slabs 3 to 10 cm thick are used for walkways in high traffic areas and must be resistant to abrasion. If used in walkways, these thin slabs must be set on a very firm base. Thicker flags of sandstone or granite may be used in walls (Fig. 4) or set on edge as curbing.

Aggregate

Uncrushed Stone: Natural aggregate is lightly processed, usually by washing or screening, yielding products suitable for decorative use. Fragments can be either rounded or angular and must be durable. Many types of decorative stone can be used for rock lawns or area covers in virtually unlimited colors. Typically, local materials are used which limits choice but lowers cost. The aggregate is placed on UV-resistant black, impermeable or semipermeable polyethylene (4 mil) covering a prepared surface treated with

Fig. 4. Thick flagstone (ashlar), capped by thin flagstone, used in wall.

Fig. 5. Rock lawn composed of small size river rock surrounding volcanic cinder in planters enclosed by rounded cobbles.

weed killer. A wide variety of sizes are used at an application rate of at least 50 kg/m^2 that varies depending on aggregate size. Some special categories of uncrushed stone are described in the following.

River Rock. A distinctive attribute of river rock is the water-rounded pebbles, cobbles, and boulders commonly used as an area cover (Fig. 5). White to gray is typically specified, but other colors are available. River rock most commonly is granite or gneiss, but any durable rock may be used. In Pennsylvania, white to buff vein quartz is a popular river rock for landscaping (Berkheiser, 1991, written communication).

Scoria and Cinder. Scoria or volcanic cinder is a lightweight, vesicular equivalent of basalt (Fig. 5) or other basic volcanic rocks. It is used primarily for desert landscaping in the southwestern United States and is less common elsewhere, but is available in most parts of the country. Scoria is sold as either red to brown or black to gray varieties, but both are otherwise similar. Reddish hues are more popular and hence, more valuable than other hues (Osburn, 1980). The color differences of cinder are a result of the presence (red) or absence (black) of oxygen during volcanic eruption and emplacement.

Fused Argillaceous Rock. Natural fires in North Dakota lignite produce fused interbedded claystone and sandstone (Murphy, 1991, written communication). In New Mexico, natural coal fires produce a similar material locally called red dog. Red baked and fused shale related to coal fires is mined in the northern Powder River Basin of Wyoming (Harris, 1991). These materials are used in the same manner as volcanic cinder.

Crushed Stone: Crushed stone is the most common decorative aggregate and can be produced from virtually any pleasing stone. It is broken mechanically and usually screened prior to use; larger sizes are often called rubble. Harris (1991) uses the term decorative aggregate to describe crushed and sized stone used for landscaping such as area cover, rock lawns, walkways, and borders around plants or gardens. Rubble, exposed aggregate, dash, and terrazzo (all defined in following segments) are included by us under crushed stone, although exposed aggregate, dash, and terrazzo are used with a binder such as cement.

Rubble. Rubble consists of large rough stone or blocks produced in quarrying (Fig. 6), often as waste, and used for retaining walls, seawalls, bridgework, and landscaping. Only landscaping rubble is considered decorative stone because it is used primarily because of color, texture, or general appearance. In New Mexico, large boulders of pegmatite are used as landscaping pieces (Austin et al., 1990).

Fig. 6. Morrison (Jurassic) sandstone rubble associated with the defunct Bishop Lamy Quarry, southwest of Santa Fe, NM.

Smaller rubble is popular as wall facing in homes and commercial buildings. The primary purpose is aesthetic—it replaces brick or other veneer—but ease of installation, weather resistance,

Fig. 7. Large exposed aggregate panel in a commercial building.

Fig. 8. Terrazzo flooring (polished) at the Dulles International Airport, Washington, DC. Bright spot at lower left is a reflection off the highly polished surface. Exposed aggregate forms the surface of the pillar and exterior walkway in the background.

light weight, and ability to bond well with mortar are also important. Rubble may be set in random patterns from about 0.05 to 4 m^2 of exposed rock. Low density rocks, such as pumice, have several advantages. Shipping costs are lower, setting is easier for the stone mason, and few, if any, anchors are required to tie the stone veneer to the wall (Power, 1983). In Minnesota, waste rock from processing granite dimension stone, called grout, is used as decorative stones including sawn, split, and even polished slabs (Johnson, 1992, written communication).

Exposed Aggregate. Exposed aggregate is one of the most common methods of using crushed stone (Fig. 7). Stith (1970) found the most important properties to be color, hardness, soundness, absorption, shape, size distribution, and impurities.

Many colors and shapes are possible making exposed aggregate compatible with almost any architectural scheme. Color should be uniform and permanent because it is the main criterion used by architects. Observation of weathered and fractured outcrops of the proposed aggregate can be very useful in determining how the stone will react (Cutcliffe and Dunn, 1967). Spalling and other forms of physical deterioration should be noted. The color should vary only slightly, if at all, between weathered and fresh outcrops. Variations in color, based upon exposure to sunlight or weather, should be noted to minimize color differences across the sides of a structure (Cutcliffe and Dunn, 1967). Color segregation of stone by quarry procedures, blasting, stockpiling, blending, batching, and due to weathering or lithology should be avoided (Cutcliffe and Dunn, 1967).

The ability to cast exposed aggregate in complex shapes and with background coloring (dash) of cement give the architect great freedom. Aggregate, mixed with white or gray cement in a 2:1 ratio, can be precast into panels, or cast in place in walls and floors or walkways, with the aggregate dispersed or concentrated in the facing layer (Stith, 1970). The aggregate is exposed by subjecting the surface of the aggregate/cement mixture to sand blasting, bush hammering, wire brushing, or acid washing (Cutcliffe and Dunn, 1967), and it is finally sealed.

Dash. Dash, either coarse (for texture) or fine (for color), is added to exposed aggregate, stucco or concrete for texture or color. Sand dash is added to stucco for color and a small scale exposed aggregate surface. Very fine dash is added to concrete or cement as a permanent pigment instead of more expensive mineral pigments which may react with the cement compounds. Non-reactive dash material, which is well mixed, avoids blotchiness or shade variation common with artificial or mineral pigments and may be used in conjunction with stucco dash or exposed aggregate.

Terrazzo. First produced by the Romans over 1500 yrs ago, terrazzo floors provide quality, at low original and maintenance cost, and very long life. Terrazzo, a mixture of sized, crushed stone, and cement, offers variety in color and design (Fig. 8). This mixture is poured into a prepared floor area and after hardening, ground smooth, sealed, and often polished (Reed, 1978). The stone aggregate has low porosity and low absorption. The portion of the terrazzo that needs protection is the portland cement matrix which is porous and absorbs stains. The primary application of terrazzo is in high traffic, public areas and buildings. Relatively soft stone, usually limestone, dolostone, or marble is preferred, but granite is also used. Quality control is paramount during quarrying, so that color can be matched through repeated batching, and during processing, so quality, color continuity, and freedom from impurities are assured. The Terrazzo Tile and Marble Association of Canada recommends a thin-gauge epoxy or polyacrylate for sealing.

Cut or Dressed Stone

Cut or dressed stone is finished on one or more sides by various methods and is used where uniform surfaces are needed. Typical

Fig. 9. Statuary, carved from tuff in Mexico, on display at the Santa Fe Stone Co. in New Mexico.

Fig. 10. Various sizes of floor tile, composed of Coconino Sandstone (Jurassic), at the Western States Stone Company yard, Ash Fork, AZ. Tiles in the foreground near the center are 0.3 x 0.3 m.

uses are in walls, monuments, sculptures, waterfalls, or other relatively small, very detailed artistic renderings. Many stones are fashioned into structural/decorative adornments such as capitals, veneers, friezes, cornices, corbels, coping, and ribbing. Non-structural statuary and art, along with stones too soft or brittle for structural use, are used in some parts of buildings, although in low volume (Perath, 1992, written communication).

Statuary and Objets d'Art: Stone is used by artists as a medium for carving, sculpting or objets d'art production (Fig. 9). Carvable stone is commonly soft and with uniform characteristics, such as marble, but commercial granite, though hard, is carved by some sculptors. Color, texture, softness, and the ability to take a polish are important for statuary stone. Carvable stone commonly commands the highest price, but accounts for the smallest sales volume of any of the stone categories. Many small firms and artisans cut and polish semi-precious gems and ornamental stone from hard stone, such as agate, quartz, jasper, chalcedony, chert, and petrified wood (Burchett and Eversoll, 1991), or softer stone, such as tuff, talc (steatite), serpentine, marble, travertine, and gypsum (alabaster).

Fireplace Rocks and Hearthstones: Many types of stone are utilized in fireplaces. Commercial use is restricted to relatively few types compared to the many used by individuals. While rough stone and aggregate are often used in rustic fireplaces, dressed stone is more typical. Moss rock, flagstone, river rock, scoria, and ashlar are often used as a decorative facing on a fireplace. Argillite or mill-stock slate is cleaved or rough-finished and used as hearthstones and mantles (Carpenter, 1983).

Ashlar: Ashlar consists of rectangular, nonuniform stones with at least two smooth parallel sides, set randomly in a wall. The exposed surface of each piece is generally less than 0.4 m^2 laid exposing the sawn or naturally smooth face, or the rough (broken) face (Fig. 4). Ashlar blocks are either prepared from natural slabs that split and fracture into usable shapes, or by sawing the required two parallel sides usually about 7 to 15 cm apart. The remaining sides have an attractive broken appearance which, when exposed, is called split-faced ashlar. Ashlar may be sawed from quarry blocks that are not suitable for finishing to dimension stone or may be made from blocks that break naturally into rectangular pieces. Natural or sawed blocks are broken in a hydraulic guillotine to assorted sizes, palletized, shipped to the job site, and laid by hand in courses similar to brick.

Monuments and Memorials: Stone is cut and polished for tombstones, historical markers, and similar monuments and memorials. A typical stone used for monuments and memorials is statuary-grade marble that is pure white. Monument stone must be hard, take a high polish, and be resistant to weathering. Tombstones are most often one of the varieties of granite.

Tiles and Paving Blocks: Tiles are cut or split stones with one very thin dimension which may be polished (Harris, 1991). Typically, they are 0.3 m^2 and 8 mm thick and can be used in many ways on all interior surfaces including floors and walls (Fig. 10). Tiles can be made from many types of stone, but most are slate, granite, limestone, basalt, or tuff.

Bluestone, slate, and argillite (Power, 1983, Carpenter, 1983), as decorative stone, are limited primarily to flagging and floor tiles, but are also used for sills, stair treads, risers, shower and toilet stalls, and exterior spandrels and facings. They are available in various shades of green, gray, purple, red, black, or variegated. Slate finishes vary from the natural split surface to knife-shaved, sand rubbed, and honed or semi-polished (Harben, 1990).

Tiles and paving blocks made of tuff generally are called cantera stone but include adoquin, sillar, tufa, tuff, calduro, and others (Kuiper, 1988). Most have been imported to the southern United States and California from Mexico, but US production has commenced in Arizona (Kuiper, 1988).

Veneers and Wall Cladding: A veneer is any rock that can be laid up on a wall. Wall cladding is composed of larger sheets, often filled and polished, of travertine, serpentine (verde antique), marble, and others. Schist (Fig. 11), which cleaves along preferred

Fig. 11. Precambrian Vadito schist used as veneer wainscoting in Santa Fe, NM.

directions yielding relatively flat stones that can be laid up in walls, is also a source.

Miscellaneous: Other non-forested areas of the United States use stone in the manner that other areas of the country use wood, such as limestone fenceposts in Kansas (Grisafe, 1991, written communication).

Manmade Stone

Stone or slag, usually crushed, is the main ingredient in the production of manufactured stone by melting and frothing. Basalt is melted and cast into various forms (Woller et al., 1992). Irregularly shaped boulder- and cobble-sized masses are used as artificial scoria or cinder. Manufactured stone can be used in floor and wall panels. Crushed marble is combined with a binding polymer, and perhaps a coloring agent, to produce marble tiles. Ground limestone is mixed with organic resins to produce cultured marble molded into a variety of shapes.

GEOLOGY AND DISTRIBUTION OF MAJOR DECORATIVE STONES

Hundreds if not thousands of types and varieties of stone are used in the decorative stone industry. The terms used in the stone industry vary widely between countries, the professions involved in the stone industry, and segments of the industry. In Scandinavia, the stone industry differentiates hard stone from soft stone. The latter includes limestones and marble, which if siliceous, may be harder than some igneous rocks like tuff which are called hard (Shadmon, 1988). Complete coverage of all local industry classifications would be nearly impossible. Consequently, we discuss only the dominant commercial rock types presented here as an overview.

Granite: Commercial granite, which has a wider meaning than geological petrographic definitions, includes all feldspathic intrusive rock with visible grains and all metamorphic rocks with gneissic texture. It includes true granite plus other intrusive igneous rock, and their metamorphic equivalents, such as syenite, monzonite, gabbro, anorthosite, amphibolite, and gneiss. Commercially, fine-grained igneous rocks, even of diabase or basalt composition, are often called black granite (Power, 1983). Commercial granite, whether light or dark, has high strength and durability, and is relatively impervious to water.

Commercial granite in the United States is quarried at over 100 sites in 20 states. The dominant companies are Cold Spring Granite, Rock of Ages Corp., and Coggins Granite Industries. Major production centers (Harben, 1990) include Georgia (Elberton), North and South Carolina, New York, Massachusetts, Vermont, New Hampshire, Minnesota (St. Cloud), South Dakota, Wisconsin, and Texas (Burnet and Llano Counties). Near Elberton, GA, about 100 companies quarry and sell granite as rough blocks or finished products such as memorial markers.

Prominent European igneous decorative stones are Scandinavian red granite (rapakivi type in Finland), norite, diabase, and larvikite (anorthoclase syenite, labradorite). Assuan red granite and red porphyry (porfido rosso antico) is produced in Egypt. Green porphyry (porfido verde antico) is produced in the Peloponnesos region of Greece (Kuzvart, 1984). Spain, India, the Republic of South Africa, Australia, and Brazil are also noted sources of granite in the world market.

Pegmatite—Most pegmatites have about the same chemical and mineralogical composition as granite, but pegmatites exhibit very large crystals. With this coarse texture, boulders are preferred. Many stone or landscaping companies stock large pegmatite boulders that are moved to the building site by flatbed truck. In Quebec, amazonite pegmatite from Saint-Ludger-de-Milot and Lac Saint-Jean is used for decorative stone aggregate and for small decorative objects (Caty, 1991, written communication). A survey of Michigan architects by Johnson (1983) showed that coarse pegmatite was highly favored. Most pegmatites have a coarsely crystalline granite composition with abundant quartz and orthoclase feldspar. Other minerals, notably muscovite or white mica, may be abundant. Although not extensively quarried in the present day, old mines operated when sheet muscovite was mined in the past are inviting to enterprising stone dealers. The texture and rough exterior make large pegmatite boulders attractive for landscaping.

Basalt and Traprock: In the western United States, Tertiary and Quaternary basaltic rocks are exposed over large areas of thousands of square kilometers. Basalt is fine-grained, hard, tough, dense, and durable. It is composed of pyroxene and calcic plagioclase and well suited for use as a decorative stone for landscaping.

Traprock is a common term for dike rock of basalt composition. Power (1983) states the correct petrologic term for traprock is diabase or dolerite. When sold as cut building stone it is commonly called black granite.

Tuff: The main uses of tuff is as cantera stone floor tiles and wall cladding (Fig. 12a). Some Mexican tuff is sculpted to statuary (Fig. 9). The tuff varies in color with pastels predominating. The stone is characterized by many inclusions of large pumice fragments (up to 30%) and phenocrysts, up to 50 mm in diameter, of quartz, biotite, feldspar, magnetite, and rock fragments such as granite or basalt (Kuiper, 1988). High porosity of tuff makes this rock suitable where low weight is a factor and as a stone with eye appeal for rustic application.

Marble and Travertine: Geologically, marble is a metamorphosed carbonate and travertine is a sedimentary carbonate deposited from flowing water such as in a spring system. Commercial marble is any crystalline rock composed dominantly (>50%) of calcite ($CaCO_3$) and/or dolomite ($CaMg(CO_3)_2$), or serpentine ($Mg_3Si_2O_5(OH)_4$), often with white calcite or dolomite stringers

Fig. 12. (a) Volcanic tuff (cantera stone), cut with a chain saw, is used as the primary building stone at The Lodge (built circa 1930) at Los Alamos, NM.

Fig. 12. (b) Although containing voids and depressions, the cantera stone hardens somewhat with time to produce a more durable veneer.

forming verde antique. Commercial marble must take a polish and may include crystalline limestone, travertine, and serpentine in addition to metamorphic carbonate. Crystals range in size from fine (easily polished) to coarse. The ability to take a polish is important, but the color and crystalline character may be more important.

Limestone, Dolostone, and Marble. Carbonate rocks, usually marine in origin, are found in many parts of the United States and many other nations. The Bloomington-Bedford District of southern Indiana has produced a well known dimension limestone (commonly called Indiana Limestone) for more than a century. Limestone and dolostone are usually gray, but range from white to buff, tan, or black. White marble is composed of nearly pure calcite. Dolostone is composed predominantly of dolomite. Mineral impurities that make limestone and dolostone darker include iron carbonates, iron oxide, iron sulfide, chert, silica, clay, graphite, and carbonaceous matter (Power, 1983). Crystallinity, bed thickness, ease of polishing (3 to 3.4 on the Mohs hardness scale), and the presence of fossils, stylolites, or other textures and structures make limestone and dolostone attractive decorative stones. Bedding may vary from thin to massive and is important in determining the end use of commercial marble and travertine. Limestone imported into the United States comes primarily from Spain and France.

The bulk of US marble production is quarried by the Georgia Marble Company in Georgia, the Vermont Marble Company in Vermont, and the Rocky Mountain Stone Company in New Mexico. Marble, particularly white, is commonly crushed and sized and sold in lots ranging from carloads to bags of less than 45 kg. White marble from Georgia (Power, 1983) and varicolored travertine from New Mexico (Austin and Barker, 1990) are premium decorative stones. The Yule quarry in Gunnison County, CO, which supplied white marble for the Lincoln and Jefferson Memorials in Washington, DC, has recently been re-opened. Marble imports come principally into the United States from Portugal, Spain, and Italy.

Travertine. Commercial travertine is hard, dense to vuggy, finely crystalline, compact, massive to concretionary or fibrous limestone that takes a polish (Austin and Barker, 1990). Impurities in travertine impact colors ranging from white to pink, red, tan, yellow, green, gold, brown, or black. Variation in impurities may cause multicolor-banding of layered travertine. Travertine may be called tufa, calcareous sinter, marble, Mexican onyx, or onyx marble.

Sandstone and Conglomerate: Sandstone suitable for cutting, flagging, and curbing has been produced worldwide for thousands of years. Commercial sandstone consists of both sandstone and siltstone. Color variations are due largely to iron oxide. Moss rock is a popular sandstone found in many areas as loose fieldstone on surface outcrops. Picture rock is sandstone, exhibiting complex color banding due to variations in weathering, and sold in slabs cut to simulate landscape paintings.

Slate, Schist, and Gneiss: Slate is a fine-grained metamorphic rock with pronounced, relatively smooth, and flat cleavage surfaces. Most slate is used for roofing, but stair treads, floor tile, flagging, wainscoting, trim, chalkboards, pool and laboratory tables, plaques, and signs are also made from it. Slate may be purple, gray or black (reduced), red to green (oxidized), or mottled. These colors can alter or bleach upon exposure (called semi-weathering), but roofing slate should have permanent colors (called unfading), or at least alter evenly to a pleasing shade and be from the same lot. Most slate is initially split by hand from quarry blocks (Sweet, 1990) with further punching and shaping by foot treadles and machines.

Slate production in the United States is concentrated in the east, specifically in New York, Vermont, Virginia, and Pennsylvania where it is centered at Pen Argyl and Slatedale. Sales of slate have been particularly strong in the southern or sunbelt United States and California (Harben, 1990).

Quartzite: Quartzite is metamorphosed quartz sandstone which breaks through the grains rather than around them. Commercial quartzite is a hard siliceous rock commonly white to light-gray to bluish-gray or pinkish. Gneiss is often marketed as quartzite.

In Virginia, thin-bedded to massive quartzite of the Cambrian Weverton Formation has been quarried since 1893 (Sweet, 1990).

Fig. 13. Palletizing ashlar from dimension stone waste at New Mexico Travertine, Belen, NM.

Fig. 14. Hand dressing (splitting) of flagstone made from Coconino Sandstone (Jurassic) at Black Cat Pit near Ash Fork, AZ.

The quartzite is marketed for flagstone, veneer facing stone, and for other decorative uses. Fern stone is a quartzite containing dendritic limonite along bedding planes. Idaho also produces a quartzite used as a building stone.

PRODUCTION

Decorative stone operations range in size from very small to very large. Many large quarries producing dimension or building stone also produce large quantities of waste rock although ideally most output should be large, sound, rectangular blocks. Often this waste material can be crushed to aggregate, screened, and sold as decorative stone or processed into ashlar (Fig. 13). Aggregate quarries produce rock that can be used as higher value landscaping stone as easily as concrete aggregate. Appearance and cost of the stone are most commonly the deciding factors.

Mining

General mining techniques in decorative stone quarries are the same as the techniques used in the production of other types of stone, but details are adjusted depending on the desired finished product. For example, blasting is minimal and is done with small charges of low-velocity explosives. Modern equipment in large dimension stone quarries, such as diamond wire saws, belt saws, chain saws, and air wedges allow rapid quarrying that is more accurate and produces thinner products than in the past. Yet great care and considerable hand dressing (Fig. 14) is necessary for high quality flagstone or slate. Accent pieces used in landscaping such as large pegmatite boulders require considerable care to preserve without damaging the coarsely crystalline nature during transport to the final site. Cranes may be needed to lift pieces and padding is used to prevent damage in transit. Contrast this to the production of crushed stone which takes little care.

Stratification in rock produces zones of weakness called rift by producers. These are important because they determine the direction in which the stone splits most easily. Spacing of the rift determines the thickness of the quarried layer. Bed seams, joints, cutters, reeds, and runs are additional terms used in the sandstone industry to describe other natural planes or directions in which the stone splits or can be cut (Bowles and Barton, 1963).

Slate production techniques have evolved little with time. Mining must be done slowly with little blasting because it would split the slate. Wire and chain saws are also used. Cold weather limits the mining season in many places. The stone is ruined if it is freeze-thawed while in large blocks, so none can be mined ahead without careful storage. The slate is still largely hand split and punched with foot-driven treadles. Because of the assistance of nature in prying weakened surfaces apart by alternating freezing and thawing, flagstone quarries tend to be small and close to the surface.

Processing

By the early 1900s muscle power had given way to steam, electricity, and compressed air. In recent years carbides and diamonds for cutting and higher grade steels for drills have allowed for more efficient cutting of stone, but the basic equipment was similar. Since the Browning Torch was developed, it has been widely used in the granite industry and has revolutionized channel cutting (Meade, 1986). Channel cutting with diamond-studded belt saws is doing the same to limestone and marble building stone production (Harben, 1990).

Automation is being rapidly introduced which allows affordable tiles and other decorative pieces such as kitchen counters, table or desk tops, bathroom counters and basins, fireplace mantels, and entry ways to be produced. The trend is toward thinner slabs, specialty surfaces (polished or flame textured), and new applica-

Fig. 15. Cutting flagstones to form square tiles.

tions, such as coordinated lines of furniture in homes, but much waste is produced.

Production of large volumes of crushed stone is assured by modern drilling equipment, blasting techniques, and crushing and processing machinery. Stone can also be treated by tumbling, leaching, scabbling, and sandblasting (Perath, 1992, written communication).

SPECIFICATIONS

Formal specifications do not exist for all categories of decorative stone, although certain individual types are covered by Power (1983, p. 163). Appearance is the primary factor in choosing a decorative stone. The color and texture are of principal importance along with the surface treatment used. The ability to be split into flat slabs is important in the production of flagstone, of less importance in fieldstone and moss rock, and of virtually no importance in river rock.

Strength, porosity, adsorption, and durability are important in some decorative stone end uses and of little importance in others. Durability and strength are significant in crushed stone used for some decorative purposes, such as terrazzo and exposed aggregate (Ault, 1989). Physical properties are covered in detail in the *Construction Materials: Aggregates* and *Construction Materials: Dimension Stones* chapters in this volume.

Flagstone

Irregularities on a flagstone wear surface must be less than 1 cm in total height (Harris, 1991). A standard test for abrasion resistance of stone subjected to foot traffic is given ASTM designation C241.

Tiles and Paving Blocks

Slate for US floor tiles is sold in four basic square or rectangular sizes from 15 × 15 to 46 × 46 cm (Fig. 15). Size is limited only by the curvature of foliation which can cause tripping in tiles over 1 m across. Tile thickness varies from 6 mm for light duty and housing to 9 mm for heavy duty use in public areas (Harben, 1990). Flexure testing of slate is covered by ASTM designation C120. A minimum abrasion resistance (ASTM C241) of 10 is required for carbonate floor tile. Where two or more marbles are combined there should be a maximum difference of 5 points of abrasion resistance. For stairways, floors, and platforms subject to heavy foot traffic, a minimum abrasion resistance of 12 is recommended.

Exposed Aggregate

In general, tests for exposed aggregate are variations of those used for cement and road building or construction aggregate end uses. Standard evaluation techniques for such end uses may have severe shortcomings when applied to exposed aggregate. Highway service records, acceptance tests, and government ratings of stone do not correlate with its quality as exposed aggregate (Cutcliffe and Dunn, 1967).

Hardness is determined by Los Angeles, ASTM, or Deval tests. The cement-aggregate mixture should be hard enough to minimize emplacement of faulty exposed aggregate because repair costs are extremely high, if repair is possible at all (Cutcliffe and Dunn, 1967, Stith, 1970). Abrasion resistance is less significant in exposed aggregate, but softness or brittleness may cause excessive fines during batching (Cutcliffe and Dunn, 1967).

Soundness is the main criterion that determines the durability of exposed aggregate during temperature and humidity cycles, weathering, and erosion. Several sides of an exposed aggregate particle often are not encased in cement making poor quality stone very susceptible to the elements. A variety of tests involving freeze-thaw cycling and exposure to solutions are applied to exposed aggregate as discussed by Stith (1970) and Cutcliffe and Dunn (1967).

Water absorption should be less than 1.5% because high absorption promotes weathering and staining (Shergold, 1954, Stith, 1970). Average pore space and percentage saturation also influence soundness. Completely saturated rocks, small pores, and capillaries are more deleterious than larger better-drained pores (Verbeck and Landgren, 1960, Yedlosky and Dean, 1961, Stith, 1970).

Particle shape and size distribution are less important factors for exposed aggregate. The two main criteria are few thin or platy particles and little dust or fines. Impurities to avoid are shale, clay, iron sulfides and pyrite, chert, gypsum, bituminous materials, dolomite, limestone, or any reactive material greater than 1% by weight (Cutcliffe and Dunn, 1967, Stith, 1970).

Dash

Dash is fine-grained (up to 9 mm) exposed aggregate and the basic principles and specifications described above for exposed aggregate apply. Stone grains as small as 1.5 mm are used as concrete block facing (Ladoo and Myers, 1951). Very fine-grained dash is commonly used as a pigment in cement and mortar in place of mineral pigments.

Terrazzo

Aggregates for terrazzo use are tested by methods applicable to aggregates for other uses. These include ASTM C33, C131, C88, Los Angeles abrasion, and sodium sulfate tests. Terrazzo is essentially exposed aggregate with only one side exposed so the criteria described above for exposed aggregate generally apply. Specifications and data on terrazzo are available from the Terrazzo, Tile, and Marble Association of Canada, Concord, Ontario, or the National Terrazzo and Veneer Association, Des Plaines, IL.

Veneer and Cladding

Modern use of decorative stone as veneers and cladding does not require the high ASTM standards for compressive or flexural strength called for in structural stone. Veneer and cladding stone is hung, using a variety of support and anchor systems, to the structural wall behind. Any stone that can support itself between anchors can be used if it meets aesthetic requirements.

Standard specification for marble dimension stone (exterior) is ASTM designation C503. Marble in this context includes calcite, dolostone, travertine, and serpentine. Marble for exterior use shall be sound, free of spall cracks, open seams, pits, or other defects that would impact its strength, durability, or appearance. Molded, cast, or artificially aggregated units are discussed under the section on manmade stone.

Cantera Stone

The criteria established by ASTM for cantera stone are as follows: bulk specific gravity (C97) 1.336 to 1.88, absorption (C97) 9.9 to 22%, modulus of rupture (C99) 435 to 1520 psi, compressive strength (C170) 1800 to 9960 psi, static coefficient of friction (C1028) 0.77 to 0.95, and an abrasive hardness (C501) of 3.3 (Kuiper, 1988).

ECONOMIC FACTORS

The decorative stone industry tends to be vertically integrated in the United States, although a dealer/distributor network is also in place. Many producers market both through their own sales operation and through various representatives. This is due in part to the need for clients to view the stone prior to purchase. Developers of large projects often visit the quarry, but it is often advantageous to have examples of the stone available regionally at distributors.

The decision to use a particular stone in a project is made by architects who are the greatest single influence on demand in the decorative stone industry. The reputation and trade name of the stone is the principal factor, rather than test results, used by architects to determine durability. Architects rated factors effecting their choice of stone (Johnson, 1983) from most important to least as follows: 1) appearance (overwhelmingly), 2) durability, 3) cost, and 4) availability. Appearance of stone depends on color, texture, and uniformity. Limestone was the most frequently selected stone with granite second, but exposed aggregate panels were preferred for small commercial buildings. Stone is specified often for government buildings, sometimes for churches, but rarely for commercial or residential structures (Johnson, 1983).

Costs

Decorative stone ranges from a moderate cost high bulk to a high cost low bulk commodity. Crushed stone for landscaping (about $65/t) is an example of the former, while stone for sculpture (perhaps $6,000 for a large block of Carrara marble) represents the latter. Decorative stone often commands a higher price than identical stone used where aesthetics are not considered.

The demand in the United States for pure white statuary-grade marble increased 850% between 1980 and 1987. World supplies have dwindled with the decreased availability at the highly subsidized deposits at Carrara, Italy, so that large blocks of this premium grade are no longer abundant. This has led to the re-opening of the Yule marble quarry in Colorado (McClean, 1990). Yule marble today sells for 700 to $1,000/m^3, but increasing demand, particularly from Japan, may drive this to $1,750/m^3 (McClean, 1990a). Marble deposits in Tennessee and Alabama have also been reactivated (Meade, written communication, 1992).

Fig. 16. Volcanic cinder in hopper cars at Santa Fe, NM, ready for shipment to eastern markets.

Transportation

All major forms of mineral transport are utilized for decorative stone. Truck transport of aggregate (86%) predominates over rail (6%) or inland/coastal water transport (barge 2% and lake <1%) while sea transport (<1%) is least common (Anon., 1988). Truck transport is very effective, though usually more costly per unit shipped, because of flexibility and orientation toward individual irregular shipments and low capital outlay (Hayes, 1991). The stone is loaded directly onto the truck at the production site and carried directly to the end use site with no intermediate handling. Delivery is very prompt by truck and damage is minimal even for polished or slabbed stone. In contrast, rail transport (Fig. 16) may cost more because it is oriented to large predictable shipments, is capital intensive, and has high fixed costs. High railcar coupling speeds may damage more stone than does road vibration during truck haulage and rail rates often are set according to the stone finish. Any transport by water, barge or ship (Fig. 17) is the lowest cost. Timing is important because most consumers need rapid delivery once a decorative stone is ordered. Distribution yards closer to consumers than production facilities are often used for flexibility and rapid delivery (Hayes, 1991).

In Michigan (Johnson, 1983), transport costs ranged from 20 to 100% more for highly finished stone than for the same stone when unfinished, rough, split, or sawn. Crushed stone transport cost was only about 40% of that for unfinished stone. In the early 1980s, shipping of decorative stone on the Great Lakes was not readily available, although its cost would have been about 10% of rail rates. Lake transport was reserved for large volumes from established producers with regular shipments, such as coal or iron ore (Johnson, 1983).

Most factors applicable to construction aggregate transport apply to decorative aggregate. Decorative stone aggregate carries significantly more value than construction aggregate, even if they are the same rock, so higher transport costs are justified. Decorative stone includes a very wide variety of products requiring transpor-

Fig. 17. Loading aggregate into barges at Portland, OR.

tation. Although much is used locally, certain segments of this market will support higher cost transport offshore and worldwide. Decorative stone is often sold in relatively small lots so bulk transport cost advantages cannot often be utilized, although they are substantial if the shipment is large.

End uses of decorative aggregate, such as landscaping rock, are the most sensitive to transport cost and thus travel the least distance, although this distance can be significant. Aggregate has been imported into the United States (Timmons and Harben, 1987) to areas accessible to ships such as the East, Gulf, and Southeast Coasts of the United States, and large portions of the Midwest by barge. This is particularly true for low volume retail sales where small lots may be shipped long distances because the per ton retail value is high. Polished or slabbed decorative stone, a high value commodity, can be shipped any distance if well protected. In contrast, decorative aggregate is resistant to exposure and shock and needs minimal protection during transport.

Substitutes

All decorative stones have competition from substitute materials and more are being developed all the time. Examples include concrete, porcelain-enameled or stainless steel, aluminum, brick, plastic, manmade stone, crushed glass and slag, and recycled materials. The aesthetic appeal, prestige, and durability of stone is exceeded by few substitutes. For this reason, substitutes simulate stone, often at lower cost. Stone substitutes are often designed as prefabricated modules or precast panels; some decorative stone, such as terrazzo or exposed aggregate, is also utilized in panels.

TARIFFS AND DEPLETION ALLOWANCES

Complex tariffs on imported dimension stone exist (Taylor, 1985) that also apply to dimension stone used as a decorative stone. The tariffs in 1990 varied from free to 7.5% for Most Favored Nation (MFN) status according to type, size, value, and degree of preparation (Taylor, 1992). Tariffs on crushed stone, including that used for decorative purposes, are 30% *ad valorem* or free for MFN (Tepordei, 1991).

The depletion allowance for decorative stone depends on the form of the stone and its end use. Crushed stone used in construction, including aggregate, has a depletion allowance of 5%. Larger blocks of decorative stone are treated as dimension stone with a depletion allowance of 14% if used in construction, or 5% otherwise.

Markets and Trends

The use of steel, glass, aluminum, plastic, and reinforced concrete made stone non-essential in construction and caused a long term decline in use of large structural blocks (Johnson, 1983). Sales of stone for exterior paving and curbing have decreased due to increased use of asphalt and concrete.

Stone acceptance and usage in the United States was growing just prior to the building slowdown in the early 1990s. In Europe, use of decorative and roughly dressed stone dropped from 50 to 0.25% relative to crushed stone during the 20th century (Kuzvart, 1984). Recently the aesthetic appeal of stone and the development of thinner stone in lightweight panels and frames have offset this to a degree (Johnson, 1983) and heightened the use of stone veneers and cladding. Thin-stone slab applications are favored by architects due to lower cost relative to glass and steel (Sweet, 1990).

Veneer style has shifted to smaller pieces of stone assembled in panels for unit construction giving a cubic look, particularly for granite buildings. Extensive engineering and testing led to precasting of stone and concrete panels, epoxy-bonding of finished stone panels prior to installation on site, and most recently, bonding of ultra-thin stone veneer to expanded aluminum backing (Meade, 1986a). Use of stone increased for decorative interiors with various shades and colors of marble in lobbies and foyers. Residential use of thin stone tiles and counter tops, particularly granite, has increased. Decorative stone panels and tiles made from agglomerated stone such as Chilean lapis bonded into tiles for high-contrast colored trim (Meade, written communication, 1992). Over time, taste in stone architecture changes so some stones fall out of favor, such as the brownstone fronts popular on older urban residences (Ladoo and Myers, 1951).

The trend in North America is for specialized use of stone for architectural purposes. The trend in stone use is away from blocks and towards veneer and cladding in larger thin panels which, depending on stone type, may need to be backed by glass fiber/epoxy or other strengtheners. Wider spans may need to be thicker or use special anchors depending on wind load or other factors. A trend in architecture is toward smaller, thinner pre-assembled stone panels set in larger panels for rapid installation onto steel frameworks or exterior walls to yield a look as if built with large stone blocks.

The broader residential market that uses a wider variety of stone is growing, particularly in the Southwest, but demand is not great enough to create demand for all quarry waste. This is in contrast to Italy where virtually all stone quarried is utilized so US costs remain higher for stone and also for fabrication. The European participation in the Marble Institute of America shows that the US market is viable for importers Multinational participation in the US stone industry is helping to establish acceptable standards for suspension systems and artificial supports. The US stone industry carries much higher product liability, environmental, safety, and health costs compared to foreign producers (Meade, 1986).

The stone industry remains labor intensive which limits markets for relatively costly US stone. Technology improvements, particularly widespread diamond cutting (Harben, 1990) or flame cutting, partly offsets this. Improvements in mechanical fasteners and cements have fostered the decorative stone panel market which is also being standardized to improve competitiveness. Increased panel sales means increased sales of exposed aggregate, often turning quarry waste into a product. Chemical treatment to increase durability of stone is making a wider variety of stone types available (Johnson, 1983).

Zoning and alternative land use are constant concerns of the stone industry. Regulation of stone quarries and sand and gravel operations is by federal agencies and sometimes state or local agencies as well. The decorative stone industry in the United States is subject to the same safety, health, and environmental pressures and regulations as the aggregate and dimension stone industries. Much decorative stone is produced as a coproduct of urban stone quarries which are under extreme pressure to relocate further from growing population centers and to impact less on wetlands. Many inoperative quarries are being converted into scenic and high value commercial and residential developments. Shortages of aggregate in most urban areas are likely because of local zoning restrictions and land development alternatives. The stone industry will continue to be concerned with environmental restrictions and safety factors.

ACKNOWLEDGMENTS

Much of the our knowledge of the various uses of decorative stone was gained though conversations and other communications with many professional and industrial personnel. Information was solicited and received from nearly every state and Canadian province, as well many governmental agencies throughout the world. It is impossible for us to thank all these individuals, but many are mentioned in the text. They gave examples which shaped our thinking on decorative stone. In particular, we thank Lance P. Meade, Cyprus Industrial Minerals Company, W. Robert Power, Frisco, CO, and Allan M. Johnson, Michigan Technological University, for reviews that greatly helped us to improve the writing and to broaden the manuscript to cover occurrences and uses that were unknown to the authors.

BIBLIOGRAPHY

Anon., 1988, "Pit & Quarry Trivia," *Pit and Quarry*, Vol. 81, No. 6, p. 23.

Ault, C.H., 1989, "Construction Materials: Crushed Stone," *Concise Encyclopedia of Mineral Resources*, D.D. Carr and N. Herz, eds., Pergamon Press, Oxford, pp. 57–64.

Austin, G.S., and Barker, J.M., 1990, "Commercial Travertine in New Mexico," *New Mexico Geology*, Vol. 12, pp. 49–58.

Austin, G.S., Barker, J.M., and Smith, E.W., 1990, "Building with Stone in Northern New Mexico," *Tectonic Development of the Southern Sangre de Cristo Mountains, New Mexico*, P.W. Bauer et al., eds., New Mexico Geological Society, Guidebook 41, pp. 405–415.

Bates, R.L., and Jackson, J.A., eds., 1987, *Glossary of Geology*, 3rd ed., American Geological Institute, Alexandria, VA, 778 pp.

Bowles, O., and Barton, W.R., 1963, "Sandstone as Dimension Stone," Information Circular 8182, US Bureau of Mines, 30 pp.

Burchett, R.R., and Eversoll, D.A., 1991, *Nebraska Mineral Operations Review, 1990*, Nebraska Geological Survey, 15 pp.

Carpenter, III, P.A., 1983, *Building Stones of North Carolina*, Department of Natural Resources and Community Development, Division of Land Resources, Geological Survey Section, 16 pp.

Cutcliffe, W.E., and Dunn, J.R., 1967, "Evaluation of Coarse Aggregate for Architectural Concrete," *Proceedings*, 3rd Forum on Geology of Industrial Minerals, E.E. Angion and R.G. Hardy, eds., State Geological Survey of Kansas, Special Distribution Publication 34, pp. 39–44.

Gary, M., et al., eds., 1972, *Glossary of Geology*, American Geological Institute, Washington, DC, 805 pp.

Hansen, W.R., 1969, "Stone," Utah Geological and Mineralogical Survey, Bulletin 73, pp. 222–228.

Harben, P., 1990, "Romancing the Dimension Stone with Diamonds," *Industrial Minerals*, Vol. 270, pp. 47–53.

Harris, R.E., 1991, "Decorative Stones of Wyoming," Geological Survey of Wyoming, Public Information Circular 31, 27 pp.

Hayes, V.L., 1991, "Product Transportation and Distribution System," *The Aggregate Handbook*, R.D. Barksdale, ed., National Stone Association, Washington, DC, pp. 10–1 to 10–20.

Johnson, A.M., 1983, "Geologic and Economic Evaluation of Building and Decorative Stone Resources in Michigan," *Proceedings*, 18th Forum on the Geology of Industrial Minerals, C.H. Ault and G.S. Woodard, eds., Indiana Department of Natural Resources, Geologic Survey, Occasional Paper 37, pp. 113–126.

Kuiper, G., 1988, "Cantera Stone Comes Home," *Dimensional Stone*, Sep., pp. 36–38.

Kuzvart, M., 1984, *Industrial Minerals and Rocks*, Vol. 18, (Developments in Economic Geology), Elsevier, Amsterdam, 454 pp.

Ladoo, R.B., and Myers, W.M., 1951, "Stone," *Nonmetallic Minerals*, McGraw-Hill, New York, pp. 506–515.

Makens J.C., et al., 1972, "Study of Technical and Economic Aspects of an Expanded Stone Industry in Michigan," Michigan Office of Economic Expansion, Research Project MOEE-23, Final Report, 124 pp.

McClean, S., 1990, "Colorado Marble was Used in Lincoln Memorial," *Paydirt*, Nov., pp. 2B-4B.

McClean, S., 1990a, "Marketing Future Looks Mighty Bright for Yule Marble," *Paydirt*, Nov., pp. 4B-6B.

Meade, L.P., 1986, "North American Quarry Economics, Selective Versus Total Extraction," *La Cava nel 2000—Tra Innovazione Technolgica e Nuove Dinamiche di Marcato*, 7·Fiera Internationale Marmi e Macchine Carrara., Carrara, Italy, pp. 56–60.

Meade, L.P., 1986a, "Dimension Stone—Surviving 2000 Plus Years of Supply/Demand Seesaw," Preprint 86–152, Society of Mining Engineers of AIME, Annual Meeting, New Orleans, 11 pp.

Osburn J.C., 1980, "Evaluation of Scoria Deposits in New Mexico," *Annual Report, July 1, 1978 to June 30, 1979*, New Mexico Bureau of Mines and Mineral Resources, pp. 75–80.

Power, W.R., 1983, "Construction Materials: Dimension and Cut Stone," *Industrial Minerals and Rocks*, 5th ed., Vol. 1, S.J. Lefond, ed., AIME, New York, pp. 161–181.

Reed, A.H., 1978, *Stone*, US Bureau of Mines, Mineral Commodity Profiles, MCP 17, 19 pp.

Shadmon, A., 1988, "Stone Technology: An Old Tradition with a New Future," *Episodes*, Vol. 11, pp. 275–282.

Shergold, F.A., 1954, "The Effect of Freezing and Thawing on the Properties of Roadmaking Aggregates," *Roads and Road Construction*, Vol. 32, pp. 274–276.

Shipley, R.M., 1945, *Dictionary of Gems and Gemology Including Ornamental, Decorative, and Curio Stones*, 2nd ed., Gemnological Institute of America, Los Angeles, 258 pp.

Stith, D.A., 1970, "Potential Use of Ohio Limestones and Dolomites for Architectural Aggregate," *Proceedings*, 6th Forum on the Geology of Industrial Minerals, W.A. Kneller, ed., Michigan Department of Natural Resources, Geological Survey Division, Miscellany 1, pp. 81–94.

Sweet, P.C., 1990, "Present and Future Dimension Stone Industry in Virginia," *Proceedings*, 24th Forum on the Geology of Industrial Minerals, A-J.W. Zupan and A.H. Maybin, III, eds., South Carolina Geological Survey, pp. 129–135.

Taylor, H.A., Jr., 1985, "Dimension Stone," *Mineral Facts and Problems*, Bulletin 675, US Bureau of Mines, pp. 769–776.

Taylor, H.A., Jr., 1991, "Dimension Stone," *Mineral Commodity Summaries, 1991*, US Bureau of Mines, pp. 156–157.

Taylor, H.A., Jr., 1992, "Dimension Stone," *Mineral Commodity Summaries, 1992*, US Bureau of Mines, pp. 168–169.

Tepordei, V.V., 1991, "Stone (Crushed)," *Mineral Commodity Summaries, 1991*, US Bureau of Mines, pp. 154–155.

Townsend, R.C, 1962, "Stone in Arizona—An Economic Study," Arizona Development Board, Phoenix, pp. 9–16.

Timmons, B.J., and Harben, P., 1987, "Have Aggregate—Will Travel," Preprint 87–131, Society for Mining, Metallurgy, and Exploration, Annual Meeting, Denver, 3 pp.

Verbeck, G., and Landgren, R., 1960, "Influence of Physical Characteristics of Aggregates on Frost Resistance of Concrete," *Proceedings*, American Society for Testing Materials, Vol. 60, pp. 1063–1079.

Woller, F., et al., 1992, "Industrial Minerals Opportunities in Czechoslovakia," Preprint 92–213, Society for Mining, Metallurgy, and Exploration, Annual Meeting, Phoenix, AZ, 11 pp.

Yedlosky, R.J., and Dean, J.R., 1961, "Petrographic Features of Sandstone That Affect Their Suitability for Road Material," *Journal of Sedimentary Petrology*, Vol. 31, pp. 372–389.

DIAMONDS, INDUSTRIAL

K. RECKLING, R.B. HOY, STANLEY J. LEFOND,
DEREK G. FULLERTON, AND UNNI H. ROWELL

In 1989 natural industrial diamonds counted for 55% of the world's natural diamond production. Australia is currently the leading producer (35%). Zaire is the second largest producer (19%), of what is primarily industrial grade rather than gem grade. Botswana (17%) is third, with the former USSR (15%) fourth, and the Republic of South Africa (8%) fifth. Diamonds are also mined in Angola, Namibia, the Ivory Coast, the Central African Republic, Ghana, Tanzania, Guinea, and other African countries. In the Western Hemisphere, Brazil is the principal producer, with some production from Venezuela and Guyana (Fig. 1). A very small output of diamonds is mined today in India, which was the first source of commercial production. In the United States, efforts at commercial diamond mining have been confined to a small area near Murfreesboro, AR. The first diamond was found in a pipe there in 1906. Small scale trial mining has not, however, proved economical.

Since diamonds were first discovered more than 2,000 years ago, only about 380 t have been mined. In order to obtain 1 g (5 metric carats) of diamonds, it is necessary to remove and process approximately 25 t of rock. Recovering this small percentage involves a combination of highly developed techniques in mining and extremely sophisticated processes in diamond recovery.

END USES

Diamonds are used for two unrelated end uses: gem diamonds are jewels of great beauty, while industrial diamonds are essential materials of modern industry. Although imitation stones are substituted for the gem diamond, none of these matches its properties sufficiently well to offer real competition. Synthetic industrial diamonds are now of a quality and size that permit them to be substituted for natural diamonds in numerous industrial applications. For example, synthetic diamonds are available today in sizes up to 100 stones per carat (1.2 to 1.4 mm). In addition, polycrystalline synthetic diamond inserts, such as De Beers Syndite®, General Electric's Compax® and Stratapax®, and Megadiamond's Megapax® have replaced natural diamonds in turning tools, mining and oil drilling bits, and dressing tool applications.

Industrial Diamonds

The diamond is by far the most important industrial abrasive. As recently as 50 years ago, consumption of industrial diamonds was less than that of gem diamonds, but since that time, industrial use has grown to a position of great dominance. During the six-year period 1929 to 1934, the material produced for industrial use amounted to about 74% by weight of the world's total output of diamonds.

In 1989 the percentage of natural industrial diamonds mined in the world was 55%. When synthetic industrial diamonds are added to the natural industrial diamond figures, this percentage becomes 87% of total world diamond production including gems, near gems, industrial, and synthetic stones. The many uses responsible for these significant increases are dependent on the properties of the diamond, including hardness, cleavage, and parting, optical characteristics, presence of sharp points and edges, and capacity for taking and maintaining a high polish.

The utilitarian role of the diamond was confined primarily to lapidary products until the industrial revolution, which created the first demand for diamond as an industrial tool. In 1777, a British inventor and instrument maker, Jesse Ramsden, used a diamond to cut a precision screw for an engine that he had invented.

The first authentic description of industrial diamonds being set in a saw was recorded in 1854 by a Frenchman, Dumain. Eight years later a Swiss watchmaker, Jean Leschot, developed the first diamond drill bit for use in a hand operated machine, which was employed to drill blastholes in rock.

In 1864, diamond bits were put to their severest test up to that time in the construction of the Mont Cenis Tunnel in the Alps. A few years later a steam-powered diamond drill with a speed of 30 rpm was able to penetrate rock at the modest rate of 0.3 m/hr.

As the industrial revolution gained momentum on both sides of the Atlantic, metal replaced wood and machines replaced people. Thus the foundation was laid for precision engineering and the recognition of diamonds as an indispensable tool of industry.

The next major demand for industrial diamonds came after World War I with the development of cemented carbide cutting tools. Diamond was found to be the most effective medium for finishing and grinding the new ultrahard metal. This discovery rapidly increased the demand for industrial diamonds.

The availability of inexpensive diamond material inspired tremendous research into applications. By 1935, the first successful phenol-resin grinding wheel containing diamond had been marketed. Soon afterward, the metal-bonded and vitrified diamond wheels were produced, and, as the matrices and bonds that held the diamond grit in place began to improve, the popularity of diamond grinding wheels grew. The outbreak of World War II, and the accompanying increase in use of hard-metal tools in the munitions industry, increased the demand for industrial diamonds.

Since about 1950, the development of ultrahard ceramics, semiconductor materials, plastics, and exotic metal alloys has further consolidated the diamond's position as an indispensable tool of industry. Only diamond is hard enough to cut these superhard materials with the precision, speed, and economy that industry demands today.

Special machines equipped with industrial diamonds are used to remove bumps from concrete runways and highways and to modify highway surfaces in order to prevent skid accidents. Many skids are caused by hydroplaning, a phenomenon that occurs when the roadway is wet. Tires mount a film of water and virtually lose contact with the road surface. Diamond machines cut neat, narrow

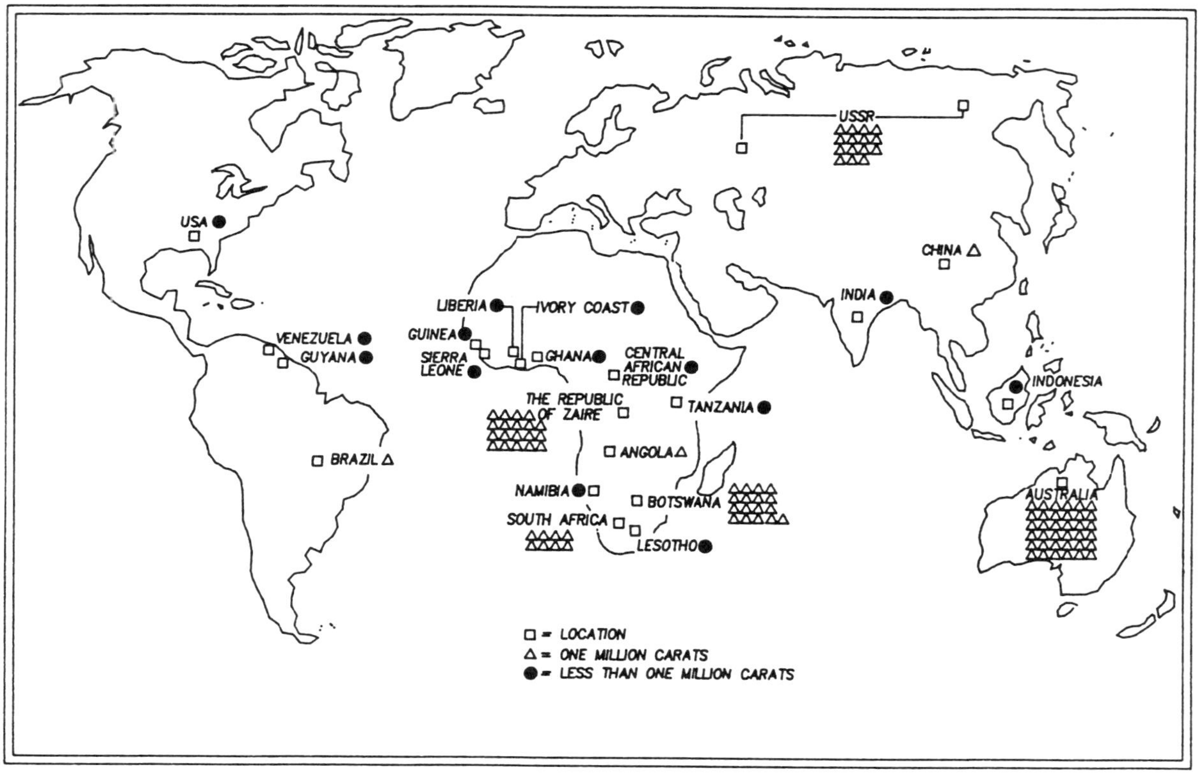

Fig. 1. World diamond production from natural sources (1990).

grooves in the pavement, thereby creating an escape route for the water.

Although dentists have used diamond drills for nearly a century, they have found a new use for diamonds as dental crowns in the form of very fine dust. The minute diamond crystals give wear and impact resistance to a dental resin employed to fabricate bridges, jackets, and other forms of dental restoration. Diamonds are also used in surgery. An ophthalmologist has developed a new technique in cataract surgery in which a perfect incision is made by a diamond knife with a microscopically perfect cutting edge. The blade's edge appears straight even when examined under 1 000-power magnification. A New York plastic surgeon specializing in the repair of peripheral nerves uses precision forceps whose tips are coated with diamond dust. The ability of the diamond-tipped forceps to grip fine tissues firmly, but without damage, is an important technical advance in the field of microsurgery.

Resembling the forceps are diamond-tipped tweezers, which have gained favor with hobbyists and collectors who work with small parts or delicate items. Because they have from three to four times the gripping power of conventional instruments, these tweezers are excellent for plucking eyebrows. Nail files coated with diamond dust are another popular item because they are almost indestructible.

In Europe, many manufacturers of pasta have switched from steel to diamond saws to slice uncooked spaghetti into desired sizes. Steel blades dull in about 2 hr, while diamond blades hold their cutting edge for several months. For similar reasons, industrial diamonds are used to cut frozen fish to package size.

Although virtually unnoticed, industrial diamonds frequently surround us. In modern oil furnaces, some nozzles through which fuel is sprayed are made of diamond, since fuel oil is abrasive and tends to enlarge metal nozzles. Fine wires used in telephones, transistor radios, and television sets owe their precision diameters to diamond dies. Eyeglasses and contact lenses are also shaped with industrial diamonds. In fact, the mirrors in your home and the glass in your automobile were most likely ground with diamonds. Because of their abrasion-resistance, diamond-tipped drills are also used for geological exploration and oil-well drilling. They also played a major role in the Mohole project in the Pacific Ocean.

Industrial diamonds have made contributions to America's space program, most notably in the production of the Gemini Apollo spacecraft. The thousands of meters of complex wiring, delicate instrumentation, ceramic fiber insulation, stabilizing gyroscope, nose cap, and reentry heat shield—all these components were produced with the aid of industrial diamond tools. Diamonds even played their part in man's spectacular step onto the moon, not only in the manufacture of the vehicles that got him there, but in the delicate instruments that helped him to probe the moon's surface to uncover the secrets that lie beneath.

Doulton, a name (Royal Doulton) synonymous with fine china for more than a century, now is a mammoth international company manufacturing not only tableware and figurines, but also a whole host of other items ranging from sonar detection equipment to bullet-resistant glass windows. The company, having a reputation for use of up-to-date methods and equipment, uses metal-band diamond wheels for grinding and Hole-O-Matic diamond drills for making holes in such items as safety doors (Herbert, 1976).

New developments in the application of industrial diamonds are described in the publication *Industrial Diamond Review*. Some of these uses are new, others are improvements in established applications. They include precision drilling of small holes (Herbert, 1976), cutting of hard, brittle materials (Forner, 1978), drilling of concrete (Seversky and Van Biljon, 1976), drilling of hard rock (Hill, 1975), trueing of tungsten carbide gages (Herbert, 1977), and

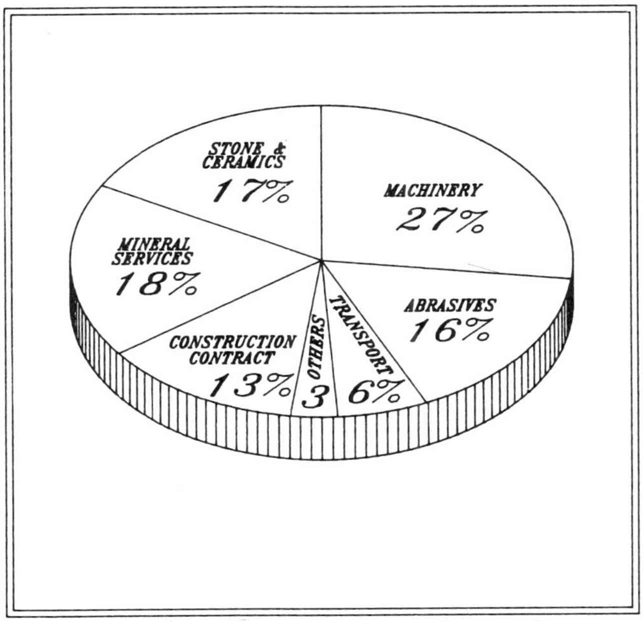

Fig. 2. US industrial diamond consumption by market sector, 1990. *Source:* US Bureau of Mines, 1991.

fabricating applications in the electronics and optical industries (Seal, 1978).

Alternate materials are available for many of the applications served by diamonds, but these materials are generally so much less efficient and adaptable that they are poor substitutes. No effective substitute has been found for use in drilling bits for hard rock or concrete.

Classification: Natural industrial diamonds come in many shapes and sizes. They are generally classified according to use in the following groups: tool and die stones, drilling material, grits, and powders. (Fig. 2 shows the pattern of industrial diamond consumption for the United States.)

Due to substitution of synthetic diamonds in drilling bits and wire drawing dies, natural diamond usage declined from 26% in 1973 to 12% in 1990. It is likely that the decline in natural industrial diamond consumption will continue into the next century with further substitution of synthetic diamond.

Industrial diamonds are further classified according to type of material, as follows:

1. *Industrial stones (fine industrials)*: Stones of large size not suited for gem use because of shape, mechanical imperfections, or undesirable color.
2. *Bort (boart, boort, bortz, bowr)*: Stones whose small size, irregular shape, content of flaws or inclusions, or occurrences in finely crystalline aggregates make them unsuitable for gem use. *Drilling bort* are those stones whose soundness permits their use in diamond drill bits. The more abundant crystals and crystal aggregates of lower grade are classed as *crushing bort*, which is suitable for being crushed into grit, powder, and dust. This latter category includes worn or broken stones from used diamond tools, and fragments formed during preparation of gems and shaped industrial stones.
3. *Carbonado (carbon, black diamond)*: Compact, opaque, dark gray to black crystalline material composed of diamond, graphite, and possibly some amorphous carbon. It has no cleavage and is extremely tough. It occurs as rounded masses of average size greater than that of gem stones.

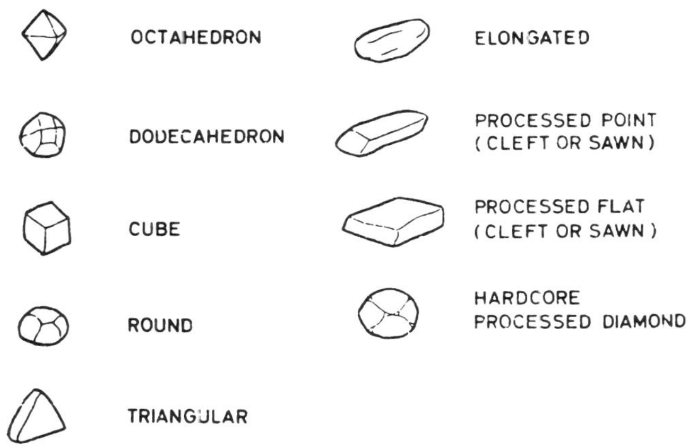

Fig. 3. Industrial diamond shapes.

4. *Ballas (bort-ballas, short-bort)*: Dense, globular aggregates of small radially oriented crystals and with extremely difficult cleavage. It is both very hard and very tough.

The average person tends to think that all rough diamonds are of the octahedron form, but many rough stones are not of this shape (Fig. 3).

Tool and Die Stones. The larger diamonds of higher quality are called tool and die stones. These are used as dressers (to true and dress conventional abrasive wheels), turning, boring, and milling tools, and wire-drawing dies.

To obtain maximum efficiency from a tool diamond, it is important to orient it so that the hard direction or the most abrasion-resistant face bears the work load. Orientation is also important in the manufacture of diamond wire-drawing dies.

Drilling Materials. Drilling accounts for 10% of all diamond use. Blasthole drill bits are usually made with small crystals of regular shape. Mineral exploration bits require diamonds ranging in size from 15 to 100 stones per carat. Masonry and concrete bits make use of smaller diamonds ranging from 50 to 200 stones per carat. Oil well bits require larger sizes, usually 12 stones per carat and sometimes stones as large as 3/4 carat.

To improve the supply of natural drilling material, stones with poor shapes are subjected to high velocity air milling with diamond dust to improve their shape and strength. In addition, surfaces of some diamonds are polished because polished surfaces reduce friction during drilling operations.

Drilling material is also used in the manufacture of multiple layer diamond dressers and rotary form dressers. These are used for dressing and trueing surfaces of precision grinding wheels, cutting glass, engraving glass, and harder materials, and for sawing and shaping carbides, corundum, dimension stone, plastics, and various metals.

Grits and Powders. Approximately 75% of all industrial diamonds—both natural and synthetic—are used in the form of grit and powder. Diamond grit is used primarily in grinding wheels and saw blades. There is a specific diamond grit type, shape, and size for each specific application. Diamond abrasive grains are sized by sieving on woven wire mesh or electroformed screens.

Diamond grit ranges in size from very coarse material (2.0 to 2.4 mm) to fine material (38 to 44 μm). Very fine diamond powder is referred to as graded micron powder and is used in the manufacture of lapping compounds and for fine finishing tools. Diamond micron powders are graded by a combination of sedimentation and centrifuging. Table 1 lists the standard grade numbers for these powders.

Table 1. Standard Grade Numbers for Diamond Powders

Grade Nos.	μ Range	Approximate Mesh Equivalent
1/2	0–1	50 000
1	0–2	14 000
3	2–4	8 000
6	4–8	3 000
9	8–12	1 800
15	12–22	1 200
30	22–36	800
45	36–54	500/600
60	54–80	400/500

Source: US Bureau of Standards, C/S 261/63.

GEOLOGY

General

Diamonds are composed of a single element, carbon, crystallized in cubic form. Diamonds range from colorless through faint tinges of blue-white, yellow, red, brown, green, blue, and gray in gem form; and from yellow-brown to dark brown and black in industrial form. It is the hardest known material, listed at 10 on the Mohs hardness scale, actually almost five times as hard as corundum listed at 9. The specific gravity is high (3.5) and index of refraction is very high (2.42). Color dispersion is exceptionally strong, producing the characteristic play of color in the gem. Table 2 lists many of the various properties of diamonds (Anon., undated).

Primary Occurrences

Kimberlite: The main primary source of diamonds is a rock called kimberlite or ''blue ground,'' which occurs in volcanic pipes and dikes. The pipes are generally vertical and range in shape from nearly circular to elliptical. They range in size from small pipes only a few feet in diameter to pipes having a surface area of hundreds of acres. (Table 3). The size of a pipe bears no relationship to its diamond content and not all kimberlites contain diamonds.

The following definition of kimberlite is taken from Kirkley et al. (1991) and augments that of Dawson (1984):

> ''Kimberlite is a hybrid, volatile-rich, potassic, ultramafic igneous rock derived from deep in the earth (>150 km below the surface) which occurs near the surface as small volcanic pipes, dikes, and sills. It is composed principally of olivine (both as phenocrysts and in groundmass), with lesser amounts of phlogopite, diopside, serpentine, calcite, garnet, ilmenite, spinel, and/or other minerals; diamond is only a rare constituent.''

The presence of pyrope garnet, ilmenite, chrome diopside, and spinel serve as indicators for geologists in their search for diamond-bearing kimberlite occurrences.

Lamproite: In 1979 the discovery in Australia of diamonds in a relatively obscure rock called lamproite led to considerable interest in lamproite as a source rock for diamonds. Since this discovery, lamproites containing diamonds have been found in several areas of the world. The possibility exists that other economic lamproite deposits will be discovered. It is of interest to note that the Prairie Creek occurrence in Arkansas, discovered in 1909, is in fact a lamproite whereas it had previously been classified as a kimberlite.

The first modern approach to defining lamproites was made by Scott-Smith and Skinner (1984) who defined lamproite as follows:

> ''Lamproite is an ultrapotassic magnesian igneous rock. It is characterized by high K_2O/Na_2O ratios, typically greater than five. Trace element concentrations are extreme with high concentrations of Cr and Ni, typical of ultrabasic rocks, as well as those more typical of highly fractionated or acid rocks, e.g., Rb, Sr, Zr, and Ba. Generally CO_2 appears to be absent.
>
> Lamproite contains, as primary phenocrystal and/or groundmass constituents, variable amounts of leucite and/or glass and usually one or more of the following minerals are prominent: phlogopite (typically titaniferous), clinopyroxene (typically diopside), amphibole (typically titaniferous potassic richterite), olivine, and sanidine. Other primary minerals may include priderite, perovskite, apatite, wadeite, spinel, and nepheline. Other minerals such as carbonate, chlorite, and zeolite, if present, may not be primary.
>
> Upper-mantle-derived xenocrysts and xenoliths may or may not be present.''

Although there has been much controversy over the origin of diamonds, present day evidence indicates that kimberlites and lamproites were the transportation vehicles by which diamonds were brought to the surface and that as such they are not genetically related to diamond (Kirkley et al., 1991). It is believed that diamonds crystallized in peridotites and eclogites at considerable depths. In the ascent of kimberlites and lamproites to the surface, eclogites and peridotites became xenoliths (i.e., foreign rocks) to the kimberlites and lamproites which incorporated them. During the transportation of these xenoliths disaggregation occurred which released the diamonds subsequently found within the kimberlite and lamproite host rock.

Alluvial Occurrences

Diamond-bearing kimberlite and lamproite in volcanic pipes served as the main source from which widely distributed alluvial diamond deposits were formed. Weathering processes disintegrated these pipes and other rocks at the earth's surface. These processes are particularly effective in warm humid climates where near surface rock formations, including kimberlites and lamproites were broken down and the products carried away by river systems and deposited over widely scattered areas, often far from their sources. Some geologists believe that a thickness of at least 900 m of rock has been eroded away in the Kimberley region of the Republic of South Africa and distributed by streams.

Diamonds and other hard, weather-resistant minerals, released from the disintegrating kimberlite and lamproite, were concentrated by flowing water and deposited in topographically favorable localities to form the alluvial deposits found today. Because of high specific gravity, diamonds tend to be concentrated at the bottom of alluvial deposits.

RESERVES

World reserves of industrial diamonds can be estimated only very roughly. Some figures have not been updated for more than a decade. The estimate in Table 4 (Austin, 1991) indicates that reserves appear to be sufficient to supply industrial stones for only 20 years at the current rates of consumption of approximately 55 million carats of industrial stones per year. Approximately half of the world's reserves of industrial diamonds are in Australia.

The former USSR estimate in Table 4 is of low reliability; recent discoveries in the USSR are likely to increase the reserve figures. With continuing exploration for diamond deposits in Australia, Africa, the USSR, North America, and South America further economic discoveries will undoubtedly be made over the next decade, thus enhancing reserves.

Table 2. Properties of Diamonds

GENERAL PROPERTIES

I Chemical composition

Diamond is composed of the single element carbon.

Major impurities: Nitrogen, up to 0.2% in natural Type I diamond (see Section II).
Nickel, iron *etc.* up to 10% as inclusions in synthetic diamond (ppm or less in natural diamond).
Aluminum, up to 100 ppm in natural Type IIb diamond and 150 ppm in special doped Type IIb synthetic diamond. Boron, between 3 and 270 atomic ppm in specially doped synthetic diamond—now thought to be responsible for semiconducting properties of Type IIb diamond.
Others generally < 100 ppm
Inclusions: 22 mineral species have been positively identified.

II Classification

Type Ia diamond: Contains nitrogen as an impurity in fairly substantial amounts (of the order of 0.1%), and which appears to have segregated into relatively large sheets or platelets within the crystal. Most natural diamonds are of this type.

Type Ib diamond: Also contains nitrogen as an impurity but in dispersed form. Almost all synthetic diamonds are of this type.

Type IIa diamond: Effectively free of nitrogen impurity. Very rare in nature, these diamonds have enhanced optical and thermal properties.

Type IIb diamond: A very pure type of diamond which has semiconducting properties: generally blue in color. Extremely rare in nature. Semiconducting properties can be imparted to synthetic crystals by the incorporation of suitable impurities.

Note: Classification is based principally on optical properties. Further subdivisions can be made almost *ad infinitum* (indeed, almost every diamond is different in some way) but the above classification is universally accepted. Some diamonds have been shown to consist of more than one type, *eg* a complex interweaving of Type I and Type II material.

III Crystal structure

Diamond type.

Unit cell: Cubic, lattice constant a_0 = between $3.56683 \pm 1 \times 10^{-5}$ Å and $3.56725 \pm 3 \times 10^{-5}$ Å (25°C).

Space group: O_h^7.

IV Density

Value: Average of 35 diamonds: 3.51524 ± 0.00005 g per cc (25°C).

MECHANICAL PROPERTIES

I Hardness

(a) Scratch hardness (Mohs' Scale)

The Mohs' hardness is a scratch hardness test and is related to the indentation hardness of the solid (see below). If the Mohs' number is M and the indentation hardness in kg per sq mm is H, the relation between these quantities is approximately

$$\log H = 0.2\,M + 1.5$$

There is reasonable equality of intervals between the first 9 integers on the Mohs' scale, but the interval between 9 (corundum) and 10 (diamond) represents a much larger difference in indentation hardness than a single unit on the Mohs' scale would suggest.

(b) Indentation hardness (Knoop Scale)

Value ((001) diamond surface): 5,700–10,400 kg per sq mm depending on crystallographic direction and with normal loads of 500 g, 1 kg, and 2 kg.

Comment: There are many other scales, some appropriate to all materials, some only to metals. A Knoop indentor produces a wedge-shaped indentation in the form of a parallelogram with one diagonal at least seven times longer than the other diagonal, and this method is generally considered to be the most accurate for crystalline solids. On any scale diamond is the hardest known material.

It has recently been shown that the measured hardness value is significantly influenced by the normal load, the indentor shape, and its crystallographic relationship with that of the indented surface. It has also been shown that the indentation hardness of diamond decreases with increasing temperature.

II Elastic moduli and compressibility

Note: Young's modulus does not vary greatly with orientation as the condition for isotropy is $\frac{2C_{44}}{C_{11} - C_{12}} = 1$ (C. S. Smith, Vol. 6, *Solid State Physics*, p. 248) and using values given below $\frac{2C_{44}}{C_{11} - C_{12}}$ is 1.54, 1.14, 1.21, and 1.30, respectively. This is not far off unity except for the first case (the ratios for copper and lead are 3.21 and 3.89 respectively). Gilman (*Fracture*, p. 204, John Wiley & Son, New York, NY, 1959) quotes values $E_{100} = 10.5$, $E_{110} = 11.6$, $E_{111} = 12.0$ (all $\times 10^{12}$ dynes cm^{-2}), confirming little variation with orientation. Since diamond specimens are small and difficult to deform, numerical values of Young's modulus from direct deflectional

(cont.)

Table 2. Properties of Diamonds (cont.)

methods are hardly obtainable. However, Young's modulus can be calculated from

$$E = \frac{(C_{11} - C_{12})(C_{11} + 2C_{12})}{C_{11} + C_{12}} = \frac{1}{S_{11}}$$

Using the values given below, $E = 7.24$, 9.48, 10.5, and 96.4, respectively (all $\times 10^{12}$ dynes cm^{-2}).

For *rough* calculations, Young's modulus can be taken as 10.5×10^{12} dynes cm^{-2} in any orientation.

(a) Elastic Moduli

Bhagavantam and Bhimasenachar:
$C_{11} = 9.5$, $C_{12} = 3.9$, $C_{44} = 4.3$
Prince and Wooster:
$C_{11} = 11.0$, $C_{12} = 3.3$, $C_{44} = 4.4$
McSkimin and Bond:
$C_{11} = 10.76$, $C_{12} = 1.25$, $C_{44} = 5.76$
Markham: $C_{11} = 10.76$, $C_{12} = 2.75$, $C_{44} = 5.19$
(all $\times 10^{12}$ dynes cm^{-2})

(b) Bulk Modulus

Calculated from elastic moduli by formula

$$K = \frac{1}{3}(C_{11} + 2C_{12})$$

Bhagavantam and Bhimasenachar: $K = 5.8 \times 10^{12}$ dynes cm^{-2}
Price and Wooster: $K = 5.9 \times 10^{12}$ dynes cm^{-2}
McSkimin and Bond: $K = 4.42 \times 10^{12}$ dynes cm^{-2}
Markham: $K = 5.42 \times 10^{12}$ dynes cm^{-2}

(c) Compressibility

Calculated from mean of Bhagavantam and Bhimasenachar and Prince and Wooster results.
Value: 1.7×10^{-7} sq cm kg^{-1}.

III Strength

(a) Tensile Strength

Diamond normally cleaves on the (111) plane, but cleavage has been observed on the (110) plane, and, to a lesser extent, some other planes (J. R. Sutton, *Diamond*, Murby & Co., London, 1928). Curved cracks can be produced by gradual loading with a spherical indentor. Ring cracks are found frequently on diamond surfaces and can be made by both slow and impact loading with indentors, not necessarily diamond.

Experimental values for strength show wide variations. This is partly a reflection of the difficulty encountered in making tests with diamonds, but also indicates that the strength of indivdual diamonds is significantly affected by the defects, inclusions and impurities which they contain. Small diamonds (or small loaded areas) usually exhibit greater strength than large diamonds (or large loaded areas), *i.e.* there is a size effect.

Value of tensile strength

4×10^{10} dynes cm^{-2} (Seal)
1.3 to 2.5×10^{10} dynes cm^{-2}
(Howes, indentation method, value depended on indentor material and size).

(b) Shear Strength

Theoretical value: 12×10^{11} dynes cm^{-2} (Tyson)
Experimental values: 8.7×10^{10} dynes cm^{-2} (Bowden & Hanwell, from a friction experiment)
3×10^{9} dynes cm^{-2} (Hull & Malloy, from a torsion test).

Note: Doubt must exist as to whether these strength values are genuine shear strengths; it is very likely that the diamonds still failed by cleavage. The variation is because of the very different method of test. In the friction experiment only a very small area of material would be stressed, while in the torsion test failure would probably start at a comparatively large defect at the specimen edge.

(ii) For synthetic diamonds with translucent inclusions or small concentrations of very fine inclusions:
Area of faces 50–75 sq mm $\times 10^{-4}$
Average 17.44×10^{10} dynes cm^{-2}
Maximum 24.50×10^{10} dynes cm^{-2}
Area of faces 660–730 $mm^2 \times 10^{-4}$
Average 6.17×10^{10} dynes cm^{-2}

(c) Compressive Strength

(i) Average value for natural octahedra containing no visible flaws or inclusions =
8.68×10^{10} dynes cm^{-2}
Maximum value =
16.53×10^{10} dynes cm^{-2}
(Area of faces over which load applied approximately 10^{-2} sq cm).

OPTICAL AND ELECTRICAL PROPERTIES

I Refractive index

Values (at H_g green 5461 Å) : $\mu = 2.4237$
(at $H\alpha$ red 6563 Å) : $\mu = 2.4099$
Near cutoff in ultraviolet at 2265 Å : $\mu = 2.7151$

(d) Plastic Flow

Recent work by Gane and Brookes suggests that plastic deformation would appear to be possible in diamonds at room temperature, although the stress required to do this must be very high and close to the crystal's theoretical strength. However, it is very difficult to achieve these stress levels in practice without fracture first occurring. Earlier it had been established that there was no detectable plastic flow in three-point bending experiments up to critical stress for cleavage at temperatures less than 1600°C.

Note: Fits empirical formula:

$$\mu^2 - 1 = \frac{\epsilon_2 \lambda^2}{\lambda^2 - \lambda_1^2} + \frac{\epsilon_2 \lambda^2}{\lambda^2 - \lambda_2^2}$$

where $\epsilon_1 = 0.3306$
$\epsilon_2 = 4.3356$
$\lambda_1 = 1750$ Å
$\lambda_2 = 1060$ Å

II Dielectric constant

Value (at 27°C and 0–3 kHz) : 5.58 ± 0.03.

Table 2. Properties of Diamonds (cont.)

III Optical transparency

Type IIa diamond : 2250 Å to 2.5μ. > 6μ
Type I diamond : 3400 Å to 2.5μ. > 10μ

Note: More restricted range in Type I diamonds due to nitrogen content.

Note: Visible region 4000–7000 Å.
Any coloration is due to absorption bands in this region, or in the case of Premier Overblue (Type IIb) to the tail of the infrared absorption spectrum associated with the acceptor center.

IV Resistivity

Value (at 20°C): Type I and most Type IIa : $> 10^{16}$ ohm cm.
Type IIb : $10–10^3$ ohm cm.

Note: Figures for diamond only apply in the dark, for there is considerable photoconductivity in ultraviolet light.

Note: Diamonds exist which would normally be classified as Type IIa, but which have resistivities intermediate between Type IIb and Types I and IIa. Hence there is really a range from $10–10^{16}$ ohm cm at 20°C. However, most diamonds are insulators with a resistivity of $> 10^{16}$ ohm cm.

Comment: Insulating diamond can have a very high resistivity, comparable with or better than many insulators in commercial use. The best insulator is probably fused silica ($> 5 \times 10^{18}$ ohm cm).

THERMAL PROPERTIES

I Thermal conductivity

Typical values (at 20°C):
Type I 9 watts deg^{-1} cm^{-1}
Type IIa 26 watts deg^{-1} cm^{-1}
Maximum at − 190°C:
Type I 24 watts deg^{-1} cm^{-1}
Type IIa 120 watts deg^{-1} cm^{-1}

Comment: Highest known value: value dominated by impurity content.

II Thermal expansion (linear coefficient)

Value (at 20°C) : $0.8 \pm 0.1 \times 10^{-6}$
(at − 100°C) : $0.4 \pm 0.1 \times 10^{-6}$
(at 100°C–900°C) : $1.5 - 4.8 \times 10^{-6}$

Note: Grüneisen's Law obeyed between 150°C and 950°C.

III Specific heat (constant volume C_v)

Value: (at 20°C : 1.478 cal g-atom^{-1} deg^{-1}
Note:
Effective Debye temperature 0–800°C
$\theta_D = 1800 \pm 10°K$
Effective Debye temperature 0°K
$\theta_D = 2220 \pm 20°K$

Source: Anon., undated.

Table 3. Largest Commercial and Near Commercial Diamondiferous Kimberlites in the World

Rank	Country	Name of Deposit	Surface Area Hectares
1	Tanzania	Mwadui	146.0
2	Botswana	Orapa	106.0
3	Angola	Camafuca Camazambo	67.5
4	Angola	Catoca	66.2
5	Botswana	Jwaneng	54.0
6	Zaire	Talala	50.0
7	Australia	Argyle	45.0
8	South Africa	Premier	32.0
9	Soviet Union	Zarnitsa	21.5
10	Soviet Union	Udatchnaya	20.0
11	Zaire	Massif I	18.6
12	South Africa	Finsch	17.9
13	Lesotho	Letseng-le-Terai	15.9
14	Botswana	Lethakane	11.6
15	South Africa	Venetia	11.4

Sources: International Diamond Annual (1971), Miller (1987), Johnson et al. (1989).

PRODUCING COUNTRIES

World production (in carats) of natural and synthetic diamonds in 1990 was as follows:

Total production of natural gem and industrial diamonds:	102 425 000
Total production of synthetic diamonds:	336 000 000
Total	438 425 000

Approximately 55 million carats of the natural diamond production are industrial diamonds. Adding this figure to the 336 million carats of synthetic diamonds, total industrial diamond production counts for 90% of the total annual diamond production, with gemstones making up the remaining 10%. Table 5 lists natural diamond production for the major producing countries in 1990.

Table 4. World Reserves of Industrial Diamonds (in Millions of Carats)

	Reserves	Reserve Base*
Australia	500	900
Botswana	125	200
Brazil	5	15
South Africa, Republic of	70	150
Zaire	150	350
Other Market Economy Countries	40	65
China	10	20
USSR	80	200
World Total	980	1 900

Adapted from Austin (1991).
*See chapter Introduction for explanation of reserves and reserve base.

Table 5. Total World Natural Diamond Production 1990

Rank	Carats
Australia	36 000 000
Zaire	19 430 000
Botswana	17 351 000
USSR	15 000 000
South Africa, Republic of	8 525 000
Brazil	1 500 000
Angola	1 100 000
China	1 000 000
Namibia	763 000
Ivory Coast	600 000
Central African Republic	447 000
Ghana	141 000
Tanzania	140 000
Venezuela	130 000
Guinea	127 000
Sierra Leone	86 000
Liberia	30 000
Indonesia	25 000
Guyana	15 000
India	15 000
Total	102 425 000

In 1986 Australia became the world's largest producer of natural diamonds, having been unknown as a diamond producer as recently as 1979. The discovery of the significantly economic Argyle lamproite occurrence in Western Australia in 1976 has led to Australia becoming the world's largest diamond producer. The grade of the occurrence is estimated at 6.8 carats/t. The USSR ranks fourth in terms of world production. Zaire, Botswana, South Africa, Angola, Namibia, the Ivory Coast, and the Central African Republic are the most important African producing countries and Brazil is the principal one in South America.

In terms of synthetic diamond production, figures (in carats) available for 1989 are as follows:

USA	90 000 000
South Africa	60 000 000
Ireland	60 000 000
USSR	41 500 000
Japan	25 000 000
Sweden	25 000 000
China	15 000 000
Czechoslovakia	5 000 000
Yugoslavia	5 000 000
Romania	4 500 000
France	4 000 000
Greece	1 000 000
Total	336 000 000

EXPLORATION

The first step in diamond exploration is selection of the locality to be explored. Naturally, favored areas are those where diamonds have been previously found in large quantities, such as countries in Africa where diamonds are abundant in both pipe and alluvial deposits and in South America in alluvial deposits. The discovery of kimberlite host rock in the USSR in 1954 led to the discovery of the Zarnitsa pipe. In 1955 the economically significant Mir pipe was discovered. In South Africa the discovery of the Venetia kimberlite in 1980 was the first significant kimberlite discovery since the Finsch pipe was found in 1961. Venetia is expected to have an open pit life of 20 years and to produce 4 million carats of diamonds per year. The diamond discoveries in the 1970s in Western Australia saw the start of Australia developing into the world's leading diamond producer. Argyle AK1 is a diamondiferous lamproite rather than a kimberlite and has a grade of 6.8 carats/t (Harben and Noetstaller, 1991). The Argyle discovery has naturally led to Australia becoming an important diamond exploration target.

In Botswana, Jwaneng, Letlhakane, and Orapa are at present the three producing diamond mines with ongoing exploration being undertaken in an effort to discover further economic kimberlite deposits. More speculative exploration has taken place in Canada and the United States where kimberlites have been found as well as alluvial diamonds. Since 1976 diamond exploration has continued unabated in the United States by companies such as Dow Chemicals, Kennecott, Cominco American, Superior, Sibeka, and others. More than 130 kimberlite occurrences have been discovered in an area extending from west of Boulder, CO, to the Iron Mountain District, Wyoming. In addition, at least eight kimberlite occurrences have been discovered in the Upper Peninsula of Michigan. To date none of the occurrences discovered has been proven to be economically viable. In Arkansas, the Crater of Diamonds discovered in 1889 has been reclassified as a lamproite occurrence. It has been intermittently mined over the past 60 years and has produced some 100 000 carats of diamonds. The last serious mining attempt was in 1956. Currently the Crater of Diamonds is a tourist attraction where tourists find up to 1 000 diamonds per year. Four mining companies are at present attempting to gain permission to drill the occurrence in order to determine its size and economic potential. While companies do not freely disclose their annual exploration budgets, it is estimated that probably 10 to $20 million is spent annually on diamond prospecting in the United States. It would be difficult to assess worldwide expenditure on diamond exploration; however, the largest consistent expenditure is by De Beers who is expected to spend $60 million on diamond exploration worldwide in 1991.

Kimberlite and lamproite are frequently intruded along zones of structural weakness in the earth's crust. For this reason, examining LANDSAT or SLAR imagery, and defining major lineaments, may help select target areas for more detailed study (Lampietti and Sutherland, 1978).

Geophysical exploration, particularly airborne and ground magnetic surveying, is adaptable to diamond exploration. The kimberlite pipes, being ultramafic rocks, usually have more magnetite than the rocks surrounding them, and they have a characteristic magnetic pattern reflecting their shapes, usually a circular or elliptical magnetic high (Gerryts, 1970). Electrical resistivity results reveal that weathered kimberlite is relatively conductive (25 to 75 Ω m) compared to intruded Precambrian granites (90 to 2 300 Ω m) in the Wyoming-Colorado area (Hausel, 1979). In areas known to contain kimberlite pipes, their locations can sometimes be pinpointed by panning alluvial material in streams. If characteristic heavy minerals of kimberlites are found (such as diamond, perovskite, ilmenite, chromite, chrome diopside, and pyrope garnet), these can be traced upstream to the source (Brown et al., 1967). The careful determination of exact composition of pyrope garnets and ilmenites may help differentiate diamond-bearing from barren kimberlites (Lampietti and Sutherland, 1978).

Geochemical exploration makes use of nickel and columbium content of stream sediments and soils. Anomalous halos have been detected in surface soils several hundred feet downslope from kimberlite outcrops, and stream sediment anomalies exist several thousand feet downstream from outcrops (Gregory and Tooms, 1969). However, geochemical methods are not widely used because heavy mineral sampling is more efficient and simpler (Lampietti and Sutherland, 1978, Hausel et al., 1979c).

In the average commercial deposit, approximately 25 t of rock must be mined to obtain 1 g of diamond—a high ratio of concentration, on the order of 1:5,000,000 or higher. Because of this extremely sparse distribution, bulk sampling methods are best for evaluating both pipe and placer deposits. Pitting and trenching is better than drilling. If holes are drilled, their diameter should be as large as practical to provide representative samples. Holes are frequently drilled in alluvial deposits merely to determine depths to, and extent of, diamond-bearing gravels.

EVALUATION OF DEPOSITS

Most diamond deposits contain both gem and industrial diamonds. Exploration should provide sufficient information to determine the quantity and quality of diamonds per unit volume, and the total volume of the deposit. Estimates are made of the cost of mining and recovery of the diamonds. A feasibility study can then be made to determine whether the deposit can be mined profitably.

PREPARATION FOR MARKETS

Mining Methods

During the second half of the 19th century before the diamond-bearing kimberlite pipes were discovered in South Africa, diamonds were produced entirely from alluvial deposits in open pits by use of very primitive tools and techniques. Picks and shovels were practically the only tools used for mining, and the hand held washing pan was used for concentration, with hand sorting to recover the diamond from the washed concentrate.

The discovery of kimberlite pipes necessitated the development of completely new methods of mining and recovery. These pipes eventually have to be mined underground with methods adaptable to treating large quantities of material.

Pipe Mining: A typical diamond pipe is roughly cylindrical in shape and from 100 to 900 m in diameter. However, the pipe at Orapa, Botswana, oval in outline, is about 1.6 km long and 900 m across (Hoppe, 1980). Pipes may be of many shapes. For example, Jwaneng in Botswana is actually two cylinders connected at depth. Operations usually begin with surface excavation of weathered peridotite or yellow ground. As mining proceeds over the entire outcrop of pipe and the pit becomes deeper, the walls begin to cave, filling the pit with rubble. To continue mining, it is necessary to mine underground. This is done by sinking shafts in the adjacent country rock and driving haulageways at 180 m intervals to connect the shafts with the diamond-bearing ground.

From the shafts, raises are driven and intermediate levels are established at about 12 m intervals. These are driven horizontally into the solid kimberlite below the open pit. This system is expanded to form a series of chambers through excavation of the diamond-bearing rock. This method combines shrinkage stoping and sublevel caving (Parkinson and Dickinson, 1952). The material is loaded into cars, hauled to the shaft, and hoisted to the surface. In some pipes, mining has reached great depth. The Kimberley mine, in South Africa, has been mined to a depth of nearly 1 200 m.

The mining method described above is called chambering. Another variation is block caving, a system in which some material is extracted leaving inverted cone-shaped openings that eventually collapse and cave the block of blue ground from above. The material is drawn away through drifts by mechanical scrapers, then hoisted to the surface. Block caving is safer, faster, and more economical than chambering.

The first attempt at block caving in diamond mining was conducted by De Beers at the Bultfontein mine in 1955. The Jagersfontein mine was completely converted to block caving in 1958, and all the Kimberley pipes are now mined by this method (Bruton, 1978).

Open Pit Mining: Open pit mining methods are relatively simple, differing from one mine to another mainly in the type of machinery used. All kimberlite mines are first operated as open pits. Only when open pit mining becomes so deep as to be inefficient and unsafe does underground mining begin. A brief description of the method used at the Finsch mine in South Africa will give a general idea of the technique used (Fig. 4).

The Finsch pipe is roughly circular, 53 by 450 m, with almost vertical sidewalls of country rock. The overburden, consisting of topsoil and banded ironstone, from 3 to 12 m thick, has to be removed to expose the weathered kimberlite, or yellow ground, which extends to a depth of about 30 m. The blue ground below continues to an indefinite depth, at least 270 m. Blastholes 14 m deep are drilled 25° from the vertical on levels. The holes are charged with explosives and blasted. The broken kimberlite is removed by mechanical shovels and trucked to the treatment plant. As mining proceeds, the waste rock from the sidewalls of the pipe is systematically cut back and removed to prevent it from falling into the pit. This waste is dumped outside the perimeter of the mine.

It is expected that open pit mining at Finsch will continue for approximately 20 years reaching a depth of 275 m before a change over to underground methods is required.

In other large scale surface mining operations in the Republic of Zaire, Botswana, Angola, South-West Africa (Namibia), and Sierra Leone other modern equipment is used, such as bucket wheel excavators, draglines, front end loaders, shovels, scraper loaders, and giant trucks, to move overburden and to mine and transport the ore.

Benching: A combination of open pit and underground mining is the method that has been used at the Premier mine in the Transvaal since 1945. The 850 by 450 m oval pipe is divided into two parts by a 60 m wide quartzite mass reaching to a depth of more than 300 m. The blue ground on both sides of the quartzite divider is blasted by benching and dropped below into inverted cone-shaped openings cut into the blue ground at a current depth of 355 m. The benches are mined to form an overall slope of 65°, steep enough to permit the ore to gravitate to the cone level below.

Placer Mining: Placer diamonds, which have been eroded from pipes, are found in both riverbed and beach deposits. At some places, diamonds are found in marine terraces elevated above the sea; elsewhere, in existing beaches or in beaches now submerged at some depth beneath the sea (Murray, 1970, Oostdam, 1970, Bruton, 1978).

The most extensive placer diamond mining is at Consolidated Diamond Mines (CDM) in South-West Africa (Namibia). CDM operates over a 80 km stretch at Orangemund, reported to be the richest gem-producing area in the world, and further south at Kleinzee.

The first step in developing an ordinary placer deposit is to sink prospecting boreholes to determine the thickness of the sand overburden and find the geological character of the diamondiferous gravel. In a typical deposit, bulldozers remove the overburden and expose the gravel. Prospecting trenches about a meter wide and 1.6 km long are then cut. These are sampled by simple hand sorting of diamonds from the gravel to determine the richness of the deposit.

If the deposit proves to be large and sufficiently rich, a formidable earth moving operation begins. Giant excavators strip away the overburden at the rate of as much as 200 tph to expose the diamond-bearing gravel, which typically averages 1 to 1.2 m thick. The gravel is either bulldozed into convenient stockpiles for mechanical loading or is loaded directly into dump trucks for delivery to the nearest screening plant. Because a high percentage of diamonds in gravel occurs on or near bedrock, it is essential to clean the rock very thoroughly. All crevices and holes in the bedrock are searched carefully either by hand labor or by mechanical devices that have been devised specifically for this purpose.

Fig. 4. Aerial view of the Finsch open pit mine and processing plant in the Republic of South Africa (*courtesy:* De Beers Consolidated Mines, Ltd.).

At the field screening plant, the diamond-bearing gravel is placed on vibrating screens, which reject 80% as waste and retain 20% as diamond-rich residue to be taken by truck or train to a central treatment plant.

Placer deposits in submerged ocean beaches off the coast of South-West Africa (Namibia) were mined for several years. The diamond-bearing gravel was hydraulically dredged aboard a ship and the diamonds were recovered. This ocean diamond mining proved uneconomic and was discontinued in 1970 (Bruton, 1978).

Processing Techniques

Lode Deposits: Material from lode deposits is crushed if necessary and concentrated by washing, winnowing, screening, and hand sorting, or by combinations of these methods. In some localities diamond-bearing material from dumps and waste heaps is frequently reworked several times.

Methods of concentrating diamonds in large-scale operations have become more mechanized with time. In the early years of shallow mining the yellow ground and the partially weathered blue ground required little crushing. The material, when exposed to the atmosphere, gradually disintegrated if allowed to remain exposed in open cuts, or if spread out in nearby areas for as many as 18 months. Thus no crushing was required before processing. However, as mining has progressed deeper into unweathered blue ground, the material is crushed in either jaw or gyratory crushers or by corrugated rolls, generally in two or more stages separated by screening. Figs. 5 and 6 (Anon., undated) show three general types of flowsheets.

Devices used for this purpose include the rotary washing pan, jigs, heavy media separators, and hydrocyclones. Each of these machines makes use of differences in specific gravity to separate diamonds and other heavy minerals from the lighter minerals. For example, in the rotary washing pan—low, circular pan of large diameter—the crushed material is mixed with mud to form a puddle of material with a density of about 1.25. Although far below the density of even the light mineral fraction of the kimberlite, the density of this puddle appears to play an important part in the process. The waste is moved off the top into a central rotating weir by rotary toothed blades. The diamonds and other heavy minerals settle to the bottom and are drawn off periodically from the outer rim of the pan. The concentrate is then classified, and either jigged or separated by heavy media to produce a rough concentrate usually amounting to about 1% of the feed from the mine. Larger, more modern plants have replaced the jigs with heavy media separators (Fig. 6). In these, a slurry of ferrosilicon powder in water with a density of 2.7 to 3.1 is used to float the light particles, separating them from the heavier minerals, including diamonds, which sink and are collected as a final gravity concentrate for further treatment.

The final concentrate was formerly processed by separating the diamonds from the other heavy minerals by hand sorting. Today most operators use one of several greased surface methods. The principle of greased surface concentration was discovered at Kimberley in 1896 when it was observed that diamonds from freshly mined kimberlite adhered tenaciously to axle grease with which they accidentally came into contact, whereas the bulk of the associated heavy minerals did not. This important discovery greatly reduced the amount of hand sorting required to separate the diamonds from other heavy minerals.

One type of concentrator is the vibrating grease table, which consists of a stepped surface made of heavy gage aluminum (Fig.

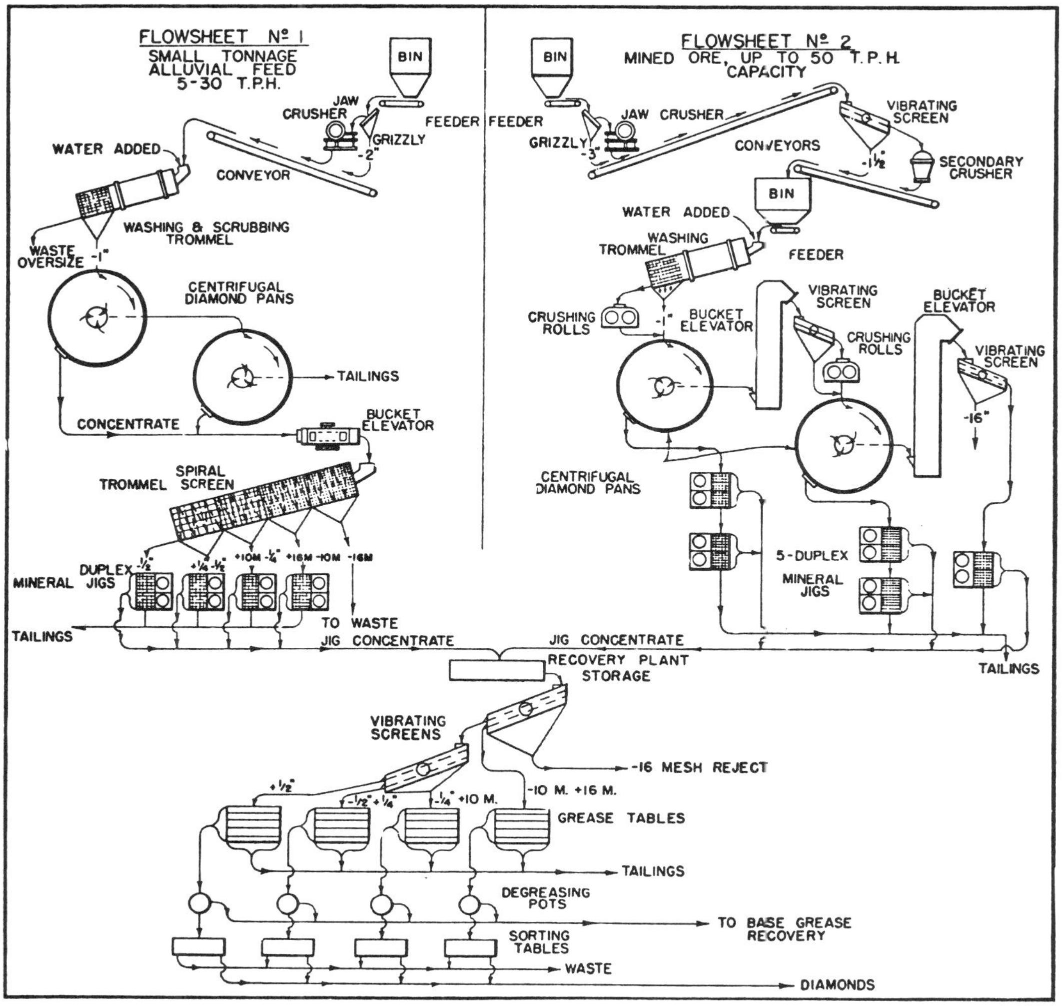

Fig. 5. General flowsheets for small to medium operations (*courtesy:* Denver Equipment Co.).

5). The surfaces of the steps are covered with a thick layer of a heavy grease. The table is vibrated while the diamond-bearing concentrate, mixed in water, is passed over the steps. The diamonds adhere to the grease; the waste minerals are washed away. The diamonds are collected by stopping the process and scraping them with some of the grease off the surface of the table. The table is regreased and the process continues. The scrapings are boiled in water to remove the grease from the diamonds.

At Orapa, Botswana, the diamonds do not adhere to the grease because of a surface coating. Instead of greased tables or belts, an X-ray unit spots the diamonds, activating a micro-second air jet that sweeps the diamond from the belt into a collector (Hoppe, 1980).

Placer Deposits: Diamonds are recovered from loose placer materials and from the weathered rock of lode deposits by washing, dry winnowing, screening, panning, jigging, tabling, and hand picking, generally in some combination. In partially consolidated gravels, some crushing is required. Usually simple devices, such as log washers, are sufficient to break up the gravels and separate the clay fractions. Other equipment used includes baskets, hand pans, grease belts, and tables.

The grease belt is the greased-surface concentration method best adapted to alluvial materials. In fact, it was developed to process diamond-bearing material from certain alluvial deposits. A modification of the normal process was necessary because alluvial diamonds, particularly those from marine terraces, frequently have a film of mineral salts on their surfaces, which renders them wettable in water. A wet diamond will not adhere to grease.

The De Beers Diamond Research Laboratory developed a method using a soap solution made of corn-acid oil and caustic soda. Treatment with this solution produces a water-repellent surface on the diamond but not on the other minerals. The method also incorporates use of a greased belt, which permits a continuous rather than a batch process (Fig. 5). Both the soap solution and the grease are added to the belt surface as the belt is in operation. The method is in use at the Consolidated Diamond Mines of South-West Africa (Namibia); at the Finsch and Williamson mines in South Africa; at the Luxilo plant in Angola; and at the State Alluvial Diggings in Cape Province.

Further treatment is required when the grease belt process is used on finer sizes of alluvial material. In order to clean surfaces and disintegrate softer materials to slimes, the degreased concentrates are processed in ball mills. The material is screened, and the clean coarser diamond particles are further concentrated by electrostatic separation.

Material from alluvial deposits is also concentrated by X-ray separation. Diamonds tend to luminesce in an X-ray beam, whereas

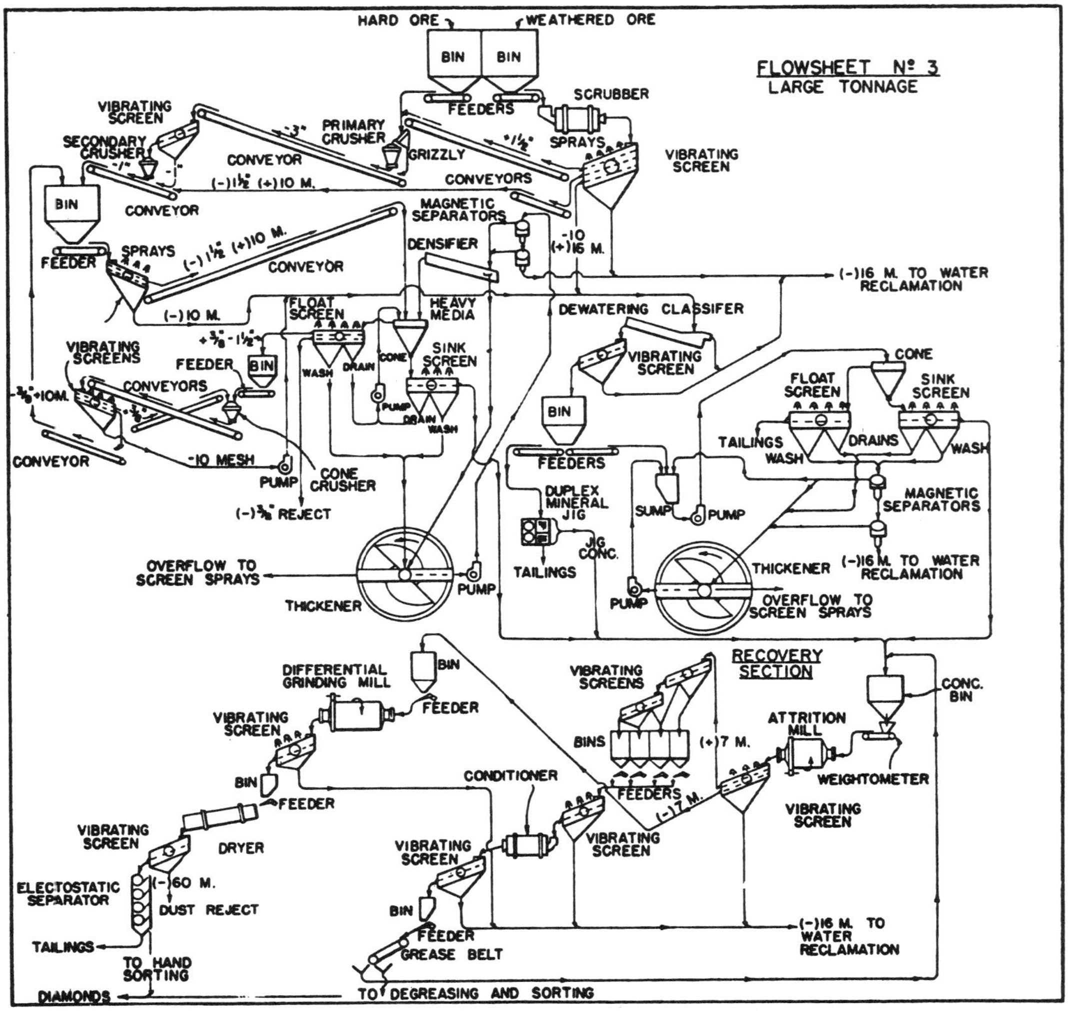

Fig. 6. General flowsheet for a large tonnage operation (*courtesy:* Denver Equipment Co.).

most of the associated minerals do not. Luminescence excites a photomultiplier that triggers a gate that diverts the diamonds from the path of the gravel passing through the machine. Calcite sometimes luminesces like diamond, but this effect is neutralized by use of a suitable filter.

TRANSPORTATION

Essentially all diamonds are shipped via air, which permits delivery almost anywhere in the world in a matter of hours. The costs are usually small compared to the value of the diamond. This favorable transportation factor also applies to grinding wheels, saws, bits, and other products.

MARKETING

Sale of rough diamonds are largely made through De Beers Central Selling Organisation (CSO). In 1990 the CSO sold US$4.17 billion worth of diamonds, or approximately 90% of world sales. Most of the world's rough diamond production not sold through the CSO finds its way to Antwerp, Belgium.

The Diamond Trading Company was founded in 1934. Together with the Diamond Producers Association, it formed the nucleus of CSO, a group of marketing companies through which all the principal producers sell diamonds in cooperative arrangement.

Prior to formation of the CSO, the diamond business was subject to slumps about every eight years. In the mid-1920s, the discovery of high quality alluvial diamonds at Lichtenburg, Namaqualand, and South-West Africa (Namibia) flooded the market with diamonds that could be produced for significantly lower costs than those from deep mines at Kimberley (Bruton, 1978), resulting in extremely low earnings from the Kimberley production.

During the world depression in the early 1930s, the situation worsened. De Beers was unable to sell enough diamonds to cover expenses. The mines were all shut down, and Kimberley was almost a ghost city.

Because of these conditions, Sir Ernest Oppenheimer, Chairman of De Beers, was able to convince a majority of producers to unite to control production and regulate sales (Bruton, 1978). Since the formation of CSO, all producers have benefited from the stability that was brought to the marketplace.

FUTURE CONSIDERATIONS AND TRENDS

Diamond Synthesis

During the 19th century many attempts were made to synthesize diamonds. Broadly speaking, three approaches were made—precipitation from saturated solutions at high temperatures and pressures, chemical reaction at high temperatures and pressures, and the application of simultaneous heat and pressure to graphite in a press (Bundy, 1962). The last of these, subjecting graphite to high temperatures and pressures in the presence of a catalyst metal is the method now in commercial use (Ringwood and Major, 1966).

Moissan and Hannay are probably the most notable of the early scientists who made these attempts, but neither was successful in manufacturing diamonds. The reason for their failure was that the essential material for successful synthesis—tungsten carbide used in pressure chambers—was not available at the time. The first successful synthesis of diamonds is attributed to a Swedish team of scientists (Liander, 1955) from Allmänna Svenska Elektriska Aktiebolaget (ASEA) in 1953. However, their process at first was not thought to be commercially viable, and no public announcement was made. Early in the 1950s, General Electric in the United States initiated a research program in superpressure reaction, which ultimately led to successful diamond synthesis in late 1954. Although this discovery was known in 1955, an order for secrecy was imposed on the General Electric process, and the apparatus patent applications restricted the publication of results until 1961.

Also in the mid-1950s, a team of scientists was assembled in the De Beers Diamond Research Laboratory in Johannesburg, South Africa, for the purpose of devising a process for diamond synthesis, and they announced success in 1958 (Legrand et al., 1980).

In the early days of diamond synthesis scientists were only interested in producing diamonds, hence, little attention was given to the quality of the product. It was soon realized that improvements in quality were necessary. Research effort for this objective has continued relentlessly. The hydraulic presses in the De Beers synthetic diamond plant in Ireland (Fig. 7) are products of such research.

A major difference between natural and synthetic diamonds is the metal content. Synthetic diamonds contain significant amounts of the metal solvent used in synthesis. The included metal is believed to exist in three forms, namely 1) macroscopic metal inclusions; 2) small platelike inclusions dominantly on the cleavage planes, and 3) substitutional metal atoms in diamonds synthesized from nickel or nickel-based alloys.

Due to differences in the coefficients of thermal expansion and the compressibilities of the metal inclusions in diamonds, the metal causes the diamonds to be stressed. The resulting strain causes birefringence which can be seen when synthetic diamonds are examined by use of a polarizing microscope. The strength of synthetic diamonds decreases with increasing metal content.

Synthetic micron-size diamonds are also produced by an explosive-shock technique first achieved by Stanford Research Institute and developed commercially by E.I. du Pont de Nemours & Co., Inc. The product is a polycrystalline type of diamond, which can, at this time, be manufactured only in micron sizes (De Carli and Jamieson, 1961).

Synthetic diamonds are being produced today in several countries throughout the world. While the De Beers Industrial Diamond Division and General Electric's Specialty Material Division are still the world's leading producers, many new companies have entered the sales fray. In the United States, there are four manufacturers of diamonds. There are also producers in the Soviet Union, Japan, Romania, People's Republic of China, Republic of Korea, India, Czechoslovakia, Germany, Greece, Ireland, Finland, Sweden, and the Republic of South Africa. While many of these producers are not able to turn out the high quality products that De Beers and General Electric make, they have caused an oversupply condition in the marketplace, resulting in substantial price reductions for most synthetic diamonds.

Fig. 7. These 10 000-t hydraulic presses in the De Beers plant in Shannon, Ireland, convert graphite into diamond by the simultaneous application of ultra high pressure and temperature.

Recent Developments

Continued research and development have improved the quality and lowered the cost of synthetic stones. Temperatures and pressures required have been lowered. Crystals can be made with better shapes. Recently, methods have been developed to recrystallize small synthetic diamonds to form gem-quality diamonds of a carat and larger. However, this has not yet been done profitably.

Sintered masses of diamond material have been produced in predetermined shapes, but only in small quantities. This process is reported to have been developed in both the USSR and the United States.

A most recent process reported in the United States is the growing of diamonds from hydrocarbon gas at atmospheric pressure and moderate temperature.

A major technological development expected and overdue in diamond synthesis is commercial production of large industrial stones to replace the use of natural stones in such applications as drilling bits (Adams, 1979).

Today production of synthetic stones far exceeds the production of natural stones. Approximately 336 million carats are manufactured by De Beers in South Africa, Ireland, and Sweden; by General

Electric in the United States; by the USSR; and by Japan. Small quantities are also produced in other countries.

Following the successful manufacture of small diamond crystals, research continued on crystallization of larger stones suitable for jewelry. Finally, in 1970 General Electric laboratories succeeded. A few stones of very high quality were produced, some as large as one carat. These can be seen on display at the Smithsonian Institution in Washington, DC. These results proved that it is possible to manufacture synthetic stones of fine quality large enough for use in jewelry. However, the process is so costly that the synthetic diamonds are far more expensive than natural stones of the same quality (Legrand et al., 1980, Bruton, 1978).

While synthetic stones can be made in sizes of 2 carats or larger, the principal market is in the 0.6 to 0.8 mm size (Austin, 1989).

Polycrystalline or composite diamond inserts are produced by sintering together small particles of diamond at temperatures above 1 400°C and at pressures on the order of 60 000 atmospheres.

The polycrystalline tool itself consists of a dense sintered diamond layer which is bonded onto a tungsten carbide substrate. In natural diamond tools chipping occurs by the diamonds dividing along naturally occurring cleavage planes which are parallel to the octahedral faces of the diamond. Wear of this nature is absent from polycrystalline diamond tools because the diamond grains, which make up the sintered diamond layers, are randomly orientated.

These polycrystalline tools are ideally suited for use in cutting tools for machining of nonferrous materials, and in abrasive wheel dressers, wear parts, wire dies, and rotary drill bits.

Chemical vapor deposition (CVD) diamonds have received a lot of publicity in recent years. The CVD technology produces a thin coating of synthetic diamond on a variety of substrates. Extensive research has been conducted not only on improving the synthesizing process itself, but also on the interaction between a material and the diamond film, on the effects of wear, tear, and heat on cutting tools, on wear surfaces, on heat sinks, and on electronic instruments.

The following chart gives the crystallographic data of the starting material and of Syndite (a trade name for an extremely hard and tough material produced by De Beers by sintering synthetic diamond micro-grit).

Material	Plane hkl	B_o [deg]	L [μm]
Synthetic diamond	111	0.13	—
	220	0.15	—
	311	0.18	0.5
Natural diamond	111	0.15	—
	220	0.18	—
	311	0.18	0.5
Syndite	111	0.18	0.088
	220	0.35	0.050
	311	0.54	0.037
	222	0.54	0.046

$$L = \frac{0.9\lambda}{B_o \cos\vartheta_B}$$

L = crystallite size
ϑ_B = Bragg angle
λ = wavelength of the radiation
B_o = half-peak width of x-ray beam

Substitutes

The same technique used in diamond synthesis has been applied to the synthesis of cubic boron nitride (CBN). Cubic boron nitride is produced by the De Beers Industrial Diamond Division under the trade name Amber Boron Nitride and by General Electric under the trade name Borazon. Today cubic boron nitride is available in several grades with varying physical properties which, when correctly applied, make it an excellent abrasive for grinding hardened and difficult-to-grind steels.

The chart below compares some properties of Amber Boron Nitride with those of other abrasive materials, using average values.

Property	Amber Boron Nitride	Diamond	Al_2O_3	SiC
Indentation Hardness (Knoop) (kg/mm^2)	4 500	8 000	2 100	2 500
Reactive (at high temperature) to	Water	Oxygen	Unreactive	Iron
Density (g/cm^3) Unclad	3.45	3.52	3.7/4.0	3.2
Density (g/cm^3) Clad	5.2 ± 0.1	—	—	—

With its excellent thermal stability, Amber Boron Nitride may be heated in air up to 1 400°C before its cubic structure reverts to hexagonal boron nitride. However, cubic boron nitride can react with water at temperatures in excess of 800°C, which can be reached easily at the point of cutting. Care should be taken, therefore, in wet grinding operations where water is used as a coolant. Reducing the water content in the grinding fluid can substantially increase the life of the grinding wheel.

Polycrystalline CBN cutting tools are manufactured by the De Beers Industrial Diamond Division under the trade name Amborite, by General Electric under the trade name BZN Compax, and by Sumitomo of Japan. The chart following lists the properties of Amborite. Results were extracted from non-standard tests described by Pipkin et al. (1980).

The CBN polycrystalline cutting tools are finding increasing application for machining hardened steels (+45 Rc), high nickel content alloys (aerospace alloys), and difficult-to-machine ferrous materials.

Commercial synthetic rutile is marketed under the name titania. In appearance, titania passes as an imitation diamond, being transparent and possessing great brilliance and fire. However, it frequently has an objectionable yellow tinge, and its hardness is less than that of quartz, so that faceted stones are subject to wear and breakage when worn in jewelry.

A relatively new imitation diamond is yttrium aluminum garnet (YAG). YAG was advertised as being only 1.5 less hard than diamond (8.5 compared to 10 on the Mohs scale). However, tests by De Beers show that YAG is only one-tenth as hard as diamond by impact tests and abrades 3000 times more readily (Caveney, 1976). YAG is produced by mixing ingredients in a platinum crucible. The seed thus formed grows as it is rotated at a temperature of approximately 1 925°C. The index of refraction is considerably less than in diamond. The dispersion is reasonably high but lower than that of diamond. The color, hardness, and freedom from impurities make the stone an inexpensive substitute for gem diamond.

Tool Material	Knoop Hardness (GPa)	Transverse Rupture Strength (GPa)	Ultimate Tensile Strength (GPa)	Young's Modulus (GPa)	Modulus of Rigidity (GPa)	Bulk Modulus (GPa)	Poisson's Ratio —
Amborite	40.5	0.57	0.45	680	279	405	0.22
Tungsten Carbide K10	17.9	1.70	1.01	630	258	375	0.22
Aluminum Oxide	17.1	0.26	0.24	365	147	234	0.24

STOCKS, PRICES AND PRODUCTION

Government Stockpile

Crushing bort, dust and powder, and industrial stones have been stockpiled by the US government since 1946. As of Dec. 31, 1989, the government stockpile contained 22.0 million carats of crushing bort, dust, and powder, 7.78 million carats of industrial stones, and 25 473 pieces of dies and small pieces (Austin, 1989).

Prices and Costs

Prices for industrial diamonds have decreased over the years (Adams, 1979). From 1954 to 1977 average prices for bort and powder decreased from 3.14 to $2.08 per carat. Average prices for stones for the same period decreased from 23.13 to $6.20 per carat. This decrease is the result of greater efficiency developed through large scale production and use of improved techniques.

US and World Production

US production, consisting of synthetic bort, powder, and dust, was 90 million carats in 1990. World production of synthetic diamond stones was 336 million carats. Total world production in 1990 for all synthetic and natural industrial diamonds was about 391 million carats.

Taxes and Tariffs

No special taxes are placed on the diamond-producing or diamond-products industries. US companies receive a 14% depletion allowance for both domestic and foreign production of natural diamonds. There is no import duty on either natural or synthetic industrial diamonds (i.e., synthetic dust, grit, and powder).

Unsorted diamonds, both gem and industrial, can be imported duty free.

A tariff rate of 4.9% *ad valorem* is in effect, as of Jan. 1, 1991, for synthetic diamond stones for industrial use, except for imports from Canada (1.9%); the USSR (30%); and free for GSP countries (including most developing countries), Israel, and Caribbean Basin countries.

SECONDARY SOURCES AND RECOVERY

The main sources of recovery for diamonds are the diamond grit and powders salvaged from industrial wastes by users of diamond grinding wheels, saws, and lapidary compounds. This amounts to about 3 million carats annually. The other major source is the substantial quantity of stones recovered from broken or worn diamond tools and drilling bits. These are returned to the manufacturer for reuse.

ECOLOGY

The diamond industry, being primarily based in countries outside the United States, has been faced with few problems regarding ecologic restraints, but in recent years concern for environmental protection has become more widespread in Africa.

FUTURE OF DIAMONDS

In 1967, the machining of cast iron and steel was a process virtually untested. De Beers DXDA-MC, a metal-clad synthetic, was specially formulated to grind certain types of steel. Wheelmakers have reported that their customers have had excellent results in grinding D series tool steels, Ferro-Tic, cast iron, titanium, stainless steels, and combination steel and carbide dies.

Scientists at the De Beers Diamond Research Laboratory feel that research indicates quite convincingly that the characteristics of natural grit can be altered in ways that will permit production of synthetic materials of virtually any type—hard or friable, octahedrons or slivers, and particles of varying sizes and qualities.

The indicated future of the industrial diamond is indeed bright. By 1976, the annual world consumption exceeded 100 million carats, an amount previously predicted as being reached by the end of the decade. It is believed that research will pave the way for an almost incredible growth.

ACKNOWLEDGMENTS

The authors wish to thank Industrial Diamonds Information Bureau, London, and Peter Silveri Associates, New York, for their help in supplying information and illustrations for this chapter. This chapter is an update based on the Diamonds chapter by K. Reckling, R.B. Hoy, and Stanley J. Lefond in the 5th edition of *Industrial Minerals and Rocks*.

BIBLIOGRAPHY AND REFERENCES

Anon., undated, "Properties of Diamonds," De Beers Industrial Diamond Div., London, 4 pp.

Anon., undated, "Concentration of Diamonds," Denver Equipment Co., Flowsheet Study, Bulletin No. M7-F74, 8 pp.

Anon., 1971, *1971 International Diamond Annual*, Johannesburg.

Anon., 1980, "Principles of a Resource/Reserve Classification for Minerals," Circular 831, US Bureau of Mines and US Geological Survey, 5 pp.

Anon., 1981, "Cominco American Will Begin Diamond Operation," *The Mining Record*, July 1, p. 4.

Anon., 1991, *Mineral Commodity Summaries,* US Bureau of Mines.

Adams, W.T., 1979, *The History and Use of Diamond*, Methuen & Co., Ltd., London, 166 pp.

Anders, E., and Lipschutz, M.E., 1966, Reply (to N.L. Carter and G.C. Kennedy's reply, 1966 to critique of "Origin of Diamonds in the Canyon Diablo and Novo Urei Meteorites"), *Journal of Geophysical Research*, Vol. 71, No. 2, pp. 673–674.

Austin, G.T., 1989, "Abrasive Materials. Industrial Diamonds," Preprint, *US Minerals Yearbook*, US Bureau of Mines, pp. 9–16.

Austin, G.T., 1991, "Diamond (Industrial)," *Mineral Commodity Summaries*, US Bureau of Mines, pp. 48–49.

Bardet, M.G., 1973, *Géologie du Diamant*, Mémoires du B.R.G.M., No. 83, 3 vol.

Berman, R., 1962, "Graphite Diamond Equilibrium Boundary," *1st International Congress on Diamonds in Industry*, Ditchling Press, Sussex, England, pp. 291–295.

Berman, R., and Simon, F., 1955, "On the Graphite-Diamond Equilibrium," *Zeitschrift fuer Electrochemie*, Vol. 59, pp. 333–338.

Bovenkerk, H.P., et al., 1959, "Preparation of Diamonds," *Nature*, Vol. 184, pp. 1094–1098.

Brown, D.D., et al., 1967, "The Source of Alluvial Kimberlite Indicator Minerals in the James Bay Lowland," Ontario Dept. of Mines Misc. Paper, MP-7, 35 pp.

Brummer, J.J., 1978, "Diamonds in Canada," *CIM Bulletin*, Oct., pp. 64–79.

Bruton, E., 1978, *Diamonds*, 2nd ed., Chilton Book Co., Radnor, PA, 532 pp.

Bundy, F.P., 1962, "Direct Conversion of Graphite to Diamond in Static Pressure Apparatus," *Science*, Vol. 137, pp. 1057–1058.

Bundy, F.P., et al., 1955, "Man-Made Diamonds," *Nature*, Vol. 176, pp. 51–55.

Carter, N.L., and Kennedy, G.C., 1966, "Origin of Diamonds in the Canyon Diablo and Novo Urei Meteorites—A Reply" (to discussion of 1964 paper by E. Andres and M.E. Lipschutz), *Journal of Geophysical Research*, Vol. 71, No. 2, pp. 663–672.

Caveney, R.J., 1976, "The Hardness of Diamond and Gem Diamond Substitutes," *Industrial Diamond Review*, Feb., pp. 45–47.

Chandler, H.P., 1964, "Industrial Diamond—A Materials Survey," Information Circular 8200, US Bureau of Mines, 150 pp.

Clarke, R.G., 1975, "Diamond-Industrial," *Mineral Facts and Problems*, Bulletin 667, US Bureau of Mines, pp. 325–338.

Collins, R.S., 1972, "Diamonds, Mining, Annual Review," *Mining Journal*, London, pp. 121–126.

Cooper, J.D., 1970, "Diamond-Industrial," *Mineral Facts and Problems*, Bulletin 650, US Bureau of Mines, pp. 953–965.

Davidson, C.F., 1964, "On Diamantiferous Diatremes," *Economic Geology*, Vol. 59, No. 7, pp. 1368–1380.

Dawson, J.B., 1960, "Comparative Study of the Geology and Petrology of the Kimberlites of the Basuto Province," PhD thesis, Department of Geology, University of Leeds, England.

Dawson, J.B., 1962, "Basutoland Kimberlites," *Geological Society of America Bulletin*, Vol. 73, pp. 545–560.

Dawson, J.B., 1964, "An Aid to Prospecting for Kimberlites," *Economic Geology*, Vol. 59, No. 7, pp. 1385–1386.

Dawson, J.B., 1968, "Recent Researches on Kimberlite and Diamond Geology," *Economic Geology*, Vol. 63, No. 5, pp. 504–511.

Dawson, J.B., 1984, "Petrogenesis of Kimberlite," *Kimberlite Occurrence and Origin: A Basis for Conceptual Models in Exploration*, J.E. Glover and P.G. Harris, eds., Geology Department and University Extension, University of Western Australia, Publication No. 8, pp. 103–111.

De Angelis, W.M., 1974, "Diamonds in the Glass Industry," *Proceedings: Diamond: A Partner in Productivity*, Symposium, Nov. 11–12, Industrial Diamond Association of America, Washington, DC.

De Carli, P.S., and Jamieson, J.S., 1961, "Formation of Diamond By Explosive Shock," *Science*, Vol. 133, pp. 1821–1822.

Dickinson, S.K., Jr., 1966, "The Chemical Erosion of Diamond by Boron—An Analogy of Graphitization (Abstract)," *American Mineralogist*, Vol. 51, Nos. 1–2, p. 262.

Fernandez, L.L., 1968, "Recent Developments in the Diamond Mining Industry in British Guiana," *Transactions*, 4th Caribbean Geologic Conference, pp. 415–417.

Forner, H., 1978, "The Precision Cutting of Hard, Brittle Materials Using Diamond Sawblades," *Industrial Diamond Review*, Apr., pp. 136–141.

Frank, F.C., 1969, "Diamonds and Deep Fluids in the Upper Mantle," *The Application of Modern Physics to the Earth and Planetary Interiors*, S.K. Runcorn, ed., Wiley-Interscience, London and New York, pp. 247–250.

Garlick, H.J., 1979, "Australian Diamond Prospects—The Story So Far," *Industrial Minerals*, No. 137, Feb., pp. 17–26.

Gerryts, E., 1970, "Diamond Prospecting by Geophysical Methods; A Review of Current Practice," *Mining and Groundwater Geophysics/1967*, Economic Geology Report No. 26, Canadian Geological Survey, pp. 439–446.

Gleason, J.J., 1974, "The Evaluation of Diamond Uses in Industry: Past, Present, and Future," *Proceedings: Diamond-Partner in Productivity*, pp. 153–157.

Gold, D.P., 1968, "Natural and Synthetic Diamonds and the North American Outlook," *Earth and Mineral Science*, Vol. 37, pp. 37–43.

Gregory, P., and Tooms, J.S., 1969, "Geochemical Prospecting for Kimberlites," International Geochemical Exploration Symposium, *Colorado School of Mines Quarterly*, Vol. 64, No. 1, pp. 265–305.

Gunn, C.B., 1967, "The Origin of Diamonds in Drift of the North Central United States—A Discussion (of paper by H.P. Schwarex, 1965)," *Journal of Geology*, Vol. 75, No. 2, pp. 232–233.

Gunn, C.B., 1968, "Provenance of Diamonds in the Glacial Drift of the Great Lakes Region, North America (Abstract)," *Bulletin*, Canadian Petroleum Geology, Vol. 16, No. 3, p. 418.

Harben, P., and Noetstaller, R., 1991, "Diamonds—Scintillating Performance in Growth and Prices," *Industrial Minerals*, Mar., 7 pp.

Hausel, W.D., et al., 1979, "Exploration for Diamond-Bearing Kimberlite in Colorado and Wyoming: An Evaluation of Exploration Techniques," Report of Investigation No. 19, Geological Survey of Wyoming, 29 pp.

Hausel, W.D., et al., 1979a, "Update on Exploration for Diamonds in Colorado-Wyoming Kimberlite Province," *AAPG Bulletin*, Vol. 63, No. 5, p. 830.

Hausel, W.D., et al., 1979b, "Preliminary Report on Exploration for Diamondiferous Kimberlites, Colorado-Wyoming," *Mining Yearbook*, Colorado Mining Association, pp. 109–120.

Hausel, W.D., et al., 1979c, "Prospecting for Kimberlites in Wyoming Using Heavy Mineral Alluvial Sampling Methods," Open File Report 79–6, Geological Survey of Wyoming, 13 pp.

Hausel, W.D., et al., 1980, "Exploration for Diamond-Bearing Kimberlite in Colorado and Wyoming—An Update," Preprint No. 80–310, SME-AIME Fall Meeting, Minneapolis, MN, 4 pp.

Herbert, S., 1976, "A New Production Plant Gets into Its Stride," *Industrial Diamond Review*, Sep., pp. 315–318.

Herbert, S., 1976a, "Superstrength Diamond Solves Small-Hole Drilling Problem," *Industrial Diamond Review*, May, pp. 170–172.

Herbert, S., 1977, "Best of Both Worlds," (Diamond Abrasive as Carbide Grinding Medium), *Industrial Diamond Review*, Dec., pp. 422–424.

Hill, B.S., 1975, "The Performance of Processed Drilling in Hard Rock," *Industrial Diamond Review*, Aug. pp. 282–288.

Hoppe, R., 1980, "Diamonds from the Kalahari," *Engineering and Mining Journal*, Vol. 181, No. 5, May, pp. 64–82.

Hunt, G., 1975, "Diamond Prospecting and Mantle Xenoliths of the Kimberlite Pipes of Elliott Co., Kentucky," *GSA Abstracts with Programs*, Vol. 7, No. 4, p. 503.

Hurlbut, P.K., 1967, "Diamonds (Abstract)," *Guidebook of Defiance-Zuni-Mt. Taylor Region, Arizona and New Mexico*, New Mexico Bureau of Mines and Mineral Resources, pp. 226–227.

Ilupin, I.P., 1977, "The Mineral Composition of Kimberlites and the Genesis from Depth in Kimberlites," *Geologiya i Geofizika*, Vol. 1, No. 12, pp. 100–105.

Johnson, C.J., Marriott, M., and von Saldera, M., 1989, "World Diamond Industry, 1970–2000," *Natural Resources Forum*, Vol. 13, No. 2, pp. 90–106.

Kennedy, G.C., and Nordlie, B.E., 1968, "The Genesis of Diamond Deposits," *Economic Geology*, Vol. 63, No. 5, pp. 495–503.

Kirkley, M.B., Gurney, J.J., and Levinson, A.A., 1991, "Age, Origin and Emplacement of Diamonds: Scientific Advances in the Last Decade," *Gems and Gemology*, Spring 1991, pp. 2–25.

Kopecky, L., 1960, "Diamond Prospects in Czech Massif," *Izvest.* Akad. Nauk SSSR, Ser. Geol., Vol. 12, pp. 46–55.

Lampietti, F.M.J., and Sutherland, D., 1978, "Prospecting for Diamonds, Some Current Aspects," *Mining Magazine*, Aug., pp. 117–123.

Lang, A.R., 1970, "Diamond Accumulation in the Upper Mantle," *Nature*, Vol. 226, No. 5243, p. 345.

Lawrence, A.L., 1967, "Field Prospecting for Diamonds in Isolated Areas," *Bulletin*, Geological Society of Liberia, Vol. 2, pp. 1–3.

Legrand, J., et al., 1980, *Diamonds: Myth, Magic and Reality*, Crown Publishers, Inc., New York, 287 pp.

Leighton, V.L., and McCallum, M.E., 1979, "Rapid Evaluation of Heavy Minerals in Stream Sediments of the Prairie Divide Area of Northern Colorado: A Tool for Kimberlite Exploration," Open File Report 79–761, US Geological Survey.

Leuria, B., 1966, "Diamond Prospecting in Lofa County," *Bulletin*, Geological Society of Liberia, Vol. 1, No. 1, pp. 27–35.

Leuria, B., 1967, "Notes on the Geology of the Lofa River South of Wea Sua, with a Comment on Diamond Dispersal and Recovery," *Bulletin*, Geological Society of Liberia, Vol. 2, pp. 4–8.

Leuria, B., and Stracke, K.J., 1966, "Diamonds and Their Occurrences in Liberia," *Bulletin*, Geological Society of Liberia, Vol. 1, No. 1, pp. 5–12.

Liander, H., 1955, "Artificial Diamonds," *ASEA Journal*, Vol. 28, p. 97.

Linari-Linholm, A.A., 1951, "Recovery of Diamonds by Electrostatic Separation," *Journal*, Chemical, Metallurgical, and Mining Society of South Africa, Vol. 52, No. 2, Aug., pp. 33–35.

Loewenstein, W., 1965, "Glaciation and the Origin and Distribution of Diamond," *Lapidary Journal*, Vol. 18, pp. 1204–1213.

Macnae, J.C., 1979, "Kimberlite and Exploration Geophysics," *Geophysics*, Vol. 44, No. 8, pp. 1395–1416.

Mathewson, W.F., et al., 1966, "An Analysis of the Coated Diamond Bond System," *Grinding and Finishing*, Vol. 12, No. 12, pp. 18–25.

McCallum, M.E., 1974, "Infrared Detection of Kimberlitic Diatremes in Northern Colorado and Southern Wyoming," *Contributions to Geology*, Vol. 13, No. 1, University of Wyoming, pp. 16–17.

McCallum, M.E., 1979, "Geochemical Prospecting for Kimberlite in the Colorado-Wyoming State Line District (Abstract)," *Abstracts for 1979*, Vol. 11, No. 6, Geological Society of America, p. 279.

McCallum, M.E., and Smith, C.B., 1978, "Minor and Trace Element Contents of Kimberlites of the Front Range, Colorado and Wyoming," Open File Report 78–1011, US Geological Survey, 20 pp.

Meyer, H.O.A., 1968, "Mineral Inclusions in Diamonds," *Yearbook 66*, Carnegie Institute, Washington, DC, pp. 446–450.

Meyer, H.O.A., and Harris, J.W., 1969, "Crystalline Inclusions in Diamonds from Ghana and Venezuela," *Symposium on Volcanoes and Their Roots*, Oxford University, p. 16.

Meyer, H.O.A., and Boyd, F.R., 1970, "Crystalline Inclusions in Natural Diamonds (Abstract)," *International Symposium on Mechanical Processes in the Mantle*, US National Academy of Sciences, p. 37.

Miller, P., 1987, *The Outlook for Diamonds: a Rejuvenated Market*, Messel, London, 210 pp.

Moore, M., 1976, "A Review of Professor Tolansky's Research on Synthetic Diamonds," *Industrial Diamond Review*, Nov., pp. 395–398.

Motti, E., 1969, "Prospecting for Diamonds and Heavy Minerals in the Alluvium of the Arabian Shield," *Mineral Resources Research*, Saudi Arabia, Mining, Petrology and Mineral Resources, Jiddah, pp. 68–69.

Murray, L.G., 1970, "Exploration and Sampling Methods Employed in Offshore Diamond Industry," *Mining and Petroleum Geology*, 9th Commonwealth Mining and Metallurgy Congress, Vol. 2, pp. 71–94.

O'Donovan, K.H., 1975, "Synthetic Diamond," *Annals of CIRP*, Vol. 24, No. 1, pp. 265–270.

O'Donovan, K.H., 1976, "Synthetic Diamond," *The South African Mechanical Engineer*, Vol. 26, Aug., pp. 311–319.

Oostdam, B.L., 1970, "Exploration for Marine Placer Deposits of Diamonds (with French abstract)," *Mining and Groundwater Geophysics/1967*, Economic Geology Report No. 26, Canada Geological Survey, pp. 447–461.

Parkinson, L.J. and Dickinson, H.T., 1952, "Practice at De Beers Diamond Mines," *Mining Engineers Handbook*, 6th ed., Vol. 1, R. Peele, ed., John Wiley, New York, pp. 10–392–10–398.

Paterson, N.R., et al., 1977, "Geophysical Exploration for Kimberlites, with Special Reference to Lesotho," *Geophysics*, Vol. 42, No. 7, p. 1531.

Pipkin, N.J., et al., 1980, "AMBORITE—A Remarkable New Cutting Material from De Beers," *Industrial Diamond Review*, June, pp. 203–206.

Pustovalov, L., and Yu. I., 1966, "Diamonds of the Russian Platform," *Mining and Minerals Engineering*, London, Vol. 2, No. 3, Mar., pp. 113–115.

Ringwood, A.E., and Major, A., 1966, "Synthesis of Diamonds," *Australian Journal of Chemistry*, Vol. 19, No. 10, p. 19.

Schwarez, H.P., 1965, 'The Origin of Diamonds in Drift of the North Central United States," *Journal of Geology*, Vol. 73, No. 4, pp. 657–663.

Scott-Smith, B.H., and Skinner, E.M.W., 1984, "A New Look at Prairie Creek, Arkansas, Kimberlites and Related Rocks," *Proceedings* of the Third International Kimberlite Conference, J. Kornprobst, ed., Vol. 1, Elsevier, Amsterdam, pp. 253–283.

Seal, M., 1978, "The Increasing Application of Diamond as an Optical Material and in the Electronics Industry," *Industrial Diamond Review*, Apr., pp. 130–134.

Sharma, S.P., 1975, "Production of Diamonds from Graphite Using Explosive-Driven Implosions," UTIAS Technical Note No. 196, Institute for Aerospace Studies, University of Toronto.

Seversky, G., and Van Biljon, J.J., 1976, "The Influence of Machine and Toll Parameters on the Drilling of Concrete with Diamond Bits," *Industrial Diamond Review*, Mar., pp. 82–89.

Tolansky, S., 1962, *The History and Use of Diamond*, Methuen and Co., Ltd., London, 166 pp.

Tolansky, S., and Komatsu, H., 1967, "Abundance of Type II Diamonds," *Science*, Vol. 157, No. 3793, pp. 1173–1175.

Weatherby, E.W., 1952, "The Principles of Diamond Sorting and Valuation," *Optima*, Johannesburg, Vol. 2, No. 2, June, pp. 5–9.

Weaver, H.B., 1964, "Development and Application of Manufactured Diamond," SME Preprint 64-H61, AIME Annual Meeting, New York, 8 pp.

Wedlake, R.J., 1977, "Diamond Synthesis," *Industrial Diamond Review*, June, pp. 196–199.

Wentdorf, R.N., and Bovenkirk, H.P., 1961, "The Origin of Natural Diamonds," *Astrophysics Journal*, Vol. 134, pp. 995–1005.

Wilks, E.M., and Wilks, J., 1978, "The Mechanical Strength and Reliability of Natural Diamonds," *Diamond Research*, pp. 2–10.

Williams, A.F., 1932, *The Genesis of the Diamond*, Vol. 2, E. Benn Ltd., London.

DIATOMITE

RICHARD O.Y. BREESE

Geologically and commercially, the term diatomite is applied to the nearly pure sedimentary accumulation of diatom frustules—the microscopic skeletons of unicellular aquatic algae belonging to the class of golden brown algae, bacillariophyceae. The sediments are fine-grained, siliceous, and consist primarily of amorphous opaline silica with only minor amounts of organic residue, secondary minerals, and co-deposited non-diatomaceous or crystalline clastic debris. In the geological sense, the name diatomite implies sedimentary accumulations that have reached appreciable thickness and when thick enough, such accumulations consequently may have possible commercial potential. Although the term diatomite is popularly and inappropriately applied to any sediment in which there is an abundance of diatom frustules, alternative terminology is more correctly employed to describe the less pure diatomaceous sediments, for example: clay-bearing diatomite or diatom-bearing clay. Synonyms in current usage include diatomaceous earth and kieselghur. More antiquated and obsolete terminology includes tripoli-powder, tripolite, and infusorial earth.

Worldwide, diatomites occur within Tertiary to Recent lacustrine and marine sedimentary facies. Although diatomite is widespread throughout the world, deposits that contain high purity, commercially versatile ore are uncommon. Physical properties of the diatom and of processed diatomite that provide unique commercial value in a broad spectrum of market end-uses include ornate fine structure, low bulk density, and high porosity and surface area. Properties of equal importance are mild abrasiveness, high absorptive capacity, insulating ability, relative inertness, and high brightness. End-use markets are diverse and range from insulating brick and absorbents through quality sensitive filter aids and premium quality functional fillers.

Notwithstanding both the economic attractiveness of the specialty markets and the commercial versatility of the high purity deposits, deposits of lesser purity are mined in nearly all parts of the world for less demanding uses. Mining costs are minimized through open pit quarrying, but in Europe, Asia, Africa, and South America underground mining methods are also employed. Blasting is not required because diatomite is soft and easily broken loose with mechanized equipment.

Following gentle crushing, the ore is dried, milled, and processed into one of three broad categories of products: naturally milled, straight calcined, and flux calcined grades. Each of these principal categories is further subdivided into additional grades through particle size adjustment of the powders. Agglomeration of particles, alteration of fine structure, and color change are achieved through calcination. Flux calcination further accentuates these changes.

Production throughout the world is dominated by the United States, followed by Romania, the former Soviet republics, and France. Other major producers include Spain, Mexico, Iceland, Korea, Japan, and Germany. Denmark is a major producer of Moler earth products, which consist of a diatomite-clay mixture.

HISTORICAL BACKGROUND

The first industrial use of diatomite can be traced back some 2000 years to the Greeks and to the use of diatomaceous earth as a component in lightweight building brick and in ceramic pottery. It was not until the mid-1800s, however, that the unique properties of diatomite were first recognized and the market end-uses explored and developed. One of the most important of the early uses followed the development of dynamite by Alfred Nobel in the mid-1860s, whereupon diatomite was used as a component of the explosive to improve stability and safety. Other early uses included low temperature insulating and refractory bricks and as a component in insulating and fireproofing construction panels. During the 1920s, processing technology underwent a very rapid evolution with the development of calcination, flux calcination, and air classification technologies. Through these technologies, innumerable different size and grade classifications could be made for market applications and for rapidly diversified end-uses.

Today, the application of processed diatomite for filter aid is the largest of the quality sensitive end-uses. Specific filtration uses include the clarification of beer, wine and liquor, vegetable oil, syrup and sugar, pharmaceuticals, and swimming pool water, to name a few. As a functional filler and extender, processed diatomite is ideally suited for application in paint, rubber, and plastic formulations. Other filler applications include use as an anti-blocking agent in plastic film, anti-caking agent for fertilizer, thermal insulating material, catalyst carrier, polish, abrasive, pesticide and fertilizer carrier, and chromatographic supports. Throughout much of the world diatomite is still used as a component of insulating brick and as an absorbent.

GEOLOGY

Composition

Diatomite is a fine-grained siliceous sedimentary rock or unlithified sediment of biogenic origin. It is chiefly composed of amorphous silica, present in the form of opaline diatom skeletons, or frustules. Kamatani (1971) notes that the silica of recent frustules resembles amorphous silica gel, whereas the composition of fossil frustules resembles an opaline silica having a lower solubility than the silica of its recent counterparts. Combined water held within the opaline structure of frustules varies between species and is noted to range as high as 9.6 wt% in some lacustrine forms to as low as 1.9 wt% in specific marine species (Tappan, 1980, Lisitsyn, 1967). In addition to silica which is the major component, alumina, iron, alkali metals, alkaline earths, as well as other minor components are present.

Table 1. Chemical Composition of Natural Diatomites (Oven-dried Basis)

Constituent	Lompoc, Calif.[1]	Nevada	Kenya Soysambu[1]	Denmark Moler	Spain Albacete[1]	Mexico Jalisco[1]	Chile	PRC Jilin
Silica (SiO_2), %	89.70	89.82	84.50	67.80	88.60	91.20	89.68	90.07
Alumina (Al_2O_3), %	3.72	2.32	3.06	10.30	0.62	3.20	2.18	1.98
Iron Oxide (Fe_2O_3), %	1.09	0.84	1.86	6.85	0.20	0.70	0.38	0.67
Titanium Oxide (TiO_2), %	0.10	0.11	0.17	1.21	0.05	0.16	0.05	0.09
Phosphate (P_2O_5), %	0.10	0.01	0.04	0.21	—	0.05	0.04	0.06
Lime (CaO), %	0.30	0.44	1.80	1.35	3.00	0.19	0.41	0.39
Magnesium (MgO), %	0.55	0.47	0.39	1.64	0.81	0.42	0.31	0.28
Sodium (Na_2O), %	0.31	0.67	1.19	0.46	0.50	0.13	0.97	0.22
Potassium (K_2O), %	0.41	0.19	0.91	1.47	0.39	0.24	0.45	0.35
Ignition Loss, %	3.70	5.40	6.08	7.91	5.20	3.60	5.90	6.30
Total	99.98	100.27	100.00	99.20	99.37	99.89	100.37	100.41

[1]*Source:* Kadey 1983.

The silica content (as SiO_2) of diatomite samples, as determined on an organic free basis, typically range from 80 to 90%, but silica contents as high as 97% have been encountered. As determined through oxide analysis, iron, calcium, aluminum, titanium, sodium, potassium, and other elements are typically present in minor amounts of several percent or less (Table 1). Many elements are present in trace amounts (Table 2). Co-deposited and secondary minerals often found in diatomites include clays, quartz, gypsum, mica, calcite, and feldspars. Less commonly, salt, pyrite, sulfur, manganese nodules, vivianite, and other phosphates are present. Chert and volcanic ash are often abundant constituents of the sediment. Common biogenic constituents include the tests of ostracods and bivalves, the siliceous remains of sponges, silicoflagellates, radiolaria, carbonized fossil leaves, and fossilized fish bones.

When chemical and mineral impurities are present in great abundance, alternative terminology is used to describe the sediment, for example: sandy diatomite, clay-bearing diatomite, diatomaceous marl, diatomaceous peat, etc. These extraneous materials can affect the properties of the final product, including pH, the solubles present, density, and abrasiveness. In such cases, the commercial suitability of the ore for some uses will be adversely affected, unless the contaminants can be removed or rendered insoluble through processing.

Thermal processing, or calcination produces fundamental changes in the composition and nature of the opaline silica frustule. Within the trade, the term calcination has been loosely and incorrectly applied to the thermal processing of diatomite with or without the addition of a fluxing agent. Technically, the thermal treatment of diatomite is a sintering or agglomeration process achieved through incipient fusion. Calcination not only dehydrates the amorphous silica but initiates conversion of the opal to cristobalite. As much as a 40 to 60% conversion to cristobalite is produced through flux calcination. As discussed later, the health issues related to the crystalline silica species present in diatomite products has come under worldwide examination during the later half of the 1980s and future promulgation of health and environment related legislation is expected. Other changes in physical properties that are induced through thermal processing are discussed in following sections.

Morphology

The diatom is a member of a diverse group of single-celled, often colonial algae. The class is subdivided into two major groupings: the Pennales forms have axial symmetry and the Centrales forms have radial symmetry. To date, more than 12 000 distinct species have been recognized (Tappan, 1980).

Each cell possesses an internal, ornate siliceous skeleton consisting of two valves which fit together much like a pill box. The morphology of the frustules varies widely among species and serves as the basis for taxonomic classification. The surface of each valve is punctated by a series of openings that comprise the complex fine structure of the frustule and impart a design that is distinctive to individual species (Fig. 1). The size of the frustules ranges from 0.75 to 1 000 μm, large enough, as in the case of *Ellerbeckia teres* (previously *Melosira teres*) and some *Coscinodiscus* forms, to be visible with a hand lens. The majority are in the 10 to 150 μm range and require a microscope to be seen. The frustule of the living cell is covered by a jelly secreted by the algae and contains organelles typical of photosynthesizing plants.

Properties

Low block density, typically ranging from 320 to 640 g/l, distinguishes diatomite from all other fine-grained sediments. High brightness, or whiteness, is an additional and distinctive attribute of nearly pure diatomites.

The fundamental chemical composition and intricate structure of the diatom frustule give diatomite unique commercial value and versatility unmatched by other natural forms of silica in filtration and filler applications. Inherently, the fine particulate structure of the diatom skeleton imparts low density and high surface area to milled powders. Ultimately, the fine structure also provides high porosity, permeability and the clarifying ability required in filtration applications. These same physical attributes give absorptive capability as well as low thermal conductivity. While the skeletal structure is rigid and serves well as an abrasive agent, the delicate particles tend to break and gently polish rather than scour. In filler and filtration applications that are sensitive to purity, diatomaceous silica remains inert and lends only its physical properties. The melting point of diatomaceous silica ranges between 1 400 and 1 750°C.

Broad adjustment of physical properties is achieved through the thermal sintering process. Calcination reduces surface area through the destruction of fine structure. Calcination and flux calcination progressively increase the specific gravity from 2.0 to 2.3, decrease the surface area of milled powders from 10 to 30 m^2/g to 0.5 to 5 m^2/g, and increase the hardness of the silica from 4½ to 5 to 5½ to 6 on the Mohs scale. The refractive index of natural powders ranges from 1.40 to 1.46 and increases through flux calcination to 1.49.

Table 2. Typical Trace Element Analysis of Celite Hyflo® Super-Cel Product

Element	Ppm	Element	Ppm	Element	Ppm	Element	Ppm
Aluminum	>1000	Gadolinium	0.47	Neodymium	2.4	Tantalum	<0.27
Antimony	0.69	Gallium	1.20	Nickel	17	Tellurium	<0.12
Arsenic	1.50	Germanium	0.37	Niobium	4.3	Terbium	0.14
		Gold	*			Thallium	<0.32
Barium	100			Osmium	*	Thorium	1.8
Beryllium	<0.18	Hafnium	<1.2			Thulium	<0.16
Bismuth	<0.70	Holmium	NR	Palladium	*	Tin	0.23
Boron	150			Phosphorus	∼2400	Titanium	∼2600
Bromine	1.20	Indium	—	Platinum	*	Tungsten	<0.46
		Iodine	<0.44	Potassium	>500		
Cadmium	0.60	Iridium	*	Praseodymium	0.48	Uranium	5.0
Calcium	∼2400	Iron	>1000				
Cerium	15			Rhenium	—	Vanadium	320
Cesium	0.49	Lanthanum	8.2	Rhodium	*		
Chlorine	110	Lead	0.82	Rubidium	11	Ytterbium	<0.59
Chromium	39	Lithium	9.2	Ruthenium	*	Yttrium	7.4
Cobalt	1.40	Lutetium	<0.13				
Copper	15			Samarium	0.83	Zinc	18
		Magnesium	∼4200	Scandium	2.7	Zirconium	14
Dysprosium	0.27	Manganese	49	Selenium	0.39		
		Molybdenum	11.0	Silicon	>1000		
Erbium	<0.50			Silver	NR		
Europium	0.27			Sodium	∼1000		
				Strontium	26		
Fluorine	690			Sulfur	100		

NR—Not Recorded.
*—<0.1 ppm.

Origin

Diatoms have successfully invaded diverse habitats including marine, brackish and lacustrine/stream environments. As well, diatoms live in moist soils and on the surfaces of rocks and vegetation (Bradbury, 1988). Species are also reported in exotic and specialized environments such as caves, snow and ice, hot springs, and in dry areas (Patrick and Reimer, 1966, Bradbury, 1988). Basic biological requirements include sufficient moisture, sufficient exposure to light, adequate physiochemical environment, and availability of nutrients, trace elements, and silica.

Within aquatic environments that provide the basic requirements, diverse communities of planktonic, attached, and benthic diatoms can flourish and contribute to the geologic sedimentary record. The species composition of a living diatom community, and eventually the sedimentary fossil community, as well as the dominance of individual species within a population can depend on environmental variables such as water chemistry, pH, nutrient abundance, turbidity, temperature, and water depth, to name a few parameters. While some species are quite tolerant of environmental variance, the presence or morphotype of others requires or depends on specific ecological conditions. Both long and short term changes in population composition, in response to environmental change through time, can be observed within individual deposits. Such changes can produce several distinct types of diatomite ore within individual deposits.

Ecological conditions vary widely between lakes, and consequently, corresponding diatom populations can show wide variation. Far from being solely of academic interest, characteristic assemblages can, in some cases, be used to fingerprint or recognize individual deposits and sources of commercial products. Several distinctive assemblages are shown in Figures 2 to 6. Moreover, recognition of assemblages and changes in diatom population can assist identification of individual beds within a single deposit, provide a useful mapping/stratigraphic location tool in extensively

Fig. 1. Scanning electron micrograph of an *actinoptychus* occurring in the Celite Corporation Lompoc, CA, deposit (*courtesy:* Celite Corporation).

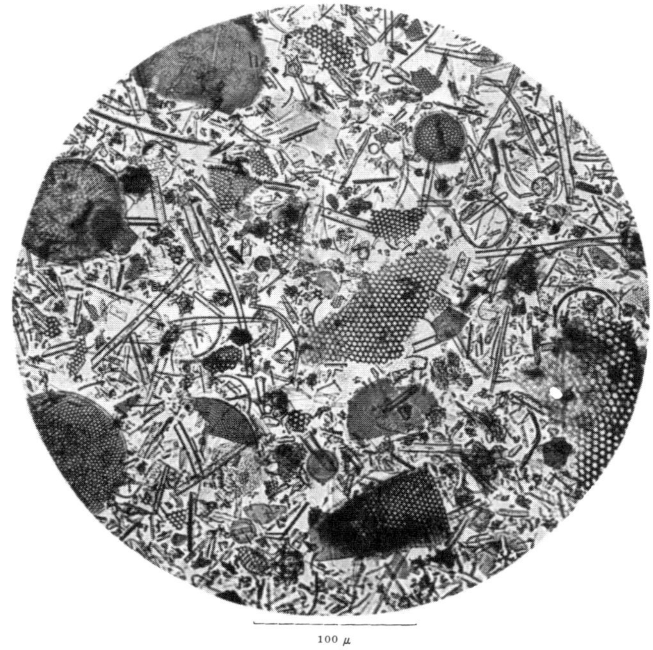

Fig. 2. Micrograph of marine diatom assemblage from Lompoc, CA.

Fig. 4. Micrograph of modern lake diatom assemblage from Myvatn, Iceland.

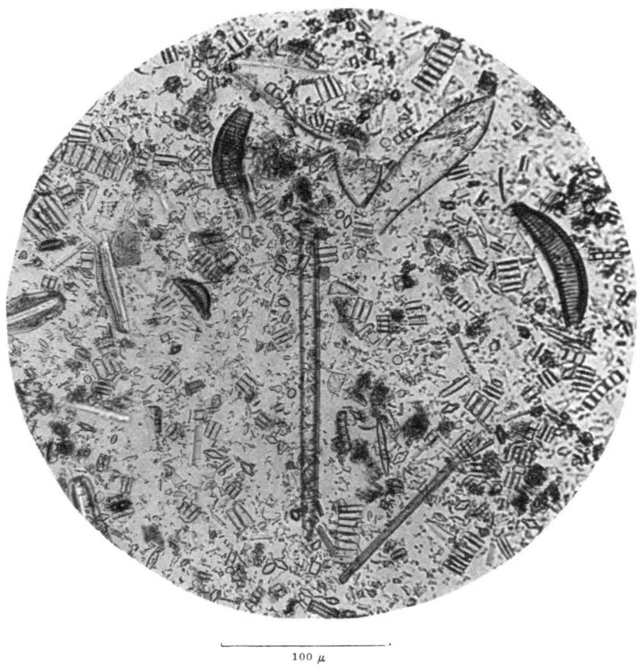

Fig. 3. Micrograph of lacustrine diatom assemblage from Carlin, NV.

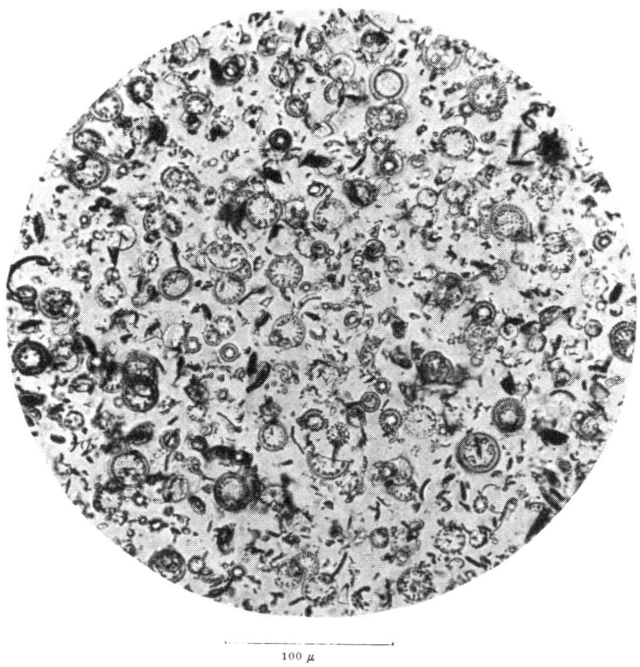

Fig. 5. Micrograph of lacustrine diatom assemblage from Elche de la Sierra, Spain.

covered or structurally complex deposits, and also assist the delineation of commercial quality orebodies.

Deposits of diatomite reflect stability of environmental and depositional conditions, as well as optimum preservation environment. While depositional rates vary, typical varve thicknesses observed in deposits suggest yearly accumulations of no more than several millimeters of compacted sediment. Strata of significant thickness and high purity, therefore, indicate prolonged environmental conditions that favored high diatom productivity and that were free of continuous and significant clastic contribution or mineral precipitation. Geologic preservation of the soft sediments requires protection from erosion, such as provided by overlying

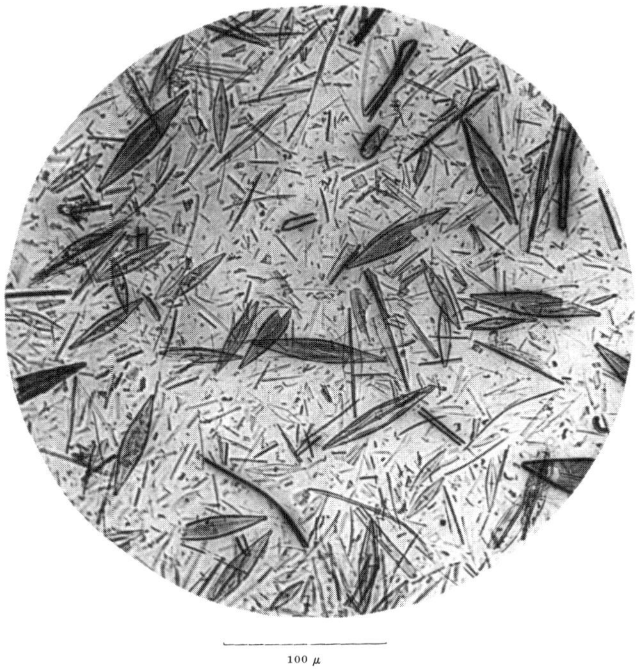

Fig. 6. Micrograph of bog diatom assemblage from Pernambuco, Brazil.

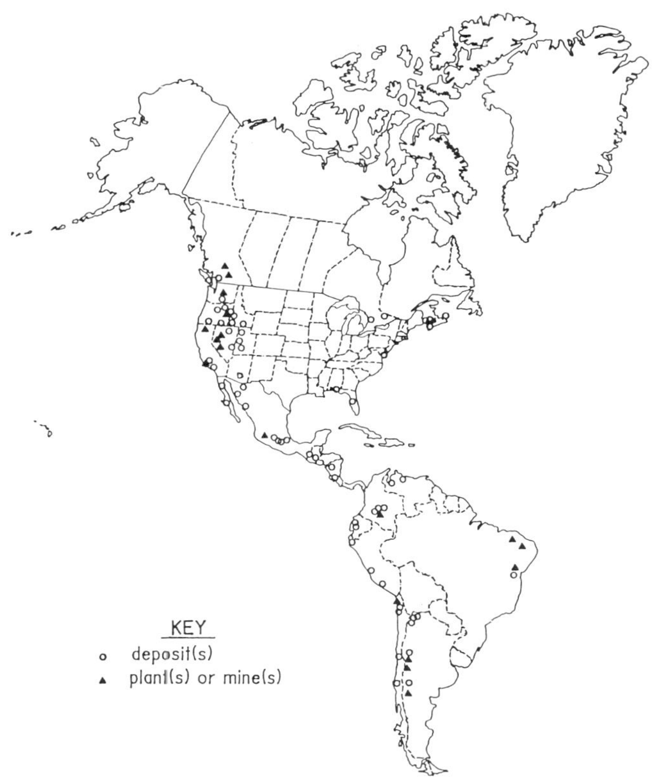

Fig. 7. Diatomite deposits and processing plants in North and South America.

volcanic rocks. Preservation also demands that the diatomite is not subjected to geological conditions which promote silica dissolution or diagenetic conversion of opaline silica to chert, porcellanite or quartz, such as through deep burial, exposure to high temperature, and through exposure to alkaline ground water. The distribution of many of the well known deposits are shown on Figs. 7 and 8.

Marine environments that are favorable for deposition of nearly pure diatomite exist in submerged coastal basin areas and shelves that are isolated from significant sources of clastic contribution and that are adjacent to areas of upwelling where waters are nutrient and silica rich, and biologically productive. Such areas occur today along the coastlines of California, Peru, and southwest Africa (Soutar et al., 1981). Seawater rich in both nutrients and silica has been observed to contain as many as one million diatom cells per milliliter (Tappan, 1980), and diatomites such as those in the Monterey Formation of California have been observed to contain as many as 1.28 x 10^6 frustules per cubic centimeter (Barron, 1987).

Tertiary marine deposits are commonly found along uplifted Pacific Rim coastlines and are associated with interbeds of volcanic ash, clay, and clastic sediments. Thick marine deposits occur in the Monterey Shales and Sisquoc Formation in Santa Barbara and Kern Counties, California, in the Miocene Tortugas Formation of Baja, Mexico, in the Miocene Pisco and Zapayal Formations along the Peruvian coast, and in the Miocene Funakawa Formation of northern Japan. Other marine diatomites and diatomaceous sediments occur in the Paleocene-Eocene Fur (Moler) Formation of Denmark, in Oligocene-Miocene sediments in Poland, in the Miocene Tellian Group (Beida Stage) of Algeria, in the Miocene Calvert Formation of Maryland and Virginia, United States, and within Tertiary sediments in Spain, South Korea, Chile, New Zealand, and Java.

Lacustrine environments typically exhibit greater variability than marine environments regarding pH, salinity, temperature, and nutrient content, but they can produce diatomite sections often tens of meters or more in thickness. Deposits occur commonly in Miocene to Recent volcanic environments and are associated or in-

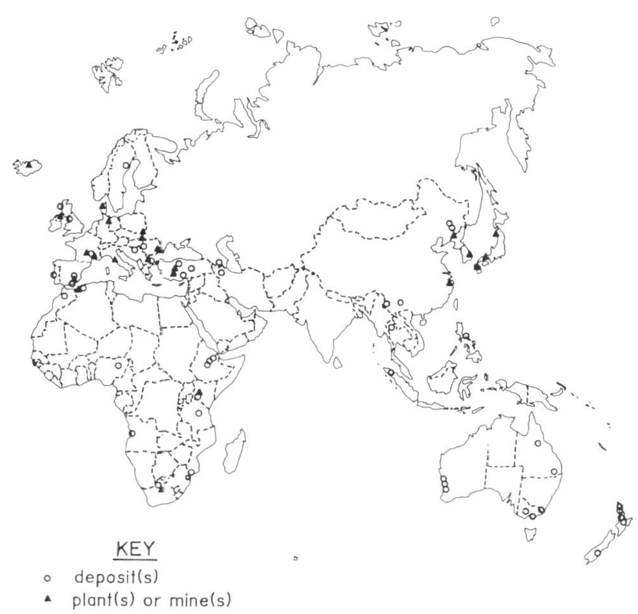

Fig. 8. Diatomite deposits and processing plants in Africa, Asia, and Oceania.

terbedded with volcanic rocks and sediments. This association is clearly demonstrated by deposits within the lake sediments of caldera moats, such as in the Trout Creek Mountains of Oregon, lava dammed rivers, e.g., Terrebonne, Oregon, in the subsiding valleys of the Nevada basin and range, and in France, within the lake sediments of volcanic maars. Throughout many countries, including the United States, Brazil, Australia, and the Philippines, bog or shallow pond deposits are found in Holocene to Recent non-volcanic, marshy lowland terrains.

The thick diatom accumulations that are common in Tertiary sediments reflect exceptional availability of silica from external sources such as runoff, ground water seepage, hot or cold springs, and/or from internal regeneration of silica released from bottom sediments (Hurley et al., 1985). Much of the silica requirement undoubtedly is provided by ground water that has percolated through silica rich volcanic rocks and sediments.

Distribution of Deposits

Diatomite deposits of both marine and lacustrine origin are commercially mined throughout the world. Although deposits occur commonly in Tertiary to Recent sediment facies, high purity deposits suitable for versatile quality sensitive products, are uncommon. The distribution of deposits and producers are shown on Figs. 7 and 8. It should be noted that classification as high or low grade is dependant on performance vis à vis ultimate application and not solely on purity. For example, although low purity diatomite is unsuitable for filter aid, it can make a high quality absorbent.

While marine deposits account for much of worldwide production, mining activities at lacustrine deposits are far more numerous. In the United States, diatomite is mined in California, Nevada, Oregon, Washington, and, until recently, Arizona. In Canada, impure diatomite is mined in British Columbia. Information regarding foreign producers and production is often lacking or available only as estimates. Notable mining operations in Europe are in France, Denmark, Spain, Germany, Iceland, and Hungary. Substantial production is reported for the former Soviet republics and Romania, but little specific information is available regarding mines and production facilities. African production is reported for Algeria, the Republic of South Africa, and Kenya. The US Bureau of Mines has reported mining in Morocco. In Asia, mining operations are known in Japan, South Korea, the People's Republic of China, and Turkey. Further south, the US Bureau of Mines has reported production in Australia. In Latin America, Mexico leads production and is followed by Brazil, Argentina, Chile, and Colombia. Production is also reported by the US Bureau of Mines for Peru and Costa Rica.

North America: North American production is dominated by the United States where eight companies operate at 14 locations in four western states. Since 1980, the industry has experienced the development of five new operations, of which one has since closed. California has continued to be the major producing state with the majority of production near Lompoc, Santa Barbara County (Fig. 9). The operations at Lompoc include the mines and mills owned by both the Celite Corporation and Grefco Inc. The marine diatomite at Lompoc is on the order of 300 m thick and is the largest center of production in the world. Other operations in California include Grefco's Lake Britton facility in Shasta County constructed during 1985 and 1986, the American Resources mine at Lake Britton that began servicing the cement industry in 1982, and the Lassinite Industry facility at Herlong.

Mines in Nevada include those operated by Grefco at Basalt, the Eagle Picher Industries Inc. quarries at Clark Station, Lovelock, and Hazen, the CR Minerals Fernley quarries and mill (previously owned by Cyprus Minerals), and the Moltan Company facility constructed during 1987 to the north of Fernley.

During 1986, Eagle Picher completed construction of the Vale, OR, facility in Malheur County. Absorbents and agricultural grades

Fig. 9. Aerial view showing extent of Celite diatomite quarries at Lompoc, CA.

of diatomite are produced by the Oil-Dri Production Co. in Lake County, Oregon, at Christmas Valley. The Celite Corporation operates several mines and two plants that produce flux calcined filter aids at Quincy, Grant County, WA.

Well known deposits occur throughout the western states in many of the Tertiary formations that contain lacustrine facies. Deposits have been known for many years in the Truckee, Desert Peak, Carlin, Esmeralda, and Panaca Formations of Nevada, within the Juntura, Drewsey, and Trout Creek Formations of Oregon, and in the Quincy and Squaw Creek portions of the Yakima Basalt in Washington. Diatomites and diatomaceous sediments are also present in the Banbury Basalt and Idaho Formation in Idaho, near Dorris and Lake Britton in northern California, and within the Quaternary age basin sediments in Utah. Along the east coast of the United States, bog type deposits have been identified in New Hampshire, New York, and Florida.

Production in Canada has historically centered on bog type deposits in Nova Scotia, and more recently on the lacustrine diatomites at Kamloops and Quesnel, British Columbia. The production in Nova Scotia has ceased. Due to the elevated clay and contaminant content of the British Columbia deposits, production has been directed towards manufacture of absorbents, fillers, carriers, and refractory/insulation bricks.

Celite Mexico (formerly Almeria SA de CV) is the largest of the Latin American producers, with mines and processing facilities southeast of Guadalajara in Jalisco State, Mexico. Unprocessed ores present in the lacustrine deposit are some of the brightest in the world and produce a full spectrum of fillers and filter aids that are exported throughout Latin America and Europe. Numerous other lacustrine deposits occur in the states of Mexico, Michoacan and Tlaxcala. Extensive, although impure, marine diatomite outcrop for many kilometers along the Baja coast.

South and Central America: Diatomite deposits in Brazil occur along the east coast in the states of Ceara, Rio Grande do Norte, Bahia, Pernambuco, Santa Catarina, and Sao Paulo. Without exception the deposits are bog or shallow pond type accumulations and contain abundant organic matter that must be burned out through an ignition process. The removal of organic matter is achieved in stationary, wood fired kilns located at or near the deposit sites. The organic free diatomite is then further processed into fillers and filter aids through size classification and/or flux calcination.

Commercial production in Argentina continues at a low level. Historically, mining of lacustrine deposits has been in Neuquen, La

Fig. 10. Aerial view of Celite quarry at Murat, France (*courtesy:* Celite Corporation).

Fig. 11. Aerial view of Celite diatomite quarry at Elche de la Sierra, Spain (*courtesy:* Celite Corporation).

Rioja, and Rio Negro Provinces. Deposits are also known in Catamarca, Chubut, Salta, and Jujuy Provinces.

Production in Chile centers on the numerous lacustrine deposits near Arica in the northernmost part of the country. Salt is a typical contaminant within the diatomites, and indeed, has been used by Sociedad Minera Condor as a natural fluxing agent in calcination. Other lacustrine deposits occur throughout Chile from the northern province of Tarapaca, south to the province of Chiloe. Marine ores were previously mined near La Serena, 400 km north of Santiago.

Elsewhere in Latin America, production is reported for Peru and at the Laguna de la Herrera deposit in Colombia. Suffice it to say that numerous lacustrine and marine occurrences are present along the west coast and within the western Andes of Peru. The most notable of these are the marine deposits near Pisco and the marine phosphate-bearing sediments at Bayovar. Undeveloped lacustrine deposits occur to the south of Quito, Ecuador, at Lake Brava in Venezuela, and near Puntarenas in Costa Rica.

Europe: Deposits in France, the former Soviet republics, and Romania are the most significant of the European diatomite sources. While production rates are nearly equivalent, fundamentally different products are made, with production in France focused on filter aid and filler, and production from the former Soviet republics and Romania primarily serving the construction and refractory product industries. Spain, Germany, and Iceland are also major producers of filter aid and filler. Denmark is the largest producer of Moler earth, a diatomite-clay mixture used in the manufacture of insulating brick, absorbent, carrier, and a variety of powders and granules.

Tertiary lacustrine deposits occur throughout the Massif Central of France. Carbonisation et Charbons Actifs SA (CECA) mines deposits at Collandres near Riom es Montagne, at Murat, and to the south near Privas at St. Bauzile. The deposit at Murat is also mined as a source for filter aids and fillers by the Celite Corporation (Fig. 10). The largest of the Spanish mines is operated by the Celite Corporation near Elche de la Sierra (Fig. 11). Although carbonate-bearing, the lacustrine ores are bright and produce a broad spectrum of premium quality filler products. Marine diatomites crop out near the coast in the south of Spain.

Germany was the first country to commercially produce diatomite, and production today remains in the Luneburg Heath area of Lower Saxony where Vereinigte Deutsche Kieselgurwerke GmbH & Co. KG (VDK) operates several mines and processing facilities. Deposits also occur in the Klieken area and in the Vogelsberg Mountain area of Bavaria. In Italy, lacustrine diatomites have been mined in the Viterbo district. Winkelmann Mineraria SpA are custom processors.

In northern Iceland, Kisilidjan HF, a joint venture between the Icelandic government and the Celite Corporation, employs dredging methods to extract the diatomaceous muds that lie beneath Lake Myvatn in northern Iceland. From the dredging barge, the muds are pumped as a slurry to landside hydrocyclones where unwanted ash and other contaminants are removed, and then to settling ponds. The resulting crude ore is dewatered and dried using geothermal steam before calcination or flux calcination.

Marine Moler earth is mined on the Isle of Mors and Fur in Denmark. The crude ore contains up to 30% plastic clay, precluding use of the ore for production of filter aids. Some lacustrine diatomite has been mined in Portugal at Couto Mineiro de Jardim, but production is minor. Another deposit is located near Rio Maior.

Although significant production is reported for Romania and the former Soviet republics, there is little information available regarding deposits and processing facilities. Reserves of marine and lacustrine diatomite in the former Soviet republics are believed to be large and to occur in the central Volga region, in Georgia, the Ukraine, Armenia, and in the St. Petersburg area. Mining in Romania has historically been at Adamclisi, Patirlapele-Buzau, and at Minis Arad. In Hungary, Orszagos Erc-es Asvanybanyak Vallalat mines and processes lacustrine diatomite at Erdobenye. Diatomite mining in Yugoslavia has centered on the Prilep and Kriva Palanka deposits. Diatomite is also present at several locations in Czechoslovakia. Impure, marine deposits occur in southeast Poland. Diatomite has also been mined in Bulgaria near Gurmen in the Blagoevgrad district.

Africa: Deposits of marine diatomite occur in Algeria at Sig and at Mostaganem. The Sig diatomite is characterized by elevated carbonate content and is mined using underground methods by the state-owned l'Enterprise Nationale des Produits Miniers Non Ferreux et des Substances Utiles. There are numerous lacustrine deposits in the Kenyan rift valley. Although most are poorly known or not of commercial significance, the reserves present at Soysambu are mined by African Diatomite Industries Ltd. and are processed at Kariandus, Kenya. Small lacustrine deposits occur at several locations in South Africa. Annual production is low and is only from the Witdraai deposit in the northern part of Cape Province. Lacustrine diatomite has been mined in Morocco at Awinat. Elsewhere in Africa, impure lacustrine deposits are known in Angola, in

Djibouti, in central Ethiopia, within a coastal belt in Mozambique, in Nigeria, and in Zimbabwe.

Asia and Oceania: The annual production rates for Japan and the People's Republic of China are only available as rough estimates. Numerous mine and processing facilities are known, however, and total production is believed to be significant. More than 70 lacustrine and marine deposits occur on the Japanese islands of Honshu, Hokkaido, and Kyushu. Showa Chemical Industry Co., Chuo Silica, Hokushu Keisodo, and Hakusan are major producers in Japan. The largest and most commercially significant of the deposits in China occurs in Jilin Province near Changbai and the border with North Korea. The lacustrine diatomites are extensive and must be mined by underground methods. Many other, lesser known lacustrine deposits occur in Jilin as well as in Heilongjiang Province. Vast reserves of clay-bearing lacustrine diatomite are present in Zhejiang Province and scattered throughout southwestern Yunnan Province.

Both marine and lacustrine deposits occur in southeastern South Korea. Filter aids, fillers, and other grades are produced at facilities located near Pohang. Although no production is reported for the Philippines, bog type, lacustrine deposits occur. Impure, clay-bearing lacustrine diatomite occurs in Viet Nam and in Thailand. Deposits are also known in Indonesia.

In Turkey, numerous deposits of lacustrine diatomite occur in the central portion of the country.

Mining of lacustrine deposits in Australia has been undertaken in New South Wales, Victoria, Queensland, and Western Australia. Marine and lacustrine diatomite occur on the North and South Islands of New Zealand, but only small scale production has been reported at several of the deposits.

World Reserves: The US Bureau of Mines estimates the world reserve base to be 2×10^9 t, and there will be no foreseeable worldwide shortage of diatomite in the future.

TECHNOLOGY

Exploration and Evaluation

Exploration and evaluation of deposits requires the integration of geology, mining feasibility, and processing feasibility with the determination of ore suitability for the products desired. During the course of these studies, the fundamental elements of an evaluation program must include identification of ore strata and determination of strata mineability, i.e., number, thickness, areal continuity, tonnage potential, and the degree of mining selectivity required. In addition, there must be a determination of processing requirements including equipment required and process flow, the determination of economic marketability, and an assessment of logistics, including proximity to utilities, transportation, and work force. A basic understanding of these elements will benefit the geologist in designing, guiding, and interpreting the sampling program.

Deposits of diatomite occur in Tertiary marine and lacustrine sedimentary facies throughout the world, and many such facies are well described in geologic literature. The review of aerial photographs can be used to augment literature study for the selection of target areas. Following the identification of a promising target, a sequence of sampling programs is initiated. A typical exploration program consists of several sampling phases that lead to the eventual determination of measured resources or proven reserves. The sampling phases are punctuated by go—no go decisions to proceed and would be followed by development and quality assurance sampling immediately prior to and following the commercialization of the deposit.

Initially, very widely spaced surface samples are collected and used to determine the general quality of the exposed diatomites, the quality variations across the deposit and between individual beds, and the tonnage potential. This also permits the identification of areas that warrant priority investigation. During this phase, representative samples are collected from well exposed outcrops or shallow trenches that have been excavated to adequately expose the near surface beds. In conjunction with surface mapping, sample testing results and data from microscopic examination of the stratigraphic sequence can be used to direct continued effort.

If the initial sampling and mapping results are encouraging, phases of progressively closer spaced sampling can be considered. Unless the sediments are well exposed naturally, subsurface potential must be determined through drilling or through the excavation of shaft-like pits. Geophysical prospecting, while unable to directly distinguish diatomite from other sediments, permits the rapid, if indirect, interpretation of subsurface geology. Consequently, geophysical methods can be used in conjunction with surface data to guide the selection of priority targets and the location of drilling sites. Both VLF—EM and seismic methods have been used to resolve geologic structure, trace beds exposed in outcrop into covered areas, and to distinguish stratigraphic units on the basis of consolidation (Ellingwood and Associates, oral communication, 1991).

Final sample spacing is dependent on the continuity of the beds and the quality variations that are observed. To permit pit design and mine planning, a final spacing pattern of 50 m or less, up to 100 m is generally used. During core drilling, care must be taken to avoid contaminating the core with drilling fluids, and the core diameter must be large enough to ensure high recovery. BQ (36.5 mm) or NQ (47.6 mm) diameter coring using triple tube wireline usually minimizes drilling problems and provides sufficient sample for testing.

If the results of sample tests indicate commercial grade ore, and mining is determined to be feasible, bulk samples are collected to determine plant-scale performance. Other special samples may be required as needed prior to the commencement of mining. While this is the final phase of exploration sampling, pit face quality assurance sampling must be undertaken during active mining.

Specialized sampling techniques can be employed when drilling is unnecessary. Where crude ores are soft and less than 5 m thick, hand augering using extension rods is effective, although the auger bucket must be emptied frequently and the cuttings laid out carefully for sampling. Mechanized augering works well in soft, wet diatomite. For sampling muds and oozes, augering can be done with a peat bog sampler.

The goal of optimum sampling is to delineate potential ore strata of uniform or nearly uniform quality. This includes the determination of the suitability of the ores for product application as well as for the delineation of waste strata. Program design thus requires identification of critical specifications and product/market requirements. The integration of these factors within sampling programs permits guidance of the program through the recognition of critical variations in ore quality. Appropriate qualitative yardsticks must be selected to judge results relative to industrial and health standards as well as to competitive commercial products. As ore quality can vary between strata in a deposit, or can vary across a deposit within a single stratum, sampling intervals and methods must be selected to maximize recognition of quality variation and to avoid sample bias.

Specialized testing of ore samples and samples of prepared test product is undertaken to examine the suitability and performance of a ore for specific product application. Testing of physical properties and milling performance can include the determination of lab scale milling recovery, density of dry milled powder, wet density, brightness, flow rate, clarity of filtrate, surface area, abrasiveness, Hegman value, and other properties related to the specific market applications. Other testing examines chemistry and composition including major and minor oxides, trace elements, heavy metals and toxic compounds, pH, soluble substances, and the determination of crystalline silica content. In addition, many countries require products that contact food and beverages to pass stringent food chemical codex specifications.

Mining, Milling, and Processing

In the United States, the cost of production can be divided as follows: 10% mining, 60% processing, and 30% packaging and shipping (Benton, 1983). Energy is the largest single element of direct cost, representing 25 to 30% of the total cost (Miles, 1990a). Freight costs, which are discussed separately, greatly impact the cost to the customer.

Mining Methods: Throughout most of the world, the lower cost of and higher recovery from open pit quarrying relative to underground mining encourages operators to use surface mining methods. Where the diatomite sequence is thick, benches from 1.5 to 15 m are developed. Benching not only facilitates the removal of ore and waste, but also permits immediate access to many ore layers and provides the opportunity to blend layers of differing quality. Notable exceptions are underground mines in Europe, South America, and Asia, and the dredging operation at Lake Myvatn, Iceland.

Prior to mining, the ore strata are carefully identified in the quarry faces through field sampling and evaluation that determine and confirm the suitability of each individual stratum. The ore grade strata can then be rated by quality and carefully segregated from the lower quality waste units using mechanized equipment. As diatomite is a soft, easily broken sediment, ores are typically dozer ripped or broken from the face by power shovel. Blasting of the crudes is neither needed nor possible due to the inherent softness of commercial quality diatomite, its porosity, and its absorptive ability. The broken material is loaded by front end loader or belt loader into haul trucks for transport to either the crude storage area or the waste piles.

Typically, the moisture content of unprocessed ore ranges from 30 to 60%. Where the climate is favorable for solar drying, significant cost savings can be achieved through reduction of moisture prior to milling. Crude grades may be blended or used separately depending on final application.

Milling Methods: The objectives of milling are twofold: to disaggregate the crude ore into individual diatoms while maintaining and preserving delicate diatom structure and to remove as much nondiatomaceous contamination as possible. Ore is gently crushed in spike rolls and hammer mills to reduce feed size to −1.27 cm. The crushed feed is then conveyed by heated air into milling fans and simultaneously milled and partially dried. Secondary milling and drying further reduces moisture content. The dried material is sized through a series of air separators and cyclones, the oversize recycled, and the coarse waste removed. Without further processing, these natural milled products consist of coarse fractions that are generally used for filter aids and finer fractions that are sold as filler grades.

Processing Methods: Diatomite is thermally processed, or calcined, to achieve fundamental adjustment of particle structure and particle size distribution. Calcination in a rotary kiln, with or without the addition of a fluxing agent, removes organic matter, reduces the percentage of fine structure through incipient fusion, and reduces surface area through the loss of fine structure (Figs. 12 and 13). Overall particle size is increased through the sintering or agglomeration of the individual diatoms. Following additional air classification, the coarser fractions or cyclone products are collected for use as filter aids (Figs. 14 and 15). The finer baghouse

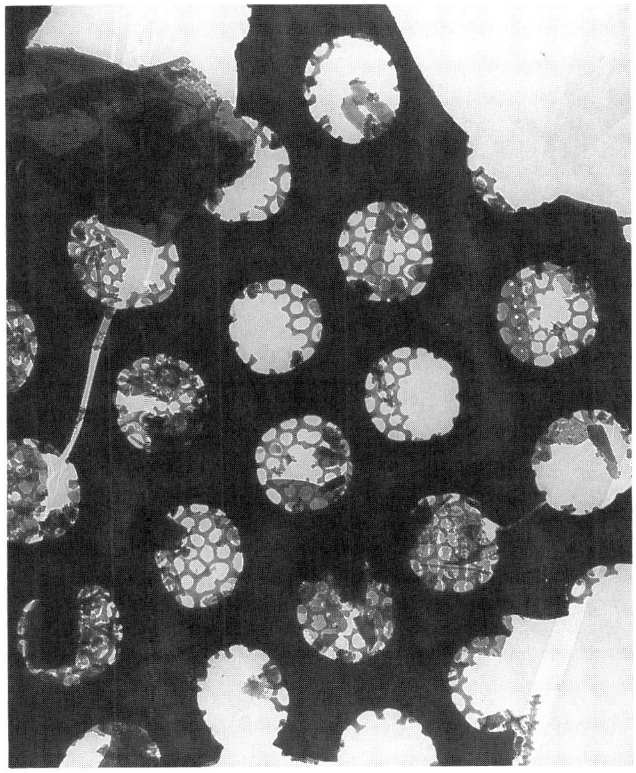

Fig. 12. Fine-structure remaining in natural processed Filter-Cel filter aid, seen at 19 500X (*courtesy:* Celite Corporation).

products are used as fillers. Permeability and flowthrough rate of the powders used for filtering are thus increased and a more diverse spectrum of products made possible through additional milling and classification. Calcination produces a color change as well as the modification of other properties described separately above.

Diatomite is calcined without the addition of a fluxing agent at temperatures that range from 870 to 1 100°C to produce pink or off-white colored straight calcined product grades. The straight calcined grades of filter aid exhibit flow rates faster than the natural grades, but achieve lower clarity of filtrate.

Calcination at 1 200°C with the addition of a fluxing agent produces white colored product and further increases the flow rate of the coarse fraction filter aid grades. The fine fraction recovered from the baghouse is sold as white fillers. In the United States, soda ash is used as the fluxing agent. Sodium chloride is commonly used elsewhere in the world. Temperature, amount of flux used, and retention time in the kiln are used to control the particle size distribution. In general, fluxing agent is added at 3 to 7 wt% of the feed.

Testing and Specifications

The factors that affect the ultimate properties of processed diatomite products are complex and interrelated. Such factors include the fundamental composition and nature of the diatomite crude ore as well as each of the handling and processing steps that follow mining. Thus, the quality of mill and kiln production can vary and the implementation of quality and process control procedures is required to assure production of uniform products of acceptable quality.

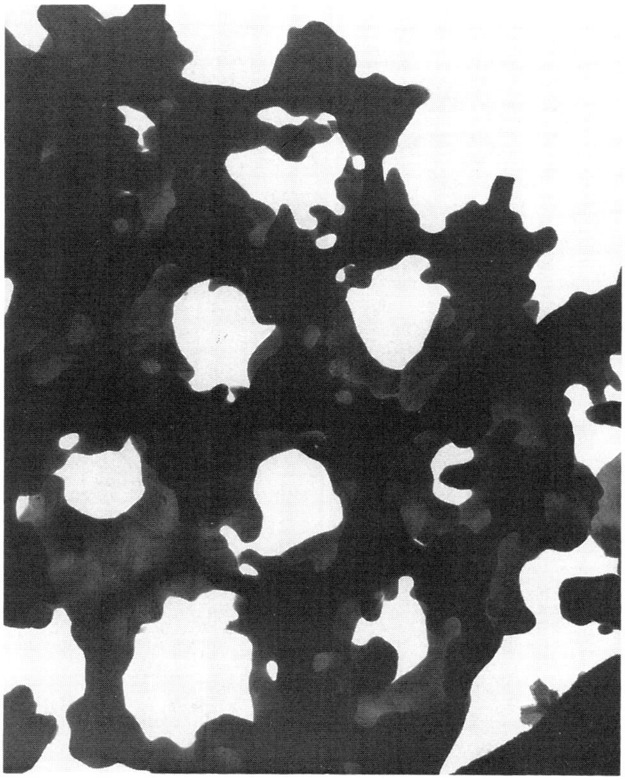

Fig. 13. Flux-calcined filter aid seen at 19 500X showing the reduction of fine-structure resulting from thermal processing (*courtesy:* Celite Corporation).

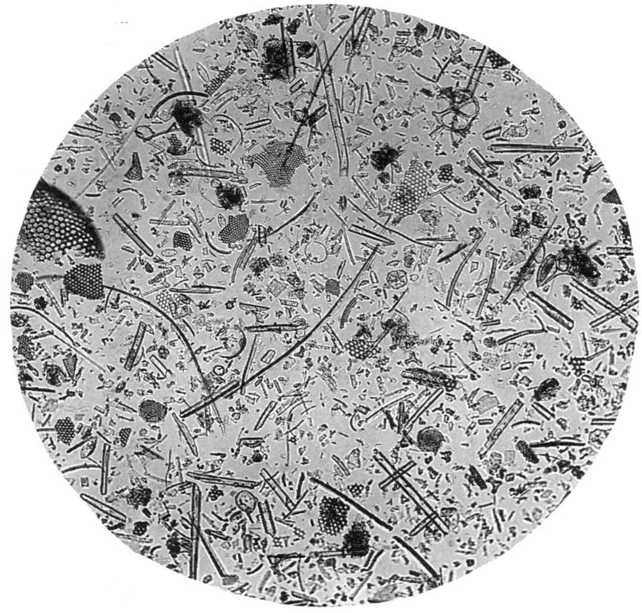

Fig. 14. Micrograph of Standard Super-Cel, a calcined filter aid (*courtesy:* Celite Corporation).

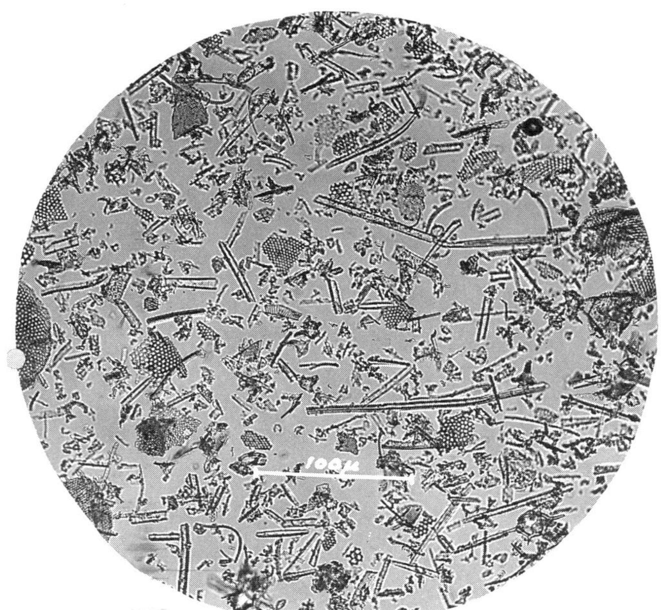

Fig. 15. Micrograph of Hyflo Super-Cel, a flux-calcined filter aid (*courtesy:* Celite Corporation).

Control of the manufacturing process can be achieved through the periodic testing of in-stream and finished product samples and through the consequent adjustment of processing variables such as kiln temperature, flux addition, fines removal, and crude blends. Laboratory testing of samples also permits the quantification of product performance in application as well as the confirmation of commercial acceptability of a product. By using standardized testing methodology, the commercial quality of a product sample can be judged relative to commercially accepted application-specific specifications and the performance of standard samples.

In filtration applications where filter aids must achieve satisfactory clarity at a reasonable flow rate, the flow rate and clarity performance of a filter aid sample can be tested under controlled laboratory conditions and measured directly. As diatomite is purchased on a weight basis and used by volume, wet cake density must also be determined. The particle and pore size distributions, which reflect ability to trap suspended particles, are also directly measured, as well as are the pH and resistivity of the powder. These latter tests indicate the presence of soluble substances that can affect the quality of the filtrate. Trace element analysis using Food Codex methods and limits must also be conducted on food, beverage, or cosmetic related products. With the advent of recent health related labeling requirements, the abundance of free crystalline silica is routinely determined to insure compliance with all regulations. Brightness testing by methods of the Technical Association of Pulp and Paper Industry (TAPPI) or the General Electric Company is used to determine the whiteness of a product relative to a magnesium oxide or barium sulfate standard. Although brightness is only of aesthetic value for filter aids, the whiteness has great importance for many filler applications.

Samples of functional filler products are typically tested for

Table 3. World Production of Diatomite[1]
(thousand ton)

Country	1980	1981	1982	1983	1984	1985	1986	1987	1988	1989[p]	1990[e]
Algeria	4.5[e]	4.5[e]	4.5[e]	4.5[e]	1.8	2.7	4[e,2]	4[e]	4[e]	4[e]	4
Argentina	6.4	4.5	6.4	10.9	5.5	8.2	14	5	7	7[e]	7
Australia	2.7	1.8	1.8	8.2	6.4	7.3	9	10	11	11[e]	10
Brazil	14.5	9.1	12.7	14.5	16.3	12.7	12[r]	16	14	16	15
Canada[e]	3.6[2]	3.6	1.8	1.8	3.6[r]	3.6[r]	4	4	4	4	4
Chile	0.9	[4]	[4]	0.9	1.8	2.7	3	3	3	3	3
Colombia[e]	0.9	0.9	0.9	0.9	0.9	0.9	—	—	—	—	—
Costa Rica	0.9	0.9	0.9	0.9[e]	0.9[e]	—	—	—	5	5[e]	5
Denmark[3]											
Diatomite	25.4[e]	3.6	3.6[r]	6.4[e]	10[e]	6.4[e]	6[e]	6[e]	6[e]	6[e]	1
Moler	125.2[e]	63.5	70.8	65.3[e]	63.5	71.7	73[e,2]	66[e]	66[e]	70[e,r]	60
France	217.8[e]	209.6[r,e]	244.1	221.4	247.7	270.4	260[e]	250[e]	250[e]	250[e]	250
Germany	52.6	42.7	42.7	44.5	49.0	48.1	49	47	47	47[r,e]	47
Iceland	18.2	20.0	25.4	25.4	27.2	29.0	23	23	25	26[r,e]	25
Italy[e]	30	25.4	20.0	25.4	28.1	30.0	27	27[r]	28	25[r]	25
Kenya	[4]	1.8	1.8	1.8	1.8	1.8	1	1	1	1	1
Korea	25.4	41.7	55.4	56.3	48.1	53.5	55	65	72	75	75
Mexico	56.3	56.3	56.3	43.6	44.5	45.4	36	35	37	45	55
Peru	7.3[e]	7.3[e]	7.3[e]	13.6	7.3	14.5	9	21	15	15[e]	16
Portugal	2.7	2.7	2.7[e]	1.8	1.8	1.8	2	3	2	3	3
Romania[e]	40.8	290.4[r]	290.4[r]	290.4	300.4	300.4	300	280	280	260	250
South Africa	0.9	0.9	0.9	0.9	[4]	0.9	2	[4]	[4]	[4]	[4]
Spain	23.6	38.1	63.5	55.4	72.6	96.2	128[e,2]	100[e]	100[e]	100[e]	100
Thailand	1.8	[4]	[4]	[4]	0.9	[4]	[4]	[4]	[4]	1	1
Turkey	9.1[e]	10.0[e]	10.0[e]	10.0	2.7	2.7[e]	6[r,e]	5	[5]	[5]	—
USSR[e]	226.9	230.5[r]	235.9[r]	235.9	240.5	245	250	255	260	260	250
United Kingdom	0.9[e]	0.9[e]	0.9[e]	[e,4]	[e,4]	[e,4]	[e,4]	[4]	[4]	[4]	[4]
United States	625.2	623.4	556.3	561.7	569.0	576.2	570[6]	596[6]	629[6]	617[6]	631[6,2]
Total	1 524.5	1 694.1	1 717	1 702.4	1752.3	1 832.1	1 843[r]	1 822	1 866	1 851	1 838

Source: Mineral Industry Surveys 1990. Minerals Yearbook 1984, 1985, 1987, US Bureau of Mines.
[e]Estimated, [p]Preliminary, [r]Revised.
[1]Table includes data through Apr. 10, 1991, and does not include production in China, Hungary, Japan, New Zealand, Yugoslavia, or recent production in Colombia.
[2]Reported figure.
[3]Data represents sales.
[4]Less than 1/2 unit.
[5]Revised to zero.
[6]Sold or used by producers.

density, TAPPI brightness, and particle size distribution. Additional testing for paint application can include the determination of Hegman value or fineness and vehicle demand expressed as oil or water absorption. The abrasiveness of the product for paper application can also be measured using the Valley Iron method.

Nearly all applications have their own unique requirements and/or formalized specifications. As a result, testing methods have been developed with which the suitability of products can be predicted. Often, evaluation can be achieved through simple, improvised testing procedures with which small samples of product are made and pertinent characteristics of the final product compared to accepted industrial norms. Similarly, comparison can be made between the performance of traditional materials used in a product formulation and the performance of potential substitutes.

The following is a partial listing of test procedures and specifications available through the ASTM and the military (Miles Industrial Mineral Research, written communication, 1991):

ASTM D 604–81 (reapproved 1989) Standard Specification of Diatomaceous Silica Pigment;

ASTM D 719–86 Standard Test Methods for Analysis of Diatomaceous Silica Pigment;

MIL-D-20550B Military Specification Diatomaceous Earth 25 July 1962;

MIL-F-52637A Military Specification Filter Aids, Water Purification, 22 June 1978;

MIL-S-15191B Military Specification Silica, Diatomaceous (flatting-extender pigment), 20 October, 1967.

ECONOMIC FACTORS

Production and Consumption

Estimates of world production vary by source, but including all quality grades, production in 1990 totalled in excess of 1.8 Mt and has shown slow, overall growth for most of the decade (*Minerals Yearbook, 1986, 1988, Minerals Industry Surveys,* 1990). With few exceptions, worldwide production of diatomite since 1980 has shown annual growth, increasing from its 1980 level to a high in 1988 of 1.87 Mt (*Minerals Yearbook, 1987, 1985, 1984, Minerals Industry Surveys,* 1990). Total production has dropped since 1988 to the 1990 level of 1.84 Mt (*Mineral Industry Survey,* 1990). Production trends are shown on Table 3. The average price per ton (Table 4), on a weighted basis by end-use volume, has followed a similar trend to production, rising to a high of $229 in 1988, followed by a drop to the 1990 level of $219 (*Minerals Yearbook,*

Table 4. Price Per Ton of Diatomite (weighted average)

	Average Value ($US)
1980	161
1981	181
1982	194
1983	204
1984	212
1985	220
1986	225
1987	225
1988	229
1989	222
1990	219

Source: US Bureau of Mines.

1981–1987, Minerals Industry Surveys, 1990). The United States remains the dominant world producer, followed by Romania, the former Soviet republics, France, Spain, and Japan (not listed by the US Bureau of Mines).

According to the US Bureau of Mines, filter aids comprised 64 to 71% of the products sold or consumed in the United States during the past decade. Sales or consumption of fillers represented 14 to 23%, insulation 1 to 4%, and other uses, including absorbents, additives, abrasives, and admixtures, 10 to 15%. The total value of diatomite products sold or used in the United States reached $136 million for 1990 (*Mineral Industry Surveys, 1990*).

Marketing and Transportation

Worldwide, production of diatomite products is reported for at least 26 countries. Five countries and approximately the same number of manufacturers dominate the world production. The major producers achieve distinct marketing advantage in the domestic and export marketplaces by providing the customer with integrated technical and sales assistance. The ability to provide such services relies largely on technical knowledge of a customer's process and product needs as well as on the ability to provide a broad spectrum of application specific products. Related services include providing reliably available products through well developed distribution systems and worldwide marketing presence and providing product development and technical support from research and development organizations. Throughout the world, the large producers have established entrenched trade or brand names that are synonymous with reliable quality. Many smaller producers compete successfully with the larger manufacturers through supplying local niche markets at lower prices. Commonly, these smaller producers supply local market requirements for which demanding quality specifications and technical sales service are not required.

Powdered diatomite products are sold in both 22.68 kg bags and in bulk form. In Europe, powdered diatomite is also sold in 1 000 to 2 000 L sacks. Bagged products are palletized and stretch or shrink wrapped for shipment.

Bags are labeled with the product and trade name and can be labeled with a customer's specific in-house identification if requested. At the request of the customer, labels may also be placed on the shrink wrap that covers palletized loads. Caution labels are placed prominently on each bag or on the palletized load advising the product handlers of the health issues associated with the crystalline silica present within diatomite products. Caution advisories may be printed in several languages on each of the labels, and in addition, special labeling may be required for different areas.

Freight costs significantly impact the overall cost of products to the customer. Diatomite products are bulky due to low density, thus, transport cost per ton is high. After palletizing and wrapping, bagged products are transported to distribution centers or directly to customers by truck or in containerized load by rail and ocean freight. Alternatively, pneumatic railcars and truck tankers are used to transport diatomite powders in bulk directly to the customer storage bin. In the United States, the majority of products are bagged and shipped by truck in palletized form.

In the United States, products destined for export are typically palletized and transported to port in containerized form. The containers are as large as 12.2 × 2.7 × 2.7 m and are easily transferable from or to ship, truck or rail systems. The major elements of export contracts that must be routinely negotiated include inland freight obligations on both ends of the transport, marine cargo insurance, and currency and fuel adjustment factors. Tariffs and taxes associated with export are considered separately in following text.

Tariffs and Taxes

Most of the major trading countries now use the Harmonized Commodity Description and Coding System to classify products for tax purposes and have replaced the Brussels Tariff designation classification system previously used. Under this system imported natural diatomite products and products calcined without sintering agents are classified under designation 25.12 as siliceous fossil meals and similar siliceous earths. This heading includes tripolite, diatomite, and moler earths as well as diatomite products which have been acid washed to eliminate impurities without altering the structure of the product. Diatomite products which are calcined with the addition of sintering agents such as sodium chloride or sodium carbonate are classified under a separate heading, 38.02, as activated products. Unfired, sawn blocks of diatomite are classified under heading 68.06 whereas otherwise they are classified under 69.01. Import duties and other taxes vary throughout the world and are set by the fiscal and customs authorities of each country.

There are no quota restrictions or tariffs applied to diatomites imported into the United States. Products are exported from the United States under general license with no restrictions. Depletion allowance is set at 14%.

Substitutes

Many minerals and materials are used in filtration applications in which a porous filter bed is used to mechanically remove suspended solids from a liquid. In addition to diatomite, other materials that are commonly used include expanded and milled perlite, sand, asbestos, cellulose, and carbon. While diatomite is acknowledged as a superior performer for a broad range of commercial and municipal filtration problems, each of the other materials can be suitable for specific applications or niche applications and can exhibit advantage regarding effectiveness, cost, durability, and availability. In the last decade, the use of ultrafiltration and synthetic membranes has become more widespread. The alternative filtration technologies, however, still require additional research.

Talc, silica sand, ground mica, clay, expanded and milled perlite, and ground calcite are a few of the many mineral fillers sold into the large and diverse filler markets shared with diatomite.

Chemically modified or synthetic fillers represent a second important group of fillers. Selection of appropriate fillers may be based on cost, availability, and/or performance. The physical properties of diatomite make it useful as a functional filler in a broad spectrum of industrial applications. While substitute fillers may be used, replacement may require reformulation of the final product.

Table 5. Typical Physical Properties[1] of Some Commercial Filter Aids

	Grade*	Color	Density, pcf		Screen Analysis % Retained, 150 Mesh	pH	Median Pore Size, μ	Perma-ability, d'Arcy	Typical Applications F ltration
			Dry	Wet					
Natural	Celite 500	Gray	7.0	16	2.0	7.0	1.5	.057	Vegetable oil
	Celite 505	Pink	8.0	23	2.0	7.0	2.5	0.16	Wine
Calcined	Standard Super-Cel	Pink	8.0	18	4	7.0	3.5	0.28	Beer and wine
	Celite 512	Pink	8.0	19	7	7.0	5.0	0.53	Beer
	Hyflo Super-Cel	White	9.0	18	6	10.0	7.0	1.2	Dry cleaning solvents, chemicals, sugar
	Celite 501	White	9.5	18	8	10.0	9.1	1.4	Grape juice
	Celite 503	White	9.5	18	9	10.0	10.0	2.0	Industrial and potable water
Flux Calcined	Celite 535	White	12.0	19	10	10.0	13.0	3.1	Industrial wastes
	Celite 545	White	12.0	19	12	10.0	17.0	4.8	Swimming pools
	Celite 580	White	18	21	30	10.0	17.0	4.8	Corn sugar
	Celite 560	White	16	20	40	10.0	22.0	30.0	Phosphoric acid

*Brands listed are Celite registered trademark, [1]English units used where appropriate.

USES

The unique physical properties of the diatom make diatomite and processed diatomite powders suitable for wide-ranging industrial applications. Early uses of diatomite blocks and powders focused on absorption, abrasive, and insulation applications. However, the use of diatomite as a filter aid in the mechanical separation of suspended solids from fluids now surpasses in importance all other specialty applications in terms of tonnage. Notwithstanding the importance of filter aid uses, specialty functional fillers and diatomite based synthetic silicates supply significant markets. Insulation, granulated or powdered absorbent, agriculture, and construction grades continue to serve high volume markets that can utilize more highly contaminated diatomite which is unsuitable for filtration purposes.

Filtration

According to the US Bureau of Mines, 71% of the diatomite sold or used in the United States is filter aid (*Mineral Industry Surveys, 1990*). Worldwide output for filtration is estimated at roughly 60% of total production.

The physical properties required of filter aid quality diatomite are demanding and must ensure the clearest possible filtrate, high flow rate of fluid through the filter cake, and a long cycle duration prior to replacement of the filter cake. Optimum permeability and flow rate-clarity characteristics of the filter cake are achieved by careful control of the particle size distribution and porosity of the filter aid cake. These properties are controlled and adjusted through thermal sintering and particle size classification to produce a broad spectrum of filter aid products and filtering characteristics. The properties of a range of products are listed in Table 5. As a rule of thumb, ability to remove small particles decreases as the permeability of the filter cake increases. In general, natural products exhibit slower flow rates but achieve the greatest clarity. Straight calcined and flux calcined products respectively exhibit faster flow rate and lower clarity than the natural products. For special applications, diatomite is mixed with cellulose fiber, and can be surface treated and acid washed.

The proper grade of filter aid is selected based on the minimum size of the suspended particle that is to be removed and on the clarity of the final filtrate that must be achieved. The optimum grade will provide the fastest flow rate while achieving acceptable filtrate clarity. Several filtration steps, each utilizing a specific grade of filter aid may be required. Nearly neutral pH and low solubles content of the filter aid powder are necessary to assure chemical inertness of the filter cake and high purity of the final filtrate.

Diatomite filter aids are used in a wide variety of food and non-food filtration applications. Brewing, vegetable sugars, fats and oils, wines, pharmaceuticals, and fruit juices represent several of the largest food contact applications. Major industrial end-users include the petroleum industry, inorganic and organic chemical manufacturers, the nonferrous metals industry, industrial waste water treatment, and dry cleaning. Swimming pool filtration is a major consumer market end-use.

Fillers

As a class of products, fillers can be divided into the two general categories of non-functional and functional, depending upon the role that the additive plays in the ultimate performance of the final product. Non-functional fillers are used as bulking agents and cost reducing agents and extenders. Functional fillers are working additives that alter and enhance the properties of the final product. Functional fillers may also improve the manufacturing process itself through conditioning of the ingredients. The physical properties of functional fillers are tailored to meet the needs of individual applications, and thus, must be tightly controlled to meet the rigorous requirements of each end-use. Many of the functional fillers command high sales price.

Most diatomite manufacturers produce filler application grades from baghouse fines. Historically, filler products were developed from the fines derived from filter aid production. Today, the filler applications are highly demanding and filler products must meet rigorous specifications and physical property requirements, and a broad spectrum of natural, calcined, and flux calcined grades are made to meet the needs of diverse end-uses. Ideally, filler grades are insoluble and inert in most chemical processes, exhibit low bulk density while maintaining small particle size, and provide suitable particle shape and size distribution. The particle size of functional fillers generally ranges from 1 to 50 μm. Many other properties may be applicable for specific applications, such as color, surface area, liquid absorption capability, abrasiveness, particle rigidity, high melting point, and porosity. While many minerals can be crushed and sized as fillers, the inherent properties of diatomite are well suited for, or can be readily modified for filler applications. The typical properties of some commercial fillers are shown in Table 6.

As the uses for diatomite fillers are numerous, only a few of the notable applications are discussed here.

Table 6. Typical Physical Properties[1] of Some Commercial Fillers

	Paper		Paint		Abrasives			Conditioning Agent		Anti-Blocking Agent	Reinforcing Agent
Application	Cylinder Machine Production	Fourdrinier Production	High Flatting	Semi-Gloss	Silver Polish	Automobile Polish	Buffing Compound	Toxicant Carrier	Fertilizer Coating	Polyethylene	Rubber
Brand*	Celite 321A	Celite 305	Celite 281	Celite 499	Snow Floss	Super Floss	Celite HSC	Celite 209	Celite 392	Celite White Mist	Celite 270
Loose wt, pcf	8.0	—	8.5	8.5	7.5	8.7	9.0	8.0	7.0	7.5	8.6
Wet density, pcf	20	24	22	22	24	24	19	24	17	23	28
Moisture content %, max	6.0	4.0	0.5	0.5	6.0	0.5	0.5	6.0	6.0	0.5	0.5
Retained on 150 mesh, %wt	0.5	0.1	Trace	Trace	0	0	6	Trace	0.5	0	0
Retained on 325 mesh, %wt	8.0	1.0	1.5	Trace	Trace	0.1	17.6	0.5	5.0	0.05	0.7
Oil absorption, %wt	210	175	110	105	175	105	185	175	210	160	150
Color	Lt. Gray	Gray	White	White	Buff	White	White	Buff	Buff	White	Pink
pH, max	7.0	7.0	10	10	7.0	9.4	9.5	7.0	7.0	10.0	7.0
Resistivity	3 000	3 000	13 000	7 400	3 000	—	—	3 000	4 000	6 500	30 000
Refractive index	1.40	1.40	1.46	1.46	1.43	1.47	1.47	1.43	1.43	1.47	1.45
Surface area m/g	10–20	10–20	0.7–3.5	0.7–3.5	10–20	0.7–3.5	0.7–3.5	10–20	10–20	0.7–3.5	4–6
Median particle size, μ	—	—	7.8	6.8	3.0	5.5	17.5	—	—	3.5	5.1

*Brands listed are Celite registered trademarks, [1]English units used where appropriate.

Paint is the most important of the filler end-uses of diatomite. Functionally, the diatomite filler serves to extend the more costly pigments and serves as flatting agents that reduce both the sheen and gloss of the dried paint surface. The fine size and irregular shape of the particles serve well to roughen the paint film, imparting a matt finish as well as tooth for better adhesion of top coats. Toughness and durability of the film are also increased. Rigorous requirements regarding brightness, pH, absorption properties, refractive index, and chemical stability must also be met.

In polyethylene film, the particulate structure as well as the optical properties of diatomite fillers makes it an effective antiblocking agent. The small and irregularly shaped particles when added to the film during manufacture, mechanically separate adjacent sheets and prevent cold welding.

In other plastics, diatomite is used to reduce cost and modify properties such as improving heat resistance and surface finish. Diatomite also reduces mould release resistance and improves water and abrasion resistance.

Diatomite filler is used as a processing aid in rubber manufacture, and as a non-scratching abrasive in polishes and cleansers. Fineness, absorptive capability, and low abrasiveness make diatomite powders and granules effective carriers and diluents for use in agrochemicals.

Diatomite provides the source of low density, heat resistent, and porous silica in catalyst carriers such as used in hydrogenation processes and in the manufacture of sulfuric acid. Similarly, diatomite is used in porous, inert chromatographic supports. Diatomite powders are also used as a silica source for the production of synthetic calcium and magnesium silicates. These unique products exhibit extremely high surface area (105 to 190 m^2/g) and are able to absorb up to 5½ times their weight in water. In fact, Celite brand synthetic silicates are able to absorb three to four times their weight in water and remain dry, free flowing powders. Natural powders and granules are used as absorbent floor sweeps and cat litters.

In concrete, mortar, plaster, stucco, and grout, diatomite filler can improve plastic and hardening properties, improve cohesiveness, homogeneity, and workability, and reduce bleeding and efflorescence. In asphalt products, diatomite can decrease cracking tendency associated with rapid temperature change and improve cohesive strength. Throughout the world, diatomite and clay-bearing diatomite are also used as pozzolans and are also formed directly into insulating brick.

Other miscellaneous applications include match head and welding rod composition, as anticaking agent on ammonium-nitrate prills, and as a stabilizer in explosives and acetylene containers.

RECENT TRENDS

US Capacity vs. Demand

During the past decade, production capacity in the United States for fillers, absorbents, and filter aids has increased through the development of the five new operations. Following the construction of the new filler and filter aid plants, US overcapacity was estimated in excess of 25% (Anon, May 1987, and Apr., 1989). While development of new products continues, no major new market applications that will greatly increase usage have been announced.

Exploration of deposits throughout the western United States has continued. Elsewhere in the world, attention has focused on the development of Chinese resources at Changbai, Jilin, and on investigation of deposits in Australia, Indonesia, and Hungary. Investigative studies were also undertaken at deposits in Chile and Mexico.

Environmental and Safety Considerations

In the United States, the diatomite industry now faces environmental issues. These include the 1987 classification of crystalline silica as a probable carcinogen by the International Agency for Research on Cancer (IARC), the product labeling requirements that followed under the Hazard Communication Standard, the addition of silica, crystalline to the State of California list of carcinogens, and the increasing cost of spent filter cake disposal in landfills.

The relationship of pneumoconiosis in workers to the inhalation of crystalline silica dust generated by mining, processing, and handling diatomite has been recognized and addressed by the diatomite industry for many years. In the United States, worker exposure to such dusts is regulated under the Occupational Safety and Health Administration (OSHA) and by the Mine Safety and Health Administration (MSHA) through mandatory occupational safety and industrial hygiene programs implemented by the manufacturer.

In 1987, the International Agency for Research on Cancer (IARC), under charter by the World Health Organization to evaluate the carcinogenic risk of chemicals to humans, classified crystalline silica as a Group 2a probable carcinogen (IARC, 1987a and b). Based on the IARC strength-of-evidence criteria used for substance classification, the IARC working group reported four results regarding silica: 1) there is sufficient evidence for the carcinogenicity of crystalline silica to experimental animals, 2) there is limited evidence for the carcinogenicity of crystalline silica to humans, 3) there is inadequate evidence for the carcinogenicity of amorphous silica to experimental animals, and 4) there is inadequate evidence for the carcinogenicity of amorphous silica to humans (Miles, 1990b). In the United States, this triggered the regulation of crystalline silica under the Hazard Communication Standard as administered by OSHA and has also triggered additional regulation under the environmental laws of several states (Miles, 1990b). In association with hazard labeling requirements imposed by the Hazard Communication Standard, analysis of crystalline silica forms as carcinogens at the 0.1% threshold is also necessary.

As crystalline silica occurs as a principal or minor accessory mineral in most mineral products, the designation of crystalline silica as a probable carcinogen affects nearly all mineral related industries. The diatomite industry formed the International Diatomite Producers Association (IDPA) in 1988 to study and report on the health, safety, and environmental issues related to the industry. In addition to sponsoring health related research, the IDPA has published educational literature regarding crystalline silica for workers and users of diatomite products. Information is available through the IDPA, San Francisco, CA. The Chemical Manufacturers Association, Crystalline Silica Panel, is sponsoring additional study of health and other industrial issues related to crystalline silica. Manufacturers are responding directly through emphasized dust reduction efforts in the workplace, worker and consumer protection, education, and training programs, and the assessment of workplace hazard through implementation of respirable dust surveys. Material Safety Data Sheets have been and will continue to be revised to meet the requirements of the US Environmental Protection Agency, and now include information concerning the probable carcinogenicity of crystalline silica. Many producers have begun labeling their product bags with appropriate hazard warnings.

The health regulation issues associated with crystalline silica are expected to evolve rapidly in the United States during the next few years.

ACKNOWLEDGMENTS

The author acknowledges the support and suggestions of J.M. Sharratt, F.L. Kadey, Jr., G. Coombs, R. Illner, J. Robertson, T. Remmers, and J. Larsen. Acknowledgment is also given to P.A. Corbetta, J. Reynolds, and B. Kelly. Grateful acknowledgment is given to Arlene Sternberg Breese and William P.S. Breese for their valuable support.

BIBLIOGRAPHY AND SELECTED READING

Anon., 1969, "Mineral and Water Resources of Oregon," Bulletin 64, State of Oregon Department of Geology and Mineral Industries, pp. 209–210.

Anon., 1986, "Acid Secrets Beneath the Lochs of Galloway," *New Scientist*, Vol. 23, p. 35.

Anon., 1987, "Diatomite-No Skeletons in the Cupboard," *Industrial Minerals*, May, pp. 22–39.

Anon., 1989, "Diatomite-Tight Competition in Saturated Market," *Industrial Minerals*, Apr., pp. 30–33.

Anon, 1987a, "IARC Monographs on the Evaluation of the Carcinogenic Risk of Chemicals to Humans: Silica and Some Silicates," Vol. 42, World Health Organization, Geneva, Switzerland.

Anon., 1987b, "IARC Monographs on the Evaluation of Carcinogenic Risks to Humans—Overall Evaluations of Carcinogenicity: An Updating of IARC Monographs Volumes 1 to 42," Supplement 7, World Health Organization, Geneva, Switzerland.

Anon., 1990, "Diatomite 1990," 6th Edition, Roskill Information Services Ltd., London.

Anderson, R.V.V., 1933, "The Diatomaceous and Fish-bearing Beida Stage of Algeria," *The Journal of Geology*, Vol. 41, No. 7, pp. 673–698.

Baldwin, E.W., 1946, "Diatomite," *The Ore-Bin*, Vol. 8, No. 1.

Barron, J.A., 1987, "Diatomite: Environmental and Geological Factors Affecting Its Distribution," *Siliceous Sedimentary Rock-hosted Ores and Petroleum*, Van Nostrand Reinhold, New York, pp. 164–178.

Benton, W.E., 1983, "Economics of Diatomite," Society of Mining Engineers Preprint No. 83–363, pp. 1–15.

Bowen, R.G., Grey, W.L., and Gregory, D.C., 1963, "General Geology of the Northern Basin," *American Philosophical Society Transactions*, Vol. 53, Part 1.

Bradbury, J.P., 1988, "Fossil Diatoms and Neogene Paleolimnology," *Paleogeography, Paleoclimatology, Paleoecology*, Vol. 62 (1988), pp. 299–316.

Breese, R.O.Y., 1989, "Diatomite: Occurrence and Evaluation," Society of Mining Engineers Preprint No. 89–111, pp. 1–8.

Brittain, R.C., 1986, "Eagle-Picher Diatomite Mine and Processing Plant, Eastern Oregon," *Oregon Geology*, Vol. 48, No. 9, pp. 108–109.

Calacal, E.L., and Whittemore, O.J., 1987, "The Sintering of Diatomite," *American Ceramic Society Bulletin*, Vol. 66, No. 5, pp. 790–793.

California Health and Safety Code Sections 25249.5 through 25249.13, referred to as Proposition 65, The Safe Drinking Water and Toxic Enforcement Act of 1986.

California Health and Safety Code Section 44300 et seq., referred to as AB 2588.

Cleveland, G.B., 1966, "Mineral Resources of California," Bulletin 191, California Division of Mines and Geology, pp. 151–158.

Colbath, G.K., and Steele, M.J., 1982, "The Geology of Economically Significant Lower Pliocene Diatomites in the Fort Rock Basin Near Christmas Valley, Lake County, Oregon," *Oregon Geology*, Vol. 44, No. 10, pp. 111–118.

Davis, L., 1989, 1990, *Mineral Industry Surveys, 1989 and 1990*, US Bureau of Mines.

Ferrante, J.G., and Parker, J.I., 1977, "Transport of Diatom Frustules by Copepod Fecal Pellets to the Sediments of Lake Michigan," *Limnology Oceanography*, Vol. 22, pp. 92–98.

Fulton, J.A., and Smith, A.M., 1932, "Nonmetallic Minerals in Nevada," University of Nevada Bulletin, Vol. 26, No. 7, pp. 6–8.

George, R., 1989, "Products From Fossil Algae," *Algae and Human Affairs*, Cambridge University Press, New York.

Gilbert, C.M., and Reynolds, M.W., 1973, "Character and Chronology of Basin Development, Western Margin of the Basin and Range Province," *Geological Society of America Bulletin*, Vol. 84, pp. 2489–2510.

Guillard, R.R.L., and Kilham, P., 1977, "The Ecology of Marine Planktonic Diatoms," *The Biology of Diatoms*, University of California Press, Berkeley, CA, pp. 372–469.

Hurley, J.P., et al., 1985, "Groundwater as a Silica Source for Diatom Production in a Precipitation-Dominated Lake," *Science*, Vol. 227, pp. 1576–1578.

Kadey, F.L., Jr., 1983, "Diatomite," *Industrial Minerals and Rocks*, 5th ed., Vol. 1, S.J. Lefond, ed., AIME, New York, pp. 677–708.

Kamatani, A., 1971, "Physical and Chemical Characteristics of Biogeneous Silica," *Marine Biology*, Vol. 8, pp. 89–95.

Kanaya, T., 1959, "Miocene Diatom Assemblages From the Onnagawa Formation and Their Distribution in Correlative Formations in Northeast Japan," The Science Reports of Tohoku University, Japan, Ser. 2, Geology, Vol. 30.

Kotlarczyk, J., 1982, "The Role of Diatoms in Sedimentation and Biostratigraphy of the Polish Flysch Carpathians," *Acta Geologica Academiae Scientiarum Hungaricae*, Vol. 25 (1–2), pp. 9–21.

Kilham, P., 1971, "A Hypothesis Concerning Silica and the Freshwater

Planktonic Diatoms,'' *Limnology and Oceanography*, Vol. 16(1), pp. 10–17.

Krebs, W., Bradbury, J.P., and Theriot, E., 1987, ''Neogene and Quaternary Lacustrine Diatom Biochronology, Western USA,'' *Palios*, 1987, Vol. 2, pp. 505–513.

Kuzvart, M., 1981, ''Industrial Minerals and Rocks in Czechoslovakia,'' *Industrial Minerals*, Mar., pp. 33.

Lisitsyn, A.P., 1967, ''Basic Relationships in Distribution of Modern Siliceous Sediments and their Connection with Climatic Zonation,'' *International Geology Review*, Vol. 9, No. 5, pp. 631–652; No. 6, pp. 842–865; No.7, pp. 980–1004; and No. 8, pp. 1114–1130.

Lohman, K.E., and Andrews, G.W., 1968, ''Late Eocene Nonmarine Diatoms From the Beaver Divide Area, Fremont County, Wyoming,'' US Geological Survey Professional Paper 593-E, pp. E1-E24.

Macdonald, 1966, ''Cascade Range and Modoc Plateau,'' Geology of Northern California, Bulletin 190, California Division of Mines and Geology, pp. 65–96.

Malde, H.E., and Powers, H.A., 1962, ''Upper Cenozoic Stratigraphy of Western Snake River Plain, Idaho,'' *Geologic Society of America Bulletin*, Vol. 73, No. 10, pp. 1197–1219.

McIntire, D.C., and Moore, W.W., 1977, ''Marine Littoral Diatoms: Ecological Considerations,'' *The Biology of Diatoms*, University of California Press, Berkeley, CA., pp. 333–371.

Meisinger, A.C., *Minerals Yearbook, 1980–1988,* US Bureau of Mines.

Miles, W.J., 1987, ''Economics of Diatomite in the 1980's,'' Society of Mining Engineers Preprint No. 87–123, pp. 1–14.

Miles, W.J., 1990a, ''Economics of Diatomite in the 1990s,'' presented National Western Mining Conference Colorado Mining Association, Feb. 8, 1990.

Miles, W.J., 1990b, ''Mining Industry Responds to Crystalline Silica Regulations,'' *Mining Engineering*, Vol. 42, No. 4, pp. 345–348.

Murata, K.J., and Larson, R.R., 1975, ''Diagenesis of Miocene Siliceous Shales, Temblor Range, California,'' *Journal Research*, US Geological Survey, Vol. 3, No. 5, pp. 553–566.

Patrick, R., 1948, ''Factors Effecting the Distribution of Diatoms,'' *The Botanical Review*, Vol. 14, No. 8, pp. 473–524.

Patrick, R., 1977, ''Ecology of Freshwater Diatoms and Diatom Communities,'' *The Biology of Diatoms*, University of California Press, Berkeley, CA., pp. 284–332.

Patrick, R., and Reimer, C.W., 1966, ''The Diatoms of the United States,'' Monographs of The Academy of Natural Sciences of Philadelphia, No. 13, The Academy of Natural Sciences of Philadelphia, Philadelphia.

Pedersen, G.K., and Surlyk, F., 1983, ''The Fur Formation, a Late Paleocene Ash-bearing Diatomite From Northern Denmark,'' Bulletin Geological Society of Denmark, Vol. 32, pp. 43–65.

Pettifer, L., 1982, ''Diatomite—Growth in the Face of Adversity,'' *Industrial Minerals*, Apr., pp. 47–69.

Plafker, G., 1956, ''Occurrence of Diatomaceous Earth Near Kenai, Alaska,'' Bulletin 1039-B, US Geological Survey.

Powers, H.A., 1947, ''Diatomite Deposits of Southwestern Idaho,'' Report No. 4, Idaho Mineral Resources.

Schrader, H.J., 1971, ''Fecal Pellets: Role in Sedimentation of Pelagic Diatoms,'' *Science*, Vol. 174, pp. 55–57.

Soutar, A., Johnson, S.R., and Baumgartner, T.R., 1981, ''In Search of Modern Depositional Analogs to the Monterey Formation,'' *The Monterey Formation and Related Siliceous Rocks of California*, Society of Economic Paleontologists and Mineralogists, pp. 123–147.

St. Stoev, 1981, ''Industrial Minerals of Bulgaria,'' *Industrial Minerals*, Oct., pp. 73–81.

Stroebel, A.T., and Goodwin, J.A., 1979, ''Diatomaceous Earth Mining and Processing,'' Society of Mining Engineers Preprint No. 79–324, pp. 1–7.

Tappan, H., 1980, ''The Paleobiology of Plant Protists,'' W.H. Freeman and Co., San Francisco.

van Bennekom, A., and van der Gaast, S., 1976, ''Possible Clay Structures in Frustules of Living Diatoms,'' *Geochimica et Cosmochimica Acta*, Vol. 40, pp. 1149–1152.

ELECTRONIC AND OPTICAL MATERIALS

JOYCE A. OBER

Minerals for electronic and optical uses divide easily into two sections: 1) quartz and 2) minerals other than quartz.

QUARTZ

The properties of quartz crystal that make it useful for radio communications were discovered in 1918. Since that time, an industry for the mining and processing of natural quartz crystal has grown, matured, and been almost entirely replaced by new technology. The new technology still involves quartz crystal, but material that is grown rather than mined.

An economic summary of the commercial growing of quartz crystals has a place in a handbook directed to the mineral engineering industry because quartz crystals have long been an important commercial mineral, and the raw material for cultured quartz—that is to say, quartz crystals grown through the ingenuity of man—is still natural quartz.

Nearly all the natural crystals that have been used for electronics and optics came from Brazil. The larger pieces which met rigorous standards of quality were used for electronic and, to a lesser extent, optical components. Smaller pieces and fragments were used for vitreous silica. The need for high quality material in quantity led to US government sponsored research and exploration programs in the 1940s. No deposits meeting the very rigid requirements for electronic-grade quartz were found, but other projects resulted in the development of a process for the factory growth of beautiful crystals of prescribed shape, size, and quality. Domestic deposits of appropriate quality were identified to use as raw materials for the quartz culturing process.

The development of the cultured quartz crystal illustrates the success that technology can have in adapting a product of the mine to increasingly sophisticated uses. A remarkable achievement perhaps, but foreshadowed by experiments by Giorgio Spezia (1908), an Italian geologist studying the relative effects of temperature and alkaline environment on the solubility of quartz.

Modern radio equipment is most often controlled as to frequency by the presence in the circuit of a separately added crystal—the 1918 discovery responsible for the existence and growth of the quartz industry. The crystal is quartz, but this component is a carefully oriented and prepared slice from a crystal, but not a crystal as recognized by a rock hound or seen in a museum.

How quartz operates to control frequencies is not a proper subject for a handbook on industrial minerals, and references should be consulted (Cady, 1964, Mason, 1964). Quartz belongs to a class of materials called dielectrics: those that do not conduct an electric current but permit electric fields to exist and act across them. Quartz shows the piezoelectric effect, which means that when a quartz plate is mechanically deformed against its natural stiffness, one of its surfaces becomes negatively charged, the other positively charged. When the plate is released quickly from the stress, the charges disappear as the plate regains its original shape, but because of mechanical momentum the plate deforms in the opposite direction (to a lesser amount) and the surfaces correspondingly become charged in the opposite direction. By thinly coating the two surfaces with metal and attaching flexible wires, these charges can be brought into an electronic circuit. If the surfaces are suddenly electrically charged by movement of current through the wires, the converse piezoelectric effect occurs and the plate deforms. Carry the thought further and it is realized that an alternating current flowing through the wires responds to the mechanical oscillation. By controlling the thickness of the plate, its mechanical vibration frequency can be varied through a wide range.

One type of quartz plate, the AT-cut, has a precisely defined orientation with respect to the crystallographic axes of the crystal and vibrates on a microscopic scale much as a book would deform when placed flat on a table and the top cover moved parallel back and forth with the hand. At least 17 other orientations have been studied, some of which have preferred uses in various applications (Cady, 1964).

The quartz crystal industry is composed of three main segments (excluding fused quartz and quartz used for optical purposes):

1. *Natural electronic-grade quartz crystals.* Mined quartz suitable for fabrication into piezoelectric units. Zlobik (1981a) estimated the waste to ore ratio at 1:1 000 to 1 000 000, depending upon the deposit.
2. *Lasca.* Mined quartz usable as feedstock in the production of cultured quartz. Approximately 0.63 kg of lasca are required to produce 0.45 kg of cultured quartz.
3. *Cultured quartz.* Cultured quartz is produced from lasca feedstock in a process of crystal growth in an autoclave under conditions of heat, pressure, and time. It is estimated that 0.45 kg of cultured quartz is equivalent to 1.4 to 4.5 kg of natural quartz crystal in yield of commercial quartz suitable for slicing into piezoelectric units.

The chronology of the development of the quartz crystal industry both natural and cultured follows:

Date	Comment
1918	Discovery of the piezoelectric effects of quartz crystal
1921	Application of the piezoelectric effects of quartz crystal in the circuitry of radios
1948	Establishment of a quartz crystal commodity stockpile by the US Government
1952	US consumption of natural quartz crystal at an all time high of 228 t
1958	First commercial production of cultured quartz crystal
1970	Cultured quartz crystal production exceeds imports of natural quartz crystal
1971	Cultured quartz crystal consumption surpasses natural quartz crystal consumption

1976 Surge in the use of quartz crystal, both natural and cultured, due to the citizen band radio sales and later decline due to change in technology in citizen band radio circuitry

1977–1990 Continued replacement of natural quartz crystal by cultured quartz crystal to the point that virtually no natural crystal is used in electronic and optical applications. New developments in the use of quartz crystal in watches and in microprocessors for automobiles and toys

Production and Consumption

In 1990, one company in the United States produced some 423 t of lasca. Specimen and jewelry material was also produced. Lasca is believed to have been imported from Brazil and Namibia, but changes in the Bureau of Census's reporting system made this information unavailable. Some 527 t of lasca were used by five companies in four states to produce an estimated 441 t of cultured quartz in 1990. The US government stockpile of natural quartz crystal was 686 t at the end of 1990, of which 577 t was classified as excess. Stockpile sales resumed in 1989 after a 3-year hiatus. Very little, if any, of the material was used for electronic or optical uses; it was used as specimen, carving, or jewelry material.

Exports and Imports

As noted previously, Brazil is the world's major producer and exporter of natural electronic-grade quartz crystal and lasca. The United States and Japan are the major importers and consumers of lasca as well as the major producers, consumers, and exporters of cultured quartz crystals. Japan, the Federal Republic of Germany, Hong Kong, and Switzerland received most of the exports of cultured quartz from the United States.

Prices and Trends

Prices of quartz material vary considerably from about \$1.00/kg for lasca to over \$100/kg for cultured quartz.

The electronic-grade quartz crystal industry ranges from fully integrated producer/consumer (involving the base-mineral production, cutting of blanks for piezoelectric unit, and fabrication into finished electronic units), to companies that are only involved in segments of the industry. Pricing in the industry reflects this segmented or captive structure depending on where the transfer of the mineral takes place.

Future trends in the industry are expected to be:

1. Some continued displacement of natural electronic-grade quartz crystal with cultured quartz crystal, although that displacement is nearly complete.
2. Further worldwide development of quartz production for lasca use as an alternative to Brazilian material.
3. Growth in the exports of US cultured quartz to countries that, due to favorable labor rates, produce piezoelectric-containing electronic units; development of cultured quartz crystal production or expansion of existing production in these locales.
4. Increased use of quartz crystal-containing electronic units as timing devices in microprocessors, for example, in automobile production.

Production of Cultured Quartz

The early history of producing large, high quality quartz crystals in the laboratory and factory, and the general manufacturing process, have been described by Waesche (1960).

Today there are five operations in the United States, and operations are reported in at least eight other countries: Belgium, Brazil, England, France, Germany, Japan, South Africa, and the former USSR.

Fig. 1. Cultured quartz crystals are grown in autoclaves. Racks of seed crystals are inserted and at the end of the growing cycle, 30 to 45 days later, racks of crystals are removed as shown here (*courtesy*: Motorola Inc.).

The manufacturing process makes use of hydrothermal growth, so named because it involves an aqueous solution at elevated temperature, and is similar to the hydrothermal action that develops ore bodies and crystal formations in the crust of the earth. The synthetic process may be briefly described as follows: a vertical steel vessel is provided with suitable internal fittings and pressure gage, safety blow-off valve, thermocouples, and a source of electrical heating. The vessel is charged with 1) nutrient material (lasca) to about $\frac{1}{4}$ of its volume, 2) dilute alkaline solution to about $\frac{3}{4}$ full, and 3) an array of seed plates. The vessel is closed and brought to a temperature of about 350°C, producing a high pressure. The heat source is adjusted to provide a temperature difference between the area of dissolution and the area of deposition and growth. Conditions of constant temperature and pressure are maintained for a period of weeks or longer. The status of the growing crystals can be monitored using radiography with cobalt-60.

Vessels up to 254 mm ID, and capable of operating at up to 275.8 MPa, are in use. Fig. 1 shows manmade quartz crystals being removed from an autoclave or reaction vessel.

The conditions of growth used by two major US producers (Sawyer Research Products, Inc. and Thermo Dynamics, Inc.) are compared here:

	Sawyer Research Products, Inc.*	Thermo Dynamics, Inc.*
Solution	0.83 molar Na_2CO_3	1.0 molar NaOH
Percent of fill†	76	80–82
Growing temperature	350°C	354.5°C
Dissolving temperature Δ	+5 to 20°C	+26.7 to 37.8°C
Pressure, MPa	82.7	158.6
Time, days	42	28–30

*Private communication.
†Percent of fill: percent of the internal volume of the vessel not occupied by solid substances when it is being charged.

Although there have been no deposits mined for natural electronic-grade quartz in the United States (Waesche, 1960), high

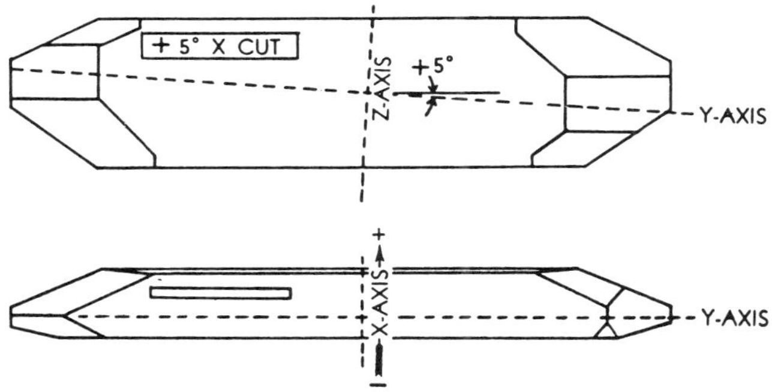

Fig. 2. Drawing of cultured quartz crystal slightly modified from standard Y-bar, the seed being cut at 85° to the c-axis, as shown, instead of 90° to the c-axis. The drawing shows elevation and plan views (Sawyer, 1972).

Fig. 3. Synthetic quartz crystals are produced in a variety of sizes, then cut to specification dependent on final end use requirements. This photograph shows two crystals as-grown and four that have been lumbered prior to slicing crystal blanks (*courtesy*: Motorola Inc.).

purity US quartzite (lasca) has been used successfully in producing cultured quartz.

Seed: The earliest seed plate that had much attention was a thin plate cut parallel to the minor rhombohedral face, often identified as *r* or as $\{01\bar{1}1\}$ by Bravais symbols, a face which makes an angle of 38° 13' with the *c*-axis. This seems a reasonable choice since the plate should grow parallel to itself, and the AT-cut plate (Cady, 1964), whose surface makes an angle of 35° 15' with the *c*-axis, is perhaps the most used for frequency-control oscillators. Indeed, this choice of seed orientation was patented (Friedman and Tuttle, 1951), particularly, however, as promoting rapid crystalline growth. The patent states that this orientation grows "much more rapidly than seed plates cut at other angles." However, it turns out that in the alkaline solutions used for growth, the minor rhombohedral face soon grows itself out of existence! The major rhombohedral face grows much slower, the prism faces grow at a negligibly slow rate, and the crystal ultimately becomes a more or less square bar (depending on how the seed was supported) elongated in the *c*-direction, lengthening slowly.

In 1954 a newer seed type called the *Y*-bar seed (Jaffe and Turobinski, 1960) was invented and studied: an elongated quartz crystal seed bar having its length extending substantially perpendicular to a crystallographic *X*-axis and substantially perpendicular to the *Z*-axis of the crystal, said seed bar being small in cross-sectional dimension perpendicular to its length. In piezoelectric engineering the *X*-, *Y*-, and *Z*-axes are mutually perpendicular, the *X*-axis being one of the *a*-axes, and the *Z*-axis being the *c*-axis, in the hexagonal system to which quartz belongs. This seed type became so successful that it is now the standard seed in the cultured quartz industry.

On this seed the growth takes place rapidly in the *c*-direction, easily 1 mm per day under good conditions, and less rapidly in the *a*-direction at right angles. The growth in the *c*-direction yields, as the mineralogist would expect, no flat face but, on the contrary, a surface composed of many rounded small mounds occasionally showing growth spirals. The *c*-face or basal face, (0001), is seen sometimes on the tiny crystals studied as micromounts, but it is not a plane.

Growth on a *Y*-bar seed proceeds in both the *c*-direction and the *a*-direction. Studies have shown that the highest quality is the *c*-direction growth; hence a modified *Y*-bar is often chosen as a seed, being long in the direction perpendicular to an *a*-axis and of moderate dimensions parallel to the *a*-axis (Fig. 2).

Growth in the *c*-direction has been found to be highly satisfactory for producing crystals destined for the relatively small AT-cut oscillator plates and various other units, but the need for longer lengths for frequency filters has led to the successful use of relatively large seeds oriented parallel to the minor rhombohedral face, $\{01\bar{1}1\}$, a cyclic sweep to the early seed (Lias and Rudd, 1969). Fig. 3 shows representative man-made quartz crystals.

The *Q* of Quartz: How does one judge the quality of a find of quartz crystal? A material with a subtle use may require a subtle test of its useful value. The *Q* of quartz, a dimensionless quotient, is the mark of electronic quality. It is often called mechanical *Q*. It is highest when energy loss is lowest in a quartz oscillator due to internal friction. Given a resonator vibrating at resonance frequency while connected to a source of ac of that frequency, the instant the current is stopped, the vibration amplitude decays. The *Q* expresses the slowness of the decay. The dying sound of a struck bell is a close analogy. A bell having a long tintinnabulation might have a *Q* of 250 000; a good quartz resonator has a *Q* of about one million.

Practically, the *Q* of a resonator can be determined as a ratio,

$$Q = \frac{2\pi fL}{R}$$

where the quantities are measured for the equivalent electrical circuit (a resonating electronic circuit of the same frequency), *f* being the resonance frequency, *L* the inductance, and *R* the resistance (Heising, 1946).

A direct method for observing *Q* displays the decay of the amplitude of a specially made 5 MHz oscillator on the screen of a cathode ray tube (King, 1959). Still another method involves the examination of the graph of resistance as a function of frequency (Adams et al., 1968).

The *Q* of the best natural quartz (measured at room temperature) lies in the range 1.0×10^6 to 3.0×10^6, and the usual good cultured quartz is about 1/10 of this range. In the early days of quartz-growth study, *Q* was much lower, and although the crystals withdrawn from the pressure vessels were clear and sparkling in their beauty, it was discouraging that the men in white coats could not improve a little on nature.

At Western Electric Co. (a former quartz crystal producer), quartz finally attained a coveted high value of *Q* by the use of $LiNO_2$, lithium nitrite, as an additive to the growth solution (Ballman et al., 1966). Lithium as LiF had been found to improve the *Q*, perhaps by decreasing the infrared absorption at 2.86 μm (or the wave number, 3 500 cm^{-1}) (King et al., 1962). The addition of the nitrite raised the *Q*, at slow growth rate, into the region of natural quartz values. The anion may be adsorbed onto the growing crystal and may decrease the presence of hydroxyl ion in the quartz.

Sawyer Research Products has developed means of growing what it labels Premium Q Cultured Quartz through slow growth in sodium carbonate solution which, in general, yields a slower rate than sodium hydroxide solutions (Rudd et al., 1966, Sawyer, 1972). Sawyer tightened quality control on all pertinent aspects of the growth process with the result that a series of runs has been produced whose 5 MHz 5th overtone Qs were consistently higher than those yielded by natural quartz.

Compared to glass, quartz has a considerably widened spectral window. Hence the use of vitreous silica for optical components and the use of quartz crystal in spectrometer prisms. However, quartz shows a number of absorption peaks in the infrared, certain of which have been identified with impurities—elements and radicals. In particular, hydrogen, which is present in great quantity in the alkaline growth solutions in the hydroxyl ion as well as in aqueous solutions in the earth's crust, is known to be responsible for a characteristic high absorption peak in the wave-number region, 3 000 to 3 700 cm^{-1} (Katz, 1962).

A relationship was determined experimentally between the infrared absorption at 3 500 cm^{-1} and the mechanical Q (Cady, 1964). Use of this information led Western Electric (Rudd and Lias, 1967) and Sawyer Research Products (Sawyer, 1972) to inspect their production of crystals for Q by measuring this absorption with an infrared spectrophotometer. This simple and rapid method has made it easier to monitor the research programs leading to improved Q as well as to provide quality control.

Defects in Quartz

For much of the product of mine and smelter, microscopic and submicroscopic details need no attention. On the contrary, quartz destined for piezoelectric oscillators must be reasonably free from various types of defects down to those affecting the crystalline structure. Defects include twinning, structural imperfections such as slip and screw dislocations, impurity ions, and microscopic inclusions. For a review of defects affecting the quality of quartz for electronic and optical use, see Waesche (1960) and National Stockpile Purchase Specification, P-43-R1, 1968 (Department of Commerce). For a more general mineralogical view see Frondel (1961).

Quartz may show two types of twinning: electrical or Dauphiné twinning and/or optical or Brazil twinning. These are beautifully displayed under reflected light on slices etched in hydrofluoric acid, a standard method for checking the quality of natural quartz during the course of oscillator manufacture (resonator plates must not show twin boundaries). The seed being carefully chosen in cultured quartz, twinning is very rarely seen.

At the start of the crystal-growing process when pressure vessel, solution, and seeds are heating up, the alkaline solution is likely to dissolve holes and tubes along dislocation lines such that when growth begins, the opening is covered and a microscopic bubble results. If, instead of continuing the process, the vessel is cooled and opened, the seed plates will show etch pits of characteristic shapes, and suggest the density of the defects.

Studies using a special double-crystal X-ray spectrometer technique have disclosed much information on structure defects. The technique produces a topography of internal structure, and assists in relating defects in structure to others such as substitutional impurities (Lang, 1967).

Considering growth on the basal plane yielding the cobble texture, the basic mechanism of growth seems to be along screw dislocations; this often results in a spiral pattern visible to the eye or under the microscope. Examination by light interference has produced pictures of great interest and beauty (Joshi and Vagh, 1964).

Quartz often darkens in certain areas when irradiated with X-rays, high voltage electrons, or gamma rays from cobalt-60. This phenomenon was traced to the presence of Al^{3+} substituting for some of the Si^{4+}. The charge compensation for the trivalent aluminum generally is provided by Na^+, H^+, or Li^+ if the latter has been added intentionally as an impurity (Ballman et al., 1966). A broad study of hydrogen in quartz has illuminated some of this behavior (Katz, 1962).

A powerful method of studying a variety of defects, substitutional, interstitial, and structural, is to place a quartz block between metal electrodes, heat the assembly, and apply a dc voltage for a period of time. Pfenninger (1961) made a detailed study using electrodes of silver, gold, copper, and platinum; voltages as much as 670; and temperatures to 800°C. Observation methods included the microscope, the electron microscope, and infrared spectrometry. A beautiful set of photographs shows veils of defects entering quartz, leaving it, and decorating lines of structural dislocations.

Regretfully this investigation did not encompass cultured quartz. However, for preparing quartz, especially stable in a radiation environment, cultured blocks are now swept of various defects by this general procedure.

A defect occasionally seen—mostly in earlier, less well controlled growing operations—is microscopic white particles identified as clumps of acmite crystals, $NaFeSi_2O_8$.

Developments in Quartz Growth and Technology

Low Optical Transmission Loss: Cultured quartz has been examined as a material for transmitting infrared laser light of 1.06 μm wavelength (from the YAG:Nd laser). It was found that growth in a silver tube to diminish iron contamination yields a quartz equivalent in low optical loss to the best commercial vitreous silica (Lias and Rudd, 1969). An interesting correlation was found between H^+ concentration in the quartz and the absorption at 1.06 μm, the hydrogen ions seemingly providing charge-compensation for the Fe^{2+} or other divalent impurity ions associated with absorption in that region.

Growth Rate 2 mm/d on Basal (0001) Plane: Stemming from an eagerness to grow the crystals faster (an economic advantage if characteristics can be maintained), growth studies at 374°C and a pressure of about 275.8 MPa led to a rate of about 2.6 mm/d and a Q of the order of 1.4×10^6 (Lias et al., 1973). The high rate due to higher expected solubility was anticipated; the excellent Q was a bonus and may signify that at the higher temperature H^+ goes less readily into the growing crystal.

Measurement of Viscoelastic Materials: A device successfully used to study the shear modulus of viscous materials employs a delay line of suitable mechanically rigid material, e.g., vitreous silica or aluminum, having mounted on the end a piezoelectric transducer to generate shear waves. This is typically quartz, providing frequencies in the range 1 to 7 MHz, but may be a piezoelectric ceramic. The delay line is inserted in the sample and the system brought to constant temperature to within 0.005°C. Pulses of waves are sent down the delay line, and the echoes are measured as to their attenuation and phase shift, from which an accurate calculation can be made of the shear modulus (Hunston et al., 1972).

An ultrasonic impedometer has been made to measure the dynamic mechanical properties of viscoelastic materials (adhesives, paints). It employs Y-cut, shear-wave resonators of 5MHz frequency attached to the ends of a 10-cm bar of vitreous silica. The bar ends are cut at 79° producing a shape suggesting a bathtub. The bulk shear wave is reflected off the top surface of the bar on which is spread the material to be studied. The reflected wave returns to the transmitting transducer and an amplified signal brought out on an oscilloscope. The impedance of a film of material can be monitored as it is drying (Myers and Schultz, 1962). For further general information on ultrasonics see Bradfield (1970).

Pressure Transducer: Quartz being piezoelectric, i.e., being a material which produces an electric field as a response to stress, the question arises how useful a pressure gage can be made from it? A gage having a range of 0 to 69 MPa and an accuracy of about 0.07 kPa is made using a periphery-stressed resonator as a pressure transducer. The transducer unit is a circular 5 MHz resonator whose edge is integral with a surrounding cylinder of quartz, the axis of the cylinder being identical with the normal through the resonator center. The cylinder and resonator are fashioned by grinding techniques from a single crystal of quartz! The resonator may be an AT-cut or other cut to provide that the temperature coefficient of frequency at zero pressure is zero at a given temperature in the range 0 to 100°C (Karrer and Leach, 1969).

Pressure Gage for High Impact Conditions: A serious study, largely at Sandia Laboratories (Albuquerque, NM), has produced a gage technique for studying the stresses due to shock-wave loading to pressures as high as 6 x 689 MPa. Based on the piezoelectricity of quartz, the device has been used to study the behavior of quartz and also other solids under great impact (Ingram and Graham, 1972). A valuable use of an outgrowth of this technique is the measurement of forces produced in underground nuclear explosions.

While stresses of 4 GPa are routinely measured, a tungsten facing on the quartz disk extends the upper stress limit of the experiment to 12 GPa. Ordinarily the stress in an oscillating quartz resonator is probably of the order of 600 kPa.

A projectile in a cylinder is faced with a disk of quartz, and another disk is provided at the closed end of the cylinder. In this Sandia Quartz Gage for impact experiments, the cylinder is a 27-m compressed gas gun which can accelerate the projectile to 510 m/s. This impact velocity can be measured to 0.01%.

Disks of *X*-cut (the face of the disk being perpendicular to an *a*-axis) cultured quartz are used, which may be 126 mm in diameter and 6.3 mm thick—these dimensions yielding about a quarter of an ampere of piezoelectric current at a stress of 1 GPa.

During the study of strains up to 4.3×10^{-2} in these experiments, the analysis of data provided unusually accurate values for the linear and nonlinear piezoelectric stress constants and longitudinal elastic constants up to the fourth order (Graham, 1972).

Quartz Thermometer: A quartz resonator having its resonance frequency unvarying with temperature has not been made, but the behavior with temperature has been thoroughly studied, so that, for a given temperature range, a resonator may be made having a reasonably low, and accurately known, variation. Advantage has been taken of this property to measure temperature difference (Smith and Spencer, 1963). A 5° *Y*-cut (Cady, 1964), operating at 5 MHz, has a temperature-frequency coefficient of about 80 ppm per °C and responds to a temperature deviation of 3.8×10^{-6}°C, indicating a very sensitive device.

A commercial development of the quartz thermometer led successfully through an analysis of plate orientations to find one yielding a linear frequency-temperature relationship. The available instrument has a range of −80 to 250°C, with a maximum resolution of 0.000 1°C. A digital readout is used, and with optional amplifiers the sensory probe may be as far as 1 372 m from the measuring equipment (Hammond et al., 1965).

Quartz Microbalance: A quartz resonator can be brought to a slightly lower frequency by loading it with a little extra mass, e.g., of evaporated metal. Hence the change in frequency on loading makes the resonator a mass-measuring device. This idea has led to the construction of microbalances reading to about 10^{-12} g for certain unusual applications, for example the measurement of gases that can be adsorbed in a layer of material coating the resonator: a hydrophilic substance for weighing water vapor. The frequency change is directly proportional to the change in mass. A 5-mm-diameter, 15 MHz resonator has a mass sensitivity of 2 600 Hz change per microgram. The resonator is usually an AT-cut. The frequency is, of course, also a function of temperature. In one method of assisting in this problem, a zigzag strip of evaporated-on nickel which serves as heater, thermocouple, and radio-frequency electrode is applied to one side of the resonator (King, Jr., 1969, Saubrey, 1964, Stockbridge, 1966).

Surface Waves on Quartz: Elastic solids may display bulk waves or surface waves. Bulk waves are found in quartz oscillators themselves and in buffer rods or other attachments for conveying a vibration from an oscillator along a length of rigid elastic solid. A use of bulk waves in such a delay line is mentioned under Measurement of Viscoelastic Materials. Bulk waves in quartz have been used also to modulate light.

Surface waves—those Lord Rayleigh waves which now find standard application in earthquake study—can be produced on various crystalline solids (Viktorov, 1968, White, 1970). They can be launched in the *X*-direction (i.e., along an *a*-axis) on a quartz bar by means of electric impulses applied to the surface through two evaporated electrodes having the appearance of two interleaved combs, or the illustrative diagram of a multiplate capacitor. They can be received and converted into electrical signals by similar electrodes at the far end. The bar or plate of quartz may be cut parallel to a plane containing an *a*- and the *c*-axis—the *Y*-cut of the piezoelectric engineer. In one application the bar is 254 mm long. The interesting advantage of using elastic waves in certain electronic circuits is the slowness with which they move: about 1.5×10^3 to 15×10^3 m/s instead of the 3×10^8 m/s which is the velocity of electric signals and light. The surface waves move in the lower third of that range. A further advantage of surface waves is that they can be sampled or modified. If they are accompanied by electric or magnetic fields (when they are produced in piezoelectric or magnetic materials), the passage of the wave and its characteristics can be picked up by a sensing device not in contact with the surface.

Another synthetic crystal is also of much interest in surface wave applications: lithium niobate (see synthetic crystal section).

Miscellaneous Uses: Further uses of quartz resonators have appeared in recent years in color television circuits and in watches. The resonator used in a watch design is not the face-shear mode most used for high frequency, 0.5 to 50 MHz, but a flexural, or wave-shaped, mode operating in the range 2 to 20 kHz (Cady, 1964).

MINERALS OTHER THAN QUARTZ

Crystals of a size to hold in the palm of your hand are of enormous interest to the technology of the 1990s because of the astonishing things they can be made to do. Not many mined mineral crystals have been among these, or are likely to be. However, a scientist in a modern physical laboratory, dreaming up a circuit (electronic, optical, magnetic, or any combination) does not ask whether an existent mineral might help, but asks whether the professional crystal grower can provide a material that the solid state physicist says might have some of the properties to make the dream realizable.

Few minerals are found having sufficiently high quality to be employed directly. There is one special exception, calcite; optical calcite is becoming so hard to find in the sizes demanded that investigations are under way to synthesize it.

Natural quartz is becoming less and less important in the frequency-control oscillator industry; mica and selenite are used in very small amounts in making light retardation plates as microscope accessories.

Table 1. Minerals for Electronic and Optical Uses

Dana No.	Mineral	Chemical Formula	Crystal System	Related or Replacing Crystal	Useful Property and Reference	Application	Device, Equipment	Commercial Growers (Keyed to Table 2)	Growth Studies, References
2613	Altaite	PbTe	Iso	PbS	Photoconductivity (Bube, 1960)	Pyrometry	Photoconductors, IR spectroscope (Bube, 1960), semiconductor laser (Harmon, 1971)		
	Almandite	$Fe_3Al_2Si_3O_{12}$	Iso	$Y_3Fe_5O_{12}$ "YIG" $Y_3Al_5O_{12}$ "YAG"	Magnetooptic Fluorescence Hardness, n=1.8 (high)	Microwaves Laser, YAG:Nd (Geusic et al., 1964) Gem, YAG	Laser, YAG:Nd (Geusic et al., 1964)	1,3,14	(Makram, 1968) (Rudness and Kebler, 1959)
	Apatite	$Ca_5F(PO_4)_3$	Hex		Fluorescence with Mn (Ryan, 1970) IR spectra (Knubovets et al., 1970)		Laser, high gain		(Tolksdorf, 1968) (Mazelsky et al., 1968)
38.4.3	Berlinite	$AlPO_4$	Hex	Quartz	Piezoel. (Stanley, 1954) Piezoel. (Cady, 1964)	Probably none; piezoel. low			
26.1.7	Boracite	$Mg_3B_7O_{13}Cl$	Iso Ort (Cady, 1964)		Pyroel. (Cady, 1964) Ferroel. (Jona and Shirane, 1962)				(Nassau and Shiever, 1972)
9.1.1.5	Bromyrite	AgBr	Iso		IR transmission 0.5–35µ		IR spectrophotometer	7	
39.2.1	Brushite	$CaHPO_4$	Mon	KH_2PO_4?	Piezoel. (Cady, 1964)				
14.1.1.1	Calcite	$CaCO_3$	Hex	$NaNO_3$	Strongly birefringent	Polarization of light. (Waesche, 1960)	Nicol prism. Glan-Thompson prism, etc.	9,15	(Hatta et al., Henisch, 1970 Liaw and Faust, Jr., 1972)
9.1.1.4	Cerargyrite	AgCl	Iso		IR transmission 0.4–30µ		IR polarizer by reflection	7	
269	Cinnabar	HgS	Hex		Optical rotation, IR transmission 1–13µ		Photoconductors		(Toudic and Aumont, 1971)
2616	Clausthalite	PbSe	Iso	PbS	Photocond. in IR to 5.6µ	Detect IR radiation.	IR spectrophotometer (Bube, 1960)		
25.1.9	Colemanite	$Ca_2B_6O_25H_2O$	Mon		Ferroel. (Cady, 1964)				
8321	Columbite	$(Fe,Mn)(Nb,Ta)_2O_2$	Rho		Piezel. Ferroel. Luminescence (Hordvik and Schlossberg, 1972)	Electroop. modulation. Hologram recording (Laudise, 1968) (Stepke, 1972)		7,8,14	(Laudise, 1968)
4411	Corundum	Al_2O_3	Hex	Sapphire, ruby	High ref. index.	Substrate for Si (Cullen, 1971) Basis for star sapphires		2,7,10,13,14	(Pollock et al., 1972)
9.2.1	Fluorite	CaF_2	Iso		Fluorescence Ref. index low Dispersion low Transmission 0.12–9.0µ	Windows, prisms, lenses.	Scintillation counter, CaF_2:Eu. Laser (Garbuny, 1965) (Nassau,1965)	7,10,11	(Nassau, 1965)
2611	Galena	PbS	Iso	Note altaite and claust-halite	Photocond. to 2.8µ	IR detection (Bube, 1960)		7,8,14	(Laudise, 1968)
2642	Greenockite	CdS	Hex		Piezoel. Photoc., Photovoltaic, Pyroel., Electrooptic, Linear compress-ability (44)	Mixing light beams.	Photoconductors, solar cells. Acousto-electric oscillator (Fochs, 1970)	5,6	(Fochs et al., 1968) (Nassau and Shiever, 1972) (Parker and Pinnell, 1968)
9.1.1.1	Halite	NaCl	Iso	NaI	Transmission, 0.2–15µ	Windows, prisms, lenses.	IR spectrometers, X-ray spectrometers.	7,10,11	
28.4.3.1	Langbeinite	$(NH_4)_2Cd_2(SO_4)_3$			Ferroel. (Jona and Shirane, 1962)				
	Leushite	$NaNbO_3$	Orth	Other niobates, see columbite	Antiferroel. (Jona and Shirane, 1962)	(See columbite)			
7216	Magnetite	Fe_3O_4	Iso	$ZnNiFe_2O_4$ $BaFe_2O_4$	Low coercive force High remanence	Ferromagnetic application; replaced by better crystals	Memory devices and permanent magnets		

	Muscovite	$KAl_3Si_3O_{10}(OH)_2$	Mon		Birefrigent, good dielectric coef. high diel. strength.	Insulation.	Capacitors, 1/4-wave wave plates.	15	
9.1.3.1	Nantokite	CuCl	Iso		Electrooptic, transmission 0.4–20μ	Electrooptic	Light modulator (Murray, 1964)		
4211	Periclase	MgO	Iso			Insulator		8	(Butler et al., 1969)
7421	Perovskite	$CaTiO_3$	Orth	$BaTiO_3$ (Tet) $PbZrTiO_3$	High diel. coef. Ferroel. (Jona and Shirane, 1962)	Modulation of light phase (Kaminow, 1965)	Phono pickups, many piezoel. devices (Wheeler, 1972)		(Nassau and Broyer, 1962) (Von Hippel, 1963)
				$SrTiO_3$ (Iso)	High ref. index	Gem			
3212	Proustite	Ag_3AsS_3	Hex		Pyroel., photocond., Electrooptic.	Electroop. modulation			(Bardsley and Jones, 1968) (Gentile and Stafsudd, 1968) (Hulme, 1967)
	Quartz	SiO_2	Hex	No equivalent crystal	Piezoel. (Cady, 1964) (Mason, 1964) Transmission, 0.15–3.5μ Birefringent. Color grown in (Crowningshield, 1972) (Pough, 1970)	Frequency control (Cady, 1964) surface waves (de Klerk, 1972) (White, 1970) cultured gems (Crowningshield, 1972) (de Klerk, 1970)	Oscillator or "crystal", delay lines, transducers, filters.	12	See Sec. 38 on "QUARTZ" IMAR, 4th ed.
4411	Ruby	Al_2O_3:Cr	Hex	Corundum	Fluorescence, high hardness	Holography, communications.	Laser, (Garbuny, 1965) (Maiman, 1960)	14	(Nassau and Broyer, 1962a)
4511	Rutile	TiO_2	Tet		High diel. coef. 173 perp. c-axis			10	(Farrell and Ling, 1972) (Nassau and Broyer, 1962) (Rosenberger and Delong, 1972)
	Sapphire—	See Corundum							
48.1.3.1	Scheelite	$CaWO_4$	Hex	Corundum	Fluorescence, electroluminescence.		Laser: $CaWO_4$:Nd (Garbuny, 1965) (Johnson and Nassau, 1961) $CaWO_4$:Sm	8	(Nassau and Broyer, 1962a)
29.6.3	Selenite	$CaSO_42H_2O$	Mon		Birefringent	Petrography	1/2-wave plates		
2621	Sphalerite	ZnS	Iso		Piezoel. (Cady, 1964) Birefringent (Yu and Cardona, 1973) Optical rotation	Light modulation (Buhrer et al., 1963)		5	(Parker and Pinnell, 1968)
7211	Spinel	$MgAl_2O_4$	Iso		Good transmission, Diel. coef.:8.4	Substrate for Si (Cullen, 1971)		10,14	(Wood and White, 1968)
9.1.1.2	Sylvite	KCl	Iso	KBr KI	Transmission 0.38–21μ	Study of color centers		7,10,11	
8322	Tantalate	$(Fe,Mn)(Ta, Nb)_2O_6$		Various tantalates $LiTaO_3$ (Jona and Shirane, 1962)	Ferroel. Low biref., transmission 0.35–4.0μ	Light modulation (Laudise, 1968)		4	
454	Tellurite	TeO_2	Tet		Piezoel., low velocity shear wave.	Acoustooptic (Bonner et al., 1972)	Light deflector, modulator.	8	(Bonner et al., 1972)
	Tourmaline	$NaFe_3B_3Al_3(Al_3Si_6O_{27})(OH)_4$ Schorl variety	Hex	Quartz	Piezoel., pyroel. (Cady, 1964) Slightly electrooptic.				
	Vaterite	$CaCO_3$	Hex		Strong pos. birefringence (Jona and Shirane, 1962)	No use noted.			
48.1.4.1	Wulfenite	$PbMoO_4$	Tet	$CaMoO_4$	High polarizability (Pinnow et al., 1969) Transmission, 0.45–3.9μ	Acoustooptic	Light deflector, modulator.	8,14	(Pinnow et al., 1969)
4221	Zincite	ZnO	Hex		Piezoel.	Acousto-electric (Kolb, 1967)		10	(Nielsen, 1968) (Kolb, 1967)

If a mineral is observed to have a useful property, the crystal grower can probably reproduce it in an improved crystal; larger, freer of optical flaws, purer or conversely doped with impurities, in either case to provide it with an effective property.

The synthetic crystals are, in any case, derived from mined ores, which undergo various refining and purifying processes—some of them rather sophisticated, leading then to the growing and cutting and polishing into usable shapes.

Many of the currently exploited properties of light and crystals have been known since early in the 18th century! Recently development techniques have been needed to study and apply them, e.g., intense light beams, high optical quality crystals, sensitive measuring equipment. The technology in optical communication is now on the edge of such an expansion as radio and television experienced in the 1940s. The frequency-control resonator, the diode, and the transistor were among those early achievements. Radio frequencies have given way, along the growing front of research, to ultra high frequencies close to and within the visible range of the spectrum.

Minerals of Technological Interest

Table 1 is a listing of minerals studied earlier, in current commercial use, or under present investigation. There are others not listed, some of only casual past interest, and some which have been unintentionally overlooked.

The list is alphabetical and includes quartz, the subject of the previous section. The mineral name is preceded in most instances by the Dana identifying number. The chemical formula and the crystal system are given. It should be noted that this is the system for the crystal at ordinary temperature; at lower temperatures a crystal may move into a less symmetric system, and vice versa. Where appropriate, related or replacing crystals are mentioned; the latter have enhanced, or developed, new and exciting properties.

The useful property, the application, and the device are not easily differentiated in every case, depending on the state-of-the-art. Sometimes they are, e.g.:

Mineral	Useful property	Application	Device
Quartz	Piezoelectricity	Frequency control	Oscillator or crystal
Ruby	Fluorescence	Light amplification	Laser

Some of the listed properties are obtained in enhanced value from water soluble crystals that are not minerals. Whereas halite and sylvite, for example, are minerals soluble in water, the crystal most phenomenal for ferroelectric and piezoelectric use is Rochelle salt, a rather complex and very soluble organic compound. Other important synthetic crystals also not found as minerals include germanium and silicon, the single crystal chemical elements from which such things as diodes, transistors, and integrated microcircuits, are made.

Almandite garnet is listed as the precursor of yttrium iron garnet and yttrium aluminum garnet. Perovskite is so listed for the much valued barium titanate and lead zirconium titanate. Columbite is listed for the family of niobates. The growers and users of optical crystals have been using niobium instead of columbium for the element of atomic number 41.

Often there are recognizable grower-suppliers for the synthetic crystals, as indicated; two listed suppliers, however, are intermediary suppliers. Other crystals may still be in the laboratory, and references to growth studies would indicate where.

GLOSSARY

The following glossary contains some of the terms applying to the use of minerals and various near mineral synthetic crystals in electronic and optical technology. Optics texts are mentioned in the general bibliography for further coverage of that limited field.

Absorbance: The value of the expression, $\ln(I_0/I)$, where I_0/I is known as the *absorbance* and is a numerical statement of the absorption of radiant energy by a transparent material: the ratio of the intensity (W/m^2) of the incident light to the intensity leaving. The constitutive equation is

$$I = I_0 e^{-\alpha d}$$

where α is the absorption coefficient and d is the length of light path in the material.

Acoustooptic. In general, a light-modulating device in which an acoustic wave, ultrasonic in frequency, passing through a suitable crystal plate sets up strained layers with its wave fronts such that an entrant light beam is deflected. Ideally the deflection can be controlled through an important angle which can be varied at frequencies up into the MHz range.

Beam Splitter. A thin sheet of glass or clear plastic having its surface treated to reflect about half the light falling on it at approximately 45° to the surface. In a more advanced type, two right-angle prisms are cemented along the hypotenuse faces with a material having selected reflection properties which may control the ratio of transmitted to reflected light and/or the proportion of color.

Delay Line. A device used to delay an electrical signal for a short time, usually of the order of milliseconds. In one method, a piezoelectric transducer is mounted on an elongated piece of rigid material, e.g., vitreous silica or aluminum. A signal is introduced by the transducer and picked up on reflection or by a transducer at the far end.

Dielectric. A substance which is an electrical insulator as contrasted with an electrical conductor. The term, of course, includes the case of no substance (a vacuum) and many gases and liquids. Many solid dielectrics, mineral and nonmineral, are opaque or absorb all nonreflected light in a very thin layer. The dielectrics useful in optical technology are transparent in the visible or near-visible range of radiation.

Ferroelectric Crystal. A member of a group characterized by being naturally electrically polarized in a restricted temperature range, and showing hysteresis in the relationship between polarization and applied electric field. If the crystal is heated to the transition temperature (Curie point), the crystal becomes nonpolar. Piezoelectric crystals, to be sure, become electrically polarized when subjected to a deforming stress.

Fluorescence. The emission of radiation from a substance which is absorbing radiation of higher energy, shorter wavelength. (But there are interesting cases of the [apparent] reverse energy relationship.) Often large crystal size is not important, and powdered fluorescent and luminescent crystals, called phosphors, are used in cathode ray tubes, television kinescopes, and fluorescent lamps. On the contrary, a large fluorescent crystal bar is used in the solid state laser: ruby, scheelite, or fluorite. Crystals do not have the property of fluorescence if they are strictly pure, and the variation in amount and type of impurity yields useful variations in behavior.

Harmonic Generation. The combining of two sets of wave motions having different, but not far removed, frequencies, so as to produce an ongoing wave whose frequency is a multiple or submultiple of the input frequency. Long known in radio technology, this effect in optics uses crystals as the medium in which the action occurs. For further information with only slight mention of minerals, see Schlossberg and Kelley (1972).

Holography. A type of whole-writing, often caught on a photographic film, involving a laser beam divided by means of a beam-splitter so that the continuing beam strikes the film and the reflected beam strikes and is reflected back by the object onto the film. The two beams, direct and twice reflected, meet, interfere, and record their interference pattern on the film. A three-dimensional view of the object can be seen on the processed film illuminated by monochromatic light. Remarkable features are: no lens is used, the view is not simply two dimensional, a piece of the film will display the same view as the whole, and a large amount of information is stored on the film. A crystal of lithium niobate has been used as the recording medium.

Laser. The device which shoots out an intense, narrow beam of coherent, monochromatic light as a result of light amplification by the stimulated emission of radiation (Maiman, 1960). Originally the laser was a ruby rod energized by an intense source of light nearby; after a few years of investigations, it can now be of glass, liquid, dye solution, or gas. The light from the ruby, at 9630Å, is at the infrared edge of vision, and now lasers can produce radiation of both longer and shorter wavelengths. The power of lasers covers the range of about 1 MW to 1 kW with the pulsed type capable of emitting at much higher levels. Through modulation at the source and demodulation at a receiving station, the laser theoretically can be used as a means of transmitting as many as 10^{10} channels each 10 kHz wide (Laudise, 1968, Nassau, 1965).

Magnetooptic. In certain crystals through which a beam of polarized light is passing, the application of a magnetic field with its lines of force parallel to the beam, causes a rotation of the plane of polarization; the Faraday effect, related to the Zeeman effect. This magnetooptic effect can be used to modulate a light beam.

Modulation. The process and the effect of adding information to a radio carrier wave or to a light beam. In radio, AM and FM are well known. Modulation of a light beam may be accomplished by altering its amplitude, frequency, phase, or plane of polarization. The modulation can take a digital form, so as to transmit simply 0 and 1, or an analog form, which includes voice and music. Modulation of the amplitude (or intensity) of a light beam can be had, for example, by applying a varying electric field to an electrooptic crystal carrying the beam. The crystal is sandwiched between crossed polarizers to provide the rest of the optical requirement.

Nonlinear. Behavior that does not follow a straight line or proportionality relationship when one variable is altered and a dependent variable is observed. In electronics, for example, the behavior of current through a diode with respect to applied voltage is nonlinear. The optical properties of a medium are linear when the intensity of the transmitted light is low, as is normally the case. Nonlinear refers to changes in the optical properties as the intensity of transmitted light is greatly increased, or when two or more strong light beams are combined in the medium. The study of nonlinear phenomena is the basis of much current progress in electronic and optical technology: solar power, transistor, laser, and holography (Schlossberg and Kelley, 1972).

Piezoelectric Effect. The reaction of a plate or bar of substance to a stress (compression, bending, or twisting) by the appearance of an electric field across two opposite faces. The converse effect applies also—the element tends to deform when an electric field is applied. Actual motion or strain is, of course, usually involved. The substance, commonly a macrocrystal, may be an assemblage of microcrystals as in a piezoelectric ceramic. A necessary crystallographic condition is that the crystal has no center of symmetry. Applications include frequency-control resonators of quartz, phonograph pickups of Rochelle salt (not a mineral), and of lead zirconium titanate (related to perovsite).

Property. Any of the principal characteristics of a substance, especially as determined by the senses, but also as determined by any variety of sophisticated sensing equipment, e.g., infrared transparency, laser capability, or semiconduction.

Pyroelectric. The property of a crystal such that electric charges appear on certain faces of it when it is heated or cooled. Strontium niobate, a synthetic crystal related to columbite, shows such a strong effect that it can serve as a practical infrared detector.

***Q*-Switch, or *Q*-Spoiler.** A fast acting light valve used with lasers to arbitrarily hold back the light emission and let it burst forth at the flick of the switch in a high energy flash. Electrooptic crystals and the Kerr cell using nitrobenzene and a high voltage pulse have been used for such shutters, and require the presence of polarizers on each side in a crossed position. Acoustooptic crystals and solutions of certain dyes are also used.

Resonator. A device or system in which a characteristic vibration may occur involving a continuous exchange between potential and kinetic energy.

A piezoelectric resonator is a rectangular plate (sometimes a disk or bar) of piezoelectric material, commonly quartz, which will vibrate characteristically when placed in an alternating voltage field having the same frequency as the natural mechanical frequency of the plate. The motion may take various types or modes: compression, shear, or flexural. Piezoelectric resonators of quartz for frequency control cover the range of about 1kHz to 55 MHz. A piezoelectric plate can be driven at frequencies different from the resonance frequency; an application of such an oscillator is found in the cutting of phonograph records.

An example of an acoustic energy cavity resonator is the organ pipe. In microwave technology a resonator is a metal box or cavity in which electromagnetic energy may be stored, the resonance frequency being determined by the geometry of the enclosure.

An example of an optical resonator is the Fabry-Perot resonator or interferometer two parallel, partially reflecting mirrors cause the light entering in one direction to oscillate before exiting. In a common application the light is passed through a converging lens to form an interference figure of highly contrasting fringes on a screen. The laser may be called a Fabry-Perot resonator.

ACKNOWLEDGMENT

The author wishes to acknowledge the work done on this chapter in the 5th edition of *Industrial Minerals and Rocks* by Danforth R. Hale. Now retired, he was a senior scientist with the Clevite Corp., and his chapter served as a base for this author's work.

QUARTZ
BIBLIOGRAPHY AND REFERENCES

Anon., 1956, *Proceedings*, Annual Frequency Control Symposium, US Army Electronics Command, Ft. Monmouth, NJ (printed from 1956, 10th Symposium and annually since).

Anon., 1971, "Premium Q Cultured Quartz," Sawyer Research Corp., advertising literature.

Adams, C., et al., 1968, "Measurement Techniques for Quartz Crystals," *Frequency*, Vol. 6, No. 8, pp. 22–25.

Ballman, A.A., and Landise, R.A., 1963, "Hydrothermal Growth," *Art and Science of Growing Crystals*, J.J. Gilman, ed., John Wiley, New York, pp. 231–251.

Ballman, A.A., et al., 1966, "Synthetic Quartz with a Mechanical Q Equivalent to Natural Quartz," *Applied Physics Letters*, Vol. 8, No. 2, pp. 53–54.

Bradfield, G., 1970, "Ultrasonic Transducers, 1. Introduction to Ultrasonic Transducers," *Ultrasonics*, Vol. 8, No. 2, Part A, Apr., pp. 112–123, Part B, July pp. 177–189.

Cady, W.G., 1964, *Pieiezoelectricity*, rev. ed., Dover Publications, New York, especially Chaps. 16, 17.

Dodd, D.W., Dodd F.D.W., and Fraser, D.B., 1965, "The 3000–3900 cm^{-1} Absorption Bands and Anelasticity in Crystalline Alpha-Quartz," *Journal of the Physics and Chemistry of Solids*, Vol. 26, pp. 673–686.

Friedman, I.I., and Tuttle, O.F., 1951, "Method of Selection of Oriented Seed Plate for Quartz Synthesis," US Patent 2,558,745, July 3.

Frondel, C., 1961, "The System of Mineralogy of J. Dana and E. Dana," Vol. 3, *Silica Minerals*, 7th ed., 6th printing, John Wiley, New York, 334 pp.

Gerber, E.A., and Sykes, R.A., 1967, "Quartz Frequency Standards," *Proceedings*, Institute of Electrical and Electronics Engineers, Vol. 55, No. 6, pp. 783–791.

Graham, R.A., 1972, "Strain Dependence of Longitudinal Piezoelectric, Elastic, and Dielectric Constants of X-Cut Quartz," *Physical Review B*, Vol. 6, No. 12, Dec. 15, pp. 4779–4792.

Hammond, D.L., et al., 1965, "A linear, Quartz-Crystal, Temperature-Sensing Element," *Transactions*, Instrument Society of America, Vol. 4, Oct., pp. 349–354.

Heising, R.A., 1946, *Quartz Crystals for Electrical Circuits*, D. Van Nostrand, New York, p. 458.

Hunston, D.L., et al., 1972, "Viscoelastic Measurements of Poly (dimethylsiloxanes) and a Polystyrene Solution Using the Layered Waveguide Technique," *Transactions*, Society of Rheology, Vol. 16, No. 1, pp. 45–57.

Ingram, G.E., and Graham, R.A., 1972, "Quartz Gauge Technique for Impact Experiments," *Proceedings of the 5th Symposium on Detonation*, Sandia Laboratories (Albuquerque, NM) Report No. SC-R-72 2722.

Jaffe, H., and Turobinski, T.J. 1960, "Method of Growing Quartz Single Crystals," US Patent 2,923,605, Feb. 2.

Joshi, M.S., and Vagh, A.S., 1964, "Growth Spirals on Prism Faces of Culture Quartz," *American Mineralogist*, Vol. 49, Nov., pp. 1771–1773.

Karrer, H.E., and Leach, J., 1969, "A Quartz Resonator Pressure Transducer," *Transactions on Industrial Electronics and Control Instrumentation*, IEEE, Vol. IECI-16, No. 1, July, pp. 44–50.

Katz, A., 1962, "Hydrogen in Alpha-Quartz," *Phillips Research Reports*, Vol. 17, N.V. Phillps Co., Eindhoven, Netherlands, pp. 133–279.

King, J.C., 1959, *Bell System Technical Journal*, Vol. 38, pp. 573.

King, J.C., et al., 1962, *Journal of Physics and Chemistry of Solids*, Vol. 2, p. 1019.

King, W.H., Jr., 1969, "Using Quartz Crystals as Sorption Detectors," *Research/Development*, Pt. 1, Apr., pp. 28–34; Part 2, May, pp. 28–33.

Kolb, E.D., et al., 1972, "Low Optical Loss Synthetic Quartz," *Materials Research Bulletin*, Vol. 7, pp. 397–406.

Lang, A.R., 1967, "Fault Surfaces in Alpha Quartz," *Crystal Growth*, supplement, *Physics and Chemistry of Solids*, Pergamon Press, pp. 833–838.

Lias, N.C., and Rudd, D.W., 1969, "Growth of Synthetic Quartz for Use in High Frequency Monolithic Crystal Filters," *Western Electric Engineer*, Vol. 13, No. 2, Apr.

Lias, N.C., et al., 1973, "The Growth of High Acoustic Q Quartz at High Grow Rates," American Association for Crystal Growth Conference, Princeton, NJ, July 30, *Journal of Crystal Growth*, Vol. 18, No. 1, pp. 1–6, Amsterdam, Netherlands.

Mason, W.P., ed., 1964, *Physical Acoustics*, Vol. 1, Pt. A, Academic Press, New York.

Myers, R.R., and Schultz, R.K., 1962, "A Kinetic Study of Film Drying via Changes in Mechanical Properties," *Official Digest*, Federation of Societies for Paint Technology, Vol. 24, 801 pp.

Ober, J.A., 1990, "Quartz Crystal Annual Report," US Bureau of Mines, 3 pp.

Pfenninger, H., 1961, "Diffusion von Kationen and Abscheidung von Metallen in Quarz unter Elektrischer Feldeinwirkung," PhD thesis, University of Zurich, 130 pp.

Rudd, D.W., et al., 1966, "Mechanical Q of Alpha-quartz Rapidly Evaluated," *Western Electric Engineer*, Vol. 10, p. 22.

Rudd, D.W., and Lias, N.C., 1967, "Recent Advances in Synthetic Quartz Production," *Chemical Engineering Progress Symposium Series*, Vol. 63, No. 76, pp. 33.

Saubrey, G., 1964, "Messung von Plattenschwingen sehr kleiner Amplitude durch Lichtstrommodulation," *Zeitschrist für Physik*, Vol. 178, p. 457; see also *Proceedings*, 15th Annual Frequency Control Symposium, Ft. Monmouth, NJ.

Sawyer, B., 1972, "Q Capability Indications from Infrared Absorption Measurements for Na_2CO_3 Process Cultured Quartz," *Transactions on Sonics and Ultrasonics*, IEEE, Vol. SU-19, No. 1, Jan., pp. 41–44.

Smith, W.L., and Spencer, W.J., 1963, "Quartz Crystal Thermometer for Measuring Temperature Deviations in $10^{-3°}$ to $10^{-6°}$ C Range," *Review of Scientific Instruments*, Vol. 34, p. 268.

Sosman, R.B., 1965, *The Phases of Silica*, Rutgers University Press, New Brunswick, NJ, pp. 148.

Spezia, G.R., 1908, "Sull'accrescimento del Quarzo," *Atti della R. Accad. delle Scienze di Torino*, Vol. 44, pp. 95–105.

Stockbridge, C.D., 1966, "Resonance Frequency Versus Mass Added to Quartz Crystals," *Vacuum Microbalance Techniques*, Vol. 5, Plenum Press, New York, pp. 193–205.

Viktorov, I.A., 1968, *Rayleigh and Lamb Waves*, Plenum Press, New York.

Waesche, H.H., 1960, "Quartz Crystal and Optical Calcite," *Industrial Minerals and Rocks, 3rd ed.*, J.L. Gillson, ed., AIME, New York, pp. 687–695.

Watson, I., 1979, "Quartz Crystal," *Industrial Minerals*, No. 140, May, pp. 21–29.

White, R.M., 1970, "Surface Elastic Waves," *Proceedings*, Institute of Electrical and Electronic Engineers, Vol. 58, No. 8, pp. 1238–1276.

Zlobik, A.B., 1981a, "Quartz Crystal," *Mineral Facts and Problems*, Bulletin 671, 1980 ed., US Bureau of Mines, pp. 729–736.

Zlobik, A.B., 1981b, "Quartz Crystal, Industrial," *Mineral Commodity Summaries, 1981*, US Bureau of Mines, pp. 120–121.

MINERALS OTHER THAN QUARTZ BIBLIOGRAPHY AND REFERENCES

Bardsley, W., and Jones, O., 1968, "On the Crystal Growth of Optical Quality Prousite and Pyrargyrite," *Journal of Crystal Growth*, Vol. 3, No. 4, pp. 268–271.

Bonner, W.A., et al., 1972, "High Quality Tellurium Dioxide for Acousto-optic and Non-linear Applicatons," *Journal of Electronic Materials*, Vol. 1, No. 1.

Bube, R.H., 1960, *Photoconductivity of Solids*, John Wiley, New York, p. 428.

Buhrer, C.F., et al., 1963, "Electro-optic Light Modulation with Cubic Crystals," *Applied Optics*, Vol. 2, Aug., p. 845.

Butler, C.T., et al., 1969, "Growth of Pure Single Crystals of MgO by Arc Furnace," American Association for Crystal Growth, Gaithersburg, MD, Aug. 11–13.

Cady, W.G., 1964, *Piezoelectricity*, Dover Publications, New York, pp. 134, 201, 229, 230, 435, 484, 704.

Crowningshield, R., 1972, *Gems and Gemology*, Spring issue.

Cullen, G.W., 1971, "The Preparation and Properties of Chemically Vapor Deposited Silicon on Sapphire and Spinel," *Journal of Crystal Growth*, Vol. 9, pp. 107–125.

de Klerk, J., 1972, "Elastic Surface Waves," *Physics Today*, Vol. 25, No. 11, pp. 32–39.

Farrell, E.F., and Ling, A., 1972, "Chemical Vapor Grown TiO_2 Single Crystals," American Association for Crystal Growth, Conference II, Princeton, NJ, July.

Fochs, P.D., et al., 1968, "Growth of Cadmium Sulfide Single Crystals of Controlled Composition from the Vapor Phase," *Journal of Crystal Growth*, Vol. 3, No. 4, pp. 122–125.

Fochs, P.D., 1970, "The Cadmium Sulfide Acousto-electric Oscillator," *Journal of Science and Technology*, Vol. 37, No. 4, pp. 171–181.

Garbuny, M., 1965, *Optical Physics*, Academic Press, New York, 371 pp.

Gentile, A.L., and Stafsudd, O.M., 1968, "Czochralski-Grown Proustite and Related Compounds," *Journal of Crystal Growth*, Vol. 3, No. 4, pp. 272–274.

Geusic, J.E., et al., 1964, *Applied Physics Letters*, Vol. 4, 182 pp.

Harmon, T.C., 1971, *Journal of the Physics and Chemistry of Solids*, Supplement, Vol. 4, 182 pp.

Hatta, J., et al., "Product Engineering for Growth of Synthetic Calcite Polarizing Material," Contract DAAB-05-71-C-2636, US Army Electronics Command, Philadelphia, PA.

Henisch, H.K., 1970, *Crystal Growth in Gels*, Pennsylvania State University Press, University Park, PA, 63 pp.

Hiroshi, Y., 1972, "Synthetic Crystals of Japan," *Japan Electronic Engineering*, June, pp. 41–47.

Hordvik, A., and Schlossberg, H., 1972, "Luminescence of Lithium Niobate," *Applied Physics Letters*, Vol. 20, No. 5, pp. 197–199.

Hulme, K.F., 1967, *Applied Physics Letters*, Vol. 10, pp. 133.

Johnson, L.F., and Nassau, K., 1961, *Proceedings*, Institute of Radio Engineers, Vol. 49, p. 1704.

Jona, F., and Shirane, G., 1962, *Ferroelectric Crystals*, Macmillan Co., New York, pp. 108, 216, 226, 338, 361, 363, 371.

Kaminow, I.P., 1965, "Barium Titanate Light Phase Modulator," *Applied Physics Letters*, Vol. 7, pp. 123.

Knubovets, R.G., et al., 1970, "Long Wave Infrared Spectra of Some Apatites," *Optics and Spectroscopy*, Vol. 29, No. 1, July, pp. 32–39.

Kolb, E.D., 1967, "Hydrothermal Growth of Low Carrier Concentration ZnO at High Water and Hydrogen Pressures," *Materials Research Bulletin*, Vol. 2, pp. 1099–1106.

Laudise, R.A., 1968, "The Search for Nonlinear Optical Materials for Laser Communication," *Bell Laboratories Record*, Vol. 46, Jan., pp. 2–7.

Liaw, H.M., and Faust, J.W., Jr., 1972, "The Growth of Calcite in Aqueous Solution," American Association for Crystal Growth, 2nd National Conference, Princeton, NJ, July 30.

Maiman, T.H., 1960, "Stimulated Optical Radiation in Ruby," *Nature*, Vol. 187, p. 493.

Makram, H., 1968, "Improvement in Yield of Yttrium Iron Garnet," *Journal of Crystal Growth*, Vol. 3, No. 4, pp. 447–451.

Mason, W.P., ed., 1964, *Physical Acoustics*, Vol. 1, Pt. A, Academic Press, New York.

Mazelsky, R., et al., 1968, "Czochralski Growth of Calcium Fluophosphate," *Journal of Crystal Growth*, Vol. 3, No. 4, pp. 360–364.

Montalvo, R.A., and Langer, D.W., 1970, "Linear Compressibilities of II-VI Compound Single Crystals," *Journal of Applied Physics*, Vol. 41, No. 10, Sept., pp. 4101–4104.

Murray, L.A., 1964, "Modulation Crystals for Lasers," *Electronic Industries*, Feb., pp. 83–85.

Nassau, K., 1964, "Growing Synthetic Crystals," *Lapidary Journal*, Vol. 18, No. 3, June, pp. 386–389.

Nassau, K., 1964a, "Growing Synthetic Crystals," *Lapidary Journal*, Vol. 18, No. 4, July, pp. 474–477, 483.

Nassau, K., 1965, "Lasers and Laser Materials," *Materials Research and Standards*, Vol. 5, Jan., pp. 3–11.

Nassau, K., and Broyer, A.M., 1962, "Calcium Tungstate: Czochralski Growth, Perfection, and Substitution," *Journal of Applied Physics*, Vol. 33, Oct., pp. 3064–3073.

Nassau, K., and Broyer, A.M., 1962a, "Application of Czochralski Crystal-Pulling Technique to High-Melting Oxides," *Journal of the American Ceramic Society*, Vol. 45, Oct., pp. 474–478.

Nassau, K., and Shiever, J.W., 1972, "Modified Technique for the Growth of Boracite Crystals," *Journal of Crystal Growth*, Vol. 16, No. 1, pp. 59–61.

Nassau, K., and Shiever, J.W., 1972a, "Vapor Growth of II-VI Compounds and the Identification of Donors and Acceptors," *Journal of Crystal Growth*, Vol. 13/14, pp. 375–379.

Nielsen, K.F., 1968, "Growth of ZnO Single Crystals by the Vapor Phase Reaction Method," *Journal of Crystal Growth*, Vol. 3, No. 4, pp. 141–145.

Parker, S.G., and Pinnell, J.E., 1968, "Molten Flux Growth of Cubic ZnS Crystals," *Journal of Crystal Growth*, Vol. 3, No. 4, pp. 490–495.

Pinnow, D.A., et al., 1969, "Lead Molybdate," *Applied Physics Letters*, Vol. 17, pp. 83–86.

Pollock, J.T.A., et al., 1972, "Edge-Defined, Film-Fed Growth of Sapphire," American Association for Crystal Growth, 2nd National Conference, Princeton, NJ, July 30.

Pough, F., 1970, "Colored Synthetic Quartz from Russia," *Lapidary Journal*, June 24, pp. 444–446.

Rosenberger, F., and Delong, M.C., 1972, "Growth of Rutile by Chemical Vapor Transport," American Association for Crystal Growth, 2nd National Conference, Princeton, NJ, July 30.

Rudness, R.G., and Kebler, R.W., 1959, "Growth of YIG by the Flame Fusion Process," *Linde Industrial Crystal Bulletin*, Linde Co., New York, Sept. 21.

Ryan, F.M., et al., 1970, "Optical Properties of Divalent Manganese in Apatite," *Physical Review B*, Vol. 2, No. 7, Oct., pp. 2341–2352.

Schlossberg, H.R., and Kelley, P.L., 1972, "Using Tunable Lasers," *Physics Today*, Vol. 25, No. 7, pp. 36–44.

Stanley, J.M., 1954, *Industrial and Engineering Chemistry*, Vol. 46, p. 1684.

Stepke, E.T., 1972, "Optical Mass Memories," *Electrooptical Systems Design*, Vol. 4, No. 11, Oct., pp. 12–17.

Tolksdorf, W., 1968, "Growth of Yttrium Iron Garnet Single Crystal," *Journal of Crystal Growth*, Vol. 3, No. 4, pp. 463–466.

Toudic, Y., and Aumont, R., 1971, "Croissance par voie Hydrothermale de Monocristaux de Cinabre de Dimensions Importances," *Journal of Crystal Growth*, Vol. 10, pp. 170–174.

Von Hippel, H., 1963, *Technical Reports*, Laboratory of Insulation Research, Cambridge, MA, Vol. 44, Mar.

Wheeler, D., 1972, "Titanates," *Ceramic Industry Magazine*, July, pp. 28–29.

White, R.M., 1970, "Surface Elastic Waves," *Proceedings*, Institute of Electrical and Electronics Engineers, Vol. 58, No. 8.

Wood, J.D.C., and White, E.A.D., 1968, "Growth of Stoichiometric Manganese Aluminate Spinel Crystal by Flux Evaporation," *Journal of Crystal Growth*, Vol. 3, No. 4, pp. 480–485.

Yu, P.Y., and Cardona, M., 1973, "Intrinsic Piezobirefringence of Several Semiconducting Chalcogenides," *Journal of the Physics and Chemistry of Solids*, Vol. 34, pp. 29–56.

EMERY

WILLIAM G. HOLROYD AND DEREK J. McCRACKEN

Emery is a rock that at one time held a large share of the abrasives market. The introduction of synthetic abrasives, however has forced it to settle for only certain niches in the marketplace. This loss of market share is due mainly to the higher amounts of impurities and weaker internal structure that naturally occurring minerals have compared to manmade materials. The use of rocks and minerals as abrasives dates back to early man, who used flint, quartz, and other natural materials to fabricate various tools. During the 19th century, with the Industrial Revolution imminent, the character of the abrasives industry was changed forever as synthetics were introduced. Frank B. Norton is credited with producing the first manmade abrasive when, in 1855, he manufactured a vitrified grinding wheel using emery grit and clay. Before Norton's accomplishment emery had been the only commercially available abrasive. The latter part of the 20th century has seen synthetics capture an increasing share of the abrasives market from natural materials.

The term emery is believed to have been derived from the words *emeri* (French), *smeriglio* (Italian), and *smiris* (Greek). The term corundum, for the principal mineral of emery, seems to have originated from *kauruntaka,* the Indian name for the mineral.

GEOLOGY

Emery is a gray to black, granular rock comprised of various minerals including corundum (Al_2O_3), an iron-bearing mineral (usually magnetite—Fe_3O_4, hematite—Fe_2O_3, or hercynite—$FeAl_2O_4$), and several trace impurities such as mullite, titania, silica, and magnesia. The hardness of emery normally ranges between 7 and 9 on the Mohs scale, depending on its mineral composition and level of impurities. Corundum, the hardest mineral in emery, typically has a hardness of 9. The specific gravity of emery ranges from 3.2 to 4.5, also depending on mineral composition and purity.

The crystallography of each mineral contained in emery differs somewhat. Corundum exhibits an hexagonal crystal structure, whereas magnetite forms within the isometric system, frequently as octahedral crystals. Hercynite, an iron spinel, is also isometric. Hematite has an hexagonal structure but often is found as botryoidal and reniform masses.

Principal deposits of commerical emery are found in the United States, Turkey, and Greece. In the United States two emery producers in the Westchester County—Peekskill area and one in the Cortlandt County area of New York closed their mines in the late 1980s. Commercial production in the United States now comes only from the western Cascade Range in Oregon.

The emery in New York has a typical Al_2O_3 content of 41.5%. Most of the rock is not true emery because it contains a mixture of spinel and magnetite but very little corundum. The deposits are of Late Ordovician-Early Silurian age and consist of veins in which emery is mixed with olivine, pyroxene, and hornblende. The veins evidently were formed by contact metamorphism and are surrounded by Precambrian schists that contain mica, sillimanite, cordierite, garnet, and quartz.

The deposits in Oregon were formed during Miocene time by the contact metamorphism of ferruginous bauxite by intrusive lamprophyric dikes and plugs. Emery is found in thin layers bordering these tabular dikes, which are usually oriented perpendicular to bedding, and in surficial placer deposits.

The emery deposits in Turkey occur as small, massive, lens-shaped bodies and as surficial placers. They are believed to have formed during Permian and Triassic times.

In Greece the principal site of emery production is at Cape Emeri on the Island of Naxos. These deposits consist of lenticular masses of emery surrounded by marble. They originated during Mesozoic time, evidently by metasomatic replacement. Smaller deposits on several other Aegean islands were formed by the metamorphism of bauxite or other alumina-rich materials.

TECHNOLOGY

All four of the deposits in Oregon that are owned by the Oregon Emery Co. can be mined by open pit methods; only two of them require drilling and blasting to free the rock. The ore is loaded by backhoes into haulage trucks for transporting to the company's processing plant at Halsey, OR. At this plant the ore is first crushed to approximately 10.2 mm in a jaw crusher and then to finer sizes in a hammer mill. The crushed emery is screened, dried, and then packaged for customers in 22.7 kg bags or 1 ton bulk bags.

Turkey's emery production is handled by Etibank in Ankara, by Ranar Mineral Industries Corp. in Istanbul, and by a small, family-owned company, Lutfullah E. Kitapci Mineral Co., Ltd. in Izmir. Etibank, founded in 1935 and the largest mining company in Turkey, produces a variety of minerals, metals, chemicals, and ferroalloys. Its emery mining at Mugla Gulluk is presently carried out by contractors. The ore is hand sorted by local villagers and is then trucked to Etibank's Milas-Mugla plant, where it is crushed and screened into three sizes: $-240 + 50$ mm, $-150 + 50$ mm, and $-50 + 10$ mm. The products are then shipped in bulk for further processing by several other companies located in Europe and North America.

Lutfullah E. Kitapci Mineral Co. supplies lump ore to Van Mannekus Universal V.O.F., which has the largest emery processing plant in Europe. This plant, a 50:50 joint venture of Rotterdam-based Van Mannekus & Co., B.V. and English Abrasives & Chemicals, Ltd. of Stafford, UK, is situated in Oudenbosch, The Netherlands, halfway between Rotterdam and Antwerp. It has the capacity to process 8 000 tpy of emery. Approximately 35 different grades are produced that range from 2.36 mm to 53 μm, plus some flour sizes. The lump ore is washed, dried, crushed, screened or air classified to size specifications and is shipped to customers in bulk or bags.

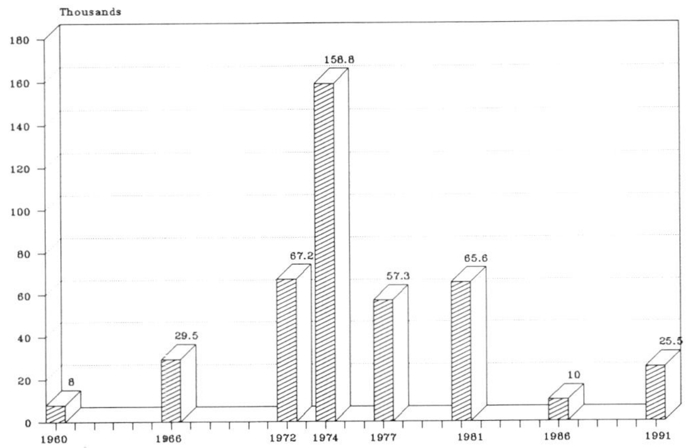

Fig. 1. Turkey's annual emery production (selected years). *Source: Industrial Minerals.*

The mines of Ranar Mineral Industries Corp. are located in the Aydin-Mugla District, the site of most of Turkey's emery deposits. Approximately 80% of Ranar Mineral's production is delivered to customers in lump form. The balance is sold in standard sizes of 2.36 mm to approximately 85 µm. Processing is carried out at the company's Aydin-Mugla plant, which has an annual capacity of 3 000 tons. It involves grinding, screening, micronizing, and air classifying. The emery is shipped in bulk or in palletized bulk bags to markets in The Netherlands, Canada, and France and in ground form to England, Pakistan, and the Far East.

In Greece the government owned emery deposits on the Island of Naxos are worked by local villagers. A number of small operations supply a central warehouse for redistribution to two domestic processors, Hellenic Corundum Ltd. and Icon Mining and Commercial Co., both in Athens. These two companies grind, crush, and classify emery into a variety of sizes in their plants at Elefinsa and Votanikos, respectively. Half of the total production in Greece is supplied to the Ministry of Industry for export in lump form.

Smaller processors of emery include Maharashita Minerals Corp., Ltd. (Bombay, India), Nigerian Mining Co. (Plateau State, Nigeria), MAF Magnesite, BV (Vlaardingen, The Netherlands), Washington Mills Electro Minerals Ltd. (North Grafton, MA), Bay State Abrasives, a subsidiary of General Abrasives (Westfield, MA), and Emeri Crete (New Castle, NH).

PRODUCTION

Oregon Emery, the newest of the major commercial emery producers, has an estimated reserve of 5 Mt of marketable emery and an annual capacity of 3 000 t. Should increased production become needed, the company can either expand its current processing plant or subcontract some of the crushing to two companies in nearby Sweet Home and Brownsville, OR. All of Oregon Emery's production is currently being sold within the United States.

Etibank controls one of the largest reserves of emery in the world, last estimated at 855 000 t. In 1991 the company produced approximately 7 690 t of emery, and maximum output was 14 800 t in 1979. Lutfullah E. Kitapci Mineral Co. has been producing 7 000 to 8 000 tpy for the past several years, while Ranar Minerals' output has been close to 10 000 tpy. Fig. 1 plots emery production in Turkey for several years beginning with 1960. Table 1 shows Turkish exports and their corresponding values. Turkey alone produced more than 158 000 t of emery in 1974, the peak year.

Table 1. Turkey's Emery Production and Value (Selected Years)

Year	Tons	Value (US$)
1966	27 000	308 000
1972	63 000	666 000
1977	50 000	1 080 000
1981	52 000	1 271 000
1984	13 000	525 000
1985	4 380	908 987
1986	6 700	250 000

Source: Industrial Minerals magazine.

Due to rapid growth in the use of synthetic abrasives, emery production in 1991 was only 40 000 t worldwide.

In 1991 Greece is believed to have produced approximately 10 000 t of emery. Half of its total production is exported, while the other half is shipped to Icon Mining and Commercial Co. and Hellenic Corundum Ltd. Deposits on the Island of Naxos are reported to contain more than 1 Mt of emery.

In other parts of the world processing is done to satisfy market requirements. For example, Nigerian Mining Corp. produces a fine grade of emery (47% + 38 µm) at an average annual rate of 2 500 t. Washington Mills Electro Minerals Ltd. of North Grafton, MA, distributes approximately 4 000 tpy of processed emery into construction and abrasives markets. MAF Magnesite in The Netherlands sells mostly to the abrasives industry. Maharashita Minerals Corp., Ltd. processes 3 000 tpy of emery in India.

USES

Emery has been replaced by synthetic materials in most markets for abrasives. Its principal uses now are in nonabrasive applications such as nonskid flooring and concrete hardening. Abrasive and nonabrasive uses for emery are outlined as follows:

- Stone sawing and metal polishing—loose grains of relatively coarse size
- Stone surfacing and rough grinding of plate glass—finer size, loose grains
- Bonded abrasives—e.g., grinding wheels
- Pressure blasting (medium to low pressure)
- Coated abrasives—grains cemented to a paper or cloth backing
- Nonskid dusting agent—for oily floors, pavements, and stair treads
- Tumbling and deburring media
- Concrete surface hardening

In the past few years Oregon Emery has concentrated its efforts on markets where emery is used as a surface hardening and nonskid agent in high strength concrete and epoxy coating and grouting systems. The end uses for these systems include steel bridgework, commerical docks, pedestrian walkways, vehicular roadways and parking ramps, steel ship decks, and airport runway repair. Oregon Emery recently was approved to supply the emery for all bridgework in a western state in the United States. The company's future business is expected to depend heavily on public works spending and, in particular, spending in connection with the building and rebuilding of infrastructures such as bridges and highways. Similarly, Emeri Crete manufactures emery for nonskid and high-wear applications.

In Turkey most of the emery produced by Etibank is exported to France for the manufacture of special alumina cement and to the abrasives market in North America. Most of Lutfullah's output is

Table 2. Specifications for Four Major Emery Producers

	Oregon Emery US	Etibank Turkey	Ranar Mineral Turkey	Lutfullah, Turkey
Al_2O_3	50–55%	54–56%	61.02%	64%
Fe_2O_3	20%	25–27%	24.95%	25%
SiO_2	28%	3–5%	7.96%	4%
TiO_2	2%	2.5%	2.60%	3%
CaO	—	1.5%	0.70%	0.2%
LOI	—	8.0%	2.25%	3%
Mohs Hardness	8–9	8–9	7.85–8.5	8–9

Source: Individual producers and processors.

exported to Van Mannekus Universal in Oudenbosch, The Netherlands. English Abrasives and Chemicals, Ltd. uses emery from the Oudenbosch facility to manufacture emery paper and cloth for which it is the largest UK producer. Van Mannekus & Co. sells into several abrasives markets.

Icon Mining & Commercial in Greece sells its processed emery for use in grinding wheels, millstones, coated abrasives, grinding and polishing pastes, wire-sawing grits, and nonskid surfaces. Hellenic Corundum, on the other hand, sells almost exclusively into the nonskid flooring market.

Maharashita Minerals Corporation Ltd. in India processes emery mainly for the abrasives industry. Nigerian Mining Corp. sells most of its fine-size product to the glass industry.

Typical specifications are shown in Table 2 for four of the major commercial emery producers.

ECONOMICS

Prices from one of the Turkish sources, Ranar, ranged from 34 to $48/t for lump and from 220 to $460/t for micronized grades in 1991. Prices were quoted FOB. Turkish Aegean ports.

Oregon Emery does not sell crude lump ore. Its processed emery ranges from 10.2 cm to 75 µm in size, and its prices start at $551.15/t FOB Halsey, OR.

Polishing-grade emery is priced between 1.75 to $2.75/kg (1,750 to $2,750/t).

The tariff system established for imported emery by the US Department of Commerce under the *Harmonized Tariff Schedule (1992)* is divided into two headings. The first (2513.21.00.00.1) is "Emery, Natural Corundum, Natural Garnet and Other Natural Abrasives; in crude or in irregular pieces." No tariff is levied on imports from any country under this heading. The other heading (2513.29.00.00.3) is "Emery, Natural Corundum, Natural Garnet and Other Abrasives; other," which covers material in processed form. The tariff levied is $0.007/kg for imports from all sources except member countries of the Caribbean Common Market (CARICOM), Canada (US-Canada Free-Trade Agreement), Israel (US-Israel Free-Trade Area), and developing countries listed under the Generalized System of Preferences (GSP), which includes Turkey.

As of January 1992 a $0.022/kg tariff was imposed on processed emery imports from fifteen former COMECON and other communist countries. This list is expected to diminish to a very few over the next several years in response to the rapidly changing political situation throughout the world.

GOVERNMENTAL MATTERS

Because of increasingly restrictive land use zoning regulations and other regulations governing drilling, blasting, hauling, etc., two emery mining companies operating in Westchester County, New York, were forced to close during the late 1980s. Their mine sites were located in the commuter belt that leads to New York City. As a consequence Washington Mills Electro Minerals, a processor of New York emery, was forced to import ore. This would have been necessary before long, because the deposits in Westchester County were almost mined out.

Between 1989 and 1990 V.S. Construction (V.S.C.) mined and processed approximately 1 000 tpd of rock in Cortlandt County, NY, to produce gravel and emery ore. However, as a consequence of operating without state mining permits, V.S.C. was fined $50,000 by the State Department of Environmental Conservation in August 1991. This fine was the highest ever in New York against a mining company. V.S.C. must also restore the 12 140 m^2 site or face further penalties.

Since the closing of these three emery mines in New York, no major land use or environmenal problems are known to have affected any of the other emery producers.

FUTURE TRENDS

Emery will probably maintain its relatively small market niche in low value abrasive, nonskid, and construction applications. Total sales of abrasive materials worldwide are expected to reach $548.2 million by 1995 based on a projected average annual growth rate of 5.5% from 1990. Of total sales in 1995, $13.2 million will be contributed by natural abrasives, which will represent an annual growth rate of 3.5% from 1990.

Because of the much better purity and uniformity of electrofused materials such as aluminum oxide, alumina zirconia, alumina chromium oxide, and of other synthetic abrasives such as silicon carbide and boron nitride, the market share held by naturally occurring abrasives such as emery will continue to erode. Since emery is comprised of a variety of minerals, its chemical composition and physical properties can vary substantially from one lot to another. This variability makes it difficult for end users to set raw material specifications within narrow ranges and to produce finished products that perform consistently in commercial application. Nevertheless, emery is likely to continue to be used in niche markets for which synthetic abrasives are too expensive and for which lower grade abrasive products such as silica, pumice, olivine, garnet, and slag are underqualified.

The balance between emery supply and demand is expected to stay relatively stable throughout the next decade. There should be no shortage of commerical emery. In fact, if emery producers and processors continue to lose ground to alternate materials, an oversupply for even the remaining niche markets could drive prices sharply lower.

BIBLIOGRAPHY

Anon., 1976, *Dictionary of Geological Terms,* Anchor Press Doubleday, Garden City, NJ, 469 pp.

Anon., 1987, *Industrial Minerals Directory,* 1st ed., G.M. Clarke, ed., Metal Bulletin Books, Ltd., London, 704 pp.

Anon., 1991, *Industrial Minerals Directory,* J. Griffiths, ed., Metal Bulletin Books Ltd., London, 557 pp.

Anon., 1991, "Abrasive Materials Adjust to Change," *Ceramic Industry,* April, pp. 16–17.

Anon., 1992, *Harmonized Tariff Schedule of the United States,* US Government Printing Office, Washington, DC.

Austin, G.T., 1990, "Abrasive Materials," *Annual Report, 1990,* US Bureau of Mines, (preprint), 22 pp.

Dickson, T., 1986, "Turkey's Minerals—Potential Still There," *Industrial Minerals,* Aug., pp. 19–26.

Endicott, W.A., 1982, "Abrasives: Key to the Industrial Revolution," *Brick & Clay Record,* Mar., p. 24.

Griffiths, J., 1985, "Hellenic Industrial Minerals," *Industrial Minerals,* Jan., pp. 19–39.

Harben, P.W., and Bates, R.L., 1984, *Geology of the Nonmetallics,* Metal Bulletin Inc., New York, 392 pp.

Harben, P.W., 1992, *The Industrial Minerals Handybook—A Guide to Markets, Specifications, and Prices,* Metal Bulletin PLC, London, 148 pp.

Harrision, D.K., 1992, "The Mineral Industry of New York in 1991," *Mineral Industry Surveys,* US Bureau of Mines, 2 pp.

Hight, R.P., 1983, "Abrasives," *Industrial Minerals and Rocks,* 5th ed., S.J. Lefond, ed., AIME, New York, pp. 11–32.

Hurlbut, C.S., and Klein, C, 1977, *Manual of Mineralogy (after James D. Dana),* John Wiley and Sons, New York, 532 pp.

Karayazici, F., 1988, "Industry Structure—Past, Present and Future," *Industrial Minerals* (special supplement), pp. 13–21.

McMichael, B., 1990, "Abrasive Minerals—Taking the Rough with Smooth," *Industrial Minerals,* Feb., pp. 19–33.

Mitchell, R.S., 1979, *Mineral Names—What Do They Mean?,* Van Nostrand Reinhold, New York, 229 pp.

Mottano, A., et al., 1978, *Guide to Rocks and Minerals,* Simon and Schuster, New York, 607 pp.

Toon, S., 1986, "Abrasive Minerals—Hard Work in Soft Markets," *Industrial Minerals,* Dec., pp. 53–73.

Wellborn, W.W., 1991, "The Expanding Role of Synthetic Minerals in Industry," *Industrial Minerals,* Apr., pp. 53–61.

ENVIRONMENTAL USES

ASBESTOS SUBSTITUTES

ROBERT L. VIRTA

An asbestos substitute is any material that replaces asbestos in a commercial product. There are many asbestos substitutes, each with its own unique physical and chemical properties. Substitutes can be inorganic or organic, fibrous or nonfibrous, and natural or synthetic. They also can be used as a direct substitute for asbestos or as a replacement product. Probably in no other field is the diversity of substitute materials so great.

The concept of asbestos substitutes is not new. The first major effort to develop asbestos substitutes was undertaken by the federal government. The United States has always been dependent on foreign sources for strategic-grade asbestos, and during World War II, the government was concerned that these supplies would be cut off. Strategic-grade asbestos was critical in many electrical and thermal insulation applications (electrical insulation, battery plate separators, flame-proof boards, insulation for rocket components, etc.). A reliable supply of nonstrategic-grade asbestos also was required for general civilian and military applications, such as in brakes, clutches, and construction materials. The government not only initiated a major search for deposits containing strategic-grade asbestos, it also funded research on synthesizing asbestos in the laboratory and developing products that did not contain strategic-grade asbestos. These efforts, as well as those conducted by the private sector, generally were unsuccessful.

In the late 1960s and early 1970s, a second major effort to develop asbestos substitutes began because the general public was concerned about the health threat posed by the release of asbestos fibers during the manufacture, use, repair, and disposal of asbestos-containing products. Strict regulations covering the mining and manufacturing sector already had been enacted because of the issue of occupational exposure. The issue then became environmental exposures. Environmental regulations were enacted, and through time they became more strict. The combination of increasingly strict environmental regulation, maturing of some asbestos markets, and liability issues prompted many manufacturers to discontinue asbestos product lines or search for asbestos substitutes. As a result, asbestos usage dramatically declined from the 1970s to the present (Table 1).

RAW MATERIALS

It is impossible to discuss the mineralogy, chemical properties, and distribution of deposits of asbestos substitutes in the manner of traditional mineral commodities because of the wide variety of materials involved. The diversity of materials required to replace asbestos is a tribute to its versatility. To understand what is required of an asbestos substitute, it is best to examine the properties of asbestos.

Asbestos

Asbestos is a term used by industry to describe six fibrous minerals. These minerals are chrysotile, anthophyllite asbestos, tremolite asbestos, actinolite asbestos, crocidolite, and amosite. Asbestos is composed of extremely fine mineral fibrils that are heat resistant, resistant to chemical attack, good electrical insulators, easily woven, and have a large surface area and tensile strength (Table 2). A full description of the properties of asbestos is given in the chapter on Asbestos.

Each type of asbestos exhibits different properties (Table 2). Because these differences affect how asbestos can be used commercially, manufacturers design products to exploit their strengths while avoiding their weaknesses. For example, anthophyllite fibers lack flexibility, so they can be used in applications where flexibility is not important, such as in filter media for acidic or alkaline solutions or for insulation board. Chrysotile is not resistant to attack by strongly acidic or alkaline solutions, but its filtration properties, flexibility, stability at high temperatures, tensile strength, etc., allow manufacturers to use it in a vast array of products (Table 3).

Although the individual properties of asbestos are important, manufacturers are just as interested in the properties that asbestos imparts to the product. Producers must be concerned about how the fiber behaves during the manufacturing process and how the fiber improves or impairs the properties of the finished product. For example, asbestos fibers make the uncured asbestos cement pipe stronger so that the pipe will not deform during manufacturing. Asbestos fibers allow water used during production to drain rapidly from the pipe after it is formed. The fibers also improve the finished pipe's impact strength and tensile strength, and they increased the pipe's life in alkali soils. Asbestos offers similar advantages in other products (Table 3).

Asbestos Substitutes

The first step taken in developing an asbestos substitute is to determine the property or properties of asbestos that are important for a particular application. The next step is to find a substitute that most closely mimics these properties, examine how the substitute affects the manufacturing process, and decide whether the properties imparted to the final product by the substitute make it competitive in cost and performance with other products.

The materials considered for use as asbestos substitutes by manufacturers can be divided into mineral-based vs. organic-based substitutes, fibrous vs. nonfibrous substitutes, and natural vs. synthetic substitutes (Tables 4 and 5). Manufacturers also consider the use of alternative products instead of asbestos or asbestos substitutes.

Table 1. US Asbestos Consumption (ton x 1 000)

Year	Quantity	Year	Quantity
1945	342 803	1970	660 550
1950	659 526	1975	551 239
1955	709 615	1980	358 700
1960	643 371	1985	162 000
1965	720 708	1990	41 348

Minerals and Rocks: Minerals that have been considered for use in place of asbestos include attapulgite, biotite, graphite, muscovite, palygorskite, serpentine, silica, talc, vermiculite, and wollastonite. The following discussion will focus on those properties that make these materials potential substitutes.

Biotite, muscovite, serpentine, talc, and vermiculite are sheet silicates and have a platy morphology. The platy shape, which results in a large contact area between the particle and the matrix, makes these minerals moderately good reinforcers, but they are not as good as is asbestos. Each asbestos fiber is composed of dozens to thousands of fibrils that are 500 to 5 000 Å in diameter and 10 to several hundred micrometers long. The fibrils have an extremely large surface area and the large number of fibrils provides a large number of contact points when properly distributed throughout the matrix.

The platy minerals are thermally stable and do not decompose at low to moderate temperatures (below approximately 750°C). Vermiculite exfoliates when heated rapidly to 300°C, but the exfoliated product is stable at moderate temperatures. Exfoliated vermiculite is an excellent thermal insulator because of the large amount of void spaces in the expanded platelets. All of the platy minerals are as resistant to chemical attack as asbestos. These minerals are less expensive than most grades of asbestos, are widely available, and are noncarcinogenic.

Attapulgite, palygorskite, and sepiolite are classified as clay minerals, although they have a lath-like structure that is fibrous in appearance. They are moderately good reinforcers because their fibrous morphology gives them a large surface area, but they are not as good as is asbestos because the fibers are not as long. Attapulgite, palygorskite, and sepiolite lose water from their structures when heated to moderate temperatures (350 to 600°C) and convert to clinoenstatite above 800°C. These minerals are less expensive than most grades of asbestos, are widely available, and are noncarcinogenic.

Silica is a framework silicate. Quartz is its most common form. Although it has been mentioned as a substitute, it has had only limited use. A major requirement of an asbestos substitute is that it provides reinforcement. Quartz is a poor reinforcer because it occurs as equidimensional grains. It is, however, resistant to chemical and thermal degradation and is a relatively inert filler. Quartz is extremely common; it is widely available and inexpensive. Some evidence points to silica being a possible human carcinogen, but this aspect has been disputed.

Graphite is a platy mineral composed of carbon. It is an excellent nonabrasive lubricant. Graphite is stable at high temperatures, makes an excellent packing in its expanded form, and is resistant to chemical attack. Graphite is an excellent electrical conductor; galvanic corrosion is a concern in valves where dissimilar metals are used. Graphite is priced competitively with asbestos. Availability is not a problem due to its limited use as a substitute.

Wollastonite is a single-chain silicate called a pyroxenoid. It possesses a modified pyroxene-like structure. Wollastonite is stable at temperatures up to 1 540°C. It occurs in a fibrous habit and is a good reinforcer, its fibers are not flexible like chrysotile fibers. Wollastonite is resistant to chemical attack and is a relatively inert mineral filler. It is priced competitively with asbestos, and it is already used commercially so availability should not be an issue. Wollastonite is considered to be noncarcinogenic.

Gypsum is a naturally hydrated calcium sulfate. It also can be synthesized easily in a partially hydrated or anhydrous form. In its partially hydrated form, it is called hemihydrate calcium sulfate. The anhydrous form is called anhydrite. Both have been used as an asbestos substitute and are sold under the trade name of Franklin Fiber. Franklin Fibers are extremely long, thin, and strong fibers, similar to ultrafine glass fiber, and are good reinforcers. The fibers are nonabrasive, stable up to a temperature of 1 000°C, and stable in mild acids and bases. It is more expensive than asbestos and less expensive than glass fiber. Franklin Fiber is readily available. No adverse health effects have been reported for Franklin Fiber.

Two other substitutes that have been examined are rocks rather than minerals. These are diatomite and perlite. Diatomite is a rock composed of the shells of diatoms, which are microscopic plants. Diatomite is stable up to 1 400°C. It is already used as a source of silica for synthesizing calcium silicate for thermal insulating applications. Diatomite is less expensive than asbestos and is widely available. Diatomite can cause silicosis; calcined forms are suspected carcinogens.

Perlite is an amorphous or glassy volcanic rock. When heated, it expands into a glassy foam. This material has a low density and is an excellent thermal insulator. Perlite already has been used in conjunction with asbestos or asbestos substitutes in thermal insulations and in some cement products. Perlite is low priced and widely available. Fine silica present in the perlite may cause silicosis.

Organic-based Substitutes: Organic-based substitutes are either natural or synthetic. The natural fiber substitutes are wood, vegetable, and animal. Wood fiber is obtained by chemically and/or thermally processing wood pulp. The resulting material, more commonly called cellulose, is the same material that is used to make paper. Vegetable fiber is derived from the bark and leaves of certain plants that grow in the tropics. Animal fiber normally refers to wool.

Cellulose is a fibrous organic material. It is a moderately good reinforcer because it is fibrous. Cellulose has a low density, almost half that of asbestos, and is suitable for making lightweight board. Cellulose fibers are resistant to alkaline attack. Cellulose fibers have a lower tensile strength than asbestos, are flammable, and absorb more water than asbestos during manufacturing. Cellulose fiber is readily available worldwide. Prices are priced competitively with asbestos.

Bamboo fiber, hemp, jute, and sisal are vegetable fibers that have been considered for use in place of asbestos. Their properties are not significantly different from those of cellulose. Their limited availability makes them less desirable than other substitutes.

Wool is the only animal fiber that has been considered as an asbestos substitute. Wool is fibrous and can be woven. It is a good thermal insulator and can withstand short term exposures to high temperatures without burning. Its heat resistance can be enhanced with an aluminized coating. Wool is widely available but has been used in only a few substitute applications.

Many different types of synthetic materials are considered for use as asbestos substitutes. Potential organic-based substitutes include aramid (PAM) fiber, nylon (PAM) fiber, polyacrylonitrile (PAN) fiber, polybenzimidazole (PBI) fiber, polyethylene (PE) fiber, polypropylene (PP) fiber, polytetrafluoroethylene (PTFE) fiber, and polyvinyl alcohol (PVA) fiber. These are all polymers or plastics consisting of long-chain organic molecules. Another organic-based substitute is a fiber composed of graphitized carbon.

Despite the formidable names of the polymers, most are recognizable in some form or another. Kevlar® or Nomex® are trade names for aramid fibers. These materials are used to make products,

Table 2. Properties of Asbestos Fibers (Badollet, 1951)

Property	Actinolite Asbestos	Amosite	Anthophyllite Asbestos	Chrysotile	Crocidolite	Tremolite Asbestos
Structure	Reticulated long prismatic crystals and fibers	Lamellar. Coarse to fine fibrous and asbestiform	Lamellar. Fibrous asbestiform	Usually highly fibrous fibers fine and easily separable	Fibrous in ironstones	Long. Prismatic and fibrous aggregates
Color	Greenish	Ash gray or brown	Grayish white. Brown-gray or green	White, gray, green	Lavender blue, metallic blue	Gray-white, greenish, yellowish, bluish
Hardness	6+	5.5–6.0	5.5–6.0	2.5–4.0	4	5.5
Specific gravity	3.0–3.2	3.1–3.25	2.85–3.1	2.4–2.6	3.2–3.3	2.9–3.2
Length	Short to long	2 to 11 in., varies	Short	Short to long	Short to long	Short to long
Texture	Harsh	Course but somewhat pliable	Harsh	Soft to harsh. Also silky	Soft to harsh	Generally harsh. Sometimes soft
Tensile strength, kPa	6 895 and less	110 316 to 620 528	27 579 and less	551 581 to 689 476	689 476 to 2 068 427	6 895 to 55 158
Temperature at maximum ignition loss, °C	—	871 to 982	982	982	648	982
Filtration properties	Medium	Fast	Medium	Slow	Fast	Medium
Electric charge	Negative	Negative	Negative	Positive	Negative	Negative
Fusion point, °C	1 393	1 399	1 468	1 521	1 229	1 316
Spinnability	Poor	Fair	Poor	Very good	Fair	Poor
Resistance to acids and alkalies	Fair	Good	Very good	Poor	Good	Good
Magnetite content	—	0	0	0–5.2	3.0–5.9	0
Mineral impurities present	Lime, iron	Iron	Iron	Iron, chrome, nickel, lime	Iron	Lime
Flexibility	Poor	Good	Poor	High	Good	Poor
Resistance to heat	—	Good. Brittle at high	Very good	Good. Brittle at high	Poor, fuses	Fair to good
Ionizable salts, μmhos (relative electrical conductance)	—	1.34	0.58	1.82	0.84	—

such as fire resistant textiles and bulletproof vests. Nylon is a synthetic fiber used in clothing. Polypropylene is a plastic that is used to produce building and household materials such as drainpipe, plastic outdoor furniture, and siding. Polyvinyl alcohol fibers are used in textiles and carpet fibers, and Teflon® is an example of a polytetrafluoroethylene fiber.

Most of the organic-based substitutes are reinforcers because of their fibrous habit. The fiber's tensile strength is dependent on the composition and structure of the organic material. The aramid (PAM) fibers are the strongest, approaching the tensile strength of asbestos, and they are the most thermally stable. The polymers are restricted to uses where temperatures do not exceed 200°C. All polymers will begin to soften or char at higher temperatures. All of the organic-based substitutes have low densities. Polymers such as PE, PP, and PTFE are resistant to attack by many acids and bases. Polymers fibers cost more than asbestos, and some are not available in sufficient quantities to permit wide scale use. Most of the polymers pose no health risk.

The carbon fiber has a composition identical to graphite, but it is a fibrous material that has been synthesized. Certain varieties of carbon fiber have an extremely high tensile strength and are excellent reinforcers. Additionally, carbon fiber can be woven and molded into intricate shapes. All varieties of carbon fiber are good electrical conductors They can be used at temperatures approaching 3 000°C in inert atmospheres. Carbon fibers are readily available, but are more expensive than asbestos.

Manufacturers also have investigated the use of nonfibrous organic-based substitutes in some applications. These have the same compositions as the fibrous organic materials and therefore the same properties. The nonfibrous organics generally are developed for specific enduse markets. Modified PP sheet for use in asbestos-cement sheet and shaped PTFE parts for gaskets are two examples of nonfibrous organic-based substitutes. Most of the nonfibrous materials are widely available.

Synthetic Inorganic Substitutes: Glass fiber, mineral wool, and ceramic (refractory) fibers have replaced asbestos in some applications. All of these forms are fibrous, with diameters ranging from 1 to approximately 25 μm. Most forms of glass fiber and mineral wool have relatively low tensile strengths, although strengths are greater for specialty glasses. Ceramic fibers have higher tensile strengths with some varieties having strengths exceeding those of asbestos. Glass fiber, mineral wool, and ceramic fibers are more brittle than asbestos fibers although some glass fiber varieties can be woven.

Table 3. End Uses for Asbestos, by Type

Application	Asbestos type	Important Properties Imparted by Asbestos
Asbestos cement pipe	Chrysotile, crocidolite	High green strength, good tensile strength, resists alkali attack, high impact strength, drains well during manufacture, heat resistant
Asbestos cement sheet	Chrysotile	High tensile strength, heat resistant, resists chemical attack, won't rot, corrode, or burn
Coatings and compounds	Chrysotile	High viscosity, resists sagging, reduced oxidation and deterioration, low cost, thixotropic, heat resistant, resistant to chemical attack
Friction products	Chrysotile	Heat resistant, low heat conductivity, durable, low cost
Packing and gaskets	Chrysotile	Flexible, heat resistant, doesn't deteriorate under pressure, corrosion resistant, resists chemical attack, low thermal conductivity, tear resistant, dimensionally stable, resists crushing
Paper	Chrysotile	Resists chemical attack, won't rot, dimensionally stable, heat resistant, won't burn or corrode, high tensile strength, relatively low price
Plastics	Chrysotile, crocidolite	Heat resistant, increased stiffness, improved dielectric strength, easy processing
Roofing	Chrysotile	High viscosity, resists sagging, reduced oxidation and deterioration, low cost, thixotropic
Textiles	Chrysotile, crocidolite	Heat resistant, flexible, durable, resists chemical attack, low thermal conductivity, high electrical resistance

Table 4. Fibrous vs. Nonfibrous Materials[1]

Fibrous	Nonfibrous
Aramid fiber	Biotite
Attapulgite	**Calcium silicate**
Carbon fiber	Diatomite
Cellulose fiber	Fibrillated polypropylene
Ceramic fiber	**Graphite**
Fiberglass	**Muscovite**
Franklin Fiber (gypsum)	Perlite
Nylon fiber	Serpentine
Palygorskite	Silica
Polyacrylonitrile fiber	Talc
Polybenzimidazole fiber	**Vermiculite**
Polyethylene fiber	
Polypropylene fiber	
Polytetrafluorethylene fiber	
Polyvinyl alcohol fiber	
Sepiolite	
Steel fiber	
Wollastonite	
Wool	

[1]Materials in bold type are the more commonly used asbestos substitutes.

Table 5. Mineral vs. Organic Materials

Mineral	Organic
Attapulgite	Aramid
Biotite	Cellulose
Franklin Fiber	Nylon
Graphite	Polyacrylonitrile
Muscovite	Polybenzimidazole
Palygorskite	Polyethylene
Sepiolite	Polypropylene
Serpentine	Polytetrafluorethylene
Silica	Polyvinyl alcohol
Talc	Wool
Vermiculite	
Wollastonite	

The resistance of glass fiber and mineral wool to attack by alkaline solutions varies with composition. Varieties with a high SiO_2 content are more susceptible to attack. Working temperatures for fiberglass and mineral wool range from 300 to 500°C. Ceramic fibers have working temperatures of 1 000 to 1 800°C, higher than those of asbestos. Glass fiber and mineral wool are more expensive than asbestos, particularly for specialty glass and ceramic fibers. Commercial production of these materials is extensive, and availability is not a factor. Some evidence suggests that glass fiber and mineral wool may be carcinogenic. Ceramic fibers are fibrogenic and carcinogenic.

To summarize, the ideal asbestos substitute would be composed of fibers that are flexible, have a high tensile strength, have a large surface area, and can be woven. The substitute also would be a good electrical insulator, resistant to degradation when exposed to high heat or chemicals, widely available, competitively priced with asbestos, and noncarcinogenic. While none of the materials discussed above (or asbestos) possess all of these qualities, many have been used to replace asbestos in commercial products (Table 4). A summary of the properties of asbestos substitutes is given in Table 6.

Alternative Products: Many manufacturers, instead of searching for an asbestos substitute, choose to replace the asbestos-containing product with a product that never contained asbestos. For example, ductile iron pipe and aluminum siding are sometimes used in place of asbestos-cement pipe and asbestos-cement shingles, respectively. Semi-metallic disc brakes are another common alternative product. Alternative products are practical for many applications because they usually are existing products, and they do not require the manufacturer to develop a new asbestos-free product (Table 7). The major concerns of customers and manufacturers when selecting an alternative product are cost and performance.

COMMERCIAL APPLICATIONS

Product Requirements

Asbestos and asbestos substitutes are used because they impart certain desirable properties in commercial products. In asbestos-cement (A/C) pipe and sheet products, well dispersed fibers allow the product to drain well in the green state, reinforce the pipe during curing, and make it more resistant to alkali attack. They also improve the tensile strength of pipe and the flexural strength of sheet, improve the bursting strength of pipes, and decrease water absorption during use. The most common substitutes are PVC for small diameter pipe and ductile iron and concrete pipe for large diameter pipe. Calcium silicate, cellulose, ceramic fiber, fiberglass, and mineral wool are used in sheet products.

Table 6. Properties of Asbestos Substitutes

Heat resistance	Reinforcing strength	Chemical resistance
good (above 400°C)	good	good
Mineral wool Refractory fiber All minerals Steel fiber Asbestos	Aramid fiber Carbon fiber PVA Glass fiber Steel fiber	PBI PAN PTFE PVA PP Carbon fiber Refractory fiber, except in alkalis Most minerals Asbestos, except in acids
moderate (200–400°C)	moderate	moderate
Aramid fiber PVA PTFE PBI PAN Carbon fiber Glass fiber	Cellulosic fiber PP PAN Refractory fiber Mineral wool	Aramid fiber Steel fiber
poor	poor	poor
Cellulosic fiber PP	All minerals PTFE	Cellulosic fiber

Modified from Hodgson (1985).

Table 7. Examples of Alternative Products

Asbestos-cement pipe	Ductile iron pipe PVC pipe Prestressed concrete Reinforced concrete pipe
Asbestos-cement sheet	Aluminum siding Cement sheet Corrugated aluminum sheet Corrugated fiberglass sheet Corrugated polyvinylchloride sheet Vinyl siding Wood siding and shakes
Coatings and pipeline wrap	Asphalt coatings Urethane tape PE films
Flooring	Vinyl tile Wood floor
Friction products	Semi-metallic disc brakes
Roofing	Single ply membranes

Coatings and compounds should be waterproof, resistant to cracking, viscous (sag resistant), have well dispersed fibers that bind with the matrix, and resist weathering. Several substitutes are cellulose, PE fiber, PP fiber, fiberglass (nonroof compounds), ceramic fiber, clay, talc, wollastonite, calcium carbonate, and silica gel. Mixtures of these substitutes often are used to obtain the desirable properties.

The asbestos substitute in flooring should provide a durable, resilient, and fire-resistant surface; it should be dimensionally stable. Flooring can be made from vinyl composition, or the matrix may contain mineral fillers such as talc, clay, wollastonite, and mica. Fiberglass and PE pulp are other substitutes.

Friction materials should be heat resistant and have low thermal conductivity, low noise coefficient, and high strength. They also should be moisture resistant, nonabrasive, and long wearing. For disc brakes, semi-metallic brakes are the most common substitute. Substitution for asbestos in drum brakes has been difficult to achieve and is limited in application. Aramid fiber, cellulose, ceramic fiber, fiberglass, Franklin Fiber, and wollastonite are used to replace asbestos.

Insulation materials (paper, paperboard, and wools) should be heat resistant, corrosion resistant, and resistant to chemical attack. Additionally, they should have a high strength, low thermal conductivity, and be moisture resistant. The most common substitute is fiberglass for moderate temperature, low cost insulation. Higher temperature applications require the use of ceramic fiber. Unspecified mineral fillers are used in some heat resistant millboards.

Packings and gaskets must be heat resistant, tear resistant, resistant to chemical attack, nonabrasive, and resilient. Fiberglass, vegetable fiber, carbon fiber, aramid fiber, expanded graphite, cork, ceramic fiber, rubber, copper, PBI, and PTFE are used in a variety of applications. Carbon fiber, expanded graphite, PBI, and fiberglass are used in packings.

Plastics fillers improve the product strength, stiffness, resistance to deformation when exposed to heat, and electrical resistance. Fiberglass, Franklin Fiber, graphite, PTFE, porcelain, and silica are the most common substitutes for asbestos in plastics. Talc, clay, and calcium carbonate also are used as inert fillers in plastics to improve strength and stiffness.

Roofing materials must be flexible yet dimensionally stable and resistant to rot, fire, and heat. Cellulose and fiberglass are used as substitutes for asbestos in roofing felts. Fiberglass or cellulose reinforced shingles, slate, metal roofing, wood shakes, and roof tile are alternatives to asbestos reinforced shingles.

Technological Problems

Although asbestos substitutes have been developed for many enduses, wholesale replacement has not occurred. Most of the products in question were developed for the particular characteristics of asbestos. The substitutes have different surface charges and textures, different fiber sizes, flexibility, and strengths, and different compatibilities with chemicals and heat. All of these properties affect the way that the fibers disperse in the matrix during manufacturing, adhere to the matrix in the final product, limit performance of the products, and restrict the processes used to manufacture asbestos products.

A few of the concerns of the manufacturers are the longevity of fiberglass-reinforced pipe, wear and performance problems with asbestos-free brakes, and the durability of cellulose-reinforced cement sheet. In some cases, the problems can be solved by using a combination of substitute material, reformulating the product to compensate for the different properties, or by using alternative products. Despite the problems, the trend is to replace asbestos as pressure is maintained on manufacturers to produce asbestos-free products. The more commonly used substitutes are shown in Table 8.

HEALTH RISKS

The switch to asbestos substitutes will not eliminate all concerns about health risks. Glass fiber has been linked with increased incidences of cancer, and ceramic fibers are proven carcinogens. Vermiculite from some deposits may contain asbestos, many other minerals will cause fibrosis of the lung when exposures to the dust are excessive. Most of the organic fibers are not carcinogens. No adverse health effects have been reported with Franklin Fibers.

Table 8. Asbestos Substitutes and Alternative Products

A/C pipe	Ductile iron, PVC pipe, prestressed concrete, reinforced concrete pipe
A/C sheet	Fibrillated PP, Ca silicate, cement, cellulose fibers, fiberglass, wood, ceramic fiber, corrugated fiberglass reinforced plastic, corrugated PVC, fiberglass reinforced cement, vinyl siding, aluminum siding
Coatings	Cellulose, PE films, ceramic fiber, clay, talc, limestone, wollastonite, fiberglass, carbon fiber, PP fiber, aramid, PE fiber
Flooring	Fiberglass, vinyl compositions, clay, PE pulp, talc, ceramic fiber
Friction	Semi-metallic brakes, fiberglass, aramid fibers, Franklin Fiber, mineral fibers, steel fibers, wollastonite, cellulose (A/T)
Insulation	Fiberglass, ceramic fiber, calcium silicate, cement
Packings	Fiberglass carbon fiber, graphite, aramid fiber, cellulose, ceramic fiber, PTFE, mica
Paper	Fiberglass, ceramic fiber, cellulose
Paper board	Nonfibrous minerals, wollastonite, ceramic fibers
Pipe wrap	Nonfibrous minerals, urethane coatings, plastic coatings
Plastics	Fiberglass, Franklin Fiber, teflon, mica, wollastonite, porcelain, fumed silica powder
Tape	Cellulose, urethane tape, carbon-based tape
Textile	Fiberglass, ceramic fiber, carbon fiber, aramid fiber, PBI

Concerns have been expressed about the risks posed during the manufacture of some of the organic substitutes and alternative products and by fumes released by some of the organics at high temperatures.

FUTURE TRENDS

Many materials are used to replace asbestos in products. Their most important properties are heat resistance, electrical resistance, resistance to chemical attack, high tensile strength, and fibrous morphology. In many cases, the higher costs and poorer performance are offset by the market demand for nonasbestos products. Some of the substitutes may encounter the same health controversy that asbestos encountered. The market for nonasbestos products should continue to grow, but it is evident from the wide variety of asbestos substitutes used that none has proven as versatile and cost effective as asbestos.

BIBLIOGRAPHY

Anon., 1986, *The Economics of Asbestos 1986*, 5th ed., Roskill Information Services Ltd., London, 181 pp.

Anon., 1987, "Silica and Some Silicates," Vol. 42, IARC Monographs on the Evaluation of Carcinogenic Risks To Humans, World Health Organization, 289 pp.

Anon., 1987a, "Recent Developments in Asbestos Substitutes," *Materials Engineering*, Vol. 104, No. 8, pp. 50–52.

Anon., 1988, *Regulatory Impact Analysis of Controls on Asbestos and Asbestos Products*, Vol. 3, Appendix F, May 6, US Environmental Protection Agency, Office of Toxic Substances, 497 pp.

Anon., 1988a, "Asbestos, A Less Favourable Filler," *Industrial Minerals*, London, Pigments, Fillers, and Extenders, pp. 160–163.

Badollet, M.S., 1951, "Asbestos, A Mineral of Unparalleled Properties," *Transactions*, Canadian Institute of Mining and Metallurgy, Vol. 54, pp. 151–160.

Corbett, R.A., 1986, "Substitutes Successfully Replacing Asbestos for Gaskets and Packings," *Oil and Gas Journal*, Vol. 84, No. 46, pp. 39–44.

Griffiths, J., 1986, "Synthetic Mineral Fibres," *Industrial Minerals*, London, No. 228, pp. 20–43.

Hodgson, A.A., 1985, *Alternatives to Asbestos and Asbestos Products*, Anjalena Publications Ltd., Berkshire, UK, 230 pp.

Hodgson, A.A., 1989, "Asbestos Alternatives," *Concise Encyclopedia of Mineral Resources*, D.D. Carr and N. Herz, eds., Pergamon Press, New York, pp. 15–20.

Parkinson, G., 1986, "Asbestos Users Step Up Search for Substitutes," *Chemical Engineering*, Vol. 93, No. 20, pp. 18–21.

Parkinson, G., 1986, "Makers of Gaskets, Packing, Try Asbestos Replacements," *Chemical Engineering*, Vol. 93, No. 20, pp. 23–26.

ENVIRONMENTAL USES

CLAY LINERS AND BARRIERS

KARAN S. KEITH AND HAYDN H. MURRAY

The sorptive and impervious nature of clay materials was recognized long before modern technology allowed us to understand the physiochemistry of clays. In Cyprus, around 5000 BC, sorbent clays were used for fulling woolen cloth. One of the earliest recorded uses of clays as an impervious liner was more than 3200 years ago in the construction of a bitumen-sand-gravel-clay liner built along the embankment of the Tigris River at Assur. Today one of the most common applications for clay minerals in the environmental industry is for use as impermeable barriers to fluid transport. Clay minerals are fine in particle size and thus have large surface areas, which, as a result of intermolecular forces and ionic substitutions, make them excellent materials for both ion exchange and ion sorption. Therefore it was logical to consider certain clay materials for applications in the environmental industry, where they can act as physical and chemical traps for potential contaminants. Certain bentonites have historically been used as sealants for fluids (Grim, 1962), and other clays have been used as sorbents for both water and oil (Haden, Jr., 1972).

Prior to the 1980s, disposal of the nation's municipal, commercial, chemical, and nuclear wastes was largely indiscriminate and poorly regulated. This created environmental pollution problems, especially with respect to groundwater contamination. Promulgation of federal regulations like the Resource Conservation and Recovery Act (RCRA) and the Comprehensive Environmental Response, Compensation and Liability Act (CERCLA—commonly known as Superfund) have forced a more careful and costly approach to waste disposal. The US Environmental Protection Agency (EPA) is the government body charged with enforcing and ensuring compliance with federal regulations governing toxic waste production, storage, and disposal. Widespread application of geotechnology to the design and construction of waste containment systems and to pollution remediation has become necessary due to such laws and because of increasing use of the subsurface environment for the disposal of hazardous wastes.

In the United States nearly 1.8 kg of trash is disposed of per person per day, which amounts to nearly 181 Mt per year. This amount is more than double the waste produced in 1960, and these numbers are expected to continue to increase. Canada produces approximately 1.68 kg of waste per person per day. The EPA estimates that in the United States 64% of our garbage is landfilled, 18% recycled, and 18% incinerated. Even with continued increases in recycling and incineration efforts, the EPA projects that the United States will still need 82% of today's landfill capacity in the year 2000. Data on the exact number of landfills in the United States and Canada is conflicting in part because of changing definitions of what constitutes a landfill and because of inadequate record-keeping by some states and provinces. An estimate of the number of municipal solid waste (MSW) landfills in the United States in 1990 to 1991 ranged from 4 462 to 10 467 with an average estimate of 6 600. In Canada the number of MSW landfills reported varied from 5 493 in a 1982 survey to 1 766 reported in a 1991 National Solid Wastes Management Association (NSWMA) survey (Repa and Sheets, 1992b). A properly operated MSW landfill on average requires 1 hectare of land per year per 25 000 people (Coates, 1981).

In addition to MSW facilities, there are also waste disposal facilities needed for hazardous waste and low level and high level radioactive waste. In 1983, approximately 266 Mt of hazardous waste was generated in the United States, almost 50% of this by industrial production. About 68% of the hazardous waste generated in the United States and Canada and almost 50% generated in Europe are still landfilled (Egger, 1987). In 1985, approximately 51 000 m^3 of commercially produced low level radioactive waste was disposed of in radioactive waste disposal sites across the United States (Jungling and Greeves, 1989). By 1987, the US defense program had produced approximately 10 000 t of high level nuclear waste. In addition, there were approximately 14 000 t of spent fuel assemblies from commercial nuclear power plants in temporary storage; the amount is expected to nearly triple by the year 2000. Presently most commercial spent fuel assemblies are stored in water-filled pools at nuclear power plants. Defense waste is stored on federal reservations in surface and underground steel tanks (Anon., 1987).

Under federal law nearly all waste disposal and containment sites are required to have a liner or barrier and a cover system design that meets federal- or state-approved hydraulic conductivity requirements for the type of waste that the site will receive. When approved materials, necessary for constructing the waste containment structure, are not available on site, one of the most economical ways to meet these requirements is through the use of clay, soil/clay, or clay/geotextile materials. In a 1991 survey by NSWMA of its members and other select private sector facilities, 81% of the surveyed landfills had some type of liner system. About 75% of these liners were clay materials (either recompacted or natural). The remaining 30% were using flexible membrane liners as part of a liner system. Clay liner materials are most commonly used in the South, Midwest, West-central, and South-central regions of the United States. About 93% of the landfill facilities had closure plans, and of these 91% were using clay materials in some way for the final cover material (Repa and Sheets, 1992a).

Waste barriers and containment structures take several basic forms, and they are used in a variety of different applications: 1) cutoff walls, 2) soil or soil/clay-admix liners, 3) geosynthetic bentonite-clay/geotextile membranes, 4) radioactive waste containment barriers, and 5) synthetic geomembranes. This chapter will discuss the first four of these categories. The field of synthetic geomembranes, although a related topic, is beyond the scope of this chapter. These containment structures serve as hydraulic barriers in applications such as landfills, nuclear waste repositories, pollution migration control, irrigation ditches, canals, cut-offs under dams, waste lagoons, mine tailings ponds, solar energy ponds, and other types of impoundments (Johnson et al., 1985). In addition to types of clay liners and waste barrier structures, this chapter will also review environmental regulations pertaining to clay liners, the important properties of clay minerals used for barrier applications, the sorptive abilities of clay minerals to potential environmental contaminants, and test methods for candidate clay liner materials.

A completely impermeable barrier does not exist. Hydrologists, geologists, and engineers should recognize this fact and anticipate the difficulties in trying to achieve a barrier that will withstand natural physical and chemical processes even through extremely short periods of geologic time, such as 10^2 to 10^4 years.

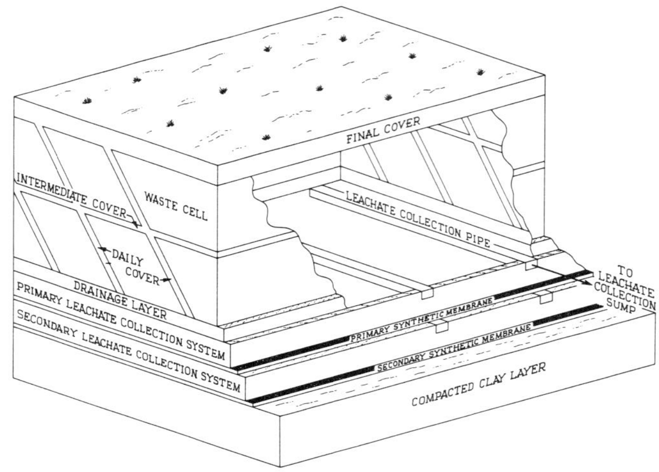

Fig. 1. Schematic cross section of a landfill with a double-liner system.

ENVIRONMENTAL REGULATIONS

Hazardous and Solid Wastes

The Resource Conservation and Recovery Act (RCRA) of 1976 was promulgated to provide guidelines for the regulation, monitoring, and disposal of hazardous and municipal wastes. The Comprehensive Environmental Response, Compensation, and Liability Act (CERCLA) was enacted to deal with already existing hazardous and municipal waste disposal problems.

Guidelines and regulations governing the disposal of solid and hazardous wastes are set forth in Subtitles D and C, respectively, of RCRA and the 1984 Hazardous and Solid Waste Amendment to this act. New RCRA Subtitle D regulations are scheduled to take effect in October 1993 and apply to all MSW landfills, except those facilities owned and operated by a state or federal government. Regulations governing the location, design, construction, operation, and maintenance of solid waste facilities are found in Parts 240–241 and 258 of Title 40 of the Code of Federal Regulations (40 CFR 240–241, 258). RCRA requirements regulating the location, design, construction, operation, and maintenance of hazardous waste facilities are found in Parts 260–265 and 122–124 of Title 40 CFR (40 CFR 260–265, 122–124). These regulations are for landfills, surface impoundments, waste piles, and land treatment facilities that generate, treat, store, or dispose hazardous wastes. Hazardous wastes (40 CFR 261.3) and solid wastes (40 CFR 258.2, 241.101) are defined in the Code of Federal Regulations.

RCRA regulations require hazardous and solid waste landfill facilities to include a double-liner system, consisting of a primary liner of synthetic material, a primary and secondary leachate collection and removal system and a final liner layer of clay, soil, or synthetic material (Fig. 1). This liner system must prevent migration of contaminant leachate during the period of facility operation, including the post-closure monitoring period. The top liner must be designed so as to prevent the migration of hazardous contaminants into such liner. The bottom liner, under the leachate collection system, must be composed of two parts. The upper part is provided to prevent migration of contaminants through the liner, and the lower component must be designed to minimize the migration of hazardous constituents if a breach were to occur in the upper component. This lower section must be constructed of at least 0.91 m of compacted soil material that has a hydraulic conductivity of no more than 1×10^{-9} m/s (40 CFR 264.221). If the liner is composed of earthen materials, data are required on the permeability, compaction density, and moisture content of the material. All liner materials, synthetic or natural, should be tested and selected for compatibility with the waste to be contained in the facility. It is also necessary to know the potential damage to the liner system that could occur during installation of the liner.

All landfill facilities are required to install a final cover system, designed to minimize infiltration and erosion, at closure of the facility (Lutton, 1980). The final cover system should include an erosion layer underlain by an infiltration layer. The erosion layer is to consist of a minimum of 0.15 m of earthen material that is capable of sustaining native plant growth. The layer designed to limit infiltration shall be a minimum of 0.46 m of earthen material with a hydraulic conductivity less than or equal to the hydraulic conductivity of any bottom liner system or natural subsoils present, or a hydraulic conductivity no greater than 1×10^{-7} m/s, whichever is less (40 CFR 258.60). State regulations must meet or exceed all federal regulations governing waste generation, transport, storage, and disposal. Many state regulations exceed those imposed by the federal government. Although many landfill facilities that are presently under construction follow these federal guidelines, the regulations are not scheduled to take effect until 1993.

Radioactive Wastes

Radioactive waste is divided into four categories depending upon origin and level of radioactivity: low level wastes, uranium mill tailings, transuranic wastes, and high level wastes. Everyday low level radioactive wastes (LLRW) are generated by industry, medical facilities, research institutions, and commercial nuclear reactor facilities and are usually disposed of through controlled shallow burial. Uranium mill tailings are the waste products of activities associated with uranium mining, and they contain low concentrations of radioactive materials that are disposed of through stabilization and burial. Most transuranic wastes are the byproducts of reprocessing of spent nuclear fuel and are characterized by elements with moderate levels of radiation and long decay times. High level radioactive wastes (HLRW) are waste products resulting from energy production at commercial nuclear power facilities and from fabrication of nuclear defense weapons. High level and transuranic wastes present long-term environmental hazards and require deep burial in a stable geologic repository (Anon., 1987, Coates, 1981, Jungling and Greeves, 1989, Peckinpaugh, 1988).

The Nuclear Regulatory Commission (NRC), the EPA, and the US Department of Transportation (DOT) regulate the activities of the Department of Energy (DOE) in the disposal and containment of radioactive wastes. The NRC is responsible for developing regulatory requirements regarding repository siting, construction, operation, and decommissioning. The EPA is responsible for the development of regulations that set standards for nuclear waste disposal designed to protect the health and safety of the public. Regulations and procedures governing transport of nuclear waste materials are the responsibility of the DOT. The DOE is the agency that is actually responsible for the permanent and safe disposal of high level nuclear waste (Anon., 1987, Rusche, 1986). Guidelines and regulations established for the disposal of nuclear wastes can be found in Parts 60, 61, and 960 of Title 10 of the CFR (10 CFR 60–61, 960) and Parts 264 and 191 of Title 40 of the CFR (40 CFR 264, 191).

The fundamental federal policies that deal with LLRW problems are the Low Level Radioactive Waste Policy Act (LLRWPA) passed by the US Congress in December 1980 and the Low Level Radioactive Waste Policy Amendments Act of 1985. This policy strongly encourages state responsibility in the management and disposal of low level radioactive wastes. The policy was developed to relieve some of the burden of disposal responsibility from what at that time were the only three commercial low level radioactive waste disposal sites in the states of Washington, South Carolina and Nevada (Peckinpaugh, 1988). Shallow burial at closely monitored sites is the most commonly accepted method of LLRW disposal. The primary requirements for land disposal of LLRW are: 1) long term, post closure site stability designed to eliminate the need for ongoing, active maintenance and 2) sufficient structural stability of the facility to prohibit infiltration of water and release of radionuclides to the environment (Jungling and Greeves, 1989, Parker, 1988).

The first major national policy to deal with the disposal of high level nuclear wastes was the Nuclear Waste Policy Act enacted in 1982 by the US Congress. Federal regulations require engineered barrier systems that will contain the radioactive wastes within the geologic repository during the first several hundred years following permanent closure of the facility. After this containment period, the engineered barrier system should work to control the release of radioactive material to the geologic setting. The function of the geologic setting in turn is to control the release of the radioactive material to the accessible environment. The transport of radionuclides should be inhibited to the degree that the amounts and concentrations of the materials entering the accessible environment are within the prescribed limits. This constitutes federally defined (10 CFR 60.110) isolation of the waste (Rusche, 1986).

IMPORTANT PROPERTIES OF CLAY MINERALS USED FOR BARRIER APPLICATIONS

Natural Untreated Clays

The common clay minerals that are used industrially include kaolinite, smectite, and palygorskite (attapulgite). These clay minerals have very different physical and chemical properties as a result of structural differences (see chapters on clay). Kaolinite has the lowest surface area (average 8 to 12 m^2/g) of the clay minerals, and has the fewest lattice substitutions, which results in kaolinite having the lowest surface charge. Palygorskite has a surface area ranging from 200 to 400 m^2/g, has a moderate charge on the surface because of lattice substitutions, and has channels in the lattice structure. Smectite also has a high surface area (up to 200 m^2/g), the smallest particle size, and the highest surface charge, making it an effective sorbent. Consequently the most commonly used clays for environmental waste disposal barriers are the smectites. Palygorskite and a similar clay, sepiolite, can be used in certain applications, particularly where sorption and flocculation are potential problems.

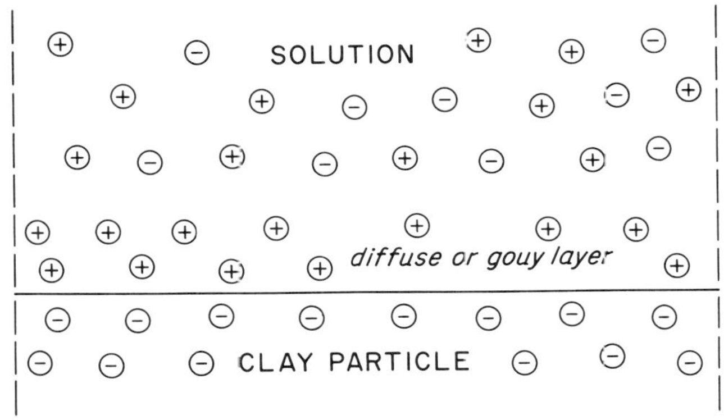

Fig. 2. Electrical double-layer (Gouy Layer) of clay minerals.

The two major processes by which clay particles interact with the chemical constituents in waste leachate are ion exchange and sorption. Clay particles are usually plate or lath shaped and are strongly anisometric The basal faces of clay particles have very large surface areas, carry negative charges, and are capable of cation exchange. Crystal imperfections, and resulting ion substitutions in the crystal lattice, give clay minerals an overall positive charge deficit. The positive charge deficit is compensated by preferentially selected counter-ions that are sorbed onto each of the layers in smectite and vermiculite, primarily on the edges and basal exterior crystal faces of illite and chlorite, and on broken-bond edges in kaolinite. The negatively charged anionic sites of the clay minerals that sorb cations are SiOH-, $AlOH_2$-, and AlOH-groups. Iron hydroxides and organics that are often associated with clay minerals are also capable of participating in cation exchange. The anionic sites on these substances are FeOH-groups in iron hydroxides, and carboxyl and phenolic OH-groups in organic substances. The intermolecular forces that are responsible for sorption phenomena in clays are electrostatic attraction alone, or electrostatic attraction augmented by hydrogen bonding and London-van der Waals attractive forces (Förstner and Wittman, 1979, Newman, 1987).

In a solution, cations necessary for charge compensation are attracted to the negatively charged basal surfaces of the clay particles and form an electrical diffuse double layer (Fig. 2). The negatively charged surface ions (co-ions) of the clay form one layer of the electrical double layer and the compensating charged cations (counter-ions) form the outer diffused portion of the double layer (Gouy layer). Competing electrostatic and diffusion forces create the diffuse nature of the double layer. The electrical double layer contains an excess of counter-ions and a deficit of co-ions. It is the sum of the charges of these two layers that give the clay particle its surface charge. The counter-ions can be exchanged for other ions and are usually referred to as exchangeable ions (Newman, 1987, Weiss, 1989). Exchange capacity in clay minerals, which is the sum of exchangeable cations expressed in milli-equivalents (meq)/100 g material, is a function of both the net negative surface charge of the clay and the particle size and related increase in surface area. The exchange capacity of clay minerals increases in order: kaolinite < chlorite < illite < smectite. Cation exchange and sorption in kaolinite occurs by sorption at broken-bond edges of the silicon-aluminum units. In smectite, illite, and chlorite substitution within the tetrahedral and octahedral layers creates a charge imbalance. Bonding of cations, primarily on the cleavage surfaces, serves to balance the charge Bonding in the tetrahedral layer position is stronger than bonding in the octahedral layer (Förstner and Wittman, 1979). In the expandable clay montmorillonite, the thickness of the double layer is a function of the cations present and this in turn controls the water sorbing capacity of the clay. Monovalent cations (e.g., Na^+) are attracted less strongly to the clay surface than are divalent cations (e.g., Ca^{2+}), resulting in a thicker diffuse double layer and consequently a greater ability to sorb more water (Hoeks et al., 1987). Smectitic clays generally swell easily in water from the hydration of interlayer cations and hydration of interlayer surfaces by sorption of water. When fully hydrated these clays usually have very low permeabilities, however, they can shrink and crack when physically or chemically dehydrated.

The edge faces of clays are broken-bond surfaces and are small in surface area. The double-layer potential on these surfaces is determined by the activity of H^+ or OH^- ions in solution. In neutral and acid solutions, the edges of clays have a positive double layer, which becomes negative at alkaline pH. Under natural conditions, hydroxide ions are usually the counter-ions surrounding the particle-edge surface, and these anions may form a diffuse anion double layer. At lower pH, the OH^- ions bond with H^+ and are replaced by other anions, which can increase both the thickness and the importance of this layer. The negative charge of the comparatively large basal faces gives the clay particle as a whole an electrostatic negative charge, but the positively charged edge surfaces play an important role in the stability behavior of clay suspensions.

Clays in the presence of water are classified as lyophobic colloidal systems. They exist in either a dispersed or flocculated state controlled by the particle interaction forces of electrical double-layer repulsions and London-van der Waals attraction. The electrochemical properties of the solution control whether the clays flocculate or disperse. When two clay particles approach each other face-to-face or edge-to-edge, electrostatic repulsion of the particles occurs in response to repulsion of similar ionic charges of the diffuse double layer. Some types of electrolytes added to the colloidal solution reduce the diffusion tendency of the ions, compressing the diffuse double layer. The range of repulsion of the particles is reduced and van der Waals attractive forces dominate. This results in decreased colloidal stability of the system and flocculation of the clay particles in a face-to-edge orientation. Under the right pH conditions, edge-to-edge attraction is possible because of the role that pH plays in influencing edge face charge.

In face-to-edge interaction the particles are attracted to each other because of their opposite signs of charge. A voluminous card-house structure is thus formed through electrostatic attraction. These card-house structured flocs have large pores, roughly the size of the diameter of the plate-shaped clay crystals, which give the overall structure a relatively high porosity and thus less effective sealing properties. In contrast, face-to-face interaction forms thick parallel particles. This association is often referred to as aggregation. Pore size in this structural arrangement is close to the thickness of the crystals and is generally smaller than pore spaces in the flocculated card-house structure. In general, edge-to-face attraction of the clay produces a more permeable structure than face-to-face attraction (van Olphen, 1977, Newman, 1987, Weiss, 1989).

The physiochemical properties of the clay and the chemical properties of the permeant are what determine the permeability and hydraulic conductivity of a clay barrier. Permeability and hydraulic conductivity are often interchangeably used to mean the velocity and ease with which a fluid, termed the permeant, passes through a solid. These two terms are not synonomous. The permeability of a material is its capacity for transmitting a fluid. Permeability is dependent on the number, size, and shape of the medium's pore space, and the size, shape, and tortuosity of the interconnections between pores. Permeability is independent of the fluid properties. Therefore permeability of a porous material is only affected when a chemical reaction, such as precipitation or leaching, occurs between the permeant and the medium altering the pore space of the material.

Hydraulic conductivity (or coefficient of permeability, as it is called in soil and rock mechanics) is dependent on properties of both the porous medium and the permeant. The size of pores or voids within the porous medium and the viscosity and specific weight of the permeant are the properties that control flow through the medium. Because hydraulic conductivity is a function of the properties of both the porous medium and the fluid, changes in either of these factors will result in changes in the hydraulic conductivity.

Organophilic Clays

In addition to the natural, untreated clays, industry produces a variety of specialty clays that have been modified in certain ways to improve one or more of their properties. One group of these treated clays, important in environmental waste management, is the organophilic clays. These specialty clays are made from natural clays that have undergone ion exchange in order to make them organophilic. The clays most commonly used in this process are smectites. Organophilic clays are prepared by exchanging the original inorganic metal exchangeable cations (e.g., Na^+, Ca^{2+}, or Mg^{2+}) of a natural clay with organic alkyl-chain bearing cations. Such treatment changes the clay surface from one that is hydrophilic to one that is organophilic, thereby allowing the clay to attract organic compounds and to swell and disperse in the presence of organic liquids.

The main attracting forces between the clays and the adsorbate are ion-dipole interactions, dipole-dipole interactions, and nonpolar interactions. The primary bonding mechanisms are: 1) hydrogen bonding to the clay surface and 2) London-van der Waals interactions between the sorbed organic compounds in the clays and the alkyl chains of the dissolved organic counter-ions. The type of interlayered organic cation and the degree of organophilic exchange control the adsorption behavior of the clay to any given adsorbate. An equilibrium is reached between adsorption on the clay and desorption to water, and it is the organophilic or hydrophilic character of the adsorbent molecules that control the position of that equilibrium. The most commonly used organic compounds in clay modification are quaternary ammonium salts (Stockmeyer, 1991a, Evans et al., 1989).

In determining the type of clay material to be used in constructing a hydraulic barrier, the physiochemical properties of the different clay minerals, the chemical nature of the leachate, and the requisite hydraulic conductivity for the barrier must all be taken into consideration. Natural or modified clays can be used alone to form a hydraulic barrier or can be mixed with local soils, clays, or shales depending on the application. It is likely that mixtures of relatively pure clay minerals will be utilized in future applications because each has some specific physical, chemical, or sorptive property that can be useful.

SORPTION ABILITIES OF CLAY MATERIALS

The mobility of hazardous waste constituents through the subsurface environment is a function of both the physical and chemical composition of the waste and the physiochemical environment of the hydrologic barrier through which the leachate must pass. Leachate generated in waste disposal sites is often complex and varies compositionally through time. The leachate composition is important because it can influence the pH and redox potential of the disposal environment which in turn affects ion exchange, solubility, precipitation, and overall mobility of the waste. Environmental regulations and subsequent industry demands for highly adsorptive and impermeable barrier materials have precipitated numerous studies, particularly in the last 15 years, on the adsorptive abilities of clays and soils for many of the anthropogenic pollutants in our environment. Most studies have focused on the group of clays known as smectites, although the least studied clays in the literature are the hormites (Helios-Rybicka, 1985, Morton and Sawyer, 1976, Singer, 1977).

Landfill leachate is derived from two sources: 1) flowable constituents of the waste itself and 2) water soluble components of the waste that are leached by water that infiltrates the waste. The four physical classes into which leachates generally fall are: 1) aqueous-inorganic, 2) aqueous-organic, 3) organic, and 4) sludges. Aqueous-inorganic and aqueous-organic leachates are wastes in which water is the liquid phase and the dissolved components are predominantly inorganic and organic, respectively. Organic leachates are liquid wastes which have an organic liquid phase and the dissolved components are other organic chemicals. Sludges are the wastes generated when a waste stream is dewatered, filtered, or treated for solvent recovery; they are characterized by a high solids content. Leachates generated from the impoundment of sludges are similar in chemical composition to the other three classes of leachates. Primary leachate is the flowable component of the waste material and includes both the liquids in the waste and the dissolved constituents. A primary leachate may be aqueous-inorganic, aqueous-organic, or organic. Secondary leachate is composed of the water percolating through the waste and its dissolved components, and is either aqueous-inorganic or aqueous-organic. Primary and secondary leachates are composed of a liquid carrier and its solutes, both of which may affect the hydraulic conductivity and permeability of a clay liner or barrier (Landreth, 1983). The composition of leachate is variable and complex, but most of its chemical constituents fall into three basic categories: 1) electrolytes, 2) metals, and 3) organics.

Electrolytes

Studies have shown that high concentrations of some electrolytes in waste leachate can compromise the integrity of clay barriers. The addition of electrolytes to a colloidal system causes compression of the diffuse ionic double layer surrounding the colloidal particles, reducing the range of repulsion of the particles, thereby decreasing colloidal stability and promoting particle flocculation. Compression of the diffuse double layer also decreases the water-adsorbing capacity of the clay resulting in an increase in hydraulic conductivity. Smectites are particularly susceptible to the effect of electrolytes. The introduction of ions with a higher valence reduces the thickness of the double-layer, thus salts with divalent cations should cause the clay to be more flocculated than monovalent cationic solutions of the same concentration. Research comparing the hydraulic conductivity of smectite clays exposed to electrolyte solutions of varying concentrations support this theory (Malik et al., 1992). A study by Alther et al. (1985) on bentonites showed an increase in hydraulic conductivity with increasing salt concentration. In addition, divalent cationic salt solutions had a greater affect on the hydraulic conductivity of these clays than monovalent solutions. Palygorskite and sepiolite, on the other hand, exhibit a very high chemical inertia and are relatively unaffected by electrolytes, particularly at moderate pH, or by acid attack. At present the authors are looking at the effects of electrolytes on the hydraulic conductivity of smectite and kaolinite blended with the hormite clays.

Metals

The mobility of inorganic metallic waste constituents through a hydraulic barrier is controlled by solubility of the metals and cation/anion exchange-sorption reactions. These two parameters are a function of pH, ionic competition, redox potential, and composition of the medium through which the inorganic constituents are passing. The large surface areas of clays, the negative charges on faces and at broken edge bonds, and the ability to exchange and sorb cations and anions make clay minerals good sorbers of many metal ions. Ionic charge, hydration behavior, electronegativity, and ionization potential are all factors that are believed to affect the selective affinity of clays for different heavy metal ions. Various sequences of affinity have been proposed for the attraction of heavy metal ions to clay minerals. In general, sorption of heavy metal cations such as lead, cadmium, zinc, copper, mercury, and chromium (III) increases with increasing pH, but sorption of the heavy metal anions such as chromium (VI), arsenic, and selenium decreases with increasing pH (Griffin et al., 1977, Förstner and Wittman, 1979).

Metal solubility, and thus ionic mobility, generally increases with decreasing pH. Under extremely acidic conditions, H^+ ions effectively outcompete metal ions for exchange sites on clays, thereby releasing the metal ions into solution. Ion exchange and sorption of most, but not all, metal cations to the clay surface increases with increasing pH. Metals have been shown to be exchanged preferentially in reverse order of their mobility. Under alkaline conditions, anions such as carbonate and hydroxide are abundant and can lead to precipitation of metal compounds (Gibb and Cartwright, 1987). Under favorable pH and redox conditions many of the heavy metal ions are preferred in cation exchange sites over the common clay interlayer cations of Na^+, Ca^{2+}, and Mg^{2+}. Ion exchange releases the clay interlayer cations into the waste stream, altering both the pH of the leachate and the relative concentration of these ions in solution. This in turn can cause precipitation of metal complexes and increased competition between the metal ion and the exchanged ion for cation exchange sites on the clay. Clays can also remove metals from leachate by acting as a mechanical substrate for the precipitation and flocculation of organics and secondary minerals, such as hydrous iron oxides, both of which have been demonstrated to have strong affinities to metals. Large amounts of organic matter and saturated conditions make reduction the more common redox state in a landfill. In addition to pH, redox state also affects metal mobility. Chromium is most mobile under oxidizing conditions, and iron and manganese are most mobile under reducing conditions (Griffin and Shimp, 1976, Griffin and Au, 1977, Förstner and Wittman, 1979).

Lead is one of the more strongly sorbed metal ions. Cation exchange-adsorption is believed to be the primary attenuation mechanism in removal of Pb^{2+} from landfill leachates by clays. Data indicate that removal of Pb^{2+} from leachate increases with increasing pH, particularly above pH 6, and is affected by competition between different ions in solution, especially di- and trivalent cations. Montmorillonite has been found to remove as much as five times more Pb^{2+} from solution than kaolinite (Griffin and Shimp, 1976, LaBauve, et al., 1988).

The mobility and removal of Cu^{2+}, Zn^{2+}, and Cd^{2+} from a landfill leachate is dependent upon pH and ionic strength of the leachate and cation exchange capacity of the soil. Removal of these metal ions from leachate increases with increasing pH. Studies indicate that below pH 7 removal of the metal ions Cu^{2+}, Zn^{2+}, and Cd^{2+} from leachate by the clay minerals kaolinite and montmorillonite is primarily a function of cation exchange, and is affected by pH and ionic strength. Low CEC values and high ionic strengths can make these metals quite mobile, particularly at lower pH. Above pH 7, metal precipitation becomes an important mechanism for removal of metal ions from the leachate (Frost and Griffin, 1977b, Griffin et al., 1977, LaBauve, Kotuby-Amacher, and Gambrell, 1988).

A study of the sorption of Co^{2+} and Ni^{2+} on montmorillonite and hectorite show that Co^{2+} exchange was greatest on hectorite and Ni^{2+} exchange greatest on montmorillonite (Davison, McWhinnie, and Hooper, 1991).

Experiments indicate that the removal of Hg^{2+} from waste leachate is pH-dependent, increases at alkaline pH, and is enhanced by clay minerals in proportion to their cation exchange capacities. Montmorillonitic clays were found to have a greater sorption capacity for Hg^{2+} than kaolinitic clays, although sorption by clays was not the primary mechanism for removal of Hg^{2+} from leachate; precipitation and volatilization are believed to be the predominant attenuation mechanisms (Griffin et al., 1977).

Sorption sites for anions on clay are associated with exposed octahedral cations at the broken edge faces of the clay particle under conditions of acidic pH. Sorption of cations is generally greater than sorption of anions due to the difference in area between basal faces and edge faces. Studies of As(V), As(III), and Se(IV) indicate that pH is a determining factor in the distribution of As and Se species present in solution and in the activation of anionic sorption sites on the clay particles. As(V) sorption reaches a maximum at about pH 5, while As(III) sorption increases over the pH range 3 to 9. Se(IV) sorption is maximized in the pH range 2 to 3 and then decreases as the pH increases. Sorption of As and Se is greater by montmorillonite than kaolinite; probably in part because of a probable larger edge surface area in montmorillonite. In general As and Se present in leachate would be expected to be relatively mobile through a clay barrier under alkaline conditions (Frost and Griffin, 1977a, Griffin et al., 1977).

Two of the most commonly occurring oxidation states of chromium are Cr(III) and Cr(VI). Cr(III) can occur as a free cation, while Cr(VI) rarely exists as a free cation and is almost always found in an anionic state with oxygen. Chromium (III) is strongly sorbed by both montmorillonite and kaolinite below pH 4. Adsorption of Cr(III) by the clays increases up to pH 4. Above pH 5 Cr(III) precipitates as hydrous chromic hydroxide and is rendered immobile. Cr(III) is extensively hydrolyzed into various species at strongly acidic pH and exchange-adsorption of hydrolyzed Cr(III) is believed to be the mechanism behind increased Cr(III) adsorption with increasing pH. The amount of Cr(VI) adsorbed from solution decreases with increasing pH, with no adsorption occurring above pH 8.5. The Cr(VI) ion most abundant between pH 2 and pH 5 is $HCrO_4^-$, which appears to be adsorbed preferentially over CrO_4^{-2}, the only species present above pH 8.5. The amount of Cr(VI) adsorbed by montmorillonite has been found to be 2 to 3 times larger than the amount adsorbed by kaolinite. Cr(III) is adsorbed 30 to 300 times more readily by clay minerals than Cr(VI) in the pH range 1.5 to 4.0. At higher pH values the ratio of Cr(III)/Cr(VI) removed is even larger. This suggests that conversion of Cr(VI) wastes to Cr(III) should be considered before landfilling wastes containing Cr(VI) (Griffin et al., 1977).

Organics

The organic constituents of leachates are another form of waste fluid that can compromise the integrity of clay barriers. Organic liquids found in leachates can be divided into four basic classes: 1) organic acids, 2) organic bases, 3) neutral, polar, water-soluble molecules, and 4) neutral nonpolar molecules. The mobility of organic wastes through the subsurface is a function of their solubility, sorptive capacity, and biodegradability.

A study by Anderson (1981) determined that organic acids affect the hydraulic conductivity of clays by acid attack, causing partial dissolution of the silica and aluminum in the structure of the clay. Acidization in turn leads to an increase in hydraulic conductivity. The other three classes of organic liquids appear to affect hydraulic conductivity via alterations in the structural fabric of the clay. Large increases in the hydraulic conductivity of clays can occur in the presence of some concentrated organic chemicals due to dehydration and shrinking of the clay. Bronsted organic bases can accept protons to become ionized cations. These cations can sorb to the clay, replacing adsorbed water and thus altering the interlayer spacing of the clay particle and its swelling and shrinking capabilities. Neutral polar organic fluids have a large dipole moment, and this allows these compounds to also compete with water for sorption sites on the clay, altering the behavior of the clay. The interlayer spacing of clay particles is affected by the dielectric constant of the surrounding liquid. Interlayer spacing controls the shrink and swell properties of the clay. When the original water solution around the clay particle is replaced by an organic compound with a lower dielectric constant than water, the interlayer spacing of the clay decreases and the clay flocculates, resulting in an increase in permeability. Neutral nonpolar organics have low water solubilities, no net charge, and almost no dipole moments, and consequently under a hydraulic gradient they tend to pass through clay without being attenuated. Some nonpolar compounds, however, are sorbed to clays by van der Waals forces (Landreth, 1983, Anderson, 1981).

A study by Brown, Green, and Thomas (1983) examined the sorptive abilities of a kaolinitic soil, a micaceous soil, and a bentonitic soil for xylene, acetone, and water. This study concluded that hydraulic conductivity values of the soils increased two to four orders of magnitude with the solvents versus the water. Xylene rapidly increased the hydraulic conductivity of the clay soils after the flow exceeded 0.2 to 0.4 of a pore volume. The acetone first caused a small decrease in hydraulic conductivity; this was then followed by a subsequent increase in hydraulic conductivity. The initial decrease in hydraulic conductivity seen in the polar acetone is believed to be the result of swelling and sorption by the clay at low acetone concentrations, followed by shrinkage of the clay structure as more water was subsequently displaced by the solvent. Anderson, Crawley, and Zabcik (1985) evaluated the hydraulic conductivity of a calcareous smectitic clay soil to water, xylene, and methanol. Again the presence of the organics caused an increase in the hydraulic conductivity of the clay over that of water.

The specialized organophilic bentonites appear to be better sorbents for organic compounds than natural clays. These clays can make good admixtures in liner systems, especially if an organophilic bentonite with a broad sorption ability is used, since the composition and concentration of landfill leachate varies with time. A study by Stockmeyer (1991b) measured the hydraulic conductivity of a silty sand that was amended with admixtures of 5% each of a natural hydrophilic bentonite, a hydrophilic/organophilic bentonite, and an organophilic bentonite. The permeants used were water and phenol. This study found that contaminant retention and hydraulic conductivity of the silt/sand soil was considerably better in the mixture containing organophilic bentonites compared with the more conventional hydrophilic bentonites. Tests also showed that no deterioration occurred in the stability and strength of the soil amended with the 5% organobentonites, except in cases of concentrated organic liquids which are unlikely in real leachates.

Another study by Stockmeyer (1991a) examined the behavior of various organophilic bentonites in contact with a variety of aqueous solutions of organic compounds. The organoclays had organic counter-ions of dimethyl-dioctadecyl ammonium and octadecyl-benzyl-dimethyl ammonium. The results showed that the adsorption performance of the different organoclays was dependent on the type of organic compounds in the permeant. Thus for waste impoundments with known chemical constituents, pollutant-specific organoclays could be used in remediation.

Leachate is rarely monochemic and generally contains both inorganic and organic compounds. Stockmeyer and Kruse (1991) evaluated the sorptive ability of both natural bentonites and organophilic bentonites to mixtures of heavy metal ions and organic compounds. As expected, they found that metal-salt solutions of Zn^{2+} and Ni^{2+} were best sorbed by the natural clays and that organic solutions of phenol and diethyl ketone were most readily sorbed by the organophilic clays. For solutions containing a mixture of both heavy metal ions and organic compounds the sorption behavior was very different. Sorption of the heavy metal ions by the organoclays was generally increased in the presence of organic compounds. The organic molecules are first attracted to the organic cations of the organoclays by van der Waals forces and then form complexes with the metal ions. During sorption the metal ions and the organic molecules enter the interlayer spaces of the clay mineral where the non-polar character of the alkyl chains of the organic cation favors formation of monodentate complexes with the metal ions.

Palygorskite and sepiolite clays have a high sorptive capacity as a result of their highly convoluted external surface, large internal pore surface, fibrous morphology, and low base exchange capacity. Sorption of organic molecules is mainly on the external surfaces of these clays. Non-polar organic molecules are sorbed to sepiolite and palygorskite primarily by van der Waals attraction. Polar molecules are sorbed on both the external surfaces of the clay and in the internal channels. Studies have shown palygorskite and sepiolite to have a specificity for organic sorbates that have a molecular size and shape that fits closely to the surface convolutions of the open channels of exposed crystal faces. Intracrystalline sorption into the structural channels is limited to compounds that have a small molecular size and high polarity because of steric hinderance and the non-expansible nature of the clays' crystal structure (Newman, 1987).

Most leachate is complex and variable through time. To address this problem, a multi-barrier waste containment system could be constructed where the clay materials used are tailored to the types of chemicals to be impounded. Because the composition and concentration of leachate varies through time, an organophilic bentonite with a broad adsorption ability would be well suited for a multi-barrier system.

TYPES OF CLAY LINERS AND BARRIERS

The main purpose of a clay barrier is to retard movement of fluids into the surrounding medium. Low permeability, barrier integrity, and chemical stability are the requisite performance criteria for clay barriers. The physical and chemical properties of clay minerals give clay-bearing strata and soils hydraulic conductivity values that often fall within acceptable EPA guidelines for hydraulic barriers. As pointed out by Johnson et al. (1985), the key issue is whether a hydraulic barrier, regardless of the material used for the barrier, will function satisfactorily over the intended lifetime of the facility. Failure of a barrier is defined as the deterioration of a slurry wall or lining material which results in the hydraulic conductivity exceeding a legally defined value (Johnson et al., 1985). The determination of hydraulic conductivity is therefore of special importance in the field of hydraulic barrier design and installation. In addition to the properties of the barrier, the properties of the permeant also affect hydraulic conductivity of the barrier. The structure and composition of a hydraulic barrier should therefore be determined by both the application for the barrier and the barrier compatibility with the type of waste fluids that the waste containment structure will contain.

Waste barriers and containment structures commonly take several basic forms: 1) cutoff walls, 2) soil or soil/clay-admix liners, 3) geosynthetic bentonite-clay/geotextile membranes, and 4) radioactive waste containment barriers.

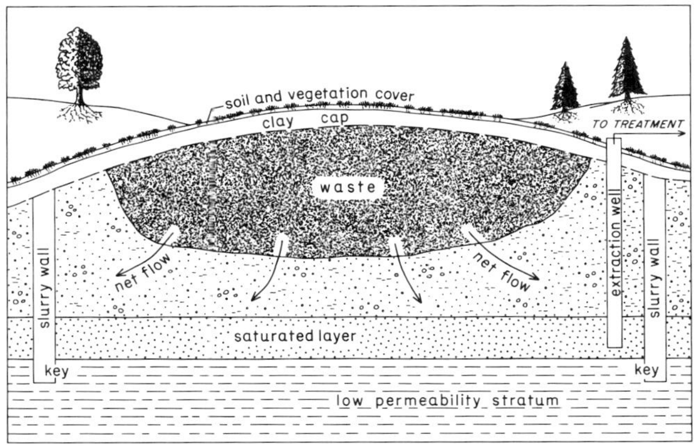

Fig. 3. Cross section of a keyed-in slurry wall containment barrier.

Cutoff Walls

A cutoff wall is a low permeability subsurface structure constructed to contain capture, or redirect groundwater flow at a given site. These walls can be composed of a variety of manmade or natural materials (Spooner et al., 1985). Some common applications for cutoff walls are: 1) retaining structures, 2) seepage barriers under dams, 3) pollution migration control, and 4) low level radioactive waste containment structures. Containment of toxic military warfare chemicals at the Denver Rocky Mountain Arsenal is one well known application of a cutoff wall in remedial clean up (Spooner et al., 1985, Wagner et al., 1986, Ryan, 1985).

There are two principal methods of cutoff wall emplacement: 1) conventional slurry trenching method and 2) vibratory beam injection (VBI). Slurry trenching construction originated in Italy and has been in use in the United States for more than 45 years, whereas VBI is a relatively new technique in use since the early 1970s and is considered by construction engineers to be a grouting method rather than a slurry trenching method. The two techniques are significantly different both in their method of construction and in the features of the finished product (Jepsen and Place, 1985, Gill and Christopher, 1985).

Slurry Trenching Method: Johnson et al. (1985) defined a slurry wall as a nonstructural wall constructed underground as a vertical barrier to the lateral flow of water or other fluids (Fig. 3). Slurry walls are vertical trenches excavated under a slurry to a relatively horizontal strata of low permeability material (Johnson et al., 1985). A nonstructural wall is one that is incapable of supporting bending moments or significant shear stress and has a strength capacity similar to that of the surrounding ground (Ryan, 1985). The function of the slurry is to hydraulically shore the walls of the trench against collapse and, through formation of a filter cake, prevent high fluid loss into the surrounding medium. Thorough testing of the compatibility between the slurry and the waste should be undertaken to ensure long term integrity of the barrier.

Conventional trenching construction of a slurry wall consists of the excavation of a narrow trench into the ground with a backhoe or other more specialized equipment. To prevent collapse of the trench wall during excavation, the trench is kept constantly filled with a slurry consisting primarily of sodium bentonite and water, similar in consistency to drilling mud. Slurry walls are often keyed-in to a low permeability confining stratum underlying the waste site (Fig. 3). The low permeability layer acts as a floor for the containment site and a good key-in is essential for proper

Table 1. Comparison of SB and CB Methods of Slurry Wall Construction

Soil-Bentonite Method	Cement-Bentonite Method
low installation cost	high installation cost (30% higher than SB walls)
wide range of chemical compatibilities (i.e., greater resistance to most pollutants)	narrower range of chemical compatibilities—more susceptible to chemical attack
lower permeability (1×10^{-9} to 1×10^{-10} m/s)	higher permeability (1×10^{-8} m/s)
backfill can be specially blended to meet specific design requirements	not dependent on availability or quality of soil available for backfill
requires a large work area for backfill mixing—when excavated material can be mixed with backfill, minimal spoil problems are created	no backfill mixing required therefore better for excavating in areas with difficult access or inadequate room for backfill mixing
trenching must be done continuously in one direction	allows for construction of trench in sections providing more flexibility to meet site constraints
least strength, highest compressibility	greater strength, less elasticity
level work site required because of flowability of slurry backfill	better for sites where topography is extreme and grading of site impractical
Wall Specifications	
width: 0.5 to 1.8 m	width: 0.3 to 1.0 m
slump value = 0.1 to 0.15 m; must stand on a slope of 5:1 to 10:1	unconfined compressive strength = 1.5×10^4 to 2.8×10^4 kg per m^2
Slurry Specifications	
4 to 7% bentonite by weight	4 to 7% bentonite by weight
93 to 97% water by weight	65 to 88% water by weight
makeup water low in total dissolved solids	8 to 25% portland cement by weight
slurry pH = 7.5 to 12.0	slurry pH = 12.0 to 13.0
density ≈ 1 050 kg per m^3	density ≈ 1 090 kg per m^3
Marsh Funnel Viscosity = 35 to 45 MFs; Apparent viscosity = 0.015 to 0.020 Pa·s (15 to 20cP)	Marsh Funnel Viscosity = 30 to 50 MFs
Backfill Specifications	
2 to 4% bentonite by weight of the total mixture	6% bentonite by weight
≈25 to 35% water by weight	≈55 to 70% water by weight
≈10 to 40% fines by weight	≈30 to 40% solids by weight ≈18% cement by weight
density ≈ 1 680 to 1 920 kg per m^3	density ≈ 1 300 kg per m^3 max.

References: Spooner et al., 1985, Jepsen and Place, 1985, Johnson et al., 1985, Ryan, 1985.

containment. Hanging slurry walls, which may only extend several meters down into the contaminated stratum, are another structural wall form used to contain contaminants that float on the upper portion of the water table.

Types of slurry walls are defined by composition and the most commonly used in environmental applications are: 1) soil-bentonite (SB) and 2) cement-bentonite (CB). After trench excavation, a SB cutoff wall is created by backfilling the excavated trench with a mixture of soil, bentonite, and water, which together displaces the slurry and forms a permanent impervious cutoff wall. This method is most economical when the on-site soil excavated from the trench is suitable material for use as backfill. Slurry is mixed with soil adjacent to the trench, worked to a smooth consistency, and then pushed into the trench, displacing the original bentonite slurry. CB slurry cutoff walls utilize a mixture of cement, bentonite, and water. The mixture is poured into the trench as it is being excavated; once excavation is completed the CB slurry is left in the trench and sets up to form a permanent seepage barrier. In the CB method there is high slurry loss due to the excavation process and seepage losses through the sides of the trench; therefore, the cost of backfill is considerably more than for the SB method (Spooner et al., 1985, Ryan, 1985). Table 1 provides a comparative list of the SB and CB methods of slurry wall construction. Individual site assessments should be made to determine which method is most appropriate for the conditions involved.

Jepsen and Place's (1985) review of slurry trenching methods identifies three areas of quality control that should be considered: 1) slurry properties and quality, 2) trench dimensions and continuity, and 3) cutoff wall composition and placement.

Slurry Properties and Quality. The clay material used in both the SB and CB methods of construction is typically a sodium (Wyoming) bentonite. Swelling and hydration, dispersability, and thixotropy are properties of bentonite that make it important in slurry trench construction. The thixotropic nature of these highly plastic clays give them excellent viscosity, density, and filter loss properties for slurry trenching. The two main functions of the bentonite slurry are: 1) maintenance of trench wall stability through hydrostatic forces and filter cake formation and 2) maintenance of slurry density. Slurries used in both the SB and CB methods must meet certain specifications in order to perform optimally (Spooner et al., 1985, Jepsen and Place, 1985). Specifications suggested for slurry used in SB and CB slurry walls are given in Table 1.

Formation of a filter cake on the walls of the trench is important in order to prevent collapse of the trench and to minimize slurry seepage out of the trench and groundwater seepage into the trench. Bentonite filter cake permeabilities can run as low as 2.3×10^{-11}

m/s. It is important to fully hydrate the slurry before it is introduced to the trench. Filtrate (water) loss is a potential problem in slurry trenching that can lead to several problems including poor filter cake formation and trench wall instability. High filtrate loss leads to an increase in the concentration of residual bentonite solids, which results in slurry thickening.

In cement-bentonite filter cakes, high calcium content in the cement can cause partial flocculation and aggregation of the bentonite leading to undesirably high filtrate losses (1.0×10^{-9} m/s). The ratio of cement to bentonite is important: too little bentonite will cause the cement to settle out causing excessive water loss; too much bentonite and the slurry will be too difficult to work with and may cause wall weakness. If too little cement is used, the slurry may not set up properly. In the CB method the cement is added to the fully hydrated bentonite-water slurry just prior to introduction to the trench, a technique that minimizes trench slurry viscosity and reduces the amount of suspended solids from the soil that might be entrained in the slurry upon hardening.

Slurry density also helps to ensure trench wall stability. The density of the bentonite helps to hold in suspension small particles that fall from the wall during excavation, creating a slurry that is denser than water. The slurry must be denser than water in order to maintain adequate force against the trench wall, especially where groundwater levels are high. Maintaining a well dispersed slurry is important because flocculation of the bentonite in the slurry can cause the bentonite to settle out, reducing trench wall stability and interfering with filter cake formation.

Contamination of the slurry by factors such as the presence of salts or extremely basic pH conditions can cause bentonite flocculation, high filtrate loss, and poor filter cake performance. Bentonite contamination can be controlled through the use of specially treated contaminant resistant bentonites, the use of chemical dispersants or bulking agents in the slurry, increased bentonite concentration in the slurry, or disposing of the contaminated slurry from the trench and replacing it with fresh slurry (Spooner et al., 1985, Jepsen and Place, 1985, Wagner et al., 1986).

Trench Dimensions and Continuity. Trench dimensions are relatively easy to control. The most important factor is to excavate a continuous trench of uniform dimensions. Trench width is controlled by the width of the excavating equipment and trench depth is a function of geology and probing to insure continuous and consistent excavation. A backhoe or a crane and clamshell are the two more common excavation machinery used in trench construction. CB slurry walls can be installed in areas with more severe topography by excavating and emplacing one section at a time (Spooner et al., 1985, Jepsen and Place, 1985).

Cutoff Wall Composition and Placement. The most economical method of construction for SB slurry walls is to create the wall using materials excavated from the trench. Imported soil materials, however, should be used when the on-site soil is too contaminated for use, when the soil is too coarse to meet necessary permeability requirements at economical bentonite application rates, or when the soil is too fine to meet the necessary strength requirements of the wall (Jepsen and Place, 1985). One of the most important properties of the SB slurry wall is its permeability. Permeability is a function of soil-particle size, soil plasticity and percentage of bentonite in the mix. The greater the percentage of soil particles in the fine range (less than 74 μm) and the more plastic those fines, the lower the wall permeability. SB slurry wall specifications generally require the backfill to consist of a minimum of 10 to 20% fines. These requirements can be achieved through careful testing and control of on-site soils or by supplementing with borrow materials. A gradation test [ASTM Particle-Size Analysis of Soils (D422–63)] and moisture content test [ASTM Laboratory Determination of Water (Moisture) Content of Soil, Rock, and Soil-Aggregate Mixtures (D 2216–80)] should be made on the backfill mixture prior to adding to the slurry or water in order to assure better quality control. Methods for mixing backfill can vary as long as a homogeneous mixture of soil and slurry is achieved. The specific weight of the backfill should be at least 250 kg/m^2 greater than that of the slurry in the trench in order to easily displace the slurry (Jepsen and Place, 1985).

Cement-bentonite cutoff walls are used where strength is a primary consideration Overall CB walls are stronger than SB walls and are therefore better suited for sites where the wall must have less elasticity, such as adjacent to roads or buildings. CB slurry walls typically have an unconfined compressive strength approximating that of the surrounding soil. They also begin to harden much faster than SB walls, which requires the use of different construction techniques from those used with SB walls. CB walls are generally thinner than SB walls because CB walls are more expensive to construct and there is no backfilling operation. In a CB slurry wall the calcium present in the cement alters the properties of the sodium bentonite, thus the viscosity, the time required for filter cake formation, and the permeability of the filter cake are all usually higher than in SB walls. Unlike SB slurry walls, suspended solids that are entrained in CB slurry walls should be kept to a minimum. At too high a concentration (above 10% by volume) interparticle contact occurs which causes an increase in wall permeability and an increase in wall strength, which can make the wall more brittle and susceptible to cracking (Spooner et al., 1985, Jepsen and Place, 1985, Wagner et al., 1986). Strongly acidic leachates can affect the permeability of cement-bentonite slurry walls because they have a tendency to dissolve the hydrated cement and flocculate the bentonite (Leonards et al., 1985). Some specifications for backfill used in SB and CB cutoff walls are given in Table 1.

Vibratory Beam Injection Method: The vibratory beam injection method was first developed in Europe more than 30 years ago. It has been improved and modified during the intervening years and today involves the injection, to a preselected depth, of a wide flange beam containing integral grout injection equipment. Consolidation and displacement of the surrounding soil occurs as the beam is vibrated into position. After reaching the desired depth, the beam is retracted leaving in its place a consolidation void that is pressure grouted with a grout that hardens in place. By overlapping each successive beam injection a continuous cutoff wall is formed. The VBI method works best in saturated, loose, granular soils. It is hard to use in stiff clays because of difficulties with penetration, and rocky soils can cause beam deflection and alignment problems. The two most important quality control parameters in the VBI method are: 1) wall composition, properties and contaminant resistance and 2) trench wall continuity (Jepsen and Place, 1985).

Wall Composition. A variety of slurries can be used in the VBI method, but some of the most common are cement-bentonite (when strength is required), fly-ash/bentonite (for reactive soils), kaolin-bentonite (in high-saline conditions), and asphalt emulsions (for highly contaminated conditions where chemical attack is probable). A thixotropic slurry is necessary with a bentonite permeability on the order of 1×10^{-8} to 1×10^{-10} m/s. Asphalt emulsions generally have permeabilities on the order of 1×10^{-14} m/s (Jepsen and Place, 1985).

In CB walls the cement to bentonite ratio is important. Cement provides strength and resistance to erosion, but it adversely affects permeability and wall flexibility. A mixture of 5 to 7% bentonite and a cement to bentonite ratio of 2:1 are usually the best to optimize the requirements for hydraulic conductivity, flexibility, and resistance to erosion. A lower cement/bentonite ratio is used if low hydraulic conductivity is the primary consideration (Leonards et al., 1985).

Trench Wall Continuity. SB and CB slurry trenching methods allow for visual inspection of the excavated material so that a qualitative assessment of all penetrated geologic materials down to

Fig. 4. Photograph showing installation of a synthetic liner over a compacted clay liner (*courtesy:* Ron Montgomery).

Fig. 5. Photograph showing sheepsfoot roller (*courtesy:* Ron Montgomery).

the required aquiclude can be confirmed. This is not the case with vibratory beam injection. The VBI method relies on indirect methods for determining both beam alignment and beam penetration into the desired aquiclude. Proper beam alignment is critical. If the beam is deflected by striking rocks or strata or if there are successive nonplumb beam penetrations, the continuity of the cutoff wall can be adversely affected. Maintaining beam verticality is essential. Each penetration of the beam must overlap the prior penetration, otherwise ungrouted voids or windows will result. These voids will act as conduits for migration of fluids through an otherwise impermeable wall (Jepsen and Place, 1985).

The type of slurry wall and method of emplacement used should be determined based upon the specific remedial goals of each site. Slurry walls are most effective when used in conjunction with other forms of remediation technology. This discussion and Table 1 summarize the comprehensive treatments of slurry walls found in Spooner et al. (1985) and Jepsen and Place (1985).

Soil or Soil/Clay Admix Liners

As defined by Johnson et al. (1985) soil or soil/clay admix liners are nonstructural liners placed on horizontal or moderately sloping surfaces as barriers primarily to the vertical flow of water or other fluids (Fig. 4). Liners serve to impede the flow of the pollutant carrier, such as water, into the subsoil and attenuate suspended or dissolved pollutants so that the existing leachate contains contaminants within acceptable ranges. Impedance of flow is attained by constructing a barrier with low permeability and hydraulic conductivity. The chemical composition and mass of the liner determine its sorptive or attenuative capacity (Eklund, 1985, Griffin et al., 1977).

Soil/clay admix liners can function as both bottom liners and cover liners in a variety of applications. RCRA and CERCLA regulations governing these applications determine the composition, properties, and installation of the clay liners. These liners consist of either compacted clay or soil/clay admixes where additives such as organoclays, cement, asphalt, or chemicals can be blended in with the clay/soil in order to enhance the liner's hydraulic properties. One of the most common uses of soil/clay admix liners is for municipal solid waste and hazardous waste landfill liners and covers. The EPA requires that clay liners used as barriers in waste disposal have a hydraulic conductivity of no more than 1×10^{-9} m/s.

Federal regulations also specify that for hazardous waste landfills, the liner materials must be compatible with the waste material being disposed. The liner must be resistant to physical forces of the disposal site, such as hydrostatic pressure and adverse climatic changes and be able to withstand the stresses of installation and the heavy equipment used in compacting the waste. Soil or soil/clay admix liner systems are required to be inspected for any imperfections, such as cracks or channels that may lead to increases in hydraulic conductivity.

Proper liner installation is required in order to meet federal regulations for hydraulic conductivity. To achieve adequate compaction of the earthen material, the material should be spread in loose lifts of 15.24 cm or less, wetted or dried to 2% or more above optimum moisture content and then compacted with a sheepsfoot-type roller (Fig. 5) or other appropriate compaction equipment to the specified relative compaction. These specified values are based upon the tested relationships between moisture content, relative compaction, and hydraulic conductivity (Wagner et al., 1986, Cope et al., 1984).

A final cover is required at the closure of a landfill facility. Cover liners serve the purpose of reducing the percolation of water through a waste storage site, which helps inhibit the generation of leachate. Bentonite makes a good material for surface capping because of its resistance to deformations caused by differential settling. Field studies have shown infiltration of rain water through properly installed bentonite caps to be negligible (Hoeks et al., 1987). The percentage of bentonite in a bottom liner mixture should be greater than that used for the cover, because of the aggressive nature of the leachate on the underlying liner (O'Leary and Walsh, 1991).

Construction of a landfill liner and the design of the landfill itself are dependent on parameters such as the geographic, geologic, and hydrologic characteristics of the available site and the regulations governing the type of waste to be disposed of at the site. The three most commonly used (sanitary) landfill designs are: 1) area method, 2) ramp method, and 3) trench method (Coates, 1981, O'Leary and Walsh, 1991).

In the area method, waste materials are placed into an excavated site. Prior to emplacement of the waste the excavated site is lined with a landfill liner material that is installed in multiple layers with compaction equipment. Solid waste is compacted into the landfill in units called cells. Each cell is a separate unit of waste surrounded by liner material. A series of adjoining cells, all the same height, is called a lift. The completed landfill consists of a series of vertical lifts, extending as high as 15 to 31 m above the original topography. Cover material is used for the top of the waste pile and the exposed sides. RCRA regulations require that landfill operators cover the solid waste with at least 15 cm of earthen material or a preapproved thickness of an alternative material at the end of each operating day, or at more frequent intervals if necessary (40 CFR 258.21).

The ramp method is similar to the area method, except that it is generally used in areas where the land is sloping. The waste is spread and compacted on a slope and the cover material is often obtained from in front of the working face.

The trench method is most often used in flat or gently sloping topography. The landfill is constructed as a series of successive parallel trenches that are excavated and filled. Trenches are at least two times the width of the excavation equipment and vary in depth, usually 2.5 to 3.5 m deep, depending on subsurface geology and hydrology. This particular method of landfilling is not conducive to the installation of landfill liners; therefore, the naturally occurring soils of the area should be cohesive and have acceptably low permeabilities.

Landfilling has become an expensive and increasingly difficult method of waste disposal as a result of growing regulatory control. Although it will continue to be needed, incineration and other treatments should be considered, and landfilling should be regarded as a last step or least attractive alternative to waste disposal. Landfilling costs vary around \$264/t for highly toxic wastes, \$132/t for ignitable materials, \$93/t for most industrial sludges, 44 to \$55/t for municipal treatment sludges, and 19 to \$22/t for municpal solid wastes (Wagner et al., 1986).

Geosynthetic Bentonite-Clay/Geotextile Membranes

Geosynthetic clay liners (or geocomposite liners) are a manufactured product developed in the early 1980s to help meet stringent federal and state regulatory requirements for various types of waste disposal, including landfilling. Geosynthetic clay liners (GCL) are composed of approximately 0.75 to 1.25 cm thick, high quality Wyoming bentonite confined between two geotextile layers (Grube, Jr., 1992, Wright, Jr., 1992). A geotextile is any of a variety of permeable synthetic textile products, generally made of woven polyester fabric, used in geotechnical engineering. Approximately 5 kg of bentonite per m^2 of liner is sandwiched between these synthetic textile sheets (Horz, 1987, Wright, Jr., 1992). GCLs are manufactured as a large sheet, generally 3.7 to 4.3 m wide and 30.5 m long and are installed with sheets overlapping and with bentonite clay used to form a continuous hydraulic seal at the sheet seams. Hydraulic conductivities of GCLs have been measured in the laboratory at less than 10^{-10} m/s and lower than 10^{-12} m/s for some products. As with most waste barriers, the hydraulic conductivity of the GCL is dependent upon the type of permeant involved. Table 2 is a comparative list of differences between compacted soil liners and geosynthetic clay liners. These types of liners provide a very low hydraulic conductivity barrier to leachate or surface infiltration seepage and are relatively inexpensive and space efficient. At present five different GCLs are available, three of which are manufactured in the United States and two in Europe (Grube, 1992).

Table 2. Comparative List of Differences Between Compacted Soil Liners and Geosynthetic Clay Liners

Compacted Soil	Geosynthetic Clay Liner
Thick (0.6—1.5 m)	Thin (< = 10 mm)
Field constructed	Manufactured
Difficult to build correctly	Easy to build (unroll and place)
Less susceptible to damage or puncture	Possible to damage or puncture
Constructed with heavy equipment	Light construction equipment
Usually requires test fill at each site and in situ percolation tests	Seam testing necessary, but repeated field testing usually not needed
Site-specific data on soils needed	Manufactured product; data available
Low to high leachate attenuation capacity, depending on soil properties	Lower leachate attenuation capacity
Relatively long containment time	Shorter containment time
Large thickness takes up space	Little space is required
Cost is highly variable	More predictable cost
Soil has low tensile strength	Higher tensile strength
Can dessicate and crack	Cannot crack until wetted
Difficult to repair	Not difficult to repair
Vulnerable to freeze/thaw damage	Freeze/thaw cycles do not change permeability
Performance is highly dependent upon quality of construction	Hydraulic properties are less sensitive to construction variables, although quality control in construction still critical
Slow construction	Much faster construction

Adapted from Grube, 1992.

Radioactive Waste Containment Barriers

As with municipal and hazardous wastes facilities, siting of radioactive waste disposal facilities is a lengthy and complicated process that involves a myriad of social, economic, political, and environmental considerations. One of the main areas of concern in disposal of these wastes is the potential for contamination of water resources from surface or near surface sites. Therefore identification of suitable locations where geological and hydrological conditions restrict surface and subsurface contaminant migration is important. Equally important is the choice of packing material that will surround the waste package.

Clays have been considered as nuclear waste packing materials in a number of studies, including a study by Allen and Wood (1988) that investigated the feasibility of a basalt/sodium bentonite mixture as packing material around individual waste containers at the Basalt Waste Isolation Project in Washington State (Albinsson et al., 1990, Baetsle et al., 1986, Beziat et al., 1992, Coulon et al., 1987, Garisto and Garisto, 1990, Manfroy et al., 1986, Marples et al., 1990, McCall and Krogness, 1986, Mertens et al., 1990, Muurinen et al., 1990, Muurinen et al., 1987, Van Iseghem et al., 1990). Clay in a nuclear waste repository serves both a mechanical and hydraulic function. The mechanical function is that of holding the waste containers in place and preventing collapse of the excavation. The plastic deformability of the clay is important in redistributing stresses from the surrounding medium to prevent excessive stresses from being transferred to the canisters. Smectite and palygorskite are the most plastic clays, illite is intermediate in plasticity, and kaolinite has the least plasticity (Bucher and Müller-Vonmoos, 1989, Lee and Tank, 1985).

Clays in nuclear waste packing material should also provide a barrier against water access to the canisters which can cause corrosion. The high swelling potential of bentonite makes it the best clay to ensure a watertight environment for the canisters. Under conditions of restrained volume change and high swelling potentials, however, the swelling stresses could exceed the allowable stress values for the canisters and the surrounding rock. Therefore it is necessary to determine the upper and lower bounds of the swelling potential for the bentonite to be used. One possible solution is to use carefully determined mixtures of clays with various swelling abilities to achieve the desired swelling capacity (Bucher and Müller-Vonmoos, 1989, Lee and Tank, 1985).

In addition to a watertight environment, bentonites provide a low permeability medium that limits radionuclide transport to diffusion and provides sorption capacity for radionuclides (Allen and Wood, 1988). Many radionuclides are metallic compounds that, when in an ionic state, are readily sorbed onto the clays by cation exchange. The sorption of radionuclides is strongly influenced by both the quantity and type of clay material (Berg et al., 1989). Anions diffuse at a much higher rate than cations, since they are most probably not sorbed to the montmorillonite surface (Bucher and Müller-Vonmoos, 1989). Sorption tests have shown that smectites favor sorption of bivalent cations due to their favorable charge compensation (Bartl and Czurda, 1991). Other sorption experiments show that sodium bentonite strongly sorbs strontium, cesium, and americium but is a less efficient sorber of uranium and neptunium. Sorption of uranium, neptunium, and technetium is increased under reducing conditions, thus indicating a correlation between Eh and sorption capacity. Desorption of uranium, neptunium, selenium, and technetium is less efficient than sorption, which is desirable in retardation of radionuclide mobility (Allen and Wood, 1988).

Allen and Wood (1988) also show that sodium montmorillonite is stable under high doses of gamma radiation. Extremely high doses of alpha radiation can cause structural damage to the clay, but this is at radiation levels exceeding that which would be expected in a nuclear repository.

In order to be effective as a containment barrier, the bentonite must retain its intercrystalline swelling potential and its cation exchange capacity over a span of several hundred thousand years. The bentonite in a nuclear waste repository is subjected to high pressures due to the overburden and high temperatures due to the radiation emitted from the waste. Studies have shown that under increasing overburden pressures and temperatures, montmorillonite is altered to illite and in this transformation its intercrystalline swelling capacity is lost. Examinations of ancient bentonite deposits with well-known pressure-temperature histories reveal that this transformation process appears to be slow enough not to affect the security of any given repository site (Bucher and Müller-Vonmoos, 1989, Kamei et al., 1990).

Studies of the clay minerals illite and chlorite have shown that the structures of these clays limit their sorption of cations primarily to their outer surface. Illites appear to sorb cations preferentially along outer surface prism faces and along exposed basal planes. Illites preferentially sorb monovalent cations such as cesium. Tests show that the bonding strengths of illite and chlorite are relatively weak allowing remobilization of cesium and strontium during desorption tests (Bartl and Czurda, 1991).

A synthesized clay, formed from mica, was developed several decades ago and has been successfully used by researchers from the Materials Research Laboratory at Pennsylvania State University as a strontium-ion sieve. The clay is a highly charged sodium fluorophlogopite mica, called by researchers Na-4-mica. The researchers have developed a method to produce this product in bulk quantities for subsequent use in the removal of strontium ions from solution. This material selectively exchanges sodium ions for strontium ions in solution at room temperature, even in the presence of large amounts of dissolved sodium. The strontium ions fill the mica's interlayer sites until the strontium content reaches a critical level (when about half of the exchange sites are occupied by strontium) at which point the layers collapse trapping the strontium, resulting in irreversible fixation. This discovery has tremendous potential in the nuclear waste field since this mica could be used to remove and isolate radioisotopic strontium from radioactive wastes or contaminated waters. It is superior to all other known strontium exchangers, including the zeolite clinoptilolite, which has been used to remove ^{90}Sr from nuclear waste solutions (Paulus et al., 1992).

Fig. 6. Photograph showing compaction-mold permeameter apparatus.

TESTING METHODS

Testing of potential clay liner materials and selection and installation of hydraulic barriers is a complicated process, and at present there is no consensus as to the methodologies involved. Some basic considerations in the design of laboratory or field testing programs are: 1) selection of the type of testing apparatus to be used, 2) determination of the effects of the permeant on the barrier or containment materials, and 3) appropriate quality control (Johnson et al., 1985).

Hydraulic conductivity of the barrier medium is one of the primary properties that must be evaluated in determining the suitability of a particular material for use as a hydraulic barrier. A good summation of terms associated with soil hydraulic conductivity and permeability is given by Johnson (1981). Permeability studies in the laboratory have used a variety of apparatus to measure hydraulic conductivity, but the most common method employs some version of a permeameter (Fig. 6) and permeameter cell. The soil or clay to be tested is emplaced in the permeameter cell at a designated unit weight or density. Then standard hydraulic conductivity tests are performed using water at 25°C or a standard aqueous permeant such as 0.01 N $CaSO_4$ solution. Laboratory studies have shown that clay hydraulic conductivities change with respect to the leachate tested, and thus it is also necessary to test the potential clay liner materials with the types of liquids they will be expected to impound.

Permeameter Cells

A permeameter is a chamber designed to measure the hydraulic conductivity of geologic materials. Permeameter cells follow two basic designs: 1) fixed-wall and 2) flexible-wall. An excellent study by Daniel, Anderson, and Boynton (1985) examined the relative advantages and disadvantages of a flexible-wall permeameter cell and four types of fixed-wall cells: consolidation-cell, fixed-cylinder, compaction-mold, and double-ring. The authors concluded that a particular type of cell is best suited for a specific application and

Table 3. Comparison Between Fixed-Wall and Flexible-Wall Permeameters

Fixed-wall Cell	Flexible-wall Cell
Low cost, easier to use	High cost
Convenient for testing compacted specimens	Convenient for testing undisturbed samples and a wide range of sample diameters
Higher potential for sidewall leakage, leading to potentially high permeability values, especially with concentrated organics. Use of a double-ring infiltrometer or consolidation cell can help minimize this problem	Flexibility of membrane conforms to the irregular shape of the sample and helps to minimize sidewall leakage
Cannot measure volume change or deformations within the specimen	Ability to measure vertical and volumetric change within the specimen
Applied effective stress is usually zero, control over stresses difficult	More control over vertical and horizontal stresses, effective stress should be less than those expected for field conditions
Harder to achieve complete sample saturation	Application of backpressure makes complete sample saturation more likely
No need to apply high confining pressures	Need to apply high confining pressures at high hydraulic gradients in order to maintain contact between the membrane and the soil
Compatibility with a wide range of permeant liquids	Flexible membrane is often not compatible with some permeant liquids
More applicable for studies of compacted clays subjected to low overburden pressure	More applicable for studies of clays subjected to high overburden pressures

Adapted from Daniel, Anderson, and Boynton, 1985.

that, overall, no individual type of cell was the best for all purposes. A comparison between the relative advantages and disadvantages of fixed-wall permeameters vs. flexible-wall permeameters are highlighted in Table 3. Specific parameters of the various types of permeameter cells are compared in Table 4. These various permeameter cells are identified below.

Consolidation cells primarily are used to test undisturbed samples. Consolidation cells consist of steel rings that can vary between 4 to 10 cm in diameter and up to 10 cm thick clamped into a consolidation loading frame that has a fluid reservoir surrounding the ring. This reservoir is filled with water (or some other liquid to be tested) and then the desired vertical pressures are applied to consolidate the sample (Daniel et al., 1985, Olson and Daniel, 1981, Boynton, 1983).

Fixed-cylinder permeameters are recommended for compacted samples, although it is possible to test undisturbed samples directly in thin-walled sampling tubes. The sampling tube acts as the permeameter cell. Fixed-cylinder permeameters are a simplified version of a compaction-mold cell and basically consist of a tube with two end plates clamped to either end (Daniel et al., 1985).

Compaction-mold permeameters are used for permeability tests

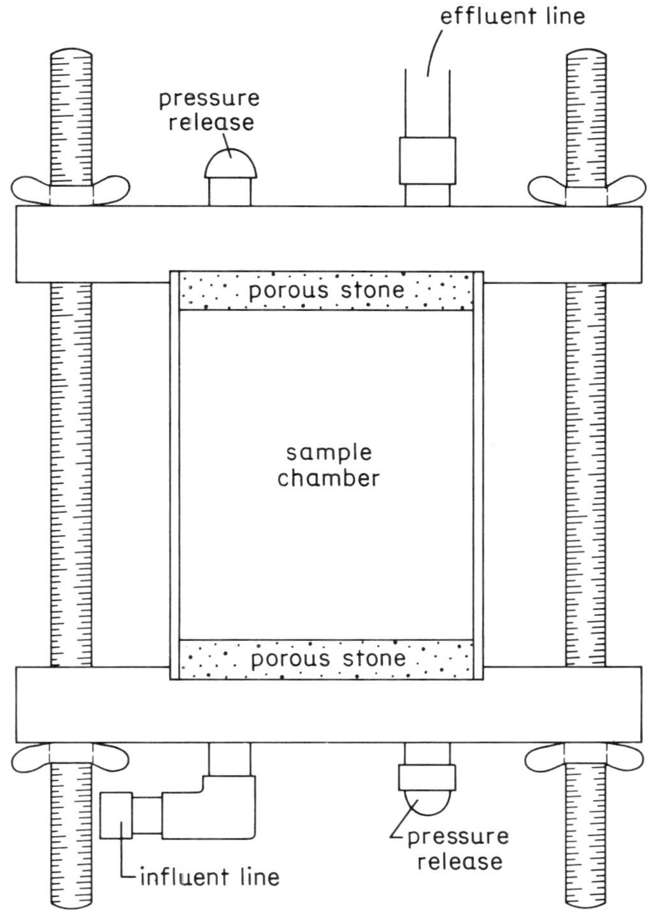

Fig. 7. Compaction-mold permeameter cell.

on compacted clay or soil specimens. This apparatus generally consists of a steel compaction cell (cylinder) 5 to 10 cm in diameter and up to 10 cm thick clamped between two steel endplates that contain inlet and outlet ports (Fig. 7). The sample is compacted into the cell, trimmed flush with the edges of the cell, and then the permeameter is assembled [ASTM Test Methods for Moisture-Density Relations of Soils and Soil-Aggregate Mixtures, Using 5.5-lb (2.49-kg) Rammer and 12-in. (305-mm) Drop (ASTM D698–78), Method A] (Daniel et al., 1985, Olson and Daniel, 1981, Gordon and Forrest, 1981, Boynton, 1983, Anderson, 1981, Brown and Anderson, 1983).

A double-ring permeameter is simply a modified compaction-mold permeameter that contains a ring built into the baseplate that separates the outflow that occurs through the center part of the sample from the outflow that occurs along the sidewall. This apparatus looks similar to a compaction-mold cell, except that it has two drainage outlets extending from the base plate, one for flow from the outer ring and one for flow from the inner ring. In this way the rate of flow from the sidewall region of the sample can be measured separately from the central part of the sample and thus the possibility of sidewall leakage can presumably be detected (Anderson et al., 1985, Daniel et al., 1985).

Flexible-wall permeameters utilize a triaxial cell or a modified triaxial cell. The flexible-wall cell is a membrane, normally made of latex, butyl, or neoprene rubber that conforms to and confines the sample. Samples can be wrapped in Teflon® before covering with

Table 4. Comparison of Various Types of Permeameter Cells

Consolidation Cell	Fixed-cylinder Permeameter	Compaction-mold Permeameter	Double-ring Permeameter	Flexible-wall Permeameter
Vertical pressures help reduce sidewall leakage by forcing sample outward against cell walls	Cannot apply vertical pressure to the sample, sidewall leakage possible	Cannot apply vertical pressure to the sample, sidewall leakage possible (especially when testing concentrated organic permeants)	Cannot apply vertical pressure to the sample, but double-ring system allows detection of sidewall leakage	Flexible membrane conforms to sample shape and helps minimize sidewall leakage
Flow through the sample is from bottom to top, which aids in complete sample saturation	Flow through the sample is from top to bottom	Flow through the sample can be from top to bottom or vice versa for better sample saturation	Flow through the sample is from top to bottom	Flow is from bottom to top of sample; applied backpressure aids in complete sample saturation
Fluid reservoir for permeant is part of and surrounds the cell	A separate reservoir is used to store the permeant	Permeant can be stored in a collar above the sample cell or in a separate reservoir	Permeant can be stored in a collar above the sample cell or in a separate reservoir	A separate reservoir is used to store the permeant
Vertical deformation of the sample can be measured with a displacement transducer	Sample deformation within the cell cannot be measured	Sample deformation within the cell cannot be measured	Sample deformation within the cell cannot be measured	Cell pressure and change in pressure across the cell can be controlled and measured; allowing control over vertical and horizontal stresses
Recommended for relatively compressible samples that will be subjected to high overburden pressures	Recommended for compacted samples	Recommended for compacted specimens that will be subjected to low overburden stresses in the field	Recommended for testing loosely packed slurry-wall backfill materials and samples subjected to low overburden stresses in the field	Recommended for samples that will be subjected to high overburden pressures in the field
Relatively inexpensive	Relatively inexpensive	Relatively inexpensive and easy to use	Relatively inexpensive	More expensive than rigid-wall permeameters

Reference: Daniel, Anderson, and Boynton, 1985.

the membrane to protect the membrane from chemical attack by the permeant. The sample and membrane are then placed between a top cap and a bottom pedestal that can be modified to accommodate samples of different diameters (Boynton, 1983, Daniel et al., 1985, Olson and Daniel, 1981).

Other Test Methods

Other laboratory studies measuring hydraulic conductivity have been done using more simplified equipment such as an API filter press and laboratory filter columns. An experiment by Alther et al. (1985) used an API filter press to form filter cakes from clay slurries. They then calculated the steady-state permeability by Darcy's Law, utilizing flow rate, area and thickness of the filter cake, and hydraulic gradient. Some studies have employed laboratory columns fitted with manometers and effluent collection systems to measure the slow, saturated, anaerobic flow of leachate through clay materials; the effluent and column contents subsequently were both analyzed to determine the concentration and vertical distribution of the chemical constituents (Griffin et al., 1976, Griffin et al., 1977).

Laboratory testing of candidate liner materials is more economical and more convenient compared to field testing. On the other hand, laboratory testing may not be representative of field conditions. Samples that have been disturbed, and then recompacted in the laboratory do not reflect the true microscopic and macroscopic features of the soil in the field. Laboratory methods should be used to conduct preliminary tests on potential clay liner materials and should simulate the expected field compactive effort of the liner material to the highest degree possible. Piezometers, infiltrometers, and field-version permeameters are all testing apparatus used to measure saturated hydraulic conductivity in the field. Several sources that comprehensively cover the topic of laboratory and field testing methods for clay barriers are: Landreth, 1983, Roberts, 1984, and Spooner et al., 1985.

In addition to tests that measure clay or soil hydraulic conductivity, it is important to evaluate more general chemical and physical properties of potential liner materials. Table 5 outlines some basic tests for chemical and physical properties of clays and the information these tests define.

SUMMARY

Clay liners/barriers are a good, relatively inexpensive method to contain municipal, hazardous, and low level radioactive wastes. The physiochemical properties of clays make them useful as bar-

Table 5. Test Methods For Clay-Liner Raw Materials

Test	Test Method	Purpose
	Physical Properties	
Particle size distribution	sieve and hydrometer [ASTM D 421–58 (1978)][a]	estimates percent sand, silt, and clay
Compaction test	compact liner at different moisture contents and measure compacted density [(ASTM D 558–82)[b] and (ASTM D 698–78)[c]]	determines the moisture content which results in maximum compacted density at a given compactive effort
Atterberg limits	measure of plastic and liquid limits [ASTM D 4318–84][d]	defines effects of varying water content on soil consistency
	Chemical Properties	
Organic matter, %	titration	organic matter in liners may attenuate organics from leachate
Soil pH	electrometric (Black, 1965)	pH affects clay structure and CEC, thus its ability to attenuate chemical contaminants
Cation exchange capacity	sodium saturation followed by ammonium extraction; analyze for sodium	estimates the extent to which the liner may exchange (remove) cations from the leachate
Major exchangeable bases	ammonium extraction; analyze for sodium, potassium, magnesium, and calcium	estimates the liner's capacity to exchange ions on base sites and identifies major ions absorbed on clay before interaction
X-ray powder diffraction	X-ray diffractometer	clay mineral identification

Adapted from Eklund, 1985.
[a]Method for Dry Preparation of Soil Samples for Particle-Size Analysis and Determination of Soil Constants.
[b]Method for Moisture-Density Relations of Soil-Cement Mixtures.
[c]Method for Moisture-Density Relations of Soils and Soil-Aggregate Mixtures.
[d]Method for Liquid Limit, Plastic Limit and Plasticity Index of Soils.

riers in waste containment. The clay minerals kaolinite, smectite, and palygorskite have very different chemical and sorptive properties so that one or an admixture of several may be best suited for a particular waste barrier structure and a particular waste disposal site. Extensive tests should be conducted to ascertain the type of clay materials that would be best suited for a particular application. The hydraulic conductivity of candidate clay liner/barrier material is an important property to ascertain in order to comply with federal regulations. Permeameters are the most commonly used apparatus to measure hydraulic conductivity. It is also important to determine the compatibility of the clay liner/barrier materials with the waste they are to contain. To meet the complex needs in waste disposal, a multi-layered barrier made up of different clay materials and different geosynthetics could be designed, where each layer would be chosen to remove selectively a specific group of constituents in the leachate.

ACKNOWLEDGMENTS

The authors wish to thank the Montgomery family of Frankfort, IN, especially Ron Montgomery, for so generously allowing access to their landfill facility and for contributing photographs to this publication. The authors also thank Rich Flores of W.W. Engineering, Bloomington, IN, Nelson Shaffer of the Indiana Geological Survey, and Ellen Crosby of the University of South Carolina, Columbia, SC, for their reviews of the manuscript.

REFERENCES

Anon., 1987, ''Answers to Your Questions on High-Level Nuclear Waste,'' US Department of Energy, Office of Civilian Radioactive Waste Management, Washington, DC, Nov., 59 pp.

Albinsson, Y., et al., 1990, ''Leaching/Migration of UO_2-Fuel in Compacted Bentonite,'' *Proceedings*, Materials Research Society Symposium—Scientific Basis for Nuclear Waste Management XIII, V.M. Oversby, and P.W. Brown, eds., Vol. 176, Pittsburgh, PA, pp. 559–565.

Albrecht, K.A., and Cartwright, K., 1989, ''Infiltration and Hydraulic Conductivity of a Compacted Earthen Liner,'' *Groundwater*, Vol. 27, No. 1, pp. 14–19.

Allen, C.C., and Wood, M.I., 1988, ''Bentonite in Nuclear Waste Disposal: A Review of Research in Support of the Basalt Waste Isolation Project,'' *Applied Clay Science*, Vol. 3, pp. 11–30.

Alther, G., et al., 1985, ''Influence of Inorganic Permeants Upon the Permeability of Bentonite,'' *Hydraulic Barriers in Soil and Rock*, ASTM STP 874, A.I. Johnson, R.K. Frobel, N.J. Cavalli, and C.B. Pettersson, eds., American Society for Testing and Materials, Philadelphia, PA, pp. 64–73.

Anderson, D.C., 1981, ''Organic Leachate Effects on the Permeability of Clay Soils,'' unpublished master's thesis, Texas A&M University, College Station, TX, 192 pp.

Anderson, D.C., Crawley, W., and Zabcik, J.D., 1985, ''Effects of Various Liquids on Clay Soil: Bentonite Slurry Mixtures,'' *Hydraulic Barriers in Soil and Rock*, ASTM STP 874, A.I. Johnson, R.K. Frobel, N.J. Cavalli, and C.B. Pettersson, eds., American Society for Testing and Materials, Philadelphia, PA, pp. 93–103.

Aughenbaugh, N.B., 1990, ''Environmental Suitability of Compacted Clay Liners,'' *Mining Engineering*, Vol. 42, No. 11, pp. 1263–1266.

Bartl, U., and Czurda, K.A., 1991, ''Migration and Retention Phenomena of Radionuclides in Clay-Barrier Systems,'' *Applied Clay Science*, Vol. 6, pp. 195–214

Baetsle, L.M., et al., 1986, ''Study of the Boom Clay Layer as a Geochemical Barrier for Long-Lived Radionuclides,'' *High-Level Nuclear Waste Disposal*, H.C. Burkholder, ed., Battelle Press, Columbus, OH, pp. 313–323.

Berg, R.C., Wehrmann, H.A., and Shafer, J.M., 1989, ''Geological and Hydrological Factors for Siting Hazardous or Low-Level Radioactive Waste Disposal Facilities,'' Illinois State Geological Survey Circular 546, 61 pp.

Beziat, A., Dardaine, M., and Mouche, E., 1992, ''Measurements of the Thermal Conductivity of Clay-Sand and Clay-Graphite Mixtures Used as Engineered Barriers for High-Level Radioactive Waste Disposal,'' *Applied Clay Science*, Vol. 6, pp. 245–263.

Bittell, J.E., and Miller, R.J., 1974, ''Lead, Cadmium, and Calcium Selectivity Coefficients on a Montmorillonite, Illite, and Kaolinite,'' *Journal of Environmental Quality*, Vol. 3, No. 3, July-Sep., pp. 250–253.

Black, C.A., ed., 1965, *Methods of Soil Analysis: Part 2-Chemical and Microbiological Properties*, American Society of Agronomy, Madison, WI, 801 pp.

Boynton, S.S., 1983, ''An Investigation of Selected Factors Affecting the Hydraulic Conductivity of Compacted Clay,'' unpublished master's thesis, University of Texas, Austin, TX, 79 pp.

Brown, K.W., and Anderson, D.C., 1983, ''Effects of Organic Solvents on the Permeability of Clay Soils,'' EPA-600/83–016, USEPA Municipal Environmental Research Laboratory, Cincinnati, OH, 6 pp.

Brown, K.W., Green, J.W., and Thomas, J.C., 1983, ''The Influence of Selected Organic Liquids on the Permeability of Clay Liners,'' *Incineration and Treatment of Hazardous Wastes*, D.W. Shultz, ed., *Proceedings*, 9th Annual Research Symposium, EPA-600/9–83–018, Ft. Mitchell, KY, pp. 114–125.

Brunelle T.M., Dell, L.R., and Meyer, C.J., 1987, ''Effect of Permeameter and Leachate on a Clay Liner,'' *Proceedings* of Geotechnical Practice for Waste Disposal '87/GT Div., American Society of Chemical Engineers, Ann Arbor, MI, 15–17 June, pp. 347–361.

Bucher, F., and Müller-Vonmoos, M., 1989, ''Bentonite as a Containment Barrier for the Disposal of Highly Radioactive Wastes,'' *Applied Clay Science*, Vol. 4, No. 2, pp. 157–177.

Cadena, F., 1989, ''Use of Tailored Bentonite for Selective Removal of Organic Pollutants,'' *Journal of Environmental Engineering*, Vol. 115, No. 4, Aug., pp. 756–767.

Coates, D.R., 1981, *Environmental Geology*, John Wiley and Sons, New York, 701 pp.

Cope, R., et al., 1984, ''Use of Liners For Containment of Hazardous Waste Landfills,'' *Pollution Engineering*, Vol. 16, No. 3, pp. 22–32.

Coulon, H., et al., 1987, ''Choice of French Clays as Engineered Barrier Components for Waste Disposal,'' *Proceedings*, Materials Research Society Symposium—Scientific Basis for Nuclear Waste Management X, J.K. Bates and W.B. Seefeldt, eds., Vol. 84, Pittsburgh, PA, pp. 813–824.

Daniel, D.E., Anderson, D.C., and Boynton, S.S., 1985, ''Fixed-Wall Versus Flexible-Wall Permeameters,'' *Hydraulic Barriers in Soil and Rock*, ASTM STP 874, A.I. Johnson, R.K. Frobel, N.J. Cavalli, and C.B. Pettersson, eds., American Society for Testing Materials, Philadelphia, PA, pp. 107–126.

Davison, N., McWhinnie, W.R., and Hooper, A., 1991, ''X-Ray Photoelectron Spectroscopic Study of Cobalt(II) and Nickel(II) Sorbed on Hectorite and Montmorillonite,'' *Clays and Clay Minerals*, Vol. 39, No. 1, pp. 22–27.

De Batist, R., Van Iseghem, P., and Casteels, F., 1986, ''Waste Form and Waste Package Characterization Studies Related to Disposal in a Geological Clay Formation,'' *High-Level Nuclear Waste Disposal*, H.C. Burkholder, ed., Battelle Press, Columbus, OH, pp. 585–596.

Eklund, A.G., 1985, ''A Laboratory Comparison of the Effects of Water and Waste Leachate on the Performance of Soil Liners,'' *Hydraulic Barriers in Soil and Rock*, ASTM STP 874 , A.I. Johnson, R.K. Frobel, N.J. Cavalli, and C.B. Pettersson, eds., American Society for Testing and Materials, Philadelphia, PA, pp. 188–202.

Egger, A.J., 1987, ''Hazardous Solid Waste Disposal in the Geological Environment,'' *Hazardous Waste Management*, Selected Papers from an International Expert Workshop Convened by United Nations Industrial Development Organization (UNIDO) in Vienna, 22–26 June 1987, S.P. Maltezou, A.K. Biswas, and H. Sutter, eds., Tycooly Publishing, London, 344 pp.

Evans, J.C., Pancoski, S.E., and Alther, G., 1989, ''Organic Waste Treatment With Organically Modified Clays,'' *Proceedings*, Third International Conference on New Frontiers for Hazardous Waste Management, Sep. 10–13, 1989, Pittsburgh, PA, USEPA, EPA\600\9–89\072, Cincinnati, OH, pp. 48–58.

Farrah, H., and Pickering, W.F., 1976, ''The Sorption of Zinc Species by Clay Minerals,'' *Australian Journal of Chemistry*, Vol. 29, pp. 1649–1656.

Förstner, U., and Wittmann, G.T.W., 1979, *Metal Pollution in the Aquatic Environment*, Springer-Verlag, New York, 486 pp.

Frost, R.R., and Griffin, R.A., 1977a, ''Effect of pH on Adsorption of Arsenic and Selenium From Landfill Leachate by Clay Minerals,'' *Soil Science Society of America Journal*, Vol. 41, No. 1, pp. 53–57.

Frost, R.R., and Griffin, R.A., 1977b, ''Effect of pH on Adsorption of Copper, Zinc, and Cadmium From Landfill Leachate by Clay Minerals,'' *Journal of Environmental Science and Health*, A12(4&5), pp. 139–156.

Garisto, N., and Garisto, F., 1990, ''The Effect of Cracks on Diffusive Mass Transport Through a Clay Barrier,'' *Proceedings*, Materials Research Society Symposium—Scientific Basis for Nuclear Waste Management XIII, V.M. Oversby, and P.W. Brown, eds., Vol. 176, Pittsburgh, PA, pp. 733–739.

Gibb, J.P., and Cartwright, K., 1987, ''Retention of Zinc, Cadmium, Copper, and Lead by Geologic Materials,'' USEPA Project Summary, EPA/600/S2–86/108, Hazardous Waste Engineering Laboratory, Cincinnati, OH, 3 pp.

Gill, S.A., and Christopher, B.R., 1985, ''Laboratory Testing of Cement-Bentonite Mix for Proposed Plastic Diaphragm Wall for Complexe LaGrange Reservoir Caniapiscau, James Bay, Canada,'' *Hydraulic Barriers in Soil and Rock*, ASTM STP 874, A.I. Johnson, R.K. Frobel, N.J. Cavalli, and C.B. Pettersson, eds., American Society for Testing and Materials, Philadelphia, PA, pp. 75–92.

Gordon, B.B., and Forrest, M., 1981, ''Permeability of Soils Using Contaminated Permeant,'' *Permeability and Groundwater Contaminant Transport*, ASTM STP 746, T.F. Zimmie and C.O. Riggs, eds., American Society for Testing and Materials, Philadelphia, PA, pp. 101–120.

Green, W.J., et al., 1983, ''Interaction of Clay Soils With Water and Organic Solvents: Implications for the Disposal of Hazardous Wastes,'' *Environmental Science and Technology*, Vol. 17, No. 5, pp. 278–282.

Griffin, R.A., and Au, A.K., 1977, ''Lead Adsorption by Montmorillonite Using a Competitive Langmuir Equation,'' *Soil Science Society of America Journal*, Vol. 41, No. 5, Sep.-Oct., pp. 880–882.

Griffin, R.A., Au, A.K., and Chian, E.S.K., 1979, ''Mobility of Polychlorinated Biphenyls and Dicamba in Soil Materials: Determination by Soil Thin-layer Chromatography,'' Illinois State Geological Survey Reprint No. 1979 G, pp. 183–189.

Griffin, R.A., Au, A.K., and Frost, R.R., 1977, ''Effect of pH on Adsorption of Chromium From Landfill-Leachate by Clay Minerals,'' *Journal of Environmental Science and Health*, A12(8), pp. 431–439.

Griffin, R.A., et al., 1976, ''Attenuation of Pollutants in Municipal Landfill Leachate by Clay Minerals: Part 1—Column Leaching and Field Verification,'' Illinois State Geological Survey, Environmental Geology Notes, No. 78, 34 pp.

Griffin, R.A., et al., 1977, ''Attenuation of Pollutants in Municipal Landfill Leachate by Clay Minerals: Part 2—Heavy-metal Adsorption,'' Illinois State Geological Survey—Environmental Geology Notes, No. 79, 47 pp.

Griffin, R.A., and Shimp, N.F., 1976, ''Effect of pH on Exchange-Adsorption of Precipitation of Lead From Landfill Leachates by Clay Minerals,'' *Environmental Science and Technology*, Vol. 10, No. 13, pp. 1256–1261.

Grim, R.E., 1962, *Applied Clay Mineralogy*, McGraw-Hill, New York, 422 pp.

Grube, W.E., Jr., 1992, ''Geosynthetic Liners Offer Cover Option,'' *Environmental Protection*, May 1992, pp. 29–33,50.

Haden, L.W., Jr., 1972, ''Attapulgite,'' SME Preprint No. 72H327, SME Fall Meeting, Birmingham, AL, Oct., 8 pp.

Helios-Rybicka, E., 1985, ''Sorption of Ni, Zn, and Cd on Sepiolite,'' *Clay Minerals*, Vol. 20, pp. 525–527.

Hoeks, J., et al., 1987, ''Bentonite Liners For Isolation of Waste Disposal Sites,'' *Waste Management and Research*, Vol. 5, pp. 93–105.

Horz, R.C., 1987, ''Geotextiles for Drainage, Gas Venting, and Erosion Control at Hazardous Waste Sites,'' USEPA Project Summary, EPA/600/S2–86/085, January, Hazardous Waste Engineering Laboratory, Cincinnati, OH, 4 pp.

Jepsen, C.P., and Place, M., 1985, ''Evaluation of Two Methods For Constructing Vertical Cutoff Walls at Waste Containment Sites,'' *Hydraulic Barriers in Soil and Rock*, ASTM STP 874, A.I. Johnson, R.K. Frobel, N.J. Cavalli, and C.B. Pettersson, eds., American Society of Testing and Materials, Philadelphia, PA, pp. 45–63.

Johnson, A.I., 1981, ''Glossary,'' *Permeability and Groundwater Contaminant Transport*, ASTM STP 746, T.F. Zimmie and C.O. Riggs, eds., American Society For Testing and Materials, Philadelphia, PA, pp. 3–17.

Johnson, R.L., Cherry, J.A., and Pankow, J.F., 1989, ''Diffusive Contaminant Transport in Natural Clay: A Field Example and Implications for

Clay-Lined Waste Disposal Sites," *Environmental Science and Technology*, Vol. 23, pp. 340–349.

Johnson, A.I., et al., eds., 1985, *Hydraulic Barriers in Soil and Rock*, ASTM STP 874, American Society For Testing and Materials, Philadelphia, PA, pp. 1–6.

Jungling, T., and Greeves, J., 1989, "Nuclear Regulatory Commission Regulations and Experience With Solidification/Stabilization Technology," *Environmental Aspects of Stabilization and Solidification of Hazardous and Radioactive Wastes*, ASTM STP 1033, P.L. Côté and T.M. Gilliam, eds., American Society for Testing and Materials, Philadelphia, PA, pp. 77–82.

Kamei, G., et al., 1990, "Estimation of Illitization Rate of Smectite From the Thermal History of Murakami Deposit, Japan," *Proceedings*, Materials Research Society Symposium—Scientific Basis for Nuclear Waste Management XIII, V.M. Oversby, and P.W. Brown, eds., Vol. 176, Pittsburgh, PA, pp. 657–663.

LaBauve, J.M., Kotuby-Amacher, J., and Gambrell, R.P., 1988, "The Effect of Soil Properties and a Synthetic Municipal Landfill Leachate on the Retention of Cd, Ni, Pb, and Zn in Soil and Sediment Materials," *Journal WPCF*, Vol. 60, No. 3, pp. 379–385.

Lagaly, G., 1989, "Principles of Flow of Kaolin and Bentonite Dispersions," *Applied Clay Science*, Vol. 4, No. 2, pp. 105–123.

Landreth, R., 1983, *Lining of Waste Impoundment and Disposal Facilities*, USEPA Technical Resource Document SW-870, USEPA, Washington, DC, 448 pp.

Lee, S.Y., and Tank, R.W., 1985, "Role of Clays in the Disposal of Nuclear Waste: A Review," *Applied Clay Science*, Vol. 1, pp. 145–162.

Leonards, G.A., et al., 1985, "Thin Slurry Cutoff Walls Installed by the Vibrated Beam Method," *Hydraulic Barriers in Soil and Rock*, ASTM STP 874, A.I. Johnson, R.K. Frobel, N.J. Cavalli, and C.B. Pettersson, eds., American Society for Testing and Materials, Philadelphia, PA, pp. 34–44.

Lutton, R.J., 1980, *Evaluating Cover Systems For Solid and Hazardous Waste*, USEPA Technical Resource Document SW-867, USEPA, Washington, DC, 57 pp.

Lutton, J.M., et al., 1986, "General Corrosion Studies of Candidate Container Materials for The Basalt Waste Isolation Project," *High-Level Nuclear Waste Disposal*, H.C. Burkholder, ed., Battelle Press, Columbus, OH, pp. 563–572.

Madsen, F.T., and Müller-Vonmoos, M., 1989, "The Swelling Behavior of Clays," *Applied Clay Science*, Vol. 4, No. 2, pp. 143–156.

Malik, M., Mustafa, M.A., and Letey, J., 1992, "Effect of Mixed Na/Ca Solutions on Swelling Dispersion and Transient Water Flow in Unsaturated Montmorillonitic Soils," *Geoderma*, Vol. 52, pp. 17–28.

Manfroy, P., Heremans, R., and Bonne, A., 1986, "Ten Years of Site-Specific Research and its Relevance to the Concept of a Waste Disposal System in Clay," *High-Level Nuclear Waste Disposal*, H.C. Burkholder, ed., Battelle Press, Columbus, OH, pp. 233–245.

Marples, J.A.C., et al., 1990, "A Comparison of the Behaviour of Vitrified HLW in Repositories in Salt, Clay and Granite: Part II: Results," *Proceedings*, Materials Research Society Symposium—Scientific Basis for Nuclear Waste Management XIII, V.M. Oversby, and P.W. Brown, eds., Vol. 176, Pittsburgh, PA, pp. 275–282.

McCall, T.B., and Krogness, J.C., 1986, "The Reference Waste Package Design for a Nuclear Waste Respository in Basalt," *High-Level Nuclear Waste Disposal*, H.C. Burkholder, ed., Battelle Press, Columbus, OH, pp. 525–531.

Mertens, L.A., et al., 1990, "A Comparison of the Behaviour of Vitrified HLW in Repositories in Salt, Clay and Granite. I: Experimental," *Proceedings*, Materials Research Society Symposium—Scientific Basis for Nuclear Waste Management XIII, V.M. Oversby, and P.W. Brown, eds., Vol. 176, Pittsburgh, PA, pp. 267–274.

Mitchell, J.K., Hooper, D.R., and Campanella, R.G., 1965, "Permeability of Compacted Clay," *Journal of Soil Mechanics and Foundries*, Vol. 91, pp. 41–65.

Morton, S.D., and Sawyer, E.W., 1976, "Clay Minerals Remove Organics, Viruses and Heavy Metals from Water," *Water & Sewage Works*, 1976 Reference Issue, Apr. 30, pp. 116–118.

Muurinen, A., Olin, M., and Uusheimo, K., 1990, "Diffusion of Sodium and Copper in Compacted Sodium Bentonite at Room Temperature," *Proceedings*, Materials Research Society Symposium—Scientific Basis for Nuclear Waste Management XIII, V.M. Oversby, and P.W. Brown, eds., Vol. 176, Pittsburgh, PA, pp. 641–647.

Muurinen, A., Penttila-Hiltunen, P., and Rantanen, J., 1987, "Diffusion Mechanisms of Strontium and Cesium in Compacted Sodium Bentonite," *Proceedings*, Materials Research Society Symposium—Scientific Basis for Nuclear Waste Management X, J.K. Bates and W.B. Seefeldt, eds., Vol. 84, Pittsburgh, PA, pp. 803–812.

Newman, A.C.D., ed., 1987, *Chemistry of Clays and Clay Minerals: Mineralogical Society Monograph No. 6*, John Wiley and Sons, New York, 480 pp.

O'Leary, P., and Walsh, P., 1991, "Landfilling Principles," *Waste Age*, Apr., pp. 109–114.

Olson, R.E., and Daniel, D.E., 1981, "Measurement of the Hydraulic Conductivity of Fine-Grained Soils," *Permeability and Groundwater Contaminant Transport*, ASTM STP 746, T.F. Zimmie and C.O. Riggs, eds., American Society For Testing and Materials, Philadelphia, PA, pp. 18–64.

Parker, F.L., 1988, *Low-Level Radioactive Waste Regulation—Science, Politics and Fear*, M.E. Burns, ed., Lewis Publishers, Inc., Chelsea, MI, pp. 85–107.

Paulus, W.J., Komarneni, S., and Roy, R., 1992, "Bulk Synthesis and Selective Exchange of Strontium Ions in $Na_4Mg_6Al_4Si_4O_{20}F_4$ Mica," *Nature*, Vol. 357, 18 June, pp. 571–573.

Peckinpaugh, T.L., 1988, "The Politics of Low-Level Radioactive Waste Disposal," *Low-Level Radioactive Waste Regulation—Science, Politics and Fear*, M.E. Burns, ed., Lewis Publishers, Inc., Chelsea, MI, pp. 45–61.

Repa, E.W., and Sheets S.K., 1992a, "Subtitle D's Effects on Private Industry," *Waste Age*, Apr., pp. 267–272.

Repa, E.W., and Sheets, S.K., 1992b, "Landfill Capacity in North America," *Waste Age*, May, pp. 18–28.

Roberts, D.W., 1984, *Soil Properties, Classification, and Hydraulic Conductivity Testing*, USEPA Technical Resource Document SW-925, USEPA, Washington, DC, 167 pp.

Robertson, R.H.S., 1985, *Fuller's Earth: A History of Calcium Montmorillonite*, Mineralogical Society Occasional Publication, Volturna Press, Hythe, Kent, UK, 421 pp.

Rusche, B.C., 1986, "Status of the United States' High-Level Nuclear Waste Disposal Program," *High-Level Nuclear Waste Disposal*, H.C. Burkholder, ed., Battelle Press, Columbus, OH, pp. 101–117.

Ryan, C.R., 1985, "Slurry Cutoff Walls: Applications in the Control of Hazardous Wastes," *Hydraulic Barriers in Soil and Rock*, ASTM STP 874, A.I. Johnson, R.K. Frobel, N.J. Cavalli, and C.B. Pettersson, eds., American Society for Testing and Materials, Philidelphia, PA, pp. 9–23.

Sai, J.O., and Anderson, D.C., 1991, "Long-Term Effect of an Aqueous Landfill Leachate on the Permeability of a Compacted Clay Liner," *Hazardous Waste and Hazardous Materials*, Vol. 8, No. 4, pp. 303–312.

Singer, A., 1977, "Dissolution of Two Australian Palygorskites in Dilute Acid," *Clays and Clay Minerals*, Vol. 25, pp. 126–130.

Spooner, P., et al., 1985, *Slurry Trench Construction For Pollution Migration Control*, Noyes Publications, Park Ridge, NJ, 237 pp.

Stockmeyer, M.R., 1991a, "Adsorption of Organic Compounds on Organophilic Bentonites," *Applied Clay Science*, Vol. 6, pp. 39–57.

Stockmeyer, M.R., 1991b, "Organobentonites and Contaminant Retention in Waste Deposit Liners," *Proceedings*, Euroclay Conference 1991, Dresden, pp. 1007–1012.

Stockmeyer, M., and Kruse, K., 1991, "Adsorption of Zn and Ni Ions and Phenol and Diethylketones by Bentonites of Different Organophilicities," *Clay Minerals*, Vol. 26, pp. 431–434.

Supak, J.R., Swoboda, A.R., and Dixon, J.B., 1978, "Adsorption of Aldicarb by Clays and Soil Organo-Clay Complexes," *Soil Science Society of America Journal*, Vol. 42, No. 2, pp. 244–247.

Van Iseghem, P., Berghman, K., and Timmermans, W., 1990, "The Interaction Between Nuclear Waste Glasses and Clay—II," *Proceedings*, Materials Research Society Symposium—Scientific Basis for Nuclear Waste Management XIII, V.M. Oversby, and P.W. Brown, eds., Vol. 176, Pittsburgh, PA, pp. 291–298.

van Olphen, H., 1977, *An Introduction to Clay Colloid Chemistry*, 2nd ed., John Wiley and Sons, New York, 317 pp.

Wagner, K., et al., 1986, *Remedial Action Technology For Waste Disposal Sites*, 2nd ed., Noyes Data Corporation, Park Ridge, NJ, 642 pp.

Weiss, A., 1989, "About Sealing of Waste Disposals by Clays With Special Consideration of Organic Compounds in Percolation Water," *Applied Clay Science*, Vol 4, No. 2, pp. 193–209.

Wright, F.B., Jr., 1992, "Bentonite Markets—Evolution and Outlook," *Industrial Minerals*, No. 298, July, pp. 43–52.

Whittle, G.P., Carlton, T.A., and Henry, H.R., 1984, "Permeability in Clay Liners of Hazardous Waste Storage Pits," *Municipal and Industrial Waste*, 7th Annual Madison Waste Conference, Madison, WI, pp. 364–372.

ENVIRONMENTAL USES

FLUE GAS DESULFURIZATION

ROBERT C. FREAS

Over the past twenty years people throughout the world, but particularly in the United States, have been taking an increasing interest in the environment, sources and causes of pollution, and pollution abatement. Air quality in the United States has been and continues to be an area of primary emphasis. In 1970, the US Congress passed the Clean Air Act and subsequently revised and strengthened it in 1977 with the issuance of the New Source Performance Standards. The Act has been further strengthened with the passage of the 1990 Amendments.

By far the largest single source of sulfur dioxide (SO_2) pollutants are coal fired electric generating power stations. Hence, a substantial portion of the flue gas desulfurization (FGD) research and equipment installation has been geared to the utility industry. Several competing FGD technologies were touted during the period 1970 to 1990 with varying degrees of success. Subsequently, the more successful technologies have been accepted, revised, and improved to the point that by 1991 third and fourth generation technology is available.

A second area of emphasis for FGD has been coal fired industrial boilers. While individually they are much smaller than any electric utility generating station, they do represent a significant source of sulfur pollution. However, due to the size of industrial boilers and their location within industrial and educational complexes, the wet scrubber technology being employed by the utilities is not appropriate for these applications. Hence, fluidized bed combustion (FBC) technology has been developed for the industrial boilers. Since the installation of the first full scale test unit at Georgetown University, several equipment vendors have come forward with an increasing array of more sophisticated FBC units. Currently, FBC technology has matured to the point that the Tennessee Valley Authority (TVA), in cooperation with the Electric Power Research Institute (EPRI), has installed an FBC system on a 250 MW test unit at the Shawnee Power Station in Kentucky.

1990 CLEAN AIR ACT AMENDMENTS

As passed, the 1990 Clean Air Act amendments are much tougher than originally anticipated and extend extensive regulatory powers to the US Environmental Protection Agency (EPA). The amendments are made up of eleven parts, or titles, several of which will impact the mining industry. However, Title IV, Acid Rain Control, is the section with the greatest impact on FGD requirements for utilities and industrial coal fired sources of sulfur emissions (Anon., 1990b).

In general, Title IV will be implemented in two phases. Phase I of the program runs from 1995 to 2000 and applies to approximately 110 electric generating stations in 22 states, (Anon. 1990a, 1991a). The basic sulfur emission allowance is calculated at a rate of 2.5 lbs (1.1 kg) of sulfur dioxide per million British thermal units (BTU) of heat input multiplied by the amount of fuel burned in that particular unit.

Phase II which begins in the year 2000 tightens the restrictions of Phase I and also imposes the Air Quality Amendment on industrial boilers. The Phase II calculation is based upon 1.2 lbs (0.5 kg) of SO_2 emissions per million BTUs and also includes restrictions on NO_x emissions. The NO_x reductions, however, have not been quantified nor their requirements clearly defined. While this level of SO_2 reduction will be both technically difficult and costly, there are permissible tradeoffs that may accrue to mitigate at least a portion of the negative economic impact (Wolfe and Lamb, 1991).

FLUE GAS DESULFURIZATION TECHNOLOGIES

There are several options available to both the utility industry and the smaller industrial boiler operations. These options or technologies include such things as wet lime or limestone scrubbing, dry scrubbing, and FBC technology, as well as several less well known and essentially untried systems. The selection of a particular technology is obviously an economic decision but will include consideration of the sulfur and BTU content of the coal which is to be burned, the delivered cost of available SO_2 sorbent, the disposal of waste products generated through sulfur removal, and local socioeconomic factors. In addition, other alternatives such as the implementation of coal cleaning or gassification technologies, solvent refined coals (SRC) and the firing with coal/oil mixtures or other coal blends will also impact on the selection of a sulfur removal system. The Electric Power Research Institute has sponsored research on all of the primary FGD methodologies as well as investigations of the availability and effectiveness of a variety of sorbents (Anon., 1991b).

Wet Scrubbing: Of the existing electric utility generating units with commercial FGD systems, over 90% employ wet calcium based lime/limestone processes. While the wet lime and limestone systems are uniquely different, the basic process chemistry is similar.

The process chemistry of a wet lime/limestone system is fairly complex and includes both the liquid phase chemistry of sulfur dioxide and the liquid-solid chemistry of lime/limestone and calcium sulfite/sulfate salts. The carbonate reagent material, either lime or limestone, provides alkalinity for SO_2 removal in the wet scrubber. This in turn generates calcium sulfite and calcium sulfate salts as products of the removal of the SO_2 from the flue gas stream. The waste products thus generated may be in the form of either a slurry or a solid material, but in either case require environmentally safe disposal. Since this chapter is not intended as a dissertation on scrubber systems, the reader is referred to three primary references which describe the preponderance of the basic work which has been done on wet scrubbing: Lowell, 1970, Struram and Morgan, 1971, Selmeczi, 1974).

Within a wet scrubber system the solubility of a calcium-based reagent material will affect alkalinity within the scrubber and therefore its removal efficiency at various stoichiometric ratios. Because of this, in very general terms, wet lime systems require less hardware and less calcium product to achieve the same sulfur removal efficiency as that of a wet limestone system. However, recent advances in limestone scrubbing utilize weak organic acids to improve SO_2 removal and limestone utilization, hence reduced volumes of limestone may be required to achieve desired removal efficiencies. Such buffering additives as adipic, acetic, citric, and sulfosuccinic acids serve to enhance limestone dissolution while maintaining a constant pH in the scrubber liquor.

Dry Scrubbing: Dry scrubbing involves contacting of sulfur containing flue gas with an alkaline reagent such that a dry waste product is generated for disposal. These systems include those that use spray dryers or injection systems with electrostatic precipitators for collection of waste products. These units have been applied to both industrial boilers and utility plants including an 860 MW unit in Craig, CO, which came on line in mid-1988.

In very simplified form the spray dryer incorporates a milk of lime slurry which is atomized and sprayed into the collection chamber at a low enough moisture concentration to allow for a fabric filter or electrostatic precipitator to control SO_2 and particulate matter simultaneously. Thus sulfur removal from the flue-gas results in a dry waste product. The very nature of the spray drying equipment dictates that the basic technology with only lime addition be restricted to low sulfur coals (less than 1.0% sulfur) and requires lime as a reagent material.

While the dry scrubber systems do not have the same liquid phase chemistry as wet scrubbing, they still are not without their problems. These include such things as inlet SO_2 concentrations, temperature drop across the spray dryer, slaking water composition, and lime reactivity. Since the first application of spray dryers, considerable work has been funded by EPA and EPRI relative to the application of this technology on higher sulfur coals (i.e., 3.0 to 4.0% sulfur). This work has included testing with calcium chloride addition for lime utilization enhancement, while other additives are presently under investigation (Anon., 1991b).

The Illinois State Geological Survey (ISGS) has taken a somewhat different approach to dry sorbent injection. This work has focused upon lime hydration and the production of a high surface area (HSA) hydrate. The ISGS research focused upon the importance of specific hydrate properties (surface area, porosity, particle size, and crystallite size) in achieving maximum SO_2 capture. Test results reported by Moran et al. (1991) indicated substantially improved SO_2 removal efficiencies with the ISGS processed lime.

Research efforts have also been carried on jointly by Joy Industrial Equipment Company and NYRO Atomizer, Inc. which developed a spray dryer process for simultaneous SO_x/NO_x removal. This process involves lime-based spray drying with sodium hydroxide (NaOH) addition on both industrial and utility scale operations (Palazzolo and Baviello, 1984).

While the initial dry scrubbing systems incorporated lime as the sorbent, dry alkali injection technology has also been developed which employs trona as a reagent material. The first commercial application of a trona based dry injection system was announced by Public Service Company of Colorado for a 500 MW unit and started up in late 1990.

Other work on dry scrubbing has focused on the utilization of nahcolite ($NaHCO_3$) as a reagent material. However, in the case of both nahcolite and trona, an important consideration in the dry injection technology is the high solubility and leachability of sodium-based waste. Thus, their acceptance as reagent materials may be limited by the environmental and cost consideration associated with the disposal of the waste product.

Fluidized Bed Combustion: This technology involves the direct mixing of the limestone with coal in the boiler during combustion. This mixed bed of coal and limestone is fluidized through the injection of air through the bed. The upward force of air equalizes the downward force of gravity such that the entire bed takes on the appearance of slowly boiling water (Remick, 1985, Roeck, 1982, Bubenick et al., 1981). Within this combustion environment the limestone is calcined. The porous lime thus produced absorbs SO_2 released from the coal in the fluidized bed thus producing calcium sulfate.

Within this technology there are two primary methodologies, atmospheric and pressurized fluidized bed combustion. This refers to the pressure within the boiler at the time of combustion, and as a consequence the thickness of the fluidized bed. Typically the bed thickness in an atmospheric system would be designed for a 1.5 to 2.0 m thick bed. This will vary with the specific boiler design, which may vary significantly (Kindya et al., 1982, Anon., 1989a).

Pressurized fluidized bed technology is less well advanced than the atmospheric, but is receiving increasing attention, particularly for larger boiler applications. In a pressurized system the bed thickness is much greater than in an atmospheric system and may exceed 5.0 m. This allows for the maintenance of the same intimate contact between the coal and the limestone, but also may require closer gradation control on coal and more precise limestone sizing (Anon., 1989b).

Other Systems: While the aforementioned technologies are the predominate systems currently finding commercial application, they are not the only systems being investigated and tried. For more information on other technologies the reader is referred to references listed in the bibliography (Anon., 1991c, Yoon and Statnick, 1989).

RAW MATERIAL REQUIREMENTS

As has already been noted, lime and limestone followed by sodium carbonate are the principal materials being used for flue gas desulfurization. Prior to publication of the 1990 Amendments, Bhagwat (1985), Freas (1980), and others offer projections as to the volume of reagent materials that could potentially be required under the New Source Performance Standards. In very general terms, these estimates have been found to be overly optimistic and thus far have not been realized. A reassessment of these estimates would now appear to be in order in light of the new regulations.

Table 1 shows the annual estimated lime or limestone consumption per megawatt that would be anticipated with a variety of different coals. These estimates are based upon an 80% load factor, 10 000 BTU per KW hour, stoichiometry of 1.05 for lime and 1.2 for limestone, and a minimum of 95% CaO or $CaCO_3$. Taking the information in Table 1 a step further, assume a 600 MW generating station using Eastern Kentucky coal. Table 1 shows that the average Eastern Kentucky steam coal would require 47 t of lime or 96 t of limestone per megawatt per year to achieve the required sulfur removal level. Therefore, a 600 MW unit would require 28 200 t of lime or 57 600 t of limestone per year.

The United States has approximately 70 000 MW of electric generating capacity fitted with FGD scrubbing equipment. This equipment utilizes approximately 1.5 Mt per year of lime and another 2 Mtpy of limestone. These are very generalized numbers, but they do give an order of magnitude as to the existing requirements imposed by FGD regulations.

Table 1. Quality Characteristics of Coal vs. Estimated Annual Lime/Limstone Consumption by Megawatt

Sources of Coal	BTU/lb	% Sulfur	Lime (t) Per Year/Per MW	Limestone (t) Per Year/Per MW
East Kentucky	11,796	1.26	47	96
West Kentucky & West Virginia	10,996	3.59	181	368
Illinois	10,775	2.92	149	302
Texas	6,601	0.66	43	89
Colorado	10,925	0.49	18	39
New Mexico	9,996	0.54	23	47
Wyoming	9,037	0.50	23	48
Pennsylvania	12,200	2.22	94	190

The size of the potential future market is, however, much more difficult to define. For example, 8 500 MW of generating capacity in the Ohio Valley utilizing a 3.4% sulfur coal would require approximately 1.2 Mt of lime or 2.4 Mt of limestone. These same 8 500 MW of generating capacity located in Colorado and burning low sulfur western coal would potentially require only 128 000 t of lime or 270 kt of limestone. Hence, the location of the power plants and the type of fuel used could potentially have a tremendous impact on the volume of lime, limestone, soda ash, etc., which might ultimately be required in meeting the new FGD requirements.

According to industry sources the period 1991 to 2000 will see an additional 47 coal fired generating stations coming on line. This represents approximately 36 000 MW of new generating capacity. Based purely on the probable geographic location of these new units, and assuming they purchase local sourced coal, this new capacity would require approximately 3 Mt of stone, plus an additional 1.3 Mt of lime. These figures assume the current distribution of approximately 70% stone and 30% lime utilized in existing and committed FGD utilities systems. While not specifically broken out, it is further assumed that soda ash will continue to command a limited segment of this market. It should be noted that these figures do not consider any additional requirements posed by new industrial boilers.

Taken to the extreme and assuming the worst case example as to the percent sulfur content of the coal utilized, one could conceivably generate numbers in excess of 20 Mt per year of additional carbonate requirement to satisfy the 1990 Amendments to the Clean Air Act. However, this is not realistic and does not take into consideration the numerous factors which will bear upon the ultimate selection of both a fuel to be consumed by a generating station and the type scrubber equipment selected for each of the individual units. This number also does not take into account retrofit requirements which may be imposed as a result of the 1990 Amendments.

One of the most basic decisions that the utility industry has to make is the trade-off between the transportation of coal from the western states to an eastern location vs. the cost of additional scrubber requirements imposed in utilizing an eastern sourced coal. This equation is further complicated by the fact that it is quite conceivable that low sulfur eastern coals will command a premium. Hence, in addition to the question of transportation, the utilities will be forced to determine the optimum coal sulfur level they can afford to purchase vs. the cost associated with scrubber installation and operation.

It should be noted that in preparing an estimate of future FGD reagent requirements, the EPA has yet to complete the writing and publication of all of the rules associated with the implementation of the 1990 Amendments. Issues yet to be answered include such things as the trading of credits between plants within a utility, or even between utilities. For example, if a newer unit of Utility A exceeds the standards in lowering its emissions, can it transfer the credits for these low emissions to another older unit less able to comply with the regulation? Could Utility A instead sell its credits to Utility B who would then use these purchased credits to bring its plants into compliance or as a minimum, reduce the degree of sulfur removal that might otherwise be required?

Other issues include the acceptability of operating a non-conforming plant at reduced capacity to bring it within the total permissible discharge limits, extension of the grandfather limits to allow new replacement capacities to be brought on line, conversion to alternate fuels, etc. If alternate fuels are selected will they be oil, gas, coal-oil mixtures, or some other modified blend still on the drawing board. A review of the legislative issues associated with the 1990 Amendments and their impact on the coal industry is presented in Platt and Hewson, Jr. (1990) as they address some of the complexities associated with implementation of the new law.

While the mining industry, and particularly the coal industry, is very much concerned with the fuel decision which will ultimately be made, the electric utility equipment manufacturers are just as interested. Estimates of the acid rain impact on coal utilization and lime/limestone consumption seldom takes into consideration the fact that equipment manufacturers are competing to secure a technical advantage. These competitive efforts have resulted in the development of a variety of technologies all aimed at achieving the highest level of sulfur removal efficiency at the lowest possible operating cost. The ultimate success of the equipment manufacturers could well change existing conventional wisdom as it relates to the trends impacting the mining industry and particularly flue gas desulfurization's impact on the mining industry.

FGD LIME/LIMESTONE SPECIFICATIONS

Lime specifications for wet scrubbing are particularly concerned with chemical quality, reactivity, and percent insoluble material, much the same as most other lime applications. The basic requirements under these specifications include a minimum of 95% total oxides, 4% maximum SO_2, 3.5% maximum loss on ignition, and maximum insoluble residue of 5%. Size specifications will vary, but in general will require 5 × 0 cm material. Because of the high efficiencies which have been achieved with Dravo's Thiosorbic® lime in wet scrubbing, several lime specifications have included provisions for a minimum of 5.0 to 5.5% MgO as a preferred alternative to a straight high calcium lime.

Limestone specifications generally include both a chemical and physical specification The chemical requirements are usually 90 to 95% $CaCO_3$, 0 to 5% $MgCO_3$, and not more than 5% inerts. The physical specification has generally been for a 0.5 × 5.0 cm material, this frequently requires a washed product and may specify a Bond work index. Several utilities have installed ball mills and grind their own product to the required −325 mesh size. This allows the utility greater flexibility in selecting a limestone vendor and maximizes their effort in achieving competitive bidding. A recent change in this trend was noted when Northern Indiana Public Service Company

(NIPSCO) recently requested bids from various limestone companies to install a grinding plant and provide NIPSCO's Bailey Station with 100% of their −325 mesh requirements. However, NIPSCO had already pre-selected the source of the stone, thus eliminating the annual competitive bidding process.

Physical specifications for FBC are frequently much more rigorous and contain a more narrow window of acceptable particle size distribution. The reactivity of the limestone may also be a factor in sorbent selection, and the user may test internal surface/pore volume and density. As a result, competitive bids frequently are more difficult to achieve and vendors may be subject to performance qualification on pilot scale units. Due to the specialized physical classification which generally does not match standard highway specifications, the vendor may have an opportunity for a premium on the sized limestone. FBC units may also require dolomite or at least a minimum level of $MgCO_3$.

BYPRODUCT MATERIALS

Any consideration of FGD would be incomplete without a discussion of the byproduct materials generated with any scrubbing system. These byproduct materials may be classified into two categories: disposable (throw away) products and regenerative products.

The waste products of most FGD systems are disposable or throw away materials. Interestingly, the US Congress and the EPA made very little effort to consider the trade-off between pollution of the air vs. the detrimental effects of land-side disposal of these captured pollutants. Nevertheless, the disposal of the waste products is a very serious consideration. Up until now the vast majority of byproduct materials generated by the utilities have been disposed of in landfills operated by the utilities solely for that purpose, or as a slurry in impoundments permitted for that purpose. These surface disposal systems may include the FGD waste products, with or without separately captured fly ash material. However, in order for the scrubber sludges to be disposed of in a dry landfill, they must be processed or treated such that they will be rendered stable over an extended period of time. This may involve the addition of additive materials which will generate a pozzolinic type reaction. This is necessitated by the fact that wet scrubber systems generate byproducts with low solids concentration which, when stabilized or partially stabilized, may be dilatant. In any case, this byproduct is purely a waste material and has no commercial significance once discharged from the utility.

Regenerative processes on the other hand are designed such that the byproduct material may be marketed and thus reduce the overall operating cost of the FGD system. To date these regenerative systems include both the recovery of sulfur and gypsum. The recovery of sulfur from FGD systems has not received a great deal of attention in the United States and is currently not considered an important byproduct regenerative process for the utility industry. However, byproduct gypsum has received a great deal of attention in Japan and other Far Eastern countries and is now receiving more serious consideration in the United States. Ellison and Makansi (1990) have authored an excellent article on byproduct gypsum utilization, and it is recommended to anyone interested in learning more about this aspect of the FGD industry.

In order for a utility to generate an acceptable gypsum product, the scrubber system must be designed such that it allows for forced oxidation to yield a principally sulfate byproduct material. The utility must also be committed to producing high purity gypsum product as well as electricity, and must understand the operational differences between the production of a marketable and non-marketable product. In earlier references to NIPSCO's Bailey Station, it was noted that they have changed the parameters in the selection of a reagent limestone for their scrubbing system. That system is designed to produce a byproduct gypsum suitable for major commodity gypsum markets, with the quality jointly being guaranteed by the scrubber vendor and the utility.

SUMMARY

The 1990 Amendments to the New Source Performance Standards will have a significant impact on the permissible emissions from both new and existing coal fired boilers. This will apply to both the utility industry and other industrial applications. Obviously, this in turn will have a major impact upon the mining industry.

First and foremost will be the impact to the coal industry. The breadth of this impact is yet to be determined, but most assuredly some degree of source switching is anticipated. This switching is expected to have a negative impact on Ohio, Indiana, and Illinois coals, while at the same time having a positive impact on the low sulfur Appalachian coal and western coal. This switch in sources and demand is expected to also carry with it pricing changes which, depending upon the sulfur of the coal, may result in either premiums or discounts.

Flue gas desulfurization represents a very real potential market for lime, limestone, dolomite, and other industrial minerals. However, the size of that market remains somewhat elusive. Nonetheless, the combined demand for reagent materials including both existing scrubbers and those which are likely to come on line during the 1990s is expected to reach at least 8 to 10 Mt per year. While this is a sizable market, it is at the same time somewhat geographically limited—particularly in the case of limestone. As has always been the case, transportation of industrial minerals is a major factor and will be no less a consideration in supplying the utility industry with lime and limestone or other products for FGD applications.

REFERENCES

Anon., 1989a, "Atmospheric Fluidized-Bed Combustion Program: Fiscal Year 1989 Summary Program Plan," NTIS, PC-A03/MF A01-OSTI, DOE/FE-0115–2, US Department of Energy, Washington, DC, 32 pp.

Anon., 1989b, "Pressurized Fluidized-Bed Combustion Program: Fiscal Year 1989, Summary Program Plan," NTIS, PC-A03/MF A01-OSTI, DOE/FE-0115–8, US Department of Energy, Washington, DC, 30 pp.

Anon., 1990a, "Title IV—Acid Rain," *Congressional Quarterly*, Nov. 24, Washington, DC, pp. 3947–3954.

Anon., 1990b, Clean Air Act Amendments of 1990, Public Law 101–549, Nov. 15, 1990, 104 Stat. 2399, U.S. Congress, Washington, DC.

Anon., 1991a, "Guideline for Methods to Measure Emissions Allowances Available at EPA," *Environmental Reporter*, Bureau of National Affairs, Inc., Washington, DC, Mar. 22, pp. 2084–2085.

Anon., 1991b, List of Current Publications, Electric Power Research Institute, Palo Alto, CA, 22 pp.

Anon., 1991c, *Current Reports, Advanced Fossil Energy Technologies*, Vol. 90, No. 1, Office of Fossil Energy, US Department of Energy, Washington, DC, selected pages.

Bhagwat, S., 1985, *The Lime and Limestone Market for Sulfur Removal: Potential for 1992*, Illinois State Geological Survey, Champaign, IL, pp. 1–13.

Bubenick, D.V., Hall, R.R., and Dirgo, J.A., 1981, "Control of Particulate Emissions from Atmospheric Fluidized Bed Combustion with Fabric Filters and Electrostatic Precipitators," Project Summary, EPA-600/F7–81–105, Research Triangle Park, NC, Dec., 6 pp.

Ellison, W., and Makansi, J., 1990, "World By-Product Gypsum Utilization," *Industrial Minerals*, London, UK, Mar., pp. 33–45.

Freas, R.C., 1980, "The Impact of Flue Gas Desulphurisation on Lime and Limestone Marketing," *Proceedings*, 4th Industrial Minerals International Congress, Atlanta, GA, 7 pp.

Kindya, R.J., et al., 1982, "Environmental Assessment: Source Test and Evaluation Report-B & W Alliance Atmospheric Fluidized Bed Combustor," Project Summary, EPA-600/F7–81–076, Research Triangle Park, NC, May, 7 pp.

Lowell, P.S., 1970, "A Theoretical Description of the Limestone Injection/Wet Scrubbing Process," Vol. 1, NTIS No. PB-183 029, June, Radion Corporation, Austin, TX.

Moran, D.L., et al., 1991, "Production and Coolside Pilot Plant Testing of High-Surface-Area Hydrated Lime," *Processing and Utilization of High-Sulfur Coals IV*, Elsevier Science Publishers B.V., Amsterdam, pp. 701–721.

Palazzolo, M.A., and Baviello, M.A., 1984, "Status of Dry SO_2 Control Systems," EPA-600/57–84–086, US Environmental Protection Agency, Research Triangle Park, NC, Sep., 6 pp.

Platt, J., and Hewson, T.A., Jr., 1990, "Coal Markets Under Acid Rain Legislation," Society for Mining, Metallurgy and Exploration, Inc., Preprint No. 90–143, Littleton, CO, Mar., 5 pp.

Remick, B.W., 1985, "A FGD Technology Primer For The Limestone Producer," *Stone Review*, Washington, DC, Dec., pp. 11–17.

Roeck, D.R., 1982, " Technology Overview: Circulating Fluidized Bed Combustion," EPA-600/F7–82–051, Research Triangle Park, NC, Sep., 6 pp.

Selmeczi, J.G., 1974, "Kinetics vs. Aqueous Chemistry of Calcium-Based Sulfur Dioxide Scrubbing Systems," 35th Annual Meeting of the International Water Conference, Pittsburgh, PA, Oct., 11 pp.

Struram, W., and Morgan, J.J., 1971, "An Introduction Emphasizing Chemical Equilibrium in Natural Waters," *Aquatic Chemistry*.

Wolfe, D.E., and Lamb, L.M., 1991, "Clean Air Act Amendments Offer Substantial Savings," *National Environmental Journal*, Sep./Oct., Campbell Publishing, Inc., Columbus, GA, pp. 32–36.

Yoon, H., and Statnick, R.M., 1989, "Flue-Gas Sorbent Injection Technologies for Retro Fit SO_2 Control," Preprint No. 89–102, Society of Mining Engineers, Littleton, CO, Feb., 15 pp.

ENVIRONMENTAL USES

NEUTRALIZATION

ARTHUR C. PINCOMB AND ALAN R. SHAPIRO

A public awareness of and commitment to abate pollution of water, air, and soil continues to stimulate demand for industrial minerals and rocks capable of neutralizing strong and weak acids. The role of acid neutralizing agents is 1) to treat acidified water and soil until the desired pH is attained and 2) to remove acid-forming compounds from industrial processes to mitigate the formation of acidic discharges. Environmental applications for these agents include: acid rain mitigation, neutralization and prevention of acidic industrial process discharges, and treatment of acid mine drainage and coal refuse. Desulfurization of industrial flue gases involves removal of sulfate compounds capable of combining with hydrogen to form acids. Details of flue gas desulfurization are covered in the previous chapter on Flue Gas Desulfurization.

Ideally, the neutralization of acids results in an upward adjustment of effluent pH 7, or neutrality. The final pH value for many acid treated wastes, especially industrial discharge streams, is commonly over neutralized to pH 8.5 to 9.5. The Clean Water Act sets the upper discharge limit at pH 9.

ACID NEUTRALIZATION REAGENTS

Acid neutralization is accomplished by adding alkali metal and alkaline earth metal compounds to acidic water or soil until the desired pH is attained. The commonly used alkali ions are calcium (Ca), magnesium (Mg), sodium (Na), and potassium (K). Sodium hydroxide, high calcium and dolomitic hydrate, high calcium and dolomitic lime (quicklime), high calcium and dolomitic limestone, soda ash, and magnesium oxide are the compounds most commonly applied to raise the pH of acidic waters or soils.

Limestone and dolomitic limestone, which are calcium and magnesium carbonates, and their equivalent oxides and hydrates formed by calcination, are the industrial materials most commonly used in the manufacture of reagents for acid neutralization. Approximately 327 kt of high calcium and dolomitic lime were consumed for acid neutralization purposes in 1988. Magnesite and brucite are industrial minerals that compete with dolomitic limestone for the magnesium-oxide market. Production of caustic calcined magnesia for neutralization purposes exceeded 6 kt in 1990. Soda ash is an industrial mineral gaining increased acceptance in the neutralization of acid waste discharges. Approximately 182 kt of soda ash were consumed in 1990 for acid waste treatment or acid prevention by flue gas desulfurization. Other mineral products either consumed in minor quantities or receiving attention as future neutralizing agents include zeolite and hectorite.

In some cases, waste products have been found to be suitable and cost effective for neutralization of acids. Common wastes gaining acceptance include basic slag, phosphate mining refuse, cement kiln dust, fly ash, and mine tailings.

POLLUTION ABATEMENT

Pollution abatement applications for treating acidified water and soil include liming of lakes and soil affected by acid rain, treatment of acidic discharges from industrial processes, and treatment of acid mine drainage and coal refuse.

Acid Rain Mitigation: The acidification of surface waters, soil, and ground water as a result of acid precipitation is recognized as a significant environmental problem in North America and Europe. According to Regens (1984), the term *acid rain* refers to the transfer of acidic materials, either wet (precipitation) or dry (deposition), from the atmosphere to the earth's surface. Concern about acid rain results largely from studies that indicate that aquatic and terrestrial ecosystems may be affected adversely by acidification. However, the extent and magnitude of the problem, its likelihood of occurrence in a given area, and the implications of the effects are not well established.

Industrial processes contribute significantly to the formation of acid rain. Primary acidic sulfates and nitrous oxides are emitted through fossil fuel combustion and a variety of industrial processes including petroleum refining, nonferrous smelting, pulp milling, and the manufacture of sulfuric acid (H_2SO_4). Sulfuric acid is produced when sulfur dioxide (SO_2) reacts with gaseous (OH) radicals, air, and oxygen (O_2), and with (H_2O_2) or ozone (O_3) in clouds or rainwater phases. Nitric acid (HNO_3) is formed primarily during daylight hours from the reaction of (NO_2) with hydroxyl radicals.

The effects of acid rain on freshwater lakes and watersheds, especially in the northeastern United States, were recognized in the late 1970s. Studies in Sweden in the 1950s indicate that the acidification of surface waters proceeds in three stages. In the first stage, the natural alkalinity of lakes in the form of bicarbonate ion (HCO_3-) neutralizes the continuous input of acid rain and the pH declines very slowly. Once the available bicarbonate is consumed, the addition of more acidic precipitation results in a rapid decline in pH, reaching values lower than pH 5. The final stage is reached when the pH stabilizes at a value of about 4.5, even though precipitation of acidic rain continues.

Congress enacted the Acid Precipitation Act of 1980 to mitigate the effects attributable to acid surface waters. The Environmental Protection Agency (EPA), in cooperation with the Interagency Task Force on Acid Precipitation, established a ten-year comprehensive program, the National Acid Precipitation Assessment Program (NAPAP), as a result of this legislation. The new source performance standards (NSPS) of the US Clean Air Act restrict SO_2 emissions from future industry and electric power plants built in the United States, especially when coal is the designated fuel. Similar requirements for nitrogen oxides (NO_x) emissions exist. Cost effectiveness is a critical factor that government and industry will use to decide which emission reduction method(s) to employ to limit the growth or further reduce SO_2 and NO_x emissions.

Several reagents, applied mainly as a slurry from boats, barges, or helicopters, have proven to be cost effective, biologically non-threatening, lake liming agents. Agricultural high calcium or dolomitic limestone and basic steel slag are the most common, biologically non-threatening agents utilized. Based on the pH of lake-bottom clays and organic content of the mud, application rates vary from ¼ ton to over 2 tons per hectare of surface area. Hydrated calcitic or dolomitic lime can be applied at a rate of about 280 to 335 kg/hectare without ecological impacts to the lake. This rate of application provides approximately one to three years of buffering.

Studies by the Ontario Ministry of the Environment in 1985 indicate that the average cost of lake liming ranges from 50 to $500 per hectare at an average application rate of 450 kg/hectare. Application rates for soils depend on the effective calcium carbonate equivalent (CCE) of the aglime agent and should not exceed 1.6 t/hectare per application.

Recent studies on lakes in Sweden indicate that soda ash, as egg-shaped briquettes, can be dropped to the bottom of lakes from floating barges. Dissolution of the briquettes provides a constant supply of sodium ions to the lake sediments, resulting in an acid neutralization effect and restoring aquatic life. Although more costly per application than other alkaline reagents, reapplication is much less frequent.

Industrial Process Waters: Sulfuric acid wastes from plating and steel plants are neutralized with alkali materials. An additional benefit of neutralizing these acid waste streams is the precipitation of heavy metal contaminants. The most commonly used neutralizing reagents are slurried high calcium and dolomitic quicklime.

Acid Mine Drainage: Acid mine drainage results from the oxidation of pyrite (FeS_2) present in mine wastes and discharges, producing sulfuric acid (H_2SO_4) and other strong acids. Acid mine drainage, a problem of both active and abandoned mines, is primarily associated with the washing of coal and the mining of coal and other sulfide minerals. Contamination of lakes, rivers, groundwater, and soil can result in negative environmental impacts for many years after mining has been completed.

Mining of high sulfur coal in West Virginia, parts of western Pennsylvania, Ohio, western Kentucky, and Illinois has created a significant acid mine drainage problem. Nearly 6 400 km of streams, 15 200 hectares of abandoned mine lands, and an additional 21 000 hectares of non-burning waste material from previous coal mining operations in West Virginia alone are affected by acid mine drainage. Most of this acreage is unvegetated because of surface toxicity.

Pennsylvania has been actively regulating coal mine discharges since the late 1960s. Treatment is required if the discharge falls below a pH of 6 and total iron exceeds 7 mg/l. Treated discharges must not have a pH greater than 9.

Lime and limestone are the primary alkaline minerals used to abate acid mine drainage. Both high calcium and dolomitic limestones and limes are used. Other materials, including caustic soda, soda ash, magnesium oxide, and phosphate are used to treat coal mine wash plant liquid wastes and acid water. Phosphate, in particular phosphate mining refuse with 25% apatite, is gaining acceptance due to the addition of phosphate as a plant nutrient which improves the physical and chemical properties of the soils.

Treatment of Coal Refuse: Spoil piles and other coal refuse remaining after mining is complete contribute sulfuric acid from the oxidation of pyrite contained in the wastes. Agricultural limestone is the preferred neutralizing agent for these wastes, at application rates exceeding 110 kg per disturbed hectare.

COMPARISON OF ACID NEUTRALIZATION MATERIALS

According to Boynton (1980), four major considerations must be evaluated in order to determine the most efficient and economical chemical reagent to employ in acid neutralization:

1. Delivered cost per ton of neutralizing agent.
2. Neutralizing power, or basicity factor, a measure of the amount of neutralizing agent required to neutralize a given acid.
3. Reactivity or rate of neutralization.
4. Sludge generation, including cost and ease of disposal. The amount of sludge generated is directly related to the solubility of the reactants.

Two additional factors in selecting neutralization materials would be:

1. Process equipment capitalization and maintenance costs.
2. Environmental impact in utilizing the neutralization agent.

A cost comparison of the most commonly used neutralizing agents is shown in Table 1. Common alkali reagents are compared with respect to cost and neutralizing power as basicity relative to high calcium lime (CaO).

High Calcium Limestone and Dolomitic Limestone: High calcium limestone and dolomitic limestone are the most widely available and least expensive of all acid neutralizing agents. They are by far the least reactive of the common neutralizing agents and the volume and density of sludge generated is usually much greater than with other materials. This requires greater plant space to neutralize a given volume of acid.

Limestone and dolomitic limestone have nearly the lowest available basicity of common acid neutralizing agents. It takes 65% more dolomitic limestone and nearly 80% more high calcium limestone than high calcium lime to neutralize the same amount of acid. Dolomitic limestone has greater basicity, but unless finely ground, generally reacts far slower than high calcium limestone.

As the pH rises during acid neutralization, the reactivity of limestone decreases, leading to under neutralization, or a final pH 6 to 7. Since most industrial processes require neutralization to a pH 8.5 to 9.5, limestone is generally used only for neutralizing weaker acids, or in combination with stronger alkalis, where over neutralization to higher pH values is required.

In addition, completeness of reaction using limestone is sometimes hampered by impervious reaction films of calcium sulfate ($CaSO_4$) coating the limestone particles. Depending on the fineness of the particles, the unreacted portion of the limestone may exceed 50%. The resulting net material costs are higher and sludge generation is greater per unit of acid neutralized. The finer the grind of the limestone or dolomitic limestone particles, the greater the reactivity and completeness of the reaction. Frothing problems may result from release of CO_2 during the neutralization reaction.

High Calcium and Dolomitic Lime and Hydrate: High calcium quicklime and high calcium hydrate are highly reactive and available at moderate cost. They are capable of neutralizing or over neutralizing to any desired pH value. Reaction with acidic waters yields a heavier, higher volume of insoluble $CaSO_4$ sludge. Calcination of limestone or dolomitic limestone results in a significantly lower delivered price per ton of high calcium or dolomitic quicklime reactant due to the volatilization of CO_2 gases.

The neutralizing power of high calcium and dolomitic quicklimes is second only to magnesium oxide. Dolomitic hydrate and quicklime have slightly more available basicity per unit weight, but have a slower rate of reaction than their high calcium counterparts. Less sludge by weight is precipitated since the main reactant, magnesium sulfate, is much more soluble than calcium sulfate. High

Table 1. Cost Comparison of Commercial Alkaline Agents, Modified from Boynton (1980)

Agent	Chemical Formula	Bulk Price $/Ton*	Equivalent Basicity Factor**	Cost $/Ton of Basicity (Rounded)
High Calcium Lime	CaO	$55	1.00	$55
High Calcium Hydrate	$Ca(OH)_2$	$56	1.32	$74
Dolomitic Lime	$CaO.MgO$	$64	0.86	$55
Dolomitic Hydrate	$Ca(OH)_2.MgO$	$75	1.02	$77
High Calcium Limestone	$CaCO_3$	$9	1.79	$16
Dolomitic Limestone	$CaCO_3.MgCO_3$	$9	1.65	$15
Sodium Hydroxide	$NaOH$	$336	1.43	$480
Soda Ash	Na_2CO_3	$110	1.89	$208
Magnesium Oxide	MgO	$258	0.72	$186

*Figures shown are approximate 1990 FOB plant prices and do not include transportation costs.
**Compares all other reagents with high calcium lime, given an arbitrary factor of 1. Thus, 1.32 kg of high calcium hydrate is equal in theoretical neutralizing power to 1 kg of high calcium lime. Calculations assume all reagents are 100% pure.

calcium lime has a higher reactivity than soda ash. High calcium and dolomitic quicklime and hydrate require processing equipment to slake the lime into a slurry, or agitation equipment necessary to maintain the slurry in storage. Undissolved lime grits cause wear on valves, pumps, and other equipment. Scaling and insoluble sludge deposits also form within the systems. Cost of replacement parts and equipment maintenance is an important consideration for the lime treatment method.

Sodium Minerals: Sodium hydroxide, known as caustic soda ($NaOH$), is a strong alkali and the rate of neutralization is nearly spontaneous. It is capable of neutralizing or over neutralizing acids to any desired pH. It is the most expensive of the reagents listed, being at least six times more costly than quicklime per ton and nearly nine times more expensive than quicklime when price per ton is adjusted to account for basicity. The reaction product of neutralization with sodium hydroxide is sodium sulfate ($NaSO_4$), which is very soluble and creates minimum sludge.

The corrosive nature of caustic soda requires special handling procedures to protect workers. The high freezing temperature necessitates installation and maintenance of insulated storage tanks and heat-tracing equipment throughout the system.

Industrial minerals used in the manufacture of chemically produced sodium hydroxide include soda ash, nahcolite, and sodium sulfate. Sodium hydroxide is also produced from salt ($NaCl$) through an electrolytic process. Interest in chemically produced caustic soda, also known as lime-soda caustic, is increasing due to concerns regarding the environmental effects of chlorine and price of coproduct electrolytic caustic soda.

Soda ash, produced from trona, is slightly less expensive, but is much less reactive and has very low available basicity than sodium hydroxide. Soda ash is twice as expensive as quicklime per ton and nearly four times more expensive than quicklime when adjusted for available basicity.

Nearly 90% more soda ash than high calcium lime is required to neutralize the same amount of acid. It is less effective for over neutralizing than sodium hydroxide or high calcium lime. Acid neutralizing reaction products using soda ash are the same as those with sodium hydroxide, resulting in a very soluble, minimum volume sludge. Generation of CO_2 gas creates frothing during the neutralization reaction.

Soda ash is obtained from underground trona mines in the Green River Basin of southwest Wyoming, solution mining of nahcolite-dawsonite deposits in northwest Colorado, and solution mining of sodium carbonate-rich brines at Searles Lake, CA.

Magnesium Minerals: Magnesium oxide (MgO) is almost five times as expensive as quicklime per ton but, due to the highest available basicity of the common reagents, is slightly more than three times as expensive than quicklime when adjusted for available basicity. More than 25% less magnesium oxide than quicklime is required to neutralize the same amount of acid. The highly soluble sludge reactant during neutralization is Epsom salt, resulting in minimum solid sludge formation. The reaction rate of magnesium oxide with acids is slower than other alkalis.

Magnesium oxide, or magnesia, is produced from the calcination of magnesite and brucite and by chemically treating seawater or magnesia-rich brines to obtain magnesium hydroxide. Magnesium hydroxide is also gaining recognition for specific acidic waste treatment applications. The magnesium hydroxide is then thermally treated to produce the magnesia. Seawater and natural brines account for about 60% of US magnesium oxide production. Magnesia is recovered from seawater in California, Delaware, Florida, and Texas and from magnesia-rich brines in Michigan. Magnesite and brucite are recovered from a mine near Gabbs, NV.

Zeolites: Natural and synthetic zeolites sometime function as neutralizers when utilized in industrial processes involving the adsorption and treatment of acidic gases and liquids. However, the neutralization which transpires is primarily a physical process and not chemical, as is the case for all the aforementioned neutralizing agents discussed in this chapter.

Their unique structural framework and chemistry provide certain zeolites with remarkable molecular sieve properties. The uniform, microporous nature of the zeolite will permit the selective adsorption of one or more molecules to the exclusion of others so that the zeolite behaves as a molecular sieve. The molecular sieving capability varies in each individual zeolite mineral according to their specific crystalline structure, void framework, and cationic composition. The zeolites are effective in acid pollution abatement due to their molecular sieving properties, which allow them to selectively remove pollutants such as sulfur dioxide (SO_2), hydrogen sulfide (H_2S), and nitrogen oxides (NO_x) from industrial effluents.

The zeolite must be activated into a dehydrated form before being utilized as a molecular sieve in an adsorption process. The adsorbed species may be removed or desorbed from the zeolite by a single or combination of physical or chemical methods involving pressure decrease, temperature increase, or displacement by another adsorbate.

An associated benefit of utilizing zeolite to extract contaminants from a process stream is that the zeolite may normally be regenerated for many years of use. Primary zeolite minerals exhibiting good resistance to attrition and acidic environments are erionite, clinoptilolite, chabazite, and mordenite. The resistance to acid attack increases as the silica to alumina ratio increases within a zeolite species.

Chabazite from Bowie, AZ, is used in a pressure-swing adsorption process to remove H_2S from acidic natural gas streams. The Linde Division of Union Carbide markets a Bowie zeolite consisting of chabazite/erionite that is reported to be stable at pHs as low as 2.5. The product can supposedly remove hydrochloric acid (HCl) from reformed hydrogen streams and purify stack gasses of SO_2 (Mumpton, 1975, Sheppard et al., 1976). The Linde Division also has a proprietary process and zeolite that is utilized for the SO_2 cleanup of tail gasses in sulfuric acid plants (Flanigen and Mumpton, 1977).

The NRG Corporation has successfully used Bowie zeolite to extract H_2S from well gas in Los Angeles, CA. The company has also used zeolite to purify methane recovered from a landfill located in Palos Verde, CA, where H_2S and other acid contaminants are eliminated through adsorption by the zeolite (Flanigen and Mumpton, 1977).

Investigative studies have shown certain mordenites to be successful in adsorbing HCl, as well as being selective in removing SO_2 and SO_3 from stack gasses. Mordenite and chabazite have both had success as NO_x adsorbents, with the additional benefit of recovering and recycling dilute nitric acid back into the industrial process (Vaughan, 1976).

The market potential for the commercial applications of zeolite functioning as an acid neutralizer or preventer appear quite promising. However, primarily the cost effectiveness, as well as the product efficiency and environmental impact of using zeolites, as opposed to other acid neutralization materials, must be examined thoroughly before the benefits and deterrents of use can be acknowledged.

OUTLOOK

Future market growth appears encouraging, but limited for the aforementioned industrial minerals and rocks utilized in acid pollution abatement, with the exception of limestone, dolomite and associated lime products, and sodium minerals used in flue gas desulfurization. Lime and limestone sales for flue gas desulfurization have doubled from 663 kt in 1980 to 1.3 Mt in 1989—an 11% average annual growth.

Liming of lakes, rivers, and soils to combat acid rain will continue to proceed, but recent biological technology involving the development of wetlands and the utilization of anaerobic bacteria to successfully treat acid mine drainage may supplant some sales of traditional acid neutralizing products. Sales should exhibit a stable to slight growth rate for neutralizers utilized in the treatment of acidic industrial wastes.

BIBLIOGRAPHIC REFERENCES

Anon., 1975, "Mineral Resources and the Environment," National Academy of Sciences, Washington, DC, pp. 52–53, 225–243, 247–255.

Anon., 1986, *Aglime Fact Book*, National Stone Association, Washington, DC, 65 pp.

Aitala, R., 1991, "U.S. Soda Ash Supply to Continue to Grow," *Glass Industry*, Dec., pp. 10–13.

Bell, A.U., 1975, "Base-Metal Mine Waste Management in Canada," *Proceedings*, The Institution of Mining and Metallurgy, M.J. Jones, ed., pp. 45–57.

Bernabo, J.C., and Ringold, P., 1985, "The Evolving Science of Acid Deposition," *Proceedings*, Illinois Energy Resources Commission and the Coal Extraction and Utilization Research Center at Southern Illinois University, D.S. Gillel and and J.H. Swisher, eds., Apr. 10, pp. 6–9.

Berner, R.A., 1971, "Worldwide Sulfur Pollution of Rivers," *Journal of Geophysical Research*, Vol. 76, No. 25, pp. 6597–6599.

Blesing, N.V., Lackey, J.A., and Spry, A.H., 1974, "Rehabilitation of an Abandoned Mine Site," *Proceedings*, The Institution of Mining and Metallurgy, M.J. Jones, ed., pp. 341–361.

Boynton, R.S., 1980, "Uses of Lime," *Chemistry and Technology of Lime and Limestone*, 2nd ed., Wiley and Sons, New York, pp. 400–424.

Boynton, R.S., and Gutschick, K.A., rev. Freas, R.C., and Thompson, J.L., 1983, "Lime," *Industrial Minerals and Rocks*, 5th ed., Vol. 2, S.J. Lefond, ed., AIME, New York, p. 821.

Britt, D.L., 1985, "The Renovation of Acid Lakes and Streams," *Proceedings*, Illinois Energy Resources Commission and the Coal Extraction and Utilization Research Center at Southern Illinois University, D.S. Gilleland and J.H. Swisher, eds., Apr. 10, pp. 19–50.

Brodie, G.A., 1991, "Engineered Wetlands for Effective Treatment of Acid Drainage—Applications, Results, and Prospects in the Tennessee Valley," *Proceedings*, Association of Engineering Geologists, 34th Annual Meeting, Sep. 29, pp. 558–568.

Charmbury, H.B., 1979, "Coal and the Environment," *Coal Surface Mining: Impacts of Reclamation*, J.E. Rowe, ed., Westview Press, Boulder, CO, pp. 71–75 (reprinted *Coal Mining and Processing*, Vol. 2, No. 1, Maclean-Hunter, Jan. 1975, pp. 6–70).

Clarke, G., 1992, "Magnesia Around the World, Hard Times in Soft Markets—Again," *Industrial Minerals*, No. 295, Apr., pp. 45–77.

Clarke, G., 1980, "Zeolites—Take Off for the Tuff Guys?", *Industrial Minerals*, No. 149, Feb., pp. 21–38.

Coope, B., 1989, "Magnesia Markets, Fit for Optimism," *Industrial Minerals*, No. 264, pp. 45–57.

Craxford, S.R., 1979, "Air Pollution from the Use of Petroleum," *The Prevention of Oil Pollution*, Graham and Trotman Ltd., London, pp. 46–59.

Crocker, T.D., ed., 1984, *Economic Perspectives on Acid Deposition Control*, Butterworth, Boston, 121 pp.

El-Shall, H., 1992, "Biotechnology, Mineral Processing, 1991," *Mining Engineering*, Vol. 44, No. 5, pp. 460–462.

Erickson, P.M., and Ladwig, K.J., 1984, "Chemical Inhibition of Acid Mine Drainage Formation," US Bureau of Mines, US Department of the Interior, Pittsburgh, PA, pp. 1–11.

Eyde, T.H., 1981, "Chemical and Physical Properties of Natural Zeolites and Their Potential Industrial Applications," *Proceedings*, TMS-AIME Process Mineralogy Committee, Feb. 22–26, pp. 645–656.

Farris, C.B., 1980, "Market Potential of the Green River Nahcolite/Dawsonite in Colorado," *Proceedings*, 15th Forum on Geology of Industrial Minerals, S.D. Schwochow, ed., Colorado Geological Survey, pp. 96–104.

Flanigen, E.M., and Mumpton, F.A., 1977, "Commercial Properties of Natural Zeolites," *Mineralogy and Geology of Natural Zeolites*, Short

Course Notes, Mineralogical Society of America, Vol. 4, F.A. Mumpton, ed., pp. 165–174.

Forster, B.A., 1985, ''Economic Impact of Acid Deposition in the Canadian Aquatic Sector,'' *Acid Deposition*, D.A. Adams and W.P. Page, eds., 1st ed., Plenum Press, New York, pp. 413–414.

Freeze, R.A., and Cherry, J.A., 1979, *Groundwater*, Prentice-Hall, Englewood Cliffs, NJ, pp. 238–239.

Guay, W.J., 1981, ''Scrubbing of Copper Reverberatory Furnace Gases with Flotation Tailings at San Manuel,'' *Proceedings*, TMS-AIME Process Mineralogy Committee, Feb. 22–26, pp. 435–440.

Hart, W., et al., 1990, ''The Use of Phosphate Refuse as a Potential AMD Ameliorant,'' *Proceedings*, 1990 Mining and Reclamation Conference and Exhibition, Apr. 23–26, pp. 43–49.

Keller, E.A., 1982, *Environmental Geology*, 3rd ed., Charles E. Merrill Publishing Co., Columbus, OH, pp. 321–323, 371, 395–397.

Kim, A.G., et al., 1982, ''Acid Mine Drainage: Control and Abatement Research,'' US Bureau of Mines, US Department of the Interior, IC 8905, 22 pp.

Kleinmann, R.L.P., and Erickson, P.M., 1986, ''Control of Acid Mine Drainage: An Overview of Recent Developments,'' *Proceedings*, Coal Extraction and Utilization Research Center at Southern Illinois University, Oct. 28–29, pp. 283–306.

Kostick, D.S., 1991, ''Soda Ash, 1990,'' US Bureau of Mines, US Department of the Interior, Washington, DC, 21 pp.

Kramer, D.A., 1991, ''Magnesium and Magnesium Compounds, 1990,'' US Bureau of Mines, US Department of the Interior, Washington, DC, 18 pp.

Lindberg, S.E., Page, A.L., and Norton, S.A., eds., 1990, *Acidic Precipitation, Volume 3: Sources, Deposition, and Canopy Interactions*, Springer-Verlag, New York, 322 pp.

Lundgren, L., 1986, *Environmental Geology*, Prentice-Hall, Englewood Cliffs, NJ, pp. 183–192.

Ma, Y.H., and Lee, T.Y., 1976, ''Sorption and Diffusion Properties of Natural Zeolites,'' *Proceedings*, Natural Zeolites: Occurrence, Properties, Use, L.B. Sand and F.A. Mumpton, eds., Pergamon Press, New York, June, pp. 373–383.

Meyers, S., 1984, ''Acid Deposition: A Search for Solutions,'' *Proceedings*, Illinois Energy Resources Commission and the Coal Extraction and Utilization Research Center at Southern Illinois University, D.S. Guilleland and J.H. Swisher, eds., Mar. 28, pp. 1–10.

Miller, M.M., 1991, ''Lime, 1990,'' US Bureau of Mines, US Department of the Interior, Washington, DC, 13 pp.

Mumpton, F.A., 1975, ''Commercial Utilization of Natural Zeolites,'' *Industrial Minerals and Rocks*, 4th ed., S.J. Lefond, ed., AIME, New York, pp. 1262–1274.

Mumpton, F.A., 1976, ''Natural Zeolites: A New Industrial Mineral Commodity,'' *Proceedings*, Natural Zeolites: Occurrence, Properties, Use, L.B. Sand and F.A. Mumpton, eds., Pergamon Press, New York, pp. 17–27.

Mumpton, F.A., 1977, ''Utilization of Natural Zeolites,'' *Mineralogy and Geology of Natural Zeolites*, Short Course Notes, Mineralogical Society of America, Vol. 4, F.A. Mumpton, ed., pp. 177–198.

Murphy, J.N., 1992, ''Research and Development, Mining 1991,'' *Mining Engineering*, Vol. 44, No. 5, p. 459.

Nielsen, I., 1980, ''Development of Nahcolite-Rich Oil Shales,'' *Proceedings*, 15th Forum on Geology of Industrial Minerals, S.D. Schwochow, ed., Colorado Geological Survey, pp. 89–95.

Palmer, K., 1974, ''Recovery of Acid and Elemental Sulfur from Copper Smelter Gases to Conform with New Environmental Standards,'' *Proceedings*, The Institution of Mining and Metallurgy, M.J. Jones, ed., pp. 435–451.

Parizek, R.R., 1986, ''Use of Waste Products to Reclaim Strip-Mined Lands,'' *Proceedings*, Coal Extraction and Utilization Research Center at Southern Illinois University, Oct. 28–29, pp. 223–281.

Parsons, W.A., 1965, *Chemical Treatment of Sewage and Industrial Wastes*, National Lime Association, Bulletin 215, Washington, DC, pp. 48–95.

Regens, J.L., 1984, ''Acid Rain: Does Science Dictate Policy or Policy Dictate Science?'' *Economic Perspectives on Acid Deposition Control*, Vol. 8, T.D. Crocker, ed., Butterworth, Boston, pp. 5–19.

Remick, B.W., 1991, ''Non-Construction Uses of Stone,'' *The Aggregate Handbook*, R.D. Barksdale, ed., National Stone Association, Washington, DC, pp. 14–18 to 14–26.

Renton, J.J., Stiller, A.H., and Rymer, T.E., 1988, ''The Use of Phosphate Materials as Ameliorants for Acid Mine Drainage,'' *Proceedings*, West Virginia Surface Mine Drainage Task Force and West Virginia Mining and Reclamation Association, June, pp. 67–75.

Ripley, E.A., Redman, R.E., and Maxwell, J., eds., 1982, ''Environmental Impact of Mining in Canada,'' Centre for Resource Studies, Queen's University, Kingston, Ontario, 284 pp.

Russell, A.M.B., 1990, ''Natural and Synthetic Soda Ash,'' *Industrial Minerals*, No. 268, Jan., pp. 19–34.

Sheppard, R.A., et al., 1976, ''Bowie Zeolite Deposit, Cochise, and Graham Counties, Arizona,'' *Proceedings*, Natural Zeolites: Occurrence, Properties, Use, L.B. Sand and F.A. Mumpton, eds., Pergamon Press, New York, June, p. 327.

Teringo, J., 1990, ''Comparing Acid Waste Stream Neutralization Methods,'' *Plant Engineering*, Nov. 8, pp. 74–77.

Vaughan, D.E.W., 1976, ''Properties of Natural Zeolites,'' *Proceedings*, Natural Zeolites: Occurrence, Properties, Use, L.B. Sand and F.A. Mumpton, eds., Pergamon Press, New York, June, pp. 353–371.

Wilmoth, R.C., and Hill, R.D., 1970, ''Neutralization of High Ferric Iron Acid Mine Drainage,'' Program 14010 ETV, Federal Water Quality Administration, US Department of the Interior, Aug., 38 pp.

ENVIRONMENTAL USES

WATER AND SEWAGE TREATMENT

KENNETH A. GUTSCHICK AND HARRY L. FRANCIS

A large number of minerals and chemicals (Table 1), some mineral-derived, are used in the treatment of both potable water and domestic sewage waste water and solids (Anon., 1990). In water treatment there are ten principal process uses as shown in Table 2 (Anon., 1990), which also lists the major chemicals in each process. Since lime (calcium oxide, calcium hydroxide, and dolomitic lime) is the largest chemical (by tonnage) used in water treatment, this chapter will deal primarily with this product. Note that it is included in five of the process uses shown in Table 2, and that indirectly it is also used in at least one other category (taste and odor control, where lime is used with potassium permanganate to provide optimum pH condition).

POTABLE WATER TREATMENT

Water Softening

The biggest use for lime in water treatment involves water softening, where it serves to reduce carbonate hardness caused by dissolved bicarbonate and carbonate compounds. Currently there are more than 1 000 domestic water softening plants (Table 3) located in most sections of the country, but particularly in Ohio, other midwestern states, Florida, and Texas, where limestone bedrock abounds (Anon., 1990). Among the other uses for lime in water treatment are coagulation, disinfection, and removal of heavy metals due to high pH softening, removal of certain radioactive and organic compounds and fluorides, neutralization of acidic water, removal of silica, particularly with dolomitic lime, and removal of color. Most of the above advantages accompany water softening, although at some plants lime is used specifically for one purpose, such as silica removal with dolomitic lime, minimizing pipeline corrosion by addition of lime to correct an aggressive water, and others.

Hardness in water comprises two types—carbonate hardness, caused by dissolved calcium and magnesium carbonates and bicarbonates, and non-carbonate hardness caused by dissolved non-carbonate salts, particularly calcium sulfate, calcium chloride, magnesium sulfate, and magnesium chloride. Total hardness is the sum of the carbonate hardness and non-carbonate hardness. Waters are classified as soft (0 to 75 mg/L as $CaCO_3$), moderate (75 to 150 mg/L), hard (150 to 300 mg/L), and very hard (above 300 mg/L). Most water plants strive to reduce total hardness to about 100 mg/L, although in recent years the tendency in some cities has been to allow a little higher value in order to reduce both chemical costs and the amount of water softening sludges produced.

There are four basic types of softening plants: lime softening, primarily to reduce carbonate hardness; lime-soda softening, for reduction of both types of hardness (in some plants caustic soda is used in place of soda ash, but it is more costly); split lime-zeolite, where a base exchange mineral is employed in place of soda ash (this system also requires salt [sodium chloride] to regenerate the zeolite mineral); and finally zeolite softening, which is generally used in small systems for private houses and small municipalities. Principal disadvantages of the latter system are its relatively high cost and the salt requirement, which introduces considerable quantities of soluble salts into the waste water discharge system.

Lime, the principal chemical used in water softening, serves to neutralize carbon dioxide, forming calcium carbonate, a precipitant, to convert soluble bicarbonate to calcium carbonate, and also to precipitate magnesium hydroxide. The following reactions occur in lime softening:

1) $CO_2 + Ca(OH)_2 \rightarrow CaCO_3 + H_2O$
2) $Ca(HCO_3)_2 + Ca(OH)_2 \rightarrow 2CaCO_3 + 2H_2O$
3) $Mg(HCO_3)_2 + Ca(OH)_2 \rightarrow MgCO_3 + CaCO_3 + 2H_2O$
4) $MgCO_3 + Ca(OH)_2 \rightarrow Mg(OH)_2 + CaCo_3$
5) $MgCl_3 + Ca(OH)_2 \rightarrow Mg(OH)_2 + CaCl_2$
6) $Mg(NO_3)_2 + Ca(OH)_2 \rightarrow Mg(OH)_2 + Ca(NO_3)_2$
7) $MgSO_4 + Ca(OH)_2 \rightarrow Mg(OH)_2 + CaSO_4$

Reactions involving soda ash are as follows:

8) $CaSO_4 + Na_2CO_3 \rightarrow CaCO_3 + Na_2SO_4$
9) $CaCl_2 + Na_2CO_3 \rightarrow CaCO_3 + 2NaCl$
10) $Ca(NO_3)_2 + Na_2CO_3 \rightarrow CaCO_3 + 2NaNO_3$

Note from these equations that the principal byproduct formed is calcium carbonate, a filterable insoluble precipitant referred to as water softening sludge. That the quantity is large is indicated by the fact that in precipitating the calcium ion (Eq. 2), two moles of calcium carbonate are formed for every mole of calcium ion removed from the water. In large plants the sludge is voluminous and often presents a disposal problem. Generally, the sludge is placed in a lagoon, or it may be dewatered with vacuum filters, belt and filter presses, or centrifuges, then disposed of on farm fields as an agricultural liming material. Its high calcium carbonate content and fineness enhance its value as a liming material. Another use is as a neutralizing chemical in industrial and sewage treatment plants. A few large municipal plants have successfully recycled the sludge by dewatering and then calcining the material in rotary or fluosolids kilns to produce quicklime for reuse at the softening facility.

After lime softening, the pH and alkalinity of the treated water must be adjusted to a level suitable for human consumption. Often a small stream of raw water is mixed with the treated water to achieve the desired alkalinity level, requiring further sterilization.

Table 1. Commonly Used Water and Sewage Treatment Chemicals

Activated carbon	Chlorinated lime	Limestone	Sodium hexametaphosphate
Aluminum oxide	Chlorine	Ozone	Sodium hydroxide
Aluminum sulfate (alum)	Chlorine dioxide	Potassium aluminum-sulfate	Sodium hypochlorite
Ammonium sulfate	Copper sulfate	Potassium permanganate	Sodium silicate
Anhydrous ammonia	Dolomitic lime	Sodium aluminate	Sodium sulfite
Bentonite	Ferric chloride	Sodium carbonate	Sulfur dioxide
Calcium fluoride	Ferric sulfate	Sodium chloride	Sulfuric acid
Calcium oxide	Ferrous sulfate	Sodium chlorite	Tetrasodium pyrophosphate
Calcium hydroxide	Hydrofluosilicic acid	Sodium fluoride	Trisodium phosphate
Calcium hypochlorite	Hydrogen fluoride		

One can also achieve the desired pH and alkalinity by re-carbonating the treated water with CO_2 gas.

The finished water must be non-aggressive (non-corrosive) to the distribution system and also must not scale. In order to achieve this balance, adjustments to pH and alkalinity must be made. Using an evaluation, such as Langelier's index, to determine the aggressive nature of the water is desirable.

Some laying down of protective scale is often desirable, but care must be taken to ensure that excessive scale and decreased flow levels are not induced. A thin calcium carbonate scale will protect the water from the dissolved salts of heavy metals, iron deposits (red water), and objectionable taste. There are several specialty chemicals used to ensure a protective coating without excessive scaling, such as glassy polyphosphates.

Coagulation

Since many water supplies, particularly river water, contain suspended solids, it is usually necessary to add coagulants for removal of turbidity. The principal coagulants are aluminum sulfate (alum), ferric chloride, ferric sulfate, and lime. Lime is also used for pH control with either of the three other coagulants. The optimum pH for alum is pH 5.5 to 7.5 (typically 7.0) and for the ferric salts, pH 5.0 to 8.5 (typically 7.5). Since many waters are acidic in nature, lime is nearly always used to raise the pH to attain the optimum coagulation efficiency; at small plants liquid caustic soda is often used instead of lime.

In recent years, various polyelectrolytes have been developed for use as coagulants, either alone or in combination with the above metal ion coagulants. In many cases, lime is also used in conjunction with certain polyelectrolytes. In the coagulation step, the lime is generally added with the coagulant chemical, although it often may be advantageous to add the lime ahead of the coagulant. Besides providing the proper pH, lime also acts as a filter aid, particularly with gelatinous-like sludges formed with certain water and coagulant combinations that are difficult to filter.

Purification

Research as far back as 1913 (Riehl, 1976) has indicated that excess lime treatment (to pH 11+ with holding times of 24 to 48 hrs) has been successful in destroying E. coli bacteria. More recent research has confirmed these earlier findings, and work has also been extended to other bacteria, virus, parasites, and parasite eggs. Sterilization is attributed to physical absorption of the pathogens on the flocs which settle out following the lime softening-coagulation reaction, and the destroying of pathogens by the high pH. According to Boynton (1980), a serious epidemic of infectious hepatitis in Bombay, India, which was traced to the potable water supply, was stopped by use of excess lime treatment to pH of 11+. However, in spite of the encouraging experiences with lime sterilization, it can never fully replace chlorine as a disinfectant because the latter's residual content protects against later contamination in the water distribution system.

Removal of Heavy Metals and Radionuclides

Table 4, based on data summarized by the US Environmental Protection Agency from various research studies involving laboratory or pilot-scale experiments, indicates that a variety of heavy metals, including arsenic, barium, cadmium, chromium, fluoride, lead, mercury, silver, and others can be removed from potable water by raising the pH 6 to 11, depending on the metal involved (Anon., 1990). These data are based on conventional treatment involving coagulation with alum or iron salts, followed by lime or lime-soda softening. Treatment steps include mixing, flocculation, sedimentation, and filtration. The elevation in pH precipitates the metals out of solution for ready removal. As noted, the pH elevation occurs during lime or lime-soda softening and/or coagulation with alum or ferric salts abetted by lime addition. The mechanism of removal for inorganic contaminants may be either precipitation of insoluble metal hydroxides, carbonation, co-precipitation with iron or aluminum hydroxides, or adsorption with natural turbidity, or on floc particles formed during the reaction.

Fig. 1 shows the extent of removal of various inorganic contaminants by lime softening, based on a pH 8 to 12 (Anon., 1990). Note in particular that lead and cadmium are completely removed at pH 8, arsenic^{+5} and chromium^{+3} at pH 11, and barium at pH 10. Other metals show substantial removal at elevated pH.

Radium, an alkaline earth material chemically similar to calcium and barium, is removed during lime softening, and generally not during coagulation. Removals of 70 to 95% have been reported, with greater removals occurring with greater hardness removals (Fig. 2) (Anon., 1990). Generally the higher the pH during softening, the greater the radium removal.

Neutralization of Acid Water: The growing concern with the presence of lead in drinking water has brought on an increase in the use of lime for neutralizing slightly acid water. By raising of the pH to >7, the lead will not dissolve from solder used in pipe fittings. Further, lime corrects corrosive waters containing excessive amounts of carbon dioxide by absorption of the CO_2, forming a thin calcium carbonate deposit on the inside of the pipes.

Silica Removal: Waters containing silica are especially harmful in high pressure boilers and turbines due to formation of troublesome scale on the pipes and turbine blades. If dolomitic lime is used in the softening process, much of the silica will be removed due to its reaction with magnesium oxide and hydroxide.

Removal of Other Impurities: Lime is also used to help remove various other impurities, especially manganese, fluorides, iron, and organic tannin, which causes undesirable colors. Generally, lime softening will remove 80 to 90% of the color, with activated carbon being used to remove the balance.

Table 2. Process Use of Chemicals

Process	Chemicals
Coagulant	Aluminum sulfate
	Bauxite
	Ferrous sulfate
	Ferric sulfate
	Ferric chloride
	Sodium aluminate
	Calcium hydroxide
	Calcium oxide
Coagulant aids	Bentonite
	Calcium carbonate
	Carbon dioxide
	Sodium silicate
Disinfection and chlorination	Anhydrous ammonia
	Ammonium hydroxide
	Ammonium sulfate
	Chlorine
	Chlorine dioxide
	Sodium chlorite
	Chlorinated lime
	Calcium hypochlorite
	Sodium hypochlorite
Dechlorination	Activated carbon
	Ion-exchange resins
	Sodium bisuflate
	Sodium sulfite
	Sulfur dioxide
pH adjustment	Calcium carbonate
	Carbon dioxide
	Hydrochloric acid
	Calcium oxide
	Calcium hydroxide
	Sodium hydroxide
	Sulfuric acid
Fluoridation and fluoride adjustment	Ammonium silicofluoride
	Calcium fluoride
	Hydrofluosilicic acid
	Sodium fluoride
	Sodium silicofluoride
	Activated alumina (aluminum oxide)
Taste and odor control	Activated carbon
	Bentonite
	Chlorine
	Chlorine dioxide
	Copper sulfate
	Potassium permanganate
Mineral oxidation	Chlorine
	Chlorine dioxide
	Ozone
	Potassium permanganate
Stabilization and corrosion control	Calcium oxide
	Calcium hydroxide
	Sodium carbonate
	Sodium hexametaphosphate
Softening	Carbon dioxide
	Calcium oxide
	Calcium hydroxide
	Sodium carbonate
	Sodium chloride

CHEMICAL TREATMENT OF MUNICIPAL WASTE WATER

Introduction

Sewage is treated similar to potable water and consists of both liquid and solid treatment processes. The liquid process mainly consists of a settling of solids from the waste stream, generating a primary sludge. The clarified waste water is then passed to a biological treatment step where bacteria and fungi remove the nutrients, generating a clear effluent and a secondary sludge. The clear effluent is then disinfected and safely returned to the environment (Fig. 3) (Christensen, 1987).

Table 3. Summary of Domestic Use Softening Plants

Softening process	Number of plants	Average daily production, mgd	Total daily production, mgd
Softening	353	0.85	299
Coagulation-softening	265	8.50	2 246
Softening-iron removal	366	0.95	348
Coagulation-softening-iron removal	45	3.60	162

However, in some treatment plants, the raw sewage is immediately treated with lime to a high pH (>12) to facilitate the primary settling of organic matter and dissolved solids (this is similar to lime softening treatment in potable water and is also effective in removing phosphates). This process produces a large volume of primary sludge. The clarified waste water is then aerated to adjust the pH to 7 to 8 and biological treatment follows as noted previously.

After secondary treatment, the clarified effluent is disinfected and discharged. The primary and secondary sludge streams (3 to 4% solids) are mixed and conditioned with polymers or ferric salts and lime and then dewatered. The resulting sludge cake, containing 25 to 30% solids, is disposed of by land application, incineration, or placed in a landfill. Fig. 4 shows the typical quantities of sludge produced from a conventional sewage treatment plant (Christensen, 1987).

Beneficial reuse of waste water treatment sludges, referred to as bio-solids, is encouraged by the EPA. The organic matter is a valuable resource, and if properly stabilized to control pathogens, bacteria, and parasites, it can be utilized in an agricultural fertilization program. Lime is being used increasingly for stabilization of sewage sludges for reuse on farm lands.

Federal Regulations on Solid Waste Disposal

Regulations governing beneficial reuse of bio-solids are contained primarily in 40 CFR 257, Criteria for Classification of Solid Waste Disposal Facilities and Practices (Anon., 1989). Many states also regulate beneficial reuse, some more strictly than the federal government. Other requirements are contained in 40 CFR 761, Solid Waste Landfills.

"In 40 CFR 257, land application is considered to be a form of solid waste disposal, as are landfilling and sludge lagooning. The same regulations, in several general sections, apply to all three practices."

Floodplains. Land application sites, landfills, and lagoons may be located in a floodplain; however, they must not "restrict the flow of the base flood, reduce the temporary water storage capacity of the floodplain, or result in washout of solid waste, so as to pose a hazard to human life, wildlife, or land or water resources." A base flood is a 100-year flood.

Surface Waters. Discharges that would violate Sections 402, 404, and 208 of the Clean Water Act are prohibited.

Table 4. Summary of Treatment Processes for Inorganic Contaminants

Contaminant	Process	pH range	Comments
Arsenic +5	Ferric sulfate coagulation	6–8	—
	Alum coagulation	6–7	—
	Lime softening	>10.5	—
Arsenic +3	Ferric sulfate coagulation	6–8	Oxidation to As^{5+} by chlorination required prior to coagulation
	Alum coagulation	6–7	
	Lime softening	>10.5	
Barium	Lime softening	10–11	—
	Ion exchange	—	Normal cationic exchange resins effective
Cadmium	Ferric sulfate coagulation	7–8	Effective over full lime-softening range
	Lime softening	—	
Copper	Stabilization and corrosion control	Site-specific	—
Chromium +3	Ferric sulfate coagulation	6–9	—
	Alum coagulation	7–9	—
	Lime softening	>10.5	—
Chromium +6	Ferric sulfate coagulation	7–9.5	—
Fluoride	Ion exchange	Neutral pH recom-mended	Ion-exchange media: activated alumina or bone char
	Lime softening	>11	Fluoride coprecipitates with magnesium hydroxide, high magnesium required
Lead	Ferric sulfate coagulation	6–9	—
	Alum coagulation	6–9	—
	Lime softening	—	Effective over full lime-softening range
Inorganic mercury	Stabilization & corrosion control/ferric sulfate coagulation	Site-specific 7–8	—
Organic mercury	Activated carbon	—	Powdered and granular carbon effective
Nitrate	Ion exchange	—	Ion-exchange media: strong base resin
Radium	Lime softening	—	Effective throughout entire softening range
	Ion exchange	—	Normal cationic exchange media effective
	Reverse osmosis	—	—
Selenium +4	Ferric sulfate coagulation	6–7	—
	Ion exchange	—	Ion-exchange media: strong base resin or activated alumina
	Reverse osmosis	—	
Selenium +6	Reverse osmosis	—	—
Silver	Ferric sulfate coagulation	7–9	—
	Alum coagulation	6–8	—
	Lime softening	—	Effective over full lime-softening range
Zinc	Stabilization and corrosion control	Site-specific	—

Groundwater. Facilities must not "contaminate an underground drinking water source beyond the solid waste boundary." States may establish an alternative boundary for a facility if such a change would not contaminate drinking water resources. The EPA is currently examining more stringent controls on land application to protect particularly vulnerable and valuable groundwater resources such as irreplaceable aquifers.

Public Health. Waste cannot cause a risk of infection by the enteric organisms which are concentrated in the sludge. For this reason, controls on pathogenic organisms are required.

To protect public health, sewage sludge or septic tank pumpings that are applied to the land or incorporated into the soil must be treated by a Process to Significantly Reduce Pathogens (PSRP) prior to application or incorporation (Table 5). The success of a PSRP can be determined by measuring the reduction in the number of organisms present. A one-log (90%) reduction in the number of pathogens present, or a two-log (99%) reduction in indicator bacteria (fecal coliforms) can be used to show that an unlisted process attains equivalent pathogen reduction. Also to protect public health, public access to the facility must be controlled for at least 12 months, and grazing by animals, whose products are consumed by humans, must be prevented for at least 1 month unless Processes to Further Reduce Pathogens (PFRP) are used (Table 6).

Food Chain Crops

Food chain crops are tobacco, crops grown for human consumption, and feed for animals whose products are consumed by humans. Land application of sludge for growth of food chain crops is subject to additional requirements and to restraints imposed by good practices and state regulations. For example, to prevent nitrate contamination of groundwater, the usual practice is to apply sludge at a rate that just satisfies the nitrogen requirement of the crop to be grown on a site. Similarly, some states protect surface waters against phosphorus contamination by limiting application rates to the phosphorus needs of the crops. Key federal regulations affecting land application to food chain crops focus on pathogen reduction, cadmium limitations, and PCB content.

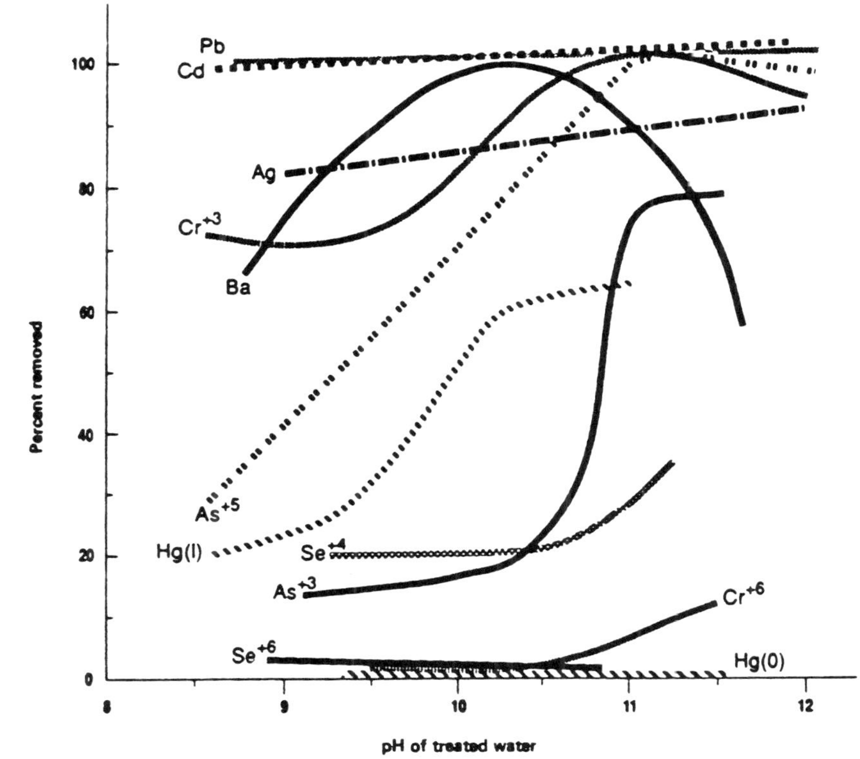

Fig. 1. Removal of inorganic contaminants by lime softening.

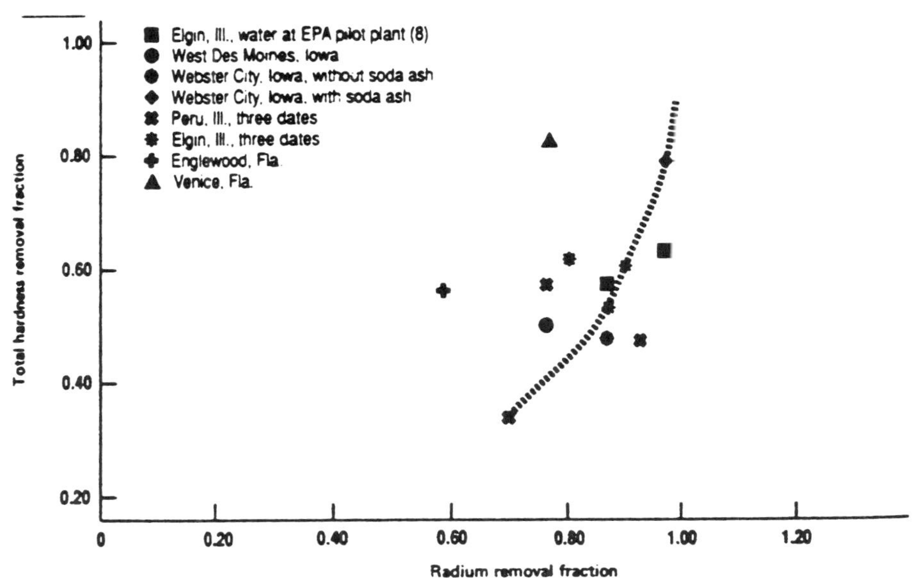

Fig. 2. Lime-soda process: total hardness removal fraction vs. radium removal fraction.

Lime Stabilization of Sewage Sludge

To meet the statutory requirements of PSRP (Table 5), lime stabilization is one of the treatment options. A sufficient quantity of chemical lime is added to the sludge so that the entire mass achieves a pH ≥ 12. It is required that this pH remains at 12 or above for a 2 hr period. This lime stabilization process can be incorporated into the conditioning/dewatering step, and also aids in plant odor control. The high pH that lime imparts also chemically fixes heavy metals into insoluble hydroxide forms.

If it is desirable to use the lime stabilized biosolids on crops for direct human consumption, it is required that treatment to a higher level be accomplished (PFRP). Again, lime can be utilized to achieve the level of safety. Sufficient amounts of quicklime can be

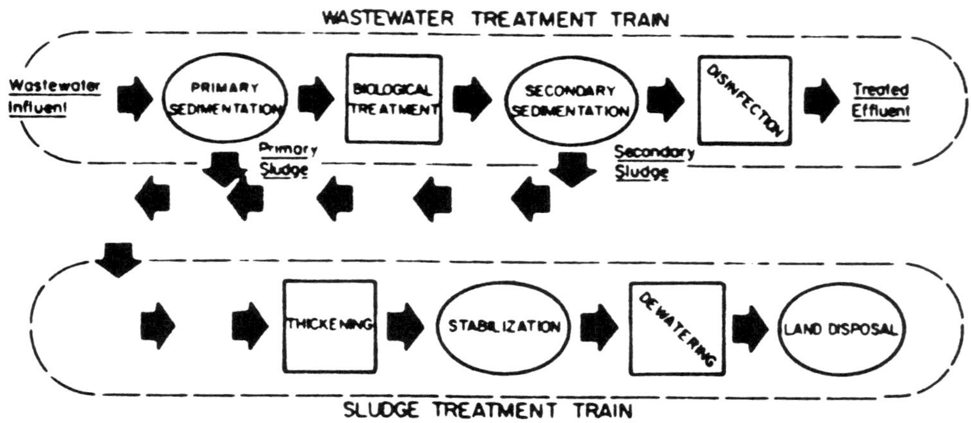

Fig. 3. Conventional wastewater treatment.

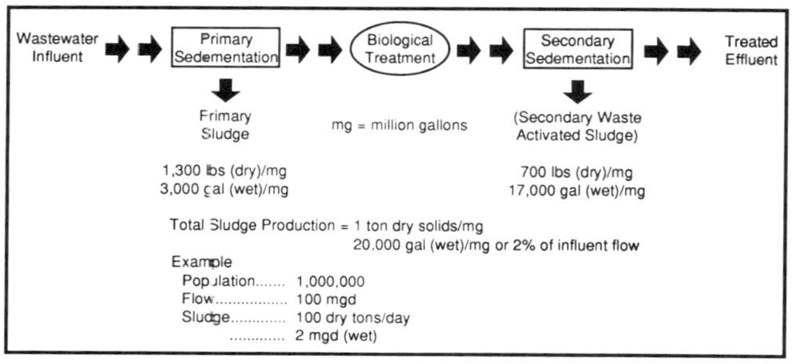

Fig. 4. Sludge production for a conventional treatment facility.

Table 5. Regulatory Definition of Processes to Significantly Reduce Pathogens

Aerobic Digestion: The process is conducted by agitating sludge with air or oxygen to maintain aerobic conditions at residence times ranging from 60 days at 15°C to 40 days at 20°C, with a volatile solids reduction of at least 38%.
Air Drying: Liquid sludge is allowed to drain and/or dry on underdrained sand beds, or on paved or unpaved basins in which the sludge depth is a maximum of 23 cm. A minimum of 3 months is needed, for 2 months of which temperatures average on a daily basis above 0°C.
Anaerobic Digestion: The process is conducted in the absence of air at residence times ranging from 60 days at 20°C to 15 days at 35 to 55°C, with a volatile solids reduction of at least 38%.
Composting: Using the within-vessel, static aerated pile, or windrow composting methods, the solid waste is maintained at minimum operating conditions of 40°C for 5 days. For 4 hours during this period the temperature exceeds 55°C.
Lime Stabilization: Sufficient lime is added to produce a pH of 12 after 2 hours of contact.
Other Methods: Other methods of operating conditions may be acceptable if pathogens and vector attraction of the waste (volatile solids) are reduced to an extent equivalent to the reduction achieved by any of the above methods.

Table 6. Regulatory Definition of Processes to Further Reduce Pathogens

Composting: Using the within-vessel composting method, the solid waste is maintained at operating conditions of 55°C or greater for three days. Using the static aerated pile composting method, the solid waste is maintained at operating conditions of 55°C or greater for three days. Using the windrow composting method, the solid waste attains a temperature of 55°C or greater for at least 15 days during the composting period. Also, during the high temperature period, there will be a minimum of five turnings of the windrow.
Heat drying: Dewatered sludge cake is dried by direct or indirect contact with hot gases, and moisture content is reduced to 10% or lower. Sludge particles reach temperatures well in excess of 80°C, or the wet bulb temperature of the gas stream in contact with the sludge at the point where it leaves the dryer is in excess of 80°C.
Heat treatment: Liquid sludge is heated to temperatures of 180°C for 30 minutes.
Thermophilic aerobic digestion: Liquid sludge is agitated with air or oxygen to maintain aerobic conditions at residence times of 10 days at 55°C to 60°C, with a volatile solids reduction of at least 38%.
Other methods: Other methods or operating conditions may be acceptable if pathogens and vector attraction of the waste (volatile solids) are reduced to an extent equivalent to the reduction achieved by any of the above methods.
Any of the processes listed below, if added to a PSRP, further reduce pathogens.
Beta ray irradiation: Sludge is irradiated with beta rays from an accelerator at dosages of at least 1.0 megarad at room temperature (ca. 20°C).
Gamma ray irradiation: Sludge is irradiated with gamma rays from certain isotopes, such as ^{60}Cobalt and ^{137}Cesium, at dosages of at least 1.0 megarad at room temperature (ca. 20°C).
Pasteurization: Sludge is maintained for at least 30 minutes at a minimum temperature of 70°C.
Other methods: Other methods or operating conditions may be acceptable if pathogens are reduced to an extent equivalent to the reduction achieved by any of the above add-on methods.

mixed with the bio-solids, utilizing the entrained water to hydrate the quicklime. The resulting high pH satisfies the stabilization requirement, eliminates offensive odors, and the heat generated by the exothermic reaction of quicklime in water pasteurizes the sludge (70°C @ 30 min). To achieve this level of treatment throughout the entire mass economically requires mixing and storage in an insulated container, so as to make the most efficient use of the heat generated, using the least amount of lime. The exothermic reaction of quicklime and water easily brings water to a boil. In fact, in the presence of water vapor, theoretical temperatures of 302°C can be achieved by this reaction with quicklime.

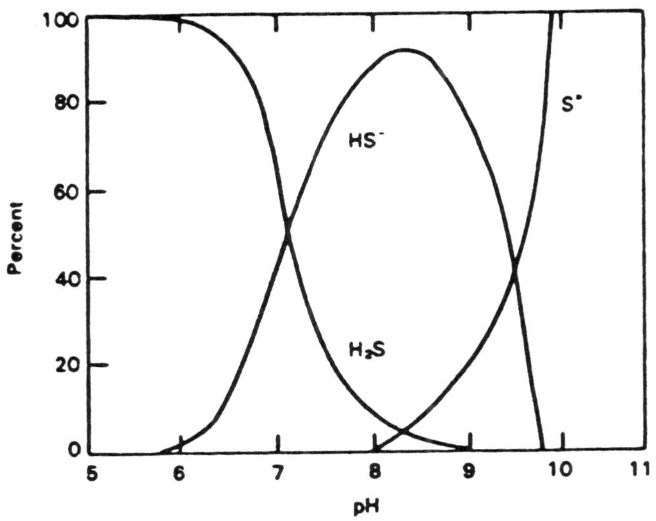

Fig. 5. Effect of pH on hydrogen sulfide equilibrium.

Odor Control

Waste water plants are plagued by offensive odors—mainly hydrogen sulfide. Hydrogen sulfide also causes considerable corrosion in the sewer pipes. Control of these odors can be accomplished if one maintains the influent pH in a range where the sulfides remain in solution and do not escape as a gas. This also prevents corrosion. Hydrated lime slurry can be utilized here (Fig. 5) to control the pH to a desirable level (Anon., 1979). Other chemicals such as potassium permanganate and hydrogen peroxide are also used to control H_2S odors and corrosion. In biological processes, small amounts of hydrated lime are added to the digesters to maintain the proper pH level that insures a suitable environment for the microbes.

REFERENCES

Anon., 1979, *Process Design Manual for Sludge Treatment and Disposal,* US Environmental Protection Agency, EPA 625/1–79–011.

Anon., 1989, "Environmental Regulations and Technology, Use and Disposal of Municipal Wastewater," US Environmental Protection Agency, EPA 625/10–84–003, 76 pp.

Anon., 1990, *Water Treatment and Design,* 2nd ed., American Society of Civil Engineers and American Water Works Association, McGraw-Hill, New York, 598 pp.

Boynton, R.S., 1980, *Chemistry and Technology of Lime and Limestone,* 2nd ed., John Wiley, New York, 578 pp.

Christensen, G.L., 1987, *Lime for Environmental Uses,* American Society for Testing and Materials, Special Technical Publication 931, 147 pp.

Riehl, M.L., 1976, "Hoover's Water Supply and Treatment," 11th ed., National Lime Association, Bulletin 211, Washington, DC, 224 pp.

FELDSPARS

ROGER A. KAUFFMAN AND DEAN VAN DYK

Feldspar is the earth's most abundant mineral group. The feldspars are aluminum silicate minerals belonging to the tectosilicate class. Feldspars are the most widespread of any mineral group and are estimated to constitute 60% of the earth's crust. They make up a large percentage of the exposed rocks as well as soils, clays, and other unconsolidated sediments. Feldspars are widely used in the glass and ceramic industry and are principal components in rock classification schemes.

Feldspar minerals and their production are important contributors to the domestic and world economies and standard of living. In 1990, the United States produced 630 kt of feldspar minerals, and world production was estimated to be more than 5 Mt. Feldspar mineral production in the United States has a long and interesting history (Feitler, 1967, Rogers Jr. et al., 1983). The first commercial production in the United States occurred in about 1825 and came from pegmatite bodies located in Connecticut, where feldspar was mined, hand sorted, and shipped to England. Feldspar production grew during the 19th and 20th centuries with important developments in production location, integrated mine and processing facilities, and the development of the flotation process for feldspar by the US Bureau of Mines and the feldspar industry. The first commercial feldspar flotation plant began operations in Kona, NC, in 1946, some six years after this process had been developed.

In 1990, the domestic feldspar industry consisted of 12 operating companies with mines and processing plants located throughout the United States (see Table 1).

During the last 25 years, the domestic feldspar industry has been dynamic, marked by new plant construction and capacity expansion at older plants. The industry has also seen plant closures, changes in ownership of existing operations, the consolidation of the Canadian nepheline syenite industry, and the increase of imported feldspar products. The industry has been affected by contractions in its major end-use markets brought about by economic cycles, product substitution, and environmental effects, such as recycling, land-use controversy, and air and water pollution. Despite problems, the industry remains mature and highly competitive.

GEOLOGY

Mineralogy

The common feldspar group of minerals form a single crystalline phase or solid solution series of minerals. Fig. 1 is a ternary phase diagram that shows the feldspar mineralogy—$KAlSi_3O_8$, $NaAlSi_3O_8$, and $CaAl_2Si_2O_8$. The end-member compositions are also referred to as potassium, sodium, and calcium feldspar. The feldspars whose chemistry ranges between the potassium and sodium end-members are known as alkali feldspars, while those between the sodium and calcium end-members are plagioclase feldspars. These feldspar groups can be further subdivided based on structural and compositional features within the alkali or plagioclase feldspar series. The distinction between and the proportion of members of the two feldspar groups are important to the classification of igneous rocks (Deer et al., 1966).

Perthites are the exsolution product of an initially homogeneous potassium-sodium feldspar solid solution which forms (in an environment of decreasing temperature) a heterogeneous intergrowth of sodium-rich feldspar lamellae in a host crystal of potassium feldspar. Antiperthites are exsolution products in which the host mineral is sodium-rich plagioclase feldspar and the intergrowths are potassium-rich feldspar.

Theoretical end-member composition of the principal feldspars are listed in the following:

	Percentages				
Feldspar	K_2O	Na_2O	CaO	Al_2O_3	SiO_2
Microcline	16.9	—	—	18.4	64.7
Orthoclase	16.9	—	—	18.4	64.7
Albite	—	11.8	—	19.4	68.8
Anorthite	—	—	20.1	36.6	43.3

The common physical properties of both feldspar series are very similar. The feldspars have a hardness of 6, a specific gravity range of 2.54 to 2.76, and a vitreous luster. Color is more specific and can range from colorless, translucent to transparent, white to gray, green, yellow, and red.

Diagnostic features of the mineral series generally require the aid of an optical microscope or the use of X-ray techniques; however, plagioclase feldspar can generally be distinguished in-hand specimen from the potassic feldspars by the presence of albite twin striations (Klein and Hurlbut, Jr., 1985).

Fig. 1 is a ternary diagram that depicts the relationship of the feldspar series and commercial feldspar products. The solid areas show homogeneous feldspars, while the hatched areas show ranges of composition for perthites and antiperthites. The Xs are data points that represent numerous domestic and foreign commercial feldspar products plotted from bulk analyses recast into theoretical molecular proportions of feldspar end-members.

The diagram shows gross compositional differences between commercial products, but does not address the modal feldspar composition of the individual products (e.g., whether each data point represents a monomineralic or mixed feldspar product). The commercial products fall within the microcline, perthite, and anorthoclase compositional ranges, and commercial aplite falls within the oligoclase range.

Origin

The major commercial feldspar deposits occur in pegmatites, granitic rocks, granitic rock types known commercially as alaskite and aplite, and certain river, dune, and beach sands.

Table 1. US Feldspar Operations

Company	Ore Body	Mining	Milling Process	Capacity (tons)	Products	Size[(a)] (μ)	Al_2O_3	Fe_2O_3	CaO	K_2O	Na_2O	Byproducts
APLITE												
Feldspar Corp. Montpelier, VA Hanover County	Pegmatite	Open pit	Flotation	160 ktpy Aplite	Aplite	600	22.00	.10	5.50	2.60	6.00	None
FELDSPAR—SILICA MIXTURE												
Arkhola Sand & Gravel Div. of APAC Arkansas, Inc. Muskogee, OK Muskogee County	River sands	Dredging	Washing, magnetic separation, leaching	650 ktpy[(b)] Feldspathic sands	Glass sand	600	5.40	.10	.08	3.04	.78	Industrial sand, construction sand
CISCO Corona, CA Riverside County	Sandstone	Open pit	Washing, spirals, magnetic separation	360 ktpy Feldspathic sand	Glass sand	600	3.80	.05	.10	2.80	.40	Construction sand, industrial sand
P. W. Gillibrand Co. Simi Valley, CA Ventura County	Deltaic conglomerates sandstone siltstone	Open pit	Washing plant with flotation	140 ktpy Feldspathic sand	Amber grade	600	9.0	.35	1.0	4.0	3.50	Construction sand, industrial sand
					Regular	600	5.0	.15	.50	2.0	1.25	
					Flint grade— High grade	600	4.0	.04	.30	1.5	1.0	
					Potash feldspar	425	18.1	.15	1.8	7.6	3.6	
Lithium Corp. of America Bessemer City, NC (mine/mill) Gaston County Pacolet, SC (Proc plant) Spartanburg County	Pegmatite	Open pit	Flotation	140 ktpy[(b)] Feldspathic sand	Lithospar®	75	11.10	.07	.20	2.50	4.05	Feldspathic sand and mica are a byproduct of lithium production
					Lithospar®	425	11.10	.07	.20	2.50	4.65	
					Lithospar® contains 0.11 Li_2O							
Manito Sand Manito, IL Mason County	Dune sand	Open pit, hydraulic	Washing, flotation, magnetic separation	180 ktpy Feldspathic sand	Glass sand[(c)]	600	—	.18	—	—	—	Construction sand, industrial sand
Unimin Corp. Bryon, CA Contra Costa County	Sandstone	Open pit	Washing, flotation	360 ktpy Feldspathic sand	Glass Sil	425	4.54	.15	.06	3.11	.34	Industrial sand, construction sand
					Gran U Sil	600	4.54	.15	.06	3.11	.34	
					Gran U Sil	425	4.54	.15	.06	3.11	.34	
Unimin Corp. Emmett, ID Gem County	Lake deposits	Open pit	Washing	90 ktpy Feldspathic sand	Glass Sil	425	7.93	.11	.75	2.82	1.11	Industrial sand
SODA FELDSPAR												
The Feldspar Corp. Middletown, CT Middlesex County	Pegmatite	Open pit	Flotation plant	90 ktpy Ore	C-6	75	18.5	.07	.90	4.10	7.20	Mica, glass sand
					C-20	600	18.5	.10	.90	4.10	7.20	
The Feldspar Corp. Spruce Pine, NC Mitchell County	Alaskite	Open pit	Flotation plant	160 ktpy Feldspar	NC-4	75	19.0	.067	1.60	4.00	7.00	Mica, industrial sand
					F-20	600	19.0	.075	1.85	3.75	7.15	
					Silo-o-spar[(d)]	90	14.0	.080	1.10	2.80	4.90	
					Felix 100	[(e)]	—	—	—	—	—	
					G-fill	[(e)]	—	—	—	—	—	
K-T Feldspar Corp. Spruce Pine, NC Mitchell County	Alaskite	Open pit	Flotation plant	110 ktpy Feldspar	Minspar 1	850	19.0	.08	1.5	4.1	6.9	Mica, industrial sand
					Minspar 200	75	19.2	.06	1.5	4.1	6.9	
					Minsilspar[(d)]	90	14.0	.08	1.1	2.8	4.9	
					Minspar	[(e)]	—	—	—	—	—	
Unimin Corp. Spruce Pine, NC Mitchell County	Alaskite	Open pit	Flotation	130 ktpy Feldspar	F-1	600	19.3	.08	1.5	4.5	6.7	Mica, high purity quartz
					F-4	90	19.3	.08	1.5	4.5	6.7	
					F-4	75	19.6	.04	1.7	6.9	4.8	
					Unispar 40	[(e)]	—	—	—	—	—	
					Unispar 50	[(e)]	—	—	—	—	—	
POTASH FELDSPAR												
The Feldspar Corp. Monticello, GA Greene & Jasper County	Granite	Open pit	Flotation	45 ktpy Feldspar	G-40	425	18.00	.10	1.02	10.50	2.85	Industrial sand
					G-200	75	18.30	.08	1.02	10.50	2.85	
KMG Minerals, Inc. Kings Mountain, NC Cleveland County	Pegmatite	Open pit	Flotation	27 ktpy Feldspar	K-200	75	18.00	.07	.14	10.10	3.60	Feldspar is a byproduct of mica production; silica sand is a byproduct of feldspar production
Pacer Corp. Custer, SD Custer County	Pegmatite	Open pit selective mining by contractors	Dry grinding	45 ktpy Feldspar	Custer potash feldspar	106	17.00	.15	.30	10.0	3.0	Mica
					Custer potash feldspar	75	17.00	.15	.30	10.0	3.0	
					Custer potash feldspar	45	17.00	.15	.30	10.0	3.0	

(a)Nominal top size, (b) Capacity estimated, (c) Detailed analysis not available, (d) Feldspar silica mixture, (e) Sub micron grind feldspar extender and filler product.

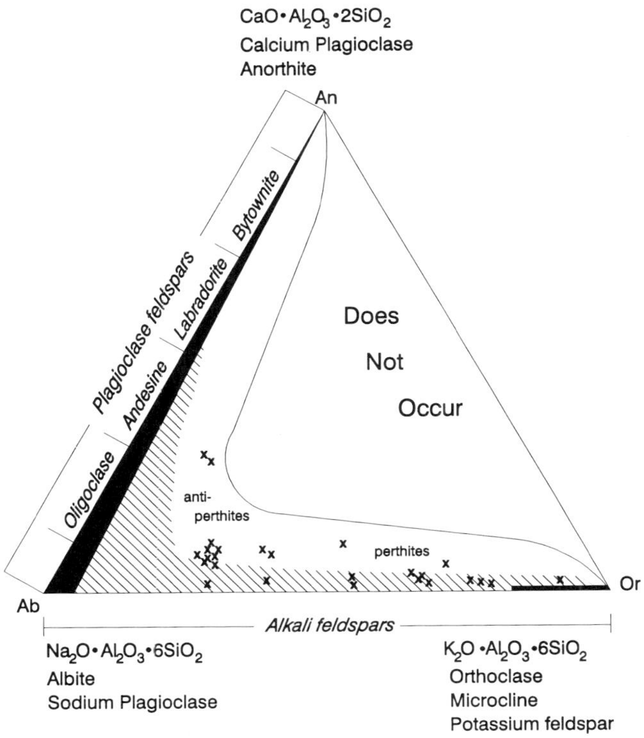

Fig. 1. Diagram representing the range of composition on nomenclature of commercial feldspar products (Smith, 1974, Bates, 1969).

Pegmatites: Granitic pegmatites are widely distributed throughout the crystalline rock areas of the United States. They occur in rocks of many geologic ages and are most abundant near the margins of granitic intrusions. Most of the important pegmatite districts are included in three broad geographic belts: the Appalachian Belt which extends from Maine to Alabama; the Rocky Mountain Belt which includes New Mexico, Colorado, Utah, Montana, Wyoming, Idaho, and in addition, South Dakota and Texas; and the Western Belt which includes Arizona, Nevada, California, Oregon, and Washington.

Most of the pegmatites from which commercial feldspar minerals have been recovered occur in metamorphic host rocks, chiefly in mica schist and gneiss, hornblende schist and gneiss, and quartzite. Some occur in igneous rocks where they are associated with numerous small pegmatites of simple mineralogy and texture. In most districts, however, the pegmatites are spatially associated with intrusions ranging in composition from granite to granodiorite. The distribution of minerals in many pegmatites has been shown to be systematically related to structural features, especially to the pegmatite-wall rock contacts. (Cameron et al., 1949)

Mineral assemblages in many pegmatite districts follow a general sequence of zones described by identity and proportions of essential minerals. These zones from the wall rock contact inward are (after Cameron et al., 1949): 1) plagioclase-quartz-muscovite; 2) plagioclase-quartz; 3) quartz-perthite-plagioclase ± muscovite and/or biotite; 4) perthite-quartz; 5) perthite-quartz-plagioclase-amblygonite-spodumene; 6) plagioclase-quartz-spodumene; 7) quartz-spodumene; 8) lepidolite-plagioclase-quartz; 9) quartz-microcline; 10) microcline-plagioclase-lithium-mica-quartz; and 11) quartz. Few pegmatites contain all eleven zones, but the zones that are present generally occur in this order.

Alaskite: Alaskite is an important feldspar ore-bearing rock that occurs in the Spruce Pine District of North Carolina. It is the major source of feldspar production in the United States. The alaskite has been described (Olson, 1944) as a coarse-grained pegmatitic granite. However, unlike pegmatites, in which irregular mineral distribution is common, the alaskite has a relatively uniform mineral composition.

Aplite: Commercial aplite was described (Brown, 1962) as a high feldspar rock produced near the Piney River area in VA. This particular rock occurrence was identified previously as a pegmatite, a syenite, and Roselond anorthosite. A description of a rock from the same locality (Castle and Gillson, 1960) suggests that the rock originally contained very coarse feldspar, which was subsequently granulated to give it a marked variability in texture. Aplite produced from Hanover County, VA, the current source of commercial aplite production, has been described (Wells, 1969) as a granitic rock of indefinite composition containing a high proportion of plagioclase feldspar. A more current description is the Montpelier Metanorthosite which was named for exposure in the Feldspar Corp. quarry near Montpelier, VA (Brice and Clement, 1982, Clement and Brice, 1982). The metanorthosite intruded the Sabot Gneiss and consists of two phases, a coarse-grained, gray, nonfoliated phase and a granulated, medium-grained foliated phase. The coarse-grained phase contains plagioclase megacrysts up to 25 to 35 cm that constitute 85 to 90% of the total rock. In general, aplite is a textural term referring to a fine-grained, hypabyssal igneous rock in which individual mineral constituents show a sugary texture with undeveloped or poorly developed crystal faces. Aplite commonly implies granitic composition, but aplite can range in composition from gabbro to granite.

Sands: Feldspar is an abundant constituent in modern sands of diverse origins (Pettijohn et al., 1973). These sands most commonly occur as beach sands, sand dunes, and river sands. River sands tend toward more feldspathic compositions than either dune or beach sands. Although relatively few petrographic analyses give a breakdown of feldspar varieties, a survey of those that do suggest that K-feldspar (orthoclase, microcline) is most abundant, followed by sodium plagioclases, and the calcium plagioclases are the least abundant.

Distribution of Major Deposits

United States: The commercially important domestic feldspar production facilities are located in the following areas by geologic occurrence:

Pegmatite. Feldspar production from pegmatites currently takes place in four states: Connecticut, North Carolina, South Dakota, and Virginia.

Pegmatites in the Middletown, CT, area have been mined almost continuously for feldspar and other minerals since about 1865. Pegmatite occurrences in the Middletown area have been recorded in the hundreds. These pegmatites are composed essentially of perthite, quartz, plagioclase, and muscovite. Approximately 42% of the pegmatites contain more than 40% plagioclase. The Middletown pegmatites are largely granitic in composition and form a vast reserve of feldspar, quartz, and scrap mica. The relative abundance of potassium and sodium feldspar-rich pegmatites is not related to the formation in which the pegmatite occurs, but rather shows a regional pattern in which potassium feldspar-rich pegmatites are more numerous to the south (Stugard Jr., 1958).

The active North Carolina pegmatites not associated with the Spruce Pine alaskites are located in the Kings Mountain Belt within the Shelby District covering Cleveland, southwestern Lincoln and western Gaston Counties. The area currently has several producing operations, but the feldspar obtained from this district is a byproduct of the lithium and mica industries (Bundy and Carpenter, III, 1969). At the south end of the district, a coarse-grained phase of the Cherryville Quartz Monzonite is mined for mica feldspar, quartz, and kaolin. The coarse-grained granite is gradational into pegmatite and is also cut by pegmatite dikes. Depending on weathering, the total feldspar content ranges from 13 to 40%, and the potassium feldspar content ranges from 6 to 7% (Horton, Jr., 1981). The active lithium pegmatites are located northwest of Kings Mountain and the Bessemer City area. The majority of the lithium pegmatites contain 20% spodumene, 32% quartz, 27% albite, 14% microline, 6% muscovite, and 1% trace minerals. The pegmatites are generally not zoned, and the fairly uniform grade of the crude ore has stimulated the recovery of feldspar and mica byproducts (Kesler, 1976).

The pegmatites of the Black Hills, South Dakota, occur in two main areas. The Harney Peak area includes the Keystone District in Pennington and Custer Counties, the Hill City District in Pennington County, and the Custer District in Custer County. The second area of occurrence is the Tinton District located in Lawrence County. The greater part of the pegmatite-bearing area is in Custer County; however, the Keystone District has a record of significant mineral production from pegmatites. The most abundant minerals of the South Dakota pegmatites are the plagioclase feldspars (oligoclase and albite), the potash feldspars (perthite and microcline), and quartz. The minerals are uniformly distributed in many pegmatites, but in others they tend to be concentrated in specific structural units. South Dakota pegmatites have been classified on the basis of mineral distribution, texture, and their structural relationships to the wall rocks (Page, 1953, Norton, 1964).

Granite. Two granitic sources in Georgia are currently being produced for feldspar. The Shadydale Granite, produced in Jasper County, has a chemical composition of 13.9% Al_2O_3, 0.9% CaO, 3.6% K_2O, and 4.7% Na_2O. The Siloam Granite, produced in Greene County, is a porphyritic granite containing perthitic microcline—47% microline, 25% plagioclase, 20% quartz, 3% perthite, and 5% biotite. The materials are blended and processed to produce a potassium-grade feldspar (Conker, 1991, private communication).

Alaskite. The Spruce Pine alaskite occurs in three principal areas, referred to as the Spruce Pine, Penland, and Crabtree alaskite bodies, which are located in Mitchell, Avery, and Yancey Counties (Olson, 1944).

The largest mass is between Spruce Pine and Chalk Mountain. The central parts of most large bodies of alaskite are relatively free of country rock inclusions, but many parallel slabs of country rock occur locally near the margins of the bodies (Brobst, 1962).

The alaskite ores in the Spruce Pine District generally have the following mineral and chemical composition: 42.9% soda feldspar (oligoclase), 28% quartz, 14.7% potash feldspar (microcline, generally perthitic), 7.5% muscovite, 6.4% calcium feldspar, and 0.5% iron minerals and garnet, SiO_2, 74.4%; Al_2O_3, 15.4%; Fe_2O_3, 0.4%; CaO, 0.9%; K_2O, 3.4%; Na_2O, 5.1%, and LOI, 0.4% (Redeker, 1981, Brobst, 1962).

Feldspar Silica Mixtures. Feldspar silica mixtures currently make up over 30% of the current feldspar market and have been a significant part of the industry for the past 25 years (Fig. 2). A major part of this production comes from the feldspar-rich silica deposits located in California, Idaho, Illinois, and Oklahoma.

The primary active California deposits are sandstone and deltaic deposits that contain quartz and a range of 10 to 35% contained feldspar (Silva, 1985). The Idaho deposit is a lacustrine sand with an estimated feldspar content of 30%. The Illinois deposit is a dune sand that is estimated to contain 15% feldspar. The Oklahoma operation processes an Arkansas River sand deposit containing 25% feldspar (Bowdish, 1978). In most cases these silica deposits have been identified as extensive.

World: Countries other than the United States that currently produce greater than 100 000 t of feldspar products annually are: Italy, Thailand, France, Brazil, Federal Republic of Germany, Republic of Korea, Mexico, Spain, Turkey, and Venezuela. Japan is also recognized as a major producer of aplite.

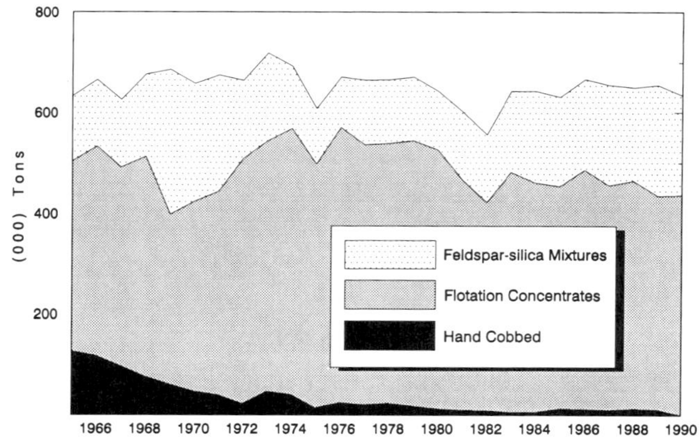

Fig. 2. Feldspar produced in the US by production type (US Bureau of Mines).

Italy, which is currently the world's largest producer of feldspar products, produces feldspar from pegmatitic, clay, sand, and aplite deposits. There are six major producers in Italy and the principal product is a sodium feldspar. Thailand and France are other major world producers of feldspar. Both countries produce potassium and sodium feldspar products. The remaining other major producing countries produce a variety of feldspar products from pegmatites and other geologic resources.

TECHNOLOGY

Mining and Processing

Commercial feldspar mining and processing in the United States involves methods ranging from hard-rock mining to dredging, and a host of mineral processing techniques (see Table 1). It is not possible, within this chapter, to detail all these various mining methods. Therefore, since the Spruce Pine alaskite ores provide and have provided the major domestic feldspar production for some time, a description of a typical alaskite feldspar operation that employs differential froth flotation for the recovery of a pure feldspar product is presented. Fig. 3 is a simplified flowsheet of such a typical feldspar flotation plant. It is intended to acquaint the reader with the concepts involved in feldspar production.

Mining: Mining is done by open pit methods, either by the mine owner or by contractors. The feldspar ore is drilled and blasted using large diameter drill holes and ammonium nitrate. Stripping ratios are nominal, less than 1:1 waste to ore. The pits are usually quite dry. Secondary breakage is with a conventional drop ball. Ore is loaded with a hydraulic shovel and hauled to the crushing plant which is adjacent to the flotation plant.

Milling: Primary crushing is done with a jaw crusher in open circuit. Secondary crushing is with a standard or shorthead gyratory crusher in two or three stages, sometimes in closed circuit with a screen. The final crushed product is –25 mm.

Grinding is normally done with rod mills in order to minimize slime generation, although one company uses ball mills without any apparent ill effects. The grinding mills are normally operated in closed circuit with a classifier in order to maximize grinding ca-

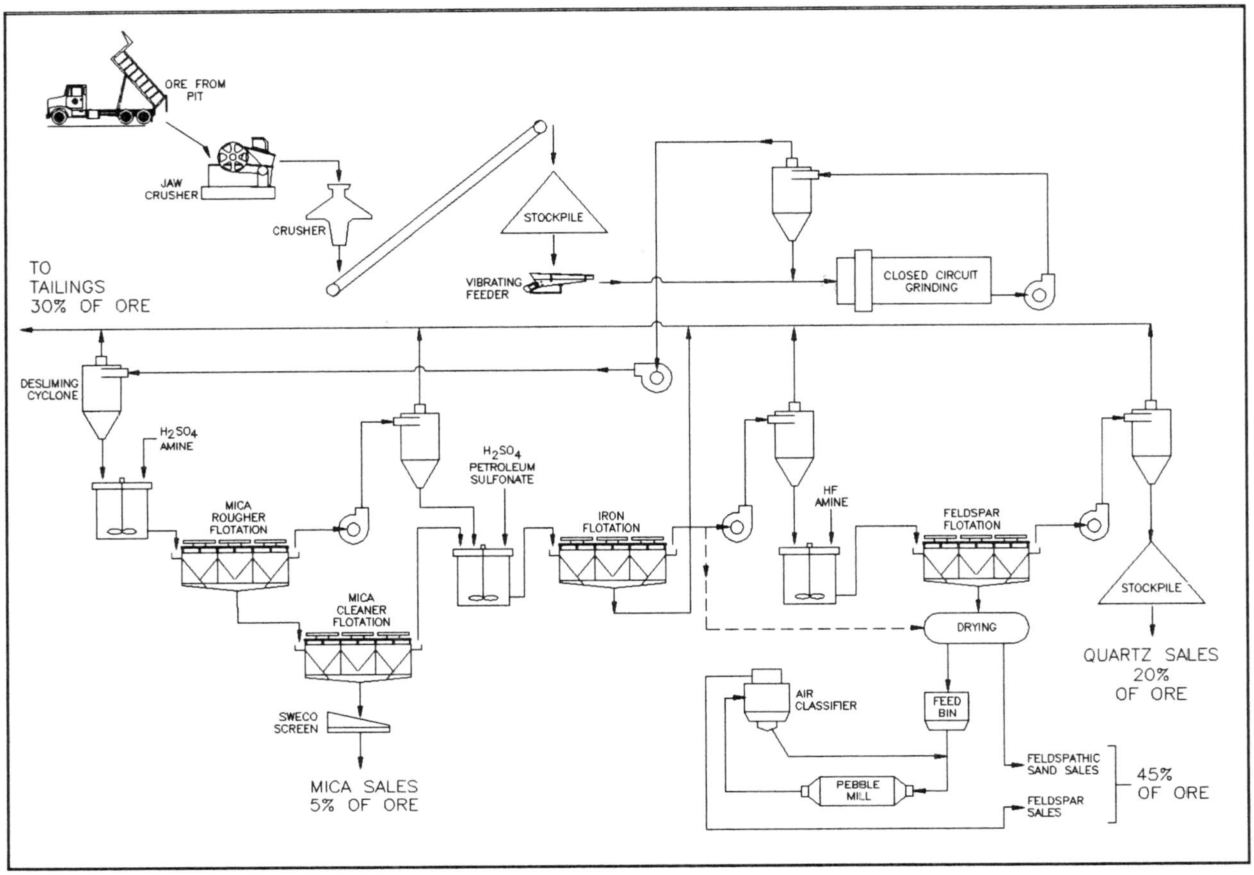

Fig. 3. Typical flowsheet for feldspar production with alaskite ore.

pacity. Although the feldspar mineral grains are liberated at 850 μ or larger, grinding is usually carried out to –600 μ in order to optimize the subsequent flotation steps. To reduce reagent consumption and produce a higher grade product, the feed to flotation must be sized to eliminate the slimes or –38 μ material. The fact that the container glass customers prefer a +75 μ product is also a factor. Desliming is normally done with hydrocyclones.

Once the feldspar is properly sized, the differential flotation process can begin. All the unit operations after crushing are operated in a continuous circuit, 24 hours a day, 5 days per week. The first flotation step is to remove mica which later can be sold as a co-product. Mica flotation is achieved by conditioning with a cationic collector (an amine) at a pH 3 using sulfuric acid. The mica is removed in a froth product and is usually cleaned once prior to being dewatered and sold. Iron-bearing minerals such as garnet and ilmenite are removed next, this time using an anionic collector, again at a pH 3. The anionic collector of choice is a petroleum sulfonate. The iron-bearing minerals are collected in the froth product and pumped to tailings.

The previous unit operations of grinding, sizing, mica flotation, and iron flotation have produced an essentially pure mixture of feldspar and quartz. In recent years it has been recognized that this makes an attractive product for ceramic producers within a certain geographic region. Almost all of the traditional feldspar producers are marketing this mixture under various trade names such as Minsilspar® or Lithospar®. The product is filtered, dried, and cleaned with high intensity magnetic separation or dry ground and sold, usually in bulk.

The production of a high grade feldspar product requires a third flotation step, again at a pH 3. The collector is the same cationic amine as used in mica flotation, but this time the pH is controlled with hydrofluoric acid to depress the quartz. The fluorine ion is a powerful depressant and the quartz-feldspar separation is nearly perfect. A high purity feldspar product is produced with good recovery in a single rougher flotation step. The feldspar containing froth product is filtered, dried, and, if necessary, cleaned with high intensity magnetic separation and stored in bulk prior to sales.

A particular plant will generally produce feldspar with a specific chemical composition. However, it will produce many different products, all with the same chemistry, but different physical specifications. The largest tonnage of these various products is ground feldspar. The dried flotation product is ground in a ceramic-lined ball mill in closed circuit with air classifiers to produce a –74 μ or a –44 μ product. As can be seen in Fig. 4, this step is one of the more expensive unit operations in the whole processing scheme and commands a higher price.

While the feldspar producers strive to sell as high a percentage of the ore as possible, tailings inevitably account for 30% or more of the head feed. Tailings disposal is a substantial portion of the total production costs. The exact method of environmentally sound tailings disposal varies greatly depending on the size and geographical location of the feldspar producer. Fig. 4 shows the relative costs of these various unit operations as compared to the total cost of producing feldspar. The data have been derived by using cost experience for the various unit operations and costs that have been published in cost estimating manuals.

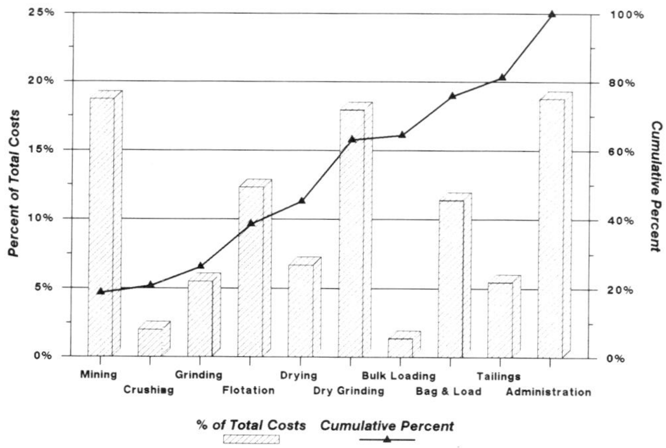

Fig. 4. Distribution of feldspar production costs.

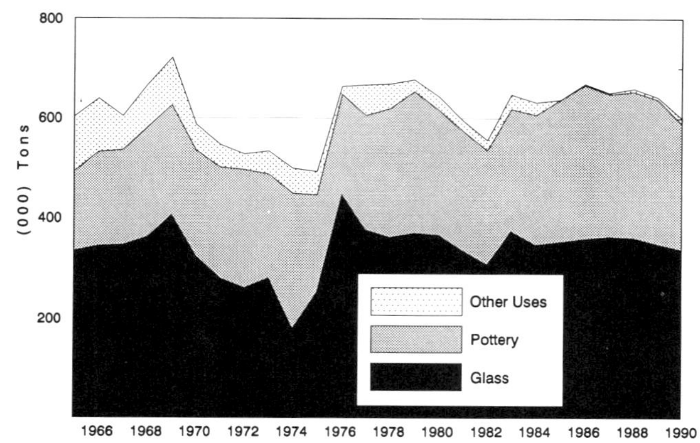

Fig. 5. Feldspar sold in the US by use (US Bureau of Mines).

USES

The US Bureau of Mines defines commercial potash spar as feldspar containing 10% or more K_2O and soda spar as feldspar containing 7% or more Na_2O.

Feldspar for glass manufacturing is usually ground to 850 to 425 µm and contains 4 to 6% K_2O, 5 to 7% Na_2O, and approximately 19% Al_2O_3 and 0.08% Fe_2O_3. Pottery grade feldspar for whiteware and similar ceramic products usually range from 5 to 14% in K_2O and are ground to 75 µm with Fe_2O_3 content in the 0.07% range (Potter, 1990).

Fig. 5 shows the consumption of feldspar by major end-use markets for 1965 to 1990. Glass accounts for the largest consumption of feldspar products and has averaged 353 kt for the years 1985 to 1990; highest consumption, 446 kt, occurred in 1976, and lowest consumption, 180 kt, was in 1974. Ceramic products consumption is the second largest market segment accounting for an average of 289 kt over 1985 to 1990. Ceramic consumption peaked in 1986 at 309 kt; the lowest, in 1965, was at 158 kt.

Other market consumption as reported by the US Bureau of Mines is currently estimated at 5 kt. This market segment has fallen, as reported, from a level of 110 kt in 1965.

Glass

Glass manufacturing is complex and encompasses an enormous range of compositions and products. Materials for glassmaking can be classified in three groups: glass formers, stabilizers, and fluxes. Fluxes are the oxides which react in the glass batch at relatively low temperatures, and stabilizers are those oxides which impart to the glass a high degree of chemical resistance. In conjunction, the fluxes and stabilizers control the working characteristics of the glass-forming operations (Tooley, 1984).

Although alumina (Al_2O_3) does not represent a large part of the composition of most glass, it is important because it increases the resistance of glass to chemical corrosion, improves the hardness and durability, and enhances the working characteristics of the glass (Pincus and Davies, 1983).

Feldspar provides both alkaline oxides (K_2O and Na_2O) for fluxing and alumina (Al_2O_3) and calcium oxide (CaO) as stabilizers. Feldspar is an important source of alumina for glassmaking. The mineral is consistent. It has a low iron and refractory mineral content, a low cost per unit of alumina, no volatile constituents and no waste. The products usually melt between 1 100 to 1 200°C and dissolve readily in the melting glassmaking batch.

As previously mentioned, nepheline syenite and aplite are also used as sources of alumina by the glass industry. Both of these minerals contain mixtures of feldspars, or feldspars and nepheline with alumina contents ranging from 22 to 26%.

The consumption and selection of glass raw materials are influenced significantly by the economics of the glass manufacturing process and the glass product market. Therefore, specifications for raw materials vary based on particular circumstances and economics. However, raw materials for the glass industry require rigid physical and chemical specifications. Table 1 is referenced again to point out the range of commercial feldspar products.

Ceramics

Ceramic and pottery products are the second largest consumer of feldspar products in the United States. The ceramic products generally consist of sanitary ware, ceramic tile, electrical porcelain, dinnerware, and ceramic glazes.

Feldspars are used in the fine ceramic industry as a flux to form a glassy phase in bodies, thus promoting vitrification and translucency. They are also used as a source of alkalies and alumina in glazes. Feldspars also provide one of the few sources of water-insoluble alkali compounds (Norton, 1970).

Consumption of feldspar varies greatly depending on the finished product. Dinnerware and various china products may contain 17 to 20% feldspar; kitchen and oven ware, 10%; vitrified plumbing fixtures, 25%; floor tile, 55 to 60%; wall tile, 0 to 11%; high tension electrical porcelains, 25 to 35%; and other special ceramic products including dental porcelain, which may require 60 to 80% feldspar (Singer and Singer, 1963).

The selection of a potash or a soda feldspar for ceramic applications has been the subject of many investigations (Norton, 1970). The feldspar fluxing differences have also been studied (Weinstein, 1985) regarding deformation and the effect of the type of alkali in feldspars on the glassy phase formed.

Fillers

Feldspar used as a functional filler and extender in the paint, plastic, and rubber industries has evolved in recent times as a non-traditional application providing improved product performance. Feldspar competes with nepheline syenite and other min-

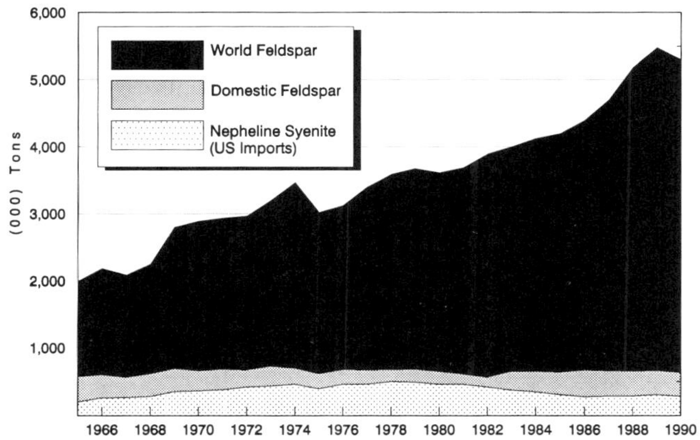

Fig. 6. World and US feldspar production and nepheline syenite imports for US consumption (US Bureau of Mines).

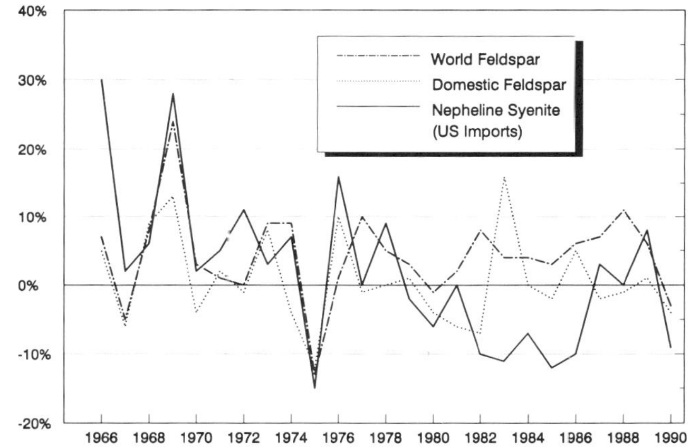

Fig. 7. Percentage annual change of feldspar production and nepheline syenite import data.

erals in these applications and requires increased levels of research and technical marketing efforts.

Feldspar products developed for these markets must comply with the following specifications: dry brightness, oil absorption, Hegman Grind, particle size distribution, surface area, and bulk density.

Feldspar provides the following favorable properties: to the paint and coatings industry, low vehicle demand and low viscosity at high pigment loading, high dry brightness and low tint strength, provides good film integrity, resistance to abrasion, chemical attack and chalking, excellent tint retention, easily dispersed, and feldspar compared to nepheline syenite has exceptional resistance to frosting, but may exhibit slightly greater health hazards due to free crystalline silica. Similar advantageous properties are found in plastic and rubber sealants and adhesives applications (Mommsen, 1988).

ECONOMIC FACTORS

Price

Current feldspar prices range between 45.36 and $61.23/t bulk FOB plant for ceramic grades. High potash ceramic grade is $74.84/t FOB plant. Glass grade ranges between 30.39 and $41.28/t FOB plant and high potash glass grade at $58.74/t FOB plant (Anon., 1992b).

Feldspar pricing, in real terms (constant dollar), over the last twenty-five years has been relatively neutral. Feldspar prices in actual and real terms increased during 1984 and 1985 and dropped in subsequent years in real terms to the 1992 levels. The price of feldspar varies over a considerable range because of the variety of grades marketed. Historical price comparisons are difficult for this reason. Feldspar grades are defined by chemical composition and particle size for different feldspar products. Prices are also affected by the quantities in which the feldspar is sold and packaged (Anon., 1990).

Markets

Domestic and world feldspar production and nepheline syenite imports for domestic consumption are shown in Fig. 6. Fig. 7 also provides an analysis of the annual percent change for the production data.

Domestic Production: Domestic feldspar production has averaged 651 000 t for the 1985 to 1990 period and approximately 10% of this production has been potash feldspar. For the 25-year period presented, feldspar production was greatest in 1973 at 713 kt and lowest during 1967 and 1982 at 558 kt. During 1990 feldspar or feldspar-silica sand was mined and produced in nine states (Table 1). North Carolina has been the dominant producer of feldspar products for the past quarter of a century and accounted for 66% of the production during 1990. Other significant feldspar-producing states include Connecticut, California, and Georgia. It is estimated that these four states have accounted for over 90% of the domestic feldspar production during the last 25 years. North Carolina and Connecticut have maintained a dominant number one and two production position, where California and Georgia have routinely swapped for the third and fourth production spots (US Bureau of Mines Mineral Yearbooks and Annual Reports, 1965 to 1990).

Fig. 2 shows the breakdown of domestic feldspar production by production type and method as classified by the US Bureau of Mines for 1965 to 1990. Flotation concentrates are the largest source of feldspar products and account for 60 to 80% of the annual feldspar production during the period. Feldspar silica mixtures make up 20 to 35% of the production and hand cobbed material has decreased from a 20% share to current estimated levels of 2% of total production.

The feldspar production mix for 1990 is estimated at 65% flotation concentrates, 33% feldspar silica mixtures, and 2% hand cobbed material.

Nepheline Syenite: Nepheline syenite imported for domestic consumption is produced at Blue Mountain, Ontario, by Unimin Corporation. Nepheline syenite, which is covered in a separate chapter, competes with feldspar products in most major market segments. The property at Blue Mountain has an estimated production capacity of 800 ktpy. Domestic nepheline syenite consumption has averaged 282 kt during 1985 to 1990. During 1965 to 1990, 1978 was the peak import year with 497 kt and 1965 was the lowest import year at 197 kt.

World Feldspar Production: World feldspar production for 1965 to 1990 presented in Fig. 6 has increased by a factor in excess of two and one-half times. In 1990 Italy was the largest world producer with an estimated 1.3 Mt, followed by the United States with 630 kt, Thailand with 520 kt, and France with 300 kt. Other countries producing greater than 100 kt of feldspar annually were: Brazil, 120 kt, Federal Republic of Germany, 315 kt, Republic of Korea, 230 kt, Mexico, 130 kt, Spain, 150 kt, Turkey, 100 kt, and

Venezuela, 100 kt. In addition, Japan produced an estimated 517 kt of aplite not previously considered in the production totals.

The Industrial Statistics Yearbook provides some insight on the increased world production of feldspar. Growth has been reported in world production of ceramic household ware, sanitary ware, and floor and wall tiles (Anon., 1986).

The major world feldspar-producing countries as noted here accounted for 78% of the world feldspar production during 1990. During 1965 to 1990, Italy's production increased from 92 kt to the current 1.3 Mt. Thailand's first record of production was in 1973 at 4.5 kt, which has expanded to the current production of 520 kt. Italy and the United States have estimated export volumes of 2 to 4% of production compared to France and Thailand with 40 to 50%.

Feldspar production from Mexico has approximately doubled from the late 1960s to the current production of 130 kt. Feldspar is produced in Mexico by several companies, but the largest producer is Materias Primas Monterrey SA which operates two mines and produces potassium and sodium feldspar products. The majority of Mexican feldspar production is consumed in domestic industries, but Mexico exported approximately 23 kt of feldspar products to the United States during the 1989 to 1990 period. The current US imports of Mexican feldspar represent an increase from the 1985 level which was less than 200 t.

Transportation

Transportation of feldspar products is an important consideration for any feldspar producer or feldspar customer because of the impact of delivered cost. The US Bureau of Mines has reported routinely the US destination of feldspar shipments. Current estimates of feldspar transportation distribution are 67% by rail, with the balance by truck.

GOVERNMENT CONSIDERATIONS

Regulation

Feldspar, due to its crystalline silica (quartz) content, has fallen under the Occupational Safety and Health Administration's (OSHA) Hazard Communication Standards, 29 CFR Section 1900.1200. The standard requires labeling and other forms of warning, material safety data sheets, and employee training for products containing identified carcinogens with concentrations greater than 0.1%.

The crystalline silica issue has created a large controversy between industry and regulators. The industry has formed the Chemical Manufacturers Association CHEMSTAR Crystalline Silica Panel to represent the industry involving crystalline silica issues (Shapiro, 1991).

Trade and Tariff

During July 1991, the Feldspar Corporation filed a petition with the US International Trade Commission alleging that an industry in the United States is materially injured or threatened with material injury by reason of Less Than Fair Value (LTFV) imports of nepheline syenite from Canada. This action has resulted in a countervailing duty margin being set for nepheline syenite on a preliminary basis. Subsequently, on Apr. 27, 1992, the Commission determined that an industry in the United States is not materially injured or threatened with material injury by reason of imports of nepheline syenite from Canada.

PROBLEMS AND FUTURE TRENDS

Plant Closure

The Feldspar Corporation announced the closure of the Middletown, CT, River Road facility during October 1991. The operation was closed due to competitive pressures and declining volumes. The plant was producing between 70 and 80% of its capacity up to 1990. But production dropped to 50% of capacity in early 1991 and suffered further reductions down to 37%. Reopening of the plant is doubtful (Stein, 1991).

Outlook

Demand for feldspar depends primarily on the glass and ceramic industries, which in turn is related to the construction industry. During the 1990 to 1995 period, new construction is expected to remain near current levels, while remodeling and repair construction will increase steadily. Despite the slowdown in the early 1990s, the demand for construction materials should rise at a modest 2 to 3% annually in the 1990s.

Although the construction and motor vehicle production markets will be weaker than in recent years, and may decline further, demand for flat glass and flat glass products should continue to be fairly high in both industries. The glass container industry is expected to grow at an average constant dollar rate of 1% annually during the 1991 to 1995 period.

Ceramic tile consumption should grow at about 3 to 4% a year over the next five years. Plumbing fixtures are expected to resume the pattern of steady growth experienced in the mid to late 1980s as construction recovers (Anon., 1991a).

ACKNOWLEDGMENT

In the organization and writing of the geology and mineralogy portions of this chapter, the assistance of James G. Clark and A. Lamar Long, geologists with Hecla Mining Company, is acknowledged with sincere thanks. I would also like to thank Linda Debo and Alan MacPhee, with Hecla Mining Company, for their assistance in the preparation of the manuscript, table, and figures.

A special acknowledgment is provided to my co-author and friend, Dean F. Van Dyk, who died September 21, 1993. Dean was more than an asset to the industry and will be greatly missed by the industry and his friends.

BIBLIOGRAPHY

Anon., 1986, "Industrial Statistics Yearbook 1984," Vol. II Commodity Production Statistics, 1975–1984, United Nations, New York, pp. 515–530.

Anon., 1990, "The Economics of Feldspar 1990," Roskill Information Services Ltd., London, 175 pp. 62–63.

Anon., 1991a, "U.S. Industrial Outlook," US Department of Commerce, Int. Trade Administration, Washington, DC, Jan., Chap. 1–53.

Anon., 1991b, "Nepheline Syenite From Canada," USITC Pub. 2415, Washington, DC, Aug., pp. 1–33.

Anon., 1992a, "Investigation No. 731-TA-525(F)," USITC, Washington, DC, Apr., 61 pp.

Anon., 1992b, "Prices," *Industrial Minerals*, Jan., No. 292, pp. 62–63.

Bates, R.L., 1969, "Feldspar," *Geology of the Industrial Rocks and Minerals*, Dover Publications, New York, pp. 241–246.

Bowdish, F.W., 1978, "Studies of the Variability of Feldspathic Sands Refined From the Arkansas River," Circular 79, Oklahoma Geological Survey, May, Norman, OK.

Brice, K.L., and Clement, S.C., 1982, "A Study of the Feldspars of The Montpelier Andesine Anorthosite, Hanover Co., VA," Abstract, Geological Society of America, Vol. 14, No. 1–2, 5 pp.

Brobst, D.A., 1962, "Geology of the Spruce Pine District, Avery, Mitchell, and Yancey Counties North Carolina," Geological Survey Bulletin 1122-A, US Department of Interior, 26 pp.

Brown, W.R., 1962, "Mica and Feldspar Deposit of Virginia," Mineral Resource Report 3, Virginia Div. of Mineral Resources, Charlottesville, VA, 195 pp.

Bundy, J.L., and Carpenter, A., III, 1969, "Feldspar Resources of North Carolina," Information Circular 20, North Carolina Department of Conservation and Development, 39 pp.

Cameron, E.N., et al., 1949, "Internal Structure of Granitic Pegmatites," Monograph 2, *Economic Geology*, Urbana, IL, 115 pp.

Castle, J.E., and Gillson, J.L., 1960, "Feldspar, Nepheline Syenite, and Aplite," *Industrial Minerals and Rocks*, 3rd ed., J.L. Gillson, ed., AIME, New York, pp. 339–362.

Clement, S.C., and Brice, K.L., 1982, "Andesine Anorthosite in the Eastern Piedmont of VA," Abstract, Geological Society of America, Vol. 14, 10 pp.

Deer, W.A., Howie, R.A., and Zussman, J., 1966, "Feldspar Group," *An Introduction to the Rock Forming Minerals*, Longman, Essex, UK, pp. 281–338.

Feitler, S.A., 1967, "Feldspar Resources and Marketing in Eastern United States," IC-8310, US Department of Interior, US Bureau of Mines, 41 pp.

Horton, J.W., Jr., 1981, "Field Guide to the Geology of the Kings Mountain Belt Between Gaffney, SC, and Lincolnton, NC," Carolina Geological Society Field Trip Guidebook, Oct. 24–25, pp. 221–222.

Kesler, T.L., 1976, "Occurrence, Development, and Long-Range Outlook of Lithium-Pegmatite Ore in the Carolinas," US Geological Survey, Professional Paper 1005, Jan. 22–24, 162 pp.

Klein, C., and Hurlbut, C.S., Jr., 1985, "Systematic Mineralogy Part IV: Silicates," *Manual of Mineralogy*, 20th ed., John Wiley and Sons, New York, pp. 445–456.

Mommsen, R.W., 1988, "Nepheline Syenite and Feldspar as Functional Filler/Extender Pigments," *8th Industrial Minerals International Congress, Industrial Minerals Div., Metals Bulletin*, Apr., pp. 270–277.

Norton, J.J., 1964, "Geology and Mineral Deposits of Some Pegmatites in the Southern Black Hills, South Dakota," US Geological Survey, 297-E, 48 pp.

Norton, F.H., 1970, *Fine Ceramics Technology and Applications*, R.E. Krieger, Malabar, FL, pp. 236–474.

Olson, J.C., 1944, "Economic Geology of the Spruce Pine Pegmatite District, North Carolina," Bulletin No. 43, Part I, US Geological Survey, US Department of Interior, 67 pp.

Page, L.R., 1953, "Pegmatite Investigations, 1942–1945, Black Hills, South Dakota," Geological Survey, Professional Paper 247, 228 pp.

Pettijohn, F.J., Potter, P.E., and Siever, R., 1973, *Sand and Sandstone*, Springer-Verlag, New York, 618 pp.

Pincus, A.G., and Davies, D.H., 1983, *Raw Materials in the Glass Industry*, Part 1, Ashlee Publishing Co., Inc., New York, 249 pp.

Potter, M.J., 1990, "Feldspar, Nepheline Syenite, and Aplite," *Annual Report*, Oct., US Department of Interior, US Bureau of Mines, 8 pp.

Redeker, I.H., 1981, "Flotation of Feldspar, Spodumene Quartz and Mica From Pegmatites in North Carolina, USA," 81–1-p, Jan., Minerals Research Laboratory, North Carolina State University, pp. 1–7.

Rogers, C.P., Jr., Neal, P.J., and Teague, K.H., 1983, "Feldspars," *Industrial Minerals and Rocks*, 5th ed., Vol. 1, S.J. Lefond, ed., AIME, New York, pp. 709–722.

Shapiro, J.C., 1991, "Chemical Manufactures Association CHEMSTAR Crystalline Silica Panel," *Proceedings*, The American Ceramic Society, Vol. 12, No. 1–2, Jan.-Feb., pp. 146–151.

Silva, M.A., 1985, "Mineral Commodity Report Silica," Special Publication 89, California Department of Conservation, Division of Mines and Geology, Sacramento, CA, 15 pp.

Singer, F., and Singer, S.J., 1963, *Industrial Ceramics*, Oxford and IBH, New Delhi, 1431 pp.

Smith, J.V., and Brown, W.L., 1988, *Feldspar Minerals*, Springer-Verlag, Berlin-Heidelberg 809 pp.

Smith, J.V., 1974, "Crystal Structures, Physical, Chemical, and Microtextural Properties," *Feldspar Minerals*, Vol. 1, Springer-Verlag, Berlin-Heidelberg, Germany, pp. 1–113.

Stein, T., 1991, "Feldspar Will End Mining Operations," *Middletown Press*, Oct., pp. 1, 6.

Stugard, F., Jr., 1958, "Pegmatites of the Middletown Area Connecticut," Geological Survey Bulletin 1042-Q, US Department of Interior, US Geological Survey, 70 pp.

Tooley, F.V., 1984, *The Handbook of Glass Manufacturers*, Vol. 1, Ashlee Publishing Co., Inc., New York, 580 pp.

Weinstein, J., 1985, "Feldspar Fluxing Differences," *Ceramic Monthly*, Sep., pp. 58–59.

Wells, R.J., 1969, "Feldspar," *Minerals Yearbook*, Vol. I-II, US Department of Interior, US Bureau of Mines, pp. 487–493.

FILLERS AND COATINGS

NIKHIL C. TRIVEDI AND ROBERT W. HAGEMEYER

Despite the apparently unflattering term filler, this group of industrial materials has become an essential component of many industrial products with varying degrees of sophistication. Mineral fillers are, by definition, inert materials included in a composition for some useful purpose. Included in this broad category are asbestos, barite, bentonite, natural calcium carbonates (limestone), clays, diatomite, feldspar, micas, silica, talc, and other industrial minerals. Also included in this category are synthetic mineral-derived materials such as aluminum hydrate, precipitated calcium carbonate, and pyrogenic and precipitated silicas and silicates.

Mineral fillers are included in compounds to accomplish a variety of final results. Inclusion of fillers modifies physical characteristics such as hardness, brittleness, impact strength, compressive strength, surface texture, fire resistance, softening point, and electrical conductivity. Depending upon the performance characteristics of the particular filler used and the quantity of the usage (i.e., percentage added), these materials can enhance the physical and optical properties, price-performance, and appearance of the finished products.

These enhancements in properties of the finished compounds are the result of properties of the fillers including chemical activity, purity, hardness, particle size, particle shape, particle size distribution, surface structure, color, density, and refractive index.

This chapter will review the general usage of fillers, the rationale for their use, trends, and economic factors influencing those trends. Major using industries will be examined in detail with extensive discussion of major fillers used in those industries. The reader is referred to the appropriate commodity chapter for specific information on deposits and processing for industrial minerals. All of these same filler materials have additional significant uses based on their chemical composition or other characteristics. Cross-references for these will be found in other use chapters and summarized in commodity chapters.

The chart in Table 1, updated from the 5th edition of *Industrial Minerals and Rocks*, illustrates the major mineral fillers (both natural and synthetic) and their more important end uses.

USE OF MINERAL FILLERS

Since ancient times man has added mineral products to compounds to improve certain performance characteristics. More than 5000 years ago, bitumin used for mortar and waterproofing contained mineral contaminants that acted as fillers. The most prevailing argument for the use of fillers has been and continues to be reduction of cost of the end product. The inclusion of a filler generally costing but a fraction of the cost of an expensive pigment, polymeric resin, pulp, or rubber will reduce the unit cost of the final product. The choice of filler, however, is not simple.

New scientific discipline is emerging because of the balancing act often required when fillers are used; the desired modification is most often achieved with some sacrifice of other qualities. Some polymeric compounds may become more brittle when cost reducing fillers are added. Some paints may show an accelerated settling, resulting in shorter shelf life, when anti-chalking fillers are added. Each using industry has found, often in collaboration with the producers of fillers, an acceptable solution to problems similar to these, usually through optimizing the degree of modification that is acceptable. Frequently, combinations of fillers are utilized to achieve a number of desirable modifications. Thus, a quality paint may contain kaolin, calcium carbonate, talc, and perhaps barite, each included to help achieve a specific favorable effect.

Depending on the sophistication of the using industry, fillers are incorporated to modify one or more of the following properties:

1. Cost
2. Physical characteristics
3. Optical properties, e.g., color, brightness, appearance, aesthetics
4. Flow characteristics, rheology
5. Fire resistance, chemical resistance
6. Density
7. Thermal properties, heat conductivity
8. Hardness, brittleness, impact strength
9. Deformity, viscosity, softening point
10. Electrical conductivity
11. Surface texture, smoothness, roughness
12. Processing problems
13. Thermal expansion characteristics
14. Abrasion resistance

The following list exemplifies the size and scope of the industrial mineral filler usage. The consuming industries and the fillers are listed here in alphabetical order.

Consuming Industries	Fillers
Adhesives	Asbestos
Asphalt	Barite
Bitumins	Bentonite
Carriers	Calcium Carbonate
Cosmetics	Clay
Drilling Mud	Diatomite
Fire Retardants	Feldspar
Insulation	Fuller's Earth
Paint	Gypsum
Paper	Kaolin
Plaster	Kyanite
Polymers	Mica
Printing Ink	Olivine
Putty and Caulks	Perlite
Rubber	Pumice
Sealants	Rock Dust
Soil	Silica
	Slate
	Talc
	Vermiculite
	Wollastonite

Table 1. The More Important Mineral Fillers

	Major Filler Uses	Distinguishing Properties	Remarks
Asbestos[a] (Chrysotile)	Floor tile, plastics, "black line" products, asb-cement composition, building and insulating compounds.	The most fibrous mineral filler. Uses based on reinforcing and viscosity control effects.	Consumption rapidly declining, possibly related to carcinogenicity concerns. Filler uses take mostly "fines" and "shorts" (7 group), but some 8, 6, and 5 group uses are akin to fillers.
Barite	Rubber, paint, and urethanes.	The highest density mineral filler; also inert.	Ground barite used for filler purpose. Consumption declining.
Bentonite	Pesticides, pellets, asphalt, detergents.	Not important as filler, except as insecticide dust diluent.	Wyoming grades growing as binder for iron ore pellets.
Diatomite	Paints, paper, plastics, pesticides, cementitious mixtures and most filler uses.	An inert or reactive opaline silica depending on use, unique microscopic structure.	Natural or calcined, milled or processed grades are used as filler. Much difference in grades.
Feldspar and nepheline syenite	Paints, plastics, foam rubber.	Translucency in plastics. Tint retention in paint. Chemical stability.	Major use is glass and ceramics.
Fuller's earth	Pesticides principally, but some in greases and paper.	Important as pesticide dust diluent; absorptive capacity relatively good; price low.	Primary uses involve its absorbent character.
Gypsum	Paints, paper, cotton goods, pesticides.	Low cost. A minor filler material used raw or calcined.	60% as calcined, 40% as uncalcined.
Kaolin	Paper, paint, adhesives, oil cloth, plastics, rubber, ink, pesticides.	Good white color, low cost, low absorption, abundant.	Paper industry consumes 70% of total production.
Limestone[b]	Asphalt, fertilizers, insecticides, paints, paper, rubber, plastics, and many others.	Abundant and cheap, particles soft, soluble in acids; has largest tonnage use of all fillers (if dolomite is included).	Large tonnages of low grade limestone and dolomite used in bitumens, fertilizers, and pesticides; quality grades employed in paints, paper, rubber, and other filler uses.
Mica (muscovite)[c]	Dry ground-roofing, joint cement and paint. Wet ground-paint, rubber, plastics, and wallpaper.	Micaeous structure unique, valuable for electrical properties, useful in specialties.	Substantial volume is byproduct of lithium and feldspar production.
Perlite	Fines used in paint, loose fill, drilling muds, plastics, and misc.	Lightweight vesicular structure. Filler uses as yet not important.	
Portland cement	Asphaltic concrete and pavements and other compositions.	Cheap and abundant; may have some bonding action.	Filler uses of portland cement greatly secondary to its uses in construction.
Pumicite	Stuccos, plasters and paint.	Natural vesicular structure, harsh. Filler uses minor.	Main uses as abrasive, concrete admixture, loose fill, and lightweight aggregate.
Pyrophyllite	Insecticides, asphalt roofing, paint, rubber, battery boxes.	Like talc, but more refractory.	Volume and usage usually reported with talc.
Rock dusts	Asphaltic compounds and cheap filler uses.	Low cost; chemical stability, strength.	Usually low-cost ground carbonates, but dust collector fines also used.
Silicas, tripoli	Paints, bitumens, foundry facings, etc.	Hardness, inertness, strength, low cost; low soluble salts.	Silica used in various related filter fields. Differentiation between amorphous and crystalline is not clear.
Slate	Roofings, sealing compounds, asphalt linoleum, paints, hard rubber.	Cheap, abundant, inert and compatible with bitumens.	
Talc	Paints, insecticides, roofing, rubber, asphalt, paper, cosmetics, textiles.	Soft platey structure, good adhesion.	Talc has important ceramic, foundry, and miscellaneous uses.
Vermiculite	Fertilizers, pesticides. Potting soils.	Filler grades are fines from expanded aggregate; filler uses not large.	

[a]Minor tonnage of anthophyllite and amphibole asbestos are used as fillers.
[b]Includes chemically precipitated calcium carbonates which are also of considerable importance as fillers.
[c]Phlogopite, biotite, and indeterminate micas are also used as fillers.

Table 2. Usage of Mineral Fillers by Industry (1990)

	Tons thousands	Dollars millions	$/t
Major Industry Segment			
Paper	5 500	660	120
Plastics	1 350	305	226
Sealants	1 125	61	54
Paint	1 000	230	230
Rubber	650	162	249
Adhesives	175	19	109
Subtotal	9 800	1 437	147
Specialty Industry Segment			
Drilling Mud	925	36	39
Carpet Backing	650	20	31
Foundry Sand	650	20	32
Ore Pelletizing	500	16	32
Subtotal	2 725	93	34
Grand Total	12 525	1 530	122

Accurate total statistics for the actual volume and dollar value of total fillers produced/used in the United States are unavailable. This is perhaps resulting from the enormously wide scope of the types of fillers and sizes of the consuming industries listed above. However, one source estimates that, in 1990, there were approximately 12.5 Mt of fillers produced or used in the United States, with an estimated value of $1.5 billion. Thus, the average price of fillers was $122/ton.

Some estimated statistics on 1990 usage of fillers in major consuming industries are given in Table 2. The paper industry is by far the most dominant user of functional fillers both in terms of volume and value. On a per ton basis, the rubber, polymers, and paint industries consumed the higher priced functional fillers. The trend in the industry is toward enhanced performance, adding value and improving functionality of fillers. The major consuming industries are technically sophisticated; research and development are pursued by the suppliers and the users, with the added emphasis on continuous improvement in cost performance ratio.

PROPERTIES OF MINERAL FILLERS

The performance of fillers in use is a result of both the inherent characteristics of the mineral itself and the characteristics imparted by processing, e.g., beneficiation, grinding, classification, calcination, slurrying, surface treatment.

Hardness affects the utilization of a filler as well as its production cost. Many silica minerals or silicates may score rolls, spreader blades, or knives, forcing the choice of an inherently softer material such as talc or calcium carbonate as alternatives. Conversely, the adverse effect of relatively high hardness may be mitigated by using fillers with finer particle sizes. High degree of hardness can often be a desirable property as it adds abrasion resistance to rubber and scrubbability to paints.

Particle size, as it decreases, improves packing, tends to fill smaller voids, imparts smoothness to surfaces, and increases viscosity. Particle size effects result from the average size, the top size, and the size distribution. In a paint system, for example, the average size of a filler may affect viscosity and binder demand; the size distribution may affect packing and resultant density and film integrity; and the top size may affect paint gloss and smoothness and cause fracture failures or cracking. Methods for reduction of particle size and manipulation of particle size distributions are commonly known techniques, such as grinding and classification, as with talc and limestones; exfoliation, as with vermiculite; and delamination, as with mica and kaolin.

Particle shapes strongly affect utilization. A prime example in recent years is the increasing use of precipitated calcium carbonates of custom designed particle shapes to achieve specific functional enhancements in the papermaking process. Similarly, acicular asbestos particles used to be most actively sought for reinforcing. Plates of mica or kaolin provide a tough, leaf-like surface film.

Color of the filler is important in any application where color of the end product is important. Obvious examples of this are the paper and paint industries. Most fillers become more acceptable as they become whiter; a distinctive blue, green, or yellow tint may result in an unacceptable gray end product. The most common techniques of process-modifying color are: 1) beneficiation, e.g., flotation, electromagnetic, electrostatic, and optical separation, to physically remove color contaminants, 2) grinding—the finer a filler is ground, the whiter it appears, or 3) classification—when objectionable coloring contaminants may concentrate in certain size ranges which can be economically removed by screening or classification. Depending on the importance of color to the consuming industry, filler producers can use techniques varying in sophistication from selective quarrying/mining and hand sorting to chemical bleaching as is used in some processing of barite, talc, and kaolin and calcination as is used for improving color of kaolin, diatomite, and talc.

Refractive index is important in some applications. If the refractive index of the filler is very nearly that of the matrix of the substance it is filling, the filler will not show as much, giving a uniform appearance—in other words, showing translucency. Prime examples of this are talc in vinyl floor tiles and nepheline syenite in many resins.

Chemical properties also affect the utilization of fillers. Because of its tendency to react with acid, calcium carbonate was not a filler of choice over kaolin in acid papermaking processes. Low water solubility is a necessity for most all fillers. Sophisticated technologies such as particle encapsulation and surface treatment tend to improve upon filler deficiencies resulting from adverse chemical properties. Because these technologies are costly, however, they are practiced only when such treatments offer other significant enhancements in desired properties, e.g., lack of agglomeration, ease of dispersion, ease of extrusion, or improved rheology.

Some properties of the important major mineral fillers can be found in Table 3 (Cummins, 1960).

APPLICATIONS

Paper

In North America, the paper industry is by far the largest consumer of filler and coating mineral products. In 1990, total volume amounted to 5.5 Mt with a value of $660 million. It is estimated that 3.6 Mt (65%) were used in coating applications, while 1.9 Mt (35%) found use as filler. For several years, growth in the coating volume has been in the range of 5% yearly, and it is expected to continue at that for an extended period. The outlook for consumption in filler applications is much more bullish as the rapid conversion from acid to alkaline papermaking has created an opportunity to incorporate more filler in the sheet without any loss in strength. The near-term outlook is an increase in filler demand of about 10% yearly, with a gradual slowing to the 5% base rate as filler loading levels are optimized.

The major reasons for coating paper are to improve its printability and appearance. Kaolin and calcium carbonate are the main

Table 3. Some Properties of the Major Mineral Fillers

	Theoretical Chemical Composition	Specific Gravity	Bulk Density kg/m³*	Hardness Mohs' Scale	Refractive Index	Reaction, pH	Oil Absorption Cc per 100 g	Particle Characteristics
Asbestos (chrysotile)	$3MgO{\cdot}2SiO_2{\cdot}2H_2O$	2.5–2.6	160–640	2.5–4.0	1.51–1.55	8.5–10.3	40–90	Fibers fine, easily separable. Fibrils hexagonally close packed 100–300Å.
Barite	$BaSO_4$	4.3–4.6	1280–2400	2.5–3.5	1.64	7	6–10	Generally equi-dimensional.
Bentonite	$(Mg,Ca)O{\cdot}Al_2O_3$ $5SiO_2{\cdot}xH_2O$	2.3–2.8	800–960	1.5+	1.55–1.56	6.2–9.0	20–30	Porous microaggregates, irregular shapes; ultimate plate structure.
Diatomite	$SiO_2{\cdot}xH_2O$	2.0–2.35	96–320	4.5–6.0	1.42–1.49	6–8.5	100–300	Unique diatom structure; micro and ultramicro porosity.
Fuller's earth	$(Mg,Ca)O{\cdot}Al_2O_3$ $5SiO_2{\cdot}xH_2O$	2.2–2.6	432–608	4	1.50	7.5–8.2	30	Apparently equi-dimensional. Electron-microscopically fibrous, lath-like.
Gypsum	$CaSO_4{\cdot}2H_2O$	2.3	400–640	1.5–2.0	1.52	6.5–7	17–25	Irregular, roughly equi-dimensional.
Kaolin	$Al_2SO_3{\cdot}2SiO_2{\cdot}2H_2O$	2.6	320–640	2.0–2.5	1.56–1.58	4.5–7	25–50	Thin, flat hexagonal plates, 0.05–2μ size and stacks of same.
Limestone	$CaCO_3$	2.7	640–1600	3	1.63–1.66	7.8–8.5	6–30	Variable size particles; ultimate rhombs.
Mica (muscovite)	$H_2KAl_3(SiO_4)_3$	2.7–3.0	192–320	2.0–3.0	1.59±	7.4–9.4	25–50	Platelike particles.
Nepheline syenite	$K_2O{\cdot}7d\text{-}_2O{\cdot}4.5Al_2O_3{\cdot}$ $20SiO_2$	2.61	800–1280	5.5–6.0	1.53	9.9	21–29	Nodular and irregular.
Perlite	Like rhyolite	2.5–2.6	64–320	5.0	1.48–1.49	9	50–275	Expanded "glass" bubbles and fragments.
Portland cement	Essentially Carbon silicates and aluminates	2.9–3.15	1440–1600	5.6	17.2±	11.0–12.6	20	Variable, smooth, rounded, angular, and flake particles.
Pumicite	A silicate, like rhylolite	2.2–2.63	640–800	5–6	1.49–1.50	7–9	30–40	Vesicular.
Pyrophyllite	$Al_2O_3{\cdot}4SiO_2{\cdot}H_2O$	2.8–2.9	400–480	1–2	1.57–1.59	6–8	40–70	Minute foliated plates or scales and extra-long particles.
Rock dusts	Variable	2.6–3.3	800–1600	4–6.5	Variable	Usually above 7	20–40	Variable.
Silicas, crystalline and microcrystalline	SiO_2	2.60–2.65	800–1280	6.5–7.0	1.53–1.54	6–7	20–50	Variable sized, angular and equi-dimensional particles, or minute particles to porous masses.
Slate	Mixture of mineral silicates	2.7–2.8	640–1280	4–6	—	6.8	20–25	Flat or wedge-shaped, or spherical grains.
Talc	$H_2Mg_3(SiO_3)_4$	2.6–3.0	416–960	1–1.5	1.57–1.59	8.1–9.0	20–50	Lamellar, foliated, or microfibrous.
Vermiculite	$(Mg,Fe)_3(Si,Al)_4$ $O_{10}(OH)_2{\cdot}4H_2O$	2.2–2.7	96–160; fines-320	1.5	1.56	Pract. neutral	—	Platelets or lamellar structure.
Wollastonite	$CaSiO_3$	2.8–3.0	320–640	4.5–5.0	1.63	9.9	25–30	Billiant white powder with acicular nature.

Modified from Cummins, 1960.
*Metric equivalent: 1 lb per ft³ x 16.018 46 = kg/m³.

constituents of the coatings, with the binders and various additives normally comprising less than 15% of the formulation. The coating provides a more uniform and receptive surface than the uncoated basestock and, in turn, both facilitates and enhances graphic reproduction. The print improvement is readily apparent, especially in image areas or when multiple colors are used.

There are several different generic species of coating minerals and numerous variations within each category. The performance of a specific material is dependent upon its inherent properties and the way in which it is used. Factors such as chemical composition, particle shape, particle size, surface area, surface charge, brightness, refractive index, specific gravity, reactivity, and absorptivity must be carefully considered in selecting a material. Comprehensive data on the physical and chemical characteristics of the commonly used materials are presented elsewhere in this chapter.

While coating of paper has been practiced for more than a century, the use of industrial minerals as fillers to enhance paper properties is a relatively recent development (circa 1940). Filler was originally added to the fiber furnish as a cost-saving diluent, with the quantity limited by the commensurate loss in strength properties. Now this use has become secondary to the selection and use of designed fillers to achieve specific improvements in paper properties. Depending upon the particular filler, its performance characteristics, and the amount used, the physical, optical, and aesthetic properties of the paper can be enhanced to a varying degree. Cost-effectiveness rather than nominal price has become the significant factor in filler selection.

Fillers are added to paper in various percentages to perform a number of different functions. The choice of material and the amount used are dependent upon the specific properties desired. The addition of fillers can contribute the following properties.

1. Better formation and sheet structure through improved fiber distribution and the fill-in of void areas.
2. Improved texture and feel resulting from a smooth surface and more uniform composition.
3. Increased opacity from both the filler per se and its impact on fiber debonding.
4. Better printability for a number of reasons related to ink transfer and setting.
5. Cost savings proportional to fiber replacement.

Kaolin: Kaolin in its various forms dominates the filler and coating materials used by the paper industry. However, it should be noted that its market share has been declining as the conversion to alkaline papermaking has permitted the increased use of calcium carbonate, especially in the wet-end of the paper machine. During the past decade, kaolin's market share has dropped from 90 to 80%.

The kaolin used by the paper industry falls into four categories: premium brightness coating, standard brightness coating, fillers, and structured pigments. Physical properties and performance parameters for representative grades in each category appear in the following table.

The premium high brightness grades are used in #1 and #2 coated papers where their high brightness, blue-white shade, and superior glossing properties are critical to clean, sharp prints with proper color tone (Table 4). High coating solids, heavy coat weights and multiple blade coating applications have become standard for these grades.

Numbers 3 and 4 coated papers are less critical in nature, but still require a good printing surface. Standard brightness clays are used in their coating formulations (Table 5). Again, blade coaters are used, but the coat weight and solids levels are often lower and single applications are common.

Number 5 coated paper is the least critical paper in terms of brightness, but is much more critical in its requirements for coverage, smoothness, and printing properties. Because this is the dominant portion of the North American coated paper market, a vast amount of research continues to go into coating development for this grade. Number 5 coated papers can be divided into two segments, Standard #5, and lightweight coated (LWC) papers; that is, sheets with a finished basis weight below 18 kg per ream. Since

Table 4. Coating Clays—Premium Brightness

	#2 High Brightness Clay	#1 High Brightness Clay	#1 Fine High Brightness Clay
TAPPI Brightness, %	89.5–90.5	89.5–91.0	90.0–92.0
Particle Size, %<2μm	80–84	90–92	96–100
Surface Area, m²/g	12	13	22
325 Mesh Screen Residue, Max., %	.01	.01	.01
Moisture, Max., % as Produced	1.0	1.0	1.0
pH (28% Solids)	6.0–7.5	6.0–7.5	6.0–7.5
Brookfield Viscosity, 20 rpm, cps	350	350	350
Solids, %	70.0	70.0	70.0

Table 5. Coating Clays—Standard Brightness

	#2 Clay	#1 Clay	#1 Fine Clay	Delaminated Clay
TAPPI Brightness, %	85.5–86.5	86.5–87.5	86.5–88.0	87.5–89.0
Particle Size, %<2μm	80–84	90–92	94–97	n/a
Surface Area, m²/g	12	13	22	14
325 Mesh Screen Residue, Max., %	.01	.01	.01	.01
Moisture, Max., % as Produced	1.0	1.0	1.0	1.0
pH (28% Solids)	6.0–7.5	6.0–7.5	6.0–7.5	6.0–7.5
Brookfield Viscosity, 20 rpm, cps	350	350	350	350
Solids, %	70.0	70.0	70.0	67–68

LWC is the grade used for magazines and catalogs, lower basis weight paper—which translates into more pages for the raw material and the mailing dollar—is the obvious choice. Such papers have base sheets which contain as much as 80% mechanical pulp, run at speeds in excess of 1 219 m per minute, and are often coated on-machine. Coat weights for lightweight papers are low, and there are ongoing attempts to reduce them further. Number 5 paper is coated with blade coaters, but in the last five years, the short dwell time applicator (SDTA) has become as important to the coating process as the blade itself. Due to the low coat weights, the pigments are required to provide good fiber coverage, good smoothness, good printing quality, and good application properties, all at a low cost.

Perhaps the property of a clay most critical to coating performance is rheology. This property is the result of many variables, including surface area, particle size and shape, contaminants and soluble salts, as well as quality of dispersion. Ideally, a clay would have Newtonian flow properties, that is, it would have the same viscosity at different rates of shear. Unfortunately, all clays are dilatant, which means that the viscosity of the slurries increases as the shear rate increases. Frequently, paper mills will run delaminated clay in combination with other clays to help alleviate the severe rheology of the delaminated clay, or to augment a particular property which might be lacking with delaminated clay alone.

In some instances, the choice of clay grade may be determined by the basestock characteristics as much as by the end use properties of the paper. It is not unusual to have paper mills producing the same grades using substantially different pigment combinations.

In general, a number of guidelines can be drawn from an in-depth study of kaolin clays in paper coatings. Some of these are:

Table 6. Filler Clays

	Air Float Filler	Fine Filler	Fine HB Filler	Delaminated Filler
TAPPI Brightness, %	80–81	81–83	84.5–85.5	86–88
Particle Size, %<2μm	62–70	82–95	82–95	n/a
Surface Area, m^2/g	15	22	22	13
325 Mesh Screen Residue, Max. %	.4	.3	.3	.01
Moisture, Max., % as Produced	2.0	slurry	slurry	1.0
pH (28% Solids)	5.0–7.5	6.0–8.0	6.0–8.0	6.0–7.5

1. As particle size becomes finer
   a) the sheet gloss increases,
   b) the difference between sheet gloss and print gloss diminishes,
   c) the binder demand of a pigment increases,
   d) opacity of the coating film decreases, and
   e) the low shear viscosity of the coating color improves.
2. As solids increase, coatings containing delaminated clays yield higher high shear viscosity than those containing non-delaminated clays.

Whereas there is basically one reason to apply coating, there are many reasons to use pigments to fill paper. Probably the most important reason is to improve the optical properties of the paper, either in brightness, opacity, or both. A second reason for filling paper is the potential cost advantage. If an inexpensive pigment can be used in place of the more expensive fiber while maintaining and/or improving the quality of the paper, the result is very beneficial for the mill. Also, pigments dewater much more quickly than paper fiber. Therefore, a filled sheet will drain more easily and require less steam in the drier section than an unfilled sheet. Some fillers are able to impart a particular property to the sheet or to the system which is otherwise not possible. Examples of this are bulking and pitch control.

Whereas the results of coating are positive, there is a trade-off when a sheet is filled. As the filler level increases, sheet strength properties decrease. This is not a linear relationship, as the drop in some of the properties can be dramatic with only a moderate change in filler level. One area of strength degradation is internal bond. The pigment particles deposit on the surface of the fibers blocking a portion of the contact area between the fibers, thus decreasing the hydrogen bonding of the cellulose. This phenomenon, referred to as debonding, is credited by many paper scientists for the increase in opacity which had previously been attributed to the addition of the mineral filler alone. Internal bond, however, is not the only area of concern. The surface strength of the paper is also reduced. If reduced too far, problems can occur on the printing press due to fiber picking, linting, and dusting. Forming paper under alkaline rather than acid conditions results in stronger fiber-to-fiber bonding, thus permitting higher filler levels.

The physical properties for the standard kaolin fillers appear in Table 6.

In the manufacture of Air Float, the crude is crushed, dried, pulverized, and air classified prior to shipment in dry form. It is the largest particle size paper clay and is only used in filling applications. East Georgia crudes, having an inherent fine particle size, are used to make the Fine and Fine HB filler grades. No drying or grinding is required so these grades in slurry form have prices comparable to Air Float with a small upcharge for the high brightness. The fine fillers have largely displaced Air Float except for geographic areas wherein the additional freight costs involved in slurry shipments make the purchase of the dry Air Float grade advantageous. The delaminated filler is a water-washed clay, processed in much the same manner as a coating clay. It has superior brightness and is more neutral in shade than the other fillers: as a result, it contributes better optical properties when added at filler levels comparable to the other grades.

Calcined Clay: As the name implies, calcined clay is produced by the thermal treatment of clay. Two grades are available: a dehydroxylated, or partially calcined clay, and a fully calcined clay. The former is produced by thermally treating clay at 650 to 700°C, thus driving off the structural hydroxyl groups. The resulting product has a specific gravity of 2.4 to 2.5. Full calcination requires a temperature of 1 000 to 1 050°C; it results in a product with a specific gravity of 2.7. The fully calcined clays are slightly brighter but slightly more abrasive than the partially calcined clays. These clays can be used as extenders for titanium dioxide.

The high viscosity of partially calcined clays limits the percentage that can be used in high-solids coating mixtures and in coatings to be applied at high speed. These clays are normally used at levels of less than 30% of the total pigment. The addition of partially calcined clay to a coating formulation provides increased bulk, increased opacity, reduction in brightness and opacity loss on calendering, increased ink receptivity, and improved printability. These properties are especially advantageous in lightweight coated papers where the added opacity and resiliency are highly beneficial.

With proper selection of the original kaolin plus careful control of calcination, a fully calcined clay equivalent in particle size and glossing characteristics to a regular No. 1 coating clay can be produced. Full calcination results in a clay with high brightness and increased light-scattering properties. Other properties, such as specific gravity, binder demand, and viscosity, and the like, are similar to those of regular clay. The addition of 10 to 15% fully calcined clay to a coating formulation can reduce ink strike-through and improve printed opacity.

Calcium Carbonate: Calcium carbonate, also referred to as chalk, whiting, and limestone, has been used as a pigmenting material for centuries. The history of its use in paper dates from the early 1900s, when the first precipitated byproduct was introduced.

The calcium carbonates used for paper fall into two general classifications: the natural products made by the grinding of chalk, marble, and limestone, or the synthetic products made by chemical precipitation. Within these two groups there are numerous grades based on differences in particle size and particle shape. Normally the natural calcium carbonates are larger in particle size and have a broader size distribution than the precipitated grades. The latter are produced with more variations in particle size and particle shape and hence are more versatile.

For many years the use of calcium carbonate in paper as a coating or filler material has been impeded by its instability in acid papermaking systems. Paper trimmings or damaged paper as a result of breaks on the paper machine and in the finishing operations, or anywhere in the process of paper manufacturing, are usually returned for reprocessing. When such material, commonly known as broke, contains calcium carbonate, it requires special treatment in the acid recycling process. The current accelerated conversion to alkaline papermaking is eliminating this drawback. As a result, the use of calcium carbonate as a coating material and as a paper filler is growing rapidly. New production concepts that have been developed to decrease the manufacturing costs and increase production capabilities are accelerating the current rate of growth.

The physical and chemical properties of calcium carbonates (Tables 7 and 8) vary according to grade, raw material source, and method of manufacture. True white color, freedom from harmful impurities, and nonabrasiveness are prime requisites for products used in paper coating and filling applications. Particle size, particle shape, and surface charge are important factors that influence performance characteristics. Most precipitated products have a chemically active surface. The relationship between particle size and

Table 7. Typical Physical Properties of Calcium Carbonate Pigments

	Natural		Precipitated	
	FGL	UFGL	Calcite	Aragonite
Specific Gravity	2.71	2.71	2.71	2.93
Index of Refraction, mean	1.58	1.58	1.58	1.63
Hardness, (Mohs scale)	3	3	3.0	3.5
Decomposition Temperature, °C	800–900°C			
Valley Abrasion, mg	25	10	5	8
Brightness, % (GE)	95	96	98	99
Oil Absorption, cc/100 g	13	23	30	55
Surface Area, m^2/g	3.2	9.6	6.8	8.5

Table 8. Typical Chemical Properties of Calcium Carbonate Pigments

	Type 1[a]	Type 2[b]	Type 3[c]	Type 4[d]
$CaCO_3$, %	96.63	98.36	98.43	98.62
$CaSO_4$, %	—	0.08	0.78	0.63
$MgCO_3$, %	2.43	0.70	0.37	0.21
Al_2O_3,	0.28	0.09	0.07	0.01
Fe_2O_3, %	0.09	0.07	0.06	0.01
SiO_2, %	0.37	0.10	0.04	0.02
NaCl, %	—	—	—	0.10
H_2O loss at 110°C, %	0.20	0.60	0.25	0.30
pH (saturated solution)	9.1	9.4	10.3	8.5

[a] Type 1 is a ground limestone. Actual values will fluctuate over a moderate range reflecting variations within and between various stone deposits.

[b] Type 2 is a calcium carbonate formed by the following reaction:

$$Ca(OH)_2 + CO_2 \rightarrow CaCO_3 + H_2O$$

[c] Type 3 is a calcium carbonate formed by the following reaction:

$$Ca(OH)_2 + Na_2CO_3 \rightarrow CaCO_3 + 2\ NaOH$$

[d] Type 4 is a calcium carbonate formed by the following reaction:

$$CaCl_2 + Na_2CO_3 \rightarrow CaCO_3 + 2\ NaCl$$

shape and surface potential and performance is complex and not fully understood. Surface area is a direct function of the physical dimensions and can be readily measured by absorption methods (Tables 7 and 8). The chemical purity of the precipitated products is normally superior to that of the ground materials. A few natural deposits may be chemically equivalent to the synthetic calcium carbonates, but these are the exception rather than the general rule. The principal chemical differences are usually in magnesium and silica content.

Calcium carbonates contribute brightness, opacity, and ink receptivity to a paper coating. With the exception of specialty coatings, calcium carbonate is rarely used as the sole pigment in a coating formulation. In standard machine, conversion, and size press coating applications, the pigment portion consists of a blend of clay and calcium carbonate, with auxiliary pigments included if necessary. The percentage of calcium carbonate in the formulation may vary from 5 to 50%, depending upon the coating properties desired.

In discussing the coating properties of a calcium carbonate, we must define the product both in terms of its particle dimensions and chemical components. The natural products perform differently than the synthetics. Physical factors of size and shape are interrelated with chemical potential and can have a pronounced effect on the flow and leveling characteristics of the coating color which, in turn, affects the orientation and packing of the pigment particles. In the same coating color formulation, the results obtained may vary quite widely, depending upon the size, shape, and surface potential of the calcium carbonate particles present. As mentioned previously, there is still much to be learned about the importance of chemical factors related to size, shape, and the method used in manufacturing the calcium carbonate. No single product is best in all applications.

Table 9. Typical Physical Properties of Calcium Carbonate Fillers[a]

	Rhombohedral Calcite PCC	Scalenohedral Calcite PCC	Orthorhombic Aragonite PCC	Fine GLS	Ultra Fine GLS
Refractive Index	1.58	1.58	1.63	1.58	1.58
Specific Gravity	2.71	2.71	2.92	2.71	2.71
TAPPI Brightness (%)	99	99	99	95	95
Surface Area (m^2/g)	6–8	9–15	9–13	5–7	10–12
Einlehner Abrasion (mg)	3	3–5	4–8	8	4
Particle Size, SediGraph					
+5 μm (%)	—	2	3	20	3
−2 μm (%)	99	45	75	70	90
Mean micrometers	0.7	1.0–3.0	0.5–1.0	2.0	0.8

[a] PCC = Precipitated Calcium Carbonate; GLS = Ground Limestone.

With the many types of calcium carbonate that are available and their significant differences in performance, any generalizations on the contribution of calcium carbonate to coating quality are of limited value. This is further complicated by the many varieties of clay that must be considered for comparison purposes. It is safe to say that calcium carbonate increases brightness, ink acceptance, and ease of finishing. The magnitude of the contribution depends upon the particular product, the percentage used, and the specific application involved.

In evaluating any data on calcium carbonates and their performance in paper coating systems, we must remember that the results depend upon the conditions under which the tests were run. While the particle size and particle shape of the calcium carbonates are important, it should be recognized that the particle size and particle shape of the other pigments in the mix will also have a pronounced influence on the coating properties. Actually, any physical, chemical, or mechanical component of the system that can influence particle arrangement and packing can influence the results.

As mentioned earlier, for many years the traditional acid papermaking systems restricted the use of calcium carbonate as a filler in the sheet. Aside from cigarette paper, where the incorporation of a rather high percentage of precipitated calcium carbonate as a filler serves to control the porosity and burning rate, other uses for calcium carbonate were quite limited. The total volume consumed was correspondingly small. Thus, though there have been many years of sustained effort directed at developing improved paper coating grades of calcium carbonate, the incentive to work on the design of filler grades is of relatively recent origin. Based on past experience, it is reasonable to expect that many new grades of calcium carbonate tailored for specific filler applications will be forthcoming within the next few years. The inherent properties of the calcium carbonate may remain the same, but changes in particle size, size distribution, and configuration should improve their functional performance.

Pigmentary calcium carbonates are very white and thus contribute brightness to the filled sheet. Also, with a refractive index of about 1.6, calcium carbonates have a significant opacifying effect when incorporated as a filler in the fiber web. The extent of the increase in opacity is dependent upon the type of calcium carbonate, i.e., precipitated or natural, the apparent particle size, and the level of addition (see Table 9). Calcium carbonates are more receptive to printing ink than are paper fibers; thus, printability improves with increasing percentages of it in the filled sheet. The effect of calcium carbonate fillers on other optical and performance properties will vary depending upon a number of factors, including the type and quantity of calcium carbonate filler, the fiber furnish, the basis weight, and the end use specifications for the finished paper. Many of the comments regarding the physical and chemical differences which affect the coating properties of calcium carbonate also apply to their relative performance as paper fillers. The precipitated calcium carbonates seem to excel in contributing to improvements in optical and printing properties. It appears that natural calcium carbonates probably in the intermediate fine ground particle size range (IFGL) provide the best potential for increasing filler levels with a minimum loss in sheet strength. Just as in coating, no one product is best for all applications.

Talc: The major use of talc in paper is as an additive in the wet-end for pitch control. Its affinity for wood pitch neutralizes the pitch particles and prevents them from coalescing into large clumps that can cause breaks on the paper machine.

Some talc is used in coatings for impact printing papers, where it imparts release characteristics to the finished sheets and prevents blocking and smudging.

Amorphous Silicates: The amorphous silicas and silicates are used in paper coating and filling to increase bulk, reduce ink strike-through, and extend the more expensive titanium dioxide pigment. There is some difference of opinion on whether the reduced requirement for titanium dioxide is the result of optimized spacing or improved dispersion; the latter is becoming the more popular theory. High viscosity of these pigments limits the amount that can be added to the coating formulations; common percentages are 5 to 10% of the total pigmentation.

Alumina Trihydrate: Alumina trihydrate pigments contribute brightness, opacity, gloss, smoothness, ink receptivity, and flame retardancy to a paper coating. They are used in coatings for no carbon required (NCR) papers where the pigment has a favorable effect on the performance of the dyes and other constituents of these coatings. The contribution to flame retardancy results from the fact that alumina trihydrate starts losing its chemically combined water at 150°C. Since the combined water represents 35% of the total weight of this pigment, the volume of water released and the energy absorbed in vaporizing it can be a significant flame retardant.

Miscellaneous: There are several other industrial minerals which have been or are being used in paper. The list of materials includes barytes, blanc fixe, calcium sulfite, calcium sulfate, diatomaceous earth, mica, nepheline syenite, zinc oxide, zinc sulfide, and various alterations of clay and talc. Total volume is quite small by comparison with the materials previously described. Regional availability and commensurate low cost is frequently the major consideration in the selection of some of these materials. In other cases, the mineral may be needed for a special property or application.

Plastics/Polymers

The North American polymers industry is by far the second most dominant user of fillers better referred to as functional inorganics. These materials contribute far more to the finished product than merely filling up space to gain cost savings in the finished product. The 1990 estimated usage of fillers in plastics is 1.35 Mt, with a value of $300 million. On the average, a functional inorganic filler modified for use in the polymers industry had an average price of $227/t.

Plastics are usually classified as thermoplastics—materials which in their final molded shape can be repeatedly softened by the application of heat and remolded into the original or another shape—and thermosetting plastics—materials which when molded are permanently shaped and cannot be softened by heat for remolding.

Plastic resins are generally formulated with filler, stabilizer, colorants, and other additives to achieve required physical properties. Resins may be dry or liquid. The dry system components may be dry mixed and fed to an extruder producing an end product. Much more commonly, both dry and wet systems are mixed, milled at a doughlike consistency with the addition of some heat (fusion), cooled, and ground to a granule or powder. This process is called compounding. This material, or compound, is then extruded, compressed, or injected to yield the desired final shape. Some extremely liquid systems may be cast, but even these usually require heat for rapid curing.

Performance tests for filler materials unique to the plastics industry are:

1. Dispersability of the filler in a fused composition.
2. Abrasiveness of filler particles as related to wear on metal extrusion dies.
3. Flow properties of the filler as related to the ease of moving it from storage hoppers to points of use.
4. Flow characteristics of filler-loaded plastics.

There are many filler materials which have wide acceptance and are used in large quantities in the plastics industry. These are outlined with some of their most important applications in the following paragraphs.

Calcium Carbonate: In 1990, almost 900 kt of natural and precipitated calcium carbonate with a value exceeding $110 million were utilized by the US plastics industry. Ground natural calcium carbonates are the most common fillers and have the greatest usage in this application. They may be wet or dry ground products produced in a 0.1 to a 20μ size category. These are particularly attractive to the plastics industry in that they are low in cost, low in oil absorption, easily and quickly dispersed in conventional mixing equipment, high in brightness range, soft in hardness value, and low in abrasion rating. The ultrafine variety may be used to impart mar resistance, produce a high gloss on finished surfaces, and help reduce the common whitening of flexed materials. In addition, when properly formulated, they may give improved dielectric properties, impact resistance, cold flow properties, low and high temperature behavior, weatherability, control of shrinkage, molded part crazing, and plasticizer migration reduction.

There is an increased use of surface-treated calcium carbonate products. The surface treatments are chemically adsorbed onto the filler and enhance various processing or end-product performance criteria.

Some of the major plastics applications for calcium carbonates are:

Thermoset molded application 20 to 200 parts $CaCO_3$ per 100 parts resin.

Auto front sections, dashboards, auto parts, bodies, pleasure and commercial boats, building panels, plumbing equipment, tubs, shower enclosures, furniture parts, electrical appliances, pipes.

Thermoplastic applications 10 to 50 parts $CaCO_3$ per 100 parts resin.

House sidings, gutters, moldings, wall covering, upholstery, floorings, carpets, structural forms, electrical wire coatings, plastic sheetings, auto parts, lumber replacements, shoe soles and heels, and housewares.

Other applications

Adhesives, sealants, caulking compounds, latexes, phonograph records, etc.

Alumina Trihydrate: In 1990, the US polymers industry used approximately $75 million worth of alumina trihydrate (ATH).

From a standpoint of volume and value, the ATH products were the second most widely used fillers for plastics. The key property enhancement provided by ATH is flame retardance and smoke suppression. ATH provides flame retardance by endothermic decomposition when exposed to high heat. Its bound water molecules are released at high temperatures (150°C) and help to dilute any combustible gases and toxic fumes that may be present. ATH also offers advantages resulting from its high brightness, low specific gravity, and ideal refractive index, which allows some translucency even at filler loadings as high as 100 parts per hundred parts resin (phr).

Reinforced polyester is by far the largest segment of the polymers industry using ATH. Typical applications include spray-up/lay-up compounds for both tubs and shower stalls, sheet molded compounds (SMC) for electronic equipment, appliances, automotive components, and cultured marbles.

Talc: Talc is processed similarly to calcium carbonate. It does not have as wide use in plastics as calcium carbonate. In 1990, the estimated use of talc in polymers applications was 70 kt, with a value of approximately $10 million. Talc has some unique physical properties which are valuable to the polyolefins industry. It is very white in color, has a high degree of softness and smoothness, providing excellent coverage, good luster, and high sheen to finished parts. Talc particles may be described as a series of very small plate-shaped particles which, when aligned, tend to make possible a finished part with improved physical properties. The nature of the particle size and shape will tend to produce slip or a form of lubricity. It is somewhat difficult to disperse in conventional plastic mixing equipment due to overall tendency for more bulking. The material is less sensitive to pH changes; therefore, it can be used in applications where minor acidity is a problem.

Talc, as a filler for plastics, has some special applications. Generally, it is used not as the sole filler in molding formulae but in combination with other filler materials. It facilitates removal from molds, helps achieve a better sheen or gloss on parts, and lowers the extrusion die wear through its lubricating effect. It is also an effective anti-blocking agent. Talc promotes better fire retardancy and aids in processing and final application of various plastic sealants and mastics. It is also useful as an aid in certain color applications, as a filler for parts in contact with acids, and as a viscosity modifier to accomplish some improvement of overall finished product physical properties.

Kaolin: Kaolin clays, especially the calcined products, are used extensively in the thermoplastic wire-coating applications. They improve the insulation properties of the coating and exhibit good dielectric properties. The nature of the particle promotes fast extrusion with little die wear. Clays have very high oil and plasticizer absorption; therefore, except in special applications, they are generally not used as a single filler in formulations. They may be combined with calcium carbonate or other mineral filler products. They are used in thermoplastic applications where density is important and good opaque qualities are required. They are economical fillers for general use and do not seriously deteriorate physical properties. Their use in sizable quantities in a formulation generally results in increased requirement for plasticizer to compensate for their high absorption rate. Clays, when combined with PVC, have a greater stiffening effect than calcium carbonate; however, this combination has greater hardness than does calcium carbonate.

Clay, combined with calcium carbonate, may be used in thermosetting polyester compounds to lower the exotherm, reduce the cost of resin, and modify the mechanical and surface properties. Clay aids the thickening mechanism of sheet molding polyester compositions. Clays are used in other thermosetting applications to lower cost where final parts are to have good electrical properties and surface appearance.

Miscellaneous: There are several other industrial mineral fillers which are used in polymers applications, including ground and synthetic silicas, barytes, and mica. Total volume is small and is predicated upon the ability of these fillers to impart special property enhancements.

Adhesives and Sealants

Adhesives are formulated compositions or chemical compounds which can bond or adhere to substrates that are similar or dissimilar. Fitting this broad definition are more commonly used terms, e.g., cement, glue, and paste. Sealants, also including putty and caulks, are widely used in consumer, construction, architectural, and engineering applications. They essentially fill gaps or joints between two or more similar or dissimilar surfaces. Common applications of sealants involve sealing, waterproofing, and weatherproofing. Sealants are highly pigmented (40 to 80% by weight) compositions based on a wide variety of polymeric vehicles. Adhesives and sealants are commonly included in a single category because of the similarities they share in chemical derivative source. However, it should be remembered that functionally they are different.

During 1990 the use of fillers in the manufacture of adhesives and sealants in the United States is estimated at 1.3 Mt, with a value of approximately $80 million. The dominant filler used in sealants is ground natural calcium carbonate (approximately 900 kt in 1990). Kaolin, air-floated and water-washed, is the filler of choice in the manufacture of adhesives. The 1990 use of fillers in adhesives is estimated at 175 kt, with a value of approximately $18 million.

Calcium Carbonate: Ground calcium carbonate is by far the most widely used filler in sealants because of its low cost. It also exhibits such desirable properties as low binder demand and comparatively narrow particle size distribution, allowing high filler loadings without adversely affecting flow. In oleoresinous putty, natural ground limestones are used at levels as high as 85% by weight. Another large application of calcium carbonate fillers is in the production of dry wall compounds used to seal and finish joints between wallboard panels.

Kaolin: Air-floated kaolin is the filler used in the largest volume in adhesives. It is added to cheaper adhesives such as reclaimed rubber. The kaolin filler improves flow properties of construction, floor tile, and paneling adhesives when troweled. Waterborne adhesives, such as dextrine, caseins, starches, and others, also use large volumes of air-floated kaolin to provide good flow and reinforcement at low cost. Water-washed kaolin is often used to raise the viscosity of certain vegetable and synthetic adhesives, thereby allowing the addition of water without preventing excess penetration. In general, kaolin addition allows cost reduction of adhesives without degrading their physical properties.

Protective and Decorative Coatings

The use of industrial mineral-based products for the *coating* application in general applies to paints and coatings and paper coating. The reader is referred to the extensive discussion in the Paper section of this chapter for details on paper coating and industrial minerals used for that application. For the more traditional use of industrial minerals in paints and decorative coatings, the 1990 volume in the United States was estimated at about 1 Mt, with an estimated value of $230 million. Ground natural calcium carbonate enjoys a dominant position in the list of industrial minerals used for paint and coatings, followed by talc, calcined clay, water-washed clay, ground silica, diatomite, mica, barytes, etc.

Mineral fillers and extender pigments are used to balance flow characteristics, control penetration, reduce requirements for high cost prime pigment such as titanium dioxide, provide flatness, improve sheen and color uniformity, aid in controlling chalking, pigment dispersion, and mildew, reduce film tack, control consistency, enhance the roughness, hardness, scrubbability, chemical resistance, or weatherability of the applied paint film, and reduce cost. The paint formulators often play an excellent game of scientific tightrope walking to derive the optimum balance of these various properties, and in so doing they find the optimum levels of use of the most important mineral fillers.

Technological developments in these fields have required increasingly specialized and precisely processed extender pigments. A broad range of coatings are carefully formulated to meet many different applications. In addition to suitable vehicles made up from oils, resins, and solvents with prime pigments, extender pigments are also a basic requirement in almost all types of coatings.

The coatings industries have come to recognize these mineral products as white extender pigments. The minerals coming within this category vary in particle characteristics and physical properties, but all are white or near white and have refractive indices substantially below 1.75. Calcium carbonates are the most widely used extender pigments. Others are kaolin, both natural and calcined, talc, pyrophyllite, diatomite, feldspar, mica, barite, silicas, and extensive quantities of the manmade fillers.

Extender pigments are employed in most types of coatings, including interior and exterior house paints of oil or water base; industrial paints, both primers and top coats; and traffic paint, plus a host of minor applications.

The practice of utilizing two or more extenders is common, since most coatings require the benefits offered. The desirable effect of one is sometimes reinforced by the presence of another of the extender pigments. The filler loading is usually stated in terms of pigment volume concentration (PVC) which is the volume percentage of the coating occupied by the pigment. This is not to be confused with the plastic PVC (polyvinyl chloride).

The durability of outside paints is related to the structure of the film and its porosity, since vapor transmission is necessary to provide blister resistance. Extender pigments also have a bearing on the chalking characteristics of the paint film. Semigloss and flat paints are formulated with extender pigments which not only aid in pigment suspension but also increase opacity and give the desired degree of flatness. Calcium carbonate is used in these paints; but diatomite and some types of talcs are the most effective flattening pigments. Emulsion paints contain a disperse phase of polymer binder and a continuous aqueous phase using mineral extenders and pigments that are as essential as in the oil base coatings.

The American Society for Testing and Materials (ASTM) rating system evaluates the performance of exterior paints through analysis of a number of factors. These include general appearance, blistering, chalking, cracking, checking, dirt retention, fading, flaking, and mildew resistance. Mineral extender pigments are a major factor in favorably meeting these criteria.

Specialties

Industrial mineral fillers have been extensively used in many traditional applications in some of the most basic industries. For example, in 1990, the US **carpet backing** industry used approximately 700 kt of fillers valued at approximately \$20 million, a major portion of which was ground calcium carbonate. The mineral filler is used in the latex or other binder used to secure the loop of tufted carpets and the secondary backing. Urethane, or other polymers used as foam backings or as a separate underlay, is often filled with up to 272 kg of filler per 100 parts of polymer. The **drilling mud** application in the United States alone consumed approximately 900 kt of mostly bentonite valued at approximately \$35 million. **Iron ore pelletizing**, a major consumer of bentonite, and the **foundry sand** industry collectively used approximately 1.2 Mt of that mineral filler in 1990, with an estimated value of \$35 million. Some **fertilizers** utilize inert fillers to adjust the desired ratio of fertilizer elements. **Textile sizings** are made up of a binder, starch, glue, or synthetic polymer together with mineral powders referred to as weight-adding agents. Kaolin, talc, and barite are the most commonly used agents. Finishing agents are employed to improve web impregnation, appearance, dye receptivity, to provide temporary sizing, and to add weight.

Fillers are extensively utilized in many facets of the **construction industry**. Asbestos cement products and specialty concrete blocks utilize ground silica to add strength and accelerate the curing process. Dry wall joint cements are heavily filled, usually with calcium carbonate. Mica is extensively used as a reinforcing medium in these compounds, and attapulgite provides sag control.

Industrial mineral fillers often find many other related uses.

Kaolin and diatomite are used as **anti-caking agents** for ammonium nitrate by coating the freshly formed prills. An anionic surfactant coating on the commonly used grades of kaolin assists in the distribution of relatively minute amounts of surfactant over all surfaces. Talc and diatomite are finding increasing usage as **anti-blocking agents** in the manufacture of linear low density polyethylene (LLDPE) and low density polyethylene (LDPE) films.

Coal mine rock dust, usually limestone, is used extensively to **dilute** the naturally occurring **coal dust** underground to mitigate the explosion hazard. It is applied wet or dry from shaft stations right up to the working face. Standard (MSHA) requirements are that the total free and combined silica content be less than 4% SiO_2 and that the dust be ground finer than 83%, passing a US Standard 200-mesh sieve.

Athletic field line marker, usually white calcium carbonate ground somewhat finer, is utilized because it presents no hazard to the athlete. Welding rod coatings, crayons, chewing gum, lubricants, match heads, cleansers, polishes, wood and concrete filling compounds, soap, papier-maché, oilcloth, window shades, oil well drill muds, cosmetics, pharmaceuticals, and animal feed nutrition also use one or more of the common mineral fillers in notable quantity. Compared to the major uses covered in detail in this chapter, these are relatively minor. However, they may present additional marketing outlets. It must be noted that many of the mineral fillers discussed here have additional applications, as insulators, soil conditioners, chemical raw materials, ceramic constituents, abrasives, bonding agents, and numerous others.

ECONOMIC FACTORS

Each of the mineral filler products reviewed has established its position in the community because of the characteristics it imparts to some end products. While a specific filler or coating material will do one job best, economic forces generally determine utilization, with the user balancing a cost-to-performance equation. Factors involved include:

1. Geographic location, with resultant transportation costs.
2. Purity of physical state, with resultant processing costs.
3. Physical characteristics, with resultant handling and processing costs.

All of these must be balanced against value of performance. Throughout this chapter, the particular characteristics that favor the utilization of a given filler in a given application have been noted. It would be impossible to equate the various fillers economically except within the framework of a specific application.

Similarly, attempting to summarize specifications would accomplish little since there are perhaps 40 grades of kaolin, 25 grades of calcium carbonate, 8 grades of barite, 10 grades of mica, 20 grades of talc, and so on, each with characteristics usually optimized for one or two industries. Coated, bleached, or treated products multiply the possibilities. One must carefully evaluate a given market to determine those specifications most important for that market. Price ranges are as broad as specifications. The *Minerals Yearbook* annually summarizes price spans for most products. A comprehensive market analysis would be required to pinpoint specific industry prices. Extreme caution must be used in the economic evaluation of deposits based on published commodity statistics, as the values shown will include substantial processing costs and may include expensive additives and packaging and, in some cases, reflect transportation factors.

TRENDS AND OPPORTUNITIES

The trend in the filler industry has been to do more processing, with the ultimate aim of making the product performance more desirable, usually in a specific application. This increased processing, while adding value, can also lead to increased prices. It should be noted that, as consumption rates increase, the economics of bulk handling become more evident. The trend toward bulk rail and truck shipments continues to increase rapidly even outside the kaolin industry, where it is already entrenched. Major quantities of water-processed kaolin are shipped as water slurries, and the shipping of ground limestone slurries is increasing. Containerized shipments have become commonplace for the overseas transport of those mineral fillers whose values support an export market.

There are three specific developments that warrant mention here. The consuming industries, e.g., ceramics, glass, polymers, paint, adhesives, sealants, and paper, are technologically quite advanced. Having recognized the importance of mineral fillers in enhancing competitive positions of their products in their markets, these consuming industries are increasingly demanding a much higher level of technical assistance from the mineral filler producers. In response to these demands, most major mineral filler producers have increased their commitments to research, development, and technical services. Usually, when all else is equal, major consumers of mineral fillers choose to work with suppliers having a higher degree of commitment to technical and customer service and quality.

Demand for consistent quality product is the other increasing trend in the industry. This, again, results from the relative sophistication of the consuming industries such as paper and polymers. The often used cliché of customer/supplier partnerships is a fact of life in the industrial mineral filler marketing, where customers insist on Statistical Process Control (SPC) and Statistical Quality Control (SQC) as essential elements for supplier qualification.

These trends are sometimes ignored by the unsophisticated mineral producers and can result in permanent barriers, keeping them from entering the high end of the market.

Last, but not the least, is the recently accelerating trend—mandated and otherwise—toward recycling. This trend can result in lower than usual growth rates of consumption for some of the mineral fillers. For example, in papermaking, as the recycle paper content increases, the use of virgin filler may be adversely affected. In building products, in glassmaking, and in the polymers and rubber industries, the trend toward recycling may well lead to technical and marketplace challenges for producers of mineral fillers.

Most of these trends, however, suggest that there are ample opportunities for innovative mineral producers to work with innovative customers to add value to their products. With their traditional low cost, mineral fillers enjoy a limited available market. At some point, freight becomes a predominant portion of the delivered cost of a filler. Innovative concepts such as on-site manufacturing or in transit processing can help increase the reach of the mineral product.

Since the consuming industries, such as paper, polymers, paint, and building products, have a favorable outlook for the future, it is safe to conclude that the future is bright for increased use of mineral fillers. The winners will be those filler producers who make strong commitments to technical service, technological development, and consistent product quality.

ACKNOWLEDGMENTS

The authors are most grateful to Nelson Severinghaus, Jr., the author of the chapter on fillers in the fifth edition of *Industrial Minerals and Rocks*, for permission to use some of the pertinent information from that chapter. We also thank Aldo F. Barsotti of the US Bureau of Mines for providing certain statistical information used in this chapter.

BIBLIOGRAPHY AND REFERENCES

Those references listed in prior editions of *Industrial Minerals and Rocks* are most important in developing the history and forecasting trends in the mineral filler industry. Some of these references are classics and have not been superseded in their field. In addition to the references, various trade associations are most helpful in providing related information. These are also listed.

General

Anon., 1985, *Mineral Facts and Problems*, Bulletin 671, US Bureau of Mines, Washington, DC.

Anon., 1986, *Fillers*, Plastics and Rubbers Institute Staff, British Plastics Federation Staff, Elsevier, Amsterdam, 200 pp.

Clark, L., Faith, W.L., and Keyes, B., 1965, *Industrial Chemicals*, 4th ed., John Wiley, New York, 914 pp.

Crull, A.W., 1980, *Fillers, Extenders and Reinforcements for Plastics*, Business Communications Co., Stamford, CT.

Cummins, A.B., 1960, "Mineral Fillers," *Industrial Minerals and Rocks*, 3rd ed., J.L. Gillson, ed., AIME, New York, pp. 567–584.

Dickson, E.M., and Griffiths, J.B., 1985, *Pigments and Extenders: Filling a Colorful Market*, Metal Bulletin Journals, London/New York, 82 pp.

Grim, R.E., 1962, *Applied Clay Mineralogy*, McGraw-Hill, New York, 422 pp.

Harrell, G.O., and Harrell, D.E., 1971, "Survey of Mineral Fillers in Selected Industries," School of Engineering, North Carolina State University, Raleigh, NC, 106 pp.

Kollonitsch, V., and Kliff, E.J., 1969, "Functional Mineral Pigments," SME Preprint 69H81, AIME Annual Meeting, Washington, DC, Feb., 24 pp.

Kulp, M.J., 1989, *Controlling the Surface Chemistry of Fillers*, Lehigh University, Bethlehem, PA.

Lewis, P.A., 1988, "Extender (Filler) Pigments," *Patton's Pigment Handbook*, 2nd ed., Vol. I, John Wiley/Interscience, New York.

Robertson, R.H.S., 1960, *Mineral Use Guide* (Robertson's Spider Webs), Cleaver-Hume Press, London, 44 leaves.

Severinghaus, N., Jr., 1985, "Fillers, Filters and Absorbants," *Industrial Minerals and Rocks*, 5th ed., S.J. Lefond ed., AIME, New York.

Solomon, D.H., and Hawthorne, D.G., 1983, *Chemistry of Pigments and Fillers*, John Wiley/Interscience, New York, 309 pp.

Solomon, D.H., and Hawthorne, D.G., 1991 (reprint), *Chemistry of Pigments and Fillers*, Krieger Publishing Co., Malabar, FL.

Skow, M.L., 1962, "Mica—A Materials Survey," Information Circular 8125, US Bureau of Mines, 241 pp.

Vogely, W.A., ed., 1985, *Economics of the Mineral Industries*, 4th ed., AIME, New York, 660 pp.

Weitz, K.F., 1970, "Applications of Mineral Fillers in the Paint and Plastics Industries," *The Canadian Mining and Metallurgical Bulletin*, Vol. 63, No. 697, May, pp. 600–606.

Uses in Bituminous Compositions

Abraham, H., 1960, *Asphalts and Allied Substances*, Vol. 3, 6th ed., Van Nostrand, Princeton, NJ, 969 pp.

Greenfeld, S.H., 1954, "Laboratory Studies of Several Factors Affecting the Behavior of Some Minerals as Stabilizers for Asphalt Roofing," Report 3616, National Bureau of Standards, 11 pp.

Greenfeld, S.H., 1956, "Effects of Mineral Additives on the Durability of Coating-Grade Roofing Asphalts," Building Materials and Structures Report 147, National Bureau of Standards, Sep., 14 pp.

Damusis, A., ed., 1967, *Sealants*, Reinhold, New York, 382 pp.

Hoiberg, A.J., 1979, *Bituminous Materials: Asphalts, Tars, and Pitches*, 3 vols. "General Aspects," Vol. 1, 446 pp., "Asphalts," Vol. 2, 716 pp., "Coal, Tars, and Pitches," Vol. 3, Krieger, Melbourne, FL, 604 pp.

Panek, J.R., and Cook, J.P., 1984, *Construction Sealants and Adhesives*, 2nd ed., John Wiley, New York, 384 pp.

The following ASTM publications are also of interest:

Symposium on Accelerated Durability Testing of Bituminous Minerals, 1949, ASTM Special Technical Publication No. 94.

Symposium on Bituminous Waterproofing and Roofing Materials, 1959, ASTM Special Technical Publication 280.

Uses in Rubber

Anon., 1961, *Compounding Ingredients for Rubber*, 3rd ed., compiled by eds. of *Rubber World*, New York, 722 pp.

Grim, R.E., 1968, *Clay Mineralogy*, 2nd ed., McGraw-Hill, New York, 384 pp.

Lamar, R.S., Mulryan, H.T., and Warner, M.F., "Reinforcing EPR and EPT with Ultra-Fine Talc," *Rubber World*, Vol. 147, No. 5, Feb., pp. 60–64.

Pollack, L., 1980, *The Rubber Redbook*, 32nd ed., Palmerton Publishing Co., New York, 778 pp.

Uses in Plastics

Hosford, R.C., 1965, "A Guide to Fillers and Reinforcements for Thermosets," *Plastics Technology*, Feb., pp. 34–44.

Hunt, R.E., 1969, "How to Select Fillers for Thermoplastics," *Plastics Technology*, Nov., pp. 38- 43.

Katz, H.S., and Milewski, J.V., eds., 1987, *Handbook of Fillers for Plastics*, Van Nostrand-Reinhold, New York.

Kollonitsch, V., and Deitsch, M., ed., 1982, *Kline Guide to the Plastics Industry*, 2nd ed., Chas. H. Kline & Co., Inc., Fairfield, NJ, 340 pp.

Miles, D.C., and Bristoh, J.H., 1965, *Polymer Technology*, Chemical Publishing Co., New York, 444 pp.

Mohr, G., and Oleesky, S.S., 1981, *SPI Handbook of Technology and Engineering of Reinforced Plastics*, 2nd ed. (reprint), Krieger, Melbourne, FL, 4[illegible]6 pp.

Simonds, H.R., and Church, J.E., 1967, *The Encyclopedia of Basic Materials for Plastics*, Van Nostrand-Reinhold, New York, 500 pp.

Thomson, R.D., 1960, "Nonmetallic Mineral Fillers in Plastics," SME Preprint 60H105, AIME Annual Meeting, New York, Feb., 18 pp.

Wells, G.W., 1980, "Barytes Market-Filler-Extender and Chemical Uses," SME Preprint 80–29, AIME Annual Meeting, Las Vegas, Feb., 4 pp.

Plastics Compounding has published (1978–1982) a series of articles on various mineral filler applications in plastics (Denver, CO).

Uses in Paints

Anon., no date, *Paint Formulary Using Wet Ground Mica*, Wet Ground Mica Association, Inc.

Bradbury, J.C., and Ehrlinger, H.P. III, 1980, "Tripoli," SME Preprint 80–389, SME-AIME Fall Meeting, Minneapolis, MN, Oct., 4 pp.

Gaynes, N.I., 1967, *Formulation of Organic Coatings*, Van Nostrand, Princeton, NJ, 386 pp.

Hall, R.F., 1959, "Effect of Particle Size Distribution of Pigment and Pigment Volume Concentration on Flat Wall Paint Properties (Solvent Type)," *Official Digest*, Federated Societies of Paint Technology, June, pp. 4–15.

Martens, C.R., ed., 1974 (reprint), *Technology of Paints, Varnishes and Lacquers*, Krieger, Melbourne, FL, 752 pp.

Parker, D.H., 1965, *Principles of Surface Coating Technology*, John Wiley/Interscience, New York, 817 pp.

Patton, T.C., 1979, *Paint Flow and Pigment Dispersion*, 2nd ed., John Wiley/Interscience, New York, 631 pp.

Payne, H.F., 1961, *Organic Coating Technology*, John Wiley, New York; Vol. 1, 1954, 674 pp.; Vol. 2, 1961, 725 pp.

Rich, S., ed., 1981, *Kline Guide to the Paint Industry*, 6th ed., Chas. H. Kline & Co., Inc., Fairfield, NJ.

Uses in Paper

Casey, J.P., 1981, *Pulp and Paper*, 3 vols., 3rd ed., John Wiley/Interscience, New York.

Delfosse, P.A., 1986, "Chalk and Marble," SME Preprint 86–8, AIME Annual Meeting, New Orleans, Mar., 11 pp.

Gill, G.J., 1984, "Talcs," *Pigments for Paper*, TAPPI Press Book, pp. 223–239.

Hagemeyer, R.W., 1984a, "Calcium Carbonate," *Pigments for Paper*, TAPPI Press Book, Atlanta, GA, TAPPI, pp 53–94.

Hagemeyer, R.W., 1984b, "Paper Coating," *Pigments for Paper*, TAPPI Press Book, Atlanta, GA, TAPPI, pp. 1–6.

Hagemeyer, R.W., 1985, "Use of Pigments in Paper," SME Preprint 85–49, AIME Annual Meeting, New York, Feb., 8 pp.

Hagemeyer, R.W., 1986, "Calcium Carbonate in Paper," SME Preprint 86–83, AIME Annual Meeting, New Orleans, Mar., 3 pp.

Hagemeyer, R.W., 1989, "Precipitated Calcium Carbonate—Fillers for Paint, Paper and Elastomers," SME Preprint 89–22, AIME Annual Meeting, Las Vegas, Feb., 14 pp.

Hanley, T., Weisler, L., and Edgett, N., 1984, "Barium Sulfate," *Pigments for Paper*, TAPPI Press Book, Atlanta, GA, TAPPI, pp. 39–52.

Harriman, E.J., 1987, "The Less Well Known Pigments In Papermaking," *Proceedings*, 8th Industrial Minerals International Congress, pp. 104–111.

Huber, J., ed., 1980, *Kline Guide to the Paper Industry*, Chas. H. Kline and Co., Inc., Fairfield, NJ, 343 pp.

Koenig, J.J., and Tauth, C.G., 1984, "Alumina Trihydrate," *Pigments for Paper*, TAPPI Press, Atlanta, GA, pp. 15–37.

Koppelman, M.H., 1984, "Paper Filling," *Pigments for Paper*, TAPPI Press Book, Atlanta, GA, TAPPI, pp. 7–13.

Lamar, R.S., and Warner, M.F., 1965, "A New Ultra Fine Talc Pigment for Paper Coating Colors," TAPPI Preprints—16th TAPPI Coating Conference, Portland, OR, May.

Lee, D.I., 1981, "Coating Structure Modification and Coating Hold-Out Mechanisms," *Proceedings*, TAPPI Coating Conference, Atlanta, GA, pp. 143–153.

Murray, H.H., 1977, "Kaolin—A Versatile Pigment, Extender and Filler," CPPA Preprint, Canadian Pulp and Paper Association Annual Meeting, Montreal, Feb., pp. A29–34.

Murray, H.H., 1984, "Clay," *Pigments for Paper*, TAPPI Press, Atlanta, GA, pp. 95–143.

Murray, H.H., and Brodhag, E.E., 1984, "Satin White," *Pigments for Paper*, TAPPI Press Book, Atlanta, GA, pp. 163–207.

Perry, C.L., and Dulin, C.E., 1984, "Amorphous Silicas and Silicates," *Pigments for Paper*, TAPPI Press, Atlanta, GA, pp. 209–222.

Rosenstock, K.H., and Thompson, R.N., 1966, "Satin White," *Paper Coating Pigments*, TAPPI Monograph 30, New York, TAPPI, pp. 141–157.

The following TAPPI (Technical Association of the Pulp and Paper Industry, Atlanta) publications are also of interest:

"Paper Coating Additives," 1963, Monograph Series No. 25.

"Paper Coating Pigments," 1966, Monograph Series No. 30.

"TAPPI Standards and Suggested Methods," 1960.

Other Uses

Fahn, R., 1979, "Acid Activated Clays and Their Adsorption Properties," SME Preprint 79–328, SME-AIME Fall Meeting, Tucson, AZ, Oct., 13 pp.

Haas, C.Y., 1971, "Attapulgus Clays for Future Industrial Minerals Market," SME Preprint 71H54, AIME Centennial Annual Meeting, New York, Mar., 13 pp.

Haden, W.L., Jr., and Schwint, I.A., 1967, "Attapulgite, Its Properties and Applications," *Industrial and Engineering Chemistry*, Vol. 59, No. 9, pp. 59–69.

Key, W.W., 1965, "Mineral Fillers for the California Pesticide Industry," Information Circular 8260, US Bureau of Mines, 39 pp.

Lang, W.J., 1971, "Bentonite: The Demand and Market of the Future," SME Preprint 71H29, AIME Centennial Annual Meeting, New York, Mar., 9 pp.

Parker, D.C., 1979, "Porters Creek Clay, Geology, Processing, and Marketing of a Unique Mineral Assemblage," SME Preprint 79–339, SME-AIME Fall Meeting, Tucson, AZ, Oct., 6 pp.

Tooker, D.J., 1980, "How a 'Tee-Hee' Mineral Got the Last Laugh," SME Preprint 80–36, AIME Annual Meeting, Las Vegas, Feb., 10 pp. (Pet Litter).

Torok, A., and Thompson, T.D., 1972, "Activated Bleaching Clay for the Future," *SME/AIME Trans.*, Vol. 252, pp. 15–17.

Industrial Minerals magazine has published many extensive articles on most of the minerals used as fillers and is helpful for reference on these minerals and their applications.

The following professional associations are also sources of information on mineral fillers:

The Adhesive and Sealant Council, Inc., Arlington, VA
American Assn. of Textile Chemists & Colorists, Research Triangle Park, NC
American Assn. for Textile Technology, Inc., New York, NY
American Forest and Paper Association, Washington, DC
American Petroleum Institute, Washington, DC
The Asphalt Institute, College Park, MD
Asphalt Roofing Manufacturers Assn., Rockville, MD
Cultured Marble Institute, Chicago, IL
Expansion Joint Manufacturers Assn., Tarrytown, NY
Federation of Societies for Coatings Technology, Philadelphia, PA
The Fertilizer Institute, Washington, DC
National Assn. of Printing Ink Manufacturers, Harrison, NY
National Paint and Coatings Assn., Washington, DC
Plastic Pipe and Fittings Assn., Glen Ellyn, IL
Rubber Manufacturers Assn., Washington, DC
The Society of the Plastics Industry, Inc., Washington, DC
Technical Assn. of the Pulp and Paper Industries, Norcross, GA
Wallcovering Manufacturers Assn., Springfield, NJ

FILTERS AND ABSORBENTS

STEVE VAN KOUTEREN

Filters and absorbents purify, clarify, and clean many types of liquids by removing solid particles. In addition, absorbents provide a convenient way to handle liquids and to clean up liquid wastes or spills. Filtration or absorptive processes are accomplished either in a mechanical way, as in absorption and filtration, or in a chemical way, as in adsorption. In filtration, minerals and rocks used as filter mediums are extensively utilized in applications requiring the filtration of high volumes of liquids. End use applications include liquid food products, waste waters, petroleum and chemicals, and drugs and pharmaceuticals.

Upon closer examination of all of the materials used as filters and absorbents, industrial minerals and rocks represent only a small proportion of the total amount, as shown in Fig. 1 (Rees, 1991). Given all the industries that use these minerals, most of the volume is concentrated in a few major end-use markets. Within these major end-use markets, a handful of minerals account for most of the total volume and value.

The focus for this chapter will be to discuss the industrial minerals and rocks used as filters and absorbents in the most important markets. The selection of the minerals and markets to review is solely based on the total volume of material used and in special cases, those where the value of the market and material is high. The major materials include diatomite, perlite, sand, gravel, and such specialty clays as montmorillonite and attapulgite. These materials are mainly used for the filtration of food, chemical, and petroleum products. In addition, they are used as absorbents for pet litter and oil and grease spills, and as carriers for pesticides and fertilizers.

RAW MATERIALS

The most common industrial rocks and minerals used as filters and absorbents are sand and gravel, diatomite, and specialty clays. Throughout this chapter the term specialty clays is used to describe clays used in such value added markets as pet litter and edible and inedible oils. Specialty clays used in filters and absorbents include calcium montmorillonite, attapulgite, and acid-activated clays.

Mineralogy

The common mineralogic trait of industrial minerals used as filters and absorbents is that these materials all belong to the silicate group. Silicates have an inherent chemical stability and unique molecular structures, which prove to be useful in filters and absorbents. Silica (SiO_2) is the dominant chemical constituent and is present in many different forms. For example, in the clay minerals, it is present in complex tetrahedral and octahedral sheets and/or chains, while in diatomite it occurs in an amorphous form.

Diatomite and Perlite: Although the mineralogy and origin of these two materials are very different, they have a physical trait that yields excellent filtration properties. Both have cell structures that form microscopic voids, which aid in filtering by allowing the passage of fluids.

Diatomaceous silica is the preferred name for the principal mineral component of which the rock diatomite is composed. Diatomite is a siliceous, sedimentary rock consisting principally of the fossilized skeletal remains of the diatom, a unicellular, microscopic aquatic plant related to algae (Kadey, Jr., 1983). The silica of the fossilized diatom skeleton closely resembles opal or hydrous silica and has the chemical formulation $SiO_2 \cdot nH_2O$.

On the other hand, perlite, in its naturally occurring form, is a dense, amorphous, hydrated volcanic glass, generally of rhyolitic composition (sodium potassium aluminum silicate). Perlite has a dual nomenclature in the industry due to its commercial applications. It is known by the same name as both the naturally occurring rock and, after processing and thermal expansion, as the lightweight aggregate (Kadey, Jr., 1983).

Specialty Clays: Compared to other minerals used as filters and absorbents, the clay minerals are much more mineralogically complex. As a general definition, clays are naturally occurring, earthy, fine-grained hydrous silicates (Patterson and Murray, 1983). Collectively, they are classified as phyllosilicates, a group of silicate minerals characterized by silica-oxygen tetrahedra linked together in two dimensional sheets (Thrush, 1968).

The phyllosilicate group of clays consists of several minerals, each having different mineralogy, geologic occurrence, technology, and uses. Although the mineralogy is complex, it is easier to think of these minerals as containing crystal structures that can be broken down into two simple building blocks: tetrahedra and octahedra. These building blocks form sheets or layers by joining at the corners (tetrahedra) or at the edges (octahedra) (Moll and Goss, 1987), as shown in Figs. 2 and 3.

Given these basic building blocks, the variations in how the blocks are arranged yield the various groups of clay minerals. The three major clay mineral groups, which account for the majority of the volume of clays used as filters and absorbents, are the smectite, hormite, and halloysite clay groups. The major structural difference between these groups are imbedded in how the building blocks are layered. Table 1 lists these clay groups, commonly used names, and mineralogical compositions.

Smectite Group. Smectite minerals have a three-layer structure. Two outer layers of silica tetrahedra are attached to a central octahedral layer of hydrous metal oxide (usually aluminum, sodium, magnesium, calcium, or lithium). The layers are combined in a way that the oxygen atoms at the tips of the tetrahedral of each silica layer are shared with the octahedral metal oxide/hydroxide layer. A small proportion of the metal cations in the central octahedral layer is replaced by cations of lower valence but similar size. This leads to an electrical charge imbalance that is corrected by the presence of cations held outside and between each of the three-layer

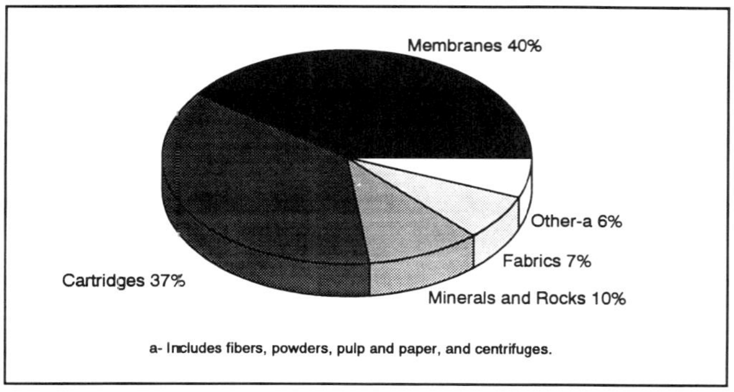

Fig. 1. The US fluid conditioning market by type of system (*courtesy*: Celite Corp.).

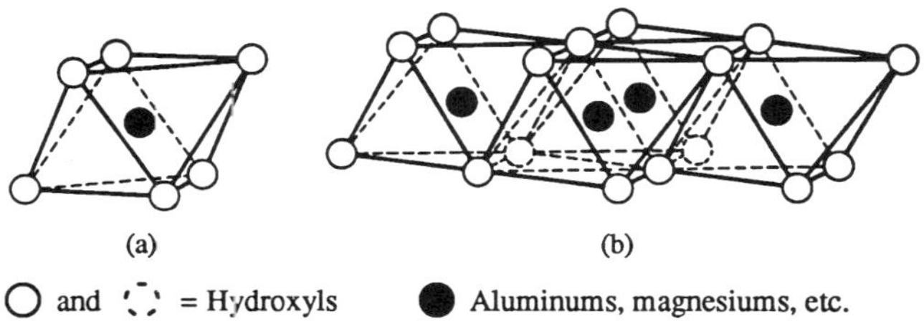

Fig. 2. Geometric representation of octahedron showing (a) a single octahedral unit and (b) the sheet structure of the octahedral units (after Grim, 1968, used with permission).

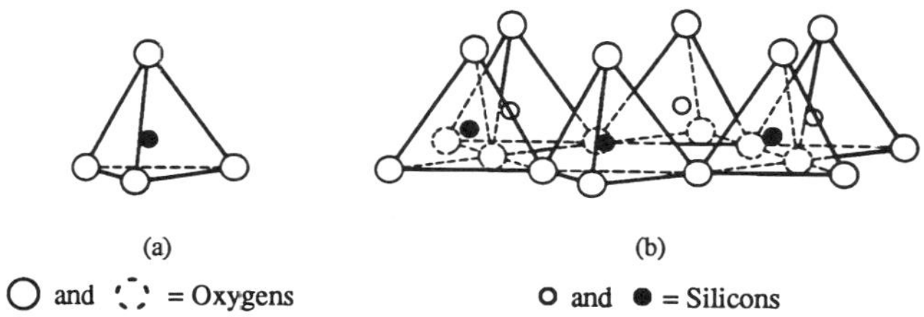

Fig. 3. Geometric representation of silica tetrahedron showing (a) a single silica tetrahedron and (b) the sheet structure of the silica tetrahedrons arranged in a hexagonal network (after Grim, R.E., Clay Mineralogy, 2nd ed., McGraw-Hill, New York, 1968, used with permission).

units. The electrical charge deficiency of the three-layer units and the presence of the exchangeable cations (commonly sodium or calcium) imparts the characteristic expansion properties to the smectite group of clays (Clarke, 1985).

Hormite Group. Hormites are markedly different in terms of their structural characteristics in that they form chain-type, rather than layered structures. The chain-type yields a fibrous-like structure (Clarke, 1985). Unlike the smectite minerals, which have sheets that extend laterally, the sheets in hormite minerals extend only a short distance and then flip up or flip down in a continuous pattern to yield a three-dimensional structure that includes long tubes instead of layers. Because of this structure, hormite minerals do not swell as do smectite minerals. However, crystal lattice substitutions, or exchange capacity, does occur at the broken bonds on the edges of the crystal (Moll and Goss, 1987).

Halloysite Group. This group of minerals, consisting principally of kaolinite, form a two-layer structure. The two layers are comprised of tetrahedra and octahedra sheets. The tetrahedra contain silicon and oxygen and the octahedra contain aluminum, oxygen, and hydroxyl, arranged in sheets. The apices of the tetrahedra join with the octahedra to form the two-layer structure with water molecules between the two layers. These sheets generally are neutral, but on the edges, where broken bonds exist, unbalanced charges exist. These sites are areas where cation exchange can occur (Moll and Goss, 1987).

Sand and Gravel: In the strictest sense, these two materials are defined by size rather than by composition or mineralogical makeup. Sand and gravel consist of separate grains or particles of detrital rock with a particular size range of $\frac{1}{16}$ to 2 mm in diameter. However, for filter and absorbent applications, sand consists mainly of pure, rounded quartz (SiO_2) grains (Davis and Tepordei, 1985) and gravel typically consists of rock fragments.

Minor Minerals: Other minerals used as filters or absorbents include vermiculite, talc, ball clay, crushed limestone, gypsum, ilmenite, and garnet. Vermiculite, talc, and ball clay are also in the phyllosilicate group. As a consequence, these minerals, especially vermiculite, are commonly used in certain absorbent applications. Such heavy minerals as garnet and ilmenite (and sometimes anthracite coal) are often used as filter media in waste water treatment plants. Calcium carbonate (especially as lime) and gypsum are two non-silicate minerals that find limited applications as filters and absorbents.

Chemical and Physical Properties

Not surprisingly, the minerals used as filters and absorbents have a number of similar chemical and physical properties. Good mineral filters and absorbents are chemically stable and have physical structures that allow liquids to either flow through or be retained in the structure. Silica or silicon dioxide, the main constituent of filters and absorbents, is responsible for the chemical stability. The common physical trait is the relatively open or porous structures of these materials which result in relative low bulk densities. Table 2 shows chemical analyses of the major materials used as filters and absorbents.

Diatomite and Perlite: Minerals used as filtration aids typically are lightweight, structurally rigid, thermally stable, chemically inert, and form high-porosity filter cakes to maintain free flow of the liquid. Both diatomite and perlite have such characteristics. However, the major difference between these minerals is in density and flow rate. Perlite is typically less dense or weighs less per unit of volume and thus it yields additional filtering capacity per gram.

Diatomite. The diatom structure, which provides high surface area, low bulk density, high absorptive capacity, and relatively low abrasion, consists of openings or pores at regular intervals, which in turn result in a very open lattice (Kadey, Jr., 1983). The pore structure can account for about 50% of the volume of a diatom, which is typically only a few microns in thickness and 50 to 150 µm in diameter (Miles, 1990).

Perlite. Perlite is known for its unique internal cellular structure, which forms upon heating and subsequent expansion. The integrity of the cellular structure is dependent on the source rock. Hard perlites, which have coarse cell walls, are relatively dense and thus are typically used as aggregate. Soft perlites, which have finer cell walls and are less dense, are typically used as filtration aids (Kadey, Jr., 1983).

Specialty Clays: The layer and chain structures, combined with cation exchange characteristics, are responsible for specialty clays' unique chemical and physical properties. Clays are platy, extremely fine grained, have flexible crystals with large surface areas, are naturally reactive, and have a house-of-cards structure (Moll, 1986). During heating, and the subsequent release of water, the clay's structure maintains its integrity. Structural integrity is responsible for the clay's low bulk density, high liquid holding capacity, and high absorptivity (Moll and Goss, 1987).

Table 1. Classification of the Major Specialty Clays Used as Filters and Absorbents

Clay group/principal mineral	Synonymous term	Theoretical chemical formula
SMECTITE (three-layer structure)		
Montmorillonite	Bentonite (Fuller's earth)	$(OH)_4M^+_{0.66}Si_8(Al_{3.3} \cdot [Mg_{3.3}$ to $Na_{0.66}])\ O_{20} + H_2O$
Calcium	Swelling bentonite	a
Sodium	Non-swelling bentonite	b
Lithium	Hectorite	$(OH)_4Si_8\ (Mg_{5.34} \cdot [Li_{0.66}$ to $Na_{0.66}])\ O_{20}$
Magnesium	Saponite	$(OH)_4(Si_{7.34} \cdot [Al_{0.66}$ to $Na_{10.66}])\ Mg_6\ O_{20}$
HORMITE (chain structure)		
Polygorskite	Attapulgite (Fuller's earth)	$(OH_2)_4\ (OH)_2\ Mg_5\ Si_8\ O_{20} \cdot 4H_2O$
	Sepiolite	$Si_4\ O_{11}\ (Mg \cdot H_2)_3\ H_2O \cdot 2(H_2O)$
HALLOYSITE (two-layer structure)		
Kaolinite	Kaolin	$Al_6\ (Si_4O_{10})\ (OH)_8$

a-Same formula as montmorillonite where M^+ is generally Ca^{2+}
b-Same formula as montmorillonite where M^+ is generally Na^+

Source: Modified from Clarke, 1985.

Table 2. Approximate Chemical Composition of the Major Industrial Minerals and Rocks Used as Filters and Absorbents

Mineral type	Sand[a]	Diatomite[b]	Perlite[c]	Attapulgite[d]	Montmorillonite[e]	Kaolinite[f]
Oxides						
$\%SiO_2$	99.45	89.7	72.1	54.0	50.2	45.2
$\%Al_2O_3$	0.05	3.7	13.5	8.6	16.2	37.0
$\%H_2O$	—	—	3.0	21.4	22.7	14.4
%Ignition loss	0.07[g]	3.7	—	—	—	—
$\%Fe_2O_3$	0.02	1.1	0.8	3.1	4.1	0.3
$\%TiO_2$	0.01	0.1	0.1	0.2	0.2	1.3
%CaO	0.04	0.3	0.9	2.0	2.2	0.5
%MgO	0.02	0.5	0.5	10.1	4.1	0.5
$\%Na_2O$ and K_2O	—	0.7	9.0	0.4	0.3	0.8
$\%P_2O_5$	—	0.2	—	—	—	—
%Other	0.34	—	0.1	0.2	—	—
%Total	100.0%	100.0%	100.0%	100.0%	100.0%	100.0%

a) LaSalle, IL (Murphy and Henderson, 1983), b) Lompoc, CA (Kadey, Jr., 1983), c) No Agua, NM (Kadey, Jr., 1983), d) Attapulgus, GA (Patterson and Murray, 1983), e) Polkville, MS (Patterson and Murray, 1983), f) Macon, GA (Patterson and Murray, 1983), g) Fusion loss.

Sand and Gravel: The important chemical and physical characteristics of sand and gravel used as filters are chemical inertness and proper particle shape and size. As filter media, the removal of particulate matter is a function of both media size and filter bed depth. Optimal packing of sand grains, which are supported by the gravel, in filtration applications occurs when a high proportion of the sand particles are rounded and tend toward a generally spherical or equidimensional shape (Anon., 1989).

Distribution of Major Deposits

United States: The United States contains major deposits of all the industrial minerals and rocks used as filters and absorbents. In fact, the diatomite, kaolin, attapulgite, and sand deposits are world class in terms of size and purity. These deposits, however, occur in select areas of the country. Diatomite and perlite deposits are in the West, the specialty clay deposits are in the South, while sand and gravel deposits occur in all regions, as shown in Fig. 4. The biggest disadvantage in regard to deposit location and market demand is that diatomite and perlite are mined in the West and the demand is concentrated in the Midwest and East.

Diatomite and Perlite. All of the commercial deposits of diatomite and perlite occur in the western United States. Diatomite is currently mined in the states of California, Oregon, Washington, Nevada, and Arizona. However, greater than 50% of US production is mined from the world class deposits at Lompoc, CA (Miles, 1990).

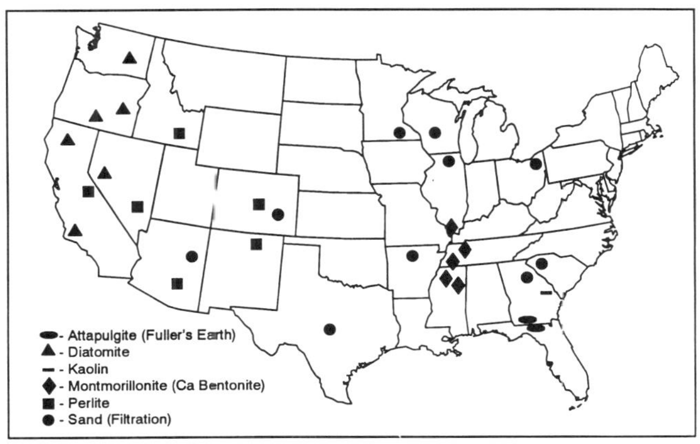

Fig. 4. Major producing locations of industrial minerals and rocks used as filters and absorbents.

New Mexico accounts for about 85% of the perlite mined in the United States with the remaining 15% mined in Arizona, California, Colorado, Idaho, and Nevada. Given the fact that expanded perlite has a very low bulk density and that the majority of perlite is consumed in the East, crushed perlite ore is typically transported to expander plants, which are located close to major end-use markets (Meisinger, 1988).

Specialty Clays. Commercial deposits of specialty clays used as filters and absorbents are, for the most part, located in the Southeast. Missouri, Mississippi, and Tennessee account for the majority of calcium montmorillonite, while Florida and Georgia account for all of the attapulgite production.

Sand and Gravel. The occurrence of sand and gravel deposits are pervasive throughout the United States. From a strictly geologic view, the quantity of sand and gravel is essentially inexhaustible (Dunn, 1983). Although this is true for gravel used in filtration, it is not true for sand. The sand that is used in filtration applications has rigid specifications as to the purity, size, and shape of the individual grains. As a consequence, the major deposits that naturally yield such material are limited to locations in northern Illinois and central Texas (Gledhill, 1991).

Rest of World: A review of the US Bureau of Mines mineral production statistics indicates that no one country dominates in producing materials used as filters and absorbents. There are, however, a number of countries that are major producers of these raw materials.

Diatomite and Perlite. The production of filter grade diatomite in the rest of world is concentrated mainly in such European countries as France, Spain, and Germany. Korea, Romania, and the former USSR are all large producers, but they are believed to have very limited, if any, production of filter grades. For perlite, nearly 80% of the production outside the United States is concentrated in Hungary, Greece, Turkey, and the USSR, but it is not known if these countries produce filter grades (Meisinger, 1988).

Specialty Clays. Rest of world production of specialty clays occurs mainly in Brazil, Canada, and Mexico. Europe and the Pacific Rim are also major producing areas with such countries as Germany, Greece, Italy, Japan, Malaysia, and Turkey accounting for most of the production (Ampian, 1988).

Sand and Gravel. Given that the geologic occurrence of sand and gravel deposits is pervasive, sand suitable for filtration applications exists in many countries. The location and production of sand from these deposits is closely correlated to the location of population and industrialized areas (Bolen, 1988).

TECHNOLOGY

Processing and Specifications

Diatomite and Perlite: The most important aspect of processing these materials is to maintain the integrity of the diatom cell walls and the expanded perlite cellular structure. After processing, these materials are subjected to filtration tests that measure flow rate through a filter cake under controlled conditions, as well as the clarity of the resulting filtrate (Kadey, Jr., 1983). Due to the difference in physical structure of perlite and diatomite, their performance varies from one grade to another. A comparison of performance and specifications for these materials is shown in Table 3.

Diatomite. The degree of processing is highly dependent on the desired physical and chemical properties needed by the end-use industry. Depending on the desired flow rates, diatomite is sold in three different forms: natural, calcined, and flux calcined. The natural products have the higher densities, smaller median pore sizes, and lower flow rates. Calcination of natural products yields less dense, larger median particle sizes, and consequently, higher flow rates.

Diatomite cannot be subjected to intense crushing or grinding, thus it is crushed in spiked rolls and hammer mills, passed through air cyclones, and then fed to air separators to yield various sizes (Miles, 1990). Material at this stage is classified as natural diatomite. For faster flow rates, the natural material is heated in rotary

Table 3. Key Properties for Diatomite and Perlite Used as Filters

Mineral/ grade	Color	Density, kg/m^3 Wet	Density, kg/m^3 Dry	Relative flow rate-a	Ignition loss, %	Medium pore size, μ
Diatomite						
Flux calcined	White	330	220	700–2 300	0.2	15.0
Calcined	Pink	375	140	100–430	0.5	3.5
Natural	Gray	260	105	<100	2.5	2.0
Perlite	White	300	120	170–930	3.0	—

a-Water permeability flow ratio, Grefco's dicalite 215 as 100.

Source: Grefco, Inc. and Celite Corp.

calciners to burn away both organic residues and the combined water that is part of the opaline structure. As a result, the diatom structure shrinks and hardens, and many of the diatom fragments are sintered into agglomerates (Kadey, Jr., 1983). This heating process yields calcined products, which are usually for medium flow rate grades and typically are pink in color due to the oxidation of iron during the heating.

Filtration applications that can not tolerate the presence of iron oxide or applications that require faster flow rates use flux calcined diatomite products (Anon., 1987). These products are produced through the addition of a flux, usually soda ash or sodium chloride, before the sintering or calcination stage. The flux allows iron oxides to enter a glassy phase where it is colorless and produces greater agglomeration of the diatom fragments. Variations in the kiln temperature, the amount and composition of the flux, and the heat time in the kiln enable suppliers to produce products with differing filtration rates (Miles, 1990).

Perlite. Perlite is used in its expanded form in filtration applications. The application for which it is intended plays a major role in the particle size range into which the material must be crushed and sized prior to expansion. Crude perlite ore is crushed with jaw crushers and impact milling and then sent to vibratory screening and air classification to separate the milled material into basic grades. The size range of the expanded perlite dictates the size of the furnace feed (Kadey, Jr., 1983).

During heating, perlite particles reach a softening range, coincident with the volatilization of combined water at about 800°C. As a result, the particles can expand or pop into a lightweight cellular aggregate that is 20 times or more its original volume. During the heating process, a good balance between the softening of the glass and the volatilization of the combined water must be maintained. If not, excess combined water will cause some perlites to explode with the subsequent production of fines. In contrast, insufficient water or an overly viscous glass will result in partially expanded, higher density perlite (Kadey, Jr., 1983).

Upon expansion, the perlite is then milled and classified to a specified size and density. The perlite bubble aggregates must be broken into curved glass platelets and bubble junctions. Filter aids require the right proportion of these with a minimum of unmilled cells. Residual unmilled cells, or floaters, are undesirable in filter applications (Kadey, Jr., 1983).

Specialty Clays: The processing of specialty clays for use in absorbent applications is fairly similar for attapulgite, montmorillonite, and kaolin. Since clay ore contains up to 60% water, the main objective for processing is to remove water and to reduce the material to the appropriate size. Before the crude clay is dried, it must first be passed through shredders to reduce it to fist sized pieces for feed to a dryer (Moll, 1986).

Proper drying techniques are essential for developing and maintaining the open structure in montmorillonite and attapulgite. At

Table 4. Key Properties for Select Specialty Clays Used as Absorbents

Grade	Color	Bulk Density kg/m^3	Surface area m^2/gm	Free moisture, % (110°C)	Base exchange capacity-a	pH
Attapulgite						
RVM	Gray	545	125	7%	20	8.0
LVM	Tan	545	125	2	20	8.0
Other[b]	Cream	275	120	1	—	8.5
Montmorillonite						
Acid activated	—	640	250	16	—	3.2[c]

a-Milli-equivalents/100 g.
b-Powdered grade.
c-25 gm of product in 75 ml D.I. H_2O.
Source: Engelhard Corp., and Floidin.

temperatures between 100 and 200°C, water in the interlayer region of montmorillonite or in the tubes of attapulgite begins to evaporate. At higher temperatures, certain tightly bound water molecules in attapulgite begin to evaporate. Between 500 and 800°C, the hydroxyl ions begin to combine to form water, which leaves the structure (Moll and Goss, 1987).

Once the clay is dried, the material is typically crushed in corrugated roller or hammer mills and passed over large shaking screens. Oversize material goes back to the crushers, and undersized material goes to the next set of smaller sized screens to produce products of increasingly smaller granule sizes. These smaller products are typically destined for the pet litter and floor absorbents markets (Moll, 1986).

Absorbents. The key properties for clays used as floor absorbents are proper granule size, low bulk density, the ability of the product to absorb liquids readily and rapidly, and durability so that the granules do not break down in normal shipment and use. Floor absorbents, mainly used for oil and grease, are sold in sizes ranging from 4.00 to 0.43 mm. The other key properties typically are a bulk density of about 625 kg/m3 and the ability to absorb about 65 to 100% of their weight in water or oil (Moll, 1986).

In cat litter, the pH and size are the important factors that affect the clay's performance. Low pH products are more effective in controlling the fermentation of urea in cat urine. The size of clay granules is important so as to avoid dustiness in use and to avoid tracking by the cat. A common size range for cat litter products is often 4.75 to 0.60 mm.

Carriers. Specialty clays used in pesticide, fertilizer, and other carrier applications often require further drying. Products that do not undergo additional heating, called regular volatile material (RVM), are typically softer and are dispersible in water. However, low volatile material (LVM) undergoes additional heating to result in a moisture content of less than 3% compared to 7% in RVM products (Ross, 1983). These products tend to be harder and far less dispersible in water due to the tightening of the interlocking crystals of the clay caused by further heating (Moll and Goss, 1987).

Another key property is surface acidity, also known as Lewis acidity. The surface acidity of the clay granules can cause degradation of some chemicals absorbed into the granule. To prevent the degradation, formulators must measure the strength of the Lewis acid sites to determine how much deactivation is required. High surface acidity clays, measured by the Hammett acidity function unit (pKa), are deactivated with alcohols or glycols to neutralize the acid cites (Moll and Goss, 1987). An example of the important specifications for carrier materials is shown in Table 4.

Filtering and Clarifying. Acid-activated montmorillonite clays used for filtering and clarifying or bleaching vegetable and petroleum oils are an important part of the market. Although the tonnage of acid-activated clay is relatively small, it represents one of the highest value products.

The treatment of calcium montmorillonite clay with hydrochloric (HCl) or sulfuric (H_2SO_4) acid yields a modified clay product with significantly greater surface area and acidity. The acid-activated clay offers enhanced absorptive and catalytic properties. These products, commonly called bleaching clays or bleaching earths, are not to be confused with fuller's earth, which are processed only by drying and grinding. Acid-activated clays are processed by predrying clay to remove excess moisture. The material is then mixed with either HCl or H_2SO_4 to form a clay-water slurry which is then transferred to a reaction vessel. Once inside the vessel, it is heated to near boiling by steam or other means until the desired degree of activation has been achieved (Taylor and Jenkins, 1986).

Sand and Gravel: The mining and processing of sand and gravel varies considerably with the type of deposit and the physical and chemical requirements of the desired product. Gravels are typically quarried or dredged and processed by screening. Sand is typically hydraulically mined due to the friable nature of the deposits (Gledhill, 1991). The sand slurry is purified by means of cyclones where clays and other impurities are removed. The material is then drained and placed into fluid bed dryers. After drying, vibratory screens, gyratory screens, mechanical, or air classifiers are used to yield various sizes and size distributions (Davis and Tepordei, 1985).

The American Water Works Association (AWWA) has prepared standard specifications for sand and gravel to cover such water filtration applications as surface water, sewage disposal, industrial waste, and swimming pools. The main specifications include freedom from organic matter, silt, minerals which fracture or deteriorate, acid-soluble components, and minimum specific gravity requirements (Severinghaus, Jr., 1983). Filter gravel should contain a high proportion of particles that are rounded and tend toward generally spherical or equidimensional shape. For sand, the grains must be rounded or angular, very durable, and have a uniformity coefficient of 1.7 or less. The required size of sand and gravel is a factor of the viscosity of the filtrate, the type of contaminant, and the flow rate desired. Sand size typically ranges from 0.35 to 0.65 mm and gravel ranges from 2 to over 60 mm (Anon., 1989).

Testing

As is common with most industrial minerals, the testing procedures by which processed minerals and aggregates are evaluated and standardized are designed to quantify an attribute required in the performance of a product (Kadey, Jr., 1983).

Filtration: The two important characteristics filter aids must produce are clarity and flow rate. A filter aid grade, then, must be subjected to a filtration test under controlled conditions. The test will measure how efficient the filter aid is in obtaining the required clarity at a reasonable flow rate.

Diatomite and perlite suppliers provide such bench-scale pressure or vacuum filtration tests as the bomb filter, Buchner funnel, and Walton filter. These tests help to determine the best balance of variables to obtain required clarity, maximum throughput, and minimum filter aid dosage. Prior to lab filtration tests, materials are tested for such physical properties as particle and pore size distribution, turbidity, and surface area. Wet and dry bulk densities are also important because diatomite and perlite are purchased on a weight basis and used on a volume basis (Severinghaus, Jr., 1983).

The use of sand and gravel in the filtration of surface and waste waters requires tests for acid solubility, sand and gravel shape, specific gravity, and particle size to ensure proper quality. The American Water Works Association (AWWA), American National Standard Institute (ANSI), and the American Society for Testing and Materials (ASTM) have established standard tests for these properties. For details on the testing procedures, contact either the AWWA or ASTM (Anon., 1989).

Absorbents: The important characteristics to measure include a clay's ability to absorb liquids easily and quickly, to have a high liquid holding capacity, to have granules that do not break down during shipment, and have relatively low bulk densities. Standard tests and techniques to measure a granule's maximum absorption and resistance to attrition are outlined by Federal Specifications P-A 1050A and P-A 1056A, respectively. In addition, agricultural carriers undergo a liquid holding capacity test. The Rhone-Poulenc Method MP-12 is a test that measures the amount of liquid the granules can absorb without sticking together (Moll, 1986).

In filtering and clarifying applications, the final color of the bleached oil is the critical property. Test methods for evaluating montmorillonite and other types of fuller's earth for bleaching edible oils are outlined in the American Oil Chemists Society (AOCS) Official Method Cc 8b-52 and AOCS Official Method Cc 8a-52. These specifications contain instructions on bench-type tests, including stirring time, heating rates and temperatures, approved equipment, quantities of clay and raw oil required, and other items (Patterson and Murray, 1983).

USES

Industrial minerals and rocks are used in many filtration and absorbent applications, as shown in Table 5. However, on a tonnage basis, the majority of the minerals, including sand, diatomite, perlite, attapulgite, and montmorillonite, are used in such end uses as municipal waters, food products, chemicals, pet litter, oil and grease, and carriers.

Table 5. End Uses for Industrial Minerals and Rocks Used as Filters and Absorbents

ABSORBENTS
- Hazardous wastes
- Oil and grease
- Pet litter
- Water
- Waxes and varnishes

CARRIERS
- Fertilizers
- Pesticides

CHEMICALS
- Acids (sulfuric, phosphoric, organic)
- Dyestuffs and intermediates
- Inorganic-general
- Organic-general
- Solvents
- Sulfur
- Titanium pigments

DESICCANTS
- Documents
- Electronic packaging
- Equipment storage
- Food packaging

DRUGS AND PHARMACEUTICALS
- Antibiotics
- Cosmetics
- Enzymes, serums, vitamins, etc.
- Pharmaceuticals

FOOD PRODUCTS
- Beverages (beer, wine, whisky)
- Corn products (sugar, glucose, oil)
- Fruit products (apple, grape, citrus)
- Miscellaneous (casein, dairy products)
- Oils and fats (vegetable, animal)
- Packing house products
- Sugars (cane, beet, plantation, etc.)
- Yeast

INDUSTRIAL
- Alginates
- Catalytic reactions (hydrogenation)
- Cellulose liquors
- Electroplating
- Dry cleaning
- Metalworking industries (coolants, cutting oils, etc.)
- Nonedible oils and greases
- Nylon and rayon liquids
- Soaps and glycerine
- Synthetic resins
- Varnishes and lacquers

METALLURGICAL
- Cyanidation
- Metal extraction (Be, W, others)
- Recovery of metal values
- Uranium processing

PETROLEUM
- Contact filtration (clay removal)
- Dewaxing
- Greases and lubricants
- Hydraulic fluid
- Jet fuel
- Oil well pressurizing
- Petrochemical
- Still residues
- Used oils (crank case, transformers)
- Waxes

WASTE DISPOSALS
- Paper mill effluents
- Sewage
- Radioactive (underground injections)
- Laundry waste waters
- Industrial wastes, generally

WATERS
- Industrial
- Military (potable, engine condensate)
- Municipal
- Super-clarified (electronics, beverages)
- Swimming pools

Source: Modified from Cummins, 1960.

Filtration

Filtration is commonly referred to as the separation of solid particles from fluids by use of a porous medium. Minerals comprise only a small part of the total market for materials used in the filtration of liquids. Such materials as membranes, cartridges, fabrics, and centrifuges account for the majority of the materials (Rees, 1991).

The main mechanisms of filtration that utilize minerals are cake filtration and depth filtration. In cake filtration, solids are separated from the liquid by forcing the liquid to flow through a porous medium, such as diatomite or perlite, and depositing the solids to form a filter cake. In depth filtration, solids are trapped within the medium using such granular media as sand (Thrush, 1968).

Depth Filtration: The major applications that utilize depth filtration are municipal water facilities that process surface water and sewage. These consumers use sand, anthracite, and, occasionally, gravel to remove impurities from water and other liquids. However, gravel is generally used, primarily in water treatment facilities, as a support media for the filtration sand (Gledhill, 1991).

In depth filtration, the filter media is the portion of the filter bed that removes particulate matter from the water. The removal of particulate matter is typically a function of both media size and filter bed depth, and removal generally improves with greater depth or with smaller media size or both. The selection of filter bed depth and media size is typically a function of the raw water conditions, plant pretreatment facilities, and the individual preferences of the designers. Therefore, the bed depths, particle sizes, materials, and the number of mediums varies widely on a plant-to-plant basis (Anon., 1989).

Most sand filters, or single-medium filters, are horizontal beds of carefully graded grains beginning with coarse particles to form the bed and becoming progressively finer toward the top layer (Severinghaus, Jr., 1983). For improved overall performance, dual- or multi-media filters have been used instead of single-medium filters in many water treatment applications, and in some cases high specific gravity filter media, usually garnet or ilmenite are used to remove more suspended solids at higher filtration rates (Anon., 1989).

Cake Filtration: Cake filtration utilizes a filter aid which controls the flow and amount of solids removed from the liquid. Filter aids form a porous layer, known as the filter cake, on the septum, which may be cloth, screen, porous stone, or metal. The filter cake on the septum acts as a filtering medium to trap solids from suspension and prevents them from blocking the septum.

Good filter aids require a suitable particle size and shape to achieve optimum cake permeability, particle retention on surface versus depth, minimum flow resistance, good cake release, and relatively low cost. Since cake filtration is a mechanical process, the structure of the particles must be such that they will not pack too closely. For example, diatomite's skeletal structure and irregular shaped particles interlock in such a way as to leave 85 to 95% voids. These voids form billions of microscopically fine interstices in which suspended particles can be trapped (Anon., 1991) and are responsible for filtration flow rate and the final filtrate clarity.

The selection of the proper grade of filter aid in cake filtration depends on the size of the suspended particles that are to be removed. As the particle size and thus the flow rate increases, the ability of the filter aid to remove small particles of suspended matter decreases. Conversely, as filter aid particle size and, therefore, flow rate decreases, the ability of the filter aid to remove small particles of suspended matter increases (Kadey, Jr., 1983).

The two major types of filters used to remove solids from food products, chemicals, petroleum, and drugs and pharmaceuticals are pressure filters and rotary filters. The most common types of pressure filters include horizontal tanks with rotating leafs, sluicing filters, tube or candle filters, horizontal plate filters, vertical tank or vertical leaf filters, and the filter press. The pressure filters are used for many different types of applications. However, for thick, difficult to filter liquids or when solids content is high, vacuum rotary filters are typically used.

Pressure Filtration. Filtration in a pressure filter is a two stage operation. First, a thin protective layer of filter aid called the precoat is built up on the filter septum by recirculating a filter aid slurry. This slurry is composed of either diatomite or perlite and the liquid. The mixture continues to circulate until the liquid becomes clear. At this point, a precoat bed of about 2 mm thick has been established and filtration is ready to begin (Anon., 1985).

The next step is called the body feed because, as the liquid is pumped through the filter, small amounts of filter aid are constantly fed into the liquid. As filtering progresses, the filter aid, mixed with the suspended solids from the unfiltered liquid is deposited on the precoat. Thus, a new filtering surface is continuously formed (Anon., 1985).

Body feed rate is increased or decreased depending upon the turbidity of the unfiltered liquid. As filtration proceeds, the volume of the cake increases until the solids trapped in the bed cause the flow to drop to an unacceptable level or the pressure to increase to an unacceptable level. At the end of the cycle, the filter is opened, cake is removed from the septum, and a fresh precoat is established for another cycle (Anon., 1985).

Rotary Vacuum Precoat Filters. As mentioned previously, these filters are generally only used where a large percentage of suspended solids is to be removed. A rotary vacuum precoat filter consists of a horizontal drum, of which 30 to 50% is submerged in a filter bowl containing the unfiltered liquid. This drum is covered with a septum capable of retaining a filter aid. Vacuum is applied to the surface of the drum by means of internal piping (Anon., 1985).

In operation, a precoat of diatomite or perlite up to 150 mm thick is built up on the drum by revolving the drum under vacuum in a filter aid slurry. After the precoat is built up to the desired thickness, unfiltered liquid is introduced into the filter bowl. As the drum rotates, a blade running across the face of the drum above the liquid level continually advances toward the drum. This results in a peeling off of the solids plus a very small amount of precoat, thereby presenting a new precoat surface for the next submergence. This continues until the knife is within 6 to 10 mm of the drum, at which time the drum is cleaned and precoated again (Anon., 1985).

Absorbents

Consumer: The use of specialty clays in the pet litter market has proven to be very effective in absorbing liquid pet waste and related odors. Specialty clays used for pet litter have the ability to absorb large quantities of liquid cat waste and retard the formation of ammonia, which is created when urea from cat urine ferments (Harris et al., 1989).

Industrial: Any user of oil and grease or machinery requiring oil and grease may ultimately require specialty clays to absorb and control spills. Slippery films of oil and grease as well as puddles of water and other chemicals often create dangerous conditions on the floors of machinery repair shops, factories, service stations, and numerous large and small industrial facilities (Harris et al., 1989). Specialty clays can absorb up to their own weight in oil, grease, water, and a variety of other chemicals. The clays also are safe, absorb rapidly, and can be cleaned up by using a broom or shovel (Moll, 1986).

Carriers: Specialty clays used as mineral carriers have a wide variety of applications in the field of agricultural chemicals. Most pesticides, including herbicides, fungicides, and insecticides, are highly toxic and typically require only a kilogram per hectare (Moll, 1986). The primary function of these carriers are therefore to dilute high potency chemicals to a spreadable field concentration and to apply them to a location where the pest can be destroyed while damage to desirable plants, wildlife, and the environment is minimized (Swayer, 1983).

The two main types of mineral carriers are absorbent and nonabsorbent. The absorbent carriers, which account for the majority of demand, provide a highly porous structure which accounts for their high absorptivity and low density. The nonabsorbent types only provide surface area to carry the chemicals and typically have high bulk densities, which limits the amount of chemicals they can carry. Montmorillonite, attapulgite, and diatomite comprise the absorbent carriers and such minerals as ball clay, calcium carbonate, quartz sand, and talc are considered nonabsorbent (Moll and Goss, 1987).

In fertilizers, clays are also used as anticaking additives in dry fertilizer mixes. The anticaking function of the clay works by coating fertilizer particles and absorbing solvents or water that may be present. This reduces the formation of crystals that often bind particles and cause caking problems (Harris et al., 1989).

Filtering and Clarifying: Acid-activated montmorillonite clays and attapulgite clays (also called bleaching clays or fuller's earth) are used to filter and clarify such liquids as edible and inedible oils, petroleum products, and beverages. Although filtration is a part of the process, selective adsorption plays a major role in clarifying these products. The bleaching or clarifying of these products is necessary to remove organic pigments from plant oils, sulfur compounds from mineral oils, and contaminants from fuel and lubrication oils (Anon., 1968).

The two main methods used for commercial bleaching are contact and percolation. The contact process is commonly used in refining edible oils and consists of mixing clay (0.5 to 3.0%) with the crude oil and heating the slurry to about 80 to 120°C for about fifteen minutes. The slurry of clay and oil is then filter pressed to remove the clay (Anon., 1968).

The percolation method, which is also a filtration process, is generally used for the clarification of lubricating oils. A granular clay is placed in a column and oil is then trickled through the clay by way of gravity. In order to obtain maximum efficiency, the viscosity of the oil is lowered by heating or dilution. This assists by allowing the oil to penetrate the minute pores of the clay. At the end of the cycle, the clay is washed with naphtha, steamed, and burned in a rotary kiln to regenerate the adsorbent, which then can be reused (Anon., 1968).

Other Markets: The most significant other absorbent market for specialty clays is desiccants. Clay desiccants are used to adsorb moisture from the air enclosed in such packaged products as military equipment and electronic components. This helps prevent corrosion of the product and retards the formation of mildew. The desiccants are made up entirely of calcium montmorillonite clays that are oven dried, crushed, and packaged in bags or canisters. These products are most effective in conditions of moderate relative humidity (15 to 40%), and are generally preferred because of their ability to adsorb more than 20% of their equivalent weight in water without any apparent change in size, shape, or texture. Furthermore, they are also available at a relatively low cost (Harris et al., 1989).

INDUSTRY STRUCTURE

As in most other mineral industries, the industry structure determines the competitive environment in which these minerals are sold. The four major factors that determine the industry structure are the concentration of the supply and the demand side of the business, the barriers to enter, the availability of substitutes that can compete on a cost performance basis, and the rivalry between the existing suppliers. These major factors, when combined with deposit quality and technology, play a significant role in the profitability of suppliers of minerals and rocks to this industry.

Industry Concentration

Supply: Supplier concentration of industrial minerals and rocks used as filters and absorbents is high due to the limited number of high quality deposits.

Diatomite and Perlite. Diatomite and perlite production in the United States is concentrated among three or four suppliers. In 1990, about 80% of diatomite production was accounted for by two companies, Celite and Eagle Picher. Grefco, which accounts for another 15% of production, is the third largest producer. The perlite industry has an even higher degree of concentration. Celite and Grefco account for about 85% of the production, with the remaining divided amongst four or five firms, of which Nord Resources is the largest (Harris et al., 1990).

Specialty Clays. Specialty clay production in the United States is highly concentrated. In 1990, more than 75% of the sales of these clays was accounted for by six companies, with only two companies selling more than one type of specialty clay. The remaining 25% of the volume was with seven to ten companies. When industry concentration is analyzed by clay type, this number increases. For example, the top four attapulgite producers accounted for nearly 90% of the sales to filtration and absorbent markets. The top five montmorillonite and top four kaolin suppliers accounted for about 80% and for nearly 100% of total sales to filtration and absorbent markets, respectively (Harris et al., 1989).

In 1990, the major producers of attapulgite were Oil-Dri, Floridin, Engelhard, and Mid-Florida Mining. These companies accounted for about 90% of the sales to absorbent and filtration markets. Golden Cat (formerly Edward Lowe Industries), Oil-Dri, Moltan, Excel Minerals, and Absorbent Clay accounted for about 80% of the montmorillonite sales. For kaolin, the top producers were Engelhard, ECC International, J.M. Huber, and Hartz Mountain (Harris et al., 1989).

Silica Sand. Although the supply of high purity silica sand is also concentrated, it is much less than diatomite, perlite, and the specialty clays. The top seven to ten companies that supply high purity sand to the filtration industry are estimated to account for about 50% of the total sales. The major suppliers of these materials are Unimin, US Silica, and Vulcan Materials (Gledhill, 1991).

Demand: On the demand side, the filtration industry is generally more concentrated than the absorbent industry.

Filtration. The demand side of the filtration market is highly concentrated. Five consumers account for about 80 to 85% of total diatomite and perlite use in such food product markets as soft drinks, juices, sugars, and beer. The concentration of consumers in the chemical, petroleum, and drugs and pharmaceutical industries is also high. Less concentrated filtration markets include industrial varnishes and lacquers, industrial water filtration, and dry cleaning. Another exception is the waste and surface water filtration market. The demand for sand and gravel in filtration is mainly from municipal water treatment plants, which exist in nearly every county in the United States (Crider, 1991).

Absorbents. The demand side of the absorbent market is unconcentrated. The pet litter and oil and grease markets consist of hundreds of brokers or wholesalers who purchase and distribute the clay products to thousands of retail stores. The carrier market is the only major end-use market for absorbent applications that is fairly concentrated. About ten firms account for more than 50% of the demand for clays in this industry (Fenn, 1991).

Barriers to Entry

Industry competition is also affected by the threat of new entrants. New entrants to an industry bring new capacity, the desire to gain market share, and often add new resources, which can result in lower profitability. The threat of entry into an industry in part depends on the barriers to entry combined with expected reaction from existing competitors.

The major entry barriers for the materials used as filters and absorbents are related to the availability of economical deposits and the location of these deposits. The limited availability of high grade deposits of diatomite, sand, and specialty clay materials provides the most significant barrier to entry in this industry. New entrants into the specialty clay markets are probably further deterred by the need to have access to distribution channels for the pet litter and oil and grease markets. In addition, brand or brand name recognition is especially important in the pet litter market.

Substitute Materials

The availability of substitute materials that are competitive on a cost-performance basis plays an important role in industry structure. Substitutes limit the potential returns of an industry by placing a ceiling on the prices firms charge. A number of good organic and inorganic substitutes for filter and absorbent applications are being used in the market today. Such materials as synthetic fabrics, ground corncobs, and ground alfalfa play a role in the absorbent markets. A discussion of these materials and others, their markets, and performance is in the Economic Factors section of this chapter.

ECONOMIC FACTORS AND GOVERNMENT REGULATIONS

Costs

The major cost factor for mining, processing, and transporting industrial minerals and rocks used as filters and absorbents is typically the energy costs associated with drying and calcining. Mining costs are relatively low because material is mined in open pits and generally does not require blasting. However, processing costs are high due to the moisture content of these deposits, which is often in the 30 to 60% range (Miles, 1990).

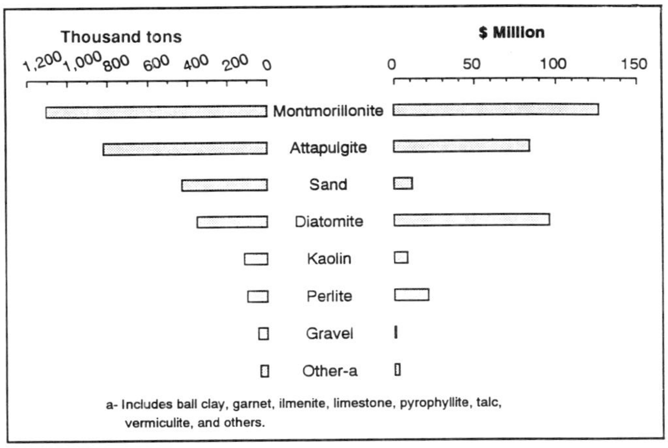

Fig. 5. Estimated demand for industrial minerals and rocks used as filters and absorbents by major mineral, 1990 (*courtesy*: Kline & Company).

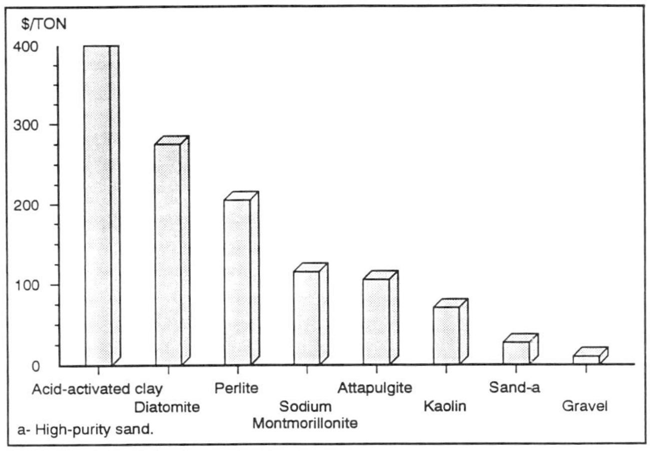

Fig. 6. Average price per ton for industrial minerals and rocks used as filters and absorbents by major mineral (*courtesy*: Kline & Company).

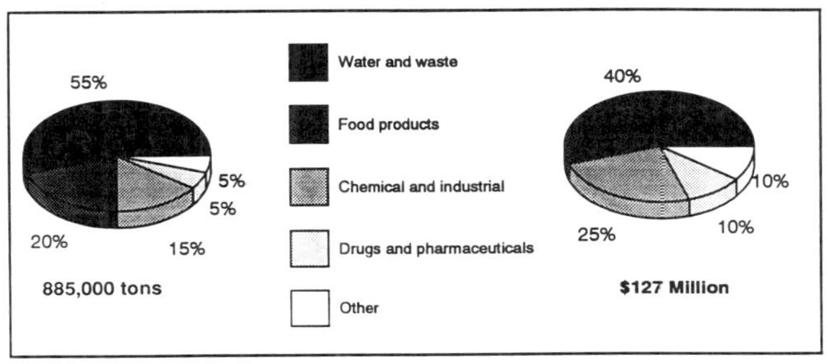

Fig. 7. Estimated demand for industrial minerals and rocks in filtration by major end use, 1990 (*courtesy*: Kline & Company).

The cost, on a tonnage basis, to transport filtration and absorbent raw materials to the market are typically higher than other industrial rocks and minerals due to their low bulk densities. As a result, transportation costs are a significant portion of the total cost of the delivered product. This is especially true for diatomite and perlite because production is in the West and the major markets are east of the Mississippi River. Furthermore, perlite must be shipped in a crude form to the major markets and then processed or expanded.

Another cost factor, although not directly related to the mining, processing, and transporting products to the market, are disposal costs. Consumers of industrial minerals and rocks are facing increasing disposal costs associated with the spent filter cakes and absorbent materials. The cost to transport the spent materials to an acceptable landfill, combined with the costs and liabilities associated with disposing the materials, are rapidly increasing.

Markets

The total demand for industrial minerals and rocks used for filtration and absorbents applications was about 3.0 Mt valued at over $350 million in 1990. The absorbents market accounted for about 70% of the total volume and about 65% of the total value, while the filtration market accounted for the remaining 30 and 35% of the volume and value, respectively. The breakdown of demand in these markets is shown in Fig. 5.

Average prices for these minerals ranged from about $10 to more than $400 a ton in 1990, as shown in Fig. 6. Sand and gravel are typically the lowest priced materials while acid-activated clays are the most expensive. Although the acid-activated clays have the highest average price, the total demand for these clays is relatively minor (Gledhill, 1991).

Filtration: The total demand for minerals used as filters was about 885 kt valued at about $127 million in 1990. High purity quartz sand, diatomite, and perlite accounted for the majority of the volume used in such major uses as waste water, food products, and chemical industries. The waste water markets account for the most tonnage due to the large volumes of municipal and industrial water and waste that require treatment, as shown in Fig. 7. In these applications, sand and gravel, along with anthracite and in some cases, such high specific gravity minerals as garnet and ilmenite are used.

Absorbents: The total demand for minerals used as absorbents was about 2.1 Mt valued at about $223 million in 1990. Clays used as pet litter accounted for more than 50% of the total volume and value with the oil and grease and carrier markets being the next largest in size and volume. Although the volume of clays used in the filtering and clarifying of oils is only 4% of the total volume, these acid-activated clays account for nearly 15% of the total value, as shown in Fig. 3.

Substitute Materials

Until recently, the threat to minerals from substitute materials or alternative technologies has been somewhat limited. However, recent environmental and health regulations combined with improvement in such alternative technologies as membrane or ultrafiltration have and will continue to adversely impact the demand for minerals.

Filtration: Competitive materials and alternative technologies are currently having an adverse impact on the demand for diatomite and perlite. Perlite, a competitive material to diatomite in filter cake applications where clarity is not so critical, has gained market share in recent years. The rivalry between the use of diatomite and perlite has intensified due to regulations on crystalline silica. Some perlite suppliers have increased their marketing efforts to exploit the fact that its crystalline silica content is much less than that of diatomite. As a result, perlite has made gains in such food product markets as juices. (Gledhill, 1991, Rees, 1991).

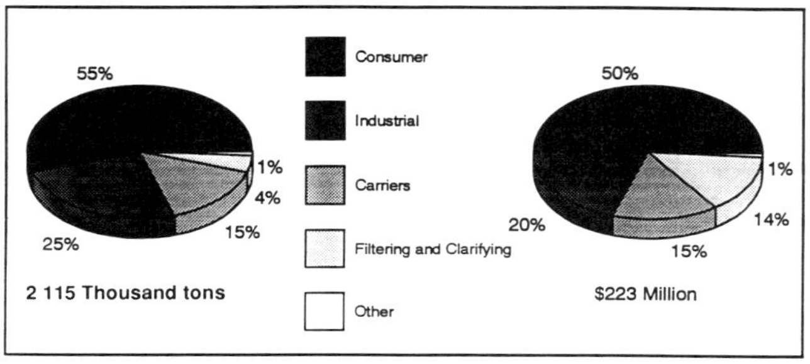

Fig. 8. Estimated demand for industrial minerals and rocks in absorbents by major end use, 1990 (*courtesy*: Kline & Company).

Ultrafiltration technology or cross-flow membrane filtration uses pressure gradients, chemical and electrical potentials, combined with synthetic membranes to separate macromolecular and colloidal solutes or particles (Clarke, 1985). In recent years, this technology has gained market share in the food products, biomedicine, and biotechnology markets. The gain is due in part to the increasing disposal costs of spent diatomite and perlite filter cake. In addition, the cost to use this technology has decreased due to advances in membrane and equipment performance and durability (Rees, 1991).

Absorbents: Substitute materials for absorbent minerals consist of such organic materials as sawdust and ground alfalfa, corn cobs, and citrus fruit grinds. Although these materials are typically less expensive, they are not as effective as specialty clays. On the other hand, such synthetic materials as molecular sieves, silica gel, precipitated silica, and nonwoven polyethylene fabrics are very effective but are too expensive. In more limited cases, such other mineral products as vermiculite, gypsum, lime, sand, and zeolites are used in specific applications where low cost and/or performance is desired. Other substitutes include diatomite and perlite, especially in the western United States due to the transportation cost advantage that they have over clays (Harris et al., 1989).

Polyethylene fabric has had the largest impact on the demand for mineral absorbents. These non-woven fabrics have a natural affinity for petroleum-based products and as a result they can absorb up to 20 times their weight of oil and grease. The use of these materials has grown due to the classification of spent clay as a hazardous material and the subsequent disposal costs associated with these clays. The big advantage of these non-woven fabrics is that they can be incinerated with less than one percent ash remaining (Harris, 1989).

Environmental and Health Regulations

Governmental regulations concerning crystalline silica have had an adverse effect on the demand for diatomite. This trend began in 1988 when the International Agency for Research on Cancer (IARC) classified crystalline silica as a probable carcinogen. The classification triggered regulation of crystalline silica at a threshold concentration of 0.1% in materials under the Hazard Communication Standard (Miles, 1990).

The diatomite and other minerals suppliers responded to this in a number of different ways including worker and user trainer programs, product labeling, and dust reduction programs. The diatomite industry also created the International Diatomite Producers Association (IDPA) to study and issue reports on health, safety, and environmental issues related to the use of diatomite. Other mineral and chemical producers formed a Chemstar Crystalline Silica Panel. The panel, formed in 1989 and sponsored by the Chemical Manufacturers Association (CMA), works with government agencies to avoid random banning of minerals suspected of being hazardous.

Another governmental regulation that has adversely impacted the demand for minerals and rocks used as filters and absorbents was the 1991 EPA ban on disposing wastes containing oil in landfills. This ruling caused some consumers of clays for oil and grease absorbents to switch to other products. The switch was mainly due to the fact that organic products could be incinerated instead of placed in landfills. Clay suppliers are currently fighting to have new tests approved for measuring the leaching or leaking of absorbed oil from clay products. Under the current regulations, a toxicity test renders the clays hazardous but if proposed liquid release tests are accepted, the absorbent minerals may prove to be less of a hazard (Fenn, 1991).

PROBLEMS AND FUTURE TRENDS

Although health and environmental regulations will continue to have an adverse impact on the demand for industrial minerals and rocks, they will not cause the demise of these materials. Overall, the minerals and rocks used as filters and absorbents do their job very effectively and economically. The regulations may increase the cost, but in most cases, the increased cost will not immediately result in the substitution of alternative materials or technology. Therefore, one must keep in perspective that despite all of the negative comments about the regulations, the use of minerals and rocks as filters and absorbents will continue to be cost effective.

Market Developments

Environmental issues will continue to affect the demand for industrial rocks and minerals. In many areas of the United States, the closure of land fills has been increasing the total cost to dump material. The total cost includes fees for material disposal and the added cost to transport the material to a more distant landfill. In addition, landfill users can incur additional liability if a site requires future cleanup due to another user's disposal practices (Miles, 1990).

The EPA ruling that spent clay oil and grease absorbents are classified as hazardous wastes also will continue to adversely impact the demand in this application. Another area that has caused problems in the past is the disposal of spent clays from edible oil filtration. The problem is that the spent clay product is highly flammable and special precautions must be taken when dumping the material. Although procedures have been developed to safely dispose of used material, some landfills still do not accept such material (Harris et al., 1989).

The end result of increasing disposal costs for spent filter cake and clay absorbents will be increased substitution or new technology to circumvent these problems. For example, diatomite and/or perlite will continue to lose market share to ultrafiltration or membrane technology. Another example of consumers switching to a different technology due to increasing spent clay disposal cost is in the petroleum refining industry. Refineries are increasing their hydrogenation capabilities, which is another way of cleaning or purifying petroleum products without incurring the additional cost and liability associated with disposing the spent clay (Harris et al., 1990).

The crystalline silica issue will continue to be a problem for the industry. The initial negative impact of the regulation appears to have occurred by 1991. Consumers have either taken protective actions and continued to use diatomite or they have switched to another material. However, the 0.1% target may not be final and could actually be changed. If it is lowered, then more problems await both the suppliers and users of any mineral product containing crystalline silica.

On the other hand, increased environmental regulations will also have a positive impact on the demand for these materials. Due to more stringent regulations on waste water, companies will be required to clean their effluents. Filtration will be one of the methods.

Specialty clays used as pet litter is another area that will show good growth. The cat population in the United States is expected to continue to increase as the population ages. Cats are, in many ways, ideal companions, especially for the elderly and apartment dwellers, because they care for themselves, require no outdoor exercise, and consume little food.

Outlook

The outlook for filtration and absorbent materials will vary considerably by end use. For the most part, however, many of the markets are mature and the growth is expected to follow the overall economy. Average annual growth of filtration and absorbent materials in the 1990s will range from 1.5 to 3.0%. However, the loss of markets to substitute materials and alternate technologies due to environmental and health problems could worsen and thus cause the average annual growth rate to be lower.

ACKNOWLEDGMENT

The author wishes to thank the American Water Works Association for permission to quote from *AWWA Standard for Filtering Material,* B100–89 (copyright © 1989).

BIBLIOGRAPHY

Anon., 1968, "Percolation Adsorptive Processes, Static Bed and Continuous," Engelhard Corp., Edison, NJ, pp. 1–22.

Anon., 1985, "Celite Filter Aids for Maximum Clarity at Lowest Cost," "Filter Aids the Winemaking Industry," Manville Corp., Denver, CO, pp. 1–28.

Anon., May 1987, "Diatomite—No Skeletons in the Cupboard," *Industrial Minerals*, No. 236, pp. 22–39.

Anon., 1989, "AWWA Standard for Filtering Material," American Water Works Association, AWWA Board of Directors, Denver, CO, pp. 1–17.

Anon., 1991, "Dicalite—High Performance Filter Aids," Filtration Bulletin B-16 (R2–90), Grefco, Inc., Torrance, CA, pp. 1–18.

Ampian, S.G., 1988, "Clays," *Minerals Yearbook, 1988*, Vol. 1, US Bureau of Mines, pp. 251–288.

Bolen, W.P., 1988, "Industrial Sand and Gravel," *Minerals Yearbook, 1988*, Vol. 1, US Bureau of Mines, pp. 833–843.

Clarke, G.M., 1985, "Special Clays," *Industrial Minerals*, No. 216, Sep. pp. 25–51.

Cummins, A.B., 1960, "Mineral Fillers," *Industrial Minerals and Rocks*, 3rd ed., J.L. Gillson, ed., AIME, New York, NY, pp. 567–584.

Davis, L.L., and Tepordei, V.V., 1985, "Sand and Gravel," *Mineral Facts and Problems*, A.W. Knoerr, ed., US Bureau of Mines, Washington DC, pp. 689–703.

Dunn, J.R., 1983, "Construction Materials, Aggregates, Sand, and Gravel," *Industrial Minerals and Rocks*, 5th ed., Vol. 1, S.J. Lefond, ed., AIME, New York, pp. 96–110.

Grim, R.E., 1968, *Clay Mineralogy*, 2nd ed., McGraw-Hill, New York.

Harris, T.S., et al., 1989, "An Analysis of the U.S. Market for Specialty Clays," Kline & Company, Fairfield, NJ, pp. 1–150.

Harris, T.S., et al., 1991, "Extender and Filler Minerals North America, 1990," Kline & Company, Fairfield, NJ, pp. 1–425.

Kadey F.L., Jr., 1983, "Perlite," *Industrial Minerals and Rocks*, 5th ed., Vol. 2, S.J. Lefond, ed., AIME, New York, pp. 997–1015.

Meisinger, A.C., 1988. "Perlite," *Minerals Yearbook, 1988*, Vol. 1, US Bureau of Mines, pp. 725–727.

Miles, W.J., 1990, "Economics of Diatomite," *Proceedings*, National Western Mining Conference—Colorado Mining Association, Feb. 8, pp. m1–10.

Moll, W.F., 1986, "Absorbent Clays," Preprint 86–334, Fall Meeting, Sep. 7–10, St. Louis, MO, Society of Mining Engineers, Littleton, CO, pp. 1–9.

Moll, W.F., and Goss, G.R., 1987, "Mineral Carriers for Pesticides-Their Characteristics and Uses," *Pesticide Formulations and Applications Systems: Sixth Volume*, ASTM STP 943, D.I.B. Vander Hooven and L.D. Spicer, eds.. American Society for Testing and Materials, Philadelphia, PA, pp 160–176.

Murphy, T.D., and Henderson, G.V., 1983, "Silica and Silicon," *Industrial Minerals and Rocks*, 5th ed., Vol. 1, S.J. Lefond, ed., AIME, New York, pp. 1167–1185.

Patterson, S.H., and Murray, H.H., 1983, "Clays," *Industrial Minerals and Rocks*, 5th ed., Vol. 1, S.J. Lefond, ed., AIME, New York, pp. 585–651.

Ross, H.W., 1983, "Available Granular Carriers—Properties and General Processing Methods," *Proceedings*, Pesticide Formulations and Application: Second Conference, A symposium sponsored by ASTM Committee E-35, K.G. Seymour, ed., Philadelphia, PA, pp. 32–44.

Severinghaus, N., Jr., 1983, "Fillers, Filters, and Absorbents," *Industrial Minerals and Rocks*, 5th ed., Vol. 1, S.J. Lefond, ed., AIME, New York, pp. 243–257.

Swayer, E.W., 1983, "Introduction to Granular Carriers, Granular Pesticide Formulations, and Processing," *Proceedings*, Pesticide Formulations and Application: Second Conference, A symposium sponsored by ASTM Committee E-35, K.G. Seymour, ed., Philadelphia, PA, pp. 26–31.

Taylor, D.R. and Jenkins, D.B., 1986, "Acid Activated Clay," Preprint 86–365, Fall Meeting, Sep. 7–10, St. Louis, MO, Society of Mining Engineers, Littleton, CO, pp. 1–25.

Thrush, P.W., ed., 1968, *A Dictionary of Mining, Mineral, and Related Terms*, US Department of the Interior, US Bureau of Mines, Washington, DC, pp. 1269.

FLUORSPAR

ROBERT B. FULTON, III AND GILL MONTGOMERY

Fluorspar is the commercial name for fluorite, a mineral that is calcium fluoride, CaF_2. The name, derived from the Latin word *fluere* (to flow), refers to its low melting point and its early use in metallurgy as a flux. It is the principal industrial source of the element fluorine.

Two other minerals, cryolite and fluorapatite, have significant fluorine content. Cryolite, sodium aluminum fluoride, Na_3AlF_6, is a rare mineral that has been found in commercial quantities only in Greenland. The natural material has been supplanted by synthetic cryolite for its principal industrial use in the manufacture of aluminum. Fluorapatite, $Ca_5F(PO_3)_2$, is a source of phosphate for fertilizer manufacture, containing a small percentage of fluorine. Commercially mined deposits of apatite have varying amounts of fluorine, chlorine, hydroxyl, and carbonate.

HISTORY

Fluorspar was used by the early Greeks and Romans for ornamental purposes as vases, drinking cups, and table tops. Various peoples, including the Chinese and the American Indians, carved ornaments and figurines from large crystals. Its usefulness as a flux was known to Agricola in 16th century Europe.

Fluorspar mining began in England about 1775 and at various places in the United States between 1820 and 1840. Production grew substantially following the development of basic open hearth steelmaking, wherein it is used as a flux. Use was stimulated by growth of the steel, aluminum, chemical, and ceramic industries, particularly during World Wars I and II. Fluorocarbons entered the picture in 1931. The use of anhydrous hydrogen fluoride (HF) as a catalyst in the manufacture of alkylate for high octane fuel began in 1942.

Differential flotation for separating fluorspar from galena, sphalerite, and common gangue minerals in the 1930s and the application of heavy media concentrating methods to the treatment of low grade ores in the 1940s were outstanding technological advances that facilitated increased production.

Pelletizing and briquetting of flotation concentrates for use in steel furnaces and the development of flotation schemes for beneficiating ores containing abundant dolomite and barite have been major improvements in the industry.

USES OF FLUORITE

Fluorspar is used to make hydrogen fluoride (HF), also called hydrofluoric acid, an intermediate for fluorocarbons, aluminum fluoride, and synthetic cryolite. It is used as a flux in the steel and ceramic industries, in iron foundry and ferroalloy practice, and has many minor specialized uses. Hydrogen fluoride is produced by reacting acid grade (97% CaF_2) fluorspar with sulfuric acid in a heated kiln or retort to produce HF gas and calcium sulfate. After purification by scrubbing, condensing, and distillation; the HF is marketed as anhydrous HF, a colorless fuming liquid, or it may be absorbed in water to form the aqueous acid, usually 70% HF.

Synthetic cryolite, organic and inorganic fluoride chemicals, and elemental fluorine are made from hydrofluoric acid. The acid itself is important in catalysis in the manufacture of alkylate, an ingredient in high-octane fuel for aircraft and automobiles; in steel pickling, enamel stripping, and glass etching and polishing; and in various electroplating operations. The manufacture of one ton of virgin aluminum requires about 12 to 29 kg of fluorine content in synthetic cryolite and aluminum fluoride. This quantity, through improved technology and recovery practices, is being lowered significantly in countries with the most advanced technology (i.e., Australia and Sweden), while others (i.e., Surinam and South Africa), remain at the high end.

Elemental fluorine is prepared from anhydrous hydrofluoric acid by electrolysis. Gaseous at room temperature and pressure, fluorine is compressed to a liquid for shipment in cylinders or in tank trucks. Elemental fluorine is used to make uranium hexafluoride, sulfur hexafluoride, and halogen fluorides. Gaseous uranium hexafluoride is used in separating U^{235} from U^{233} by the diffusion process. Sulfur hexafluoride is a stable high dielectric gas used in coaxial cables, transformers, and radar wave guides. Halogen fluorides have important applications, mostly as substitutes for elemental fluorine, which is more difficult to handle.

Emulsified perfluorochemicals, organic compounds in which all hydrogen atoms have been replaced by fluorine, are undergoing investigation as promising blood substitutes. They transport oxygen and, in conjunction with a simulated blood serum, perform many functions of whole blood. With further development, these organic compounds may ultimately, in emergencies, be useful in saving lives of animals and humans during periods of acute shortages of natural blood.

Inorganic fluorides are used as insecticides, preservatives, antiseptics, ceramic additives, and fluxes and in electroplating solutions, antioxidants, and many other products. Boron trifluoride is an important catalyst.

Organic fluorides are volume leaders in the fluorine chemical industry. Fluorinated chlorocarbons and fluorocarbons are prepared by the interaction of anhydrous HF with chloroform, perchlorethylene and carbon tetrachloride, and are characterized by low toxicity and notable chemical stability. They perform outstandingly as refrigerants, aerosol propellants, solvents, and cleaning agents and as intermediates for polymers such as fluorocarbon resins and elastomers. Fluorocarbon resins are inert compounds that have unusually low coefficients of friction and have found a number of applications as lubricants for parts that cannot be oiled, e.g., bearings for window raising equipment located inside of automobile doors, in small electronic equipment, for the manufacture of chemical resistant gaskets and valve parts, as pipe and tank linings, in flexible tubing and containers, and on non-stick cookware.

In the steel industry, fluorspar is used as a flux in basic open hearth, basic oxygen, and electric furnaces where it is added to the heats in amounts ranging from 1 to 10 kg/t of steel produced. The average in the United States is currently about 1.4 kg/t but is declining as some steel plants discontinue using it. Fluorspar promotes fluidity of the slag and thus facilitates removal of sulfur and phosphorus from the steel into the slag. It serves the same purpose in iron foundries, where it is added to the cupola charge in the proportion of around 7 to 9 kg/t of metal melted.

In the ceramic industries, fluorspar is used to make flint glass, white or colored opal glasses, and enamels. Flint glass mixtures commonly contain 3% fluorspar. Opal glasses, containing 10 to 20% fluorspar, are used in containers for foods, drugs and toiletries, and in ornamental glassware and lavatory and restaurant fixtures. Opaque enamels are used to cover steel stoves, refrigerators, cabinets, bathtubs, and cooking ware and for facings on brick, tile, and other structural materials. Fluorspar makes up to 3 to 10% of the weight of the enamel. Many types of welding rod coatings incorporate fluorspar or fluorspar mixtures. Ceramic grade fluorspar is used in the manufacture of magnesium and calcium metals, and in the preparation of some manganese chemicals.

Lower grades of ceramic spar are used in making fiberglass insulation, in zinc smelting, as a clinkering aid in the manufacture of portland cement, as an inhibitor of vanadium green scumming in the manufacture of buff-faced brick, and as an abrasive on certain types of sandpapers. Various grades of fluorspar are used in electric furnace manufacture of calcium carbide, in making electrodes for arc lamps, and as a bonding material for abrasive wheels, among numerous minor uses.

Product Specifications

The three principal market grades of fluorspar are acid, ceramic, and metallurgical. Specifications for these are fairly well defined although the requirements set by individual consumers may vary in detail.

Acid grade fluorspar is defined as not less than 97% CaF_2, but some manufacturers of hydrofluoric acid, both in the United States and in Europe, can use 96%, or slightly lower, if the remaining impurities are acceptable. The Russian specification for acid spar from the Mongolian flotation mill at Bor Ondor is not less than 92% CaF_2. The US tariff structure formerly favored fluorspar of 97% or over, but in 1990 regulations imposing a 13.5% ad valorem duty on less than 97% material was suspended through December 1992, making it duty free, while keeping a duty of $2.10/t on acid grade. Users specify limits for silica, calcium carbonate, arsenic, lead, sulphide sulfur, phosphorous, and other deleterious constituents. Moisture content of dried concentrates for acid and ceramic grades is usually specified to be not more than 0.10% H_2O, but in some ceramic uses up to 1 or 2% may be required to relieve dust problems in handling bulk materials.

There are two standard ceramic grades: No. 1 ceramic containing 95 to 96% CaF_2 and No. 2 ceramic which customarily contains 85 to over 90% CaF_2. A medium grade of about 93 to 94% CaF_2 is specified by some users. In some instances, buyers of ceramic grades may specify not over 2.5 or 3.0% silica, limited amounts of calcite and ferric oxide, and only traces of lead and zinc sulfides. Practically every ceramic grade user has his own specifications, and producers tailor their products accordingly.

In the United States, metallurgical grade, called metspar, contains a minimum of 60 effective percent of fluorspar and generally not over 0.30% sulfide sulfur and 0.25 to 0.50% lead. The effective percentage is calculated by multiplying the silica percentage in the analysis by 2.5 and subtracting this number from the calcium fluoride percentage in the analysis; the result is termed effective units. Thus, a concentrate with 85% calcium fluoride and 6% silica would be 70% effective grade. To qualify as lump or gravel spar, it must pass a 2.5 to 3.8 cm screen and contain less than 15% of material passing a 1.6 mm screen.

In markets outside the United States, metspar typically contains a minimum of 80% CaF_2 and a maximum of 15% silica.

In fluorspar's use as a flux in the steel industry, the trend is toward briquettes and away from metspar due to growing scarcity of satisfactory gravel fluorspar. Briquettes are made from the fines accumulated during metspar preparation and from flotation concentrates. Most briquettes, from peach seed sizes up to 5 cm^2, are made on roll presses. The most popular binders used are molasses and lime, the use of which does not require baking ovens. Pellets have been made on balling machines using sodium silicate binder, and extruded shapes have been fashioned. There is a strong trend toward making diluted briquette containing steel mill wastes, such as mill scale, flue dust, shredded scrap, iron ore fines, and manganese ore fines. Some diluents have fluxing values. Briquettes that contain as little as 25% CaF_2 are in common use, but those made from acid grade fluorspar grading above 90% are in more demand. Users of the diluted briquette claim fluorspar economy plus a way to make use of steel mill wastes. Fluorine from fluorspar can have a serious adverse effect in steelmaking, causing erosion of refractory linings of the furnace. This has caused some steel manufacturers to severely reduce or even eliminate the use of fluorspar.

GEOLOGY

Composition and Properties

Theoretically pure fluorite contains 51.1% calcium and 48.9% fluorine. Substitution of small percentages of cerium and yttrium for calcium has been noted. Inclusions of gases and fluids, such as petroleum and water, and of solid minerals such as pyrite, marcasite, and other sulfides are common. Free fluorine is present in some crystals. The fluorspar of commerce contains attached and admixed mineral impurities, such as calcite, quartz, barite, celestite, various sulfides, or phosphates.

Fluorite tends to occur in well formed isometric crystals, forming cubes and octahedrons. It also occurs in both massive and earthy forms, and as crusts or globular aggregates with radial fibrous texture. Crystalline fluorspar exhibits a great range of colors, from colorless and waterclear to yellow, blue, purple, green, rose, red, bluish and purplish black, and brown. The colors may occur in alternating bands parallel to cube faces. Coloration may be altered by exposure to X-rays, heat, ultraviolet light, and pressure. Colors are caused by a variety of factors including the presence of trace impurities and displaced ions in the lattice. Long exposure to sunlight, such as occurs on mine dumps, frequently results in the fading of the original coloration.

Some varieties fluoresce blue or violet under ultraviolet light or cathode rays. Some specimens phosphoresce when heated or after exposure to sunlight or ultraviolet light, and some exhibit triboluminescence.

The mineral has a hardness of 4, and is the type mineral of that hardness on the Mohs scale. Its specific gravity is 3.18 when crystalline but ranges from 3.01 to 3.6 in various forms. The luster is vitreous. The mineral has perfect octahedral cleavage, and diamonds made by cleaving off the corners of cubic crystals are seen in many collections.

When pulverized and treated with sulfuric acid, fluorite decomposes into gaseous hydrogen fluoride and calcium sulfate, which is the fundamental reaction in the production of hydrofluoric acid. When added to metallurgical slags, it imparts greater fluidity at lower temperatures making it valuable in steelmaking and ferrous and nonferrous foundry practice. In ceramic mixes, it promotes crystallization around individual centers and thus is useful in making opal glasses. Crystalline fluorite has a very low index of refraction (n = 1.4339), low dispersion, is isotropic, and has an unusual ability to transmit ultraviolet light. These are the properties that make it useful as prisms in optical systems and as components of high quality, special purpose lenses. Synthetic fluorite has replaced the natural mineral for optical uses.

Modes of Occurrence

Fluorite occurs in a wide variety of geological environments, evidencing deposition under an extended range of physical and chemical conditions. At one extreme, it is present as an accessory mineral in granites and related igneous rocks; at the other extreme, it is sometimes found as crystals in geodes and as botryoidal linings in limestone caves. From an economic standpoint, the most important modes of occurrence of the mineral are:

1. Fissure veins in igneous, metamorphic, and sedimentary rocks.
2. Stratiform replacement deposits in carbonate rocks.
3. Replacements in carbonate rocks along contacts with acid igneous intrusives.
4. Stockworks and fillings in shattered zones.
5. Deposits at the margins of carbonatite and alkalic rock complexes.
6. Residual concentrations resulting from the weathering of primary deposits.
7. Occurrences as recoverable gangue in base metal deposits.

Less common modes, in some places commercially important, include:

8. Fillings in breccia pipes.
9. Fillings in open spaces.
10. Pegmatites.

Least common but of potential importance are:

11. Deposits in lake sediments.

Fissure Veins: Fissure veins, usually along faults or shear zones, are the most readily recognized form in which fluorspar deposits occur. Silica, calcite or other carbonates, iron, lead and zinc sulfides, and barite are the typical associated minerals. In some vein deposits, such as those in the Rosiclare District of southern Illinois, fluorite appears to have replaced a prior vein filling of calcite. Along some veins in carbonate rocks, fluorspar has replaced the wall rock at intersections with favorable beds, providing large minable tonnages. Although vein structures are remarkably persistent, the fluorspar commonly occurs as lenses or ore shoots separated by barren or poorly mineralized portions of veins. Ore shoot widths of 0.5 to 10 m and lengths of 60 to 300 m are common, but there is great variation from deposit to deposit. Vein systems may be up to several kilometers in length, and ore may be present to depths of 300 m or more below the surface. The CaF_2 content of minable portions of veins normally ranges from 25 to 80%, although grades above 90% occur in limited areas.

Important vein deposits have been developed recently in China, adding to the world's great vein deposits that include the Osor deposit in northeastern Spain, the Torgola deposit in northern Italy, the Muscadroxiu-Genna Tres Montis vein system in Sardinia, Hucklow Edge and Longstone Edge-Sallet Hole vein systems in England, the Rosiclare-Goodhope vein system in Illinois, and the El Hammam vein system in Morocco.

Stratiform Deposits: Stratiform, manto, or bedded deposits occur in carbonate rocks. Certain beds are replaced along or adjacent to structural breaks such as joints or faults. This relationship to structural features is very clear in some deposits but obscure in others. Frequently there is a sandstone, shale, or clay capping. Typically, there is evidence of loss of volume in the replaced zones with attendant development of gentle synclinal structures in overlying strata or of collapse structures, the latter sometimes pipelike in shape. In some districts, no connection is recognized between the mineralization and any igneous activity. In southern Illinois, mafic dikes, offset by the younger mineralized structures, intrude host rocks superjacent to a rift complex (Baxter and Bradbury, 1989). Post-dike igneous activity, related to rift reactivation is invoked as a deep seated source of fluorine. In other districts, such as the Encantada District in northern Coahuila, Mexico, the presence of rhyolite plugs and sills in the general vicinity of the spar deposits, and the association of spar with rhyolite injections along bedding planes, suggests a more direct association.

Stratiform deposits are known in many parts of the world and are particularly well developed in the Cave in Rock District of southern Illinois, in the northern part of the State of Coahuila in Mexico, and in the Ottoshoop (Zeerust) District of the Province of Transvaal in the Republic of South Africa. In Illinois, the deposits range from a few centimeters to 6 m thick, 15 to 150 m wide, and are up to 7 km long. In Coahuila, the individual ore bodies are smaller, but relatively more numerous and widespread. In the Ottoshoop District, bedded deposits occur in an area 16 km long and 10 km wide in a dolomite facies that underlies a prominent chert bed.

In stratiform deposits, textural features, such as sedimentary banding of the parent rock, are commonly preserved. A massive, crystalline ore-type, frequently associated with the banded ore, appears to have filled open spaces left from the dissolution of limestone by the orebearing solutions or their precursors. Minerals accompanying fluorspar in stratiform deposits are calcite, dolomite, quartz, galena, sphalerite, pyrite, marcasite, barite, and celestite. CaF_2 content in minable deposits ranges from 15% upward. A few ore bodies in Illinois yield direct shipping metallurgical grade fluorspar containing 85% or more CaF_2.

Replacement Deposits: Replacement deposits in carbonate rocks are well developed along contacts with intrusive rhyolite bodies in the Rio Verde, San Luis Potosi, and Aguachile districts in Mexico. They include some of the largest and highest grade fluorspar deposits. The fluorspar, not thought to be contact metamorphic in origin, was apparently introduced later by ore solutions that followed conduits along the contact zone and, either massively or selectively along certain beds, replaced the limestone outward from the contact. At Aguachile, cross sections show ore shoots resembling one side of a Christmas tree.

Stockworks: Fluorspar often occurs as stockworks and fillings in shear and breccia zones. Many occurrences in the American West are of the stockwork type and, though wide, usually have low overall CaF_2 content. Deposits in the Zuni Mountains of New Mexico and near Jamestown, CO, are examples. The Zwartkloof deposit, in the Transvaal Province of South Africa, consists of three vertical breccia zones containing stockworks of fluorite-carbonate veins. The mineralized breccia zones are in an east-west line and developed in felsite. The largest zone is 60 by 180 m in plan and persists to 900 m below the surface. The fluorspar grade is about 14%. The Buffalo deposit, near Naboomspruit in the Transvaal, consists of a network of fluorspar veinlets that occur in sill-like bodies of fine-grained, pink granite in coarse, red granite of the Bushveld complex.

Carbonatite and Alkalic Rock Complexes: Fluorspar is a common mineral in carbonatite and alkalic rock complexes, rarely in sufficient abundance to be economic. The Okorusu deposit in Namibia is of this type and consists of a number of bodies of fluorspar in limestones, quartzites, and related rocks that have been intruded and metamorphosed by an alkaline igneous rock complex that includes a nepheline syenite stock. The fluorspar replaces bedded and brecciated limestone, marble, and quartzite, forming large lenticular masses. Apatite and quartz are abundant accessory minerals. At Amba Dongar, India, veins and replacement bodies of fluorspar occur in carbonate rocks bordering ankeritic carbonatite intrusives. Fluorite is also present in the carbonatite itself.

Residual Deposits: Concentrations of fluorspar in clayey and sandy residuum, from surficial weathering of fluorspar veins and replacement deposits are, principally, sources of metallurgical spar. This category includes detrital deposits blanketing the apex of veins, as well as deeply weathered upper portions of the veins themselves. Such deposits, extending to depths of 30 m or more, were of major importance in Illinois and Kentucky. Similar deeply weathered ore has been mined in England, Thailand, and the Asturias District of northwestern Spain. Weathered residuum, called *kokoman*, is mined in the Marico district in South Africa, where it is the result of the weathering of gently dipping replacement bodies.

Gangue Mineral: Fluorspar occurs as a major gangue mineral in lead-zinc veins in many parts of the world, and in some, averaging 10 to 20%, is economically recoverable. Acid-grade fluorspar has been produced on a large scale from lead-zinc mill tailings near Parral in Mexico.

Breccia Pipes: Fluorite occurs in breccia pipes in the Thomas Range in Utah, and near Beatty, NV. Pipes in the Thomas Range are circular to oval in plan, up to 45 m in diameter and over 60 m in depth. They formed in dolomite by replacement along shattered zones associated with faults and intrusive breccias. The fluorspar occurs as soft, friable masses and in boxworks of fine-grained more resistant veinlets, and is nearly unrecognizable. At the Daisy mine near Beatty, fluorspar replaces brecciated dolomite in pipe-like bodies bounded by gouge zones along two sets of intersecting faults.

Fillings in Open Spaces: Fluorspar occasionally partially fills open spaces, both in veins and stratiform deposits. Spectacular examples of this type occur in the San Vicente District of northern Coahuila, where fluorspar occurs in veins and mantos as pure massive incrustations of mammillary, stalactitic, and stalagmitic forms. The Fluorspar-Gero-Penber vein system of the Northgate mine in Colorado is a similar occurrence where fluorspar occurs in botryoidal layers on the walls of open fissures and as concretionary coatings surrounding loose fragments of country rock. The lower parts of open areas of these fissures are partly filled with concretionary pebblelike masses, in places cemented into porous, rubbly aggregates.

Pegmatites: Many pegmatites contain minor amounts of fluorspar. Grade in three ore bodies was high enough to support a mining operation at the Crystal Mountain occurrence in Montana. These large, tabular bodies of massive fluorspar also contain minor amounts of biotite, quartz, feldspar, and other igneous-type accessory minerals and occur in coarse-grained biotite granite. This deposit is of particular interest for the scandium content of its tailings.

Lake Sediments: Fluorspar occurs in unconsolidated clayey and sandy pyroclastic sediments in the beds of former lakes near Castel Giuliano, about 40 km north of Rome, Italy. Fluorine of volcanic origin permeated the lake sediments resulting in deposition of minute disseminated crystals of fluorspar, which make up as much as 50 to 60% of the clayey parts and 15% of the sandy parts of the deposits. It is accompanied by barite, apatite, calcite, dolomite, and opal.

Distribution of Deposits

Fluorspar deposits occur on every continent. Major producers have been the United States, Mexico, Spain, England, France, Italy, South Africa, China, Thailand, Mongolia, and the former USSR. Lesser producers include Argentina, Brazil, Bulgaria, Canada, Czechoslovakia, Kenya, Tunisia, Morocco, Germany, Romania, India, and Korea. Countries now most active in exporting are China, Kenya, Mongolia, Morocco, and South Africa.

North America: *Canada.* The only large production in Canada was from the St. Lawrence area of the Burin Peninsula in southern Newfoundland. Over 3 Mt of ore was produced from 1940 until operation ceased in 1978. Production resumed in 1987, then stopped in 1990. Lesser deposits occur at Madoc, in southern Ontario and near Wilberforce and Cobden, also in Ontario. Other occurrences in Canada are at Rock Candy and Birch Island, near Kamloops and Quesnel Lake, both in British Columbia; along the Lower Laird River in northern British Columbia; and at Lake Ainslie on Cape Breton Island in Nova Scotia.

On the Burin Peninsula, fluorspar occurs in veins in granite and rhyolite porphyry along steeply dipping faults. Average thickness of higher grade veins is 1 to 2 m and of lower grade veins is 4 to 6 m. Some veins have been traced for more than a 1 600 m.

Veins at Madoc are small and have been worked principally in wartime. The veins at Lake Ainslie are low grade fluorspar with associated abundant barite. A deposit at Birch Island remains undeveloped with reserves said to be 1.5 Mt of 29% CaF_2 grade. Bedded deposits on the Lower Laird River near Mile 497 on the Alaska Highway in northern British Columbia have not proved to be of current commercial interest. Two low grade fluorite deposits near Likely, BC, were reported to be under exploration early in 1981.

Mexico. Mexico vies with China for being the world's largest producer of fluorspar. In 1981, production was 1.1 Mt, having held at that approximate level steadily since 1975. After 1981, production declined sharply to levels in the range of 750 to 820 kt and, in 1990, to a low of about 630 kt. Important deposits occur in San Luis Potosi, Coahuila, Chihuahua, and Guanajuato.

Many types of deposits are mined. In the Musquiz District of northern Coahuila, mantos in Cretaceous limestones are most important. Northwestern Coahuila has vein deposits and replacement deposits in limestone associated with rhyolite sills and plugs. Deposits in the Paila District of central Coahuila are veins. In San Luis Potosi and adjoining Guanajuato, there are several large replacement deposits in limestone, associated with rhyolite intrusives. Numerous vein deposits occur in Chihuahua, but the principal production is acid grade concentrate recovered from sulfide mill tailings near Parral.

At the height of production, eight flotation plants were in operation in Mexico with annual production capacity of about 600 kt of acid grade. A plant with a nominal capacity of 90 ktpy, formerly located at Eagle Pass, TX, and based on Mexican ore, has been moved into Mexico. After several mill closures and curtailment of production by others, total acid grade output was reportedly about 400 kt in 1990. Metspar production was about 200 kt, and ceramic output was a nominal 30 kt. While much is exported, Mexican consumption of acid grade has increased substantially, resulting in domestic acid grade sales of about 200 kt in 1990, when four HF acid plants were in operation.

The flotation mills continuing in operation derive their ore mainly from manto deposits of Coahuila and the large Las Cuevas replacement ore body in San Luis Potosi.

United States. ALASKA—The Lost River deposit, located in the western portion of the Seward Peninsula 140 k northwest of Nome, was worked intermittently for tin from 1904 to 1955 and has since been explored for fluorspar, beryllium, and tin.

Fluorspar-tin-tungsten-beryllium mineralization occurs in acidic dikes, in skarn deposits along granite-limestone contacts, and in brecciated limestone adjacent to thrust faults. Mineralization in one zone was reported to be 27 Mt grading 16.35% CaF_2, 0.15% tin, and 0.03% tungsten. Fluorspar also occurs in six other zones in the district (Anon., 1972), but there has been no production.

ARIZONA—Although 56 occurrences of fluorspar were listed in one summary (Van Alstine, 1969) and 96 in another (Elevatorski, 1971), cumulative production to date has probably been less than 50 kt. Deposits are widely scattered in the southern half of the state, principally as small epithermal veins in fissures and brecciated zones along faults in intrusive, metamorphic, and sedimentary rocks. No large replacement deposits have been found. Fluorite occurs mostly with calcite and quartz in nonmetaliferous veins. In the Castle Dome District, it is a gangue mineral along with barite, quartz, and calcite in veins which were mined on a small scale for their lead-silver content. The Castle Dome, Duncan, Sierrita Mountain, Tonto Basin, and Whetstone Mountain districts have been producers. The Lone Star mine, in the Whetstone Mountain District, yielded 10 kt of metallurgical spar in 1967 but then closed down. Vein ore from the Packard and other nearby mines was processed into acid grade fluorspar at a flotation mill in the Tonto Basin area until the veins were exhausted. Ore from the Turkey mine on McFadden Peak, north of Globe, was processed at that mill before it was decommissioned in 1979.

CALIFORNIA—Although there are many occurrences of fluorspar in California, only small amounts have been produced for ceramic and metallurgical purposes. Metspar was produced on a small scale from the Clark Mountain area of San Bernardino County in the 1950s and 1960s. California occurrences have been described by Chesterman (1966) and Elevatorski (1968).

COLORADO—There has not been production from Colorado in recent years. However, cumulative production from that state is estimated at about 2.5 Mt. During some past years, Colorado ranked second in annual production, after Illinois.

A total of 63 occurrences have been identified in the state (Van Alstine, 1964), including five major districts, all now idle. Northgate and Jamestown districts were the largest and most recently closed. The others are Browns Canyon, Poncha Springs, and Wagon Wheel Gap.

In the Jamestown District, fluorspar occurs in pipelike bodies and breccia zones in granite and granodiorite and in associated veins. Output of the Burlington mine was treated at the former Valmont Mill near Boulder. Other mines include the Argo, Emmett, and Blue Jay, ore from which was concentrated to acid grade in a mill located at Jamestown. Neither mill exists now.

At Northgate, fluorspar occurs in veins in shear and breccia zones in granite and schist and has been mined open cut and underground. During World War II, the district produced metspar. After the war, the ore was concentrated to acid grade in a mill built in 1952, now idle.

The Browns Canyon District included a number of mines developed on veins in rhyolite and in granite. Two flotation mills were operated at various times. In the Poncha Springs District a flotation mill operated for several years in the mid-1950s on low grade ore consisting of fluorspar veinlets in fractured and brecciated zones in granite and granite gneiss. The Wagon Wheel Gap mine, on a vein in a sheeted zone in rhyolite, yielded metspar.

IDAHO—The principal deposits of fluorspar in Idaho are in the Bayhorse-Keystone Mt. area of Custer County, southwest of Challis, and in the Meyers Cove, Big Squaw Creek, and Stanley areas (Anderson and Van Alstine, 1964).

The Bayhorse dolomite replacement deposits, as outlined by core drilling and partially developed for mining, may exceed 1.5 Mt of 36% CaF_2, with a 25% grade cutoff. Other deposits, both vein and replacement, occur in the Bayhorse District and on adjoining Keystone Mt.

A flotation mill operated briefly at Meyers Cove in the fifties, producing about 13 kt of acid grade concentrate from fluorspar mined in brecciated volcanics and fissure fillings. Later, a heavy media plant was built to make metallurgical gravel spar, but this venture was unsuccessful due to excessive silica locked in the ore. This mill was moved to the Garden Creek area west of Challis to be used on free-milling vein ores but is not in operation.

At Big Squaw Creek, a vein is exposed in the northeast canyon wall of the Salmon river west of Shoup. At Stanley, fluorspar occurs in shear zones in granite and volcanics.

ILLINOIS-KENTUCKY—Current production is limited to one flotation mill situated in an area of 1 800 km^2 in southern Illinois and adjacent Kentucky. For a number of years, the Illinois-Kentucky district was the world's largest producer (Baxter, Bradbury, and Hester, 1973) and is now the sole, significant US producing district. Shipments of fluorspar concentrates of all grades during the period 1880 through 1990 are estimated to have totaled 11.2 Mt, which is probably equivalent to mine production of approximately 33 Mt of crude ore.

Vein deposits are more numerous, but since 1950 the bedded deposits near Cave-in-Rock, IL, have been the source of most of the district's output and outrank the vein deposits in terms of total, all time, district production. At various times, as new vein deposits have been discovered and mined, production from vein deposits has regained some of its former importance, especially in Kentucky.

In addition to the major district, small production has been recorded from veins southwest of Lexington, in central Kentucky.

Vein deposits of the Illinois-Kentucky District occur along an extensive and intricate system of faults in sedimentary rocks of Mississippian age (Baxter, Potter, and Doyle, 1963, Baxter and Desborough, 1965, Baxter, Desborough and Shaw, 1967). Most of these faults are the steeply-dipping normal type, trending northeast, with displacement ranging from a few meters to over 300 m. The most important deposits are on faults of 15 to 150 m displacement. Veins generally range in width up to 4 m and to as much as 18 m in exceptional cases. They have been mined for as much as 270 m vertically and for as much as 3 km along faults. Veins pinch and swell horizontally and vertically with the proportion of fluorspar to total vein material ranging from 0 to 100% within short distances. Near the surface, and to depths as great as 75 m, the veins contain much clay owing to the dissolution of vein calcite and limestone wall rock by circulating ground water, leaving behind the argillaceous material originally in the limestone. Strong walls are found at depth in most places. A large inflow of water at high pressure was characteristic of the deeper mines, particularly those at Rosiclare.

Major replacement deposits occur in Illinois adjacent to a major structure, the Rock Creek Graben (Baxter, Potter, and Doyle, 1963). These include deposits of the well known Cave-in-Rock District, located on the southeast side of the graben, northwest and north of the village of Cave-in-Rock, and the relatively recently discovered Harris Creek District, approximately 6 km northwest of the Cave In Rock ore trend and on the northwest side of the graben (Baxter and Bradbury, 1989). Other replacement deposits are in Kentucky, south of Carrsville, and near Hampton, in Livingston County.

The Cave-in-Rock group of mines occupies a zone of minor structural disturbance 1.6 km wide and over 9 km long, paralleling a major northeast trending fault which has maximum displacement of about 300 m (Baxter, Potter, and Doyle, 1963). The deposits, characteristically long and narrow, lie along one or both sides of minor fractures and small faults. Major deposits occur along fracture-fault systems with less than 0.3 m of displacement. Widths of 15 to 45 m and thicknesses of 1 to 5 m are typical. Maximum widths of 100 to 110 m have been found. Lengths are mostly 60 to 450 m.

One ore body is known to extend continuously for at least 7 km. The deposits occur through a vertical range of about 55 m in four preferred sets of limestone beds and follow those beds' down dip from the outcrops (which were discovered originally in a prominent bluff) to depths well over 300 m. Mining in the former Minerva #1 Mine (now owned and operated by Ozark-Mahoning) currently is below 400 m and drilling discloses ore at least 36.6 m deeper in lower horizons. Three to four bedded horizons often exist superimposed along the same fracture system, and within a vertical interval of 50 m.

Structural and stratigraphic relationships of deposits in the Harris Creek District, north of the Rock Creek Graben, are similar to those of the Cave-in-Rock District.

Fluorite, accompanied by sphalerite and galena in many ore bodies, is the principal mineral in replacement deposits. Quartz, barite, and calcite are common, along with pyrite, marcasite, witherite, and strontianite. Banded ore, known locally as coontail ore, is characteristic of some deposits and may be a relic from the replaced limestone or may result from rhythmic metasomatic deposition. Crude ores have 15 to 90% calcium fluoride; some contain up to 3½% zinc and 5% lead.

Residual deposits have been less important producers but have yielded high quality fluorspar. These gravel spar deposits are as much as 20 m wide and extend to depths of a 30 m or more. The fluorspar occurs as weathered fragments, which range from boulder-size down to that of sand grains, dispersed in a clayey matrix.

MONTANA—Most of the fluorspar production has come from the Crystal Mountain deposits near Darby in Ravalli County. The deposits of massive fluorite in granite and metasediments were discovered in 1951 and developed into one of the nation's major metallurgical grade fluorspar mines. The mine was a side-hill benched open pit, at the 2 100 m elevation on Sapphire Mt. Ore was mined, hand-sorted, and sized at the mine continuously, from 1952 to 1973. This roughly concentrated ore was hauled 40 km by truck to Darby, where finished concentration was accomplished in a 60-tph heavy media cone plant. The metspar was shipped by rail to steel plant consumers and government stockpiles.

Numerous other fluorspar occurrences have been recorded in Montana, but have no production history. Fluorspar occurs as gangue in metalliferous veins in marginal portions of the Boulder batholith in west-central Montana and in the north-central part of the state is present as veins and replacement seams in intrusive syenite. In several other north-central areas, fluorspar occurs in limestone within terrane characterized by alkalic intrusives.

The presence of fluorspar has been used as a guide to some types of gold deposits. In the Old Glory deposit in the Pryor Mountains, fluorspar occurs with the uranium mineral, tyuyamunite, replacing limestone in brecciated and cavernous portions of the Madison limestone.

NEVADA—Fluorspar became commercially important in Nevada in 1919 when the Daisy mine near Beatty began its long productive life. Relatively small production has, in recent years, placed Nevada second in domestic production, behind Illinois.

Nevada deposits are described in a publication by Papke (1979). Most of the 62 deposits he describes lie in a broad, arcuate band across the southern part of the state and northward through the central part. Total production has been about 500 kt, most from the Daisy (Crowell) mine in southern Nye County, the Baxter (Kaiser) mine in northeastern Mineral County, the Goldspar mine near the Daisy, and the Carp (Wells Cargo) mine in southeastern Lincoln County. About 19 other deposits have had some production, but only the Daisy mine is active. More than a fourth of the deposits are in the Quinn Canyon and Adaven District in the southeastern part of the state, but the small size of the deposits and remoteness of this area has hindered their development. Small metspar shipments have been made from Caliente.

Deposits are hydrothermal or pyrometasomatic. They occur as replacement bodies, veins, replacement and/or filling in jasperoid bodies, stockworks, and breccia pipes. The Daisy deposit is mined underground, with total past production of about 200 kt. It consists of irregular, generally steeply dipping, structurally complex, replacement bodies in dolomite. The ore is soft, friable, and relatively low in silica. Most has been used as metspar. Some has been shipped to a southern California cement producer. The Baxter deposit, a vein in andesitic volcanic rocks, developed to a depth of 180 m, had a total mine production of 160 kt, of which two-thirds was made into acid-grade concentrate in a mill operated from 1952 to 1957 by Kaiser Aluminum and Chemical Corporation at Fallon, NV. The Goldspar and nearby Mary deposits are breccia pipe and replacement bodies in dolomite. All the fluorspar from these open pit mines was used in making cement. At the Carp mine an estimated 40 kt of metspar was produced in four open pits. The ore came from nearly flat, manto-like replacement bodies, generally concordant with bedding of the dolomite host rock.

NEW MEXICO—There are numerous occurrences of fluorspar in New Mexico, but production has been small. Van Alstine (1965) listed 65 areas in the central, north-central, and southwestern parts of the state. The fluorspar map MR-60 of the USGS and accompanying tabulation (Worl, Van Alstine, and Heyl, 1974) classified some 51 locations as to type, tonnage, and associated minerals.

Eight areas each produced at least 20 kt of crude ore and total, cumulative, statewide production has probably been about 700 kt. Production peaked in World War II when New Mexico was the fourth largest producing state after Illinois, Kentucky, and Colorado.

Most production came from the Zuni Mountains, Sierra Caballos Mountains, Gila District, Burro Mountains, Cooks Peak, and from Fluorite Ridge and Tortugas near Las Cruces. Flotation plants have operated at various times at Chise, Silver City, Gila, Lordsburg, Deming, Los Lunas, San Antonio, and near Hachita.

In the Zuni Mountains, the "27" vein deposits were the most productive, with the Mirabal deposit contributing. Until closed in 1953, these had total production of about 200 kt of ore that was concentrated to acid grade at the Zuni Milling Co. flotation mill at Los Lunas (Goddard, 1966).

Important producers in New Mexico were the White Eagle and Greenleaf deposits, near Deming, in the Cooks Peak District; mines in the Fluorite Ridge District; and Burro Chief and Shrine deposits in the Burro Mountains. All are vein deposits found in brecciated or fractured zones along faults in igneous rocks. The Lyda K mine, south of Truth or Consequences, is a siliceous vein in Precambrian granite. The Ruby mine, north of Las Cruces, contains veins in limestone near an igneous contact. At Chise, small manto replacement bodies occur in limestone beds. The Bishops Cap deposit is a vein in limestone, with some adjacent replacement ore. In the Hansonburg District, east of San Antonio, fluorite occurs with lead and barite as vug-filling in brecciated limestone.

The Salado deposit, southwest of Truth or Consequences, consists of tabular masses of fractured and brecciated jasperoid with the fluorspar forming a filling and cement. Between 2.5 and 3 Mt of low grade mineralization have been estimated from drilling. A thickness of over 12 m in places would permit open pit mining, with little overburden. Difficult flotation metallurgy, low grade, and remoteness from market have inhibited development.

OREGON—Near Rome in the southeastern part of the state, fluorspar occurs as nearly spherical grains in volcanic tuff, tuffaceous mudstone, and mudstone of Tertiary lake deposits (Sheppard and Gude, 1970). The fluorspar content reaches 16% in some zones, but is generally much lower. The amount of contained fluorspar is estimated at 11 Mt. Because of very fine particle size, concentration to acceptable grade has not proved practical.

TENNESSEE—A small amount of metspar was produced from a small fissure vein in limestone in the central part of the state, where it occurs with barite and sulfides (Jewell, 1947).

Fluorite has been noted as a gangue mineral in zinc deposits near Carthage, locally comprising 12 to 15% of the ore, but averaging much less.

A drilling program in the Sweetwater area indicated fluorspar-zinc-barite mineralization several hundred meters below the surface in buried karst breccia zones similar to the Mascot District, TN, zinc deposits. Exploratory drifting was done from a 180 m shaft sunk in 1979, now plugged.

TEXAS—Metallurgical grade fluorspar production began in 1971 in the Christmas Mountains, south of Alpine, from bedded limestone replacement mantos adjacent to intrusive rhyolite bodies. In late 1980, a 270 m decline was sunk beneath a laccolithic structure, but results were disappointing and the work stopped in about 1986.

In the Eagle Mountains southwest of Van Horn, fluorspar occurs as replacement of limestone and as fissure deposits in rhyolite. In the period 1942 to 1950, about 15 kt of metallurgical gravel and acid grade were produced.

UTAH—Most of the deposits are located in the west-central part of the state in the Spor Mountain, Indian Peak, Pine Grove, and Star Districts. The Spor Mountain District, in the Thomas Range northwest of Delta, was the most consistent producer. The deposits are classified as pipes, veins, and disseminated bodies in dolomite and volcanic rocks. Most of the production has come from the pipes. The most productive mines were the Lost Sheep, Fluorine Queen, and Bell Hill.

In the Indian Peak District, veins occur along faults in volcanic rocks. The Cougar Spar and Blue Bell mines were World War II producers of metspar. Small production has also been recorded from the Pine Grove and Star districts, and from the Rain Bow mine in Millard County and the Silver Queen mine in Tooele County. Cumulative Utah production of crude ore to date is estimated to have been 230 kt.

WASHINGTON—A small production of metspar has come from the Mitchem mine near Keller, in Ferry County.

WYOMING—There are reports of fluorspar in Crook and Albany counties.

Central America: Occurrences have been reported in Honduras and Guatemala, but details are not available.

South America: *Argentina.* Vein deposits have been mined intermittently in the Sierra Comechigones, 50 km northwest of Cordoba, and some have been developed for production in the southern part of Rio Negro Province and the adjoining northern part of Chubut. In recent years metspar has been produced from a group of veins in the Sierra Grande District of Rio Negro, including the Delta vein, reportedly with 4.0 Mt of drill-indicated ore of 51% CaF_2 grade. A small flotation mill at Valcheta formerly produced concentrates for domestic use. In Chubut, a number of veins were discovered west of Puerto Madryn. The ore was used to supply a small 10 ktpy flotation mill that was operated in Chubut from 1984 through 1990.

Brazil. Principal deposits are in the vicinity of Criciuma in Santa Catarina Province of southern Brazil and in Parana Province near the town of Cerro Azul where a new 50 ktpy flotation mill started operation in 1989, based on a large alkalic replacement deposit. Small scale production has been reported from deposits in the provinces of Rio Grande do Norte, Paraiba, and Bahia.

In the Criciuma District, three flotation mills have an aggregate capacity of about 35 ktpy of acid grade concentrate. Known deposits are not large and continued output will depend on the discovery of new reserves and the resolution of water problems.

Chile. Small tonnage of fluorspar has been produced from veins in granite in the Province of Coquimbo. One deposit, the Mercedes mine, is near the village of Paihwano.

Europe: *France.* Important deposits occur in the Departments of Haute-Loire, Pyrenees-Orientales, Var, Tarn, Saone-et-Loire, and Puyde-Dome. Details have been described in a series of articles by Chermette (1972–1973, 1979).

Epithermal vein deposits occur in the Morvan, Auvergne, Limousin, and Albi Districts of the Massif Central, the Maures and Esterel Districts of the Mediterranean coastal area northeast of Toulon, in the eastern Pyrenees west of Perpignan, and in the Vosges. The vein deposits contain about 50% CaF_2, 20% silica, 5 to 10% barite, and 3 to 5% sulfides.

Stratiform deposits with about 35% CaF_2 and 15% barite occur in the Morvan District. Deposits having stratiform characteristics are important producers in the Escaro District in the Pyrenees southwest of Perpignan. Deposits at Le Beix, Chavaniac, and Le Maine in the Massif Central are exhausted. Fonte Sante, La Charbonniere and Langeac have also closed. Le Burc and Montroc (near Albi) and Escaro continue in operation. Fonte Sante was noteworthy for the economic occurrence of sellite (MgF_2) which is blended with the fluorspar product and is used in the Lyons hydrofluoric acid plant.

Germany. The former East Germany normally produced about 100 ktpy of acid grade, in a market insulated from outside competition. Its two principal flotation mills were located at Ilmenau-Gehren in Thuringia and at Lengenfeld in Saxony, each deriving its feed from nearby mines exploiting vein deposits. Following reunification in 1989 and the freeing of markets, these operations were noncompetitive and production ceased in late 1990 and early 1991.

Production in the former West Germany is limited to two surviving mills; one at Wohlfach and the other at Pforzheim, both in the Schwarzwald, with total 1990 acid grade production of about 100 kt. In both cases, the ore is derived from immediately adjacent vein deposits.

Greece. Fluorspar occurs as a gangue mineral in the silver ores of Larium, but Greece imports its requirements.

Italy. Italy produced about 80 kt in 1990. Mines in the Gerrei District of Sardinia are the only significant producers, producing ore from one of the world's longest and most persistent vein systems on which four underground mines have been linked along strike into a single producing unit. Ore production was about 290 kt in 1990, grading about 33% CaF_2 and 3.3% PbS. Smaller outputs have come in the past from Brescia, Trento, Bolzano, and Bergamo in northern Italy. The Genna Tres Montis-Muscadroxiu-S'Acqua Frida vein system in southern Sardinia, near the village of Silius, is credited with 8 Mt of ore reserve averaging 40 to 45% CaF_2, making it one of the world's largest deposits. Vein widths are 5 to 8 m. Barite and galena are recoverable accessory minerals. Underground workings have demonstrated a strike length of about 3 km. Numerous other vein deposits and a skarn deposit occur in southern Sardinia.

One flotation plant at Assemini, near Cagliari, is active with a nominal capacity of 200 ktpy of acid grade concentrate and metspar briquettes, plus barite and galena concentrates. Actual capacity is said to be about 150 ktpy due to restrictive labor regulations. An older mill with acid grade capacity of 40 ktpy is idle. In 1990, Sardinia produced about 80 kt of acid grade from 290 kt of ore.

In northern Italy, important deposits include Prestavel and Vallarsa near Bolzano, Torgola near Brescia, and Camerata Cornello near Bergamo, all of which are now closed.

The Pianciano deposits in the Castel Giuliano area, 40 km northwest of Rome, consists of fluorspar impregnating lake beds of volcanic ash (Spada, 1969). The fluorspar content, ranging from 20 to 55%, was derived from volcanic emanations along with barium, strontium, and phosphorous. The fluorspar is too fine-grained to be recovered economically.

Norway. Veins are reported in the Buskerud, Telemark, Vest Agder, and Drammen regions west and southwest of Oslo. The Lassedal mine near Kongsberg was a former producer.

Spain. Spain is estimated to have produced about 140 kt of metgrade and acid grade in 1990, 20 kt metgrade and 120 kt acid grade, all from the Asturias region.

Principal deposits are in the Asturias region of northern Spain, near the ports of Ribadella and Aviles; in the southeast Spain, near the port of Almeria; and in south-central Spain, near Seville and Cordoba. Deposits are both vein and replacement bodies. Acid-grade material was formerly derived by the retreatment of old lead mine wastes in the Almeria-Berja area. In 1980, seven companies produced acid grade, four in Asturias, two in the Seville-Cordoba region, and one near Berja. Grades of ore range from 40 to 45% for some of the vein mines, down to 15 to 20% in the case of the old lead mine waste. The large Osor vein in Gerona is exhausted. Weakening market conditions in 1982 led to closure, at least temporarily, of all but three mills. Two of the remaining mills, one in Asturias and the other in Almeria, belonged to Minersa. The third mill, near Berja, belonged to Minas de Orgiva. By 1990, only Minersa's Minas de Ana mill in Asturias remained in operation.

Sweden. The sole producer in Sweden, the government-owned Yxsoberg mine near Ludvika, 200 km northwest of Stockholm, where fluorspar occurs with tungsten and copper, closed in 1977. Other deposits have been reported between Branteriks and Onslunda in Skane in southern Sweden. The Osterlen District was formerly a small producer.

Switzerland. Occurrences, none of commercial importance, are known.

United Kingdom. Among numerous fluorite occurrences in the United Kingdom, present operations are located in the Southern Pennine Orefield of north Derbyshire and in the Northern Pennine Orefield of west Durham. Production comes from fissure veins and bedded replacement deposits, together with associated old lead mine waste and tailings dumps. The deposits occur in rocks of the Carboniferous Limestone Series (Mississippian). Ore is mined both underground and open pit. The deposits contain variable amounts of galena, sphalerite, calcite, barite, quartz, and iron sulfides. Some were originally mined for lead, even before written record.

Installed capacity for the production of all grades of fluorspar is now about 200 ktpy, of which about 160 kt is acid-grade capacity. Two mills are in operation, one in the south Pennines at Stoney Middleton in northern Derbyshire with capacity output of 100 ktpy, operated by Laporte Minerals; the other in the north Pennines at Frosterly in County Durham with a capacity of 60 ktpy, operated by Sherburn Stone Company.

Former Soviet Bloc. The former USSR is estimated by the US Bureau of Mines to have produced about 560 kt of fluorspar in 1979, and the same in 1980. Details regarding the sources of this output are not available, but deposits are reported in the Transbaikal, Tashkent, and Yaroslav regions. Czechoslovakia produces about 95 ktpy from vein deposits, mainly in the Krusne Hory (Erzgebirge) Mountains. Mongolia produced about 120 kt of acid grade in 1990 at Mongolsovtsvetmet flotation mill at Bor Ondor. Operations at Bor Ondor, supplanting an older operation at Berh, were started in 1982 as a joint venture of the Mongolian Peoples Republic and the USSR. Mining, which began in 1982, was followed with the mill operation in 1985. A separate joint venture between Mongolia and Czechoslovakia operates an open pit mine at Chulut Cagan Del, producing about 75 ktpy run-of-mine ore and shipping screened products to Czechoslovakia for flotation mill feed and for metallurgical flux. Total crude ore production in Mongolia in 1990 was possibly as much as 750 kt from ten identified mines. Romania produces about 20 ktpy, Bulgaria about 25 ktpy, and Yugoslavia has a small production. The republics of the former USSR import about 100 ktpy of acid grade and about 400 ktpy of met grade to supply their requirements of more than 1 Mtpy. Principal sources include China and Mongolia. A supply contract with Kenya terminated at the end of 1985 and was not renewed.

Africa: *Angola.* Deposits are reported in Angola, but details are lacking.

Kenya. A 100% government owned acid grade plant, formerly jointly owned with International Minerals and Chemical Company (IMC) and Bamburi Portland Cement, produces concentrates relatively high in P_2O_5 from epithermal vein and replacement deposits in the Kerio Valley, about 130 km northwest of Nairobi. Ore ranging from 30 to 45% CaF_2 is mined open pit. The acid grade mill's output capacity is rated at 9 kt per month. Concentrates are trucked to rail at Kaptagat, near Eldoret, and loaded on vessels at Mombasa. Other similar deposits are reported at other points along the rift structure.

Malawi. Exploration for fluorspar deposits related to carbonatite structures has been reported; five such occurrences are noted by Van Alstine and Schruben (1980).

Morocco. The El Hammam deposit near Meknes is a nearly vertical vein on an east-west fault in sericite schist, grading about 45% CaF_2. A flotation plant with a capacity of 110 kt/a of acid grade concentrate is operated by Societe Anonyme d'Entreprises Minier (Samine), owned 58% by Omnium Nord Africain and 42% by the Moroccan Government through Bureau de Recherches et de Participations Minieres (BRPM). Three other deposits are recognized in this region. Other veins occur in the Djebel Tirremi area, 10 km northeast of Taourirt, in eastern Morocco (Van Alstine and Schruben, 1980).

Mozambique. Large vein deposits occur approximately 150 km southeast of Tete. Remote from transportation and from sources of power, this area has been a minor producer of metspar. Further development may result as the area is opened up as a result of the construction of the Cabora Bassa dam on the Zambesi River. Six localities are listed by Van Alstine and Schruben (1980).

Nigeria. Deposits containing fluorspar with galena are reported in limestones in the Arufu District of Gongola State near the border with Benue State, in the Benue Trough about 80 km east of Makurdi, and fluorspar occurs in greisens near Jos (Van Alstine and Schruben, 1980).

South Africa. Acid grade output of the three operating flotation plants was about 260 kt in 1990. Other production consisted of 7.5 kt of ceramic and 41.5 kt of metspar. Acid grade is exported to Australia, Europe, Japan and North America through the port at Durban. In 1980, total production of all grades was almost 523 kt.

Domestically, AECI consumes acid grade to make hydrofluoric acid and the steel industry consumes essentially all of the metgrade as well as about 20 kt/a of 80% CaF_2 briquettes.

The Buffalo open pit mine of Transvaal Mining and Finance, a subsidiary of General Mining and Finance near Naboomspruit, exploits 14 to 16% grade deposits consisting of a network of fluorspar veinlets in sill-like bodies of fine-grained, pink granite in coarse, red Bushveld granite. Similar deposits occur nearby.

Near Warmbad, the Zwartkloof deposit consists of three breccia zones containing a stockwork of fluorspar veinlets in felsite. Operating difficulties and the low grade of the ore for underground operation, approximately 14% CaF_2, caused operations to close in 1973.

At the open pit Vergenoeg mine of Chrome Chemicals, a subsidiary of Bayer, near Pienaars-rivier, fluorspar ore grading 40% occurs with abundant iron oxides (up to 50%) in a massive deposit in felsite.

In the Marico-Zeerust District, fluorspar occurs as replacements in dolomite and in associated residual deposits. Much of the ore in dolomite is disseminated along intersecting fractures. For many years, metspar was produced by screening coarse fragments from residual surface deposits. Exploration has disclosed the existence of large reserves of primary ore, probably on the order of 60 Mt, containing 15 to 30% CaF_2. One open pit mine, belonging to Phelps Dodge Corp., is operating with an annual acid grade capacity of about 85 kt, produced from about 450 kt of ore grading 16 to 20% CaF_2. The former Marico mine of Barlow Rand is closed. The formerly projected operation of Armco Bronne failed to materialize.

Namibia. The principal fluorspar deposits are on the Marburg Farm, with lesser deposits on the adjacent Okorusu, Belvedere and Brandenburg Farms, located about 20 km northwest of Ojiwarango, which is 425 km northeast of the port of Walvis Bay. These fluorspar deposits are part of an alkalic complex that forms an erosionally broken, ring structure about 5 km in diameter. The fluorspar bodies lie mainly in the southern part of the complex where they occur, with accompanying thin magnetite cappings, as late metasomatic replacements of syenite, fenitic rocks, and carbonatite. Reserves are 6 Mt, grading about 60% CaF_2 and about 5% P_2O_5 in, the latter in fluorapatite. A flotation mill, rated at 50 ktpy, began operation in 1988 and has made sales to the United States, United Kingdom, and Germany, despite the drawback of relatively high P_2O_5 content (0.20%) in the acid grade product. A paved highway and rail provide alternative means of transport from Ojiwarango to the port at Walvis Bay.

Sudan. Fluorspar deposits occur at Jebel Sameih and Jebel Dumbeir approximately 320 km southwest of Khartoum and 1 330 km by rail from Port Sudan. At Jebel Sameih, a highly siliceous fluorspar vein occurs in a granitic host rock. At Jebel Dumbeir, fluorspar occurs in thin quartz-fluorite veins in metasediments, replacing bands of marble, and in bodies of fluorite associated with red feldspathic rock. No production is reported.

Tunisia. A flotation mill located at the Hamman Zriba deposit near Zaghouan, 54 km south of Tunis, produces about 30 ktpy of acid grade concentrate from a bedded deposit in Jurassic limestone. Proven reserves are said to be 4 Mt of ore containing 35% CaF_2 and considerable barite. In the same general area, replacement vein deposits occur at Djebel Staa, Hammam Djedida, Djebel Oust, and Djebel el Kohol. Other deposits are at Bourchiba, Cap Bon, and Bou Jaber.

Turkey. Fluorspar deposits in Turkey have been described (Anon., 1965); there is a small annual production.

Asia: *China.* This country is emerging as a major producer with an annual production of about 1 500 ktpy, mainly from deposits in the central and northern parts of Zhejiang Province, northeastern Hebei Province, eastern and southern Hunan, northern Guangdong Province, northeastern Anhui Province, eastern Shandong Province, Inner Mongolia, and Liaoning Province. In all, fluorspar occurrences are reported in 17 provinces.

China exported about 170 kt of acid grade to Japan and about 125 kt to the United States in 1990. Acid grade production has mushroomed with as many as 40 flotation mills now in operation, some obtained by conversion of pre-existing mills and others by new construction. Most of the fluorspar deposits now being mined in China are veins in acidic terranes. One, De An in Jiangxi Province, is known to be a series of large replacement ore bodies.

India. At Amba Dongar, 560 km north of Bombay in Gujarat State, veins and replacement bodies occur in the carbonate wall rock around ankeritic carbonatite intrusives, similar to the occurrences at Okurusu in Namibia (Deans et al., 1972). A heavy media and flotation concentration plant has been constructed nearby at Kadipani, with a nominal capacity of 37.5 ktpy of product of which 30% is acid grade and 70% metgrade. Reserves are reported to be 11.6 Mt of 30% fluorite content, of which proved reserves are said to be 4.7 Mt.

Other mineralized areas are reported at Chandri-Dungri in the Drug District of Madhya-Pradesh and at Mando-ki-Pal in the Dungarpu District of Rajasthan.

Japan. Japan formerly produced less than 10 ktpy of fluorspar from the Island of Hokkaido, but ceased production in 1972 and imports its requirements.

Korea, North. North Korea has an annual production of about 40 kt.

Korea, South. Fluorspar deposits occur in the northern, central, and southern parts of the country. Approximately 70 mines have been producers, formerly with annual production of about 50 kt largely for export (Kim et al., 1972), but production has been negligible since that time. Both vein and replacement deposits occur. Minable ore reserves have been estimated at 660 kt (Jee and Kye, 1971). Producers have included the Choonchon Shinpo, Dojon, Kumi, Busang, and Kumbo mines.

Pakistan. Minor production has been reported from the Kalat district 70 to 90 km south of Quetta in Baluchistan. There are also deposits south of Quetta.

Thailand. Fluorspar deposits occur in a region extending from the northern border with Burma southward for over 1 280 km to a point about 640 km southwest of Bangkok. Active mines are concentrated in three areas: Chiangmai, Lumphun, and Mai Hongsun Provinces in the north; Kanchanaburi, Petchaburi, and Ratchaburi Provinces southwest of Bangkok; and Krabi Province in the south (Gardner and Smith, 1965, Hodge, private communication, 1981).

The northern deposits are mainly replacements in Paleozoic limestones which, so far, have been found suitable only for the production of metspar, derived by hand sorting. As a result, there are large waste dumps containing low grade ore. Residual deposits are widespread. A heavy media concentrating plant, with a 60 ktpy production capacity originally installed at Ban Hong in Lamphun Province in the northern area, 515 km north of Bangkok, has been moved a short distance to Mae Tha. The product is trucked to river barges at Bank Pa In. In the southern area, the deposits are veins in metamorphosed, fine-grained, fractured Paleozoic clastics and in granite. Two flotation plants have been operating intermittently, one at Ban Lard in Petchburi Province, about 160 km south of Bangkok, with an annual production of 50 kt of acid grade concentrate, and another in Khlong Thom District, Krabi Province, about 100 km miles northwest of Trang, with an annual production of about 15 kt of acid grade.

Australia: Deposits of fluorspar have been found in all the Australian states except the Northern Territory (Liddy, 1971), but are undeveloped because they are small and remote. There have been investigations in the Chillagoe-Mungana-Almaden and Forsayth areas of northern Queensland, and in Western Australia in the Pilbara district 217 km southeast of Port Hedland, the Yinnietharra area 320 km from Carnarvon, and the Speewah Valley area southwest of Wyndham.

World Activity in Fluorspar

Reported reserves of fluorspar, along with annual world production, have been increasing for several decades as producer interest in fluorspar has grown, either to fill a need for captive raw material or in response to specific marketing opportunities. Most notable are increases in the People's Republic of China where fluorspar is widespread and has become a significant item of export for individual provinces. Provincial and even local governments have built flotation plants, many with capacities of only 10 to 12 ktpy. As a result, Chinese production has become a very significant factor in world markets. This flood of new production has come at a time of general market weakening due to restrictions on ozone depleting substances. Production in other countries has, as a result, had to be curtailed through temporary closure or, in some cases, complete shutdown of many mines and mills. The combination of weak market and substantial new production hit Mexican and European producers particularly hard, allowing only the most efficient to compete for the present. It is possible that the market may revive substantially if new products, replacing the ozone depleting substances, turn out to be ones with higher fluorine content than the products they are replacing.

EXPLORATION

Because fluorspar resists chemical weathering, it can be traced in the soils overlying weathered veins. Cleavage fragments, washed clean by rainwater or exposed in anthills or spoil piles from animal burrows, are useful clues. Silicified veins, resisting erosion, may stand up like a reef. Bedded deposits are less discernable, but, in areas of sufficient topographic relief, the outcropping edges of deposits or slumped fragments of ore may be found. Owing to its softness and cleavability, fluorspar does not survive in the beds of streams and ordinarily cannot be traced by panning.

The search for bedded deposits usually involves locating a mineralized horizon and following its downdip with vertical drillholes, drilled to intersect the favorable beds. Geophysical prospecting methods are not applicable. Photogeology can be of assistance in locating and tracing structures and outcrops, and as an aid in geologic mapping to determine the areal distribution of stratigraphic units known to be favorable for veins or replacement deposits.

Geochemical methods have been employed with varying degrees of success. Fluorine anomalies have been found in ground water and surface streams, in soil samples, and in stream sediments, but using such anomalies as a prospecting tool for the discovery of ore deposits has not been very successful. Various nonfluorine elements that can be detected geochemically can be guides to fluorspar deposits and insoluble residues of potential host rock limestones can serve as a sample medium (Erickson et al., 1987). Once mineralization has been detected, core or churn drillhole, prospect shafts, and drifts may be used to probe the structures.

PREPARATION FOR MARKETS

Mining Methods

In some areas, such as in the Illinois-Kentucky District and the Southern Pennine District of Derbyshire in England, veins may be weathered to depths of as much as 75 m. Such weathered ore, a mixture of clay and fragments of fluorspar and detached wall rock, may be mined open pit with draglines, scrapers, or power shovels to depths of as much as 50 m. Below that, underground mining methods, involving modified top slicing or overhead shrinkage stoping, are used.

Until the late 1960s, most vein mining in the Illinois-Kentucky District was done by shrinkage stoping and open stoping where strong walls occurred. Closely spaced shrinkage stope bins have given way to widely spaced bins, with electric and air slushers being used in the tops of stopes to transport the overbreak to the ore pass. Air-operated, rubber-tired, muckhaul units have been adapted to in-stope work. Where shrinkage stoping is used, broken ore is commonly moved to the shaft by track haulage using battery-powered locomotives and one- or two-ton side dump cars.

With the development of diesel haul units of less than 1.5 m width, one Illinois company changed from shrinkage stope mining to ramp subleveling in veins. Loaders, which traversed the sublevels and ramps, were served by small diesel trucks carrying three to four tons. Output per manshift improved by the elimination of labor intensive timbering, chute building, and track work. Ventilation for the diesel equipment is usually handled by lines of woven plastic tubing. In shafts, bucket hoisting has been supplanted by skips with guides in all instances. In larger mines, crushers are installed over skip loading pockets at the shaft bottom, improving skip loading.

In bedded deposits, room and pillar patterns are used, with the widths of rooms governed by roof conditions. Roof bolting has supplanted wooden supports. Most equipment is rubber-tired and diesel-operated, including the muckhaul units which have buckets ranging in size from 0.9- to 4.6-m capacity and rubber-tired diesel trucks with 3- to 18-ton capacity. Drilling is done by diesel-propelled jumbos in the bedded ore mines, but the jackleg drills are still used in narrower working places and drifts. In multileveled ore bodies, haulage ramps on 12 to 15% grades connect the levels. Vertical raises are used to facilitate ventilation requirements. Most drilling uses tungsten carbide bits, or the throwaway type hardened steel bits. In blasting the trend is increasingly toward the use of ammonium nitrate-fuel oil mixtures. Mine crews, supervisors, and mechanics are usually provided with diesel personnel vehicles to facilitate mobility.

Wherever widths of ore bodies, depths of overburden, and the strength of the sidewalls permit, it is common practice to mine by open pit methods, as is the case for all mines now producing in South Africa. This optimizes the obvious advantages to be gained by using lower cost explosives, rubber-tired earth moving equipment, and economical quarry bench mining methods.

Beneficiation

Most fluorspar must be upgraded for marketing. It is sold in a range of calcium fluoride contents, and the beneficiation techniques used are tailored to meet the character of the ores and the specifications for particular uses.

In countries with abundant low cost labor, metallurgical spar with a calcium fluoride content in the range of 75 to 85% is often produced by the hand sorting of high grade lump crude ore, followed by crushing and screening to remove most of the minus 10-mesh fraction. In the case of ores of lower grade, and/or ores with relatively coarse interlocking of minerals, gravity processes of concentration are used for producing metspar based on the specific gravity of 3.2 for spar and less than 2.80 for most gangue minerals.

Heavy media cone and drum separators are particularly effective in the size range of 3.8 × 0.5 cm, either for producing metallurgical gravel or for preconcentrating the crude ore for flotation feed. For the finer sizes the heavy media cyclone process is frequently used. The high tonnage capacity and low operating costs of heavy media methods give very satisfactory results. The sink can be sold as metspar, the sands for flotation feed, and the float frequently is marketed for road gravel and concrete aggregate. Ores as low as 14% calcium fluoride are being preconcentrated to yield a flotation feed of 40% CaF_2 or more. Lead and zinc sulfides and barite, by reason of their high gravity, concentrate with the fluorspar to enrich the flotation feed in these valuable minerals. Many ores which would otherwise be too low grade to warrant mining and processing can be preconcentrated to acceptable flotation feed grade with a tailings loss of less than 12% of the CaF_2 in the crude ore. Washing plants are also used ahead of flotation to remove clay or wad (manganese oxides) in some areas.

Barite, a fairly common accessory mineral in fluorspar deposits, in most instances, can be effectively separated from the fluorspar by flotation and made into a salable product.

As a consequence of the decreasing supply of natural metspar gravel, the agglomeration of fluorspar products by pelletizing and briquetting has expanded. Feed for agglomeration covers a variety of products such as gravel fines or screenings and various grades of flotation concentrates.

Ceramic and acid grades of fluorspar are produced by the froth flotation process, which has reached a high level of metallurgical efficiency. In many cases, ceramic grades with a calcium fluoride content in the range of 90 to 96% can be made with good recovery from ores that otherwise would be difficult to concentrate to 97% CaF_2 acid-grade spar. Ore from the mine is crushed and ground to proper size, care being taken not to overgrind and cause fluorspar to be lost in the slimes. If sulfides are present, they are floated off with xanthate collector in separate circuits ahead of the fluorspar, first the lead sulfide and then the zinc sulfide. Then all the easy floating fluorspar is removed in a quick pass through a rougher flotation circuit and sent on to the cleaner circuit; the rougher tailing is then discarded. The middling product is reground to separate the more finely interlocked grains of fluorspar and gangue, and these are then separated in one or more cleaner circuits. The final products generally comprise an acid-grade concentrate and in some cases one or more concentrates of lower grade which are sold as ceramic grade or pelletized and sold as metallurgical grade. Fatty acids are used as collectors for the fluorspar. Quebracho or tannin is used to depress calcite and dolomite, sodium silicate to depress iron oxides and silica, and chromates, starch, and dextrin to reject barite. Cyanide is used to depress sulfides. Lime, caustic, or soda ash can be used for pH control. Flotation temperatures range from ambient to 25 to 80°C with heating.

Tests on the use of column flotation have been performed by operators at Deng Feng Mining Company in Zhejiang Province in China and by Ozark Mahoning Company, a subsidiary of Atochem North America, at Rosiclare, IL. Promising results, obtained in both cases, may lead to commercial scale installations when economic conditions permit.

A standard method for determination of calcium fluoride in fluorspar is being promulgated by ASTM under Designation E 815–81. The sample is digested with nitric and perchloric acid and the fluorine is expelled by fuming. The residue is dissolved in dilute HCl, the solution made alkaline, and the calcium titrated with standard EDTA solution. Calcium present as carbonate is determined in a separate sample after extraction with dilute acetic acid. A correction for calcium fluoride, solubilized by dilute acetic acid digestion, is applied by determining the fluoride in the acetic acid extract by fluoride ion-selective electrode. The CaF_2 content is then calculated.

Neutron activation is used as an analytical method in determining calcium fluoride in ores and mill products having a content of less than 95%. The method is rapid, and is a particularly useful tool in analyzing ore samples and for mill control. The X-ray fluorescence analyzer is used to determine the abundance of other minerals occurring with the fluorspar such as barite, sphalerite, and galena.

Transportation

Most acid grade fluorspar in world markets is shipped in bulk by boat, rail, or truck in the form of damp filter cake containing about 8 to 10% moisture. Damp spar is easier to load and unload, and dust losses are minimized. This moisture is removed by heating in rotary drying kilns, or other types of dryers, before treating with sulfuric acid to make hydrofluoric acid. The fluorspar has to be practically bone dry when it goes into the acid reactors. Some acid spar is dried at the point of origin and shipped in covered hopper cars or tank trucks. A relatively small amount of acid grade spar is dried and bagged for shipment to purchasers of small quantities for specialty uses.

Metallurgical grade fluorspar is shipped as lump or gravel, in barges, ships, or railroad cars. Ceramic grades of fluorspar are usually marketed in bags; larger ceramic users, however, buy dried concentrates shipped in bulk by rail.

Markets, Production, and Consumption

The top ten producing countries in 1990 were as follows (approximate production in kt):

China	1 500
Mexico	634
Mongolia	400
USSR	380
South Africa	310
France	200
Germany	190
Spain	135
Italy	130
UK	120

The total world production in the five years from 1986 through 1990, according to US Bureau of Mines figures, was as follows (in kt):

Year	Quantity
1986	4 850
1987	4 817
1988	5 207
1989	5 586
1990	5 107

Principal consuming countries are the Russian Republic of the former USSR, the United States, Japan, China, Germany, Mexico, Italy and France. Because of disparities in production vs. consumption, the principal importing countries in 1990 were the United States, Japan, USSR, and Germany. Major exporting countries were China, Mongolia, Mexico, South Africa, Kenya and Morocco. Total market-economy production was about 2.8 Mt. World production of all grades approached 4.6 Mt.

The United States relies heavily on imports of acid grade and metspar and does not produce enough ceramic grade to meet full requirements. In 1990, figures compiled by the US Bureau of Mines show estimated domestic shipments of 63 kt, compared to imports of 513 921 t and reported consumption of 564 545 t. The Bureau's estimate of world production in 1990 was 5 107 kt.

Producing Facilities in the United States: In 1990, about 95% of US production came from southern Illinois where Ozark-Mahoning Company, a subsidiary of Atochem North America, Inc., operated underground mines and a flotation mill and imported filter cake for drying. Drying of imported fluorspar was also done by Seaforth Mineral & Ore Co. at East Liverpool, OH. Hastie Mining Company has a small metallurgical gravel operation near Cave-in-Rock, IL, that produces intermittently. Ozark-Mahoning's mill, located at Rosiclare, IL, has an annual treatment capacity of 210 kt. It is currently supplied by three shaft mines located immediately north of Cave-in-Rock, IL. The ore bodies are bedded replacements occurring mainly in Downeys Bluff Limestone and in the Ste. Genevieve Limestone. The main ore levels are commonly called Bethel ore horizon and the Rosiclare ore horizon in reference to the overlying sandstone formations. Other ore levels known as Levias ore, lower Rosiclare ore, and sub-Rosiclare ore have been productive in some mines and minor occurrences have been reported in a lower zone that miners generally refer to as St. Louis ore.

Ore body widths range from 8 to 45 m, with thicknesses from 1 to 6 m. Length exceeds 2 100 m in the former Minerva No. 1 mine in the Cave-in-Rock District, now owned and operated by Ozark-Mahoning, and 1 200 m in the Annabel Lee mine in the Harris Creek District. Ore depth exceeds 450 m in northeasternmost workings, as the host rock dips in that direction. Grade of the ore, as mined, is 30 to 55% CaF_2, 3 to 11% ZnS and 10 to 45% $BaSO_4$. All three are recovered and marketed.

Exploration is being continued in this district by Ozark-Mahoning and, as a result, that company has one new property ready for development when market conditions warrant.

In Kentucky, Cerro Corp. constructed a heavy media and flotation plant in 1974, with a stated capacity of 54 ktpy of acid grade flotation concentrate, at the Babb-Barnes property north of Salem, KY. After a series of ownerships, none of which resulted in an on-going operation, Triple M Mining Company operated the mill briefly in early 1990, using feed obtained from nearby small open pits after initially upgrading it in a logwasher. Triple M ceased these operations in Third Quarter, 1990. The flotation mill of Pennwalt Corp., near Mexico, KY, with a capacity of about 32 ktpy of flotation products, was idled in 1973 following mine exhaustion and was subsequently briefly leased to Armco, but is now idle. There are several small mines in both Kentucky and Illinois that, operating intermittently, sell ore to established mills.

In Colorado, Ozark-Mahoning Co. operated a flotation mill until late 1973 adjacent to its Northgate mine, one of the nation's largest, near Cowdrey. This mill, with a capacity of 45 ktpy of acid grade spar, is now idle. At various times, Ozark has operated other vein mines in the area to supplement Northgate production.

Roberts Mining Co. formerly processed ore from its Crystal Mountain mine near Darby, MT, in a heavy media plant, producing only metspar. Reynolds Metal Co. once operated a flotation plant, since dismantled and moved to Mexico, at Eagle Pass, TX. For mill feed it used ores and metallurgical screenings of Mexican origin. Small flotation mills, with short operating histories, have been operated in Arizona, New Mexico, Colorado, Utah, and Nevada. J. Irving Crowell at Beatty, NV, still produces metspar.

Briquetting presses are operated by Cametco in Chicago, IL, and Duquesne, PA, Ogelbay-Norton in Brownsville, TX, the Delhi Foundry Sand Company in Youngstown, OH, and the National Briquette Company in Chicago.

Flux bricks containing varying amounts of fluorspar and other additives, usually patented formulations, are produced by the Cleveland Flux Co. in Cleveland, OH, and Allen Park, MI, and by Glen-Gerry Shale Brick Corp. in Reading, PA.

Most of the metspar screenings used in briquettes and flux brick are imported from Mexico. The imported flotation concentrates used for briquetting, now coming mainly from China, are usually filter cake of medium moisture content. The small amount of domestic flotation concentrates used in metspar briquettes is normally below acid grade and contain only enough CaF_2 to achieve acceptable grades.

Binders used in making roll-compacted briquettes include molasses-lime mixtures (with self-curing properties), tall oil derivatives, and lignin paper mill byproducts, the latter two requiring oven curing. Other binders have practical applications, both in the United States and foreign areas. Sodium silicate was used in making disk pellets. Most of the flux bricks use portland cement as a binder.

FUTURE CONSIDERATIONS AND TRENDS

Fluorspar Substitutes in Metallurgical Use

A number of substitutes for the traditional metspar gravel of natural origin have been proposed and investigated from time to time (Bauxton and Sandaluk, 1972). From the worldwide viewpoint, high grade sources of natural metspar are declining and most of the crude ores currently available tend to be of lower CaF_2 content and require flotation to separate the fluorspar from the other minerals in the ore. Rising fluorspar prices and shipping costs have stimulated the search for substitutes, as has the need to reduce fluoride air and water pollution at points of consumption. Investigations have taken three general courses: 1) making artificial metallurgical gravel by forming flotation concentrates into pellets and briquettes, with or without the admixture of natural gravel spar screenings; 2) developing substitute nonfluoride fluxes; and 3) making pellets from mixtures of fluorspar and substitute fluxes.

Mixtures of iron, aluminum, and calcium oxides are now used as extenders for fluorspar, to reduce the poundage of fluorspar required per ton of steel produced. Steel company research groups and governmental agencies, seeking substitutes for fluorspar, have examined various nonfluoride substances including aluminum-bearing materials such as bauxite and the red mud residue from the manufacture of aluminum. Other materials receiving attention include dolomite, olivine, serpentine, manganese ores, ferromanganese slags and dusts, titanium ores such as ilmenite and natural ilmenite-hematite mixtures, borates (including borax, colemanite, and rasorite), soda ash, and various iron residues such as mill scale and dust. In the event that a serious shortage of fluorspar did develop, the work done to date indicates that substitutes would be available.

Phosphate-derived Fluorides

Natural phosphate rock contains approximately 3.5% fluorine as a component of the mineral fluorapatite. In phosphate processing, the fluorine volatilizes as silicon tetrafluoride, which hydrolizes to fluosilicic acid when the gas is passed through water scrubbers. Currently most of this acid is neutralized with lime. However, some fluosilicic acid is converted to aluminum fluoride and cryolite (Steininger, 1972) and several plants in the United States are recovering fluorine values in this fashion.

In the past, several attempts have been made to develop processes to produce hydrofluoric acid (HF) from phosphate rock in fertilizer manufacturing. A recently developed process, showing particular promise for commercialization, is described in United States Patent Number 5,180,569, titled "Process for the Production of Phosphoric Acid and Hydrogen Fluoride from Phosphate Rock and Fluosilicic Acid," dated Jan. 19, 1993, issued to inventors William R. Erickson and Leif E. Bouffard and assigned to Phosphate Engineering and Construction Company, Lakeland, FL.

The quantity of fluorine present in phosphate rock deposits is enormous (Worl et al., 1973). Current US identified reserves of phosphate rock are estimated to be 10.5 Gt, which at 3.5% would contain approximately 370 Mt of fluorine, equivalent to about 740 Mt of fluorspar.

Depletion and Duties

The depletion allowance is 22% for domestically mined fluorspar and 14% for foreign operations of US companies. The duty on imported fluorspar is \$2.10/t on fluorspar which assays 97.0% CaF_2 or more (acid grade). A 13.5% ad valorem for all lower grades and ores, most of which are metallurgical grade, has been suspended through December 1992.

The Defense Logistics Agency maintains a strategic stockpile of fluorspar. National Defense Stockpile objectives as of 1990, converted to metric, were 816 kt for acid grade and 281 kt for metallurgical grade; 1990 stockpile inventories of 809 178 t of acid grade and 266 756 t of metallurgical, are approximately in balance with current goals.

Changes in Patterns of Fluorspar Consumption

In recent years there has been a trend in the aluminum industry toward lower consumption of fluorine salts per ton of metal produced. Aluminum companies are recovering and recycling fluorine values from pot linings and fumes, larger pots are being used which absorb proportionately less fluoride electrolyte, more aluminum fluoride and cryolite are being manufactured from fluosilicic acid, and greater overall operating efficiency is being achieved.

In steel metallurgy, there has been a substantial decrease in consumption of fluorspar used in steel production. Current average use in the United States, about 1.4 kg of CaF_2 per ton of steel, is declining.

A recent development in the steel industry is the bottom oxygen-blowing process which has been applied to both open hearth furnaces and oxygen converters. In both applications, provision is made for injecting hydrocarbon fuel, oxygen, and pulverized fluxes such as fluorspar and lime into the base of the steelmaking vessel below the surface of the molten metal. It has been claimed that fluorspar consumption per ton of steel will be reduced by as much as 50%, in addition to various other benefits. As this method becomes widely adopted, the requirement for lump metallurgical spar will continue to drop and specifications will call for a finely ground spar, either ceramic or acid grade.

Increasing industrialization around the world may, over the years ahead, lead to increased fluorspar production and consumption, but, in the short term, it is expected that changes in unit consumption will occur in accordance with new developments, the desire to cut costs, and in response to environmental concerns. The latter will have a profound effect under existing international protocols and the increasingly stringent controls scheduled to take effect in the near future.

Ecology

Environmental restrictions and controls are severely affecting the fluorspar industry. Meeting official standards will require additional investment and cause significant increases in cost of production of fluorspar and its derivatives.

The disposal of mill tailings creates a problem because, even after impoundment and clarification, the liquid effluents may contain traces of chemicals that are in violation of environmental standards. The trace amount of fluoride ion concentration resulting from the natural solubility of calcium fluoride is, of itself, a concern.

Dust emissions from drying and dry-handling equipment can be controlled. Since the loss of valuable product is involved, it is desirable to reduce the generation and loss of fines from a purely economic standpoint, as well as for environmental reasons.

CRYOLITE

Natural cryolite (Na_3AlF_6) is a colorless mineral containing 54.3% fluorine, 12.8% aluminum, and 32.9% sodium. It has a hardness of 2.5, a specific gravity of 2.96, and crystallizes in the monoclinic system. Its mean index of refraction is about 1.339, which is close to that of water.

Cryolite has been found in commercial quantities only in one place in the world, Ivigtut, West Greenland. Small economically unimportant occurrences have been found in the St. Peter's Dome District near Pike's Peak in Colorado, and in the USSR, Spain, Canada, and Nigeria.

The Greenland deposit is a pegmatite in porphyritic granite in which the cryolite is accompanied by fluorspar, siderite, pyrite, arsenopyrite, galena, topaz, molybdenite, and a number of rare aluminum fluoride minerals. Approximately 90 different minerals have been found in the deposit, several of which have not been found elsewhere. The deposit lies on the shore of the Arsuk Fiord on the southwest coast of Greenland. It was worked by open pit and was mined out in 1962 (Maersk, 1973) after approximately a century of operation. Cryolite was stockpiled at Ivigtut for shipment to a concentrating plant in Copenhagen operated by Kryolitselskabet Oresund, which is partly owned by the Danish government. In this plant, the crude ore was concentrated by flotation to a product containing 98 to 99% cryolite. For 100 years, crude ore also was shipped to the United States for processing by Pennsalt Chemicals Corp., but this arrangement terminated in 1965.

Originally cryolite was used mainly in the production of soda, alum, and aluminum sulfate, but this eventually became unprofitable, and in 1889 the mineral assumed a new and important role as the molten bath in which alumina was dissolved for electrolytic reduction to aluminum metal by the Hall process. Demand for cryolite in the aluminum industry outstripped the natural supply and the gap was filled by synthetic cryolite manufactured from hydrofluoric acid, sodium carbonate, and aluminum hydrate. In the ceramic industry, cryolite is used as a whitener for enamels and an opacifier in glasses. It is also used in bonding grinding wheels and abrasives, and as an ingredient in welding rod coatings. Very finely ground cryolite is an active agent in some insecticide mixtures.

ACKNOWLEDGMENT

The authors appreciate valuable contributions to this paper by Ernest P. Cox of Alcoa and Victor A. Evans of Ozark-Mahoning.

BIBLIOGRAPHY AND REFERENCES

Anon., 1965, "Barytes and Fluorite Deposits of Turkey," Maden Tetkik Arama Enstit. Yayinl, No. 126, Ankara, 11 pp.

Anon., 1972, Annual Report, Lost River Mining Corp., Alaska.

Anon., 1985, "The Economics of Fluorspar," Roskill Information Services, Ltd., London, 365 pp.

Anon., 1991, "The Future of Refrigerants Begins Today," Dupont Chemicals, Wilmington, DE, 6 pp.

Anderson, A.L., and Van Alstine, R.E., 1964, "Fluorspar," *Mineral and Water Resources of Idaho*, 88th Congress, 2nd Session. US Government Printing Office, Washington, DC, pp. 79–84.

Argall, G.O., Jr., 1949, "Fluorspar," "Industrial Minerals of Colorado," *Quarterly of the Colorado School of Mines*, Vol. 44, No. 2, pp. 179–208.

Bauxton, F.M., and Sandaluk, P.A., 1972, "Fluorspar Substitutes in Steelmaking," reprint, American Iron and Steel Institute, Regional Technical Meeting, Pittsburgh, PA, Nov. 9.

Baxter, J.W. and Bradbury, J.C., 1989, "Illinois-Kentucky Fluorspar Mining District," *Precambrian and Paleozoic Geology and Ore Deposits in the Midcontinent Region*, J.W. Baxter, et al., 28th International Geological Congress, Field Trip T147, American Geophysical Union, pp. 3–36.

Baxter, J.W., Bradbury, J.C., and Hester, N.C., 1973, "A Geologic Excursion to Fluorspar Mines in Hardin and Pope Counties, Illinois," Guidebook Series II, Illinois State Geological Survey, 28 pp.

Baxter, J.W., and Desborough, G.H., 1965, "Areal Geology of the Illinois Fluorspar District Pt. 2.—Karbers Ridge and Rosiclare Quadrangles," Circular 385, Illinois State Geological Survey, 40 pp.

Baxter, J.W., Desborough, G.H., and Shaw, C.W., 1967, "Areal Geology of the Illinois Fluorspar District Pt. 3—Herod and Shetlerville Quadrangles," Circular 413, Illinois State Geological Survey, 41 pp.

Baxter, J.W., Potter, P.E., and Doyle, F.L., 1963, "Areal Geology of the Illinois Fluorspar District Pt. 1.—Saline Mines, Cave-in-Rock, Dekoven and Repton Quadrangles," Circular 342, Illinois State Geological Survey, 43 pp.

Chermette, A., 1950, "L'Exploitation du Spath-Fluor en France de 1938 á 1946," *L'Echo des Mines et de la Metallurges*, Vol. 427, No. 3, pp. 547–549.

Chermette, A., 1964–65, ''Les Ressources du Mexique en Spath-Fluor,'' *Mines et Metallurgie*, Oct., pp. 467–469; Nov., pp. 519–521; Dec., pp. 571–574; Jan. pp. 25–27; Feb., pp. 77–79.

Chermette, A., 1972–1973, ''Un Demi-Siècle de Spath-Fluor Français (1922–1972),'' *Mines et Metallurgie*, Oct., pp. 179–182; Nov./Dec., pp. 201–204. 213; Jan./Feb., pp. 9–12, 14; Mar., pp. 8, 11; Apr., pp. 10–13; May, pp. 9–16.

Chermette, A., 1979, ''Le Spath Fluor est a un Tournant,'' *Mines et Metallurgie*, No. 145, pp. 14–19; No. 146, pp. 20–24.

Chesterman, C.W., 1966, ''Fluorspar,'' *Mineral Resources of California*, Bulletin 191, California Division of Mines and Geology, pp. 165–168.

Coope, B., 1978, ''Fluorspar—Down But Not Out,'' *Industrial Minerals*, No. 125, Feb., pp. 39–66.

Cornwall, H.R., 1972, ''Geology and Mineral Resources of Southern Nye County, Nevada,'' Bulletin 77, Nevada Bureau of Mines and Geology, 49 pp.

Cox, E.P., 1989, ''Use of Fluorspar in the Aluminum Industry,'' Metal Bulletin Conference, Monterey, Mexico.

Dasch, M.D., 1964, ''Fluorine,'' *Mineral and Water Resources of Utah*, Bulletin 74, Utah Geological and Mineralogical Survey, pp. 162–168.

Deans, T., et al., 1972, ''Metasomatic Feldspar Rocks (Potash Fenites) Associated with the Fluoride Deposits and Carbonatites of Amba Dongar, Gujarat, India,'' *Transactions, Sec. B.*, Institution of Mining and Metallurgy, Vol. 81, Bulletin 783, p. B6.

Dunham, K.C., 1952, ''Fluorspar,'' Memoir, Special Report on Mineral Research, Geological Survey of Great Britain, Vol. 4, 143 pp.

Elevatorski, E.A., 1968, ''California Fluorspar,'' *Mineral Information Service*, State of California, Vol. 21, No. 9, pp. 127–130.

Elevatorski, E.A., 1971, ''Arizona Fluorspar,'' Arizona Dept. of Mineral Resources, 51 pp.

Erickson, R.L., et al., 1987, ''Subsurface Geochemical Investigation in Western and Southern Illinois: a Pilot Study,'' Illinois Mineral Notes 98, Illinois State Geological Survey, 45 pp.

Fulton, R.B., 1991, ''Fluorspar,'' *Mining Engineering*, Vol. 43, No. 6, p. 599.

Funnell, J.E., and Wolff, E.J., 1964, ''Fluorspar (Fluorite),'' *Compendium on Nonmetallic Minerals of Arizona*, prepared for Arizona Public Service Co. by Southwest Research Institute, pp. 105–114.

Gardner, L.S., and Smith, R.M., 1965, ''Fluorspar Deposits of Thailand,'' Report of Investigation, No. 10, Department of Minerals Resources, Bangkok, 42 pp.

Gillerman, E., 1953, ''Fluorspar Deposits of the Eagle Mountains, Trans-Pecos Texas,'' Bulletin 987, US Geological Survey, 98 pp.

Goddard, E.N., 1966, ''Geologic Map and Sections of the Zuni Mountains Fluorspar District, Valencia County, New Mexico,'' Map I-454, Miscellaneous Geologic Investigations, US Geological Survey.

Gossling, H.H., 1972, ''A Review of the World's Fluorspar Industry, with Particular Reference to South Africa,'' Report No. 1424, National Institute of Metallurgy, Johannesburg, 30 pp.

Gossling, H.H., 1978, ''Fluorspar, 1973–1980: A Commodity Profile,'' Report No. 3, Mineral Bureau, Department of Mines, Republic of South Africa, p. 36.

Grogan, R.M., and Bradbury, J.C., 1967, ''Origin of the Stratiform Fluorite Deposits in Southern Illinois,'' *Genesis of Stratiform Lead-Zinc-Barite-Fluorite Deposits*, Monograph 3, The Economic Geology Publishing Co., Lancaster, PA, pp. 40–51.

Grogan, R.M., and Bradbury, J.C., 1968, ''Fluorite-Zinc-Lead Deposits of the Illinois-Kentucky Mining District,'' *Ore Deposits of the United States, 1933–1967*, Vol. 1, J.D. Ridge, ed., AIME, New York, pp. 370–399.

Hodge, B.L., 1991, ''Fluorspar,'' *Mining Annual Review, 1991*, pp. 116–117.

Horton, R.C., 1961, ''An Inventory of Fluorspar Occurrences in Nevada,'' Report 1, Nevada Bureau of Mines, 31 pp.

Horton, R.C., 1962, ''Fluorspar Occurrences in Nevada,'' Nevada Bureau of Mines, Map 3.

Horton, R.C., 1964, ''Fluorspar,'' *Mineral and Water Resources of Nevada*, 88th Congress, 2nd Session, US Government Printing Office, Washington, DC, pp. 198–202.

Jee, J.M, and Kye, J., 1971, ''Report on the Fluorite Deposits of Korea,'' Bulletin No. 13, Geological Survey of Korea, pp. 7–368 (English Abstract).

Jewell, W.B., 1947, ''Barite, Fluorite, Galena, Sphalerite Veins of Middle Tennessee,'' Bulletin 51, Tennessee Division of Geology, 114 pp.

Kilgore, C.C., Kraemer, S.R., and Bekkala, J.A., 1985, ''Fluorspar Availability—Market Economy Countries and China,'' Information Circular IC9060, US Bureau of Mines, 57 pp.

Kim, H., John, Y., and Kang, W., 1972, ''Mineral Requirements for Korea's Industrialization: Patterns of Requirements and Availabilities,'' Korean Institute of Mineral and Mining Engineers, 153 pp.

Liddy, J.C., 1971, ''Fluorspar in Australia,'' *Australian Mining*, Jan., pp. 34–38.

Maersk, B., 1973, ''Cryolite Concentrator in Copenhagen,'' *World Mining*, Vol. 26, No. 3, Mar., pp. 60–63.

Miller, M.M., 1991, ''Fluorspar—1990,'' Annual Report, US Bureau of Mines, 12 pp.

Papke, K.G., 1979, ''Fluorspar in Nevada,'' Bulletin 93, Nevada Bureau of Mines and Geology, 77 pp.

Parkinson, G., 1990, ''HF's Future is up in the Air,'' *Chemical Engineering*, May, 1990.

Peters, W.C., 1958, ''Geologic Characteristics of Fluorspar Deposits in the Western United States,'' *Economic Geology*, Vol. 53, pp. 663–688.

Peterson, M.R., 1990, ''Column Flotation of Multiple Products from a Fluorite Mine,'' Report of Investigations 9309, US Bureau of Mines.

Prud'homme, M., 1987, ''Fluorspar,'' Energy, Mines and Resources Canada Yearbook, pp. 29–1 to 297.

Rothrock, H.E., Johnson, C.H., and Hahn, A.D., 1946, ''Fluorspar Resources of New Mexico,'' Bulletin 21, New Mexico Bureau of Mines and Mineral Resources, 245 pp.

Sahinen, U.M., 1962, ''Fluorspar Deposits in Montana,'' Bulletin 28, Montana Bureau of Mines and Geology, 38 pp.

Sheppard, R.A., and Gude, A.J., 1970, ''Authigenic Fluorite in Pliocene Lacustrine Rocks Near Rome, Malheur County, Oregon,'' *Geological Survey Research 1969*, Professional Paper 650-D, US Geological Survey, pp. D69-D74.

Snyder, K.D., 1978, ''Geology of the Bayhorse Fluorite Deposits, Custer County, Idaho,'' *Economic Geology*, Vol. 73, No. 2, Mar.-Apr., pp. 207–214.

Spada, A., 1969, ''Il Giacimento de Fluorite e Bartina Esalativo-sedimentario in 'Facies' Lacustre, Intercalato nei Sedimenti Piroclastici della Zona di Castel Giuliano, in Province di Roma,'' *Industria Min.*, Vol. 20, pp. 501–518.

Staatz, M.H., and Osterwalld, F.W., 1959, ''Geology of the Thomas Range Fluorspar District, Juab County, Utah,'' Bulletin 1069, US Geological Survey, 97 pp.

Steininger, E., 1972, ''Making Fluorine Compounds from Waste Fluosilicic Acid,'' *Engineering & Mining Journal*, Vol. 173, No. 12, pp. 73–75.

Tabor, J.W., 1953, ''Montana's Crystal Mountain Fluorspar Deposit is Big and High Grade,'' *Mining World*, Vol. 15, No. 7, pp. 43–46.

Thurston, W.R., et al., 1954, ''Fluorspar Deposits of Utah,'' Bulletin 1005, US Geological Survey, 53 pp.

Van Alstine, R.E., 1964, ''Fluorspar,'' *Mineral and Water Resources of Colorado*, 88th Congress, 2nd Session, US Government Printing Office, Washington, DC, pp. 159–165.

Van Alstine, R.E., 1965, ''Fluorspar,'' *Mineral and Water Resources of New Mexico*, Bulletin 87, New Mexico Bureau of Mines and Mineral Resources, pp. 260–267.

Van Alstine, R.E., 1969, ''Fluorspar,'' *Mineral and Water Resources of Arizona*, Bulletin 180, Arizona Bureau of Mines, pp. 348–357.

Van Alstine, R.E., and Schruben, P.G., 1980, ''Fluorspar Resources of Africa,'' Bulletin 1487, US Geological Survey, 25 pp.

van Zijl, P.J., 1962, ''The Geology, Structure, and Petrology of the Alkaline Intrusions of Kalkfeld and Okorusu and the Invaded Damara Rocks,'' *Aunale Univ. Stellenbosch*, Ser. A., Vol. 37, No. 4, pp. 237–346.

Weller, J.M., Grogan, R.M., and Tippie, F.E., 1952, ''Geology of the Fluorspar Deposits of Illinois,'' Bulletin 76, Illinois State Geological Survey, 147 pp.

Worl, R.G., Van Alstine, R.E., and Shawe, D.R., 1973, ''Fluorine,'' *United States Mineral Resources*, Professional Paper 820, US Geological Survey, pp. 223–235.

Worl, R.G., Van Alstine, R.E., and Heyl, A.V., 1974, ''Fluorite in the United States: Exclusive of Hawaii,'' Minerals Investigation Resource Map, MR-60, US Geological Survey.

Zurowski, M., 1972, ''Barite-Fluorite Deposits of Lake Ainslie—An Appraisal From an Economic Viewpoint,'' *Transactions*, Canadian Institute of Mining and Metallurgy, Vol. 75, pp. 318–321.

GARNET

GORDON T. AUSTIN

Garnet is the general name for a family of complex silicate minerals having similar physical properties and crystallizing in the isometric (cubic) system. All garnets have the same general chemical formula but vary greatly in chemical composition. The name garnet is derived from the Latin word *granatus,* meaning like a grain, which refers to the mode of occurrence wherein crystals resemble grains or seeds embedded in the matrix.

GEOLOGY

Mineralogy

Chemical Properties: The general formula for garnet is $A_3B_2(SiO_4)_3$, where A can be calcium, magnesium, ferrous iron, or manganese, and B can be aluminum, ferric iron, or chromium, or rarely titanium. The formulas and names of common garnet species are:

Uvarovite:	$Ca_3Cr_2Si_3O_{12}$	Pyrope:	$Mg_3Al_2Si_3O_{12}$
Grossularite:	$Ca_3Al_2Si_3O_{12}$	Almandite:	$Fe_3Al_2Si_3O_{12}$
Andradite:	$Ca_3Fe_2Si_3O_{12}$	Spessartite:	$Mn_3Al_2Si_3O_{12}$

Almandite and almandite-pyrope solid solution garnets are the best abrasive types, but andradite, grossularite, and pyrope also are used. All species of garnet have been used as gemstones.

The structure of garnet consists essentially of isolated SiO_4 tetrahedra connected by oxygen-cation-oxygen bonds through the distinct A and B group cation sites. Within this structure magnesium, ferrous iron, and manganese easily interchange and substitute for each other in the A cation position, and calcium does so less readily. Additionally, aluminum, ferric iron, and chromium substitute for each other to a limited extent in the B cation position.

This ability to substitute or exchange ions without changing the crystal structure is called isomorphism, and garnet is one of the finest examples of a isomorphous series. Because of this isomorphism there is complete solid solution between certain garnet species but not between others. Fig. 1 illustrates these solid solution relationships.

Some rare species of garnet are known that illustrate the wide range of cation substitution that the garnet crystal structure can accommodate. They include:

Hydrogrossular:	$Ca_3Al_2(SiO_4)_{3-x}(OH)_4$
Henritermierite:	$Ca_3(Mn,Al)(SiO_4)_2(OH)_4$
Goldmanite:	$Ca_3V_2Si_3O_{12}$
Kimzeyite:	$Ca_3(Zr,Ti)_2(Al,Si)_3O_{12}$
Knorringite:	$Mg_3Cr_2Si_3O_{12}$
Majorite:	$Mg_3(Fe,Al,Si,)_2Si_3O_{12}$
Schorlomite:	$Ca_3(Fe,Ti)_2((Si,Ti)O_4)_3$
Yamatoite:	$Mn_3V_2Si_3O_{12}$

These rare species are not of interest for industrial applications, but can be of interest to mineralogists and the gem industry.

Physical Properties: Garnet displays the greatest variety of color of any industrial mineral. Garnets have been found in all colors except blue. For example, grossularite can be colorless, white, gray, yellow, yellowish green, various shades of green, brown, pink, reddish, or black. Andradite garnet can be yellow-green, green, greenish brown, orangy yellow, brown, grayish black or black. Pyrope is commonly purplish red, pinkish red, orangy red, crimson, or dark red; and almandite is deep red, brownish red, brownish black or violet-red. Spessartite garnet can be red, reddish orange, orange, yellow-brown, reddish brown, or blackish brown.

A few garnets exhibit a color-change phenomenon. They are one color when viewed in natural light and another color when viewed in incandescent light. Because of the great variation in color of garnet within each species and similarities in color between garnets of different species it is recommended that garnet identification not be based on color alone.

The Mohs hardness of garnet varies from 6.5 to 9.0. Grossularite and uvarovite have a hardness of 6.5 to 7.5; andradite is 6.5 to 7.0; and pyrope, almandite, and spessartite are 7.0 to 7.5 in hardness. There are reports of almandite having a hardness of between 8.0 and 9.0

As with hardness, the specific gravity of garnet varies considerably. The specific gravity may be as low as 3.2 or as high as 4.3 depending on chemical composition.

Garnet crystallizes in the isometric system with rhombic dodecahedra and trapezohedra the most common forms. Crystals also form in combinations of dodecahedra and trapezohedra or either of these in combination with hexoctahedra. Crystals can have cubic or octahedral faces, but these are rare. Under favorable conditions of formation garnet will crystalize in nearly perfect forms that rival study models. Garnet also forms as irregular blebs, grains, knots, or masses, with or without distinguishable crystal faces, and as coarse- or fine-grained granular masses that appear to be totally lacking in crystal form.

The fracture of garnet also shows great variation. In some garnets, particularly those that are well crystallized and glassy in appearance, the fracture is subconchoidal to conchoidal. Other, more poorly crystallized garnets exhibit a fracture that can only be described as uneven.

Garnet occasionally has an indistinct dodecahedral cleavage. Certain species of garnet from specific locations have a pronounced laminated structure consisting of planes of weakness along which parting takes place. This parting may resemble cleavage, but since it is mechanical and not related to crystal structure it is not true cleavage.

The optical properties of garnet are sensitive to even small changes in chemical composition or strain in the crystal structure. For this reason each species does not have a single index of refraction (IR) but has a range of IRs. Uvarovite in its pure form should have an IR of 1.870, but in nature samples vary from 1.74

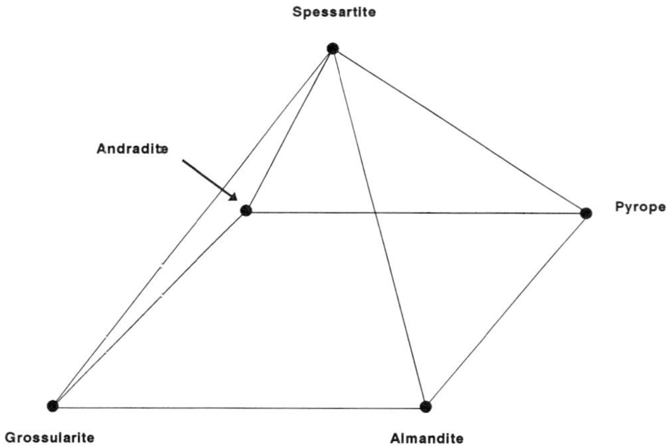

Fig. 1. Phase diagram of garnet end-members.

to 1.87. Grossularite should have an IR of 1.735, yet specimens vary from 1.72 to 1.80, although most often the range is from 1.73 to 1.76. Pure andradite has an IR of 1.895, although varieties are found with IRs from 1.88 to 1.94. Pyrope should have an IR of 1.705, but tests have shown that in most samples it varies from 1.730 to 1.766. Almandite theoretically has an IR of 1.830, however measurements on most samples range from 1.75 to 1.83. Spessartite's IR is 1.800 if the crystal is pure and unstrained; but most samples range between 1.79 and 1.81, and the majority have IRs greater than 1.78.

The fusibility of garnet is a function of its chemical composition, and when fused by itself garnet disassociates into other minerals. Almandite, spessartite, and grossularite easily fuse to light brown or black glass at about 1 050°C. Andradite and pyrope fuse to the same types of glass at about 1 125°C, whereas uvarovite is almost infusible.

Tenacity, the resistance of a mineral to break, crush, bend or tear, is a function of its hardness, structure, and chemical composition. Because of variations between different species of garnet and within the same species the tenacity of garnet is not consistent. Aggregates of garnet that are comprised of small, individual, irregular blebs, grains, or masses are brittle and break or crush readily. In comparison, garnet formed as coarse- or fine-grained granular masses or well-formed crystals is very tough and shatters with difficulty. Also, certain species of garnet are more brittle than others. Almandite and pyrope tend to be less brittle than other species.

The clarity of different species of garnet can vary from opaque to transparent. Their luster can range from vitreous to resinous or even dull.

Mode of Occurrence

Garnet is found as an accessory mineral in a wide variety of rocks. Most commonly it is found in gneisses and schists, but it also is found in contact metamorphic rocks, crystalline limestones, pegmatites, volcanic rocks, eclogites, and serpentinites. Additionally, garnet is found in the gangue of veins formed at high temperatures. Since garnet is resistant to chemical and mechanical weathering, it is common in the sands of current or ancestral beaches, streams, and other types of alluvial deposits.

Garnets are usually small in size. Grossularite crystals range up to about 15 cm in diameter. Almandite approximates the same range in size except at Gore Mountain and North Mountain, New York, where crystals range up to 60 cm in diameter. Spessartite crystals from Brazil can weigh as much as several kilograms. Most garnet crystals are less than 2.54 cm in size.

Table 1. Production of Garnet in the United States

	Crude Garnet Concentrate		Refined Garnet	
Year	Production t × 10^3	Value $ × 10^6	Sold or Used t × 10^3	Value $ × 10^6
1972	17.2	2.0	17.2	2.0
1973	20.7	2.4	20.7	2.4
1974	22.4	2.6	22.4	2.6
1975	15.6	1.7	15.6	1.7
1976	23.3	3.0	22.3	3.6
1977[a]	19.9	1.3	18.2	3.3
1978	18.9	1.3	20.0	3.9
1979	19.3	1.5	21.1	4.6
1980	24.4	1.9	24.1	4.9
1981	23.1	2.1	23.2	5.2
1982	24.8	2.3	24.2	5.5
1983	27.0	2.8	27.5	6.0
1984	26.9	2.7	25.1	5.7
1985	33.3	3.0	27.8	6.1
1986	24.3	2.6	28.9	6.8
1987	38.4	4.3	35.8	7.7
1988	42.5	4.7	41.9	11.1
1989	42.6	4.4	41.3	9.8
1990	47.0	6.9	45.9	11.8

[a] Crude garnet concentrate reported as sold and used up to this date
Source: US Bureau of Mines

The majority of world garnet production is from alluvial deposits in the United States, Australia, China, and India. Possibly the world's highest quality garnet is mined, however, from almandite-bearing igneous/metamorphic rock of uncertain origin in the Adirondack Mountains of New York. Garnet production as published by the US Bureau of Mines is shown in Table 1.

Distribution of Major Deposits

Deposits of garnet are found in many countries. Members of the garnet group are common accessory minerals in a wide variety of rocks, particularly granites, gneisses, and schists. They also are found in serpentinites, peridotites, and crystalline limestones. Since most varieties of garnet are resistant to chemical and mechanical weathering, they often are found in alluvial deposits of sand and gravel. Yet, an estimated 95% of the world's production of the highest quality abrasive garnet comes from a small area in the Adirondack Mountains of New York, although total production of garnet from New York is less than from Idaho. Gem quality garnet has been produced in the United States from deposits in Arizona, Arkansas, California, Colorado, Connecticut, Idaho, Maine, Michigan, Nevada, New Jersey, New Mexico, New York, North Carolina, Pennsylvania, South Dakota, Utah, Vermont, and Virginia.

United States: During the past 10 years industrial garnet production in the United States has been confined to three states, Idaho, Maine, and New York, with production since 1988 coming exclusively from Idaho and New York. However, deposits in several other states are notable.

CALIFORNIA—Several skarn-type tungsten deposits have been mined in the past, and the processing of their ore has led to significant accumulations of tailings. One of the major gangue minerals in these deposits was garnet, so the tailings contain between 5 and 40% andradite garnet. These tailings are already ground and ready for processing to recover the garnet, but because the species of garnet present is andradite, it is questionable whether they are an economic source of garnet. In many instances andradite has not proved satisfactory for use as an air-blasting medium or in other abrasive applications.

CONNECTICUT—Garnet in mica schist near Roxbury and Roxbury Falls in Litchfield County was once mined, although not recently, for use in abrasive applications. The andradite garnet was found as well-formed dodecahedral crystals from less than 7 mm to more than 500 mm in size. The crystals separate easily from the host schist (Ladoo and Myers, 1951).

IDAHO—Alluvial deposits of almandite garnet were discovered in the early 1880s near Fernwood in Benewah County, but commercial mining did not start until the early 1940s. The deposits are on Emerald, Carpenter, and Meadow Creeks about 6.4 km from a mica-garnet schist formation. The source schist is estimated to be about 1 200 m thick. The garnet-bearing gravels are from nearly 1 m to about 1.2 m thick and contain 8 to 15% garnet. Garnets from these gravels range up to 40 mm in diameter, although about 95% of the garnet is smaller than 7 mm in diameter.

MAINE—The Wing Hill garnet deposit, located near Rangeley in western Maine, is one of the largest and highest grade garnet deposits in the world. The host rock is a homogeneous garnet granofels that consists of 50 to 70% almandite garnet. It is medium grained and consists essentially of garnet and andesine plagioclase along with biotite and minor quartz. The garnet is present as discrete, well-formed crystals from less than 1 mm to about 10 mm in diameter, but averaging 1.5 to 2 mm. The granofels is an east-west trending, tabular body more than 1 000 m in length and as much as 700 m thick. The garnet ranges in Mohs hardness from 7 to 8 and in specific gravity from 3.6 to 4.3. When crushed the garnet breaks into blocky grains, a shape that persists down to the finest fragment. The garnet is amenable to gravity and magnetic concentration when the ore is crushed to -3.36 mm. Reserves are estimated to be several million tons.

MONTANA—Several alluvial deposits of almandite-pyrope garnet are located on the drainages of the Ruby River in Madison County. One such deposit, the Alder Gulch deposit, is in the alluvial fan formed where Alder Gulch joins the Ruby River valley. It is at elevations of 1 550 to 1 615 m above mean sea level. The deposit contains about 40 Mt of old placer gold tailings that grade 4.5% almandite-pyrope garnet and 0.03 g/t of gold. The alluvium consists of high energy, fluvial, well rounded material, approximately 50% of which is +9 mm in size. The garnet-bearing tailings are as much as 15 m thick and average more than 11 m thick. The source of garnet along the Ruby River is highly metamorphosed Archean rocks in the Tobacco Root and Greenhorn Mountain Ranges to the east (Tansley et al., 1933, Vitaliano et al., 1979). The garnets are present mostly as broken fragments of crystals which were originally as much as several centimeters in diameter. The particle size distribution of garnet in this deposit is 43% $-3.36 + 1.0$ mm, 15% $-1.0 + 0.59$ mm, 30% $-0.59 + 0.25$ mm, and 12% -0.25 mm.

NEVADA—An alluvial deposit of almandite garnet is found along Hampton Creek Canyon in White Pine County about 3 km from the mouth of the canyon. No production history is available for the deposit. It is about 1 to 30 m in width, from 5 cm to 10 m in thickness, and extends for about 800 m along the canyon floor. The source of the garnet is quartz-garnet-mica-staurolite schist that forms a portion of the walls of the canyon. The schist averages about 15 m in thickness and contains between 8 and 12% garnet. The garnets range from about 0.25 to 3 mm in size.

NEW HAMPSHIRE—A garnet deposit is located near North Wilmont in Merrimack County, where small almandite crystals are found in a biotite, quartz, and albite feldspar matrix. The crystals range up to about 19 mm in diameter but average only 6 to 10 mm in diameter. The grade ranges up to 66% garnet but averages much less. The garnet contains quartz and biotite inclusions and must be finely ground to liberate these impurities (Myers and Anderson, 1925a, Ladoo and Myers, 1951).

NEW YORK—Deposits of industrial garnet are found at two locations in New York. In the Gore Mountain area garnet is mined as the primary product, and near the town of Willsboro, it is recovered as a byproduct of wollastonite mining.

The deposit near Gore Mountain is an almandite-bearing diorite of uncertain, igneous or metamorphic origin. The ore body is approximately 2 km long and 15 to 90 m wide. The ore grade varies from less than 5% to about 20% garnet and averages slightly less than 10%. Hornblende and plagioclase feldspar make up 40 to 80% of the host rock, while hypersthene, magnetite, biotite, apatite, and pyrite are present in minor amounts. The garnet is present as imperfectly developed crystals surrounded by a rim of coarsely crystalline hornblende. The crystals range from about 1 mm to almost 1 m in diameter but average about 100 mm in diameter. The garnet has a pronounced laminar structure, which enables it to naturally break into thin plates from about 2 to 6 mm in thickness. Garnet fragments maintain this platy particle shape even as they are crushed smaller and smaller.

Garnet is recovered as a byproduct from a wollastonite deposit on the western side of Lake Champlain near Willsboro. The deposit is in a belt of wollastonite-bearing rocks that is about 10 km long and 0.4 km wide. The largest zone of wollastonite rock within the belt is from about 9 m to more than 20 m thick but averages about 13 m thick. The wollastonite is interbanded with almandite garnet and iron-rich diopside. The garnet and diopside together comprise 10 to 20% of the ore. Both the garnet and diopside are feebly magnetic and are easily separated from the wollastonite by magnetic separators.

NORTH CAROLINA—Large deposits of andradite and rhodolite garnet of good abrasive quality are known in Clay, Jackson, Macon, Madison, and Burke counties. Garnet was produced from some of these deposits from 1900 to about 1926, but no production has been recorded recently. The deposit of almandite garnet in Clay County is in a hornblende gneiss at Penland Bald on Buck Creek.

PENNSYLVANIA—Almandite garnet crystals are found in a quartzose mica schist about 1.6 km west of Chelsea in Delaware County. Near the surface the schist is badly weathered and the garnets are easily recovered. At about 6 m below the surface the schist is unweathered and requires crushing, grinding, and processing to recover the garnet, which comprises as much as 75% of the rock locally but averages much less. Garnet was produced from this deposit prior to 1900, but no production has been recorded since the turn of the century.

Small trapezohedral crystals of almandite garnet are scattered throughout a badly decomposed gneiss near Chester Heights, also in Delaware County. An attempt was made to mine this deposit years ago by means of a shaft and underground workings.

VIRGINIA—A deposit of garnet is located on a steep bluff on the northeast side of the Tyle River about 6.5 km south of Arrington in Nelson County. The small, dull-red garnets are found in a sericitic schist. Attempts were made to mine the deposit in the past, but there is no record of commercial production.

Other Countries: Garnet is mined or has been commercially mined from both hard rock and alluvial deposits in several other countries. Currently, a number of garnet deposits are being evaluated for production.

Australia. A large, alluvial beach sand deposit of garnet is located about 100 km north of Geraldton in Western Australia. The deposit grades about 35% garnet, is approximately 12 km long, and has a mining width of more than 100 m. Ore reserves are estimated to be at least 6 Mt.

Canada. Nearly 100 garnet deposits have been identified in Canada, but only 10 or 12 of them have a history of production or have economic potential.

BRITISH COLUMBIA—According to one mining company's reports a high grade andradite garnet deposit in a skarn host rock near Penticton contains an estimated 40 Mt of ore with a grade of 80% garnet. The outcrop of the deposit measures approximately 800 by 300 m.

Almandite garnet schists, similar to the famous garnet schists found along the Stikine River near Wrangell, Alaska, have been reported farther up the river in British Columbia. The garnets are well formed, nearly transparent crystals. Many of them are more than 24 mm in diameter, but they average about 12 mm in diameter. The garnets are of excellent abrasive quality and are easily freed from the host rock. Commercial development of the deposit is unlikely because of its remoteness (Eardley-Wilmot, 1927).

NOVA SCOTIA—Near Chegoggin Point there is a belt of garnetiferous hornblende schist approximately 8 km in length that strikes northeast-southwest and dips steeply to the northwest. The schist contains an average of at least 40% garnet over a width of more than 9 m. The garnet is present as well-formed, opaque, red-brown crystals that vary in size from about 5 to 19 mm (Eardley-Wilmot, 1927).

ONTARIO—On the northern shore of Fishtail Lake, north of Mumford Station, there is a 9 m wide band of garnetiferous amphibolite associated with granite gneiss. The band averages about 30% garnet and is exposed for an east-west strike length of about 450 m. The garnets are large, pink or clear red crystals, many of which are nearly 40 mm in diameter. Tests have shown that the garnet can be concentrated easily from the ore by gravity methods, and the garnet appears to be of good abrasive quality (Eardley-Wilmot, 1927).

West of Actinolite near the Black River on the road to Madoc is a 4.5 m wide zone of garnets in mica schist. The garnet-mica schist is exposed for a northwest-southeast strike length of about 0.8 km. The garnets are dark brown to almost black, well-formed crystals (Eardley-Wilmot, 1927).

South of Hardwood Lake Post Office in the extreme northwest corner of Lennox and Addington County there is a garnetiferous, quartzose hornblende-biotite gneiss. The garnet also is found in the fibrous hornblende and quartz-mica schist. The deposit strikes approximately east-west, but its strike length is unknown. It is about 45 m wide where it was stripped years ago. The average ore grade is 30% garnet, but some zones contains as much as 40% garnet. In the mica schist the garnet crystals are well formed, clear deep red, range in size from 3 to 5 mm, and are uniformly distributed. In hornblende gneiss the crystals are sightly smaller and pale pink. The total extent of the deposit is not known. Garnet produced from it in the 1920s was well received by the abrasives industry and commanded a price equal to that of the best garnet on the market at that time (Eardley-Wilmot, 1927).

About 3 km northeast of Addington Station and 0.8 km south of the Canadian Pacific Railroad there is a band of garnetiferous mica schist about 15 m wide. The schist strikes approximately northeast-southwest and has an exposed length of more than 3 km. Garnet is present in three zones that comprise the deposit. On the south side is of muscovite mica schist about 9 m and containing densely clustered garnets 3 to 5 mm in size. The central zone is 1.8 m thick, black biotite mica schist that contains evenly scattered, clear deep red to almost black garnets 13 to 25 mm in size. On the north side is a band of muscovite mica schist that contains smaller garnet crystals (Eardley-Wilmot, 1927).

Near Depot Harbor on the north side of Parry Island are wide bands of garnetiferous hornblende gneiss alternating with mica schist. The bands strike east-west and can be traced for several kilometers inland from the coast. The garnet is found in zones within these bands. Several zones are 2 to 3 m wide and contain garnet crystals 25 to 50 mm in diameter. However, the garnets average only 3 to 5 mm in size. Tests have shown that the garnet has a good, sharp fracture and is of good abrasive quality (Eardley-Wilmot, 1927).

About 6.5 km north of Markstay Station on the Canadian Pacific Railroad there is a belt of garnetiferous biotite gneiss. The belt is about 30 m wide and is exposed for about 800 m along its northeast-southwest strike. It can be divided roughly into three zones, each of which is about 10 m wide and contains a particular size of garnet that ranges from very large to medium or small. The zone nearest the eastern contact contains the largest garnet crystals, which average more than 35 mm in diameter and make up about 25% of the rock. The crystals are nearly perfect dodecahedra but have a granular appearance on fractured surfaces. In the zone that contains the medium-size garnet, 3 to 5 mm in diameter, the garnet comprises about 33% of the rock. These crystals are much clearer than the larger crystals and appear to be of better abrasive quality (Eardley-Wilmot, 1927).

QUEBEC—In the extreme southwest corner of Cathcart Township there is a belt of garnetiferous quartzose gneiss about 10 m wide that can be traced for about 225 m along its strike. Garnet makes up more than 30% of the gneiss and is present as crystals that range in size from 6 to 25 mm. The garnet appears to be of good abrasive quality (Eardley-Wilmot, 1927).

Northwest of Saint Felix, near where the Black River turns south, horizontal bands of weathered sillimanite gneiss contain large garnet crystals. The gneiss is about 30 m wide, is exposed for about 400 m along its strike, and contains large red-brown garnets that are extensively fractured and weathered. Near this same location is a 10 m wide band of unaltered hornblende gneiss that contains about 25% garnet. The garnet is present as well-formed, deep-red crystals and is of good abrasive quality (Eardley-Wilmot, 1927).

On the east side of a small lake about 3 km southwest of Labelle Station on the Mont-Laurier branch of the Canadian Pacific Railway, there is a belt of garnetiferous biotite-quartz gneisses that contain veins of pyrrhotite and massive, crystalline garnet. Four parallel veins strike N 30^{0} E for an exposed length of 400 m and are about 15, 20, and 60 m apart. For a width of about 2 m on each side of the veins the host gneiss contains a high percentage of pale red garnets that vary in size from about 1 to 6 mm. The first vein is about 2 m wide, consists of pyrrhotite and large garnet crystals, and has highly garnetiferous gneiss at both contacts. The second vein is about 1 m wide and consists of solid pyrrhotite with a 50 to 70 mm wide zone of massive, crystalline garnet at both contacts. The total minable width of ore, including garnetiferous gneiss wall rock, is about 8 m. The third vein, about 1 m wide, consists of pyrrhotite with massive garnet along both contacts and throughout the pyrrhotite. The estimated total minable width of ore, including garnetiferous gneiss, is about 7 m. The fourth vein is about 5 m wide and consists of weathered gneiss containing large garnet crystals. Although the quantity of garnet in any one vein may be small, the combination of veins together could contain a significant tonnage of high quality abrasive garnet (Eardley-Wilmot, 1927).

China. Ten provinces in China are known to have deposits of abrasive-grade garnet. The deposits include both alluvial and primary, hardrock types. The garnet reserves in Hubei Province are the largest, but the deposit in the eastern part of the Zhongtiao mine in Shaanxi Province is extremely rich, averaging about 83% garnet. Only the most significant deposits are discussed below.

NEI MONGGOL—The Ming Xin garnet deposit at Wu La Te Hou Qi is a large metamorphic deposit in sericite-garnet-quartz schist. The ore consists of 45% sericite, 35% quartz, 15% almandite, and 5% minor metallic minerals. The garnet crystals average 8 mm in size and exhibit some laminar parting. There are some instances where other minerals are pseudomorphs of the garnet crystals, which limits the quality of the garnet product. There is a mill located at the deposit that has the capacity to produce 10 000 t of refined garnet per year (Changyin, 1990).

JIANGSU—The Dong Hai mine, located about 15 km southwest of the town of Dong Hai in northeast Jiangsu, is in eclogite that contains an average 55% almandite garnet. The eclogite body is 2 200 m long and averages 170 m wide. Within it a total of ten garnet ore zones have been delineated based on a cutoff grade of 30% garnet, resulting in an average grade of at least 40% garnet. The largest ore zone averages 55% garnet and is 1 300 m long, an average 170 m in width, and 300 m thick. The highest grade ore zone averages 90% garnet but is only 50 m long, 40 m wide, and 20 m thick. The garnet is dark red, orange-red, and pink. Crystals range in size from 0.1 to 10 mm and average about 0.8 mm. Their Mohs hardness varies from 7.0 to 7.3, and their specific gravity ranges from 3.52 to 4.04. The garnet has a glassy luster and an uneven fracture (Changyin, 1990).

Near the commune of Fanshan in Donghai County the alluvial Zhimafang Ruby deposit is mined for gemstone garnet. The ore body is stripped periodically by a government team. Local miners produce about 100 kg per year of high grade, gem quality material. Although the mine is called a ruby mine, all of the production is precious garnet (Changyin, 1990).

Near the commune of Anfeng in Donghai County the alluvial Anfeng Ruby deposit also is mined for gemstone garnet. The deposit is stripped periodically by a government team, and miners in the commune recover, cut, and polish precious garnet and fabricate it into jewelry. Annual production is about 20 kg of high quality gem garnet (Changyin, 1990).

SHANGXI—In Danfeng County a garnet-biotite-sillimanite-plagioclase gneiss is mined, and the garnet is recovered at the Shanyan Garnet plant. The ore averages only about 7% almandite garnet, which ranges in size from 5 to 15 mm. The plant's annual output is about 1 500 t, and its major product is a 38 μm abrasive-grade garnet powder (Changyin, 1990).

SICHUANG—Many small mines in the Leshan Mountain area supply garnet ore to the Bauzhuwan Emery plant, which is a garnet production facility even though it is called an emery plant. It produces a complete range of high quality garnet products for domestic use. The products range in size from 2.38 mm grains to technical grade powders (Changyin, 1990).

HEBEI—Independent miners supply garnet ore to the Xingtai Emery plant, which is a garnet, not emery, processing facility. It produces three basic abrasive grades—coarse ($-2.38 + 0.71$ mm), medium ($-0.59 + 0.25$ mm), and fine ($-117 + 45$ μm). Annual output is about 10 000 t and is sold to domestic and export markets (Changyin, 1990).

India. Garnet deposts have been identified in several states in India, the most important of them being located in Andhra Pradesh, Kerala, Madras, Mysore, Rajasthan, and Tamil Nadu.

TAMIL NADU—The modern beach sand deposit at Manavalakurichi extends over 6 km from the mouth of Pelliyar River to Colachel. The sands contain an average of about 5% almandite garnet, and ore reserves amount to hundreds of millions of tons. The garnet is recovered as a byproduct during ilmenite, zircon, and rare earth minerals production.

Another beach sand deposit on the Thanjavur coast extends for 12 to 14 km from the mouth of the Cuauvery River near Kaveripattinam to Sirkali. Reserves have not been measured, but it is thought that they are at least equal to those at Manavalakurichi. The deposit contains a high percentage of ilmenite, zircon, and garnet, along with monazite.

KERALA—Large deposits of beach sand on the west coast near Chavara are more than 7 m thick, 250 to 800 m wide, and about 22 km long. Reserves amount to hundreds of millions of tons having an average grade of 0.5% garnet. The almandite garnet is present as coarse, pale pink, nonangular grains that have a vitreous luster. It is recovered as a byproduct during the production of ilmenite, zircon, and monazite.

ORISSA—A Quaternary eolian sand dune deposit, 18 km long and 1.5 km wide, is found on the coast near Chatrapur. It contains reserves estimated at 240 Mt averaging 6.8% garnet. The dunes have a maximum minable thickness of about 15 m. Current plans do not call for the recovery of garnet, but recovery should be simple as part of the mineral sands operation. The mining rate being planned for the deposit suggests that garnet production easily could be 100 ktpy.

Norway. The area from Romsdal to Sunnfjord in western Norway contains many deposits of almandite-pyrope garnet-bearing rock known as eclogite. Some of the deposits are only small lenses, whereas others contain hundreds of millions of tons of garnet-rich rock. The Tverrberget deposit is near the towns of Kvalsund and Fosnavåg and is located on Nerlandsöy Island off the southwestern coast of Norway. The 110 m thick ore body is altered eclogite that strikes roughly east-west and dips at about 75^0 to the north. Garnet crystals are more abundant, larger, and less fractured in the essentially unaltered core of the deposit than in the transitional zone between the core and the enclosing hornblende gneiss (Alsobrook, 1991).

The Tverrberget deposit contains an estimated 28 Mt above mean sea level. On a volume percent basis the ore contains 38% garnet, 28% plagioclase, 20% biotite, 11% pyroxene, 3% amphibole, and essentially no quartz. On a weight percent basis the garnet content of the orebody varies from 11 to 66%, is usually 41 to 46%, and averages about 44%. The almandite-pyrope crystals range in size from 0.2 to 5 mm but are mostly from 0.5 to 3 mm in diameter (Alsobrook, 1991).

USSR. Garnet mining has been reported in Russia, but details regarding the deposits are not available.

TECHNOLOGY

Exploration Techniques

Successful exploration for garnet deposits begins with selection of target areas based on the presence of known favorable host rocks. Geologic studies and maps, topographic maps, and aerial photographs are used to identify favorable metamorphic, igneous, alluvial, or eolian geological formations.

The second step is field examination of the selected targets. This may include geologic mapping and limited sampling, but in many cases reconnaissance studies are sufficient. The results of the examinations are used to determine whether additional sampling and data collection are warranted.

The next step, if warranted, is to sample the deposit in detail, the purpose being to measure the physical parameters of the deposit, specifically its grade and size. The method of sampling used depends upon the type of deposit. Hardrock deposits, igneous or metamorphic, are sampled differently than alluvial or eolian deposits.

Hardrock deposits can be either diamond core drilled or trenched. Data from field investigations are used to select the spacing, pattern, and attitude (vertical or directional) of drill holes. Trenches should be oriented perpendicular to the strike of the formation and can be dug by hand, with a dozer, or with a backhoe. Backhoe trenching causes the least land disturbance for the volume of sample collected and leaves free-standing walls that can be examined.

Placer deposits, alluvial or eolian, can be sampled by drilling, trenching, or by excavating pits or shafts. The physical nature of the deposit, its thickness, hardness, grain size, and whether it is above or below the water table influence the method of sampling chosen. If the deposit is 10 m or less in thickness and does not contain boulders or cobbles, hand augers or power augers can be used to drill sample holes. Hand augering with a two man crew using a jointed auger is fast and inexpensive. In deposits that are too hard to hand auger because of the presence of clay, iron oxide cement, or mild calcium carbonate cement, two-man motor-driven augers or vehicle-mounted augers (such as post hole diggers or telephone pole hole diggers) work well. However, augering does not always provide uncontaminated samples and is not effective below the water table.

Truck-mounted water-jet drills and rotary hammer drills are used to test thick deposits and deposits that are too hard to drill with other methods. A water-jet drill uses flush-jointed drill pipe with perforations near the bit to direct jets of water forward and downward from the bit. A hammer advances the drill bit and casing at the same time. Water from the jets flushes out the hole and returns cuttings to the surface in the annulus between the drill pipe and casing. Also, truck-mounted rotary drills equipped with double-pipe drill string and downhole hydraulic hammers can be used. Compressed air is forced down the inner pipe and returns samples to the surface in the annulus between the inner pipe and outer casing. Bits are selected depending on the type of material to be drilled.

Trenching can be carried out using either a dozer or a backhoe, depending on the size and depth of the deposit. In unconsolidated sediments it is difficult to maintain the stability of the walls of trenches, and samples can be contaminated by material sloughing from above. Bulk samples can be collected from 1 m^2 pits or shafts. However, here again, wall stability can be a problem unless some form of shoring is used.

Evaluation of Deposits

The evaluation of garnet deposits to determine their suitability for commercial production can be divided into five major factors: 1) size and grade of the reserves; 2) mining conditions; 3) quality of the garnet; 4) location of the deposit relative to markets and infrastructure; and 5) milling costs. The relative importance of each of these factors will vary from one deposit to another, but the final determination of economic viability will depend upon all of them.

Size and Grade of Reserves: An economic garnet deposit has reserves that are large enough and high enough in grade to provide full project capital recovery, meet all operating costs, and furnish a reasonable profit over the depreciable life of the project, which is normally at least 10 to 20 years. The physical size required will be a function of the anticipated annual production rate, but generally a deposit should contain a minimum of 2 Mt of ore. The cutoff grade used in calculating reserves will depend on the economics of the specific deposit, but cutoff grades for most deposits average about 20% garnet. Commonly, using a cutoff of 20% garnet will result in an average grade of between 30 and 35% for the deposit. The Chinese use a 20% cutoff, or boundary grade, and their deposits have an overall average grade of about 40% garnet.

Mining Conditions: Mining conditions greatly affect the economics of a deposit and can even control whether the deposit can be developed. Early field examinations should identify any conditions, such as nearby houses, historical sites, national monuments, archaeological or paleontological sites, wildlife refuges, municipal water sheds, etc., that could preclude mining the deposit. Local, county, and state zoning regulations, environmental regulations and other constraints should be checked to establish their effects on mining the deposit.

Hardrock garnet deposits are mined in benched open pits using conventional drill-and-blast methods. Some placer deposits are mined with dredges. Data collected during the exploration stage should include the dimensions, attitude, and structure of the orebody, the topography of the area, and the slope stability characteristics of the ore and overburden. The configuration and structure of the deposit should be such that there are no blocks of waste rock within the orebody and that the garnet is, ideally, uniformally distributed throughout the ore. Data on potential surface and/or subsurface water problems and other physical features of the ore and overburden that would affect mining costs also must be collected and evaluated. All of these factors will influence whether the deposit is amenable to open pit mining and will impact the total mining cost per ton of ore and overburden.

Quality of the Garnet: The garnet in a deposit must be tested to evaluate its performance and marketability as an abrasive. It should be recognized in this regard that most garnet used in abrasive applications is either the almandite or almandite-pyrope type. The garnet should be present in the deposit as discrete crystals that are at least 6 to 13 mm in size and that are liberated when the ore is crushed to -2 mm. The crystals should be free of any inclusions, such as quartz, mica, hornblende and feldspar, and also should be free of alteration. The garnet should be relatively uniform in hardness and density throughout the deposit. It should not be badly weathered or highly friable. If an incipient parting is present in the garnet crystals, the effect this parting has on the quality of the final product and the distribution of such crystals within the deposit should be determined.

Location Relative to Markets and Infrastructure: The location of a garnet deposit must be evaluated relative to its proximity to the markets for garnet abrasives and filter media. If the deposit itself is not near major market areas, then it must be near adequate and inexpensive means of transportation, especially by rail or water. Its location should be such that a skilled or trainable work force and adequate sources of electric power, fuel, and both process and potable water are available in the area. Deposits should be reasonably accessible from existing roads.

Milling Costs: The type of milling required and the cost of processing the ore must be analyzed carefully. Ideal ore is that from which the garnet can be liberated with minimum crushing and can be recovered using gravity/density-based methods of mechanical concentration, and where size classification of the concentrate yields a complete range of product sizes. At most garnet operations primary, secondary, and third stage crushing are required to completely liberate the garnet from the ore. Concentration methods used include gravity/density, flotation, and magnetic and electrostatic separation. Concentrates are screened or otherwise size classified to obtain finished products that have different particle size distributions. In some cases portions of the concentrate will be ground to achieve proper particle size distribution. To facilitate recovery of high quality garnet concentrates the ore should not contain minerals that have densities equal to or higher than garnet or that have similarly low magnetic susceptibility.

Mining and Processing

Mining: At the Barton Mine in New York, the Ruby Mountain ore body covers approximately 41 000 m^2, which represents nearly the entire mountain. The ore is mined by open pit methods with benches 9 m high. The ore is drilled and blasted, a drop ball provides secondary breakage, and power shovels load ore into trucks for haulage to the mill about 0.4 km from the mine.

At Emerald Creek and Carpenter Creek in Benewah County, Idaho, backhoes are used to strip 1 to 2 m of overburden and expose the garnetiferous gravels. Mining is then carried out with draglines and ore is trucked to a mill for processing.

Wollastonite production at Willsboro, NY, yields a byproduct concentrate containing garnet, diopside-hedenbergite, and minor

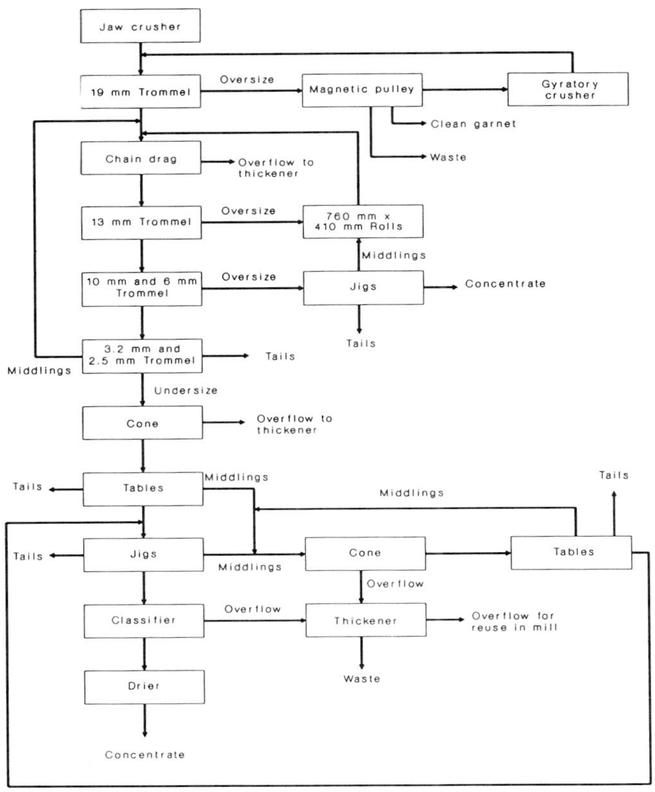

Fig. 2. Generalized flowsheet for a garnet processing plant (Myers and Anderson, 1925).

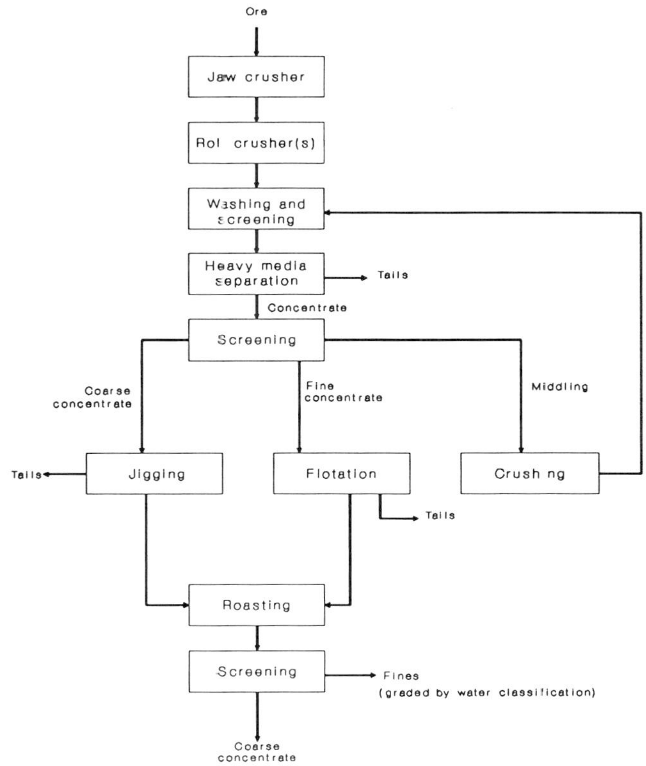

Fig. 3. Estimated flowsheet for Barton Mines Corp. garnet processing plant.

amounts of wollastonite. After drilling, blasting, and loading, the wollastonite ore is transported 22 km to a plant for separation of the wollastonite from garnet-diopside. For additional information on the mining operation see the chapter on Wollastonite.

At the Garnet Millers Associates' mine in Australia beach sands containing garnet are mined by wheel loaders working in conjunction with articulated trucks. The loaders and trucks work in tandem at the mining face and also for overburden stripping.

Processing: The primary factors controlling the behavior of garnet ore during processing are the identity and characteristics of the minerals present, the percentage of each mineral present, and the size of the garnet grains. The major minerals associated with garnet in most deposits are feldspar, mica, hornblende, pyroxene, and quartz. Magnetite, pyrite, ilmenite, limonite, pyrrhotite, and occasionally chalcopyrite, rutile, zoisite, chlorite, and corundum may be present in much smaller amounts. Sodium feldspar is the most common feldspar, biotite is the most common mica, and hypersthene and diopside are the most common varieties of pyroxene present. The most important physical property controlling the behavior of the major and minor minerals during concentration of the garnet is their specific gravity. Fig. 2 illustrates a generalized flowsheet for processing garnet.

At Barton Mines Corp.'s plant in New York garnet is liberated, concentrated, and further processed by combinations of crushing and grinding, screening, tabling, flotation, magnetic separation, water sedimentation, drying, heat treating, and air separation. Occasionally a single process will accomplish both sizing and the removal of gangue minerals, but this usually requires several processes. Particle size control can be difficult but is important in order for the concentration processes to work effectively.

The process used by Barton Mines to recover garnet has evolved over time. It began in 1924 with construction of the first modern plant where separation was accomplished using jigs. In 1941 heavy media separation was added to the system, and in 1945 flotation was added. The current state-of-the-art mill is a combination of heavy media and flotation circuits. Ore crushed to -7 mm is screened to 3.36 to 2.38 mm with the coarse fraction going to a heavy media circuit using a density of 3.02 for primary separation and a density of 3.20 to 3.40 for secondary separation. The fine fraction from the screen goes to a flotation circuit, where it is first ground to −0.3 mm in ball mills and then fed to float cells. The concentrate from the heavy media circuit is roll crushed, combined with the float concentrate, and then dried and heat treated in a rotary kiln. Fig. 3 illustrates a conceptualized flowsheet for Barton Mines' mill. The flowsheet was prepared from a written description of the milling circuit (Oskam, 1980).

At Emerald Creek Garnet Milling's operation in Idaho the ore is first processed through trommels to remove the gravel and is then concentrated in jigs and on shaking tables. The garnet concentrate is dried, crushed, and screened. The maximum grain size of finished products is 4.6 mm. Fig. 4 shows a process flowsheet using trommels, jigs, and tables.

At NYCO Minerals' mill in New York beneficiation of the wollastonite-garnet ore requires two stages. The first stage involves drying, crushing, screening, and recrushing the feed to −1.19 mm. In the second stage the −1.19 mm ore passes over a series of screens to attain −1.19 + 0.84, −0.84 + 0.42, −0.42 + 0.25 and −0.25 mm splits. A battery of magnetic separators then removes the garnet from each of these fractions. Four sizes of final product are bagged for shipment.

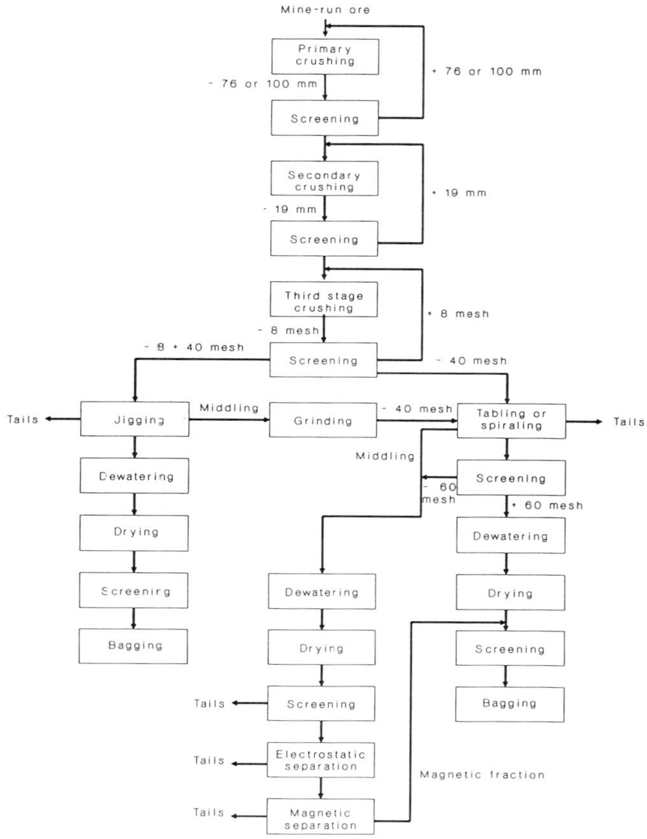

Fig. 4. Flowsheet for a proposed garnet processing plant in Europe (Alsobrook, 1991).

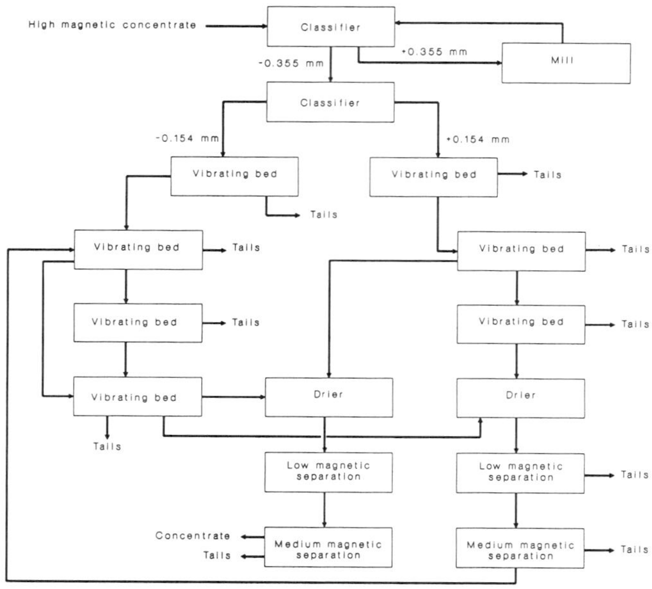

Fig. 5. Flowsheet for a typical chinese garnet processing plant (Changyin, 1990).

At Garnet Millers Associates' operation a wet gravity separation circuit at the mine recovers garnet from the beach sand and shell fragments. The concentrate is trucked to the company's dry processing plant for drying, screening, and packaging.

Fig. 5 shows the flowsheet for a typical Chinese garnet processing plant.

Specifications

Consumers purchase garnet that has been crushed, ground, and graded to different particle sizes for different end uses. Descriptions of testing methods for materials to be used in the manufacturing of coated abrasives are contained in American National Standards Institute (ANSI) Specifications B74.18–1977, "Specifications for Grading of Certain Abrasive Grains on Coated Abrasive Products." Other specifications are contained in ANSI Specification B74.12–1977, "Specification for Size of Abrasive Grains-Grinding Wheel, Polishing, and General Industrial Uses."

If the garnet is to be used as a sandblasting medium at a US Navy shipyard or a private shipyard working on US Navy vessels, then it must meet MIL-A-22262(SH) specifications that qualify it for product list QPL-22262. These specifications set forth limits on the content of total metals, soluble metals, and radiation, specify the acceptable level of chemical conductivity, and require that the material be certified by the California Air Resources Board.

The California Air Resources Board (ARB) requires all loose grain abrasives used in the state to be certified at the ARB testing facility. The testing is conducted according to "Methods of Test of Abrasive Media Evaluation," Test Method No. Ca 371-A and as required under ARB's Title 17, Subchapter 6 entitled "Abrasive Blasting." Copies of these documents are available from the Compliance Division of the California Air Resources Board, Sacramento, CA. In addition, air pollution control agencies in Utah, Minnesota, and Louisiana have adopted the ARB tests and require that only ARB-certified abrasives be used on state projects requiring sandblasting. The ARB's testing procedures and standards are focused on the environmental impact of sandblasting abrasives, not on their qualities or cutting performance/efficiency.

The Steel Structures Painting Council, a technical organization in Pittsburgh that has a committee on abrasives, is currently drafting abrasive specification SSPC-XAB1X, "Mineral and Slag Abrasives," which will address garnet specifically. The specification defines the requirements for selecting and evaluating mineral and slag abrasives used in blast cleaning steel and other surfaces for painting and other purposes.

The American Water Works Association (AWWA) has established specification B100–89, "Standards for Filtering Materials," but this specification does not cover high-density media such as garnet and ilmenite. The AWWA plans to include garnet and ilmenite in the updated specification B100–94. In general, specifications for filtration involve garnet's particle shape, specific gravity, effective grain size, coefficient of grain size uniformity, content of acid-soluble impurities, and content of radionuclides and heavy metals.

The sale of garnet for sandblasting, water filtration, and abrasive cutting is by screen size. In US sandblasting the most widely used domestically produced garnet is $-1.0 + 0.3$ mm, followed by $-0.59 + 0.21$ mm. The most popular grade of imported garnet in the US market is the $-0.59 + 0.25$ mm grade from Australia. In US multimedia water filtration systems two grades of garnet are used, a coarse grade that is $-4.75 + 1.41$ mm, and a fine grade that is either $-0.84 + 0.3$ or $-0.71 + 0.25$ mm. Abrasive cutting applications' most popular grade of garnet is $-0.25 + 0.177$ mm, which may account for as much as 90% of the garnet used.

Testing

There is no definitive method for testing the quality of garnet or any other loose-grain abrasive except by actual application. There are tests and examinations that can indicate a garnet product's probable abrasive performance in use. The fracture, sharpness, shape, and structure of the garnet and the presence of inclusions all can be studied with a microscope. The hardness and friability of the garnet can be evaluated by placing a sample between two glass microscope slides and rubbing them together. The relative scratch hardness of the grains and their degree of breakdown (attrition) can be measured in this manner. Additionally, ANSI standard B74.8–1965, "American Standard Test for Ball Mill Test for Friability of Abrasive Grain," can be used to determine the friability of garnet grains (Anon., 1965).

The US National Institute for Standards and Technology developed an apparatus for evaluating the abrasive quality of corundum. This tester can be adapted for use with any loose-grain abrasive. Testing of loose-grain abrasives also can be performed with various production-scale and laboratory machines where the removal of glass from flat plates is evaluated. Data from such tests can be correlated with the anticipated performance of any abrasive grain relative to the other material to be abraded. In these tests abrasive performance is determined by the amount of stock removed, as found either by weighing or by dimensional measurements.

In judging the abrasive quality of garnet, stock removal is not the only important factor. Surface finish is just as important. If an abrasive is too hard or contains too coarse grains for use on a certain material it can cause deep scratches in the surface, which cannot be removed in subsequent finishing operations. Additionally, if an abrasive does not have the proper grain shape or does not break down in such a manner as to resharpen itself, it will tend to burn or gouge the material being abraded, thus defeating the purpose of its use.

The California Air Resources Board's testing facility tests loose-grain abrasives for certification for use in the state. The test is focused on the environmental impact of sandblasting abrasives, not on their abrasive quality.

Other firms or laboratories that test the quality of loose abrasive grains include three firms involved in the abrasive cutting industry, Flow International Corp., Ingersoll-Rand Corp., and Jet Edge, Inc., who have laboratories for testing and evaluating garnet for use in their systems. Most of the evaluations in these laboratories involve the comparative testing of various abrasive grits, including measurements of particle size distribution and mechanical properties. KTA-Tator, Inc. of Los Angeles, CA, is a major consulting/engineering/testing firm whose laboratory tests and evaluates the quality of garnet and other abrasives. Much of its work is oriented to the Steel Structures Painting Council (SSPC) specifications. The SSPC also can carry out comparative testing of garnet and other loose-grain abrasives to determine whether they meet military specifications, how clean the material is, how well the material cleans, the amount of degradation during use, and other quality factors. SSPC is part of Carnegie-Mellon Research Institute.

Byproducts and Coproducts

Mica and quartz are byproducts of garnet production at some deposits. More often, garnet is a byproduct during the production of other mineral commodities. Garnet from India, for example, is a byproduct at ilmenite and other mineral sand deposits. In New York it is a byproduct of wollastonite production, and in Turkey a small amount of garnet is recovered as a byproduct during feldspar production.

USES

The major industrial uses of garnet are for sandblasting, the manufacture of coated abrasive products, the grinding and lapping of glass, metals, ceramics, and plastics, and as a filter medium. Minor industrial uses are as jewel bearings, in the manufacture of grinding wheels, as a tumbling medium, in various nonskid applications, and as a grit in abrasive cutting and slabbing of marble dimension stone. As a gemstone garnet is used in manufacturing a wide price range of different types of jewelry.

Abrasive Paper and Cloth

Coated papers and cloths made from technical-grade garnet are used primarily for sanding wood and also for finishing leather, hard rubber, plastics, glass, and soft metals. Coated abrasive products continue to improve in quality as new manufacturing techniques and adhesives are developed.

Garnet-coated products make up about 10% of the total volume of abrasive paper and cloth produced in the United States each year. This percentage has remained fairly constant in recent years despite the replacement of abrasive paper and cloth by other surface finishing materials. A unique use of garnet-coated paper is the removal of the red hulls that cover peanuts.

Glass Grinding

Garnet was first used commercially as a powder in the grinding of optical lenses during World War II. By the time the war ended the abrasive qualities of garnet were widely know, and it was well established in the glass surfacing industry. By decreasing the amount of scratching the use of garnet improved the quality of lenses and reduced the frequency of rejects. Garnet also produced a finish that required less polishing time. More recently, garnet has become the medium of preference for grinding television face plates. The same properties that made it a competitive abrasive in the lens and television industries continue to generate new applications for it in the glass, plastics, ceramics, and soft metals industries.

Loose Grains

Major nontechnical uses of garnet in loose grain form are sandblasting, water filtration, and abrasive jet cutting. Garnet has the advantage of containing essentially no crystalline silica, so its use does not cause silicosis and does not fall under the jurisdiction of the Occupational Safety and Health Administration's (OSHA) Hazard Communication Standard. In 1987 International Agency for Research on Cancer (IARC) determined that crystalline silica is a probable carcinogen for humans. As a result OSHA is required to regulate materials containing more than 0.1% crystalline silica. Also, garnet contains no leachable heavy metals. It has a higher specific gravity than quartz, so when used as a sandblasting or abrasive jet cutting agent it deliver more impact than quartz grains of the same size. It is estimated that in similar sandblasting applications 3 to 4 times as much silica sand is required to complete the work and that garnet has 5 to 6 times the useful life of silica sand.

In abrasive jet cutting, garnet grains, usually 0.18 mm in size, are entrained in a stream of high velocity water. The supersonic abrasive slurry is impinged on the material to be cut at a pressure of $5.5 \times 10^{+8}$ Pa. Abrasive jet cutting is used to shape-cut materials such as marble, granite, artificial stone, concrete, aluminum, titanium, high strength steel and steel bridge decking, automotive glass, textiles, corrugated boxboard, plastics laminates, and aerospace composites.

When used as a water filtration medium garnet resets the filter bed faster than quartz after the bed is backflushed. It also has the advantages of being more recyclable, essentially chemically inert, and of fracturing less readily.

ECONOMIC FACTORS

Price

In 1990 the average value of crude garnet concentrates produced in the United States was estimated by the US Bureau of Mines to be $178/t. This value represents the total cost of the garnet concentrate at the end of the milling process, but does not include any packaging, transportation, or other costs beyond the last processing stage. The average value during 1981 to 1990 was $105/t, with a high of $178 in 1990 and a low of $88 in 1986. In 1990 the average value per ton of garnet sold or used by US producers was estimated to be $255. The average value during 1981 to 1990 was $232 per ton, with a high of $266 in 1988 and a low of $216 in 1987. Market prices for garnet vary widely depending on the type of garnet, purity, particle size, size of order, and packaging. The value of a ton of garnet can range from as low a $60/t for low quality sandblasting material to as much as $2,000/t for top quality, micronized, technical-grade material.

Markets

The demand for garnet in sandblasting is growing because of environmental concerns related to the use of silica sand and slags. Silica sand is made up entirely of free, crystalline silica, the particles of which that are small enough to be inhaled can cause silicosis. The silicosis health risk in using silica sand for sandblasting has been known for many years. However, in 1987 it was determined that crystalline silica may pose an additional health hazard as a probable carcinogen. These health risks associated with crystalline silica have improved the demand for garnet as a sandblasting medium in certain markets, and it is anticipated that they will continue to encourage its use in the future. Many of the slags used in sandblasting contain leachable heavy metals, which can pose both environmental and health risks. Because garnet is environmentally less hazardous and has a price advantage over most of the other environmentally acceptable sandblasting agents, its market share should continue to grow.

The demand for garnet in abrasive jet cutting applications is growing because the number of installations using the technology is increasing rapidly. This technology is only 10 years old, and the market is dominated by three firms, Flow International Corp., Ingersoll-Rand Corp., and Jet Edge, Inc. Garnet demand will be a function of the number of installations using the technology, the number of hours per day each installation is actually operating, and the g/min consumption rate at each installation during operation.

The US Bureau of Mines estimates that 30% of annual demand for garnet in the United States is for use as a filtration medium, mostly for water filtration. The worldwide demand pattern appears to be the same as that in the United States. Demand was stable during 1986 to 1990 but can be expected to increase in the future as populations grow and require additional water supplies and as aging water filtration plants have to be replaced.

Alternative Materials

Garnet competes with a number of natural and manufactured abrasives. In coated abrasive paper and cloth garnet competes with aluminum oxide, silicon carbide, flint, emery, and silica sand. In the sandblasting market for loose-grain abrasives competitive substitutes include aluminum oxide, silicon carbide, silica sand, slags, staurolite, and olivine. In the loose-grain filter media market garnet competes with quartz sand, ilmenite, magnetite, and plastics. In the grinding and polishing of glass other abrasives used include diamond, aluminum oxide, silicon carbide, and natural corundum. Nonskid flooring and coatings can use silicon carbide, quartz sand, and emery instead of garnet.

GOVERNMENTAL CONSIDERATIONS

Tariffs and Depletion Provisions

Tariffs: In 1992 the US tariff rates on imported garnet and items made from garnet are shown below for countries with Most Favored Nation (MFN) status and for countries without MFN status.

Item	MFN	Non-MFN
Crude garnet	duty free	duty free
Garnet other than crude	$0.007/kg	$0.022 /kg
Garnet on woven textile	2.5% ad valorem	20% ad valorem
Garnet on paper or paperboard	2.5% ad valorem	20% ad valorem
Garnet sheet, strip, disk, belt, sleeve, or similar form	2.5% ad valorem	20% ad valorem

Depletion Provisions: The percentage depletion allowance against US federal income tax is 14% of gross income from both domestic and foreign garnet production.

PROBLEMS AND FUTURE TRENDS

It is the opinion of the US Bureau of Mines that US production and consumption of industrial garnet during the next 5 years will depend largely on producers' installed production capacity. Currently the garnet industry in the United States is operating at or near capacity. The Bureau also believes that if capacity is added, production and consumption could increase to 60 ktpy by 1997 or 1998. If additional garnet is not available, other abrasives will have to be used instead, but in many cases at some sacrifice in terms of productivity, efficiency, finished product quality, environmental and health protection, and/or cost savings.

BIBLIOGRAPHY AND REFERENCES

Anon., 1964a, "American Standard Test for Bulk Density of Abrasive Grain," Specification B74.4, American National Standards Institute.

Anon., 1964b, "Garnet," *Indian Minerals Yearbook, 1964*, Government of India, pp. 457–461.

Anon., 1965, "American Standard Test of Ball Mill Test for Friability of Abrasive Grain," Specification B74.8, American National Standards Institute.

Anon., 1976, "Specifications for Size of Abrasive Grains-Grinding Wheel, Polishing, and General Industrial Uses," Specification B74.12, American National Standards Institute.

Anon., 1977, "Specifications for Grading of Certain Abrasive Grains on Coated Abrasive Products," Specification B74.18, American National Standards Institute.

Anon., 1985, "Abrasive Materials," *Minerals Yearbook 1985*, US Bureau of Mines, Department of the Interior, Vol.1, pp. 95–108.

Anon., 1989, "Mineral and Slag Abrasives," Specification SSPC-XAB1X (Draft No. 10), Steel Structures Painting Council.

Anon., 1991, "Open Pit Garnet Mining in W. Australia," *Mining Magazine*, Vol. 165, No. 4, Oct., p. 188.

Adams, W.T., 1976, "Abrasive Materials," *Minerals Yearbook 1976*, US Bureau of Mines, US Department of the Interior, Vol. 1, pp. 137–147.

Arem, J.E., 1987, *Color Encyclopedia of Gemstones,* 2nd ed., Van Nostrand Reinhold, New York, 343 pp.

Austin, G.T., 1989, "Abrasive Materials," *Minerals Yearbook 1989*, US Bureau of Mines, Department of the Interior, Vol. 1, pp. 77–96.

Austin, G.T., 1990, "Abrasive Materials," *Annual Report 1990*, US Bureau of Mines, Department of the Interior, 22 pp. (preprint).

Changyin, L., 1990, *Perspective of Production and Development for Abrasive Garnet in China*, Editorial House of Non-Metallic Mines, Suzhou Jiangsu, 38 pp.

Eardley-Wilmot, V.L., 1927, "Garnet," *Abrasives*, Part 3, Canada Department of Mines, Mines Branch, 69 pp.

Ford, W.E., 1957, "Descriptive Mineralogy," *Dana's Textbook of Mineralogy*, 4th ed., John Wiley and Sons, London, pp. 591–597.

Hight, R.P., 1983, "Abrasives," *Industrial Minerals and Rocks*, 5th ed., Vol. 1, S.J. Lefond, ed., AIME, New York, pp. 11–32.

Hurlbut, C.S., 1971, "Descriptive Mineralogy," *Dana's Manual of Mineralogy*, 18th ed., John Wiley and Sons, London, pp. 377–380.

Hutchins, J.P., 1907, "Gold-Dredging Practice at Ruby, Montana," *Engineering and Mining Journal*, Vol 83, No. 26, 4 pp.

Ladoo, R.B., 1922, "Garnet," Reports of Investigations No. 2347, US Bureau of Mines, 16 pp.

Ladoo, R.B., and Myers, W.M., 1951, "Garnet," *Nonmetallic Minerals*, 2nd ed., McGraw-Hill, New York, pp. 241–249.

Myers, W.M., and Anderson, C.O., 1925a, *Garnet: Its Mining, Milling, and Utilization*, US Bureau of Mines, 54 pp.

Myers, W.M., and Anderson, C.O., 1925b, "Recent Developments in the Production and Consumption of Abrasive Garnet," Reports of Investigations No. 2691, US Bureau of Mines, 11 pp.

O'Connor, M.P., 1985, "The Wing Hill Garnet Deposit of Rangeley, Maine and Its Relationship to World Markets," Preprint 85–29, SME-AIME Annual Meeting, New York, Feb., 5 pp.

O'Driscoll, M., 1990, "Wollastonite Production Tempo Rises As Markets Grow," *Industrial Minerals*, No. 279, Dec., pp. 15–23.

Oskam, J., 1980, "Garnet—Abrasive and Gems," Preprint 80–382, SME-AIME Fall Meeting, Minneapolis, MN, Oct., 5 pp.

Smoak, J.F., 1981, "Abrasive Materials," *Minerals Yearbook 1981*, US Bureau of Mines, Department of the Interior, Vol. 1, pp 57–73.

Tansley, W., et al., 1933, "A Geological Reconnaissance of the Tobacco Root Mountains, Madison County, Montana," Memoir No. 9, Montana Bureau of Mines and Geology.

Vitaliano, C.J., et al., 1979, "Geologic Map of Southern Tobacco Root Mountains, Madison County, Montana," Map MC-31, Geological Society of America.

Winchell, A.N., 1914, "Mining Districts of the Dillon Quadrangle, Montana," Bulletin No. 574, US Geological Survey, pp. 57–61.

GILSONITE

HARRY D. LEWIS

Gilsonite is a lustrous black material that can be easily crushed into a chocolate-colored powder. It is a solid but relatively soft, lightweight hydrocarbon that visually resembles the volcanic glass obsidian. It was once thought to be a type of coal, but because it is thermoplastic it melts rather than burns.

Gilsonite is found in commercial quantities only in the Uinta Basin of eastern Utah and western Colorado, where it forms vertical, nearly parallel veins of varying depth, width, and length. Gilsonite from different veins has different properties, most notably variations in softening point and solubility. These differences result in its use in a variety of industrial applications.

Gilsonite's earliest commercial uses were in varnishes, buggy paints, and beer vat linings. Eventually it came to be used in more than 160 products ranging from inks to explosives. It is exported from the United States to most industrialized countries throughout the world. Today, due to competition from synthetics and other hydrocarbon products, its use has narrowed to several specific industries. Although sold mostly as a commodity, current research is focused on applications for gilsonite as a value added product.

Historical Overview

The discovery of gold in California and the dawn of the mining era in Nevada, Colorado, and Utah around 1850 brought a wave of prospectors to the western United States. One of those pioneers, Samuel H. Gilson, is generally credited with a major role in the discovery of gilsonite.

The material now known as gilsonite was discovered several times from the late 1860s through the mid-1880s. It was William P. Blake who, after receiving samples of the material, wrote a paper in 1885 describing "a new variety of asphaltum from the Uinta Mountains." He termed the material uintaite. Although Sam Gilson was not the first to discover the mineral, he was an inventor who used gilsonite in products that he patented. It was his conviction that the substance was of great value, which caused some to refer to it as gilsonite. The term *Gilsonite* became the registered trademark of the American Gilsonite Co., and it continues to be in use today. When the first company was formed to mine this substance in 1886 it was named the Gilsonite Manufacturing Co. This firm acquired many gilsonite mining claims and marketed a varnish made from the material.

It was the unusual properties of gilsonite varnish that attracted the attention of Adolphus Busch of the Anheuser-Busch Brewing Co. In 1889 Anheuser-Busch bought out Gilson's interests and formed the Gilson Asphaltum Co. Busch had been importing asphalt from Sicily for use in lining beer barrels and saw gilsonite as being more readily available and cheaper. Although gilsonite did not work well in this application, Busch did acquire a large number of mining claims in the Uinta Basin.

In 1903 the parent firm of Gilson Asphaltum financed the building of an 85-km railroad from near Grand Junction, CO, to Dragon, UT, to haul gilsonite ore to market. Underground mining was initially centered near Dragon and then later at Bonanza, UT, as different veins were mined. The railway was abandoned in 1937 in favor of truck transportation.

In 1938 Gilson Asphaltum changed its name to Barber Asphalt Co., the predecessor of Barber Oil Corp. In 1946 Barber Oil teamed with Standard Oil of California to form a partnership, American Gilsonite Co., which today has its operational headquarters at Bonanza, UT.

Barber Oil investigated many new uses for gilsonite, including the making of petroleum derivatives such as gasoline. This work led to the construction of a 116-km pipeline to carry slurried gilsonite to a refinery in Colorado for the production of gasoline, fuel oil, and coke. Pipeline operations began in 1957 but stopped when the economics of competing with petroleum became unfavorable, and the refinery was sold in 1973.

In 1981 Standard Oil (later Chevron) became the sole owner of American Gilsonite and remained so until 1991, when it sold its interest to Stratford Enterprises Co. of Tulsa, OK. The companies have since been merged and now operate under the name American Gilsonite Co. with headquarters in Salt Lake City, UT. American Gilsonite continues to be the world's largest miner and supplier of the mineral.

The American Asphalt Association was formed in 1902 by former stockholders in Gilson Asphaltum. This company mined various gilsonite claims until it was dissolved in 1953 and its properties were sold to American Gilsonite and Gordon S. Ziegler and Associates. The Utah Gilsonite Co. was established in 1920 and mined several claims until 1952, when its holdings were sold to Gordon Ziegler. Several other gilsonite mining ventures have been pursued in Utah, although they were significantly smaller in scale than those described above. There are currently three companies which mine gilsonite resin in Utah. They are American Gilsonite Co., Ziegler Chemical & Mineral Corp., and Lexco.

GEOLOGY

Mineralogy

Gilsonite is a natural hydrocarbon also known as uintaite. It is a brittle, glossy, solid bitumen with a lustrous, conchoidal fracture. It is jet black when mined, but brown when ground to a powder. It resembles obsidian and belongs to the group of bituminous materials called asphaltites, natural asphalt-like substances characterized by high softening points (e,g, + 110°C). Glance pitch and grahamite are other members of this group. Asphaltites are soluble in carbon disulfide and are fusible with difficulty. A comparison of the properties of gilsonite, glance pitch, and grahamite is shown in Table 1.

Fig. 1. An early gilsonite mine hoist located north of Bonanza, UT.

Properties

Although gilsonite is chemically similar to asphalt, there are several important differences that make it unique. Gilsonite is a solid that has a softening point much higher than that of petroleum-derived asphalt. Its softening point generally ranges between 149 and 204°C depending on grade. Gilsonite contains a high concentration of nitrogen, generally around 3%, which appears from FTIR analysis to be present in heterocyclic compounds of the pyrrolic or quinolinic type. It is believed that the presence of these polar molecules is what gives gilsonite its dispersing ability. Also present in smaller concentrations are hydrogen-bonding moieties such as amide, carbonyl, and hydroxyl groups.

A number of analytical procedures have been carried out to characterize general purpose (GP) grade gilsonite, which has a softening point of about 177°C. Chemical analyses of gilsonite give the following elemental composition (in weight percent): carbon, 83.95; hydrogen, 10.03; nitrogen, 3.26; sulfur, 0.27; and oxygen, 1.38. The atomic ratio of hydrogen to carbon is 1.423. Carbon 13 NMR analyses give the following results (in weight percent): aliphatic carbon, 68.3, and aromatic carbon, 31.7. Average molecular weight, as determined by vapor pressure osmometry (VPO) in two different solvents is 3 172 in toluene and 2 976 in pyridine. VPO molecular weights in toluene are believed to be affected by intermolecular interaction, so the results in pyridine are probably more accurate.

Origin and Distribution of Major Deposits

The only area in the world where gilsonite is presently mined commercially is the Uinta Basin of northeastern Utah and northwestern Colorado. From a point 6 to 8 km within Rio Blanco County, CO, an elliptical area in which gilsonite veins are found extends westward about 100 km into Utah and is about 10 km wide.

Not all agree on how gilsonite was formed, but the consensus is that it was derived from the kerogen in oil shale. Gilsonite and shale oil have similar compositions and a nitrogen content much higher than that of petroleum. Gilsonite veins are found in a parallel series of near-vertical fractures cutting the Uinta Formation, which overlies the Green River Formation in the Uinta Basin. The Green River Formation contains rich oil shales, which are thought to have been the source of the gilsonite. Regional upwarping of the shale squeezed heavy, viscous hydrocarbons from it into tension fractures formed in overlying beds of Uinta sandstone. This organic material subsequently lost most of its volatile constituents and gradually solidified into gilsonite over time. These vertical veins are the deposits from which gilsonite is mined today.

The veins usually extend to about 365 m in depth, vary in width from a few centimeters to 7 m, and continue in length for as many as 40 km. They average 1.8 m in width, although veins as narrow as 0.46 m have been mined. In almost all cases they outcrop at the surface. They strike northwest-southeast through the Uinta Basin. The principal deposits are found within shales of the upper Green River Formation and massive sandstone and shale beds of the lower and middle Uinta Formation. These beds dip about 2.5° to the west.

The walls of gilsonite veins are usually quite regular; but locally the bitumen penetrated the wall rock, leaving large, dish-shaped sections of sandstone separated from the formation by a thin gilsonite film. Horizontal and near-horizontal offsets found along the veins vary from a few centimeters to the entire width of a vein. Both the upper and lower portions of these offsets are weakly attached to the wall rock, so stulling and roof bolting are required to hold them in place after the ore is removed.

Inclusions of wall rock are occasionally found within the ore. These rock masses are normally large and may span the entire width of a vein. They are most common near an offset. A corresponding cavity in the wall usually can be found not more that 15 m above an inclusion. This suggests that the gilsonite was quite viscous at the time the rock was dislodged. Veins that extend down into the Green River shales in the Bonanza area divide into narrow, irregular stringers that are presently uneconomical to mine.

TECHNOLOGY

Exploration Techniques

The primary tools used in exploring for gilsonite are the mapping of outcrops and the angle-drilling of a vein along its strike. Drilling targets are identified based upon surface mapping and previous mining locations. Once a target is established, a drill rig is offset from the vertical plane and a drill hole is angled, usually at 45°, toward the vein. The vein and wall rock immediately adjacent to it are cored to collect samples and to determine the vein's width.

Each mining region or deposit is evaluated prior to and during early phases of development. Evaluation factors include height, depth, width, accessibility to the top of the vein, softening point, ash content, water table, topography, location, construction costs, and projected mining costs. These factors influence the mine development plan and are included in a long range (10-year) mining schedule.

Mining Methods

Mine development begins with the boring of 213 cm diameter shafts spaced 305 m apart along a vein through about 46 m of overburden. Total shaft depth varies, depending on the vein, but can reach 457 m. Shafts are connected by horizontal drifts in the ore, which provide a second escapeway. The shafts are equipped with 9.1 m long steel inserts that are comprised of four 25.4 cm diameter pipes equally spaced around the perimeter of the shaft. These pipes serve as the main support for the guides and ladderway and carry conduit, air hose, electrical cables, and pipelines from the headframe to the working areas.

Once a shaft is bored, the 9.1 m long inserts are lowered into the borehole by the same rig that was used to raise and lower the drill bit during shaft sinking. As each 9.1 m section is lowered into the shaft the next section is lined up with it, and the two are welded

Table 1. Properties of Gilsonite, Glance Pitch, and Grahamite

Mineral	Streak	Specific gravity at 25°C	Softening pcint, ring and ball method,°C	Fixed carbon[a] %
Gilsonite or uintaite	brown	1.03–1.10	132–190	10–20
Glance pitch or manjak[b]	black	1.10–1.15	132–190	20–35
Grahamite[b]	black	1.15–1.29	138–329	35–55

[a]By proximate analysis, as for coal.
[b]When substantially free of impurities.
Source: Kirk-Othmer, 1980.

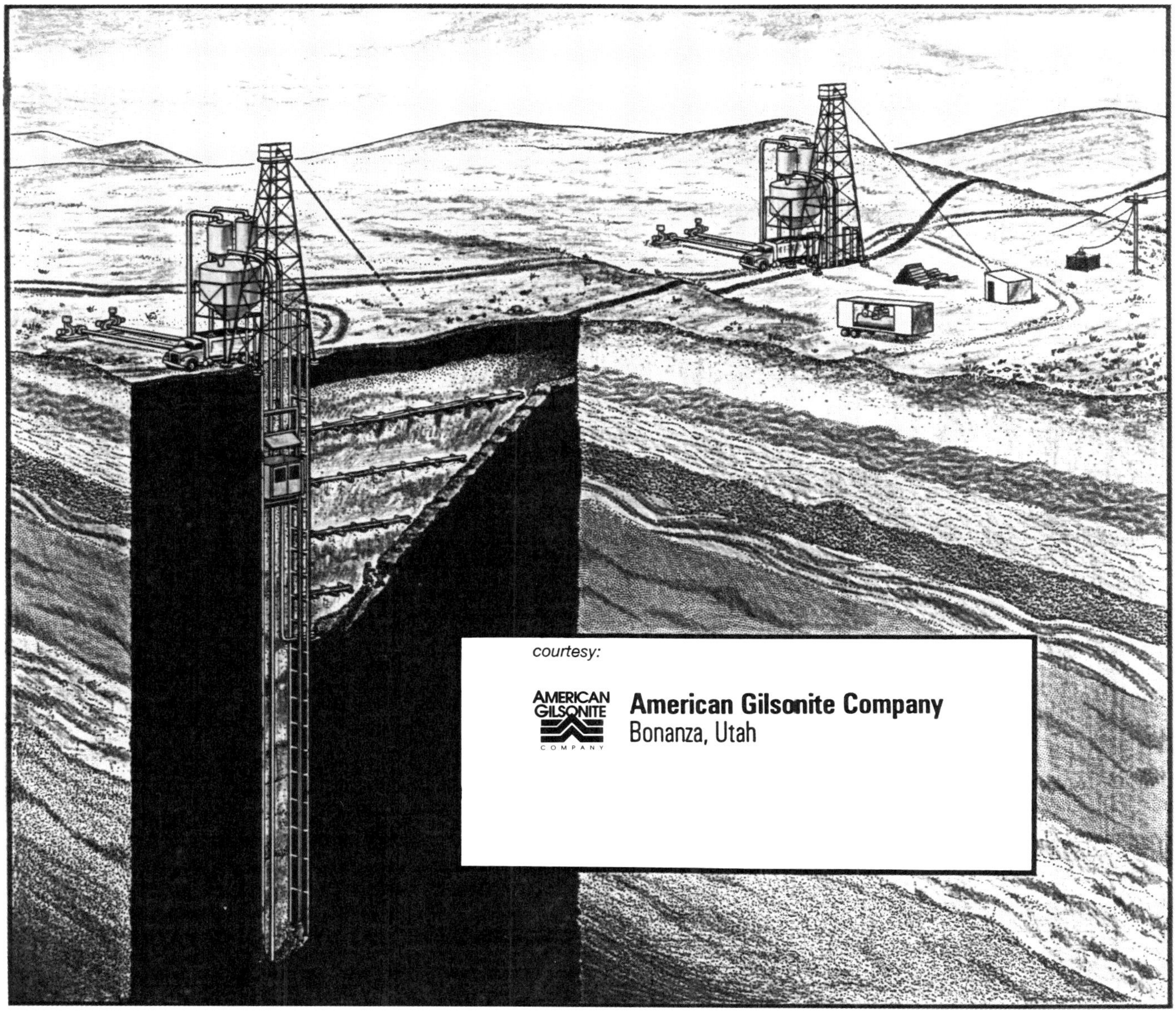

Fig. 2. Typical gilsonite mining operation.

together at the surface. This procedure is repeated until the inserts line the shaft from top to bottom. When a mine is worked out the liner assembly is pulled in 9.1 m sections and reused.

Drifts are started near the edge of the gilsonite vein on both sides of a shaft. Hand-held pneumatic chipping hammers with moilpoint bits are used to break out the ore in 1.5 m wide slices that slope upward from the floor of the drift at a 40 to 45° angle. Drifts extend as far as 153 m from a shaft, measured horizontally. Com-

Fig. 3. Hand mining of gilsonite with a pneumatic hammer.

Fig. 4. Cross-section of a gilsonite mine.

Fig. 5. Headframe and storage bin at an underground gilsonite mine.

pressed air is supplied by a 112 kW compressor on the surface and is delivered underground through a 7.6 cm diameter pipe.

Broken ore flows by gravity to the toe of the sloping face at the floor of the drift. From there it is air-lifted to the surface through a 35.6 cm diameter pipe. Vacuum for the air lift is produced by a 75 kW motor capable of producing a pressure of 14.9 kPa.

Drifts are mined to a length of 153 m, the halfway point between shafts. As mining progresses downward in successive drifts, horizontal pillars are left at vertical intervals of 92 m for the full strike length of the vein. Pillars vary from 4.6 to 7.6 m in thickness depending on the width of the vein. Wooden stulls to support the walls are installed in a horizontal line perpendicular to the shaft. The stulls are placed in hitches cut into the walls with air hammers and are secured, if necessary, with wedges.

Access to working areas is provided by walkways that are extended as mining advances toward the 153 m horizontal limit of a drift. These walkways are installed on another set of stulls, also perpendicular to the shaft. Wooden lagging and chain link fencing form the floor. The fencing prevents rock from falling into the working area but does not cut off visual contact between successive drifts. Careful hand mining keeps the ash content of run-of-mine ore to less than 0.25%.

When the air-lifted ore reaches the surface it enters a baghouse. The larger pieces drop first, and the rest is collected in filter bags. All solids are discharged into 136-t elevated storage bins and are then transferred by truck to a processing plant.

Processing

American Gilsonite Co. has a 110-ktpy processing plant that consists of concrete storage silos, truck receiving bins, a vibrating bed dryer, pulverizing machinery, and packaging equipment. Pneumatic conveying systems and sophisticated dust control equipment are state-of-the-art and allow for the handling of gilsonite in large quantities.

Before entering the processing plant gilsonite ore is segregated by grade in receiving bins. From these bins the ore is processed through a vibrating bed dryer, where excess moisture is removed. It is then passed over a double-deck screen, where it is classified according to particle size for storage in silos. From the silos ore is fed to product bins from which it is either loaded directly as bulk product, fed to a bagging machine, or fed into a pulverizer. Pulverized product is segregated into product bins from which it can be loaded directly as bulk product or packaged at a second bagging machine.

Product Specifications

Gilsonite products are marketed on the basis of specific requirements for each end use. The primary specifications are softening point, ash content, and particle size. Processing facilities sort, size, and dry incoming ores to meet these specifications. Careful attention to mining methods and plant screening to ensure

Fig. 6. Processing plant of American Gilsonite Co. at Bonanza, UT.

removal of pieces of sandstone are required in order to control the level of ash contained in the final product.

USES

Gilsonite was first used commercially in varnishes and paints for horse-drawn wagons and buggies. It was later used in a process known as japanning, which provided automobiles with a long lasting, deep, rich black, satin finish. Since 1904 gilsonite has been used in battery boxes, floor tiles, thermoplastic molding compounds, and linoleum and rubber products. One of the revolutionary developments was the establishment near Grand Junction, CO, of a major refinery that produced high octane gasoline, electrode-grade coke, industrial fuels, and road oils from gilsonite. This operation lasted from 1957 to 1973, when production ceased for economic reasons.

The challenge for producers has always been to develop new markets based on the three unique properties of gilsonite. These properties are: 1) high softening point (132 to 204°C), 2) very high purity (gilsonite is at least 99% pure as it comes from the mine), and 3) relatively high nitrogen content (about 3%).

Historical Uses

Gilsonite has been used in the manufacture of more than 160 products for the automotive, building materials, printing inks, highway products, oil well drilling, paint, rubber, steel, and foundry industries. A chronology of the major industrial uses of gilsonite over the past 100 years follows.

1890–1900 Wire and cable compounds, varnish
1900–1910 Roofing materials, paints, varnish
1910–1920 Paints (automotive), canvas saturants
1920–1930 Molded products (electrical), paints, inks
1930–1940 Battery cases, mineral wool binder
1950–1960 Asphaltic floor tile, printing inks, paints & protective coatings
1960–1970 Auto sealers, printing inks, oil well cements, asphaltic floor tile, refinery feedstock
1980–1984 Oil well cements, nuclear coke, board binders, foundry sand binder, printing inks
1984-Present Oil well mud additives and cements, printing inks, foundry sand binder, asphalt modification.

Current Uses

Printing Inks: Gilsonite is used as the primary carbon black wetting agent for dark-colored newspaper and magazine inks. Higher concentrations of gilsonite are used in the formulation of low rub-off newsprint inks.

Paints and Protective Coatings: Gilsonite is used as an additive in asphaltic paints and varnishes, where it enhances hardness, gloss, and weather and chemical resistance. It is easy to incorporate in these products, because it is a free-flowing, granular solid and is highly compatible with the oils and resins used in the manufacturing of coatings.

Oil Wells: The addition of gilsonite reduces the slurry weight of oil well cement without reducing its compressive strength. Because of gilsonite's plastic flow characteristics, it is also an effective bridging and plugging agent for sealing fractures in weak formations during cementing. In oil well drilling operations the addition of gilsonite to water-based drilling fluids helps minimize hole washout by stabilizing troublesome shale formations. It also seals off highly permeable sands while at the same time reducing torque and drag on the drill stem.

Road Paving: Gilsonite is used as a reinforcing agent in asphalt pavement hot mixes. Gilsonite-modified paving mixes have higher stability, reduced deformation, reduced temperature susceptibility, and increased resistance to water stripping. Gilsonite is also used in the manufacture of solvent-based and emulsion pavement sealers that have superior appearance and weathering properties.

Foundries: Gilsonite is blended with sea coal and used as a special binder for foundry sand to reduce defects in finished metal castings. It also improves mold release and the overall finish of castings. Gilsonite's value in this application derives from its high temperature volatility and relatively high lustrous carbon content.

Building Products: Gilsonite is mixed with wood fibers or chips in the manufacturing of high quality, composite wood sheet products, such as exterior siding and insulation board for homes and commercial buildings. Process conditions generate sufficient heat and pressure to allow it to flow and wet the wood particles.

ECONOMIC FACTORS

Prices

Depending on the industrial application, the physical and chemical properties required of gilsonite by consumers, and therefore the prices, vary. Prices depend on the following factors: softening point, percent insoluble (ash) content, packaging, particle sizing, moisture content, degree of refinement, quantity ordered, competitive pressure from other gilsonite producers, and competitive pressure from other substitute materials. The relative importance of these factors to an end user will partly determine the price he is willing to pay. American Gilsonite Company's 1992 prices, FOB its plant site at Bonanza, UT, ranged from 220 to $995/t depending on grade.

Transportation

Gilsonite is sold in the United States and is exported to approximately 40 countries throughout the world. It is usually packaged in 22.7 kg bags in either lump or pulverized form. A recent trend has been to package the product in one ton bulk bags that can be reused. Domestic shipments can be made by truck in bagged and palletized form or in bulk for certain particle sizes if requested. Transportation by rail, previously used for large-volume shipments, is not currently being used.

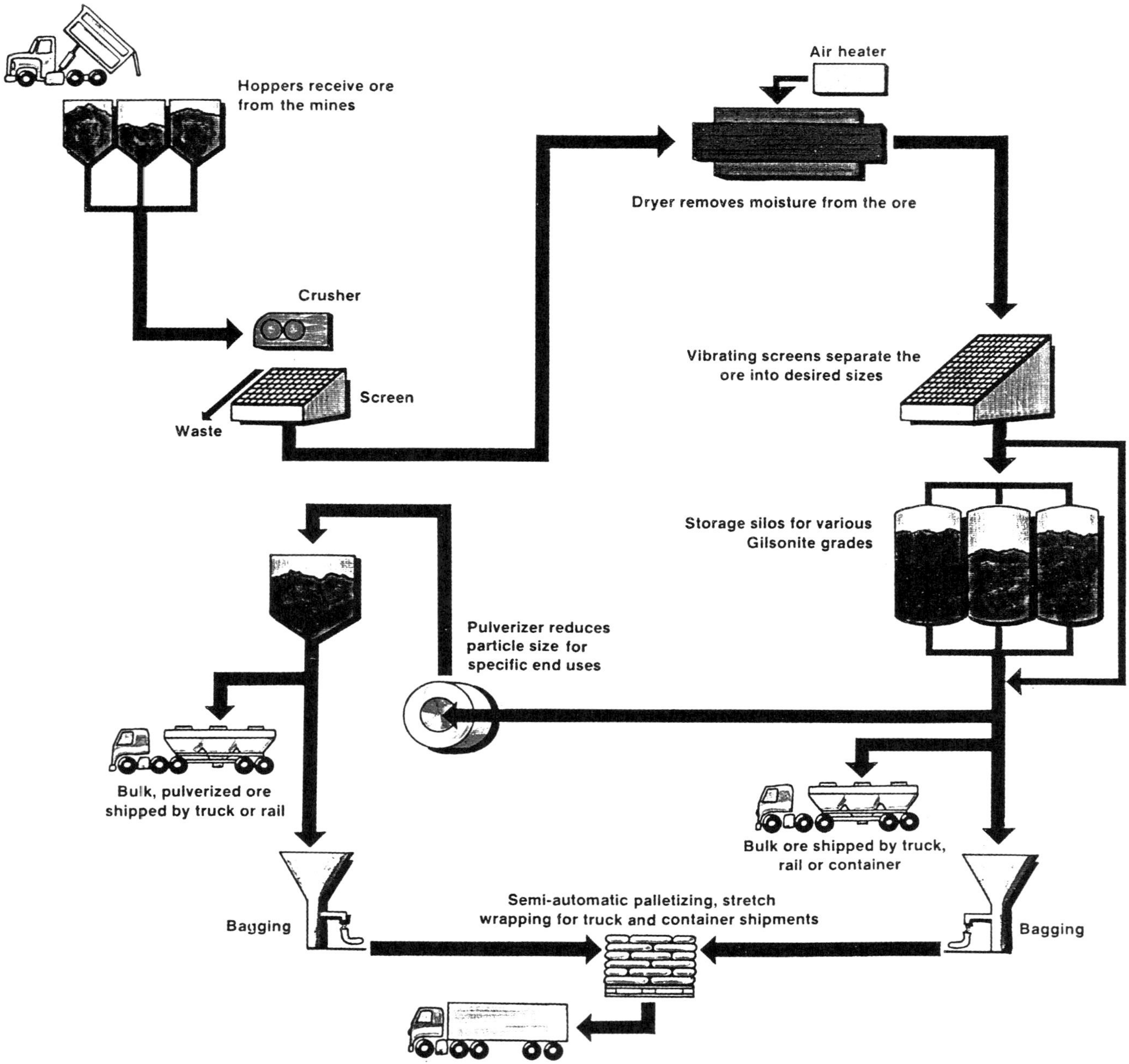

Fig. 7. Schematic flowsheet of gilsonite processing (*courtesy:* American Gilsonite Company, Bonanza, UT).

For the export market loose bags can be shipped in 6.1 m containers. Bagged and palletized product or bulk bags also can be shipped in 12.2 m containers for transportation by ocean-going vessels. Break-bulk shipments (i.e., less than a container load) are rarely made because of their high unit cost for transportation. The export foundry industry has received bulk shipments in open-top containers and LASH barges holding more than 340 tons.

Transportation rates in the United States are assessed on a per-mile basis up to a maximum allowable weight. Gilsonite is sold on an FOB plant-site basis. Export transportation costs include truck and/or rail haulage to and from both ports, terminal handling charges at both ports, ocean freight (including currency adjustments and fuel surcharges), insurance, and an import tax in some countries. All of these costs influence the price of gilsonite delivered to an export customer compared to the price of locally competitive products.

Technical Considerations and Alternate Materials

Gilsonite is a unique natural hydrocarbon resource. It would probably be impossible and certainly impractical to exactly match its chemical composition and properties with substitutes in all applications. Competitive products and formulations for certain uses may be acceptable, but they are not equal to gilsonite in performance.

Fig. 8. Gilsonite, a glossy, black, solid hydrocarbon resin.

Gilsonite has been tested extensively as a possible environmental, health, and toxicological hazard. Detailed information is contained in American Gilsonite Company's Material Safety Data Sheet 002542 (dated Sept. 6, 1991). Gilsonite in its natural state has been found to have a low level of toxicity and to be noncarcinogenic. Accordingly, gilsonite products do not have to carry the warnings that most petroleum and coal products must carry. In some cases gilsonite is used instead of products of comparable cost that do have health/environmental warnings associated with them.

Specific Uses: *Printing Inks.* Gilsonite's main use is as an additive for dispersing the carbon black particles that give dark printing inks a black color. It is normally used at dosages of 2 to 5% in ink formulations. It is sometimes used at 10 to 20% loadings as a low-cost substitute for other resins in dark ink formulations, where it substitutes for phenolic and other resins. In Europe high-nitrogen asphalt has been used as a substitute for gilsonite as a dispersing agent, but this has not been practiced in the United States because of health implications.

Oil Field Products. WELL CEMENTING—Gilsonite is used in oil well cementing for two purposes: 1) as a lightweight additive to reduce cement density, and 2) to prevent loss of circulation by sealing the formation. Products that compete with gilsonite as a lightweight aggregate are coal, diatomaceous earth, perlite, and synthetic microspheres. Products that compete with it in controlling circulation loss are perlite, walnut shells, coal, cellophane, portland cement, and synthetic microspheres.

DRILLING MUD—Gilsonite's major function as a drilling mud additive is for sealing microfractures in shale and plugging sand zones. Its secondary functions are high-temperature and high-pressure fluid loss control, torque reduction, and drill bit life improvement. Its major competition includes sulfonated asphalts, hard-blown asphalts, and proprietary products made with gilsonite. Many asphaltic products have environmental and health warnings associated with their use. They are slowly losing market share in the United States, because gilsonite does not have these risks associated with its use.

Foundries. Gilsonite is used as a carbonaceous additive in foundry sands to produce smoother surfaces on gray iron castings. It volatilizes at a lower temperature than does coal, which promotes peel from the casting and produces smoother, cleaner casting surfaces. It serves as a supplement to or substitute for sea coal, the most popular carbonaceous additive for gray iron castings. Sea coal is normally used as the primary additive, and other materials are used as supplements. Gilsonite is added at 1.5% of the mold mix to ensure superior casting quality. Other materials that can substitute for gilsonite are coal tar pitch, petroleum asphalt, blown asphalt, aruba pitch, rose asphalt, asphaltene chips, asphaltic emulsions, and hydrocarbon resins. All of these alternatives to gilsonite have health and/or environmental warnings associated with their use.

One of the benefits obtained by using carbonaceous additives at foundries is derived from their lustrous carbon content. The higher this content the better; but there is a cost per unit of lustrous carbon associated with each material. The use of such carbonaceous additives must weigh environmental and health considerations against cost and performance.

Asphalt Modification. Gilsonite's basic function in this application is to improve the stability of pavements and their resistance to deformation. It increases load-carrying ability and is used in high stress areas where high performance pavements are required. About 10% of the asphalt paved roads in the United States are considered high stress areas. Modifiers such as gilsonite are being used in only about 1% of these roads, however. Competing modifiers for road paving include hard asphalt, blown asphalt, rubbers, polymers, carbon black, and fibers. There are about 35 different substitutes, but gilsonite's cost/benefits ratio is among the lowest of them all.

Other Uses. Gilsonite can be used in a variety of other applications that are able to accept a wide range in product specifications. Examples include explosives, briquette or pellet binding, and asphalt pavement sealing.

GOVERNMENTAL CONSIDERATIONS

Gilsonite production in Utah is subject to all normal county, state, and federal taxes, but is exempt from Utah's mineral severance tax. The gilsonite veins are located in rural, remote areas of northeastern Utah, and mining takes place mainly on federal fee lands. Zoning and land use matters are both addressed in mining permits and in local county government ordinances. Future reserves are located on disputed federal lands within the original Ute indian reservation boundaries and on state lands.

Environmental regulations are monitored and enforced principally by the state of Utah. Permits are required for mining operations, water discharge, landfills, and solid waste facilities. Safety is monitored and enforced by the federal Mine Safety and Health Administration (MSHA). Safe practices are of utmost importance in gilsonite mining and require the attention of all gilsonite workers.

BIBLIOGRAPHY

Kirk, R.E. and Othmer, D.F., eds., 1980, *Encyclopedia of Chemical Technology*, 3rd ed., Vol. II, John Wiley & Sons, New York, pp. 802–806.

GLASS RAW MATERIALS

H. LYN BOURNE

Daily everyone depends on the great variety of glass products, so much so that glass is often taken for granted. In fact most people do not realize how versatile glass has become. Consider the various uses and then try to imagine a day in which we are not influenced by glass. Common uses include container ware, table ware, window glass, lead crystal, automobile glass, and fiber glass. Several less common, but important, uses include laboratory ware, pharmaceutical, TV bulbs, light bulbs, glass ceramics, optical glass, fiber optics, and laser glass. Corning, Inc., a leader in specialty products, uses nearly 1 000 different compositions to manufacture about 60 000 different products (Edwards and Copley, 1977).

Glass is such a complex product that definitions vary and exceptions can be found for most definitions. Glass is an inorganic amorphous (non-crystalline) solid. Most glasses are produced by melting of a mixture of oxide raw materials, and then cooled to room temperature. Soda-lime-silica compositions account for about 90% of all glasses melted (Anon, 1973).

The properties of the glass product come mainly from its chemical composition. All of the different glasses require melting a combination of raw materials and forming the molten material into the desired shape.

Both the melting and the forming processes use sophisticated technology and these technologies require experts to manage these production systems. The manufacturing process is continuous and takes place in tonnage quantities, so adjustments in the batch to achieve the desired finished product requires a great deal of expertise. Raw materials are fed to the batch mixing area in very large quantities (tons in most cases). As a result, impurities in the range of 0.1% result in addition of that impurity within the molten glass in kilogram amounts.

More than twenty different industrial minerals are consumed in the manufacture of various kinds of glass (O'Driscoll, 1990). This chapter describes the major and minor ingredients of the various glass batches. It discusses the roles of the various oxides in the glass batch and most importantly considers the mineral raw materials which supply the glass industry. Each of the raw materials is described in detail in other chapters so the geology and mineralogy sections are kept brief here.

Container glass, by far, accounts for the most production; followed by flat glass, fiber glass, and specialty glass of which table ware accounts for the greatest tonnage. Table 1 shows the general production data for 1987 through 1990. Statistics for many of the uses do not appear because production volumes are small compared to the major uses.

The glass industry is organized in four categories: containers, flat glass, fiber glass and specialty glass. The US Department of Commerce, Bureau of Census, publishes production data about the glass industry in three different categories: 1) glass containers, 2) consumer, scientific, technical and industrial glassware, and 3) flat glass. The Bureau has very complete statistics about the glass industry in these three categories but they report production data in different units according to industry standards. Therefore, Table 1 gives the production data in dissimilar units.

The production of most glass articles follows similar steps. The raw materials are mixed and the resulting batch is fed into the furnace. In soda-lime-silica glasses melting begins between 600 and 900°C. At these temperatures CO_2 and other gasses are released which create bubbles in the molten glass. To remove the bubbles and insure complete melting the temperature is raised to between 1 500 and 1 600°C. This is the melting-refining stage during which the refining agents in the glass batch serve to aid in the release of gas bubbles, homogenize the melt, and prevent the formation of scum on the surface of the molten liquid. At the conclusion of the melting-refining stage the glass is too fluid for working and the melt is cooled to about 1 100°C to attain the proper viscosity for working and forming to begin. After the glass article has been made, it must undergo annealing (slowly and uniformly reheated and cooled) to remove thermal stresses that were created during the forming process.

Production Methods

As mentioned, the industry groups glass products into one of four categories: glass containers, fiber glass, flat glass, and specialty glass. Some of the glass production methods are briefly described.

Glass Containers: Currently, bottles and jars are made on a continuous machine by pressing and/or blowing hot glass in a mold. Individual gobs of glass are formed in the molds on the container machines. Lamp bulbs come from a similar process, but the glass is fed from the melting furnace as a ribbon, not as a gob (Doremus, 1989).

Fiber Glass: Insulating fiber glass (glass wool), reinforcing glass fibers and optical fibers represent the three types of fiber glass and the optical fibers rightfully belong with specialty glasses. In the production of glass wool, controlled streams of glass flow into spinners (rotating dishes) which have thousands of tiny holes. The molten glass is 1) forced through the tiny holes and "fiberized," 2) sprayed with a binder, 3) collected on a forming conveyor, and 4) annealed and trimmed to the final dimension. Reinforcing fibers are produced by drawing up to 500 fibers from small orifices in a bushing, spraying the fibers with a sizing and then winding all of these onto a spool. Bushings can be fabricated with several thousand holes (Trott, personal communication, 1992). These fibers are subsequently chopped and added to various materials to add strength to the composite material. Fiber glass for reinforced plastics accounts for about 90% of fiber glass production (Russell, 1991).

Flat Glass: Until the 1950s most high quality plate glass was produced by flowing the glass from the furnace through rollers. The glass from this process had rough surfaces which required grinding and polishing. In the 1950s the Pilkington Glass Company of England, developed the float process in which the glass from the

Table 1. Glass Production Data

Glass sector	1987	1988	1989	1990
Containers (m.g.u.)*	281.6	280.4	285.6	285.0
Flat glass (Mm^2)	434.2	437	455.1	440
Glass fiber (kt)	1 600			1 800
Consumer glassware** (m.g.u.)			5.2	6.8

Source: Anon., Dept Commerce, Bureau of Census (1991)
* Millions of gross units, ** comprises tumblers, stemware and tableware
Statistics for other glasses give only value, not quantity. This includes the following glass products: lighting, electronic, tubing, scientific, ophthalmic, TV tubes.

melting furnace is floated onto a surface of molten metal (usually tin). The temperature and viscosity of the glass are carefully controlled which allows it to be pulled along the surface of the molten metal until it has cooled enough to be handled by rollers. This process results in sheet glass with smooth, undistorted surfaces that do not require grinding or polishing (Doremus, 1989).

Table Ware: Plates, tumblers, vases and similar glass pieces result from pressing the hot glass in a mold. This kind of glass became popular in the United States because it was cheaper than cut-crystal (Doremus, 1989).

RAW MATERIALS

The geology and mineralogy of individual raw material components appear in separate chapters in this book. This section of this chapter describes the role of the major and minor raw materials used in the production of various glasses.

Role of Oxides

The oxides found in glass compositions play one of four roles (Holscher, 1965):

1. Glass formers
2. Glass modifiers
3. Glass stabilizers
4. Glass colorants

Glass Formers: These constitute the major part of the glass composition and give the ability to become stable in the glassy state. Silica is the major oxide in this category; others include boric oxide, phosphorous oxide and germanium oxide.

Glass Modifiers: This group of compounds helps lower the fusion temperature during the melting process. Soda is the major oxide in this category and other alkali oxides also serve as glass modifiers.

Glass Stabilizers: Sodium silicate is water soluble as are most combinations of glass formers and glass modifiers. Other oxides must be added to provide an insoluble vitreous product. Lime is the most common compound used to provide this property (insolubility) to the glass composition. Other alkaline earths also qualify as stabilizers. Alumina is intermediate between glass formers and stabilizers. It increases viscosity of the molten material, improves durability in the finished product, and retards devitrification (the act of crystallizing out of the glassy state).

Glass Colorants: This category includes those compounds used both to add color and increase transparency (decolorize). These are minor components of the glass composition, but can have a major influence on the end product. Oxides of iron, manganese, chrome, copper and transition elements impart color. The elements have two or more oxidation states and the color produced in the glass depends on the oxidation-reduction conditions. Most compounds that add color in glass do so in solution, but silver and gold can exist in glass as colloidal particles whereby absorption and reflection produce unusual colors.

There are both major and minor oxide compounds in glass and the descriptions that follow describe the oxide and the raw material(s) that provide it.

Major

Silica (SiO_2): This is the most important oxide and constitutes the most tonnage of all the glass making raw materials. Glass makers depend on a relatively small number of sources. Silica is abundant but glass grade material is relatively scarce. Optical and specialty glasses require even more pure materials which are scarcer. Quartz sand supplies most of the silica to the glass batch. Glass grade quartz sand deposits occur in several areas within the United States. In Europe glass sand comes from Austria, Belgium, France, Italy, Spain, Sweden, and the United Kingdom.

Soda Ash (Na_2CO_3): This represents the second largest commodity used for glass manufacture. Soda ash represents about 50% of the raw material costs for glass containers and glass technologists try to substitute or minimize the amount of this raw material (clear glass has 13 to 15%, amber has 13.8 to 15.6%, and green glass 13.6 to 15.7% Na_2O from soda ash). The Solvay process yields synthetic soda ash after salt, limestone and ammonia are reacted. The process has been abandoned in the United States for energy conservation and environmental reasons; the US glass industry now exclusively uses soda ash from natural deposits, mainly trona $[Na_2(CO_3){\cdot}Na(HCO_3){\cdot}2H_2O]$ from Wyoming. Turkey also has large deposits of trona. In the United States soda also comes from brine operations at Searles Lake, CA. Other natural soda deposits are brines at Lake Magadi, Kenya, at Sua Pan in the Makgadikgadi Basin, Botswana, and Mexico. In many parts of the world the Solvay process still provides soda for the glass industry.

Limestone and Dolomite: These commodities combined form the third largest raw material used in glass. They supply the lime (CaO) in soda-lime-silica compositions and also serve as a flux. Magnesia (MgO), in small quantities, helps the melting process and enhances the chemical resistance of the finished glass. Limestone and dolomite deposits of adequate purity for container and flat glass occur widespread, but higher grades for specialty glasses are less common. These oxides may be added as natural stone or as the kiln fired products lime or dolime. There are advantages and disadvantages to either the natural or burned state of the raw material (O'Driscoll, 1990).

Minor

Sodium Sulfate ($NaSO_4$): Known as salt cake, it too supplies soda to the soda-lime-silica compositions. However, unlike soda ash, the primary role of salt cake is as a refining agent. It helps remove gas bubbles, homogenizes the melt, and prevents scum from forming on the surface of the molten glass during the refining stage. The addition of salt cake to the glass batch often requires that some reducing agent such as carbon (anthracite or coke) be added to serve as an aid to decomposition of the sulfate (Anon, 1973). To counteract reducing agents in the batch salt cake or gypsum is added and to counteract high levels of tramp SO_3, carbon is added (Trott, 1992, written communication). Salt cake comes both from bedded deposits, and natural brines plus synthetic processes. Sodium sulfate occurs as bedded mirabilite or in brines beneath saline lakes or playas (Harben and Bates, 1990). The United States, Canada, and

Mexico produce natural sodium sulfate. Other countries which produce natural sodium sulfate include Argentina, Belgium, Chile, Spain, Turkey, and parts of the former Soviet Union.

Potash (K_2O): Potash replaces or supplements soda, but it makes the glass more viscous, plus it is more corrosive to furnace refractories. Decorative and illuminating glassware use potash in their composition (Coope et al., 1977). Potash comes from evaporite deposits where sylvite (KCl) is the major mineral. Processing of KCl yields KCO_3 or KNO_3 and some potash in the batch comes from feldspars or nepheline syenite. Major evaporite deposits which contain KCl occur in eastern and western Europe plus North America.

Alumina (Al_2O_3): Feldspar, kaolin, and nepheline syenite supply alumina to glasses that need it. Alumina is a major component in alumino-silicate glasses such as fiber glass. It is a minor component in boro-silicate glasses (e.g., laboratory ware), TV tubes and soda-lime glasses. When used, alumina adds strength, chemical durability and toughness to the finished glass. Some specialty glasses need purer Al_2O_3 and use calcined alumina or alumina hydrate. The best glass grade feldspar comes from the United States, Norway, and Finland. Canada and Norway produce the rarer nepheline syenite as an alumina source for the glass industry; and some alumina comes from blast furnace slag. Kaolin provides alumina and other oxides for fiber glass compositions. The southeast part of the United States supplies kaolin for the glass industry; in Europe Belgium, France, Germany, and Spain have suitable deposits of kaolin. Pyrophyllite is a common source of alumina in southeast Asia.

Boron Compounds: In borosilicate glasses like laboratory ware boron is a major component. Boron in glass reduces thermal expansion, improves chemical durability, and increases resistance to thermal shock. Containers, tableware, and fiber glass also use boron in their batch. Borax ($Na_2O{\cdot}B_4O_7{\cdot}10H_2O$) causes frothing and anhydrous borax or colemanite ($Ca_2B_6O_{11}{\cdot}5H_2O$) is preferred for soda-lime glasses. Borosilicate glass producers prefer boric acid because it provides the boron free from other compounds. Borates are produced in Argentina, Chile, China, Peru, Turkey, the United States, and the former Soviet Union.

Lead: High quality table ware (lead crystal known for its ring) uses lead oxide as part of the lead-alkali-silica composition as does some optical and illuminating glasses. Red lead (Pb_3O_4) is preferred over litharge (PbO), because litharge is susceptible to reduction to metallic lead in the glass melt. Lead adds brilliance and density to the finished glass product.

Barium and Strontium: Picture tubes for television sets contain barium and strontium in the glass batch. The presence of these elements in the finished product shields the viewer from radiation. Strontium is more effective than barium in color television sets. The compounds are usually added as a carbonate which is manufactured from the sulfates; barite (barytes) ($BaSO_4$) or celestite ($SrSO_4$). In addition barium oxide (BaO) is sometimes substituted for lead in crystal ware. Significant concentrations of celestite occur in some sedimentary deposits and the major producing countries are Turkey, Spain, Mexico, Iran, and the United Kingdom. Barite deposits are bedded, residual, or vein and cavity filling. Most of the barite in the United States comes from Nevada; some other producing countries are Mexico, Turkey, Spain, the United Kingdom, Argentina, and China.

Lithia (Li_2O): Specialty glasses such as heat resistant borosilicate, alumino-silicate glasses, glass ceramics, opal glass and TV tube compositions use lithia. Lithium is usually added as lithium minerals and to a lesser degree as the carbonate (Edwards, 1992, written communication) and acts as a flux in these compositions which are difficult to melt. Lithia also provides other benefits; it reduces browning from radiation in cathode ray tubes, provides development of desired crystal species in glass ceramics, and adds strength and durability to cosmetic containers (Shore and Haigh, 1990). Commercial production comes from pegmatite deposits and lithium rich brines. In pegmatites naturally occurring lithium-aluminum-silicate minerals include spodumene ($LiAlSi_2O_6$), lepidolite and petalite ($LiAlSi_4O_{10}$). Major pegmatite deposits occur in Zimbabwe, Brazil, the United States, Canada, Australia, China, Zaire, and South-West Africa. Lithium produced from brines comes from Salar de Atacama in Chile. Bolivia, Israel, and the United States have brine deposits which contain lithium, but at concentrations far lower than those in Chile (Harben and Bates, 1990).

Fluorspar and Cryolite: Soda-lime glasses use small amounts of these as a fluxing agent, plus F_2 reduces surface tension and controls foam. Opal glasses use more because it is the major opacifying component. About thirty countries produce fluorspar (CaF_2) and the major producers are Mexico, the Republic of South Africa, Mongolia, China, and the former Soviet Union. However, Greenland is the only source of natural cryolite (Na_3AlF_6) and production ceased in 1962 (Harben and Bates, 1984). So, synthetic cryolite is often used; produced as an intermediate product in the manufacture of aluminum. The US glass industry currently uses less fluorspar than in the past due to environmental concerns.

Miscellaneous: Several other oxides and minerals are added for various minor roles. Salt cake is the main refining agent, but other refining agents include calcium sulfate (gypsum or anhydrite), sodium nitrate, antimony oxide, arsenic oxide, and zinc oxide. Fluorspar and cryolite can add opacity and other opacifiers are zinc oxide and titanium dioxide. Oxides of arsenic, selenium and cobalt are often used singly or in combination as decolorizing agents. Color can be obtained by adding oxides of chrome, cobalt, copper, iron, manganese, and rare earths. The specific color depends on which are present, the oxidation state, and the amount of the oxide added to the batch. Rare earths are used in small quantities in special high tech glasses. For example neodymium oxide is added to compositions used to make television filters; cerium is added for radiation control in cathode ray tubes; and neodymium and samarium are used in laser glass compositions.

INDUSTRY STRUCTURE

Four categories define the glass industry: container glass, fiber glass, flat glass, and specialty glass. Some companies compete in a single category and some compete in two or more categories. Likewise some companies are international in scope and others do not produce and sell outside their country.

Occasionally international politics and crisis affect the glass industry. The conflict in Lebanon has had a major impact on the glass industry in that country (Khawlie and Khalaf, 1989). Glass makers in Lebanon produced containers, flat glass, and fiber glass (both insulation and reinforcing) for internal and export markets. They exported to Arab countries, Europe, the United States, Cyprus, and Senegal. Lebanon has good high purity silica deposits and an abundance of cullet, but the turmoil has diminished its manufacturing capabilities.

Container Glass

In the United States, Owens Illinois acquired Brockway in 1988 (Harben, 1991). In 1990 the company, Owens Brockway Glass Containers, operated 24 glass plants. Anchor Glass Container Corporation, headquartered in Tampa, FL, operated 22 plants; Ball-Incon Glass Packaging Corporation had 14 plants and acquired four more from Kerr (Anon, 1992). The other major glass container manufacturer in the United States is Foster-Forbes Glass Division with 8 plants.

Fiber Glass

There are three categories of fiber glass; insulation (glass wool), reinforcement, and optical. Most fiber glass manufacturers produce both glass wool and reinforcing fiber glass. Optical fibers are a specialty and require more sophisticated production methods and very special glass compositions and high purity raw materials.

Four companies make up the major producers of fiber glass in the United States: Owens-Corning Fiberglas® Corporation, PPG Industries, CertainTeed Corporation, and Manville Sales Corporation. Owens-Corning has the most diverse production facilities, but all four produce fiber glass both for reinforcement and insulation uses. Owens Corning is the largest of the fiber glass producers with 80 plants worldwide (Russell, 1991). It is headquartered in Toledo, OH, and has plants in Belgium, France, Scandinavia, Canada, England, and Brazil. In addition they have licensees in Japan, Saudi Arabia, and Taiwan, plus an affiliate in Mexico.

The French company, Saint Gobain, owns CertainTeed in the United States, which produces both glass wool and reinforcing fiber glass. PPG Industries likewise is international, as is Manville Sales. Other European fiber glass producers include Bayer AG of Germany and Ahlstrom of Finland.

Flat Glass

Within this category, 57% of the product goes to the construction industry, 25% to the automotive industry, and the remainder goes for such products as mirrors, solar panels, store displays, furniture, and advertising signs. The six major flat glass producers in the United States all use the float glass process. PPG Industries has ten float tanks; AFG Industries Inc has seven and three other companies Guardian Industries, Ford Motor Company (Glass Division) and Libbey-Owens-Ford Company (LOF), each have six.

Pilkington PLC of the United Kingdom bought LOF in 1986 and Nippon Sheet Glass bought 20% of LOF in 1989. Asahi Glass Company of Japan has two plants in the United States which produce automotive glass; plus Asahi has an interest in AFG (Harben, 1990). In Europe Pilkington PLC produces flat glass in the UK; in France BSN Gervais Danone and Saint-Gobain have flat glass operations; and in Germany Flachglass AG (a member of the Pilkington Group) and Vereinigte Glasswerke Gmbh dominate the production of flat glass (Toon, 1986).

Specialty Glass

Specialty glass only makes up 1 to 2% of the total glass produced but accounts for 15 to 16% of the value (Mennemann, 1987). There are numerous specialty glass companies in the United States, but Corning, Inc. dominates this segment of the industry and Owens Illinois probably ranks second. Specialty glass markets are much smaller than those of the other categories, but this category has more diversity than the other three combined. Included among these specialty products are tumblers, table ware, lead crystal, scientific and laboratory ware, glass tubing, TV glass, cookware, pharmaceutical, optical glass, ophthalmic glass, lasers, optical waveguides, glass ceramics, lighting glass, fiber optics, glass spheres, and numerous others.

TECHNOLOGY

Technological advances have come in both the mineral processing industry and the glass manufacturing industry. As easy to mine and high purity deposits were depleted, the raw material supplier had to improvise and beneficiate to meet the glass makers' requirements. Likewise, the glass maker found ways to improve old products and develop new ones. Higher speed manufacturing processes for glass products usually has meant changes in the raw material specifications: smaller grain size distributions to give faster melting, plus considerations about grain shape, and tolerance limits for impurities. Specifications can never be static for glass making raw materials. The glass maker must find the best available raw materials at the lowest possible delivered cost to make a competitive product.

Specifications

The production of glass articles results from a continuous process and any error in raw material purity or in the melting and forming appears first in the finished product. The demand for adherence to specifications reflects this fact. Faster melting rates in modern furnaces demand increased quality in raw materials. Subtle differences in composition can effect melting, viscosity, and the forming that takes place after melting is complete. Impurities can also diminish the properties of the finished product (index of refraction, tensile strength, appearance, etc.).

Some glass making raw materials, like silica, are produced and consumed within a limited geographic area. Therefore, glass manufacturers have accepted regional differences in quality and price. Within the regional variations, certain minimum standards must be met with respect both to chemical and physical specifications. Most glass manufacturers recognize that raw material deposits have variations in chemistry from deposit to deposit. They often issue working specifications for each raw material location. These specifications give the tolerance limits for each raw material deposit.

Chemical: Glass makers have two considerations relative to chemical specifications—purity (or lack of contaminants) and consistency (or lack of variation). Their specifications reflect these requirements either as a minimum percentage for a single oxide ingredient, or a maximum percentage when describing an impurity, or a narrow range of percent variation usually in reference to raw materials that contribute more than one component to the batch. Table 2 shows chemical specifications for some of the more common components. However, some specialty glasses require raw materials of such purity that **no** minerals can meet the required specifications; and glass production of these specialty items is done with highly purified chemicals (Edwards and Copley, 1981), e.g., optical waveguides.

Physical: The physical specifications deal exclusively with particle size. The grain size of batch materials strongly affects the amount of energy required for melting. Glass makers prefer a near uniform size to the batch ingredients to insure efficient melting; they typically use material whose grains range from 0.59 to 0.149 mm. However, in textile and reinforcing fiber glass more than 90% of the raw material grains are smaller than 0.045 mm (45μm). There is consideration to tighten these limits from 0.3 to 0.1 mm, especially for sand since it is the most difficult component to melt. Grain shape of the sand also affects melting. Angular grains offer more surface area and faster melting than rounded grains. Likewise, frosted or pitted grains offer an increase in surface area that can enhance melting.

Uniform size among all of the ingredients also speeds melting, minimizes segregation during the batch handling, and homogenizes the melt. If very much of the batch is coarser than the specified range, incomplete melting often occurs, which results in a poor quality product. If very much of the batch is finer, dusting occurs which creates housekeeping problems outside the furnace. Inside the furnace fines can harm the furnace refractories and heat exchangers.

Table 2. Chemical Specifications

	Silica Sand				Limestone		Dolomite	
Compound	Flat glass	Container glass	Soda Ash	Feldspathic content	Flat glass	Container glass	Flat glass	Container glass
SiO_2	≥99.5%	≥99.5%		±0.5%		±0.5%		±0.5%
Fe_2O_3	≤0.04%	≤0.03%	≤20 ppm	≤0.1%	≤0.075	≤0.25%	≤0.1%	≤0.1%
Al_2O_3	≤0.3%	±0.01%		±0.5%	≤0.35%	≤0.4%	±0.5%	±0.5%
TiO_2	≤0.1%	≤0.03%						
Cr_2O_3	≤2 ppm	≤10 ppm		≤10 ppm			≤10 ppm	≤10 ppm
Co_3O_4	≤2 ppm							
MnO_2	≤20 ppm							
CaO-MgO		±0.1%						
ZrO_2		≤0.01%						
Na_2O-K_2O		±0.1%		±0.5%				
Moisture		≤0.1%						
Na_2CO_3			≥99.2%					
Na_2O			≥58%					
Na_2SO_4			≤0.2%					
NaCl			≤0.2%					
CaO				±0.5%	≥54.8%	±0.3%	≥29.5%	±0.3%
MgO					≤0.8%	±0.3%	≥21.4%	±0.3%

Processing

Refractory heavy minerals (RHM) represent the raw material impurity most avoided in most glass compositions. RHM include kyanite-sillimanite-andalusite, zircon, corundum, chrome and other spinels, rutile-leucoxene, and staurolite. These minerals do not melt in most glass compositions at normal glass making temperatures and wind up, unmelted, in the finished product as a solid inclusion or stone defect. Specifications for RHM usually relate to particle size and quantity. Particles larger than 0.25 mm are the greatest concern and the quantity may be specified either as a weight percent or on a particle count basis (Mills, 1983). Typical working specifications for a glass sand limit the RHM content to 0.0003% (3 ppm) for particles larger than 0.25 mm. Fortunately, when present in a deposit, these minerals can usually be removed by magnetic separation and/or through flotation.

The other major impurities often depend on the glass product. Iron in the glass raw material imparts color to the finished glass. Therefore, clear glass has a lower tolerance limit for iron in its raw materials than amber or green glass. Conversely, optical grade glass has even more stringent specifications regarding iron than clear container ware. If iron is present in a raw material commodity as a discrete mineral, removal is usually accommodated through magnetic separation. If the iron is present as a surface coating on the grains (often the case with silica sand or feldspathic sand), a surface treatment may be necessary. If the iron is structurally locked within the mineral, it may respond to magnetic separation—or it may render the deposit unsuitable.

Size segregation represents the other major processing function. Each glass maker may have slightly different requirements for their gradation, but generally they want raw materials to range from 0.59 to 0.149 mm. After processing to remove impurities and classifying the product to its proper size range, the final step is drying. Glass producers generally want the raw materials to contain less than 0.1% moisture.

USES

The uses of finished glass articles depend on the composition of the glass; Table 3 gives the nominal compositions for several glass products. Sometimes specific properties are sought and glass technologists experiment until they achieve a glass which provides the desired property. Table 4 identifies some of the more common glass types by their composition and their range of uses. This table shows that soda lime compositions dominate the range of use. The table also reveals some of the less common glasses and their uses. In addition to those listed in the table there are glasses used for some following purposes: solder glasses for connection of glass to glass and glass to metal seals, laser glasses, infrared glasses, fiber optics, ophthalmic glasses for eyeglass wear and optical glass for optical instruments.

Glass spheres represent a specialty use and are produced in both hollow and solid form from compositions of soda lime and borosilicate (Harben, 1988). None of the major glass producers make spheres as they represent a small volume specialty, but they are produced in the United States, the United Kingdom, and Sweden and marketed for use as fillers.

Optical glass has two different compositions: crown glass and flint glass (Scholes, 1952). Crown glass has compositions which contain lime and flint contains considerable lead. Ordinary spectacle glass is crown glass similar in composition to plate glass. Some of the glass used in bifocals contains 40 to 50% lead.

ECONOMIC FACTORS

The raw material costs reflect location of the deposit, the purity of the finished product, and the amount of processing that was required to bring the raw material to that level of purity. Costs often reflect, also, the supply and demand conditions for the various raw materials. For instance, how scarce is the material, who and where are competitive operations. Table 5 gives the f.o.b. price range for some common glass making raw materials. The prices range from about \$15/t for limestone to \$600/t for barite, but these costs do not reveal the true story relative to the effect on the finished glass product. In reality most of the expensive ingredients are only used in small amounts and have a small effect on the glass production costs. Soda ash is the ingredient that has the major effect on the cost of the glass batch for three reasons—it has a high unit cost, it is *not* a regional commodity so it has high transportation costs, and a large amount is used in the glass batch. For fiber glass and borosilicate glasses, the borate component of the batch drives up the raw material costs.

Table 3. Nominal Glass Compositions

Oxide	Container glass	Flat glass	Pressed ware	Fiber glass "E"	Laboratory glass	Lead crystal	Glass ceramics	Fiber optics
SiO_2	73.0%	73.0%	72.2%	54.5%	80.2%	55.7%	67.4%	61.0%
CaO	10.0	8.2	6.2	21.5	0.1	0.2	2.7	
Na_2O	14.0	14.0	16.9		4.2			14.0
K_2O				0.8		12.1		
Al_2O_3	2.0		0.3	14.0	2.4	0.3	20.9	3.0
MgO		4.0	4.4	1.0				
B_2O_3				7.0	12.9			22.0
F_2				0.4				
Fe_2O_3				0.3				
TiO_2				0.5			1.8	
ZrO_2							2.0	
Li_2O							3.9	
Fe		0.1 to 0.6						
SO_3	1.0							
PbO						31.4		
ZnO							1.3	
	Kephart et al., 1986	Poley, 1992	Holscher, 1968	Watkins, 1986	Holscher, 1968	Holscher, 1968	Grubb, 1991	Grubb 1991

Table 4. Uses for Common Glass Compositions

Glass type	Uses
Soda lime	Blown and pressed ware including containers, lamp bulbs, dishes, tumblers, flat glass, and clear window glass.
"E" lime alumino-silicate	Reinforcing fiber glass
Borosilicate	Laboratory ware, headlamps, and cookware
Alkali lead	Lead crystal, lamp tubing, sealing glass and some TV glass
Lithia-alumino-silicate	Glass ceramics, high temperature glasses, astronomical mirrors.
Vitreous silica	Optical components, semi-conductors, crucibles, specialty lamps, and fiber optics.
Alkaline earth	High temperature thermometers, electrical resistance.
Alkali-lead-silicate	TV tubes

Table 5. Price Range of Common Glass Making Raw Materials (FOB)

Silica sand	8 to $25/t
Soda ash	93 to $123/t
Limestone/dolomite	15 to $25/t
Sodium sulfate	115 to $135/t
Feldspar	35 to $65/t
Nepheline syenite	22 to $28/t
Borax	270 to $300/t
Lithium (spodumene concentrate)	200 to $250/t
Fluorspar	95 to $135/t
Celestite	$0.82/kg
Barite	70 to $95/t

Source: Harben, 1992.

Transportation

Like most industrial minerals, transportation costs usually exceed the f.o.b. price of the raw material. Most glass plants demand frequent deliveries because melting and production takes place around the clock. They must add batch every day and they lack storage for large quantities of raw materials. The recent trend is toward more raw materials delivered by truck and less by rail. However, the transportation cost of the raw material affects the choice of delivery system and rail is usually less expensive than truck.

Alternate Sources and Materials

Most glass plants use more than one supplier for a given raw material. Because they need new raw materials daily, they cannot afford to rely on a single producer for key raw materials. In the event of a labor strike or any natural disaster or equipment failure any single source raw material could cause a shut down in the glass plant.

Substitutions of raw materials are common in the batch either to reduce costs or to improve melting. For example, most of the silica in the glass batch comes from quartz sand, but those compositions that require alumina often use a feldspathic material (feldspar, nepheline syenite, feldspathic sand, etc.) which supplies part of the silica to the batch in addition to the alumina. Likewise, some iron furnace slag is used for glass making, and this brings several oxides the batch, e.g., CaO, Mgo, Al_2O_3. Several examples of raw material substitutions in the glass batch appear in the literature (e.g., Reynolds, 1986) and one reference (Ansems, 1990) describes a computer model for analyzing the effects of raw material substitutions. However, Statistical Process Control (SPC) methods used in production of glass making raw materials is a new trend that is forming closer relationships between supplier and customer.

Cullet: This term applies to recycled glass that is added to the melting cycle. Cullet has a lower melting temperature than the rest of the glass batch and affords energy savings as well as extending the life of the furnace refractories. In 1991, cullet in the United States accounted for about 20% of the container glass melted (Freas, oral communication, 1992). This use of cullet translates into a similar reduction of raw materials needed. However, delivery of recycled glass containers can be disrupted due to holiday interruption at collection centers. Therefore, delivery of cullet often is not as reliable as other raw materials.

Color segregation of containers is critical to preserve the proper end result (clear, amber, or green). Recycled glass normally is used only in production of glass containers and so only glass from containers can be used. As can be seen from the table of compositions (Table 3), each category of glass products comes from a different and distinct chemical composition. Random substitution of glass types to the melting would destroy the compositional integrity of the melt and yield a lack of control in the glass forming area. Specifications for cullet:

≤ 0.05% ferrous metal
≤ 0.5 % non-ferrous metal
≤ 0.5 % organic (paper labels, etc)

These specifications assume the glass containers have been segregated by color and have been crushed to the same size as the other raw materials.

Blast Furnace Slag: This byproduct of iron making contains many of the oxides used in making glass, plus slag melts at relatively low temperatures. While slag provides alumina and sulfur plus silica and lime, its chief role is as a melting aid and refining agent (Mills, 1975). Iron and chrome are the major contaminants but can be controlled by proper selection and through magnetic separation. Blast furnace slag has been used in the manufacture of containers, fiber glass, and flat glass. The Calumite Company began producing slag for the glass industry in the United States, and its process has been extended to the United Kingdom and Germany (Toon, 1986).

GOVERNMENT CONSIDERATIONS

US Environmental Regulations

In the United States both raw material producers and glass makers are affected by environmental regulations. The federal Environmental Protection Agency (EPA) has standards for asbestos and free silica. The EPA regulations still have not settled completely on definitions or tolerance levels. Too often, it seems that detection limits are equated to health risks. The regulations affect the raw materials through all stages of production and at the glass manufacturer's operation until the material begins to melt.

Glass producers also are affected by the US Clean Air Act of 1977 as it relates to exhaust gases from the furnaces. The regulation of particulates and nitrogen compounds (NO_x) released required the installation of electrostatic precipitators and special exhaust systems. Glass makers also sought better batch control, especially the sulfate content, to reduce sulfur emissions and the fluorine content to reduce HF emissions.

Safety Regulations

In 1978 the US Mine Safety and Health Act brought industrial mineral miners and producers under the jurisdiction of the Mine Safety and Health Administration (MSHA). Similarly the Occupational Safety and Health Administration (OSHA) has dominion over workers outside of the mining industry. Both of these acts and administrations intend to protect workers. However, the interpretation of standards varies among inspectors such that compliance can be difficult.

Glass recycling in the United States has a major impact on the demand for glass making raw materials. Due to differences in composition between glass types (see Table 3 for compositional differences) only like glass uses can be used as a substitute for raw materials. In the United States some states have laws that require recycling and require a deposit on beverage containers, only to minimize litter and relieve some pressure on landfills. Glass containers make up an estimated 7% of the recyclable waste stream and glass manufacturers encourage recycling because the use of cullet conserves natural raw material deposits and reduces the energy required for melting. However, cullet needs clean segregation to avoid contamination in the glass furnaces for glass container manufacturers.

PROBLEMS AND FUTURE TRENDS

This section discusses the problems and trends for the glass industry which have a direct relationship to the glass making raw materials. Recycling has already been mentioned and accounts for about 20% of the US raw material needs for the glass container industry. The container industry has also been challenged to develop lighter weight containers to conserve fuel in shipping finished products. The container segment also faces increased competition from metal and plastic containers. All of the trends (weight reduction, recycling, and alternate packaging materials) have the affect of reducing raw material needs.

In the 1980s containers saw a 25 to 40% reduction in weight and these lightweight programs occurred in Australia, Europe, Japan, and the United States (Reynolds, 1988). Recycling of glass containers varies from 10 to 50% in Europe depending on the country, and, as mentioned, about 20% in the United States.

The demand for flat glass depends heavily on the construction and automotive markets which account for 57 and 25%, respectively, of flat glass production. These markets each have a cyclical history which can be expected to continue.

Higher quality and more consistent raw materials are taking priority over batch cost in the glass industry. Statistical Process Control (SPC) programs and methods among raw material suppliers are leading to closer relationships with the glass manufacturers. A consistent set of batch ingredients (with less substitution for reducing the cost of the batch) leads to more stability in the melting and in the finished glass products. There is a growing trend toward an attitude of teamwork between the supplier and the glass company as opposed to a buyer-seller relationship.

As glass makers look to reduce fuel costs, they will seek finer sizes in the difficult to melt raw materials e.g., silica, feldspar, nepheline syenite. The finer sizes also benefit the production of lightweight containers. The glass companies need to reduce energy costs *and* speed up the manufacturing process. Both of which have led toward the need for finer sizes. The old specifications of raw materials size 0.105 by 0.84 mm often reflected the grain size most common in natural deposits (Dickson, 1985). Glass producers may approve more agglomerated raw materials and/or pelletized batch which could influence the specifications for raw materials.

Glass manufacturing demands a lot from raw materials. While glass is an everyday material we often take for granted, the industry and its suppliers face changes in raw material demands.

ACKNOWLEDGMENTS

Everett Grubb was a good friend who supplied a great deal of information about the container glass industry. Four people agreed to provide a critical review and each added important facts, ideas, and clarifications. George Edwards of Corning, Inc. represented the specialty glass segment of the industry; Harry Mills, formerly with Owens Illinois, now a consultant to the industry, used his expertise in the container industry; Joseph Poley with Libbey Owens Ford added his comments relative to the flat glass industry; and Teresa Trott with Owens-Corning Fiberglas® provided her comments with the background of the fiber glass industry. Thanks to all of them the chapter is more complete.

BIBLIOGRAPHY

Anon, 1973, "An Introduction to the Glass Industry," *Industrial Minerals*, Nov., No. 74, pp. 9–23.

Anon, 1991, Glass Industry Statistics, US Dept of Commerce, Bureau of Census.

Anon, 1992, *Industrial Minerals*, Feb., No. 293, p.62.

Ansems, R., 1990, "Economic Evaluation of Glass Raw Materials," *Industrial Minerals*, June, No. 273, pp. 57–63.

Bates, R.L., 1969, *Geology of the Industrial Rocks and Minerals*, Dover, New York, 459 pp.

Coope, B.M., et al., 1977, "Raw Materials for the Glass Industry. An Industrial Minerals Consumer Survey," Metal Bulletin Plc., London, 131 pp.

Dickson, T., 1985, "The US Glass Industry, A Mixed Batch," *Industrial Minerals*, Sep., No. 216, pp. 91–103.

Doremus, R.H., 1989, "Glass: An Overview," *Concise Encyclopedia of Mineral Resources*, D.D. Carr and N. Herz, eds., Pergamon Press, Oxford, pp. 151–155.

Edwards, G.H., and Copley, J.G., 1977, "Specialty Glass," *Raw Materials for the Glass Industry. An Industrial Minerals Consumer Survey*, Metal Bulletin, Plc., London, 131 pp.

Edwards, G.H., and Copley, J.G., 1981, "Strategic Minerals and Chemicals for Specialty Glass & Ceramics," *Proceedings of Minerals and Chemicals in Glass and Ceramics—the next decade,* Metal Bulletin, Plc., London, 99 pp.

Harben, P.W., ed., 1981, "Minerals and Chemicals in Glass and Ceramics—the Next Decade," *Proceedings*, An Industrial Minerals Meeting at Corning, NY, Metal Bulletin, 99 p.

Harben, P.W., 1988, "Glass and Ceramic Spheres," *Industrial Minerals*, May, No. 248, pp. 53–57.

Harben, P.W., 1991, "Glass Raw Materials—Aspects of Quality, Quantity, and Prices," *Industrial Minerals*, July, No. 286, pp. 31–43.

Harben, P.W., 1992, *The Industrial Minerals HandyBook, A Guide to Markets, Specifications & Prices*, Metal Bulletin, Plc., London, 148 pp.

Harben, P.W., and Bates, R.L., 1984, *Geology of the Nonmetallics*, Metal Bulletin, Plc., London, 392 pp.

Harben, P.W., and Bates, R. L., 1990, *Industrial Minerals Geology and World Deposits*, Metal Bulletin Plc., London, 312 pp.

Holscher, H.H., 1965, "Hollow and Specialty Glass: Background and Challenge," *The Glass Industry*, Vol. 46, pp. 42–43.

Holscher, H.H., 1968, "The Relationship of Viscosity to the Processing of Glass," Owens Illinois special publication.

Huntley, F.D. and Snow, R.R., 1986, "Mineral Raw Materials for Flat Glass Manufacturing," Preprint 86–40, SME Annual Meeting, New Orleans, Society of Mining Engineers, Littleton, CO, 9 pp.

Kephart, W.W., Poole, J.P., and Ryder, R.J., 1986, "Raw Materials for Container Glass Manufacturing," Preprint 86–131, SME Annual Meeting, New Orleans, Society of Mining Engineers, Littleton, CO, 9 pp.

Khawlie, M.R., and Khalaf, L., 1989, "Lebanese Glass Industry During Crisis," *Industrial Minerals*, Sep., No. 264, pp. 71–79.

Mennemann, K., 1987, "Special Glasses, The Importance of Raw Materials," *Industrial Minerals*, Apr., No. 235, pp. 53–59.

Mills, H.N., 1975, "Glass Raw Materials," *Industrial Minerals and Rocks*, 4th Ed., S.J. Lefond, ed., AIME, New York, pp. 327–334.

Mills, H.N., 1983, "Glass Raw Materials," *Industrial Minerals and Rocks*, 5th Ed., Vol. 1, S.J. Lefond, ed., AIME, New York, pp. 339–347.

O'Driscoll, M., 1990, "US Glass Industry—Over Capacity Shadows Strong Demand," *Industrial Minerals*, Feb., No. 269, pp. 34–53.

Poley, J., 1992, written communication.

Reynolds, A., 1986, "Seen Through Glass Darkly," *Industrial Minerals*, Mar., Number 222, pp. 97–102.

Reynolds, A., 1988, "Lightweighting and Recycling, a Knockout Combination for Raw Material Suppliers?" *Industrial Minerals*, Mar., No. 246, pp. 50–57.

Russell, A., 1991, "Minerals in Fibre Glass—Current Malaise in a Growth Industry," *Industrial Minerals*, Nov., No. 290, pp. 27–41.

Scholes, S.R., 1952, *Modern Glass Practice*, Industrial Publications, Inc., Chicago, IL, 312 pp.

Shore, I., and Haigh, M., 1991, "Lithium Minerals Review—1990," *Industrial Minerals*, July, No. 286, pp. 44–45.

Toon, S., 1986, "European Glass, Seeking Batch Cost-Efficiency," *Industrial Minerals*, Aug., No. 227, pp. 39–58.

Watkins, E.C., 1986, "Mineral Raw Materials for Fiber Glass Manufacturing," Preprint 86–98, SME Annual Meeting, New Orleans, Society of Mining Engineers, Littleton, CO, 5 pp.

GLAUCONITE

NENAD SPOLJARIC

Greensand, greensand marl, and green earth are names given to sediments rich in the bluish green to greenish black mineral known as glauconite. The word glauconite is derived from the Greek word *glaukos*, meaning bluish green. The term "greensand" as a rock name for a glauconite-bearing sediment is more appropriate than "greensand marl," a term that has been doggedly perpetuated in the literature.

Because of its potash and phosphate content, greensand was mined and marketed as a natural fertilizer and soil conditioner for more than 100 years. The advent of manufactured fertilizers with adjustable nutrient ratios led to a decline in the use of greensand in agriculture. The material has since been recognized as useful in water treatment. Unfortunately, despite large reserves and worldwide distribution, glauconite has not been utilized to any significant commercial extent because no major application has been found for a substance with its chemical composition and properties. This is probably due mostly to a paucity of research on its potential commercial uses.

Extraction of potash received considerable attention during and just after World War I. Because of relatively high extraction costs and a generally low potash content (viz., less than 8%), glauconite lost its appeal as a source of this commodity.

Historical Background

Greensand was used as a fertilizer in New Jersey in the latter part of the 1700s. During the early 1800s its use became more common; applications of as much as 22.5 kg/m^2 were sometimes made, although recommendations for agricultural use suggested 4.5 to 11 kg/m^2 (Tedrow, 1957). Many crops, especially the forage type, were said to improve with greensand application; however, because of its slow release of potash, large quantities were required. Certain greensands that contain sulfur and sulfide minerals are harmful to plant growth, and these were classified as poison, burning, or black marls. The availability of higher grade potash salts from other mineral sources and the manufacture of prepared fertilizers displaced the agricultural use of greensand during the latter 1800s.

During the mid-1800s the greensand industry, centered in a small section of the eastern United States, grossed more than $500,000/y. Toward the end of the century, however, annual production had dwindled to less than $100,000 in value. By 1910 there were only six or eight greensand producers grossing less than $5,000/y each (Tyler, 1934). There was a brief revival of the US industry during World War I because of the curtailment of foreign potash, especially from Germany.

During the latter 1940s and early 1950s greensand was again recommended as a food nutrient for plants and farm crops. Agronomic studies discussed its potential as a soil additive that gradually releases potash and many trace element nutrients essential for plant growth (Tedrow, 1957). Greensand was sold with the idea that it would condition soil and absorb and hold water while its base exchange properties would release trace elements.

For a short time glauconite was used in certain parts of New Jersey as a binding additive in the brick industry, and in the 1800s it was used for making green glass (Cook, 1868).

In the early 1900s the base exchange properties of glauconite were recognized for water treatment and the mineral gained acceptance as a water softener. Mansfield (1922) does not mention base exchange even though this phenomenon was known in 1916 or earlier. From 1916 through 1922 several patents for the use of glauconite as a water softening agent were granted. A method was also patented for treating greensand to improve it for water softening and ready regeneration with common sodium chloride brine (Borrowman, 1920, Spencer, 1924, Kriegsheim and Vaughan, 1930).

Treated glauconite, on contact with water containing magnesia or lime, takes up magnesium or calcium ions and releases sodium ions. This exchange is limited to the outer surface of glauconite grains, and when all the surfaces have absorbed their capacity, the grains must be regenerated. Regeneration, simply stated, consists of treating or backwashing the glauconite with a sodium chloride solution, which replaces the hard water elements with sodium, thus reviving the glauconite. The process has become more sophisticated due to competition among companies in the water softening business.

Greensand products for water softening generally consisted of several different grades distinguished by the particular treatment the glauconite was given during processing. The standard greensand water softener was produced from natural glauconite that was only washed and classified. Its characteristics for water softening are given in Table 1.

To increase its water softening capacity glauconite was treated by a process that made it more porous. Although most of the specifications remained the same as those for the washed greensand shown in Table 1, the porosity changed from 2 to 3% to 11 to 13%, and the exchange capacity (as $CaCO_3$) increased from 6.4 to 6.9 kg/m^3 to 8.6 to 10.2 kg/m^3. Maximum permissible raw water turbidity dropped from 10 ppm to 5 to 7 ppm, and the maximum permissible iron in raw water decreased from 2.5 to 1.5 ppm. By increasing the porosity of glauconite even more, the exchange capacity was also further increased, with a corresponding change in some of the other specifications. This special porosity process increased the salt requirement for regeneration, for example, from 22.4 kg/m^3 to 40 kg/m^3. Another method of processing greensand involved heating it to a temperature of approximately 200 to 400°C to drive off water. After cooling, it was treated with a salt brine, which made the greensand product more efficient because softened water did not become cloudy.

During the latter 1940s greensand was displaced as a water softening agent by phenol formaldehyde resin, which has twice the softening capacity, is not affected by water temperatures above 45°C, and requires the same amount of salt for regeneration. After

Table 1. Greensand in Water Softening

Porosity, %	2 to 3
Effective size, mm	0.30 to 0.33
Uniformity coefficient	1.40 to 1.55
Screen analysis, mm	1.0 to 0.25
Attrition loss per annum, %	2
Weight per m^3, air-dried, kg	1 360
Recommended pH operating range	6.6 to 8.3
Maximum permissible temperature of raw water, °C	46
Maximum permissible turbidity in raw water, ppm	10
Maximum permissible iron in raw water as Fe, ppm	2.5
Bed expansion during backwash, %	30 to 35
Exchange capacity as $CaCO_3$, kg per m^3	6.4 to 6.9
Salt requirement, kg per m^3	22.4
Time required for regeneration in water softener, min	40
Recommended maximum softening rate, lpm per m^2 bed area	204
Recommended wash rate, lpm per m^2 bed area	285
Recommended minimum bed depth, m	0.75
Recommended maximum bed depth, m	1.85

Source: New Jersey Geological Survey.

1950 phenol formaldehyde resin was displaced by styrene resins, which have an even higher water softening capacity but require no more salt for regeneration.

GEOLOGY

Glauconitic greensands are unconsolidated sandy, silty, and clayey sediments of marine origin found in formations ranging in age from Precambrian to Recent. Although greensands are generally moderately sorted due to the presence of fine matrix, the glauconite fractions of these sediments are usually well sorted because glauconite pellets are remarkably uniform in size.

The contacts with overlying and underlying sediments are either gradational or unconformable. Primary sedimentary structures are generally lacking, but cross bedding and massive bedding have been locally reported.

No specific studies of the matrix of glauconitic greensands have been carried out. Nevertheless, preliminary examinations suggest that the clay mineral composition is quite variable and can include any or all of the following: glauconite, illite, smectite, and mixed-layer clays.

Most greensand deposits contain, in addition to glauconite, materials such as quartz, mica, pyrite, some heavy minerals, and calcium carbonate in the form of shell or cement derived from marine organisms. Iron oxides, phosphatic nodules, and fecal pellets are locally common, and skeletal remains of marine vertebrates are found occasionally.

Mineralogy

Glauconite grains range in size from 0.84 to 0.149 mm and develop different shapes (Figs. 1–4). The largest percentage by weight is 0.84 to 0.42 mm in size. Grains are usually smooth and round. Clay-size glauconite occasionally fills the interstices of coarse-grained sediments. True authigenic glauconite sometimes forms the cement.

Glauconite is a hydrous iron potassium silicate containing varying amounts of aluminum, calcium, magnesium, sodium, and numerous trace elements such as beryllium, cobalt, chromium, nickel, molybdenum, vanadium, titanium, and uranium. It has the chemical formula $3X_2(Fe^{+3}, Fe^{+2}, Y)_6(Si_4O_{10})(OH)_4 \cdot nH_2O$, where X is K^+, Na^+, or Ca^{+2} and Y is Al or Mg. The chemical composition of selected glauconites from Delaware is shown in Table 2.

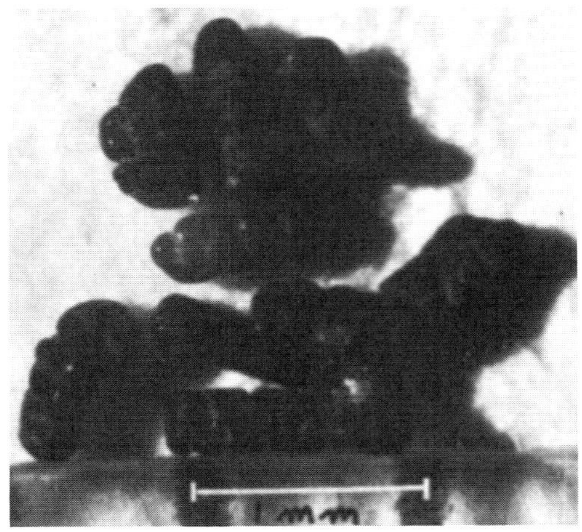

Fig. 1. Vermiform pellets of glauconite. Note the elongated and segmented shape. X-ray diffraction and thin-section studies indicate that the segments are books of glauconite layers (see Fig. 3). The pellet at the far right has a longitudinal line dividing a more solid, perhaps more recently crystallized portion from a rougher, perhaps older portion, which suggests epitaxial crystal growth (Tapper and Fanning, 1968).

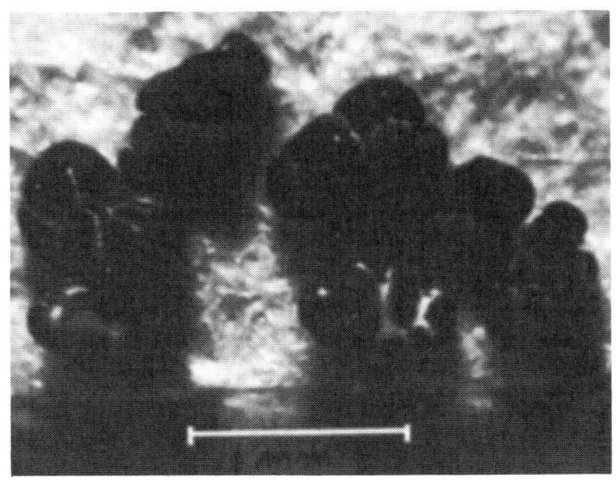

Fig. 2. Lobate pellets exhibiting the typically rounded, botryoidal shape. Lobate pellets in the Maryland deposts are more numerous and generally larger than the vermiform variety and also have a different internal morphology (Tapper and Fanning, 1968).

Greensand analyses from ten localities in the United States are listed in Table 3.

The composition of glauconites was investigated by Foster (1969), who computed atomic ratios and layer charges for 32 samples from 12 countries. Table 4 shows the compositional range of these glauconites.

In the structure of mica, according to Grim (1968), two silica tetrahedral sheets face one central octahedral sheet. The tips of the tetrahedra point toward the central unit, and the three sheets are

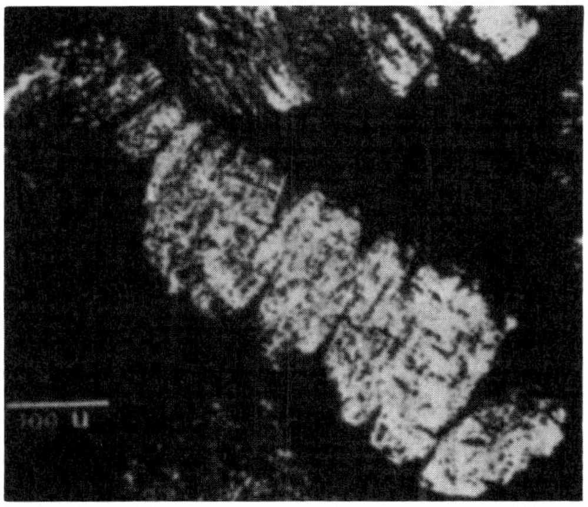

Fig. 3. Longitudinal section of a vermiform pellet showing glauconite layers crossing the pellet. (Thin-section viewed with crossed nicols.) Note the appearance of a rectangular crystal outline within the pellet just above the center of the photograph. A portion of another pellet with a gross micaceous internal morphology is at the top (Tapper and Fanning, 1968).

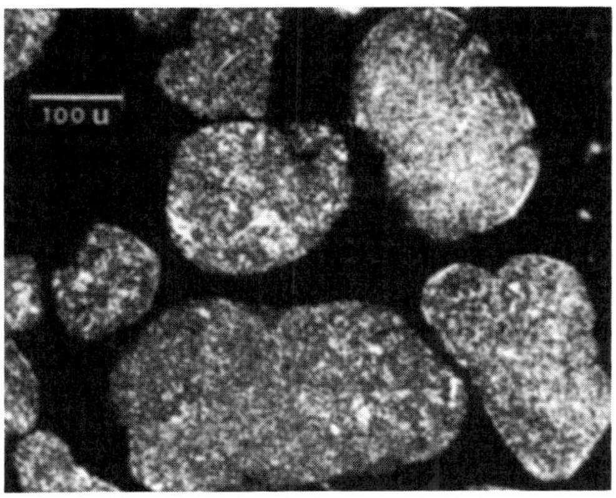

Fig. 4. Thin-section, as viewed under crossed nicols, showing the grainy internal morphology that appears to be characteristic of lobate pellets (Tapper and Fanning, 1968).

combined into a single layer, with a suitable replacement of OH and O. As much as 25% of the silicon is replaced by aluminum, and the resulting charge deficiency is balanced by potassium ions. In glauconite aluminum can substitute for 4.2 to 14.3% of the silicon in tetrahedral sheets. Some of the charge deficiency is balanced by Fe^{+2} and Mg^{+2} ions replacing Fe^{+3} and Al^{+3} ions in the octahedral sheet. Most glauconites are not pure dioctahedral micas but a mixture of 10Å micas and 14Å expandable-lattice clays (Burst, 1958, Cimbalnikova, 1971).

Like other layer minerals, glauconite has the capacity to absorb ions from solution. Its exchangeable sites are on the outside of the silica-alumina-iron framework, and the exchange reaction generally does not affect the glauconite structure. Both cations and anions are exchangeable, but not much information is available on the latter. Ion exchange is important in chemical separations, water softening, and the properties of soil for both agriculture and engineering. It is usually expressed in milliequivalents (meq) per 100 g, and a typical cation exchange capacity for glauconite is 20 meq/100 g.

Table 2. Chemical Composition of Glauconites from the Delaware Coastal Plain

	1	2	3
Oxide		Weight %	
SiO_2	49.6	47.3	47.3
TiO_2	0.84	0.28	0.27
Al_2O_3	9.6	7.1	7.2
Fe_2O_3	14.0	17.6	17.5
FeO	2.26	1.89	2.30
MnO	0.04	0.03	0.03
MgO	3.1	3.7	3.6
CaO	trace	2.1	2.2
Na_2O	0.11	0.11	0.05
K_2O	6.2	8.0	8.0
Cr_2O_3	0.045	0.034	0.056
H_2O_{total}	10.13	8.99	8.2
CO_2	—	1.0	1.1
Total	95.9	98.1	97.8
(Fe_{total} as Fe_2O_3)	(16.5)	(19.9)	(20.1)

Source: Spoljaric and Crawford, 1978. (Published with permission of Springer-Verlag, Inc., Heidelberg.)

The compositional relationship between glauconite and other layer silicates is illustrated in the ternary plot by Yoder and Eugster (1955) shown as Fig. 5.

The shapes of typical glauconite grains are shown in Figs. 1 and 2, and their internal morphology is illustrated in Figs. 3 and 4.

Properties

Magnetic Susceptibility: Glauconite has marked magnetic susceptibility. Magnetic separation is, therefore, a commonly used method for separating it from impurities. With a Frantz Isodynamic separator the following settings give a clean separation of glauconite from quartz, coprolites, pyrite, and calcite particles:

side slope	20°
forward slope	15°
current	0.50 A (ampere)
mass magnetic susceptibility	$(K_m) = 25 \pm 5 \times 10^{-6}$

Three steps of medium-intensity magnetic separation, preceded by washing and removal of +0.59 and −0.074 mm fractions, results in purified glauconite.

Specific Gravity: The specific gravity of glauconites ranges widely depending on the abundance of mixed-layer minerals and K_2O, the degree of substitution in the tetrahedral and octahedral sheets, and the extent of weathering. Shutov et al. (1972) reported the results of careful fractionation of purified glauconites by means of a density column, which permitted separation in heavy liquids in steps of 0.020 ± 0.001 g/cm^3. The specific gravity of glauconites of various geologic ages and provenances ranged between 2.32 and 2.92.

The wide range in specific gravity of glauconites limits the effectiveness of heavy liquid separation as a purification method. It can be used on some greensands with good results to separate mixtures of glauconite of different provenances and different stages of diagenesis.

Thermal Characteristics: Thermal curves for glauconites show weight losses between 80 and 255°C and between 520 and 825°C. The first temperature range corresponds to dehydration, the second to actual decomposition of the mineral (Smykatz-Kloss, 1974). Prolonged heating at 1 000°C results in the formation of

Table 3. Greensand Analyses (%)

	1	2	3	4	5	6	7	8	9	10
Silica (SiO_2)	51.83	46.28	63.55	77.93	77.80	73.00	35.18	52.61	30.00	36.44
Alumina (Al_2O_3)	6.23	5.42	4.41	7.48	7.86	9.60	5.30	9.56	14.11	29.02
Iron oxide (Fe_2O_3)	20.08	23.64	15.74	3.54	5.42	7.20	17.35	23.04	25.09	13.36
Lime (CaO)	0.52	1.10	0.48	2.39	2.12	2.88	16.00	1.39	10.80	2.72
Magnesia (MgO)	3.66	3.57	2.35	1.01	0.82	1.55	trace	2.87	3.46	1.88
Potash (K_2O)	6.60	7.50	6.24	2.30	3.60	3.30	1.69	3.49	0.80	—
Soda (Na_2O)	0.76	0.83	0.03	0.12	—	—	1.39	0.42	4.41	—
Phosphorus (P_2O_5)	0.31	0.52	0.19	0.22	—	—	3.30	—	0.44	—
Carbonic acid	—	—	—	—	—	—	8.00	—	—	—
Ignition loss	10.34	10.75	6.91	4.83	2.26	2.40	10.10	5.96	—	15.59

Source: Markewicz and Lodding, 1983.
[1]Hornerstown greensand, Sewell, NJ (NJ Geological Survey).
[2]Navesink greensand, Sewell, NJ (NJ Geological Survey).
[3]Nanjemoy Formation, MD (Grim, 1968).
[4]Aquia Formation, MD (Glasser, 1971).
[5]Aquia Formation, Hop Yard, Rappahannock River, VA (Clark, 1912, Gildersleeve, 1942).
[6]Nanjemoy Formation, Woodstock, VA (Clark, 1912, Gildersleeve, 1942).
[7]Lower Eocene greensand, Gas Ridge anticline, TX (Jones, 1936).
[8]Glauconite, McNairy County, TN.
[9]Weches Formation, Houston County, TX (Fisher, 1965).
[10]Winona Formation, Montgomery County, MS.

Table 4. Compositional Range of Glauconites (%)

SiO_2	46.9–52.9
Al_2O_3	5.8–15.2
Fe_2O_3	9.3—24.1
FeO	1.0–6.3
MgO	2.3–4.6
CaO	Traces–1.9
Na_2O	Traces–1.6
K_2O	5.1–9.3
TiO_2	0–1.8
H_2O	4.3–7.2

Source: Foster, 1969.

maghemite, which inverts into hematite (Spoljaric and Crawford, 1980).

Mode of Occurrence and Origin

Glauconite is found in rocks ranging in age from Precambrian to Recent; however, when the terms glauconite greensand or greensand are used, a Tertiary or Cretaceous age is generally implied. The term greensand denotes an unconsolidated glauconite-rich sand. Glauconite frequently forms the major constituent in many sandy, silty, or even clayey formations of Tertiary or Cretaceous age. Greensand can be present as a solid massive bed, as disseminated grains, as glauconite-rich masses and aggregates, or as fillings in fossil worm tubes or foraminifera. Brown, ovoid, phosphatic fecal pellets are common as accessory grains in greensand deposits.

Fresh glauconite is typically green to dark green in color and polylobate to round or tabular in grain shape. Glauconite imparts a greenish color to the formation in which it occurs, depending on the amount present.

The genesis of glauconite has been puzzling for a long time, but facts are emerging that permit establishing several workable theories. The process still accepted by some (Hower, 1961, Ehlmann et al., 1963, Triplehorn, 1966, Grim, 1968, Bailey and Atherton, 1969, Zumpe, 1971, Odin, 1971, Odin and Giresse, 1972, Robert, 1972, Shutov et al., 1972) requires the presence of montmorillonite (smectite) in a shallow marine environment (e.g., water depths of

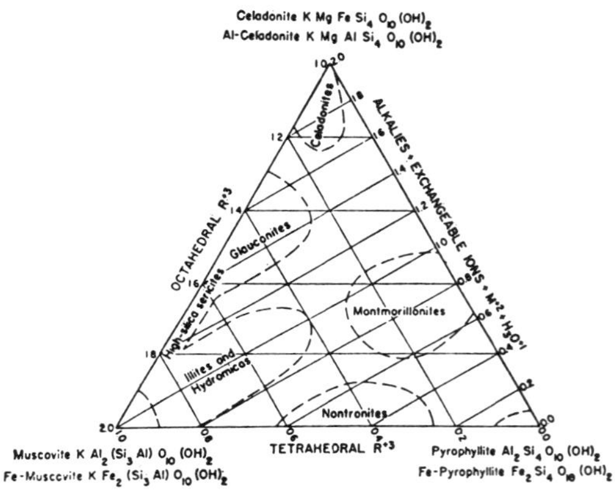

Fig. 5. The relationships of dioctahedral mica-type phases, with fields indicated in which natural clay mineral compositions fall (Yoder and Eugster, 1955).

100 to 300 m), and very slow rates of sedimentation. Montmorillonite is converted to glauconite by substitution of Fe^{+3}, Fe^{+2}, and Mg^{+2} for Al^{+3} in the octahedral layer and the consequent fixation of K^{+}. The resulting mineral is a mixed-layer clay. The potassium-rich layers are those which have attained a sufficiently high negative charge (because of the substitution of Mg^{+2} and Fe^{+2} for Al^{+3}) to fix K^{+} and dehydrate. Aluminum ions moving out of the octahedral layer can form new montmorillonite on the surface of the potassium- and iron-enriched globules, continuing the process of accretion. Most of the abundant ferric iron and silica carried in solution by freshwater is chemically precipitated as hydroxide gels in a brackish environment (Jeans, 1971) and becomes available for glauconitization.

A suite showing alteration from 1) bleached-clear biotite mica containing dark cores and swollen edges to 2) well-expanded, light olive, altered hydromica grains to 3) light green, vermiform, peripherally sutured, early-stage glauconite has been observed in drill cuttings from several New Jersey coastal plain formations. The mica-glauconite diagenesis reported by Galliher (1939), Takahashi

(1939), and Takahashi and Yagi (1929) finds support in these observations. Growth of the globules is limited by ocean currents, which move the newly formed glauconite pellets from areas of formation to those of deposition and burial. Diagenesis continues, with the general substitution of hydrated layers by additional potassium, thus reducing the amount of mixed-layer material and increasing the 10Å mica component of mature glauconite.

In recent years new theories on the origin of glauconite have been proposed. According to Odin and Matter (1981), the mineral forms a series having glauconitic smectite and glauconitic mica as its end members. The parent material has been found to include a variety of particles ranging from foraminiferal tests to rock fragments. Glauconitization appears to take place under very specific conditions in a microenvironment and to be controlled by the ion exchange processes in open marine seawater. Regionally, the formation of glauconite seems to depend to a large extent on the availability of iron and potassium.

Kemper (1982) determined that a transgressive sea and slow rate of sedimentation are not necessary for the formation of glauconite. According to the results of his study the requirements are pelagic conditions involving the upwelling of cold water from great depth into marginal areas of even cooler seas. It appears that siliceous organisms, glauconite, and phosphorite are genetically related and that a close genetic relationship also exists among glauconite, phosphorite, and montmorillonite.

In an earlier study in Hungary (Bondor, 1982, private communication), low sedimentation rate, grain size, and calcium carbonate were found to be unimportant in the formation of glauconite. An investigation of Eocene and Oligocene glauconites also clearly established the genetic relationship of glauconite to pyroclastic biotite, amphibole, and volcanic glass through the process of halmyrolysis.

The formation of glauconite in cool water was also suggested for high-alumina (e.g., more than 10% Al_2O_3), Middle Cambrian glauconites in Sweden (Berg-Madsen, 1983). A process characterized by gradual transition from illitic minerals was proposed as the means of glauconite formation.

A detailed Mossbauer investigation carried out in New Zealand (McConchie et al., 1979) gives insight into one of several possible processes of actual glauconite formation. Apparently the formation starts with the defective or degraded phyllosilicate structure of an authigenic or detrital mineral decomposed in part by marine biogenic activity. The structure is highly charged but is temporarily neutralized by potassium, sodium, calcium, and particularly magnesium cations. Iron enters the structure as ferrous iron replacing neutralized cations and is partially oxidized to octahedral ferric iron. Ferrous iron continues entering the structure while the system proceeds toward equilibrium. Not all ferrous iron is oxidized, and the glauconitization process may be temporarily or permanently stopped by change in Eh, diminished supply of iron, or the filling of all octahedral sites.

Whether these theories can adequately explain the formation of glauconite in a marine environment and the accumulation of greensand strata of considerable thickness and extent remains to be seen.

Distribution of Major Deposits

United States: The most important glauconitic greensands are developed in the Mid-Atlantic Coastal Plain. Glauconite is found as an accessory mineral in many coastal plain sediments of New Jersey, Delaware, and Maryland, but highly glauconitic greensand is best developed in Upper Cretaceous and Lower Tertiary formations.

Greensand crops out in a narrow belt that trends northeast-southwest across New Jersey, Delaware, Maryland and beyond, but in general the glauconite content and thickness of the strata and the potash content of glauconite in them decrease southward. Chemical compositions of glauconites from Delaware are shown in Table 2.

Table 5. Major Greensand Formations in the US Mid-Atlantic Region

State	Formation	Age	Thickness (m)
New Jersey	Hornerstown	Paleocene-Late Cretaceous	10+
	Navesink	Late Cretaceous	15+
Delaware	Hornerstown	Paleocene-Late Cretaceous	35+
	Marshalltown	Late Cretaceous	10+
Maryland	Aquia	Eocene-Paleocene	35+

Sources: State Geological Surveys.

The Marshalltown Formation of Late Cretaceous age is composed of dark greenish gray, massive, highly glauconitic and silty fine sand. Its thickness ranges from 3 to more than 10 m.

The Hornerstown Formation of Paleocene age is a green, medium-grained glauconitic sand containing bright green clay cement and a little quartz sand. Locally it contains a few thin, brown to reddish brown, clayey or indurated limonitic bands. Its thickness ranges from 5 to more than 15 m.

The Navesink Formation of Late Cretaceous age is a massive, dark greenish gray, medium-grained, poorly sorted glauconitic sand containing varying amounts of quartz. It contains calcareous clay toward the top and limonitic bands or masses locally. Its thickness ranges from 7 to more than 15 m.

The Aquia Formation is composed of fine- to medium-grained, highly glauconitic sand with calcite-cemented layers. Its maximum thickness exceeds 35 m.

Other Countries: Glauconitic greensands are developed on all of the continents and have been geologically mapped in many countries. At some locations glauconite is the most abundant constituent of the formation. In most cases, however, it is present in relatively small amounts or as a minor component of the sediment.

Great Britain. The Upper Greensand Formation is composed of highly glauconitic siltstones and sandstones and is developed in the southern and southeastern parts of England. The formation is well exposed at Ballard Cliff (Bennison and Wright, 1969).

Hungary. The Lower Cretaceous Nana Formation in the Transdanubian region is composed of highly glauconitic marls. The largest concentration of glauconite is in the lower part of the formation.

In the southern part of the Buda Mountains clays and sands rich in glauconite are well developed in the Torokbalint Formation of Oligocene age (Bondor, 1982, private communication).

Switzerland. The Altmann Schist of Early Cretaceous age contains highly glauconitic layers, reaches a maximum thickness of 40 m, and is best developed in the Helvetic Alps, especially near Santis and Pilatus (Funk, 1971).

USSR. Glauconitic sandstone is widespread in the northwestern region of the European USSR. It occurs in the Ukraine, in the Volga region, on the eastern slope of the Ural Mountains, in Kazakhstan and in Uzbekistan, and also in the Far East. These glauconitic deposits are mainly in Jurassic, Cretaceous, and Paleocene sands and sandstones. Their glauconite content commonly reaches 50 to 80% (Tatarinov, 1972, private communication).

Israel. The Cretaceous glauconitic sediments in the Negev area first appear in the Hatira Formation and continue for several meters into the overlying Hazera Formation. In some beds glauconite makes up 80% of the rock (Bentor, 1966).

The Pliocene Abu Hareira Sandstone is hard, cross-bedded sandstone rich in glauconite. With a maximum thickness of 30 m the formation is best developed between Gaza and Beersheba (Picard and Solomonica, 1936).

Pakistan. The lower member of the Chichali Formation in the western Salt Ranges, Trans-Indus Ranges, and parts of Kohat is composed of sandy glauconitic shale that passes upward into dark green glauconitic and fossiliferous sandstone (Danilchik and Shah, 1967).

Thailand. The green sandstones of the Phra Vihan Formation are rich in glauconite. The formation is of Early Jurassic age and is well developed on the Khorat Plateau (Ward and Bunnag, 1964).

Egypt. In the Western Desert the upper part of the Nubian Sandstone is composed of a 380 m thick sequence of gray-green, very glauconitic shale alternating with sandstone. It is of Early Cretaceous age (Said, 1962).

South Africa. The major part of the Igoda Formation, which is of Late Cretaceous age, is composed of highly glauconitic sandy limestone and glauconitic calcareous sandstones. The formation is exposed in cliffs on the southwestern side of the Igoda River (Klinger and Lock, 1978).

New Zealand. All major glauconitic deposits in New Zealand are found on South Island.

The Iron Creek Greensand is a highly glauconitic, massive sand in the Iron Creek area. The formation reaches as much as 250 m in thickness and is of Late Cretaceous to Eocene age (Gage, 1970).

Waipara Greensand, which is Paleocene-Eocene in age, consists of alternating soft and hard, green, richly glauconitic sandstones. It is exposed in banks of the Waipara River and the South Branch of the Waipara River.

Kokoamu Greensand is pure, highly glauconitic sand. The formation is of Late Oligocene age and is developed and exposed in the Duntroon area.

Gee Greensand is exposed in the coastal Kakanui area. The formation shows distinct stratification and is very rich in glauconite. It is at least 15 m thick and is of Oligocene-Early Miocene age (Gage, 1957).

Australia. North of Perth in Western Australia there are three glauconitic formations that have a cumulative thickness of 90 m: the Molecap Greensand at the base, the Gingin Chalk, and the Poison Hill Greensand at the top. In some beds the glauconite content reaches 50% of the sediment (McLeod, 1965). The entire sequence is of Late Cretaceous age (Brown et al., 1968).

West New Guinea (Irian Jaya). The Kembelangan Formation, which is Jurassic to Paleocene in age, is well developed in the Vogelkop region. Its Upper Cretaceous portion is very glauconitic. Glauconite usually makes up more than 50% of the rock (Hermes, 1981, private communication).

Venezuela. The Cogollo Group includes a sequence of sediments containing as much as 80% glauconite. The Group reaches 610 m in thickness, is of Early-Late Cretaceous age, and is well developed in the central Lake Maracaibo area (Bartok et al., 1981).

TECHNOLOGY

Exploration

Standard exploration procedures consist of outcrop examination, trenching, split-spoon and auger drilling, and sampling. Drilling is essential for determining the extent of a deposit and its thickness. Exploration data are used in constructing geologic, isopach, isograde, and other maps, in calculating the volume of a deposit, and to determine minable reserves. Only color, grain size range, and clay matrix content are examined in the field to make certain that the deposit meets required criteria.

Mining

Most greensand deposits are unconsolidated sediments. Various methods of open pit extraction have been used ranging from pick and shovel in the earliest days to power shovels, draglines, and hydraulic mining. At Inversand Company's operation at Sewell, NJ, which is presently the only commercial greensand producer in the United States, hydraulic mining methods are used in the Hornerstown and underlying Navesink formations. The pit has been opened to a depth in excess of 15 m. The overburden, consisting of sandy units of the Kirkwood Formation, is about 7.5 m thick and is removed prior to greensand extraction. The lower part of the Hornerstown Formation, which comprises the upper part of the deposit, is 5 to 6 m thick. Below it are 3 to 4.5 m of nearly pure glauconite belonging to the Navesink Formation. After the beds are slurried by a jet of high pressure water, the greensand is pumped from a sump through a pipeline to a tank at the nearby processing plant.

Processing and Purification

Processing at Sewell, NJ, involves first the screening of crude, mine-run greensand. Approximately two-thirds of the crude is recovered as −1.0 +0.25 mm feed for the production of manganese greensand. The remaining one-third consists of -0.25 mm material that is too fine for use as a water filtration medium. The −0.25 +0.088 mm fraction is stockpiled and sold as a soil conditioner and natural fertilizer, and the −0.088 mm fraction is pumped as waste to settling ponds.

After screening, the −1.0 +0.25 mm fraction is washed for two hours with clean water to remove silt, dirt, etc. This is followed by several chemical treatments. Treatment with each chemical solution is followed by a fresh water rinse before the next chemical is used. The treatments each take between $\frac{1}{2}$ hour and $1\frac{3}{4}$ hours. The first four are with solutions of sodium aluminate, sodium silicate (twice), and aluminum sulfate. These chemicals serve to harden the glauconite particles. The next eight treatments use manganese sulfate and potassium permanganate to build a manganese dioxide coating on the glauconite. This coating is the active agent in removal of iron, manganese, and hydrogen sulfide from well water.

The total time required for all chemical treatments is approximately 14 hours. Afterwards the material is slurry-pumped to a drying building, drained of most of its free water in holding pans, and dried in gas-fired ovens. The drying time for each batch is approximately 20 hours. The material is then packed in $0.03 m^3$ bags or one-ton super-sacks, palletized, and stretch wrapped for shipment.

Evaluation of Deposits

Whether a deposit of glauconitic greensand is suitable for commercial exploitation is determined largely by the physical and chemical properties of the glauconite itself. Both field and laboratory tests are employed in this determination. Important physical properties include color, grain size, degree of sorting, and the amount of glauconite in the sediment. Economically acceptable greensands contain at least 90% glauconite. For use in filters glauconite must be green, dark green to almost black in color and must be present in sand-size pellets. Deposits should contain not more than 2 to 3% clay matrix and must show no evidence of weathering.

Chemical properties are determined by analytical means and include analyses for major elements (Si, Al, Fe, Mg, Ca, Na, K, and P) as well as some trace elements. Chemical composition is especially important for greensand used in agriculture. Most important is the K content, which should be at least 6% (as K_2O). Agricultural

Table 6. Greensand Production in New Jersey

Years	Tons	Value, $
1922–23	16 780	188,000
124–25	24 490	330,000
1926–27	21 770	337,000
1928–29	22 680	460,000
1930	11 520	225,000
1947–51	27 884	—
1967–71, average	3 118	—

Sources: New Jersey Geological Survey and US Bureau of Mines.

Table 7. Greensand Production in New Jersey by Use (Tons)

Year	Water Treatment	Agricultural Use
1959	1 815	4 175
1960	1 725	3 400
1961	1 360	2 630
1962	1 725	2 040
1963	1 815	not reported

Source: New Jersey Geological Survey.

greensand deposits must be fine-grained, preferably clayey. Once a glauconitic greensand is determined to be of commercial quality, the reserves available in the deposit must satisfy the producer's production requirements.

PRODUCTION

United States

In the late 1860s almost a million tons of greensand were mined each year for use as fertilizer. In 1855 greensand brought $0.07/bushel ($2.00/m^3), and in 1877 it was selling for $0.65/t at the pit. Reports indicate that production was principally from New Jersey. By the 1870s annual production had dropped to less than 100 ktpy. Toward the turn of the century demand for greensand improved, but in 1908 less than 10 000 t were produced, for a total value of only $3,500. The number of producers dropped from 13 in 1908 to 4 in 1912. The use of greensand for water softening in the early 1920s revived the industry. Table 6 shows the tonnages produced from 1922 through 1971.

During 1926 Inversand Co. started operations near Sewell in Gloucester County, New Jersey. In the late 1940s National Soil Conservation Co. started operations at Medford in Burlington County, New Jersey, and produced greensand at that location until the early 1960s. Production of New Jersey greensand for water treatment and agricultural uses is shown in Table 7.

USES

Present day use of glauconite is limited to two applications, water treatment and soil conditioning. Production is concentrated in the coastal plain of New Jersey, where greensand is extracted from Cretaceous and Tertiary sediments. A small amount of greensand for soil conditioning was previously mined in Maryland, but that operation closed in 1970, leaving New Jersey the sole producer in the United States.

Processed greensand, also called manganese greensand, is produced in the United States by Inversand Co., a Division of Hungerford & Terry, Inc. In water treatment it is used as a filter medium for removing soluble iron and/or manganese salts and hydrogen sulfide from well water. It is employed in vertical and horizontal pressure filters as well as open gravity filter systems by almost all water conditioning equipment manufacturers. Purple to almost black manganese greensand is used in two water treatment processes. In the older, intermittent regeneration (IR) process it is regenerated intermittently with a weak solution of potassium permanganate ($KMnO_4$).

A newer process, developed and patented by Hungerford & Terry around 1960, has replaced the older one and is called continuous regeneration (CR). Potassium permanganate is injected directly into the raw water upstream of the unit containing manganese greensand and may be preceded by chlorine injection. In this process the greensand does not require intermittent regeneration, only an occasional wash. The CR process uses potassium permanganate more efficiently than does the IR process and has gained favor because of its ability to reduce very high iron levels in water to extremely minute amounts, frequently to less than 10 ppb. When a well water supply contains very little iron but significant quantities of manganese, then the older IR process is more appropriately used.

For home and garden use greensand is recommended as a mulch, top dressing, and conditioning additive to soil for gardens, potted plants, and vegetable starting plots. It slowly releases nutrients, loosens clayey soil, is non-burning, holds moisture, and contributes potassium to plants to stimulate photosynthesis. The usual application is 0.05 to 0.49 kg/m^2, although larger amounts may be applied depending on the nature of the soil.

A number of other uses have been suggested or tried locally. Finely ground glauconite has been employed as a green paint pigment, but its low refractive index and tendency to weather to red iron oxide inhibit its suitability for this application. Pure glauconite has been used to a limited extent as a glass polishing agent. Glauconite from the Weches Formation in Texas contains clayey beds that were formerly cut or sawed and dried to form a tough, reasonably durable building stone (Jones, 1936, Fisher, 1965).

ECONOMIC FACTORS

Pricing

The list prices effective January 1, 1991, for manganese greensand sold by Inversand Co., the only producer of greensand in the United States, ranged from $740 to $990/m^3 depending on the quantity purchased. All prices were FOB the point of shipment in New Jersey.

Standard packaging consists of: 0.03 m^3 bags stretch wrapped on pallets holding 25 bags each; an octagonal box strapped to a pallet and holding 0.74 m^3 of product; or super-sacks holding one ton each.

Marketing and Distribution

There are two avenues of distribution for manganese greensand, dealers and original equipment manufacturers. Dealers are primarily small- to medium-size water treatment companies that resell tanks, chemicals, etc. and provide service for domestic and other small water systems. Original equipment manufacturers design and manufacture equipment for large scale municipal and industrial installations, as well as smaller commercial and domestic systems, and provide service for these water treatment facilities. They receive a small price incentive for making large single orders and provide promotion for the product. Many dealers are located in the midwestern United States. Large dealers in Europe distribute greensand from ports of entry in Sweden and Italy. Inversand Co. has also shipped greensand to Indonesia, South Korea, Trinidad, Chile, and Thailand in the past few years. A large manganese greensand water treatment system, possibly the largest ever, is being built in Saudi Arabia.

Alternate Materials

Unusually high levels of dissolved iron in water are commonly due to low pH. This problem can be corrected with a neutralizing calcite filter, a water softener, raising the pH to a minimum of 6.7, or oxidation with chlorine bleach followed by filtration. Manganese can be removed from water by reverse osmosis and by oxidation with chlorine bleach followed by filtration. Small concentrations of manganese can be removed with a water softener. Hydrogen sulfide in concentrations exceeding 6 ppm is treated with constant chlorination followed by filtration/dechlorination (McGowan, 1984, Anon., 1978).

GOVERNMENTAL CONSIDERATIONS

Zoning and Land Use

The area of southern New Jersey where glauconitic greensand is being produced is heavily populated. It is likely, therefore, that any new mining permits or changes to present land use zoning to accommodate expansion of the greensand mines will be difficult to obtain. However, in 1988 Inversand Co. had reserves estimated to be sufficient for at least 20 to 25 years of production.

Environmental Regulation

Inversand Co. is required to comply with a number of environmental regulations and to operate a wastewater treatment plant under the jurisdiction of the New Jersey Department of Environmental Protection (New Jersey Pollution Discharge Elimination System or NJPDS). The Company must have permits for both surface water discharge and groundwater discharge. Its operations are inspected periodically by the Mine Safety and Health Administration (MSHA), which checks workers for exposure to crystalline silica dust and enforces regulations regarding safety practices.

FUTURE TRENDS

Potential New Uses

In light of declining demand for greensand, efforts have been made recently to develop new uses. Glauconite's catalytic properties have been investigated in some detail in the United States, the former USSR, Japan, Germany, and France (Hartough and Kosak, 1947, Kvirikashvili, 1962, 1964, Gornak, 1963). Acylating reactions, dehydration of cyclohexane to benzene, and condensation of aldehydes have been carried out with glauconite-based catalysts. The relationship between catalytic activity and cation exchange capacity (CEC) has encouraged efforts to enhance the former by increasing the latter through heat treatment and chemical methods. Since CEC increases with the amount of mixed-layer component present, these efforts also have focused on replacing more of the potassium in the glauconite structure with hydroxyl ions. Heating of glauconite in a reducing atmosphere leads to the formation of metaglauconite, the sorptive properties of which are being studied.

An entirely different approach is the destructive dissolution of glauconite and recovery of at least three marketable products, high-purity silica, potash, and iron oxides. This process was first proposed by Turrentine (1925). At present, efforts are underway to prove its economic feasibility with the help of up-to-date chemical engineering practice.

Some extensive greensand deposits are situated near population centers, so glauconite's capacity to absorb hazardous waste fluids is being investigated.

Recent studies in Germany and the United States suggest that glauconite can be used as a filter for radioactive wastes (Kohler, 1976) and heavy metals from landfill leachates and industrial effluents (Spoljaric and Crawford, 1978, 1979). An increasing demand for this mineral in solving pressing pollution problems can therefore be anticipated. Food shortages in several parts of the world may also bring about renewed interest in glauconite as an inexpensive fertilizer.

ACKNOWLEDGMENTS

I wish to express my appreciation to Raymond W. Ellis of the Inversand Company, who kindly provided timely information and data on manganese greensand and Inversand's activities. I also thank A.F. Alsobrook, Associate Editor, for his patient guidance and advice in preparation of this chapter. The author is further indebted to Frank J. Markewicz (Acting State Geologist, New Jersey Geological Survey, Trenton, NJ, now retired) and William Lodding (formerly at Rutgers University, New Brunswick, NJ, now deceased) who wrote this chapter for the 4th edition of *Industrial Minerals and Rocks*, and to John Hower (University of Illinois, Urbana, IL, now deceased) who wrote the revised chapter for the 5th edition of *Industrial Minerals and Rocks*. Several parts from the 5th edition, some with only minor modifications, are included in the present edition.

BIBLIOGRAPHY

Anon., 1978, "Water Treatment: Iron, Boiler Water and Water Analyses," *Ground Water Age*, Vol. 13, pp. 19–23 and 53 (Reprinted with permission of DE/Domestic Engineering).

Bailey, R.J., and Atherton, M.P., 1969, "The Petrology of a Glauconitic Sandy Chalk," *Journal of Sedimentary Petrology*, Vol. 39, No. 4, Dec., pp. 1420–1431.

Bartok, P., Reijers, T.J.A., and Juhasz, J., 1981, "Lower Cretaceous Cogollo Group, Maracaibo Basin, Venezuela: Sedimentology, Diagenesis, and Petrophysics," *American Association of Petroleum Geologists Bulletin*, Vol. 65–6, pp. 1100–1134.

Bennison, M.G., and Wright, A.E., 1969, *The Geological History of The British Isles*, Edward Arnold Ltd., 406 pp.

Bentor, Y.K., 1966, "Clays of Israel," *Guidebook*, The International Clay Conference, Jerusalem, Israel Program of Scientific Translations, 121 pp.

Berg-Madsen, V., 1983, "High-Alumina Glaucony from the Middle Cambrian of Oland and Bornholm, Southern Baltoscandia," *Journal of Sedimentary Petrology*, Vol. 53, pp. 875–893.

Borrowman, G.L., 1920, "Water-Softening Compound and Method of Producing Same," US Patent 1 348 977, Aug. 10.

Brown, D.A., Campbell, K.S.W., and Crook, K.A.W., 1968, *Geological Evolution of Australia and New Zealand*, Pergamon Press, Oxford, 409 pp.

Burst, J.F., 1958, "Glauconite Pellets: Their Mineral Nature and Applications to Stratigraphic Interpretations," *Bulletin of American Association of Petroleum Geologists*, Vol. 42, No. 2, pp. 310–327.

Cimbalnikova, A., 1971, "Chemical Variability and Structural Heterogeneity of Glauconites," *American Mineralogist*, Vol. 56, No. 7–8, pp. 1385–1392.

Clark, W.B., 1912, "The Physiography and Geology of the Coastal Plain Province of Virginia," Bulletin 4, Virginia Geological Survey, 274 pp.

Cook, G.H., 1868, ''Geology of New Jersey,'' State Geological Survey, New Brunswick, 899 pp.

Danilchik, W., and Shah, S.M.I., 1967, ''Stratigraphic Nomenclature of Formations in Trans-Indus Mountains, Mianwali District, West Pakistan,'' US Geological Survey, Project Report (IR) PK-33, 45 pp.

Ehlmann, A.J., et al., 1963, ''Stages of Glauconite Formation in Modern Foraminiferal Sediments,'' *Journal of Sedimentary Petrology*, Vol. 33, No. 1, Mar., pp. 87–96.

Fisher, W.L., 1965, ''Rock and Mineral Resources of East Texas,'' Report of Investigations 54, Texas Bureau of Economic Geology, 439 pp.

Foster, M.D., 1969, ''Studies of Celadonite and Glauconite,'' Professional Paper 614-F, US Geological Survey, 15 pp.

Funk, H., 1971, ''Zur Stratigraphie und Lithologie des Helvetischen Kieselkalkes und der Altmannschichten in der Santis-Churfirsten-Gruppe (Nordostschweiz),'' *Eclogae Geologicae Helvetiae*, Vol. 64, 433 pp.

Gage, M., 1957, ''Geology of the Waitaki Subdivision,'' Bulletin 55, New Zealand Geological Survey, pp. 53–56.

Gage, M., 1970, ''Late Cretaceous and Tertiary Rocks of Broken River, Canterbury,'' *New Zealand Journal of Geology and Geophysics*, Vol. 13, pp. 507–559.

Galliher, E.W., 1939, ''Biotite-Glauconite Transformation and Associated Minerals,'' *Recent Marine Sediments*, P. Trask, ed., pp. 513–515.

Gildersleeve, B., 1942, ''Eocene of Virginia,'' Bulletin 57, Virginia Geological Survey, 43 pp.

Glaser, J.D., 1971, ''Geology and Mineral Resources of Southern Maryland,'' Report of Investigations 15, Maryland Geological Survey, 85 pp.

Gornak, A.I., 1963, ''Influence of Preliminary Treatment on the Exchange Capacity of Glauconite from the Loevsk Deposit of White Russia,'' SSR Ionobmen i Sorbtsiya iz Rastvorov, Akademii Nauk Belorussk, SSR, Inst. Obshch. i Neorgan. Khim,; Chemical Abstracts 61: 67h.

Grim, R.E., 1968, *Clay Mineralogy*, 2nd ed., McGraw-Hill, New York, 596 pp.

Hartough, H.U., and Kosak, S., 1947, ''Acylation Studies in the Tiophene and Furan Series, III, Natural and Synthetic Silica-Metal Oxide Catalysts,'' *Journal of American Chemical Society*, Vol. 69, pp. 1014–1016.

Hower, J., 1961, ''Some Factors Concerning the Nature and Origin of Glauconite,'' *American Mineralogist*, Vol. 46, No. 3–4, pp. 313–334.

Jeans, C.V., 1971, ''The Neoformation of Clay Minerals in Brackish and Marine Environments,'' *Clay Minerals*, Vol. 9, London, pp. 209–217.

Jones, C.A., 1936, ''Report on a Mineral Resource Survey of Bexar County, Texas,'' Mineral Resource Survey Series, Circular 2, Texas Bureau of Economic Geology, 6 pp.

Kemper, E., ed., 1982, ''Das Spate Apt und fruhe Alb Nordwestdeutschlands, Versuch der umfassenden Analyse einer Schichtenfolge,'' *Geologisches Jahrbuch*, Hrft 65, 679 pp.

Klinger, H.C., and Lock, B.E., 1978, ''Upper Cretaceous Sediments From the Igoda River Mouth, East London, South Africa,'' Annual Summary, African Museum, Vol. 77, pp. 71–83.

Kohler, E.E., 1976, ''Zur moglichen Verwendung von Glaukonit als Ionenaustauscher bei der Aufbereitung radioaktiver Abwasser,'' Zutschrift fur Wasser und Abwasser Forschung 9, Jahrgang Nr. 3, pp. 76–82.

Kriegsheim, H., and Vaughan, W., 1930, ''Water-Purifying Material and Process of Making the Same,'' US Patent 1 757 372, May 6.

Kvirikashvili, V.L., 1962, ''The Conversion of n-Heptane in the Presence of Glauconite,'' *Transactions*, Inst. Kh8m. Akademii Nauk Gruz., Vol. 16, pp. 89–98; Chemical Abstracts, 60: 9131D.

Kvirikashvili. V.L., 1964, ''Contact Conversions of Methylcylopentane in the Presence of a Natural Glauconite Catalyst,'' *Transactions*, Inst. Khim. Akademii Nauk Gruz., Vol. 17, pp. 201–205; Chemical Abstracts, 62:11697h.

Mansfield, G.R., 1922, ''Potash in the Greensands of New Jersey,'' Bulletin 27, US Geological Survey, 146 pp.

Markewicz, F.J., and Lodding, W., 1983, ''Glauconite,'' as revised by J. Hower, *Industrial Minerals and Rocks*, 5th ed., Vol. 2, S.J. Lefond, ed., AIME, New York, pp. 745–756.

McConchie, D.W., et al., 1979, ''A Mossbauer Investigation of Glauconite and Its Geological Significance,'' *Clays and Clay Minerals*, Vol. 27, pp. 339- 348.

McGowan, W., 1984, ''The Key to Water Treatment Skills, Using Your Senses,'' *Ground Water Age*, Vol. 18, pp. 15–19.

McLeod, I.R., 1965, ''Australian Mineral Industry: The Mineral Deposits,'' Bulletin 72, Bureau of Mineral Resources, 690 pp.

Oakley, K.P., 1943, ''Glauconite Sand of Brackesham Beds,'' Geological Survey of England & Wales, Vol. 33, 28 pp.

Odin, G.S., 1968, ''Glauconie, Glauconite et phyllosilicate verts,'' *Bulletin*, Gr. Franc. Argiles, Vol. 20, pp. 11–12.

Odin, G.S., 1971, ''Sur la genese des glauconies et leur signification sedimentologique d'apres l'etude detaillee du sondage de Mont Cassel (Nord),'' *Comptes Rendus*, Academy of Sciences, Paris, 272-D, pp. 697–699.

Odin, G.S., and Giresse, P., 1972, ''Formation des mineraux phylliteux (berthierine, smectites ferriferes, glauconite ouverte) dans les sediments du Golfe de Guinee,'' *Comptes Rendus*, Academy of Sciences, Paris, 275-D, pp. 177–180.

Odin, G.S., and Matter, A., 1981, ''De Glauconiarum Origine,'' *Sedimentology*, Vol. 28, pp. 611–641.

Picard, L., and Solomonica, P., 1936, ''On the Geology of the Gaza-Beersheeba District,'' Bulletin, Vol. 1, Geology Department, Hebrew University, Jerusalem.

Robert, M., 1972, ''Transformation experimentale de glauconites et d'illites en smectites,'' *Comptes Rendus*, Academy of Sciences, Paris, 275-D, pp. 1319–1322.

Said, R., 1962, *Geology of Egypt*, Elsevier, Amsterdam, 377 pp.

Shutov, V.D., et al., 1972, ''Crystallochemical Heterogeneity of Glauconite as Depending on the Conditions of its Formation and Post-Sedimentary History,'' *International Clay Mineral Conference*, Madrid, pp. 327–339.

Smykatz-Kloss, W., 1974, *Differential Thermal Analysis*, Springer-Verlag, Heidelberg, 185 pp.

Spencer, A.C., 1924, ''Process of Improving Glauconite,'' US Patent 1 491 561, Apr. 22.

Spoljaric, N., and Crawford, W.A., 1978, ''Glauconitic Greensand: A Possible Filter of Heavy Metal Cations From Polluted Waters,'' *Environmental Geology*, Vol. 2, pp. 215–221.

Spoljaric, N., and Crawford, W.A., 1979, ''Removal of Contaminants From Landfill Leachates by Filtration Through Glauconitic Greensands,'' *Environmental Geology*, Vol. 2, pp. 356–363.

Spoljaric, N., and Crawford, W.A., 1980, ''Glauconitic Greensand: Parent of Some Sedimentary Iron Ores?,'' *Proceedings*, 5th Quadrennial IAGOD Symposium, pp. 581–590.

Takahashi, J.I., 1939, ''Synopsis of Glauconitization,'' *Recent Marine Sediments*, P. Trask, ed., pp. 502–512.

Takahashi, J.I., and Yagi, T., 1929, ''The Peculiar Mud-Grains in the Recent Littoral and Estuarine Deposits with Special Reference to the Origin of Glauconite,'' *Economic Geology*, Vol. 24, No. 8, Dec., pp. 838–852.

Tapper, M., and Fanning, D.S., 1968, ''Glauconite Pellets: Similar X-Ray Patterns from Individual Pellets of Lobate and Vermiform Morphology,'' *Clays and Clay Minerals*, Vol. 16, pp. 275–283.

Tedrow, C.F., 1957, ''Greensand Soils Subject of Study by Station Scientists,'' *New Jersey Agriculture*, Nov.-Dec., 3 pp.

Triplehorn, D.M., 1966, ''Morphology, Internal Structure, and Origin of Glauconite Pellets,'' *Sedimentology*, Vol. 6, pp. 247–266.

Turrentine, J.W., 1925, ''Potash from Greensand,'' *Industrial & Engineering Chemistry*, Vol 17, Nov., pp. 1177–1181.

Tyler, P.M., 1934, ''Greensand,'' Information Circular 6782, US Bureau of Mines, 8 pp.

Ward, D.E., and Bunnag, D., 1964, ''Stratigraphy of the Mesozoic Khorat Group in Northeastern Thailand,'' Report of Investigation No. 6, Department of Mineral Resources, Thailand, 95 pp.

Yoder, H.S., and Eugster, H.P., 1955, ''Synthetic and Natural Muscovites,'' *Geochemica et Cosmochimica Acta*, Vol. 8, pp. 225–280.

Zumpe, H.H., 1971, ''Microstructure in Cenomanian Glauconite,'' *Mineral Magazine*, Vol. 38, pp. 215–244.

GRAPHITE

HAROLD A. TAYLOR, JR.

The first use of graphite is lost in the mists of time. It was used by primitive man to make drawings on the walls of caves and by the Egyptians to decorate pottery. As early as 1400 AD graphite crucibles were being made in the Haffnerzell District of Bavaria.

Through the Middle Ages graphite was confused with other minerals, especially galena and molybdenite. Two common names that are still being used for the mineral are plumbago, meaning lead-silver, and black lead. The latter name implied that graphite is either composed of lead or at least contains a large percentage of it. Gessner is credited with having recognized it as a separate mineral in 1565, but its composition was not determined until 1779, when Scheele demonstrated that graphite oxidized to carbon dioxide, thus proving its carbon constitution. In 1789, Werner named it graphite from the Greek word *graphein*, meaning to write.

The United States has long relied on foreign countries for its graphite supplies. This reliance has been due more to the lack of suitable domestic graphites than to any particular preference for buying minerals overseas. Sri Lankan graphites were imported as early as 1820, and Madagascar graphites have been imported since the properties on that island were opened up about a hundred years later. Mexico has long supplied the United States with most of its requirements for amorphous graphite.

PHYSICAL PROPERTIES AND CLASSIFICATION

Graphite is found in laminated, flaky aggregates disseminated in schistose rocks. It also occurs in veins and exhibits a foliated or fibrous structure. Graphite is a black lustrous mineral that crystallizes in the hexagonal system, with rhombohedral symmetry. The crystals have tabular form and are six-sided; the faces commonly are striated. The flakes have perfect basal cleavage and are opaque. When well crystallized, the flakes have a black metallic luster whereas the amorphous material is black and earthy, with a microcrystalline compactness. The flakes feel greasy. Graphite is an excellent conductor of heat and electricity. It melts at approximately 3 550°C at a triple point under 88 kg/cm^2. It sublimes between 3 300 and 3 500°C at 1.033 kg/cm^2. At 3 726°C a second triple point occurs at approximately 100,000 atmospheres. Thermal oxidation in the presence of oxygen begins at 300°C and the rate increases with temperature.

The three forms of carbon (charcoal, graphite, and diamond) are distinguished by chemical and physical tests. The specific gravity of charcoal is 1.3 to 1.9, of graphite 2.266 g/cc (crystal density), and of diamond 3.5. Graphite has a hardness of 1 to 2 (Mohs scale).

Natural graphite can be divided into three classes: disseminated flake, crystalline vein (fibrous or columnar), and amorphous. Flake graphite is a lamellar form found in metamorphic rocks, such as marble, gneiss, and schist. Each flake is separate, having crystallized as such in the rock. Crystalline vein graphite (also called lump or high crystalline graphite) is found in well-defined veins or pocket accumulations along the intrusive contacts of pegmatites with limestones. The graphite of such deposits is of two types, foliated and columnar. The important Sri Lankan deposits are vein type.

Amorphous graphite is commonly found as minute microcrystalline particles more or less uniformly distributed in weakly metamorphosed rocks, such as slates and shales, or in beds consisting almost entirely of graphite. The latter usually are metamorphosed coal seams with as much as 80 to 85% graphitic carbon, whereas the former, being altered carbonaceous sediments, commonly range from 25 to 60% carbon. The graphite content of such amorphous deposits is dependent on the amount of carbon originally present in the sediments. There is no evidence of enrichment by the intrusive rocks. Certain amorphous graphite deposits have undoubtedly been formed by contact (thermal) metamorphism, while others are probably the result of dynamic (regional) metamorphism.

Manufactured (artificial) graphite is produced in electric furnaces from petroleum coke.

DEFINITION OF TERMS, GRADES, AND SPECIFICATIONS

Through its long history of use, graphite has been marketed as graphite, plumbago, or black lead. Furthermore, industrial usage of terms may be confusing to one not intimately associated with the industry. For example, even though natural and manufactured graphite is used extensively by industry, commercially the two main categories of natural graphite are crystalline and amorphous. Inasmuch as all graphite is truly crystalline and the distinction is only one of crystal size, the term amorphous is actually a misnomer.

"Amorphous," as applied to graphite, has been further complicated by longstanding industrial application of the term to very fine particles of crystalline flake graphite that can be sold only for low value uses, such as foundry facings. This very fine flake is also called crystalline dust. Also, fine-grained varieties of lump graphite that can be reduced easily to fine particle size by grinding are called amorphous lump to distinguish them from tough, platy, and acicular varieties, known as crystalline lump, that can be reduced in particle size with difficulty.

Crystalline flake graphite is well defined in paragraph 213 of the Tariff Act of 1930 as follows:

> The term "crystalline flake" means graphite or plumbago which occurs disseminated as a relatively thin flake throughout its containing rock, decomposed or not, and which may be or has been separated therefrom by ordinary crushing, pulverizing, screening, or mechanical concentration process, such flake being made up of a number of parallel laminae, which may be separated by mechanical means.

Under the foregoing definition finely divided particles of crystalline flake graphite would be classified as crystalline graphite.

However, the Court of Customs Appeals has repeatedly held that commercial designations and not scientific terms must govern classification and that when a commercial meaning differs from the technical meaning the commercial designation must govern. Thus, large quantities of fine-crystalline flake graphite are imported under the amorphous classification.

Sri Lankan lump graphite, one of the most important, is classified either as amorphous or crystalline. Each type is divided into a number of grades, depending upon the particle size (lump, ranging from the size of walnuts to that of peas; chip, from the size of peas to about that of wheat grains; dust, finer than 60 mesh), graphite carbon content, and degree of consolidation.

Amorphous graphite is graded primarily on graphitic carbon content. Commercial products contain about 50 to 94% carbon, depending on their source.

Crystalline flake graphite from the Malagasy Republic is divided into two main grades, flake (coarse flake) and fines (fine flake). Malagasy crucible flake must have a minimum of 85% carbon and be essentially all −8 to +60 mesh. Other crystalline flake graphite is graded according to graphitic carbon content and particle size.

The synthetic graphites generally are classified as electrodes, anodes, molded shapes, and powders or scrap, and they are made from petroleum coke and pitch. Graphite fiber is made from textile threads or extruded pitch fibers.

MODE OF OCCURRENCE

Graphite is widely distributed throughout the world. It occurs in many types of igneous, sedimentary, and metamorphic rocks. Many occurrences are, however, of little economic importance. The more important are those found in metasomatic-hydrothermal deposits and in sedimentary rocks that have been subjected to regional or thermal metamorphism.

Most, if not all, of the world's deposits of flake and crystalline graphite occur in metamorphic rocks of Precambrian age. Marble, gneiss, and schist are the most common types of rock in which economic deposits of flake graphite occur. In many cases, the rocks have been intruded by pegmatitic veins. Vein graphite is normally found in rocks similar to these, but the enclosing wall rock is not necessarily graphitic. This type of deposit assumes the character of a true lode.

Economic deposits of graphite include five main geological types:

1. Flake graphite disseminated in metamorphosed, silica-rich sedimentary rocks.
2. Flake graphite disseminated in marble.
3. Amorphous deposits formed by metamorphism of coal or carbon-rich sediments.
4. Veins filling fractures, fissures, and cavities in country rock.
5. Contact metasomatic or hydrothermal deposits in metamorphosed, calcareous sedimentary rocks.

Natural graphite in these deposits varies widely in physical appearance. Klar (1958) classified graphite occurrences on the basis of crystal characteristics into microcrystalline dense graphites and macrocrystalline silvery-bright graphites. Microcrystalline graphite occur in deposits of high carbon content as exceptionally small crystals discernable only by high powered microscopes, whereas the macrocrystalline type generally occur in lower concentrations and in a larger crystal form.

Following are descriptions of the five main geological types, and Table 4 shows world graphite resources by country.

Flake Graphite Disseminated in Metamorphosed Silica-rich Sedimentary Rocks

A large part of the world's total production of graphite is derived from rocks such as quartz-mica schist, feldspathic or micaceous quartzite, and gneiss. The graphite flakes in these rocks are oriented so that they lie parallel to the plane of foliation. The graphite varies widely in physical appearance and characteristics. The flakes range in size from a fraction of a millimeter to an average of 40 cm, and may vary from one deposit to another in thickness, toughness, density, and shape.

The principle deposits of flake graphite are in lenses or layers as much as 33 m thick and some thousands of meters long. The lenses have variable graphite content both within themselves and from one lens to another. The content of an average deposit is about 10 to 12%. Some deposits containing as little as 2% graphite have been worked. In Madagascar one rich lens contains 60% graphite. Crystals in each deposit vary in size, usually reflecting the grain size in the parent rock. In a moderately good deposit, crystals as much as $\frac{3}{4}$ cm across are common.

Description of Deposits: *Alabama.* The deposits of northeastern Alabama, in Chilton, Coosa, and Clay counties, occur in two northeast-trending belts $1\frac{1}{2}$ to 3 km wide with a combined length of about 90 km. Worked intermittently since the late 1800s, the mines reached their peak production during World War I. The deposits, called leads, are groups of parallel layers and lenses in the Ashland quartz-mica schist that contain 1 to 5% disseminated flake graphite. The leads mined range from 6 to more than 35 m in thickness, dip gently to steeply, and have been traced up to 1 200 m along strike. Some are essentially constant in strike and dip over long distances; others are folded, faulted, or both. The deposits are weathered to depths as great as 35 m. In general, the flake ranges from less than 1.25 to 5 mm in diameter, but most of it is less than 1.25 mm in diameter. Assays of samples from 32 properties in the northeastern belt by the US Bureau of Mines gave the following results:

Size	Kg of Flake per Ton of Ore
+50 mesh	3.0 to 7.0
−50 +100 mesh	3.5 to 19.0
−100 mesh	4.0 to 35.0
Total flake	17.5 to about 65.0

How much of each of the commercial grades that could be produced from the deposits, however, is uncertain. Most of the graphite produced in Alabama during and since World War II has been stockpiled, but Alabama graphite has been marketed at one time or another for most of the uses of graphite.

Texas. The flake graphite deposits of Texas occur in the Precambrian Packsaddle Schist in Llano and Burnet Counties. The deposits show a wide range of graphite content and flake size. The most important deposit was one in Burnet County operated by the Southwestern Graphite Co.

Canada. The Lac Knife graphite deposit at Fermont, Quebec, has unusually rich ore; quite atypical of the other commercially important Canadian deposits. The graphite deposit and its surrounding host rock, a quartz-feldspar (biotite) gneiss, usually migmatised, represent the southern extension of the Menihek shales and argillites of the Churchill Province into the Grenville Province. Garnet, kyanite, and pyrrhotite are associated minerals. The graphite is uniformly disseminated when the ore has less than 10% graphite. Graphite blebs appear in ever-increasing size as the ore increases in richness above 10%. When the ore is above 25% graphite, veins and massive zones of graphite are observed. The graphite occurs as independent grains, specifically medium to coarse flakes or masses of fine grains, or as inclusions in other minerals, including graphite interlain with mica.

Norway. In Europe the principal productive graphite deposits of Norway, Germany (Bavaria), and Czechoslovakia, consist of flake graphite disseminated in mica schist or mica gneiss or closely associated with these rocks. The productive deposits of Norway (A/S Skaland Grafitverk) are on Senja Island, North Norway (latitude 60°39′N; longitude, 17°E) and are lenses of graphitic rock enclosed in mica schists. The richest ore contains 25 to 30% graphite. The ore is hard and must be crushed to recover the graphite.

Germany. Deposits east of Passau, Bavaria, near the Austrian border, were known to prehistoric man and have been worked since the Middle Ages. They consist of crystalline graphite in seams, lenses, and disseminated flakes in gneiss and schist. Individual seams and lenses as much as ten meters or so in thickness are mined for a few hundred meters. At Kropfmuhl, the only mine now operating, a very heavily folded bed of graphite is mined at eight successive depths at some places. This bed of graphite is interbedded with marble and micaceous gneiss in a zone that is 130 m in stratigraphic thickness. Exposures 0.5 to 1 m thick are mined. Graphite forms 10 to 30% of the lenses. Associated minerals are mainly feldspar and calcite, and less quartz, pyrrhotite, pyrite, biotite, hornblende, sphalerite, and galena. Both weathered and unweathered rock have been mined. Two grades of crucible flake (92 to 97%), three of pencil flake (96 to 97%), and five of fines are produced. The deposits are syngenetic and formed by metamorphism of organic carbonaceous sediments.

Madagascar. Probably the largest resources of high grade, crystalline flake graphite in the world are on the island of Madagascar. The deposits occur in belts of micaceous gneiss and schist over a distance of more than 800 km in the eastern half of the island from the latitude of the city of Tamatave to the southern end. The principal producing area extends about 110 km along the east coast from Tamatave to Marovintsy. Individual deposits are graphite-rich layers that range from 3 to more than 35 m in thickness; some can be followed for a few thousand meters. The graphite content reportedly ranges up to 60%, but ore averaging 4 to 11% disseminated graphite has been worked in recent years. Operations are confined to the weathered upper parts of the deposits, and reserves of weathered ore are considered ample for many years at current rates of production. Flake graphite from Madagascar is noted for its high proportion of coarse flake, uniform thickness, toughness, and cleanness, and for the care with which it is graded for export. It sets the world standard for high quality flake.

Flake Graphite Disseminated in Marble

Flake graphite disseminated in marble probably derives from carbonaceous impurities. Graphite commonly is less than 1% of the rock, although in some localities it is as much as 10%. Deposits of this type are structurally complex and exhibit much variation in grade and kind of accessory minerals over short distances. Their contribution to world production is consequently much less than that from deposits in schists and gneisses.

Description of Deposits: *Canada.* The Lac-des-Iles crystalline flake graphite deposit is presently the producer of the great majority of Canadian graphite; it is the most significant graphite deposit of its type in the world. The host rock is a jumble of Grenville biotite-garnet gneiss, quartzite, and graphite-containing marble. Although most of the graphite is found in the higher-grade marble core of a gneissic body, some larger-flake graphite is found in the quartzite ring surrounding the core. The entire sequence is cut by gabbro-diorite dikes and pegmatite bodies and is extensively folded and faulted.

Amorphous Deposits formed by Metamorphism of Coal or Carbon-rich Sediments

The graphite in deposits formed by the metamorphism of coal or carbon-rich sediments is almost invariably the microcrystalline variety known to the trade as amorphous graphite. A substantial part of the world production of amorphous graphite is from such deposits. The graphite occurs in seams, often distorted by folding and faulting, and is commonly intimately mixed with ungraphitized material. The ratio of graphite to ungraphitized material can vary widely. For example, some high grade Mexican varieties contain as much as 95% graphite, whereas low grade Korean graphites are commonly burned for fuel because of high nongraphitic carbon content.

Description of Deposits: *United States.* In the United States, coals in the Narragansett Basin of Rhode Island have locally been metamorphosed to graphitic anthracite.

Mexico. The most important deposits of amorphous graphite in the western hemisphere occur in the State of Sonora, Mexico, in an area 30 km long by 15 km wide, approximately 60 km southeast of Hermosillo. The deposits were discovered in 1867. Six beds of graphite are reported, the thickest averaging 3 m, but locally swelling to 7 m because of folding. They were earlier described as beds of coal interlayed with sandstone, folded, and then intruded by granite, the coal thereby being converted to graphite. As the result of recent detailed examinations, however, the deposits at the Moradillas mine are now believed to be hydrothermal veins; no evidence has been found to indicate that they are metamorphosed coal beds. The veins are said to occupy fissures that cross the bedding of the country rocks at low angles. In the district as a whole, the graphite deposits are reported to be steeply dipping and are said to range from mere stringers to seams that are locally 4 to 5 m wide. High grade Mexican graphite averages 80% carbon, but some contains as much as 95%.

The mineral suite for amorphous ore includes mica, clay minerals, tourmaline, and hematite with occasional pyrite and gypsum locally. The ore is prepared for shipment by hand sorting the admixed rock that comes from the hanging walls; then it is crushed, screened, and dried.

Austria. In southeastern Austria, large resources of graphite are present in a 50 km-long belt of folded metasediments extending from the towns of Leoben to Rottenmann in the Styrian Alps. There are two producing mines. At Kaiserberg, the principle mine, a series of graphitic beds with an aggregate thickness of 12 m are enclosed in graphitic schists and quartzites. Crude ore contains 30 to 80% carbon and averages 50 to 60%. Refined graphites ranging from 66 to 95% carbon are produced by grinding and flotation. The products are amorphous graphite and are used for foundry facings, pencils, paint lacquer pigments, and lubricants.

South Korea. The important amorphous graphite deposits of southern Korea, which have yielded about 2 Mt, occur as irregular lenses parallel to the structure of enclosing schists, and phyllites of sedimentary origin. In the Kyeng-Sang District three seams having minable widths of 24, 6, and 15 m have been reported. Neither the nature of the source rocks of these seams nor the origin of the seams is clear from published data. Deposits formed by dynamic and contact metamorphism of coal beds occur in southern Korea, but Overstreet (1947) reports that these are impure. Korean graphite is low in carbon content and is mostly exported to Japan for foundry facings and pigment; some of it is burned for fuel. Reserves of graphite in southern Korea are large.

Veins Filling Fractures and Cavities

Veins of graphite are typically fillings of fractures or cavities in the country rock. In the thinner veins, graphite occurs as crust or layers of closely packed, coarse, elongate plates (needle lump) oriented perpendicular to the vein walls. Plates vary in length with vein thickness and may be as much as 10 cm long. Central portions of thicker veins may consist of platy layers with the crystals parallel to the vein walls.

The origin of vein deposits of graphite is uncertain. Winchell (1916), discussed various hypotheses and concluded that the most probable mode of origin involved deoxidation of oxides of carbon.

Description of Deposits: *United States.* Deposits of this type at Sturbridge, MA, were worked in the 1640s, the first graphite mining in the United States. The only sizable domestic vein deposits, however, are those at the Crystal Graphite mine near Dillon, MT. There graphite fills networks of fractures in gneiss and pegmatite around the northwesterly plunging nose of an isoclinal fold, a structure that involves granite gneiss, schist, marble, and quartzite of the Precambrian Cherry Creek Series. The veins are a few millimeters to 60 cm thick and 1 to 15 m long. The formation of the veins has been related to the development of the pegmatites, but Ford (1954) has later shown that fracturing and vein formation took place after the formation of these rocks.

Sri Lanka. The vein graphite deposits of Sri Lanka have been worked systematically for more than 140 years, yielding more than 1.2 Mt of graphite since 1880. The main deposits are in Southern, Western, and Sabaragamuwa provinces. The deposits occur in a complex series of Archean granulites, quartzites, garnet-sillimanite gneisses, and marbles in narrow, structurally controlled belts that persist for many miles along strike. The productive deposits range from simple fissure veins to branching veins or stockworks. Veins commonly are a few millimeters to almost a meter thick. Some of the larger deposits are tabular bodies parallel to foliation of the enclosing rocks. Graphite filled vugs as much as 24 m long and 2 × 1½ m in cross section are reported. The highest grades of Sri Lanka graphite carry 97 to 98% carbon. A variety of minerals, chiefly quartz, pyrite, calcite, apatite, and pyroxene, and lesser rutile, allanite, magnetite, and various lime-magnesia silicates, has been reported to be associated with graphite in the veins, but it is not clear whether these minerals were deposited with the graphite or merely minerals of the wall rocks accidentally enclosed in the veins. Resources of graphite in Sri Lanka are probably very large.

Mexico. The most important deposits of amorphous graphite in the western hemisphere occur in the State of Sonora, Mexico, in an area south and east of Hermosillo and another south and east of Navajoa. In addition to these strictly amorphous deposits, there are a few deposits like the one at the Lourdes mine which has supplied vein graphite. Substantially more than half of the total state production formerly came from the Lourdes mine at Moradillas, now closed. These vein deposits, interlayed with sandstone, consist of steeply dipping north-south striking graphite veins ranging from small stringers to seams that at depth average 3 to 4 m wide. Some veins more than 6 m wide were mined. The graphite content averaged 85 to 90%.

Other. Vein deposits have also been reported from southern India (e.g., in the Madras Province).

Contact Metasomatic or Hydrothermal Deposits in Marble

Concentrations of graphite occur in various parts of the world in silicified carbonate rocks. Some are clearly contact metasomatic; others show gradations to typical hydrothermal deposits. As a group, they are characterized by variable, but generally small size, variability in flake size and content, and irregularity of form. The average grade of most deposits is low, and, consequently, they have accounted for only a minor part of total world production.

In the United States and Canada, deposits of this type occur at a number of places in the Grenville Series in northern New York, southwestern Quebec, and southeastern Ontario. Small deposits west and northwest of Ticonderoga, NY, were worked between 1850 and 1900. They consist of fine-to-coarse flake graphite irregularly concentrated in silicified marble adjacent to bodies of pegmatite.

REGIONAL OCCURRENCE, PRODUCTION, AND USE

Asia

Republic of Korea. The Republic of Korea is one of the world's largest producers of natural graphite and produces both amorphous and flake. In 1961, the ratio was about 7 of amorphous to 1 of flake, steadily increasing to the current of 100 amorphous to 1 flake.

The principal mining areas are in the provinces of Chung Nam, Kyong Gi, Lyung Pak, and Kang Won. All mines are fairly near the Seoul area. Almost all the deposits are in the vicinity of anthracite coal mines, which lends support to the idea that the amorphous graphite is derived from coal. Some of the graphite produced has a fairly high carbon content. The refined amorphous graphite is 80% carbon or higher. One sample of amorphous lump was 87.86% carbon. Material with 90% carbon remaining on a 60-mesh sieve has been mined.

The Korean government sometimes restricts the exportation of anthracite coal, and it may be shipped as graphite. Therefore, the actual tonnage of graphite exported is probably less than the reported figure.

Sri Lanka. Sri Lanka contains the largest known deposits of vein graphite. Very large deposits are located in the western and southwestern sections of the island.

The extent of the deposits has inspired much study, but, although the resulting literature is extensive, little is known about the origin of the vein graphite. According to Wadia (1943):

> The vein graphite deposits are believed (by some) to have been formed by reduction, at high temperatures and pressures, of carbon dioxide liberated during the assimilation of limestones and dolomites by charnockitic and related igneous magmas.

The structures of these ore bodies contrast markedly with those of other forms of graphite and are typical of ores deposited as vein, dike, pegmatitic, or pneumatolytic occurrences.

Almost all of the high quality vein graphite comes from Sri Lanka. Despite rising labor costs, a considerable proportion of the production is still prepared by hand cobbing and hand sorting to satisfy buyers who still retain a preference for lump grades. Because most graphites are utilized in industry as a powder, the present operators anticipate a trend toward the direct supply of the powder from the mine to the importer. The requirements for hand sorting would be reduced progressively.

Sri Lankan graphite is highly graphitized; the best grades are completely graphitized. It is favored for lubrication, pencils, electromotive brushes, and other uses requiring quality graphite. The State Graphite Corp. of Sri Lanka is in charge of the mining operations and the sale of the Sri Lanka graphite.

China. Graphite is produced in a number of provinces, of which two are by far the most significant. Shandong Province is the largest flake graphite producer, with a producing capacity around 60 kt. Its major mines are the Nan Shu (largest), Bai Shu, and Dong Guan. Bai Shu also has an exfoliated graphite plant. Heilongjiang Province is the other major producer, from three mines east of Harbin at Jixxi: the Liu Mao, Lin, and Moshuan mines. The productive capacity of Heilongjiang Province is around 35 kt. Inner Mongolia produces graphite, mostly from the Xing Ho mine. Shanxi Province produces graphite from the Guodian and Deisheng mines. Hunan and Guandong Provinces also produce graphite. China has very low production costs, which allowed it to carve out a major niche in the world market in the 1980s. While domestic consumption of graphite has increased rapidly and infrastructure problems have appeared, Chinese graphite exports are likely to continue to increase.

India. The major graphite-producing states are Orissa and Rajasthan, but Andra Pradesh, Bihar, Gujarat, and Tamil Nadu also produce. Orissa usually accounts for 65 to 75% of all Indian production. Major firms include Agrawal Graphite Industries, with mines near Ganjaudar and Temrimal in Orissa State, and TP Minerals Pvt. Ltd., with mines near Phulbani, Madagudarf, and Sargipali, all in Orissa State. Both firms produce a variety of crystalline flake and powder graphites, which mostly goes to a wide variety of domestic end uses. The most important Indian end uses are in crucibles and foundries.

Indian graphite deposits usually occur in garnet-sillimanite-muscovite gneisses of the Khondalite Group. Of the total reserves of recoverable graphite, 35% are located in Kerala State, 31% in Bihar State, 13% in Orissa State, and the balance elsewhere.

Europe

Austria. A mineralogical region in Styria makes Austria a major producer of natural graphite. The graphite is thought to have originated by the metamorphism of carboniferous and bituminous substances under high pressure and temperature, during deformation. The deposits, therefore, occur in comparatively small lenses and, with one exception, must be mined underground.

The major mines in Styria, some of them two centuries old, are near Kaiserberg and Triben. Geologically, these Alpine deposits are embedded in a zone of fine-grained, sçaly schists in lenses and individual beds as much as 10 m thick. The ore is black, soft, fine-grained, and dense, with carbon content of 40 to 88%. It is low in sulfur (0.2 to 0.3%) and iron and practically free of carbonate and phosphate.

The first and only flotation plant for amorphous graphite is at Kaiserberg. The process was developed in the laboratories of Franz Mayr-Melnhof & Co's Grafitbergbau Kaisersberg, where fine (2μ average particle size) graphite with 92% carbon is produced, suitable for lubrication, pencils, pigment, and drilling mud.

Germany. The Passau district (Bavaria) has long produced flake graphite suitable for crucibles. Old documents frequently refer to this graphite, which probably furnished the graphite in the Middle Ages for alchemists' crucibles.

Today, Graphitewerk Kropfmuhl A.G. produces flake graphite exceeding 96/97% graphitic carbon suitable for crucibles, pencil leads, and lubricants. This company also chemically purifies graphite to +99.9% carbon in several size grades. The country rock is a part of the "kristallines Grundgebirge," the old gneissic and schistose rocks of the Bohemian massif; the gneissic rocks are believed to have been metamorphosed during the Carboniferous period.

The firm mines one 1.5-m thick graphite bed that is so heavily folded that it can be intersected eight different times in depth at many locations. Thicker layers occur in places because of folding. The ore runs 20 to 25% graphitic carbon. It is beneficiated and processed by flotation, grinding, and screening.

Czechoslovakia. Graphite is mined in two areas, one in Bohemia near the Austrian border and one in Moravia near the Polish border. Both mine fine-grained crystalline flake, or more usually, crystalline dust from metamorphic host rocks, mostly gneiss or schist. In Bohemia, the ore contains 15% graphite; its benefication by flotation creates a product containing 80 to 96% graphite, and a final leaching yields a 99.99% graphite product. In Moravia, the ore contains 30 to 40% graphite.

USSR. Graphite occurs in a number of places in this country. Locations of particular significance are the Urals, Siberia, Uzbekistan, and particularly the Ukraine. Reliable sources indicate that 75 to 90% of the graphite produced is crystalline flake, almost all fine flake. Until the early 1980s the country was a significant exporter to most world markets, including the United States.

The biggest producer, probably accounting for 75% of total production or 60 kt, is the Zavalyevsky complex in the Ukraine, near Gayvorgon. The mine is an open pit operation on steeply dipping ore bodies that range from 3 to 35 m thick and produces ore grading from 3 to 7% crystalline flake graphite. The benefication plant also produces colloidal graphite products, lubricants, and high purity graphite containing between 99 and 99.9% graphite.

Norway. A/S Skaland Grafitverk has been the only operating mine in Norway; currently closed, it is expected to be reopened. The deposit is on Senja Island in northern Norway. Flotation brings the carbon content up to 80 to 93%.

Spain. Although Spain produces little or no graphite, there are large deposits in the province of Jaen and near Toledo.

North America

United States. Graphite is no longer produced in the United States. The one flake producer, Southwestern Graphite Co., in Burnet, TX, closed its mine operation in 1980. The graphite plants in the United States process imported graphites that are beneficiated, ground, and blended as required to maintain the individual company grades.

Flake graphite deposits in both Alabama and Pennsylvania have been worked intermittently; the last Alabama mine closed in 1953 and the last Pennsylvania mine in 1961. Production of amorphous graphite ceased in Rhode Island in 1959. An amorphous graphite-like material was produced in Montana in the 1980s.

Canada. Canada is endowed with a sizable number of flake graphite deposits, almost all in Quebec or Ontario. These deposits are associated with Grenville Series marbles and gneisses. The most important producer by far is the mine and plant of Stratmin Inc. at Lac-des-Iles, Quebec. The Cal Graphite Corp. operation at Huntsville, Ontario, is producing at a low level; the present graphite oversupply will make it very difficult for it to reach the large capacity the firm was hoping for. The Exploration Graphicor Inc. operation next to Stratmin's produced very small amounts of graphite and is now closed. Societe d'Exploration Miniere Mazarin Inc. has a good deposit at Fermont, Quebec, presently underdeveloped.

Stratmin Inc. operates an open pit mine on reserves totalling 23.7 Mt averaging 7.5% graphite. With this feed, the firm is able to produce as much as 25 kt of concentrate running 92 to 98% graphite. Stratmin operated the Asbury Graphite Inc. mine and mill nearby for several years before closure. In November 1988, Stratmin leased the Asbury Graphite operation and contracted to supply Asbury with at least 10 kt of graphite each year for five years.

Mazarin has extensively drilled its Fermont deposit and estimates that its reserves for the first 20 years of mining total 3 Mt averaging 17% graphite. Additional reserves total 5.8 Mt averaging 14.2% graphite. The firm is looking for partners or financing for an operation with a capacity of over 20 kt. The deposit is near the railroad to the iron ore towns of Quebec from Sept Iles and has good access.

Mexico. The state of Sonora contains extensive deposits of amorphous graphite and most of Mexico's graphite mines and mills are located here. The Sonoran deposits are homogeneous, contain about 80% graphitic carbon, and are an important source of amorphous graphite. They were discovered in 1867 and exploitation began in 1891. There are said to be as many as seven distinct beds of graphite within alternating layers of metamorphosed andalusite-bearing Triassic rocks.

The operating mines are located about 400 km southeast of the US border in the region of Moradillas. This region is arid, and water for the operations is scarce. Other amorphous graphite deposits are reported in the states of Baja California and Coahuila in northern Mexico, and Hidalgo and Guerrero in central Mexico.

Several crystalline flake deposits are known in Oaxaca State, one of which is worked at Telixtlahuaca. The owner, Grafito de Mexico SA de CV, beneficiates the ore nearby by flotation.

Africa

Malagasy Republic (Madagascar). The island of Madagascar is one of the most important producers of graphite because of the size and quality of its flake. Occurrences are widespread along the eastern coast near Tamatave, in the central section near Tananarive, and in a southern section between Betroka and Bekily. Present production comes from the deposits along the eastern coast. The main producers are Societe Miniere de la Grande Ile, Establissments Gallois, Establissments Izouard, Societe Louys, and Establissments Rostaing (Pettifer, 1980). Each company markets its own graphite; graphites from different deposits are not blended.

The deposits are in a region where lateritic deposits of iron and bauxite also are found. The graphite, being resistant to weathering, is found among the weathered residue. The graphite occurs in highly metamorphosed schists and gneisses that have weathered deeply to soft, ferruginous clays. The graphitic content of the original rock has been raised through the natural leaching process.

The crude ore contains 3 to 10% graphite, of which about two-thirds is large flake and one-third fine flake. Commercial interest is primarily in the large flake. The flake is large, strong, and flexible and is the best graphite for refractories. It is also favored for crucibles. The pyrometric cone equivalent (PCE) of the ash is cone 16 to 20 and is higher than the ash of other flake graphites.

The nongraphitic portion of the flake product is uniform and consists (after oxidizing) chiefly of a clayey residue with some quartz along with a sprinkling of accessory minerals such as mica, zircon, rutile, and epidote. Sulfur compounds are absent and iron is low.

Zimbabwe. This country's only producer is the Lynx mine near Karoi, although there are other nearby deposits. It is operated by its part-owner, a subsidiary of Graphitwerk Kropfmuhl A.G. of Germany. The ore is mined underground from a lenticular ore body running 20 to 35% crystalline flake graphite. The ore is beneficiated by flotation to make products containing 80 to 83% and 90 to 94% graphite.

Other African Sources. Namibia produced an 85% carbon graphite in the District of Bethanien, and has deposits in other locations.

Deposits are present in the Transvaal at Carolina, Mopani, and Steamboat.

Kenyan graphite is mined in open pits near Kanziku. The ore runs about 10% graphitic carbon.

Two mines, now closed, formerly operated in the Republic of South Africa, at Cumbu and Mutali. The ore is high in flake graphite, on the order of 20%.

Other Sources of Graphite

Brazil. The two major crystalline flake graphite mines are in the State of Minas Gerais, and both belong to Nacional de Grafite Ltda. The mines are at Itapecerica, near Belo Horizonte, and Pedra Azul, about 16 km from the Minas Gerais-Bahia boundary. Pedra Azul is by far the largest producer of the two. The ores at both mines are graphite-containing soils, the results of very extensive weathering of the original host rock. Therefore, no blasting is required before the ore is removed by front end loader or shovel. The ore at Itapecerica averages 20% graphite and occurs in numerous small pockets spread over an area of 130 sq km. Usually only a small number of these pockets are mined at one time. The ore at Pedra Azul averages 5 to 12% graphite, is mined at a number of faces and usually blended to a mill feed running 6 to 7% graphite. Beneficiation plants at each of these two locations use flotation to make a product. A chemical leaching plant that produces high purity (99.6%) graphite is located at Itapecerica. A sizeable newly discovered graphite deposit is about 50 km from Pedra Azul.

A handful of very small mines located in the States of Minas Gerais and Bahia produce unbeneficiated product sold as is.

Australia and New Zealand. There are some medium-sized deposits in Australia. Several occurrences are also reported in New Zealand.

EXPLORATION AND DEVELOPMENT

Prospecting has consisted primarily of outcrop examination, trenching, and sampling. This more or less standard procedure has from time to time been augmented by either churn drilling or diamond core drilling. However, the nature of disseminated deposits in particular makes accurate evaluation of grade and tonnage difficult to achieve by drilling alone.

During World War II several more integrated programs involved geological mapping of specific areas in Alabama and Pennsylvania in addition to trenching, pitting, and sampling. In 1942 there was some small production at one of the areas, the Benjamin Franklin Mine, in Chester County, PA; the ore was of low grade (3%), however, and mining ceased in 1943.

Because of the high electrical conductivity of graphite, geophysical methods such as resistivity and self-potential have been employed in the search for deposits in Japan, Australia, India, and North America.

MINING METHODS

During the period 1890 to 1920, underground mining of graphite was practiced in New York and Pennsylvania. From 1942 until the end of World War II, open pit operation was the only method used, primarily because of the comparative ease of working in weathered rock. Graphite was mined underground at Dillon, MT, during World War II and shortly thereafter until extraction of the ore became too costly to compete with Sri Lankan graphite and the operation ceased.

Madagascar operations are entirely open pit, but in Sri Lanka, Korea, Mexico, and Bavaria, underground mining is usually essential due to the depth and physical characteristics of the deposits.

Mexican underground mining operations are 100 to 400 m below the surface, measured on the angle of the vein. Some of the older mines in Sri Lanka reached depths in excess of 450 m on a vertical plane. For many years mining operations in Sri Lanka were primitive and ore extraction slow and cumbersome. The mines were mechanized after World War II.

Madagascar operations also were primitive because of the low labor cost. After 1938 the mines began to use mechanical equipment to remove the overburden, and the graphite-bearing schists were easily removed by bulldozers and tractors.

PREPARATION FOR MARKET

In Sri Lanka the graphite ore is extracted in lumps and hand cobbed on a sorting patio to remove quartz inclusions. The finished product is graded as follows: large lump, 97 to 99% carbon; ordinary lump, 94 to 96% carbon; ordinary lump, 90% carbon; chips, 85 to 90% carbon; and chippy dust, 80% carbon. However, the chippy dust and flying dust grades are not produced in any quantity today.

In Madagascar the crude ore averages 10 to 12% carbon. It is crushed by a primary crusher and then conveyed to a series of roll crushers and classifiers to remove the oversizes and gangue. The −6.35 mm product is then passed to flotation rougher cells, which separates the graphite flakes from the clay and sand. The product of the rougher cells passes to the finishing cells where each particle of graphite is recovered by means of a pine oil frothing process. This yields an 80 to 84% carbon concentrate, which is dried; any residual pine oil is volatized. The run-of-mill product is upgraded by mechanical means, such as rod mill grinding, to 0.3 mm sieve with 85 to 90% carbon. Higher carbon concentrates (up to 95%) are obtained by regrinding and flotation.

At the mill of Graphitwerk Kropfmuhl, S.A., in Bavaria, crude ores carrying 15 to 20% carbon are crushed, passed through a grinding system, and then a rod mill, which reduces the ore to 40 mesh. This product is fed to a series of flotation and grinding steps and the graphite concentrate is filtered and kiln dried with the final product averaging 90 to 93% carbon. For further refinement this product is milled and screened, producing grades with 95 to 96/97% carbon.

The above operations are fairly typical of those found in other countries. Some firms make high purity graphite by leaching concentrate with strong acids or alkalis.

USES

Batteries

Natural graphite is suitable for batteries that are required to furnish small currents over long periods of time. However, it is not particularly efficient in the type of battery currently used in flashlights and has been replaced by a form of carbon black. Since carbon black is satisfactory and cheap, there seems to be little opportunity for natural graphite to reverse this downward trend through product improvement.

Bearings

Graphite is used extensively in bearings but the potential for increased use varies with the type of bearings being produced.

Brake Linings

Graphite consumption as an asbestos substitute increased greatly in the 1980s. As long as organic friction materials impregnated with graphite are used and accepted as satisfactory, the consumption of graphite should increase gradually in accordance with the increased number of vehicles on the road. Graphite is particularly useful in brake linings for heavy duty vehicles as opposed to passenger vehicles.

Carbon Brushes and Other Electrical Uses

Natural graphite encounters considerable competition from artificial graphite for this use. It seems that wherever artificial graphite has advantages it has already captured the market. For example, in the production of sliding contacts the variation in the ash content of natural graphite has proved unacceptable; now artificial graphite in which the abrasive materials can be controlled is being used.

Crucibles

Crystalline flake graphite consumption to make crucibles has been steadily declining in market share over the last decade. Changes in crucible manufacturing technology now allow much greater flexibility in type and flake size of the graphite utilized.

Expanded Graphite

This use has recently been growing rapidly. Expanded graphite is used to make high performance gasket material for use at high temperatures or pressures and with radioactive or corrosive fluids. Expanded graphite can be made into a foil laminate that is used in valve packings and as a gasket material. Expanded graphite is being used as a hot topping compound to insulate hot ingots and decrease heat loss. Expanded graphite is made from crystalline flake graphite by treating it with chromic acid followed by sulfuric acid to weaken the bonds between the graphite flakes. The graphite is then heated to drive the water off, causing expansion, and giving a product analogous to that observed in vermiculite.

Foundries

One of the larger uses for graphite is in foundries. Graphite has a number of uses in foundries, but the largest single one is as a foundry facing or mold wash that allows the metal casting to be easily removed from the mold on cooling. Another significant use in foundries is as a carbon raiser, a carbonaceous material added to hot metal to raise its carbon content.

Lubricants

This market is tremendously complex because of the variety of lubricating jobs that must be done and the imposing array of types of lubricants that are available. Graphite has been an important lubricant and will probably continue to be, but competition is stiff. Graphite added to a soapy lubricant is considered to be a good die lubricant, particularly for galvanized wire, though the discoloration it produces is not a desirable feature. Natural graphite is a good anti-seize agent in steel mill and railroad applications because it creates a solid lubricant film that retains its properties under pressure and high temperature. In this particular use the graphite content is specified by the customer; hence its use may not be based on strictly technical grounds. Natural graphite is used in grease where the application is such that the siliceous portion of the natural graphite is overshadowed by the indigenous contamination of dirt and sand, i.e., gear lubrication for mining machinery.

Graphite of colloidal size dispersed in kerosene, oil, or another carrier, can be used when a film with a quick-drying characteristic is desired in high-temperature lubricating jobs, i.e., conveyors passing through ovens.

Packings

Fine flake graphite used with oils or greases is very satisfactory because of its resistance to heat up to 310°C. Closely controlled particle size is not too important for packing use. Use of graphite will tend to remain relatively steady, but it may decline because equipment being manufactured today doesn't utilize old style pack-

Table 1. Salient Natural Graphite Statistics

		1986	1987	1988	1989	1990
United States:						
Production	tons	—	—	W	W	—
Apparent consumption	do	31 784	31 634	42 799	50 867	W
Exports	do	7 034	11 700	11 068	11 452	11 537
Value	thousands	$3,416	$6,218	$5,815	$7,421	$9,481
Imports for consumption	tons	38 818	43 334	53 867	62 319	50 213
Value	thousands	$15,758	$17,654	$23,238	$33,707	$35 222
World: Production	tons	624 718	648 156	660 168	648 827[p]	660 600[e]

[e]estimated, [p]preliminary, W withheld to avoid disclosing company proprietary data.

Table 2. US Consumption of Natural Graphite By Use

Use	Crystalline		Amorphous[1]		Total[2]	
	Quantity (tons)	Value (thousands $)	Quantity (tons)	Value (thousands $)	Quantity (tons)	Value (thousands $)
1988[3]	18 897[r]	$23,871[r]	17 708[r]	$12,978[r]	36 605[r]	$36,848[r]
1989:						
Batteries	W	W	W	W	1 243	1,924
Brake linings	2 281	2,436	2 427	5,450	4 708	7,886
Carbon products[4]	340	886	167	242	507	1,128
Crucibles, retorts, stoppers, sleeves, nozzles	W	W	W	W	1 497	1,697
Foundries[5]	408	481	4 847	1,412	5 255	1,893
Lubricants[6]	3 391	5,033	3 125	1,068	6 516	6,101
Pencils	1 510[r]	1,876	248	150	1 758[r]	2,026
Powdered metals	1 401	2,551	58	111	1 459	2,662
Refractories	W	W	W	W	12 599[r]	11,985[r]
Rubber	72	108	406	314	478	422
Steelmaking	218	122	1 035	484	1 253	606
Other[7]	2 108	7,542	189	284	2 297	7,826
Withheld uses	7 516[r]	9,260[r]	7 823	6,346	—	—
Total[2]	19 245[r]	30,295	20 325[r]	15,861[r]	39 570[r]	46,156[r]
1990:						
Batteries	W	W	W	W	319	505
Brake linings	1 520	2,195	3 808	4,331	5 328	6,526
Carbon products[4]	243	647	114	142	358	789
Crucibles, retorts, stoppers, sleeves, nozzles	W	W	W	W	1 391	1,716
Foundries[5]	825	878	3 348	1,024	4 174	1,902
Lubricants[6]	3 999	6,326	3 540	1,397	7 539	7,723
Pencils	964	1,339	237	128	1 201	1,467
Powdered metals	1 278	2,691	37	66	1 315	2,757
Refractories	4 024	5,823	7 610	5,594	11 634	11,417
Rubber	W	W	W	W	421	329
Steelmaking	W	W	W	W	1 174	657
Other[7]	736	1,845	744	740	1 480	2,585
Withheld uses	1 839	2,333	1 467	874	—	—
Total[2]	15 428	24,077	20 905	14,296	36 334	38,373

[r]revised, W Withheld to avoid disclosing company proprietary data; included with "Withheld uses," [1]Includes mixtures of natural and manufactured graphite, [2]Data may not add to totals shown because of independent rounding, [3]Revision reflects a data correction in the refractories and other use categories, [4]Includes bearings and carbon brushes, [5]Includes foundry facings, [6]Includes ammunition, packings, and seed coating, [7]Includes paints and polishes, antiknock and other compounds, soldering and/or welding, electrical and electronic products, mechanical products, magnetic tape, small packages, industrial diamonds and drilling mud.

ings. One consumer has mentioned that very small quantities of molybdenum offer excellent lubricating qualities.

Pencils

There has been some shift from Sri Lankan graphite to Mexican amorphous graphite because of the grayish cast of the Sri Lanka variety. Crystalline flake use has also grown. The pencil industry seeks to make improvements that would lead to greater jetness, higher carbon content, smaller particle sizes, and more unctuousness and opaqueness in the graphite. However, such improvements must not add to the cost of graphite to the pencil manufacturer.

Refractories

Graphite is used in both plastic (castable) refractories, principally in ramming and gunning mixes, and shaped refractories, principally carbon-magnesite brick and alumina graphite continuous casting ware. Over the last ten years, graphite-containing refractories have gained marketshare at the expense of other refrac-

Table 3. Graphite: World Production, By Country[1] (tons)

Country[2]	1986	1987	1988	1989[p]	1990[e]
Argentina	40	216	24	100	100
Austria	36 167	39 391	7 557	15 307	17 500
Brazil (marketable)[3]	28 586	31 404	34 520	31 700	32 000
Burma[4]	722	—	—	—	—
China[e]	185 000	185 000	200 000	200 000	200 000
Czechoslovakia[e]	25 254[5]	25 000	25 000	25 000	25 000
Germany, former Federal Republic of	13 233	9 891	9 666	7 000[e]	8 000
India (mine)[6]	38 412	42 589	57 325	47 731	50 000
Korea, North[e]	25 000	25 000	25 000	35 000[r]	35 000
Korea, Republic of:					
Amorphous	96 577	106 507	107 767	100 282	100 000
Crystalline flake	641	838	678	1 186	1 000
Madagascar	16 187	13 169	14 706	15 863	16 000
Mexico:					
Amorphous	36 018	36 674	47 871	38 304	35 000
Crystalline flake	1 838	1 787	1 735	1 942	2 000
Norway	—	—	—	1 800	5 000
Romania[e]	12 000	12 000	12 000	10 000	20 000
Sri Lanka	7 453	9 400	8 547	4 163	4 000
Turkey (mine)	3 586	11 760	12 911	11 302	12 000
USSR[e]	83 000	84 000	84 000	84 000	80 000
United States	—	—	W	W	—
Zimbabwe	15 004	13 530	11 441	18 147	18 000
Total	624 718	648 156	660 168	648 827	660 600

[e]estimated, [p] preliminary, [r]revised, W withheld to avoid disclosing company proprietary data, [1]Table includes data available through May 9, 1991, [2]In addition to the countries listed, Namibia may have produced graphite during the period covered by this table, but output is unreported and available general information is inadequate for formulation of reliable estimates of output levels, [3]Does not include the following quantities sold directly without beneficiation, in tons: 1986—19 074; 1987—10 505; 1988—18 269, 1989—20 000 (estimated); and 1990—20 000 (estimated). [4]Data are for fiscal year beginning Apr. 1 of that stated, [5]Reported figure, [6]Indian marketable production is 10 to 20% of mine production.

Table 4. World Graphite Resources (thousand tons of recoverable flake or crystalline concentrates or of amorphous graphite)

Country	Type	Reserves	Reserves base[1]	Inferred reserve base
Austria	amorphous	50	50	1 100
Brazil	flake	500	1 000	2 800
Canada	flake	1 500	2 700	3 000
China	amorphous	3 000	300 000	
	flake	2 500	7 500	4 000
Czechoslovakia	flake	900	2 300	
Germany, Federal Republic of	flake	130	220	600
India	flake	735	735	10 400
Korea, North	amorphous	1 000	11 000	20 000
	flake	700	1 000	1 100
Korea, Republic of	amorphous	3 000	20 000	25 000
	flake	160	214	620
Madagascar	flake	980	980	180 000
Mexico	amorphous	3 000	3 000	10 000
	flake	106	106	320
Norway	flake	200	200	40
Romania	amorphous	300	300	—
Sri Lanka	crystalline[2]	50	50	150
USSR	amorphous	1 000	21 000	540 000
	flake	1 800	3 800	5 000
United States[3]	amorphous	—	1 000	20 000
	flake	—	750	400
Zimbabwe	flake	600	600	1 200
Other	flake	280	300	900
	Total amorphous	11 350	356 350	616 100
	Total flake	11 090	22 405	210 380
	Total crystalline	50	50	150

[1]The reserve base includes reserves; the inferred reserve base does not include the reserve base, [2]Known as high crystalline or lump, comprising crystalline lump and amorphous lump, [3]Reserve base is Alabama only, inferred base is Alaska plus New York.

Table 5. Representative Yearend Graphite Prices ($ per ton)

	1990	October 1991
Industrial minerals:		
Crystalline large flake, 85 to 90% carbon	820–1,300	650–1,200
Crystalline medium flake, 85 to 90% carbon	770–1,120	450–1,000
Crystalline small flake, 80 to 90% carbon	540–900	400–600
Powder (200 mesh), 95 to 97% carbon	770–1,000	770–1,000
Powder (200 mesh), 97 to 99% carbon	1,000–1,300	1,000–1,300
Amorphous powder, 80 to 85% carbon	220–440	220–440
Custom value, at foreign ports:		
Flake	935	617
Lump and chip, Sri Lankan	1,216	1,458
Amorphous, Mexican	116	119

Source: Industrial Minerals, No. 267, Dec. 1989, p. 82, and No. 279, Dec 1990, p. 78.

tories as the need for high-performance ware has grown. Such ware is principally needed for high temperature, thermal shock, and corrosion resistance. The steel industry consumes around 70% of refractories worldwide.

Gunning and ramming mixes, used where the refractory must be deformable or flexible, commonly use amorphous graphite. Carbon-magnesite mixes are made from crystalline flake graphite.

Carbon-magnesite brick is used to line basic oxygen converters, in the slaglines and water-cooled sidewalls of electric arc furnaces, and to make steel ladles. Graphite consumption in the first two uses grew rapidly in the 1980s, and consumption in the third use is still growing significantly as ladles have to hold hot metal for longer times and at higher temperatures. Use in basic oxygen converters is still growing in some countries.

Continuous casting ware made of alumina graphite comprises items such as shrouds, sub-entry tubes, and stopper rods. Alumina graphite use could grow rapidly in torpedo ladles.

BASIC DATA

Tables 1 through 5 present data on graphite for production, consumption, imports, and exports and prices for the United States, plus world production and reserves.

ACKNOWLEDGMENT

The author has drawn heavily from the chapter on Graphite by George D. Graffin in the 5th edition of *Industrial Minerals and Rocks.*

BIBLIOGRAPHY AND REFERENCES

Alling, H.L., 1917, "The Adirondack Graphite Deposits," Bulletin 199, New York State Museum, 150 pp.

Bastin, E.S., 1912, "The Graphite Deposits to Ceylon, with a Description of a Similar Graphite Deposit near Dillon Montana," *Economic Geology,* Vol. 7, No. 5 Aug., pp. 419–443.

Bonneau, J., and Raby, R. 1990, "The Lake Knife Graphite Deposit," *Mining Magazine,* Vol. 163, No. 1, July, pp. 12–18.

Cameron, E.N., 1960, "Graphite," *Industrial Minerals and Rocks,* 3rd ed., J.L. Gillson, ed., AIME, New York, pp. 455–468.

Cameron, E.N., and Weis, P.L., 1960, "Strategic Graphite, a Survey," Bulletin 1082-E, US Geological Survey, pp. 250–252.

Clark, T.H., 1921, "The Origin Of Graphite," *Economic Geology,* Vol. 16, No. 3, Apr.–May, pp. 167–183.

Ford, R.B., 1954, "Occurrence and Origin of the Graphite Deposits near Dillon, Montana," *Economic Geology,* Vol. 49, No. 1, Jan.–Feb., pp. 31–43.

Goffaux, D.P., 1992, "Stratmin Graphite Inc., A Graphite Mine Study," SME Preprint #92-104, AIME Annual Meeting, Phoenix, AZ, Feb., 21 pp.

Gupta, B.C., 1949, "Graphite," *Indian Minerals,* Vol. 3, No. 1, pp. 17–31.

Jones, W.B., 1929, "Summary Report on Graphite in Alabama," Circular 9, Alabama Geological Survey, 29 pp.

Kemberling, S.M., and Walker, P.L., Jr., 1961, "Progress Report on Natural Graphite Studies," Fuel Technology Dept., The Pennsylvania State University, University Park, PA, pp. 1–6.

Klar, G., 1958, "The Important Graphite Deposits of the World," *Mining Magazine,* pp. 98, 137–142.

Murdock, T.M., 1963, "Mineral Resources of the Malagasy Republic," Information Circular 8196, US Bureau of Mines, pp. 112–121.

Needham, A.B., 1946, "Mining and Milling Operations of Southwestern Graphite Co., Burnet County, Texas," Information Circular 7339, US Bureau of Mines, 3 pp.

Overstreet, W.C., 1947, "Graphite Deposits of Southern Korea," *Economic Geology,* Vol. 42, No. 4, June–July, pp. 424–425.

Pallister, H.D., and Thoenin, J.R., 1948, "Flake Graphite and Vanadium Investigations in Clay, Coosa, and Chilton Counties, Alabama," Report of Investigation 4366, US Bureau of Mines, 84 pp.

Pettifer, L., 1980, "Natural Graphite—The Dawn of Tight Markets," *Industrial Minerals,* No. 156, Sep., pp. 19–39.

Pettifer, L., 1980a, "Synthetic Graphite-Electrodes Continue to Lead," *Industrial Minerals,* No. 158, Nov., pp. 19–31.

Russell, A., 1988, "Graphite-Current Shortfalls in Flake Supply," *Industrial Minerals,* No. 255, Dec., pp. 23–43.

Schanz, J.J., Jr., 1960, "Natural Graphite Market Research," Progress Reports Nos. 1, 2, Jan. 1–Nov. 30, College of Mineral Industries, The Pennsylvania State University, University Park, PA.

Shobert, E.I., II, 1964, "Natural and Synthetic Graphites," SME Preprint #64-H56, AIME Annual Meeting, New York, Feb., 14 pp.

Seeley, S.B., 1963, "Natural Graphite," *Encyclopedia of Chemical Technology* Vol. 4, 2nd ed., Kirk-Othmer, ed., Interscience Publishers, New York, pp. 304–335.

Taylor, H.A., Jr., 1986, "Graphite," *Mineral Facts and Problems, 1985,* Bulletin 675, US Bureau of Mines, pp. 334–348.

Taylor, H.A., Jr., 1989, "Graphite," *Minerals Yearbook, 1988,* US Bureau of Mines, pp. 1–11.

Taylor, H.A., Jr., 1991, "Graphite," Annual Report, 1990, US Bureau of Mines, pp. 1–10.

Tron, A.R., 1964. *The Production and Uses of Natural Graphite,* Her Majesty's Stationery Office, London, Feb., 83 pp.

Wadia, D.N., 1943, "Origin of the Graphite Deposits of Ceylon," Professional Paper 1, Records, Dept. of Mineralogy, Ceylon, Pt. 2, pp. 15–24.

Weis, P.L., 1973, "Graphite," *United States Mineral Resources,* Professional Paper 820, US Geological Survey, pp. 277–283.

Weis, P.L., and Salsa, G.A., 1978, "Estimating Reserves of Amorphous Graphite in Sonora, Mexico," *Engineering and Mining Journal,* Vol. 174, No. 10, pp. 123–128.

Winchell, A.N., 1916, "A Theory for the Origin of Graphite as Exemplified in the Graphite Deposits near Dillon, Montana," *Economic Geology,* Vol. 6, pp. 218–230.

GYPSUM AND ANHYDRITE

DUANE B. JORGENSEN

The two calcium sulfate minerals, gypsum and anhydrite, occur in many parts of the world, and gypsum has long been of economic importance in the family of industrial minerals. Gypsum, the dihydrate form of calcium sulfate ($CaSO_4 \cdot 2H_2O$) and anhydrite, the anhydrous form ($CaSO_4$) are frequently found in close association, and it is seldom that a calcium sulfate deposit will consist exclusively of one mineral or the other.

Although known gypsum deposits are extensive, anhydrite makes up the larger part of total calcium sulfate resources. Although it has very minor economic use, it is important to the understanding of gypsum deposits and is referred to frequently in the following discussion.

Calcium sulfate is one of the principal constituents of evaporite deposits, and when pure, has the following composition:

	Lime (CaO)	Sulfur Trioxide (SO_3)	Combined Water (H_2O)
Gypsum	32.6%	46.5%	20.9%
Anhydrite	41.2%	58.8%	0.0%

Deposits of pure gypsum or of pure anhydrite which are large enough to be considered commercial have not yet been found because of both the metastable relationship between the two minerals, and the presence of impurities such as calcium or magnesium carbonates, chlorides, other sulfate minerals, clay minerals, or silica. Most gypsum produced ranges from 80 to 95% pure and beneficiation is normally limited to selective mining or quarrying or upgrading by crushing and screening during mill processing.

END USES

The largest use for gypsum is based upon the unique property which calcium sulfate has of readily giving up, or taking on, water of crystallization. With the application of a moderate amount of heat in a process known as calcining, gypsum is converted to plaster of paris, the hemihydrate of calcium sulfate ($CaSO_4 \cdot \frac{1}{2} H_2O$), which when mixed with water will set or harden as the calcium sulfate returns to the stable dihydrate form. This semifinished product, usually called stucco, is then manufactured into a large variety of plasters, wallboard, and block for construction use, or into plasters for industrial applications. About 75% of the gypsum used in the United States is calcined for these purposes.

Uses of uncalcined gypsum are principally as a retarder for portland cement, as a soil conditioner, as a mineral filler, and other minor industrial applications. About 25% of the gypsum mined in the United States goes into these markets; however, in other countries with building practices different from those in the United States and Canada, relative usage varies widely.

Calcium sulfate deposits are the world's largest sulfur resources; minor quantities of gypsum and anhydrite have been used to produce sulfur or sulfur compounds. This use is accompanied by a unique site-specific set of economics because sulfur is generally available from non-gypsum sources at lower cost.

MINERALOGY

Anhydrite, $CaSO_4$, because of its geologically rapid conversion to gypsum and relatively high solubility (about 0.2 g per 100 g H_2O), is not often found outcropping in climates wet enough to support abundant vegetation where gypsum, the hydration product, is continuously removed. Anhydrite rock is most often light to bluish gray in color, and under the microscope, varies from granoblastic (ver Planck, 1952) to felty-lath crystal aggregates. Fig. 1 shows felty-lath anhydrite crystals from Shoals, IN.

Bassanite, $CaSO_4 \cdot \frac{1}{2} H_2O$, is a distinct phase intermediate between anhydrite and gypsum, but is identifiable only by X-ray diffraction or petrographic techniques using very carefully prepared samples. Bassanite is metastable under ordinary conditions (Wood and Wolfe, 1969); however, its occurrence in amounts under 1% is suspected to be widespread in calcium sulfate mineral deposits. Bassanite is natural stucco.

Gypsum, $CaSO_4 \cdot 2H_2O$, is the commonly observed calcium sulfate mineral and has diverse origins. It is easily distinguished from anhydrite by its inferior hardness (2.0 vs. 3 to 3.5) and lower specific gravity (2.2 to 2.4 vs. 2.7 to 3.0). Most gypsum is white to grayish white although the impurities in any given deposit frequently determine the color of the rock and resulting products. Petrographically, most rock gypsum is granoblastic (Goodman et al., 1957). Commercial deposits show a great variety of crystal sizes and textures, both from deposit to deposit and within a single deposit. Relatively undisturbed deposits often contain texturally consistent stratigraphic units, but in deposits that have been deformed the gypsum is often recrystallized and exhibit porphyroblastic textures wherein two distinct sizes and ages of crystals occur (Holliday, 1970), and in some deposits fibrous gypsum similar to felty-lath anhydrite makes up the mass (Van't Hoff et al., 1903).

Alabaster is a compact, very fine grained variety of rock gypsum prized by sculptors for its uniform workability and occasionally is found within commercial deposits. Fibrous gypsum composed of needle-shaped crystals in orientation parallel to the C-axis (Schmidt, 1914), is known as **satin spar** and is a type of stress mineral indicative of deformation. It is universally a secondary mineral and occurs widely as a fracture filling wherein the needle lengths are perpendicular to the fracture walls, and less commonly in shear zones wherein the needles are parallel to the direction of movement. Large euhedral gypsum crystals and large cleavages commonly known as **selenite** form in fluid-filled spaces or in an easily deformable host material, and are sometimes mistakenly identified as mica. Both satin spar and selenite are accessory

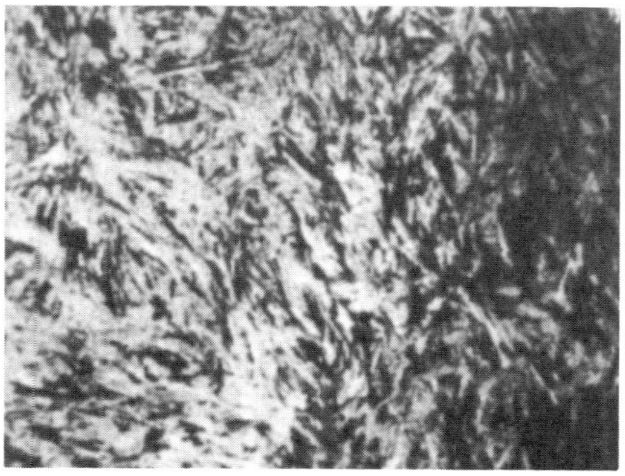

Fig. 1. Felty-lath anhydrite from Shoals, IN, × 39 m polarized light.

Fig. 2. Selenite crystal from Sperry, IA.

minerals in rock gypsum deposits and do not commonly form the bulk of workable deposits. Notable exceptions are the selenitic deposits of the San Joaquin Valley, California and the huge selenite deposits in Spain. Fig. 2 shows selenite from Sperry, IA.

ORIGIN AND OCCURRENCE

The calcium sulfate minerals are deposited by precipitation from aqueous solution when the concentration of the components and the physical conditions are suitable, with the majority of the deposits originating from evaporation and concentration of marine brines in a dry climate, i.e., evaporite conditions. That this is the basic mechanism of formation of economic calcium sulfate deposits is unquestioned; however, there is considerable discussion in the literature as to the physical environment in which precipitation takes place and as to whether the originally precipitated mineral is gypsum or anhydrite.

The older, well established concept of deposit formation idealized a brine-filled basin having restricted circulation (Branson, 1915, King, 1971) which permits limited replenishment of the brine as it evaporates. Increasing the brine concentration leads to precipitation of the contained salts in the inverse order of their solubility, and substantial thickness of essentially monomineralic deposits (e.g., calcium sulfate) are visualized as being derived from this type regime, an example of which is the A subzone of the Mississippian Windsor series of southeastern Canada.

A more recent concept of evaporite deposition has resulted from considerable work done in the Trucial Coast of Arabia (Butler, 1970, Kinsman, 1966, 1969) which describes the sabkha supratidal environment wherein calcium sulfate minerals are currently being deposited (*sabkha,* an Arabic word denoting salt-flat). These deposits are characterized by a distinctive sequence of sediments including lagoonal limestones, intertidal algal mat limestones, and precipitation of gypsum crystals and nodular anhydrite in the preexisting host sediments, which if originally calcium carbonate, become dolomitized. Clastic sediments, if present, are unaffected chemically, but are contorted and displaced as the gypsum crystals and nodular anhydrite grow in them. In the sabkha regime a basin is not required for deposition, although some concentration of the original seawater brine does take place in alongshore lagoons. Thick chloride accumulations would not be expected (Schroeder, 1970). The Mississippian evaporites of southwestern Indiana illustrates this type of deposit (Jorgensen and Carr, 1972). Most North American gypsum deposits were deposited in sabkha or similar depositional environments. The gypsum deposits of southwestern Virginia and south central Utah are examples of mud rich sabkha-type deposits.

The classical restricted-basin accumulations of sulfates and chlorides might be expected to have sabkha deposits along the margin of the basin, and the Silurian gypsum of Ohio, Ontario, and New York is possibly of this origin. The individual sulfate beds in sabkhas do not commonly exceed 7 to 13 m in thickness, although 70 m accumulations have been described in Jamaica, WI (Holliday, 1971).

Gypsum may also precipitate along fractures, bedding planes or in other available spaces where ground water carrying sulfate ions from the oxidation of sulfides comes into contact with carbonate rocks. Gypsum also occurs in minor amounts in some igneous and hydrothermal deposits. Such deposits have no influence on gypsum production economics.

Because gypsum will precipitate wherever brine concentration is appropriate, and recrystallizes or dissolves readily as the metastable or temporary environments change, interpretation of geologic evidence to determine exact depositional environments is difficult at best. Several recent works (Dean and Schreiber, 1978, Kendall, 1978, Raup, 1991, and others) are helpful in determining depositional environments.

Gypsite deposits usually occur in semiarid and arid climates, and result from the solution of existing gypsum deposits by ground water, which in turn is drawn toward the surface by capillary action where it evaporates leaving the dissolved sulfate precipitated as a porous aggregate of impure gypsum. Reworking of existing deposits by wind erosion has concentrated gypsum in sand deposits, as at White Sands, NM, or Cuatrocienegas, Coahuila, Mexico.

Of totally different origin are those calcium sulfate deposits which constitute part of the cap rock of salt domes in the Gulf Coast basin. The exact origin of these domes is not fully understood but most prevailing views involve flowage and upward movement of plastically deformed salt in response to overburden pressure from the overlying sediments (Martinez, 1974). Once the upward movement is initiated, it is possible for the top of the lighter salt mass to be injected into and through the overlying sediments.

The most widely accepted hypothesis for the origin of the cap rock is that it represents a residue of anhydrite and other relatively insoluble minerals which accumulate at the upper surface of the salt dome as it rises through water-bearing sediments and the salt is dissolved. The anhydrite crystals, which make up about 99% of the water insoluble residue (Walker, 1974), are then recrystallized and compacted to form a massive rock which may be locally hydrated to form gypsum.

The total height of a salt dome can exceed 3 000 m, with the cap rock being well over 300 m thick. Most domes are roughly circular in plan view, and range in diameter from about 1 km to more than 10 km, although there is great diversity in dome geometry and dome-to-cap-rock relationships. Most domes do not reach the present surface, although occasionally cap rock is found at or near the surface and might be of economic significance.

Edinger has compiled a very extensive bibliography and reference that includes brief abstracts of studies of and references to gypsum from ancient Greek times to about 1975.

DISTRIBUTION OF DEPOSITS

Calcium sulfate minerals have been found within every geologic system from Silurian through Quanternary. The largest production in North America is from Mississippian rocks in southeastern Canada, Michigan, Indiana, and Virginia, followed by that from Permian rocks in Texas, Oklahoma, Kansas, Colorado, Wyoming, Nevada, and Arizona. Production from Tertiary rocks is obtained in California, Jamaica, and Arizona; from Jurassic rocks in Iowa, New Mexico, Colorado, Utah, Wyoming, and Montana; and from Silurian rocks in New York, Ontario, and Ohio. Elsewhere, Tertiary rocks in Australia, France, Spain, and Italy; Permian and Jurassic rocks in England; and Triassic rocks in France, Germany, and England contain large commercial deposits. Commercial deposits range in thickness from 1 to 30 m and reserves of individual deposits are usually measured in millions of tons. Production rates commonly average over 300 ktpy from each operation, but also range from a few thousand tons to well over 2 Mtpy.

Although gypsum is produced from rocks of many ages, many deposits are Quaternary in age due to the recent and continuing conversion of anhydrite to gypsum that occurs under certain conditions. This conversion of anhydrite to gypsum, as well as the reverse reaction, has been extensively studied because of its importance in the original sequence of deposition from brines, as well as in diagenetic changes that occur in calcium sulfate deposits after deposition. Several reviews have been made (Braitsch, 1971, Hardie, 1967, MacDonald, 1953, Wood and Wolfe, 1969) of such variables as brine composition, temperature, water pressure, and overburden pressure. From these it is evident that no one set of conditions applies to all deposits. On the contrary, all these variables, plus structural, lithological, and topographical considerations, which might affect water migration, must be examined in order to understand the degree of hydration of any specific calcium sulfate deposit. To date, however, only limited attempts have been carried out in an effort to integrate theoretical models with field evidence of rock textures and structures.

Most commercial gypsum deposits are believed to have resulted from the action of surface and ground waters upon anhydrite, and hydration has been found at depths ranging from zero to more than 700 m below the surface. In many deposits, the degree of hydration is readily predictable once the variables controlling it are understood; however, in others the degree of hydration can vary widely, with corresponding reduction in the amount of available reserves, and increase in the cost of rock production. In all cases, the operator is well advised to thoroughly explore and understand the anhydrite-gypsum relationship prior to development of a mine or quarry.

IMPURITIES IN CALCIUM SULFATE DEPOSITS

The number of minerals that can occur in evaporite deposits is quite large (Braitsch, 1971), and many can be found as impurities in calcium sulfate deposits. Their occurrence is often dependent upon the mode of formation of the deposit. In varying degrees, most gypsum and anhydrite deposits contain terrigenous and calcareous clastic sediments, commonly clay minerals and fine sands, as well as chemical precipitates such as carbonates. In sabkha-type deposits, these preexisting sediments fill internodular spaces of the gypsum and anhydrite, or form crude layers, and often appear to be replaced when they actually are displaced. Relatively insoluble evaporite minerals such as celestite, certain borates, some carbonate minerals, and silica may occur in calcium sulfate deposits as discrete crystals, crystal aggregates or nodules, and probably are in many cases original depositional features. However, strontium or boron, which may be present in only trace amounts in anhydrite, may migrate during gypsification to result in strontium and boron mineral accumulations.

Soluble evaporite minerals, e.g., halite, sylvite, mirabilite, epsomite, and others are also frequently found in calcium sulfate deposits. These minerals components are expelled from anhydrite during gypsification, and most commonly become associated with the clay minerals that may be present, probably being adsorbed on the clay mineral surfaces. They also are associated with the carbonate impurities where they may be adsorbed onto mineral surfaces, held as fluid inclusions, or included in a disordered lattice. In addition, these soluble minerals can occur as microfracture filling. The hydration halo of mixed gypsum-anhydrite commonly contains many times the soluble salt content of either unhydrated anhydrite or hydrated gypsum. Individual deposits show great variation in the quantity and type of soluble salts present, i.e., sulfates or chlorides, and it has been noted that, in general, chloride salts predominate in gypsum deposits of eastern North America, whereas sulfate salts exceed chlorides in western North American deposits.

In addition to the foregoing evaporite mineral impurities which origin is generally a feature of original calcium sulfate deposition, impurities in any given mining operation may also be encountered as a result of structural features such as tight folding or faulting, or from recent erosional activity. These often are sands or gravels and occur most often in near surface deposits, although structural deformation can result in severe impurity problems at considerable depth. More often than not, the greatest quantity of impurity in a gypsum deposit is anhydrite (and vice versa), a result of the metastable relationship of these two minerals as previously discussed.

As utilized, most gypsum contains from 10 to 15% of impurities, although some deposits may be exceptionally pure (i.e., +95%) or somewhat impure (i.e., 80%). In general, the amount of impurity which can be tolerated depends upon: 1) the type of impurity, 2) the product being manufactured, and 3) the competitive situation.

Based upon the effect impurities may have upon the manufacturing process and finished products, impurities may be separated into three categories:

1. Insoluble or relatively insoluble minerals such as limestones, dolomite, anhydrite, anhydrous clays, silica minerals, etc.
2. Soluble evaporite minerals including chlorides: halite, sylvite, etc., and sulfates: mirabilite, epsomite, etc.
3. Hydrous, but insoluble minerals, e.g., the montmorillonite group of clays.

The first category, to the extent that the minerals replace gypsum, reduce the strength of rehydrated stucco and increase weights of the finished plaster or wallboard, i.e., more kilograms of an impure stucco are required to obtain a given strength. These minerals occasionally act as hydration accelerators and many commercial gypsum deposits contain as much as 10 to 15% of these insoluble impurities. The presence of the second category affects calcining temperature and stucco slurry fludity, set or rehydration time and paper-core bonding. These minerals are usually limited to no more than 0.02 to 0.03% by weight of the gypsum rock and strongly affect vapor pressure which in turn affects the gypsum hydration-dehydration phenomenon. The principal impact of the third category is in moisture pickup of the finished product and on bonding characteristics of the gypsum stucco core of wallboard to its paper covering. Hydrous clays up to 1.0 to 2.0% may be tolerated.

MAJOR NORTH AMERICAN PRODUCING AREAS

The principal gypsum producing areas in North America are shown on Fig. 3. Table 1 lists the North American gypsum production. In addition, there are many more known, but undeveloped gypsum deposits, especially in the midwestern and western states, southeastern Canada, and Mexico. Domestic resources are geographically distributed in 24 states and have been estimated to contain billions of tons or hundreds of years supply at current usage rates. The US Bureau of Mines and the Canadian Department of Energy, Mines, and Resources gather and distribute information about worldwide resources, reserves and usage. Deposits are widespread and resources are inferred to be essentially unlimited.

A brief description of the major North American producing districts follows:

Southeastern Canada

Marine evaporites of Mississippian (Visean) age underlie much of southern New Brunswick, central and northern Nova Scotia, and southwestern Newfoundland. The first production of gypsum in North America was recorded in Nova Scotia about 1735, and it has grown to make Nova Scotia the largest gypsum mining region in the world, with an annual production between 5 and 7 Mt.

Calcium sulfate is found in, or is correlative with, the Windsor Group which has been divided into five subzones: A, B, C, D, and E in ascending order. Although calcium sulfate is present in each of these subzones, the economic deposits of gypsum and anhydrite are confined to the Lower Windsor, i.e., subzones A and B. The most extensive is the A subzone, consisting of up to 300 m of anhydrite with occasional thin dolomite beds. Where topographic conditions are favorable, surface hydration, which rarely exceeds 10 m has produced several deposits of dense, white, high purity gypsum.

The B subzone, which varies in thickness from 75 to 150 m consists of interbedded calcium sulfate, limestones, and siltstones. The calcium sulfate, representing about 80% of the section, has been hydrated to depths varying from 15 to 70 m, although occasional remnant anhydrite lenses with associated soluble salts are encountered. Unlike most gypsum deposits which have only minor structural features, if any, the rocks of the B subzone are frequently deformed by complex plastic folding and by occasional faults of substantial displacement, which make for difficult mining conditions. Nevertheless, the B subzone accounts for approximately 75% of the region's total production.

The southeastern Canada gypsum deposits are usually covered by glacial till which may be as thick as 70 m, but averages 12 to 15 m. All mining is by quarry methods, and in 1990, four major companies operated in the region.

Iowa

The largest single gypsum producing district in the United States is at Fort Dodge, IA, where mining began in 1872. Long considered to be an outlier of the Permian Basin, this deposit is now thought to be Jurassic age (Cross, 1966). It consists of a single bed of gypsum up to a maximum of 10 m in thickness, but which probably averages nearer 4 to 5 m. Its areal extent of some 40 km^2 appears to be limited by post-depositional erosion, particularly during the period between Late Cretaceous and the Pleistocene. Its continuity is interrupted by wide channels apparently formed prior to glaciation and subsequently modified.

Overburden consisting of glacial till, varies from 7 to 25 m and the rock surface is commonly cut with mud-filled erosion crevices which must be cleaned out to retain purity; however, gypsum purity as used is usually plus 90%. Mining is by quarry methods, with four companies operating in the district. Total production in 1990 was 1 973 000 tons.

Another gypsum operation in Iowa is at Sperry in southeastern Iowa. It began production in 1961, and at about 200 m below the surface, is the deepest currently operating gypsum room and pillar mine in North America. The gypsum mining horizon is reached through a 4.8 m diameter concrete lined shaft and is developed using a room and pillar mining geometry in an average 3 m mining height. The gypsum occurs in the Wapsipinicon Formation of Devonian age. This calcium sulfate was apparently originally deposited as anhydrite. Maximum hydration is found along the deposits solution margin and beneath pre-glacial valleys, and reserves are limited to these areas.

Michigan

The second most productive gypsum source in the Midwest is the Michigan Basin, with two producing districts: Grand Rapids in the western portion of the basin, and Alabaster-National City in the eastern part and near Lake Huron. These deposits are in the Michigan Formation of Mississippian age, and consist of multiple units of gypsum from 1.5 to 13 m thick, separated by beds of shale varying from a fraction of a meter to 15 m thick. Overburden is glacial till ranging in thickness from 13 to 25 m.

Three operations, all quarries, are active in the Alabaster-National City area. Three gypsum beds (separated by 1.5 m of shale) are recovered at Alabaster, and a single bed is recovered at National City and Turner. This production serves five manufacturing plants: one in National City, and after water shipment via the Great Lakes, plants at Waukegan, IL, Detroit, MI, Lorain and Gypsum, OH, as well as numerous cement plants. At Alabaster, where the gypsum grades into anhydrite as the overburden thickness increases, a minor amount of anhydrite is also produced.

Two operations are also active in the Grand Rapids district, both are underground mines employing the room and pillar system. Two different beds, 2.5 to 4 m thick, are being mined with the production utilized by manufacturing plants at the mine sites. A recent paper (Hart et al., 1991) describes the local geologic conditions and methods of shaft construction at one of the mines.

New York—Ontario

Gypsum is found in a belt of Upper Silurian rocks which extends eastward from the Niagara Peninsula in Ontario across New York to the vicinity of Utica. The first recorded mining of gypsum in the United States took place in this belt in 1808 near Syracuse. In recent years, however, all production has come from the Oakfield area east of Buffalo in the western part of the state, where one operation currently is active. The economic gypsum here consists of a single bed approximately a meter thick in the Salina Group of Upper Silurian age, overlain by 3 to 13 m of thinly bedded shales and dolomites. Glacial till ranging up to 15 m in thickness covers bedrock. Mining is by underground methods using the room and pillar system, and the gypsum averages about 85% purity.

In the westerly end of this belt, two underground room and pillar mines in the Caledonia-Hagersville area of Ontario are im-

Fig. 3. North American gypsum producing districts.

Table 1. Gypsum Production, North America

Country	Thousand Tons 1980	1989	1990
Canada[1]			
Nova Scotia	4 941	5 926	6 213
Newfoundland	666	X	X
New Brunswick	43	0	0
Ontario	860	1 345	1 086
Manitoba	136	X	X
British Columbia	750	358	396
Subtotal	7 396	8 196	8 202
United States			
Arizona, New Mexico	355	510	707
Arkansas, Kansas, Louisiana	*	1 669	1 462
California, Nevada, Utah	2 527	3 205	2 917
Colorado, Montana, South Dakota, Washington, Wyoming	493[2]	612	475
Indiana, New York, Virginia, Ohio	1 624	1 953	1 860
Iowa	1 335	2 066	1 993
Michigan	1 256	1 899	1 818
Oklahoma	1 205	2 294	1 985
Texas	1 518	1 812	1 698
Others	2 431	—	—
Subtotal	11 251	16 022	14 915
Mexico	1 713	4 813	4 808
Dominican Republic	173	188	165
Jamaica	60	86	91
Total	20 593	29 305	28 181

X—confidential. [1]From O. Vagt, "Gypsum and Anhydrite" chapter in the *1990 Canadian Minerals Yearbook,* does not include gypsum produced by or shipped for use by Canadian Portland Cement Procedures. All other data from US Bureau of Mines. [2]Colorado and Wyoming only, *included in others.

portant producers. The Hagersville mine also operates in a bed approximately a meter thick which is correlative with the gypsum being mined in Oakfield, NY, giving these operations the dubious distinction of having the lowest mining heights in the gypsum industry of North America. However, the Caledonia mine is in a bed 2.5 to 3 m thick that stratigraphically is approximately 70 m below the Hagersville bed.

In addition to the Caledonia-Hagersville mines, an operation came into production approximately 48 km northwest of this area in 1978 near the town of Drumbo, mining a bed averaging 1.8 m thick about 120 m below the surface. This mine also had a room and pillar geometry. The operation closed down in 1993.

Ohio

On the shores of Sandusky Bay in northern Ohio, gypsum has been continuously mined since 1849, with one quarry currently active. Numerous beds of calcium sulfate are found in the Tymochtee Formation of Upper Silurian age, all of which are relatively impure due to inclusion of dolomite stringers and nodules, shale, anhydrite, and chert. As quarried, this gypsum is probably the lowest grade of any operating deposit in the United States (with the exception of gypsite in California) being in the 60 to 70% range; however, quarry production is beneficiated by heavy media sink-float methods to produce an approximately 80 to 81% gypsum product.

Indiana

In 1952, gypsum was discovered in southern Indiana near the town of Shoals, and this has become one of the most important gypsum producing districts in the United States. In part of the Illinois Basin, gypsum occurs in the St. Louis Formation of Mississippian age, and lies 100 to 250 m below the surface of this hilly area. The bed averages 4 to 5 m thick, with room and pillar mining methods being employed by the two mines operating in the district.

The St. Louis Limestone overlying the gypsum has been subjected to solution along bedding and joint planes resulting in a network of water-filled cavities of wide extent. This cavity system presents a potential flooding hazard to the mine workings; however, careful study of all the factors affecting the hydrology of the area, plus the adoption of precautionary operating measures has enabled mining to proceed safely.

Virginia

In the western tip of Virginia near Saltville, gypsum has been produced continuously since 1835, first from shallow pits or quarries and currently from one underground mine.

The calcium sulfate originated in the McCrady Formation of Mississippian age, but the deposits are structurally very complex as a result of Appalachian mountain building, and in particular, the Saltville fault, a major northeast-southwest trending thrust fault. The calcium sulfate is found as deformed masses in several stratigraphic horizons. The host rocks are soft shales, brecciated limestones, and rubbly classics of local origin deposited in a rapidly changing sabkha environment. The mine is multi-level, currently 230 m deep with potential extension to about 700 m. A combination of sublevel and cut-and-fill mining methods are used to extract the gypsum. Sharpe (1985) describes the deposits, history, and mining methods in detail.

Permian Basin

One of the major geological features of the United States is the Permian Basin extending over parts of Texas, New Mexico, Oklahoma, and Kansas, and calcium sulfate is found extensively over this area. Where hydration has occurred, the resulting gypsum is usually of high (+92%) purity, and in fact most of the specialty gypsum products which require high quality raw material are made from deposits of Permian age. The best purity and whitest color gypsum being commercially exploited in the United States is found in the Permian age Blaine Formation at Medicine Lodge, KS, Southard, OK, and the Easly Creek Formation at Blue Rapids, KS. Several formations contain numerous gypsum beds up to 10 m thick. The quarryable gypsum occurs near outcrop and anhydrite is found where shale and sandstone overburden thickens.

Currently, there are six operating quarries in Texas, six in Oklahoma, and two in Kansas. All of the Texas and Oklahoma deposits are worked by quarry methods. One of the two Kansas operations also has an underground mine in addition to a quarry.

Although considerable Permian calcium sulfate is found in southeastern New Mexico, none currently is being mined. There are, however, three operating quarries in the north central part of the state working gypsum of the Todilto Formation of Jurassic age.

Rocky Mountain and Inter-mountain States

This area is particularly rich in calcium sulfate deposits, most of which are at, or just below, the surface and lend themselves to quarrying. Structural features play an important part in limiting the size of the deposits, and in quarry design. In general, purity is high, but the extent of hydration is frequently highly variable. A major consideration to development is the high cost of transportation to major markets that are far from the deposits.

Gypsum is currently produced in the Powder River Basin area of Wyoming from rocks of Upper Permian to Middle Jurassic age, and from the Front Range of Colorado in Larimer and Fremont Counties. A new operation is working in structurally complicated deposits in Eagle County, in central Colorado. Production continues from four operations in the Arapien Shale and its equivalent Jurassic age formations in Sevier, Juab, and Emery Counties, Utah. Several deposits in Washington County, Utah, are worked intermittently. Small deposits near Raynesford, MT, supply local cement plants.

In Nevada, five deposits are quarried in Clark County and one each in Pershing and Lyon counties, with most of the product from these operations being shipped to California markets. In Arizona, a deposit at Winkleman provides gypsum for a calcining and manufacturing operation in Phoenix, and three smaller operations in the state supply portland cement retarder and agricultural markets. Small production comes from deposits near Weiser, ID, and from gypsite deposits in Okanogan County, Washington.

California

One gypsum products manufacturing facility in California uses locally quarried rock. This facility at Plaster City, Imperial County, annually quarries over 1 Mt, making it the largest operating gypsum quarry in the United States. This deposit, of Miocene age, occurs as a remnant of a calcium sulfate bed well over 30 m thick whose present extent is limited by structure and erosion. Little or no overburden is present, and quarrying is carried out by multi-bench methods (ver Planck, 1952). Other operations manufacturing gypsum products use rock from Mexico, Nevada, or other places as market conditions require.

For many years, several low grade gypsite deposits have been worked in San Joaquin Valley to produce a soil conditioner that contains 45 to 75% gypsum. These deposits are Holocene age in occurrence and have limited extent. Although too low a purity for portland cement retarder or plaster products, they are important in treating the alkaline soils which predominate in the great agricultural valleys of the state.

Western Canada

Gypsum is produced from five quarries, two in Manitoba, and three in British Columbia. These operations provide rock for six calcining and manufacturing plants in the western provinces. Also, rock from Mexico and Spain has been used in Vancouver wallboard and cement plants.

Production from these deposits is used almost entirely in the manufacture of building products and as a portland cement retarder. In all cases, only rock production and primary crushing operations are carried on at the quarry site, with the crushed gypsum rock being shipped to plants located in or near the major cities of Winnipeg, Calgary, Edmonton, and Vancouver.

Mexico

The primary use for gypsum in Mexico is as a retarder for portland cement and for building plaster. A small wallboard plant in Puebla and a plant near Monterrey produce more than enough wallboard to satisfy the Mexican market. Considerable gypsum is also exported to the West Coast of the United States and, on occasion, to Japan.

Many large deposits are known to exist; however, at the present time the major producers are located on San Marcos Island in the Gulf of Baja California; a few kilometers north of Santa Rosalia on the Baja Peninsula; and in the states of Puebla, San Luis Potosi, and Nuevo Leon. Smaller production comes from Sonora and Coahuila. The gypsum is found in rocks of Mesozoic and Tertiary age.

The largest gypsum production in Mexico is from San Marcos Island where a quarry and a deep-water ship loading facilities are operated to produce and ship a crushed gypsum rock. A large deposit a few kilometers north of Santa Rosalia, Baja California Sur, started production in 1990. Both deposits are of Miocene age and are related to basins and vulcanism associated with movement along the San Andreas fault.

In the state of Puebla, from the town of Izucar de Matamoros on the east to the village of Axochiapan on the west, a zone of gypsum bearing rocks of Miocene-Pliocene age occurs in which several small quarries are operated. These provide gypsum for eight or nine cement plants in the Mexico City-Puebla area, for the gypsum board plant at Puebla, and for several small calcining operations making gypsum plaster. The highest purity gypsum presently mined in Mexico comes from near Monterrey, Nuevo Leon, where at least two calcining plants operate to produce plaster for both construction and industrial uses.

Caribbean Area

Two deposits are being worked in the Caribbean area; one at Bull Bay, Jamaica, about 16 km east of the capital, Kingston, and the other near Barahona, Dominican Republic, some 210 km west of the capital city, Santo Domingo. In both countries only small domestic gypsum markets have been developed, primarily for portland cement retarder, and the major part of the production serves the export trade.

In Jamaica calcium sulfate is of Lower Eocene age, and occurs as several separate bodies of relatively small size in a northwest-southeast trending basin known as the Wagwater Trough (Holliday, 1971). Gypsum is currently being mined from two different deposits, utilizing multi-bench quarrying methods.

The Dominican Republic gypsum is probably Upper Miocene age, and the largest outcrop is found in the Cerro de Sal, a 19 km long hogback ridge trending northwest-southeast. Calcium sulfate is interbedded with limestones, claystones, sandstones, and salt of the Cerro de Sal Formation with a total thickness of 700 to 750 m. Bentonite clays and soluble salts are a potential problem associated with some beds. The geological structure has been complicated by faulting, resulting in several repetitions of the stratigraphic column. This large deposit potentially contains hundreds of millions of tons of economic gypsum, recoverable by quarrying methods. At least three other gypsum deposits are known in this general area, which to date have not been developed for production.

GYPSUM RESOURCES OUTSIDE NORTH AMERICA

Reflecting its wide geologic occurrence and its basic use in the construction industry, gypsum is produced in over 70 countries, the major ones being listed in Table 2. The US Bureau of Mines and the Canadian Department of Energy, Mines, and Resources have additional data providing details about production in countries not listed in Table 2. In 1990, the world total estimated production was 98.3 Mt with the United States consuming an estimated 16.3 Mt or 16.6%. No statistics are available as to the potential world reserves of calcium sulfate, but it can be safely predicted that they are enormous both in terms of total tons and of years at the present rate of consumption.

Table 2. Gypsum: World Production, Summary (thousand tons)

	1980	1989	1990
United States	11 251	15 988	16 329
Iran	7 000	8 437	8 437
Canada	7 225	8 196	8 202
People's Republic of China	3 364	8 074	8 074
Japan	6 118	6 260	6 260
Spain	5 234	5 498	5 535
France	6 000	5 684	5 715
USSR	5 455	4 808	4 808
Mexico	1 713	4 813	4 808
United Kingdom	3 407	3 992	3 992
Federal Republic of Germany	2 345	1 851	1 814
Australia	1 297	1 796	1 814
Other Countries	18 055	22 642	22 589
Total World	78 464	98 039	98 377

Figures are preliminary estimates and are subject to revision. *Sources:* Canadian Department of Energy, Mines and Resources, US Bureau of Mines.

In many of the countries reporting production, gypsum was used only for portland cement retarder, this being by far the most common use of the mineral on a worldwide basis. Generally, the calcining of gypsum to make wallboard or gypsum panels is limited to the United States, Canada, Europe, Japan, and other highly developed countries, although a few wallboard plants have recently been built or are being considered in certain South American and Asian countries. In these countries, interior building walls and ceilings are often open structural frames, made of wood or formed metal. Some type of covering material for walls and ceilings is required, and gypsum products have gained wide acceptance for this use because they are fireproof and are low cost. About three-quarters of the gypsum used in the United States and Canada is used to manufacture wallboard.

In countries where the usual methods of interior building construction rely heavily on masonry products, such as concrete, concrete block, tile, stone, and brick (including adobe), the demand for gypsum wallboard is much less, and the main uses of gypsum are for plasters and portland cement manufacture. The production and use of gypsum in a given country, therefore, is usually a result of its mode of use rather than the availability of gypsum deposits.

EXPLORATION METHODS

As for most minerals, a thorough understanding of the origin and occurrence of gypsum is essential before undertaking an exploration program. This may be obtained by a review of published literature, and a basic reference would be *Selected Annotated Bibliography of Gypsum and Anhydrite in the United States and Puerto Rico* by C.F. Withington and M.C. Jaster (1960).

By studying the stratigraphic column in regions where sedimentary rocks occur, an indication can usually be obtained from the lithologies whether or not an evaporite depositional environment may have existed, this being, of course, a requisite for the deposition of calcium sulfate. If evaporite-type rocks are indicated, further details can be obtained from examination of outcrops of the particular formations, or if outcrops are not available, from logs of oil or gas wells, or from water wells which may have been drilled in the vicinity. Perhaps drill cuttings, or even core, might be available for inspection. As stratigraphic and lithologic details of an area are thus developed, structure should also be mapped so as to provide as detailed a picture of the area as the basic data will permit. From this, the possibility of calcium sulfate occurrence usually can be predicted, and also, it may be possible to judge the potential for hydration of the sulfate zones.

Sources of this type of information are published literature, unpublished university theses, geological surveys, geological departments of colleges and universities, oil companies, and water-well drillers.

Geophysical work is only partially applicable to the search for calcium sulfate. The right combination of logs may indicate gypsum or anhydrite, but no log has yet been devised that will replace accurately a laboratory analysis. Geophysical logs are useful for determining structural details and can be used, in conjunction with other tests, to help determine geological environment conditions. Geochemical methods are rarely applicable to calcium sulfate, although heavy concentrations of sulfate in ground waters might serve as a useful clue, and in a few instances differences in plant species may indicate changes in rock types that can aid in surface mapping.

Even though the extent of gypsum can be mapped from surface outcrops, drilling is necessary to determine the extent of hydration, i.e., how much of the calcium sulfate is gypsum, and how much is only partially hydrated or is anhydrite. For deeper deposits, additional drilling data is almost always required to map projections of lithologies and/or structures. Core drilling is usually preferred, taking BX (or larger) size core to provide adequate material for sampling and to permit good core recovery. In some cases where it is only necessary to determine the thickness of overburden, or to locate the contact between gypsum and anhydrite, rotary drilling is sufficient and is much less expensive. Also geophysical methods can be successfully used under some conditions to determine the thickness of overburden.

Although calcium sulfate beds may sometimes be continuous over tens or even hundreds of kilometers, the exploration geologist should keep in mind that they may also occur as discontinuous lenses. Also, the extent of hydration is even less predictable, as is the occurrence of minor, but detrimental, amounts of impurities. Hence, a successful exploration program requires knowledgeable consideration and interpretation of all the factors involved.

EVALUATION OF DEPOSITS

The singlemost important factor in the evaluation of a gypsum deposit is its location with respect to markets, i.e., the logistical realities of the industry. The term place value is often used to describe this characteristic, which results because a local market demand will make nearby deposits more valuable even though they may not be as pure nor as easily mined as more remote deposits. Thus, the factor which more often than any other determines the value of one deposit as against another is the cost of transportation from the mine or quarry to the major market areas.

Also basic to the exploitation of gypsum is the fact that it is a relatively low priced commodity. This is due in part to the low cost inherent to extraction of the mineral from the ground and its preparation for market, and in part to the fact that in its major product areas, it must compete with other relatively low cost materials. With this highly competitive pattern, which as a practical matter results from differing manufacturing and delivery costs from one operation to another, and the constantly varying market demand, there is no set market price to measure against, and each deposit must be evaluated within the market area it might logically serve. Most gypsum produced is converted to a manufactured product. Because of the high cost of building a manufacturing facility, assured rock supplies are an industry characteristic. Effectively this means most gypsum produced is captive production with mines and quarries owned by the manufacturing companies. Hence, early in the evaluation of any gypsum deposit, an investigation of marketing factors is essential.

The geologist and mining engineer should carefully study the geological environment of a deposit, looking at such features as overburden ratio if a shallow deposit or roof conditions if an underground mine is indicated. Also, thickness of the bed, structural features, and the possibility of unusual water conditions are important, as these translate into capital investment and operating costs. There is no simple rule as to how much overburden one can afford to strip for a quarry, or how much depth one might consider for underground mining. In general, the economics of the industry are such that mining costs must be relatively low, and many operating deposits today mine and deliver rock to the mill (on site) for a cost between 3 and $11/t.

The grade of the deposit must also be assessed, particularly the type and amounts of impurities. Impurities should be carefully analyzed, and if they are outside the limits, steps to eliminate or reduce them by beneficiation should be investigated. In such a case, the cost increase due beneficiation may well eliminate the deposit from further consideration.

A separate chapter on utilization, discussing processing methods, markets, and price trends, provides further background for the evaluation of gypsum deposits.

EXTRACTION METHODS

Gypsum is recovered both from quarries and underground mines. Quarrying of gypsum employs conventional methods and equipment, such as draglines and scrapers to remove overburden, and shovels or front end loaders to load rock. Rotary and/or auger drilling is almost universally used because of the relative softness of the rock, and penetration rates of up to 7 m/min are possible. Drill holes are usually from 50.8 to 101.6 mm in diameter, and spaced relatively close together in order to distribute the explosive throughout the rock mass in a somewhat dense pattern. This is because gypsum tends to absorb explosive force without readily fracturing. A poor blasting pattern commonly results in irregular fragmentation, excessive fines, etc. Ammonium nitrate-fuel oil explosive is used where water conditions permit, and a ratio of 1 t of rock broken per 1 kg of explosive is about average.

Because little or no beneficiation is usually performed in the processing steps following extraction, extra effort is required to remove all overburden, or to selectively recover a face in order to eliminate impurities or to blend varying grades of rock. Quality control is most important in the mining or quarrying step, and may often result in somewhat higher per ton costs.

Underground mining commonly employs room and pillar methods with pillar geometry based upon the conditions of each mine; however, recovery of total rock is in the 65 to 80% range. Auger drilling is standard practice, and loading is done with front end loaders or load-haul-dump units. Underground haulage may be by rail, or by electric or diesel powered trucks. In a few cases, conveyor belts are used. The use of diesel power has become quite common in gypsum mines, in part because explosive gas conditions do not exist, and in part because mine design usually can readily incorporate good ventilating conditions. Continuous miners are only beginning to be used. Underground crushing is quite common and haulage distances from face to crusher are minimized by using moveable feeder-breaker type crushers.

In general, gypsum extraction operations are highly mechanized, although the limited volume of production, usually 500 to 3 000 t/day, does restrict the size of equipment which can be utilized economically. The physical environment of most gypsum deposits favors good working and safety conditions, a major factor in contributing to low cost.

Table 3. US Imports of Crude Gypsum (thousand tons)

Country	Year 1980	1989	1990
Canada	5 463	5 714	5 722
Mexico	1 565	2 139	1 544
Jamaica	11	0	22
Dominican Republic	69	*	*
Spain	250	562	560
Others	8	44	96
Total Imports	7 366	8 458	7 933
Domestic Production	13 039	16 022	14 915
Total	20 405	24 480	22 848
% Imports	36	35	35

*Under 1, *Source:* US Bureau of Mines.

FUTURE CONSIDERATIONS

As indicated throughout this discussion, there is no foreseeable shortage of either gypsum or anhydrite in the United States or in the world. Paradoxically, however, there are instances where it may be difficult to find gypsum that can be considered economic at a given time and location, a problem that has its roots in place value. The best evidence of this situation is that the United States historically has imported 33 to 36% of its needs from Canada, Mexico, Jamaica, the Dominican Republic, and Spain, as shown in Table 3.

Almost all of the imported gypsum (with the exception of a few thousand tons of special grade rock) is used by wallboard plants on the Atlantic, Pacific, and Gulf Coasts of the United States. The basic reasons for this situation are: 1) large markets for gypsum products are concentrated in and around the coastal cities, 2) there are no developed and producing gypsum deposits along any of the three United States coastlines, and 3) large gypsum deposits exist on or near deep water of the exporting countries from which rock can be shipped to these major markets at lower cost than from inland deposits in the United States.

It is likely that these conditions will persist into the future. The industry has been built around these facts and has made large investments in offshore deposits, shipping facilities, and domestic port locations with appropriate marine rock handling equipment. There is no import duty on crude gypsum rock, but duties are placed upon processed gypsum, although continuing free trade agreements and laws may change the duties.

Because of its widespread occurrence and huge potential reserves and also because its uses are such that it is not basic to survival in the event of a national emergency, gypsum is not considered to be a strategic mineral. This has permitted natural economic factors to prevail in the development of the mineral, literally on a worldwide basis, which overall is a healthy situation that hopefully will continue to prevail.

FUTURE TRENDS AND PROBLEMS

Competition between manufacturers of gypsum building products, as well as from manufacturers of substitute materials, has resulted in a continuing pressure to improve the quality of gypsum products. In turn, mine operators are being asked to reduce the average quantity of impurities, i.e., upgrade the gypsum purity, and maintain a uniform level of impurities. To date, these objectives are being accomplished by selective extraction techniques and screening during primary and secondary crushing stages to reject a size fraction in which impurities tend to concentrate. Other beneficiation techniques can be and are considered, but are slow to be adopted because of adverse economic impact.

Environmental concern in the mining and quarrying of gypsum includes a full range of concerns from disturbance of old growth forests by underground mining to storm water run-off in the desert. Land reclamation of quarries has become a requirement as new laws and regulations take effect, and considerable ingenuity is needed to do a satisfactory job at a reasonable cost. It is believed that gypsum miners have as good, and in some cases, a better opportunity of doing satisfactory reclamation than miners of many other types of minerals because there seldom are associated minerals which lead to such problems as acid waters. Also, the configuration of many gypsum quarries is such that surface restoration is readily feasible. In general, the water pumped from either quarry or underground gypsum mines is of a quality not considered as a pollutant; however, it is necessary to sample and analyze the pump discharge from time to time to develop adequate background data and to ensure that requirements are being met. In general the mood of increasing regulatory control over every aspect of business activity already requires and will continue to require increased planning from pre-permit to post-closure.

Processing of gypsum is energy-intensive; energy requirements represent the largest single cost in the production of gypsum wallboard. Conservation of energy meshes well with utilization of synthetic gypsum from power plant desulfurization processes, as waste heat from the power plant can be utilized in the processing of gypsum to make commercial products.

SYNTHETIC OR BYPRODUCT GYPSUM

Synthetic gypsum as a byproduct of various chemical operations has been available for some time in various parts of the world. Whether or not it might be substituted for natural gypsum requires analysis of the same factors that one would investigate in evaluating a calcium sulfate deposit, the most important being logistics, quality, and cost. In most cases, byproduct gypsum is of unacceptable quality for the majority of gypsum uses because of the type and amounts of impurities, or crystal size and shape. In all cases, it is a relatively fine, wet material which comes out of the chemical process as a filter cake or after settling in ponds or tanks.

Both the chemical and physical characteristics of byproduct gypsum are such that further treatment is necessary to make it into a useful product. The cost of such treatment varies widely with the type of byproduct and the use for which it is intended. By far the major source of byproduct gypsum at present is the manufacture of phosphoric acid. The provisions of the Clean Air Act of 1990 are having a very strong impact on coal-fired electricity-generating plants. These plants have the potential to produce much more gypsum than the United States uses annually, and many alternatives are being considered to lessen the economic impact of reducing SO_2 emissions. Availability of widespread wallboard quality synthetic gypsum has had the potential to change the industry significantly; however, in the past, alternatives to using synthetic gypsum for wallboard have provided better economics, and utilization of flue gas desulfurization gypsum to date is small. Synthetic gypsum from phosphoric acid manufacture is not used, although production is millions of tons per year and over 100 Mt are available. Use of phosphogypsum in California ceased after closure of the phosphate fertilizer manufacturing facilities.

Japan makes greater use of synthetic gypsum for wallboard than the United States does, in part because Japan has almost no natural deposits. Other countries make greater use of synthetic gypsum than the United States as well.

ACKNOWLEDGMENT

The author wishes to acknowledge his debt to Frank C. Appleyard, the author of this chapter in the 5th edition of *Industrial Minerals and Rocks*. A former Vice President with US Gypsum in Chicago, he is now retired and living in Arizona. His work was outstanding and provided a base for this author's contribution to the 6th edition.

BIBLIOGRAPHY AND REFERENCES

Adams, J.E., 1971, "Upper Permian Ochoa Series Of Delaware Basin, West Texas and Southeastern New Mexico," *Origin of Evaporites*, American Association of Petroleum Geologists, Tulsa, OK, Reprint Series No. 2. pp. 60–89.

Borchert, H., and Muir, R.O., 1964, *Salt Deposits, The Origin, Metamorphism, and Deformation of Evaporites*, D. Van Nostrand, London, 338 pp.

Braitsch, O., 1971, *Salt Deposits, Their Origin and Composition*, Springer-Verlag, New York, 297 pp.

Branson, E.B., 1915, "Origin of Thick Gypsum and Salt Deposits," Geological Society of America Bulletin, Vol. 26, pp. 231–242.

Briggs, L.I., 1970, "Geology of Gypsum in the Lower Peninsula, Michigan," *Proceedings*, 6th Forum on Geology of Industrial Minerals, Michigan Geological Survey, Misc. 1, pp. 66–76.

Butler, G.P., 1970, "Holocene Gypsum and Anhydrite of the Abu Dhabi Sabkha, Trucial Coast: An Alternative Explanation of Origin," *Proceedings*, 3rd Symposium on Salt, Northern Ohio Geological Society, Cleveland, OH, Vol. 1, pp. 120–152.

Buzzalini, A.D., ed., 1969, "Evaporites and Petroleum," American Association of Petroleum Geologists Bulletin, Vol. 53, No. 4, pp. 775–1011.

Cole, L.H., 1930, "The Gypsum Industry of Canada," Canadian Dept. of Mines, Publication 714, pp. 1–164.

Cross, A.I., 1966, "Palynologic Evidence of Mid-Mesozoic Age of Fort Dodge (Iowa) Gypsum," Abstract, San Francisco Meeting, Geological Society of America, Nov. 14–16.

Davis, L.L., 1990, 1991 "Gypsum," Monthly and Annual Mineral Industry Surveys, US Bureau of Mines, Washington, DC.

Dean, W.E., and Schreiber, B.C., eds., 1978, "Marine Evaporites," Mineral Short Course, Notes, No. 4, Society of Economic Paleontologists, 188 pp.

Edinger, S.E., 1971–1975 "The Chemistry of Gypsum and its Dehydration Products," NTIS, US Department of Commerce Report No. Steed 72-1-1-3-4, 3 Vol.

Goodman, N.R., et al., 1957, "Gypsum," The Geology of Canadian Industrial Mineral Deposits, 6th Commonwealth Mining and Metallurgical Congress, pp. 111–137.

Ham, W.E., 1962, "Economic Geology and Petrology of Gypsum and Anhydrite in Blaine County," Oklahoma Geological Survey Bulletin, Vol. 89, pp. 100–151.

Hardie, L.A., 1967, "The Gypsum-Anhydrite Equilibrium at One Atmosphere Pressure," *American Mineralogist*, Vol. 52, Jan.–Feb., pp. 171–200.

Holliday, D.W., 1970, "The Petrology of Secondary Gypsum Rocks: A Review," *Journal of Sedimentary Petrology*, Vol. 40, No. 2, June, pp. 734–744.

Holliday, D.W., 1971, "Origin of Lower Eocene Gypsum-Anhydrite Rocks, Southeast St. Andrew Jamaica," *Transactions*, Sec. B, Institution of Mining and Metallurgy, Vol. 80, pp. B305–B315.

Jorgensen, D.B., and Carr, D.D., 1972, "Influence of Cyclic Deposition, Structural Features, and Hydrologic Controls on Evaporite Deposits in The St. Louis Limestone in Southwestern Indiana," *Proceedings*, 8th Forum on Geology of Industrial Minerals, Iowa Geological Survey, 195 pp.

Keith, S.B., 1969, "Gypsum and Anhydrite," Bulletin 180, Arizona Bureau of Mines, pp. 371–382.

Kendall, A.C., 1978, "Calcium Sulfate Diagenesis, Mississippean of Eastern Williston Basin," Abstract, American Association of Petroleum Geologists Bulletin, Vol. 62, No. 3, p. 530.

Kerr, S.D., and Thompson, A., 1963 "Origin of Nodular and Bedded

Anhydrite in Permian Shelf Sediments, Texas and New Mexico,'' American Association of Petroleum Geologists Bulletin, Vol. 47, pp. 1726–1732.

King, R.H., 1971, ''Sedimentation in Permian Castile Sea,'' *Origin of Evaporites,* American Association of Petroleum Geologists, Reprint Series No. 2, Tulsa, OK, pp. 90–97.

Kinsman, D.J.J., 1966, ''Gypsum and Anhydrite of Recent Age, Trucial Coast, Persian Gulf,'' 2nd Symposium on Salt Northern Ohio Geological Society, Vol. 1, Cleveland, OH, pp. 302–326.

Kinsman, D.J.J., 1969, ''Modes of Formation, Sedimentary Associations, and Diagenetic Features of Shallow Water and Supratidal Evaporites,'' American Association of Petroleum Geologists Bulletin, Vol. 53, No. 4, pp. 830–840.

Lewis, W.L., and Holleman, M., 1983 ''Gypsum in Atlantic Canada,'' *Proceedings,* 19th Forum on Geology of Industrial Minerals, Ontario Geological Survey Miscellaneous Paper 114, 216 pp.

MacDonald, G.F.J., 1953, ''Anhydrite-Gypsum Equilibrium Relations,'' *American Journal of Science,* Vol. 251, pp. 883–898.

Martinez, J.D., 1974, ''Tectonic Behavior of Evaporites,'' *Proceedings,* Vol. 1, 4th Symposium on Salt, Northern Ohio Geological Society, Cleveland, OH, pp. 155–168.

Murray, R.C., 1964, ''Origin and Diagenesis of Gypsum and Anhydrite,'' *Journal of Sedimentary Petrology,* Vol. 34, No. 3, Sep., pp. 512–523.

Newland, D.H., and Leighton, H., 1910, ''Gypsum Deposits of New York,'' New York State Museum Bulletin 143, 94 pp.

Posnjak, E., 1940, ''Deposition of Calcium Sulfate From Sea Water,'' *American Journal of Science,* Vol. 238, pp. 559–568.

Pressler, J., 1981, ''Gypsum,'' *Minerals Yearbook 1980,* US Bureau of Mines, pp. 385–396.

Raup, O.B., 1991, ''Descriptive Model of Bedded Gypsum; Deposit Subtype, Marine Evaporite Gypsum,'' US Geological Survey, OFR 91-0011-A, pp. 34–35.

Riley, C.M., and Byrne, J.V., 1961, ''Genesis of Primary Structures in Anhydrite,'' *Journal of Sedimentary Petrology,* Vol. 31, pp. 553–559.

Schenk, P.E., 1969, ''Carbonate-Sulfate-Redbed Facies and Cyclic Sedimentation of the Windsorian Stage (Middle Carboniferous), Maritime Provinces,'' *Canadian Journal of Earth Sciences,* Vol. 6, pp. 1037–1066.

Schmidt, R., 1914, ''Uber di Beschaffenheit und Entstehung Parallelfaseriger Aggregate von Steinsalz und Gips,'' *Kali.* 8, pp. 161, 218, 239.

Schroeder, H.J., 1970, ''Gypsum,'' Mineral Facts and Problems, Bulletin 650, US Bureau of Mines, pp. 1039–1048.

Sharpe, R.D., 1985, ''Geology and Mining of Gypsum in Virginia,'' *Proceedings,* 20th Forum on Geology of Industrial Minerals, Maryland Geological Survey Special Publication No. 2, pp. 41–49.

Shearman, D.J., 1965, ''Origin of Marine Evaporites by Diagenesis,'' *Transactions,* Sec. B, Institution of Mining and Metallurgy, Vol. 75, pp. B208–B215.

Stone, R.W., et al., 1920, ''Gypsum Deposits of the United States,'' Bulletin 697, US Geological Survey, 326 pp.

Stonehouse, D.H., 1970, ''Gypsum and Anhydrite,'' *Canadian Minerals Yearbook,* Mineral Resources Branch, Dept. of Energy, Mines, and Resources, Ottawa, pp. 775–1011.

Stonehouse, D.H., 1978, ''Gypsum in Canadian Mineral Industry,'' Dept. of Energy, Mines and Resources, Ottawa, pp. 211–216.

Stonehouse, D.H., 1979, ''Gypsum in Canadian Mineral Industry,'' Dept. of Energy, Mines and Resources, Ottawa, pp. 221–228.

Stonehouse, D.H., 1980, ''Gypsum in Canadian Mineral Industry,'' Dept. of Energy, Mines and Resources, Ottawa, preprint, 8 pp.

Van't Hoff, J.H., et al., 1903, ''Gips und Anhydrit,'' *Zeitschrift fuer Physik Chem.,* Vol. 45, pp. 257–306.

ver Planck, W.E., 1952, ''Gypsum in California,'' Bulletin 163, California Division of Mines, 151 pp.

Walker, C.W., 1974, ''Nature and Origin of Cap Rock Overlying Gulf Coast Salt Domes,'' *Proceedings,* 4th Symposium on Salt, Northern Ohio Geological Society, Vol. 1, Cleveland, OH, pp. 169–175.

Walsh, A.R., Hart, D.E., and Maishman, D., 1991 ''Shaft Construction by Raise Boring Through Artificially Frozen Ground,'' *Proceedings,* Rapid Excavation and Tunnelling Conference, Seattle, WA, June 17–19, SME, Littleton, CO, pp. 705–719.

Withington, C.F., 1962, ''Gypsum and Anhydrite in the U.S.,'' Mineral Investigation Research, Map MR-33 and text, US Geological Survey, 18 pp.

Withington, C.F., and Jaster, M.C., 1960, ''Selected Annotated Bibliography of Gypsum and Anhydrite in the United States and Puerto Rico,'' Bulletin 1105, US Geological Survey, 126 pp.

Wood, G.V., and Wolfe, M.J., 1969, ''Sabkha Cycles in the Arab-Darb Formation of Trucial Coast of Arabia,'' *Sedimentology,* Vol. 12, pp. 165–191.

IODINE

KENNETH S. JOHNSON

Iodine, a grayish-black nonmetallic element, with a density of 4.9 g/cm^3, is a solid at ordinary temperatures. It is a member of the halogen family, along with fluorine, chlorine, bromine, and astatine. Iodine melts at 114°C, and at 184°C it is volatilized to a blue-violet gas that has an irritating odor. It does not occur as an element in nature, but occurs as iodates, iodides, or other combined forms. It is the 47th most abundant element in the earth's crust.

Iodine was discovered by Bernard Courtois in 1811. He observed an unknown substance in the crude soda ash that results from the burning of seaweed. Samples of this unknown substance were identified to be a new element, and in 1813 Gay-Lussac named the substance *iode*, from the Greek word for violet color.

The world production during 1989 and 1990 (Table 1) is estimated to be about 16 million kg per year (Lyday, 1991), of which about 30% is consumed in the United States.

GEOLOGY AND MINERALOGY

Compounds of iodine are minor constituents in seawater and brines, in certain marine organisms, and in minerals of the Chilean nitrate deposits. Seawater contains approximately 0.05 ppm iodine, and certain marine organisms, such as seaweed, sponges, fish, and some brown algae, are able to further concentrate iodine (Lyday, 1989a). Some seaweed can extract and accumulate iodine up to 0.45% of their weight, on a dry basis. The northern Chilean nitrate deposits, in the Atacama Desert, contain the following iodine minerals: lautarite, $Ca(IO_3)_2$ (calcium iodate); dietzeite, $Ca_2(IO_3)_2 \cdot (CrO_4)$ (calcium iodate-chromate); and bruggenite, $Ca(IO_3)_2 \cdot H_2O$ (Erickson, 1981).

Various subsurface brines also contain iodine compounds. Some gas-field brines in the United States and Japan locally contain 30 to 1 300 ppm iodine. Several coals in Germany also contain iodine compounds.

Iodine has been recovered from brines mainly in Japan and the United States, but also in Java, Indonesia, Italy, England, and the former USSR. Iodine has also been recovered from seaweed in Ireland, Scotland, France, Japan, Norway, and the USSR. Seaweed was a major source of iodine for the world in the first half of this century, and it remains as a large resource. The reserves and future resources of iodine are large, even excluding the resources in seaweed and seawater, and are shown in Table 1.

Analysis

Iodine as the free element can be detected by the characteristic blue color it gives to a starch solution. Quantitatively it is determined as the free element by titration with standard thiosulfate solutions using starch as an indicator. Colorimetric methods are also applicable.

PRINCIPAL PRODUCING COUNTRIES

Major iodine-producing nations are Japan, Chile, the USSR, and the United States, with lesser amounts being produced in China and Indonesia (Fig. 1; Table 1). Annual world production in 1989 and 1990, respectively, is estimated at 15.6 and 16 million kg. In Japan and Chile, the production of iodine depends on production of other materials, such as natural gas or nitrates, respectively, whereas in US operations (in Oklahoma) iodine is the major product recovered from natural brines. Chile was for a long time the principal world producer of iodine from its nitrate-fertilizer operations, but in recent years Japan has become the world leader with increased production of natural gas and associated iodine-rich brines.

Japan

Iodine production in Japan, which began in 1934, comes from iodine-bearing natural brines that are recovered along with natural gas produced in three different regions of the country (Fukuta and Fujii, 1982, Fukuta et al., 1982, Fukuta, 1985, Wu, 1985). Production of the natural gas entails co-production of the associated brines, and the iodine is then removed from the brine before disposal of the brine. The three iodine-producing regions are the southern Kanto gas field (near Tokyo in Chiba and Kanagawa prefectures), Niigata and Nakajo gas fields (about 250 km north of Tokyo in Niigata Prefecture), and the Sadowara gas field (Miyazaki Prefecture in southern Kyushu). The southern Kanto gas field produces more iodine than any other region in the world. In Japan, a total of 18 plants, operated by 6 companies, produced 7.3 million kg of iodine in 1984, and their combined capacity was 9.22 million kg (Table 2). The maximum iodine content of the brines is about 160 ppm. In addition, iodine has been confirmed in the southern Okinawa gas field, on Okinawa, and in the Oshamanbe and Ishikari gas fields, on Hokkaido (Fukuta, 1985).

Iodine-rich brines in Japan are found in Miocene, Pliocene, and Pleistocene sands and silts. Most of these sediments are of marine origin, deposited in bathyal and/or outer-neritic environments, although those in the Niigata region are sands and gravels of shallow-marine to estuarine or lagoonal origin (Fukuta, 1985). The iodine-rich brines of Japan are regarded as being primary in origin, are derived from seawater, and are limited to the younger marine sediments. The iodine-rich brines are saturated or supersaturated with natural gas (mainly methane); however, some Japanese brines saturated with natural gas may not contain iodine, because methane can be derived from the remains of a variety of life forms (including land species) that contain little or no iodine (Fukuta, 1985).

Table 1. Crude Iodine: World Production and Reserves (thousand kilograms)

	Mine Production		
Country	1989	1990[e]	Reserves
Japan	7 600	7 600	4 000 000
Chile	4 000	4 000	900 000
United States	1 500	1 860	550 000
USSR[e]	2 000	2 000	400 000
China[e]	500	500	400 000
Indonesia	14	14	100 000
World Total[1]	15 600	16 000	6 350 000

[e]Estimated, [1]figures are rounded; excludes other countries.
Source: Lyday (1991).

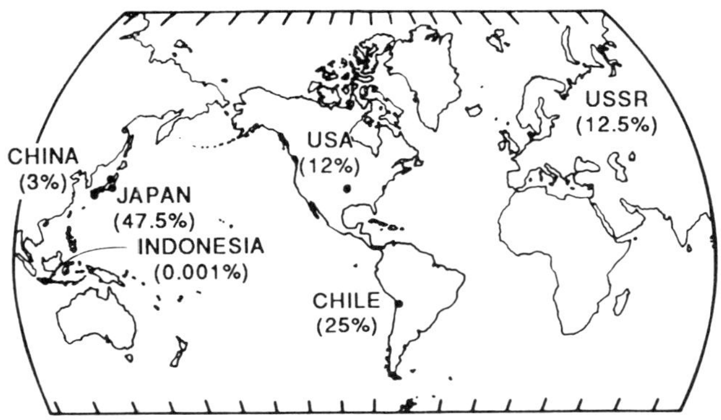

Fig. 1. Countries producing iodine and percentage of estimated world production during 1990.

Chile

Chile was the world's leading producer of iodine for many years, but now it ranks second, behind Japan. Iodine is obtained mainly as a byproduct during production of sodium nitrate fertilizer from caliche (salitre) deposits in the Atacama Desert of northern Chile. Lately, however, it also has been produced as the primary product of a newly developed process to extract iodine from abandoned nitrate tailings (Velasco and Gurmendi, 1988). The nitrate fields are mainly in the Central Valley, called the Pampa del Tamarugal, that lies between the relatively low Coastal Range on the west and the high Andes Mountains on the east. They are mainly in Tarapaca and Antofagasta provinces.

The iodine-bearing nitrate ores, described by Erickson (1981), Garrett (1985), and Lyday (1985, 1989a, 1989b), are extensive in a belt several hundred kilometers long and tens of kilometers wide. Caliche ore layers typically are 1 to 3 m thick, flat lying or gently dipping, and near the land surface. They contain evaporite minerals as a cement in unconsolidated surface materials or as veins and impregnations in bedrock. The caliche is typically 5 to 30% sodium nitrate, and it contains iodine as calcium iodate (lautarite and dietzeite). Crushed caliche is leached to produce a solution containing sodium nitrate and calcium iodate (Lyday, 1989a). After precipitation of the sodium nitrate, the remaining liquid is then stripped of iodine.

Most of the iodine is produced by Sociedad Química y Minera de Chile (SOQUIMICH), the recently privatized state nitrate company. Other producers and/or potential producers include: Amsterdam Chemical Pharmaceutical Minera; Amsterdam Pharmaceutical Co. of the Netherlands; Compañía de Salitre y Yoda de Chile; Compañía Minera del Alba; Taltal Joint Venture of USA and Canada; Cia Minera Chilbras; Merck Pharmaceutical Co. of Germany; Compañía Minera Sierras de Tarapaca; and Banco Bilboa Vizcaya S.A. of Spain (Velasco and Gurmendi, 1988; Lyday, 1989b). Chile exports its iodine primarily to Europe and the United States.

United States

Iodine was first produced in the United States between 1917 and 1921, from seaweed harvested in California; after distillation to produce acetic acid, the residue was processed for potash fertilizers and iodine (Lyday, 1986). The first US commercial production of iodine from brines was in Louisiana between 1928 and 1932; some of the Louisiana oil-field brines contained about 35 ppm iodine (Lyday, 1986). Oil-field brines in parts of California contain 30 to 70 ppm iodine (Fig. 2) in the Monterey Formation (Miocene) and Repetto Formation (Pliocene); these were produced at various times between 1928 and 1966. In Michigan, natural brines containing 15 to 30 ppm iodine are present in the Sylvania Formation (Devonian) at a depth of about 1 300 m. The Michigan brines were processed mainly for bromine, and byproduct iodine, by Dow Chemical Company until the wells were plugged and abandoned in 1987.

Iodine production in the United States now comes totally from iodine-rich (300 ppm iodine) natural brines in the deep subsurface of the Anadarko basin of northwestern Oklahoma (Fig. 3). Oklahoma production began in 1977, and at present three companies operate three major plants and one miniplant for the recovery of iodine. Total production from the three companies in 1989 was 1 500 000 kg, and in 1990 was estimated to be 1 860 000 kg (Table 1).

Oklahoma's first iodine operation was the Woodward Iodine Corporation. It opened early in 1977 as a joint venture between Amoco Production Company and Houston Chemicals, a subsidiary of Pittsburg Plate Glass Industries (Cotten, 1978); in 1984 the company was purchased by Asahi Glass Co. of Japan (Lyday, 1986). Discovery of iodine-rich brines at Woodward resulted from a 12-year program of analyzing brine samples collected by Amoco (Cotten, 1978). Amoco noted unusually high concentrations of iodine in the Woodward area, with concentrations as high as 1 560 ppm in Chesterian (Mississippian) limestones and 700 ppm in Morrowan (basal Pennsylvanian) sandstones. Morrowan sandstones in the area are as much as 100 m thick and are preserved as channel sands in a south-trending paleovalley (the Woodward trench) cut into the Chesterian surface (Fig. 4). Although iodine concentrations are higher in the Chesterian strata, Chesterian limestones have low permeability and do not yield large volumes of water. On the other hand, Morrowan sandstones here have higher permeability and yield large volumes of brine averaging about 300 ppm iodine.

The Woodward trench has an average width of 1.6 km and a known length of about 115 km (Cotten, 1978). Woodward Iodine Corporation operates 22 production wells and injects the waste brine back into the trench through 10 injection wells. Brine-production wells and injection wells are 2 130 to 2 290 m deep at Woodward.

A second iodine plant in Oklahoma was started in late 1987 by Iochem Corporation of Japan. The Iochem plant, located near Vici, also extracts iodine-rich brines from Morrowan sandstones, at depths of 3 000 to 3 183 m, in a southern extension of the Woodward trench. The company has 9 production wells and 4 injection wells.

North American Brine Resources operates the remaining iodine plants in Oklahoma. At a miniplant near Dover, the company processes oil-field brines collected from a number of producing oil and gas wells in nearby parts of northwestern Oklahoma. The company also operates a major plant, constructed about 35 km north

Table 2. Iodine Production in Japan, 1984

Area	Company	Number of plants	Production, thousand kg	Capacity, thousand kg
Chiba	Ise Chemical Industry Co.	5	2 445	3 000
	United Resources Industry Co.	2	1 610	1 800
	Nippon Natural Gas Industry Co.	4	1 040	1 320
	Kanto Natural Gas Development Co.	3	880	1 030
	Nippon Chemicals Co.	1	290	450
Niigata	Ise Chemical Industry Co.	1	575	900
	Nippon Halogen Chemicals Co.	1	280	360
Miyazaki	Ise Chemical Industry Co.	1	180	360
TOTAL		18	7 300	9 220

Source: Wu (1985).

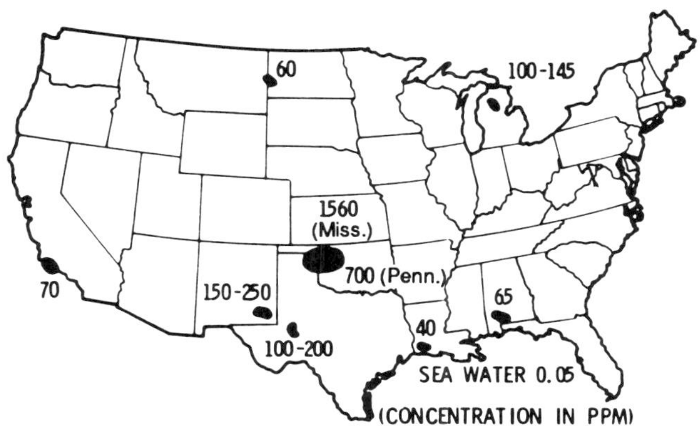

Fig. 2. Areas of known iodine concentrations in the United States (from Cotten, 1978). Exceptionally high iodine concentrations in northwest Oklahoma are in Mississippian (Miss.) and Pennsylvanian (Penn.) strata.

of Woodward in 1989, where they produce iodine-rich brine from Morrowan sandstones in a northern extension of the Woodward trench. Operations north of Woodward include two production wells and three injection wells, all about 1 800 m deep. North American Brine Resources is a joint venture of Beard Oil Co. (40% share) and two Japanese firms (Godoe USA, Inc., 50% share, and Inorgchem Development, Inc., 10% share) (Ohl and Arndt, 1988).

Other Countries

Data are sparse on other iodine-producing nations, but an overview is provided by Lyday (1986) and many of those data are summarized here.

The USSR has produced iodine since 1926, when borax springs at Baku began yielding iodine as a byproduct. Later, production came from oil-field brines that contained 67 ppm iodine at Lake Bejuk-Schor, near Baku. Brines in the Slavyansko-Troitsko area are reported to contain 80 to 120 ppm iodine in the Neftechinski oil field near the Black Sea, and iodine was also produced from brines at the Nebit-Dag iodine/bromine plant in Turkmenistan (Lyday, 1986).

China has produced iodine from salt lakes at Chaerham, Yuncheng, and elsewhere, but data on this industry are not available (Lyday, 1986).

In Indonesia, iodine is present with trace amounts of bromine in oil-field brines in Pliocene sandstones and diatomaceous marls in the Gujangon anticline (Lyday, 1985). The sole producer is P.T. Kimia Farma, a state-owned pharmaceutical firm with facilities near Mojokerto, East Java (Lyday, 1988).

Italy produced iodine from treated mineral water at Salsomaggiore between 1925 and 1963, and during World War II iodine was extracted from coal dust in Germany and Czechoslovakia (Lyday, 1986). Iodine is now produced from mineral waters at a depth of 1 800 m at Rabasomjen, Hungary, and as a byproduct of the largest seaweed-drying plant in the world located in Iceland (Lyday, 1988).

USES

Iodine products were first used commercially in the early part of the 20th century as a remedy for goiter, as a disinfectant for cuts and abrasions, and for sanitation (Lyday, 1989a, 1989b). Since then, iodine has been used in a variety of specialty chemicals for many commercial applications. Major uses include catalysts, stabilizers, animal feeds, disinfectants, pharmaceuticals, photography, and colorants. The US consumption (by principal end use) is given in Table 3. The following description of uses is from Lyday (1989a, 1989b).

Catalysts

A major use of iodine is in catalysts. With development of the Monsanto process for producing acetic acid (using an iodide-promoted rhodium complex as catalyst) in the 1960s, about 90% of new acetic-acid capacity used this process. Iodide catalysts also are used in the dehydrogenation of butane and butene to butadiene, and in the preparation of stereo-regular polymers.

Stabilizers

A second major use is as a stabilizer in the manufacture of nylon for tire cord and carpets, and for converting rosins, tall oil, and other wood products to more stable forms.

Animal Feeds

Another major use of iodine is in the form of additives for animal feeds, with about 25% of US domestic consumption going to that end. Cattle and sheep are fed iodized salt and other iodine compounds to reduce certain livestock ailments, such as soft-tissue lumpy jaw and goiter.

Disinfectants

A fourth major use of iodine is the manufacture of disinfectants. Iodine is an effective germicide for many microorganisms, and its

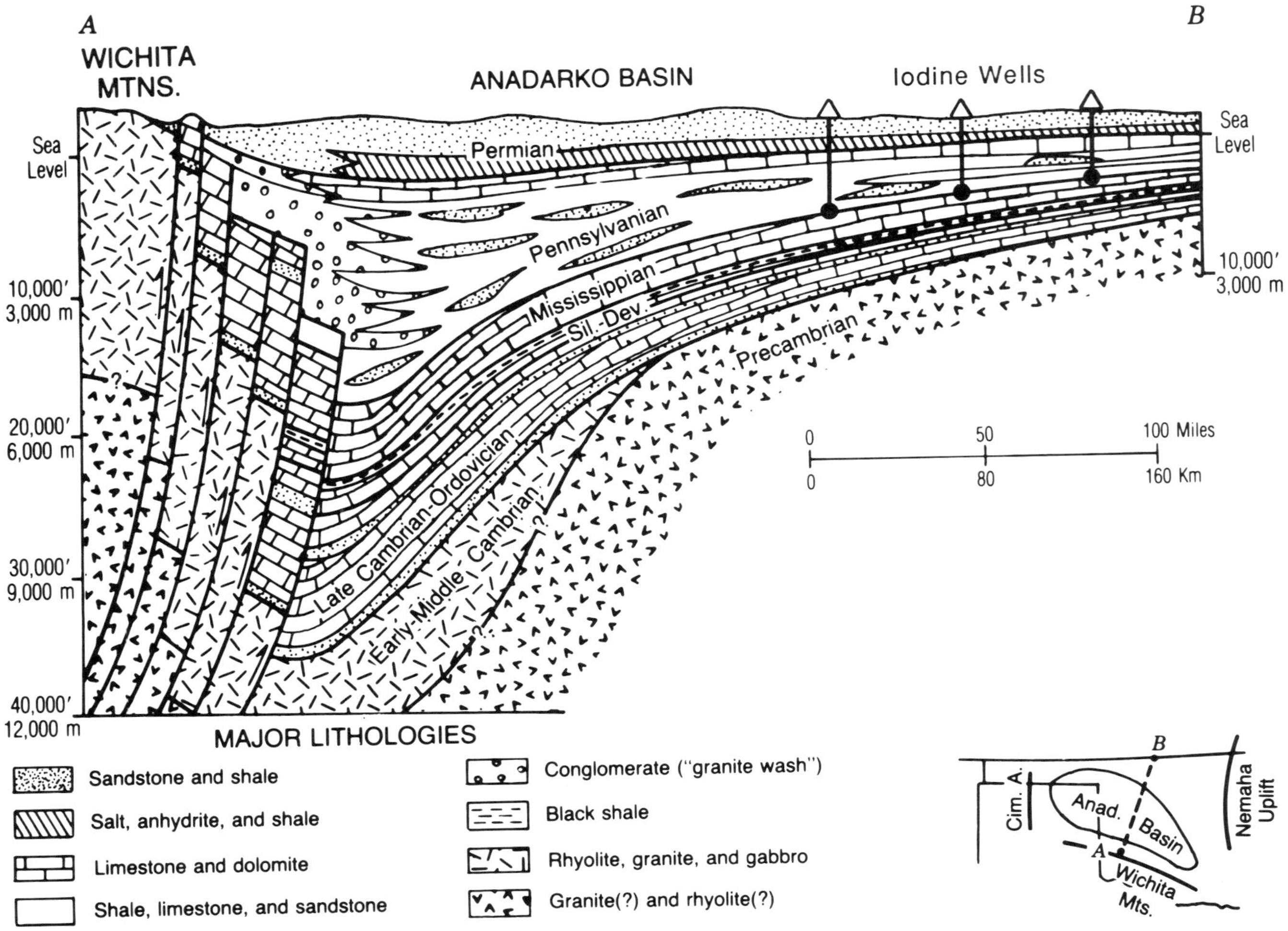

Fig. 3. Generalized north-south structural cross section through the Anadarko basin of western Oklahoma (modified from Johnson, 1989). Iodine-rich brines are produced from basal Pennsylvanian sandstones on north flank of the basin.

compounds are used in dairies, laboratories, food processing plants, and dish washing in restaurants. Iodine also is used as a disinfectant in swimming pools and in water supplies.

Pharmaceuticals

Pharmaceuticals are another major use of iodine, mainly as radiopaque media that are injected into the blood stream for producing X-ray photographs of the internal structure or blood circulation of the body. Potassium iodide is widely used in cough medicines as an expectorant, and iodine compounds are used in the synthesis of amphetamine, methamphetamine, and ethylamphetamine.

Photography

Photography is one of the oldest industrial uses of iodine. As much as 7% of the silver salt in the emulsions of negatives is iodide, and ammonium iodide is used as a photographic developer.

Colorants

Iodine is a coloring agent in some dyes used in foods and on materials. Red food coloring has been used in carbonated soft drinks, powdered drinks, gelatin deserts, icings, and pet foods, whereas red dyes are used in dyeing and/or printing of cotton, half-silk, jute, and straw products.

Other Uses

Iodine is also used in: the modification of selenium to make semiconductors; the manufacture of high purity metals such as titanium, zirconium, boron, and hafnium; the production of motor fuels; additives in rechargeable dry cells; smog inhibitors; and cloud seeders to induce rainfall.

ECONOMIC FACTORS

Iodine and its compounds are generally marketed in the form of crude iodine, resublimed iodine, calcium iodates, calcium iodide, potassium iodide, sodium iodide, and numerous organic compounds.

Product Specifications

Commercial-grade crude iodine normally has a minimum purity of 99.5% I_2, with the main impurities being water, sulfuric acid,

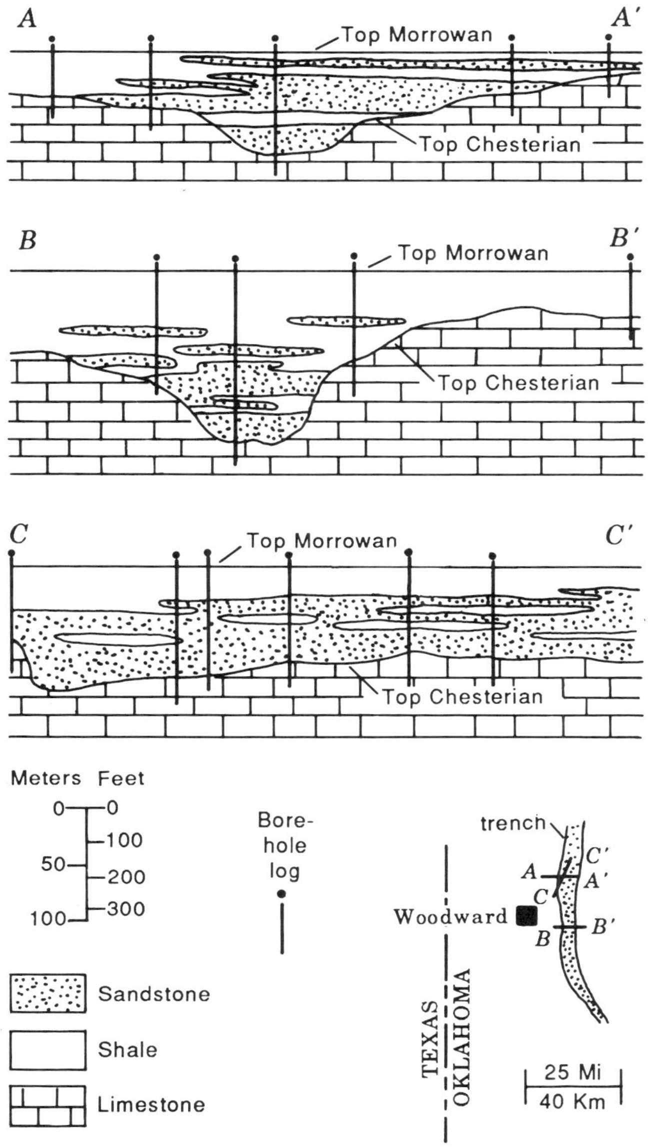

Fig. 4. Cross sections showing iodine-bearing Morrowan channel sandstones preserved in the Woodward trench that has been cut into Chesterian limestones (modified from Cotten, 1978). Datum is top of Morrowan strata. Length of A-A′ and B-B′ is about 10 km; length of C-C′ is about 15 km.

iron, and insoluble materials (Lyday, 1989a). Specifications of US Pharmacopoeia XVII call for crude iodine to be not less than 99.8% I_2.

Resublimed iodine is usually 99.9% pure, and ACS specifications call for not more than 0.005% total bromine and 0.020% nonvolatile materials (Anon., 1971).

Prices

United States prices for iodine (including cost, insurance, and freight charges for imported iodine) have been steadily increasing over the past two decades, rising from $3.48 per kg of crude in 1970 to an average of $16.98 per kg in 1989 (Lyday, 1989b). The highest average price during this 19-year span was $17.46 per kg in 1988.

Table 3. US Consumption of Iodine in 1985

Use	Thousand kg	% of Total
Catalysts	816	23
Stabilizers	680	19
Animal Feeds	499	14
Disinfectants	544	15
Pharmaceuticals	363	10
Photography	272	8
Colorants	181	5
Other	181	5
Total	3 536	99*

*Percentages do not total 100 because of independent rounding.
Source: Lyday (1989a).

Table 4. Tariffs on Iodine, 1991

Item	Number	Most Favored Nation (MFN) 1/1/91	Non-MFN 1/1/91
Iodine, crude	2801.20.00	Free	Free
Iodide, calcium and cuprous	2827.60.10	Free	25% ad valorem
Iodide, potassium	2827.60.20	2.8% ad valorem	7.5% ad valorem

Source: Lyday (1991).

US Consumption, Imports, Exports

Reported US consumption of iodine has increased steadily from 2 330 000 kg in 1986 to 2 900 000 kg in 1989 (Lyday, 1991). In 1989, about 3 300 000 kg was imported to the United States and about 2 100 000 kg was exported. US imports, during the years 1986 to 1989, were from Japan (71%), Chile (28%), and other countries (1%).

Transportation

Crude iodine is shipped in double polyethylene plastic-lined fiber drums holding 45 to 90 kg each. Resublimed iodine is also shipped in the same type of container, as well as in 11.3-kg glass jars and smaller bottles. Freight classification is "chemicals, not otherwise indexed by name" (NOIBN) and requires no special label.

Stockpiles, Depletion, Tariffs

On Sept. 30, 1990, the US Government strategic stockpile contained 2 750 000 kg of iodine, which was 122 000 kg in excess of the inventory goal (Lyday, 1991). The stockpile also contained 609 kg of nonstockpile-grade material.

Depletion allowances for iodine from both domestic and foreign production is 14% on solid material and 5% on brine (Lyday, 1992).

Tariffs on iodine are reported in Table 4.

REFERENCES CITED

Anon., 1971, "Iodine in 1971," *Mineral Industry Surveys,* US Bureau of Mines, annual preliminary survey.

Cotten, H.M., 1978, Iodine in Northwestern Oklahoma, *13th Annual Forum on the Geology of Industrial Minerals,* K.S. Johnson and J.A. Russell, eds., Circular 79, Oklahoma Geological Survey, pp. 89–94.

Erickson, G.E., 1981, "Geology and Origin of the Chilean Nitrate Deposits," Professional Paper 1188, US Geological Survey, 37 pp.

Fukuta, O., 1985, 'Japanese Iodine—Geology and Geochemistry," *Salts and Brines '85,* W.J. Schlitt, ed., AIME, New York, pp. 149–168.

Fukuta, O., and Fujii, N., 1982, "Japanese Iodine—Production, Geology, and Geochemistry," *Industrial Minerals,* No. 175, Apr., pp. 101–115.

Fukuta, O., Nagata, S., and Fujii, N., 1982, "Iodine from the Niigata Dissolved Dry Gas Producing Area, Japan," *Industrial Minerals, No. 180,* Sep., pp. 69–79.

Garrett, D.E., 1985 "Chemistry and Origin of the Chilean Nitrate Deposits," *6th International Symposium on Salt, 1983,* Vol. 1, B.C. Schreiber, ed., Salt Institute, Alexandria, VA, pp. 285–302.

Johnson, K.S., 1989, "Geologic Evolution of the Anadarko Basin," *Anadarko Basin Symposium, 1988,* K.S. Johnson, ed., Circular 90, Oklahoma Geological Survey, pp. 3–12.

Lyday, P.A., 1985, 'Iodine," *Mineral Facts and Problems, 1985,* Bulletin 675, US Bureau of Mines, pp. 377–384.

Lyday, P.A., 1986, "Crude Iodine Production," *Industrial Minerals,* No. 222, Mar., pp. 65–76.

Lyday, P.A., 1988, "Iodine," *Minerals Yearbook,* Vol. 1, Metals and Minerals, US Bureau of Mines, pp. 479–482.

Lyday, P.A., 1989a, "Iodine and Iodine Compounds," *Ullman's Encyclopedia of Industrial Chemistry,* 5th ed., Vol. A 14, VCH Publishers, New York, pp. 381–391.

Lyday, P.A., 1989b, "Iodine," *Minerals Yearbook,* Vol. 1, Metals and Minerals, US Bureau of Mines, pp. 515–520.

Lyday, P.A., 1991, "Iodine," *Mineral Commodity Summaries 1991,* US Bureau of Mines, pp. 78–79.

Lyday, P.A., 1992, "Iodine," *Mineral Commodity Summaries 1992,* US Bureau of Mines.

Ohl, J.P., and Arndt, R.H., 1988, "The Mineral Industry of Oklahoma," *Minerals Yearbook,* Vol. 2, Area Reports: Domestic, US Bureau of Mines, pp. 395–404.

Velasco, P., and Gurmendi, A.C., 1988, "The Mineral Industries of Southern South America, 1988 International Review," US Bureau of Mines, 38 pp.

Wu, J.C., 1985, "The Mineral Industry of Japan," *Minerals Yearbook,* Vol. 3, Area Reports: International, US Bureau of Mines, pp. 491–516.

LIME

ROBERT C. FREAS

Lime, the versatile chemical is, generally speaking, a calcined or burned form of limestone commonly known as quicklime, calcium oxide or, when water is added, calcium hydroxide or slaked lime. Almost 40 different lime products are available, a fact which has contributed to the rather loose use of the term *lime* as well as to confusion and misunderstanding. The term is frequently, albeit erroneously, used to denote almost any kind of calcareous material or finely ground form of limestone or dolomite.

Lime is made from high calcium or high magnesium limestone, generally having a minimum of 97% combined carbonate content. Normally, high calcium lime has less than 5% MgO. When the lime is produced from a high magnesium limestone, the product is referred to as dolomitic lime.

CALCINATION/HYDRATION

Calcination, or the production of lime, has its origins in the earliest days of alchemy, with the general reaction class having been identified in an Arabic text printed in 1000 AD. It was much later, however, in the mid-1700s to the mid-1800s before this basic reaction became understood from scientific perspective. The production of lime has been so basic and simple that its underlying scientific principles have over the years received only intermittent investigation. Rather, much of the thought and inquiry has been directed toward the development of kilns. Thus, it has only been in the last 25 to 30 years that lime has received any concentrated scientific investigation relative to the thermodynamics and kinetics of the calcination and hydration reactions. Particular emphasis has been focused upon energy consumption and fuel efficiency.

Calcination refers to a broad class of reactions, of which the lime/limestone reaction is just one, wherein a substance is heated to less than its melting point, resulting is a weight gain or weight loss. In the calcination of limestone to produce lime, the basic chemical reaction is as follows:

$$\frac{CaCO_3}{100}\text{(limestone)} + \text{heat (1 000° to 1 300°C)} = \frac{CaO}{56}\text{(quicklime)} + \frac{CO_2}{44}\uparrow$$

$$\frac{CaCO_3}{100}\cdot\frac{MgCO_3}{84}\text{(limestone)} + \text{heat (900°C to 1 200°C)} = \frac{CaO}{56}\cdot\frac{MgO}{40}\text{(quicklime)} + \frac{2CO_2}{88}\uparrow$$

While there is nearly universal agreement about the equilibrium conditions related to the limestone/lime reaction above, there have been numerous and varied calcination models developed for the reaction. Recent investigations have resulted in the development of the model shown in Fig. 1. From this model it can be seen that calcination is a function of both temperature and CO_2 pressure. It does not, however, provide any indication of the rate at which the reaction takes place.

Calcination is strongly time variant with different limestones. In a very broad sense this relates to the fact that the calcination reaction starts on the exterior surface of the limestone and then proceeds toward the center. As the calcination reaction takes place, the CO_2 released at the interface must make its way through the lime to the exterior surface. Since calcination is limited by gas diffusion to the surface of the partially calcined limestone, the natural impurities in the stone, differences in crystallinity, grain boundary chemistry, density variations, and imperfections in the atomic lattice all play a significant role in calcination rate. Therefore, the suitability of a given limestone as a source material for lime production can be determined only after completion of adequate burn tests designed to evaluate the various limiting factors. When a coal-fired kiln system is considered, the entire process of calcination is made even more complex by the introduction of additional chemical constituents into the calcination environment. The reader is referred to the references for a more detailed description of the complete calcination reaction and the differences inherent between differing kiln systems.

In the foregoing discussion it was noted that CO_2 is released during the reaction. This release results in a 44% weight loss during the complete calcination of a high calcium limestone, or a 48% weight loss for a highly dolomitic limestone. The trade term for this weight percent loss is *loss on ignition*, or LOI. This weight loss is frequently used as a measure of the completeness of calcination.

Because the calcination reaction is chemically reversible, *quicklime* or *burned lime* is frequently referred to a being highly reactive, or unstable. The more stable form of lime, *hydrated lime*, is commonly preferred and generally specified by the user. Hydrated lime is obtained by adding water to quicklime to produce a dry, fine powder. Quicklime's affinity for moisture is then satisfied, although it still retains a strong affinity for CO_2.

Hydration is the combining of calcium oxide with water in a reversible reaction to form calcium hydroxide:

$$CaO + H_2O \rightarrow Ca(OH)_2$$

As a standard condition the reaction is exothermic for the formation of hydrate, with the accepted value for the heat of hydration being 15.455 kcal/g mol of CaO. This is a three-step reaction, with the first being the dissociation of CaO into a Ca^{++} ion and 0 = ion:

$$CaO \rightarrow Ca^{++} + O^{=}$$

Next, the water ionizes into two hydroxyl ions due to the strong attraction that an oxygen anion exerts on a water molecule:

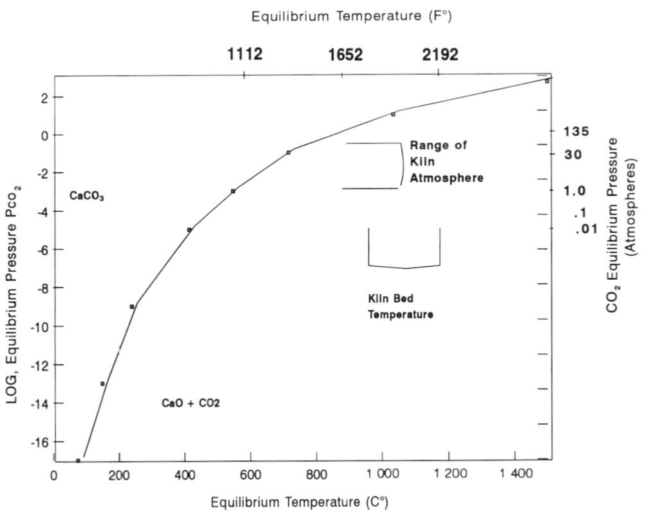

Fig. 1. Equilibrium graph for $CaCO_3 \leftrightharpoons CaO + CO_2$ as a function of temperature and CO_2 pressure in atmospheres. Conditions below the curve cause the reaction to go toward lime (CaO), while above the curve the reaction proceeds toward calcite. Only on the curve is the equilibrium maintained and the reaction does not proceed in either direction. Most lime kilns have production bed temperatures of 1 000 to 1 150°C (Thompson, 1978).

$$H_2O + O^{=} \rightarrow 2(OH)^-$$

Finally, the calcium and hydroxyl ions combine to form calcium hydroxide:

$$Ca^{++} + 2(OH)^- = Ca(OH)_2 + \text{heat}$$

As noted this is a reversible reaction, and therefore hydrated lime can, when heated, revert to the original oxide form.

The success of calcination, or lime production, is measured in the available lime content (CaO). In the case of dolomitic lime this is expressed as total oxides. This, in effect, is the purity of the lime. Typical chemical analyses of commercial quicklimes are presented in Table 1. The purity of lime is influenced primarily by the purity of the limestone, and secondarily by its manufacture. The major impurities are silica, iron, alumina, and sulfur. During calcination these impurities react with some of the calcium oxide, yielding mineral forms (silicates, aluminites, ferrites, etc.) that reduce the available CaO by 3 to 5% or more. Even the purest calcium limes assay less than 95% in available CaO, several percentage points less that the total oxide content. In addition, uncalcined limestone particles may remain as core in the individual lime pieces.

Impurities and core also affect the hydration reaction. As an example, sulfur in the form of calcium sulfate inhibits the hydration reaction out of all proportion to its actual weight percent concentration (perhaps more than 100 times) (Boynton, 1980). If the parent stone contains dolomite, this will also inhibit the reaction as its hydration may be nearly three orders of magnitude slower than that of calcium oxide. Early in the hydration reaction dolomite appears practically inert, but given sufficient time it will hydrate. The heat of reaction is nearly the same as for calcium oxide, but the time necessary for complete hydration may be hours, whereas for calcium oxide it is only seconds.

Table 1. Typical Analysis of Commercial Quicklimes

Component	High Calcium Quicklimes Range, *%	Dolomitic Quicklimes Range, *%
CaO	93.25–98.00	55.5–57.50
MgO	0.30–2.50	37.60–40.80
SiO_2	0.20–1.50	0.10–1.50
Fe_2O_3	0.10–0.40	0.05–0.4
Al_2O_3	0.10–0.40	0.05–0.40
H_2O	0.10–0.90	0.10–0.90
CO_2	0.40–1.50	0.40–1.50

Source: National Lime Association (1989).
*The values given in this range do not necessarily represent minima and maxima percentages.

Important physical properties of lime are listed in Table 2. Lime is white, although with brightness of varying intensities, some limes have a light cream, buff, or gray cast, depending upon the impurities present. The severity of calcination (time and temperature) determines the degree of porosity and chemical reactivity of the lime. When a lime is soft-burned, very little or no shrinkage occurs and a porous, softer, very reactive lime is produced. Boynton (1980) presents an excellent discussion of the physical properties of both soft and hard burned lime.

When a lime is hard burned, a denser, physically stronger, less reactive lime results. In either case the lime will readily hydrate in water although the rapidity of hydration is much greater with the former type than with the latter. In any event the heat of hydration is appreciable: 1 134 and 886 kJ/kg of quicklime for high calcium and dolomitic types, respectively. This strong exothermic reaction will boil water easily and, under certain hydration conditions, temperatures of 290 to 315°C have been reached, causing dehydration of the freshly slaked lime. Quicklime can be made so reactive that it will literally explode on contact with water.

Although limestone is almost totally insoluble, lime is slightly soluble in water in the range of 0.54 to 1.4 mg/l, depending on the temperature. As water temperature rises, lime's solubility diminishes. The addition of sugar will increase these solubilities manyfold.

On slaking into a slurry or milk-of-lime, the saturated solution ionizes immediately into Ca^{++}, Mg^{++}, and OH^- ions, one of the strongest bases. Even a trace of lime will yield a pH of 11.2 and on up to nearly pH of 13 at saturated solution and low temperature.

As an alkali, unlike other strong bases, (NaOH, KOH), $Ca(OH)_2$, and $Mg(OH)_2$ are diacid bases. Therefore, one molecule of lime will neutralize two molecules of acid (sulfuric, hydrochloric, etc.). For example, at equal weights CaO has 30% more neutralizing power than caustic soda (sodium hydroxide). However, dolomitic quicklime, because of its MgO content has 16% greater neutralizing power than pure CaO.

LIME MANUFACTURE

Production of lime involves three main processes—stone preparation, calcination, and hydration—although the latter is not necessary for several lime uses. The simplified flowsheet (Fig. 2) outlines the various production steps. Details of stone preparation, which comprises quarrying or mining, crushing, and screening to produce kiln-size stone, are covered in the chapter on limestone and dolomite. However, because lime must meet existing chemical specifications, the various steps in stone production are carefully

Table 2. Properties of Typical Commercial Lime Products

QUICKLIMES		
	High Calcium	Dolomitic
Primary Constituents	CaO	CaO and MgO
Specific Gravity	3.2–3.4	3.2–3.4
Bulk Density (pebble lime), g/cm^3	0.88–0.95	0.88–0.96
Specific Heat at 38°C kJ/kG	0.4	0.94
Angle of Repose	55°*	55°*
HYDRATES		
	High Calcium	Normal Dolomitic
Primary Constituents	$Ca(OH)_2$	$Ca(OH)_2$ + MgO
Specific Gravity	2.3–2.4	2.7–2.9
Bulk Density, g/cm^3	0.4–0.56†	0.4–0.56
Specific Heat at 38°C kJ/kg	0.62	0.62
Angle of Repose	70°*	70°*

Source: National Lime Association (1988).

*The angle of repose for both types of lime (hydrate in particular) varies considerably with mesh, moisture content, degree of aeration, and physical characteristics of the lime; e.g., for quicklime it generally varies from 10 to 13°C and for hydrated lime it may range as much as 2 to 26°C.

†In some instances these values may be extended. The Scott method is used for determining the density values. In calculating bin volumes the lower figure should be used.

controlled to prevent contamination with undesirable impurities, particularly silica, alumina, and iron oxide. Accordingly, at several lime plants practices such as selective quarrying and washing are carried out.

Calcining

Although lime calcining (or burning) is a relatively simple operation chemically, a wide variety of kiln systems are used today. These include vertical, rotary, rotary with preheaters, multiple-shaft regenerative, and a variety of other types of kilns, in striking contrast to the portland cement industry which utilizes rotary kilns with preheaters almost exclusively.

Lime kiln selection depends upon several factors, the most important one being burning characteristics of the stone, fuel consumption, and capital equipment costs. Other factors include market requirements, fuel availability and air pollution regulations. The discussion which follows describes the various calcining systems and their relative advantages and disadvantages.

Vertical Kilns

The US Bureau of Mines (1991) reports that of the 113 currently active lime plants in the United States, 38 (33.6%) are utilizing vertical kilns. Of these 38 lime plants, 34 have the older, smaller capacity vertical shaft kilns while 4 have the new multi-shaft regenerative kiln. Interestingly, 25 of the 34 older style plants are operated by sugar companies processing sugar beets. These kilns tend to be lower volume, less than 250 tpd and are captive to their respective sugar plants.

Vertical kilns have traditionally burned only larger stone, 75 to 300 mm, with a size range of approximately 1:2. However, the newer designs can handle stone sizes as small as 20 mm and a widened size range to as much as 1:4. Fuel consumption has been a major advantage of shaft kilns, normally requiring less than 5.2 GJ/t of lime produced. This has been furthered improved upon with the newer designs which reportedly operate as low as 3.3 to 3.9 GJ/t of lime. A disadvantage of traditional vertical kilns in today's energy environment is their fuel requirement of oil, natural gas, or coke. However, several recent developments in burner technology have come to the point where pulverized coal can now be utilized in modern shaft kilns.

Vertical kilns may be of stone masonry, reinforced concrete, or boiler plate construction. The most widely used kiln has a refractory-lined steel shell and is usually circular in cross section. These kilns may be 2.7 to 7.3 m diameter and 15 to 48.5 m high. Capacities vary from as low as 9 tpd to in excess of 635 tpd with the larger tonnages being restricted to the newer generation of kilns.

Among the newer vertical kilns are the center burner, gas fired (Azbe, Union Carbide), oil fired annular (Catagas & Beckenback), and the Beckenback double-incline (Cascade) kilns. Also, there are multiple shaft, parallel flow regenerative kilns of which five are now in operation in the United States.

Multiple Shaft Kilns

Multiple shaft kilns were developed to burn small stone. This type kiln utilizes the parallel flow calcining principle in double and triple shaft units. The shafts are interconnected in the burning zone and while one shaft is being fired the other is preheated (Fig. 3). Fuel and combustion air are supplied to the burning shaft from above, ignite at the upper end of the burning zone, and calcine the lime in uniflow. The exhaust gases then pass into the second shaft preheating the stone in counter flow. After a 10 to 15 minute interval the shaft firing is reversed. Cooling air is blown into both shafts simultaneously. Because of the novel heat regenerative system, fuel consumption is 3.8 to 4.0 KJ/t of lime produced. These type kilns vary in capacity from 90 to 550 tpd. While normally gas or oil fired, recent developments in burner technology have lead to several installations now being equipped to fire pulverize coal (Fig. 4). Kilns of this type have found worldwide acceptance and an increasing number of them are being installed in the United States. One of the newest ones belongs to Chemstar Inc., in Soda Springs, ID. This is a 550 tpd Mertz vertical shaft kiln and was in production the second quarter of 1992. Chemstar, also installed a second kiln similar to the Soda Springs, ID, facility at Cosgrave, NV, bringing it on line in early 1993.

Annular Shaft Kilns

The burning process in this type kiln is based upon counter flow for preheating and calcining and uniflow for residual calcination. Partitions provided by an inner cylinder and by staggered bridges in the burning zone permit even distribution of heat and uniform flow of material down the kiln. Stone as small as 25 × 75 mm can be calcined to produce a soft burned lime using either oil or gas. Capacities vary from 90 to 270 tpd with fuel consumption reported at under 6.2 KJ/t.

Rotary Kiln

Unlike vertical kilns, which operate fully charged, the rotary kiln has about 90% of its volume filled with flame and hot gases. As the kiln slowly rotates new surfaces of stone are exposed to the hot gases but there is little passage of gases through the solids. Hence, radiation interchange between gas solids and refractory wall is an important part in the overall heat transfer mechanism. Because the area of solids exposed is relatively small, a rotary kiln is less efficient than a shaft kiln. However, the rotary has the advantage of burning stone from 6.5 to 57 mm. Generally though the size ratio is 1:3 in order to minimize segregation and to promote thorough calcination. This sizing minimizes the residual core left in the individual lime fractions, while at the same time allowing a thorough soft burn of all of the limestone feed. Also, when using these

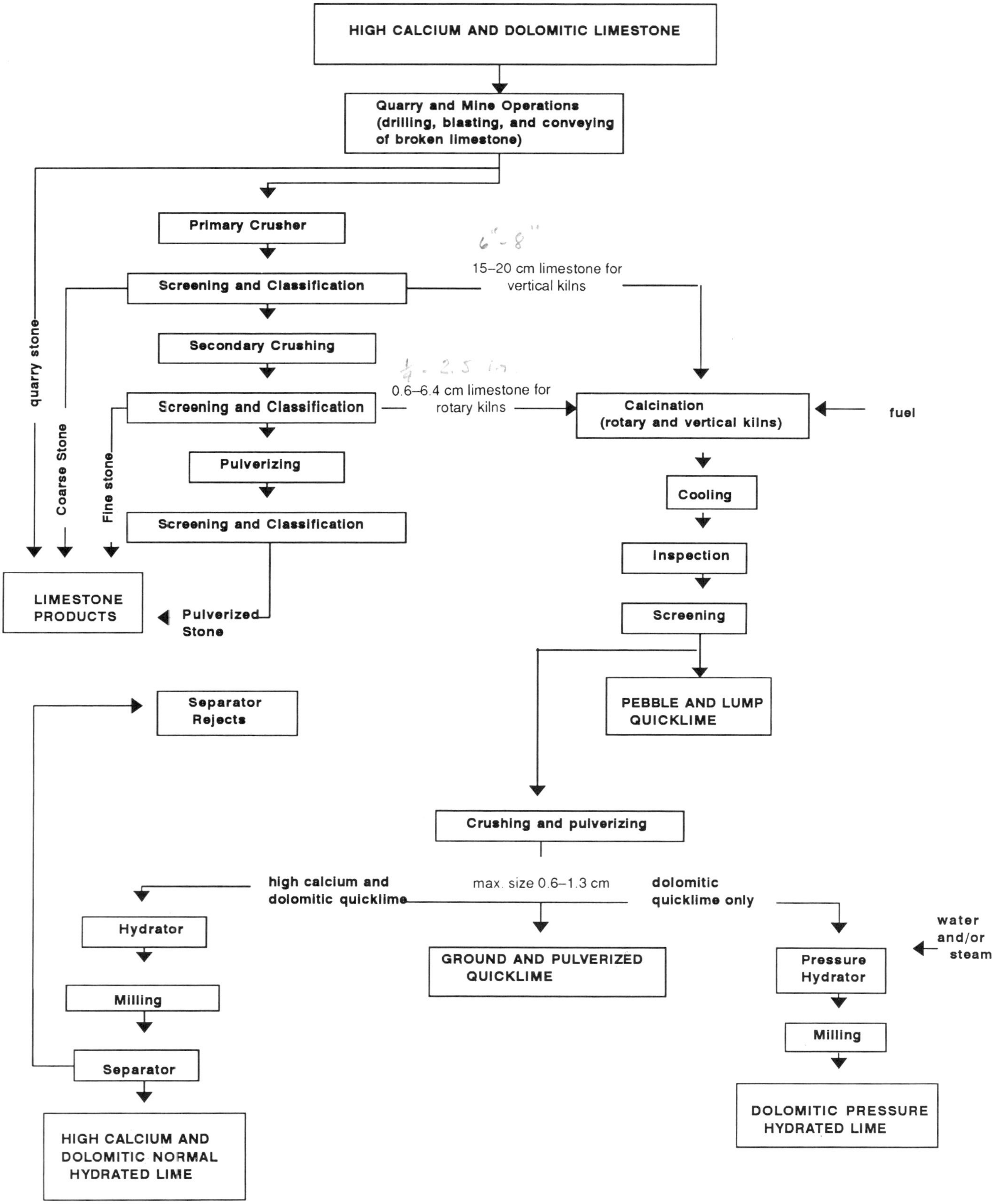

Fig. 2. Simplified flowsheet for lime and limestone products (Boynton, 1980).

Fig. 3. Multiple shaft parallel flow regenerative kiln.

Fig. 4. Fuel lances on a parallel flow shaft kiln.

type kilns it generally is not necessary to crush the final product as it comes out in an acceptable pebble lime size.

Rotary kilns vary greatly in size ranging from 2 × 25 m to 5.5 × 190 m with capacities of 45 to 945 tpd. Because of the relatively high length to diameter ratio these kilns have come to be termed long kilns. In North America these long kilns are typically 2.5 to 4.0 m × 45 to 105 m long producing 181 to 410 tpd of lime. The largest rotary kilns are 1 000 tpd capacity. While rotary kilns can burn a wide range of fuels, a major drawback of these long kilns has been their lack of fuel efficiency. Many of the earlier rotaries require in excess of 12.6 KJ/t of lime. This high fuel requirement has been reduced to 7.4 KJ/t through the use of internal heat exchangers and dams, recuperative type coolers, more sophisticated and complex instrumentation, and improved refractories.

Rotary Preheater Kilns

Of the several advances which have been made in improving the heat efficiency of rotary kilns, probably the most pronounced has been the trend away from long kilns to medium length kilns with external preheaters. These vertical, cylindrical, or shaft type preheater kilns are available through Kennedy Van Saun, Fuller Company, and Allis-Chalmers. The preheater is a refractory-lined chamber located below the raw feed bin and ahead of the kiln. Exhaust gases from the kiln are drawn counter-current through the stone preheating it to 540 to 900°C. Retention time within the preheater is 1.0 to 1.5 hours depending upon the system design, burning characteristics of the stone, etc. During preheating approximately 30% calcination is achieved before the stone is discharged into the kiln. Following preheating the partially calcined stone is sent to the kiln at predetermined rate by means of hydraulically actuated plungers. This type preheater has been adapted to the new large volume kilns which typically may be 5.2 m in diameter by 62.5 m long with a capacity of 907 tpd.

Allis-Chalmers has also developed a second type of preheater kiln which is known as a grate-kiln. This is a single pass, down draft, enclosed traveling grate on which the finer stone sizes, placed on top of the coarser stone sizes are virtually calcined while the coarser sizes are preheated. Then in the rotary kiln which follows the coarse sizes are calcined while the finer sizes are protected from over burning because they sink into the kiln load through a sifting action. Grate-kiln systems of 270 to 540 tpd capacity are in operation in North America.

Heat exchangers and coolers which recuperate waste heat are also utilized to affect improved fuel efficiency. The heat exchangers, principally trefoil or quadrant sections, are mounted at the feed end. Their purpose is to turn over the stone more effectively and increase the turbulence of the hot gases. Dams and lifters may be used to turn over the stone, create gas turbulence, or retard the passage of the stone. Coolers on the other hand, are located at the discharge end of the kiln and serve to return hot gases to the kiln as secondary air. Three types of coolers are generally used with the rotary kilns—the shaft or contact, the planetary, and the rotary. The most prevalent type is the Niem's contact (counter flow), which is refractory lined and mounted below the kiln hood. Cooling air is provided by a fan of sufficient size to reduce the lime temperature to 65°C or less. The planetary cooler consists of eight or more tubes, approximately 4.2 × 6.2 m in size mounted at the kiln discharge end. The lime is moved in each tube by conveyor flights counter flow to the cooling air. The rotary cooler is generally not as effective as the other two and has the additional disadvantage of attrition loss during rotation.

The net effect of use of both shaft and grate preheaters, and improved heat exchange recovery within the kilns has been to reduce fuel ratios in rotary kilns from 6.9 to 8.5 KJ/t of lime produced. In addition, advances in instrumentation and monitoring have reduced manpower requirements and increased product quality and uniformity. Taken together, these changes have helped the large capacity rotary kiln with preheater become the most commonly selected system today for new or expanded lime production facilities.

Calcimatic Kilns

The Rotary Hot Calciner, or Calcimatic kiln, consists of a preheater, circular hearth and cooler all of which are refractory lined. Like the rotary this kiln burns small stone which is typically sized at a 1:3 ratio. The stone is carried on the hearth in a thin layer and one revolution of the hearth constitutes the calcining cycle. Numerous burners located inside and outside of the hearth are used for firing, utilizing gas, fuel oil, coke oven gas, and most recently, pulverized coal. Fuel requirements approximate 7.7 KJ/t and can be adjusted to accommodate a wide range of calcination. A significant advantage of this kiln is the fact that mechanical attrition loss is negligible. Stone rests virtually motionless on the hearth, therefore permitting the utilization of soft limestones which would otherwise be subject to high mechanical degradation during calcination. At present there is only one active calcimatic kiln in North America, although there are several 90 to 270 tpd capacity kilns in use in Europe.

FluoSolids Kilns

For limestone that is friable or decrepitates during calcining the Dorr-Oliver FluoSolids kiln has been used to produce a highly reactive lime. In this fluid bed process the material is maintained in suspension in a rising current of hot gases until calcined. Kiln feed is normally 60 to 65 mesh in size. The system which can be fired with gas or oil incorporates a FluoDry unit (preheater) and FluoSolids reactor (calciner) the latter resembling a rotary kiln. Fuel consumption is approximately 7.8 KJ/t. Fluid bed calciners are available in size ranges of 45 to 225 tpd, however, the only kilns of this design in North America are operated by Pfizer at their North Adams, MA, plant.

Flash Calciners

The Flash Calciner is one of the most recent developments in calcination and in reality is not a kiln. This system developed by Fuller Company is designed to calcine the undersized material or fines remaining after the primary kiln feed has been sized. The system includes preheater, flash calciner, and cooler. Preheating is achieved in a series of cyclones in which the heat of rising hot gases is absorbed in the counter flowing material which is fed from the top and heated to 980°C. As the preheated feed drops into the furnace or calciner, a forced air vortex flow pattern pulls the particles into the center of the furnace and then up the sides before discharge. Within the furnace the temperature is 1 200°C, but retention time is only a few seconds. A suspension type cooler recovers heat from the calcined product before it is conveyed to the product bin. The system is designed in two modes depending upon what size feed is to be used. Where stone fines are to be calcined, 60 to 80% of the calcination occurs in the furnace. The final calcination takes place in a fluid bed located between the furnace and the cooler. In those systems where −28 mesh material is fed to the calciner, no fluid bed reactor is necessary. This system can utilize gas, oil or coal, and will normally require 6.2 to 8.5 KJ/t of product.

MILLING AND HYDRATING

Following cooling, quicklime is carefully inspected to avoid core, or pieces of uncalcined rock. This is particularly true for shaft kilns burning large stone. The quicklime is then crushed and sized for shipment. Generally, the fines are converted to hydrated lime.

Hydrated lime shipped in 23 kg bags or in bulk is available in two principal forms—standard or 85% passing #200 sieve and super fine or spray hydrate where 98% or more passes the #325 sieve. These in turn are manufactured in two types N (normal) and S (special, see ASTM designation C-207–80), each including high calcium or dolomitic hydrates.

Type S limes are highly hydrated generally by processes involving pressure hydration. They contain less than 8% unhydrated oxides and develop high and quick (15 minutes) plasticity and high water retention. Since World War II the demand for type S lime has grown rapidly until currently it is the chief lime for structural uses. Some lime manufacturers now add entraining catalysts in the lime to increase plasticity and durability (called S.A. and N.A. lime).

Mechanical hydrators used for type N lime are made in various designs including batch hydrators, like the Clyde, and continuous hydrators like the Kritzer, Schaffer, Knibbs, Hardinge, and Kennedy Van Saun. The hydration process consists of adding water slowly to quicklime (usually in crushed or ground form), followed by agitation to produce intimate contact of lime with water. Accurate proportioning is essential and this is accomplished by using flow meters for water and weighing feeders for lime. More than the theoretical amount of water required should be used for hydration in order to allow for the water that is lost as steam from the heat of hydration. Following hydration the product is generally fed to a pulverizer or tube mill equipped with an air separator. The core together with most of the silica and aluminum is rejected by the air separator. By adjusting the separator, hydrates of varying degrees of fineness can be made.

Type S dust hydrates are manufactured by various processes, the principal one being the Corson Explosion Method of continuous pressure hydration. In this process ground dolomitic quicklime and water are automatically fed in constant proportions to a high speed slurry mixer. Hydration then takes place in an insulated autoclave operating at about 5.3 kg/cm^2 at 150°C. After about thirty minutes retention time the product is discharged continuously (or exploded) through a small pipe at the rate of 3 050 m/sec into a special cyclone collector at atmospheric pressure. There the hydrate and water are separated by the spray drying principle. The product is then ground in a tube mill to increase the plasticity. Several US plants use the Corson Process marketing their product under the registered tradename Miracle Lime®.

CONSUMPTION AND USES

Lime consumption in the United States has recovered somewhat from the lows of 1982 and 1986, yet still remains well below the high of 10 to 15 years ago (Fig. 5). Current consumption levels reflect both current recession economics as well as changing consumption patterns. Table 3 compares 1990 consumption by end use to similar applications ten years earlier, in 1980. These figures show that in 1990 the major end uses included 66% chemical and industrial, 24% environmental uses, 8% construction uses, and 2% refractory dolomite. Captive lime included in these figures was used mainly in sugar refinery and the production of steel.

Steel

Over the past thirty years the steel industry has been the major consuming market for lime. While this situation continues today, there has been a well defined decline in sales to the steel industry. Since 1980 this market has dropped from 6 Mt or 38.1% of US lime

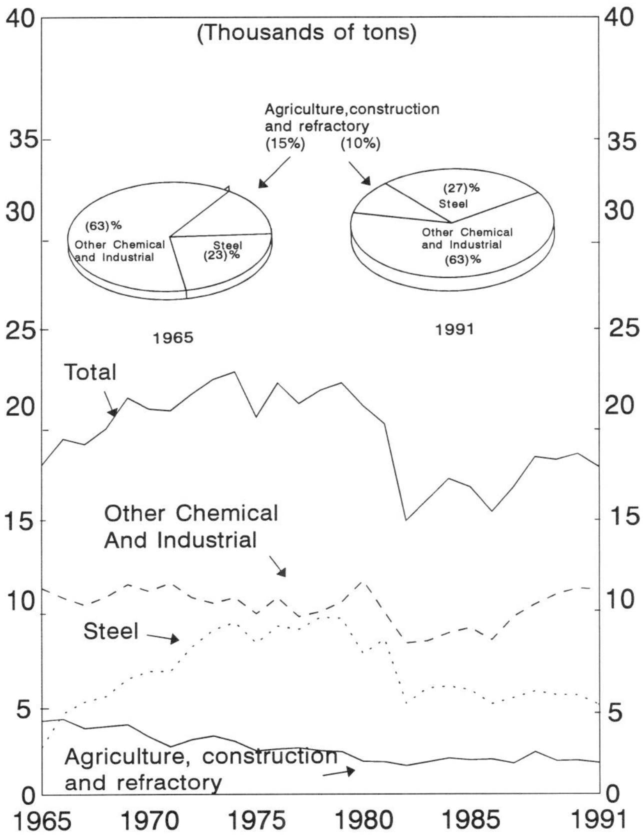

Fig. 5. Lime consumption in the US.

production to 4.6 Mt or 29.4% in 1990. While both the importation of steel and automobiles have impacted the steel industry requirements for lime, the reduction is also a result of a shift to more polymer composites in car and truck design and manufacturing as well as the greater role being played by mini-mills in steel production.

Nevertheless, the basic oxygen furnace (BOF) is still the dominant process, and continues to utilize approximately 80% of the lime sold for steel production. A typical BOF process required 60 to 65 kg/t of steel versus an average 30 kg/t of steel produced in an electric furnace or mini-mill. Both of these applications use lime as a scavenger or flux to remove impurities such as phosphorous, silica, alumina, and sulfur. While quicklime is the product of choice for a BOF furnace, 10 to 30% dolomitic quicklime is used with the high calcium lime to extend refractory life. Dead-burned dolomite, or refractory lime, is also used in open hearth steel furnaces for the same reason.

Lime used in steel production is normally a highly reactive, 90 to 93% available lime, and should have less than 0.06% sulfur. Pebble lime, 1.25 to 5 cm, or pelletized lime of the same size is normally required in US steel processes. However, the USX developed Q-BOP process in some European mills requires a ground or pulverized lime. The steel industry also uses small volumes of lime in wire drawing steel plants, neutralization of sulfuric pickle liquors, water softening, and coating molds for pig casting. There has also been some experimental use of hydrated lime in place of pulverized limestone in making self-fluxing ferrous sinter.

Non-Ferrous Metallurgy

Non-ferrous metallurgy accounts for the utilization of approximately 1.36 Mt of lime per year, or about 8.7% total US lime consumption. Non-ferrous metal ore beneficiation normally involves the recovery of the metals through some form of flotation process. Here, lime is used for pH control, may be used to neutralize sulfuric acid waste liquors, or in processes employing cyanide, lime may be used to curtail the loss of cyanide. Lime is also employed as a flux in the sintering of low carbon chrome and in the recovery of nickel by precipitation.

Copper: Most domestic copper mining companies use lime as the principal reagent in beneficiating copper ores during flotation. Lime is used to neutralize the acidic effects of pyrites and maintain the proper pH. There is limited use of lime as a flux in smelting concentrated copper ores and to absorb SO_2 from the stack gases of copper smelters.

Alumina: The manufacturer of alumina by the Bayer process frequently requires large quantities of lime to be used to causticize sodium carbonate solutions to regenerate sodium hydroxide for recycle use. Nearly all plants will also use some lime for secondary desilication in refining alumina. Some of the experimental processes developed by the US Bureau of Mines (USBM) for making alumina from clay or kaolin instead of bauxite or ore also require lime.

Magnesia: Most methods for the manufacture of magnesia, such as the Dow seawater and natural brine processes, ferro silicon processes, and seawater magnesia processes require lime. Dolomitic lime, because of its magnesium content is generally preferred but high calcium lime has been successfully used. Dolomitic lime, however, cannot be used in the manufacture of magnesium metal. This process requires a very high calcium, high purity lime.

Pulp and Paper

On the surface, the use of lime in pulp and paper production would appear to have been fairly stable since 1980. However, there has been a significant shift in US paper manufacturing from clay to alkaline sizing. This has resulted in an increased use of precipitated calcium carbonate and a corresponding doubling in lime usage for PCC production. Thus, lime consumption directed at the end use of paper manufacturing has increased by about 100 ktpy.

When lime is used in the manufacture of sulfate (craft) paper pulp it is used to causticize the waste sodium carbonate (black liquor) to regenerate sodium hydroxide for reuse in the process. In spite of the fact that pulp mills in turn recover 90 to 96% of the lime employed by recalcining the dewatered calcium carbonate sludge a substantial quantity of lime continues to be required as makeup material.

A secondary use of lime in the paper industry is the making of calcium hypochlorite bleach which is used to bleach paper pulp to the desired degree of whiteness. Lime is also used in the clarification of process water and in color removal of waste water effluents.

Chemicals Manufacture

Alkalies: In the manufacture of soda ash and bicarbonate of soda by the solvay process lime is employed to recover ammonia for recycle use in the process. This process requires 635 kg of quicklime per ton of soda ash produced. In the process the lime reacts with the chloride ion to form calcium chloride much of which is recovered and sold commercially. In addition, the process also yields byproduct precipitated calcium carbonate. Also, lime is used to causticize sodium carbonate solutions to produce sodium hydroxide (caustic soda). Lime usage for this application has dropped tenfold over the last ten years as the solvay process has lost most of its market share to natural soda ash (trona) and electrolytic caustic soda. However, this loss of lime market in chemical manufacturing

Table 3. Lime Sold or Used by Producers in the United States by Use[1] (thousand tons and thousand dollars)

	1980				1990			
Use	Sold	Used	Total[2]	Value	Sold	Used	Total[2]	Value
Agriculture	65	—	65	3,286	45	—	45	2,898
Chemical and industrial:								
Alkalies	5	1 138	1 144	58,804	W	W	96	5,170
Aluminum and bauxite	145	104	250	12,036	141	—	141	8,122
Copper ore concentration	309	289	598	28,859	W	W	339	17,540
Food products, animal or human	34	—	34	1,602	19	—	19	1,171
Glass	164	—	164	7,910	90	—	90	5,022
Magnesia from seawater or brine	—	—	589	38,414	W	W	594	31,987
Oil well drilling	35	—	25	1,689	11	—	11	693
Oil and grease	29	—	29	1,395	W	W	594	31,987
Ore concentration, other	16	—	16	773	301	—	301	15,940
Paper and pulp	945	105	1 051	50,658	W	W	1 150	64,901
Precipitated calcium carbonate	59	43	102	4,905	W	W	231	11,718
Steel, BOF	4 008	1 310	5 318	256,469	W	W	3 667	193,254
Steel, electric	686	31	717	34,556	W	W	885	50,698
Steel, open-hearth	513	35	547	26,407	111	—	111	5,856
Sugar refining	53	826	879	42,414	27	5 710	597	35,596
Tanning	25	—	25	1,243	24	—	24	1,548
Other[3]	895	710	1 775	49,067	8 108	1 071	2 199	131,599
Total[2]	7 922	4 591	4 184	107,201	8 832	1 644	10 474	582,100
Construction:								
Soil Stabilization	658	—	658	35,071	611	—	611	36,188
Finishing lime	90	—	90	4,777	145	—	145	14,708
Mason's lime	262	36	298	15,916	W	W	231	17,932
Other[4]	15	25	40	2,111	727	—	268	12,217
Total[2]	1 024	62	1 085	57,872	W	W	1 255	81,046
Environmental:								
Acid water, mine or plant	381	64	445	21,467	269	—	259	16,120
Sewage treatment	771	11	782	29,905	475	—	425	22,014
Flue gas sulfur removal	675	—	675	32,566	1 521	—	1 521	89,694
Water purification	1 352	8	1 360	65,603	975	—	975	50,271
Other[5]	—	—	—	—	588	9	597	33,900
Total	3 179	83	3 263	157,341	3 766	9	3 776	211,999
Refractory Lime (dead-burned dolomite)	382	68	449	28,308	W	W	343	26,988
Grand Total[2]	12 571	4 804	18 136	854,008	14 069	1 822	15 892	905,032

Source: Miller, 1990.
W Withheld to avoid disclosing proprietary data.
[1]Excludes regenerated lime. Includes Puerto Rico.
[2]Data may not add to totals shown because of independent rounding.
[3]Includes briquetting, brokers, calcium carbide, chrome, citric acid, commercial hydrators, desiccants, environmental uses (1989), explosives, ferroalloys, fiber glass, glue, insecticides, ladle desulfurizing, magnesium metal, manganese, metallurgy, pelletizing, pharmaceuticals, petrochemicals, precipitated calcium carbonate, (1989), rubber, silica, brick, soap, wire drawing, and uses indicated by symbol W in "Chemical and Industrial" lime only.
[4]Includes asphalt antistripping.
[5]Includes industrial solid waste treatment, industrial wastewater treatment, scrubber sludge solidification, and other environmental uses.

may be about to turn around as Dow Chemical is currently considering a switch from caustic soda in its US propylene oxide production. The company already has a propylene oxide facility in Brazil that uses lime and thus, this switch is technically feasible. If carried to its fruition the additional lime potentially required by this new market could represent as much as 3% of the total US lime production.

Inorganic Chemicals: Inorganic chemicals made from lime include mono-, di-, and tricalcium phosphates, chrome chemicals, purifying salt brines used to produce USP food grade salt, pesticides such as lime-sulfur sprays and powders, and calcium arsenate. Hydrated lime is also used in proprietary water paints which are used for masonry applications as well as in white portland cement and some paint pigments. Lime also has minor uses in varnish manufacture, casein paints, and certain performance pigments.

Organic Chemicals: Lime is required in the manufacture of ethylene and propylene glycols in the chlorohydrin process, in calcium based organic salts, such as calcium stearate, acetate, lactate, and lignosulfonate. Lime is also used in the manufacture and refinement of citric acid and glucose. Within the petroleum industry lime has limited use for neutralizing organic sulfur compounds, and SO_2 emissions control in petroleum refineries. Lime is also used in the manufacture of certain lubricating greases and drilling muds.

ENVIRONMENTAL

Water and Waste Treatment

Over 1.36 Mt of lime are used annually in the treatment of municipal potable water and industrial waste water. Lime is used either by itself or in combination with soda ash to soften water in systems where ion exchange processes are not employed. Lime is required to remove the temporary bicarbonate hardness from the water. The high pH of 11.5 induced by the lime serves as a secondary sterilization agent to chlorine since retention for 3 to 10 hours at this pH will kill 99% of the bacteria and most viruses. By introducing CO_2 into the lime treated water the pH is lowered to acceptable levels and most of the lime in solution is precipitated as a carbonate sludge. An important application is in coagulation where lime is employed with such coagulants as alum or iron salt to remove turbidity from river waters as they are purified for potable purposes. Here lime is used for pH control to negate the acidic effect of coagulants and to achieve optimum coagulation efficiency. Subsidiary to its role in water softening, sterilization and coagulation, is its absorption of iron, magnesium, and organic tannins from untreated water.

For several decades sewage disposal plants have used lime for a variety of functions including the chemical precipitation of sewage, pH control in the sludge digester in biological treatment plants, and for conditioning sludge when dewatering with ferric chloride.

As a result of the growing problem of eutrophication of surface waters, caused to some extent by sewage plant effluents, more lime will be used in the future. Sewage effluents will require much more treatment to remove dissolved and suspended solids which contain plant nutrients. Lime is useful in precipitating phosphates directly and when alum or iron salts are used as a coagulant, lime is generally required for pH control. It appears that a high pH of 10 to 11 inherent in lime treatment is necessary for efficient removal of nitrogen. As federal and state EPA requirements become more stringent, there will be a pronounced swing in sewage treatment from biological to chemical treatment processes and as a consequence it is anticipated that lime requirements will increase.

In addition to the treatment of potable water and sewage, lime is also used in a variety of processes directed at treating industrial waste water. As an example, lime is used to neutralize acid mine drainage, and to precipitate iron in coal washing plants. Waste from sulfuric acid pickling plants and plating waste are neutralized with lime as are chemical and pharmaceutical plant waste.

Flue Gas Desulfurization

The passage of new source performance standards in 1971 made it apparent that coal fired electric generating stations were going to have to scrub their stack gases with either lime or limestone systems in order to achieve new air quality control standards. Subsequent to the passage of those air quality control standards lime scrubbers began to appear throughout the utility industry. The development of the market was greatly enhanced by advances in scrubber technology including both wet and dry processes, as well as the development of Dravo Corporation's Thiosorbic® lime. Lime scrubbing was proved to be both an effective and viable means of meeting the US Environmental Protection Agency air pollution control requirements. As a result during the period of 1980 to 1990 the use of lime for this particular application has more than doubled.

In 1990, Clean Air Act Amendments were passed to the new source performance standards which require a much higher level of attainment for sulfur removal. These new standards phase in compliance requirements with the first phase being implemented by 1995 and the second phase by the year 2000. A more complete description of the impact of the 1990 amendments, whether lime or limestone scrubbing or other technology is involved, remains to be seen and will be an individual power plant decision. However, it is anticipated that the requirement for additional sulfur removal will mandate that the market for lime used in flue gas desulfurization will increase over the next decade. The full magnitude of that increase is yet to be determined, and is discussed in more detail in the Environmental Uses section of this publication.

CONSTRUCTION

Soil Stabilization

Soil stabilization is a process wherein lime is mixed with clay soils to react with available silica and alumina in the soil to form complex cementing compounds which bind the soil into a hard, stable mass that is not sensitive to water saturation. The plasticity of the soil, as well as shrinkage and swell, are markedly reduced. The amount of lime used for soil stabilization generally ranges between 3 to 5% of the weight of the dry soil. The majority of lime used in this application has been hydrated lime which is used in slurry form. This contrasts with slaked lime slurries made with pebble quicklime and has lead to discussion within the construction industry as to which process is most appropriate. In using a quicklime it is critically important that it be a soft burned high calcium lime with not more than 10% residue upon slaking. Recently, the California Department of Transportation (CALTRANS) compared the unconfined compressive strengths of a high and low residue lime. The low residue lime showed seven day cure strengths which were as much as 26% higher than the higher residue limes. Since quicklime slurries tend to have finer particles than hydrated lime slurries and appears that there may be a gradual shift from hydrated to pebble quicklime for this particular application. Regardless of whether pebble or hydrated lime is used, soil stabilization with lime is recognized by all federal agencies and most state highway departments as a worthwhile practice. This type system has been used on at least 3 200 km of interstate highway construction in 26 states. In addition, several major construction projects have utilized soil stabilization, particularly in the states of Texas, Oklahoma, and Louisiana. Lime is also used by many contractors to expedite construction by drying up muddy subgrades and to construct all weather haul roads.

Lime-Fly Ash

Lime-fly ash mixes were first introduced by the Corson Company in a proprietary process called POZ-O-PAC. Since the introduction of this system, lime-fly ash pozzolan mixes have achieved increasing success and acceptance. The initial construction of the Newark, NJ, airport used lime-fly ash mixtures throughout its pavement system and subsequently, again used lime-fly ash mixtures in the expansion of the airport. These mixtures were also used during the recent reconstruction of Interstate 80 across northern New Jersey, and have also found acceptance in construction projects in the states of Tennessee, North and South Carolina, Georgia, and Kentucky.

In these systems lime is mixed with fly ash and aggregate in the presence of a small amount of water, generally with the use of a pug mill, in a central mix plant. Moisture control is critical to achieving the proper pozzolinic reaction and adequate strength in the subgrade. The product may then be laid down either with standard paving equipment or utilizing a method compaction. Lime consumption for this particular use has increased from approximately 44 kt in 1980 to 250 kt in 1990.

Mortar

Lime has been used for thousands of years in mortar materials and continues to be used even today. Before the advent of portland cement in the late nineteenth century, mortars were composed only of lime and sand. Presently lime-cement mortars of varying proportions are used, ranging from 3 to 1/4 parts of lime by volume to 1 part cement. Lime is still the most dependable plasticizer for mortar, making it workable and more trowelable. In the cement-lime mixes, the cement provides the rapid set. These type masonry mixes however, encounter severe competition from masonry cements made by cement companies.

A similar application for lime is found in plaster. While at one time it was used in base plaster coats, these generally have been replaced with gypsum boards. However, in interior plastering today, lime is still used in the finish or white coat and is usually mixed with gypsum. Also, lime is used in exterior plaster or stucco, particularly in California and Florida and again is mixed with portland cement and sand in varying proportions.

Asphalt

Increasingly, 1 to 2% of hydrated lime is added to asphalt hot mixes in central mix plants as a paving material. The lime acts as an antistripping agent preventing the aggregate from raveling from the bituminous binder. In remote areas, it permits the use of marginal, unwashed aggregates, and where needed, serves as a filler. Based upon the immersion compression test with some asphalts and aggregates it markedly increases the wet/dry strength ratio and reduces volume change. The application of lime in asphalt mixes is relatively new and over the last ten years has been receiving increasing acceptance. ASTM's committee on lime, C-7, has adopted specialized specifications for the use of hydrated lime in asphalt and asphaltic concrete mixtures. Nevertheless, there continues to be some resistance within the construction industry to use lime in asphalt.

AGRICULTURE

Sugar Refining

While sugar refining is normally included with most chemical and industrial lime use statistics, it is in reality an agricultural application. Lime is essential for sugar making in both beets and cane sugar refinery processes. However, sugar beet refining requires a significantly higher volume of lime than does cane sugar. On the average beet sugar averages 200 kg of quicklime per ton of sugar, whereas cane sugar would require only 1.9 to 4.9 kg/t of lime is required. In sugar refining a milk of lime slurry is used to raise the pH of the product stream and thus precipitate colloidal impurities. These impurities are normally phosphatic and organic acid compounds and are removed from the stream through filtration. Most sugar beet plants have captive lime plants on site as each of these processing plants also has a demand for CO_2 which is obtained from the lime kiln stack gas.

Agricultural Lime

Although most agricultural liming today is achieved with pulverized limestone, some farming still uses ground quicklime or hydrated lime. This is particularly true in truck farming where lime is able to neutralize soil acidity more rapidly and maintain a pH level that allows for three crops to be grown per year. The added cost for lime is easy to justify in higher value crops. In some sandy soil areas dolomitic lime only is used due to persistent magnesium deficiencies.

Other

Lime is used as a CO_2 absorbent in controlled atmospheric storages for fresh fruit and certain vegetables to extend the freshness of the produce. This prevents CO_2 from building to abnormally high levels which would accelerate the rotting of produce. It is used to recover glue and gelatin from packing house byproducts. In citrus fruit processing lime is used to treat the waste pulp which after grinding and drying is sold as cattle feed.

ECONOMICS

As with most other mineral products, the past twenty years have witnessed intense pressure on lime producers to achieve economic efficiencies. During the period 1970 to 1980 worldwide petroleum shortages and the emergence of oil cartels forced producers to look for more fuel efficient calcining equipment. It was during this period that a significant switch was made away from long rotating kilns to short preheater rotary kilns and parallel-flow regenerative kilns. This switch included increased emphasis on coal firing, including advanced coal lance firing systems on some parallel-flow shaft kilns.

Fuel efficiency has been only one part of the equation in achieving economic efficiency. There has also been a significant focus on kiln design with the emphasis being on higher capacity and positive control of the calcining environment. These newer kilns have ranged from 544 to 907/tpd for the short preheater kilns. Typical of these are Dravo Lime's three 907/tpd kilns installed at its Maysfield, KY, plant in the late 1970s (Fig. 6). The world's largest kiln, brought on line by Marblehead Lime in 1981, is a 4.6 × 5.2 × 4.5 × 149 m long rotary kiln rated at 1 089/tpd (Levine, 1981). Programmable controllers and centralized process control are an integral part of these installations, as are innovations in coolers, heat exchangers, and heat recovery systems.

The third part of the efficiency equation has been consolidation of production capacity. The number of lime plants in the United States has steadily declined, from 1 200 in 1910 to 154 in 1980, to the 114 plants in production in 1991. These 114 plants represented 70 individual lime companies; however, 26 plants owned by 10 companies produced 57% of the lime consumed in the United States in 1990 (Miller, 1991). The impact of consolidation is even more pronounced when one considers that in 1991 the lime industry is estimated to have worked at only 79% capacity. In these slow production periods it is the larger lime plants and multi-plant companies which have been able to cut operating costs sufficiently to survive.

Pricing

Traditionally, lime has been considered as a commodity material and as a consequence, its pricing levels have been relatively low. During the period 1910 to 1970, lime pricing was relatively constant and rarely exceeded $15 per ton FOB the lime production plant. However, beginning in 1970, lime prices increased steadily due to fuel costs and government mandated environmental controls. In 1974 the oil embargo initiated a substantial increase in energy costs which were clearly reflected in the costs of production and the cost of lime.

Table 4 shows the price relationship for lime over time in both actual price and constant dollars. In looking at this table it is quite apparent that the actual price of lime continued to escalate through the mid and late 1970s to 1985 or 1986, and has steadily declined since then. However, on a constant dollar basis lime pricing has steadily decreased since 1980 and in 1990 reflected an actual constant dollar decrease of nearly 21%.

Cost

While lime pricing has continued to decline through the 1980s and is only now experiencing a modest increase in the actual dollars

Fig. 6. Modern preheater rotary kilns.

realized for lime, the cost of production continues to escalate. In general terms, production costs can be divided into the cost of energy, the cost of kiln feed, environmental, and direct labor. Of these only kiln feed costs and labor held relatively constant over the last two or three years. However, it is anticipated that these costs will continue to escalate on a modest scale over the next five to ten years as employee benefit costs, particularly those associated with health benefits and workman's compensation continue to escalate. The fuel costs have been depressed for the last several years and at this point in time there is every indication that they will remain relatively constant with only moderate increases paralleling the increases which the lime manufacturer is likely to receive in his cost of material. The major impact on lime costs for the next several years will the be the continuing implementation of environmental and regulatory costs which escalate with no end in sight. The Clean Air Act Amendments of 1990 while potentially representing additional markets for the lime producer also include additional regulatory requirements and increased compliance regulations for the lime producer.

DISTRIBUTION AND TRANSPORTATION

The destination of shipments of lime sold or used by producers in the United States is reported annually by the US Bureau of Mines. In 1990 the top six destination states were Pennsylvania, Ohio, Texas, Indiana, Michigan, and Illinois. Together these six states represent 49.1% of the consumption of US produced lime capacity (Miller, 1991).

For many years railroads were the primary carrier for bulk lime; however, the convenience of truck delivery via the US interstate highway system has significantly reduced the volume of bulk lime moved by railroad. The trucks used for this type of movement are primarily pneumatic carriers which can both provide an enclosed container in which to ship the lime and at the same time afford a safe handling system wherein the operator does not have to come in direct contact with the product. While truck and rail shipments continue to dominate an increasing tonnage of lime is moved by barge on the nation's inland waterways. Today over 15% of the lime produced in the United States is moved via barge. This volume is

Table 4. Time-price Relationships for Lime[1] (US dollars per ton)

Year	Actual Price[2]	Based on Constant 1990 Dollars
1971	15.78	47.76
1972	16.78	47.53
1973	17.42	46.28
1974	22.02	53.62
1975	27.46	60.89
1976	30.19	62.92
1977	33.50	65.46
1978	36.76	66.95
1979	41.26	69.03
1980	44.50	68.28
1981	47.01	65.76
1982	49.47	65.05
1983	51.10	64.67
1984	51.12	62.42
1985	51.69	61.29
1986	52.50	60.67
1987	50.11	56.13
1988	48.12	52.17
1989	49.83	51.88
1990	51.77	51.77

[1]Includes Puerto Rico, [2]average value of lime sold or used by producers FOB plant excluding cost of containers.
Source: Miller, 1990.

Table 5. World Lime Production (in descending order) (thousand tons)

Country	1990
USSR	34 008
China	20 607
United States	19 264
Germany (East & West)	12 386
Japan	10 359
Mexico	7 284
Brazil	6 921
Poland	5 345
Italy	4 672
Czechoslovakia	4 066
Rumania	3 890
France	3 637
United Kingdom	3 163
Canada	3 042
Yugoslavia	2 424
South Africa	2 224
Belgium	2 182
Turkey	1 984
Austria	1 873
Australia	1 818
Bulgaria	1 697
Chile	1 576
Colombia	1 576
Spain	1 455
Other	8 013
Total	165 466

Source: USBM, Miller, 1990.

expected to increase even more in the decade ahead as an increasing number of utilities located on navigable waterways are required to utilize lime in their flue gas desulferization systems.

Intermodel shipments of lime are also being used in a number of locations to take advantage of long distance, low unit cost barge and rail rates. In these type systems a silo is installed at the barge or rail unloading terminal and is equipped with loading chutes suitable for the transfer of the product to trucks. Bulk pneumatic trucks then deliver the product to the end use customer. These systems are quite effective in allowing off-rail or off-water small volume users to take advantage of the bulk transportation rates available to larger volume movements.

While most lime is moved via bulk transportation, hydrated product frequently is handled in a bag or semi-bulk container. When packaged in paper bags they normally contain a polyethylene liner to protect the lime from moisture and insure its satisfactory end use performance. Semi-bulk containers are normally sized to handle one ton of product. However, this type of container does have the flexibility of being sized to customer requirements and may be either larger or smaller than one ton capacity. There are a variety of bulk containers on the market and include both one-way and returnable bulk bags, as well as cardboard "Gaylords" and reusable metal containers.

FOREIGN PRODUCTION

On a worldwide basis 165 466 000 tons of lime were produced in 1990 (Miller, 1991). Table 5 shows the world lime production by country in descending order. Clearly the USSR has been the largest worldwide producer of lime in 1990, a position which they have held for many years. However, the events of 1991 will be reflected in future year's statistics as the newly independent, individual countries of the former USSR begin to report their own production numbers. Of these new countries it is anticipated that the Ukraine will dominate. Traditionally lime usage in the Eastern Bloc nations and Russia has been focused on steel production. This fact is reflected in the production numbers for the countries of Poland, Czechoslovakia, Rumania, and Yugoslavia, all of which were part of the industrialized segment of the Eastern Bloc countries. While steel production is anticipated to continue to demand large volumes of lime in each of these countries, the implementation of environmental clean up throughout Eastern Europe and the countries which formally made up Russia is anticipated to require an increasing volume of lime.

European production is largely dedicated to steel production and is generally consumed within the European trading community. Within the European steel industry there has been a move toward greater use of dolomitic lime as well as an increased use of powdered lime rather than pebble lime. Powdered lime is used in synthetic slags and ladle furnaces during secondary steel making. Future trends in Europe are expected to see increasing demand for flue gas desulfurization (FGD), industrial waste and sewage treatment, and domestic water treatment.

Within the Pacific Rim, Japan is clearly the dominant producer. As in most other industrialized areas the predominant market for lime has been the steel industry which traditionally accounts for 50 to 55% of the lime produced in Japan. The Japanese steel industry has made a strong move toward higher performance steels and particular attention has been drawn to significantly reducing phosphorous contents within the steel to 0.005% P. At Nippon Steel Corporation, a mixture of 18 kgt powdered lime, 28 kgt of scale, 2 to 5 kgt of feldspar, and 2.5 kgt calcium chloride has been used to reduce hot metal phosphorous levels to < 0.005% (O'Driscoll, 1988). Other Japanese companies have developed similar procedures all of which are targeted at reducing phosphorous and sulfur levels in these high performance steels. Interestingly, unlike most other modern industrialized nations, Japan's utility FGD market does not require a significant volume of lime. Here wet limestone processes which generate byproduct gypsum are much more in vogue since Japan has no natural resources for gypsum and little land space is available for byproduct disposal. A small but growing market for lime is found in a unique application in Japan wherein lime is used in the production of non-explosive blasting agents. This type blasting agent is required in Japan to meet their restrictive permissible noise and vibration standards.

The leading Latin American producer of lime is Mexico where there are about 130 lime plants of which only 15 are classified as modern. These 15 plants produce about 50% of Mexico's annual production, a part of which is exported to the United States, particularly from those lime plants located in the Sonora Province and in the Yucatan Peninsula. Approximately 80% of Mexico's production is consumed by the steel industry with the remainder almost equally divided between sugar production and construction. Brazil is the other major Latin American producer and enjoys the benefit of rich deposits of both high quality limestone and dolomite. The Brazilian lime industry has been expanding over recent years with new capacity being added and projections for an expanded environmental and construction market both within Brazil and in surrounding countries.

Canadian lime production is concentrated in the provinces of Quebec, Ontario, Alberta, and British Columbia. The predominate markets for Canadian produced lime are metallurgical and glass industries as well as export sales to the United States. Canada was the number one trading partner for the United States in lime and received 80% of US exports, while shipping 80% of US imports (Miller, 1991).

FUTURE TRENDS

Lime is a commodity chemical finding a vast array of uses in a variety of industries. As a consequence the long term prospects for lime are excellent despite the fact that at present there is some over capacity in the US lime industry. The key to the success of the future of any lime producer will be the ability to develop new markets.

As has been noted previously, the major growth area for lime over the next decade will be environmental applications. Certainly the focus here has been on flue gas desulferization and anticipated use of lime based wet and dry scrubbers. While the 1990 Clean Air Act Amendments are expected to provide a real boost to this industry, the fact remains that many of the environmental systems which are now being studied and which will be employed in the future may or may not utilize lime as the primary reagent material. Nevertheless, environmental applications are not restricted to FGD. The Clean Water Act and enforcement efforts by both federal and state environmental protection agencies will bring additional pressure to bear on municipal and industrial sewage treatment plants. This should lead to expanding lime markets for this application. Additionally, the Electric Power Research Institute (EPRI), Little Lake, Inc., and the EPA have been working on the restoration of lake fisheries (Emler, 1990). In this program lakes are treated with lime to raise the pH levels to the point that both lake and brook trout can both spawn and survive on a long term basis.

On the production side, lime producers increasingly are being required to implement statistical process control (SPC) in their manufacturing procedures. The move toward SPC has been initiated by the US automobile industry in its attempts to improve quality and be competitive on a worldwide basis. One of the major customers of the auto industry has been the steel producers who in turn are now applying pressure to the lime manufacturers for SPC. Rukavina (1990) noted that Marblehead Lime Company has implemented this technology on a corporate wide basis and has now moved ahead with an integrated process control (IPC) program. While Marblehead was one of the earliest lime producers to implement SPC no major producer of lime can long avoid the implementation of SPC technology.

As noted earlier, the competitive nature of the lime industry dictates that cost control will continue to be a major issue with all producers. While kiln sizing seems to be settling in at 544 to 726 tpd for most new units, advances in computerized process control will continue to be critical in achieving the most effective cost controls. The installation of programmable controllers and centralized process control units represent the leading edge of this technology. Continuing refinements in software packages should lead to increased energy efficiency, reduced manpower, and a more consistent end product.

Finally, it can also be anticipated that there will be additional consolidation within the lime industry and potentially, expanded foreign ownership of US lime producers. The trends which developed in the 1980s are expected to continue through the 1990s, albeit at a somewhat slower pace. Developments in both production and market technology which are taking place in Europe and Japan should translate to increased profitability for the large lime producers.

SPECIFICATIONS AND TESTING

Because of the number of limes available, their many uses, and their wide variation in physical and chemical characteristics, there is no one overall lime specification. Numerous specifications and buyer requirements for lime exist not only for most of the major users, but even among individual companies or plants.

The National Lime Association with its member companies and a majority of the lime consumers subscribe to the American Society for Testing and Materials (ASTM) specifications on lime, promulgated by Committee C-7. These specifications which cover many uses and testing procedures include the following:

- C25 Chemical analysis of limestone, quicklime and hydrated lime
- C50 Sampling, inspection, packing, and marking of quicklime and lime products
- C51 Terms relating to lime
- C110 Physical testing of quicklime, hydrated lime, and limestone
- C400 Testing quicklime and hydrated lime for neutralization of waste acid
- C593 Fly ash and other pozzolins for use with lime
- C602 Agricultural liming materials
- C706 Limestone for animal feed use
- C737 Limestone for dusting of coal mines
- C821 Lime for use with pozzolins
- C911 Quicklime, hydrated lime, and limestone for chemical uses

Although these are the primary tests and specifications for lime, this is not a complete or comprehensive list. For a complete listing the reader is referred to ASTM in Philadelphia, PA.

Another important lime specification, promulgated by the American Waterworks Association (AWWA), is the Standard for Quicklime and Hydrated Lime (B202). This specification includes a bonus and penalty arrangement based on available lime content, slaking rate, and slaking residue. It is also applicable to sewage and industrial waste treatment plants.

While ASTM procedures are used in testing limes intended for flue gas desulferization, many of the specifications relating to a specific utility are predicated upon EPA performance standards applicable to the specific utility in question. As a consequence, individual utility customers may have specifications which go beyond normal ASTM standards.

BIBLIOGRAPHY AND REFERENCES

Because lime is one of mankind's oldest materials and its end use pattern is diverse, the literature on lime is so extensive that it would be impossible to give a complete bibliography. The following list of sources presents the more general references on lime with emphasis on manufacture and major uses. Among the principal sources of information are R.S. Boynton's textbook *Chemistry and Technology of Lime and Limestone*, (2nd ed., 1980), publications of the National Lime Association, articles from *Pit and Quarry*, *Rock Products*, and *Zement-Kalk Gipps* magazine, and publications of the US Bureau of Mines.

A selected bibliography follows:

Anon., 1955, "Whitewash and Cold Water Paints," Bulletin 304-G, National Lime Association, Washington, DC, 15 pp.

Anon., 1960, "Exterior Masonry Construction," 2nd ed., Bulletin 324, National Lime Association, Washington, DC, 70 pp.

Anon., 1962, "Hydrated Lime is Asphalt Paving," Bulletin 325, National Lime Association, Washington, DC, 15 pp.

Anon., 1965, "Chemical Treatment of Sewage & Industrial Wastes," Bulletin 215, National Lime Association, Washington, DC, 139 pp.

Anon., 1966, "Specifications for Lime and Its Uses in Building," Bulletin 322, National Lime Association, Washington, DC, 16 pp.

Anon., 1967, "Lime for Steelmaking: Tailored to Fit New Demands," *Metals Producing Magazine*, Vol. 33, Jan., 18 pp.

Anon., 1968, "Phosphate Extraction Process," Dorr-Oliver, Inc., 22 pp.

Anon., 1969, "Masons Lime," National Lime Association, Washington, DC, 12 pp.

Anon., 1969a, *Reaction Parameters of Lime*, American Society for Testing and Materials, Symposium, Atlantic City, NJ.

Anon., 1969b, *Water Treatment Plant Design*, American Water Works Association, Denver, CO, 359 pp.

Anon., 1970, "Lime," Bulletin T60.350–1, Wallace and Tiernan.

Anon., 1970a, "Lime Dries Up Mud," National Lime Association, Washington, DC, 4 pp.

Anon., 1970b, "Lime Industry Safety Manual," 2nd ed., Bulletin 400, National Lime Association, Washington, DC, 43 pp.

Anon., 1971, *Process Design Manual for Phosphorous Removal*, Black & Veatch, EPA Tech. transfer, Oct.

Anon., 1972, "Flexible Pavement Design Guide," Bulletin 327, National Lime Association, Washington, DC, 43 pp.

Anon., 1972a, "Lime Stabilization Construction Manual," 5th ed., Bulletin 326, National Lime Association, Washington, DC, 47 pp.

Anon., 1976, "Acid Neutralization with Lime," Bulletin 216, National Lime Association, Washington, DC.

Anon., 1976a, "Lime Handling, Application, and Storage in Treatment Processes," Bulletin 213, National Lime Association, Washington, DC, 73 pp.

Anon., 1976b, "Water Supply & Treatment," 11th ed., Bulletin 211, National Lime Association, Washington, DC, 220 pp.

Anon., 1977, "A Long Range Durability Study of Lime Stabilized Bases," Bulletin 328, National Lime Association, Washington, DC.

Anon., 1977a, "Standard for Quicklime and Hydrated Lime," Standard B 202–77, American Water Works Association, Denver, CO, 12 pp.

Anon., 1978, "Lime," *The Fuller Engineer*, Vol. 28, No. 1, Fuller Company, Bethlehem, PA, 15 pp.

Anon., 1980, "Lime in Municipal Sludge Processing," Bulletin 217, National Lime Association, Washington, DC.

Anon., 1980a, lime brochure, National Lime Association, Washington, DC.

Anon., 1981, map, "Commercial Plants in U.S. and Canada," National Lime Association, Washington, DC.

Anon., 1988, "Chemical Lime Facts," 5th ed., rev., Bulletin 214, National Lime Association, Washington, DC, 44 pp.

Anon., 1991, ASTM Standard, Sec. 4, Vol. 04.03, American Society for Testing and Materials (ASTM), Philadelphia, PA.

Anon., 1991a, "High Performance Hydrate Lime," *Geonews*, Winter, Illinois State Geological Survey, 8 pp.

Ahlberg, H.L., and Barenber, E.J, 1965, "Pozzolanic Pavements," Engineering Equipment Station Bulletin 473, University of Illinois, Champaign-Urbana, IL, 130 pp.

Ary, T.S., 1992, *Lime Plants in the United States in 1990*, US Department of the Interior, US Bureau of Mines, Washington, DC, 18 pp.

Azbe, V.J., 1946, *Theory and Practice of Lime Manufacture*, St. Louis, MO, 423 pp.

Azbem, V.J., 1954, "Rotary Kiln Performance, Evaluation and Development," National Lime Association, Washington, DC, pp 89.

Bishop, D.L., 1939, "Particle Size and Plasticity of Lime," *Journal of Research*, Vol. 23, No. 2, pp. 285–292.

Boynton, R.S., 1980, *Chemistry and Technology of Lime and Limestone*, John Wiley, New York, 578 pp.

Boynton, R.S., and Gutschick, K.A., 1968, *Building Lime, Its Properties, Uses, and Manufacture*, United Nations Ind. Dev. Organization, New York, July, 104 pp.

Boynton, R.S., and Jander, F.K., 1952, "Lime and Limestone," *Encyclopedia of Chemical Technology*, R.E. Kirk et al., eds., Interscience, New York.

Clarke, G., 1981, "Burnt Lime in Western Europe," *Industrial Minerals*, No. 164, London, May, pp. 25–51.

Davison, J.L., 1969, "Curing of Lime Cement Mortars," STP 472, American Society for Testing and Materials, Phildelphia, PA, pp. 193–208.

Derge, G., and Shegog, J.R., 1969, "Observations on the Solution of Calcium Oxide in Slag Systems," STP 472, American Society for Testing and Materials, Philadelphia, PA, pp. 132–142.

Dixon, T., 1982, "North American Lime," *Industrial Minerals*, No. 177, London, pp. 51–63.

Eades, J.L., and Sandbert, P.A., 1969, "Characterization of the Properties of Commercial Limes by Surface Area Measurements and Scanning Electron Microscopy," STP 472, American Society for Testing and Materials, Philadelphia, PA, pp. 3–4.

Emler, P., 1990, "Lime: A New 'Old' Tool for Fisheries Management," *Stone Review*, National Stone Association, Washington, DC, Apr., pp. 8–10.

Falk, L., and Hauber, D., 1990, "The Eastern German Industrial Minerals Contribution," *Industrial Minerals*, No. 279, London, Dec., pp. 24–49.

Freas, R.C., 1990, "Lime," *Encyclopedia of Materials Science and Engineering*, Vol. 2, Pergamon Press, Oxford, UK, pp. 1038–1042.

Grancher, R.A., 1970, "Past, Present and Future of Lime in the United States," International Lime Conference, Salzberg, Austria, May, 34 pp.

Gutschick, K.A., 1987, *Lime for Industrial Uses*, STP 931, American Society for Testing and Materials, Philadelphia, PA, 31 pp.

Hauson, D., 1991, "Chemical Method for Destroying PCB's," *Chemical & Engineering News*, Vol. 69, No. 11, New York, pp. 5–6.

Hedin, R., 1961, "Structural Processes in the Dissociation of Calcium Carbonate," Azbe Award No. 2, National Lime Association, Washington, DC, 12 pp.

Hedin, R., 1962, "Plasticity of Lime Mortars," Azbe Award No. 3, National Lime Association, Washington, DC, 11 pp.

Lamar, J.R., 1961, "Uses of Limestone and Dolomite," Circular 321, Illinois Geological Survey, 119 pp.

Levine, S., 1981, "World's Largest Lime Kiln Onstream in Chicago," *Pit and Quarry*, Chicago, IL, May, pp. 58–63.

Levine, S., 1977, "Lime for Flue Gas Desulferizing," *Pit & Quarry*, Chicago, IL, May, 8 pp.

Limes, R., and Russell, R.O., 1969, "Crucible Test for Lime Reactivity in Slags," STP 4632, American Society for Testing and Materials, Philadelphia, PA, pp. 161–172.

Mayer, R.P., and Stowe, R.A., 1964, "Physical Characterization of Limestone and Lime," Azbe Award No. 4, National Lime Association, Washington, DC, Aug., p. 26.

McDonald, E.B., 1971, *Lime Stabilization Training Manual*, South Dakota Department of Highways, 79 pp.

McDowell, C., 1972, "Flexible Pavement Design Guide," Bulletin 327, National Lime Association, Washington, DC, 43 pp.

Miller, M.M., 1991, *Lime, 1990 Annual Report*, US Department of the Interior, US Bureau of Mines, Washington, DC, 13 pp.

Minnick, L.J., 1971, "Control of Particulate Emissions from Lime Plants—A Survey," *Journal of Air Pollution Control Association*, Apr., pp 195–200.

Mullins, R.C., and Hatfield, J.D., 1969, "Effects of Calcination Conditions on Properties of Lime," STP 462, American Society of Testing and Materials, Philadelphia, PA, pp. 173–192.

Obst, K., et al., 1969, "Present Status and Technical Advances in Steelworks

Lime for Basic Oxygen Furnaces in Germany,'' STP 462, American Society for Testing and Materials, Philadelphia, PA, pp. 173–192.

O'Driscoll, M., 1988, ''Burnt Lime/Dolime,'' *Industrial Minerals*, London, May, pp.23–51.

Parsons, W.A., 1965, ''Chemical Treatment of Sewage and Industrial Wastes,'' Bulletin 215, National Lime Association, Washington, DC, 139 pp.

Remick, B., 1990, ''Lake and Stream Liming: An Overview,'' *Stone Review*, Washington, DC, Apr., p. 11.

Riehl, M.L., 1970, ''Water Supply and Treatment,'' 10th ed., National Lime Association, Washington, DC, 220 pp.

Ritchie, T., 1969, ''Influence of Lime in Mortar on the Expansion of Brick Masonry,'' STP 462, American Society for Testing and Materials, Philadelphia, PA, pp. 67–81.

Rooney, L.F., and Carr, D.D., 1971, ''Applied Geology of Industrial Limestone and Dolomite,'' Geological Survey Bulletin 46, Indiana Department of Natural Resources, 59 pp.

Rukavina, M., 1990, ''Lime Producers Accept Challenge,'' *Rock Products*, Chicago, IL, Feb., pp. 139–147.

Schlitt, W.J., and Healy, G.W., 1969, ''A Comparison and Scaling Down of Coarse Grain Tritration and ASTM Slaking Rate Test,'' STP 462, American Society for Testing and Materials, Philadelphia, PA, pp. 143–160.

Schwarzkopf, F., 1970, ''A Comparison of Modern Lime Calcining Systems,'' *Rock Products*, July, pp. 68–71.

Thompson, J.L., 1978, ''Pebble Lime From Preheater Kilns,'' Preprint 78 H362, SME-AIME Fall Meeting, Sep., Lake Buena Vista, FL, pp. 2041–2044.

Thompson, M.R., 1968, ''Lime Stabilization of Soils for Highway Purposes,'' Series No. 25, Illinois Highway Engineering, Dec., 25 pp.

Timucin, M., and Morris, A.E., 1969, ''Phase Equilibria and Thermodynamic Properties of Lime-Iron Oxide Melts,'' STP 462, American Society for Testing and Materials, Philadelphia, PA, pp. 25–31.

Voss, W.C., 1960, ''Exterior Masonry Construction,'' Bulletin 324, National Lime Association, Washington, DC, 70 pp.

Walker, D.D., et al, 1991, ''Innovations and Uses for Lime,'' STP 1135, American Society for Testing and Materials, Philadelphia, PA, p. 151.

LIMESTONE AND DOLOMITE

DONALD D. CARR, LAWRENCE F. ROONEY, AND ROBERT C. FREAS

Perhaps no other mineral commodities in this volume have as many uses as limestone and dolomite. These carbonate rocks are the basic building blocks of the construction industry, the material from which aggregate, cement, lime, and building stone are made. Carbonate rocks, and their derived products, are used as aggregates, fluxes, glass raw material, refractories, fillers, reactive agents in sulfur-oxide removal, abrasives, soil conditioners, ingredients in a host of chemical processes, and much more. Although the use of limestone and dolomite goes back before recorded history, new uses and products are continually being found.

Carbonate rocks form about 15% of the earth's sedimentary crust and are widely available for exploitation. Found extensively on all continents, they are quarried and mined from formations that range in age from Precambrian to Holocene. Reserves of carbonate rock are large and will last indefinitely, although high-purity deposits may be absent or have limited availability in certain areas.

Limestone and dolomite are so useful and so abundant that in 1989 about 798 Mt were produced in the United States. In fact, about 75% of all stone quarried or mined in the United States was carbonate rock. Sand and gravel was the only mineral commodity produced in greater quantity.

In this chapter we have sought to provide the general reader an overview of carbonate rocks, their composition, distribution, production, and uses. The literature on the limestone and dolomite resources of the United States and Canada is so vast and diverse that one can easily become mired in print. We tried to list those references that are easily accessible and that provide the economic geologist with a starting point for more intensive study. The table listing selected references on limestone and dolomite resources in Canada and the United States reflects our attempt to be selective rather than exhaustive and to direct the reader to the best sources of detailed information.

GEOLOGY

Mineralogy

The principal carbonate rocks used by industry are limestone and dolomite. Limestones are sedimentary rocks composed mostly of the mineral calcite ($CaCO_3$), and dolomites are sedimentary rocks composed mostly of the mineral dolomite ($CaCO_3 \cdot MgCO_3$). Aragonite ($CaCO_3$), which has the same chemical composition as calcite but a different crystal structure, is economically important only in modern deposits, such as oyster shells and oolites. Aragonite is metastable and alters to calcite in time. Some other carbonate minerals, notably siderite ($FeCO_3$), ankerite ($Ca_2MgFe(CO_3)_4$), and magnesite ($MgCO_3$), are commonly found associated with limestones and dolomites but generally in minor amounts.

Because of their similar physical properties, the carbonate minerals are not easily distinguished one from the other. Specific gravity, color, crystal form, and other physical properties (Table 1) are aids to mineral identification if the rock is relatively monomineralic and compact. The interested geologist can find further information on the chemical, mineralogical, and physical properties of carbonate minerals in publications by Graf and Lamar (1955) and by Tucker and Wright (1990).

The rate of solubility of the different minerals in dilute hydrochloric acid is perhaps the most useful technique to identify them in the field. Calcite is much more soluble in dilute acid than dolomite; hence, if a fresh rock surface is etched, the amount of dolomite left standing in relief can be estimated by the use of a hand lens. Some staining techniques are based on differences in solubility of the carbonate minerals (in decreasing order of solubility: aragonite, calcite, and dolomite). These techniques are useful in the laboratory and have limited application in the field (Friedman, 1959, Högberg, 1971, Warne, 1962). Staining of thin sections is particularly effective in the laboratory (Dickson, 1965, 1966).

The X-ray diffractometer is commonly used in the laboratory for determining carbonate mineralogy of bulk samples. Techniques of determining calcite and dolomite ratios or the percentage of these minerals based on comparison of their diffraction intensities with those of known standards can be found in Azároff and Buerger (1958), Cullity (1956), Graf and Goldsmith (1963), Gulbrandsen (1960), Müller (1967), and Runnells (1970).

Thin section analysis may be a helpful adjunct to binocular examination of carbonate rocks. Although it is difficult to distinguish between calcite, dolomite, and ankerite in thin section unless staining techniques are used, identification of some other minerals, types of carbonate grains, fabrics, textures, and structures are sometimes facilitated by this method (see also "Physical Properties"). Appropriate parts of "Microscopic Sedimentary Petrography" (Carozzi, 1960) and "Petrology of Sedimentary Rocks" (Hatch and Rastall, 1971) provide introductions to microscopic investigation of carbonate rocks, as does the review on that subject by Gubler et al. (1967). Good places for assistance in the identification of fossils and fossil fragments in thin section are works by Horowitz and Potter (1971) and by Scholle (1978).

Color, an important property of carbonate rocks, can be a rough guide to purity, but it can also be misleading. Only a small amount of noncarbonate material is necessary to produce a marked change in color. The famous building stone called Indiana Limestone, for example, with its distinct gray and buff colors, commonly contains less than 0.2% Fe_2O_3; "Carthage Marble," a fossiliferous dimension limestone from Missouri in shades of gray, generally has a total iron and aluminum oxide content less than 0.2%. Most high-purity limestones are shades of light brown and gray to white. Limestones in shades of gray or green generally indicate the presence of minerals containing ferrous iron oxides and/or carbonaceous matter. As the state of oxidation increases, the colors change to yellows, browns, or reds. A color reference chart is useful in maintaining uniformity of rock descriptions. One of the most useful has been

Table 1. Physical Properties of Some Common Carbonate Minerals

Calcite: $CaCO_3$, hexagonal crystal system, commonly good rhombohedral cleavage. Mohs hardness, 3. Specific gravity, 2.72. Commonly colorless or white but may be other colors due to impurities.
Dolomite: $CaCO_3 \cdot MgCO_3$, hexagonal crystal system, commonly good rhombohedral crystals with curved faces. Mohs hardness, 3.5 to 4. Specific gravity, 2.87. Commonly white or pink.
Aragonite: $CaCO_3$, orthorhombic crystal system. Mohs hardness 3.5 to 4. Specific gravity, 2.93–2.95. Commonly colorless, white, or yellow, but may be other colors due to impurities.
Siderite: $FeCO_3$, hexagonal crystal system, commonly distorted rhombohedral crystals. Mohs hardness, 3.5 to 4. Specific gravity, 3.7 to 3.9. Commonly brown or black.
Ankerite: $Ca_2MgFe(CO_3)_4$, hexagonal system, commonly rhombohedral crystals. Mohs hardness, 3.5 to 4. Specific gravity, 2.9. Commonly white, pink, or gray.
Magnesite: $MgCO_3$, hexagonal, usually in granular or earthy masses. Mohs hardness, 3.5 to 4.5. Specific gravity, 2.96 to 3.1. Commonly white or yellowish but may be other colors due to impurities.

made available by the Geological Society of America (Goddard et al., 1963).

Impurities in carbonate rock vary considerably in type and amount but are important from an economic standpoint only if they affect the usefulness of the rock. Generally the two most important considerations of each impurity are how much is present and how it is distributed. A considerable amount of some impurities is tolerable in carbonate rock for some uses if the impurity is disseminated throughout the rock. On the other hand, if the impurity is concentrated in laminae, it may form planes of weakness that seriously affect the performance of the rock.

Perhaps the most common impurity in carbonate rocks is clay. The clay minerals—mainly kaolinite, illite, chlorite, smectite, and mixed-lattice types—may be either disseminated throughout the rock or concentrated in laminae or thin partings. The basic molecular building blocks of clay minerals are silica tetrahedra (a silicon atom and four oxygen atoms) and alumina and/or magnesium octahedra (an aluminum or magnesium atom and six hydroxyl ions). Other chemical elements are incorporated into the structure, however, so that it is difficult to determine the type of clay mineral by chemical analysis alone. If clay mineral identification is important, other analysis techniques, such as X-ray diffraction, differential thermal analysis, and electron microscopy, may be employed. A comprehensive review of clay minerals can be found in the clay chapters of this volume.

Chert is another common impurity in carbonate rocks and may be disseminated as grains throughout the rock or concentrated in nodules, lenses, or beds. It is composed mainly of very fine grained quartz (SiO_2) that may appear under the microscope as minute subequant crystals, usually 1 to 10 μ in diameter (microcrystalline quartz), or as radiating fibers (chalcedonic quartz). Chert easily incorporates impurities, including water, into its structure so that it may be found in almost all colors; its surface texture may range from dense or porcelaneous to porous or earthy. Dense cherts have a Mohs hardness of 7 and high impact toughness, which make them particularly abrasive to crushers and other processing equipment. Porous cherts, mainly because of their large surface area available for chemical reaction and moderate solubility in alkalies, are considered as deleterious components in aggregates used in concrete. Comprehensive coverage of the deleterious properties of cherts can be found in Dunn and Ozol (1962) and Kneller et al. (1968), the latter of which contains a useful bibliography with 600 entries. Good places to start for further information on the origin, texture, and composition of chert are papers by Folk and Weaver (1952), Calvert (1974), and Wise, Jr. and Weaver (1974).

Silica is also found in carbonate rocks as discrete silt- or sand-size grains of the mineral quartz. These grains may be disseminated throughout the rock or concentrated in laminae and beds. Detrital limestone especially may contain a considerable percentage of quartz silt and sand. These grains may act as the nuclei for coated carbonate grains, such as some ooliths and pisoliths.

Finely disseminated organic matter is a common constituent of limestones and dolomites and may give the rock a pronounced brown or black color. Bituminous material, an organic derivative of petroleum and residue of former pore fluids, may be present in sufficient quantity to make the rock undesirable for some uses.

Thin-section and insoluble-residue studies reveal trace amounts of a wide variety of other minerals in most carbonate rocks. Although these trace minerals may affect the economic usefulness of rocks used for chemical purposes, such as glass manufacture, they have little effect on rocks used principally for physical properties, such as construction materials.

Origin

Most limestones of economic importance were partly or wholly biologically derived from seawater and accumulated in a relatively shallow marine environment. The obvious skeletal material in limestones speaks of a biologic origin, but even nondescript, fine-grained material may derive from the life of organisms. Pellets in many instances are fecal material, and silt- and clay-size particles may be aragonitic-sheath crystals released upon the death of algae. Ooliths, which in the past have largely been thought to be the inorganic accumulation of calcium carbonate around a nucleus, may also depend in part on algal activity.

In some places, lime-secreting organisms, such as corals, calcareous algae, and mollusks, are able to erect large, wave-resistant structures called reefs. The biologically active parts of these structures are generally near the edge of shallow marine shelves where upwelling currents supply nutrients for the growth of the organisms. In other places along such shelves, small skeletal particles or other material may become coated by concentric layers of calcium carbonate to form ooliths. Oolites develop best in the high energy zone of shelves where water currents agitate the grains, and as the oolites build up they form elongate lenticular bars that nearly reach the water surface. Very fine grained carbonate muds (micritic limestone) derived from the comminution of coarser skeletal material or precipitated directly from seawater accumulate in low-energy environments, such as lagoons or deep water.

Environment of deposition is significant to the economic geologist because it determines the size, shape, purity, and other economically significant characteristics of the carbonate rock deposit. Limestones that form in high-energy zones generally contain little noncarbonate material and hence may be the source of high-purity carbonate material. Micrite, which accumulates in zones of low energy, is more likely to be diluted by clay and silt-size noncarbonate material.

Carbonate sediments are highly susceptible to postdepositional alteration and modification. The origin of dolomite is especially significant to the economic geologist. Although some dolomite may be precipitated directly from seawater, most dolomite is a result of the alteration of calcium carbonate sediments or rocks by hypersaline brines. Good examples are the almost-pure dolomite Silurian reefs in northern Illinois, Indiana, and Ohio, and in southern Michigan.

To understand the depositional environment and postdepositional history of carbonate rocks, economic geologists are well

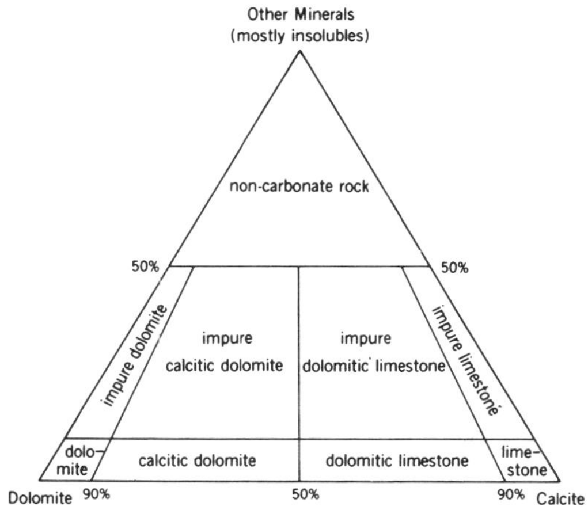

Fig. 1. Mineralogical classification of carbonate rocks.

advised to study modern carbonate deposition. Good places to start for this information are the general reviews by Baars (1963), Ginsburg et al. (1963), and Milliman (1974). For further information on locations and distributions of modern carbonate sediments, the summaries by Graf (1960a) and Taft (1967) are useful; the comprehensive discussions of the origin and occurrence of carbonate rock in the papers compiled by Jordan (1978) and Scholle, Bebout, and Moore (1983), in the book by Wilson (1975), and the reviews of limestone by Sanders and Friedman (1967) and of dolomites by Friedman and Sanders (1967) fill the gaps of the other coverages. A good coverage of carbonate sedimentology is provided by Tucker and Wright (1990).

Classification

The explosive growth in the study of modern carbonate sedimentation from the late 1950s to the present has had a notable effect on carbonate rock classification. Numerous classification schemes have emerged based on this newfound information with the result that carbonate rock descriptions are now more explicit and more conducive to genetic interpretation than before. It is unlikely that the ultimate objective of the industrial geologist is the origin of the rock; still a classification is useful in providing a written record that can be used to interpret the rock's physical and chemical properties.

Many aspects of carbonate rocks can be used as the basis of a classification scheme, but perhaps the most useful are composition and texture. Composition can be thought of in terms of mineralogy, types of fossils or grains, or chemical constituents. Texture refers to both depositional and post-depositional features, such as relative proportions of framework grains and lime mud, grain size, cement, and pores.

Carbonate rocks are rarely monomineralic in nature; thus, a mineralogical classification of these rocks needs to take into consideration the variation in amounts of calcite, dolomite, and noncarbonate materials (Fig. 1). Such a classification is useful in rock descriptions, especially when combined with textural parameters, but it commonly is not sufficient for industrial purposes. Although limestone and dolomite can be used equally well for a large number of uses, certain uses have special chemical requirements. These special requirements are stated in terms of chemical composition rather than mineralogical composition and specify the quantity of $CaCO_3$ (or CaO) and $MgCO_3$ (or MgO) or both in the rock along with the maximum percentage of impurities that can be tolerated. A practical chemical classification considers that ultrahigh calcium limestone is more than 97.5% $CaCO_3$, high-calcium limestone is more than 95% $CaCO_3$, high-purity carbonate rock is more than 95% combined $CaCO_3$ and $MgCO_3$, and high-magnesium dolomite is more than 43% $MgCO_3$ (theoretically pure dolomite is 45.7% $MgCO_3$).

A textural classification, along with a mineralogical classification, is fundamental to geologic studies aimed at determining the origin of carbonate rocks. One such classification by Leighton and Pendexter (1962) considered that most limestones can be characterized by the types and relative amounts of four textural components: grains, lime mud (micrite), cement, and pores. The ratio of the relative proportions of grains to micritic material, which is the basis of their nomenclatural system (Fig. 2), gives some indication of water turbulence because muds cannot be deposited in areas of strong bottom currents. Other classifications, such as the ones by Folk (1962) (Fig. 3) and Dunham (1962) (Fig. 4) make use of framework grains to mud ratios and have practical applications.

Dolomite presents a special problem in classification and may require separate handling from limestone. The textural classification can be used for secondary dolomite if the original depositional texture is preserved. Some dolomites, however, show only faint traces of original texture, called ghosts or relics, and others may

Grain/ Micrite Ratio	% Grains	Grain Type					Organic Framebuilders	No Organic Framebuilders
		Detrital Grains	Skeletal Grains	Pellets	Lumps	Coated Grains		
		Detrital Ls.	Skeletal Ls.	Pellet Ls.	Lump Ls.	Oolitic Ls. Pisolitic Ls. Algal Encrusted Ls.	Coralline Ls. Algal Ls. Etc.	Caliche Travertine Tufa
9:1	~90%							
		Detrital-Micritic Ls.	Skeletal-Micritic Ls.	Pellet-Micritic Ls.	Lump-Micritic Ls.	Oolitic-(Pisolitic-Etc.) Micritic Ls.	Coralline-Micritic Ls. Algal-Micritic Ls. Etc.	
1:1	~50%							
		Micritic-Detrital Ls.	Micritic-Skeletal Ls.	Micritic-Pellet Ls.	Micritic-Lump Ls.	Micritic-Oolitic (Pisolitic Etc.) Ls.	Micritic-Coralline Ls. Micritic-Algal Ls Etc.	
1.9	~10%							
		← Micritic Limestone →						

Fig. 2. Classification of limestones according to the scheme of Leighton and Pendexter (1962).

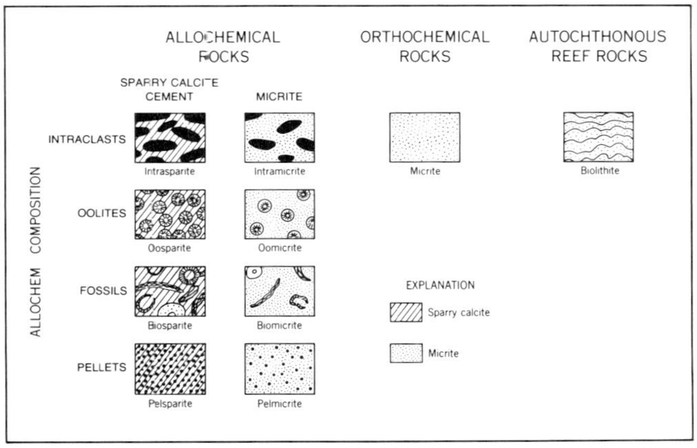

Fig. 3. Classification of limestones according to the scheme of Folk (1962).

have had their original texture completely obliterated. For these cases, and for dolomite of primary origin, a classification based on crystal size may be required.

Distribution of Deposits

Carbonate rocks have been deposited from Precambrian to Holocene time, and, although they compose only about 0.25% of the volume of the crust of the earth (Parker, 1967), they comprise about 15% of the sedimentary rocks. In 1991, limestone and/or dolomite were quarried or mined in all of the 50 states except Delaware, Louisiana, New Hampshire, and North Dakota; they were mined in all Canadian provinces except Saskatchewan.

Because carbonate rocks are widely distributed and differ in their geologic characteristics, each deposit must be considered on its own attributes. The best source of information for carbonate rock deposits is the state geological surveys, or their equivalents. Most state publications are oriented toward areal geology, commodities, or the two combined, rather than uses or methods. The US Bureau of Mines (USBM) and the US Geological Survey (USGS) also publish information on limestone and dolomite. The Survey's 7½-min geologic quadrangle maps are especially useful, as are the Bureau's circulars that contain cost analyses of limestone production (Eilertsen, 1964, Evans and Eilertsen, 1957, Kline, 1962, Kusler and Corre, 1968, Marshall, 1963, Riley and Schroeder, 1963). Table 2 lists the most useful publications on the limestone and dolomite resources of the different states and provinces. The annotated bibliographies of Gazdik and Tagg (1957) on high-calcium limestone deposits and of Davis (1957) on some high-magnesium dolomite deposits are helpful guides to the older literature, as is the report on high-grade dolomite deposits by Weitz (1942).

Although high-purity carbonate rock deposits are not overly abundant, they are by no means rare. All of the states, except Delaware, Hawaii, Louisiana, Mississippi, New Hampshire, and Rhode Island, and all of the Canadian provinces report potentially commercial deposits of high-calcium limestone and/or high-magnesium dolomite (Table 2). Twenty-one Mexican states produced high-calcium limestone or high-magnesium dolomite in 1992 (Fig. 5). Because deposits are present, however, does not necessarily mean that they can be exploited. In many areas, competition is intense for potential mineral lands for construction sites, recreation areas, nature preserves, and highways; even nature has her own requirements for flowing streams and soil development. Society has also imposed environmental controls, now firmly ensconced in state and federal statutes, which prohibit or restrict mineral production in areas where it might significantly affect the quality of man's environment.

EXPLORATION

Exploration for limestone and dolomite in North America is largely the detailed examination of known deposits. Because most limestone is a sedimentary rock, it occurs in strata generally of considerable extent. Some data as to chemical composition and physical character are available for most such strata in the published reports or files of state, provincial, and national geological surveys. Exploration for a new limestone deposit, therefore, in most instances begins with a search of these records to find the locations of deposits that satisfy the various economic factors and proceeds to a sampling program of favorable deposits. Geophysical techniques, if used at all, are used to determine the thickness of overburden. Geochemical techniques are not used.

All aspects of exploration are important, but the one most likely to be slighted is sampling. Yet it determines the validity of further study and may become the basis for hundreds of thousands and sometimes millions of dollars worth of development work. Its goal must be the accurate representation of the limestone deposit. Coring, rock bitting, and surface (ledge) sampling are the most common sampling methods, and the choice among these depends on such matters as the geology of the deposit, the proposed use of the material, and the availability of equipment.

<table>
<tr><th colspan="5">DEPOSITIONAL TEXTURE</th></tr>
<tr><td colspan="4">Original components not bound together during deposition</td><td rowspan="4">Original components were bound together during deposition, as shown by intergrown skeletal matter, lamination contrary to gravity, or sediment-floored cavities that are roofed over by organic or questionably organic matter and are too large to be interstices.</td></tr>
<tr><td colspan="3">Contains mud
(particles of clay and fine silt size)</td><td rowspan="3">Lacks mud
and is
grain supported</td></tr>
<tr><td colspan="2">Mud supported</td><td rowspan="2">Grain supported</td></tr>
<tr><td>Less than
10 percent grains</td><td>More than
10 percent grains</td></tr>
<tr><td>Mudstone</td><td>Wackestone</td><td>Packstone</td><td>Grainstone</td><td>Boundstone</td></tr>
</table>

Fig. 4. Classification of limestones according to the scheme of Dunham (1962).

Table 2. Limestone and Dolomite Resources in Canada and United States—Information Supplied by State and Provincial Surveys in 1991

	Limestone and Dolomite Currently Being Produced	High-Calcium Limestone and/or High-Magnesium Dolomite Currently Being Produced	Potentially Commercial Deposits of High-Calcium Limestone and/or High-Magnesium Dolomite are Present	Selected References (Number Keyed to Bibliography) on Limestone and Dolomite Resources
CANADA				
Alberta	x	x	x	159, 173, 201, 202, 203
British Columbia	x	x	x	102, 128, 159, 271
Labrador	x	x	x	70, 106
Manitoba	x	x	x	29, 154, 159, 169, 170
New Brunswick	x	x	x	155
Newfoundland	x	x	x	70, 106
Nova Scotia	x	x	x	155, 296, 358
Ontario	x	x	x	154, 158, 167, 191, 192, 346
Quebec	x	x	x	154, 156, 215, 216, 243
Saskatachewan			x	140, 159, 241
UNITED STATES				
Alabama	x	x	x	3, 105, 251, 312, 323
Alaska	x	x	x	64, 66, 218, 319, 332, 345
Arizona	x	x	x	228
Arkansas	x	x	x	60, 200
California	x	x	x	54, 55, 56, 179
Colorado	x	x	x	15, 16, 288, 391, 408
Connecticut	x	x	x	6, 7, 90, 184, 292
Delaware			x	
Florida	x	x	x	205, 325, 353, 393, 417, 418
Georgia	x	x	x	61, 277, 318, 321
Hawaii	x			403
Idaho	x	x	x	349, 350
Illinois	x	x	x	151, 152, 153, 210, 244, 246, 247, 248, 409
Indiana	x	x	x	23, 74, 78, 79, 135, 211, 275, 276, 317, 333, 334, 335, 399
Iowa	x	x	x	115
Kansas	x		x	141, 165, 214, 227, 330, 341, 342, 343, 344, 376
Kentucky	x	x	x	107, 108, 109, 110, 230, 231, 274, 375, 376
Louisiana				
Maine	x	x	x	4
Maryland	x	x	x	9, 267, 270
Massachusetts	x	x	x	14, 83, 127, 316, 327
Michigan	x	x	x	222, 251, 252, 365
Minnesota	x		x	255, 301, 303, 354, 383
Mississippi	x			47, 48, 256, 279, 294
Missouri	x	x	x	11, 197, 198, 208, 265, 266, 289
Montana	x	x	x	82, 98, 272, 353, 366
Nebraska	x	x	x	67, 68, 89
Nevada	x	x	x	80, 129, 308, 314
New Hampshire				
New Jersey	x	x	x	32, 239, 240, 263, 415
New Mexico	x	x	x	20, 237, 238, 360
New York	x	x	x	27, 220, 221, 307, 322, 386, 395, 419, 420
North Carolina	x	x	x	2, 41, 46, 91
North Dakota			x	12, 69, 72, 174
Ohio	x	x	x	217, 250, 370, 371, 372, 373, 374, 377
Oklahoma	x	x	x	10, 40, 42, 124, 172, 223, 337, 379, 398
Oregon	x	x	x	5, 62
Pennsylvania	x	x	x	45, 285, 309, 311
Rhode Island	x			316, 326, 327, 406
South Carolina	x	x	x	28, 87, 189
South Dakota	x		x	164
Tennessee	x	x	x	190, 381
Texas	x	x	x	31, 85, 122, 143, 259, 300, 331, 355, 369
Utah	x	x	x	293, 387, 388
Vermont	x	x	x	134, 278, 392, 402
Virginia	x	x	x	33, 34, 92, 93, 120, 121, 145
Washington	x		x	44, 101, 287, 290
West Virginia	x	x	x	234, 242, 273, 405
Wisconsin	x		x	63, 207, 368
Wyoming	x	x	x	176, 177, 178, 313

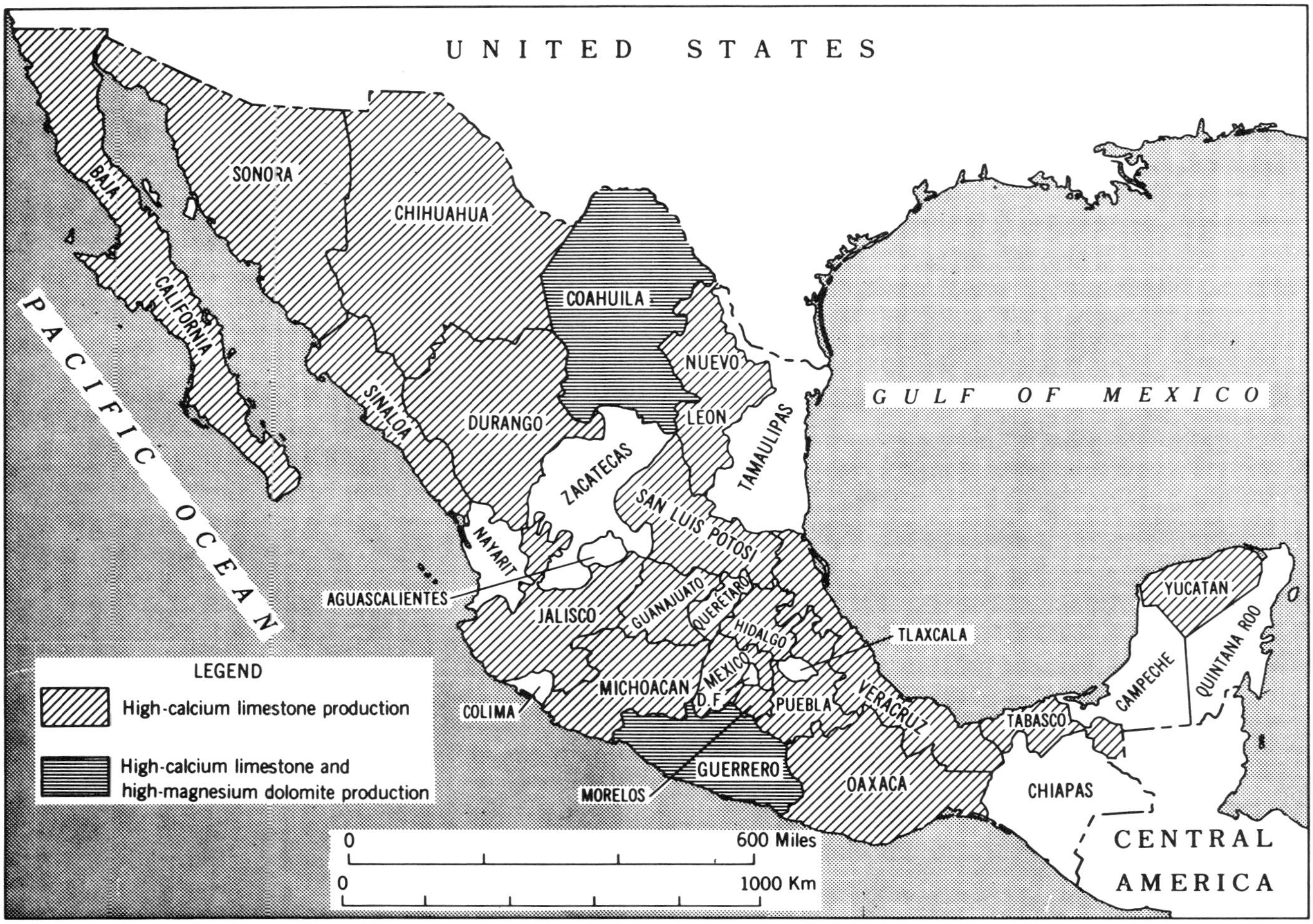

Fig. 5. Map showing states in Mexico producing high-calcium limestone and high-magnesium dolomite in 1992. Data from Cosejo de Recursos Minerales.

Special care should be taken when sampling weathered outcrops. In humid regions the surface layer of a carbonate rock may be leached of calcite and dolomite and hence be less pure than the rest of the unit. On the other hand, in arid and semiarid regions, where evaporation exceeds precipitation for long periods of time, the surface layer may be enriched in calcite and dolomite. Thus, if surface samples must be taken under these conditions, the geologist should be aware of a potential bias.

Coring

Coring is generally the best method of exploring a new deposit. A single core may not always be more representative than a section sampled in a quarry or in a natural exposure, where some judgment may be used as to what is representative, but cores taken on a grid pattern constitute a more representative and unbiased sample of a deposit than an equal number of surface sections. Coring avoids contamination by soil and weathered material and disproportionate sampling of different parts of a single unit, yet retains the surface material that may have worked down in solution cavities.

Initial drilling is generally widely spaced both to locate a potential deposit and to determine its potential size. Once an apparently large and suitable deposit is discovered, it should be drilled in a more or less regular pattern. The core grid and type of drilling depends largely on the proposed use of the limestone, although other factors, such as the homogeneity of the deposit, geologic complexity, topography, and cost of drilling, are significant. In areas of steeply dipping strata, the grid spacing must take into account whether vertical or inclined drilling will be used. If stone is to be used as cement raw material and the magnesium content of the rocks is believed to be marginal and unpredictable, no greater than 30-m centers would be required. If the deposit is relatively homogeneous in one direction and not in another, a rectangular rather than a square grid might be used. If the stone is to be used as aggregate and has proved relatively uniform in other deposits, generally only a few cores need to be taken in several hundred hectares. We recommend that for most deposits of flat-lying strata in which the chemical composition of the rocks is important, cores be spaced on 30-m centers until a pattern of uniformity indicates that the spacing can be increased safely. We know of one lime company, quarrying in a structurally complex area of Pennsylvania, that found coring on 15-m centers was necessary for adequate quality control.

For most limestone exploration a BX core (4.13 cm in diam.) suffices, but, if physical tests are required, a core of larger diameter must be used. NX core (5.4 cm in diam.) is adequate for chemical analyses and limited physical testing, but too small for some physical tests.

Each common physical test of rock used for aggregate requires about 5 kg of rock. If tests for absorption, abrasion, and soundness are performed, more than 14 kg of rock is required. If stone is to be tested for use as highway materials (ASTM D75C and AASHOTO T2), a minimum of 23 kg of rock is required. Highway commissions in some states do not perform physical testing of cores and accept only ledge samples. Testing agencies in other states will run physical tests on cores. For example, in prospecting for a quarry site near Skaneateles, NY, the General Crushed Stone Co. took a core 31 cm in diameter and had the stone tested for quality by the New York State Department of Public Works before opening the quarry (Moore et al., 1968).

Rock Bitting

If used alone, drill cuttings are probably the least reliable samples in exploration, but, if used to supplement cores or information obtained from nearby quarries or outcrops, they are a cheap, rapid method of acquiring much information. If the drilling is done in a carefully cased hole so as to prevent contamination by overburden, if the cuttings are collected carefully, and if the geologist is experienced in interpreting well cuttings, these samples may be as reliable as core and surface sampling for some purposes. In fact, the percussion air drill is probably the fastest and least expensive method for preliminary sampling.

Because of the large amount of drilling for oil and gas during the past few decades, cuttings of thousands of wells are on file in government and other sample libraries. Although not intentionally so, these cuttings constitute the largest exploration program for industrial minerals that has ever been undertaken. Cuttings cannot be interpreted properly unless the interpreter understands drilling techniques, and the best interpretation of the well cuttings requires considerable well site experience.

Surface Sampling

Chip samples taken carefully on a quarry face can provide a good representation of the limestone deposit. The geologist should first inspect the quarry face and divide the face into units of uniform lithology and should mark tops of units by paint or flagging. He/she should subdivide thick homogeneous beds arbitrarily, so that no sample is more than about 1.5 m thick. Starting at the base of the unit, one should take chips of uniform size as nearly as possible along a selected vertical line to the top of each unit. (The geologist might start at the top and work down, but it is easier to work up than down.) This method of sampling is sometimes called channel sampling. Chip samples should be taken from unweathered surfaces even if the weathered rind must be chipped away. Samples should be washed to remove some contaminants, such as lichens or soil, from the samples, but care should be taken not to wash out thin, interbedded shales.

Many channel or core samples weigh 5 to 10 kg, and this amount must be crushed, thoroughly mixed, and quartered several times to reduce the sample to the few grams needed for chemical determination.

EVALUATION AND TESTING

If given sufficient cores or exposures of rock, an experienced industrial minerals geologist can evaluate a limestone deposit largely just by looking at it, using a hand lens, hammer, and weak hydrochloric acid as the only tools. The geologist can generally appraise whether a stone will make class A aggregate and/or cement raw material. One may not be able to determine whether it is of sufficient purity for some other uses, but much stone can be eliminated that is not of sufficient purity for those uses. Final evaluation of rocks suitable for aggregate, dimension stone, and similar uses requires physical testing. Rocks used for making lime, cement, or other products that depend on chemical purity should be chemically analyzed.

A list of laboratories capable of making chemical and physical determinations of rocks can be found in the *Directory of Testing Laboratories, Commercial-Institutional*, available from the American Society for Testing and Materials.

Physical Properties

Procedures for physical testing can be obtained from two main sources: the American Society for Testing and Materials (ASTM) and the American Association of State Highway and Transportation Officials (AASHTO). Both organizations describe explicit procedures, in cookbook fashion, for testing limestone and dolomite. State highway commissions may specify slightly different procedures and should be consulted if stone is to be used as aggregate. In earlier years, several federal organizations developed physical tests for building stone, such as those reported by the US Census Office (1884) and the US Bureau of Standards (Kessler and Sligh, 1927), but now building-stone testing generally follows procedures outlined by ASTM. Testing methods and procedures developed by the USBM for testing diamond drill cores of mine rock are described in a report by Obert et al. (1946).

Physical tests are designed to test how well a rock will perform for a particular use. As might be expected, a great number of tests have been developed in response to the many uses to which rock is put. The selected physical properties of carbonate rock shown in Table 3 are from reports of a lengthy study of the physical properties of mine rock by the USBM. In addition to the physical properties listed in Table 3, many other properties have been reported in this same study, which include tensile strength, abrasive hardness, scleroscope hardness, Young's modulus, modulus of rigidity, Poisson's ratio, specific damping capacity, and longitudinal bar velocity. A large number of physical properties of limestones and other rocks have been collected and published in the *Handbook of Physical Constants* (Birch et al., 1950). This handbook contains data on bulk density, compressive strength, compressibility at high and low temperature, dielectric constant, electrical resistivity, porosity, thermal conductivity, thermal expansion, and other properties that are especially useful to the geophysicist. Considerable research on the properties of rock that have application to blasting and mining have been undertaken by the USBM (Atchison et al., 1964, Atchison and Pugliese, 1964, Bur et al., 1969, Dick et al., 1973, Merrill, 1956, Pugliese, 1972, Thill et al., 1969, Willard and McWilliams, 1969). A tabulation of porosity and bulk density determinations of carbonate rock, as reported in the more accessible American, British, German, and Swiss literature, was published by Manger (1963).

A great many physical tests of carbonate rock used for aggregate have been performed in connection with state and federal road building programs, but much of this information is unpublished. In 1953, Woolf (1953) compiled results of physical testing of road-building aggregate as performed by the US Bureau of Public Roads until 1951. Table 4 shows average values of physical properties for selected rock types from this compilation. Unfortunately data from later testing of aggregate in the federal highway program have not been tabulated. The Highway Research Board of the National Research Council has sponsored much research dealing with the relationships of the physical properties of carbonate rock to its serviceability as an aggregate. A report by Renninger and Nichols (1970) reviewed some of this work and gave the current status of aggregate research. Marek (1991) gives a useful overview of sampling and testing principles.

Table 3. Physical Properties of Some Carbonate Rocks

Age	Formation	Location	Rock Type	Apparent Specific Gravity	Apparent Porosity, %	Compressive Strength, MPa	Modulus of Impact Rupture, MPa	Toughness cm/cm^2	Reference
Miocene	Unspecified	Eniwetok	Ls.(fossil.)	2.39	4.0	97.2	16.5	0.4	Blair (1956)
Eocene	Unspecified	Eniwetok	Ls.(dense)	2.53		122.0	22.0	0.8	Blair (1956)
Cretaceous	Niobrara	S. Dakota	Ls.(chalky)	1.81	8.3	25.5	4.1	0.5	Blair (1955)
Mississippian	St. Louis	Missouri	Ls.(oolitic, fossil.)	2.56		115.8	15.9	1.1	Blair (1956)
Mississippian	Maxville	Ohio	Ls.(fn.-grained)	2.41		108.9	17.2	1.2	Blair (1956)
Mississippian	Salem (Spergen)	Indiana	Ls.(fossil.)	2.37	11	75.2	11.0	0.7	Windes (1949)
Devonian	Columbus	Ohio	Ls.(v. fn-grained)	2.60	5.4	123.4		1.4	Blair (1956)
Devonian	Columbus	Ohio	Ls.(fn.-grained)	2.69	0.7	196.5	20.0	3.4	Windes (1949)
Silurian	Brassfield	Ohio	Ls.(dolomitic)	2.8	1.3	179.2	193.1	1.6	Windes (1950)
Silurian	Niagara	Ohio	Dol.(fn.-grained)	2.4	8.6	89.6	75.8	0.7	Windes (1950)
Silurian	Clinton	Alabama	Ls.(cs.-grained)	2.83	0.9	165.5		2.6	Windes (1949)
Ordovician	Chickamauga	Tennessee	Ls.	2.73	3.4	>173.0	5.5	2.2	Blair (1956)
Ordovician	Lenoir	W.Virginia	Ls.(siliceous)	2.68	6.	158.6	13.1	1.0	Windes (1950)
Ordovician	Knox	Tennessee	Dol.(fn.-grained)	2.84	0.7	322.0	26.2	2.3	Windes (1949)
Cambrian	Bonne Terre	Missouri	Ls.(dolomitic)	2.66	3.3	175.1	12.4	1.9	Blair (1955)
Cambrian	Oro Grande	California	Marble	2.72	0.2	165.5	16.5	0.7	Blair (1955)
Precambrian	Cockeysville	Maryland	Marble	2.87	0.6	212.4	19.3	1.1	Windes (1949)

Table 4. Average Values for the Physical Properties of Rocks Used for Aggregate*

Rock Type	Bulk Specific Gravity	Absorption,%	Loss of Abrasion		Toughness†
			Deval Test,%	Los Angeles Test,%	
Basalt	(229) 2.86	(228) 0.5	(203) 3.1	(24) 14	(203) 19
Chert	(74) 2.50	(74) 1.6	(78) 8.5	(6) 26	(29) 12
Diabase	(332) 2.96	(309) 0.3	(340) 2.6	(63) 18	(285) 20
Dolomite	(668) 2.70	(667) 1.1	(708) 5.5	(134) 25	(612) 9
Gneiss	(419) 2.74	(424) 0.3	(602) 5.9	(293) 45	(386) 9
Granite	(662) 2.65	(666) 0.3	(718) 4.3	(174) 38	(703) 9
Limestone	(1695) 2.66	(1673) 0.9	(1677) 5.7	(350) 26	(1315) 8
Marble	(184) 2.63	(162) 0.2	(175) 6.3	(41) 47	(188) 6
Quartzite	(208) 2.69	(204) 0.3	(233) 3.3	(119) 28	(161) 16
Sandstone	(716) 2.54	(707) 1.8	(699) 7.0	(95) 38	(681) 11
Schist	(297) 2.85	(296) 0.4	(314) 5.5	(136) 38	(212) 12

*Number of samples tested in parenthesis.
†Height in cm for a 2 kg steel plunger to cause a rock core about 2.5 cm diameter and 2.5 cm high to fail.
Source: Woolf, 1953.

Through the years, the US Army Corps of Engineers has tested carbonate rock from many quarries in the continental United States in order to evaluate its potential use in constructing locks, dams, and other structures. A summary of this testing, which includes the standard tests for sulfate soundness, abrasion resistance, specific gravity, and absorption, can be found in Technical Memorandum No. 6–370, "Test Data—Concrete Aggregates in Continental United States" of the US Army Corps of Engineers, Waterways Experiment Station. This compilation, which occupies nearly a meter of shelf space, cannot be purchased but is available for inspection at the Corps of Engineers' libraries. Other more accessible publications that give physical test values and results of the Corps' extensive research relating to the use of carbonate rock as aggregates in portland cement concrete are those by Curry and Buck (1966) and Stowe (1969).

Some manufacturers of pulverized limestone, defined as limestone or dolomite having a minimum fineness of 97% passing a 325 mesh sieve, have established standards and test methods through the Pulverized Limestone Division, National Stone Association. Test methods that are recommended include particle size, pH, dry brightness, water demand, and calcium and magnesium carbonate content.

A petrographic examination of much carbonate rock can foretell its physical properties and subsequent suitability for certain uses, but too few people make use of this method. The definitive reference for this type of applied petrography is ASTM Standard C 295, "Petrographic Examination of Aggregates for Concrete," which is based on the work of Mather and Mather (1950). Techniques of petrographic examination and some applications of this kind of work can be found in Mather (1953), Mielenz (1954, 1955), Newlon, Jr. et al., (1972), and Scholle (1978).

Rippability is a physical property of rock that concerns mainly the construction industry. Limestones that are thinly bedded, low in compressive strength, or sufficiently inhomogeneous so as to contain horizontal planes of weakness can generally be ripped easily. Empirical testing by the Caterpillar Tractor Co. (1972) found a relationship between seismic wave velocities and rock rippability: the higher the wave velocity, the more difficult it is to rip the rock.

Various research workers have examined the relationships between different physical as well as chemical properties of carbonate rocks. These attempts have not met with much success because of the inhomogeneity of the rocks. Further information can be found in Judd and Huber (1961), D'Andrea et al., (1965), West and Johnson (1965), Mutmansky and Singh (1968), Baxter and Harvey (1969), and Carr et al., (1970).

Further information on physical requirements of carbonate rock can be found in the chapters of this book on Stone, Crushed and Dimension, and Cement and Cement Raw Materials.

Chemical Properties

The techniques for making chemical analyses are beyond the scope of this chapter. Instrumental methods of analysis have become popular in recent years, and as a result a wide selection of instructional material is available on theory and techniques. Some of the most useful are those on atomic absorption spectroscopy by Angino and Billings (1967), Galle and Angino (1967), and Slavin (1968); and one on rapid chemical methods by Shapiro and Brannock (1962). Details of standard gravimetric and complexometric (EDTA titration) methods can be found in Hillebrand and Lundell (1953) and Diehl (1964), respectively.

The chemical and physical properties of carbonate rocks are interdependent. Pure calcite in the form of poorly cemented chalk is not only unique in its low strength and high absorption among the carbonate rocks, but it is also highly reactive chemically due to the large surface area of the component grains. Pure calcitic marble of the same chemical composition as the chalk is relatively strong, unabsorptive, and unreactive. Dolomite that contains quartz sand grains may have the same overall composition as dolomite that contains chert, but their suitability as aggregate differs widely due to the difference in reactivity of the two forms of silica. Processing also affects the degree of fracturing of stone and thus the surface area and chemical reactivity. Therefore, physical and mineralogic descriptions of carbonate rocks are of considerable use in predicting the chemical properties of the product from a deposit.

For some uses of carbonate rocks, chemical analysis may be of little or no help in estimating the suitability of a rock unit; for other uses various chemical determinations must be available to compare with specific optimum compositions. For example, in stone used for chemical purposes, such as glass raw material, flux, or cement, the percent of certain elements must fall within specified limits or ranges. On the other hand, the chemical content may or may not be important for stone that is used because of its physical properties, such as aggregate, building stone, or riprap. A thick section of rock that is almost pure dolomite is likely to be well-cemented reefal dolomite, and it can be predicted generally to make good aggregate (Ault, 1989). Rock that is pure calcite, however, may be a skeletal or oolitic limestone that is either well cemented or poorly cemented and thus might make excellent or only fair aggregate.

The proportion of alumina (Al_2O_3) and silica (SiO_2) in the rock may be helpful in determining the value of a carbonate rock for a use in which physical properties are important. Most of the silica in a carbonate rock is likely to be present as clay, silt- and sand-size quartz, or chert. For the common clay minerals found in limestones and dolomites, as much as 2% silica may be present for each 1% of alumina. The higher the alumina content, the more argillaceous the sample is likely to be. The alumina content can be multiplied by two to obtain an estimate of the amount of silica tied up in the clay. The way the remaining silica occurs can be determined only by reference to the sample description or by examining the sample itself.

The chemical analysis of a carbonate rock is essential for estimating the neutralizing value of agricultural limestone, which is usually expressed in terms of calcium carbonate equivalent (CCE). Pure calcite ($CaCO_3$) is assigned a CCE value of 100. Pure dolomite ($CaCO_3.MgCO_3$) has a theoretical CCE value of 108.6; that is, it is 8.6% more effective than pure limestone as a neutralizer. Such interpretation assumes that a molecule of $MgCO_3$ with a molecular weight of 84.32, is as effective a neutralizer as 1 molecule of $CaCO_3$, with a molecular weight of 100.09. Thus, a given weight of $MgCO_3$ is 1.19 times as effective as the same weight of $CaCO_3$. Because of differences in solubility, however, a dolomitic liming material will take longer to neutralize a given amount of acid than a limestone one, even though the CCE of the two is the same. A useful guide to selecting agricultural limestone materials has been prepared by Goodwin (1979).

Many chemical analyses of carbonate rocks have been published to show the varied compositions of limestones and dolomites (Clarke, 1924, Gillson et al., 1960, Graf, 1960b, Siegel, 1967, Wedepohl, 1970), but these may not always be typical of the formation from which they were taken. Most sedimentary carbonate rocks vary in their impurities—including clay minerals, resistant minerals such as quartz, and organic material—because they were deposited in different environments. In addition, the rocks have evolved chemically as well as physically during compaction, dehydration, and lithification. Subsequent processes, including burial and exposure to percolation of water, provide for synthesis of authigenic minerals and alteration, such as oxidation of organic matter. Because of the limited extent of identical environments of deposition and subsequent postdepositional conditions, composition of any given rock unit is likely to be variable. Analyses of many samples taken at different sites are required to reveal the approximate composition of a particular unit of rock.

Good sources of chemical data on carbonate rocks are the state geological surveys or their equivalents. Many state surveys have files of chemical data obtained from quarry sampling and coring programs. Many of the references listed in Table 2 contain extensive lists of chemical analyses. The USGS published 1 131 analyses of carbonate rocks from Montana, Wyoming, North Dakota, South Dakota, Nebraska, Colorado, and Kansas (Hill et al., 1967) and 3 585 analyses of carbonate rocks from Alaska, Idaho, Oregon, and Washington (Hill and Werner, 1972).

Specifications

Limestone specifications vary with the end use of the stone. These specifications may be either physical or chemical, but frequently they include both parameters. Physical specifications, such as durability and gradation, are more important if the stone is to be used as is, such as for construction aggregate. Chemical properties are more important if the stone is to be subject to pyro-processing as in the production of lime or cement. Most industrial and agricultural applications require adherence to both a physical and a chemical specification. For example, glass-batch raw materials might have both a rigid chemical specification and a narrow gradation requirement.

Physical specifications focus on both the natural properties of the rock as well as properties imparted during processing. Natural properties are intrinsic, such as hardness, composition, texture, color, porosity, and density, and processing properties are derived from the physical gradations and result from crushing, screening, washing, and air classification. Obviously the results of processing are directly related to the natural characteristics of the stone.

Physical specifications relating to gradation and durability are by far the most common when the rock is to be used as an aggregate material. These specifications are based upon standardized testing and practices developed through ASTM. ASTM specified Grading Requirements for Coarse Aggregate are shown in Table 5. Other related ASTM physical specifications include (ASTM, 1991):

11.1.1 Sampling—Practice D 75, and Practice D 3665
11.1.2 Grading and Fineness Modulus—Method C 136
11.1.3 Amount of Material Finer Than No. 200 Sieve—Test Method C 117
11.1.4 Organic Impurities—Test Method C 40
11.1.5 Effect of Organic Impurities on Strength—Test Method C 87
11.1.6 Soundness—Test Method C 88

Table 5. Grading Requirements for Coarse Aggregates (ASTM 1991)

Size Number	Nominal Size (Sieves with Square Openings)	Amounts Finer than Each Laboratory Sieve (Square Openings), Weight %									
		100 mm	90 mm	75 mm	63 mm	50 mm	37.5 mm	25.0 mm	19.0 mm	12.5 mm	9.5 mm)
1	90 to 37.5 mm	100	90 to 100	—	25 to 60	—	0 to 15	—	0 to 5	—	—
2	63 to 37.5 mm	—	—	100	90 to 100	35 to 70	0 to 15	—	0 to 5	—	—
3	50 to 37.5 mm	—	—	—	100	90 to 100	35 to 70	0 to 15	—	0 to 5	—
357	50 to 4.75 mm	—	—	—	100	95 to 100	—	35 to 70	—	10 to 30	—
4	37.5 to 19.0 mm	—	—	—	—	100	90 to 100	20 to 55	0 to 15	—	0 to 5
467	37.5 to 4.75 mm	—	—	—	—	100	95 to 100	—	35 to 70	—	10 to 30
5	25.0 to 9.55 mm	—	—	—	—	—	100	90 to 100	20 to 55	0 to 10	0 to 5
56	25.0 to 9.5 mm	—	—	—	—	—	100	90 to 100	40 to 85	10 to 40	0 to 15
57	25.0 to 4.75 mm	—	—	—	—	—	100	95 to 100	—	25 to 60	—
6	19.0 to 9.5 mm	—	—	—	—	—	—	100	90 to 100	20 to 55	0 to 15
67	19.0 to 4.75 mm	—	—	—	—	—	—	100	90 to 100	—	20 to 55
7	12.5 to 4.75 mm	—	—	—	—	—	—	—	100	90 to 100	40 to 70
8	9.5 to 2.36 mm	—	—	—	—	—	—	—	—	100	85 to 100

11.1.7 Clay Lumps and Friable Particles—Test Method C 142
11.1.8 Coal and Lignite—Test Method C 123
11.1.9 Weight of Slag—Test Method C 29
11.1.10 Abrasion of Coarse Aggregate—Test Method C 131 or C 535
11.1.11 Reactive Aggregates—Appendix XI
11.1.12 Freezing and Thawing—Test Method C 666
11.1.13 Chert—Test Method C 123

ASTM Standards provide a guideline both for the required gradations and the test methods employed in evaluating materials for use as construction aggregates. The specific gradation requirements may vary to some degree by individual states as a result of specific regional application or generally accepted practices. As an example, Table 6 compares gradations for an ASTM No. 8 specification with gradation for a No. 8 stone from Ohio, Indiana, and Kentucky.

Table 6. Comparison of Size Gradations No. 8 Stone Screen Size % Passing

State Specifications	1″	3/4″	1/2″	3/8″	No. 4	No. 8	No. 10	No. 16
ASTM	—	—	100	85–100	10–30	0–10	—	0–5
Ohio	—	—	100	85–100	10–30	0–10	—	0–5
Indiana	100	85–100	20–60	—	0–5	0–2	—	—
Kentucky	—	100	85–95	40–65	5–20	0–5	—	—

While there may be state to state modifications or variations in the physical gradations of specific stone sizes, most states generally have durability requirements that are fairly consistent across the country for aggregates for use as highway surface materials. The most common specifications for durability relate to abrasion, soundness, and/or freeze-thaw, all of which are covered by an ASTM standard procedure. These tests are a measure of the rocks ability to withstand repeated cycles of weathering and continued impact of traffic. Several states have also included skid resistance for aggregates to be used in surface-course paving and bridge decks. As a consequence, carbonate rocks have been excluded for these uses in several states because they polish easily.

Frequently the specifications for rock used in non-state- and federal-funded projects are modified for economic reasons, such as lower transportation costs. This type of project could include non-sensitive uses such as subgrade material for driveway or other domestic, non-load-bearing purposes. Conversely, a higher transportation cost might be justified in order to achieve a specific appearance as in some exposed-aggregate applications where architectural considerations are more important than economic considerations.

Chemical specifications for carbonate rocks generally are industry or application specific. In any consideration of chemical specifications, or the suitability of a limestone or dolomite for a particular application, it is important to understand that chemical properties are not necessarily related to physical properties. For example, the high-purity limestone from the Mosheim Formation is very hard, but the equally pure limestone from the Ocala Formation is very soft. The marble from the Franklin Formation has both high-brightness and high-carbonate values, yet the dark gray, almost black, marble from the Dundee Formation has a higher calcium-carbonate value.

Limestone and dolomite specifications may be developed in a variety of ways, including rigid testing and evaluation, mutual consent between the buyer and seller, common practice in an area, or even as a compromise to offset some other item, such as high freight costs. A specification may also be written to describe a particular limestone source, such as when the buyer is not sure which components are important for a particular application. In these instances of uncertainty, the final specification results from testing several different limestones; selection is based upon best performance.

Some specifications may be unique to a particular industry and require testing properties based upon an industry standard or procedure. Examples of such properties are thermal decrepitation, odor, taste, and crystallinity. In some instances, these requirements may be related to a specific company and be unique to the specific application. For instance, when a limestone is to be used in rice polishing, it may be subjected to a taste test to be sure that the limestone does not impart an unpleasant taste to the rice during polishing.

As noted above, it is not at all uncommon for a limestone specification to incorporate both physical and chemical requirements. This is particularly true of industrial uses, such as glass manufacturing, where limestone is one component of the glass-batch raw materials. In this example, particle size, gradation, chemical degradation, and chemistry are equally important. Additionally, batch-to-batch uniformity of composition, both in chemistry and physical properties, are important and may require statistical process control (SPC). The importance of SPC should not be underestimated. For example, Ford Motor Company requires all glass-batch suppliers of raw materials to employ audited SPC in their process technology in order to be qualified as an acceptable vendor.

Table 7. Physical and Chemical Specifications for Glass-Grade Limestone

	Typical Sieve Analysis		
Size (mm)	% retained	% retained cumulative	% passing cumulative
1.68 (12 MESH)	0.00	0.00	100.00
1.19 (16 MESH)	0.35	0.17	99.83
0.84 (20 MESH)	5.06	5.20	94.80
0.30 (50 MESH)	57.05	62.25	37.75
0.15 (100 MESH)	26.26	88.90	11.10
0.07 (200 MESH)	9.98	98.40	1.60
PAN	1.60	100.00	0.00

Moisture Content 0.09%

	Typical Chemical Analysis	
Chemical	Reported as	%
Calcium carbonate	$CaCO_3$	97.80
Magnesium carbonate	$MgCO_3$	1.25
Iron oxide	Fe_2O_3	.095
Silica	SiO_2	.56
Alumina	Al_2O_3	.23
Nickel	Ni	<.002
Chromium	Cr_2O_3	<.001
Strontium oxide	SrO	.03
Manganese oxide	MnO	<.01

Typical specifications for limestone to be used in glass-batch raw materials are shown in Table 7.

Industry-specific specifications, such as those required for limestone or dolomite used in the production of lime or cement, in flue gas desulfurization, and filler and extender specifications, are included in the chapters on these materials. The point is made here, however, that any industrial or agricultural application or use may have a unique specification that is predicated upon specific performance requirements. Individual publications on specifications, uses, and other aspects of the crushed stone industry are available from the National Stone Association (1991).

PRODUCTION AND USES

The USBM reports that 798 Mt of limestone, dolomite, and related materials were sold or used in the United States in 1989 (Table 8).

Table 8. Crushed Carbonate Rock Sold or Used in the United States, by Kind in 1989 (Tepordei, 1991)

Kind	Number of quarries	Quantity (1 000 t)	Value (1,000 $)	Unit Value
Limestone	2,533	731 748	$3,234,317	$ 4.42
Dolomite	136	44 725	$ 216,957	$ 4.85
Marble	65	11 088	$ 137,481	$12.40
Calcareous marl	15	4 643	$ 11,614	$ 2.50
Shell	18	6 281	$ 34,794	$ 5.54
TOTALS Weighted (AVERAGE)	2,767	798 485	$3,635,163	$ 4.55

In 1989, carbonate rock was produced in 47 of the 50 states, with only Delaware, Louisiana, and North Dakota not reporting any limestone or dolomite production. Louisiana, however, did report shell production. The top eight producing states in descending order were Florida, Texas, Pennsylvania, Illinois, Tennessee, Missouri, Kentucky, and Ohio. These eight states accounted for 55.7% of the total US production.

Whereas this tonnage is consumed in hundreds of applications, the predominate markets for limestone and dolomite can be divided into 10 major categories and 38 primary uses (Table 9). The first five major categories are related to the use of carbonate rocks as construction materials. These five categories, a total of 302 Mt or 38% of all crushed carbonate rock produced in the United States in 1989, was produced for aggregate markets. Limestone and dolomite thus represent 54% of the total 562 Mt of aggregate material sold or used.

Agricultural applications consumed 17.7 Mt of limestone and dolomite (Table 9), plus another 3.8 Mt of shell materials. Agricultural liming materials made up 84% of this category. Chemical and metallurgical uses account about 103 Mt or about 13% of the limestone and dolomite produced each year.

Special uses cover a host of applications, the largest of which is fillers, extenders, and whiting materials. Because of the unique applications and value of these products, they are covered in a separate chapter. Mine dusting and acid-water treatment are a unique category unto themselves. High-calcium limestone, in excess of 95% calcium carbonate, is used as rock dust in coal mines for explosion suppressant. This product must contain less than 2% silica and is normally ground to 95% less than 325 mesh. Limestone or dolomite used in acid-water treatment may be specified with a variety of sizes depending both upon the pH of the water to be treated as well as the available residence time. In some instances it might be injected as a 325-mesh powder into a low-pH effluent stream or it might be used as boulders in acid streams not suitable to continuous treatment with finer-grained materials.

Dimension stone is the one category not included in Table 9. Dimension stone production is closely tied to commercial construction; in 1989 about 417 kt were produced. Limestone and marble together account for approximately 24% of total US dimension stone production. Indiana was the number one state in limestone dimension stone production, and Georgia, followed by Vermont, was the leading state for marble production.

When granite production is included with limestone and marble production, the United States is the fourth largest producer of dimension stone in the world. Nevertheless, total US dimension stone production is one-fourth that of Italy and one-half that of Spain (Harries-Rees, 199[illegible]). Italy, Spain, and Greece produce more dimension stone than the United States; each is a net exporter. The United States is a net importer of dimension stone.

Interestingly, dimension stone production is reported in tons, but sales and prices are normally based upon the cost per cubic foot. These prices have remained fairly constant over the last several years; increases in profitability realized by producers have come from increases in productivity.

PREPARATION FOR MARKET

Mining

Limestone and dolomite are high volume, low-value commodities. This segment of the industrial minerals industry is highly competitive and is characterized by thousands of operations serving local and/or regional markets. Thus competitive environment dictates that production cost control is a critical element in any stone operation.

Most limestone and dolomite is mined from open quarries, although in many areas economic and/or environmental considerations favor large scale production by underground mining. The

Table 9. Crushed Limestone and Dolomite Sold or Used by Producers in the United States in 1989, by Use (Tepordei, 1991)

	Limestone		Dolomite	
Use	Quantity, 1 000 t	Value, $1,000	Quantity, 1 000 t	Value, $1,000
Coarse Aggregate (+38 mm):				
Macadam	6 457	26,970	809	3,437
Riprap and jetty stone	11 159	48,612	547	2,213
Filter stone	5 309	22,760	57	284
Other coarse aggregate	1 329	4,134	W	W
Coarse Aggregate, graded:				
Concrete aggregate, coarse	69 672	340,740	5 762	23,230
Bituminous aggregate, coarse	39 771	190,816	3 574	14,832
Bituminous surface-treatment aggregate	13 127	63,848	1 244	5,937
Railroad ballast	3 518	14,246	645	2,596
Other graded coarse aggregate	1 201	3,399	—	—
Fine aggregate (−9.7 mm):				
Stone sand, concrete	11 659	59,622	346	1,666
Stone sand, bituminous mix or seal	11 015	48,175	1 109	4,311
Screening, undesignated	17 526	77,773	386	1,762
Other fine aggregate	565	2,481	W	W
Coarse and fine aggregates:				
Graded road base or subbase	99 815	392,120	6 474	24,852
Unpaved road surfacing	18 384	76,917	719	2,661
Terrazzo and exposed aggregate	2 607	16,284	[1]	[1]
Crusher run or fill or waste	28 460	116,119	1 051	4,617
Other coarse and fine aggregates[2]	481	1,589	37	129
Other construction materials	9 693	38,502	232	1,065
Agricultural:				
Agricultural limestone	14 775	73,036	2 402	16,247
Poultry grit and mineral food	2 143	21,115	[3]	[3]
Other agricultural uses	737	3,911	92	853
Chemical and metallurgical:				
Cement manufacture	83 042	264,028	547	1,359
Lime manufacture	14 358	83,641	309	1,393
Dead-burned dolomite manufacture	65	278	1 174	7,162
Flux stone	4 276	19,636	641	2,556
Chemical stone	415	1,911	—	—
Glass manufacture	241	3,063	W	W
Sulfur oxide removal	1 051	4,536	—	—
Special:				
Mine dusting or acid water treatment[4]	611	7,433	W	W
Asphalt fillers or extenders	1 487	11,727	—	—
Whiting or whiting substitute	603	11,344	—	—
Other fillers or extenders[5]	1 892	50,304	930	10,136
Roofing granules	106	3,483	—	—
Other miscellaneous uses:				
Waste material	1 828	7,024		
Other uses not listed	346	5,750	687	6,208
Unspecified:[6]				
Actual	203 532	900,919	11 687	63,676
Estimated	48 531	216,070	3 269	13,774
Total	731 748	3,234,300	44 725	217,000

W Withheld to avoid disclosing proprietary data; included with "Other uses not listed"
[1] Included with other construction materials, [2]Includes stone used in drain fields, [3] Included with other agricultural uses, [4] Includes stone used in other coarse and fine aggregates, and mine dusting or acid water treatment, [5]Includes stone used in abrasives, paper manufacture, and pipe bedding, [6]Includes production reported without a breakdown by use and estimates for no respondents.

only carbonate materials not consistently recovered by surface or underground mining are shell products that are dredged from parts of US coastal waterways.

Surface Mining: The basic elements of surface mining include overburden removal, drilling, blasting, and hauling to the crushing and processing plant. Ultimately the surface mine must also contend with reclamation requirements. The selection of surface mining equipment varies with the particular requirements at each individual operation, production capacity required, size and shape of the deposit, haul distances, estimated life of the operation, location relative to urban centers, and other social/economic factors. Other factors that must be considered in undertaking surface mining are the value of the products that are to be produced, location of competitive operations, and environmental and safety requirements associated with a particular deposit.

Frequently a surface mine or quarry will contract for the drilling and/or blasting part of the production cycle. The specialized nature of this aspect of production, combined with the unique regulatory

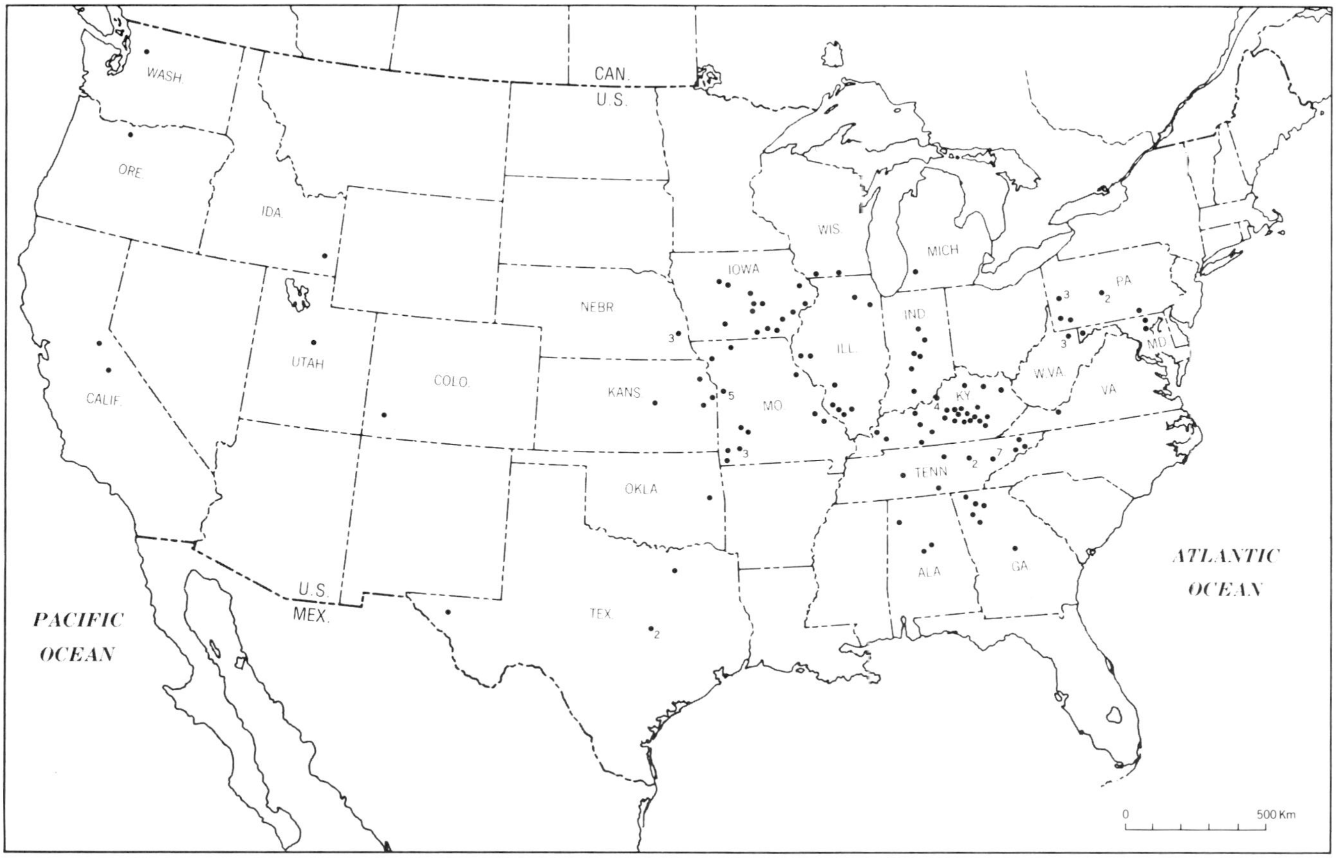

Fig. 6. Map showing locations of active underground limestone mines in the United States. The figure beside a location symbol indicates the number of mines represented by the symbol. Information supplied by the US Bureau of Mines in 1991.

and safety requirements associated with the handling and use of explosives, may make the use of an outside contractor economically attractive. Particularly for quarries in close proximity to densely populated urban environments. The contractor is responsible for the maintenance of an explosives magazine and all record and reporting requirements and may arrange for seismic monitoring of each blast.

Overburden removal is a key element in the cost of any surface mining operation and its ability to compete with other mines in its market area. If the overburden ratios for all surface mines in a given area are relatively low, for example, if 3 m of overburden must be removed to recover 30 m of limestone, a mine with 15 m of overburden for the same 30 m of rock will not be competitive. If the mine is unique to its particular area, however, high overburden ratio may be economically justified. Additionally, the production of high value end products may support mining operations that would not be economical for low-value products. For example, in East Tennessee as much as 90 m of overburden has been removed from a 30-m-thick, steeply dipping deposit of chemical-grade limestone for use in the production of lime. If this were a conventional aggregate operation, this overburden ratio would probably not be justified.

Underground Mining: The most important fact about underground mining of limestone is that it is done. In fact, 127 underground limestone mines were operating in the United States in 1989. In 1900, total production of limestone from all underground mines was 95 kt, or 0.37% of the total limestone production (Thoenen, 1926). In 1924, production was 5.2 Mt, or 4.5% of the total production. In 1965, it was 31.2 Mt or about 6% of the total limestone production. In 1989, it was 57 Mt or about 7.5% of the total production. Many of the mines are in the midwest (Fig. 6), but their concentration reflects the abundance of high-quality limestone in horizontal beds in that part of the country as much as a concentration of population. This distribution pattern will change gradually as more mines are built to serve urban centers (Carr and Ault, 1983, and Baxter, 1989).

The basic operations in underground mining include drilling, blasting, loading and hauling, scaling, and roof bolting. Drilling equipment includes horizontal drill jumbos as well as downhole track drills. This equipment is generally quite different from that used for surface mining and results in the drilling of much smaller blast holes and a lower volume of rock pulled in each individual blast. Other equipment required in the underground mine includes powder loaders, which are used to blow ammonium nitrate-fuel oil mixtures (ANFO) into the blast holes. Scaling rigs, which are used to remove loose rocks from the ribs and roof of the mine, and roof-bolting equipment may also be required in an underground mine.

Most underground limestone and dolomite mines are room and pillar type operations and many recover rock from both headings and benches. It is not uncommon for an underground limestone mine to have several benches and an overall mine height that may reach 30 m. Whereas the thickness of the deposit being mined is directly controlled by the thickness of the rock and related roof conditions, it is not at all uncommon for an individual heading to

Fig. 7. Underground mining of limestone. Photo courtesy of Dravo Lime Company.

be 7.5 to 10.5 m high and in some instances to reach as high as 15 m. Rooms are generally 13.5 to 15 m in width, which, depending on the type drilling jumbo used, can normally be excavated in one- or two-drill set ups. A V type drill pattern is commonly employed to maximize the amount of rock pulled with each individual shot and reduce the number of boots or unbroken rock in the shot face. Scaling is normally required as a safety measure; roof bolting may or may not be required depending upon roof conditions at the individual mine. Loading and hauling equipment may include standard 22 to 45 t haul trucks and correspondingly sized front end loaders. In some mines the loading equipment may be more typical of that used in underground hard rock operations and include load-haul-dump units or other type of tramming equipment (Fig. 7). For more detailed information on both surface and underground mines, the reader is referred to Kennedy, 1990, Hall, 1981, Bise, 1986, and Cummins and Given, 1973.

Processing

Limestone processing varies with the end product and target industry but of necessity includes several basic similarities. These similarities relate primarily to sizing and include crushing and screening. Depending upon such factors as the volume of rock processed, type of mine, haul distance, and surface topography, primary crushing may be done either at or near the mining face or at the primary processing plant, which is usually adjacent to the loading and shipping point. When primary crushing is done in the quarry or mine, the rock is normally reduced to a -15 cm size and is moved via conveyor to the main processing plant.

Crushing equipment selection is dependent upon the plant size, physical properties of the rock, and the product to be produced. In general, primary crushing is by jaw or gyratory crusher, but impact crusher or other specialized equipment may be used. Several underground limestone mines use Stamler feeder-breakers as the primary crusher, which is a low-profile piece of equipment. Secondary crushing is usually by cone and/or gyratory crushers, although impact and roll crushers and hammer mills are commonly used.

Screening is one of the most critical steps in the process cycle, particularly in the production of crushed stone. The selection from a wide variety of screen types and screen cloth depends upon the requirements for the end product. Most screens incorporate some form of inclined vibratory equipment. Screen cloth may range from a rod deck or punctured steel plates for larger product sizes, down through woven wire, welded wire cloth, rubber, plastic, and polyurethane for smaller product sizes. Screens have to be checked regularly for holes and tears because oversized material can quickly contaminate a significant volume of product.

Quarried stone usually has some moisture associated with it, and, as a consequence, fines may adhere to larger pieces of rock. Many crushed stone facilities include washing equipment in the production line to remove these fines and thus assure performance with gradation specifications. Where washing facilities are included in the process cycle, settling ponds are also required in order to eliminate the discharge of silt-laden water to neighboring streams. Where stone washing is not practical and when moisture cannot be tolerated in either subsequent processing steps or the final product, a dryer may be required. The dryer is normally installed in the process ahead of the screening equipment. These dryers are usually counter-flow rotary dryers that operate at temperatures of 120 to 150°C. Fluidized-bed dryers may also be used and are particularly useful where the flash calcination of fines might be a problem.

Limestone and dolomite processing for industrial and agricultural applications may also require air classification and milling equipment, particularly when products are smaller than 10 mesh and must have a clean bottom cut, usually a 100 or 200 mesh. Air classifiers are easily adjusted and have the flexibility to make a wide range of products. Cage mills, roller mills, and ball mills are used in the production of a variety of particle sizes ranging from 20 mesh down to 2 μm in size. Wet processing is usually required to produce products smaller than 2 μm.

Once processed, the stone products must be stored prior to shipment. Most crushed stone products are stored in open areas or a combination of open areas and loading bins. The volume of the material produced and the variety of products generally dictate the design of the storage area. In many larger aggregate operations, a series of radial stackers radiate out from a screening plant. Other operations may also have several stone bins to facilitate truck and/or rail loadout. Storage of industrial and agricultural products requires enclosed silos and bins to protect them from moisture and contamination by other products. These bins and associated loadout spouts or equipment require positive dust control, including adequately sized baghouses. They may also include pneumatic conveying equipment and/or bucket elevators in the product-handling system.

Programmable controllers and/or computers are finding increased use in the processing of the limestone and dolomite. As process plants become larger and more complex, these computer-controlled process systems are an invaluable tool in achieving improved production efficiencies and improved product quality. This equipment may be incorporated into the statistical process-control program of the plant and may be interactive with loadout and scale computers. Where interactive systems are employed, they may provide on-line inventory control, bills of lading, and input directly from loadout scales to computer-generated invoicing. Several software packages currently available are particularly well adapted to aggregate operations. A useful reference is the National Stone Association (1990) publication, *Guidelines for the Successful Automation in the Crushed Stone Industry.*

Transportation

Transportation is a major factor in the delivered price of limestone and dolomite products. In very general terms, the lower the value of the processed product the shorter the distance it can be transported. Stated another way, construction aggregates are less likely to move long distances to the marketplace than are 2 μm carbonate filler/extender materials. In many places, the cost of transportation equals or exceeds the FOB plant value of the stone.

Thus, limestone and dolomite aggregates are generally marketed locally.

In the limestone and dolomite chapter in the 5th Edition of *Industrial Rocks and Minerals*, we reported that trucking accounted for 82% of the transportation of limestone and dolomite products. The remainder of the material was transported about equally by rail or waterway. This distribution of transportation has not changed appreciably in the last ten years. However, many factors related to the cost of this transportation have changed. Both rail and trucking rates have essentially been deregulated, and, with the exception of the State of Texas, where the Texas Railroad Commission sets freight rates, competition has been allowed to function openly in the marketplace. As a consequence, it is frequently possible for the stone processing company to negotiate rates for specific jobs or contracts. This competition has also fostered a number of intermodal transportation arrangements that have allowed stone products to be delivered to more remote locations than might otherwise be thought possible.

In looking at the cost of trucking limestone and dolomite products, we must consider the truck costs associated with aggregate and crushed stone materials separately from the cost of transporting industrial and agricultural products. Whereas the specifics of trucking vary from area to area, in very general terms, approximately half of the trucking is done by the contractor purchasing and using the aggregate as opposed to owner/operators or commercial haulers. Many state and local governments also have their own trucking equipment and can be grouped with those contractors hauling for their own purposes or jobs. The actual cost of trucking in the case of contractor- or government-owned equipment is extremely difficult to determine because each company or governmental unit values and costs this transportation somewhat differently. However, with regard to owner/operators or contract haulers, using open dump trucks, for hauls of less than 40 km, a charge of 5 to 6¢ per ton-km is not uncommon. These rates, however, are subject to negotiation and may be cut significantly on very large construction projects.

In many places, trucking of industrial and agricultural products is handled differently from the hauling of limestone and dolomite aggregates. The equipment used normally includes pneumatic tankers, bottom-dump grain trucks, and vans and flatbeds for bagged products. Where bottom-dump grain haulers can be used, back-haul freight rates may be possible, allowing the producer to reach customers whose geographical location is far more remote than would otherwise be possible with normal tariff rates. In the case of industrial products, the potential for contamination by foreign matter is a major concern. As a consequence, it is not at all uncommon for the producing company to have its own equipment or to contract with a specific hauler for dedicated equipment. Even then it may be necessary to employ rigorous clean out procedures in order to preclude any possibility of product contamination, particularly when the product includes micron-sized materials. Because of the specialized nature of equipment used for industrial and agricultural purposes, trucking rates tend to be slightly higher than those for aggregate products and may be in the 5 to 8¢ per ton-km range. However, even these rates may be subject to negotiation when a consistent haul can be established and several units can be dedicated to a specific movement.

Railroads continue to be a major factor in the transportation of limestone and dolomite products, whether they be aggregate materials or industrial and agricultural products. For aggregates, rail movement may employ intermodal facilities at the receiving end in which the railed stone is either transferred to a stockpile or directly to a truck for delivery to the customer. Such facilities have been used successfully in selected markets and most recently has been employed by one Pennsylvania producer to move aggregates into the Baltimore-Washington market area. On a single line haul with mini-unit trains and a minimum ten-car movement, it is possible to ship aggregate products at a cost of 1 to 2¢ per ton-km. In many places, agricultural and industrial applications are large-volume users and require the use of covered-hopper equipment. Nevertheless, again it is possible to negotiate very favorable rail freight rates if a single-line haul can be used. Where average rail tariffs might be in the range of 2 to 3¢ per ton-km, it is possible to obtain negotiated contract rates at 70 to 80% of tariff rates. The negotiation of favorable rail rates, however, becomes far more difficult when two or more rail lines are involved and when switching charges are imposed by short line rail carriers. Nevertheless, even in multi-line hauls, when the competition is a straight truck movement, the volumes of stone to be transported may justify a very competitive bid from the combined rail effort.

Waterway transportation continues to hold its own and has continued to maintain about the same volume of stone products as it has for the past 10 years. Most of the stone moved via waterway is either construction aggregate or chemical-grade limestone or dolomite used in the production of lime or cement. When the movement is by barge, it normally requires a shipment of 1.4 kt per barge and generally is not attempted on movements of less than 480 km. Freight quotations of ½ to 1¢ per ton-km are not uncommon depending upon the specifics of the movement in question. Barge movement, however, is not the only means of waterway transportation. Self-unloading ore carriers on the Great Lakes, ocean going barges, and 27- to 54-kt bulk ship movements on the ocean are also being employed to move crushed stone. Ocean freight and Great Lakes shipment rates can be negotiated at less than 1¢ per ton-km. The utilization of this type of transportation, however, is clearly restricted to large volume movements.

LEGISLATIVE AND ENVIRONMENTAL ISSUES

Limestone and dolomite producers, along with the rest of the mining industry, are faced with an ever-increasing array of regulatory and environmental legislation. Today, the average US citizen is far more environmentally conscious than just ten years ago. As a consequence, the public expects a positive and responsible commitment from the mining industry to environmental, health, and safety concerns. These concerns divide into two basic groups: local community sensitivities and state and federal mandates.

At the local level, the limestone industry is faced with zoning and community relations requirements, the response to which will dictate the success or failure of each individual quarry in the years ahead. The response by the stone producer must include the development of long- and short-range reclamation plans, aggressive dust and noise control, attention to the aesthetics of plant entrances and other street-appeal items, and a willingness to listen to community concerns. In some instances, restrictions may be imposed on the hours that blasting can take place or on the hours a quarry might be able to operate. One of the major concerns of many communities is the traffic associated with the quarry's or mine's operations. In some places, trucks moving to and from the limestone producer's operations must travel through neighboring residential areas. As a consequence, additional driver safety and courtesy training may be required and speed restrictions must be aggressively enforced.

State and federal regulations include a number of items of critical interest to the limestone and dolomite industry. As previously noted, reclamation plans may be a critical part of the permit process and should be developed well in advance of mining. In the past, it has not been at all uncommon for a limestone producer to view reclamation requirements as being a future item not needing current input or concern. Many local communities and some states, however, are now requiring aggressive reclamation plans that include the recovery of abandoned areas as a quarry proceeds throughout its entire mine life. In the development of these reclamation plans, several items need to be specifically addressed. These include long term land use decisions, health and safety of the community, and elimination of hazardous wastes and other potentially toxic discharges.

State and federal legislation that is currently under debate and that will have a significant impact on the mining industry is the *Federal Wetlands Delineation Manual* and associated legislation. Given the current federal thinking of no net wetlands lost and the location of many of our limestone and dolomite resources, conflict appears to be almost inevitable. Industrial, environmental, and political factions are currently debating the definition of wetlands, and it may be some time before a clear delineation and definition of what is and what is not a wetland can be determined. It is an issue, however, that has the potential of having a tremendously negative impact on the limestone and dolomite industry.

Other critical issues currently receiving public attention include crystalline silica, storm-water runoff, underground storage tanks, and the 1990 Clean Air Act Amendments. The crystalline-silica issue is of particular concern due to the vagueness of the definition of crystalline silica and to the fact that currently its regulation is only under discussion. At present crystalline silica is not regulated by the Occupational Safety and Health Administration (OSHA) or the Mine Safety and Health Administration (MSHA), but it is included on the list of possible carcinogens by the International Agency for Research Against Cancer (IARC). As a consequence, debate continues on whether respirable crystalline silica is a hazardous or toxic substance, the hazards of silicosis, and the possible carcinogenicity of crystalline silica. As yet, there is no clear direction from congress or the Environmental Protection Agency (EPA). The problem is further complicated by the fact that silica is present in almost every mineral material, and it will be virtually impossible for limestone and dolomite producers and other mining and mineral companies to avoid litigation related to crystalline silica.

Storm-water runoff is another of the current environmental concerns of the quarry operator and mineral processor. This program, mandated by the federal government, requires storm-water discharge permits for all of the mining industry. This requirement is of particular concern to the limestone and dolomite producer because most of its plant areas are broad and cover several acres of unpaved ground. As a consequence, it is quite conceivable that storm waters may pick up sizeable sediment loads and discharge these to nearby streams. Under the regulations, these discharges would be illegal because they do not meet clean water requirements. The National Stone Association has been working extensively with both state and federal legislative committees and stone producers to promote and assist compliance with these new requirements.

Federal mandates to eliminate underground storage tanks also demand the stone producers' attention. These tanks could potentially contaminate ground water as they age and develop leaks. Because it is extremely difficult to monitor the integrity of these underground tanks, their removal has been mandated. All of these regulations and environmental issues, along with clean air and water requirements, are costly and make it imperative that today's stone producer take a proactive approach to environmental compliance. For further information the reader is referred to Lootens, Greenslade, and Barker (1991).

FUTURE TRENDS

Looking ahead to the next decade, we expect the demand for limestone and dolomite to grow at an average annual rate of about 2 to 2.5% per year. Limestone is the primary raw material for crushed stone, and the demand therefore is expected to match approximately the demand for new construction. In recent years, considerable attention has been focused on the nation's deteriorating infrastructure and the need for repair and replacement. The 102nd US Congress passed the Inter-Modal Surface Transportation Infrastructure Act of 1991 in late 1991. The bill provides the basic framework for the US highway and mass transportation programs for the next six years. The bill also provides total spending authority of $151 billion divided between highways ($119 billion) and mass transit ($32 billion).

The demand for chemical-grade carbonate rock used in industrial and agricultural applications is expected, at a minimum, to equal the growth of the US population. In addition, the 1990 Clean Air Act Amendment could greatly expand the demand for both lime and limestone. Lime is a primary reagent material for both wet and dry scrubbing; limestone is used in both wet scrubbing and fluidized-bed combustion. These uses are covered in more detail elsewhere in this volume.

Although the demand for limestone and dolomite is expected to remain strong throughout the 1990s and into the 21st century, the structure of the stone industry is undergoing change. The most pronounced changes relate to consolidation within the industry and the increasing role being played by foreign owners. In the 1980s and early 1990s, many mergers, acquisitions, and joint ventures have been effected by foreign and domestic companies. These have resulted in a tremendous change in the names on the front gates of many limestone and dolomite operations. Benbow (1988) presents an excellent overview of several of the developments in the aggregate industry wherein foreign ownership has come to dominate many of the larger limestone quarry operations in the United States. In 1983, ARC, Ltd. acquired the Mule Pen limestone quarry in Florida, and, in 1987, it acquired American Aggregates. These acquisitions gave ARC a presence in Florida, Ohio, Indiana, and Michigan; later ARC strengthened its position with the additional acquisition of Ernst Aggregates of Dayton, OH. Another British company, Tarmac, Plc., acquired Florida-based Pennsuco Limestone quarry from Lone Star Industries in 1984, and then went on to secure a position in Texas with the acquisition of Lampasas Quarry and Chico Crushed Stone. In 1982, Redland, Plc., a diversified UK building products corporation, merged with Texas-based McDonough Brothers, Inc., to form Redland-Worth Corporation. With an infusion of several million dollars into the San Antonio limestone quarry, Redland-Worth enlarged the quarry to where it is now one of the five largest plants in the United States. In 1985, Redland aquired Genstar Stone Products Company with its limestone operations in Maryland and Texas. Another UK building company, C.H. Beazer, also entered the US market with its acquisition of Gifford-Hill and its Texas quarry operations. Later, after months of negotiations, Beazer also acquired Koppers Company, which gave it control of limestone operations in Ohio, Michigan, Indiana, Kentucky, Pennsylvania, and New York.

While the British have been and continue to be involved in these and many other acquisitions in the United States, they are not the only companies actively looking for US investments. CSR, Ltd., a diversified Australian corporation, acquired Rinker Materials Corporation of Florida, which has sizable holdings in that state. Other companies making acquisitions in the United States included Australia's Pioneer Construction Services, Ltd., and Japan's Onoda Cement Company, Ltd. These have not been the only companies or countries involved, but certainly they have been among those that have been more visible. This trend that started in the 1980s is expected to continue and will effectively consolidate more operations under common ownership in the future.

Acquisition and consolidation, however, has not been limited to foreign companies. US-based Vulcan Materials Corporation,

Florida Rock Industries, and Dravo Corporation all continue to expand their operations within the United States. Each company has made limestone and/or dolomite acquisitions, which have served to further increase their presence in the crushed stone industry as well as their dominance in selected market areas. Companies producing limestone for construction aggregates have not been the only ones undergoing consolidation, acquisition, and merger in the United States. Industrial and agricultural producers of limestone products have also had face changes in recent years. Georgia Marble, English China Clay, Franklin Industrial Minerals, and JM Huber have all made significant acquisitions that have increased their market share and presence in the production of industrial- and agricultural-grade limestone.

A second trend that is having an increasing impact on US limestone production is the increasing role of imported crushed stone materials. Offshore limestone operations located in Newfoundland, the Bahamas, and Mexico have all been able to take advantage of low cost ocean freight to reach markets along the East and Gulf Coasts of the United States. Newfoundland Resources and Mining Company, Ltd., initiated its first shipments of limestone to the United States in May 1990, and it has targeted full production to be 3.5 Mt (Rukavina, 1990). The $30 million dollar plant was designed specifically for exporting crushed stone by ship along the eastern coast of the United States. Canada is only one of the countries currently being used as an off-shore source of crushed limestone for domestic US markets. Dravo Corporation opened its Bahamas quarry near Freeport, Bahamas, in 1989, and targeted production at 1.4 Mtpy (Rukavina, 1989). Prior to the start up of Dravo's Bahamian operations, the company already operated distribution terminals in Tampa, FL, and Mobile, AL. These were natural ports of entry for the Bahamas stone and have been the initial distribution points for that rock. The Yucatan Peninsula of Mexico has also been a source area for offshore stone. Vulcan Materials started up an operation west of Cancun, with a capacity greater than 3.6 Mtpy, and it is bringing much of that rock into a port facility in Houston, TX. While the efforts of Vulcan Materials, Dravo, and Newfoundland Resources and Mining Company have been directed at the importation of crushed stone for aggregate materials, they have not been the only operations bringing limestone into the United States. On a much smaller scale, white, chemically pure industrial-grade limestone has also been imported from the Caribbean for use in filler/extender materials. These products, however, have thus far represented only a fraction of 1% of the total high-calcium filler/extender material produced and sold in the United States.

In summary, limestone and dolomite have been and continue to be one of the most important raw materials in the United States. Demand will increase, albeit at a modest rate, and new uses and applications will be found. But, limestone and dolomite will continue to be commodities produced and sold in a highly competitive marketplace. Competitive pressure will dictate that limestone and dolomite producers be attentive to cost control and apply improvements in technology to increase productivity. At the same time, they will have to adapt to a social and political climate that will demand no deterioration in the environment and thus drive up costs.

BIBLIOGRAPHY AND REFERENCES

1. Aase, J.H., 1970, "Transportation of Iron Ore, Limestone, and Bituminous Coal on the Great Lakes Waterway System, With Projections to 1995," Information Circular 8461, US Bureau of Mines, 61 pp.
2. Adair, R.B., et al., 1947, "Evaluation of North Carolina Raw Materials for the Production of Portland Cement, Part 1, Preliminary Laboratory Investigation," Bulletin 35, North Carolina State College, Dept. Eng. Research, 31 pp.
3. Alabama Geological Survey, see Publications List of County Reports and Maps.
4. Allen, H.W., 1953, "Progress Report of Limestone Survey, Knox County," Report of the State Geologist for 1951–1952, Maine Development Commission, pp. 11–30.
5. Allen, J.E., 1946, "Reconnaissance Geology of Limestone Deposits in the Willamette Valley, Oregon," Short Paper 15, Oregon Dept. of Geology and Mineral Industries, 15 pp.
6. Altamura, R.J., 1987, "Bedrock Mines and Quarries of Connecticut (Map)," Map, 1:125,000 scale, Connecticut Geological and Natural History Survey, Natural Resources Series.
7. Altamura, R.J., 1987, "Citation List and References Cited for Bedrock Mines and Quarries of Connecticut," Connecticut Geological and Natural History Survey, 41 pp.
8. Ames, J.A., 1966, "Limestone Deposits vs. Beneficiation," *Ohio Journal of Science*, Vol. 66, No. 2, pp. 131–136.
9. Amsden, T.W., 1954, "Geology of Garrett County," *Geology and Water Resources of Garrett County*, Bulletin 13, Maryland Dept. of Geology, Mines and Water Resources, pp. 1–95.
10. Amsden, T.W., and Rowland, T.L., 1965, "Silurian Stratigraphy of Northeastern Oklahoma," Bulletin 105, Oklahoma Geological Survey, 174 pp.
11. Anderson, K.H., et al., 1979, "Geologic Map of Missouri," Scale: 1 in. = 8 miles, Missouri Geological Survey and Water Resources.
12. Anderson, S.B., and Haraldson, H.C., 1969, "Cement Rock Possibilities in Paleozoic Rocks of Eastern North Dakota," Report of Investigation 48, North Dakota Geological Survey, 62 pp.
13. Angino, E.E., and Billings, G.K., 1967, *Atomic Absorption Spectrometry in Geology*, Elsevier, New York, 144 pp.
14. Apfel, E.T., 1944, "Dolomite Marble in the Vicinity of Lee, Massachusetts, as an Available Source of Metallic Magnesium," Open File Report, US Geological Survey, 34 pp.
15. Argall, G.O., Jr., 1949, "Dolomite, Industrial Minerals of Colorado," *Colorado School Mines Quarterly*, Vol. 44, No. 2, pp. 144–149.
16. Argall, G.O., Jr, 1949, "Limestone, Industrial Minerals of Colorado," *Colorado School Mines Quarterly*, Vol. 44, No. 2, pp. 254–274.
17. ASTM, 1991, "Concrete and Aggregates," *Annual Book of ASTM Standards*, Vol. 04.02, Philadelphia, PA, Oct., 840 pp.
18. Atchison, T.C., et al., 1964, "Effect of Decoupling on Explosion-Generated Strain Pulses in Rock," Report of Investigations 6333, US Bureau of Mines, 49 pp.
19. Atchison, T.C., and Pugliese, J.M., 1964, "Comparative Studies of Explosives in Limestone," Report of Investigations 6395, US Bureau of Mines, 25 pp.
20. Austin, G.S., 1990, "Commercial Travertine in New Mexico," *New Mexico Geology*, Vol. 12, pp. 49–58.
21. Ault, C.H., 1989, "Construction Materials: Crushed Stone," *Concise Encyclopedia of Mineral Resources*, D.D. Carr and N. Herz, eds., Pergamon Press, Oxford, pp. 57–64.
22. Ault, C.H., 1992, "Geologic Factors Affecting the Underground Limestone and Dolomite Mines of Indiana," *Proceedings*, 26th Forum on Geology of Industrial Minerals, P.C. Sweet, ed., Virginia Geological Survey Publication 119, pp. 83–94.
23. Ault, C.H., and Carr, D.D., 1981, "Search for High-Calcium Limestone in Silurian Reefs of Northern Indiana," *Geological Society of America Bulletin*, Vol. 92, pp. 641–647.
24. Ault, C.H., and Haumesser, A.F., 1990, "A Central Indiana Model for Predicting Jointing Characteristics in Underground Limestone Mines," *Proceedings*, 24th Forum on Geology of Industrial Minerals, South Carolina Geological Survey, pp. 1–8.
25. Azároff, L.V., and Buerger, M.J., 1958, *The Powder Method in X-Ray Crystallography*, McGraw-Hill, New York, 342 pp.
26. Baars, D.L., 1963, "Petrology of Carbonate Rocks," *Shelf Carbonates of the Paradox Basin*, Four Corners Geological Society 4th Field Conference, pp. 101–129.
27. Baird, G.C., 1979, "Sedimentary Relationships of Portland Point and Associated Middle Devonian Rocks in Central and Western New York," New York State Museum & Science Service Bulletin 433, 24 pp.
28. Banks, R.S., 1977, "Stratigraphy of Eocene Santee Limestone in Three Quarries of the Coastal Plain of South Carolina," Geologic Notes, Vol. 21, No. 3, South Carolina Division of Geology, pp. 85–149.

29. Bannatyne, B.B. 1975, "High-Calcium Limestone Deposits of Manitoba," Man. Mines Br., Pub. 75–1, Manitoba Mines Branch, 103 pp.
30. Barksdale, R.D., ed., 1991, *The Aggregate Handbook*, National Stone Association, Washington, DC, 775 pp.
31. Barnes, V.E., and Bell, W.C., 1977, "The Moore Hollow Group of Central Texas," Report of Investigations 88, Bureau of Economic Geology, University of Texas, Austin, 169 pp.
32. Barnett, S.G., 1970, "Upper Cayugan and Helderbergian Stratigraphy of Southeastern New York and Northern New Jersey," *Geological Society America*, Bulletin, Vol. 81, pp. 2375–2402.
33. Bartlett, C.S., Jr., and Biggs, T.H., 1980, "Geology of the Abingdon, Wyndale, Holston Valley and Shady Valley Quadrangles, Virginia," Publication 16, Virginia Division of Mineral Resources, 39 pp.
34. Bassler, R.S., 1905, "Cement Materials of the Valley of Virginia," Bulletin 260, US Geological Survey, pp. 531–544.
35. Bates, R.L., 1966, "Geology of Cement Raw Materials in Ohio: A Summary," *Proceedings*, 2nd Forum on Geology, Industrial Minerals, Indiana Univ., Bloomington, Indiana Geological Survey, pp. 23–29.
36. Bates, R.L., 1972, "Limestone Quarry in a Gravel Pit," *Proceedings*, 8th Forum on Geology of Industrial Minerals, Public Information Series 5, Iowa Geological Survey, pp. 5–10.
37. Baxter, J.W., 1980, "Factors Favoring Expanded Underground Mining of Limestone in Illinois," *Mining Engineering*, Vol. 32, No. 4, pp. 1497–1504.
38. Baxter, J.W., 1989, "Possible Underground Mining of Limestone and Dolomite in Central Illinois," *Proceedings*, 23rd Forum on Geology of Industrial Minerals, R.E. Hughes and J.C. Bradbury, eds., Illinois Minerals 102, Illinois State Geological Survey, pp. 21–33.
39. Baxter, J.W., and Harvey, R.D., 1969, "Alumina Content of Carbonate Rocks as an Index to Sodium Sulfate Soundness," Industrial Minerals Notes 39, Illinois Geological Survey, pp. 5–10.
40. Beach, J.O., and English, S.G., 1940, "Dolomite and Magnesium Limestone," Mineral Report 6, Oklahoma Geological Survey, 19 pp.
41. Beatly, K.O., and Adair, R.B., 1948, "Evaluation of North Carolina Raw Materials for the Production of Portland Cement, Part 2. Small-Scale Production of Portland Cement," Bulletin 42, North Carolina State College Engineering School, 51 pp.
42. Bellis, W.H., and Rowland, T.L., 1976, "Shale and Carbonate-Rock Resources of Osage County, Oklahoma," Circular 76, Oklahoma Geological Survey, 50 pp.
43. Benbow, J., 1988, "Developments in Aggregates—A Changing Corporate Landscape," *Industrial Minerals*, No. 254, London, Nov., pp. 57–69.
44. Bennett, W.A.G., 1944, "Dolomite Resources of Washington, Part 1," Report of Investigations 13, Washington Division of Mines and Geology, 35 pp.
45. Berkheiser, S.W., Jr., 1983, "Reconnaissance Survey of Potential Carbonate Whiting Sources in Pennsylvania," Mineral Resources Report 83, Pennsylvania Geological Survey, 4th Series, 53 pp.
46. Berry, E.W., 1947, "Marls and Limestones of Eastern North Carolina," Bulletin 54, North Carolina Division of Mineral Resources, 16 pp.
47. Bicker, A.R., Jr., 1970, "Economic Minerals of Mississippi," Bulletin 112, Mississippi Geological Survey, 80 pp.
48. Bicker, A.R., Jr., and May, J.H., 1976, "Agricultural Lime in Central Mississippi," Information Series MGS-77-1, Mississippi Geological Survey, 55 pp.
49. Birch, F., et al., 1942, "Handbook of Physical Constants," Special Paper 36, Geological Society of America, 325 pp.
50. Bise, C.J., 1986, "Mining Engineering Analysis," Society of Mining Engineers, Inc., Littleton, CO, 160 pp.
51. Bissell, H.J., and Chilingar, G.V., 1967, "Classification of Sedimentary Carbonate Rocks," *Developments in Sedimentology, Carbonate Rocks*, Vol. 9A, G.V. Chilingar, H.J., Bissell, R.W. Fairbridge, eds., Elsevier, New York, pp. 87–168.
52. Blair, B.E., 1955, "Physical Properties of Mine Rock, Part III," Report of Investigations 5130, US Bureau of Mines, 69 pp.
53. Blair, B.E., 1956, "Physical Properties of Mine Rock, Part IV," Report of Investigations 5244, US Bureau of Mines, 69 pp.
54. Bowen, O.E., Jr., 1957, "Limestone, Dolomite, and Lime Products," *Mineral Commodities of California*, Bulletin 176, California Division of Mines and Geology, pp. 293–306.
55. Bowen, O.E., Jr., 1966, "Limestone, Dolomite, and Lime Products, *Mineral Resources of California*, Bulletin 191, California Division of Mines and Geology, pp. 221–233.
56. Bowen, O.E., et al., 1973, "The Mineral Economics of the Carbonate Rocks," *Limestone and Dolomite Resources of California*, O.E. Bowen, ed., Bulletin 194, California Division of Mines and Geology, 60 pp.
57. Bowles, O., 1956, "Limestone and Dolomite," Information Circular 7738, US Bureau of Mines, 29 pp.
58. Bowles, O., and Jensen, M.S., 1941, "Limestone and Dolomite in the Chemical Processing Industries," Information Circular 7169, US Bureau of Mines, 15 pp.
59. Bowles, O., and Jensen, M.S., 1947, "Industrial Uses of Limestone and Dolomite," Information Circular 7402, US Bureau of Mines, 19 pp.
60. Branner, G.C., 1941, "Limestones of Northern Arkansas," Miscellaneous Publication, Arkansas Geological Commission, 24 pp.
61. Brantley, J.E., 1916, "Limestones and Marls of the Coastal Plain of Georgia," Bulletin 21, Georgia Geological Survey, 300 pp.
62. Brooks, H.C., 1989, "Limestone Deposits in Oregon," Special Paper 19, Oregon Department of Geology and Mineral Industries, 72 pp.
63. Buckley, E.R., 1898, "On the Building and Ornamental Stones of Wisconsin," Bulletin 4, Wisconsin Geological Survey, 544 pp.
64. Bundtzen, T.K., et al., 1990, "Alaska's Mineral Industry, 1989," Special Report 44, Alaska Geological and Geophysical Surveys, 100 pp.
65. Bur, T.R., et al., 1969, "An Ultrasonic Method for Determining the Elastic Symmetry of Materials," Report of Investigations 7333, US Bureau of Mines, 23 pp.
66. Burchard, E.F., and Chapin, T., 1920, "Marble Resources of Southeastern Alaska," Bulletin 682, US Geological Survey, 118 pp.
67. Burchett, R.R., 1980, "Mineral Resource Map of Nebraska," Scale, 1:1,000,000, Nebraska Geological Survey.
68. Burchett, R.R., and Eversoll, D.A., 1991, "Nebraska Mineral Operations Review, 1990," Nebraska Geological Survey, 15 pp.
69. Burr, A.C., et al., 1950, "Raw Materials for the Manufacture of Portland Cement in North Dakota," *1949 Proceedings*, North Dakota Academy of Science, Vol. 3, pp. 15–18.
70. Butler, J., and Bartlette, G., 1969, "Bibliography of the Geology of Newfoundland and Labrador, 1814 through 1968," Bulletin 38, Newfoundland and Labrador Mineral Resources Division, 273 pp.
71. Calvert, S.E., 1974, "Deposition and Diagenesis of Silica in Marine Sediments," *Pelagic Sediments: on Land and Under the Sea*, K.J. Hsü and H.C. Jenkyns, eds., International Association of Sedimentologists, Special Publication 1, pp. 273–299.
72. Carlson, C.G., 1964, "The Niobrara Formation of Eastern North Dakota; Its Possibilities for Use as a Cement Rock," Report of Investigation 41, North Dakota Geological Survey, 56 pp.
73. Carozzi, A.V., 1960, *Microscopic Sedimentary Petrography*, John Wiley, New York, 485 pp.
74. Carr, D.D., 1973, "Geometry and Origin of Oolite Bodies in the Ste. Genevieve Limestone (Mississippian) in the Illinois Basin," Bulletin 48, Indiana Geological Survey, 81 pp.
75. Carr, D.D., and Ault, C.H., 1980, "Potential for Deep Underground Limestone Mining in Indiana," Preprint 80–334, SME-AIME Fall Meeting, Minneapolis, MN, Oct., 11 pp.
76. Carr, D.D., and Ault, C.H., 1983, "Potential for Deep Underground Limestone Mining in Indiana," *Trans. SME-AIME*, Vol. 272, pp. 1912–1916.
77. Carr, D.D., et al., 1970, "Relationships Between Physical and Chemical Properties of the Brassfield Limestone (Silurian) in Indiana, Ohio, and Kentucky," *Proceedings*, 6th Forum on Geology of Industrial Minerals, Miscellany 1, Michigan Geological Survey, pp. 127–137.
78. Carr, D.D., et al., 1971, "Crushed Stone Aggregate Resources of Indiana," Bulletin 42-H, Indiana Geological Survey, 38 pp.
79. Carr, D.D., et al., 1978, "Crushed Stone Resources of the Blue River Group (Mississippian) in Indiana," Bulletin 52, Indiana Geological Survey, 225 pp.

80. Castor, S.B., 1990, ''Industrial Minerals,'' Special Publication 89, Nevada Bureau of Mines and Geology, pp. 26–29.
81. Caterpillar Tractor Co., 1972, *Handbook of Ripping, A Guide to Greater Profits*, 4th ed., Caterpillar Tractor Co., Peoria, IL, 44 pp.
82. Chelini, J.M., 1965, ''Limestone, Dolomite, and Travertine in Montana,'' Bulletin 44, Montana Bureau of Mines and Geology, 53 pp.
83. Chute, N.E., 1967, ''Structural and Mineralogical Features of the Stockbridge Marble and Berkshire Schist Near Adams, Massachusetts,'' *Economic Geology in Massachusetts*, O.C. Farquhar, ed., University of Massachusetts Graduate School, Amherst, pp. 169–180.
84. Clarke, F.W., 1924, ''The Data of Geochemistry,'' Bulletin 770, US Geological Survey, 841 pp.
85. Cloud, P.E., Jr., and Barnes, V.E., 1946, ''The Ellenburger Group of Central Texas,'' Publication 4621, University of Texas, Austin, 473 pp.
86. Colby, S.F., 1941, ''Occurrences and Uses of Dolomite in the United States,'' Information Circular 7192, US Bureau of Mines, 21 pp.
87. Colquourn, D.J., and Duncan, D., 1966, ''Geology of the Eutawville Quadrangle, South Carolina,'' MS-12, map with text, South Carolina Division of Geology.
88. Comstock, H.B., 1963, ''Magnesium and Magnesium Compounds,'' Information Circular 8201, US Bureau of Mines, 128 pp.
89. Condra, G.E., and Reed, E.C., 1959, ''The Geological Section of Nebraska,'' Bulletin 14A, Nebraska Geological Survey, 82 pp.
90. Connecticut Geological and Natural History Survey, see Publications List of Geologic Quadrangle Maps and Quadrangle Reports.
91. Conrad, S.G., 1960, ''Crystalline Limestones of the Piedmont and Mountain Regions of North Carolina,'' Bulletin 74, North Carolina Division of Mineral Resources, 56 pp.
92. Cooper, B.N., 1944, ''Industrial Limestones and Dolomites in Virginia: New River-Roanoke River District,'' Bulletin 62, Virginia Division of Mineral Resources, 98 pp.
93. Cooper, B.N., 1945, ''Industrial Limestones and Dolomites in Virginia: Clinch Valley District,'' Bulletin 66, Virginia Division of Mineral Resources, 259 pp.
94. Cooper, J.D., 1970, ''Stone,'' *Mineral Facts and Problems*, Bulletin 650, US Bureau of Mines, pp. 1219–1235.
95. Cullity, B.D., 1956, *Elements of X-Ray Diffraction*, Addison-Wesley, Reading, MA, 514 pp.
96. Cummins, A.G., and Given, I.A., eds., 1973, ''SME Mining Engineering Handbook,'' Society of Mining Engineers, Inc., Littleton, CO, 2 Vols., 2,200 pp.
97. Curry, R.L., and Buck, A.D., 1966, ''Development of Methods of Testing Aggregates Larger than 1½ Inch,'' Technical Report 6–747, US Army Engineer Waterways Experiment Station, 44 pp.
98. Czehura, S.J., and Berg, R.B., 1977, ''Index of Graduate Theses on Montana Geology,'' Special Publication 77, Montana Bureau of Mines and Geology, 104 pp.
99. Dale, T.N., 1914, ''The Commercial Marbles of Western Vermont,'' Report of the State Geologist on the Mineral Industries and Geology of Vermont, 1913–1914, Vermont Geological Survey, pp. 1–160.
100. D'Andrea, D.V., et al., 1965, ''Prediction of Compressive Strength from Other Rock Properties,'' Report of Investigations 6702, US Bureau of Mines, 23 pp.
101. Danner, W.R., 1966, ''Limestone Resources of Western Washington,'' Bulletin 52, Washington Division of Mines and Geology, 474 pp.
102. Danner, W.R., 1976, ''Limestone Resources of Southwestern British Columbia,'' *Proceedings*, 11th Forum on Geology of Industrial Minerals, Special Publication 74, Montana Bureau of Mines and Geology, pp. 171–186.
103. Davis, R.E., 1957, ''Magnesium Resources of the United States—A Geologic Summary and Annotated Bibliography to 1953,'' Bulletin 1019-E, US Geological Survey, pp. 373–515.
104. Dean, T.J., et al., 1969, ''Underground Mining in the Kansas City Area,'' *Missouri Mineral Industry News*, Vol. 9, Apr., Missouri Geological Survey, pp. 37–56.
105. Dean, L.S., ed., 1991, ''Minerals in Alabama 1990,'' Information Series 64I, Alabama Geological Survey, 47 pp.
106. Degrace, J.R., 1974, ''Limestone Resources of Newfoundland and Labrador,'' Report 74–2, Newfoundland Department Mines Energy, 117 pp.
107. Dever, G.R., Jr., 1980, ''High-carbonate and Low-Silica Stone in the High Bridge Group (Middle Ordovician), Fayette County, Central Kentucky,'' Information Circular 4, Series 11, Kentucky Geological Survey, 45 pp.
108. Dever, G.R., Jr., 1990, ''Use of Limestone, Lime, and Dolomite for SO_2 Emission Control in Kentucky,'' Information Circular 31, Series 11, Kentucky Geological Survey, 14 pp.
109. Dever, G.R., Jr., and McGrain, P., 1969, ''High-Calcium and Low-Magnesium Limestone Resources in the Region of the Lower Cumberland, Tennessee, and Ohio Valleys, Western Kentucky,'' Bulletin 5, Series 10, Kentucky Geological Survey, 192 pp.
110. Dever, G.R., Jr., et al., 1978, ''Industrial Limestone Resources Along the Ohio River Valley of Kentucky,'' *Mining Engineering*, Vol. 30, No. 4, pp. 396–401.
111. Dick, R.A., et al., 1973, ''A Study of Fragmentation from Bench Blasting in Limestone at a Reduced Scale,'' Report of Investigations 7704, US Bureau of Mines, 24 pp.
112. Dickson, J.A.D., 1965, ''A Modified Staining Technique for Carbonates in Thin Section,'' *Nature*, No. 4971, Feb. 6, pp. 587.
113. Dickson, J.A.D., 1966, ''Carbonate Identification and Genesis as Revealed by Staining,'' *Journal of Sedimentary Petrology*, Vol. 36, pp. 491–505.
114. Diehl, H., 1964, ''Calcein, Calmagite, and o,o'-Dihydroxyazobenzene Titrimetric, Colorimetric and Fluorometric Reagents for Calcium and Magnesium,'' G. Frederick Smith Chemical Co., Columbus, OH, 124 pp.
115. Dorheim, F.H., 1970, ''Mineral Resources of Iowa,'' Map with Text, Iowa Geological Survey.
116. Drehmel, D.C., and Harvey, R.D., 1974, ''Carbonate Rock Properties Required by Desulfurization Processes,'' *Proceedings*, 10th Forum on Geology of Industrial Minerals, Miscellaneous Report 1, Ohio Geological Survey, pp. 59–66.
117. Dunham, R.J., 1962, ''Classification of Carbonate Rocks According to Depositional Texture,'' *Classification of Carbonate Rocks*, W.E. Hamm, ed., American Association of Petroleum Geologists, Memoir 1, pp. 108–121.
118. Dunn, J.R., 1991, ''Geology and Exploration,'' *The Aggregates Handbook*, R.D. Barksdale, ed., National Stone Association, Washington, DC, pp. 4–1 to 4–46.
119. Dunn, J.R., and Ozol, M.A., 1962, ''Deleterious Properties of Chert,'' Physical Research Report No. 12, New York Dept. of Public Works, 121 pp.
120. Edmundson, R.S., 1945, ''Industrial Limestones and Dolomites in Virginia: Northern and Central Parts of Shenandoah Valley,'' Bulletin 65, Virginia Division of Mineral Resources, 195 pp.
121. Edmundson, R.S., 1958, ''Industrial Limestones and Dolomites in Virginia: James River District West of the Blue Ridge,'' Bulletin 73, Virginia Division of Mineral Resources, 137 pp.
122. Eifler, G.K., Jr., 1968, ''Industrial Carbonates of the Texas Gulf Coastal Plain,'' *Proceedings*, 4th Forum on Geology of Industrial Minerals, Texas Bureau of Economic Geology, pp. 45–56.
123. Eilertsen, N.A., 1964, ''Mining Methods and Costs, Kimballton Limestone Mine, Standard Lime and Cement Co., Gile County, Va.,'' Information Circular 8214, US Bureau of Mines, 50 pp.
124. English, S.G., et al., 1940, ''Limestone Analyses,'' Mineral Report 5, Oklahoma Geological Survey, 28 pp.
125. Ericksen, G.E., and Cox, D.P., 1968, ''Limestone and Dolomite,'' US Geological Survey and US Bureau of Mines, Mineral Resources of the Appalachian Region, Professional Paper 580, US Geological Survey, pp. 227–252.
126. Evans, T.B., and Eilertsen, N.A., 1957, ''Mining Methods and Costs at the Sunbright Limestone Mine, Foote Mineral Co., Sunbright, Virginia,'' Information Circular 7793, US Bureau of Mines, 44 pp.
127. Farquhar, O.C., 1967, ''Drill Records of Stone and Basalt in Western Massachusetts,'' *Economic Geology in Massachusetts*, O.C. Farquhar, ed., University of Massachusetts Graduate School, Amherst, pp. 75–82.
128. Fischl, P., 1991, ''Limestone and Dolomite Resources of British Columbia,'' Open File Report, British Columbia Ministry of Energy, Mines, and Petroleum Resources, 150 pp.
129. Fleming, K.L., 1990, ''Directory of Mining and Milling Operations,''

Special Publication M1–89, Nevada Bureau Mines and Geology, pp. 30–43.

130. Folk, R.L., 1959, "Practical Petrographic Classification of Limestones," *Bulletin*, American Association of Petroleum Geologists, Vol. 43, pp. 1–38.

131. Folk, R.L., 1962, "Spectral Subdivision of Limestone Types," *Classification of Carbonate Rocks*, W.E. Hamm, ed., American Association of Petroleum Geologists, Memoir 1, pp. 62–84.

132. Folk, R.L., and Weaver, C.E., 1952, "A Study of the Texture and Composition of Chert," *American Journal of Science*, Vol. 250, pp. 498–510.

133. Ford, C.T., 1974, "Selection of Limestones as Neutralizing Agents for Coal-Mine Water," *Proceedings*, 10th Forum on Geology of Industrial Minerals, Miscellaneous Report 1, Ohio Geological Survey, pp. 30–42.

134. Fowler, P., 1950, "Stratigraphy and Structure of the Castleton Area, Vermont," Bulletin 2, Vermont Geological Survey, 83 pp.

135. French, R.R., 1967, "Crushed Stone Resources of the Devonian and Silurian Carbonate Rocks of Indiana," Bulletin 37, Indiana Geological Survey, 127 pp.

136. French, R.R., 1969, "Transportation of Mineral Aggregates in Indiana," *Proceedings for 1968*, Indiana Academy of Science, Vol. 78, pp. 348–354.

137. Friedman, G.M., 1959, "Identification of Carbonate Minerals by Staining Methods," *Journal of Sedimentary Petrology*, Vol. 29, pp. 87–97.

138. Friedman, G.M., and Sanders, J.E., 1967, "Origin and Occurrence of Dolostones," *Developments in Sedimentology, Carbonate Rocks*, Vol. 9A, G.V. Chilingar, H.J. Bissell, R.W. Fairbridge, eds., Elsevier, New York, pp. 267–348.

139. Fulkerson, F.B., 1969, "Transportation of Mineral Commodities on the Inland Waterways of the South-Central States," Information Circular 8431, US Bureau of Mines, 88 pp.

140. Fuzesy, L.M., 1980, "Geology of the Deadwood (Cambrian), Meadow Lake and Winnepegosis (Devonian) Formations in West-Central Saskatchewan," Department Mineral Resources Report 210, Saskatchewan Geological Survey, 64 pp.

141. Galle, O.K., 1967, "The Geochemistry of the Oread Limestone," *Proceedings*, 3rd Forum on Geology of Industrial Minerals, Special Distribution Publication 34, Kansas Geological Survey, pp. 97–109.

142. Galle, O.K., and Angino, E.E., 1967, "Determination of Calcium and Magnesium in Carbonate and Silicate Rocks by Atomic Absorption," Bulletin 187, Pt. 1, Kansas Geological Survey, pp. 9–11.

143. Garner, L.E., et al. (compilers), 1979, "The Mineral Resources of Texas," Map, Scale 1:1,000,000, Bureau of Economic Geology, University of Texas, Austin.

144. Gazdik, G.C., and Tagg, K.M., 1957, "Annotated Bibliography of High-Calcium Limestone Deposits in the United States Including Alaska, to April 1956," Bulletin 1019-I, US Geological Survey, pp. 675–713.

145. Gianni, W.F., 1991, "Analyses of Carbonate Rocks—Northern Virginia," Virginia Division of Mineral Resources, 104 pp.

146. Gillson, J.L., et al., 1960, "The Carbonate Rocks," *Industrial Minerals and Rocks*, 3rd ed., J.L. Gillson, ed., AIME, New York, 1960, pp. 123–201.

147. Ginsburg, R.N., et al., 1963, "Shallow-Water Carbonate Sediments," *The Sea*, Vol. 3, John Wiley, New York, pp. 554–582.

148. Goddard, E.N., et al., 1963, *Rock-Color Chart*, Geological Society of America, 16 pp.

149. Goodridge, E., 1973, "Status Report: the AGA/OCR Coal Gasification Program," *Coal Age*, Vol. 78, No. 1, pp. 54–59.

150. Goodwin, J.H., 1979, "A Guide to Selecting Agricultural Limestone Products," Illinois Mineral Note 73, Illinois Geological Survey, 7 pp.

151. Goodwin, J.H., 1983, "Geology of Carbonate Aggregate Resources of Illinois," Industrial Minerals Notes 87, Illinois State Geological Survey, 12 pp.

152. Goodwin, J.H., and Baxter, J.W., 1981, "High-Calcium, High-Reflectance Limestone Resources of Illinois," *Geological Society of America Bulletin*, Vol. 92, pp. 621–628.

153. Goodwin, J.H., Mikulic, D.G., and Baxter, J.W., 1990, "Industrial Minerals and Metals Publications of the Illinois State Geological Survey through December 1989," Illinois Minerals 104, Illinois State Geological Survey, 33 p.

154. Goudge, M.F., 1933, "Canadian Limestones for Building Purposes," Report 733, Canadian Bureau of Mines, 196 pp.

155. Goudge, M.F., 1934, "Limestones of Canada, Their Occurrence and Characteristics; Part 2, Maritime Provinces," Report 742, Canadian Bureau of Mines, 186 pp.

156. Goudge, M.F., 1935, "Limestones of Canada, Their Occurrence and Characteristics; Part 3, Quebec," Report 755, Canadian Bureau of Mines, 274 pp.

157. Goudge, M.F., 1937, "Limestone and Lime—Their Industrial Uses," *Mining and Metallurgy*, Vol. 18, No. 368, pp. 371–374.

158. Goudge, M.F., 1938, "Limestones of Canada, Their Occurrence and Characteristics; Part 4, Ontario," Report 781, Canadian Bureau of Mines, 362 pp.

159. Goudge, M.F., 1944, "Limestones of Canada, Their Occurrence and Characteristics; Part 5, Western Canada," Report 811, Canadian Bureau of Mines, 233 pp.

160. Graf, D.L., 1960a, "Geochemistry of Carbonate Sediments and Sedimentary Carbonate Rocks, Part I, Carbonate Mineralogy, Carbonate Sediments," Circular 297, Illinois Geological Survey, 39 pp.

161. Graf, D.L., 1960b, "Geochemistry of Carbonate Sediments and Sedimentary Carbonate Rocks, Part IV-A, Isotopic Composition, Chemical Analyses," Circular 308, Illinois Geological Survey, 42 pp.

162. Graf, D.L., and Goldsmith, J.R., 1963, "Carbonate Mineralogy," *Subsurface Geology of Eniwetok Atoll*, by S.O. Schlanger, Professional Paper 260 BB, US Geological Survey, pp. 1048–1053.

163. Graf, D.L., and Lamar, J.E., 1955, "Properties of Calcium and Magnesium Carbonates and Their Bearing on Some Uses of Carbonate Rocks," *Economic Geology*, 50th Anniversary Vol., pp. 639–713.

164. Gries, J.P., 1964, "Limestone," *Mineral and Water Resources of South Dakota*, Bulletin 16, South Dakota Geological Survey, pp. 96–102.

165. Grisafe, D.A., 1977, "Kansas Building Limestones," Mineral Resource Series 4, Kansas Geological Survey, 42 pp.

166. Gubler, Y., et al., 1967, "Petrology and Petrography of Carbonate Rocks," *Developments in Sedimentology, Carbonate Rocks*, Vol. 9A, G.V. Chilingar, H.J. Bissell, R.W. Fairbridge, eds., Elsevier, New York, pp. 51–86.

167. Guillet, G.R., and Kriens, J., 1984, "Ontario and the Mineral Filler Industry," Industrial Mineral Background Paper 5, Ontario Ministry of Natural Resources, 175 pp.

168. Gulbrandsen, R.A., 1960, "A Method of X-Ray Analyses for Determining the Ratio of Calcite to Dolomite in Mineral Mixtures," Bulletin 1111-D, US Geological Survey, pp. 147–152.

169. Gunter, W.R., 1989, "Industrial Minerals Mapping, Swan River-Mafeking Area," Report of Field Activities in 1989, Manitoba Energy and Mines, Minerals Division, pp. 127–131.

170. Gunter, W.R., 1990, "Industrial Minerals Investigations in the Gypsumville—Steep Rock Area," Report of Activities in 1990, Manitoba Energy and Mines, Minerals Division, pp. 152–154.

171. Hall, C.J., 1981, "Mine Ventilation Engineering," Society of Mining Engineers, Inc., Littleton, CO, 334 pp.

172. Ham, W.E., 1949, "Geology and Dolomite Resources, Mill Creek-Ravina Area, Johnson County, Oklahoma," Circular 24, Oklahoma Geological Survey, 104 pp.

173. Hamilton, W.N., 1987, "Field Evaluation of Prospective Filler-Grade Limestones in Alberta," Open File Report 1989–14, Alberta Research Council, 51 pp.

174. Hansen, M., 1953, "Geologic Report on Limestone Deposits in Stark County and Hettinger County, North Dakota," Report of Investigation 8, North Dakota Geological Survey, 71 pp.

175. Harries-Rees, K., 1991, "Dimension Stone Review, The New Stone Age," *Industrial Minerals*, No. 290, London, Nov., pp. 43–49.

176. Harris, R.E., 1987, "Geology and Economic Potential of a High-Calcium Limestone and Dolomitic Limestone Deposit Near Manville, Niobrara County, Wyoming," Report of Investigations 39, Wyoming Geological Survey, 16 pp.

177. Harris, R.E., and King, J.K., 1989, "Industrial Minerals and Construction Materials Map of the Powder River Basin and Adjacent

Uplifts, Wyoming,'' Map MS-30, Wyoming Geological Survey, 1:500,000.

178. Harris, R.E., and Meyers, J.E., 1986, ''Construction Materials Map of Wyoming,'' Map MS-21, Wyoming Geological Survey, 1:500,000.

179. Hart, E.W., 1978, ''Limestone, Dolomite, and Shell Resources of the Coast Ranges Province, California,'' Bulletin 197, California Division of Mines and Geology, 103 pp.

180. Harvey, R.D., and Steinmetz, J.C., 1971, ''Petrographic Properties of Carbonate Rocks Related to Their Sorption of Sulfur Dioxide,'' Environmental Geology Notes 50, Illinois Geological Survey, 37 pp.

181. Harvey, R.D., et al., 1974, ''Lake Marls, Chalks, and Other Carbonate Rocks with High Dissolution Rates in SO_2-Scrubbing Liquors,'' *Proceedings*, 10th Forum on Geology of Industrial Minerals, Miscellaneous Report 1, Ohio Geological Survey, pp. 67–80.

182. Harvey, R.D., et al., 1976, ''Behavior of Dolomite in Absorption of H_2S from Fuel Gas,'' *World Mining and Metals Technology*, AIME, New York, pp. 163–188.

183. Hatch, F.H., and Rastall, R.H., (rev. by J.T. Greensmith), 1971, *Petrology of the Sedimentary Rocks*, Vol. 2, Hafner, New York, 502 pp.

184. Hatch, N.C., Jr., and Stanley, R.S., 1988, ''Post-Taconian Structural Geology of the Rowe-Hawley Zone and the Connecticut Valley Belt West of the Mesozoic Basins,'' US Geological Survey Professional Paper 1366 C, pp. C1-C36.

185. Hatmaker, P., 1931, ''Utilization of Dolomite and High Magnesium Limestone,'' Information Circular 6524, US Bureau of Mines, 18 pp.

186. Herod, B.C., 1969, ''Pennsylvania Producer Handles Tough Recovery Problem,'' *Pit & Quarry*, Vol. 62, No. 4, Oct., pp. 125–127, 129.

187. Herod, B.C., editor-in-chief, 1980–81, *Pit and Quarry Handbook*, 73rd ed., Pit and Quarry Publications, Chicago, 686 pp.

188. Herod, B.C., 1972, ''Rock Removed = Valuable Space,'' *Pit & Quarry*, Vol. 64, No. 11, May, pp. 72–75, 82.

189. Heron, S.D., Jr., 1962, ''Limestone Resources of the Coastal Plain of South Carolina,'' Bulletin 28, South Carolina Division of Geology, 128 pp.

190. Hershey, R.E., and Maher, S.W., 1985, ''Limestone and Dolomite Resources of Tennessee,'' Bulletin 65, 2nd ed., Tennessee Division of Geology, 252 pp.

191. Hewitt, D.F., 1960, ''The Limestone Industries of Ontario,'' Industrial Mineral Circular 5, Ontario Division of Mines, 177 pp.

192. Hewitt, D.F., and Vos, M.A., 1972, ''Limestone Industries of Ontario,'' Industrial Mineral Report 39, Ontario Division of Mines, 79 pp.

193. Hill, R.D., 1974, ''Overview of Use of Carbonate Rocks for Controlling Acid Mine Drainage,'' *Proceedings*, 10th Forum on Geology of Industrial Minerals, Miscellaneous Report 1, Ohio Geological Survey, pp. 25–29.

194. Hill, T.P., and Werner, M.A., 1972, ''Chemical Composition of Sedimentary Rocks in Alaska, Idaho, Oregon, and Washington,'' Professional Paper 771, US Geological Survey, 319 pp.

195. Hill, T.P., et al., 1967, ''Chemical Composition of Sedimentary Rocks in Colorado, Kansas, Montana, Nebraska, North Dakota, South Dakota, and Wyoming,'' Professional Paper 561, US Geological Survey, 241 pp.

196. Hillebrand, W.F., and Lundell, G.E.F., 1953, (rev. by G.E.F. Lundell, H.A. Bright, J.I. Hoffman), *Applied Inorganic Analysis, with Special Reference to the Analysis of Metals, Minerals, and Rocks*, 2nd ed., John Wiley, New York, 1034 pp.

197. Hinchey, N.S., 1946, ''Missouri Marble,'' Report of Investigation 3, Missouri Geological Survey, 47 pp.

198. Hinchey, N.S., et al., 1947, ''Limestones and Dolomites in the St. Louis Area,'' Report of Investigation 5, Missouri Geological Survey, 80 pp.

199. Högberg, E., 1971, ''Staining Method for Examination of Siliceous, Cretaceous Limestone II,'' *Geol. Fören. Stockholm Förh.*, Vol. 93, Part 4, pp. 707–713.

200. Holbrook, D.F., 1965, ''High-Calcium Limestones in Independence and Izard Counties, Arkansas,'' Miscellaneous Publication, Arkansas Geological Commission, 21 pp.

201. Holter, M.E., 1976, ''Limestone Resources of Alberta,'' Alberta Research Council Economic Geology Report 4, Edmonton, 91 pp.

202. Holter, M.E., 1976, ''Limestone Resources of Alberta,'' *Proceedings*, 11th Forum on Geology of Industrial Minerals, Special Publication 74, Montana Bureau of Mines and Geology, pp. 37–50.

203. Holter, M.E., and Hamilton, W.N., 1989, ''An Evaluation of Alberta Limestones for Use as Paper Filler Materials,'' Open File Report 1990–11, Alberta Research Council, 89 pp.

204. Hopkins, D.A., 1985, ''Refractory Dolomite Production in a Geologically Complex Area,'' *Proceedings*, 20th Forum on Geology of Industrial Minerals, Special Publication 2, Maryland Geological Survey, pp. 117–124.

205. Hopkins, R.H., 1942, ''The Dolomitic Limestones of Florida,'' Report of Investigation 3, Florida Geological Survey, 105 pp.

206. Horowitz, A.S., and Potter, P.E., 1971, *Introductory Petrography of Fossils*, Springer-Verlag, New York, 302 pp.

207. Hotchkiss, W.O., and Steidtmann, E., 1914, ''Limestone Road Materials of Wisconsin,'' Bulletin 34, Wisconsin Geological Survey, 137 pp.

208. Howe, W.B., et al, 1961, ''The Stratigraphic Succession in Missouri,'' Missouri Geological Survey and Water Resources, Series 2, Vol. 40, 185 pp.

209. Huntley, F.D., and Snow, R.R., 1988, ''Mineral Raw Materials for Flat Glass Manufacturing,'' *Proceedings*, 22nd Forum on Geology of Industrial Minerals, C.W. Colton, ed., Miscellaneous Publication 21, Arkansas Geological Commission, pp. 11–16.

210. Illinois Geological Survey, see Publications List of Road Materials Resources Maps, Regional and County Resource Reports, and Research Reports on Various Uses of Limestone.

211. Indiana Geological Survey, see Publications List of Regional Geologic Maps.

212. Institution of Mining & Metallurgy, 1965, *Opencast Mining, Quarrying and Alluvial Mining*, American Elsevier, New York, 772 pp.

213. Interstate Commerce Commission, 1968 ''Carload Waybill Statistics, 1966, State-to-State Distribution, No. 32, Stone, Clay, and Glass, Products,'' Interstate Commerce Commission, Washington, DC, 58 pp.

214. Ives, W., and Runnels, R.T., 1960, ''Lime Raw-Materials in Kansas,'' Bulletin 142, Part 3, Kansas Geological Survey, pp. 123–148.

215. Jacob, H.L., and Petry, K.A.A., 1989, ''Sommaire des Principaux Gîtes de Dolomie au Québec,'' MB89–49, Ministry of Energy and Resources, Québec, 20 pp.

216. Jacob, H.L., 1990, ''Les gîtes de marbre de Québec,'' MB90–14, Ministry of Energy and Resources, Québec, 46 pp.

217. Janssens, A., 1977, ''Silurian Rocks in the Subsurface of Northwestern Ohio,'' Report of Investigation 100, Ohio Geological Survey, 96 pp.

218. Jasper, M.W., and Mihelich, M., 1961, ''Kings River Limestone Deposits, Anchorage Quadrangle,'' Report for the Year 1961, Alaska Division of Mines and Minerals, pp. 40–48.

219. Jewett, J.M., et al., 1967, ''Inland Underground Facilities,'' *Guidebook for 3rd National Forum on the Geology of Industrial Minerals*, Kansas Geological Survey, University of Kansas, Apr., 13 pp.

220. Johnsen, J.H., 1958, ''Preliminary Report on the Limestones of Albany County, New York,'' Special Publication, New York State Museum and Science Service, 43 pp.

221. Johnsen, J.H., 1971, ''Limestones (Middle Ordovician) of Jefferson County,'' Map and Chart Series 13, New York State Museum and Science Service, 88 pp.

222. Johnson, A.M., and Sorensen, H.O., 1978, ''Drill Core Investigation of the Fiborn Limestone Member in Schoolcraft, Mackinac, and Chippewa Counties, Michigan,'' Report of Investigation 18, Michigan Geological Survey, 51 pp.

223. Johnson, K.S., 1969, ''Mineral Map of Oklahoma (Exclusive of Oil and Gas Fields),'' Map GM-15, Scale 1:750,000, Oklahoma Geological Survey.

224. Johnstone, S.J., and Johnstone, M.G., 1961, ''Limestone, Chalk and Whiting,'' *Minerals for the Chemical and Allied Industries*, 2nd ed., John Wiley and Son, New York, pp. 293–326.

225. Jordan, C., (compiler), 1978, ''Sedimentology Processes: Carbonate Sedimentology,'' Reprint Series 5, Society of Economic Paleontologists and Mineralogists, 235 pp.

226. Judd, W.R., and Huber, C., 1961, "Correlation of Rock Properties by Statistical Methods," *Proceedings,* Symposium on Mining Research, Vol. 2, Pergamon Press, New York, pp. 621–648.

227. Kansas Geological Survey, see Publications List of Regional and County Reports.

228. Keith, S.B., 1969, "Limestone, Dolomite, and Marble," *Mineral and Water Resources of Arizona*, Bulletin 180, Arizona Bureau of Mines, pp. 385–398.

229. Kennedy, B.A., ed., 1990, *Surface Mining*, 2nd ed., Society for Mining, Metallurgy, and Exploration, Inc., Littleton, CO, 126 pp.

230. Kentucky Department of Commerce and Kentucky Geological Survey, 1962, "Mineral Resources and Mineral Industry Map of Kentucky."

231. Kentucky Geological Survey, see Index to Geologic Quadrangle Maps.

232. Kephart, W.W., 1972, "Limestone and Dolomite Requirements for Glass Containers," *Proceedings*, 8th Forum on Geology of Industrial Minerals, Public Information Series 5, Iowa Geological Survey, pp. 11–15.

233. Kessler, D.W., and Sligh, W.H., 1927, "Physical Properties of the Principal Commercial Limestones Used for Building Construction in the United States," Technical Paper 349, Vol. 21, National Bureau of Standards, pp. 497–590.

234. King, H.M., and Kirstein, D.S., 1987, "Mineral Resources of West Virginia," Map WV24, West Virginia Geological Survey, 1:500,000.

235. Kline, H.D., 1962, "Methods and Costs of Mining and Crushing Limestone at Three Quarries, Anderson-Oxandale Rock Co., Kansas," Information Circular 8084, US Bureau of Mines, 15 pp.

236. Kneller, W.A., et al., 1968, "The Properties and Recognition of Deleterious Cherts which Occur in Aggregate Used by Ohio Concrete Producers," Final Report No. 1014, University of Toledo Research Foundation, 201 pp.

237. Kottlowski, F.E., 1962, "Reconnaissance of Commercial High-Calcium Limestones in New Mexico," Circular 60, New Mexico Bureau of Mines and Mineral Resources, 77 pp.

238. Kottlowski, F.E., 1965, "Limestone and Dolomite," *Mineral and Water Resources of New Mexico*, Bulletin 87, New Mexico Bureau of Mines and Mineral Resources, pp. 345–353.

239. Kümmel, H.B., 1901, "Report on Portland Cement Industry," Annual Report of the State Geologist for the Year 1900, New Jersey Geological Survey, pp. 9–101.

240. Kümmel, H.B. 1940, "The Geology of New Jersey," Bulletin 50, New Jersey Geological Survey, 203 pp.

241. Kupsch, W.O., 1953, "Ordovician and Silurian Stratigraphy of East Central Saskatchewan," Report 10, Saskatchewan Geological Survey, Department of Natural Resources, 62 pp.

242. Kusler, D.J., and Corre, H.A., 1968, "Limestone Resources in Western West Virginia," Information Circular 8369, US Bureau of Mines, 15 pp.

243. Lachambre, G., and Biron, S., 1987, "Calcaires de l'est du Québec," ET83–12, Ministry of Energy and Resources, Québec, 180 pp.

244. Lamar, J.E., 1957, "Chemical Analyses of Illinois Limestones and Dolomites," Report of Investigations 200, Illinois Geological Survey, 33 pp

245. Lamar, J.E., 1961, "Uses of Limestone and Dolomite," Circular 321, Illinois Geological Survey, 44 pp.

246. Lamar, J.E., 1966, "High-Purity Limestones in Illinois," Industrial Minerals Notes 27, Illinois Geological Survey, 20 pp.

247. Lamar, J.E., 1967, "Handbook on Limestone and Dolomite for Illinois Quarry Operators," Bulletin 91, Illinois Geological Survey, 119 pp.

248. Lamar, J.E., et al., 1956, "Preliminary Report on Portland Cement Materials in Illinois," Report of Investigations 195, Illinois Geological Survey, 34 pp.

249. Lamar, J.E., and Willman, H.B., 1938, "A Summary of the Uses of Limestone and Dolomite," Report of Investigations 49, Illinois Geological Survey, 48 pp.

250. Lamborn, R.E., 1951, "Limestones of Eastern Ohio," Bulletin 49, Series 4, Ohio Geological Survey, 377 pp.

251. Landes, K.K., 1963, "Metallurgical Limestone Reserves in the United States," 2nd ed., National Lime Association, Washington, DC, 31 pp.

252. Landes, K.K., et al., 1945, "Geology of the Mackinac Straits Region," Publication 44, Michigan Geological Survey, 204 pp.

253. Landreth, J.O., 1968, "High Calcium Limestone Deposit in the Rattler Gulch Area, Granite County, Montana," Special Publication 44, Montana Bureau of Mines and Geology, 10 pp.

254. Leighton, M.W., and Pendexter, C., 1962, "Carbonate Rock Types," *Classification of Carbonate Rocks*, W.E. Hamm, ed., American Association of Petroleum Geologists, Memoir 1, pp. 62–84.

255. Lipp, R.J., 1987, "Minnesota Industrial Minerals Directory, 1987," University of Minnesota, Mineral Resources Research Center, Minneapolis, MN, 50 pp.

256. Logan, W.N., 1916, "Preliminary Report on the Marls and Limestones of Mississippi," Bulletin 13, Mississippi Geological Survey, 82 pp.

257. Loofbourow, R.L., 1966, "Will Aggregate Shortages Drive Stone Producers Underground?" *Rock Products*, Vol. 69, No. 10, Oct., pp. 67–75.

258. Lootens, D.J., Greenslade, W.M., and Barker, J.M., eds., 1991, "Environmental Management for the 1990's," Society for Mining, Metallurgy, and Exploration, Inc., Littleton, CO, 379 pp.

259. Lozo, F.E., 1959, "Stratigraphic Relations of the Edwards Limestone and Associated Formations in North-Central Texas," *Symposium on Edwards Limestone in Central Texas*, Publication 5905, University of Texas, Austin, pp. 1–20.

260. Manger, G.E., 1963, "Porosity and Bulk Density of Sedimentary Rocks," Bulletin 1144-E, US Geological Survey, 55 pp.

261. Mantell, C.L., 1931, *Industrial Electrochemistry*, McGraw-Hill, New York, 528 pp.

262. Marek, C.R., 1991, "Sampling and Testing Principles," *The Aggregates Handbook*, R.D. Barksdale, ed., National Stone Association, Washington, DC, pp. 16–1 to 16–37.

263. Markewicz, F.J., and Dalton, R., 1977, "Stratigraphy and Applied Geology of the Lower Paleozoic Carbonates in Northwestern New Jersey," 42nd Annual Field Conference, Pennsylvania Geologists, 117 pp.

264. Marshall, L.G., 1963, "Mining and Beneficiating Costs at Two Crushed Limestone Operations, Madison County, Iowa," Information Circular 8199, US Bureau of Mines, 18 pp.

265. Martin, J.A., 1967, "Limestone and Dolomite," *Mineral and Water Resources of Missouri*, Missouri Geological Survey and Water Resources, Vol. 43, Series 2, pp. 126–133.

266. Martin, J.A., 1967, "Cement and Cement Raw Materials," *Mineral and Water Resources of Missouri*, Missouri Geological Survey and Water Resources, Vol. 43, Series 2, pp. 172–183.

267. Maryland Geological Survey, see Publications List of County Reports.

268. Mather, K., 1953, "Crushed Limestone Aggregates for Concrete," *Mining Engineering*, Vol. 5, No. 10, pp. 1022–1028; *Trans. AIME* 1953, Vol. 196, 1954.

269. Mather, K., and Mather, B., 1950, "Method of Petrographic Examination of Aggregates for Concrete," *Proceedings*, American Society for Testing and Materials, Vol. 50, pp. 1288–1312.

270. Mathews, E.B., and Grasty, J.S., 1909, "The Limestones of Maryland," Vol. 8, Part 3, Maryland Geological Survey, pp. 227–484.

271. Mathews, W.H., and McCammon, J.W., 1957, "Calcareous Deposits of Southwestern British Columbia," Bulletin 40, British Columbia Department of Mines, 105 pp.

272. McCulloch, R., 1988, "Directory of Montana Mining Enterprises for 1988," Special Publication 97, Montana Bureau of Mines and Geology, 28 pp.

273. McCue, J.B., et al., 1939, "Limestones of West Virginia," West Virginia Geological Survey, Vol. 12, 560 pp.

274. McGrain, P., and Dever, G.R., Jr., 1967, "Limestone Resources in the Appalachian Region of Kentucky," Bulletin 4, Series 10, Kentucky Geological Survey, 12 pp.

275. McGregor, D.J., 1958, "Cement Raw Materials in Indiana," Bulletin 15, Indiana Geological Survey, 88 pp.

276. McGregor, D.J., 1963, High-Calcium Limestone and Dolomite in Indiana," Bulletin 27, Indiana Geological Survey, 76 pp.

277. McLemore, W.H., and Hurst, V.J., 1970, "The Carbonate Rocks in the Coosa Valley Area, Georgia," Geology Department, University of Georgia, Athens, 170 pp.

278. Meade, L.P., 1980, "The Vermont Marble Belt and the Economics of Current Operations," *Proceedings*, 14 Annual Forum on Geology

of Industrial Minerals, J.R. Dunn et al., eds., Bulletin 436, New York State Museum, pp. 54–56.

279. Mellen, F.F., 1942, "Mississippi Agricultural Limestone," Bulletin 46, Mississippi Geological Survey, 20 pp.

280. Merrill, R.H., 1956, "Roof-Span Studies in Limestone," Report of Investigations 5348, US Bureau of Mines, 38 pp.

281. Mielenz, R.C., 1954, "Petrographic Examination of Concrete Aggregate," *Proceedings*, American Society for Testing and Materials, Vol. 54, pp. 1188–1218.

282. Mielenz, R.C., 1955, "Petrographic Examination," *Significance of Tests and Properties of Concrete and Concrete Aggregates*, American Society for Testing and Materials, Special Technical Publication 169, pp. 253–273.

283. Mihok, E.A., 1968, "Mine Water Research-The Limestone Neutralization Process," Report of Investigations 7191, US Bureau of Mines, 20 pp.

284. Mihok, E.A., 1970, "Mine Water Research-Plant Design and Cost Estimates for Limestone Treatment," Report of Investigations 7368, US Bureau of Mines, 13 pp.

285. Miller, B.L., 1934, "Limestones of Pennsylvania," Mineral Resource Report M20, Pennsylvania Geological Survey, 729 pp.

286. Milliman, J.D., 1974, "Marine Carbonates," Part 1, Springer-Verlag, New York, 375 pp.

287. Mills, J.W., 1962, "High-Calcium Limestones of Eastern Washington," Bulletin 48, Washington Division of Mines and Geology, 268 pp.

288. Mineral and Water Resources of Colorado, 1964, *Report of the US Geological Survey*, US Government Printing Office, Washington, DC 302 pp.

289. Missouri Geological Survey, 1988, "Mineral Resources and Industry Map of Missouri," Scale 1:500,000.

290. Moen, W.S., 1966, "Stone, Including Limestone and Dolomite," *Mineral and Water Resources of Washington*, Reprint 9, Washington Division of Mines and Geology, pp. 265–273.

291. Moore, F.C., et al., 1968, "Problems Encountered in the Design and Construction of a New Aggregate Plant," SME Preprint 68-H-324, 24 pp.

292. Moore, F.H., 1935, "Marbles and Limestones of Connecticut," Bulletin 56, Connecticut State Geological and Natural History Survey, 56 pp.

293. Morris, H.T., 1969, "Limestone and Dolomite," *Mineral and Water Resources of Utah*, Bulletin 73, Utah Geological and Mineral Survey, pp. 188–194.

294. Morse, W.C., 1930, "Paleozoic Rocks," Bulletin 23, Mississippi Geological Survey, 212 pp.

295. Müller, G., 1967, *Methods in Sedimentary Petrology*, Part 1 (English trans. by Hans-Ulrich Schmincke), Hafner, New York, 283 pp.

296. Murray, D.A., 1975, "Limestones and Dolomites of Nova Scotia, Part II, Antigonish, Guysborough, Picton and Cumberland Counties," Bulletin 2, Nova Scotia Department of Mines, 155 pp.

297. Mutmansky, J.M., and Singh, M.M., 1968, "A Statistical Study of Relationships Between Rock Properties," *Status of Practical Rock Mechanics*, AIME, New York, pp. 161–177.

298. National Stone Association, 1990, "Guideline for the Successful Automation in the Crushed Stone Industry," 1990 Conferencey Proceedings, National Stone Association, Washington, DC.

299. National Stone Association, 1991 NSA Publications, Washington, DC, 3 pp.

300. Nelson, H.F., 1959, "Deposition and Alteration of the Edwards Limestone, Central Texas," *Symposium on Edwards Limestone in Central Texas*, Publication 5905, University of Texas, Austin, pp. 21–95.

301. Nelson, S.L., Oberhelman, M.W., and Olson, D.J., 1990, "Inventory of Industrial Mineral Pits and Quarries in Minnesota," Report 282, Minnesota Department of Natural Resources, Division of Minerals, Vol. 1, 108 pp; Vol. 2, 415 pp.

302. Newlon, H.H., Jr., et al., 1972, "An Evaluation of Several Methods for Detecting Alkali-Carbonate Reaction," Progress Report 5, Virginia Highway Research Council, 73 pp.

303. Niles, H.R., and Mossler, J.H., 1990, "Evaluation of the Carbonate Resources of Southern Minnesota," Technical Report NRR1/TR-90/08, Minnesota Natural Resources Research Institute, 200 pp.

304. Obert, L., et al., 1946, "Standardized Tests for Determining the Physical Properties of Mine Rock," Report of Investigations 3891, US Bureau of Mines, 67 pp.

305. Odgen, D.G., 1967, "Considerations of Structural Geology in the Search for White, High-Calcium Marble as a Source of Industrial Fillers," O C. Farquhar, ed., University of Massachusetts Graduate School, Amherst, pp. 181–184.

306. O'Donnell, J.J., and Sliger, A.G., 1972, "Availability of Limestones and Dolomites," Task 1 Final Report, M.W. Kellogg Co., National Technical Information Service PB-206-963, Feb. 1, 191 pp.

307. Oliver, W.A., Jr., 1956, "Biostromes and Bioherms of the Onondaga Limestone in Eastern New York," New York State Museum and Science Service Circular 45, 23 pp.

308. Olson, R.H., 1964, "Limestone," *Mineral and Water Resources of Nevada*, Bulletin 65, Nevada Bureau of Mines, pp. 217–220.

309. O'Neill, B.J., Jr., 1964, "Limestones and Dolomites of Pennsylvania," *Atlas of Pennsylvania's Mineral Resources*, Part 1, Mineral Resource Report M50, Pennsylvania Geological Survey, 40 pp.

310. O'Neill, B.J., 1975, "Potential High-Calcium Limestone Resources in the Mount Joy Area, Lancaster County," Information Circular 76, Pennsylvania Geological Survey, 18 pp.

311. O'Neill, B.J., Jr., 1976, "The Distribution of Limestones Containing At Least 90 Percent $CaCO_3$ in Pennsylvania," *Atlas of Mineral Resources*, Part 4, Mineral Resources Report M50, Scale 1:500,000, Pennsylvania Geological Survey.

312. Osborne, W.E., et al., compilers, 1988, "Geologic Map of Alabama, Special Map 220, Scale 1:250,000, Alabama Geological Survey.

313. Osterwald, F.W., et al., 1966, "Mineral Resources of Wyoming," Bulletin 50, Wyoming Geological Survey, pp. 180–184.

314. Papke, K.G., 1973, "Industrial Mineral Deposits of Nevada," Map 46, Nevada Bureau Mines and Geology.

315. Parker, R.L., 1967, "Composition of the Earth's Crush," *Data of Geochemistry*, 6th ed., M. Fleischer, ed., Professional Paper 440-D, US Geological Survey, 19 pp.

316. Pearre, N.C., 1956, "Mineral Deposits and Occurrences in Massachusetts and Rhode Island, Exclusive of Clay, Sand and Gravel, and Peat," Mineral Investigations Resources Map MR 4, US Geological Survey.

317. Perry, T.G., and Smith, N.M., 1958, "The Meramec-Chester and Intra-Chester Boundaries and Associated Strata in Indiana," Bulletin 12, Indiana Geological Survey, 110 pp.

318. Pickering, S.M., Jr., 1970, "Stratigraphy, Paleontology, and Economic Geology of Portions of Perry and Cochran Quadrangles, Georgia," Bulletin 81, Georgia Geological Survey, 67 pp.

319. Plafker, G., and Berg, H.C., 1964, "Limestone and Marble," *Mineral and Water Resources of Alaska*, US Geological Survey Report, pp. 135–138.

320. Potter, P.E., 1968, "A Selective, Annotated Bibliography on Carbonate Rocks," *Bulletin of Canadian Petroleum Geology*, Vol. 16, No. 1, pp. 87–103.

321. Power, W.R., 1978, "Economic Geology of the Georgia Marble District," *Proceedings*, 12th Forum on Geology of Industrial Minerals, Information Circular 49, Georgia Geological Survey, pp. 59–68.

322. Prucha, J.J., 1953, "The White Crystal Dolomite Deposit near Gouverneur, New York," Report of Investigations 9, New York State Museum and Science Service, 13 pp.

323. Puckett, T.M., et al., 1990, "High-Calcium Deposits in the Newala Limestone," Circular 149, Alabama Geological Survey, 29 pp.

324. Pugliese, J.M., 1972, "Designing Blast Patterns Using Empirical Formulas, A Comparison of Calculated Patterns with Plans Used in Quarrying Limestone and Dolomite, with Geologic Considerations," Information Circular 8550, US Bureau of Mines, 33 pp.

325. Puri, H.S., et al., 1967, "Geology of Dixie and Gilchrist Counties, Florida," Bulletin 49, Florida Geological Survey, 155 pp.

326. Quinn, A.W., 1971, "Bedrock Geology of Rhode Island," Bulletin 1295, US Geological Survey, 68 pp.

327. Quinn, A.W., et al., 1949, "Bedrock Geology of the Pawtucket Quadrangle, Rhode Island-Massachusetts," Geological Quadrangle Map 1, Scale 1:31,680, US Geological Survey.

328. Renninger, F.A., and Nichols, F.P., Jr., 1970, "Aggregate Technology—Wider Horizons Through Research," *Trans. SME/AIME*, Vol. 247, pp. 38–42.

329. Riley, H.C., and Schroeder, H.J., 1963, "Crushed Limestone Oper-

ations, Watauga Quarry, Watauga Stone Co., Carter County, Tenn.,'' Information Circular 8198, US Bureau of Mines, 21 pp.
330. Risser, H.T., 1960, ''Kansas Building Stone,'' Bulletin 142, Part 2, Kansas Geological Survey, pp. 53–122.
331. Rodda, P.V., et al., 1966, ''Limestone and Dolomite Resources, Lower Cretaceous Rocks, Texas,'' Report of Investigations 56, Texas Bureau of Economic Geology, 286 pp.
332. Roehm, J.C., 1946, ''Some High Calcium Limestone Deposits in Southeastern Alaska,'' Pamphlet 6, Alaska Division of Mines and Geology, 85 pp.
333. Rooney, L.F., 1970, ''High-Calcium Limestone and High-Magnesium Dolomite Resources of Indiana,'' Bulletin 42-B, Indiana Geological Survey, 20 pp.
334. Rooney, L.F., 1970, ''Dimension Limestone Resources of Indiana,'' Bulletin 42-C, Indiana Geological Survey, 29 pp.
335. Rooney, L.F., and Ault, C.H., 1970, ''Potential Limestone and Dolomite Resources of Northern Indiana,'' *Proceedings*, 5th Forum on Geology of Industrial Minerals, Mineral Resources Report M64, Pennsylvania Geological Survey, pp. 179–224.
336. Rooney, L.F., and Carr, D.D., 1971, ''Applied Geology of Industrial Limestone and Dolomite,'' Bulletin 46, Indiana Geological Survey, 59 pp.
337. Rowland, T.L., 1972, ''General Survey of Carbonate Mineral Deposits in Oklahoma,'' *Oklahoma Geology Notes*, Vol. 32, No. 3, pp. 73–89.
338. Rukavina, M., 1989, ''Dravo Steps up Oceanic Shipment,'' Rock Products, Chicago, IL, June, pp. 34–37.
339. Rukavina, M., 1990, ''A Model Plant Opens in Canada,'' Rock Products, Chicago, IL, Aug. pp. 31–35.
340. Runnells, D.D., 1970, ''Errors in X-Ray Analysis of Carbonates Due to Solid-Solution Variation in Composition of Component Minerals,'' *Journal of Sedimentary Petrology*, Vol., 40, pp. 1158–1166.
341. Runnels, R.T., 1951, ''Some High-Calcium Limestones in Kansas,'' Bulletin 90, Part 5, Kansas Geological Survey, pp. 77–104.
342. Runnels, R.T., 1959, ''Cement Raw Materials in Kansas,'' *1959 Reports of Studies*, Part 2, Bulletin 134, Kansas Geological Survey, pp. 105–124.
343. Runnels, R.T., and Dubins, I.M., 1949, ''Chemical and Petrographic Studies of the Fort Hays Chalk in Kansas,'' Bulletin 82, Part 1, Kansas Geological Survey, 36 pp.
344. Runnels, R.T., and Schleicher, J.A., 1956, ''Chemical Composition of Eastern Kansas Limestone,'' Bulletin 119, Part 3, Kansas Geological Survey, pp. 81–103.
345. Rutledge, F.A., et al., 1953, ''Preliminary Report—Nonmetallic Deposits Accessible to the Alaska Railroad as Possible Sources of Raw Materials for the Construction Industry,'' Report of Investigations 4932, US Bureau of Mines, 129 pp.
346. Sage, R.P., 1987, ''Geology of Carbonatite—Alkalic Rock Complexes in Ontario: Prairie Lake Carbonatite Complex, District of Thunder Bay,'' Study 46, Ontario Geological Survey, 91 pp.
347. Salotti, C.A., et al., 1970, ''Limestone as a Raw Material for Hydrocarbon Fuels,'' *Proceedings*, 6th Forum on Geology of Industrial Minerals, Miscellany 1, Michigan Geological Survey, pp. 48–55.
348. Sanders, J.E., and Friedman, G.M., 1967, ''Origin and Occurrence of Limestone,' *Developments in Sedimentology, Carbonate Rocks*, Vol. 9A, G.V. Chilingar, H.J. Bissell, and R.W. Fairbridge, eds., Elsevier, New York, pp. 169–265.
349. Savage, C.N., 1965, ''Economic Geology of Carbonate Rocks Adjacent to the Snake River South of Lewiston, Idaho,'' Mineral Resources Report 10, Idaho Bureau of Mines and Geology, 26 pp.
350. Savage, C.N., 1969, ''Distribution and Economic Potential of Idaho Carbonate Rocks,'' Bulletin 23, Idaho Bureau of Mines and Geology, 93 pp.
351. Scholle, P.A., 1978, ''A Color Illustrated Guide to Carbonate Rock Constituents, Textures, Cements, and Porosities,'' American Association of Petroleum Geologists, Memoir 27, 241 pp.
352. Scholle, P.A., Bebout, D.G., and Moore, C.H., eds., 1983, ''Carbonate Depositional Environments'' American Association of Petroleum Geologists, Tulsa, OK, 708 pp.
353. Schmidt, W., et al., 1979, ''The Limestone Dolomite and Coquina Resources of Florida,'' Report of Investigations 88, Florida Bureau of Geology, 54 pp.
354. Schwartz, G.M., et al., 1959, ''Investigation of the Commercial Possibilities of Marl in Minnesota,'' Office of Iron Range Resources and Rehabilitation, 190 pp.
355. Scott, A.J., 1968, ''Environmental Factors Controlling Oyster Shell Deposits, Texas Coasts,'' *Proceedings*, 4th Forum on Geology of Industrial Minerals, Texas Bureau of Economic Geology, pp. 129–150.
356. Searle, A.B., 1935, *Limestone and Its Products; Their Nature, Production, and Uses*, Ernest Benn, London, 709 pp.
357. Shapiro, L., and Brannock, W.W., 1962, ''Rapid Analysis of Silicate, Carbonate and Phosphate Rocks,'' Bulletin 1144-A, US Geological Survey, 56 pp.
358. Shea, F.S., and Murray, D.A., 1967, ''Limestones and Dolomites of Nova Scotia, Part 1, Cape Breton Island,'' Bulletin 2, Nova Scotia Department of Mines, 300 pp.
359. Siegel, F.R., 1967, ''Properties and Uses of the Carbonates,'' *Developments in Sedimentology, Carbonate Rocks*, Vol. 9B, G.V. Chilingar, H.J. Bissell, R.W. Fairbridge, eds., Elsevier, New York, pp. 343–393.
360. Siemers, W.T., 1982, ''New Mexico Limestones—Geology, Uses, and Economic Potential of High-Calcium Limestone,'' Circular 182, New Mexico Bureau of Mines and Mineral Resources, pp. 39–42.
361. Sinclair, J., 1969, *Quarrying, Opencast and Alluvial Mining*, Elsevier, New York, 375 pp.
362. Singleton, R.H., 1980, ''Stone,'' Preprint, Bulletin 671, *Mineral Facts and Problems*, US Bureau of Mines, 16 pp.
363. Slack, A.V., 1973, ''Removing SO_2 from Stack Gases,'' *Environmental Science & Technology*, Vol. 7, No. 2, pp. 110–119.
364. Slavin, W., 1968, *Atomic Absorption Spectroscopy*, Interscience Publishers, New York, 307 pp.
365. Smith, R.A., 1915, ''Limestones in Michigan,'' Publication 21, Part 2, Michigan Geological and Biological Survey, pp. 103–402.
366. Smith, D.L., and Gilmore, E.H., 1980, ''Mississippian and Pennsylvanian (Carboniferous) Systems in Montana,'' Memoir 46, Montana Bureau Mines and Geology, 32 pp.
367. Stearn, E.W., 1965, ''Underground Storage: How Does It Look Today?'' *Rock Products*, Vol. 68, No. 12, pp. 86–87, 96.
368. Steidtmann, E., 1924, ''Limestones and Marls of Wisconsin,'' Bulletin 66, Wisconsin Geological Survey, 208 pp.
369. Stenzel, H.B., et al., 1948, ''Geological Resources of the Trinity River Tributary Area in Oklahoma and Texas,'' Publication 4824, University of Texas, Austin, 252 pp.
370. Stith, D.A., 1969, ''Potential Use of Ohio Limestones and Dolomites for Architectural Aggregate,'' Report of Investigations 73, Ohio Geological Survey, 14 pp.
371. Stith, D.A., 1972, ''High-Calcium Limestone Facies of the Devonian Dundee Limestone,'' Report of Investigations 86, Ohio Geological Survey, 14 pp.
372. Stith, D.A., 1979, ''Chemical Composition, Stratigraphy, and Depositional Environments of the Black River Group (Middle Ordovician), Southwestern Ohio,'' Report of Investigations 113, Ohio Geological Survey, 36 pp.
373. Stith, D.A., and Stieglitz, R.D., 1979, ''An Evaluation of Newberry Analysis Data on the Brassfield Formation (Silurian), Southwestern Ohio,'' Report of Investigations 108, Ohio Geological Survey, 25 pp.
374. Stith, D.A., and Van Horn, R.G., 1983, ''A Geological Sketch—Construction Materials in Ohio,'' *Proceedings*, 18th Forum on Geology of Industrial Minerals, Occasional Paper 37, Indiana Geological Survey, pp. 143–155.
375. Stokley, J.A., and McFarlan, A.C., 1952, ''Industrial Limestones of Kentucky No. 2,'' Report of Investigations 4, Series 9, Kentucky Geological Survey, 95 pp.
376. Stokley, J.A., and Walker, F.H., 1953, ''Industrial Limestones of Kentucky No. 3,'' Report of Investigations 8, Series 9, Kentucky Geological Survey, 62 pp.
377. Stout, W., 1941, ''Dolomites and Limestones of Western Ohio,'' Bulletin 42, Ohio Geological Survey, 468 pp.
378. Stowe, R.L., 1969, ''Strength and Deformation Properties of Granite, Basalt, Limestone and Tuff at Various Loading Rates,'' Miscellaneous Paper C-69–1, US Army Engineer Waterways Experiment Station, 168 pp.

379. Suffel, G.G., 1930, ''Dolomites of Western Oklahoma,'' Bulletin 49, Oklahoma Geological Survey, 155 pp.
380. Taft, W.H., 1967, ''Modern Carbonate Sediments,'' *Developments in Sedimentology, Carbonate Rocks*, Vol. 9A, G.V. Chilingar, H.J. Bissell, R.W. Fairbridge, eds., Elsevier, New York, pp. 29–50.
381. Tennessee Division of Geology, see Publications List of Geologic Quadrangle Maps.
382. Tepordei, V.V., 1991, ''Crushed Stone 1989,'' *Minerals Yearbook*, US Bureau of Mines, Washington, DC, Apr., 28 pp.
383. Thiel, G.A., and Dutton, C.E., 1935, ''The Architectural, Structural and Monumental Stones of Minnesota,'' Bulletin 25, Minnesota Geological Survey, 160 pp.
384. Thill, R.E., et al., 1969, ''Correlation of Longitudinal Velocity Variation with Rock Fabric,'' *Journal of Geophysical Research*, Vol. 74, No. 20, pp. 4897–4909.
385. Thoenen, J.R., 1926, ''Underground Limestone Mining,'' Bulletin 262, US Bureau of Mines, 100 pp.
386. Trainer, D.W., Jr., 1932, ''The Tully Limestone of Central New York,'' New York State Museum and Science Service Bulletin 291, 23 pp.
387. Tripp, B.T., 1991, ''High-Calcium Limestone Resources of Utah,'' Map Series, Utah Geological and Mineral Survey.
388. Tripp, B.T., 1991, ''The Industrial Rock and Mineral Industry in Utah, 1990,'' Circular 82, Utah Geological and Mineral Survey, 31 pp.
389. Tucker, M.E., and Wright, V.P., 1990, *Carbonate Sedimentology*, Blackwell Scientific Publications, Oxford, 482 pp.
390. US Census Office, 1884, ''Report on the Building Stones of the United States and Statistics of the Quarry Industry for 1880,'' *10th Census*, Government Printing Office, Washington, DC, 410 pp.
391. Vanderwilt, J.W., 1947, ''Mineral Resources of Colorado,'' Colorado State Mineral Resources Board, Denver, 547 pp.
392. Vermont Geological Survey, see Publications List of Quadrangle Reports.
393. Vernon, R.O., 1951, ''Geology of Citrus and Levy Counties, Florida,'' Bulletin 33, Florida Geological Survey, 256 pp.
394. Vineyard, J.T., ed., 1969, *Missouri Mineral Industry News*, Missouri Geological Survey, Vol. 9, July, pp. 102.
395. Walker, K.R., 1973, ''Stratigraphy and Environmental Sedimentology of Middle Ordovician Black River Group in the Type Area—New York State,'' New York State Museum and Science Service Bulletin 419, 43 pp.
396. Warne, S.S.J., 1962, ''A Quick Field or Laboratory Staining Scheme for the Differentiation of the Major Carbonate Minerals,'' *Journal of Sedimentary Petrology*, Vol. 32, pp. 29–38.
397. Waugh, T.C., 1966, ''Analyses of High Calcium Chert-Free Beds in the Keokuk Limestone, Cherokee County, Kansas,'' Bulletin 180, Part 2, Kansas Geological Survey, 7 pp.
398. Wayland, J.R., and Ham, W.E., 1955, ''General and Economic Geology of the Baum Limestone, Ravia-Mannsville Area, Oklahoma,'' Circular 33, Oklahoma Geological Survey, 44 pp.
399. Wayne, W.J., 1971, ''Marl Resources of Indiana,'' Bulletin 42-G, Indiana Geological Survey, 16 pp.
400. Wedepohl, K.H., 1970, ''Geochemische Daten von Sedimentären Karbonaten und Karbonatgesteinen in Ihrem Faziellen und Petrogenetischen Aussagewert,'' *Verhandl. Geol. Bundesanst*, (Austria), Vol. 4, pp. 692–705.
401. Weitz, J.H., 1942, ''High-Grade Dolomite Deposits in the United States,'' Information Circular 7226, US Bureau of Mines, 86 pp.
402. Welby, C.W., 1961, ''Bedrock Geology of the Central Champlain Valley of Vermont,'' Bulletin 14, Vermont Geological Survey, 296 pp.
403. Wentworth, C.K., 1939, ''Physical Geography and Geology [of the Hawaiian Islands],'' Hawaii Territorial Planning Board, 1st Progress Report, Honolulu, pp. 13–20.
404. West, T.R., and Johnson, R.B., 1965, ''Analysis of Textural and Physical Factors Contributing to the Abrasion Resistance of Some Indiana Carbonate Aggregates,'' *Indiana Academy of Science Proceedings for 1964*, Vol. 75, pp. 153–162.
405. West Virginia Geological and Economic Survey, see Publications List of County Reports.
406. Willard, B., 1931, ''Commercial Limestones of Rhode Island,'' *Pan-American Geologist*, Vol. 56, No. 2, pp. 116–122.
407. Willard, R.J., and McWilliams, J.R., 1969, ''Microstructural Techniques in the Study of Physical Properties of Rock,'' *International Journal of Rock Mechanics and Mining Sciences*, Vol. 6, No. 1, pp. 1–12.
408. Williamson, D.R., and Burgin, L., 1960, ''Limestone Occurrences in Colorado,'' *Mineral Industries Bulletin*, Vol. 3, No. 1, Colorado School of Mines, Golden, CO, 12 pp.
409. Willman, H.B., 1943, ''High-Purity Dolomite in Illinois,'' Report of Investigations 90, Illinois Geological Survey, 89 pp.
410. Wilmoth, R.C., and Hill, R.D., 1973, ''Mine Drainage Pollution Control Via Reverse Osmosis,'' *Mining Engineering*, Vol. 25, No. 3, pp. 45–47.
411. Wilson, J.L., 1975, *Carbonate Facies in Geologic History*, Springer-Verlag, New York, 471 pp.
412. Windes, S.L., 1949, ''Physical Properties of Mine Rock, Part I,'' Report of Investigations 4459, US Bureau of Mines, 79 pp.
413. Windes, S.L., 1950, ''Physical Properties of Mine Rock, Part II,'' Report of Investigations 4727, US Bureau of Mines, 37 pp.
414. Wise, S.W., Jr., and Weaver, F.M., 1974, ''Chertification of Oceanic Sediments,'' *Pelagic Sediments: on Land and Under the Sea*, K.J. Hsü and H.C. Jenkyns, eds., International Association of Sedimentologists, Special Publication 1, pp. 301–326.
415. Wolfe, P.E., 1948, ''Agricultural Mineral Resources of New Jersey,'' Bulletin 2, Bureau of Mineral Research, Rutgers University, New Brunswick, NJ, pp. 18–50.
416. Woolf, D.O., 1953, ''Results of Physical Tests of Road-Building Aggregate,'' US Department of Commerce, Bureau of Public Roads, Washington, DC, 225 pp.
417. Yon, J.W., Jr., and Hendry, C.W., Jr., 1969, ''Mineral Resource Study of Holmes, Walton and Washington Counties,'' Bulletin 50, Florida Geological Survey, 161 pp.
418. Yon, J.W., Jr., and Hendry, C.W., Jr., 1972, ''Suwannee Limestone in Hernando and Pasco Counties, Florida,'' Bulletin 54, Part 1, Florida Geological Survey, pp. 1–42.
419. Zenger, D.H., 1965, ''Stratigraphy of the Lockport Formation (Middle Silurian) in New York State,'' New York State Museum and Science Service Bulletin 419, 210 pp.
420. Zenger, D.H., 1981, ''Stratigraphy and Petrology of the Little Falls Dolostone (Upper Cambrian), East-Central New York,'' New York State Museum and Science Service Map and Chart Series 34, 138 pp.

LITHIUM RESOURCES

IHOR A. KUNASZ

Lithium minerals occur predominantly in pegmatites that contain mineral assemblages derived from the crystallization of post-magmatic fluids or from the metasomatic action of residual pegmatitic fluids. They have been the traditional sources of raw material for ceramic and chemical industries. With the discovery of the Searles Lake deposits, brines became new sources of lithium, and, today, two brine deposits satisfy nearly one-half of the free world demand for lithium chemicals.

Indeed, lithium production from pegmatites has been superseded by brines sources. In the United States, the brines of Clayton Valley, NV, have been exploited since 1964 by Cyprus Foote Mineral Company, a subsidiary of Cyprus Minerals Company. In Chile, Sociedad Chilena de Litio Ltda., an affiliate of Cyprus Minerals Company, is now producing lithium carbonate from the brines of the Salar de Atacama. Recently identified brine deposits suggest that these may well become the prime sources of lithium chemicals. In Bolivia, the government is negotiating the rights to exploit the brines of the Salar de Uyuni. The Salar del Hombre Muerto in Argentina is being evaluated as a future center of chemical production. The brines of the Great Salt Lake of Utah, the Smackover Formation, and the Imperial Valley geothermal field have been defined as important resources of lithium. In Tibet, high concentrations of lithium have been reported in several saline lakes. In the People's Republic of China, lithium may be produced from the brines of the Dabusan playa in the Qinghai Basin.

The 1950s saw a great number of lithium companies develop. However, following the termination of the Atomic Energy Commission purchase program in 1959, several companies, such as Maywood Chemical Co., American Lithium Chemicals, and Quebec Lithium Corp., were forced to close their operations. Nevertheless, with the development of new applications in glass ceramics, air conditioning systems, synthetic rubber, and metallurgy (aluminum potlines), the lithium industry grew steadily. With the potential application of lithium metal in batteries, nuclear reactors, and fuel cells, a steady growth can be foreseen. New products and new applications are likely to stimulate exploration for expanded reserves as well as the search for new sources from which lithium could be extracted economically. Presently identified reserves and production capacity are sufficient to satisfy the conventional market growth of lithium.

GEOCHEMISTRY OF LITHIUM

Lithium is the third element in the periodic table. It is the lightest of all the metals, having an atomic weight of 6.938, an ionic radius of 0.68A°, and a charge of +1.

The geochemistry of lithium has been extensively studied and has been summarized by Rankama and Sahama (1950), Goldschmidt (1937), Hortsman (1957), and Cerny (1991).

The distribution of lithium in igneous rocks is controlled by its size and its charge and by the $(MgO+FeO)/Li_2O$ ratio. In the early stages of crystallization of a magma that ratio is very large. Consequently, both magnesium and iron are removed by ferromagnesian minerals in preference to lithium, which is then concentrated in the residual magma. The result is an enrichment of lithium in silicic rocks and pegmatites (Strock, 1936).

Pegmatites are coarse-grained igneous rocks formed by the crystallization of postmagmatic fluids. Minerals within pegmatites may also form by metasomatism (Jahns, 1955). Genetically, the pegmatites are associated with neighboring intrusives. Mineralogically, granitic pegmatites contain feldspar, quartz, and mica as the main constituents and a variety of exotic elements such as lithium, beryllium, tantalum, tin, and cesium, which may or may not occur in economically significant concentrations.

Detailed studies by numerous investigators (Cameron et al., 1949, 1954, Hanley et al., 1950, Jahns, 1952, 1955, Page et al., 1953, Norton and Schlegel, 1955, Cerny, 1991) indicate that many pegmatites exhibit an internal zonal arrangement, with each zone containing a specific suite of minerals. The lithium minerals are usually found in the intermediate zones, and, although as many as 13 zones have been recognized (Cameron et al., 1949), a complete zonal arrangement is rarely found. Zoning of pegmatite bodies has also been observed on a regional basis. The regionally zoned pegmatite sequences exhibit mineral assemblages and complexity according to their respective distance from the granitic bodies to which they are genetically related. Although various theories have been proposed for the genesis of pegmatites, Cerny (1991) offered convincing evidence that rare metal pegmatites are essentially magmatic phenomena. Although pegmatites exhibit a broad diversity of paragenetic, geochemical, and structural styles, pegmatites have crystallized from a volatile-rich melt enriched to various degrees in lithophile elements. From a practical exploration standpoint, Cerny concluded that Late Archean and Early Proterozoic fields are possibly the most productive, and that lower amphibole facies of volcano-sedimentary rocks are the main hosts for pegmatites.

Lithium is also found in small proportions in a variety of rocks. The average lithium content of igneous rocks is estimated at about 28 ppm Li. Sedimentary rocks contain an average of 53 ppm Li, the highest being recorded in shales (Hortsman, 1957).

Unusual amounts of lithium are found in the clay mineral hectorite, which is expandable and belongs to the magnesian end member of the smectite group. Recent work by the US Geological Survey suggest that lithium-rich clays are ubiquitous in a number of clay environments. A large lithium resource has been identified in the McDermitt Caldera (Nevada-Oregon), where concentrations of 0.35% Li have been recorded (Glanzman et al., 1978).

Lithium is also present in significant amounts in waters associated with geothermal areas (White, 1957), in oil well brines (Mayhew and Heylmun, 1966), and certain brines of California (Searles Lake), Nevada (Clayton Valley), Utah (Great Salt Lake), and a number of salars in Chile, of which the Salar de Atacama is the richest, Bolivia (Salar de Uyuni), Argentina (Salar del Hombre Muerto, El Rincon), as well as Tibet and the People's Republic of China. Lithium also occurs in certain geothermal waters, such as the ones presently exploited for energy in the Imperial Valley, California.

LITHIUM MINERALS

Although lithium occurs in some 145 minerals, only spodumene, lepidolite, petalite, amblygonite, and eucryptite have been commercial sources of lithium. Today, the principal sources of lithium ores and chemicals are spodumene and petalite.

Spodumene

Spodumene, a lithium aluminum silicate ($LiAlSi_2O_6$), is a monoclinic member of the pyroxene group. It has a very pronounced cleavage plane (110), which results in typically lath-shaped particles upon breaking. The color of spodumene is variable, being nearly white in the low iron variety and dark green in iron-rich crystals.

When clear, spodumene is considered a gemstone. Three varieties are known: hiddenite, the green variety from Alexander County, North Carolina, first discovered in Brazil; triphane, the yellow variety also from Alexander County; and the lilac-colored kunzite from the Pala district, California, Brazil, and Afghanistan.

Spodumene undergoes pseudomorphic alteration to a variety of minerals. Norton and Schlegel (1955) have described spodumene replacement by quartz, albite, perthite, muscovite, beryl, amblygonite, apatite, and tourmaline. Weathering commonly alters spodumene to kaolinite and to montmorillonite.

Spodumene constitutes the most abundant commercial source of lithium. Theoretically, it may contain up to 3.7% Li, but the actual lithium concentrations range from 1.35 to 3.56%, probably as a result of sodium and potassium substitution for lithium. Spodumene concentrates typically contain 1.9 to 3.3% Li. Spodumene is the traditional source of lithium concentrates and chemicals in the United States (North Carolina), Brazil (Minas Gerais), Canada (Manitoba), Zimbabwe (Bikita), Australia (Greenbushes), the former USSR (Chita Region), and the People's Republic of China (Altai Mountains).

Lepidolite

Lepidolite is a phyllosilicate with the general formula $K_2(Li,Al)_{5-6}\{Si_{6-7}Al_{2-1}O_{20}\}(OH,F)_4$. The chemical variability expressed in the formula stems from a structural complexity attributed to a mixture of polymorphs, which include muscovite, lithium muscovite, and polylithionite (Winchell, 1942). On the other hand, Foster (1960) and Deer et al. (1962) suggested that there is a continuous series between muscovite with a $2M_1$ structure to lepidolite with 1M, $2M_2$, and 3T structures. The structural transition takes place when the lithia content in the mica reaches 1.53%.

The lithium concentration in lepidolite ranges from 1.53% to a possible theoretical maximum of 3.6%. In commercial deposits, the concentrations are more normally 1.4 to 1.9% Li. In addition to lithium, lepidolites also carry substantial concentrations of rubidium and cesium (Deer et al., 1962).

The major commercial deposits of lepidolite are located in Zimbabwe (Bikita), Namibia (Karibib), Canada (Bernic Lake, Manitoba), and Brazil (Minas Gerais).

Petalite

Petalite, $LiAlSi_4O_{10}$, is a monoclinic mineral with a framework silicate structure. Its color is grayish white and more rarely pinkish. It has two cleavage directions, which form an angle of 38.5°. The basal cleavage is perfect.

The theoretical lithium content of petalite is 2.27%. In commercial deposits, it ranges from 1.6 to 2.1% Li. Sizable deposits of petalite occur with lepidolite in Zimbabwe (Bikita), Namibia (Karibib), Brazil (Aracuai), Australia (Londonderry), in the former USSR (eastern Transbaikalia), and Sweden (Utö).

In certain pegmatites, there is evidence that petalite alters to a mixture of spodumene and quartz. In the Bernic Lake pegmatites of Manitoba, Cerny and Turnock (1971) have described pseudomorphs of spodumene and quartz after petalite.

Eucryptite

Eucryptite is also a lithium aluminum silicate that is deficient in silica. It has a formula $LiAlSiO_4$ and may contain 5.53% Li. The only large deposit of eucryptite is found in Zimbabwe (Bikita), where its occurrence with quartz suggests spodumene origin (Westernberger, 1963). The grade of the eucryptite is 2.34% Li.

Amblygonite

Amblygonite, with the generalized formula $LiAl(PO_4)(F,OH)$, is the fluorine-rich end member of a phosphate series, whereas montebrasite represents the hydroxyl-rich end member. It occurs in white to gray masses. Most cleavage planes are pearly, others are vitreous. Amblygonite weathers to earthy apatite, wavellite, and other lithium-deficient phosphates. Although amblygonite may contain as much as 4.74% Li, most commercial ores carry 3.5 to 4.2%. Amblygonite has been mined in Canada, Brazil, Surinam, Zimbabwe, Ruanda, Mozambique, Namibia, and the Republic of South Africa.

Hectorite

Lithium also occurs in significant concentrations in the mineral hectorite, a trioctahedral smectite. The purest deposit is found at Hector, CA, where the white clay is exploited for its swelling characteristics in cosmetic applications. The lithium concentration in the hectorite is 0.53% Li. Hectorite has also been identified in Clayton Valley, NV (Kunasz, 1970), and McDermitt, NV (Glanzman et al., 1978).

Brines

Lithium is found in commercial quantities in certain brine deposits. The brines, volcanic in origin, are present in desert areas and occur in playas and saline lakes where dilute lithium solutions have been concentrated by solar evaporation. In Searles Lake, where production of di-lithium phosphate began in 1938, the lithium concentration is 70 ppm Li. In Clayton Valley, Nevada, the lithium concentration in brines varies from 100 to 300 ppm. Lesser concentrations of lithium (28 to 60 ppm Li) are found in the Great Salt Lake. High concentrations of lithium have been recorded in several salares in Argentina (Salar del Hombre Muerto: 100 to 700 ppm Li), Bolivia (Salar de Uyuni: 100 to 500 ppm Li), Chile (Salar de Atacama: 1 000 to 5 000 ppm Li), Tibet (Lake Zabuye: 700 to 1 000 ppm Li), and the People's Republic of China (Quinghai: 100 ppm Li).

RESERVES

Projections of lithium metal requirements to satisfy the needs of the automotive and storage battery system as well as the potential

Table 1. Reserves and Resources of Lithium in the United States and Other Western Countries (Penner, 1978)

	Tons Lithium[2]	
	United States	Other Western Countries
CLASS A[1]		
From pegmatites	329 100	366 400
From brines	40 500	1 290 000
From stockpile[3]	6 100	—
	375 700	1 656 400
CLASS B		
From pegmatites	47 300	456 800
From brines	—	—
	47 300	456 800
CLASS C		
From pegmatites	2 780 900	1 969 000
From brines	77 300	3 000 000
	2 858 200	4 969 000
CLASS D		
From brines	283 700	—
Total, CLASS A	375 700	1 656 400
Total, CLASSES A and B	423 000	2 113 200
Total, CLASSES A, B, and C	3 281 200	7 082 200
Total, CLASSES A, B, C, and D	3 564 900	7 082 200
Combined Totals:		
CLASS A	2 032 100	
CLASS A and B	2 536 200	
CLASS A, B, and C	10 363 400	
CLASS A, B, C, and D	10 647 100	

In the absence of data indicating otherwise, mine losses are assumed at 25% for open pit and 50% for underground operations. Brine totals are for recoverable brine.

[1]Classification system

A. Reserves proved by systematic exploration

B. Reserves indicated by limited exposures and/or exploration

C. Resources inferred on geological evidence

D. Quantities largely known, but economic lithium extraction probably dependent upon marketing of coproducts.

[2]Tonnages to beneficiation and process.

[3]Approximate Li content of lithium hydroxide currently available for purchase from US Government stockpile.

use in thermal nuclear fusion energy generation resulted in concern regarding the availability of sufficient lithium. The US Geological Survey created the Lithium Exploration Group, which sponsored two symposia, one held in Denver, CO, and another in Corning, NY. Concurrently, the Energy Research and Development Administration requested the National Academies of Sciences and Engineering to form a National Research Council Committee on Nuclear and Alternative Energy Systems (CONAES). One of the groups was the Lithium Subpanel consisting of experts from the industry and the government, formed to evaluate the lithium availability in the western world. The results of the Subpanel study were presented at the second lithium symposium held at Corning, NY, in 1977. This document is considered the best reference regarding reserves and resources, and the serious student of lithium should have this particular publication on hand (Penner, 1978). The study showed that known reserves and resources of lithium in the western world were 10.6 Mt of Li (Table 1). Many of the figures quoted in this review have been taken from the Subpanel study. Following the beginning of lithium production in Chile, a lithium symposium was held in Santiago, Chile, in 1986, at which time estimates of world reserves were updated by Evans (1986), who included the resources identified at the Salar de Uyuni (5.5 Mt Li), the Imperial Valley geothermal field (4 Mt Li), and the lithium contained in hectorite clays of the western United States (15.1 Mt Li).

In the 1975 edition of *Industrial Minerals and Rocks*, the importance of brines as potential sources of lithium was predicted. Today, it is clear that, with the discovery of several important lithium brine deposits, brines are displacing pegmatites as sources of lithium chemicals. In fact, the market applications will determine the future balance between brines and high-purity concentrates as preferred sources of lithium. Whereas in 1975 Foote Mineral Company operated a brine and a pegmatite deposit, today Cyprus Foote Mineral Company operates exclusively from two brine deposits. FMC, Greenbushes, Hudson Bay, Bikita Pty., Arqueana de Minerios e Metais, the Russian Federation, and the People's Republic of China all produce from pegmatites, but there are strong indications that the search for brine resources is proceeding at an accelerated pace. FMC is involved in active negotiations with the Bolivian, Argentinean, and Chilean governments for the potential developments of the Salares of Uyuni, Hombre Muerto, and Atacama. The Chinese have announced the anticipated production of lithium chemicals from the brines of the Dabusan playa in the Quinghai Basin.

PRODUCTION

At the present time, the United States, Chile, Australia, Zimbabwe, Canada, and Namibia supply most of the free world's demand for lithium raw materials (Fig. 1), and the United States is the leading producer. Within the communist bloc, the Russian Federation and the People's Republic of China are the major producers; their production figures, however, are not available. The People's Republic of China is reported to produce lithium compounds from pegmatites in the northwestern part of the country, whereas the bulk of the lithium in the Russian Federation is produced in the Chita oblast, east of Lake Baikal.

United States

Lithium ores and lithium chemicals in the United States are supplied by three producing areas. Two of the areas are in the Tin-Spodumene Belt of North Carolina, and the third is in Clayton Valley (Silver Peak), Nevada.

The Tin-Spodumene Belt of North Carolina constitutes one of the largest developed hard rock reserves in the free world. The area is characterized by numerous pegmatites that intrude amphibolites and schists along the eastern margin of the Devonian Cherryville quartz monzonite (Kesler, 1961).

The pegmatites are typically unzoned, and spodumene is distributed throughout the bodies. The contact zones and aplitic layers contain substantially lower percentages of spodumene. On the average, the pegmatites consist of 20% spodumene, 32% quartz, 6% muscovite, 41% feldspar (including albite and microcline), and about 1% trace minerals, such as beryl, lithiophilite, columbite, tantalite, pyrite, sphalerite, apatite, and rhodochrosite (Kesler, 1961).

Two companies operate in the Tin-Spodumene Belt: Cyprus Foote Mineral Co., a subsidiary of Cyprus Minerals Company, and the Lithium Division of the FMC Corporation. Spodumene has been mined for decades from a pegmatite cluster located about 1.6 km southwest of the town of Kings Mountain, NC. At the Cyprus Foote Mineral Company property, extensive drilling has outlined ore reserves of 29 Mt averaging 0.7% Li. At the present time, only the northern pegmatites are included in this estimate. Approximately

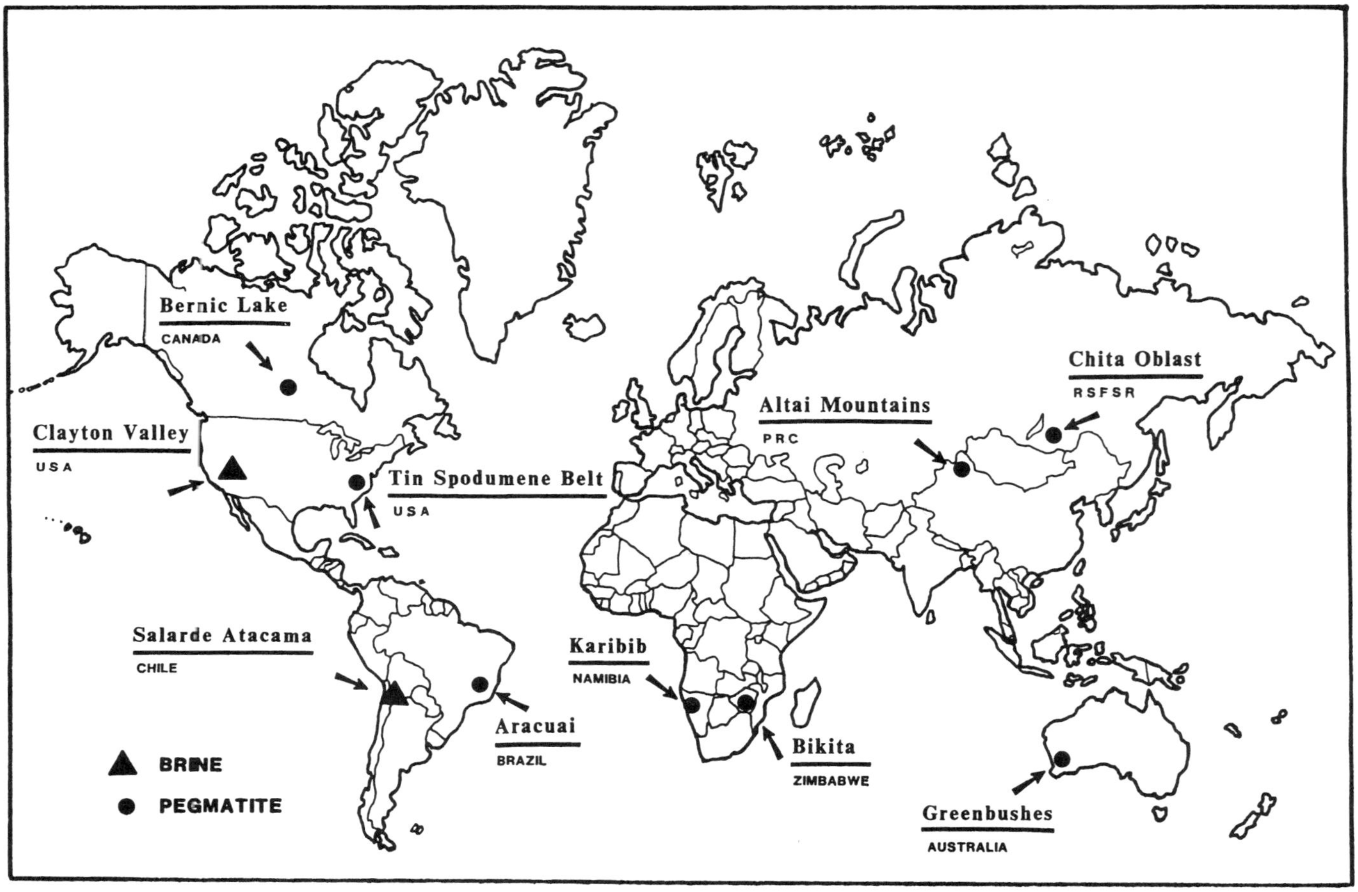

Fig. 1. Lithium production centers (1991).

one-half of the property, where surface outcrops of spodumene-bearing pegmatites exist, has not been evaluated by drilling. Geological inference permits an estimate of some additional 14 Mt of spodumene-bearing pegmatites for the unexplored portion of the property. However, as a result of the development of the low cost brine reserve at the Salar de Atacama in Chile, the Kings Mountain, NC, pegmatite operation is idle.

Lithium Corp. of America, a wholly-owned subsidiary of FMC, has developed an open pit in the Long Creek area in western Gaston County, North Carolina. Proven and probable reserves on the Hallman-Beam property are reported at 26 Mt of spodumene-bearing pegmatite grading 0.64% Li.

The Tin-Spodumene Belt is about 50 km long, and the exploration activity has been restricted to 300 m depth. The geological setting and distribution of pegmatites suggest that pegmatites may well occur at greater depth. Kesler, in a study of the potential lithium contained in the total belt, has suggested the potential existence of a resource of 754 Mt at a grade of 0.7% Li typical for the area (Kesler, 1978).

Other known pegmatite sources in the United States are not of great significance. These include the Black Hills of South Dakota, the Pala District of southwestern California, the White Picacho District of Arizona, the Harding and Pidlite mines in New Mexico, and several smaller areas in Colorado, Wyoming, Utah, and New England.

In addition to pegmatitic sources, significant quantities of lithium are produced by Cyprus Foote Mineral Company from a brine deposit located in Clayton Valley (Silver Peak), Esmerelda County, Nevada. Lithium was also produced for a number of years from the brines of Searles Lake, where brines are presently beneficiated only for potash, sodium carbonate, and borate salts by the Kerr-McGee Corporation.

In Clayton Valley, lithium-bearing brines occur in an undrained structural depression filled with Quaternary sediments composed mainly of clay minerals, including hectorite, volcanic sands, alluvial gravels, and saline minerals consisting of gypsum and halite (Kunasz, 1970). The brine that saturates the sediments is chemically simple. It is a concentrated sodium chloride solution containing subordinate amounts of potassium and minor amounts of magnesium and calcium. The lithium concentration is variable and shows a decrease in concentration as a function of pumping; the lithium concentration in the brine varies from 100 to 300 ppm Li.

The lithium resources, to a depth of 300 m, are estimated at 115 kt Li. Recent exploration has identified new lithium-bearing aquifers, and has resulted in an increase in average lithium concentration pumped to the pond system as well as an increase in the recoverable-lithium reserve category.

Chile

The first published reference on the occurrence of lithium in Chile was made in 1969 by the Instituto de Investigaciones Geologicas (IIG) when it undertook an extensive survey of the Salar de

Atacama. The Institute published its findings in 1974 (Moraga et al., 1974). The Salar came into particular focus in 1970 when the *Christian Science Monitor* (June 22, 1970) and the *Mining Journal of London* (June 26, 1970) reported a lithium content of 0.2% and resources of 1.2 Mt. Subsequent studies by CORFO showed that the salt nucleus contains a resource of 4.3 Mt of lithium (Penner, 1978).

In April 1974, Foote Mineral Company (Cyprus Foote Mineral Company) verified the high lithium concentrations in the shallow brines below the saline crust. In January 1975, an agreement was signed with CORFO (Corporacion de Fomento—Chilean State Development Agency) to initiate a four-year feasibility study in order to assess the potential of producing lithium carbonate from the brine. Following the successful completion of the feasibility study in 1979, a joint venture between CORFO (45%) and Foote Mineral Company (55%)—the Sociedad Chilena de Litio, Ltd. (SCL)—was created. SCL started commercial production of lithium carbonate in 1984. In 1988, Cyprus Minerals Company acquired Foote Mineral Company, which survived as Cyprus Foote Mineral Company. In 1989, Cyprus Foote Mineral Company acquired full control of SCL.

The Salar de Atacama is located in the east-central portion of the Second (Antofagasta) Region. It straddles the Tropic of Capricorn, extending from latitude S23° to S24°35' and from longitude W68°05' to W68°30'. The floor of the salar lies at an elevation of 2 300 m. It is bounded on the east side by the Andean Cordillera and on the west side by the Domeyko Cordillera.

The Salar de Atacama Basin covers approximately 3 000 sq km, and the salt nucleus proper about 1 100 sq km. It is a tectonic basin filled with clastic sedimentary material and evaporites. The nucleus consists predominantly of halite, with some narrow sulfate and carbonate facies at the eastern and southern margins. The deltaic sediments of the Rio San Pedro bound the salt nucleus to the north. The surface of the salar is inaccessible due to extensive polygonal cracking. Preliminary drilling performed by CORFO over various parts of the basin indicated a minimum thickness of 360 m of halite near the center of the basin, diminishing to about 40 m near the southern margin. However, recent drilling by the Hunt Oil Company indicates that the salt thickness exceeds 1 000 m (oral communication, 1989). Salt cores show that only the near-surface portion of the halite crust has high porosity and permeability. A 10-km-long seismic survey revealed that the highest porosity extends to a depth of 20 to 25 m and that some additional lower porosity halite may exist at depths from 25 to 35 m. Below this depth, salt cores show complete recrystallization of the halite into a solid mass, devoid of any pores. The yield characteristics of the upper halite layer were tested by completing shallow wells to 30 m and 60 m. The wells, pumped for a period of 3 months at a pumping rate of 64 liters/second, stabilized at drawdowns of 20 cm and 7.9 m respectively, corroborating that only the upper 30 m have a high transmissivity.

The origin of the lithium at the Salar de Atacama must be viewed from several perspectives. The leaching of the volcanic rocks was the principal source of lithium, which was discharged in the salar by stream and ground water flow. A second source of lithium was the weathering and leaching of the lacustrine sediments predating the formation of the salar. High-lithium sediments (to 490 ppm Li) have been identified in a numbers of formations (Ide and Kunasz, 1989). Geothermal fluids were a third source. Solutions from the El Tatio geothermal field, located to the north of the Salar de Atacama, contain 28 to 40 ppm Li.

As in most others salares and playas, the high concentration of elements such as lithium and potassium is caused primarily by evaporation. Three basic requirements are needed for the development of such systems:

1. They occur near volcanic systems, Tertiary or younger in age.

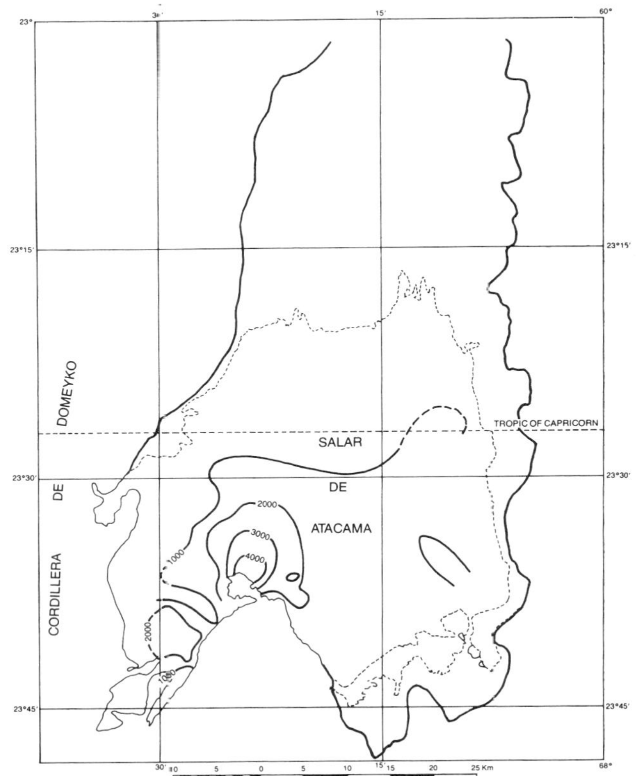

Fig. 2. Map showing lithium concentration (ppm) isopleths in the southern part of the Salar de Atacama (Ide and Kunasz, 1989).

2. Geothermal and other fluids must drain into structurally enclosed basins from which they cannot escape.
3. They are found only in desert areas of the world, where solar evaporation exceeds rainfall.

Although these conditions have been met by a number of basins, each basin exhibits a particular geochemical signature reflecting specific structural, climatic, and hydrological conditions. At the Salar de Atacama, conditions were such that it resulted in the formation of the richest lithium concentrations known to date in the world. Whereas the average lithium for the salar is 1 350 ppm Li, lithium concentrations in certain areas reach 5 000 ppm, and values as high as 7 000 ppm have been recorded in the southern area presently exploited by the Sociedad Chilena de Litio, Ltd. (Fig. 2).

Australia

A significant source of spodumene was identified by Greenbushes in Western Australia during drilling to outline tin and tantalum reserves. The deposit is located about 200 km south of Perth. The north-south striking pegmatites, which intrude metasediments and meta-basic intrusives of the Balingup Metamorphic Belt, have been traditional sources of alluvial tin and tantalum. The pegmatites contain a high lithium-low iron spodumene that is quite suitable for the glass and ceramic industry. Greenbushes began the production of two types of spodumene concentrate: a high grade concentrate (3.49% Li and 0.10% Fe_2O_3) and a run of mine glass-grade concentrate (1.86% Li) (Hatcher and Elliott, 1988). The company has also considered the conversion of spodumene concentrates into lithium carbonate. Gwalia Consolidated, the parent company, is preparing a feasibility study on an expanded operation that contemplates the production of 150 kt of spodumene products. The reserves of the spodumene ore body, to a depth of 220 m, have been calculated at 40 Mt averaging 1.35% Li. Contained in these reserves is a 6 Mt block averaging 1.86% Li (Anon., 1991b).

In the Coolgardie District, small quantities of petalite and minor amounts of spodumene and amblygonite were produced at Londonderry.

Other lithium mineral occurrences include spodumene near Ravensthrope, lepidolite near Wodgina, and minor amblygonite near Euriowie, New South Wales (McLeod, 1965).

Zimbabwe

The largest lithium-bearing area in Zimbabwe is that of the Bikita tinfields, located about 75 km east of Fort Victoria. Exploration has indicated the presence of an important mineralized zone within the Al Hayat, Bikita, and Southern Sectors. The pegmatite is about 1 700 m long and its width varies from 30 to 70 m. It strikes north-northeast and dips from 14 to 45° east. The property is operated by Bikita Minerals (Pvt.) Ltd., controlled by B.P. Minerals.

The pegmatite is asymmetrically zoned and contains a variety of commercially important lithium minerals as well as beryl and pollucite.

Drilling and development work have proved a reserve of 12 Mt of quarry ore grading 1.4% Li. Mining is presently done in the Al Hayat sector. The plant produces about 2 200 tons per month of petalite, 800 tons per month of lepidolite and 800 tons per month of spodumene, In the reduction plant, feldspar is separated from petalite and ground for sale (Wegener, 1980).

Other minor lithium-bearing occurrences are known in the Wankie, Salisbury, Umtali, Mtoko, Insizia, Matobo, and Mazoe districts (Toombs, 1962).

Canada

In the Bernic Lake area of southeastern Manitoba, the lithium-bearing pegmatites are part of a zoned complex, previously exploited for tantalum by the Tantalum Mining Corp., subsequently acquired by the Hudson Bay Mining and Smelting Corp. The property is located in the Cat Lake-Winnipeg River District, which lies some 130 km northeast of Winnipeg. The pegmatite is complexly zoned and contains several lithium-bearing units. It is a lenticular body about 1 200 m long and 500 m wide, with a maximum thickness of 90 m (Cerny and Turnock, 1971). Two lithium-rich zones are characterized by the occurrence of spodumene, petalite, and amblygonite. The spodumene reserves have been measured at 7.4 Mt grading 1.34% Li. The pegmatite complex also contains three lepidolite zones, with lithium concentrations ranging from 0.87 to 1.31% Li (Crouse et al., 1979). Tanco is producing 12 kt of lithium ores annually, and some 160 tons of tantalum ore.

The pegmatite also contains 300 kt of pollucite averaging 23% Cs_2O, 1.1 Mt of tantalite with a grade 0.16% Ta_2O_5, and 1 Mt of beryl ore averaging 0.22% BeO.

South America

Only two countries produce small amounts of lithium minerals in South America: Brazil and Argentina. The amblygonite deposit in Surinam has been essentially depleted.

In Brazil, lithium-bearing pegmatites are known in the districts of Minas Gerais and in the northeastern part of the country, which includes the states of Paraiba, Rio Grande do Norte, and Ceará (Afghouni, 1978).

In the State of Minas Gerais, near Aracuai, several pegmatites have been exploited on a sporadic basis. The pegmatites, which carry spodumene, amblygonite, petalite, and lepidolite, have been traditionally mined for cassiterite, tantalite, and beryl, while lithium minerals have been sporadically recovered. As a result of an increase in demand for lithium minerals (petalite), exploration activities resulted in the discovery of important petalite pegmatites, reported to contain 100 kt of petalite grading 2% Li. Spodumene reserves have been estimated at 300 kt, whereas lepidolite reserves are considered to be nearly exhausted.

The most important producer of lithium minerals is Arqueana de Minerios e Metais (Sao Paulo). The company mines spodumene, petalite, lepidolite, amblygonite, beryl, and cassiterite from the pegmatite bodies located near Aracuai in Minas Gerais. The company also supplies spodumene concentrates to Companhia Brasileira do Litio (CBL), a small producer of lithium carbonate. The processing facilities have been constructed in an economically depressed region several hundred kilometers north of the mining district in order to benefit from government incentives. The plant is estimated to produce about 1 500 tons of lithium carbonate (oral communication, 1991).

In Argentina, lithium-bearing pegmatites occur in the western part of the state in the Sierras Pampaneas, which include the productive districts of San Luis, Cordoba, and Catamarca. The pegmatites are zoned and contain spodumene. The reserves, considered to be small, total about 18 kt as spodumene (Angelleli and Rinaldi, 1963, 1965). The occurrence of brine has been reported in several salars of northwestern Argentina (Nicolli et al., 1980), where exploration has been conducted by several foreign companies.

Namibia

Following a low production period, mining in the Karibib District of Namibia was stimulated by the trade sanctions imposed in 1964 on Southern Rhodesia (now Zimbabwe). The district, located approximately 190 km from Walvis Bay on the Atlantic Ocean, contains several strongly zoned pegmatites, which contain lepidolite, petalite, and small amounts of amblygonite. In 1968, Klockner & Co. A.G. acquired the Rubicon and Helicon mines previously operated by S.W.A. Lithium Mines (Pty.) Ltd. Production has been expanded by the installation of a lepidolite flotation plant. Lepidolite reserves suitable as flotation feed material have been estimated at 1 Mt. Petalite reserves consist of some 200 kt of crude ore (Anon., 1968).

Other African Deposits

Smaller productive areas occur in Mozambique, Uganda, and Ruanda. In the Alto Ligonha area of Mozambique, lepidolite, amblygonite, and spodumene occur in zoned pegmatite bodies. Only lepidolite is exploited commercially. Amblygonite is rare, and spodumene is completely altered to kaolinite (Hutchinson and Claus, 1956).

Large amblygonite masses are found in pegmatite districts located west of the capital city of Kigal, in Ruanda (Valarmoff, 1954), and north of Kampala in Uganda (Roberts, 1948).

Other occurrences of lithium minerals include the Sudan, the Malagasy Republic, the Ivory Coast, and the Republic of South Africa (Nel, 1968).

Others

In addition to the producing areas described here, lithium minerals are known to occur in a number of other countries. A new and significant discovery of spodumene pegmatites has been made in the Koralpe in southern Austria. Minerex, the Austrian State Exploration Company, conducted underground testing and pilot metallurgical testwork. Lithium occurrences are also known in Ireland, India, Korea, and Japan. Sizable deposits exist in the Russian Federation (Chita Oblast, Ural Mountains, Karelia). Although a number of deposits have been described (Ginzburg and Lugovskoi, 1977), actual reserve data are not available. In the People's Republic of China, lithium minerals and chemicals are produced from pegmatite deposits located some 600 km north of the city of Urumchi in the Altai Mountains.

POTENTIAL PRODUCING DISTRICTS

Zaire

Probably the largest hard rock lithium resources in the world are contained in the Manono and Kittolo pegmatites located in Zaire. At the present time, only cassiterite and columbite are mined by Congo-Etain from the Manono pegmatite, which is 5 km long and from 120 to 425 m wide. The adjoining Kittolo pegmatite has similar dimensions. Although the pegmatites are apparently zoned, their dimensions imply spodumene reserves that dwarf the presently known world reserves. However, the deposit may not have an economic value for years because of the very poor transportation facilities. The deposit is located about 2 200 km from the Angolan port of Lobito.

Canada

In the Preissac-Lacorne district of western Quebec, spodumene-bearing pegmatites intrude Precambrian metamorphic rocks. In addition to spodumene, they contain a variety of other minerals such as lepidolite, beryl, pollucite, molybdenite, and minor amounts of columbite and tantalite. The largest reserves occur on the property of the Quebec Lithium Corp., where drilling has indicated 15 Mt of ore grading 0.56% Li (Mulligan, 1965). Although the company produced spodumene concentrates for a few years, the operation is presently inactive. Several other companies control spodumene-bearing properties in the district.

Lithium-bearing pegmatites are known in the Herb Lake and East Braintree-West Hawk districts of Manitoba (Williams and Trueman, 1978); the Yellowknife-Beaulieu districts of the Northwest Territories (Lasmanis, 1978); and the Nipigon, Dryden, Lac La Croix, Root Lake, Falcon Lake, and O'Sullivan Lake districts of Ontario (Mulligan, 1965). These are believed to contain significant additional lithium reserves.

Chile

In addition to the southern part exploited by the Sociedad Chilena de Litio, Ltd., the northern part of the Salar de Atacama constitutes a potential future source of lithium.

The interstitial brine contains an average lithium concentration of 1 500 ppm. Lithium resources are reported at 4.3 Mt Li. In 1988, Amax Corporation signed a contract with CORFO to initiate a three year study to evaluate the feasibility of producing potassium chloride (400 kt), potassium sulfate (150 kt), lithium carbonate (15 kt), and boric acid (20 kt). The capital costs for this project were estimated at US$400 million. However, in late 1991, Amax announced that its interest in the Minsal project was for sale. The partners of Amax in the Minsal project are CORFO (25%), the Chilean government development agency, and Molymet (11.25%), a Chilean molybdenum roaster.

FUTURE RAW MATERIALS

Several areas of the world carry potential lithium raw materials in the form of brines, geothermal brines, oil well brines, and clays.

Among the existing brine deposits, the Great Salt Lake of Utah constitutes an important potential source of lithium chemicals. The lake holds a brine formed as a result of concentration by solar evaporation of waters contained in a much larger and more dilute Pleistocene lake—Lake Bonneville. In the Bear River Basin, the Great Salt Lake Minerals and Chemicals Corp. has constructed 5.7 km^2 of ponds and is producing sodium chloride and potassium and sodium sulfates. The recoverable lithium resources are estimated at 286 kt (Evans, 1978).

A potential geothermal area lies within the Imperial Valley of southern California. The chemically complex brine contains large concentrations of sodium, calcium, and potassium. It also carries about 0.020% Li. Koenig (1970) indicated that the volume on the brine under an area of 31 to 62 km^2 is in excess of 4 km^3. This suggests a potential reserve exceeding 840 kt Li. However, a subsequent analysis indicated that potential exploitation of lithium will depend upon whether or not some complex chemical engineering problems are solved.

Other geothermal areas include the Reykanes Field of Iceland, where the visible active sector covers an area of about 5 km^2. Feasibility studies (Ludviksson and Hermannsson, 1970) indicate that a yearly production of 500 tons of lithium compounds is possible. The brine contains 8 ppm Li. Additional thermal areas are known at Reykir, Hveragardi, and Drysuvik (Karlsson, 1961, Prast, 1972).

A similar potential source of lithium is the geothermal waters at Wairakei, New Zealand. Potential production from bore waters containing 13 ppm Li are estimated at 2 400 tpy as Li_2CO_3, assuming a discharge rate of 3 785 000 L per hr (Kennedy, 1964).

Lithium-bearing clays are known to occur in several localities in the United States. The highest lithium value of 0.53% has been recorded at Hector, CA, where the Baroid Division of National Lead Co. mines hectorite for its swelling characteristics (Ames et al., 1958). Hectorite has been identified in Clayton Valley, where it occurs as an alteration product of volcanic ash along a fault zone and is one of the clay minerals in Tertiary and Quaternary lake sediments (Kunasz 1970). The hectorite contains 0.24 to 0.35% Li. At Spor Mountain, UT, 0.11% Li has been reported in beryllium-bearing tuffs (Shawe et al., 1964, Shawe, 1968), where lithium probably also occurs in hectorite. An important discovery of lithium-bearing clays has been made in the McDermitt caldera in northern Nevada. The occurrence of lithium-bearing clays is also reported (Norton, 1965) in Yavapai County, AZ (0.10% Li), in Tertiary clays at Kramer, CA (0.19% Li) (Muessig, 1966), and in the Turilari playa (0.14% Li) in Argentina (Muessig, 1966). An extensive survey by the Lithium Exploration Group of the US Geological Survey has indicated that vast resources of lithium lie in a great number of sedimentary clays (Vine, 1976). Extensive research on the extraction of lithium from clay has been conducted by the US Bureau of Mines.

Although the clays contain lithium concentrations of the same order of magnitude as commercial pegmatites, they are not likely to become lithium sources in the near future because of technological problems associated with the extraction of lithium from the crystal lattice, and also because they cannot compete against the low cost brine operations in the United States and Chile.

In the 1970s, following the discovery of the high lithium brines at the Salar de Atacama, much exploration activity was directed at identifying other such potential deposits. In South America, the Chilean CORFO organization identified a number of salares with substantial lithium concentrations. In Argentina, the efforts of Nicolli at the Commision Nacional de Investigaciones Espaciales

(CNIE) and Poppi at the University of Salta identified anomalous concentrations of lithium in a number of salares in the Argentinean Altiplano (Nicolli et al., 1980, Poppi, 1981). The Salar del Hombre Muerto has received considerable attention by the Lithium Division of FMC (Anon., 1991a). Preliminary exploration indicates lithium concentrations varying from 100 to 700 ppm. The Salar is located on the Argentinean Altiplano at an elevation exceeding 4 000 m and is quite isolated. FMC announced that the deposit may yield between 7 to 15 kt of lithium carbonate per year, requiring an investment ranging from 40 to $55 million. The company expects that it will take five to eight years to bring the deposit into production (Anon., 1991a).

In Bolivia, the Salar de Uyuni became the focus of a joint USGS-GEOBOL program, followed by extensive additional studies by companies such as the Lithium Division of FMC. The Salar de Uyuni is reported to contain resources estimated at 5.5 Mt of lithium, 110 Mt of potassium, and 3 Mt of boron under an area of 18 000 km^2 (Ballivian and Risacher, 1981). FMC has announced plans to produce 18 kt of lithium carbonate per year.

In the People's Republic of China and Tibet, lithium has been identified in several salt flats and lakes. In the Quinghai Basin, in north-central China, the first phase of a potash plant has been completed near the town of Golmud. Although potassium salts constitute the primary products, lithium salt production is also contemplated.

In Tibet, on the Qinghai-Xizang Plateau, saline lakes contain highly concentrated alkaline chloride-sulfate brines (Zheng, 1989). Lake Zabuye, one of 352 salt lakes, has been intensively studied during the last decade including, in 1988, a study by a Sino-American expedition sponsored by the US National Academy of Sciences. Lithium concentrations in the saline lake waters ranges from 700 to 1 000 ppm (Holland et al., 1991).

BENEFICIATION

Most of the world's lithium is manufactured by two companies: Cyprus Foote Mineral Co., at Silver Peak, NV, Kings Mountain, NC, Sunbright, VA, New Johnsonville, TN, and Exton, PA, and the Sociedad Chilena de Litio, Ltd.—a Cyprus Mineral Company affiliate—in Northern Chile; and the Lithium Division of FMC at Bessemer City, NC, and in the United Kingdom. In Europe, an important chemical producer is Chemetall, a wholly owned subsidiary of Metallgesellschaft AG, in Germany. In addition, lithium chemicals are produced in Japan by Honjo Chemical and several other companies. Brazil is reported to have started minor production of lithium carbonate. Commercial quantities of lithium ores are produced in the United States, Zimbabwe, Australia, Canada, Brazil, and Namibia. The USSR produces sizable quantities of chemicals. Lithium chemicals are also supplied by the People's Republic of China. Production statistics for these two latter countries are, however, not available.

Mining of lithium ores can be divided into three categories: underground, open pit, and the mining of brines.

Where practical, such as in the Tin-Spodumene Belt, United States, at Greenbushes, Australia, and Bikita, Zimbabwe, the open pit is the most economical method. Higher cost underground methods are used in Canada, Brazil, and Namibia.

Although hand cobbing and screening are still used in Brazil and Africa, flotation is the main beneficiation method of lithium ores. In the production of spodumene concentrates, the mined ore undergoes several stages of crushing, grinding, and classification, followed by flotation with organic reagents. At the FMC operation in North Carolina, the spodumene concentrate collected from the flotation cells is further processed into lithium chemicals. A low iron spodumene variety was produced by the chlorination of B-spodumene (Fishwick, 1967, Richardson and Fishwick, 1968) and used as a substitute for petalite in glass-ceramic applications. Today, a low iron spodumene concentrate from Greenbushes, Australia, competes with both petalite and lithium carbonate in ceramic and glass applications.

In addition to spodumene concentrates, the beneficiation of pegmatite ore includes several byproducts such as quartz, feldspar, and mica. Other potential byproducts are beryl, columbite, tantalite, cassiterite, and wolframite.

The extraction of lithium from spodumene can be achieved by several methods. Presently, the process used by Cyprus Minerals Company and FMC involves calcination to B-spodumene, which is then reacted with sulfuric acid to produce lithium sulfate. Conversion to lithium carbonate is accomplished by soda ash addition (Ellestad and Leute, 1950). Other processes have included the calcination of a spodumene-calcium carbonate mixture, hot water leaching, and evaporation-crystallization to produce lithium hydroxide as a final product. Another process, used in the past by the Quebec Lithium Corp., involved the calcination of spodumene and its reaction with an aqueous solution of sodium carbonate. Purification involved reaction with carbon dioxide, and then heating to produce a final lithium carbonate product (Archambault, 1963).

Lithium is produced from brines at two locations in the world—Clayton Valley, Nevada, and the Salar de Atacama, Chile.

In Nevada, at the Silver Peak operation, Cyprus Foote Mineral Co. ''mines'' by pumping saline solutions contained in sediment and salt aquifers into large solar evaporation ponds (Barrett and O'Neill, Jr., 1970). Cyprus Foote Mineral Company presently operates between 40 to 60 wells ranging from 150 to 300 m in depth. The company has been recovering lithium since 1965. The brine is pumped through PVC pipelines into 1 600 hectares of clay-lined evaporating ponds, where it is concentrated by solar evaporation. Following a 40-fold concentration, the final lithium chloride solution is converted to lithium carbonate by addition of soda ash. Small impurities are removed within the pond system by the addition of lime (Dillard and McLean, 1991).

At the Salar de Atacama, Chile, operation of the Sociedad Chilena de Litio, Ltd. brine is pumped from shallow production wells into a series of solar evaporation ponds. The brine has an initial strength of 2 500 ppm Li and undergoes a 25-fold concentration. An extensive evaluation of pond designs at the salar resulted in the construction of a 130-hectare pond system lined with polyvinyl-chloride (PVC).

In the first series of ponds, halite and sylvinite precipitate successively, followed by the precipitation of carnallite, and then bischofite. In the last ponds, the lithium solution is concentrated to its final design strength of 6% Li. The precipitated salts are harvested annually and stored in waste piles near the ponds. Recently, SCL has begun the recovery of potash from the sylvinite stockpiles.

The concentrated brine is delivered by tank trucks to the Salta-Antofagasta railhead, located some 80 km south of the salar operations, from where it is delivered to the chemical processing plant, located near the port city of Antofagasta (Fig. 3). SCL has expanded its capacity to 11.8 kt of lithium carbonate, and considers the installation of a lithium chloride plant.

USES

Lithium is marketed and used in three basic forms: as an ore and concentrate, as a metal, and as chemical compounds. As shown in Table 2, there are numerous applications.

Ores and concentrates are consumed by the glass, ceramic, and porcelain enamel industries. Petalite, lepidolite, and amblygonite

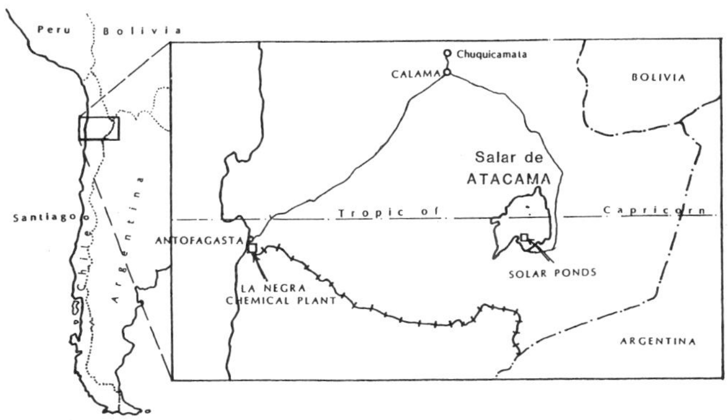

Fig. 3. Operational setup of SCL at the Salar de Atacama.

can be used without prior beneficiation, except handcobbing, whereas spodumene must be concentrated and treated for removal of iron. Lithium is useful because it creates favorable internal nucleation conditions. By comparison, sodium favors external crystallization. Because of its small ionic radius and high field strength, lithium imparts high mechanical strength and thermal shock resistance, as well as good chemical resistance, to the product (Fishwick, 1966). In these applications, lithium is also introduced in the form of lithium carbonate.

In metal form, lithium is the lightest solid element, having an atomic weight of 6.94 and a specific gravity of 0.534 (20°C). It is manufactured by the electrolysis of a molten salt mixture of lithium and potassium chlorides. Lithium metal is used in the synthesis of butyllithium. In nonferrous metallurgy, the high reactivity of lithium with gases is used for scavenging oxygen and sulfur, converting them into stable compounds. Lithium is also used in lithium-aluminum and lithium-magnesium alloys, where it imparts high temperature strength, improves elasticity, and increases the tensile strength (Bach et al., 1967). Demand for aluminum-lithium alloys in commercial and military aircraft industries, while encouraging, has not met earlier growth forecasts (Alexander, 1992).

In addition to being the lightest metal, lithium is also the most electronegative metal and, therefore, is ideal for use in many battery applications. Lithium battery characteristics include high energy density, high operating voltage, wide operating temperature range, and long shelf life (Grady, 1980). Present applications include heart pacemakers, military hardware, cameras, computer memory backup, watches, and measuring equipment in the oil drilling industry. Future growth in this area may include use in rechargeable lithium batteries for hand held power tools and electronic and communications equipment, and even as a power supply for electric vehicles and the total artificial heart.

Among the various chemicals produced, the most important are: lithium carbonate, lithium hydroxide, lithium chloride, lithium bromide, and butyllithium. The remaining compounds constitute only a small proportion of the lithium market.

Lithium carbonate, in addition to its consumption by the ceramic industry, is being used at an increasing rate in the aluminum reduction cells, where it improves the conductivity of the molten bath, reduces operating temperature, and results in higher production. It is also very effective in reducing fluorine emissions by retaining the fluorine as lithium fluoride in the bath. In purified form, lithium carbonate is being used in the chemotherapeutic treatment of manic depression. In the United States, Japan, and Europe, development programs have been initiated on molten carbonate fuel cells. Due to the energy efficiency and environmental cleanliness of all types of fuel cells, worldwide financial support for the development of this technology has increased. Molten carbonate fuel cells are expected to utilize various lithium compounds, principally lithium aluminate and lithium carbonate (Alexander, 1992).

Lithium hydroxide was first employed as an ingredient in alkaline storage batteries where its presence increased the life of the cells (Bach et al., 1981). In the 1940s, a new application was discovered by Clarence Earl, who developed a special grease by reacting lithium hydroxide with fatty acids. The resulting grease was found to retain its viscosity over a wide temperature range and remain stable in the presence of water. Today, lithium-based greases have become a standard product in military, industrial, and automotive lubrication fields. Anhydrous lithium hydroxide is capable of absorbing large quantities of carbon dioxide, a property that has been utilized in the air-regeneration system of the Apollo command

Table 2. Applications of Lithium

Ores				
Spodumene	Petalite	Lepidolite	Amblygonite	Metal
Ceramics Glass Ceramics	Ceramics Glasses Glazes Enamels	Glasses	Li-Chemicals	Deoxidation Degasification Alloys Organic compounds Synthetic rubber Vitamins Nuclear energy Battery

Chemicals						
Lithium Carbonate	Lithium Hydroxide	Lithium Chloride	Lithium Fluoride	Lithium Bromide	Butyl Lithium	Other Lithium Comp.
Ceramics Li-Chemicals Fluxing Pharmaceuticals Aluminum electrolysis cells Fuel cells	Greases Li-Salts CO_2 absorption	Li-Metal Fluxing Dehumidification Synthetic fibers Tracer	Porcelains Enamels Glazes Welding Brazing Optics	Absorption Refrigeration	Polymerization Pharmaceuticals Organic synthesis	Electrolytes Catalysis

and lunar modules, as well as in new applications in the mining industry for rescue-breathing apparatus.

Lithium chloride and lithium bromide brines have low vapor pressures and are used in absorption refrigeration systems. Lithium chloride also constitutes the feed material for the production of lithium metal. Lithium fluoride is used mainly as a flux in enamels, glasses, glazes, and in welding and brazing.

The most important organic compound is butyllithium. It serves as a catalyst in the polymerization of butadiene, isoprene, and styrene to produce polymers with special properties. Butyllithium has applications in the synthesis of pharmaceutical and agricultural intermediates.

Numerous miscellaneous applications of lithium chemicals include sanitation and bleaching, hydrogen generation, oxygen generation, catalysis, and vitamin synthesis.

PRICE

The 1992 prices of various ores (in dollars per ton, FOB Amsterdam) are as follows:

Ceramic Spodumene—7.25% Li_2O : $385
Glass Grade Spodumene—5% Li_2O : $175
Petalite—3.5 to 4.5% Li_2O : $230

Prices for lithium chemicals in 1992 are shown in Table 3 (*Chemical Marketing*, Mar. 23, 1992).

Table 3. Typical Prices of Lithium Compounds in 1992

	$ per kg
Lithium carbonate, powder	4.20
Lithium hydroxide, monohydrate	5.39
Lithium chloride, anhydrous	10.01
Lithium fluoride	14.83
Lithium metal, (454-kg) lots	69.96
n-Butyllithium, 15% solution	41.47
Lithium bromide, anhydrous	15.29
Lithium bromide, brine	11.68
Lithium chloride, brine	8.43

FUTURE TRENDS

Lithium is still a developing commodity. Its applications in various phases of the metallurgical and chemical industries have been rapidly diversifying and expanding in the past few years. In 1990, the estimated world consumption of lithium metal and chemicals was reported at 29.5 kt of lithium carbonate equivalents, while the installed production capacity of the western world is 40.4 kt of lithium carbonate equivalents (Alexander, 1991) In addition, it is also estimated that the USSR consumes probably about 4.5 kt of lithium carbonate equivalents. At the present time, the industry has sufficient production capacity to supply the existing market, and the announced increases for new capacity appear to be sufficient to meet future growth.

A potential application for lithium is in nuclear power as an absorption blanket in fusion generation. In thermonuclear reactions, the most likely fuel will be a deuterium-tritium mixture because of its very large reaction rate (Bogart, 1980). Although tritium is naturally scarce, it can be artificially generated within the reactor itself by bombarding the lithium-6 isotope in the blanket with the neutrons created by the fusion reaction.

A promising application for lithium metal is in batteries. In the past decade, large efforts have been made to find a battery with a high energy-weight ratio. Because of its low weight and high electrochemical potential, lithium is a prime candidate. Primary lithium batteries have moved from the development laboratory into pilot plant production. Commercial utilization could require many tons of lithium metal.

In the 1970s, projections made by several agencies concluded that the known lithium reserves were not sufficient to satisfy the large demand potential required by the introduction of massive lithium battery application in the automotive industry and thermonuclear fusion programs (Penner, 1978). This concern gave rise to the formation of the Lithium Exploration Group of the US Geological Survey, which undertook an ambitious exploration program for lithium in the western United States and participated in overseas programs as well. Their work resulted in a comprehensive inventory of an element that had not been evaluated in detail. Concurrently, the National Academy of Sciences commisioned a study on the availability of lithium resources in the world. In 1976, the Lithium Subpanel completed its study, in which the resources of the western world were estimated at 10.7 Mt of Li (Evans, 1986). An analysis of the lithium demand against the known lithium resources indicates that, while the lithium demand, since 1963 when the lithium industry completed its rationalization, has grown at an annual rate of 3 to 5%, lithium reserves have increased by more than ten times. Market projections, even those including battery and fusion scenarios, show that known reserves and resources are sufficient to support such requirements beyond 2050. With the discovery of additional reserves in Australia and with new estimates of resources in Canada, at the Salar de Uyuni in Bolivia, in the geothermal brines in Imperial Valley, and in the lithium-bearing clays identified in the western United States, known resources have more than trebled to 36.7 Mt (Evans, 1986).

Presently known reserves exceed the cumulative conventional demands by about two orders of magnitude. Thus the amount of exploration has been justifiably limited. Should increasing demands be imposed on the lithium industry, vast resources can be tapped within a short period of time (Kunasz, 1980).

BIBLIOGRAPHY

Anon., 1968, "Klockner Buys SWA Lithium," *Metal Bulletin*, No. 5288, Apr. 5, p. 26.

Anon., 1976, "Lithium by the Year 2000," *Industrial Minerals*, No. 108, Sep., pp. 45–46.

Anon., 1977, "Lithium Chemical Expansion Insures Supply," *Industrial Minerals*, No. 122, Nov., pp. 67–73

Anon., 1991a, "FMC Unit Ups Lithium List Prices," *American Metal Market*, Oct. 24, p. 12.

Anon., 1991b, "Greenbushes Planning for the Future," *Mining Journal of London*, Vol. 316, No. 8123, p. 393.

Afghouni, K., 1978, "Lithium Ores in Brazil," *Energy*, S.S. Penner, ed., Vol. 3, No. 3, pp. 247–253.

Alexander, J.H., 1981, Lithium Annual Review, *Engineering and Mining Journal*, pp. 111–113.

Alexander, J., 1992, "Lithium," *Mining Engineering*, Vol. 44, No. 3, p. 564.

Ames, L.L., et al., 1958, "A Contribution on the Hector, California, Bentonite Deposit," *Economic Geology*, Vol. 53, No. 1, Jan.-Feb., pp. 22–37.

Angelleli, V., and Rinaldi, C.A., 1963, "Yacimientos de Minerales de Litio de las Provincias de San Luis y Cordoba," Info. No. 91, Argentina, Com. Nac. Energia Atomica, 79 pp.

Angelleli, V., and Rinaldi, C.A., 1965, "Resena Acerca de la Estructura, Mineralizacion, y Aprovechamento de Nuestras Pegmatita Portadoras de Minerales de Litio," *Acta Geologica Lilloana*, Vol. 5, pp. 1–18.

Archambault, M., 1963, US Patent 3,112,171, Nov. 26.

Bach, R.O., et al., 1981, "Lithium and Lithium Compounds," *Encyclopedia of Chemical Technology*, A. Standen, ed., Vol. 12, 2nd ed., Interscience, New York, pp. 448–476.

Ballivian, O., and Risacher, F., 1981, "Les Bassins a Evaporites de l'Altiplano Bolivien (Les Salars)," *ORSTOM*, Paris, pp. 109–135.

Barrett, W.T., and O'Neill, B.J., Jr., 1970, "Recovery of Lithium from Saline Brines Using Solar Evaporation," *3rd Salt Symposium*, J.L. Rau and L.F. Dellwig, eds., Vol. 2, Northern Ohio Geological Society, Inc., Cleveland, pp. 47–50.

Bogart, L.S., 1980, "Potential Lithium Requirement for Fusion Electric Power," *Foote Prints*, Vol. 43, No. 1, pp. 2–11.

Buckley, S., 1983, "Lithium—Recession Delays Market Lift Off," *Industrial Minerals*, No. 185, Feb., pp. 25–35.

Caldwell, H.C., et al., 1971, "A Pharmaco-Kinetic Analysis of Lithium Carbonate Absorption from Several Formulations in Man," *Journal of Clinical Pharmacology*, Vol. 11, No. 5, pp. 349–356.

Cameron, E.N., et al., 1949, *Internal Structure of Granitic Pegmatites*, Monograph 2, Economic Geology Publishing Co., Urbana, IL, 115 pp.

Cameron, E.N., et al., 1954, "Pegmatite Investigation, 1942–1945, New England," Professional Paper 255, US Geological Survey, 352 pp.

Cerny, P., 1982, "Short Course in Granitic Pegmatites in Science and Industry," P. Cerny, ed., Mining Association of Canada, Friesen Printers, 555 pp.

Cerny, P., 1991, "Rare Elements in Granitic Pegmatites, Part I; Anatomy and Internal Evolution of Pegmatite Deposits," *Geoscience Canada*, Vol. 18, No. 2, pp. 49–67.

Cerny, P., and Turnock, A.C., 1971, "Pegmatites of Southeastern Manitoba," Special Paper No. 9, Geological Association of Canada, pp. 119–127.

Crouse, R.A., et al., 1979, "The Tanco Pegmatite, Southeastern Manitoba," *CIM Bulletin*, pp. 142–150.

Deer, W.A., et al., 1962, "Lepidolite," *Rock-Forming Minerals*, Vol. 3, Wiley and Sons, New York, pp. 85–91.

Dillard, G., and McLean, S., 1991, "Cyprus Foote Minerals Taps a Unique Treasure," *Pay Dirt*, Sep., pp. 4A-7A.

Eardley, A.J., 1970, "Salt Economy of the Great Salt Lake," *3rd Symposium on Salt*, J.L. Rau and L.F. Dellwig, eds., Vol. 1, Northern Ohio Geological Society, Inc., Cleveland, pp. 78–105.

Ellestad, R.B., and Leute, K.M., 1950, US Patent 2,516,109, July 25.

Evans, R.K., 1978, "Lithium Reserves and Resources," *Energy*, Vol. 3, No. 3, S.S. Penner, ed., Pergamon Press, New York, pp. 379–385.

Evans, R.K., 1986, "Reservas y Recursos de Litio en el Mundo Occidental," *El Litio, Un Nuevo Recurso Para Chile*, G. Lagos, ed., Editorial U. de Chile, Santiago, Chile, pp. 45–52.

Fishwick, J.H., 1966, "Glass Ceramics," *Foote Prints*, Vol. 35, No. 1, pp. 20–23.

Fishwick, J.H., 1967, US Patent 3,394,988, Oct. 31.

Foster, M.D., 1960, "Interpretation of the Composition of Lithium Micas," Professional Paper 354-B, US Geological Survey, 146 pp.

Ginzburg, A.I., and Lugovskoi, G.P., 1977, "Deposits of Lithium," *Ore Deposits of the USSR*, V.I. Smirnov, ed., Pittman Publishing, London, pp. 295–311.

Glanzman, R.K., McCarthy, H.J., and Rytuba, J.J., 1978, "Lithium in the McDermitt Caldera, Nevada and Oregon," *Energy*, Vol. 3, S.S. Penner, ed., Pergamon Press, New York, pp. 347–353.

Goldschmidt, V.M., 1937, "The Principles of Distribution of Chemical Elements in Minerals and Rocks," *Journal of the Chemical Society*, pp. 655–673.

Grady, H.R., 1980, "Primary Lithium Batteries—A Status Report," *Foote Prints*, Vol. 43, No. 1, pp. 19–22.

Hanley, J.B., et al., 1950, "Pegmatite Investigations in Colorado, Wyoming, and Utah, 1942–1944," Professional Paper 227, US Geological Survey, 125 pp.

Hatcher, M.I., and Elliott, A., 1988, "Greenbushes, a New Source of Lithium," *Industrial Minerals International Congress*, G.M. Clarke, ed., Metal Bulletin PLC, London, pp. 217–232.

Hegelson, H.C., 1968, "Geological and Thermodynamic Characteristics of the Salton Sea Geothermal System," *American Journal of Science*, Vol. 266, Mar., pp. 129–166.

Heinrich, E.W., 1964, "The Tin-Tantalum-Lithium Pegmatites of Sao Joao Del Rei District, Minas Gerais, Brazil," *Economic Geology*, Vol. 59, No. 6, Sep.-Oct., pp. 982–1002.

Holdren, J.P., 1971, "Adequacy of Lithium Supplies as Fusion Energy Source," Hearing Before the Joint Committee on Atomic Energy, Congress of the United States, Nov. 10 and 11.

Holland, H.D., et al., 1991, "Lake Zabuye and the Climatic History of the Tibetan Plateau," *Die Geowissenschaften*, Vol. 9, No. 2, pp. 37–44.

Hortsman, E.L, 1957, "The Distribution of Lithium, Rubidium and Cesium in Igneous and Sedimentary Rocks," *Geochimica et Cosmochimica Acta*, Vol. 12, pp. 1–28.

Hutchinson, R.W., and Claus, R.J., 1956, "Pegmatite Deposits, Alto Ligonha, Portuguese East Africa," *Economic Geology*, Vol. 51, pp. 757–779.

Ide F.Y., and Kunasz, I.A., 1989, "Origin of Lithium in Sales de Atacama, Northern Chile," *Geology of the Andes and its Relation to Hydrocarbon and Mineral Resources*, Houston, Texas, Circum-Pacific Council for Energy and Mineral Resources Earth Science Series, Vol. 11, pp. 165–172.

Jahns, R.H., 1952, "Pegmatite Deposits of the White Picacho District, Maricopa and Yavapai Counties, Arizona," *Bulletin*, Vol. 23, No. 5, University of Arizona, 105 pp.

Jahns, R.H., 1955, "The Study of Pegmatites," *Economic Geology*, 50th Anniversary Volume, Part II, pp. 1025–1130.

Karlsson, T., 1961, "Drilling for Natural Steam and Hot Water in Iceland," *Proceedings*, the United Nations Conference on New Sources of Energy, Rome, Vol. 3, pp. 215–221.

Kennedy, A.M., 1964, "Recovery of Lithium and Other Minerals from Geothermal Water at Warakei," *Proceedings*, the United Nations Conference on New Sources of Energy, Rome, Vol. 3, pp. 502–511.

Kesler, T.L., 1961, "Exploration of the Kings Mountain Pegmatites," *Mining Engineering*, Vol. 13, No. 9, pp. 1062–1068.

Koenig, J.B., 1970, "Geological Setting of the Imperial Valley and its Geothermal Resources," Compendium of Papers Presented at the Imperial Valley—Salton Sea Area Geothermal Hearing, Oct. 22–23, Sacramento, CA, pp. E1-E5.

Kunasz, I.A., 1970, "Geology and Geochemistry of the Lithium Deposit in Clayton Valley, Esmeralda County, Nevada," PhD thesis, The Pennsylvania State University, University Park, PA, 114 pp.

Kunasz, I.A., 1975, "Lithium Raw Materials," *Industrial Minerals and Rocks*, 4th ed., S.J. Lefond, ed., AIME, New York, pp. 791–803.

Kunasz, I.A., 1980, "Lithium, How Much?" *Foote Prints*, Vol. 48, No. 1, pp. 23–27.

Kunasz, I.A., 1986, "Reservas Mundiales de Litio," *El Litio, Un Nuevo Resurso Para Chile*, G. Lagos, ed., Editorial U. de Chile, pp. 33–43.

Lasmanis, R., 1978, "Lithium Resources in the Yellowknife Areas, Northwest Territories," *Energy*, Vol. 3, No. 3, S.S. Penner, ed., Pergamon Press, New York, pp. 399–407.

Ludviksson, V., and Hermannsson, S., 1970, "Some Possibilities for Major New Industries in Iceland," National Research Council, Reykjavik, 26 pp.

Mayhew, E.J., and Heylmun, E.B., 1966, "Complex Salt and Brines of the Paradox Basin," *2nd Symposium on Salt*, J.L. Rau, ed., Northern Ohio Geological Society, Inc., Cleveland, pp. 221–235.

McLeod, I.R., 1965, *Atlas of Australian Mineral Resources Deposits*, 2nd ed., Department of National Development, Canberra.

Moraga, B.A., et al, 1974, "Estutio Geologico del Sales de Atacama, Provincia de Antofagasta," *Instituto de Investigaciones Geologicas de Chile*, Bulletin 29, 56 pp.

Muessig, S., 1966, "Recent South American Borate Deposits," *2nd Symposium on Salt*, J.L. Rau, ed., Northern Ohio Geological Society, Inc., Cleveland, pp. 151–159.

Mulligan, R., 1965, "The Geology of Canadian Lithium Pegmatites," Economic Report No. 21, Geological Survey of Canada, 131 pp.

Nel, L.T., 1968, "Ore Deposits of Lithium in the Republic of South Africa," Atomic Energy Board, Republic of South Africa, PEL 23, 23 pp.

Nicolli, H.B., et al., 1980, "Caracteristicas Geoquimicas de Aguas y Salmueras de la Puna Argentina," Miscelana No. 63, Academiia Nacional de Cinencas, Cordoba, Argentina, pp. 3–38.

Norton, J.J., 1965, "Lithium-Bearing Bentonite Deposit, Yavapai County, Arizona," Professional Paper 525-D, US Geological Survey, pp. D163-D166.

Norton, J.J., and Schlegel, D.M., 1955, "Lithium Resources of North America," Bulletin 1027-G, US Geological Survey, pp. 325–350.

Page, L.R., et al., 1953, "Pegmatite Investigations, 1942–1945, Black Hills, South Dakota," Professional Paper 247, US Geological Survey, 228 pp.

Penner, S.S., 1978, "The Lithium Symposium 1977," *Energy*, Vol. 3, No. 3, 413 pp.

Poppi, R.F., 1981, "Los Salares de la Puna Argentina," *Revista Processos*, No. 92, pp. 6–13.

Prast, W.G., 1972, "Scope for Expanded Use of Iceland's Resource," *Mining Magazine*, Nov., pp. 465–468.

Rankama, K., and Sahama, T.G., 1950, "The Alkali Metals: Lithium, Sodium, Potassium, Rubidium, Cesium," *Geochemistry*, University of Chicago Press, Chicago, IL, pp. 422–442.

Richardson, L.S., and Fishwick, J.H., 1968, "Removal of Iron from Spodumene Concentrates by Chlorination," *Glass Industry*, July, pp. 377–381.

Roberts, R.O., 1948, "Amblygonite and Associated Minerals from the Mbale Mine, Uganda," *Imperial Institute Bulletin*, Vol. 46, pp. 342–347.

Russell, A., 1988, "Bikita Minerals, 35 Years On and Still Further Potential," *Industrial Minerals*, pp. 63–71.

Shawe, R.D., 1968, "Geology of the Spor Mountain Beryllium District, Utah," *Ore Deposits in the US, 1933/1967*, J. Ridge, ed., Vol. 2, AIME, New York, pp. 1143–1161.

Shawe, R.D., et al., 1964, "Lithium Associated with Beryllium in Rhyolitic Tuff at Spor Mountain, Western Juab County, Utah," Professional Paper 501-C, US Geological Survey, pp. 86–87.

Singleton, R.H., 1979, "Lithium," *Mineral Commodity Profiles*, US Bureau of Mines, 25 pp.

Strock, L.W., 1936, "Zur Geochemie des Lithiums," Nachrichten von der Gesellschaft der Wissenschaften zu Goeltingen, Mathematisch-Physikalische Klasse, IV, N.F. 1, No. 15, 171 pp.

Toombs, R.B., 1962, "A Survey of the Mineral Industry of Southern Africa," Mineral Information Bulletin MR 58, Department of Mines and Technical Survey, Ottawa, 275 pp.

Valarmoff, N., 1954, "Matériaux pour l'Etude des Pegmatites du Congo Belge et du Ruanda-Urundi," *Annales de la Société Géologique Belgique*, Vol. 78, pp. 1–25.

Vine, J.D., 1976, "Lithium Resources and Requirements by the Year 2000," Professional Paper 1005, US Geological Survey, 162 pp.

Wegener, J.E., 1980, "Lithium Minerals in Zimbabwe with Special Reference to the Operations at Bikita Minerals (Pvt) Ltd.," International Conference on Zimbabwe, Sep. 1–5, Salisbury.

Wegener, J.E., 1981, "Profile on Bikita—Processed Petalite, the New Priority," *Industrial Minerals*, No. 165, June, pp. 51–53.

Westenberger, H., 1963, "Lithium Minerals, Their Formation and Occurrence," *Review of Activities*, No. 6, Metallgesellschaft A.G., Frankfurt/Main, pp. 16–26.

White, D.E., 1957, "Thermal Waters of Volcanic Origin," *Geological Society of America Bulletin*, Vol. 68, pp. 1637–1658.

Williams, C.T., and Trueman, D.L., 1978, "An Estimation of Lithium Reserves and Potential in Northwestern Ontario, Manitoba, and Saskatchewan, Canada," *Energy*, Vol. 3, No. 3, S.S. Penner, ed., Pergamon Press, New York, pp. 409–413.

Winchell, A.N., 1942, "Further Studies of the Lepidolite System," *American Mineralogist*, Vol. 27, pp. 114–130.

Zheng, M., 1989, "Saline Lakes on the Quinghai-Xizang (Tibet) Plateau," *Beijing Scientific and Technical Publishing House*, Beijing, 431 pp.

MAGNESITE AND MAGNESIA

L.R. DUNCAN AND W.H. McCRACKEN

Magnesium is the eighth most abundant element in the earth's crust and the third most plentiful element in seawater. It is found in more than sixty minerals and in brines and seawater as a magnesium salt. One of the more important magnesium minerals is magnesium carbonate, $MgCO_3$, with a theoretical maximum magnesia, MgO, content of 47.6%; this carbonate form represents the world's largest source of magnesia. The next most used sources for magnesia are magnesia-rich brines and seawater, where it occurs as magnesium sulfate and magnesium chloride. Other commercially important magnesium-bearing minerals are dolomite, $CaMg(CO_3)_2$, which serves the aggregate industry as well as the chemical industry; brucite, $Mg(OH)_2$, which is used in the production of both caustic and sintered magnesia; olivine, ($[MgFe]_2$-SiO_2), which serves the refractory and heat storage industries; talc, ($H_2Mg_3(SiO_3)_4$), which serves several industries as a filler and as an ingredient in cosmetics; and serpentine, ($H_4Mg_3SiO_9$). All these minerals are of commercial value because of the desirable characteristics given to them by magnesia and by its placement in the crystal structure.

These minerals are the starting raw materials for a wide range of products. These include the production of magnesium metal, and several grades of magnesia used for the production of both sintered magnesia for refractory manufacture and lighter fired caustic magnesia, the latter is used in fluxes, fillers, insulation, cements, decolorants, fertilizers, chemicals, in the treatment of waste water including pH control, and in the removal of sulfur compounds from exhaust stacks.

Magnesite is one of several major minerals possessing a dual character, i.e., non-metallic industrial mineral because of its principal end use as a calcined/dead burned raw material in non-metallic end products and magnesium metal. A large percentage of the magnesia used to produce magnesium metal comes from seawater/brines. The total primary metal production worldwide is about 350 ktpy. This report reviews magnesite solely as an industrial non-metallic raw material.

Magnesite for most refractory purposes must have a content of sintered magnesia above 85%, with a preference for those sources that have the potential for 90 to 98% magnesia values. Dolomite, used in large tonnages because of its physical properties in the construction industry, also has a use in the chemical and the refractory industry. In its calcined form, dolomite is used as a slag addition in the making of steel, and it is used in a high-purity calcine to precipitate magnesium hydroxide from brines or seawater. The precipitated magnesium hydroxide is intermediate for the production of deadburned clinker for refractories and the production of magnesium metal. In many cases the high-purity magnesium hydroxide obtained by precipitation with calcined dolomite, or in some cases limestone, has supplanted much of the world's production of magnesia from the natural mineral magnesite. This is especially true of the United States and in major industrialized areas of Europe and Japan.

The best known of the minerals directly and widely exploited for its magnesia content is magnesite ($MgCO_3$), one of the calcite group of rhombohedral carbonates, which includes calcite ($CaCO_3$), siderite (Fe_2CO_3), and rhodocrosite ($MnCO_3$), among others. The members of this group enter into a wide range of substitutional solid solutions when the positive ions have similar radii. The radii of magnesium and iron ions are within 6% of each other; hence magnesite and siderite form a complete series, of which the mineral breunnerite (ferroan magnesite) is a well known end member. The radii of calcium ion is 36% larger than that of magnesium ion and only limited substitution exists at each end of the $MgCO_3$-$CaCO_3$ series.

Dolomite is not a member of the calcite group, but it occurs when calcium and magnesium ions alternate in equal number in an ordered structure among carbonate ions. The result of this relationship is that calcite and dolomite are commonly found intermixed with magnesite. They occur, commonly, as identifiable crystal entities, which can be separated to a varying degree from the magnesite by benefication techniques.

Magnesite, when pure, contains 47.8% MgO and 52.2% CO_2. The pure mineral is found occasionally as transparent crystals resembling calcite. Magnesite principally contains variable amounts of the carbonates, oxides, and the silicates of iron, calcium, manganese, and aluminum.

Although the genesis of natural magnesite deposits can be complex, it is distinguished in nature in two distinct physical forms, namely crystalline, (with a wide range of visible crystal sizes) and cryptocrystalline, sometimes referred to as amorphous, where the crystal size is not detectable to the eye and will range from 1 to 10 µm. The two types not only differ in crystal structure but in the sizes of the deposits and in modes of formations.

The crystalline form has a Mohs hardness of 3.5 to 4.0. The color range is from white to black with shades of yellow, blue, red, or gray. The color is not a significant indicator of purity but in a given deposit an experienced person can often roughly grade the magnesite by observing color and crystallinity. Macrocrystalline deposits occur in relatively few, but generally large, deposits on the order of several million tons. The ore shows a marble-like crystal structure and belongs to the sedimentary or metasomatic groups of origin.

The cryptocrystalline variety of magnesite frequently occurs in many small deposits, although there are exceptions. Cryptocrystalline magnesite is typically massive with no cleavage and is sometimes descriptively called bone magnesite. The fracture is usually conchoidal, with a hardness of 3.5 to 5.0. The color is mainly white, but it can have tints of yellow, orange, or buff. Accessory silaceous minerals such as serpentine, quartz, or chalcedony are usually present. Calcium minerals are normally absent or in low concentration, this contrasting with the macrocrystalline form where calcium minerals are almost invariably present and sometimes in high concentration. The specific gravity of cryptocrystalline magnesite ranges between 2.90 and 3.00, whereas the value of pure crystalline magnesite is 3.02. In actuality, the normal specific gravity of magnesite is usually higher than 3.02 because of the presence of iron carbonate.

Magnesite ($MgCO_3$) dissociates on heating to form magnesia (MgO) and carbon dioxide (CO_2). When heated sufficiently, the magnesia develops a crystal structure identified with that of the natural mineral periclase. Periclase occurs only rarely in nature and not in any known commercially workable deposit.

Brucite ($Mg[OH]_2$) is another mineral that can be directly reduced to periclase through the application of heat (Comstock, 1963). Brucite ore has been exploited in the past for the production of magnesia but is no longer an important source. Theoretically it contains 69.1% MgO and 20.9% H_2O. The mineral is often associated with limestone as well as magnesite, is translucent, and is relatively soft and lightweight. Its hardness is 2.5 and specific gravity about 2.4. The color may be white, but blue or green with a gray cast is the more common coloration.

Large crystalline magnesite deposits are exploited in Austria and Czechoslovakia of the high-iron type containing a substantial percentage of brunnerite ($MgCO_3$). The iron is of the isomorphous type, substituting for magnesium ions in the carbonate lattice of the natural magnesite. This iron does not lend itself to a physical separation.

Total current world productive capacity of all commercial magnesium compounds is approximately 11 Mt, roughly 20 to 25% of which is produced from seawater and brines.

NOMENCLATURE OF PRODUCTS

The word *magnesite* literally refers only to the natural mineral, but common usage applies this name, plus a prefatory word, to two other types of materials, dead-burned magnesite and caustic-calcined magnesite. For the most part these are commercial products of magnesia, differing mainly in density and crystal development that results from different levels of heat application. When magnesia produced from seawater or brines first made its appearance on the world scene, the products also received the appellations dead-burned or caustic-calcined magnesite, but later the technical literature has increasingly referred to the materials as refractory magnesia or refractory grade magnesia and caustic-calcined magnesia. These terms are now also increasingly being applied to magnesias derived from the natural mineral magnesite, especially for those materials with high MgO content. The older terms still persist in the trade.

The terms dead-burned magnesite or refractory magnesia refer to the granular product produced by firing magnesite, magnesium hydroxide, or other material reducible to magnesia at temperatures in excess of 1 450°C. The heat treatment must be of sufficient duration to produce a dense, reasonably weather-stable granule for use in the manufacture of refractory materials.

The use of the terms describing magnesia products is somewhat confusing to those not closely connected with the industry. The vagueness of the terminology can lead to misreading of statistics to such a degree that simultaneous shortage and surplus of certain magnesia products can be indicated. Confusion has been further amplified by use of such terms as high-purity or super high-purity magnesia; in most cases these terms have been used with magnesia supplied to the refractory industry. They usually refer to the content and relationship of the accessory oxides and density of the material as they affect the refractoriness of the final product rather than any specific magnesia (MgO) content.

The terms high grade and high-purity generally refer to a refractory magnesia containing over 96% MgO, a density greater than 3.30, preferably 3.40, and a proper relationship of auxiliary oxides.

Caustic-calcined magnesia is the resultant of 800 to 1 000°C heat being applied, frequently in a rotary kiln, to a magnesite or other material reducible by heat to magnesia. It is heated to such a degree that less than 10% ignition loss remains and the product displays absorptive capacity or activity. The caustic magnesia readily absorbs water vapor or carbon dioxide, an action which is referred to as activity. This activity feature then facilitates the production of high density periclase. This is generally done through briquetting and the application of additional high temperature heat. The production of caustic magnesia involves less heat energy than is required to produce dead-burned magnesite.

Fused magnesia is produced by heating high grade magnesite to a molten state for up to six hours in electric arc furnaces to temperatures aprroximating 3 000°C. The resultant product, at 96 to 98% MgO, has a favorable high density of 3.50 g/cc and relatively high chemical stabilty, strength, and resistance to abrasion.

In addition to Baymag Canada's current production of 14 ktpy, fused magnesia is also produced in Australia, China, France, Israel, Japan, the UK, and the US. World production is some 160 ktpy, used both in refractories and in electrical products.

ORIGIN AND OCCURRENCE OF MAGNESITE

Crystalline Magnesite

Crystalline magnesite deposits are generally of a large size—on the order of several million tons of ore. Most of the deposits have a striking resemblance to each other, indicating a common mechanism of formation. The deposits are usually associated with dolomite as a host rock, but some, such as those in Brazil, are in limestone formations. In these cases the magnesite is generally not in direct contact with the limestone but is separated by a dolomitized zone.

Magnesite is closely associated both spatially and chronologically with intrusive activity (Bain, 1924, Bodenlos, 1954, Siegfus, 1927, Spencer, 1972). It has been suggested that igneous activity has been the source of the initial CO_2-bearing solutions. The mineralogy of magnesite deposits suggests a high temperature of formation. The world's major deposits of crystalline magnesite occur in Austria, Brazil, Canada, Australia (Tasmania), the former USSR, North Korea, China, Nepal, Czechoslavakia, Spain, and the United States.

The crystalline deposits supply about two-thirds of the magnesia used commercially. A large portion of this is used in the Eastern economic countries where the demand for low iron magnesia has not kept pace with the Western countries. Most refractory magnesia is used for the manufacture of refractory products for the steel industry. Greatest usage for low iron magnesia has been to satisfy the refractory requirements of basic oxygen steel furnaces.

Dolomite geologic measures are widely viewed as being sedimentary (Dunbar and Rogers 1957), and some investigations have also suggested a sedimentary origin for both the magnesite and the enclosing dolomite (Paone 1970, Schroeder 1948), but consensus favors secondary placement of magnesite in preexisting dolomite by hydrothermal action on a volume-for-volume basis (Bain 1924, Bennett, 1943, Campbell and Loofbourow 1962). The immense size of the magnesite deposit in Liaoning Province of China, which extends for some 60 km and reportedly contains some 130 Gt has led to the proposal that the magnesite was precipitated with the dolomite in a Precambrian lagoon or shallow sea (Nishihara, 1956).

At Kilmar, Quebec, there is a rather unique occurrence of magnesite where the intermixed rock of magnesite-dolomite is thought to be the product of hydrothermal dolomitization of the limestone (Bray and Hilchey, 1957).

A large deposit of low lime crystalline magnesite occurs at Timmins, Ontario, intermixed with talc and quartz. The iron content is on the order of 3% Fe_2O_3. Since dolomite is absent it is suggested that this magnesite formed from magnesium-rich solutions in ultrabasic rocks.

The multi-million ton essentially undeveloped deposit of magnesite at Lamasangu, Nepal, is also lime-poor with a fairly high iron content and considerable associated talc.

Igneous rocks associated with magnesite vary widely in composition. Bodenlos (1954) compiled a list which includes amphibolite, pyroxenite, diabase, peridotite, rhyolite, quartz, porphyry, basalt, granite, and others. This mineral association suggests a relatively high temperature of formation (Niinomy, 1925). Olivine, pyrrhotite and magnetite have also been recognized in the deposit at Chewelah, WA. Bodenlos has suggested a temperature of formation of 300°C or higher, and Schroeder (1948) a temperature within the hydrothermal range of 300 to 500°C.

Although the field evidence is clear and constant, the specific action of the hydrothermal solutions and the mechanism of replacement is not yet fully understood. Mechanisms generally suggested for the hydrothermal implacement of magnesite in dolomite involves a reaction with Mg-rich, CO_2-bearing solutions. Although these CO_2-bearing solutions may be conveniently derived from nearby intrusive bodies, it is not easy to account for the required Mg-enrichment of such solutions.

An experimental investigation (Rosenberg and Holland 1964, Rosenberg and Mills, 1966) threw some new light on the matter. The investigation was carried out on the basis of observed field relationships, but under the conditions that obviated the requirement for Mg-rich solutions. The results were consistent with the presence of high temperature mineral assemblages associated with magnesite. The suggested mechanism may be summarized as follows:

1. Attack on preexisting dolomite by CO_2-bearing solutions at temperatures below 200°C, resulting in solutions with Ca:Mg ratio of unity.
2. Heating of these solutions to temperatures in excess of 200°C, resulting in the replacement of host rock dolomite by magnesite with the production of Ca-rich solutions.

To explain some of the deposit anomalies such as the Liaoning (Nishihara, 1956) and Timmins (Ontario) deposits and specifically the Himalayan deposits of India and Nepal, the role of algae has been presented (Valdiya, 1968). Valdiya discusses a chain of major lens-like deposits of coarsely crystalline magnesite extending for 130 km northwest from the Kali Valley in Nepal to the Alaknanda Valley in India. The deposits are uniform in thickness and have similar chemical and mineralogical composition. The magnesite is restricted to a narrow stratigraphic interval in the upper part of the Gangolihat dolomites and is usually associated with the stromatolite-bearing dolomite. Reference is made to the Coorong Lagoon and its associated ephemeral lakes of South Australia. Development of barred basins are attributed to barrier-building algae. The Mg:Ca ratio of the basin waters progressively increased with the passage of time as a cumulative result of biogenic and organic precipitation of $CaCO_3$ during the aqueous stages of the sedimentary process. Accumulation of algoid debris further increased the concentration of magnesium. The Mg:Ca ratio became so high that during periods of high pH attributable to prolific growth of the algae, the carbonate assemblages were converted into carbonates of higher magnesium content. The coarse granularity of the magnesite is attributed to later recrystallization during regional metamorphism. Australia has large beds of crystalline magnesite in northwestern Tasmania. This magnesite is associated with dolomite and limestone but it has not yet been commercially developed.

Cryptocrystalline Magnesite

Cryptocrystalline or amorphous magnesite is an alteration of serpentine or allied magnesium-bearing rocks which have been subjected to the action of carbonate water. The serpentine which lies near or surrounds the magnesite is itself an alteration product of ultrabasic rocks. The mode of formation of the magnesite limits the amounts of impurities to relatively small quantities of iron, lime, and silica.

Occurrences of this type of magnesite are fairly common throughout the world, but because of their usually limited size, few, with the notable exceptions of those in Greece, India, Turkey, and Australia, are worked commercially. The deposits are of two types. The first is closely associated with the serpentine mass where the magnesite occurs as veins of various thicknesses, and as massive bodies and lenticular masses or as stockwork with irregular veins from a few centimeters to several meters in thickness. The second is also associated with ultrabasic rocks as the magnesium source, but here the host rock was weathered, eroded, transported, and deposited in a lacustrine environment. At this point decomposition was completed and magnesite was precipitated in a mud matrix together with impurities. Subsequently, a recrystallization took place and magnesite was accumulated to form nodules and boulders (lumps) on the shores of the mud-bearing lakes, thus resulting in large scale deposits of secondary magnesite. The deposits have benefited from a natural benefication process and are normally of a high purity.

In both types of cryptocrystalline deposits reaction of water on the serpentine host was a necessity. The relationship between temperature and carbon dioxide pressure was critical to the formation of magnesite. The zones nearer to the earth's surface are more favorably situated for this reaction. Therefore, most of the deposits of this type of magnesite are found near the surface and have limited extent in depth.

DISTRIBUTION OF DEPOSITS/PRODUCTION CENTERS

The world production of magnesite is in a constant state of flux. With the Eastern economic countries moving towards a free market economy, their demand for the lower grades of magnesite will decrease. At the same time the world steel industry, the largest consumer of magnesite and magnesite refractories, is under pressure for market share from competitive metals and plastics. A previously-projected production estimate of 21 Mtpy of crude beneficiated magnesite could be inaccurate, with more conservative estimates being as low as 15 Mt. Changing steel practices can result in lower magnesite usage. Chemically-produced magnesia from seawater or brines amounts to some 2.5 Mt worldwide.

Prior to 1938 when the first large scale seawater magnesia plant was started at Hartlepool, UK, the principal supply of magnesia for the refractory industry came from the natural deposits located in central Europe and Asia. The magnesia was brown in color from its iron content and was the accepted color for refractory grade sinter. When low iron magnesia first appeared, the refractory industry was slow to accept it because the color was now different, white or offshade white.

Even today the natural magnesites of the CIS, China, North Korea, and Czechoslovakia dominate the refractory market in the eastern European countries.

With the increased cost of fuel, there is a distinct advantage to the conversion of natural magnesite rather than producing magnesia from seawater or brine. After the first synthetic magnesite plant was started in the 1940s, fuel oil cost has increased some tenfold. This is a considerable impact because it requires three times as much energy to produce sinter from the chemical process. The offsetting advantage is the high-purity and uniformly high quality of the synthetic magnesias.

The major world deposits of natural magnesite that are commercially exploited are located in the CIS, North Korea, China, Czechoslovakia, Yugoslavia, Austria, Greece, Brazil, Spain, Turkey, India, and Australia; the CIS, North Korea, and China represent more than half of world capacity. Smaller operations are in Nepal and the Republic of South Africa.

Although they are not natural magnesite deposits, the sources of magnesium contained in seawater and natural brines are an important contributor to world magnesia production. The current production of dead-burned magnesite from these synthetic sources is approximately 2.5 Mtpy, principally in the Western economic countries. Seawater/brines magnesite production can be carefully controlled to give magnesia (MgO) content as desired up to 99.4%, together with high density and low boron (B_2O_3) content. The major synthetic magnesia plants are located in the United States, Mexico, Japan, Italy, the Netherlands, Ireland, the United Kingdom, and Israel. Other operations are in South Korea and the former USSR.

A brief description of some of the world's major natural magnesite deposits are described below:

United States

In the United States the principal crystalline magnesite deposits are near Chewelah, WA (Bennett 1941, Siegfus 1927). Commercial operations were discontinued there in 1969. Another natural magnesite property in the United States is the brucitic magnesite at Gabbs, NV, still being worked on a reduced scale.

In addition, there is a potentially commercial deposit of brucite ore located in the state of Texas. It has been reported (Walper, 1964) that in the Marble Canyon Area, Culberson Country, TX, lying within the frontal scarp of the Sierra Diablo, 48 km north of Van Horn, there occurs a large deposit of brucitic marble, reportedly containing more than 2 Mt with an average brucite content of 38%. Samples of this ore have shown brucite proportions ranging from 23 to 100%. The reason for this wide range of values is that while the brucite generally occurs as relatively small nodules in the marble, there are also larger veins and pockets of pure brucite. The brucite in the marble could be a consequence of both an enrichment from outside sources as well as the thermal metamorphism of dolomite. If the brucite had been formed solely from the thermal metaphorism of dolomite, the brucite content in the host ore could not exceed 36.7%. This occurence of brucite ore is not currently being exploited but there have been recent reports of a renewed interest in its development.

Canada

In Canada the brucite operation at Wakefield, PQ, has ceased production. The crystalline deposit at Kilmar, PQ, ceased operations in 1992. The iron-bearing crystalline magnesite located near Timmins, ON, is being reevaluated but there is no commercial production. There are several deposits in British Columbia, but the only one being commercially operated now is the Brussiloff Mountain occurrence. This open pit deposit is being operated by Baymag with a capacity to produce 100 ktpy of caustic magnesite in two rotary kilns, and 14 ktpy of fused magnesia in two electric arc furnaces. Other deposits in British Columbia are the East Kootenays beds of crystalline magnesite occurring interlayered with sedimentary rocks, the cryptocrystalline magnesite in the Clinton and Bridge River areas of central British Columbia, and the vein deposits in the fault zones along the Yalakom River and the Pinchi Mountains.

Commonwealth of Independent States

The identifiable reserves of crystalline magnesite in the former Soviet Republics are reported to be on the order of 400 Mt, located in the southern Ural Mountains, and 2 000 Mt in the Savan Mountain area of eastern Siberia. Most of the CIS production comes from the Ural Mountain deposits, especially those near Satka in Ufa Province. The Satka deposits alone are said to contain more than 250 Mt of reserves. The crude ore has an average MgO content of 46% and is similar in appearance to the Austrian magnesite, possessing a somewhat lower iron content, but similar lime values. With a total capacity exceeding 3 Mt, CIS ore production in the past has reached some 3 Mt annually. There was a plan to increase this to 5.5 Mt, but this was shelved. Average production today is approximately 2 Mtpy. To meet Western standards for BOF steel process refractories, the Russian magnesite would require benefication. The deposits in the Savan Mountain have not been exploited to the same degree as in the Ural Mountains because of their remote location and somewhat lower quality.

North Korea

North Korea has a reported reserve of 300 Mt of crystalline magnesite. The main deposits are located in Kankyo Province and average on a crude basis 45% MgO. The annual production of crude magnesite has been reported to be in excess of 1 Mt. They have added rotary kiln and benefication facilities, thus permitting the production of a sinter with 95% MgO, a density in the range of 3.25 gm/cc, and Fe_2O_3 content slightly over 1.2%. A substantial portion of the North Korean dead-burned magnesia and caustic-calcined magnesia is exported. Productive capacity is reported to exceed 2 Mtpy, including caustic.

South Korea

In South Korea at Pohang there is a 50 ktpy dead-burned seawater magnesite operation.

China

The largest magnesite deposits in China are located approximately 65 km south and southeast of the steel-producing city of Anshan in Liaoning Province (a district of the former Manchuria). The deposit occurs over a 60 km long section northeast from Da Shi-chiao. The high relief and hills of the local terrain facilitates mining. The geological formation of the magnesite zone corresponds to the Precambrian-Algonkian system. The reserves are reported to be in excess of 30 Gt. The colors in the ore range from white to pink and gray. The current mine production is on the order of 3 Mtpy. Only relatively small sections of the total deposit are being worked, by open pit methods. The run-of-mine ore, after crushing, hand sorting, and screening, generally produces a sinter containing 90 to 95% MgO. A benefication plant plus a calciner and shaft furnace have recently been installed to produce a 98% MgO product; sintered density is currently at a 3.30 to 3.35 level. Most of the Chinese magnesite exports have been of the regular grade of sinter. In addition, crude Liaoning magnesite is being sold to the Norsk Hydro magnesium metal operation in Quebec. The ore as mined has a specific gravity of 2.99 to 3.02 gm/cc. The proportion of waste and overburden requires the mining of approximately 2.5 t for each ton of raw magnesite ore. Total productive capacity is on the order of 1.25 Mt of dead burned magnesia, 300 kt of caustic.

Czechoslovakia

Czechoslovakian natural magnesite reserves are estimated at 100 Mt. This is a breunnerite type and is extensively mined in Slovakia. Operations are in Kosice, Jelsava, and Lobinobana. Benefication facilities have recently been installed, together with a calciner and a modern shaft furnace designed to produce an improved 99% MgO grade of sinter for the domestic steel industry. Major producer SMZ has dead-burned magnesia capacity of approximately 500 kt.

Yugoslavia

Yugoslavia has deposits of magnesite in Serbia and Bosnia. The material is mined in both underground operations and in open pit quarries; reduced production precludes any significant exports. The Yugoslavian magnesite deposits are hydrothermal sedimentary and hydrothermal vein types. Mines are located in the area of Kraljevo at Bela Stena and Ilinjaca and at Beli Kamen and Evati. Vein type deposits are located in Sumadija, Goles, Dubovac and Ibar. All the ore requires beneficiation, in this case heavy media separation, in order to produce commercial quality. Magnochrom, the major producer, has a productive capacity of dead burned magnesia reported to be on the order of 200 ktpy; caustic 50 kt.

Austria

Austria, one of the oldest magnesite producers in Western Europe has a narrow belt of breunnerite magnesite extending westward from Semmering. The most important deposits are located at Semmering, Veitsch, Breitenau Trieben, Radentheim, and Dientin. Consolidation and acquisition have given the RHI Vienna group total dead burned magnesite capacity of more than 500 ktpy, including 80 kt for Radex at Radentheim and 350 kt at Veitscher. Magindag reportedly produces 40 kt of caustic magnesite. While these are important commercial deposits with approximately 6% Fe_2O_3, they have lost much of their value with the introduction of the new steelmaking methods using the oxygen process. In addition to the steel industry, large quantities of caustic magnesite were supplied for the manufacture of wallboard. This business has also declined. Some chemical grade, with 99%+ MgO has been produced.

Greece

Greece has deposits of cryptocrystalline magnesite. A deposit located at Vavdos in Khalkidiki region in the southwest Thessalonika was first worked in 1957. Other deposits which have been commercial are located on the island of Euboea and in Ormylia in Khalkidiki. The Euboea operations, once producing 300 ktpy, have been closed. Greek magnesite production has been important to the refractory production in Western Europe.

The Grecian deposits are low in iron and boron and produce a grade of sinter suitable for refractories used in the BOF steel furnaces. The magnesite is of high purity but tends to be intermixed with some siliceous gangue; beneficiation of all ore is necessary for the production of commercial sinter. Large amounts of caustic magnesia have been produced, some calcined with wood to enhance purity. Most of the magnesite occurs as veins or small lenses, resulting in high mining cost. The removal of about 80 t of waste and overburden are required to obtain one ton of magnesite. The development of new competitive world magnesite sources, and the high costs of Grecian production, have made profitable operations difficult in recent years. With a reported capacity exceeding 300 kt, current annual production is probably less than 100 kt of magnesite including caustic.

Brazil

Brazil has substantial reserves of crystalline magnesite. The two major areas are located in the Iguatu area of Ceara Province and in the Eguas range near the town of Brumado in southwestern Bahia. Resources are reported to be in excess of 500 Mt. Magnesita SA is the largest producer of magnesite sinter in Brazil, with an annual production of approximately 300 kt of crude magnesite, 140 kt of dead-burned. They produce two main grades, M_{10} and M_{20}, both having similiar chemistry, with MgO of 94 to 95%, but differing in the degree of sinter. A third grade M_{40}, with $CaO:SiO_2$ ratio of 3:1, has been produced on a limited basis, with higher density and an MgO content of some 97%. The ore is generally white in color, sometimes dark gray to black, and is mined open pit. The color generally does not distinguish significant differences in the grade of ore. Productive capacity of other producers in Bahia and Ceara is a reported 60 ktpy including caustic.

Spain

Spain's magnesite deposits are principally located in Navarra Province, near the French border. There are also smaller deposits in Northwest Spain. The total production is approximately 400 ktpy of crude magnesite. A major part of this is calcined and used as a caustic magnesia addition to animal feeds and agricultural fertilizers. The company Magnesitas de Rubian SA has undertaken research to better define the role of different grades of magnesite, as well to evaluate optimum burning temperatures needed to produce the most effective grades for animal feed and fertilizer. Their productive capacity is reported to be 75 ktpy of dead-burned, 120 kt of caustic.

Turkey

Turkey has both crystalline and cryptocrystalline deposits of magnesite. The principal crystalline deposit is in the Black Sea region. The more important cryptocrystalline deposits are located in the Eskisehir and Kutaya regions. These deposits are of the vein-filling type. Another cryptocrystalline magnesite deposit is at Salda Lake in southern Turkey, approximately 150 km northwest of the port of Antalia and near the town of Yesilova. The magnesite is mixed with mud and occurs as dunes and as lumps on the lake shore. The magnesite generally is not compacted, and has a specific gravity of 1.60 to 1.80 gm/cc. The total annual production of magnesite ore in Turkey is of the order of 1 Mt, including approximately 250 kt of dead-burned.

India

India, a traditional supplier of crude and caustic-calcined magnesite, now produces sintered magnesia for its own refractory industry. The deposits in the southern Salem district of Madras State, with three suppliers having a productive capacity of 100 kt, are cryptocrystalline. These magnesites are of high quality but are mixed with serpentine and require benefication to obtain optimum quality. The deposits in the North in the Almora district in Uttar Pradesh are crystalline. The ore is white to yellowish in color and produces a sinter in the range of 90% MgO. The annual production of Salem is about 200 kt of crude magnesite. The deposit is becoming depleted and benefication has become more necessary; hand sorting alone does not produce commercial tonnages or quality. The Uttar Pradesh area produces some 35 to 40 kt of low quality magnesite sinter. Production capacity is 90 ktpy.

Republic of South Africa

South Africa has been a producer of magnesite for more than twenty years, but the output has been somewhat irregular. The problem is with the mining; the Transvaal ore occurs in pockets at irregular intervals. A deposit is presently being worked in Zimbabwe near the border with South Africa.

Nepal

The Nepal crystalline deposit at Lamasangu has been under commercial development since 1983. A production facility to produce 50 ktpy of sinter has been installed. The ore body is lime-poor, but contains Fe_2O_3 in the order of 5%. The MgO content of the sinter ranges between 90 to 94%. The reserves of ore are substantial.

Iran

In eastern Iran more than 1 Mt of cryptocrystalline magnesite have been reportedly identified in a 20 km long series of deposits located 50 km southeast of Birjand.

Saudia Arabia

In Saudia Arabia a crystalline high lime deposit of magnesite is located at Jabal Al Rokhan, north of Jiddah. This is a hydrothermal replacement of dolomite. Reserves are reportedly 40 Mt. Another cryptocrystalline deposit, reported by BRGM to contain several million tons, is located in the region of Zarghat and Jabal Wask in the Hijaz Mountains. There is no current commercial activity.

Australia

Australia has identified cryptocrystalline magnesite deposits in Queensland, New South Wales (Fifield, Young), and in Western Australia at Ravensthorpe. The total annual production from these deposits, excluding Queensland, is approximately 16 kt. What is reported to be one of the world's largest sedimentary deposits of nodular, low iron cryptocrystalline magnesite was discovered in 1985 at a point 55 km northwest of Rockhampton in central Queensland. This became the Queensland Kunwarara Magnesite Project. Reserves are reported to be in excess of 750 Mt, approximately one-half as low iron. The ore occurs as nodules and lumps and is white in color. The calcines have a pink cast. The ore nodules occasionally contain inclusions of dolomite. Nickel content is somewhat higher than for most cryptocrystalline deposits. The magnesite zone rests on tertiary sands and gravels which in turn rest on a granitic basement. The nodules range in size from several millimeters in diameter to over one meter. They are surrounded by a soft mudstone/sandstone/earthy matrix. The specific gravity of the ore is 2.5 to 2.8 g/cc. The ore body is now being developed to annually produce approximately 150 kt of sintered magnesite with a density of 3.4 g/cc, plus and additional production of caustic magnesia. There are parallel studies to produce 25 ktpy of fusted magnesia and magnesium metal as a long term part of the project.

Other

Other apparently minor magnesite deposits have been reported to occur in Bolivia, Bulgaria, Mindanao in the Philippines, Pakistan, Colombia, Guatemala, and Zimbabwe.

SYNTHETIC MAGNESITE

Prior to 1938 the natural-occurring mineral magnesite was essentially the source of all refractory magnesia. Since that time the manufacture of refractory grade magnesia and caustic-calcined magnesia from other than natural magnesite has steadily increased in importance. Today approximately two-thirds of refractory grade magnesia used in the western economic countries comes from synthetic magnesite sources. Several basically different methods for production of magnesia from non-magnesite sources have been studied and at least four have operated commercially. All involve the calcination of either magnesium hydroxide or magnesium chloride. The magnesium hydroxide may be obtained as a precipitate from magnesium-rich solutions or as a residue remaining after the lime fraction of calcined dolomite is removed as a soluble compound by reaction with solutions of ammonium chloride or hydrogen sulfide. The magnesium chloride may also be obtained as an end liquor after solar concentration of solutions of natural brines for production of salt or potash, or from well brines.

There are over 20 plants around the world producing refractory magnesia and caustic magnesia from magnesium hydroxide. In these plants, magnesium hydroxide is obtained by precipitation from solutions such as seawater, seawater bitterns, inland well brines, or end liquors from other chemical processes. The plant in Israel produces magnesia by pyrolytic decomposition of magnesium chloride.

The commercial production of magnesia from seawater dates back to around 1885 when a small operation existed for a short time on the Mediterranean coast of France (Egleston, 1886). In this operation milk of lime and seawater were combined in large masonry tanks which overflowed continuously into two similar tanks in series. From the last tank, the mixture of spent seawater and precipitated magnesium hydroxide was discharged into shallow troughs about 5 m wide and 305 m long excavated in the beach sand. The sand acted as a filter for removal of the liquid. When the troughs became full of magnesium hydroxide mud, the inflow was stopped and after a few weeks in the sun, the material became sufficiently dry to allow removal. The broken lumps were calcined. This early French operation eventually closed down, presumably because it could not compete with the world's developing natural magnesite operations. The technical principles used in this original French process are similar to those now practiced in seawater-magnesia plants.

All operations which make chemical magnesia by calcining magnesium hydroxide follow a basically similar process (Wicken, 1949). The soluble magnesium salts are converted to insoluble magnesium hydroxide by treating with calcium hydroxide. The calcium source can be oystershells, limestone, or dolomite. Any of the common magnesium-bearing solutions such as seawater or brines may be used provided they can be treated to reduce carbonate and sulfate levels sufficiently to prevent the precipitation of calcium compounds with the magnesium hydroxide. The main concern in the selection of a reactant is the degree of purity and economic availability. After the reactants are established for any given plant, the limitations of pumping and thickening equipment will prevent that plant from changing to more dilute reactants without loss of capacity. A plant using seawater bitterns cannot change to using raw seawater, nor can one using dolomite change to using limestone without sacrificing production. However, the opposite changes are feasible.

United Kingdom

In 1938 a commercial plant for refractory magnesia was built at Hartlepool, Durham (Gilbert and Gilpin, 1951). The plant used calcined dolomite to treat the seawater. The dolomite is quarried, crushed and washed before the sized stone is calcined in a rotary kiln. The calcined dolomite is then slaked in hydrators, with the hydrate being slurried and classified to remove impurities before being mixed with softened seawater.

The incoming seawater is treated with sulfuric acid to remove carbon dioxide (CO_2). The softened seawater is reacted with a slurry of calcined dolomite precipitating magnesium hydroxide. The dilute suspension of magnesium hydroxide is pumped to large primary settling vessels. The underflow from the thickener is repulped with treated seawater or fresh water and then rethickened. The underflow from the secondary thickener is filtered on rotary disk filters giving a cake of 50% solids. This cake is fed to a rotary kiln for dead-burning or to a hearth furnace for the production of caustic magnesia. This caustic is pelletized and dead-burned in separate shaft kilns at temperatures in excess of 1 800°C.

To increase efficiency and improve grade, high pressure filters have been installed to produce a cake with approximately 70% solids. This UK plant produces some 190 ktpy of product.

United States

In the United States the first commercial production of synthetic magnesite was by the Marine Magnesium Products Corp., a company that produced non-refractory magnesium compounds at its South San Francisco plant.

Several commercial US plants treating seawater for magnesite have ceased to operate, notably the Harbison-Walker plant of Dresser Industries at Cape May, NJ, and the Corhart plant at Pascagoula, MS. Currently operating plants in the US are National Refractories at Moss Landing, CA, with a capacity of 120 ktpy; a plant at Port St. Joe, FL, capacity 45 kt; Dow Chemical at Freeport, TX, for producing magnesium metal; with a small production also reported at South San Francisco, CA, and at Lewes, DE. In comparison with the magnesium content of seawater, brines generally have much higher quantities of the magnesium ion. A typical analysis for a Michigan brine is NaCl, 5.8%; $CaCl_2$, 17.0%; $MgCl_2$, 9.9%; KCl, 2.5%; Br_2, 0.1%; specific gravity 1.29; and pH 5 ± 0.5. The brine does not contain carbonate or sulfate ions and therefore requires no pretreatment ahead of precipitation of magnesium hydroxide. It is pumped to the surface from depths of 762 to 1 067 m. A plant using brines is Harbison-Walker at Ludington, MI, with a productive capacity of 200 ktpy of dead-burned magnesite; Harbison-Walker receives magnesium hydroxide from the Dow Chemical plant at Ludington; Martin-Marietta has a brines plant at Manistee, MI, producing magnesium hydroxide. This feeds a plant with a magnesia capacity of 275 ktpy.

Lake brines are used to produce metal and chemical grade magnesia in Utah.

The California seawater bitterns at South San Francisco are an end liquor from a solar salt operation. The seawater was concentrated by solar evaporation until most of the sodium chloride was crystallized, leaving a bittern with a nominal composition of NaCl 12 to 16%; $MgCl_2$, 6 to 9%; $MgSO_4$, 4 to 5%; KCl, 1 to 2% and $MgBr_2$, 0.16%. The bittern must be treated with calcium chloride solution to remove the sulfate ions as calcium sulfate before precipitation of the magnesium hydroxide.

In all these operations the processes are fundamentally alike in that seawater or brine is treated to remove the soluble bicarbonates via the addition of lime or treated with acid to liberate the carbon dioxide. The softened liquid is mixed with dry or slaked lime or dolime to precipitate magnesium hydroxide. The magnesium hydroxide is thickened and washed to remove any calcium chloride. The washed magnesium hydroxide is calcined, normally briquetted, and dead-burned. All produce refractory grade magnesia with a high percentage of MgO, controlled chemistry, and with a high sintered density.

Annual refractory grade synthetic dead-burned magnesia production in the United States is on the order of 400 kt; total capacity is nearly 950 kt.

Japan

In Japan, magnesium hydroxide is produced from the treatment of seawater with lime. The largest plants are those of Ube Chemical Industries Ltd. at Ube Yamaguchi with a capacity of 400 kt, and Shin Nihon Chemical Ind. Co. at Minamata, Kyushu, capacity 100 ktpy. Ube has recently reduced productive capacity to approximately 300 ktpy by closing their smaller plants.

Mexico

There are two chemical magnesia plants in Mexico, both owned by Industrial Penoles. The plant Quimica del Mar, Ciudad Madero, Tampico, operates on seawater and the plant Quimica del Rey, Laguna del Rey, Coahuila, uses the end liquor from the sodium sulfate operation. The total Mexico refractory magnesia productive capacity is approximately 160 ktpy.

Italy

Italy has two plants, both based on seawater and calcined limestone, operated by Sardamag; one is at Sant' Antioco and the other at Priolo near Siracusa. Productive capacity of refractory grade magnesia is some 170 ktpy with grades ranging from 92 to 98.5% MgO.

Ireland

In Ireland the plant at Dungarvan, County Waterford, has been closed, leaving the plant of Premier Periclase at Drogheda, County Louch, as the only operating unit. Premier Periclase has recently installed two modern shaft furnaces to produce 100 ktpy. They are offering a high grade refractory grade product with MgO contents of 98%, low boron, high density, and crystal sizing in the 150 μm range.

Commonwealth of Independent States

A plant has been reported in the Crimea with a capacity of some 100 ktpy. The finished product quality is reportedly not up to standards because of the poor dolomite quality and brackish seawater.

Norway

In Norway, Norsk Hydro AS has a plant near Porsgrunn operating on seawater and dolomite. The magnesium hydroxide is for nonrefractory use going to the production of magnesium metal and caustic.

South Korea

In South Korea a plant operated by Sam Hwa Chemicals at Pohang has been upgraded to produce 50 ktpy of sintered magnesia.

Netherlands

A synthetic magnesia plant is operated by Billiton Minerals Division of the Royal Dutch Shell Group at Veendam in northern Netherlands. The project is based on using the magnesium values contained in the large salt beds located in the Groningen region. Extraction is by solution mining. Natural gas is available from the nearby giant Slochteren field. The salt bed is part of the Zechstein basin which extends from Poland to eastern Scotland. The quality of the $MgCl_2$ brine produced from solution mining of the carnalite deposits is excellent, with a low boron content. High quality calcined dolomite is obtained from Belgium. The Veendam plant has a capacity of approximately 100 ktpy using multiple-hearth calciners and two modern shaft furnaces to produce a high quality refractory grade magnesia. Approximately 1.25 t of $MgCl_2$ are required to produce one ton of dead-burned magnesite.

Israel

The Dead Sea Periclase Ltd., subsidary of Israel Chemicals Ltd., operates a plant at Mishor Rotem. The principal raw material is the magnesium chloride brine obtained from the Dead Sea. Reserves have been given as 22 Gt. The waters of the Dead Sea are concentrated to a 30% solution of the chlorides of magnesium, potash, sodium, calcium, and, to a lesser extent, bromides of various elements. Bromine and potash are recovered. The brine is further concentrated to increase the magnesium chloride content and decrease the other salts. The brine is pumped some 25 km to Mishor Rotem where it is treated by the Aman process to give a fine magnesium oxide powder and a hydrochloric acid content of 18 to 20%. The crude magnesium oxide is hydrated to form insoluble magnesium hydroxide, which is washed with pure water to remove soluble salts and upgrade the magnesium hydroxide to over 99% purity. The purified magnesium hydroxide is calcined, briquetted, and sintered in a shaft furnace at 2 000°C to produce a product with a density of 3.4 and MgO of 99.4%. The dead-burned magnesite capacity of the plant has recently been increased to 77 ktpy; there is a parallel capacity to produce 6 ktpy of high grade speciality magnesia products. Some magnesium hydroxide is obtained from magnesium chloride using sodium hydroxide as the precipitant.

Mexico

In the Quimica del Rey operation of Penoles, the brine is pumped to the surface from the dry lakes of the Laguna del Rey. It is rich in sulfates, chlorides, magnesium, and sodium ions. The brine has the average ion content of SO_4, 221 g/l; Cl, 66 g/l; Mg, 33 g/l; Na, 86 g/l; H_2O, 896 g/l, and a specific gravity of 1.3. Current capacity on a 97% MgO dead-burned basis is approximately 70 ktpy. The Quimica del Mar operation of Penoles near Tampico, has a current capacity of 70 ktpy of an upgraded 97% MgO material.

Typical seawater contains about 34 g of dissolved salts per kg; of this, 23 g is sodium chloride, and the balance mainly magnesium and potassium compounds. The amount of magnesium ion is equivalent to about 2.1 g of magnesia (MgO) per kg of seawater, or over 8 kg per 3 785 l.

Choice of suitable sites for seawater magnesia plants are limited. Large land masses can cause the seawater to be diluted because of freshwater runoff and tide action, with the seawater becoming merely brackish. Other factors such as coastal topography, availability of high purity limestone or dolomite, fuel source, and market location must be closely studied before a site is selected.

PROSPECTING AND EXPLORATION

The search for deposits of any type of magnesite should be based on considerations involving the probable origin of magnesites. In the case of crystalline magnesite, possible locations of deposits are areas of limestone or dolomite terrain that have been subjected to folding or igneous activity. In the case of cryptocrystalline magnesite, possible locations are areas of ultrabasic rocks which show extensive alteration to serpentine. Thus, in either case, the search can be confined within the boundaries of areas that show evidence of dynamic geologic activity; areas such as the midwestern United States would not be fruitful in spite of the large deposits of limestone and dolomite.

Magnesite formations are generally more resistant to weathering than associated formations; bold outcrops frequently characterize magnesite deposits. This is very evident in the Baja California Peninsula where the magnesite stands out as caps on the mountains. Outcrops are sampled by surface chipping and shallow trenching. Field differentiation between commercial and noncommercial magnesite can be aided by the use of cold, dilute hydrochloric acid; limestone will effervesce when treated with the acid whereas magnesite will not. Sometimes the Jolly balance is used for the field identification of magnesite on the basis of specific gravity of the minerals (magnesite, 3.0+, dolomite, 2.85).

Following preliminary investigations, a diamond drill program may be carried out to assess the commercial potential of the deposit. The formal final evaluation is based on the size and location of the deposit, the quantity and distribution of undesired minerals containing silica, lime, iron oxide, and alumina. Tests to determine whether the impurities can be removed by ore dressing is part of the evaluation. Magnesite, to be commercially acceptable, should contain at least 95% magnesium carbonate, an amount which, depending on the accessory oxides, would result in a dead-burned product containing 90 to 94% MgO. Refractory grade magnesia of that quality would have some demand at the lower end of the price scale. The growing demand is for even higher MgO content with this placing increased emphasis on the need to beneficiate the ore.

MINING AND PROCESSING

The type of the deposit dictates the method of mining. Large massive near surface deposits are usually worked by open pit methods; narrow and deep deposits by underground drifts and stopes.

The mined ore is rarely shipped or used in crude form. It is processed near the mine site to yield magnesia products. Invariably some degree of sorting or benefication is applied to the ore prior to heat treatment.

Magnesite mining and processing practices of varying forms and complexity are found in many places around the world.

North America

United States, Nevada: The discovery of magnesite in the Gabbs area in 1927 followed the finding of deposits of brucite in terrain originally prospected for tungsten. Exploration delineated total magnesite reserves in excess of 25 Mt. The ore bodies occur in the folded beds of Upper Triassic dolomites, being faulted and intruded by a variety of dikes and granodiorite stocks.

The deposits were worked by Basic Inc. and its successor company to produce many diversified magnesite products. The plant is located at 1 524 m elevation and the open pit at 1 829 m in the foothills of the Paradise Range in Nye County.

Mining at Gabbs has been by open pit method with 3 m benches, the desired height to maintain proper ore control. Blastholes were drilled 4 m deep on a 2 m square pattern to give a good pit floor for the benches and good ore fragmentation (Willard and Gates, 1963). Blasting was with ammonium nitrate sensitized in a proportion of seven pints of diesel fuel for every 36 kg of nitrate. Charges are primed with one stick of dynamite and an electric blasting cap. Usually 400 to 1 000 holes are shot at one time. Secondary blasting is seldom necessary. The ore is transported 3 km to the crushing plant.

The different grades of raw ore cannot be readily distinguished visually, and sorting into grades is controlled by chemical analysis. Mapping allows selective mining to be carried out. The complex mineralogical nature of the ore body gives a wide range of both lime and silica. Control of quality at the pit is very important.

The ore is crushed in three stages. The minus 9.5 mm material is sent to waste.

Ore from the stockpile containing low grade ore high in lime and silica is treated by flotation to produce a concentrate.

About one-third of the concentrate is calcined in a hearth furnace. Some of the calcines were combined with the flotation concentrates. The mixture is briquetted and fired.

United States, Washington: The magnesite occurs as massive lenses in dolomite. Stringers of siliceous and calcareous impurities are present. The magnesite deposits are found within the Stensgar dolomite formation. While the magnesite outcrops were numerous, only a few of the deposits were worked. Operations ceased in 1968.

The mine had a series of 12 m benches. Primary rock breakage was obtained by drilling 15 cm diameter holes and blasting. This was followed by some secondary breakage. The broken ore was put through a 107 cm gyratory crusher and loaded into aerial tramway buckets and transported some 13 km to the mill.

The quarry rock was crushed, scrubbed and screened into two fractions. The coarse 38 x 3.2 mm size was sent to the heavy media separation plant while the minus 3.2 mm material was stockpiled for flotation. The ore from the heavy media plant was fired in a rotary kiln at 1 593°C. The flotation product was thickened, filtered, dried and sent for calcination.

Canada, Quebec: A deposit of high lime magnesite at Kilmar consists of a series of veins dipping about 85°. The footwall is quartz monzonite and the hanging wall consists of serpentine, diopside, and quartzite. It is mined underground with drifts and cross cuts driven into the vein at suitable levels.

The ore is crushed, washed and screened into two fractions. The coarse size 38 to 4.8 mm is treated in a heavy media separation plant and the minus 4.8 mm size is used as a blending material at the calcination plant. The floats are sent to waste. After blending at the calcination plant the ore is ballmilled, homogenized and dead-burned in a rotary kiln.

Europe, Greece: While the importance of the deposits in Greece have diminished because of the high production costs, the mining methods applied to this type of deposits are classic.

The northern part of the island of Euboea consists of schists-hornstone and serpentine of the oriental Greece zone. The magnesite is of the cryptocrystalline variety located in ruptures, stratified zones, and stockwork. The three types of magnesite present can be classified as:

1. Vein type which may be from 1 cm to 4 m in width and of varying length.
2. Stockwork, which is a network of veins varying from 5 to 60 cm in thickness.
3. Mass type, forming lens-like concentrations of magnesite, with volumes ranging up to several thousand cubic meters. This type is a local anomaly of vein or stockwork magnesite.

The largest operator on the island was FIMISCO who mined the ore principally by open pit methods. Benches have heights ranging from 10 to 20 m. The broken ore is loaded by shovel or front end loader into off-the-road trucks for transport to the ore dressing plants. The ratio of waste and overburden to one ton of magnesia is 80 to 100 t.

At the dressing plant the ore is washed, crushed, and beneficiated. The beneficiation is a combination of optical, magnetic, and heavy media separation. The waste ore, which contains about 12% magnesia, is treated in a flotation plant.

An extensive program of sampling and chemical analysis facilitates the separation of different grades of ore at the calcining plant. The ore is separated into storage piles and reclaimed as required by a front end loader. The regular beneficiated ore is dead-burned in one of four rotary kilns. The flotation concentrate is calcined in a hearth furnace, then briquetted and dead-burned in a modern pressurized shaft furnace.

There are also deposits of ore on mainland Greece which are cryptocrystalline magnesite requiring beneficiation.

Asia, Turkey: Magnesite, mainly cryptocrystalline, is located in the Eskisehir and Kutaya region in the west-central part of the country. The ore is a vein-filling type and is hand sorted to meet specifications before calcination. One company has a shaft furnace but the others use rotary kilns.

Asia, Nepal: The deposit is crystalline magnesite with stringers of pyrite and talc. The ore is mined by open pit methods and selectively mined and hand sorted. Productive capacity of dead-burned magnesite is 50 ktpy.

Australia

The Queensland Kunwarara natural magnesite ore body occupies an arc-shaped 22 km long shallow depression with a granite bottom. There are essentially no outcrops, and the deposit is overlain by a humus black clay overburden with a thickness ranging from 0.30 to 12.0 m.

The magnesite ore, contained within a continuous flat-lying zone ranging in thickness from 2.5 to 20 m, is of the low iron nodular cryptocrystalline type. The nodules range in size from 1 mm to 50 cm in diameter. Origin, geology, and mineralogy here and at Young, Fifield, New South Wales, are similar to the Salda Lake deposits in Turkey.

Mining is done by open pit, and reserves are on the order of several million tons.

OTHER MAGNESIUM MINERALS

Brucite

The deposits of Nevada and Quebec have lost their commercial importance and are being replaced by more economical magnesite deposits. The deposit in Quebec is not operational, having been shut down in the 1950s.

Olivine

This mineral, a magnesium iron silicate, is composed of forsterite ($2MgO\ SiO_2$) and fayalite ($2FeO\ SiO_2$) in solid solution; in common olivine, the latter, fayalite, is the minor constituent present, in an approximate proportion up to about 15%. Olivine is a common constituent of igneous rocks but deposits of sufficient purity to be commercially attractive are not numerous. Olivine is inert, chemically stable, requires no calcination prior to use in making refractory products, has a low heat conductivity that is nearly one-third of magnesite, does not hydrate, is resistant to most metals and slags, and is compatible with most other refractory products.

The North Carolina and Georgia reserves in the United States of olivine have been estimated at over 200 Mt containing approximately 48% magnesia (Hunter, 1941). In Washington State the olivine occurs in dunite rock, a peridotite in which olivine is the dominant mineral. Reserves are estimated at 50 Mt on Cypress Island and several million tons in the Twin Sister Region.

Olivine is mined in Norway for refractory purposes as well as for the manufacture of night storage heaters, oil platform ballast, and as a slag conditioner in iron/steel making. In Norway, the two major producers of olivine are A/S Olivin which operates a large quarry at Aaheim in the Sunmore district producing nearly 3 Mtpy from a 2.5 Gt reserve and Norddal Olivin which has a quarry at Norddal near Alesund.

A Swedish company, Handols Toljstens, produces olivine from a quarry near the Norwegian border.

Austrian olivine production comes from a quarry at St. Lorenzen in Styria. Dunite, from which olivine is derived, is used on a smaller scale in some refractory operations. Mineable quantities of dunite are found in Austria, Sweden, Spain, New Zealand, the CIS, Japan, Zimbabwe, and New Caledonia.

Magnesium Sulfate

This product has been made by leaching serpentine or olivine with sulfuric acid, or by the extraction from well brines and lakes. Most comes from treating magnesia with sulfuric acid.

POLITICAL AND COMMERCIAL CONTROL

The use of magnesia as a refractory material has a relatively short history when compared to that of other heat-resisting materials used in metallurgical processes since antiquity.

The use of magnesia for metallurgical operations was mentioned in 1866 and the first magnesia-based bricks were manufactured in 1868. Prior to World War I magnesia was not produced regularly in the United States; industrial needs were met by imports from Central Europe. A small amount of magnesite was mined in California as early as 1886, with output reaching 10 kt by 1910. Under the impetus of World War I, US production of magnesite reached an annual production level of 21 kt in 1917. The first production from the state of Washington was in 1916, and it soon became the leading producer in the United States, eventually reaching a level of 200 ktpy. Production in Washington continued until the need for higher purity magnesia, plus the depletion of high grade ore, forced closure in 1968.

DEPLETION, TAXES, AND TARIFFS

Depletion allowance for magnesite, dolomite, and brucite mined in the United States has remained at 14% and for olivine 22%.

Post World War I high import tariffs on magnesite imported into the United States ($12.68/t) began to be reduced in 1951, reaching the current minimum levels shown in Table 1.

Table 1. US Import Duties

	Harmonized Tariff Schedule	MFN*	Non-MFN
Dead-burned/fused	2519.90.1000	0.4¢/kilo	1.7¢/kilo
Caustic-calcined magnesia	2519.90.2000	$2.07/ton	$20.70/ton
Calcined dolomite	25.18.20	6% ad valorem	30% ad valorem

*Most Favored Nation

PRODUCTION AND CONSUMPTION

About 75% of all the refractory magnesia is used in one form or another by the steel industry. Thus any change in technology or rate of production in the steel industry results in a parallel change in the market for magnesite. Trends in the total industrial use of magnesia can be conveniently expressed in terms relating to steel production. On that basis, the total apparent consumption of refractory grade magnesia by all industries in 1930 was equivalent to 1.9 kg per ton of steel. By 1958 the rate increased to 5.2 kg as magnesia refractories supplanted other materials in the production of steel. The European advent of the basic oxygen process in 1954 began a new era of steelmaking technology. The process, not fully exploited in the United States until the 1960s, requires a different class of refractories than from the open hearth. Basic oxygen increased production and decreased the quantity of refractories required per ton of steel. The introduction of water-cooled panels in the electric furnace also decreased the overall use of refractories. A current figure for magnesia is now once again 2 kg per ton of steel.

There will probably be no large change in the proportionate quantities of steel produced in the near future; it is not expected that plastics and other metals such as aluminum and magnesium will significantly alter the world per capita consumption of steel.

The new processes for steel making also created a market for high purity magnesia with a closely controlled chemistry. The market for intermediate grade magnesia has diminished considerably. Use of 90% MgO magnesite and 50 to 70% MgO olivine in tundish spray applications is growing, with the proportion of olivine gradually increasing.

There is a growing market for caustic magnesia in water treatment and other environmental projects.

MARKETING, USES AND PRICES

Crude Magnesite

Commercial movement of crude magnesite is relatively minor. Small quantities are used for the production of chemicals as Epsom salts. Crude sales in Europe and the Middle East are made by producers who do not have their own processing facilities for converting crude magnesite to calcined products or chemicals. Norsk Hydro in Becancour, Quebec, takes large quantities of crude ore form China for the production of magnesium metal. The selling price for crude ore ranges from 50 to $70 per ton delivered.

Dead-burned Magnesite

Dead-burned magnesite is used almost entirely for the preparation of refractories. It is used directly as a sintered grain to manufacture ramming mixes, gunning mixes, or brick. The refractories mainly are used in the steel industry but also serve the cement, glass and copper industries. The great merit of magnesia-based refractories is their ability to resist basic slags at high temperatures (Spencer, 1972).

Refractory magnesia is freely sold on the world market, but many refractory producers operate their own source of supply. The total capacity production of refractory magnesia is generally adequate to meet market demands, although it is possible that shortages in certain special grades of high purity magnesia may occur periodically.

The production costs and prices of refractory magnesia are determined principally by the physical and chemical properties of the product. The magnesia oxide (MgO) content alone is not the sole factor in establishing the price. The amounts and proportions of accessory oxides and the physical characteristics of the finished product are equally influential on price. Two different magnesites with identical magnesium oxide contents may be priced at different levels. The range of price for high purity (98%+) refractory grade magnesia is generally in the range 330 to $360 per ton. The price for maintenance grade (90%) generally ranges from 100 to $165 per ton.

Magnesite consumption in US refractory manufacture dropped approximately 25% between 1980 and 1990, with this somewhat related to a similar drop in other major industrial countries. It is anticipated that the consumption level will now stabilize and remain relatively constant always directly subject to steel consumption levels, which in turn will be affected by the replacement of steel by other materials. However, the consumption of magnesite-based refractories will increase in some regions as the per capita consumption of steel significantly increases among the developing countries of the world. Nevertheless, the result could be an overall net sustained level world consumption of dead-burned magnesite over the next ten years. It is anticipated that the availability of high density fused magnesia will preclude the use of dead-burned magnesite containing more than 99% MgO.

Caustic-calcined Magnesia

Caustic-calcined magnesia has many uses. Among these are: In the building industry in the form of lightweight board and flooring; the paper industry in as a binder for grinding wheels and discs; the rubber, ceramic, glass and resin industries; the petrochemical and pharmaceutical industries; the treatment of water; the removal of sulfur oxide; the production of chemicals; and in fertilizer and animal feeds.

Specifications, including sizing, vary greatly and are set by each of the many consuming industries. The price for caustic magnesia varies greatly depending on grade and end use. Natural industrial grades sell for 100 to $125 per ton. Special caustics from seawater brines range from 180 to $260 per ton. Very high purity can sell for 500 to $600 per ton.

Fused Magnesite

Fused magnesite's high density favors its use in mag-carbon and other speciality refractories for the steel industry. Pricing is in the range 350 to $650.

Energy is the major cost factor in producing caustic, dead-burned, and fused magnesite.

Olivine

Processed olivine is used in limited, but growing quantities for refractories, principally in tundishes and ladles. It is also used for the production of magnesium phosphates. In addition to foundry sand, slag control, soil conditioning, and sand blasting, olivine is also used greatly for heat storage blocks where a dense, nonfriable, fairly pure grade of olivine is required.

One of the world's largest manufacturers of olivine-based refractory products is A/S OLIVIN located in western Norway, producing approximately 25 ktpy of a full line of fired and unfired shapes and monolithic refractories for the European market.

Refractory grade olivine pricing is in the range 75 to $120 Mt.

TESTS AND SPECIFICATIONS

Crude Magnesite

Crude magnesite is used mainly to produce dead-burned or caustic-calcined magnesia. Thus the chemical composition is most important; in some cases it is important to know whether the magnesite decrepitates with the application of heat.

Refractory Magnesia

Refractory magnesia (dead-burned magnesia) is subjected to both physical and chemical tests. The nature of the tests and the product specifications are somewhat dependent upon the end use of the material. If the magnesite is to be used for the manufacture of monolithic refractories certain characteristics are needed to assist sintering in actual industrial service. Generally, these same characteristics would not be suitable for brick-making mixes. However, both uses usually require dense, well burned grain with a predominance of large grains. Tests for free lime and hydration are common. Clinkers or sinters of magnesia containing lime are tested for resistance to hydration. Lime, if present in large quantities, may combine with silica to form dicalcium silicate (2CaO SiO_2) which under certain conditions undergoes a rapid inversion that can reduce the magnesia to a dust by hydration. Complete stabilization of dicalcium silicate is possible by additions such as phosphates.

In synthetic high-purity magnesia the size of the periclase crystal is of special importance. The presence and relationship of small quantities of lime and silica is important for the production of resin (carbon-bonded) refractories; this should be at or above a dicalcium silicate ratio. The density requirement for first quality refractory magnesia has been established at a minimum level of 3.40 gm/cc. Boron levels of B_2O_3 should be preferably below 0.02%; some specs require 0.005%.

Caustic-calcined Magnesia

Calcined at about 850°C, most caustic magnesia is sold on a tonnage basis in relatively large lots to chemical, paper, and fertilizer manufacturers.

The specifications are varied. In general, specific requirements as to particle size, color, ignition loss, chemical analysis, bulk density, and activity can be readily satisfied.

Chemical Grade

A class of magnesia differing from the regular commercial grades is the material sold to specific technical and United States Pharmacopeia (USP) standards. They meet rigorous physical and chemistry specifications. The products are used for medicinal purposes, specialized fillers, and various chemical needs. They may be referred to as light or heavy calcined magnesia. When calcined at higher temperatures, 1 000 to 1 200°C, the caustic is generally used in electrical applications. These products are sold by the kilo rather than by the ton as is the case of the refractory grades.

ENVIRONMENTAL CONSIDERATIONS

Because magnesite mining operations are largely in mountainous regions, they have little interface with urban society. The major problems are noise, dust, and disposal of overburden and processing waste.

Mine drainage from underground or open pit operations does

not contribute significantly to the pollution of streams because the drained area contains no significant amounts of soluble or oxidizable minerals/chemicals.

Magnesite calcining plants emit dust in substantial quantities, but this can be controlled by the installation of modern dust collectors and precipitators.

In the production of synthetic magnesia the return of spent seawater with excess contents of lime needs to be controlled to prevent the formation of a white colored area in the water body and a change of pH in the immediate discharge area.

Proper dust collection is also required on the dead-burning equipment. Wet scrubbers and electrostatic precipitators are generally installed to take care of the problem. With proper remedial measures the cosmetic quality of the discharges can meet acceptable levels, as well as meet all environmental codes.

Actually, magnesite itself appears in medicine as Milk of Magnesia and is used in water treatment and waste gas desulfurizing.

BIBLIOGRAPHY AND REFERENCES

Anon., 1956, ''Magnesium Chemicals,'' *Chemical Engineering*, Vol. 63, pp. 346–349.

Anon., 1987, *Minerals Yearbook,* US Bureau of Mines.

Anon., 1990a, ''Metals and Minerals Annual Review,'' *Industrial Minerals.*

Anon., 1990b, *Magnesium and Magnesium Compounds,* US Bureau of Mines.

Anon., 1991, *Ceramic Bulletin,* Vol. 70, No. 5.

Bain, G.W., 1924, ''Types of Magnesium Deposits and Their Origin,'' *Economic Geology,* Vol. 19, No. 5, pp. 412–423.

Bennett, W.A.G., 1941, ''Preliminary Report on Magnesite Deposits of Stevens County, Washington,'' Report of Investigation No. 5, Washington State Division of Geology, p. 25.

Bennett, W.A.G., 1943, ''Character and Tonnage of the Turk Magnesite Deposits,'' Report of Investigation No. 7, Washington State Division of Geology, 22 pp.

Bodenlos, A.J., 1954, ''Magnesite Deposits in Serra Das Equas, Brumado, Bahia, Brazil,'' Bulletin 975-C, US Geological Survey, pp. 87–170.

Bray, W.T., and Hilchey, G.R., 1957, ''Magnesite Deposits of Kilmar, Quebec,'' Geology of Canadian Industrial Mineral Deposits, Canadian Institute of Mining and Metallurgy, Montreal, pp. 164–166.

Clark, G., 1992, ''Magnesia Around The World,'' *Industrial Minerals,* Apr., pp. 45–77.

Comstock, H.B., 1963, ''Magnesium and Magnesium Compounds,'' Information Circular 8201, US Bureau of Mines, 128 pp.

Coope, B., 1981, ''Caustic Magnesia Markets,'' *Industrial Minerals,* Feb., pp. 43–51.

Coope, B., 1989, ''Magnesia Markets,'' *Industrial Minerals,* Sep., pp. 45–57.

Campbell, I., and Loofbourow, J.S., 1962, ''Geology of Magnesite Belt of Steven County,'' Bulletin 1142-F, US Geological Survey, pp. F1–50.

Dunbar, C.O., and Rogers, J., 1957, ''Principles of Stratigraphy,'' John Wiley and Sons, New York.

Duncan, L.R., and McCracken, W.H., 1981, ''The Impact of Energy on Refractory Raw Materials,'' ILAFA-ALAFAR 1980, Lima, Peru, *Industrial Minerals,* Jan., pp. 19–27.

Egleston, T., 1886, ''Basic Refractory Materials,'' *Trans. AIME,* Vol. 14, pp. 460–492.

Frost, M.T., McCallum, D., and Jones, M.H., 1988, ''The Kunwarara Magnesite Project,'' 8th Industrial Minerals International Congress, CSIRO, London.

Gilbert, F.C., and Gilpin, W.C., 1951, ''Production of Magnesite from Seawater and Dolomite,'' *Research Magazine,* No. 8, pp. 348–353.

Gouge, M.F., 1944, ''Brucite Magnesia,'' *Journal of the American Ceramic Society,* Vol. 27, No. 1, pp. 8–10.

Hunter, C.E., 1941, ''Forsterite Olivine Deposits of North Carolina,'' Bulletin 41, Department of Conservation and Development, Division of Minerals Research, pp. 114–116.

Kato, T., 1929, ''The Origin of the Manchurian Magnesite Deposits,'' *Economic Geology,* Vol. 24, pp. 90–93.

Mortl, G., and Leopold, F., 1978, ''Caustic Magnesia for the Building Industry,'' *Proceedings,* 3rd Meeting of Industrial Minerals, Paris.

Niinomy, K., 1925, ''The Magnesite Deposits of Manchuria,'' *Economic Geology,* Vol. 20, No. 1, Jan.-Feb., pp. 25–53.

Nishihara, H., 1956, ''Origin of the Bedded Magnesite Deposits of Manchuria,'' *Economic Geology,* Vol. 51, No. 7, pp. 698–711.

Paone, J.O., 1970, ''Magnesium,'' Mineral Facts and Problems Bulletin No. 650, US Bureau of Mines, pp. 621–638.

Rosenberg, P.E., and Holland, H.D., 1964, ''Calcite-Dolomite-Magnesite Stability Relations in Solutions at Elevated Temperatures,'' *Science,* Vol. 145, pp. 700–701.

Rosenberg, P.E., and Mills, J.W., 1966, ''A Mechanism for the Replacement of Magnesite in Dolomite,'' *Economic Geology,* Vol. 61, No. 3, May, pp. 582–586.

Schroeder, N.C., 1948, ''The Genesis of the Turk Magnesite Deposits, Stevens County Washington,'' *Compass,* Vol. 26, No. 1, Nov., pp. 37–46.

Schmid, H., 1989, ''Turkey's Salda Lake,'' *Industrial Minerals,* pp. 19–30.

Schmid, H., 1984, ''China—The Magnesite Giant,'' *Industrial Minerals,* Aug., pp. 27–45.

Siegfus, S.S., 1927, ''Some Geological Features of the Washington Magnesite Deposits,'' *Engineering Mining Journal,* Vol. 124, No. 22, Nov., pp. 853–856.

Spencer, D.R.F., 1972, ''Basic Refractory Raw Materials,'' *Transactions and Journal of British Ceramic Society,* Vol. 71, No. 5, pp. 123–134.

Valdiya, K.S., 1968, ''Origin of the Magnesite Deposit of South Pithoragarh, Kumaun Himalyaya, India,'' *Economic Geology,* Vol. 63, No. 8, Dec., pp. 924–934.

Van den Assem, L., 1981, ''Planning for New Industrial Projects,'' Billiton Industrial Minerals, Jan.

Walper, J.L., 1964, ''Geological Report on The Marble Canyon Area, Culbertson County, Texas,'' Texas Bureau of Economic Geology, Austin, TX, pp. 1–27.

Wicken, O.M., 1949, ''Production of Seawater Magnesite,'' *Electrical Furnace Steel Conference Proceedings,* Vol. 7, AIME, New York, pp. 212–217.

Willard, H.P., and Gales R.W., 1963, ''Selective Open Pit Mining Featured at Gabbs,'' *Mining Engineering,* Vol. 15, No. 10, pp. 45–46.

MANGANESE MINERALS

R.A. HOLMES

Although manganese is a metallic element and is widely dispersed in nature, it never occurs except as a compound in combination with other elements. Use of such compounds in the production of glass is known to have occurred in early Egypt.

The dioxide of manganese was considered a compound of iron until 1774 when C.W. Schule first recognized it as an element. In the same year, a Swedish mining engineer, J.G. Gahn, became the first to isolate manganese. In 1856 development of the Bessemer process of steelmaking gave economic importance to manganese, and later, in 1882 Robert Hadfield discovered the benefits of high manganese steels.

GEOLOGY

Physical Properties

Elemental manganese is a silver-gray metal, resembling iron but harder and more brittle and used primarily in alloys (both ferrous and non-ferrous) and a wide variety of chemical compounds. Some of the physical properties of manganese include: melting point—1 245°C; boiling point—2 150°C; density at 20°C—7.43 g/cm^3; specific heat at 25.2°C—0.115 cal/g; latent heat of fusion—63.7 cal/g; hardness on Mohs scale—5.0; linear coefficient of thermal expansion from 0 to 100°C—22×10^{-6}.

Mineralogy

There are over one hundred minerals that contain manganese. These minerals vary from those with compositions that are predominantly manganese to those having only minor percentages of manganese.

Mineral	Chemical Composition	% Mn
Hausmanite	Mn_3O_4	72
Polianite	MnO_2	63.1
Pyrolusite	MnO_2	60–63
Cryptomelane	$KR_8O_{16}R$ (R=Mn)	Variable
Psilomelane	$BaMn\ Mn_8O_{16}(OH)_4$	45–60
Coronadite	$PbR_8O_{16}R$ (R=Mn)	Variable
Hollandite	$BaR_8O_{16}R$ (R=Mn)	Variable
Manganite	$Mn_2O_3 \cdot H_2O$	62
Braunite	$3Mn_2O_3 \cdot MnSiO_3$	62
Tephroite	$2MnO \cdot SiO_2$	54.3
Rhodochrosite	$MnCO_3$	47
Rhodonite	$MnSiO_3$	42
Spessartite	$3MnO{\cdot}Al_2O_3 \cdot 3SiO_2$	33.3
WAD	Hydrous Mn Oxides	Variable
Franklinite	$(Fe,Mn,Zn)O(Fe,Mn)_2O_3$	Variable
Asbolan	Cobaltiferous WAD	Variable
Alabandite	MnS	63.14

Distribution of Deposits

Manganese ore deposits are found worldwide and were formed in various geological environments, but only a rather limited number of deposits have high grade manganese ore in sufficient quantities to be mined and utilized economically on an industrial scale. It is worth noting the fact that almost all of the significant deposits can be classified into two types of deposits: marine chemical sedimentary deposits, and residual (secondary) enrichment deposits. There are, however, a much larger number of geological types but these are not of commercial significance at this time.

Sedimentary deposits are the most common and are usually stratiform or lenticular. Manganese minerals were formed by a chemical process during the deposition of marine sediments. They usually contain manganese oxides and carbonate minerals, sometimes interbedded together or with other sedimentary rocks such as limestone or shale Examples of this type of deposit are the Russian ore bodies of Nikopol and Tchiatoura, as well as the Kalahari deposits in South Africa and deposits of Groote Elyandt in Australia.

Residual deposits were formed in a different way: by alteration of existing manganese deposits or by concentration of the manganese minerals when other minerals were washed away by weathering or ground water processes. The Nsuta deposit in Ghana, the Amapa deposit in Brazil, the Moanda deposit in Gabon, and nodules in the residual clays of the US Southern Appalachians are examples of this type of geological process. In the case of the Ghana and the Amapa deposits, this is only true for the outer layers of the deposit containing oxide minerals, the inner part being comprised of carbonate minerals including manganese carbonate, probably from marine origin. Some sedimentary and residual-type deposits have been metamorphosed, giving rise to small high grade ore bodies. These deposits are regionally metamorphosed, occurring in marbles, slates, quartzites, schists, and gneisses. Some of these deposits, such as the Franklin, NJ, deposit are rich enough to be commercial without secondary enrichment; however, most of the exploitable deposits have been secondarily enriched.

Due to the diversity and complexity of manganese deposits, both with respect to deposition and chemistry, a wide range of impurities are almost invariably present in the ores.

Table 1 shows that the reserves (i.e., a measured resource that can be economically and legally extracted) are estimated at 814×10^6 tons of contained Mn, which equates to more than a 100 year supply at the current level of production. The reserve base is made up of the marginally economic reserves and sub-economic resources and are some 4.5 times greater than the proven reserves.

Also apparent from study of Table 1 is the fact that some 75% of manganese reserves are found in two countries, the USSR and South Africa. On the other hand, North America (i.e., the United States and Canada) have few significant deposits.

The largest deposit, or at least one of the largest, in the United States is located at Chamberlain, SD. This deposit is sedimentary

Table 1. World Manganese Ore Reserves and Reserve Base (million tons of manganese content)

	Reserves	Reserve Base
North America		
Mexico	4	9
South America		
Brazil	21	59
Europe: USSR	295	454
Africa		
Gabon	53	163
Ghana	1	5
Republic of South Africa	369	2630
Asia		
China	14	29
India	17	25
Oceania: Australia	40	154
WORLD TOTAL (Rounded)	814	3528

Source: "Manganese," Thomas S. Jones, US Department of the Interior Bureau of Mines, 1990.

in origin and consists of manganiferous concretions in the Pierre shale, which is Upper Carboniferous in age.

The Batesville, AR, deposit contains a sporadic scattering of pyrolusite and wad-type residual materials in clay.

The Cuyuna iron range in Minnesota contains iron ore carrying varying quantities of manganese. The manganese content ranges from less than 1 to more than 17%.

Southern Appalachian deposits begin in the area of the Shenandoah Valley and extend south to the area of Cartersville, GA, and on into Alabama. While a wide variety of manganese minerals occur in lenses, beds, and residual pockets, only the oxides in residual pockets have been of any economic importance.

The deposit at Artillery Mountain in Arizona has been operated from time to time during national emergencies. It has been estimated that the deposit contains approximately 175 Mt of ore, averaging from 3 to 4% manganese.

The Three Kids property near Henderson, NV, was operated by Manganese, Inc., during World War II. At that time, it was the largest producer of manganese in the United States. The primary ore, containing about 18% manganese, was beneficiated to produce a concentrate averaging 45% manganese.

The occurrence of large tonnages of manganese nodules in the three major oceans has been well established. The major elements are manganese and iron with manganese values of 8.2 to 50.1% Mn (median value 24.2%) determined. On the same samples, nickel values of 0.16 to 2.0% were found.

It is highly probable that if sea nodules were to be mined, manganese and cobalt would be considered as byproducts with mining depending mostly on the world demand for nickel.

TECHNOLOGY

Exploration and Evaluation

Manganese deposits are one of the more difficult ores to evaluate. In general, the deposits are small and scattered and expensive geophysical means are not usually economically feasible. Also, the geophysical exploration costs are often beyond the financial capacity of the small operator. However, since manganese has relatively high solubility, geochemical techniques can be an effective tool in an exploration program.

One of the main difficulties is securing representative samples of an ore body for proper evaluation and analysis. In the past, a large percentage of the holes drilled for samples were by cable or churn drill methods. These methods require extreme care in order to secure valid data. In most cases, the data have been inaccurate.

Rotary drilling techniques are somewhat better, but again, extreme care must be exercised in order to secure representative samples. Where the deposit contains wad or soft oxides in clay, the particle size of the manganese mineralization ranges from large boulders to −200 mesh material. This makes sampling procedures difficult to employ. In general, the hard and deeper deposits are easier to sample with some degree of accuracy. There are many factors to consider in the economic evaluation of manganese deposits. Among these are deposit size, concentration of manganese in the deposit, impurities associated with the manganese mineralization, and the amenability of the ore to beneficiation.

Due to the diversity and complexity of manganese formations, the impurities are many in number and complex in nature. Following is a broad generalization of the types of impurities:

1. Metallic impurities: iron, lead, zinc, nickel, copper, cobalt, arsenic, and silver minerals
2. Non-metallic impurities: sulfur, phosphorus, alkali, and alkaline earth minerals
3. Gangue: silica and alumina
4. Volatiles: water, carbon dioxide, and organic matter

Prior to World War II, few manganese operations had milling plants which were more complicated than washing, screening, jigs, tables, and, in a few instances, flotation. Now it is necessary to increase the manganese content and decrease impurities through increasingly complex beneficiation practices. Because each deposit is distinctly different from most other deposits, no single process is applicable to all.

Manganese Ore Production

It is worth noting that, except for 1983, manganese ore production worldwide has remained at a fairly constant level. The production in India is almost entirely devoted to the indigenous metallurgical production (steel and ferroalloys) with only some 250 kt exported. Chinese production is important but almost entirely constituted of low grade manganese ore (Mn less than 30%) and exports are insignificant.

Table 3 shows that the four main western world producers of manganese ore are South Africa, Gabon, Australia, and Brazil. Table 4 provides statistics for 1989 for the main manganese producing countries, including production share, average grade, and reserves of the relevant mine areas.

The manganese industry in the United States relies almost entirely on foreign ores containing 35 to 55% manganese. The small quantities of manganese ore produced in the United States are in general used as a pigment in the manufacture of brick.

The Federal Emergency Management Agency (FEMA) continues to maintain a stockpile of manganese ore and manganese products. The stockpile goals in 1990 were the same as those established in 1980: natural battery ore—23 kt, synthetic manganese dioxide—23 kt, chemical ore—154 kt, metallurgical ore—2 450 kt, and high-carbon ferromanganese—398 kt. Beginning in 1987 and to continue for seven years, FEMA established a basis for a ferroalloy upgrading program. The goal of the program was to add not less than 61 kt of high-carbon ferromanganese to the National Defense Stockpile each year. The program calls for the Administrator of General Services to obtain bids from domestic producers of high carbon ferromanganese and award contracts for the conversion of manganese ores held in the National Defense Stockpile into high carbon ferromanganese.

Table 2. Manganese Ore Production 1981–1990 (thousand tons per year)

	1981	1982	1983	1984	1985	1986	1987	1988	1989	1990
Mexico	578	590	350	476	396	459	385	444	395	406
Brazil	2 042	2 340	2 092	2 693	2 523	2 697	2 067	1 990	2 080	1 996
Chile	—	—	—	—	—	32	32	44	44	40
Gabon	1 488	1 512	1 857	2 119	2 340	2 510	2 403	2 254	2 593	2 631
Ghana	223	160	173	269	318	304	275	259	279	254
Morocco	110	96	73	59	44	40	43	30	32	32
South Africa	5 039	5 216	2 886	3 049	3 601	3 719	2 892	3 454	3 623	3 672
USSR	9 153	9 825	9 825	9 879	10 070	9 888	9 344	9 435	9 072	9 072
Australia	1 411	1 123	1 370	1 849	2 003	1 649	2 180	1 986	2 124	1 988
China	1 597	1 597	1 597	2 903	2 631	2 722	2 631	3 175	3 175	3 175
India	1 526	1 490	1 281	1 130	1 240	1 213	1 302	1 333	1 334	1 361
Other	389	355	348	434	369	278	276	288	312	314
TOTAL	23 556	24 223	21 907	25 051	25 353	24 967	23 921	24 328	25 063	24 667
Mn Content (Ave.)	35.6%	35.4%	35.5%	34.3%	34.3%	35.4%	34.6%	34.6%	34.9%	34.9%

Source: Department of the Interior, US Bureau of Mines.

Table 3. Production Share (%) of Main Western World Manganese Ore Producing Countries (excluding Russia and China)

Country	1982	1983	1984	1985	1986	1987	1988	1989
South Africa	42	29	27	30	30	27	31	32
Gabon	14	22	21	22	23	25	22	23
Australia	11	15	17	17	15	17	19	19
Brazil	17	17	19	17	19	16	14	14
India	9	10	8	8	7	8	7	7
Mexico	3	3	4	3	4	4	3	4
Ghana	1	2	2	2	2	2	2	2
Others	2	2	2	1	1	1	1	1

Note: rounded figures.

Specifications

Typical ore analyses for some of the higher grades ores available are shown in the following table. All analyses are on a dry basis.

Ferromanganese is supplied in three categories: standard (high-carbon) ferromanganese, medium-carbon ferromanganese, and low-carbon ferromanganese. Several additional grades are designated within these three categories. Chemical specifications for these grades, as designated by The American Society for Testing and Materials (ASTM) A 99–76, are shown in Table 6.

The grade or type of alloy to be used is selected on the basis of the most cost effective addition which conforms with the end use specifications.

Chemical specifications for silicomanganese as defined by ASTM A 483–64 (reapproved 1980) are as follows:

	Composition, %		
Element	Grade A	Grade B	Grade C
Manganese %	65.0–68.0	65.0–68.0	65.0–68.0
Silicon %	18.5–21.0	16.0–18.5	12.5–16.0
Carbon, max %	1.5	2.0	3.0
Phosphorus, max %	0.20	0.20	0.20
Sulfur, max %	0.04	0.04	0.04

USES

The preponderance (95%) of all manganese ore produced is consumed in the manufacture of steel primarily as ferromanganese and silicomanganese, and other minor alloy related industries. The remaining 5% is consumed by the non-alloying industries; i.e., chemical industry, paints, fertilizers, and batteries, and the production of electrolytic manganese metal (EMM).

Although the non-steel segment is small in comparison to the total, it is constituted of a wide variety of products and applications and primarily employs high grade manganese ores (48 to 52% Mn). Basically, this segment can be broken down into three general market groups.

1. Bulk commodity market—In this grouping, final pricing is of extreme importance and quality generally of a secondary nature. Products required undergo minimal transformation which may include drying and grinding when natural manganese ore is required as in colorant usage, and chemical oxidations to reduction of MN^{+4} to Mn^{+2} for use in general animal feed and fertilizer industry or as head feed for the production of powder and granular manganese sulfate/oxy-sulfate, which also is used primarily in the feed and fertilizer industry.
2. Commodity products—Manganese ore used in this segment goes into the production of potassium permanganate, electrolytic manganese metal (EMM), and battery products, including both naturally occurring manganese dioxide (NMD) and synthetically produced manganese dioxide (SMD). The synthetic product is produced using chemical processing (CMD) or more commonly the electrolysis of manganese sulfate to generate $Mn0_2$ (EMD). Product pricing in this group must be competitive, but quality and performance are of prime interest.
3. Specialty chemicals—This segment is characterized by moderate to small volume, value-added products relying on high quality standards, strong R & D and customer service inputs, quality-oriented production teams, and a close customer-supplier coordination.

It is estimated that some 40% of all manganese ore not used for alloying operations is consumed by the primary battery industry, either as natural battery-grade manganese dioxide or in production of synthetically produced manganese dioxide.

In metallurgical applications, manganese is essential in steel manufacture. In most grades of steel, higher levels of manganese are required than would normally be present in the steel. In some cases, the manganese may be introduced as ore. However, the additional requirement is most commonly obtained by the controlled addition of ferromanganese usually produced in a submerged arc furnace with electrical energy. It may also be produced in a ferromanganese blast furnace utilizing coke.

Manganese in steel serves three basic purposes: alloys with the

Table 4. Comparison of Manganese Ore Producing Countries (1989)

Producing Country	Production (1 000 t)	Average Mn%	Average Content Mn (1 000 t)	Mining Reserves (1 000 t)	Production Share %
Australia	2 124	45.98	977	170 000	8.6
Gabon	2 593	47.66	1 236	180 000	10.5
South Africa	3 623	42.51	1 540	2 900 000	14.7
Brazil	2 080	39.00	811	65 000	8.4
Ghana	279	31.07	87	5 000	1.2
India	1 334	30.00	400	28 000	5.4
Mexico	395	39.06	154	10 000	1.6
USSR	9 072	30.77	2 791	500 000	36.7
China	3 175	28.00	889	32 000	12.9

Table 5. Ore Source

Compound Element %	Amapa (Brazil)	Comilog (Gabon)	GEPF (Australia)	CVRD (Brazil)	Ghana
Mn	53.0	52.6	52.3	48.8	48.6
MnO_2	79.0	78.5	77.8	72.5	71.8
Chem.H_2O	3.20	4.75	4.2	5.43	—
Fe	3.1	2.1	2.6	3.7	3.7
SiO_2	2.0	1.8	2.65	2.4	8.5
Al_2O_3	4.3	5.6	3.12	7.0	3.5
Phos.	0.08	0.11	0.07	0.08	0.12
As	0.18	0.004	0.007	0.015	0.03
BaO	Trace	0.21	2.3	0.23	0.09
CaO	—	0.10	0.10	0.20	0.11
Na_2O	0.25	0.04	0.15	0.05	0.24
K_2O	1.95	0.80	0.80	0.94	0.58
C	0.15	0.03	0.013	—	—
Ni	Trace	0.04	0.024	0.05	—
Cu	0.12	0.03	0.007	0.05	—
Co	—	0.07	0.012	0.04	—
Cr	—	<.002	0.003	—	—
Pb	Trace	0.005	0.005	0.02	—
Zn	0.05	0.03	0.001	0.07	—
Mg	—	0.11	0.23	0.32	—
Mo	—	0.003	0.008	—	—
TiO_2	—	0.04	0.05	0.04	0.11
S	0.017	0.017	0.02	—	0.011

Source: Chemetals, Inc.

Table 6. Ferromanganese Categories

Standard Ferromanganese

	Grade A	Grade B	Grade C
Manganese %	78.0 to 82.0	76.0 to 78.0	74.0 to 76.0
Carbon, max %	7.5	7.5	7.5
Silicon, max %	1.2	1.2	1.2
Phosphorus, max %	0.35	0.35	0.35
Sulfur, max %	0.05	0.05	0.05

Medium Carbon Ferromanganese

	Grade A	Grade B	Grade C	Grade D
Manganese %	80.0 to 85.0	80.0 to 85.0	80.0 to 85.0	80.0 to 85.0
Carbon, max %	1.5	1.5	1.5	1.5
Silicon, max %	1.0	1.5	0.70	0.35
Phosphorus, max %	0.30	0.30	0.30	0.30
Sulfur, max %	0.02	0.02	0.02	0.02

Low Carbon Ferromanganese

	Grade A	Grade B
Manganese %	85.0 to 90.0	80.0 to 85.0
Carbon, max %	as specified	0.75
Silicon, max %	2.0	5.0 to 7.0
Phosphorus, max %	0.20	0.30
Sulfur, max %	0.02	0.02

steel to improve strength, toughness, and improvement during heat treatment; combines with sulfur yielding better rolling and forging properties; and acts as a deoxidizer.

Table 7 shows consumption of manganese ferroalloys in the United States

Table 8 shows the unit consumption of manganese in iron and steel making stated in kilograms of manganese per ton of raw steel.

ECONOMIC FACTORS

Manganese Ore Pricing

Metallurgical manganese ore containing minimum 48% Mn underwent a substantial increase in cost starting in 1988. Annual prices for metallurgical manganese ore delivered CIF US ports for the years 1982–90 are shown in Table 9. Manganese ore of this quality is sold on a dry ton unit (DMTU); the cost per DMTU being the price for 10 kg of contained Mn.

INDUSTRY STRUCTURE

Several changes have occurred in recent years that have had effects on the chemical manganese industry in the United States. Foote Mineral Corporation ceased production of EMM at its facility in New Johnsonville, TN. This facility was converted to the production of EMD and the New Johnsonville operation was subsequently acquired by Chemetals, Inc., in 1987. Tennessee-Eastman Company, long a supplier of manganese sulfate (Techmangan R) to the agricultural and animal feed industry, changed its method of hydroquinone manufacture such that manganese dioxide was no longer required as an oxidant, thus eliminating manganese sulfate as a byproduct. This market is now served primarily by imports from Mexico and China and domestic production.

Forward integration occurred when BHP initiated production from an EMD facility in Australia and with Comilog's acquisition of the facilities of Sadacem S.A., which included Sadaci, Ghent, Belgium, Sedema, Tertre, Belgium, and Chemetals, Inc., USA. Hoechst (Germany) and Daichi (Japan) withdrew from the EMD market.

EMD manufacturers have continued to improve quality by continually decreasing levels of impurities in the final product. This has been necessary to meet the environmental requirements of primary batteries to decrease the mercury content required for dry cells and to eventually produce mercury-free batteries.

Table 7. US Consumption, by End Use, and Industry Stocks of Manganese Ferroalloys and Metals in 1990 (tons, gross weight)

	Ferromanganese		
End Use	High Carbon	Medium and Low Carbon	Total
Steel:			
Carbon	232 747	79 229	311 976
Stainless and heat-resisting	12 634	[1]	12 634
Full alloy	30 912	7 829	39 540
High-strength, low alloy	24 876	4 729	29 605
Electric	[1]	[1]	[1]
Tool	235	2.7	238
Unspecified	122	581	702
Total steel[2]	302 326	92 371	394 697
Cast irons	11 760	909	12 671
Superalloys	W	—	W
Alloys (excluding alloy steels and superalloys	864	W	864
Miscellaneous and unspecified	5 515	211	5 727
TOTAL CONSUMPTION	320 467	93 467	413 959
TOTAL MANGANESE CONTENT[5]	250 000	74 545	324 545
Stocks, Dec. 31: Consumers and producers	47,592	13,949	61,541

End Use	Silico-manganese	Manganese metal
Steel:		
Carbon	67 383	[1]
Stainless and heat-resisting	4 652	2 819
Full alloy	16 375	629
High-strength, low alloy	5 426	[1]
Electric	[1]	[1]
Tool	[1]	42
Unspecified	210	1 650
Total steel[2]	94 646	5 170
Cast irons	1 575	—
Superalloys	—	164
Alloys (excluding alloy steels and superalloys)	W	17 315[3]
Miscellaneous and unspecified	2 515	464
TOTAL CONSUMPTION	98 136[4]	23 113
TOTAL MANGANESE CONTENT[5]	64 546	22 727
Stocks, Dec. 31: Consumers and producers	8,328	3,807

[W] Withheld to avoid disclosing proprietary data; included with "Miscellaneous and Unspecified," [1]Withheld to avoid disclosing company proprietary data; included with "Steel: Unspecified," [2]Includes estimates, [3]Approximately 90% of this subtotal was for consumption in aluminum alloys, [4]Internal evaluation indicates that silicomanganese consumption is considerably understated, [5]Estimated based on typical percent manganese content (rounded).
Source: "Manganese," 1990, T.S. Jones, US Department of the Interior, US Bureau of Mines.

Table 8. US Unit Consumption of Manganese in Ironmaking and Steelmaking

Form Used	1989	1990
Ironmaking		
Basis: Production of Raw Steel, Mt[1]	44.4	44.9
Ore[2]	0.5	0.3
Steelmaking		
Basis: Production of Raw Steel, Mt	44.9	45.4[c]
Ore[2]	[3]	[3]
Ferromanganese	3.0	3.1
Silicomanganese	0.6	0.6
Manganese metal	.05	.04
TOTAL, Steelmaking	3.7	3.8

[c]Estimated. [1]As ingots, continuous or pressure cast blooms, billets, slabs, etc. [2]Containing 35% or more manganese. [3]None reported.
Source: "Manganese," 1990, T.S. Jones, US Department of the Interior, US Bureau of Mines.

Table 9. Yearly Pricing—Manganese Ore—Metallurgical Grade—CIF US Ports

Year	Price (DMTU) in US Dollars
1982	$1.56
1983	1.36
1984	1.40
1985	1.41
1986	1.32
1987	1.27
1988	1.75
1989	2.80
1990	3.85

Source: Chemetals, Inc.

PROBLEMS AND FUTURE TRENDS

Manganese ore production will continue to depend primarily on the production of steel. Growth rates are estimated to be about 1% per year. Prices for manganese ore, which escalated rapidly in 1989 and 1990, are expected to stabilize or possibly soften a little in the immediate future.

Efforts will continue to find a process for upgrading low grade domestic deposits to compete with imported ores. This work will occur primarily through the efforts and auspices of the US Bureau of Mines.

ACKNOWLEDGMENT

Appreciation is expressed to Chemetals, Inc., for allowing access to internal reports prepared for and by Chemetals and for general information supplied by their personnel.

BIBLIOGRAPHY AND REFERENCES

Anon., ASTM Specifications A99–76; A483–64.

Anon., 1984, *The Processing and Marketing of Manganese; Areas for International Cooperation*, United Nations Conference on Trade and Development, Geneva.

Anon., 1985a, *Exploring Our New Ocean Frontier*, Marine Minerals, Office of Technology Assessment, US Government Printing Office, Washington, DC.

Anon., 1985b, *Mineral Facts and Problems*, US Department of the Interior, US Bureau of Mines.

Anon., 1986–1989, *Mineral Year Book*, Vol. 1, Metals and Minerals, US Department of the Interior, US Bureau of Mines.

Anon., 1987, *Stockpile Report to the Congress*, Apr.-Sep., Federal Emergency Management Agency.

Jacoby, C.H., 1983, "Manganese," *Industrial Minerals and Rocks*, 5th ed., S.J. Lefond, ed., Society of Mining Engineers, pp. 897–906.

Jones, T.S., 1990, *Manganese*, US Department of the Interior, US Bureau of Mines.

METALLURGICAL USES

FLUXES FOR METALLURGY

HAROLD R. KOKAL AND MADHU G. RANADE

A metallurgical *flux* is a substance that is added to combine with gangue (unwanted minerals) during ore smelting, with impurities in a molten metal, or with other additives in metal refining processes to form a *slag* that can be separated from the metal. Because slags are immiscible with the metallic melt and are of lower density, a separation of the slag and metal occurs if the viscosity and surface tension are of the proper values. The chemistries of slags are adjusted to provide the proper melting point, viscosity, surface tension, conductivity, specific heat, density, or chemical properties to effect the desired task. In addition to absorbing impurities from the metal, the purposes of the slag are to thermally insulate the metal bath, protect the molten metal from the atmosphere, and control the chemical potential of the system. Several excellent references on slags and metallurgical fluxes are available (Boynton, 1980, Lankford, Jr., et al., 1985, Turkdogan, 1983, Rosenqvist, 1974, Fine and Gaskell, 1984).

The function of the slag might vary at different stages within a process prior to final melting. For example, the slag composition and behavior will change as materials descend in the iron blast furnace, or as they melt in the early stages of formation of steelmaking slags. Selection of the chemistry for a slag might be influenced by factors outside the primary function as when blast furnace slags are used to make cement, rock wool insulation, or fertilizer supplements. Sometimes slags contain a sufficient amount of valuable recoverable elements to be sold as raw materials for other processes.

Fluxes are often referred to as *acid*, *basic*, or *neutral*. Acid fluxes are those that generally form acids in water and bases are those that would generally form bases in water. Typical acid fluxes are silica, alumina, and phosphorus, although alumina can function as either an acid or base. Typical basic fluxes are lime and magnesia. Fluorspar, or calcium fluoride, is a neutral substance because it can be viewed as the reaction product of a base and an acid. The degree of acidity (ratio of acids to bases) or basicity (ratio of bases to acids) is often specified to characterize the slag chemistry for a particular system. The system might also be referred to as acid or basic depending on the choice of slag, for example, basic steelmaking uses slags with more bases (lime and magnesia) than acids (silica and alumina).

Many forms of minerals and compounds have been used as fluxes depending on the process requirements, availability, costs, requirements for recycling of intermediate products, and environmental concerns. Because slags are most often mixtures of oxides and silicates, fluxes are usually oxides, carbonates (which decompose to oxides) and silicates. Slags comprising phosphates, borates, sulfides, carbides, or halides have also been used (Rosenqvist, 1974, Moore, 1981, Szekely et al., 1989). In ironmaking and steelmaking, slags are basic, and the primary flux added is lime-bearing, although some acidic components are sometimes added. In nonferrous processes slags and fluxes are acidic with silica as the primary component, and although the use of basic flux such as limestone or lime is not extensive, some amount is used to modify slag properties and some refining slags are lime based.

The production of iron is accomplished by smelting of ores, pellets, and sinter in blast furnaces with subsequent refining of molten iron and scrap in oxygen-blown processes. The electric furnace is used to make steel, stainless steels, ferroalloys, and special alloys, whereas ferromanganese is often made in blast furnaces. Fluxes are used in all of these processes, and either limestone or lime is the major flux component with some dolomite, dolomitic lime, silica, alumina, and fluorspar being used. In ironmaking, the flux is added either by direct charging or through sinter and fluxed pellets. In steel refining by the basic oxygen process, flux is added as lime and dolomitic lime that are either charged as lumps or injected as fines. Some small amount of limestone is sometimes used. Fluorspar is often added as lumps or mixed with other fine materials in briquettes. Alumina is sometimes added to blast furnaces either by direct charging of sized lumps or through sinter.

Typical acid fluxes are sand, gravel, quartz rock, used silica brick, or raw siliceous ore (Lankford, Jr., et al., 1985). Olivine, which is a magnesium-silicate mineral, has been added to iron blast furnaces to enhance removal of alkalis (Lankford, Jr., et al., 1985). Alumina in the form of bauxite, aluminiferous clay, and recycled alumina brick has been used as flux. Alumina can function amphoterically as either an acid or base, for example it can form aluminum silicate in high silica slags or calcium aluminate in lime-bearing slags (Lankford, Jr., et al., 1985).

Limestone (calcium carbonate) and dolomite (magnesium-calcium carbonate), or fluxstones as they are sometimes called (Boynton, 1980), and their calcined forms, lime and dolomitic lime are the major basic fluxes. In the latter cases, the carbonates are decomposed in a kiln-type process to drive off carbon dioxide that might otherwise interfere with the subsequent smelting or refining process or require expensive forms of energy to effect the decomposition (e.g., by combustion of coke in the iron blast furnace). Limestone is considerably less expensive than lime because it is not calcined; however, lime is far more difficult to handle and reacts readily with water. Dolomite and dolomitic lime are used for their magnesia content.

Fluorspar is a neutral flux often used as an additive in steelmaking slags to improve fluidity, but it has also been used in combination with lime as a primary slag in electroslag refining (Duckworth and Hoyle, 1969). Fluorspar is available in several grades, with the lowest grade often used for steelmaking. When the price of fluorspar has risen, substitutes including aluminum smelting dross, borax, manganese ore, titania, iron oxides, silica sand, and calcium chloride have been used (Peters, 1982). Briquettes containing various combinations of low grade manganese ore fines, mill scale, precipitator dust, recycled slag, and fine fluorspar have been used as an addition to the steelmaking process to form a fluid slag.

Some flux materials are added to repair smelting and refining vessel refractory linings. The selection of the flux to be added depends on the type of process and type of refractory. In the basic oxygen process for steel, lime from the flux combines with silica formed by oxidation of silicon from the metal to form a slag. The steelmaking vessel is lined with basic magnesia brick that will be attacked by the slag, so common practice includes addition of 5 to 10% magnesia (magnesium oxide) to help prevent attack on the refractory in the vessel (Boynton, 1980, Lankford, Jr., et al., 1985). In copper smelting, an acid slag practice is used, and silica is added for repair of refractories in some parts of the smelting furnaces (Anderson, 1961, Fowler, 1961).

The slag formed in one operation can also serve as a flux in another. Slags formed in refining operations are often recycled to a smelting operation to provide valuable flux materials and also to recover elements that otherwise might be lost. For example, steelmaking slags are recycled to the blast furnace smelter to recover unreacted, pre-calcined lime, iron, and manganese oxides. The limitation on the recycle of steelmaking slag to the iron blast furnace is typically the impurity content, particularly the sulfur, phosphorus, and zinc. In copper smelting, converter slags can be recycled to recover copper and flux. The amount recycled is affected by the minor elements such as bismuth, antimony, and arsenic.

Since about 1980, ironmakers and steelmakers have been using synthetic slags for external treatment of hot metal from the blast furnaces and for treatment of steels in ladles subsequent to basic oxygen and electric furnace processing. These treatments include reagent additions for removal of silicon, sulfur, and phosphorus. There have been applications in nonferrous metallurgy, the ferroalloy industries, and foundries as well (Szekely et al., 1989). Synthetic slags are made through additions of magnesium or aluminum metal, magnesium-aluminum or calcium-silicon alloys, calcium carbide, prefused calcium aluminate, and mixtures of some of the above with coke, lime, and fluorspar (Stubbles, 1984). Sodium carbonate (soda ash) can be used for desulfurization and dephosphorization of steel (Szekely et al., 1989, Yamamoto et al., 1980). Special synthetic compounds are also used as mold fluxes in the continuous casting of steel (Branion, 1987).

INDUSTRY STRUCTURE

About 90% of all metal consumed in the United States and in the world is iron and steel (Anon., 1991a, Habashi, 1986), and consequently, use of flux is preponderantly in the iron and steel industry. In the United States, metallurgical flux required by both the ferrous and nonferrous industries is abundant, and most of the demand is satisfied by domestic production with the exceptions of fluorspar and alumina. Fortunately, the major flux minerals limestone, dolomite, and silica are widespread throughout the world, although there are examples where an adequate flux source was not available and this affected the process selection.

Limestone and Dolomite

Limestone and dolomite fluxes used for ironmaking and steelmaking in the United States originate primarily from Michigan. Historically, these fluxstones were found to be of the type and purity useful for ironmaking and could be readily and economically transported to iron and steel producing centers by lake boats and rail. Fluxstone is moved by barge where feasible. Fluxstone is also obtained in Ohio, Pennsylvania, and Illinois (Lankford, Jr., et al., 1985, Lawrence, Jr., 1975). Until recently, major iron and steel producers owned limestone and dolomite quarries, but most have sold the operations and now recycle slags or purchase flux as needed.

Because limestone is very readily available around the world, a suitable source can almost always be found. Long distance transportation of stone is not common due to the low value and widespread occurrence. However, higher valued lime might be transported greater distances. In some special cases, due to the purity required, flux might be moved long distances. For example, low impurity flux has been transported from Japan to Australia to supplement local sources for use in fluxed pellets (Murray, 1988, personal communication).

For many years in the United States, limestone and dolomite were transported to the steel producing centers for use in blast furnaces, lime plants, and sinter plants. The sinter plants were often located at or near steel mills to take advantage of the availability of a variety of iron-bearing raw materials, coke fines, and slag. Recently, some major iron producers have incorporated flux into pellets, and the fluxed pellets are transported to iron producing centers.

The trend towards the use of fluxed pellets in blast furnaces has required a change in the transportation and distribution patterns of limestone and dolomite. In the United States, fluxed pellets are not made near the steel mills, but instead are made at the mines in Minnesota and Michigan that are located farther from the steel mills than the limestone and dolomite quarries. Therefore, the limestone and dolomite must first be transported to the mines before being returned to the steel mills in the form of fluxed pellets. Fluxed pellets made in Minnesota with flux from Michigan are transported as far as Alabama and Utah.

Extensive processing of limestone and dolomite materials is generally avoided; however, the economical production of materials for flux in the sizes desired by the customers requires that producers make a wide range of products for multiple markets. Thus, the limestone and dolomite in various sizes might also be sold for land erosion control, construction, aggregate, and agricultural purposes as well as for flux. The beneficiation process generally comprises only crushing, sizing, and washing. The sizes of stone used for flux are generally less than 15 cm but greater than 150 mesh (0.104 mm). The coarsest sizes are used for direct charging to furnaces, and intermediate sizes are used for lime burning, production of flux for fluxed pellets, and for sinter production. The finest sizes, often contain impurities and are disposed of as waste.

In the manufacture of fluxed pellets, the flux stone is crushed and ground to 60 to 80% passing 200 mesh (0.074 mm) thus with appropriate facilities for crushing, grinding, and sizing, there is some flexibility in the stone sizes that can be used. In fluxed pellets, a specific chemistry is usually targeted in terms of both lime and magnesia in relation to silica, and this requires a blend of limestone and dolomite. In some instances, the blend can be made from quarries in close proximity to one another, and this is the most economical choice. In other cases, the blend must be made with stones from separate locations, and blending is done while loading boats or at a blending site.

The US Bureau of Mines (USBM) publishes information on the production and consumption of crushed stone which includes flux (Tepordei, 1991). A directory of principal producers of crushed stone is also published annually (Anon., 1990). The major sources of limestone and dolomite in the United States are domestic, and the quantity used in the iron and steel industry is about 1% or less of total crushed stone consumption (Anon., 1987, Tepordei, 1991). The trend in the production and consumption of limestone and dolomite in the iron and steel industry is reported as Crushed Stone by the USBM as shown in Table 1. During the period 1975 to 1985, the amount of limestone and dolomite used by the iron and steel industry was reduced from about 22 to 9 Mt. During the same period, the value of stone increased from about \$2.20 to \$4.46/t, but

Table 1. Demand (Mt) and Prices ($/t) for Crushed Stone, Lime, and Fluorspar in Iron and Steel Production, 1975–1985

	1975	1976	1977	1978	1979	1980	1981	1982	1983	1984	1985
a. Crushed Stone (Limestone and Dolomite)											
Demand	22	22	21	22	20	15	14	NA	8	NA	9
Total US	817	816	865	952	998	892	792	717E	782R	868E	909
Avg. Price, $/t[a]	2.20	2.47	2.59	2.73	2.91	3.30	3.66	3.95	4.27	4.33	4.46
1985 Price, $/t[b]	4.64	4.57	4.52	4.51	4.67	4.76	4.68	4.54	4.59	4.47	4.46
b. Lime (Calcined Limestone and Dolomite)											
Demand	7.0	7.9	7.6	8.2	8.2	6.5	7.2	4.4	4.9	5.1	4.9
Total US	17.4	18.4	18.1	18.6	19.0	17.3	17.1	12.8	13.5	14.5	14.2
Avg. Price, $/t[a]	30.26	33.27	36.92	40.51	45.47	49.04	51.80	54.52	56.31	56.33	56.96
1985 Price, $/t[b]	56.28	58.88	61.23	62.70	64.59	63.94	61.52	60.91	60.62	58.20	56.96
c. Fluorspar (Demand in Crude Steel Production Only)											
Demand	0.47	0.55	0.49	0.51	0.47	0.32	0.33	0.18	0.16	0.17	0.12
Total U.S.	1.13	1.15	1.05	1.09	1.03	0.89	0.84	0.48	0.51	0.68	0.52
Avg. Price, $/t[a]	86	95	106	112	122	150	176	191	181	W	NA
1985 Price, $/t[b]	162	168	175	174	174	195	209	213	195	W	NA

Adapted from USBM (Anon., 1987), NA = not available E = estimated R = revised W = withheld to avoid disclosing company proprietary data.
[a]Average annual price—FOB quarry, actual dollars/t.
[b]Average annual price—FOB quarry, based on constant 1985 dollars.

was about $4.40/t in constant 1985 dollars. From 1985 to 1989, the consumption of limestone and dolomite in the steel industry decreased further. In 1989, the consumption of limestone for flux was on the order of 4.3 Mt, and the consumption of dolomite was about 0.64 Mt (Table 2).

Lime

Lime and dolomitic lime produced by calcination of limestone and dolomite are widely used as fluxes in steelmaking. Some steel producers operate lime plants, and others purchase lime. Lime plants are sometimes located near steel consuming operations, and in some cases are connected by conveyor belts to the steel shops. In other situations, lime is transported by truck, barge, or rail. The sizes of stone required for lime burning are fairly well fixed and must be integrated with the needs of the entire steel operation when the limestone and dolomite operations are captive. Some nonferrous operations have operated captive lime plants, but in the nonferrous industries lime has uses in addition to smelting and refining such as treatment of acid wastes and control of acidity in flotation plants. The USBM has published a directory of lime plants in the United States (Anon., 1991b).

The USBM publishes data related to production and consumption of lime (Ober, 1989, Miller, 1990). Most lime is domestically produced, and about one-third of total lime demand arises from the iron and steel industry. The trends for the period 1975 to 1985 are shown in Table 1 and for 1986 to 1989 in Table 3. Consumption of lime for iron and steel use decreased by about 50% between 1975 and 1989. There was a decrease from a high of 8.2 Mt in 1978 and 1979 as shown in Table 1 to about 4.5 Mt for the period 1986 to 1989 as shown in Table 3. The cost of lime increased from about $30/t in 1975 to about $57/t in 1985, but remained fairly constant at about $60/t in constant 1985 dollars. In 1989, the value of metallurgical lime was about 50 to $55/t.

Other Flux Materials

Other major flux materials used in ironmaking and steelmaking include fluorspar, alumina, and siliceous sources. Fluorspar is largely imported to the United States from Mexico, China, and the Republic of South Africa. The USBM publishes trends in fluorspar production and consumption as shown in Table 1 (Anon., 1987).

Table 2. Quantity (kt) and Value ($1 000) of Crushed Limestone and Dolomite Sold or Used Within the Chemical and Metallurgical Industries in the United States in 1989, by Use

	Limestone		Dolomite	
Use	Quantity	Value	Quantity	Value
Lime manufacture	14 330	83 600	308	1 400
Flux stone	4 262	19 600	635	2 600
Total, all uses	731 586	3 234 300	44 715	217 000

Adapted from USBM, Tepordei (1991).

About 25% of total fluorspar demand was for crude steel production. Between 1975 to 1983, fluorspar demand for crude steel production decreased from about 470 ktpy to 120 ktpy as shown in Table 1. The price doubled from about 86 to $181/t during the same period; the price rose from 162 to $195/t in constant 1985 dollars.

In ironmaking, small amounts of alumina-bearing and siliceous materials are used. In the United States there is a shortage of high-grade, domestic alumina sources except for aluminiferous clays which contain silica. Thus, bauxite must be imported from Greece, Turkey, or Jamaica. In contrast, siliceous materials are plentiful. Silica sources for the blast furnaces include crude ore (taconite) from the iron mines and local gravel. Local sands and other fine siliceous materials can be used to increase the silica content of sinter. In the nonferrous industries, fluxes are largely siliceous, and often lean ores or nearby rock can be used as flux if the impurities are low. In some instances, high grade silica rock is used. Gypsum has been used as a source of lime and sulfur in nickel smelting (Habashi, 1986).

RAW MATERIALS

The selection of flux for a particular process requires specification of the chemistry of the flux materials required to form a slag of the required composition to attain a desired effect. Ideally, only pure substances would be selected, however, these are seldom available in nature. In general, the level of impurities in the flux should be as low as possible because the greater the amount of impurities, the less effective will be the flux, and the greater the amount that will be required. In addition, some minor impurities can

Table 3. Quantity (kt) and Value ($1 000) of Lime Sold or Used in the Metallurgical Industries in the United States, by Use

	1986		1987		1988		1989	
Use	Quantity	Value	Quantity	Value	Quantity	Value	Quantity	Value
Metallurgy	9	467	4.5	218	W	316	W	W
Steel, BOF	3 338	175 300	3 701	189 700	3 845	182 900	3 563	180 720
Steel, Electric	801	41 530	689	35 240	860	44 530	888	46 294
Steel, Open Hearth	111	5 940	131	6 733	45	2 190	58	2 877
Other[a]	822	50 490	2 511	125 000	2 127	116 000	3 211	178 958
Total steel	4 250	222 770	4 521	231 673	4 750	229 620	4 509	229 891
Total (Chemical & Industrial)	11 455	637 800	12 789	682 400	13 424	686 900	13 759	730 800
Total, all uses	13 151	761 200	14 294	789 700	15 492	821 700	15 582	855 900

Adapted from USBM, Ober (1989), Miller (1990).
W = withheld to avoid disclosing company proprietary data.
[a]Includes metallurgical applications including briquetting, chrome, ferroalloys, ladle desulfurizing, magnesium metal, manganese, pelletizing, chemical and industrial uses specified as W above, as well as many other non-metallurgical uses (see original tables for detail).

have significant detrimental effects on the process or process equipment. After the chemistry is specified, the size of flux will be determined largely by the handling and gas flow requirements of the particular process, and to a lesser extent, on the size of flux that is available. Geographic location and economics might dictate that a flux from a particular source and with a particular chemistry be selected. If the desired chemical or physical specifications important for a given process can not be attained from the most economical source, a new source might have to be found, or an alternate process might have to be developed.

Since 90% of all metal production is iron and steel, the major fluxes consumed will be those used in ironmaking and steelmaking, that is, limestone, dolomite, lime, fluorspar, and to a small extent, siliceous sources. Many of the same considerations that apply to the fluxes used in ferrous smelting apply to fluxes for nonferrous smelting. Although limestone and dolomite are used largely in ironmaking and steelmaking, some small amounts are used in nonferrous smelting, particularly in smelting and refining of copper and lead ores where the ores are generally acidic. Foundry cupolas also use some stone (Boynton, 1980).

Chemistry of Limestone and Dolomite

Limestone and dolomite are used in smelting iron ores specifically to supply lime (CaO) and magnesia (MgO) to form a slag with the acid components, particularly silica (SiO_2) and alumina (Al_2O_3). Typical chemical analyses for limestone, dolomite, and their respective calcined products are shown in Table 4. The ideal flux will be low in acids (silica, alumina and sometimes titania and iron oxide) as well as low in sulfur (S), phosphorus (P), alkali metals (sodium and potassium), and halides (chlorides and fluorides). All phosphorus in ironmaking raw materials including iron ores or agglomerates, coke, and fluxstone is reduced to metal in the blast furnace and appears in the iron product, therefore, phosphorus must be minimized in the flux. This is particularly true with recycled slags, and it is the phosphorus content that determines the amount of steelmaking slag that can be recycled to ironmaking. A major purpose of the ironmaking slag is to remove sulfur, therefore sulfur in flux as well as other materials should be low. Alkali metals are generally detrimental to blast furnace operation often forming deposits on the furnace walls, therefore alkali inputs from raw materials should be minimized. During the production of fluxed pellets or sinter, the alkali metals in combination with sulfur, chlorides, and fluorides have been found to be extremely detrimental to process equipment in some pellet and sinter plants.

Table 4. Typical Analyses of Limestone, Dolomite, and Lime Fluxes

	Limestone		Dolomite		Burnt Lime	
Component	Blast Furnace[a]	Sinter Plant	Blast Furnace[a]	Sinter Plant	High-Calcium	Dolomitic
	(one year average, % by weight)				(calculated)	
$CaCO_3$	95.3	93.8	54.5	52.4		
$MgCO_3$	3.1	3.6	42.0	40.0		
CaO	53.4	52.5	30.6	29.4	88.5	56.0
MgO	1.5	1.7	20.1	19.1	2.5	36.8
R_2O_3[b]	0.3	0.4	0.3	0.4	0.5	0.5
SiO_2	0.7	1.8	2.6	6.8	1.2	4.7
	(typical, % by weight)					
Fe_2O_3	0.20	0.30	0.2	0.3		
Al_2O_3	0.30	0.20	0.3	0.3		
Mn	0.01	0.02	0.01	0.03		
P	0.01	0.01	0.01	0.01		
S	0.03	0.04	0.05	0.03		
K_2O	0.10	0.10	0.10	0.10		
Na_2O	0.02	0.02	0.02	0.02		
LOI[c]	43.40	46.10	6.00	1.00	1.0	2.0

[a]Also typical of flux used in fluxed pellets and as stone for lime production.
[b]$R_2O_3 = Fe_2O_3 + Al_2O_3 + Cr_2O_3 + TiO_2$
[c]Loss-on-Ignition

Silica addition to the blast furnace is often required to obtain the proper ratio of basic to acid components in the slag, and addition of a siliceous source enables accurate control of this ratio. Historically, in traditional blast furnace practice, up to 5% silica in the flux was tolerated and 1 to 3% was common (Gault and Ames, 1960). Today, the levels are lower and the amount tolerated depends on whether the flux is used for sinter or for direct charging, and what the levels of silica are in other burden materials. The allowable level depends on the desired slag chemistry and slag quantity required for control of the furnace and the sulfur and alkali.

The silica content in flux used for fluxed pellets should be as low as possible because the total silica content of the pellets is carefully monitored. A silica content of less than 1% in the flux is commonly desired. Furthermore, uniformity of composition over time is important. The silica content of the flux adds to the silica content in the fluxed agglomerates, and therefore more flux is required, and more energy is required to calcine the additional flux.

High silica is less of a problem in flux used for sintering, as the sinter can be used as a vehicle for introducing small amounts of silica to the blast furnace in a controllable way, and energy is available from combustion of coke fines and from oxidation of mill scale (ferrous oxide) in the sintering process.

The magnesia content of limestone added to the blast furnace is often not critical because both magnesia and lime are added to form a slag. The amount that can be tolerated depends on the magnesia contents of all burden materials and the target slag chemistry. Generally, a higher magnesia content keeps slag fluid over a wider temperature range (Gault and Ames, 1960, Rosenqvist, 1974). Dolomite might be added to provide sufficient magnesia to produce slag of a certain specification, for example, for use as aggregate. Magnesia added in sufficient amount increases the melting point, and this effect is used to advantage when making fluxed pellets for the blast furnace. The magnesia content of the limestone must be known with certainty when producing blends of limestone and dolomite for use in fluxed agglomerates.

Physical Properties of Limestone and Dolomite

Traditional ironmaking practice has relied on direct charging of the iron blast furnace with flux materials of up to 15 cm in size. However, more recent practice that recognizes the need for a uniformly sized and permeable burden requires a size more compatible with other burden materials, and stone as fine as 5 by 1.25 cm has been used. The location in the blast furnace where complete calcination occurs can be regulated by changing the size of stone (Lankford, Jr., et al., 1985). Typically, flux added to sinter is in the range of −6.35 to −3.17 mm in size with 90% coarser than 150 mesh (0.104 mm). Some iron producers have adopted the use of fluxed pellets, which requires that the stone be ground to 60 to 80% passing 200 mesh (0.074 mm), a size more compatible with the formation of pellets from fine concentrates. The size ranges of limestone and dolomite required for the blast furnace, lime plant, sinter plant, and fluxed pellet plant are shown in Table 5.

Thermal decrepitation and physical degradation are not desirable because the permeability of the furnace burden is reduced by the presence of fines. Experience has shown that fine-grained stone decrepitates less than coarse-grained stone in the blast furnace (Gault and Ames, 1960). Resistance of the stone to physical degradation during handling and transportation is also important to minimize the amount of fines formed. When the fines will be deleterious to the process, they must be removed by screening. If the fines do not have an alternate use, the cost of the flux is effectively increased. A drop test from heights approximating those in handling of the stone can be used to estimate the expected amount of degradation during handling and transportation (Fagerberg and Sanberg, 1973).

Many standard tests for characterizing lime and limestone (as well as dolomitic lime and dolomite) have been developed by the American Society for Testing Materials (ASTM) (Anon., 1988). A list of standard ASTM tests commonly applied to limestone and dolomite are listed in Table 6. Accurate sampling of the stone is often a problem due to the large size or wide range of sizes. Sampling of stone is described in ASTM tests C-50 and D-75. Chemical analysis of limestone and dolomite are easily accomplished according to standard procedures defined in ASTM C-25 and by published or proprietary methods using modern analytical techniques. Physical testing is defined in ASTM C-110. The physical test most often done is sizing or screening of the stone to determine the size distribution as described in ASTM E-11, C-136, E-276, and E-389. Additional tests that might be applied to determine the degradation characteristics are the Los Angeles test (ASTM C-131, C-535) and the ASTM tumble test (ASTM E-279).

Table 5. Size Analyses of Typical Flux Stone for Blast Furnace Sinter Plant, Lime Plant, and Fluxed Pellet Plant

	(% passing)					
	Blast Furnace or Lime plant stone		Mixed Blend	Sinter Flux		Pellet Flux
Size	Limestone	Dolomite	50:50[a]	Limestone	Dolomite	50:50[b]
cm						
7.620			99.6			
6.350	87.6		99.0			
5.080	46.2	88.7	91.0			
4.445	18.7	72.7				
3.810	9.1	50.5	64.6			
3.175	4.4	26.4				
2.540	2.6	14.8	15.7			
1.905	1.7	2.0				
1.270			2.5			
0.952				99.8	100.0	
0.635			1.2			
mm						
4.699				97.5	97.8	
3.327				90.6	86.1	
1.651				66.0	58.6	
0.589				20.3	20.0	
0.295				8.2	10.0	99.9
0.208						99.3
0.147				2.4	4.7	90.4
0.074				0.9	2.4	65.9
0.053						58.2
0.043						52.2
0.038						47.2
0.026						42.0

[a]Feed to crusher at fluxed pellet plant (50% limestone, 50% dolomite), [b]Ball Mill Grinding.

Chemical Properties of Lime

Limestone and dolomite are calcined or heated to drive off the carbon dioxide prior to use in steelmaking. Calcination is usually done in rotary kilns or rotary kiln-shaft combinations. Typically, the temperatures are 900 to 1 100°C. Calcination is strongly endothermic and requires energy input. The reaction proceeds according to:

$$CaCO_3(s) = CaO(s) + CO_2(g) \text{ for limestone, and}$$

$$(Mg,Ca)CO_3(s) = (Mg,Ca)O(s) + CO_2(g) \text{ for dolomite}$$

where (s) is a solid, and (g) is a gas. In the blast furnace, calcination of limestone begins at temperatures over 800°C and dolomite decomposition typically begins at 700°C or lower (Ricketts, 1992). Thermal requirements are discussed in the literature (Boynton, 1980).

High-calcium lime, and to a lesser extent dolomitic lime, are typically used as fluxes in steelmaking and in some nonferrous smelting and refining processes. The chemistry of lime is specified more closely in steelmaking than in ironmaking. Typical analyses of lime and dolomitic lime are shown in Table 4. In steelmaking, high calcium lime with a low silica content is specified because the lime is used to combine with the silica formed by oxidation of silicon in the steel bath. Dolomitic lime is added in steelmaking to control the MgC content to minimize wear of the basic refractory lining.

The available base for reaction with acidic components is im-

Table 6. ASTM Standard Physical and Chemical Tests and Definitions Applicable to Limestone, Dolomite, and Fluorspar

ASTM No.	Description
C 25	Standard Test Methods for Chemical Analysis of Limestone, Quicklime, and Hydrated Lime.
C 29	Unit Weight and Voids in Aggregate.
C 50	Sampling, Inspection, Packing and Marking of Lime and Limestone Products.
C 51	Standard Definition of Terms Relating to Lime and Limestone.
C 70	Surface Moisture in Fine Aggregate.
C 110	Physical Testing of Quicklime, Hydrated Lime, and Limestone.
C 125	Definition of Terms Relating to Concrete and Concrete Aggregates.
C 127	Specific Gravity and Absorption of Coarse Aggregate.
C 128	Specific Gravity and Absorption of Fine Aggregate.
C 131	Standard Test Method for Resistance to Degradation of Small-Size Coarse Aggregate by Abrasion and Impact in the Los Angeles Machine.
C 136	Sieve or Screen Analysis of Fine and Coarse Aggregates.
C 184	Fineness of Hydraulic Cement by the 150 μm (No. 100) and 75 μm (No. 200) Sieves.
C 204	Fineness of Portland Cement by Air Permeability Apparatus.
C 294	Descriptive Nomenclature of Constituents of Natural Mineral Aggregates.
C 295	Petrographic Examination of Aggregates for Concrete.
C 430	Fineness of Hydraulic Cement by the 45 μm (No. 325) Sieve.
C 535	Standard Test Method for Resistance to Degradation of Large-Size Coarse Aggregate by Abrasion and Impact in the Los Angeles Machine.
C 566	Total Moisture Content of Aggregate by Drying.
C 702	Reducing Field Samples of Aggregates to Testing Size.
C 786	Fineness of Hydraulic Cement and Raw Materials by the 300 μm (No. 50), 150 μm (No. 100), and 75 μm (No. 200) Sieves by Wet Methods.
D 75	Sampling Aggregates.
E 11	Wire-Cloth Sieves for Testing Purposes.
E 12	Standard Definitions of Terms Related to Density and Specific Gravity of Solids, Liquids, and Gases.
E 276	Particle Size or Screen Analysis at No. 4 (4.75 mm) Sieve and Finer for Metal Bearing Ores and Related Materials (including wet methods).
E 279	Tumbler Test for Iron Ore Pellets and Sinter.
E 382	Determination of Crushing Strength of Iron Ore Pellets.
E 389	Particle Size or Screen Analysis at No. 4 (4.75 mm) Sieve and Coarser for Metal Bearing Ores and Related Materials.
E 463	Silica in Fluorspar by the Silico-Molybdate (Photometric) Method.
E 815	Test Methods For Calcium Fluoride in Fluorspar by Complexometric Titration.

From ASTM (Anon., 1988).

portant (Lankford, Jr., et al., 1985, Peters, 1982). The available base (AB) is calculated as:

$$CaO/(SiO_2 + R_2O_3) \text{ or } CaO + MgO\text{-}SiO_2\text{-}R_2O_3\text{-}LOI$$

where R_2O_3 is the sum of the Fe_2O_3, Al_2O_3, and TiO_2.

The sulfur content of the lime should be low, and absorption of sulfur originating with both the stone and fuel during calcining can be a major problem. The sulfur level in the calcined lime is controlled by regulating the oxygen content in the calcining system exit gas. Gypsum in the flux stone is not desirable because it is difficult to decompose, but sulfur in pyrite or pyrrhotite is generally removable. Sulfur also occurs as organic sulfur in graphite within the stone. Sulfur levels of 0.03 to 0.06% in the lime are typical.

After calcining, lime is hygroscopic which means it readily absorbs water from the air and slakes to form calcium hydroxide. In addition, some uptake of carbon dioxide and sulfur dioxide can occur thereby forming a dense layer of calcium carbonate and/or gypsum on the surface. This layer can adversely affect the subsequent reaction rate of the lime. In addition, if the lime is not completely calcined, a core of unreacted limestone (or dolomite) will remain in the center of the larger lime (or dolomitic lime) pieces. The unreacted core and the adsorbed water and carbon dioxide will result in a ''loss-on-ignition'' or LOI when the lime is heated. The lime for steelmaking generally requires a low LOI of 2 to 4% or less, and some plants require an LOI not exceeding 0.5% (Boynton, 1980).

Physical Properties of Lime

The size of lime used in steelmaking is typically 37 by 6.35 mm The particle size of the lime and dololime fluxes added to steelmaking vessels is important because if particles are too fine, they will be removed by the intense gas flow above the steelmaking vessel before they can react to form slag. Typically, all particles 3 mm and less in size will be removed. Therefore, considerable effort is made to achieve a low fines content in the lime used. However, lime degrades by physical handling and by degradation due to hydration and carbonation in storage. Lime placed in storage might contain as little as 5% to as much as 20% of material finer than 3.17 mm, yet after several days of storage, and depending on the relative humidity and natural characteristics of the lime, fully one third of the lime can degrade, and up to one half will exist as fines after handling and transfers before the steelmaking vessel. Some stones characteristically produce significant dust during calcining resulting in plugging of baghouse filters.

Several steelmaking processes use bottom blowing and some use combined top and bottom blowing so that much finer flux can be used as an injectant. Fine lime can be pneumatically conveyed by oxygen or a carrier gas and injected through tuyeres (openings) in the vessel bottom. The high surface area of the fine lime promotes rapid slag-metal reactions. The sizes of lime typically used in the injection process are about the same as those shown for Pellet Flux in Table 5, or finer (Koros, 1991).

The reactivity of the lime added to the steelmaking vessel is important. One test often used to estimate the reactivity of the lime in steelmaking is the ASTM water reactivity test (ASTM C-110) (Schlitt and Healy, 1970). In this test, the lime is broken to −6 mesh (3.33 mm), a prescribed amount is stirred into water held in an insulated bottle to retain the exothermic heat of reaction in the system, and the temperature rise of the water with time is recorded. The time of reaction and the temperature rise are used to indicate the reactivity of the lime. The American Water Works Association Standard B2–2 is similar. Although some correlation of water reactivity with slag formation has been reported, a crucible test is probably more directly related to lime behavior in steelmaking (Limes and Russel, 1970).

Requirements for Fluorspar

Fluorspar or calcium fluoride (CaF_2) is used as a flux along with lime to improve the fluidity of slag in steelmaking and subsequent ladle metallurgy processing. Typical consumption of fluorspar in steelmaking is 2 to 10 kg/t or 5 to 10% of the lime added (Peters, 1982). Fluorspar is also used with lime in electroslag

Table 7. Composition of Metallurgical Fluorspar (Dry Basis)

Component	(% by weight) High-Grade	Medium-Grade[a]	Low-Grade[b]
CaF_2	92.7	89.1	82.0
SiO_2	3.5	3.9	8.9
Fe	0.09[c]		
S	0.05[c]		
$CaCO_3$	3.5	4.6	4.2
Effective CaF_2	83.9	79.3	59.7

[a]Gravel Size; minus 6.35 cm, plus 0.95 cm.
[b]Screen Fines; 78% minus 6.35 mm and plus 100 mesh (0.147 mm); 11% -325 mesh (.043 mm)
[c]Oxides of iron and aluminum and sulfur can sometimes reach 1% in lower grade products.

refining in a ratio of about 70 parts fluorspar to 30 parts lime (Duckworth and Hoyle, 1969). The purpose in electroslag refining is to form a very conductive slag with the required melting point.

The specification for metallurgical-grade fluorspar is 72% minimum effective fluorspar where the effective fluorspar is calculated from:

$$\text{Effective } CaF_2 = CaF_2 - (2.5 \times SiO_2)$$

The main impurity in fluorspar is silica, the balance of impurities generally being calcium carbonate, barite, lead, and zinc. Analysis is done by ASTM E815 and E463 (Anon., 1988) or by other generally accepted analytical methods. The chemical analyses of typical fluorspars are shown in Table 7.

The fluorspar can be added as lumps, gravel-sized material, or as fines incorporated in briquettes. Sizes of gravel and screen fines are shown in the footnotes to Table 7. Typically, 5 to 15% fluorspar fines would be added to briquettes made with recycled slag fines or other steelmaking ingredients. The latter form is often preferred to achieve better mixing of fluxes and better reaction with the slag. As with lime, any loose fines present will be lost during the blow.

Other Materials

Silica-bearing materials are sometimes added to the blast furnace to control the process basicity and counteract over addition of basic fluxes. Additions are generally kept to a minimum. The chemistry of the source should be compatible with the process, and deleterious elements such as alkali metals, phosphorus, and sulfur should be avoided. In ironmaking, typical silica sources are low grade taconite ore, flint clay containing alumina, sandstone, gravel, or used brick. Specifying the material as low grade with respect to silica allows a larger, measurable, and more controllable quantity to be added. For example, addition of taconite ore containing 50% silica and 35% iron allows additional iron units to be added to the furnace while adding a larger volume of material than if adding pure silica. In copper smelting, low grade ores or rock are used, but sometimes high grade silica is used. In any process, the criteria for size is that the material should be of a size compatible with the process and fine enough to react rapidly, but not so fine that it will cause permeability problems or be lost as dust.

TECHNOLOGY

The successful application of a slag in a metallurgical process depends on the slag chemistry and properties of the molten slag which are obtained by selecting the proper fluxes. The properties of a slag that are important are melting point, viscosity, surface tension, density and chemical activities of the components present. In electric smelting practices, conductivity is also important. These parameters all depend on a knowledge of the composition. The effects of composition on melting temperature and other properties are often described through temperature-composition diagrams (phase diagrams) and property-composition diagrams. In practice, it has been found convenient to use index values to describe composition requirements for a given process. Thorough discussions of the properties of metallurgical slags can be found in the literature (Derge and Tenenbaum, 1951, Turkdogan, 1983, Rosenqvist, 1974, Lankford, Jr., et al., 1985, Elliott, 1984).

Properties of Slags

Melting Point: The most common fluxes used in smelting and refining are lime (or limestone) and silica. The melting point of pure lime is higher than the process temperatures during smelting of iron ores or refining of steel. However, it has been found that when lime is added to other components such as silica, the melting point is lowered and a liquid phase forms. This phenomenon can be extended to multi-component systems. In most systems, there are more than two components, and interactions between components become complex. It has been found convenient to represent the interactions on ternary or 3-component composition diagrams. Phase diagrams for many systems including those in ironmaking, steelmaking, and nonferrous smelting and refining can be found in a series of publications entitled *Phase Diagrams for Ceramists* (Levin et al., 1956, 1969).

Iron blast furnace slags are generally represented in the CaO-SiO_2-Al_2O_3 ternary system. Typical iron blast furnace slag contains 32 to 42% silica (SiO_2), 7 to 16% alumina (Al_2O_3), 32 to 45% calcia (CaO), and 5 to 15% magnesia (MgO) (Lankford, Jr., et al., 1985). The composition of iron blast furnace slag is generally controlled to maintain a composition with about 10% alumina (some practices contain greater amounts) to achieve a relatively constant melting point and liquid slag over a wide range of lime and silica variations. The melting point is typically 1 400 to 1 500°C. Ferrous iron will react with silica at much lower temperatures than calcia and alumina, and local melting can occur in the lower part of the shaft of the iron blast furnace. One purpose of flux in fluxed pellets is to avoid early melting that results from the formation of iron silicate (Fe_2SiO_4 or fayalite) by preferentially reacting silica in the pellets with lime and magnesia (Taguchi et al., 1980, Ranade, 1992). In non-ferrous smelting, iron silicate slags that melt near 1 200°C are formed.

Viscosity: A major component of many slags is silica which can form complex silicate networks in the molten state. These networks comprise chains of silica molecules, and this results in very viscous flow conditions. Silica also melts at a high temperature, therefore, separation and movement of silica is difficult. It has been found that the addition of lime breaks up the silicate networks, lowers the melting point, and enhances the movement of the slag. Therefore, the viscosity of iron blast furnace slags decreases with increasing lime addition and increases with increasing silica and alumina (Rosenqvist, 1974). Addition of a basic oxide such as magnesia (MgO) has only a small effect on viscosity and is used to minimize variability over a range of compositions. If sufficient lime is present to form dicalcium silicate, viscosity will increase sharply. In basic slags, a small decrease in temperature will increase viscosity a small amount, however, as solid phases such as dicalcium silicate precipitate, viscosity will increase rapidly (Rosenqvist, 1974).

In steelmaking slags, calcined lime is added as a flux to react with silica and phosphorus oxidized from the steel bath. The reaction of the lime is sometimes impeded due to formation of a dicalcium silicate or calcium ferrite shell on the lime particles. Fluorspar (CaF_2) is often added to the steelmaking vessel to aid in dissolution of the lime and to reduce the viscosity of the slag. The mechanism is not precisely known, however, it is thought that the fluorspar attacks the coating on the lime (Peters, 1982). Fluorspar melts at 1 386°C and dicalcium silicate melts at over 2 000°C, but when approximately equal portions of fluorspar and di-calcium silicate are mixed, the melting point decreases to below 1 200°C. (Levin et al., 1956, 1969).

Surface Tension: Surface tension of slag is important, but is not often used to characterize slags, probably due to the difficulty of measurement. However, surface tension affects the wetting of both metals and refractories by slags and this affects the rates at which various components dissolve in the slag and the ease of separation of slag and metal. Foaming during blowing in basic oxygen steelmaking and also in the electric furnace is affected by surface tension. The surface tension decreases with increasing silica content, with addition of alkali oxides, calcium sulfide, and phosphoric oxide, and with an increase in temperature. Lime, iron oxide, and alumina increase the surface tension of slag (Rosenqvist, 1974).

Density: Density can be controlled to achieve separation of metal and slag by adjusting the proportions of components of differing densities. In ironmaking, the slag density increases with increasing lime and iron contents, but is not specifically controlled. In copper smelting, where the product is a matte of lower density than metal, the separation of slag and matte is enhanced by decreasing the density of the iron-rich slag by addition of lime (Rosenqvist, 1974).

Activity: The activity of a species in a slag is a parameter that is a function of both concentration and temperature and determines how a particular species will react with other components in a given system. Values of thermodynamic activity can be used to predict the equilibrium in the process. Activity (a_i) is related to the more familiar concentration term expressed as the mole fraction (X_i) through the activity coefficient (γ_i):

$$a_i = \gamma_i \cdot X_i$$

For ideal gases, the activity is the partial pressure (p_i). A thorough discussion of thermodynamics of slags can be found in the literature (Derge and Tenenbaum, 1951, Coudurier et al., 1985, Lankford, Jr., et al., 1985, Turkdogan, 1983, Elliott, 1984). In general, activity can't be predicted, and the activity or the activity coefficient is usually measured experimentally. Values for particular systems can be found in the literature.

Index Values

The majority of slags can be described as either basic slags as used in ironmaking and steelmaking or acid slags as used in nonferrous smelting. The desired composition of a slag in a metallurgical system is often defined by index values that express the acidity, basicity, or the capacity of a slag to perform a desired function. In practice, the composition of the slag is controlled by adjusting the proportions of flux added to combine with impurities or gangue to yield a desired index value. Common index values are basicity or total basicity (TB), V-ratio, optical basicity (OB), and sulfide capacity (C_s).

Basicity: The general description of basicity is the ratio of basic oxide components to acid oxide components in the slag. Care must be exercized in specifying which basicity definition is being used. Because lime and silica are the most prevalent species in basic slags, the definition of basicity (B) is often given as the lime-to-silica ratio:

$$B = CaO/SiO_2$$

where the weight percent of the chemical species is used to compute the ratio. This ratio is also called the C/S ratio in the manufacture of fluxed pellets. The magnesia-to-silica ratio or M/S ratio is sometimes also specified for fluxed pellets.

The most often used definition of basicity in blast furnace practice is the four-component basicity or total basicity (TB) defined as:

$$TB = (CaO + MgO)/(SiO_2 + Al_2O_3)$$

The total basicity is sometimes modified to include titania (TiO_2) as an acid.

V-Ratio: In steelmaking, the basicity is often referred to as the V-ratio which expresses the ability of a slag to desulfurize or dephosphorize the steel bath. The V-ratio is usually defined as one of the following:

$$CaO/SiO_2;\ (CaO + MgO)/SiO_2;\ \text{or}\ (CaO - 4P_2O_5)/SiO_2$$

Sometimes, the V-ratio is defined the same as total basicity given above.

Optical Basicity: The optical basicity is sometimes used to characterize a slag for correlation with desulfurization (Sosinsky and Sommerville, 1986, Blattner, 1992). The optical basicity (OB) is defined as:

$$OB = \frac{(\%CaO + 1.11\%MgO + 0.915\%SiO_2 + 1.03\%Al_2O_3)}{(\%CaO + 1.42\%MgO + 1.910\%SiO_2 + 1.69\%Al_2O_3)}$$

Sulfide Capacity: Slags are desulfurized primarily by reaction with lime to form calcium sulfide which dissolves in the slag. The ability of the slag to dissolve sulfides is expressed as the sulfide capacity (C_s) which is:

$$C_s = (\%S)(P_{O_2}/P_{S_2})^{0.5}$$

where P_i is the partial pressure of component i. Sulfide capacity can be expressed in terms of optical basicity (OB) and temperature as follows:

$$\log C_s = [(22690 - 54640OB)/T] + 43.6OB - 25.2$$

where T is the temperature in degrees Kelvin (°K) over the range 1 400 to 1 700°C. The sulfide capacity of a slag increases with increasing lime (CaO), and a low silica activity (a_{SiO_2}) is desirable.

USES

Fluxes are used at various stages in the production of iron and steel. Iron used for steel production is produced in the iron blast furnace largely by traditional practices described in the literature (Lankford, Jr., et al., 1985). As a result of fairly recent studies of iron-bearing raw materials and dissections of the burdens of iron blast furnaces, the high temperature properties of the iron-bearing burden materials have been improved (Taguchi et al., 1980, Yamoaka et al., 1980). This has resulted in modifications to fluxing practice at some furnaces wherein flux is introduced to the furnaces through fluxed sinter and fluxed pellets rather than by direct charging. There has also been an increase in the use of external desulfurization of hot metal prior to steelmaking which allows greater flexibility in selection of furnace burden materials with respect to sulfur content, and greater flexibility in furnace operating practices for control of other furnace parameters (Stubbles, 1984). Steelmaking is largely accomplished in basic oxygen furnaces with the virtual phasing out of the open hearth process, although electric furnace steelmaking continues to increase. Modern steelmaking includes the treatment of steel in ladles for improved cleanliness, greater throughput in the steelmaking vessels, and shape control of inclusions in the continuous casting process which now dominates over ingot processing (Szekely et al., 1989). A variety of powder mixtures and prefused slags are used as mold fluxes in the continuous casting of steel (Branion, 1987). Fluxes are also used in smelting and refining of nonferrous metals.

Iron Blast Furnace

The modern iron blast furnace uses iron-bearing agglomerates, coke, and flux to produce hot metal. The flux, which is largely limestone and dolomite, provides basic elements (calcium oxide and magnesium oxide) to the furnace charge to form a slag with acid flux elements (silicon oxide and aluminum oxide) present in the iron-bearing agglomerates and coke. This slag provides a vehicle to remove the silica and alumina gangue that was introduced with the ore and coke, and sulfur that was introduced with the coke. Hot metal produced from the blast furnaces is further treated in transport vessels and in subsequent steelmaking operations to reduce impurity levels. These operations also require slags.

The limestone and dolomite flux used in the blast furnace can be added as lumps 2.5 to 7.5 cm in size, or can be added to iron-bearing agglomerates and charged to the furnaces as fluxed pellets or fluxed sinter. Agglomerates are large particles formed from smaller particles and fused at high temperatures to form a material that is readily handled and transported without formation of fines. The agglomerates are also designed to react in a predictable manner inside the blast furnaces. In the agglomerates, limestone and dolomite are ratioed to the silica and alumina. The ratio can be expressed as the basicity ratio discussed earlier, or as in fluxed pellets, the basicity might be calculated as the lime-to-silica ratio (C/S ratio) to reflect only the major components. A comprehensive overview of the use of raw materials in the blast furnace was recently published (Ranade, 1992).

The total basicity of iron blast furnace slags is typically between 1.05 and 1.15 to achieve a fluid slag with adequate capacity to remove sulfides, although ratios as low as 0.9 and as high as 1.4 are reported (Anon., 1989). Typically, fluxed pellet basicities will be between 0.7 and 1.7, and sinter basicities will be between 1.7 and 4.0. When the basicity of the agglomerates is higher than the basicity of the furnace, siliceous materials must be added to reach the required furnace slag basicity. Coke ash furnishes silica to the furnace, but additional siliceous material might be required. The basicity of the agglomerates is selected to obtain specific reducing and melting properties to improve furnace operation. The higher basicities for sinter are used when the primary function of the sinter is to provide calcined flux to the furnaces. At very high basicities, melting of the sinter can be a problem due to formation of dicalcium silicate.

The amount of flux charged directly to blast furnaces varies widely depending upon the amounts charged in fluxed agglomerates. The majority of blast furnaces operating in the United States with fluxed pellets and/or sinter charge 10 to 50 kg/t of hot metal. Those not using any fluxed burden materials are charging 100 to 250 kg/t (Anon., 1989). A typical blast furnace charge could contain 900 kg of pellets, 650 kg of fluxed sinter, 25 to 100 kg of recycled steelmaking slag, and 50 kg of dolomite depending upon the chemistries of the individual materials (Lankford, Jr., et al., 1985). Typical blast furnace slag contains 32 to 42% silica (SiO_2), 7 to 16% alumina (Al_2O_3), 32 to 45% lime (CaO), and 5 to 15% magnesia (MgO) (Lankford, Jr., et al., 1985). The slag composition varies significantly between countries depending upon the chemistries of the raw materials available (Blattner, 1992).

With emphasis on consistent quality, high productivity, and lower cost hot metal from blast furnaces, improvements have been made in the type of burden and the characteristics of the agglomerates used. Furthermore, to minimize the heat lost in slag removed from the furnaces which requires expensive coke to be burned, a trend to lower slag volumes (actually, slag mass per unit of hot metal produced) has occurred. The lower slag volumes result from using agglomerates with low gangue contents and coke with low ash content. With lower gangue contents, lower limestone and dolomite consumption has resulted. Typically, 225 to 275 kg/t is normal practice, but many modern furnaces are approaching 200 kg/t. Some furnaces have reached 150 kg/t under special circumstances, while some are much higher (Anon., 1989, Blattner, 1992). With increased use of external desulfurization (described here), the lower limit is a function of furnace operating practice. Because slag is removed with the hot metal during tapping, slag removal is not independently controlled, and factors such as furnace size and design can affect the lowest possible slag volumes attainable. Also, there can be some economic tradeoffs, for example, less expensive high sulfur coals might be available for making coke, and this might offset the cost of using more flux.

The presence of alkali (sodium oxide and particularly potassium oxide) in the blast furnace can lead to operating problems such as degradation of coke and formation of scaffolds on furnace walls. These scaffolds can greatly reduce furnace permeability and can cause damage if they break loose. Alkali metal content in the furnace is regulated through selection of raw materials and through control of the basicity of the slag (Sciulli, 1992). Alkali removal is greatest with slags of low basicity. Typically, 60 to 80% of the alkali input will be removed with the slag.

External Treatment of Hot Metal

In the United States, external treatment of hot metal or pretreatment prior to steelmaking is most often done for desulfurization. This treatment, done in transfer cars or ladles, allows control of sulfur input to steelmaking, reduces the slag requirement in the steelmaking vessel, allows the use of higher sulfur raw materials in the blast furnace, and permits production of low-silicon hot metal. Thus, blast furnace performance can be improved (Stubbles, 1984, Szekely et al., 1989). Some common reagents used are magnesium, lime, and fluorspar. Limestone, calcium carbide, coke, sodium carbonate, salt-coated magnesium, and aluminum are sometimes used. The consumption of these reagents is on the order of 4 to 7 kg/t except for the salt-coated magnesium which is added at 0.5 kg/t (Szekely et al., 1989, Cameron, 1992). The thermodynamics of the process and practice at several plants are described in the literature (Pehlke, 1984, D'Orazio, 1984, Dukelow, 1984).

Simultaneous desulfurization and dephosphorization can also be accomplished with soda ash (sodium carbonate) or lime if the metal is low in silicon (Szekely et al., 1989). The sodium carbonate (Na_2CO_3) reacts with sulfur, silica, and phosphorus to form sodium sulfide (Na_2S), sodium silicate ($Na_2O{\cdot}SiO_2$), and sodium phosphate ($3Na_2O{\cdot}P_2O_5$) in the slag. An advantage of this process is that the sodium carbonate can be recovered and recycled, however, high refractory wear is typical (Yamamoto et al., 1980).

Oxygen Steelmaking

Steel is produced by refining hot metal for removal of carbon, silicon, and other impurities. Most steel is now produced by the basic oxygen process (BOP) in which hot metal is contacted with oxygen in a basic oxygen furnace (BOF) to oxidize impurities (Lankford, Jr., et al., 1985). The oxidized impurities rise to the top of the hot metal and are absorbed in a slag. The slag is formed from lime and dolomitic lime with addition of other compounds such as fluorspar to improve fluidity. Fluorspar (calcium fluoride) is added in an amount of 2 to 10 kg/ton (Peters, 1982), either as lump or gravel-sized material, or as briquettes made with about 15% fluorspar fines combined with slag residues from the BOF. The amount of flux added is usually proportioned to the expected silicon content in a ratio often referred to as the V ratio described earlier. Typically, the V ratio is 2.5 to 4.0 (Peters, 1982). The use of lime in steelmaking has been described in the literature (Boynton, 1980).

The amount of flux added depends on the silicon, sulfur, and phosphorus contents of the metal, and the purity of the flux (Lankford, Jr., et al., 1985). Typically, 40 to 100 kg of lime/t of hot metal and 10 to 30 kg of dolomitic lime/t of hot metal are added to the BOF (Lankford, Jr., et al., 1985, Peters, 1982). The amount of lime required in steelmaking has been steadily declining due to the trend towards decreasing silicon content of hot metal, and lower manganese contents. The increased use of scrap and the use of external desulfurization have also resulted in lower flux consumption in steelmaking. A typical steelmaking slag will contain about 10 to 20% silica (SiO_2), 40 to 50% lime (CaO), 10% magnesia (MgO), 15 to 20% iron (mixed oxides), 2 to 5% alumina (Al_2O_3), and 10% manganese oxide (MnO) and other oxides. The iron oxide content will depend upon the carbon level attained in the metal, generally increasing with lower carbon levels.

A second function of flux in steelmaking is to provide a slag that is neutral with respect to the lining of the steelmaking vessel. The lining in basic steelmaking is typically magnesia brick. To help prevent erosion and dissolution of the brick, a level of magnesia, typically 5 to 12%, is maintained in the steelmaking slag (Boynton, 1980, Lankford, Jr., et al., 1985, Peters, 1982). This is also useful for controlling the viscosity of the slag. The solubility of magnesia is greater in acid slags, that is, as the V ratio decreases. The effect of magnesia on phosphorus removal is often debated.

Ladle Metallurgy

Reagents can be added to liquid steel in the transfer ladles for additional refining to remove sulfur, reduce metallic oxides, and condition the slag prior to casting. Additions can also be made for alloying and for shape control of inclusions in the cast product. Reagents similar to those for hot metal desulfurization are used, and are added at the top of the ladle or are injected. Metallic elements can be added in the form of wire. Typical reagents are magnesium, lime, alumina, aluminum, fluorspar, calcium carbide, calcium-silicon alloy, and prefused di-calcium aluminate (calcium-aluminate). Magnesium is often added in combination with lime. Typical addition rates are 1 to 5 kg/t (Szekely et al., 1989). Calcium-aluminate increases the extent of deoxidation of steel and has a high capacity for sulfides (Turkdogan, 1985). An exothermic synthetic calcium-aluminate mixture comprising 58% dry burnt lime, 30% hematite, and 12% aluminum powder might be added at tap (Gilbert et al., 1988).

Electric Furnace Steelmaking

Electric furnaces are used to make steel and ferroalloys from scrap, other ferroalloys, and alloying elements. Oxidation is accomplished by injection of oxygen or by the addition of iron oxide either as mill scale or iron ore. Carbon is added as coke, used carbon electrodes, or coal, and lime is added as flux. In the electric furnace route to making steel, a two stage process with both oxidation and reduction is used. This is required to achieve necessary purification of the metal and high retention of alloying elements. In alloy steels, many components are readily oxidized and will report to the slag in an oxidizing environment, but can be recovered in a subsequent reducing environment. Therefore, double-slag, slag-and-a-half, and single slag practices followed by ladle metallurgy have evolved (Reebel, 1950, Sharp, 1967, Peters, 1982, Lankford, Jr., et al., 1985). In a single slag practice, after the melt down the oxidizing slag is held in the furnace and the steel is removed and treated further in a ladle. In a double-slag practice, the slag and metal are held in the furnace and the slag composition is adjusted. In either practice, alloy additions are made prior to the second stage. In slag-and-a-half practice, the first slag is removed and a lime-fluorspar slag is added to the electric furnace (Lankford, Jr., et al., 1985)

For oxidizing slags, limestone can be added, but calcined lime is most often used for slag formation. Limestone and coke are added in some practices to help create a foaming slag to protect refractories and to improve power efficiency. Lime and dolomitic lime are also added to protect the furnace refractories. Lime addition is 15 to 50 kg/t (Boynton, 1980). Other slag-forming ingredients are fluorspar, silica sand, crushed coke, and aluminum. During melting, oxidation of carbon occurs, and oxidation of phosphorus, silicon, manganese, and chromium results in removal to the slag. The V-ratio is controlled to over 2.5 in the melt down stage and a range of 2 to 4 is typical (Lankford, Jr., et al., 1985). The basicity depends on the silicon in the charge and in the flux added. The iron oxide in the slag is inversely related to the carbon in the bath as in oxygen steelmaking. An oxidizing melt down slag might contain 40 to 50% lime (CaO), 12 to 20% silica (SiO_2), 5 to 15% iron oxide (FeO), 3 to 10% alumina (Al_2O_3), 2 to 12% magnesia (MgO), and 5 to 15% manganese oxide (MnO) (Peters, 1982, Lankford, Jr., et al., 1985). Generally, 40 to 50% lime (CaO) is required for good sulfur removal (Lankford, Jr., et al., 1985).

The second stage can be done in a mildly reducing slag in a ladle in which a basic, non-oxidizing CaO-Al_2O_3-SiO_2 slag is built up from additions of lime, dicalcium aluminate and aluminum. During the second process stage, metal oxides in the slag are reduced and sulfur enters the slag as calcium sulfide, particularly in low iron oxide slags. In the double-slag electric furnace practice, a finishing or deoxidizing slag is formed by adding coke to react with lime to form a calcium carbide slag. High purity of both the coke and lime are desirable. Fluorspar equal to about 10% of the lime is added to increase reaction with lime and increase fluidity (Sharp, 1967). Silica is sometimes added, and alumina is sometimes added in place of fluorspar. The slag quantity is generally about 2 to 3% of the metal. A typical reducing slag might contain 55 to 70% lime (CaO), 15 to 25% silica (SiO_2), 1 to 3% alumina (Al_2O_3), 3 to 12% magnesia (MgO), and 0.5 to 2% FeO and MnO (Peters, 1982).

Continuous Casting

Fluxes are used during the continuous casting of steel and during bottom pouring of ingots. These fluxes are often referred to as *mold powders.* The fluxes are either mixtures of compounds or fused synthetic slags formulated to provide thermal insulation and prevent solidification in the mold, prevent oxidation, absorb oxide inclusions, provide lubrication at high temperature, and maintain uniform heat transfer. Important physical properties are flowability, melting rate, viscosity, crystallization temperature, inclusion absorption capacity, and insulation characteristics. The mold fluxes are largely silica based with additions of lime, alumina, fluorspar, alkali metal oxides and carbon. The fluxes are used both as raw mixtures (synthetic flux) and in premelted and sintered forms. The particle sizes range from −10 mesh (1.5 mm) granules to 90% −200 mesh (0.074 mm) powders. Typical mold fluxes contain 17 to 56% silica (SiO_2), 2 to 45% lime (CaO), 0 to 25% alumina (Al_2O_3), 0 to 15% fluorine (F_2), 0 to 10% iron oxide (Fe_2O_3), 0 to 20% borate (B_2O_3), 0 to 25% alkali metal oxides (Na_2O, K_2O, Li_2O), and 0 to 30% carbon (C). The amounts vary widely to attain the desired function. The basicity (CaO/SiO_2) is typically 0.65 to 1.67. Consumption of mold fluxes is 0.3 to 0.8 kg/t in continuous casting and 1 to 3 kg/t in ingot casting (Branion, 1987, McCauley, 1987).

Nonferrous Metals

The use of fluxes in nonferrous smelting is described in depth in the literature (Turkdogan, 1983, Rosenqvist, 1974, Habashi, 1986, Moore, 1981). Nonferrous slags differ from ironmaking and steelmaking slags because they are based largely on formation of iron silicate rather than calcium-aluminum silicate. A typical nonferrous slag will contain 30 to 55% iron oxide (FeO), 30 to 45% silica (SiO_2), and less than 10% lime (CaO), although there will be variation around these numbers depending upon the system. Phase diagrams of typical nonferrous systems showing the compositional ranges of slags can be found in the literature (Coudurier et al., 1985, Rosenqvist, 1974). In copper smelting, silica is the flux used in the largest amount, but lime is also added to modify slag viscosity and density. Limestone is also used in small amounts to aid in refining metallic copper, lead, zinc, tin, and antimony (Boynton, 1980). Limestone purity and particle size are generally similar to those used in ironmaking and steelmaking, but high-calcium limestone is most often used. Both high grade siliceous materials and lower grade siliceous ores are used as fluxes along with limestone. An overview of the technology applicable to smelting and refining of copper, lead, nickel, and other nonferrous metals was recently published (Peacey, 1989).

Copper: Copper occurs as both oxide and sulfide minerals, and smelting is used to recover copper from the sulfide ores. Many copper sulfide ores contain iron that must be fluxed, and typically, silica is used to flux the iron and form an iron silicate slag (fayalite). During smelting, the sulfur combines with copper and some iron to form a copper-iron sulfide matte which is separated from the slag. The matte is converted to metallic copper by oxidation to remove the iron and sulfur, and the copper is further refined to remove impurities (Turkdogan, 1983, Moore, 1981).

Copper smelting is done in bath processes and flash smelting furnaces with oxygen enrichment, but flash smelting is the predominant method practiced today (Peacey, 1989). The older reverberatory furnace practice has been described in the literature (Anderson, 1961, Fowler, 1961, Nicholson et al., 1961). Electric furnace smelting of copper sulfides has also been reported, and other smelting processes are being developed (Barth, 1961, Floyd et al, 1980, Phelps, 1991). In general, the slags in reverberatory furnace smelting, flash smelting, and other processes are similar. The Mitsubishi continuous smelting and refining process differs in that a calcium ferrite slag is used in the converting furnace.

Copper-bearing concentrates often contain 5 to 10% silica. About 5% additional silica is added as pure silica or ore, and the remainder of the silica is obtained from slag recycled from the converters used to make copper metal from the matte. Generally, lime in some form is added to break up silicates, take up excess silica from aluminosilicates, and reduce the viscosity of the silica-rich slag. Typically, 3 to 10% limestone or lime-bearing mix is added. A typical smelting slag contains about 30 to 45% FeO, 30 to 40% SiO_2 , 5 to 10% Al_2O_3, 2 to 6% CaO, and 2 to 4% MgO (Watanabe et al., 1980). The ratio of slag to metal varies but is often in the range 0.6 to 1.5.

One problem in recovering copper is the tendency to form magnetite in the converters, although a certain amount of magnetite is required to attain high copper recoveries. Magnetite melts at a temperature higher than the temperature in either smelting or converting, has a low solubility in slags and matte, and doesn't form a silicate with the silica usually present in smelting because much of the silica is tied up in other forms. Typically, some free silica is added to tie up magnetite as iron silicate, but excess will react with basic furnace brick. The silica used is often of very high purity—greater than 90% SiO_2 (Fowler, 1961, Watanabe et al., 1980, Floyd et al., 1980). Although siliceous slags are usually used, it has been reported that lime-based slags compared to siliceous slags have a lower viscosity, a higher solubility for magnetite, separate easily from the matte, and are more efficient in removing arsenic, antimony, and bismuth (Turkdogan, 1983, Chaubal et al., 1989).

The size of flux must be compatible with the process, but in general the sizes used are the same as in ironmaking and steelmaking. In converters, the flux is in the range of 2.5 by 0.6 cm. Flux coarser than 2.5 cm is difficult to inject and to dissolve, and flux finer than 0.6 cm tends to float on the bath and is also a source of dust (Anderson, 1961). In the Sirosmelt process, the limestone used was 0.5 to 1.0 cm in size (Floyd, 1980).

Lead: Lead is processed by a number of older processes that include sintering followed by smelting in the lead blast furnace, electric furnace, or Imperial Smelting Process (ISP), and several newer direct and flash smelting methods such as Kivcet, Outokumpu, and QSL are being tried (Castle et al., 1989, Pickard and Crawford, 1989). In the blast furnace and ISP processes, lead sulfide ores are roasted to form lead oxide sinter which is charged with limestone and coke to the lead blast furnace or the Imperial Smelting Furnace (Moore, 1981). In the latter, lead is collected in the furnace bottom where it separates from slag, and zinc vapor exiting the top is condensed. In the direct and flash processing methods, the lead is partially or totally oxidized into the slag and the slag is treated in an electric furnace with carbon as coal or coke to reduce the oxidized lead to lead bullion (Turkdogan, 1983). Copper is removed from the lead by adding sulfur and by several other processes. Further refining is accomplished by oxidizing impurities with air, lead oxide, sodium hydroxide, or sodium nitrate flux and by electrorefining (Moore, 1981, Peacey, 1989).

Slags formed are complex. A typical lead blast furnace slag composition is: 2 to 12% ZnO, 20 to 25% FeO, 1 to 2% Fe_3O_4, 0 to 2% PbO, 16 to 19% CaO, 26 to 31% SiO_2, 5% MnO, 10% Al_2O_3, 1 to 2% CaS, and 5 to 10% MgO (Grant, 1980).

Nickel: Nickel ores occur as both sulfides and oxides. The sulfide ores often contain copper-iron sulfides in association with nickel-iron sulfides. For sulfide ores, the processing is accomplished by reverberatory or flash smelting furnaces and horizontal converters (Pickard and Crawford, 1989, Moore, 1981). Flash converting is being introduced (Phelps, 1991). Separation of copper and nickel sulfides from the matte is accomplished by crystallization, grinding, and froth flotation, and the copper and nickel sulfides are refined further. In nickel smelting from some ores, slag formation can be complex due to the presence of magnesia (MgO) in various silicate minerals. The slag is typically iron and silica rich, but with elevated magnesia levels. The magnesia will increase the melting point, and olivine $(Mg,Fe)SiO_2$ can crystallize out. This can be countered with silica and lime addition wherein lime decreases the viscosity of the melt (Davey et al., 1980).

In New Caledonia, an oxide ore comprising a nickel-bearing hydrated silicate of iron and magnesia containing 37% SiO_2, 2.2% Al_2O_3, 23% MgO, 20% Fe_2O_3, 2% Cr_2O_3, 0.3% CaO, and 2.2% Ni was sintered with gypsum ($CaSO_4 \cdot nH_2O$) and smelted in a blast furnace to an iron-nickel matte. This was further treated in a converter to remove iron (Thurneyssen et al., 1961, Habashi, 1986). At the now closed smelter at Riddle, OR, garnierite, a magnesium silicate containing alumina was smelted. The ore contained 25 to 38% MgO, 45 to 55% SiO_2, 1 to 3% Al_2O_3, 1 to 2% CaO, 8 to 15%

Fe, and 1.5% nickel (Ni). The Ugine process was used in which silicon as ferrosilicon (FeSi) was used as the reductant, and silica was formed in the process. Phosphorus was removed by oxidizing the bath with iron ore in the presence of a high lime slag. The metal was further refined with a slag containing lime and fluorspar (Coleman and Vedensky, 1961).

Tin: High grade tin oxide concentrate is smelted in reverberatory and electric furnaces, and low grade concentrate is smelted in blast furnaces, kilns, and horizontal furnaces (Castle et al., 1989). In the first stage of a two stage process, part of the oxide is reduced to metal with formation of a ferrous silicate slag containing a small amount of lime and unreduced tin as a tin silicate. First stage slags typically contain 30 to 40% SiO_2, 15 to 25% FeO, 5 to 15% CaO, and 5 to 25% SnO_2. In the second stage, the tin silicate is reduced by iron metal to form iron silicate and an impure tin which is recycled. Second stage slags from southeast asia typically contain 24 to 28% SiO_2, 9 to 11% Al_2O_3, 20 to 22% CaO, 12 to 20% FeO, 2–6% MgO, and impurity metals (Habashi, 1986).

ECONOMIC FACTORS

In industrialized nations, metal consumption per capita is thirty times that of underdeveloped nations (Anon., 1991a) and thus, the worldwide potential for increased metal production and the attendant flux consumption is very large. However, the rate at which demand will increase is not clear. World demand for metals and fluxes would be expected to be a function of population growth and economic growth, however, population growth is greatest in underdeveloped countries where economic resources are scarce. Whether reserves of metals will be sufficiently large to meet this potential demand is a subject of debate, however, it is clear that previously projected shortages are not yet developing (Castle et al., 1989). Fortunately flux minerals, and particularly limestone and silica, that would be required to support growth in metals consumption are abundant and widespread.

An excellent review of the state of the metals industry was recently published (Davis, 1989). This review suggests that consumption of both ferrous and nonferrous metals ceased growing around 1960, and in fact, has been negative recently. This is attributed to both the decreasing growth rates of the economies in industrialized countries and the decreasing intensity of metal use in both service industries and manufacturing. It has been proposed by several economists that the demand for raw materials has become uncoupled from economic growth (Davis, 1989). This is particularly true in Japan and the United States, and the impact has been greatest on the steel industry, the largest flux user. The implication of this trend is that economic recovery will not necessarily result in increased demand for metals (and hence fluxes). It is suggested that a period of new demand creation and market development will be required to stimulate further growth.

Recent changes in the Republic of South Africa and the new Commonwealth of Independent States (formerly the Union of Soviet Socialist Republics) and increased trade with China, will alter the supply and demand situation of certain minerals which could affect flux consumption. Greater imports of metals from these countries to the United States will impact domestic flux production and consumption. Even with increased imports of metals and other mineral commodities, it is not likely that any large scale trade in stone or silica flux will occur due to the abundance of sources, low value, and high transportation costs.

However, fluorspar, which is used at at rate of about 10 kg/t of steel is almost entirely imported to the United States because of high domestic production and transportation costs (Anon., 1991a). Metallurgical fluorspar imports and domestic consumption were about 110 kt in 1990. These imports were 70% from Mexico, 28% from China, and 2% from South Africa (Hodge, 1991). The latter is largely a low grade flotation concentrate for agglomeration into briquettes. The imports from China are increasing because Chinese fluorspar costs 50 to $60/t FOB port compared to Mexican fluorspar which is $90/t FOB port (Hodge, 1991). The demand for metallurgical-grade fluorspar is decreasing in steel and allied industries.

In both ferrous and nonferrous smelting processes, the trend to lower consumption of flux per unit of metal is due to economics resulting from an intense effort to reduce the quantity of slag formed which requires energy and to changes in smelting practices. The present deficiency of capital in the capital intensive metal industries has resulted in plant closures and consolidation in both the metals industries and the attendant flux suppliers. Alternative uses for limestone and dolomite previously used for fluxstone are being sought by producers.

GOVERNMENT CONSIDERATIONS

Government influences mineral and metal markets through labor laws, land use and environmental regulations, trade and energy policies, and funded programs and research. In the United States, the flux industry can be affected by actions of the Department of the Interior, Department of Energy, the Environmental Protection Agency, Occupational Safety and Health Administration, Mine Safety and Health Administration, and the Bureau of Land Management, to name a few.

The 1990 Amendment to the Clean Air Act will affect the domestic manufacture of coke which might drive the steel industry to conversion to scrap-based and premetallized burden electric furnace steel processing or coal-based direct ironmaking processes. Electric furnaces would require lower gangue feed stock, and hence would require less flux. Coal-based processes could require a change in the size requirement of flux used. Sinter plants presently making agglomerates from ores and steel plant wastes might be affected by these changes or environmental regulations in a way that will alter slightly the way in which flux is introduced to the blast furnaces.

Fluorspar imports and consumption were impacted by several recent actions. One was the September 1987 Montreal Protocol and the United Nations Environmental Program (revision), London, June 1990, in which chlorinated fluorocarbons (CFCs) and halons are to be restricted by the year 1995 and banned by the year 2000 (Hodge, 1991). In 1990, Mexican acid-grade fluorspar became eligible for preferential treatment, and a 13.5% tariff on imports of metallurgical-grade fluorspar was suspended through December 1992 (Anon., 1991a).

Mining and quarrying are industries that affect the environment, particularly the use of land. The stone industry in some instances has developed near populated areas which causes problems with dust, noise, blasting, trucking, etc. Little emphasis was placed on the environment in years past, however that has changed due both to increased awareness of the effects on the environment and to state and federal regulations. Environmental regulations that affect mining and quarrying are those that affect reclamation of process water and solid waste, reduction of dust and noise, and control of air and ground shock from blasting.

The Bureau of Land Management works under the Federal Land Policy and Management Act of 1976. This Act mandates that public lands be managed for multiple use (Anon., 1991a). This also requires that, following mining, land will be reclaimed for other uses. Other regulations of importance are those that affect use of wetlands and dunelands, revegetation of waste, and gradation of pit slopes. The Bureau of Land Management also administers mining on lands governed by the General Mining Law of 1872 (Anon.,

1991a). Recently, a study of the potential economic impact of repealing the General Mining Law was released (Anon., 1992). The study predicts that 30,000 jobs would be affected, and the cost to the federal government in lost revenues and expenditures would be several hundred million dollars annually.

For updates on minerals and land use planning and the General Mining Law, readers should contact the Bureau of Land Management. For further information on mineral statistics, minerals and the environment, recycling, or mining, readers should contact the United States Bureau of Mines.

PROBLEMS AND FUTURE TRENDS

In a low growth economic scenario, and with the demand for saving energy and improving the environment, the consumption of flux for metallurgical purposes is likely to continue downward. Metal production and flux consumption will be influenced by new process technologies, recycling, and substitution (Castle et al., 1989). Identifiable trends are the use of composite materials and lightweight materials in transportation vehicles, the increased use of high technology devices and miniaturization. Substitution of plastics, composites, and light alloys including aluminum will impact the use of steel and the attendant flux consumption; however, any increased use of aluminum will require some increase in fluorspar consumption. Space required for slag disposal and methods for containment of effluents from slag dumps to prevent transport of undesirable elements into the environment also drive lower flux usage.

Ironmaking and Steelmaking

Current blast furnace practice using domestic iron ore pellets with a nominal silica content of 5% requires a certain blast furnace slag practice that dictates flux consumption. Therefore, further reductions of flux consumed per ton of hot metal are not likely. Recently, there has been a trend to using fluxed pellets in the blast furnaces. In 1990, fluxed pellets comprised 34% of domestic production and the trend is increasing (Ranade, 1992). The use of fluxed pellets with flux from Michigan has displaced some flux used in sinter and, at some blast furnaces, flux from other sources. Presently, recycled steelmaking slag provides the additional flux needed by many furnaces. Sinter plants, although useful for recyling of steel plant wastes, could be closed for environmental reasons. There is also a trend to production of lower silicon hot metal and lower hot metal manganese requirements which results in lower flux consumption in steelmaking. Decreases in the consumption of lime and use of fluorspar in BOF steelmaking are also likely to continue.

Presently, there is overcapacity in the steel industry, and prices are weak (Anon., 1991a). Some steel mills are downsizing, and some are looking to alternative modes of operation. To achieve greater process throughputs, and higher quality steel, additional installations for external treatment of hot metal, particularly desulfurization will be built, more ladle metallurgy stations and vacuum degassing facilities will be installed, and scrap preheating for electric furnaces will be increased. The success of mini-mills, a desire by steel producers to decrease coke production to meet environmental regulations, and increased recycling of scrap is resulting in greater use of electric furnaces. According to a recent article, Karlis Kirsis of Paine Webber's World Steel Dynamics has projected that mini-mills will capture 14% of the domestic steel market in the year 2000, up from 6% today (Ashby, 1992). If major steel producers adopt mini-mill and electric furnace practices, the demand for lower silica iron sources is likely to increase, and a decrease in flux consumption is likely. Many iron and steel producers that once owned captive limestone and dolomite operations have sold them.

Over the last several years, there has been renewed interest in direct ironmaking and direct steelmaking processes. Direct ironmaking will very likely require a different slag practice than blast furnace ironmaking. New slag functions such as foaming to provide a blanket over the bath to achieve post combustion of gases and allow adequate heat transfer might redefine flux requirements. Both limestone and lime might be used in the same smelting vessel. The direct ironmaking processes will use coal in preference to coke which will increase the sulfur burden on the process. Use of coals of lower grade than present metallurgical coals might require alteration of flux practices.

It is well known that there is a trend to more recycling and reuse of wastes to protect the environment, and the steel industry is the single largest recycler of materials (Anon., 1991a). The reuse of steel saves iron ore, coal, limestone and dolomite flux, and energy. Byproduct slags have been used for landfill, ballast, aggregate, cement, and rock wool, and more uses for slag are needed.

Nonferrous Industry

Excellent reviews of the nonferrous industries in Europe and North America have been published recently (Castle et al., 1989, Pickard and Crawford, 1989). In the copper smelting industry, there has been considerable consolidation which has brought supply and demand more closely into balance, however, the relationship is tight due to disruptions in supply from foreign sources (Castle et al., 1989, Pickard and Crawford, 1989, Anon., 1991a). In copper smelting, there has been a trend to new technology development, with flash smelting and bath processes largely replacing reverberatory furnaces. Direct smelting using a lime flux has been proposed (Peacey, 1989). Flash converting is gaining in importance (Carter, 1990).

Copper is the third most consumed metal after iron and aluminum, and the greatest use for copper is in electrical applications (Anon., 1991a). Copper is affected by substitution, for example the use of plastic piping in domestic plumbing and the substitution of fiber optic cable for copper wire. The trend towards computerization and miniaturization, the development of lower voltage lighting systems allowing finer wire, lower power requirements of solid state devices, introduction of better magnet steels with lower power losses, and the development of high energy permanent magnets that allows development of smaller motors all could diminish the demand for copper. The effect of superconductivity on copper use is uncertain (Rieber, 1992). The known superconductors all contain copper, but wire will be smaller. However, the technology might spur some new industries and increase the demand for copper. There is an increasing demand for electronic shielding materials, and manufacture of all electric vehicles could increase demand for copper. Currently, 26% of copper production is from scrap (Phelps, 1991).

Lead is the fifth most produced metal after zinc, which is fourth (Castle et al., 1989). Lead smelting is likely to be adversely affected by the push for a lead-free environment which might lead to some decrease in flux consumption, however, demand for all electric vehicles could increase the demand for lead-acid batteries, the largest use for lead. A large portion of demand for lead is met by recycling, and this could increase in the future. Hydrometallurgical processing of lead ores might prove attractive in the future (Peacey, 1989).

ACKNOWLEDGMENTS

The authors wish to thank their colleagues Pinakin Chaubal, Howard Pielet, William Heitmann, Mark Keilman, and Dan Kruse for their helpful comments and the Research Librarian, Barbara Banek for her assistance.

BIBLIOGRAPHY AND REFERENCES

Anon., 1987, "Industrial Minerals Supply/Demand Data, 1975–1985," *Mineral Industry Surveys*, US Department of the Interior, US Bureau of Mines, June 22, pp. 29, 43, 75.

Anon., 1988, *Annual Book of ASTM Standards*, Vol. 4.01 Cement, Lime, Gypsum; Vol. 4.02 Concrete and Aggregates; Vol. 3.05 Chemical Analysis of Metals and Metal Bearing Ores, ASTM, Philadelphia, PA.

Anon., 1989, "Technical Committee on Blast Furnace Practice, Operating Report, Fourth Quarter, 1989," American Iron and Steel Institute, Washington, DC.

Anon., 1990, "Directory of Principal Crushed Stone Producers in the United States in 1989," *Mineral Industry Surveys*, US Department of the Interior, US Bureau of Mines.

Anon., 1991a, *Minerals in 1991*, US Department of the Interior, US Bureau of Mines, pp. 19, 59–68.

Anon., 1991b, "Lime Plants in the United States in 1990," *Mineral Industry Surveys*, US Department of the Interior, US Bureau of Mines.

Anon., 1992, "The Economics of Changing the Mining Law," *Mining Engineering*, Vol. 44, no. 5, p. 395.

Anderson, J.N., 1961, "Reverberatory Furnace and Converter Practice at the Noranda and Gaspé Smelters," *Extractive Metallurgy of Copper, Nickel, and Cobalt*, P. Queneau, ed., Interscience Publishers, New York, p. 133f.

Ashby, R.B., 1992, "Industrial Heating—Washington Bureau," *Industrial Heating*, May, p. 6.

Barth, O., 1961, "Electric Smelting of Sulfide Ores," *Extractive Metallurgy of Copper, Nickel, and Cobalt*," P. Queneau, ed., Interscience Publishers, New York, p. 241f.

Blattner, J.L., 1992, "Blast Furnace Slag," *Blast Furnace Ironmaking*, Vol. 1, Principles and Design, McMaster University, Hamilton, Ontario, May, p. 4–1.

Boynton, R.S., 1980, *Chemistry and Technology of Lime and Limestone*, 2nd ed., John Wiley & Sons, New York, pp. 95–104, 163, 220–321, 380–385.

Branion, R.V., 1987, "Mold Fluxes for Continuous Casting," *Mold Powders for Continuous Casting and Bottom Pour Teeming*, Iron and Steel Society, Warrendale, PA, p. 3ff.

Cameron, I.A., 1992, "External Treatment of Hot Metal," *Blast Furnace Ironmaking*, Vol.3, Practice, McMaster University, Hamilton, Ontario, May, p. 17–1.

Carter, R., and Suttill, K., 1992, "New Smelter for Utah Copper," *Engineering and Mining Journal*, June, p. 33ff.

Castle, J., et al., 1989, "Non-Ferrous Metals—An Industrial/Economic Overview from a European Viewpoint," *Productivity and Technology in the Metallurgical Industries*, Proceedings of Symposium, Cologne, Germany, Sep. 17–22, 1989, The Minerals, Metals, and Materials Society, Warrendale, PA, pp. 3–67.

Chaubal, P.C., et al., 1989, "Mathematical Modeling of Minor-Element Behavior in Flash Smelting of Copper Concentrates and Flash Converting of Copper Mattes," *Met. Trans*, B, Vol. 20B, Feb., p. 39.

Coleman, E.E., and Vedensky, D.N., 1961, "Production of Ferronickel at Riddle, Oregon," *Extractive Metallurgy of Copper, Nickel, and Cobalt*, P. Queneau, ed., Interscience Publishers, New York, p. 263.

Coudurier, L., Hopkins, D.W., and Wilkomirsky, I., 1985, *Fundamentals of Metallurgical Processes*, 2nd ed., Pergamon Press, New York, p. 242.

D'Orazio, J.J., 1984, "Desulfurization at Inland Steel," *Proceedings*, Low Sulfur Steel Symposium, AMAX Materials Research Center, Ann Arbor, MI, p. 111.

Davey, T.R.A., Segnit, E.R., and Boow, J., 1980, "Viscosity and Composition Relationships in Kalgoorlie Nickel Smelter Slags," *Proceedings*, Australia/Japan Extractive Metallurgy Symposium, Australasian Institute of Mining and Metallurgy, Parkville, Vic., p. 115.

Davis, G., 1989, "The Declining Influence of Economic Growth on Metals Demand," *Productivity and Technology in the Metallurgical Industries*, Proceedings of Symposium, Cologne, Germany, Sep. 17–22, The Minerals, Metals, and Materials Society, Warrendale, PA, pp. 159–186.

Derge, G., and Tenenbaum, M., 1951, "Physical Chemistry of Slag-Metal Reactions," Chap. 17, *Basic Open Hearth Steelmaking*, W.O. Philbrook et al., eds., AIME, New York, p. 691f.

Duckworth, W.E., and Hoyle, G., 1969, *Electro-slag Refining*, Chapman & Hall, London, p. 17f.

Dukelow, D.A., 1984, "Desulfurization at National Steel Corporation," *Proceedings*, Low Sulfur Steel Symposium, AMAX Materials Research Center, Ann Arbor, MI, p. 107f.

Elliott, J.R., 1984, "Slags for Metallurgical Process," *Metallurgical Slags and Fluxes*, 2nd International Symposium, Metallurgical Society-AIME, Warrendale, PA, p. 45ff.

Fagerberg, B., and Sanberg, N., 1973, "Degradation of Lump Ores in Transport," Vol. 2, *Mineral Transportation*, Proceedings of 2nd International Symposium on Transport and Handling of Minerals, Rotterdam, Oct., p. 128f.

Fine, H.A., and Gaskell, D.R., eds., 1984, *Metallurgical Slags and Fluxes*, 2nd International Symposium, Metallurgical Society-AIME, Warrendale, PA.

Floyd, J.M., Grave, N.C., and Lightfoot, B.W., 1980, "Small Pilot Plant Trials of Sirosmelt Copper Refining," *Proceedings*, Australia/Japan Extractive Metallurgy Symposium, Australasian Institute of Mining and Metallurgy, Parkville, Vic., p. 63f.

Fowler, M.G., 1961, "Smelting Practices of Phelps Dodge in Arizona," *Extractive Metallurgy of Copper, Nickel, and Cobalt*, P. Queneau, ed., Interscience Publishers, New York, p. 159f.

Gault, H.R., and Ames, J., 1960, "Fluxing Stone," *Industrial Minerals and Rocks*, 3rd ed., J.L. Gillson, ed., AIME, New York, p. 191.

Gilbert, S., Monos, G.G., and Turkdogan, E.T., 1988, "Ladle Refining of Steel Using an Exothermic Synthetic Slag," *Proceedings*, Steelmaking Conference, Iron and Steel Society, Warrendale, PA, p. 291f.

Grant, R.M., 1980, "The Derivation of Thermodynamic Properties of Slags from Slag Fuming Plant Data," *Proceedings*, Australia/Japan Extractive Metallurgy Symposium, Australasian Institute of Mining and Metallurgy, Parkville, Vic., p. 75f.

Habashi, F., 1986, *Principles of Extractive Metallurgy*, Vol 3., Pyrometallurgy, Gordon and Breach, New York, pp. 313–332.

Hodge, B.L, 1991, "Fluorspar," Industrial Minerals, Metals and Minerals Annual Review, *The Mining Journal*, p. 116.

Koros, P.J., Petrushka, R.G., and George, T.R., 1991, "Flowability of Powder Materials," *Injection in Process Metallurgy*, Proceedings, International Symposium, T. Lehner, P.J. Koros, and V. Ramachandran, eds., Minerals, Metals, and Materials Society, Warrendale, PA, p. 81ff.

Lankford, W.T., Jr., et al., eds., 1985, *The Making, Shaping and Treating of Steel*, 10th ed., Association of Iron and Steel Engineers, Pittsburgh, PA, pp. 325f, 333f, 539f, 599f, 627f.

Lawrence, F.V., Jr., 1975, "Fluxes," Industrial Minerals and Rocks, 4th ed., S.J. Lefond, ed., p. 260.

Levin, E.M., et al., 1956, 1969, *Phase Diagrams for Ceramists*, (continuing series), The American Ceramic Society, Columbus, OH.

Limes, R.W., and Russel, R.O., 1970, "Crucible Test for Lime Reactivity in Slags," *The Reaction Parameters of Lime*, Symposium, 72nd Annual Meeting, American Society for Testing and Materials, Atlantic City, NJ, June 22–27, 1969, Special Technical Publication 472, ASTM, Philadelphia, PA, p. 161f.

McCauley, W.L., and Koul, M.K., 1987, "Mold Powders Technology: Bottom Pouring vs. Continuous Casting," *Mold Powders for Continuous Casting and Bottom Pour Teeming*, Iron and Steel Society, Warrendale, PA, p. 111ff.

Miller, M.M., 1990, "Lime," Preprint for Minerals Yearbook, US Department of the Interior, US Bureau of Mines, Oct., p. 7.

Moore, J.J., 1981, *Chemical Metallurgy*, Butterworths, London, p. 150f.

Nicholson, M.K., Lockridge, P.L., and Beals, G.C., 1961, "Copper Converting Practice at Chino Mines Division of Kennecott Copper Corporation," *Extractive Metallurgy of Copper, Nickel, and Cobalt*, P. Queneau, ed., Interscience Publishers, New York, p. 225f.

Ober, J.A., 1989, "Lime," *Minerals Yearbook 1987*, Vol. 1, US Department of the Interior, US Bureau of Mines, p. 575.

Peacey, J.G., 1989, "Non-Ferrous Extractive Metallurgy: A Technical Overview," *Productivity and Technology in the Metallurgical Industries*, Proceedings of Symposium, Cologne, Germany, Sep. 17–22, The Minerals, Metals, and Materials Society, Warrendale, PA, p. 85f.

Pehlke, R.D., 1984, "Thermochemistry of Desulfurization," *Proceedings*, Low Sulfur Steel Symposium, AMAX Materials Research Center, Ann Arbor, MI, p. 81.

Peters, A.T., 1982, *Ferrous Production Metallurgy*, John Wiley & Sons, New York, pp. 20–24, 135–137, 148–167.

Pickard, F.G.T., and Crawford, G.A., 1989, "Productivity and Technology Changes in the North American Copper, Zinc, and Nickel Extractive Industries," *Productivity and Technology in the Metallurgical Industries*, Proceedings of Symposium, Cologne, Germany, Sep. 17–22, 1989, The Minerals, Metals, and Materials Society, Warrendale, PA, pp. 69–84.

Ranade, M.G., 1992, "Iron-Bearing Burden Materials," *Blast Furnace Ironmaking*, Vol. 2, Raw Materials, McMaster University, Hamilton, Ontario, May, p. 2–1.

Reebel, D., 1950, *ABC of Iron and Steel*, 6th ed., Penton Publishing Co., Cleveland, OH, pp. 133–147.

Ricketts, J.A., 1992, "Historical Development and Principles of the Iron Blast Furnace," *Blast Furnace Ironmaking*, Vol.1, Principles and Design, McMaster University, Hamilton, Ontario, May, p. 1–1.

Rieber, M., 1992, "High-T_c Superconductivity's Potential Impact on Copper," *Journal of Metals*, Mar., p. 16.

Rosenqvist, T., 1974, *Principles of Extractive Metallurgy,* McGraw-Hill, New York, p. 324f.

Schlitt, W.J., and Healy, G.W., 1970, "Characterization of Lime: A Comparison and Scaling Down of the Coarse Grain Titration Test and the ASTM Slaking Rate Test," *The Reaction Parameters of Lime* Symposium, 72nd Annual Meeting, American Society for Testing and Materials, Atlantic City, NJ, June 22–27, 1969, Special Technical Publication 472, ASTM, Philadelphia, PA, p. 143.

Sciulli, C.M., 1992, "Blast Furnace Reactions," *Blast Furnace Ironmaking*, Vol. 1, Principles and Design, McMaster University, Hamilton, Ontario, May, p. 3–1.

Sharp, J.D., 1967, *Electric Steelmaking*, CRC Press, Cleveland, OH, pp. 81–85.

Sosinsky, D.J., and Sommerville, I.D., 1986, "The Composition and Temperature Dependence of the Sulfide Capacity of Metallurgical Slags," *Met. Trans*, B, Vol. 17B, June, pp. 331–337.

Stubbles, J.R., 1984, "Desulfurization Technology, State of the Art," *Proceedings*, Low Sulfur Steel Symposium, AMAX Materials Research Center, Ann Arbor, MI, p. 99.

Szekely, J., Carlsson, G., and Helle, L., 1989, *Ladle Metallurgy*, Springer-Verlag, New York, pp. 1–26.

Taguchi, K., Hanaoka, K., and Ikeda, K., 1980, "Production of Self-Fluxed Pellets and Improvement of Quality," *Proceedings*, Australia/Japan Extractive Metallurgy Symposium, Australasian Institute of Mining and Metallurgy, Parkville, Vic., p. 245.

Tepordei, V.V., 1991, "Crushed Stone (1989)," Preprint, *Minerals Yearbook*, US Department of the Interior, US Bureau of Mines, p. 16.

Thurneyssen, C.G., Szczeniowski, J., and Michel, F., 1961, "Ferronickel Smelting in New Caledonia," *Extractive Metallurgy of Copper, Nickel, and Cobalt*, P. Queneau, ed., Interscience Publishers, New York, p. 287.

Turkdogan, E.T., 1983, *Physicochemical Properties of Molten Slags and Glasses*, The Metals Society, London, pp. 1–68, 89–179, 425–439.

Turkdogan, E.T., 1985, "Slags and Fluxes for Ferrous Ladle Metallurgy," *Ironmaking and Steelmaking*, Vol. 12, No. 2, pp. 64–78.

Watanabe, T., Okada, S., and Muto, S., 1980, "Behavior of Magnetite in the Flash Smelting Furnace with Furnace Electrodes at Tamano Smelter," *Proceedings*, Australia/Japan Extractive Metallurgy Symposium, Australasian Institute of Mining and Metallurgy, Parkville, Vic., p. 31f.

Yamamoto, S., Kajioka, H., and Nakamura, Y., 1980, "Refining of Hot Metal with Na_2CO_3," *Proceedings*, Australia/Japan Extractive Metallurgy Symposium, Australasian Institute of Mining and Metallurgy, Parkville, Vic., p. 363.

Yamoaka, Y., et al., 1980, "On the High Temperature Properties of Blast Furnace Burdens," *Proceedings*, Australia/Japan Extractive Metallurgy Symposium, Australasian Institute of Mining and Metallurgy, Parkville, Vic., p. 245.

METALLURGICAL USES

FLUXES FOR SOLDERING, BRAZING, AND WELDING

JAMES WATSON BAXTER

Fluxes are used to promote pyrometallurgical processes that rely on adhesion (soldering or brazing) or fusion (gas and arc welding) to join metallic surfaces. In the adhesive processes, the metal surfaces to be joined are not melted; the join is formed using a filler metal with lower melting point than the base metal. Fusion welding involves use of heat in excess of the melting point of the base metal. The fused joint may be achieved either by simply fusing together metal surfaces brought in contact with each other or by introducing additional molten metal of similar composition to form a fused joint.

ADHESIVE PROCESSES—SOLDERING AND BRAZING

In order for molten filler metal, solder or braze, to spread in a manner that creates a successful join; the work surfaces on the base metal must be thoroughly cleansed. Fluxes remove stubborn oxide films and other surface contaminants, promote wetting of the work surfaces, add fluidity to the solder or braze, and enhance workability and ease of spreading. Brazing processes involve higher temperatures than those reached in soldering. Brazing fluxes, which must remain active and effective at the higher temperatures, differ from those employed in soldering. Some common fluxes used in adhesive processes are rosin for soldering tin and electrical connections, hydrochloric acid for use in soldering galvanized iron and other zinc surfaces, and borax for brazing.

Soldering and brazing are similar processes, the primary difference being the temperature at which the joining operation is carried out. Soldered joints, produced with low-melting-point fillers (solders) that melt and flow at temperature less that 450°C (Althouse et al., 1988) can sustain loads of 1 to 1.7 MPa for extended periods of time (Anon., 1966). Brazing involves the use of filler materials with melting points commonly above 500°C and generally provides stronger joints than those obtained with solder. Both processes require local application of heat to melt and spread the filler so that the molten filler can wet (adhere to) the base metals by alloying and diffusion.

Soldering

Soldering is a means of joining metals by adhesion using a metallic bonding alloy as the filler, commonly a mixture of lead and tin. However, the adhesion of solder depends more on its ability to be keyed into minute surface irregularities than on alloying. The most familiar application is to provide and secure electrical connections. Soft solders can range from 1 to 70% tin with the remainder mostly lead. However, for general-purpose, soft-solder work, the alloy is commonly 50% lead-50% tin. Higher lead contents provide a wider range in the melting temperature and, for this reason, a 60% lead-40% tin alloy, which yields a mushy mixture, is used for wiped joints in lead sheet and pipe work. Conversely, 40% lead-60% tin alloys are used in soldering tin and other low-melting-point materials for which a narrower range of melting temperature is required. There are numerous other solder compositions such as tin-silver, 95% tin-5% silver and antimony-tin, 95% tin and 5% antimony (Carlin, Jr., 1992).

Heat needed to melt and spread the solders is commonly provided by electrically heated, copper-tipped soldering irons or by means of torches; the solder is applied by hand, usually face-fed by means of wire. For wiped joints in plumbing and lead-cable splicing, the solder is manipulated with cloth pads. The molten solder wets the joint surfaces and is drawn, by surface tension, into minute fissures and capillary openings. Other applications involve use of induction heaters and furnaces with pre-shaped solder appropriately placed prior to fluxing and heating. In some processes, the joints are immersed in molten solder.

Constituents and Role of Soldering Fluxes: Soldering fluxes generally fall into one of three categories: highly corrosive fluxes, intermediate fluxes, and noncorrosive fluxes. These same categories are sometimes designated inorganic, organic and rosin-based respectively (Althouse et al., 1988). Common constituents of each group are discussed briefly below.

Corrosive Fluxes (Inorganic). Work with aluminum, magnesium, stainless steel, high alloy steel, aluminum bronzes, and silicon bronzes is carried out at temperatures in the upper portion of the range for solder operations. Soldering these materials requires use of highly active, corrosive fluxes to remove and prevent the formation of the especially stubborn, hard, oxide films that form on these materials upon exposure to the atmosphere. The corrosive fluxes consist of inorganic acids and salts that are applied either as pastes or dry. They are active at elevated temperatures and, since they remain active after the soldering is completed, must be completely removed.

The main constituent of most corrosive fluxes is zinc chloride with a melting temperature well above the solidus temperature of most commercial tin-lead solders. It is made by the action of hydrochloric acid on zinc. When zinc chloride is used alone, unmelted particles of this corrosive salt get caught up in the joint and weaken it. For this reason, other inorganic salts such as ammonium chloride (NH_4Cl) or sodium chloride ($NaCl$) may be added to lower the melting temperature. A mixture of zinc chloride and ammonium chloride is very effective because the excellent oxide reducing properties of ammonium chloride and the protective action of the molten zinc chloride combine to produce a fluxing action superior to that achieved when either is used alone. In addition to zinc chloride, ammonium chloride, and sodium chloride; common constituents of corrosive soldering fluxes that can be used alone or in combination with other materials include but are not limited to stannous chloride, potassium chloride and fluorides, lithium and aluminum chlorides and fluorides, hydrochloric acid, hydrofluoric acid, and orthophosphatic acid.

Intermediate Fluxes (Organic). Intermediate fluxes include certain organic acids and bases that are most active at soldering temperatures but decompose at higher temperatures, limiting their usefulness. The residues formed by this decomposition are easily removed. Some ingredients that, combined with water and wetting agents, form typical intermediate fluxes are 1) glutamic acid hydrochloride 2) urea, 3) hydrazine monohydrobromide, and 4) lactic acid.

Noncorrosive Fluxes (Rosin-based). Noncorrosive fluxes are rosin-based. All types of electrical and electronic soldering is best done with a rosin-based flux. Rosin is not, of itself, a highly active flux therefore additives are commonly employed, either as stabilizers or activators. Depending on the use of stabilizers and/or activators, rosin-based fluxes are classified as rosin flux, stabilized rosin flux, or activated flux. Alternative designations for these same categories are non-active (R), mildly active (RMA), or fully active (RA)(Althouse et al., 1988).

ROSIN FLUXES—Water white rosin dissolved in an organic solvent such as alcohol, turpentine or petroleum spirits is a *non-active (R)* flux used only on highly solderable surfaces. Rosin fluxes are active at soldering temperatures up to 316°C, after which they begin to decompose. Rosin-based fluxes may be incorporated into the core of the wire solders or, dissolved in a solvent, may be applied directly to the work.

STABILIZED ROSIN FLUXES—These *mildly active (RMA)* fluxes are those in which, by addition of stabilizers, the anhydrite structure of the rosin has been made fully effective. They clean more thoroughly and faster than non-active fluxes.

ACTIVATED ROSIN FLUXES—*Fully active (RA)* fluxes depend on additives for increased activity. They are the most commonly used and clean best. They may be corrosive but only during the soldering operation.

COMMON ADDITIVES—Some common compounds, added in amounts of 0.2 to 5%, are hydrazine hydrohalite, glumatic and biguanide hydrochlorides, benzoic acid, succinic acid, cetyl pyridinium bromide, and aniline compounds.

Brazing

Brazing refers to a family of metallurgical joining processes in which the filler metal is a nonferrous metal or alloy with melting point above 500°C but below the melting point of the base metal. The fillers range in composition from pure metals such as copper, nickel, or silver to complex alloys. Closely fitted parts are required to assure a good brazed joint. Heat is applied locally usually by means of a torch fueled by a gas (acetylene, propane, or natural gas) burned with oxygen, air, or compressed air. As the flame melts the filler and warms the base metal, a thin film of molten filler metal is drawn into the joint by capillary action. The filler metal is either face-fed manually or, when the metal surfaces have reached appropriate temperatures, preplaced in the joint. Low temperature brazing techniques, such as silver brazing, are similar to soldering and are sometimes referred to as hard soldering. Silver brazing, also called silver soldering, utilizes a filler alloy composed of 10 to 80% silver with the remainder primarily copper and zinc in various proportions.

Constituents and Role of Brazing Fluxes: Chemical fluxes are employed in common brazing practice. However, some brazing operations are carried out in an artificial cleansing or protective environment (combusted fuel gas, dissociated ammonia, hydrogen, inorganic vapors, or inert gases) maintained in a furnace with dew point control (Anon., 1966). In some cases, as for alloys with volatile constituents, chemical fluxes may be required in addition to brazing atmospheres. Chemical brazing fluxes commonly fall in one of three general categories: borax, fluoride, and alkali halide fluxes.

Borax Fluxes. Borax fluxes, American Welding Society (AWS) type 5, are useful at temperatures ranging from 760 to 1 200°C for brazing operations involving steel, stainless steel, cast iron and iron-based alloys, copper and copper-based alloys, and nickel and nickel-based alloys as the base metal. They are not satisfactory for metals with refractory oxides such as aluminum and titanium. The major flux constituents are borax, boric acid, borates, and wetting agents. Such formulations cannot contain fluorine in any form.

A popular mixture of borax and boric acid used in brazing and braze welding fluxes, formulated for use with iron and steel, consists of 75% powdered borax and 25% boric acid (Althouse et al., 1988). To obtain more fluidity, the borax content is lowered to as little as 25% with the rest boric acid.

Fluoride Fluxes. Typical fluoride fluxes are AWS Type 3, which must contain a fluorine compound. Fluoride based fluxes are useful with filler metals that melt below about 800°C. The addition of alkali fluorides and bifluoride to borax-based fluxes lower their melting point and decrease their viscosity. Fluoride-based formulations may consist of a mixture of potassium borate, potassium fluoride, and boric acids. Alkaline bifluoride is used as a flux for brazing or braze welding stainless steel, silicon bronzes, aluminum, or beryllium copper alloys (Althouse et al., 1988).

Alkali Halide Fluxes. Fluoride-based fluxes are capable of the removal of most refractory oxides but special formulations with more aggressive fluxes are required for aluminum, aluminum alloys, titanium, and zirconium. The preferred fluxes are essentially mixtures of alkali metal chlorides and potassium or lithium fluoride. When it is not possible to thoroughly clean an aluminum surface prior to brazing or when the joint will be slowly heated, use of a *reaction flux* containing 7 to 12 wt% zinc chloride is beneficial.

SOURCES OF FLUXES FOR SOLDERING AND BRAZING

Inorganic Fluxes

Chlorides: Practically all the chlorine and chlorine compounds produced in the United States are produced from sodium chloride solution by electrolysis. Sodium chloride, mineral halite (NaCl), occurs in seawater, salt beds, salt lakes, and brines. In the United States, it is produced in New York, Michigan, Louisiana, Utah, and other places. Potassium chloride, mineral sylvite (KCl), is found in salt lake deposits with halite and carnalite ($KMgCl_36H_2O$) and in brines.

Chlorine reacts readily with hydrogen to form hydrochloric acid (HCl); chlorine and HCl both react with most metals (zinc, tin, etc.) to form chlorides. Hydrous aluminum chloride ($AlCl_3 \cdot 6H_2O$) is formed by the reaction of aluminum hydroxide and HCl. Ammonium chloride (NH_4Cl) is formed as a sublimate from the reaction of ammonia gas with hydrogen chloride gas or by the reaction of ammonium hydroxide with hydrochloric acid.

The manufacture of lithium compounds such as lithium chloride (LiCl) is, in the United Sates, based on spodumene deposits in North Carolina and geothermal brines in Nevada. Lithium carbonate is the primary product of lithium chemical plants and the compound from which all other downstream lithium products are obtained.

Borax and Boric Acid: The mineral borax, hydrous sodium borate ($Na_2B_4O_7 \cdot 10H_2O$), boric Acid (H_3BO_3), and potassium borate, are produced in California, the only domestic source of boron minerals. The majority of that production comes from Kern County, with the balance from San Bernardino and Inyo Counties. Minor amounts of colmanite, ulexite, and some boric acid are imported from Turkey.

United States Borax and Chemical Corporation mines and processes crude and refined sodium borates, anhydrous derivatives, and anhydrous boric acid from tincal deposits at Boron in Kern County. US Borax also makes technical grade boric acid from extensive kernite deposits and produces potassium and other borates at a plant in Wilmington, CA.

Kerr-McGee Chemical Corporation, at their Trona Plant at Searles Lake in San Bernardino County, produces borax with byproduct potash from brines, using a differential evaporation process, and boric acid by solvent extraction. Sodium borates are produced by a carbonation process at their Westend plant, also at Searles Lake.

The total amount of boron minerals and compounds used in metallurgy, in terms of boron oxide content, amounted to 2 554 t in 1990, compared to 3 040 t in 1989 (Lyday, 1990). The proportion finding use in welding flux formulations is not revealed.

Fluorides: Prominent fluoride compounds used in soldering and/or brazing fluxes include aluminum fluorides, potassium fluoride, and lithium fluoride. Major sources of fluorine are fluorite (CaF_2) in fluorspar deposits and the mineral fluorapatite in natural phosphate rock (Fulton and Montgomery, 1993). Hydrofluoric acid (HF) is made by reacting acid-grade (97% CaF_2) fluorspar with sulfuric acid in a heated kiln or retort. Elemental fluorine, prepared from anhydrous HF by electrolysis, reacts with all metals to form various fluorides including artificial cryolite (Na_3AlF_6), aluminum fluoride, potassium fluoride, and other inorganic fluoride chemicals. Fluosilicic acid is produced in the processing of phosphate rock and some of it is converted to aluminum fluoride and cryolite.

Organic Compounds

Organic compounds, utilized mostly as intermediate fluxes in soldering operations and as additives in noncorrosive fluxes, include a variety of organic acids, organic hydrohalides (chlorides, bromides etc.) and amines.

Organic Acids: Some organic acids used in fluxes and their method of preparation are:

- Lactic acid—fermentation of corn starch, molasses, potatoes, etc., or synthesized
- Citric acid—fermentation
- Oleic acid—from animal tallow or vegetable oils
- Glutamic acid—hydrolysis of wheat gluten or sugar beets
- Benzoic acid—from benzoin or other balsams
- Oxalic acid—reacting carbon monoxide with sodium hydroxide or certain carbohydrates with acids or alkalis
- Abietic acid—from the resin of a species of pine
- Phthalic acid—by oxidation of naphthalene

Organic Hydrohalides: Hydrochlorides and hydrobromides are salts, especially of alkaloids, formed by the direct union of hydrochloric acid and hydrobromic acid with an organic base, making the base more soluble. Typical organic salts used in fluxes include:

- Glutamic acid hydrochloride
- Aniline hydrochloride or phosphate
- Hydrazine hydrobromide or hydrochloride
- Cetyl trimethyl ammonium bromide
- Ethyl dimethyl detyl ammonium bromide
- Cetyl pyridinium bromide
- Biguanide hydrochlorides

Amines and Others: Amines are derivatives of ammonia (NH_3) of which one or more hydrogen atoms have been replaced by an alkyl, e.g., ethyl (-CH_3), methyl (-C_2H_5), etc. Typical examples of amine and other constituents are:

- Urea—formed by the reaction of liquid ammonia and liquid carbon dioxide
- Diethylene di- or triamine
- Glycerol—usually obtained by the saponification of natural fats and oils.

FUSION WELDING

Fusion welding processes involve melting of the edges of the metal parts to be joined. A clean surface and intimate contact is required to assure a successful weld. Ideally, intimate contact is achieved when the atomic spacing at the interface of the welded surfaces is equal to that of the parent material (Anon., 1968). Gas welding, in which heat is supplied by a flame, and electric arc welding are the most common types of fusion welding. In arc welding an arc, struck between the welding electrode and the base metal, melts the base metal to form the weld.

A variety of materials used to promote the joining of metals by fusion welding are generally considered as fluxes although they mainly function as deoxidants, slag formers and arc stabilizers rather than true fluxes. The term flux, in this chapter, is applied to mineral formulations used in welding procedures in which their primary and/or secondary function (Lawrence, Jr., 1983, Griffiths, 1985) may be as :

- Shielding gas formers—volatile compounds and mineral substances that, when heated give off gases that shield the arc and weld pool from the atmosphere
- Slag formers—minerals that, at temperatures reached in the operation, form a molten/solid slag that protects the slag pool and solidified weld metal
- Slag modifiers—minerals that control the physical properties of the slag (melting point, freezing range, conductivity, wettability, and fluidity)
- Arc stabilizers—mineral compounds that, with heat, evolve elements with low ionization potential which stabilize the arc
- Deposition rate augmenters—metallic additions that increase the rate of deposition of the weld metal
- Deoxidizers—metallic additions, including iron powder and ferroalloys, that are added to remove free or combined oxygen from the molten metal in the weld pool
- Alloying agents—metallic substances added to flux formulations to enhance the properties of the weld metal
- Fluxing agents—minerals that are active in the removal of impurities from the weld

In addition, flux formulations may contain materials that serve as binders and extrusion aids in the manufacture of coated welding rods or promote agglomeration and extrusion processes in the manufacture of granular flux.

Fluxes and flux formulation used in pyrometallurgical processes (and the slags produced) are generally referred to as *acid*, *basic*, or *neutral*. As with smelting fluxes (Kokal and Renade, 1993), acid fluxes (silica, for example) generally form acids in water, basic fluxes (calcium carbonate and lime) form bases in water, and fluorspar, or calcium fluoride, is a neutral substance because it can be viewed as the reaction product of a base and an acid.

Gas Welding

Gas welding utilizes mixtures of oxygen and acetylene or, primarily with aluminum and metals with low melting point such as lead, employs oxygen and hydrogen (Anon., 1966). Welding with oxygen and acetylene is also known as the oxyacetylene welding (OAW) process. Other gases that may be employed as fuel include liquified petroleum (LP), methylacetylene propadiene (MAPP), and natural gas (Althouse et al., 1988).

The gas and oxygen, stored separately in pressurized tanks, are fed through a series of valves to the tip of a torch where, at reduced pressure, they mix and are ignited. The torch mixture can be controlled to produce a flame that is oxidizing, reducing, or neutral. Adjusted to proper conditions, size, and temperature, the flame is applied to the work surfaces and the temperature raised until metal from the edges of the juxtaposed surfaces melt, the molten metals pool together, and on cooling form the weld. Any additional metal needed to complete the weld, is supplied either in the form of wire or rod. Flux is not used in welding ordinary steel but is employed with stainless steel and nonferrous metals, particularly aluminum. Gas welding continues to find applications in welding thin materials and for work requiring low gradients of temperature, as in welding cast iron. However, a far greater volume of welding is done using arc welding processes.

Arc Welding

Arc welding employs an electric arc, either ac or dc, as a heat source. Arc welding processes differ mainly in terms of 1) electrode type, 2) means of providing a suitable gaseous atmosphere for arc and weld pool protection, 3) manner of introduction of chemical fluxes, and 4) degree of automation.

Electrode Type: Welding electrodes are either consumable, fabricated from metal that will melt in the arc and provide a filler, or nonconsumable, fashioned out of metal that will not melt at welding temperatures. In most processes, the electrode is composed of a consumable metal or metal alloy and constitutes one pole of the arc; the other pole is the work. When the arc is struck between the base metal and the filler metal, both are melted to form the weld pool. The filler metal is generally a wire or rod of the same composition as the base metal.

Nonconsumable electrodes, fashioned from tungsten or carbon, have high melting points, are not melted, and therefore do not contribute to the weld pool. Carbon electrodes, are less commonly used because they vaporize more rapidly than tungsten.

Introduction of Fluxes: Depending on the process and equipment used, chemical fluxes are introduced with the filler metal, by means of a flux-coated welding rod (stick electrode) or continuous flux-cored wire, or by means of a layer of granular flux, laid over the joint prior to welding.

Well Pool Protection: Exposure of the well pool to atmospheric nitrogen causes porosity in the weld; exposure to oxygen leads to oxidation (Griffiths, 1985). Providing a slag cover and/or a vapor shield to protect the well pool and the arc, minimize these problems.

Shielding Gases. Gases (CO_2, H_2O, F_2), which evolve when certain minerals used in the manufacture of coated electrodes and flux-cored wires are heated, form a gas shield that protects the well pool from the air and provides a suitable gaseous atmosphere for the arc. Some arc welding processes do not involve use of mineral fluxes and others exclude gas-forming minerals from the flux mixture used. The vapor shield is, in such techniques, achieved by the introduction of carbon dioxide or an inert gas (argon, helium). In another application, the shielding gas supplied by gas formers in the flux is supplemented by auxiliary gas.

Role of Slag. Some constituents in fluxes melt to produce a slag that spreads over the work and further protects the weld and deoxidizes the metal in the slag pool. In one process that utilizes a solid-wire electrode submerged in a blanket of flux, the arc and weld pool are protected by the flux and by the slag formed when the flux is melted.

Automation: Arc welding processes vary in the degree of automation. In a manual process, as the stick electrode is consumed, it is fed into the weld by the operator. Semiautomatic welding employs a continuous solid or flux-coated wire that is fed mechanically into the weld but the operator guides the arc manually along the joint. In automatic welding the arc is guided mechanically.

ARC WELDING PROCESSES

Welding Processes Not Utilizing Fluxes

Two common welding processes, gas tungsten arc welding (GTAW) and basic gas metal arc welding (GMAW), require no flux and produce welds without a slag cover.

Gas Tungsten Arc Welding: Gas tungsten arc welding (GTAW) (Althouse et al., 1988) employs a nonconsumable, tungsten electrode that does not melt but does constitute one pole of the arc, usually the negative electron-emitting pole (Anon., 1966). A shielding gas is supplied from a heat-resistant gas flow cup or nozzle that surrounds the electrode. Additional metal, when needed to fill a joint, is provided by means of filler wire that is feed into the arc. Inert gas (argon or helium) is essential in the provision of fluxing atmosphere. The process, also known as Tungsten Inert Gas (TIG) welding (Griffiths, 1985) is well adapted for welding stainless steel, aluminum, titanium, and other nonferrous metals especially when working with thin materials and for the first (root) pass in pipes.

Gas Metal Arc Welding: Solid Wire Electrode: Gas metal arc welding (GMAW) (Althouse et al., 1988) is carried out with continuous, consumable, solid-wire electrodes in an atmosphere of argon or helium supplied from a gas cup on the torch. This process is also known as Metal Inert Gas (MIG) welding (Griffiths, 1985). Inert gas welding is particularly suited for reactive metals such as titanium that react with siliceous fluxes and is used with aluminum and stainless steel to avoid loss of corrosion resistance experienced in the presence of flux residues. As a variation on this process, carbon dioxide is sometimes used instead of an inert gas as a low cost shielding gas, but use of CO_2 is restricted to steel. To avoid porosity in the weld metal, caused by carbon monoxide (CO) that can result from reaction of CO_2 with the carbon in the steel, electrodes for CO_2 applications contain such deoxidizers as manganese, silicon, aluminum, titanium, or zirconium (Anon., 1966).

Welding Processes Utilizing Fluxes

Welding processes and variations that involve fluxes include:

- Shielded metal arc welding (SMAW)
- Self-shielding flux-cored arc welding (FCAW)
- Gas metal arc welding (GMAW): self-shielding flux-cored electrode
- Gas metal arc welding (GMAW): gas-shielded, flux-cored electrode
- Submerged arc welding (SAW)
- Electroslag welding (ESW)
- Electrogas (EGW) processes.

Shielded Metal Arc Welding. Shielded Metal Arc Welding (SMAW) (Althouse et al., 1988) is a welding method also know as Manual Metal Arc (MMA) Welding (Griffiths, 1985). A consumable stick electrode, coated with a complex mixture of minerals, is maneuvered by hand in a holder. Each of the minerals in the coating (flux) performs a specific function in the welding process, in the manufacture of the welding rod, or in its preservation in storage prior to use.

In this process the arc is struck between filler metal exposed at the end of the coated electrode which is self-shielded. The heat from the arc melts the electrode and the base metal forming a weld pool that is protected from atmospheric contaminants by a shield formed by gases that evolve from constituents of the coating. The coating also includes appropriate slag-forming, arc-stabilizing, and alloying agents. The consumable electrode is hand-fed into the weld. The SMAW process is commonly used for maintenance and small production welding and for almost all heavy pipe welding (Althouse et al., 1988).

Materials for electrode coatings, including extrusion aids and binders, are prepared in many different compositions but the mixtures are primarily of 4 types: 1) acid, 2) basic, 3) rutile, and 4) cellulosic.

Acid types, with significant proportions of silica, manganese oxides, and/or iron oxide, are rarely used and basic type electrodes have found increased popularity over rutile (Griffiths, 1985). Basic fluxes, contain significantly larger proportions of limestone and fluorspar, constituents that are particularly active. Use of basic electrodes results in a cleaner weld metal that is low in hydrogen and therefore has lesser susceptibility to cracking. Rutile type electrodes are easier to work with than basic types, produce a nicer shaped bead, and make a slag that is easily removed (Griffiths, 1985). Rutile electrodes are especially well adapted to welding across gaps up to ¼ in. (Anon., 1966).

Cellulose electrodes can be used advantageously to weld flat, vertical, and overhead surfaces (Anon., 1966) and where sections cannot be rotated (Griffiths, 1985).

Flux-cored Arc Welding: In flux-cored arc welding (FCAW), the arc is struck between the end of a flux-cored wire and the work. Flux cored arc welding combines the benefits of shielded metal arc welding (SMAW), submerged arc welding (SAW) and gas metal arc welding (GMAW) (Althouse et al., 1988). Fluxing agents dissolve and remove oxides, a slag layer protects the weld pool, and a gas shield is provided either by ingredients in the flux or by auxiliary gas. Flux-cored arc welding is generally semiautomatic; the operator guides the torch along the joint. However, the process can be used on fully automatic machines with good results.

Flux-cored electrodes are of two general types: self-shielding and gas-shielded. Self-shielding wires, with more carbonates and fluorspar, produce a stronger gas shield than those designed for gas shielding. There are four types of self-shielding flux-cored wires:

- Fluorspar-alumina (CaF_2-Al_2O_3)
- Fluorspar-titania (CaF_2-TiO_2)
- Fluorspar-lime-titania (CaF_2-$CaCO_3$-TiO_2)
- Fluorspar-lime (CaF_2-$CaCO_3$)

The first and second types listed are the ones most commonly used and the last type listed can be employed either as a self-shielding electrode or with an auxiliary gas.

Self-shielding Flux-cored Arc Welding. A semi-automatic welding method, related to shielded metal arc MMA welding (Griffiths, 1985, Althouse et al., 1988), employs a continuous, self-shielding, flux-cored wire. On heating, gases that evolve from the flux form the gas shield and other ingredients in the flux mixture promote the welding process in much the same manner as the coverings on coated electrodes.

Gas Metal Arc Welding (GMAW): Self-shielding Flux-cored Electrode: This process (Althouse et al., 1988) is a variation on the solid wire GMAW or MIG method. It differs in that it involves the use of a continuous, self-shielded, flux-cored electrode. The carbon dioxide or inert gas that is introduced supplements the shielding gas formed when fluxing agents within the core vaporize. Carbon dioxide, being less expensive, is preferred over inert gas in some applications.

Gas Metal Arc Welding (GMAW): Gas-shielded Flux-cored Electrode: This variation on the GMAW process (Althouse et al., 1988) is related to MIG welding in that the gas shield is provided by introduction of an inert gas. Fluxes used in gas-shielded flux-cored welding electrodes, designed for work that relies on introduction of auxiliary shielding gas, are of three basic types: 1) high titania (acid), 2) lime-titania (neutral), and 3) lime (basic).

Arc welding with high titania type gas-shielded electrodes produces a smooth weld bead with little splatter and has good penetration. Lime-titania electrodes provides good arc stability and fairly low splatter. The lime type flux produces a basic slag.

Submerged Arc Welding: Submerged arc welding (SAW) (Griffiths, 1985 and Althouse et al., 1988) is, in some operations, a fully automatic method performed with a continuous electrode and a layer of granular flux. The flux, composed of silicates with or without deoxidizers and alloying agents, is fed automatically from a hopper. The arc, struck between bare filler metal wire and the base metal, melts some of the flux and is submerged in molten slag that forms a protective layer over the weld pool. Unmelted flux can be recovered by means of vacuum and recycled.

Submerged arc welding can be done manually with a combination electrode and wire feed cable attached to an electrode holder with a small hopper for flux built into the holder (Althouse et al., 1988). This process is similar to GMAW but the electrode is manually submerged beneath a layer of granular flux. The operator guides the electrode along the joint. Submerged arc fluxes are generally neutral and must not evolve large amounts of gas. They are supplied in three forms: prefused, bonded, and agglomerated.

Prefused Fluxes. Prefused fluxes are produced by melting the ingredients in an electric furnace. After melting the flux is quenched and crushed. The resulting product is a glass which has good chemical homogeneity. The fine particles can be reused without changing chemical composition. Prefused fluxes are nonhygroscopic, unfused portions of the flux can be reused without changing the chemical composition, and they are suitable for high welding speeds. The largest disadvantage of prefused fluxes is the segregation and/or loss of the ferroalloys and deoxidizers which occur during the high temperature prefusing operation.

Bonded Fluxes. Bonded fluxes are produced by grinding the constituents together with a potassium silicate or sodium silicate binder, which results in the pelletization of the flux. The advantages of this method over use of prefused flux are: the deoxidizers and ferroalloys are retained; a lower density flux allows use of thicker flux blankets; and the resulting slag is more easily removed after welding. The disadvantages of bonded flux are that the fines cannot be removed without altering its composition and that the flux is more hygroscopic.

Agglomerated Fluxes. Agglomerated fluxes are similar to bonded fluxes except that a ceramic binding agent is used and higher processing temperatures are required.

Electroslag Welding: Electroslag welding (ESW) (Griffiths, 1985, Althouse et al., 1988) is used to weld butt joints on thick metal. The weld, done vertically, moves upward. Before the weld is started a conductive flux material several inches thick is laid down between the two base metal. An arc is struck between one or more continuous, consumable electrodes, solid wire or flux-cored, and the base metal. When the flux becomes molten, the arc is stopped, but the electricity continues to flow from the electrodes, through the slag, to the base metal. The flux (slag) remains molten due to its resistance to the flow of electricity. The molten flux melts the base metal and the continuously fed filler to form the weld. Movable water-cooled copper shoes, one on each side of the joint, confine the molten flux, filler metal, and weld metal to the gap and are moved up as the weld progresses upward. The electrode can be automatically oscillated when very thick plates are welded.

Electrogas Welding: Electrogas Welding (EGW) (Althouse et al., 1988) resembles ESW in that the weld is done vertically and moves upward confined by movable copper shoes. It differs from ESW in that a shielding gas is used, a flux-cored wire is automatically fed into the weld joint, and an electric arc is maintained continuously between the electrode and the weld pool.

MATERIALS FOR WELDING FLUXES

A great variety of rock and mineral products are used in the formulation of fluxes used in welding processes. For all commodities, the flux market represents only a minor proportion of the total market, and the number of raw material suppliers and/or flux manufacturers are limited in number. As a consequence, statistical data related to use in fluxes are often concealed or included in miscellaneous categories.

The following is a brief review of the major welding flux materials. For more detailed information on sources of materials and specific aspects of deposits, the reader is referred to the commodity chapters of this publication.

Specifications

Griffiths (1985) has discussed the specifications for flux formulations, emphasizing the proprietary nature of such information and the secrecy that prevails. Of the number of common physical and chemical requirements, consistency in particle size distribution and chemical composition are said to be the most important.

Particle Size Distribution: Particle size is important in the manufacture and use of welding fluxes, particularly with respect to ease of fabrication, durability in storage, and performance of coated stick electrodes. Oversized materials and excessive fines each pose their own set of problems. According to Griffiths (1985, p. 29):

> "In general, for the manufacture of fluxes, the particle size does not exceed 250 µm while the average size is about 60 µm for standard size electrodes (350 mm long, metal diameter <6 mm). For larger electrodes (450 mm long, metal diameter >6 mm), the average particle size is 100 to 120 µm. The particle size for the manufacture of agglomerated fluxes should be less than 60 µm to ensure even distribution of the constituents. For flux-cored wires the particle size is smaller."

Chemical Specification: As with size distribution, consistency is the most important factor with respect to the chemical composition of welding flux materials. However, tolerance for phosphorus and sulfur is low and minimum levels for these elements are specified, generally less than 0.03%. However, if the deleterious element is contained in a mineral making up a large proportion of the formulation (30 to 55% rutile in ASW E6012 coverings), the critical minimum is lower (perhaps 0.01%) (Griffiths, 1985). Lead, zinc, arsenic, antimony, boron, cadmium, and vanadium have adverse effects on the ductility of the weld metal and the fluidity of the slag.

Moisture content, especially with respect to formulations for basic electrodes which must be dried before use, and potential for chemical reactivity with the alkaline silicate binder are other problem areas.

Limestone and Calcium Carbonate

Limestone, composed of the mineral calcite (calcium carbonate ($CaCO_3$), is one of the most abundant constituents in basic electrodes and fluxes. When heated to 830°C, limestone converts to lime which fuses with silica to form a slag at about 1 427°C. In addition to its function as a slag-former, limestone also serves as a shielding gas-former, and an active fluxing agent. As in most other chemical and metallurgical uses of limestone, a high-calcium, high-purity product may be preferred to maintain consistency in formulations. High-calcium, high-reflectance calcium carbonate of a sort known in the trade as whiting may be used to lighten the color of the slag or, may be added to flux mixes for coated electrodes, to provide a uniform light-colored base prior to the addition of pigments (Griffiths, 1985).

High-calcium limestone is generally defined as containing greater than 95% $CaCO_3$, but for welding fluxes higher contents are preferred. For example, a limestone used in the United Kingdom (Griffiths, 1985) is a standard 200 mesh grade with a typical chemical analysis of $CaCO_3$, 98.5%; Fe_2O_3, 0.03%; Al_2O_3, 0.3% (maximum); and SiO_2, 1.0% (maximum). In many instances, it may be assumed that trade-offs can may be made to accommodate somewhat less pure but nearby sources and avoid the costs involved in long haulage distances.

Fluorspar

Fluorspar, mineral name fluorite (CaF_2), functions primarily as a slag former (modifier) and secondarily as a true fluxing agent (weld metal cleaner). When fluorspar is incorporated in coated electrodes or with self-shielding, flux-cored wires, fluorine gas evolves and along with other gases provides a shielding gas atmosphere that protects the arc and weld pool.

The United States is highly dependent on the foreign imports of fluorspar, especially for acid-grade (more than 97.0% CaF_2). Essentially all US production is from the Illinois-Kentucky Fluorspar District (Fulton and Montgomery, 1993). Estimated domestic production in 1990 for all uses was 63.5 kt compared to imports of 513.9 kt. Major exporters of acid-grade fluorspar to the United States in 1990 were Mexico, China, and South Africa. Mexico accounted for about 66% of all lower grades. Although ceramic-grade products meet the requirements for most welding flux formulations, most of the material imported for that use has, in recent years prior to mid-1991, been acid-grade because of duty differential. With a $2.10/t import duty on acid-grade fluorspar compared to an ad valorem of 13.5% on lower grades, importers were forced to sell acid-grade without regard to customer requirements. Effective July 1, 1990, acid-grade from Mexico received differential tariff treatment and was exempt from duty upon entering the United States. Effective August 20, 1990, the 13.5% ad valorem tariff on subacid-grade was suspended through December 1992.

Titanium minerals

The titanium minerals rutile, ilmenite, and leucoxene are incorporated in fluxes to serve as slag-formers and slag-modifiers. The major sources of titanium minerals are the heavy mineral sands of Australia (east and west coasts), the United States (North Central Florida), Sri Lanka, Malaysia, Brazil, India, Sierra Leone, and the Republic of South Africa. The heavy mineral suites commonly consist of such minerals as ilmenite, leucoxene, rutile, tourmaline, spinel, kyanite, sillimanite, corundum, topaz, zircon, and staurolite. Other than the heavy minerals sands in Florida, US resources of titanium minerals include ilmenite in California, Minnesota, New York and Wyoming, and ilmenite and/or rutile from sands in Georgia, Idaho, New Jersey, North Carolina, Oregon, and South Carolina.

Ilmenite ($FeTiO_2$), theoretically containing 52.7% TiO_2 and 47.3% FeO, is the most abundant titanium mineral in nature and in heavy mineral suites from beach sands. The term ilmenite is also applied to oxidized and leached material, containing up to 70% TiO_2, 25 to 30% iron oxides and about 5% other oxides (Kramer, 1992). Extremely altered material, essentially composed of amorphous to finely crystalline TiO_2 is called leucoxene. Rutile (TiO_2) is essentially crystalline and as sold usually contains about 95% TiO_2 (Kramer, 1992).

Griffiths (1985) summarized the world market. Ilmenite has not been widely accepted for use in welding fluxes but does have a market in Japan and Southeast Asia. Japan has developed specifications based on ilmenite formulations. Rutile has found application worldwide because of its consistently high TiO_2 content, but in recent years use of rutile-based fluxes has decreased in favor of basic formulations used in coated electrodes and other applications. Leucoxene, in some quarters thought to be too variable to compete in the market, has increased in popularity as a low cost substitute for rutile.

TiO_2 pigments, like calcium carbonate whiting, may be added to lighten the base-color of flux mixes to which pigments are to be added.

Welding flux represents such a small fraction of the titanium mineral market and involves so few producers that data on production and consumption are largely confidential. However, US consumption in 1985 was probably about equal to the 19 000 t predicted by Griffiths (1985). US consumption of rutile, natural and synthetic in 1990 for use in welding rod coatings and other fluxes (Kramer, 1992), amounted to 3 807 t gross weight with a TiO_2 content of 3 603 t (94.4% of gross weight). Similar data for ilmenite and leucoxene are concealed.

Silica

Silica, silicon dioxide (SiO_2), generally incorporated in acid flux formulations as silica flour, serves primarily as a slag-former/modifier and secondarily as a fluxing agent. Silica deposit are wide-spread and common to most regions. Any one of several natural sources of silica—silicate minerals, quartz sand, diatomaceous earth, or tripoli (microcrystalline silica)—may be used. In addition, ground glass or ferrosilicon can suffice. A natural mixture of quartz (70%) and kaolin (30%) known as Neuburg silica, which provides the benefits of both materials, is used in Europe where sales of 4 to 5 ktpy have been reported (Griffiths, 1985).

Clay and Clay Minerals

Clay is an extremely fine-grained, unconsolidated rock composed predominantly of clay minerals, mostly hydrous aluminum silicates. Some clay also contain appreciable amounts of iron, alkalies, and alkaline earths. Clay deposits classified as china clay (kaolin), ball clay (predominantly kaolin), and bentonite (sodium smectite) have found applications in welding fluxes (Griffiths, 1985). However, ball clays vary widely in composition and their behavior in flux formulations is so unpredictable that they have not been widely used. Much of the kaolin produced in the United States comes from Georgia; Wyoming bentonite is the standard material for sodium smectite.

In welding fluxes, clays function as slag-formers and by the evolution of elements with low ionization potential (potassium and sodium) help to stabilize the arc. In addition, gaseous H_2O that is given off during the heating and dehydroxylation of clay minerals acts as a shielding gas. Clays such as Wyoming bentonite, added to coated electrode formulations in amounts up to 5%, serve as an extrusion aid in the manufacture of coated electrodes.

Griffiths (1985) reported that china clay (kaolin), 100% below 90 μm and with the bulk less than 63 μm, was then being supplied to flux fabricators, and that a standard 200 mesh foundry grade Wyoming bentonite (75 to 90% passing 200 mesh) was in use.

Mica

Mica is a group of monoclinic, hydrous potassium silicate minerals which, due to strong basal cleavage, break off into thin, tough but flexible cleavage flakes. In fluxes, mica performs in much the same manner as clay minerals producing a shielding gas by dehydroxylation. It also serves as a slag former and contained potassium helps to stabilize the arc. The platy shape of cleavage flakes makes the mineral a valuable extrusion aid.

Some data on size and chemical specifications have been provided by Griffiths (1985). Flux manufacturers set their own size specification. Chemical specifications require that phosphorus and sulfur content be as low as possible (generally less than 0.02%) and a K_2O content of 8 to 12%. SiO_2 content is usually set at a level low enough to assures material relatively devoid of free silica. Fluorine must generally be less than 0.02% and moisture, 4 to 5.5% (Griffiths, 1985).

Most of the ground mica produced in the United States, 73%, comes from three producers in North Carolina with lessor amounts from single producers in New Mexico and South Carolina (Davis, 1991). Information on the US market for mica used in welding flux applications are not available but Griffiths (1985) reported that the UK market dwindled from about 1 500 t in the late 1970s to around 200 to 400 t in 1984, reflecting the decrease in demand for rutile type electrodes. Griffiths estimated a European market of 1 500 to 2 000 t.

Manganese Minerals

Manganese, incorporated in welding fluxes in the oxide form, serves primarily as a slag former and secondarily as an alloying agent; $MnCO_3$ serves as a gas-former, deoxidizer, active fluxing agent, and slag former; ferromanganese, as deoxidant, slag former, and alloying agent; and natural ores containing MnO and MnO_2 are constituents of fluxes for submerged arc and electroslag welding (Griffiths, 1985).

US production and shipments in 1989 consists only of an undisclosed amount of manganiferous materials having a manganese content of 5 to 10% mined in Cherokee County, South Carolina, for use in coloring brick (Jones, 1990). The United States is completely dependent on foreign imports of chemical and metallurgical grade ores; primarily from Gabon, Australia, and Brazil. According to Griffiths (1985), one grade commonly supplied to welding flux manufacturers is 80%, another 84%, MnO_2. Specifications require low sulfur, phosphorus, lead, arsenic, and vanadium.

Feldspars and Nepheline Syenite

Feldspar and nepheline syenite are used as slag formers in fluxes. Feldspar is the name applied to a complex group of aluminum silicate minerals that contain varying amounts of the alkali metals potassium, sodium, and calcium. Theoretically the use of the potassic feldspars, orthoclase and microcline (both $KAlSi_3O_8$), helps to stabilize the arc. Anorthite ($CaAl_2Si_2O_8$) and albite, a sodic feldspar ($NaAlSi_3O_8$), are end members of a compositional series known as plagioclase feldspars. Feldspar in flux formulations on low hydrogen basic electrodes beneficially slows down deleterious hygroscopic activity (Griffiths, 1985).

Potash feldspar is fairly abundant throughout Europe especially in Italy which accounts for more than a third of the world production. Potash feldspar is not widely available in Asia. A welding rod market in Japan, Korea, Taiwan, and areas of Southeast Asia, are largely met by Australian imports (Potter, 1991). In the United States, feldspar or feldspar-silica sands are mined in seven states, led, in 1990, by North Carolina with 66% of the total; nepheline syenite is imported from Ontario. Norway serves syenite markets in Western Europe (Potter, 1991).

FUTURE TRENDS

The future for all welding consumables is directly related to metals and is particularly dependent on the steel industry. Consumption of both ferrous and non-ferrous metals ceased to grow in about 1960 due to decreasing economic growth rates in the industrialized countries (Davis, 1989). US production of pig iron and steel remained fairly constant from 1988 to 1990 and growth in demand is expected to be slow during the remainder of the century (Peters, 1992). However, it has been proposed (Drucker, 1986) that the demand for raw materials in the United States has been uncoupled from economic growth, implying that economic recovery may not, of itself, result in increased demand for iron and steel and welding flux products. Future US auto production, augmented by plants being constructed in the United States by foreign automakers, may outweigh such negative factors as increasing use of composites and the development of other lightweight, nonmetallic materials. Improved international relations and the expected receipt of a peace dividend will be reflected in lesser need for metals used in the manufacture of big ticket defense items: naval vessels, aircraft, and armored vehicles.

ACKNOWLEDGMENT

The author wishes to note his gratitude to Frederick V. Lawrence, Jr., author of the Fluxes chapter in the 5th edition of *Industrial Minerals and Rocks*. This is a revised and updated version, with the omission of the section on Fluxes Used in Smelting.

BIBLIOGRAPHY AND REFERENCES

Anon., 1955, *Brazing Manual*, American Welding Society, Reinhold, London, pp. 33–47.

Anon., 1959, *Soldering Manual*, American Welding Society, New York, pp. 25–42.

Anon., 1966, ''Welding and Cutting of Metals,'' *McGraw-Hill Encyclopedia of Science*, Vol. 14, McGraw-Hill New York, pp. 465–469.

Anon., 1968, ''Welding,'' *Van Nostrand's Encyclopedia of Science*, 4th ed., Van Nostrand Company, Princeton, NJ, pp. 1970–1973.

Anon., 1971, ''Welding and Brazing,'' *Metals Handbook*, 8th ed., Vol. 6, American Society for Metals, Metals Park, Aug., pp. 6–77.

Althouse, A.D., et al., 1988, *Modern Welding*, Goodhart-Willcox, South Holland, IL, 736 pp.

Brooker, H.R., and Beatson, E.V., 1953, *Industrial Brazing*, Iliffe and Sons, London, pp. 117–126.

Carlin, J.F., Jr., 1992, ''Tin,'' *Annual Report, 1990*, US Bureau of Mines, 36 pp.

Davis, G., 1989, ''The Declining Influence of Economic Growth on Metals Demand,'' *Productivity and Technology in the Metallurgical Industries*, Proceedings of Symposium, Cologne, Germany, Sep. 17–22, 1989, The Minerals, Metals, and Materials Society, Warrendale, PA, pp. 159–186.

Davis, L.L., 1991, ''Mica,'' *Annual Report, 1990*, US Bureau of Mines, 9 pp.

Drucker, P.F., 1986, *The Frontiers of Management: Where Tomorrow's Decisions Are Being Made Today*, Heinemann, London, pp. 21–30.

Fulton, R.B., III, and Montgomery, G., 1993, ''Fluorspar,'' *Industrial Minerals and Rocks*, 6th ed., D.L. Carr, ed., SME, Littleton, CO, pp. 509–522.

Griffiths, J., 1985, ''Minerals in Welding Fluxes—The Whys and Wherefores,'' *Industrial Minerals*, No. 210, pp 19–41.

Jones, T.J., 1990, ''Manganese,'' *Minerals Yearbook*, US Bureau of Mines, 13 pp.

Kramer, D.A., 1992, ''Titanium,'' *Annual Report, 1990*, US Bureau of Mines, 22 pp.

Kokal, H.R., and Renade, M.G., 1993, ''Fluxes for Metallurgy,'' *Industrial Minerals and Rocks*, 6th ed., D.L. Carr, ed., SME, Littleton, CO, pp. 661–675.

Lyday, P.A., 1991, ''Boron,'' *Annual Report 1990*, US Bureau of Mines, 9 pp.

Lawrence, F.V., Jr., 1983, ''Fluxes,'' *Industrial Minerals and Rocks*, 5th ed., S.J Lefond, ed., AIME, New York, pp. 259–270.

Lancaster, J.F., 1965, *The Metallurgy of Welding, Brazing and Soldering*, American Elsevier, New York, pp. 130–134.

Peters, A.T., 1992, ''Iron and Steel,'' *Annual Report 1990*, US Bureau of Mines, 27 pp.

Potter, M.J., 1991, ''Feldspar, Nepheline Syenite, and Aplite,'' *Annual Report 1990*, US Bureau of Mines, 8 pp.

METALLURGICAL USES

FOUNDRY SANDS

EZRA L. KOTZIN

Foundry sands used to make molds and cores have been very important to foundrymen since metal casting began hundreds of years ago. They are now used in two basic ways, either in an uncured green state, grain-bonded with clay, or in a set cured state achieved with sand that, mixed with resins or oils, is cured by baking or by chemical reaction.

The term sand, as generally used in foundry applications, may best be defined as material composed of granular particles of mineral matter, ranging in size from 0.5 to 2 mm in diameter. The particulate nature of the material may be the result of natural disaggregation or result from crushing and screening of rock or ceramic materials; crushed and screened products are sometimes referred to as "manufactured sand." Based on their origin, the raw materials for foundry sands will vary in grain shape, grain composition, relative surface, grain size, and grain distribution patterns. These properties, in addition to their chemical analyses, sintering point and expansion characteristics play an important part in the choice of sands used as the base molding or core aggregate in metal casting. Sands with different origins may be blended to produce specific compositions and grain size distributions.

Early raw material requirements based on silica sand were simple, but as sand/core/mold technology became more sophisticated and binder technology continued to be developed, the physical properties of foundry sands became more critical. Silica sand is composed of quartz, along with, in most cases, small amounts of feldspar, mica, clay, and other common minerals. Although silica sand is still by far the most widely used base material in production of molds and cores used in metal casting, other natural mineral sands have their own unique characteristics and fill an important niche in special applications. As a result, the term sand, as applied to raw materials used in metal casting, has been logically extended to include granular materials composed of a group of minerals other than quartz. With increased use in advanced foundry technology, natural sands composed predominantly of zircon, chromite, staurolite, or aluminum silicate minerals and of granularized chromite or olivine, have become more than just alternate materials and are now commonly classified as sands when the grains are sand-sized. Foundry sand may therefore be considered under two categories: silica sands and non-silica sands.

SILICA SANDS

Silica is the common term applied to silicon dioxide (SiO_2); the mineral name is quartz. The term *sand* has two different meanings, depending on the context in which it is used. In the context of rock type or mineralogical composition, the term is usually understood to refer to unconsolidated quartz grains within a certain range of grain size. In ancient rocks, the granular particles have become cemented together to form a consolidated sedimentary rock referred to as sandstone. However, the term sand is also used as a size term in a continuum which, listed in order from finest to coarsest, includes other terms with recognizable size connotations: clay, silt, sand, granule, cobble, and boulder.

The most common foundry sand used over the years has been silica sand. This is not surprising because it is the most abundant and one of the most easily mined mineral resources at the earth's surface. Silica sands are produced from rocks which formed under a variety of geologic conditions and are used in a number of industrial applications other than foundry sands: in aggregates, sand blast abrasives, fillers, glass manufacture, chemicals, refractories, and as a source of silicon.

Modes of Occurrence

Silica sand deposits occur naturally as a result of such geologic processes as weathering and erosion of preexisting quartz-bearing igneous and sedimentary rocks, glacial gouging, and transportation to the site of deposition. Disaggregated rock and mineral particles may be carried hundreds of kilometers from their ultimate source by fluvial, glacial, glaciofluvial, marine, and eolian processes. The composition of minable deposits depends on the nature of the rocks that were eroded, the intensity and time-scale of weathering, the distance and agent(s) of transportation, the manner of deposition, and post-deposition history. Deposits may be reworked by natural processes and redeposited under similar or entirely different depositional conditions.

Silica sands accumulate along channels and flood plains of rivers and streams, in lake and lake shore deposits, and are deposited under marine conditions in seacoast environments. Geologically young deposits are unconsolidated or semi-consolidated. Ancient sand bodies, which became buried under great thicknesses of younger sediments and consolidated, are termed sandstone. Subsequent uplift, due to the heaving and shifting of the earth's surface, can result in complete cessation of sedimentation and the initiation of widespread erosion and exhumation of previously deeply buried formations. Those that either crop out at the surface or have only a thin layer of overburden can be readily mined. Repeated cycles of weathering, erosion, transportation, size-sorting, and deposition results in well-sorted, high-purity sandstone deposits from which practically all minerals heavier than quartz have been winnowed out.

Advantages and Disadvantages

Silica sand has the advantages of abundance, ease of bonding with organic or inorganic binders, low cost, and ability to be re-

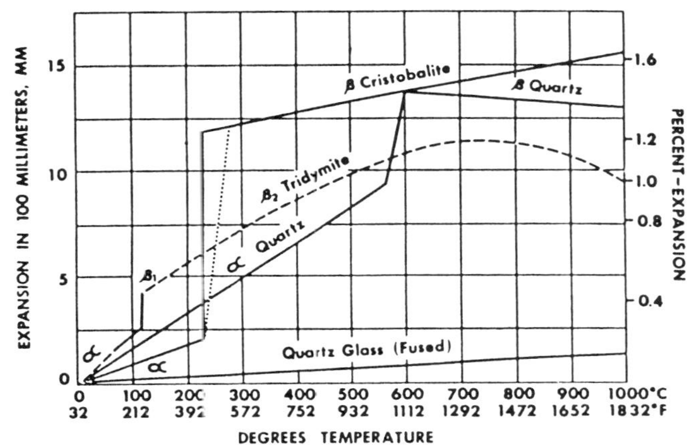

Fig. 1. Thermal expansion of silica materials.

claimed for reuse by wet, dry, or thermal methods. However, it also possesses certain disadvantages when used in the production of metal castings. The major disadvantage of silica sand is its characteristic high thermal expansion. This expansion behavior causes casting quality problems, e.g., rat-tails, buckles, and scabs and contributes to other expansion-type defects. Fig. 1 illustrates the typical thermal expansion of various silica materials and depicts alpha quartz, the polymorph characteristic of silica sand, expanding at a constant rate until a temperature of approximately 573°C is reached. As the temperature increases above this point, a sudden expansion takes place due to the change in crystal form, from alpha to beta quartz. This high thermal expansion requires carefully controlled additions of cushioning materials (e.g., cellulose additives) to minimize the deformation and rupture of mold surfaces in contact with molten metal. In addition, silica sand is unable to resist metal penetration and reactions that occur in quartz when in contact with casting surfaces. These problems arise where there are reentrant angles in hot spot areas of large iron and steel castings and, in Hadfield's austenitic manganese steel castings, when the steel contains high amounts of manganese that attack (wet) the silica sand mold surfaces. In addition, continuing environmental concern about respirable silica as a potential cause of silicosis works as a disadvantage.

Classification of Silica Sands

Silica sands used in foundry practice vary in grain size distribution (Table 1), purity, structure, grain shape, and refractoriness. However, two broad classifications are generally applied to foundry molding sands, naturally bonded sands and unbonded sands.

Naturally Bonded Sand: Naturally bonded molding sands are water deposited sands that are naturally combined with clay that serves as a bonding agent. The well known Albany naturally bonded sand, from the Albany, NY, area, was deposited in a glacial lake from glacial melt-water. Naturally bonded sands occur with a wide range of grain fineness and clay content, depending upon their source and geologic history. Typical size analysis of a fine grade naturally bonded Albany sand and a coarse naturally bonded sand or gravel are shown in Table 1.

Processing. The processing of naturally bonded molding sands consists primarily of screening to remove oversized particles, roots, and other deleterious material. For some sands, this is all that is required to meet specifications. Other sands are dried to approximately 6.0% moisture and may then be passed through mullers, cage mills, or dry pans to mill the sand and distribute the natural clay bond. Sands with dissimilar compositions, from different pits or from different strata in the same pit, are often blended to produce a range of sands having consistent, uniform properties. Sand with a wide range of properties can be produced and specifications are negotiated between producer and consumer.

Foundry Use. Fine grade naturally bonded sands are generally used as shipped in the casting of aluminum, light copper-based alloys, and light gray iron. Use is becoming more and more restricted to smaller, less mechanized foundries. Heavy, naturally bonded sands and gravels are used in the molding of large iron castings (Henderson, 1983). Applications for all naturally bonded products have declined in favor of synthetic formulations.

Unbonded Sands:

Silica Sand. Unbonded sands that are often referred to as washed and dried silica sands are found in many areas of the United States. Analyses of typical silica sands are shown in the top row on Table 1. Major production in the Midwest comes from the St. Peter Sandstone and medium- and fine-grade washed sands are produced in the East, particularly in Cumberland County, New Jersey. Silica sand from Nevada has been shipped to West Coast foundries.

PROCESSING—Washed, graded, and dried (clay free) silica sands are prepared for use in casting production by being bonded with the required amounts of binders, additives, and the necessary amounts of temper water. The total sand mixture is prepared in the sand muller in order to gain uniform distribution and dispersion of the clay and additives over and around the individual sand grains.

FOUNDRY USE—Medium-grade sands are used predominantly for casting of steel and heavy, gray and ductile iron; fine-grade eastern silica sand is used in making precision castings (Henderson, 1983). The St. Peter Sandstone is a popular washed and dried product because its roundness of grain promotes flowability, and, due to smoothness of grain, this sand requires less bonding material.

Bank Sands. Unconsolidated deposits with a low clay content (less than 5%), of a sort that at one time received little processing beyond drying and scalping, are still referred to as bank sands although they are now usually subjected to considerable processing and blending before shipment. These bank sands, of sedimentary origin, vary in purity depending on the amount of foreign materials and minerals with which they are mixed. In many areas, they are of high purity and suitable for foundry use. A typical analysis of a fine bank sand from New York is shown in Table 1.

PROCESSING—In the processing of bank sands, after stripping, the sand is hauled to the plant. A front end loader dumps the sand into a hopper from which it is conveyed to a rotary screen with 12.7 mm openings to remove roots. Some screened sand is dried and shipped without any further processing and meets the specifications as a system sand addition and as a synthetic sand base; some is washed to produce a cleaner and more consistent product.

The sand to be washed is dropped into a surge bin to assure a uniform feed. The sand is then conveyed to a tank where it is mixed with water and then pumped to a scalping screen where fine roots and organic materials are removed. The scalped sand is pumped to a deslimer to remove the clay and silt, freshwater is added, and the sand slurry is either pumped to a second deslimer and then to ground storage, or it may be fed to banks of screens and separated into a fine particle size and a course particle size. Each sand grade is then collected in a sump, mixed with water, pumped to a deslimer and dewatering cyclone, and piled for drainage and storage. The sand is then dried, cooled, and passed over a final scalping screen after which it is conveyed to storage bins prior to shipment.

FOUNDRY USE.—Fine bank sand (Table 1) is used as a base for molding sand in medium- to small-sized, gray iron castings typical of those used in casting hardware and hand tools and for casting aluminum and copper-based alloys (Henderson, 1983).

Lake Sands. The term lake sands has been applied to sands of

Table 1. Typical Analyses of Various Foundry Sands

Medium Grade Washed Eastern Silica Sand		Fine Grade Washed Eastern Silica Sand		Midwestern Rounded Grain Silica Sand, St. Peter's Sandstone		Silica Sand from Nevada	
US Series	% Retained	US Series	% Retained	US Series	% Retained	US Series	% Retained
20	0.4	50	0.4	40	2.0	30	0.5
30	1.6	70	5.2	50	24.0	40	7.5
40	6.4	100	14.6	70	40.0	50	19.5
50	22.6	140	23.4	100	22.0	70	42.0
70	40.4	200	25.8	140	8.0	100	18.5
100	23.6	270	16.2	200	3.0	140	10.0
140	4.6	Pan	14.4	270	1.0	200	1.5
200	0.6			Pan	0.0	270	0.5
270	0.0					Pan	0.0
Pan	0.0						
AFS Fineness #53		AFS Fineness #148		AFS Fineness #60		AFS Fineness #57	

Fine Bank Sand— New York State		Fine-grade Naturally Bonded Albany Sand		Naturally Bonded Molding Gravel		Lake Sand	
US Series	% Retained	US Series	% Retained	US Series	% Retained	US Series	% Retained
30	0.2	12	Trace	6	3.0	30	Trace
40	0.2	20	Trace	12	8.8	40	2.4
50	1.0	30	0.2	20	16.0	50	22.6
70	4.4	40	0.2	30	13.8	70	44.4
100	12.4	50	1.0	40	17.6	100	24.0
140	20.6	70	3.4	50	10.8	140	6.0
200	24.8	100	10.2	70	5.6	200	0.6
270	18.2	140	12.0	100	2.0	270	0.0
Pan	18.0	200	13.6	140	0.8	Pan	0.0
		270	13.0	200	0.4		
		Pan	33.4	270	0.2		
				Pan	1.0		
		AFS clay 13.4		AFS clay 20.0			
AFS Fineness #157		AFS Fineness #191		AFS Fineness #29		AFS Fineness #55	

Source: Henderson, 1983

the type dredged from Lake Michigan or dug from the dunes bordering that lake; the dune sands have been shifted by the wind but are still a part of the lake sand deposit. During the depositional process, some natural classification by grain size, grain composition, and specific gravity can take place. A typical size analysis of a lake sand is shown in Table 1. In general, the degree of purity of these materials is the result of their geologic history and processing.

PROCESSING—In a typical lake sand processing operation, the overburden is stripped and a front end loader is used to mine the sand and feed a hopper that discharges to a portable screen by means of which deleterious material is removed. The hopper and screen can be moved along a field conveyor as required. Conveyor belts transport the sand to the plant site for further processing and/or blending.

FOUNDRY USE—Lake sands in large quantities are used in the production of automotive and farm machinery castings (Henderson, 1983).

Consumption of Silica Sands

About 25.4 Mt of industrial sand and gravel were sold or used in the United States in 1990 (Bolen, 1991). Of this total, more than 22% (5.65 Mt) was molding and core sand and refractory sand. Refractory sand accounted for only 2% of the total foundry sand sold or used. Data for ground molding and core facing sand were not disclosed. Three-fourths of the foundry sand was produced in the Midwest.

NON-SILICA MATERIALS

The use of silica sands for metal casting has been the subject of studies by many investigators and improved techniques have been developed. However, the continued demands for improved cast surfaces and closer dimensional tolerances have emphasized some of the physical and chemical limitations of silica sands and their inability to provide all of the properties required to meet these quality levels. A number of specialty sands composed of material other than silica possess properties which make them superior to silica sand for certain applications.

The major non-silica aggregates (zircon sand, chromite sand, and olivine sand) have gained wide recognition and use throughout the metal casting world for their individual properties. Other non-silica molding media such as aluminum silicate sand and staurolite sand also possess unique properties, but their use in casting production has been on a much smaller scale. Table 2 shows some of the typical properties of the non-silica sands in comparison with silica sand.

Non-silica aggregates have a much lower rate of thermal expansion than that of silica sand (Fig. 2) and the rate of expansion on heating is not affected by phase change relationships like those exhibited by quartz. Experience in the use of the major non-silica minerals has shown that expansion type defects are therefore greatly reduced and, in many cases, eliminated.

Zircon Sand

Sources of Zircon Sand: The major sources of zircon ($ZrSiO_4$) sand are the heavy mineral sands of Australia (east and west coasts), the United States (North Central Florida), Sri Lanka, Malaysia, and the Republic of South Africa. The heavy mineral suites commonly consist of ilmenite, leucoxene, tourmaline, spinel, kyanite, sillimanite, corundum, topaz, zircon, and staurolite. The

Table 2. A Comparison of Silica and Non-Silica Sand Properties

	Silica	Olivine	Chromite	Zircon	Zircon/Aluminum Silicate	Staurolite
Origin	USA	USA (Washington, N. Carolina), Norway	Republic of South Africa	USA Australia	USA (Florida)	USA (Florida)
color	white-light brown	greenish gray	black	white-brown	salt & pepper	dark brown
hardness	6.0–7.0	6.5–7.0	5.5–7.0	7.0–7.5	6.5–7.0	6.5–7.0
dry bulk density (kg/m^3)	1 362–1 602	1 602–2 002	2 483–2 643	2 563–2 964	2 483–2 691	2 291–2 339
specific gravity	2.2–2.6	3.2–3.6	4.3–4.5	4.4–4.7	3.2–4.0	3.1–3.8
grain shape	angular/rounded	angular	angular	rounded/angular	rounded	rounded
thermal expansion (cm/cm)	0.018	0.0083	0.005	0.003	.005[a]	.007[a]
apparent heat transfer	average	low	very high	high	high	high
fusion point	1 427°C–1 760°C	1 538°C–1 760°C	1 760°C–1 982°C	2 038°C–2 204°C	1 815°C–1 982°C	1 371°C–1 538°C
high temperature reaction	acid	basic	basic	acid	slightly acid	slightly acid
wettability with molten metal	easily	not generally	resistant	resistant	resistant	resistant
chemical reaction	acid-neutral	basic	neutral-basic	acid-neutral	neutral	neutral
grain distribution[b]	2–5 screens	3–4 screens	4–5 screens	2–3 screens	3 screens	3–4 screens
AFS grain fineness number ranges	25–180	40–160	50–90	95–160	approximately 80	approximately 70

[a] Clay bonded sand mixture, [b] 10 % or more retained on USA standard sieve
Source: Kotzin, 1989.

primary purpose for mining is to recover titanium bearing heavy minerals. However, zircon, staurolite, and aluminum silicate minerals are valuable byproducts used in foundry practices.

Mining and Concentration: The following is a description of procedures followed in a typical marine, dunal shoreline deposit in the southeastern United States. After stripping, a suction-type dredge digs sand at rates up to 1 000 t/h. The sand, containing about 4% heavy minerals is pumped to the wet mill which is on a floating barge where three stages of gravity concentration are made, using spirals. The wet mill produces a concentrate averaging 85% heavy minerals with a recovery of 80%. This concentrate is then pumped to the land-based dry mill where the scrubbed ore is dried and subjected to various magnetic and electrostatic separations to remove titanium-bearing minerals. The tailings from the titanium separation are fed to high intensity magnets for separation of byproducts. Zircon is the major constituent of the non-magnetic, non-conductive fraction produced by the procedure.

PROPERTIES—Zircon sand is chemically and thermally stable and possesses most of the desirable properties for a foundry sand. Its major advantages are:

- Lowest thermal expansion of any foundry sand
- High thermal conductivity and bulk density giving approximately four times the cooling rate of quartz
- Completely unwetted by molten metal
- Chemically unreactive
- Requires less binder than any of the other sands when using chemical binders
- Compatible with all known binder systems (organic or inorganic)
- Excellent dimensional and thermal stability characteristics at elevated temperature
- pH is neutral or slightly acid

Foundry Use: Zircon sand is used as a molding medium, as a core sand, and, when finely ground and combined with a suspending agent and a binder, zircon is widely employed as a refractory in the manufacture of core and mold washes.

Chromite Sand

Sources of chromite Sand. The element chromium (Cr) combines with iron (Fe), magnesium (Mg), calcium (Ca), lead (Pb), copper (Cu), phosphorus (P), and sulfur (S) to form a variety of minerals. Of these, chromite ($FeCr_2O_4$) is the only ore mineral of metallic chromium and the only source of chromium compounds, chemicals, and foundry sands. About 60% of the world production of chromite is consumed in metallurgy, 20% in refractories, 12% in chemicals, and 8% in foundry sand.

Most of the world's production of chromite comes from South Africa (the Steelpurt and Rustenburg areas of the Transvaal), the former USSR, and Zimbabwe. Albania, Turkey, Finland, and the Philippines also produce small amounts of chromite. The United States has a small amount of chromite reserves, but over 92% of the material has relatively high iron (Fe) and low chromium (Cr) contents.

Mining and Milling: Chromite deposits are associated with the basic igneous peridotites and pyroxenites or with serpentine resulting from the alteration of these rocks. Most of the South African mines extend underground. The ore is drilled, blasted, loaded onto conveyors or mine cars, and carried to the processing plant. Hand sorting and/or mechanical screening is used to separate lumps and chips from the sand, followed in some cases by bar or rod mills to reduce the sand to AFS GFN 80 and release the gangue minerals (primarily serpentine and talc). The natural grain size of approximately AFS GFN 55 is generally preferred for metal casting purposes. The chromite sand is further beneficiated by the use of auger type washers to remove the slimes, and by gravity concentration using riffling tables or spirals. The sand is then stored on concrete pads to drain. At this point, the sand, containing a minimum of 45% Cr_2O_3, is assayed. A turbidity test has been developed to control the washing processes and assure the production of clean sand. The sand is trucked 8 to 32 km to a rail siding and then transported 400 to 650 km a shipping port where it is loaded onto vessels for shipment to the United States and other points.

Finishing: At finishing plants in the United States, chromite is dried, cooled, and screened to produce grades ranging from AFS GFN 55 to 80. The sand may also be further washed to below 150 ppm turbidity for certain core operations. Quality control tests include the turbidity test, acid demand values, and screen and chemical analyses.

Properties: Chromite sand was introduced to the foundry in-

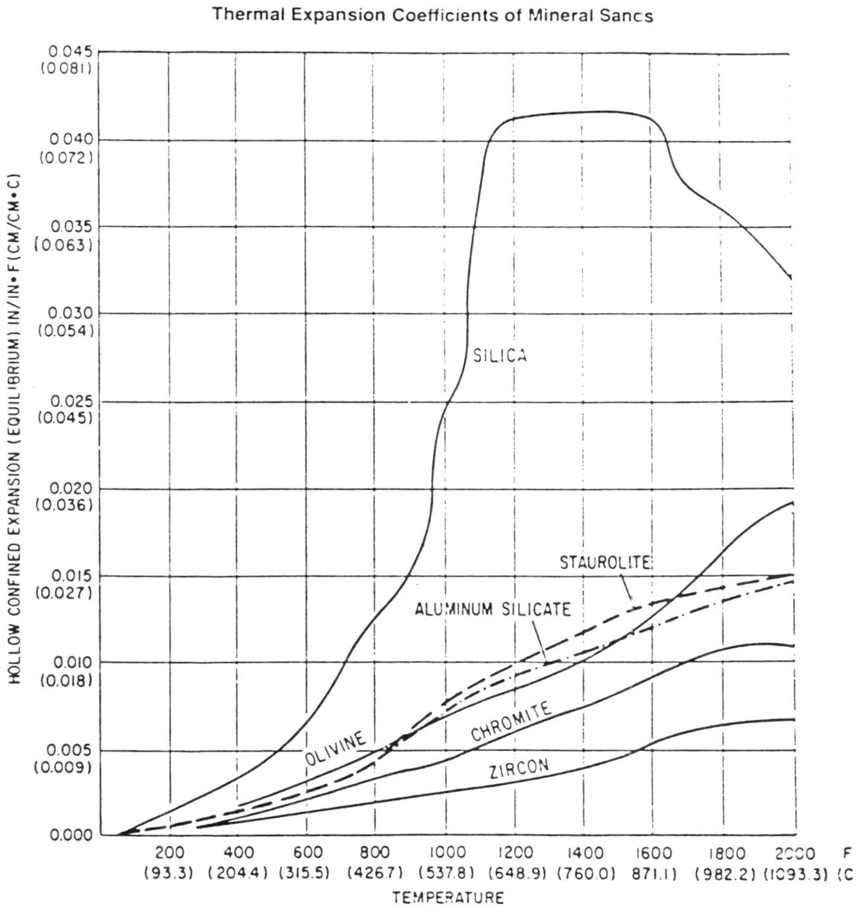

Fig. 2. Typical thermal expansion data of non-silica aggregates as compared with silica sand.

dustry in the early 1960s when zircon sand was in short supply. Later, in the face of rising zircon sand prices, foundrymen adopted chromite sand as a substitute for zircon sand. It has been successful as a non-silica sand because of the following properties:

- Excellent thermal stability
- Good heat diffusivity characteristics
- Not easily wetted by molten metals
- Highly refractory
- Chemically unreactive

One of the major disadvantages in using chromite sands is the presence of hydrous impurities which can contribute to pinholing and blows.

Foundry Use: The unusual properties of chromite sand (Table 2) lend themselves ideally to fulfilling some of the extreme thermal requirements of heavy sectioned ferrous castings. In addition, the basic nature of chromite sand also makes its use highly desirable when producing castings of Hadfield's austenitic manganese steel.

Olivine Sand

Sources of Olivine Sand: Olivine is a magnesium-iron silicate mineral $(MgFe)_2SiO_4$. Olivine sand is produced from dunite, a rock composed predominantly of olivine. Large deposits of olivine-bearing dunite are found in Norway, Sweden, the former USSR, Austria, Zimbabwe, South Africa and the United States. Norway furnishes much of the olivine sand consumed in Europe, while dunite deposits in the states of Washington and North Carolina furnish the major amounts of olivine sand for foundry use in the United States.

Mining and Milling: A typical operation in the United States follows fairly simple procedures. After blasting, a drop ball is used to break the large lumps to two feet or smaller. Ore is passed through a primary jaw crusher, reduced to 15 cm, and dumped down hill for storage to feed the processing plant. At the mill, the rock is further reduced to 12.7 mm particles by jaw and gyratory crushers. It then passes through a screw classifier to wash out clays and slimes, to a rod mill for reduction to sand size, and is then pumped to scalping screens. Oversized material is returned to the rod mill, and undersized material is fed to riffling tables to make a gravity separation of the olivine from the gangue which is predominantly serpentine and talc. Tables are fed a slurry of 25% solids and water is added. The amount of water, the tilt of the tables, and reciprocating speed control the quality of the product produced. The olivine concentrate, representing 60% of the table feed, is then pumped to a dewatering screen and then to drainage tanks. The sand is dried before screening to produce four basic grades. Each grade is separately stored.

Properties: Olivine sand is less stable under thermal shock than zircon sand or chromite sand, but its thermal expansion is much less than silica sand. However, the presence of quantities of hydrous magnesium silicates (serpentine) may contribute to a pinholing or pockmarking condition when olivine sand (uncalcined) is used in the production of low carbon steel castings. Compared to other non-silica sands, olivine sand particles are less durable although of comparable hardness. This may be attributable to the mineral's cleavages and other physical or chemical characteristics.

Foundry Use: Olivine sand has been very successful in non-ferrous casting as well as ferrous casting production where high thermal requirements are not necessary. Basic characteristics also makes olivine an ideal aggregate for the production of castings of Hadfield's austenitic manganese steel.

Aluminum Silicate Sand

Sources of Aluminum Silicate Sand: Along with zircon sand, aluminum silicate is a byproduct of the mining of titaniferous heavy mineral beach sands. Those being mined today in Florida contain significant quantities of kyanite, sillimanite, and andalusite.

Mining and Concentration: Mining and concentration of Florida beach sands and the processing of the various heavy mineral products were discussed briefly in the section on zircon sands. The aluminum silicate concentrate is further processed by means of scrubbing and washing the sand being processed to remove contaminants, dust, and fines. The product from intensive magnetic and electrostatic separation is a rounded-grain sand much like the zircon sand with which it is associated in the non-magnetic, non-conductive product.

Properties: The aluminum silicate minerals are mineralogically similar. Kyanite, sillimanite, and andalusite all have the same chemical composition (Al_2SiO_5). In addition, all three minerals are thermally stable up to a temperature at which they break down to yield 88% mullite and 12% free silica (cristobalite). The mullite, stable to 1 810°C, imparts the properties of high refractoriness, low thermal expansion, resistance to thermal shock, intermediate thermal conductivity, and resistance to chemical erosion.

Foundry Use: The aluminum silicate sand, being of a neutral characteristic, has been successful in its use in the production of Hadfield's austenitic manganese steel and other ferrous castings. This non-silica sand is also compatible with all known binder systems.

Staurolite Sand

Source of Staurolite Sand: Staurolite is a naturally occurring mineral with the general formula ($FeAl_5Si_2O_{12}$)(OH). Staurolite sand is produced as a byproduct of the same southeastern United States heavy mineral operation that also yields zircon sand and aluminum silicate sands.

Mining and Concentration: In its processing, staurolite and the other heavy minerals are removed from the deposit by use of a floating dredge and concentrated in three stages of spiral separators on a floating wet mill. The heavy mineral concentrates are pumped from the wet mill to a conditioning unit where they are scrubbed with caustic to remove organic and clay coatings from the grain surfaces. After scrubbing, the minerals are rinsed and dried, and the staurolite is removed by means of electrostatic and magnetic separation. The magnetic, nonconductive product used as a foundry sand consists principally of staurolite with small percentages of tourmaline, spinel, and silicates with magnetic inclusions.

Properties: Like other non-silica minerals, the key to the use of staurolite sand in casting production is its low thermal expansion, its refractoriness, durability and compatibility with the various binder systems in use. Although, as noted in Fig. 2, its thermal expansion is slightly greater than that exhibited by zircon sand or chromite sand, it is approximately equal to the thermal expansion properties of aluminum silicate sand and olivine sand.

Foundry Use: Staurolite sands have been used singly as a base aggregate or blended with silica sands in the production of non-ferrous castings and ferrous castings of light to medium section thickness although, because of its lower melting point of approximately 1 538°C, staurolite is not recommended for use in steel casting production.

Overall Benefits from Non-silica Sands

In the production of increasingly complex castings, problems of design and geometry have presented the foundryman with the challenge of preplanning the use of special aggregates for molds and cores. The judicious use of the various non-silica sands has minimized casting quality problems. Their use is reflected in savings in finishing costs while achieving improved quality of the casting surface.

ECONOMIC FACTORS

Most of the economic factors bearing on the production and processing of materials destined for use in foundry sand applications are covered in the commodity chapters of this book. However, some aspects of these factors, mainly with respect to silica sand, will be discussed here.

Demand

Statistical data for foundry silica sands sold or used by US producers during the 10-year period 1981 to 1990 (Bolen, 1991) show a high in 1981 (8.88 Mt), followed by decreases in 1982 and 1983, an increase in 1984, a low in 1985 (4.59 Mt), followed by annual increases during the period 1986 to 1989, and decreases in 1989 and 1990. The amount for 1990 was 5.7 Mt, a decrease of 35.8% from 1981 levels. Demand, production, and use of silica sands reflect the economic vagaries of the 1980s, but the overall decrease is no doubt offset, to some extent, by increased use of non-silica foundry sands.

Price

Along with a decrease in demand, silica sand produced for traditional end uses, including foundry sand, has experienced price decreases (Bolen, 1991). In 1990, molding and core sand sold at an average value of $11.91/t; refractory at an average value of $16.92/t. The total value of foundry sands sold or used in 1990 was approximately $68 million; less than 3% of that value was attributed to refractory sands.

Transportation

Transportation costs, to point of first sale or use, has a pronounced impact on the pricing structure of sand products and the viability of deposits. As previously noted by Henderson (1983), because of higher shipping costs, the market price of zircon sand from Florida heavy mineral deposits, in which zircon concentration runs about 1.5%, is comparable to that from Australian sands, some of which run up to 15% zircon.

Of the total industrial sand and gravel produced in 1990, 56% was moved by truck, 40% by rail, 3% by waterway, and 1% was not transported (Bolen, 1991). Such statistics for foundry silica sand would, perhaps, be similar. The trend continues to be toward truck haulage and use of pressurized tank trucks.

REGULATORY CONSIDERATIONS

Zoning and Land Uses

Certain permits are required for the installation and operation of sand pits and dredging operations. Provisions for land use and mining permits vary from state to state and differ with respect to private as opposed to public land ownership. The following requirements are generic in nature and may be more closely related to silica sand operations in midwestern states.

Special Use Permit: Many pits and quarries operate in lands zoned for agriculture under a special use permit issued by local zoning agencies. State agencies are not necessarily involved in this permitting process. Applications submitted to and reviewed by a zoning board may then be referred to the county board of supervisors or similar agency for approval. Some counties require that a copy of the application also be submitted to a soil and water conservation district from which a response may be required before the application can proceed.

Mining Permit: A mining permit may be required if the operation is to disturb more than some specified acreage annually or involve the removal of more than a specified thickness of overburden. The application ordinarily goes to a state regulatory agency with jurisdiction over mining. An Environmental Impact Statement (EIS) is usually required but may be prepared by the state agency from information furnished by the applicant on a questionnaire that must be completed and submitted with the application. In some states, the application for a mining permit must, at the same time, be filed with the appropriate county clerk. The county, in this case, has some specified length of time in which to respond. If no timely objections are lodged, the permit is issued within some specified time interval. If there are objections and call for public hearings, the time period can be extended indefinitely. Permits are for a few years and extendable. In some states, if no development has occurred during the original life of the permit on the property, the extension may be limited to a fraction of the original acreage.

Construction and Operating Permits: Construction and operating permits are generally needed. Applications, in some cases submitted to the state environmental protection agency, may require an EIS prepared by the applicant.

Dredging Permits: Dredge operations require a permit from the US Army Corps of Engineers.

ENVIRONMENTAL CONSIDERATIONS

In 1987, the International Agency for Research on Cancer (IARC) expressed the opinion that crystalline silica (quartz) is probably carcinogenic to humans (Anon., 1987a, 1987b). Citing IARC as a source to be used in recognizing whether or not a chemical is a carcinogen or a potential carcinogen, the Occupational Safety and Health Administration (OSHA) requires that a product containing greater than 0.1% of a material designated as a carcinogen or potential carcinogen must be labelled with a warning regarding this substance (Anon., 1987). The question about the carcinogenicity of crystalline silica has been further reviewed (Shapiro, 1990) and is still under study. Meanwhile, producers and consumers of crystalline silica products are taking measures to restrict the amount of respirable, fugitive dusts (including silica) that is released in the workplace and into the atmosphere. They must be constantly aware of current requirements and comply with all regulations.

Water Pollution

Many sand pits and other sources of foundry sand are located in riparian areas or, in the case of dredge operations, the materials may be derived directly from a wetland area or water body. This supplies a source for the large amounts of water required for thorough washing and/or classification of sand products. Sand producers in some western and southwestern states rely on dug wells.

Process water often becomes laden with fines and must eventually be discharged back into water courses. Although the process water is not generally affected by use of chemicals, the turbidity of the water leaving a plant becomes a matter of regulatory scrutiny. Different methods of water clarification are used and in some cases the quality of water leaving a plant is better than that entering.

Air Pollution

Dust can be a nuisance to neighbors and under certain conditions contribute to more widespread air pollution. By the very nature of mining and processing sand, a certain amount of dust is produced. Stockpiles, roadways, and sand dryers pose particular problems. Sand washing and the clarification of process water produces a residue of fines (clay and fine quartz) that, if discarded dry or left to dry in a manner that exposes the fine material to winds, can be a major source of air pollution. The unclear carcinogenic status of crystalline silica adds to the dilemma posed by dispersal of respirable dust. Watering of roadways, dampening of stockpiles, use of dust collectors, and other dust control techniques are used extensively in the industry.

FUTURE PROSPECTS

Supply-Demand Considerations

The future for molding sand is of course tied to metals. It has been suggested that the consumption of both ferrous and nonferrous metals ceased to grow in about 1960, due to decreasing economic growth rates in the industrialized countries (Davis, 1989). Conventional wisdom may indicate that, with economic recovery in the 1990s, better days are ahead. However, it has been proposed (Drucker, 1986) that the demand for raw materials in the United States has been uncoupled from economic growth, implying that economic recovery may not, of itself, result in increased demand for cast metal products and molding sand.

Although the sand used in both green sand (clay bonded) and bonded sand molds and cores is reclaimed for reuse, the normal foundry procedure of constructing a mold of sand, clay and other binders, and additives for a single casting can be viewed as extremely inefficient. Permanent ceramic molds, with cordierite or other refractory material as the primary raw material, could possibly reduce molding sand use in some applications (Burst, 1991). However, maintenance of precise dimensions may be a problem as the permanent mold is used.

Future US auto production, augmented by plants being constructed in the United States by foreign automakers, may outweigh such negative factors as increasing use of composites and the development of other lightweight, nonmetallic materials. The nature of the final resolution of the environmental safety issue with respect to crystalline silica will have an immense effect on the future supply-demand picture for the various types of foundry sands.

New Technology and Practices

Despite environmental pressures, silica sand will continue to dominate foundry sand practice in the foreseeable future, but it is anticipated that advancements in metallurgy, increased use of light metals, and requirements for more uniform and higher finishes will require increased use of non-silica sands and the implementation of new technology, some not yet on the drawing board.

PROJECTIONS AND FORECASTS

The US Bureau of Mines forecasts that use of foundry sand in the year 2000 will fall within a range of 8 to 10 Mt and probably be about 8.5 Mt. This would involve an average increase of 2.6% per year and is based on the assumption that, as foreign automakers continue to open plants in the United States, a greater percentage of cars that are sold in the United States will be built domestically. (Bolen, 1991).

BIBLIOGRAPHY AND REFERENCES

Anon., 1987, "Hazard Communication," *Federal Register,* Vol. 52, No. 163, Aug. 24, pp. 31852–31886.

Anon., 1987a, "Silica and Some Silicates," *The Evaluation of the Carcinogenic Risk of Chemicals to Humans*, Monographs Vol. 42, International Agency for Research on Cancer, Lyon, France, 289 pp.

Anon., 1987b, "Overall Evaluations of Carcinogenicity: An Updating of International Agency for Research on Cancer, Monographs Vols. 1–42," *The Evaluation of the Carcinogenic Risk of Chemicals to Humans,* Monographs Supplement 7, International Agency for Research on Cancer, Lyon, France, 440 pp.

Ammen, C.W., 1954, "Preparation of Synthetic Sand," *Foundry*, Sep., pp. 202, 204–206, 208.

Asanti, P., 1963, "The Wetting of Quartz and Olivine Sand by Iron at High Temperature," International Foundry Congress, Prague, pp. 83–101.

Asanti, P., 1966, "On the Interface Reactions of Chromite, Olivine and Quartz Sands with Molten Steel," International Foundry Congress, India, Paper No. 4.

Beckius, K., 1967, "Chromite Sand in the Steel Foundry," *Foundry Trade Journal*, Vol. 10, pp. 562–566.

Beckius, K., 1973, "Olivine Sand—Recent Research and Experience," 40th International Foundry Congress, Moscow, Paper No. 11.

Bolen, W.P., 1991, "Industrial Sand and Gravel," Annual Report, US Bureau of Mines, 15 pp.

Burst, J.F., 1991, "The Application of Clay Minerals in Ceramics," *Journal of Applied Clay Science*, Vol. 5, pp. 421–443.

Davis, G., 1989, "The Declining Influence of Economic Growth on Metals Demand," *Productivity and Technology in the Metallurgical Industries*, Proceedings of Symposium, Cologne, Germany, Sep. 17–22, The Minerals, Metals, and Materials Society, Warrendale, PA, pp. 159–186.

Drucker, P.F., 1986, *The Frontiers of Management: Where Tomorrow's Decisions Are Being Made Today*, Heinemann, London, pp. 21–30.

Feldcamp, R.H., 1970, "Properties of Chromite Sand and Its Use in the Foundry Industry," *Giesserei*, Vol. 57, Dec. 17, pp. 815–825.

Garnar, T.E., Jr., 1977, "Mineralogy of Foundry Sand and Its Effects on Performance and Properties," *AFS Transactions*, Vol. 85, pp. 399–415.

Goettman, F.P., 1947, "Preparation of Foundry Sands for Market," *AFS Transactions*, Vol. 55, pp. 490–492.

Goettman, F.B., 1975, "Production of Sands for the Foundry Industry," *AFS Transactions*, Vol. 83, pp. 15–24.

Gould, H., 1977, "Olivine for Ferrous and Non-Ferrous Castings," AFS California Regional Conference, AFS Current Awareness No. CA-79–284.

Hardy, T.H., 1954, "Natural Sands of Mid-America," *Foundry*, Nov., pp. 114–115.

Henderson, G.V., 1983, "Foundry Sand," *Industrial Minerals and Rocks*, 5th ed., S.J. Lefond, ed., SME-AIME, New York, pp. 272–278.

Kotzin, E.L., 1981, "The Application of Non-Silica Aggregates for Steel and High Alloy Steel Casting Production," *AFS Transactions*, Vol. 89, pp. 15–26.

Kotzin, E.L., 1989, "Foundry Sands: Silica Sand and Nonsilica Minerals," *Concise Encyclopedia of Mineral Resources,* D.D. Carr and N. Herz, eds., Pergamon, Oxford, pp. 127–131.

Locke, C., Brings, C.W., and Ashbrook, R.L., 1954, "Heat Transfer of Various Molding Materials for Steel Castings," *AFS Transactions*, Vol. 62, pp. 589–599.

Marais, J.J., 1966, "Foundry Techniques Using Chromite Sands," *Modern Casting*, Oct., pp. 45–47.

Middleton, J.M., and Bownes, F.F., 1970, "Properties of Chromite and its Application in the Foundry," *The British Foundryman*, Vol. 3, No. 8, pp. 248–260.

Petro, A., and Flinn, R.R., 1978, "Mold Metal Interactions Between Chromite and Cast Steel," *AFS Transactions*, Vol. 86, pp. 357–364.

Rassenfoss, J.A., 1977, "Mold Materials for Ferrous Castings," *AFS Transactions*, Vol. 85, pp. 583–596.

Schaller, G.S., and Snyder, W.S., 1954, "Olivine-Silica Molding Sands," *AFS Transactions*, Vol. 62, pp. 413–419.

Shapiro, J.C., 1990, "Chemical Manufacturers Association Chemstar Crystalline Silica Panel—Update," *Preprint*, Regulation of Crystalline Silica Meeting, Washington, DC, SME, Apr. 9, 1990.

Winter, W.P., 1960, "Molding Sand," *Foundry*, Nov., pp. 103–105.

MICA

JAMES T. TANNER, JR.

Mica is a generic term that applies to a group of complex aluminosilicate minerals that have a sheet or plate-like structure and with different chemical compositions and physical properties. All micas form flat six-sided monoclinic crystals with remarkable cleavage parallel to the direction of the large surfaces, which permits them to be split into optically flat films; as thin as 1 μm in thickness in some cases (Rajgarhia, 1987). Some micas are more easily split than others. For example, muscovite mica found in pegmatites is more easily split into thin films than is phlogopite. Mica found in mica schist is not as easily delaminated as mica found in weathered pegmatites.

According to Dawson (1949), the commercially important mica minerals are:

Muscovite—potassium mica (colorless to pale green and ruby)
Biotite—magnesium-iron mica (dark brown to black)
Phlogopite—magnesium mica (pale yellow to dark brown)
Vermiculite—hydrated biotite or magnesium-iron-mica (bronze yellow flakes)
Lepidolite—potassium-lithium mica (pale lilac to deep purple)

Some micas of less importance are:
Zinnwaldite—lithium-iron mica (gray to smoky brown)
Roscoelite—vanadium-potassium-magnesium mica (various shades of green and brown)

Mica, principally muscovite, is widely used for industrial applications because of its exceptional physical, chemical, electrical, thermal, and mechanical properties (Table 1). Phlogopite and vermiculite follow muscovite in importance. Biotite finds very little commercial application at the present time.

Commercially, mica is used in a number of forms. These forms and their description follow:

Sheet mica refers to books of mica either mined from hard rock (pegmatites) or from loosely consolidated clayey material formed from the weathering of pegmatites and alaskites. These books can be readily split into thin film or splittings with specific thicknesses ranging from 0.0031 to 0.10 mm (Davis, 1985). Quality sheet mica is graded into 10 quality classifications (Tables 2 and 3). Books of mica that are flawed with excess inclusions, cracks, or folds (A mica) are called *scrap mica* and are either ground into commercial products by a dry or wet process.

Built-up mica was developed in 1894 as a result of the cost and scarcity of sheet mica, and because many satisfactory electrical parts could be made from it (Preston, 1971). This type mica is formed by layering pieces of mica splittings upon one another and binding them together with inorganic or organic binders. These sheets are then pressed together under high temperature. Mica for this process is produced from pieces of sheet mica too small to be punched into electrical parts (Davis, 1985).

Reconstituted mica is mica paper produced by forming a mat of very thin delaminated flakes of scrap mica. This mat is usually impregnated with an organic binder, but is also available unimpregnated. It is dried at 110°C. The mica for this process is either produced from the trimmings of a sheet mica operation or from blocks of A type mica (Davis, 1985, Chapman, 1983).

Ground mica constitutes by far the largest use of mica, both quantitatively and application wise. It is produced by grinding and sizing scrap mica (also mica concentrate) obtained from five sources, 1) trimming from sheet mica, 2) A type mica blocks, 3) as a byproduct of spodumene and feldspar mining, 4) weathered pegmatites and alaskites, and 5) metamorphic schists.

Synthetic mica is produced by crystal growth in a slowly cooled melt of accurately proportioned chemical oxides. Extended cooling times at a specific viscosity result in larger crystals. Many different micas have been formed by this process; however, fluorophlogopite is the most common.

MINERALOGY

Bailey (1984) classified mica into three groups: 1) the mica group proper, 2) the clintonite group, or the brittle micas, and 3) the chlorite group. Supplementary to these are the vermiculites, which are hydrated compounds, that can result from the alteration of any one of the micas, but usually biotite. All minerals in these groups belong to the monoclinic crystal system, and all of them show plane angles of 60 and 120° on the basal section. The crystals usually form in hexagonal or rhomb-shaped scales, prisms, or plates.

The structure of mica is composed of two silica tetrahedral sheets with a central octahedral sheet. The tips of the unjoined tetrahedrons in each silica sheet point toward the center of the unit and are shared with the octahedral sheet in a single layer with suitable replacement of OH by O. The unit is the same as that for the smectite group except some of the silicons are always replaced by aluminums and the resultant charge deficiency is balanced by potassium ions. Varying amounts of other minor elements can also be present depending upon the mica (Grim, 1968). These elements are calcium, magnesium, iron, and lithium. The general formula describing the composition of micas is (Chapman, 1983):

Where:
$X_2Y_{4-6}\,Z_8O_{20}(OH,F)_4$
X is mainly K, Na, or Ca
Y is mainly Al, Mg, or Fe
Z is mainly Si or Al

In many of the well-crystallized micas one-fourth of the silicons are replaced by aluminum ions (Grim, 1968). The unit layers extend indefinitely in the *a* and *b* dimension and are stacked in the *c* dimension. The potassium ions occur between unit layers, where they just fit into perforations in the surface layers. Adjacent layers are stacked in such a way that the potassium ion is equidistant from 12 oxygens, 6 in each layer.

Muscovite is dioctahedral. Two-thirds of the possible octahedral positions are filled, and the octahedral sheet is populated only

Table 1. Chemistry and Selected Properties of Various Micas

Chemical Constituents	% by Weight		
	Muscovite	Phlogopite	Biotite
SiO_2	46.5	40.0	37.0
Al_2O_3	34.0	17.0	18.0
K_2O	10.0	10.0	9.0
Na_2O	0.8	0.5	1.0
MgO	0.5	26.0	8.0
CaO	0.3	—	—
Fe_2O_3	2.5	0.2	2.0
FeO	1.0	2.8	21.0
Minor Elements	—	0.5	1.0
H_2O	4.5	3.0	3.0
Total	100.0	100.0	100.0
PROPERTIES			
Specific gravity	2.77–2.88	2.76–2.90	2.70–3.30
Mohs hardness	2.8–3.2	2.5–3.0	2.5–4.0
Shore hardness	80–150	70–100	—
Specific heat (at 25°C)	0.207	0.207	—
Volume resistivity (ohms/cm^3)	2×10^{13}–1×10^{17}	less	—
Modulus of elasticity (Pa)	172×10^9	172×10^9	—
Compression strength (Pa)	221×10^6	221×10^6	—
Volume resistivity in ohms/cm^3 @ 25°C	5×10^{13}	—	—
Optical axial angle (2V)	38°–47°	0–10°	0–25°
Temperature of decomposition	400–500°C	850–1000°C	—
Dielectric constant	6.5–9.0	5.0–6.0	—
Linear coefficient of expansion $\times 10^{-6}$ cms/°C; range 20°–600°C	58–79	79–97	—
Coefficient of expansion per °C perpendicular to cleavage 20°–100°C (10^{-6})	15–25	1–1×10^{-3}	—
Coefficient of expansion per °C parallel to cleavage 0°—200°C (10^{-6})	8–9	13–14.5	—
Tensile strength Pa $\times 10^6$	225–297	255–297	—
Modulus of elasticity Pa $\times 10^9$	172	172	—
Dielectric strength (0.025—0.030 mm thick) volts/cm $\times 10^6$	2.4–11.2	1.7–0.8	—
Resistivity ohms—cm	10^{12}–10^{15}	10^{10}–10^{13}	—
Thermal conductivity perpendicular to cleavage @ 100°C (K cal/m^2/hr/°C)	0.57	0.57	—

by aluminum ions. The structural form is $(OH)_4K_2(Si_6Al_2)Al_4O_{20}$; and the theoretical composition is 11.8% K_20, 45.2% SiO_2, 38.5% Al_2O_3, and 4.5% H_2O (Grim, 1968).

Biotite micas are trioctahedral with octahedral positions completely filled, mostly by Mg^{+2}, Fe^{+2}, and/or Fe^{+3}. Examples are biotite $(OH_4K_2(Si_6Al_2)(Mg{\cdot}Fe)_6O_{20})$ with the proportions of iron and magnesium varying widely, and phlogopite $(OH)_4K_2(Si_6Al_2)Mg_6O_{20}$. Hendricks and Jefferson (1939) have shown the existence of at least six polymorphic variations of biotite. The variations occur due to the number of silica-alumina-silica units per unit cell and the manner in which the unit cells are stacked. Unit cells composed of 1, 2, 3, 6, and 24 silica-alumina-silica units are known, with stackings yielding monoclinic, rhombohedral, or triclinic forms.

Certain clays are commonly described as *illites* or *hydrated micas*. These differ from macro-crystalline mica with respect to both chemical composition and X-ray diffraction data. Chemical analysis indicates the presence of slightly less K_2O and more H_2O than in muscovite mica. K_2O is probably present, in part at least, between the mica-type layers and this gives rise to irregularities which are indicated by the characteristically different basal reflections, particularly the *101* reflection. Minerals similar to smectite in structure are essentially mica-type minerals with regards to the structure of the silicate layer. They differ from illites and hydrated micas in that they have fewer interlayer cations and larger and variable amounts of interlayer water. Smectites are characterized by a large cation exchange capacity and by swelling properties. Vermiculites are mica minerals of the smectite group. Vermiculites have a large cation exchange capacity, interlamellar adsorption properties, and are normally trioctahedral minerals. Vermiculite occurs in both coarse and fine crystalline forms.

The muscovite member of the mica group, formed as a primary mineral in pegmatites and alaskites is readily identifiable by its distinctive physical properties of being flexible and easily delaminated. The ability to be split or delaminated into infinitely thin layers is due to its excellent basal cleavage. Mineralogical properties of the four major micas are shown in Table 4. Muscovite micas, formed by secondary alteration, found in schist and gneiss, do not exhibit these properties to the same degree as well formed muscovite. In fact, these micas tend to be brittle.

HISTORICAL BACKGROUND

Sheet Mica

The name mica is believed to have been derived from the Latin word *micare,* which means to shine or glitter. The word muscovite,

Table 2. Quality Classification of Muscovite Block Mica, 0.018 cm Minimum Thickness, Based on Visual Properties

ASTM Visual Quality Classification	Crystall. Discolor.	Air Inclusions: Very[a] Slight	Slight[b]	Medium	Heavy	Cloudy Stains	Lt. Blk.[f] and Red Date (Mnrl)	Black Stains (Mnrl)[h]	Red Stains (Mnrl)	Bk. & Red[i] Stains (Mineral)[c]	Green Stains (Veg. Type)	Clay Stains	Waviness: Nearly Flat	Slight	Med.	Heavy
V-1, Clear	X	X	X	X	X	X	X	X	X	X	X	X	√	X	X	X
V-2, Clear and slightly stained	√[d]	√	X	X	X	X	X	X	X	X	X	X	√	X	X	X
V-3, Fair stained	√[d]	√	√	X	X	X	X	X	X	X	X	X	√	X	X	X
V-4, Good stained	√[d]	√	√	√[e]	X	X	X	X	X	X	X	X	√	√	√	X
V-5, Stained A quality	√	√	√	√[f]	X	X	X	X	X	X	√[d]	X	√	√	√	X
V-6, Stained B quality	√	√	√	√	X	√	X[d]	X	X	X	√[g]	√	√	√	√	√
V-7, Heavy stained	√	√	√	√	√	√[g]	√[d]	X	X	X[g]	√	√[d]	√	√	√	√
V-7A, Densely stained	√	√	√	√	√[h]	√[h]	√[h]	X[g]	√[g]	X	√	√[g]	√√	√	√	
V-8, Black dotted	√	√	√	√	√	√	√	X	X	X	√	X	√	√	√	X
V-9, Black spotted	√	√	√	√	√	√	√	√[d]	X	X	√	X	√	√	√	X
V-10, Black stained	√	√	√	√	√	√	√	√[g]	√[d]	X	√	√[d]	√	√	√	X
V-10A, Densely black and red stained	√	√	√	√	√	√	√	√[h]	√[d]	√	√	√[d]	√	√	√	√

ASTM Visual Quality Classification	Hardness: Hard	Soft	Stones and Holes[k]	Buckles	Reeves[l]	Ridges	Tears	Cracks	Hairline Cracks	Wedge	Tangle Sheet	Herringbone[j]	Sand Blast
V-1, Clear	√	X	X	X	X	X	X	X	X	X	X	X	X
V-2, Clear and slightly stained	√	X	X	X	X	X	X	X	X	X	X	X	X
V-3, Fair stained	√	X	X	X	X	X	X	X	X	X	X	X	X
V-4, Good stained	√	X	X	X	X	X	X	X	X	X	X	X	X
V-5, Stained A quality	√	X	X	X	X	X	X	X	X	X	X	X	X
V-8, Stained B quality	√	X	X	X	X	X	X	X	X	X	X	X	X
V-7, Heavy stained	√	S	X	S	X	S	X	X	X	X	X	X	S
V-7A, Densely stained	√	√	X	√[g]	X	√[g]	X	X	X	X	X	√	√
V-8, Black dotted	√	X	X	X	X	X	X	X	X	X	X	X	X
V-9, Black spotted	√	X	X	X	X	X	X	X	X	X	X	X	√
V-10, Black stained	√	X	X	X	X	X	X	X	X	X	X	X	√
V-10A, Densely black and red stained	√	S	X	X	X	X	X	X	X	X	X	X	√

Symbols: √ = Permissible X = Not permissible S = Permissible only if specified
[a]Few and tiny in one-fourth of usable area (must not contain air chains, etc.)
[b]In one-half usable area (must not contain air chains, etc.)
[c]Very dense
[d]Slight amount of black mineral stain
[e]In two-thirds usable area (must not contain air chains, etc.)
[f]Uniformly distributed
[g]Medium
[h]Heavy
[i]Black and red
[j]Numerous rulings that intersect to form a series of v's
[k]perforation through the laminae
[l]North Carolina term for cross grain
[m]The terms slight, medium, heavy, etc., are arbitrary subjective terms intended for use by experienced personnel only
[n]Air chains, a term used for a series of air inclusions forming a chain pattern.

the most commonly commercially used mica mineral, was derived from muscovia, the district in Russia where de Boodt found and identified the mineral in 1609 (Chapman, 1983).

Evidence indicates that the ancient Hindus mined mica as early as 2000 BC for use in medicine, window glazing, for decorative purposes, and as a surface for painting mythological scenes. They believed that mica crystals were preserved lighting flashes (Preston, 1971). The Romans used mica or mica-bearing rock to cover the floor of the Circus Maximus and other arenas and as a window covering. The use of mica in North America by the American Indians of the southern Appalachian region as an ornamentation at grave sites can be traced back to the 14th century. The ancient Indians also used mica as a medium of exchange. Mica has been used in Europe for the last two to three centuries, principally as

Table 3. Quality Classification of Muscovite Block Mica, Based on Size

ASTM[a] grade number	ASTM size (area of rectangle cm^2)	Minimum dimension of one side (cm)
A-1 Special	232–310	8.89
1	155–232	7.62
2	97–155	5.08
3	65–97	5.08
4	39–65	3.81
5	19–39	2.54
5 1/2	14–19	2.22
6	6–15	1.91

[a]ASTM Specification D 351.

Table 4. Mineralogical Properties of Selected Micas

Properties	Muscovite	Biotite	Lepidolite	Phlogopite
Hardness (Mohs)	2–2.25	2.5–3	2.5–4	2–3
Specific gravity	2.7–3	2.7–3.1	2.8–3.3	2.8–2.9
Luster	Vitreous to pearly	Splendent sometimes submetallic	Pearly	Pearly
Optical sign	Negative	Negative	Negative	Negative
Crystal system	Monoclinic	Monoclinic	Monoclinic	Monoclinic
Crystal symmetry	Rhombic or hexagonal	Pseudo-rhombohedral	Hexagonal	Hexagonal
Colors	Gray, brown, pale green, violet, yellow, dark olive-green, and ruby	Green, black, and yellow	Rose-red, violet-gray, lilac, yellowish, grayish white, and white	Yellow-white gray to green, pearly
Streak	Colorless	Colorless	Colorless	Colorless
Other properties	Transparent to translucent	Transparent to opaque	Twining common, translucent	

Source: Dana, 1945, Chesterman et al., 1989.

glazing for windows, in lanterns, in medicines, and for ornaments.

Mining in the United States started in 1803 at the Ruggles mine in Grafton, MA. Only sheet mica was recovered since there was no known use for ground or scrap mica at that time. Prior to 1890, the bulk of sheet mica was used in the production of stove windows, shades for open flame lights, and for furnace viewing glass (Preston, 1971). Originally, the United States outstripped other countries in the production of sheet mica because of the popular reliance on stoves rather than open fire places. However, by 1885, India had become a major supplier of muscovite sheet.

In 1884, imports began to effect the small domestic industry and tariffs were imposed beginning in 1890. Growth of the US electrical industry opened up many uses for sheet mica. This was due to the unique electrical and heat insulating properties of mica. Mica proved to be the best insulator for use in commutator segments of large electric generators and motors. Sheet mica also found use in electric irons, toasters, fuse plugs, and later in radio tubes, airplane spark plugs, condensers, compactors, radios, radar, and telephone equipment. The US electrical industry grew so fast that domestic reserves of large blocks of sheet mica were rapidly depleted (Preston, 1971).

In 1892, a patent was issued for the first built-up mica. This mica, which was made from mica splittings bonded together with inorganic and organic binders, mimicked the dielectric and mechanical properties of good quality sheet mica to such a degree that it could be used in transformers, armatures, and commutators. The invention of builtup mica largely overcame one of the limitations of sheet mica in that relatively inexpensive sheets of any size could be produced.

Mica was used as spacers and insulators in the diode vacuum tube developed in 1904, and in the triode which followed in 1906. The vacuum tube industry grew in importance with the outbreak of World War I. Large quantities of mica were required to produce the tube and other items that supported the war effort. From 1938 to 1944, Germany put a great deal of effort into growing synthetic mica sheets. However, few crystals were produced over 10 cm^2, and the process never became commercial.

During World War II, mica demand was spurred by the great increase in sophisticated electronic gear. The United States worked together with the United Kingdom to stimulate production in India and other producing countries. The US Government also started a program to increase domestic mica production.

After World War II, domestic sheet mica production dropped drastically due to the ending of the government subsidy program. In 1952, the United States began a new domestic mica purchase program for the strategic materials stockpile. Elevated prices paid by the government for domestic sheet mica brought about the resumption of mica mining. This program was terminated in 1962 (Zlobix, 1980).

Sheet mica consumption has decreased substantially from 5 487 t in 1953 to 56 t in 1989. The tremendous decline is due to the advancement in solid state electronic technology, which resulted in a dramatic decrease in the use of vacuum tubes and capacitors, and to the discontinuance of government-supported purchases.

Splittings now represent the largest part of the sheet mica industry in the United States. Both muscovite and phlogopite splittings are used to make builtup mica. Ninety-six percent of the splittings were imported from India (Davis, 1990).

Wet Ground Mica

In 1890, J.S. Williams, of Richmond, VA, experimented with making an axle grease for wagons from wet ground mica. It worked so well that he built a wet grinding mill. This mill was named Richmond Mica Co. In Plumtree, NC, young D.T. Vance and his older brother, T.B. Vance, heard of the new use for mica and thought it would be a way to market the trimmings or waste from their sheet mica operation (Preston, 1971).

D.T. Vance made a trip to Richmond and returned to North Carolina to start the first wet ground mica operation. It was based on grinding the mica in a chaser mill or muller. Power to operate the mill was furnished by a water wheel. The ground mica was dried on a table composed of steam heated pipes. After drying, the mica was sized in silk bolting racks much the way flour was sifted. Thus, the first pigment was made from mica at the Vance Mill where production started on Christmas Day 1891.

In 1908, the English Mica Company was founded in Spruce Pine, NC, by T.A. English. R.R. Dent of Pennsylvania became associated with the English Mica Company and arranged for financing to build a wet grinding mill. The plant was located on the banks of the North Toe River where plenty of water power was available, adjacent to the railroad, and centrally located in the mica mining area.

In the 1920s, the Brown family, which owned extensive lands and mineral rights in North Carolina, as well as the Asheville Mica Company that fabricated electrical parts from sheet mica, purchased the majority interest in the Richmond Mica Company from the Williams family. Also in the 1920s, the J.B. Preston Company purchased an interest in the English Mica Company from R.R. Dent and later merged it with the Biotite Mica Company and the Roofing Mica Company started by C. Gunter, who built the first wet grinding biotite mill in Spruce Pine.

In Franklin, NC, J. Davenport started the Franklin Mineral Products Company, and R. Hayden started the Hayden Mica Company. This was the second largest mica producing center next to the Spruce Pine District.

In New Hampshire, the predecessor of the Concord Mica Company was built by E. Moore. Two new companies, the Diamond Mica Company and the Hayden Mica Company, were started after World War II. The Diamond Mica Company was started in Spruce Pine, NC, by R.T. Dent, son of R.R. Dent. It was located a short distance from the English Mica Company.

In the 1930s, wet ground mica production began in Derby, England. Wet grinding plants were also established in Norway, France, and South Africa (Chapman, 1983).

Only two companies in the United States produce wet ground mica at the present time. These are: 1) KMG Minerals, Inc., a successor to the English Mica Company, and 2) Mearl Corporation, which owns Franklin Mineral Products Company, Inc. J.M. Huber recently closed its operations in Spruce Pine and Kings Mountain, NC. The wet grinding process has changed very little since the beginning despite many attempts to improve upon the batch grinding Muller system. Improvements, however, have been made in the screening and drying processes.

Dry Ground Mica

Dry ground mica production started in the early 1920s as a result of the need for an economical mineral filler, particularly in the roofing industry. This mica product was first produced as a byproduct from primary kaolin operations and later from the feldspar flotation process. It was also found that many weathered granitic and schist ore bodies contained commercial quality and quantities of mica, suitable for roofing, that could be easily extracted by simple (wet) crushing and screening (jigging—term used locally for a hydraulic mica mining operation; equipment is called a jig) methods. As a result of the demand by the roofing industry and later by the oil well drilling industry, many small mica jig operations sprang up in the mountains of western North Carolina and in Georgia.

Some of the original producers of dry ground mica were: 1) The Harris Clay Company, 2) The Deneen Mica company, 3) Southern Mica Company, 4) The English Mica Company, 5) Asheville Mica Company, 6) Funkhouser, 7) United States Gypsum Company, and 8) Western Mica Company. Mearl Corporation and MICA are late arrivals in the production of dry-ground mica.

A large growth in dry ground mica was realized from the 1940s to the mid-1960s due to the increase in housing which required roofing felts and shingles containing mica. Mixtures of talc and silica sand then began to replace mica for this purpose because of their lower cost. However, the use for mica in oil well drilling as a loss-drilling-fluid preventative increased. The mica market for wallboard compounds also increased. These increases more than made up for the lost roofing mica business.

In recent years, the oil well mica market has declined. However, the use of mica in automotive plastics continues to increase, and promises to more than offset losses in other markets.

GEOLOGIC FEATURES OF MICA DEPOSITS

Sheet Mica

Large books of muscovite mica are found in coarse-grained pegmatites. These pegmatites intrude into the surrounding country rock in the form of sills or dikes and they range in color from white to pink, depending upon which type of feldspar predominates (orthoclase or plagioclase). This granitic type of material is also found intruded into alaskite ore bodies. The alaskite ore is of similar composition to that of the pegmatite but generally contains less potassium feldspar and is somewhat finer in grain size. Sheet mica is also found in the alaskite ore bodies. The mica crystals found in these deposits range from less than 2.5 cm to many meters in length. Generally mica sizes vary widely within the same deposit.

Pegmatites can range in size from stringers to bodies which may be several hundred meters wide and long. Depths also vary from a few meters to hundreds of meters. Alaskites are usually larger and less well defined. Both ore bodies are composed of feldspar (albite, orthoclase, and plagioclase in various proportions), quartz, mica, and garnet. Some ore bodies also contain tourmaline and beryl as well.

Phlogopite mica is found in areas of metamorphosed sedimentary rocks into which pegmatite-rich granite rocks have intruded. Deposits that contain large books of phlogopite mica that can be economically mined are found in only a few places. Phlogopite ore bodies may be classified into vein, pocket, and contact deposits. Vein deposits, which are narrow, are enclosed in fine-to-medium-grained pyroxenite. This later type deposit is the major source of phlogopite sheet mica.

Scrap and Flake Mica

Waste mica trimmings from sheet mica operations was the original source of scrap mica. Later, as the demand for this type mica increased, methods were developed to extract scrap and flake mica from weathered pegmatites and alaskite ore bodies, where it is found in abundance. Mica schist deposits were also found to be an excellent source for this type mica. The weathered pegmatite and alaskites have the appearance of light-colored loosely consolidated soil sprinkled with shiny mica particles. They vary in size from stringers to several meters square. The mica schist ore bodies are generally large, covering several meters square, and have a higher percentage of mica than do pegmatites and alaskites. Flake or scrap mica is also extracted from hard rock alaskite as a byproduct from spodumene and feldspar flotation.

COUNTRIES OF ORIGIN

A brief description of the important mica producing locations in major mica producing countries follows. A summary of the mica production and reserves of countries of the world is shown in Tables 5 and 6.

United States

Approximately 60% of the flake mica produced and used in the United States is produced in North Carolina (Table 7). The remaining 40% is produced in Georgia, Alabama, Connecticut, New Mexico, and South Carolina. This figure has held relatively steady for a number of years. A large part of this mica is produced as byproducts from the feldspar, kaolin, and spodumene operations. The remaining portion is produced by two mining operations whose primary product is mica. These are Deneen Mica Company, located in Micaville, NC, and MICA, located near Española, NM. Nearly all of the mica produced is ground in mills located near the mine site. There were two exceptions where scrap mica or mica ore is transported over a considerable distance to the mill. These are MICA, and the J.M. Huber operations in North Carolina (presently closed).

USSR

Muscovite mica is found in the region of Ekaterinburg, at Alabaschka near Mursinsk in the Ural Mountains, and in the Ilmen

Table 5. MICA: World Mica Production by Country (Tons)[a,b]

Country[c]	1985	1986	1987	1988	1989[d]
Argentina:					
Sheet	347	234	340	330	330
Waste, scrap, etc.	374	317[e]	451	630	500
Brazil	2 881	2 060[e]	2 400	2 520	2 500
Canada[d]	11 500	12 000	13 500	12 500	12 000
France	10 084	10 834	11 000[d]	11 000[d]	11 200
India:[d]					
Exports:					
Block	1 200	1 200	1 000	1 000	1 000
Film and disk	250	250	200	200	200
Splittings	4 000	4 000	3 000	3 000	3 000
Scrap	11 000	11 000	10 000	10 000	12 000
Powder	4 700	4 700	3 800	3 800	4 000
Manufactured	1 000	1 000	2 000	2 000	2 000
Indian domestic production for all forms of consumption	3 500	3 500	4 000	4 000	4 000
TOTAL (India only)	25 630	25 650	24 000	24 000	26 200
Iran[d,f]	820[g]	820	820	820	800
Korea, Republic of (all grades)	20 044	41 997	31 938	30 000[d]	30 000
Madagascar (phlogopite)	589	1 594	402	617	600
Mexico (all grades)	1 446	1 748	3 419	2 584	2 500
Morocco[d]	1 440[g]	1 500	1 500	1 500	1 500
Mozambique[d]	300	300	([h])	([h])	—
Namibia	—	—	—	—	5
Peru[d]	550	550	550	550	500
South Africa, Republic of:					
Sheet	81	—	—	—	—
Scrap	2 072	2 509	970	1 669	1 700
Spain	727	325	250[d]	200	300
Sri Lanka (scrap)[d]	200	200	200	200	200
Sudan (all grades)[d]	10	10	10	10	10
Taiwan	114[d]	774	787	4 387	4 000
Tanzania (sheet)	(5)	(5)	(5)	(5)	(5)
USSR (all grades)[d]	50 000	50 000	50 000	50 000	50 000
United States (scrap and flake)[i]	124 800	134 400	145 650	129 900	119 142[g]
Yugoslavia	382[d]	674	1 043	768	800
Zimbabwe	582	1 340	800	800[d]	800
WORLD TOTAL	254 993[e]	289 836[e]	290 030	274 585	265 387

[a]Davis, 1989, [b]Table includes data available through April 25, 1990, [c]In addition to countries listed, China, Norway, Pakistan, Romania, and Sweden known to produce mica, but available information is inadequate to make reliable estimates of output levels, [d]estimated, [e]revised, [f]data are for year beginning March 21 of that stated, [g]less than 1/2 unit, [h]excludes US production of low quality sericite and sheet mica, if any, [i]1989 last year statistics available.

Mountains. Most of the production comes from the pegmatite deposits of Karelia and Maura River regions.

India

India represents the largest producer of mica outside the USSR and the United States. The Indian mica producing industry consists of a very large number of operators producing small volumes for further processing and manufacturing by other countries. The major consumers of Indian block mica are the United States and Czechoslovakia.

Although no reliable figures are available for mica consumption in India, the Indian Bureau of Mines estimates annual consumption to be of the order of R550m. Domestic consumption is understood to be growing rapidly, with the communication industry being the most important consumer. The most important consumer of buildup mica (micanite) for domestic as well as export markets is BHEL, of Blopal in Madya Pradesh. The company either buys micanite or produces its own material, using up to 227 tpy.

Seventy mica fabricators are located in the towns of Giridih, Kodarma, and Gudur. Many of these are very small. Two plants located at Giridih and Santacruz, near Bombay, produce mica.

Table 6. World Mine Production, Reserves, and Reserve Base

	Mine Production			
	1988	1989[a]	Reserves[b]	Reserves Base[b]
United States[c]	119	113	Large	Large
Canada	12	12	Large	Large
India	20	20	Large	Large
Korea, Republic of	30	30	Large	Large
USSR	45	45	Large	Large
Others	27	30	Large	Large
World Total	253	250	Large	Large

[a]estimated, [b]measured, indicated, and inferred reserves, [c]depletion allowance: 22% (domestic), 14% (foreign)
Source: Davis, 1990.

Table 7. Major US Producers of Mica—1991

Company	Parent Company	Location	Capacity (tpy)	Product	Remarks
North Carolina					
Asheville Mica Co.	—	Asheville, NC	5 442	Dry	Purchased scrap from Indusmin, Feldspar Corp., Harris, etc.
Lithium Corp. of America	FMC Corp.	Cherryville, NC	9 070	Scrap	Supplies oilwell use, its subsidiary, Spartan, and other grinders
Franklin Mineral Products	Mearl Corp.	Franklin, NC	NA	Wet	Has another plant in Hartwell, GA
KMG Minerals, Inc.	—	Kings Mountain, NC	680 680 2 270	Wet, dry, and micronized	Mica is main product; up to 22 675 mtpy
Foote Mineral Co.	Cyprus Minerals	Kings Mountain, NC	Small	Wet	Sold for tape joint cement; mill presently closed
US Gypsum Corp.	—	Spruce Pine, NC	NA	Dry	Believed to have about 31 745 mtpa
Harris Mining Co.	Unimin Corp.	Spruce Pine, NC	22 675	Dry	Presently operating well below capacity; about 60% of output is dry ground
The Feldspar Corp.	Corona Mining	Spruce Pine, NC	10 884	Scrap	Sold under contract to US Gypsum
Kentucky-Tennessee Feldspar and Clay Co.	Hecla Mining Co.	Spruce Pine, NC	7 619	Scrap	Sold locally for further processing
Deneen Mica Co.	—	Spruce Pine, NC	22 675	Dry	Sold for oilwell use; presently closed
J.M. Huber Corp.	—	Spruce Pine, NC	v. smal	Wet	Absorbed Mitchell Mining and Manufacturing in December, 1982; presently closed
Connecticut					
Feldspar Corp.	Pacific Tin Consolidated Corp.	Middletown, CT	4 535	Flake	Sold under contract to US Gypsum; presently closed
Georgia					
Franklin Mineral Products	Mearl Corp.	Hartwell, GA	NA	Wet	
New Hampshire					
Concord Mica	—		1 633	Wet	For paint and lining rubber molds
New Mexico					
MICA	Franklin Limestone Co.	Taos County, NM	18 140	Dry	Many improvements in dry and wet processing; new flotation plant
South Carolina					
Spartan Minerals Corp.	Lithium Corp. of America	Pacolet, SC	NA	Dry	Sold into the joint compound market
Mineral Mining Corp.	—	Kershaw, SC	22 675	Dry	Product is sericite for paint industry
South Dakota					
Pacer Corp.	—	Custer, SD	Unknown	Flake	A 70% muscovite mica for oil well use
Texas					
Alamo Mining Corp.	—	Van Horn, TX	16 326	Flake	Sold for oil well use; presently closed

Some 63 mica-powder plants operate on a custom milling basis to produce dry ground mica. One plant in Gudur has produced wet ground mica since 1979. Two micronized mica plants are located in the cities of Madras and Jhumeritalaiya. Each of these plants can produce 50 t of micronized mica per month.

Arun Mica Industries Ltd., has a 2 400 tpy capacity and is located in the village of Taswaria in the Gehsil-Hirda District of Bhilwara. This company is able to produce mica to a fineness of 0.25 μm for export.

Based in Calcutta, Mica Manufacturing Co. Pvt., Ltd. is one of the principal manufacturers of mica products in India, using mined mica mainly from Kodarma in Bihar. United India Minerals Ltd. and Gunpatroy Pvt. Ltd. are also important India processors of mica products.

Norway

At one time A/S Norwegian talc produced its own mica from the Rednolsvik mine in northern Norway. Since the mine's closure, the company's mica grinding operation has relied solely on imported material, mostly from India. Imports of waste mica to Norway in 1981 were 1 577 t.

Spain

In the 1980s, mica production in Spain has hovered around 4 535 tpy. The principal producer of Spanish mica is Ciaminera Comba from its 1 814 t per month kaolin operation at Santa Comba in the La Corona District of Galicia in northwestern Spain.

Morocco, Tanzania, Zimbabwe, Mozambique, Sudan, Angola

Small amounts of mica are mined in the following African countries: Morocco, Tanzania, Zimbabwe, Mozambique, and the Sudan. At one time Angola produced some of the finest quality sheet mica. Exploitation of mica in Morocco is undertaken by Mines de Zenaga from deposits located about 300 km inland from Agadir in the Tazenakt area of the Plane of Zenaga. Tanzania is a supplier of high quality sheet mica for European markets. Mica is mined in the Rubatab area of Sudan. In Mozambique, Minas Gerais de Mozambique Lia (Magma) produces up to 500 tpy of ground mica. Mica for this operation is mined in the providence of Zambezie, and is exported to the UK, the Republic of South Africa, and Czechoslovakia.

France

In Brittany, the English China Clay Company's subsidiary, Micarec, has the capacity to produce up to 13 000 tpy from its China clay operation at Ploemeur. Another English China clay subsidiary, Kaolins du Finistére, recovers 4 to 6% (by weight) mica by flotation from its Berrien China clay deposit. The plant has the capacity to produce up to 4 000 tpy.

Madagascar

The island of Madagascar is one of the few places in the world where phlogopite mica is mined. The deposits are located in the south of Madagascar at Ampandrandava and Sakamasy. These deposits are located 250 and 80 km from Taoláñaro (Fort Dauphin) on the South Coast. Ste des Mines d'Amprandrandava is the sole company exploiting the deposits. Annual production capacity from underground mining operations is around 1 814 tpy of which 150 t is exported in the form of blocks, splittings, and books, while in excess of 907 t is exported as scrap phlogopite.

Republic of South Africa

The major source of ground muscovite mica on the African continent is the Republic of South Africa. A wet processing plant (Gelletich Mining Industries Pty., Ltd.) is located in the Transvaal. Another grinding plant, Garieb Mineral Pty., Ltd., is located in the northwest Cape Province, and a third plant, Pegmin Pty., Ltd., is located in the city of Mica in the northern Transvaal.

Australia

In Western Australia, Pilbara Mica Corporation Pty., Ltd., began producing dry ground mica from the Pippingarra pegmatite deposit some 40 km southeast of Port Hedland in 1979. This pegmatite was intruded into Archaen granitoid rocks and is up to 30 m thick. It contains up to 92% muscovite, the other minerals are feldspar and quartz. Some 50 000 t of mica to a depth of 68 m had been proven as of June 1983.

China

Mica is produced in a number of regions from pegmatites or as byproducts of feldspar processing. The most important area is the Tientsin Province.

Canada

Most of the Canadian output of mica consists of phlogopite mines and is processed by LaCana Mining (formerly Martin-Marietta) in the Quebec Province of Lac Letondal near Suzor Township. This company has a capacity of about 31 745 tpy.

EXPLORATION

Sheet Mica

The spotty occurrence of sheet mica in pegmatites, and the unpredictable nature of the size, shape, and attitudes of pegmatites precludes the development of a systematic exploratory drilling program. Drilling can only serve to delineate the pegmatite, and will not yield much information about the size, quality, or amount of mica present. The reason for this is that sheet mica occurs in pockets within the pegmatite. These pockets are randomly scattered in the ore body, making it very difficult to detect their presence by drilling and impossible to determine a pattern. The common method of exploring for sheet mica is to sink a test pit in an effort to locate the mica in the pegmatite. This is costly and the results are problematical at best. However, the quality of mica present can be determined with a sufficient degree of certainty that gives the miner reason to continue or abort the exploratory program. The size, shape, and attitude of the pegmatite is often determined by stripping and trenching. Test pit sites may be selected in this fashion or information may be obtained to supplement the data from test pits. Another common method of exploring for sheet mica is to sink a shaft along the dip of the vein or sink a vertical shaft to cut the vein at the desired depth. Any large mica crystals found are examined by experienced workers for quality (Chapman, 1983).

Flake Mica

Mica ranging in size from thumbnail to subsieve size is found in weathered pegmatites, alaskites, schist (both weathered and hard rock), gneiss, and clay deposits. The amount and quality of this mica is evaluated by conventional techniques. This usually involves core drilling at 61 m intervals to establish the extent of the deposit. Once it is determined that there is the possibility of a viable deposit, a closer drilling pattern is used. In pegmatites, it has sometimes been found to be necessary to drill on 7.6 m centers in order to determine the presence of xenoliths or roof pendant and thereby more accurately determine the actual size of the ore body. Drill cores are usually 3.8 to 7.6 cm in diameter and 15 to 31 m long. The cores or drill cuttings are normally removed in 1.5 m segments. Auger drilling is sometimes used in soft material. However, it is very difficult to obtain a representative sample of a particular section when using this procedure because of the probability of intermixing foreign material with the vein material.

The evaluation of the cores is difficult and time-consuming. The determination of the mica content by chemical analysis is of little help because the accessory minerals have the same elements present. X-ray diffraction can be used, but is time consuming and very expensive. Petrographic analysis is also time consuming and its accuracy is often questionable. A combination of flotation and vanning have been found to be the most effective and least time consuming method for determining the amount of mica present.

MINING

Flake Mica

At the present time, flake mica is recovered in the United States from weathered pegmatites, weathered alaskites, weathered mica schist, and unweathered alaskite and pegmatites. It is mined by conventional open pit methods. In soft residual material, dozers, shovels, scrapers, and front-end loaders are used in the mining process. Trucks transport the ore to the processing plant. Often kaolin, quartz, and feldspar are recovered along with the mica.

Hard rock mining of mica-bearing ore requires drilling and blasting. After blasting, the ore is reduced in size with a drop ball and then loaded on a truck with a shovel for transportation to the processing plant, where mica, quartz, and feldspar are separated and recovered.

In early times, hydraulic mining was common. Numerous mica washing plants or jigs dotted the sides of the mountains in western North Carolina, Georgia, and Alabama. The soft weathered mica-bearing ore was pushed up into a pile with a dozer and then washed with a hydraulic hose into the processing plant.

Sheet Mica

Sheet mica is recovered by either sinking a shaft along the strike and dip of a pegmatite or by open pit surface mining of semi-hard pegmatite ore. In either case, it is a very economically risky mining procedure because of the cost involved in locating the unpredictable vein, and the unpredictability of the quantity and quality of the mica that might be recovered when the vein is located and worked (Chapman, 1983).

In underground mining, the main shaft is driven through the pegmatite at suitable angles to the dip and strike. This is accomplished with air drills, hoists, and explosives. Crosscuts and raises are developed to follow promising exposures of mica. When a pocket of mica is found, extreme care is exercised in removing it in order to minimize the damage to the mica crystals. Small charges of 40 to 60% strength explosive are carefully placed around the pocket of mica and care is exercised in the drilling procedure so that the mica will not be penetrated. The charge is just sufficient to shake the mica free from the host rock. After blasting, the mica is hand-picked and placed in boxes or bags for transporting to the trimming shed where it is graded, split, and cut to various specified sizes for sale.

Sheet mica is no longer mined in the United States because of the high cost of mining, the small market, and the high capital risk. Most sheet mica is mined in India, where the labor costs are comparatively low, and where there are very few environmental regulations.

BENEFICIATION PROCESS FOR FLAKE MICA

Flake Mica

In past eras, a high percentage of flake or scrap mica produced from soft weathered ores was washed hydraulically into a series of roll crushers and trommel screens with different size openings in order to remove mica flakes. Coarse mica, generally from 7.5 to 10 cm down to 8 to 14 mesh, was recovered in this fashion. The rolls would preferentially crush the quartz and feldspar down to a size that passed the screen, thus leaving a heavy mica concentrate on the screen. After each crushing and screening operation, abundant amounts of mica of a finer size than the openings of the screen reported to the gangue minerals. This mica, along with the gangue minerals, was conveyed to the next roll crusher which was gapped to crush finer than the previous roll crusher. Crushed material was then fed from this crusher to a screen with a smaller opening than the previous screen to remove the next size mica fraction. This process continued until mica down to about 14 mesh was removed. The −14 mesh mica was either discarded as waste or in some cases (in later years) was ground in a small rod mill and fed to Humphrey spirals to recover finer mica particles, thereby improving recovery. This jig process was inefficient from both recovery and grade standpoints. A high percentage of mica reported to waste, and the grade was usually around 75% mica concentrate.

This washing or jig operation disappeared with the arrival of stricter environmental laws. Many of these plants were located on hillsides above rivers or creeks, and there were no provisions for treating the wastewater. This waste was simply discharged into these nearby streams, from which this type of operation was dependent upon for the large amount of water necessary to operate.

Plants using rod mills, Humphrey spirals, and flotation (Fig. 1) gradually replaced the mica jig operation because of the improvement in recovery and quality that could be obtained. Because of the economics resulting from the improved recovery, operations could be profitably located away from steep mountain slopes, thus, waste impoundment ponds could be safely built, providing waste storage areas as required by the environmental laws.

This type of operation requires the use of trucks to haul the ore to the plant where it is either belt fed into a rod mill in an as mined state or hydraulically washed from a bin or stockpile into the rod mill. In some cases the feed is slurried, and then fed to a cyclone that removes slime. The deslimed feed is stockpiled, and fed from the stockpile to the rod mill with a belt feeder. Better control of the plant feed is obtained by belt feeding the ore to the rod mill. The proper amount of dispersant is added at this point to a mill charge of about 40% solids. The rod mill serves to blunge the ore into a slurry and to grind the quartz, feldspar, and mica to liberation. A rake classifier, hydraulic classifier, cyclones, or some other type of classifier equipment is used at this point to produce coarse and fine fractions.

The coarse mica is separated from coarse quartz on spirals. The quartz tailings from the spiralling process are sometimes upgraded further by scrubbing, desliming, and flotation. This quartz can be used in glass sand or in some cases can be further processed to produce high-purity quartz. The coarse mica fraction is sometimes further upgraded by additional spiralling steps or flotation. This cleaner coarse mica can be used to produce high-quality dry-ground or wet-ground mica products. For example, coarse mica further concentrated to 90 to 95% purity by repeated spiralling steps is used for wet-grinding products that specify a zero chemical residue.

The fine material from the classifier (−30 mesh) is deslimed, usually on 80 or 100 mesh Trommel screens, or with hydrocyclones. The pulp is then fed to a flotation circuit where it is floated either in an acid circuit with rosin amine and sulfuric acid (2.5 to 4.0 pH), or an alkaline circuit (7.5 to 9.0 pH) with tall oil amine, goulac, rosin amine, and caustic soda. The slime is then either further processed to recover commercial-grade clay or pumped to a settling pond for disposal. Both coarse and fine mica concentrate are dewatered in preparation for drying and grinding using either drip bins, centrifuges, or vacuum belt filters.

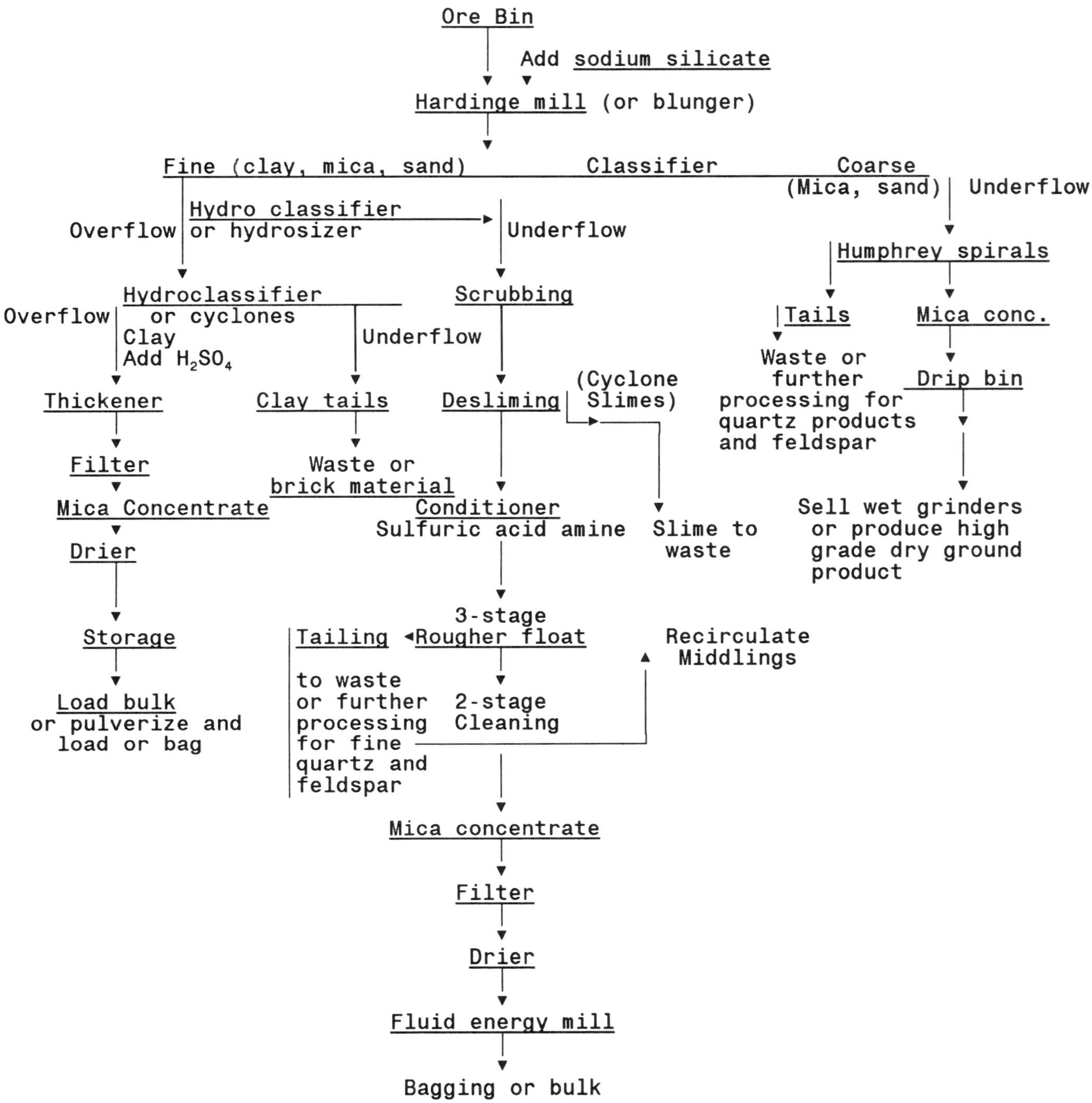

Fig. 1. Flowsheet for processing and recovery of mica from weathered alaskite ore (acid circuit). This flowsheet can be converted to an alkaline-cationic circuit by inserting a second conditioner; the pH is adjusted to 7.5 to 8.0 with NaOH in first conditioner and DRL fatty acid is added; lignin sulfonate is added in second conditioner.

Mica is also extracted from hard alaskite ore as a byproduct from the beneficiation of feldspar or spodumene. In this process, the ore is crushed to 2.5 to 3 cm with a jaw crusher, and cone crusher to 1.3 cm. It is then further ground with a rod mill to 0.6 mm. The mica is removed by flotation in an acid circuit. The mica concentrate is sold as flake mica to the dry grinders where it is processed into saleable products.

Pneumatic Processes

The US Bureau of Mines has developed a dry pneumatic process for extracting mica from ore, and it has been reported that similar processes are being successfully operated in Europe and Canada. Although several dry processes have been attempted in the United States, none of them have proven successful. The recovery and quality of mica have been poor and the economics prohibitive (Chapman, 1983).

PREPARATION OF FINISHED MICA PRODUCTS

Sheet Mica

The preparation of sheet mica for sale is strictly a labor intensive hand-cobbing operation. The mica crystals are first cobbed to remove adhering dirt, rock, and defective mica. The cobbed mica is then split into suitable thicknesses, and any mica with major flaws is removed. The mica produced in this manner is broadly classified as untrimmed mica and scrap mica. The next step in the operation is to trim and bevel the edges of the split-mica sheets with a knife. Sheets with broken and ragged edges and other major imperfections are carefully trimmed into usable pieces so that a minimum of waste is realized. Finished mica sheets are graded according to size and quality. They are either sold as block mica or split further for film or splittings (Zlobix, 1980).

Film is split from block mica and sheets that are too thin to be classified as block. The skilled hand labor required must develop a touch to split mica to narrow thickness ranges.

Splittings are processed from lower quality block and from sheets that are too thin for block and unsatisfactory for producing film. The splittings are packed in three ways: 1) book form—are laminae that are split to the desired thickness from the same book of mica, then dusted with mica dust and restacked in the same form as the original book, 2) pan packed—splittings are put in a pan in even layers with each layer separated by a thin sheet of paper and pressed together, and 3) loose packed—splittings are sized with screens and then packed loosely in a case for shipment.

The splittings processed by the methods described are formed into larger sheets of builtup mica by overlapping the irregularly-shaped pieces by hand or with a machine, and binding them with an organic or inorganic binder. The layered mica is then heat-treated to complete the bonding process. The sheets thus formed may be built up to any thickness by being bound together with heat and pressure. The sheets may also be bound to fiberglass or asbestos. After forming, the builtup mica sheets are cut to the desired shape and design.

High quality sheet mica (see Tables 2 and 3) is fabricated by punch pressing into various sizes and shapes for application as insulators in electronic parts. The higher the quality and grade of the mica; the more sophisticated the application.

Considerable waste results from the trimming operation required to prepare sheet mica for fabrication. This waste is often used as feed stock in the wet ground mica grinding process.

Grinding Processes for Flake Mica

Four pieces of equipment are used in grinding mica concentrate into salable mica products. These are hammer mills, fluid energy mills, Chaser or Muller mills for wet grinding, and Raymond imp mills. All of these mills are used in conjunction with screens, and all but the Hammer mill incorporate air classifiers as a part of the circuit. These processes will be described in more detail elsewhere in this text.

Ground Mica

The relation of chemisty to physical properties of mica is shown in Table 8, and Table 9 shows the particle size distribution for various mica products.

Table 8. Typical Chemical Analysis and Physical Properties of Different Forms of Ground Muscovite Mica (British standards)

Chemical Analysis (approx.)	Dry Ground wt %	Micronized wt %	Wet Ground wt %
SiO_2	45.57	46.27	48.65
Al_2O_3	36.10	35.24	32.04
K_2O	9.87	9.87	9.87
Fe_2O_3	2.48	2.48	3.68
Na_2O	0.62	0.60	0.28
TiO_2	0.20	0.20	0.20
CaO	0.21	0.21	0.21
MgO	0.15	0.16	0.16
H_2O	0.10	0.20	0.20
P_2O_5	0.03	0.02	0.018
S	0.01	0.01	0.010
C	0.44	0.44	0.44
LOI	4.30	4.30	4.30
TOTAL	100.08	100.00	100.05
Physical Properties			
Index of refraction	1.58	1.58	1.58
Hardness (Mohs scale)	2.5	2.5	2.5
Ph in distilled water	6.2	5.2	5.2
Oil absorption (B.S. 3483)	60.75%	60.75%	60.75%
Water soluble (B.S.1765)	<0.3%	<0.3%	<0.3%
Brightness	66–75	75	75
Effect by common acids	Slight	Slight	Slight
Phericity factor	0.01	0.01	0.01
Softening point	1 538°C	1 538°C	1 538°C
Apparent density (kg/m^3)	1 920–256	160–224	160–224

Source: Rajarhia, 1987.

Water ground mica: Approximately 15% of the total mica market is produced by the wet grinding process. This process, which involves batch grinding of mica containing 25 to 35% moisture in Chaser mills or Muller mills, has not changed significantly in almost a century (Preston, 1971, Zlobix, 1980). Steel-spoked wheels are used at the present instead of the wooden wheels made from the black gum tree that were originally used, and two wheels or rollers are used instead of four that were originally used. A 1.13 t charge of mica is placed in the mill which consists of a 4.57 m diameter tub containing two 1.22 m diameter wheels or rollers. These wheels roll over the mica for a period of from six to eight hours, causing the mica flakes to slide over each other. This action causes the mica to delaminate. This process not only preserves but enhances the sheen and slip of the mica. At the end of the grinding period, the mica is washed from the mills down a flume and into settling tanks. Recesses in the flume line serve as traps for quartz or other coarse impurities that might be present in the mica. These traps are periodically cleaned out manually. After allowing the mica to settle for a period of time, the settled portion is pumped from the settling tanks to a holding tank. The unsettled portion is decanted into additional settling tanks. The settlings from these tanks are mixed with the first settlings. The combined settlings are dewatered on plate and frame type filter presses.

The unsettled material from the second set of tanks is decanted into a third set of tanks where it is allowed to settle over a period of time. These settlings are eventually mixed with feed from the first and second tank. The total feed is then pumped to a leaf-type filter press. The decanted material from the third set of settling tanks goes to waste. The filter cakes (62% solids) are then dried in a steam-tube-drier. The dry mica is screened over sieves containing 170 to 200 mesh stainless steel cloth. Oversized mica from these sieves is returned to the feed bin while product or undersize mica reports to a bagging machine.

Table 9. Particle Size Distribution of Ground Mica Products in Weight Percent Retained[a]

Sieve Analysis Mesh Size	Size Microns	Water-Ground Mica		Micronized Mica	Dry Ground Mica						
					Oil Well		Joint Cement	High K_2O	Welding Rod	No. 160	Special[b] Plastics
		170 Mesh	325 Mesh		Fine	Coarse					
6	3 360				—	0–10	—	—			
16	1 190				—	25–55	—	—		—	
20	850				trace	—	—	—		—	
28	600				—	—	—			—	
32	500				10–30	—	—	—		—	
35	475				—	—	—	trace		trace	
60	250				10–50	—	trace	5–0		—	1.0
80	180				—	—	—	—		—	
100	150	0.0	0.0	0.00	10–70	25–65	1.5 max	10.0		10.0 max	30.0
140	106	C.5–1.0			10–30[c]	10–20[c]	—	—		—	50.0
200	74	5.0 max	1.0 min				10.0 min	10.0 max		65.0	
270	53	—	—	0.00			—	—			80.0
325	44	7.0–14.0	3.0–10.0	—			—	—		75.0 max[d]	15.0
−325	44	75.0–85.0	90% min	99.0 min			50.0 min[e]	40–75		—	5.0
Bulk Density (gm/cm^3)		0.2 max	0.2 max	0.2 max	(0.4–0.6)		0.18–0.26		0.2–0.26	0.29–0.45	

[a]information furnished by consumers, [b]micronized mica and wet-ground mica both used in certain types of plastics, [c]−100 mesh, [d]retained on 325 mesh sieve, [e]usually run 70–75%.

Wet ground mica is packaged in 22.72 kg bags and is usually pelletized in 1.13 t lots for shipment. Wet ground mica has a much higher aspect ratio than does dry ground mica, and it also possesses much better slip and sheen.

Dry ground micas: Dry ground mica concentrate is processed by four different methods. The relatively coarse particle sizes (10 to 100 mesh), used in oil well drilling, some types of welding rod coatings, roofing, and for some other types of fillers, are hammer milled and screened to size. Some roofing mica are also produced using a Raymond mill in closed circuit with an air classifier. These products are either packaged in 22.72 kg bags and pelletized in 1.13 t lots for shipment or in some cases shipped in tank trucks in bulk.

The finer particle size micas (−100 to −325 mesh) are used chiefly in textured paints and joint cement compounds and are ground on a variety of fluid energy mills. Most of the fine dry ground mica produced is ground with superheated air. The very finest dry ground mica product, down in the 10 to 20 μm range, is ground with superheated steam, and is labeled micronized mica by one producer. Micronized mica is produced in a special type of dry grinding machine that has no moving parts. This machine depends on jets of high pressure superheated steam to reduce the mica to ultrafine particle sizes. The jets of steam are placed around the periphery of the grinding chamber and are set at an angle to the wall. They keep the mica rotating at high speed. Grinding is accomplished by innumerable collisions of mica particles whirling at high speed (Zlobix, 1980).

In the most commonly used fluid-energy process, the wet flotation mica concentrate is fed into the mill at 8 to 35% moisture. Pre-drying the scrap mica to 8% moisture greatly increases the production of these mills. As the mica feed is conveyed by superheated air pressure to an air classifier, it is flash dried from an air heater attachment. The oversize from the classifier is returned to a grinding chamber where high velocity superheated air jets cause the mica particles to collide with each other and thereby grind themselves. Ground mica leaving the grinding chamber combines with the incoming feed and returns to the air classifier.

The fines, or product from the air classifier, are conveyed in an airstream to a baghouse. Mica from the baghouse is fed onto multiple deck sieves containing 105 to 120 mesh stainless steel sieves. (Sometimes finer sieves are used, depending upon the product.) Oversize mica from the sieves returns to the mill's grinding chamber. Product from the sieves is conveyed to storage bins from which it is bagged in 22.72 kg units and pelletized in 1.13 t lots for shipment, or in some cases it is conveyed to bulk trucks or railroad hopper cars for shipment.

Testing

Some of the conventional tests required by consumers of mica follow:

Screening: A 100-gm sample of mica is usually used for this test, plus a rack of six Tyler sieves, and a pan. The stack of six Tyler sieves containing the sample is placed in a Tyler Ro-tap for 10 min. for coarse mica, and 20 min. for fine mica. After screening, the mica remaining on each screen is weighed, and the percentage retained is calculated.

A combination of wet and dry screening is often preferred for fine mica. The mica is first wet screened on a 325 mesh Tyler sieve. The +325 mesh fraction is dried and then dry screened. The −325 mesh mica resulting from dry screening is added to the −325 mesh mica obtained from wet screening.

Bulk Density: Bulk density of fine mica is determined with a Scott-Schaeffer-White volumeter using a 2.54-cm cube. A 7.6-cm diameter cylinder that is 12.7 cm long is used to determine the bulk density of coarse mica such as oil well grade mica.

True Specific Gravity: True specific gravity is determined with an air comparison pycnometer or can be determined with a conventional glass pycnometer using distilled water as the displacement liquid.

Chemical Analysis: Chemical analysis is determined with conventional atomic absorption methods. Reliable wet techniques are sometimes employed as in the past.

Moisture: Moisture is determined by weighing a 50-gm sample, and then drying it at 110°C. It is reweighed after drying. The difference between the two weights, divided by the wet weight, and multiplying by 100 equals the percentage of moisture.

Free Silica: Free silica down to 1% can be determined with XRD (X-ray diffraction techniques).

Refraction Index: Refraction index is determined with a petrographic microscope using emersion oils of known refractive index.

Oil Absorption: The Gardner rub-out (ASTM-D-281) method is used. A 5-gm sample of mica is placed on a smooth glass plate, and then linseed oil is added drop-wise from a buret. The mica is constantly mixed with the oil with a spatula while the oil is being added. When the mica becomes saturated with oil to a point where it begins to curl when the spatula is scraped across the mica-oil mix, the end point has been reached and the pounds (0.4545 kg) of oil per 45.45 kilos of mica is calculated as follows:

20 times the milliliters of oil times specific gravity of linseed oil = pounds of oil/45.45 kg of mica.

Brightness: The brightness of mica is determined with a Photovoltmeter (instrument made by the Photovolt Co., a division of Soragen) or other suitable reflectance meters using a green (550 µ) filter. The mica sample is prepared for the brightness reading by pressing it into a smooth layer on a smooth surface.

Surface Area: Surface area can be measured with any one of several good surface area analyzers now on the market.

Grit Content: The grit content is determined for fine mica by thoroughly mixing a 50-gm sample with 1 L of water. After mixing the mixture is allowed to stand for 15 sec, after which ½ of the contents is carefully decanted. The volume is again increased to 1 L and again thoroughly mixed. The slurry is again allowed to stand for 15 sec before one-half of the contents is decanted. This procedure is continued until the slurry begins to become transparent. When this occurs the slurry is allowed to stand for 10 sec before being decanted. This is repeated until the water is perfectly clear. At this point the water is carefully decanted as much as possible without pouring out grit. The grit is then washed into a casserole and dried at 110°C. After drying, it is weighed and reported as percent grit. In some cases, the grit is screened on a 70 mesh sieve and the percentages of +65 mesh and −65 mesh grit are reported as separate percentages. Another method of determining the grit content of fine mica is with a Frantz Isodynamic Separator. This apparatus provides a magnetic field that will remove magnetic mica from non-magnetic grit.

Coarse Mica: The grit content of coarse-mica is determined by the vanning method. A 25-gm sample of dry mica is placed on a vanning plaque or a 21.6 by 28 cm piece of rough cardboard. The cardboard is cupped and the mica is rolled to and fro so that the sand becomes separated from the mica. A good separation is noted by inspection. The sand is then scrapped from the plaque or cardboard and the percentage of grit is calculated by dividing the weight of the grit by the weight of the sample and multiplying by 100.

Aspect Ratio: The aspect ratio of mica is determined with electronmicroscopic image analysis techniques.

Byproducts

The main byproducts of mica processing plants are kaolin, quartz, and feldspar. Some plants produce all of these products while others produce only one or two of these products.

Glass-grade silica can be produced from most mica operations with additional beneficiation of the quartz. Ceramic-grade kaolin can also be produced from some mica flotation plants by selective mining and by additional processing of the clay slime that was removed prior to mica flotation.

MICA MARKET AND CONSUMPTION DATA

Sheet Mica

Good quality sheet mica, principally muscovite, is widely used for many industrial applications, particularly in the electrical and electronic industries because of its high dielectric strength, uniform dielectric constant, low power loss (high Q factor), high electrical resistivity, uniform dielectric constant, and low temperature coefficient. In addition to these properties, mica resists temperatures of 600 to 900°C, and can be easily machined into structurally strong complicated parts of different sizes and shapes (Rajgarhia, 1987).

Table 10. US Sheet Mica Imports for Consumption (tons and thousands of dollars)

	Unworked Sheet Mica		Split Block Consumption		Splitting	
Year	Quantity	Value ($)	Quantity	Value ($)	Quantity	Value ($)
1985	761	1,080	25	112	737	957
1986	846	653	5	61	827	580
1987	1 116	1,230	43	235	1 072	991
1988	1 603	2,083	120	403	1 452	1,587
1989	1 616	2,054	158	235	1 376	1,355
1990	1 615	2,051	380	364	1 201	1,419

Source: Davis, 1989, 1990.

Commercially, flat books of sheet mica are delaminated into thin sheets, and cut to size (no less than 2.54 by 2.54 cm). These thin sheets of mica are graded into 10 different qualities according to size and the number of impurities present (Tables 10 and 11).

Some of parts fabricated from sheet mica follow:

Microwave Windows: Windows having excellent mechanical strength, and low power loss can be readily assembled by low temperature methods.

Condenser Mica: Mica has an average dielectric constant of 7, which makes it one of the most dependable types of insulators for all types of electronic applications.

Transistor Mica: The excellent mechanical and electrical properties of natural mica coupled with its high surface leakage resistance, low moisture absorption, and ability to dissipate heat make mica ideally suited for transistor mounting washers.

Interlayer Insulator: Mica 0.10 mm thick is used as insulation for coil workings in small Class H transformers. The use of mica in this manner helps reduce the size and weight of the transformers which is very important in the electronic industry.

Resistance and Potentiometer Cards: The properties of high heat resistance, low bulk density, and high mechanical strength make natural mica suitable for winding non-inductive resistance cards. In potentiometers, it offers a high temperature material that can be wound and later bent into a circle.

Vacuum Tube Mica: At one time this application represented the largest use of sheet mica. The high mechanical strength, electrical properties, and ability to be precisely machined make mica ideally suited for this application.

Natural Mica Bushings and Tubes: Insulation for electrical components requiring a round completely inorganic tube can be completely formed from sheet mica. These tubes (12.7 to 15.2 cm in length) are rolled cigarette fashion on a mandrel.

Target and Mosaic Mica: Target and mosaic micas are used in the telecasting industry in image orthacons and in computers requiring optically flat mica selected from the best V-1, V-2, or V-3 quality. This mica, which is coated with gold or silver, is split to a thickness of 25 to 38 µm.

Guided Missile Micas: Most of the uses for natural mica in guided missile components are secret. However, the electrical, physical, and heat endurance properties of natural mica make it an excellent material for the missile engineer.

Other Uses: High quality natural sheet mica is used in helium neon lasers, special optical filters, lining for glasses of high pressure steam heaters, diaphragms for oxygen-breathing equipment, washer

Table 11. Builtup Mica[a] Sold or Used in the United States by Product (tons and thousand dollars)

	1985		1986		1987		1988		1989		1990	
Product	Quantity	Value ($)	Quantity	Value ($)	Quantity	Value ($)	Quantity	Value ($)	Quantity	Value ($)	Quantity	Value ($)
Flexible (cold)	88	180	77	80	381	626	85	499	85	545	96	588
Heater plate	32	182	29	132	24	88	21	76	18	74	W[a]	W
Molding plate	297	2,397	283	2,326	308	2,397	328	1,898	316	1,011	291	1,623
Segment plate	328	2,974	311	2,940	259	2,211	325	2,741	378	2,216	284	1,777
Tape[b]	121	1,675	115	1,559	75	1,197	37	305	51	277	W	W
Other	181	1,796	178	1,741	195	1,758	195	1,810	141	1,607	190	2,223
TOTAL[c]	1 047	9,835	993	9,501	942	8,277	991	6,829	989	5,730	861	6,211

[a]withheld to avoid disclosing company proprietary data; included in other, [b]consists of alternative layers of binder and irregularly arranged and partly overlapped splittings, [c]data do not add to total shown because of independent rounding.
Source: Davis, 1986, 1987, 1988, 1989, 1990.

dials for navigator compasses, quarter-wave plates of optical instruments, pyrometers, and thermal regulators. The heart of a CAT scan is a piece of high quality mica coated with gold.

Builtup Mica

Builtup mica made from splittings serves as a substitute for natural sheet mica when the primary property needed is insulation. Principal end uses for builtup mica are segment plate, molding plate, flexible plate, heater plate, and tape (Zlobix, 1980). Over the years, segment plate has been the largest use for builtup mica (Table 11). The segment plate acts as insulation between copper commutator segments on direct current universal motors and generators. Phlogopite builtup mica is preferred for these segments since it will wear at the same rate as the copper segments.

Molding plate is the sheet from which V-rings are cut and stamped for use in insulating the copper segments from the steel shaft at the ends of a commutator. Molding plate is also fabricated into tubes and rings for insulation in transformers, armatures, and motor starters.

Flexible plate is used in electric motor and generator armatures, field coil insulation, and magnet and commutator core insulation. Heater plate is used where high insulation strength at high temperature is required.

Some types of builtup mica are bonded to special paper, silk, linen, muslin, glass cloth, or plastic. These products are very flexible and are produced in continuous wide sheets. These sheets are either shipped in rolls or cut into ribbons, tapes, or any other desired shape.

Wet Ground Mica

Wet ground mica (Table 12) is used in the following products because of one or more of its unique properties of luster, slip and sheen, and high aspect ratio (Preston, 1971, Rajgarhia, 1987).

Wallpaper and Coated Paper: The shiny particles of mica give a silky or pearly luster when applied to paper. The effect is pleasing to the eye, and in some cases simulates fabric.

Nacreous Pigments: Mica is used as a substrate for coatings of various metal oxides to obtain a pearlescent effect. Mica coated in this fashion is used as filler and coloring agents in certain types of plastics.

Rubber: A thin coating of mica acts as a mold release compound in the priming of rubber goods such as tires. It prevents the migration of sulfur from the tire to the airbag when the tire is being vulcanized. Mica is also dusted on rubber inner tubes to prevent sticking.

Outside House Paint: Mica acts as a reinforcing pigment to reduce checking and cracking while at the same time controlling chalking in outside latex, oleoresinous, alkyd, and alkyl-modified latex exterior paints. It also reduced penetration into porous surfaces while providing excellent adhesion. In addition, mica reduces running, sagging, improves weatherability, and improves brushability.

Table 12. Ground Mica[a] Sold or Used by Producers in the United States, by Method of Grinding (thousands of tons and thousands of dollars)

	Dry Ground		Wet Ground		Total	
Year	Quantity	Value ($)	Quantity	Value ($)	Quantity	Value ($)
1985	112	15,993	12	5,263	124	21,256
1986	99	14,682	13	7,190	112	21,872
1987	101	15,140	12	7,237	113	22,376
1988	95	14,570	14	9,117	109	23,687
1989	88	13,409	13	9,034	101	22,443
1990	W[b]	W	W	W	97	21,610

[a]domestic and some imported scrap combined; low quality sericite is not included, [b]withheld to avoid disclosing company proprietary data.
Source: Davis, 1989, 1990.

Aluminum Paints: Mica is substituted for up to 25% of the aluminum in this type paint as an economic measure. In addition, mica is inert, which tends to protect the more reactive aluminum from corrosive atmospheres; thus, helping the paint to maintain its luster.

Sealers: Mica is used in all types of sealers for porous surfaces, such as wallboard masonry and concrete blocks, to reduce penetration and improve holdout. It permits a thicker film to be applied and at the same time reduces sagging. Cracking is reduced by the reinforcing action of the flakes, and gaps and holes in rough masonry are bridged by the mica flakes.

Plastics: Mica is used as a filler in plastics to improve its electrical and thermal resistance and its insulating qualities. Although dry ground mica is also used as a filler in plastics, wet ground enjoys the higher value market and is used because of its extrusion properties, and because it can replace the higher priced glassfiber in some cases.

Dry Ground Mica

Hammer Mill Mica: Dry ground mica produced from hammer mills and screens is used in oil well drilling, coating roofing shingles, roofing felt, and for some types of welding rod flux.

Mica is mixed with the drilling mud to prevent loss of circulation and seepage in loose rock formations in oil well drilling operations. Mica mixed with mud additives acts as a sealant because of its platy structure. It also helps keep the other solids in the mud in suspension. Different particle size distributions of mica are used

Table 13. Ground Mica Sold or Used by Producers in the United States by End Uses (tons and thousands of dollars)

	1985[a]		1986[a,b]		1987[a,b]		1988		1989		1990	
End Use	Quantity	Value ($)	Quantity	Value ($)	Quantity	Value ($)	Quantity	Value ($)	Quantity	Value ($)	Quantity	Value ($)
Joint cement	62	10,372	57	10,499	64	11,352	60	10,936	58	11,198	53	8,984
Paint	26	3,451	20	3,676	14	3,012	17	3,536	10	1,512	14	2,894
Plastics	2	498	1	252	2	338	1	366	3	876	2	675
Well drilling mud	15	1,730	5	576	6	797	7	741	7	816	7	1,114
Other[a]	26	5,205	33	8,868	25	6,877	24	8,108	23	8,041	21	7,944
TOTAL[b]	131	21,256	116	23,872	113	22,376	109	23,687	101	22,443	97	21,610

[a]includes mica used for molded electrical insulation, rubber, textile, and decorative coatings, welding rods, and miscellaneous, [b]data may not add to totals shown because of independent rounding.
Source: Davis, 1986, 1987, 1988, 1990.

to prevent specific types of circulation losses that are associated with particular rock formations (Preston, 1971, Rajgarhia, 1987).

Certain sizes of hammer-milled and screened mica is used as a flux coating on the metal wire of welding rods. Fine particle size, fluid-energy ground mica is also used for this purpose. High heat resistance, excellent mechanical and thermal strength, low moisture absorption, high surface leakage resistance, unusual chemical inertness and non-hydroscopic properties are some of the distinctive qualities which make mica an indispensable ingredient for use in welding electrodes.

Fluid-Energy-Produced Mica: The largest use at the present time for fine, dry ground mica is in the manufacture of wallboard joint cements (Table 13). Ground mica that is essentially −100 mesh (see Table 9) and approximately 70% passing through a 325 mesh Tyler sieve (45 μm) is used in the joint compound mixture as a filler and extender. These compounds are used to fill the joints between panels of gypsum plaster board. In this particular use, mica contributes to making a non-absorbing smooth surface that reduces shrinkage and eliminates cracks.

It is also used in the finished coating on ceilings and to improve ceiling tile and prefabricated concrete blocks and slabs, where thermal insulation and acoustical qualities are of particular importance. Slabs prepared from ground mica and portland cement mixed in different proportions possess far superior tensile strength and conductivity values, as well as lower bulk densities than other types of slabs.

Fine particle size dry ground mica is also used as an extender and filler in certain texture and traffic paints where sheen and slip are not as important as they are in high quality paints. Not only does its inclusion mean a great savings in cost by replacing more expensive pigments, but it also enhances certain essential paint properties. Mica particles are stronger than iron and are not brittle like other inert material. Finely ground particles reinforce a protective coating like paint in almost the same way that steel rods reinforce concrete or glass is reinforced by wire. It is an anti-friction, anti-fouling, anti-settling, anti-corrosive, anti-tarnish and anti-siege agent. It is a superior reinforcing pigment that acts as a sealer over porous surfaces and reduces penetration and flushing; moreover, it improves the moisture resistance of protective coatings and their adhesion to all types of surfaces.

Mica is one of the most inert substances; thus it helps protect the paint coating against acid and alkali attacks, from deterioration due to heat or sunrays, from atmospheric gases, and from moisture. It also decreases blistering and staining from the substrates.

Micronized Mica: Micronized mica is a trade name for a very fine particle size dry ground product. It is usually ground with superheated steam in a special fluid energy mill and is used as a replacement for wet ground mica in certain types of paints. Micronized mica, preferably calcined, is also used in cosmetic applications, i.e., nail varnishes, lipsticks, eyeshadows, and barrier cream. It is used in cosmetics because it has the advantage of high ultraviolet light stability, excellent lubricity, skin adhesion, and compressibility. Some of these micas are coated with oxides like titanium and iron.

New Applications

Some of the latest research and testing results have shown that mica can be used successfully in a wide range of new applications. Some of these are air conditioner fan blades, dashboard panels, head lamp assemblies, fan shrouds, and floor panels for automobiles. Mica can also be substituted for more expensive glass flakes to strengthen lightweight plastic seat backs, load floors, grill panels, ignition system parts, and air conditioning and heater valve housings. Both American and Japanese automakers are incorporating ground mica in place of asbestos in acoustic compounds that change vibrations and eliminate road and engine noise. Mica is also an environmentally accepted replacement for asbestos in brake linings.

MICA PRODUCTION AND CONSUMPTION (ECONOMIC FACTORS)

The data described in this section is taken from the US Bureau of Mines *Mineral Industry Survey 1990* (Davis, 1990). This includes the price per ton of types and end uses (Table 14).

North Carolina remained the major producing state with 60% of the total production. The remainder was produced in Connecticut, Georgia, New Mexico, Pennsylvania, South Carolina, and South Dakota. The five largest producers were 1) The Feldspar Corporation, Spruce Pine, NC, 2) Franklin Mineral Products, Division of Mearl Corporation, Hartwell, GA, 3) KMG Minerals, Inc., Kings Mountain, NC, 4) Lithium Corporation of America, Gastonia, NC, and 5) Unimin Corporation, Spruce Pine, NC. These five companies produced 67% of the nation's total output.

Eleven companies operated 14 grinding plants to produce 97 kt of ground mica. Ten produced dry ground and four produced wet ground mica. The five largest producers accounted for 73% of the total. They were: 1) KMG Minerals of Kings Mountain, NC, 2) MICA (Franklin Limestone) near Española, NM, 3) Spartan Minerals Corporation (FMC Corp.) of Pacolet, SC, 4) Unimin Corp. of Spruce Pine, NC, and 5) United States Gypsum Corporation, Spruce Pine, NC. The major end uses were joint cement 55%, paint 14%, and oil well drilling muds 7%. Other end uses were molded electrical insulation, plastics, roofing, rubber, and welding rods.

The United States imported most of its manufactured sheet mica and paper-quality scrap mica from India (Tables 15 and 16). Con-

Table 14. Average Price per Ton Reported by US Producers of Wet Ground and Dry Ground Mica in 1990

Type	Price ($)
Wet Ground	663
Dry Ground	151
By End Uses	
Joint Cement	168
Paint	212
Plastics	288
Well Drilling Mud	162
Other[a]	381

[a]Includes mica used for molded electrical insulation, roofing, rubber, textile, and decorative coating, welding rods, and miscellaneous.
Source: Davis, 1990.

Table 15. US Imports for Consumption of Mica Powder and Waste in 1989 by Country (tons and thousands of dollars)

	Powder		Waste	
Country	Quantity	Value ($)	Quantity	Value ($)
Canada	8 641	3,927	—	—
China	—	—	72	32
India	72	29	1 078	474
Japan	72	931	—	—
Norway	100	68	—	—
Other[a]	18	16	39	37
TOTAL[b]	8 902	4,971	1 189	542

[a]includes Brazil, Finland, France, the Federal Republic of Germany, and the United Kingdom, [b]data may not add to totals shown because of independent rounding.
Source: Davis, 1989.

sumption of muscovite block (ruby and non-ruby)[1] totaled 6.0 t in 1990, a 20% decrease from that of 1989 (Davis, 1990). Five companies consumed muscovite block and film in five plants in four states, two in North Carolina and one each in New Jersey, Ohio, and Virginia. Consumption of mica splittings decreased 24% to 842 t in 1990 from 1989. The splittings were fabricated into various builtup mica products by nine companies operating nine plants in seven states. Total production sold or used of builtup mica decreased 13% from that of 1989. Molding plates and segment plates were the major end products, accounting respectively for 35 and 33% of the total. The combined value of all mica imports increased 4% to $15.6 million, while the combined value of all mica exports increased 10% to $10.5 million. Sheet mica prices ranged from approximately 4.40 to $1,100/kg, depending upon the grade, while splittings were bought for 1.10 to $3.30/kg. Wet ground micas selling prices ranged from 440 to $800/t, while some of the very finest cosmetic type mica brought up to $6,600 to $9,000/t. Dry ground mica sold for 110 to $220/t FOB the plant site. See Table 17 for comparative quantities and value of sheet and scrap mica from 1986 to 1990, and Table 14 for average selling prices per ton of dry and wet ground mica products in 1990. Table 12 shows comparative quantities and values of ground mica for the years 1985 to 1990. Exports of worked mica, sometimes to countries supplying raw mica, is an important consideration in the economic picture of mica (Table 18).

[1]Refers to a specific type of muscovite mica denoted by its color.

Transportation

Mica is shipped by truck or railroad. Maximum truckload shipments are considered to be 20.4 t while maximum rail shipments are 27.2 t. Bulk truckloads and bulk rail shipments are also made to avoid extra costs, such as palletizing, bagging, and shrink or stretch wrapping. Mica is usually palletized and shrink wrapped or stretch wrapped for shipment.

Alternate Materials

Mica is usually used in an application because of its unique properties and not for a simple filler clone. In these cases, there is no substitute for it. However, in some cases, such as roofing shingles, substitutes of sand and talc have been made. Although these shingles were inferior, the price of the raw ingredients was materially less. Fiberglass and mica are interchangeable in certain types of plastics, and in these cases, mica is usually substituted for fiberglass because of the price differential.

Talc is used in the place of mica in certain low priced paints. It can be used as a mica replacement in some mold release compounds.

GOVERNMENTAL CONSIDERATIONS

Depletion Provisions

According to Section 1289 of the 1991 Standard Federal Tax Report, depletion is defined as the exhaustion of natural resources, such as mines, wells, and timberlands, as a result of severance or production. The right to a depletion allowance is based upon the taxpayer's economic interest in the property. An economic interest exists if he: 1) has acquired by investment any interest in minerals in place or in the standing timber, and 2) looks to the income from the extraction of the minerals or severance of the timber for a return on his investment.

The basis upon which a deduction is allowed is the adjusted basis of the property (Reg. No. 1.612–1). Determination of cost depletion first requires an estimate of the number of units (tons, barrels, etc.) that make up a deposit, then that part of the cost or other adjusted basis of the property that is allocable to the depletable reserves is divided by the number of units. The quotient is the cost depletion per unit. This amount, multiplied by the number of units extracted and sold during the year, determines the cost depletion deductible for the year.

A special method of computing depletion, known as percentage depletion, applies to all depletable property except timber. Under this method, a flat percentage of gross income from the property is taken as the depletion deduction. The percentage depletion may not exceed 50% of the taxable income from the property computed without regard to the depletion allowance. In computing taxable income (for the 50% limitation), the deductible mining expenses must be decreased by the amount of Section 1245 gains allocable to the property [Code Section 613 (a)]. If cost depletion results in a greater deduction, cost depletion must be used (Reg. 1.613–1).

Depletion is subject to recapture as ordinary income upon the sale or other disposition of an oil, gas, geothermal, or other mineral property at a gain generally for disposition of property placed in service after 1986 (Code Section 1254). Recapture is limited to the amount for which the depletion deduction reduced the adjusted basis of the disposed property. Percentage depletion is allowed under Code Sec. 613 (b), at the following percentages of "gross income from property:" 22% for mica, olivine, quartz crystals,

Table 16. US Imports for Consumption of Crude and Rifted[a] Mica in 1989 by Country (tons and thousands of dollars)

	Split block		Splittings		Other			
					Less than $0.55/kg		Greater than $0.55/kg	
Country	Quantity	Value ($)	Quantity	Value ($)	Quantity	Value ($)	Quantity	Value ($)
Canada	—	—	—	—	54	22	—	—
France	2	34	[b]	12	—	—	42	56
India	155	177	1 348	1,264	2 992	685	21	312
Madagascar	1	24	15	69	—	—	15	—
Other[c]	—	—	16	10	1	27	—	—
TOTAL[d]	158	235	1 379	1,355	2 996	714	79	436

[a]crudely split block mica from mine, [b]less than 1/2 unit, [c]includes Brazil, China, Japan, the Republic of Korea, and the United Kingdom, [d]data may not add to totals because of independent rounding.
Source: Davis, 1989.

Table 17. Summation of US Mica Trade Data (tons and thousands of dollars)

	Scrap and Flake Mica				Sheet Mica			
	Powder		Waste		Unworked		Worked	
Year	Quantity	Value ($)	Quantity	Value ($)	Quantity	Value ($)	Quantity	Value ($)
Exports:								
1986	5 300	1,758	1 454	472	44	196	NA	4,502
1987	4 235	1,275	824	259	77	145	NA	4,748
1988	5 202	1,623	1 088	367	63	228	NA	6,148
1989	3 628	1,634	1 224	555	60	156	415	7,227
1990	4 319	2,050	580	646	148	272	612	7,568
Imports for Consumption:								
1986	5 450	2,326	4 510	1,225	846	653	955	4,859
1987	5 672	2,685	3 916	1,243	1 116	1,230	746	5,125
1988	7 778	4,222	4 345	1,339	1 603	2,083	751	5,679
1989	8 902	4,971	4 185	1,256	1 616	2,054	1 229	6,711
1990	9 142	5,133	4 034	987	1 615	2,051	1 085	7,431

Source: Davis, 1990.

rutile, block steatite, and many other minerals (References are to paragraphs 1289, 1291, and 1298 of the 1991 Standard Tax Reports).

Zoning and Land Use

Mica mining is subjected to local land zoning regulations, and state regulations where applicable. After mining the land must be reclaimed.

Environmental Regulations

Mica mining like all mining operations is subjected to local, state, and federal laws. Mining, Safety and Health Administration (MSHA) regularly monitors mica mining operations for safety violations. Most states have labor departments that more or less mirror the efforts of MSHA.

Both state air and water environmental departments together with the Environmental Protection Agency (EPA) regulate and oversee air and water quality associated with mica mining operations. Most states work under the auspice of the EPA.

Most states also have land management departments that regulate dam safety, erosion, sedimentation, and reclamation. The mica mines, like other mines, must control erosion and sedimentation and must restore the mined out areas. This is accomplished by either backfilling, contouring and seeding operations, or in cases where this is impractical or undesirable, lakes for water related recreation may be built. The Corps of Engineers have jurisdiction over laws governing wet lands.

Health provisions are also maintained by county and state health departments. Some states require employees to receive an annual mine employee chest X-ray to detect silicosis and other lung diseases.

PROBLEMS AND FUTURE TRENDS

The number one problem in the United States is to educate the public on what mining entails, and how dependent the economy and our standard of living is on having a viable mining industry. This education program must start at the elementary school level so that the student achieves a balanced view of the world and how things work.

The future for mica is in the specialty plastic market. It is hoped that mica will find more use as a substitute for the more expensive glassfiber in plastic automobile bodies, and as a molecular barrier in plastic containers. Giant strides are being made toward these goals and, if finally brought to fruition, will offer a promising future for mica.

ACKNOWLEDGMENTS

The author acknowledges the special contributions made by several individuals in writing this chapter on mica for the 6th edition

Table 18. US Exports of Worked Mica in 1989 by Country (tons and value in thousands)

	Plates and Sheets		Other	
Country	Quantity	Value ($)	Quantity	Value ($)
Australia	7	118	([a])	2
Austria	5	47	—	—
Bahamas	4	21	6	24
Bolivia	—	—	3	10
Brazil	41	901	2	68
Canada	66	1,414	35	898
Colombia	2	10	1	25
Egypt	18	59	—	—
France	1	4	3	55
Germany (West)	23	557	2	12
India	1	34	8	77
Italy	15	373	([a])	14
Japan	([a])	5	3	105
Mexico	46	500	26	520
New Zealand	4	17	—	—
Peru	—	—	12	395
Philippines	8	30	—	—
Spain	5	110	—	—
Taiwan	9	30	40	502
United Kingdom	—	—	2	29
Other[b]	13	198	2	61
TOTAL[c]	268	4,430	147	2,797

[a]less than 1/2 unit, [b]includes Argentina, Barbados, Belgium, Bermuda, Chile, China, the Dominican Republic, Ecuador, Guatemala, Honduras, Hong Kong, Ireland, the Republic of Korea, Kuwait, the Netherlands Antilles, Nigeria, Pakistan, Panama, Saudi Arabia, Sierra Leone, Singapore, the Republic of South Africa, Surinam, Sweden, and Switzerland, [c]data may nct add to totals shown because of independent rounding.
Source: Davis, 1989.

of *Industrial Minerals and Rocks*. These contributors were Louis Schlesinger, Engineer at MRL, who spent several hours proofreading the manuscript; Hugh Lancaster, General Manager of Kings Mountain Minerals, Inc., who also thoroughly proofread the manuscript and provided the author with valuable information; and Bill Wilkins, retired Sales Manager of Deneen Mica Company, who provided valuable historic information.

BIBLIOGRAPHY

Anon., N.D., "Mica Fabricators," The Tar Heel Mica Company, Brochure, Plumtree, NC, 3 pp.

Anon., 1990, "National Mica for Industry," The Tar Heel Mica Company, Buffalino, 3 pp.

Bailey, S.W., 1984, "Classification and Structures of the Micas," Reviews in Mineralogy, Vol. 13, Mineralogical Society of America, Washington, DC, pp. 1–12.

Chapman, G.P., 1983, "Mica," *Industrial Minerals and Rocks*, 5th ed., Vol. 2, S.J. Lefond, ed., AIME, New York, pp. 915–929.

Chesterman, C.W., et al., 1989, *The Audubon Society Field Guide to North American Rocks and Minerals*, Alfred A. Knorf, New York, pp. 531–536.

Clarke, G., 1983 "Mica—A Review of World Developments," *Industrial Minerals*, No. 189, pp. 27–50.

Dana, E.J., 1945, *A Text Book of Mineralogy*, John Wiley and Sons, New York, pp. 659–665.

Davis, L.L., 1985, "Mica," *Mineral Facts and Problems*, US Bureau of Mines, Preprint from Bulletin 675, 12 pp.

Davis, L.L., 1986, "Mica," Mineral Industry Survey, US Bureau of Mines, p. 8.

Davis, L.L., 1987, "Mica," Mineral Industry Survey, US Bureau of Mines, p. 8.

Davis, L.L., 1989, "Mica," Mineral Industry Survey, US Bureau of Mines, p. 9.

Davis, L.L., 1990, "Mica," Mineral Industry Survey, US Bureau of Mines, p. 4.

Dawson, E.H., 1949, "Mica," *Industrial Minerals and Rocks*, 2nd ed., S.H. Dolbear et al. eds., AIME, New York, pp. 551–566.

Grim, R.E., 1968, *Clay Mineralogy*, McGraw-Hill, New York, 596 pp.

Hendricks, S.B., and Jefferson, M., 1939, "Polymorphism of Micas with Optical Measurements," *American Mineralogist*, Vol. 24, pp. 729–771.

Preston, J.B., 1971, "Mica," Draft of Report for *Pigment Handbook*, 30 pp.

Rajgarhia, M.L., 1987, "Ground Mica," Mica Manufacturing Company, Pvt. Ltd., 30 pp.

Zlobix, A.B., 1980, "Mica," US Bureau of Mines, Preprint from Bulletin G71, 16 pp.

NEPHELINE SYENITE

G. ROBERT GUILLET

Nepheline syenite is a light-colored feldspathic rock that is used primarily as an addition to the glass batch. It is also used in ceramics and as an industrial filler. Composed essentially of the soda and potash feldspars and nepheline, nepheline syenite contributes needed alkalis and alumina to the glass composition. And because it contains a higher proportion of these oxides per unit weight, it is a formidable competitor for feldspar in northeastern North America and parts of western Europe.

Nepheline-bearing rocks are not uncommon on a world scale, but few are able to be beneficiated to a composition sufficiently low in iron to be used in the making of glass. Nepheline can only form in an environment deficient in silica. Hence free silica, as quartz, is always absent in the primary mineral assemblage. Biotite, hornblende and magnetite are usually present as minor accessory minerals, and zircon, garnet, muscovite, and apatite may be present as trace minerals.

Nepheline syenite is produced for glass, ceramics, and filler markets by Unimin Canada Limited at Blue Mountain, in southeastern Ontario. Unimin acquired the former Indusmin and IMC properties from Falconbridge Ltd. in 1990. Elkem Nefelin A/S commenced production of similar products from Stjernoy Island in northern Norway in 1961. Unimin acquired this company in 1992.

Nepheline has also been a byproduct of apatite mining on the Kola Peninsula for many years, and in 1951 the production of alumina from nepheline was commenced in a plant near St. Petersburg (Leningrad). There are now four plants in the Russian Republic, two smaller ones based on nepheline tailings from apatite mining and two larger and more recent plants in central Siberia using other nepheline resources. All four of these plants are integrated complexes that produce alumina for the subsequent production of aluminum metal, sodium and potassium carbonates, and portland cement.

Roofing granules are produced by the 3M Company from a nepheline rock in Arkansas. In 1980 and 1989, trial shipments of nepheline syenite were recorded from new sources in Brazil and western Turkey respectively. Recent interest is also recorded for deposits in China, Saudi Arabia, and British Columbia.

GEOLOGY

Mineralogy

Nepheline syenites and nepheline pegmatites are igneous rocks characterized by the absence of quartz and the presence of the feldspathoid mineral, nepheline. In an alkali-rich rock magma deficient in silica, nepheline will form instead of albite feldspar. However, some nepheline gneisses and other nepheline rocks may also have formed by metasomatic nephelinization processes.

Nepheline rocks are composed essentially of nepheline, sodic plagioclase (usually albite or oligoclase), and microcline, but the proportions of these minerals can vary widely. Nepheline typically comprises 20 to 30% of the rock, but segregations of nearly pure nepheline are known. Biotite, hornblende (usually hastingsite), magnetite, and pyroxene are the usual accessory minerals. Other minerals occurring in trace amounts include muscovite, sodalite, cancrinite, hydronephelite, garnet, zircon, apatite, corundum, titanite, ilmenite, scapolite, calcite, tourmaline, pyrite, vesuvianite, chlorite, and zeolites (Hewitt, 1961).

Nepheline is the most common of the feldspathoid minerals, a group that also includes leucite, sodalite, and cancrinite. It crystallizes in the hexagonal system, and in vuggy occurrences can sometimes be found in simple hexagonal prisms. Most commonly it occurs in anhedral grains, interstitial to feldspars. It has a Mohs hardness of about 6, and a specific gravity of 2.6. Nepheline is usually white or slightly off-white, and sometimes pale salmon pink. It has a vitreous or greasy lustre. In the field it is more easily recognized on weathered surfaces where its easier solubility results in a pitted surface. Nepheline alters readily to a variety of poorly known secondary products, usually more or less micaceous and variously referred to as hydronephelite and giesekite. And on weathered rock surfaces these result in a coloration that varies from dove-gray to distinctly bluish. Indeed, it is these colorations and the pitted surfaces that are the most distinctive characteristics for recognizing nepheline in the field. Nepheline also alters to sodalite, cancrinite, and zeolites.

Chemical Composition

Nepheline has the theoretical composition $NaAlSiO_4$, but potassium substitutes for sodium in the amount of about 1 to 4; silica is usually slightly in excess and calcium is sometimes present. Table 1 shows the theoretical composition of pure nepheline and pure albite. While both minerals in nature fall a little short of these compositions (Table 2), principally in the substitution of potassium for sodium, and in excess silica, it is obvious that nepheline offers a distinct advantage to the glass batch, where it is the content of alumina and alkalis that count. Similarly, with a scarcity of domestic bauxite, Russia has found it viable to use nepheline concentrates in the making of aluminum metal, the economics being aided by the simultaneous production of alumina, sodium and potassium carbonates, and portland cement. Table 2 gives the actual compositions of nepheline, albite, and microcline from Unimin's Blue Mountain deposit.

Table 3 is a typical analysis of raw nepheline syenite from Blue Mountain, Ontario. Commercial nepheline syenite products are essentially mixtures of nepheline and feldspars from which any dark minerals have been removed by dry magnetic separation. In both the Canadian and Norwegian products, which are the principal sources of glass-grade nepheline syenite, nepheline comprises 20 to 30%. Typical analyses of these products are given in Table 4.

Table 1. Theoretical Composition of Nepheline and Albite

Oxides	Nepheline $NaAlSiO_4$ %	Albite $NaAlSi_3O_8$ %
Soda	21.8	11.8
Alumina	35.9	19.5
Silica	42.3	68.7

Table 2. Composition of the Primary Minerals at Blue Mountain, Ontario

	Nepheline %	Albite %	Microcline %
SiO_2	43.01	67.90	64.50
Al_2O_3	34.01	20.60	19.31
MgO	0.04	0.06	0.42
CaO	0.36	0.24	0.07
Na_2O	15.42	10.78	2.03
K_2O	6.15	0.15	13.59
H_2O+	0.99	0.22	0.28
Fe_2O_3	0.14	nil	nil
FeO	nil	nil	nil
Total	100.12	99.95	100.20
Specific gravity	2.62	2.59	2.55

Source: Keith, 1939.

Table 3. Typical Raw Composition of Blue Mountain Nepheline Syenite

	%
SiO_2	59.18
Al_2O_3	23.06
Fe_2O_3	2.15
Na_2O	10.48
K_2O	3.94
CaO	0.76
MgO	0.17
TiO_2	0.064
ZrO_2	0.05
P_2O_5	0.021
Loss on ignition	0.40
	100.28

Source: Derry and Phipps, 1957.

Table 4. Nepheline Syenite: Typical Glass-grade Compositions

Chemical composition:	Unimin Canada Ltd. A Grade %	Unimin Canada Ltd. B Grade %	Elkem Nefelin North Cape %
SiO_2	60.2	60.1	55.9
Al_2O_3	23.5	23.4	24.2
Fe_2O_3	0.08	0.35	0.1
CaO	0.3	0.3	1.3
MgO	trace	trace	trace
Na_2O	10.6	10.5	7.9
K_2O	5.1	4.9	9.0
BaO	—	—	0.3
SrO	—	—	0.3
P_2O_5	—	—	0.1
Loss on ignition	0.4	0.3	1.0
Sieve analyses:	US Sieve No. %		Tyler Sieve No. %
On 25 mesh	0.0	0.0	On 28 mesh 0.0
30	0.1	0.1	32 0.1
40	14.5	14.0	35 4.9
50	48.0	46.0	48 30.0
100	86.0	84.0	65 52.0
200	98.0	97.2	200 89.0
Pan	2.0	2.8	Pan 11.0

Source: modified from Minnes et al., 1983.

Physical Properties

Nepheline syenite generally occurs in irregular-shaped intrusive bodies of small to medium size, either as individual plutons, or as rudely concentric sheets arranged in ring structures; it also occurs in elongated gneissic units that may be the result of nephelinization. Texture is typically granitic to gneissic, medium to coarse grained, but massively crystalline nepheline pegmatites are locally common.

An open texture, free of mineral intergrowths, is typical of commercial deposits where good liberation of the mafic minerals (biotite, hornblende, and magnetite) is essential for their removal by a dry magnetic separation process. In general, nepheline rocks associated with alkaline ring complexes are not readily beneficiated to a low iron glass grade because of intergrowths of dark iron-bearing minerals.

Commercial deposits of nepheline syenite are white with a high dry brightness, and they contain a minimum of dark minerals. Specific gravity of nepheline syenite from the Blue Mountain deposit is 2.61, and its average refractive index is 1.53. Average hardness on Mohs scale is 5.5. Acidity measured in terms of pH value (5 g in 6 cc water) is 9.6.

Origin and Classification of Deposits

There is a general preference to refer to most nepheline rocks as nepheline syenites. However, because they are noted for great variability in their mineral content, a plethora of names have been coined in an attempt to categorize them. Some of the more common are:

Litchfieldite: A nepheline syenite composed essentially of nepheline, albite, and potash feldspar, with minor biotite.
Foyaite: A nepheline syenite containing about equal amounts of nepheline and potash feldspar, with minor dark minerals.
Theralite: A nepheline gabbro composed essentially of andesine, nepheline, and pyroxene.
Urtite: A rock containing more than 70% nepheline with minor aegirine and apatite.
Ijolite: A rock containing less than 70% nepheline with aegirine and minor apatite.
Jacuperangite: A nepheline gabbro comprising aegirine, biotite, and magnetite, with subordinate nepheline.
Phonolite: The volcanic equivalent of nepheline syenite.

Nepheline syenite has various modes of occurrence, as follows (Moorhouse, 1959):

1. Feldspathoidal rocks associated with undersaturated volcanics.
2. Differentiated ring complexes, often associated with carbonatite and usually characterized by metasomatism around their borders.
3. Layered intrusives.
4. Borders of satellitic stocks associated with syenites or granites.
5. Nephelinized gneisses, usually associated with nepheline pegmatites.

Distribution of Major Deposits

Because of their unusual compositions, nepheline rocks have received a lot of academic interest worldwide, and for that reason they appear to be more numerous than in fact they are. Those deposits which might otherwise be suitable in terms of size and uniformity invariably contain too much iron for glass and ceramic uses. And while many have been subjected to beneficiation testing, such information is often proprietary. Some of the more interesting occurrences are discussed on the following pages.

Ontario: The Precambrian Grenville rocks of southeastern Ontario contain a variety of nepheline occurrences in Peterborough, Haliburton, and Hastings counties. They have been the subject of numerous studies, notably Adams and Barlow (1910) and Hewitt (1961). The Blue Mountain occurrence is the most important of these. However, Ontario's Grenville Series also contains an isolated occurrence near the French River, south of Sudbury. The Blue Mountain and French River occurrences are described in the following notes.

Blue Mountain. In September 1990, Unimin Canada Ltd. acquired the industrial mineral operations of the Indusmin Division of Falconbridge Limited. This included several silica operations in Ontario and Quebec, and the nepheline syenite operations at Blue Mountain, Ontario. Unimin Canada Ltd. is a subsidiary of Unimin Corporation of New Canaan, CT.

Although a metasomatic replacement origin has been put forward by some authors, notably Derry (1951) and Derry and Phipps (1957), it is more generally believed that the Blue Mountain deposit is a stock of nepheline syenite intruded into the trough of a southwest-pitching syncline of metasedimentary rocks. Authors who have held to an intrusive view include Keith (1939), Hewitt (1961), Guillet (1962) and Payne (1966). The body is about 9 km long overall, the northern third being 2 km wide; a long sill 300 m wide extends to the southwest (Fig. 1).

Blue Mountain nepheline syenite is a remarkably uniform rock consisting of 54% albite, 20% microcline, 22% nepheline, and 4% mafic minerals (MacGregor, 1983). As such it is a leucocratic rock in which biotite and magnetite tend to occur on the fringes of the deposit, and hastingsite and magnetite (with minor aegirine and garnet) in the interior parts of the intrusive. A foliation due to the alignment of dark minerals is characteristic of the border facies. The interior is more massive.

The rock is typically white, medium-grained, with an open granitic texture. Dark minerals are normally liberated by crushing to about 20 mesh, a feature which has made possible effective beneficiation by dry magnetic separation. Commercially it has been found that the biotite-magnetite facies can be more consistently beneficiated to low iron products suitable for the glass industry. Minute inclusions or intergrowths of iron-bearing minerals make this more difficult in rock containing hastingsite.

French River. Two lenticular bodies of nepheline syenite outcrop in Bigwood Township, 60 km south of Sudbury, Ontario. They have been described by Hewitt (1961). Both are enclosed by pink alkaline syenite and are intrusive into a regional granitic terrain (Fig. 2).

The northern deposit, known as the Rutter body, is exposed over an area of about 5 000 by 500 m. It outcrops in a series of low glaciated ridges in rather low swampy terrain. The rock is coarse-grained, slightly gneissic to almost massive, and contains numerous small patch pegmatite segregations. It is composed approximately of 40% white albite, 30% salmon pink nepheline, 20% pink microcline, and less than 10% hornblende with minor zircon, magnetite, biotite, and aegirine.

The southern deposit, known as the French River body, is poorly exposed over an area of 4 000 m by 700 m. It is cut by the main channel of the French River near its south end. The rock is strongly foliated in a northwesterly direction, dipping moderately to the east. Its mineral composition is about 40% albite, 30% microcline, 20% nepheline, and 10% biotite with minor graphite, zircon, pyrite, magnetite, and hornblende. The feldspathic minerals are all white.

Between 1980 and 1984, Steep Rock Resources, Inc. carried out detailed studies of the deposits, including laboratory beneficiation testing of split diamond drill cores. Dry magnetic separation and flotation testing were attempted, but optimum results were achieved by the former. Table 5 shows typical analyses for both the raw nepheline syenite in the two bodies, and the beneficiated products. Product recoveries by dry magnetic separation were about 70% for the Rutter deposit, but only 55% for the French River deposit. An A-grade product (in terms of residual iron content) was only possible for the French River body.

British Columbia: An extensive alkaline igneous province occurs in the Canadian Cordillera of eastern British Columbia. It includes carbonatites, nepheline and sodalite syenite gneisses, and related alkaline rocks, all in a broad zone which parallels and includes the Rocky Mountain Trench. The major occurrences have been described by Pell (1987), and some of them are summarized in the following notes.

Ice River. The Ice River complex is an alkaline ultramafic intrusion located 23 km south of Field. It is an arcuate mass, 18 km in length, intruding Cambrian and Ordovician shales and carbonates (Fig. 3). Much of the complex lies in Kootenay and Yoho National Parks, and access is difficult due to steep mountain topography.

An early layered mafic rock series is intruded by a feldspathic series of syenites. The early series includes nepheline-rich ijolite and urtite, and is characterized by a repetitive sequence of graded layers, 10 to 200 m thick, in which the nepheline content increases progressively towards the top of each layer. Ijolites predominate in the layered sequence, and consist of (in decreasing order of abundance) clinopyroxene, nepheline, phlogopite, magnetite, and sphene. Urtites are more interesting because nepheline predominates, with lesser aegirine. They are coarse-grained, foliated, and may contain minor amounts of wollastonite, kaersutite, melanite, albite, sphene, calcite, apatite, and nepheline alteration products.

The syenite series is unlayered, but forms an elliptical pipe-like mass, zoned from a greenish sodalite syenite core through pale gray nepheline syenite to darker colored mafic-rich rocks at the margin. Microcline predominates in these rocks, with subordinate amounts of perthite and albite. Nepheline may comprise up to 30% of the rock, and sodalite up to 20%. Mafic minerals—pyroxene, kaersutite, and biotite—are usually present in minor amounts. Accessory minerals may include sphene, apatite, cancrinite, and various opaques. Typical compositions of some of these rocks are given in Table 6.

Trident Mountain. Trident Mountain is located in the Big Bend of the Columbia River, about 85 km northeast of Revelstoke. The area is very rugged, and the syenites occur as a concordant lenticular mass high on the mountain slopes. The best exposures are in cliff faces at elevations between 2 200 and 3 000 m, adjacent to large icefields (Fig. 4). Access is by helicopter from Revelstoke. The occurrence is described by Pell (1987) and MacLean and White (1991).

The syenite gneisses are medium-grained and foliated. They are composed of white to pink microcline (25 to 50%), and white albite (10 to 30%) and nepheline (10 to 40%). Nepheline generally has an irregular poikiloblastic texture, and it may be partly altered. Biotite varies from trace amounts to more than 30%, and accessory minerals may include sodalite, cancrinite, calcite, apatite, sphene, ilmenite, pyrochlore, and zircon. Local pegmatitic patches may be present.

The syenites are intruded into psammitic and kyanite-bearing pelitic schists. Compositional layering is typical of the syenites,

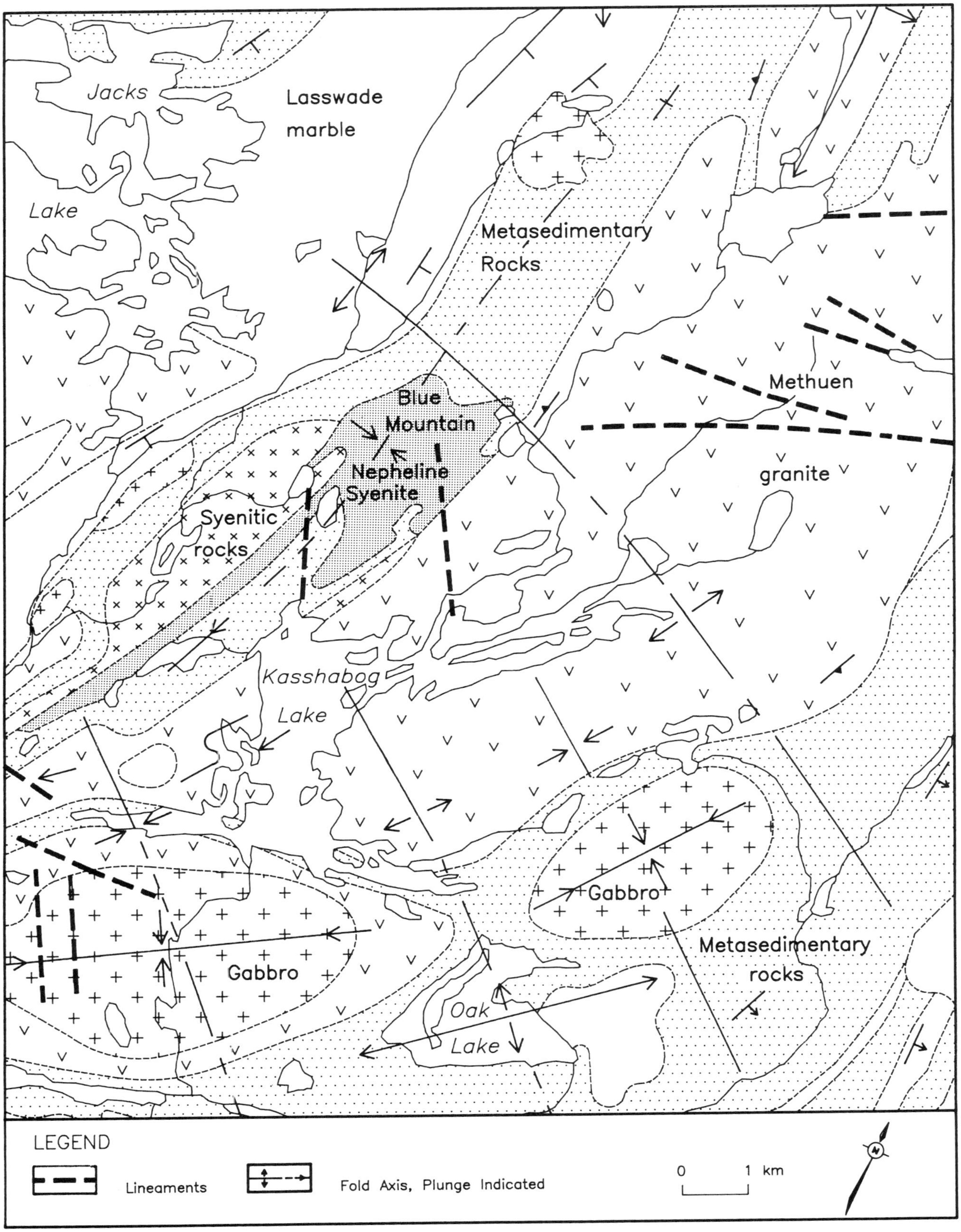

Fig. 1. Geological setting for the Blue Mountain nepheline syenite deposit (modified from Hewitt, 1961).

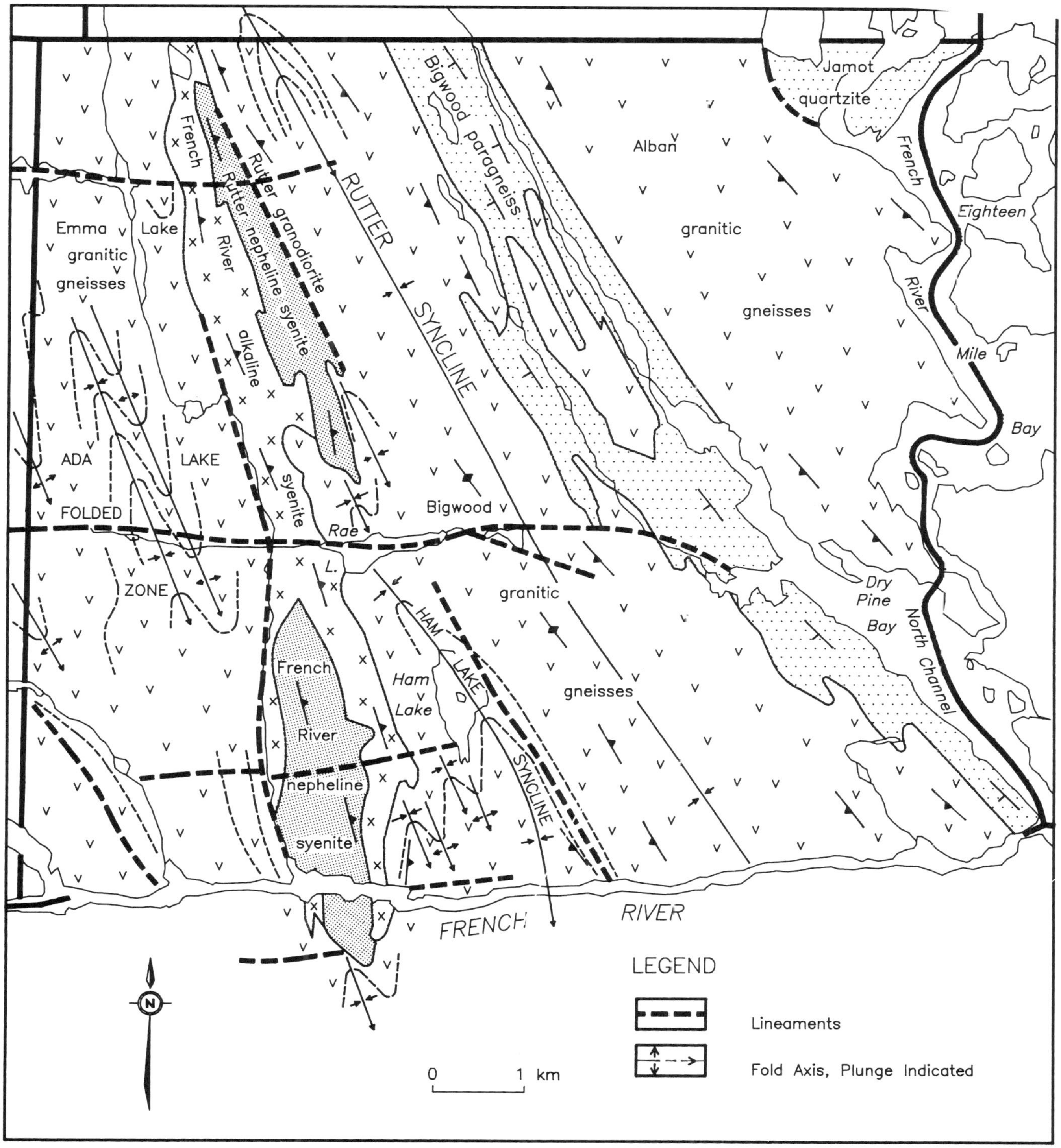

Fig. 2. Geological setting for the French River nepheline syenite deposits (modified from Hewitt, 1961).

defined by alternating biotite-rich and biotite-poor layers. Foliation is parallel to the margins of the folded syenite body and also to the axial plane of the antiform and the bedding of the countryrocks.

Two analyses of the raw nepheline syenite are reproduced in Table 7. Beneficiation tests by CANMET (Canadian Government testing laboratories in Ottawa) were successful in producing low iron products by dry magnetic separation, with good product recoveries (MacLean and White, 1991).

Mount Copeland. Nepheline syenite and syenite gneisses are exposed over a distance of 6 km in the Jordan River area on the southern flank of Mount Copeland, 15 km northwest of Revelstoke. The area is reasonably accessible via an old mining road. The

Table 5. Nepheline Syenite Deposits at French River, Ontario

	French River Deposit		Rutter Deposit	
	Raw %	Product %	Raw %	Product %
SiO_2	56.35	62.2	59.03	61.7
Al_2O_3	20.31	22.0	21.81	22.2
Fe_2O_3	2.66	0.08	2.34	0.18
FeO	4.09		1.34	
TiO_2	0.16	0.01	0.03	0.01
CaO	1.69	0.33	0.53	0.44
MgO	0.11	0.08	0.06	0.07
MnO	0.18	0.00	0.03	0.00
Na_2O	8.31	10.2	10.78	10.5
K_2O	4.86	4.45	4.36	4.42
P_2O_5	0.04	0.03	0.02	0.02
LOI	1.36	0.74	0.15	0.59
	100.12	100.12	100.48	100.13

Sources: Raw compositions, Hewitt (1961); product compositions, Steep Rock Resources, Inc.

occurrence is described by Currie (1976), Pell (1987), and Maclean and White (1991).

There are three main alkaline rock units, consisting of a basal nepheline syenite gneiss overlain successively by alkaline amphibolite and saturated syenites. These rocks intrude micaceous quartzites and calcsilicate gneisses. The nepheline syenite unit is itself roughly zoned into a weakly foliated core consisting of nepheline, potash feldspar, and albite, with minor aegirine and sphene. Accessory minerals include calcite, fluorite, garnet, and zircon, and traces of biotite, muscovite, and cancrinite.

The highest nepheline content is found in an augen gneiss unit which is faintly purple in color and contains an abundance of nepheline porphyroblasts up to 2 cm in diameter in a fine-grained groundmass. In addition to nepheline, this unit is composed of alkali feldspar, perthite, calcite, biotite, and fluorite. Accessory minerals include cancrinite, tourmaline, apatite, sphene, riebeckite, and aegirine.

A third type of nepheline gneiss is moderately foliated due to the parallel alignment of biotite which, however, rarely exceeds 5%. Microcline, perthite, and nepheline, commonly replaced by cancrinite and thompsonite, are the major minerals. Minor aegirine and hastingsite may be present; sphene, apatite, calcite, and specularite are accessories.

Typical analyses of the three nepheline rocks are given in Table 8. Test samples were beneficiated by CANMET using dry magnetic separation and a mica-iron flotation, but iron levels could not be reduced to glass-grade specifications.

Norway:

North Cape. The North Cape nepheline syenite deposit of Elkem Nefelin was opened in 1961. It is located on the island of Stjernoy in the Alta fjord in the Norwegian Arctic. The deposit is described by Geis (1975) as lenticular in plan, 1 800 by 300 m, outcroping between 250 and 700 m above sea level. It is enclosed by carbonatite and hornblendite on the northeast side, and by gabbro gneiss on the southwest. Typical composition is 56% perthite, and 34% nepheline, with biotite or hornblende and minor amounts of albite, magnetite, pyroxene, calcite, and sphene. The rock is well foliated and consists of two principal facies: a hornblende type and a biotite type. Biotite nepheline syenite tends to be concentrated at the two ends of the lens, while hornblende nepheline syenite occupies the central portion (Fig. 5). Typical chemical compositions of the two types are given in Table 9. A number of diabase dikes up to a few tens of centimetres thick cut the body. Mining is underground and largely within the hornblende nepheline syenite facies. The North Cape deposit was purchased by Unimin in 1992.

Republic of Russia:

Kola Peninsula. Russia has utilized its nepheline rocks longer than any other country, and for more diverse uses. The unique apatite-nepheline deposits of the Khibiny Complex on the Kola Peninsula in the extreme north of the Russian Republic were first developed in 1929. These deposits have continued to be not only a source of very high grade phosphate, but a major source of aluminum and cement derived from nepheline concentrates from the apatite tailings. The deposits are briefly described by Notholt (1983) and summarized in the following notes.

The Khibiny alkaline igneous complex is remarkable not only for its unusual composition but also for its size; it is about 40 km in diameter. The apatite-nepheline deposits are restricted to an arcuate lenticular zone of layered ijolite and urtite rocks about 100 m thick and 11 km long. The zone parallels the southwestern boundary, and it dips towards the center of the intrusion at a moderate angle. A number of mines have been developed near Kirovsk, and compositions of their deposits range from 15 to 75% apatite, 10 to 80% nepheline, up to 25% aegirine and aegirine-augite, 5 to 12% sphene, and minor feldspar and titanomagnetite. Typically the upper parts of the deposits are richer in apatite than the lower.

The Russian deposits are the richest source of igneous phosphate in the world, and the phosphate products are of superior grade, typically 86% BPL (bone phosphate of lime, equivalent to 39.4% P_2O_5). Igneous phosphate production, largely from this source, was 17.7 Mt in 1982. Byproduct nepheline concentrates are used in the making of aluminum metal, alumina, sodium and potassium carbonates, and portland cement. Production methods are described in a later section. Large reserves of nepheline rock at Goryachegorsk in central Siberia are also used for the recovery of alumina and the making of aluminum metal (Minnes, 1975).

Arkansas:

Fourche Mountain. The nepheline syenite deposits in Pulaski, Saline, and Hot Springs Counties in central Arkansas have been known for many years. Open pit mining of bauxite deposits derived from the nepheline rocks began about 1895, and only since World War II has it declined in favor of foreign sources. Since 1947 the 3M Company has produced roofing granules from the dense uniform gray nepheline syenite. Construction aggregates are also produced on a large scale from the Fourche Mountain deposit in Pulaski County.

The nepheline syenite of Fourche Mountain was described by Smothers et al. (1952), and both a blue syenite containing hornblende, and a gray biotite-bearing syenite, were recognized. Beneficiation testing was not successful in lowering the iron and titanium content to acceptable levels for glass and ceramic applications (Table 10).

Brazil:

Canaan Deposit. The Canaan deposit of nepheline syenite gneiss has been described by de Ferran (1983). It is located 40 km from Rio de Janeiro in the municipality of Duque de Caxias. In 1978 it was investigated by Austral Mineracao & Servicos Ltda., and 2.5 kt of glass-grade product was produced in a pilot plant in 1980. Initial startup for commercial production was expected to be 1982, targeting the glass and ceramic industries of the Rio-Sao Paulo area. An annual market of 134 kt for glassmaking and 36 kt for ceramics had been identified, primarily to replace hand-cobbed potash feldspar from pegmatites located 1 000 km north of Rio. Exports to overseas destinations were also anticipated, through the port of Rio de Janeiro. However, no further development has been reported.

Alkaline rocks are common near Rio de Janeiro, and a number of Cretaceous-Tertiary ring complexes are known. Nepheline rock from one of these in the Tingua area was unsuccessfully tested for

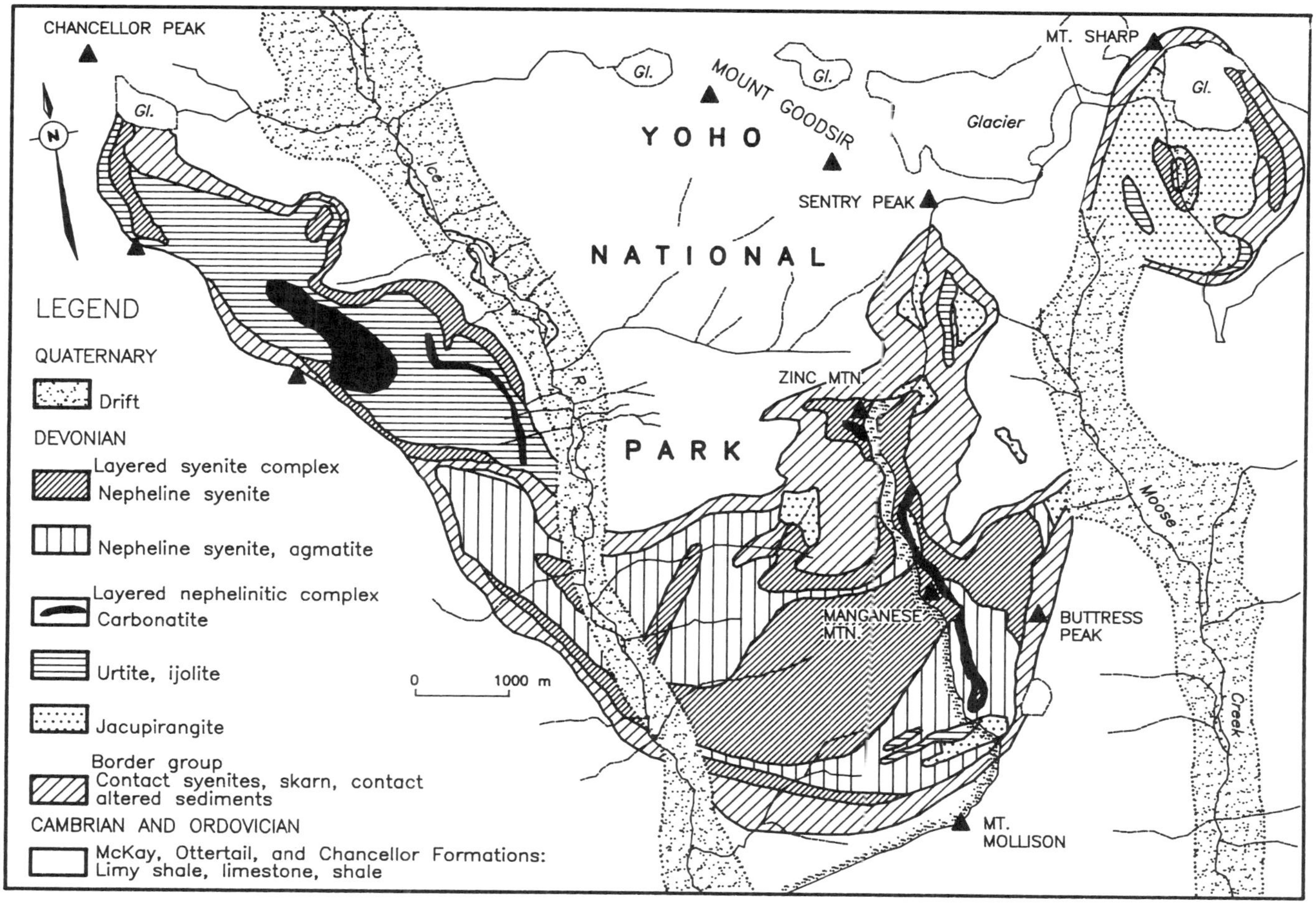

Fig. 3. Geology of the Ice River complex (from Currie, 1976).

Table 6. Nepheline Rocks of the Ice River Complex, British Columbia

	Ijolite %	Urtite %	Nepheline Syenite %	Sodalite Syenite %
SiO_2	43.7	43.5	53.9	54.4
Al_2O_3	25.8	26.2	23.1	21.9
Fe_2O_3	1.6	1.1	0.8	2.0
FeO	3.7	2.1	0.8	0.9
TiO_2	1.20	0.19	0.11	0.20
CaO	5.4	9.7	1.6	1.3
MgO	2.7	0.4	0.1	0.8
MnO	0.15	0.50	0.05	0.11
Na_2O	9.9	10.4	7.8	10.7
K_2O	4.5	4.9	8.8	6.3
CO_2	0.8	0.3	0.1	0.2
P_2O_5	0.09	0.05	0.00	0.01
	100.4	100.2	98.2	99.8

Source: Pell, 1987.

the removal of iron-bearing minerals; others occur in the Sao Goncalo, Itatiaia, and Rio Bonito areas. However, the Canaan deposit does not appear to be related to a ring complex, but rather is a stock of nepheline gneiss about 1 km in diameter. The deposit forms a series of hills at the foot of the Petropolis Range. Overburden is heavy and there are few outcrops.

The rock is classed as a litchfieldite, and consists approximately of 55% microcline and microcline perthite, 20% nepheline, 15% albite, and 10% biotite with minor pyrite, pyrrhotite, magnetite, and zircon. Raw and product compositions are given in Table 11. Beneficiation in the pilot plant was by dry magnetic separation.

China:

Sichuan Deposit. Testwork on a nepheline rock in Sichuan Province has been successful in yielding a low iron product by magnetic separation. The product is said to be characterized by high contents of alumina, soda, and potash, and to be suitable for glass-making. Fourteen other nepheline syenite bodies in southwestern China have been found to be too high in iron (Potter, 1989).

Turkey:

Bursa Orhaneli Deposit. In 1989, a trial production of 6 kt of nepheline syenite was made from the Bursa Orhaneli region of western Turkey. The material was apparently not beneficiated and was used in the making of floor tiles. However, there are said to be plans for a commercial plant that may include beneficiation (Burger, 1990).

Other Deposits: Numerous references can be found in the literature to alkaline rocks in many parts of the world, many of which are nepheline-bearing. In many cases there is little information on their possible economic potential, other than perhaps a chemical analysis. Table 12 reproduces some of these analyses. However, in some cases it is not clear whether the analysis represents a phonolite, a unit associated with an alkaline ring complex, or a nepheline syenite gneiss.

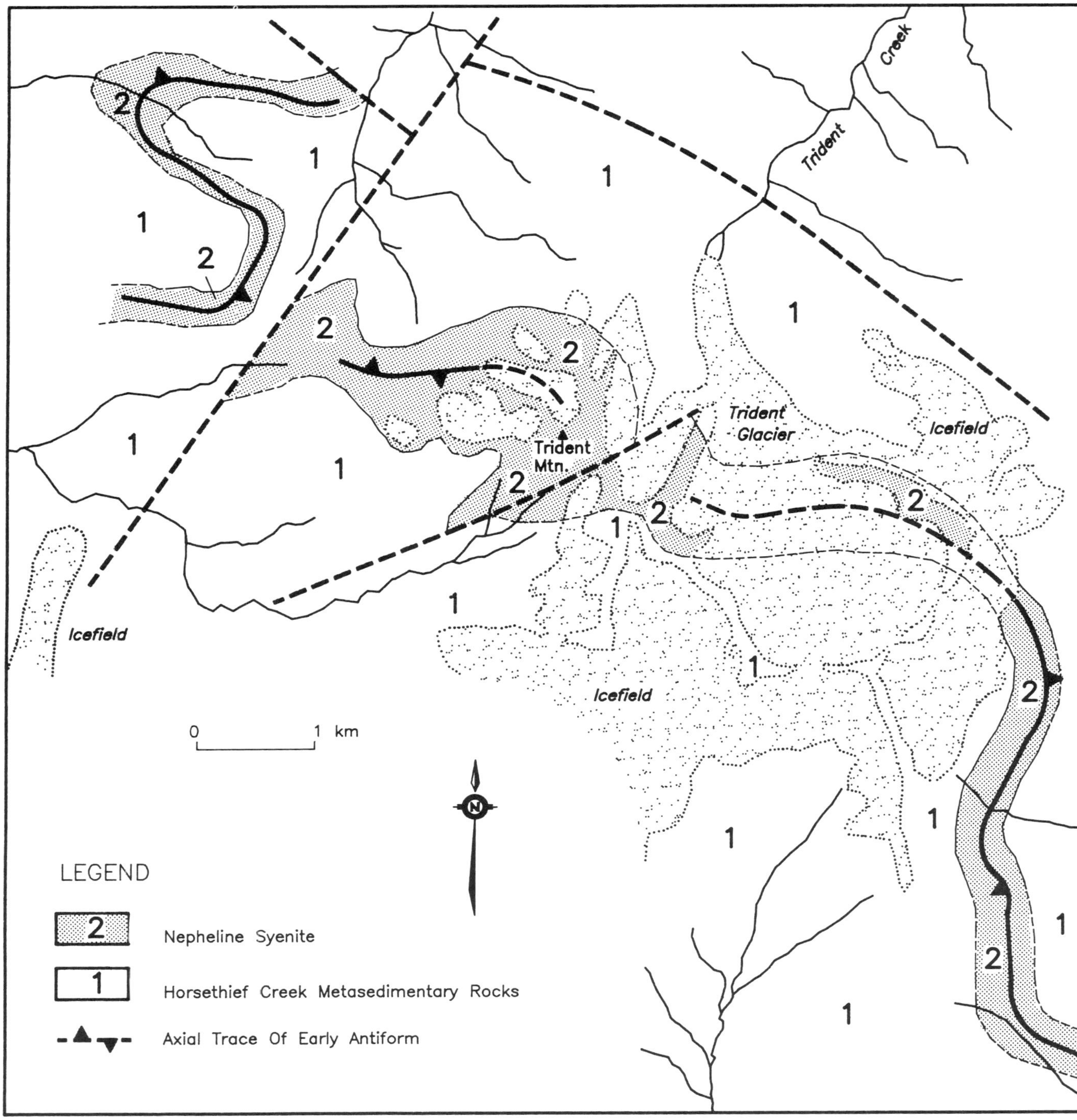

Fig. 4. Geology of the Trident Mountain nepheline syenite area (from Pell, 1987, Perkins, 1983).

USES AND SPECIFICATIONS

The beneficiated nepheline syenite products from Canada and Norway are primarily used for glassmaking, and to a lesser extent ceramics. A growing market for the finely-ground Canadian product is for industrial fillers. And because it contains no free silica as quartz, there is a steady demand for its use as sandblasting sand. In Russia, nepheline concentrates are used for the production of alumina, sodium and potassium carbonates, and portland cement in integrated industrial complexes. Near St. Petersburg, several of these production facilities are associated with large-scale recovery of apatite from nepheline rocks on the Kola Peninsula. Two larger complexes are based on other nepheline resources in central Siberia. In the United States, nepheline syenite has been used for years in Arkansas for the production of roofing granules and construction aggregates.

Table 7. Nepheline Syenite Gneiss, Trident Mountain, British Columbia

	Sample 3A %	Sample 3G %
SiO_2	56.66	55.59
Al_2O_3	24.36	24.69
Fe_2O_3	0.17	0.59
TiO_2	0.02	0.04
CaO	0.59	0.56
MgO	0.05	0.07
MnO	0.01	0.01
Na_2O	8.16	8.39
K_2O	8.22	7.98
P_2O_5	0.00	0.02
LOI	0.99	1.55
	99.23	99.49

Source: Pell, 1986.

Table 8. Nepheline Syenite Gneisses, Mount Copeland, British Columbia

	Nepheline Augen Syenite %	Nepheline Syenite %	Nepheline-biotite Gneiss %
SiO_2	48.78	54.92	54.11
Al_2O_3	25.22	23.01	21.69
Fe_2O_3	3.46	2.57	2.48
FeO	0.23	2.62	0.62
TiO_2	0.73	0.55	0.49
CaO	3.08	1.53	3.01
MgO	0.32	0.67	0.19
MnO	0.12	0.22	0.11
Na_2O	9.89	6.06	5.85
K_2O	7.21	7.47	9.37
H_2O	0.30	0.49	1.26
CO_2	0.12	0.23	0.24
P_2O_5	0.01	0.02	0.01
	99.47	100.36	99.43

Source: Pell, 1987.

Glass

The major use for nepheline syenite in Western Europe and North America is for glassmaking, where it is an inexpensive source of necessary alumina, soda, and potash for the glass batch. Alumina increases the chemical resistance and physical strength of the glass, and the alkali content reduces the requirement for more expensive soda ash. Nepheline syenite improves the workability of the glass by lowering the vicosity of the batch. Its low fusion point lowers the melting temperature, promoting faster melting, higher productivity, and fuel savings. Nepheline syenite typically comprises 4 to 8% by weight of a soda-lime container glass batch. About 70% of Canadian production and 80% of Norwegian production is used for glass, particularly for container glass, tableware glass, and fiberglass. More than 75% of Canadian production is exported to the United States; more than 90% of Norwegian production is exported to Western Europe, particularly Germany, the United Kingdom, France, and Belgium.

Nepheline syenite used for glassmaking is a sand-sized product falling within the range of 40 to 200-mesh. Its iron content must be very low, not exceeding 0.1% as Fe_2O_3, since iron contributes a green color to the glass. However, amber glass and fiberglass can tolerate somewhat higher levels of iron. Content of alumina and alkali should be as high as possible, typically at least 23 and 14% respectively. Refractory minerals must be absent since they may form unmelted stones in the glass, and other contaminants must be very low. Table 4 gives compositions of typical glass-grade products from Canada and Norway.

Ceramics

Finely-ground nepheline syenite is also used in the manufacture of whiteware ceramic products, particularly sanitaryware, dinnerware, floor, and wall tile, and various porcelain products for electrical, dental, and chemical uses. A low fusibility temperature and high fluxing capacity allows nepheline syenite to act as a vitrifying agent by contributing an early glassy phase that binds other constituents of the mix. This permits a lower flux content in the ceramic body, lower firing temperatures, and faster firing schedules. Its long firing range results in higher physical strength of the finished body. According to Taylor (1990) sanitaryware bodies can be commercially formulated to fire at cone 4, resulting in much lower fuel and refractory costs and faster firing time. Pressed vitreous floor tiles, where nepheline syenite comprises 50 to 55% with accessory flint and clay, enjoy lower firing temperatures and faster firing cycles. Vitreous china can be formulated solely from clays and nepheline syenite without an auxiliary flux, maturing at cones 3 to 6, and exhibiting high translucency. Ultrafine ceramic grades are used successfully in electrical porcelain, especially in alumina bodies, to provide faster and lower firing temperatures, increased firing range, increased strength, increased shrinkage, and decreased absorption over comparable bodies using potash feldspar. Nepheline syenite is also useful in formulating practically all ceramic glazes, particularly because of its fluxing ability in both fast-fire and conventional firing cycles.

About 15% of Canadian and Norwegian production is used in ceramic applications.

Nepheline syenite for the ceramics industry is finely-ground, typically into products of 200, 270, and 400 mesh. Such products produced at Blue Mountain, Ontario, are 95, 98, and 100% finer than 325-mesh, and they have pyrometric cone equivalents of 6, 5, and 3 respectively. Ceramic-grade nepheline syenite must be free of dark minerals and must fuse on firing to a uniformly white body without specking. Typically the nepheline syenite is mixed with ball clays, kaolin, flint, and talc, comprising as much as 60% of the body, depending on the ceramic product.

Extender Pigments and Fillers

Applications for nepheline syenite as extender pigments and fillers has been pioneered in Canada, and now represents about 15% of total sales. Finely-ground nepheline syenite is especially used as an inert filler in paints, both latex and alkalid systems, for use in high traffic areas, as metal primers, wood stains, sealers, and undercoats. Nepheline syenite contributes a high dry brightness, inertness, high bulking value, low vehicle demand, and easy wetting and dispersion in paint formulations.

In plastics, nepheline syenite is used as an inert, low cost mineral filler in polyvinyl chloride (PVC), epoxy, and polyester resin systems. Because it exhibits a low resin demand, high filler loadings are possible, permitting reduced requirements for more expensive components. In PVC resins it exhibits a low tinting strength, has a refractive index close to that of vinyl resin, and has a very low optical dispersion, all of which allows it to be used in nearly transparent stock.

Nepheline syenite is also used as an inert filler in the manufacture of foam carpet-backing. It has a specific gravity lower than either calcium carbonate or talc.

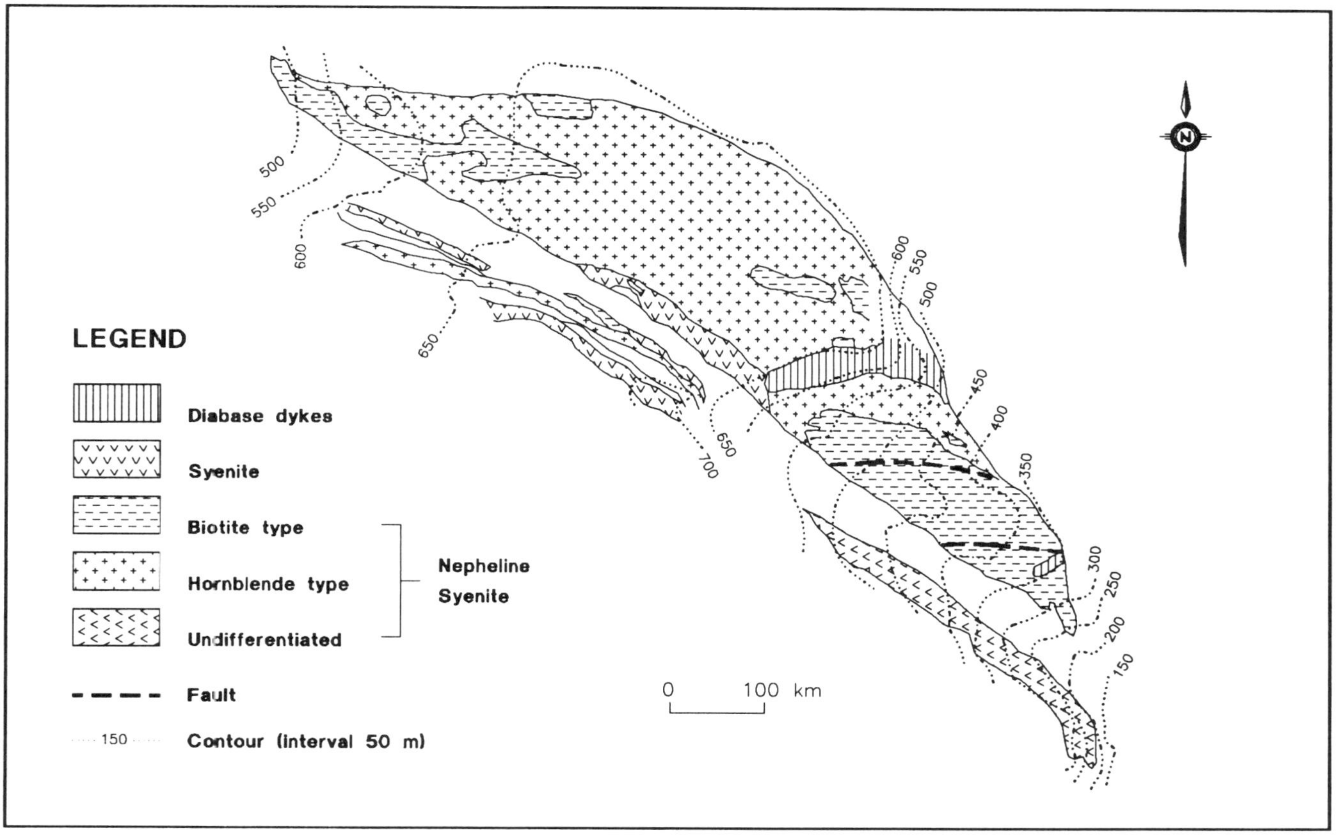

Fig. 5. Surface geology of the Stjernoy Island nepheline syenite deposit (modified from Geis, 1983).

Table 9. North Cape Nepheline Syenite, Norway

	Raw Composition		Product Quality	
	Hornblende type %	Biotite type %	Glass-grade %	Ceramic-grade %
SiO_2	52.73	52.37	57.0	57.0
Al_2O_3	23.71	23.22	23.8	23.8
Fe_2O_3	1.9	1.1	0.10	0.12
FeO	1.89	1.14		
TiO_2	0.51	0.61		
CaO	2.54	3.11	1.3	1.1
MgO	0.24	0.25		
MnO	0.06	0.09		
Na_2O	7.78	6.87	7.9	7.8
K_2O	8.08	8.30	9.0	9.1
P_2O_5	0.05	0.09		
CO_2	0.77	1.88		
H_2O	0.26	0.26		
BaO	0.40	0.47		
	100.92	99.76	99.1	98.9

Sources: Anon., 1987, for product compositions; Heier, 1966, for raw compositions.

For filler uses, nepheline syenite must be very finely ground and have a high dry brightness (96 to 98). Products produced at Blue Mountain, Ontario, have mean particle sizes as small as 4.5 µm, an average refractive index of 1.53, and oil absorptions from 22 to 29%.

Table 10. Nepheline Syenite, Fourche Mountain, Arkansas

	Raw Composition		Beneficiated Product	
	Blue Syenite, % Hornblende type	Gray Syenite, % Biotite type	Blue Syenite %	Gray Syenite %
SiO_2	60.3	60.8	61.2	61.9
Al_2O_3	19.9	19.6	21.5	21.1
Fe_2O_3	4.7	4.3	0.9	1.1
TiO_2	1.1	1.2	0.3	0.2
CaO	1.3	0.8		
MgO	1.2	0.9		
Na_2O	6.3	6.7		
K_2O	5.3	5.9		
LOI	0.1	1.3		

Source: Smothers et al., 1952.

Alumina, Alkali Carbonates, Portland Cement

Nepheline concentrates are used in the Republic of Russia for the extraction of alumina, most of which is used in the making of aluminum metal. The process also includes the recovery of sodium and potassium carbonates, and the dicalcium silicate residue is used with a nearly equal amount of limestone in the making of portland cement. Several such manufacturing complexes were originally established near St. Petersburg, using a nepheline-rich tailing from apatite mining on the Kola Peninsula. Newer facilities have since been established in Central Siberia, based on other large nepheline

Table 11. The Canaan Deposit, Brazil

	Raw Composition %	Beneficiated Product %
SiO_2	55.6	58.8
Al_2O_3	22.9	23.0
Fe_2O_3	3.5	0.10
CaO	0.8	0.6
Na_2O	9.1	10.3
K_2O	6.7	6.3
LOI	1.0	0.9
	99.6	100.0

Source: de Ferran, 1983.

resources. Plants at Volkhov, Pikalevo, and Achinsk are together believed to consume more than 3 Mt of nepheline rock annually.

While there has been much research along these lines in the United States (Archibald and Nicholson, 1948, Lundquist and Singleton, 1962, Johnson and Peters, 1968) and in Canada (Archibald, 1942), the process has not proved economic as long as bauxite resources are reasonably available. However, during World War II inventories were made of resources of nepheline syenite, anorthosites, and kaolin clays in North America as strategic alternatives to scarce domestic bauxite resources, but they were never utilized for this purpose.

TECHNOLOGY

Exploration Techniques

Nepheline syenite is no different than any other bulk commodity when it comes to evaluating its potential for development. Consistent product uniformity is mandatory, and because haulage costs are a major factor, accessibilty to markets must be carefully evaluated. Some of the more important aspects are discussed in the following notes.

Field Studies: The first requirement for a glass-grade product is that the nepheline rock be reasonably uniform, light-colored, medium to coarse-grained, with a clean, fresh, open texture with minimum mineral intergrowths. Needless to say the deposit should have sufficient size for long term utilization and preferably be massive or not more than moderately gneissic. Dikes and veins of nepheline syenite will normally be too small, and phonolites will be too fine-grained for effective beneficiation. Experience has shown too that rocks associated with alkaline ring complexes tend to suffer from mineral intergrowths and/or fine dark mineral inclusions, such that complete liberation is not possible during crushing, and hence effective beneficiation is difficult. Experience has also shown that white massive to gneissic deposits, in which biotite and magnetite are the principal dark minerals, together constituting not more than 5% of the rock, have the best chance for making low-iron products for glass and ceramic uses by a dry magnetic beneficiation process. In the Bancroft area of Ontario there has also been small production from six nepheline pegmatites during the period 1937–42; these were crudely beneficiated by hand-cobbing.

Needless to say, the higher the nepheline content the better, as higher nepheline contributes desirable alumina and alkalis to the glass-grade product. It also provides a richer raw material for the chemical industry (for the recovery of alumina, soda, and potash) or as an alternative to bauxite for the manufacture of aluminum. Any iron-bearing mineral is objectionable if it is not readily removed by beneficiation. Even trace amounts of iron will contribute a green color to the glass In fact most metal ions, particularly iron, titanium, chromium, manganese, copper, and cobalt are strong coloring agents, and they should be virtually absent from the glass-grade product. Refractory minerals such as corundum, zircon, and graphite are objectionable even in trace amounts as they may not melt in the glass batch, remaining as stones in the finished glass. Likewise, any dark mineral not removed by beneficiation may cause undesirable specking in filler applications or on ceramic wares. A high level of whiteness, or brightness, is particularly important for industrial fillers. In the production of roofing granules, however, a dense opaque rock of uniform gray color, free of sulfide minerals, and capable of yielding well shaped granules on crushing, free of slabby or splintery particles, is desirable.

Laboratory Testing: Since it would be most unusual to find a nepheline rock that could be used commercially for glass and ceramic purposes without beneficiation, it is important to test its amenability by one or more techniques. Initial sampling need only involve a few character samples taken from surface. avoiding heavy weathering by finding suitable ledge exposures, or if necessary, exposing fresh rock by drilling and lightly blasting.

Samples of perhaps 5 kg should be crushed through 30 mesh and tested first in a lab-scale electromagnetic separator, of which numerous makes are available: Carpco, Dings, Exolon, Jones, Rapid, and Stearnes. These are induced-roll, high intensity, electromagnetic machines, and samples are normally given several passes at increasing intensity levels. It should be possible to adjust the intensity levels and the splitter settings on these machines, based largely on visual evaluation of the nonmagnetic fractions. Dark minerals such as magnetite, biotite, hornblende, aegirine, garnet, pyrite, and tourmaline are normally readily removed. However, proof lies in careful analysis for total iron as Fe_2O_3, which ideally should be less than 0.10% for potential use in glass-making.

If the result by dry magnetic separation is close to being acceptable, some additional improvement may be possible by further experimentation with the dry process, or by using a wet process with special Carpco or Jones magnetic separators, or by flotation. Needless to say, a dry process is to be preferred from the point of view of cost. Nevertheless, flotation may be necessary if, for example, certain nonmagnetic refractory minerals are a problem. In general, flotation products will be somewhat improved over those produced by dry magnetic separation, but at the expense of higher losses and additional cost.

Optimum development of a beneficiation process will necessarily involve much experimentation and chemical analysis, perhaps checking different screen fractions separately in the event that impurities are concentrated in the finer fractions, for example, which might then be removed from the circuit by screening.

Following beneficiation it is important to check a number of samples for heavy minerals. This is accomplished using acetylene tetrabromide, which has a specific gravity of 2.96, as the separating medium. It effectively allows the feldspathic minerals to be skimmed off while heavy minerals such as zircon, corundum, chromite, ilmenite, and sphene are recovered in the sink fraction. Not more than a few grains of any of these minerals can be tolerated in the glass batch, and if any should approach several tenths of a percent, much further study will be necessary to try and isolate the source. If uniformly distributed throughout the deposit, flotation or other techniques may be necessary for their removal.

For ceramic applications, the relatively coarse glass-grade products are further ground to 200 or 325 mesh in pebble mills. Nepheline syenite provides important fluxing properties in ceramic mixtures, so it is important to check the fusion point (PCE, pyrometric cone equivalent) and the whiteness of the fused product. Typical ceramic grades from Blue Mountain have PCEs from 3 to 6, depending on the fineness of grind. They start sintering at cone 08 and have a sintering range of 13 cones.

Table 12. Chemical Compositions of Some Other Nepheline Deposits, %

	Kirumba[a] The Congo	Litchfield[b] Maine	Lulwe Hill[c] Malawi	Pleasant Valley[d] New Mexico	Brathagen[e] Norway	York River[f] Ontario	Algarve[g] Portugal	Krasnoyarsk[h] Russia	Aqaba[i] Saudi Arabia	Norra Karr[j] Sweden
SiO_2	55.44	60.39	55.77	56.24	55.50	45.42	55.22	54.57	50.3	56.37
Al_2O_3	23.59	22.51	22.26	21.43	22.45	25.60	22.59	22.56	25.8	24.95
Fe_2O_3	0.44	0.42	1.33	2.01	1.03	0.55	1.14	3.44	1.39	0.31
FeO	1.42	2.26	2.50	0.55	1.32	7.16	1.17		0.46	0.20
TiO_2	0.20		0.50	0.26	0.50	0.30	0.59	0.15	0.07	0.01
CaO	1.56	0.32	1.25	1.38	1.60	4.75	2.12	1.61	1.46	0.25
MgO	0.14	0.13	0.58	0.15	0.47	0.27	0.28	0.19	0.18	0.07
MnO	0.15	0.08	0.19	0.08		0.18	0.13		0.06	0.05
Na_2O	10.20	8.44	8.05	10.53	10.74	8.15	8.76	9.69	9.38	11.73
K_2O	6.26	4.77	6.66	5.74	5.48	4.04	5.59	5.51	4.48	5.20
P_2O_5	0.18		0.21	0.06		0.02		0.02	0.03	0.01
H_2O	1.07	0.57	0.43	0.86	0.9	1.53	2.16		5.93	0.1

Sources: [a] Nepheline syenite, Kirumba, Republic of The Congo (Allen and Charsley, 1968), [b]nepheline syenite, Litchfield, ME (Allen and Charsley, 1968), [c]biotite-hastingsite nepheline syenite, Lulwe Hill, Port Herald District, Malawi (Tilley, 1961), [d]aegirine phonolite, Pleasant Valley, Colfax Co., New Mexico (Clarke, 1910), [e]nepheline syenite, Brathagen, Lougendal, Norway (Brogger, 1890), [f]biotite nepheline plagioclase gneiss, Egan Chute, York River, Ontario (Tilley, 1958), [g]Algarve Province, Portugal (Allen and Charsley, 1968), [h]leucocratic nepheline syenite, Central Tartar Massif, Krasnoyarsk Territory, Siberia, Republic of Russia (Allen and Charsley, 1968), [i]leucocratic nepheline syenite, Jabal Sawda, Aqaba, Saudi Arabia (Liddicoat et al., 1985), [j]nepheline syenite (grennaite), Norra Karr, Sweden (Tilley, 1953).

To test its fusion characteristics, a specially shaped cone is made solely from the ceramic-grade product. It is fired at a controlled rate along with several standard cones having PCEs of 6, 7, and 8, and changes in the height and diameter of the sample are compared with the standards. Another test of fluxing character is given by the "inclined plane viscosity test," wherein a given weight of sample is packed into a specially designed wedge and fused under controlled conditions, noting the relative flowage characteristics of the sample against several known standards.

In addition to the fusion tests, the +200 mesh residue from a wet sieve analysis may be fused to check the fired color and the presence of any speck-producing impurities, the assumption being that these will be most obvious in the coarser fractions. Obviously, it is important that the fused color be very white.

Grain size distribution, and its uniformity from shipment to shipment, is a particularly important control feature once into production. While screen-testing of coarse products is fairly straight forward, the evaluation of subsieve particles requires the use of the Andreasen or Bahco methods. These are summarized by Minnes et al. (1983).

Geological Mapping: A reconnaissance level of geological examination will typically be a part of the surface sampling program. But once a sufficient number of surface samples have been tested to indicate a deposit amenable to beneficiation—one which is of sufficient size and of reasonable uniformity—a program of outcrop mapping should be undertaken. There is no substitute for the detailed mapping of surface outcrops; the information obtained on mineral variations, mineral and rock contaminants, and structural features can be obtained in no other way. Mapping should be on a scale of 1:2 000, and this can be based on a conventional grid pattern of marker stakes, or on suitably enlarged aerial photographs. The latter is perhaps the easiest, particularly if there is not too much tree cover. Outcrops should be visible on the enlarged photographs, and mapping of various geological features can be marked directly on them or on transparent overlays.

Drilling: Mapping and surface sampling is logically followed by diamond drilling. Small diameter core drilling is adequate. Holes should be collared initially at 100 m intervals, with later fill-in drilling at 50 m if necessary. Where possible, the holes should be angled to intersect the foliation, gneissosity, or layering of the rock at right angles.

The core must be split longitudinally to provide samples for laboratory testing, retaining one half for permanent record. Sample intervals should be based on stratigraphy, the aim being to isolate units or zones of similar appearance and lithology. Unusually wide zones of visually uniform rock should be sampled in units not exceeding 10 m. After carefully prepared written logs have been completed, and the cut-off points for individual samples have been decided, the entire split half of each core interval is separately bagged and labeled for delivery to the testing laboratory.

Laboratory testing of each sample will duplicate the optimum crushing, grinding, and beneficiation sequence that will have been determined by the earlier testwork. Chemical analyses for all major oxides should ideally be performed on each beneficiated product. At the very least, total iron content as Fe_2O_3 must be determined for each beneficiated sample, with complete analyses on appropriate composites. Of major importance are the contents of alumina, soda, and potash, as these have a significant bearing on the value of the commercial nepheline syenite product.

MAJOR PRODUCERS

Canada

Unimin Canada Limited: An extensive deposit of nepheline syenite near Peterborough, Ontario, has been commercially operated since 1935. The deposit, known locally as Blue Mountain, was originally opened by Canadian Nepheline Limited, with a quarry near the south end of the deposit and a 50 tpd mill at railhead in the village of Lakefield, 20 km to the southwest. Canadian Nepheline Ltd. pioneered the use of nepheline syenite as an alternative to feldspar for the glass and ceramics markets.

In 1937 a subsidiary, American Nepheline Corp., was formed to process and distribute the nepheline syenite product from a mill in Rochester, NY. The crude rock was barged across Stoney Lake to Lakefield, then railed to Cobourg on the north shore of Lake Ontario and transported across the lake by car ferry to Rochester. The townsite of Nephton was subsequently created at the quarry, and in 1946, with a name change to American Nepheline Limited, the primary milling facility was transferred to Nephton and fine-grinding facilities were established in Lakefield.

In 1951 all of the milling was consolidated at Nephton, and in 1954 the Canadian Pacific Railway completed a spur line to Nephton from Havelock, 20 km to the south. In 1956 a new mill was built

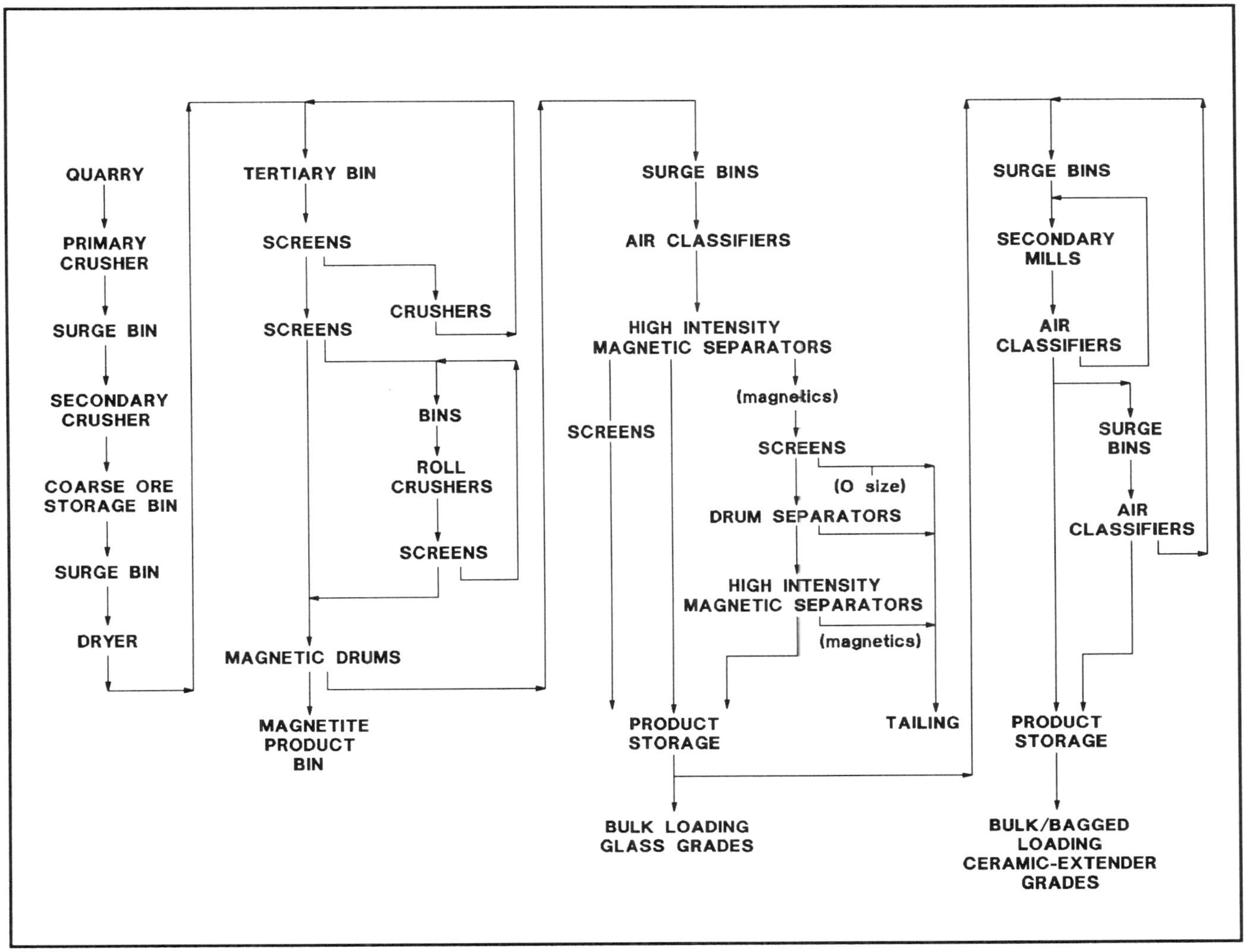

Fig. 6. Production flowsheets for the two milling plants at Blue Mountain, Ontario (from Minnes et al., 1983).

at Nephton, and with several subsequent stages of expansion, that mill now produces 550 ktpy of finished product. In 1961 the company's name was changed to Indusmin Ltd., and the company subsequently became the Indusmin Division of Falconbridge Ltd.

In 1956 American Nepheline Ltd. was joined at the north end of the Blue Mountain deposit by a second producer, Canadian Flint and Spar, a division of International Minerals and Chemical Corporation (Canada) Limited. Indusmin purchased the IMC operation in June 1985, and the combined operation was subsequently purchased by Unimin Corporation in September 1990. Unimin Canada Ltd. is a wholly owned subsidiary of Unimin Corporation of New Canaan, CT.

Mining. Although a number of quarries have been opened in the deposit over the years, Unimin presently operates only one quarry, located near the north end of the deposit. There is little overburden on the Blue Mountain ridge because of the scouring action of the glacial ice sheets. Gradall and other mine equipment is used for cleaning the rock surface prior to quarrying.

The quarrying operation uses a crawler-mounted Tamrock hydraulic drill on a 3 by 3 m pattern. The holes are about 11 cm in diameter, and bench heights are as much as 25 m. Ammonium nitrate and emulsion are used as blasting agents, detonated by a non-electric sequential blasting technique. Up to 20 kt of rock are broken at a time. A drop ball provides secondary breaking where necessary. Broken rock is loaded by front end loaders into off-highway straight-frame trucks of 32 t capacity and 60 t tractor trailer units. Drill cuttings are analysed prior to blasting so that blending can be effected where necessary.

Milling and Processing. Unimin operates both the Indusmin and IMC milling plants, which are closely similar. A typical flowchart is illustrated in Fig. 6, and described in the following notes.

Broken rock from the quarry is discharged directly into an Allis-Chalmers jaw crusher with an opening of about 1 by 1.2 m. Discharge from the primary crusher is −18 cm. Secondary crushing is accomplished in a 1.4 m Symons standard cone discharging at −5 cm. After secondary crushing, the rock is held in a crude rock storage building of 4 kt capacity. A constantly travelling overhead discharge point provides a degree of blending in the storage building, in the event of any minor quality fluctuations in the quarried rock. The primary and secondary part of the plant is operated on a single shift per day, while the rest of the milling process is continuous.

Crushed rock is recovered from storage from sublevel draw points and is dried in 1.8 by 15 m parallel flow oil-fired rotary dryers, then conveyed to a 270 t bin. From this bin it is passed over a series of double deck mechanical screens in closed circuit with two 1 m Symons short head cone crushers or one impact crusher. Material finer than 0.5 cm is fed to six 1.2 by 3 m Tyler Hummer double deck screens, and the fraction coarser than 30 mesh is further ground in vertical shaft impact crushers.

Material finer than 30 mesh is passed over low intensity magnetic drums and other screens before being directed to banks of high intensity electromagnetic separators of the Lurgi and Exolon types. The white sand-sized product of these separators is sold for glassmaking, having a residual iron content not greater than 0.08% as Fe_2O_3. Both plants have four lines comprising feed bins, autogenous crushers, screens, magnetic drums, high intensity electromagnetic separators, and scalping screens to achieve the necessary capacity. Rejects from the magnetic circuits are mixed with water and pumped to a tailings pond. Magnetite collected from the magnetic drums is sold as iron ore. Overall recovery from the quarry, milling and beneficiation processes is better than 75%.

For ceramic and filler markets, the glass-grade products are further ground in six pebble mills operating in closed circuit with air classifiers. Ultrafine grades are produced in pebble mills coupled with air classifiers.

Bulk storage for the final products include four concrete silos of 700 t capacity, four of 500 t, and three of 135 t. Warehousing is also available for several hundred tons of bagged products. Shipments are made by truck and rail, mostly in bulk except for very fine ceramic and filler products which are bagged. While the major market is the United States, more than twenty countries have been served. About 70% of the production goes for glassmaking, and 15% each for ceramics and fillers.

Particular attention is paid to dust collection throughout the entire mill, and bag collectors are attached to the exhaust stacks.

Quality Control. Quality and uniformity of the products is maintained by a rigid system of control that starts with original core testing from the diamond drilling program and continues with the testing of rock cuttings from blasthole drilling in the quarry. Chip samples from quarry faces are sometimes separately checked, and there are also numerous check points in the mill and on the product loading conveyors. Routine analyses include determinations for alumina, soda, potash, lime, magnesia, and iron; also heavy mineral evaluations, screen analyses, and subsieve particle-size analyses; also cone fusion and dry brightness determinations as necessary.

Norway

Elkem Nefelin S/A: In 1961, commercial production was commenced from an underground mine on the Island of Stjernoy in the Alta Fjord, Norwegian Arctic, near the northern tip of Norway. Rugged topography and a heavy annual snowfall contributed towards the decision to mine underground. Nabberen Mountain is steep, rising to more than 700 m above sea level, and all surface facilities are protected by barriers designed to deflect avalanches. Fig. 7 is a schematic diagram of the mine workings. The mining and milling process has been described by Minnes et al. (1983) and Wyllie (1973).

Mining. Access to the mine workings is gained by a portal 100 m above the sea and is reached by a switchback road cut into the side of the mountain. From the portal a paved haulage road leads 500 m underground along the 100 m level to the primary crusher, and from here a spiral ramp ascends to the drawpoint level 330 m above. A further system of ramps and inclines gives access to the stope blocks. Only rock above the portal level is mined, using a room and pillar method.

The stopes are oriented across the strike of the body. A crosscut is first driven from the footwall ramp to the hanging wall. A typical crosscut will be 120 m long, and this will open two stopes that will each be 50 m in length separated by a strike pillar 25 m long and 12 m wide. The stope sections along the crosscut are mined out 5 m high and 18 m wide, and a slot is created by longhole drilling across the full width of the new stope down to the open stope below. In 1973, the company was experimenting with stopes 45 m high, by drilling 30 m down and 15 m up from each level. Development headings are drilled by a Gardner-Denver three boom jumbo drilling holes 5 cm in diameter. For stoping, a twin boom Gardner-Denver jumbo is used to drill holes 6 cm in diameter. Each stope yields about 470 kt at a daily rate of about 1.0 kt.

A Caterpillar 966 front end loader is used to load two Kiruna trucks of 25 and 33 t capacity. This equipment is fitted with filters to reduce gas emissions within the mine. Broken rock is delivered by truck to an ore chute on the 200 m level above the primary crusher, a Blake No. 9 jaw crusher with an opening of 70 by 100 cm. A storage silo beneath the jaw collects the −10 cm discharge, from whence it is trucked 1 460 m along the 100 m level to the secondary crusher. The secondary is a 1.2 m Symons standard cone that reduces the rock to minus 3 cm and discharges it into either a 2 500 t underground storage area or a 250 t steel bin located on the mountain slope above the milling plant. All crushing and handling to this point is underground. A percentage breakdown of mining costs follows:

	Percent
Explosives	16
Transport	23
Loading	13
Drilling	20
Other (crushing, ventilation, administration and capital)	28

Milling and Processing. The mill building is located near the base of Nabbaren Mountain. Crushed rock is retrieved from storage by conveyor and is screened on a 2 by 6 m single deck screen in closed circuit with a 1.7 m Symons shorthead cone crusher. The −1 cm discharge is dried in an oil-fired rotary Newell Dunford cascade dryer 2.3 m in diameter and 14 m long, and is then distributed to one of five 180 t steel bins. These bins provide feed for further grinding and screening circuits that comprise four 1.2 by 3 m double deck Hummer screens, sixteen double deck Rotex screens, eighteen rolls crushers, two magnetic drums and one 60 cm by 6 m screen. At this point the material has been ground to 90% −30-mesh and has been reduced in iron content from run-of-mine 2.75% Fe_2O_3 to 1.5%.

The rest of the beneficiation process includes a series of high intensity magnetic separators: three single roll Lurgi machines followed by double roll Lurgi separators and fourteen three roll Exolon separators. These reduce the iron oxide level from about 1.5% to 0.1% Fe_2O_3. Rejects from the beneficiating circuit are reprocessed to make an amber glass-grade product containing 0.4% Fe_2O_3.

Beneficiated material coarser than 30 mesh is removed on twelve Sweco screens 1.2 m in diameter, in closed circuit with a Unidan pebble mill 2 by 12 m in size. Material finer than 30 mesh passes through a Wedag air classifier 2.4 m in diameter, from which the oversize is directed to glass-grade bins, and the undersize is further ground to a −325 mesh ceramic-grade in two flint-lined pebble mills.

Quality control is accomplished by complete chemical analyses of blasthole drill cuttings, chip samples across stope faces, and diamond drill cores. Mill samples are taken every second hour around the clock, analyzing them for Fe_2O_3 and grain-size distribution.

Glass-grade material is stored in bulk silos at the company's

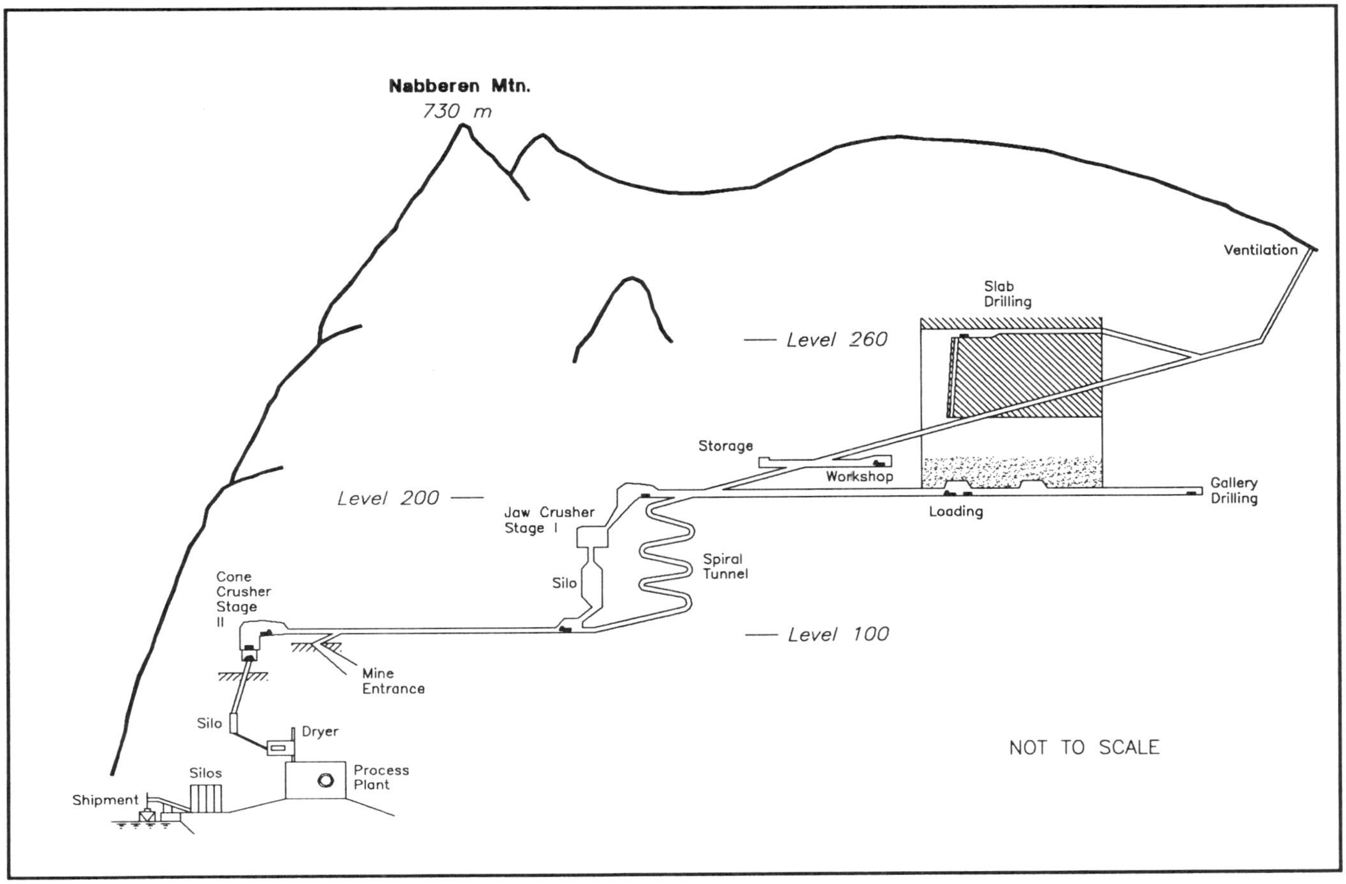

Fig. 7. Schematic sketch of the mine workings, Stjernoy Island, Norway (from Minnes et al., 1983).

own ice-free harbour. About 80% of the company's output is for the European glass industry, two-thirds for container glass and one-third for tableware. The ceramic-grade is shipped either in bulk or in bags and is largely consumed in sanitaryware (Anon., 1987). All shipments are necessarily made by water for markets in seventeen countries, notably Germany, the United Kingdom, France, and Belgium. The company maintains storage silos in some European ports such as Birkenhead near Liverpool, England, Terneuzen in the Netherlands, and Duisburg in Germany. Stockpiles are also maintained in France and Spain. Distribution of both ceramic and glass grades from stockpiles in Europe is normally by truck or covered hopper.

Originally designed for an annual capacity of 45 kt in 1961, the plant has been progressively expanded to its present production level of about 300 kt. Of the feed to the mill, 55% is recovered as glass-grade product, 10% as ceramic-grade, and 35% goes to waste. Alta Fjord is ice-free all year round due to the moderating effect of the Gulf Stream. There are no permanent residents on Stjernoy Island, and the workers commute by motor launch from the nearby community of Alta.

Republic of Russia

The process for the conversion of nepheline concentrates to alumina, alkali carbonates, and portland cement was pioneered in Russia and was based on the large apatite-bearing nepheline syenite deposits of the Kola Peninsula. Minnes et al. (1983) has given a thorough review of the use and treatment of nepheline in the Russian Republic, based on the works of a number of authors who are referenced. The following notes are a summary.

In 1938 a cement plant using a nepheline slurry tailing from the recovery of apatite from urtite and ijolite rocks on the Kola Peninsula was constructed at Volkhov, near St. Petersburg, and in 1941 a pilot plant to produce alumina using the same slurry was established. Commercial production of alumina from nepheline began in 1951, and in 1959 a second plant was established at Pikalevo, not far from Volkhov. A third and larger alumina facility was operational in 1970 at Achinsk in central Siberia, using nepheline concentrates from the large Goryachegorsk deposits 150 km to the southwest. A fourth and even larger facility was planned near the Sayano-Shushenskaya Dam (Minnes et al, 1983). In each case these plants are associated with portland cement facilities which use the dicalcium silicate residue from the alumina and alkali carbonate extraction process, and mix it with limestone before burning it in rotary cement kilns.

Extraction of apatite from the rich ijolite and urtite deposits of the alkaline igneous Khibiny Complex on the Kola Peninsula began in 1929. While these deposits contain as much as 75% apatite, they also yield a tailing rich in nepheline. Typical composition of the tailing ranges from 65 to 75% nepheline, 6 to 10% apatite, and 1 to 3% sphene. This is further concentrated by flotation to yield a product containing 90 to 93% nepheline, as well as a secondary apatite product. The nepheline product grades 29.3% alumina. The concentrate is shipped by rail to plants at Volkhov and Pikalevo where it is mixed with high-calcium limestone and is wet ground to 95% −170-mesh.

Alumina Extraction. The nepheline-limestone pulp is fed through a coal-fired sinter kiln where intake temperatures are as much as 1 600°C. Reaction between nepheline and limestone takes place at 1 300°C, and the product is a sinter cake of essentially beta dicalcium silicate and sodium and potassium aluminate. The chemical reaction is:

$$4CaCO_3 + (Na,K)_2O \cdot Al_2O_3 \cdot 2SiO_2$$
$$= (Na,K)_2O \cdot Al_2O_3 + 2Ca_2SiO_4 + 4CO_2$$

The cake is cooled, crushed in a cone crusher and wet ground in a ball mill to 97% −10 mesh, 30% −200 mesh. The ground material is then leached in a tube digestor using a solution of caustic soda and mother liquor derived from other circuits in the plant, and it is subjected to a special filter-condenser to remove the alkali aluminates. The dicalcium silicate residue is washed and thickened for use in making portland cement.

After a desilication process in pressurized autoclaves, where minor contaminating silica is removed by filtration, the aluminate solutions are transferred to carbonization towers where the alkalis remain in solution as soluble carbonates and the alumina is precipitated as aluminum hydroxide.

$$2(Na_2K)AlO_2 + CO_2 + 3H_2O = 2Al(OH)_3 + (Na,K)_2CO_3$$

The hydroxide is calcined to release alumina, and the carbonates are individually separated by selective crystallization.

Portland Cement Production. Dicalcium silicate from the alumina plant can be used as a major constituent in the making of portland cement. Known as bellite, a typical composition at the Volkhov plant in 1961 was:

	%
SiO_2	29.85
Al_2O_3	3.57
Fe_2O_3	2.7
CaO	56.58
MgO	1.27
Na_2O	2.21
SO_3	0.34
LOI	3.61
	100.13

Bellite comprises about 40% of the batch. It is mixed in slurry form with additional limestone, low-grade bauxite, and pyrite clinker. The ground mixture is burned in a coal-fired rotary kiln at about 1 600°C, and the clinker is mixed with about 15% dry bellite and 5% gypsum, and further ground to the required fineness.

Construction of a portland cement plant at Achinsk, in central Siberia, is described in a paper by Comte (1959). Planned to start in 1960, the plant was believed to have the largest cement kilns in the world at that time. It was to have two identical rotary kilns 175 m long, with inside diameters of 5.8, 4.9 and 5.3 m corresponding to the drying zone, the decarbonation zone, and the burning zone respectively. Feed to the kilns was to be a slurry comprising 48% bellite and 52% limestone. Hard North Siberian lignite was to be the fuel, with consumption estimated at 25 t per hour. Designed capacity for the plant was 12.5 million bbl per year.

Economics. Russia is unique in the extensive use it makes of its nepheline rocks, stimulated, of course, by the unusual and very large deposits on the Kola Peninsula. Starting with the recovery of phosphate from the rich apatite-nepheline rocks of the Khibiny alkaline igneous complex, then extracting alumina, sodium and potassium carbonates from the nepheline tailing, and finally making portland cement from the byproduct dicalcium silicate, sounds like any miner's dream: there is very little waste. Table 13 shows the approximate tonnages of coproducts obtained during the production of 1 t of alumina and the raw materials required for their production. Fig. 8 is a flowsheet for the process.

Table 13. Materials List for the making of Alumina and Coproducts

	Tons
Raw Materials:	
Nepheline concentrate	3.9 to 4.3
Limestone: for alumina manufacture	6.0 to 7.8
for portland cement	6.0 to 6.0
Coal (min. 7 000 kcal/kg):	
for alumina manufacture	1.67 to 1.70
for cement	1.3 to 1.6
Steam heat	4.68 to 4.12 g-cal
Electric power: for alumina	1 050 to 1 190 kW-hr
for cement	700 to 860 kW-hr
Products:	
Alumina	1.0
Sodium carbonate	0.62 to 0.76
Potassium carbonate	0.18 to 0.28
Portland cement	9.0 to 11.0

Source: Minnes et al., 1983.

It is believed that operating costs for an integrated plant producing these products saves perhaps 10 to 15% over their individual production by conventional methods. However, the capital cost is somewhat higher.

MARKETING

Production

After more than half a century of continuous production in Canada, nepheline syenite is well established as an alternative to feldspar in the making of glass and ceramics. Production from the large Blue Mountain deposit in southeastern Ontario began in 1935. Norway became the second producer of similar low iron products for glass and ceramics in 1961. Annual capacities of the Canadian and Norwegian plants are estimated at 800 and 300 kt respectively. The record of production from these two producers is shown in Table 14.

The first producer of nepheline, and still much the largest, was the Republic of Russia. Commencing about 1930, nepheline-rich tailings from apatite mining on the Kola Peninsula were found to be useful, with additions of limestone, for making portland cement. Subsequently, alumina and alkali carbonates were extracted from a nepheline concentrate derived from the tailing, and a residual dicalcium silicate became the feed to the cement plant. Recovery of alumina, and the production of metallic aluminum, has now become the dominant use for nepheline, and several large plants are based on nepheline syenite deposits in central Siberia. In all, the four known alumina-alkali-portland cement complexes are believed to consume in excess of 3 Mt of nepheline syenite annually. However, relatively minor amounts are used in the production of colored glass and ceramics, because of an apparent lack of material suitable for making low iron products.

Consumption

Glassmaking is the major use for nepheline syenite both in North America and Europe, accounting for about 80% of Norwe-

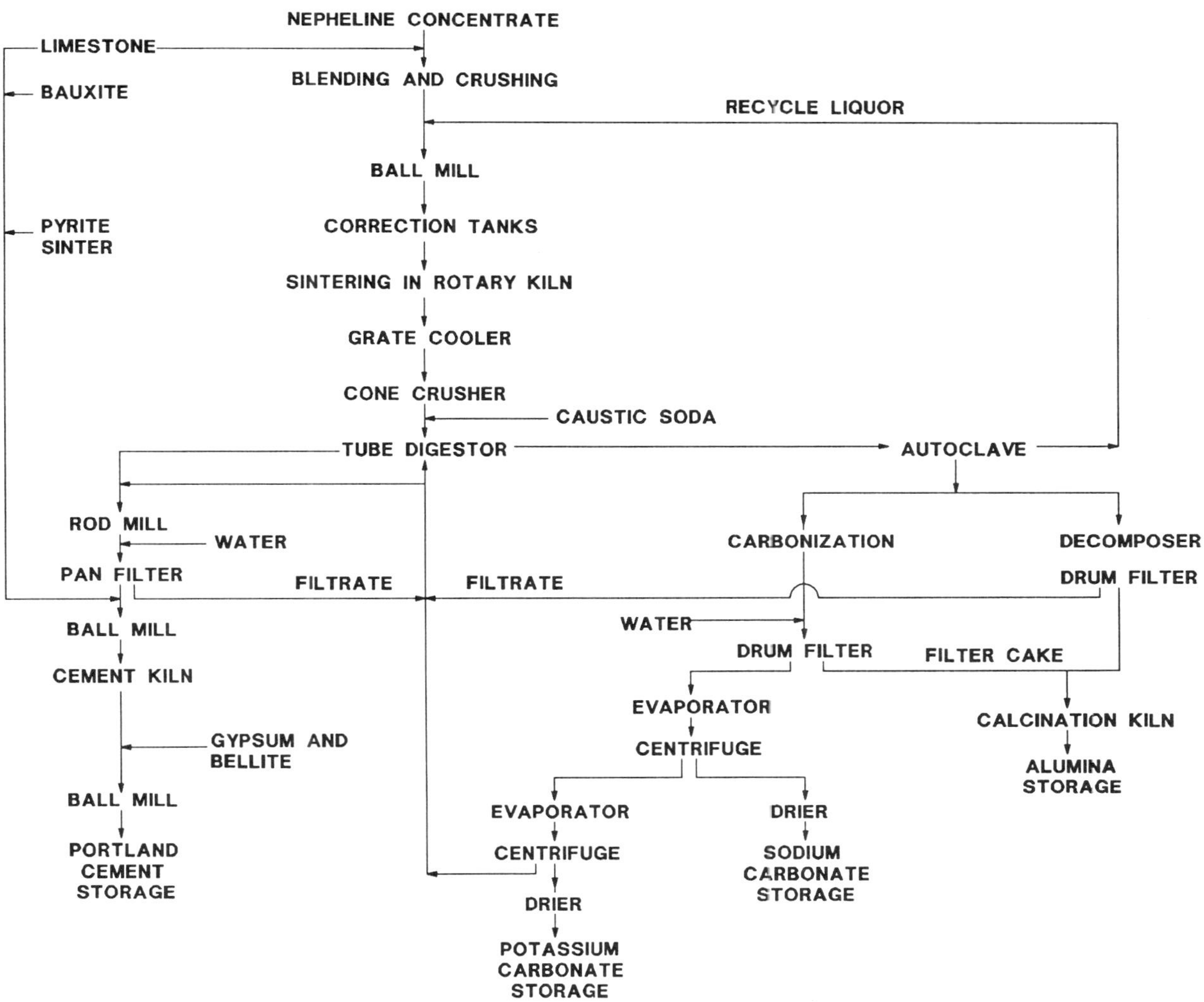

Fig. 8. Generalized flowsheet for the production of alumina and coproducts from nepheline (from Minnes et al., 1983).

Table 14. Nepheline Syenite Production in Canada and Norway (tons)

	Canada	Norway
1980	599 699	217 000
1981	587 565	223 000
1982	552 838	224 000
1983	523 249	219 565
1984	520 640	225 731
1985	467 186	227 465
1986	467 491	218 421
1987	506 415	240 000
1988	539 835	287 000
1989	555 728	
1990		306 000
1991	486 000	
1992	566 000	340 000

Sources: Anon., 1987, 1988, 1993, Boucher, 1990, Burger, 1990, Olerud, 1993.

Table 15. Canadian Consumption of Nepheline Syenite by Use (tons)

	1986	1987	1988
Primary glass and containers	58 265	62 731	49 913
Ceramic products	13 750	14 592	19 225
Paints	6 095	6 876	6 968
Fiberglass, plastics, sandblasting sand, structural clay products, and minor uses.	16 294	15 452	14 902
	94 404	99 651	91 008

Source: Statistics Canada.

gian production and 70% of Canadian. Canadian consumption of nepheline syenite in ceramics is about 15%, while pigment and filler applications, and others, also total about 15%. Table 15 gives a breakdown of Canadian consumption for the years 1986 to 1988 inclusive.

Because glass containers typically contain from 1 to 3% alumina, which gives the glass a degree of toughness, feldspathic materials have become the preferred source for this oxide. Hence, container glass is the largest single application for nepheline syenite, followed by other types of glass and fiberglass. A typical batch formulation for a soda-lime container glass is given in Table 16, where nepheline syenite comprises 5.9% of the batch, and the resulting glass contains 1.5% alumina.

Nepheline syenite is also used with clay, flint and other materials in the making of a variety of ceramic products, but partic-

Table 16. Typical Soda-lime Container Glass Batch Formulation

	Weight %
Silica sand	62.7
Nepheline syenite	5.9
Soda ash	21.5
Burnt lime	8.8
Barite	0.6
Gypsum	0.5
Cullet and decolorizers	as required

Source: Taylor, 1990.

ularly sanitaryware where it typically comprises 25 to 35% of the mix. In hotel china, and semi-vitreous china, finely ground nepheline syenite typically comprises 15 to 30% of the mix. For chemical and electrical porcelain, nepheline syenite may comprise 25 to 35%.

The use of nepheline syenite as an extender pigment in paint, and as an inert filler in plastics and rubber, is increasing gradually. So also is its use for sandblasting, where its lack of free quartz poses less of a health hazard than silica sand.

Exports

Both Canada and Norway are major exporting countries for nepheline syenite. Canada exports about 75% of its production, principally to the United States, but also to The Netherlands and Italy. Norway exports more than 90% of its production, mainly to Germany, but also to the United Kingdom, France, and Belgium. There is no record of exports of nepheline syenite materials from the Russian Republic.

Table 17 is a record of Canadian exports of nepheline syenite for recent years, showing also the average unit value of these exports. The sharp decline in sales to the United States after 1989 reflects primarily a declining demand for new glass containers and competition from feldspar. Both the feldspar producers in the United States, and the nepheline syenite producer in Canada, increased their glass-grade prices about 10% in 1990. About 70% of nepheline syenite exports to the United States is for glassmaking.

Out of a total 1988 feldspar production of 657 kt in the United States, about 350 kt valued at $15.5 million was used for glass (O'Driscoll, 1990). This is equivalent to about $44.30 per ton, closely similar to the apparent value of nepheline syenite exports to the United States that year, viz. $39.44.

Prices

Prices for Canadian nepheline syenite are quoted in the April 1992 issue of *Industrial Minerals* as follows: Low iron, 30 mesh, glass-grade material is sold in bulk at Cdn $32.75 per short ton (approximately equivalent to US$30.70 per ton) in carload and truckload lots FOB mine. A higher iron product for colored glass is quoted at Cdn $25.50 per short ton (US$23.90 per ton). Ceramic-grade, 200 mesh, bagged in 1-t lots is quoted at Cdn $78 (US$73.00 per ton). Filler and extender grades, bagged, are listed at Cdn $92 to 106 (US $86 to 99 per ton).

Norwegian nepheline syenite, glass-grade, 32 mesh, is quoted in the same source at £64 per ton (US$112 per ton) in bulk at a port in the United Kingdom; £85 (US$149) in bags. Ceramic-grade, 325 mesh, is quoted at £75 (US$131 per ton) in bulk, £101 (US$177) in bags.

Table 17. Canadian Exports of Nepheline Syenite (tons)

	1988	1989	1990	1991[a]
Australia	88 478	573	2 958	2 091
France	1 101	361	465	477
Italy	8 773	14 265	24 767	12 576
Japan	1 758	659	1 854	1 221
Korea	36	36	601	5 130
Netherlands	52 437	32 647	55 858	20 978
Spain	4 194	1 773	6 525	4 403
Taiwan	2 017	798	3 867	4 107
United Kingdom	164	236	6 353	400
United States	325 950	360 272	275 682	267 734
Other	1 300	675	3 123	2 321
	486 208	412 295	382 053	321 438
Average value per ton (US $)	39.92	39.38	58.87	55.00

[a] Preliminary. Figures are for January to November inclusive, *Source*: Statistics Canada.

Transportation

Glass-grade nepheline syenite is largely sold in bulk and transported in rail hopper cars. Unimin Canada Ltd. leases a fleet of hopper cars for the exclusive use of their nepheline syenite to ensure shipments are free of contamination. Finely ground ceramic and filler grades are usually bagged and shipped by boxcar or truck. Container loads are shipped by rail and boat to European destinations.

Shipping costs play a major role in determining the market range of nepheline syenite. For markets in Canada and the United States, where numerous rate increases have been a part of the inflating economies of the 1970s and 1980s, increased rail costs have favored greater haulage by truck. So also has the trend towards lower inventories at consuming plants, and the greater use of just-in-time truck deliveries. Truck haulage is now favored for destinations within a 24-hour trucking range. Cost increments at border crossings have always worked against shipments of nepheline syenite from Canada.

In general, boat shipping has been less affected by increases than truck and rail. Hence, the transport of Norwegian nepheline syenite, which is all by water over most of the distance, has some advantage over other European feldspathic materials that are moved mostly by land carriers.

Alternative materials

For most applications, feldspar can be used as a replacement for nepheline syenite. For glassmaking in particular, feldspar, aplite, and some quartz-feldspar sand mixtures can be used with suitable adjustment to the batch proportions. Feldspar sources in Georgia, and North Carolina, and aplite in Virginia, compete with each other and with Canadian nepheline syenite in the large industrial markets of the eastern United States.

Because the various feldspathic materials that may be available differ in respect to their contributions of desirable oxides, glass manufacturers use a weighting system to determine the material that gives them the best value (Minnes et al., 1983). Tooley (1953) and Feitler (1967) give detailed descriptions of the calculations, and the the latter method results in the determination that 41 kg of feldspar, 36 kg of aplite, and 33 kg of Canadian nepheline syenite have equal value. Table 18 gives a comparison of the important oxides in the three principal feldspathic materials used in eastern North America and illustrates the method used in calculating the weight of the

Table 18. Chemical Value of Competing Feldspathic Materials

Oxides	Chemical Composition, %		
	Nepheline Syenite	Aplite	Feldspar
SiO_2	60.0	63.5	67.9
Al_2O_3	23.4	22.4	19.2
$Na_2O + K_2O$	15.6	8.58	11.2
CaO	0.03	5.2	0.01
Fe_2O_3	0.08	0.1	0.07
Raw material needed per unit of alumina	4.27	4.46	5.21
Weight of contained oxides			
Al_2O_3	1.00	1.00	1.00
SiO_2	2.56	2.84	3.54
$Na_2O + K_2O$	0.67	0.38	0.58
CaO	—	0.23	—
Total Weight	4.23	4.45	5.12

Source: modified from Minnes et al., 1983.

feldspathic alternatives for equivalent chemical contribution. Knowing the delivered costs of silica sand, the various feldspathic materials, soda ash, limestone, and alumina, and by crediting each raw material with its SiO_2, $Na_2O + K_2O$, and CaO content, the consumer can determine which feldspathic material gives the lowest net cost of alumina and hence is the best buy. The value of its higher alumina and alkali contents is apparent in calculations such as these, and indicates why nepheline syenite has successfully competed with other feldspathic sources.

In much the same way, feldspar competes with nepheline syenite in ceramic applications. For pigment and filler uses there are a variety of alternative minerals: calcium carbonate, kaolin, feldspar, silica, and talc, all of which, however, have somewhat different physical properties and may not be suitable alternatives in specific cases.

Norwegian nepheline syenite competes with feldspars from other European sources, with Cornish Stone from the United Kingdom, and with Canadian nepheline syenite in certain areas.

GOVERNMENTAL MATTERS

Tariffs

Nepheline syenite, processed or raw, enters the United States duty free from any source country. The United Kingdom levies no duty on nepheline syenite from Commonwealth and European Trade Area countries. Nepheline syenite enters the European Economic Community free of duty, but member countries may level value added taxes. Exports of Canadian nepheline syenite are permitted to all countries (Minnes et al., 1983).

Environmental

Although glass containers represent only about 6.5% of municipal waste (Minnes et al., 1983), it is a highly visible and undesirable form of litter, and an unnecessary pressure on limited refuse disposal sites. Glass has therefore joined metals, certain plastics, and newspapers in municipal recycling programs. Waste glass, or cullet, is a desirable addition to the glass batch, although it must be carefully sorted for color and contaminants before use. Additionally, in Canada, most glass beverage containers are reusable and have had a refundable deposit put on them. Both recycling of waste glass and refundable deposits on beverage containers have had a depressing effect on new glass manufacturing and hence on nepheline syenite consumption.

Both the Canadian and Norwegian nepheline syenite operations pose no threat to widespread concerns for the environment. Their tailings are minimal in volume and non-toxic in character. However, being dry milling processes, there is a significant cost involved in maintaining effective dust collection systems.

PROBLEMS AND TRENDS

Consumption of nepheline syenite and other feldspathic materials has been relatively flat for more than a decade, largely due to recycling of waste glass, reuse of beverage containers, a trend towards thinner and lighter bottles, increased use of metal and plastic containers, and more recently a depressed housing market for whiteware bathroom fixtures and fiberglass insulation. However, glass is perceived as easier to recycle than plastics, and with technical advances in producing bottles of lighter weight and greater strength, new glass containers are providing formidable competition for metal. Additionally, glass beverage bottles can be reused numerous times, an advantage not yet shared by metal and plastic containers to any appreciable extent. But while glass has a number of advantages in competition with other containers, the outlook for the near future is nevertheless not likely to see significant growth in nepheline syenite consumption.

Most producers of feldspathic materials are now attempting to diversify their markets to include the pigment extender and filler industries, a market long enjoyed by Canadian nepheline syenite. However, tight specification standards, particularly the requirement for extreme fineness and whiteness, puts a relatively high cost on production, increasing their vulnerability to competition from other mineral commodities such as calcium carbonate, kaolin, silica, and talc.

Prospects for the use of nepheline syenite for other than glass, ceramics, and fillers are not great. The key to the establishment of new commercial operations will continue to be the ability to produce low iron products. The use of nepheline syenite for roofing granules and construction aggregates is a function solely of the physical properties of the Arkansas rock, which are not characteristics unique to nepheline syenites.

It is also doubtful that the production of alumina, and its byproducts alkali carbonates and portland cement, will ever be economic except as a strategic alternative to bauxite. In the Russian case, a nepheline-rich tailing grading nearly 30% alumina was available in large volumes from the extraction of apatite. And in the absence of bauxite, this proved to be a viable source of alumina, to the point where several large industrial complexes were subsequently established in central Siberia based on nepheline resources unassociated with the recovery of other minerals.

However, the unique chemical properties of nepheline syenite continue to attract attention the world over, and it seems likely that new sources of low iron products may ultimately be developed. In recent years there has been interest in such possible developments from Brazil, China, Turkey, Saudi Arabia, and British Columbia.

BIBLIOGRAPHY

Anon., 1987, "The Economics of Feldspar 1987," 5th ed., Roskill Information Services, London, 132 pp.

Anon., 1988, "Volume and Value of Production of Nepheline Syenite in Ontario," *1988 Ontario Mineral Score*, Ontario Ministry of Northern Development and Mines, pp.76–78.

Anon., 1993, "Mineral Production of Canada in 1991 with Estimates for 1992," *The Northern Miner*, Mar., Toronto, pp. 20–21.

Adams, F.D., and Barlow, A.E., 1910, *Geology of the Haliburton and Bancroft Areas, Province of Ontario*, Memoir No. 6, No. 1082, Canada Department of Mines, Geological Survey Branch, Ottawa, 419 pp.

Allen, J.B., and Charsley, T.J., 1968, *Nepheline Syenite and Phonolite*, Institute of Geological Sciences, Minerals Resources Division, London, 169 pp.

Archibald, F.R., 1942, "Report of Investigation on Production of Alumina and Alkalis from Nepheline Syenite," unpublished report to the Bureau of Mines, Ottawa, 40 pp.

Archibald, F.R., and Nicholson, C.M., 1948, "Alumina from Clay by the Lime-Sinter Method II," Technical Publication 2390, AIME, New York, 25 pp.

Boucher, M.A., 1989, "Nepheline Syenite and Feldspar," *Canadian Minerals Yearbook*, Annual Report of the Mineral Resources Division, Energy, Mines and Resources, Canada, Ottawa, pp. 43.1–43.5.

Brogger, W.C., 1890, "Die Mineralien der Syenitpegmatitgange der sudnorwegischen Augit und Nephelinsyenite," Zeits. fur Kryst. 16, Oslo, p. 41.

Burger, J., 1990, "Feldspar & Nepheline Syenite, At the Mercy of Glass Markets," *Industrial Minerals*, No. 275, Aug., pp. 21–33.

Clarke, F.W., 1910, "Analyses of Rocks and Minerals," US Geological Survey Bulletin 419, p. 123.

Comte, J.M.A., 1959, "Russia gets World's Largest Cement Kilns," *Rock Products*, May, pp. 128–131.

Currie, K.L., 1976, "The Alkaline Rocks of Canada," Bulletin 239, Geological Survey of Canada, 228 pp.

de Ferran, A., 1983, "The Canaan Nepheline Syenite Venture," *Proceedings*, 5th Industrial Minerals International Congress, B.M. Coope and G.M. Clarke, eds., Metal Bulletin Plc., London, pp. 101–104.

Derry, D.R., 1951, "The Lakefield Nepheline Syenite: Evidence of a Non-intrusive Origin," *Transactions*, The Royal Society of Canada, Vol. 45, pp. 31–40.

Derry, D.R., and Phipps, C.V.G., 1957, "Nepheline Syenite Deposit, Blue Mountain, Ontario," *The Geology of Canadian Industrial Mineral Deposits*, Congress Volume, Canadian Institute of Mining and Metallurgy, Stovel-Advocate Press, Winnipeg, pp. 190–195.

Feitler, S.A., 1967, "Feldspar Resources and Marketing in Eastern United States," Information Circular 8310, US Bureau of Mines, 41 pp.

Geis, H-P., 1975, "The Stjernoy Island Nepheline Syenite Deposit," Elkem Spigerverket, private communication.

Guillet, G.R., 1962, "A Chemical and Inclusion Study of Nepheline Syenite for Petrogenetic Criteria," MA thesis, University of Toronto, Toronto, 75 pp.

Heier, K.S., 1966, "Some Crystallo-Chemical Relations of Nepheline and Feldspars on Stjernoy, North Norway," *Journal of Petrology*, Vol. 7, Pt. 1, pp. 95–113.

Hewitt, D.F., 1961, "Nepheline Syenite Deposits of Southern Ontario," Vol. 49, Pt. 8, 1960, Ontario Department of Mines, Toronto, 194 pp.

Johnson, P.W., and Peters, F.A., 1968, "Methods for Producing Alumina from Anorthosite, and Evaluation of a Lime-Soda Sinter Process," Report of Investigation 7068, US Bureau of Mines, 42 pp.

Keith, M.L., 1939, "Petrology of the Alkaline Intrusive at Blue Mountain, Ontario," *Bulletin*, Geological Society of America, Vol. 50, No. 12, Pt. 1, pp. 1795–1826.

Liddicoat, W.K., Ramsay, C.R., and Hedge, C.E., 1985, "Cambrian Nepheline Syenite Complex at Jabal Sawda, Midyan Region, Kingdom of Saudi Arabia," *Felsic Plutonic Rocks and Associated Mineralization of the Kingdom of Saudi Arabia*, A.R. Drysdale, C.R. Ramsay, and D.B. Stoeser, eds., Mineral Resources Bulletin 29, Deputy Minister for Mineral Resoures, Jiddah, pp. 139–150.

Lundquist, R.V., and Singleton, E.L., 1962, "Some Characteristics of Iron in the Lime-Soda Sinter Process for Recovering Alumina from Anorthosite," Report of Investigation 6090, US Bureau of Mines, 16 pp.

MacGregor, D.D., 1983, "Geology, Mining and Processing of Nepheline Syenite," *Proceedings*, 19th Forum on the Geology of Industrial Minerals, S.E. Yundt, ed., Miscellaneous Paper 114, Ontario Geological Survey, pp. 49–53.

MacLean, M.E., and White, G.V., 1991, "Feldspathic Mineral Occurrences in British Columbia," Open File 1991–10, British Columbia Ministry of Energy, Mines and Petroleum Resources, 87 pp.

Minnes, D.G., 1975, "Nepheline Syenite," *Industrial Minerals and Rocks*, 4th ed., S.J. Lefond, ed., AIME, New York, pp. 861–894.

Minnes, D.G., Lefond, S.J., and Blair, R., 1983, "Nepheline Syenite," *Industrial Minerals and Rocks*, 5th ed., S.J. Lefond, ed., AIME, New York, pp. 931–960.

Moorhouse, W.W., 1959, *The Study of Rocks in Thin Section*, Harper and Brothers, New York, 514 pp.

Notholt, A.J.G., 1983, "Igneous Phosphate Resources: Their Growing Contribution to World Markets," *Phosphates: What Prospects for Growth*, Metal Bulletin Plc., London, pp. 43–58.

O'Driscoll, M., 1990, "US Glass Industry: Over Capacity Shadows Strong Demand," *Industrial Minerals*, No. 269, Feb., pp. 34–53.

Olerud, S., 1993, "Norway's Industrial Minerals, Production and Development Trends," *Industrial Minerals*, No. 307, Apr., pp. 55–63.

Payne, J.G., 1966, "Geology and Geochemistry of the Blue Mountain Nepheline Syenite Body," PhD thesis, McMaster University, Hamilton, Ontario, 183 pp.

Pell, J., 1986, "Nepheline Syenite Gneiss Complexes in British Columbia," Paper 1986–1, Geological Field Work, British Columbia Ministry of Energy, Mines and Petroleum Resources, pp. 255–260.

Pell, J., 1987, "Alkaline Ultrabasic Rocks in British Columbia: Carbonatites, Nepheline Syenites, Kimberlites, Ultramafic Lamprophyres and Related Rocks," Open File 1987–17, British Columbia Ministry of Energy, Mines and Petroleum Resources, 109 pp.

Perkins, M.J., 1983, "Structural Geology and Stratigraphy of the Northern Big Bend of the Columbia River, Selkirk Mountains, Southeastern British Columbia," PhD thesis, Carleton University, Ottawa, 238 pp.

Potter, M.J., 1989, "Feldspar, Nepheline Syenite, and Aplite," *Minerals Yearbook*, US Bureau of Mines, pp. 389–396.

Smothers, W.J., et al, 1952, "Ceramic Evaluation of Arkansas Nepheline Syenite," Research Series No. 24, Arkansas Resources and Development Commission, 21 pp.

Taylor, G.H., 1990, "Nepheline Syenite," *Ceramic Bulletin*, Vol. 69, No. 5, pp. 872–873.

Tilley, C.E., 1953, "The Nephelinite of Etinde, Cameroons West Africa," *Geological Magazine*, Vol. 90, No. 3, May-June, pp. 145–151.

Tilley, C.E., 1958, "Problems of Alkali Rock Genesis," *Quarterly Journal of the Geological Society of London*, Vol. 113, May, pp. 323–360.

Tilley, C.E., 1961, "The Nepheline Rocks of the Port Herald Hills, Southern Nyasaland," *Overseas Geology and Mineral Resources*, Vol. 8, No. 3, pp. 255–259.

Tooley, F.V., 1953, *Handbook of Glass Manufacture*, Vol. 1, Ogden Publishing, New York, pp. 75–79.

Wyllie, R.J.M., 1973, "Mining Nepheline Syenite on Norwegian Arctic Island," *World Mining*, Aug., pp. 36–41.

OLIVINE AND DUNITE

ROBERT J. HENNING

The term olivine was first used in 1790 by J. Werner, and the corresponding mineral was so named because of its olive-green color (Hunter, 1941). Olivine is the principal component of the rock known as dunite. While by classic definition dunite is composed of 90% or more olivine, commercial use of these terms has resulted in more specific definitions. Commercially, the term olivine is used to designate a material that consists of 85% or more forsterite and contains 45 to 50% MgO, 39 to 42% SiO_2, 5 to 8% Fe_2O_3, and has an ignition loss (LOI) of 1 to 2%. The term dunite designates a rock that contains 36 to 45% MgO, 36 to 39% SiO_2, and has an ignition loss of as much as 10% (Teague, 1983).

GEOLOGY

Olivine is found in basic igneous and basic metamorphic rocks whose age varies from pre-Ordovician to Tertiary. Economic deposits are of magmatic origin and are restricted to the dunite variety of the peridotite group of ultrabasic igneous rocks. Olivine also occurs in basic igneous rocks such as gabbro, basalt, and diabase and more rarely as an accessory mineral in granite and metamorphosed crystalline limestones.

Mineralogy

The term olivine refers to a group of orthosilicate minerals in isomorphous series with magnesium-rich forsterite (Mg_2SiO_4) as one end member and iron-rich fayalite (Fe_2SiO_4) as the other. In most olivines the ratio of magnesium to iron varies from 16:1 to 2:1. Commercial olivine deposits contain a mixture of forsterite and fayalite with the fayalite content restricted to less than 15% (Reed, 1959; Teague, 1983). Accessory constituents may include primary minerals such as ilmenite, magnetite, diopside, chromite, and garnet and secondary minerals such as chlorite, serpentine, talc, and vermiculite as alteration products.

Chemical and Physical Properties

The chemical composition of olivine can be expressed as either $(Mg, Fe)_2\ SiO_4$ or $2(Mg, Fe)O{\cdot}SiO_2$. Table 1 presents chemical analyses for some commercial olivines.

The physical properties of olivine are shown in Table 2.

Distribution of Major Deposits

Commercial olivine and dunite deposits are found in Norway, Japan, Spain, India, Italy, the United States, Sweden, Austria, Pakistan, and Mexico. Noncommercial deposits have been reported in New Zealand, Zimbabwe, South Africa, New Caledonia, Greece, Brazil, and Canada (Teague, 1983). The present markets for olivine are characterized by low unit value, high volume, and high sensitivity to freight costs, so few deposits are commercially viable.

Literature from the USSR indicates that considerable research on olivine refractories is being done. (Afansa'ev, 1988), lending credibility to the conjecture that commercial deposits are being mined there.

In eastern Canada synthetic olivine is being produced by calcining chrysotile asbestos mine tailings.

In the United States numerous lenses of olivine extend in a belt from northeast Georgia northeastward across western North Carolina. Total reserves in North Carolina and Georgia are estimated at between 125 and 200 Mt. The largest US deposit is Twin Sisters Mountain, located in Whatcom and Skagit Counties in the state of Washington. This deposit, which has an outcrop area of 93 km^2 and a relief of about 1 525 m, is estimated to contain 1 800 Mt of fresh, unaltered olivine (Teague, 1983). In addition, the US Bureau of Mines reports reserves of 50 Mt on Cypress Island in the state of Washington, and the literature reports eight complexes in southeastern Alaska, some containing cores of olivine as large as 1.6 km in diameter (Taylor, Jr., 1967).

The world's largest deposit of olivine occurs in the Aaheim area of Norway, where reserves are estimated at a 2 000 Mt. Several other Norwegian deposits contain as much as 5 Mt each.

In Japan the production of ultrabasic rocks is currently about 6 Mtpy, but the bulk of this output is serpentinite. Olivine/dunite accounts for only 17% of this production. The major olivine deposit in Japan is located at Horoman Hill on the northern island of Hokkaido. Reserves at this deposit are estimated at 100 Mt (Anon., 1990).

Spain is one of the largest producers of dunite. Reserves in the La Coruna District in northwestern Spain are estimated at 100 Mt. India, also a major dunite producer, has a large deposit at Salem in Tamil Nadu.

In Italy olivine/dunite is mined from a deposit near Turin. In Sweden the largest deposit is at Arutats and is reported to be comparable in size to the Twin Sisters deposit in the United States. At present the only production in Sweden is from a deposit 30 km south of the town of Gallivare. Reserves at this deposit are reported to be between 1 and 2 Mt.

Other significant deposits are: 1) in Austria at St. Stefan near Leoben, 2) in the Swat District of the North West Frontier Province in Pakistan, and 3) in Mexico at Cuidad Victoria in Tamaulipas (Anon., 1990).

TECHNOLOGY

Mining

In all but one case olivine/dunite is mined by open pit methods. The exception is the Franzefoss Bruk operation in Norway. Fran-

Table 1. Chemical Analyses of Some Commercial Olivines/Dunites (%)

	1	2	3	4	5	6
MgO	45–49	45–48	48–50	46.5–50	38–40	41–43
SiO_2	39–42	39–42	42–43	41.5–43	36–40	42–44
Fe_2O_3	6–8	6–8	6.8–7.3	6.8–7.2	8–12	6–8
Other	1.5–2.4	1.3–2.4	1.4–1.9	0.1–2.1	1–5	3.5–6.1
LOI	0.5–1.0	0.8–1.4	0.7–1.0	0.8–1.2	2 max	?

1. Hamilton, WA, 2. Burnsville, NC, 3. Aaheim, Norway, 4. Briggja, Norway, 5. Salem, India, 6. Vidracco, Italy.
Note: Care should be taken when comparing chemical analyses and LOI (Wilds, Jr., 1988) to assure that identical testing procedures were used.
Source: Producers' literature.

Table 2. Physical Properties of Olivine

Hardness (Mohs Scale)	6.5–7
Specific Gravity	3.27–3.37
Bulk Density	1.5–2.0 g/cm^3
Fusion Temperature	1 665–1 743°C

zefoss Bruk operates an underground mine at Briggja on the north shore of the Nord Fjord.

When using the open pit method bench heights generally range from 3 to 9 m depending upon the hardness of the olivine and the extent of fractures and joints present in the deposit. At most olivine deposits blasting is required. Depending upon the hardness and degree of alteration, however, some deposits can be mined by ripping. When blasting, holes are usually drilled on a 1 to 2 m pattern, which normally results in reasonably good breakage. Some secondary drop ball work may be required at times. Blasting charges are usually relatively small, generally yielding less than 10 kt of crude olivine. Blasting charge consumption per ton of crude is relatively high, however, since much of the explosive force is dissipated through fractures and joints in the deposit.

In the Franzefoss Bruk underground mine the ore is extracted using room and pillar methods. This mine has some of the largest unsupported rooms in the world, extending up to 10 m in height. The rooms are mined in two lifts, and the pillars are elongated to ensure that fractures in the rock do not cross completely through the thickness of a pillar.

Processing

At the mine −0.5 m olivine rock is loaded into trucks with either shovels or front end loaders and is hauled to a primary crusher. The primary crusher, usually a jaw or cone, is located either at the mine site or at the processing plant. Primary crushing reduces the ore to −40 mm. Depending upon the degree of alteration of the rock and the product quality required, the −6 mm fines may be discarded at this point. Since altered material is softer than the olivine grains, a disproportionate amount of it is found in the fines generated by primary crushing.

The −40 mm material may be processed further at the plant or screened into a −40 + 10 mm product ready for the market. When further processed to obtain a finer grain size, the material is crushed and ground using cone, impact, or roll crushers or ball mills and is then screened. Dry processing is used wherever possible, such as at Aimcor's operations in the states of Washington and Indiana and at the A/S Olivin and Franzefoss Bruk operations in Norway.

In some cases the level of impurities present requires the ore to be beneficiated in order to yield product of the desired quality. Aimcor's operation in North Carolina uses wet processing and gravity separation. Magnolithe GmbH in Austria calcines its dunite in rotary kilns at 1 650°C to obtain high quality olivine for refractory and foundry applications (Anon., 1990).

Dalmia Magnesite Corporation recovers dunite as a byproduct of magnesite production in India. The dunite is the host rock for magnesite in a deposit that consists of 10:1, dunite to magnesite (Anon., 1990).

Canada produces synthetic olivine from the tailings of a chrysotile asbestos mine at Thetford Mines in Quebec (Mathieu et al., 1992). The company, Les Sables Olimag, Inc., calcines the tailings in a 4 ktpy rotary kiln at a temperature of 850°C. The product formed is a mixture of forsterite and silica. This company has experimented with producing a relatively pure forsterite by adding magnesia during the calcination. In these tests the magnesia and tailings were pelletized, calcined in a rotary kiln at 800°C, and then further heated at 1 350°C in a static furnace. At this latter temperature the added magnesia combines with the silica to form forsterite. Total retention time was two hours. Various magnesia addition levels were tested to produce forsterite/magnesia mixtures, some of which gave fusion temperatures above 1 680°C.

Specifications

The critical specifications for olivine are usually 1) its chemistry, especially its MgO content, 2) the presence of accessory minerals such as diopside, magnetite, and aluminosilicates, 3) loss on ignition, which reflects the presence of hydrated minerals such as chlorite, serpentine, and vermiculite, and 4) grain size distribution. Generally, there is no universally accepted standard sizing for olivine. Each producer screens to its own specifications to suit individual market requirements.

USES

Almost 80% of the worldwide consumption of olivine is dependent directly or indirectly on the production and forming of metals. Its use as a slag conditioner, foundry sand, and as a raw material in the manufacture of refractories and olivine-fluxed pellets make it highly dependent upon the strength of especially the iron and steel industry.

Slag Conditioning

By far, the largest use of olivine is for conditioning blast furnace slag. The olivine can be added to a furnace either as lump, as a component in the sinter feed, or mixed with iron ore when making pellets. Another slag conditioning application is in electric arc furnaces, where olivine is added to the slag to provide basicity control and thereby prevent slag attack on the refractory lining. When olivine is added to a blast furnace in lump form it replaces some of the limestone or dolomite used as a flux in the reduction of iron ore. Its MgO content and other properties provide olivine with the following benefits (Currier, private communication):

- Reduces alkali recirculation within the furnace by forming magnesium alkali silicates, which are stable at blast furnace temperatures.
- Olivine, because of its higher reaction temperature, reduces low temperature breakdown and swelling of the burden, thus maintaining permeability and reducing coke consumption.
- Because of olivine's high MgO content, less material is required to achieve a specific MgO level than when dolomite is used.
- MgO:SiO_2 ratio of olivine permits raising the MgO level without changing the basicity of the slag.

- Olivine has a low loss on ignition, so energy is not consumed in driving off the unwanted carbon dioxide from dolomite or water from serpentinite.

There are additional benefits when olivine is added as a component in sinter. Along with the benefits achieved in a blast furnace, olivine increases a sinter furnace's capacity by 10 to 15%. This is because the addition of olivine reduces the sintering temperature by as much as 100°C and produces a tougher sinter, which generates less fines that have to be recycled.

The addition of olivine to iron ore during the production of pellets results in several advantages (Currier and Green, 1986). Olivine-fluxed pellets swell less, reduce more quickly, and have a narrow melting range. These features all contribute to lower melting costs. The addition of olivine is limited, however, to iron ores that contain low levels of silica. Because olivine also contains silica, high-silica ores result in excessive slag levels in the furnace. At present only LKAB in Sweden and Hoogovens B.V. in Holland are producing olivine-fluxed pellets. Hoogovens is producing only for its own consumption.

Refractories

Olivine's use in refractories has undergone radical change during the last ten years. Changes in technology and processing have almost eliminated many old uses but have also opened up new applications.

Olivine's traditional applications in the checker works of open hearth furnaces, as a protective cover for soaking pit bottoms, for heat storage, and as backing sand for tundish boards have almost disappeared. Changes in steel mill processing and technology have eliminated the open hearth furnace, have nearly eliminated the need for soaking pits by conversions to continuous casting, and have replaced the use of tundish boards with the use of monolithic gunable linings in continuous casting tundishes. Magnesia refractories are now preferred for heat storage in Europe.

The steel industry, by far the largest consumer of refractories, is demanding more specialized slag-resistant refractories in its continuing drive to reduce costs. Refractories that are MgO-based are mechanically very stable and resistant to slag attack and therefore are now being used more by the steel industry.

Olivine is a low cost raw material for MgO-based refractories, costing about half as much as magnesite. It is now being used extensively in forsterite bricks (Koltermann, 1988) and in the gunable working linings of continuous casting tundishes.

Another growing market for olivine is in precast refractory linings for incinerators (Smith, Jr., 1992), where its heat storage properties result in more complete combustion. Olivine's low expansion rate also permits heat cycling without spalling of the lining.

Despite its higher cost, olivine is finding greater acceptance as a substitute for silica flour in refractory washes for foundry cores and molds. This is primarily because the use of olivine does not create a health hazard (viz., silicosis) in the workplace.

Foundry Sands

Olivine was first used as a foundry sand in the middle 1930s in Norway. The Norwegian organization, Jernknontoret, undertook an extensive program to identify a better molding material than silica sand and an alternative that would eliminate the silicosis hazard in Norwegian foundries (Anon., 1992). Olivine, with its high fusion temperature (approximately 1 800°C), low thermal expansion rate, and absence of free silica, was the recommended alternative.

Today olivine is the molding sand of choice at manganese steel, brass, bronze, and aluminum foundries. It is also used at gray iron and alloy steel foundries where an extremely fine finish is required or where the foundry is distant from a source of silica sand. In the United States many states will lower a foundry's workmen's compensation payments if olivine is used exclusively. This reduction is in recognition of the greatly reduced risk of silicosis.

While its price is usually four to five times that of silica sand, olivine does provide some economic benefits that help offset this difference. When used in a bentonite-bonded green sand system, the basicity of olivine enhances the bonding properties of the bentonite, so less bentonite is required. In addition, the low and uniform expansion rate of olivine requires that little or no cushioning agent be added (Smith, Jr., 1991). This not only saves on the cost of these additives but allows the sand to be reused many more times. An olivine-based system requires 40 to 70% less makeup sand than a silica-based system. This not only saves on the cost of the makeup sand but on the disposal costs for discarded sand as well.

When using olivine in making manganese steel castings there is an additional benefit. If manganese steel is poured into silica sand molds a low melting point $FeO\text{-}MnO\text{-}SiO_2$ eutectic is formed, generating a liquid-liquid interface at the surface of the mold. This liquid interface penetrates into the mold, resulting in a fused layer of manganese steel, eutectic, and silica sand on the surface of the casting. Olivine does not form this eutectic, providing instead a solid mold wall against which the manganese steel can solidify.

Very fine olivine sands (e.g., American Foundrymen's Society, or AFS, grain fineness numbers 100, 140, and 180) are used in brass, bronze, and aluminum foundries. Because of olivine's non-wetting properties and its angular grain shape, using these fine sizes results in castings that have an extremely smooth surface finish. The angularity of olivine grains provides good permeability, thus allowing use of these fine sizes.

Olivine's usage in foundries is somewhat restricted, however. Because of its basicity it cannot be used with acid-catalyzed furan and phenolic no bake binders. In addition, because of its angular grain shape olivine requires approximately 50% more resin than round-grain silica sand when used with compatible binder systems.

Abrasives

Although olivine is not used in bonded or coated abrasive applications, its physical properties make it well suited as a blasting abrasive. Its hardness, high specific gravity, and conchoidal fracture provide efficient cleaning rates and scouring ability. Olivine's hardness of 6.5 to 7 on the Mohs scale indicates a grain that has minimum shatter on impact. Its high specific gravity of 3.2 to 3.4 gives a high force of impact, because the higher the specific gravity the greater the impact for grains of the same size. Its conchoidal fracture produces sharp edges that cut readily through paint and scale. Olivine has an advantage over smelter and power plant slags in that it does not leave a black, dusty surface. Its advantage over silica sand is that it contains no free silica and therefore poses no risk of silicosis. It is expected that stricter environmental regulations will lead to increasing use of olivine in blasting applications.

Olivine is also used as an abrasive grit in waterjet cutting applications. The worldwide consumption of olivine in this market is estimated at 6 to 8 ktpy.

Other Uses

In Japan 74 μm olivine is reacted with conventional fertilizers and pelletized for use in agriculture (Nakagome, private communication). The addition of olivine provides both magnesium and iron. Recent tests at North Carolina State University (Warren and Shelton, 1989) have shown that raw olivine, unprocessed except for

Table 3. Estimated Worldwide Olivine/Dunite Consumption, 1990

Application	kt
Abrasives	114
Asphalt filler	143
Construction/ballast	1 015
Fertilizer	31
Foundry sand	221
Iron ore pellets	125
Refractories	262
Slag conditioning	4 165
Tiles	23
Other	20
Total	6 119

sizing, will dissolve in acidic soils, releasing both magnesium and iron in forms that can be absorbed by plants.

In Italy olivine is added to roofing tile formulations to render the tiles opaque to ultraviolet light (Smith, Jr., 1991).

Other uses for olivine/dunite include asphalt fillers, construction aggregates, ballast, and as water filtration and fluid bed furnace media. An estimate of worldwide olivine consumption by application is given in Table 3.

ECONOMIC FACTORS

Costs

The cost of mining crude olivine/dunite varies from 7 to $15/t depending upon conditions encountered at the minesite. The cost per ton of olivine to be processed can be considerably higher, however, depending upon the cost of transporting the crude to a processing plant and the amount of crude material discarded in order to meet quality requirements. Processing costs also vary widely, depending upon whether the ore can be dry processed or whether wet processing or calcination is required.

Markets

As would be expected from its uses, the major markets for olivine are in Japan, western Europe, and North America. In all of these areas alternative materials exist that compete with olivine.

In its use as a slag conditioner olivine competes with dolomite and serpentinite. In refractories it competes with magnesite. In its use as a foundry sand olivine competes with zircon, chromite, and silica sand. In abrasive applications olivine competes with slags, garnet, staurolite, and silica sand. Being a relatively low priced material, olivine's success in competing with these materials is dependent upon reaching its markets at the lowest possible cost. This requires that deposits be located close to consumers or very close to means of low cost transportation. The deposits in Norway meet this latter requirement. All three commercial Norwegian operations are located on deepwater fjords. This enables them to take advantage of low cost ocean shipping in delivering their products to markets worldwide.

GOVERNMENTAL CONSIDERATIONS

In the United States olivine is permitted a 22% depletion allowance for federal income tax purposes. Olivine imported into the United States must pay a duty of 4.9% ad valorem.

State and federal regulations in the United States now require that all open pit mining operations be reclaimed. These and other possible future regulations pertaining to environmental protection and land use must be taken into consideration when evaluating the economics of producing olivine. The costs of compliance with existing governmental guidelines have rendered uneconomic several deposits once thought to be economic.

FUTURE TRENDS

The outlook for olivine is one of cyclical demand because of its heavy dependence upon the iron and steel industry. Sufficient capacity presently exists to meet expected demand for the next few years. The price of olivine is expected to change very little in the near future, because it is already a low priced commodity and its markets have ready access to alternative materials.

As consumers of olivine strive for quality improvements in their products, pressure for improvements in consistency and quality will be passed on to olivine suppliers. Olivine producers will be expected to develop total quality management (TQM) systems utilizing statistical process control (SPC) techniques, failure modes and effects analyses (FMEAs), and the goal of continuous quality improvement.

BIBLIOGRAPHY AND REFERENCES

Anon., 1990. *The Economics of Olivine,* 2nd ed., Roskill Information Services, 90 pp.

Anon., 1992, "Olivine," Aimcor corporate literature, 2 pp.

Afansa'ev, A.P. 1988, "Olivinites From Overburden of The Kovorsk Phlogopite Deposits—A New Source of Raw Materials for Refractories Production," translated from *Ogneupory,* No. 12, Plenum Publishing, pp. 23–27.

Currier, J.W., and Green, R.A., 1986, "The Function of Magnesia in Iron Oxide Agglomerates," *Proceedings,* 45th Iron Making Conference, pp. 23–30.

Hunter, C.E., 1941, "Forsterite Olivine Deposits of North Carolina and Georgia," Bulletin 41, North Carolina Department of Conservation and Development, Division of Mineral Resources, 117 pp.

Koltermann, M., 1988, "Olivine $(Mg,Fe)_2$ SiO_4 as a Refractory Material in the Steel Industry," *Taikabutsu Overseas,* Vol. 8, No. 1, pp. 3–8.

Mathieu, G.I., et al., 1992, "Technology to Produce Synthetic Forsterite for Applications in Steel Mills," SME Preprint No. 92–82, SME Annual Meeting, Phoenix, AZ.

Reed, J.J., 1959, "Chemical and Modal Composition of Dunite From Dun Mountain, Nelson," *New Zealand Journal of Geology and Geography,* Vol. 2, No. 5, pp. 916–919.

Smith, C.S., Jr., 1991, "Olivine," *Mining Engineering,* Vol. 43, No. 6, pp. 608–609.

Smith, C.S., Jr., 1992, "Olivine, the Industrial Mineral," SME Preprint No. 92–102, SME Annual Meeting, Phoenix, AZ.

Taylor, H.P., Jr., 1967, "The Zones of Ultramafic Complexes of Southeastern Alaska," *Ultramafic and Related Rocks,* Wiley, New York, 464 pp.

Teague, K.H., 1983, "Olivine," *Industrial Minerals and Rocks,* 5th ed., Vol. 2, S.J. Lefond, ed., AIME, New York, pp. 989–996.

Warren, S.L., and Shelton, J.E. 1989, "Olivine: A Slow-Release Magnesium Source for Ornamental Container Production, 1988–89," Research Report, North Carolina State University, Asheville, NC.

Wilds, A.A., Jr., 1988, "Determination of Olivine Loss on Ignition," Report No. 88–21-P, North Carolina State University Minerals Research Laboratory, Asheville, NC.

PERLITE

RICHARD O.Y. BREESE AND JAMES M. BARKER

The recognition of perlite as a distinct volcanic glass, and of the arcuate fractures that often characterize these natural glasses, dates back at least to the late 19th century and perhaps into the 18th century (Howell, 1974). Some authors suggest that recognition of perlite dates considerably further into antiquity, perhaps to the 3rd century BC (Caley and Richards, 1956).

Most recent definitions identify perlite by the presence of vitreous, pearly luster and by the presence of characteristic concentric or arcuate perlitic fractures. As discussed more fully below, arcuate fractures in perlite need not be megascopic. The textures of perlite which commonly occur in deposits range from dense to highly vesiculated to pumice-like. Thus, the classical definition of perlite is descriptively restrictive compared to the natural range of textures and neglects to address the issues of genetic origin and geologic occurrence.

In spite of the petrological usage of the term perlite, the introduction of expanded perlite aggregate into industrial markets caused the term to be applied as well to the lightweight, cellular aggregate that is produced through the rapid thermal expansion of milled perlite ore.

Perlite is distinguished petrologically from other natural glasses by a silicic or rhyolitic composition by the presence of 2 to 5 wt% total chemical water held within the glass structure and often by the presence of pearly luster and onionskin-like perlitic fractures. Worldwide, occurrences are associated with Tertiary through early to middle Quaternary volcanics and with the glassy portions of silicic domes and flows, with vitric tephra, with the glassy chill margins of dikes and sills, and with the glassy portions of welded ash flow tuffs. Less commonly, perlite is reported in association with the glassy portions of volcanic plugs and laccoliths. The perlite carapaces that partially or fully comprise extrusive domes and flows are often thick and areally extensive, and thus provide commercially attractive targets for low cost open pit mining. Perlite tuffs and tephras are also important commercial sources.

Mining costs are minimized by open pit quarrying and by either bulldozer ripping or blasting. Crushing and sizing facilities are generally close to the pits. Due to the low weight and large volume of expanded perlite, unexpanded perlite that has been crushed and sized is usually shipped nationwide directly to local markets where it is expanded and processed for distribution to end users. The vast majority of international trade is in unexpanded perlite.

In the United States, the évolution of the expanded perlite industry can be traced to the late 1930s and early 1940s when expanded perlite aggregate was introduced into gypsum plaster and concrete markets as a substitute for vermiculite (Schundler, Jr., oral communication, 1991) and was promoted for use in a much broader spectrum of products (Howell, 1974). By the early 1950s, end user markets were well established and small expanders were numerous east of the Mississippi River. Today, mining and expansion is undertaken throughout the world. Worldwide production of perlite is dominated by the former Soviet republics, the United States, and Greece. Total worldwide production of processed perlite during 1990 is estimated at roughly 1 778 000 tons (Davis, 1991).

The physical properties of expanded perlite that give this product unique commercial value include: low bulk density, chemical inertness, high insulating ability, nonflammability and fire resistance, and its ability to retain water. These and other physical properties give expanded perlite value in such uses as: lightweight acoustical and thermal insulating aggregate; component in plaster, insulating cements and lightweight concrete; loose fill insulation; as aggregate in horticultural applications; and in milled form as filter aid. In the United States and Europe, the largest end use is within building construction products.

GEOLOGY

Composition and Origin

Perlite is a member of the family of water-bearing natural glasses that includes hydrous and low water obsidian, hydrated volcanic ash or pumicite, and perlitic and non-perlitic pitchstone. This general classification system differentiates perlitic and non-perlitic glasses of varying origin but of similar composition based on water content and physical characteristics. The commonly accepted, classical definition of perlite distinguishes it from other water bearing glasses by 2 to 5 wt% chemical water content, a pearly luster and by the presence of onionskin perlitic fractures. This definition is descriptively restrictive. Geologically, the term must be expanded to include the less dense textures of perlite which are genetically related with the classical variety but which rarely, or only, exhibit microscopic perlitic fractures, and which may not exhibit pearly luster. Textures of perlite commonly found in deposits range from the dense and classically fractured variety to vesiculated and pumice-like varieties which do not necessarily exhibit pearly luster or megascopic perlitic fracture.

Most silicic obsidians contain from 0.1 to 1.0 wt% total water bound as silanol groups (SiOH) within the silica framework. This water is thought to represent the original magmatic component present in the magma upon chilling. Few recent dense glasses are hydrated, although older glasses occur extensively as perlite. Perlite is a hydrated volcanic glass formed through the secondary alteration of obsidians by the incorporation of water into the glass silica structure. While several theories describing the timing and water sources of hydration have been proposed during the past few decades, clearly the slow, post-emplacement hydration of cooled glasses by meteoric water is of primary geological importance in the formation process. Although mechanisms such as hydration by magmatic or meteoric waters during or shortly after emplacement, or hydration by water exsolved during cooling cannot be discounted, their significance has not yet been demonstrated. Rather, the hydration of obsidians and the formation of perlite deposits is

Table 1. Some Typical Perlite Chemical Compositions (%)

	No Agua, NM	Superior, AZ	Pioche, NV	Big Pine, CA	Milos, Greece	Akita, Japan	Bulgaria	Argentina	Hungary
SiO_2	72.1	73.6	73.1	73.6	74.2	74.2	73.8	72.3	73.5
Al_2O_3	13.5	12.7	12.8	13.2	12.3	12.9	12.8	13.4	13.0
Fe_2O_3	0.8	0.7	0.7	0.8	0.95	0.68	0.56	1.0	1.8
TiO_2	0.06	0.1	0.08	0.07	0.08	0.06	0.07	0.08	—
CaO	0.89	0.6	0.9	0.6	0.85	0.45	0.50	0.59	1.5
MgO	0.50	0.2	0.2	0.1	0.13	0.05	0.03	0.3	0.4
Na_2O	4.6	3.2	3.0	4.1	4.0	4.1	3.0	3.4	3.5
K_2O	4.4	5.0	4.7	4.1	4.4	4.0	4.9	4.7	3.8
H_2O+	3.0	3.8	3.9	3.3	2.8	3.3	4.0	3.7	3.0

Source: Kadey, Jr. (1983).

a gradual process, coincident with the inward migration of meteoric water into glass selvages and its incorporation into the glass structure as molecular water (H_2O) and as silanol groups. Jezek and Noble (1978) describe the process as being associated with the progressive incorporation of water and ion exchange. From empirical studies, composition of the glass phase and effective temperature have been shown to be important factors that influence the rate of hydration (Friedman and Long, 1976). Hydration rate slows with decreasing temperature and with increasing calcium and magnesium content. The rate increases with increasing silica content. Some typical perlite compositions are shown in Table 1. The arcuate or concentric fractures which characterize the denser textural varieties of perlite are the result of tensional strain associated with the hydration of glass and its consequent increase in volume (Friedman and Smith, 1958, Jezek and Noble, 1978). Much study yet remains to understand and describe the process of hydration, much less its application to the geologic formation of deposits.

Perlitized volcanic glasses occur within the rapidly quenched selvages of silicic domes and flow lavas, with vitric or glassy tephra, and with the glassy zones of welded ash-flow tuffs. While perlitized glassy plugs, laccoliths, and dikes are also reported, descriptions of these occurrences are either rare or poorly documented. The extrusion of viscous silica-rich lavas often produces steep-sided domes formed through the external distension and expansion of the lava mass as it accumulates above its vent. Peripheral flows that can reach several square kilometers or more in area are commonly associated with the dome structures and complexes. The eruption of silicic lavas with lower emplacement viscosities has produced flow and dome complexes of even greater areal extent. Glass zones are frequently present on the domes and flows as exterior selvages, as mantles, and as associated tephras. When hydrated, such occurrences can represent important commercial sources of perlite.

A typical zonation observed in many silicic flows and domes consists of a texturally zoned exterior glass unit which encases interior, partially devitrified glasses and felsite core (Fig. 1). This basic concentric zonation is produced by the rapid quenching of exterior surfaces and crystallization within the more slowly cooled interior environments. Glass mantle can occur as basal, marginal, and/or upper selvage, and can be tens of meters thick. Exterior glasses are typically pumiceous and encase progressively denser obsidians. These grade into and often interfinger with more highly crystalline, interior core rocks. The glasses exposed to hydration processes, including associated vitric tephras, alter to perlite through the incorporation of water. Hydration and the resulting perlitic fractures are superimposed on the primary textures and zones of texture. Primary texture remains intact and is evident as the textural varieties of perlite that are described below. While no single model can be applied to all occurrences and this basic expression of zonation is dependent on many factors, the recognizable textures and lithological zonation provide a useful basis for geologic mapping and for the uniform comparative description and modeling

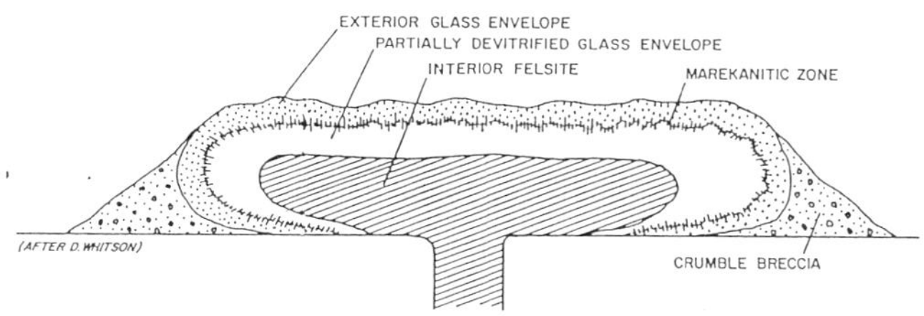

Fig. 1. Generalized stratigraphic section present at the No Agua, New Mexico deposit (*courtesy:* D. Whitson, "Geology of the Perlite Deposit at No Agua Peaks, New Mexico," *Industrial Rocks and Minerals of the Southwest*, New Mexico Bureau of Mines and Mineral Resources, Circular 182, 1982, 111 pp.).

of deposits. The following lithologic and textural features that are exhibited at the No Agua Peaks, NM, deposit clearly illustrate features that can be observed in deposits worldwide and that can be broadly applied to modeling perlite deposits.

Progressing towards the interior of the flows and domes, the textural subunits of the exterior glass mantle are distinguished as admixtures of pumiceous (shardy), granular and classical textural types. The textural types are recognized by the degree and distortion of primary vesiculation and by the presence of distinctive fracturing. These subunits are gradational in nature, and are recognized by the predominance of individual textural types. At the base of the exterior perlite mantle and below the zone of perlite that is commercially important, the hydration of obsidian to perlite is incomplete and the distinctive marekanitic zone is present, typified by numerous remnant obsidian nodules. The remnant obsidian nodules, or Apache tears, occur as isolated or interlocking grains that form pods of varying dimension. This unit extends as a gradational zone into the outer portion of the interior devitrified glass envelope, which, in turn, encases the crystalline, felsite core. As the stratigraphic position of these units exhibits predictability, the zones can be used as markers for placement within the section, and consequently, can be used to guide exploration and mine planning.

Common Textures of Perlite

Pumiceous Perlite: Perlite at the exterior of the deposit is typified by frothy, open vesicles which can show flattening and distortion due to compression and flowage shear (Fig. 2). Remnant obsidian cores and perlitic fractures, if present, are generally restricted to dense vesicle wall junctions and may not be readily evident megascopically. The friable nature of pumiceous textures can reduce core recovery during exploration drilling unless care is

Fig. 2. Hand specimen of shardy perlite showing open vesicles (*courtesy:* Celite Corporation).

Fig. 3. Hand specimen of granular perlite.

Fig. 4. Hand specimen of classical perlite. Note the concentric onionskin-like cracks.

taken by the drilling crew and the coring is sufficiently large in diameter.

Granular Perlite: The granular texture is characterized by saccharoidal or sugary appearance and by blocky fracture (Fig. 3). The gradational nature of this texture with dense pumiceous texture is evident from highly collapsed vesicles that impart a silky luster to surfaces that are nearly parallel with flow foliation. Well developed perlitic fracture and obsidian nodules are rarely observed in this type of perlite. As a denser textural variety, granular perlites typically mill well and are more successfully cored during drilling.

Classical Perlite: Well developed perlitic fracture and pearly to resinous luster are only displayed in the classical textural variety, and it is these perlites that initially served as the basis for the definition of perlite (Fig. 4). As viewed in thin section, this is the densest of the perlite types. Widely spaced vesicles are present, but are highly collapsed. Remnant obsidian nodules can be quite abundant and can occur as isolated grains or as pods of varying dimension. The initial appearance of obsidian grains in the classical zone, although gradational, can be abrupt, and the abundance can rapidly increase towards the interior of the unit or laterally into subzones in which hydration of the glasses has been incomplete.

WORLD PRODUCTION AND DISTRIBUTION OF DEPOSITS

Because hydrated natural glasses are thermodynamically unstable and alter with time to secondary minerals, most perlite deposits are associated with Tertiary or younger silicic volcanics. Patterns of worldwide distribution have been broadly described by Kadey, Jr. (1983) and Chesterman (1975) as being contained within circum-Pacific and Euro-African rhyolite belts. Clearly, distribution patterns generally correspond with island arc and continental volcanism related to plate boundary movement, or as in Iceland, are associated with volcanism at oceanic-spreading centers. The locations of mines and many reported occurrences are described here and are shown on Fig. 5. For clarity, many deposits have not been shown. In addition, there are many smaller perlite mining operations which have not yet been adequately identified with the worldwide perlite community and so have not been included. Occurrences in the western United States are too numerous to be shown adequately on a world scale and are identified on Fig. 6 to illustrate their abundance.

During 1990, worldwide production of perlite totalled roughly 1 778 000 tons (Bolen, 1991) with the major producing countries including the former Soviet republics, the United States, Greece, Turkey, and Hungary. Greece is the leading exporter of crude and processed perlite. Estimates of worldwide production dating back to 1980 are listed in Table 2.

In the United States, perlite mining operations are in New

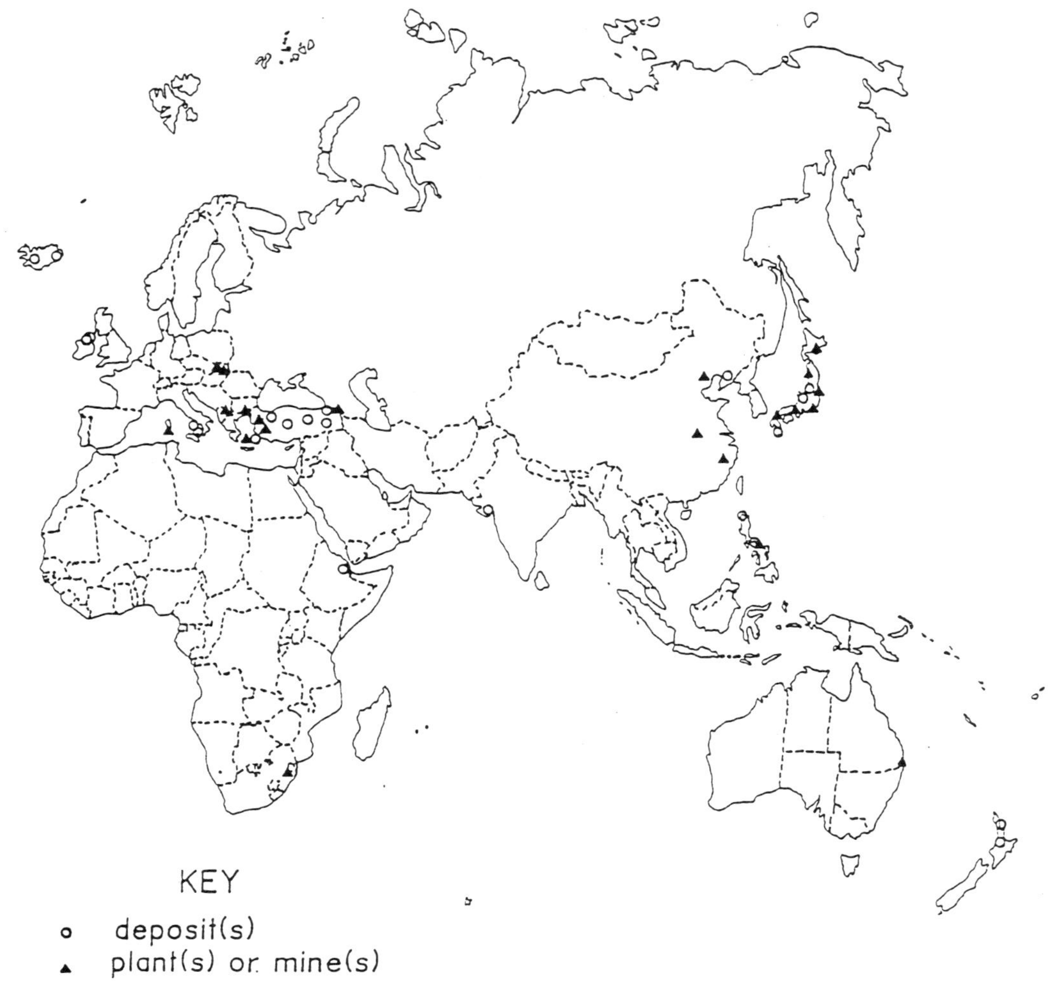

Fig. 5. Perlite deposits and processing plants, Africa, Asia, and Oceania.

Mexico, Arizona, California, Nevada, Idaho, Colorado, and Oregon. Perlite ore is expanded in 33 states at 61 plants. Information regarding foreign producers and production is often lacking or available only as estimates, but notable European mining operations are in Hungary, Bulgaria, Czechoslovakia, Yugoslavia, Armenia (USSR), Sardinia, Italy, and on the island of Milos, Greece. African production is only reported for South Africa. In Asia, mining operations are in western Turkey, Japan, the Philippines, and the People's Republic of China. Further south, mining has been undertaken in Australia, and until 1982/1983, also in New Zealand. Mexico and Argentina are the primary countries in Latin America mining perlite. Undoubtedly there are many additional small producers of perlite ore, as well as producers of expansible obsidian.

North America

North American production is dominated by the United States where, as of 1990, 10 companies operated 11 mines located in 7 western states. New Mexico continues to be the major producing state with the majority of production centered at No Agua Peaks in Taos County. The operations at No Agua include mines and mills owned by both the Celite Corporation and Grefco, Inc. Milled and sized perlite is trucked from No Agua to railheads at Antonito, CO, where Grefco also maintains an expansion facility. The No Agua deposit is the largest producer in the United States. Grefco also quarries and mills perlite at its deposit located in Socorro County, NM. The US Gypsum Corporation operates New Mexico's fourth perlite mine near Grants. Numerous other deposits in Sandoval, Hidalgo, and Grant Counties have attracted prospectors and entrepreneurs since the 1950s, but none has been mined for sustained periods of time.

Elsewhere in the United States, production centers on the large Pickett Post Mountain deposit near Superior, AZ, where, as of 1990, two companies operated mining and milling operations: Harborlite and Nord-Sil Flo, Inc. Nord Resources Corporation acquired Sil Flo during 1986. The perlite exhibits a full range of textures and is particularly well suited for filter aids. In Custer County, CO, Persolite Co. both mines and expands perlite quarried from the Rosita Hills deposit near Florence. Paramount Perlite, Inc. produces from a deposit at Caliente, NV, where the perlite is mined under contract by Wilkens Mining and Trucking Co., and is crushed onsite for transport to the expansion facility in Paramount, CA. Perlite deposits near Fish Springs, CA, and near Malad City, ID, are respectively mined and processed by the American Perlite Co. and by National Perlite Products (Oneida) which was acquired in 1986 by Oglebay Norton. Supreme Perlite operates an expander near Portland, OR, and mines the Dooley Mountain deposit in Baker County, OR.

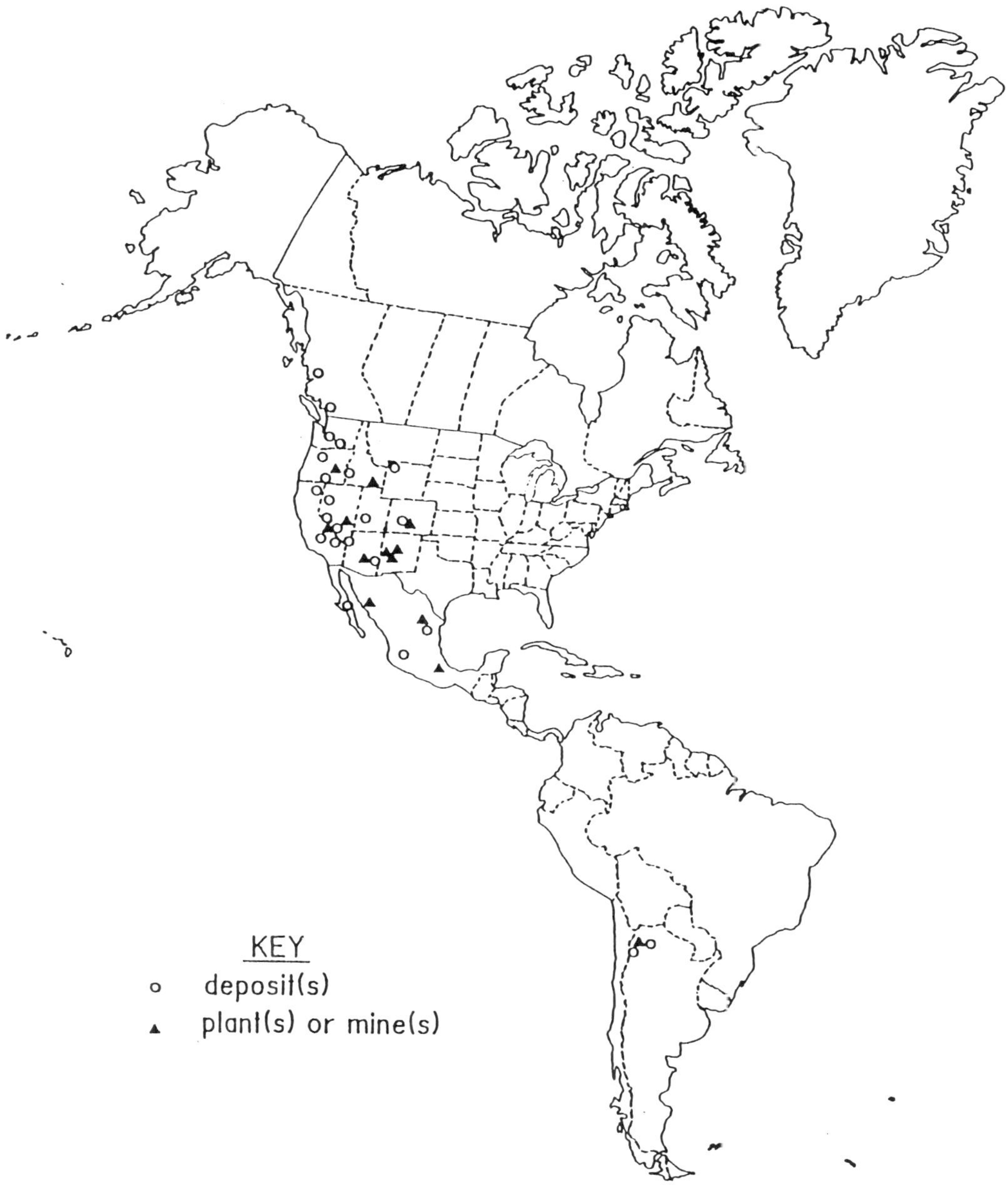

Fig. 6. Perlite deposits and processing plants—North and South America.

Until closure, Aurun Mines Ltd. operated the first perlite mine in Canada at Frenier, British Columbia. Perlite mined at Frenier was delivered to the Fraser crushing and expansion plant roughly 400 km from the mine. Further to the north in British Columbia, perlite is also reported near Burns Lake.

Latin America

Compania Minera Oriental SA is the largest of the Mexican producers. The company, which is partly owned by Grefco, Inc., bases its mining operation at the Dimesa deposit near Oriental in Pueblo. After milling, the perlite is expanded by Dicalite de Mexico SA de CV. Other producers include Termolita SA, based in the state of Durango with an expander located 600 km to the east near Monterrey, Neuvo Leon, and Aislantes del Pacifico SA based in Sonora near Ciudad Obregon.

Perfiltra SA operates two open pit mines and four plants in Argentina. The mining operations, Salar de Pocitos and San Antonio, are in western Salta Province at high elevation. The Perfiltra plants include a screening operation in Salta Province, screening and expansion facilities in Tucuman, and an expansion facility in Rosario, Santa Fe Province. In 1983, a filter aid plant was constructed in Paulinias, Sao Paulo. The facility is operated by Perfiltra do Brasil and utilizes perlite ore from the Perfiltra mines in Argentina. Several perlite occurrences are near San Antonio de los Cobras to the east of the Perfiltra mines and to the south in Catamarca Province.

Table 2. World Production[1] (in thousands of tons)

Country	1980	1981	1982	1983	1984	1985	1986	1987	1988	1989[p]	1990[e]
Australia[2]	1.8	1.8	0.9	2.7	3.6	2.7	3.6	5.4	4.5	5.4[e]	5.4
Czechoslovakia	39.9	42.7	45.4	44.5	44.5[e]	44.5[e]	41.7	41.7	43.5	44.5[e]	43.6
Greece	197.8	131.6	151.5	151.5	177.9	161.5	184.2	208.7	208.7[e]	157.9	159.7
Hungary[2]	98.9	95.3	89.8	93.5	94.4[r]	94.4	109.8	112.5	120.7	108.9	109.8
Italy[e]	90.7	85.3	79.9	75.3	79.9	79.9	73.5	69.9	69.9[r]	70.8	70.8
Japan[e]	77.1	75.3	75.3	75.3	75.3	75.3	75.3	75.3	75.3	77.1	77.1
Mexico[2]	44.5	57.2	32.7	41.7	31.8	37.2	46.3	39.0	39.0	37.2	36.3
New Zealand[2]	0.9	0.9	1.8	0.9	—	—	—	—	—	—	—
Philippines	8.2	7.3	3.6	1.8	15.4	3.6[3]	3.6	5.4[r]	7.3[r]	0.9[r]	5.4
Turkey	25.4	45.4	121.6	29.0	60.8	59.9	103.5[r]	144.3	154.3	143.4	145.2
USSR[e]	363.0	598.9	598.9	598.9	598.9	598.9	598.9	598.9	598.9	598.9	544.5
United States[4]	578.9	536.3	459.2	430.1	451.9	470.1	460.1	483.7	522.7	545.4	579.9[3]
TOTAL	1 527.1	1 677.9	1 660.6	1 545.2	1 634.3	1 628.0	1 700.5[r]	1 784.9[r]	1 844.8	1 790.4	1 777.7

[1]Mined and/or processed; does not include production in Algeria, Bulgaria, China, Iceland, Mozambique, the Republic of South Africa, South America, and Yugoslavia, [2]crude ore, [3]reported figure, [4]processed ore sold or used by producers, [e]estimated, [r]revised, [p]preliminary.
Source: Meisinger, 1980–1988, Bolen, 1989–1991.

Europe

Greece is the largest producer and exporter of perlite in western Europe. Deposits are reported on many of the Aegean islands, including Kos, Lesbos, and Kimolos, as well as on the mainland, but production centers on the island of Milos. The major producers include Silver and Baryte Ores Mining Co., Otavi Minen Hellas SA, Milopan SA, and Peletico Minerals Ltd, a subsidiary Cypriot Peletico Ltd. All the companies operate milling and shipping facilities on the island. Peletico ships to the crushing and expansion facility operated by Peletico Plaster in Lanarca, Cyprus. Little is known about N. Bouvas and Co. which reportedly mines perlite on the island of Kos.

Italy is a major Mediterranean producer and exporter. While perlite is reported on the mainland as well as on the islands of Ponza and Lipari, all production comes from Sardinia. Perlite SpA owned in part by British Gypsum, is the largest of the operations. Its deposits are near Uras in Oristano Province in west central Sardinia and are situated to the east of their processing facility at Torre Grande. Loading facilities and a pierhead are maintained at Torre Grande. Ceca Italiana SpA, a subsidiary of Ceca SA of France, also mines perlite in Sardinia and in addition maintains expansion plants in Salerno and Pioltello.

Turkey is a major perlite supplier and is considered to have vast reserves. Numerous deposits occur throughout eastern, central, and western Anatolia. Production is mainly from occurrences in the western portion of the country. Major producers include Perlisan Mining and Construction Ltd. Co. and its associate Perlitas group with a mine and milling facility at Maruflar southeast of Bergama; Ege Industrial Minerals Industries, Inc., a member of the Kiska Construction Corporation Group, which has mines and a mill in the Bergama vicinity; Pabalk Madencilik Sanayi AS operates mines in the Biga vicinity east of Canakkale and maintains a mill near Karabiga and an expansion facility at Tepebasi on the outskirts of Istanbul; and Etibank, the state mining concern, with mines near Manisa and a processing plant near Cumaovasi.

Numerous deposits and significant production is reported from eastern European countries and the former Soviet republics, but information regarding producers is not readily available. As the largest world producer, production in the former Soviet republics is believed to center at Aragats in Armenia and at Beregovsk in the Ukraine. Large reserves are reported in Buryat, ASSR, near Mongolia, and in Kazakhstan Province. In Hungary, numerous deposits occur in the northeast portion of the country (Ilkey-Perlaki, 1987). The National Ore and Mineral Mining Co. mines and mills perlite at Palhaza. Mineralimpex markets Hungarian production. The only reported production in Czechoslovakia is from the Lehotka pod Brehmi deposit in the northwestern part of the Stiavnicke hory Mountains in the eastern portion of the country. The Jastraba deposit and other deposits are reported in the eastern portion of the country. Large perlite deposits are reported within the Eastern Rhodope region of Bulgaria near Greece, but mining is only reported at the Broken Mountain deposit (Yanev, 1987), at Dzebel, and at Svetoslav. In Yugoslavia, perlite has been mined in the Mariovo District of Macedonia.

Elsewhere in the European community, perlite occurs at Sandy Braes in County Antrim, Northern Ireland. The deposit has been dated at 65 million years old and was mined until 1968. Two perlite deposits in Iceland have repeatedly attracted attention since the 1960s. The foremost occurs at Priest Mountain, 100 km from Reykjavik. During the 1970s the Icelandic government developed a small pilot classification and expansion plant to evaluate the Priest Mountain ore, however, the facility has since closed. The perlites are reported to be friable and to contain obsidian. Perlite at Lodmundafjordur is described in a dike occurrence located near Seydisfjordur along the east coast. Although perlite occurrences are reported in France and Germany, the deposits are not considered economically viable (Anon., 1977).

Africa

The Republic of South Africa is the only known producer of perlite, with two operations, Chemserve Perlite Ltd. and Prately Perlite Mining Co. Chemserve operates a surface mine in the Lebombo Mountains and a mill and expansion facility at Olifantsfontein (Anon., 1989a). Prately also produces perlite in northern Zululand and processes ore at Krugersdorp, Transvaal; the plant operated by their wholly-owned subsidiary, Prately Perlite Products (Pty.) Ltd. (Anon., 1989a). Deposits are known in Djibouti and Morocco, but production has not been reported. The US Bureau of Mines has previously reported production in Mozambique and Algeria.

Asia and Oceania

Numerous mines operating on the islands of Honshu, Hokkaido, and Kyushu have been developed in Japan. The major producers include Mitsui Metal, Asano Perlite, and Toko Perlite (Showa Chemical). In addition, several companies produce expanded obsidian. Production in China is not widely known, but numerous mining operations undoubtedly exist. Production from over 100 quarries is reported to exceed 450 ktpy (Anon., 1987a). The better known mining areas are located near Jianping in Liaoning Province, the Jinyun mining area in Zhezjiang Province, and in the Xinyang vicinity of Henan Province. Perlite deposits also occur within the provinces of Heilongjiang, Anhui, Jilin, and Shanxi. Many of the Chinese expanders are coal-fired, although several are known to use fuel oil. Further south, perlite has been mined in the Philippines at several locations on the northern island of Luzon. Induplex, Trinity Lodge Mining Corp., and Asaphil Construction and Development Corp. mine in the Legaspi City vicinity. Asaphil also mines at Bagumbayan, Baao, and Sun Chemicals Corp. mines at Bicol.

Production in Australia has centered on the Australian Perlite Pty. Ltd. Numinbah deposit located in Queensland. Ore is trucked south to the expansion plant in the Sydney vicinity. The Kenlieth deposit on the North Island of New Zealand historically has been the center of production there and most recently has been mined by New Zealand Perlite Ltd. Substantial reserves are reported on Great Barrier Island off the coast, northeast of Auckland (Kadey, Jr., 1983).

Reserves: The US Bureau of Mines estimates the worldwide reserve base at in excess of 1.8 Gt (1990), and given the numerous deposits worldwide, there will be no shortage of perlite in the foreseeable future.

EXPLORATION

Deposits of perlite occur commonly throughout the world and geologically the outlook is excellent for discovery and development of new sources of crude perlite. Economically, the competitive pressure exerted by existing producers within market areas will impair the viability of new sources which would supply the same market and end users.

Fundamental strategy for the exploration for perlite requires an integrated approach incorporating the recognition of geologic occurrence and lithologic and temporal modeling. The association of perlite with Tertiary to Early to Middle Quaternary silicic volcanics, and the attractiveness of flow/dome-type occurrences for commercial mining provide general parameters to guide regional reconnaissance programs for new, undeveloped perlite sources. Rhyolitic complexes of Oligocene to Early to Middle Quaternary age have proven to be successful target areas for initial field reconnaissance. The glasses within this age range are old enough to have hydrated extensively and are young enough to have escaped extensive erosion and devitrification. More recent glasses are likely to be incompletely hydrated and to have less commercial potential as sources for high quality perlite. As the hydrated glass selvages of flows and extrusive domes are often extensive and can be easily mined using low cost open pit methods, this occurrence type is a particularly attractive target for reconnaissance. Moreover, the perlite tuffs and breccias, which are often associated with domes and flows, can offer additional reserve potential that can be easily recovered at low cost.

Through the study of many deposits, versatile lithologic and textural models have been developed that are used in the exploration and evaluation of deposits. While the presence and the development of textures and zonation are variable among deposits, the features exhibited at the No Agua deposit in New Mexico are commonly observed in flow/dome occurrences throughout the world. No single model can be applied to all occurrences. However, the rock types, textures, and zonation as described at No Agua, provide useful bases for geologic mapping and for the uniform comparative description and interpretation of deposits. Moreover, the basic zonation of lithology and texture provides a model to assist in the identification of potential ore bodies and in the prediction of ore zone location and lithologic trends. Through the prediction of lithology, the models serve to guide the placement (and termination) of drill holes, the elimination of unnecessary or unadvisable drilling sites, and eventual mine planning.

Traditional methods used for exploration of orebodies include deposit mapping to identify source vents and flows, detailed surface mapping, core drilling, and trenching. Detailed mapping is employed to delineate basic rock types, perlite textures, and the abundance of contaminants such as clay, felsite, and obsidian. Mapping and modeling can be used in conjunction with geophysical techniques for the prediction of trends and the location of concealed orebodies. Core drilling and trenching are necessarily used to assess subsurface and lateral continuity of orebody lithology and to permit evaluation sampling. Typically, drill hole spacing of 60 m is used to delineate ore bodies and to provide the basis for ore reserve calculation. Optimum spacing and placement of drill holes, however, must reflect the variability of lithology and ore quality within individual deposits. The core diameter must be large enough to ensure high core recovery and to provide sufficient sample for evaluation testing. Where perlite quality is uniform and, where the geology is well known, percussion or reverse-circulation air drilling can be used.

MINING, MILLING, AND PROCESSING

Mining Methods

Throughout most of the world the lower cost of surface mining relative to underground mining drives producers to use open pit methods. Most perlite mines use either dozer ripping or blasting, or a combination of the two. Where perlite ores are soft and friable, brecciated, or extensively jointed, dozer ripping can be employed with significant cost savings. The development of benches and blasting is required where perlite is hard and competent and cannot be readily broken using rippers, but care must be taken to achieve fragmentation without production of excessive fines or oversized material. Once broken, the perlite is loaded into trucks by front end loader, or alternatively, where dish shaped pits can be developed, the loosened crude can be collected by scrapers for transport to the mill. Routinely, selective mining is used where the quality of perlite or the abundance of contaminants is variable. Where ore bodies contain perlites of varying texture, the ores can be blended to produce a mill feed with consistent milling characteristics and with consistent suitability for market end users.

Milling Methods

In general, the ultimate size of the expanded perlite particle is determined by the particle size of the initial furnace feed. Consequently, the applications for which the crushed and sized perlite is to be used in its expanded form will determine the particle sizes into which the perlite ore must be milled. While the eventual particle size distribution achieved by mills will vary among producers, basic milling and classifying processes can be described as including primary and secondary crushing, drying, classification, tertiary crushing, and dust collection systems. Run-of-mine perlite ore, which has been stockpiled, is reduced in size through jaw crushing. Finer size reduction is accomplished by secondary jaw crushing or by cone crusher prior to drying. Following drying, particle size is further reduced through additional crushing, and the product is classified by vibratory screens and air classifiers. Variants on this general procedure include the use of gyratory crushers for primary

Table 3. Typical Size Distribution For Some Perlite Ores

	Celite Grade							
	PA-4000	PA-3000	PA-1000	PA-116	PA-130	PA-220	PA-420	PA-610
	Application							
Sieve Size	Silica Source Micro Bubbles		Plastic and Resin Filler	Cryogenics & Acoustical Tile	Roof Insulation Board & Filter Aid	Plaster Aggregate	Concrete Aggregate, Masonry, & Loose Fill	Horticulture & Foundry Slag
+8							0–2	10–25
8–16							22–30	
8–12								40–60
12–16								25–42
+20						7–16		
16–20								0–8
+30				Trace	0–9			
16–30							53–65	
20–30						26–35		
−20								0–4
30–50				4–12	48–75	40–52	10–21	
50–100				45–75	20–42	6–13		
−50							0–4	
−100					0–6	0–4		
+100		0–1	0–8					
100–200	0–15	0–15	20–47	20–40				
−200	85–100	85–100	50–75	0–10				

Source: Manville Bulletin FF 391, September, 1988.

crushing, and impact mills, rod mills, hammer mills, gyrodisks, and roll crushers in secondary and tertiary crushing. The basic mill size grades are stored in silos and final blends are made to meet the specialized requirements of customers (Table 3).

Expansion and Product Processing

When heated to softening, perlite can expand to as much as 20 times its original volume. Expansion is generally achieved at temperatures that range from 870 to 1 100°C in one of two basic furnace types: the stationary vertical expander and the rotary horizontal expander. Both operate on the same principle. Crushed and sized perlite ore is rapidly introduced into a hot zone, where volatilization of the combined water within the glass is coincident with the softening of the glass. The resulting frothy particle (Fig. 7) exhibits extremely low density, high surface area and light or white color.

The stationary vertical furnace is the more widely used of the two furnace types (Jackson, 1986). Sized perlite, which may or may not be preheated, is gravity fed through a side port into the hot zone where expansion occurs. The expanded particles are carried by updraft into the cyclone collection system. Nonexpandable particles pass through the hot zone and drop from the system, while shattered fines are carried upward and are collected in the baghouse.

Within a rotary horizontal furnace, the feed is introduced into the hot end such that the particles fall into the hot zone. As in the vertical furnace, the feed may or may not be preheated before introduction into the hot zone. As the furnace rotates, the feed travels by lifters and spirals down the length of the tube, and the expanded product is drawn by draft up into the cyclone and baghouse system. Following the cyclone sizing, the product may be milled and classified for use as filter aid or may be surface treated for other applications. Final products are either packaged in bags or shipped to customers in bulk rail and truck containers.

The densities and particle strengths of expanded perlite required by the end user varies by application. The final density and recovery of the product as well as the fuel efficiency of the expander can be improved through preheating the furnace feed (Jackson, 1986). For vertical furnaces, preheating is achieved in an external, separate heater. In horizontal furnaces, the feed is preheated as it is transported between the inner and outer shells of the furnace to the hot zone where expansion occurs. Preheating the coarser grades can be important due to the relationship between the ratio of volume to

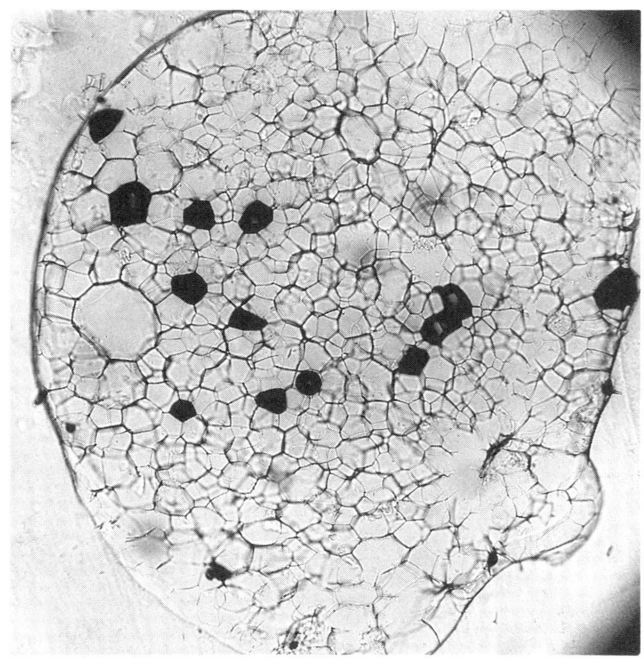

Fig. 7. A kernel of expanded perlite seen in thin section at 200X magnification (*courtesy* Celite Corporation).

surface area of the particle, heat transfer, and residence time in the hot zone of the furnace (Jackson, 1986). Although some flexibility of finished product properties can be achieved through adjustment of furnace conditions and through preheating, it is important to note that all perlites do not expand alike. Kadey, Jr., (1983) has written that perlites containing excessive combined water are likely to shatter upon expansion, whereas perlites which contain insufficient water are likely to produce higher density products. The ultimate suitability of crude for market end uses must be measured by the inherent ability of each perlite ore to meet end use specifications.

TESTING AND SPECIFICATIONS

Perlite deposits are evaluated for suitability in four main end uses: horticulture, construction, filteraids, and fillers/extenders. Basic testing of expanded perlite is done for exploration. Some additional tests of perlite aggregate are done for horticulture and construction end uses. A special grinding regime and chemical tests are added to filteraid-grade perlite testing owing to its use in the food industry. Use of microspheres of perlite as a filler/extender is relatively new and testing procedures and specifications are often kept confidential. One of the most important tests for all end uses is for the crystalline silica content so compliance requirements related to health regulations can be determined.

Coincident with the growth of the perlite industry, the need for product development and common standards and specifications was recognized. In 1949, the Perlite Institute was formed to provide an international industry-based forum chartered to develop and provide technical and engineering knowledge regarding all aspects of the perlite industry. To maintain quality and performance standards, the Perlite Institute has assembled a comprehensive testing procedure and specification manual which includes tests developed by the Perlite Institute, as well as specifications established by the American Society for Testing and Materials (ASTM). The Institute is also involved with the International Congress of Building Officials, the International Standards Organization (ISO), and the European Committee for Standardization (CEN). The *Manual of Test Methods* is available from the offices of the Perlite Institute in New York City. A partial listing of Perlite Institute tests and ASTM specifications that apply to perlite follows:

PI 108–77 Bulk Box Car Sampling of Sized Perlite Ore
PI 109–77 Sampling Sized Perlite Ore from Hopper Cars
PI 110–77 Sampling of Perlite
PI 111–77 Preparation of Perlite Sample for Testing by Coning and Quartering Method
PI 112–77 Preparation of Perlite Sample for Testing by Splitter or Riffle Method
PI 113–77 Sieve Analysis of Fine Materials
PI 114–77 Sieve Analysis of Expanded Perlite Volume Basis
PI 115–77 Sieve Analysis of Expanded Perlite Weight Basis
PI 116–77 Fractional Density of Expanded Perlite
PI 117–77 Wet Screen Analysis
PI 118–77 Determination of Free and Combined Moisture in Perlite
PI 200–77 Loose Weight Determination of Expanded Perlite
PI 201–77 Compacted Density
PI 202–77 Test for pH Value of Expanded Perlite
PI 300–77 Determination of Floats and Sinks of Expanded Perlite
PI 301–80 Percent Non-expansibles in Perlite
PI 303–85 Water Repellency Test
PI 305–77 Yield Test for Perlite Aggregate
PI 306–80 Compaction Resistance Test
PI 307–77 Perlite Ore Expansibility Test
PI 400–77 Setting Time of Gypsum Plaster
PI 401–77 Plaster Coverage Test
PI 501–83 Method for Determination of Wet Density and Relative Water Permeability of Powder Type Filter Aids
C 28–80 Specification for Gypsum Plaster
C 29–78 Unit Weight and Voids in Aggregate
C 35–76 Inorganic Aggregates for Use in Gypsum Plaster
C 136–83 Sieve or Screen Analysis of Fine and Coarse Aggregates
C 331–81 Lightweight Aggregates for Concrete Masonry Units
C 332–82 Lightweight Aggregates for Insulating Concrete
C 472–79 Physical Testing of Gypsum Plasters and Gypsum Concrete
C 495–80 Compressive Strength of Lightweight Insulating Concrete
C 513–82 Securing, Preparing, and Testing Specimens from Hardened Lightweight Insulating Concrete for Compressive Strength
C 520–81 Density of Granular Loose Fill Insulations
C 549–81 Perlite Loose Fill Insulation
C 926–81 Application of Portland Cement-Based Plaster
C 11–81 Wire-Cloth Sieves for Testing Purposes

Discussion of testing procedures for expanded and unexpanded perlite grades follows.

Crystalline Silica Testing

Quartz has been of interest in occupational health for many years because it can cause pulmonary fibrosis (silicosis) if exposure is prolonged. Although there are three common polymorphs of crystalline silica, quartz is by far the most abundant form found in perlite, so initial research has concentrated on trace analysis or quartz in both ore and expanded perlite (Hamilton and Peletis, 1988, 1990, Miles and Hamilton, 1991).

Traditional XRD analysis of quartz in industrial dusts uses well established procedures (Klug and Alexander, 1974) that are sensitive to quartz only to the level of 1 wt%. In order to meet the new regulations, the development of an XRD method to determine quartz in perlite at the 0.1 wt% level was initiated (Hamilton and Peletis, 1988). Their method emphasized careful preparation of samples and standards in conjunction with very long counting times to obtain a satisfactory signal to noise ratio.

Subsequent study has confirmed and extended procedural methods (Barker and McKee, 1990, McKee et al., 1990, Renault et al., 1991, 1992). Recent work has concentrated on four methods of analysis for trace quartz; three using XRD, a fourth using XRF, and all utilized deconvolution to enhance the signal to noise ratio (Renault et al., 1992) The XRD methods used: 1) natural perlites as standards, 2) artificial mixtures of quartz and glass as standards, and 3) a new method of additions. The XRF method uses chemical analysis and a mass balance calculation of quartz in glass mixtures.

The scatter of data from natural perlite standards is high. Inhomogeneity due to small samples is a serious problem partly addressed by the method of additions. The best calibration curves are from mixtures of quartz and window glass. Those curves have a slope near unity and give estimates of quartz in the perlite standards which are closest to those of the new method of additions than the calibration obtained from the natural perlites.

The new method of additions is particularly useful for analysis of materials in which the mass absorption coefficient of the matrix differs from that of the analyte. This is important in the analysis of industrial mineral products because mines and quarries usually have variable bulk compositions. The method is more useful for determining trace concentrations than high concentrations because it is

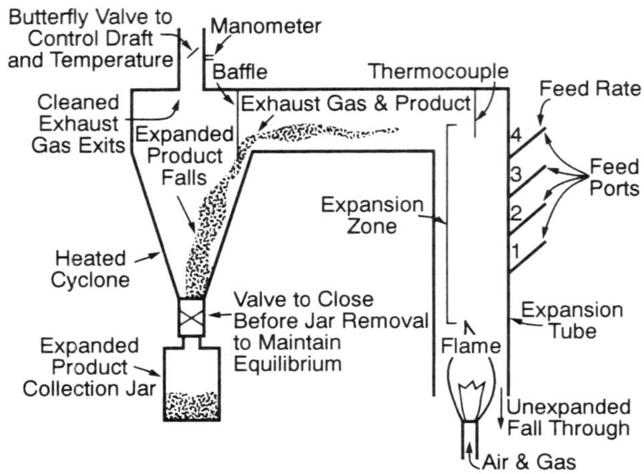

Fig. 8. Schematic diagram of a laboratory perlite furnace showing material flow and major control points (after Barker and Harris, 1990).

sensitive to the difference in XRD intensities of the analyte in the spiked and unspiked samples.

The chemical mass-balance method was disappointing at low quartz concentrations because the CaO, and possibly the SiO_2, analyses could not be made with sufficient accuracy and precision. At concentrations above 2.5 wt%, the results show very close correspondence with the amounts of quartz as mixed.

Laboratory Testing

Facilities for testing perlite, both unexpanded and expanded, are sparse on a commercial, cost-effective, or readily accessible basis. Little information or data are accessible regarding perlite testing other than from the methods and standards published by the Perlite Institute (Anon., 1984) and several related tests of ASTM.

Perlite testing is not an exact science, interpretation of results requires experience and scaling up from laboratory results demands caution. Scaleup problems abound in designing full-scale plants based on bench-scale expansion tests. No correlation has been demonstrated between bench-scale expansion tests using horizontal furnaces and similar tests in commercial plants. Scaling up from vertical furnace bench testing to full-scale testing has better correlation, but full-scale testing must be done prior to commitment of resources to investment, mining, or processing of perlite (Kadey, Jr., 1983).

Furnace Operation: Efficient expansion of perlite particles is highly dependent on their surface area to volume ratios. A narrow, unimodal size distribution is preferred for testing. The small size of the laboratory expansion furnace (Fig. 8) makes it sensitive to feed size. Samples must be crushed and sized prior to expansion in a vertical expansion furnace. Expansion is best using the −50+100-mesh fraction, although the −30+50-mesh fraction can also be used.

Perlite expansion is a function of its composition (especially water content), softening temperature, furnace temperature, and particle residence time in the flame. Additional factors affecting perlite expansion are particle fall-through time, humidity, barometric pressure, and ambient air temperature (Fig. 9). Temperatures as low as 510°C can be used in the laboratory furnace.

Perlite can be classified as lively or dead based on its water content and softening temperature (Murdock and Stein, 1950). Dead perlites generally have less combined water and higher softening temperatures than lively perlites. More heat must penetrate dead

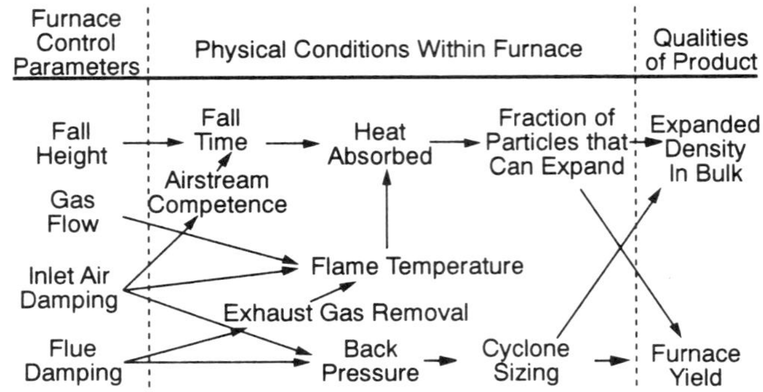

Fig. 9. Synergistic interactions between control settings, furnace conditions, and product quality in a perlite expansion furnace (after Barker and Harris, 1990).

perlite particles to achieve good expansion. This can be accomplished by preheating the perlite feed, increasing the furnace temperature and/or the particle residence time in the furnace.

Testing Expanded Perlite: The Perlite Institute test manual (Anon., 1984) and Barker et al. (1987) give descriptions of testing equipment, procedures, and calculations. Expansion and testing of perlite over time unavoidably is done under varying conditions so a standard must be used. The higher the expanded density of an unknown perlite sample compared to the standard, the more energy-intensive perlite production will be. Test results can also be reduced to the ratio between an unknown sample and the standard to dampen some of the variability, or can be normalized on the standard.

Loose Weight Determination of Expanded Perlite (Expanded Density): Laboratory expanded perlite is handled less than commercially expanded perlite. Laboratory perlite, because it is less milled, has lower expanded density for otherwise identical perlite. Commercial feed, and hence product, often is not as well sized as the material tested in laboratories. These relationships must be considered when scaling up to commercial production. The density to which a perlite expands affects many of its commercial properties so the expanded density measurement is crucial to assigning a given perlite to its most favorable end uses.

Furnace Yield: Furnace yield is the weight of expanded perlite recovered from the furnace relative to the weight of perlite fed into the furnace. Commercial perlite expanded to near its optimum expansion temperature has high furnace yields, often in excess of 90%.

Compacted Density: The Perlite Institute test procedure for compacted density of expanded perlite requires that a compacting machine be custom-built. Measurement of the density of expanded perlite allowed to settle into minimum space is of technical and economic significance for insulation. The thermal conductivity of expanded perlite insulation under atmospheric pressure correlates directly with its density and cost in place (Anon., 1984).

Compaction Resistance: Compaction resistance determines the resistance of perlite to crushing under a standard compressive load. The standard for commercial grade ore is 355 kg/m^3. Perlite testing less than 97 kg/m^3 is sometimes sold as lightweight aggregate. Compaction resistance is directly proportional to the expanded density of the sample. The log of compaction resistance vs. expanded density is nearly a straight line. Compaction resistance can be estimated over the linear range of densities for all practical expanded densities, based on a few measurements at varying densities (Barker and Harris, 1990).

Percent Nonexpansibles in Expanded Perlite: The nonexpansibles or sinkers are composed of inclusions, partially expanded perlite, or shattered perlite (bubble-wall shards). Common inclu-

Table 4. Perlite Filteraid Tests Recommended by the Perlite Institute and the US Food Chemical Codex (USFCC)

Test	Expanded	Expanded & Milled
Material balance		
Expansion	x	
Milling		x
Loose weight	x	x
Shattering	x	
Color	x	x
Compaction	x	
Particle size distribution	x	x
Sinks, sinks density	x	
Wet density		x
Floaters content		x
Filtration rate		x
Acidity (pH)		x
USFCC		x
Crystalline SiO_2		x

Source: Bodycomb, 1987, written communication.

sions are microlites, crystallites, spherulites, and phenocrysts of quartz, feldspar, biotite, hematite, hornblende, magnetite, sphene (Weber and Austin, 1982), clay, and obsidian. The petrography of the nonexpansible fraction should be described.

The proportion of sinkers in expanded perlite is partly dependent upon its expanded density as set by the furnace expansion conditions. Perlite expanded near its optimum expansion temperature range will have fewer sinkers than if it were underexpanded or overexpanded. If a perlite is underexpanded, particles which could be fully expanded are not expanded or are only partially expanded. Overexpansion may yield higher expanded density and high sinkers due to the destruction of intact coherent bubbles producing bubble-wall shards that sink readily.

Relative Brightness: Perlite used in paints, plaster, plastics, concrete, and insulation may be required to meet strict brightness or whiteness specifications. Brightness of expanded perlite is influenced by particle size, packing, and void space. Coarser particles are less bright by a few percent owing to light lost into the larger intergranular pores. Closely packed particles will have less void space, and therefore, higher brightness. Brightness is indirectly influenced by expanded density, which itself is partially controlled by the amount of impurities which tend to be darker than expanded perlite.

Sieve Analysis: Sieve analysis of perlite before and after expansion are compared. The crossover point on a single plot of both curves is used, particularly in Europe. Wide variance between the amount of expanded perlite in each size fraction is undesirable if a particular size range is targeted for production, as other size fractions are costly waste. Expanded perlite particle-size distribution is set mainly by feed size. An increase in the coarser size fraction upon expansion is desirable for certain applications such as horticulture where the material acts partly as a propant for soil conditioning. Abundant fines, some of which may result from milling of the particles during sieving, is beneficial to filteraids. Expanded density measurements of individual size fractions may indicate inhomogeneity between size fractions as impurities may be partitioned by size. Recognition of this partitioning aids in matching specific size fractions of the perlite to the most suitable end uses and may reveal size fractions which could be eliminated to produce a higher grade product.

Testing Filteraids: The tests recommended for expanded perlite or expanded and milled perlite are in Table 4. The test methods are from the appropriate sections of the Perlite Institute *Handbook* (Anon., 1984), the US Food Chemical Codex (USFCC), and F.M. Bodycomb (written communication, 1987). These data generally cannot be interpreted separately and scaling up from laboratory results must be done with caution.

Expansion and Yield. A sample of −50+100-mesh ore is expanded to a loose weight of 0.03 to 0.04 kg/m^3 while minimizing shattering. The percent recovery (yield) is also calculated.

Milling. The milling of expanded perlite is the central filteraid test. Insufficient milling produces excessive floaters. Aggressive milling yields filteraids with excessive fines, giving slow filtration rates. An overmilled filteraid has high density, negating the economy of usage of low density filteraids.

In the laboratory, expanded perlite is progressively reduced in particle size to a filteraid product through hammer milling and sieving to a −100 mesh fraction. Any oversize, +100 mesh perlite, is recycled through the mill. All −100 mesh fractions are thoroughly, but gently, recombined to produce the filteraid used in subsequent tests. The particle size distribution of the recombined filteraid is determined by sieving using 16, 30, 50, 100, and 200 mesh screens. The particle size distributions of the three sieve fractions, compared to a standard, are an indication of the ease of milling the perlite (Bodycomb, written communication, 1987).

Loose Weight. Loose weight of expanded perlite measures the value of perlite ore, expansion efficiency, and suitability for filteraids. It is run on expanded perlite both before and after milling. The test is operator sensitive so careful adherence to procedures is mandatory for repeatability.

Shattering. A sieve result on expanded, unmilled perlite in excess of 10% −100 mesh indicates friable particles or a lively ore of questionable use as a filler, aggregate, or in horticulture. A high −100 mesh content indicates the ore has a tendency to shatter.

Color. The color of a filteraid should be white and free from visible dark particles. A visual evaluation is recommended. While this is a cosmetic property that does not affect utility, the notable presence of dark particles can affect sales into markets where appearance is important.

Compaction Resistance. A high resistance to compaction indicates stronger particles, possible greater difficulty in milling, and higher loose weights. This test, run on expanded perlite before milling, is useful for evaluation of filteraids but is of less importance than other tests.

Particle Size. Expanded perlite should be sieved before milling and at various stages of the milling process including filteraid product. This aids in evaluating the mill feed, ease and efficiency of milling, and characterization of the filteraids.

Sinkers and Sinkers Density. These tests are run on expanded perlite before milling to determine the quantity of unexpanded or shattered material and the density of the sinks. Low sinks density indicates shattered perlite and high density, unexpanded contaminants. A minimum of high density sinks is desirable but is not always objectionable.

Wet Density. In part, wet density measures economy of usage, the density of wet filtercake under pressure or vacuum, and the clarifying ability of the filteraid.

Filtration Rate. Filtration rate in conjunction with wet density is a highly valuable yardstick for comparing filteraid samples. Using water which is pulled under vacuum through a carefully prepared column of filteraid, the relative permeability of filteraids can be directly measured and compared with other filteraid products.

Floaters Content. The percent floaters has a profound effect on all filteraid test results. As the presence of non-sinking intact bubbles disrupts formation of efficient filter cake, floaters should be minimized during milling, taking care not to overmill the product.

Table 5. Expanded Perlite Sold and Used by Producers in the United States, by Use (% of Total)

Use	1987	1988	1989	1990
Concrete Aggregate	1.6	2.2	2.2	2.2
Fillers	6.2	5.5	5.5	5.0
Filter aid	14.2	13.1	13.7	15.8
Formed products[1]	59.5	59.1	56.3	55.5
Horticultural products[2]	9.0	10.3	9.6	9.4
Low temperature insulation	0.8	1.3	1.2	0.9
Masonry and cavity fill insulation	2.9	2.2	2.0	2.5
Plaster aggregate	1.7	2.3	3.5	1.9
Other[3]	4.1	3.9	6.2	6.6
Total in tons	422 868	434 664	469 147	478 221

[1]Includes acoustic ceiling tile, pipe insulation, roof insulation board, and unspecified formed products, [2]includes fertilizer carriers, [3]includes fines, high temperature insulation, paint texturizer, refractories, and various nonspecified industrial uses.
Source: Meisinger, 1980–1988, Bolen, 1989–1991.

Table 6. Average Price in $ Per Ton in the United States, FOB Mine

1980	28.50
1981	32.48
1982	34.94
1983	36.42
1984	36.82
1985	37.30
1986	37.31
1987	34.11
1988	34.05
1989	34.23
1990[e]	33.79

[e]estimated
Source: Meisinger, 1980–1988, Bolen, 1989–1991.

Acidity (pH). The pH of a perlite filteraid extract indicates the calcium content for certain food and beverage applications.

Food Chemical Codex. Perlite filteraids should be tested for compliance with worldwide food codex standards. By example, USFCC standards establishes limits as follows: loss on drying (3% powdered form), LOI (7%, glassy form), pH (5 to 9 in filtrate from 10% suspension), and, importantly, no more than 10 ppm As or Pb. The mercury content should be measured, but is not required under the USFCC. The testing is often by inductively coupled plasma (ICP) methods.

Crystalline Silica. Testing for crystalline silica content of perlite filteraid should be done as described previously.

TRADE, PRODUCTION, AND MARKETING

Due to the cost of transporting ultralight expanded perlite, most perlite enters international trade in unexpanded form. Greece, Italy, the former Soviet republics, Turkey, and Hungary are the largest exporters, and trade primarily into Europe. Imports to the large European customers from sources in Turkey, Greece, and Sardinia are made by boatload to ports with convenient road or rail access to the expanding plants. Ocean freight exports from Greece to the United States have continued from 1983 to the present. In the United States and Canada, shipping to customers is generally by rail, and, as in Europe, the perlite is expanded locally for redistribution within nearby geographic markets.

Although unexpanded perlite is used as a slag coagulant, abrasive, metal cleanser, and silica source, expanded perlite comprises the majority of worldwide consumption. Most of the market for expanded perlite is in construction-related products such as roof insulation board, acoustical ceiling tile, pipe insulation, concrete aggregate, and plaster aggregate (Table 5). The market as a whole is sensitive to the impact of economic trends on the construction and industrial sectors. Following level production from 1981 through 1986, world production has shown steady but modest growth (Bolen, 1991, Meisinger, 1985, Table 2). Average price per ton in the United States (FOB mine) peaked during 1985 and 1986 at roughly $37.31 and has since declined to its 1990 level of $33.79 (Bolen, 1990, Meisinger, 1985, Table 6).

The physical properties of expanded perlite that give this product unique commercial value include low bulk density, physical resilience, chemical inertness, low thermal and acoustical conductivity, fire resistance and nonflammability, and water retention capability.

In the building construction industry, perlite is mixed with mortar, concrete, and plaster to impart thermal and acoustic properties, lightweight, and fire resistance. Applications include lightweight concrete used for roof decks, floors and roofing tiles, and precast components. In formed construction products, perlite is used in acoustic ceiling tile, pipe insulation, and roof insulation boards. These products are lightweight and provide excellent insulation. As a loose-fill insulation, applications include lofts, floors, and wall cavities. For loose-fill applications, perlite can be treated with silicone to increase water repellence and to improve the insulating properties. Inorganic foamed perlite products have been developed that can be injected as infill. Perlite is also used as a cryogenic insulation.

Milled perlite finds wide application in the filtration of industrial effluents, sugar, fruit juices, oils, and other fluids. It has particular suitability for application in rotary precoat filtration.

In horticultural applications, perlite is used to condition soils by adding loft and reducing compaction, and through facilitating water drainage and moisture retention. Perlite is also used as a propagating medium for seedlings and for the packaging and storage of bulbs and plants. In the United States, the horticultural market has grown steadily since 1985 (Larsen, oral communication, 1990).

Perlite is used as a filler in textured paints and as an extender in plastics. A sampling of expanded perlite grades and applications is as follows:

Type	Application
Superfine	Silica source (unexpanded perlite) and microbubbles
Fine	Manufacture of PVC and nylon Plastic, paint, and resin filler
Medium	Cryogenics, acoustical tile, filter aid, mortars, plasters, roof board, and general insulation
Coarse	Concrete aggregate, masonry, and loose fill
Very coarse	Horticultural applications, oil absorbents, and foundry slag (unexpanded perlite)

HEALTH AND ECOLOGY

Perlite, while mostly composed of amorphous volcanic glass, does contain trace amounts of crystalline materials such as quartz and feldspar. The crystalline silica in perlite occurs primarily as both alpha and beta quartz and rarely as trace amounts of cristobalite or tridymite, which are about twice as reactive in the body as quartz (Sax and Lewis, 1988). While quartz is present in low amounts within perlite, generally ranging from less than 1 to 5%, it comprises about 12% of the earth's crust and 17% of surficial materials. Little of this occurs naturally as respirable dust of less than 10 μm in diameter. However, when perlite is mined, processed or utilized, some of the crystalline silica may become airborne. In the past, this has been treated and regulated as a nuisance dust with appropriate exposure limits enforced. The relationship of pulmonary fibrosis in workers exposed to crystalline silica dust for widely varying lengths of time is well documented (Sax and Lewis, 1988) and the perlite industry has addressed these health issues for many years.

In 1987 and 1988, the International Agency for Research on Cancer (IARC) under charter by the World Health Organization to evaluate carcinogenic risk of chemicals to humans reviewed available literature regarding the health affects of crystalline silica and classified crystalline silica as a Class 2A (probable carcinogen) in humans. This finding was based on limited evidence in humans and sufficient evidence in animals (Kelse, 1991) and only applies to inhaled (respirable) crystalline silica, not to dust that is ingested or contacted. Health regulation has ensued as many nations, including US federal and state agencies, consider the findings of IARC when writing health related legislation.

In the United States, the regulatory situation is complex, and the reader is referred to Miles and Harben (1991) and Miles (1990) for summary. Relevant programs and agencies that used IARC data when promulgating current regulations include the National Toxicology Program (NTP), the Occupational Safety and Health Administration (OSHA), and the Mine Health and Safety Administration (MSHA). Research sources for current regulations are:

- the National Toxicology Program Annual Report on Carcinogens (crystalline silica listed 1989, Chemical Abstracts Service CAS Nos. 14808–60–7, 144654–46–1, 15468–32–3)
- the OSHA-administered Hazard Communication Standard (HCS) of 1986 29 Code of Federal Register (CFR) 1910.1200 (amended 1987, 1988)
- OSHA 29, Code of Federal Register (CFR) 1910, subpart Z (1989)
- MSHA is creating a Hazard Communication Standard (HCS) for mine workers

The Clean Air Act of 1990 does not specifically list or regulate crystalline silica dust, however, dusts in general are regulated as PM 10 particulates. Under the HCS, workers exposed to crystalline silica-bearing dusts must be advised of the hazards and protected if the permissible exposure limit (PEL) established by OSHA is exceeded. MSHA is using the 1974 National Institute for Occupational Safety and Health (NIOSH) levels that are essentially the same as the OSHA PELs until the NIOSH PELs are established in 1992.

As the result of health concerns and regulation, labeling of perlite is now undertaken in the United States and several states, including California (Roberts, 1989, Miles and Harben, 1991) and New Jersey (Miles, 1990, Miles and Harben, 1991), are regulating silica.

The US regulations presuppose that untested mixtures of products and crystalline silica present the same hazard as any component of the mixture comprising more than 1% by weight or volume. If a component is a suspected carcinogen then the threshold becomes 0.1% by weight of volume and a carcinogenic warning label is required if this value is exceeded. Dilution of the mixture to below the threshold level of 0.1% can eliminate the need for labeling.

The industry has responded to silica regulations by mobilizing several organizations, mandating protective equipment wear by workers, updating Mineral Safety Data Sheets (MSDS), dust reduction and monitoring, development of test methods suitable to the 0.1 and 1.0% thresholds, and training. The leading organizations are the CHEMSTAR Crystalline Silica Panel, the International Diatomite Producers Association (IDPA), the Perlite Institute (PI), the Refractories Institute (TRI), and the National Industrial Sand Association (NISA). The analysis for silica at the 0.1% level is difficult. The detection method often used is X-ray diffraction (Hamilton and Pelites, 1990, Miles and Hamilton, 1991) because of its speed, availability, relative insensitivity to material, ability to detect below the 0.1% level, and cost effectiveness. Other methods are optical petrography, electron microscopy, wet chemistry, heavy liquid separation, solubility, and infrared absorption.

FURTHER CONSIDERATIONS AND TRENDS

Substitutes

Expanded perlite shares many of its markets as an ultra-lightweight aggregate, thermal and acoustical insulator, and bulking agent with vermiculite. Regardless of the physical properties which are shared, the worldwide availability of perlite deposits and expansion facilities and its lower price offer a distinct advantage over vermiculite for applications where physical resilience or brightness of particles is of key importance. In concrete, where degradation can occur in wet conditions, the resilience of perlite is preferred. In cryogenic applications, where the insulating material must have low thermal conductivity as well as low moisture absorption, perlite clearly dominates the market. Aesthetically, the white color of expanded perlite offers cosmetic advantage in light colored, visible surface coatings. In horticultural applications, perlite and vermiculite can be used selectively in different areas of the market. The greater hardness of perlite, its chemical resistance, and inherent porosity, are important properties of use where soil conditions are wet for prolonged periods of time, and where good drainage, aeration, and water retention are required. Conversely, in dry soils, the addition of perlite facilitates the retention of water. With its higher brightness, finely ground perlite applied to the surfaces of seed blocks reflects light to the underside of seedlings and promotes rapid and sturdy growth.

In loose-fill insulation, polystyrene beads are now a competitive product.

The filter aid applications have long been dominated by diatomite. As a milled product, the irregular and platy particles of perlite lend themselves well in rotary vacuum filters and as a filter aid in applications where high flow rate is required and clarity is not of utmost importance.

Tariffs and Taxes

No tariff is applied on perlite imported into the United States. The depletion allowance for perlite is 10%.

ACKNOWLEDGMENTS

The authors wish to acknowledge the contributions of F.L. Kadey, Jr., J.M. Sharratt, J. Larsen, G. Coombs, and P. Corbetta. Acknowledgment is also given to B. Kelly and J. Reynolds. Our very special thanks to Robert Illner, now deceased.

REFERENCES AND SELECTED READING

Anon., 1977, "Perlite—Market Patterns and Future Potential," *Industrial Minerals*, Vol. 116, pp. 17–37.

Anon., 1984, *Test Methods and Related Standards*, Perlite Institute Inc., New York, p. 95.

Anon., 1987a, ''A Window on Perlite,'' *Industrial Minerals*, Vol. 238, pp. 9–10.

Anon., 1987b, ''Silica and Some Silicates,'' *IARC Monographs on the Evaluation of the Carcinogenic Risks of Chemicals to Humans*, Vol. 42, International Agency for Research on Cancer, pp. 1–144.

Anon., 1988a, ''Overall Evaluations of Carcinogenicity, An Update of IARC Monographs Volumes 1 to 42,'' IARC Monographs on the Evaluation of Carcinogenic Risks to Humans, Supplement 7, International Agency for Research on Cancer, pp. 29–74, 341–343.

Anon., 1988b Manville Bulletin FF391, Sep.

Anon., 1989a, ''Perlite,'' *Industrial Minerals,* Vol. 263, p. 53.

Anon., 1989b, ''Air Contaminants, Final Rule; 29 CFR Part 1910,'' *Code of Federal Register*, Vol. 54, Book 2, pp. 2330–2983.

Anon., 1989c, ''Monographs on the Evaluation of Carcinogenic Risk of Chemicals to Humans,'' Vol. 47, International Agency for Research on Cancer.

Austin, G.S., and Barker, J.M., 1994, ''Production and Marketability of Perlite in the Western United States,'' *29th Forum on the Geology of Industrial Minerals,* California Division of Mines and Geology.

Bacon, C.R., et al., 1981, ''Pleistocene High-Silica Rhyolites of the Coso Volcanic Field, Inyo County, California,'' *Journal of Geophysical Research*, Vol. 86, No. B11, pp. 10223–10241.

Barker, J.M., and Harris, M.J., 1990, ''Practice and Problems of Perlite Expansion and Testing at the New Mexico Bureau of Mines and Mineral Resources, Socorro, New Mexico,'' *Industrial Rocks and Minerals of the Pacific Northwest*, R.P. Geitgey and B.F. Vogt, eds., Oregon Department of Geology and Mineral Industries, Special Paper 23, pp. 63–68.

Barker, J.M., et al., 1987, ''Perlite Testing at the New Mexico Bureau of Mines and Mineral Resources, Socorro, New Mexico,'' *Mining Engineering*, Vol. 39, No. 9, pp. 877–882.

Barker, J.M., and McKee, C., 1990, ''Overview of the Impact on the Perlite Industry of the IARC Classification of Crystalline Silica as a Group 2A (Probable) Carcinogen in Humans,'' *Industrial Rocks and Minerals of the Pacific Northwest*, R.P. Geitgey and B.F. Vogt, eds., Oregon Department of Geology and Mineral Industries, Special Paper 23, pp. 51–62.

Benbow, J., 1987, ''Minerals in Fire Protection,'' *Industrial Minerals*, Vol. 240, p. 61.

Benton, W.E., 1984, ''Economics of Perlite,'' Preprint, Society of Mining Engineers, Littleton, CO.

Bolen, W.P., 1990–1991, ''Perlite,'' *Mineral Industry Surveys 1990 and 1991*, US Bureau of Mines.

Breese, R.O.Y., 1984, ''Rhyolite Domes and Flows at No Agua Peaks, New Mexico,'' *Rio Grande Rift: Northern New Mexico*, New Mexico Geological Society, 35th Annual Field Conference, Oct. 11–13, pp. 373–374.

Breese, R.O.Y., 1985, ''Occurrence and Origin of Perlite,'' Preprint No. 85-359, Society of Mining Engineers, Littleton, CO, p. 12.

Caley, E.R., and Richards, J.F., 1956, ''Theophrastus on Stones: Introduction, Greek Text, English Translation and Commentary,'' Ohio State University, Columbus, OH.

Chesterman, C.W., 1975, ''Perlite,'' *Industrial Minerals and Rocks*, 4th ed., S.J. Lefond, ed., AIME, New York, pp. 927–934.

Christiansen, R.L., and Lipman, P.W., 1966, ''Emplacement and Thermal History of a Rhyolite Flow near Fortymile Canyon, Southern Nevada,'' *Bulletin*, Geological Society of America, Vol. 77, pp. 671–684.

Eichelberger, J.C., et al., 1985, ''Research Drilling at Inyo Domes, California: 1984 Results,'' *EOS*, Apr. 23, pp. 186–187.

Eichelberger, J.C., and Westrich, H.R., 1981, ''Magmatic Volatiles in Explosive Rhyolitic Eruptions,'' *Geophysical Research Letters,* Vol. 8, pp. 757–760.

Epel'baum, M.B., and Salova, T.P., 1987, ''Experimental Study of Volcanic Glasses: Water Specification, Application to Perlite Genesis and Vesiculation.'' *2nd International Conference on Natural Glasses*, J. Konta, ed., Charles University, Prague, pp. 73–80.

Fediuk, F., 1987, ''Glass in Czechoslovak Volcanics,'' *2nd International Conference on Natural Glasses*, J. Konta, ed., Charles University, Prague, pp. 25–33.

Friedman, I., 1977, ''Hydration Dating of Volcanism at Newberry Crater, Oregon,'' *Journal Research US Geological Survey*, Vol. 5, pp. 337–342.

Friedman, I., and Long, W., 1976, ''Hydration Rate of Obsidian,'' *Science*, Vol. 191, pp. 347–352.

Friedman, I., and Obradovich, J., 1981, ''Obsidian Hydration Dating of Volcanic Events,'' *Quaternary Research*, Vol. 16, pp. 37–47.

Friedman, I., and Peterson, N., 1971, ''Obsidian Hydration Dating Applied to Basaltic Volcanic Activity,'' *Ore Bin*, Vol. 33, p. 158.

Friedman, I., et al., ''Hydration of Natural Glass and Formation of Perlite,'' *Bulletin*, Geological Society of America, Vol. 77, pp. 323–327.

Friedman, I., and Smith, R.L., 1958, ''The Deuterium Content of Water in Some Volcanic Glasses,'' *Geochimica et Cosmochimica Acta*, Vol. 15, pp. 218–228.

Friedman, I., and Trembour, F.W., 1978, ''Obsidian: The Dating Stone,'' *American Scientist*, Vol. 66, pp. 44–51.

Hamilton, R.D., and Peletis, N.G., 1988, ''The Determination of Quartz in Perlite by X-ray Diffraction,'' Company Report, Corporate Research and Engineering Center, Manville Sales Corporation, Littleton, CO, p. 11.

Hamilton, R.D., and Peletis, N.G., 1990, ''The Determination of Quartz in Perlite by X-ray Diffraction,'' *Advances in X-ray Analysis*, Vol. 23, C.S. Barrett et al., eds., Plenum Press, New York, pp. 493–497.

Harris, C.R., and Wilson, R.K., 1991, ''Challenge 2000: The Effect of 'The Right To Know' Regulations,'' *Environmental Management for the 1990's*, D.J. Lootens et al., eds., Society for Mining, Metallurgy, and Exploration, Littleton, CO, pp. 321–326.

Howell, W.R., 1974, ''The Early History of the Perlite Industry,'' Unpublished Address to the Perlite Institute Inc., Colorado Springs, CO, Apr. 21.

Ilkey-Perlaki, E., 1987, ''Volcanic Glass and Its Relation to Tertiary Acid Volcanism in the Tojak Mountains,'' *2nd International Conference on Natural Glasses*, J. Konta, ed., Charles University, Prague, pp. 111–119.

Jezek, P.A., and Noble, D.C., 1978, ''Natural Hydration and Ion Exchange of Obsidian: An Electron Microprobe Study,'' *American Mineralogist*, Vol. 63, pp. 266–273.

Jackson, F.L., 1986, ''Processing Perlite for Use in Insulation Applications,'' *Mining Engineering*, Vol. 38, No. 1, pp. 40–45.

Kadey, F.L., Jr., 1983, ''Perlite,'' *Industrial Minerals and Rocks*, 5th ed., S.J. Lefond, ed., AIME, New York, pp. 997–1015.

Keith, H.D., and Padden, F.J., Jr., 1963, ''A Phenomenological Theory of Spherulitic Crystallization,'' *Journal of Applied Physics*, Vol. 35, pp. 2409–2421.

Keller, W.D., and Pickett, E.E., 1954, ''Hydroxyl Water in Perlite from Superior, Arizona,'' *American Journal of Science*, Vol. 252, pp. 87–98.

Kelse, J.W., 1991, ''Cancer Warnings: How Justified Are They?'' *Environmental Management for the 1990's*, D.J. Lootens et al., eds., Society for Mining, Metallurgy, and Exploration, Littleton, CO, pp. 313–320.

Klug, H.P., and Alexander, L.E., 1974, *X-ray Diffraction Procedures*, John Wiley and Sons, New York, p. 966.

Lajcakova, A., and Kraus, I., 1987, ''Acid Volcanic Glasses in the Western Carpathians,'' *2nd International Conference on Natural Glasses*, J. Konta, ed., Charles University, Prague, pp. 121–127.

Langford, R.L., 1979, ''Perlite,'' Mineral Dossier No. 21, Institute of Geological Sciences, London, p. 28.

Laursen, T., and Lanford, W.A., 1978, ''Hydration of Obsidian,'' *Nature*, Vol. 276, pp. 153–156.

Lipman, P.W., 1965, ''Chemical Comparison of Glassy and Crystalline Volcanic Rocks,'' Bulletin 1201, US Geological Survey, pp. D1-D24.

Lipman, P.W., et al., 1969, ''Retention of Alkalis by Calc-Alkalic Rhyolites during Crystallization and Hydration,'' *American Mineralogist*, Vol. 54, pp. 286–291.

Lipman, P.W., et al., 1978, ''Pleistocene Rhyolite of the Mineral Mountains, Utah: Geothermal and Archeological Significance,'' *Journal of Research US Geological Survey*, Vol. 6, pp. 133–147.

Lofgren, G., 1971, ''Experimentally Produced Devitrification Textures in Natural Rhyolitic Glass,'' Bulletin, Geological Society of America, Vol. 82, pp. 111–124.

Marakushev, A.A., and Yakoleva, E.B., 1980, ''On the Origin of Perlites,'' *Geologiya*, Vol. 35, Vestnik Moskovskogo Universiteta, Moscow, pp. 1–14.

Marakushev, A.A., et al., 1987, ''Endogenic Nature of Perlites,'' *2nd In-*

ternational Conference on Natural Glasses, J. Konta, ed., Charles University, Prague, pp. 89–96.

McKee, C., et al., 1990, ''Quantitative Analysis of Quartz in Perlite by X-ray Diffraction I,'' Open-File Report 364, New Mexico Bureau of Mines and Mineral Resources, p. 16.

McMichael, B., 1989, ''Pyroprocessing Techniques, the Fiery World of Industrial Minerals,'' *Industrial Minerals*, Vol. 263, pp. 54–69.

Meighan, C.W., et al., 1968, ''Obsidian Dating in West Mexican Archeology,'' *Science*, Vol. 160, pp. 1069–1075.

Meisinger, A.C., 1980–1988, ''Perlite,'' *Minerals Yearbook 1980–1988*, US Bureau of Mines.

Miles, W.J., 1990, ''Mining Industry Responds to Crystalline Silica Regulations,'' *Mining Engineering*, Vol. 42, No. 4, pp. 345–348.

Miles, W.J., and Hamilton, R.D., 1991, ''Detection and Measurement of Crystalline Silica in Minerals,'' *Environmental Management for the 1990's*, D.J. Lootens et al., eds., Society for Mining, Metallurgy, and Exploration, Littleton, CO, pp. 329–333.

Miles, W.J., and Harben, P.W., 1991, ''US Crystalline Silica Regulations—Approaching the Detection Limits?,'' *Industrial Minerals*, No. 291, pp. 21–22, 25, 27.

Murdock, J.B., and Stein, H.A., 1950, ''Comparative Furnace Designs for the Expansion of Perlite,'' *Trans. AIME*, Vol. 187, pp. 111–116.

Nasedkin, V.V., 1987, ''Hydration Types, Minerals and Geology of Volcanic Glasses,'' *2nd International Conference on Volcanic Glasses*, J. Konta, ed., Charles University, Prague, pp. 65–71.

Noble, D.C., 1967, ''Sodium, Potassium, and Ferrous Iron Contents of Some Secondarily Hydrated Natural Silicic Glasses,'' *American Mineralogist*, Vol. 52, pp. 280–286.

Noble, D.C., 1968, ''Stress-Corrosion Failure and the Hydration of Glassy Silicic Rocks,'' *American Mineralogist*, Vol. 53, pp. 1756–1759.

Otvos, E.G., Jr., 1961, ''Pumice and Perlite in a Rhyolite Complex in Hungary,'' *Economic Geology*, Vol. 56, pp. 1112–1122.

Palm, S.K., 1991, ''The Economics of Product Labeling and Hazard Communications,'' *Environmental Management for the 1990's*, D.J. Lootens et al., eds., Society for Mining, Metallurgy, and Exploration, Littleton, CO, pp. 307–312.

Pavlovski, B., et al., 1987, ''Chemical and Physical Characteristics of Volcanic Glasses of Macedonia, Yugoslavia,'' *2nd International Conference on Natural Glasses*, J. Konta, ed., Charles University, Prague, pp. 163–168.

Pettifer, L., 1981, ''Perlite—Diversification the Key to Overall Expansion,'' *Industrial Minerals*, Vol. 171, pp. 55–75.

Power, T., 1986, ''Perlite & Vermiculite, the Market Overlap,'' *Industrial Minerals*, Vol. 230, pp. 39–49.

Renault, J., McKee, C., and Barker, J., 1991, ''Quantitative X-ray Diffraction Analysis of Trace Quartz in Selected Mineral Products: Standardization II,'' *Environmental Management for the 1990's*, D.J. Lootens et al., eds., Society for Mining, Metallurgy and Exploration, Littleton, CO, pp. 361–362.

Renault, J., McKee, C., and Barker, J., 1992, ''Calibrating for X-ray Analysis of Trace Quartz,'' *Advances in X-ray Analysis*, Vol. 35, C.S. Barrett et al., eds., Plenum Press, New York, pp. 363–373.

Roberts, L., 1989, ''A Corrosive Fight Over California's Toxics Law,'' *Science*, Vol. 243, pp. 306–309.

Roedder, E., and Smith, R.L., 1964, ''Liquid Water in Pumice Vesicles, a Crude but Useful Dating Method,'' Special Paper #82, *Abstracts for 1964*, Geological Society of America, p. 164.

Ross, C.S., and Smith, R.L , 1955, ''Water and Other Volatiles in Volcanic Glasses,'' *American Mineralogist*, Vol. 40, pp. 1071–1089.

Ross, C.S., and Smith, R.L., 1961, ''Ash Flow Tuffs: Their Origin, Geologic Relations and Identification,'' Professional Paper 366, US Geological Survey, pp. 1–81.

Sargent, K.A., 1969, ''Petrography and Heavy Minerals of Three Groups of Rhyolitic Lavas, Pahute Mesa, Nevada Test Site,'' Professional Paper 650-C, US Geological Survey, pp. C18-C24.

Sax, N.I., and Lewis, R.J., eds., 1988, *Dangerous Properties of Industrial Materials*, 7th ed., Vol. III, Van Nostrand Reinhold, New York, pp. 3020–3023.

Shackley, D., 1988, ''Characterization and Expansion of Perlite,'' PhD thesis, University of Nottingham, Nottingham, UK.

Soloninko, I.S., and Chernyavskii, P.N., 1962, ''Transcarpathian Perlites,'' *Perlite and Vermiculite (Geology, Exploration and Production Technology)*, Gosudarstvennoe Nauchno-Tekhnicheskoe Izdatel'stvo Literatury Po Geologii I Okhrana Nedr., Moscow, pp. 33–46.

Steen-McIntyre, V., 1975, ''Hydration and Superhydration of Tephra Glass: A Potential Tool for Estimating Age of Holocene and Pleistocene Ash Beds,'' *Quaternary Studies*, Royal Society of New Zealand, Wellington, pp. 271–278.

Stein, H.A., and Murdock, J.B., 1955, ''The Processing of Perlite,'' *California Journal of Mines and Geology*, Vol. 51, pp. 105–116.

Taylor, B.E., et al., 1983. ''Hydrogen Isotopic Evidence of Rhyolite Magma Degassing during Shallow Intrusion and Eruption,'' *Nature*, Vol. 306, pp. 541–545.

Thompson, B.N., and Reed, J.J., 1954, ''Perlite Deposits in New Zealand,'' *New Zealand Journal of Science and Technology*, Sec. B, Vol. 36, pp. 208–226.

Turner, L.M., 1985, ''Perlite Processing and Quality Control,'' Preprint No. 85–345, Society of Mining Engineers, Littleton, CO, p. 6.

Weber, R.H., and Austin, G.S., 1982, ''Perlite in New Mexico,'' *Industrial Rocks and Minerals of the Southwest*, G.S. Austin, compiler, Circular 182, New Mexico Bureau of Mines and Mineral Resources, pp. 97–101.

Whitson, D., 1982, ''Geology of the Perlite Deposit at No Agua Peaks, New Mexico,'' *Industrial Rocks and Minerals of the Southwest*, Circular 182, New Mexico Bureau of Mines and Mineral Resources, pp. 89–95.

Wu, C-K, 1980, ''Nature of Incorporated Water in Hydrated Silicate Glasses,'' *Journal of the American Ceramic Society*, Vol. 63, pp. 453–457.

Yanev, Y., 1987, ''Characterization of Volcanic Glasses from the Eastern Rhodopes, Bulgaria,'' *2nd International Conference on Natural Glasses*, J. Konta, ed., Charles University, Prague, pp. 129–138.

Zielinski, R.A., et al., 1977, ''Minor-element Abundances in Obsidian, Perlite, and Felsite of Calc-Alkalic Rhyolites,'' *American Mineralogist*, Vol. 62, pp. 426–437.

PHOSPHATE ROCK

JAMES J. BARTELS AND THEODORE M. GURR

Phosphate produced fertilizers provide the phosphorous nutrients required by plant life to sustain significant growth, thus improving the production of food for the world's population. Phosphorous replenishment of soil currently can be obtained efficiently by no other means than by direct application of phosphate fertilizer to the soil. Phosphate products are also utilized in animal feeds, detergent, and various industrial processes. Thirty-three countries are presently mining phosphate rock for the production of fertilizers, and its process byproducts. The major phosphate rock producers are the United States, the former Soviet Union, Morocco, and China (Table 1).

Phosphate rock is generally defined as a rock material that contains phosphate minerals which are sufficient for commercial usage. Phosphate minerals are found in sedimentary igneous and metamorphic rocks. Economic resources of phosphates are primarily developed from sedimentary rock sources.

The apatite mineral family comprises the majority of phosphate constituents from both sedimentary and igneous rocks. Igneous phosphate rocks are generally composed of fluroapatite whereas sedimentary rocks are generally composed of carbonate fluroapatite, or chlorapatite.

Phosphate rock is mined by a multitude of processes, including surface and underground mining. The mined rock requires mechanical and chemical processing to liberate the phosphate for utilization as fertilizer. Approximately 85% of the world's phosphate production uses sulfuric acid, with the remaining processes utilizing nitric and phosphoric acids. Ammonia is also introduced in the manufacture of liquid fertilizer, ammonium phosphate and ammonium polyphosphate.

GEOLOGY

Mineralogy and Chemical Properties

Phosphate most commonly is derived from the mineral apatite, which is chemically described as $CA_5(F,Cl,OH)(PO_4)_3$. Apatite is also broken up into the composite apatite minerals which are fluroapatite $Ca_5F(PO_4)^3$, chlorapatite $Ca_5C1(PO_4)_3$ and hydroxylapatite $Ca_5(OH)(PO_4)_3$. Carbonate CO_3 can substitute for PO_4 forming carbonate apatite, called francolite. The guano phosphate (bird excrement) mineralogy commonly occurs as brushite $CaHPO_4 \cdot 2H_2O$, monetite $CaHPO_4$, whitlockite $Ca_3(PO_4)$, and dahlite $Ca_{10}(PO_4)_6(OH)_2$.

Apatite minerals form from igneous sources where they are developed deep in the earth's crust; and the cooling processes are slow, producing long, hexagonal prismatic to tabular crystals. The termination of the crystal can be basal plans or pyramids. In some instances, the crystals are bipyramids. Faster cooling, or sedimentary derived apatites are more commonly cryptocrystalline. Collophane is a name given to cryptocrystalline apatite found in phosphate rock in fossil form.

Apatite minerals are relatively soft and are used to describe the 5 hardness on the Mohs scale. Cleavage is poor and develops along the C(0001). The color of apatite minerals varies from colorless, to violet, and blue, but are predominantly green or brown; luster is vitreous to subresinous. Collophane's physical appearance is often opaline, dense, with colloform structure and sometimes concretionary, nodular or pulverulent.

Phosphate Bearing Materials Origins

Phosphate deposits of igneous and metamorphic origins have been well defined. Sedimentary origins have been defined, but with a greater degree of uncertainty. Research in the 1980s proved out some of these processes through studies of real time deposition in Chile, Peru, and Australia. Igneous intrusive alkali rock and associated contact metamorphic rocks, provide approximately 20% of the world's phosphate. Fluorine containing apatite minerals are the most common materials containing phosphorous. Fluorapatite occurs in most igneous rocks. Fluorapatite and fluorine hydroxylapatite, together with carbonate varieties of these, are important members of the group. Where an essentially pure chlorapatite, carbonate apatite, and hydroxylapatite are rare and restricted in occurrence, the fluorine containing types occur in most all igneous rocks as early formed accessory mineral, usually in microscopic crystals, and may occur as extremely large bodies as magmatic segregations from alkalic igneous rocks. Apatite is also found crystallized in pegmatitic faces of both acidic and basic types of igneous rocks. More specifically, apatite is associated with magnetite deposits, in hydrothermal veins, especially those formed at relatively high temperatures, and in veins of the Alpine type. Apatite is common in both regionally and contact metamorphosed rocks, especially in the crystalline limestones where it is associated with sphene, zircon, pyroxine amphibole, spinel vesuvianite, phlogopite, talc, chloride schists, and as a contact metamorphic mineral.

Sedimentary Phosphate Deposition

Sedimentary phosphate deposits are comprised of sedimentary rock in which phosphate is contained as pellets, nodules, grains, replaced calcium in animal skeletal remains, and animal excrement. Sedimentary deposition of phosphate occurred throughout geologic time and continues today.

Economically viable phosphate deposits are from the following ages: Proterozoic, Cambrian, Ordovician, Devonian, Permian, Cretaceous, Tertiary, Miocene, and Eocene.

"Marine deposition of pelletal phosphorites occurs in regionally restricted basin areas along continental shelves. The richest accumulations of phosphate occur where clastic sedimentation is at a minimum away from cranton, either on stable continental shelves

Table 1. Phosphate Rock Production for the World's Major Producers (thousand tons)

Africa	**1980**	**1981**	**1982**	**1983**	**1984**	**1985**	**1986**	**1987**	**1988**	**1989**	**1990**
Algeria	1,025	916	947	893	1,000 *	1,207	1,203	1,073	1,332	1,223 *	1,102
Egypt	658	720	711	647	1,043	1,074	1,271	1,167	1,146	1,146	1,143
Morocco	18,824	18,562	17,754	20,106	21,245	20,737	21,178	20,955	25,015	17,988	21,189
Senegal	1,632	1,699	1,182	1,521	1,912	1,814	1,850	1,874	2,326	2,273	2,147
S. Africa	3,185	2,718	3,161	2,887	2,585	2,433	2,920	2,623	2,850	2,963	2,164
Tanzania				20	15	15	10	18	4	5	5 *
Togo	2,933	2,215	2,800	2,081	2,696	2,452	2,314	2,644	3,464	3,356	3,314
Tunisia	4,582	4,596	4,196	5,924	5,346	4,530	5,951	6,390	6,103	6,621	6,566
Zimbabwe	130	125	120	133	134	135	136	155	125	135	140 *
Subtotal	**32,969**	**31,551**	**30,871**	**34,212**	**35,976**	**34,397**	**36,833**	**36,899**	**42,365**	**35,710**	**37,770**

N. America	**1980**	**1981**	**1982**	**1983**	**1984**	**1985**	**1986**	**1987**	**1988**	**1989**	**1990**
Mexico	397	252	379	389	653	787	747	689	842	851	636
USA	54,415	53,624	37,414	42,573	49,197	50,835	40,320	40,954	45,389	49,817	46,343
Subtotal	**54,812**	**53,876**	**37,793**	**42,962**	**49,850**	**51,622**	**41,067**	**41,643**	**46,231**	**50,668**	**46,979**

S. America	**1980**	**1981**	**1982**	**1983**	**1984**	**1985**	**1986**	**1987**	**1988**	**1989**	**1990**
Brazil	2,612	3,238	2,732	3,208	3,855	4,214	4,509	4,777	4,672	3,580	2,967
Chile					5	7	7	10	9	10	10 *
Columbia	6	17	20	17	11	24	27	34	35 *	30	30 *
Peru	14	12	29	3	13	12	5	61	13	13	15 *
Subtotal	**2,632**	**3,267**	**2,781**	**3,228**	**3,884**	**4,257**	**4,548**	**4,882**	**4,729**	**3,633**	**3,022**

Europe	**1980**	**1981**	**1982**	**1983**	**1984**	**1985**	**1986**	**1987**	**1988**	**1989**	**1990**
Finland	138	201	233	381	477	512	527	553 *	584	560	546
Sweden	88	124	128	107	133	187	191	221 *	142	160	6
Subtotal	**226**	**325**	**361**	**488**	**610**	**699**	**718**	**774**	**726**	**720**	**552**

Asia	**1980**	**1981**	**1982**	**1983**	**1984**	**1985**	**1986**	**1987**	**1988**	**1989**	**1990**
China	10,726	11,500	11,720	12,500	11,800	6,970	6,700	9,000 *	15,000	17,000	17,300
India	541	565	631	688	892	929	667	679	678	700	674
Iraq		50	363	1,199	1,000 *	1,000 *	1,000 *	1,500 *	1,273	1,300	600
Israel	2,307	1,919	2,148	2,969	3,312	4,076	3,673	3,798	3,479	3,922	3,516
Jordan	3,911	4,244	4,390	4,749	6,263	6,067	6,249	6,800	6,611	6,675	5,925
Pakistan							50	5	5	8	9
Korea DR*	500	500	500	500	500	500	500	500	500	500	500
Sri Lanka	5	15	20	16	14	14	15	21 *	23	25	25 *
Syria	1319	1321	1455	1229	1514	1270	1606	1986	2342	2250	1670
Thailand	6	6	4	5	3	4	5	5	8	9	10 *
Turkey	21	43	26	50	96	37 *	3	19	74	84	85 *
Vietnam DR	90	181	110 *	200 *	200 *	516	530	300	330	400	400 *
Subtotal	**19,426**	**20,344**	**21,367**	**24,105**	**25,594**	**21,383**	**20,998**	**24,613**	**30,323**	**32,873**	**30,714**

Oceania	**1980**	**1981**	**1982**	**1983**	**1984**	**1985**	**1986**	**1987**	**1988**	**1989**	**1990**
Australia	7	22	11	5	15	23	34	11	13	8	10 *
Christmas Is.	1,713	1,423	1,328	1,094	1,259	1,187	880	842			
Nauru Is.	2,087	1,480	1,359	1,684	1,358	1,508	1,494	1,376	1,540	1,500	926
Phillipines	17	8	6	4	7	6	6	8 *	7	7	7 *
Subtotal	**3,824**	**2,933**	**2,704**	**2,787**	**2,639**	**2,724**	**2,414**	**2,237**	**1,560**	**1,515**	**943**

	1980	**1981**	**1982**	**1983**	**1984**	**1985**	**1986**	**1987**	**1988**	**1989**	**1990**
USSR*	25,300	30,700	31,300	33,100	33,300	33,750	33,900	34,100	34,400	39,000	36,868

World Total	**139,189**	**142,996**	**127,177**	**140,882**	**151,853**	**148,832**	**140,478**	**145,148**	**160,334**	**164,119**	**156,848**

*estimated, *source*: US Department of Interior, Bureau of Mines *Minerals Yearbook*, Phosphate Rock chapter (1984, 1988, and 1990).

or platforms in shallow, warm water, and at low latitudes'' (Sheldon, 1964). The mechanism for deposition is generally believed to be the movement of oceanic currents over long periods of time. Phosphate deposition is theorized to occur through upwelling of nutrient-rich zones of ocean currents. Anoxic conditions commonly arise in the bottom waters beneath the nutrient-rich upwelled layer. Upwelling of nutrient rich oceanic waters is only a segment of phosphorite deposition.

The origin of phosphorites was studied extensively during the 1980s including offshore areas in Africa, Chile-Peru, and Australia where phosphorite deposition is presently occurring. The studies also included the effects of palaeoceanography and palaeoclimatology, sea level changes, plate tectonics, phosphogenesis and the role of bacteria, evolution, petrography, minerology, and geochemistry. A clear and concise picture of phosphate sedimentary deposition has not been developed. Continued research in the above areas is ongoing. Greater emphasis is now placed on interstitial water microenvironment vs. water precipitation as the primary source of sedimentary apatite minerals.

Island guano deposits occur through the accumulation of bird and mammal excreta. The excreta is an organic compound with its major constituents being uric acid, calcium, phosphate and potassium. The excreta reacts with the islands substrata of limestone resulting in the formation of the minerals brushite, monetite, whitlockite, and dahlite. Additional mineral alterations and recrystallization occurs with exposure to sea water through spray, wash, and wave action.

Weathering

The reworking of phosphate rich sediment is a primary factor in the development of a economically viable phosphate deposit. Variation in oceanic current intensities and directions causes removal of matrix material, resulting in the exposure of phosphatized clasts. Subsequent movement of phosphate clasts through currents capable of transport in effect beneficiates and redeposits the phosphate clasts. The transgressing sea can cause multiple cycles of phosphate pelletal burial and subsequent erosion. For this reason, phosphate rich material is found in many sedimentary faces including glauconitic sands, anorobic shales and carbonates.

DISTRIBUTION OF MAJOR DEPOSITS

North America

United States: The majority of United States phosphate rock resources are marine origin and are distributed along the eastern coastline and Rocky Mountain areas of the United States (Fig. 1). The primary phosphate rock producing area is the Florida-North Carolina land pebble field (Fig. 2).

Atlantic Coastal Plain phosphate deposits are generally composed of carbonate fluorapatite silt to pebble sized particles in a matrix of quartz, sand and clays. The majority of deposition occurred during the Miocene age with subsequent reworking and redeposition occurring in Pliocene through Holocene times.

Florida: The majority of the currently producing mines in the Florida land pebble phosphate are located near the center of the state (Fig. 2). The phosphate occurs in the Miocene Hawthorne Formation and the Pliocene to Holocene Bone Valley Formation (Fig. 3). The phosphate sand sized material ranges from 10 to 12.8% P_2O_5 and the phosphate pebble material ranges from 33.0 to 35.7% P_2O_5 (Gurr, 1977).

In the period from 1980 to 1991 a drastic decline in the number of of phosphate companies operating in Florida occurred, reducing

Fig. 1. Major phosphate deposits in the Western Hemisphere (altered after Cathcart, 1983).

the total number of companies operating mines from over 15 to 9. However, the total number of mines open remains fairly constant at 17. The recessional effects forced several companies to remove themselves from the market in 1985. Operating companies were reduced from 15 down to 10 as discussed in the following.

Considerable restructuring of the Florida producers occurred during the 1980s. Beker Industries commenced operations of the Wingate Creek mine in 1981. A joint Beker venture between W.R. Grace and the International Minerals and Chemical Corporation (IMC) brought the Four Corners mine on stream in 1984. Noranda also opened a mine in Hopewell, FL, in 1984. These additions increased Florida's production of phosphate rock by approximately 10%. However, in 1985 Beker filed Chapter 11, while Grace and IMC closed the newly opened Four Corners mine. IMC purchased the Haynsworth and Lonesome Mine reserves from Brewster Phosphates in 1985. Florida Crushed Stone (FCS) purchased Amax's Big Four mine in 1986 and subsequently sold it to Mobil Mining and Chemical Company for their expansion needs due to the anticipated closing of the Mobil Ft. Meade mine. Nu West Industries purchased the Wingate Creek, FL, mine from Beker in 1988 and started operating it in 1989. Grace sold its share of the Four Corners mine to IMC and its remaining phosphate properties to Seminole Fertilizer Corporation (SFC) in 1988. IMC closed the Lonesome and Haynsworth mines in 1989. Nu West Industries sold the Wingate Creek mine to Gulf Atlantic Company early in 1990. Mobil mined out and closed the Ft. Meade mine in 1990 due to reopening

Fig. 2. Location of major phosphate deposits in the Southeastern United States (*source*: Gurr, 1977).

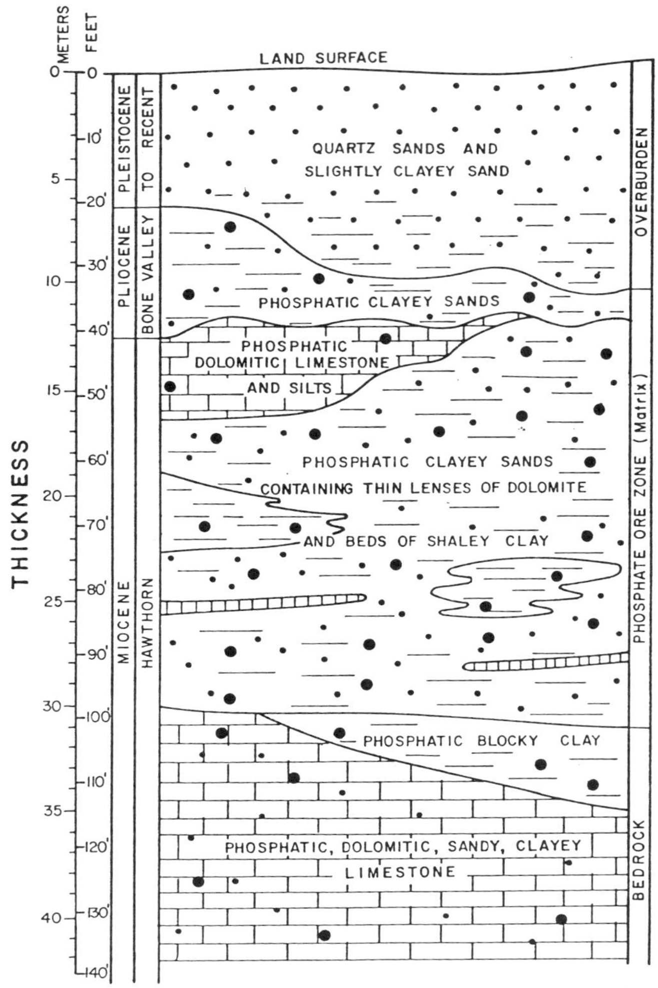

Fig. 3. Generalized geologic section in the South Florida and the North/South Central Georgia areas (*source*: Gurr, 1977).

the Big Four mine and future prospects for opening the South Ft. Meade reserve. The US Agri-Chemicals' phosphate operation was sold by USX Corporation to Sinochem (part of the Chinese National Import Export Corporation) in 1989.

Throughout the turmoil of the 1980s, the Florida phosphate rock industry remains the major producer of phosphate in the United States and continued to be a major producer on the world market.

North Carolina: Mining in North Carolina began in 1966 when Texasgulf, Inc. opened the Lee Creek mine along the coastal area of North Carolina. Texasgulf, Inc. mines the phosphatic Miocene Age sands of the Pungo River Formation. The sand is fine to medium grained and is comprised of 70 to 80% phosphate ore. The flotation concentrate produces 29 to 30% P_2O_5 (Gurr, 1977). Texasgulf, Inc., continues to operate the mine today (Fig. 4).

Tennessee: Middle Ordovician sedimentary phosphate rocks in the Nashville Group have been mined in Alabama, Tennessee and Kentucky. The phosphate rock is called brown rock and is comprised of phosphatized fossil fragments and pellets in a calcite and quartz silt matrix. The major brown rock phosphate rock formation is the Bigby Limestone, Cannon, and Leipers Formations. The mined phosphate ore ranged from 17 to 27% P_2O_5 and was upgraded to 27 to 32% P_2O_5 (Gurr 1977). The last two Tennessee phosphate rock mining companies closed in early 1991.

Western States: Phosphate rock is mined in Idaho, Utah, and Montana (Fig. 4). The phosphate bearing rock is Permian age and consists of the Phosphoric Formation with its members being the Meade Park phosphatic shale which is overlain by the Rex Chert member. Minable quantities of phosphate are primarily found in the Meade Park member. The phosphate rock is mostly carbonate fluorapatite and averages 24 to 26% P_2O_5 (Gurr, 1977).

Utah has one open pit operated by Chevron Resources near Vernal. Idaho ranks third in the nation in phosphate rock production. FMC is developing a phosphate mine in Dry Valley near Soda Springs, Idaho. The Dry Valley mine will replace the reserves of the Gay Mine site set to close in 1994. The western United States phosphate resources are vast, but are presently at depths uneconomical to mine.

Alaska: Alaska's Brooks Mountain Range contains phosphate deposits of the Mississippian age Lisbourne Formation and Triassic age Shublik Formation. The phosphate occurs with black shale, carbonate rock, and some chert. The pelletal phosphate beds range from 3 to 7 m in thickness. Average P_2O_5 content is estimated at approximately 12%.

Offshore: Both of the Atlantic and Pacific Coasts of the United States have offshore deposits of phosphate. The ocean floor deposits are Holocene age. Reserve estimation and further purity analyses are required before production possibilities are realistic. Most of the phosphate is reworked from deposits on land, but some are being formed today (Burnett, 1977).

Mexico: Mexico has three phosphate resource areas; Zacatecas, Baja California Sur, and offshore along the west coast of Mexico. The Zacatecas deposit is Jurassic age and consists of La

Fig. 4. Major phosphate mining areas of the United States (*source*: Gurr, 1977).

Caja Formation's phosphate nodules and pellets in a limestone matrix. The phosphate beds are up to approximately 5 m thick. Reserves are reported to be 150 Mt in rock averaging 13% P_2O_5, and resources may be several times greater than the reserves (Rogers et al., 1961). The Baja California Sur deposit is Miocene age. The phosphate is pelletal and is associated with diatomaceous earth and chert in unconsolidated sand and clay. Volcanics of the Pacific Coast has provided mechanisms for some reworking of these unconsolidated sediments.

The ocean floor phosphate deposits off the coast of Baja California is Holocene age. Reserves are estimated high and may prove to be economically feasible to mine and process. Further exploration and resource evaluation are required, but the deposits have potential in the future.

South America

Brazil: Brazil is the major phosphate rock producer in South America. Brazil's present phosphate mines are centrally located in the southern half of the country. Phosphate resources are found along the north-south edge of the Parana sedimentary basin.

The Barreiro Deposit of the state Minas Gerias is the largest deposit producing phosphate in Brazil. The Tapira and Araxá area contains weathered carbonate complexes composed mainly of magnetite and barytes. In situ P_2O_5 is reported from 8 to 12%. Additionally, precambrian age metamorphosed shale containing 11% P_2O_5 is mined at Patos de Minas.

Brazil's State of San Paulo has two significant phosphate deposits. Jacapiranga and Ipanema. The Jacapiranga deposit is the first phosphate mine to operate in Brazil. Originally the mine exploited the weathered apatite bearing carbonate. Presently, the mine has used the reserve of weathered material and has been forced to mine the unaltered apatite bearing carbonate. The in situ Upper Jurassic—Late Cretaceous age rock contains approximately 5% P_2O_5. Ipanema's phosphate resource is presently not utilized. The deposit is alkaline rock with glimerite and fenites mineralized with apatite and magnetite. Ipanema's reserves are estimated at 120 Mt at approximately 7% P_2O_5.

Several other small phosphate resources have been determined to have economic significance to be mined, including deposits in the states of Maranhão, Parnambuco, and Santa Catarina. The state of Santa Catarina has phosphate rock reserves of 320 Mt at 8% P_2O_5. Maranhão and Parnambuco have an estimated combined reserves of 90 Mt at an average 21.5% P_2O_5.

Columbia: Columbia has steadily increased phosphate rock production through the 1980s. The Upper Cretaceous age La Luna Formation is mined, and its phosphate content estimated at 15% P_2O_5. Reserves are estimated at 50 Mt with an estimated unproven resource at 600 Mt.

Venezuela-Ecuador: Venezuela also has phosphate reserves of 50 Mt generated from the La Luna Formation. Lithologically, the Napo Formation of Ecuador correlates with the La Luna Formation of Columbia and Venezuela. No reserve or purity estimates were attained. The La Luna and Napo Formations have two marine facies. Black shale and calcareous phosphorite are characteristics of the deep water facies and the sandstone, shale and quartzose phosphorite comprises of the shallow shelf sediments.

Peru: Peru has two phosphate rock resources to exploit. The vast Sechura deposit is comprised of Miocene age unconsolidated pelletal phosphate and sandy carbonate sediments. The Sechura deposit reserves are estimated at 500 Mt with possible phosphate resource estimated at 1 000 Mt.

The second phosphate rock resource is the Jurassic Age Aramachay Formation located in northern Peru. This deposit also is estimated to have vast reserves. Aramachay Formation sediments are comprised of phosphatic shales and phosphorite that are folded and faulted. The upper 15 m of the formation produces an estimated 10% P_2O_5.

Offshore: The coastlines of Venezuela, Chile, and Peru offer significant phosphate resources. Cost for exploration and capital investment to exploit these deposits will be tremendous.

Africa

The African continent contains one of the largest phosphate resources in the world. Morocco is the third largest producer of phosphate rock in the world, next to the United States and the former Soviet Union. Other significant African phosphate producers or countries with phosphate resources are: Algeria, Angola, Burundi, Egypt, Kenya, Senegal, the Republic of South Africa, Tanzania, Togo, Tunisia, Uganda, Zaire, and Zimbabwe (Fig. 5).

Morocco: Moroccan phosphate rock formations are Late Cretaceous to Eocene age. As many as 23 phosphate beds separated by limestone and chert are known in Morocco (Northolt, 1980). The phosphate is found in 1 to 40 mm particle fractions. The particles are carbonate fluorapatite associated with minerals such as calcite, dolomite, quartz, and clay. There are four developed phosphate deposits in Morocco: Oulad Andoun, Gantour, Meskala, and Sahara.

Morocco's phosphate rock largely comes from the Khouribga area of the Ouland Abdoun deposits. Khouribga is located in west central Morocco. Sidi Daoui and the Recette mines of the Khouribga area are approaching the exhaustion of reserves during the 1990s. Mera El Arech and Sidi Chenane mines will provide phosphate to compensate the upcoming closure of the Sidi Daoui and Recette mines. According to the Office Chêrifién des Phosphates (OCP), there are six to ten phosphate bearing beds, of which seven are presently mined. The P_2O_5 content for the Khouribga area is estimated to range from 25.2% to 34.3% per the OCP. Reserves are calculated to be 26.8 Gt utilizing 36% of the area's phosphate resource.

Phosphate rock mining of the Gantour deposit, in the area of Youssoufia, is an additional vast phosphate resource for Morocco. The reserve is calculated at 8.02 Gt, incorporating only 18% of the area's phosphate resource. Phosphate content ranges from 23 to 33% P_2O_5, slightly lower than the Oulad Abdoun deposits. The Ben Guerir mine located on the east edge of the Youssoufia Plateau was expanded in 1989 to offset production losses from other areas.

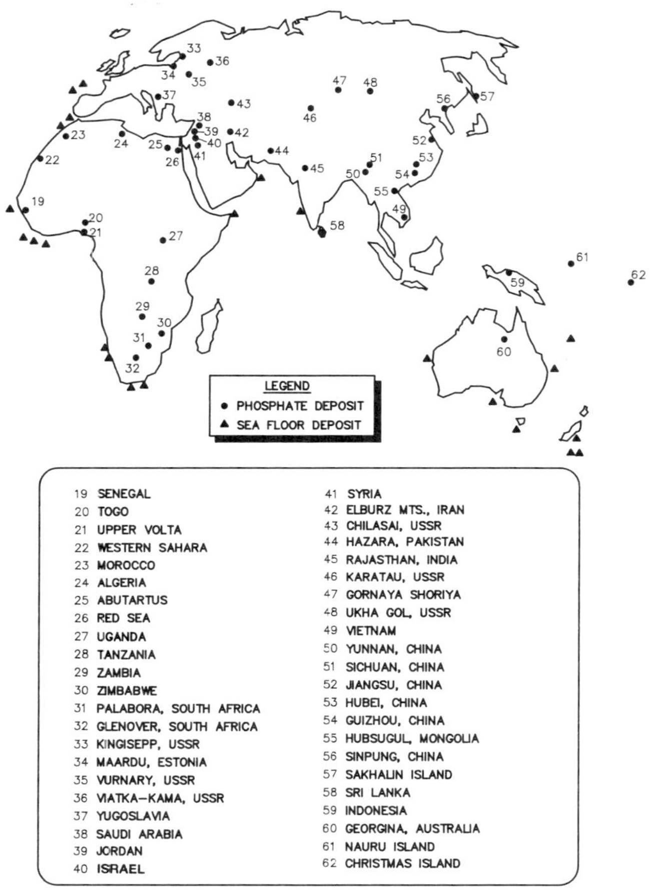

Fig. 5. Major phosphate deposits of the Eastern Hemisphere (altered after Cathart, 1983).

A third deposit, called the Meskala Deposit, holds reserves of 20.5 Gt and is presently mined south of Essaouira area. P_2O_5 content is reported by the OCP to range from 20.6% to 29.7% from an ore zone of approximately 18 m thick.

Morocco's fourth phosphate deposit is the Laayoune Boucraa area of the Sahara Desert. In 1977, the Boucraa mine was closed, only to be reopened in 1982. Mining exploits two to six beds of phosphate bearing ore resulting in 0.95 Gt of reserves. Additional phosphate resources are available in the Laayoune Boucraa area. The Boucraa ore's P_2O_5 content ranges from 31 to 34%.

Northern Africa is blanketed with the Upper Cretaceous to Eocene phosphate bearing sediments. These sediments are not only exploited in Morocco, but also in

	Gt		Gt
Angola	0.03	Toga	0.1
Algeria	0.6	Tunisia	0.2
Egypt	0.7	Tanzania	0.01
Senegal	0.5		

Phosphate reserve amounts for these countries have yet to be estimated.

Igneous and metamorphic phosphate deposits are located in Central to South Africa. These countries are Burundi, Kenya, Nigeria, South Africa, Tanzia, Uganda, Zaire, and Zimbabwe. Burundi's phosphate reserve is calculated at 15 Mt graded at 11% P_2O_5. No mining of the Burundi phosphate has occurred to date. Kenya also possesses an unmined phosphate resource estimated at 10 Mt of 4% P_2O_5 ore. Nigeria's Upper Volta Basin contains Pre-cambrian phosphate resources in excess of 300 Mt, with reported P_2O_5 content ranging from 15 to 32%. South Africa's Pre-cambrian Palabora Igneous Complex's P_2O_5 content ranges from 5 to 9%. Tanzania's phosphate deposits have been mined off and on for many years. The Mijingu Phosphate Company began mining southwest of Arusha in 1982. The phosphate resource is estimated at 10 Mt. An additional Tanzania phosphate resource is the Panda Hill complex in the southeast section of the county. Uganda's phosphate region consists of ten apatite bearing deposits stretching north to south. The Sukulu mine, in the Sukulu phosphate deposit, began operation in 1983. Phosphate reserves of the Sukulu deposit are estimated at 230 Mt at an average P_2O_5 concentration of 12.8%. Exploration of Zaire's carbonate complexes indicate possible phosphate resources are available in the Bingu and Lueshe areas of the country. Zimbabwe's Dorowa Complex steadily produced phosphate rock yielding approximately 38% P_2O_5.

Africa possesses offshore phosphate resources along the coasts of Morocco, Senegal, Togo, South Africa, and Somalia. The offshore deposits are derived from Miocene or older rocks, except for the west coast of South Africa, which is Holocene age. African offshore deposits require further study of resources to prove if they are viable for mining.

Asia

Asian nations produced approximately 17% of the world's phosphate rock from 1980 through 1990. Twelve Asian countries produced phosphate rock in 1990, with several others possessing significant phosphate resources.

China: China produces the majority of phosphate rock in Asia and is the fourth largest producer in the world. Production is primarily from the Hunnan, Jingshan, Zhongxiang, Hubei, and Gaiyang deposits. The Jingshan, Zhongxiang, Hubei, and Gaiyang deposits are comprised of the Pre-cambrian Doushanto Formation shale and dolomite, where the Hunnan deposit is comprised of precambrian Chayuanpo Group limestone, shale, and sandstone. Phosphate resources are estimated to be 30% P_2O_5, with reserves approximately 100 Mt. Additional Cambrian aged phosphate deposits are located in the Guizhou, Jiangsu, Sichuan, and Yunnan provinces of China. China's Cambrian aged phosphate deposits are associated with shale and silicious carbonates. Cambrian phosphate reserves are estimated at 100 Mt of 25% P_2O_5 material.

Jordan: Jordan produces phosphate from the Late Cretaceous sediments of carbonate and shales. Jordan utilized four mines through most of the 1980s, but is presently operating only three mines. The Jordan Phosphate Mines Company is producing 6 Mt of phosphate rock from the Wadi Al Abyad, Al Hasa, and the Al Shidiyah mines. The Al Shidiyah mine opened in 1989 and is located about 120 km east of the port of Aqaba. The Al Shidiyah mine replaced the closed Ruseifa mine and will replace the Al Hasa and Al Abyad mines around the year 2000. Jordan's phosphate reserves are reported to be 1.2 Gt, with an additional 1.0 Gt of phosphate resources.

Israel: Israel's phosphate rock source is Late Cretaceous sediments which are also mined throughout North Africa and the Middle East. Israel has three active phosphate quarries in the Oran and Arad areas of the country. Developing a fourth phosphate mine near Arad has been discussed for the Zohar deposit. Israel's phosphate reserves and resources are each estimated at 300 Mt.

Egypt, Iraq, Syria, and Saudi Arabia: Egypt, Iraq, and Syria also produce phosphate from the Late Cretaceous sediments similar to many nations of the Middle East and North Africa. Egypt's phosphate reserves and resources are reported to be approximately 700 Mt and 3 Gt respectively. Iraq entered into phosphate production in 1981 by commissioning the Akashat mine. Iraq's phosphate reserves and resources are estimated at 60 and 400 Mt respectively. Syria's phosphate rock reserves and resources are estimated at 400 Mt each. Saudi Arabia's Upper Cretaceous and Eocene aged Al Jalamid phosphate deposits are located approximately 1 126 km from the Arabian Gulf. The deposit contains an estimated 900 Mt of reserves. Development studies are presently in progress.

Iran: Iran contains an Devonian age phosphate deposit in the Elburz Mountains. Reserves are estimated at 40 Gt of rock at 11 to 22% P_2O_5. The phosphate consists of pelletal apatite grains and bone fragments in the Geirud Formation quartzite-shale sequence.

India: India's phosphate rock resources are located in four main areas: Mussoorie, Udaipur, Jhabua, and Harapur-Lalitpur. All of the phosphate deposits are precambrian age and are comprised of phosphatic carbonates and quartzite units. The phosphorite of the Jhabua area is associated with Stromatolites. The Jhamarkotra deposit near Udaipur is the largest high grade deposit in India. Resources are estimated at about 76 Mt of 10% P_2O_5 and 17 Mt of 30% P_2O_5. The small quarry operation reserves at Maton and Udaipur are estimated at 9 Mt at 25% P_2O_5.

The Harapur-Lalitpur area contains two basins of phosphate deposits. The Harapur basin consists of quartzites, shales and dolomites and has been exploited since 1978. Harapur's phosphate resources are estimated to range from 15 to 35% P_2O_5. The Lalitpur deposit is similar to Harapur, except for brecciated zones and a broader phosphate resource range of 5 to 25% P_2O_5.

Jhabua phosphate deposit lies about 200 km south of Udaipur and consists of phyllite, quartzite, and carbonates. Average P_2O_5 content is about 28%.

Mussoorie Syncline area is located in the northern section of India. The Mussoorie deposit is estimated to contain 45 Mt of 16 to 18% P_2O_5 material. Production from the area began in 1970.

Sri Lanka: Sri Lanka increased phosphate rock production from 5 Mt in 1980 to 25 Mt in 1990. Sri Lanka's phosphate rock is produced from the precambrian age apatite bearing carbonatite at Eppawala. Reserves are estimated at 20 Mt averaging 25 to 30% P_2O_5.

Pakistan: Pakistan is one of four countries that entered into phosphate rock production in the 1980s. Pakistan opened the Kalkul mine in 1986. Cambrian age carbonate fluorapatite mixed with impurities produces 10 to 30% P_2O_5. The deposit's reserves are estimated at 14 Mt. Additional deposits at Kalkul and Lagarban are being investigated. Pakistan also contains an Igneous phosphate resource at Loe Shilman, estimated to have 200 Mt of reserves with 5% P_2O_5 content.

Turkey and Greece: Turkish and Greek phosphate deposits belong to the same phosphogenic province as North Africa and the Mediterranean countries. The deposits are Late Cretaceous age and relatively small in comparison with deposits of neighboring countries. Turkey's phosphate resources are estimated at 250 Mt.

People's Republic of Mongolia: Mongolia's phosphate deposits in the area of Lake Hubsugal are substantial with reserves estimated at 1 Gt of rock with 20% P_2O_5. The phosphate bearing carbonate rocks are Cambrian in age.

Korea: Korea's phosphate deposits are located at Sinpung, Korea. The Sinpung phosphate deposit is precambrian age and is comprised of metamorphosed apatite-bearing marble. Phosphate reserves are estimated at 90 Mt containing 12.5% P_2O_5.

Vietnam: Vietnam has dramatically increased mining phosphate rock from 90 Mt in 1980 to an estimated 400 Mt in 1990. Cambrian age apatite bearing dolomite is utilized containing an estimated 23% P_2O_5 content. Reserves are reported at 100 Mt.

Thailand: Thailand produced an average of 6.5 kt of guano phosphate throughout the 1980s. Phosphate rock production is developed from the Ban Sob Moli deposit in the Lamphun province. Total phosphate reserves are estimated at approximately 150 kt.

Oceania

Islands in Oceania are the largest producers of guano phosphate in the world. Oceania's phosphate production is proportional to local demand.

Nauru Island: Nauru's phosphate rock production has steadily declined through the 1980s. Guano has been mined on the island since 1910. Reserves are estimated at 25 to 30 Mt of approximately 40% P_2O_5 ore. Nauru's present production of approximately 925 kt of phosphate rock continues to be the largest in the region.

Christmas Island: Christmas Island ceased production of phosphate rock in 1987. The island's maximum production year was 1980 with 1.7 Mt of guano phosphate being produced. The mine has since been sold and is planned to be reopened early in the 1990s.

Philippines: Guano phosphate resources are also exploited in the Philippines. The guano is mined on Negros Island. The Negros reserves are estimated at 50 kt. The Philippine islands of Bantique and Leyte have also been discovered to contain significant quantities of guano.

Australia: Australia produced an average of 14 kt per year of phosphate rock through the 1980s. Australian phosphate is produced from cambrian pelletal phosphorite of the Battle Creek Formation. Phosphate reserves are reported at 1.2 Gt of phosphorite averaging 18% P_2O_5.

Indonesia: Indonesian guano phosphate resources are estimated at 1.5 Mt containing an estimated 28 to 30% P_2O_5. Mining of the guano deposits is small, at 5 ktpy.

Fiji: Fiji's phosphate deposit is located on Lau Island, estimated resources are at 2.2 Mt of phosphate ore. The phosphate is associated with Miocene carbonates and clays.

French Polynesia: Phosphate on the Mataiva and Tuomoto Islands was discovered in the 1980s and is reported to contain in excess of 1 Mt of greater than 20% P_2O_5 material.

Europe

Europe's viable phosphate deposits are located along the northern and eastern margins of the continent, in Estonia, Yugoslavia, Finland, and Sweden. Europe's annual total phosphate rock production is approximately 0.7 Mt.

Finland: Finland's phosphate rock is mined at two deposits totalling an estimated 580 Mt of reserves. The Siilinjarvi and Sokli deposits utilize Late precambrian carbontite sediments. The Siilinjarvi deposit reserves are estimated at 470 Mt averaging 3.8% P_2O_5. The Siilinjarvi carbontite deposit mine opened in 1979.

Sweden: Late precambrian carbontite is exploited at the Alno deposit in Sweden. The deposit reserves are estimated at 60 Mt of ore containing 4% P_2O_5. Additional deposits are located at Grängesberg and Kiruna.

Yugoslavia and Estonia: Ordovician age phosphate deposits, comprised of phosphorized shells, are located in southeast Yugoslavia and in Estonia. Estonia's phosphate deposit is estimated to contain 300 Mt.

Former USSR

The former USSR is the second largest producer of phosphate rock in the world. Production of phosphate rock is mainly from the

Kola Peninsula. The Kola Peninsula's Khibiny Complex has produced rock since 1929. Phosphate is extracted from the complex's apatite-nepheline rock which on average contains 18% P_2O_5. The Kovdor mine located on the west side of the Kola Peninsula utilizes Apatite-Forsterite averaging 6.6% P_2O_5 content. A new mine was opened in 1981 along the east side of the Kola Peninsula at Vostochny. Total reserves for the peninsula are estimated at 2.0 Gt.

One hundred km southwest of the city of St. Petersburg is the Kingisepp phosphate deposit. The deposit is comprised of Ordovician carbonates and phosphorized shells. Reserves are estimated at 230 Mt containing 6 to 10% P_2O_5.

Forty-five Cambrian age phosphate deposits are located at Karatav, Kazakhstan. The Cambrian Age phosphate is associated with sandstone and dolomite. Reserves are estimated at 1.7 Gt at 26% P_2O_5 content. Southern Kazakhstan Chilisai deposit is comprised of Upper Cretaceous phosphate nodules in glauconitic sands and clay. Chilisai's reserves are estimated at 250 Mt.

Central Siberia phosphate deposits are located at Slyudyank, Irkutsk. The deposit's phosphate is contained in stromatolites generating a range of 4 to 20% P_2O_5. Reserves are estimated at 500 Mt.

Precambrian deposits of phosphate are located at Gornaya Shoriya at Kamorousk and Ukha Gol in Buryat. Both deposits are carbonate sequences with phosphate content of approximately 6 to 11% P_2O_5. Reserves are estimated for Gornaya Shoriya at 270 Mt and Ukha Gol at 2.6 Gt.

Three igneous phosphate deposits are located at Seligdar and Belaya Zima-Kruchinino. Reserves are estimated at 300 Mt for each deposit. Reserve P_2O_5 content ranges from approximately 7 to 10%.

EXPLORATION TECHNIQUES

Phosphate deposits of economic interest, both igneous and sedimentary, are difficult to recognize in the field. Commonly, even the discrete apatite crystal, or the small phosphorite pellet, cannot be positively identified. But a simple field test is possible by placing a drop of acid (sulfuric, nitric, or hydrochloric, not weaker than approximately 20% concentrated acid in water by volume) on the questionable area and then adding a small crystal, or a small amount of powder, of ammonium molybdate off the end of a knife blade. The rapid development of a bright yellow color indicates phosphate.

Exploration is generally directed toward: 1) finding new deposit areas and 2) searching for new deposits associated with known deposits. In the case of the latter, exploration will, of course, be guided by the practical knowledge acquired in that field before the new exploration begins.

Igneous Deposits

Although the world's great phosphate deposits are of the pellet phosphorite type, igneous deposits are also important as they yield about 20% of the present world production of phosphate rock.

Before field work begins, there must be a critical search for possible clues in the regional literature, maps, files, and rocks and mineral collections of all the geological departments in each country so that specific areas may be selected for reconnaissance inspection.

Having selected localities or small districts for examination on the basis of the more promising clues found in this search, a round of geological field inspections can begin. The objective is now to locate carbonatites, promising alkaline or ultrabasic rocks. Collecting material for identification or confirmation in the laboratory is the major task, and in addition to rock specimens, samples of residual soils are most important. This is because they are often more representative of the chemistry and mineralogy of the bedrock as a whole than are the resistant and often heterogeneous rocks which form outcrops. Normally they will be enriched relative to bedrock in some of the diagnostic elements of carbonatites, notably in phosphorus, barium, rare earths, and niobium (strontium is leached out).

Having located carbonatitic alkaline or ultrabasic complexes, the apatite-rich material may be quite nondescript and variable in appearance, especially when fine-grained or weathered, and chemical or microscopic checks should be applied to all doubtful or suspect rocks, coupled with soil sampling. In general the most deeply eroded complexes appear to be most favorable for apatite deposits of the phosphorite or pyroxenite type. In these, carbonatites may be relatively minor constituents of the complex, so a complex should not be underrated just because the area of carbonite is small. The maximum development of carbonatites appears at a somewhat higher level of intrusion in the form of large carbonatite plugs, and these may yield important residual deposits of apatite, provided climatic and erosional factors are favorable. In the less eroded volcanic complexes one may find apatite associated with the highly potassic feldspathic breccias and trachytic rock, as exemplified by the Tundulu deposits and several minor occurrences. The grade of deposit which may be regarded as of possible practical interest depends on many factors, but present practice suggests that a recoverable apatite content of about 15% (6% P_2O_5) may prove economic provided there are adequate reserves, and location and working costs are favorable.

Phosphate reserves are determined through core samples. Cores are sectionally sampled in lengths of 2.5 to 3 m and analyzed for P_2O_5 and Al_2O_3. Other chemical analyses are performed on groupings of samples or individual samples.

Pellet Phosphorites

The exploration and development of pellet phosphorites in new areas might be separated into four phases as follows:

Phase 1: The search for areas of marine sedimentation.

Phase 2: Within areas of marine sedimentation, search for beds of chert, or marine diatomite, and/or black shales.

Phase 3: Search for beds of phosphoric in or near the black shales or cherts. Also search for residual phosphatic soil areas associated with beds of phosphatic dolomite or limestone.

Phase 4: Determine thickness, grade, and extent of weathering by trenching and drilling.

The thickness of chert beds, so commonly associated with pellet phosphorites, may be from a few centimeters to over 60 m. Pellet phosphorite commonly contains from 0.001 to 0.02% U_3O_8.

Commonly, an outcrop of phosphorite will have patches of a thin skin of phosphate bloom where exposed at the surface. Its characteristic light gray-blue color is readily visible up to distances of several feet. A 10 to 20-power hand lens is indispensable to aid in identifying the small ovoid-shaped phosphate pellets.

In areas of marine sedimentation which appear to be a favorable locale for phosphorite, but where no outcrops are present, previously drilled holes, if present, can prove a valuable aid. For example, water wells in North Carolina when flushed out revealed phosphate pellets. Later, preliminary and successful exploration of the field was conducted by gamma-ray testing of about 100 existing holes drilled for water.

The important pellet phosphorites of the Georgina Basin in the Queensland-Northern Territory of northeastern north central Australia resulted from a study of cores of holes drilled in a search for oil. It was then established that phosphate beds could be identified on the gamma-ray logs of the holes. The phosphate beds so identified were then found in outcrops some distance away (Howard, 1971, 1972).

Chemical testing of sludge obtained from many holes drilled for oil in Turkey showed a few holes with enough phosphate to merit interest. Later gamma-ray logs of such holes were examined for phosphate. the results obtained from the drill holes led to the discovery of pellet phosphorite outcrops.

Other than gamma-ray measuring techniques, there has been to date very little application of geophysical methods in the exploration of pellet phosphorites. Cooksley, Jr. (1967) describes some work on the Phosphoria Formation in southeastern Idaho made in connection with obtaining additional information on previously discovered phosphorite deposits. Road cuts and an occasional gully will reveal the underlying phosphorite. Even at such places the inexperienced eye probably would not distinguish between the soil and the phosphorite.

Phosphorite beds vary laterally in thickness and phosphate content. In exploration therefore, extensive lateral exploration should be made even though preliminary exploration may detect only low grade ore. The economically important detrital (reworked residual) phosphorites of the Bone Valley Formation in Florida occur as unevenly distributed beds over large areas. Their lateral variation in thickness and grade requires extensive drilling to ascertain the tortuous outlines of minable material.

After a bedded primary pellet phosphorite has been found, it is commonly difficult to follow the bed in outcrop because weathering has softened the formation releasing loose soil-like material that masks the beds. Even when the exact location is known, pellets or small pieces of phosphorite are difficult to find or recognize. Therefore, harder marker beds stratigraphically above or below the phosphorite often can be better used to map for location of a phosphorite bed than the phosphate bed itself.

In certain areas, shallow drilling has proven itself where the objectives are near surface deposits. Residual phosphate deposits of Florida-Georgia-South Carolina-North Carolina and Tennessee are representative. Drilling in those areas is generally limited to depths that could be open pit mined. For the past several years, exploration drilling for phosphate deposits in selected areas of Florida and North Carolina has reached depths of 90 m or greater.

In wet places, tests can be made by washing cuttings from the hole. Cuttings are caught in tubs on the surface. This type drilling commonly results in loss of slimes and, consequently, under-reporting of the quantity of slimes in the final laboratory tests.

The discovery of the economically important Bu-Craa phosphate in the Sahara is an example of drilling being used in the exploration phase. As described by ENMINSA (Anon., 1967), geological reconnaissance work started at the already known low grade phosphate horizon in the Izic district. These studies, in a flat desert terrain with poor rock exposures, indicated a richer phosphate concentration 96 km away. Drilling in this area on a 1-km grid found the important Bu-Craa deposit. This drilling was done with a rotary rig using a trichion bit. The diameter was 11.34 cm with cuttings air-blown to the surface where they were collected as samples.

Phase 4 in the exploration and development of pellet phosphorites involved the determination of the thickness, grade, and extent of weathering by surface work which may include one or more of the following: trenching, shafts, or drilling. Trenching can be by hand or with mechanical equipment—dozers or backhoes. Its purpose is to expose the phosphatic beds for measurement and sampling. Depths of 4.5 m or more may be necessary to get through loose overburden and minimize the effects of surface creep. Downslope surface creep can significantly string out and thin the phosphatic beds and change the apparent dips of the beds. Surface slopes as low as 5° can result in significant creep. Where the material moved is dry and relatively in cohesive, deep trenches pose a real danger to men and equipment from sudden slumping of the vertical walls.

Diamond drilling is used as well as rotary drilling with air return of cuttings. When the rotary drilled hole has reached the phosphate zones (determined visually), the cuttings are collected as samples. Coring can, of course, also be done with a rotary drill rig, but this is not common nor is drilling with water common with these rigs. Air return of cuttings works very well, and without casing, up to depths of about 100 m; however, encountering wet material usually prevents recovery of cuttings.

Diamond drilling has certain advantages, i.e., ability to make deeper holes needed to determine the limits of alteration, wet ground does not stop drilling, holes can be angled to better intersect dipping beds, and a core is a more accurate sample for geological study.

Additional information is required for the purpose of better defining local structures such as faulting and change of dip of the beds. These local structural features dictate the amount of additional drilling.

In the Florida fields, the matrix (phosphate ore) is commonly separated every 2 m into an individual sample based upon changes in phosphate content, texture or color of the matrix. Each sample is processed in laboratories by bench tests to simulate plant operations, i.e., a weighed sample is pulped with water and deslimed; slime is any material less than 150 mesh. The deslimed material is then screened at 14 or 16 mesh to obtain coarse phosphate aggregate called pebble, and the undersize $-14+150$ mesh portion is floated to recover a phosphate concentrate. The pebble and flotation feed portions furnish the sample for assaying and not the original matrix.

In the United States, analyses on samples are performed by the established wet chemical procedures or by X-ray fluorescence. The X-ray process, however, is not always accurate. It depends on the establishment of standard curves, but between deposits or even within one deposit variations frequently do not fit these curves. For exploration and development, therefore, the X-ray procedure should not be used. A titration process is used for P_2O_5, Fe_2O_3 and CaO, but SiO_2 and Al_2O_3 analyses are determined gravimetrically.

Even before exploration begins, a search for phosphate deposits has an end use objective. Objectives include export of rock, export of fertilizer, export of phosphorus, or satisfying domestic markets for fertilizer or industrial chemicals.

Where igneous apatite is contained in unweathered alkalic rocks, such as in the Kola Peninsula area of the USSR and in South Africa, the absence of carbonates permits flotation of apatite after fine grinding. Where the igneous apatite is in a carbonatite, the presence of the large quantities of carbonates in the unweathered rock has prevented the recovery of rock by flotation because flotation reagents work on the calcium end of the mineral molecule and calcium carbonate will float along with calcium phosphate.

With pelletal phosphorites, impurities contained in the phosphate ore may influence end use of the rock. Such impurities include detrital silica grains, cryptocrystalline silica derived from siliceous skeletons of small organisms, clay, iron oxide, calcite, dolomite, and hydrocarbons. Most of these impurities occur in the ore as discrete particles separate from the phosphate; however, a certain amount is present in the individual phosphate pellet or particle.

Impurities present as discrete particles in weathered phosphorites can be partially, or nearly wholly, removed by simple washing or by washing followed by flotation. However, the small particles of impurities present in a phosphate pellet generally cannot be economically removed, except that calcining, generally in the range of 900 to 950°C, removes the CO_2 of carbonates and the hydrocarbons. The remaining CaO can be removed by slaking and screening.

Because under normal conditions phosphate rock for any use has a relatively low sales value, the evaluation of phosphate deposits from the start is involved with limiting cost factors. Among these are the need for low cost open pit mining, large volume operations with their cost benefits, low cost large supplies of water for beneficiation, and low cost transportation to markets.

The mining and milling techniques for the preparation of rock are essentially the same whether the rock is used to make phosphorus or fertilizer. Rock for phosphorus production can be as low grade as 24% P_2O_5; therefore, it may require no beneficiation, or, if beneficiated, a low grade concentrate is acceptable. On the other hand, for fertilizer use, the rock (concentrate) should be as high grade as economically possible.

Mining and Processing

Underground mining is utilized in the China, Russia, Morocco, Tunisia, and the United States. In Morocco and Tunisia underground mining employs room and pillar methods. The relatively soft ore does not require drilling and blasting but large quantities of timber posts are used to prevent roof slabs from falling. Underground mining of the Russian Kola ore is done by sublevel open stoping (Anon., 1968a, 1979, 1979a, Woodrooffe, 1972).

The bulk of the world's phosphate rock is mined open pit for the simple reason that it is the least expensive way to move large volumes of material. Unit costs are largely affected by the thickness and hardness of the overburden, thus larger equipment with its lower unit cost of operation permits removal of greater thicknesses of overburden. Both overburden and phosphate ore are moved by 1) shovels and trucks, 2) draglines and trucks, 3) rubber-tired scrapers, 4) draglines and slurry pipelines, 5) bucket wheel excavators, and 6) grabbing cranes (guano).

Where the overburden is hard, such as the thin-bedded carbonate formations in Morocco and Utah and the chert bed in Idaho, drilling with rotary rigs and shooting with ammonium nitrate fuel oil (ANFO) may be required. Rippers on heavy duty dozers are also useful.

Generally, more efficient use of equipment, and therefore lower overall costs, results from using the same equipment to both strip and mine. This is the practice in Florida where the mining operations use draglines. The same machine strips the sandy overburden, then picks up the matrix (ore) and dumps it into a pump-sump from where it is pumped to beneficiation plants.

At some open pit operations in Morocco, larger draglines are used on the overburden matrix to beneficiation plants. The overburden consisting of thin-bedded carbonate rocks is drilled and blasted, then removed with draglines up to 50 m^3 in size. The matrix draglines are about 7.5 m^3 in size. Matrix is mined with a 600 tph bucket wheel excavator.

At the Fosfago mine at Catalão, Brazil, mining is conducted by open pit techniques using rippers, loaders, and trucks on an altered carbonate. Beneficiation is similar to Jacupiranga except the major contaminate, magnetite, is removed by magnetic separation prior to fine grinding and flotation.

The examples given here of equipment used in surface mining of phosphate rock do not, of course, cover all world phosphate mining. They do indicate types and sizes commonly used. In soft material the large draglines move material at the lowest cost. With harder material and smaller machines, of whatever type used, costs are much greater.

Beneficiation

Igneous Deposits

At present, there are four areas producing substantial quantities of apatite—Russia, South Africa, Brazil, and Zimbabwe.

Russia: In 1972 production of Kola concentrates was about 12 Mt containing 39% P_2O_5 (Woodrooffe, 1972). By 1980 this had increased to about 20 Mt.

South Africa: In South Africa the unweathered apatite-bearing ore of the Phalaborwa alkalic igneous complex (pyroxenite and foskorite) is crushed, ground, and floated. The flotation concentrate contains about 36% P_2O_5. Production in 1980 was about 3.5 Mt.

Brazil: At the Jacupiranga mine in Brazil mining operations are on an unweathered carbonatite containing 5% P_2O_5. Beneficiation consists of crushing, grinding, and floating an apatite concentrate containing 36 to 38% P_2O_5.

With the exception of Jacupiranga all apatite is on soft residual ore resulting from weathering. Grade of ore is 7 to 15% P_2O_5 and beneficiation is crushing, grinding, magnetic separation, desliming, and flotation. Brazil's rapidly developing phosphate industry is described in Anon., 1978, and Breathitt and Finch, 1977.

Zimbabwe: At the Dorowa alkalic complex in Zimbabwe, weathered surface material is mined at about 7% P_2O_5. The beneficiation plant grinds, washes, and floats the ore to produce apatite concentrate.

Pellet Phosphorite Deposits

Basically, the beneficiation flowsheets are about the same. The matrix is pumped several miles as a water slurry containing about 35% solid from the mine pit to the plants in steel pipes. At the plants the slurry passes over coarse screens to remove trash and waste rock, then over about 14 mesh screens to remove pebble rock. The −14 mesh material is deslimed at 150 mesh. The −14 +150 mesh material is called flotation feed, and its grade is 6.4 to 11.4% P_2O_5. It is separated into two sizes and each is floated to recover the concentrate. References for more details on beneficiation practices are: Beall and Merritt (1966), Aparo (1969), Anon. (1971a), and Lawver et al. (1978). Beall and Merritt not only describe beneficiation operations in Florida, but mining as well. Additionally, they describe mining and beneficiation in North Carolina. Beneficiation of the North Carolina pellet phosphorite is described also by Wright (1971) and Breathitt and Finch (1977).

Israel

The production of the newest mine in Israel, Nahal Zin, was about 1.7 Mt of concentrates in 1980. The surface mined ore is crushed and wet-screened to remove +1 mm waste. Hydrocycloning removes −0.1 mm slimes. The pulp is washed with fresh water to reduce chlorides, then dried. A description of Israel's deposits and mining/beneficiation operations is given in Anon. (1980b).

Jordan

As an additional beneficiation process, calcining of pellet phosphate rock, is being increasingly used. When the objective is elimination of hydrocarbons, temperature on the order of 800°C is adequate, but removal of CO_2 from carbonates requires higher temperatures, from 900 to 950°C.

There are a number of advantages in using calcined rock in the production of wet process phosphoric acid. Among them is the ability to produce clear acid, which is a more acceptable component of liquid fertilizers.

Products

The United States is the world's largest producer and user of fertilizer and is the leader in developing new types of fertilizers and

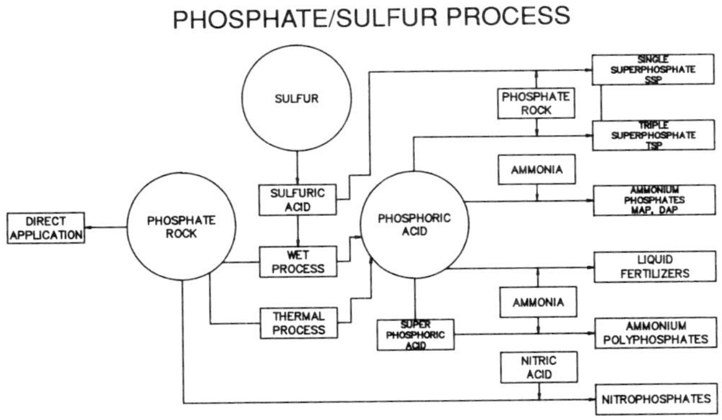

Fig. 6. Flowchart illustrating various fertilizer products and their source materials.

their uses. There has been a rapid switch to mixed and higher grade fertilizers, both granular and liquid types.

Most world fertilizer manufacturing processes use sulfuric acid, although some, mostly in Europe, use nitric acid. Where acid soil conditions favor its use, some finely ground rock is applied directly to the soil, but the quantity used is minor and in the United States continues to decline (Fig. 6).

A continuing trend is to produce phosphoric acid (52 to 54% P_2O_5) which is then used in the formulation of final fertilizers. Some superphosphoric acid (67 to 76% P_2O_5) is used for the same purpose.

The grade of a fertilizer is denoted by the use of three numbers such as (18–46–0). The number defines the percent elemental nitrogen (N), the second, P_2O_5, and the last is K_2O. In recent years there has been a trend to using numbers referring to the element only as (N-P-K). Chief among the types of phosphate fertilizers are the following:

- normal superphosphate (0–18–0) to (0–20–0) (solid)
- concentrated superphosphate (0–42–0) to (0–50–0) (solid)
- diammonium phosphate (18–46–0) (solid)
- monoammonium phosphate (16–48–0) and (11–48–0) (solid)
- granular mixed fertilizers (combination of foregoing types and K_2O where needed).

Liquid mixed fertilizers (phosphoric acid neutralized with ammonia and nitrogen solution and potash added where needed).

As pointed out earlier, any phosphate rock, igneous or sedimentary, can be used to meet any of man's required end products; however, economics help govern the selection. Rock grindability and reactivity with acid influence costs. For example, the fluorapatite crystals in igneous apatite are denser than the francolite crystals of pellet phosphorites. Igneous apatite is, therefore, not as easy to grind nor is it as reactive to acid. Conversely, igneous apatite does not have the hydrocarbon content which generally occurs in all pellet phosphorites and which may impose a penalty during acidification. Robinson (1972), Hill (1960), Everhart (1971), and Pacl et al. (1978) describe the physical and chemical characteristics of phosphate rock which influence its reaction to acidulation.

Phosphorus (Industrial Chemicals) Industry

Phosphorus is produced in electric furnaces charged with calcined phosphate rock, silica, and coke. It is used to produce high purity compounds—mostly sodium, calcium, potassium, and ammonium phosphates. The manufacturing steps begin with burning the phosphorus in air and dissolving the resultant oxide in water to make a relatively pure phosphoric acid suitable for the manufacture of industrial phosphates. They include detergents, industrial cleaners, and fire retardants. Further purification with hydrogen sulfide to remove traces of arsenic yields a very poor food grade acid. Phosphorus is also used in the manufacture of organophosphorus compounds for such uses as flame-retardant additives for fabrics and plastics, and insecticides.

Phosphorus electric furnaces (enclosed) operate in a reducing atmosphere so that the vaporized phosphorus can be pulled off in the gases without contacting oxygen in the air. The gases are mainly CO and vaporized phosphorus. They are first cleaned of entrained particulate matter, then sprayed with water which cools them, causing condensation of the phosphorus into droplets which settle and are collected underwater.

In the phosphorus furnace coke is the reducing agent. Silica in the melt reacts with the calcium end of the phosphate mineral to produce a calcium silicate slag. Any other calcium, magnesium, or aluminum mineral, such as calcite, dolomite, or clay also reacts with silica and adds to the slag. Part of any iron oxide present in the phosphate rock is reduced to molten iron which settles to the bottom of the furnace under the silicate melt. As the iron is formed it reacts with phosphorus at a ratio of about three parts iron to one part phosphorus to make an alloy called ferrophosphorus. It is tapped separately from tapholes located below the silicate-melt tapholes.

Calcium, magnesium, and clay minerals, therefore, are undesirable in the phosphate rock, as they consume electricity. Iron is undesirable for the same reason, but, even worse, it ties up phosphorus in a low value end product—ferrophosphorus.

The calcined phosphate rock charged to the furnace cannot be fine-grained because a porous bed must be maintained in the furnace to permit circulation of gases. The rock is charged as prepared material about an inch in size. This prepared material is made by one of two routes—nodulizing and calcining at about 1 350°C in nearly horizontal kilns or calcining rolled flakes, or round pellets in grate kilns.

Phosphate rock used to produce phosphorus can be as low grade as 24% P_2O_5. the CaO/P ratio should be in the range of 3.30 to 3.60—preferably the lower figure. An acceptable CaO/P ratio is, however, directly related to the cost of power. About 6 kWh is required in a phosphorous plant for each pound of phosphorus produced. The iron content of phosphate rock should not exceed a few percent Fe_2O_3.

In the United States phosphorus produced at an electric furnace plant is shipped in closed and insulated railroad tank cars to manufacturing plants located near markets for the end product. these tank cars, with capacities of 30 300 to 83 300 L, are jacketed so that if the temperature of the phosphorus drops to 44°C (when the liquid phosphorus becomes a solid), the contents can be remelted by stream at destination for pumping to the burning towers. In the burning tower oxygen in air reacts with the phosphorus to make P_2O_5 powder, which is mixed with water to make a very pure phosphoric acid containing 54% P_2O_5.

Phosphorus is not inexpensive. The principal production cost factors are electricity, coke, and rock coupled with heavy capital investment. Its production is justified by its high purity which is required to meet strict specifications of end products.

Byproducts from the Use of Phosphate Rock

Gypsum: Gypsum, $CaSO_4 \cdot 2H_2O$, is produced as a byproduct in making wet-process phosphoric acid. About 5 000 kg of gypsum results from the production of a ton of P_2O_5—equivalent to roughly 1 500 kg of gypsum for each ton of rock used. Commonly, the gypsum is stockpiled near the fertilizer plant. Because of the increasing use of rock, with the trend to wet-process phosphoric acid production, together with the trend to larger plants, the quantity of gypsum accumulating in stockpiles is becoming large.

Much attention has been given to putting this byproduct gypsum into the market for plasterboard and road base materials, but for a number of reasons, both economic and social, efforts to use any significant quantity of byproduct gypsum have not been successful. Processes for the recovery of the sulfur content from the gypsum are not currently economic.

Uranium: Uranium occurs in minor quantities in the mineral apatite $3Ca_3(PO_4)_2CaF_2$, which is the source of phosphorous in phosphate ore. After mining, phosphate ore is crushed and ground, leaving an apatite concentration with a grade of 30 to 34 % P_2O_5. If the apatite is derived from a marine deposit, as most are, this concentrate generally carries 50 to 200 ppm U_3O_8; and lower if the apatite is derived from an igneous deposit.

There was, during the 1980s, much interest in the United States in the recovery of uranium from wet-acid by solvent extraction methods. However, current market prices for yellow cake, with the static nature of the nuclear power industry, leaves uranium recovery economically unattractive.

Because uranium is recovered as a byproduct from existing operations, the environmental difficulties are significantly less troublesome than those associated with new uranium mining and milling facilities.

Vanadium: Pellet phosphorites commonly contain significant quantities of other metals besides uranium. Those of the Phosphoria Formation of the western United States in places contain considerable vanadium. In the 1930s and 1940s, rock mined in Idaho was processed into wet-process acid at a plant in Montana and red cake, sodium vanadate recovered. The ferrophosphorus from an electric furnace operation in Idaho is processed in a separate plant for recovery of vanadium.

Fluorides: The world's phosphate rock deposits represent the largest known resources of the all important element fluorine. Much of the fluorine in rock is evolved in gaseous form in the production of fertilizer. Plants in Florida have long recovered H_2SiF_6, used by municipalities to fluorinate water and used to produce synthetic cryolite for aluminum manufacturing processes. Other plants around the world either in production or planned to recover fluorides from rock are located in Canada, Mexico, Russia, India, Japan, western Germany, China, and Australia. In all likelihood the future will see a growing recovery of fluorine from this source.

Other Byproducts: Nepheline syenite tailings from the Kola Peninsula apatite operation in Russia are utilized in the production of aluminum and ceramics.

In Brazil, a plant makes cement from calcite tailings resulting from the flotation of apatite from a carbonatite phosphate deposit.

ECONOMIC FACTORS

Domestic and export phosphate rock prices are annually published by the US Bureau of Mines. Average domestic phosphate rock prices in 1991 was reported to be $0.25 per kg. The price of phosphate rock has only increased 8.7% from the average 1982 price.

The world supply/demand situation continues to be controlled by excess production. Late in the 1980s companies reduced production to control glutted stockpiles (Fig. 7). The amount of companies operating in the United States continues to decline. Exports for the United States will decline with the advent of Third World phosphate rock production, in addition to present foreign competition.

Transportation continues to be by conventional truck, rail and ship modes. Several facilities in the world use slurry pipelines to transport phosphate from the mines to processing sites.

Government Considerations

The United States federal income tax system includes provisions for mineral depletion deductions. Either cost depletion or

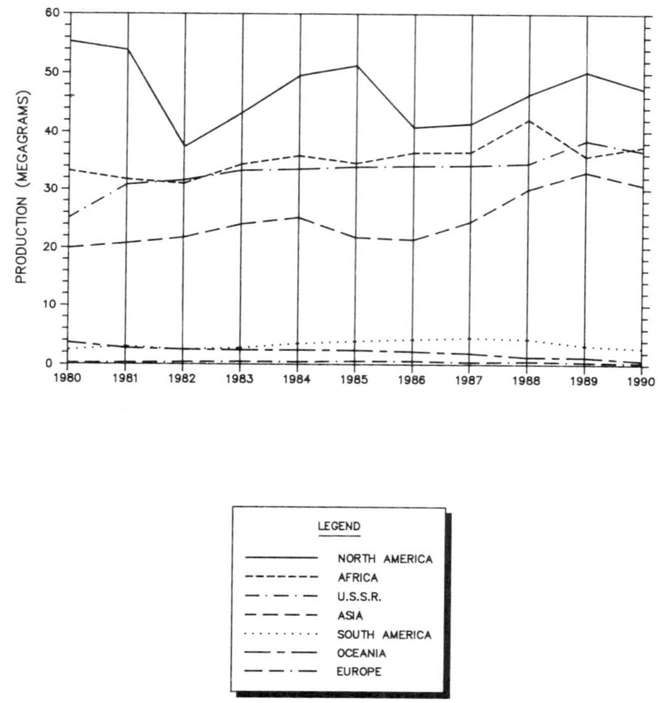

Fig. 7. World production of phosphate rock from 1980 to 1990 (*source*: US Department of the Interior, Bureau of Mines Minerals Yearbook, Phosphate Rock chapter (1984, 1988, and 1990).

percentage depletion can be elected, but is limited to 50% of the net income. Phosphate industry profits are also subject to applicable state sales taxes, state motor fuel taxes, county and local property taxes and state mineral severance taxes, where applicable.

Zoning and land use permitting has become increasingly difficult. During the 1970s, states produced legislation that provided for environmental agencies and regulations. Florida mining operations are required to have Development of Regional Impact (DRI) studies prepared and approved prior to expansion into new mineral blocks, or for any major construction. Environmental and occupational health regulations also increased through the 1980s. Gypsum stacks, potential ground water pollution, low level radiation and wetland conservation, water conservation and other environmental concerns are just a few of the issues facing the US phosphate rock producers.

The United States does not place tariffs or restrictions on phosphate rock shipments. However, the European Economic Community imposes duties on phosphoric acid and diammonium phosphate.

Future Trends of the Phosphate Rock Industry

Phosphate rock production fluctuated significantly through the 1980s. In general, production increased from 1980 through 1990 by approximately 11% (Fig. 7). Fluctuations of production can be correlated to the world's economic climate.

Future phosphate rock production will also increase with the world's population demand for food. Third world countries will become more self-reliant for their fertilizer requirements during the 1990s. However, the high demands of capital investment will hinder major development in these countries. Major phosphate rock or fertilizer exporters will continue relatively at the present rate of increase.

Improving phosphate processing and deposit exploration methods in conjunction with the advent of offshore phosphate deposits, ensures the future of the phosphate industry.

BIBLIOGRAPHY AND REFERENCES

Anon., 1965, "Phosphate Rock Mining and Treatment Plant at Phalaborwa," *Phosphorous & Potassium*, No. 19, The British Sulphur Corp., London, Oct., pp. 12–18.

Anon., 1966, "Elemental Phosphorus in the United States," *Phosphorus & Potassium*, No. 24, The British Sulphur Corp., London, Aug.-Sep., pp. 12–24.

Anon., 1966a, "World Elemental Phosphorus-Part II," *Phosphorus & Potassium*, No. 25, The British Sulphur Corp., London, Oct.-Nov., pp. 10–16.

Anon., 1967, *A Study of the Bu-Craa Phosphate Deposit*, Empresa Nacional Minera del Sahara, S.A. (ENMINSA), Madrid, 28 pp. (English translation).

Anon., 1968, "Elemental Phosphorus Developments," *Phosphorus & Potassium*, No. 36, The British Sulphur Corp., London, July-Aug., pp. 16–25, 33.

Anon., 1968a, "Kola," *Phosphorus & Potassium*, No. 37, The British Sulphur Corp., London, Sep.-Oct., pp. 13–20.

Anon., 1968c, "Phosphate Rock Calcination in Israel," *Phosphorus & Potassium*, No. 35, The British Sulphur Corp., London, May-June, pp. 12–15, 21.

Anon., 1969, "Calcined Phosphate Rock," *Phosphorus & Potassium*, No. 41, British Sulphur Corp., London, May-June, pp. 13–17, 19.

Anon., 1971, *Fluorides*, Committee on Biologic Effects of Atmospheric Pollutants, National Academy of Sciences, Washington, DC, 295 pp.

Anon., 1971a, "The Phosphate Fertilizer Industry of the U.S.," *Phosphorus & Potassium*, No. 55, The British Sulphur Corp., London, Sep.-Oct., pp. 21–28.

Anon., 1972, "Developments in Morocco's Calcined Rock Capacity," *Phosphorus & Potassium*, No. 62, The British Sulphur Corp., London, Nov.-Dec., pp. 24–26.

Anon., 1975, "Morocco Plans for the Future," *Phosphorus & Potassium*, The British Sulphur Corp., Ltd., London, Mar.-Apr., pp. 25–36.

Anon., 1978, "Brazil's Phosphate Rock Industry," *Phosphorus & Potassium*, The British Sulphur Corp., Ltd., London, Jan.-Feb., pp. 21–24.

Anon., 1978a, "Phosphorus Acid Trend in 1977," *Phosphorus & Potassium*, The British Sulphur Corp., Ltd., London, July-Aug., pp. 17–21.

Anon., 1978b, "Does Banning Detergent Phosphate Make Any Significant Difference?" *Phosphorus & Potassium*, The British Sulphur Corp., Ltd., London, Jan.-Feb., p. 42.

Anon., 1978c, "Evaluation of the Phosphate Deposits of Florida Using the Minerals Availability System," Open File Report 112–78, US Bureau of Mines, 199 pp. Contract No. 14–08–001–17501, Zeller-Williams, Inc., available from TIS PB 286 648/AS.

Anon., 1979, "Soviet Phosphates in the Tenth FYP-I (Five Year Plan)," *Phosphorus & Potassium*, The British Sulphur Corp., Ltd., London, July-Aug., pp. 16–20.

Anon., 1979a, "Soviet Phosphates in the Tenth FYP-II (Five-Year Plan)," *Phosphorus & Potassium*, The British Sulphur Corp., Ltd., London, Sep.-Oct., pp. 21–23, 43.

Anon., 1979b, "Uraniferous Phosphate Resources and Technology and Economics of Uranium Recovery from Phosphate Resources, United States and Free World," Open File Report GJBX-110 (79), Earth Sciences, Inc., US Dept. of Energy, Grand Junction, CO.

Anon., 1980, "OCP Unveils Its Plans for the Future, Part I," *Phosphorus & Potassium*, The British Sulphur Corp., Ltd., London, Jan.-Feb., pp. 20–26.

Anon., 1980a, "OCP Unveils Its Plans for the 1980s, Part II," *Phosphorus & Potassium*, The British Sulphur Corp., Ltd., London, Mar.-Apr., pp. 34–39.

Anon., 1980b, "Negev Phosphates Comes of Age," *Phosphorus & Potassium*, The British Sulphur Corp., Ltd., London, May-June, pp. 27–32.

Altschuler, Z.S., 1973, "The Weathering of Phosphate Deposits, Geochemical and Environmental Aspects," *Environmental Phosphorus Handbook*, John Wiley-Interscience, New York, pp. 33–96.

Aparo, S.J., 1969, "Current Flotation Practices in the Beneficiation of Florida Phosphate," Preprint 69-H-319, SME-AIME, New York, 13 pp.

Barber, J.C., and Marks, E.C., 1962, "Phosphorus Furnace Operations," *Journal of Metals*, Dec., pp. 1–5.

Barr, J.A., 1960, "Phosphate Rock," *Industrial Minerals and Rocks*, 3rd ed., J.L. Gillson, ed., AIME, New York, pp. 649–668.

Barrie, J., 1967, "The Geology of Christmas Island," Australia Department of National Development, Bureau of Mineral Resources, Geology and Geophysics, Canberra, 46 pp.

Beall, J.V., and Merritt, P.C., 1966, "Phosphate and Potash," *Mining Engineering*, Vol. 18, No. 10, pp. 83–89.

Benchekroun, A., 1983, "Prospects for Morocco's Position in World of Phosphate Rock and Derived Products by the Year 1990," Industrial Mineral Conference Proceedings, *Phosphate—What Prospects for Growth?* Orlando, FL, pp. 78–79.

Breathitt, H.W., and Finch, E.P., 1977, "Phosphate Rock Beneficiation," Preprint 77-H-134, SME-AIME Annual Meeting, Atlanta, GA, 31 pp.

Burnett, W.C., 1977, "Geochemistry and Origin of Phosphorite Deposits from Off Peru and Chile," Geological Society of American Bulletin, Vol. 88, pp. 813–823.

Burt, R.B., and Barber, J.C., 1952, *Production of Elemental Phosphorus by the Electric-Furnace Method*, Chemical Engineering Report No. 3, Tennessee Valley Authority, Muscle Shoals, AL, (republished 1967), 312 pp.

Cathcart, J.B., 1983, "Sedimentary Phosphate Deposits of the World—Present Status and Outlook for the Future," Industrial Minerals Conference Proceedings, *Phosphate—What Prospects for Growth?* Orlando, FL, pp. 78–79.

Clitheroe, J.B., and Mele, S., 1967, "Beneficiation Techniques for Western Phosphate Rock," *Anatomy of the Western Phosphate Field*, L.A. Hale, ed., Intermountain Association of Geologists, Salt Lake City, UT, pp. 225–235.

Cook, P.J., 1972, "Petrology and Geochemistry of the Phosphate Deposits of Northwest Queensland, Australia," *Economic Geology*, Vol. 67, No. 8, pp. 1193–1213.

Cook, P.J., 1990, "Phosphate Research: A Historical Overview," Geologic Society Special Publication No. 52, pp. 8–13.

Cooksley, J.W., Jr., 1967, "Application of Geophysical Methods in Phosphate Exploration Southeastern Idaho," *Anatomy of the Western Phosphate Field*, L.A Hale, ed., Intermountain Association of Geologists, Salt Lake City, UT, pp. 161–166.

Deans, T., 1968, "Exploration of Apatite Deposits Associated with Carbonatites and Pyroxenites," *Proceedings of the Seminar on Sources of Mineral Raw Materials for the Fertilizer Industry in Asia and the Far East*, United Nations Mineral Resources Development Series, No. 32, pp. 109–119.

Douglas, J.R., 1979, "World Phosphate Supply Plagued with Questions," 7th Phosphate-Sulphur Symposium, Innisbrook, FL, Jan. 18–19, TVA National Fertilizer Development Center, Muscle Shoals, AL, 10 pp.

Emigh, G.D., 1967, "Petrology and Origin of Phosphorites," *Anatomy of the Western Phosphate Field*, L.A. Hale, ed., Intermountain Association of Geologists, Salt Lake City, UT, pp. 103–114.

Emigh, G.D., 1972, "World Phosphate Reserves—Are There Really Enough," *Engineering and Mining Journal*, Vol. 173, No. 4, pp. 90–95.

Emigh, G.D., 1973, "Economic Phosphate Deposits," *Environmental Phosphorus Handbook*, John Wiley-Interscience, New York, pp. 97–116.

Emigh, G.D., 1975, "Phosphate Rock," *Industrial Minerals and Rocks*, 4th ed., S.J. Lefond, ed., AIME, New York, pp. 935–962.

Everhart, D.L., 1971, "Evaluation of Phosphate Rock Deposits," Preprint No. 71-H-52, SME-AIME, New York, 7 pp.

Gurr, T.M., 1977, "Geology of Phosphate Deposits," SME preprint 77-H-84, pp. 684–690.

Hale, L.A., 1967 "Phosphate Exploration Using Gamma-Radiation Logs, Dry Valley, Idaho," *Anatomy of the Western Phosphate Field*, L.A. Hale, ed., Intermountain Association of Geologists, Salt Lake City, UT, pp. 147–159.

Harben, P., 1980, "Where is Florida's Phosphate Industry Going?" *Industrial, Minerals*, No. 148, London, Jan., pp. 48–55.

Harben, P., 1983, "Phosphates in Brazil, Imports Slashed as Production

Rises,'' Industrial Mineral Conference Proceedings, *Phosphate—What Prospects for Growth?* Orlando, FL, pp. 93–56.

Harre, E.A., et al., 1972, *Estimated World Fertilizer as Related to Future Demand*, Tennessee Valley Authority, Muscle Shoals, AL, Aug., 25 pp.

Harris, G.E., and Harre, E.A., 1979, ''World Fertilizer Situation and Outlook,'' International Fertilizer Development Center and National Fertilizer Development Center, TVA, Muscle Shoals, AL, Mar., 27 pp.

Heinrich, E.W., 1966, *The Geology of Carbonatites*, Rand McNally and Co., Chicago, IL, 607 pp.

Hignett, T.P., and Striplin, M.M., Jr., 1967, ''Elemental Phosphorus in Fertilizer Manufacture,'' *Chemical Engineering Progress*, Vol. 63, No. 5, pp. 85–92.

Hill, W.L., 1960, ''Phosphate Rock: Reactivity Scales for Ground Rock,'' *Chemistry and Technology of Fertilizers*, V. Sauchelli, ed., Reinhold Publishing Corp., New York, pp. 116–138.

Howard, P.F., 1971, ''The Discovery of Phosphorites in Australia—A Case History,'' Preprint 71-H-97, SME-AIME Annual Meeting, 11 pp.

Howard, P.F., 1972, ''Exploration for Phosphorite in Australia—A Case History,'' *Economic Geology*, Vol. 67, No. 8, pp. 1180–1192.

Hutchinson, G.E., 1950, ''Survey of Contemporary Knowledge of Biogeochemistry: 3. The Biogeochemistry of Vertebrate Excretion,'' *Bulletin*, American Museum of Natural History, Vol. 95, 554 pp.

Jayawardena, D., 1979, ''The Occurrence of Rock Phosphate Deposits in Sri Lanka in Relation to the Precambrian Basement,'' Paper presented at I.G.C.P. Project 156, East-West Center, Hawaii, Aug. 9.

Kurtanjek, M.P., 1983, ''The Phosphate Situation in Central Africa: A Regional Assessment of Fertilizer Demand and Possible Future Supply Sources of Phosphate Rock,'' Industrial Mineral Conference Proceedings, *Phosphates—What Prospects for Growth?* Orlando, FL, pp. 103–108.

Lawver, J.E., et al., 1978, ''New Techniques in Beneficiation of Phosphate Rock,'' Preprint 78-B-331, SME-AIME Fall Meeting, Orlando, FL, 36 pp.

Lehr, J.R., 1983, ''Impact of Phosphate Rock Quality on Fertilizer Market Uses,'' Industrial Mineral Conference Proceedings, *Phosphate—What Prospects for Growth?* Orlando, FL, pp. 276, 277.

Mayberry R.C., 1981, ''Phosphate Reserves, Supply and Demand—Southeastern Atlantic Coastal States, 1980–2000 A.D.,'' Preprint 81-409, SME-AIME Fall Meeting, Denver, Nov., 39 pp.

McClellan, G.H., and Hill, J.M., 1983, ''The Far East as a Phosphate Producer and Consumer,'' Industrial Mineral Conference Proceedings, *Phosphate—What Prospects for Growth?* Orlando, FL, pp. 110, 113–115.

McKelvey, V.E., 1973, ''Abundance and Distribution of Phosphorus in the Lithosphere,'' *Environmental Phosphorus Handbook*, John Wiley-Interscience, New York, pp. 13–31.

Merriman, G.M., and Hobbs, C.S., 1962, ''Bovine Fluorosis from Soil and Water Sources,'' *Tennessee Agricultural Experiment Station*, Bulletin 347, University of Tennessee, Knoxville, 46 pp.

Northolt, A.J.G., 1968, ''Phosphate Exploration Techniques,'' *Proceedings of the Seminar on Sources of Mineral Raw Materials for the Fertilizer Industry in Asia and the Far East*, United Nations Mineral Resources Development Series, No. 32, pp. 214–228.

Northolt, A.J.G., 1980, ''Economic Phosphatic Sediments: Mode of Occurrence and Stratigraphical Distribution,'' *Journal of the Geologic Society*, Vol. 137, Part 6, pp. 793–805.

Northolt, A.J.G., 1980a, ''Igneous Apatite Deposits,'' *Mode of Occurrence, Economic Development and World Resources*, pp. 263–284.

Northolt, A.J.G., 1982, ''Igneous Phosphate Resources: Their Growing Contribution to World Markets,'' Industrial Mineral Conference Proceedings, *Phosphate—What Prospects for Growth?* Orlando, FL, pp. 43–48, 50–52.

Nyiri, K.F., 1991, ''Phosphate in the Nineties,'' *Phosphorus and Potassium*, No. 176, Nov.-Dec. 1991, The British Sulphur Corp., Ltd., London, pp. 14–18.

Pacl, J., et al., 1978, ''The Reactivity of Natural Phosphates,'' *Phosphorus & Potassium*, The British Sulphur Corp., Ltd., London, Nov.-Dec., pp. 20–25.

Pfleider, E.P., ed., 1968, *Surface Mining*, AIME, New York, 1061 pp.

Robinson, N., 1972, ''The Assessment of Phosphate Rocks for Phosphoric Acid Ammonium Phosphate Manufacture,'' *Phosporus & Potassium*, No. 57, The British Sulphur Corp., London, Jan.-Feb., pp. 31–36.

Rogers, C.L., et al., 1961, ''Reconocimiento geologico y depositos de fosfatos del norte de Zacatecas y areas adyacentes en Coahuila, Nuevo Leon, y San Luis potosi,'' Boletin 56, Consejo de Recursos Naturales no Renovables, Mexico, 322 pp.

Sandvik, P.O., 1979, ''US Phosphates—Abundant Resources Will Last for Hundreds of Years,'' *Engineering & Mining Journal*, Oct., pp. 99–101.

Sheldon, R.P., 1964, ''Paleolatitudinal and Paleogeographic Distribution of Phosphorite,'' US Geological Survey Professional Paper 501-c, pp. 106–113.

Sheldon, R.P., et al., 1967, ''Sedimentation of Rocks of Leonard (Permian) Age in Wyoming and Adjacent States,'' *Anatomy of the Western Phosphate Field*, L.A. Hale, ed., Intermountain Association of Geologists, Salt Lake City, UT, pp. 1–13.

Stowasser, W.F., 1979, ''Phosphate,'' Preprint, *Mineral Commodity Profile 1/79*, pp. 9–12, US Bureau of Mines, 20 pp.

Stowasser, W.F., 1982, ''Phosphate Rock,'' *Mineral Commodity Summaries 1982*, US Bureau of Mines, pp. 112–113.

Stowasser, W.F., 1984, ''Phosphate Rock,'' US Bureau of Mines, *Minerals Yearbook*, pp. 15–16.

Stowasser, W.F., 1988, ''Phosphate Rock,'' US Bureau of Mines, *Minerals Yearbook*, pp. 15–16.

Stowasser, W.F., 1990, ''Phosphate Rock,'' US Bureau of Mines, *Minerals Yearbook*.

Sweeney, J.W., and Timmons, B.J., 1972, ''Availability and Potential Utilization of Byproduct Gypsum in Florida Phosphate Operations,'' *Proceedings,* 8th Forum on Geology of Industrial Minerals, Iowa City, IA, Apr. 13, pp. 89–97.

Tuttle, O.F., and Gittins, J., eds., 1966, *Carbonatites*, John Wiley-Interscience, New York, 610 pp.

Waggaman, W.H., 1952, *Phosphoric Acid, Phosphates and Phosphoric Acid*, 2nd ed., Reinhold Publishing Corp., New York, 683 pp.

Warin, O.N., 1968, ''Deposits of Phosphate Rocks in Oceania,'' *Proceedings*, Seminar on Sources of Mineral Raw Materials for the Fertilizer Industry in Asia and the Far East, United Nations Mineral Resources Development Series, No. 32, pp. 124–132.

Woodrooffe, H.M., 1972, ''Phosphate in the Kola Peninsula, USSR,'' *Mining Engineering*, Vol. 24, No. 12, pp. 54–56.

Wright, T.J., 1971, ''Texas Gulf Sulphur Beneficiation of North Carolina Phosphate,'' Preprint No. 71-H-116, SME-AIME, New York.

PIGMENTS, IRON OXIDE

GEORGE PODOLSKY AND DAVID P. KELLER

Iron oxides are unique in that they are the only significant *colored* mineral found in a natural state suitable for use as a pigment after being pulverized to pigmentary size. The world production of iron oxide pigments is estimated by the authors to be between 350 and 400 kt. The fact that highly colored natural deposits of iron ore are found throughout the world accounts for its use in prehistoric man's artistic cave paintings. These early artists did not realize that they had established man's first paint test which now has some 20,000 years of exposure.

In addition to abundance, iron oxides (about 7% of the earth's crust) have the advantage of low cost, permanency, and are non-toxic. Through the centuries, succeeding civilizations have utilized iron oxides as a major source for decoration and protection when these properties were desired. In the last century the chemical industry has improved on nature by developing a complete range of synthetic iron oxide pigments which surpass the pigments produced from natural iron ores in uniformity, color quality, and chemical purity. In the United States, the combined production of natural and synthetic iron oxides produced sales of over $97 million in 1980 (Spinrad, Jr., 1980), increasing to $138 million by 1990 (Templeton, 1990).

NOMENCLATURE

With the increasing importance of the synthetic iron oxides it is necessary to make a distinction between the natural or mineral pigments and the synthetic pigments. Natural pigments are those products that are derived from selected ores and should not be confused with iron ores mined for steel production. Iron ores that are mined for steel must be capable of being mined and reduced to iron on a competitive commercial basis. These ores are selected on the basis of iron content and processing economics. It is therefore unusual when iron ores for steel production are suitable for use as mineral pigments. Natural pigment ore sources are selected for their special physical-chemical properties and are able to command a premium price over iron ores used for steel manufacturing.

Synthetic pigments, and in this instance iron oxides, are those pigments produced from basic chemicals. Through chemical synthesis pigmentary particles are produced as opposed to the use of comminution, the major procedure common to all natural iron oxide pigments.

CLASSIFICATION

An important characteristic of a pigment is its color. Therefore, a logical separation can be made on this basis.

A. Yellow Iron Oxide Pigments
  1. Natural mineral origin
    a) Goethite
    b) Lepidocrocite
    c) Ochres
    d) Siennas
    e) Limonite
  2. Synthetic pigments
    a) Goethite
    b) Lepidocrocite
    c) Akagenite

B. Red Iron Oxide Pigments
  1. Natural mineral origin
    a) Hematite
    b) Siderite (calcined)
    c) Pyrites (calcined)
  2. Synthetic pigments
    a) Hematite

C. Brown Iron Oxide Pigments
  1. Natural mineral origin
    a) Umbers
    b) Limonite (calcined)
    c) Siderite (calcined)
    d) Goethite (bog ore or sulfur mud)
  2. Synthetic pigments
    a) Blends of hematite, goethite, and magnetite
    b) Coprecipitated hematite-magnetite
    c) Maghemite

D. Black Iron Oxide Pigments
  1. Natural mineral pigments
    a) Magnetite
    b) Slate (mixed minerals)
  2. Synthetic pigments
    a) Magnetite

Pigmentary minerals, natural or synthetic, have physical-chemical properties that are important and distinguish them from metallurgical, cementitious, construction, agricultural, and other types of industrial minerals.

The iron oxide content of natural minerals is important. Also important is the amount, the type and the form of the companion minerals. For example, an iron ore possessing good pigmentary qualities but associated with appreciable quantities of quartz would be undesirable due to the difficulty in grinding quartz. High percentages of organic matter in the ore would require calcination to decompose the organics. This step would normally be followed by grinding. Synthetic oxides, on the other hand, exhibit a high iron content due to chemical synthesis. They also contain minimal impurities.

Chemical Properties

Most of the pigmentary iron ores are relatively non-reactive (Casey, 1966, Mellor, 1953) and contain only traces of heavy metals and toxic elements, usually chemically bound in the form of complex silicates. Synthetic pigments produced from basic raw materials have the advantage that purification steps can be employed to reduce the heavy metal content (e.g., lead, antimony, arsenic, cadmium, mercury, and selenium) of the starting raw materials. If subsequent precautions are followed avoiding processing contamination, the heavy metal impurities in synthetic oxides will be about one-tenth that of the best naturals.

Iron oxide pigments have a high degree of light-fastness. This is a term which refers to the ability of a pigmented film to resist the exposure to radiant energy with minimal physical and/or chemical change.

Physical Property Definitions

The physical properties of iron oxide pigments are usually more important than the chemical properties in view of their inertness. Several physical properties require definition to provide a clear understanding of a pigment's merit and to distinguish between naturals and synthetics or pigments of other families.

Oil Absorption: Oil absorption is defined as that weight of vehicle required to wet a definite weight of pigment to form a paste. There are numerous methods (Anon., 1972a, 1983, Stewart, 1942, Von Fisher, 1948) and devices for determining oil absorption. Values given in this article will be those from standards of the American Society for Testing and Materials (ASTM), spatula rub-up procedure. Oil absorption, or binder demand, is an important property since it relates to particle size, shape, and surface characteristics and places certain limitations on paint formulation or other end uses where solid-liquid interactions are important.

Surface Area: Two pigments of essentially the same mean particle size may have widely divergent surface areas. This may be due to shape and/or surface characteristics such as porosity, pore size, smoothness, and convolutions. Since spectral and surface energy properties depend on the surface, its measurement becomes an important factor. The surface area of a particle may be defined as the area of the total external surfaces including pores but excluding *internal* (isolated) pores. When external pores are smaller than the molecule of the media being absorbed, they will not be measured. The most widely accepted and fundamental method for determining surface area is by the adsorption of a gas, usually nitrogen (Brunauer et al., 1938, Herdan, 1953) and is designated as the BET method. The value obtained expresses surface area in square meters per gram.

Particle Size: No single method or instrument currently in existence will adequately measure particle sizes in the range from 100 to 0.01μ. Complete books have been devoted to this subject, which is probably the most important physical property of pigments. It is not sufficient to state an average particle size since the size distribution of particles is equally important. Unless otherwise specified, particle diameter will be the diameter of a sphere having the same terminal velocity as the particle. This is commonly called a Stokes' diameter (Cadle, 1955) and is derived from Stokes' equation for particles settling under the action of gravity through an infinite fluid.

Terms that are often confused in particle technology are:

1) *Discrete Particle:* A single unit of matter which can change in size only by the breakdown or fracture of chemical bonds within its structure.

2) *Crystallite:* A minute single crystalline unit, i.e., that part of a particle that has its atoms and molecules arranged in a perfect crystal lattice. A *discrete* particle is composed of an assemblage of a large number of crystallites.

3) *Aggregate:* A group of discrete particles bound together by strong chemical bonds.

4) *Agglomerate:* A group of discrete particles and/or aggregates bound together by weak physical bonds.

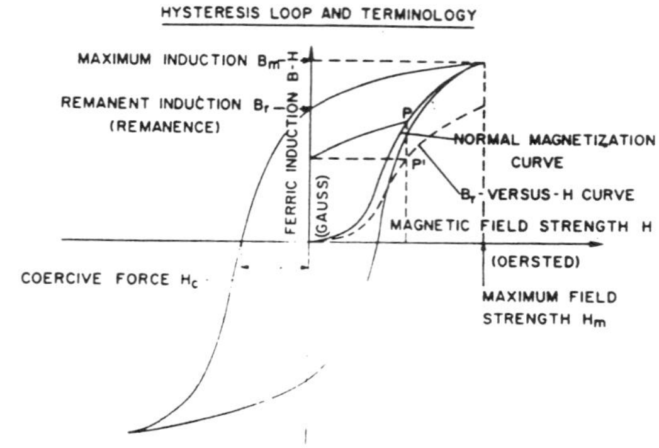

Fig. 1. Hysteresis loop used to characterize magnetic iron oxides.

5) *Flocculate:* Groups of aggregates or particles usually formed in suspensions and readily dispersed with minimal forces.

Shape: In some applications particle shape can have a more profound effect than size. Micaceous iron oxide (lamellar platelets) and acicular particles contribute completely different characteristics to a pigmented system than their equivalent spherical counterpart. Various techniques have been devised to apply numerical values for shape factors (Cadle, 1955, Herdan, 1953). Pigments may be classified according to shape into five categories: 1) spheroidal, 2) cubical, 3) nodular, 4) acicular, and 5) lamellar.

Optical Properties: The optical properties of pigments are dependent on the molecular and crystal structure and are influenced by the size and shape of the discrete particles. Since pigments are rarely used in their powder state, the ultimate effect depends on the system. Light entering a pigmented system may be absorbed, reflected, or refracted through particle-to-particle interactions. The manner and intensity of the light reflected results in spectral reflectance which can be measured by a spectrophotometer.

Magnetic Properties: Iron oxides are usually antiferromagnetic (Schieber, 1967), but they can also be ferromagnetic as produced synthetically and in certain natural mineral species. The magnetic properties of ferromagnetic materials may be determined by various techniques. A few of the more common are: magnetic moment balance, vibrating sample magnetometer, 60-Hz B-H meter, dc hysteresisgraph, and permeameters.

Magnetic materials are best described by measuring the response to a magnetic field of varying intensity. This produces a magnetization curve or BH curve as shown in Fig. 1. Ferromagnetic materials may then be characterized with values of coercive force (H_c), remanence (B_r), and saturation (B_m) at a specified field strength in oersteds (H).

NATURAL IRON OXIDE

Natural iron oxide minerals as mined for pigment applications normally require more than simple grinding to attain commercial status. The major minerals are discussed along with their chemical and physical properties. It should be noted that the chemical compositions, especially those of goethite and hematite, do not represent the pure mineral but rather the range as received by the pigment manufacturer.

Hematite

This important iron-bearing mineral, besides providing a source for steel manufacture, is a major source for mineral pigments.

Occurrence: Major iron ore sources are sedimentary as exemplified by the Lake Superior deposits. Hematite also occurs as igneous, volcanic, hydrothermal, metamorphic, and weathered deposits. Several deposits are unique for their high color value, such as those on the Island of Ormuz in the Persian Gulf or Spanish deposits near Malaga. Micaceous hematite, unique for its lamellar structure, is mined in Austria and Australia. Synthetic processes for the manufacture of micaceous iron oxide have been developed.

Physical-Chemical Properties:

1) *Chemical Formula*—Fe_2O_3
2) *Chemical Composition of Ore*—50 to 95% Fe_2O_3, associated with other iron or silicate minerals.
3) *Crystal Class*—hexagonal; scalenohedral.
4) *Habit*—thick to thin tabular, as rosettes, rhombohedral, sometimes platy, also earthy.
5) *Physical*—no cleavage, fracture subconchoidal to uneven. hardness—5 to 6.
6) *Other properties:*
   Melting point—1 475° to 1 565°C.
   Boiling point—decomposes
   Solubility—insoluble in water and common organic solvents
   Index of refraction—3.042
   Specific gravity—5.26
   Magnetic—usually antiferromagnetic, exists as ferromagnetic synthetically and naturally
7) *Reaction Mechanism*—will form by solid state reactions a wide variety of ferrites of the composition $M \cdot Fe_2O_4$ where M represents one or more metals such as manganese, zinc, magnesium, nickel, barium, and yttrium. These ferrites have important electrical and magnetic properties and to a lesser extent pigmentary properties.

Goethite

The second most common iron-bearing mineral is goethite. This includes other hydrated mineral iron oxides such as the limonites, ochres, and siennas.

Occurrence: Goethite is typically formed under oxidizing conditions as a weathering product of ferriferous minerals, siderite, pyrite, and magnetite, and is widespread as deposits in bogs. It is commonly associated with hematite, pyrolusite, manganite, calcite, quartz, lepidocrocite, clay minerals, and limonite. Hydrous iron oxide is also called limonite. By X-ray examination most limonites are actually cryptocrystalline goethite with absorbed or capillary water (Berman et al., 1946). Several excellent pigment ore sources are found in India. Domestic ochre is mined in Georgia and Virginia.

Physical-Chemical Properties:

1) *Chemical Formula*—$\alpha FeOOH \cdot xH_2O$
2) *Chemical Composition of Ore*—50 to 88% Fe_2O_3, 9 to 10% H_2O, balance silicate or clay minerals
3) *Crystal Class*—orthorhombic, dipyramidal
4) *Habit*—prismatic and striated, also in tablets or scales, grading into acicular and long prismatic forms; also earthy
5) *Physical*—cleavage [010] perfect, [100] less so, fracture uneven, hardness 5 to 5 1/2
6) *Other Properties*:
   Melting Point—decomposes: naturals (finely ground) start to dehydroxylate at 200°C and is essentially dehydroxylated at 320°C, synthetics start at 200°C and are completely dehydroxylated at 300°C.
   Boiling point—decomposes
   Solubility—considered insoluble in water and common organic solvents, 0.0015 gpl cold water
   Index of refraction—2.26 to 2.27
   Specific gravity—3.3 to 4.0
   Magnetic—antiferromagnetic
7) *Reaction Mechanisms*—will form ferrites through solid state reactions as noted under hematite.

Can also be converted from $\alpha FeOOH \cdot xH_2O$ to αFe_2O_3 by heat and to ferromagnetic γFe_2O_3 by reduction and oxidation.

Lepidocrocite

This natural mineral has essentially the same chemical composition as goethite, but has different crystallographic constants.

Occurrence: Formed under similar circumstances as goethite and often associated with it.

Physical-Chemical Properties:

1) *Chemical Formula*—$\gamma FeOOH \cdot xH_2O$
2) *Chemical Composition*—50 to 89% Fe_2O_3, 10 to 11% H_2O; minor amounts of silicates
3) *Crystal Class*—orthorhombic; dipyramidal
4) *Habit*—as flattened elongated scales sometimes striated; usually as isolated crystals or plumose groups on a matrix
5) *Physical*—cleavage [010] perfect, [100] less perfect, [001] good; hardness—5.0
6) *Other Properties:*
   Melting Point—decomposes, similar to goethite
   Boiling point—decomposes
   Solubility—considered insoluble in water and common organic solvents
   Index of refraction—1.94
   Specific gravity—4.05 to 4.10
   Magnetic—antiferromagnetic
7) *Reaction Mechanisms*—lepidocrocite converts to ferromagnetic γFe_2O_3 on heating at low temperature and to αFe_2O_3 at high temperatures; will enter into other reactions similar to goethite.

Magnetite

The mineral magnetite occurs abundantly and is found worldwide in many locations. The magnetic nature of this mineral gives it considerable historic interest going back into antiquity.

Occurrence: Magnetite deposits are found as a magmatic segregation deposit, as an accessory mineral in igneous rocks, in metamorphic deposits in limestones, in chlorite schists with pyrite, in sulfide vein deposits, and in meteorites. Commercial deposits are found in the United States, Norway, and Sweden.

Physical-Chemical Properties:

1) *Chemical Formula*—Fe_3O_4
2) *Chemical Composition*—84 to 99% Fe_3O_4 associated with minor amounts of silicate minerals; purity can be high for a natural mineral
3) *Crystal Class*—isometric (cubic); hexoctahedral
4) *Habit*—usually octahedral, well-formed nearly perfect crystals are common; occurs in coarse to fine grained masses
5) *Physical*—octahedral parting on [111] especially good; conchoidal to uneven fracture. [001], [011], [138] also reported as parting planes in magnetite; hardness 5.5 to 6.5
6) *Other Properties:*
   Melting point—decomposes
   Boiling point—decomposes
   Solubility—insoluble in water and common organic solvents
   Index of refraction—2.42
   Specific Gravity—5.16 to 5.17
   Magnetic—ferromagnetic

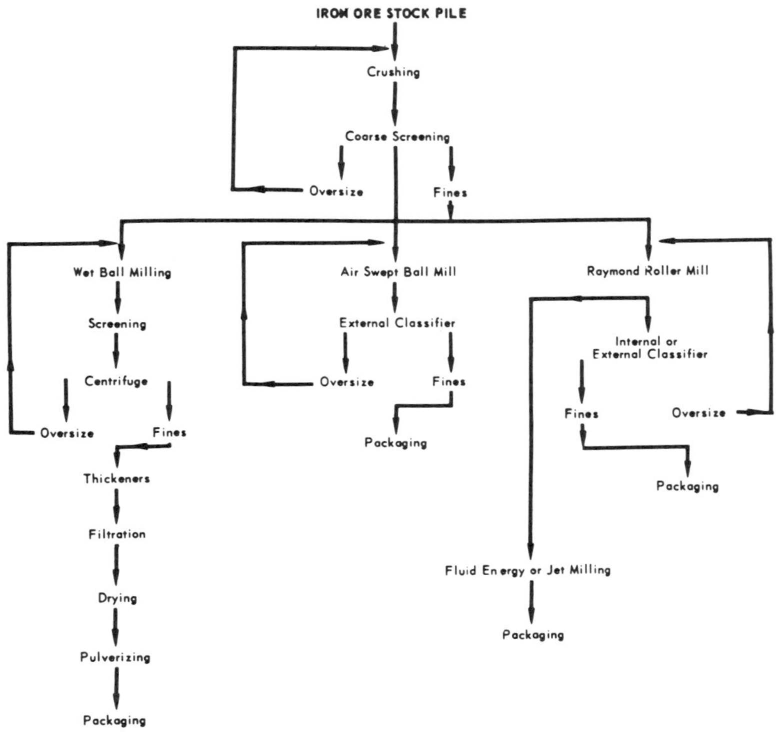

Fig. 2. Simplified flowsheet showing alternate processing routes for natural iron oxide pigment ores requiring grinding.

7) *Reaction Mechanisms*—synthetic magnetites start to decompose to γFe_2O_3 at 105°C and are completely oxidized at 385°C, respectively. Natural magnetite, if finely ground, will start to decompose at 124°C and be completely oxidized at 550°C. It will form spinel compounds of the general formula $M{\cdot}Fe_2O_3$ where M represents metals such as manganese, zinc, and barium.

Pyrites and Siderites

These minerals do not fall into the class of iron oxides since they require transformation by heat to convert to the oxide. As such, they are potential raw materials rather than end products.

Manufacturing

The manufacturing procedures for the naturals must be designed around the starting ore base. For example, siderite ($FeCO_3$) requires calcination to convert to the oxide whereas pyrites cinder (calcined pyrite), although essentially Fe_2O_3, must be freed of its acidic residues. Also, certain ores, although suitable for use by simple grinding, develop more desirable colors if calcined. Other natural ores, siennas, and umbers are produced in raw and calcined form. Typical flowsheets for both ore types are presented in Figs. 2 and 3.

Some of the hematites and the magnetites are magnetically beneficiated to increase the chemical purity and remove undesirable siliceous minerals. This is particularly true if the end use has chemical purity requirements such as ferrites (Anon., 1965). Magnetic beneficiation is most efficient if the mesh size is controlled within fairly narrow limits. Therefore, wet magnetic beneficiation usually follows the crushing and screening operation and prior to fine grinding or calcination. The critical variables are: magnetic field strength, pole and plate designs, solids concentration and feed rate, demagnetizing field strength, and wash water feed rates. The process produces a super-concentrate with purities in excess of 99.5% Fe_2O_3 or Fe_3O_4 from lower grade concentrates.

The use of fluid energy mills, micronizers, and improved methods of classification has enabled the manufacturer to produce natural iron oxide pigments for many applications where chemical purity and the ultimate discrete particle size are not factors.

Products

The broad selection of manufacturing procedures coupled with the wide variety of natural ore results in about 100 different products being marketed in the United States alone. Some of the major ore sources are tabulated in Table 1. Data for a *single* typical product prepared from the ore are listed. Selling price and end use considerations result in a range of products being produced from a single ore body. For example, a natural iron oxide for a paint primer must be finely ground but has wide latitude in its color specifications. Concrete coloring pigments, however, have stringent color requirements but moderate fineness requirements.

All natural products, without exception, are subjected to some type of final grinding. Comminution, as opposed to particle synthesis, produces characteristic fragmented particles covering a wide size distribution. In Fig. 4, a synthetic iron oxide is compared with a natural ground iron ore showing a major difference in size and fragmented nature of the natural ore.

The difference in particle size distribution between a synthetic and a natural ground product is shown in Fig. 5.

Spectral reflectance curves are shown in Figs. 6 and 7 which compare a synthetic red and a synthetic yellow with similar natural ground products. The spectral reflectance curves show the naturals to have weaker tinting strengths and lower color purity.

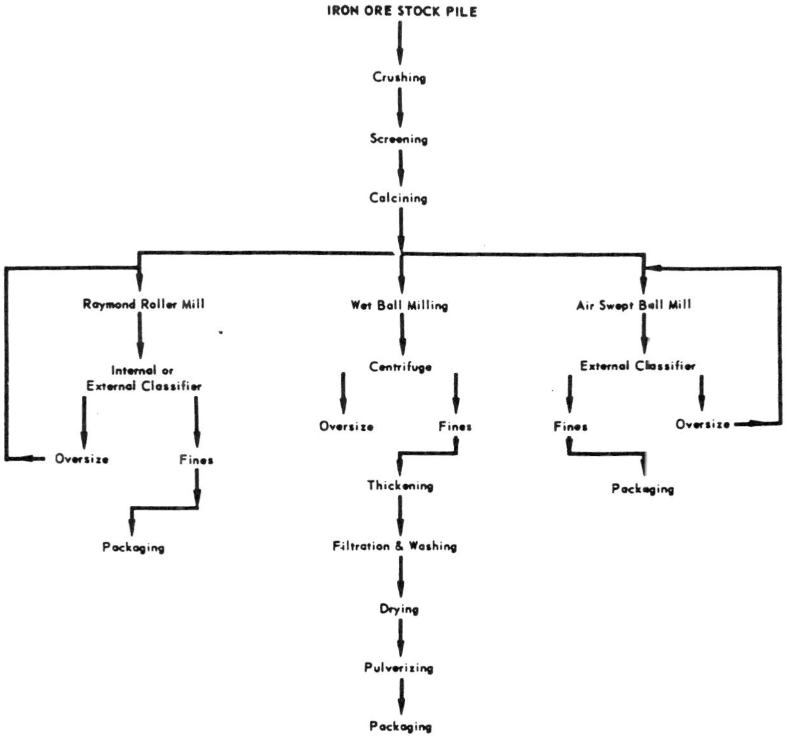

Fig. 3. Simplified flowsheet showing alternate processing routes for natural iron oxide pigment ores requiring calcining.

The price range of natural iron oxide pigments marketed in the United States and the appropriate varieties available within the various categories are listed in Table 2.

Uses

Many of the applications for natural iron oxides are based on low cost and permanency (Anon., 1991). There is some overlap in use of the naturals and synthetics; both, for example, are used extensively in paints and coatings.

Paints and Coatings: The variety and types of industrial coatings are too numerous to mention individually. There are several general types of paints wherein natural iron oxides provide excellent properties and represent major uses. *Primers* vary in the degree of pigment loading and function as a base for various top coats. The primer pigments are usually hematites and can be highly loaded due to high density, fairly large particle size, and low oil absorption.

Industrial maintenance finishes for railway equipment and metal structures largely employ hematites. Micaceous iron oxide (Anon., 1972a), due to its platelike structure, has found extensive use in Europe in railway maintenance coatings, bridge paints, industrial maintenance finishes, and other exterior applications.

Micaceous iron oxide is not a primer or a rust inhibiting pigment but is used as part of a corrosion control system. To capitalize on its platelike structure, the paint formulator must grind the paint properly to secure a lamellar film structure (Anon., 1971).

Natural umbers and siennas have unique color characteristics which are not easily matched with synthetics and hence are the major pigments in stains. With the increased use of prestained wall panels, this is an important use.

The naturals are also used in general interior and exterior house paints, but to a lesser extent than synthetics.

Electronic: Natural iron oxides are used to some extent in the preparation of barium ferrite ($BaFe_{12}O_{19}$) and to a lesser extent in the preparation of manganese zinc ferrite ($Mn_xZn_y{\cdot}Fe_2O_4$). The magnetic properties of manganese zinc ferrites are very sensitive to impurities and hence the use of naturals has been limited. Highly beneficiated magnetite or hematite can be used to prepare barium ferrites. Beneficiated magnetite when finely ground is used in magnetic inks (Love, 1961). The magnetic energy (remanence) of an ink is proportional to the volume of iron oxide in the ink. Due to the low binder demand of the dense natural magnetites, high loadings of 60 to 70% by weight of magnetite pigment can be obtained.

Rubber, Plastics, and Related Compounds: Natural pigments are used to a limited extent in rubber and many of the more common plastics such as vinyls, phenolics, polyurethanes, and epoxies (Woernle, 1967). Umbers contain a high percentage of manganese which can cause undesirable effects such as accelerating the aging of rubber and retarding the curing of polyesters. The general trend has been toward the synthetics based on color value.

Concrete Products and Building Materials: Natural oxides are suitable for use in coloring concrete, mortar, concrete roof tile, concrete pavers, and similar building materials. In the coloring of concrete the tint is the important characteristic and synthetic pigments have replaced most of the naturals because of their higher tinting strength.

Rouge: Prior to the introduction of the float glass process, large tonnages of natural iron oxides were used in polishing plate (flat) glass. This use has declined dramatically due to the float glass process. Ophthalmic rouge and metal polishing rouge represent a small market. Specially processed grit-free hematites having high glass removal rates are required. Another market is in coloring glass and is referred to a batch rouge.

Miscellaneous: Natural iron oxides are used in numerous other industrial applications, such as foundry-core washes, ceramics, paper, fertilizer, and as the iron source in glassmaking.

Table 1. Iron Oxide Pigments From Natural Ores—Typical Physical and Chemical Properties

Mineral Class:	Hematite	Hematite	Hematite	Hematite	Hematite (Specular-) (Micaceous)	Hematite	Goethite (Ochre)	Goethite (Ochre)	Goethite (Sienna)	Goethite (Sienna)	Umber	Umber	Magnetite
Ore Source:	US	Venezuela (Calcined)	Persian Gulf	Spanish Oxide	Austria	Pyrites Cinder	Georgia	France	Italy	Italy (Calcined)	Cyprus	Cyprus (Calcined)	US
Chemical Properties													
% Fe_2O_3	63.0	96.5	73.0	85.0	90.0	82.0	50.0	20.0	62.0	68	45.0	55.0	98.5*
% SiO_2	12.6	1.50	21.0	6.5	3.5	Balance—Mainly Al Silicates	14.0	50.0	Balance—Mainly Al Silicates	Balance—Mainly Al Silicates	Balance—Mainly Al Silicates	Balance—Mainly Al Silicates	Balance—Mainly Al Silicates
% Al_2O_3	4.8	1.50	3.0	1.0	2.4		6.00	18.0					
% MgO	3.2	0.02	0.33	1.0	1.0		0.20	0.25					
% CaO	7.5	0.03	0.80	3.0	0.5		0.20	0.25					
% MnO_2	—	0.02	0.06	—	—		0.62	—			11.0	10.0	
% Loss on ignition													
@ 1000° C	8.2	0.50	2.50	3.0	0.15	—	9.00	9.00	15-20	—	10-15	—	—
% H_2O soluble salts	0.8	0.20	<3.0	1-2	0.08	3-3.5	0.50	—	0.45	—	0.8	0.4	0.10
Other Properties													
Color	Brownish red	Reddish brown	Bright red	Bright red	Brownish red	Red	Brownish yellow	Dull yellow	Brownish yellow	Reddish brown	Greenish brown	Brown	Black
Oil absorption†	15	14	13	17	11	21	24	32	36	42	44	54	15
Particle size‡													
Range (10-90%)	0.3-7	0.3-8	0.3-7	0.25-15	8-65¶	0.2-6	0.2-15	0.2-40	0.1-8	0.05-6	0.6-7	0.13-3	1-10§
median	0.8	3.0	0.6	1.5	25¶	2.0	1.0	4.0	0.6	0.7	3.0	0.6	—
% retention +325	<0.20	<0.01	<0.20	<0.50	30	<0.01	<2.0	<3.0	<1.0	<0.20	<0.20	<0.20	<0.10
Surface area#	8.0	6.1	4.7	5.8	0.08	8.73	15.4	—	33.4	36.0	101.7	86.9	4.3
Specific gravity	3.78	4.90	4.35	4.67	4.8	4.48	3.47	2.88	3.58	4.03	3.24	3.74	5.03

Source: Technical bulletins and reports of domestic iron oxide pigment producers.

* Contains 30% FeO.

† Oil absorption, pounds of oil per 100 lb of pigment.

‡ In microns as determined by Andreasen pipette or Micromeritics sedigraph, except as noted. Range is the percentage by weight—between 10% to 90%. 50% by weight point is median size.

¶ Values by microscopy due to lamellar structure.

§ Magnetic—estimated from electron micrographs.

\# Square meters per gram—by BET.

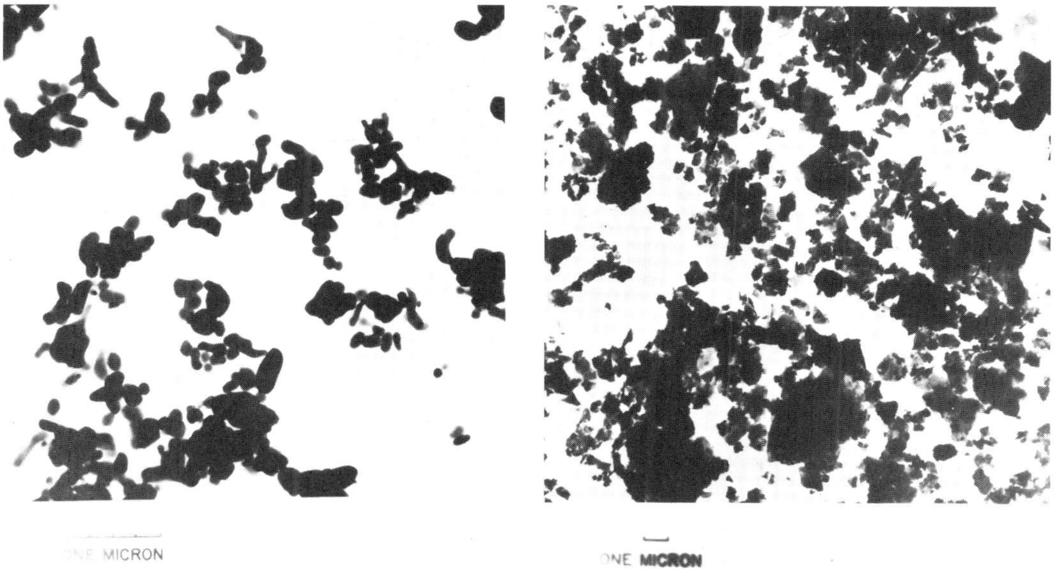

Fig. 4. Comparison of particle size between synthetic iron oxide (left) and ground natural iron ore (right).

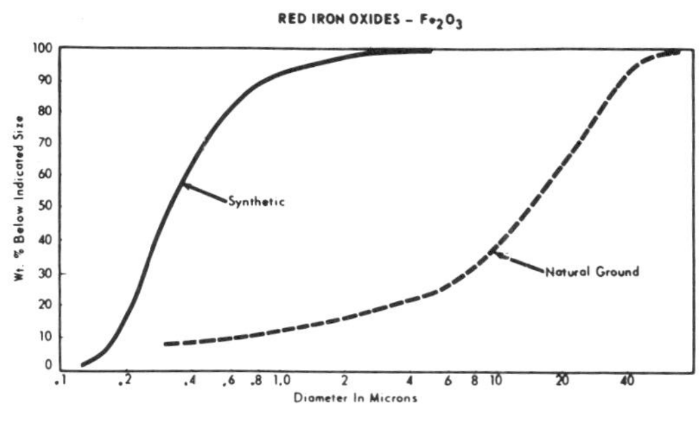

Fig. 5. Particle size distribution comparing a fine particle synthetic with a natural iron oxide.

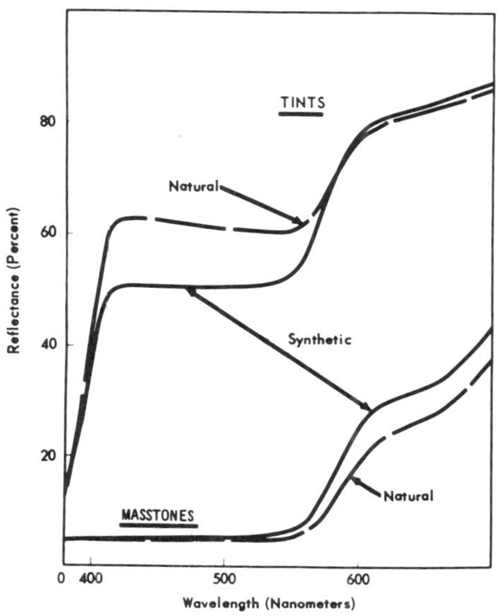

Fig. 6. Spectral reflectance curves of masstone and 1:50 TiO_2 tint of a synthetic copperas red vs. a natural red oxide.

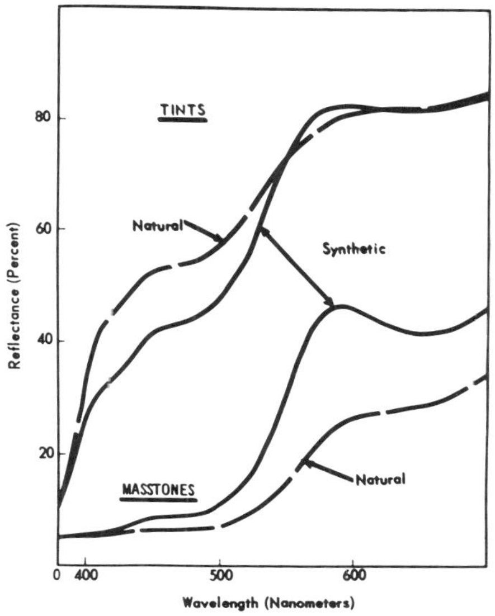

Fig. 7. Spectral reflectance curves of masstone and 1:10 TiO_2 tint of a synthetic yellow oxide vs. a natural domestic ochre.

SYNTHETIC IRON OXIDES

There are four major types of synthetic iron oxides: reds, yellows, browns, and blacks. All can be prepared from basic chemicals, usually iron salts and alkalis. The advantage of chemical synthesis is that it provides precise control over three important pigment parameters: particle size, particle shape, and chemical purity.

Raw Materials

Iron Salts: There are only two iron salts of commercial significance, ferrous sulfate and ferrous chloride, that are used in

Table 2. Typical 1991 Price Range for Natural Iron Oxide Pigments Marketed in the United States (value rounded to nearest dollars per kg)

	FOB # Selling Price	Products Available
Micaceous iron oxide	1.96	1
Vandyke brown	1.30	2
Yellow ochre	.73–1.25	6
Sienna (raw)	1.19–1.72	9
Sienna (burnt)	1.32–1.91	10
Umber (raw)	1.14–1.50	14
Umber (burnt)	1.14–1.56	28
Black iron oxide (magnetite)	.97–1.03	2
Natural hematites	.33–1.47	23
Price range and total available products	.73–1.96$/kg	95

Source: Suppliers published price and other industry sources.
#Truckload price in 22.7 kg multi-wall bags. Prices about 23 to 33¢ per kg higher for small quantities, minimum weight depends on supplier. Special milling, palletizing, and packaging carry a premium.

the manufacture of synthetic iron oxides. Ferrous sulfate ($FeSO_4 \cdot 7H_2O$), or copperas, as it is commonly called, was the historical base for red oxides produced by calcination. Subsequently, ferrous sulfate was used as the iron salt for the production of yellows, reds, browns, and black by precipitation. The increased use of hydrochloric acid for the pickling of steel resulted in an excess of ferrous chloride. The quantities of both iron sources far exceed the need of synthetic iron oxide producers.

Ferrous sulfate for pigment production is available from three sources. The first consists of neutralizing scrap iron with sulfuric acid. A second is byproduct sulfate pickle liquor from the steel mills. This is a mixture of about 15% $FeSO_4$ and 10% H_2SO_4, the balance water. The excess acid is converted to $FeSO_4$ by neutralization with iron. The third major source is the titanium pigment industry which produces byproduct $FeSO_4 \cdot 7H_2O$. This iron salt must be redissolved and purified. The subsequent processing is the same as steel mill pickle liquor.

Ferrous chloride has rarely been utilized for the production of color pigments by domestic pigment manufacturers. Its use limitations are more fully discussed under *Byproducts*.

Iron: The iron source is generally stampings and punchings of light gage soft black steel. Pigment producers avoid scrap that is plated (galvanized) or contains alloying elements due to the adverse effect of these impurities.

Alkalis: The major alkali used in the United States is sodium hydroxide (NaOH) followed by ammonium hydroxide (NH_4OH). Lime (CaO) is used to a limited extent but could become more important with increased regulation of the use of ammonium hydroxide. Both sodium hydroxide and liquid NH_3 can be purchased with a high degree of chemical purity, which is less true of commercial limes.

Byproducts: Most of the byproducts require further processing either as raw materials or end products. The spray roasting of ferrous chloride pickle liquor produces Fe_2O_3 and HCl. The spray roasted oxides can be returned to the steel making process but usually are landfilled. However, some steelmakers divert a portion of the Fe_2O_3 to other applications where color is not important. The spray roasted oxides are generally poor in color quality. Even poorer in quality are the dust collector recovered oxides produced from the basic oxygen steel lancing process. The thousands of tons of these two waste products have depressed pigment prices in a few areas where color quality is not a major consideration.

Mill scale, although abundant, has found only very minor usage in the iron oxide industry. One of the major contaminants of this product is refractory material due to spalling of furnaces and linings. Also, mill scale is not a completely oxidized product and cannot be used for pigmentary purposes without further processing. It is projected that the use of this type of iron oxide will continue to diminish and level off at about 3% of the iron oxide raw materials for ferrites (Anon., 1991a).

Other waste products that have attracted attention over the years but have not proven suitable for commercial pigments are: red mud from bauxite processing, sponge iron from the cement copper mining operation, and sulfur mud or yellow boy, a precipitate formed from acid coal mine waters.

Manufacturing Processes

There are three major methods for the manufacture of synthetic iron oxides: thermal decomposition of iron salts or iron compounds, precipitation of iron salts usually accompanied by oxidation, and organic reduction processes utilizing iron.

Thermal Decomposition of Iron Salts

There are numerous combinations of processing techniques depending on the starting basis. Only one flowsheet (Fig. 8) will be outlined, starting with scrap iron and sulfuric acid (Ayers, 1946). If the starting base is steel mill pickle liquor or titanium copperas, the process is essentially the same with the exception that steps may be introduced to remove undesirable impurities.

The production of iron oxides by the calcination of ferrous sulfate is illustrated in Figure 8. These products are commonly known as copperas reds. The major chemical reactions that are involved in this process are (Hoffman, 1972):

Chemical Reactions Involved In The Thermal Decomposition of Ferrous Sulfate

Neutralization

$$H_2SO_4 + Fe + H_2O \rightarrow FeSO_4 + H_2\uparrow + H_2O$$

Evaporation

Crystallization

$$FeSO_4 + 7\ H_2O \rightarrow FeSO_4 \cdot 7H_2O$$

Dehydration

Decomposes above 649°C

$$FeSO_4 \cdot 7H_2O \xrightarrow{\Delta} FeSO_4 \cdot H_2O + 6H_2O$$

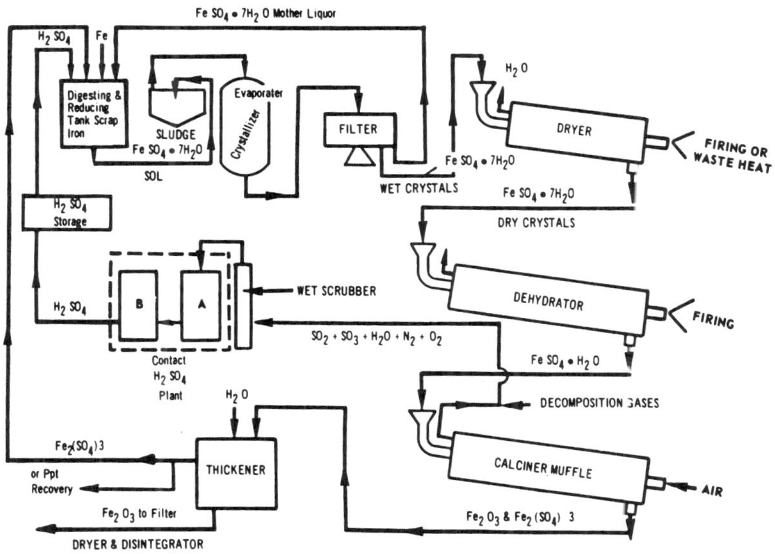

Fig. 8. Simplified flow diagram for synthetic iron oxide (copperas red type).

Without air

$$6FeSO_4{\cdot}H_2O \xrightarrow{\Delta} 2Fe_2O_3 + Fe_2(SO_4)_3 + 6H_2O + 3SO_2$$

$$Fe_2(SO_4)_3 \xrightarrow{\Delta} Fe_2O_3 + 3SO_3$$

$$3SO_3 \xrightarrow{\Delta} 3SO_2 + 1\tfrac{1}{2}\ O_2$$

With air

$$6FeSO_4{\cdot}H_2O + 1\tfrac{1}{2}O_2 \xrightarrow{\Delta} Fe_2O_3 + 2Fe_2(SO_4)_3 + 6H_2O$$

$$2Fe_2(SO_4)_3 \overset{\Delta}{\rightleftharpoons} 2Fe_2O_3 + 6SO_3$$

$$6SO_3 \overset{\Delta}{\rightleftharpoons} 6SO_2 + 3O_2$$

The last two chemical reactions with and without air are reversible at temperatures below 650°C down to about 316°C.

One major advantage of this thermal decomposition process is that the products of combustion are reclaimed and reused.

One of the earliest synthetic iron oxides and a variation of the calcination of ferrous sulfate are the Venetian reds. These oxides are produced by calcining an intimate mixture of ferrous sulfate and lime or preferably hydrated lime ($Ca(OH)_2$). The chemical reactions may be expressed as follows (Love and Ayers, 1942):

$$4FeSO_4{\cdot}7H_2O + 4CaO + O_2 \xrightarrow{\Delta} 2Fe_2O_3 + 4CaSO_4 + 28H_2O \tag{1}$$

Or

$$2FeSO_4{\cdot}7H_2O + CaO \xrightarrow{\Delta} Fe_2O_3 + CaSO_4 + SO_2 + 14H_2O \tag{2}$$

The difference between Reactions 1 and 2 is the amount of oxygen present in the calcining operation. Generally speaking, both reactions are involved and an end product of about 40% Fe_2O_3 is obtained. The calcination product must be ground to be suitable for pigmentary applications. The process has one advantage in that the kiln product does not require washing. Venetian reds have been supplanted by other synthetic reds.

Most steel mills have converted from sulfuric acid to hydrochloric acid in their pickling lines. This facilitates the recovery of the acid and generates iron oxide as a byproduct. The amount of byproduct iron oxide generated greatly exceeds the total present production of iron oxide pigments. The byproduct oxide may be pelletized and recycled into the steel process. Some of the oxide is also utilized as a raw material for the ferrite industry.

Precipitation Reactions: Precipitation processes were developed in the early 1900s (Penniman and Zoph, 1921), but did not achieve commercial importance until the 1930s to 1940s. Since that time the major growth in US synthetic iron oxide manufacture has been by precipitation technology. The precipitation process will produce yellows, reds, browns, and blacks. Additional products can be produced by subsequent procedures, such as calcination, reduction, etc.

In the manufacture of yellow oxides by the Penniman-Zoph process (often referred to as the scrap iron process) the initial reaction is one of nucleation, wherein a seed nucleus is formed by the reaction of an iron salt and an alkali. The nucleation reaction is normally conducted in an excess of $FeSO_4$, and can be described by the following equations:

$$4NaOH + 2FeSO_4 \rightarrow 2Fe(OH)_2 + 2Na_2SO_4$$

$$2Fe(OH)_2 + \tfrac{1}{2}O_2 \rightarrow + \underset{\text{nucleus}}{\underline{2FeOOH}} + H_2O$$

The nucleus is transferred to a precipitator containing $FeSO_4$, H_2O, and iron, usually in the form of thin-gage stampings. The seed particles are continuously circulated over the bed of iron in an oxidizing environment at elevated temperatures [60 to 87.8°C] and, as the iron oxidizes or dissolves, the reaction products precipitate on the nuclei causing the particles to grow. The chemical reactions involved are:

$$2\ Fe^{++} + \tfrac{1}{2}\ O_2 + 3H_2O \rightleftharpoons 2FeOOH + 4H^+$$

also

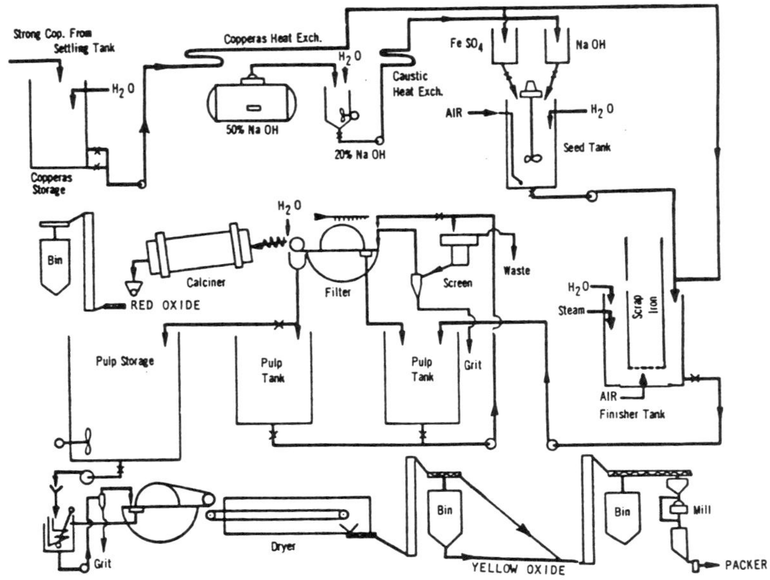

Fig. 9. Simplified flow diagram of a synthetic yellow iron oxide plant.

$$2FeSO_4 + 3H_2O + \tfrac{1}{2} O_2 \rightarrow 2FeOOH + 2H_2SO_4$$

$$H_2SO_4 + Fe \rightarrow FeSO_4 + H_2\uparrow$$

From the simplified chemical reactions, the precipitation process would not appear to be complex but the actual gas-liquid-solid chemical interactions are not completely understood. A host of variables affect the reaction and the products. Some of the more important variables are: rate of circulation, quantity of oxygen, temperature, quantity of initial seed nuclei, size and shape of the seed nucleus (Hancock, 1970). A typical precipitation flowsheet that will produce a wide color range of yellow iron oxides is shown in Fig. 9.

A class of pigments that has attracted considerable attention in recent years is the transparent or low opacity pigments. These pigments using nucleation are produced by precipitation techniques (Anon., 1971a) and are characterized by very fine particle size, usually less than 0.10μ.

Precipitated reds can be produced by utilizing equipment that is essentially the same as that employed for yellow iron oxide manufacturing. The difference lies in the preparation of the nucleus which contains ultrafine hematite (Fe_2O_3) nuclei (Ayers, 1960, Bennetch, 1957, 1961, Marsh, 1955, Toxby, 1952).

Subsequent processing conditions are also much more critical since normal precipitation reactions tend to produce FeOOH rather than Fe_2O_3.

The two preceding processes employ scrap iron as the major source of iron in the growth of the nuclei. Both yellow and red iron oxide can be produced using nucleation by simultaneously precipitating alkali and iron salts in the presence of the nuclei (Martin, 1960). The process is commonly referred to as the direct precipitation or DP process. If conducted properly, the nuclei will grow to pigmentary size in the same manner as in the Penniman-Zoph process.

Another class of pigments produced by precipitation are the browns and blacks. The blacks are Fe_3O_4 and browns are χ FeO·y Fe_2O_3. Commercially, they are prepared by reacting ferrous sulfate with an alkali at temperatures close to boiling at an alkaline pH (Ayers, 1938). Another method of producing black iron oxide is the conversion process. This process consists of the conversion of yellow or red iron oxide in the presence of an alkali and a ferrous salt.

The precipitation processes have a further advantage in that their end products are suitable for additional processing to produce different pigments. The yellows, blacks, and browns all may be calcined to produce red and brown pigments. If the temperature is below 649°C, the shape of the precursor is usually retained. Synthetic browns are also made by producing intimate blends of synthetic reds, yellows, and blacks.

One of the most sophisticated products to be manufactured by a combination of precipitation and controlled calcination techniques is γFe_2O_3 (maghemite) from acicular FeOOH (Camras, 1954). In this process the γFe_2O_3 particle size is determined by the size of the yellow oxide. The yellow is processed in a rotary kiln at temperatures between 315 and 538°C. The reactions are:

$$\underset{\text{(yellow)}}{FeOOH} \xrightarrow{\Delta} \underset{\text{(red-nonmagnetic)}}{\alpha Fe_2O_3 + H_2O}$$

$$\underset{\text{(red)}}{3Fe_2O_3 + H_2\uparrow} \xrightarrow{\Delta} \underset{\text{(black-magnetic)}}{2Fe_3O_4 + H_2O}$$

$$\underset{\text{(black)}}{2Fe_3O_4 + \tfrac{1}{2} O_2} \xrightarrow{\Delta} \underset{\text{(brown-magnetic)}}{3\gamma Fe_2O_3}$$

Reduction of Organic Compounds: The reduction of organic compounds, using iron as the reducing agent, resulting in iron oxides has been practiced in Germany for over 50 years. Most organic chemical manufacturers, however, use different reducing agents or, if they use iron, are more concerned with the organic compound than with the iron oxides produced. Consequently, most iron oxide sludges from the reduction of organic compounds are an impure waste product.

If nitrobenzene is reduced to aniline in the presence of various chemicals, it will produce specific colors of iron oxides (Love, 1946). $AlCl_3$ may be used to produce yellow oxides but is omitted

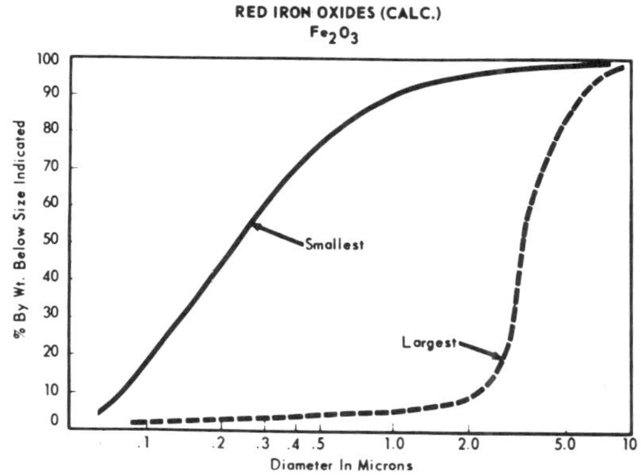

Fig. 10. The range in particle size distribution for synthetic red iron oxides of the copperas red type.

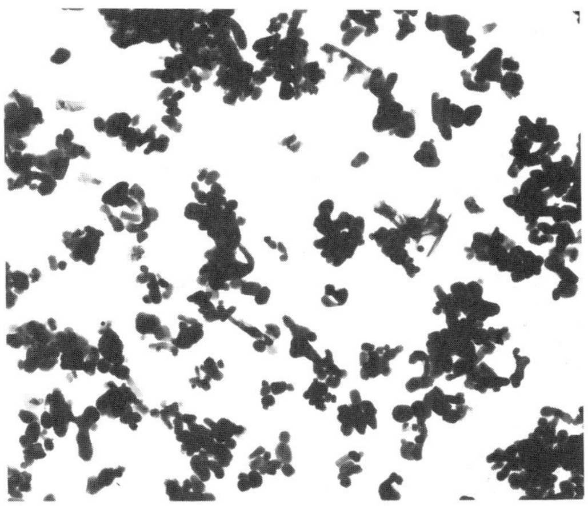

Fig. 11. Light synthetic red iron oxide, copperas red type.

Fig. 12. Light synthetic red iron oxide, precipitated type.

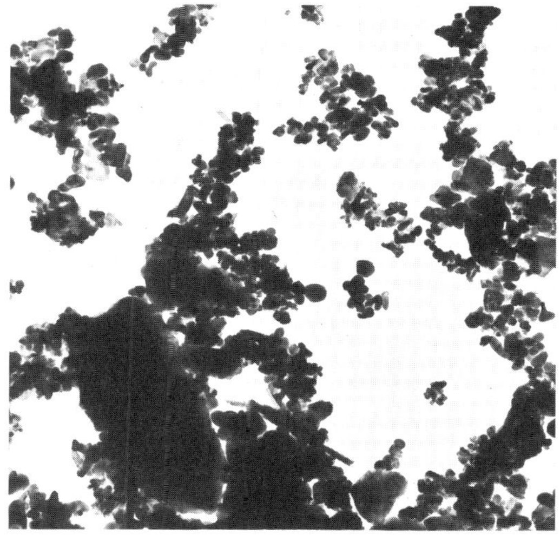

Fig. 13. Light synthetic red iron oxide, Venetian red type.

when blacks are desired. Reds can be produced by calcining either yellow or blacks. In each the choice of iron source is carefully controlled and usually consists of iron filing or iron powder.

Miscellaneous Manufacturing Processes: Another related product line consists of tan pigments. One major domestic pigment producer manufactures tan pigments that are ferrites (Downs and Martin 1950, 1959). These are formed by interaction of iron oxides with other metallic oxides to form ferrites. The reactions are as follows:

$$ZnO + Fe_2O_3 \xrightarrow{\triangle} Zn{\cdot}Fe_2O_4$$

$$MgO + Fe_2O_3 \xrightarrow{\triangle} Mg{\cdot}Fe_2O_4$$

A final example of a product which is between a synthetic and a natural, with respect to its variable nature and very high impurity level, is the byproduct oxide from the basic oxygen furnace (BOF) steel production. The byproduct oxide is collected from the gases evolved during the lancing process.

Products and Properties

The *synthetic* manufacturing approach opens up an almost infinite number of distinct product possibilities. An iron oxide producer wishing to cover all major *synthetic* marketing applications should have more than 100 iron oxide products available for sale.

In Table 3 a listing is made of synthetic reds, yellows, browns, blacks, intermediate colors, and some specialty items such as the transparent iron oxides, magnetic recording oxides, and a few that are not color oxides but still are classed as synthetics (byproduct steel mill oxide).

In the major classifications, such as the synthetic yellows and reds, those products representing the smallest and largest discrete particles are tabulated. Within any one classification, for example, synthetic yellow oxides, there can be as many as 10 to 12 intermediate colors.

Table 3. Synthetic Iron Oxide Pigments—Typical Physical and Chemical Properties

Chemical Formula:	Fe_2O_3	Fe_2O_3	Fe_2O_3	Fe_2O_3	Fe_2O_3	Fe_2O_3	Fe_2O_3	$CaSO_4$	Fe_2O_3 $CaSO_4$	Fe_2O_3
Mineral Class:	Hematite	Hematite	Hematite	Hematite	Hematite	Hematite	Hematite	Hematite	Mixed	Hematite
Method of Mfg:	Decomp. $FeSO_4$	Decomp. $FeSO_4$	Ppt.	Ppt.	Calc. of FeOOH	Calc. of Fe_3O_4	Pickle Liq. Decomp. (HCl)	BOF Flue Dust	Decomp $FeSO_4$ & $Ca(OH)_2$	Organic Reduction (Calcined)
Chemical Properties										
% Fe_2O_3	99.5	99.5	97.0	98.0	97.5	99.3	98.4	82.2	40.0	96.5
% FeO	—	—	—	—	—	—	—	—	—	—
% SiO_2	0.03	0.03	0.05	0.06	0.10	0.01	0.07	9.10	—	
% Al_2O_3	0.02	0.03	0.01	0.01	0.01	0.06	0.006	3.10	—	2.5
% MgO	0.01	0.006	0.01	0.01	0.01	0.015	0.03	1.00	—	—
% CaO	0.02	0.03	0.10	0.10	0.06	0.006	0.06	0.40	25.0	—
% Mn	0.02	0.25	0.02	0.02	0.02	0.10	0.30	0.36	—	0.10
% SO_3	0.06	0.02	0.80	0.80	0.60	0.10	—	—	15.0	—
% LOI @ 1 000°C	0.20	0.06	2.50	1.50	0.70	0.60	—	2.50	—	0.50
% Soluble Salts	0.10	0.10	0.10	0.10	1.00	0.12	0.10	0.22	—	0.30
Other Properties										
Color	Light Red	Dark Red	Light Red	Dark Red	Medium Yellow Red	Red	Dark red (poor color)	No Value	Light Red	Light Red
Oil absorption*	24	13	24	22	48	—	24	—	20	26
Particle size†										
Range (10–90%)	0.1–1	2–6	0.1–.6	0.5–1.5	0.5–1	0.3–1.2	0.3–7	Varies widely	0.3–5	—
Median, μ	0.25	3.0	0.35	0.90	0.75	0.50	1.5	Varies	1.2	0.3
Particle shape	Nodular	Nodular	Spheroidal	Spheroidal	Acicular	Cubic	Spheroidal	Spheroidal	Mixed system	Spheroidal
% Retention +325	<0.10	<0.10	<0.05	<0.10	<0.10	—	<0.10	20.0	<0.50	0.05
Surface area¶	10.0	2.2	9.4	3.7	9.1	5.4	6.3	—	8.8	—
Specific gravity	5.15	5.15	4.90	4.9	4.9	5.15	5.18	—	3.56	5.2

(cont.)

The particle size range of the copperas red family is shown in Fig. 10. The typical shape of a calcined copperas red (Fig. 11) may be compared to a precipitated red oxide (Fig. 12). The broad particle size distribution of a Venetian red is shown in Fig. 13. The very large particles are the $CaSO_4$.

In Fig. 14, a yellow iron oxide is shown before and after being subjected to high temperatures of about 1 038°C.

A synthetic, precipitated black iron oxide is shown in Fig. 15. A magnetic recording gamma iron oxide is shown in Fig. 16. Note that the product retains the acicular shape of its yellow precursor.

In Table 4, the price range and number of products are listed.

Specifications

Specifications vary widely based on the end use application. A paint manufacturer will be most concerned with color and top size whereas a manufacturer of ferrites requires chemical purity.

Typical Color Pigment Specifications:

Mass color	3△E FMC-2 units (max.)*
Tint tone	3△E FMC-2 units (max.)
Tint strength	±5% of standard (max.)
% Fe_2O_3	Min. 99.0% (copperas red type) Min. 96.0% (precipitated red type)
325 mesh sieve retention	0.10% max.
Moisture at 110°C	0.25% max.
H_2O soluble salts	0.15% max.
Dispersion for specialty milled reds	6.0 min. Hegman fineness (ASTM D1210)
Oil absorption	±10% of standard (max.)

*FMC-2 color difference method as described by Chickering, 1971.

Natural pigments will have much wider tolerances and in some cases are not controlled for color and tinting strength.

Applications

As previously noted there is an overlap between natural and synthetic oxides. There are some applications, either by custom or necessity, where only synthetics are used.

Paint: Synthetic oxides have higher chroma or color purity and greater tinting strength than natural oxides. The color purity variation from lot to lot can be closely controlled, which is not always true of a natural. They have excellent dispersion properties. Consequently they are ideally suited for universal tinting systems, powder coatings, automotive finishes, appliance enamels, interior and exterior paints (Anon., 1968, Parfitt, 1969, Parker, 1965, Patterson, 1967). They are nontoxic and meet existing federal regulations.

A group of ultra fine iron oxide pigments are known as transparent or low opacity products. These pigments have found application in the automotive industry in the manufacture of metallized finishes in the copper and orange color range. These iron oxide

Table 3. Synthetic Iron Oxide Pigments (cont.)

$Fe_2O_3{\bullet}H_2O$	FeOOH	FeOOH	FeOOH	FeOOH	$ZnFe_3O_4$	$xFeO{\cdot}yFe_2O_3$	—	$\gamma\, Fe_2O_3$	Fe_3O_4
—	Goethite	Goethite	Goethite	—	Zinc Ferrite	Hematite Magnetite	Goethite Hematite Magnetite	Magnetite	Magnetite
Ppt.	Ppt.	Ppt.	Organic Reduction	Ppt.	Calcination	Ppt.	Blend	Reduction & Oxidation	Ppt.
94.0	87.0	87.5	86.5	83.0	99.6 ($ZnO{\cdot}Fe_2O_3$)	93.3#	94.0	98.5	98.8#
—	—	—	—	—		9.0		—	23.2
0.10	0.05	0.05		—	0.15	0.01		0.06	0.12
—	0.04	0.04	{2.0}	—	0.002	0.01	Balance	0.01	0.07
—	0.02	0.02	—	—	—	1.12	depends	0.03	0.03
—	—	—	—	—	—	0.04	on	—	—
0.14	0.05	0.05	—	0.18	0.01	0.19	blended	0.06	0.20
—	1.00	1.00	—	—	0.03	0.20	components	1.00	0.60
5.50	11.0	11.0	12.5	16.2	0.10	—		1.00	—
—	0.15	0.10	0.30	—	0.07	—		0.40	0.15
Transparent Red	Light Yellow	Dark Yellow	Yellow	Transparent Yellow	Tan	Brown	Brown	Brown	Black
84	55	52	42	55	34	28	26	60	27
—	0.1–0.4	1–2.5	—	—	0.4–1.2	0.2–0.8	—	—	0.2–0.6
<0.10	0.20	1.5	0.2 x .7	<0.10	—	0.4	0.5§	0.2§	0.4§
Acicular	Acicular	Acicular	Acicular	Acicular	Acicular	Cubic	Mixed systems	Acicular	Cubical
0.20	<0.10	<0.10	0.10	0.16	<0.10	0.20	<0.20	<1	<0.10
—	15.2	7.4	—	—	—	11.8	—	15.4	7.2
4.56	4.03	4.03	3.97	3.88	5.24	4.77	4.70	4.60	4.96

Source: Technical bulletins and reports of pigment producers. Federal Regulations covering paints and coatings: ANSI (American Natural Standards Institute) and FDA 1973 and 1974. *Source:* "Coatings," National Paint and Coatings Assn., Washington, DC, Vol. 24, No. 17, Sep. 1972.

* Oil Absorption—kg. of oil per 100 kg. of pigment.

† As determined by Andreasen pipette or Micromeritics sedigraph, except as noted. Range is the weight percentage between 10% and 90%. 50% weight point is median size.

By electron microscopy, acicular, width x length.

¶ Square meters per gram—by B.E.T.

§ Magnetic, by electron microscopy.

\# Total iron expressed as Fe_3O_4.

pigments are used with other metal pigments (e.g., aluminum) to produce striking automobile finishes. The present market for these iron oxide pigments is estimated at 15 to 25 kt per year.

Electronic Applications (Ferrites): The traditional synthetic reds had the major share of the ferrite raw material market until about 1970. In recent years highly beneficiated naturals and the steel mill oxides derived from ferrous chloride have cut into this market primarily in the less critical barium ferrite market. The soft ferrite market remains a consumer of synthetic reds. Copperas reds and calcined yellows are preferred due to chemical purity, high density, and minimum volatiles (Stephens, 1959). The 1990 western world production of permanent magnetics is estimated to be about $2 billion. The iron oxide portion of the total ferrite dollar sales is less than 5% (Kline, 1981).

Magnetic Recording: Magnetic tape recording and magnetic ink oxides represented a US market in excess of 14 ktpy in 1990 for all applications and has continued to show healthy growth. Most magnetic recording applications utilize gamma ferric oxide or cobalt treated gamma ferric oxide products. CrO_2 has made some inroads, but the major part of the market is gamma ferric oxide.

The requirements for audio, video, and data recording vary and special oxides are produced for these applications. Slow recording speeds and high fidelity require ultrafine oxides. Cobalt modification is required to elevate the intrinsic coercivity of gamma ferric oxide to the higher levels required by video and the higher density applications.

Magnetic inks for check data processing, credit cards, and railway tickets use synthetic magnetite or other ferrites which provide high pigment concentrations in the binder system.

Concrete Products and Building Materials: Synthetics have the advantage of greater tinting strength and uniformity and have replaced naturals in most applications. This is a growing market consuming between 30 and 35 kt per year domestically. A new iron oxide product line is iron oxide slurry (liquid color). This product line has found wide application in the pigmentation of concrete products.

Plastics, Floor Tile, Linoleum, Rubber: High chemical purity, high chroma masstones, and strong tints result in a low cost per color unit, accounting for large usage of synthetics. Many plastics require stabilization which is more easily accomplished with the consistent chemical purity of the synthetic iron oxide.

Miscellaneous: Synthetic iron oxides are used singly or in

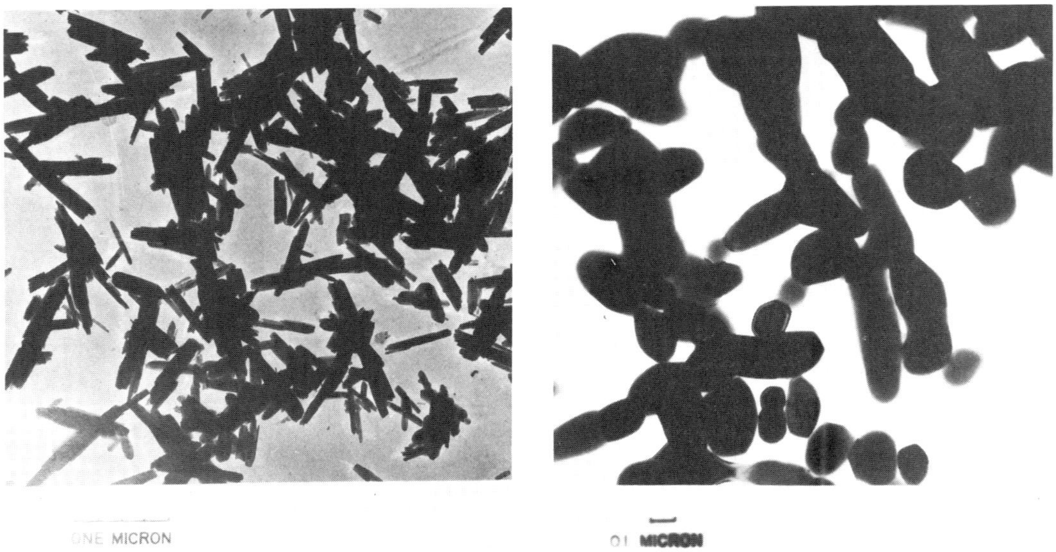

Fig. 14. Change in morphology in synthetic yellow iron oxide (FeOOH) by heat treatment to red oxide (Fe_2O_3).

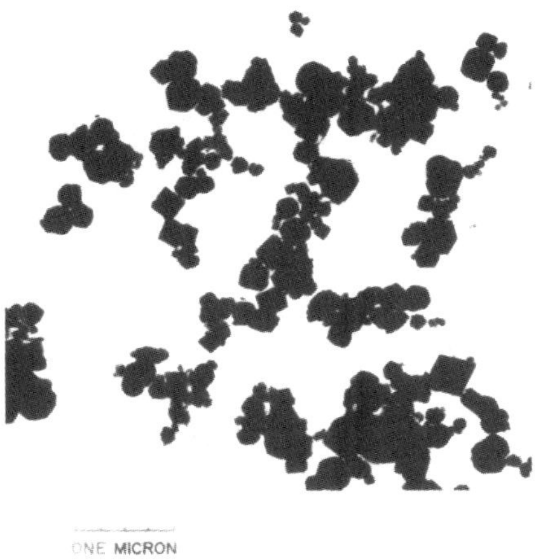

Fig. 15. Synthetic black iron oxide (precipitated magnetite).

Fig. 16. Magnetic iron oxide ($\gamma\ Fe_2O_3$) retains shape of yellow oxide precursor.

combination with other elements as catalysts in various conversions for ammonia, styrene, butadiene, hydrogen, and formaldehyde (Mills, 1964). Rocket and airbag propellants also utilize iron oxides.

Iron oxide is added to animal feeds as a source of iron nutrient and also as a coloring agent for pet foods.

Iron oxides are considered to have a generally recognized as safe (GRAS) rating by the Food and Drug Administration (FDA). Certain grades that meet the maximum of 3 ppm arsenic, 10 ppm lead, and 3 ppm mercury are approved for drug and cosmetic use. Iron oxides for pet foods must not exceed 5 ppm arsenic, 20 ppm lead, and 3 ppm mercury. High-purity iron oxides meeting these heavy metal requirements carry a premium of about 22 to 26¢ per kg.

Synthetic oxides are used for ophthalmic polishing and as the iron source for coloring glass and providing ultraviolet absorptive properties.

An interesting application—but as yet of no commercial significance—is in the field of medicine. Magnetic oxides have the ability to absorb myxoviruses which can then be magnetically separated from the system (Warren, 1969).

MARKETING

The demand for iron oxides is closely related to domestic and worldwide requirements of the construction, coatings, automotive, plastics, electronics, and chemicals industries.

Iron oxide prices were very stable from 1960 through 1970. The inflationary costs of labor and supplies plus high energy and environmental regulation costs necessitated significant price increases during the period 1970 to 1990. The twenty-year price trend of a selected range of natural and synthetic iron oxides is found in Table 5; they continue to be the most cost effective color pigment.

The transportation of iron oxides within the United States is accomplished primarily by truck and rail. In 1968 as much as 50% of iron oxides moved via rail. This figure was drastically reduced over the next several years due to damage in transit, shorter lead

Table 4. Typical 1991 Price Range for Synthetic Iron Oxides Marketed in the United States (Values rounded to closest dollars per kg)

Synthetic Oxide Type	FOB Selling Price	Products Available
Red Iron Oxides		
Copperas reds	1.96–2.27	20
Precipitated reds	1.94–2.09	7
Calcined yellows	1.76–3.96	13
Organic reduction (calcined)	1.72–2.02	20
Yellow Iron Oxides		
Regular	1.65–1.89	21
Low Oil	1.85–2.13	16
Tans		
Browns and Tans		
Synthetic blends	1.85–1.98	200**
Blacks		
Precipitated	1.96–2.09	7
Organic reduction	1.80–2.02	5
Magnetic Oxides		
Gamma ferric oxide	2.75–17.60	19
Magnetite	5.50–11.00	4
Transparent		
Yellows	5.80–8.58	6
Reds	6.05–9.24	5
Electronic (Ferrite) Grades		
Red iron oxides	2.24–2.42	10
Pet Food and Cosmetic Grades		
Reds	2.22–2.40	10
Blacks	2.38–2.49	2
Price range and total available products	1.65–17.60	365

Source: Suppliers published prices and other industry sources. *Truckload price in 22.7 kg multiwall bags. Price about 13 to 33¢ per kg higher for smaller quantities. Minimum weight depends on supplier. Special milling, palletizing, and packaging carry a premium. **estimated.

time on orders, lower average inventories, and the increase in freight rates. The amount of oxide moved by rail bottomed out in the late 1970s at between 5 and 15% of the total.

IRON OXIDE STATISTICS

The statistics for natural iron oxides are compared with the total natural/synthetic market as shown in Table 6.

This table shows that the consumption of naturals in volume (tons) has remained at about the same level, over the 1970–1990 interval. A small surge occurred in 1980. However, it returned to the 1970 level in 1990. The volume percent of the naturals declined mildly from about 41 to about 38% during this 20 year period.

This was not the case for naturals in value. Although the value of naturals increased from $6 million in 1970 to about $14 million in 1990, its market dollar share diminished significantly from about 21 to about 10%.

The average price of naturals derived from the above data has increased from 6.0 cents per pound in 1970 to 30 cents per kg in 1990. This contrasts with a much larger increase from 33 cents

Table 5. Typical Iron Oxide Prices 1970–1991

	1970 ¢ per kg*	1980 ¢ per kg	1991 ¢ per kg
Blacks:			
Synthetic	.35–.42	.95–1.08	1.80–2.09
Micaceous	—	1.21	1.96
Browns:			
Ground iron ore	.15–.22	.22–.29	.33
Metallic	.22	.40–.44	.44–.81
Pure synthetic	.42–.48	1.03–1.21	1.85–1.98
Burnt sienna (Italy)	.29–.62	1.54	1.32–1.91
Burnt umber (Turkey)	.24–.33	.68–.79	1.14–1.56
Red:			
Domestic primers	.15–.22	.48–.64	—
Pure synthetic	0.40	.99–1.14	1.94–2.27
Spanish	.26	.62–.70	—
Yellow:			
Synthetic	.35–0.40	1.14	1.65–2.13
Ochre (French)	.22–.26	0.51	—
Ochre (Domestic)	.13	.29	.73–1.25

Source: American Paint Journal; Tables 5 & 7 (1991) *Prices are FOB per kg.

Table 6. US Natural and Synthetic Iron Oxide Production and Price Statistics 1970–1990

Year	Naturals	Synthetics	Total	% Naturals
		Thousand Tons		
1970	45 638	66 824	112 462	40.6%
1980	52 958	70 703	123 661	42.8%
1990	46 526	76 373	122 899	37.9%
		Thousand Dollars		
Year	Naturals	Synthetics	Total	% Naturals
1970	6,012	22,185	28,197	21.3%
1980	11,143	86,126	97,269	11.5%
1990	13,951	124,392	138,343	10.1%

Chart data: Iron Oxide Pigments, US Bureau of Mines, Stepp 1970, Spinrad 1980, Templeton 1990.

Table 7. Salient US Iron Oxide Pigments Statistics

	1986	1987	1988	1989	1990
Mine production tons	30 744	31 816	30 747	36 661[r]	32,421
Crude pigments sold or used, tons	37 183	38 803	39 711	40 202[r]	37 071
Value $ thousands	$2,908	$3,596	$3,815	$4,561[r]	$4,615
Finished pigments sold, tons	116 444	124 293	126 640	127 665	122 921
Value $ thousands	$126,388	$136,427	$134,325	$134,567	$138,343
Exports tons	26 164	20 184	21 966	9 966[1]	9 510
Value $ thousands	$30,830	$31,689	$33,0141	$15,700	$18,533
Imports for consumption, tons	33 360	38 394	38 929	36 669[1]	34 000
Value $ thousands	$21,517	$20,680	$27,128	$30,540	$36,909

[r]Revised, [1]data reflects conversion to the Harmonized Tariff Schedule, effective Jan. 1, 1989, Public Law 100–418, Subtitle B.

Table 8. Finished Iron Oxide Pigments Sold by Processors in the United States, by Kind

Kind	1989 Quantity (tons)	1989 Value (thousand $)	1990 Quantity (tons)	1990 Value (thousand $)
Natural:				
Black: Magnetite	10 253	2,129	W	W
Brown:				
Iron Oxide	W	W	W	W
Umbers:				
Burnt	W	W	W	W
Raw	W	W	W	W
Red:				
Iron Oxide[1]	26 395	4,710	25 219	4,540
Sienna, burnt	W	W	W	W
Yellow:				
Ocher	W	W	W	W
Sienna, raw	W	W	W	W
Undistributed	12 725	7,218	21 315	9,411
Total	49 373	14,057	46 534	13,951
Synthetic:				
Brown: Iron oxide[2]	23 621	36,048	22 766	35,579
Red: Iron oxide	32 748	50,833	32 489	54,238
Yellow: Iron oxide[3]	21 923	33,629	21 132	34,575
Other: Specialty oxides	[4]	[4]	[4]	[4]
Mixtures of natural and synthetic iron oxides	[4]	[4]	[4]	[4]
Total	78 292	120,510	76 387	124,392
Grand Total	127 665	134,567	122 921	138,343

W Withheld to avoid disclosing company proprietary data; included with Undistributed, [1]Includes pyrite cinder, [2]Includes synthetic black iron oxide, [3]Includes other specialty oxides and mixtures of natural and synthetic iron oxides, [4]Included with Synthetic Yellow Iron Oxide to avoid disclosing company proprietary data.

per kg to \$1.63 per kg for synthetic iron oxide over the same period.

Additional data is presented on natural and synthetic iron oxides in Tables 7 and 8.

ENVIRONMENTAL CONSIDERATIONS

One of the major factors affecting iron oxide manufacturing today has been federal and state regulations regarding the environment. Whereas the major steel and titanium processors were disposing of their iron salts by dumping, they are now required to recover and process these salts prior to disposal. This has resulted in large quantities of cheap byproduct iron oxides. The iron oxide manufacturers have also been required to recover and process any soluble iron salts which were previously discharged, resulting in increased manufacturing costs.

The heavy metal content in several iron oxide products, both natural and synthetic is shown in Table 9. The synthetic iron oxides are generally low in these elements as can be expected by the chemical control provided in their manufacture. Certain grades of copperas reds show particularly low values of arsenic, lead and mercury. Precipitated reds, synthetic yellows, and synthetic blacks are somewhat higher in the analysis of these heavy metals but are low compared with the naturals. Table 9 shows this natural red to be particularly high in arsenic, lead and barium. It should be noted that natural iron oxides can show considerable variation in these heavy metal contents in different areas of the world and sometimes in the same ore deposit.

FUTURE OUTLOOK FOR IRON OXIDES

Two major markets for iron oxides are coatings and concrete products. Sales of coatings products increased from 9.9 to \$11.6 billion from 1985 to 1990 (NPCA data). The concrete masonry market increased from 2.7 billion units in 1984 to 4.1 billion units in 1990. The value increased from 2.1 to \$3.8 billion during this same period. About 12% of these concrete masonry units (eg., concrete block, brick, pavers, roof tile) are pigmented. Almost all of the pigments used in concrete masonry units are iron oxides.

The last 30 years produced a few iron oxide innovations—transparent iron oxides, high-purity ferrite oxides, grades meeting special FDA requirements for heavy metals, magnetic ink pigments, new magnetic iron oxides, and new types of surface treated magnetic oxides. Due to the high cost of energy and a shift to water-based coatings systems, iron oxides have recently become available in a high concentration slurry form (liquid color). This simplifies handling and provides for accurate metering of the pigment. Additionally it provides for more efficient use of the iron oxide pigmentary value. This product line has found wide use in the coloring of concrete products.

The number of iron oxide pigments available as easy dispersing types increased during the last ten years. The use of spectrophotometers coupled with computers to monitor and control iron oxide pigments is now common practice. With the projected population

Table 9. Concentrations of Heavy Metals in Light or Medium Shade Iron Oxides in Parts Per Million

Element	Copperas Reds	Precipitated Reds	Precipitated Yellow	Synthetic Black	Natural Red	Raw Sienna	Burnt Umber
Arsenic	<3	20–60	5–40	3–15	50–100	70–120	100–200
Antimony	5–15	5–20	5–20	10	40–80	†	†
Cadmium	<10	<20	<20	<20	<50	<50	<50
Selenium	<20	<20	<20	<20	<100	<100	<100
Mercury	<5	<5	<5	<5	<5	<5	<5
Lead	1–15	10–140	10–50	50–100	6000+	100–500	100–300 400–1100
Barium (acid sol)	<10	30–120	100–1000	100–1000	100–2000	NA	

Code < indicates if the element is present at all, it is below the detection limit of analytical method used. Hg determined by flameless atomic absorption as by USP colorimetry method. Remaining elements by x-ray fluorescence.

NA—Data not available.

†Potassium spectral interference.

growth and its attendant demands for construction, wood products, appliances, paints, plastics, chemicals, and related industrial products, the future for the permanent nontoxic iron oxides is very good.

BIBLIOGRAPHY AND REFERENCES

Anon., 1965, ''Wet Magnetic Separation Comes of Age,'' *Canadian Chemical Processing*, Feb.

Anon., 1966, ''Trends in Steel Pickling and Waste Treatment,'' *The Magazine of Metals Producing*, Mar., pp. 65–76.

Anon., 1968, ''Inorganic Color Pigments—Federation of Societies for Paint Technology,'' Unit 8, Sec. 2, Mar.

Anon., 1971, ''Iron Oxide Pigments,'' *Industrial Minerals*, No. 50, Nov., pp. 9–19.

Anon., 1971a, ''Trans-Oxide Pigments,'' Technical Bulletin, SC22—3/71, Hilton Davis Chemical Co., Cincinnati, OH.

Anon., 1972, ''Oil Absorption by Spatula Rub Out,'' *ASTM Annual Book of Standards*, Vol. 20, p. 170.

Anon., 1972a, ''Technical Report '316' MIOX,'' Reichard-Coulston Inc., Aug. 18, Bethlehem, PA.

Anon., 1983, ''Determination of Oil Absorption by Absorti-corder,'' Harcros Test No. PP-OIL5060.

Anon., 1984, 1990, ''Market Planning Statistical Survey,'' National Concrete Masonry Association, Herndon, VA.

Anon., 1985, ''Management Information Report,'' US Bureau of Census Data, National Paint and Coatings, Washington, D.C.

Anon., 1990, ''Current Industrial Reports—Paint, Varnish & Lacquer,'' Dec., Bureau of Census, US Dept. of Commerce.

Anon., 1991, ''Metallic Browns,'' ''Umbers,'' Harcros Technical Data, Easton, PA.

Anon., 1991a, ''Iron Oxides,'' Falmouth Associates, Portland, ME.

Ayers, J.W., 1938, ''Manufacture of Black Iron Oxide,'' US Patent 2.133.267 (assigned to Pfizer Inc.), Oct. 18.

Ayers, J.W., 1946, ''Process for the Production of Iron Oxide,'' US Patent 2.394.579 (assigned to Pfizer Inc.), Feb. 12.

Ayers, J.W., 1960, ''Production of Red Oxide of Iron Pigments,'' US Patent 2.937.927 (assigned to Pfizer Inc.), May 24.

Bennetch, L., 1957, ''Preparation of Red Iron Oxide,'' US Patent 2.785.991 (assigned to Pfizer Inc.), Mar. 19.

Bennetch, L., 1961, ''Red Oxide of Iron,'' US Patent 3.009.821 (assigned to Pfizer Inc.), Nov. 21.

Berman, H., et al., 1946, *Dana's System of Minerology*, Vol. 1, 7th ed., John Wiley, New York, pp. 685–687.

Brunauer, S., et al., 1938, ''Adsorption of Gases in Multi-Molecular Layers,'' *Journal*, American Chemical Society, Vol. 60, p. 309.

Buckley, J.A., 1967, ''Liquor Regeneration Slashes Cost of Steel Pickling,'' *Chemical Engineering*, Jan. 2, pp. 56–58.

Cadle, R.D., 1955, *Particle Size Determination*, Interscience Publishers Inc., New York, pp. 23–25, 198.

Camras, M., 1954, ''Magnetic Impulse Record Member,'' US Patent 2.694.656 (assigned to Armour Research Foundation), Nov. 16.

Casey, R., 1966, ''Iron Compounds, Iron (III) Oxide,'' *Encyclopedia of Chemical Technology*, Kirk-Othmer, eds., Vol. 12, 2nd ed., Interscience, New York, p. 39.

Chickering, K.D., 1971, *Journal of Optical Society of America,* Vol. 61, No. 118.

Downs, C.D., and Martin, J., 1950, ''Preparation of Magnesium Ferrite Pigment,'' US Patent 2.502.130 (assigned to Cities Services Inc.), Mar. 28.

Downs, C.D., and Martin, J., 1959, ''Method for Producing Zinc Ferrite Pigment,'' US Patent 2.904.395 (assigned to Cities Services Inc.), Sep. 15.

Hancock, K.R., 1970, ''Preparation of Ultrafine Particles of Metal Oxides,'' *Ultrafine Grain Ceramics*, Syracuse University Press, pp. 39–59.

Herdan, G., 1953, *Small Particle Statistics*, Elsevier Publishing, New York, pp. 37–39, 230–240, 267–269.

Hoffman, M., 1972, ''Iron Oxide Technology,'' Videotape Presentation, Pfizer Inc.

Jolly, J.L.W., and Collins, C.T., 1980, ''Natural Iron Pigments,'' Information Circular 8813, US Bureau of Mines, pp. 79.

Kline, C., 1981, ''Ceramic Raw Materials,'' Charles Kline & Co. Report, Fairfield, NJ.

Love, C.H., 1946, ''German Production of Some of the Most Important Inorganic Pigments'' FIAT Final Report 814, June 24, Office of Military Government for Germany (US), pp. 7–19.

Love, C.H., 1961, ''Magnetism, Magnetic Recording and Magnetic Inks,'' *American Ink Maker*, Vol. 39, No. 5, pp. 50–52, 55–57.

Love, C.H., and Ayers, J.W., 1942, ''The Iron Oxide Pigments,'' *Protective and Decorative Coatings*, J.J. Mattiello, ed., Vol. 2, John Wiley and Sons, New York, pp. 317–318.

Marsh, B., 1955, ''Manufacture of Red Hydrous Ferric Oxide,'' US Patent 2,716,595 (assigned to Pfizer Inc.), Aug. 30.

Martin, J., 1960, ''Manufacture of Iron Oxides,'' US Patent 2,939,767 (assigned to Cities Services Inc.), June 7.

Mattiello, J.J., 1942, ''Protective and Decorative Coatings,'' *The Iron Oxide Pigments*, Vol. II, John Wiley and Sons, New York, pp. 317–318.

McGannon, H.E., 1968, ''Steel Oxygen Steel Making Process,'' *Encyclopedia of Chemical Technology*, Kirk-Othmer, ed., Vol. 18, 2nd ed., Interscience, New York, pp. 741–747.

Mellor, J.W., 1953, *A Comprehensive Treatise on Inorganic and Theoretical Chemistry*, Vol. 13, Lowe and Brydone Ltd., London, pp. 805–817.

Mills, G.A., 1964, ''Catalysis,'' *Encyclopedia of Chemical Technology*, Kirk-Othmer, ed., Vol. 4, 2nd ed. Interscience, New York, pp. 534–586.

Parfitt, G.D., 1969, *Dispersion of Powders in Liquids*, Elsevier Publishing Co. Ltd., pp. 81–121, 285–324.

Parker, D., 1965, *Principles of Surface Coating Technology*, Interscience, New York, p. 817.

Patterson, D., 1967, *Pigments, an introduction to their Physical Chemistry*, Elsevier Publishing Co. Ltd., pp. 50–65.

Penniman, R., and Zoph, N., 1921, ''Process of Manufacturing Iron Compounds and Product,'' US Patent 1,368,748 (assigned to National Ferrite Co.), Feb. 15.

Schieber, M.M., 1967, *Experimental Magneto-chemistry*, E.P. Wohlfarth, ed., Chap. 4, Vol. 8, North Holland Publishing Co., Amsterdam, pp. 157–243.

Spinrad, W.I., Jr., 1980, ''Iron Oxide Pigments,'' *Minerals Yearbook, 1980*, US Bureau of Mines, pp. 429–436.

Stephens, R.A., 1959, ''Iron Oxides as Raw Materials for the Manufacture of Magnetic Ferrites,'' *American Ceramic Society Bulletin*, Vol. 38, No. 3, Mar. 15, pp. 106–109.

Stephenson, W., 1972, ''Project Cope,'' presentation to Federated Societies of Paint Technology, May 2.

Stewart, J.R., 1942, *The National Paint Dictionary*, 2nd ed., Stewart Research Laboratory, Washington, DC, p. 95.

Taylor, A.H., 1966, ''Oxygen Uses,'' *Encyclopedia of Chemical Technology*, Kirk-Othmer, ed., Vol. 14, 2nd ed., Interscience, New York, pp. 402–403.

Templeton, D.A., 1990, ''Iron Oxide Pigments in 1990,'' *Mineral Industry Surveys*, US Bureau of Mines, pp. 4–5.

Toxby, T., 1952, ''Method of Making Iron Oxide Pigment,'' US Patent 2,620,261 (assigned to Pfizer Inc.), Dec. 2.

Von Fisher, W., 1948 *Paint and Varnish Technology*, Reinhold Publishing Corp., New York, pp. 56–57.

Warren, J., 1969, ''The Adjuvant Effect of Powdered Ferric Oxide: Enhancement of Response to Mycoplasma Pneumonia in Respiratory Synctal Virus Vaccines,'' *The Journal of Immunology*, Vol. 102, No. 5, pp. 1200, 1308.

Woernle, A.K., 1967, ''Iron Oxide Pigments,'' *The Encyclopedia of Basic Materials for Plastics*, Reinhold Publishing Co., New York, pp. 261–264.

The following company technical bulletins are also of interest:

Columbian Chemicals, Technical Bulletin, ''Low Opacity Colors.''

Harcros Pigments Inc., Technical Bulletins on Natural and Synthetic Iron Oxides.

POTASH RESOURCES

SHERILYN C. WILLIAMS-STROUD, JAMES P. SEARLS, AND ROBERT J. HITE

Potash is a generic term that includes potassium chloride, potassium magnesium sulfate, potassium sulfate, potassium nitrate, and sodium-potassium nitrate mixtures. In the ceramics industry, potash is also used to refer to potassium oxide. Potash, primarily in the form of potassium carbonate, was the first industrial mineral produced in the United States, and the first US patent issued was for an apparatus and process developed in 1790 for its production (Paynter, 1990). Prior to the 1860s, potash was primarily sold as an impure form of potassium carbonate produced by burning hardwood trees and leaching the potassium salts from the ashes. The major early uses of potash include soap and glass making, dyeing fabrics, baking, and saltpeter for gunpowder.

In 1859, the development of a purification process to remove the sodium and magnesium chlorides was developed for the carnallite found at Stassfurt, Germany, and mined potash became available. With the appearance of mined potash and the earlier (1840) discovery in Germany by Justus von Liebig that potash was a nutrient for crops, potash started to be used for high valued crops such as cotton and vegetables. The German potash companies quickly developed a manufacturing process for producing potassium sulfate for tobacco fertilization. German potash supplied nearly all American needs until the embargo of the First World War when imports from Germany were interrupted (Bateman, 1918). With the discovery of potash deposits in New Mexico in 1931, the United States became self-sufficient in potash. In 1962, the United States began importing potash from Canada, and two years later domestic apparent consumption began to exceed domestic production.

Along with nitrogen and phosphorus, potassium is one of the three essential plant nutrients, the "K" of NPK terminology. As a result, 95% of potash production is used as plant fertilizer. In all plants, inadequate potassium diminishes growth, causes increased disease, stalk and stem breakage, and susceptibility to other stress conditions. Plants take up large quantities of potassium from the soil, and potash fertilization replaces this loss so that each new crop can be grown with the same vigor and productivity as the previous year's crop. The potassium depletion of the soil from growing repeated cotton and tobacco crops is well known in the history of southern agriculture in America. George Washington was known to have studied alternative crops that could be grown on soil that had been depleted by repeated tobacco crops.

Most of the remaining 5% of potash consumption is by the chemical industry, as potassium hydroxide to produce soaps and detergents, glass and ceramic products, dyes, explosives, alkaline batteries, and medicines. Potash as chemical is used in oil field drilling mud, the aluminum recycling industry, and the electroplating industry. Additional minor uses for potassium chloride include water softener regeneration, sidewalk deicing, and salt substitution for human consumption. Potash is used in the food industry as potassium phosphate, and in production of glass products as potassium carbonate or nitrate.

GEOLOGY

Potassium is the seventh most abundant element in the earth's crust and the sixth most abundant element in seawater. It is found in silicate minerals of igneous, metamorphic, and sedimentary rocks and is also a major constituent of many surface and subsurface brines. The majority of world potash resources are found in subsurface bedded salt deposits which yield high grade, large tonnage ore bodies and are amenable to low cost mining and beneficiation. Because of the relatively high solubility of potassium minerals, potash from salt deposits is ideal for use as fertilizers. Some potash production is from evaporation of naturally occurring brines, but the vast majority of current domestic and international production is from bedded salt deposits.

Sylvite, carnallite, kainite, and langbeinite are some of the more important potassium minerals (Table 1). Sylvinite, a mixture of KCl and NaCl is the highest grade potash ore. Carnallite can be considered a potash ore when removal of magnesium chloride is included in the beneficiation, but it can also be considered a contaminant when mining for sylvite. Potassium sulfate and potassium nitrate are typically manufactured products. Potassium sulfate is produced from mined minerals through conversion processes in Italy, Germany, and Carlsbad, NM, and from brines in southern California and at the Great Salt Lake in Utah. Natural deposits of potassium nitrate occur only in small amounts in Chile.

The majority of potash-bearing bedded salt deposits are believed to have originated from the evaporation of seawater or mixtures of seawater and other brines in restricted marine basins (Schmalz, 1969). The reflux depositional model for evaporite deposition was first described in the literature in 1888 by Ochsenius. A shallow bar, or sill, across the mouth of a basin lets in a restricted flow of seawater which evaporates into a salt-precipitating brine (Fig. 1). The density of the brine at the distal end increases with increased salinity, sinks to the bottom, and sets up a reflux current of higher density brine back toward the ocean. The sill, which restricts the inflow of seawater, allows inhibited flow of evaporation-concentrated brines back to the ocean. The least soluble salts are precipitated nearer the sill, and the most soluble components come out of solution in the deeper parts of the basin. The result is a lateral facies change in a tabular-shaped deposit that is due to the salinity gradients in the brine (Fig. 2A). The asymmetrical facies distribution of the Paradox Formation (Middle Pennsylvanian) Utah (Hite, 1970), the Prairie Formation (Middle Devonian) in Saskatchewan (Holter, 1972), and the Salado Formation (Upper Permian) in New Mexico (Lowenstein, 1988), might prompt explanation by such a model. Other deposits, such as the Salina Formation (Upper Silurian) in Michigan (Matthews and Egleson, 1974), show a facies distribution that could be described as a bull's eye pattern. Although some small subbasins of high grade sylvite are found near the margins, the potash is generally located in a central part of the basin surrounded by successively less soluble facies (Fig. 2B). The sparse

Table 1. Some Minerals, Ores, and Products of Potash Deposits

Name	Formula	K_2O, Equivalent Wt%	Remarks
MINERALS			
Chlorides			
Sylvite	KCl	63.17	Principle ore mineral
Carnallite	$KCl{\cdot}MgCl_2{\cdot}6H_2O$	16.95	Ore mineral and contaminant
Kainite	$4(KCl{\cdot}MgSO_4){\cdot}H_2O$	19.26	Important ore mineral
Bischofite	$2MgCl_2{\cdot}12H_2O$	0	Accessory contaminant
Halite	$NaCl$	0	Principle ore contaminant
Sulfates			
Polyhalite	$K_2Ca_2Mg(SO_4)_4{\cdot}2H_2O$	15.62	Ore contaminant
Langbeinite	$K_2SO_4{\cdot}2MgSO_4$	22.69	Important ore mineral
Leonite	$K_2SO_4{\cdot}MgSO_4{\cdot}4H_2O$	25.68	Ore contaminant
Schoenite (Picromerite)	$K_2SO_4{\cdot}MgSO_4{\cdot}6H_2O$	23.39	Accessory
Glaserite (Aphthitalite)	$3K_2SO_4{\cdot}Na_2SO_4$	42.51	Accessory
Syngenite	$K_2SO_4{\cdot}CaSO_4{\cdot}H_2O$	28.68	Accessory
Bloedite	$Na_2SO_4{\cdot}MgSO_4{\cdot}4H_2O$	0	Accessory
Loeweite	$6Na_2SO_4{\cdot}7MgSO_4{\cdot}15H_2O$	0	Accessory
Vanthoffite	$3Na_2SO_4{\cdot}MgSO_4$	0	Accessory
Kieserite	$MgSO_4{\cdot}H_2O$	0	Common ore contaminant
Hexahydrite	$MgSO_4{\cdot}6H_2O$	0	Accessory
Epsomite	$MgSO_4{\cdot}7H_2O$	0	Accessory
Anhydrite	$CaSO_4$	0	Common ore contaminant
ORES			
Sylvinite	$KCl + NaCl$	10–35	Canada, US, USSR, Brazil, Congo, Thailand
Hartsalz	$KCl + NaCl + CaSO_4 + (MgSO_4{\cdot}H_2O)$	10–20	Germany
Carnallitite	$KCl{\cdot}MgCl_2{\cdot}6H_2O + NaCl$	10–16	Germany, Spain, Thailand
Langbeinite	$K_2SO_4{\cdot}2MgSO_4 + NaCl$	7–12	US, USSR
Mischsalz	Hartsalz + Carnallite	8–20	Germany
Kainite	$4KCl{\cdot}4MgSO_4{\cdot}11H_2O + NaCl$	13–18	Italy, Ethiopia, USSR
PRODUCTS			
Potassium chloride	KCl	63.17	Principle potassium product
Potassium sulfate	K_2SO_4	54.05	Artificial product
Potassium-magnesium-sulfate (langbeinite)	$K_2SO_4{\cdot}2MgSO_4$	22.69	Natural product (New Mexico)
Manure salts	$KCl + NaCl$	40–60	USSR and Germany
OTHER SOURCES			
Alunite	$K_2Al_6(SO_4)_4(OH)_{12}$	11.4	
Niter	$KNO_3 + NaNO_3 + Na_2SO_4 + NaCl$	0.6–1.9	Chile
Mixed Potassium Nitrate and Sodium Nitrate	$KNO_3 + NaNO_3$	10–14	

Source: Adams and Hite, 1983, modified after Adams, 1968.

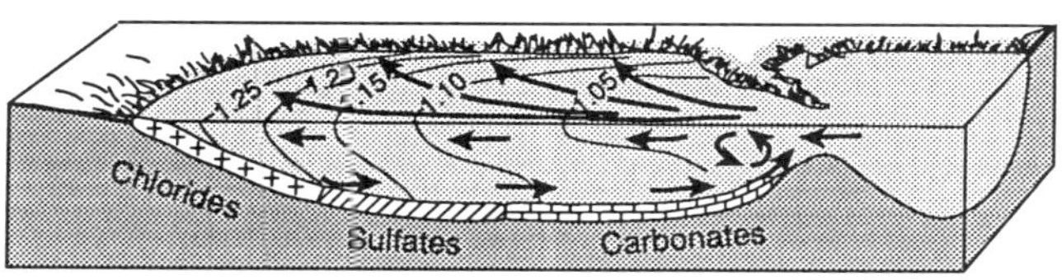

Fig. 1. The reflux model of evaporite deposition. Evaporation of seawater entering the basin over a restricting sill, increases in salinity at the distal end of the basin. Contours represent density of the brine. The least soluble minerals precipitate near the inlet, and halite precipitation occurs furthest from the inlet (after Briggs, 1957).

data available on the Miocene Mediterranean Evaporite (Hsu, 1972) suggest it also has a bull's eye facies distribution.

Sedimentary features in the marginal carbonates of some evaporite basins suggest subaerial exposure (Cercone, 1988, Peterson and Hite, 1969). This could have resulted in a basin which underwent evaporative drawdown as was suggested by Hsu (1972) for the Mediterranean Evaporite of Miocene Age, the Permian Upper Zechstein in Europe, the Silurian Salina evaporite deposit in Michigan, and by Maiklem (1971) for the Devonian Prairie Formation in Canada. The brine depth in any ancient evaporite basin could have undergone fluctuations related to sea level, groundwater inflow, precipitation and meteoric runoff, so that saline minerals could have been deposited in deep or shallow water, and may even have experienced subaerial exposure before burial. Depositional and petrographic features in evaporite deposits show evidence of the entire range of these environments. However, it is critical that interpretations take into regard whether observations were made at basin margins or center. It is possible that many evaporite deposits which show evidence of being disconnected from free inflow of seawater were derived from mixtures of seawater and meteoric or other waters.

The evaporation of seawater results in the precipitation of alkaline earth carbonate minerals, followed by calcium sulfates, halite, magnesium sulfates, and then magnesium and potassium chlorides. The ratio of sodium to potassium in seawater is 27:1, and, in general, minable potash beds are accompanied by thicker halite deposits. Often, the potash ore zone is located near the tops of halite beds in relatively thin layers because the potash is precipitated from brines of higher salinities occurring near the end of the evaporation sequence. The potash salt precipitated from evaporation of seawater

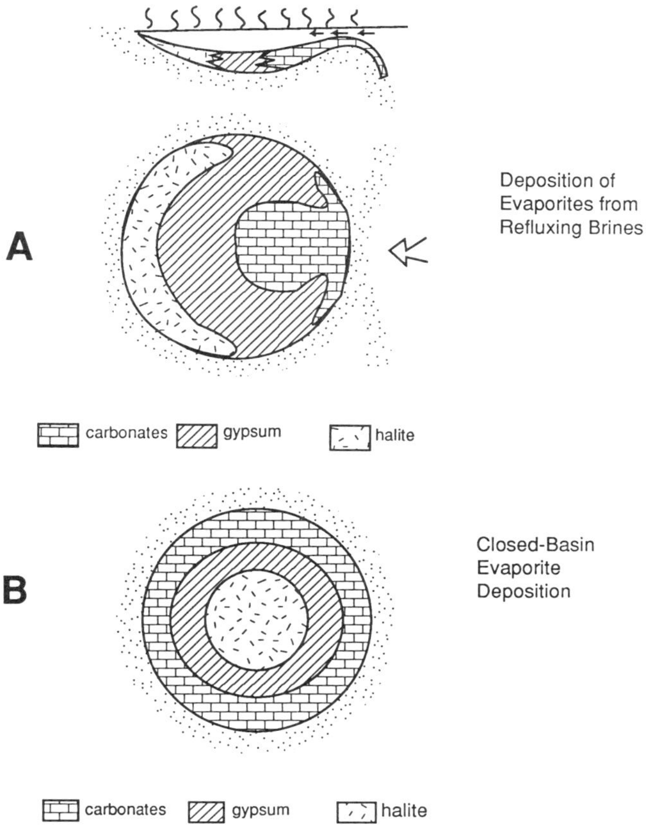

Fig. 2. A) Asymmetrical facies distribution of evaporite deposit formed by reflux model deposition. B) Bull's eye pattern of evaporite facies resulting from deposition in a basin drawdown model (after Hsu, 1972).

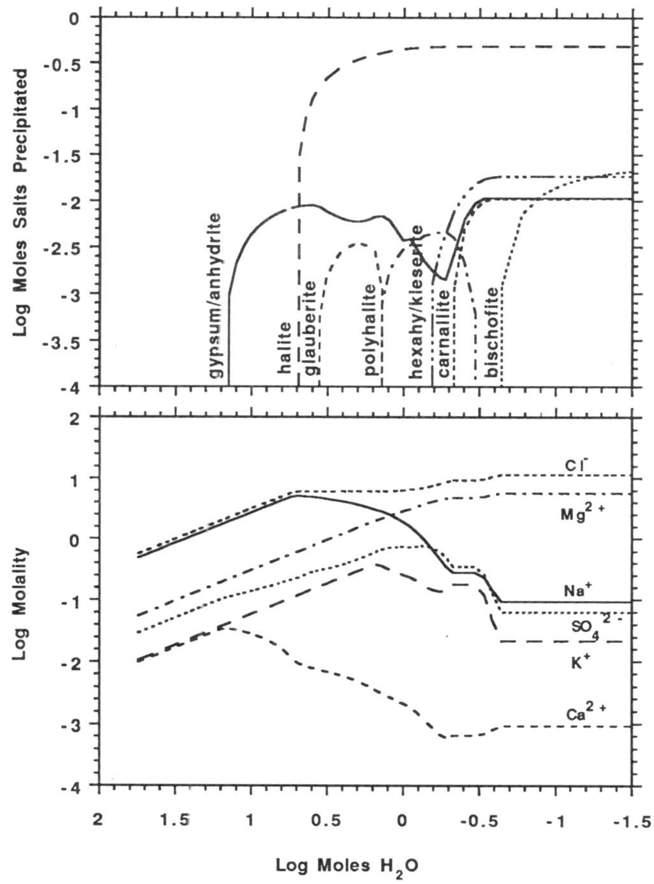

Fig. 3. Computer simulation of evaporation of modern seawater, using the program of Harvie, et al., 1980. Bottom graph shows change in concentration of major ionic species with evaporation, top shows masses of minerals precipitated.

after precipitation of magnesium sulfates is carnallite ($KCl{\cdot}MgCl_2{\cdot}6H_2O$) rather than sylvite (KCl) due to the high concentration of magnesium in seawater (Fig. 3).

The origin of evaporites believed to be marine derived which lack magnesium sulfate and magnesium chloride minerals is often explained by post-depositional dissolution processes. Carnallite in the salt could have been altered to sylvite by reaction with a magnesium-poor brine (Garrett, 1970, Braitsch, 1971, Borchert and Muir, 1964), or meteoric water (Hite, 1982). In the sylvinite deposits of Thailand, the Congo, and the Elk Point Basin of Canada there is evidence that point source ingress of meteoric, or sea water of nearly normal salinity, brought about local destruction of a precursor deposit of carnallite leaving behind sylvinite or only barren halite. This diagenetic process in some instances began shortly after the deposition of the carnallite layer, as in the case of the Thailand deposits, and may even be taking place today (Hite, 1982). However, recognition that non-marine evaporite deposits are not necessarily restricted to small, terrestrial saline lake basins, has led some workers to reassess the importance of mixing of brines of seawater origin with brines of other origin and with meteoric water (Hardie, 1984, Williams-Stroud, 1993). Some modern evaporites, such as the Basque Lakes in British Columbia, Canada (Jones, 1966), and those which occur in the Qaidam Basin of Western China (Lowenstein et al., 1989), which are clearly non-marine in origin, precipitate upon evaporation a suite of minerals very similar to that calculated to precipitate from seawater.

Theoretical and experimental salt assemblages calculated by D'Ans (1933) bore close resemblance to naturally occurring assemblages, and the presence or absence of some of the accessory minerals found in naturally occurring evaporite deposits was explained by temperature changes in precipitating brines, mixing of brines of different compositions, and reaction of brines with previously precipitated salt. The theoretical assemblage of primary precipitates from seawater at 25°C by Braitsch (1971) is given in Table 2. The sequence and amounts of minerals calculated with more recent computer simulations of the evaporation of seawater (Harvie et al., 1980) in Table 3 shows a slightly different mineral sequence from that calculated by Braitsch (1971). The computer generated sequence predicts the occurrence of polyhalite and glauberite ($CaSO_4{\cdot}Na_2SO_4$), which was not predicted in the earlier theoretical models, and which does occur in many evaporite deposits believed to be of marine origin, although in many of these deposits polyhalite appears to be secondary after anhydrite or gypsum. For deposits such as the Stassfurt horizon of the Permian Zechstein potash in Germany, the computer-generated sequence suggests the natural mineral assemblage could be primary and does not necessarily require alteration by post-depositional processes (Table 3). The initial brine composition is critical to the sequence of minerals which precipitate upon evaporation, but the mixing of waters during evaporation can also have a profound effect on the resulting precipitates. Computer simulations of evaporation show that it is pos-

Table 2. Theoretical Primary Precipitation of Stable Chloride Assemblages from Normal Seawater + Static Evaporation at 25°C (without reaction at transition points)

Layer	Mineral	%	Thickness of Layer (m)
A	Halite	100	100.0
B_1	Halite	72	6.3
	Bloedite	28	
B_2	Halite	20	4.5
	Epsomite	80	
C_1	Halite	29	6.3
	Epsomite	30	
	Kainite	41	
C_2	Halite	21	5.7
	Hexahydrite	3	
	Kainite	76	
C_3	Halite	11	1.7
	Kieserite	4	
	Kainite	85	
D_1	Halite	12	3.6
	Kieserite	40	
	Carnallite	48	
E	Halite	0.5	38.0
	Kieserite	1	
	Carnallite	0.5	
	Bischofite	98	

Source: Adams and Hite, 1983, adapted from Braitsch, 1971.

Table 3. Calculated Mineral Sequence from Evaporation of Modern Seawater and Observed Sequence from Zechstein 2 Potash Deposit[a,b]

Equilibrium calculated	Observed Zechstein II
G	G
A	A
A + H	A + H
Gl + A + H	Gl + A + H
Po + A + H	Po + A + H
Ep + Po + A + H	
Hx + Po + A + H	
Ki + Po + A + H	Ki + Po + A + H
Car + Ki + Po + A + H	Car + Ki + Po+ H
Car + Ki + A + H	Car + Ki + A + H
Bi + Car + Ki + A + H	

[a]Equilibrium evaporation in the first column contains some minerals not found in the observed sequence from the Zechstein 2. Most of the differences can be explained by dehydration reactions.
[b]Abbreviations: G, gypsum; A, anhydrite; H, halite; Gl, glauberite; Po, polyhalite; Ep, epsomite; Hx, hexahydrite; Ki, kieserite; Car, carnallite; Bi, bischofite.

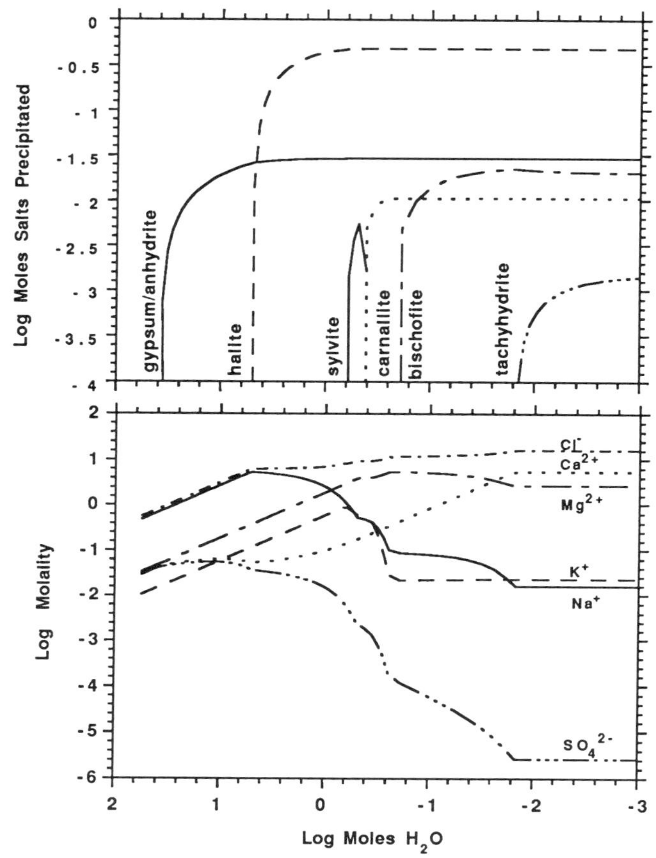

Fig. 4. Computer simulation of evaporation of altered seawater where the ratio of magnesium to calcium is approximately 1:1 instead of the seawater ratio of 5:1.

sible to produce widely varying mineral suites by altering the relative amounts of only one or two of the initial aqueous species (Fig. 4).

Petrographic and textural relationships and chemical analysis of fluid inclusions of associated halites in potash evaporites suggest that sylvite has occurred in many instances as a primary mineral (Lowenstein and Spencer, 1990, Wardlaw, 1972, Williams-Stroud, 1993). Enrichment of seawater with respect to calcium will result in early depletion of sulfate with gypsum/anhydrite precipitation, and prevent deposition of magnesium sulfates. If some mechanism for removal of magnesium exists, then carnallite also may not precipitate, and the primary potash mineral laid down will be sylvite. Mechanisms by which calcium enrichment and magnesium depletion can occur include, dolomitization of associated limestones or of a limestone precursor (Hite, 1983, Shearman, 1966), interactions with hydrothermal waters (Hardie, 1990), and clay mineral cation exchange (Hite, 1983). Sulfate depletion, which also will lead to non-deposition of magnesium sulfates, can be accomplished through brine mixing (Raup, 1970). When additional calcium is brought into the system by other brines, sulfate can be consumed by precipitation of gypsum and anhydrite before saturation with respect to magnesium sulfate minerals is reached.

Potash-bearing evaporites can be divided into two different types: Magnesium sulfate-rich deposits (containing the minerals polyhalite, hexahydrite, kieserite, and epsomite), and magnesium sulfate-poor deposits (Hardie, 1990). In the former type, the primary potash mineral is carnallite, while the latter is characterized by sylvite and carnallite. The magnesium sulfate-rich evaporites, such as the Stassfurt horizon of the Permian Zechstein 2, exhibit the same mineralogy expected to result from evaporation of modern seawater (Fig. 5). The magnesium sulfate-poor potash deposits probably precipitated from brines which were high in calcium and constitute 60% or more of known potash basins (Hardie, 1991). The fact that these two types of potash deposits also display a distinctive temporal relationship, with magnesium sulfate-rich types occurring only in the Permian, Miocene, and Quaternary (Fig. 6) may suggest changes in the composition of seawater throughout Phanerozoic time, or that many potash evaporite deposits did not precipitate from

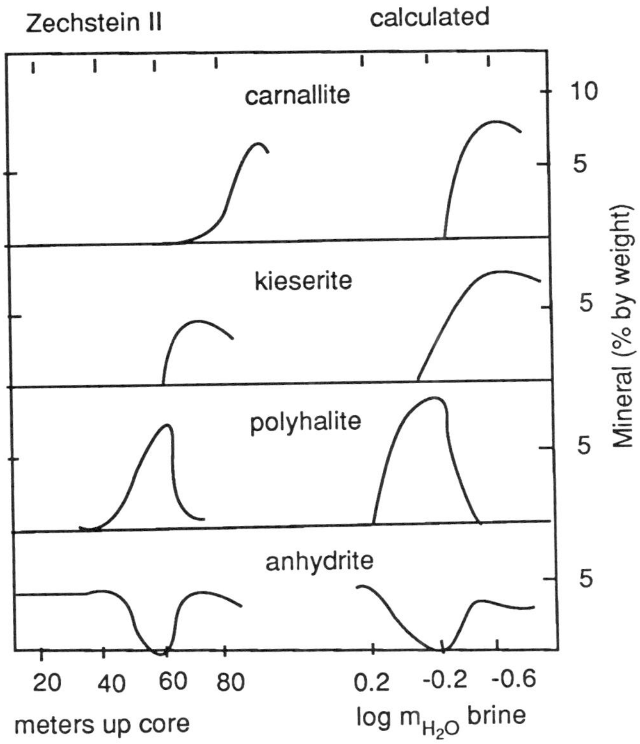

Fig. 5. Mineral sequence measured in the magnesium sulfate-rich Zechstein II Stassfurt potash zone and the sequence calculated with computer simulation of equilibrium evaporation of modern seawater (Harvie et al., 1980).

Fig. 6. Distribution of magnesium sulfate rich evaporite deposits which occur only in the Permian, Miocene, and Quaternary.

seawater, but are the result of evaporation of nonmarine or mixed marine-nonmarine waters (Hardie, 1991). Fig. 7 shows the distribution of potash deposits which contain little or no associated magnesium sulfate salts. Many of the basins are smaller than the magnesium sulfate-rich evaporite basins, and some show an elongated shape which could suggest formation in a graben or rift valley setting.

EXPLORATION

Evaporite deposits which are found in cratonic basins are usually associated with biogenic deposits such as shelf carbonates and

Fig. 7. Distribution of magnesium sulfate-poor evaporites. Occurrences are throughout geologic time and not restricted to any particular geologic period.

fringing reefs, and with clastic redbed deposits. Some evaporite deposits are found at continental margins which were sites of tectonic extension during the Mesozoic and Cenozoic Eras. The high solubility of evaporite minerals precludes the formation of salt outcrops in most parts of the world, and their occurrence is generally limited to the subsurface. The proximity of evaporite deposits may be inferred from saline waters in springs and wells, solution breccias, or thick outcrops of gypsum or anhydrite. Domal or collapse structures at the surface may indicate salt domes, or piercement structures, such as those found under and around the Gulf of Mexico (Halbouty, 1967), under the North Sea, and in the Zechstein Basin (Brunstrom and Walmsley, 1969, Trusheim, 1960). Domes or anticlines that show on outcrop several periods of upward growth during a broad range of geologic time can generally be expected to have a central core of evaporites. The distribution and shape of subsurface salt structures can also be determined by geophysical techniques such as gravity and seismic methods (Sawatzky et al., 1960).

The origin of salt domes is believed to be due to the effect of gravity on layered rock where a lower density material (salt) is overlain by higher density material (sandstones, compacted shales, and carbonates). The shear strength of halite and potash salts is very low compared to other geologic materials, and their low compressibility maintains their low density even under deep burial conditions. Low shear strength is further abetted by high temperatures associated with deep burial. Over geologic time, salt can be squeezed upward due to the effect of gravity on a denser overburden with or without the addition of lateral tectonic compression and its flow can be modeled as a viscous fluid (Ramberg, 1981). Points of instability, brought about by intersections of fault systems or differential loading, are sites for nucleation of salt pillows, or salt hills in the subsurface. As the salt masses continue to rise, anticlines formed in the overlying layers can be pierced by upward salt movement and result in the development of mushroom-shaped salt diapir structures which, in extreme cases, can become completely detached from the rest of the salt deposit (Jackson and Talbot, 1986). In the arid climate of northern Iran, the salt diapirs have pierced the surface and flow downhill as salt glaciers. The structure of the original depositional layering in the salt inside of a diapir is highly deformed and chaotic (Talbot and Jackson, 1987), and may not be suitable for conventional potash mining techniques.

The vast majority of potash deposits were discovered as a result of drilling for petroleum. Evaluation of oil and gas exploration geophysical logs, such as gamma-ray, neutron, sonic, caliper, density, and resistance logs are useful in interpretation of evaporite

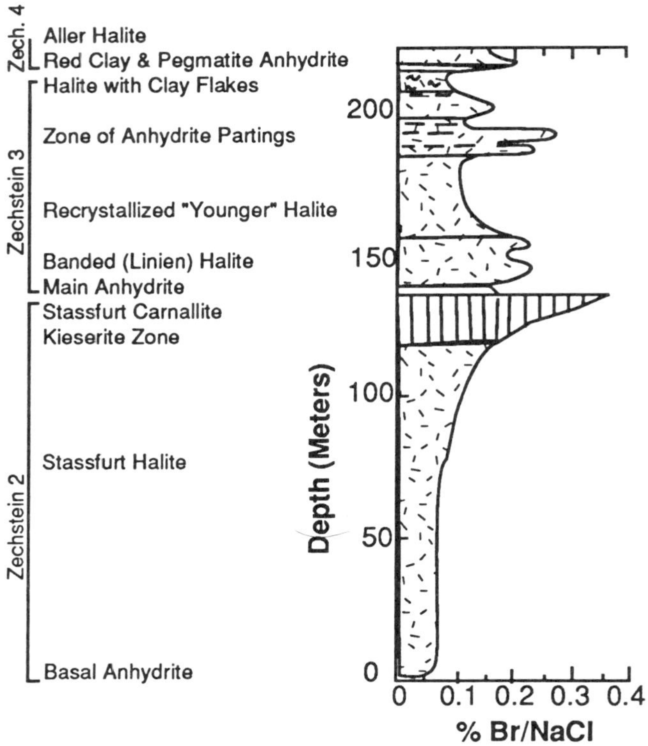

Fig. 8. Bromine profile from halite in the Zechstein 2 showing regular increase in the amount of bromine which substitutes for chlorine in the halite crystal lattice.

lithology, stratigraphy, and mineralogy (Alger and Crain, 1968). The gamma-ray log detects the gamma radiation from the natural potassium isotope K^{40}, and is particularly useful for providing a continuous measure of the potassium content of the salt section (Adams and Hite, 1983). Bromine content of salt layers has been used to evaluate potash potential, and is useful as an indication of the stage of evaporation, especially in waters which were predominantly marine in origin. During evaporation of a halite precipitating brine, the bromine incorporated into the halite crystal lattice increases as the bromine content of the brine increases. Since precipitation of potash occurs in highly concentrated brines, the halite in and near potash beds should contain more bromide. A bromine concentration profile from seawater would show a regularly increasing bromine content toward the top of the halite bed (Fig. 8), while a salt precipitated from mixtures of seawater and other waters or brines or a salt which has undergone extensive post-depositional recrystallization by groundwaters may show an irregular profile, or one that shows no increase in bromine concentration. In northeastern Thailand, economic potash deposits were discovered by the use of bromine geochemistry (Hite and Japakasetr, 1979).

Some general guidelines for evaluating the features peculiar to an evaporite deposit which may determine the success of a mining operation were given by Adams and Hite (1983) and are listed in Table 4. The initial appraisal of a deposit is based on drill cores spaced on 400 to 1 600 m centers, depending on the depth and complexity of the deposit. The salt cores are logged with attention to all mappable stratigraphic, lithologic, and mineralogic features, and the potash horizons are sampled by individual geological units for chemical analysis. It is generally sufficient to analyze for K^+, Na^+, Ca^{2+}, Mg^{2+}, Cl_2^-, SO_4^{2-}, and H_2O. The chemical analysis and the megascopic mineralogy estimated from hand specimen and thin section are used together with the stratigraphic data to develop a geologic interpretation to be evaluated in terms of the factors listed in Table 4 to determine the potential for a minable deposit or the need for additional drilling or metallurgical tests.

WORLD POTASH OCCURRENCE AND RESOURCES

The reserves of potash recoverable with existing technology from known deposits are estimated to be approximately 9.3 Gt K_2O equivalent. Canada and the Commonwealth of Independent States (former USSR) have approximately 47 and 32% of the total respectively. The United States has only about 1% of the world's potash reserves. The resource estimates compiled in Table 5 are based largely on evaluation of published information. For the most part, the reserves may best be referred to as probable ore, or ore whose occurrence is to all essential purposes reasonably assured but not absolutely certain. There is some question, for example, as to the reliability of the reserves reported for the Commonwealth of Independent States (CIS), as well as the portion of the Canadian solution mining reserve that will actually be recoverable. A less conservative estimate of world reserves, using a US Bureau of Mines (Sullivan and Michael, 1986) estimate of the 1986 market economy countries, yields about 11.7 Gt K_2O to which must be added the former East Germany and CIS reserves. With the change in economic philosophy in the latter countries, some deposits may be dropped from production in the near future, making estimates of actual reserves difficult.

North America

Canada: The largest high grade sylvinite ore reserves in the world are found in Canada. Reserves recoverable by underground mining are estimated to be 4.7 Gt (Table 5), based on 14 to 32% K_2O and 1 100 m maximum depth. The zones may be as thick as 10 m, but grade from approximately 3% K_2O at the top and bottom to 27% in the center.

Saskatchewan. Eleven mines currently produce potash from the Middle Devonian Prairie Formation of the Elk Point Basin in Saskatchewan (Fig. 9). There are three potash horizons which occur in the upper part of the evaporite sequence, the Esterhazy, Belle Plaine, and Patience Lake Members. The potash zones are composed primarily of halite, sylvite, and carnallite in zones 20 to 25 m thick (Holter, 1972). Varying amounts of insolubles consisting of anhydrite and gypsum, carbonates, quartz, hematite, and clays, form distinct layers or finely disseminated bands in the potash salt. The entire evaporite sequence dips gently to the south at 1 to 8 m/km, with the potash horizons occurring from 600 to 2 500 m below the surface. The lateral continuity of the beds and the presence of only gentle warping permits the use of continuous mining machinery in a room-and-pillar method, which minimizes rock instability and ventilation problems. International Minerals & Chemical (Canada) Ltd. may lose two mines to flooding but may probably open up another mine in the same reserves to replace it, using the same beneficiation plant.

New Brunswick. Two mines produce potash from the Windsor Group of Mississippian Age in New Brunswick. The high grade ore (21 to 25% K_2O) was discovered in 1970 in ore zones located 300 to 1 000 m deep (Miller, 1982). The Windsor Group evaporites are mostly sylvinite with minor local occurrences of carnallite. The evaporites are located in a fault-bounded synclinorium and are structurally more complex than the Prairie Formation. In some

Table 4. Guides to the Evaluation of Potash Deposits

Factor to be considered	Potential Problem	Observation to be made
Structure of ore horizon	Faulting, folding, or flowage of ore horizon complicating or precluding room-and-pillar mining method	a) Correlation of stratigraphy between drill holes b) Examination of core for evidence of deformation and recrystallization
Stratigraphy	Incorrect correlation of ore horizons; salt thickness above ore insufficient to protect deposit from water; slabbing and ore dilution from clay beds above ore horizon	Detailed core logging and correlation of beds between drill holes
Ore depth	Low mine extraction or excessive development and mining costs	Mining method and size of mine openings designed on basis of rock mechanics and geology
Ore thickness	Local systematic or random thinning of ore zone	a) Isopach maps of stratigraphic units and mineralization in ore zone b) Inspection of core for evidence of metasomatism and potash leaching, i.e., recrystallized halite, clay in intergranular clots, and horizontal mineral zoning.
Ore mineralogy	Assemblages incompatible with proposed mill circuit; certain assemblages reduce stability of mine openings; high insoluble content requires extra desliming capacity	Determinative mineralogy of core
Ore grade	Grade may be too low or erratic to maintain acceptable mill feed; recoverable K_2O may be significantly less than total K_2O	Correlation and interpretation of drill hole data; comparison of chemical analyses with determinative mineralogy
Ore continuity	Erratic barren or thin areas in ore zone	Correlation of stratigraphic units and mineralization in ore zones; inspect core for evidence of recrystallization and horizontal mineral zoning
Ore crystallinity	Inclusion of contaminants within potash minerals; fine-grained assemblages requiring finer grind	Correlation of microscopic examination of ore samples with chemical analyses of mineral separates
Ore reserves	Insufficient reserves or excessive mining and milling costs	Accurate geological description of deposit for mining and metallurgical studies.

Source: Adams and Hite, 1983.

places the evaporites are diapiric in character. The two mines produced about 1 Mt K_2O in 1988.

United States: In the United States, several mines and facilities producing potash from near-surface brines and lake brines have at least 35 years of reserves, but some mines are approaching depletion in ten years or less.

New Mexico. Since 1931, the majority of potash production has come from the Carlsbad mining district in the Permian Delaware basin of southeast New Mexico (Fig. 10). In this district the hosting Salado Formation contains sylvite, langbeinite, and lesser amounts of sulfate minerals in 12 potash horizons that occur over an area of approximately 5 000 km^2 (Lowenstein, 1988, Jones, 1972). Several mines are producing from five of the 10 potash horizons (Searls, 1985). The potash ore area is confined to a relatively small interval of the Salado Formation called the McNutt Zone, in which local barren patches of almost pure halite, called salt horses complicate mining operations.

Utah. The Middle Pennsylvanian Paradox Formation has large reserves of potash in the Paradox Basin of southeast Utah and southwest Colorado. These deposits are found in a thick sequence of evaporite cycles. Thirty-one cycles have thus far been identified in this sequence and 18 of these are known to contain potash, although only 11 cycles contain potentially valuable deposits (Hite, 1964). The potash deposits are found at depths suitable for conventional shaft mining only along a series of northwest-trending salt anticlines. In 1964, a shaft was sunk on the Cane Creek anticline to a depth of about 850 m by Texas Gulf Company to commence mining of a high grade sylvinite deposit which occurs in the upper part of the evaporite sequence. After a period of about four years, the company ceased mining by room and pillar methods and converted the underground openings to a solution mine.

Forty kilometers northwest of Moab, on the Ten Mile Wash structure, Buttes Oil and Gas Company has been exploring a high grade sylvinite deposit at depths that indicate a potential solution mining target. Currently the company is seeking a joint venture partner to assist in development of this deposit.

Potash is also produced in Utah by solar evaporation of the brines of the North Arm of the Great Salt Lake, and from shallow brines in the Great Salt Lake Desert at Wendover.

Michigan. The Salina Formation of Late Silurian Age hosts a large potash resource. These resources consist of a sylvinite deposit in the A-1 Salt which underlies an area of about 33 800 km^2 in the Michigan Basin (Matthews and Egleson, 1974). The total zone of mineralization is over 30 m thick in the central part of the basin. The vertical concentration of sylvinite is somewhat erratic in this thick deposit but some intervals show K_2O values as high as 40%. Depths to the deposit in basin center are as much as 2 550 m, and although these depths decrease towards the basin margins, the deposit can only be exploited by solution mining. Currently Kalium Corporation is developing a solution mine in this deposit near the northern pinnacle reef trend.

Arizona. In east-central Arizona, the Supai Formation of Pennsylvanian and Early Permian Age includes thick halite deposits which also host potash. This area, known as the Supai Salt Basin, was extensively explored for potash by several companies through the period of 1963 to 1966. This drilling outlined a northeast trending sylvinite deposit underlying an area of about 800 km^2 (Pierce, 1969). The deposit proved to be highly discontinuous with high clay content. Depths to the deposit range from 200 to 600 m, and it is apparently relatively undeformed.

California. Potash production takes place in California by mechanical evaporation of subsurface brines.

Texas. The Palangana salt dome in southeast Texas is the only salt dome in the Gulf Coast region which hosts a significant concentration of potash. This occurrence, which consists of sylvinite, was explored by 25 drill holes in the early 1960s (Hofrichter, 1968). The internal structure of this salt diapir is very complex and resulted in the abandonment of the project by the exploration company.

Table 5. World Potash Resources

Country	Potash Mineralogy	Grade (% K_2O)	Depth (m)	Resource (t)	Reserves (t)	Remarks
NORTH AMERICA						
Canada						
Saskatchewan	Sylvinite	14–32	1 000–1 800	11 000	4 700	
New Brunswick	Sylvinite	14–28	550–900	770	440	Region has not been completely drilled
United States						Much of the resources in Utah are too deep
New Mexico	Sylvinite	14–18	240–2 400	3 300	80	
New Mexico	Langbeinite	9	260–430	6	4	Estimated
California	Brines		Surface	30	10	
SOUTH AMERICA						
Argentina	Sylvinite	20	1 100	170	30	
Bolivia	Brines		Surface	130	70	
Brazil	Sylvinite	14–20	1 000	1 200	380	
Chile	Caliche	1	Surface	13	11	
	Brines		Surface	30	NA	
EUROPE						
France	Sylvinite	16.7	550–1 000	20	18	
Germany	Various	10–25	760–1 100	NA	770	
Italy	Kainite	10–17	140–730	NA	20	
Spain	Sylvinite	12–17	440–980	40	30	
United Kingdom	Sylvinite	18–25	1 100	130	28	
Former USSR	Sylvinite	15–20	270–400	5 000	4 000	
	Carnallite	14	270–400	20 000	NA	
	Various	9–12	60–210	NA	NA	Western Ukraine
AFRICA						
Congo	Sylvinite	15–20	260–400	70	48	
	Carnallite	15	260–400	5 000	NA	
Egypt	Sylvinite					Currently being explored by BHP-Utah International
Ethiopia	Sylvinite	19	80–330		70	Drilled by Parsons in 1966, hot springs in the area
Tunisia	Brines		Surface			
ASIA						
China	Sylvinite		Surface	350	220	
	Brine		Surface			
Thailand	Sylvinite	14–18	120–240	30	NA	
	Carnallite	18	60–430	300	NA	
MIDDLE EAST						
Israel	Carnallite			160	60	An underground deposit near Sdom
	Sylvinite					
	Brine		Surface			
Jordan	Carnallite		Surface	160	60	
Saudi Arabia	Sylvinite					Four locations are being explored by the Saudis
	Brine					

South America

The South American continent has only a few potash evaporite deposits, and so far no extensive, high grade, continental craton type evaporites, such as those found in North America and Europe have been discovered.

Argentina: A possible reserve exists high in the Argentinean Andes provinces of Neuquén and Mendoza in evaporites of Middle Cretaceous Age. This reserve is nearly 1 000 m deep and very high formation temperatures (>50°C) may preclude underground mining so solution mining methods are being investigated. The ore is sylvinite and is divided into two zones. The upper zone averages about 3 m thick and the lower about 11 m. Grade of the ore averages 20 to 25% K_2O.

Brazil: The large Amazonas Basin of Brazil contains sylvite-bearing evaporite deposits of Permian Age (Szatmari et al., 1979). Two possible reserves near Manaus in the Amazon basin are about 1 000 m deep with an average thickness of 2.7 m and average K_2O content of 16.5% K_2O.

A minor proven potash reserve in South America occurs in the Ibura Member of the Muribeca Formation of Middle Cretaceous Age in the state of Sergipe. The deposits probably formed as a rift valley evaporite sequence which resulted from alteration of seawater brines by the addition of calcium-rich hydrothermal fluids during the separation of the African and South American plates in Jurassic to Cretaceous times (Wardlaw, 1972). The mineralogy of the deposit includes halite, carnallite, sylvite, and very thick (up to 90 m) and nearly pure deposits of tachyhydrite ($CaCl_2{\cdot}2MgCl_2{\cdot}12H_2O$) in a bedded sequence displaying primary depositional growth fabrics of carnallite crystals in halite, but with no bischofite and no magnesium sulfate minerals.

The production of potash from the only mine is from the Ibura member, and it is complicated by complex structure in the ore, rapid and frequent lateral variation of sylvite content, and a thick underlying deposit of metastable tachyhydrite. The sylvinite deposit is divided into two seams by a thin halite layer which is sometimes missing. Because the ore directly overlies tachyhydrite, 3 m of sylvinite must be left in the floor to prevent floor heave. The extreme variations in ore grade have required mining large tonnages of barren salt. The geologic complexities in the mine also require an extensive program of underground horizontal drilling.

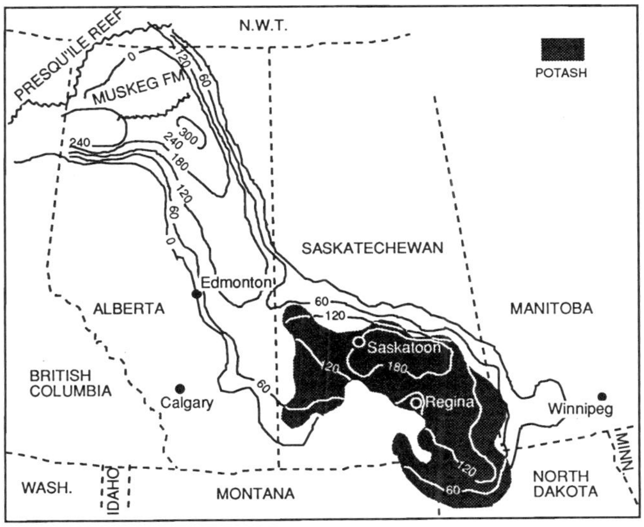

Fig. 9. Middle Devonian Prairie Formation in Canada. Hatched area represents the areal distribution of potash. Isopach contours in meters (after Zharkov, 1984).

Chile: Significant reserves occur in the several salars surrounding the area where the corners of Chile, Bolivia, and Argentina form a triple junction. The Salar de Atacama in Chile is producing less than 50 kt of KCl per year as a byproduct of lithium production. The potash is sold to Soquimich for conversion to high grade potassium nitrate. Part of the caliche deposits in Chile have a minor amount of potassium nitrate in the ore. This can be purified up to 14% K_2O as a mixed sodium nitrate-potassium nitrate fertilizer which has a niche market in the United States. MINSAL (Sociedad Minera Salar de Atacama, Ltd.) is developing an evaporation operation through AMAX in the Salar. Sociedad Quimica y Minera de Chile (SQM) purchased AMAX's interest in this development in 1992. Studies indicate production output after startup to be in the range of 530 ktpy.

Europe

France: Two mines in the Mulhouse area in Alsace are currently producing potash. The principal potash reserves of France are in evaporites which were deposited in the Oligocene Rhine Graben. The same deposits were mined where they extend into Germany. The potash salts occur in the subsurface at depths from 420 to 1 100 m in two laterally continuous zones in the upper third of the evaporite section. The lower potash interval containing sylvite with some carnallite is up to 5 m thick. It is separated from the upper sylvite zone which is up to 2 m thick. The ore contains several bands of claystone which imparts a high insoluble content. Bedding dips of more than 10° are common, and mining by the longwall, rather than room and pillar method, is used. The last French Alsatian mine is scheduled to close in 2004.

Germany: The largest production comes from the Permian Zechstein, where the world's first subsurface occurrences of potash were discovered (Fig. 11). The Zechstein contains four evaporite cycles with the lower three containing five potash horizons, the Thuringen, Hessen, Stassfurt, Ronnenberg, and Riedel horizons. The Hessen and Thuringen horizons are mined for carnallite and hartsalz. The latter is a potash ore containing sylvite and the relatively hard mineral, kieserite. In the Werra-Fulda district, these deposits are mined at depths of 400 to 1 000 m. The beds are flat-lying and relatively undeformed with uniform potash grade and distribution permitting the use of the room and pillar mining method. Potash deposits in the Hannover district are on steeply dipping flanks of conformable and piercement salt structures. Mining is by stoping and back-filling and is concentrated between depths of 300 and 800 m. The ore is irregular and discontinuous and requires underground exploration drilling.

Mining in the Stassfurt district of what was formerly East Germany also exploited potash reserves of the Zechstein salt. The Stassfurt mines shut down in December 1972. One mine in the former East Germany shut down in 1990 and four mines shut down in 1991 due to low ore grade or problems disposing of waste. The shutdowns in East Germany in early 1991 accounted for about 30 Mt K_2O of reserves removed but there was plenty of unused reserves and capacity that could pick up the demand.

England: The Boulby mine in England has produced potash from sylvinite beds in the English Zechstein basin since 1973. Two sylvinite beds averaging 6 to 8 m thick occur at depths between 1 000 and 2 700 m. Original plans had been to mine to the west from the shaft, but a fault zone was discovered to the west and the mine is now mining to the east under the English channel. Reportedly the great depths of the ore, bad roof conditions, erratic nature of the ore, and methane gas have caused difficult mining conditions.

Spain: Several potash mines in northern Spain produce potash from evaporites of late Eocene to early Oligocene Age. In the Navarra mining district, potash occurs in two beds, a lower sylvinite bed and an upper bed which is principally carnallite, at depths of 120 to 600 m. The carnallite and sylvinite deposits in the Catalonia mining district occur in four beds, which are folded and deformed. Mining in beds dipping 30° is common.

Italy: Italy has several potash mines on the island of Sicily in which kainite is mined from the Solfifera Series of Miocene Age. The only other known kainite reserves in the world are found in Ethiopia and in the Ukraine. Other minerals present in the Solfifera include carnallite with some sylvite, kieserite, and bischofite in beds 2 to 30 m thick with dips up to 60°.

Poland: In Poland, potash occurrences are known in the Zechstein in three separate areas (Orska and Werner, 1978). One of these includes the Klowada mine which is located on a large salt dome about 125 km west of Warsaw. The principal production from the Klowada mine is sodium chloride but some potash is also produced from an associated carnallite deposit. In the late 1960s polyhalite deposits were found in the Puck Gulf region along the Baltic Sea. According to Werner (1975), one drill hole intersected 6.5 m of nearly pure polyhalite at a depth of 788 m. At present no development of this deposit is under way. Another deposit in the Zechstein occurs in the Nowa Sol area 60 km east of the border with Germany (Orska and Werner, 1990). This deposit, which consists of carnallite, has not yet been developed.

CIS: The Commonwealth of Independent States contains 32% of the world's potash reserves in several sedimentary basins (Fig. 12). In the Pripyat depression and the Dnieper-Donets depression, potash occurs in Devonian age salt. Several potash-bearing salt deposits of Permian age occur in the East European Basin (Zharkov, 1984). In addition to the Devonian salt described above, the Dnieper-Donets depression contains Lower Permian Asselian and Sakmarian deposits (Fig. 12). Sakmarian salt is also found northeast of Moscow in the Moscow Basin. The Caspian depression contains Kungurian deposits which also occur in the Belsk and Solikamsk depressions of the Cis-Uralian trough. The thickest salt strata with potash salt horizons occur in the Kungurian deposits. One mine was lost to flooding in the CIS, but ore from another mine has kept the beneficiation plant operating.

Byelorussia. In the Pripyat Depression, four potash horizons

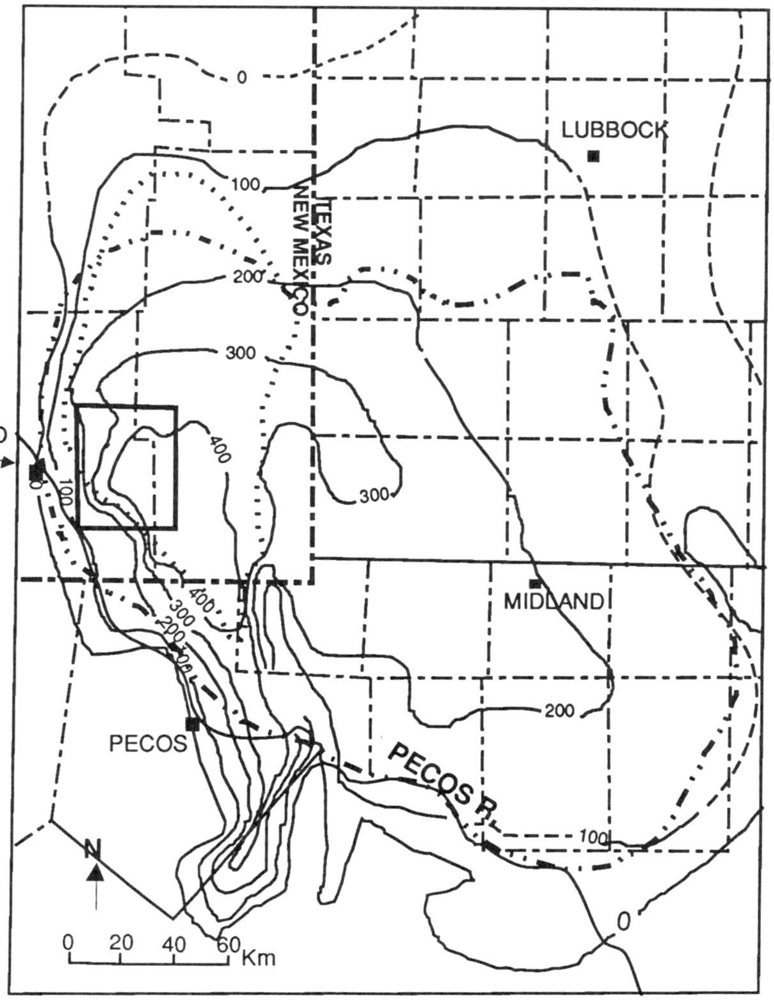

Fig. 10. Salado Formation (Upper Permian) in the Permian Basin of New Mexico and Texas. Isopach contours in meters.
–••– = Extent of polyhalite facies, ––– = extent of soluble potash salt facies (from Lowenstein, 1988).

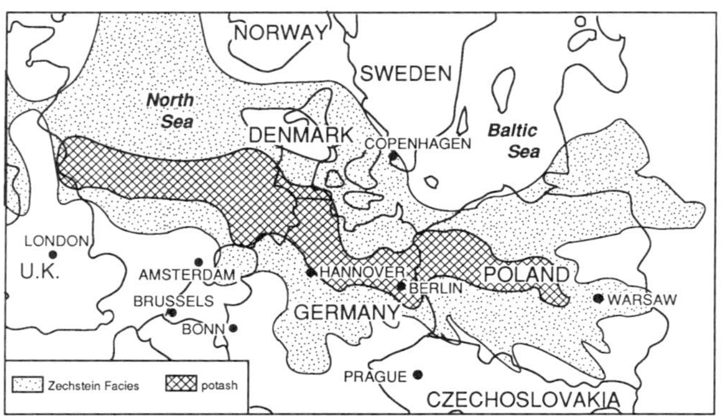

Fig. 11. Lithologic map of Permian Zechstein 3 evaporite deposit (after Zharkov, 1984).

have been identified in the lower salt sequence of the Liven Horizon of Devonian (Frasnian) age (Zharkov, 1984). The areal extent of the potash horizons ranges from about 1 500 to 130 km^2, and they contain beds of sylvite with minor carnallite in beds 4 cm to 1.5 m thick which are interbedded with muddy halite. Above the salt of the Liven Horizon, 60 potash horizons or potash occurrences have been identified in the upper salt sequence in the Elets Horizon (Famennian). The potash occurs under an area covering about 5 000 km^2 at depths varying from 3 000 to 200 m on local uplifts. Potash-bearing halite occurs in the upper part of the sequence in zones of sylvite interbedded with halite which are as thick as 70 m. Major potash production takes place from the Elets Horizon.

Russia. The Permian salt in the Solikamsk depression is found in the Iren Horizon of the Kungurian Stage. The potash member is divided into two intervals, the lower sylvinite and upper sylvinite-carnallite. The average thickness of the lower interval is about 21 m thick, and the upper interval averages 60 m thick. The deposits are generally medium grade (8 to 20% K_2O), occur at depths of about 250 m, and are undeformed.

Kazakhstan. Kungurian (Lower Permian) salt in the Precaspian depression contains polyhalite-halite, bischofite-carnallite, carnallite-halite, and sylvite-halite rocks. The salt sequence has been studied in more detail on the flanks of the Precaspian depression. In the center of the depression where the salt is strongly deformed, several salt domes were drilled for potash salts or other

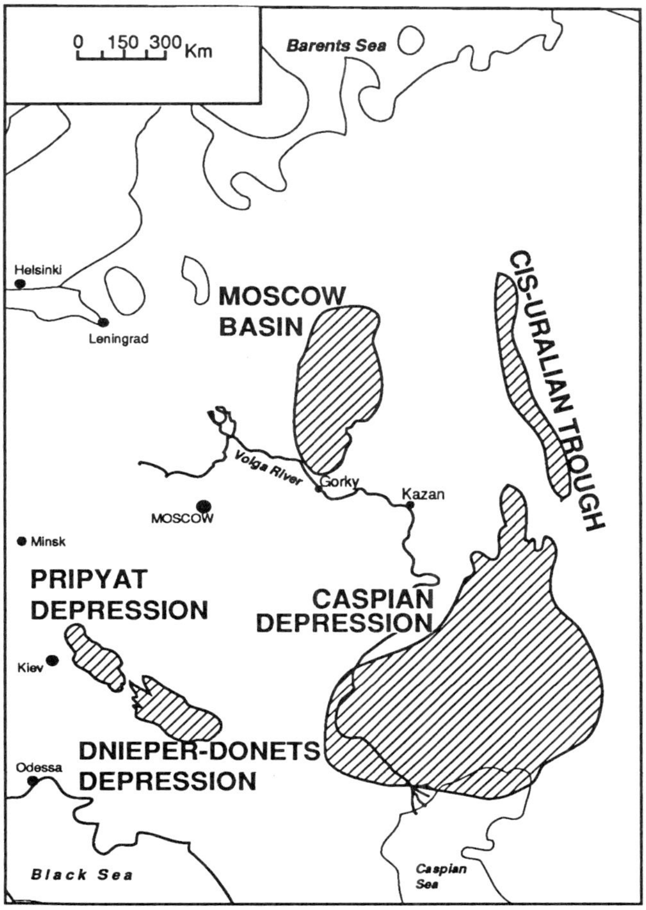

Fig. 12. Distribution of Permian potash-bearing salt deposits in the East European Basin (after Zharkov, 1984).

mineral resources. Potash is produced from three beds of predominantly sylvinite and carnallite in the Elton Salt Dome.

Ukraine. In the western Ukraine, lower grade potash deposits are being mined from lower and middle Miocene rocks near the Stebnik and Kalush-Golyn regions (Kopnin, 1963). The first three of four evaporite cycles contain individual beds of langbeinite, kainite, and mixed langbeinite-kainite with minor chlorides, as well as beds that contain sylvinite, carnallite, kainite, langbeinite, schoenite, and polyhalite. The percentage of insoluble residues is high, averaging 9 to 14% K_2O in the langbeinite-kainite ores to as much as 33% in the sylvinite ores. The fourth evaporite cycle contains polyhalite, but no soluble potassium salts. The deposits are deformed and mining in steeply dipping beds is common.

Potash is also produced from the Kramatorsk Formation of probable Sakmarian Age in the Dnieper-Donets Basin. Interbeds of sylvinite, carnallite, carnallite-halite, and sylvite-carnallite occur in the upper part of the formation. The potash salts form two horizons, the Privoleno sylvinite horizon and the upper, Chasov-Yar sylvinite-carnallite horizon.

Siberia. Cambrian salt covers an extensive area in east Siberia. Potassium salts are found in the Chara Horizon (Lower Cambrian), where three types of sections are distinguished: carnallite, sylvite-carnallite, and sylvinite. The potash horizon lies at depths of 600 to 900 m, and contains some sylvite with as much as 30 to 50% K_2O. No production is currently taking place in Siberia.

Africa and the Middle East

Congo: Brazzaville began potash production in 1969 with completion of the facilities at Holle, 40 km northeast of Point Noire. The Cretaceous Age potash beds contain sylvite and carnallite and are probably equivalent to the Cretaceous coastal basins of Brazil. Irregularities in the potash beds have reportedly caused initial difficulties for the boring machines. In 1977, the Holle potash mine was abruptly flooded and lost when a mining machine opened a brecciated collapse column with connection to an aquifer. Since then, there have been a few studies to renew production from this district. Although sylvinite mining grades are reportedly high, the latest study concerned mining carnallite; it may be inferred from this that the reserves of sylvinite are small and erratic in grade. In contrast, carnallite grade is more consistent and the deposits occur in beds 3 to 12 m thick. Highly deformed beds of carnallite which are also Cretaceous in age are found in the Gabon basin of West Africa.

Ethiopia: The potash deposits in the Danakil Depression of Ethiopia are young (Quaternary) and contain extensive kainite beds. Potash occurs in three members in the Houston Formation. The uppermost sylvinite member is up to 10 m thick, the intermediate member is 3 to 24 m thick and contains carnallite throughout with sylvite at the top and kainite at the base, and the Lower Kainite Member is 4 to 13 m thick. The deposit is in a salt pan in a rift valley depression characterized by recent volcanism and high heat flow (Holwerda and Hutchinson, 1968). Several hot springs and brine pools are found in the salt pan, and groundwater within the sediments interbedded with the evaporites are potential hazards for underground mines. The deposit was explored by several companies in the 1960s and 1970s. From 1968 to 1973 exploratory drilling and mining was performed, but flooding of the mine, via a hydrologic connection with an overlying water-bearing alluvial fan, stopped work. Feasibility and data appraisal studies were conducted during the 1970s and 1980s, but the site still remains to be developed.

Morocco: The Moroccan Meseta basins contain Late Triassic Age sylvite-carnallite deposits with bischofite and rinneite ($FeCl_2 \cdot NaCl \cdot 3KCl$). The potash salt deposits are interbedded with basalt flows (Tortochaux, 1968) and occur at depths between 490 to 790 m. The sylvinite bodies are reportedly low grade and erratic.

Egypt: The Red Sea rift zone contains a thick fill of Miocene evaporites which extends northwest into the Gulf of Suez region. In the Gulf of Suez, potash minerals, including sylvite, polyhalite, and langbeinite, have been identified in side wall cores from petroleum exploration wells (Hite and Wassef, 1983). Correlations in the thick sequence of evaporites in the Zeit and South Garib Formations show that there are millions of tons of polyhalite in the Suez area (Hite, unpublished data). Currently BHP-Utah International has obtained an exploration concession for the area and will begin exploration drilling in early 1992.

Saudi Arabia: The thick (up to 3 000 m in cores) and extensive evaporites of Miocene Age in the Red Sea rift zone rise near the surface in a group of salt domes in the Farasan Islands along the west coast of Saudi Arabia. The Saudi government has recently implemented an exploration program for potash in these islands due to identification of potash deposits on petroleum exploration wells drilled in the area. Mineral identification thus far includes bischofite and only minor carnallite.

Israel and Jordan: The only major producers of potash in the Middle East are Jordan and Israel. The brine of the Dead Sea, which is highly concentrated and contains 980 Mt of K_2O, is the source of this production (Neev and Emery, 1967). Both countries use a system of solar evaporation ponds along the margins of the Dead Sea to produce carnallite.

The Dead Sea rift zone also hosts several salt beds beneath the floor of the sea. These beds which were originally discovered by petroleum exploration wells are reported to contain potash deposits.

Israel was reported to be contracting an exploration program for these deposits in 1990. Once the Dead Sea brines are exhausted as a potash source it is possible that production from the bedded deposits beneath the sea might begin.

Asia

Pakistan: The Salt Range in West Pakistan has a thick sequence of Lower Cambrian evaporites which include some potash occurrences. These evaporites are near the surface at several localities along the Salt Range where they are mined for their halite content. The potash occurrences are known in rock salt mines at Khowra and elsewhere, where they form ten discrete zones or deposits (Jones, 1970). Potash minerals present include langbeinite, kainite, sylvite, and polyhalite. The potash deposits are thought to extend over a very large area along the Salt Range and offer promise for commercial development (Jones, 1970). As yet these deposits have not been explored by drilling.

Thailand: Potash deposits are present in the Maha Sarakham Formation of Cretaceous age on the Khorat Plateau of northeast Thailand and eastern Laos. These deposits are in the lowermost of three thick halite deposits. The deposits are divided into a southern basin (Khorat) and a northern basin (Sakon Nakhon) by a northwest-trending uplift (Pu Phan). This uplift is interpreted to have formed after the deposition of the evaporites because the evaporite stratigraphy is identical in both basins. The base of the evaporite sequence in both basins is undeformed, but the top is folded into numerous salt anticlines and domes. Most of the potash consists of a carnallite deposit, which is as much as 95 m thick (Hite and Japakasetr, 1979). This carnallite deposit, which is thought to be primary, also contains the highly soluble mineral tachyhydrite. Locally around the anticlines and domes, the carnallite deposit has been altered by meteoric water into lenticular bodies of high grade sylvinite (Hite, 1982). Most of the exploration has been done by the Thailand Department of Mineral Resources although some drilling by concessionaires has taken place in recent years. The government of Thailand is currently planning to exploit the carnallite deposit near the village of Banmet Narong. The first trial mine started by the government was lost to flooding. However, plans are now underway to begin a second mine. The carnallite deposit in the Khorat and Sakhon Nakhon Basins represents an enormous potash resource. The sylvite deposits are much smaller and their extent is not well known at this time.

China: Although China has been exploiting and exploring evaporite deposits for nearly 2000 years, only minor amounts of potash have been discovered. The most extensively explored basin is in the Sichuan where brines have been produced from Triassic evaporites for many centuries. These brines are utilized primarily for their sodium chloride content but other small amounts of byproducts, including potash, are recovered. The Triassic halite deposits in this basin locally have bromine values as high as 300 ppm and as a result the Ministry of Geology and Mineral Resources (MGMR) has concentrated most of their exploration efforts on this basin. Unfortunately, other than minor amounts of polyhalite, no economic concentrations of potash have been discovered.

The most recent exploration effort by MGMR is centered on the Orthos Basin which covers about 1.0 Mkm^2 in Shaaxi Province. Halite-bearing evaporites in this basin are of Ordovician age and include cycles with beds of halite up to 174 m thick. Most of these halite-bearing cycles are at depths of 2 000 m or more so that if potash is discovered here, it will have to be recovered by solution mining.

In western China, sylvite deposits have been found in rock salt mines in the Simao Basin in Yunnan Province. The evaporites here are found in rocks of Middle Jurassic, Late Cretaceous, and Paleocene age. Although their precise age is unknown, it seems likely that the Simao Basin evaporites are the same age (Cretaceous) and a northerly extension of the potash-bearing evaporites of the Khorat Plateau of Thailand and Laos. The Yunnan sylvite deposits are lenticular and associated with a high degree of deformation. Grade of the ore seldom exceeds 15% K_2O. Because of their lenticular deformed nature, low grade, and extremely remote location, the government does not consider this area to merit further exploration.

Currently the principal production of potash in China comes from a large salt lake in the Qaidam Basin in Qinghai Province. About 30 ktpy of K_2O are extracted by solar evaporation from the lake brine.

China's needs for potash are acute. The soils of the principal farming areas of the country have been steadily depleted in potash for centuries and agricultural crisis is impending until this depletion can be made up. Current potash consumption (1986 to 1987) was only 545 kt K_2O. The global ratio of N:P:K applied to agricultural lands is 1:0.45:0.36 but in China the current ratio is 1:0.19:0.03. The government plans to consume 3 Mtpy K_2O by the year 2000, but according to agronomists in China, the actual needs will be closer to 20 Mt. Thus it is apparent that the country must either discover more potash deposits or work out some trade agreements with other producing countries.

Australia

Australia has several large salt basins but as yet no economic occurrences of potash have been discovered there. In the Carnarvan basin, minor occurrences of sylvite have been noted in petroleum test holes in a halite deposit of Silurian age. In addition, three petroleum tests drilled in the Adavale Basin found potassium minerals, mainly sylvite, in the Devonian age Etonvale Formation. Some of the Australian basins have been the sites of minor potash exploration efforts but the results of this drilling are unknown at the present time.

PRODUCTION TECHNOLOGY

Mining

A rule of thumb for underground mining of sylvinite is that the ore should be 14% K_2O or above, and 1.2 m thick, with a salt back (or roof) of about 8 m thick. The 14% K_2O rule has been used since it was established in Carlsbad, NM, in the early 1930s by the US Geological Survey. It is still a reasonably good rule of thumb, but the East Germans and the Russians have apparently mined down to 11% K_2O while operating under centrally planned economics and receiving subsidies to provide jobs for the populace. The Carlsbad sylvite potash producers have operated with 80 to 85% recovery in the mill (with some producers operating at 92% recovery by pulling pillars). The French and the British have decided that potash is intrinsically valuable, and they have added crystallizers to the back end of their mills to capture the potash from the waste liquor before it is sent to the tailings pile with the waste salt. They have raised the recovery of their plants to as high as 92% with this additional step. However, it is not clear if the marginal cost of this additional step of recovery is worth the marginal profit from the additional potash.

The amount of ore to be proved for a new mine should be enough for 20 years of production given the mill size, mill recovery, and the underground recovery factor likely for the ore depth. In this market, plant or mill size ought to be at least 300 kt K_2O to compete in a market with many plants in the 1 Mt K_2O size. This is the economist's returns to scale concept. Another factor to consider is the existence of salt horses in the ore zone (Jones and Prugger, 1982). In Saskatchewan, these barren patches of almost pure halite can constitute as much as 10% of the ore, which requires additional monitoring of the mining operation.

Two extremes for percent of underground recovery are illustrated by North American conventional mining methods. The Carlsbad, NM, mines operate at about 270 m, with one mine sloping down to a depth of 430 m. These Carlsbad mines recover about 75% of the ore on the first pass and attain 90%, sometimes with surface subsidence, on the retreating pass. In contrast, the mines in Saskatchewan are normally operated between 1 000 and 1 100 m with a 35 to 45% recovery. The mines in the Patience Lake Member recover about 35 to 40% of the ore while the mines in the Esterhazy Member, which has at least 18 m of competent salt back above the ore zone, recover about 45%. The Canadian operations do not have a retreating pass but have started studying the feasibility of returning waste halite from the mill back to the galleries of the mine. After the original pillars collapse enough for the waste heaps to take up the weight of the back (roof), the miners plan to return and mine the pillars. This provides a higher underground recovery but the cost and complexity of returning mill tailings (waste halite) to the galleries has not been fully assessed. Another example of a mine which operates at a lower recovery rate, is the Boulby mine, near Whitby, England. This mine is at a depth of about 1 100 m and achieves around 35% underground recovery.

Much sylvinite mining is now mechanized, utilizing either borers, drum miners, or longwall miners and occasionally road headers. Boring machines in Canadian mines can continuously produce 12 t/min, occasionally ranging up to 17 t/min for as long as eight hours. The mining machines take the central three meters of high grade ore. Mine depths of more than 1 000 m in a sedimentary sequence require sophisticated techniques which sometimes involve controlled collapse of panels. All borers feed directly onto conveyor belts connected to the main haulage ways or over bridge conveyors that are directly connected to the conveyor belt system (Jones and Prugger, 1982). Drum miners are becoming continuous miners with the addtion of bridge conveyors for direct connection to the main conveyor belts. The French mines use longwall mining for higher underground recovery and continuous mining. The longwall conveyor belt discharges to a ladder-chain conveyor lying on the floor of a development roadway at one end of the longwall panel. The ladder-chain conveyor discharges onto the conveyor belt leading to the skips of the ore hoist. Road headers typically use shuttle cars to connect to the conveyor belts but can be directly connected. The Carlsbad mines that recover langbeinite still must use jumbo drills, undercutters, blasting, mucking, and shuttle cars because of the hardness of the ore.

Borers mine the ore in very long galleries to take advantage of their cutting method, rather than room and pillar which is the method used with drum mining. Some of the Canadian mines produce 1 200 meter-long galleries with long walls in between. Other Canadian mines use panels having five entries per panel with the outer two entries collapsing in one or two days. Sylvinite and langbeinite mining in Carlsbad, NM, is room and pillar. The Boulby mine on the eastern coast of England drives five panels with cross-overs in a V shape to allow the stress to redistribute.

Where the ore zones are not flat-lying, other methods of underground mining are used. The Potash Corporation of America mine at Sussex in New Brunswick, Canada, mines with road headers in a cut-and-fill technique up one side of an anticline. The environmental interests in the province require that no tailings pile be on the surface so all the tailings are dried and returned to the mine using a second conveyor belt system. Potacan Mining Company's Clover Hill mine in New Brunswick uses a complex system involving boring machines, drill and blast, and road headers. The environmental requirements are met at this operation by returning the mill tailings to the mine as a slurry. In some of the German potash mines, the potash ore is on the side of salt domes and the block-caving (bench-mining and funnel-stoping) method is used.

Solution mining can substitute for conventional shaft mining in potash deposits that occur below approximately 1 100 m. However, for some potash zones which are too deep for conventional mining, it is not the automatic answer. Thickness, mineralogy, and structure of the deposit, as well as technical considerations of the mining operation, must be evaluated to determine the suitability of solution mining. There are two forms of solution mining now in practice. A site which was initially mined by conventional underground methods can be converted to solution mining to recover the ore remaining in pillars. Pipes are drilled to meet the former underground galleries to provide communication between an inlet pipe which supplies undersaturated brine to the ore zone and a return pipe which brings saturated brine to the surface. This method is effective with relatively thin ore zones of slightly more than one meter. The Texasgulf mine at Moab, UT, was the first to employ this type of solution mining in 1971. Bore holes were drilled to the lowest point in the mine and to all the highest points in the branch galleries. The boreholes in the branch galleries supplied fresh water and the lowest point in the main gallery became the return pipe for the dense brine. If seen from above, the galleries appear something like a Christmas tree with the butt of the trunk at the lowest point. The lower elevation of the return pipe allows only saturated brine to come up from the ore zone, where as in a flat-lying ore zone it would be difficult to prevent the less dense, undersaturated brine from returning to the surface. The Potash Company of America's Patience Lake mine was converted to a solution mine in 1988 after it was flooded by water coming from a breccia occurring down dip from the ore zone. The return pipe inlet was located near the breccia zone, and brine was pumped into the mine until a hydrologic head was established sufficient to prevent further entry of water from the breccia zone. Heated brine, which was nearly saturated with respect to NaCl, was then introduced into the mine. The brine dissolved only potash for return to the surface ponds. The imposed hydrologic head prevented most of the less concentrated brine from the intruding aquifer from entering the mine.

Solution mining with concentric pipes solves the problem of establishing communication between the inlet and the return pipe. Three concentric pipes are required, the inner most and deepest for the return flow and the middle for the inlet flow of hot NaCl-saturated brine. The outermost concentric pipe injects oil, which floats above the brine, to force the injected water to mine outward instead up upwards around the pipe. A large diameter for the cavity must be established, and it is critical that the potash zone or sequence of zones be sufficiently thick. Development of mine production by this technique generally requires that the cavity have enough potash to pay for the drilling of the well, the equipment in the well field, the pipes leading to the crystallizers, and all the recovery plant costs.

Solution mining can also be employed in shallow high grade deposits where the structural complexity precludes successful underground mining by conventional methods. The conversion to a solution technology at the Texasgulf mine in Moab, UT, was dictated by the high degree of structural deformation of the ore layer; a problem that was not indicated by the original exploration drilling (Evans and Linn, 1970). Solution mining was quite successful initially because of the large surface area of potash ore presented by the original room and pillar structure. The K_2O content of the brine pumped to the surface has continuously diminished with removal of the mine pillars and there is lttle indication of active dissolution of ore beyond the original mine workings.

Other solution mining techniques have received some attention. Davis and Shock (1970) experimented with hydrofracturing to provide connection between wells 60 m apart at a depth of 350 m near Carlsbad, NM. The experiment was a technical success but apparently was not repeated or put into commercial operation.

Processing and Beneficiation

In operations where the source of potash reserves is surface or subsurface brines or the ore is mined by solution technology, the potash must be recovered with technologies which are specific to brines. The three processes currently used for performing this recovery are crystallization of the potash by solar evaporation, by reduction of the temperature of the brine, and by artificial processing through crystallization-evaporation cycles. Each of the processes utilizes selective dissolution and precipitation of sylvite from the brine. Where brine is mutually saturated with with KCl and NaCl, the solubility of NaCl decreases slightly with increasing temperature up to 90° C, above which the solubility of NaCl increases with temperature. Since the solubility of KCl increases dramatically with increasing temperature up to 90°C, a warm brine would dissolve much less NaCl if there were sufficient sylvite available to saturate the solution with respect to KCl. When the brine is cooled, sylvite will precipitate more rapidly than halite, and at certain temperatures, a solution which is saturated with respect to both KCl and NaCl will precipitate only sylvite, and NaCl will remain in solution.

For effective solar evaporation from surface ponds climate is extremely critical. The brine brought to the surface at the Texasgulf mine at Moab, UT, is totally evaporated in single ponds to precipitate a mixture of sylvite and halite. Dyes are added to the pond brines to increase solar energy adsorption and thus increase the brine temperature and evaporation rates. The operations at Ogden and Wendover use lake water or near-surface brines and multiple ponds. The multiple ponds are used to establish brine densities for crystallization of various minerals at different stages. Efficient solar evaporation occurs during only three months of the year (June, July, and August), with over half of the total evaporation occuring in July. During the three-month evaporation period, low rainfall must combine with sufficient wind to keep the brine mixed and to enhance evaporation. At Moab and Wendover, both sylvite and halite are precipitated in the potash-harvesting ponds and the harvest is sent to a standard flotation-type beneficiation plant for separating the two salts.

The two potash operations at the south end of the Dead Sea benefit from a climate that promotes evaporation throughout the year. Upon evaporation the Dead Sea brine precipitates carnallite and halite in the harvesting ponds. Magnesium chloride is washed away with precious freshwater, leaving a mixture of sylvite and halite. On the Israeli side the sylvite is separated from the halite in three different processing plants. The oldest processing plant uses flotation technology. A capacity expansion added a process which uses dissolution-recrystallization. The newest plant uses a proprietary cold crystallization process which uses a low-heat crystallizer to produce KCl.

The new solution mine of Potash Company of America in Saskatchewan, Canada, employs a selective precipitation process that takes advantage of natural cooling. One large pond operates from September through May. As the warm brine from the mine flows into the pond, only sylvite crystallizes upon exposure to the cold winter air. The pond is then dredged and the precipitated sylvite dried for sale. The process is predicated on the different temperature dependance of the solubility of NaCl and KCl with temperature which is described above.

An additional method of recovery is used by Kalium Chemical in Belle Plaine, Saskatchewan, Canada. The brine is processed using two trains of evaporators and crystallizers. The NaCl is removed in the evaporators, and KCl precipitates in the crystallizers.

Ore beneficiation methods also differ according to the type of ore being recovered. For less soluble ores such as the langbeinite mined in the Carlsbad, NM, area, leaching is an effective beneficiation method for about 9% K_2O ore grade. One of the two processes involves preferential dissolution of associated salts (mostly halite with a little sylvite) from the ore. The other process uses heavy media separation and flotation in addition to leaching.

The potash industry began using flotation technology for sylvinite in the early 1930s in Carlsbad, NM. The use of flotation has gradually spread to France, England, Germany, the CIS, and Israel. After the slimes (clay and hematite) are washed away, the pulp (the suspension of crushed sylvinite in saturated brine) is treated with a long chain amine (C-18 to C-22) and a frother (such as pine oil) and then is passed to an agitation cell where compressed air is introduced from the bottom. A froth of amine and entrained air bubbles carries the potash to the surface where it is mechanically swept over the side of the cell into a trough. Carnallite is avoided during mining because it has a detrimental effect on flotation cell recovery (Jones and Prugger, 1982).

Heavy media technology is practiced by International Minerals and Chemical (Canada) Ltd. in their Canadian plant. The sylvite and halite crystals of the Esterhazy zone are crushed to 1 cm and separated by heavy media utilizing the small density difference between halite and sylvite. In Carlsbad, NM, sylvinite and langbeinite are separated in one heavy media circuit, then the sylvite and halite are separated in a second heavy media circuit. The French apparently tried this technology in the early 1960s, but did not continue its use (Anon., 1963).

In the early 1910s, the French developed dissolution-recrystallization technology for processing of potash ore and continue its use today. The French potash ore has a high-insoluble content and the beneficiation methods originally developed in Germany did not work. The first Carlsbad, NM, potash mill also used this technology but converted to flotation when it became obvious that the latter would produce adequate beneficiation a clean sylvinite ore (less than 3.5% slimes). One Carlsbad mine continues to use the dissolution-recrystallization technology for ores with a greater than 5% slimes content.

Electrostatic separation of potash was first investigated in the Carlsbad, NM, area after the Second World War but was not implemented. It has been resurrected in Germany and, after years of development, has proved to be a low cost, high volume process (Singewald and Neitzel, 1983). The waste product is dry, there are no waste brine disposal problems, and little heat input is required of the operation. The process involves moderate heat, reagent coating, a source of electric power, and careful control of the relative humidity in the separation chamber. Separations can be serialized in a multi-stage process line.

Potash Products

Within the fertilizer trade potassium chloride is known as muriate of potash (MOP), potassium sulfate is known as sulfate of potash (SOP), and potassium magnesium sulfate is known as sulfate of potash-magnesia (SOPM). About 94% of the potash used for fertilizer is in the form of potassium chloride (MOP) with the remainder as sulfates and nitrates (Searls, 1985). Lixivation is used for production of potassium sulfate. The two major practitioners are Italy and Great Salt Lake Minerals & Chemicals. The process starts with kainite, either mined or crystallized from lake brines. The kainite is converted to schoenite ($MgSO_4 \cdot K_2SO_4 \cdot H_2O$), and potassium sulfate is crystallized from the schoenite. Another method of beneficiation of potassium sulfate is by ion exchange. This method is being considered for production in Canada where lakes with high concentrations of sodium sulfate are located near the large supply

Table 6. Product Specifications

Grade	Minimum K_2O Equivalent	Approximate particle size range[a] Mesh[b]	mm	Type of potash
Granular	60, 50, 22	6–30	3.34–0.85	Muriate and sulfates
Coarse	60	8–28	2.4–0.6	Muriate
Standard	60, 50, 22	14–65	1.2–0.21	Muriate and sulfates
Special Std.	60	35–150	0.4–0.11	Muriate and sulfates
Soluble	62	35–150	0.4–0.11	Muriate
Chemical	63	NA	NA	Muriate

[a]from approximately 2 to 98% by weight percent cumulative, [b]Tyler standard, NA not applicable.

of potassium chloride mined from underground. It is not clear if this is a profitable technology at current market prices.

Sulfate of potash does not occur naturally as a mineral. Most of the commercially produced sulfate of potash is synthesized by heating muriate of potash and sulfuric acid to approximately 650°C in large furnaces. Some of the German potash mines have produced sulfate of potash by combining kieserite ($MgSO_4 \cdot H_2O$) and sylvite (Singewald and Neitzel, 1983). The IMC Fertilizers Inc. plant in Carlsbad, NM, produces sulfate of potash by mining sylvinite and langbeinite and performing a base exchange (Anon., 1968). Great Salt Lake Minerals & Chemicals Corp. uses solar evaporation to produce kainite ($KCl \cdot MgSO_4 \cdot {}^{11}/_4H_2O$) from the Great Salt Lake brines, and follows with the previously mentioned lixivation process in beneficiation. Great Salt Lake Minerals finds that it has enough leftover sulfate ions in the brine that they can buy muriate of potash and produce sulfate of potash profitably. The potash mines of Sicily also use the lixivation process.

Potassium chloride (MOP), which constitutes 90% to 95% of the world's production of potash, has a minimum K_2O content requirement of 60% K_2O except for minor uses in Eastern Europe. Potassium sulfate (SOP) has a 50% minimum K_2O content requirement and potassium-magnesium sulfate is 19% minimum K_2O content. Product specifications, including the various grain sizes of the different grades of potash are shown in Table 6. Coarse MOP is the most popular size sold in the United States, but granular muriate of potash for use in machine-spread fertilization is a close second. Standard muriate and the even smaller special standard muriate of potash are preferred in countries where fertilizers are still hand cast. The particle sizes listed in Table 6 do not fully describe the current expectations of farmers. Current potash producers claim to match the current particle size distribution of monoammonium phosphate (MAP) or diammonium phosphate (DAP) particles.

Byproducts and Coproducts

The typical underground sylvinite mine and muriate of potash mill beneficiates with flotation technology and produces a dirty NaCl as waste. Some waste has clay, hematite, and even bits of shale in it. Other potash sources and beneficiation techniques can produce a variety of byproducts. The German mines, by using electrostatic separation, can produce magnesium sulfate (kieserite) and rock salt that is pure enough for road deicing. The German mills that use dissolution-recrystallization produce magnesium chloride, salt, and bromine. The French and the Israelis also produce bromine as a byproduct. Potash produced from brines of Searles Lake in California has soda ash and salt cake coproducts. The surface brine at Wendover, UT, produces salt and magnesium chloride as byproducts. Great Salt Lake Minerals & Chemicals Co. produces salt for road de-icing, Glauber's salt ($NaSO_4 \cdot 10H_2O$), and magnesium chloride. The solution mine at Belle Plaine, Saskatchewan, produces large amounts of clean salt, only a small portion of which is sold to a salt producer. The dissolution-recrystallization mill in Carlsbad, NM, occasionally processes ore which contains some sulfate. The sulfate plugs up the circuit as glaserite (K_3NaSO_4), which is recovered and sold in small amounts as a low priced substitute for sulfate of potash if the crop can tolerate the slight amount of sodium.

ECONOMIC FACTORS

Potash company cost data is proprietary but some estimate of costs can be obtained from profit or loss statements and from annual reports. Mines de Potasse d'Alsace in France, working about 14% K_2O ore with a 3 900-member workforce produced approximately 1.3 Mt K_2O of potash in 1990 while operating at about 85% capacity, but lost about $30 million on total sales of about $200 million. Potash Corporation of Saskatchewan, working about 23% K_2O with a 1,090 member workforce produced approximately 2.1 Mt K_2O of potash in 1990 while operating at about 40% capacity and earned about $23 million on sales of $250 million. The Dead Sea Works in Israel, working with 0.7% concentration of K^+ ions in the Dead Sea, produced 1.3 Mt K_2O of potash (nearly full capacity), and earned $65 million on sales of $567 million. The average price for standard muriate of potash at Vancouver for much of 1990 was $97/t KCl ($161.67/t K_2O), with a price rise to $102/t in November.

A recent presentation at the joint International Fertilizer Industry Association-State (former USSR) Agrochemical Association meeting in Moscow (Bingham, 1991) placed Saskatchewan (Canada) and Middle East production costs at an index of about 75. The United States and New Brunswick (Canada) production costs were placed at about 110. European and the CIS production costs were placed at an index of about 220. The absolute costs were not given.

United States Historical Consumption

The United States passed through an S-shaped consumption curve between 1920 and 1990 (Fig. 13) as it learned to feed an ever-growing population. The decline in potash consumption after 1980 appears to reflect a more efficient use of fertilizers and possibly some decline in crop exports as other countries have begun their own green revolutions. Generally, the need for fertilizer increases as the population increases but populations do not buy fertilizer until they can afford it (Fig. 14). Many poor countries around the world need fertilizers (in addition to good weather and modern farming practices) but will not be able to purchase the fertilizers until the wealth (GNP per capita) of the population increases. By grouping the countries and taking averages of the groups one can clearly see the progression of potash consumption as a function of wealth in Fig. 14.

A worldwide over-capacity of potash forced most of the potash producers to produce under capacity during 1990 (Table 7). Potash Corporation of Saskatchewan led the way by producing at about 40% capacity. The Carlsbad, NM, mines produced at about 79% capacity. Only the Israeli and Jordanian producers worked at full capacity. United States potash companies cannot supply the current demand in the United States at today's prices (Fig. 15). The potash mines in the Province of Saskatchewan provided about 70% of the US apparent consumption in 1990 (Table 8). However, the US potash producers exported about 27% of their sales to the Caribbean and the Pacific Basin. Transportation costs govern much of this routing of exports although the Canadians do not produce any sulfate of potash, which constitutes the larger portion of the US exports to the Pacific Basin.

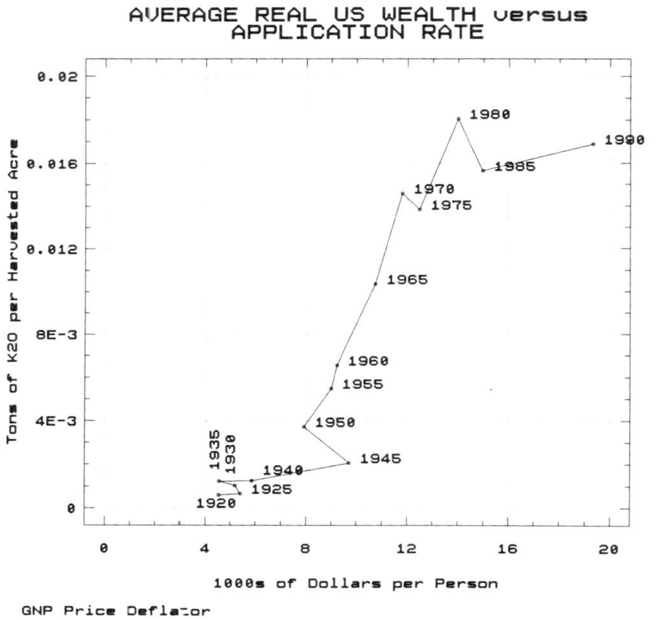

Fig. 13. History of US potash consumption and growth of wealth.

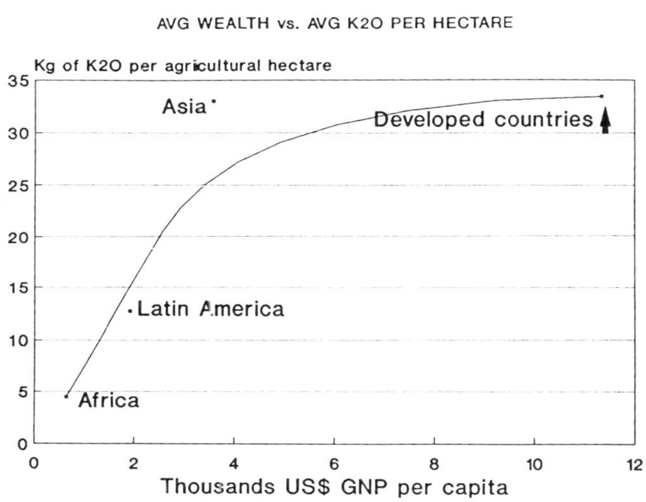

Fig. 14. Relationship of wealth to consumption of potash (*source*: World Bank and UN FAO).

United States Prices: Prices for a commodity are rather hard to define since many firms sell at slightly lower prices to large buyers. Large buyers may also receive extended terms, the period allowed for the buyer to pay the bill without interest penalty. Terms may be 30, 90, or even (rarely) 180 days. The prices presented in Fig. 16 represent the average annual prices received by the US potash producers, FOB at the plant. Annual prices are computed by dividing the total dollars received sold by the tonnage sold. The average for all US producers includes all types of potash sold (MOP, SOP, and SOPM), and all grades (sizes). Large size particles command a premium price in the United States.

Canadian Prices: The price (FOB plant) for equivalent potash in Saskatchewan is approximately three to five dollars less than the US producer's price. United States and Canadian potash pro-

Table 7. Estimated World Potash Production, 1990 (kt of K_2O equivalent)

Brazil	60
Canada	7 372
Chile	25
China	40
France	1 300
Germany, Federal Republic of	4 900
Israel	2 200
Italy	60
Jordan	790
Spain	690
USSR	9 500
United Kingdom	490
United States	1 713
Total	28 290

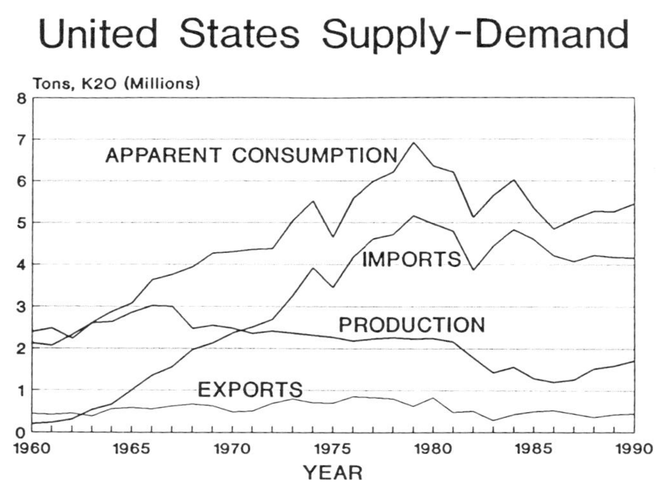

Fig. 15. Components of US supply and demand for potash.

Table 8. World Consumption of Potash, Crop Year 1987–88 (kt K_2O equivalent)

Africa	446
North and Central America	5 350
South America	1 907
Asia	4 149
Europe	8 396
Oceania	229
USSR	7 052
Totals	27 529

Source: *Food and Agricultural Organization Yearbook*, "Fertilizer," 1988.

ducers compete at the fertilizer retailer level. This price is the sum of the price at the mine plus the cost of transportation to the retailer's location.

Anti-dumping petitions were filed against the Canadian potash producers with the International Trade Commission and the International Trade Agency of the US Department of Commerce by potash producers in the United States in 1988. In Saskatchewan in the late 1960s six mines opened within three years, and the price of potash dropped to its lowest level. The Province of Saskatchewan set a floor price for Saskatchewan-produced potash and allocated the market among the producers. Several US potash producers, who also had potash mines in Saskatchewan, were sued in a Chicago

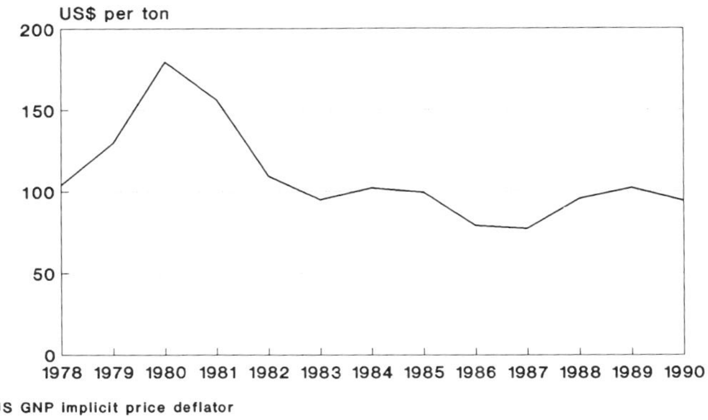

Fig. 16. Real price of US-produced potash, all type and grades, FOB plant, 1990 dollars.

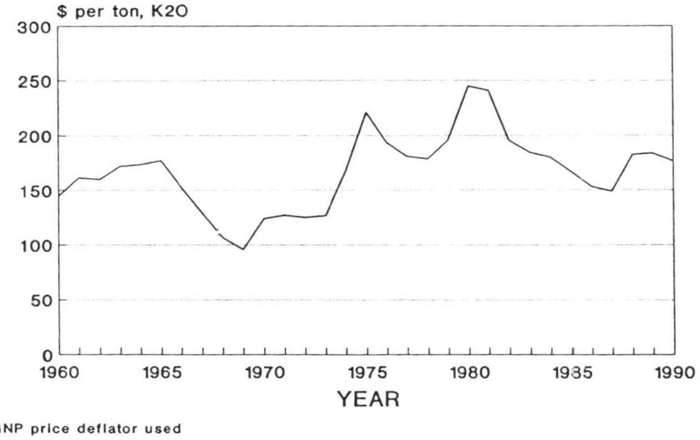

Fig. 17. Real average world annual potash price, FOB Vancouver, BC, standard grade KCl (*source*: Potash Corp. of Saskatchewan).

federal court for joining in a price setting and market allocating cartel.

World Prices: World prices are much more difficult to chart for there is no data published on producer's prices other than in the United States. Generally the world price is set between Canpotex of Canada, Zen-Noh of Japan, Sinochem of China, and MMTC of India. Canpotex is the Canadian Potash Exporting Association, an exporting association of the Saskatchewan potash producers and the largest single potash exporter in the world. Prices can be quoted for standard muriate of potash, FOB Vancouver, British Columbia, Canada. Prices around the world tend to be set by other firms or countries by summing the Vancouver price and the transportation cost from Vancouver to a port of competition, then backing out their own transportation cost to that port of competition (Fig. 17). The price at Vancouver tends to rise when the North American producer's stock of potash decreases to 1 Mt K_2O and falls when the same stocks increase to 2 Mt K_2O. Sulfate of potash prices tend to be one and one-half to two times the price of MOP, stated in dollars per ton product (not K_2O) because the largest producer of SOP (Belgium) manufactures SOP from MOP and sulfuric acid.

Transportation

The cost of railroad transportation from Saskatoon to Vancouver is approximately \$30/t. The potash tariff from Carlsbad to Houston is also about \$30/t, but private contracts for lower rates handle most of the movements. In 1990, the equal point between Saskatoon and Carlsbad for MOP was somewhere in Missouri, as evidenced by the nearly equal sales by both producing areas into that state. United States SOP and SOPM travel into Canada because Canadian potash producers do not produce a SOP. Deepwater transportation fixtures rise and fall with the overall supply-demand for all types of ocean shipping but as an example, during the summer of 1991, potash moved from Hamburg, Germany, to the west coast of India for \$39/t. Ocean shipping is much less expensive than shipping overland by rail or truck. Certain potash ocean shipments move at lower costs because they are backhauls. A backhaul is made on a ship that must return to an originating port whether or not it carries a cargo. The ship owners desire to cover some costs of returning to the home port and will charge less for the shipment while returning. There are regular shipments of grain from the Gulf of Mexico to Israel and the return shipments of potash from Israel to the eastern and southeastern coast of the United States sometimes are at lower rates. Because there is a large movement of iron ore from Venezuela and Brazil to Japan, potash from Vancouver can sometimes make a shipment back to South America at a lower rate.

Substitutes

The K^+ ion is essential to plant and animal life and neither Li^+, Na^+, Rb^+, nor Cs^+ is a substitute. The K^+ is essential to protein synthesis in plants, and it specifically activates the enzyme responsible for starch formation. In humans, K^+ ions and Na^+ ions in the nerve sheaths comprise the essential sodium pump that allows signals to travel along nerve pathways. Potash is the low priced, water-soluble form of K^+ ions available at this time. For plants, green manure is a substitute but only near farms that fatten beef and raise pigs, chicken, sheep, and dairy cattle. Green manure does not contain enough potash to transport over distances greater than about 80 km. Muriate of potash is concentrated enough to ship for a few thousands of kilometers. Greensand (glauconite) is available and is used by organic farmers. Greensand contains up to 7% K_2O but is shipped long distances because it contains some phosphorus and because it meets the requirement of organic farmers for a natural fertilizer for which they are willing to pay extra.

For industrial uses, NaOH (caustic soda) is a cheaper substitute for KOH (caustic potash). Because of its wide availability and lower cost, NaOH is used in the range of 12 Mt (100% basis) per year vs. about 250 kt (90% basis) per year for caustic potash. For the glass industry, potassium nitrate or potassium carbonate provide different optical, density, and luster qualities to the glass, making their use desirable for certain glasses. Potassium carbonate is used in chocolate and cocoa manufacture, liquid soaps, welding rods, vat dying, photographic chemicals, electroplating compounds, fire extinguishers, and rubber antioxidant applications. In small amounts it is used in many consumer endproducts.

GOVERNMENTAL CONSIDERATIONS

Environmental Regulations

The mining and production of potash, as with mining of other commodities, is not without environmental impact. Potash tailings in the Carlsbad, NM, area do not appear to pose a serious environmental problem because the land is so arid that very little salt enters the surface water or groundwater. In Canada, the groundwater around the tailings piles is monitored for contamination using a network of surrounding wells. Some mines are considering returning the tailings to the mines as a strategy to increase underground recovery factors, which would also lessen the impact of the salt on the local aquifers. The Belle Plaine solution mine has started returning waste salt to the caverns by piping salt suspensions to the caverns and letting the particles settle out. Both mines in the Canadian Province of New Brunswick must return the mill tails to the mines. One mine returns the tails as dry solids which requires another conveyor belt, and two men operating load-haul-dump vehicles to move the tailings from the end of the conveyor belt to the working area. In this mine the road header rides up on the tailings to make the next cut off the roof of the gallery. The other mine in New Brunswick returns the tailings as a slurry which is largely unattended once the pipe is placed in the old galleries. The gallery floors dip down to the center of the gallery and the brine is recovered from the lowest point in the old gallery, leaving a tailings pile hard enough in a week for the mining machines to cross. In Germany, the tailings piles are on the surface and mountainous (after many years of low grade ore production and beneficiation) and have to be monitored for ground water contamination. The French mines have dumped a portion of their waste into the Rhine River, which the Dutch have been protesting for many years. The Alsatian mines were estimated to introduce 40% of the salt load in the Rhine River.

Safety and Health Considerations

Potash mines are in sedimentary deposits, along with oil and gas and most potash mines have been discovered through oil and gas exploration. Some potash mines in the world contain high amounts of methane gas. Recovery in these mines require special precautions and expensive equipment but, as the Alsatian and Boulby mines demonstrate, they are minable with good safety records. Scaling of the roof of the mine is a constant concern in some potash mines. Often mud seams are associated with or right above potash zones; often the level of the cut is gauged by the mud seam. In the Esterhazy ore zone in Canada, the mud is present as a thin band in the middle of the highest quality ore. The mud-bearing band can only be seen by holding a miners lamp to the face of the ore and watching for the disappearance of the halo around the lamp. In other potash ore zones the mud bands are too few or too thin to be recognized during mining and are insufficient to provide additional support to the potash beds. There is little adhesion between sylvinite and clay, and roof falls can be injurious. In some of the mines in Carlsbad relief holes are drilled at intersections to bleed off any nitrogen that may lie about one meter above the potash zone. Roof bolts are installed at the intersections for support. The health effects of potash dust were studied by the University of Saskatchewan (Markham and Tan, 1981). The investigators noted that chronic phlegm production, mild shortness of breath, and chronic cough were more common in high exposure groups such as mining crews and screening operators, but severe respiratory symptoms were rare.

Problems and Future Trends

Supply: The market for potash throughout the world could be in oversupply for the next 10 to 20 years (Searls, 1991). The present price of potash is maintained at a profitable level only by the major Saskatchewan potash producer operating below capacity by 50% or more, which results in approximately 3.6 Mtpy of potash being kept off the market. In the CIS, modernization could result in higher production levels from presently underutilized capacity. Several smaller potash production areas in the United States and Europe are operating at about 80% capacity. Conversely, at least five mines have shut down in the former East Germany. Several smaller potash sources may start up in the next five to 15 years. In the United States, a mine which is located very close to a large consuming area, such as Michigan, would be able to carve out a market for itself by taking advantage of low production and transportation costs. In other areas, such as Argentina, Egypt, Israel, and Saudi Arabia, transportation costs from present potash production areas will allow a higher cost mine to start up and run a profitable operation.

Demand: The developed countries around the world have shown a general pattern of decreasing potash consumption since the 1979 oil shock. These countries are the world's largest potash consumers, and the downward trend in potash consumption is seen despite a slightly growing population. Although the populations of the poorest countries are growing at a faster rate, they may not be able to purchase potash in amounts reflecting the actual needs of the population. A growing demand for potash is likely to occur in the Pacific Basin and Latin America, in countries with a developing industrial base resulting from the increase in wealth of the population. A demand for more and higher quality foodstuffs in these countries may produce strong growth in potash demand which could cancel out the slight decline in the demand of the developed countries for potash.

Forecasting the demand curve for potash is greatly complicated by possible changes in consumption trends in Eastern Europe and the CIS, which have historically been areas of major potash consumption. Potash consumption trends in these countries will be largely determined by such factors as the availability of funds to purchase potash, an efficient system for distribution of fertilizers, and the rate of conversion to mechanized farming from the present labor-intensive farming practices now used.

BIBLIOGRAPHY AND REFERENCES

Anon., 1963, "Potash Mining in Alsace," *The Mining Journal*, Vol. 260, Mar. 29, p. 295.

Anon., 1968, "Potassium Compounds," *Encyclopedia of Chemical Technology*, Vol. 16, John Wiley & Sons, New York, pp. 381–383.

Adams, S.S., and Hite, R.J., 1983, "Potash," *Industrial Minerals and Rocks*, 5th ed., S.J. Lefond, ed., AIME, New York, pp. 1049–1077.

Alger, R.P., and Crain, E.R., 1968, "Defining Evaporite Deposits with Electrical Well Logs," *2nd Symposium on Salt*, Vol. 2, J.L. Rau, ed., Northern Ohio Geological Society, Cleveland, pp. 116–130.

Bateman, E., 1918, "Wood Ashes and Production of Potash," R1049, US Department of Agriculture, Forest Service, Forest Products Laboratory, 21 pp.

Bingham, J., 1991, "Potash Production Costs: A Discussion of Factors Affecting Production Costs," IFA-AgroChem Meeting, Moscow (USSR), Oct. 8–10, 1991, 11 pp.

Borchert, H., and Muir, R.O., 1964, *Salt Deposits: The Origin, Metamorphism and Deformation of Evaporites,* Van Nostrand, London, 338 pp.

Braitsch, O., 1971, *Salt Deposits: Their Origin and Composition*, Springer-Verlag, New York, 297 pp.

Briggs, L.I., 1957, "Quantitative Aspects of Evaporite Precipitation," *Michigan Academician*, Vol. 42, University of Michigan, Ann Arbor, MI, pp. 115–123.

Brunstrom, R.G.W., and Walmsley, P.J., 1969, "Permian Evaporites in North Sea Basin," *Bulletin*, American Association of Petroleum Geologists, Vol. 53, pp. 870–883.

Cercone, K.R., 1988, "Evaporative Sea-Level Drawdown in the Silurian Michigan Basin," *Geology*, Vol. 16, pp. 387–390.

D'Ans, J., 1933, *Die Losungsleichgewichte de System der Salze Oceanischer Salzablagerungen*, Hrsq. von Kali-Forschungs-Anstalt, Verlagsgellschaft fur Ackerbau, Berlin, 254 pp.

Davis, J.G., and Shock, D.A., 1970, "Solution Mining of Thin-Bedded Potash," *Trans. SME-AIME*, Vol. 247, New York, pp. 93–96.

Evans, R., and Linn, K.O., 1970, "Fold Relationships Within Evaporites of the Cane Creek Anticline, Utah," *3rd Symposium on Salt*, Vol. 1, J.L. Rau and L.F. Dellwig, eds., Northern Ohio Geological Society, Cleveland, pp. 286–297.

Garrett, D.E., 1970, "The Chemistry and Origin of Potash Deposits," *3rd Symposium on Salt*, Vol. 1, J.L. Rau and L.F. Dellwig, eds., Northern Ohio Geological Society, Cleveland, pp. 211–222.

Halbouty, M.T., 1967, *Salt Domes, Gulf Region—United States and Mexico*, Gulf Publishing Company, Houston, TX, 425 pp.

Hardie, L.A., 1984, "Marine or Non-Marine," *American Journal of Science*, Vol. 284, pp. 193–240.

Hardie, L.A., 1990, "The Roles of Rifting and Hydrothermal $CaCl_2$ Brines in the Origin of Potash Evaporites: An Hypothesis," *American Journal of Science*, Vol. 290, pp. 43–106.

Hardie, L.A., 1991, "On the Significance of Evaporites," *Annual Reviews of Earth and Planetary Sciences*, Vol. 19, pp. 131–168.

Harvie, C.E., et al., 1980, "Evaporation of Seawater: Calculated Mineral Sequences," *Science*, Vol. 208, pp. 498–500.

Hite, R.J., 1964, "Salines," *Mineral and Water Resources of Utah*, Report of US Geological Survey for Use of Senate Committee on Interior and Insular Affairs, 88th Congress, 2nd Session, US Government Printing Office, Washington, DC, pp. 206–215.

Hite, R.J., 1970, "Shelf Carbonate Sedimentation Controlled by Salinity in the Paradox Basin, Southeast Utah," *3rd Symposium on Salt*, Vol. 1., J.L. Rau and L.F. Dellwig, eds., Northern Ohio Geological Society, Cleveland, pp. 48–66.

Hite, R.J., 1982, "Progress Report on the Potash Deposits of the Khorat Plateau, Thailand," US Geological Survey Open-File Report 82–1096, 70 pp.

Hite, R.J., 1983, "The Sulfate Problem in Marine Evaporites," *6th International Symposium on Salt*, Vol. 1, B.C. Schreiber, ed., Alexandria, VA, pp. 217–230.

Hite, R.J., and Japakasetr, T., 1979, "Potash Deposits of the Khorat Plateau, Thailand and Laos," *Economic Geology*, Vol. 74, pp. 448–458.

Hite, R.J., and Wassef, A.S., 1983, "Potential Potash Deposits in the Gulf of Suez, Egypt," *Annals of the Geological Survey of Egypt*, Vol. 13, pp. 39–54.

Hofrichter, E., 1968, "Stratigraphy and Structure of the Palangana Salt Dome, Duval County, Texas," Special Paper 88, Geological Society of America, pp. 365–379.

Holter, M., 1972, "Geology of the Prairie Evaporite Formation of Saskatchewan, Canada," *Geology of Saline Deposits*, G. Richter-Bernberg, ed., UNESCO, Paris, pp. 183–189.

Holwerda, J.G., and Hutchinson, R.W., 1968, "Potash-Bearing Evaporites in the Danakil Area, Ethiopia," *Economic Geology*, Vol. 63, pp. 124–150.

Hsu, K.J., 1972, "Origin of Saline Giants: A Critical Review after the Discovery of the Mediterranean Evaporite," *Earth-Science Reviews*, Vol. 8, pp. 371–396.

Jackson, C.J., and Talbot, M.P.A., 1986, "External Shapes, Strain Rates, and Dynamics of Salt Structures," *Bulletin*, Geological Society of America, Vol. 97, pp. 305–323.

Jones, B.F., 1966, "Geochemical Evolution of Closed Basin Waters in the Western Great Basin," *2nd Symposium on Salt*, Vol. 1, J.L. Rau, ed., Northern Ohio Geological Society, Cleveland, pp. 217–230.

Jones, C.L., 1970, "Potash in Halitic Evaporites, Salt Range, West Pakistan," US Geological Survey, Professional Paper 700-D, pp. 140–145.

Jones, C.L., 1972, "Permian Basin Potash Deposits, Southwestern United States," *Geology of Saline Deposits*, G. Richter-Bernberg, ed., UNESCO, Paris, pp. 103–110.

Jones, P.R., and Prugger, F.F., 1982, "Underground Mining in Saskatchewan Potash," *Mining Engineering*, Vol. 34, No. 12, pp. 1677–1683.

Kopnin, V.I., 1963, "Areal Distribution of Potassium Chloride in the Sylvinites of the Upper Kama Region," *Doklady*, Akademiia Nauk SSSR, Vol. 149, pp. 172–175.

Lowenstein, T.K., 1988, "Origin of Depositional Cycles in a Permian Saline Giant: The Salado (McNutt Zone) Evaporites of New Mexico and Texas," *Bulletin*, Geological Society of America, Vol. 100, pp. 592–608.

Lowenstein, T.K., and Spencer, R.J., 1990, "Syndepositional Origin of Potash Evaporites: Petrographic and Fluid Inclusion Evidence," *American Journal of Science*, Vol. 290, pp. 1–42.

Lowenstein, T.K., Spencer, R.J., and Pengxi, Z., 1989, "Origin of Ancient Potash Evaporites: Clues from the Modern Nonmarine Qaidam Basin of Western China," *Science*, Vol. 245, pp. 1090–1092.

Maiklem, W.R., 1971, "Evaporative Drawdown—A Mechanism for Water-Level Lowering and Diagenesis in the Elk Point Basin," *Bulletin*, Canadian Petroleum Geology, Vol. 19, pp. 487–503.

Markham, J.W., and Tan, L.K.T., 1981, "Concentrations and Health Effects of Potash Dust," *Journal*, American Industrial Hygiene Association, Vol. 42, Sep., pp. 671–674.

Matthews, R.D., and Egleson, G.C., 1974, "Origin and Implications of a Mid-Basin Potash Facies in the Salina Salt of Michigan," *4th Symposium on Salt*, A.H. Coogan, ed., Northern Ohio Geological Society, Cleveland, pp. 15–34.

Miller, G.C., 1982, "Potash: A Proposed Strategy," Mineral Bulletin MR 194, Energy, Mines and Resources, Canada, 54 pp.

Neev, D., and Emery, K.C., 1967, "The Dead Sea: Depositional Processes and Environments of Evaporites," Bulletin 41, State of Israel Ministry of Development Geological Survey, 147 pp.

Ochsenius, K., 1888, "On the Formation of Rock Salt Beds and Mother Liquor Salts," *Proceedings*, Philadephia Academic Society. Part 2, pp. 181–187.

Orska, J., and Werner, Z , 1978, "The Zechstein Salt-Bearing Formation within the Leba Elevation," *Proceedings*, International Symposium on the Central European Permian, T.S. Piatkowski, J. Porkorski, and R. Wagner, eds., Jablonna, Poland, pp. 649.

Orska, J., and Werner, Z., 1990, "Rock and Potassium Salt," *Mineral Deposits*, Geology of Poland, Vol. 4, R. Osika, ed., Publishing House Wydawnictwa Geologiczne, Warsaw, pp. 197–201.

Paynter, H.M, 1990, "The First US Patent," *Invention and Technology, Fall,* American Heritage, New York, pp. 19–22.

Peirce, H.W., 1969, "Salines," *Mineral and Water Resources of Arizona*, Report of US Geological Survey for Use of Senate Committee on Interior and Insular Affairs, 90th Congress, 2nd session, US Government Printing Office, Washington, DC, pp. 417–424.

Peterson, J.A., and Hite, R.J., 1969, "Pennsylvanian Evaporite-Carbonate Cycles and their Relation to Petroleum Occurrence, Southern Rocky Mountains," *Bulletin*, American Association of Petroleum Geologists, Vol. 53, pp. 884–908.

Ramberg, H., 1981, *Gravity, Deformation and the Earth's Crust*, 2nd ed., Academic Press Inc., London, 452 pp.

Raup, O.B., 1970, "Brine Mixing: An Additional Mechanism for Formation of Basin Evaporites," *Bulletin*, American Association of Petroleum Geologists, Vol. 54, pp. 2246–2259.

Sawatzky, H.B., Agarwal, R.G., and Wilson, W., 1960, "Helium Prospects of Southwest Saskatchewan," Report No. 49, Saskatchewan Department of Mineral Resources, 26 pp.

Schmalz, R.F., 1969, "Deep-Water Evaporite Deposition: A Genetic Model," *Bulletin*, American Association of Petroleum Geologists, Vol. 53, pp. 798–823.

Searls, J.P., 1985, "Potash," *Mineral Facts and Problems, 1985*, US Bureau of Mines, Bulletin 675, pp. 617–633.

Searls, J.P., 1991, "Potash," *1991 Annual Report*, US Bureau of Mines, pp. 11–14.

Shearman, D.J., 1966, "Origin of Marine Evaporites by Diagenesis," *Transactions*, Institute of Mining and Metallurgy, Vol. 379, pp. 155–162.

Singewald, A., and Neitzel, U., 1983, "Electrostatic Separating Procedures for Raw Ore Materials," *Proceedings,* 1st International Potash Technology Conference, R.M. McKercher, ed., Pergamon, Ontario, pp. 589–95.

Sullivan, D.E., and Michael, N., 1986, "Potash Availability—Market Economy Countries," Information Circular 9084, US Bureau of Mines, 32 pp.

Szatmari, P., et al., 1979, "A Comparison of Evaporite Facies in the Late Paleozoic Amazon and Middle Cretaceous South Atlantic Salt Basins," *Economic Geology*, Vol. 74, pp. 432–447.

Talbot, C.J., and Jackson, M.P.A., 1987, "Internal Kinematics of Salt Diapirs," *Bulletin*, American Association of Petroleum Geologists, Vol. 71, pp. 1068–1093.

Tortochaux, R., 1968, "Occurrence and Structure of Evaporites in North Africa," Special Paper 88, Geological Society of America, pp. 107–138.

Trusheim, F., 1960, "Mechanism of Salt Migration in Northern Germany," *Bulletin*, American Association of Petroleum Geologists, Vol. 44, pp. 1519–1540.

Wardlaw, N.C., 1972, "Unusual Marine Evaporites with Salts of Calcium and Magnesium Chloride in Cretaceous Basins of Sergipe, Brazil,'' *Economic Geology*, Vol. 67, pp. 156–168.

Werner, Z., 1975, ''Chemical Mineral Resources in the Baltic,'' *Studia i Materialy Oceanologiczne*, Vol. 11, pp. 27–37.

Williams-Stroud, S.C., 1993, ''Evolution of Brines from Marine to Non-Marine during Deposition of Potash Evaporites in the Pennsylvanian Paradox Basin,'' *SEPM Special Publication 50.*

Zharkov, M.A., 1984, *Paleozoic Salt-Bearing Formations of the World*, Springer-Verlag, New York, 427 pp.

PUMICE AND VOLCANIC CINDER

RONALD P. GEITGEY

Pumice and volcanic cinder are volcanic rocks characterized by a cellular structure. They are formed as gases (primarily water) dissolved in molten rock, or magma, exsolve to generate a froth that cools and solidifies into rigid foam. The cells or bubbles are referred to as vesicles and they range in size from a few thousandth of a millimeter to several centimeters. Due to their vesicular character, both pumice and cinder have lower density and higher porosity than most other rock types, and, as the vesicle walls are broken, sharp cutting edges are continually generated. These properties are the basis for their commercial value as lightweight aggregates, insulators, absorbents, and abrasives.

There is a certain imprecision in the terminology applied to volcanic products, reflecting to varying degrees the diversity of volcanic rocks, the relative youth of volcanology as a distinct field of study, and customs and practices of commercial trade. In general terms, pumice is light colored (white, gray, pink, pale yellow, or brown) and highly vesicular, usually with vesicle walls that are visibly glassy. Many pumice fragments are light enough to float on water. Pumice is readily cut by steel tools and some can be crushed by hand. Individual pumice fragments may range from 1 or 2 m in one dimension to <1 mm. Commercially, finer fragments are called pumicite or volcanic ash. In volcanology the term ash refers to any fragment <2 mm in size regardless of its composition. Cinder, or scoria, is dark colored (black, red, brown) with thick vesicle walls that appear dull or stony and that may exhibit iridescence. Cinder typically is heavier than pumice and has a higher crushing strength. In volcanology the term scoria refers to dark vesicular material throughout a wide size range while it is common commercial practice to use scoria for larger fragments (usually greater than 2.5 cm) and cinder for smaller fragments.

GEOLOGY

Composition and Physical Characteristics

Most pumice is silicic ranging from 60 to 70% SiO_2, that is, dacitic to rhyolitic in composition. Pumice of less silicic composition, including basaltic, occurs but it is less common and does not have extensive commercial use. Scoria and cinder are typically basaltic to andesitic with approximately 50 to 60% SiO_2.

Pumice and cinder may contain phenocrysts of feldspars and various ferromagnesian minerals that crystallized in the magma prior to eruption. Fragments of rock through which the magma has passed may be entrained in the melt and wall rock may be fragmented and admixed during an explosive eruption. Some scoria particles have lithic fragment cores. Phenocrysts and lithic fragments are relatively unimportant in cinder end uses but their presence may be detrimental in certain pumice products.

Pumice typically has vesicles <1 mm in size separated by thin walls. Vesicle shapes include irregular, spherical, elliptical, and elongate to the point of being tubular with a silky appearance (Fig. 1). In some pumices, the vesicles are interconnected making the fragments permeable and highly absorbent; in others the vesicles are isolated forming a highly porous but very impermeable pumice. Floating masses of pumice fragments from the eruption of Krakatoa in 1883 were reported in the Indian Ocean for up to two years after the eruption. Vesicles in cinder are larger and range from spherical to highly irregular in shape with much thicker walls and abundant interconnections. Scoria and cinder are typically heavier, porous, and do not float.

Density and hardness of pumice must be expressed precisely to avoid confusion. Density may refer to the glass itself, the apparent density of the vesicular pumice particle, or the bulk density of pumice in a deposit or in a pumice product. Pumice glass has a specific gravity of 2.5 or more depending on its composition. Pumice fragments typically have a specific gravity of <1.0 and so will float on water, at least until enough vesicles are saturated to cause them to sink. Expressed as density, pumice fragments typically weigh less than 1 g/cm^3. Depending on moisture content, particle density, and particle size distribution, the bulk density of pumice typically ranges from 500 to 700 kg/m^3. Typical bulk densities of cinder range from 700 to 900 kg/m^3 for material used as lightweight aggregate while the bulk density of cinder used for highways and railroad ballast may be considerably higher.

Hardness in pumice may refer to the glass vesicle walls, possibly including crystals or lithic fragments, or to the apparent hardness of the particle as a whole, which is more a measure of the strength of the vesicle structure. Pumice glass typically has a Mohs hardness of 5 to 5.5. A pumice particle may have a much lower apparent hardness and be easily cut with a knife or steel saw because the vesicle walls break readily.

Pumice and cinder are primarily pyroclastic deposits formed as fragmental products of volcanic eruptions. Several classification systems differing in various details have been proposed for pyroclasts and pyroclastic deposits based on particle size (Fisher, 1961, Schmid, 1981, Cas and Wright, 1988). Generally, particles >64 mm are referred to as blocks or bombs depending on their shape. Particles between 64 mm and 2 mm (4 mm in some schemes) are called lapilli and smaller particles are called ash. These names may be modified by compositional terms, for example pumice lapilli or scoria lapilli. Most commercial deposits are composed of fragments in the lapilli size range. In the United States, pumice fragments having one dimension of 5 cm or more are legally defined as block pumice. The procedures for acquiring pumice from federal land are determined in part by this definition and are described in a later section.

Origin

The science of volcanology has grown rapidly, particularly in the last 20 years. Both field and laboratory studies have greatly

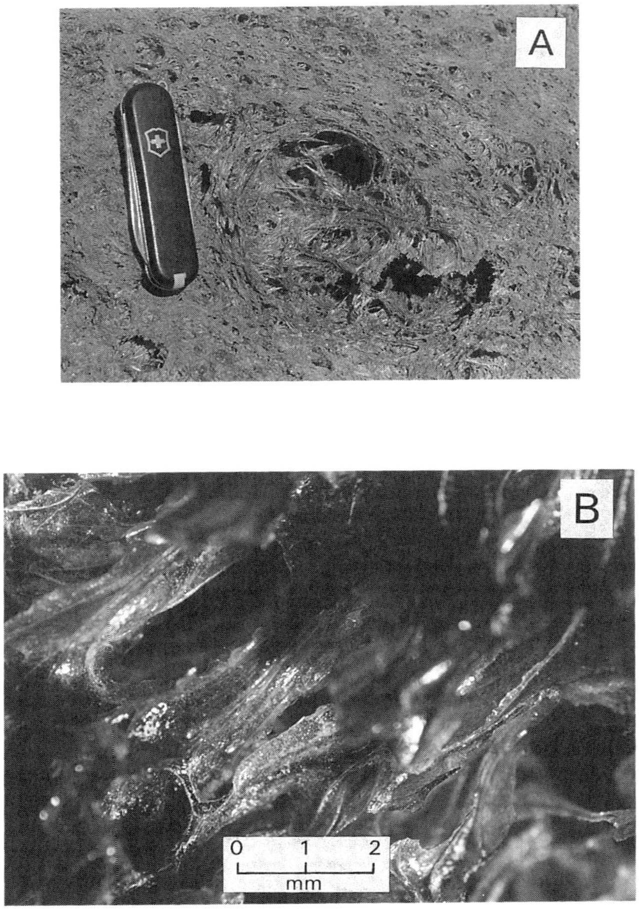

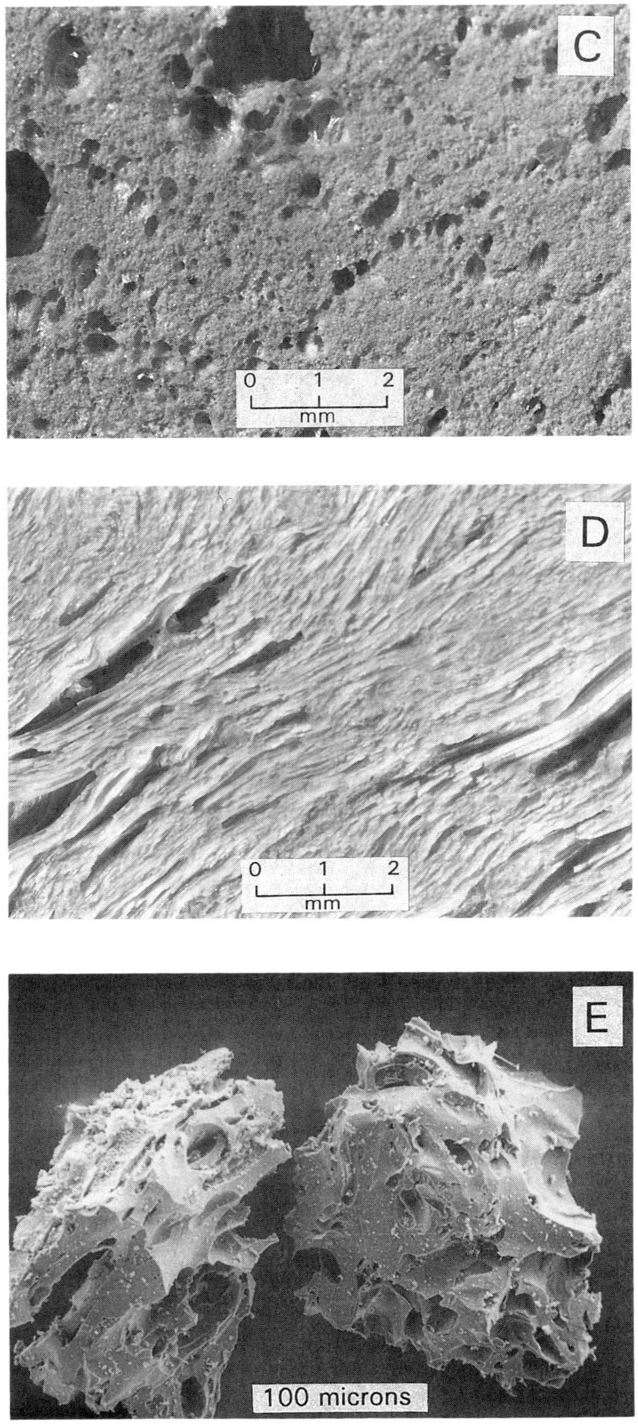

Fig. 1. Variations in vesicle shape and size. A) Large vesicle cavity in the Mount Mazama (Crater Lake) climactic pumice, Oregon. B) Coarsely vesicular pumice from Mono Craters, California, marketed as Featherock. C) Bend pumice, Bend, OR, used for lightweight aggregate. D) Highly elongated vesicles from air-fall deposit near Burns, OR. Individual strands are tubular vesicles. E) Finely vesicular pumice from Hess Pumice Products pit near Malad City, ID. Even small particles are vesicular and friable making the material suitable for fine abrasives and polishes.

expanded the understanding of eruptive rocks and eruptive processes. The reader is referred to Cas and Wright (1988) and their extensive list of references for details of this recent work.

The formation and preservation of highly vesicular rocks involves the interplay of a large number of variables including temperature, viscosity, gas pressure, and diffusion rates within the erupting magma and external conditions such as wall rock permeability, water influx, vent blockage, wind conditions, and whether the eruption is subaerial or subaqueous. None of these is constant and may change significantly even during a single eruptive event.

Basaltic magmas tend to be more fluid than more silicic types and, as they near the surface, gases may escape quickly without generating a stable vesicular structure. Even after vesicle formation, scoria particles may be fluid enough to collapse on impact or to adhere to each other forming a dense mass rather than a friable deposit.

Silicic magmas are less fluid and offer less opportunity for gas escape and collapse of vesicle walls. Conditions that reduce gas escape, such as high magma viscosity, impermeable wall rock, or a blocked vent, may lead to eruptions explosive enough to shatter the vesicle walls generating a fine ash rather than a vesicular particle. Pumice particles entrained in a pyroclastic flow may be reheated enough to soften and collapse into nonvesicular glass.

The unconsolidated nature and low density of pumice and scoria deposits make them readily susceptible to erosion, and the extremely large surface area of their vesicular particles promotes rapid chemical weathering. As a result, most deposits of pumice and scoria are quite young geologically, rarely more than a few million years old, and often only a few thousand years old.

Types of Pumice Deposits

Pumice deposits can be broadly classified into four major types: flows and domes, air-fall deposits, pyroclastic flows, and epiclastic or reworked deposits. There are gradations between types and deposits may be a mixture of types reflecting variations in eruption conditions. Changes in wind direction, blocking and clearing of the vent, increase in vent diameter, influx of water into the magma chamber, influx of a different magma, and rapid gas exsolution by unloading and depressurization may contribute to the character of the resulting ash and pumice deposits.

Flows and Domes: Viscous magmas may extrude with little explosive activity forming lava flows and domal mounds. A vesicular rind, or carapace, can develop on their rapidly cooled outer surfaces, often forming a surface of blocky pumice rubble (Fig. 2) underlain by obsidian or by nonvesicular lava. Continued cooling and crystallization of the flow increases the volatile content of the remaining liquid which can then cause more vesiculation within the flow, often interlayered with nonvesicular rock. Buoyant masses of this pumice can forcibly pierce the overlying flow to the surface as pumice diapirs. Increased volatile content and resultant increased internal gas pressures can also form explosive craters on the surface of the flow and generate pyroclastic flows long after eruption has ceased. Pumice flows and domes typically are only a few square kilometers in areal extent. Rock Mesa, a pumice flow in the central Cascade Range of Oregon, is about 6.5 km^2 (Fink and Manley, 1987) and Mono Craters, a multiple dome and flow complex in east central California, covers about 30 km^2 (Chesterman, 1956). Pumices associated with domes and lava flows often have larger vesicles than other pumice types and may form blocks up to about 1 m in dimension. The erratic nature of vesiculation can make exploration and development difficult.

Fig. 2. Undisturbed surface of a pumice flow with rubble blocks up to 1.2 m in size on Rock Mesa, in the Cascade Mountains of Oregon. The flow lies within a wilderness area and is unavailable for commercial development.

Air-fall Deposits: Explosive eruptions eject fragments of dense magma, vesiculated magma (pumice), and country rock in various proportions and with various velocities and degrees of fragmentation. These fragments, and in many cases the deposits formed by them, are broadly referred to as pyroclastic.

In a Plinian-type eruption, named after Pliny the Younger's description of the 79 AD eruption of Mount Vesuvius, pyroclastic material is explosively ejected upward from the volcanic vent. The pyroclastics are boosted into the atmosphere by the explosion and maintained in suspension by convection giving the appearance of a vertical column that flattens and spreads out at its top. Pliny the Younger compared its shape to that of a pine tree with a tall trunk (Bullard, 1976). Based on grain size and dispersal characteristics, Walker (1980) has concluded that some Plinian columns, or eruptive columns, may exceed 50 km in height.

Dense or large fragments leave the vent ballistically and fall at distances dependent on their exit velocity. Smaller or less dense particles such as pumice lapilli and ash may be removed from the Plinian column by winds and cooled and deposited over wide areas depending on column height, wind velocity and direction, and the particle size and density.

Air-fall deposits typically decrease in thickness and particle size with increasing distance from the vent and commonly contain pumice particles ranging in size up to several centimeters. Air-fall deposits may blanket the pre-existing topography over very large areas. Thicknesses may range from tens of meters near the vent to only millimeters several hundred kilometers away from the vent. Most commercial pumice operations utilize air-fall deposits.

Pyroclastic Flows: Decreases in eruption activity or overloading by continued eruption can cause the eruptive column to collapse either continuously or sporadically. The hot pyroclastic material falls from the column and flows outward from the vent following topography and may flow over areas of tens or hundreds of square kilometers. Such pyroclastic flows can retain enough heat to fuse or weld the particles together after movement stops. The names applied to these rocks historically have included tuffs, welded tuffs, ash flows, and ignimbrites. Pumice fragments may survive intact on upper and lower surfaces which cool before welding occurs but pumice fragments in the flow interior soften and collapse. Pumice deposits associated with pyroclastic flows tend to be thinner than air-fall deposits and since unwelded portions are either easily eroded flow tops or are overlain by welded portions, pyroclastic flow pumices are less frequently exploited commercially than air-fall pumice deposits.

Epiclastic Deposits: Epiclastic processes include erosion, transportation, and redeposition by such mechanisms as water, wind, and mass movement. From a commercial standpoint these processes can either degrade or enhance a pumice deposit. All could reduce pumice particle size and possibly intermix non-pumice material, but wind and water can also sort by size and density and beneficiate the deposit by separating pumice from crystals and rock fragments. Minable thicknesses of pumicite were produced in the central United States when surface drainage washed thin blankets of airborne ash into lake basins.

Scoria and Cinder Deposits

Scoriaceous material can be associated with many eruptive styles but those deposits exploited commercially are nearly all cinder cones. Cinder cones are unconsolidated piles of scoria formed by subaerial eruptions. Fresh cinder cones are typically conical, steep-sided, and often with a central crater. They are rather small volcanic landforms with diameters ranging from 100 m to

Table 1. Uses of Pumice

Use	Product form and processing	Essential properties
lightweight aggregate		
decorative and structural concrete blocks; cast concrete; lightweight structural members, wall panels, floor decking; stucco and plaster mixes; pozzolan in cement; civil engineering, lightweight fill	granular crushing, screening, blending	low density, good crushing strength, thermal insulator; acoustical insulator, fire resistance; moisture resistance
abrasives		
grill cleaners; scouring sticks for porcelain, tile, swimming pools; buffing wheel cleaners; cosmetic skin removal	blocks sawing irregular lumps as mined	broken vesicle (bubble) walls form sharp edged particles; wear continues to generate fresh cutting edges
stonewashing: water, pumice, and garments tumbled together in laundry machine, pumice must float, abrades and softens textile fibers.	coarse granular, ± 1.9 cm crushing, screening	broken vesicle (bubble) walls form sharp edged particles; wear continues to generate fresh cutting edges
hand soaps; scouring compounds; rubber erasers; polishing compounds for glass, metal, plastics; dental cleaners; wood finishing; nonskid paints; cleaning printed circuit boards; tumble polishing; leather finishing; matches and striking surfaces	granular drying, milling, screening, air flotation, blending	broken vesicle (bubble) walls form sharp edged particles; wear continues to generate fresh cutting edges
absorbents		
potting soils, hydroponic media, pet litter, floor sweep, turf aeration	granular crushing, screening	high porosity, large surface area, low chemical reactivity
acid washing: impregnated with bleaching agents, tumbled dry with garments, requires high absorption rate; gas "charcoal" grills, absorbs fat and grease drippings	coarse granular, ± 1.9 cm crushing, screening	
catalyst carriers; carriers for pesticide, herbicides, fungicides	granular drying, crushing, milling, screening, blending	high porosity, large surface area, low chemical reactivity
architectural		
loose fill insulation; roofing granules; textured coatings; ground cover	granular crushing, screening	low density, thermal insulator, acoustical insulator, fire resistance, moisture resistance
landscaping; decorative interior and exterior veneer	blocks boulders as mined; sawn slabs	low density, easily shaped, low maintenance
fillers		
in rubber, paints and plastics; mold release compounds; hot asphalt mixes; brake linings	granular crushing, drying, milling, screening, blending	particle shape, cost
filter media		
both expanded and unexpanded forms used to filter animal, vegetable, and mineral oils	granular crushing, drying, milling, screening, firing, air flotation	particle shape, expandability

about 2.5 km and heights from 30 to 500 m with slopes to about 30° (Wood, 1980a, 1980b, Osburn, 1982). Heights and slope angles are degraded with age and erosion and composite cones of cinder and lava flows may have lower angles than cones of only cinders.

Cinder cones formed from a single vent can be nearly circular but elongate forms may develop under strong prevailing wind conditions or from fissure eruptions or multiple vent eruptions. Variations in color from red to black are often concentric around the vent. Cinders are black initially but those deposited near the vent are subject to heating after deposition causing oxidation of iron to shades of brown and red.

USES AND TECHNOLOGY OF PUMICE AND CINDER

Pumice and cinder at first appear to be interchangeable in many applications, but in fact each has distinct qualities that preclude direct substitution without some change in performance. Both are used extensively as lightweight aggregates in concrete products but pumice produces lighter weight and lighter color while cinder produces greater strength. Pumice is used in abrasives ranging from screened material for stonewashing to highly processed abrasives for polishing optical glass. The abrasive action of cinder is much too harsh for such applications. Cinder is used for road and highway aggregate and surfacing but pumice often lacks sufficient mechanical strength to be similarly useful. Both are used as landscaping material but with obvious differences in color, weight, and texture.

Pumice

The uses, processing and essential properties of pumice are summarized in Table 1. Lightweight aggregate in concrete is by far the largest volume end use in the United States, followed by stonewashing and abrasives (Anon., 1992). Due to the method used to collect production data the lesser volume uses cannot be accurately ranked.

Lightweight Aggregate: Pumice is used as a lightweight aggregate in both cast concrete and concrete block units. Lightweight concrete reduces the total weight of the structure and reduces

Fig. 3. Structural concrete block unit, 0.2 × 0.2 × 0.4 m. Such a block with a 50/50 mix of pumice aggregate and sand and gravel aggregate weighs about 13 kg compared to about 17 kg for the same block with only sand and gravel aggregate.

Fig. 4. An installation of pumice aggregate concrete blocks used as both structural and decorative elements. The smaller blocks are 0.1 × 0.4 m and molded with smooth faces. The larger blocks are 0.2 × 0.4 m with split faces formed by breaking apart a double block unit before the concrete has completely cured.

bearing strength requirements of the supporting members while providing thermal and acoustical insulating qualities. Similar advantages are realized in plaster and stucco applications.

Decorative and structural concrete block units with pumice aggregate are easily handled reducing construction time and worker fatigue. A typical 0.2 × 0.2 × 0.4 m structural block with sand and gravel aggregate weighs about 17 kg. Commonly a mixture of the two aggregate types is used to produce a block that retains the advantages of lightweight aggregate with an increased strength while sacrificing some of the weight advantage. A block unit with a 50/50 mix of pumice aggregate and sand and gravel aggregate weighs about 13 kg (Fig. 3).

The light color of pumice is also advantageous in producing colored decorative blocks using mineral pigments. Blocks produced with broken surfaces, or split faces, exhibit a color contrast between the pumice particles and the pigmented cement matrix (Fig. 4).

Pumice for lightweight aggregate is typically sold in sizes from 8 mm downward. Grading requirements (particle size distribution) for concrete masonry units are given in American Society for Testing and Materials (ASTM) Standard C 331, but these may be waived for special applications where particular texture, strength, weight, acoustical, or thermal properties are required. Grading is critical to the shrinkage characteristics of masonry units and it is controlled in processing by crushing and dry screening.

ASTM Standard C 331 also specifies unit weights of oven-dry material from 880 kg/m^3 for fine aggregate to 1 120 kg/m^3 for coarse aggregate. A limited amount of lithic fragments can be tolerated because their principal effect is to increase weight, but obsidian fragments are deleterious because they may hydrate, expand, and weaken the concrete. Organic debris and particle coatings of clay, iron oxide, or fine ash are also detrimental.

Suitable crushing strength of pumice particles is often tested qualitatively by hand or under foot while quantitative measurements are made on masonry units. Ranges of compressive strengths and various other properties of concrete with pumice aggregate are reported by Schmidt (1956).

Pumice has also been used as a lightweight aggregate to mitigate highway damage from landsliding. Removing a portion of the slide mass and backfilling with pumice to road level helped stabilize the slope by reducing loading and promoting drainage (Valdez, 1992, private communication).

Pumicite and finely ground pumice are also added to concrete mixtures as pozzolanic material, an application distinct from lightweight aggregate. Pozzolans form cementitious compounds by reacting with calcium hydroxide that is formed as portland cement hardens. Without the addition of pozzolanic material, calcium hydroxide can be leached from the concrete causing it to deteriorate. The US Bureau of Mines (Anon., 1992) reports production of large volumes of pumice for pozzolan from Europe and from Latin America.

Abrasives: Pumice abrasive products include sawn and shaped blocks, granules and powders, and bonded and reconstituted forms. The glassy vesicle walls are brittle, exhibit conchoidal fracture forming thin, sharp cutting edges, and have a Mohs hardness of 5 to 5.5. Fresh cutting edges are continuously generated as the friable pumice particles wear down during use.

Sawn and shaped blocks and irregular lumps of pumice are used in applications requiring hand held abrasives for scouring various surfaces. Sawn blocks are produced by the United States Pumice Corporation from their block pumice deposits on Mono Craters near Lee Vining, CA. Large lumps and blocks of uniformly vesicular pumice, free of lithic fragments, crystals, and obsidian stringers, are sawn into rectangular blocks of various sizes. The 7.6 × 7.6 × 15.2 cm blocks are sold under the brand name of Grillmaster for cleaning restaurant grills (Fig. 5). The pumice gently scours the grill while neither scratching the metal surface nor removing the thin seasoned layer that prevents food from adhering. Smaller blocks are sold for home use to clean porcelain and tile surfaces and for cosmetic skin removal. Cosmetic blocks hand shaped to a mouse-like form are manufactured in Europe but their high unit cost has generated a competitive product formed by compressing and baking granular pumice (McMichael, 1990). Pumice blocks and lumps are still used to clean buffing wheels but demand has dropped with the decreasing use of chrome plating by the automotive industry.

The most rapidly growing segment of the pumice industry is stonewashing or laundry use. Two distinct processes are used: abrasion, in which coarse granular pumice abrades the garment surface, and absorption, in which pumice acts as a carrier for bleaching chemicals. The term stonewashing is applied both to laundry use in general and specifically to the abrasive process in

Fig. 5. Scouring block produced by United States Pumice from its Mono Craters deposit near Lee Vining, CA. With an aluminum handle attached the block is used to clean restaurant grills. Smaller blocks are also produced for other commercial and household scouring applications.

which finished garments, water, and pumice are tumbled together in large washing machines. The term acid washing refers specifically to the treatment process using pumice as a chemical carrier. Both processes utilize pumice particles from 2 cm to over 8 cm in size. Smaller particles produce a more uniformly worn effect while larger particles impart a more irregular effect (Hoffer, 1991). Each process requires pumice of different characteristics and acid washing is discussed in the section on absorbents.

Stonewashing, the abrasive process, uses pumice particles with a low enough density and a low enough permeability to float in the laundry machine. The pumice must be hard enough to survive at least one washing cycle, but not so hard that it unduly damages the cloth fibers. Clean, light colored pumice with few lithic fragments is preferred. The industry is young and highly competitive and there are no standardized specifications. Often a new source of pumice is tested simply by using it, a potentially wasteful approach. Testing procedures to evaluate pumice for laundry use have been developed by Hoffer (1991; in preparation) that measure particle density, absorption, surface fines, impregnation rate, and abrasion loss. Particle density and abrasion loss are the most important properties for stonewashing pumice.

The popularity of stonewashed garments has led to the development of many new pumice deposits, particularly in Turkey, Indonesia, and several Latin American countries. Traditional producers of aggregate pumice supply some laundry pumice but many were unable to expand quickly enough to meet demand, some do not have appropriate particle sizes or other characteristics, and some are reluctant to make extensive modifications to their current operation and product mix to serve what may be an ephemeral market driven by clothing fashions.

Pumicite and ground pumice are used in a large number of abrasive products as loose granules and compounded in such products as hand soaps, rubber erasers, and polishing compounds for metal, plastics, and glass. Particle sizes range from a few millimeters for non-skid coatings to less than 100 μm for fine polishing compounds. Consumer perception determines the particle size in some applications. Pumice is used in bar soaps, soft soaps, and liquid soaps but products marketed for heavy duty hand cleaning are more acceptable if they contain coarser pumice giving a rougher texture in use. There may be differences between producers and particle size distributions of respective grades but abrasive pumice is still described by the system reported by Hatmaker (1932). Letters F through FFFF are used for progressively finer particles from about 150 μm downward and whole and fractional numbers from 0 through 10 for progressively coarser particles up to about 5 mm.

Uniformly finely vesicular pumice is required for most granular abrasives. Typically the vesicle diameters are several times smaller than the finished product particle size (Fig. 1E). Crystal, obsidian, and lithic fragment content must be very low to produce an abrasive with uniform hardness and wear characteristics. For example, pumice is used as an intermediate abrasive in polishing television tubes, optical components, and certain types of flat glass. It is used following grinding by coarser abrasives and is followed by final polishing with metal oxides. In a recirculating slurry system, any harder or tougher particles will accumulate to intolerable levels and produce scratches. Less than 1% non-pumice particles is required and less than 0.5% is preferred (Hess, private communication, 1992). Beneficiation to remove crystals, obsidian, and lithics is difficult, especially in fine particles, and few deposits have such a low non-pumice content.

Processing pumice and pumicite for abrasives includes drying, crushing, milling, air flotation, screening, and blending. Sizing is often done with wire or cloth screens and equipment designed for flour mills. Screening can be a difficult and time-consuming operation because the density, shape and surface texture of pumice particles often causes clogging, or blinding, of the screens. Product particle size distribution is further controlled by blending. Abrasive pumice prices of up to 100 times aggregate-pumice prices reflect the rarity of suitable deposits, the higher capital and operating costs of processing plants, and the higher product tolerances required for pumice abrasives.

Absorbents: The high porosity, large surface area, and low chemical reactivity of pumice, make it suitable for many absorbent applications ranging from floor sweep products to catalyst carriers. Processing ranges from simple screening to the same sophisticated techniques used to produce fine pumice abrasives.

Coarse granular pumice, +1.9 cm in size, is used to acid wash garments. The pumice is impregnated with bleaching chemicals by various immersion and vacuum techniques and then tumbled dry with finished garments. The chemicals are released as the pumice contacts the fabric. Unlike pumice for abrasive stonewashing, acid washing pumice does not have to float and it must have sufficient porosity and permeability to absorb and release the chemicals. Surface coatings of fine ash or clays reduce the absorption rate and the clays release the chemicals at a more rapid rate than pumice causing uneven bleaching. Surface fines should be less than 5% by weight (Hoffer, 1991). Iron oxides are also detrimental because they can be deposited on the fabric producing a yellow or brown stain.

Lumps of pumice about 5 cm in diameter are used in gas grills to absorb grease drippings and reduce flaming. The pumice is marketed both as the manufacturer's original equipment in the grill and as a replacement product through retail outlets. These markets are shared with scoria.

Finer granular pumice is used in potting soils and as a hydroponic growth medium. Both the absorption and abrasive non-skid characteristics of pumice are valuable in its use as a floor sweep material on machine shop or meat-packing plant floors. Finely ground pumice is also used as a catalyst support and carrier and it is compounded with pesticides, herbicides, and fungicides to permit uniform application of small amounts of chemicals.

Architectural Applications: Pumice has architectural applications in addition to its use as lightweight concrete aggregate. Its low density, fire and moisture resistance, and thermal and acoustical insulating qualities make it useful in textured plasters and coatings, as loose-fill insulation, and as roofing granules. The high reflectivity of white pumice is also advantageous in roofing applications.

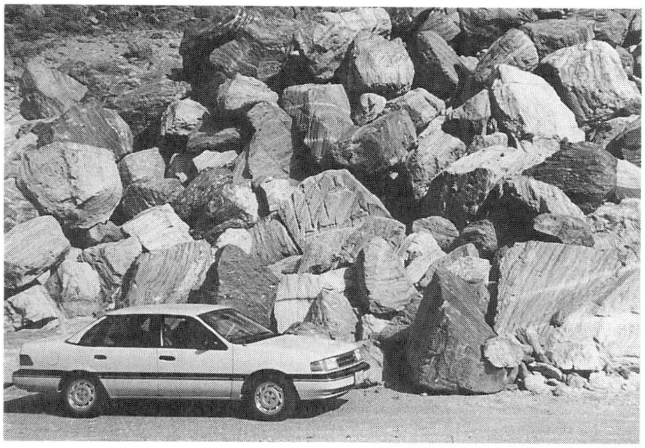

Fig. 6. Pumice landscaping boulders quarried from the Mono Craters deposit near Lee Vining, CA, by Featherock.

Fig. 7. Pumice veneer slabbed and palletized for delivery, Lee Vining, CA.

Coarse granules and pebbles are used as ground cover to reduce the water otherwise required to maintain plant ground covers and pumice is distinctively colored and easily handled.

Large blocks and boulders of pumice are used for decorative landscaping. They are easily handled and can be readily fitted together or sculpted to specific shapes such as planters and lanterns. Pumice can be bored and hollowed for the installation of electrical and plumbing systems for fountains and waterfalls (Fig. 6).

Pumice blocks are sawn on one surface and used as decorative veneer on both interior and exterior walls producing a highly textured fire resistant surface. The slabs can be shaped easily with hand tools for rapid and precise installation (Fig. 7).

Density, pattern, and color are important for both landscaping and veneer products but the most specific requirement is large block size. Few pumice deposits, primarily flows and domes, contain blocks a meter or more in size and only one is currently in production in North America by Featherock, at Lee Vining, CA.

Fillers: Mineral fillers are rock and mineral materials added to formulations to give products specific characteristics such as tensile strength in plastics, gloss (or lack of gloss) in paints, or weight and surface finish in paper. Typically fillers are added as very fine powders and many factors affect the choice of materials used.

The important attributes of pumice and pumicite as fillers include cost, hardness, particle shape, particle size, absorbency, vesicularity, density, and low chemical reactivity. Permeable, vesicular, pumice particles are absorbent and can be used to vary viscosity, impermeable vesicular pumice can be used to decrease the density of the product, and platey vesicle wall fragments can increase tensile strength. The hardness of pumice can increase the abrasion resistance of the finished product but it also increases wear on processing equipment. Very finely vesicular pumicite from Kern County, California, has been tested successfully as an opacifying agent to partially replace titanium dioxide in paints (Simonson, private communication, 1992). Grading or particle size distribution is critical to the performance of asphalt and pumicite can be used to provide the proper ratio of fines in asphalt mixes.

Filter Media: Liquids, ranging from water, juices, and beverages to oils and fuels are clarified by filtration by mechanically removing suspended particles. Filters may be beds of granular material (filter media) or granular material supported by porous membranes. Filter media may also be mixed into the liquid to remove more efficiently particles and to maintain porosity of the filter. Finer filter-media particles are required to remove finer particles from the liquid.

Pumice is useful as a filter aid because of its fine particle size, particle shape, low chemical reactivity, and in some applications, its expandibility by pyroprocessing. Pumicite in Kansas and adjoining states is composed of broken vesicle walls, which are used to form a filter bed of overlapping plates to remove fine particulates (Fig. 8A). Some of the pumicites can be fired and expanded into hollow spheres similar to those formed by expanding perlite (Fig. 8B). The expanded pumicite is then milled to produce filter media particles of specific sizes and flow characteristics.

Volcanic Cinder

The uses of volcanic cinder and pumice overlap in some applications but cinder is not processed into as wide a variety of products as pumice Volcanic cinder usually has thicker vesicle walls than pumice and produces a heavier but stronger lightweight concrete with insulating qualities less than those of pumice concrete of comparable strength and weight (Schmidt, 1956). Block units have a gray or red color depending on the cinder color.

Cinder and scoria are also used in the same construction applications as sand and gravel and crushed rock, including fill, base courses, road surfacing, asphalt aggregate, and railroad ballast. Cinder for these purposes must meet the same specifications as any other aggregate material including specifications such as abrasion resistance, immersion disintegration, and aggregate degradation. Cinders that qualify have a higher density than those used for lightweight concrete. The friable, fragmental nature of cinder deposits results in much lower production costs than from non-vesicular rock that must be crushed.

Cinders are also used to sand highways to improve traction on icy surfaces. Cinders are typically crushed and screened to about −16+4 mm. Coarser particles can be damaging to vehicle windshields and finer particles can retain moisture and freeze in stockpiles or in transit (Iverson, private communication, 1992).

Cinders and scoria are used in landscaping as both placed boulders and coarse granular ground cover. Scoria blocks are also used as building stones. Uniformity of color is of importance in landscaping and decorative uses.

Scoria lapilli is used as an absorbent bed in gas cooking grills.

DISTRIBUTION OF DEPOSITS

The ephemeral nature of pumice and cinder deposits limits their occurrence to areas of geologically recent volcanism such as the

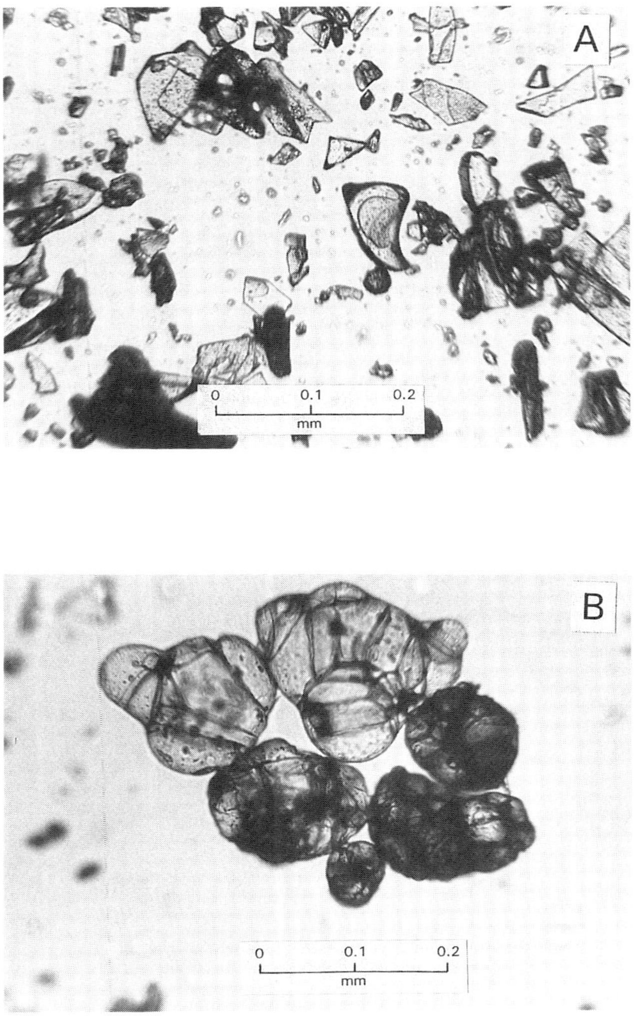

Fig. 8. Pumicite from deposit near Mankato, KS. A) Pit run material composed of vesicle wall fragments. B) Pumicite expanded by pyroprocessing for use in filtering, Kansas Minerals, Mankato, KS.

Table 2. Sources of Information on Pumice and Pumicite Deposits

World	
	Anon., 1977
	Harben and Bates, 1984
	McMichael, 1990
	Robbins, 1984
	US Bureau of Mines Yearbook, published annually
United States	
Arizona	Hoffer, 1991
California	Chesterman, 1956
Colorado	Williamson and Burgin, 1960
Idaho	Asher, 1965
Kansas	Bauleke, 1962,
	Hardy et al., 1965
Nebraska	Burchett, 1990
Nevada	Horton, 1964
New Mexico	Clippinger and Gay, 1947
	Hoffer, in preparation
Oklahoma	Burwell, 1949
Oregon	Moore, 1937
	Geitgey, 1992
Utah	Van Horn, 1964
Washington	Carithers, 1946
Wyoming	Harris and King, 1986

Table 3. Pumice and Pumicite Mines Active in 1991, Based on US Bureau of Mines Data

	Deposit Location	
Company	County	State
Arizona Tufflite, Inc.	Coconino	Arizona
California Industrial Minerals Co.	Madera	California
Featherock Inc. and US Pumice Co.	Mono	California
Glass Mountain Pumice Inc.	Siskiyou	California
Hess Pumice Products	Oneida	Idaho
Producers Pumice	Bonneville	Idaho
Calvert Corporation	Norton	Kansas
Kansas Minerals Inc.	Jewell	Kansas
Copar Pumice Co.	Santa Fe	New Mexico
General Pumice Corp.	Rio Arriba	New Mexico
Cascade Pumice Corp.	Deschutes	Oregon
Central Oregon Pumice Co.	Deschutes	Oregon

Circum-Pacific belt, the Mediterranean region, and oceanic ridges. Sources of information on pumice and pumicite are listed in Table 2. Most of those tabulated have extensive lists of references to specific deposits and to more detailed descriptions.

Pumice lapilli deposits in the United States are largely confined to the area from the Rocky Mountains westward while airborn ash from this area has formed pumicite deposits throughout the Great Plains states. Pumice has been produced in the past for various purposes from nearly every western state including Hawaii but currently mines in only six are active. Pumice mines active in 1991 are listed in Table 3.

The principal pumice deposits in Arizona are in the San Francisco volcanic field near Flagstaff in Coconino County in the north-central part of the state. Several air-fall pumice lapilli and ash units occur in the area that have been mined for lightweight aggregate, horticultural, and laundry use. In several other counties, epiclastic deposits of pumice and pumicite have been mined for lightweight aggregate and pozzolonic material (Hoffer, 1991).

Chesterman (1956) listed and described over 60 deposits of pumice and pumicite in California. All types of occurrences are represented including flows, domes, air-fall, pyroclastic flow, and epiclastic lacustrine and fluvial deposits. Since the first recorded production in 1883, numerous companies have been active throughout the state producing pumice and pumicite for nearly every end use. Production from the three currently active areas illustrates this wide range of markets.

Pumice in the Glass Mountain area of Siskiyou County in northern California occurs as air-fall deposits and block pumice flows. Glass Mountain Pumice produces stonewashing pumice and lightweight aggregate from an air-fall deposit. Pumice for scouring blocks has been mined from Glass Mountain in the past but the block pumice deposits are not being mined at present.

The Mono Craters deposit near Lee Vining in Mono County is a multiple dome and flow complex. The surface of the domes is a jumbled mass of pumice rubble at least 45 m thick through which spires and pinnacles of pumice protrude (Loney, 1968). The flow surfaces are covered with loose angular blocks and exhibit complex movement pattern and the flows terminate in steep talus slopes. Blocks of up to 2 m are mined by Featherock for landscaping and decorative veneers. The company produces large blocks of coarsely vesicular pumice in a variety of colors and textures (Figs. 1B and 6). Smaller blocks are sawed into thin slabs for decorative veneer (Fig. 7). United States Pumice, a wholly owned subsidiary of Featherock, utilizes pumice of finer texture from the same deposit to produce abrasive blocks of several sizes for cleaning grills and ceramic surfaces, and for cosmetic purposes. Small trimmings are sold for use in gas cooking grills so there is very little waste in the processing operation.

Water-laid pumicite in Madera County, central California, is mined by California Industrial Minerals, and processed for filler applications. Most of the pumicite particles in the deposit are −325 mesh (Chesterman, 1956).

Pumice and cinder occurrences in Idaho are associated with volcanics of the Snake River region in the southern part of Idaho (Asher, 1965). The active pumice mines are Producers Pumice near Idaho Falls and Hess Pumice Products north of Malad City in Oneida County, both of which supply graded material for lightweight aggregate. The Hess deposit is partially indurated, and the pumice is finely vesicular with a very low lithic and crystal content. From it, Hess produces high quality abrasives for a wide range of markets including glass polishing, metal finishing, soaps, and erasers.

Numerous beds of volcanic ash, Pliocene and Pleistocene in age, occur throughout the Great Plains states. The Valles Caldera, site of the Jemez Mountains in northern New Mexico, has been identified as the source for much of the airborne material. The ash occurs as thin air-fall blankets and as reworked fluvial and lacustrine deposits with varying amounts of admixed detrital minerals. Some beds are altered to clays, others are composed of clean, unaltered glass fragments nearly all of which are less than 100 μm in size.

Deposits in several states have been mined for abrasives, polishes, asphalt filler, absorbents, and ceramics. Some of the pumicites bloat into hollow spheres under proper heating conditions and extensive testing has been done to produce filter aids and cellular glass blocks (Burwell, 1949, Bauleke, 1962, Hardy et al., 1965). Currently two companies are mining two different beds of expandable pumicite in north central Kansas. Kansas Minerals, near Mankato, produces filter aids by drying, air classifying, and expanding the pumicite and milling the expanded fragments to the desired size ranges. Calvert Corporation, near the village of Calvert, sells a dried and screened product which is used by its customers in both expanded and unexpanded forms to filter vegetable oils.

Pumice occurs in several areas in New Mexico but those deposits currently being mined are air-fall units associated with the Valles Caldera volcanics in the Jemez Mountains in the area of Santa Fe and Los Alamos. Copar Pumice and General Pumice produce lightweight aggregate, stonewashing pumice, and material processed into granular abrasives. Several other areas of the state have produced pumice in the past, including East Grants Ridge near Grants in Cibola County. Development work is in progress on a small epiclastic pumice deposit near San Antonio in Socorro County (Hoffer, in preparation).

Pumice production in Oregon is limited primarily to the Bend pumice, an air-fall lapilli unit around the city of Bend in Deschutes County. Cascade Pumice and Central Oregon Pumice produce lightweight aggregate, horticultural pumice, and landscaping material. Less than 5% of the fragments in the Bend pumice is greater than 2.5 cm in diameter but Cascade Pumice screens this coarser fraction for stonewashing material.

A limited amount of pumice has also been produced from an extensive air-fall and pyroclastic flow unit formed when Mount Mazama collapsed forming Crater Lake 6845 ±50 years b.p. (Bacon, 1983). The climactic air-fall lapilli unit now blankets an area of over 5 200 km^2 with a thickness greater than 1 m and an area of over 900 km^2 at a thickness greater than 3 m, with little or no overburden. To date the Mazama pumice has not proven as satisfactory as the Bend pumice for lightweight aggregate but some of it has been sold for stonewashing material (Geitgey, 1992).

Pumice deposits are distributed in other volcanic areas throughout the world. Historically the relatively low value of lightweight aggregate has limited large scale production to deposits such as those in Germany, Greece, and Italy that are near markets or near bulk transportation facilities. Increased industrialization has created greater markets for lightweight aggregates and the higher value of stonewashing pumice has led to the development of formerly uneconomic deposits. The following summary of international pumice occurrences is based on articles published in *Industrial Minerals* (Anon., 1977, Robbins, 1984, and McMichael, 1990), Harben and Bates (1984), and various editions of the US Bureau of Mines Minerals Yearbook.

The Italian peninsula is well endowed with volcanics but the country's most important commercial pumice deposit is on the island of Lipari, about 35 km off the north coast of Sicily. An unconsolidated deposit of high quality pumice has been the source of aggregate and industrial pumice for Europe and North America for many decades. Pumex SpA, Italy's largest producer, supplies lightweight aggregate, powders and granular material for various industrial uses, and cosmetic stones. Italpomice SpA produces aggregate and stonewashing pumice. Both companies can load directly from their operations onto ocean-going vessels.

Many of the islands of Greece, particularly the Cyclades and Dodecanese groups in the southern Aegean Sea, have pumice deposits. Several have been mined in the past, including Thera (Santorini), but production is now only on the island of Yali where Lava Mining and Quarrying mines aggregate grade pumice for markets throughout Europe, North America, North Africa, and the Middle East. Yali is the principal source for lightweight aggregate pumice for markets on the east coast of the United States.

Pumice exports from Turkey have increased over the last decade from almost none to several hundred thousand tons per year. Nearly all of the production is stonewashing pumice, primarily for European markets The deposits are located in relatively remote areas of central and eastern Turkey and the pumice must be transported long distances by truck to ports on the Mediterranean Sea. Small amounts are used domestically for construction and some companies are now producing granular and powdered material. Pumice mines are operated by about a dozen companies.

Pumice is produced in many other countries for both export and domestic use. Spain has active mines in the province of Ciudad Real and on Tenerife in the Canary Islands. Pumice production is diminishing in Germany and has ceased in France. Both Iceland and New Zealand are mining in excess of local needs and are now supplying various export markets. Stonewashing pumice is being shipped from several countries including Ecuador, Guatemala, Mexico, and Indonesia.

ALTERNATIVE MATERIALS

For nearly every end use of pumice and cinder there are natural or manufactured alternative materials that have similar performance characteristics. Lightweight concrete is produced with blast furnace slag, expanded perlite, expanded clays, and air-entraining agents. Numerous minerals are used as granular abrasives. Clays, diatomite, and zeolites function as various types of absorbents and carriers. Other rocks and various manufactured materials serve for landscaping and veneer uses. Hollow microspheres of glass and plastic are used as low density fillers, and diatomite and expanded perlite are used extensively as filter media.

The choice of pumice and cinder is often a function of costs, both production costs and transportation costs. Pumice and cinder deposits are typically not indurated and relatively inexpensive to mine. Unlike perlite, expandable clays, and diatomite, many pumice uses do not require expensive pyroprocessing.

Pumice offers unique characteristics in some applications, such as lightweight landscaping boulders and stonewashing, that are not precisely matched by either manufactured materials or other natural rocks. Similarly block and granular pumice occupy unique niches in the abrasive field because of their texture and friability.

PROBLEMS AND FUTURE TRENDS

Caution should be exercised in interpreting pumice and cinder statistics. Clear distinctions are not always made among various volcanic materials or their end uses, either currently or historically. Some reports cite only volcanic materials, without specifying pumice, pumicite, cinder or scoria, pozzolans, or other volcanic rocks. Volcanics produced for construction materials may include pumice or cinder for lightweight aggregate, pumice or pumicite ground for pozzolan, and pozzolanic materials other than pumice. Uniform methods of data collection are impossible, particularly when production is sporadic, mines are in remote areas, or production is only for local use.

In the United States mineral production statistics are reported to the US Bureau of Mines only on a voluntary basis. Through 1980, the US Bureau of Mines reported combined statistics of pumice, pumicite, and volcanic cinder. Beginning in 1981, volcanic cinder statistics have been reported instead with crushed stone, a more accurate reflection of the principal market of cinder. The apparent abrupt decrease in production and number of producers in volcanic materials reflects a change in reporting procedures rather than a decline of the industry. The Bureau of Mines has also modified its pumice questionnaire to receive more detailed data on specific products and markets.

Problems have arisen in the United States concerning acquisition of mineral rights to pumice on federal land. A large number of pumice deposits occur on federal land, administered by the US Bureau of Land Management or by the US Forest Service, on which access to minerals is obtained by leasing, by staking mining claims, or by purchase, depending on the mineral. Leases and mining claims grant exclusive development right while saleable commodities are available to any purchaser. Leases require royalty payments, mining claims do not. Federal law has been modified over the years to specify which minerals are acquired by which methods.

Prior to 1955 pumice on federal land could be acquired by mining claims and if these claims have been maintained they remain valid. The Surface Resources Act of 1955 (US Code, Title 30, Section 611) names sand, stone, gravel, pumice, pumicite, and cinders as common variety materials. As such they can only be purchased, they cannot be acquired by staking mining claims. But the Act also expressly excludes block pumice from the list of common variety materials, defining block pumice as pumice "which occurs in nature in pieces having one dimension of two inches or more." Also excluded are "deposits of such minerals which are valuable because the deposit has some property giving it distinct and special value." Many pumice deposits have some particles larger than 5.1 cm, but that in itself may not necessarily qualify the deposit as block pumice. Stonewashing pumice has a much higher value than aggregate pumice but particles as small as 2 cm may be suitable. Various, sometimes conflicting, interpretations have been given by the courts and boards of appeal and it is still unclear under what conditions, if any, exclusive mineral rights can be obtained to pumice on federal land. Continued litigation is probable.

Future trends in pumice production will depend principally on building construction and fashion. Aggregate pumice remains the largest volume market and sales fluctuate with construction activity which in turn is a reflection of general economic conditions. The appearance and softness of stonewashed garments remain in high demand and as yet no satisfactory alternative to pumice has been developed. The higher value of laundry pumice has led to the development of new sources throughout the world which may in turn lead to a more diversified pumice industry in those areas.

REFERENCES

Anon., 1977, "Pumice—A Dual Role in Industry," *Industrial Minerals*, Sep., pp. 15–29.

Anon., 1992, "Pumice and Pumicite in 1991," *Mineral Industry Surveys*, US Bureau of Mines, 5 pp.

Asher, R.R., 1965, "Volcanic Construction Materials in Idaho," Pamphlet 135, Idaho Bureau of Mines and Geology, 150 pp.

Bacon, C.R., 1983, "Eruptive History of Mount Mazama and Crater Lake Caldera, Cascade Range, U.S.A.," *Journal of Volcanology and Geothermal Research*, Vol. 18, pp. 57–115.

Bauleke, M.P., 1962, "What's New in Volcanic Ash for Industry?" Bulletin 157, Part 3, State Geological Survey of Kansas, 19 pp.

Bullard, F.M., 1976, *Volcanoes of the Earth*, University of Texas Press, Austin, TX, 579 pp.

Burchett, R.R., 1990, "Volcanic Ash Deposits in Nebraska," *Geonotes*, Nebraska Geological Survey, Dec., 3 pp.

Burwell, A.L., 1949, "Cellular Products from Oklahoma Volcanic Ash," section on geology and petrology by W.E. Ham, Circular No. 27, Oklahoma Geological Survey, 89 pp.

Carithers, W., 1946, "Pumice and Pumicite Occurrences of Washington," Report of Investigations No. 15, Washington Division of Mines and Geology, 78 pp.

Cas, R.A.F., and Wright, J.V., 1988, *Volcanic Successions*, Unwin Hyman, London, 487 pp.

Chesterman, C.W., 1956, "Pumice, Pumicite, and Volcanic Cinders in California," Bulletin 174, California Division of Mines, 119 pp.

Clippinger, D.M., and Gay, W.E., 1947, "Pumice Aggregate in New Mexico, Its Use and Potentialities," Bulletin 28, New Mexico Bureau of Mines and Mineral Resources, 56 pp.

Fink, J.H., and Manley, C.R., 1987, "Origin of Pumiceous and Glassy Textures in Rhyolite Flows and Domes," *The Emplacement of Silicic Domes and Lava Flows*, J.H. Fink, ed., Special Paper 212, Geological Society of America, pp. 77–88.

Fisher, R.V., 1961, "Proposed Classification of Volcanoclastic Sediments and Rocks," Bulletin, Geological Society of America, Vol. 72, pp. 1409–1414.

Geitgey, R.P., 1992, "Pumice in Oregon," Special Paper 25, Oregon Department of Geology and Mineral Industries, 26 pp.

Harben, P.W., and Bates, R.L., 1984, "Pumice and Scoria," *Geology of the Nonmetallics*, Metal Bulletin, New York, pp. 65–68.

Hardy, R.G., et al., 1965, "A Feasibility Study for the Production of Filter Aids from Kansas Volcanic Ash," Special Distribution Publication 19, Kansas Geological Survey, 34 pp.

Harris, R.E., and King, J.K., 1986, "Pumice, Scoria and Pumicite in Wyoming," Open File Report 86–17, Geological Survey of Wyoming, 20 pp.

Hatmaker, P., 1932, "Pumice and Pumicite," Information Circular 6560, US Bureau of Mines, 23 pp.

Hoffer, J.M., 1991, "Pumice and Pumicite in Arizona," Open File Report 91–8, Arizona Department of Mines and Mineral Resources, 67 pp., 1 plate.

Hoffer, J.M., in preparation, "Pumice and Pumicite in New Mexico," Bulletin 140, New Mexico Bureau of Mines and Mineral Resources.

Horton, R.C., 1964, "Pumice, Pumicite, and Volcanic Cinders," *Mineral and Water Resources of Nevada*, Bulletin 65, Nevada Bureau of Mines, pp. 237–240.

Loney, R.A., 1968, "Flow Structure and Composition of the Southern Coulee, Mono Craters, California—A Pumiceous Rhyolite Flow,"

Studies in Volcanology, R.O. Coats et al., eds., Memoir 116, Geological Society of America, pp. 415–440.

McMichael, B., 1990, "Pumice Markets—Volcanic Rise of Stonewashing," *Industrial Minerals*, May, pp. 22–37.

Moore, B.N., 1937, "Nonmetallic Mineral Resources of Eastern Oregon," Bulletin 875, US Geological Survey, 180 pp.

Osburn, J.C., 1982, "Scoria Exploration and Utilization in New Mexico," *Industrial Rocks and Minerals of the Southwest*, G.S. Austin, compiler, Circular 182, New Mexico Bureau of Mines and Mineral Resources, pp. 57–59.

Robbins, J., 1984, "Pumice—The Ins and Outs of a Reinforced Market," *Industrial Minerals*, May, pp. 31–51.

Schmid, R., 1981, "Descriptive Nomenclature and Classification of Pyroclastic Deposits and Fragments: Recommendations of the IUGS Subcommission on the Systematics of Igneous Rocks," *Geology*, Vol. 9, pp. 41–43.

Schmidt, F.S., 1956, "Technology of Pumice, Pumicite, and Volcanic Cinders," *Pumice, Pumicite, and Volcanic Cinders in California*, C.W. Chesterman, ed., Bulletin 174, California Division of Mines. pp. 99–117.

Van Horn, R., 1964, "Lightweight Aggregate," *Mineral and Water Resources of Utah*, L.S. Hilpert, compiler, Bulletin 73, Utah Geological Survey, pp. 185–188.

Walker, G.P.L., 1980, "The Taupo Pumice: Product of the Most Powerful Known (Ultraplinian) Eruption?" *Journal of Volcanology and Geothermal Research*, Vol. 8, pp. 69–84.

Williamson, D.R., and Burgin, L., 1960, "Pumice and Pumicite," *Mineral Industries Bulletin*, Vol. 3, No. 3, Colorado School of Mines, Golden, CO, 12 pp.

Wood, C.A., 1980a, "Morphometric Evolution of Cinder Cones," *Journal of Volcanology and Geothermal Research*, Vol. 7, pp. 387–413.

Wood, C.A., 1980b, "Morphometric Analysis of Cinder Cone Degradation," *Journal of Volcanology and Geothermal Research*, Vol. 8, pp. 137–160.

PYROPHYLLITE

P.A. CIULLO AND C.S. THOMPSON

Pyrophyllite is a naturally occurring hydrous aluminum silicate with the molecular formula $SiO_{10}Al_{12}(OH)_2$ or, as more commonly expressed, the oxide formula $Al_2O_3 \cdot 4SiO_2 \cdot H_2O$.

Pyrophyllite is rarely found in deposits as the pure mineral and is often intimately associated with quartz, sericite, kaolinite, and diaspore. These associated minerals, to a large degree, help determine the most appropriate commercial application for a given pyrophyllite ore.

Structural similarities plus the generally soft platy nature of pyrophyllite has in the past caused it to be included with talc in applications and production data. *The Minerals Yearbook*, published annually by the US Bureau of Mines (USBM), still reports on talc and pyrophyllite together, although admitting that they do not share the same uses. Pyrophyllite has also at times been considered a clay mineral because of its structure and chemistry.

The refractory properties of pyrophyllite have been known for nearly 200 years, but it has only been in the past 60 to 70 years that a variety of applications have been developed to exploit the unique characteristics of this mineral. Worldwide, pyrophyllite is a component of refractories, whiteware, foundry mold dressings, pesticides, paint, plastics, rubber, cement, fiberglass, and soap.

Historical Overview

The most salient refractory features of pyrophyllite are its low coefficient of thermal expansion, high thermal conductivity, low moisture expansion, and good resistance to corrosion by molten metals and basic slags. One of the first recorded uses of this mineral was as firebrick in Japan in the early 1800s. This followed its discovery in 1797 on Mount Omotoyama at the site of the present Mitsuishi mine. In addition to firebrick, sawn blocks of ore were used for carvings and in the manufacture of slate pencils and signature seals. The Mitsuishi mine is still one of Japan's largest. By the mid-1800s, refractories were being made in Japan from roseki (a pyrophyllite-bearing ore) for the construction of metal melting furnaces.

The mineral did not get its name until 1829 when R. Herman, testing a sample of supposed talc from the gold-quartz veins at the Berezovo deposit in the Urals, showed that it was actually an aluminum silicate and designated it pyrophyllite. This name was derived from the Greek *pur* for fire and *phullon* for leaf, in allusion to the effect of heat in causing separation of the laminae in foliated varieties (Mellor, 1927).

The exact date of the discovery of pyrophyllite in the United States is uncertain, but it is believed that the talc and soapstone deposits noted in an 1827 report on the geology of the slate belt of North Carolina were actually pyrophyllite occurrences (Olmstead, 1827). Pyrophyllite was first used in North Carolina on a local basis in the early 1800s. Blocks were sawn and cut for use as stove linings, mantels, chimneys, fireplaces, and gravestones. This was the material of choice because it did not shrink when heated, but cut easily and hardened as it weathered. Pyrophyllite crayons were sold commercially by local gold miners for steel marking and as tailor's chalk from about 1800 until 1922.

It was during this period that pyrophyllite was discovered near the town of Hemp (later renamed Robbins), NC. The large quantity of scrap resulting from the sawing of crayons at this deposit prompted the construction in 1921 of a grinding plant. The ore scrap was ground to fine granules and sold to asphalt producers to prevent shingles from sticking together. By the mid-1930s, milled pyrophyllite from North Carolina was also being successfully utilized by the ceramics industry in whiteware for its low cost and ability to shorten firing time. Also during the 1930s, pyrophyllite was first used in kilncar furniture in the United States.

Canada was another early entrant in the pyrophyllite business. Mining at the Foxtrap deposit on the Avalon Peninsula of Newfoundland began in 1904, with ore shipped to the United States. Production from this deposit was mostly sporadic in the ensuing years until 1956 when mining began for shipment and captive use in US tile manufacture. This production has remained continuous and still essentially captive to the present. The Foxtrap deposit is currently the sole commercial source of pyrophyllite in Canada.

GEOLOGY

Crystal Structure and Chemical Properties

Pyrophyllite has a 2:1 phyllosilicate layer structure composed of an octahedral alumina (gibbsite) sheet sandwiched between two silica tetrahedral sheets. These sheets are combined so that the tips of the silica tetrahedra and two-thirds of the hydroxyls of the alumina sheet become shared oxygens. The alumina sheet is dioctahedral in structure, with two thirds of the available sites occupied by aluminum and one third empty. The pyrophyllite lattice is continuous in the a and b directions and stacked in the c direction, accounting for the usual flake or platelet particle morphology. A schematic representation of the prophyllite lattice is presented in Fig. 1.

In ideal form, the pyrophyllite structure is electrically neutral. In nature, very small substitutions of aluminum for silicon are found and very minor replacement of aluminum with iron, magnesium, and titanium has been determined. The predominant interlayer bonding forces are believed to be van der Waals, although minor electrostatic attraction and hydrogen bonding have been suggested (Evans and Guggenheim, 1988). These weak interlayer forces account for pyrophyllite's soft nature and often soapy or slippery feel. They also explain the difficulty in finding samples with well developed crystallinity for mineralogical study, since in nature the pyrophyllite structure often exhibits stacking disorder of the 2:1 layers.

The three common forms of pyrophyllite are:

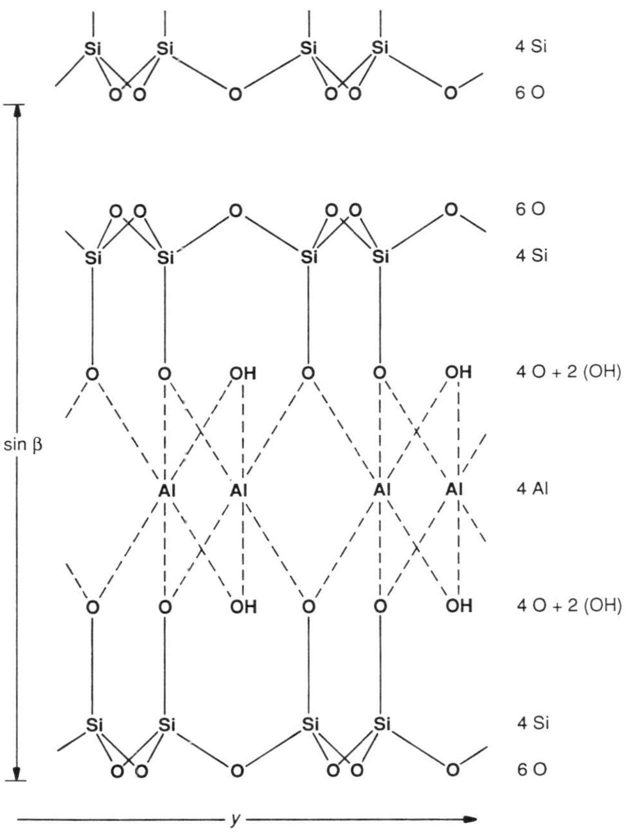

Fig. 1. Schematic diagram of the structure of pyrophyllite as viewed along the x axis.

1. Fine-grained foliated laminae with platy cleavage. This type is commonly used for whiteware and filler applications.
2. Massive spherulitic aggregates of small crystals. This type is preferred for use in refractories.
3. Radiating large needle-like crystals. This type is used as a filler and in whiteware and finds minor use in refractories.

The most common polymorph of pyrophyllite is the two layer monoclinic form. Other polymorphs found in nature are a one layer triclinic form, a mixture of the triclinic and monoclinic forms and a disordered form (Brindley and Wardle, 1970).

The pyrophyllite crystal structure is readily distorted by mechanical means. Studies have shown that dry grinding of pyrophyllite, as with many minerals, can readily introduce disorder into its structure. Detailed in a recent work (Perez-Rodriguez et al., 1988), an initial grinding period produced delamination, particle size reduction, and a large surface area increase. Additional grinding caused an abrupt increase in the degree of structural disorder as observed by X-ray diffraction (XRD). Continued grinding further increased disorder, and decreased the specific surface area due to agglomeration.

This mechanical instability stands in contrast to the chemical stability of pyrophyllite. This mineral is generally considered inert, although this description is more in conformance with a chemical perspective rather than a mineral one since the surface chemistry of even a low charge phyllosilicate such as this cannot be discounted. Pyrophyllite is highly resistant to strong alkalis and acids but can be decomposed to gelatinous silica with sulfuric acid.

Because high purity pyrophyllite is uncommon both in nature

Table 1. Chemical Analysis of Selected Pryophyllite Samples

	SiO_2	Al_2O_3	Fe_2O_3	MgO	CaO	Na_2O	K_2O	LOI
Theoretical	66.7	28.3						5.0
USA								
North Carolina[a]	75.0	19.3	0.8	0.1	0.1	0.1	0.2	3.9
North Carolina[b]	67.2	27.0	0.7	0.02		tr	0.01	4.6
California[c]	66.8	28.2	0.1	tr	tr	0.05	tr	5.0
CANADA								
Newfoundland[d]	63.1	27.4	0.8			0.2	3.1	4.9
AUSTRALIA								
Pambula[c]	78.5	15.8	0.4	0.1	0.3	0.1	0.2	3.9
JAPAN								
Okayama[e]	75.8	17.7	0.6	0.02	0.2	0.2	1.0	4.4
Shokozan[e]	77.4	18.2	0.2	0.1	0.2	0.1	0.3	3.5
REP. of KOREA								
Tongnae[e]	67.6	26.3	0.8	0.3	0.2	0.7	0.3	3.9
BRAZIL								
Matheus Leme[e]	63.6	25.6		0.01		0.3	2.5	4.8
INDIA								
Madhya Pradesh[f]	61.6	29.6	0.5	tr	1.3	0.1	1.4	5.3
SOUTH AFRICA								
Transvaal[c]	56.7	33.3	1.5	0.3	0.1			6.2
TURKEY								
Malatya[c]	65.9	26.4	0.6	0.2	0.7	0.7	0.3	4.3
THAILAND								
Changwat[c]	67.8	22.8	0.1			0.2	0.2	6.7
CHINA								
Quingtian[g]	64.8	29.6	0.4	0.04	0.1	0.02		4.0

[a](Anon., 1985)
[b](Sykes and Moody, 1978)
[c](Cornish, 1983)
[d](Papezik, Keats, and Vahtra, 1978)
[e](Benbow, 1988)
[f](Prasad, Mukhopadhyay, and Maiti, 1989)
[g](Huifen, Damei, and Zechun, 1989)

and commerce, reported chemical analyses vary greatly and are of limited use without corresponding mineralogical characterization. Chemical analyses of pyrophyllite samples from eleven countries are presented in Table 1. Analyses of several samples from India, Canada, and North Carolina are rationalized with their mineralogy in Table 2.

Physical Properties

Pyrophyllite is generally characterized as soft, slippery, and sallow. Its color in nature actually ranges from white, to yellowish, to pale green, to light gray. Its luster varies from dull matte to pearly. Pyrophyllite in pure or near pure form is rare in nature and in commerce. In fact, the majority of current uses for pyrophyllite exploit the desirable features of the mineralogical blends that constitute most commercial grades. The impurities in pyrophyllite can contribute to physical properties and uses to a substantial degree. Quartz content in pyrophyllite products, for example, can range from 5 to 70%, with corresponding variations in product abrasiveness and refractoriness. Sericitic pyrophyllite products are used in whiteware applications for improvement in craze resistance and thermal shock resistance due to their low thermal expansion.

Wonderstone, a South African ore, is atypical among high pyrophyllite rocks. With a pyrophyllite content approaching 90%, this massive block form is highly compacted. Associated minerals are chloritoid, rutile, and epidote. End uses take advantage of its characteristic acid and abrasion resistance.

Thermal Properties

On heating, pyrophyllite forms the metastable dehydroxylate at about 700°C. Dehydration occurs in the internal plane common to

Table 2. Chemical Analysis vs. Mineralogy of Selected Pyrophyllite Samples

SAMPLE	A	B	C	D	E	F
Chemical Analysis						
SiO_2	61.6	57.7	63.1	63.5	76.3	83.3
Al_2O_3	29.6	31.6	27.4	29.6	19.8	13.9
Fe_2O_3	0.5	0.7	0.8	0.1	0.2	0.1
MgO	tr	0.1		0.2		
CaO	1.3	1.1		nd	0.1	0.4
Na_2O	0.1	0.2	0.2	0.2	0.1	0.2
K_2O	1.4	1.0	3.1	1.1	0.3	nd
LOI	5.3	7.2	4.9	5.0	3.4	2.2
Mineralogy						
Pyrophyllite	79	61	60	89	66	48
Quartz			12		31	51
Sericite			27	11	3	1
Kaolin	4	26				
Muscovite	12	8				
Paragonite	2	2				
Calcite	2	2				

A—Madhya Pradesh, India (Prasad, Mukhopadhyay, and Maiti, 1989)
B—Uttar Pradesh, India (Prasad, Mukhopadhyay, and Maiti, 1989)
C—Newfoundland, Canada (Papezik, Keats, and Vahtra, 1978)
D—Newfoundland, Canada (Papezik, Keats, and Vahtra, 1978)
E—North Carolina, USA (Cornish, 1983)
F—North Carolina, USA (Cornish, 1983)

the octahedral and tetrahedral sheets by the loss of one OH- group, and H+ from the neighboring OH group. Overall disruption of the pyrophyllite structure is minimal (Wardle and Brindley, 1972). Decomposition to cristobalite and mullite occurs at about 1 200°C (Deer, Howie, and Zussman 1967). At this temperature, Mohs hardness has increased from 1–2 to 7–8 due to the formation of these minerals. Pyrophysical properties are greatly influenced by mineralogy and chemistry, which can vary widely depending on the type of pyrophyllite deposit.

Classification and Origin of Deposits

Historically, pyrophyllite deposits have been classified as either hydrothermal deposits, generally associated with acid-volcanic complexes, or metamorphic deposits, less common and associated with metamorphosed volcanic ash. The following expanded classification scheme has been proposed, with five types of deposits, each relating to geological setting and condition of formation (Zaykov, Udachin, and Sinyakovskaya, 1988).

Type 1: Hydrothermal deposits in metasomatites in continental and island-arc volcanic zones, in Precambrian platforms and Paleozoic and Mesozoic-Cenozoic folded systems.

Type 2: Deposits in metamorphosed metasomatites in submarine Paleozoic volcanic zones enclosing sulfide ores.

Type 3: Hydrothermal deposits in metasomatites in the wall rock of hydrothermal quartz veins among Archean and Paleozoic granitoids and in metamorphosed Paleozoic clastic suites.

Type 4: Stratiform deposits in metamorphosed Paleozoic and Mesozoic clastic-clay suites containing pyroclastic material and coal seams.

Type 5: Deposits in clays formed by weathering.

Examples, according to this scheme, of deposit type, composition and location are given in Table 3.

Commercial Deposits

Commercially exploited deposits of pyrophyllite are found in Japan, the Republic of Korea, Brazil, India, the United States,

Table 3. Pyrophyllite Deposits by Classification Type

Deposit Type	Predominant Ore Types	Example
1	pyrophyllite pyrophyllite-quartz sericite-pyrophyllite	North Carolina, USA
1	pyrophyllite diaspore-pyrophyllite kaolinite-pyrophyllite chlorite-pyrophyllite	Pambula, Australia
1	pyrophyllite-quartz quartz-pyrophyllite	Okayama, Japan
2	pyrophyllite-quartz quartz-pyrophyllite diaspore-pyrophyllite sericite-pyrophyllite	Ural, USSR
3	kaolinite-pyrophyllite muscovite-pyrophyllite	Madhya Pradesh, India
4	pyrophyllite kaolinite-pyrophyllite sericite-kaolinite-pyrophyllite	Las Aguilas, Argentina
5	kaolinite-pyrophyllite illite-montmorillonite-pyrophyllite	Utah, USA

Canada, Pakistan, Thailand, Australia, the Republic of South Africa, Argentina, and Peru. In 1990, these twelve countries produced approximately 2 4 Mt of pyrophyllite. Of these, the top five producing countries, and their respective percentage of total production, are as follows (Virta, 1991):

1) Japan (51%)
2) Republic of Korea (31%)
3) Brazil (7%)
4) India (4%)
5) United States (3%)

Japan: Pyrophyllite production is widespread throughout Japan, with scores of mines from Toya in the north to Goto in the southwest. The most important pyrophyllite areas have been exploited within the Cretaceous rocks of southern Honshu. The major producing region is Wake-gun in Okayama Prefecture near the south coast of southwest Honshu, about 145 km west of Osaka.

In Okayama Prefecture, pyrophyllite occurs as veins and irregular masses within sedimentary and volcanic rocks. Typically, kaolin accompanies pyrophyllite within the upper parts of the ore bodies, with nodules of diaspore occurring in the central part of the ore mass, and pyrite and marcasite in the lower parts. These deposits were likely formed by the replacement of andesites, high alumina shales, and rhyolites, which were permeated by hot hydrothermal solutions associated with volcanism.

Major production in Okayama comes from the Ohira, Kawabe, Kato, and Mitsuishi mines. Of these, the oldest and largest is the Mitsuishi mine which has a current output of about 100 ktpy. The pyrophyllite lodes here are irregular or lenticular. Accessory minerals include quartz, sericite, kaolinite, and diaspore.

To the west of Okayama and northeast of the city of Hiroshima, pyrophyllite is mined from the Cretaceous rock formations in the Shokozan area of Hiroshima Prefecture. Principle production is from the Yanoshokozan mine (100 ktpy). In Hiroshima Prefecture the massive hydrothermal pyrophyllite deposits are in rhyolite and porphyrite, and include the accessory minerals diaspore, kaolinite, corundum, pyrite, quartz, and alunite.

To the northeast of Okayama lie the largest pyrophyllite reserves in Japan, according to the Imperial Geological Survey of 1947. These occur in the Kinki region of Hyogo Prefecture, mainly in Kinosaki-gun and Kanzaki-gun. This survey listed probable reserves in these two areas at 50 Mt, or about 69% of Japan's total. However, little is known and less reported concerning the present level of mining activity in these regions, although it is believed to be relatively minor. The geology and mineralogy of these deposits are similar to those in Hiroshima Prefecture.

The third major producing region is on Kyushu Island in Minaminatsura-gun in Nagasaki Prefecture. To the west of Nagasaki city, the Goto mine produces 110 ktpy from a high grade pyrophyllite deposit hosted by Miocene rocks. In the deposits of Nagasaki Prefecture, lodes are lenticular or irregular, typically with quartz, sericite, kaolinite, and diaspore as associated minerals.

Much of the pyrophyllite produced in Japan is mined from many small deposits and by numerous operators of widely varying size and resources. A significant portion of production is captive, principally by refractories companies. In 1988, the ten largest producers, with output ranging from 130 down to 15 ktpy, accounted for less than 50% of Japan's total production (Benbow, 1988).

Just what constitutes the pyrophyllite of commercial statistics is not always clear. The colloquial name roseki (waxstone) is applied to ores of pyrophyllite and associated minerals such as kaolinite and sericite. Roseki can be used to describe both high and low grade pyrophyllite ore, as well as sericitic and kaolinitic ores containing little or no pyrophyllite. Although very high grade deposits of pyrophyllite are known and mined, most of the pyrophyllite currently mined is believed to be of the roseki type and subject to beneficiation. Again, the significant captive use of mine output makes it very difficult to delineate the consumption of high vs. lower quality pyrophyllite, whether obtained naturally or by selective classification.

Pyrophyllite mining by the major producers is both open pit and underground in accordance with the usual geological and economic criteria. Beneficiation is wet or dry depending on the ore quality and end use. Roseki can be crushed and wet separated to rosekiclay of higher pyrophyllite content. This in turn can be further processed to high purity pyrophyllite. Each industry dictates the balance of pyrophyllite quality and cost that is suitable. A low grade ground roseki stone might be suitable for a general filler application, while rosekiclay could be required for fiberglass or paper, and high grade pyrophyllite for refractories and whiteware.

Republic of Korea: The pyrophyllite-bearing rocks of western Japan can be traced northward to the Republic of Korea. The major producing areas in this country are all near the southern coast—the Haenam area of Chaollanam-do Province and the Tongnae area of Kyongsangham-do Province. Most of the country's 50+ operating mines are in this region. As in Japan, these mines vary greatly in size and output extracting pyrophyllite formed by the alteration of acid volcanic rocks. The largest producers are the Wan-Do mine and the Nohwa-Do mine, both in Haenam. The larger Wan-Do operation mines lenticular ore bodies by open pit. The ore is composed of pyrophyllite, quartz, kaolin, and alunite, with accessory diaspore and disseminated pyrite. Output from the Wan-Do and Nohwa-Do mines is geared to the production of pyrophyllite for the whiteware and refractories markets, the latter of which accounts for about 80% of domestic consumption.

The Haenam pyrophyllite deposits occur as fissure filling in tuffs and rhyolites. In addition to Wan-Do and Nohwa-Do, other major operations in Haenam are the Sungsan, Hwansan, Okmesan, Dae-Do, and Chin-Do mines. The country's major pyrophyllite reserves are in Haenam, primarily at Wan-Do where inferred ore reserves total approximately 30 Mt.

The second major pyrophyllite producing area in South Korea is at Tongnae where the Nilyang, Yangsan, Kimhae, Pusan, and Kyongnam mines are worked. The deposits in the Tongnae area were formed by the alteration of intermediate volcanic rocks including andesite, trachyte, trachytic andesite, and quartz porphyry. The host rocks of the Pusan and Kyongnam mines consist of quartz porphyry and feldspar porphyry, while trachytic andesite is the host rock in the Tongnae, Imgi, and Nilyang mines. Pyrophyllite bodies in the Tongnae area occur along fissures in tuffs and other volcanics. They are believed to have formed by alteration of granite intrusives of Cretaceous age by hydrothermal solutions. The Yangsan and Un-Yang granite are thought to be the principal mineralizers in the Tongnae area.

Much of the ore in this country is of the massive variety, which is locally referred to as agalmatolite. In the major producing regions the three common mineralogical suites are pyrophyllite with accessory quartz, kaolin, and alunite; kaolin-pyrophyllite with minor diaspore; and pyrophyllite-sericite with minor quartz.

The general commercial situation with pyrophyllite in Korea is similar to that in Japan. The overwhelming bulk of mine output is devoted to the ceramics industry. In Korea most of the consumption is domestic, but about one-third is for export to Japan and China, as well as the developing whiteware industries in Sri Lanka and Indonesia. The higher grades of pyrophyllite are sold for refractory (principally firebrick), whiteware, and wall tile production, but, as in Japan, the source of these grades is not clearly delineated (e.g., high grade ore vs. beneficiated agalmatolite) because of the largely captive use.

Production statistics are likewise subject to interpretation, as suggested by a recent study on beneficiation of pyrophyllite (Shin, Kook, and Choi, 1990). The purpose of this work was to determine the potential of flotation to upgrade the pyrophyllite in Chun-Nam Province, which accounts for 70% of the country's production. One of the ores studied was from the Poo-kok mine, which produces a range of products from wet screened refractory and 1st Grade pyrophyllites to dry screened Low Grade pyrophyllites, differentiated largely on the relative predominance (by XRD) of quartz. However, the XRD evaluation provided showed that neither the raw ore, nor any of the products contain pyrophyllite. Other than quartz, the only other constituents are kaolinite and dickite. Rather than any basis in mineralogy, the use of the term pyrophyllite for this ore and resultant products apparently relates to its intended applications, or possibly to its original discovery and uses, much the same way that pyrophyllite was often mistaken for talc prior to the development of modern techniques of mineralogical characterization. Nevertheless, it does provide one more indication of the difficulties that can accompany the collection of commercial statistics when production of pyrophyllite is predominately for one industrial sector and captive.

Brazil: The production of pyrophyllite in Brazil is concentrated in the states of Minas Gerais, Sao Paolo, and Parana. Most deposits occur in Minas Gerais in a 64 km long belt spanning the counties of Pitangui, Para de Minas, Onca do Pitangui, Matheus Leme, Conceicao do Para, and Betim. Proven reserves in this state total about 5.7 Mt, according to Brazil's national mineral production department (1986). The deposits are lenticular or podiform and are hosted by sericite and sericite-chlorite schists.

Brazilian pyrophyllite production figures are reported in terms of both pyrophyllite and agalmatolite. The latter is an ore of intimately mixed pyrophyllite, sericite, diaspore, kyanite, and quartz. Agalmatolite is subdivided into three further categories: sericite agalmatolite, diaspore-pyrophyllite agalmatolite, and quartz-kyanite agalmatolite. The distinction in production figures between pyrophyllite mined as such and pyrophyllite derived from agalmatolite is not clear, although the USBM figures indicate an almost 300% increase in pyrophyllite production from 1987 to 1990 (60 to 170 kt), while estimates of agalmatolite over this period have remained static at about 120 kt (Virta, 1991).

Most pyrophyllite/agalmatolite mining is open pit and much of this production is captive for use in refractories. The main consuming industries besides refractories are whiteware, paint, and soap. There is also a wide variety of lesser uses, such as in paper, cardboard, animal feeds, rubber, plastics, bitumen, cement, fertilizers, insecticides and cosmetics. Ore beneficiation to accomodate each application concentrates on the most appropriate control of color, particle size, and abrasivity. Typically, ore is selectively mined, washed, crushed, hand sorted to control color, screened, and ground or micronized as required.

The use of pyrophyllite in such a variety of applications is due largely to its relatively low cost and easy accessibility within the heavily industrialized Belo Horizante—Rio de Janeiro—Sao Paolo triangle in southeastern Brazil. This has allowed it to capture some use in applications like paper and paint that would otherwise have been the sole province of kaolin and talc.

India: Significant known reserves of Indian pyrophyllite have been identified in the states of Uttar Pradesh (3 Mt) and Madhya Pradesh (1.1 Mt), with the bulk (46%) of annual production coming from mines at Tikamgarh and Shrivpuri in Madhya Pradesh. Although its reserves position is less well defined, the state of Orissa provides about 20% of India's production from the Konjhar District. Additional important production occurs near Udaipur in the state of Rajasthan and in Hamirpur in the state of Uttar Pradesh (Russell, 1991). The USBM reports an estimate of 95 kt for 1990 production in India.

Indian pyrophyllite ore is generally classed as either muscovite-diaspore pyrophyllite or kaolinite-pyrophyllite. However, some interpretation of commercial statistics must be allowed in terms of the actual amount or presence of pyrophyllite in material that is produced as such. Recent work on the suitability of Indian pyrophyllites for ceramics applications involved twelve samples from mines in the states of Madhya Pradesh, Uttar Pradesh, and Rajasthan (Prasad et al., 1989). Of these, two contained no pyrophyllite, three less than 10% pyrophyllite, and five greater than 50% pyrophyllite. The pyrophyllite content from six mines in Madhya Pradesh ranged from 5 to 79%. The three mines in Uttar Pradesh produced samples ranging in pyrophyllite content from 26 to 92%. The three mines in Rajasthan (all in Udaipur) produced two samples with no pyrophyllite and one with 4% pyrophyllite. All twelve samples contained muscovite, with content ranging from 4 to 83%.

Unlike the other major pyrophyllite producing countries, India's major domestic uses are outside the ceramics industry. Most pyrophyllite is sold for filler use in rubber, paper, roofing, soap, cosmetics, asbestos products, and pesticides. Pyrophyllite exports are currently limited by the government, although in the past significant quantites, as much as 5 kt, have been exported to Sri Lanka for use in ceramic wall and floor tiles.

United States: The production of pyrophyllite in the US is limited to the states of North Carolina and California, with 1991 annual production at approximately 80 kt. Pyrophyllite occurs in considerable quantities in Moore and Orange counties in North Carolina, with statewide reserves reported to exceed 12 Mt. The pyrophyllite deposits of North Carolina occur in the Carolina Slate Belt, a series of volcanic pyroclastic and epiclastic rocks that have undergone greenschist-facies metamorphism. The slate belt, a part of the Avalon Zone belt of Precambrian sedimentary and volcanic rocks, starts near the Virginia border and extends across North Carolina and through South Carolina to Graves Mountain, GA. Acid tuffs and breccias are the host rocks for irregular oval or lenticular pyrophyllite bodies 140 to 460 m wide, up to 1 800 m long, and often more than 450 m deep. The pyrophyllite is predominately foliated, with quartz, sericite, muscovite, pyrite, chloritoid, feldspars, and iron oxides as accessory minerals.

Open pit mines at the Glendon and Robbins deposits in Moore County account for about half of US production. The ore selectively mined from these deposits is laminar to semimassive, and after crushing, screening, grinding and blending, results in two major product lines. One is high pyrophyllite/sericite, high alkali whiteware grades, while the other is high pyrophyllite, low alkali refractory grades.

The Hillsborough deposit in Orange County is a lenticular pyrophyllite/andalusite/quartz body, 205 m long and 15 to 60 m wide hosted by andesitic and dacitic metavolcanic rocks. Accessory minerals include sericite and kaolinite. The ore selectively mined from this deposit is crushed, screened, ground, and blended to produce products mainly for refractories and whiteware applications. Refractory grades are low alkali/high andalusite, while whiteware grades are high alkali/high pyrophyllite.

The only other commercial pyrophyllite mine in the United States is in the White Mountains of Mono County to the north of Laws, CA. The lenticular pyrophyllite-quartz ore is hosted by Precretaceous volcanic and sedimentary rocks. Mining is open pit, and after crushing, screening and grinding, the resultant products are sold mainly for regional use in pesticides and construction products. Current production is estimated at about 1 ktpy from estimated reserves of 100 kt.

Other Producing Countries: The other seven minor producing countries account in aggregate for approximately 5% of world pyrophyllite production. At about 40 ktpy each, Canada, Pakistan, and Thailand make up the overwhelming bulk of this volume.

The sole producing pyrophyllite mine in Canada exploits the Foxtrap deposit on the Avalon Peninsula of Newfoundland. The ore consists of finely grained pyrophyllite, sericite, diaspore, and quartz. The deposit is hosted by the rhyolitic flows and pyroclastics that predominate in the Harbor Main volcanics, which also contain a minor sedimentary component. The open pit ore is selectively mined and subject only to primary crushing before shipment to the United States for captive use in tile production. A minor percentage of output is shipped to Germany for further processing and sale primarily to refractories producers.

The USBM reports an estimated 40 ktpy production of pyrophyllite in Pakistan. This is believed to be the soapstone extensively mined in North West Frontier Province, likely for primary use in whiteware and refractories. Little has been reported about these deposits, their geology, or their mineralogy.

Thailand appears to be exploiting the growth in the Southeast Asian ceramics industry with pyrophyllite mined in Changwat Province. The ore is a mixture of pyrophyllite, kaolinite, and diaspore. Three classes of product are produced based on alumina content: refractories/sanitaryware (35 to 40% alumina), whiteware (28 to 35% alumina), and cement (19 to 24% alumina).

The main pyrophyllite deposit in Australia is located at Pambula near the deepwater port of Eden on the southeast coast of New South Wales. The ore is the result of metamorphism and hydrothermal alteration of Upper Devonian volcanics. The core of the ore body is nearly pure pyrophyllite, with local concentrations of diaspore and cookeite (an aluminum lithium chlorite). Away from the core the ore grades into chalcedonic and chalcedonic-sericitic pyrophyllite. Selective open pit mining is followed by crushing, screening, and blending. About 80% of production is of chalcedonic pyrophyllite for refractories applications, principally zircon-pyrophyllite ladle bricks. The balance of production is about equally split between whiteware and filler applications. The latter uses a chloritic pyrophyllite as an alternative to talc in paint, plastics, and rubber. Mine production in recent years has paralleled the declining use of ladle bricks in favor of continuous casting in steel manufacture. Current production is estimated at approximately 5 ktpy.

In the Republic of South Africa, pyrophyllite, known as wonderstone, is mined near Ottosdal in the Lichtenburg District. The ore body is lenticular and hosted by the volcanic and rhyolitic rocks of the Dominion Reef series of the Witwatersrand System. Wonderstone is a massive block sedimentary rock composed approximately 90% of pyrophyllite, with chloritoid, epidote, and rutile. This highly compacted pyrophyllite is used in submerged pump bushings, abrasion/acid resistant pipe linings, and coal chute linings. It also finds use in high performance refractories for synthetic diamond manufacture. Current yearly production is about 3 kt.

Argentina currently produces about 1.3 kt of pyrophyllite, all from La Rioja Province, and primarily for use in ceramics. The smallest production reported by the USBM as pyrophyllite is that of Peru, at about 0.75 ktpy. Notable quantities in the past have been extracted from Junin, Lima, and Ica, but the origin of present production is not clear, nor are current uses. Minor production of pyrophyllite is inferred from production statistics of Egypt and Colombia. Both countries report combined soapstone/pyrophyllite production (plus steatite in Egypt).

A similar situation but on a much grander scale, exists with reported production from China. This country currently reports approximately 1 Mtpy production of talc and related materials (steatite, soapstone, pyrophyllite). Many hydrothermal pyrophyllite deposits have been identified along the coastal area of southeast China. In 1989, the United Nations Fund for Natural Resources Exploration entered into an agreement with China for the exploration of pyrophyllite in Fujian Province. This was based on the discovery in 1987 of an estimated 18 Mt of reserves at Dongzi and 1 Mt at Lingli, both in Fujian. China is believed to be a large pyrophyllite producer, with consumption primarily in whiteware and refractories.

Non-Commercial Deposits

All of the preceeding pyrophyllite producing countries also contain deposits that are not commercially exploited at the present time. For all of its distinct functionality in ceramics applications, pyrophyllite is considered basically an inexpensive industrial mineral. As such it can be subject to the problem of high transportation costs relative to its nominal price. It is also vulnerable in certain uses to substitution by even lower priced, locally available bulk fillers like calcium carbonate, or, on the other hand, higher priced but mineralogically purer clays and talcs. Japan's historical predominance in pyrophyllite production and consumption has been due in large part to its lack of talc and clay reserves. Alternatively, the presence of such reserves is likely to contribute to the lack of emphasis on development of pyrophyllite in other countries.

Turkey has in the past been reported to produce as much as 3 ktpy of pyrophyllite for captive use in ceramics manufacture. Mining was from the estimated 20 Mt deposit at Poturge Township in Malatya Province of eastern Turkey. This lenticular ore body grades from nearly pure pyrophyllite to pyrophyllite-sericite and is hosted by hydrothermally altered dacite tuffs. No pyrophyllite production in Turkey has been reported in recent years.

In the former Soviet Union, minor production of pyrophyllite reportedly occurred at the Kur'yanos deposit near Ovruch, Ukraine, up until 1980. Many deposits have been identified and described in the provinces of Ovruch, Caucasus, Ural, Kazakhstan, and Altai-Sayan (Zaykov, Udachin, and Sinyakovskaya, 1988). Production of Yugoslavian pyrophyllite has in the past been reported from deposits near Konjic in Hercegovina, although there has been no output recorded recently. Milled products from these deposits have been evaluated for general filler and ceramics applications and recently for paper fillers (Hanus et al., 1988), but their present status is unknown.

TECHNOLOGY

Exploration

Exploration programs for pyrophyllite are usually conducted in areas where geological evidence indicates that underlying rocks belong to acid-volcanic complexes. Pyrophyllite is generally found in hydrothermal alteration zones associated with faults or shears that transect these felsic-igneous rocks.

In semiarid to arid climates, aerial photographs are useful in reducing the area to be explored to a manageable size. This is achieved by plotting on the photographs linear features which may represent faults or shear zones. The geologist then examines these features on the ground and undertakes reconnaissance work along favorable zones. Progress has recently been reported in the use of airborn high resolution imaging spectrometers to map exposed hydrothermal alteration patterns and identify certain minerals, including pyrophyllite (Carrere, 1989, Akiyama et al., 1989).

Pyrophyllite is generally identified in the field by its translucent green color and waxy feel. Outcrops of chalcedonic pyrophyllite tend to be bolder than the enclosing rocks and display a characteristic weathering pattern, whereas outcrops of chloritic pyrophyllite are typically subdued and feature jagged boulders. When a discovery has been made, drilling or trenching is performed to provide fresh samples for preliminary laboratory evaluation.

Geochemical Evaluation: The pattern of alteration in pyrophyllite deposits is typically zoned. A central core of pure pyrophyllite or chloritic pyrophyllite may be surrounded by imperfectly segregated zones of chalcedonic pyrophyllite and sericitic pyrophyllite which grade into unaltered host rocks. Geochemical techniques have been found to be successful in delineating the zonal pattern and defining the margins of the ore body. In areas of good outcrop conditions, representative samples can be obtained from surface exposures without difficulty. In areas of poor outcrop conditions, shallow drilling or trenching may be necessary. Contoured plots of alumina and total-alkali values can then be prepared and superimposed on a geological map. The results not only define the zonal pattern but also provide a good basis for designing a detailed drilling program.

Drilling: Two kinds of drilling operations generally apply to pyrophyllite deposit development, rockbit (diamond or tungsten carbide) drilling and percussion drilling.

Rockbit drilling provides core samples that enable the geologist to predict the subsurface behavior of the ore body and to find and correlate the zonal pattern determined from surface observations. Drill holes are usually inclined at low angles perpendicular to the ore zone so as to penetrate the true width of steeply dipping ore bodies.

Percussion drilling is carried out as forerunner to production operations for detailed ore grade evaluation. Vertical holes are typically drilled on a square grid spacing, with chip samples collected at desired intervals. Compressed air raises drill chips to the surface and into a cyclone where they are removed from the airstream and discharged into a sample collecting tube.

Deposit Evaluation: Drilling samples are first graded according to the relative abundance of pyrophyllite vs. accessory minerals. Mineralogy is determined indirectly by chemical analysis, principally for Al_2O_3, K_2O, and Fe_2O_3, and directly by XRD. Depending upon anticipated end uses, additional testing, such as firing trials for refractoriness (pyrometric cone equivalent, or PCE) and fired color might be undertaken. A variety of grades is generally feasible for a given deposit. Evaluation of drilling samples determines reserves of exploitable ore grades, and the most suitable mining plan. The pyrophyllite grades produced are the most favorable balance of ore mineralogy and market requirements.

Mining Methods

Pyrophyllite is usually mined by conventional open pit methods, employing modern mining, earth moving, and bulk handling equipment. Open pit operations start with clearing of topsoil, which in the United States is generally stockpiled for use during site restoration. Standard drilling and blasting techniques are used to establish and develop pit faces according to a predetermined mine design. Ore blasted onto the pit floor is sorted according to size (oversize rock must be further broken), and according to grade from preliminary visual sorting. Lump pyrophyllite ore is then transported to the processing plant.

Underground mining of pyrophyllite is much less common, and is generally disfavored unless ore quality and market conditions can support the higher costs. Historically, nevertheless, deep deposits were often first mined underground. For example, earlier this century the Robbins mine in North Carolina was an extensive underground operation. The mine consisted of shafts, drifts, and crosscuts to reach the ore, which was mined from open square-set stopes. The underground workings were eventually abandoned in favor of open pit extraction, which continues to this day.

Similarly, the Gerhardt mine at Staley, NC, was started with a tunnel and raise. The top of the raise was widened into a glory hole and as extraction progressed, the operations grew into a large open pit.

Processing

Many processing methods apply to pyrophyllite and each operation has a processing plant layout based on its individual ore characteristics and market requirements. Nonetheless, the principal processing methods comprise crushing and screening to provide a series of size fractions for further grinding and/or beneficiation.

Strict quality control is essential during processing because pyrophyllite ore commonly consists of a mixture of soft pyrophyllite and relatively harder quartz. During crushing and screening, the softer pyrophyllite tends to become concentrated in the finer size fractions. This can be used to advantage in the production of higher grade, air floated products. A representative flowsheet of a pyrophyllite processing plant is shown in Fig. 2.

Wet beneficiation of pyrophyllite ore has on occasion been reported and may still be carried out in some countries. Heavy media separation of Hillsborough, NC, pyrophyllite/andalusite to obtain a purified andalusite product has been reported in the past.

Product Test Methods

The many uses of pyrophyllite have spawned correspondingly appropriate test methods that are used in both product and process quality control. Since the most important markets worldwide are in whiteware and refractories, most testing is geared toward these applications. Chemical analysis is concerned primarily with aluminum and alkali content and is complemented by XRD analyses for mineralogical composition. Other properties of interest for whiteware and refratory grades are screen residue and particle size distribution. These latter two are of common interest in nearly all applications. In addition, the assorted aqueous (e.g., paint, mastics) and non-aqueous (e.g., plastics, pesticide dust diluent) filler applications also variously require testing of BET surface area, moisture, water absorption, oil absorption, abrasivity, pH, and bulking values.

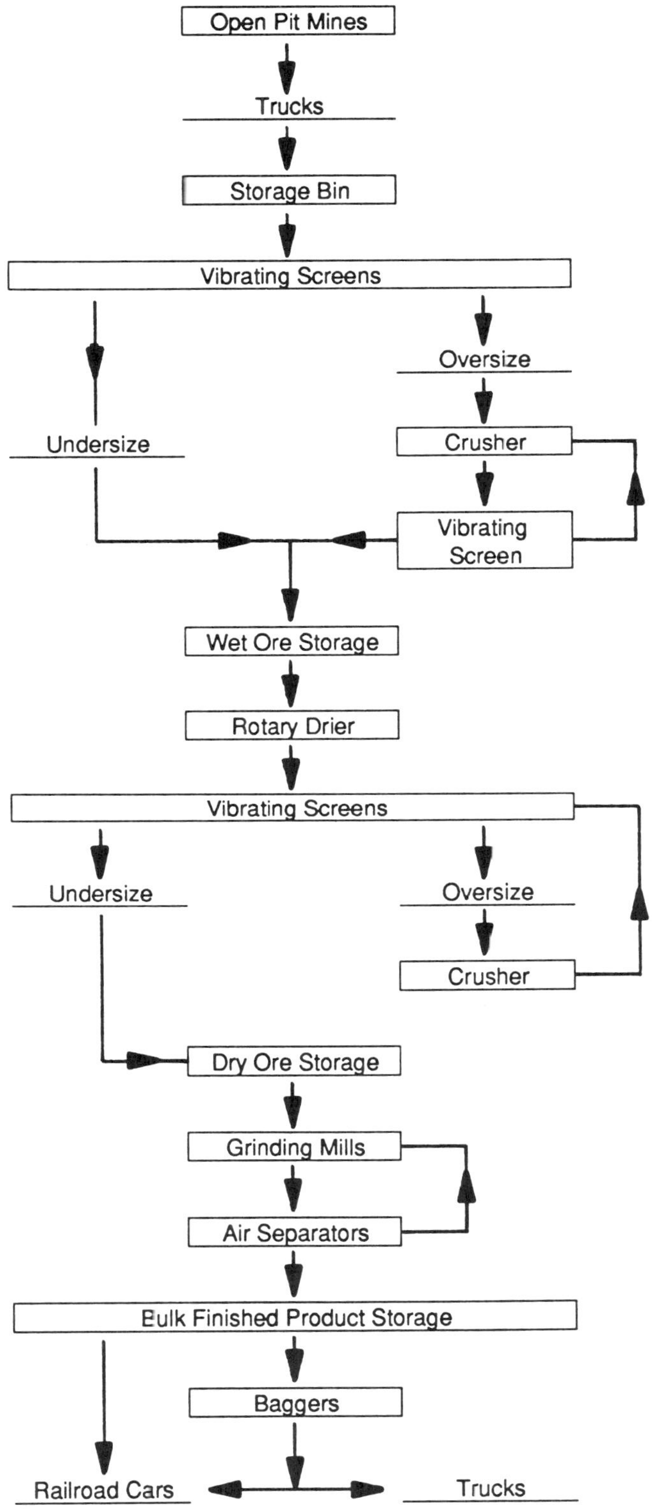

Fig. 2. Representative flowsheet of a pyrophyllite processing plant.

USES

Worldwide, the highest volume uses for pyrophyllite are in refractories (including foundry applications), whiteware, and pesticides. In the United States, the order of the first two is reversed. Lesser quantities are used in the manufacture of paint, cement, paper (Japan), soap (Brazil), soda glass, and fiberglass (Japan), construction materials (joint compounds, wallboard), rubber, animal feeds, plastics, synthetic diamonds (wonderstone), and road aggregate.

Refractories

Applications for refractory grade pyrophyllites include metal pouring refractories (ladle, tundish, cupola liners), kiln car furniture, insulating firebrick, monolithics (ramming mixes, gunning mixes), foundry mold coatings, and refractory mortars. The properties of pyrophyllite that are exploited for these applications are its permanent expansion on firing, excellent reheat stability, low hot load deformation, good hot creep resistance, low reversible thermal expansion, low thermal conductivity, and high resistance to corrosion by molten metals and basic slags. The largest refractory use is in pyrophyllite and pyrophyllite/zircon ladle bricks, although this market has seen some erosion due to new continuous casting techniques.

For ladle liners, and refractories in general, pyrophyllites with low sericite and low alkali content are required. Sericite and alkalis reduce the melting point of pyrophyllite. Alkali content should not exceed 1% by weight. Content of coloring oxides (Fe_2O_3 + FeO + TiO_2) is also preferably less than 1%. High pyrophyllite, or pyrophyllite-quartz products are generally used, with control of refractoriness (PCE), chemistry, mineralogy and particle size the main support for commercial specifications. Table 4 shows the typical properties of a refractory grade pyrophyllite available in the United States.

Whiteware

Applications for whiteware grade pyrophyllite include wall tile, floor tile, sanitaryware, pottery, chinaware, electrical porcelain, and glazes. Pyrophyllite has been used to advantage in whiteware in Japan and the United States for some time, and more recently in Australia and Southeast Asia, paralleling the growth in tile manufacture in those areas. Whiteware grade pyrophyllite offers increased firing strength in vitreous bodies, improved thermal shock resistance due to its low coefficient of thermal expansion, the ability to produce bodies with little or no shrinkage and reduced warpage, and the ability to produce low moisture expansion bodies with excellent craze resistance.

The major use of pyrophyllite is in wall tile, for which higher sericite grades are widely preferred. The high sericite content contributes to the formation of mullite at relatively low temperatures. The interlocking grain structure of mullite contributes to greatly increased fired strength in vitreous bodies, where the pyrophyllite can substitute for all the flint, or all the flint and some of the feldspar, without affecting shrinkage and water absorption. High sericite pyrophyllite is experiencing increasing use in monoporosa (porous, rapid single fire) tile bodies.

Whiteware grades of pyrophyllite generally benefit from low iron, and the relatively high alkali content that promotes early mullite formation. As with refractory grades, production of whiteware grades requires control of mineralogy, chemistry, and particle size. Table 5 shows the typical properties of a siliceous pyrophyllite-sericite widely used in the United States.

Pesticides

Pyrophyllite rose to prominence as the pesticide dust diluent of choice after World War II due to the widespread use at that time of DDT, with which it is particularly compatible. Although the use of DDT has been long since proscribed, and the use of pesticidal dusts has diminished in favor of more environmentally suitable (and effective) application forms, finely ground pyrophyllite is still a preferred diluent, extender, vehicle, and carrier. Its practically neutral pH, inertness, good flowability, and nonhygroscopic and fluffy characteristics make it well suited for this purpose. It is compatible with both acid and alkaline poisons. It allows longer storage of prepared insecticides without loss of effectiveness. It readily settles and covers the foliage of the plants. In passing through the blowers of dusting machinery it picks up an electrostatic charge causing its attraction and adherence to the underside of leaves in addition to coating the upper exposed plant surfaces.

The requirements for agricultural grade pyrophyllite are few compared to refractory and whiteware grades. Market acceptance is based mainly on fineness, pH, bulking values, moisture, and performance, rather than on chemistry, mineralogy, or color. Because of this, and the relatively low market value, agricultural grades are generally byproducts of deposits exploited for higher value materials.

Besides use in pesticides, pyrophyllite of similar quality is used in fertilizers, and in animal feed, where it can be used as a blending agent, anticaking agent, pelleting aid, or carrier.

Miscellaneous Uses

Paint applications use a finely ground, high brightness pyrophyllite for low cost pigment extension. The mineral's laminar nature also promotes good dispersion by inhibiting settling, helps film-dry, and increases resistance to film cracking. Worldwide, paint is the largest filler market for pyrophyllite. Use in paints is minor in the United States, generally utilizing relatively coarse grades for imparting mud-crack resistance to high build coatings (texture paints, block fillers), and checking/ cracking/ frosting resistance to exterior latex paints. In the Far East and in Brazil, however, very finely ground products have traditionally been used in place of less readily available talc and kaolin. In Brazil, paint is the largest single application for beneficiated agalmatolite, primarily in PVA-based paints where loadings are in the 20 to 25% range.

The second largest market for beneficiated agalmatolite in Brazil is soap manufacture, although this is not a common application elsewhere in the world. The mineral is used as a filler in bar and powdered soaps where it can replace the soda ash at loading levels up to 30%. The diaspore-pyrophyllite variety is generally used for its good color, dispersion, and bulk.

Fine ground, laminar pyrophyllite is commonly used in producing countries for the manufacture of wallboard, wallboard joint cements, and mastics. In all cases, good color and fine particle size are required. For joint cements in particular, water absorption is an important property, because this affects the critical cement properties of consistency, spreading, shrinkage on drying, and resistance to cracking.

Low iron pyrophyllite is used in the production of substantial tonnages of white and off-white cement in Japan. Compared to silica and clay mixtures, better color and lower clinker sintering temperatures due to pyrophyllite's higher reactivity are obtained.

In countries like Japan and Brazil, where supplies of talc and kaolin have traditionally been less accessible, fine ground high brightness, low abrasivity pyrophyllite has been commonly used as a paper and plastics filler. Similar grades with less stringent color requirements have been used as a rubber filler as well. In recent years, however, the difficulty in minimizing abrasivity (from quartz content), coupled with the improved availabilty of talc and/or kaolin, have caused some contraction in the use of pyrophyllite for these applications.

The reaction vessel used for the manufacture of synthetic diamonds is usually composed of South African wonderstone. This pyrophyllite variant is easily machined to accurate dimensions, remains inert during the reaction, is an excellent electrical insulator, and forms a superior seal because it flows during the reaction.

Table 4. Typical Properties of US Refractory Grade Pyrophyllite

Refractory grade (RG) pyrophyllite is used in:
1) Insulating firebrick
2) Metal pouring refractories
3) Alumina-silica monolithic refractories
   - Ramming mixes
   - Gunning mixes
   - Castable mixes
4) Kiln car refractories

General Description:

Description	Pyrophyllite ore with low alkali content
Crystalinity	Laminar to semimassive
Hardness	1 to 2 (Mohs scale)
Workability	Nonplastic
Density	2.8 to 2.9 Mg/m^3
PCE, ASTM C-24 (Anon., 1990)	27 to 28

Typical Chemical Analysis:
(Calculated as oxides)

SiO_2	75.0%
Al_2O_3	19.3
Fe_2O_3	0.8
K_2O	0.2
Na_2O	0.1
CaO	0.1
MgO	0.1
LOI	3.9

X-ray and Petrographic Analysis:

Mineral:	Composition:	Amount %
Pyrophyllite	$Al_2O_3 \cdot 4SiO_2 \cdot H2O$	40 to 50
Quartz	SiO_2	30 to 45
Kaolin	$Al_2O_3 \cdot 2SiO_2 \cdot 2H_2O$	5 to 15
Muscovite	$K_2O \cdot 3Al_2O_3 \cdot 6SiO_2 \cdot 2H_2O$	1 to 3

Typical Screen Analysis and Bulk Densities of RG Pyrophyllite Grades:

Screen Nos.	Cumulative Percent Retained				
US Series	RG 3/8	RG 1/4	RG 16	RG 140	RG 200
4	17.0	2.2			
8	40.0	25.1			
16	57.8	45.9	1.0		
30	72.5	61.2	14.5		
50	80.8	72.0	41.0		
100	84.4	78.7	62.4	1.0	trace
200	88.6	84.9	74.9	14.0	1.0
325	92.0	92.0	81.9	21.5	4.6
Compact Bulk Kg/m^3	1617	1585	1537	977	945

Properties:
RG pyrophyllite has the following properties of particular interest for refractory applications:
1) Permanent expansion
2) Excellent reheat stability
3) Low hot load deformation
4) Low reversible thermal expansion
5) Low bulk density
6) Low thermal conductivity
7) Highly resistant to corrosion by molten metals and basic slags

The permanent expansion of RG pyrophyllite is due to physical changes taking place during dehydration. The greatest expansion occurs between 650 and 870°C which corresponds approximately to the range of greatest weight loss. The amount of expansion is dependent on the particle size. The reversible thermal expansion of RG pyrophyllite is low for specimens fired below C/13 (~1 330°C).
This characteristic is dependent on heat treatment, for at higher temperatures the silica is gradually converted to cristobalite and there is corresponding increase in the expansion.
The dimensional stability of RB pyrophyllite is excellent

Initial Firing Temperature	% Linear Change After 3 Cycles of Reheat at 1 350°C	
°C	RG 1/4	RG 16
1 349	+0.10	+0.08
1 399	+0.10	+0.07
1 449	+0.08	+0.07

Compositions containing RG pyrophyllite show superior hot load bearing ability. Resistance to thermal shock (spalling) is high by virtue of its high thermal conductivity and low coefficient of thermal expansion. Above 1 050°C a transformation of pyrophyllite to mullite corresponds to significant increases in mechanical properties, such as compressive strength, flexural strength, and hardness.
RG Pyrophyllite Applications:

	RG 1/4	RG 3/8	RG 16	RG 140	RG 200
Insulating Firebrick				X	X
Metal Pouring Refractories				X	
Foundry Mold Dressings				X	
Investment Casting Molds			X		
Castables	X				
Stiff Plastic Refractories	X	X			
Gunning Mixes	X	X			
Kiln Car Refractories	X	X	X		

Table 5. Typical Properties of US Whiteware Grade Pyrophyllite

High sericite (HS) grade pyrophyllite for whiteware applications is used in:

- Wall tile
- Floor tile
- Sanitaryware
- Electrical porcelain
- Stoneware
- Earthenware
- Glazes
- Refractories

General Description:

Description:	Pyrophyllite ore with high alkali content:
Crystallinity	Laminar to semimassive
Hardness	1 to 2 (Moh's scale)
Workability	Nonplastic
Density	2.8 to 2.9 Mg/m^3
PCE, ASTM C-24 (Anon., 1990)	20 to 21

Typical Chemical Analysis:
(Calculated as oxides)

SiO_2	80.9%
Al_2O_3	13.8
Fe_2O_3	0.2
TiO_2	0.1
Na_2O	0.4
K_2O	2.3
LOI	2.3

X-ray and Petrographic Analysis:

Mineral:	Composition:	Amount %
Pyrophyllite	$Al_2O_3 \cdot 4SiO_2 \cdot H_2O$	20 to 30
Quartz	SiO_2	50 to 60
Sericite	$K_2O \cdot 3Al_2O_3 \cdot 6SiO_2 \cdot 2H_2O$	20 to 25
Kaolin	$Al_2O_3 \cdot 2SiO_2 \cdot H_2O$	5 to 10

Particle Size Distribution:
Coulter Counter

Particle Size (μm)	%< Indicated Size
40	90
30	78
20	58
10	28
5	10.6
2	1.6
1	nd

Compact Bulk:

Kg/m^3	1137

Properties:

HS pyrophyllite has the following properties of particular interest for whiteware applications:

1) Lowers the firing temperature.
2) Produces low moisture expansion bodies resulting in good craze resistance.
3) Produces bodies with little or no shrinkage and reduced warpage. HS pyrophyllite shows some 3.5% expansion from cone 016 (~777°C) to cone 01 (~1 127°C), and cone 8 (~1 249°C) needs to be reached before any shrinkage from the original die size occurs.
4) Increases thermal shock resistance due to its high thermal conductivity and low coefficient of thermal expansion.
5) Greatly increases firing strength in vitreous bodies.

HS pyrophyllite aids in the formation of mullite at a comparatively low temperature. The mullite, with its interlocking grain structure, results in greatly increased fire strength in vitreous bodies, such as sanitaryware, where the HS pyrophyllite can substitute for flint and some of the feldspar. This reformulation will result in reduced pyroplastic deformation, increased thermal shock resistance, and faster firing capabilities.

In semivitreous ware, such as wall tile, the use of HS pyrophyllite results in low moisture expansion bodies with good craze resistance. In addition, the nature of pyrophyllite to expand on heating results in more uniform size and less warpage.

Typical body formulas using HS pyrophyllite:

	Wall Tile	Wall Tile	Wall Tile	Floor Tile	Electrical Porcelain
Flint	10 %	17 %			
HS Pyrophyllite	40	16	7 %	16 %	19 %
Ball Clay	18	24	28	7	16
China Clay	18			25	30
Tremolitic Talc, 200 m	12	35	61		
Whiting	2	8	1		
Feldspar			3	52	35
	100 %	100 %	100 %	100 %	100 %
Firing Cone	6 to 9	3 to 5	1	9	11
% Absorption	16.0	13.0	13.5	0.0	0.1
% Shrinkage	1.6	0.4	1.2	11.2	9.0

THE FUTURE

In recent years, a strong market has developed for the use of low iron (<0.5% Fe_2O_3) kaolin clay in the manufacture of reinforcement grade fiberglass. Its main function is to replace the more expensive alumina. In Japan, pyrophyllite has made inroads into this market because of its lower cost and ready availability vs. kaolin. Elsewhere, the potential for pyrophyllite to compete with kaolin in this application is small.

In the mainstay ceramics markets, little overall growth is foreseen for pyrophyllite outside of Japan, although the mineral might find expanded use in certain niches. For example, it has earned considerable attention, and lately growing use, in the manufacture of foundry mold dressings. Iron and steel cast in conventional sand molds can form a thin skin from reaction between the carbon in the metal and the steam evolved from the mold. This skin blunts cutting and finishing tools and adversely affects the frictional properties of the casting surface. Skin formation is prevented by treating the mold surface with a dressing containing pyrophyllite. The pyrophyllite penetrates the surface layer of the mold and seals the pores. At casting temperatures, the pyrophyllite expands and forms a hard impervious refractory coating with excellent heat insulating properties. This barrier prevents metal-mold reaction, eliminates skin formation, and generally improves the surface finish of the casting (Reisner, 1962).

Over the years, the use and growth of pyrophyllite in world markets has been variously promoted and inhibited by the nearly ubiquitous association of pyrophyllite with quartz in both ore and commercial products. In most applications, the presence of quartz is either beneficial or innocuous. In other uses, the abrasivity imparted by the quartz has been a traditional detriment. Since the opinion by the International Agency for Research on Cancer (IARC) in 1987 that crystalline silica is a potential carcinogen, health and liability issues have been raised that affect many industrial minerals, including pyrophyllite. Such considerations would not seem to greatly impact pyrophyllite's future in its core ceramics and foundry applications. These industries are well acquainted with the proper handling and use of quartz, routinely dealing as they do with silica per se as a basic raw material. Considering that the performance advantages of pyrophyllite are not readily replaced, and that quartz impurites are generally of no technical detriment, the future of pyrophyllite in these applications should be reasonably secure.

The use of fine ground pyrophyllite in pesticide dusts has been in decline in the United States over the past decade. Part of this has been due to the move toward more sophisticated and effective application forms. But cost and quartz content have also worked against pyrophyllite, and in favor of less effective but less expensive and perceptually safer minerals like calcium carbonate.

Judging by present political and economic conditions, the 1990s will likely see the development and maturation of a true world economy. This will entail world class production techniques and quality control practices. Pyrophyllite is at something of a disadvantage as a raw material in this scenario because of its relative rarity, its perceived low inherent value vs. transportation costs, and because of variations in mineralogy, properties, and performance from producer to producer and country to country. The challenge to the pyrophyllite producer is to take advantage of these disadvantages. Capitalizing on cost/performance benefits for applications where transportation costs do not figure prominently will always be important. Essential, however, will be the need to develop and exploit the inherent technical superiority of pyrophyllite in specific applications, so that cost becomes of secondary importance and allows for upgrading or modification of the pyrophyllite so that performance can be optimized.

ACKNOWLEDGMENTS

The authors are grateful to Mr. Konrad Rieger of the R.T. Vanderbilt Co. for his suggestions and comments regarding the ceramic applications of pyrophyllite. Many thanks also to *Industrial Minerals* magazine for the many fine articles over the years dealing with pyrophyllite in particular and industrial minerals in general, which have provided the context of our presentation on the world-wide production and use of pyrophyllite.

REFERENCES

Anon., 1985, "PYRAX RG," Technical Data Sheet, R.T. Vanderbilt Co., Norwalk, CT.

Anon., 1990, "Designation C-24," Annual Book of ASTM Standards, 1990, Vol. 15.01, American Society for Testing and Materials, Philadelphia, PA, pp. 27–29.

Akiyama, Y., et al., 1989, "Digital Processing and Analysis of Airborne Multispectral Data for Mapping Hydrothermal Alteration at Yerington, Nevada," *Proceedings*, 7th Thematic Conference on Remote Sensing for Exploration Geology: Methods Integration Solutions, Report No. 718, Vol. 2, Environmental Research Institute of Michigan, Ann Arbor, MI, pp. 969–980.

Benbow, J., 1988, "Pyrophyllite: Far East Steels the Market," *Industrial Minerals*, No. 249, June, pp. 37–49.

Brindley, G.W., and Wardle, R., 1970, "Monoclinic and Triclinic Forms of Pyrophyllite and Pyrophyllite Anhydride," *American Mineralogist*, Vol. 55, pp. 1259–1272.

Carrere, V., 1989, "Mapping Alteration in the Goldfield Mining District, Nevada, with the Airborne Visible/Infrared Imaging Spectrometer (AVIRIS)," *Proceedings*, 7th Thematic Conference on Remote Sensing of Exploration Geology: Methods Integration Solutions, Report No. 725, Vol. 1, Environmental Research Institute of Michigan, Ann Arbor, MI, pp. 365–378.

Cornish, B., 1983, "Pyrophyllite," *Industrial Minerals and Rocks*, 5th ed., Vol. 2, S.J. Lefond, ed., AIME, New York, pp. 1085–1108.

Deer, W.A., Howie R.A., and Zussman, J., 1967, "Pyrophyllite," *Rock Forming Minerals*, Vol. 3, Longmans, London, pp. 115–120.

Evans, B.W., and Guggenheim, S., 1988, "Talc, Pyrophyllite, and Related Minerals," *Reviews in Mineralogy—Hydrous Phyllosilicates*, Vol. 19, Mineralogical Society of America, Washington, DC, pp. 225–294.

Hanus, V., Malechova, V., and Discantiny, P., 1988, "Konjic Pyrophyllite as a Raw Material for Production of Paper Fillers," *Papir a Celulosa*, Vol. 43, No. 12, pp. 248–250.

Huifen, Z., Damei, C., and Zechun, J., 1989, "A Study on Mineralogy and Mineral Physics of Chinese Pyrophyllite," *Developments in Geoscience: Contributions to the 28th International Geological Congress*, Science Press, Beijing, pp. 137–145.

Mellor, J.W., 1927, *A Comprehensive Treatise on Inorganic Chemistry*, Vol. 16, Longmans, London, pp. 498.

Olmstead, D., 1827, "A Report on the Geology of North Carolina," Conducted under the Board of Agriculture, Raleigh, NC, 1825–1827, J. Gales and Son.

Papezik, V.S., Keats, H.F., and Vahtra, J., 1978, "Geology of the Foxtrap Pyrophyllite Deposit, Avalon Peninsula, Newfoundland," *CIM Bulletin*, Mar., pp. 152–160.

Perez-Rodriguez, J.L., Del Villar, L.M.S., and Sanchez-Soto, P.J., 1988, "Effects of Dry Grinding on Pyrophyllite," *Clay Minerals*, Vol. 23, pp. 399–410.

Prasad, C.S., Mukhopadhyay, T.K., and Maiti, K.N., 1989, "Characterization and Utilization of Some Indian Pyrophyllites," *Transactions*, Indian Ceramics Society, Vol. 48, No. 5, pp. 95–104.

Reisner, H.O., 1962, "Some Aspects of the Formation and Structure of a Skin on Iron Castings and a Method Used to Obviate Its Occurrence," *British Foundryman*, Sep., pp. 362–369.

Russell, A., 1991, "India's Industrial Minerals," *Industrial Minerals*, No. 288, Sep., pp. 45–55.

Shin, B.J., Kook, N.P., and Choi, J.M., 1990, "A Study on the Improvement of Value Added of Pyrophyllite by Mineral Processing," *Han'guk Chawon Konghak Hoechi*, Vol. 27, pp. 20–30.

Sykes, M., and Moody, J., 1978, ''Pyrophyllite and Metamorphism in the Carolina Slate Belt,'' *American Mineralogist*, Vol. 63, pp. 96–108.

Virta, R.L., 1991, ''Talc and Pyrophyllite: 1990 Annual Report,'' US Bureau of Mines, Washington, DC, pp. 1–7.

Wardle, R., and Brindley, G.W., 1972, ''The Crystal Structures of Pyrophyllite, 1TC, and of Its Dehydroxylate,'' *American Mineralogist*, Vol. 57, pp. 732–750.

Zaykov, V.V., Udachin, V.N., and Sinyakovskaya, I.V., 1988, ''Pyrophyllite Deposits,'' *International Geology Review*, Vol. 30, pp. 90–103.

RARE EARTH MINERALS

STEPHEN B. CASTOR

The rare earth elements (REE) which include the 15 lanthanide elements (Z = 57 through 71) and yttrium (Z = 39) are so called because the elements were originally isolated in the late 18th and early 19th centuries as oxides from rare minerals. Most REE are not as uncommon in nature as the name implies. Cerium, the most abundant REE (Table 1), comprises more of the earth's crust than copper or lead. Many REE are more common than tin and molybdenum, and all but promethium are more common than silver or mercury (Taylor, 1964). Promethium (Z = 61) is best known as an artificial element, but has been reported in very minute quantities in natural materials. Lanthanide elements with low atomic numbers are generally more abundant in the earth's crust than those with high atomic numbers. In addition, lanthanide elements with even atomic numbers are two to seven times more abundant than adjacent lanthanides (Table 1) with odd atomic numbers.

The lanthanide elements traditionally have been divided into two groups: the light rare earths (LREE), lanthanum through europium (Z = 57 through 63); and the heavy rare earths (HREE), gadolinium through lutetium (Z = 64 through 71). Although yttrium is the lightest REE, it is usually grouped with the HREE to which it is chemically and physically similar.

The REE are lithophile elements (elements enriched in the earth's crust) that invariably occur together naturally because all are trivalent (except for Ce^{+4} and Eu^{+2} in some environments) and have similar ionic radii. Increase in atomic number in the lanthanide group is accompanied by addition of electrons to an inner level rather than the outer shell. Consequently, there is no change in valence with change in atomic number, and the lanthanide elements all fall into the same cell of the periodic table. The chemical and physical differences that do exist within the REE group are caused by small differences in ionic radius, and generally result in segregation of REE into deposits enriched in either light lanthanides or heavy lanthanides plus yttrium.

The relative abundance of individual lanthanide elements has been found useful in the modelling of rock-forming processes. Comparisons are generally made using a logarithmic plot of lanthanide abundances normalized to abundances in chondritic (stony) meteorites. The use of this method eliminates the abundance variation between lanthanides of odd and even atomic number, and allows determination of the extent of fractionation between the lanthanides because such fractionation is not considered to have taken place during chondrite formation. The method is also useful because chondrites are thought to be compositionally similar to the original earth's mantle. Europium anomalies (positive or negative departures of europium from chondrite-normalized plots) have been found to be particularly effective for petrogenetic modelling.

REE were originally produced in minor amounts from small deposits in granite pegmatite, the geologic environment in which they were discovered. During the second half of the 19th century and the first half of the 20th century, REE came mainly from placer deposits. With the exception of the most abundant lanthanide elements (cerium, lanthanum, and neodymium), individual REE were not commercially available until the 1940s. Since 1965, most of the world's REE have come from two hard rock deposits: Mountain Pass, United States, and Bayan Obo, China.

GEOGRAPHIC DISTRIBUTION OF REE DEPOSITS

More than 70% of the world's REE raw materials come from three countries: China, the United States, and Australia. China emerged as a major producer of REE raw materials during the 1980s, while Australian and United States market share decreased dramatically (Fig. 1). Table 2 gives recent annual production figures along with estimated reserves by country, and Fig. 2 shows locations of significant REE mining.

MINERALS THAT CONTAIN REE

Although REE comprise significant amounts of many minerals, almost all production has come from less than ten minerals. Table 3 lists minerals that have yielded REE commercially or have potential for production in the future. Extraction from a potentially economic REE resource is strongly dependant on its REE mineralogy. Minerals that are easily broken down, such as bastnasite, are more desirable than those that are difficult to dissociate, such as allanite. In general, producing deposits contain REE-bearing minerals that are relatively easy to concentrate because of coarse grain size or other attributes. For more thorough discussions of REE-bearing minerals see Mariano (1989a) and Cesbron (1989).

GEOLOGY OF REE DEPOSITS

Carbonatite Deposits

The Mountain Pass, CA, carbonatite deposit has been in production continuously since 1954 by Molycorp, Inc. (originally Molybdenum Corp of America) and is the only deposit mined exclusively for REE. In 1987, proven and probable reserves totalled 29 Mt, with an average grade of 8.9% rare earth oxide (REO) based on a 5% cutoff grade (Castor, 1991). Bastnasite is the only mineral processed, although nine other REE minerals occur at Mountain Pass. The ore typically contains 10 to 15 % bastnasite, 65 % calcite and/or dolomite, and 20 to 25 % barite. Other gangue minerals, such as strontianite and talc, are present in significant amounts locally. Galena is locally abundant, but other sulfide minerals are rare. LREE dominate the Mountain Pass bastnasite concentrate (Fig. 3).

The Mountain Pass carbonatite, which has been dated at about 1.4 b.y. (DeWitt et al., 1987), is a moderately dipping, tabular

Table 1. Rare Earth Elements, Atomic Numbers, and Abundances

ELEMENT	SYMBOL	ATOMIC NUMBER	CRUSTAL ABUNDANCE	CHONDRITE ABUNDANCE
Yttrium	Y	39	33	—
Lanthanum	La	57	30	0.34
Cerium	Ce	58	60	0.91
Praseodymium	Pr	59	8.2	0.121
Neodymium	Nd	60	28	0.64
Promethium	Pm	61	—	—
Samarium	Sm	62	6	0.195
Europium	Eu	63	1.2	0.073
Gadolinium	Gd	64	5.4	0.26
Terbium	Te	65	0.9	0.047
Dysprosium	Dy	66	3	0.30
Holmium	Ho	67	1.2	0.078
Erbium	Er	68	2.8	0.20
Thulium	Tm	69	0.48	0.032
Ytterbium	Yb	70	3	0.22
Lutetium	Lu	71	0.5	0.034

Crustal abundance data in ppm from Taylor (1964). Chondrite abundance data in ppm from Wakita, Rey, and Schmitt (1971).

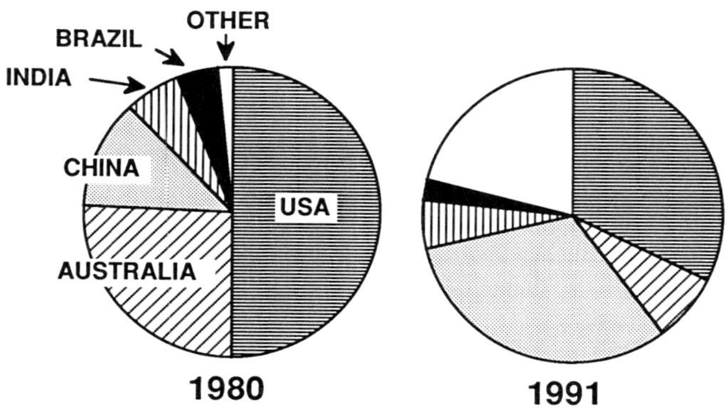

Fig. 1. Production of rare earths by country in 1980 and 1991. *Source*: Hedrick (1993). Data for 1980 do not include USSR production, while those for 1991 do, explaining the large change in other.

Table 2. Annual Rare Earth Oxide Production from 1987 through 1990 and Estimated Present Reserves by Country in Tons

COUNTRY	1988	1989	1990	1991[a]	RESERVES
United States[c]	11 533	20 787	22 713	16 465[b]	12 600 000
China	29 640	25 220	16 480	16 150[b]	43 000 000
USSR	8 500[a]	8 500[a]	8 500[a]	8 500	450 000
Australia	6 530	7 150	6 050	3 850	5 100 000
India	2 200	2 365	1 925	2 750	2 300 000
Brazil	1 690	1 900	1 100	1 100	280 000
Malaysia	1 630	1 646	1 825	1 050	30 000
Thailand	375	365	385	360	1 000
Canada	100	100	—	—	940 000
Other	862	866	962	890	19 600 000

Source: Hedrick (1991a, 1993). [a]estimated, [b]reported, [c]REO from bastnasite only.

intrusion (Fig. 4) into granulite-grade gneiss. It is associated with potassic alkaline plutons of similar age, size and orientation, as well as with abundant carbonatite and alkaline dikes. The Mountain Pass carbonatite/alkaline complex is part of a narrow north-trending zone of potassic alkaline igneous rocks at least 130 km long (Castor and Gleason, 1989).

Table 3. Minerals that Contain Rare Earth Elements and Occur in Economic or Potentially Economic Deposits

Mineral	Formula	REO wt%
Aeschynite	$LnTiNbO_6$	36
Allanite (orthite)	$(Ca,Ln)_2Fe\ Al_2Si_3O_{12}(OH)$	30
Anatase	TiO_2	3
Ancylite	$Sr\ Ln(CO_3)_2(OH)H_2O$	46
Apatite	$CaPO_4$	19
Bastnasite	$Ln\ CO_3\ F$	75
Brannerite	$(U,Ca,Fe,Ln,Th)_3Ti_5O_{16}$	6
Britholite	$(Ln,Ca)_5(SiO_4,PO4)_3(OH,F)$	62
Cerianite	CeO_2	81[a]
Churchite	YPO_4H_2O	44[a]
Eudialyte	$Na(Ca,Ln)_2(Fe,Mn,Y)ZrSi_8O_{22}(OH,Cl)_2$	10
Euxenite	$(Ln,Ca,U,Th)(Nb,Ta,Ti)_2O_6$	<40[a]
Fergusonite	$(Ln,Fe)(Ta,Nb,Ti)O_4$	47
Florencite	$Ln\ Al_3(PO_4)_2(OH)_6$	32[a]
Gadolinite	$Be_2Ln_4FeSi_2O_{13}$	52
Huanghoite	$BaLn(CO_3)_2F$	38
Kainosite	$Ca_2(Y,Ln)_2(Si_4O_{12})CO_3H_2O$	38
Loparite	$(Ln,Na,Ca)(Ti,Nb)O_3$	30
Monazite	$LnPO_4$	71
Mosandrite	$(Ca,Na,Ln)_{12}(Ti,Zr)_2Si_7O_{31}H_6F_4$	<65[a]
Parisite	$Ca\ Ln_2(CO_3)_3F_2$	64
Samarskite	$(Ln,U,Ca,Fe,Pb,Th)(Nb,Ta,Ti,Sn)_2O_6$	12
Synchisite	$Ca\ Ln(CO_3)_2F$	51[a]
Thalenite	$Y_4Si_4O_{14}\ H_2O$	63[a]
Xenotime	$Y\ PO_4$	61[a]
Yttrotantalite	$(Fe,Y,U,Ca)(Nb,Ta,Ti,Sn)O_4$	<24[a]
Zirconolite	$CaZrTi_2O_7$	6

Sources: Kapustin, 1980, Overstreet, 1967, Frondel, 1958, Joint Committee on Powder Diffraction Standards, (Anon., 1980), Mariano, 1989a, Mazzi and Munno, 1983. Where more than one analysis is available, the analysis with the highest REO content is reported (e.g., REO for monazite from the Mountain Pass carbonatite is reported; monazite from pegmatites and metamorphic rocks generally has low REO). Ln = lanthanide elements, [a] stoichiometric calculation of REO content.

The Mountain Pass deposit shares features with other carbonatites in the world. It has textural and structural features supporting igneous intrusion (Olson et al., 1954); it has associated alkali-rich fenitic alteration typified by development of sodic amphibole and potash feldspar; and it has low $^{87}Sr/^{86}Sr$ (Powell et al., 1966). However, the mineralogy and chemistry of the Mountain Pass ore body, in addition to its association with potassic alkaline plutons, serve to distinguish it from the more common niobium- and phosphorus-rich carbonatites in nephelinite or nepheline syenite complexes. In addition, the Mountain Pass ore body is the only large carbonatite mass that contains significant amounts of a REE mineral as a primary phase.

Elsewhere in the western United States, younger carbonatites have been investigated as possible sources of REE. Most have high LREE/HREE ratios, but none rival the Mountain Pass deposit as a REE resource. The large Cambrian Iron Hill carbonatite in the Powderhorn district of Colorado has been considered a possible source of REE, but Armbrustmacher (1980) reported an average grade of about 0.5% REE. Carbonatite dikes in the Powderhorn district that contain up to 3% REO constitute a resource of unknown size (Olson and Hedlund, 1981), and dikes in the nearby Wet Mountains constitute a small resource with an average grade of about 2.5% REO (Armbrustmacher, 1988). Phanerozoic carbonatite dikes that contain as much as 21% REO (Crowley, 1960), mainly in monazite, allanite, and ancylite (Heinrich and Levinson, 1961) are present in the North Fork area on the Idaho-Montana border, but

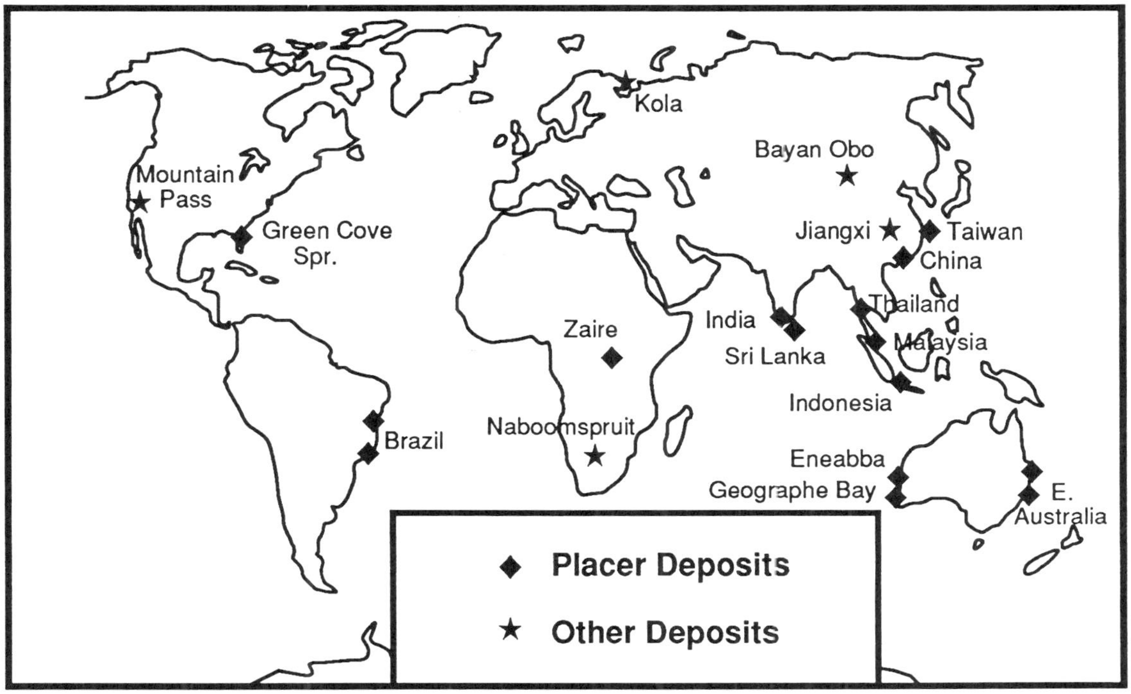

Fig. 2. Locations of the world's rare earth mines.

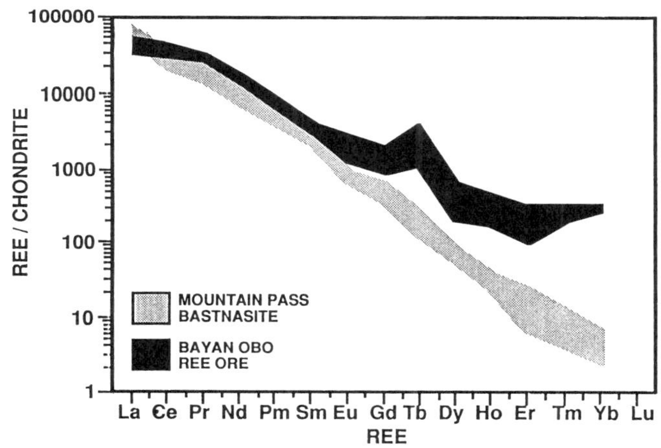

Fig. 3. Rare earth element/chondrite distributions for the Mountain Pass and Bayan Obo REE deposits. Mountain Pass curve constructed from values obtained by dividing bastnasite concentrate values by 10. Bayan Obo data from Qiu et al. (1983).

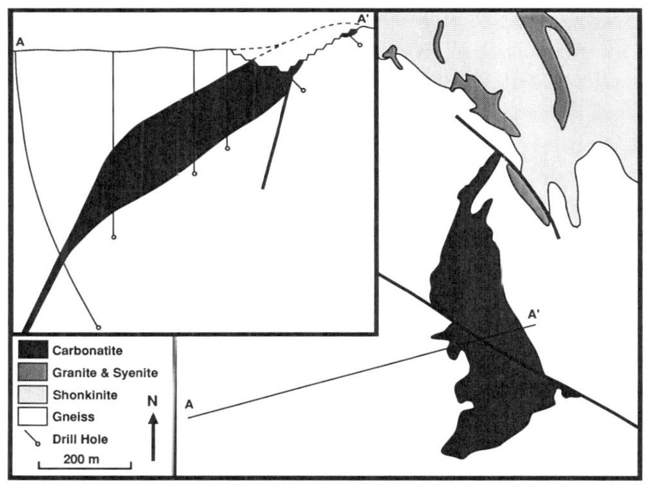

Fig. 4. Simplified geologic map and cross section of the Mountain Pass ore body. Map modified from Olson et al. (1954). Cross section modified from Castor (1991).

the size of this resource is probably small because dikes with high REO contents are generally less than 1 m thick.

Carbonatite complexes in eastern Canada also contain REE mineralization. A britholite-bearing carbonatite dike in Mesozoic carbonatite at Oka, Quebec, is a REE resource of unknown size and grade (Mariano, 1989a). Bastnasite and monazite zones containing up to 4.5% LREE have been identified in the Late Proterozoic St. Honore complex in Quebec, with one 80-m drill intercept averaging over 3% REO (Vallee and Dubuc, 1970).

The Mount Weld deposit in southern Western Australia is in the laterized cap of a large circular carbonatite dated at 2.0 b.y. Although LREE/HREE ratios are generally high, the laterite is locally enriched in HREE and yttrium. Reserves include 15.4 Mt with 11.2% REO + Y_2O_3 based on a cutoff grade of 5.0% and 4.0 Mt containing 0.36% Y_2O_3 based on a 0.20% cutoff grade, and the laterite is also a niobium and tantalum resource (Duncan and Willett, 1990). Unweathered carbonatite at depth contains REE and niobium in amounts that are common in carbonatites. REE enrichment at Mount Weld is thought to have been caused by long-term leaching and redeposition by groundwater; it mainly consists of REE phosphates (Mariano, 1989a) including the yttrium-rich mineral churchite (Lottermoser, 1988). The Yangibana REE prospect in northern Western Australia is gossan containing as much as 11% REO (O'Driscoll, 1988). The deposit is thought to have formed from a carbonatite, and is associated with Early Proterozoic alkaline rocks.

Brazilian carbonatites contain significant REE resources, but have yielded only minor production to date. Niobium ore in the Barreiro complex at Araxá, Minas Gerais, which comprises 450+ Mt averaging 2.5% Nb_2O_5, contains about 4.4% REO (Issa Filho et al., 1984). In addition, the complex contains 800 kt of supergene-enriched laterite that averages 13.5% REO, mainly in the phosphate minerals gorceixite and goyazite (Mariano, 1989a). In the Catalao I complex in Goiás about 300 km southeast of Araxá, phosphatic laterite comprises a resource of about 2 Mt averaging 12% REO. Laterized carbonatite in the Maicuru complex, Pará, northern Brazil, contains phosphatic material with as much as 17% REE (Costa et al., 1991). Cerro Impacto, a large circular structure thought to be a deeply weathered carbonatite that is located in the Amazon region of Venezuela, contains as much as 11% REE in thick laterite (O'Driscoll, 1988). Other poorly explored carbonatites in the Amazon region of South America are known to host supergene REE mineralization (Mariano, 1989b).

Deeply weathered Brazilian carbonatites that contain large titanium resources in the form of anatase with associated REE phosphate minerals are also a potential source of by product REE (Mariano, 1989a). Titanium concentrates from a pilot plant contain more than 3 % REE which must be removed to produce commercial titania. Reserves are known to be substantial; the largest deposit, at Serra Negra, Minas Gerais, contains 200 Mt averaging 27.7% TiO_2 (Anon., 1984). These resources have high LREE/HREE ratios.

African carbonatites also contain REE resources with high LREE/HREE ratios that are largely unexploited. Kangankunde Hill, Malawi, is underlain by carbonatite dikes in which the main REE phase is monazite that is almost thorium-free and is associated with florencite and bastnasite (Garson, 1966). The dikes are thought to contain a resource of several hundred thousand tons of monazite at an average grade of more than 5% monazite (Deans, 1966). Mariano (1989a) gave evidence that suggests that monazite in the Kangankunde Hill carbonatite is of primary igneous origin (like the bastnasite at Mountain Pass). Carbonatite at Mrima Hill, Kenya, is overlain by monazite-rich laterite including 6 Mt of material that contains about 5% REO, and carbonatite dikes at Wigu Hill, Tanzania, contain up to 20% REO as monazite and REE fluorocarbonate minerals of possible hydrothermal origin (Deans, 1966). The Phalaborwa carbonatite in South Africa, which occurs in a large Early Proterozoic potassic alkaline complex, has significant copper and apatite production and has been evaluated as a source of REE. Potential byproduct from apatite concentrates that contain 0.4 to 0.9% REO has been estimated at 4 250 tpy REO (Russell, 1977, as cited in Notholt et al., 1990).

Sill-like lenses of bastnasite-fluorite-barite rock as much as 30 m thick with about 5% REE are associated with carbonatite at Kizilçaören, Turkey (Hatzl et al., 1990). The carbonatite occurs in dikes associated with intrusive Tertiary phonolites and trachytes along with alkaline pyroclastic rocks. The Kizilçaören resource is dominantly LREE (Morteani, private communication, 1991).

Iron-REE Deposits

Iron-LREE-niobium deposits at Bayan Obo in Inner Mongolia, China, are said to constitute the largest REE resource in the world, at 48 Mt REO (Drew et al., 1990). REE are currently recovered as a byproduct of iron production from two large ore bodies (Fig. 5) that contain 33 to 35% iron and 6% REE (Argall, 1980). The iron ore consists of several types, including massive magnetite ore, banded magnetite-hematite-fluorite-REE mineral-sulfide ore, magnetite-dolomite ore, and riebeckite- and aegerine-bearing ores (Qiu, private communication, 1984). REE are mainly in bastnasite and monazite, but aeschynite and barium-REE fluorocarbonates such as huanghoite are also present. The mineralization is in dolomite in a Middle Proterozoic sedimentary sequence (Qiu et al., 1983), but the age of mineralization is not well defined. Isotopic dates of mineralized rocks range between 1400 and 240 m.y. (Chao et al., 1993; Drew et al., 1990). The Bayan Obo ore is similar to the Mountain Pass carbonatite ore in having extreme LREE enrichment with no europium anomalies (Fig. 3) and low $^{87}Sr/^{86}Sr$ (Philpotts et al., 1989). As at Mountain Pass, alkali-rich alteration is associated with mineralization at Bayan Obo, where alteration assemblages include sodic amphibole and potash feldspar (Drew et al., 1990).

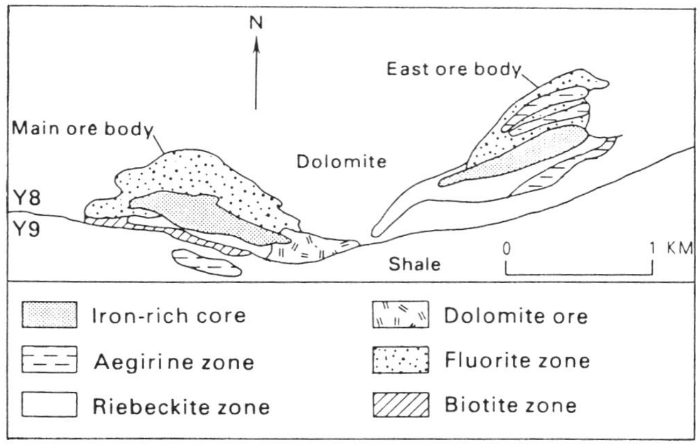

Fig. 5. Simplified geologic map of Bayan Obo iron-REE ore bodies (Drew et al., 1990).

Although there are some mineral and chemical similarities between the Mountain Pass and Bayan Obo deposits, there are also some major differences: notably high phosphorus, iron, and niobium contents and relative lack of barium in Bayan Obo ore; and the absence of alkaline intrusions at Bayan Obo. In addition, Bayan Obo ore contains nearly equal amounts of bastnasite and monazite (Qiu, private communication, 1984), whereas Mountain Pass ore contains about ten times as much bastnasite as monazite. REE/chondrite patterns for Bayan Obo show enrichment in HREE relative to Mountain Pass (Fig. 3).

The Pea Ridge iron deposit in Missouri contains a high grade REE resource of unknown size in breccia pipes associated with 1.48-billion-year-old granite and syenite (Kisvarsanyi et al., 1989). Bulk samples of REE-rich breccia, which contains monazite and xenotime, average about 12% REE (Nuelle et al., 1991). Although the Pea Ridge resource is LREE-dominated, it contains significant HREE.

Mill tailings from processing magnetite deposits in Precambrian gabbro and syenite at Mineville, New York, are a REE resource of unknown size. They contain apatite that averages more than 11% REO, including about 2% Y_2O_3 (McKeown and Klemic, 1956).

The huge Middle Proterozoic Olympic Dam copper-uranium-silver-gold deposit, South Australia, may also fall into this category because most of the ore is breccia that contains 40 to 90% hematite along with quartz, sericite, fluorite, barite, sulfides, and REE minerals (Oreskes and Einaudi, 1990). The deposit is a potential source of byproduct REE; typical content is about 0.5% total REE. Extremely fine-grained monazite and bastnasite are the most abundant REE minerals, and the LREE/HREE ratio is high in Olympic Dam ore.

Placer Deposits

Nearly all placer accumulations with significant amounts of REE minerals are Tertiary or Quaternary deposits derived from source areas that include granitic rocks or high-grade metamorphic rocks. Most commercial deposits are in sands of marine origin along or near present coastlines, and consist of titanium mineral placers with by-product zircon and monazite. Some contain xenotime. Monazite and xenotime from titania-zircon paleobeach placers in Australia constitute the third most important source of REE in the world, about 7 kt of REO per year.

Placer rutile-zircon-ilmenite deposits at Eneabba on Australia's west coast north of Perth, which produce about 2.5 kt of monazite annually, were brought into production in the late 1970s. Heavy minerals comprise about 6% of the sand mined at Eneabba, and monazite makes up about 0.5 to 7.0% of the heavy minerals (Shepherd, 1990). The heavy mineral sands lie 30 m or more above present sea level, and were deposited in late Tertiary or early Pleistocene time (Lissiman and Oxenford, 1975). Subparallel high grade zones within the deposits mark strandline accumulations on progressively lower wave-cut benches. Archean metamorphic rocks in the interior provided heavy minerals to Mesozoic sedimentary rocks that were reworked to form the deposits. In the Geographe Bay area near Capel, ilmenite-zircon paleobeach deposits have been mined since the mid-1950s. The deposits occur along a strike length of about 50 km in Pleistocene to Recent sands up to 75 m above the present sea level (Welch et al., 1975). The setting and source are similar to those described for Eneabba. Although the average heavy mineral grade of Geographe Bay deposits is somewhat higher than at Eneabba, the monazite content of heavy mineral concentrates is lower, at 0.04 to 1.1% (Welch et al., 1975).

Less than 10% of Australia's monazite is produced from placers on the east coast. Rutile and zircon are the main mineral products from these operations, which exploit Quaternary beach placers and sand dunes. Heavy mineral content in some of the deposits is less than 0.5% (Griffiths, S. J., 1984) and monazite comprises less than 1% of the heavy mineral concentrate (McKellar, 1975).

Byproduct monazite is also extracted from beach deposits in other countries. In approximate order of decreasing production, these countries are: Brazil, India, Malaysia, Thailand, China, Taiwan, the United States, New Zealand, Sri Lanka, Indonesia, Zaire, and Korea. In Brazil and India, monazite production from titania-zircon placers is government controlled, and both countries have plants that process monazite concentrate into REE products. In Malaysia, Thailand, and Indonesia monazite and xenotime are won from concentrates produced during placer tin mining. Prior to 1988 when China began to dominate yttrium production, xenotime from Malaysia was the largest source of yttrium in the world. Monazite has been produced since 1952 as a byproduct of titania-zircon production from Pleistocene sands near Green Cove Springs in Florida. Data are not available for recent years, but production levels in the past were approximately 500 t of monazite per year. Other countries with potential for placer REE production include Madagascar, Guyana, and the Republic of South Africa.

Prior to large scale exploitation of the Mountain Pass deposit, REE production in the United States came almost exclusively from non-marine placers, although little or none now comes from such sources. The Carolina monazite belt, from which a total of about 5 kt of monazite was produced between 1885 and 1917, has considerable placer reserves that average about 0.25 kg/m^3 of monazite (Overstreet, 1967). Recent developments with possible byproduct REE potential in the southeastern United States include a gold placer in North Carolina and a titania beach placer in Virginia.

REE minerals were produced in the 1950s from placers in Idaho. Dredging near Cascade is thought to have produced about 7 kt of monazite from gravels containing about 1 kg/m^3 and monazite reserves are estimated at 38 kt (Overstreet, 1967). Bear Valley, ID, where monazite and yttrium-bearing euxenite were mined by dredging, contains an estimated 10 kt of REO along with significant niobium and tantalum on the basis of data from Kline et al. (1953). Some Idaho placers contain xenotime, particularly those in northern Idaho, and REE are a possible byproduct of placer gold mining in that area.

Future REE production may come from the WIM 150 deposit, a placer in an interior basin of Victoria, Australia, that contains large REE mineral resources. The deposit consists of an extensive area amenable to dredging that is underlain by a 14-m-thick sequence of sand enriched in titanium and zirconium minerals. Estimated reserves are 4 900 Mt averaging 2.8% heavy minerals, with monazite and xenotime comprising 1.4 and 0.4% respectively of the heavy mineral fraction (Williams, 1990). The heavy minerals are finer in grain size than those in coastal deposits; the fine size may cause recovery problems.

Metamorphosed Early Proterozoic conglomerate mined for uranium at Blind River, Ontario, Canada, contains REEs in monazite, uraninite, and brannerite (Roscoe, 1969) and was the source of minor REE production from the 1950s to 1970s. The heavy minerals in this ore are believed to be primarily of placer origin with possible enrichment during metamorphism. Of the approximately 230 Mt of ore averaging about 0.1% U_3O_8 estimated to be present in the district in 1973 (Robertson, 1981), perhaps half remains unmined. In 1986 technology was reinstalled at a uranium mill to recover yttrium from Blind River ore, but this operation was closed in 1990.

At Music Valley in southern California, xenotime-rich zones with as much as 16% REO (including 6% Y_2O_3) in Precambrian gneiss are thought to be placer accumulations (Evans, 1964). At Bald Mountain, WY, a significant paleoplacer monazite resource has been identified in Cambrian conglomerate (Borrowman and Rosenbaum, 1962).

HREE Deposits in Peralkaline Igneous Rocks

During the 1980s when yttrium demand exceeded supply, exploration for yttrium-rich REE deposits led to the discovery of several zirconium and HREE-dominated resources that are associated with peralkaline syenitic and granitic rocks. These resources are large, but have lower REE content than the LREE-dominated deposits now being mined, and none have been exploited. REE/chondrite patterns for HREE deposits are nearly flat with pronounced negative europium anomalies (Fig. 6) and differ distinctly from patterns for LREE deposits such as Mountain Pass and Bayan Obo (Fig. 3).

A HREE-zirconium deposit at Pajarito Mountain in New Mexico containing 2.4 Mt of rock with 0.18% Y_2O_3 and 1.2% ZrO_2 (Sherer, 1990) consists of a 10-km^2 dome-shaped syenitic intrusion (Kelley, 1968). The main zirconium-REE phase was found to be eudialyte (Mariano, 1989a). A foliated syenite complex at Kipawa Lake, ON, includes a 1 300 by 100 m zone of yttrium- and zirconium-rich rock that contains eudialyte, mosandrite, and britholite (Allan, private communication, 1990). The Y_2O_3 content of this deposit is comparable to that at Pajarito Mountain, but grade varies widely within the deposit, which was intensely deformed during the Grenville orogeny.

The Strange Lake zirconium-HREE-niobium-beryllium deposit in Quebec-Labrador is in a circular peralkaline granite complex about 6 km in diameter (Currie, 1985). Reserves are reported at 52 Mt containing 0.31% Y_2O_3, 2.9% ZrO_2, and 0.54% REE, along with beryllium, and niobium credits (Miller, 1988). Total reserves in the complex, which has not been thoroughly explored, could be much larger. Most of the REE are in gadolinite, bastnasite,

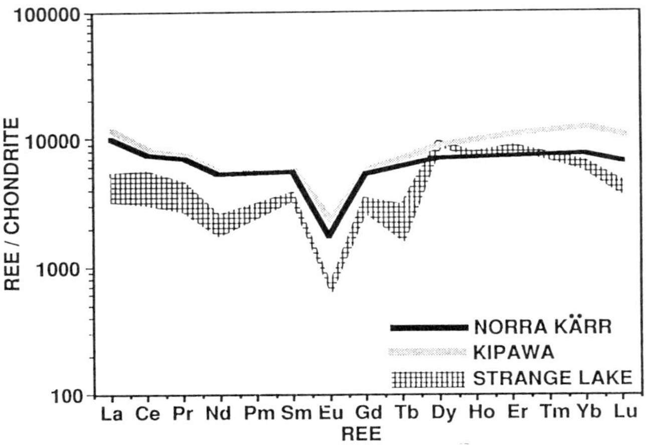

Fig. 6. Rare earth element/chondrite distributions for peralkaline HREE deposits. *Source*: Fryer and Edgar (1977).

and kainosite (Mariano, private communication, 1989). Alkaline complexes in the Shallow Lake and Letitia Lake areas in Labrador, about 250 km southeast of Strange Lake, also include rocks with high yttrium content (Miller, 1988). The 1.35-billion-year-old Red Wine complex in the Shallow Lake area contains foliated eudialyte-bearing syenite that has been investigated as a potential source of REE (Currie, 1976).

A peralkaline syenite complex at Ilimaussaq in southern Greenland dated at 1170 m.y. contains layers with as much as 27% eudialyte (Ferguson, 1970); the discovery of ore bodies containing 0.12% Y_2O_3 and 1.2% ZrO_2 has been reported (O'Driscoll, 1988). The richest layer is 3.5 m thick and contains about 1 Mt of rock that averages 4% ZrO_2 (Steenfelt, 1991). At Norra Kärr, Sweden, 1.54-billion-year-old eudialyte-bearing peralkaline syenite is associated with anorthosite and rapakivi granite (Lundqvist, 1979). Eudialyte from Norra Kärr contains 1.3% yttrium (Fryer and Edgar, 1977). The grade and tonnage of the deposit are not known, but it has been evaluated as a zirconium source.

A HREE-dominated resource has been identified at Bokan Mountain in southeastern Alaska, where dikes averaging about 0.5% REO (including more than 0.15% Y_2O_3) are associated with Jurassic peralkaline granite (Barker and Mardock, 1988). The dikes contain many REE minerals including silicates (e.g., thalenite), fluorocarbonates (e.g., bastnasite), and fluorides. Because the REE-rich dikes at Bokan Mountain are generally less than 1 m thick and are mineralogically complex, the potential for large scale exploitation may be low. A relatively small deposit at Thor Lake in the Northwest Territories, Canada, contains 0.85% BeO and 0.26% Y_2O_3 (Trueman et al., 1988) in a 2.1-billion-year-old syenite-granite complex. Because of its remoteness, near-term development of this deposit is unlikely.

Four Saudi Arabian deposits carrying REE occur in alkaline granites of probable Late Proterozoic age. Zirconium is abundant in these deposits, with zircon as the only zirconium phase reported; REE minerals that have been identified are bastnasite, monazite, and xenotime (Drysdall et al., 1984). The deposits include both LREE- and HREE-rich types, and estimated sizes range from 6 to 440 Mt with yttrium contents from 0.13 to 0.52% (Drysdall et al., 1984).

The Brockman deposit in northern Australia is being evaluated for zirconium, yttrium, HREE, niobium, and tantalum. The REE are in very fine-grained minerals disseminated in a metamorphosed lower Proterozoic rhyolite tuff (Buckovic, private communication, 1990). Announced reserves are 9 Mt containing 1.3% zirconium, 0.15% Y_2O_3 and 0.12% HREE (O'Driscoll, 1988).

On the Kola Peninsula, in the former USSR, Paleozoic peralkaline syenite complexes at Khibina and Lovozero contain large amounts of eudialyte-bearing rock (Gerasimovsky et al., 1974). According to Hedrick and Templeton (1991) REE production is from apatite in the Khibina complex and from loparite in the Lovozero complex. Data on grade, production, and reserves are not available. A. Notholt (private communication, 1992) believes that REE production from the Kola Peninsula has only been on a laboratory/pilot-plant scale. In South Africa, small REE reserves have been estimated for britholite-bearing veins in the huge Proterozoic Pilanesberg peralkaline complex, which also contains eudialyte syenite (Lurie, 1986). At Pocos de Caldas, Brazil, a circular peralkaline syenite complex 30 km in diameter that has yielded 53 to 90 m.y. dates (Woolley, 1987) contains local concentrations of eudialyte.

Vein Deposits

In South Africa, more than 50 kt of monazite were produced in the 1950s and 1960s at Steenkampskraal (Highley, 1982) from a monazite-apatite-quartz vein in Proterozoic granite. The vein, which occupies a shear zone, is about 300 m long and as much as 2 m thick. Monazite comprised 60 to 75% of the vein material (Overstreet, 1967). A similar, but smaller, monazite-apatite vein cuts Precambrian granite near Crescent Peak in Nevada about 30 km east of the Mountain Pass REE deposit. Samples from this vein contain about 10% REO (Castor, 1991).

At Karonge, Burundi, in east Africa, quartz-barite-bastnasite-monazite stockwork veins that cut Precambrian quartzite and schist have produced small amounts of REE (Notholt et al., 1990). The bastnasite and monazite have high LREE typical of carbonatites, but secondary cerianite and rhabdophane have abnormally high or low cerium contents due to oxidation (Van Wambeke, 1977).

REE- and thorium-bearing quartz veins scattered over an area of about 250 km^2 at Lemhi Pass on the Idaho-Montana border in the United States comprise a REE resource of 370 kt (Staatz et al., 1979). The mineralization is mostly LREE-dominated and averages less than 1% REO, but some veins contain up to 0.3% Y_2O_3 and 2.0% total REO. A small area at Diamond Creek about 50 km northwest of Lemhi Pass contains veins with similar mineralogy, but with low LREE/HREE ratios. The Snowbird occurrence in northwestern Montana is a REE resource of unknown size with 0.1 to 0.2% Y_2O_3. The REE minerals are in a Cretaceous quartz-carbonate-fluorite-parisite vein considered to be of hydrothermal origin (Metz et al., 1985). Thorite-bearing quartz-carbonate-fluorite veins at Hall Mountain near Porthill in northernmost Idaho contain as much as 0.2% Y_2O_3. The Tertiary Bear Lodge trachyte-phonolite complex in Wyoming contains a minor REE resource in veins that average about 3% REO (Staatz, 1983), but the veins are narrow and commercial development is unlikely.

Other Deposits

Some REE accumulations that do not fit into the catergories discussed above have had REE production or have been evaluated as possible sources. For instance, REE have been produced in increasing quantities in recent years from the Jiangxi province in southern China, and this source has had a strong impact on yttrium supplies since 1988. The raw material is said to be REE-bearing clay from two sites, one yielding HREE- and yttrium-rich material and the other LREE-rich material (O'Driscoll, 1988). Little data are available on the geology, grade, or tonnage. Individual REE distributions suggest supergene enrichment similar to that in deeply weathered carbonatites; however Ren (1985) reported that Jiangxi REE comes from a weathering crust over granite. Additionally,

Hedrick and Templeton (1991) listed several other REE-bearing clay deposits in China as REE producers.

Marine phosphorites have been proposed as a potential source of REE (Altschuler et al., 1967). Certain members of the Permian Phosphoria Formation, which are mined for phosphate in large quantities in Montana and Idaho, contain significant amounts of REEs including as much as 0.1% yttrium (Gulbrandsen, 1966).

Near-economic HREE-uranium mineralization occurs as xenotime accumulations in sandstone of the late Precambrian Athabasca Group near Wheeler River in Alberta, Canada. Similar accumulations of HREE and uranium are in quartzite in Western Australia (Mariano, 1989a).

Some fluorspar deposits offer potential for REE production from associated REE minerals or from REE that substitutes for calcium in fluorite. The only deposit that presently yields REE is one at Naboomspruit in South Africa where a small amount of monazite is produced (Hedrick and Templeton, 1991). About 65 t of bastnasite concentrate were produced from a fluorspar deposit in the Gallinas Mountains, NM, in the 1950s (Adams, 1965).

REE deposits in Vietnam are reported to contain several million tons with grades that range from 1.4 to 5% REO with LREE- to moderate HREE-enriched distributions. The deposits are described as crushed zones in Paleozoic limestone affected by metasomatic processes (Premoli, 1989).

Pegmatite REE deposits are common but generally are too small or too low in grade to be commercially exploited. However, production of REE from pegmatite mined for other minerals, such as feldspar or mica, is possible. Allanite-bearing pegmatites are relatively common, generally with high LREE/HREE ratios. In Australia's Northern Territory a pegmatite deposit containing 1 Mt with 4% allanite has been investigated as a source of REE (O'Driscoll, 1988).

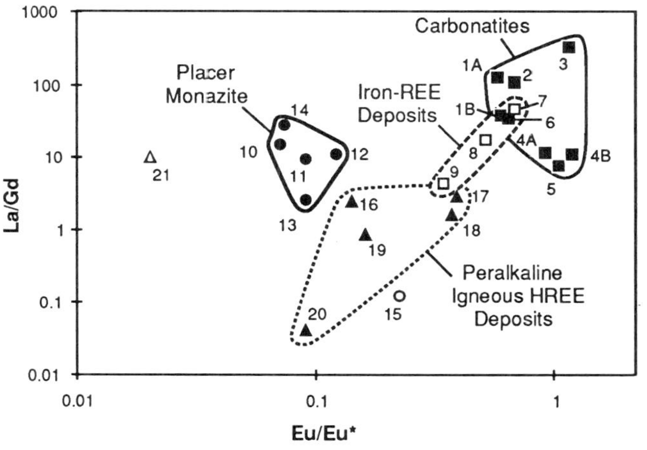

Fig. 7. Plot of La/Gd against Eu/Eu* for some REE-rich mineral concentrate and rock samples. Carbonatite rock or minerals samples (■): 1A, Mountain Pass bastnasite; 1B, Mountain Pass monazite; 2, Kangankunde REE-rich carbonatite (Nelson et al., 1988); 3, Kizilcaoren REE-rich carbonatite (Morteani, 1991, private communication); 4A and 4B, Salitre II perovskite and anatase (calc. from Mariano, 1989a); 5, Mount Weld carbonatite; 6, Maicuru REE-rich carbonatite (Costa et al., 1991). Iron-REE deposits (□): 7, Bayan Obo ore (Qiu et al., 1983); 8, Pea Ridge breccia (Tucker, private communication, 1991); 9, Mineville apatite (Roeder et al., 1987). Placer monazite (●): 10 and 11, W. and E. Australia (Hedrick, 1985); 12, S. China (Hedrick, 1985); 13, Florida (Hedrick, 1985); 14, Brazil (Deans, 1966). Placer xenotime (○): 15, Malaysia (Hedrick, 1985). Peralkaline igneous HREE deposits (▲): 16, Strange Lake; 17, Kipawa Lake eudialyte (Fryer and Edgar, 1977); 18, Bokan Mountain (Barker and Mardock, 1988); 19 and 20, Jabal Sa'id and Jabal Tawlah (Drysdall et al., 1984). Vein deposit (△): 21, Crescent Peak.

ORIGIN OF DEPOSITS

The distribution of REE in nature is dependent upon several petrogenetic processes, including enrichment and complexing in late stage magmatic or hydrothermal fluids, fractionation into mineral phases, oxidation or reduction, and redistribution during weathering. LREE enrichment in igneous rocks generally is ascribed to fractionation of HREE into minerals such as garnet and pyroxene during partial melting of source materials or during fractional crystallization. In addition, anomalously low Eu/chondrite contents, a common feature in highly evolved igneous rocks of crustal derivation, are considered to be due to Eu fractionation into plagioclase. Eu/Eu* (measured europium divided by europium calculated from interpolation between samarium and gadolinium: Henderson, 1984) generally is near unity in mantle-derived rocks such as carbonatites and low in granitic rocks thought to have been derived by partial melting of crustal materials. La/Gd, as a measure of LREE/HREE ratio, plotted against Eu/Eu* is instructive in the interpretation of the origin of REE deposits (Fig. 7). Whole rock and mineral plots for carbonatite REE deposits fall in a field of high La/Gd and Eu/Eu*. By comparison, monazite concentrates from Australia, China, and Florida placer deposits plot in a tight cluster with high La/Gd and low Eu/Eu*, probably reflecting derivation from rocks that contained plagioclase or were produced by melting of crustal rocks. HREE deposits in peralkaline rocks plot in a diffuse field with relatively low La/Gd and Eu/Eu*, probably due to crustal derivation, and placer xenotime from Malaysia plots near the peralkaline field. Iron-REE deposits occupy a linear field that appears to connect the carbonatite and peralkaline HREE deposit fields, and ore from the Bayan Obo deposit plots within the carbonatite field.

The chemistry of the carbonatite and associated alkaline rocks at Mountain Pass suggests that they were derived from primitive or depleted mantle mixed with an enriched mantle or crustal component; but REE distributions in these rocks suggest that crustal contribution was negligible (Castor, 1991). The Mountain Pass carbonatite shares features with other carbonatite complexes in the world, including mineral assemblage, relative REE distribution, and low $^{87}Sr/^{86}Sr$ (Powell et al., 1966). However, the shape of the carbonatite orebody, its potassic alkaline association (most carbonatites are associated with alkaline rocks that are dominantly sodic), and chemistry (it is unusually low in iron, phosphorus, and niobium for REE-rich carbonatite), suggest that its source was different than that of other carbonatites. In addition, the geologic setting of the Mountain Pass LREE deposit is similar to that of Proterozoic HREE deposits emplaced in the North American continent (Castor, 1993). A northeasterly-trending belt of anorogenic Middle Proterozoic plutonism extending across North America contains four HREE deposits in 1.2- to 1.4-billion-year-old sodic alkaline complexes. The LREE-dominated Proterozoic Mountain Pass, Pea Ridge, and Mineville deposits also occur in this belt, which probably extended through Greenland to northern Europe in Precambrian time (Fig. 8), includes the Ilimaussaq and Norra Kärr deposits. Alkaline rocks containing the HREE deposits are enriched in some of the same elements as alkaline rocks at Mountain Pass (e.g., zirconium and LREE), but contain much lower contents of some elements (e.g., barium and chromium) and much higher amounts of others (e.g., beryllium). The HREE deposits are asso-

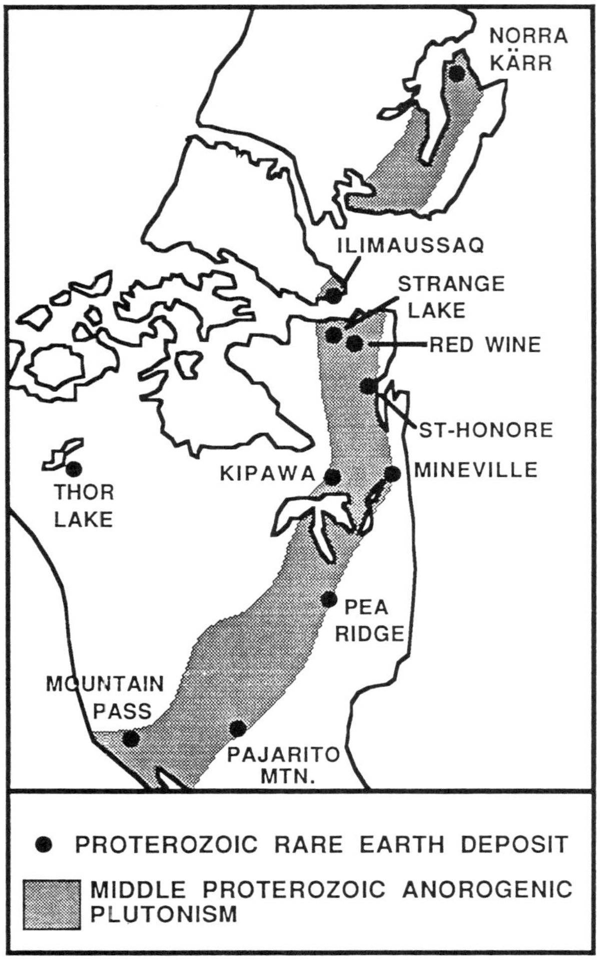

Fig. 8. Location of REE deposits on a portion of the late Proterozoic Laurussian continent. Reconstruction taken from Piper (1983).

ciated with alkaline rocks that are dominantly sodic, in contrast to the potassic alkaline rocks at Mountain Pass.

Proterozoic REE deposits such as those at Bayan Obo, Pea Ridge, and Mineville differ from Mountain Pass in their association with large amounts of iron. REE/chondrite patterns for Bayan Obo ore (Fig. 3) and low initial $^{87}Sr/^{86}Sr$ support a mantle source (Philpotts et al., 1989). The Bayan Obo deposits have been compared to banded-iron formations (Qiu et al., 1983) or to carbonatite deposits (Bai and Yuan, 1985), and have most recently been interpreted as replacements of dolomite by hydrothermal fluids derived from unexposed alkaline-carbonatite intrusion (Drew et al., 1990).

The chemical compositions and mineral assemblages of most vein-type REE deposits suggest a genetic relationship with carbonatite-alkaline complexes. Such a relationship has been proposed for the Karonge, Burundi, veins (Van Wambeke, 1977). However, the monazite-apatite vein at Crescent Peak, NV, has REE distribution that includes a prominent negative europium anomaly that suggests a different origin than the Mountain Pass deposit 30 km away.

TECHNOLOGY

Exploration

Geologic conditions favorable for REE deposits are widespread. The world's non-placer REE deposits are mainly in areas of Precambrian rock. Most carbonatite REE deposits and many HREE deposits occur in, or are associated with, alkaline rocks in circular complexes. Such complexes may be identified using aerial photography, even in deeply weathered and highly vegetated areas. Although isolated alkaline-carbonatite complexes such as Mount Weld in Australia do occur, most are in clusters or linear belts like those associated with rift zones in eastern Africa and the Araxá-Catalao belt in Brazil.

Sources for placer REE deposits include granitic rocks or high grade metamorphic rocks, or both. Coastal monazite-bearing placers in Western Australia, Brazil, India, Sri Lanka, and the southeastern United States were derived from highly metamorphosed Precambrian shield rocks, whereas monazite and xenotime-producing placers in Eastern Australia, Malaysia, Indonesia, China, and Korea were eroded from Phanerozoic granites, including tin-rich granites. Sources of monazite, euxenite, and xenotime in Idaho placers include both highly metamorphosed Precambrian rocks and Mesozoic granitic rocks.

Because REE are associated with thorium and uranium, radiometric exploration techniques are extremely useful in REE exploration. REE-rich carbonatite at Mountain Pass was discovered during surface prospecting for uranium using a Geiger counter. Many other REE deposits were found by surface or airborne radiometric surveying. Although most placer REE deposits yield only subtle radiometric signatures, careful data collection and analysis can be used to locate favorable intrabasin areas or horizons.

Geochemical prospecting using heavy mineral concentration of active stream sediments can be very effective because most REE minerals are relatively heavy and resistant. Concentrates made by hand panning, mechanized gravity concentration, or heavy media separation can be analyzed for REE using inexpensive multi-element analytical techniques. Because many REE minerals are resistant to chemical breakdown, care must be taken to use analytical methods that do not require dissolution, such as X-ray fluorescence or neutron activation analysis, or to use methods that assure total dissolution when using a technique such as ICP emission spectroscopy. During prospecting for REE placer deposits in Idaho, the writer found favorable basins to yield heavy mineral concentrates with 1% or more cerium and/or 0.1% or more yttrium. Splits of heavy mineral concentrates taken during exploration should be saved so mineral determinations can be performed using microscopic or other techniques.

Normal surface exploration techniques, such as trenching and pitting by heavy equipment, are used to evaluate poorly exposed REE deposits. Exploratory dredging is used to locate beach placer deposits. In countries with low labor costs, hand pitting may be used to evaluate placer deposits or unconsolidated deposits such as those in laterite.

Drilling techniques utilized for hard rock REE exploration and prospect evaluation are mainly core drilling and dual-tube rotary or hammer drilling. The method of choice for most heavy mineral placer evaluations is dual-tube rotary drilling. The collection of large samples, such as are necessary to evaluate gold placers, generally is not needed because the nugget effect is not a factor in most heavy mineral sands that contain REE. Sample collection by a dual-tube technique (air or foam) helps to assure heavy mineral sample integrity. Down hole natural gamma probes can be used to detect concentrations of REE minerals.

Because of the economic importance of differences in processing REE-bearing minerals, identification of the REE-bearing phases in prospective deposits is important. Light-colored REE minerals, such as bastnasite and monazite, may generally be distinguished by a green luminescence when illuminated by a mercury vapor light (Murata and Bastron, 1956). A conventional black light with the purple filter removed may be used for this test. This luminescence, which is caused by absorption and indicates the presence of significant amounts of neodymium, has been used successfully for preliminary REE grade determination during core logging at Mountain Pass. Because the test is based on neodymium content, HREE minerals may not respond. Absorption bands in the visible spectrum may also be used to distinguish REE minerals with the aid of a hand spectroscope (Mertie, 1960).

Geophysical methods that can be used to search for buried iron-REE deposits and some REE carbonatite deposits include gravity surveys and airborne or surface magnetic surveys. The effectiveness of such methods is based on anomalous density and magnetic susceptibility of a deposit and/or the associated rocks. Because many carbonatite complexes have a central carbonatite mass surrounded by mafic alkaline rocks, a common signature is a magnetic bull's-eye with high central values combined with a centrally located gravity low and a concentric gravity high. However, the Mountain Pass carbonatite does not follow this pattern because of its high density (due to abundant barite and bastnasite) and lack of magnetite. Other geophysical techniques include seismic surveys, which may be used in conjunction with drilling to determine the depth to bedrock of placer deposits.

Table 4. Prices in $US/kg of Selected Rare Earth Products

PRODUCT	PURITY (%)	QUANTITY (kg)	PRICE ($/kg)
Bastnasite-unleached	60[a]	907	.59[b]
Bastnasite-leached	70[a]	907	.60[b]
Lanthanum concentrate	80[a]	91	.63[b]
Cerium concentrate	60[a]	907	1.04[b]
Lanthanum oxide	99.99	136	3.94
Lanthanum carbonate	99.90	79	2.66
Cerium oxide	99.00	91	4.05
Cerium carbonate	99.00	68	2.36
Cerium fluoride	Tech Grade	113	1.58
Cerium acetate	96.00	45	1.51
Praseodymium oxide	96.00	136	7.56
Neodymium oxide	96.00	136	3.04
Neodymium carbonate	96.00	136	2.03
Samarium oxide	96.00	25	24.75
Europium oxide	99.99	11	335.25
Gadolinium oxide	99.99	25	24.75
Terbium oxide	99.90	20	168.75
Dysprosium oxide	99.00	23	36.00
Erbium oxide	98.00	23	29.25
Yttrium oxide	99.99	23	23.63
Yttrium nitrate	99.90	25	8.10

Source: Molycorp, Inc. Prices, effective Oct. 15, 1991, are FOB Mountain Pass, CA; Louviers, CO; or York, PA. Prices for smaller quantities than those shown are 4 to 12% higher. [a] typical REO content, [b] price per kg contained REO.

Mining

At Mountain Pass REE ore is mined in an open pit approximately 100 m deep using 80-t haul trucks and a 10-m^3 shovel (30-t trucks and front end loaders were used until late in 1991). Blast holes drilled at 3 to 4 m spacing are routinely assayed for total REO and other elements by X-ray fluorescence methods. Approximately 300 ktpy are mined with a stripping ratio of 4 or 5:1. The ore is crushed and blended in stockpiles that contain about 40 kt and fed to a mill located less than 100 m from the pit.

At Bayan Obo, iron-REE ore is mined from two large open pits at a rate of at least 15 ktpd using electric shovels and rail haulage. Due to lack of water near the mine, crushed ore is hauled about 150 km by railroad to milling facilities (Argall, 1980).

REE placer mining is done by either dry land mining or dredging techniques. In Australia, dredges with capacities of as much as 2.8 ktph (Anon., 1987) are used to mine beach and dune sand deposits, particularly on the east coast. Dry land techniques, including loader and truck, scraper, drag line, or bucket wheel mining, are used for most of the heavy mineral sand mining in Western Australia (Griffiths, S. J., 1984).

Processing

Flotation is used at Mountain Pass to make a bastnasite concentrate containing about 60% REO. This concentrate is either used on site as feed for chemical separation of REE, leached to produce a 70% REO concentrate, or shipped as is. Iron ore tailings that contain up to 40% REE from the Bayan Obo deposit are the feed for one or more REE treatment plants. A concentrate containing about 60% REO can be prepared from these tailings by flotation.

REE minerals are separated commercially from associated minerals in placer deposits using a combination of gravity, magnetic, and electrostatic techniques (Griffiths, S. J., 1984). Gravity methods include the use of jigs, spiral and cone concentrators, and shaking tables. Sizing and preconcentration commonly is performed at the mine site by trommels, shaking screens, and gravity separation. Many dredges have such facilities on board or utilize floating preconcentration plants.

REE extraction from monazite and xenotime is accomplished by dissolution in hot concentrated base or acid solutions. At Rhône-Poulenc Inc. plants, which process most of the world's monazite, REE is extracted in a concentrated solution of sodium hydroxide at 140 to 150° C (Kaczmarek, 1980). After cooling, hydroxides of REE and thorium are recovered by filtration, and thorium is separated by selective precipitation or dissolution. At Mountain Pass, bastnasite is calcined to drive off CO_2 and fluorine and leached with hydrochloric acid to dissolve most of the trivalent REE. The residue, which consists mostly of CeO_2, is sold as a polishing abrasive.

Because of their chemical similarity, the trivalent REE are difficult to separate. Although fractional crystallization and ion-exchange techniques are used to separate them in small amounts, commercial separation generally is done using liquid-liquid solvent extraction (Kaczmarek, 1980). This process consists of addition of a solvent composed of a mixture of organic compounds to the pregnant aqueous solution in a series of mixing/settling cells which allow repetitive fractionation during a more-or-less continuously flowing process. Following precipitation and drying, specific REE compounds at purities in excess of 99.99% can be produced by this process. A REE fraction produced during solvent extraction can be treated to separate europium by reduction to the divalent state.

Specifications and Testing

Specifications for REE mineral concentrates, compounds, and metals vary depending on use. Purity specifications for some compounds may be determined by reference to Table 4. Testing of REE products mainly consists of purity determinations. In addition to specific wet chemical analytical procedures, X-ray fluorescence, ICP emission spectroscopy and instrumental neutron activation analyses are the most effective techniques used to test products for individual REE content. Impurities, such as phosphate, silica, lead

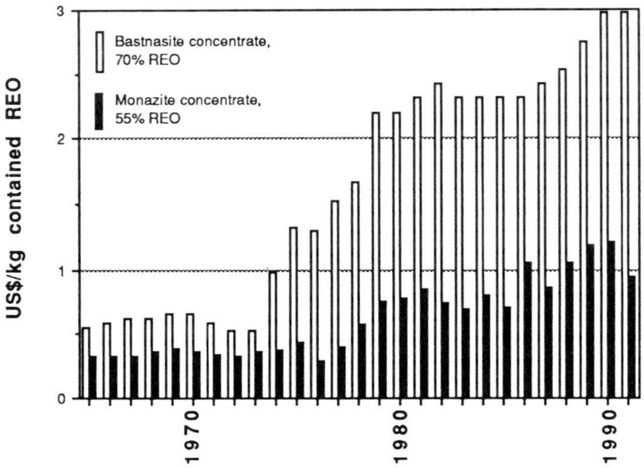

Fig. 9. Prices of rare earth raw materials, 1965 through 1991. Data from: Molycorp, Inc., *Minerals Yearbook* (USBM), and *Industrial Minerals*.

and thorium, may be deleterious to product utilization or the environment. The need to limit such impurities depends on use. Physical specifications, such as particle size distribution, color, and moisture content, also are important for some products.

ECONOMIC FACTORS

Prices

REE raw materials have had relatively steady price increases (Fig. 9), as well as production increases, since the 1960s when markets were created in response to new technologies. Yttrium raw materials varied greatly in price during the 1980s, selling for as much as \$50/kg for contained Y_2O_3 in a 60% concentrate in the mid 1980s, and retreating to \$32/kg since then. Individual REE compounds vary widely in value (Table 4), mostly due to relative abundance and production costs. The most expensive REE is the heaviest, lutetium, which sold as high-purity oxide for as much as \$9,000/kg in 1989 (Hedrick, 1991a). Metallic REE commodities bring higher prices than oxides or other compounds. Mischmetal, a mixture of metallic REE, was quoted at \$12.35/kg at the end of 1989 (Hedrick, 1991a). During the 1980s, individual REE prices have generally been static or increased slightly, but price decreases for individual REE commodities have been caused by market changes. Samarium oxide, for which demand often exceeded supply during the late 1980s, sold for \$187/kg in 1988, but currently is available at \$121/kg, and europium oxide has decreased in price from about \$2,000/kg in 1980 to about \$1,700/kg in 1991.

RARE EARTH USES AND MARKETS

Older REE uses in mischmetal, lighter flints, arc-lighting carbons, and iron and steel additives have declined from 75% of the overall market in the 1950s to about 30% in the 1980s. At present, the glass industry uses the most REE by weight, mainly cerium oxide as a decolorizing agent and an abrasive. Recent overall REE production increases probably are due mostly to the use of cerium in automotive catalytic converters (O'Driscoll, 1988). Sales of lanthanum-rich mixed REE compounds for use in catalysts for petroleum refining (Griffiths, J., 1984) peaked in 1984, but have declined substantially since then. REE uses in electronics, laser technology, and ceramic glazes provide relatively minor markets.

Beginning in the 1960s, color television manufacture spurred demand for high value REE phosphors based on yttrium, europium, gadolinium, and terbium. More recently, increased production of REE phosphors has resulted from their use for color control in fluorescent lighting (Griffiths, J., 1984).

The introduction of high-strength cobalt-samarium permanent magnets in the 1970s, particularly for use in small electric motors, created a dynamic new REE market. Neodymium-iron-boron permanent magnets are now used as substitutes for the more expensive cobalt-samarium magnets in some applications.

Proposed new uses have caused increased awareness of potential profits in REE raw materials. Superconducting substances containing lanthanum or yttrium have been in the spotlight since 1986. However, volume markets for these substances await further technological breakthroughs. New methods have been devised to use lanthanum in hydrogen-based refrigeration and energy storage. Expansion of markets for yttrium may provide incentive for production from HREE deposits. In addition to use in phosphors, yttrium has found increased use in zirconia ceramic; potentially this is a large market.

Although REE raw material production is dominated by three countries, worldwide REE commodity markets are complex because of the variety of commodities, and because raw materials are processed in some countries without raw material production. High purity REE products may then be shipped back to countries that produce REE raw material. France and Japan, which have no internal REE raw material sources, are major producers of processed REE commodities, exporting approximately 27 and \$16 million worth of REE compounds, respectively, to the United States in 1991 (Hedrick, 1993). The French production is by a single company, Rhône-Poulenc, which utilizes monazite and xenotime that come mainly from Australia, and has REE processing plants in both France and the United States. Six other European countries exported REE commodities to the United States in 1991 (Hedrick, 1993). At least 20 companies in Japan produce REE commodities that are used by that country's steel, auto, and electronics industries. Chinese raw materials are being used increasingly by Japanese REE processors.

GOVERNMENTAL CONSIDERATIONS

Tariff Rates and Depletion Provisions

The classification of China as a Most Favored Nation in 1978 by the United States has had a strong impact on REE markets. Tariff rates are shown in Table 5.

Environmental Regulations

In the United States, REE mines and processing plants are required to meet the same environmental quality guidelines as other mining and chemical facilities. The US Environmental Protection Agency is responsible for setting air-, water-, and soil-quality standards for such operations nationwide, but standards for some states are more stringent. Reclamation requirements for mining are under state authority, and differ from state to state; however, reclamation on federally-owned land is also overseen by agencies such as the US Forest Service.

REE ores and commodities, as well as byproducts and waste materials from REE processing, are naturally radioactive, mostly due to contained thorium. The thorium content of monazite typically ranges from 4 to 10% ThO_2, whereas thorium comprises less

Table 5. Tariff Rates for Rare Earth Commodities Imported into the United States

Commodity	Most Favored Nation	Non-MFN
Rare earth metals	70.5¢/kg	$4.41/kg
Cerium compounds	7.2% ad valorem	35% ad valorem
Rare earth oxides except cerium	7.2% ad valorem	25% ad valorem
Other rare earth compounds	7.2% ad valorem	35% ad valorem
Other mixtures of rare earth oxides or other rare earth chlorides	3.7% ad valorem	25% ad valorem
Ferrocerium and other pyrophoric alloys	48.5¢/kg + 2.5% ad valorem	$4.41/kg + 25% ad valorem
Yttrium-bearing materials and compounds	Free	25% ad valorem

Source: US International Trade Commission (Anon., 1991).

than 1 000 ppm in bastnasite from Mountain Pass (Mariano, 1989a). Handling, shipping, storage, and disposal of such materials are controlled by federal and state agencies. Because only about 50 t of thorium commodities, all from imported or stockpiled material, are consumed in the United States annually (Hedrick, 1991b), thorium produced during REE processing presents an expensive disposal problem for the industry.

Environmental concerns centered around the disposal of radioactive waste have precluded the development of REE extraction plants in Australia, where environmental policies are mainly under state control. A processing facility in Western Australia designed to produce REE from 15 kt of monazite annually, originally slated to begin production in 1989, may be in jeopardy because of lack of approval from the Environmental Protection Authority of Western Australia. Groups concerned about radioactive contamination of the environment have become active worldwide as shown by protests of plans to begin producing REE from monazite in Malaysia.

FUTURE TRENDS

Future increases in demand for REE are partially dependent upon the development of significant new markets. Examples include widespread use of REE-based superconducting compounds, lanthanide hydride in refrigeration or energy storage, and yttrium stabilized zirconia ceramic in automobile engines. Commercial development of all three examples will be based upon the world's needs for more efficient and environmentally safe use of energy resources. Long term outlook for the development of markets is therefore optimistic.

World supplies of LREE are expected to be adequate to meet short term demand. Because LREE availability is, in part, dependent upon byproduct production, oversupply is effectively buffered during worldwide economic downturns. HREE elements may be another matter. At the 1991 International Conference on Rare Earth Minerals for Electronic Uses, which brought together delegates from all sectors of the REE industry, a potential shortage of HREE was forecast by the mid-1990s unless new deposits are brought into production.

REFERENCES

Anon., 1980, *Mineral Powder Diffraction File,* Joint Committee on Powder Diffraction Standards, Swathmore, PA, 1168 pp.

Anon., 1984, *Carbonatitic Complexes of Brazil: Geology*, Companhia Brasileira de Metalurgia e Mineraçao, Sao Paulo, 44 pp.

Anon., 1987, "Mineral Sands, the Big Scale Time," *Industrial Minerals*, No. 231, pp. 47–61.

Anon., 1991, "Inorganic Chemicals, Organic or Inorganic Compounds of Precious Metals, of Rare Earth Metals, of Radioactive Elements, or of Isotopes," *Harmonized Tariff Schedule of the United States,* USITC publication 2333, US International Trade Commission, US Government Printing Office, Washington, DC, pp. 28–1 to 28–14.

Adams, J.W., 1965, "Rare Earths," Mineral and Water Resources of New Mexico, New Mexico Bureau of Mines and Mineral Resources, Bulletin 87, pp. 234–237.

Altschuler, Z.S., Berman, S., and Cuttitta, F., 1967, "Rare Earths in Phosphorite—Geochemistry and Potential Recovery," Professional Paper 575-B, US Geological Survey, 9 pp.

Argall, G.O., 1980, "Three Iron Ore Bodies of Bayan Obo," *World Mining*, Jan., pp. 38–41.

Armbrustmacher, T.J., 1980, "Abundance and Distribution of Thorium in the Carbonatite Stock at Iron Hill, Powderhorn District, Gunnison County, Colorado," Professional Paper 1049-B, US Geological Survey, 11 pp.

Armbrustmacher, T.J., 1988, "Geology and Resources of Thorium and Associated Elements in the Wet Mountains Area, Fremont and Custer Counties, Colorado," Professional Paper 1049-F, US Geological Survey, 34 pp.

Bai, G., and Yuan, Z., 1985, "On the REE-Rich Carbonatites," *New Frontiers in REE Science and Applications*, Xu Quangxian, ed., Science Press, Beijing, China, pp. 126–136.

Barker, J.C., and Mardock, C.L., 1988, "Lithophile Metal, REE-Y-Nb Deposits on Southern Prince of Wales Island, Alaska," D.J.T. Carson and A.H. Vassiliou, eds., *Process Mineralogy VIII*, Minerals, Metals and Materials Society, Warrendale, PA, pp. 139–158.

Borrowman, S.R., and Rosenbaum, J.D., 1962, "Recovery of Thorium from a Wyoming Ore," Report of Investigation 5917, US Bureau of Mines 8 pp.

Castor, S.B., 1991 "Rare Earth Deposits in the Southern Great Basin," *Geology and Ore Deposits of the Great Basin*, Geological Society of Nevada, Reno, pp. 523–528.

Castor, S.B., 1993, "Rare Earth Deposits and Proterozoic Anorogenic Magmatism," *Proceedings*, Y.T. Maurice, ed., 8th IAGOD Symposium, 1990, Ottawa, Canada, 1990, Schweizerbart'sche, Stuttgart, pp. 331–340.

Castor, S.B., and Gleason, J.D., 1989, "Proterozoic Ultrapotassic Intrusive Rocks in Southeastern California," *Abstracts with Programs*, Vol. 21, No. 5, Geological Society of America, p. 64.

Cesbron, F.P., 1989, "Mineralogy of the Rare Earth Elements," *Lanthanides, Tantalum and Niobium; Mineralogy, Geochemistry, Characteristics of Primary Ore Deposits, Prospecting, Processing and Applications*, P. Möller, P. Cerny, and F. Saupé, eds., Springer-Verlag, Berlin, pp. 3–26.

Costa, M.L., et al., 1991, "Geochemical Exploration of the Maicuru Alkaline-Ultramafic-Carbonatite Complex Northern Brazil," *Journal of Geochemical Exploration*, Vol. 40, pp. 193–204.

Chao, E.C.T., et al., 1989, "The H8 Dolomite Host Rock of the Bayan Obo Iron-Niobium-Rare-Earth-Element Ore Deposit of Inner Mongolia, China: Origin, Episodic Mineralization, and Implications," Circular 1035, US Geological Survey, p. 8.

Chao, E.C.T., et al., 1993, "Multiple lines of evidence for establishing the mineral paragenetic sequence of the Bayan Obo rare ore deposit of Inner Mongolia, China," *Proceedings,* Y.T. Maurice, ed., 8th Quadrennial IAGOD Symposium, Ottawa, Canada, 1990, Schweizerbart'sche, Stuttgart, pp. 55–73.

Crowley, F.A., 1960, "Columbium—Rare Earth Deposits, Southern Ravalli County, Montana," Bulletin 18, Montana Bureau of Mines, 47 pp.

Currie, K.L., 1976, "The Alkaline Rocks of Canada," Bulletin No. 239, Geological Survey of Canada, 228 pp.

Currie, K.L., 1985, "An Unusual Peralkaline Granite near Lac Brisson, Quebec-Labrador," *Current Research, Part A*, Geological Survey of Canada Paper 85–1A, pp. 73–80.

Deans, T., 1966, "Economic Geology of African Carbonatites," *Carbonatites*, O.F. Tuttle and J. Gittins, eds., Wiley Interscience, New York, pp. 385–413.

Dewitt, E., Kwak, L.M., and Zartman, R.E., 1987, "U-Th-Pb and Ar/Ar dating of the Mountain Pass Carbonatite and Alkalic Igneous Rocks, S.E. Cal.,'' *Abstracts with Programs*, Vol. 19, No. 7, Geological Society of America, p. 642.

Drew, L.J., Meng, Q., and Sun, W., 1990, ''The Bayan Obo Iron-Rare Earth-Niobium Deposits, Inner Mongolia, China,'' *Lithos*, Vol. 26, pp. 43–65.

Drysdall, A.R., et al., 1984, ''Rare Element Mineralization Related to Precambrian Alkali Granites in the Arabian Shield,'' *Economic Geology*, Vol. 79, pp. 1366–1377.

Duncan, R.K., and Willett, G.C., 1990, ''Mount Weld Carbonatite,'' *Geology and Mineral Deposits of Australia and Papua New Guinea*, F.E. Hughes, ed., Australasian Institute of Mining and Metallurgy, Parkville, Vic., pp. 591–597.

Evans, J.R., 1964, ''Xenotime Mineralization in the Southern Music Valley area, Riverside County, California,'' Special Report 79, California Division of Mines and Geology, 24 pp.

Ferguson, J., 1970, ''The Differentiation of Agpaitic Magmas: The Ilimaussaq Intrusion, South Greenland,'' *Canadian Mineralogist*, Vol. 10, pp. 335–49.

Frondel, C., 1958, ''Systematic Mineralogy of Uranium and Thorium,'' Bulletin 1064, US Geological Survey, 400 pp.

Fryer, B.J., and Edgar, A.D., 1977, ''Significance of Rare Earth Distributions in Coexisting Minerals of Peralkaline Undersaturated Rocks,'' *Contributions to Mineralogy and Petrology*, Vol. 61, pp. 35–48.

Garson, M.S., 1966, ''Carbonatites in Malawi,'' *Carbonatites*, O.F. Tuttle and J. Gittins, eds., Wiley Interscience, New York, pp. 33–72.

Gerasimovsky, V.I., et al., 1974, ''Alkaline Provinces: Kola Peninsula,'' *The Alkaline Rocks*, H. Sorensen, ed., John Wiley, London, pp. 206–221.

Griffiths, J., 1984, ''Rare-earths—Attracting Increasing Attention,'' *Industrial Minerals*, No. 199, pp. 18–36.

Griffiths, S.J., 1984, ''Review of Placer Deposits in Australia,'' Placer Exploration and Mining, Short Course, C. McKay and P. Mousset-Jones, eds., University of Nevada Division of Continuing Education, Reno, pp. 265–307.

Gulbrandsen, R.A., 1966, ''Chemical Compositions of Phosphorites of the Phosphoria Formation,'' *Geochemica et Cosmochimica Acta*, Vol. 30, No. 8, pp. 769–778.

Hatzl, T., et al., 1990, ''Petrology and Geochemistry of the Bastnaesite-Fluorite-Barite Mineralization Near Kizilçaören, Province Eskisehir,'' Abstracts, IESCA, International Earth Sciences Congress on Aegean Regions, Oct. 1–6, 1990, Izmir, Turkey, pp. 83–84.

Hedrick, J.B., 1985, ''Rare-Earth Elements and Yttrium,'' *Mineral Facts and Problems*, Bulletin 675, US Department of the Interior Bureau of Mines, pp. 647–664.

Hedrick, J.B., 1991a, ''Rare-Earth Metals,'' *Mineral Commodity Summaries, 1991*, US Department of the Interior Bureau of Mines, pp. 128–129.

Hedrick, J.B., 1991b, ''Thorium,'' *Mineral Commodity Summaries, 1991*, US Department of the Interior Bureau of Mines, pp. 128–129.

Hedrick, J.B., 1993, ''The Rare Earths, Yttrium, and Scandium, 1991,'' US Department of the Interior Bureau of Mines, 27 pp.

Hedrick, J.B., and Templeton, D.A., 1991, ''Rare-Earth Minerals and Metals, 1989,'' US Department of the Interior Bureau of Mines, 20 pp.

Heinrich, E.W., and Levinson, A.A., 1961, ''Carbonatite Rare Earth Deposits, Ravalli County, Montana,'' *American Mineralogist*, Vol. 46, pp. 1424–1447.

Henderson, P., 1984, ''General Geochemical Properties and Abundances of the Rare Earth Elements,'' *Rare Earth Element Geochemistry*, P. Henderson, ed., Elsevier, pp 1–32.

Highley, D.E., 1982, ''Rare Earths: World Resources and Exploitation,'' Metals Society and Chinese Society of Metals, Book 1, Paper 5, 10 pp.

Issa Filho, A., Lima, P.R.A., and Souza, O. M., 1984, ''Aspects of the Geology of the Barreiro Carbonatitic Complex, Araxá, MG, Brazil,'' *Carbonatitic Complexes of Brazil: Geology*, Companhia Brasileira de Metalurgia e Mineraçao, Sao Paulo, pp. 21–44.

Kaczmarek, J., 1980, ''Rare Earths: Discovery and Commercial Separations,'' Text of address to the American Chemical Society, Rhône-Poulenc Inc., Monmouth Junction, NJ, 17 pp.

Kapustin, Y.L., 1980, *Mineralogy of Carbonatites*, Amerind Publishing Co. Pvt. Ltd., New Delhi, 259 pp.

Kelley, V.C., 1968, ''Geology of Alkali Precambrian Rocks at Pajarito Mountain, Otero County, New Mexico,'' Bulletin, Vol. 79, Geological Society of America, pp. 1565–1572.

Kisvarsanyi, E.B., Sims, P.K., and Kisvarsanyi, G., 1989, ''Proterozoic Ore-forming Systems: Examples of Iron Metallogenesis from the United States Midcontinent,'' *Abstracts with Programs*, Vol. 21, No. 6, Geological Society of America, p. A34.

Kline, M.H., et al., 1953, ''Bear Valley Radioactive Mineral Placers, Valley County, Idaho,'' US Bureau of Mines Report of Mineral Evaluation 3130, 23 pp.

Lissiman, J.C., and Oxenford, R.J., 1975, ''Eneabba Rutile-Zircon-Ilmenite Sand Deposit, W.A.,'' *Economic Geology of Australia and Papua New Guinea, I: Metals*, Monograph, Series 5, Australasian Institute of Mining and Metallurgy, Parkville, Vic., pp. 1062–1070.

Lottermoser, B.G., 1988, ''Supergene, Secondary Monazite from the Mt Weld Carbonatite Laterite, Western Australia,'' *Neues Yahrbuch Mineralogie*, Vol. 2, pp. 67–70.

Lundqvist, T., 1979, ''The Precambrian of Sweden,'' *Geology of the European Countries Denmark, Finland, Iceland, Norway, Sweden*, 26th International Geologic Congress, Paris, France, pp. 211–271.

Lurie, J., 1986, ''Mineralization of the Pilanesberg Alkaline Complex,'' *Mineral Deposits of South Africa, II*, Geological Society of South Africa, pp. 2215–2228.

Mariano, A.N., 1989a, ''Economic Geology of Rare Earth Elements,'' *Reviews in Mineralogy*, Vol. 21, B.R. Lipin and G.A. McKay, eds., Mineralogical Society of America, pp. 309–337.

Mariano, A.N., 1989b, ''Nature of Economic Mineralization in Carbonatites and Related Rocks,'' *Carbonatites—Genesis and Evolution*, K. Bell, ed., Unwin Hyman, London.

McKeown, F.A., and Klemic, H., 1956, ''Rare-Earth-Bearing Apatite at Mineville, Essex County, New York,'' Bulletin 1046-B, US Geological Survey, pp. 9–23.

Mazzi, F., and Munno, R., 1983, ''Calciobetafite (New Member of the Pyrochlore Group) and Related Minerals from Campi Klegrei, Italy: Crystal Structures of Polymignyte and Zirkelite: Comparison with Pyrochlore and Zirconolite,'' *American Mineralogist*, Vol. 68, p. 262.

McKellar, J.B., 1975, ''The Eastern Australian Rutile Province,'' *Economic Geology of Australia and Papua New Guinea, I: Metals*, Monograph, Series 5, Australasian Institute of Mining and Metallurgy, Parkville, Vic., pp. 1055–1062.

Mertie, J.B., 1960, ''Monazite and Related Minerals,'' *Industrial Minerals and Rocks*, 3rd ed., J.L. Gillson, ed., AIME, New York, pp. 623–629.

Metz, M.C., et al., 1985, ''Geology and Geochemistry of the Snowbird Deposit, Mineral County, Montana,'' *Economic Geology*, Vol. 80, pp. 394–409.

Miller, R.R., 1988, ''Yttrium (Y) and Other Rare Metals (Be, Nb, REE, Ta, Zr) in Labrador,'' Current Research (1988), Newfoundland Department of Mines, Mineral Development Division, Report 88–1, pp. 229–245.

Murata, K.J., and Bastron, H., 1956, ''Convenient Method for Recognizing Nonopague Cerium-earth Minerals,'' *Science*, Vol. 123, pp. 888–889.

Nelson, D.R., et al., 1988, ''Geochemical and Isotopic Systematics in Carbonatites and Implications for the Evolution of Ocean-Island Sources,'' *Geochimica et Cosmochimica Acta*, Vol. 52, pp. 1–17.

Notholt, A.J.G., Highley, D.E., and Deans, T., 1990, ''Economic Minerals in Carbonatites and Associated Alkaline Igneous Rocks,'' *Transactions*, Institution of Mining and Metallurgy, Vol. 99, pp. B59-B80.

Nuelle, L.M., et al., 1991, ''Rare Earth Element- and Gold-Bearing Breccia Pipes of the Pea Ridge Iron Ore Mine, Washington County, Missouri,'' AIME, Preprint 91–109, 14 pp.

O'Driscoll, M., 1988, ''Rare Earths—Enter the Dragon,'' *Industrial Minerals*, No. 254, pp. 21–54.

Olson, J.C., et al., 1954, ''Rare-earth Mineral Deposits of the Mountain Pass District, San Bernardino County, California,'' Professional Paper 261, US Geological Survey, 75 pp.

Olson, J.C., and Hedlund, D.C., 1981, ''Alkalic Rocks and Resources of Thorium and Associated Elements in the Powderhorn District, Gunnison County, Colorado,'' Professional Paper 1049-C, US Geological Survey, 34 pp.

Oreskes, N., and Einaudi, M.T., 1990, ''Origin of Rare Earth Element-Enriched Hematite Breccias at the Olympic Dam Cu-U-Au-Ag Deposit, Roxby Downs, South Australia,'' *Economic Geology*, Vol. 85, pp. 1–28.

Overstreet, W.C., 1967, ''The Geologic Occurrence of Monazite,'' Professional Paper 530, US Geological Survey, 327 pp.

Philpotts, J.A., et al., 1989, ''Petrography, Chemistry, Age, and Origin of the REE-Iron Deposit at Bayan Obo, China, and Implications of Proterozoic Iron Ores in Earth Evolution,'' Circular 1035, US Geological Survey, pp. 53–55.

Piper, J.D.A., 1983, ''Dynamics of the Continental Crust in Proterozoic Times,'' *Proterozoic Geology: Selected Papers from an International Proterozoic Symposium*, L.G. Medaris, Jr. et al., eds., Memoir 161, Geological Society of America, pp. 11–34.

Powell, J.L., Hurley, P.M., and Fairbairn, H.W., 1966, ''The Strontium Isotopic Composition and Origin of Carbonatites,'' *Carbonatites*, O.F. Tuttle and J. Gittins, eds., Wiley, New York, pp. 365–378.

Premoli, C., 1989, ''Vietnamese Metallic Mineral Resources,'' *Engineering and Mining Journal*, Vol. 190, No. 12, pp. 31–35.

Qiu, Y., Wang, Z., and Zhao, Z., 1983, ''Preliminary Study of REE Iron Formations in China,'' *Geochemistry*, Vol. 2, pp. 185–200.

Ren, X., 1985, ''A Brief Account on Rare Earth Mineralization in China,'' *New Frontiers in REE Science and Applications*, Q. Xu, ed., Science Press, Beijing, China, pp. 39–41.

Robertson, J.A., 1981, ''The Blind River Uranium Deposits: the Ores and their Setting,'' *Genesis of U- and Au-bearing Quartz-pebble Conglomerates*, F.C. Armstrong, ed., Professional Paper 1161-A-BB, US Geological Survey, pp. U1-U23.

Roeder, P.K., et al., 1987, ''Cathodoluminescence and Microprobe Study of Rare Earth Elements in Apatite,'' *American Mineralogist*, Vol. 72, pp. 801–811.

Roscoe, S.M., 1969, ''Huronian Rocks and Uraniferous Conglomerates,'' Geological Survey of Canada Paper 68–40, 205 pp.

Shepherd, M.S., 1990, ''Eneabba Heavy Mineral Sand Placers,'' *Geology and Mineral Deposits of Australia and Papua New Guinea*, F.E. Hughes, ed., Australasian Institute of Mining and Metallurgy, Parkville, Vic., pp. 1591–1594.

Sherer, R.L., 1990, ''Pajarito Yttrium-Zirconium Deposit, Otero County, New Mexico,'' *New Mexico Geology*, New Mexico Bureau of Mines and Mineral Resources, Vol. 12, p. 21.

Staatz, M.H., 1983, ''Geology and Description of Thorium and Rare-earth Deposits in the Southern Bear Lodge Mountains, Northeastern Wyoming,'' Professional Paper 1049-D, US Geological Survey, 51 pp.

Staatz, M.H., Sharp, B.T., and Hetland, D.L., 1979, ''Geology and Thorium Resources of the Lemhi Pass Thorium District, Idaho and Montana,'' Professional Paper 1049-A, US Geological Survey, 90 pp.

Steenfelt, A., 1991, ''High-Technology Metals in Alkaline and Carbonatitic Rocks in Greenland: Recognition and Exploration,'' *Journal of Geochemical Exploration*, Vol. 40, pp. 263–279.

Taylor, S.R., 1964, ''Abundance of Chemical Elements in the Continental Crust—a New Table,'' *Geochimica et Cosmochimica Acta*, Vol. 28, pp. 1273–1285.

Trueman, D.L., et al., 1988, ''The Thor Lake rare-metal deposits, Northwest Territories,'' *Recent Advances in the Geology of Granite-related Mineral Deposits*, Special Vol., 39, R.P. Taylor and D.F. Strong, eds., Canadian Institute of Mining and Metallurgy, pp. 280–290.

Vallee, M., and Dubuc, F., 1970, ''The St-Honore Carbonatite Deposit, Quebec,'' *Transactions*, Canadian Institute of Mining and Metallurgy, Vol. 73, pp. 346–356.

Van Wambeke, L., 1977, ''The Karonge Rare Earth Deposits, Republic of Burundi: New Mineralogical-Geochemical Data and Origin of the Mineralization,'' *Mineralium Deposita*, Vol. 12, pp. 373–380.

Wakita, H., Rey, P., and Schmitt, R.A., 1971, ''Abundances of the 14 Rare Earth Elements and 12 Other Elements in Apollo 12 Samples,'' *Proceedings*, 2nd Lunar Science Conference, A.A. Levinson, ed., MIT Press, Cambridge, MA, p. 1319–1329.

Welch, B.K., Sofoulis, J., and Fitzgerald, A.C.F., 1975, ''Mineral Sand Deposits of the Capel Area, W.A.,'' *Economic Geology of Australia and Papua New Guinea, I: Metals*, Monograph, Series 5, Monograph, Australasian Institute of Mining and Metallurgy, Parkville, Vic., pp. 1070–1088.

Williams, V.A., 1990, ''WIM 150 Detrital Heavy Mineral Deposit,'' *Geology and Mineral Deposits of Australia and Papua New Guinea*, F.E. Hughes, ed., Australasian Institute of Mining and Metallurgy, Parkville, Vic., pp. 1609–1614.

Woolley, A.R., 1987, Alkaline Rocks and Carbonatites of the World, Part 1: North and South America, University of Texas Press, Austin, 216 pp.

REFRACTORIES

LOUIS J. TROSTEL, JR.

Refractories, called the Hidden Industry by The Refractories Institute, provide the temperature and chemically resistant linings for the multitude of vessels that are used today in high temperature processes ranging from giant iron-making blast furnaces to industrial incinerators to petrochemical cracking units to the smallest furnaces used by the jewelry trade. While these refractory linings are absolutely essential to the proper functioning of these processes, the linings are almost always on the inside of the vessels and thus hidden from the eyes of the public.

Refractories are defined in Standard C71 of the American Society for Testing and Materials (ASTM), titled "Standard Definitions of Terms Relating to Refractories," as "nonmetallic materials having those chemical and physical properties that make them applicable for structures, or as components of systems, that are exposed to environments above 1 000°F (538°C)." This criterion of ability to withstand exposure to environments above 538°C is the critical distinction separating refractories from other ceramics, fibers, and coatings applicable only at lower temperatures.

In addition to resisting temperature, the refractories must withstand abrasion, chemical and slag attack, resist thermal shock, and carry sustained structural stresses at high operating temperatures for as long as years.

The most common methods of classifying refractories are by forming method and composition. First are the traditional formed and fired bricks, batts, and the associated multitude of special shapes. These refractories are usually formed at room temperature, dried, and then fired to high temperatures to develop the final properties desired. Other formed refractories are chemically bonded or resin or tar bonded. To complete the formed refractory matrix, some fired refractory bricks and shapes are subsequently impregnated with resins or tars.

Physical dimensions of these formed refractories are classified into two general categories: standard sizes and special shapes. The standard sizes are those in common usage in the United States described in ASTM Standard C909 "Dimensions of a Modular Series of Refractory Brick and Shapes" based on the 38 mm basic module as described in ASTM Standard C861 "Determining Metric Dimensions of Standard Series Refractory Brick and Shapes." The most common size is the "9-inch straight" 228 × 114 × 64 mm (9 × 4½ × 2½ in.) of the "2½-inch" series or 228 × 114 × 76 mm (9 × 4½ × 3 in.) of the "3-inch" series.

Bricks of nonstandard sizes are referred to as special shapes. Their use is required in specific furnaces for particular requirements or applications.

In addition there is the large and growing group of refractories sold as unshaped mixtures that may be formed into their final shapes at the application site. These unshaped mixtures are sometimes referred to as specialties. The mixtures are prepared to be hydraulically cast, plastic or wet rammed, dry rammed, or dry vibrated into their final shapes. Some are designed to be projected through nozzles (or gunned) onto their forms. The refractory mortars used to lay up the shaped refractories fall into this group referred to as specialties.

Prepared refractory grains also are sold. These are usually calcined or fused materials that are subsequently crushed and sized to a variety of specific carefully controlled grain sizes or size distributions. These grains include calcined fire clay, dead burned magnesite, periclase (fused magnesia), dead burned dolomite, kyanite, tabular and fused aluminas, fused zirconia, fused mullite, and calcined alumina-magnesia spinel.

The primary composition groups of refractories are the 1) alumino-silicate refractories made largely from fire clay and high-alumina materials, 2) silica refractories made from quartzites and ganisters, and 3) basic refractories from magnesia (or magnesite) and chromia (also referred to as chrome ore), alone and in combination, and dolomite. Special refractories include carbon, zirconia, zircon, high purity alumina, mullite, silicon carbide, silicon nitride, and some borides. In addition there are groups of special insulating refractories.

TYPES AND CLASSES OF REFRACTORIES

Fireclay Refractories

Classification: Fireclay refractories are composed mainly of fire clay and calcined fire clay in varying proportions. ASTM C27 "Classification of Fireclay and High Alumina Brick" gives five classes of fireclay brick: super-duty, high-duty, semi-silica, medium-duty, and low-duty. The semi-silica class is based on its hot load-bearing ability and high silica content. The other four classes use their refractoriness or pyrometric cone equivalent (PCE) as their basis for classification. Table 1 illustrates this classification system.

The class of super-duty fireclay refractories is characterized by its significantly improved hot load-bearing properties. This enhanced property is based on the use of raw materials of extremely low alkali contents.

Another development is the pitch or resin impregnated fireclay refractory. The resulting carbon in the refractories' pores makes these refractories more resistant to wetting and penetration by metals and slags.

Pouring pit refractories are refractories used in the handling of molten iron and steel from the furnace to the ingot molds or casting machines. These refractories include ladle brick, stopper rod sleeves, hot tops, teeming nozzles, runner brick, tuyeres, ladle-gate parts and certain other shapes, such as ingot-mold plugs. These pouring pit refractories account for a significant fraction of the fireclay brick and shapes sold.

Acid resistant brick and tile are generally classified as fireclay refractories as most are made from special fire clays. The bulk of these refractories fall into the high and super-duty classes.

Raw Materials: Fire clays are considered to include all clays having a fusion point above 1 425°C and are not white burning. They mainly are composed of hydrated aluminum silicates with

Table 1. Fireclay and High Alumina Refractory Brick Classified According to Classes and Subdivided into Types

Class	Type	PCE, min	Panel Spalling Loss, Max, %	Hot Load Subsidence, max, %	Reheat Shrinkage, max, %	Modulus of Rupture, min, MPa	Other Test Requirements
Fireclay Brick							
Super-duty	Regular	33	8 at 1 650°C	—	1.0 at 1 600°C	4.14	—
	Spall-resistant	33	4 at 1 650°C	—	1.0 at 1 600°C	4.14	—
	Slag-resistant	33	—	—	—	6.89	Bulk density, min, 2.24 g/cm³
High-duty	Regular	31½	—	—	—	—	—
	Spall-resistant	31½	10 at 1 600°C	—	—	3.45	—
	Slag-resistant	31½	—	—	—	8.27	Bulk density, min, 2.19 g/cm³ or porosity, max, 15%
Semi-silica		—	—	1.5 at 1 350°C	—	2.07	Silica (SiO_2) content, min, 72%
Medium-duty		29	—	—	—	3.45	—
Low-duty		15	—	—	—	4.14	—
High-Alumina Brick:							Alumina (Al_2O_3) Content, %
Alumina, %							
50		34	—	—	—	—	50 ± 2.5
60		35	—	—	—	—	60 ± 2.5
70		36	—	—	—	—	70 ± 2.5
80		37	—	—	—	—	80 ± 2.5
85		—	—	—	—	—	85 ± 2.0
90		—	—	—	—	—	90 ± 2.0
99		—	—	—	—	—	min, 97

Source: ASTM C27 "Classification of Fireclay and High-Alumina Refractory Brick."

minor amounts of accessory minerals. The clays of greatest importance as refractory raw materials are flint and semiflint fire clays, plastic and semiplastic fire clays, and kaolins.

Flint and semiflint clays are found in sedimentary deposits from the Mississippian and Pennsylvanian ages. They may be derived from the weathering of pegmatites or as the result of a clay mineral weathering in place. In the United States, they occur in rather limited areas of Pennsylvania, Ohio, Kentucky, Maryland, Missouri, and Colorado. Physically, these flint clays are hard dense clays that exhibit a concoidal fracture, will not slake in water, and are resistant weathering. They are essentially pure, extremely fine-grained kaolinite ($Al_2O_3 \cdot 2SiO_2 \cdot 2H_2O$). The best grades of refractory flint clays have a PCE of cone 33 to 35 and are low in impurities, such as iron and alkalies. The drying and firing shrinkages of flint clays are quite variable, ranging from medium to high.

The plastic and semiplastic clays are found in sedimentary deposits of the Mississippian to Tertiary ages. Deposits of these clays occur in many states, but the principle ones occur in Pennsylvania, Ohio, Kentucky, Maryland, West Virginia, Alabama, Missouri, Texas, and Colorado. These clays vary considerably in plasticity, refractoriness, and bond strength. They will work up readily with water into a plastic mass showing high drying and firing shrinkages. Top quality plastic fire clays have a PCE of cone 26 to 33 and generally relatively low amounts of iron and alkali impurities.

Kaolins are found in two types of deposits: residual and sedimentary. The residual deposits are those where the kaolin is in the location where it was altered from the parent rock. Kaolins for use in refractories, however, are generally from sedimentary deposits. The majority of the kaolin deposits in the United States are located below the fall line in North Carolina, South Carolina, Georgia, and Alabama. Refractory kaolins have a PCE of cone 28 to 35. They are generally moderately plastic with low green strength and high firing shrinkage. The best refractory grades of kaolin have low iron and alkali contents and exhibit high fired density. The extremely low alkali content of some kaolins, particularly from Alabama, make them valuable in the production of super-duty firebrick with enhanced hot load-bearing properties.

In addition to the above raw materials, some fireclay brick, especially those of the low-duty class of pouring-pit refractories, are made of clays with special properties, such as the ability to expand or bloat when exposed to high temperatures.

Manufacture: The raw clays and calcined clays or grogs are crushed by various conventional means. The crushed materials are prepared for forming by either of two procedures. In the first, the various raw materials are weighted out together, ground to the desired grain size, then mixed with water to achieve the right consistency for forming. In the second, the crushed ingredients are crushed and sized separately, then individually weighed out to form the batch. Next, the correct amount of water is blended into the mix to achieve the right consistency for forming.

There are several forming methods used for the production of fireclay brick and tile. The mixes for brick and tile vary from 10% raw clay and 90% calcined fire clay or crushed brick to 100% raw clay. The grain size distribution of each mix is optimized for the forming method used and the desired properties of the fired refractory.

Most of the fireclay brick and tile are made by the dry press process. The mix batch, tempered with the correct amount of water, is loaded into the die cavity of the press and pressed to shape at high pressure, either hydraulic or mechanical.

Some brick shapes are formed on vibratory presses in which the pressure is less but vibration aids in the compaction.

To give special properties or to form brick or tile whose shape is difficult to dry press, some bricks are made by the stiff-mud process. In this process the mix is tempered with water to a stiff plastic consistency and extruded through a die to form a column. The column is cut into individual units that may be repressed to give sharp edges and corners or formed into intricate shapes by hydraulic or runner presses.

Large and/or intricate shapes are sometimes made by hand molding. More often these shapes are made by hand ramming the mix into a steel mold with an air hammer. The air hammer makes the use of stiffer mixes possible forming refractories that more closely duplicate the properties of machine-pressed bricks.

Slip casting is used to make some tile for special purposes. For these, the ground mix is made into a slurry or slip with water and a small amount of electrolyte, which gives the slip the correct flow properties with a minimum amount of water. Plaster of paris molds are generally used in the casting process. The plaster quickly absorbs water from the mix, permitting early removal of the refractory shape from the mold.

The damp unfired (green) refractory brick and tile from all of these processes are dried in either periodic or continuous tunnel driers. The more difficult shapes are dried under closely controlled humidity and temperature conditions. The dried refractory shapes are then fired in periodic or continuous tunnel kilns using an appropriate firing schedule (time and temperature) to give the desired properties in the fired products. The firing schedules for different products will vary widely depending on initial mix and the properties desired in the finished product.

For instance, pouring-pit refractories generally are fired to a low temperature on a relatively long schedule to give the desired reheat expansion of the refractories in service.

During firing, mechanical and chemically combined water are driven off. Iron, sulfur, and organic impurities are oxidized. Mineral transformations and volume changes occur and refractory bonds develop.

Refractory Uses: Fireclay refractory brick and tile are used because of their generally good properties and relatively low cost. The largest consumers of fireclay brick are blast furnaces and blast-furnace stoves in which the blast furnace combustion air is heated to improve blast furnace efficiency. Other large uses of fireclay brick are ladles and accessory equipment used to transfer molten metal between furnaces and between finishing furnaces and ingot molds or casting machines.

Fireclay brick and tile also are widely used in furnaces for baking the carbon electrodes used in the aluminum industry, as general purpose steel mill brick, fireplace brick, ceramic kiln walls and roofs, ceramic kiln furniture, glass tank regenerator walls, checker support brick, and linings of heat treating furnaces.

High-Alumina Refractories

Classification: High-alumina brick and tile are those alumina-silica refractories containing from 50 to 99⁺% alumina. These refractories are classified according to their alumina content as shown in Table 1, from ASTM C27. In addition to these groups of high alumina bricks, there is a special group of mullite refractories described in ASTM C467 "Classification of Mullite Refractories." These have an alumina content of 56 to 79%, a maximum amount of metal oxides other than alumina and silica of 5%, and a maximum deformation of 5% under a compressive load of 172 kPa for 90 minutes at 1 593°C. It is the high hot load bearing ability of these refractories in this chemistry range that distinguishes them as containing mullite. The mullite content itself is hard to determine quantitatively.

Raw Materials: The more common raw materials used in high-alumina refractories are bauxite [$Al(OH)_3$], kyanite ($Al_2O_3 \cdot SiO_2$), sillimanite ($Al_2O_3 \cdot SiO_2$), andalusite ($Al_2O_3 \cdot SiO_2$), and fused, calcined, and sintered aluminas (Al_2O_3). Bauxite for refractories is considered to be a mixture of gibbsite [$Al(OH)_3$] and kaolinite ($Al_2O_3 \cdot 2SiO_2 \cdot 2H_2O$). Most of the bauxite used in refractories in the United States originates in China (PRC) with some also coming from Guyana. There are commercial-size deposits in Arkansas, Alabama, and Georgia. The Alabama material has been particularly suited to the manufacture of high quality, high-alumina brick because of its high purity and high fired density. Bauxite is nearly always calcined before use in refractory mixes because of its high shrinkage when fired.

Diaspore ($Al_2O_3 \cdot H_2O$) has lost its previous widespread use because of the depletion of reserves.

Kyanite, sillimanite, and andalusite all form mullite and silica when heated to 1 200 to 1 600°C. India is a source of kyanite and sillimanite. Large reserves of kyanite occur in Virginia, Georgia, and South Carolina.

Mullite grain or grog may be a raw material for high alumina refractories. It is made by calcining at high temperatures or fusing mixtures of high-purity alumina and silica or mixtures of alumina-silica minerals with additional alumina. The resulting material consists predominately of the mineral mullite.

Manufacture: Most of the principle ingredients used in high-alumina brick and tile mixes have been calcined or fused to reduce subsequent firing shrinkage and to convert them to all corundum, corundum and mullite, or mullite and a refractory glass depending on the silica content of the mix. The calcined or fused-grain materials in the mix are crushed or ground and sized prior to batching with the more reactive uncalcined clay or alumina bonding portion of the mix. The mix is tempered in a suitable mixer with the correct amount of water to give a good consistency for forming.

High-alumina brick and tile are made predominately by the dry-press method However some large or intricate shapes are formed by air hammer or hand ramming. For these processes the mixes are prepared by the same process as used for dry pressing with an appropriate adjustment in the amount of tempering water to give the right consistency for the ramming process. In addition, some special shapes and tiles are made by the stiff-mud and the slip-casting processes.

The formed brick and tile are dried in periodic or continuous dryers and then fired in continuous-fired tunnel kilns or periodic kilns. Firing programs for the various high-alumina refractories are designed for the specific refractory's particular needs with time-temperature schedules that will give the desired properties in the final fired products.

Uses: High-alumina refractories can withstand higher temperatures and a number of more severe chemical environments than fireclay refractories. With the continuing trend towards higher operating temperatures and more severe service conditions, there has been a steady move toward the use of high-alumina refractories to replace fireclay brick and tile.

Among the large consumers of high-alumina refractories are rotary kilns for the production of cement and lime products, some roofs for electric steel-melting furnaces, blast furnace linings and stove brick, ladle brick and aluminum contact refractories in aluminum-melting furnaces. Their applications include furnaces operating as high as 1 900°C, dry hydrogen-atmospheres above 1 100°C, special kiln furniture and some thermal-shock-prone applications.

Silica Brick

Classification: ASTM C416 "Classification of Silica Brick" covers all silica brick with an alumina content of less than 1.50%, a titania (TiO_2) content of less than 0.20%, an iron oxide (Fe_2O_3) content of less than 2.50%, a calcia (CaO) content of less than 4.00%, and an average modulus of rupture of not less than 3.45 MPa.

Within this general group of silica brick, there are two types based on the refractory's flux factor. The flux factor equals the percent of alumina plus twice the percent of total alkalies. Type A silica brick are those with a flux factor of 0.50 or less. Type B brick are all other silica brick falling within the scope of this classification.

Raw Materials: Gannister, a true quartzite, is commonly used for the manufacture of silica brick. The principal deposits in the United States are the Medina Sandstone of Pennsylvania, the Baraboo Quartzite of Devil's Lake Region of Wisconsin, and the quartzites of Alabama and Colorado. Pebble rock, such as the Sharon Conglomerate of Ohio, is also used for brick. Not all silica raw materials of these types are suitable for silica-brick manufacture. The raw materials must be high in chemical purity and have firing characteristics that yield refractories with sufficiently high strength.

Manufacture: To remove the harmful alumina and alkalies, some silica raw materials are crushed and washed free of clay and other impurities. The washed silica is ground in either a muller type mill or a rod mill, and the coarser fractions of the mix are screened to specific sizes. A portion of the silica is ball milled to make the finer size fraction of the brick mix.

The several grain size portions are added together to give a mix with a desired specific size distribution. This is mixed in a muller-type mixer with about 3% hydrated lime, water, and an organic binder.

Brick and tile are either dry pressed at high pressure on a hydraulic or mechanical press or, at a slightly lower pressure, formed on a vibrating-impact press where the vibration aids the consolidation of the mix. Silica shapes are also hand molded by ramming into molds with an air gun or a heavy hammer.

The formed brick and tile are dried in periodic or continuous driers before firing to approximately 1 480°C in continuous tunnel or periodic kilns. A rigid firing schedule must be followed when firing silica brick to accommodate the several mineral changes (with accompanying rapid growth or shrinkage) the refractory undergoes when heated up to 1 480°C. The cooling rate also must be carefully controlled in the lower temperature portion of the cooling-firing curve to accommodate the cristobalite inversion in the 120 to 290°C range, which is accompanied by a large volume expansion (several percent) and a density reduction.

Uses: Silica brick and tile are relatively inexpensive, but their areas of application are limited by a rapid change in size as the refractory passes through the cristobalite inversion range (see previous section). Cracking can occur when they are heated rapidly through this low temperature range. Historically, silica bricks were used for the roofs of open hearth and electric furnaces. Changes in steelmaking technology have drastically reduced the number of open hearths and raised the operating temperatures of electric furnaces above the service temperature of silica refractories. As a consequence, the amount of silica brick sold has decreased more than eight-fold since the middle 1950s.

Because of excellent hot load-bearing ability at operating temperatures close to its melting point, silica brick and tile continue to be used in coke ovens, ceramic kilns, and glass tank crowns. Some are used as checkers in the regenerators in glass tanks. In some European and Far Eastern steel mills, silica brick are used as blast furnace checkers.

Insulating Refractories

Classification: Insulating refractories based on alumina-silica mixes (insulating firebrick also referred to as IFB) are characterized by low thermal conductivity, low heat capacity, low density, and high porosity. They are classified in ASTM C155 "Classification of Insulating Firebrick" as summarized in Table 2. This table shows these refractories broken down into eight groups based on maximum operating temperatures. The group numbers were derived by dividing the maximum use temperatures in degrees Fahrenheit by 100. Each group has a specified maximum bulk density as well.

In addition to the above group of alumina-silica insulating refractories, other lightweight refractory shapes of high-purity silica, alumina, zirconia, and zircon are available commercially.

Raw Materials: Insulating refractory bricks are generally made from standard alumina and fireclay raw materials, such as calcined alumina, bauxite, kyanite, kaolin, and fire clay. In some instances special raw materials are used, such as pure silica, alumina, or fused zirconia bubbles.

Table 2. Grouping of Insulating Firebrick

Group No.	Reheat Change, Not More Than 2% When Tested at °C	Bulk Density Not Greater Than g/cm^3
16	845	0.54
20	1 065	0.64
23	1 230	0.77
26	1 400	0.86
28	1 510	0.96
30	1 620	1.09
32	1 730	1.52
33	1 790	1.52

Source: ASTM C155 "Classification of Insulating Firebrick."

Special ingredients are added to the mixes to produce the high porosity, lightweight structures desired. These materials include (a) combustible materials, such as sawdust, coal, coke, paper, naphthalene, anthracene, and plastic beads that leave voids after burning; (b) lightweight inorganic aggregates or fibers such as diatomaceous earth, expanded perlite, expanded vermiculite, expanded clay, foamed aggregate, mineral wool, and synthetic alumina-silica fibers; and (c) foaming agents to form a myriad of small voids throughout the brick shape through the release of many small gas bubbles.

Manufacture: Insulating brick is generally made by either extrusion, pressing, or casting. The extrusion process normally uses combustible materials to achieve high porosity and light weight although lightweight aggregates are also used in this process. The pressing process uses some of the combustible materials that will not re-expand after the forming pressure is released. Lightweight aggregates and hollow fused bubbles (for the high-purity alumina and zirconia refractories) are commonly used in the pressing process.

The casting process frequently uses foaming agents to induce porosity in the structures. Some other pore-inducing materials may be used as well.

After forming, drying and firing are accomplished under closely controlled conditions. Most of the insulating brick requires some cutting and grinding after firing to provide uniform sizes of bricks. Some pressed brick made with lightweight aggregate or fused hollow spheres do not require final finishing to size after firing.

Uses: Insulating refractory bricks are used to reduce the heat flow through furnace walls and to reduce the thermal mass of refractory structures. Insulated walls reduce the fuel consumption necessary to hold a furnace at temperature. Lower thermal mass enables the structure being heated to be brought up to temperature more rapidly and with less energy than if dense refractories of the same composition were used.

Special Refractories

This group of refractories includes very high alumina, zirconia, silicon carbide, fusion-cast products, zircon, chrome oxide, carbon, and graphite.

Zircon: Zircon is a highly refractory zirconium silicate ($ZrO_2 \cdot SiO_2$), quite resistant to thermal shock but susceptible to attack by basic slags and iron oxide. ASTM C545 "Classification of Zircon Refractories" describes the class of zircon refractories as having not less than 60% zirconia (ZrO_2) and not less than 30% silica (SiO_2). Within this group of zircon refractories are two types:

Type A whose bulk density is less than 3.84 g/cm^3 and Type B whose bulk density is 3.84 g/cm^3 or greater.

Zircon brick is normally made by dry pressing, drying, and firing to high temperatures. In addition to dry pressing, relatively large and small special shapes are made by isostatic pressing. For this process, very fine size zircon powder is mixed as a slip and spray dried to form granules. These granules are then formed under very high pressure in a flexible mold in an isostatic chamber and fired to the required high temperature. The isostatic process provides products with very uniform density. Slip casting is also used to make large or intricate shapes.

Zircon bricks and shapes are used in the glass industry, generally limiting the applications to the low soda, borosilicate glasses. Type E fiberglass glass tanks and forehearths use this refractory extensively.

Zirconia: Zirconia (ZrO_2) is very refractory with a high melting point of 2 710°C. Bricks and shapes of zirconia are strong, hard, and resistant to chemical attack, particularly by molten silicates. Because pure zirconia undergoes severe size changes during crystallographic inversions around 1 000°C during heating or cooling, it must be stabilized by the addition of roughly 5 to 15% calcia, magnesia, or yttria. These additions reduce the size changes during heating or cooling by reducing or eliminating the phase inversions of the zirconia. Thus the manufacture and use of larger zirconia parts are commercially feasible. The manufacture of zirconia bricks, blocks, and tiles is normally by dry pressing, ramming, isostatic pressing, or slip casting.

Very High Alumina: The very high alumina products are characterized by purities above 98% Al_2O_3. Compositions consist of essentially all-alumina raw materials; calcined, tabular, and fused white aluminas. Only a very small amount of a sintering aid or inorganic plasticizer may be added. Particle sizing of these products ranges from normal coarse grain (6 mesh and finer) compositions used to make bricks, tile, and large blocks and shapes to all fine-grain bodies. The high melting point, 2 050°C, makes structural applications possible as high as 1 900°C. Applications use very high alumina refractories for their extreme resistance to abrasion, resistance to chemical attack, and compatibility with a number of glasses.

A wide variety of processes are used for forming very high purity alumina refractories. These range from the standard processes of dry pressing and ramming to slip casting, isostatic pressing, and even hot pressing. Densities achieved with some fired very high alumina parts approach the theoretical density of corundum.

Silicon Carbide: Silicon carbide grain used in the manufacture of refractories is made in large electric resistance furnaces by a process invented near the turn of the century. In this operation a mixture of coke, sand, sawdust, and a small amount of salt is heated to around 2 200°C by a large graphite electrode that runs through the middle of the mix.

Silicon carbide grain does not soften or melt but dissociates at 2 510°C. It is not attacked by most acids, but it reacts with basic slags and can be dissolved into molten iron that is not saturated with carbon. Silicon carbide oxidizes when heated in air. An important characteristic of a number of bond systems for silicon carbide refractories is the ability of the bond to impart good oxidation resistance to the fired refractory for a particular application.

Silicon carbide refractories are formed by dry pressing, air hammer ramming, vibration (jolting), and slip casting. Some advanced ceramics high tech parts are hot pressed to ultimate theoretical density and very high strength.

Refractory shapes of silicon carbide have good thermal shock resistance, low thermal expansion, good abrasion resistance, high thermal conductivity, and excellent hot strength. These silicon carbide refractories are used as ceramic setter tile, kiln furniture structural beams and supports, and muffles. Bricks are used in boilers and incinerators, as well as in heat treating, forging, and annealing furnaces. Self bonded silicon carbide is used as electrical resistance-heating elements and in a number of high temperature, high stress and wear resistance applications.

Nitride-bonded bricks retaining excellent strength up to 1 550°C have been used successfully in linings of blast furnaces in Japan, Europe, and the United States. The use of ramming mixes containing significant amounts of silicon carbide in blast furnace troughs, tapholes, and runners has found significant application.

Fusion-cast Products: A number of refractories are formed by fusing or melting the refractory mixture and casting the melt into the desired forms. These products include alpha and beta alumina, basic compositions, and a variety of alumina-zirconia-silica compositions. Applications include glass tanks and other glass melting holding and forming locations where a low rate of corrosion of the refractories is required. Other applications of fused-cast refractories make use of high wear resistance and strength. Fused-cast refractories also find application in limited areas of high vulnerability in the steel industry.

Chromic Oxide: Refractories made of chromic oxide (Cr_2O_3) predominantly are formed by slip casting or isostatic pressing. The fine-grained, extremely dense bodies are used primarily in glass-fiber tanks because the refractories have extremely low rates of corrosion or attack by the glass. Chromic oxide is a strong colorant in glasses, but the rate of solution of these refractories is low enough to permit application in some glass tanks.

Carbon and Graphite: The largest application of carbon refractories is in the hearth, bosh, and sidewalls of blast furnaces. The primary raw material for most carbon brick and blocks is anthracite-based carbon. Graphite is used in the manufacture of crucibles and ladle stopper heads. Some graphite is used in plastic firebrick. Oxygen, alkalies, water vapor, and carbon dioxide all attack carbon brick at high temperatures, but under reducing conditions, carbon shapes can withstand loads at temperatures above 3 315°C without melting or softening.

Refractory Specialties

Refractory specialties or monolithics are those refractory products that are sold as unformed products. They are usually formed in place by the user usually at the application site. These refractories include mortars, castables, plastics, ramming mixes, and dry vibratable products.

Mortars are generally used for bonding together brick or shapes. The mortar joints used in refractory construction are quite thin, so the refractory mortar must be a fine-grained mix capable of making a thin joint. Some mortars are used as a coating on refractory brick. For this application the mortar is mixed to a thicker than normal consistency.

Many castables are used in the same manner as portland cement concretes to form large refractory shapes. A trend toward gunning castables, both cold and hot, is occurring as this type of application takes less time, molds, and forming labor than the casting process. The installation economies, speed, and good properties of installed castables and gunnables have lead to wide use in the petrochemical industry, in large incinerators, and power generating stations as well as in lining and repairing of a number of different process units in today's steel plants.

Plastics and ramming mixes are used to repair refractory installations where the repairs can be installed against the existing refractory structure. Because these mixtures are capable of being molded in place, they also are used for refractory installations that would otherwise require intricate, expensive preformed shapes.

Mortars: Refractory mortars are of two general types. The first, air-set mortars, have a bond that sets the mortar when exposed

Table 3. Regular Castable Refractories

Test Requirements	Classes of Alumina-Silica Base Castable Refractories						
	Class A	Class B	Class C	Class D	Class E	Class F	Class G
Permanent linear shrinkage, not more than 1.5% when fired for 5 h at	1 095°C	1 260°C	1 370°C	1 480°C	1 595°C	1 705°C	1 760°C

Insulating Castable Refractors

Test Requirements	Classes of Insulating Castable Refractories				
	Class N	Class O	Class P	Class Q	Class R
Permanent linear shrinkage, not more than 1.5% when fired for 5 h at	925°C	1 040°C	1 150°C	1 260°C	1 370°C
Maximum bulk density, g/cm^3 after drying at 105 to 110°C	0.88	1.04	1.20	1.44	1.52
	Class S	Class T	Class U	Class V	
Permanent linear shrinkage, not more than 1.5% when fired for 5 h at	1 480°C	1 595°C	1 650°C	1 760°C	
Maximum bulk density, g/cm^3 after drying at 105 to 110°C	1.52	1.60	1.68	1.68	

Source: ASTM C401 "Classification of Alumina and Alumina-Silicate Castable Refractories."

to the air. When heated, in some cases, the original strength from the air setting bond decreases while the ceramic bond that develops during heating increases. The second type, heat-set mortars, have only moderate strength when dried, developing a high-strength ceramic bond when heated. Refractory mortars vary greatly in composition but generally are similar to the composition of the brick they are being used with. The mortars are of fine-grain size and use plasticizing agents to give the necessary workability.

The manufacture of refractory mortars is relatively simple. Raw materials are crushed or milled, sized, batched with the necessary bonds and plasticizing agents, thoroughly mixed either wet or dry, and packaged.

Castables: Refractory castables are made from refractory grains or aggregates and bonds that will set in a reasonable time due to a chemical reaction when mixed with water. Castables are generally mixed at the application site with the water to the correct consistency for placement and poured into a mold to set. Most refractory castables are bonded with hydraulic-setting calcium aluminate cement.

In ASTM C401, "Classification of Alumina and Alumina-Silicate Castable Refractories," these castables are divided into two types: regular or dense and insulating. Each of these two types is further subdivided according to service temperature. Each class of insulating castable has maximum bulk density criteria as well (Table 3).

The majority of castables produced are of the fire clay (alumina-silicate) and high-alumina types. Other aggregates may be used, such as chrome ore and chrome magnesite mixtures. When intended for gunning applications, special additives are used.

Manufacture of refractory castables is straightforward. Coarse aggregate is sized after crushing, or simply purchased presized by the raw material supplier in the case of tabular or fused refractory grains. Coarser materials are mixed thoroughly with the fines, bonds, and any additives, and then packaged in bags.

Plastics: Plastics are supplied in a stiff mud or plastic condition, which enables them to be formed into shape or placed at the installation site. Various types of plastic firebrick are made depending primarily on the type of coarse aggregate used. These include fireclay grog, crushed fireclay brick, bauxite, tabular alumina, and chrome ore. These products are divided into air-setting and heat-setting types. The quality or refractoriness of a plastic depends on the aggregate, as well as the type and amount of bond and plasticizers, used.

The manufacture of plastics consists of crushing and sizing the coarse aggregate, then mixing it with the fines, plasticizers, and sufficient water. The wet mix is generally extruded as a solid column, which is sliced into blocks for ease of handling and installation. Plastics are usually packaged in cartons with a protective plastic wrap to prevent premature drying. Some are sold without extrusion, simply packaged in drums.

Ramming Mixes: Ramming mixes may be either the air-setting or the heat-setting type. When rammed into place, they harden to form well-bonded monolithic refractories. These mixes consist primarily of ground and sized refractory aggregates with minor amounts of other materials added for workability and for a bond. The refractory aggregates used in these products include silica, tabular alumina, calcined fireclay, calcined bauxite, silicon carbide grain, zircon, periclase, and chrome ore. The amounts of clay, bentonite, or other plasticizers and bonds in ramming mixes are normally less than that used in refractory plastics. Some ramming mixes are supplied dry for preparation with water at the installation site, but others are supplied already mixed with water to give the proper ramming consistency. When supplied already wet, the ramming mixes are usually shipped in drums. When shipped dry, they are usually supplied in bags.

Vibration Mixes: Vibration mixes, supplied dry, are monolithic mixtures that can be dumped directly out of the bag at the installation site into a form, vibrated or dry rammed to density, and fired immediately to maturity without the normal drying out period required after installation with wet monolithics. This speed of installation and quick turn around in furnace equipment during relines compensates for the more complex forms required with the dry vibratables versus the wet rams.

The bonds in a dry vibratable frequently are of two types: first, a relatively low temperature (red heat) bond that develops before the forms are removed (dismantled after a low fire or burned out or melted out), and second, a higher temperature sintered or refractory glassy bond providing the integrity to the refractory lining during its operation. Behind this well-bonded hot face refractory, in the cooler section near the furnace shell, is loose unbonded backup refractory material that will prevent hot-face cracks in the refractory, and the resulting metal fines, penetrating completely through the refractory lining to the furnace shell.

These mixes consist of carefully sized materials that will quickly vibrate to a dense packed condition. The coarse grains used may be high aluminas, such as brown-fused alumina or tabular

Table 4. Classification of Chrome Brick,[A] Chrome-Magnesite Brick, Magnesite-Chrome Brick, and Magnesite Brick According to MgO Content

Class	MgO Content, %[B] Nominal	Minimum
Chrome-Magnesite and Magnesite-Chrome Brick[C]		
30	30	25
40	40	35
50	50	45
60	60	55
70	70	65
80	80	75
Magnesite Brick[C]		
90	90	86
95	95	91
98	98	96

[A]Chrome brick is a refractory brick manufactured substantially or entirely of chrome ore.
[B]Ignited basis.
[C]This classification applies to both chemically bonded and burned brick.
Source: ASTM C455 "Classification of Chrome Brick, Chrome-Magnesite Brick, Magnesite-Chrome Brick and Magnesite Brick."

alumina; magnesias, such as dead-burned magnesite or fused magnesia; or a number of alumino-silicates. The fines incorporate a variety of materials matching the chemistry of the coarse fractions and developing the bonding desired for the particular application. The dry-vibration installation technique demands that the sizing of the fine fractions of these mixes be carefully chosen and well controlled.

Manufacture of the compositions is simple and straightforward. The raw materials, which are generally received pre-sized, are carefully weighed out, mixed in a pan or intensive-type mixer, and bagged for shipment to the customer. Care must be taken during mixing to not change the particle size of the raw materials and to develop a completely homogeneous mix.

Basic Brick

Classification: Basic refractories are defined in ASTM C71 as "Refractories whose major constituent is lime, magnesia or both and may react chemically with acid refractories, acid slags or acid fluxes at high temperatures." In an added note the definition also says "Commercial use of this term also includes refractories made of chrome ore and combinations of chrome ore and dead-burned magnesite." For this chapter, the broad definition is used.

ASTM C455 "Classification of Chrome Brick, Chrome-Magnesite Brick, Magnesite-Chrome Brick, and Magnesite Brick," classifies basic brick, composed of chrome ore and dead-burned magnesite as end members, and combinations of these minerals on the basis of MgO content (Table 4). These basic bricks may be supplied either as burned (or fired) brick or as chemically-bonded brick. Burned bricks are supplied as either burned at the normal temperature or burned at high temperatures, generally exceeding 1 650°C which gives direct-bonded brick with enhanced strength at high temperatures. These refractories may be resin or pitch impregnated, and they may have metal additions in the magnesia brick. Both burned and chemically bonded brick may be metal encased with steel plates that act to bond the refractory structure together like mortar would when the refractories are exposed to high application temperatures.

The development and rapid adoption of basic-oxygen steel making in the last decade or so has led to the rapid development of the resin and pitch-bonded and impregnated refractories. This group includes pitch-bonded periclase, pitch-bonded dolomite, pitch-bonded periclase dolomite, and pitch-impregnated periclase brick. Pitch-bonded brick are sometimes subjected to a low temperature heat treatment and are then referred to as pitch-bonded tempered brick. These refractories are characterized by resistance to peeling or slumping during the burn in of a basic-oxygen furnace (BOF) lining. The pitch-bonded and the pitch-bonded tempered periclase brick are generally of two types: one, of high purity (about 96% MgO and above) and two, of lower purity (nominally 90% MgO). Dolomite bricks are of two types: one, a regular dead-burned dolomite and two, a low-flux dolomite characterized by lower amounts of oxides other than CaO and MgO. The pitch-impregnated periclase bricks are also of two types: one, a regular-burned brick and two, a direct-bond (high-fired) type.

Raw Materials:

Chrome Ore. Chromite has a theoretical composition of $FeO.Cr_2O_3$ but commercial refractory chrome ores contain substantial amounts of alumina, magnesia, and silica. Typically they contain at least 60% chromic oxide (Cr_2O_3) plus alumina (Al_2O_3) and less than 6% silica (SiO_2). They generally range from 30 to 50% Cr_2O_3, 12 to 16% FeO, 13 to 30% Al_2O_3, 14 to 20% MgO, 1 to 6% SiO_2, and up to 1% CaO. High quality refractory ores have about 32% Cr_2O_3 and 30% Al_2O_3. Other excellent ores have as much as 42% Cr_2O_3 with correspondingly lower amounts of Al_2O_3. Sources of chrome ore include the Philippines, Turkey, Zimbabwe, and the Republic of South Africa. Chromite is also mined in Albania and Russia.

Magnesite and Periclase. In ASTM C71, magnesite grain is defined as "dead-burned magnesia in granular form of size suitable for refractory purposes." Seawater magnesite is defined as "dead-burned magnesia made by a chemical process using seawater or other saline solutions (brines)." These refractory terms differ from the mineralogical term "magnesite—a mineral composed of $MgCO_3$." The term periclase is often used in refractories to describe dead-burned magnesia derived from seawater or brines while the mineralogical use of the term periclase refers to the high-temperature form of magnesia.

Most of the dead-burned magnesite grain derived from natural deposits occurs as magnesium carbonate, although some brucite [$Mg(OH)_2$] deposits have been used. The largest deposits of magnesite in North America occur in Washington, Nevada, western Canada, and Quebec. Also, natural deposits occurring in China (PRC), Greece, Austria, and Yugoslavia are used as raw materials to derive dead-burned magnesite in the United States.

Some natural deposits provide grain of excellent quality and high purity. However, the need for periclase grain of high purity and closely controlled properties has lead to significant growth of production of periclase grain from seawater and brines. This seawater or brine process involves precipitation of magnesium hydroxide from solutions using slaked lime, slaked dolomite, or caustic, followed by conversion of the magnesium hydroxide to periclase by calcining in rotary or shaft kilns at extremely high temperatures.

This highly calcined grain derived from either natural deposits or seawater/brine solutions is rendered relatively inert to reactions with moisture or carbon dioxide and so is termed dead-burned.

The chemical compositions of periclase grain derived from seawater and brine solutions can be modified during its production. Grains having purities over 98% MgO are commercially available.

Dolomite. Dolomite grain is obtained by treating the mineral dolomite [$Ca,Mg(CO_3)_2$] at high temperatures, driving off the CO_2 and densifying the resulting material. Extensive deposits of high-

Table 5. Shipments of Refractories in the United States (millions of dollars)

	Year										
	1980	1981	1982	1983	1984	1985	1986	1987	1988	1989	1990
Total Refractories	1662	1790	1297	1398	1632	1582	1521	1682	1965	2011	2026
Total Clay	**697**	**789**	**601**	**664**	**755**	**694**	**670**	**740**	**824**	**823**	**803**
Brick & Shapes	421	438	316	343	384	349	307	278	328	422	427
Fire Clay	179	175	111	102	110	97	79	47	48	75	81
High Alumina	143	151	116	130	148	139	123	173	198	164	169
IFB	36	41	29	25	29	27	25	24	44	58	54
Unshaped	256	328	260	277	323	338	360	397	418	402	376
Mortars, Plastics & Rams	99	134	111	126	144	132	114	123	129	136	142
Castables	101	121	108	108	130	122	119	162	180	179	161
Total Non-Clay	**965**	**1001**	**694**	**734**	**878**	**887**	**851**	**943**	**1140**	**1188**	**1224**
Brick & Shapes	536	553	292	402	492	463	404	550	686	659	672
Silica	d	d	d	14	16	15	14	20	23	24	27
Basic	278	309	115	206	248	223	176	208	252	233	229
Extra High Alumina	41	45	41	39	50	65	63	57	85	103	97
Unshaped	429	448	403	333	386	424	447	393	454	528	552
% Change from Last Year in Total Refractories		8%	−28%	8%	17%	−3%	−4%	11%	17%	2%	1%
in Clay Refractories		13%	−24%	11%	14%	−8%	−4%	10%	11%	0%	−3%
in Non-Clay Refractories		4%	−31%	6%	20%	1%	−4%	11%	21%	4%	3%
% of Total Clay Production											
as Bricks & Shapes		56%	53%	52%	51%	50%	46%	38%	40%	51%	53%
as Castables		15%	18%	16%	17%	18%	18%	22%	22%	22%	20%
% of Non-Clay Production											
as Bricks & Shapes		31%	17%	28%	28%	25%	21%	22%	22%	20%	19%

Source: United States Department of Commerce, Bureau of Census, Report MA32C.

purity dolomite are readily available. Two types of grain are produced for refractory use: one, a regular grain and two, a low flux grain of higher purity.

Manufacture: When manufacturing basic brick, chrome ore and periclase or magnesite are crushed, ground, and carefully sized into several fractions. Generally two separate systems are used to minimize the contamination of one material with the other during the crushing and sizing operations. Because the size fractions of the chrome ore and the periclase have such a strong influence on the properties of these brick, it is important that the grain size of each component be controlled within specific limits. The grain size fractions of each ingredient are carefully weighed out and then mixed in a muller-type mixer with the correct amounts of water and additives. Most basic brick are pressed on hydraulic or mechanical presses at high pressures, but difficult shapes are formed by ramming with an air hammer. The chemically-bonded brick are dried at a temperature high enough that the chemical bond will develop. The burned brick are dried and fired at a high temperature to develop the ceramic bond. Direct bonded brick is fired at about 1 760°C to develop the crystal-to-crystal bond. Both the chemically bonded and the burned bricks may be metal cased with soft steel plates as mentioned earlier.

For the basic oxygen furnace (BOF), manufacture of the pitch-bonded brick is begun by crushing, grinding, and sizing the periclase grain into fractions. The coarse-size fractions are weighed out and heated to about 205°C. This heated coarse-grain fraction is then mixed with the pulverized (fine-grained) portion of the periclase, heated pitch, and carbon particles in a blade- or a pan-type mixer. The hot mix is quickly conveyed to a press and the brick shapes are pressed at high pressure. After cooling, the pitch-bonded brick can be shipped. The tempered equivalents of this brick are heated to 315°C to remove some volatiles and develop carbon bonding. Tempered brick show better strength during early heat up of the BOF than do the regular pitch-bonded type.

The resin and pitch-bonded and the pitch-bonded tempered dolomite and dolomite-periclase brick are made in the same manner, substituting dolomite grains in the coarse-size fractions.

Burned-impregnated periclase brick are made by the same process as the conventional basic brick with the added steps of heating the fired brick to about 205°C, subjecting it to a vacuum, and then impregnating it with pitch in a chamber held at several atmospheres of pressure.

Uses: Basic brick find their greatest use in the steel industry, including electric furnaces. They are also used widely in cement kilns, copper and nickel processing furnaces, and in glass-tank regenerators. Most of these applications use the burned type of brick. In the steel, cement, and nonferrous industries, direct-bonded brick is favored. The brick used in glass tank regenerators is of the ceramic- and the direct-bonded types, but in the glass tank's basic checkers, chemically-bonded brick of various qualities is used. Pitch-bonded, pitch-bonded tempered, and pitch-impregnated brick find their principal use in the basic oxygen furnace. The trend is toward zoned linings in these furnaces, for instance, pitch-impregnated high fired periclase brick used in the charge-pad area where the scrap and hot metal impinge.

ECONOMICS AND FUTURE TRENDS

Table 5 shows data for US refractories shipments during the most recent decade (1980–1990) for which these statistics are available. These data indicate that overall growth in the refractory industry has been minimal. Extra high alumina brick and specialties (unshaped refractories) in fire clay and nonclay have shown increases. Basic refractory products (brick, shapes, and the unshaped mixes) remain the largest dollar category throughout these years. Shipments of fire-clay bricks and shapes diminished throughout the decade.

Table 6. Refractory Consumption by Industry in 1985

	$ (Millions)	% Total
US Total Production	1,521	100
Iron and Steel	746	49
Nonferrous	115	8
Glass	98	6
Ceramic	126	8
Mineral Processes	62	4
Chem and Petrocher	54	4
Utilities	13	1
Engr Contractors	112	7
Other	61	4
Not Specified	59	4
Total Domestic	1,444	95
Exports	77	5

Notes:
Mineral products include cement, lime, dolomite, clay bauxite, and magnesia.
Ceramic includes brick, tile, whiteware, and refractories.

Source: United State Department of Commerce, Bureau of Census, Report MQ32C(86)-5 Summary of 1986.

These changes are driven by the refractories users' push toward lower refractory cost per unit of product. Increased life and less down time justify the use of higher performance refractories although they may be more expensive per unit than fireclay refractories as purchased. Minimizing down time is also a factor in the increased use of unshaped refractory specialties used for patching and partial rebuilds. Finally, refractory consumption changes are driven by changes in the technology of the user industries.

The consumption of refractories by specific industries is shown in Table 6 and illustrates the large proportion of refractories consumed by the iron and steel industry (49%). Thus one of the most important changes driving refractory developments and consumption changes is the shift in the steel industry to the basic oxygen process. These smaller units produce more steel annually than the older furnaces, and they use considerably less refractories per ton of steel produced.

Refractories are constantly upgraded to yield longer life, or more product per refractory unit consumed or even to allow the commercialization of new processes.

All the trends are toward higher performance, higher purity, higher cost, and lower unit volumes of refractories. Future pressures on the technology should continue these directions.

Most of the fireclays and their calcines are produced domestically with many of the refractory producers mining and processing their own fireclays and calcines. With the increasing quality requirements for refractories and the decreasing availability of high quality fireclays, beneficiation of the clays that are available may become more of a requirement. This could significantly increase the cost of fireclay refractories, which may hasten the switch to other types of refractories, a trend already in progress.

Some refractory raw materials are entirely imported from abroad. There are no commercial refractory-grade deposits of chrome ore in the United States, so all the chrome ore used must be imported. Bauxite also is largely imported into the United States, although there are domestic deposits. Both bauxite and chrome ore are found in areas of the world that are politically sensitive. Bauxite is largely imported from China (PRC). Chrome ore is found in such areas as Turkey, Greece, Korea, the Philippines, and South Africa.

Magnesite, dolomite, and periclase are found in domestic natural deposits, and they can be extracted in almost limitless quantities from seawater and brines as well.

Factors having a strong influence on future refractory usage patterns and changes in these patterns are 1) increasing environmental considerations, 2) changes in refractory consuming manufacturing practices, particularly in the field of metallurgy, and 3) changing sources and costs of available energy.

In the late 1970s and early 1980s, energy costs seemed to be the major factor in driving refractory usage changes. More recently, however, energy costs have stabilized and are less of a factor. Currently, much stricter environmental considerations, many newly mandated by state and federal regulations, are changing refractory usage. The difficulties arising in the use and disposal of chrome oxides are an obvious example, reducing the demand and production of any chrome containing refractory.

The use of electric furnaces in place of fuel-fired furnaces is an example of a change in technology that will result in improved emission controls from manufacturing plants as well as requiring different refractories for optimized furnace construction.

New metallurgical practices would result in changes in refractory usage. Direct reduction of iron ore is one that has been tried for several years. The direct steel making processes, being developed on a number of fronts, will have profound effect on refractory consumption (both type and quantity) if successful.

ACKNOWLEDGMENTS

The author would like to acknowledge the help of The Refractories Institute, Pittsburgh, PA, in supplying copies of the MA32C report ''Industrial Reports—Refractories'' published by the Census Bureau of the United States Department of Commerce and the excellent chapter on ''Refractories'' by James A. Crookson and William D. Fitzpatrick in *Industrial Minerals and Rocks*, 5th edition.

BIBLIOGRAPHY AND REFERENCES

Anon., 1961, *Modern Refractory Practices*, 4th ed., Harbison Walker Refractory Company, Pittsburgh, PA, 607 pp.

Anon., 1990, *Phase Diagrams for Ceramists*, Vol I-VII, 1964–1989, The American Ceramic Society, Westerville, OH.

Anon., 1992a, *UNITECR'91, Proceedings*, 2nd ed., 2nd Biannual Unified Technical Conference on Refractories, The German Refractories Association, Verlag Stahleisen mbH, Dusseldorf, 610 pp.

Anon., 1992b, *1992 Annual Book of Standards*, Vol. 15.01, American Society for Testing and Materials, Philadelphia, PA, 502 pp.

Budnikov, P.P., 1964, *The Technology of Ceramic Refractories*, MIT Press, Cambridge, MA, 647 pp.

Chesters, J.H., 1963, *Steel Plant Refractories, Testing, Research and Development*, 2nd ed., United Steel Co., Sheffield, UK.

Fisher, R.E., 1989, ''Advances in Refractories Technology,'' *Ceramic Transactions*, Vol. 4, The American Ceramic Society, Westerville, OH, 620 pp.

Fisher, R.E., 1985, ''New Developments in Monolithic Refractories,'' *Advances in Ceramics*, Vol. 13, The American Ceramic Society, Westerville, OH, 424 pp.

Krause, C.A., 1987, *Refractories: The Hidden Industry—A History of Refractories in the United States, 1860–1985*, The Refractories Institute, Pittsburgh, PA, 272 pp.

Newsome, J.W., et al, 1960, *Alumina Properties*, Alcoa, Pittsburgh, PA, 88 pp.

Norton, F.H., 1968, *Refractories*, 4th ed., McGraw-Hill, New York, 450 pp.

Searle, A.B., 1948, *Refractories for Furnaces, Kilns, Retorts, Etc.*, 2nd ed. rev., C. Lockwood, London.

Sosman, R.B., 1960, *Phases of Silica*, Rutgers University Press, New Brunswick, NJ.

Trostel, L.J., Jr., ed, 1989, *UNITECR'89, Proceedings*, 1st Unified International Technical Conference on Refractories, The American Ceramic Society, 2 vols., Westerville, OH, 1868 pp.

SALT

DENNIS S. KOSTICK

Sodium is a silver-colored metal that is so unstable that it reacts violently in the presence of water, and chlorine is a greenish-colored gas that is dangerous and lethal. Yet the combination of these two elements forms sodium chloride, commonly known as salt, which is a white-colored compound essential to life itself. Salt is perhaps the only mineral in the world that is used virtually by every human being in every country. Although its importance has diminished over time, salt has had a profound influence on history. Sources of salt have determined the location of cities, such as Salzburg, Austria, Salzgitter, Germany, Salt Lake City, UT, and Saltville, VA. Populations have migrated in search of salt, and wars have been fought to obtain or to protect it.

Although there is no written record of when prehistoric man first discovered salt, he probably instantly appreciated its unique and distinctive taste and sought more of it. He noticed animals were attracted to salt springs and salt licks, where they satisfied their innate cravings. Man used this knowledge to hunt for food; where he found the salt, he would find the animals. Because prehistoric man was not aware of his physiologic need for salt, he automatically received his salt requirements from the meat of the animals he ate. A Mesopotamian tale tells of a wounded pig that ran into the ocean and drowned. After being recovered from the ocean, where it was saturated in the salt brine, the pork was found to taste better than unsalted meat. It was not until man graduated from a nomadic to an agricultural society when he introduced vegetables and cereals into his diet that he realized he needed additional supplements of salt. The Sumerians ate salted meat and used salt to preserve food about 3500 BC. In approximately 1000 BC, the Trojans learned to use salt for preserving fish. The maritime civilizations, such as that of the Phoenicians, engaged in an extensive salt trade throughout the Mediterranean.

Many superstitions, customs, and traditions surround the religious, social, economic, and political aspects of civilization. Although some commodities have provided the basis of some ageless metaphors and cliches (i.e., heart of gold, strong as steel, etc.), none are more commonly used than those pertaining to salt. Being ''the salt of the earth'' referred to a person's worthiness, as referenced in Matthew 5:13 of the *Bible*. Someone ''not worth his salt'' reportedly was used in ancient Greece in bartering salt for slaves. When a laborer was worth his salt, he was paid a salary (*sal* is Latin for salt), a term still in use today that dates back to Roman times when part of a soldier's pay was salt rations. Fearing those around him, the Roman leader Pompey was known to add a grain of salt to his drinks as a supposed antidote to poison. Hence, ''taking it with a grain of salt'' is to regard something with suspicion.

The earliest known record of salt production was in China about 2000 BC when Emperor Yu decreed that salt had to be supplied to the court as a tribute. Because salt was a commodity that had universal demand, governments often intervened to control salt production and distribution to generate revenue for major national projects. State-run salt monopolies were common in history, such as in Rome about 506 BC. Revenue was also derived from taxes levied on salt sales. The word *gabelle* refers to the direct salt tax used in France. This tax was so despised that it was a major cause of the French Revolution. Although the tax was repealed in 1790, it was reinstated in 1805 to raise revenue for war. France continued to used the gabelle until 1945.

Salt making was an important economic activity in many areas of the world. Archaeological evidence indicates that salt was quarried from hills near the Dead Sea before the Bronze Age. At Hallstatt in the eastern Alps of Austria, rock salt was mined as early as 1400 BC. Solar salt was being produced in the northern Yucatan peninsula by the Mayan civilization from about 50 BC to AD 300. Other than the prehistoric use of salt in the southwestern part of the United States by the Native Americans, the first production of salt by American colonists began in June 1614 on Smith's Island, VA, seven years after the founding of the Jamestown colony. The short-lived activity was in response to the high price of salt being charged by ship's captains who brought salt over as ballast from England.

Although Nathaniel Loomis is credited with starting the first commercial inland salt production in the United States at Onondaga Lake in New York in 1790, salt had been made by individuals for private use years earlier. Indians had long known of the salt springs in the Kanawha Valley of Virginia (now known as West Virginia). In 1755, Mary Ingles, a white settler, was kidnapped by Indians and was forced to make salt from the saline springs. After her escape, she told other settlers how to make salt. At Saltville, VA, Arthur Campbell began making salt in 1782 by boiling brine obtained from salt springs. In 1799, William King sank a shaft about 3 m in diameter down about 61 m to the underlying Saltville salt deposits. The shaft flooded before he could extract any rock salt; however, King has the distinction of digging the shaft of the first salt mine in the United States. Fig. 1 shows the King site as it appears today in Saltville, VA. Production of brine from the shaft continued until 1892. Settlers moved into the Kanawha Valley in 1785, but it wasn't until 1806 that Elisha Brooks established the first commercial salt works in Kanawha.

GEOLOGY

The sequence of precipitation of chemical compounds from seawater was shown by the Italian scientist J. Usiglio in 1849 using water collected from the Mediterranean Sea. His order of precipitation was iron hydroxide, calcium carbonate, calcium sulfate, sodium chloride, and the soluble salts of potassium and magnesium in the mother liquors. Although the deposit at Strassfurt, Germany, is a classic example of this sequence, this order of precipitation is normally the exception to the rule.

Mineralogy

Pure sodium chloride contains by weight 39.34% sodium and 60.66% chlorine. The mineralogical name for salt is halite, from the

Fig. 1. First salt mine in the United States, Saltville, VA. Shaft was filled in many years ago for safety reasons. Note rebuilt salt-boiling block at right, used during Civil War to produce salt for the Confederacy (*courtesy:* author).

Greek word *hals* meaning salt. Because of its simple crystal structure, it was the first structure to be analyzed by X-ray. Salt has the following properties:

Formula: NaCl
Crystallography: isometric, hexoctahedral, 4/m 32/m
Cell contents: Na_4Cl_4
Habit: usually cubic, rarely octahedral; massive, granular to compact
Physical properties:
  Refractive index: 1.554
  Twinning: observed on {111}
  Cleavage: {001} perfect
  Fracture: conchoidal, brittle
  Hardness: Mohs hardness of 2
  Specific gravity: 2.168
  Melting point: 804°C
  Boiling point: 1 413°C
  Luster: vitreous, normally colorless to white, but occasionally red, yellow, blue, and purple
  Solubility: 0°C: 35.7 parts per 100 parts of water
    100°C: 39.8 parts per 100 parts of water

Classification of Deposits

Salt deposits are found in solution or in a solid state. It is the method of recovery from these two forms that categorizes salt as rock salt, solar salt, salt brine, or vacuum pan salt.

Solution: Solutions of salt are found in the oceans, lakes, natural brines, and ground water.

Oceans. The largest resource of salt in the world is the oceans. The world's oceans contain about 530 Mkm^3 of water, in which is dissolved about 42 Pt of salt. One cubic kilometer of seawater contains 93 Mt of dissolved matter, of which 78 Mt is salt. To visualize what this quantity of salt looks like, it would take about 18 Vertical Assembly Buildings (the world's largest building in terms of volume) like the one at the Kennedy Space Center at Cape Canaveral, FL, to hold all the salt contained in 1 km^3 of seawater.

The salt content of the oceans varies from 1 to 5% depending on the location, depth, and external influences, but it averages about 3.5%. In coastal areas, where evaporation rates exceed precipitation rates, solar evaporation of seawater may be the more advantageous method to produce salt. In certain countries that have limited technology and markets, the operations may be crude and relatively unchanged from methods their predecessors used. Some facilities in arid regions take advantage of natural shallow depressions along the coastlines. The seawater is introduced during high tides or storms and evaporated during dry periods, leaving a dry crust of salt on the surface. Most of these operations are small and the salt is marketed for local consumption. In other nations, modern operations that utilize more advanced scientific techniques to improve product quality and quantity while maintaining low production costs become major competitors in world markets.

Lakes. Lakes are the result of natural accumulations of water in topographical depressions. Usually water carries dissolved material whose composition depends on the type and solubility of the rocks and material transversed by the ground and surface waters. Clarke (1924) classified mineralized lakes as follows:

1) Sodium chloride lakes
2) Natural bittern lakes—magnesium salts dominate
3) Sulfate water lakes
4) Sulfate-chloride lakes
5) Alkaline lakes; carbonate lakes, carbonate-chloride lakes, chloride-sulfate-carbonate lakes

Sodium and calcium cations that dominate most mineralized lake deposits give rise to deposits rich in sodium sulfate (both mirabilite and thenardite), burkeite, gaylussite, trona, and halite. Landes (1960) reported that the sodium chloride lakes are "either separated bodies of ocean water or they owe their composition to the presence of salt within the rocks that floor the watershed." Land-locked bodies of saline water provide a natural source of salt to develop, such as the Great Salt Lake in Utah, where several solar salt evaporation facilities have been constructed. Alkaline lakes, such as Searles Lake and Owens Lake, both in California, are more noted for their sodium carbonate content than their sodium chloride potential.

Ground Water. Ground waters are classified as either connate or meteoric water. Connate waters are those that are trapped in the country rock at the time of formation, and reflect the composition of the water at the time. Most sedimentary rocks were deposited at the bottom of the sea, so a substantial quantity of connate water could have been entrapped during the process. However, sedimentary rocks are also formed in freshwater lakes and interior salt lakes.

Meteoric, or rain water, is fresh as it is precipitated; however, it absorbs soluble material as it descends through the air, and into the soil, rocks, etc. Depending on the environment, the composition and the salinity of ground water varies. The salinity of the ground water can increase on contact with sodium chloride-rich strata. Clarke (1924) classified ground waters into the following groups: chloride, sulfate, carbonate, mixed (chloride-sulfate, chloride-carbonate, sulfate-carbonate, chloride-sulfate-carbonate), siliceous, borate, nitrate, phosphate, and acid.

Ground waters range in chemical content from practically nil to as much as 254 000 ppm. Many ground waters that were saturated with salt provided the source of surface salt springs that were utilized by animals and prehistoric man. The salt-bearing ground waters in New York, Virginia, and West Virginia were commercialized to take advantage of this resource. Some brines are not processed for the salt but for the bromine or magnesium contents.

Solid Salt: Crystalline salt is found in playa lake deposits, bedded salt deposits, and salt domes. Halite may be the dominant mineral; however, it also may be commingled with other sedimentary minerals.

Playa Salts. A playa is a naturally occurring sandy, salty, mud-covered floor of a desert basin, according to Grabau (1913). During an influx of rain, the playa may be covered with rainwater, which dissolves part of the soluble playa surface. After the water evaporates, a soluble evaporite remains that may be concentrated in one part of the playa. With subsequent periods of additional precipitation and evaporation, the crystal's surface promotes the continued growth of the size of the crystalline deposit. Searles Lake in California is considered an example of a crystal body in a playa lake.

Playa lake deposits generally are common in arid regions. Saline playas are formed from the evaporation of a mineralized lake that was created by the leaching of the rocks in the area surrounding the enclosed basin. Sodium chloride-bearing solutions contribute to the salt content of the playa. The presence of certain other minerals may indicate that other geologic processes occurred in the area. For example, the presence of borates would indicate recent volcanism in the area.

Salt is not the only mineral found in playa deposits. The mineralogic composition of a playa may vary considerably. Several playa lakes in California illustrate this clearly. Cadiz Lake is predominantly composed of calcium chloride, while Bristal Lake is predominantly salt, and Owens Lake is chiefly sodium carbonate. Searles Lake contains a mixture of many compounds, primarily salt, sodium carbonate, sodium sulfate, potash, and borates.

Bedded Salt Deposits. Salt is also found as sedimentary bedded deposits of varying thicknesses. It is usually associated with other sedimentary rocks, such as shales, limestones, dolomites, gypsum, anhydrite, etc. They are found in every geologic period from the Cambrian to the Tertiary, and evidence indicates that some deposits are Precambrian. Table 1 lists examples of salt deposits in the different geologic periods.

Although bedded salt deposits are found in many countries throughout the world, geologists have been intrigued with the origin of the very thick deposits, some of which exceed several hundreds of meters in thickness and cover many hundreds of square kilometers. The concept of the time span and the amount of evaporation required to produce that magnitude of salt overpowers the imagination.

Various theories have been developed to explain the formation of bedded salt deposits. None of the ideas adequately account for the extremely thick beds, or for the concept of how salt may be the dominant mineral in one basin with gypsum-anhydrite in another. The generally accepted theory of formation involves the precipitation of salt from an evaporating body of salt water. Landes (1960) outlined the classification of the various theories as follows:

I. Terrestrial type: evaporating interior sea becomes saline due to:
  A. Primary sodium chloride produced as a weathering product.
  B. Flushing of connate water form sedimentary rocks cropping out above water level in drainage basin.
  C. Leaching of salt from sedimentary rocks cropping out in drainage basin.

II. Marine type: Concentration of ocean water in cutoff sea:
  A. Without further enrichment
  B. With enrichment by
    1. Wind-blown salt
    2. Ocean water
      a. marginal salt pans
      b. marine salinas
      c. barred lagoons
      d. silled downwarping basins.

Landes (1960) also suggests the following stages for the formation of a major bedded salt deposit:

Generalized Chronological History Proposed for Thick Salt Deposits. The following stages are suggested in the natural history of a major bedded salt deposit:

Table 1. Ages of Salt Deposits

CENOZOIC ERA

Recent. Mexico, United States, (solar salt operations and playa lakes).
Pleistocene. Israel, Mexico, United Sates (CA, NV), USSR.
Pliocene. Italy, Jordan United States (NV, UT).
Miocene. Algeria, Cyprus, Czechoslovakia, Dominican Republic, Egypt, Iran, Iraq, Morocco, Poland, Spain, Sudan, Syria, Trucial States, Turkey, USSR.
Oligocene. France, Germany, Spain, Iran, Iraq, Spain, Turkey.
Eocene. Belize, Iran, Morocco, Pakistan (?), United States (Green River Basin, WY).

MESOZOIC ERA

Cretaceous. Angola, Bolivia, Brazil, Colombia, Congo, Gabon, Libya, Mexico, Morocco, Nigeria, Peru, Senegal, United States (FL), USSR, Zaire.
Jurassic. Aden, Cuba, Chile, Germany, Kuwait, Tanzania, United States (ID, Gulf Coast area).
Triassic. Algeria, Bolivia, Bulgaria (?), Ethiopia, France, Germany, Greece (?), Libya, Mexico (Isthmus of Tehuantepec), Morocco, Netherlands, Peru, Portugal, Spain, Switzerland, Tunisia, United Kingdom.

PALEOZOIC ERA

Permian. Australia (?), Brazil, Denmark, Germany, Greece (?), Netherlands, Mexico, Peru, Poland, United Kingdom, United States (Permian Basin, Supai Basin, Williston Basin), USSR.
Pennsylvanian. Brazil, United States (CO, Paradox Basin).
Mississipian. Canada (New Brunswick, Nova Scotia), United States (VA, Williston Basin).
Devonian. Australia, Canada, United States (Williston Basin, WY), USSR.
Silurian. Canada, United States (Salina Basin).
Ordovician. Bolivia (?), United States (Williston Basin, WY).
Cambrian. Australia, Canada (Northwest Territories), Iran, Pakistan (?), USSR.

PRECAMBRIAN

Australia, Iran, Pakistan (?).

(?) unconfirmed, see Fig. 2 for more information regarding US deposits. *Source*: Modified from Lefond, 1969.

1. Peneplanation of the continental platform so that it is a nearly flat land of very little elevation above the sea.
2. Submergence of wide areas beneath the sea. Shallow ocean waters spread over a considerable part of the continent.
3. Beginning or renewed downwarping of a basin marginal to this epicontinental sea.
4. Sedimentation of both basin and epicontinental sea. Mainly clastic and organic (reef) deposits.
5. Lowering of sea to approximately sill level. Period of evaporite deposition. Enrichment by seawater flowing across sill. Because of the shallowness of the water, and the distance from the open ocean in such areas as the Michigan Basin, the brine crossing the sill may have been abnormally strong. Some enrichment could also take place by wind-blown salt, for the flat topography and scantiness of land vegetation no doubt resulted in incessant winds of considerable magnitude.

As more salt and/or sediments are deposited onto the layer of salt crystals on the sea floor, the water is squeezed out and the salt crystals become compacted or recrystallized into massive salt beds. Other theories of salt formation suggest that salt may have been deposited from deep density-layered brine contained in basins that did not dry up. This theory raises more questions that it resolves.

Salt Domes. One of salt's properties is that it will flow plastically when compressional forces are exerted on competent strata

Fig. 2. Salt resources of the United States.

that contain salt beds. The forces may be due to the static weight of the overlying sediments, or other tectonic factors. If the pressure increases, the salt will flow into zones of decreased pressure, usually vertically upward through the overlying sediments. The salt will resemble domes or diapiric folds. Salt domes are found in the Gulf Coast of the United States, and in the Paradox Basin of Colorado-Utah. They are also found in Algeria, Germany, Iran, Mexico, Romania, Tunisia, and elsewhere. Sometimes the salt dome reaches the surface and appears as a circular expression, such as the Avery Island, LA, salt dome. A salt dome can also flow like a glacier onto the surface, such as the Kuh-i Namak salt plug in the Persian Gulf area.

Distribution of Major Deposits

Solid deposits of salt or salt-bearing solutions are found in virtually every country in the world. Although the world contains abundant reserves of salt, the location and size of the deposits in relation to the location and size of the salt markets are important factors to consider when developing a commercial salt operation.

The United States: The domestic salt resources are located primarily in four depositional basins covering an area totaling about 1.3 Mkm^2 in 18 states. Fig. 2 shows the distribution of domestic salt resources. Although large regions of the country are underlain with bedded, dome, or brine salt deposits, the northwest and most of the eastern states are devoid of salt. For consumers located in these regions, salt must be imported. The total identified resource of salt in the United States is estimated at 55 Tt. At the present rate of mining, only 36 Mt are being depleted annually from the resource.

The four basins are: the Gulf Coast of Permian-Jurassic age, which covers part of Alabama, Arkansas, Florida, Mississippi, eastern Texas, and all of Louisiana; the Permian of Permian age, which includes part of Colorado, Kansas, New Mexico, Oklahoma, and western Texas; the Salina of Silurian age, which covers part of Michigan, New York, Ohio, Pennsylvania, and West Virginia; and the Williston of Ordovician-Permian age, which includes portions of North and South Dakota, Montana, and Wyoming. The Salina and Williston basins extend into Canada while the Gulf Coast and Permian basins reach into Mexico.

There are also a few other minor basins that contain salt. They are the Supai of Permian age, in part of eastern Arizona and western New Mexico; the Paradox of Pennsylvanian age, in southeast Utah and southwest Colorado; the Sevier, in central Utah; and the Green River of Eocene age, in southwest Wyoming.

World: Salt is produced in Canada in the provinces of Alberta, New Brunswick, Nova Scotia, Ontario, Quebec, and Saskatchewan. Although Canada has large reserves of salt, about 62% of Canadian salt production is from Ontario. Part of the Salina basin and the Williston basin extends northward into Canada.

Rock salt deposits in Mexico occur near Monterrey, Coatzacoalcos, and Punta Gorda. Salt domes and salt anticlines are associated with the deposit at Coatzacoalcos. At Monterrey, the salt is solution mined and used as feedstock for the one synthetic soda ash plant located locally. Although there were extensive solar works in the northern Yucatan Peninsula that had been used by the early Mayan civilization, the major solar salt produced in Mexico today is at Guerro Negro along the Pacific coast of Baja.

Brazil has large reserves of salt in the Amazon basin, Sergipe, and offshore salt domes. Salt is also found in Argentina, Chile, and Colombia. Except for the underground salt mine at Zipaquira, Colombia, the majority of salt produced in South America is from playa lake deposits or solar salt pans.

The salt industry of Europe can be traced back to early Roman times. The Romans introduced open pan evaporation of seawater to England. Salt was produced in Austria since the 8th century, and the rock salt mine at Wieliczka, Poland has been in continuous operation since the 13th century. The USSR, France, Germany, Italy, England, the Netherlands, Poland, and Romania are the major salt-producing countries of Europe. Approximately one-fourth of the salt produced is rock salt, and most of the remainder is salt brine for use by the European chemical industry.

One of the major salt deposits in Europe is the Zechstein (Permian) salt formation that extends from eastern England to western Poland. In Germany, the Keuper (Triassic) salt beds are well developed. Salt beds of the Miocene, Oligocene, and Jurassic periods are found in several European countries. Solar salt is produced along the coasts of France, Greece, Italy, Portugal, and Spain.

Although rock salt deposits are found in Africa, most salt is produced by solar evaporation from seawater and saline lakes. The countries that are known to produce salt include Algeria, Benin, Burkina Faso, Egypt, Ghana, Kenya, Libya, Madagascar, Mali, Mauritania, Mauritius, Morocco, Mozambique, Namibia, Niger, Senegal, Sierra Leone, Somalia, the Republic of South Africa, Sudan, Tanzania, Tunisia, and Uganda.

China is the largest salt-producing nation in Asia, and the second largest in the world behind the United States. Coastal solar salt works and inland drilling for salt brines have been conducted since about 2000 BC. Japan has no indigenous rock salt or salt spring deposits; therefore, must depend on imports to satisfy most of its salt requirements. It does, however, produce food-grade salt by processing seawater using ion exchange electrodialysis technology, which is very energy intensive. The improper climate and lack of available coastal land makes solar salt production impractical. Rock salt deposits in Asia are found in China, India, Pakistan, and Thailand.

TECHNOLOGY

Exploration and Development

Many salt deposits have been discovered while exploring for oil and gas and potash. For example, salt was found in Morocco's Mohammedia region while searching for commercial-grade potash in the Khemmiset-Berrechid basins by a United Nations exploration team. Accidental discoveries of salt have occurred in Brazil, France, Morocco, Portugal, and Thailand.

Although many reconnaissance projects have located salt resources, it is the task of the exploration geologist to delineate the ore zone and attempt to quantify the salt reserves. The way the exploration is conducted is governed to a large extent by the proposed method for recovering the salt—that is by either conventional underground room and pillar mining or by solution mining, including the projected use of the resultant cavity.

If the salt body is to be dry mined, the exploration project must be more extensive and greater care taken in detailing the results than if the deposit were to be mined using solution mining techniques. The extra care is necessary because of the greater capital investment incurred with underground dry mining than in a brine field.

Evaluation of Geologic Factors: When evaluating a salt deposit for dry mining, the geologic conditons surrounding the deposit are almost as important as the deposit itself. These conditions include:

1. Ground Water. The quantity of ground water and its hydrostatic head will have a direct influence on the method and cost of shaft sinking. If the ground water is a potable water supply for a community, the protection of the aquifer requires careful hydrological investigation. Shaft sinking difficulties will increase if an aquifer contains a brine or significant amounts of hydrogen sulfide, ammonia, or hydrocarbon gases or if the hydrostatic pressure is very high.

2. Overburden Characteristics. Considering the overburden as the alluvium plus the rock sequences above the salt deposit, it is essential that the physical characteristics of these horizons be known as fully as possible. It is particularly important that this information is gathered from the area in which the shaft or shafts are to be sunk. The data should include the thickness and hardness of the beds, drillability, fracture pattern, friability, and tendency to oxidize. Contractors that sink the shafts occasionally make physical examinations of the drill cores themselves.

3. Physical Characteristics of the Rocks Associated With the Salt. The structural integrity of the sedimentary rocks underlying and overlying the salt bed are particularly important. The roof spans, pillar size, percentage of extraction, mine layout, and the roof bolting program are some of the mine features which will be dictated by the rocks surrounding the deposit.

4. Faults. Fault systems which are common to many salt formations can create numerous serious problems and should be defined as carefully as possible during the exploration project (Jacoby, 1969). Salt domes in the gulf area of the United States have many faults associated with the spines that characterize their upper limits. These vertical or near vertical fractures often contain brine, oil and/or gases. Failure to develop mining patterns to avoid creating differential stresses across these features often results in the loss of expenditures of money and energy to seal off the liquid intrusions.

5. Facies Changes. Major and minor facies changes may occur in salt beds over relative short distances. This has been documented in the salina salts of Michigan, the potash salts of the Khemmiset Basin of Morocco, and the salts of Realmonte, Sicily. In Michigan, the A-1 salt changes from a pure anhydrite on the basin's edge near Detroit to an almost pure salt in East China Township, MI, and finally to a heavily contaminated potash salt at Midland. In the Khemmiset Basin, salts grade from a lean potash ore at Berrechid to a 98.3% sodium chloride in Mohammedia. Although the salt at Realmonte, Sicily is metamorphosed, it shows a similar gradational change from lean potash mineralization to a pure halite. Most salt consumers prefer a product of uniform chemical composition; therefore, the exploration program must define the limits of the various salt grades laterally and vertically within the bed and the reserves of a uniform or standard product.

6. Attitude and Position. The attitude and position of bedded salt and the flowage lines in salt domes influence the mining system, the percentage of extraction, the depth of the mining level, the selection of a shaft versus an inclined entry, and the rock mechanics of superimposing pillars where multilevel systems are contemplated. Whenever the internal features of the deposit or the continuity of the salt quality is questionable, a mill sample should be obtained by a bored shaft or inclined entry.

When solution mining is considered, exploration is conducted concurrently with the expansion or replacement of the current brining facilities. Evaluating the regional geology leads to the selection of the brine field site. Based on previously published information, a decision may be made on whether or not geophysical information would be beneficial in guiding subsequent drilling. Where domes are involved, or where the beds are known to have been altered by diapirism or faulting, geophysical studies may have distinct value (Mattox, 1968).

7. Chemical Composition. The deposit generally should have a minimum sodium chloride content of 95% or greater to be commercially acceptable. The northeastern section of the United States has an average sodium chloride content of 97%. Compared with salt obtained from southern salt domes or from solar evaporation, the sodium chloride content must be 99% in order to be competitive. Depending on the end use of the salt, various trace elements such as boron, chromium, copper, iron, and vanadium, as well as gypsum, ammonia, and water insolubles must be considered.

Drilling and Coring: If seismic or other geophysical data show a uniform elevation of the top of the salt, the drilling fluid can be changed from water to saturated brine just above the top of the salt formation. Where zones of lost circulation have been encountered up the hole, a casing point is usually selected above the top of the salt and a string of protective casing installed. Depending on the design of the well and the economics, the casing is either set temporarily on a packer or permanently cemented in the hole. When drill holes expose shales which swell and spall or slough in the presence of freshwater, early conversion to a saturated salt brine drilling fluid often is justified.

A clean and fully saturated or supersaturated brine drilling fluid is very important so that the salt is not mechanically etched or leached. The solubility of salt at different temperatures is also very important. Once the drill rig drills and recovers the cores, the cores are cut into one-third meter lengths and halved. One section is halved again to form a quarter which is crushed and ground for chemical tests. Physical testing is usually done on a second quarter of the core.

Once the core hole is completed, it should be logged with gamma-neutron, sonic, or 3 dimensional equipment. The logs are then correlated with the physical sections of core and the geolograph that records the rate of drill penetration. Because faults, especially low-angle thrust faults, are difficult to recognize in cores of salt beds, the correlation with logs will often show repeat sequences of rock strata. Subsequently, a reexamination of the core may indicate a fine-grained granular halite section of core that was previously unrecorded. Sonic logging differentiates anhydritic-dolomite from dolomitic-anhydrite, while gamma-neutron will clearly depict salt from anhydrite (Leroy, 1951).

Evaluation of Deposits and Plant Siting: Jacoby in 1972 outlined a good procedure for evaluating a salt deposit with respect to its development using solution mining.

1. Preliminary Considerations
   A. Plant requirements
      1. Quantity of brine
      2. Quality of brine
      3. Degree of saturation required
   B. Subsurface factors
      1. Geology
      2. Rock mechanics
      3. Hydrology
      4. Previous extractive operations and environment
   C. Surface factors
      1. Environment
      2. Topography

3. Property boundaries
4. Plant-brinefield relationships

D. Feasibility study
1. Preliminary brinefield layout
2. Capital investment
3. Cost of capital
4. Operating cost
5. Brine treatment cost
6. Taxes and tax incentives
7. Return on investment

2. Preliminary Engineering
A. Property
1. Acquisition
a. Option of surface and mineral rights
b. Rights of well sites
c. Ingress and egress rights
d. Right-of-way for roads, pipelines, and power lines
e. Storage rights
f. Conditions of abandonment
B. Water supply
1. Environmental considerations
a. Ground water table
b. Salt water encroachment
c. Effect on local agriculture, industry, and domestic use
2. Supply characteristics
a. Volume (continuity and cyclic variations)
b. Quality
C. Salt supply
1. Exploration
a. Geophysical surveys
1) Seismic
2) Gravity
b. Drill holes
1) Cores
2) Geophysical logs
3) Hydrological tests
2. Design data development
a. Analyses of well logs and cores
b. Physical testing of cores
c. Water supply evaluation
d. Equilibrium studies

3. Design and Cost Analysis
A. Brinefield Layout and Cost
1. Salt wells design and cost
2. Subsidence grid system design
3. Roadways and well sites
4. Pipeline and power line
5. Cathodic protection
6. Pump and pump facilities
7. Feed water treatment facilities
8. Reservoirs and aeration
9. Security
B. Supervision costs
C. Operating costs
D. Maintenance costs
E. Taxes and insurance
F. Depletion and depreciation
G. Cost of capital
H. Return on investment.

Methods of Recovery

Data pertaining to commercially produced salt in the United States has been reported since 1797. Table 2 lists domestic salt production by type from 1970 through 1992. Fig. 3 shows historical salt production data from 1880–1992.

Table 2. Salt Production[1] in the United States (thousand tons)

	Vacuum pans and open pans[2]	Solar[2]	Rock[3]	Brine[4]	Total[5]
1970	3 391	1 598	12 855	23 794	41 636
1971	3 537	1 841	12 428	22 180	39 986
1972	3 590	1 717	13 094	22 441	40 843
1973	3 490	1 848	11 201	23 297	39 835
1974	3 540	1 833	13 458	23 386	42 217
1975	3 315	1 535	12 957	19 415	37 222
1976	3 405	1 683	14 214	20 790	40 089
1977	3 422	1 728	13 570	20 663	39 383
1978	3 401	1 901	13 325	20 263	38 890
1979	3 646	1 998	13 509	22 390	41 543
1980	3 522	2 206	10 710	20 168	36 607
1981	3 450	2 176	10 769	18 901	35 296
1982	3 376	2 581	12 033	16 180	34 169
1983	3 354	1 862	8 572	16 124	29 913
1984	3 292	2 454	12 386	17 413	35 544
1985	3 278	2 312	12 692	17 295	39 217
1986	3 299	2 430	12 096	15 996	33 822
1987	3 426	2 830	11 095	16 163	33 514
1988	3 469	3 065	11 567	17 434	35 534
1989	3 606	2 849	12 682	16 496	35 632
1990	3 662	2 985	12 772	17 374	36 794
1991	3 654	2 813	11 188	18 660	36 316
1992	3 811	3 221	11 411	17 574	36 016

[1]Production data for 1970 through 1981 represent sold or used because actual production statistics were not collected by type. Production data for 1982 through 1992 represent actual production data. [2]Includes bulk, packaged, compressed pellets, and pressed blocks. [3]Includes bulk, packaged, and pressed blocks, [4]Tons of anhydrous salt contained on a bulk basis. [5]Data may not add to totals shown because of independent rounding.

Underground Mining: Mining underground bedded salt or salt dome deposits includes the construction and development of underground rock salt mines and the solution mining of the subsurface solid salt or natural saline brines.

Rock Salt Mining. All the US and Canadian rock salt mines use the room and pillar method of underground mining. The height of the pillars is limited to the thickness of the salt bed and the vertical section of salt that can be economically extracted. Usually a certain amount of salt is left behind in the floor and the roof to avoid mining into the underlying and overlying non-salt beds that would contaminate the quality of the salt being mined. Bedded salt deposits generally are vertically restrictive but laterally unrestricted while salt dome deposits are laterally restricted but vertically unrestricted. Room heights can range from about 3 m in a bed to more than 30 m in a dome. The height of pillars in salt domes, therefore, can be substantially higher than those in bedded salt operations. The room widths are basically a function of the thickness of the overlying incompetent rocks and the physical characteristics and thickness of the salt that forms the roof. Fig. 4 shows a wide angle view of Akzo Salt Company's Avery Island rock salt mine with its 30-m height rooms.

The pillar widths are controlled by the percentage of extraction permissible at the various depths and room widths. Most room and pillar operations recover about 45 to 65% of the resource, with the remainder left behind as pillar supports. Most mines in bedded deposits use roof bolts to support their haulageways, shaft entries, and permanent work areas. The length, diameter, and spacing of the

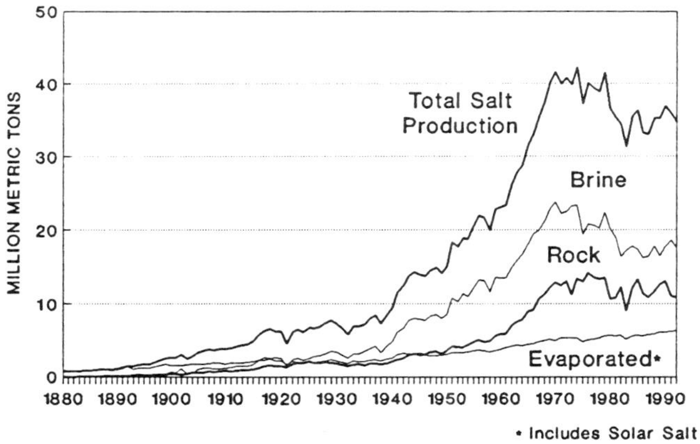

Fig. 3. Historical US salt production, 1880–1992. *Source:* US Bureau of Mines.

Fig. 4. Wide-angle view of Akzo Salt (Company's Avery Island rock salt mine in Louisiana) (*courtesy:* National Geographic Society).

bolts vary with the circumstances. Most mines use bolts 1.1 to 3 m long on 1.1-m spacing centers. Very little roof bolting is done in salt dome mines because of the problems encountered in bolting in salt. Bolt holes in salt have a smooth slick surface that forms poor seats for bolt anchors and which deform plastically over time due to the pressures. Wedge-type slotted bolts are impossible to set.

Because of the physical properties of salt, almost all domestic rock salt mining uses conventional mining equipment. Continuous mining equipment, which use rotating drum heads fitted with tungsten-carbide cutting tips, had been tested but it tended to pulverize a lot more of the salt into useless fine particles. In conventional mining, the working face is first horizontally undercut about 3 m into the face to produce a kerf. This allows a smoother floor to work from later on. Next, a drill rig is used to drill a series of horizontal (or slightly angled) holes about 3 m into the salt. Explosives, usually ammonium nitrate and fuel oil (ANFO), are pneumatically blown into the drill holes and set off electrically later after the working shift is over.

After the salt is blasted free, front end loaders pick up the loose rock salt and transport it to a breaker-feeder unit that crushes and conveys the salt on belt conveyors to other underground crushing and screening stations. The salt can be stockpiled underground or hoisted to the surface and stored in silos or bins for processing at a later time. Some operations prefer to crush and screen the salt to different sizes at the surface.

In 1990, 10 companies operated 14 rock salt facilities in the United States. The plants were located in Kansas, Louisiana, Nevada, New York, Ohio, and Texas. Rock salt mining accounted for 12.8 Mt, or 35% of total US salt production in 1990.

Solution Mining. The practice of solution mining is more than 2200 years old; however, the concept remains the same today although there have been a few improvements made since 1965. Solution mining involves the injection of a solvent to dissolve and recover underground soluble minerals. Prior to 1955, all solution mining used either annulus injection, that used a pair of concentric pipes (one carried the solvent downward and the other contained the brine upward), or tubing injection, that introduced the solvent at the bottom of the tube.

Hydraulic fracturing was used in oil and gas industry and was adopted by the salt industry. This technology was preferred because:

1) It altered the cavity configuration from the morning glory shape to hemispheres and ellipsoids, thus increasing the stability of openings and reducing the likelihood of subsidence.

2) It increased the percentage of extraction from less than 5 to something in excess of 40%.

3) It reduced brining costs in most operations by 50%.

4) It increased the productive capacity of wells from approximately 50 to about 500 gpm or more.

5) It reduced well maintenance costs to less than 5% of their former amount.

Solution mining is used to obtain a sodium chloride feedstock for vacuum pan salt production and for synthetic soda ash, chlorine, and caustic soda manufacture. The quantity of underground salt dissolved and recovered as brine to make vacuum pan salt usually is not reported as primary salt production, only the quantity of vacuum pan salt manufactured is. The quantity of brine used to make chloralkali chemicals is reported as either the amount of captive brine used or brine sold. The chemical industry is the largest consumer of salt brine in the world; almost 88% of the total domestic and imported salt consumed by the US chemical industry is as brine. Approximately 93% of brine production in the United States is for chemical uses. In 1992, 14.6 Mt of brine were produced and consumed for chloralkali manufacture.

Most salt and chemical companies use contract drillers to drill brine wells. Almost all new wells are constructed with rotary drilling equipment. Steel-cased pipes are inserted in the drill holes and cemented on the outside back to the surface. Most companies use a series of well pairs to establish a brine field. One well pair consists of an injection well, in which the water solvent is introduced, and a production well, in which the salt-bearing solution is pumped to the surface for processing. Some chemical companies have turned over the routine responsibility of operating and maintaining the brine field and pipelines to brine supply companies. The chemical producers are charged on the meter for how much brine they receive.

The cavities created from solution mining have been utilized by companies for storing a variety of products underground. The empty caverns provide an inert and sealed environment to store hydrocarbons, ethylene dichloride, and compressed air.

Solar and Mechanical Evaporation: Other than recovering salt from underground deposits, salt can be obtained from seawater along coastal margins, and from landlocked bodies of natural saline water and artificial brines. Salt production uses the wind and the sun to evaporate the water, leaving behind relatively pure crystals of salt. Solar salt production is restricted to areas of the world that have high evaporation rates and low precipitation. The practice of solar salt production from seawater can be traced back many hundreds of years and has not changed much throughout history. Laborers in some countries today, such as India, Thailand, Jordan, and Colombia, continue to produce solar salt in the same manner as their ancestors once did.

Seawater Evaporation. Seawater contains various dissolved salts that will separate depending on their relative solubilities. Calcium carbonate, which is the least soluble, will separate out first. Highly soluble magnesium salts tend to separate last. The order of separation of mineral salts from seawater from first to last are: calcite, gypsum, halite, astrakainite, epsomite, kainite, hexahydrite, kieserite, carnallite, and bishofite.

In the United States, solar salt production from seawater is restricted to the San Francisco Bay area and southern California. Seawater from the Pacific Ocean is collected and allowed to evaporate in specially constructed concentration and evaporation ponds. The initial step concentrates the brine to raise the salinity and to allow various calcium, magnesium, and iron compounds to precipitate from solution. The brine circulates among a network of interconnecting gravity-fed ponds, with salinity increasing with each transfer. It takes approximately five years from the time seawater is initially introduced before the first salt is ready for harvest.

The brine is treated with lime to remove excess calcium sulfate and then pumped to evaporation ponds and harvesting ponds to permit the salt to crystallize. After about 85% of the salt has crystallized, the remaining supernatant liquid, called bitterns, is drained to adjacent ponds for subsequent extraction of magnesium, potassium, bromine, and sodium compounds. Every year, between August and November, solar salt is harvested by mechanical harvesting equipment.

Inland Solar Evaporation. Solar salt is produced from the Great Salt Lake in Utah, which is an example of a landlocked body of saline water. The principles of solar salt concentration and production are similar to those along coastal margins except that salinity in inland lakes usually is greater than that of seawater. As water flows over or beneath the surface, it dissolves minerals from underlying soils and rocks. The salt lakes are topographically lower than most of the surrounding areas and therefore, become excellent sumps for mineral accumulation. For example, the Great Salt Lake receives about 2 Mt of new minerals annually by this process. The percentage of salt in the lake varies, depending on the influx of freshwater.

A second form of inland solar evaporation includes evaporating artificial brines that were discharged as waste products into containment ponds. One example are the solar salt operations in New Mexico that are associated with potash mining. Underground commingled deposits of halite and sylvite are mined for the potash and the salt is discharged during potash processing as a brine to tailing ponds and left to evaporate. Over the years, many meters of salt have accumulated that have attracted various companies to recover and sell the salt. Although the network of concentration and evaporation ponds are not required in this case, the harvesting equipment is similar. The arid location permits virtually year-round harvesting.

The total quantity of solar salt produced in 1992 in the United States was 3.2 Mt, or 9% of US salt production. Twelve salt companies operated 15 solar evaporation facilities in the United States. Solar salt from seawater was harvested in California. Solar salt from inland solar evaporation was produced in Arizona, California, New Mexico, Oklahoma, and Utah.

Vacuum Pan Salt: Vacuum pan salt is not mined but is a type of salt produced using mechanical evaporation technology. Although any of the other three types of salt may be dissolved to make vacuum pan salt (i.e. rock salt, solar salt, and salt brine), virtually all domestic vacuum pan salt is obtained from solution mining underground salt formations.

The salt is obtained by dehydrating the incoming brine using heat alone or in combination with a vacuum. The vacuum pan process conserves energy by using multiple-effect evaporators connected to vacuum pumps. A saturated salt solution will boil at a higher temperature than freshwater; however, when a vacuum is applied, the brine boils at a lower temperature, enabling the superheated vapor that is generated to act as the heating medium for the next evaporator.

Another mechanical evaporation process that is used in the United States on a small scale is the open pan process. Rectangular pans with steam-heated immersion coils are used to evaporate the water in the brine. Rotating rakes scrape the salt that has accumulated on the bottom of the pan into a sump, or up a ramp depending on the method, and onto conveyors for debrining and drying. The open pan process produces a flake-shaped crystal rather than the typical cubic form. Flake salt is preferred for production of cheese, butter, and baked goods. The Alberger process is a modified open pan technique that produces cubic salt with some flake salt.

Five companies operated 17 vacuum pan salt plants accounted for 11% of total US salt production in 1992. Vacuum pan salt is produced in California, Kansas, Louisiana, Michigan, New York, and Ohio.

Processing

After rock salt is mined and solar salt is harvested, the material undergoes some basic mechanical processing that usually consists of crushing and screening to reduce and separate the particles into different marketable size fractions. Rock salt mines have underground stations consisting of a combination of primary crusher and feeder units that load the salt onto conveyor belts for transport to secondary crushers elsewhere in the mine. Some mines have underground screening equipment to size the fragments; other operations conduct the screening at the surface. After screening, the salt is shipped to either underground or surface storage bins, each containing a different size fraction. From this point, the salt is ready for packaging and shipping or loaded as bulk salt into railcars, trucks, barges, or ships.

Solar salt processing consists of hauling the salt to wash plants or directly to a stockpile. There are different variations of wash plants but they should be designed to fit the local conditions and quality requirements. After the solar salt is dumped into a pit at a typical wash plant, the salt is fed to a vibrating feeder or screw conveyor where it gets washed with saturated salt brine to minimize dissolving the salt crystals. The washed salt goes to a classifier and then on for dewatering before being conveyed to a storage site. Some solar salt operations dry the salt in rotary dryers and screen the salt before sending it to the stockpile.

Vacuum pan salt is produced by processing salt brine in large vertical evaporators. Three to five evaporators are placed in series with each one in line under a greater vacuum than the preceding one. A unit consisting of three evaporators is also known as a triple-effect evaporator system. When pressure is lowered in an evaporator, the temperature at which water will boil also is reduced. Steam is fed to the first pan causing the brine to boil. The steam liberated by the boiling brine is used to heat the brine in the second pan. The pressure in the second pan is less, thereby allowing the steam made by the boiling action in the first pan to boil the brine in the second evaporator. The pressure continues to be reduced in each succeeding evaporator. Although the boiling process could be done using only one vacuum pan, more salt can be made per pound of steam using several pans in a row and saves energy.

The processing of brines usually consists of removing impurities dissolved with the salt in the solution. Calcium and magnesium are two common ions associated with salt that are removed during brine purification treatment. Soda ash, also known as sodium carbonate, is used to remove calcium ions, which precipitates as calcium carbonate. Magnesium ions are removed by adding caustic soda or lime. Barium carbonate is used to remove calcium sulfate from brines. Sometimes hydrogen sulfide is dissolved in the brine but can be eliminated by reacting it with sulfuric acid and neutralizing the acid with lime.

Product Forms: Salt is marketed and shipped in various forms, the most common being bulk. Most large chemical and industrial consumers of salt buy in bulk quantities to reduce costs. Purchasers of bulk salt should have adequate storage space to accommodate the volume of material. Shipping costs are usually less than those for packaged salt.

Packaged salt consists of salt sold in round cans, boxes, shakers, single-serve packets, and bags. Most vacuum pan salt for human consumption is packaged in one or more type of container. Because of the packaging and handling costs, the unit price is higher than for bulk salt. The majority of packaged salt is sold through grocery, food service, institutional, and related outlets, or to food processors or manufacturers.

Pressed blocks includes salt bricks, spools, and other large compressed forms used by the water conditioning and agricultural sectors. Twenty-three kilogram blocks are the most common pressed block sold. Pressures of 550 to 560 t per block are used to form the block. Blocks sold for animal nutrition often have trace elements added to them, based on veterinary advise, to supplement what is missing from the normal diet of the animal.

Compressed pellets are briquette-shaped pieces of salt that have been compressed together for ease of handling. Most compressed pellets are for water softening applications. Salt tablets used for heat relief and for canning applications are types of compressed pellets.

Additives: Temperature and moisture affects salt in storage or in use. For these reasons, free-flowing agents are added to help the salt pour freely. The agent should have very good adsorptive, covering, and adhering power, be nontoxic, odorless, colorless, dustless, tasteless, inexpensive, and soluble enough or fine enough to show no sediment in brine (Kaufmann, 1960). The agent also should not cover the salt grain so completely as to slow up the recognition of the salt taste when preparing food.

Between 0.5 to 1.0% of free-flowing additives are normally added to table salt. The agents could include magnesium carbonate, calcium silicate, dicalcium and tricalcium phosphate, and calcium carbonate.

Salt is an excellent carrier of potassium iodide, which is used to prevent Iodine Deficiency Disorders (IDD). About one billion people are at risk of contracting IDD throughout the world, mainly in the underdeveloped nations. IDD includes several abnormalities, such as goiters, mental retardation, deaf mutism, squint, difficulties in standing and walking, and stunted limbs. Without treating the population by simply using iodized salt, the social and economic progress of countries becomes adversely affected. The cost to prevent IDD is $0.05 per person per year. The International Council for Control of Iodine Deficiency Disorders has been working with the World Health Organization and UNICEF of the United Nations to overcome this international treatable problem. In developed countries, iodized salt contains 0.01% potassium iodide, with 0.2% of a stabilizing mixture of sodium thiosulfate and calcium hydrate.

Specifications

Salt is sold to various end use markets based on specific chemical and physical specifications formulated by reputable scientific and medical organizations. The specifications are important in order to ensure that consumers are obtaining a safe, consistent, and reliable grade of salt for their requirements. The specifications of salt for human consumption are different than those for highway deicing. The quality of salt also varies depending on the type of salt; rock salt is generally less pure than more refined vacuum pan salt. Table 3 lists the specifications that apply to salt sold in the United States.

USES

World salt consumption is equal to world salt production, assuming that inventories on a global basis are discounted. Therefore, the total quantity of salt consumed in the world in 1990 was 193 Mt. The pattern of end use markets is very similar among developed nations in the world with the chemical industry as the largest international consumer of salt. Most of the markets are mature and show minimal growth rates for salt consumption. Lesser developed countries tend to use more salt for human and agricultural applications. Table 4 lists the trends of the top 10 salt-producing countries of the world since 1970.

Salt consumption in the United States has increased during the past two centuries as noted in Fig. 5. Various events have had effects on domestic per capita consumption. Prior to the beginnings of the chemical industry in the United States in about 1884, most of the salt consumed was for food and animal hide preservation and in refrigeration. New chemicals made from salt and new applications using salt contributed to the significant growth of salt use during the early twentieth century. It should be noted that the chemical and highway deicing sectors were responsible for the greatest increases in salt usage but also have contributed to some of the environmental issues that caused consumption to begin declining after the mid-1970s.

It has been reported that there are approximately 14 000 direct and indirect uses of salt. It would be virtually impossible to collect data for all these uses of salt. The US Bureau of Mines and the Salt Institute, the trade association of the North American salt industry, surveys end use consumption information on 28 individual uses in 8 major categories that include chemical, food processing, general industrial, agricultural, water treatment, ice control, distributors, and miscellaneous. These industries are the first-tier consumers of salt; information on most of the other downstream customers (such as who all the distributors sell salt to) is unavailable. Fig. 6 shows a comparison of the end use patterns in 1940, when end-use data was first collected by the US Bureau of Mines, with those of 1992. During the 52-year span, changes in the quantity of salt consumed by the chemical industry and for highway deicing changed substantially.

Chemical Industry

The US chemical industry is the largest consumer of salt, representing 44% of reported consumption in 1992. This category is distinguished by two important subsectors, the chloralkali industry and other chemical manufacturers.

Chloralkali: The chloralkali industry is the largest segment of the chemical sector that uses salt. Traditionally, this end use included salt consumed for chlorine, coproduct sodium hydroxide (also known as caustic soda and lye), and synthetic soda ash; however, since 1986 no synthetic soda ash has been manufactured in the United States. The majority of the rest of the world continues to use salt for synthetic soda ash production.

Salt is used as the primary raw material in chlorine manufacture because it is the primary source of chlorine ions. For sodium hydroxide production, salt is the main source of the sodium ions. Approximately 97% of the domestic chlorine and sodium hydroxide produced is obtained from the electrolysis of salt brine feedstock using three different cell technologies. The types of cells, and percent chlorine manufactured by them, are diaphragm, 75%; mercury, 16%; and membrane, 6%. It takes about 1.75 t of salt to make

Table 3. Specifications Applying to the Salt Industry

Type specification	Title	Designation	Jurisdiction
Analysis	Standard Methods for Chemical Analysis of Sodium Chloride	ASTME-534-81[1]	American Society for Testing and Materials
Food[2]	Sodium Chloride	FCC III Food Chemicals Codex, 3rd ed.	Food and Nutrition Board, National Academy of Sciences
Highway	Standard Specification for Sodium Chloride	ASTM-D-632-89	American Society for Testing and Materials
Do	Do	AASHTO-M-143-86	American Association of State Highway Transportation Officials
Medical[3]	Official Monograph on Sodium Chloride	United States Pharmacopeia XXI, 1985, 21st rev.	United States Pharmacopeia Convention, Inc.
Packaging	Salt packages	ANSI/Z353.1-1983[4]	American National Standards Institute Inc.
Do	Federal Specification-Sacks, Shipping, Paper	UU-S-48F[5]	General Services Administration
Reagent	Sodium Chloride	Reagent Chemicals 7th ed., 1986	American Chemical Society
Soil stabilization	Standard Test Methods for Water-Soluble Chlorides Present as Admixes in Graded Aggregate Road Mixes	ASTM-D-1411-82	American Society for Testing and Materials
Table salt[2]	Proposed Federal Specification—Salt, Table	SS-S-31K	US Army Natick Laboratories
Water	Federal Specification Commercial Item, Description, Sodium Chloride, Technical (Water Conditioning Grade)	A-AZ-694	General Services Administration
Do	AWWA Standard for Sodium Chloride	ANSI/AWWA-B200-88	American Water Works Association

[1]E-534-81 is an analytical standard that is necessary for a meaningful salt specification. [2]Input regarding food and table salt specification is necessary to maintain quality products in these areas. [3]Work has been done with medical salt standards to obtain reasonable specifications. [4]The ANSI Packaging Standard is a voluntary standard for primary salt containers, secondary containers, and pelletized shipping units. [5]UU-S-48F gives Federal Sack Specifications. An effort has been made in this area to obtain a reasonable standard. *Source:* Salt Institute.

Table 4. World Salt Production, 1970–1990 (thousand tons)

	1970	1975	1980	1985	1990
Australia	3 071	4 659	5 665	5 835	7 227
Canada	4 862	4 835	7 029	10 085	11 097
China[e]	15 966	29 937	17 280	14 446	20 000
France	5 664	5 537	7 104	7 113	6 450
Germany, Federal Republic of[1]	12 627	11 746	14 524	16 219	13 258
India	5 588	4 856	8 009	9 878	9 503
Mexico	4 153	5 354	6 575	6 467	7 135
USSR	12 428	13 699	14 600	16 100	14 700
United Kingdom	9 188	7 629	7 155	7 146	6 434
United States	28 781	24 264	25 896	22 114	36 794
Other	38 133	43 612	47 936	50 419	51 464
World Total	146 130	161 667	168 878	172 936	193 493

[e]estimated, [1]includes the former German Democratic Republic.

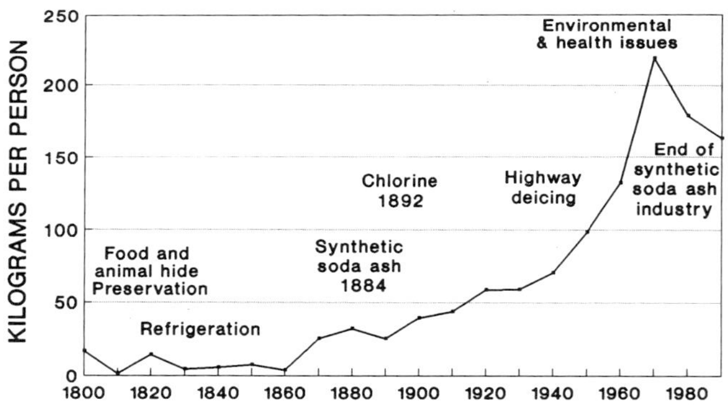

Fig. 5. Factors that influences per capita consumption of salt in the United States, 1800–1980.

1.0 t of chlorine and 1.1 tons of coproduct caustic soda. The electrolytic process splits the sodium chloride compound into ions that selectively pass through special membranes. Chlorine gas forms at the anode while sodium ions migrate through the membranes toward the cathode where sodium hydroxide forms and hydrogen gas evolves. In 1992, 10.6 Mt of chlorine and 11.2 Mt of caustic soda were produced in the United States.

Making synthetic soda ash had been the primary method to obtain soda ash in the United States until 1971, when natural soda ash production exceeded that of synthetic. The main process was, and is still the main technique used by the rest of the world, the Solvay process, which was developed in the early 1860s. Carbon dioxide gas is percolated through a carbonation tower containing a nearly saturated salt brine to which ammonia has been added. Ammonium bicarbonate forms, which reacts with the brine to form sodium bicarbonate and ammonium chloride. The sodium bicarbonate precipitate is collected, filtered, and calcined into soda ash. Residual sodium chloride is discharged as plant effluent. The chloralkali subsector is classified under Standard Industrial Classification (SIC) code 2812.

Other Chemicals: The other subsector that uses salt includes chemical establishments that make sodium chlorate, metallic sodium, and other downstream chemical operations that use salt as feedstock. In powdered soaps and detergents, salt is used as a bulking agent, and as a coagulant for colloidal dispersion after

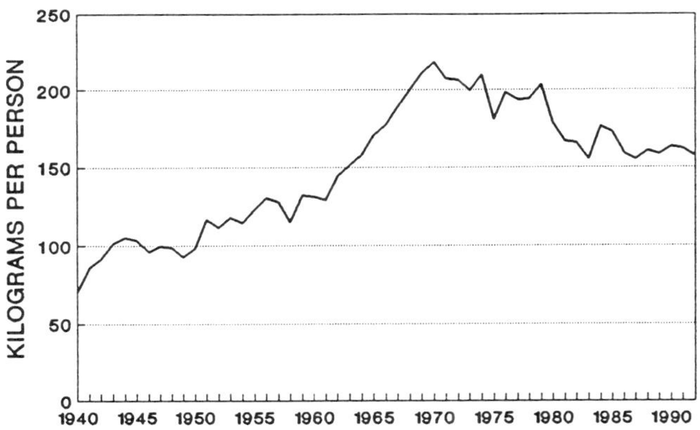

Fig. 6. US per capita consumption of salt, 1940–1992.

saponification. In pharmaceuticals, salt is a chemical reagent and used as the electrolyte in saline solutions. It is also used as a co-feedstock with sulfuric acid to produce sodium sulfate and hydrochloric acid. This subsector is relatively small, comprising only 10% of domestic salt sales for the entire chemical sector and only 4% of total domestic salt consumption. These chemicals are classified under SIC 28, excluding SIC 2812—alkalies and chlorine.

Food Processing Industry

Although every person uses some quantity of salt in their food, the entire category only represented 5% of reported consumption in 1992. Salt is added to food as a flavor enhancer, preservative, binder, fermentation control additive, texture aid, and color developer. This major category is subdivided into six subsectors, in descending order of salt consumption: meat packers, canning, other food processing, baking, dairy, and grain mill products. Additional salt for human consumption is also included in grocery wholesalers. It is estimated that about 75% of the grocery store category is salt for food processing; the remainder is for other household applications.

Meat Packers: Salt added to processed meats promotes the color development in bacon, ham, and other processed meat products. As a preservative, salt inhibits the growth of bacteria, which would lead to spoilage of the product. Salt acts as a binder in sausages to form a binding gel comprised of meat, fat, and moisture. Salt also acts as a flavor enhancer and a tenderizer. This subsector is engaged in slaughtering and processing meat and poultry products and in sausage and other prepared meat manufacturing. It includes SIC codes 2011, 2013, 2016, and 2017.

Canning: Salt is primarily added as a flavor enhancer and preservative. It also is used as a dehydrating agent, tenderizer, enzyme inhibitor, and as a carrier for other ingredients. Establishments that can vegetables, vegetable juices, jams, jellies, pickled fruits, and specialty products such as baby foods and seafood are in this subsector. It includes SIC 2032, 2033, 2035, and 2091.

Other Food Processing: Salt is used mainly as a seasoning agent in this end use. Other food processing includes miscellaneous establishments that make food for human consumption (i.e., potato chips, pretzels) and domestic pet consumption (i.e., dog and cat food). They are included in SIC 206, 207, 208, 2047, 2099, and any other food category not included in aforementioned SIC codes.

Baking: Salt is added to control the rate of fermentation in bread dough. It also is used to strengthen the gluten (the elastic protein-water complex in certain doughs), and as a flavor enhancer, such as a topping on baked goods. This subsector includes companies making bread, cakes, cookies, and other perishable bakery products. It includes SIC codes 205, 2051, and 2052.

Dairy: Salt is added to cheese as a fermentation control agent, and as a color and texture control agent. The dairy subsector includes companies that manufacture creamery butter, natural and processed cheese, condensed and evaporated milk, ice cream, frozen desserts, and specialty dairy products. It incorporates SIC codes 202, 2021, 2022, 2023, 2024, and 2026.

Grain Mill Products: This subsector includes establishments engaged in milling flour and rice, and manufacturing cereal breakfast food, and blended or prepared flour. SIC codes 204 (excluding 2047), 2041, 2043, 2044, and 2045.

General Industrial

The industrial uses of salt are diverse. This major sector of salt consumption includes, in descending order of salt usage, oil and gas exploration, metal processing, other industrial, pulp and paper, textiles and dyeing, tanning and leather treatment, and rubber manufacture. In 1992, general industrial applications consumed 7% of total reported salt consumption.

Oil and Gas Exploration: Salt is an important component of drilling fluids in well drilling. It is used as a flocculent and as a compound to increase the density of the drilling fluid in order to overcome high down-well gas pressures. Whenever drilling activities encounter salt formations, salt is added to the drilling fluid to saturate the solution and minimize the dissolution within the salt strata. Salt is also used to increase the set rate of concrete in cement casings. This subsector includes companies engaged in oil, gas, crude petroleum, and refining and compounding lubricating oil. It includes SIC major groups 13 and 29.

Metal Processing: Salt is used in concentrating uranium ore into uranium oxide, or yellow cake. It is also used in processing aluminum, beryllium, copper, steel, and vanadium. Establishments engaged in the smelting, refining, and fabrication of ferrous and nonferrous metals represent this subsector, which includes SIC 33, 34, 35, and 37.

Other Industrial: This subsector is represented by any group not included in any of the general industrial groups in this major category.

Pulp and Paper: Salt is used to bleach wood pulp. It also is used to make sodium chlorate, which is added with sulfuric acid and water to manufacture chlorine dioxide—an excellent oxygen-based bleaching chemical. Although the chlorine dioxide process originated in Germany after World War I, it is becoming more popular because of environmental pressures to reduce or eliminate chlorinated bleaching compounds. Companies that manufacture wood pulp, cellulose fibers, pulp from rags, paper and paperboard, and paper and paperboard converted products are included in this subsector, which includes SIC group 24.

Textiles and Dyeing: In commercial dye manufacture, salt is used as a brine rinse to separate organic contaminants, promote salting out of dyestuff precipitates, and blending with dyes to standardize concentrated dyes. One of its main roles is to provide the positive ion charge to promote the absorption of negatively-charged ions of dyes. The textile and dyeing industry includes businesses involved in the preparation of fibers and subsequent manufacturing of yarn and woven fabrics, dyeing and finishing of fibers. This includes SIC group 22.

Tanning and Leather Treatment: Salt is added to animal hides to inhibit microbial activity on the underside of the hides and to replace some of the moisture in the hides. This subsector includes establishments curing and/or tanning animal hides. It includes SIC 311.

Rubber Manufacture: Salt is used to make neoprene rubber, white rubber, and buna rubber. Salt brine and sulfuric acid are used to coagulate an emulsified latex made from chlorinated butadiene. Businesses that make rubber and rubber products, such as synthetic rubber, tires and inner tubes, rubber and plastic footwear, reclaimed rubber, rubber and plastic hose and belting, and other fabricated rubber products comprise this subsector. It includes SIC codes 2822, 30 (excluding 3079), 3011, 3021, 3031, 3041, and 3069.

Agricultural Industry

Since prehistoric times, man has noticed animals satisfied their salt hunger by locating salt springs, salt licks, or playa lake salt crusts. Barnyard and grazing livestock need supplementary salt rations to maintain proper nutrition. Veterinarians have advocated adding salt in commercially mixed feeds or in block forms sold to farmers and ranchers. Salt also acts as an excellent carrier for trace elements not found in the vegetation consumed by grazing livestock. Sulfur, selenium, and other essential elements are commonly added to salt licks, or salt blocks, for free-choice feeding.

The agricultural end use sector contains 3 subsectors—feed retailers and/or dealers and mixers, feed manufacturers, and direct buying end users. The entire sector represents 4% of total reported consumption in 1992; however, agricultural distribution (contained under distributors) represented an additional quantity, raising the total agricultural end use to 5%.

Feed Retailers/Dealers/Retailers-Mixers: Establishments engaged in the retail distribution of animal feeds, fertilizers, agricultural chemicals, pesticides, seeds and other farm supplies represent this subsector. It incorporates feed retailers who mix on-site some or all of the feed which they sell and whose primary customers are feed end-users, not other feed retailers. There are no suitable SIC classifications for this subsector.

Feed Manufacturers: This subsector includes establishments engaged in preparing feeds and feed ingredients and adjuncts for animals and fowl, including poultry and livestock feed and feed ingredients such as feed supplements and concentrates and premixes. It does not include the local dealer-mixer, whose primary business is the resale of prepackaged feed and feed ingredients to end-users, but who does a moderate amount of custom-mixing secondarily. The end use includes SIC 2048.

Direct-buying End Users: Salt used in this subsector is for consumption by farms, ranches, dairies, feed lots, etc., buying on a direct basis from salt producers. It includes SIC major group 2.

Agricultural Distribution: This subsector of distributors is a secondary, or second-tier consumer of salt, includes companies involved in the wholesale distribution of animal feed products, fertilizers, agricultural chemicals, pesticides, seeds, and other farm supplies to retail feed dealers and some large end-users who buy on a wholesale basis. It includes co-op (except mixing operations) and independent wholesale/distributors. SIC 5159 is the subsector category classification.

Water Treatment

Approximately 1.2 TL of water are used daily in the United States for residential and commercial uses. Many areas of the United States have hard water, which contain excessive calcium and magnesium ions that contribute to the buildup of a scale, or film of alkaline mineral deposits in household and industrial equipment. Commercial and residential water softening units use salt to remove the ions causing water hardness. The sodium ions captured on a resin bed are exchanged for the calcium and magnesium ones. Periodically the water softening units must be recharged because the sodium ions become depleted. Salt is added, dissolved, and converted to brine which replenishes the lost sodium ions. In 1992, water treatment represented 4% of total domestic reported consumption.

Government and commercial water treatment are the primary subsectors of water treatment and water treatment distribution is the secondary, which is the second-tier market for salt sales.

Government: This subsector includes local, state, and federal customers purchasing directly from salt producers for water treatment usage. It includes SIC 2899.

Commercial: All commercial establishments that purchase salt primarily for water treatment, and not classified elsewhere, are in this category. It includes car washes, laundries, and utilities. It also is included in SIC 2899.

Water Treatment Distributors: This subsector is represented by water conditioning salt route operators and distributors whose exclusive or dominant business is the distribution of water conditioning products to home or industry. It is included in SIC 7399.

Ice Control and Road Stabilization

The second-largest end use of salt is for highway deicing. Anyone who has ever driven on snow covered, icy roads can appreciate the benefits that salt provides. Gabriel Daniel Fahrenheit, the developer of the Fahrenheit temperature scale, discovered that salt mixed with ice (at a temperature below the freezing point) creates a solution with a lower freezing point than water by itself. The brine forms below the surface of the ice and snow and prevents the water from freezing into ice and bonding to the road surface. Therefore, salt causes snow and ice to melt. Salt is an inexpensive, widely available, and effective ice control agent. It does, however, become less effective as the temperature decreases below about −6°C. At lower temperatures more salt would have to be applied to maintain higher brine concentrations to provide the same degree of melting. Most winter snowstorms and ice storms occur between −4°C and 0°C, a range in which salt is most effective.

Salt is also added to stabilize the soil and provide firmness on which highways are built. The salt acts to minimize the effects of shifting due to changes in humidity and traffic load in the subsurface. In highway deicing, salt has been associated with corrosion to motor vehicles, bridge decks, unprotected steel structures, and reinforcement bar and wire used in road construction. Surface runoff, vehicle spraying, and windblown actions also affect roadside vegetation, soil, and local surface and ground water supplies. Although there is evidence of environmental loading of salt during peak usage, the spring rains and thaws usually provide sufficient water to dilute the concentrations of sodium in the area.

The deicing category includes government (state, federal, and local) and commercial subsectors of consumption. Approximately 21% of total reported consumption was used in this category during 1992.

Government Deicing: More than 95% of this major category is for consumption by separate local, state, and federal agencies to ensure safe and navigable roads during adverse climatic conditions.

Commercial Deicing: This subsector includes companies that purchase salt for ice control, such as transportation companies, shopping centers, ice control service contractors, etc. It includes any distributors whose primary or exclusive business is the distribution of salt for ice control purposes to business and/or government. It does not include institutional purchasers. There are no suitable SIC codes for this section.

Other Uses of Salt

The other uses of salt include any consumer that is not included in any of the aforementioned categories. It also includes sales by distributors, excluding those by agricultural and water conditioning

Table 5. Time-Value Relationships for Various Types of Salt[1]

	Average annual value and 1992 constant dollars,[2] dollars per ton							
	Salt in brine		Rock salt		Solar salt		Vacuum pan and open pan salt	
Year	Annual value	Based on constant 1992 dollars	Annual value	Based on constant 1992 dollars	Annual value	Based on constant 1992 dollars	Annual value	Based on constant 1992 dollars
1970	4.05	13.95	7.28	25.08	7.30	25.14	28.12	96.86
1971	3.94	12.87	7.07	23.10	12.15	39.70	28.97	94.66
1972	3.63	11.31	6.82	21.25	9.26	28.85	29.50	91.92
1973	3.94	11.53	6.82	19.96	9.91	29.01	32.68	95.67
1974	4.07	10.96	7.87	21.19	10.25	27.60	38.03	102.40
1975	4.31	10.59	9.85	24.20	15.66	38.48	47.41	116.50
1976	4.19	9.69	8.62	19.93	15.83	36.59	55.92	129.27
1977	4.31	9.32	9.85	21.30	15.66	33.87	57.33	123.99
1978	4.67	9.36	11.13	22.32	16.17	32.42	64.88	130.08
1979	4.97	9.17	11.02	20.34	13.40	24.73	67.95	125.42
1980	7.17	12.09	16.15	27.23	17.25	29.09	84.26	142.08
1981	6.51	9.98	15.17	23.25	20.23	31.00	87.83	134.45
1982	6.85	9.88	15.31	22.09	19.72	28.45	95.59	137.91
1983	5.75	7.97	14.80	20.52	23.67	32.82	96.33	133.56
1984	5.57	7.40	15.19	20.18	21.68	28.80	102.27	135.87
1985	6.77	8.67	16.70	21.39	25.46	32.61	102.14	130.81
1986	5.68	7.09	15.99	19.95	26.19	32.68	100.61	125.53
1987	5.43	6.56	15.81	19.11	28.00	33.85	103.85	125.55
1988	3.95	4.60	15.94	18.55	29.31	34.11	107.71	125.33
1989	6.25	6.97	18.06	20.14	30.73	34.27	102.22	114.01
1990	5.36	5.74	18.52	19.83	34.83	37.30	110.58	118.42
1991	5.45	5.63	19.25	19.89	27.78	28.71	114.75	118.58
1992	4.35	4.35	19.63	19.63	32.56	32.56	112.53	112.53

[1]Values are based on the average of all salt producers' valuations reported to the US Bureau of Mines of the finished salt in bulk, compressed pellets, and packages, FOB plant, and includes all processing costs, depreciation of equipment, taxes and profit. [2]From final 1992 implicit price deflators for gross domestic product, by the Council of Economic Advisors. Based on 1987=100.

distributors that are included as subsectors under agriculture and water treatment. It includes part of grocery wholesale and retail, institutional wholesale distributors, US Government resale, and other wholesale/retailers.

Institutional Wholesale Distributors: This subsector includes companies engaged in the distribution of food and other salt products to restaurants, hospitals, schools, and other institutional establishments. These are included in SIC groups 58 and 70.

US Government Resale: Included in this category are government units, such as military commissaries, that purchase salt for resale to armed forces personnel and dependents at the retail level. It also includes purchases by the government of foodgrade salt used by armed forces/government foodservices facilities, as well as all other government salt purchases not classified elsewhere. SIC 9199 pertains to this subsector.

Other Wholesale and Retail: This subsector includes wholesalers, not included elsewhere, primarily engaged in distributing merchandise for personal and household consumption to retailers, such as hardware stores, home centers, filling stations, and discount centers. It also includes all retail outlets that sell merchandise for personal or household consumption that purchase directly from salt producers. The SIC category is 5251.

ECONOMIC FACTORS

Prices

The price of salt depends on the type of salt, location, product form, and type of sale. Salt brine is the least expensive type of salt because it is not expensive to produce and processing is minimal. Vacuum pan salt is the most expensive because of the energy input and degree of purity of the final product. The price of salt also can vary because of regional differences in labor rates, energy costs, and other operating factors. Large consumers of salt generally prefer to purchase salt in bulk quantities. When salt is packaged in bags, cylinders, or single-serve packets, or sold in compressed pellets or pressed blocks, the sales price increases compared with the bulk price. The price of salt can also vary if the consumer is buying based on bid contracts or spot purchases.

There is a difference between quoted list prices that are published in trade journals and issued by the salt producers, and salt values which are aggregated national averages of similar salt sold on the same basis. Table 5 lists the average values of all types of salt sold since 1970.

Foreign Trade

Salt is one of the largest internationally traded mineral commodities in the world. It is estimated that total world trade is about 15% of total world salt production. Because salt is a low valued material, it is uneconomic to transport long distances. Overland transportation rates often are more expensive than ocean freight rates; therefore, some countries tend to import salt to certain regions where rail or truck rates from domestic salt production sites are cost-prohibitive.

Imports: Some countries import salt to satisfy demand requirements when indigenous salt resources are minimal or production capacity is inadequate. Japan is an example of a nation that imports the majority of its salt because it does not have any rock

Table 6. US Imports and Exports of Salt (thousand tons and thousand dollars).

	Imports								Exports[4]	
	Salt in brine[1]		In bags, sacks, barrels, or other packages		Bulk		Total		Total	
	Quantity	Value	Quantity	Value	Quantity	Value	Quantity	Value	Quantity	Value
1970	—	—	41	$ 625	3 167	$12,704	3 208	$13,329	384	$ 3,657
1971	27	$199	24	574	3 445	13,656	3 496	14,429	608	4,182
1972	—	—	24	535	3 118	11,444	3 142	11,979	788	5,544
1973	—	—	24	559	2 885	11,995	2 909	12,554	552	4,400
1974	8	5	25	746	3 013	13,677	3 046	14,428	473	4,276
1975	[2]	4	9	580	2 908	14,688	2 917	15,272	1 208	9,070
1976	[2]	12	17	691	3 931	22,762	3 948	23,465	914	10 326
1977	33	187	21	883	4 055	25,624	4 109	26,694	914	10,881
1978	[2]	72	1	1,209	4 880	32,966	4 881	34,247	704	9,795
1979	[2]	51	1	1,760	4 784	39,049	4 785	40,860	632	9,025
1980	11	63	1	1,478	4 763	42,530	4 775	44,071	754	12,829
1981	3	53	24	1,483	3 891	42,987	3 918	44,523	949	17,429
1982	[2]	50	43	1,613	4 902	54,521	4 945	56,184	908	16,647
1983	[2]	4	27	1,826	4 513	58,365	5 440	60,195	469	12,368
1984	[2]	5	64	2,386	6 781	71,522	6 845	73,913	744	12,299
1985	4	117	60	3,794	5 567	61,682	5 631	65,593	820	15,988
1986	[2]	34	64	3,170	5 983	76,505	6 046	79,709	1 057	16,928
1987	1	67	40	5,122	5 145	61,747	5 185	66,936	491	8,217
1988	2	125	71	6,682	4 893	70,550	4 966	77,357	802	10,858
1989	[3]	[3]	[3]	[3]	5 519	74,474	5 519	74,474	1 422	20,211
1990	[3]	[3]	[3]	[3]	5 969	88,419	5 969	88,419	2 266	32,944
1991	[3]	[3]	[3]	[3]	6 188	87,380	6 188	87,380	1 777	29,875
1992	[3]	[3]	[3]	[3]	5 394	87,714	5 394	87,714	992	32,193

[1]Anhydrous salt content, [2]Less than 1/2 unit, [3]Included in bulk because categories were merged under HTS code No. 2501000000 in 1989. Before 1989, salt imports were listed under TSUS code Nos. 4209200 for salt in brine, 4209400 for bulk salt, and 4209600 for other salt, [4]Prior to 1989, the Schedule B export code was 4209500. In 1989, it was changed to HTS System code 2501000000. *Source:* Bureau of the Census.

salt deposits and the climate and lack of available coastal land are not favorable for solar salt production. Other countries import salt because of economics. Although the United States is the largest salt producer in the world, it imported almost five times what it exported in 1992. The data would indicate that the United States is import reliant and that it's salt industry cannot keep up with demand. Less expensive labor costs, transportation costs, and currency exchange rates are the main cost considerations why the United States imports large quantities of salt. Most of the imported salt is solar salt from Mexico and the Caribbean that is consumed by the chemical industry to manufacture chlorine and caustic soda in areas of the country where there are no local salt deposits, such as the Pacific Northwest and Southeast. In addition, a large quantity of rock salt, mainly from Canada and Chile, is imported for highway deicing purposes.

The majority of salt imports are in bulk form, but it comes also in bags, sacks, barrels, and other forms. Prior to 1989, all imported salt was classified under the Tariff Schedule of the United States, Annotated (TSUS). The TSUS codes for salt in brine, bulk salt, and other salt were numbers 4209200, 4209400, and 4209600, respectively. After 1989 with the passage of the Omnibus Trade and Competitiveness Act of 1988, the United States adopted the internationally-used Harmonized Tariff System (HTS) that aggregated the three classifications into one named ''Salt (including table and denatured salt) and pure sodium chloride, whether or not in aqueous solution, seawater.'' This HTS code number is 2501000000. Table 6 lists some of the recent US import statistics.

Exports: Although the United States imports most of its salt from Canada, the majority of US salt exports are to the Canadian Provinces of Ontario and Quebec. Canada represents about 87% of total US exports. The remainder of export sales are small shipments mainly to other countries in the Western Hemisphere. Table 6 lists the trend in US salt exports since 1970.

Transportation and Storage

Overland and waterborne transportation rates of salt are important economic factors to salt consumers. Transportation charges generally are always higher than the sales price of the salt being carried. Rock salt, solar salt, and vacuum pan salt is shipped in bulk or in some packaged form, such as shrink-wrapped bags on pallets, and one-ton superbags. Because salt brine is a solution, it is transported in underground or surface pipelines.

Overland: Overland movements include salt moved by rail, truck, and pipeline from salt production localities to regional consumers. Because overland transportation rates are more costly than the salt, most salt shipped overland is not transported economically greater than about a 500-km-radius from the production site. Competition from other salt producers is usually encountered at distances beyond that radius.

Rail Movement. Boxcars are used to convey packaged salt loaded on pallets. Conveyor belts and forklifts assist the loading of the salt into the boxcars. The majority of all railcars used to transport bulk salt are covered hopper cars. The protected cars prevent any inclement weather from affecting the salt. To avoid caking due to moisture and humidity, deicing salt is treated with 50 to 100 ppm of ferric ferrocyanide (known as Prussian Blue) or sodium ferrocyanide (yellow prussiate of soda, or YPS).

To reduce freight rates, most producers ship in unit trains that contain up to 12 railcars. The loading and unloading of the covered hopper cars includes pneumatic equipment using flexible hoses and pipes. Many cars are equipped with bottom-dump doors or gates that allows faster unloading. Conveyors beneath the railcars take the salt to truck-transfer stations or to warehouses or stockpiles.

Truck Movement. Trucks are used in shorter hauls from production sites to customers' locations, or from distribution centers and warehouses to consumers. Bulk salt is carried in dump trucks or specially built pneumatic and tank trucks. Dump trucks should be covered properly with tarps to prevent caking or losses from being blown or shaken off the vehicle. Some trucks are owned and operated by special trucking companies. Other trucks are operated by independent truckers who often will carry salt as a backhaul to their point of origination and at competitive prices.

Dump trucks can drop their loads directly into hoppers or onto a surface pad so that the salt can be picked up by front-end loaders. The trucks can also unload into pits, which are convenient for companies that convert dry salt into brine.

Pipeline. Some salt brine is transported by pipeline between production sites and consumers locations. The chemical industry is the largest consumer of brine but some chloralkali operations are not located adjacent to salt deposits which could be solution mined. Several chemical companies in Louisiana and New York rely on brine supply companies to provide a continual supply of brine feedstock. Some of the pipelines are only a couple of kilometers in length; however, others can be up to 100 km long or more.

Waterborne: Waterborne movements of salt include salt shipped in barges and freighters. This type of transportation is the most economical method to move salt great distances. Although rock salt and solar salt are the two major types of salt shipped on the water, bulk vacuum pan salt from the Netherlands is also shipped on water.

Barge. The Erie Canal and the Chesapeake & Ohio Canal owe part of their origin to the early salt industry of the eastern United States. Barges were the easiest way to transport salt and other goods to customers far away. Barges are used today in shipping salt along the Intracoastal Waterway of the Gulf Coast, up the Mississippi River, the Florida coast, and inland waterways. They are also used to transport salt across the Great Lakes to and from Canada, and from Mexico to the Pacific Northwest. Several barges are linked together and towed inexpensively to distant locations. When freezing winter conditions in the Midwest affect the major rivers and their tributaries, salt movements by barge ceases until the ice is broken or thaws.

Salt is usually loaded onto barges using conveyor belts with moveable swing arms to evenly distribute the salt in the hold. A conveyor system is also found in the bottom of the holds of many barges to unload the salt at the final destination. Barge-mounted or shore-mounted cranes are also used to unload salt.

Freighter. Bulk freighters are the dominant method of transporting waterborne salt throughout the world. Salt may represent a partial cargo or the entire cargo on a ship. When the Jamestown and Plymouth colonial settlers arrived in the New World, imported salt was carried as ballast in the hulls of sailing ships arriving from England. Today, the bulk freighters may range in size from 6 to more than 100 Kt.

Some freighters have self-unloading equipment to discharge the salt from the cargo hold. The most common method of unloading bulk salt is by clamshell bucket and crane. The ship is brought alongside the dock and the salt is transferred to a dockside storage site or to a transportation system.

Storage: Bulk salt may be stored outside on specially constructed pads that reduce the amount of runoff that threatens surface and ground waters. The storage sites are usually located near areas accessible by water, rail, or truck. Salt is stored in cone-shaped piles or long windrows. Plastic, canvass, or burlap covers are often used to protect the salt from the wind and rain.

Inside storage bins, silos, and warehouses are also used to store bulk salt. The site is usually located near the rail, truck, or ship loading facilities. Packaged salt is normally stored in warehouses adjacent to where the salt is bagged or containerized.

ENVIRONMENTAL AND HEALTH CONSIDERATIONS

Since the early 1970s, people throughout the world have become more conscious about the environmental effects of pollution caused by industrialized societies. In pursuit of better living conditions and longer lifespans, many people have changed their lifestyles because of health considerations. Both of these topics have a bearing on salt consumption, and therefore, salt demand.

The national energy crisis that began in 1973 adversely affected several energy-intensive industries, one being the synthetic soda ash industry that used tremendous quantities of salt. Higher energy costs, environmental problems pertaining to discharges of sodium chloride and calcium chloride to groundwater supplies, and competition from natural soda ash sources, contributed to the closure by 1979 of seven of the eight synthetic soda ash plants that operated during the 1970s. The last plant closed in 1986, which ironically was the first plant constructed in 1881 and commenced production in 1884. The demise of this sector of the domestic chemical industry resulted in the significant loss of salt production. The decline is noted in Figs. 3 and 5.

The application of salt for highway deicing has contributed to rising levels of sodium and chloride concentrations in the Nation's ground water and surface water supplies. Water-quality trends were taken from two nationwide water sampling networks for the period 1974 through 1981 (Smith et al, 1987). The results indicated that river salinities have been increasing, and are attributed to highway deicing and human waste. Salt has been used for highway deicing since about 1940, which was before the major national highway construction projects began. In 1940, about 149 Kt of salt were used for deicing whereas in 1990, 10.3 Mt were used.

The US salt industry through its Salt Institute initiated a sensible salting program in 1972 to educate transportation departments on the proper ways to store, handle, and apply deicing salt to protect the environment while maintaining winter mobility and safety. The practice of knowing when to use salt, what alternative ice control agents to use, and how much salt to apply have contributed to some decline in deicing salt sales over the last decade. Although there are environmental implications using deicing salt, salt is the fastest, least expensive, and most effective agents to control the hazards associated with ice and snow. Without salt, there would be significant increases in the number of traffic accidents, loss of life, and productivity.

Population increases have caused increases in municipal waste treatment that has contributed to elevated salinities in waterways. As salt is assimilated by individuals, excess sodium is excreted by the body and disbursed to the environment through municipal sewage systems. The cities with growing populations are experiencing these increases. Although runoff from irrigated agriculture influences salinity, especially in the West, chloride trends were not appreciably correlated with changes in irrigated acreage nationwide.

In the late 1960s—early 1970s, health authorities attributed salt as the alleged cause of hypertension, or high blood pressure. People began restricting their salt intake and measuring the sodium levels in ingested food. It was not until the mid-1980s that other medical research suggested that there were other factors besides salt that contributed to hypertension. In 1988, the US Environmental Protection Agency (EPA) issued the National Primary Drinking Water Regulations that eliminated sodium as a regulated drinking water contaminant. The EPA rationale was that evidence was building that salt did not cause hypertension, and that the quantity of sodium in drinking water was very low compared with what is contained in food and other beverages. Further medical studies using control groups indicated that only a small segment of the population was salt sensitive, and that a large reduction in sodium intake would only result in a minor reduction in systolic and diastolic blood pressure (Johnson, 1989). Unfortunately, there were no positive tests to indicate who was or was not sensitive to salt. Using the analogy that an ounce of prevention is worth a pound of cure, most doctors recommended a general restriction of salt usage rather than attempting to diagnose who may be salt sensitive or address other contributing factors leading to high blood pressure, such as obesity, smoking, alcohol intake, stress, and lack of exercise. People today are not as concerned about their salt intake as they were in the 1970s.

FUTURE TRENDS

Despite the environmental concerns regarding deicing salt, rock salt will continue as the deicing agent of choice. Other deicing products, such as calcium magnesium acetate, are commercially available but are more expensive, and more must be used to have the same effect as salt. Because the markets for salt are mature, domestic and world consumption will grow proportional to the rise of population and the gross domestic product of countries.

The United States will remain among the largest salt-producing nations of the world. Since the middle of the twentieth century, the US salt industry has restructured itself through acquisitions, name changes, and consolidations to become a more integrated North American industry competing in the Western Hemisphere.

The world will continue to mine, harvest, process, transport, and sell salt among neighboring countries. New deposits will be developed in the future that will probably compete with existing facilities operating today. The operations that have the lowest production costs, offer a high quality product, and have low transportation costs, will be the leaders of the world salt industry.

BIBLIOGRAPHY

Anon., 1978a, ''New Problems Arise for Nuclear Waste Storage,'' *Chemical and Engineering News*, Vol. 56, No. 24, June 12, p. 28.

Anon., 1978b, ''Pakistan Opens New Salt Mine,'' *Mining Journal*, Vol. 290, No. 7446, May p. 337.

Anon., 1987, *The Economics of Salt*, 5th ed., Roskill Information Services, 284 pp.

Adams, J.E., 1968, ''Permian Salt Deposits, West Texas and Eastern New Mexico,'' Special Paper No. 88, Geological Society of America, p. 407.

Ahlbrecht, H., 1980, ''Auftausalz—Ja Oder Nein? (Deicing Salt—Yes or No?),'' 5th International Symposium on Salt, Vol. 2, Northern Ohio Geological Society, Cleveland, pp. 411–415.

Al Naquib, N.M., 1970, ''Geology of Jabal Sanam, South Iraq,'' *Journal*, Vol. 3, No. 1, Geological Society of America, pp. 9–36.

Aleksin, A.G., Kozhevnikov, I.I., and Sokolin, K.G., 1968, ''Geologicheskoye Stroyeniye i Neftegazonosnost; Solyanokupol'nykh Basseynov Zapadnoy Afriki (Gas and Oil-Bearing Salt Domes of West Africa),'' *Geologiya Nefti i Gaza*, Vol. 12, No. 12, pp. 47–51.

Andreyeva, R.I., and Kabyshev, V.P., 1970, ''Kharakyernyye Osobennosti Solyanoy Tektoniki Dneprovsko-Donetskoy Vpadiny (Characteristics of Salt Tectonics of the Dnieper-Donets Basin),'' *Sovetskaya Geologiya*, No. 2, pp. 135–142.

Antoine, J.W., and Dryant, W.R., 1969, ''Distribution of Salt and Salt Structures in Gulf of Mexico,'' Bulletin, Vol. 53, No. 12, American Association of Petroleum Geologists, pp. 2543–2550.

Atwater, G.I., 1968, ''Gulf Coast Salt Dome Field Area,'' Special Paper No. 88, Geological Society of America, pp. 29–40.

Auzende, J.M., et al., 1971, ''Upper Miocene Salt Layer in the Western Mediterranean Basin,'' *Nature*, Vol. 230, No. 12, pp. 82–84.

Ballard, J.A., and Feden, R.H., 1970, ''Diapiric Structures on the Campeche Shelf and Slope, Western Gulf of Mexico,'' Bulletin, Vol. 81, No. 2, Geological Society of America, pp. 505–512.

Batterson, M., and Boddie, W.W., 1972, *Salt—The Mysterious Necessity*, The Dow Chemical Co., Dallas, TX, 112 pp.

Bays, C.A., 1963, ''Significant Uncertainties in Current Salt Solution Extraction Operations,'' *1st Symposium on Salt*, Northern Ohio Geological Society, Cleveland, pp. 467–481.

Bays, C.A., et al., 1960, ''Solution Extraction of Salt Using Wells Connected by Haudraulic Fracturing,'' *AIME-Transactions*, Vol. 217, pp. 266–277.

Benavides, V., 1968, ''Saline Deposits of South America,'' Special Paper No. 88, Geological Society of America, pp. 249–290.

Benavides, G.L., and Sansores, E., 1968, ''Salt Deposits of Southern Mexico,'' Special Paper No. 88, Geological Society of America, pp. 407–408.

Benbow, J., 1990, ''Salt—A Status Report,'' *Industrial Minerals*, No. 271, pp. 19–31.

Bentor, Y.K., 1968, ''Salt Deposits of the Dead Sea Region,'' Special Paper No. 88, Geological Society of America, pp. 139–156.

Bertram, B.M., 1993, ''The Future of De-Icing Salt in North America,'' *7th International Symposium on Salt*, Vol. 2, Elsevier, The Netherlands, pp. 479–488.

Bleimeister, W.C., 1970, ''Mining the Salar Grande,'' *3rd Symposium on Salt*, Vol. 2, Northern Ohio Geological Society, Cleveland, pp. 479–486.

Bloch, M.R., 1963, ''The Social Influence of Salt,'' *Scientific American*, Vol. 209, No. 1, pp. 89–98.

Braitsch, O., 1971, ''Minerals, Rocks and Inorganic Materials,'' *Salt Deposits, Their Origin and Composition*, Vol. 4, Springer-Verlag, New York-London-Heidelberg, 297 pp.

Briggs, L.T., 1968, ''Evaporite Deposits of the United States,'' Abstract, Special Paper No. 88, Geological Society of America, p. 408.

Brunstrom, R.G.W., 1967, ''Origin of the Keuper Salt in Britain,'' *Nature*, Vol. 215, No. 5109, p. 1474.

Carosella, M.E., 1980, ''The Use of Salt Domes for the Strategic Petroleum Reserve,'' *5th International Symposium on Salt*, Northern Ohio Geological Society, pp. 69–75.

Cater, F.W., and Craig, L.C., 1970, ''Geology of the Salt Anticline Region in Southwestern Colorado,'' Professional Paper No. 637, US Geological Survey, 80 pp.

Clarke, F.W., 1924, ''The Data of Geochemistry,'' Bulletin 770, US Geological Suvey, 841 pp.

Cunha, T.J., 1974, ''Value of Salt and Trace Minerals in Animal Feeding,'' *4th International Symposium on Salt*, Vol. 1, Northern Ohio Geological Society, Cleveland, pp. 451–455.

DeMille, G., and Schouldice, J.R., 1968, ''Saline Deposits of Alberta and Saskatchewan,'' Special Paper No. 88, Geological Society of America, pp. 408–409.

Denton, D., 1982, *The Hunger For Salt*, Springer-Verlag, Berlin-Heidelberg-New York-Tokyo, 633 pp.

Dickinson, W.E., 1983, ''Salt Products in North America and Outlook for Future,'' *6th International Symposium on Salt*, Vol. 2, The Salt Institute, Alexandria, VA, pp. 657–659.

Duckworth, D., 1987, *The Economics of Salt*, Roskill Information Services, London, 284 pp.

Eskew, G.L., 1948, *Salt—The Fifth Element*, J.G. Ferguson and Assoc., New York, 239 pp.

Foster, R.J., 1980, ''Salt,'' Preprint, US Bureau of Mines, 12 pp.

Frey, M.G., and Grimes, W.H., 1969, ''Bay Marchand-Timbalier, Bay-Caillou Island Salt Complex, Louisiana,'' *Transactions*, Vol. 19, Geology of the American Mediterranean, Gulf Coast Association of Geological Societies, p. 266.

Gawel, A., 1967, ''Salt Deposit at Wieliczka,'' Bulletin, No. 211, Pt. 2, Polish Institute of Geology, pp. 325–328, 337–339.

Grabau, A.W., 1913, *Principles of Salt Deposition*, McGraw-Hill, New York, 435 pp.

Grott, G.J., 1988, ''High Purity Salt From Solar Ponds,'' Preprint No. 88–148, SME Annual Meeting, Phoenix, AZ, 5 pp.

Halbouty, M.J., 1967, *Salt Domes*, Gulf Publishing Co., 425 pp.

Hamilton, W.N., 1971, "Salt in East-Central Alberta," Bulletin No. 29, Research Council of Alberta, 53 pp.

Hanneman, R.L., 1993, "The Politics of Sodium Restriction in the US," *7th International Symposium on Salt*, Vol. 2, Elsevier, The Netherlands, pp. 231–239.

Harben, R.W., and Bates, R.L., 1984, "Salt," *Geology of the Nonmetallics*, Metal Bulletin, Inc., New York, pp. 215–224.

Hardy Jr., H.R., 1980, "Development of Design Criteria for Salt Cavity Storage of Natural Gas," *5th International Symposium on Salt*, Northern Ohio Geological Society, Cleveland, pp. 13–20.

Hardy Jr., H.R., and Langer, M., eds., 1988, *The Mechanical Behavior of Salt*, Trans Tech, F.R. Germany, 781 pp.

Harris, F.E., 1939, "Marketing of Salt," Information Circular 7062, US Bureau of Mines, 52 pp.

Hawkins, M.E., and Jirik, C.J., 1966, "Salt Domes in Texas, Louisiana, Mississippi, Alabama, and Offshore Tidelands: A Survey," Information Circular 8313, US Bureau of Mines, 78 pp.

Hetzel, B.S., 1989, *The Story of Iodine Deficiency—An International Challenge in Nutrition*, Oxford University Press, New York, 236 pp.

Hills, J.M., 1968, "Permian Basin Field Area, West Texas and Southeastern New Mexico," Special Paper No. 88, Geological Society of America, pp. 17–27.

Hite, R.J., 1968 "Salt Deposits of the Paradox Basin, Southeast Utah and Southwest Colorado," Special Paper No. 88, Geological Society of America, pp. 319–330.

Holser, W.T., 1971, "Geologic Significance of the Minor-Element Composition of Marine Salt Deposits," *Economic Geology*, Vol. 66, No. 5, p. 813.

Hou, Te-P., 1942, *Manufacture of Soda With Special Reference to the Ammonia Process*, Reinhold, New York, 590 pp.

Hutchinson, F.E., 1974, "Effect of Deicing Salts Applied to Highways on the Contiguous Environment," *4th International Symposium on Salt*, Vol. 1, Northern Ohio Geological Society, Cleveland, pp. 427–434.

Istvan, J.A., and Querio, C.W., 1983, "Storage of Natural Gas in Salt Caverns," *6th International Symposium on Salt*, Vol. 2, The Salt Institute, Alexandria, VA, pp. 183–190.

Jacoby, C.H., 1969, "Faults and Their Impact on (Salt) Mining," *3rd Symposium on Salt*, Vol. 2, Northern Ohio Geological Society, Cleveland, pp. 447–452.

Jacoby, C.H., 1972, "Evaluation of Salt Deposits," *Trans. SME-AIME*, Vol. 252, pp. 118–122.

Jacoby, C.H., 1977, "Geology-Hydrology of Avery Island Salt Dome," July, Federal Energy Administration Report, Y/Office of Waste Isolation, /SUB.

Johnson, G., 1989, "High Blood Pressure: A Silent Killer—Hypertension," American Council on Science and Health, New York, 40 pp.

Johnson, K.S., 1970, "Salt Produced by Solar Evaporation on Big Salt Plain, Woods County, Oklahoma," *Geological Notes*, Vol. 30, No. 3, Geological Survey of Oklahoma, pp. 47–54.

Johnson, K.S., and Gonzalez, S., 1978, "Salt Deposits in the United States and Regional Geologic Characteristics Important for Storage of Radioactive Wastes," March, Y/Office of Waste Isolation, /SUB-74141.

Jones, P.H., and Hutchon, H., 1983, "Road Salt in the Environment," *6th International Symposium on Salt*, Vol. 2, The Salt Institute, Alexandria, VA, pp. 615–619.

Joshi, B.S., and Bhatt, R.B., 1983, "Design and Layout of Solar Salt Works," *6th International Symposium on Salt*, The Salt Institute, Alexandria, VA, pp. 281–287.

Kaufman, D.W., 1960, *Sodium Chloride*, Reinhold, New York, 743 pp.

Kawahara, T., Suzuki, K., and Asada, K., 1983, "Concentration of Seawater by New Electrodialysis Process," *6th International Symposium on Salt*, Vol. 2, The Salt Institute, Alexandria, VA, pp. 499–513.

Kirkland, D.W., and Evans, R., 1973, *Marine Evaporites, Origin, and Diagenesis and Geochemistry*, Dowden, Hutchinson, and Ross, Inc., Stroudsburg, PA, 426 pp.

Klingman, C.L., 1975, "Salt," *Mineral Facts and Problems*, Bulletin 667, US Bureau of Mines, pp. 921–932.

Kostick, D.S., 1980, "Salt," *Mineral Facts and Problems*, Bulletin 671, US Bureau of Mines, pp. 769–780.

Kostick, D.S., 1981, "Salt," *Minerals Yearbook*, 1980, US Bureau of Mines, pp. 683–694.

Kostick, D.S., 1989, "Salt," *Minerals Yearbook*, 1989, US Bureau of Mines, pp. 849–872.

Kostick, D.S., 1982, 'Salt," *Mineral Industry Surveys*, US Bureau of Mines, pp. 132–133.

Kostick, D.S., 1983, 'The Future Supply-Demand Balance of Salt—A View of the US Bureau of Mines," *6th International Symposium on Salt*, Vol. 2, The Salt Institute, Alexandria, VA, pp. 661–666.

Kostick, D.S., 1991, "Directory of Companies and Plants Producing Salt in the United States—1990," *Mineral Industry Surveys*, US Bureau of Mines, 28 pp.

Kostick, D.S., 1992, 'Salt," *Mineral Commodity Summaries, 1992*, US Bureau of Mines, pp. 148–149.

Kostick, D.S., 1993a, "The US Salt Industry—A Heritage of Progress for the 21st Century," *7th International Symposium on Salt*, Vol. 1, Elsevier, The Netherlands, pp. 21–28.

Kostick, D.S., 1993b, "The Evolution of the US Salt Industry as a Partner in World Trade." *7th International Symposium on Salt*, Vol. 2, Elsevier, The Netherlands, pp. 453–460.

Kuhajek, E.J., and Fiedelman, H.W., 1974, "Nutritional Iodine in Processed Foods," *4th International Symposium on Salt*, Vol. 1, Northern Ohio Geological Society, Cleveland, pp. 447–449.

Landes, K.K., 1960, "Geology of Salt Deposits," *Sodium Chloride*, D.W. Kaufman, ed., Reinhold, New York, pp. 28–69.

Lefond, S.J., 1965, "Salt—The Universal Mineral," *The Explorer*, Vol. 7, No. 4, Cleveland Natural Science Museum, pp. 11–15, 34–35.

Lefond, S.J., 1969, *Handbook of World Salt Resources*, Monographs in Geoscience, Plenum Press, New York, 384 pp.

Leroy, L.E., 1951, *Subsurface Geologic Methods*, 3rd ed., Colorado School of Mines, Golden, 1166 pp.

Lotze, F., 1968, "Salt Deposits of Europe (Including the USSR)," Special Paper No. 88, Geological Society of America, pp. 411–412.

MacMillan, R.T., 1972, "Salt," *Minerals Yearbook, 1972*, US Bureau of Mines, pp. 1093–1102.

McArthur, J.N., 1980, "An Approach to Process and Quality Control Relevant to Solar Salt Field Operations in the Northwest of Western Australia," *5th International Symposium on Salt*, Vol. 2, Northern Ohio Geological Society, Cleveland, pp. 325–334.

Mannar, M.G.V., and Bradley, H.L., 1984, "Guidelines for the Establishment of Solar Salt Facilities," *Seawater, Underground Brines, and Salted Lakes*, United Nations Industrial Development Organization, UNIDO/IS.330/rev.1, 173 pp.

Martinez, J.D., 1971, "Environmental Significance of Salt," Bulletin, Vol. 55, No. 6, American Association of Petroleum Geologists, pp. 810–825.

Mattox, R.B., ed., 1968, "Saline Deposits," Special Paper No. 88, Geological Society of America, 701 pp.

Multhauf, R.P., 1978, *Neptune's Gift*, The Johns Hopkins Press, Baltimore and London, 325 pp.

Notholt, A.J.G., and Highley, D.E., 1973, "Salt," Mineral Dossier No. 7, Mineral Resources Division, Institute of Geological Sciences, London, 36 pp.

Phalen, W.C., 1917, "Technology of Salt Making in the United States," Bulletin, No. 146, Mineral Technology 20, US Bureau of Mines, pp. 149.

Phalen, W.C., 1919, "Salt Resources of the United States," Bulletin, No. 669, US Bureau of Mines, pp. 284.

Pierce, W.G., and Rich, E.I., 1962, "Summary of Rock Salt Deposits in the United States as Possible Storage Sites for Radioactive Waste Materials, Geological Survey Bulletin 1148, US Geological Survey, 91 pp.

Read, L.E., and Jacoby, C.H., 1957, "Exploring and Mining for Salt," *Mining Engineering*, Vol. 9, No. 5, pp. 538–541.

Severy, C.L., 1946, "Salt Cored Successfully with Oil Field Equipment," *Engineering and Mining Journal*, Vol. 147, No. 8, pp. 85–87.

Sheflin, M.J.E., 1983, "Road Salt: Friend or Foe," *6th International Symposium on Salt*, Vol. 2, The Salt Institute, Alexandria, VA, pp. 621–623.

Smith, D.A., and Reeve, F.A.E., 1970, "Salt Piercement in Shallow Gulf Coast Salt Structures," Bulletin, Vol. 54, No. 7, American Association of Petroleum Geologists, pp. 1271–1289.

Smith, R.A., Alexander, R.B., and Wolman, M.G., 1987, "Water-Quality Trends in the Nation's Rivers," *Science*, Vol. 235. pp. 1607–1615.

Standiford, F.C., 1974, "Sea Water Desalting and Salt Manufacture," *4th International Symposium on Salt*, Vol. 2, Northern Ohio Geological Society, Cleveland, pp. 481–487.

Turk, L.J., 1973, " Hydrogeology of the Bonneville Salt Flats, Utah," Water Resources Bulletin 19, Utah Geological and Mineral Survey, 80 pp.

ver Planck, W.E., 1957, "Salt in California," Bulletin 175, California Division of Mines and Geology, 168 pp.

White, R.M., and Spiers, C.A., 1983, "Characterization of Salt Domes for Storage and Waste Disposal," *6th International Symposium on Salt*, Vol. 1, The Salt Institute, Alexandria, VA, pp. 511–551.

Will, R.K., Leder, A.E., and Yoshida, Y., 1992, "Sodium Chloride," *Chemical Economics Handbook*, pp. 770.8000A-770.8002I.

Yang, S.T., 1991, "A Novel Method to Produce Road Deicer From Cheese Whey," *Proceedings,* Waste Stream Minimization and Utilization Innovative Concepts, An Experimental Technology Exchange, Vol. 2, Battelle Corp., Cleveland, OH, pp. 9.2–9.13.

SAND AND GRAVEL

HAROLD B. GOLDMAN

On the basis of tonnage, the sand and gravel industry is the second largest nonfuel mineral industry in the United States. In 1990, the production of sand and gravel was 927 Mt valued at $3.4 billion. California, which leads the nation with more than 126 Mt, together with Texas, Washington, Michigan, and Ohio, account for 36% of the total production in the nation (Table 1).

In commercial usage, sand applies to rock or mineral fragments ranging in size from particles retained on a No. 200 Sieve (0.074 mm openings) to those passing a No. 4 Sieve (4.76 mm openings). Gravel consists of rock or mineral fragments larger than 4.76 mm, ranging up to 88.9 mm maximum size.

The construction industry consumes 97% of the sand and gravel produced; the remainder is sand used for specialized products such as glass (see chapter on Industrial Sand and Sandstone and Glass Raw Materials).

Utilization

The building industry uses sand and gravel chiefly as aggregate in portland cement concrete, mortar, and plaster; the paving industry uses sand and gravel in both asphaltic mixtures and portland cement concrete. Aggregate is commonly designated as the inert fragmental material that is bound into a conglomerate mass by a cementing material such as portland cement, asphalt, or gypsum plaster. Sand and gravel is also used as construction fill, road base and subbase, and decorative material.

Portland Cement Concrete Aggregates: Portland cement concrete consists of sand and gravel surrounded and held together by hardened portland cement paste. Concrete mixes commonly contain 15 to 20% water, 7 to 14% cement, and 66 to 78% aggregate. Sand and gravel used as concrete aggregate have to meet many requirements (Goldman and Reining, 1983). Premature deterioration of concrete has been traced in many instances to the use of unsuitable aggregates.

Asphaltic Aggregate: Asphaltic mixtures used predominantly for paving consist of combinations of sand, gravel, and mineral filler (material finer than 0.076 mm), uniformly coated and mixed with asphalt produced in the refining of petroleum. Sand and gravel used as asphaltic aggregate must meet the same general physical requirements as materials used for portland cement aggregate.

GEOLOGY

General Requirements of Aggregates

Construction aggregate has many requirements that are difficult to meet if only unprocessed material from natural deposits is used. Suitable material is composed of clean, uncoated, properly shaped particles that are sound and durable. Soundness and durability are terms used to denote the ability of aggregates to retain a uniform physical and chemical state over a long period of time so as not to disintegrate when exposed to weathering and other destructive processes. Individual particles must be tough and firm, possessing the strength to resist physical stresses and chemical and physical changes, that may cause swelling, cracking, softening, and leaching. The aggregate should not be contaminated by excessive clayey material, silt, mica, organic matter, chemical salts, and surface coatings.

Physical Properties: The quality of aggregate depends upon its physical and chemical properties. These, in turn, may be inherent mineralogical and textural features of the rock or may be the effects of later changes such as tectonics, mechanical or chemical weathering, or incrustations.

The physical properties most significant for concrete use are: 1) abundance and nature of fractures and pores, 2) particle shape and surface texture, and 3) volume changes which may occur because of freezing and thawing or wetting and drying. An aggregate is considered to be physically sound if it is adequately strong and capable of resisting the agencies of weathering without disruption or decomposition. Minerals or rock particles that are physically weak, extremely water absorptive, and easily cleavable are susceptible to breakdown. The use of such materials in concrete reduces strength or leads to early deterioration by promoting weak bond between cement and aggregate, or by inducing cracking, spalling, or popouts. Severely weathered, soft, micaceous, or porous materials may cause localized stresses to develop in concrete by swelling and shrinking during wetting and drying or freezing and thawing cycles.

Physical Suitability of the Various Rock Types. Sedimentary rocks have a wide range in physical and chemical qualities. Sandstones and limestones, if hard and dense, are ordinarily satisfactory, but many sandstones are friable and excessively porous and commonly are clay-bearing. Shales generally make poor aggregate material, being soft, weak, and absorptive. Most igneous rocks are satisfactory, being normally hard, tough, and dense. Tuffs and certain flow rocks may be extremely porous and have high water absorption and low strength. Metamorphic rocks differ in character. Most quartzites are massive, tough, and dense. Fine-grained marbles are usually durable, but coarse-grained marbles have low abrasion resistance. Gneisses are ordinarily very tough and durable. Some schists contain micaceous minerals that are undesirable because they are soft, laminated, and absorptive. Micaceous minerals are susceptible to splitting along cleavage planes and thereby impair particle strength and durability. Some schists and slates in particular are thinly laminated and tend to assume flat slabby shapes that lack strength and do not pack well.

Any or all of these rock types may be rendered undesirable because of harmful exterior coatings. Weathering processes, particularly the action of ground waters, deposit these coatings. The most common coatings are calcium carbonate, clay, silt, opal, chalcedony, iron oxide, manganese oxide, and gypsum. Particles with these coatings are generally undesirable as aggregates because the bond between particle and coating may be weak, and decreasing the strength of the aggregate-cement bond.

Table 1. Construction Sand and Gravel Sold or Used by Producers in the United States, by State (thousand tons and thousand dollars)

State	1988		1989[e]	
	Quantity	Value	Quantity	Value
Alabama	10 652	41,417	9 435	36,500
Alaska	15 604	48,749	15 422	48,500
Arizona	29 392	123,854	30 754	133,900
Arkansas	7 005	26,201	6 804	25,500
California	128 773	622,074	125 466	670,800
Colorado	19 565	69,882	22 952	104,000
Connecticut	7 507	32,102	5 262	24,700
Delaware	1 754	5,988	1 724	6,200
Florida	16 923	53,083	16 239	55,500
Georgia	8 642	30,185	5 534	18,900
Hawaii	591	3,173	544	3,200
Idaho	6 273	19,897	5 262	18,900
Illinois	27 305	93,504	29 938	108,900
Indiana	23 517	79,985	26 853	99,200
Iowa	10 778	36,087	11 612	37,800
Kansas	9 761	25,329	11 794	33,200
Kentucky	5 738	15,243	4 990	15,100
Louisiana	12 912	52,820	12 338	54,400
Maine	9 238	33,007	7 802	30,100
Maryland	17 478	95,169	15 332	84,500
Massachusetts	20 111	79,364	12 610	57,000
Michigan	48 542	138,171	43 546	132,000
Minnesota	30 635	72,678	30 573	82,600
Mississippi	12 078	38,806	14 152	51,500
Missouri	10 176	32,941	9 072	32,500
Montana	7 243	20,225	5 262	13,900
Nebraska	10 187	28,928	13 789	41,800
Nevada	14 269	50,928	18 144	70,000
New Hampshire	8 246	32,614	5 443	20,400
New Jersey	16 618	74,183	13 789	68,400
New Mexico	7 972	31,367	10 705	45,400
New York	30 740	124,341	28 668	118,500
North Carolina	10 048	38,459	10 161	43,700
North Dakota	3 422	8,079	3 266	8,100
Ohio	41 826	156,318	40 280	148,700
Oklahoma	8 412	22,654	7 711	20,000
Oregon	13 499	52,657	13 064	49,700
Pennsylvania	17 986	91,966	17 690	94,600
Rhode Island	1 681	7,847	998	3,900
South Carolina	6 830	20,751	6 804	23,300
South Dakota	7 193	18,681	5 806	20,800
Tennessee	6 202	23,343	5 534	21,900
Texas	45 696	171,167	39 826	155,800
Utah	16 187	49,796	12 973	41,500
Vermont	5 486	17,478	6 260	20,400
Virginia	11 386	42,573	11 703	49,700
Washington	28 277	94,402	34 292	124,700

[e]estimated, *source:* Tepordei, 1991.

Chemical Properties: The chemical properties that may affect service life of concrete are reaction of certain rocks and minerals with high-alkali cement (alkali-aggregate reactivity), leaching of water soluble substances, solution of certain secondary minerals, such as the zeolites, to release sodium and potassium which accelerates deterioration of susceptible aggregate particles, and oxidation by weathering to produce compounds that may retard cement hydration, discolor pavement surfaces, or weaken the concrete.

Alkali-aggregate reactivity has been discussed at length in many publications. A reactive aggregate is any rock, gravel, or sand that contains one or more constituents that react chemically with the alkalies (sodium and potassium) in portland cement with more than 0.6% alkali by weight. This reaction causes expansion, cracking, and deterioration of concrete and arises from osmotic pressures produced by the formation and hydration of alkali-silica gels. The gels are formed through interaction between reactive silica in the mineral aggregate and the alkalies which are liberated by the cement during hydration. Opal (amorphous hydrous silica) is the most conspicuous aggregate material reacting in this manner. Other rocks and minerals known to be reactive are: glassy volcanic rocks of medium to high silica content (andesite and rhyolite), chalcedonic rocks, certain rhyolites that contain a hydromica, and the minerals tridymite and heulandite. Silt-size quartz and highly strained quartz may also be reactive, as will some argillaceous carbonate rocks. Any rock containing a significant proportion of reactive substances may be deleteriously reactive; thus normally nonreactive sandstone, shale, basalt, granite, and other rock types may be reactive if impregnated or coated with opal, chalcedony, or other reactive substance.

Certain sulfide minerals, such as the iron sulfides, pyrite, and marcasite, oxidize and cause unsightly rust stains or generate acidic compounds injurious to the concrete matrix and may cause popouts. Chemical salts such as sulfates, chlorides, carbonates, and phosphates also occur in some aggregates. Some of these substances dissolve easily or react to impede the setting of the cement.

A chemical alteration contributing to physical unsoundness is hydration. Shales, clays, and some rock nodules are examples of materials that expand when they absorb water and shrink as they dry. This absorptive character decreases the strength of concrete and increases the concrete's susceptibility to disruption by weathering.

The potential quality of sand and gravel for aggregate use often can be observed in the field, but most generally laboratory test procedures are used to make final evaluations of the engineering properties of the aggregate.

Origin

In the United States, sand and gravel is obtained commercially from sedimentary units of many types and ages in open pit or dredging operations. The principal sources are from existing or ancient river channel deposits, glacial deposits, and alluvial fans; marine and lake deposits and older geologic formations are significant but of lesser importance.

Stream Deposits: *Stream Channel Deposits.* Stream channel deposits consist of sand and gravel deposited in stream beds along present or former stream courses. Most channel deposits are generally easily accessible and easily mined.

Materials in these deposits are desirable as aggregate for many reasons. The natural abrasive action of stream transport has removed most of the weak rocks, leaving only the harder and firmer particles. These latter particles have undergone some degree of rounding and are subrounded to well-rounded, a desirable aspect for use in concrete.

Most sand and gravel deposits in active stream channels are replenished by material carried by seasonal floods, except in portions of the streams downstream from dams. Overburden is rarely present, but high flood waters may leave silt, clay, and wood debris covering parts of some channel deposits. The size of gravel gradually decreases downstream in the streams with long reaches. Commercial production is concentrated in the deposits where a proper blend of sand and gravel can be obtained. Fortunately, the favorable portions of many streams occur in flat-lying areas near population centers.

Channel deposits normally are free of excessive amounts of silt and clay and most deposits contain sand and gravel in the size gradations necessary for concrete use. Mining operations may be relatively simple, consisting of no more than washing and screening to obtain aggregate in suitable size gradations. In spite of these qualities, some deposits are unsuitable because they contain harmful ingredients such as physically unsound or chemically reactive rocks. The nature of the material in the stream channel is determined in large part by the nature of the source rocks within its drainage area. Different geological formations in a drainage basin contribute a variety of rock types that show a wide range in chemical composition and degree of weathering.

An example of the influence of source rocks on the quality of materials in stream deposits is provided by the Colorado River between Hoover and Parker Dams. The Colorado River in the vicinity of Hoover Dam contains a complex assemblage of rock types of which only a small proportion are chemically reactive. Hoover Dam concrete made with this aggregate shows no sign of harmful chemical reactivity after 40 years of service.

Below Hoover Dam the Colorado River flows over predominantly volcanic rocks and the content of reactive volcanic rock increases progressively downstream. At Davis Dam, 108 km below Hoover Dam, the content of reactive ingredients increases to the point where it was unwise to use the aggregate unless special precautionary measures were taken to forestall alkali-aggregate reaction. Farther downstream, the tributary Bill Williams River contributes large amounts of highly reactive andesites and rhyolites to the main stream. Parker Dam built with this aggregate exhibited signs of alkali-aggregate reaction within two years after completion.

Flood Plain Deposits. True flood plain deposits consist of material deposited on plains bordering streams by periodic overflow of the streams from their channels. Normally, these sediments deposited are composed of silt and sand grains. However, fine materials may mantle buried usable deposits of sand and gravel, particularly in areas where, in the geologic past, the streams were more vigorous and transported greater volumes of coarser material. The sand and gravel in these older, deeper flood plain deposits is similar to that in channel deposits.

Terrace Deposits. Stream terrace deposits are bench-like deposits of sand and gravel that border a stream but lie above the level of the present flood plain. These deposits are remnants of older flood plains or channels through which the stream has cut. Terrace deposits, because they are above the stream level, may be more desirable than stream channel deposits where water may make operations difficult. However, there is little possibility of replenishment of terrace deposits. The materials in these deposits have the general properties of stream channel materials, but weathering processes may have diminished the quality of some of their constituents by converting certain minerals to clay or by coating some grains with deleterious substances. Not all terrace deposits can be used for aggregate. Some deposits are only thin veneers of sand and gravel deposited on a stream-cut bedrock surface. The materials in the deposit may not be suitable if they have undergone post-depositional weathering or if the grains have become coated with harmful substances due to the action of ground waters.

Alluvial Fans: An alluvial fan is formed when streams carrying large volumes of sand and gravel down a steep mountain slope enter an adjacent valley or plain. The abrupt change in slope results in a decrease in the speed of the stream causing the stream to deposit the sand and gravel it has been carrying. The deposited matter spreads in a gently sloping fan-shaped mass from the mouth of the canyon onto the valley floor. The heavier, coarser material is deposited near the mouth of the valley, while the finer material is carried out toward the edges of the fan. Alluvial fan deposits have been found to reach depths of more than 60.8 m.

Fan deposits, because of the frequent shifting of the stream channel, ordinarily contain lenticular beds or tongues of poorly sorted sand and gravel interbedded with varying proportions of silt and clay. The particles are generally subangular to angular in the fans built by streams with small drainage areas. The particles in large fans, covering several square kilometers, are generally subangular to subrounded. Suitable aggregate is obtained from alluvial fans that are free from thick clay lenses.

Glacial and Fluvial-Glacial Deposits: Sand and gravel is produced commercially from glacial deposits in the midwestern and northeastern parts of the United States. The bulk of the sand and gravel was deposited directly or indirectly from continental ice sheets that originated in Canada and spread south into the United States. Glacial deposits that have been reworked by stream action are called fluvial-glacial deposits.

The glaciation periods consisted of advances and recessions of the continental ice sheets that were hundreds of meters thick and moved roughly north to south. Fragments of the rocks in the areas over which the glaciers passed, such as the igneous and metamorphic rocks in Canada, were picked up and incorporated in the ice. Debris consisting of clay, silt, sand, pebbles, and boulders was deposited directly from ice as lodgement till. When the advance and melting of the ice were in balance the debris deposited by water along the margin of the glacier formed irregular deposits termed moraines. As the water flowed away from the melting glaciers, the glacial debris was transported and deposited in outwash plains or elongate valley trains that are now major sources of sand and gravel (Ekblaw and Lamar, 1964). In addition, kames (hills and valley terraces) and eskers (ridges) of sand and gravel were deposited by melt waters within and alongside ice sheets.

The fluvial-glacial deposits exhibit a better degree of sorting and contain less clay than lodgement till and are often of high quality.

Outwash Deposits. Outwash debris varies in size with the distance from the former glacial ice. The coarsest materials, pebbles, cobbles, and boulders, were deposited near the glacier, and the finer materials, sand, silt, and clay, were carried successively farther away from it. Since the water velocities fluctuated, the deposits are characterized by great variation in particle sizes. Instances are cited where layers of gravel abruptly terminate to be replaced by thick accumulations of silt or clay.

Kames and Eskers. Kames are hills or terraces composed of material deposited by glacial melt water in vertical crevasses or holes in the glacier, or by melt water running alongside ice in a valley. Eskers consist of material deposited in streams that flowed under or in the glacier. When the ice melted, the kames remained as rounded hills or valley terraces, and the eskers were left as ridges. These deposits, which show a wide variation in particle size and physical characteristics, are a common source of sand and gravel.

Beach and Dune Deposits. During the last stage of glaciation, large lakes such as one in the Lake Michigan basin had extensive beach deposits at various elevations reflecting water levels. Each former water level is marked by beach ridges of gravelly sand. In places the beach sand has been blown into dunes. These deposits may be suitable as sources of sand.

Lodgement Till. Some glacial deposits such as lodgement till have been deposited directly from glacial ice with little or no reworking by streams. Such debris contains material of diverse size, shape, and quality ranging from pulverized rock flour to erratic boulders and rarely is usable due to the high cost of processing.

Dredge Tailings: As a result of bucket dredge operations for gold, huge furrow-like ridges of gravel have been accumulated as tailings along many streams in western states. A normal gold-dredging operation involves washing and screening the stream deposit. The oversize gravel, which ordinarily consists of everything over 12.7 or 19.1 mm, is washed and sent to tailing piles behind or alongside the dredge. The undersize fine gravel and sand is then processed to remove the gold and returned to the dredge pond and subsequently covered by other oversize tailings. This results in segregated deposits of fine gravel and sand that underlie coarser gravels, requiring considerable material be processed by aggregate producers to obtain the desired gradings of sand and gravel. Most commercial operations employ equipment to crush the oversize tailings. Tailing piles that have not been subjected to excessive weathering contain material as suitable as the related stream deposit. Considerable scrubbing and washing are necessary when abundant clay is present.

Older Geologic Formations: Pre-Quaternary beds of sand and gravel that are not well cemented may be usable sources of aggregate. While outcrops of such deposits may be well cemented, suitable unconsolidated material may be present behind the outcrop zone. Such pre-Quaternary deposits may be as old as Paleozoic. Ordinarily, these formations have been subjected to long periods of weathering and are either too well-cemented or contain too much clayey material to be processed economically. Accessibility and depth of overburden may present problems.

Beach Deposits: Sand and gravel formed by the winnowing action of currents and waves on a beach make excellent aggregate. The gravel and coarse sand size particles are generally well-rounded, hard, and firm and the sands are commonly composed predominantly of resistant quartz and feldspar. Many beach deposits, however, are thin and lack proper size gradation.

Regional Distribution

In the northern states, the principal sand and gravel resources are various types of glacial-related deposits. Fluvial and marine terraces, both ancient and recent geologically, are the major sources of sand and gravel in the Atlantic and Gulf Coastal Plains. River deposits are the prime sources in several of the southeastern and south-central states. In the Great Plains, sand and gravel are mainly stream deposits. On the West Coast, deposits are principally alluvial fans, stream deposits, beach and dune sand, and fluvial-glacial deposits.

Reserves: There are enormous undeveloped deposits in the United States that are uneconomical to mine today because of the high cost of transportation to market. These would constitute potential reserves if some means can be found to set aside these lands for future use and avoid the conflicts between urbanization and the mineral industry (Goldman, 1959a).

The heavily populated and urbanized Middle Atlantic, New England, and Pacific Coast states now are facing rapid depletion of deposits because of loss of mineral lands to urbanization, and because of opposition from environmental groups to development of new sources.

Supplies in the rest of the world may be considerable, but as is the case with the United States, many deposits may not be located in the areas of greatest demand. The more highly developed countries also are faced with the problem of loss of resources to urban encroachment (Pajalich, 1976).

TECHNOLOGY

Exploration

A deposit should meet certain criteria in order to be of economic interest for a commercial producer. Typical criteria are proximity to market, ease of obtaining the necessary mining permit, and quality and quantity.

As sand and gravel is a low cost, high volume commodity, it cannot be transported far before the cost of transportation equals or exceeds the FOB plant cost. The target area for exploration is largely defined by the cost of transportation from new sites relative to transportation costs from competitive operations. In aggregate-deficient parts of the country where materials are transported hundreds of kilometers, the target area naturally would be much larger.

Another important criteria to be considered before field exploration begins is the socioeconomic restrictions on a potential mining operation. If the governing bodies have adopted a policy of refusing to grant operating permits, there would be no purpose to exploring in that jurisdiction. Also, there should be an awareness of the opposition that might be generated by conservation and ecology-minded groups to granting of mining permits.

The size of a deposit is critical and generally should be sufficiently large to amortize the cost of the processing plant.

Field Techniques: Preliminary field evaluation by a geologist or by an engineer with some geologic training may often suffice as a first phase evaluation of a deposit as a possible source of aggregate material. The characteristics that bear upon the usefulness of a deposit as a source of aggregate material have been determined by the geologic processes by which the deposit was formed and subsequently modified. A basic understanding of the processes, therefore, is helpful to the person evaluating it. The important characteristics are types and physical condition of the rock, grading, rounding and degree of uniformity, particle size and shape, location, thickness and type of overburden and ground water level.

By the judicious use of geologic maps a geologist can determine the type of materials to be encountered in a certain drainage area and predict the type of alluvial material to be found in a given stream channel. The geologist thereby can recognize likely sources of aggregate and often can rule out areas unlikely to contain suitable materials. The geologist uses local commercial sand and gravel deposits as a yardstick to guide him in evaluating new deposits. By comparing service records and laboratory test results with the petrographic character of commercially proven aggregates, the geologist builds up a background of data from which he can extrapolate to predict the behavior of untested deposits of similar petrographic character. This technique has been used successfully by the US Bureau of Reclamation in preliminary planning studies for large structures that require large amounts of suitable aggregates. In many instances field evaluation has obviated the need of expensive and time-consuming tests.

Target Areas: Preliminary investigations, which include determinng such features as crude estimates of reserves, thickness and type of overburden, water and power availability and access are based primarily upon visual examination either by a geologist or someone with experience in aggregate production. As outlined in the section on geology, the areas to be prospected for sand and gravel are predominantly those associated with either the present stream drainage or ancient drainages. Clues to buried or older deposits may be gleaned from well-drilling logs, geophysical data, and from geologic reports.

Aerial photographs, coupled with topographic maps, are valuable tools. Any pertinent geologic information in the form of maps or reports should be researched. In most instances, the state geological surveys can provide the necessary maps. In certain situations, initial exploration can be greatly expedited with the use of airplanes or helicopters to observe promising geologic features.

Detailed exploration: The method of exploration will depend upon the particular set of conditions, the objective being to obtain representative samples and to delineate the dimensions of the deposit. No matter what method of exploration, ultimately representative samples must be obtained so that standard laboratory tests can be performed.

Where the depth of the deposit is less than 9 m, test pits may be excavated with a bulldozer, backhoe, clamshell, or dragline. The location of the water table often will determine the type of test pit. For thicker deposits, truck-mounted drill rigs should be used. Ordinarily, truck-mounted bucket augers, which use buckets from 0.5 to 0.9 m in diameter are satisfactory when large boulders are not present in the deposit. When drilling beneath the water table, steel or concrete casing is used to prevent the hole from caving. Samples may also be taken by rigs, such as clamshells, reverse circulation drills, or split spoon samplers. Choice of drilling method will depend upon the nature of the deposit and upon the availability of competent drillers in the area.

Geophysical methods can be used in conjunction with a drilling program. Geophysics may be used to check continuity of geologic units between drill sites or beyond a tested deposit. For example, electrical resistivity techniques may be used to locate thick zones of clay interbedded in the sand and gravel. The depth of a sand and gravel deposit resting on a hard bedrock surface can be determined by means of refraction seismic methods. The cost of geophysical methods equals the cost of drilling, on an hourly basis. However, much more geometric information is obtained within the same time interval through geophysics.

Obviously the geophysical methods are useful exploration tools but cannot substitute for the other methods when samples must be obtained for laboratory testing. Standard laboratory tests are used to determine the quality of a deposit. Equally important as laboratory testing is the sampling in the field. Where possible, large bulk samples should be obtained and quartered in the field. In some cases, material should be stockpiled and trucked or railed to an existing plant to determine size gradation and quality of the finished product. A more thorough description of sampling and testing of aggregate is presented in the US Bureau of Reclamation Concrete Manual (Anon., 1975).

As an integral part of the exploration program, cross sections should be drawn of the deposit and estimates made of the quantity of material available. This figure should be reduced by the percentage of waste material, by the material in setbacks, and in finished slopes required by mining permits.

Mining Methods

Sand and gravel deposits are mined with power shovels, draglines, slackline cable draglines, front end loaders, or dredges. The choice of excavating equipment depends upon whether the deposit is mined dry or wet, and other considerations.

In a dry-pit operation, shovels, loaders, or draglines load the sand and gravel into trucks or onto conveyor belts for transfer to the processing plant.

A wet-pit operation recovers sand and gravel from deposits that are below the water table. The sand and gravel may be excavated by a land-based dragline or by a floating dredge on a manmade or natural body of water.

Ocean Mining: Ocean mining may supply future aggregate needs in metropolitan coastal areas as local deficiencies develop in construction aggregate supplies from nearby land deposits. Sand and gravel probably constitute the most widespread resources of the sea floor of the continental shelf. Although no large scale commercial exploitation of offshore sand and gravel deposits is known to be underway at present on the continental shelf proper in the United States, tidal rivers and estuaries have been dredged commercially for many years.

Marine technology today is essentially an extension of conventional dredging systems that have been used successfully in rivers and estuaries. These include clamshell, dragline, and dipper dredges for relatively shallow water (generally less than 30.4 m), bucket ladder types (45.6 to 60.8 m depth), and hydraulic systems (generally equipped with a rotary cutter) capable of pumping dredged material in slurry form.

Processing Techniques

Commercial operations are located at deposits where the proper blend of sizes of sand and gravel can be obtained in order to produce a variety of products. Production by highway contractors generally is obtained from portions of deposits where there is an abundance of +38.1 mm gravel to provide crushed materials for use as road base or bituminous aggregate.

Ideally a commercial deposit would contain about 60% gravel and 40% sand (−4.76 mm), enough +38.1 mm gravel to crush for use as road base or bituminous aggregate, and sand in the correct sizes and proportions for use in concrete. Few deposits meet these requirements and plants are designed to cope with natural deficiencies. Plant capacities can range from 30 to 1 000 tph of processed material.

Most plants are departmentalized to produce different products. There is usually a dry side, where pit run material is screened and crushed for use as road base, or in some instances as bituminous aggregate, and a wet side where sand and gravel are washed and screened for use as concrete aggregate. Various stages of crushing are used to reduce the gravel fraction to the necessary sizes and/or the required number of broken faces. Commonly, a jaw crusher is used for primary reduction, particularly where boulders greater than 76.2 mm are present. Gyratory, roll or impact crushers further reduce the gravel to −25.4 and −19.1 mm. Rod mills are used to manufacture sand from 6.4 to 9.5 mm (pea gravel) to supplement natural sand that is deficient in fine sizes.

The sand fraction ordinarily is washed and classified in spiral classifiers (sand screws). Plants processing sand deficient in fine sizes usually employ wet cyclone-type separators to recover fine sand which may be present in the overflow from a classifier. In a number of plants, hydraulic settling tanks are used to separate the sand into various sizes. The desired blend of sand sizes is drawn off through gates in the bottom of the tank and dewatered in spiral classifiers.

Plants processing clayey material use revolving scrubbers or log washers to clean the pit run material before further processing. Standard screens that are mechanically vibrated, separate the gravel into the necessary sizes.

Some plants beneficiate their raw material by removing soft, porous particles, such as shale, in a heavy media separation that utilizes the sink-float principle. The desired material sinks and the undesirable lighter fraction is floated off. The heavy liquid is made up of a mixture of finely ground magnetite and ferrosilicon, with a specific gravity that can be varied from 1.0 to 3.4. Jigs are also used to remove light particles in the granule size range (Goldman, 1962).

General Specifications

The study of the results of laboratory tests of aggregates that have good service records in concrete has led to the establishment of certain minimum requirements or specifications to which aggregates are expected to conform. Laboratory tests and specifications are designed so that completely serviceable concrete may be made using any aggregate that meets the requirements. Most specifications are written by government agencies, engineering societies, and concrete technologists, and generally attempt to conform to one standard set of specifications, those set up by the American Society for Testing and Materials (ASTM). These ASTM standards (summarized in Table 2) are frequently modified to meet specific requirements of certain regions or types of construction and to accomodate to new testing technology. Some deposits that do not meet certain required specifications may still have to be utilized because of other factors, such as the greater expense of importing a more suitable aggregate.

In general, aggregate will be satisfactory for most uses if it meets the following minimum standards. These specifications are a compilation of the typical requirements recommended by the ASTM, California Division of Highways, US Army Corps of-

Table 2. Laboratory Methods of Determining Suitability of Sand and Gravel for Use as Aggregate in Portland Cement Concrete

Property of Aggregate	Importance in Concrete	Test Methods	Reference to Test
Hardness and durability (resistance to abrasion)	Affects strength, resistance to wear	"Los Angeles Rattler," measure proportion of fine material produced by abrasion in revolving metal drum after 100 and 500 turns	ASTM Test C131
Soundness (lack of fissures in particles)	Affects strength, susceptibility to frost damage from expansion of absorbed water	Alternately soak in sodium or magnesium sulfate solution and dry; crystallization of absorbed solution forces open invisible cracks. Subject test beams made with aggregate to alternate cycles of freezing and thawing	ASTM Test C88 ASTM Test C666 C682
Specific gravity—dry and with absorbed liquid	Determine mass (specific gravity commonly specified 2.5 or more); absorption affects bond of cement paste to particles	Compare oven-dry weight with immersed weight, and weight after surface re-dried	ASTM Tests C127 and C128
Size grading characteristics	Affects flowability, residual void spaces, strength	Standard sieve analysis; screen in standard size screens; weigh various fractions; plot on appropriate graphs	ASTM Test C136
General characteristics: 1. Particle Shape 2. Character of surface 3. Grain size 4. Texture (e.g., pore space, grain packing cementation) 5. Color 6. Mineral composition 7. General physical condition (e.g., weathering) 8. Presence of potentially deleterious chemical substances (e g., gypsum, zeolite, pyrite, opal, chalcedony, volcanic glass)	Different effects on strength, hardness, color, and permanence of concrete	Examine by naked eye, hand lens, and under petrographic microscope	ASTM Test C295
Potential chemical reactivity	Affects permanence of concrete; reactive substances cause "popouts" and failures due to expansion	Weigh silica dissolved in sodium hydroxide solution; measure reduction in alkalinity caused by immersion of sample in standard sodium hydroxide solution Measure expansion of mortar bars made with aggregate over lengthy periods; 1–2 years	ASTM Test C289 ASTM Test C227
Lack of organic matter (coal, lignite, organic impurities)	Affects strength, resistance to wear	Separate material lighter than 2.0 specific gravity in heavy liquid, and weigh; compare color of sample with standard color solution—dark color assumed due to organic material	ASTM Tests C123 and C40
Cleanness (lack of dirt, clay, or silt finer than 200 mesh)	Determines quality of bond with cement	Measure material passing 200 mesh sieve; measure suspended material after shaking in water	ASTM Test C117
Unit weight	Determines mass	Weigh aggregate contained in standard cubic foot measure	ASTM Test C29
Lack of soft or friable fragments	Affects strength, resistance to wear	Scratch test using brass rod of Rockwell hardness B65 to B75; rock softer than rod is unsatisfactory	ASTM Test C235
Toughness	Affects strength, resistance to wear	Impact; measure distance a standard-size hammer drops on specimen to fracture it	ASTM Test D3-18

Source: Annual Book of ASTM Standards, Sec. 4, Vol. .04.02 (concrete aggregate) Vol. .04.03 (road and paving materials) 1991.

Engineers, and the US Bureau of Reclamation. However, other tests or other test limits may be used to meet specific local needs.

Abrasion: The Los Angeles abrasion test loss should be less than 30% (ASTM C131).

Soundness: The loss in the sodium sulfate soundness test should be less than 10% for coarse aggregate (ASTM C88).

Specific Gravity: The specific gravity should be greater than 2.55 (ASTM C127).

Size and Grading: (ASTM C136)

a) The product has proper grading so that the fine aggregate should contain no more than 45% of the material between two consecutive sieve sizes.

b) The fineness modulus should be between 2.3 to 3.1.

c) No more than 5% of the material should pass the No. 200 sieve (.074 mm).

Reactivity: A mortar bar containing the aggregate should have an expansion less than 0.10% in one year with a 0.8% alkali content cement (ASTM C289).

Absorption: The absorption should not exceed 3% (ASTM C127).

Durability: The concrete containing the aggregate should not have a loss in the modulus of elasticity exceeding 50% in the freeze-thaw test (ASTM C666).

Sand Equivalent: The fine aggregate should have a sand equivalent of not less than 75 (ASTM D2419).

Testing and Evaluation

Petrographic Examination: Petrographic examination is a laboratory method used for determining physical and chemical properties of aggregates primarily by visual appraisal (ASTM C295). Standard techniques for mineral and rock identification, which commonly involve the use of a binocular or polarized light microscope and microchemical tests, are employed. Representative samples usually consisting of 300 or more particles are carefully examined and classified as to 1) rock and/or mineral type, 2) physical condition, 3) chemical stability—presence of reactive or otherwise deleterious ingredients, 4) shape of particles, and 5) nature of particle coatings. Reports are tabulated to show percentage composition, condition of the rock types present, and percentage of reactive ingredients, as well as diagnostic features of the particular constituents and the relative qualities of the different size fractions. Examination by petrographers experienced in concrete work permits interpretation of routine laboratory tests and aids in comparing new aggregate deposits with other deposits for which service data are available. On the basis of petrographic observations, supplementary tests other than the usual laboratory tests may sometimes be recommended.

USES

Within the construction industry, highway construction or road building consumes the major volume of sand and gravel used as aggregate for both concrete and asphalt pavements; in concrete for bridges, tunnels, and appurtenant structures; in treated and untreated road base material; and in both structural and nonstructural fill. The next largest segment of use is in building construction, where practically all of the volume is used in the manufacture of concrete for both cast-in-place and precast structural and cladding elements. The remaining quantities go into fill, septic fields, and other construction-related uses.

Specialty sand uses demand more rigid specifications as to types and gradings. Such specialty sands are classified by the US Bureau of Mines as "industrial sand" and are discussed in the chapter on Silica Sand and Sandstone.

ECONOMIC FACTORS

Marketing

Commercial interests produce approximately 85% of the sand and gravel supply, with the remaining 15% produced by government and/or contractors. In the latter category, it is apparent that such production operations are set up to handle extremely large projects (such as dams, rural highways, etc.), or in remote or rural areas where no commercial operations exist. The work is done with semi-mobile plants that remain only until completion of a specific project and then relocate for the next job.

In or around metropolitan market areas where transportation networks (rail, water, highways) are adequate, most production is done by commercial companies with permanent production facilities. Since transportation of sand and gravel is a disproportionate share of selling cost, it is common for asphaltic concrete and ready-mix concrete plants to be located on the sand and gravel production site. Indeed, some producers are in the ready-mix and/or asphaltic concrete business, and some are also construction contractors.

On the basis of tonnage, the construction sand and gravel industry is the second largest nonfuel mineral industry in the United States. According to the US Bureau of Mines, there were 4 175 companies producing construction sand and gravel operating 5 687 pits in 1988. The individual operations ranged in size from those producing millions of tons annually to those reporting less than 25 ktpy. Most operations are small, turning out one product or a limited range of products, but most of the tonnage comes from large operations. For example, the 26 operations reporting more than 2.0 Mt each in 1988 represented only 0.4% of the total number of operations, but produced 9.5% of the total tonnage, while 95 operations reporting between 1 and 2 Mt each, representing only 1.6% of the total number of operations, produced 13.6% of the total tonnage. For the same year, most of the construction sand and gravel was produced by operations reporting between 0.2 and 1 Mt each—1 109 operations, representing 19.5%, produced 50.3% of the total tonnage (Tepordei, 1991).

Price Trends

The price of sand and gravel at its point of use is a function of two elements: the FOB plant cost and the cost of transport to the using site. These two elements must be considered when evaluating industry trends. Tepordei (1991) revealed that although the unit price of sand and gravel increased since 1970 from $1.11 per ton, FOB plant to $3.62 in 1989, the constant 1970 dollar value fluctuated between $1.11 and $1.25 per ton in that interval (Fig. 1).

The low price has been maintained by more efficient mining and processing equipment in spite of rising labor costs, higher land values, and increasing restrictions. Such improvements do have limits and it is expected that prices will continue to rise slowly during the 1990s.

Major technological developments that have been instrumental in maintaining adequate production of sand and gravel at stable or slightly declining real cost include: use of larger operating units, more efficient portable and semi-portable plants, unitized plants for versatility of plant capacity, new prospecting methods utilizing aerial and geophysical surveying methods, and greatly increased rehabilitation and higher resale value of depleted mining sites.

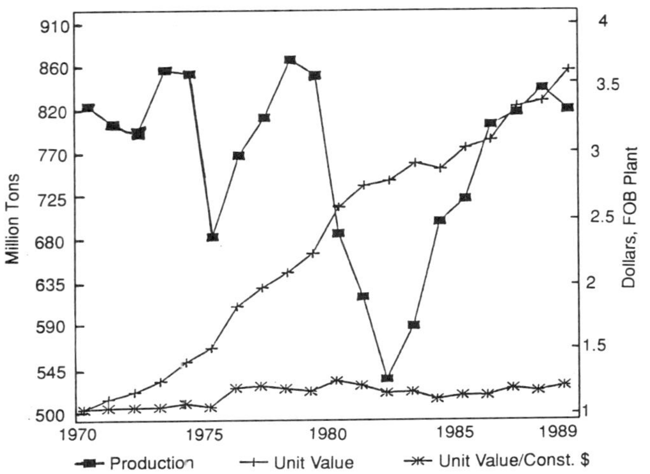

Fig. 1. Production and unit value in actual and 1970 constant dollars of construction sand and gravel in the United States for 1970–1989.

Consumption

The demand for sand and gravel in the early 1990s is expected to remain at the same level as in 1989. The slowdown in the residential building market may be offset by increased demand for aggregate in road construction and maintenance. Tepordei (1989) estimated that demand for sand and gravel will reach 1 Gt in 1995, primarily due to activity in the public construction sector.

A report by the Department of Transportation indicates that a significant increase in the volume of work on the infrastructure should be expected in the next 5 to 10 years. The value of public construction work as part of the gross national product (GNP) has declined constantly from about 3% in the 1950s and 1960s to less than 2% in the 1980s. If this trend begins to reverse in the 1990s, a significant increase in spending for public construction work could be expected that will dramatically alter the above forecasts.

The average annual growth rate of domestic demand for sand and gravel since 1954 has been about 2.8%. The estimated annual growth rate for the balance of the century, obtained by contingency forecasting based on anticipated requirements for individual end use, is expected to be between 2.5 and 4.7%. The forecast demand for sand and gravel in the United States in the year 2000, corresponding to this growth range, covers a range from 2.0 to 2.6 Gt (Pajalich, 1976).

Transportation

The proximity of market to the resource to date has resulted in a relatively stable delivered price for sand and gravel for most regions. The cost of transporting sand and gravel is a function of the equipment used and road conditions. A good rule-of-thumb estimate on rates for transporting sand and gravel is approximately 6 to 10¢ per ton for each kilometer hauled, although in some congested areas costs of 12 to 15¢ or more per ton-kilometer are realized.

The single factor that most escalates the delivery price of sand and gravel is the distance of haul. At 6¢ a ton-kilometer, an increase of 32 km in the hauling distance would increase the price of aggregate $2 per ton. This could easily represent an increase of 50 to 100% on the delivered cost of sand and gravel to the consumer. When aggregates are moved great distances, the final cost for the finished product may be mostly transportation cost.

In some metropolitan areas, public utility commission truck zone rates apply; in other more outlying areas truck rental rates apply. The rate is highly variable depending on method of transportation, governmental regulation, labor contracts, and area. Other factors include haul distance, legal load limits, on-road or off-road operation, and competition.

Truck haulage is the predominant form of transportation used by the industry, both as the only method or often in conjunction with other methods. Rail hauls are used in many areas where spurs are available, quantities are large, and distances from production to destination are long. In coastal areas and areas on inland waterways, barge hauling is a readily accepted method. Together, rail and barge transportation account for approximately 10% of the total sand and gravel movement. Rehandling of material is usually required when rail or water transportation are used, adding to cost, and sometimes causing breakdown of the aggregate.

Depletion Allowance

The depletion allowance for common sand or gravel is 5%. There are no import tariffs.

Alternate Materials

The most widely used substitute or alternate material for sand and gravel is crushed stone, which competes in almost every category, and is produced in a wide variety of types, sizes, and gradation. Natural sand has a considerable FOB price advantage over sand manufactured from crushed stone because of the high cost of crushing. Coarser gradations of crushed stone are also more expensive to produce, but is competitive usually because there is insufficient gravel to fill the coarse aggregate needs.

A growing competitor to sand and gravel in the concrete aggregate market area is lightweight aggregate, which is manufactured by a sintering process in which appropriate clays, shales, slate, or fly ash are passed through rotary kilns or traveling grate furnaces to produce a bloated aggregate product. Lightweight aggregate is used extensively in cast-in-place structures, in precast elements, and in concrete blocks. Other lightweight materials for concrete aggregate include natural pumice, expanded perlite, expanded (or exfoliate) vermiculite, and iron blast furnace slag, which find use in insulating fills, roof decks, and similar basically nonstructural uses.

Recycling of concrete and asphalt used mainly in pavements has been increasing in areas where aggregate is scarce or costly. Recycled concrete also has been used as road base material.

GOVERNMENTAL CONSIDERATIONS

Zoning and Land Use

The sand and gravel industry, more than any other major segment of the mineral industry, is beset with land use conflicts and environmental problems associated with rapid urban expansion. The problems are complicated because efficient urban growth depends upon centrally located sources of low cost construction materials.

As a city or county expands, the growth process itself may cut off the vital supply of industrial minerals that are needed for this growth. This usually happens in the following sequence. As a community or county begins to grow, a demand or market is created for industrial minerals such as sand, gravel, crushed rock, and clay for brick products. Since sand and gravel form a high-bulk, low cost product, the extraction operations begin as close as possible to the consuming areas—the population centers. As the community grows, more and more homes are built in the outlying or rural areas. This expansion often encroaches upon and then surrounds the established gravel pits. New homes may cover sand and gravel deposits making it impossible to expand the sand and gravel reserves. In addition to covering potential reserves, the new homeowners may resent the existing extraction operations. Complaints of excessive noise, dust, and hazards to children are often made to the local government authorities. Continued pressure may result in ordinances that restrict operating conditions, truck routes, etc. Finally, as a result of depleted reserves and/or harassment, a pit operator may be forced to move to more remote locations, or to shut down. Ultimately, this reduces the available supply and increases construction costs (Goldman, 1959a).

Rapid population growth and urbanization have had a decided impact upon the sand and gravel industry. Their ability to continue operations and supply the needed aggregates to maintain the present and projected construction needs of our society is imperiled. Often local government imposes planning and zoning restrictions that are inconsistent and usually only the zoning needs for that particular area are considered. Ordinarily, little thought is given to those areas in the total region served by the mineral resources and prime mineral resources are often lost to development and urban growth.

In California, under the Surface Mining and Reclamation Act passed in 1975, the state and local planners are charged with the responsibility of designating those mineral resources that are of economic significance to the state. Initial studies have focused on sand and gravel deposits.

Environmental and Safety Regulations

Closely associated with land use conflicts are problems of environmental contamination, such as water pollution, dust, solid waste, noise, and unsightly conditions that sometimes result from sand and gravel operations.

A more recent concern to operators are the regulatory requirements of the Occupational Safety and Health Administration (OSHA) concerning exposure to forms of crystalline silica and asbestos. Under these regulations, all products containing a specified concentration of asbestos and non asbestiform varieties of the amphibole minerals anthophyllite, tremolite, and actinolite must be labeled as having a constituent that is a carcinogen. This entails examination of sand and gravel deposits in the United States at a minimum cost of $1,000 per deposit (Tepordei, 1991).

BIBLIOGRAPHY AND REFERENCES

Anon., 1957, *Standard Specifications for Construction of Roads and Bridges on Federal Highway Projects*, FP-57, US Department of Commerce, Washington, DC, pp. 363.

Anon., 1958, "Bibliography on Mineral Aggregates," No. 23, Highway Research Board, Washington, DC, p. 111.

Anon., 1962, "Engineering Problems of Sand and Gravel Production," NSGA Circular No. 88, National Sand and Gravel Assn., Silver Spring, MD, pp. 34.

Anon., 1975, *Concrete Manual*, 7th ed., US Bureau of Reclamation, Denver, CO, p. 642.

Anon., 1991, *Annual Book of ASTM Standards*, Sec. 4, American Society for Testing and Materials, Philadelphia, PA, p. 10.

Ahearn, V.P., Jr., 1964, "Land Use Planning and the Sand and Gravel Producers," National Sand and Gravel Association, Silver Spring, MD, p. 30.

Barksdale, R.D., ed. 1991, *The Aggregate Handbook*, National Stone Association, Washington, DC.

Cooper, J.D., 1970, "Sand and Gravel," *Mineral Facts and Problems*, Bulletin 650, US Bureau of Mines, pp. 1185–1199.

Drake, H.J., 1972, "Stone," *Minerals Yearbook, 1972*, US Bureau of Mines, pp. 1153–1173.

Dunn, J.R., and Cutcliff, W.E., 1971, "Selecting Aggregate Deposits: A Geologic View," *Rock Products*, Mar., pp. 75–79.

Dunn, J.R., and Hudec, P.P., 1972, "Frost and Sorption Effects in Argillaceous Rocks," No. 393, Highway Research Board, Washington, DC, pp. 65–78.

Ekblaw, G.E., and Lamar, J.E., 1964, "Sand and Gravel Resources of Northeastern Illinois," Circular 359, Illinois State Geological Survey, p. 8.

Goldman, H.B., 1956 "Sand and Gravel for Concrete Aggregate," *California Journal of Mines and Geology*, Vol. 52, No. 1, pp. 79–104.

Goldman, H.B., 1959a, "Franciscan Chert in California Concrete Aggregate," Special Report 55, California Division of Mines and Geology, p. 28.

Goldman, H.B., 1959b, "Urbanization and the Mineral Industry," *Mineral Information Service*, California Division of Mines and Geology, Vol. 12, No. 12, pp. 1–5.

Goldman, H.B., 1961a, " Sand and Gravel in Northern California," Bulletin 180 A, California Division of Mines and Geology, p. 38.

Goldman, H.B., 1961b, "Urbanization—Impetus and Detriment to Mineral Industry," *Mining Engineering*, Vol. 13, No. 7, p. 717.

Goldman, H.B., 1962, "Aggregate from 'Fossils," *Rock Products*, Vol. 65, No. 11, pp. 65–68.

Goldman, H.B., 1964, "Sand and Gravel in Central California," Bulletin 180 B, California Division of Mines and Geology, p. 58.

Goldman, H.B., 1968, "Sand and Gravel," *Mineral and Water Resources of California*, Bulletin 191, California Division of Mines and Geology, pp. 361–369.

Goldman, H.B., 1969, "Sand and Gravel in Southern California," Bulletin 180C, California Division of Mines and Geology, p. 56.

Goldman, H.B., and Klein, I.E., 1958, "Sand and Gravel Resources of Cache Creek in Lake, Colusa, and Yolo Counties, California," *California Journal of Mines and Geology*, Vol. 54, No. 2, Apr., pp. 237–296.

Goldman, H.B., and Klein, I.E., 1961, "Sand and Gravel Resources of the Kern River, near Bakersfield, California," Special Report No. 70, California Division of Mines and Geology, pp. 33.

Goldman, H.B., and Reining, D., 1983, "Sand and Gravel," *Industrial Minerals and Rocks*, 5th ed., Vol. 2, S.J. Lefond, ed., AIME, New York, pp. 1151–1166.

Meininger, R.C., 1988, *Technical Information Letter*, No. 395, National Aggregates Association, Silver Spring, MD, p. 7.

Pajalich, W., 1972. "Sand and Gravel," *Minerals Yearbook, 1972*, US Bureau of Mines, pp. 1103–1121.

Pajalich, W., 1976, 'Sand and Gravel," *Mineral Facts and Problems, 1975*, US Bureau of Mines, pp. 933–946.

Schellie, K.L., and Rogier, D.A., 1963, " Site Utilization and Rehabilitation Practices for Sand and Gravel Operations," National Sand and Gravel Association, Silver Spring, MD, p. 80.

Tepordei, V.V., 1991, "Construction Sand and Gravel," *Minerals Yearbook, 1989*, US Bureau of Mines, p. 26.

Yeend, W., 1973, "Sand and Gravel," *US Mineral Resources*, Professional Paper 820, US Geological Survey, pp. 561–565.

SILICA

INDUSTRIAL SAND AND SANDSTONE

MARK J. ZDUNCZYK AND MICHAEL A. LINKOUS

Silica in the form of sand and sandstone is one of the most common, and at the same time, unique industrial minerals. Found in every rock type of every geologic age and virtually everywhere in the world, silica is used in products that touch just about every aspect of daily life. Imagine a world in the 1990s without computer chips, fiber optics, or glass, and you have just begun to understand how important silica is to the quality of life we enjoy.

The elements silicon (Si) and oxygen (O) comprise roughly 60% of the earth's lithosphere to a depth of about 16 km. The crystal structure of silicon dioxide consists of one atom of silicon bonded to four surrounding atoms of oxygen to form a three-dimensional network of SiO_4 tetrahedra. This network makes up the mineral quartz (Murphy and Henderson, 1983), the most common detrital mineral in most sandstones. Quartz is also a major constituent of many igneous and metamorphic rocks and is widespread as a siliceous cementing agent in various rock types.

Although quartz is common, sandstones, quartzites, and pegmatites and the unconsolidated sediments derived from them that have a silica content high enough and pure enough to meet today's market demands for quality and consistency are not common.

USES AND SPECIFICATIONS

Silica sand that is mined and processed for industrial uses must conform to the chemical and physical specifications set by customers. In the United States almost half of the silica sand produced is used in the manufacture of glass. Other important products include foundry sand, ground silica, blasting sand, and fracturing sand.

Glass Sand

Silica is the principal glass-forming oxide in a glass batch. Glass manufacturers develop model specifications for each source of silica sand used. These specifications broadly define the limits and ranges for chemical and physical properties of the sand and are used by the manufacturer in calculating the desired batch mix or formula. Some specifications may be critical to a glassmaker and require very stringent limits on the quantity of impurities in the sand. For example, the total iron oxide content of a batch is extremely crucial when making white or flint glasses (Mills, 1983). Iron is present in almost every raw material used in a glass batch and must be carefully controlled in order to obtain a consistent color in the finished product.

It is difficult, however, for a raw material supplier to tightly control the chemistry of a naturally occurring material such as silica sand. To a great extent the commercial quality of a sand is determined by its geologic history. Realizing this, glass producers tailor their model specifications to each source of approved material. In general, a glass company is concerned most about the consistency of raw materials on a day-to-day basis.

Soda-lime-silica glass was the earliest type of manmade glass (Baker-Carr, 1967) and still accounts for most of the glass manufactured for commercial use today (Mills, 1983). It is relatively easy to melt and shape and is less expensive per ton to produce than most other types of glass (Baker-Carr, 1967). Soda-lime-silica glass is used in fabricating containers, flat glass products, incandescent and fluorescent lamps, glass fiber, and many other products.

Heavy minerals such as ilmenite, leucoxene, kyanite, and zircon are impurities on which strict limits are placed for a glass batch. Because of their refractory nature they either do not melt or only partially melt, which results in stones or feathers in finished glass. Aluminosilicates such as kyanite also contribute unwanted alumina to the batch as they partially melt. Limits are especially rigid for refractory mineral grains coarser than 0.60 to 0.425 mm (30 to 40 mesh).

Tables 1 and 2 present typical specifications for silica sand used in flat glass and container glass products. The percentages shown represent an average of many companies' specifications.

Foundry Sand

Foundry molds and cores are the second largest industrial use of silica sand in terms of tons consumed. Foundry sand must generally exceed 98% SiO_2 in purity, and limits are placed on the amounts of CaO and MgO present. The greater the amount of alkaline earth oxides a sand contains, especially CaO, the more synthetic binder is required in making a mold or a core. The amount of binder needed is called the acid demand value (ADV) of a sand.

Although specialty sands such as chromite, olivine, fused silica, and zircon are being increasingly used in foundry applications, silica still represents the largest tonnage consumed (Hoyt, 1987). Silica sand's primary advantage is its abundance, which translates into a lower delivered cost per ton compared to specialty sands. Its performance is also well understood by the foundryman.

In the past many foundries used naturally bonded silica sands for molding. The sand was held together to form a specific shape by its natural clay content, and little or no additional binder was needed. Most foundries today use synthetic sand, silica sand that is free of clay and other deleterious material and has a very high SiO_2 content.

The properties that are important for a commercial foundry sand include its grain shape and size distribution and its base permeability, sintering point, clay content, and mineralogical composition (Dietert, 1966).

Silica sand is used at foundries in the manufacture of molds and cores for casting metals such as steel, ductile iron, gray iron, and aluminum- and copper-based alloys (Wilborg and Henderson,

Table 1. Glass Sand Specifications for Flat Glass

Chemical Specifications	
SiO_2	99.5% min.
Fe_2O_3	0.04% max.
Al_2O_3	0.30% max.
TiO_2	0.1% max.
Cr_2O_3	2 ppm max.
Co_3O_4	2 ppm max.
MnO_2	0.002 ppm max.
H_2O	0.05% max.
Physical Specifications	
Sizes (mm)	Cum. Retained %
1.18	0.0
0.85	0.01 max.
0.425	0.10 max.
0.106	92.0 min.
0.075	99.5 min.

Table 2. Glass Sand Specifications for Flint Container Glass

Chemical Specifications	
SiO_2	98.5% min.
Fe_2O_3	0.035% max.
Al_2O_3	0.5% max.
CaO + MgO	0.2% max.
TiO_2	0.03% max.
ZrO_2	0.01% max.
H_2O	0.1% max.
Cr_2O_3	0.001% max.
Physical Specifications	
Sizes (mm)	Cum. Retained %
1.18	0.0
0.850	0.0
0.600	4 max.
0.425	25 max.
0.106	95 min.

1983). In molds, fine- to medium-grained sand is generally preferred, although in molds for large castings relatively coarse-grained sand may be needed. Core sands are used for filling voids within a mold and are generally medium to coarse grained (Whitlatch, 1939).

The use of blends of silica sand and specialty sands such as zircon is increasing. Such blends not only cost less than specialty sands alone but can also provide a superior mold and, therefore, a superior casting.

The effect of grain size on the permeability of sand is critical to a foundryman. After ramming, the permeability of a sand must remain high enough to allow gases, which are generated when the molten metal is poured, to escape. Even a small amount of fine, −0.106 mm material in a poorly graded sand will greatly reduce its permeability compared to that of an open sand (Dietert, 1966).

The American Foundrymen's Society (AFS) grain fineness number is used in describing the results of a sieve analysis and is also a means of calculating the mean diameter of sand particles. This calculation can be misleading, however, because two sands may have the same AFS grain fineness number but very different grain size distribution patterns. To further characterize the grain size profile of a sand the number of adjacent screens having 10% or more of the material retained on them can be noted (Hoyt, 1987). The AFS grain fineness number is approximately the number of openings per inch corresponding to a sieve that would just pass a sand sample if its grains were of uniform size, i.e., the average size of grains in the sample. It is approximately proportional to the surface area per unit weight of sand exclusive of any clay binder or matrix (Anon., 1978).

Table 3. Glass Sand Specifications for Fiberglass

Chemical Specifications	
SiO_2	99% min.
Al_2O_3	0.30% max.
Fe_2O_3	0.50% max.
Na_2O	0.10% max.
K_2O	0.10% max.
LOI + H_2O	0.50% max.
Physical Specifications	
Sizes (mm)	Cum. Retained %
0.250	0.01 max.
0.075	0.6 max.
0.045	3.0 max.

In addition to their size, the shape of sand grains also affects the porosity of a mold. High porosity promotes high green strength and allows the use of less binder. The more angular the sand grains, the lower the porosity and green strength of a mold and the more binder required to form it. Angular grains also resist compaction because of their interlocking nature, so they produce mold densities 8 to 10% lower than round grains.

Ground Silica

Ground silica is used as a functional filler and pigment extender in many industrial products including paint, plastics, sealants and adhesives, and rubber. It is also the major batch component in the production of continuous strand fiberglass.

The term *silica flour* connotes lower value filler applications. The term *ground silica* denotes a higher value functional filler that modifies the end product in which it is used.

Finely ground crystalline silica, both chemically treated and uncoated, is used as a functional filler to modify the physical properties of a final product in ways not readily obtained with other mineral fillers. In paint it is used as a flatting agent and dispersant. Ground silica is also used to improve toughness, increase scrubbability, improve weatherability and prevent chalking, and as a pigment extender in paint. In plastics it enhances flexural strength and provides dielectric properties. In other applications it is used to extend higher priced pigments, such as titanium dioxide, and as an inert filler (Guillet and Kriens, 1984).

Tables 3 and 4 present general chemical and physical specifications for glass fiber and ground silica products.

A number of additional specifications can apply to the use of ground silica in various applications. Acceptable values vary depending on the source of the silica and the specific product in which it is to be used.

Reflectance: This is the percentage of light reflected from blue, amber, and green light sources of known wavelengths. Most commercial ground silica reflects well in excess of 80% of the light from each such source.

Brightness: This is the percentage of light reflected by a pressed sample of dry, sized material. The wavelength of light used is 457 nm, and values range from 1 to 100%. To be commercial a ground silica's brightness should be +89%.

Yellowness: Yellowness is usually tested by a lab along with reflectance and brightness. It is an indication of the amount of iron oxide present in a sample. A value of less than 2 is most desirable.

Hegman Grind: The paint industry in particular uses this test to define the largest particle in a dispersion. Values are usually

Table 4. Specifications for Ground Silica

Chemical Specifications	
Fe_2O_3	0.10% max.
Al_2O_3	0.38% max.
Na_2O	0.10% max.
K_2O	0.10% max.
Physical Specifications	
Sizes (mm)	Cum. Retained %
0.250	0
0.075	1 max.
0.045	3 max.

reported between 0 (100 μm) and 8 (0 μm). Values of 3 to 4 are required for flat paints, whereas gloss and semigloss paints require a Hegman number of 7 (Phillips, 1989).

Oil Absorption: This is a measure of the amount of raw linseed oil absorbed on a pigment's surface. Test values indicate the kilograms of oil required to wet 45.4 kg of pigment. Each application has its own specifications.

Fusion Point: The fusion point of silica ranges from 1 605 to 1 980°C depending on the purity of the sand.

Filtration Sand

Specifications for silica sand used as a water filtration medium in the United States are governed largely by the American Water Works Association. The sand must be relatively pure and free of clay, dust, and micaceous or organic matter. There are no strict specifications defining grain shape except that grains should not be elongate or flat. Angular and round grains appear to work equally well. Filtration sand must be uniform in size, however, with a narrow grain size distribution. An excess of fine material reduces the efficiency of a filter bed by reducing the rate of water flow and increasing the amount of friction within the bed. Therefore, the effective size of sand grains within a filter bed is determined by fine particles rather than coarser material. Filtration sand must meet two specifications that are used to control its average grain size and uniformity. These specifications are termed effective size and uniformity coefficient.

A sand's effective size is the size of a sieve opening that will just pass 10% by weight of a representative sample. For example, if the size distribution of a sand is such that 10% by weight is finer than 0.45 mm, the sand has an effective size of 0.45 mm (Anon., 1989). The uniformity coefficient of a filter sand is a ratio calculated by dividing the size of a sieve opening that will just pass 60% of a representative sample of the sand by the size of a sieve opening that will just pass 10% (the effective size) of the same sample (Anon., 1989).

Table 5 presents typical specifications for a commercial filtration sand product.

Blasting Sand

Silica sand is used as an abrasive to clean metal, concrete, brick, and other surfaces. It can also be used as a grinding or polishing medium and in stone sawing.

Specifications focus on a desired angular shape and particle size grading. An abrasive sand must also be free of impurities such as mica, organics, carbonates, and clay. Finally, it should be both hard and durable. In the grading of sand a smaller grit number (e.g., G10) represents coarse material, while a larger number (e.g., G120) denotes finer sand (Shufflebarger, Jr., 1982). Coarse blasting sand is used on large steel or iron structures, while finer sand is used in cleaning brass, aluminum, and even wood.

Silica sand for use in blasting applications has always competed with ground slag, alumina grit, and garnet. However, the health risk associated with breathing fine crystalline silica dust has made these alternative products much more attractive.

Table 5. Typical Grain Size and Chemical Composition of Filtration Sand

Sizes (mm)	Uniformity Coefficient
0.40 to 0.50	1.60 max.
0.50 to 0.60	1.60 max.
0.60 to 0.70	1.60 max.
0.70 to 0.80	1.60 max.
0.80 to 1.00	1.65 max.
1.00 to 1.50	1.70 max.
Chemical Analysis	
SiO_2	99.39
Fe_2O_3	0.24
Al_2O_3	0.19
TiO_2	0.12
CaO	0.01
MgO	0.004
LOI	0.046

Acid solubility (1:1 HCL) 0.08 to 0.11%. Specific gravity 2.64 to 2.66

Proppant Sand

Proppant or hydraulic fracturing sand is used to prop open fissures and voids in bedrock in order to increase the flow of gas, oil, and other fluids toward a well. The sand should consist predominantly of durable, well-rounded silica grains with only minor amounts of impurities such as clay, feldspar, and calcite. The most common screen sizes of proppant sand are as follows:

3.35 × 1.70 mm
2.36 × 1.18 mm
1.70 × 0.850 mm
0.425 × 0.212 mm
0.212 × 0.106 mm.

For each sand product 90% of the material should be within the size range specified. Within each major size classification is a series of sieves on which specified percentages of material must be retained. For example, less than 1% of the total sample should be retained on the first sieve or passed by the last sieve in the series (Hoagberg and Koerner-Moore, 1981).

GEOLOGY AND DISTRIBUTION OF MAJOR DEPOSITS IN NORTH AMERICA

The silica sand and sandstone deposits that are being mined commercially in the United States vary greatly in terms of both physical and chemical characteristics. The highest quality industrial sands are those produced from material that has been upgraded by several cycles of erosion and deposition. They include the eastern and mid-continent marine quartz arenites containing less than 10% matrix material and less than 10% rock particles or feldspar (Pettijohn et al., 1987).

Commercial sands produced in the southwestern United States have not gone through cycles of beneficiation and are therefore generally of lower quality. The primary sources of production are

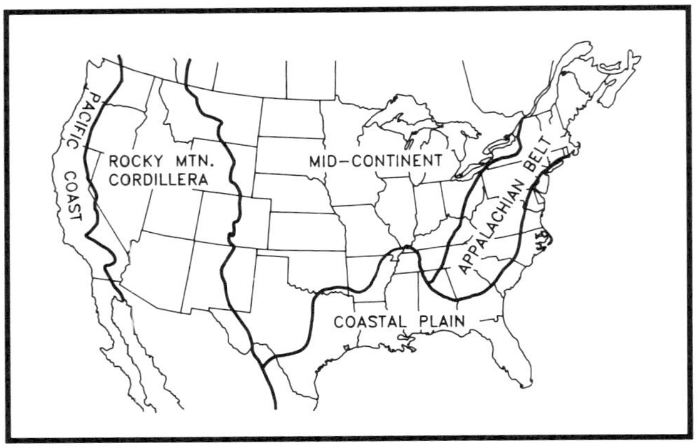

Fig. 1. General geologic provinces for silica sand and sandstone resources in the United States.

marine, deltaic, and near-shore nonmarine feldspathic, clayey sand formations in southern and central California. Because of the lower quality of these feldspathic sands, higher grade material must be shipped in from Nevada or even the mid-continent area. This material is needed to blend with and upgrade feldspathic sand for glass production on the West Coast (Harben and Bates, 1990). In addition, none of the commercial sands in the western United States are pure enough or bright enough to produce ground silica suitable for use as a functional filler. Therefore, this product must also be shipped in from the mid-continent area.

As shown in Fig. 1, silica sand deposits in the United States can be grouped into five geologic provinces (Heinrich, 1981). Each province has a distinctive geological history that has influenced the sand's grain shape and composition, grain size and distribution, surface area, and accessory mineral content (Pettijohn et al., 1987). These parameters largely determine the commercial value of a silica sand as well as the markets for which it is best suited. However, since silica sand is a low cost, high volume commodity that cannot be economically transported great distances (with some special exceptions as mentioned previously), consumers may use lower quality sand located close to their plants. Local sand may be less than ideal for use in a manufacturing process (e.g., foundry casting) or end product (e.g., glass), but it can save a manufacturer considerable expense for transportation as well as for the material itself. These savings must be balanced against any increase in process costs that result from using lower quality sand.

Each of the five geologic provinces shown in Fig. 1 is briefly discussed below along with examples of high-quality silica sand produced from each area. Figs. 2, 3, and their legend depict the distribution of known high-quality silica sand resources in the United States.

Coastal Plain

Along the Atlantic Coastal Plain unconsolidated, nearly flat-lying sediments of Late Cretaceous to Holocene age are mined for nearly all industrial uses. Four important areas are southern New Jersey, the Sand Hills District of the Carolinas, central Florida, and West Tennessee.

Cohansey Formation: Tertiary sediments that make up the New Jersey Coastal Plain are thought to have been deposited in an environment that ranged from continental to mid-shelf or outer shelf marine. The sources of these sediments were the Piedmont Lowland, Valley and Ridge, and New Jersey Highlands provinces, as well as Cretaceous and Tertiary beach sands that were being eroded and reworked. These sediments were carried and deposited on a submerged coastal shelf by southeastward flowing streams and were then reworked and upgraded by marine action (Wolfe, 1977).

In the coastal plain of southern New Jersey silica sand of the Miocene Cohansey Formation is mined extensively for use in the manufacture of glass containers, specialty glass, borosilicate glass, ceramic sanitaryware, and electrical porcelain insulators. It is also used as construction, foundry, filtration, and golf trap sand. The Cohansey Formation consists of a 30 to 75 m thick section of poorly sorted, subangular, fine- to medium-grained sand containing minor amounts of clay and coarse sand. Heavy minerals in the Cohansey include rutile, leucoxene, and zircon, which have been recovered commercially by dredge mining near Lakehurst, NJ.

The source of the Cohansey may have been the underlying Kirkwood Formation, also Miocene in age. It is a well-sorted, subangular to well-rounded quartz arenite (Pettijohn et al., 1987). The Kirkwood was deposited in Middle Miocene time as the sea transgressed over southern New Jersey. After its deposition there was evidently a brief period of emergence when the Kirkwood was exposed and eroded. When the sea covered this area again the sand, gravel, and clay of the Cohansey were deposited disconformably over the Kirkwood in the shoreward facies of a marine environment (Wolfe, 1977).

Pinehurst Formation: The Pinehurst Formation makes up the Sand Hills District of the Carolinas and is an important source of very high quality eolian quartz sands. The age of the Pinehurst is probably Late Miocene to Early Pliocene based on its stratigraphic position. A windblown origin has been postulated for the Pinehurst Formation based on the presence of well-defined dune topography, the inverse grading in some crossbeds, and the lack of clay, silt, or fossils in the sand (Nystrom et al., 1991).

Where mined in central South Carolina and south-central North Carolina the Pinehurst ranges from a thin veneer to more than 10 m in thickness. The sand varies from nearly white to yellow-orange and consists of about 90% quartz. The angular to subrounded sand is generally poorly sorted and medium to very coarse in grain size.

Florida: Glass-grade sand is produced near Davenport, Plant City, and Edgar, FL, from marine terraces of Pleistocene age (Olson, 1970) and from the closely associated Citronelle Formation of Pliocene/Pleistocene age. Four marine shorelines have been recognized in the Gulf Coast terraces representing the inland limits of transgressive seas. These marine terraces consist of fine- to medium-grained, well-sorted, unconsolidated quartz sand. They usually contain trace amounts of heavy minerals that commonly include apatite, but there is essentially no clay present in the deposits. The quartz grains, generally clear to milky, are often stained with iron oxides or organic material (Cathcart, 1963).

The Citronelle Formation is a fluvial blanket deposit of highly crossbedded sands, clays, and gravel extending across western Florida, Alabama, and Mississippi. It attains a maximum thickness of about 50 m in western Alabama at elevations above 130 m. At lower elevations the Citronelle lies beneath Pleistocene terraces, but it forms a terrace surface itself at elevations higher than the highest estimated Pleistocene sea level. The character and distribution of the Citronelle do not suggest a definite age for the formation other than Pliocene/Pleistocene. It is believed that the Citronelle sediments were deposited as coalescing deltas by several rivers that emptied into the Gulf of Mexico (MacNeil, 1950).

McNairy Sand: The McNairy Sand is mined for high quality silica sand in western Tennessee. The McNairy is part of the Gulf Coastal Plain-Upper Mississippi Embayment. It is considered to have been deposited in a cyclic near-shore geologic environment. The outcrop belt generally parallels the Tennessee River and varies from about 19 km wide in McNairy County to about 13 km wide in Benton County (Russell and Parks, 1975).

The McNairy is best described as: 1) a regressive, near-shore,

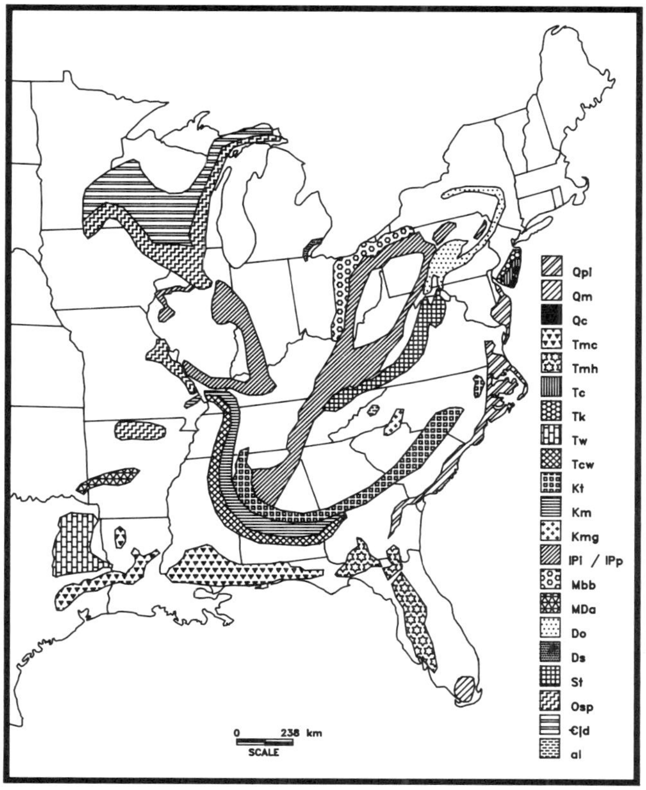

Fig. 2. Eastern United States, major silica sand resources by geologic formation.

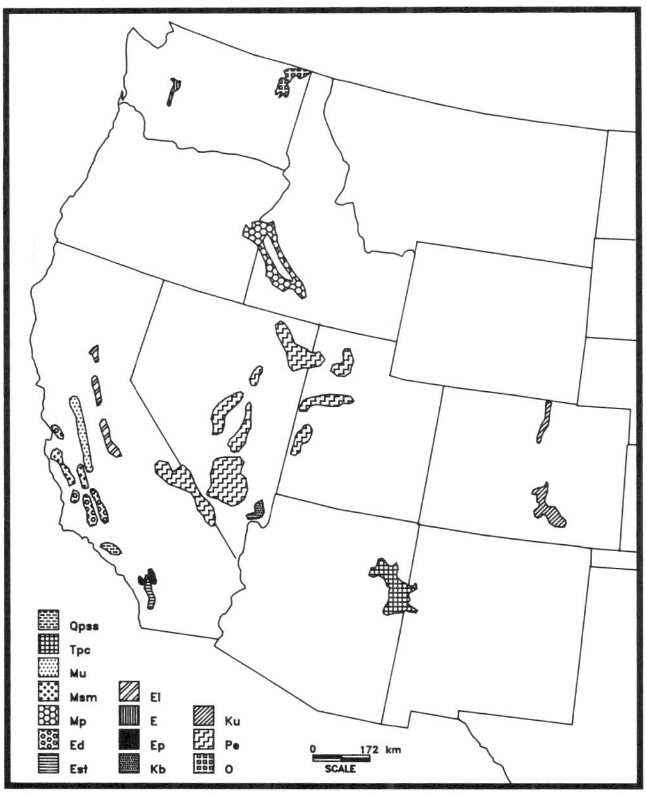

Fig. 3. Western United States, major silica sand resources by geologic formation.

basal sand that is very fine grained and contains heavy minerals; 2) a middle wedge of coarser grained sand that appears to be fluvial, deltaic, and partly estuarine in origin; and 3) an upper transgressive marine sand, which is present only in southwestern McNairy County.

Appalachian Belt

The high quality quartz sandstones mined within the Appalachian Belt resulted from the deposition of thick wedges of nonmarine clastics during the Taconic Orogeny in Late Ordovician time. As erosional unloading occurred following large scale tectonic events, terrigenous material shed from the fold-thrust mountain belt was deposited westward into an adjacent foreland basin. Following the Orogeny, the Silurian and Early Devonian were periods of relative quiescence. With erosion and subsidence slowed, the siliclastic sediments that accumulated in the foreland basin were reworked into pure quartz sandstones (Smosna, 1992). Two of these sandstone formations, the Silurian Clinch and the Lower Devonian Oriskany, are important sources of high quality industrial silica sand.

LEGEND

EASTERN UNITED STATES MAJOR SILICA SAND RESOURCES

Symbol	Age	Formation
Qpl	Quaternary	Undivided Recent Sands
Qm	Quaternary	Miami Fm.
Qc	Quaternary	Cape May Fm.
Tmc	Tertiary	Citronelle Fm.
Tmh	Tertiary	Hawthorne Fm.
Tc	Tertiary	Cohansey Fm.
Tk	Tertiary	Kirkwood Fm.
Tw	Tertiary	Wilcox Fm.
Tcw	Tertiary	Claiborne Fm.
Kt	Cretaceous	Tuscaloosa Fm.
Km	Cretaceous	McNairy Fm.
Kmg	Cretaceous	Kings Mountain Group
IPl / IPp	Pennsylvanian	Pottsville Group Includes: Lee & Lower Breathitt – KY Sewanee & Bonair – TN Lookout – GA Homewood and Connoquenessing – WV Connoquenessing, Massillon, & Sharon – OH Sharon & Olean – PA
Mbb	Mississippian	Black Hand & Berea Fms.
MDa	Mississippian	Arkansas Novaculite
Do	Devonian	Ridgeley Fm.
Ds	Devonian	Sylvania Fm.
St	Silurian	Tuscarora Fm. (Clinch in S.W. VA & E. TN)
Osp	Ordovician	St. Peter Fm.
€ld	Cambrian	Jordan / Dresbach Fm.
al	alaskite	alaskite

WESTERN UNITED STATES MAJOR SILICA SAND RESOURCES

Symbol	Age	Formation
Qpss	Plio–Pleistocene	San Pedro / Saugus Fm.
Tpc	Pliocene	Bidahochi Fm.
Mu	Upper Miocene	Undivided
Msm	Miocene	Santa Margarita Fm.
Mp	Miocene	Payette Fm.
Ed	Eocene	Domengene Fm.
Est	Eocene	Santiago Fm.
El	Eocene	Ione Fm.
E	Eocene	Puget Group
Ep	Paleocene	Silverado Fm.
Kb	Cretaceous	Baseline Fm.
Ku	Cretaceous	Dakota Sandstone
Pe	Permian	Eureka Quartzite
O	Ordovician	Addy Quartzite

In the northern Valley and Ridge area of Tennessee the Clinch Sandstone, an equivalent of the Tuscarora Sandstone to the north, is mined at one location for glass-grade sand. Its history of deposition includes alluvial fan, coastal plain, and tidal river environments as well as near-shore shelf sand wave complexes (De Witt, Jr. and Milici, 1989). The Oriskany, widely recognized throughout the Appalachians as a fossiliferous, predominantly marine, calcite-cemented sandstone, is probably reworked Tuscarora.

Oriskany Group: The Lower Devonian Oriskany, or more precisely the Ridgeley Member of the Old Port Formation, has been a major source of glass sand and other high quality industrial silica sands for many years (Berkheiser, 1985). Because the term Oriskany is more widely recognized it is commonly applied to the commercial unit.

The commercial sandstone unit of the Oriskany is an extremely hard, white to bluish gray, clean marine quartz arenite that extends from New York into Virginia. Current economic occurrences, however, are limited to a folded belt of the Central Appalachians in the Virginia, West Virginia, and Pennsylvania tri-state area. Here the Oriskany is well sorted and averages about 90 m in thickness. Fracturing and jointing have allowed ground water to leach out the calcareous matrix, resulting in a more friable sandstone or even loose sand (Harben and Bates, 1990).

About 95% of the silica grains in the Oriskany are 28 × 65 mesh, or −0.60 + 0.212 mm in size (Berkheiser, 1985). The grain size becomes finer to the west, and to the east the sandstone consists of rounded and cemented coarser grains and pebbles (Patchen, 1968). Minor amounts of altered feldspar, hornblende, and tourmaline are also present (Harben and Bates, 1990).

South of New York the most likely source of Oriskany sand was the Silurian and Cambrian quartzites in southeastern Pennsylvania (Berkheiser, 1985). The Oriskany is, therefore, at least a second generation orthoquartzite. The predominance of quartz and the presence of various marine fossils and conglomerate lenses indicate that it was deposited under tectonically stable, near-shore, shallow water conditions (Bates, 1960). The Oriskany is mined primarily for glass, foundry, and construction grades of sand and for the production of ground silica for use as a functional white filler.

Clinch Formation: The Clinch or Tuscarora quartz arenite has been sporadically mined in Virginia and Tennessee for glass and foundry sand. It is currently mined near Rogersville, TN, for the production of glass, for use on golf courses, and for construction sand. The shape of the sand grains varies from subangular to subrounded. The sandstone is moderately to firmly cemented by silica, although in some areas ground water has almost totally removed this matrix, leaving an extremely soft and friable sand.

Mid-Continent

In the mid-continent area (Fig. 1) of the United States some of the highest quality industrial silica sands in the world are produced. These sands occur as mature to supermature, multicycle sheet sands that are Early Paleozoic in age, flat lying, and moderately consolidated. Their quality has been enhanced by repeated cycles of erosion, winnowing, and redeposition. They include the St. Peter, Sylvania, Jordan, Oil Creek, and McLish Formations.

St. Peter Sandstone: The Ordovician St. Peter Sandstone is a remarkably uniform, widespread, quartz arenite sheet sand that is used in virtually all industrial applications. It is exposed in southeastern Minnesota and crops out across the southern margin of the Wisconsin Dome into eastern Iowa. The St. Peter continues in outcrop southward into Illinois and across the flanks of the Ozark Dome in eastern Missouri and northern Arkansas (Bates, 1960). It is also present in the subsurface to the west in South Dakota, where it joins with basal clastics of the Winnipeg Formation (Bunker et al., 1989).

A marine origin is widely accepted for the St. Peter Sandstone. It is a multicycled sand whose source is believed to have been Precambrian quartzites of the Canadian Shield as well as Cambrian and Early Ordovician sandstones (Lamar, 1927). Regional studies have shown that the St. Peter is a transgressive sheet sand of reworked beach deposits formed along the edge of a northwestward advancing sea (Bates, 1960). The St. Peter is best known for its homogeneous, round, and inclusion-free silica sand grains. The sand is well sorted with almost 80% of the grains sized to 40 × 140 mesh, or −0.425 + 0.106 mm (Greenburg, 1960). In the Ottawa District quarry-run rock averages more than 97% SiO_2 (Harben and Bates, 1990). It is white to buff and contains only a minor amount of clay matrix.

The Ottawa District in northern Illinois is the best known area for production from the St. Peter Sandstone. The formation was deposited on an erosional surface of the underlying Shakopee Dolomite and attains a maximum thickness of 130 m. Post-depositional erosion, however, has reduced the sandstone to about 70 m in thickness within the district. Most producers restrict their mining to the upper half of the formation, because its iron, alumina, and carbonate content generally increases below 30 m in depth.

The St. Peter's exceptionally high silica content and uniformity have made it the most successfully exploited silica sand resource in North America. It is not, however, the most expensive silica sand on the market. In fact, despite its unique purity and grain shape and homogeneous nature, St. Peter sandstone from the Ottawa District is one of the lowest priced glass sands in North America. This is primarily because of the formation's uniformity and great lateral extent in outcrop, which allows more than a dozen companies to mine, process, and sell nearly identical, high quality silica sands within the same market area. More than six mines are located within a 60 km radius of each other in the district. The availability of such a large volume of product for a highly competitive market is the reason prices for this high quality silica sand are unusually low. This is also why sand from the Ottawa District can be economically transported to the West Coast, bypassing other, higher cost producers as well as numerous good, undeveloped silica sand deposits along the way.

Sylvania Formation: The Sylvania Formation in southeastern Michigan is a basal blanket quartz arenite of Middle Devonian age. It is the lowermost unit of the Detroit River Group. It has been influenced structurally by the Michigan Basin, which dips gently to the northwest, and by the Findlay Arch in Ohio, where the Sylvania rests on the northwestern flank (Bates, 1960). The Sylvania is one of the premier glass sand resources in North America. It is a littoral eolian sand that was deposited in a near-shore environment. Its source was outcroppings of the St. Peter sandstone to the northwest (Heinrich, 1977). The Sylvania represents one more cycle of reworking, making it a third or fourth generation, supermature sandstone.

The Sylvania is extremely pure, friable, and highly quartzose. The major contaminant is dolomite with minor carbonaceous matter, which weakly cement the silica grains. Overall, the sand grains are well sorted and rounded, but the grain size frequency distribution varies both horizontally and vertically. Although the larger grains are well rounded, the more common smaller grains are quite angular as a result of secondary silica overgrowths. These secondary overgrowths serve as a binder, causing the sandstone to be denser than expected and to be resistant to weathering (Heinrich, 1977).

Although the Sylvania is considered by many to be the best glass-grade silica sand in North America, its economic significance is limited because it is one of the least exposed high quality silica sand units. It crops out only in a narrow, northeast-trending belt across the southeastern corner of Michigan and to a lesser extent in northern Ohio. Products from the Sylvania have the lowest iron content of any commercial sands sold in North America. For this reason silica from the Sylvania is used in many specialty applications and is shipped well outside a normal marketing radius.

Jordan Sandstone: During Late Cambrian time most of the central mid-continent area was covered by an epicontinental sea characterized by carbonate deposition. The regression of this sea was followed by the prograding of sandstone units eroding from the Wisconsin Dome across much of Minnesota, Iowa, and northeastern Missouri. These Upper Cambrian sandstone units included the Lower Jordan Sandstone in Minnesota. During Early Ordovician time sandy carbonates and sandstones were deposited across most of the central mid-continent. These units included the Upper Jordan Sandstone (Bunker et al., 1989).

In Minnesota the Lower Jordan Sandstone is mined primarily for hydrofracturing or proppant sand. It is a white to yellow, well-rounded and well-sorted, medium- to coarse-grained sandstone. The Jordan actually consists of three members (Austin, 1972). From the top down they are: 1) the Norwalk Member, a fine-grained, silty, yellow quartzose sandstone; 2) the Van Oser Member, a coarse- to medium-grained, white or yellow orthoquartzite that is mined for fracturing sand; and 3) the Sunset Point Member, a dolomitic and argillaceous quartz sandstone containing pebble-size clasts of dolomitic sandstone, which suggest a near-shore depositional environment (Ojakangas and Matsch, 1982).

Oil Creek and McLish Formations: In southwestern Oklahoma the Ordovician Simpson Group consists of 300 to 700 m of sandstone and limestone with interbedded green shales (Buttram, 1913). The Oil Creek and McLish Formations within this Group have been considered lithologically equivalent to and of the same age as the St. Peter (Ham, 1945). However, during early Middle Ordovician time, when deposition was taking place in the southern mid-continent (e.g., the Oil Creek Formation), erosion was the dominant process in the remainder of the central mid-continent. It was not until later, when Middle Ordovician seas onlapped the central mid-continent, that the St. Peter Sandstone was deposited (Bunker et al., 1989).

Both the Oil Creek and the McLish yield high quality glass sand from beds that were folded into anticlines and synclines during the Early Pennsylvanian period. Subsequent faulting, folding, and overthrusting followed by erosion have produced a complex structural picture. Although potential reserves are enormous, economic deposits are restricted to locations accessible to rail transportation. The sand is generally poorly consolidated and contains minor amounts of clay. The grains are well rounded, frosted, and considerably finer than those of the St. Peter in Illinois (Ham, 1945).

Rocky Mountain Cordillera

Immense deposits of Ordovician orthoquartzites extend from British Columbia to southern California. The beds are strongly indurated and dip at varying attitudes. These orthoquartzites have not been extensively developed commercially because of their distance from large industrial markets along the Pacific Coast.

Ketner (1968) proposes that the silica source for the Cordilleran quartzites was Cambrian sandstones on the Peace River-Athabaska Arch in Northern Alberta. The sand was deposited in a geosyncline as far away as southern California. Ketner cites five lines of evidence supporting this hypothesis: 1) a lack of adequate sources of sand except in northern Alberta, 2) the progressively younger age of the basal quartzite from north to south, 3) an increasing width and decreasing thickness of the deposits toward the south, 4) a fining of the grain size toward the south, and 5) improved sorting toward the south.

Currently only one of these orthoquartzites, the Addy Quartzite, is being mined commercially in the United States. The operation is located in northeastern Washington and supplies markets in northern Oregon, Washington, and western Canada.

Addy Quartzite: A structurally-controlled, friable portion of the Cambrian-Ordovician Addy Quartzite is mined for glass sand in Stevens County, WA. The rock is crushed, attrition scrubbed, floated, dried, screened, and subjected to magnetic separation to produce several grades of high quality silica sand (Joseph, 1988).

The Addy Quartzite varies from unconsolidated sand to true quartzite. It appears to be a tabular mass 60 to 90 m thick and 5 km in length along strike. Where mined the material is leached to a highly friable condition. It is buff to white in color (Reed, 1974). Individual grains range from angular to subangular and from clear to milky white with some inclusions. Most of the sand produced is sold to container glass manufacturers in western Washington, Oregon, and British Columbia (Reed, 1974).

Swan Peak and Eureka Quartzites: The Swan Peak and Eureka Quartzites of Utah and Nevada have been mined in the past and are promising future sources of industrial sand within the Cordilleran region. Although a number of high grade prospects and short term operations have been reported, none of the deposits have been friable enough or located close enough to transportation facilities or markets to sustain a long term commercial operation.

Pacific Coast

Immature, feldspathic, clayey sands of Paleocene to Pliocene/Pleistocene age are mined in California for the glass industry. They include the Ione, Saugus, Domengine, and Silverado Formations. Most flat glass and container glass manufacturers upgrade these feldspathic sands by blending them with more expensive, higher quality sand mined from the Baseline Formation in Nevada or with sand shipped in by rail from the mid-continent area. No sands on the Pacific Coast are high enough in quality or brightness to produce ground silica.

Ione Formation: In central and southern California, along the lower foothills of the Sierra Nevada Mountains, the Eocene Ione Formation crops out in a belt of discontinuous, trough-like depressions from Butte County south to Madera County. Near the town of Ione in Amador County moderately pure sand is mined for use in both flat and container glass. Ceramic clays and lignite are also recovered as byproducts from the same sediment. The lignite is sold as a low grade fuel and for use in manufacturing montan wax.

The source of the Ione sediments was the deeply weathered granitic and metamorphic rocks of the Sierra Nevada Mountains, which were uplifted beginning in Late Jurassic time during the Nevadan Orogeny. By Eocene time the plutonic core of the Sierra Nevadas was exposed and eroded to the level of low mountains (Norris and Webb, 1990). The crystalline rocks were dissected by freshwater streams, and the resulting sediments were deposited in lagoons and shallow basins along a broad, swampy coastal plain bordering the Eocene sea to the west. The sands and clays were then upgraded by chemical leaching in what evidently was a near-tropical environment. Later, Miocene and Pliocene volcanism buried the Ione sediments under rhyolitic ash and andesite mudflows (Anon., 1956). Finally, erosion accompanying renewed uplift of the Sierra Nevadas re-exposed the Ione Formation.

The Ione is typically a light-colored, clay-bearing sandstone comprised of white sand, pallid kaolin, and black lignite. The lower portion of the formation is a major source of glass-quality silica sand and ceramic- and filler-grade kaolinitic clay. Where mined the Ione consists of interbedded lenses of clayey sand and sandy clay. Ore-grade material usually consists of about 60% quartz sand, 35% clay, and minor heavy minerals. The glass sand produced from the Ione is white to buff, quite angular, fine grained, and relatively well sorted. Impurities may include shale chips, anauxite, feldspar, and heavy minerals.

Saugus Formation: The Pliocene-Pleistocene Saugus Formation is mined in Ventura County, southern California, for construction and container glass grades of feldspathic silica sand. It is a locally conglomeratic, light-colored, feldspathic sandstone (Sharp, 1978). The mining area is part of the east-west trending Transverse Ranges, which extend from the coastline of Santa Barbara County east to the mountains of the Joshua Tree National Monument at the edge of the Mojave Desert. This east-west trend is an exception to the generally north-south structural trends in California and is not fully understood.

Within the Transverse Ranges a 190 km long synclinal structure includes the Ventura Basin to the west and the Soledad Basin to the east. The Soledad Basin is a nonmarine depression that was cut off from the sea until subsidence during Miocene time allowed the sea to fill the western part. The Saugus Formation represents the resumption of freshwater deposition as this sea regressed at the end of the Pliocene epoch (Norris and Webb, 1990).

Domengine Formation: In Contra Costra County, central California, a feldspathic sand in the Eocene Domengine Formation is mined for container glass and construction uses. The Domengine is part of a thick sedimentary unit ranging in age from Paleocene to Pleistocene. It is widely distributed around the margins of the Diablo Range, which extend along the western side of the San Joaquin Valley. The Domengine is considered to be of shallow marine origin, although nonmarine beds east of Mount Diablo consist of container glass grades of feldspathic silica sand, ceramic grades of clay, and coal. Because the clay and quartz sand appear to have been affected by intense chemical weathering, these nonmarine beds are believed to indicate that almost tropical conditions existed at the time of deposition (Norris and Webb, 1990).

Silverado Formation: The basal member of the Paleocene Silverado Formation is currently being mined in the northern Santa Ana Mountains south of the city of Corona in Riverside County, CA. The Silverado makes up part of the Peninsular Ranges in southern California. This geologic unit, one of the longest in outcrop extent in western North America, consists primarily of a complex of Cretaceous igneous rocks situated in northwest-southeast fault blocks. Intense chemical weathering and erosion of this igneous basement (Sharp, 1978) provided the major source of sediments for Early Tertiary formations (Norris and Webb, 1990), including the Silverado. The Early Tertiary units are a veneer of marine, nonmarine, and volcanic sediments that lies unconformably on Cretaceous sediments and crystalline basement rocks and is restricted to coastal margins.

The Silverado Formation consists of about 425 m of Paleocene nonmarine sediments located in the Elsinore-Temecula trough. The lower section, which was deposited in brackish water, contains the units of glass-grade sand currently mined in Riverside County. Only one other location is known where a quartz-rich facies occurs in the Silverado. It is near Trabuco Creek in Orange County along the western slope of the Santa Ana Mountains. This area has been developed with single-family homes, however, which makes the likelihood of mining very remote. The history of clay and sand mining from the Silverado in this area goes back to the early 1900s (Norris and Webb, 1990).

Baseline Formation: The principal area in which the Cretaceous Baseline crops out is in Clark County, NV, where it is mined for the glass markets in California. The Baseline is a friable sandstone weakly cemented by clay, calcium carbonate, and some iron oxide. Commercial material is mined from the lower and central portions of the formation, where the sand is uniformly fine to medium grained and white to light yellow. Iron oxide is the major contaminant, and the heavy mineral content is variable. Laminated clays and clay pockets are locally present (Papke, private communication, 1987).

The Baseline is limited in outcrop to an area of about 6 km^2 southwest of Overton, NV. This indicates that the source of the sand was probably the Jurassic Aztec Sandstone (Stewart, 1980). The Baseline is the major source of high quality silica sand used by southwestern US glass manufacturers for blending with the lower quality feldspathic sands produced in southern and central California.

Canada

Bar River Formation: On Badgeley Island in northern Georgian Bay, ON, the Precambrian Bar River Formation is quarried for high-purity lump silica, glass sand, and ground silica. The Bar River is the youngest and purest of a series of three quartzite units at the top of the Cobalt Group in the Huronian Supergroup of Middle to Late Precambrian age (Guillet and Kriens, 1984). It crops out on a chain of islands located in the North Channel of Georgian Bay.

The Bar River quartzite is commonly light gray to white. Where completely recrystallized by metamorphism it consists of interlocking grains of quartz, which causes it to be extremely hard and to break across grains with a conchoidal fracture (Hewitt, 1963). Where metamorphism has been less intense it is easier to see that the quartzite was originally made up of intergradational units of well-sorted and -rounded, fine- to medium-size sand grains (Guillet and Kriens, 1984). Accessory minerals include minor amounts of muscovite, zircon, hematite, monazite, and tourmaline. The nearly vertical quartzite beds strike northeast-southwest and are cut by numerous, narrow diabase dikes (Hewitt, 1963).

Potsdam Formation: Near St. Canut, PQ, the Cambrian Potsdam Formation is mined for high-purity industrial sand. The Potsdam sandstone can vary in color from white to red depending on whether the matrix is primarily ferruginous, calcareous, or siliceous. Where mined, the material is a white, massive, saccharoidal sandstone with a siliceous matrix or cement (Keith, 1949). Secondary overgrowths of silica are common and affect both the friability of the rock and the shape of the grains. Keith showed in 1949 that an original matrix of carbonate, clay, and/or iron oxides was replaced by silica. Where this replacement was incomplete, a porous, friable, and fairly pure sandstone made up of poorly cemented aggregates of individual grains resulted. When crushed, this sandstone easily breaks into angular to sub-rounded grains, more than 80% of which are between 0.60 and 0.15 mm in size (Hewitt, 1963).

In southeastern Ontario the Potsdam is the lowermost Paleozoic formation present, and it is assumed to be equivalent to the Potsdam in New York. It is not folded but shows evidence of regional metamorphism (Keith, 1949). Both deepwater and near-shore facies have been identified. The near-shore facies seems to contain the purest sandstone (Guillet and Kriens, 1984).

Mt. Wilson Formation: In the Rocky Mountains near Golden, BC, a zone of friable quartzite within the Ordovician Mt. Wilson Formation is mined for industrial sand. The rock is fine grained, white, and massive. Its quartz grains are well-rounded and average 0.25 to 1 mm in diameter. The friable zone that is mined lies along the crest of an anticline at the most northwestern exposure of the formation. Most of the sand produced is sold to glass manufacturers, but some of it is sold for a variety of other uses including specialty ceramics, filtration, and golf courses (Hancock, 1991).

Mexico

The largest producer of high quality industrial silica sand in Mexico operates three mines, two of which are located in the Gulf Coastal Plain. The other mine is situated in the Central Mesa

Province, which is equivalent to the Basin and Range Province of the western United States. The coastal plain sediments of eastern Mexico are a Late Tertiary molasse of clastic material deposited by several major river systems that emptied into the Gulf of Mexico. The source of these sediments was the eroded eastern side of the Sierra Madre Oriental fold belt, which emerged following the Early Eocene Hidalgoan Orogeny (De Cserna, 1974).

Near the town of Canoitas, Coahuila, in northeast Mexico, coastal plain sands of probable Pliocene age are mined. They are upgraded by flotation, magnetic and gravity separation, and even by acid leaching to produce glass grades of white, angular sand that is also ground for sale as a functional filler. In Veracruz near the city of Jaltipan, Miocene coastal plain sands containing lenses of kaolin are mined by dredging. They are upgraded by gravity separation and froth flotation for industrial uses. The ore averages about 45% sand and 55% clay and contains a significant amount of fine-grained heavy minerals.

The third operation is located in the Central Mesa Province at San Jose Iturbide, Guanajuanto. Industrial grades of silica sand along with potassic feldspar are recovered from a Miocene crystalline tuff unit that grades about 45% quartz and 55% potash feldspar. Clay and heavy minerals account for as much as 1% of the ore (Griffiths, 1988). The sand product is comprised of clear, angular, quartz grains and can be used in glassmaking and for ground silica applications.

TECHNOLOGY

Exploration

As is true for most industrial minerals, exploration for silica sand deposits leads almost immediately into marketing and transportation studies. Identifying the locations of other producers as well as consumers allows one to define exploration areas that are closer to markets and in which equal or perhaps superior products can be produced. Rarely, if ever, is a new, "me-too" operation commercially successful. Even if a prospect is closer to markets, success is not guaranteed. Beneficiation, transportation, and land costs, as well as local environmental opposition to a project, can have an immense negative impact. All of these factors, therefore, must be evaluated as part of the exploration program.

Areas of interest may be as large as a province or state or restricted to the immediate vicinity of a single consumer. In any case, the market to be supplied defines the quality and quantity of material that must be found. It is wasteful to spend exploration dollars and time evaluating deposits of opportunity without first fully understanding the requirements of prospective consumers.

After characterizing the markets and competition, the next step is a detailed literature research to identify potential geologic targets and sites of previous prospecting or mining. In some states, such as Wyoming, commodity studies designed to promote new industrial development are available from universities and government agencies. Most state geological agencies are well aware of local silica sand resources and can be helpful by providing information. State publications often provide chemical analyses for known sand prospects or formations within a study area. Frequently, however, these analyses were performed on field samples and may not reflect the potential quality of a sand after even simple benefication.

Most silica sand deposits in the United States are known and have been investigated many times. As market and economic conditions change, however, some of these deposits can become viable. New opportunities also arise as older deposits become less profitable because of increasing mining costs or decreasing quality of the sand. Environmental regulations may result in costly operating restrictions and can speed the demise of a floundering operation. Any of these situations can create an exploration opportunity. In addition, as overall industrial activity expands new markets appear, so exploration opportunities may develop in areas largely ignored in the past.

Other, perhaps more entrepreneurial situations are represented by holes in the market, for example, where a large consumer is supplied by producers located more than 120 km away. A silica sand source located substantially closer to its plant can save a consumer many hundreds of thousands of dollars per year by reducing transportation costs by only a few dollars per ton. A good example of this is a successful new operation in eastern Tennessee, which supplies a major silica sand consumer nearby that formerly bought sand from suppliers several hundred kilometers away. However, in many cases there are good reasons why such close-to-market deposits are not in production, the most common being the need for costly beneficaticn of the sand.

Once a prospect has been located, reconnaissance of the site is conducted. It is important to collect unoxidized samples, if possible, during the initial investigation. The minimum lab evaluation of exploration samples should include grain size analysis and chemical analysis for Fe_2O_3 and Al_2O_3 after magnetic separation, wet scrubbing, and flotation. Subsequent geologic mapping and more detailed surface sampling are relied upon to determine the number, location, and pattern of drill holes and/or test pits.

Economic silica sand and sandstone deposits are found in a wide variety of geologic environments that require different types of exploration drilling techniques. In unconsolidated deposits, such as those found in the Atlantic Coastal Plain where the water table is shallow, using a rotary drill rig with a 7.6 cm (outside diameter) split-spoon sampler is recommended. The spoon is driven approximately 80 cm into the sand for each sample interval by a 136 kg weight. A fishtail or drag bit is used for cleaning the hole prior to sampling the next interval. Bentonite-based drilling mud can be used to prevent the hole from collapsing, but if the mud contaminates the sand it will elevate the sample's Fe_2O_3 and Al_2O_3 content. Using synthetic mud such as Revert (trade name) can prevent such contamination. Care must be taken to keep the sample as clean as possible, and the lab must be instructed to thoroughly wash and scrub the sample before analyzing it.

Auger drilling also can be used in unconsolidated deposits. Hollow-stem augers recover uncontaminated samples with a split-spoon sampler that is rotated instead of hammered into the deposit. Solid-stem augers are used when quick and dirty sampling will suffice. In this type of drilling the sample tends to become contaminated by material falling in from the wall of the hole above the sample interval. It is an inexpensive method but is recommended only for obtaining general information during the reconnaissance stage.

Reverse circulation or water jet drilling is a rapid exploration technique but also provides mixed samples that can be difficult to interpret. This type of drilling also tends to wash the matrix (e.g., clay) from the sand, which can lead to overly high estimates of the sand's quality. Excavating test pits or trenches with a backhoe can be useful for shallow deposits, although good reclamation of the sample sites is necessary.

In consolidated silica resources such as sandstone and quartzite core drilling is required. A drill rig equipped with NX size (54.9 mm diameter) wireline coring capability is generally adequate. If the rock is loosely cemented, the use of bentonite-based drilling mud can improve core recovery. Larger diameter coring can also improve sample recovery in highly fractured or friable rock.

Percussion drilling, such as is used for blast holes at quarries, is comparable to using solid-stem augers in unconsolidated sands. The crushed and mixed samples recovered provide only a rough estimate of the quality and quantity of material present. Percussion drilling is not recommended for use in detailed investigations.

The cuttings or cores obtained by drilling are logged in terms of the type and amount of matrix material or cement present and the degree of iron staining on the silica grains. Opaque mineral impurities, staining and inclusions within the quartz grains, and all accessory minerals likewise should be noted. In the case of consolidated deposits it is also important to note whether the rock breaks around or across the grains. Quartzites and sandstones that are cemented by silica tend to break across grains, which results in an angular product that is less desirable in most industrial applications. Grain shape should also be noted in drilling logs.

As in the case of all minerals, accurately calculating the reserves in a silica sand deposit depends on defining its dimensions and grade. Methods for making these determinations are well documented in the literature. One key variable in reserve calculations for silica sand and sandstone is the tonnage factor or in-place bulk density. Unconsolidated sand can have a bulk density as low as 380 kg/m^3 if uncompacted, whereas indurated or consolidated sandstone and quartzite can weigh as much as 700 kg/m^3. Generally, until a more precise tonnage factor is determined, a conservative figure of 485 kg/m^3 can be assumed.

Laboratory Testing

Glass producers, whether they specialize in containers, pressed ware, blown ware, or flat glass, all establish rigid chemical and particle size distribution specifications that must be met from one shipment to the next by a supplier. One reason for this in the case of quartz sand is that it usually represents 60 to 75% by weight of a total batch mix, which can contain as many as 6 to 13% (or even more) other raw materials. Most if not all of the raw materials used in the batch contain contaminants; and there are few, if any, silica deposits that do not contain detrital particles of refractory minerals (Murphy and Henderson, 1983).

Even the smallest silica producer today must have some type of quality control laboratory to ensure that the products shipped meet consumers' specifications. The testing done in-house before a product is shipped usually includes one or more of the following: sieve analysis, X-ray diffraction and/or fluorescence (for mineral and chemical composition), heavy media separation, magnetic or electrostatic separation, and microscopic analysis. Sieve and X-ray diffraction analyses are the two most commonly used quality control procedures.

Other silica products may require a variety of specialized tests in addition to those listed above. For example, filtration media require detailed sieve analyses to determine the uniformity coefficient and effective size. Other tests may include an evaluation of organic impurities and acid solubility. Foundry sand testing includes sieve analyses, acid demand, and compressive strength tests. Quality control for blasting sand generally involves sieve analyses and determining the sand's particle shape as well as the amount and type of deleterious material present. Similarly, proppant sand generally requires particle size, compressive strength, grain shape, and deleterious material testing.

Quality control labs also are used in evaluating exploration samples. However, since most labs regularly deal with finished product samples and not raw material, their ability to wet scrub, bench float, and run magnetic separation tests on an exploration sample (or on hundreds of samples from a drilling program) may be limited.

Mining

Industrial silica sand is mined by different methods from different types of deposits. In unconsolidated sand deposits, such as those found in the Atlantic Coastal Plain where the water table is shallow, mining is accomplished in manmade lakes with floating dredges. A hydraulic dredge uses a suction pipe to excavate the sand, which is pumped through a pipeline to surge piles on shore or directly into a processing plant. In some cases, such as where a deposit contains hard clay lenses, the suction pipe is equipped with a cutter head to facilitate excavation of material from a mining bank below water. Draglines can also be used at such deposits instead of dredges, but they are usually more expensive to operate and maintain and are less efficient.

In unconsolidated or loosely consolidated deposits above the water table a front end loader, hydraulic excavator, or power shovel can extract material directly from a mining face. The material is then stockpiled as feed to a processing plant. Another option in loosely consolidated material is to use a high pressure water monitor. Material washed from the working face is collected in a sump, from which it is pumped to the processing plant.

At well-consolidated deposits open pit mining methods are preferred. The rock is drilled and blasted. It is then loaded by either front end loaders or power shovels into trucks for transport to the processing facilities.

Mining underground in well-consolidated material uses standard drill-and-blast methods. The material is loaded by power equipment and hauled to a processing plant. Underground mining is usually more expensive than surface mining and is currently employed at few operations in the United States.

Processing

Beneficiation is the key in producing high-quality glass or foundry sand from most deposits. To remove clay coatings from the sand grains a sand-water slurry is pumped to a primary desliming circuit in which hydrocyclones remove −0.01 mm material, which is pumped to settling ponds. The clean sand is then sized, either by hydrosizing in hindered settling tanks and/or on vibrating screens, to remove the coarse fraction, which usually consists of +0.6 mm particles. (Some producers use air separators instead of hydrosizers to size their ore.) The sand is then dewatered in hydrocyclones to 70% solids prior to attrition scrubbing, which is carried out in tanks equipped with propeller-type blades. Scrubbing removes the iron oxides, clay, and other materials that coat the sand grain surfaces. It also produces fresh surfaces on non-quartz minerals (e.g., heavy minerals) to enhance reagent attachment in the next step, which is flotation. Two types of flotation are used, cationic and anionic. Anionic flotation, currently the method of choice in the silica sand industry, employs petroleum sulphonate as the collector in an acid circuit to remove heavy minerals.

For most applications the finished sand product must be dried. This is normally done in rotary or fluid-bed dryers equipped with wet scrubbers for removing dust from the exhaust gases. Many glass sand producers use fluid-bed dryers because of their high reliability and thermal efficiency. After drying, the sand is usually cooled and conveyed to another set of screens for final sizing.

The following list (adapted from Shufflebarger, Jr., 1983) outlines the processing steps that can be used in producing high quality industrial silica sand. The combination of processes actually required is a function of the deposit to be mined, products to be offered, and markets to be served.

1. Drying and screening only, which is rarely sufficient
2. Washing (primarily for the removal of clay), drying, and screening
3. Washing, scrubbing, flotation, drying, and screening
4. Washing, scrubbing (high pulp density attrition), drying, screening, and sizing in air separators
5. Iron removal (magnetic, wet or dry)

6. Grinding (rod mill or ball mill), regardless of other processes used, if necessitated by the grain size distribution of plant feed
7. Acid leaching for high-purity products
8. Bagging facilities for some or all products
9. Cooling facilities for foundry sand.

ECONOMIC FACTORS

Market Trends

Market conditions and competition dictate companies' decisions to expand facilities, to acquire other operations, or to explore for new deposits and open new silica sand mines and processing plants. Logically, an in-depth market study is prepared before any exploration or investigations of a prospect are begun. Such a study will help determine the profitability and growth potential for a new operation.

For example, the supply of silica sand to consumers in the midwestern United States has always been plentiful. Long established companies produce from the Jordan and St. Peter Sandstones, dune sands in Michigan, the Sylvania Formation, and the Sharon Conglomerate. Because of an oversupply of material, price wars occur, especially in the Illinois-Indiana District. The odds are against a new supplier building a successful business in this area. In the early 1980s a large construction aggregates producer built a plant in Minnesota for the production of proppant sand which failed. A large silica sand producer nearby subsequently bought the plant but had to close it because of a lack of sales. Although the downturn in oil exploration did not help this plant (which produced mostly proppant sand), the oversupply of product on the market proved to be the key reason for the failure of the business.

Markets for container-grade glass sand are being adversely affected by strong competition from aluminum cans, plastic bottles, and cardboard cartons. The use of nonglass containers is expected to grow in the 1990s. Overcapacity in the silica sand industry and the threat of conversion to alternative packaging materials by container consumers should keep price increases for glass sand at a low rate. One segment where modest growth is expected is clear glass containers for specialty food products such as sauces, isotonic drinks, and flavored sparkling water (Anon., 1992).

The recycling laws enacted in many areas have greatly increased the supply of post-consumer cullet (broken glass). In order to achieve cost savings, especially in the form of reduced fuel consumption, container glass producers have turned to cullet as a batch raw material. Cullet has become an important ingredient of the manufacturing process, because less energy is required to melt it than the same weight of raw sand. Cullet can easily comprise 15% if not 25% or more of a glass batch, thereby also reducing the glass producer's consumption of silica sand and other raw materials. The cullet used in a container glass plant is obtained both from in-house sources and from outside recycling programs. The "bottle bills" passed in many areas require the use of returnable bottles and have created a large supply of cullet. Major beer and soft drink distributors have built recycling plants to separate cullet and clean it for shipment, usually by rail, to glass container plants. Recycling is expected to expand, further reducing the consumption of silica sand.

The annual volume of flat glass produced in the United States varies with the strength of the economy. This is due to its major role in the automotive and construction industries. With the trend to smaller, lighter weight automobiles, the glass parts used as windows and windshields are also becoming smaller and thinner. Although demand for glass used in automotive applications is not expected to show any growth, flat glass demand in construction applications is projected to increase modestly by 3.8% per year until 1995.

Growth in silica sand sales may also come from increasing demand for specialty sand products, although the continued use of these products is questionable in light of the health hazard issues surrounding crystalline silica.

Proppant sand (frac sand) consumption depends on the level of oil well drilling, which as been depressed since 1982. Foundry sand demand is also cyclical, because it depends heavily on economic conditions in the auto, heavy machinery, and other capital goods industries. Foundry sand consumption has also been reduced in markets where plastics have replaced iron and steel products.

A new supplier of silica sand can expect to have difficulty entering the market. For one thing, consumers always test a new product on a trial basis, sometimes for many months before accepting the material for use in commercial production. In addition, many consumers consider a new producer as a secondary or minor source of supply. However, serious consideration may be given by a consumer to a long term supply of a good quality material for a fixed price that is usually lower than the rates from established sand producers.

During the 1980s the silica sand industry in the United States consolidated by means of acquisitions into several large companies producing mainly glass sand. Acquisitions not only supply the buyer with reserves and production facilities, but also market share.

In addition to the two or three large producers in North America there are also several dozen viable smaller companies producing such products as foundry, filtration, and blasting sand, but only limited amounts of glass sand. Some of these smaller producers have carved out a niche in the market by capturing one or two large customers near their plants, thereby displacing one or more of the major producers.

Overall growth in the use of silica sand between 1980 and 1990 was relatively weak, as evidenced by the fact that only six new production facilities were built in the United States during that period. Under these market conditions a commitment to maintain high and consistent product quality, tight controls to reduce operating costs, and effective product marketing are essential.

Transportation

The main modes of transportation to markets are rail, truck, barge, and ship. The availability of any one of these methods of transportation near a silica sand deposit is essential for a profitable operation. Industrial grades of silica sand, unlike common construction grades, can be shipped a few hundred kilometers to customers. However, transporting silica sand is often complex. High labor and equipment costs are involved in making a large number of trucks or railcars available to customers to ensure that delivery schedules are met (Peddicord and Regis, 1981). Truck transportation is usually more reliable but more costly than rail transportation. Scheduling so as to provide customers with reliable rail service is also sometimes difficult to accomplish. Freight contracts with transport companies often must include a guaranteed minimum annual tonnage in order for rates to be favorable. The availability of railcars must be established well in advance of the date of shipment. The mode of transportation selected is also affected by regulations, competition, economic conditions, and insurance costs. Silica products are usually shipped bulk in covered trucks, covered railroad hopper or tank cars, or pneumatic tank trucks. Some products are bagged and palletized for shipment.

Production and Pricing

The silica sand industry in the United States was relatively static between 1981 and 1990. In 1981, according to the US Bureau of Mines, a total of 27.2 Mt of industrial sand and gravel valued at $332 millon FOB plant, were produced by 91 companies at 141 active operations. Most of this production came from operations that had annual output in excess of 272 kt. Forty operations (28%) accounted for 71% of the total tonnage.

In 1990 a total of 25.8 Mt of industrial sand and gravel valued at $436 million FOB plant, were produced by 83 companies at 154 active operations. Most of this production came from operations that had annual output in excess of 181 kt. Forty-seven operations (31%) accounted for 44% of the total tonnage.

These figures indicate that US production decreased by 5%, or 0.5% per year, from 1981 through 1990. However, the average price per ton, including bulk and bagged sand, increased from about $11.00 in 1981 to $15.35 in 1990. This represents an increase of nearly 40% over 10 years, or 3.4% per year.

Silica sand prices vary according to geographic region, the amount of product sold to a consumer, general economic conditions, product purity, and the terms of sales contracts. Producers are reluctant to quote prices to nonconsumers, but the price ranges shown below were representative of the last quarter of 1991.

	Price (dollars) Per Ton FOB Plant	
Region	Regular Melting Sand	Foundry Sand
East	14.00–16.00	15.00–16.00
Illinois/Indiana District	6.00–8.00	11.00–20.00
Midwest, Ohio, Michigan	15.00–17.00	14.00–16.00
West Coast and Northwest	14.00–25.00	20.00–25.00

Government Controls and Restrictions

In the last two editions of *Industrial Minerals and Rocks* this portion of the Silica and Silicon chapter began with the following statement: "Restrictions on silica producers from the environmental and ecological viewpoints are proliferating yearly" (Murphy and Henderson, 1983). This statement is certainly still true today. State and local government agencies as well as the federal Mine Safety and Health Administration (MSHA) and Occupational Safety and Health Administration (OSHA) in the United States have developed and enforce regulations covering the mining industry. Environmental issues that are generally addressed in connection with applications for mining permits including the following:

- restrictions on hours of daily operating time
- water well monitoring of ground water quality
- noise and seismic monitoring related to blasting and trucking
- fugitive dust monitoring and emissions controls
- reclamation plans
- detailed site and mine plans
- water discharge permits, including process water and storm water runoff
- materials safety determinations (data sheets)
- wildlife surveys
- archaeological surveys.

By far the most significant environmental issue facing silica sand producers today is the proposed regulation of crystalline silica as a carcinogen. In 1987 the International Agency for Research on Cancer (IARC), a unit of the World Health Organization, concluded that there is sufficient evidence for the carcinogenicity of crystalline silica to experimental animals and limited evidence for the carcinogenicity of crystalline silica to humans. This was followed shortly by IARC's publication of its Supplement 7 in which crystalline silica was classified in Group 2A, meaning probably carcinogenic to humans (Shapiro, 1990). This issue has ramifications throughout virtually all of the mining industry but clearly causes the greatest concern to silica sand producers.

OSHA has established 0.1% of a suspected carcinogen as the maximum amount any product may contain before its sale must be accompanied by a health hazard warning. As mentioned at the beginning of this chapter, silicon dioxide (quartz) is common in nearly all types of rocks, and the industrial uses of silica are encountered daily by everyone. In many of its commercial applications there are no suitable substitutes for silica. So, although actions taken by regulatory authorities are meant to protect the well-being of the environment and the public, satisfactory compliance can be not only difficult but sometimes impractical.

BIBLIOGRAPHIC REFERENCES

Anon., 1956, "The Mineral Resources of the Ione Formation," *Mineral Information Service,* Vol. 9, No. 8, California Department of Natural Resources, pp. 1–5.

Anon., 1978, *Mold and Core Test Handbook,* American Foundryman's Society, Des Plaines, IL, pp. 4–11.

Anon., 1989, "AWWA Standard for Filtering Material," American Water Works Association, p. 1.

Anon., 1992, "US Flat Glass Outlook—A Repeat Performance of 1991 Is Predicted," Glass Industry, Jan. 10, pp. 8–9.

Austin, G.S., 1972, "Paleozoic Lithostratigraphy of Southeastern Minnesota," *Geology of Minnesota: A Centennial Volume,* P.K. Sims and G.B. Morey, eds., Minnesota Geological Survey, pp. 459–473.

Baker-Carr, C.D.T., 1967, *Sand and Glass,* Nelson Doubleday, New York, 64 pp.

Bates, R.L., 1960, "Sandstone," *Geology of the Industrial Rocks and Minerals,* Harper & Row, New York, pp. 98–117.

Berkheiser, S.W., 1985, "High-Purity Silica Occurrences in Pennsylvania," MR Report 88, Pennsylvania Bureau of Topographic and Geologic Survey, 67 pp.

Buie, B.F., and Robinson, G.C., 1958, "Silica for Glass Manufacture in South Carolina," Bulletin 23, South Carolina Division of Geology, 41 pp.

Bunker, B.J., et al., 1989, "Phanerozoic History of the Central Midcontinent, United States," in "Sedimentary Cover-North American Craton: U.S.," L.L. Sloss, ed., *The Geology of North America,* Vol. D-2, Geological Society of America, pp. 243–260.

Buttram, F., 1913, "The Glass Sands of Oklahoma," Bulletin 10, Oklahoma Geological Survey, 71 pp.

Carter, G.J., Kelly, H.J., and Parsons, E.W., 1962, "Industrial Silica Deposits of the Pacific Northwest," IC 8112, US Bureau of Mines, 57 pp.

Cathcart, J.B., 1963, "Economic Geology of the Keysville Quadrangle, Florida," Bulletin 1128, US Geological Survey, pp. 1–80.

De Cserna, Z., 1974, "El Senério Geográfico; Introducción Ecologica," *Mexico-Panorama de Histórico y Cultural,* Vol. 1, Instituto Nacional de Antropologia y Historia, 306 pp.

De Witt, W., Jr., and Milici, R.C., 1989, "Energy Resources of the Appalachian Orogen: The Appalachian-Ouachita Orogen in the United States," R.D. Hatcher, W.A. Thomas, and G.W. Viele, eds., *The Geology of North America,* Vol. F-2, Geological Society of America, pp. 495–510.

Dietert, H.W., 1966, *Foundry Core Practice,* American Foundryman's Society, Des Plaines, IL, pp. 2–16.

Guillet, G.R., and Kriens, J., 1984, "Ontario and the Mineral Filler Industry," Industrial Mineral Background Paper No. 5, Ontario Ministry of Natural Resources, pp. 99–124.

Greenberg, S.S., 1960, "Petrography Of Indiana Sandstones Collected for High-Silica Evaluation," Bulletin 17, Indiana Department of Conservation, Geologic Survey, 64 pp.

Griffiths, J., 1988, "Mexico's Industrial Minerals, Meeting Mañana's Challenge," *Industrial Minerals,* No. 250, July, pp. 19–41.

Ham, W.E., 1945, "Geology and Glass Sand Resources, Central Arbuckle Mountains, Oklahoma," Bulletin 65, Oklahoma Geological Survey, 103 pp.

Hancock, K., 1991, "Some Significant Industrial Mineral Producers of the Southern Rocky Mountains," *Guidebook for Fieldtrips 2 and 3,* 27th Forum on the Geology of Industrial Minerals, Banff, AB, 34 pp.

Harben, P.W., and Bates, R.L., 1990, "Sandstone and Related Materials,"

Industrial Minerals Geology and World Deposits, Metal Bulletin, London, pp. 238–245.

Hatcher, R.D., 1989, ''Appalachians Introduction: The Appalachian-Ouachita Orogen in the United States,'' R.D. Hatcher, W.A. Thomas, and G.W. Viele, eds., *The Geology of North America,* Vol. F-2, Geological Society of America, pp. 1–6.

Heinrich, E.W., 1977, ''Sylvania Sand Deposit, Monroe County, Michigan,'' unpublished report.

Heinrich, E.W., 1981, ''Geologic Types of Glass-Sand Deposits and Some North American Representatives,'' Bulletin, Vol. 92, Geological Society of America, pp. 611–613.

Hershey, R.E., 1960, ''The High-Silica Resources of Tennessee,'' RI No. 10, Tennessee Division of Geology, 62 pp.

Hewitt, D.F., 1963, ''Silica in Ontario,'' Industrial Mineral Report No. 9, Ontario Department of Mines, pp. 15–25.

Hoagberg, R.K., and Koerner-Moore, J., 1981, ''Silica-Sand Proppants Used in Well Stimulation,'' 16th Forum on the Geology of Industrial Minerals, St. Louis, MO, pp. 15.

Hoyt, D., 1987, ''Back to the Basics of Silica Sand,'' *Modern Casting,* Vol. 77, Sep., pp. 24–27.

Huntley, F.D., and Snow, R.R., 1986, ''Mineral Raw Materials for Flat Glass Manufacturing,'' SME Preprint 86–40, SME-AIME Annual Meeting, New Orleans, LA, 9 pp.

Jaster, M.C., 1957, ''Selected Annotated Bibliography of High-Grade Silica of the United States and Canada Through December 1954,'' Bulletin 1019-H, US Geological Survey, pp. 609–673.

Joseph, N.L., 1988, ''The Mineral Industry in Washington, 1987,'' *Washington Geologic Newsletter,* Vol. 16, No. 2, pp. 3–20.

Keith, M.L., 1949, ''Sandstone as a Source of Silica Sands in Southeastern Ontario,'' 55th Annual Report, Vol. 55, Part 5, Ontario Department of Mines, 33 pp.

Kephart, W.W., Poole, J.P., and Ryder, R.J., 1986, ''Raw Materials for Container Glass Manufacturing,'' SME Preprint 86–131, SME-AIME Annual Meeting, New Orleans, LA, 9 pp.

Ketner, K.B., 1968, ''Origin of Ordovician Quartzite in the Cordilleran Miogeosyncline,'' Professional Paper 600-B, US Geological Survey, pp. 169–177.

Lamar, J.E., 1927, ''Geology and Economic Resources of the St. Peter Sandstone of Illinois,'' Bulletin 53, Illinois State Geological Survey, 175 pp.

Lamey, C.A., 1966, *Metallic and Industrial Mineral Deposits,* McGraw-Hill, New York, 567 pp.

Larson, C.E., 1986, ''Mineral Raw Materials and Their Effect on the Properties of Glass,'' SME Preprint 86–124, SME-AIME Annual Meeting, New Orleans, LA, 8 pp.

MacNeil, F.S., 1950, ''Pleistocene Shore Lines in Florida and Georgia,'' PP 221-F, US Geological Survey, pp. 95–106.

McMaster, R.L., 1954, ''Petrography and Genesis of the New Jersey Beach Sands,'' Geologic Series Bulletin 63, New Jersey Department of Conservation and Economic Development, 239 pp.

Meckel, L.D., 1970, ''Paleozoic Alluvial Deposition in the Central Appalachians: A Summary,'' *Studies of Appalachian Geology: Central and Southern,* G.W. Fisher et al., eds., Interscience Publishers, New York, pp. 49–68.

Mills, H.N., 1983, ''Glass Raw Materials,'' *Industrial Minerals and Rocks,* 5th ed., Vol. 1, S.J. Lefond, ed., AIME, New York, pp. 339–347.

Murphy, T.D., 1954, ''Silica Resources of Clark County, Nevada,'' Bulletin 55, Nevada Bureau of Mines, 33 pp.

Murphy, T.D., 1960, ''Distribution of Silica Resources in Eastern United States,'' Bulletin 1072-L, US Geological Survey, pp. 657–664.

Murphy, T.D., and Henderson, G.V., 1983, ''Silica and Silicon,'' *Industrial Minerals and Rocks,* 5th ed., Vol. 2, S.J. Lefond, ed., AIME, New York, pp. 1167–1186.

Norris, R.M., and Webb, R.W., 1990, *Geology of California,* 2nd ed., John Wiley & Sons, New York, 541 pp.

Nystrom, P.G., Jr., Willoughby, R.H., and Price, L.K., 1991, ''Cretaceous and Tertiary Stratigraphy of the Upper Coastal Plain, South Carolina,'' *The Geology of the Carolinas,* J.W. Horton, Jr. and V.A. Zullo, eds., Carolina Geological Society 50th Anniversary Volume, University of Tennessee Press, Knoxville, pp. 221–240.

Olson, N.K., 1970, ''Silica Sand for Glass Production in a Portion of the Atlantic Coastal Plain,'' *Proceedings,* 5th Forum on the Geology of Industrial Minerals, K.V. Hoover, ed.. Pennsylvania Geologic Survey, pp. 3–22.

Ojakangas, R.W., and Matsch, C.L., 1982, *Minnesota's Geology,* University of Minnesota Press, Minneapolis, MN, 255 pp.

Patchen, D.G., 1968, ''Oriskany Sandstone-Huntersville Chert Gas Production in the Eastern Half of West Virginia,'' Circular Series No. 9, West Virginia Geological and Economic Survey, 47 pp.

Peddicord, R.C., and Regis, A.J., 1981, ''The Effect of Environmental, Regulatory, Technological, and Marketing Constraints on the Cost of Producing Glass Sands,'' *Proceedings,* Minerals and Chemicals in Glass and Ceramics—The Next Decade, Conference held in Corning, NY, Oct., pp. 86–88.

Pettijohn, F.J., Potter. P.E., and Siever, R., 1987, *Sand and Sandstone,* 2nd ed., Springer-Verlag, New York, 544 pp.

Pettke, C.R., 1919, ''Glass Manufacture and the Glass Sand Industry of Pennsylvania,'' Report No. 12, Topographic and Geologic Survey of Pennsylvania, 278 pp.

Phillips, K.A., 1989, ''Industrial Minerals in Arizona's Paint Industry,'' O-F Report 89–1, Arizona Department of Mines and Mineral Resources, 13 pp.

Rast, N., 1989, ''The Evolution of the Appalachian Chain: The Geology of North America-An Overview,'' A.W. Bally and A.R. Palmer, eds., *The Geology of North America,* Vol. A, Geological Society of America, pp. 323–348.

Reed, H.E., 1974, ''Silica Sand Production in Washington,'' Report No. 6, Energy and Minerals Department, Billings, MT, 24 pp.

Russell, E.E., and Parks, W.S., 1975, ''Statigraphy of the Outcropping Upper Cretaceous, Paleocene, and Lower Eocene in Western Tennessee, (Including Descriptions of Younger Fluvial Deposits),'' Bulletin 75, Tennessee Department of Conservation, Division of Geology, p. A41.

Shapiro, J.C., 1990, ''Chemical Manufacturers Association, Chemstar Crystalline Silica Panel-Update,'' Conference on Regulation of Crystalline Silica, Society for Mining, Metallurgy, and Exploration, p. 1.

Sharp, R.P., 1978, *Field Guide Coastal Southern California,* Kendall/Hunt Publishing, Dubuque, IA, 268 pp.

Shufflebarger, T.E., Jr., 1978, ''Economics of Glass-Sand Prospects,'' Circular 79, Oklahoma Geological Survey, pp. 25–35.

Shufflebarger, T.E., Jr., 1982, ''Systems and Specifications,'' *Steel Structures Painting Manual,* Steel Structures Painting Council, Pittsburgh, PA, pp. 16.

Shufflebarger, T.E., Jr., 1983, ''Glass Sand Prospects and Exploration,'' Preprint 83–100, AIME Annual Meeting, Atlanta, GA, 9 pp.

Smosna, R., 1992, ''Geological History of West Virginia's Industrial Mineral Resources,'' *Program with Abstracts,* 28th Forum on the Geology of Industrial Minerals, West Virginia Geologic and Economic Survey, pp. 61–62.

Stewart, J.H., 1980, ''Geology of Nevada,'' Spec. Publ. 4, Nevada Bureau of Mines and Geology, pp. 136.

Watkins, E.C., 1986, ''Mineral Raw Materials for Fiber Glass Manufacture,'' SME Preprint 86–98, SME-AIME Annual Meeting, New Orleans, LA, 5 pp.

Weidman, S., 1907, ''The Geology of North Central Wisconsin,'' Bulletin 16, Series 4, Wisconsin Geological and Natural History Survey, pp. 396–408.

Weigel, W.M., 1927, ''Technology and Uses of Silica Sand,'' Bulletin 266, US Bureau of Mines, pp. 138–147.

Whitlatch, G.I., 1939, ''Foundry Sands,'' Markets Circular No. 9, Tennessee Division of Geology, 20 pp.

Wilborg, H.E., and Henderson, G.V., 1983, ''Foundry Sand,'' *Industrial Minerals and Rocks,* 5th ed., Vol. 1, S.J. Lefond, ed., AIME, New York, pp. 271–278.

Wolfe, P.E., 1977, *The Geology and Landscapes of New Jersey,* Crane, Russak & Company, New York, 351 pp.

Zdunczyk, M.J., 1990, ''Quartz/Silica Sand,'' *Surface Mining,* 2nd ed., B.A. Kennedy, ed., Society for Mining, Metallurgy, and Exploration, Littleton, CO, pp. 235–236.

SILICA

SPECIALTY MATERIALS

A.F. ALSOBROOK

Specialty silica materials, as defined in this chapter, are rocks that consist predominantly of silica (SiO_2) and are produced as chunks, round pebbles, and sawed or trimmed blocks and other shapes for nonconstruction applications. The rock types include quartzite, sandstone, novaculite, flint, chert, and vein quartz. Other terms used commercially include jasper, silex, lascas, and rock crystal.

In its publication, *Minerals Yearbook*, the US Bureau of Mines (USBM) reports the production and consumption of some specialty silica materials under several headings. Quartzite, sandstone, and novaculite used in abrasive applications are addressed by the USBM under Abrasive Materials. Lascas, a type of vein quartz used in growing synthetic quartz crystals and in fusing applications, is discussed under Quartz Crystal. Silica gravel used in manufacturing silicon metal and ferrosilicon, as metallurgical fluxstone, and as a filtration medium is covered under Industrial Sand and Gravel. Information on these and other specialty silica materials is consolidated in this overview as a complement to the preceding chapter on Silica Sand and Sandstone.

GEOLOGY

Sandstone

Sandstone is a sedimentary rock comprised predominantly of quartz grains cemented together by argillaceous, calcareous, ferruginous, or siliceous material. Common detrital impurities include feldspar, mica, and a wide variety of heavy minerals such as ilmenite and kyanite. The purest quartzose sandstones, also termed quartz arenites, consist almost entirely (e.g., +95%) of fine to medium, well-sorted, and well-rounded quartz grains. They are more common in the near-shore facies of a formation than in its deepwater facies. The degree of induration varies from tightly cemented, hard, and tough sandstone to loosely consolidated, friable rock. Induration usually increases as grain size decreases. Color depends largely on the type of interstitial cementing material present. Most sandstones that have been used in specialty applications in the United States, England, Canada, and Germany are Cambrian, Silurian, or Pennsylvanian (Carboniferous) in age and range from less than 3 to more than 85 m in thickness. Those of Cambrian age usually rest unconformably on an eroded Precambrian surface.

Formations of medium- to coarse-grained sandstone that yield rock for specialty applications include those that contain lenses of round quartz pebbles or thicker beds of quartz pebble conglomerate. The pebbly zones commonly are found as long, narrow bands formed by deposition in channels cut by streams into the underlying rock. The pebbles range from 0.3 to 8 cm in diameter, and the conglomeratic beds range from 15 to 90 m in thickness.

Quartzite

Quartzite is a metamorphic rock derived from sandstone and is distinguished from sandstone by it fracture. Sandstone breaks along boundaries between constituent grains, whereas quartzite is so tightly cemented and indurated that it breaks across the grains. Orthoquartzite, the metamorphic equivalent of quartz arenite, consists of quartz grains cemented by silica and commonly exhibits relict sedimentary textures (Murphy and Brown, 1985). In metaquartzite the grains of quartz have lost their individual identity through metamorphism-induced recrystallization, and the rock no longer has the granular texture of sandstone.

Quartzite formations range from 100 to more than 1 200 m in thickness and mostly from Precambrian to Silurian in age. The rock can vary widely in texture, friability, color, and chemical purity depending on the type of cementing material present, the grain size and mineral composition of the original sandstone, and the intensity of metamorphism. Quartzite derived from argillaceous sandstone often contains sericitic partings, which may raise its alumina and alkali content above acceptable limits and cause it to crush into undesirable slabby pieces. Fine- to medium-grained, massive quartzite that contains little or no feldspar, mica, or carbonate mineral impurities tends to be dense, hard, and tough and to crush into sharp, brittle, angular fragments. Its chemical purity can range from 96 to 99.8% SiO_2. Coarser grained quartzite tends to be less pure and more friable, particularly if the cementing material is not siliceous but argillaceous or calcareous. Colors range from white and gray-white to shades of red, pink, purple, blue, and orange. Compressive strength can vary from 1.52×10^5 to 4.96×10^5 kPa, and the specific gravity of most quartzites is between 2.65 and 2.70.

Novaculite

The term novaculite is derived from the Latin word *novacula*, meaning razor hone (Steuart, et al., 1985). Novaculite is a very dense, hard, fine-grained, homogeneous, and highly siliceous sedimentary rock that has a Mohs hardness of 7 and a specific gravity of 2.63 to 2.64. It varies in color from white to gray, light brown, bluish, or black, is translucent on thin edges, and breaks with an uneven conchoidal or subconchoidal fracture (Holbrook and Stone, 1979). It resembles chert, although its fracture surfaces are finely granular or gritty in comparison to the smooth fracture surfaces of chert and flint (Sholes, 1977). The rock's luster varies from waxy to dull, the latter resembling that of unglazed porcelain. Its chemical purity can exceed 99% SiO_2 in the form of microcrystalline or cryptocrystalline quartz.

Perhaps the most distinguishing characteristic of novaculite is its texture. Rock that has been little affected by thermal metamorphism is comprised of anhedral quartz grains. Rock that has been thermally metamorphosed, by comparison, such as near Hot Springs, AR, and Broken Bow, OK, consists of coarser, polygonal or polyhedral, triple-point crystals or sharp-edged blocks of quartz 5 to 20 μm in diameter (Steuart, et al., 1984, Keller, et al., 1984). The porosity of novaculite from near Hot Springs ranges from 0.07 to 0.25% for the Hard Arkansas Stone variety, which has a waxy luster, and from 5 to 16% for the Washita Stone (Ouachita Stone, or Soft Arkansas) variety, which has a porcelanic appearance (Harben, 1983). Washita Stone contains rhombic cavities from which calcite or other carbonate minerals have dissolved by weathering. Broken surfaces of Hard Arkansas Stone are smooth, whereas those of Ouachita Stone are rough. Beds of Ouachita Stone contain fewer joints and quartz veins, and the stone becomes tougher and less easily fractured when exposed to weathering.

The Arkansas Novaculite Formation lies conformably between the older Missouri Mountain Shale and the younger Stanley Shale. It attains a maximum thickness of 290 m and consists of upper and lower novaculite members interbedded with three chert and shale units. The upper member has a maximum thickness of 55 m, is Mississippian in age, and is a white to light gray or blue-black, massive calcareous rock that contains as much as 30% carbonate minerals where unweathered. It weathers to beds of soft, powdery tripoli that reach 12 m in total thickness. The purer lower member ranges from 15 to 135 m in thickness, is Devonian in age, and is a white to light gray or light brown massive rock that is cut extensively by joints oriented normal to bedding. It is slightly calcareous near the base, where dissolution of calcite rhombs has produced a distinctive whetstone texture (Steuart, et al., 1984).

Vein Quartz

Massive quartz for specialty applications is obtained from the central core of zoned pegmatites and from veins, dikes, and plugs within sedimentary, igneous, or metamorphic host rocks. Host rock formations usually range in age from Precambrian to Jurassic. Most deposits are believed to have formed by hydrothermal processes, although not all are located near granitic intrusions. They occur as fissure and cavity fillings and less commonly as metasomatic replacement bodies.

Pegmatites associated with granitic intrusions generally yield very pure quartz, especially in the case of microcline pegmatites that are devoid of muscovite and rare earth oxides but have a well developed quartz core with druse cavities. The core of such pegmatites commonly contains abundant fluid inclusions, however, which render the quartz unsuitable for fusing into clear quartz glass. Massive quartz veins of hydrothermal or metasomatic origin may have a zonal structure consisting of several generations of quartz. The oldest generation is coarse grained and usually contains abundant inclusions; the younger generation is finer grained and purer; and the youngest generation consists of crystals in druse cavities. The wall rocks of such veins are usually leached, sericitized, or chloritized, and the size of cavities containing quartz crystals is directly proportional to the intensity of wall rock alteration. In general, the more chlorite a vein contains, the purer is the quartz in it. Hydrothermal veins that are metamorphosed or were segregated from country rocks during regional metamorphism tend to be free of fluid inclusions and therefore to yield fusing-grade quartz (Kuzvart, 1984).

Primary quartz deposits in Brazil occur in the form of veins, pipes, pockets, stockworks, and composite lodes confined to siliceous host rocks. They are found mostly within crossbedded sandstone and quartzite of Late Precambrian age and along the contact between Precambrian granite and Silurian shale. The veins and pockets were emplaced in early Paleozoic, probably Silurian, time (Bates, 1960). They are comprised mostly of very large crystals of milky or gray bull quartz that grew inward from the walls of druse cavities to form a comb structure. Vugs are abundant, some lined with well-formed quartz crystals and filled with clay. Some of the crystals, especially the smaller ones that weigh less than 250 g, are clear and of electronic quality. Most piezoelectric-grade quartz, however, occurs as clear or translucent pyramidal terminations on milky bull quartz crystals that can exceed one ton in weight.

Primary vein deposits in bedrock in Brazil are overlain by a residual soil blanket containing elluvial and colluvial deposits of clear, unworn quartz crystals. They are also associated with nearby alluvial placer deposits in present or former stream channels that contain rounded and frosted but usable crystal fragments. Some alluvial deposits consist of several beds of quartz gravel separated by lenses of clay. They are commonly enriched in high-quality quartz, because fractured or twinned crystals break down more readily during weathering and stream transportation. Elluvial placer deposits are elliptical in shape with their long axis parallel to the strike of the vein, whereas colluvial placers are fan-like in shape (Kuzvart, 1984).

Silica Pebble

In addition to the quartz pebble facies of sandstone formations, geologic resources that yield silica pebbles for specialty applications include sedimentary chalk and marl beds and fluvial deposits. These are generally much younger than other specialty silica rock resources. They usually range in age from Cretaceous to Holocene. The high-purity silica pebbles they contain may be flint, quartzite, or massive, vein-sourced quartz.

DISTRIBUTION OF MAJOR DEPOSITS

Sandstone, quartzite, vein quartz, and other silica-rich rocks are common and widely distributed around the world, but at few locations does the rock have the physical and chemical properties required for use in specialty applications.

Sandstone

United States: The Pottsville Group of Lower Pennsylvanian age contains quartz pebble conglomerate facies in sandstones of the Sharon and Olean Formations. These formations have been mined in New York, Pennsylvania, Tennessee, and Ohio as sources of ganister for refractory brick and of metallurgical gravel for silicon metal and ferrosilicon production (Murphy and Brown, 1985). The Sharon Conglomerate, the lowermost member of the Pottsville Group, varies in thickness from 15 m in the northeastern Ohio counties of Geauga, Portage, and Summit to 70 m in the southern Ohio counties of Pike and Jackson. It is a friable, medium- to coarse-grained sandstone grading 95 to 99% SiO_2 and containing pebbly zones that occur as long, narrow, north-south bands deposited in channels cut into underlying Mississippian rocks. Metallurgical-grade quartz pebbles have also been mined in Franklin County on the Cumberland Plateau of east-central Tennessee, where the Sewanee Conglomerate Member of the Pennsylvanian Crab Orchard Formation contains 99% SiO_2.

At Keck Center in Fulton County, NY, a white, granular sandstone containing as much as 99% SiO_2 was mined years ago for use in ferrosilicon production. Sandstone from the Homewood Formation has been quarried for ganister near Upper Hillville and Foxburg, PA. Cambrian sandstone has been quarried for whetstone manufacturing near Allentown in Lehigh County, for ganister and millstone production in Lancaster County, and for ganister in Montgomery County, all in Pennsylvania. Fine-grained quartzites and quartzitic sandstones of the Pennsylvanian Stanley and Jackfork Formations in the Ouachita Mountains of Oklahoma are suitable for the fabrication of millstones and buhrstones. In Montana a sandstone quarry at Columbus in Stillwater County has produced grindstones, and another at Dillon in Beaverhead County has produced metallurgical fluxstone and refractory ganister. High quality sandstone suitable for specialty applications has been quarried near Denison in Spokane County, WA.

The Shawangunk Conglomerate crops out in a belt that extends from near Kingston, NY, southwestward into New Jersey and Pennsylvania. It has been quarried extensively for millstones at Shawangunk Mountain and near Kyserike, St. Josen, Granite, and Kerhonkson, NY. It is a well cemented and hard, light gray, pebbly sandstone that ranges from 15 to 90 m in thickness. Tough, fine-grained sandstone near Rawlins in Carbon County and on Baldwin Creek near Lander in Fremont County, WY, was once used for grindstones, as was the Dakota Sandstone northeast of Edgemont, NE, near the South Dakota border. More recently the Dakota Sandstone has been mined in Colorado for use as ganister in refractory applications.

Pulpstones have been fabricated from sandstone quarried in Pierce, Wilkerson, and Skagit Counties, Washington, and whetstones have been produced from a 2.5 m thick bed of sandstone in Skamania County, Washington, between White Salmon and Stevenson. Good pulpstones, grindstones, and millstones also have been produced from the Dunkard and Berea Sandstones in Ohio, West Virginia, and Michigan. Hard sandstone, such as the Hindostan stone from Indiana and the Queen Creek stone from Ohio, have been used as waterstones (Bateman, 1959). Until 1989 sandstone quarried at Orleans in Orange County, Indiana, was fabricated into cuticle removers, sharpening stones, and table coasters.

One of the few remaining US producers of sandstone for non-construction purposes is Cleveland Quarries Co., which extracts and saws the Berea Sandstone into refractory bricks and blocks at South Amherst in Lorain County, OH. In this part of northern Ohio the Berea, which is part of the Waverly Series of Mississippian age, can exceed 68 m in thickness, although it is only 15 m thick elsewhere. Cleveland Quarries cuts Buckeye Silica Stone into beams measuring 1.6 to 3.2 cm^2 in cross-section and into slabs normally 1.2 cm in thickness. These refractory firestone shapes are used in fabricating acid tanks, steel pickling tanks, acid wash trays, soaking pits, ladle linings, and for related applications in the steel and chemical industries. The company also has produced grindstones, pulpstones, and deburring media as coproducts at its dimension stone operations.

Canada: The Potsdam (Nepean) Sandstone has been quarried in eastern Ontario and southern Quebec. It is Late Cambrian to Early Ordovician in age, rests unconformably on an eroded Precambrian surface, and ranges from 20 to 85 m thick in the eastern Ontario counties of Frontenac, Leeds, and Lanark. Where exposed between Kingston, Brockville, Perth, and Smiths Falls in Ontario, this sandstone consists of well-indurated grains 75 to 850 µm in size (Guillet and Kriens, 1984). It is mined and processed at St. Canut north of Montreal, PQ, and is sold for use in glassmaking, foundries, ceramics, and silicon carbide manufacturing.

The Medina Sandstone of Early Silurian age crops out in several counties west of the Niagara Escarpment in Ontario. It has been quarried principally for dimension stone but has been used also in ganister mixes for lining cupolas and ladles in the steel industry. A quartzitic sandstone is mined at Melocheville in Beauharnois County, PQ, and is used at two ferrosilicon plants at Beauharnois.

The Winnipeg Formation of Cambro-Ordovician age crops out on three islands and two promontories in Lake Winnipeg, MB. It consists of nearly white, poorly consolidated, round-grain sandstone and is mined principally on the southeastern shore of Black Island. After being washed and screened, the silica product is barged 140 km south to a plant at Selkirk, MB, for further processing (Guillet and Kriens, 1984). The plant at Selkirk, on the Red River 40 km north of Winnipeg, has a production capacity of 115 ktpy.

Grindstones, millstones, and pulpstones have been produced from Carboniferous and Permo-carboniferous sandstones in New Brunswick, Nova Scotia, and British Columbia.

England: One of the principal sources of grindstones, pulpstones, and millstones in England has been the fine-grained Newcastle Stone quarried from Carboniferous sandstone associated with the Coal Measures near Newcastle (Bateman, 1959). Derbyshire (Peak) and Yorkshire Stones also have been quarried for this purpose from the Millstone Grits, which yield some of the best pulpstones in the world.

Carboniferous sandstone for grindstones, millstones, and pulpstones also has been quarried in Germany and Italy.

New Zealand: At Pebbly Hills, 50 km northeast of Invercargill on South Island, a quartz pebble conglomerate deposit contains more than 200 Mt of high quality gravel suitable for use in ferrosilicon production. The +12 mm gravel averages more than 98.75% SiO_2 (Benbow, 1990).

Quartzite

United States: In the eastern United States massive quartzite of the Lower Cambrian Erwin Formation is more than 100 m thick and is well exposed in an anticline in the Lick Mountain District, which is part of the Appalachian Valley in Virginia. It contains 97.63 to 99.37% SiO_2 and has been quarried as metallurgical-grade stone for silicon metal and ferrosilicon manufacture in Virginia (e.g., at Sand Mountain), West Virginia, and east-central Tennessee.

Belts of high-purity quartzite extend across central and southeastern Pennsylvania and across Calhoun, Talladega, Cherokee, Clay, and Cleburne Counties in Alabama. Quartzite (actually quartzitic sandstone) from the Lower Silurian Tuscarora (Medina) Formation has been mined in Carbon, Bedford, Blair, and Huntingdon Counties, Pennsylvania, for use as ganister in refractory bricks for furnace linings. It also has been quarried in Virginia and near Minnehaha Springs in Pocohontas County, West Virginia, for use as metallurgical stone. It is light gray to white and 120 to 180 m thick in central Pennsylvania, such as at Lock Mountain near Point View. In addition to the Tuscarora, quartzite from the Chickies Formation in southeastern Pennsylvania has been used as refractory ganister and for tube mill linings. In Cherokee County, Alabama, quartzite containing 97.75 to 98% SiO_2 can be obtained from the Lower Cambrian Weisner Formation.

In Cherokee County, North Carolina, the Tusquitee Quartzite is of metallurgical quality. The hard, tough, Poughquag quartzite near Highland, NY, is suitable for chemical uses, as metallurgical stone, and for cutting into blocks for lining grinding mills. Buhrstone has been quarried from the Cambrian Cheshire quartzite between Dalton and Washington in Berkshire County, MA, where the formation is nearly 55 m thick at Coltsville. The Carboniferous Quabin quartzite is a pure, white, sugary but locally flaggy rock that forms part of the Quabin and Felton Mountains in Rhode Island and Massachusetts. Many years ago near Stafford, CT, it was quarried and sawed into hearthstone and firestone for iron ore smelting.

In the central United States one of the most important sources of specialty silica stone is the Sioux Quartzite, also known as the Sioux Falls Granite, of Late Huronian (Middle Precambrian) age. A fine-grained quartzite that ranges from 460 to 1 220 m in thickness, the Sioux crops out in a belt more than 95 km wide that extends over 15 540 km^2 from the junction of the Cottonwood and Minnesota Rivers at Redstone, MN, westward to Mitchell on the James River in South Dakota. This belt includes parts of the southwestern Minnesota counties of Nicollet, Pipestone, Cottonwood, Rock, and Watonwan, part of northwestern Iowa, and parts of the southeastern South Dakota counties of Turner, McCook, Hanson, and Minnehaha. Outcrops are most prevalent in river valleys, such as those of the Big Sioux at Dell Rapids and South Falls, the Vermillion at Parker, the James south of Mitchell, Wolf Creek at Salem, and

Pierre Creek at Alexandria, all in southeastern South Dakota. The color of Sioux quartzite ranges from pink to orange, purple, and blue; its silica content varies from 96% to nearly 99%; its compressive strength ranges from 1.52×10^5 to 4.96×10^5 kPa; and crushed pieces vary from equidimensional to slabby. The stone's softening point commonly exceeds 1 800°C, and its density is about 1 682 kg/m^3.

One of the most productive portions of the outcrop belt of the Sioux Quartzite extends from Jasper in the southwest corner of Minnesota westward to Sioux Falls, Spencer, and Mitchell, SD. Within this district several companies quarry and crush quartzite for construction applications, and one also ships metallurgical-grade material to a ferrosilicon producer in Iowa. Another firm, Jasper Stone Co. of Jasper, MN, is primarily a dimension stone producer for nonconstruction markets. The company produces some quartzite memorials or monuments, but its chief products are split and sawed blocks for lining tube-type grinding mills, acid tanks, and chutes. As a byproduct of its dimension stone operations Jasper Stone produces quartzite cubes that measure 1.0, 1.2, or 1.6 cm on a side and are pretumbled for use as grinding media in pebble or ball mills.

Artificially rounded grinding pebbles have also been produced from the Sioux quartzite near Sioux Falls, SD. In addition to Jasper, other producing sites in Minnesota include Pipestone, Luverne, and New Ulm, but the stone from those locations is sold mostly for construction applications. Some blocks for lining tube mills and chunks as grinding media have been produced by New Ulm Quartzite Quarries, Inc. in Nicollet County, south-central Minnesota. This company also has sold crushed Sioux quartzite as ganister for steel furnace refractories, as filtration rock, and as fluxstone for copper smelters. The rock is excellent for the manufacture of refractory silica brick.

Other important resources of specialty silica stone in the central United States are the Baraboo Quartzite in Sauk and Columbia Counties, south-central Wisconsin, and the Nesnard Quartzite in Michigan, both of which are Precambrian (Huronian) in age. The Baraboo Quartzite District is 45 km long, 3 to 19 km wide, and encompasses nearly 583 km^2. The formation is about 1 220 m thick and consists of massive beds of hard, very tough quartzite comprised of round quartz grains cemented by secondary quartz. Where free of thin slate or schist partings along bedding planes the stone has a compressive strength of nearly 5.0×10^5 kPa. Until it ceased operations in 1988, Baraboo Quartzite Co. quarried stone from the Netzebaum Quarry near Devil's Lake, WI, and marketed crushed, sized, and pretumbled deburring and burnishing media to foundries and metal stamping plants worldwide. Grinding media and liner blocks for pebble mills and ganister for refractory applications also have been produced from the Baraboo Quartzite. In north-central Wisconsin the Rib Hill and Powers Bluff Formations are medium-grained quartzites that contain 99.07% SiO_2 and are suitable for specialty applications.

High-purity quartzite is widespread in the western United States, especially in eastern Washington. The Addy Quartzite of Cambrian age crops out over an area of 648 km^2 in Stevens County, WA (Murphy and Brown, 1985), and most of its exposures are within 20 km of a railroad. Until 1991 Northwest Alloys, Inc., a subsidiary of Aluminum Company of America (Alcoa), produced 22 ktpy of Addy quartzite from the Blue Creek Mine in Stevens County and used the output captively in the production of silicon metal and ferrosilicon at Addy, WA, where ferrosilicon capacity was 27.5 ktpy. Elsewhere in Stevens County at Valley, 64 km northeast of Spokane, Lane Mountain Silica Co. mines a part of the Addy Quartzite that is complexly faulted and consists of alternating hard and soft, friable zones. Products contain less than 0.005% Fe_2O_3. Addy quartzite also has been quarried at Kettle Falls in Stevens County. Other counties in Washington that have produced or have the potential to produce quartzite for specialty markets are Ferry, Pend Oreille, Spokane, and Whitman.

The Eureka Quartzite of Middle Ordovician age crops out over more than 260 km^2 from Idaho into southern California (Murphy and Brown, 1985). It ranges up to 99.9% SiO_2 in purity and has been mined in Clark County, NV, for refractory, metallurgical, and other uses. Quartzite is produced at the Canyon Creek Quarry near Maiden Rock in Beaverhead County, MT, for use as fluxstone at an elemental phosphorus and ferrophosphorus plant in Silver Bow County west of Butte, MT (Rice, et al., 1989).

Quartzite for silicon metal production has been quarried at Rogue River near Gold Hill in Jackson County, OR. It also has been produced from the Precambrian Dripping Spring Formation at Oracle, AZ. Quartzite from the Tintinnic Formation has been mined in Tooele County, UT, primarily for the manufacture of refractory coke oven shapes. Quartzite for furnace-lining refractories also has been produced west of Douglas, AZ. The Sawatch Quartzite of Late Cambrian age is a hard, massive, gray-white formation that lies on an eroded Precambrian surface, is as much as 40 m thick in the Mosquito Range of Colorado, and may be suitable for specialty markets.

Metallurgical-grade quartzite from other sources in the US Pacific Northwest has been used in the production of elemental phosphorus and ferrophosphorus at electric furnace facilities. Examples in Idaho include stone quarried in Caribou County for a plant at Soda Springs and the Kit Creek mines in Bannock County, which supply plants at Pocatello.

Canada: In eastern Canada three quartzite formations comprise the top of the Cobalt Group of Middle to Late Precambrian (Middle Huronian) age. The purest is the Bar River Formation at the top of the series, which is underlain by the Gordon Lake Formation and then the Lorrain Formation. The Bar River is dense, massive, white, high-purity quartzite composed of fine to medium, round and well-sorted quartz grains cemented by secondary quartz (Guillet and Kriens, 1984). It is at least 900 m thick and is exposed on a chain of islands in the North Channel of Georgian Bay (Lake Huron) including Pine Island north of Killarney, ON, as well as Badgeley, Centre, Partridge, and Heywood Islands.

The Bar River Formation has been quarried at Shequiandah on Manitoulin Island south of Little Current, where lump quartzite was produced for ferrosilicon manufacturing. It is now being quarried on Badgeley Island, 6 km west of Killarney, where the beds dip nearly vertically and contain a few lenses of diabase and amphibolite. Crushed quartzite is shipped to ferrosilicon and silicon metal producers at Welland, Thorold, and Niagara Falls, ON, at Bécancour, PQ, and at Niagara Falls, NY. Fines are shipped 190 km south to a plant at Midland, ON, for further processing. Production capacity is 1 Mtpy at Badgeley Island and 0.5 Mtpy at Midland. The chemical purity of the Bar River and Lorrain quartzites is illustrated in Table 1.

The Lorrain Formation approaches the purity of the Bar River but contains more alumina. It is exposed along the northern shore of Lake Huron between Sault Ste. Marie and Killarney, ON. The Bellevue Quarry north of Sault Ste. Marie formerly supplied quartzite as ganister for refractory brick, and a quarry at Killarney produced lump quartzite for ferrosilicon. Some lump quartzite from the Bar River Formation has been shipped to the Sudbury area in Ontario and the Noranda area in Quebec for use as metallurgical fluxstone in smelting nickel and copper matte. The Lawson Quarry at Whitefish Falls, ON, was a captive operation that supplied silica fluxstone to nickel smelters at Sudbury (Guillet and Kriens, 1984). Quartzite is also supplied to a nickel smelter at Thompson, MB.

A friable orthoquartzite of Grenville (Precambrian) age is quarried near St. Donat, PQ. It is processed at St. Canut, PQ, 80 km to the southeast, where the Potsdam (Nepean) Sandstone is quarried. Several quartzite mines in Beauharnois County, PQ, supply met-

Table 1. Typical Composition (percentage) of Precambrian Quartzites in Ontario, Canada

	Bar River Formation		Lorrain Formation	Grenville Quartzite
	Badgeley Island %	Shequiandah %	McGregor Point %	Alban %
SiO_2	98.50	97.90	94.10	96.90
Al_2O_3	0.80	0.80	3.46	0.68
Fe_2O_3	0.04	0.08	0.02	0.22
MgO	0.03	0.07	0.13	0.03
CaO		0.04	0.01	0.00
Na_2O	0.01	0.10	0.13	0.01
K_2O	0.14	0.30	0.38	0.09
LOI	0.16	0.24	0.43	0.12
TiO_2	0.02	0.05	0.06	0.06
P_2O_5	0.01	0.03	0.01	0.01
MnO		0.01	0.00	0.00
	99.71	99.62	98.73	98.12

Source: Guillet and Kriens, 1984.

allurgical-grade stone to local ferrosilicon manufacturers. Elsewhere in eastern Canada high-purity quartzite is found near Labrador City, LB. In Newfoundland 100 ktpy of quartzite have been quarried at Villa Marie on the Avalon Peninsula in Placentia Bay and shipped to Long Harbour as fluxstone for the electric furnace production of elemental phosphorus. In Nova Scotia a fine-grained, brittle quartzite of George River (Precambrian) age has been quarried in Cape Breton County as ganister for the refractory brick linings of open hearth steel furnaces at Sydney. The stone is highly fractured and contains serpentine and chlorite impurities.

In western Canada the Mount Wilson Formation of Ordovician age consists of massive, uniform, white, medium-grained orthoquartzite or silica-cemented quartz arenite. The formation ranges in thickness from more than 330 m near Golden, BC, to as little as 20 m at Canal Flats 110 km to the southeast. It is quarried at Nicholson, 11 km from Golden in southeastern British Columbia, from which 30 to 60 ktpy are shipped as metallurgical lump for silicon metal and ferrosilicon production at Rock Island near Wenatchee, WA. It has supplied lump for the manufacture of polycrystalline silicon at Moses Lake, WA. The Mount Wilson quartzite grades as high as 99.85 to 99.9% SiO_2 with only 0.10% Al_2O_3 and 0.04% Fe_2O_3 (Simandl, et al., 1992). Reserves of lump silica at Nicholson are estimated at 3 Mt. At Mount Moberly, BC, a friable variety of Mount Wilson quartzite is mined and used in conventional markets for industrial silica sand. It contains 99.67% SiO_2, 0.06% Al_2O_3, and 0.02% Fe_2O_3 (Simandl, et al., 1992).

In the vicinity of Longworth east of Prince George, BC, the Nonda Quartzite of Silurian age crops out for a strike length of 25 km, may be as much as 250 m thick, and is reportedly even purer than the Mount Wilson stone at Golden (Hamilton and Hora, 1987). Quartzite for ferrosilicon and silicon carbide manufacture was formerly quarried near Oliver in south-central British Columbia near the US border.

Australia: In Western Australia high-purity quartzite is quarried at the rate of 60 ktpy at Picton Junction near Moora, 185 km north of Perth. It is hauled 350 km to a 26.8 ktpy silicon metal plant at Kemerton, 20 km north of the port of Bunbury and 165 km south of Perth. The quarry has capacity to produce 75 ktpy of stone, and the Kemerton plant produces about 24.3 ktpy of chemical-grade silicon metal.

Republic of South Africa: Iscor Ltd. produces 61 ktpy of lump quartzite at Donkerhoek, 20 km east of Pretoria, and uses the stone as blast furnace flux at its steel mills at Vanderbijlpark and Newcastle. The stone contains more than 95% SiO_2 and less than 0.18% alkalis and 1.7% Al_2O_3. At Delmas, 85 km from Johannesburg, a subsidiary of Samancor Ltd. quarries 480 ktpy of quartzite for use by domestic ferroalloys and refractories producers. The −180+100, −100+76, and −20+12 mm grades average 98.8% SiO_2, 0.5% Fe_2O_3, and less than 0.5% Al_2O_3, 0.04% S, and 0.01% P. Metallurgical-grade silica is also produced at Letaba, Bronkhorstspruit, and Lydenburg (Griffiths, 1989).

Norway: At Tana in northern Norway quartzite is quarried by Elkem, AS and is used by the company in manufacturing 50% ferrosilicon. Lump quartzite from Tana has been shipped to Iceland, 2 575 km away, for use in a ferrosilicon plant at Grundartangi near Reykjavik.

Sweden: At Åmal in Dalsland, approximately 150 kt of quartzite are quarried annually. Of this total some 35 to 45 kt of high-purity rock are used 15% in ferrosilicon production and 85% in the manufacture of monolithic refractories (Anon., 1987, Russell, 1990).

Italy: An operation at Valsinni in Matera can produce as much as 50 ktpy of lump quartzite for local steel and chemical markets (Robbins, 1986).

Pakistan: Lump quartzite for use as a steel mill flux is produced at the rate of 20 to 40 ktpy in the Lasbela region of Baluchistan (Griffiths, 1987b).

Czechoslovakia: In central Europe quartzites of Early Tertiary age, especially freshwater quartzites, are used as ganister for refractories and in the manufacture of ferrosilicon and silicon metal (Kuzvart, 1984). Since depletion of deposits in the Most district, acid refractories (*dinas*) in Czechoslovakia have been made with fine-grained quartzite from the Lahost-Jeníkov deposit near Teplice in North Bohemia. This deposit is 20 m thick, and the stone contains less than 1% Al_2O_3 and more than 97% SiO_2 (Hruska, 1991).

Novaculite

United States: The Arkansas Novaculite Formation crops out in several narrow belts that extend along the borders of the Benton-Broken Bow Uplift portion of the Ouachita Mountains between Little Rock, AR, and Atoka in southeastern Oklahoma. This portion of the Ouachita Fold Belt ranges from 25 to 65 km in width and extends for 320 to 360 km east-west. Mining operations that produce novaculite for commercial use as oilstones, hones, files, whetstones, and for other specialty applications are concentrated in Garland County just northeast of Hot Springs, AR. A few deposits are also mined in nearby Pike and Montgomery Counties, AR. In 1990 there were nine companies actively engaged in novaculite production in this area. Five additional companies, four of which had nonintegrated finishing operations, were inactive in 1989. Nearly all commercial whetstone mines are confined to the massive lower member of the formation.

In Oklahoma the outcrop belts of novaculite continue from the Potato Hills east of Talihina through Latimer, Pushmataha, and McCurtain Counties and westward to near Atoka in Atoka County. Silica rock that has been called novaculite comprises part of the Caballos Formation in Brewster County, southwestern Texas. It is intensely folded and fractured but crops out prominently and extensively in the Marathon Uplift area (Jaster, 1957). In the counties of Alexander and Union in southern Illinois widespread deposits of primary novaculite and alluvial novaculite gravel are present in the same area where major deposits of tripoli are being mined. A novaculite locality near Marquette, MI, has also been reported (Jaster, 1957); but deposits outside the United States have not been well documented.

Vein Quartz

United States: Hydrothermal quartz veins that cut sandstone and shale are the most common source of electronic-grade quartz crystals and lascas around the world. In the United States minor quartz crystal deposits have been found in California, Idaho, Oklahoma, and Virginia, but the major deposits are near Hot Springs, AR, in the foothills of the Ouachita Mountains. The Arkansas deposits are being mined mostly for quartz crystal specimens, however, not for electronic-grade crystal. World renowned crystals with exceptionally smooth, clean faces are extracted from veins and cavities in sandstone at two small open pits in Garland County 24 km north and 32 km northwest of Hot Springs. The best quality, water-clear specimens having highly polished faces are found in cavities filled with reddish brown, sticky, rouge-like clay.

The only producer of lascas in the United States is Coleman Quartz, Inc. At an open pit near Paron, AR, 48 km northeast of Hot Springs in Saline County, this company mines one of two massive bull quartz veins that lie mostly conformably between units of thinly bedded and deeply weathered shale. The veins range from 2.4 to 3.7 m in width and extend for 0.4 km in strike length. Annual production varies from 450 to 570 t. Undrilled reserves are estimated at 200 kt. Mine-run material is hauled by trucks to the company's processing plant 48 km away. The quality of finished lascas produced in Arkansas is equal to or slightly superior to that of lascas produced in Brazil, and it is usually less variable. Total impurity levels range from 25 ppm for grades #1 (90% clear) and #2 (50 to 60% clear) to 44 ppm for grade #3 (translucent) and 54 ppm for grade #4 (opaque). Impurities include 15 to 20 ppm Al, 2 to 25 ppm Na, 2 to 10 ppm K, 2 to 5 ppm Fe, and 1 ppm each of Ca, Mg, Li, and Ti (Griffiths, 1987). The relatively high sodium level is associated with fluid bubble inclusions.

In Connecticut between the towns of Ledyard and North Stonington the massive Lantern Hill vein quartz deposit is being mined. It measures 2 000 m long and 300 m wide and yields products that grade 99 to 99.4% SiO_2. A similar deposit is located just north of Glasgo about 16 km north of Lantern Hill. Vein quartz and high-purity quartzite deposits have been mined in the past in Adams County, PA, and at Port Henry and Fort Ann, NY. Bull quartz dikes have been found west of Wirtz in Franklin County, VA, and numerous vein quartz deposits are known in the North Carolina counties of Anson, Moore, Montgomery, Harnett, Buncombe, and Transylvania. In Texas massive quartz veins occur in Precambrian rocks of the Llano Uplift and in the Carrizo Mountains of Culberson and Hudspeth Counties.

High-purity, metallurgical-grade silica resources are abundant in the western United States, particularly the Pacific Northwest. Silicon metal and ferrosilicon plants at Rock Island, WA, and Springfield, OR, now use mostly quartzite but were supplied in the past with lump quartz from massive vein deposits located 1) north of Weiser, ID, 2) at Basin and west of Galen, MT, and 3) near Hawthorne, NV. The deposit near Weiser is a quartz plug within granodiorite and yielded material averaging 99.78% SiO_2. The Basin deposit is a quartz plug within quartz monzonite and yielded metallurgical lumps grading more than 99.5% SiO_2. Near Galen a massive quartz body averaging 99.82% SiO_2 is present within Precambrian limestone and quartzite on the southwestern slope of the Flint Creek Range. The vein near Hawthorne, NV, measures nearly 850 m long and 95 m wide, is associated with Jurassic granite, and yielded stone that averaged 99.89% SiO_2 (Peterson, 1976). Similar deposits in western Montana are located at Deer Lodge, Haines Point, and Quartz Creek.

The Quartz Mountain deposit 32 km from the town of Mead in Spokane County, WA, is a large vein of massive quartz within granitic host rocks. It has been quarried as a source of lump quartz for silicon metal and ferrosilicon producers. Other deposits in eastern Washington include 1) the Latshaw deposit, a 200 m^2 plug of quartz 20 km north of Spokane that grades 98% SiO_2 and has been mined mostly for ganister and 2) massive quartz veins at Mount Spokane and near Mica Peak. In central Washington a dike of white quartz 15 m wide and averaging 96.6% SiO_2 is found near Wenatchee, and another large vein is located near Merritt in Chelan County. In western Washington three large quartz veins are located near Marblemount in Skagit County, and another has been mined at Rockport 13 km to the southwest. Other deposits of massive quartz have been quarried for ganister in King and Pierce Counties. Altogether there are 36 known massive quartz deposits in 13 counties in Washington, most of which are found as veins and pods in metamorphic rocks or as the cores of pegmatites.

Elsewhere in the western United States a pure quartz dike 300 m long and 60 m wide has been mined in Jackson County, OR. Lenses of massive quartz or quartzite more than 60 m thick are found in metamorphic rocks of Paleozoic age 30 km northeast of Oro Grande and 15 km east of Victorville in San Bernardino County, CA. They have been quarried mostly for ganister. The Veta Grande Mine near Carson City, NV, is another massive vein deposit that has yielded metallurgical-grade lump quartz.

Brazil: Brazil has the largest known reserves of natural, piezoelectric-grade quartz crystal and lascas in the world, with the total estimated at 26.8 Mt (Smith, 1984). On the Brazilian altiplano, an extensive plateau underlain by Precambrian and Lower Paleozoic rocks, there are hundreds of small, deeply weathered vein and alluvial deposits scattered over thousands of square kilometers, an area twice the size of the state of Texas (Bates, 1960). This area is located along the southeastern margin of the Brazilian Shield. Production comes mostly from the states of Minas Gerais, Santa Catarina, São Paulo, and Bahía, although significant reserves are also known in Espírito Santo, Paraiba, Paraná, Rio de Janeiro, and Rio Grande do Sul. The richest deposts are in Minas Gerais.

Some piezoelectric-grade crystal is produced in Brazil, but shipments from the official export cities of Rio de Janeiro, Bahía, and San Salvador consist predominantly of lascas for growing cultured quartz crystals and in Europe, after extensive chemical refinement, for the manufacture of fused quartz preforms. One Brazilian-Japanese joint venture has operated a 60 tpm plant to produce quartz crystal, lascas, and ultrapure quartz powder from material mined from six pits at Pequeri in Minas Gerais that are owned by one of the Brazilian partners. Of the total quantity of quartz produced in Brazil approximately 97% is lump for industrial applications such as silicon metal and silicon alloys manufacture, and the remaining 3% consists of piezoelectric crystal, lascas, and decorative specimens. Quartz for silicon metal smelting is produced at Gouveia in Minas Gerais.

Namibia: Quartz for the electronics industry is recovered as a byproduct of lithium ore production at the Rubicon Mine of SWA Lithium Mines near Karibib. Another large quartz deposit is located at Marbad near Windhoek (Murray, 1989).

Angola: Production of lascas from the Pocarica vein deposits near Conda in Angola's Cuanza Sul Province was nationalized in 1979 with the establishment of Miniquartzo to manage operations. Angola has exported no lascas since 1980, compared to a level of 3 900 t in 1975; but the recent peace treaty and cessation of military action could lead to a resumption of exports. One cultured quartz producer at Libramont, Belgium, used to purchase nutrient quartz from Angola but switched to Namibian lascas in 1989.

Republic of South Africa: A dike of massive, high-purity quartz within Precambrian host rocks near Pietersburg in the northern Transvaal region is known as the Witkop Deposit and is quarried by Samancor Ltd. The open pit mine, located 380 km north of Johannesburg, has a capacity of 150 ktpy and supplies −38+25 mm lump to a silicon metal smelter nearby (Griffiths, 1989).

Malagasy: The chief source of lascas in Malagasy is vein and associated elluvial placer deposits in the northeastern coastal and central plateau regions.

India: India produces lump quartz containing 99.8% SiO_2 for domestic metallurgical uses as well as for export mostly (97%) to

Japan, where it is used predominantly in the manufacture of fused quartzware for the electronics industry. There are more than 100 vein quartz mines in India, 40 of which produce in excess of 1 000 tpy. One producer exported 21.8 kt of high-purity quartz in 1990 (Russell, 1991). Andhra Pradesh leads in production, followed by Karnataka, Rajasthan, Gujarat, Tamil Nadu, Haryana, and Madhya Pradesh (Smith, 1984). Production also comes from the Tiruchirapalli District in Orissa.

Sri Lanka: Veins of massive high-purity quartz (+99.8% SiO_2) have been found at more than 50 locations in the areas of Openaike, Pelmadulla, Pusella, Rattota, and Ratnapura (Harben, 1991). The deposit at Galaha (Ambalamana) contains inferred reserves of 6 Mt (Herath, 1990).

Venezuela: Quartz crystal for cultured quartz growing and silicon metal and alloys manufacture is produced at Santa Elena, Bolivar.

Guyana: Quartz crystal of possible optical and electronic quality is associated with diamonds in alluvial deposits.

Chile: Lump quartz is produced from vein deposits for use as fluxstone at copper smelters. The largest producer has capacity for 200 ktpy, but the +0.635 cm stone is not of exceptional chemical purity. A pegmatite deposit containing electronic- and possibly optical-grade quartz has been found in Chile, however.

Canada: The Mount Rose deposit of high-purity, lascas-grade vein quartz is located near Armstrong, BC. The Fs quartz vein, situated 40 km northeast of Kamloops, BC, has supplied milky quartz for silicon carbide manufacture in Oregon. It is at least 400 m long and 3.5 to 15 m wide. Near Oliver in southern British Columbia quartz veins 0.3 to 4 m wide cut the Oliver Plutonic Complex of Late Jurassic age. They are widest in the porphyritic quartz monzonite phase of the complex and have been quarried for fluxstone and ferrosilicon manufacture. The Gypo deposit near Oliver reportedly produces less than 10 ktpy of lump quartz (Simandl, et al., 1992).

Spain: This country supplies most of the quartz used for silicon metal and ferrosilicon production in continental Europe. Production of metallurgical-grade lump quartz in northwestern Spain is centered at La Coruna, Galicia, where the rock contains less than 0.05% Fe_2O_3, 0.02% Al_2O_3, and 0.01% CaO. Output in 1988 amounted to 966.5 kt, of which about 435 kt were exported principally to Iceland, Norway, and Sweden (Griffiths, 1985b, 1991). For silicon metal and high-purity ferrosilicon alloys (e.g., 75% and 95% Si) Norwegian and Swedish companies use quartz imported mostly from Spain, while their domestic quartz is used for ferrosilicon of lower silicon content (e.g., 50%).

In northwestern Spain Explotacion Rocas Industriales y Minerales SA (Erimsa) produces 350 ktpy of quartz near La Coruna, one-third for sale to domestic producers of silicon and ferrosilicon and two-thirds for export to similar producers in Norway and Sweden. Cuarzos Industriales SA mines a vein of quartz or quartzite at Vivero and exports 80% of its annual output of 250 kt to silicon metal and ferrosilicon producers in Norway, Sweden, France, and Germany (Griffiths, 1985b).

Greece: In eastern Macedonia about 25 ktpy of metallurgical-grade lump quartz are mined from vertical pegmatite veins 1 to 8 m wide within schist at Paranesti, Thessaloniki. The quartz contains 99.5% SiO_2, 0.04% Al_2O_3, 0.03% Fe_2O_3, and 0.01% each of CaO, K_2O, and Na_2O. Most of the output is exported to silicon metal manufacturers (Griffiths, 1985a, Anon., 1987, Georgiades, 1988).

Italy: Lump quartz is produced at Montioni (100 ktpy capacity), Sanfront (80 ktpy capacity), Martiniana Po (50 ktpy capacity), and Sondalo. Markets include a ferrosilicon plant at Montioni and a silicon metal plant at Sondalo as well as refractories, filtration, abrasives, chemicals, and blast furnace steelmaking (Robbins, 1986). One company produces about 130 ktpy of lump quartz, including 70 ktpy at Sondalo in Lombardy. It sells 30 ktpy for silicon metal manufacture and grinds 100 ktpy into potter's flint for the ceramics industry (Anon., 1987).

Portugal: Quartz crystal and lump quartz are mined at three sites in Guardia, where 2 500 tpy have been produced for sale to silicon metal and ferrosilicon manufacturers (Anon., 1987).

Czechoslovakia and Other Countries: Chamber-type pegmatites that contain quartz crystals are of Precambrian age in Brazil, Malagasy, and the Ukraine but are of Late Paleozoic age in central Kazakhstan, China, Mongolia, the northern Ural Mountains, and the Plateau Central in France. Hydrothermal quartz veins that yield piezoelectric crystals in the Urals are found in Proterozoic mica schists and lower Ordovician quartzites. Inclusion-free quartz from hydrothermal veins at Detkovice in Czechoslovakia contains 99.43% SiO_2 and 0.07% Fe_2O_3 after hand sorting. Similar deposits at Grapa, Czechoslovakia, contain 99.86% SiO_2 and 0.062% Fe_2O_3 after hand sorting and only 0.003% Fe_2O_3 after further washing, calcination, magnetic separation, and chemical refining (Kuzvart, 1984). Elluvial placer deposits of quartz crystals in the Ural Mountains of Ukraine may contain gold in economically recoverable amounts.

Other producers of high-purity lump quartz for silicon metal and silicon alloys manufacture as well as specialty glass and refractory products include Finland, France, and Yugoslavia. Silicon and ferrosilicon producers in Japan use quartz imported from China, Korea, and India. Elluvial placer deposits of quartz crystals containing no gold or diamond coproduct values have been found in China. Other vein deposits of quartz crystal and/or lascas are known in Uganda, South Africa, and Australia.

Silica Pebble

United States: One of the most important sources of metallurgical-grade silica gravel for silicon metal and high-purity ferrosilicon production in the eastern United States is high level terrace deposits of the Citronelle (Lafayette) Formation of Pliocene or Holocene age. These are fluvial channel and bar deposits found in raised terraces that roughly parallel the present course of the Pee Dee River in North Carolina and of other major rivers in the coastal plain of South Carolina, Georgia, Florida, and Alabama. They consist of unconsolidated sand and gravel that lie unconformably upon Late Cretaceous sediments. The most productive area encompasses 65 km^2 and is centered at Lilesville in Anson County, NC. The Pee Dee terraces near Lilesville contain a well developed basal gravel zone 1.5 to 6 m thick overlain by sand and a few lenses of gravel and clay. They consist of 46% sand and nearly 54% gravel, and the gravel is 75 to 83% quartzite and 17 to 25% vein quartz (Callahan and Craig, 1990).

Two companies near Lilesville are together currently producing about 11 ktpd of sand and gravel from the Pee Dee River terraces (Callahan and Craig, 1990). Products include construction sand, metallurgical-grade pebble, and clear or white pebbles exceeding 5 cm in size for packing acid towers used in the chemical industry. The pebbles have been used as grinding mill media in the past. One of the Lilesville producers has recently begun a new dredging operation on Beech Island in the Savannah River floodplain near Augusta, GA. A third company produces metallurgical gravel from similar deposits in Harnett County, NC.

Lower purity ferrosilicon (e.g., 50% Si) is manufactured in the southeastern United States from silica gravel dredged from the Coosa River north of Montgomery, AL, and from the Alabama River between Montgomery and Selma, AL. In Florida subrounded quartz and quartzite pebbles have been dredged from the Escambia, Apalachicola, Flint, and Chattahoochee Rivers, and one company still produces river-bed gravel at Lecanto and Crystal River, FL. Blocks for lining grinding mills and artificially rounded grinding pebbles were formerly fabricated from silica pebbles recovered from beds of Ocala limestone in Florida. Quartz and quartzite pebbles in the Bryn Mawr Formation of Pliocene age have been mined for metallurgical and refractory applications in Harford County and to a lesser extent in nearby Baltimore, Carroll, and Howard Counties, MD. Grinding pebbles and blocks for lining grinding mills have been produced in the past from residual deposits of chert derived from the Fort Payne Formation near Iron City in Wayne County, TN.

Numerous alluvial deposits containing laterally extensive but thin beds and lenses of flint pebbles occur in interstream divides, river terraces, and major stream valleys in the Gulf of Mexico Coastal Plain of Texas. These deposits extend west from the Guadalupe River to the Rio Grande. Grinding pebbles that can be milled to a quality equal to that of similar pebbles produced in Europe have been mined in Texas from deposits in Frio, Bastrop, Llano, and Travis Counties, but pebbles in deposits associated with the Colorado and Brazos Rivers generally are unsuitable for use as grinding media. Some flint as stock from which blocks for lining grinding mills are cut has been produced from the Edwards Formation in western Travis County, TX. One company now mines the Austin Chalk Formation in south-central Texas, from which flint nodules are recovered for captive use as an abrasive medium in the production of ground silica flour.

Silica pebbles, apparently derived from the Sioux Quartzite, are abundant along the shores of Lake Superior between Grand Portage and Pigeon Point, MN. As grinding mill media they compare favorably in performance with flint pebbles produced in Europe. Many years ago silica pebbles recovered from the beach between Oceanside and Encinitas, CA, were used as grinding mill media. They were harder than flint pebbles from Denmark, and they were coarser grained and rougher in texture and therefore faster cutting when used in grinding hard materials. Round silica gravel for use in water filtration is produced at several locations in the United States including Muscatine, IA, Eau Claire and Fairwater, WI, and Thompson, OH.

European Countries: *Silex* is a term used commercially for flint or chert, or chalcedonic silica, that is produced in several European countries including France, Belgium, England, and Yugoslavia. Silex is used in the form of round pebbles as grinding mill media and as cut blocks, known as *Belgian block*, for iron-free grinding mill linings. The silex from Belgium is an extremely fine-grained, tough, chalcedonic form of silica containing secondary calcite and radial chalcedony (Griffiths, 1987a). Silex from Yugoslavia is somewhat coarser grained and contains tightly packed quartz crystals. Flint pebbles produced in France, Belgium, England, and Denmark (actually Greenland) are available in two grades and in seven to twelve sizes ranging from 1.3 × 1.6 to 13.3 × 15.2 cm. Some pebbles from Denmark are as much as 20.3 cm in diameter. Pebbles of "Select" quality from Dieppe in France contain 98.43% SiO_2, 0.02% Fe_2O_3, and 0.10% Al_2O_3.

Blocks of silex, or buhrstone, are produced by at least two companies in Belgium from deposits located just east, north, and west of the town of Mons. Blocks range from 60 to 160 mm in thickness and are rectangular or conical in shape for lining cylindrical, tube-type pebble and ball mills. They are cut and fabricated by hand from flint nodules recovered in a marl quarry at Eben-Emael. The best quality flint is obtained from the lowest stratigraphic intervals at a depth of more than 30 m. The average chemical composition is 94.0% SiO_2, 4.9% CaO + $CaCO_3$, 0.9% MgO, and 0.1% Fe_2O_3. One company in Belgium produces approximately 1 000 tpy of finished blocks (Benbow, 1989), and another in Jaen, Spain, produces about 500 tpy (Anon., 1987). Silex blocks and crushed silex also are produced from flint pebbles mined at Dieppe, France.

Flint nodules as grinding mill media are produced from Cretaceous chalk formations in western and northern Europe. Perhaps the largest resource is in the Paris Basin, where a 400 m thick section of chalk contains flint nodules (Kuzvart, 1984). Flint pebbles also are produced from chalk deposits as a byproduct at cement operations in southeastern England. At Northfleet and Shoreham pebble capacity approximates 1 000 tpy, although output in 1984 was only 500 to 600 tpy. The pebbles are washed to remove chalk and iron oxides and are used mostly in the ceramics industry. The three sizes produced are −10.16+5.08, −5.08+2.54, and −2.54 cm (Clarke, 1984).

Canada: A large deposit of well rounded quartzite pebbles is found near Cypress Hills, AB. As grinding media the pebbles from this deposit compare favorably with flint pebbles produced in Denmark (Greenland). Lump quartz has been produced from a deposit on the southwestern shore of Lake Baskatong at Charlevoix near St. Ludger, PQ. The material quarried was used mostly in silicon metal manufacturing and to a much less extent as grinding pebbles.

TECHNOLOGY

Exploration

Exploration for deposits of specialty silica materials is guided first by knowledge of which geologic formations are likely to contain rock of suitable quality and then by geologic reconnaissance and mapping in areas where those formations are known to be present. At hardrock prospects, such as for sandstone, quartzite, novaculite, and vein quartz, the evaluation of commercial potential normally requires drilling and drill sample testing to establish the extent and reserves of suitable stone. Percussion drilling is seldom adequate, because physical properties of the stone are as important, or even more important, than its chemical purity. Chip samples from percussion drilling provide some indication of chemical purity but are unsuitable for many physical property measurements. Diamond-bit core drilling is therefore usually necessary at some stage in the evaluation process. Problems of poor core recovery are commonly encountered when holes are drilled at angles between horizontal and vertical and especially where the silica rock is highly fractured, jointed, or friable. Where conglomeratic sandstones are sought for their silica gravel content, core drilling must be close spaced in order to delineate pebble-rich zones.

At prospects where silica gravel is present in unconsolidated sediments, or where the host rocks enclosing quartz veins are deeply weathered, drilling can be impractical. In such cases trenches are dug with dozers or backhoes to locate and sample the desired material, and the samples are washed and screened to determine the yield of product per cubic meter excavated. Exploration for deposits of river gravel, to the extent that it is carried out at all, is done by dredging.

Mining

Deposits of vein quartz in Brazil are small and irregular in shape and are mined mostly by hand in open pits seldom more than 20 m deep. Dozers are used for overburden removal, and pumps are required for pit dewatering. Clear, piezoelectric-grade crystals of small size are common, but those that weigh 250 g or more are rare, accounting for only 1/10 000 (0.01%) of the total mass of a vein filling (Kuzvart, 1984). The ratio of clear crystal to total material mined ranges from 1:1 000 to 1:1 000 000, so enormous quantities of milky quartz are extracted to recover only a few kilograms of piezoelectric crystal (Lamey, 1966). In exceptionally rich pockets the yield of such crystal can be as much as 0.4 or even 1.6%.

During World War II 40 000 m^3 of vein quartz mined in

Arkansas yielded 200 t of rock crystal, but only about 2 t of the crystal were suitable for piezoelectric applications (Kuzvart, 1984). Three deposits of vein quartz currently are mined for quartz crystal specimens and lascas in Arkansas. At the Paron Mine in Saline County there is little or no overburden, and very little drilling and blasting is required. Massive bull quartz for lascas production is mined with a backhoe during one or two months in the summer, and a stockpile containing about 2 000 t is built to sustain processing operations for a year. Mining methods used at the two crystal specimen operations in Garland County, Arkansas, are similar to those at Paron except that the pits are much deeper, no drilling and blasting is used, and the pockets of crystals are mined by hand. Bull quartz extracted in reaching the pockets is used as feed for lascas production, supplementing that from the Paron mine.

Sandstone and quartzite for nonconstruction dimension stone applications are quarried in a manner that uses joints and bedding planes to advantage. Primary vertical cuts along quarry walls or for separating key blocks can be made with channeling machines, although drilling and broaching is more standard practice. In drilling and broaching, rows of 51 mm diameter vertical holes are drilled and the channel cut is made with pneumatic or hydraulic wedging hammers or by detonating very light charges of black powder. Quarry blocks are freed from a bench or ledge by drilling rows of horizontal holes and then either detonating low velocity explosive charges or using the plug-and-feather method of manual wedging. Blocks are moved from quarries to finishing plants with fixed or mobile derricks, front end loaders, or trucks. Similar methods are used for quarrying whetstone-grade novaculite in Arkansas, where the yield of usable stone from the total amount quarried or within a deposit ranges from a maximum of 10% for hard, translucent grades to a maximum of 25% for softer grades (Kirschman, private communication, 1992).

Metallurgical-grade quartzite deposits in Oregon and Washington are quarried by conventional drilling and blasting in open pits with benches 12 m high and 6 m wide (Peterson, 1976). Similar methods are used in extracting quartzite on Badgeley Island, Ontario, where the quarry measured 700 m long and 200 m wide in 1984 (Guillet and Kriens, 1984). In Stevens County, Washington, the friable portion of the Addy Quartzite is mined in two open pits with benches as high as 45 m. About 50% of the rock can be ripped with a dozer, but the remainder requires drilling and blasting. Conventional drilling and blasting is also used in the open pit mining of conglomeratic sandstone that contains metallurgical-grade pebbles, such as from the Sharon Formation in Ohio.

Metallurgical-grade quartz gravel from the Alabama River between Selma and Montgomery, AL, and at Beech Island, SC, near the Savannah River is recovered by dredging. Suction dredges are suitable for deposits consisting of loose sand and gravel, whereas bucket-ladder or suction cutterhead dredges are used for deposits that contain beds or lenses of clay. Raised fluvial terrace deposits of metallurgical gravel or pebble, such as those along the Pee Dee River in the Carolinas, are mined with front end loaders, draglines, and hydraulic excavators.

Processing

Brazilian lascas is quartz that is defective for piezoelectric (e.g., radio component) or optical applications because it is cloudy or milky instead of clear or contains optical or electrical twinning, cracks, fluid bubbles, or other inclusions. It is produced by hand sorting and trimming such material from higher quality, electronic-grade quartz crystals.

In Arkansas mine-run vein quartz is crushed and wet screened to −3.175+1.27 cm, washed, leached in vats of hot oxalic acid, and then rinsed several times with water. It is then hand sorted, dried, and finally examined on a light table for defects before being packaged in 45.36 kg bags. Defective rejects, −5.08 cm mine-run rock, and the -1.27 cm screenings represent waste that averages about 50% of the rock mined, so two tons of vein quartz are mined for each ton of finished lascas sold.

Processing of metallurgical-grade quartzite and novaculite consists only of crushing, washing, and screening to sizes ranging from −10.16+3.81 to −3.81+0.32 cm. Metallurgical-grade quartz and quartzite gravels from river channel and raised terrace deposits are processed similarly, except that no crushing is required.

Quartzite, sandstone, and jasper slabs for nonconstruction dimension stone applications used to be split by hand. Because they were therefore rough on all six sides, more mortar was required for their installation. Quarry blocks today are cut with wire saws and are split hydraulically. Acid blocks have either two or four sides sawed smooth with the other sides hydraulically split and rough. They consist of beams and slabs of sandstone that are sawed to a thickness of 10.16 to 20.32 cm. Grinding mill and chute liner blocks are usually sawed on all six sides. The red and gray jasper (quartzite) in Minnesota is still cut, trimmed, and shaped by hand into rectangular adamant silica blocks measuring 5.08 to 12.7 cm thick, 10.16 to 15.24 cm wide, and 20.32 to 25.4 cm long for lining pebble mills. The unevenness or irregularity of the surfaces of these blocks is advantageous, because it reduces the need to install lifters in a mill.

As a means of increasing the yield of salable product from each quarry block one producer in Minnesota cuts and shapes cubes of adamant silica by hand for use as grinding media in competition with flint pebbles produced in Europe. The cubes are pretumbled in a mill to round off their edges and reduce their high initial wear, which results in a weight loss of about 20%. The smallest cube that can be hand cut economically measures about 5.08 to 6.35 cm on a side, so the cubes are too large to be used as deburring media. Crushed quartzite has been pretumbled near Baraboo, WI, and sold as a deburring medium in eight sizes ranging from −19.05+14.29 to −3.175+2.38 mm.

Grindstones made mostly of sandstone are produced in Europe as blocks that are cut and shaped and measure 0.6 to 1.5 m across. From the same sandstone beds pulpstones are cut and shaped into blocks 1.5 m in diameter and 0.9 to 3.0 m wide. Used for grinding logs into paper pulp the stone is often seasoned for one to two years after being quarried but before being cut (Bateman, 1959).

Novaculite oilstones, files, and whetstones are produced in Arkansas by sawing quarry blocks with diamond-tipped saws and then shaping and lapping the cut stones by hand. The yield of finished product from carefully selected quarry blocks ranges from about 10% for hard, translucent grades to 25% for softer grades and a maximum of 30 to 40% for perfect stone, which is seldom found (Kirschman, private communication, 1992). This means that the overall yield of finished product from all of the novaculite quarried or within a deposit ranges from only 1% for translucent grades to 6 or 7% for softer grades. At least one company in the district crushes, screens, and pretumbles the waste or scrap novaculite generated at whetstone operations and markets 11 different sizes from 5.08 to 0.635 cm and finer as deburring media for barrel tumbling and vibratory finishing.

Specifications

Metallurgical Applications: Crushed quartzite and novaculite lumps and round quartz and quartzite pebbles from fluvial deposits and conglomeratic sandstones are used in the electric furnace production of silicon metal, ferrosilicon (FeSi) and other silicon alloys. Most FeSi producers prefer quartzite, even though its Al_2O_3 and Fe_2O_3 content tends to be higher, because vein quartz is more brittle and therefore usually generates excessive fines during handling. Vein quartz also often contains fluid inclusions, which cause it to explode into additional fines when exposed to furnace temperatures. For metallurgical applications silica rock should be tough (i.e., not friable), should have a high thermal shock resistance, should be free of fines when delivered, and should not crumble or decrepitate into fines when heated.

Fines cannot be tolerated in submerged arc electric furnaces because they reduce the porosity of the charge, which blocks the flow of gases, causes premature fusion and crusting, and results in a buildup of gas that produces furnace blowouts. Blowouts are sudden releases of high-temperature gas consisting of CO and SiO, so they result in a loss of silicon and, therefore, reduced efficiency. High-purity novaculite has been used in silicon metal and FeSi production, but the fines it generates when heated create a foam that accumulates and requires the furnace to be shut down and cleaned out periodically, which means costly downtime.

Silicon metal producers prefer to use quartz or quartzite lumps that exceed 2.54 cm in diameter, have a minimum softening point of 1 700°C, and do not decrepitate below 950°C. The lumps can range from 1.27 to 10.16 or 12.7 cm in average diameter, but −10.16+3.81 and −7.62+2.54 cm sizes are most commonly used. The rock should contain at least 98.5 or 99% SiO_2 (typically 99.3 to 99.8% SiO_2), and less than 0.1% Fe_2O_3, 0.15% Al_2O_3, 0.2% CaO, 0.2% MgO, and 0.2% loss on ignition (LOI). LOI is directly proportional to the rock's tendency to decrepitate upon heating, and Fe_2O_3 and CaO are especially critical impurities for metallurgical-grade silicon.

For chemical-grade silicon production the silica feed should have high reactivity and a very low alumina content. Some grades of silicon metal require the feed to contain less than 0.05% Fe_2O_3, 0.10% Al_2O_3, 0.005% CaO, and 0.002% TiO_2. In all cases the rock should contain essentially no phosphorus, sulfur, or arsenic impurities, which are particularly objectionable because of the poisonous gases they form in the furnace. Metallurgical-grade quartz in Brazil is typically high in titanium, so the silicon metal produced from it is suitable for secondary aluminum refining but usually not suitable for manufacturing chemicals such as silicones and silanes.

Ferrosilicon producers can accommodate smaller lumps of silica rock ranging from 0.32 to 10.16 cm in diameter, with the size varying in direct proportion to the silicon content of the alloy produced. The rock certainly should contain more than 96% and preferably more than 98% SiO_2. It also should contain less than 0.4% Al_2O_3 (usually less than 0.25%) and 0.2% Fe_2O_3. Its Al_2O_3 content influences the consumption of electric energy during smelting, which can amount to as much as 40 or 45% of total manufacturing costs. The rock's sulfur content should be very low, although its phosphorus content can range up to 0.1%. Its iron and carbon content is usually not as critical as its alumina, alkali, and alkaline earth metals content. One US producer's maximum specifications are 0.03% Fe_2O_3, 0.20% Al_2O_3, 0.01% CaO, and 0.02% TiO_2. Silica rock for silicomanganese (75% Si, 35% Mn) does not have to be of such high purity. Typical analyses of quartz gravels used in FeSi production in the southeastern United States are shown in Table 2.

The Alabama river gravel mentioned in Table 2 is produced in various sizes including −15.24+2.54 and −7.62+2.54 cm. The quartz gravel from North Carolina is produced in sizes ranging from +5.08 to −0.635+0.3175 cm.

In furnaces used for the production of silicon carbide a bed of −10.16+2.54 cm silica rock, through which gas is drawn, filtered, and cleaned, is placed on the floor. Although angular rock has more surface area, round river rock is preferred because it compacts into a looser, more permeable bed.

In the smelting of iron, nickel, zinc, lead, and copper ores the price of an acid flux is more important than its chemical purity, so local sources of −4.445+1.905 cm stone are normally used. The charge to blast furnaces and basic open hearth furnaces used in steelmaking are trimmed with washed, −1.59+0.635 or −2.54+0.79 cm river gravel that contains +99% SiO_2 and is usually locally available and inexpensive.

In the electric furnace production of elemental phosphorus −2.54+0.3125 cm lump quartz or −3.175+0.635 cm quartz pebble from river channel deposits is used. In general, quartz or quartzite used as fluxstone in metal and phosphorus smelting should contain more than 90% SiO_2 and less than 1.5% Al_2O_3, 1.5% Fe_2O_3, and 0.2% combined CaO and MgO.

Table 2. Typical Composition of Quartz Gravel for Ferrosilicon Manufacture

	Wt %					
	SiO_2	Al_2O_3	Fe_2O_3	CaO	P	Ti
North Carolina terrace gravel	99.2	0.068	0.41	0.031	0.007	0.006
Alabama river gravel	98.7	0.27	0.53	0.039	0.009	0.018

Source: R.A. Falconer, private communication, 1989.

Abrasive Applications: For use as abrasive media in grinding mills and deburring equipment quartzite, flint, jasper, and novaculite must have high crushing strength, high specific gravity, toughness or durability, high purity, homogeneity, high hardness, and resistance to staining, fracturing, chipping, and high temperatures. The rock should have low porosity and must contain no toxic substances or dark colored metallic impurities such as iron and manganese. It must be tightly cemented and chemically inert in the presence of strong alkalis and acids. The media should be round in shape and closely sized and should produce a white powder when abraded. The size and shape of constituent grains and their cleavage and fracture characteristics are also important.

High density grinding media are preferred, because their greater impact and grinding force mean that less mill volume is needed for a given load of media or that a faster grind can be obtained without reducing the media load. In either case production capacity is greater than with less-dense media. Novaculite is not usually suitable as a grinding medium because of its comparatively low specific gravity, friability, and low resistance to breaking upon impact during milling. Crushed quartzite is also not usually qualified for high impact applications, because crushing can produce microfractures in the rock that cause it to break apart during use. Properties that determine the suitability of silica rock for use as a grinding medium also determine its suitability as a grinding mill lining.

Silica rock that qualifies as a grinding mill medium and lining material is also normally suitable as a deburring abrasive, where closely controlled sizing is required in order to minimize lodging. Certification tests are carried out in a 0.28 m^3 tub-type vibrator that is loaded with 90.7 kg of media and 4.5 kg of angle iron and is operated at 1 800 vibrations per minute for 24 hours. What is desired is a steady, uniform rate of cutting and wear and the development of a fine, high luster finish on the deburred part. Performance of the abrasive medium is evaluated in terms of its cutting rate, wear rate, cutting efficiency, and the quality of finish produced on the metal part, which should be free of scratches. The cutting rate is the weight percent of metal removed; the wear rate or chipping factor is the weight percent of media lost as nonrecyclable fines; and the cutting efficiency factor is the ratio of cutting rate to wear rate, with high numbers preferred.

Very few sandstones qualify for use as grindstones, for which uniform hardness and sharp and even grains are required. The rock must be cemented tightly enough to be sound and tough but not so tightly cemented that it glazes during use. Coarser grains grind faster but produce a rougher surface. Pulpstones also require that the cement be weak enough to wear and permit the harder quartz grains to protrude. Coarse-grained sandstone produces wood fiber that can be too coarse (Bateman, 1959).

Electronics: For use in electronics and optics quartz crystals should weigh no less than 50 to 100 g (Kuzvart, 1984). Piezoelectric-grade crystals should contain at least 16.4 cm^3 of flawless material (Lamey, 1966). Even clear crystal is unfit if it is optically or electrically twinned or contains phantoms, cracks, liquid bubble inclusions, rutile fibers, or intergrowths with other minerals. Phantoms are growth outlines within a crystal that are commonly marked by very fine clay particles (Bates, 1960).

Pieces of quartz crystal that weigh 10 to 30 g and defective material trimmed from piezoelectric-grade crystals can be used as the nutrient in growing cultured (synthetic) quartz crystals if their iron and aluminum content is low enough. Crystal growers prefer to use a nutrient (i.e., lascas) that contains less than 50 ppm total impurities, although material containing 50 to 100 ppm impurities can be used where requirements are less critical. Iron and aluminum are the most important contaminants, because they govern the amount of undesirable sodium and lithium that enters the crystal during the growth process.

Nonpiezoelectric-grade pieces of ultrapure quartz that weigh 10 to 30 g also can be fused and fabricated into optical fiber preforms (e.g., rods and tubing), precision optical and electro-optical blanks, billets, and prisms, specialty lamp tubing, and quartzware apparatus (e.g., crucibles) for semiconductor electronics and labware applications (Kuzvart, 1984). In order to manufacture fused quartz (vitreous silica, or quartz glass) products that are transparent, a feed that contains less than 30 ppm (0.003%) total impurities is required, which means a silica content of at least 99.997%. The minimum silica purity required in manufacturing opaque fused quartz glass is 99.8%, versus 99.97% for translucent products (Griffiths, 1987a).

Acid Tanks and Towers: Blocks of sandstone and quartzite for lining acid tanks, towers, and trays are resistant to hydrochloric and sulfuric acids but not to hydrofluoric or high-temperature phosphoric acids. The blocks should have good abrasion resistance, a crushing strength of 55×10^3 to 70×10^3 kPa, and a transverse strength of 8.3×10^3 to 9.0×10^3 kPa. Quartz and quartzite gravel used as a packing medium in acid towers must have low porosity (e.g., less than 3%), a high crushing strength, high degree of size uniformity, and high surface area per cubic meter. The gravel should be resistant to dissolution by sulfuric acid and should have a neutral or slightly acid pH, since a pH of 7.5 or higher can cause problems. Low porosity is required for low acid absorption, and a high crushing strength minimizes the generation of fines that can plug the gas orifices and other equipment at the base of a tower. The packed bed of gravel should contain a large percentage of void space per cubic meter.

Air Preheaters: High-purity quartz river gravel is used in three of the four layers that serve as a heat transfer medium in the chambers of air preheaters. Gravel 12 to 15 mm in diameter is used for the bottom (first) layer, which accounts for 5 wt% of the bed. Gravel 3 to 5 mm in diameter is used for the second and topmost (fourth) layers, which together account for about 70 wt% of the bed. The gravel must contain at least 99% SiO_2 and should be round in order for the bed to have sufficient void volume and void size to provide adequate heat transfer and pressure drop. Angular fragments restrict air flow. The gravel also should have high thermal conductivity, high heat storage capacity, and high thermal efficiency (e.g., 98%), and it must be closely sized. Some quartz gravels work well while others do not.

Filtration: Rapid gravity filters at municipal water treatment plants contain quartz gravel 2.54 cm in diameter at the base, grading upward to 3.2 mm gravel at the top, for a total bed thickness of 30.5 to 45.7 cm. According to American Water Works Association specifications only uncrushed media can be approved for municipal filtration installations. The media should contain less than 25% flat or elongate fragments having length:width ratios of 5:1 or more. Crushed rock is angular in shape, so it and flat or elongate fragments tend to interlock in a filter bed. When the bed is backwashed periodically to flush out entrained sediment, locked-together media fragments will explode or erupt. Round quartz gravel does not interlock and provides the 35 to 40% void space required in a bed. The gravel need not be of high chemical purity, but it should have a high specific gravity, very low acid solubility, high coefficient of size uniformity, and an effective size that meets filter specifications.

At trickle filter installations silica rock that is −11.43+3.81 cm in size is used as the support medium on which bacteria are grown for digesting raw sewage sludge. The rock should be of high chemical purity but not necessarily high density. It should be angular in shape in order to have a high surface area. Filter beds used in drying sewage sludge for compost applications require silica gravel having a high coefficient of size uniformity and a particle size of between 3.2 and 12.7 mm. Gravel that is +15.9 mm in diameter is used in septic leach fields.

Refractories and Other Uses: The quartzite (ganister), quartz gravel, and vein quartz used in manufacturing refractory products is usually crushed to between −3.81 cm and −2.38 mm and must be low in iron, alkali, and phosphorus impurities. The stone must have low porosity after being fired at 1 500°C. Fine-grained quartzites usually have lower porosity than coarse-grained quartzites after firing. Even more stringent chemical specifications apply to novaculite that is used in producing special vitreous silica refractory materials. For this purpose the best quality novaculite from Arkansas contains 99.19% SiO_2, 0.02 to 0.03% Fe_2O_3, 0.10 to 0.25% Al_2O_3, and 0.043% Na_2O+K_2O and is crushed to −3.81+1.27 cm.

For use as aggregate in furan polymer concrete, where it is loaded at 90 wt%, silica rock must be angular in shape but can range from 6.35 mm to 44 μm in size.

USES

Electrometallurgy

Specialty silica gravel and lump, metallurgical-quality quartz are used in the production of silicon metal and silicon alloys and as a flux in the smelting of elemental phosphorus, iron, nickel, zinc, copper, and lead.

Depending largely on the purity of the silica feed, silicon metal is produced in electronic grades for semiconductor markets, chemical grades for manufacturing silicones and silanes, and metallurgical grades for secondary aluminum refining. The consumption of quartz raw material ranges from 2.6 to 3.5 per finished ton of 96.0 to 99.7% Si metal. Assuming an average ratio of 2.9, producers of silicon in the United States consumed about 423 kt of silica feed during 1990, excluding that which was used in electronic-grade metal production.

Although ferrosilicon (FeSi) is produced in at least six different grades ranging from 20 to 95% Si, the standard grades are 50% Si (typically 48%) and 75% Si (typically 76%). In addition to silica rock, the charge to an electric furnace consists of coal, coke, or charcoal as the reductant, wood chips to provide porosity, and steel scrap or iron ore as the source of iron. The consumption of quartz raw material per finished ton of alloy ranges from 1.05 to 1.1 t for 45 to 50% FeSi, from 1.7 to 2.0 t for 75% FeSi, and from 2.3 to 2.5 t for 90% FeSi. Assuming an average ratio of 1:9, ferrosilicon producers in the United States consumed approximately 957 kt of silica feed during 1990, raising total consumption in manufacturing all silicon metal and alloys to about 1.38 Mt.

Steelmaking consists of electric furnaces that remelt scrap, blast furnaces that reduce iron ore pellets to pig iron, and basic oxygen furnaces (BOFs) that convert iron into steel. The principal use of silica rock in steelmaking is as an acid flux and slag conditioner at blast furnace operations. Its functions are: 1) to correct, balance, or otherwise adjust the basicity ratio ($CaO:SiO_2$) of the burden when too much lime has been charged, 2) to raise the volume of slag sufficiently to ensure complete removal of sulfur and alkalis, and 3) occasionally to clean the furnace. Electric arc furnace minimills are operated with a basic, not acid, furnace environment, so they use lime or dolomitic lime and fluorspar instead of silica as fluxes. In remelting scrap they do use alloys such as 75% FeSi and silicomanganese (SiMn), however.

Elemental phosphorus is still produced in the United States in Idaho, although production in western Tennessee has ceased. It is manufactured by first nodularizing beneficiated phosphate sand in rotary kilns and then smelting the nodules to elemental phosphorus in electric furnaces charged with fine coke and quartz pebbles.

Abrasives

Novaculite from Arkansas and silex from Belgium are cut into whetstones, oilstones, hones, files, and other special shapes for residential, industrial, leisure, and craft uses. Residential uses include the sharpening of kitchen knives and other cutlery such as scissors, shears, and lawn and garden tools. Industrial uses include the sharpening and honing of cutting surfaces and the polishing of metal surfaces in watch and clock manufacture and repair, gun and reciprocating jet engine manufacture and repair, and tool and die work. Leisure applications include the sharpening of sports knives, arrowheads, fishhooks, spear points, etc. Craft applications include the sharpening of woodcarving tools, gun engraving, jewelry making, and other engraving (Austin, 1991). Hard Arkansas whetstones are used for fine finishing and edging work, whereas Ouachita Stone-based whetstones are used in the coarse sharpening of large tools such as scythes.

One Arkansas novaculite producer's price list consists of four grades, twelve shapes, and more than 300 different sizes of cut stone products. The grades are Soft Arkansas for medium work, Hard Select Arkansas for fine work, and True Hard Arkansas and Translucent for extra-fine work. The most popular grade is Soft Arkansas, which is used by woodcarvers, sportsmen, butchers, commercial knife sharpeners, and housewives. The Translucent grade is used mostly for obtaining an extremely fine polish on industrial and surgical instruments. The shapes include 1) bench stones, 2) triangular and square files, 3) special round, flat, bevel, diamond, and oval files, 4) slip stones, 5) carving tools, 6) pocket stones, 7) penknife pieces, and 8) hones (Kirschman, private communication, 1992).

Oilstones for fine sharpening have been made from fine-grained sandstone quarried in Ohio. Waste from whetstone operations in Arkansas is used as an abrasive medium in wet (vapor) pressure blasting, as a grinding and deburring medium, as a -74 µm loose-grain abrasive for cleaning printing press plates, and in the manufacture of cuticle stone. The largest US producer of cuticle and coaster stones in Bedford, IN, uses a patented process to bond scrap novaculite, which has been ground to 149 and 74 µm, with epoxy resins.

Pretumbled nonwhetstone-grade novaculite and hand cut quartzite cubes are used in barrel tumbling and vibratory machines for light burr, rust, and scale removal, radius formation, cleaning, and degreasing. Applications include the refinement, burnishing, and high luster finishing of cast and punched parts made of mild steel, hardened steel, yellow brass, aluminum, and stainless steel and the deflashing, degating, and deburring of cast and extruded Teflon rings and other plastic parts. Rough, untumbled, and unrounded media are seldom used.

The deburring market is highly diverse and fragmented, ranging from numerous small tool and die shops to huge metal casting foundries and stamping operations in the automotive and aerospace industries. In the United States this market consumes well in excess of 45 Mkg of abrasive media annually and consists predominently (+99%) of wet (vs. dry) deburring operations. A typical 0.57 m^3 vibratory machine uses a 907 kg load of abrasive media and requires nearly 23 kg per day of new media, so the daily makeup represents an average 2.5% of the media load. Novaculite is not suitable for use with complicated, intricate castings where abraded particles can lodge in blind holes and tight spots, but it can be used with lighter weight, simpler castings.

Quartzite (jasper) quarried in Minnesota and silex (chalcedonic silica) produced in France are cut and shaped mechanically and by hand into rectangular blocks that are used in lining tube or cylindrical, tile-type ball and pebble mills. Grouted into place with high-alumina cement or epoxy resin loaded with alumina or silica, jasper linings seem to last indefinitely in mills used for grinding pyrophyllite, feldspar, high-purity quartz, industrial silica flour, talc, gypsum, clays, and wollastonite, for example. They are used primarily in applications where iron contamination from metal alloy liners or alumina contamination from ceramic liners cannot be tolerated. The same quartzite is cut into cubes that are prerounded and used as the grinding medium in cylindrical or conical ball and pebble mills, as are flint pebbles and silex (buhrstone) cubes from Texas, France, Belgium, England, and Yugoslavia.

Natural silica grinding media have been replaced to a large extent by high density, high-alumina manufactured ceramic media but are still used for some nonmetallic minerals and ores, such as those mentioned above, as well as some metallic ores. They are also preferred for fine grinding materials that are sensitive to contamination by discoloration, such as specialty coatings (paints), certain chemicals, ceramic glazes and enamels, organic pigments, white portland cement clinker, and technical ceramic raw materials (e.g., zircon). A major use for silica media is in grinding porcelain glazes and lacquers, because they do not cause the product to become cloudy during grinding.

Although largely replaced now by artificial/synthetic abrasives, grindstones made from fine-grained sandstone are still used in sharpening edge tools such as saws, machine knives, scythes, shears, and other harvesting and die-making equipment. Hard sandstone, quartzite, and quartz pebble conglomerate have been used as millstones, which are large circular stones that include buhrstone and chaser stone. True buhrstone is chalcedonic silica and is used in grinding grain, paints, fertilizers, and graphite. Chaser stone has been used for grinding feldspar, quartz, barite, and other nonmetallic minerals (Bateman, 1959). Sandstone is also cut into pulpstones for grinding logs into pulp for paper manufacture.

Acid Tanks and Towers

At South Amherst, OH, sandstone (firestone) is quarried and cut into blocks or bricks used at steel mills for lining soaking pits, acid tanks, acid wash trays, and steel pickling tanks. The blocks are used also at chemical plants and foundries, by oilfield service companies, and to line acid towers. Acid blocks cut from quartzite at Jasper, MN, are used mostly in the chemical industry.

Acid towers are usually packed with ceramic media of various sizes and shapes but can be packed with quartz or quartzite gravel. There are hundreds of applications for such towers, including the drying (dehydration) and purification of gases such as chlorine at chemical processing plants. Sulfuric acid or another acid is introduced at the top of a tower containing a packed bed of gravel, and it coats the surfaces of the packing as it trickles downward. Wet chlorine gas is then introduced at the bottom of the tower, and as it migrates upward through the packing its water content is stripped by the gravel's acid coating.

Refractories

Sandstone, quartzite, quartz pebbles, conglomerate, and nonwhetstone-grade novaculite are crushed and used as ganister in manufacturing refractory bricks, tiles, kiln furniture, and other shapes as well as ramming mixes, mortars, patching compounds, and other monolithic specialties. Such products are used in lining cupolas, ladles, Bessemer converters, metal-pouring shrouds and nozzles, crucibles, mixer cars, byproduct coke oven doors, and blast furnace hot stoves and checkers at steel mills. They are used also in lining kilns for firing ceramic products and in the glass industry for lining glass tank crowns and regenerator checkers. High-purity but nonwhetstone-grade novaculite also is being fused in a high-temperature rotary kiln by one refractories manufacturer in Missouri. The water quenched chunks of vitreous silica produced are crushed, closely sized, and sold as shell and core sand to investment casting foundries. Vitreous silica has an extremely low coefficient of thermal expansion and good dimensional stability at high temperatures.

Electronics and Optics

Natural electronic-grade quartz crystal was the principal form of quartz used in electronics and optics until 1971, when cultured (synthetic) quartz crystal took over. Cultured crystal has now displaced natural crystal in nearly all applications. Very small quantities of natural crystal from Brazil are consumed in the United States today, mostly for cutting the seed plates used in growing cultured quartz and in a few direct applications, such as deep-well pressure transformers (Ober, 1991).

Five companies in the United States grow synthetic quartz crystals in autoclaves by means of a hydrothermal process. In 1990 they used a total of 527 t of high-purity lascas as the nutrient. By comparison, less than 0.4 t of natural quartz crystal was consumed in the United States in 1990 (Ober, 1991). For producing 1 kg of as-grown cultured quartz between 1.3 and 1.4 kg of lascas are required.

The worldwide consumption of further purified, lascas-grade quartz in the manufacture of specialty lamp tubing, fiberoptic preforms, precision optical and electro-optical devices, and fused quartzware for the electronics (e.g., semiconductor) and laboratory markets is easily ten times the consumption of lascas in the cultured quartz industry. In the manufacture of vitreous silica or fused quartz products the ratio of quartz feed to finished product ranges from 1:1 to 3:1.

Other Uses

Air preheaters operate continuously at temperatures of 1 165 to 1 294°C and are used in the thermal destruction of volatile organic compounds (VOCs) present in the offgases from industrial facilities such as painting booths. Their heat transfer chambers are equipped with perforated plates at the bottom, above which are placed four layers of heat transfer media. The bottom (first) layer consists of coarse quartz gravel; the second and top (fourth) layers consist of finer quartz; and the third layer, where the heating elements are installed, consists of synthetic mullite aggregate. Altogether the four layers comprise a bed averaging approximately 30 t of media in each preheater. One major US manufacturer of air preheaters purchased a total of 1 545 t of 5 mm quartz gravel and 440 to 550 t of 12 to 15 mm gravel in 1989.

Water filtration applications for specialty silica stone range from rapid sand or rapid gravity filters for the purification of municipal water supplies to trickle filters and drying beds for municipal sewage sludge used as compost at mushroom farms and for other agricultural purposes. They also include commercial and industrial water treatment installations and septic leach fields. The second and fourth quarters usually are the annual peaks for filter media sales. Another filtration application for quartz gravel is in the cleaning of gases during the manufacture of silicon carbide, where it is placed as a lining on the floor of electric furnaces.

Other commercial applications for specialty silica rock, often produced as a byproduct at operations that serve principally the markets mentioned above, include the following construction products: roofing granules and chips; decorative, exposed, semilightweight aggregate in precast concrete panels and other white, architectural concrete products; aggregate in polymer concrete used for the protection of concrete flooring from acids; aggregate bonded with epoxy resins in the manufacture of curtain walls; aggregate for subbase road construction; nonpolishing, skid-resistant aggregate for asphalt highways; rip-rap and railroad ballast; and nonskid aggregate in the top coating of concrete flooring that is subjected to heavy wear, such as from forklift traffic.

Nonwhetstone-grade novaculite is ground and used as a filler/extender in latex paints, silicone rubber, and plastic casting, molding, and potting compounds, such as for large electrical insulators.

Miscellaneous uses include: poultry, pigeon, and parakeet grit (mixed with crushed oyster shells), sandblasting sand, decorative landscaping and garden stone, sodium silicate manufacture, the polishing of electronic microcircuit dice pedestals, aquarium chips, aggregate for swimming pool plaster, and fine aggregate in abrasion resistant linings.

ECONOMIC FACTORS

Production

According to the US Bureau of Mines 99% of the value and 97% of the total quantity of specialty silica stone sold or used by US producers in 1990 were accounted for by novaculite from Arkansas. The whetstone and oilstone industry in Arkansas consists of four integrated companies that quarry crude novaculite and cut finished products, three companies that cut finished shapes from purchased crude stone, one that quarries and sells crude novaculite, and one that produces pretumbled deburring media from purchased scrap. Production of crude silica stone for grindstones (in Ohio), oilstones, whetstones, and deburring media in 1990 amounted to 454 t valued at $230,000, excluding grinding mill pebbles and liners. Finished specialty silica stone products sold or used by US producers in 1990 amounted to 450 t valued at $6.33 million, including only grindstones, oilstones, and whetstones (Austin, 1991).

Production of lascas in the United States amounted to an estimated 423 t in 1990, and US producers of synthetic quartz crystals consumed about 527 t of lascas and less than 400 kg of natural, electronic-grade quartz crystal. US production and reported consumption of lascas in 1991 were 425 and 530 t, respectively, although apparent consumption was 564 t in 1990 and 550 t in 1991 (Ober, 1992). US production of as-grown and lumbered cultured quartz in 1990 was 324 and 33 t, respectively (Ober, 1991).

US production of industrial silica gravel for electrometallurgical, filtration, and other specialty applications in 1990 amounted

Table 3. Specialty Silica Stone Production in the United States

	1986	1987	1988	1989	1990
Special silica stone, crude[a]					
Tons produced	973	1 378	1 892	898	454
Value (thousands)	$ 501	$ 489	$ 566	$ 147	$ 230
Special silica stone, finished[b]					
Tons sold or used	463	598	371	377	450
Value (thousands)	$6,520	$7,367	$ 5,415	$ 5,459	$ 6,328
Lascas					
Mine production (tons)	544	—	272	454	423
Consumption[c] (tons)	277	499	451	572	527
Silica gravel[d]					
Tons sold or used	439	572	1 154	1 256	1 315
Value (thousands)	$4,853	$6,424	$11,796	$14,388	$15,284

[a] Includes grindstones, oilstones, whetstones, and deburring media but excludes grinding pebbles and tube mill liners.
[b] Includes grindstones, oilstones, and whetstones but excludes grinding pebbles, tube mill liners, and deburring media.
[c] Excludes quartz consumed in the manufacture of fused quartzware.
[d] For silicon metal and ferrosilicon manufacture, filtration, flux, and other uses.
Sources: Anon, 1989, Austin, 1991, Bolen, 1989, 1990, 1991, Ober, 1990, 1991, Tepordei, 1988, 1989.

to 1.32 Mt valued at $15.28 million (Bolen, 1991). Table 3 summarizes US production statistics for several forms of specialty silica materials for the period 1986–1990. Similar statistics for other countries are seldom published.

Silica stone for which US production statistics are unavailable include: 1) Arkansas novaculite used in manufacturing silicon metal and alloys, vitreous silica refractories, oilfield proppants, deburring media, and fillers, 2) quartzite quarried in Minnesota and South Dakota for ferrosilicon production, grinding mill pebbles and liners, and fluxstone, 3) high-purity quartz mined from pegmatite and vein deposits in South Dakota and New Mexico for the manufacture of specialty glass products, 4) sandstone quarried in Ohio for grindstones and lining acid tanks and towers, and 5) highly refined quartz grain produced from alaskite deposits in North Carolina for the manufacture of fused quartz products. In 1990 the author estimated that annual production of lascas and fusing-grade quartz grain in the United States was about 15.7 kt valued at $25.5 million and that annual consumption of these materials worldwide was about 28.1 kt valued at $34.7 million.

Pricing

The average unit values of specialty silica materials produced in the United States during 1986–90 are shown in Table 4. The values shown are heavily skewed toward whetstone-grade novaculite and do not represent all specialty silica materials as defined in this chapter.

The price of Arkansas lascas ranges from 0.77 to $0.99/kg, depending on the volume purchased, and averaged $0.93/kg at the end of 1991. In 1988 the average customs value of imported lascas at US ports was $2.78/kg for Brazilian material (crude Brazilian pebble) and $0.64/kg for quartzite from Namibia, and the overall average price for the 97 640 kg imported was $1.84/kg (Ober, 1990). The average price of all lascas consumed by the five US cultured quartz crystal growers was $.85/kg in 1990 and 1991 (Ober, 1992). The average value of cultured quartz produced in the United States during 1990 was $28.35/kg for as-grown crystal and $107.48/kg (vs. $120/kg in 1991) for lumbered quartz, which is as-grown crystal that has been further processed by sawing and grinding (Ober, 1991).

As of August 1991 the price of fusing-grade lump quartz exported from India to Japan was only about $0.05/kg (Russell, 1991). Fused quartz ingots are sold on the merchant market for 66 to $88/kg at the low end in terms of purity, whereas synthetic fused silica is priced at 5 to 10 times that much.

Silica gravel produced in the United States during 1990 was valued at 11.02 to $11.70/t for silicon metal and ferrosilicon manufacture and 9.65 to $9.71/t for flux applications. Its unit value for filtration ranged from $11.14/t in the South to $55.13/t in the Midwest and for other uses ranged from $9.57/t in the West to $18.69/t in the Midwest (Bolen, 1991).

Producers of silicon metal and silicon alloys in the eastern United States currently pay delivered prices of 23 to $25/t for high-purity quartz gravel from North Carolina, 15 to $20/t for lower quality river gravel from Alabama, and 17 to $24/t for high-purity novaculite from Arkansas. The cost of transportation generally doubles the FOB prices quoted by silica producers, which range from 10 to $13/t for North Carolina gravel, 9 to $12/t for Arkansas novaculite, and 7 to $9/t for Alabama gravel. Prices are 2 to $3/t higher for the larger sizes (e.g., −10.16+3.81 and −7.62+2.54 cm) than for the smaller sizes (e.g., −3.81+1.27 and −3.81+0.95 cm). Higher purity and larger sizes are required for silicon metal and 75% FeSi manufacture, whereas less pure and smaller stone, such as inexpensive river gravel, can be used in producing 50% FeSi, silicomanganese (SiMn), ferrochromium silicon, and magnesium ferrosilicon. Delivered prices as low as $8/t have been quoted for locally sourced gravel for SiMn manufacture.

Blast furnace operations in the steel industry do not require high-purity silica rock as an acid flux, so they normally trim with river gravel that is dredge-mined locally and delivered for prices of less than $10/t and commonly only $5 to $6/t. Nonferrous metal smelters likewise are extremely sensitive to raw material pricing and do not require high-purity fluxstone. Before they closed in 1991 the two elemental phosphorus smelters in western Tennessee purchased local −3.175+0.635 cm river gravel for less than $10/t delivered.

The value of crude Arkansas novaculite suitable for cutting into finished whetstones and related products in 1990 ranged from $0.33/kg for the Washita grade to $2.75/kg for the Black Hard Arkansas grade. The average value of finished oilstones, hones, whetstones, and grindstones sold or used by US producers in 1990 was $13.86/kg (Austin, 1991). Industrial files and special shapes are priced higher than sporting goods products.

The value of nonwhetstone-grade novaculite produced in Ar-

Table 4. Average Unit Values of Specialty Silica Materials Produced in the United States

	$/Ton				
	1986	1987	1988	1989	1990
Special silica stone:					
Crude	514.90	354.86	299.15	163.70	506.61
Finished products	14,082.07	12,319.40	14,595.69	14,480.11	14,062.22
Silica gravel					
Silicon and ferro-silicon	10.19	10.77	11.45	12.06	11.60
Filtration	25.72	18.58	5.56	13.57	16.26
Flux	—	—	8.82	9.36	9.70
Other uses	16.25	11.50	10.66	—	—
Overall average	11.06	11.22	10.22	11.45	11.62

Modified from Austin, 1991, Bolen, 1989, 1990, 1991, Tepordei, 1988, 1989.

kansas in 1990 ranged from $39 to more than $300/t (Austin, 1991). As much as 1 Mtpy of such rock has been quarried and sold as construction aggregate for only 3.50 to $5.50/t (Steuart, et al., 1985). Novaculite of high purity but not whetstone quality is being purchased by a manufacturer of refractory vitreous silica grain for a delivered price of 50 to $60/t for −3.81+1.27 cm stone. Novaculite unsuitable for cutting into whetstones is crushed, sized, pretumbled, and sold as a deburring medium for 0.44 to $0.55/kg wholesale and 0.66 to $0.88/kg through distributors. It is screened into 11 sizes ranging from 5.08 to −0.635 cm, packaged in 22.68 kg plastic bags, and sold in lots containing from 45 to as much as 18 150 kg. Novaculite ground to −74 µm for cleaning printing press plates is priced at 0.26 to $0.33/kg.

Rounded flint pebbles from France, Belgium, England, and Denmark are available in as many as 12 sizes for use as grinding media in pebble mills. In the United States they are priced at 1.10 to $1.98/kg for small orders and 0.66 to $1.32/kg in bulk quantities. The smallest pebbles command the highest prices because they are scarce but provide the fastest grind. US prices for French pebbles generally range from $1.15/kg for largest, nominally 10.16 to 15.24 cm size to $1.96/kg for the smallest, −1.59+1.27 cm size. Pricing also depends on whether the pebbles are of Standard or Select quality. The most popular pebble is Standard #1, which is −6.35+3.81 cm in size and sells for about $1.48/kg in the United States.

Cut and pretumbled cubes of jasper (quartzite) from Minnesota are priced at about $1.10/kg but are now rarely used as grinding media in pebble mills. Cut blocks of Minnesota jasper for ball and pebble mill linings are priced at 0.83 to $1.98/kg if purchased through distributors or 0.44 to $1.10/kg if purchased directly from the producer. Before the operation closed in 1988, pretumbled chunks of quartzite from a producer in Wisconsin were sold as deburring media for prices that averaged about $0.66/kg but ranged as low as 0.33 to $0.44/kg. Prices varied not by the size of stone but by the size of order, namely, less than 1 815 kg, between 1 815 and 9 075 kg, and in excess of 9 075 kg.

Round river gravel for use in water filtration is packed in 22.68 kg, 45.36 kg, and one-ton bags and is shipped in lots ranging from 36 t to 36 kt. Prices of filter gravel vary with the size of stone and whether it is sold direct to municipal waterworks, to contractors who install water treatment plants, or to the manufacturers of filter plants. Round, high-purity quartz gravel for use in air preheaters is priced in the United States at 72 to $110/t packaged in 1 360 kg bags. Crushed quartz and quartzite aggregate for furan polymer concrete is priced at 66 to $110/t.

Transportation

The place value of specialty silica materials varies inversely with their market value. High priced products, such as whetstone-grade novaculite, lascas, fusing-grade lump quartz, flint grinding pebbles, silex and quartzite grinding mill linings, electronic-grade quartz crystal, and some metallurgical-grade lump quartz and quartzite, are items of international trade. Some quartz gravel for water filtration is also exported. Lower priced materials, such as are used in refractories, as fluxstone at ferrous and nonferrous metal smelters, and in the manufacture of some silicon alloys, are shipped only to local or regional markets.

Metallurgical-grade silica raw materials for silicon metal and alloys production used to be obtained with 500 km of the smelter. Rising power costs have forced many electric furnace operations to be relocated closer to supplies of low cost energy, and rising transportation costs have forced others to be relocated closer to steel mills. In both cases the consequence is that silica rock is now hauled from more distant sources (Murphy and Brown, 1985). The centers of ferroalloy production in Canada, for example, are in British Columbia, near Welland. Thorold, and Niagara Falls, ON, and near Beauharnois, PQ, where hydroelectric power is available and comparatively inexpensive. The major centers of ferroalloy production in the United States are also where hydroelectric power is available, such as in the Tennessee Valley Authority's service territory (e.g., Alabama, Kentucky, Ohio, Tennessee, and West Virginia), the Niagara Falls, NY, area, and the Pacific Northwest (e.g., Washington and Oregon).

Some US manufacturers of silicon metal and alloys purchase 1) quartz gravel from North Carolina delivered in open-top, gondola-type railcars holding payloads averaging 68 t (Peterson, 1976), 2) crushed Sioux quartzite delivered by rail or river barge from South Dakota, and/or 3) high-purity novaculite delivered by low cost river barge in 1 360 t payloads from Arkansas. Other manufacturers, particularly those that have operations located close to their silica source, prefer to receive just in time shipments by truck in order to minimize inventory costs. In any case electric power costs generally represent at least 33% of the total direct operating cost for silicon metal production, and silica raw materials account for about 11 and 8% of the direct operating costs for ferrosilicon and silicon metal manufacturing, respectively. Lump quartz from Spain is used in producing 75% FeSi and silicon metal in Norway, where its delivered price of $60/t includes an estimated $30/t in transportation cost.

Alternative Materials

Grinding Mill Linings: Pink granite from Salisbury and Faith, NC, was once used in lining pebble mills but has been replaced by Minnesota jasper (quartzite), which is harder and tougher and can be shaped more accurately. Blocks of Minnesota jasper are still used for lining tile mills at feldspar grinding operations. Hardinge mills for grinding talc and wollastonite are lined occasionally with the lightest, BB grade of silex (buhrstone) blocks measuring 25.4 to 27.94 cm long, 10.16 cm wide, and 10.16 to 12.7 cm thick. Silex blocks from France and Belgium are not readily available and are very expensive, however, so they are seldom used today.

Most ball and pebble mills are lined with ceramic materials, chromium-molybdenum steel alloys, aluminum alloys, fused alumina, or special rubber. Where an iron-free lining is required, natural stone has been replaced largely by single-piece porcelain ceramic liners that contain 27 to 74% Al_2O_3, are 5.08 cm thick, and cost 2.40 to $2.90/kg. Ceramic liners are also constructed with precast porcelain blocks set with special cements or epoxy resins and are used in grinding paint, ceramic, and chemical materials.

Grinding Mill Media: As the grinding medium in ball and pebble mills flint pebbles, silex, and jasper chunks and cubes have been widely replaced by manufactured abrasive media including carbon steel and chrome alloy steel balls, glass beads, high-alumina porcelain balls, and zirconia beads. High density, high-alumina porcelain grinding media are manufactured as balls, discs, cylinders, and rods ranging from 1.59 mm to 7.62 cm in diameter and from 87 to 99% in alumina content. They are made in various grades of hardness, which average about 8, and they have a specific gravity of 3.4, which is 40 to 50% that of forged steel. Their bulk density averages about 2 210 kg/m^3, compared to 1 680 kg/m^3 for Minnesota jasper.

The prices of high-alumina ceramic grinding media depend on their size, shape, and Al_2O_3 content and range from 0.66 to $2.20/kg for large diameter sizes and from 2.40 to $4.00/kg for small diameter sizes. The smaller sizes are used for really fine grinding, such as of inks, paints, and chemicals.

Factors in the choice of grinding abrasives include the price of the media, grinding speed, media wear or loss, product contamination, and mill lining wear. In most applications high-alumina ceramic materials are the most cost effective choice, which explains why they have captured most of the grinding media market. Their high specific gravity provides high grinding speed, and their toughness minimizes fracturing, wear, and loss. They are longer lasting than Belgian silex pebbles, and their use usually extends the life of grinding mill linings, although their high density can cause strain on mill bearings. Feldspar producers typically experience a media loss of only 0.55 kg per ton of ground feldspar. Some ground silica flour producers use pretumbled jasper (quartzite) cubes or flint pebbles in combination with high density porcelain ceramic balls as a blended medium in order to reduce costs and the load on mill bearings.

Forged steel and stainless steel balls from 1.27 to 15.24 cm in diameter are priced between 4.35 and $17.65/kg and are often used where iron contamination of the product is not of concern. Zirconia grinding media have a high specific gravity and wear resistance similar to ceramic materials. Glass beads range in specific gravity from 2.5 for lead-free glass to 2.94 for leaded glass. They vary in diameter from less than 1 to 15 mm and cost nearly $5.00/kg. Chunks of recycled high-silica refractory brick priced at only $0.25/kg have been used as a grinding medium, but they cause excessive wear to a mill's lining at its discharge end. In some instances, such as nepheline syenite, the grinding medium has been fabricated from the same rock as that being ground.

Deburring Media: As in the case of grinding mill media, natural stone abrasives once used as deburring media have been largely displaced by manufactured products, including glass beads, fused aluminum oxide, high-alumina ceramic materials, stainless steel, and even plastics. Such media are used in vibratory and tumbling types of machines for the deburring and surface finishing of stamped metal parts and foundry castings and for the deflashing of cast plastic parts.

High-alumina ceramic and fused or sintered aluminum oxide tumbling media are manufactured in a variety of shapes, called preforms, including cones, tubes, pyramids, and stars. They are harder, tougher, and more abrasive than natural stone media and do not represent a silicosis hazard. They also are produced in crushed and rounded form, called random shapes or nuggets. Their bulk density ranges from a low of 1 360 kg/m^3 for ceramic media to a high of 2 080 to 2 400 kg/m^3 for fused aluminum oxide. Pricing depends on the size of medium, its Al_2O_3 content, and the quantity ordered. Prices range from 0.65 to $2.00/kg for ceramic media, 1.75 to $6.60/kg for fused alumina preforms, and 1.55 to $3.30/kg for random-shaped fused alumina media. Small sizes are priced 7.5 to 8.5% higher than large sizes. The higher the Al_2O_3 content, the faster the product's cutting ability and the higher its price; so the highest priced media are small, complex preforms having a very high alumina content. Preforms having the highest Al_2O_3 content cut fastest but have the lowest efficiency factor due to high wear rate, so they are used in applications where an extra-fast cutting rate is more important than media loss and makeup.

The total US market for ceramic preforms is well in excess of 45 Mkg per year, and random shapes hold 7 to 10% of the total US market for ceramic deburring media. Individual customer consumption ranges from 45 kg per year to 2 270 kg per month. Ceramic media are too aggressive in some applications but not aggressive enough in others. Less aggressive polyester plastic preforms average 1 040 kg/m^3 in bulk density and cost between 2.00 and $4.40/kg. Prices for stainless steel deburring media range from 4.35 to $17.65/kg. Emery and corundum also have been used as random-shaped, natural abrasive deburring media.

Acid Tanks and Towers: As a lining for acid towers, extruded bricks can be used instead of silica pebble. The bricks are composed of coarse, −0.84 mm silica sand plus ball clay or an organic binder. As the packing medium for such towers special ceramic, plastic, or metal materials can be used instead of quartz gravel. Plastic media are the cheapest, but they are not easily wetted by acids and can be used only where operating temperatures are low. Ceramic and special metal packing media are used where operating temperatures are high. Ceramic materials are more easily wetted by acids and are less expensive than metallic media.

Air Preheaters: Instead of high-purity quartz gravel, most manufacturers of air preheaters use high-alumina ceramic pieces to fill the metal chambers. The ceramic media are heated to high temperatures and serve as heat exchangers in stripping VOCs from industrial exhaust gases blown through the chambers.

Silicon Metal: At East Selkirk, MB, silicon metal is being smelted from silica sand instead of traditional metallurgical-grade quartzite. The sand is mined on Black Island in Lake Winnipeg and is smelted in a plasma furnace to produce silicon metal for the manufacture of silicone fluids, resins, and elastomers.

Cultured Quartz: As the seed for growing synthetic quartz crystals, plates cut from specially grown cultured seed crystals have nearly replaced those cut from natural crystal. A slow deterioration of the properties of crystal grown from synthetic seeds, however, requires that producers periodically start the process over with high-purity natural seed plates. High-purity silica sand represents the major threat to lascas as the nutrient in growing cultured quartz crystals.

Whetstones, Hones, and Grindstones: Natural abrasives have been displaced by manufactured abrasive products made of such materials as fused aluminum oxide and silicon carbide. In most applications synthetic materials are longer lasting and more cost-effective, although whetstones, hones, and oilstones made from Arkansas novaculite continue to be used in niches of the abrasives market. Some pulpstones still are fabricated from granite and other types of stone.

GOVERNMENTAL CONSIDERATIONS

Policies adopted by national governments can impact the availability, pricing, and consumption of some specialty silica materials. The Brazilian government, for example, establishes the minimum prices for first, second, and third quality Brazilian lascas in the export market. In 1974 it embargoed shipments of lascas from Brazil, and as a result prices increased nearly 6-fold that year and nearly 10-fold subsequently. This action was what led to the initiation of lascas production in the United States, Angola, and Namibia. Nationalization of lascas operations in Angola in 1979 and the civil war that followed have interrupted supplies of Angolan lascas since 1980.

In past years the government of India has followed policies that restricted the exporting of certain minerals. One of the objectives of its current five-year plan, however, is to promote mineral exports in order to generate foreign exchange earnings (Russell, 1991).

Import Tariffs

Lascas for growing cultured quartz crystals and manufacturing fused quartzware products are imported into the United States free of tariffs in the case of most-favored nation (MFN) and non-MFN sources. This applies to lascas designated as crude Brazilian pebble from Brazil and as quartzite from Namibia. No tariffs are levied on US imports of other types of special silica stone, such as flint or silex pebbles from Europe and metallurgical-grade quartzite from Canada.

Depletion

The depletion allowance for federal income tax purposes in the United States is 14% of gross income for domestic as well as foreign production of metallurgical-grade quartzite and river gravel, lascas from vein quartz deposits, nonconstruction dimension stone, novaculite for industrial applications, and other industrial silica gravel. The rate allowed for natural electronic-grade (radio-grade) quartz crystal is 22% for domestic production and 14% for foreign production.

US Stockpile

The United States designated natural electronic-grade quartz crystal as a strategic and critical material after World War II and proceeded to purchase crystal from Brazil for the National Defense Stockpile (NDS). Since that time the technology for growing synthetic quartz crystals has been developed, and cultured quartz has now displaced natural crystals in essentially all applications. As a consequence, the NDS inventory has been reduced periodically by sales of excess material on the open market, primarily for specimen and gemstone uses. Sales during 1989 and 1990 amounted to about 74 800 kg valued at $368,000 and 172 000 kg valued at $444,000, respectively. At the end of 1990 the NDS contained about 686 000 kg of natural quartz crystal valued at $9 million, which represented an excess of 576 000 kg above the 110 000 kg goal now set for the stockpile. An additional 36 000 kg were sold during the first three quarters of 1991, at the end of which the NDS inventory contained 650 000 kg, or 540 000 kg in excess of the goal. Approximately 25% of the material sold in 1990 was used in upgrading certain ferroalloys in the stockpile. Some of the material is suitable as a source of seed plates for cultured quartz crystal production, but virtually none of it is suitable as a substitute for cultured quartz itself (Ober, 1990, 1991).

Environmental Regulations

Most of the types of specialty silica materials discussed here are comprised essentially entirely of crystalline silica, which in respirable form has long been known to cause silicosis, a disabling pulmonary disorder. Ultrafine silica particles about 0.5 μm in size are believed to constitute the greatest health hazard, because their surfaces exercise a strong oxidizing effect comparable to that of ozone. This effect is especially pronounced for quartz particles that have been recently ground, because freshly fractured and exposed surfaces are more chemically reactive than aged or weathered surfaces.

The silicosis risk associated with crystalline silica materials has caused consumers in some markets to abandon them in favor of alternative materials. Since 1988 this abandonment has accelerated, due in large part to a monograph issued that year by the International Agency for Research on Cancer (IARC). In that monograph IARC classified crystalline silica as a Group 2A (i.e., probable) carcinogen in humans. The subsequent IARC data sheet listing suggested that all Material Safety Data Sheets for products containing 0.1% or more crystalline silica should contain a cancer warning. A number of federal and state agencies in the United States have adopted that suggestion and now require appropriate product labeling.

The facts and issues involved in the current crystalline silica debate are too many and too complex to be analyzed meaningfully here. They are addressed in more detail in other chapters of this book, and they have had an adverse impact on demand for silica stone products. Because of the ongoing controversy, for example, one major US manufacturer of abrasive media recently refused to introduce and market a new line of novaculite-based deburring media, despite the fact that more than 99% of all deburring operations are carried out in an aqueous environment where essentially no dust is generated.

PROBLEMS AND FUTURE TRENDS

The outlook for specialty silica materials, with the exception of ultrapure quartz grain and lascas, is not particularly attractive. Manufactured materials have already displaced natural silica in most abrasive and other industrial applications and can be expected to slowly penetrate the entrenched niches that remain, especially at the high end of the price range. This applies to novaculite for whetstones, hones, oilstones, and deburring media, sandstone as grindstones and blocks for acid tank linings, quartz pebbles for acid tower linings and packing media and as the heat transfer medium in air preheaters, quartzite and silex as blocks for grinding mill linings and as pretumbled cubes for grinding and deburring, and pebbles of flint and silex as grinding mill media.

At the low end of the price range quartz gravel should continue to experience modest growth in demand for use in water filtration, partly because it faces little competition from as cost effective substitutes. There are likewise few substitutes for quartz gravel, quartzite, and perhaps high quality novaculite for use as fluxes in ferrous and nonferrous metal smelting and as raw materials in the manufacture of silicon metal and various silicon alloys, although these end-use industries are typically highly cyclical.

Until means are devised for using high-purity silica sand instead, lascas should continue to be the uncontested feedstock in producing cultured quartz crystals for electronics and optics. The markets for cultured quartz, and therefore lascas, are somewhat cyclical but should experience a higher rate of growth than most other markets for naturally occurring, specialty silica materials.

The strongest rate of growth in demand should be experienced by ultrapure, fusing-grade quartz grain for specialty glass and fiberoptic, precision optical, semiconductor, high intensity lighting, electro-optical, and labware applications. Competition from synthetic silicas will intensify in these high tech markets, however, as the need for raw materials of higher and higher chemical purity intensifies. This need is being driven largely by the trend to higher capacity memory devices in the semiconductor industry but also by developments in fiberoptic communications. Ultrapure synthetic silica is now being produced by new sol-gel technology, and commercial production of synthetic fused silica containing less than 2 ppm total impurities (including hydroxyl and chlorine ions) was recently begun by a new company in the United States.

Although few of the commercial applications for specialty silica materials generate significant amounts of crystalline silica dust, the silicosis and now carcinogenicity issues will undoubtedly continue to adversely impact the future market demand for these products. Those would be producers that are willing to accept this situation and perhaps turn adversity into opportunity should focus on developing resources of lascas, ultrapure quartz grain, filtration-grade quartz gravel, and metallurgical-grade quartzite that are located close to geographic concentrations of major consumers.

BIBLIOGRAPHY

Anon., 1976, "Silica: World Production, Consumption and Trade," *Industrial Minerals*, May, pp. 31–47.

Anon., 1985, "Quartz for the National Defense Stockpile," Report No. NMAB-424, National Materials Advisory Board, National Research Council, National Academy Press, Washington, DC, 99 pp.

Anon., 1987, *Industrial Minerals Directory*, 1st ed., G.M. Clarke, ed., Metal Bulletin Books, Surrey, UK, 704 pp.

Anon., 1989, "Other Industrial Minerals," *Minerals Yearbook 1987*, Vol. 1, US Bureau of Mines, pp. 965–972.

Alsobrook, A.F., 1992, "Coleman Quartz Inc.—Novaculite for the Ferrosilicon Industry," SME Preprint 92–119, SME Annual Meeting, Phoenix, AZ, Feb., 5 pp.

Austin, G.T., 1988, "Abrasive Materials," *Minerals Yearbook 1988*, US Bureau of Mines, 20 pp.

Austin, G.T., 1989, "Abrasive Materials," *Minerals Yearbook 1987*, Vol. 1, US Bureau of Mines, pp. 71–84.

Austin, G.T., 1990, "Abrasive Materials," *Minerals Yearbook 1989*, US Bureau of Mines, 20 pp.

Austin, G.T., 1991, "Abrasive Materials," 1990 Annual Report, US Bureau of Mines, 22 pp.

Bateman, A.M., 1959, *Economic Mineral Deposits*, 2nd ed., John Wiley, New York, 916 pp. (See chapter 23, "Abrasives and Abrasive Minerals," pp. 821–833.)

Bates, R.L., 1960, *Geology of the Industrial Rocks and Minerals*, Harper and Brothers, New York, 441 pp. (See chapter 8, "Vein and Replacement Minerals," pp. 269–305.)

Benbow, J., 1989, "Belgium—An Industry Profile," *Industrial Minerals*, May, pp. 23–40.

Benbow, J., 1990, "New Zealand's Minerals—For Domestic Consumption," *Industrial Minerals*, June, pp. 19–35.

Bolen, W.P., 1989, "Industrial Sand and Gravel," *Minerals Yearbook 1988*, US Bureau of Mines, 11 pp.

Bolen, W.P., 1990, "Industrial Sand and Gravel," *Minerals Yearbook 1989*, US Bureau of Mines, 15 pp.

Bolen, W.P., 1991, "Industrial Sand and Gravel," 1990 Annual Report, US Bureau of Mines, 15 pp.

Callahan, J.E., and Craig, J.R., 1990, "The Potential for Heavy Minerals in the Lilesville, North Carolina Sand and Gravel Deposits," *Proceedings*, A-J.W. Zupan and A.H. Maybin, III, eds., 24th Forum on the Geology of Industrial Minerals, South Carolina Geological Survey, pp. 39–44.

Clarke, G., 1984, "Industrial Minerals in the UK—An Overview of Recent Developments," *Industrial Minerals*, Nov., pp. 21–47.

Ferrell, J.E., 1985, "Quartz Crystal," *Mineral Facts and Problems*, A.W. Knoerr, ed., Bulletin 675, US Bureau of Mines, pp. 641–646.

Gambogi, J., 1991, "Silicon," 1990 Annual Report, US Bureau of Mines, 12 pp.

Georgiades, G., 1988, "Greek Raw Materials for the Glass and Ceramics Industry," *Industrial Minerals*, Apr., pp. 135–141.

Griffiths, J., 1985a, "Hellenic Industrial Minerals," *Industrial Minerals*, Jan., pp. 19–39.

Griffiths, J., 1985b, "Spain's Industrial Minerals," *Industrial Minerals*, Oct., pp. 23–63.

Griffiths, J., 1987a, "Silica—Is the Choice Crystal Clear?" *Industrial Minerals*, Apr., pp. 25–43.

Griffiths, J., 1987b, "Pakistan's Mineral Potential—Prince or Pauper?" *Industrial Minerals*, July, pp. 20–43.

Griffiths, J., 1989, "South Africa's Minerals—Diversity in Adversity," *Industrial Minerals*, Aug., pp. 18–53.

Griffiths, J., 1991, "Spain's Minerals—Mixed Fortunes," *Industrial Minerals*, June, pp. 23–47.

Guillet, G.R., and Kriens, J., 1984, "Silica," *Ontario and the Mineral Filler Industry*, Industrial Mineral Background Paper No. 5, Ontario Ministry of Natural Resources, pp. 99–124.

Hale, D.R., and Blair, R.E., 1983, "Electronic and Optical Uses," *Industrial Minerals and Rocks*, 5th ed., Vol. 1, S.J. Lefond, ed., AIME, New York, pp. 213–231.

Hamilton, W.N., and Hora, Z.D., 1987, "Opportunities Exist for Industrial Minerals Development in British Columbia and Alberta," *Mining Engineering*, Vol. 39, No. 12, pp. 1061–1067.

Harben, P., 1983, "Tripoli and Novaculite—The Little Known Relations," *Industrial Minerals*, Jan., pp. 28–32.

Harben, P., 1991, "Sri Lanka—Freeing Its Mineral Potential," *Industrial Minerals*, Jan., pp. 45–51.

Herath, J.W., 1990, "Sri Lanka's Minerals & National Development," *Industrial Minerals*, Mar., pp. 133–135.

Hight, R.P., 1983, "Abrasives," *Industrial Minerals and Rocks*, 5th ed., Vol. 1, S.J. Lefond, ed., AIME, New York, pp. 11–32.

Holbrook, D.F., and Stone, C.G., 1979, "Arkansas Novaculite—A Silica Resource," Reprint, Arkansas Geological Commission, 8 pp; from 13th Annual Forum on the Geology of Industrial Minerals, Circular 79 (1978), Oklahoma Geological Survey.

Hruska, J., 1991, "Czechoslovakia's Minerals—Production, Reserves and Exploration," *Industrial Minerals*, Oct., pp. 49–53.

Jaster, M.C., 1957, "Selected Annotated Bibliography of High-Grade Silica of the United States and Canada Through December 1954," Bulletin 1019-H, US Geological Survey, 673 pp.

Keller, W.D., Stone, C.G., and Hoersch, A.L., 1984, "The Geologic Significance of Textures of Paleozoic Chert and Novaculite in the Quachita Mountains of Arkansas and Oklahoma," Contributions to the Geology of Arkansas, Vol. 2, Miscellaneous Publication 18-B, Arkansas Geological Commission, pp. 87–89.

Kuzvart, M., 1984, "Deposits of Industrial Minerals," and "Deposits of Industrial Rocks," *Industrial Minerals and Rocks*, Elsevier, Amsterdam, pp. 122–352.

Lamey, C.A., 1966, *Metallic and Industrial Mineral Deposits*, McGraw-Hill, New York, 567 pp. (See chapter 17, "Mica, Feldspar, Nepheline Syenite, and Quartz," pp. 454–466.)

Murphy, G.F., and Brown, R.E., 1985, "Silicon," *Mineral Facts and Problems*, A.W. Knoerr, ed., Bulletin 675, US Bureau of Mines, pp. 713–728.

Murphy, T.D., and Henderson, G.F., 1983, "Silica and Silicon," *Industrial Minerals and Rocks*, 5th ed., Vol. 2, S.J. Lefond, ed., AIME, New York, pp. 1167–1185.

Murray, R., 1989, "Namibia's Mineral Potential—Independence Brightens Development Prospects," *Industrial Minerals*, Nov., pp. 85–89.

Murray, R., 1991, "Namibia's Industrial Minerals," *Industrial Minerals*, Sep., pp. 97–103.

Ober, J.A., 1990, "Quartz Crystal," *Minerals Yearbook 1989*, US Bureau of Mines, 3 pp.

Ober, J.A., 1991, "Quartz Crystal," *1990 Annual Report*, US Bureau of Mines, 3 pp.

Ober, J.A., 1992, "Quartz Crystal in 1991," *Mineral Industry Surveys*, Annual, Preliminary Preprint, US Bureau of Mines, Jan., 3 pp.

O'Driscoll, M., 1989, "Brazil's Industrial Minerals—Seeking Success Against All Odds," *Industrial Minerals*, Nov., pp. 23–69.

Pearse, G.H.K., 1972, "Silica," Preprint from *Canadian Minerals Yearbook 1971*, Department of Energy, Mines, and Resources, 5 pp.

Peterson, P.M., 1976, ''Silica Rock in the Northwest,'' 11th Industrial Minerals Forum, Special Publication 74, Montana Bureau of Mines and Geology, pp. 93–97.

Prud'homme, M., and Francis, D., 1986, ''The Refractories Industry,'' Mineral Policy Sector Internal Report MRI 86/4, Department of Energy, Mines and Resources, Canada, 18 pp.

Rice, W.L., Lawson, D.C., and Berg, R.B., 1989, ''The Mineral Industry of Montana,'' *Minerals Yearbook 1987*, Vol. 2, US Bureau of Mines, pp. 239–248.

Robbins, J., 1986, ''Italy's Industrial Minerals,'' *Industrial Minerals*, Dec., pp. 19–50.

Russell, A., 1990, ''The Swedish Minerals Industry—Seeking Fresh Horizons,'' *Industrial Minerals*, Apr., pp. 35–55.

Russell, A., 1991, ''India: A Treasure Trove of Minerals—Lifting the Lid on Exports,'' *Industrial Minerals*, Aug., pp. 17–33.

Sholes, M.A., 1977, ''Arkansas Novaculite Stratigraphy,'' Reprint, Arkansas Geological Commission; from Symposium on the Geology of the Ouachita Mountains, Vol. 1, pp. 139–145.

Simandl, G.J., Jakobsen, D., and Fischl, P.S., 1992, ''Refractory Mineral Resources in British Columbia, Canada: A Fresh Look,'' *Industrial Minerals*, May, pp. 67–77.

Smith, M., 1984a, ''Quartz Crystal—Currently Sparkling,'' *Industrial Minerals*, Aug., pp. 19–25.

Smith, M., 1984b, ''Quartz in Brazil,'' *Industrial Minerals*, Sep., p. 111.

Steuart, C.T., Holbrook, D.F., and Stone, C.G., 1984, ''Arkansas Novaculite: Indians, Whetstones, Plastics and Beyond,'' Contributions to the Geology of Arkansas, Vol. 2, Miscellaneous Publication 18-B, Arkansas Geological Commission, pp. 119–134.

Steuart, C.T., Holbrook, D.F., and Stone, C.G., 1985, ''Arkansas Novaculite: Indians, Whetstones, Plastics and Beyond,'' SME Preprint 85-87, SME-AIME Annual Meeting, New York, Feb., 12 pp.

Tepordei, V.V., 1988, ''Sand and Gravel,'' *Minerals Yearbook 1986*, Vol. 1, US Bureau of Mines, pp. 801–823.

Tepordei, V.V., 1989, ''Sand and Gravel,'' *Minerals Yearbook 1987*, Vol. 1, US Bureau of Mines, pp. 745–758.

SILLIMANITE GROUP

ANDALUSITE

A.P. GROBBELAAR

Andalusite is a member of the sillimanite group of minerals consisting of andalusite, kyanite, and sillimanite. These minerals are the structural polymorphs of Al_2SiO_5 (the anhydrous silicate of aluminium) and although chemically similar, they differ in physical properties due to their different crystal structures. For further information on the sillimanite group, see the chapter on Kyanite and Related Minerals that follows.

The minerals are generally the products of the metamorphism of alumina-rich sediments and are found in schists, gneisses, or hornfelses, which have undergone either regional or thermal metamorphism from their original form. Sedimentary deposits may also form since the minerals are chemically and physically resistant to breakdown.

When heated to a sufficiently high temperature the sillimanite group minerals convert to mullite with a resultant volume expansion and the liberation of silica. Due to the stability of the mullite formed, its high hot strength and resistance to chemical and physical erosion, they are widely used in the manufacture of refractories and special ceramics (Table 1).

The reaction can be represented as follows:

$$3(Al_2O_3{\cdot}SiO_2) + \text{heat} = 3Al_2O_3{\cdot}2SiO_2 + SiO_2$$
$$\text{andalusite} + \text{heat} = \text{mullite} + \text{free silica}$$

The worldwide use of these minerals depends to a large degree on their availability (Fig. 1). Andalusite, with its lower conversion temperature and the lowest volume expansion on being heated is the natural choice as a raw material to produce energy-saving refractories without prior calcination.

The Republic of South Africa holds by far the largest minable andalusite deposits of the world, estimated at some 70 Mt of known in situ reserves. The only other large commercial production of andalusite comes from France, with known reserves of about 5 Mt. Potential new deposits are known from China, Ireland, Australia, and South America.

World consumption of andalusite in special performance refractories has risen from less than 10 ktpy in 1960 to nearly 300 ktpy in 1990, which is due to cost effectiveness in many industrial applications compared to other conventional refractory products.

Andalusite-based refractories have traditionally been used in Europe, South Africa, and the Far East; usage in America was limited in the past due to relatively high delivery costs and availability of other local refractory minerals. However, there is a significant trend towards greater use of andalusite in North America, thus opening new markets and replacing other traditional refractory raw materials.

GEOLOGY

Andalusite occurs in metamorphosed rocks of clay-like composition, as in the andalusite-hornfelses in thermal aureoles (contact metamorphism), formed under conditions of high temperatures and low stress, and in regional metamorphic rocks, such as the andalusite-schists. These schists are mostly unconnected with definite igneous intrusions but were presumably also formed at high temperatures and under low stress. Andalusite can also appear in alluvial deposits associated with the primary deposits.

At depths of 10 to 20 km, temperatures of up to 450°C might be expected. Another result of increasing depth is an increase of pressure due to the load of the rock cover. The load stresses involved in metamorphism is generally believed to lie in the range 0 to 10 000 bars.

During contact metamorphism the increase in temperature is brought about by the injection of magma. The temperature reached in the country rock at some distance from the igneous body will always be less than the initial temperature of the intrusion, which is presumably less than 1 200°C, and will vary with time and distance from the contact.

A simple metamorphic reaction involving dehydration of an aluminous sediment can be:

$$KAl_2(AlSi_3)O_{10}(OH)_2 + SiO_2 \rightarrow KAlSi_3O_8 + Al_2SiO_5 + H_2O$$
$$\text{muscovite} + \text{quartz} \rightarrow \text{K-feldspar} + \text{andalusite} + \text{water}$$

At higher temperatures sillimanite would form at the expense of andalusite, while kyanite would be the mineral formed at higher pressures, as can be seen in the phase diagram of Al_2SiO_5 (Fig. 2).

Mineralogy

The mineral andalusite occurs in the form of transparent to opaque, usually subtranslucent prismatic orthorhombic crystals with an uneven, subconchoidal fracture (Fig. 3). The color can be whitish, rose-red, flesh-red, violet, pearl-gray, reddish brown, beige-brown to olive-green, with a hardness of about 7.5 on the Mohs scale and a relative density of 3.10 to 3.30 gm/cc.

Most of the andalusite occurs as the variety chiastolite. The crystals when cut or broken across exhibit definite cruciform growth patterns which are due to impurities enclosed in the crystals during their formation.

The impurities in the chiastolite cross can consist of quartz, iron-rich ilmenite and graphite not completely expelled during crystal formation (Fig. 4). Further typical impurities that can be found are muscovite, biotite and chlorite. During weathering an-

Table 1. Mullitization Data of KSA Minerals

Mineral	Relative Density	Temperature of Mullitization	Volume Change on Conversion
Andalusite	3.20	1 380°C	+ 5%
Sillimanite	3.25	1 550°C	+ 7%
Kyanite	3.60	1 350°C	+18%
Mullite	3.05	—	—

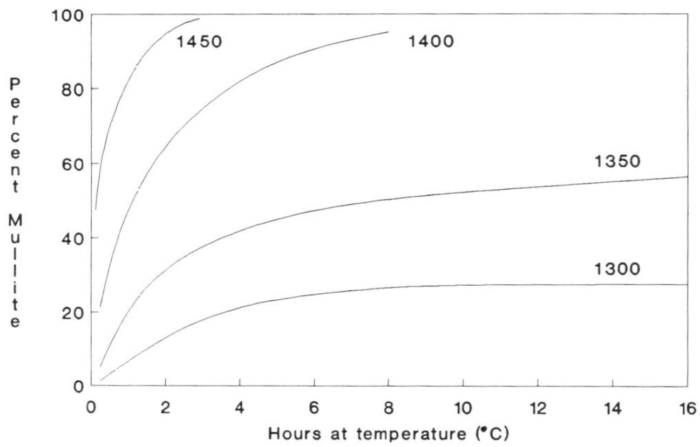

Fig. 1. Formation of mullite from andalusite (Anon., 1985).

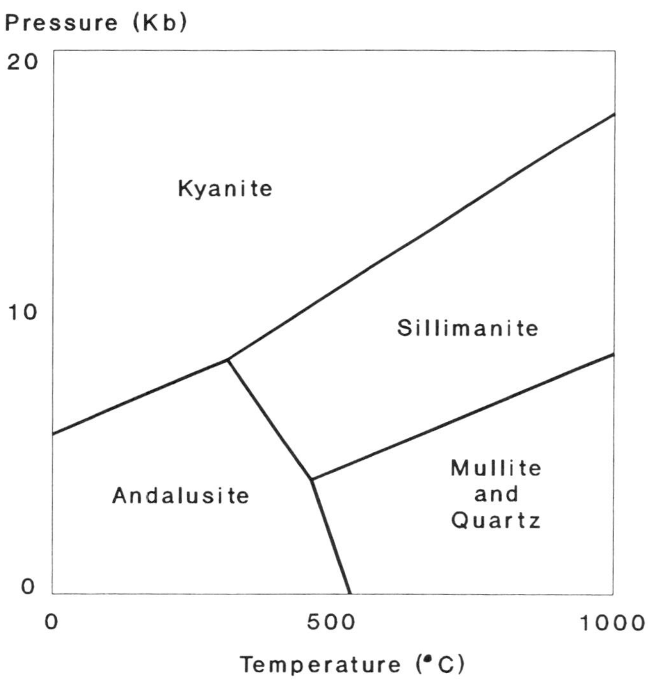

Fig. 2. Mineral phases within the $Al_2O_3 \cdot SiO_2$ (Anon., 1985).

dalusite alters progressively to fine-grained sericite. It is particularly liable to alteration to sericite near the inclusions.

Certain andalusite products suffer from iron spotting owing to the iron being present in localized concentrations in the deposits, as well as in the crystal. This form of contamination is unacceptable

Fig. 3. Andalusite crystals.

Fig. 4. Photomicrograph of a cross-section of an andalusite crystal showing the chiastolite cross.

in certain refractory applications. Generally, however, these spots have not been found to have any measurable effect on the performance of the refractory material.

Chemical Properties

Andalusite has the chemical formula Al_2SiO_5 or more commonly $Al_2O_3{\cdot}SiO_2$ and theoretically contains 36.8% silica and 63.2% alumina.

Commercially available andalusites vary in alumina content from about 53 to 60% with Fe_2O_3 being the major impurity varying between 0.8 and 1.5%. Small amounts of alkalis and alkaline-earths are invariably also present in the crystal lattice lowering the refractory properties of the material.

Physical Properties

As will be discussed later, there are major differences in the physical properties of the andalusite and host rock from the various locations. These differences affect to some degree the mining and beneficiation procedures that are used to recover the andalusite, although the basic principles are similar. Weathering and alteration have a marked effect on the quality of the crystals, their relative density, the size of the crystals and the particle-size distributions of concentrates.

Origin

Andalusite was first found in 1798, and named after the location in Andalusia, Spain. It typically occurs as a result of low- to medium-grade contact metamorphism of argillaceous rocks. Andalusite is rarely found in a pure form in nature, and commercial concentrates generally contain some impurities resulting from alteration which must be regarded as part of the andalusite crystal.

Distribution of Major Deposits

Although the name andalusite was derived from the province of Andalusia, it is now generally associated with the Republic of South Africa.

South Africa holds the largest known minable andalusite deposits in the world, estimated at some 70 Mt of known in situ reserves, of which approximately 35 Mt are economically recoverable using current methods (Hammerbeck, 1986). Approximately 284 kt of andalusite were produced in 1990 of which nearly 150 kt were exported.

The only other large commercial production outside of South Africa comes from France, with known reserves of 5 Mt, where about 65 ktpy are mined, and chiefly sold locally or elsewhere in Europe.

Potential new deposits are known in China, Ireland, Spain, and South America, but limiting factors for new production are remote locations, lack of infrastructure, as well as metallurgical problems. As will be shown later, water is essential in the beneficiation of andalusite and is generally scarce in these locations.

The original location in Andalusia is of insignificant commercial interest. In Brazil, an outstanding precious stone-grade andalusite is mined, polished, faceted, and sold by the carat.

South Africa: The development of andalusite resulted mainly from the metamorphism and the chemistry of the pelitic rocks of the Pretoria Series of the Transvaal System (Later Precambrian) due to the emplacement of the Bushveld Igneous Complex.

The Bushveld Igneous Complex is regarded as consisting essentially of four phases. Phase 1 consisted of an extrusive lava phase resting on the Transvaal system. Phase 2 was intrusive where numerous sills were injected into the sediments of the Transvaal system, principally the shales of the Pretoria Series. This was followed by a basic intrusive phase where successive stages of rocks of gabbroidal composition were emplaced within the upper part of the Pretoria Series. An acid intrusive phase then followed, which intruded mainly into the lavas of Phase 1, but also into the basic rocks of the Complex. It is estimated that the Complex was formed about 1.95 billion years ago.

The metamorphic aureole in the Pretoria series approximates in outline the shape of the main basic phase of the Bushveld Igneous Complex with the grade of metamorphism increasing in a northerly direction as the thickness of this phase increases.

The larger part of the rim rocks in the metamorphic aureole show alterations typical of thermal or contact metamorphism with the characteristic minerals like chiastolite, cordierite, and biotite. Because of the vast size of the metamorphic zone it is obvious that many factors existed that could have affected the minerals formed at any particular place. The main factors being the chemical nature and the thickness of the original sedimentary layers and their respective distances from the intrusive phases of the Bushveld Igneous Complex. Temperatures of 550 to 600°C and low pressures

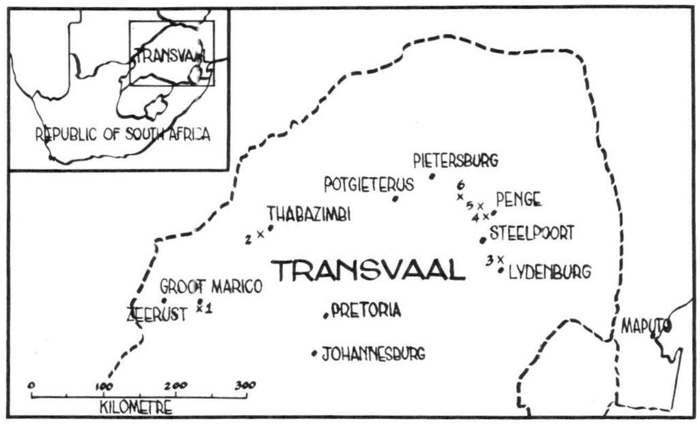

Fig. 5. Location of operating andalusite mines in the Republic of South Africa: 1) Andafrax, 2) Rhino, 3) Krugerspost, 4) Annesley, 5) Havercroft, and 6) Hoogenoeg.

of about two kilobars are regarded as ideal conditions for the formation of andalusite.

Although the metamorphic aureole extends for more than a thousand kilometers around the Complex, and andalusite can be found in most areas, only a few areas have satisfied the necessary conditions of metamorphism to produce good ore deposits of andalusite. These areas are in the southwest, northwest, and the east of the Transvaal province.

Andalusite is developed at various levels in the stratigraphic succession. Economic deposits of andalusite are found in the lower pelitic unit of the Timeball Hill Stage at Thabazimbi, the upper Timeball Hill Stage at Groot Marico and in the Eastern Transvaal in the upper Timeball Hill Stage as well as in the Boven and Lydenburg members of the Silverton Stage.

Fig. 5 shows the location of the six main productive mines in the Republic of South Africa.

Southwestern Transvaal: Andalusite is developed extensively in the metamorphosed shales of both the Timeball Hill and Daspoort Stages. The andalusite crystals occur as conspicuous porphyroblasts 5 to 40 mm in length and 1 to 4 mm in diameter. The color of the crystals varies from dark red to black. In many cases the color is obscured by the formation of a white micaceous alteration product. The recoverable andalusite is in the order of 5 to 10%. The andalusite occurs in association with biotite, iron oxides, quartz, cordierite and epidote. Staurolite is absent.

Alluvial andalusite rich sand is found in the drainage system of the Klein Marico River and its tributaries containing 10 to 50% andalusite.

Only one major producer still operates in this area. The Andafrax mine, originally owned by George Alexander, regarded by many as one of the fathers of the andalusite industry, was acquired by AC Nesbitt (Trading)(Pty) Ltd. in 1986. The Andafrax mine is operated as Purity Minerals (Pty) Ltd. It was sold to Damrec of France in 1991 and is now trading as Samrec (Pty) Ltd. The mine is located on a farm near Groot Marico, a town about 200 km west of Johannesburg.

The andalusite occurs in the metamorphosed shales of the Daspoort Stage which are over 60 m thick and contains 5 to 20% andalusite crystals. Since the overburden is thin and the shales relatively soft, extraction is easy without recourse to blasting. The ore then undergoes an initial crushing and screening operation before passing through scrubbers and a two stage heavy media plant. This removes the shale and free silica. After drying the andalusite is passed over high intensity magnets to reduce the iron content. The mine produces about 36 ktpy and present proven reserves of andalusite are close to 1 Mt.

Andalusite product typically contains:

$$59.3\%\ Al_2O_3$$

$$0.8\%\ Fe_2O_3$$

and a maximum grain size of 3 mm.

Northwestern Transvaal: Andalusite is well developed in the Time Ball Hill shale near Thabazimbi, approximately 240 km north of Johannesburg. The shale is generally very iron rich and the chiastolite can be iron stained. The chiastolite crystals are very well developed in a highly weathered mica-rich hornfels.

The hornfels can be readily crumbled by hand and the andalusite is consequently easily liberated. The rock is highly mineralized and contains up to 15% andalusite.

The largest andalusite mine in the world is situated on the Grootfontein farm some 12 km west of Thabazimbi. The deposit was discovered and developed by Weedon Minerals (Pty) Ltd, and it was sold to AngloVaal Industrial in 1991. The company is now trading as Rhino Andalusite Ltd.

Production capacity is an estimated 140 ktpy of andalusite with an alumina content of up to 60.0% and Fe_2O_3 0.8 to 0.9%. Maximum grain size is 4 mm, although the bulk of the product is below 3.35 mm. At present the ore is mined from four open quarries over a distance of about 4 km. Near the surface the ore is relatively soft and can be mined by machine, but the ore grows progressively harder with depth and mining has to revert to drilling and blasting. The ore is then crushed, washed, screened, and subjected to heavy media separation. The dried material is then split into fractions before treating the main product with high intensity magnets.

The andalusite from this area is light brown, frequently with a pinkish tint. The crystals are also fairly brittle and tend to break down more than those from other areas.

Eastern Transvaal: In the Eastern Transvaal generally a higher degree of metamorphism is found with added structural disruptions such as faults and dikes; a larger part of the Pretoria Series was affected. Andalusite is found throughout the Pretoria Series, but because of this higher degree of metamorphism, it appears in lower formations towards the north and northeast. Minerals such as staurolite, epidote, and garnet also appear. It is therefore possible to divide the area into two parts: the northern part containing staurolite and having a large crystal size, and a southern area, close to Lydenburg, containing finer grained andalusite and without the association of staurolite.

The four andalusite producers in this area primarily mine in situ ore, although some do mine the alluvial deposits associated with the shale as well.

In 1977 Cullinan Minerals (Pty) Ltd. started production of its new mine at Krugerspost just north of Lydenburg. The mine was sold to Damrec in 1992 and is trading as Samrec (Pty) Ltd.

Andalusite occurs in a chiastolite-muscovite-biotite hornfels at the base of the Lydenburg member of the Silverton Stage and is staurolite free. Extensive weathering rendered the ore body soft allowing it to be mined without blasting. The deposit extends for over 6 km and has an average width of 20 m. The overburden is less than 2 m thick and the andalusite content in the shale varies between 8 and 12%.

Mining is done by means of a bulldozer. The shale is then crushed, washed, and passed through cyclones to remove the fines. The solids are then treated in tube mills to liberate and clean the andalusite crystals. After washing and screening the concentrate is treated in a heavy media plant before passing over high intensity magnets.

Current capacity is about 30 ktpy of product with an alumina content of 57.5%, iron content of less than 0.9% and a maximum grain size of 4 mm. Again the bulk of the product is less than 3.35 mm.

In the Penge area, 250 km east of Johannesburg, two companies mine from essentially the same ore body. The Havercroft Andalusite mine, previously owned by Verref Mining (Pty) Ltd., was sold to Anglo Vaal in 1992, and is about 10 km northwest of Penge—this mine is now dormant. Annersley Andalusite is situated about 5 km from Penge and is trading as Samrec (Pty) Ltd.

The andalusite occurs in the upper pelitic unit of the Time Ball Hill Stage. This formation has been metamorphosed into a gray carbonaceous, shaly hornfels showing porphyroblasts of coarse-grained chiastolite. This is the most important andalusite-bearing zone in the Eastern Transvaal.

The economic zone being mined occurs within a sequence of andalusite-staurolite-hornfels and andalusite-hornfels layers rich in andalusite. The 50 to 60 m-thick zone which is mined for andalusite is a relatively soft unit between hard unweathered beds. Most of this zone is unfortunately covered by thick overburden of talus and alluvium which has to be removed before the ore body can be mined. The andalusite is of the large crystal variety with dimensions of up to 15 mm in cross section being common. A thin layer or coating of sericite has formed around the andalusite crystals due to weathering or alteration of the hornfelses. Dark inclusions of carbonaceous matter forming the chiastolite cross and microscopically small crystals of stauralite and chloritoid are also present.

The Havercroft Andalusite mine produces high grade large-crystal andalusite. The 50 m-thick andalusite-hornfels zone is subdivided into the Giant crystal zone (with crystals up to 60 mm in length) and the upper zone.

The deposit is situated on the Havercroft and Streathham farms. Current output is about 50 ktpy with less than 10% from the associated alluvium.

The recoverable andalusite in the ore body averages between 7 and 8%, with the concentrate containing 58 to 59% Al_2O_3 and about 1.3% Fe_2O_3. This product was further beneficiated by Vereeniging Refractories to give a final product with an alumina content of 60.5 to 61% and an Fe_2O_3 content of 0.6%.

The deposit which is now known as Annesley Andalusite (Pty) Ltd. has had a checkered career. It was the first of the large crystal deposits to be discovered in the early 1960s, and initially a semi-concentrate was supplied to the local refractory brick producers for upgrading to a raw material for the production of ladle bricks. At that time only the alluvial deposits were mined and beneficiation was by means of relatively simple screening, washing, and heavy medium separation. In the early 1970s the demand for andalusite exceeded the then available alluvial deposits, and equipment was installed to win andalusite from the main ore body. In July 1981 the deposit was acquired by Rand London Corporation Ltd. from ZIMRO (Pty) Ltd. The plant was upgraded and new plant was added to increase output. More recently ownership was transferred to Annesley Andalusite (Pty) Ltd. part of the mining division of Zimco Industries (Pty) Ltd., ultimately owned by ZIMRO (55%) and the Cookson Group PLC (45%). The mine is presently owned and operated by Damrec of France.

Production is running at about 35 ktpy from both the ore body and alluvial deposits associated with the deposit. Individual crystals from this deposit can attain a length of up to 10 mm and product is sold in two grades. The 59/12 grade contains 59% Al_2O_3 and 1.2% Fe_2O_3 while the 60/09 grade has 60% Al_2O_3 and 0.9% Fe_2O_3. The maximum grain size is 8 mm. A premium grade is now also available with an Fe_2O_3 content of 0.75%.

In the northern part of the region metamorphism progressed so far as to include the lower pelitic unit of the Time Ball Hill Stage. It is exposed on the farm Hooggenoeg where it consists of a gray, carbonaceous chiastolite-staurolite-biotite hornfels over and underlain by a diabase. This ore body is between 20 and 50 m thick.

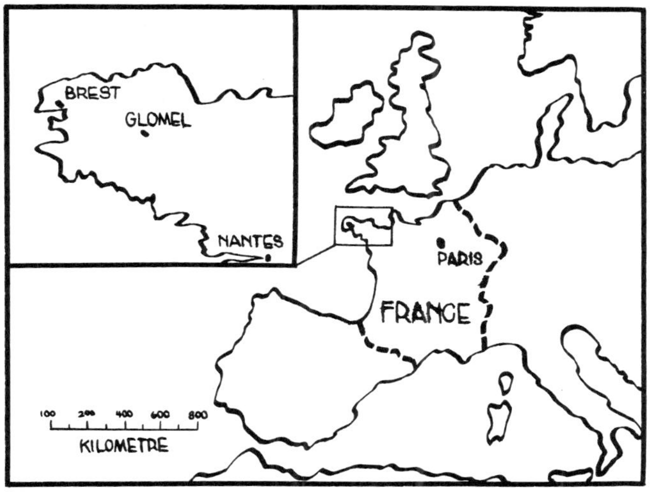

Fig. 6. Location of the Glomel deposit in France.

The massive ore in the middle of the reef, contains large crystals which are randomly orientated. Weathering of the hornfels is not constant, and the rock often becomes so hard that the recovery of andalusite is not economically viable.

The Hoogenoeg mine is situated halfway between Pietersburg and Burgersfort, approximately 300 km from Johannesburg in the northeastern part of the Transvaal. Andalusite mining at Hooggenoeg started in 1974 when George Alexander erected a small washing plant as a simple heavy medium concentrator for the treatment of the alluvial gravels and weathered bedrock. The deposit was worked intermittently between 1975 and 1982. In 1981 the deposit was taken over by the French andalusite producer Denain-Anzin Mineraux and another French company Intersteel. In 1985 Hoogenoeg Andalusite (Pty) Ltd. was acquired from Cheminerals (Pty) Ltd. by ENRO (Energie-Rohstoff-und Anlagen GmbH) West Germany.

The old plant was modified and upgraded to produce about 20 kt of final product. In late 1989 a new and redesigned plant was started up which increased production to 35 ktpy. The ore contains between 8 and 12% andalusite and recovery ranges from 50 to 60%.

The mining operation is straightforward. The overburden and ore body are drilled, blasted, and the ore transported by truck to the crushers. The ore dressing plant consists mainly of three stage crushing, screening and scrubbing, two stage heavy medium separation, drying and multi-stage magnetic separation. The andalusite product has a maximum grain size of 8 mm and a typical alumina content of 59.5% and maximum iron content of 1.0%.

France: During 1965 Denain-Anzin Minéraux Refractaire Ceramique (Damrec), a French company specializing in industrial minerals, undertook a study of all possible French and European deposits of high-alumina raw materials. The investigation led to the discovery of many deposits in France and Spain, with a fine to extremely fine-grained andalusite mineralization, most of which would pose problems with exploitation.

The most promising deposit was found in a zone of schists near Angers of Ordovician age and strongly metamorphosed by the granitic intrusion of Rostrenen. The deposit is in the region of Glomel, in Brittany, near the town Kerphales from which the mineral takes its name (Fig. 6).

The deposit is fairly homogeneous with interleaving of quartzites and dolerites and contains approximately 15% andalusite crystals, commonly 1 to 4 mm thick and several centimeters in length. The ore body consists of a succession of schists some 400 m thick with a low overburden ratio but a large number of big boulders within the formation; metamorphic events complicate the mining process. Quarrying is by drilling and blasting in benches of 15 m wide.

Because of fairly tight crystal interval, the firm and hard ore has to be broken down and milled in the processing plant by autogenous grinding to a size of −1.6 mm and +0.3 mm in order to provide suitable liberation of the andalusite from the other minerals. The chlorite and mica are then removed by high intensity magnetic separators remaining quartz and andalusite is separated after the treatment in a double circuit heavy media separation unit. A very high grade but fine-grained product is also produced by froth flotation.

Products are sold under the tradename Kerphalite. Kerphalite KA contains a minimum of 59% Al_2O_3, while KB has a minimum of 53%. The flotation product KF has a minimum of 60% Al_2O_3.

Proven and possible reserves for the deposit were quoted in the region of 10 Mt. Present production is approximately 65 ktpy of which nearly 50 kt are exported. This ranks the French operation second in size of the world's andalusite producers.

Rest of the World: Andalusite is most common as a contact mineral in clay slates and argillaceous schists, especially in connection with granitc intrusions. Commercially viable deposits are not common.

Confusion in the reporting of production and export of minerals of the sillimanite group creates the impression that a number of other countries commercially exploit deposits of andalusite.

Andalusite has been mined and produced in the United States, India, Spain, Korea, China, and Czechoslovakia. Various reports of new deposits are also recieved from time to time, but no other major supplier of andalusite can be foreseen in the immediate future.

United States: Although there are known deposits of andalusite in the western United States, difficulties in beneficiation and remoteness from major markets have restricted their economic viability.

In 1919 a high grade deposit was discovered at White Mountain, CA. Andalusite from this deposit was initially used in the manufacture of spark plug insulators for aircraft during World War II. This deposit is not being mined in 1992.

Another deposit at Hillsborough, NC, is owned and operated by Piedmont Minerals and represents potential for many applications. This deposit, located in the slate belt, stretches about 457 m long by 91 m wide. It is made up of several lenticular pods of a fine-grained quartzite with about 25 to 30% andalusite and a minor constituent of massive pyrophyllite blended within it. At this time, crushing, screening, and further beneficiation yield a maximum 35% alumina product, containing about 45% andalusite, with 0.7% iron. Although this deposit consists of the finer grained andalusite crystals (3.2 mm maximum, bluish-to-gray), the refractory grades liberate generously at −5 mm and even better at −3 mm to give the best product currently available from this location. Reserves are sufficient and will extend production well into the next century.

The technology exists to render a 56% alumina product from the ores mined by Piedmont, but grinding to −212 μm would be required for proper liberation. No plans are being made at this time, however, to produce a concentrated andalusite by this method.

Uses for the current Piedmont andalusite/pyrophyllite products include soaking pit refractories in the form of plastics, gunning mixes, and castables. Kiln car blocks mortars, and foundry washes are among other refractory uses incorporating this mineral product. Some low iron ore found at Hillsborough has been useful over the years in wall tiles, electrical porcelain, and other ceramic products.

China: Andalusite from the provinces of Henan and Liaoning has been reported to have reached the market in the United States. A uniform commercial product of a low iron content has yet to be identified. An alumina content between 55 and 60%, together with an iron oxide content of 1 to 2%, has been reported. There are also indications that the iron is locked up within the large andalusite crystals, making beneficiation very difficult.

India: Andalusite is mined intermittently from the state of Uttar Pradesh. The last production figures reported were on the order of a few hundred tons per year.

Ireland: A major andalusite deposit was discovered at Tomduff, in County Carlow in southeastern Ireland. The deposit is situated 7 km northeast of the small town of Borris and adjacent to the hamlet of Killedmond.

The geology of the area is dominated by two of the five constituent granite plutons which comprise the Leinster Batholith. This batholith is of late Caledonian age (400 M.Y.) and intruded into Lower Paleozoic, predominantly greywacke, with minor basic volcanic units. Contact metamorphism and local metasomatism is well developed.

The Tomduff deposit consists of a series of steeply dipping interlayered andalusite-mica schists and biotite staurolite-mica shists. A program of detailed mapping and trenching has revealed that the zone of mineralization is 5 km by 300 m. Preliminary sampling indicated that the content of the ore body varies from 9 to 25%. Processing testwork has indicated that conventional crushing and grinding could liberate up to 98% of the andalusite. Current research focuses on the use of heavy media, and magnetic separation is used for upgrading the andalusite.

Navan Resources PLC and its Swiss partner Realstate AG believe that the deposit could support a production capacity of 50 ktpy.

Australia: Ramsgate Resources Ltd. is planning to explore and develop an andalusite deposit at the old gold mining center of Spargoville, west of Kambalda in Western Australia's Eastern Goldfield, according to Resource Information Unit Ltd., Australia.

The property covers 337 hectares and is reported to contain several occurrences of outcropping andalusite-bearing chlorite-biotite sc 350 kg of the andalusite-bearing chlorite-biotite schist. Some 350 kg of the schist has been sent to laboratories to determine the ore characteristics. It is present as discrete crystals within metamorphosed Archean aluminous metasedimentary rocks.

Work carried out by Ramsgate Resources showed the crystals range in size from 0.5 to 20 cm in length. The company also reported that the small number of samples analyzed indicated that the Spargoville andalusite is of very high quality.

TECHNOLOGY

Exploration Techniques

It is known that andalusite principally occurs in areas of metamorphism, where masses of sedimentary rocks have been altered by heat from large intrusions of igneous masses or have been extensively distorted, crumpled, and overturned by geological processes. It has also happened that the discovery of alluvial deposits led to the discovery of the andalusite-bearing formations. Identification of such areas is then followed up by regional geological surveys and mapping.

Trenching and drilling are used to collect samples for evaluation. Pilot plant tests will include determination of the grade and an estimation of the yield of recoverable andalusite. Liberation of the andalusite from the host rock by milling and washing and recovery by heavy media separation and high intensity magnets are but a few of the methods being used to evaluate a deposit.

Mining and Processing

Because andalusite is a typical large volume and fairly low cost commodity, mining and milling processes must be relatively simple and cost effective.

Ore bodies are commonly fairly thick, and reef thicknesses of more than 50 m are not uncommon.

Mining is essentially by open pit methods and, depending on the prevailing dip of the ore body, mining or quarrying may be as contour mining planned around a specific ore/overburden ratio depending on the cost structure of the overall operation.

The mining operation as such is straightforward. All the deposits currently exploited are mined by opencast methods, using both face development and bench mining procedures. The overburden can vary in thickness between a few meters to over 60 m.

Once the ore has been exposed, the mining basically consists of loosening the ore by means of rippers, scrapers, or shovels. Blasting can also be used in deposits where the ore is too hard. The ore body also increases in hardness with depth. The mines, treating the coarse crystal type of andalusite, all use controlled blasting to loosen the ore.

At some of the mines where the overburden is thick and the ore body extends to a considerable depth, underground mining may be practiced in the future.

In contrast to the South African deposits, the French operation loosens the ore by vertical blasting and then breaks the ore by hydraulic-percussion lump breakers. Several stages of autogenous grinding reduce the size of the ore to −1.6 mm in order to liberate the andalusite from the other minerals, particularly the micas.

After mining of the ore it is then transported to the processing plant. The beneficiation process normally depends on the physical characteristics of the ore, the most important being:

1) friability of the host rock and the ease of liberation of the andalusite crystals from the host rock,

2) difference in relative density between the andalusite and the other minerals present, and

3) differences in magnetic susceptibility between the andalusite (non-magnetic) and the dense iron-bearing impurities, which are usually only feebly magnetic.

Impure andalusite has a relative density of 3.05 to 3.10 gm/cc compared with a relative density of 2.70 to 2.85 gm/cc for the shale component. This difference in relative density allows separation by means of gravity, but is not sufficiently great for simple procedures such as jigging, tabling, or spiralling to be effective. Shape is also a negative factor when jigs are considered, and the crystal size rules out the use of tabling or other forms of thin film concentration.

Heavy-medium separation has been found to be a very efficient means of concentration, and it is the only method of concentration currently used in the production of andalusite. Before dense-medium separation can be applied, however, the material has to be properly prepared and it is in the preparation of the feed to this separation that variations are found.

A standard beneficiation procedure consists of the stages shown in Fig. 7. Since no two ores behave similarly when treated by the standard procedure, producers have incorporated variations in these procedures. These variations are indicative of differences in approach to the problems of beneficiation by producers.

In the application of the various unit operations it must be borne in mind that the object is, first, to release the andalusite crystal with minimum breakage and second, to clean the liberated crystal from the adhering impurities while at the same time dispersing the particles of soft shale that can adversely affect the subsequent concentration processes. The final object is the production of a pure concentrate that will conform to the specifications.

Ore dressing processes consist commonly of one to three stages of crushing, extensive screening and scrubbing, two stages of heavy medium separation, drying in a rotary kiln, and multi-stage dry magnetic separation (Fig. 7).

Both induced roll and Permroll high intensity magnets are used. One producer has introduced a roasting stage prior to magnetic separation. The roasting under controlled conditions increases the

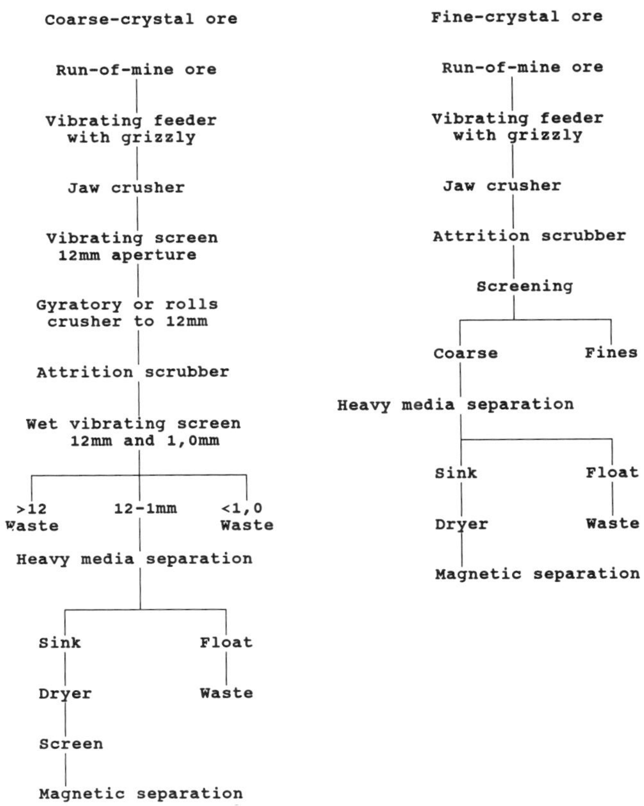

Fig. 7. Simplified flowsheet for a beneficiation plant operating on coarse- and fine-crystal andalusite (Overbeek, 1989).

magnetic susceptibility of feebly magnetic mineral particles and thereby decreases the iron content.

The standard ore dressing procedures do not recover any andalusite smaller than 0.6 mm (0.3 mm in the case of the French operation). The demand for fine andalusite down to 0.1 mm caused various producers to investigate the recovery of the fines from their slimes dams, which represents a vast source of high grade andalusite. Heavy medium separation is being investigated as well as wet-drum magnetic separation. In France a flotation process is already used to recover the fine andalusite up to a grain size of 0.3 mm.

Product Data

Tables 2 and 3 give data on the different andalusite qualities available on the market.

Byproducts

Staurolite and almandine garnet frequently accompany the occurrence of andalusite and, where present in commercial quantities, they are recovered and sold as abrasive aggregates.

USE OF ANDALUSITE

Although the first use of andalusite was in the manufacture of ceramic spark plugs, the bulk of the andalusite production is now used in the refractories industry, with relatively small quantities used in other applications such as ceramics, abrasives, and filler materials. The bulk of the refractories are high-alumina bricks and monolithics for the steel industry, and the consumption of andalusite is therefore closely related to developments in the steel industry.

The refractory characteristics of andalusite depend largely on its ability to form the mineral mullite at high temperatures. Mullite combines high strength at high temperatures with resistance to chemical and physical erosion. The resultant mullite consists of interlocking crystals which remain dimensionally and chemically stable to temperatures exceeding 1 800°C.

The superiority of andalusite over other natural refractory raw materials has been extensively reported in the literature (see bibliography) and can be summarized as follows:

Andalusite converts to mullite at relatively low temperatures and the associated volume change is so small that no prior calcination is required before manufacturing refractory or ceramic shapes. This leads to lower energy requirements and better controlled dimensions. The small volume change is also beneficial in the application of andalusite in monolithic materials and unfired bricks and shapes since it will seal the joint and form a monolithic structure.

The high temperature strength of a refractory material depends largely on the amount of liquid that is formed at elevated temperatures. The low impurity content of andalusite and the interlocking mullite crystal network results in better creep resistance than materials made from other natural raw materials. Lower thermal expansion, higher thermal conductivity, and low glassy phase content also results in exceptional thermal shock resistance.

The ability of andalusite to be converted to dense, low porosity shapes also enhances its resistance to penetration and attack by deleterious slags and gases. The low-flux content also enhances resistance to attack of the glassy phase by slags and volatile alkalies.

The inherent hardness of the andalusite crystals and the dense materials that can be manufactured from it result in a highly abrasion resistant material.

Some of the applications where andalusite refractories and ceramics are used include: stove bricks and checkers; direct reduction kilns; blast furnace stack and runners; torpedo, transfer, and casting ladles; hot metal mixers; and induction furnaces in the steel industry.

Other applications include cement kilns, incinerators, glass furnaces, carbon-baking furnaces, copper-roasting furnaces, kiln furniture for the ceramic industry, and foundry sands.

Product can be supplied as fired, chemically bonded, pitch bonded or resin bonded shapes or as unformed materials in the form of castables, ramming materials, mortars, or graded materials.

ECONOMIC FACTORS

Worldwide use of andalusite has risen from 10 kt in 1960 to 320 kt in 1990. Since the major use of andalusite is in refractories, it is a clear indication of its cost effectiveness—especially in the steel industry where there is a steady decline in the specific consumption of refractories used per ton of steel produced.

More than half the andalusite produced in South Africa is exported to other countries. It is therefore very clear that transport is a very big cost component in the price of andalusite. The bulk of andalusite exported comes from South Africa where the mines are far from the coast, making road and rail transport compulsory resulting in a considerable difference between the ex-mine and FOB prices. Shipping to markets in the United States, Europe, and the Far East further adds to cost. The greater availability of shipping to

Table 2. Properties of South African Andalusite

	Samrec					Anglovaal		Hoogenoeg
	Andafrax 3000	Annersley Premium	Annersley Randalusite 60/09	Krugerspost K57	Krugerspost K55	Rhino Purusite	Havercroft	Enrosit A
LOI	0.55	0.5	0.5	0.47	0.55	0.4	n.d.	0.4
SiO_2	38.5	38.0	38.0	40.8	41.80	39.0	37.5	39.0
Al_2O_3	59.5	60.0	59.8	57.2	54.4	59.5	60.5	59.5
Fe_2O_3	0.8	0.75	0.85	0.91	1.2	0.9	0.87	0.9
TiO_2	0.2	0.14	0.17	0.26	0.33	0.20	0.15	0.20
CaO	0.05	0.1	0.15	0.1	0.15	0.20	0.10	0.20
MgO	0.1	0.06	0.06	0.11	0.27	0.10	0.05	0.10
Na_2O	0.01	0.04	0.04	0.10	0.15	0.10	0.05	0.10
K_2O	0.2	0.19	0.26	0.20	0.4	0.25	0.15	0.25
Grain Size (mm)	0.5–3	3–8	2–8 1–4	0.6–3.5	0.6–3.5	0.6–3.5	2–8	0.6–3.5

Table 3. Properties of French Kerphalite

	Kerphalite France		
	KA	KB	KF
LOI	0.70	0.85	
SiO_2	38.3	44.3	
Al_2O_3	59.0	53.0	60.0
Fe_2O_3	0.90	1.0	0.65
TiO_2	0.25	0.25	
CaO	0.20	0.2	
MgO	0.15	0.15	
Na_2O	0.1	0.2	
K_2O	0.3	0.35	
Grain size (mm)	Unground 0.3–1.6		<0.3

Europe results in making andalusite more cost effective in Europe than in other parts of the world.

Until 1990 the andalusite market was reasonably healthy, but demand started to fall in 1991 due to the recession in most of the major consuming countries. The Rand price of South African andalusite has been rising steadily due to inflation and rising costs. The devaluation of the Rand has helped to cancel out these effects on the export price of the material to some extent. The fall in the value of the Rand has slowed and an increase in price could be possible in the near future.

Andalusite is a unique mineral and no direct alternatives exist. Bauxite with a higher alumina content and very competitive prices from China can replace andalusite in certain applications. Synthetic materials with approximately the same chemical composition can be used in many of the applications, but cost effectiveness will be a problem. In specific applications other materials such as chamotte, alumina, zircon, calcined topaz, magnesia, and doloma could be used, but in the end availability and cost of the raw material are important factors in selecting refractory products. This can be clearly seen from the fact that South Africa is also the largest producer of andalusite-based refractory materials.

Future growth in the demand for andalusite will depend on substitution of other refractories because of the continuous decline in the market and on the price and availability of andalusite relative to other raw materials.

BIBLIOGRAPHY

Anon., 1985, "Sillimanite Minerals—Europe Places Demand on Andalusite," *Industrial Minerals*, Jan., pp. 41–59.

Blain, M.R., Coetzee, C.B., and Lombaard, B.V., 1976, "Andalusite, Sillimanite and Kyanite," *Mineral Resources of South Africa*, 5th ed., Geological Survey, Pretoria, pp. 255–260.

Botha, J.C., 1989, "Andalusite—Trends in its Supply and Demand," Bulletin, Minerals Bureau of South Africa, B11/89, 7 pp.

Carrol, J., and Grobbelaar, A.P., 1981, "Refractories for Anode Baking Furnaces," *Proceedings*, 110th AIME Annual Meeting, TMS Light Metals Committee, Feb., pp. 543–567.

Carrol, J., and Mathews, G.W., 1982, "The Move Towards Andalusite for Stove Checkers and Ladle Linings," *Proceedings*, Ilafa-Alafar Congress, Lima, Peru.

Carrol, J., and Mathews, G.W., 1983, "Sillimanite Minerals as Raw Materials for Refractories," *Industrial Minerals, Refractories Supplement*, London.

Dana, E.S., 1958, *A Textbook of Mineralogy*, 4th ed., W.E. Ford, ed., Wiley & Sons, New York, pp. 615–617.

Dickson, T., 1991, "Sillimanite Minerals," *Metals and Minerals Annual Review*, p. 104.

Hammerbeck, E.C.I., 1986, "Andalusite in the Metamorphic Aureole of the Bushveld Complex," *Mineral Deposits of Southern Africa*, Geological Society of South Africa, Johannesburg, RSA, pp. 993–1004.

Human, D.R., and Collins, L.A., 1986, "The Havercroft-Streatham Andalusite Deposit, Eastern Transvaal Complex," *Mineral Deposits of Southern Africa*, Geological Society of South Africa, Johannesburg, RSA, pp. 1005–1008.

Kriek, H.J.S., 1988, "Recent Developments with Andalusite as a Refractory Material in Steelmaking," *Interceram*, No. 4, pp. 11–15.

Lister, R.R.,1984, "Randalusite for Unfired Refractory Products," *Interceram*, pp. 38–39.

McCracken, W.H., 1988, "Refractory Raw Materials in 1992," *Ceramic Bulletin*, Vol. 67, No. 7, pp. 1155–1157.

McCracken, W.H., 1991, "Chinese Refractory Raw Materials in the 1990's," *Interceram*, Vol. 40, No. 8, pp. 583–586.

Muchart, H., 1976, "Kerphalite-Europe's Own Source of High-Alumina Andalusite for Refractories," *Proceedings*, 2nd Industrial Minerals International Congress, May, 8 pp.

Overbeek, P.W., 1989, "Andalusite in South Africa," *Journal of the South African Institute of Mining and Metallurgy*, Vol. 89, No. 6, June, pp. 157–171.

Power, T., 1986, "Sillimanite Minerals—Europe Places Demands on Andalusite," *Industrial Minerals Refractories Survey*, pp. 77–91.

Russell A., 1988, "Industrial Minerals of France," *Industrial Minerals*, June, pp. 19.

Siegers, A., and Lange, S.P., 1991, "The Hoogenoeg Mine in NE Transvaal, RSA—A Producer of High Grade Andalusite," *Erzmetall*, Vol. 44, No. 7–8, pp. 384–390.

Stowers, R., 1991, "Andalusite," *Ceramic Bulletin*, Vol. 70, No. 5, pp. 850–851.

SILLIMANITE GROUP

KYANITE AND RELATED MINERALS

PALMER C. SWEET

Kyanite, andalusite, and sillimanite are anhydrous aluminum silicate minerals, and they have the same chemical formula, Al_2SiO_5. Dumortierite and topaz are also classified as aluminum silicates although they contain boron and fluorine respectively. Mullite (a rare mineral named for its discovery site, the Scottish island of Mull) is produced when kyanite, andalusite, or sillimanite are heated. Synthetic mullite, also classified as an aluminum silicate, is formed by melting alumina and silica, bauxite, and kaolin or alumina, kaolin, and kyanite, depending on the type of synthetic mullite desired. These materials are either sintered in rotary, periodic, or tunnel kilns, or fused in an electric arc furnace.

The world's largest producer of kyanite is the United States. Domestic kyanite is presently produced from one company at two sites in Virginia; Kyanite Mining Corporation mines kyanite from Willis Mountain and adjacent East Ridge in Buckingham County in the central Piedmont province of Virginia. Internationally, kyanite is produced in Australia, Brazil, India, Kenya, Sweden, and Zimbabwe; the former USSR also produces kyanite (Fig. 1). Andalusite is produced by Piedmont Minerals Company, Inc. in the Piedmont province near Hillsborough, NC. Andalusite is also produced in France, India, the Republic of South Africa, and Spain (Fig. 1). Sillimanite is produced in Australia, Brazil, India, Spain, and the Republic of South Africa; the former USSR is also thought to be a major producer (Fig. 1). Some of the countries produce only for domestic use. Synthetic mullite, made by sintering and by electric arc fusion, is produced by Mulcoa (Mullite Company of America) in Andersonville, GA, and by Washington Mills Electro Minerals Corporation in Niagara Falls, NY. Annual world production of kyanite, andalusite, sillimanite, and synthetic mullite is approximately 672 727 tons (Potter, 1988).

GEOLOGY

Mineralogy

Kyanite is a blue to light-green, triclinic mineral that occurs in long, thin-bladed crystals and crystalline aggregates in gneiss, schist, and pegmatites. The mineral has a hardness of 5 parallel to the long axis of the crystal and a hardness of 7 across the crystal.

Andalusite is a yellow, brown, green, or red orthorhombic mineral that occurs in thick, almost square prisms in schist, gneiss, and hornfels. Gem quality andalusite exhibits a strong pleochroism; brownish-green in one direction and brownish-red perpendicular to it. Andalusite also occurs in placer concentrations where it has weathered out of rock.

Sillimanite is a brown, pale-green, gray, or white orthorhombic mineral that occurs in long, slender, needlelike crystals in schist and gneiss.

Chemical and Physical Properties

Chemical and physical properties of kyanite, andalusite, and sillimanite are noted in Table 1. Kyanite is the only mineral that irreversibly expands when heated (calcined to produce mullite) and may offset the shrinkage of most ceramic or refractory ingredients when heated.

Distribution of Major Deposits

Kyanite:

Kyanite Quartzite. Kyanite quartzite occurs in the Piedmont region of the Appalachians and trends from Georgia through South and North Carolina into central Virginia. The kyanite quartzite is in the Arvonia Formation of Ordovician age in Virginia and is enclosed in a sequence of metamorphosed felsic and mafic flows and volcanoclastic rocks on the southeast limb of the Blue Ridge Anticlinorium. The kyanite deposits, presently being mined at Willis Mountain and East Ridge in Virginia, were originally deposited as basal quartz gravel or coarse sand overlain by silt and clay that was later metamorphosed to form kyanite. Hydrothermal fluids were later introduced (Marr, 1992). Kyanite quartzites are rocks containing 15 to 40% kyanite and usually about 5% of other minerals such as pyrite, rutile, and mica.

The kyanite quartzite deposits in the Georgia-South Carolina-McCormick Districts would seem to be similarily formed as the mineralogy, etc., are virtually the same as in Virginia. The metavolcanics and sediments are known as the Little River Series in Georgia, but they are considered to be in rocks of the Slate Belt in Virginia.

There are at least 13 distinct deposits of kyanite quartzite within the Appalachians. All of them are known as mountains, owing to their resistance to weathering which results in their prominence above the surrounding countryside. The operation at Baker Mountain, Virginia, was closed in 1979 and that at Graves Mountain, Georgia, was closed in 1986. From 1948 through 1969, Commercialores, Inc.. a subsidiary of Combustion Engineering, Inc. worked the Henry Knob deposit in York County, South Carolina.

There is a zone of metavolcanics whose composition and lithology resembles that of the deposits in the Appalachians in southeastern California and extending into southwestern Arizona that contains some deposits of kyanite quartzite. However, the kyanite in these deposits, while averaging 25 to 35% of the rock in apparently large and recoverable crystals, has proven to be contaminated with extremely fine inclusions of quartz. As of 1991, it has not been

Fig. 1. Kyanite, andalusite, sillimanite, and synthetic mullite production.

economically feasible to produce a competitive kyanite product from these western kyanite quartzite deposits.

Kyanite quartzite has been described in several other places in the world, and usually seems to occur within a geologic framework similar to that of the deposits of the southeastern United States. Such deposits have been described and explored in Surinam, Norway, Kenya, and Austria. It is reported that a kyanite deposit (Lac Croche property), located 35.4 km south-southeast of Fermont in Quebec, Canada, was being evaluated. Exploration indicated a zone with 20 to 40% kyanite in large bluish crystals, 1 to 2 cm in size and with an early estimate of 4 090 909 tons to a depth of 32.8 m (Potter, 1988).

Kyanite Schist and Gneisses. Kyanite is very common in the highly metamorphosed schists and gneisses of the metamorphic regions of the world. Typically, the kyanite occurs in quantities ranging from a percent or two to as much as 25% in a gangue of biotite, feldspar, muscovite, garnet, and occasionally hornblende and other common rock-forming minerals. Rocks containing a few percent of kyanite are extremely abundant and widespread. They are exposed over hundreds of square kilometers in the eastern and western metamorphic areas of North America and in the metamorphic rocks of other continents. Repeated attempts to recover the kyanite from such rocks have been made and have been described in earlier editions of *Industrial Minerals and Rocks*. The most recent attempt was in the Timiskaming District in western Quebec by North American Refractories Co. of Cleveland, OH. An earlier effort was made in the late 1950s near Burnsville, NC. As of 1991, no kyanite schists are being mined for kyanite.

In order for such kyanite to be economic it is necessary that the region be deeply weathered and of gentle relief so that a mantle or segregation of resistant kyanite nodules, cobbles, and boulders have accumulated at the surface. It is further necessary for labor to be abundant and cheap.

A great deal of literature has been published by the US Bureau of Mines, the US Geological Survey, and agencies of other governments describing investigations of kyanite-bearing schists. An exhaustive study was conducted by the US Bureau of Mines in Idaho on the huge deposits of kyanite, andalusite, and sillimanite at Goat Mountain and the kyanite deposit on Woodrat Mountain near Kemiah, ID (Van Noy, 1970).

Massive Kyanite. Kyanite is found locally as nodules, knots, and huge boulder-sized segregations in very highly metamorphosed areas of aluminous sediments. This has been the principal source of kyanite from India for the past 40 years. Production is estimated to be 34 545 tons in 1988 (Potter, 1988). Similar segregations are the basis for the kyanite production in Kenya.

Table 1. Chemical and Physical Properties of Kyanite, Andalusite, and Sillimanite

	Kyanite	Andalusite	Sillimanite
Chemical formula	Al_2SiO_5	Al_2SiO_5	Al_2SiO_5
Composition	Al_2O_3 63.2% SiO_2 36.8%	Al_2O_3 63.2% SiO_2 36.8%	Al_2O_3 63.2% SiO_2 36.8%
Type	Aluminum silicate	Aluminum silicate	Aluminum silicate
Crystal system	Triclinic	Orthorhombic	Orthorhombic
Cleavage	{100} perfect {010} good	{110} good {100} poor	{010} good
Specific gravity	3.56 to 3.66	3.16 to 3.20	3.23
Hardness	5 to 7	7.5	6 to 7
Temp. at which mullite is formed	1 100 to 1 480°C	1 450 to 1 500°C	1 550 to 1 650°C
Volume change during calcination	Sizeable increase	Very slight increase	Slight increase
Density after calcining at 1 445°C	3.05	3.04	3.10

Source: in part, Klinefelter and Cooper (1961).

In several counties in Georgia, similar lumps of massive kyanite are found (Furcron and Teague, 1945). While the abundance and purity of the Georgia massive kyanite meet the requirements for commercial exploitation, high labor cost makes economic production by hand sorting and gathering unfeasible and production has been limited to a few carloads during World War II.

Kyanite mineral segregations are probably the result of local pneumatolytic migration of silica and alumina during the late stages of regional metamorphism. Introduction of alumina does not seem to have been a factor since the overall composition of the segregations are similar to that of the country rock if the sample area considered is large enough to include the barren quartz segregations which invariably accompany the kyanite-corundum segregations. Aside from the size and abundance of the segregations in the Lapsa Bura deposits of India, they do not seem mineralogically dissimilar from kyanite segregations found occasionally in all kyanite schists.

Massive kyanite typically contains corundum and minor amounts of rutile. The kyanite is often felty and acicular and occurs in tightly interlocking aggregates. Kyanite from India is usually produced in lumps large enough to be hand sorted according to kyanite and corundum content, and three grades are offered. Massive Indian kyanite has properties quite unlike that of large coarse kyanite crystals; it is essentially volume stable and calcines to a dense white aggregate that is much prized by European refractory manufacturers. On the other hand, the coarse kyanite produced from Georgia placer deposits in the 1940s, and more recently in Kenya, crumbles and loses much of its density and physical strength upon calcining. Apparently the interlocking acicular crystal mode of ''Indian'' kyanite prevents such expansion and consequent breakdown. From 1968–1973, a kyanite-sillimanite concentrate was recovered as a byproduct from the processing of Pleistocene beach sands in Florida.

Andalusite:

Constituent of Metamorphic Rocks. Andalusite is a frequent constituent of metamorphic rocks, although it is not as abundant or common as kyanite or sillimanite. It is found in argillaceous and micaceous slates, in schists and gneisses and as crystals resulting form the contact metamorphism of intrusive rocks. Andalusite will readily incorporate foreign matter in its crystals, and it frequently grows around preexisting materials, including carbon. One variety, chiastolite, is so named for the crosslike inclusions of carbon oriented normal to the axis of the crystal.

In France, near Glomel in Brittany, an extensive, deeply weathered body of andalusite schist is being mined at the present time by Denain-Anzin Mineraux Refractaire Ceramique (Damrec), a branch of the Imetal Group. This company has been active at this site since 1969. The quarry is about 966 m^2 by 40 m deep. About 681 818 tons of material is moved annually, and of that amount 59 091 tons is commercial kerphalite (two grades of andalusite concentrates). Of the two grades of kerphalite (andalusite concentrates), grade KA contains a minimum of 59% alumina and grade KB contains a minimum of 52% alumina (Anon., 1989). Near Glomel the andalusite occurs as matchstick-sized crystals embedded in a fine-grained black groundmass composed of biotite, hornblende, muscovite, and feldspar. The andalusite is evenly disseminated and constitutes about 20% of the rock. The rocks in the area are very poorly exposed, and the geology of this occurrence is not well understood.

Near Canso, Nova Scotia, there is an extensive deposit of andalusite schist. Here the andalusite makes up about 15% of the rock and is evenly disseminated as large porphyroblasts averaging about 1.3 to 2.5 cm in cross section. The groundmass is principally muscovite, garnet, and feldspar. The crystals of black andalusite incorporate about 10% of finely disseminated magnetite and muscovite, however, and effective beneficiation is not practical.

Near Hillsborough, NC, a monadnock of andalusite-pyrophyllite-sericite rock is being mined by the Piedmont Minerals Company, Inc. The ore consists principally of pyrophyllite and quartz and contains 15 to 20% of disseminated pink andalusite. The andalusite is mined along with the pyrophyllite, and being calcined in part, by nature, exhibits refractory characteristics usually found only with prepared calcined materials. The product is consumed by Resco Products, Inc. in the manufacture of kiln car refractories, plastic and castable refractories, fire brick, and ramming and gunning mixes.

In the Goat Mountain deposit in Idaho, andalusite coexists with kyanite and sillimanite. No attempt has been made to separate it from the aluminum silicate minerals in the tests made so far. The andalusite in this deposit typically incorporates a great many deleterious impurities. The Goat Mountain deposit is extremely large, but to date no commercial exploitation of the deposit seems feasible owing to beneficiation difficulties (Abbott and Prater, 1954).

In early 1991 exploration work by Ramsgate Resources Ltd. at Spargoville, in Western Australia's Eastern Goldfield, has uncovered several occurrences of andalusite-bearing chlorite-biotite schist. The andalusite occurs as discrete crystals, from 0.5 to 20 cm in length, in metamorphosed aluminous metasedimentary rocks. Reportedly the andalusite is of very high quality, equal if not better than South African material (Anon., 1991).

In the Vihorlat Mountains of East Slovakia, Czechoslovakia, there are new findings of secondary quartz with corundum, mullite, and andalusite (Hruska, 1991).

Residual Andalusite. Alluvial deposits of andalusite sand occur in the Republic of South Africa. The andalusite has been weathered from the parent rock and subsequently concentrated by the action of wind and water; apparently concentration is still going on. Preliminary figures indicate that production was 259 556 tons in 1988. In 1964 a heavy media separation plant was installed and since then the production of high grade concentrates has increased substantially. Concentrate grades averaging almost 60% alumina have been obtained. Total reserves are reported to be 45.91 Mt where the andalusite is found in metamorphosed shales of the Strubenskop Formation near Zeerust, in the weathered shales of the Timeball Hill Formation at Thabazimbi, and in the hornfels and schists of the eastern Transvaal near Penge; all of these units are of the Pretoria Group, that have been intruded and metamorphosed by the Bushveld Complex. Operational mines include the Annesley mine of Annesley Andalusite (Pty.), the Krugerpost mine of Cullinan Refractories (Pty.) Ltd., and the Timeball Hill mine of Weedon Corp. (Morgan, 1990).

Crystals. In 1986, exceptionally large reddish to flesh-colored andalusite crystals were discovered in Campbell County, VA. Some of the single crystals measured 25.4 cm long and up to 11.7 cm wide. They were formed through contact metamorphism, along both sides of a quartz vein which has intruded a staurolite-rich muscovite schist (Giannini and Penick, 1986).

Sillimanite:

Sillimanite Schist. Sillimanite is a very common rock-forming mineral in metamorphic rocks of relatively high rank. It is common in a series of metamorphosed rocks to find sillimanite, kyanite, and andalusite interchanging occurrences in given strata as the local conditions of temperature vary, as in the proximity to intrusives. For this reason the aluminum silicate minerals are often used to identify parameters of metamorphic intensity. Sillimanite, while common, seldom occurs as potentially exploitable crystals. The typical mode is what is often called fibrolite which is a felty aggregate of extremely fine whiskers of acicular sillimanite interlaced and interlocked with quartz, mica, and other minerals. Beneficiation is usually impossible. In some areas the sillimanite occurs as nodules and buttons which are marginally potential as in the Peltzer area of South Carolina (Espenshade and Potter, 1960).

In Hart County, Georgia, there is a north-easterly trending zone about ten miles long in which matchstick sillimanite occurs. Beneficiation tests have been conducted, from which it appears that a limited production of sillimanite in the 35 mesh range could be accomplished. However, the deposits are narrow and limited, and there is presently no existing market or incentive for production of sillimanite concentrates in the United States (Furcron and Teague, 1945).

The Geological Survey states that a 30 cm wide vein of sillimanite in Namaqualand schist "might be economically mined on a small scale." The sillimanite vein is located in the Bethanien district of Namibia, Africa (Murray, 1991).

Massive Sillimanite. Massive sillimanite has been produced for many years from the state of Assam, India. The deposits of Assam consist of huge segregations of sillimanite and corundum, commonly in intimate association, and weighing several tons. Considerable hand effort is employed to recover the boulders in a form suitable for sawing to refractory shapes, particularly for the English glass industry. Production of sillimanite in 1988 was estimated to be 15 kt (Potter, 1988).

The most important deposits of massive sillimanite-corundum occur in the Republic of South Africa, in the Pella District near Pofadder, Namaqualand. Production reached a high of 50 909 tons in 1963, but was down to 781 tons in 1988 (Morgan, 1990).

In the vicinity of Adelaide, South Australia, late-stage metasomatism resulted in a mixture of kaolin and included boulders and nodules of sillimanite. The sillimanite is recovered as a byproduct in the process of manufacturing refractory clays. Other deposits of residual boulders of massive sillimanite have been exploited in this region. Beneficiation has been tried with some success on the sillimanite itself and on the byproducts too low in grade to use directly. The Australian domestic market for such concentrates is limited, however, and thus far no large scale production has been described; reported production in Australia in 1988 was 73 tons of sillimanite and 7 273 tons of kaolinized sillimanite (Potter, 1988).

TECHNOLOGY

Synthetic Mullite

Synthetic mullite results from three main heating processes: 1) melting of alumina and silica or bauxite and kaolin in an electric furnace at 1 914°C produces a fused synthetic mullite, 2) sintering alumina and kaolin, bauxite and kaolin, or alumina, kaolin, and kyanite above 1 763°C produces a high-temperature synthetic mullite, 3) low temperature synthetic mullite results by sintering a siliceous bauxite or mixtures of bauxite and kaolin at about 1 561°C. United States production of synthetic mullite in 1985 was estimated to be 24 546 tons (Potter, 1988).

C-E Minerals/Mulcoa (Mullite Corporation of America) of Andersonville, GA, produces three grades of synthetic mullite at various temperatures, by sintering bauxite with kaolin in their two plants. The company mines bauxitic kaolin from pits in Sumter County, near Andersonville, and also from properties near Eufaula, AL. Washington Mills Electro Minerals Corporation of Niagara Falls, NY, produces a standard and electric grade of synthetic mullite by fusing alumina and glass-grade silica. They also produce a zirconium mullite which offers better resistance to thermal shock for ceramic products.

At Mulcoa, GA, the three grades produced are: Mulcoa 47 containing 47% alumina produced from the kaolin component of the ore, Mulcoa 60 containing 60% alumina produced from the bauxitic kaolin, and Mulcoa 70 containing 70% alumina produced from the bauxite lenses. The two plants have a total capacity of 545 455 tpy. Steady growth has occurred at Mulcoa since the downturn in the early eighties. All seven of the kilns were in operation during 1989, while there were only five operations in 1985 and one in 1981 (Sikich and O'Connor, 1991).

Potter (1985) notes 1984 estimated production figures of synthetic mullite as follows: Federal Republic of Germany and Italy, 9 091 tons each; United Kingdom, 13 636 tons; and Japan, 36 364 tons.

Exploration and Evaluation

Kyanite and sillimanite are fairly common minerals and are found in metamorphosed rocks on every continent. Andalusite is less abundant but still common. Kyanite is consumed mainly by highly developed industrialized countries and thus its consumption is restricted to specific areas. The bulk of American production is consumed in a relatively small region east of the Mississippi River. A similar pattern exists in Europe, where consumption is concentrated in those parts of England, France, and Germany where the producers of refractories are concentrated. An example is noted in *Industrie Minerale* (Anon., 1989), where the andalusite near Glomel, France, is said to be in a very favorable location. This is because of its qualities for refractory products and its geographical position, which permits direct deliveries to European markets, even in small tonnages. Kyanite and related minerals must be delivered to the consuming areas at competitive prices. An important preliminary consideration in any exploration project is the prospective cost of delivering the kyanite to the market. Typical freight charges will range from 0.03 to $0.08 t/km. A small market on the west coast of the United States would pay dearly for product from the eastern part of the country.

A salable product must also be produced. Consumers become very used to specific materials used by them in the manufacture of refractory products. Any deviation from the commonly accepted specifications imposes a considerable burden on products attempting to enter the market. One of the most important considerations in this regard when evaluating a prospect is the liberation size of the kyanite crystals. There are some attractive kyanite-quartzite deposits in the western United States, but the material cannot be beneficiated to the required specifications without grinding the ore to around 200 mesh, which is quite fine considering that most existing kyanite production is in concentrates in the size range of −28 to +35 mesh.

After initial investigation has indicated that both freight costs and the quality of the product can be competitive, normal techniques of surface sampling, geologic mapping, trenching, and diamond drilling are used to determine the quantity of potential reserves. Since the kyanite deposits are produced by regional metamorphism, their composition is fairly consistent, although there may be variation of grade in zones. A pilot plant production of the material would be the next step to determine if a salable product could be produced economically. It is very important to produce, by reproducible pilot plant techniques, a quantity of concentrates from a representative selection of ores sufficient to permit prospective consumers to test the new material extensively in their plants; there is no laboratory substitute for this kind of testing. Prospective producers of aluminum silicates have found, in many cases, that consumers may not be willing to use a new material from a new area. Competitive testing and specific approval is the only sure way that the marketability of a given kyanite product can be assured.

Mining and Processing

Kyanite: In the United States, for almost the last 40 years, kyanite concentrates have been produced by flotation and magnetic separation. The two kyanite quartzite deposits presently being mined (at Willis Mountain and East Ridge in Virginia) are monadnock-type features whose resistant ridges are exposed with little or no overburden. The deposits lie on the opposing limbs of the Whispering Creek anticline. Minerals other than quartz and kyanite associated with the deposit include muscovite, pyrite, rutile, topaz, magnetite, feldspar, and apatite. The rare mineral trolleite $Al_4(PO_4)_3(OH)_3$, a hydrous aluminum phosphate, has also been noted in the kyanite-quartzite (Giannini et al., 1986).

The kyanite quartzite is drilled and blasted; secondary breaking is sometimes done with a drop ball. The ore is picked up with diesel powered shovels, loaded into trucks (up to 22.7 ton loads), and hauled to the primary (jaw) crusher. At the primary crusher, the ore is reduced to −4 cm and passed by conveyor belt to the rod mill, in a closed circuit with a classifier, to grind the ore to −35 mesh. Water is added which creates a slurry from which the −200 mesh is removed (deslimed). The new slurry is conditioned with several ingredients and passed through a series of flotation cells which remove the pyrite and micaceous contaminants from the slurry. The tailings from the pyrite circuit are again deslimed and conditioned with sulfuric acid and petroleum sulfonate and passed through a section of rougher flotation cells. The rougher concentrate goes to a two-stage recleaning circuit; tails of the rougher circuit go to waste.

The flotation concentrate consists of about 91% kyanite, 2 to 5% iron oxides, and the balance quartz. The concentrates are dewatered by draining in open stockpiles or in drainage silos and are subsequently conveyed to a dryer where the moisture is reduced from almost 8 to near 0%. A reducing roasting technique in a rotary kiln elevates the temperature during the drying process to 482°C, followed by cooling in an oxygen-deficient atmosphere to render the iron oxide highly magnetic; ferric iron (Fe_2O_3) is converted to ferrous iron (Fe_3O_4). The dried concentrates are passed over high intensity magnetic separators, which reduce the iron content of the product to less than 1%. The result is a raw 35 mesh kyanite product, ready for shipment. Five mesh sizes (35, 48, 100, 200, and 325) are available. A typical analysis of the product is as follows:

Al_2O_3	54 to 60.06%
SiO_2	37.64 to 43.70%
Fe_2O_3	0.40 to 1.16%
TiO_2	0.67%

The magnetite (pyrite with magnetite rims) is stockpiled, with no present market available. The byproduct quartz is marketed for golf courses, masonry and concrete sand, and for other applications.

Kyanite deposits of India have traditionally been exploited by labor-intensive techniques. In the beginning it was sufficient to simply gather the cobbles and boulders of kyanite which had been exposed by weathering. Later, primitive mining methods were initiated and the large segregations were drilled and blasted and hand-cobbed in preparation for shipment. Dozens of these small operations employed numerous men and women. As time passed and the availability of readily hand-picked material decreased, the beneficiation methods became more sophisticated. The picking belt is still an important beneficiation technique, however. High grade kyanite nodules, and segregations too low in grade to be shipped directly, are now crushed and subjected to more sophisticated beneficiation techniques including flotation and magnetic separation. At the present time there are no large-scale integrated beneficiation plants in India of the kind employed in America.

Andalusite: In the Republic of South Africa, andalusite sands in the Marico District of the Transvaal are beneficiated with heavy media separators and high intensity magnetic separators. A typical analysis of the product is as follows:

Al_2O_3	52 to 57%
SiO_2	35 to 44%
Fe_2O_3	1 to 4.5%
TiO_2	0.04 to 4%

In France, Damrec produces andalusite concentrates near Glomel, in Brittany. The black andalusite schist contains about 20% coarse euhedral andalusite porphyroblasts. The crystals are about the diameter of matchsticks and range from 1.9 to 2.5 cm in length. The schist is deeply weathered and friable to at least a depth of 9.1 m. The ore is drilled with a wagon drill, blasted, and transported with a rubber tired front end loader with a 2.29 to 2.74 m bucket. It is fed into a surge hopper through an 20.3 x 20.3 cm grizzly and discharged onto a 61 cm conveyor belt by a reciprocating feeder. Oversized material is reduced by hand with a sledge hammer to pass the 20.3 cm grizzly.

The conveyor belt carries the broken ore about 45.7 m to an autogenous mill, which consists of a drum 4.6 m in diameter revolving on eight truck tires, and which reduces the ore to 100% −0.3 cm. It then travels to the furnace which is air-swept with heated air in closed circuit with two air classifiers. The sized ore is fed into a hopper, from which it is passed over a four-roll, high intensity magnetic separator. The nonmagnetic fraction is conveyed to the heavy media separator and the magnetic particles are discarded.

The heavy-media separator circuit consists of a hopper in which the andalusite ore is blended with a ferrosilicon slurry and then pumped through a hydrocyclone. The specific gravity of the slurry is controlled automatically by a water-metering device ahead of the blending hopper. The first stage of the two-stage heavy media separator circuit is controlled at about 2.3 specific gravity. The final effective specific grvity is adjusted by regulating the velocity of the slurry through the cyclone. The second stage is controlled in similar fashion. The heavy (sink) product from the heavy media separator is screened at about 35 mesh to recover the ferrosilicon media. The finished andalusite product consists of dark brown to black, roughly equidimensional grains ranging in size from minus 8 mesh to +35 mesh, and contains about 59.2% Al_2O_3 and 1.0% Fe_2O_3.

In 1988, the Council for Mineral Technology (Mintek) published a review on the development of the andalusite industry in the Republic of South Africa. The publication discussed Mintek's role in early benefication methods and techniques developed since 1962; Mintek was also instrumental in introducing heavy medium separation as a standard benefication operation. The report also discusses treatment of both hard shales and andalusite fines (Potter, 1988).

Sillimanite: Segregations of sillimanite-corundum rock in the Pellar District in the northwestern part of Cape Province, South Africa, have been produced for many years by labor-intensive techniques similar to those employed in India.

USES

Today it is estimated that 90% of the kyanite-mullite output is used in refractories: 55% for smelting and processing of ferrous metals, 20% for nonferrous metals, and 15% for glassmaking and ceramics. The remaining 10% is used for nonrefractory uses (Potter, 1988).

The main uses of kyanite are in high temperature (refractory) cements/mortars, ramming mixes, and castables. With clay or other ingredients, the kyanite is used in fire brick, insulating brick (brick for rotary kilns and furnaces), and other kiln furniture. Expansion characteristics of kyanite are valuable because they can offset shrinkage of other ingredients and produce a stable refractory body. Kyanite in mortar will expand and form a tight lock in a furnace or kiln. Finely ground kyanite, in the 200-mesh range, is used in body mixes for sanitary porcelains, wall tile, casting molds, and special purpose ceramics.

Before being used in some refractories, kyanite is converted to mullite by heating at high temperature. The addition of mullite to refractory and porcelain bodies will increase the fired strength, increase resistance to deformation under load, increase dielectric properties at both normal and high temperatures, and increase thermal resistivity; mullite will also decrease the coefficient of expansion and the tendency to spall. The synthetic mullite, produced in Georgia, is used in investment castings and moldings and sold to the refractories industry to make refractory bricks. Almost half of the 1989 production was exported through the Port of Savannah (Sikich and O'Connor, 1991).

The main end uses for andalusite are refractory and ceramic such as tile, foundry mould washes, and electrical porcelain.

The largest market for sillimanite minerals is for the manufacture of refractory products, mainly for the iron and steel and nonferrous metal smelting industry.

ECONOMIC FACTORS

Prices

United States kyanite, FOB, in carload lots, is valued at 70 to $137 raw per ton and 123 to $172 calcined per ton. Price of andalusite, Transvaal, South Africa, CIF (cost, insurance, and freight) a main European port, ranges up to $200/t. South African sillimanite (70% Al_2O_3) in bags, CIF a main European port is $293/t (Potter, 1988). Synthetic mullite, 165 to $1,650/t, depending on grade and type. A sharp increase in the price of 60% alumina of South African andalusite was caused by the value of money relative to that of other countries.

Reserves

Cooper (1965) notes that reserves of kyanite-bearing quartzose rock containing 10 to 30% kyanite in the United States are reported to be about 91 Mt. A US Bureau of Mines drilling program, in 1949 on Willis Mountain in Buckingham County, indicated that 59.1 Mt of 25% kyanite were exploitable by open pit mining (Jones and Eiletsen, 1954). Potential reserves are probably much larger. Reserves of the East Ridge deposit, 8 km distant, are not included in this figure. Other deposits in Virginia include Woods Mountain in Buckingham County as well as properties to the south in Prince Edward County. With annual mining at 363 636 to 454 546 t of ore, the deposits of kyanite in Virginia will be adequate for many years. At Graves Mountain, Georgia, where the deposit is no longer mined, previous geologic mapping and diamond drilling indicated a reserve of 27.3 Mt of 25% ore. In Van Noy (1970), the study on Goat Mountain and Woodrat Mountain, near Kemiah, ID, indicated that hundreds of millions of tons of 25% aluminum silicate ore are available for open cut mining. However, remoteness from the major markets and the difficulty of beneficiation have stalled development. If the schists and gneisses, which contain approximately 10% kyanite, are noted as resources in North America, the resource potential is probably very large and indeed may not even be recognized or discovered yet (Potter, 1985).

According to the Minerals Bureau of South Africa, reserves of andalusite and sillimanite are 104.6 Mt. Kyanite reserves in Brazil are given at about 2.7 Mt. Andalusite occurs in France where reserves of andalusite content amount to about 2.7 Mt. Sillimanite deposits are located in Australia, India, and the USSR (Potter, 1985).

GOVERNMENT CONSIDERATIONS

Depletion allowance rates for kyanite were unchanged through 1988. They are 22% for domestic production and 14% for foreign operations.

For kyanite, andalusite, and sillimanite, as of Jan. 1, 1991, there were no tariffs for most favored nations (MFN) or for other nations. On the other hand, mullite holds a 4.9% tariff ad valorem (according to the value) for MFN and for non-MFN, the tariff is 30% for the commodity.

From Aug. 23, 1951, through Mar. 4, 1970, the eligibility for stockpiling of lots of kyanite ore and concentrates, or of fused synthetic mullite, was determined according to criteria in the National Stockpile Purchase Specification P-27-R (Feb. 29, 1960, superseding issue of Aug. 23, 1951). The statement listed as chemically acceptable three types of kyanite if they contained not less than 59% alumina, not more than 39% silica, and not more than specified proportions of iron, titanium, lime, magnesia, alkalies, and total fluxes. Chemical requirements for fused synthetic mullite called for alumina within the range of 75 to 80%; silica 20 to 25%; and total impurities not exceeding 1%. Additionally, both kyanite and mullite for stockpiling were required to conform to a catalog of specific physical properties and pyrometric cone equivalent ratings. Now, sales of kyanite-group minerals are based on bargaining of the consumer and the seller. Chemical analyses, physical characteristics and service testing required are usually not available. The stockpile of lump kyanite as of Sep. 30, 1990 was 100 t; there are also 863.6 t of nonstrategic grade material retained in government stockpiles.

FUTURE TRENDS

Today the overall use of kyanite and mullite has increased throughout the world, due mainly to consistent high quality. Potter (1985) noted the forecast range for demand of kyanite and synthetic mullite in the year 2000 to range from 109 to 182 kt. However, production from 1989 to 1990 is believed to have decreased due to the turndown in the economy and subsequent lack of production in the iron and steel and other industries in which refractories are utilized. Substitutes are possible for all end-uses, but kyanite and mullite have competed very favorably in refractories. With the increasing price of zircon (from $209/t in 1986 to an estimated $422/t in 1990), the increased use of mullite as a direct substitute is being used in many zircon applications, especially refractories. Both fused and sintered synthetic mullite are substitutes for kyanite in refractories. Garner (1991) notes that demand for zircon, even with increasing price may continue to grow at a rate of 4 to 5% a year over the next decade. A forecast for demand of kyanite-group minerals and synthetic mullite in the next 10 years may indeed be in the range of 909 kt.

BIBLIOGRAPHY

Anon., 1989, "La Kerphalite Bretonne," *Industrie Minerale*, Jan., pp. 66–69.

Anon., 1991, "Andalusite Deposit Found by Gold Explorer," *Industrial Minerals Magazine*, Apr., p. 9.

Abbott, A.T., and Prater, L.S., 1954, "The Geology of Kyanite-Andalusite Deposits, Goat Mountain, Idaho, and Preliminary Beneficiation Tests on the Ore," Pamphlet 100, Idaho Bureau of Mines and Geology, 27 pp.

Cooper, J.D., 1965, "Kyanite and Related Minerals," Mineral Facts and Problems, Bulletin 630, US Bureau of Mines, pp. 481–488.

Espenshade, G.H., and Potter, D.B., 1960, "Kyanite, Sillimanite, and Andalusite Deposits of the Southeastern States," Professional Paper 336, US Geological Survey, 121 pp.

Furcron, A.S., and Teague, K.H., 1945, "Sillimanite and Massive Kyanite Deposits in Georgia," Bulletin 51, Georgia Department of Natural Resources, 76 pp.

Garner, T.E., Jr., 1991, "Zirconium," *Mining Engineeering*, June, Vol. 43, No. 6, p. 619.

Giannini, W.F., and Penick, D.A., Jr., 1986, "Mineral Update-Large Andalusite Crystals from Virginia," Virginia Division of Mineral Resources, Virginia Minerals, Vol. 32, No. 4, Nov., pp. 43–44.

Giannini, W.F., Penick, D.A., Jr., and Fordham, O.M., Jr., 1986, "Mineral Update-Virginia (Trolleite)," Virginia Division of Mineral Resources, Virginia Minerals, Vol. 32, No. 1, Feb., p. 11.

Hruska, J., 1991, "Czechoslovakia's Minerals—Production, Reserves, and Exploration," *Industrial Minerals Magazine*, Oct., pp. 49–53.

Jones, J.O., and Eiletsen, N.A., 1954, "Investigation of the Willis Mountain Kyanite Deposit, Buckingham County, Virginia," Report of Investigations No. 5075, US Bureau of Mines, 54 pp.

Klinefelter, T.A., and Cooper, J.D., 1961, "Kyanite: A Materials Survey," Information Circular 8040, US Bureau of Mines, 55 pp.

Marr, J.D., 1992, "Geology of the Kyanite Deposits at Willis Mountain, Virginia," *Proceedings*, 26th Forum on the Geology of Industrial Minerals, P.C. Sweet, ed., Publication 119, Virginia Division of Mineral Resources, pp. 129–134.

Morgan, G.A., 1990, "The Mineral Industry of the Republic of South Africa," *Minerals Yearbook: Area Reports: International*, Vol. 3, US Bureau of Mines, pp. 127–158.

Murray, R., 1991, "Namibia," *Industrial Minerals Magazine*, Sep., pp. 97–103.

Potter, M.J., 1985, "Kyanite and Related Minerals," *Mineral Facts and Problems*, Bulletin 675, US Bureau of Mines, pp. 425–432.

Potter, M.J., 1988, "Kyanite and Related Materials," *Minerals Yearbook: Metals and Minerals*, Vol. 1, US Bureau of Mines, pp. 577–580.

Sikich, S.W., and O'Connor, B.J., 1991, "Georgia," *Minerals Yearbook*, US Bureau of Mines, 13 pp.

Van Noy, R.M., 1970, "Kyanite Resources in the Northwestern United States," Report of Investigations 7426, US Bureau of Mines, 81 pp.

SODA ASH

DENNIS S. KOSTICK

Soda ash is the common name for sodium carbonate (Na_2CO_3), an alkali compound that is the 11th largest inorganic chemical in terms of production of all domestic inorganic and organic chemicals, excluding petrochemical feedstocks. Despite that most people have never heard of soda ash, it is an important industrial compound used to manufacture glass, chemicals, soaps and detergents, pulp and paper, and many other familiar consumer products. Natural alkalis have been used by mankind for thousands of years; however, their industrial manufacture dates back to only the latter half of the 18th century.

Natural deposits of sodium carbonate have been known as early as about 3500 BC, when the Egyptians used natural soda ash in making glass. They also mixed lime and soda ash to make sodium hydroxide (caustic soda) that was combined with silicate minerals found in the Sinai desert. This made a soluble silica, which when added to aluminum-rich silt from the Nile River, produced a silica-aluminate cement mortar with excellent bonding properties for construction. The Romans in the first century AD also used natural soda ash in glass manufacture but expanded its use medically for the treatment of colic pains and skin eruptions, and in the making of bread.

Elsewhere, people used the ashes of wood to obtain their source of alkali. People learned over time that different types of wood yielded various ashes with different properties; therefore, various plants were burned specifically for their ashes. Ash from plants grown in salt-bearing soils (such as saltwort) and from kelp and other seaweeds, especially Spanish barilla, were so different from ashes obtained from other vegetation that it became known as soda ash (due to its sodium content) versus pot ash, a potassium-based alkali ash. About 13 t of barilla ashes produced 1 t of sodium carbonate and 14 kg of iodine as a byproduct. The difference between these two ashes was relatively obscure until Duhamel Dumonceau made the distinction clear in 1736. The word soda ash developed from common usage and is perhaps more well known today than its synonym sodium carbonate. One of the primary ores of soda ash is trona, a sodium carbonate-bearing mineral that's name is traced back to Arabic origin. Trona also is known internationally by other names such as urao (Venezuela), kaum (Nigeria), natron (from Greek nitron and Latin natrium), and szekso (Hungary).

The beginning of the Industrial Revolution in Western Europe in the late 18th century soon put a strain on the availability of raw materials to meet consumer demands. One of these scarce materials was soda ash. Because of the Seven Years War with England and the Napoleonic wars with other parts of Europe, France could not obtain sufficient quantities of Spanish barilla or other supplies of vegetable alkali to meet the growing demand. Efforts were needed to synthesize soda ash. In 1775, the French Academy of Sciences offered a large prize of 2,400 livres to someone who could find an inexpensive method to make soda ash. In September 1791 at St. Denis, Nicolas Leblanc (1742–1806), a French chemist, developed such a technique using salt, sulfuric acid, coal, and limestone. The French Revolution interfered with its development, and his patent and factory were confiscated with Leblanc receiving only token compensation. Napoleon returned his factory to him; however, Leblanc was not able to raise enough capital to reopen it, and he committed suicide in 1806. A small, but not particularly successful, Leblanc plant was established in England in 1814. It was not until 1823 when the process first became commercially successful in Liverpool, England. The process was introduced in Germany in 1843 and in Austria in 1851 (Harness and Coons, 1942). Soda ash production by the Leblanc process reached its peak of about 599 500 t in 1880, after which it began to decline as the Solvay process became more popular. The Leblanc process was used to a limited extent in Europe during World War I but had disappeared by World War II.

The Solvay process, also known as the ammonia-soda process, was developed by Alfred and Ernest (1838–1922) Solvay in 1861 based on a concept by Fresnel that had been known since 1811. For the next 50 years, the implementation of the idea evaded industrial chemists because no large-scale and economic means could be found to commercialize the concept. Although Ernest Solvay was unaware of the existance of an ammonia-soda concept, he solved the problems by utilizing carbonating towers. The Solvay process produced soda ash from salt, limestone, and coke, with ammonia as a catalyst. With a capital of 136,000 francs, the first plant was built by the Solvay brothers at Couillet, Belgium, in 1863 with production commencing in 1865. Synthetic soda ash production amounted to 1.5 tpd in 1866, but reached 10 tpd by 1872. They patented their use of carbonating towers in 1872 that had made ammonia-soda manufacturing a successful continuous process. Their second plant was built in 1872 at Dombasle, France. In 1874, the first Solvay plant in England was built at Northwich by Ludwig Mond, the namesake of Brunner Mond and Co. (formerly Imperial Chemical Industries), the company that currently operates the two English Solvay plants.

In the United States, William B. Cogswell in February 1879 heard a presentation on the Solvay process and sailed to Europe to meet with the Solvay brothers to seek their support for using their process in New York. At first, the brothers were not interested; however, Mr. Cogswell was persistent and ultimately gained their support. The Solvay Process Co. was formed Sep. 21, 1881, and construction began immediately on the first Solvay plant at Geddes, near Syracuse, NY. The plant came onstream Jan. 10, 1884, and produced 11 180 t in its first year of operation. In 1910, rotary calciners were installed that increased capacity to 1 000 tpd. By 1930, plant capacity was up to 2 400 tpd. The facility remained in continuous operation for more than 100 years. By 1939, 10 Solvay soda ash plants were in operation in 6 States throughout the nation. The Syracuse plant was officially closed Jan. 6, 1986, with total shutdown completed by Feb. 1986. Ironically, this facility was the first Solvay plant as well as the last. Other than the first Solvay plant beginning in 1884, the majority of US soda ash supplies during the 19th century came from imports. The remainder of the US supply was obtained from small natural soda ash deposits in Nevada and California that served consumers in the western part of the country.

The first commercial natural soda ash operation in the United States began in Nevada in 1868 at the Ragtown Soda Lakes, better known as Big Soda Lake and Little Soda Lake, which are the remnants of old volcanic craters located about 7 km west of the present town of Fallon. Asa L. Kenyon, a pioneer who settled in the area in 1855 to sell provisions to the western settlers passing through, acquired the lake properties and sold the smaller lake in 1868 to Higgins and Duffy, who sold it to a San Francisco group known as Doe and Dowd. Little Soda Lake was primarily dry most of the year and reportedly had a mass of crystallized sodium carbonate on its surface that was more than 10 m thick, whereas Big Soda Lake was filled with water that was saturated with many salts, including sodium carbonate and borax.

Two types of solar processes were used to produce the soda ash—a summer process where trona precipitated in evaporation ponds on the perimeter of the lakes and soda ash was the final product, and a winter process where natron precipitated and thermonatrite was the final product. Soda ash production from Little Soda Lake averaged about 300 tpy while output from Big Soda Lake, which began operation in 1875, averaged about 450 tpy. The majority of the soda ash was sold to miners in Virginia City, NV, to treat silver ore and to soap manufacturers in Reno, NV. Soda ash also was bagged and transported to Hazen, NV, where it was shipped by rail for $19 per ton to San Francisco for further refining where it sold for between 55 to $65 per ton.

With the beginning of the Newlands Reclamation Project in 1902, the lake brines became increasingly diluted by infiltrating ground waters from the Truckee-Carson irrigation project that conveyed water to California. By the end of 1907, the brines were so dilute that soda ash production became impractical and the rising water level in the lakes began covering the evaporation ponds and building structures. The soda ash producers brought claims against the US Government on June 17, 1917, citing that they believed there was a direct correlation between the flooding of their operations and the irrigation project, and that the Government should be responsible for damages of $205,000. On Apr. 7, 1919, the United States Court of Claims decided in favor of the Government. After years of appeals and investigations, which ultimately did prove that the canals were responsible for the rising lake levels, the case was resolved on Nov. 21, 1921, by the Supreme Court of the United States ruling in favor of the Government. The case was a landmark decision because it set a new legal precedent that ". . . the government is not liable for all actual damages resulting from public improvements, but only for such damages as inevitable to result and which with due diligence and expert study could have been anticipated" (Anon., 1921). The fact that the soda ash producers gave a right-of-way to the Government prior to 1907 was the main reason why they lost their case (Taylor, 1987–1988). The right-of-way stated "that in consideration of the premises, the first party hereby releases the second party form all claims and damages for entry, survey, or construction of said works." (US Supreme Court Case Nos. 26 and 32: John Horstmann Co. and Natron Soda Co. vs. the United States, Oct. Term, 1921).

More than 85 years have passed since Big and Little Soda Lakes were flooded. Today, two lakes lie quietly in the virtually barren landscape of western Nevada. There are no traces of the first natural soda ash operation in the United States anymore, except about 10 m down at the bottom of Big Soda Lake, which sometimes is explored by local skin divers that see the remains of the wooden buildings, rusting steam tractors, and beehive ovens that once produced soda ash. The site is nominated for the National Register of Historic Places, a fitting tribute to this historic operation.

Natural sodium carbonate was also observed in other areas besides Nevada. In 1835, the Reverend Samuel Parker crossed the Rocky Mountains and noticed epsom salts on the surface near Independence Rock, WY. The Mormon pioneers are believed to have used the natural sodium bicarbonate occurring along the Sweetwater River in Wyoming as early as 1849 for laundry and medicinal purposes. Commercial development of sodium carbonate brines began in 1907 by the Western Alkali Corp., which produced sal soda, or washing soda, in solar evaporation ponds from 13 wells about 100 m deep. The operation ceased in 1920 because of high operating costs (Milani, Jr., 1983).

The largest natural deposit of trona known in the world is in Green River, WY. The deposit was accidentally discovered in January 1938 when the Mountain Fuel Supply Co. was drilling for oil and gas (Lindeman, 1954). The company abandoned the well as a dry hole, but the company's geologist, William T. Nightingale, observed sodium carbonate in the drill cores at the well site. Core samples were sent to Roger C. Wells and Joseph J. Fahey of the US Geological Survey, who verified the presence of trona. Robert D. Pike, a consulting chemical engineer, recognized the economic potential of refining soda ash from the trona in 1940 but not until he convinced the Westvaco Chemical Corp. (which merged with FMC Corp. in the fall of 1948) of this did commercial interest begin. The first mine shaft was sunk about 460 m deep in 1947 and active mining began on a developmental basis later that year. Calcined trona was sold from 1950 (38 814 t mined and sold the first year) to 1953, when the first refined soda ash plant came onstream with an annual capacity of about 272 kt. Today there are five natural soda ash producers in Wyoming with a combined annual nameplate capacity of more than 10.6 Mt. Table 1 lists the natural and synthetic soda ash plants that operated in the United States from 1868 to the present.

Beginning with the first world energy crisis in 1973 and the growing international concern for the environment, synthetic soda ash output in the world has been decreasing as illustrated in Fig. 1. The majority of the world continues to use the Solvay process, which is energy, capital, and labor intensive, and generates byproduct calcium chloride and sodium chloride that cause environmental problems when discharged. In 1971, synthetic soda ash represented 85% of total world production; in 1991, it amounted to 70% with natural soda ash comprising the balance of 30%. Of the more than 31 Mt of total soda ash produced in the world today, 9 Mt or 29% originates from the United States, with Botswana, China, Kenya, and Mexico accounting for the remaining 1%.

GEOLOGY

Mineralogy

There are several sodium carbonate-bearing minerals found in deposits throughout the world. The most important economic mineral is trona.

Trona ($Na_2CO_3 \cdot NaHCO_3 \cdot 2H_2O$):

System—monoclinic

Appearance—prismatic, tabular, colorless spar-variety of crystals. Efflorescences are earthy crusts or white to yellow-white. Massive trona is buff-colored to shades of tan-brown, due to organic material content.

Physical properties—soft to semi-hard; 2.5 to 3.0 Mohs hardness. Fragile with perfect cleavage. Transparent to translucent with dull, earthy luster. White streak. Soluble in hydrochloric acid. When heated, it loses water and then carbon dioxide.

Specific gravity—2.17 gm/c^3.

Soda ash (Na_2CO_3):

Molecular weight—105.98895

Table 1. History of US Natural and Synthetic Soda Ash Plants, 1868–1992

Company	Location	Deposit	Startup Date	Closure Date	Capacity (Mt) 1950	1960	1970	1980	1990
Natural Soda Ash Plants									
American Potash and Chemical Corp.[1]	Trona, CA	W, Searles Lake	1926	1979	.150	.150	.150	—	—
California Alkali Co.[2]	Cartago, CA	S, Owens Lake	1917	1932	—	—	—	—	—
FMC Wyoming Corp.	Green River, WY	Green River trona	1948	—	.300	.500	1.250	2.550	2.850
General Chemical (Soda Ash) Partners[3]	Green River, WY	Green River trona	1968	—	—	—	.550	2.200	2.400
Inyo Development Co.[4]	Keeler, CA	ne, Owens Lake	1987	1920	—	—	—	—	—
Natron Soda Co.[5]	Ragtown, NV	Big Soda Lake	1885	1907	—	—	—	—	—
Natural Soda Products Co.[6]	Keeler, CA	ne, Owens Lake	1912	1952	.020	—	—	—	—
Nevada Soda Co.[7]	Ragtown, NV	Little Soda Lake	1868	1907	—	—	—	—	—
North American Chemical Co.[8]	Trona, CA	nw, Searles Lake	1978	—	—	—	—	1.300	1.300
Occidental Alkali Co.	Schurz, NV	Double Springs Marsh	1886	1917	—	—	—	—	—
Pacific Alkali Co.[9]	Bartlett, CA	nw, Owens Lake	1926	1968	.009	.070	—	—	—
Permanente Metals Corp.[10]	Bartlett, CA	w, Owens Lake	1947	1950	.022	—	—	—	—
Rhone-Poulenc of Wyoming, L.P.[11]	Green River, WY	Green River trona	1962	—	—	—	.750	1.520	2.200
Sodium Products Co.[12]	Wilson Creek, WA	Mitchell Lake	1936	1941					
Solvay Minerals, Inc.[13]	Green River, WY	Green River trona	1982	—	—	—	—	—	1.850
Tg Soda Ash, Inc.[14]	Granger, WY	Green River trona	1976	—	—	—	—	1.300	1.300
Union Pacific Alkali Works[15]	Laramie, WY	Union Pacific Lakes	1895	NA	—	—	—	—	—
Washington Sodium Products Co.	Warden, WA	Carbonate Lake	1927	1941	—	—	—	—	—
West End Chemical Co.[16]	Westend, CA	sw, Searles Lake	1926	1988	.150	.190	.160	.150	—
Western Alkali Corp.	Green River, WY	Green River brines	1907	1918	—	—	—	—	—
Total natural capacity					.651	.910	2.860	8.720	11.900
Synthetic Soda Ash Plants									
Ammonia-Soda Process:									
Diamond Alkali Co.[17]	Painsville, OH	—	1912	1976	.727	.800	.800	—	—
Mathieson Alkali Works, Inc.[18]	Lake Charles, LA	—	1935	1975	.365	.365	.375	—	—
Mathieson Alkali Works, Inc.[18]	Saltville, VA	—	1893	1971	.200	.410	.400	—	—
Michigan Alkali Co.	Wyandotte, MI[19]	—	1892	1978	.835	.700	.800	—	—
Pittsburgh Plate Glass Co.	Barberton, OH	—	1900	1973	.350	.600	.600	—	—
Solvay Process Co.[3]	Hutchinson, KS[20]	—	1908	1918	—	—	—	—	—
Solvay Process Co.[3]	Baton Rouge, LA	—	1935	1975	.750	.750	.785	—	—
Solvay Process Co.[3]	Detroit, MI	—	1898	1969	.800	.800	—	—	—
Solvay Process Co.[3]	Syracuse, NY	—	1881	1986	.850	.900	1.000	.900	—
Southern Alkali Corp.[21]	Corpus Christi, TX	—	1934	1978	.200	.240	.270	—	—
Caustic Carbonation Process:									
Dow Chemical Co.	Freeport, TX	—	1955	1970	—	.100	—	—	—
Total synthetic capacity					5.077	5.665	5.030	.900	.000
Total industry capacity					5.728	6.575	7.890	9.620	11.900

NA Not available. Note: This table does not include plants that used the electrolytic soda process.

[1]Began as California Trona Co. in Feb. 17, 1908, which was acquired by American Trona Corp. on June 13, 1913. American Potash formed in 1926 and soda ash production beginning in 1931. Bought by Kerr-McGee Chemical Corp. in 1967; ceased production in 1979.

[2]Owned by Great Western Electro-Chemical Co., subsidiary of Dow Chemical Co. Bought Inyo Development Co. in 1918 and acquired in May 1924 by Inyo Chemical Co. Operated until January 1932.

[3]Formerly part of Allied Corp.; separated in 1985. Joint venture ACI of Australia (1986-25%) and Tosoh of Japan (1992-24%).

[4]Sold to California Alkali Co. in 1918; property leased to Natural Soda Products Co. in 1922.

[5]Formed by Griswold and Epperson Sept. 22, 1885. Capacity about 1.5 ktpy but only about 450 tons produced annually.

[6]Bankrupt in 1927; sold to receivership and kept operating until 1933. Natural Soda Products reorganized as subsidiary of Michigan Alkali Co. (part of Wyandotte Chemical Corp.). Resumed full production by 1942; sold to independent group in 1950; ended in June 1952.

[7]Sold by A. Kenyon in 1868 to Higgins & Duffy, who sold lake to San Francisco group Doe & Dowd. Nevada Soda Co. probably their company until about 1879 when John Horstmann Co. acquired facility and operated until 1907. Produced about 300 tons annually.

[8]Plant began July 1978. Acquired Argus plant of Kerr-McGee Chemical Corp. Dec. 1, 1990. OCI of South Korea is 27% partner.

[9]Operated until 1944 when acquired by Columbia Southern Chemical Corp. New plant built and onstream in late-1958. Became subsidiary of Pittsburgh Plate Glass Co. on Jan. 1, 1961. PPG closed plant in 1968.

[10]Acquired as a division of Kaiser Aluminum and Chemical Corp.

[11]Formerly Stauffer Chemical Co. Joint venture with Union Pacific Resources (49%) and Rhone Poulenc of France (1987-51%).

[12]Reorganized into Sodac Chemical Co. in 1941.

[13]Formerly Tenneco Minerals. Joint venture with Asahi Glass of Japan (1990-20%) and Solvay of Belgium (1992-80%).

[14]Bought in 1987 by Societe Nationale Elf Aquitaine of France (100%).

[15]Acquired by American Soda, Glass and Chemical Co. in 1899. The American Development Co. and American Native Soda, Glass and Chemical Co. incorporated in 1898 to develop the soda deposits in Wyoming.

[16]Formed as the Westend Chemical Co., subsidiary of Westend Consolidated Mining Co., in 1920. Bought by Stauffer Chemical Co., in 1956, which was sold to Kerr-McGee Chemical Corp. in 1974.

[17]Later became Diamond Shamrock Corp.

[18]Became Olin-Mathieson in 1954, and later Olin Corp.

[19]Facility had two plants; North plant built in 1892 and South plant built in 1895.

[20]During World War I, the plant was leased from Kansas Chemical Co. and sold to the Solvay Process Co.

[21]Production began September 1934. Owned by Pittsburgh Plate Glass Co. and American Cyanamid Co.

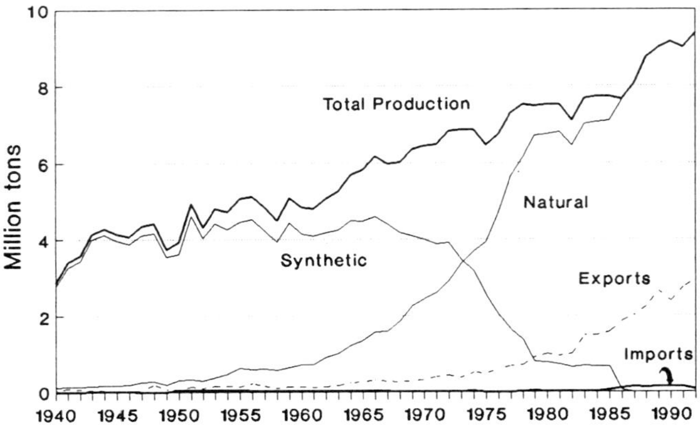

Fig. 1. US soda ash statistics, 1940–1992. *Source:* US Bureau of Mines.

Specific gravity—2.533 gm/c[3]
Bulk density—0.59 to 1.04 gm/ml
Melting point—825°C
Heat of formation (ΔH_f)—1 131 kJ/mol
Solubility—5.9 wt% @ 0°C; 30.8 wt% @ 100°C
pH—1% solution @ 25°C to 11.4

Origin

Various theories have been proposed to explain the origin of sodium carbonate found in natural deposits. Different deposits may have been formed from a combination of one or more different physical and chemical processes. These include: 1) the evaporation of sodium carbonate-rich thermal springs, 2) carbonation of sodium sulfide (from bacterial decomposition of sodium sulfate) to sodium carbonate, 3) ion exchange in sodium-bearing soils, 4) concentration-temperature-dependent equilibrium relationship among carbon dioxide, sodium bicarbonate, and carbonate that converts carbonate solutions to sodium bicarbonate or carbon dioxide removed from sodium bicarbonate solutions to form carbonates, and 5) leaching of alkaline carbonatites or basic-ultrabasic rocks. The sodium may have been derived from the leaching of sodic feldspars or volcanic ash deposits, and the carbon dioxide from the atmosphere and biogenic processes.

Most ground waters in metamorphic, volcanic, or igneous terrains have produced alkaline solutions upon evaporation (Hardie and Eugster, 1970). The relative absence of chloride and sulfate in these types of rocks permits solutions to become predominantly sodium-bearing and carbon dioxide-bearing. The chemical fractionation of inflowing waters and brines within closed depositional basins can produce different minerals accumulating in separate areas. The mechanisms may be mineral precipitation, selective dissolution of efflorescent crusts, degassing of brines, and/or redox reactions (Eugster and Jones, 1979). Ground water percolating through siliceous soils, such as those developed on granite or windblown sands, can become highly alkaline and enriched in sodium carbonate (Garrels and Mackenzie, 1967). Very high carbon dioxide concentrations (up to 10% of soil gas) are known to occur in soils as a result of biological activity. This can produce more than 200 mg/L of carbon dioxide in the soil water (Jakucs, 1977). The Pretoria pan in the Republic of South Africa, the sandhill regions of Nebraska, and the Sind of Pakistan are examples of sodium carbonate in soil waters (Barbour, 1916, Cotter, 1923, Dutoit, 1948). In lakes, deposition of monomineralic beds of trona may be aided by the continual generation of CO_2 by algae and bacteria in

Table 2. Sodium Carbonate-Bearing Minerals

Mineral	Composition	Percent Na_2CO_3[1]
Thermonatrite (monohydrate)	$Na_2CO_3{\cdot}H_2O$	85.5
Wegscheiderite	$Na_2CO_3{\cdot}3NaHCO_3$	74.0
Trona (sesquicarbonate)	$Na_2CO_3{\cdot}NaHCO_3{\cdot}2H_2O$	70.4
Nahcolite		
(sodium bicarbonate)	$NaHCO_3$	63.1
Bradleyite	$Na_2PO_4{\cdot}MgCO_3$	47.1
Pirssonite	$Na_2CO_3{\cdot}CaCO_3{\cdot}2H_2O$	43.8
Tychite	$2MgCO_3{\cdot}2Na_2CO_3{\cdot}Na_2SO_4$	42.6
Northupite	$Na_2CO_3{\cdot}NaCl{\cdot}MgCO_3$	40.6
Natron (sal soda		
or washing soda)	$Na_2CO_3{\cdot}10H_2O$	37.1
Dawsonite	$NaAl(CO_3)(OH)_2$	35.8
Gaylussite	$Na_2CO_3{\cdot}CaCO_3{\cdot}5H_2O$	35.8
Shortite	$Na_2CO_3{\cdot}2CaCO_3$	34.6
Burkeite	$Na_2CO_3{\cdot}2Na_2SO_4$	27.2
Hanksite	$2Na_2CO_3{\cdot}9Na_2SO_4{\cdot}KCl$	13.6

[1] includes bicarbonate converted to carbonate.

the lake water (Eugster, 1971). Semiarid climates, rather than arid, are most favorable for producing alkaline waters because the evaporation rate is too rapid to stabilize the soda solutions in very arid regions (Perelman, 1967).

In sedimentary rock terrains, the drainage usually contains too much chloride, sulfate, and calcium to produce either brine or solid sodium carbonate salts. Although sodium carbonate solutions originate from the weathering of igneous and volcanic rocks, these same waters may interact with other waters or with other rock and soil types in the area of accumulation so that mixed chloride and sulfate salts are the final products (Risacher et al., 1978).

Mode of Occurrence

Sodium carbonate resources commonly occur as precipitates in shallow, nonmarine alkaline lakes and marshes and usually coexist with various chloride and sulfate salts. For this reason, it is not uncommon to find different sodium carbonate-bearing minerals within the same deposit. Table 2 lists some minerals that contain sodium carbonate in varying concentrations.

Because sodium carbonate generally was deposited in shallow bodies of water, many of the deposits were subjected to vast seasonal changes in area and mineral concentration, and were susceptible to destruction by mechanical and chemical erosion due to the relative softness and solubility of the minerals. Therefore, large economic sodium carbonate deposits are relatively rare and are considered geologically young, ranging from the Eocene to Pleistocene Epochs. Pretertiary deposits are virtually absent from the geologic record.

Mannion (1975, 1983) presented a list of the more common forms of sodium carbonate occurrences that include:

1) natron crystals in cool, wet environments on the bottom of ponds or at shallow depths in salt marshes, as at Lake Goodenough, BC (Cummings, 1949), and Owens Lake, CA (Smith and Friedman, 1975); 2) powdery surface efflorescences of thermonatrite around alkaline lakes or marshes, as in the alkaline soils of the lower Danube River valley and at Lake Nyasa, Tanzania (Orr and Grantham, 1931); 3) relatively hard but porous deposits of trona, either around the shores or at the bottom of shallow alkaline lakes or playas, as at Lake Magadi, Kenya (Baker, 1958, Eugster, 1970); Owens Lake, CA (Dub, 1947), and Lake Chad basin, Africa (Eugster and Maglione, 1979); 4) older beds of buried trona, nahcolite, or other sodium

carbonate-bearing minerals, as the Green River formation in Wyoming (Bradley and Eugster, 1969, Culbertson, 1971, Deardorff and Mannion, 1971) and in Colorado (Smith and Milton, 1966, Anon., 1972) and at Railroad Valley, NV (Free, 1913); and 5) in solution as a constituent of brine, either surface or subsurface, as at Searles Lake, CA (Smith, 1979), and Makgadikgadi, Botswana (Massey, 1973, Lefond, 1981).

Although certain deposits are made up of combinations of the five forms of occurrences, the economic and subeconomic sodium carbonate deposits in the world can be classified by their economic importance (Kostick, 1983). The five forms, in terms of decreasing economic significance, are buried, surface or subsurface brines, crystalline shoreline or bottom crusts, shallow lake bottom crusts, and surface efflorescences.

Occurrences of Sodium Carbonate

Natural deposits containing sodium carbonate-bearing minerals are located in various countries throughout the world. Some of the deposits are large while others are small. Most of the deposits contain solid salts, solutions, or both. Mannion (1975, 1983) prepared an excellent compilation of the world sodium carbonate occurrences. It is reproduced as Table 3 and includes some updated information on some of the previously known deposits as well as on a few of the newer deposits recently evaluated and described.

The majority of the deposits listed are geologic occurrences and are not large enough or of sufficient sodium carbonate content to be of economic significance. Many small scale deposits, such as those in Chad and the Sudan, have been mined strictly for local consumption.

Africa: Africa has numerous sodium carbonate occurrences scattered throughout the continent. The arid to semi-arid climates in parts of Africa prevent the dissolution and transport of the soluble carbonate minerals. Some of the occurrences are small isolated lake deposits while others are major deposits found in a chain of deposits in the East African Rift Valley. Garrett (1992) has provided an excellent description of all the major and minor soda occurrences in the Rift Valley.

A $504 million natural soda ash facility commenced production in April 1991 on the Sua Pan in the Makgadikgadi Basin in northeast Botswana. The project is operated by Soda Ash Botswana (SAB), a joint venture between the Government (48%) and the African Explosives and Chemical Industries (AECI) consortium (52%) with its partners AECI (25.5%), the Anglo American Corp. (12.75%), and De Beers Holdings (12.75%). The facility produces about 300 kt of refined soda ash from underground brines and 600 kt of sodium chloride. The majority of the soda ash is marketed in South Africa, which had relied heavily on soda ash imports from the United States. SAB also exports soda ash to Zambia and Zimbabwe, which both use about 15 kt annually. The soda ash reserves at Sua Pan exceed 360 Mt.

Impure sodium carbonate in the form of natron mixed with halite has been obtained from crystallizing pans among sand dunes in the northeast section of Lake Chad (Eugster and Maglione, 1979). Most of the crude material has been for local consumption; however, it had been reported that a few thousand tons of crude trona had been exported to Nigeria previously. Some of the reported uses include: 1) local remedy for stomach aches and constipation, 2) women mix the natron with ashes of burnt wheatstalks to make a crude soap, 3) Chadian herders give it to cattle as a saltlick as a preventative medicine, 4) it is mixed with raw tobacco to flavor it for chewing, and 5) tanners use it as an additive for certain tree gums to tan hides and skins.

Wadi Natron in Egypt, about 80 km northwest of Cairo and 35 km west of the Nile River, consists of up to 20 individual lakes, some of which dry up completely during the summer (Lucas, 1912, Atia et al., 1970). Trona obtained from these lakes has been utilized for more than 5500 years, since early Egyptian times. Soda ash was used to make ornamental glassware, and was used in the mummification process. Samples of trona have been found in one of the Pharaoh's tombs at Thebes, suggesting the importance of its unique value. Other occurrences have been reported at El Barnugi in Lower Egypt, Mahamid, and at Wadi Tumilat in upper Egypt. Trona has been identified in cores taken from the outer Egyptian shelf of the Nile Cone in the Mediterranean Sea. The trona probably was eroded from the continental margin 18,000 to 23,000 years ago and buried very quickly in the deep marine environment (Stanley and Sheng, 1979). Ethiopia has several sodium carbonate-bearing lakes. The lakes in the southern part of the Rift Valley are richer in sodium carbonate than those in the north.

The second major natural soda ash facility in Africa operates at Lake Magadi in Kenya. The 65 km^2 deposit is probably the largest of all those in the African Rift Valley. Trona has been mined since 1914 by the Magadi Soda Co., Ltd., owned by Brunner Mond Holdings, Ltd. (formerly Imperial Chemical Industries). The trona is excavated by bucket dredgers mounted on a floating pontoon. The ore is crushed and made into a slurry so that it can be pumped through a floating pipeline to the refinery on the shore. The slurry is screened, washed, re-screened, and de-watered to remove salt and mud. The trona is calcined to remove the CO_2 and converted into soda ash. The operation has an annual production capacity of about 300 kt.

The lake bed is underlain by about 45 m of relatively pure trona with some interbedded trona and trona-bearing muds. Brine is found within a meter of the surface. The alkaline spring water that delivers the sodium carbonate to Lake Magadi is considered to be the result of weathering and leaching of igneous silicate minerals (Baker, 1958). Dilute ground water from the igneous highland regions recharges a deep reservoir, which in turn supplies the warm springs with concentrated carbonate waters (Eugster, 1970). The same process appears to be responsible for the alkaline waters at Lake Natron, Lake Nakuru, Lake Elmenteita, and Lake Bogoria (Orr and Grantham, 1931). Based on the flow rates of the rivers and springs and the rate of evaporation, it is estimated that about 200 kt of new trona crystallizes annually in Lake Magadi (Garrett, 1992).

An alkali lake is located in Malawi in a downfaulted rift basin. Alkaline hot springs and alkaline igneous instrusives in the area were the presumed source of the sodium carbonate in Lake Chilwa. Trona, or "kaum," is found in many small lakes and in alkaline soil in Nigeria. It also accumulates as effervescent crusts on the surface of a sandy area in Dallol Bosso in western Nigeria (Garrett, 1992).

In Namibia, several pans contain sodium carbonate that have been used by the local population for centuries. In the Owambo Basin, the two Otjiwalunda pans (No. 1 and No. 2) have been known since 1915, and contain trona (78.65%), salt (0.31%), thenardite (0.94%), and burkeite on the surface (Dutoit, 1948, Schneider and Genis, 1991). The deposit grades from almost pure trona to trona mixed with other salts. A rare sodium carbonate mineral, pirssonite ($CaCO_3 \cdot Na_2CO_3 \cdot 2H_2O$) was reported to occur with the thenardite from the No. 2 Otjiwalund pan. Also in the Basin are the Ngandjela Pan and the Ngandjela East pan, which also contain halite and soda ash intermixed with thenardite. The Naruunda and Onanzi pans that are also located within the region contain brines enriched in various sodium-bearing minerals. Production in 1975 amounted to 4 189 t, and only 70 t in 1976.

The Pretoria Salt Pan in the Republic of South Africa is an extinct volcanic crater containing sodium carbonate. The first commercial development began in 1912 when the Government leased to P.A. Ogilvie the rights to mine crude trona from the deposit. The lease was ceded later to the South African Alkali Ltd., which mined about 400 tons in 1914. Calcined trona was produced between 1915

Table 3. Occurrences of Sodium Carbonate in Various Parts of the World

Name and Location	Physical Character	Geologic Character	Chemical Composition	Remarks and References
ARGENTINA				
Santa Cruz area in southern Patagonia	Seven lakes in 4 000 km^2 area of Andes, dry climate	Natron, trona, and other carbonates in bottom muds	Solid salts dried Na_2CO_3 56.3% $NaHCO_3$ 8.3 NaCl 11.3 Na_2SO_4 19.3 SiO_2 2.2 H_2O 2.0 Insol 0.7	Bernasconi and Cangioli, 1982
La Ernestina lake in Utracan area	Several lakes in region also located	Salts and brine	Salts (summer) Na_2CO_3 11.5% $NaHCO_3$ 3.0 NaCl 45.9 Na_2SO_4 39.6 Insol 1.1 Salts (winter) Na_2CO_3 3.7% $NaHCO_3$ 6.4 NaCl 1.8 Na_2SO_4 88.1 Insol 0.9	Cordini, 1967
AUSTRALIA				
Bowen Basin, Queensland		Dawsonite in mudstone, brines	Brine Na_2CO_3 566–16 700.0 ppm	
Denison Trough, Comet, Queensland	600 to 1 100 m deep	Sodium carbonate brine	Brine $NaHCO_3$ 1.6–2.5%	
Sydney Basin, Queensland		Dawsonite		
BOLIVIA				
Collpe Laguna, southwest corner of Bolivia (Lipez) just east of Salar de Chalvira	Shallow lake 800 m in diameter	Salt crust and subsurface brine in mud	Brine Na_2CO_3 6.0% NaCl 2.6 Organic 0.5	Ahlfeld and Schneider-Scherbena, 1964, Risacher et al., 1978
Cochi Laguna, southwest Bolivia in northern Lipez, km south of Chiguana	Shallow lake 1 000 m across, located south of the volcano Tapoquilaha	Brine in mud, also 6 to 10 cm crust with trona and thermonatrite	Solid salts dried at 80° Na_2CO_3 47.3% $NaHCO_3$ 11.0 NaCl 7.1 Na_2SO_4 11.0 Insol 1.0	Soda ash produced from brine in past, not economic; Ahlfeld and Schneider-Scherbena, 1964, Risacher, et al., 1978
BOTSWANA				
Sua Pan, Makgadikgadi basin in northeast Botswana	Low area in a vast drainage area of flat terrain; Sua Pan is eastern of two	Brine in sand aquifer overlain by sand and clay layers totaling 30 m thick; aquifer underlies 900 km^2	Avg. composition of brine Na_2CO_3 26.0 g/L NaCl 140.0 Na_2SO_4 14.5	Investigation and test work on possible commercial use since 1963; Pelletier, 1964, Lefond, 1981
BRAZIL				
Upper Sao Francisco drainage, northeast of Brasilia	Numerous salt pans up to 175 000 m^2 area in plateau region 900 to 1 100 m in elevation	Salt crusts and shallow brine in area of alkaline igneous rocks	Range of 5 crust analyses Na_2CO_3 23–41.0% NaCl 1.5–10 Na_2SO_4 1–7 $MgSO_4$ 0.4–2.2 H_2O 40–50	Freise, 1932
CANADA				
Eighty-Three Mile Lake, Green Timber Plateau, south-central BC, north of Clinton	Area 0.1 km^2; max depth brine 0.6 m; in level terrain	Alkaline lake with a mud bottom on glacial drift and basalt. Crystals and brine present	Brine Na_2CO_3 94.0 g/L NaCl 1.3 Na_2SO_4 Trace	Many other small lakes and ponds in region also contain high carbonate content; Cummings, 1949, Livingston, 1963
Goodenough Lake, Green Timber Plateau, south-central BC, north of Clinton	Area 0.5 km^2 dries up	Crust of salines 2 cm thick on natron containing mud 3 m thick	Brine Na_2CO_3 208.0 g/L NaCl 11.0 Na_2SO_4 71.0 KCl 8.7	Solid Crust Na_2CO_3 97.0% $NaHCO_3$ 0.7 NaCl 0.8 Na_2SO_4 0.5 $MgCo_3$ 0.35 Cummings, 1949
Last Chance Lake, Green Timer Plateau south-central BC, north of Clinton	Area 0.1 km^2; mostly dry	Lake contains permanent crystal layers in numerous bowl-shaped depressions separated by mud	Brine Na_2CO_3 180.0 g/L NaCl 12.0 Na_2SO_4 37.0 KCl 3.3	Cummings, 1949

(cont.)

Table 3. Occurrences of Sodium Carbonate in Various Parts of the World (cont.)

Name and Location	Physical Character	Geologic Character	Chemical Composition	Remarks and References
CHAD				
Lake Chad, near Bo, Chad	Lake highly variable in area 10 000 to 26 000 km^2; depth is 308 m; vast drainage area	Lake held in by sand dunes northeast shores; layers of natron with halite found 1.5 m below surface	Solids as mined are impure sodium carbonate mixed with sodium chloride	Long continued export of 5 to 7 ktpy to Nigeria; also Lake Orari 60 km northeast of Bedao, and Lake Tellie near Ounianga Kebir, northern Chad; Eugster and Maglione, 1979, Maglione, 1972
CHINA				
Heilungkiang Province Tsitshar (Ch'ich'haerh)	Soda lakes in steppe in valley of Nonni River			Torgasheff, 1929; Hou, 1942
Henan Province				
Anpeng, Tongbai County	Area of 13 km^2	Buried Eocene trona; two ore zones 1 288 to 1 686 m and 2 055 to 2 408 m deep	Trona and brine Na_2CO_3 10.0% $NaHCO_3$ 70–80 $CaCO_3$ 0.31 $MgCO_3$ 0.08 NaCl 0.61 B_2O_3 400–1 400.0 ppm K 40–1 200.0 ppm	Discovered in 1976, 120 Mt resource; Wang, 1987
Henan Province				
Wucheng	Elliptical basin with 260 km^2 area	Buried late-Eocene deposit, 650 to 970 m deep. Has 36 trona and halite beds 0.5 to 3 m thick	Solids Trona 77% Nahcolite 23	About 130 Mt soda ash resource; Youxun, 1985
Kiangsi Province				
Kian-Hsien	Soda lakes in steppe		Natural solid salts Na_2CO_3 81.0% $NaHCO_3$ 7.0	Other soda lakes in region; Hou, 1942
Kirin Province				
Taboos Nor, 320 km north of Shenyang, near Fuyu	Elliptical lake, 8 km across in flat plain surrounded by sand dunes	Generally dry lake with 1 cm crust of thermonatrite on 30 cm mud layer; mud contains natron 0 to 30%, which lies on gaylussite-bearing muds up to 2 m depth.	Solids Na_2CO_3 35.0% H_2O 30.0 Insol 0.27	Possibly a million tons of sodium salts; numerous other lakes including Fu-U-Hsien; Hou, 1942, Niinomy, 1930
Nei Mongol Province				
Xilin Gol League	Alkali lakes	Contain surface and bottom trona layers	Solids Sodium sesquicarbonate Sodium chloride Sodium sulfate	Effort to construct 150 ktpy soda ash plant began in 1971; suspended plans in 1979; resumed in 1986 as the Changannuoer alkali mine; Anon., 1986
Nei Mongol Province Wulan Buh desert			Solids	Planned soda ash project
Nei Mongol Province				
Alashan and Ehtingol areas	Alkali lakes abundant			Produced soda ash; Torgasheff, 1929
Nei Mongol Province				
Chalar Region, Dolon Nor, 250 km north of Beijing	Alkali lakes abundant			Produced soda ash; Torgasheff, 1929
Hailar Area, 150 km east of Manchouli	Two groups of lakes north and south of Hailar; aggregate 500 km^2	Salt crust and brine in steppe	Brine Na_2CO_3 115.0 g/L NaCl 213.0 Na_2SO_4 71.0 Salt crust Na_2CO_3 29.0% Na_2SO_4 12.0 H_2O 58.0	Include Chassun-Nor and Bain-Tsugan-Nor; production from south group; Hou, 1942, Torgasheff, 1929
Nei Mongol Province				
Polishan (Lake Liao-Ning)	One of large group alkali lakes		Salt crust Na_2CO_3 12.0% NaCl 1.5 Na_2SO_4 1.0	Also Tafus Lake; Torgasheff, 1929

(cont.)

Table 3. Occurrences of Sodium Carbonate in Various Parts of the World (cont.)

Name and Location	Physical Character	Geologic Character	Chemical Composition	Remarks and References
CHINA *(cont.)*				
Nei Mongol Province Ih Ju League	Alkali lakes in arid to semi-arid regions	Sodium carbonate found in lake bottom black ooze mud		Youxun, 1985
Shansi Province U-Tsu-Hsien	Dry Lake		Natural solid Salts Na_2CO_3 54.0% $NaHCO_3$ 16.0 NaCl 3.5 Na_2SO_4 17.0 H_2O 12.0 Insol 0.21	Hou, 1942
Shensi Province Shen-Mu-Hsien	Soda lakes in steppe		Natural solid salts Na_2CO_3 78.3% $NaHCO_3$ 0.5 NaCl 5.0 Na_2SO_4 0.9 H_2O 15.0	Other soda lakes in region; Ordos area in adjacent Inner Mongolia incl. Cha-Han-Nor, Nalin-Nor, and Pa-Yen-Nor; Hou, 1942
Xinjiang Province Hami	Alkali lake in arid to semi-arid area	Trona found in lake bottom black ooze mud		Youxun, 1985
Xizang (Tibet) Bangkog Cuo	Three alkaline lakes	Tibetan plateau	Surface Brine (mmol/L) Na 1 131–2 696 CO_3 77–222 HCO_3 27–202 Cl 675–1 119	Zhang et al., 1989
Guogaling Cuo	Alkaline lake	Tibetan plateau	Surface brine Na 1 761	Zhang et al., 1989
Zabuye Caka	Alkaline lake	Tibetan plateau	Brine (wt%) Na 7.29% CO_3 3.27 SO_4 2.71 Cl 9.53	Zhang et al., 1989
EGYPT Wadi El-Natron, 97 km northwest of Cairo	Chain of 12 small, partly ephemeral lakes; central ones carry natron, others sodium sulfate. Area of 12 km^2	Brine of varying composition; bottom and bordering solid salts. Some lakes are dry in summer; others remain wet	Brine Na_2CO_3 52.3 g/L Na_2SO_4 67.1 NaCl 221.2 Product impure carbonate-bicarbonate with mud, NaCl, and Na_2SO_4	Probably oldest known soda sources; probably fed by seepage from Nile River; other occurrences at El Barnougi in Baheira Province, 50 km north of Wadi El-Natron; and in small lakes at El-Mahamid, near Edfo; Lucas, 1912, Said, 1962
ETHIOPIA Lake Magado and El Soda, near Mega, 500 km south of Addis Ababa	Salt pans in volcanic craters	Brine plus solid salts	Brine a mixture of sodium chloride and sodium carbonate	Exploited on local scale; Murdock, 1949
Galla Lakes southern Ethopia	Four lakes (Abijata, Langalo, Shala, & Zwai) near Lake Shala	Fed by hot springs and rivers in East Rift Valley	Sodium carbonate-bicarbonate brines and lake bottom crystals	Baumann et al., 1975
INDIA Sambhar Lake in northwest Rajasthan (Rajputana), 310 km southwest of New Delhi	Variable area, max 230 km^2 and 4 m deep	Brine with salt layers and saline muds to 85 m depth	Brines predominantly NaCl with lesser Na_2SO_4 and Na_2CO_3	A significant salt producer, remaining bittern contains Na_2CO_3 3.8% $NaHCO_3$ 1.0 Na_2SO_4 7.7 Roy, 1959, Sahni, 1951
Lonar Lake, Buldana District of Berar, Lat. 20°N, Long. 76½°E	Circular shallow lake 150 m deep, 1 km^2, variable amounts of water	Impact crater in basalt plateau; crusts of trona around edge, mixed trona and halite on bottom; spring-fed	Brine Na_2CO_3 25.0 g/L NaCL 55.0 Na_2SO_4 2.0 K 0.2	Christie, 1912, Fredrikson et al., 1973

(cont.)

Table 3. Occurrences of Sodium Carbonate in Various Parts of the World (cont.)

Name and Location	Physical Character	Geologic Character	Chemical Composition		Remarks and References
KENYA					
Lake Amboseli	Dry lake with 38 km^2 area	Fed by springs and rivers in volcanic area	Gaylussite bed 0.3 to 3.5 m thick and 1 to 3 m deep		About 15 Mt of gaylussite containing 5 Mt soda ash; Williams, 1972
Lake Bogoria, southeastern Kenya	Alkaline lake in Rift Valley that is 17 km long, 500 to 3 500 m wide and 9 to 12 m deep	Fed by hot springs, geysers, and fumeroles	Brine $NaHCO_3$ NaCl Na_2SO_4	 48.0 g/L 5.8 0.3	Livingston, 1963
Lake Elmenteita	Shallow lake about 4 m deep	Trona crusts along shore	Lake Na HCO_3 Cl SO_4 F	 23 000 ppm 41 000 ppm 11 000 ppm 1 000 ppm 2 000 ppm	McCall, 1967
Lake Magadi, southeastern Kenya	Trona salt flat 65 km^2, in a low part of Rift Valley	Trona surface with subsurface trona layers and brine, in alkalic volcanic terrane	Brine, variable Na_2CO_3 $NaHCO_3$ NaCl Na_2SO_4 NaF	 100.0 g/L 20 60 1–2 0.01–0.2	Trona dredged as raw material for soda ash production since 1914; Baker, 1958, Eugster, 1970, 1980a
Lake Nakuru, southeastern Kenya, northwest of Nairobi	Shallow brine lake in Rift Valley, modest relief	Brine with shore crusts of trona until lake level rose	Brine $NaHCO_3$ NaCl Na_2SO_4	 17.0 g/L 2.0 0.4	Abundant organisms in water; Livingston, 1963
Lake Turkana	Largest lake in Rift Valley is 250 km long, 15 to 30 km wide, 35 to 125 m deep	Receives drainage from Ethiopian Plateau alkali lakes	Lake Na HCO_3 K Ca Mg Cl	 753 ppm 1 160 ppm 17 ppm 5 ppm 2 ppm 514 ppm	Yuretich, 1979
MALAWI					
Lake Chilwa	Alkali lake	Alkaline intrusives including carbonatites with regional hot springs			Garrett, 1992
MEXICO					
Bahia de Adair, 36 km northwest of Puerto Penasco in Senora	Area of 7.5 km^2		Solids and brines Na_2CO_3 $NaHCO_3$ NaCl Na_2SO_4 KCl MgO Insol(SiO_2)	 30.9% 9.6 8.4 15.5 2.7 8.4 23.6	Contains 250 to 400 kt soda ash; Garrett, 1992
Chihuahua, north-central area, 70 km southwest of Ciudad Juarez	5 000 km^2 area with playa lakes	Alkaline soil	Efflorescences 1 mm thick containing sodium carbonate, chloride, and sulfate Na HCO_3 Cl SO_4 Ca Mg	 54 000 ppm 1 700 ppm 67 000 ppm 18 500 ppm 10.8 ppm 3.4 ppm	Garrett, 1992
Durango, 40 km northeast of Durango	Central basin	Alkaline springs	Brine Na_2CO_3 $NaHCO_3$ NaCl Na_2SO_4	 4.12 g/L 2.19 2.46 0.74	Schultze, 1953
Lake Texcoco, west of Mexico City	Low point in Valley of Mexico, subsurface brines	Residual waters from drying up of former Lake Texcoco; now subsurface brines dilute	Water Na_2CO_3 NaCl Na_2SO_4 $Na_2B_4O_7$ KCl	 8.0 g/L 8.6 0.1 0.1 0.8	Solar evaporation in spiral-shaped ponds; much of soda ash production converted to caustic soda; Durand-Chastel, 1968, Gonzalez-Reyna, 1956

(cont.)

Table 3. Occurrences of Sodium Carbonate in Various Parts of the World (cont.)

Name and Location	Physical Character	Geologic Character	Chemical Composition	Remarks and References
NAMIBIA				
Otjiwalunda Pan just west of Etosha Pan in northern Namibia	Four alkali pans in nearly flat terrian (Otjiwalunda Nos. 1 and 2, Ngandjela, and Ngandjela East pans)	Trona and halite precipitated from brines	Solid sodium salts Na_2Co_3 and $NaHCO_3$ 79% NaCl 2 Na_2SO_4 3 Brine Na_2CO_3 and $NaHCO_3$ 3.4 g/100 cm^3 NaCl 24.7 Na_2SO_4 4.5	Possible production of soda investigated in 1964; Pelletier, 1964, Schneider and Genis 1991
NIGER				
Bilma oasis in central Sahara, 350 km north of Lake Chad	Shallow saline lakes containing mainly sodium chloride	Trona precipitated from salt lakes	Trona crystallizes from brine	Significant trade in salt
NIGERIA				
Various locations	Small lakes; alkaline soil	Trona, or "Kaum" is found in lakes		Makanjuola and Beetlestone, 1973
Dallol Bosso, western Nigeria	Area of 3 400 km^2 sodium-bearing soil 1 m deep		Brine and trona crusts Na 100.0 ppm HCO_3 252 Ca 14.5 K 9.8 Mg 3.8 Cl 17.8 SO_4 27 SiO_2 13.5	Bui and Wilding, 1988
PAKISTAN				
In southern Sind, 100 to 110 km northeast of Hyderabad	Hundreds of shallow lakes mostly a kilometer or less in max dimension in sand dunes, level terrain	Brine lakes with clay bottom, fed by ground water; trona crusts deposited with scarce natron	Solids Na_2CO_3 24.0% $NaHCO_3$ 15.0 NaCl 3.0 Na_2SO_4 7.5	Both saline and carbonate lakes present; much organic matter in carbonate lakes; past production at several thousand tpy; Cotter, 1923
SOUTH AFRICA				
Konga Pan	Salt pan	Sodium carbonate found with burkeite and halite		Sutherland et al., 1987
Pretoria salt pan, 40 km north of Pretoria	Crater 1 km diam, 130 m deep in granitic terrane, salt pan in bottom	Salts including trona and halite, underlain by laminated clays containing brine; volcanic rocks occur in crater	Brine Na_2CO_3 37.4 g/L NaCl 56.9 SO_4 0.30 SiO_2 0.07 CaO 0.02 MgO Trace	Worked in past; proportion of Na_2CO_3 to NaCl is greater in outer parts of pan; brine likely represents highly concentrated normal ground water; Bond, 1946, Dutoit, 1948
SUDAN				
Bir El Natron		Trona found with burkeite and halite	Trona is 35% of orebody	Farag et al., 1975
TANZANIA				
Bahi Swamp, central Tanzania	Salt pan underlain by 110 m of clays and sandy clays	Alkaline waters contained in sediments. Drill holes had some with high concentrations sodium carbonate	Variable brine composition Na_2CO_3 20.0 g/L $NaHCO_3$ 17.5 Na_2SO_4 9.0 NaCl 28.0 F_2 0.15	Wells can be pumped at rates in excess of 2 500 L/min; Fawley, 1958
Lake Balangida (Balangda), north-central Tanzania, 80 km south of Lake Eyasi	Salt pan, 30 km^2, at foot of escarpment in Rift Zone	Spring-fed lake; crust forms only 10 cm thick on mud; volcanic terrane	Specimen of crust Na_2CO_3 22.3% $NaHCO_3$ 13.8 NaCl 28.2 Na_2CO_4 21.1 Insol. 10.0	Dries up in summer; only 140 kt sodium carbonate and bicarbonate in crust; Orr and Grantham, 1931
Lake Eyasi (Nyarasa), north-central Tanzania	Saline, sandy pan 1 000 km^2, at foot of escarpment in Rift Zone	Principal salts occur as thin crusts and in brine pools along Sibiti River bed in southwest part of pan	Variable composition, both brine and crust; a supersaturated brine from pool shows: TDS 432.0 g/L Na_2CO_3 3.6% $NaHCO_3$ 1.5 NaCl 51.2 Na_2SO_4 9.1	Sample of alkali salt a top crust of NaCl Na_2CO_3 32.5% $NaHCO_3$ 25.7 NaCl 15.8 Na_2SO_4 10.5 Moisture 11.1 Insol. 1.3 Organic matter 3.1 Orr and Grantham, 1931

(cont.)

Table 3. Occurrences of Sodium Carbonate in Various Parts of the World (cont.)

Name and Location	Physical Character	Geologic Character	Chemical Composition		Remarks and References
TANZANIA (*cont.*)					
Lake Lagarja, Olduvai Valley, north-central Tanzania	Rift Valley Lake	Mostly dry salt pan in volcanic terrane	Crust Na_2CO_3 $NaHCO_3$ NaCl Na_2SO_4	 44.0% 17.0 13.9 9.8	Guest, 1956
Lake Manyara	Area of 570 km^2 Lake is 41 km long and 15 km wide		Brine Na_2CO_3 $NaHCO_3$ NaCl Na_2SO_4 KCl $MgCl_2$ SiO_2	 0.40% 0.33 0.23 0.04 19.0 ppm 10 16	Contains 70 Mt soda ash. Stoffers and Holdship 1975
Lake Natron, northeast Tanzania on Kenya border	570 km^2, large shallow lake in Rift Valley; northwest part contains salts	At times as much as 200 km^2 are underlain by a meter of salts	Salt crusts contain Na_2CO_3 NaCl Na_2SO_4	 53–60.0% 20–20.0 1.5–2.0	Estimated 65 Mt of total salts in mud layers; Orr and Grantham, 1931
TURKEY					
Beypazari, 30 km west of Ankara in central Turkey	Area of 8 km^2	Middle-upper Miocene age buried trona deposit. Total of 38 m of trona found in several beds ranging from 2 cm to 11.5 m thick	Trona Na_2CO_3 $NaHCO_3$ NaCl Na_2SO_4 Fe H_2O Insol	 45.0% 37 0.08 0.18 0.05 16.5 0.36	Reserves of 235 Mt. Dundar, 1988
Lake Van, southeast Turkey	3 730 km^2, avg. depth 55 m, appears on 1 700 m high plateau	Dilute alkaline water; may be density stratified	Water Na_2CO_3 NaCl Na_2SO_4 K_2SO_4	 8.6 g/L 9.4 3.3 0.5	Also some lakes in the Araxes plain are carbonate; Lake Urmi is a sodium lake; Karajian, 1920, Tulus, 1944
UGANDA					
Lake Albert	Lake with hot springs. Other saline lakes occur in region	Rift valley igneous origin	Lake Na HCO_3 Mg K Ca Cl SO_4 SiO_2	 94.0 ppm 479 32 66 11 26 31 2	Talling and Talling, 1965
Lake Katwe, southwest Uganda	In a volcanic crater, near Lake Edward, salt lake 0.6 km in diameter		Salt from evaporation of brine Na_2CO_3 Na_2SO_4 NaCl K_2SO_4 Fe_2O_3	 2.46% 5.3 81.7 8.4 0.15	Investigation made toward commercial extraction from brine similar to Searles Lake, CA; Barnes, 1961
USSR					
Altay Region (M. Khailor Lakes)					
Petukhov Lakes (Siberia), 210 km from Slavgorod	One large, several small, interconnected lakes in Barnaul region of Siberia	Brine lake with alkali crusts in steppe region; when temperature drops below 0°C, natron precipitates in 8 to 15 cm layers	Solid salts Na_2CO_3 NaCl Na_2SO_4 H_2O	 36.4% 0.39 Trace 63.21	Estimate Na_2CO_3 at 2.4 Mt in lake water, 4.5 Mt precipitated in muds; natrum mined and dehydrated to 99% Na_2CO_3; Niinomy, 1930
Lake Tanatar	Lake in semiarid steppe of Siberia	Trona and brine	Brine Na_2CO_3 $NaHCO_3$ NaCl Na_2SO_4	 5.11% 1.2 3.7 1.3	Past production from trona at Mikhaylovskiy; Bukshtein and Zakina, 1949
Chita Region					
Doronin Lake, 400 km east of Lake Baikal	Alkaline lakes in semiarid steppe		High in sulfate as well as carbonate		Produced soda ash; crystalline salts amount to 350 to 450 kt; Torgasheff, 1929
Kazakhstan Region	Several soda lakes of Neogene epoch				

(*cont.*)

Table 3. Occurrences of Sodium Carbonate in Various Parts of the World (cont.)

Name and Location	Physical Character	Geologic Character	Chemical Composition	Remarks and References
USSR (*cont.*)				
Krasnaya Polyana Region	Several alkali lakes		Brine $NaHCO_3$ 19–29.0 g/L	
Krasnoyarsk Region				
Tura salt lakes, Lat. 56° N, Long. 90° E	Various lakes in steppe	Brines and solid sulfate salts	Brines containing sulfate and some with high carbonate and bicarbonate content	These are mostly sulfate lakes; some have large quantities of thenardite and mirabilite
UNITED STATES				
Alaska				
Yukon Flats basin, east-central Alaska	Shallow saline lakes in a basin 240 km long long by 38 km wide in northeast trend	Surface trona (25%) with nahcolite, thermonatrite, and calcite/dolomite	Trona Crusts Na 12.04% Cl 0.48 CO_3 15.91 Ca 2.73 Mg 0.62 K 0.83 Ti 0.30 Al 3.78 Fe 1.47 F 348.0 ppm B 47.0 ppm	Clautice and Mowatt, 1981
California				
Black Lake, north of Benton, Mono County	Small lake 1.6 km long, up to 150 m wide and up to 20 m deep	Carbonate-sulfate lake colored with organic matter	Dilute lake water Na_2CO_3 12.0 g/L NaCl 2.0 Na_2SO_4 3.4 KCl 0.7	Clarke, 1924
Borax Lake, southeast of Clear Lake, Lake County	0.8 to 1.2 km^2 in small basin adjacent to Clear Lake	Carbonate water with notable borate content; recent volcanic rocks adjacent	Variable composition Na_2CO_3 18.5 g/L NaCl 1.7 KCl 1.7 $Na_2B_40_7$ 0.5 Na_2SO_4 0.01 Misc. and Organic 0.5	1.4 Mt of Na_2CO_3; lake supports a notable amount of life; Averill, 1947
Deep Springs Lake, northern Inyo County	Mountain valley, 21 × 7 km, contains saline playa 2.6 km^2 in area	Shallow brine pond underlain and bordered by saline crust with thenardite, burkeite, trona, and nahcolite	Solid crush contains mainly sodium carbonate, sulfate, and chloride	Jones, 1965
Middle Alkali Lake, western Modoc County	One of three shallow lakes in downfaulted valley	Dilute brine in region dominated by Cenozoic volcanic rocks	TDS 9.3 g/L Na_2CO_3 3.5 NaCl 5.4 Na_2SO_4 0.9	Livingston, 1963
Mono Lake, Mono County	Lake area variable 190 to 222 km^2, mean depth 18 m	Rather complex saline water in volcanic terrane	Dilute lake water Na_2CO_3 46.0 g/L NaCl 23.0 Na_2SO_4 10.0 $Na_2B_4O_7$ 1.0 KCl 2.1	84 Mt salts; Livingston 1963
Owens Lake, west-central Inyo County	Up to 260 km^2, very shallow	Brine on surface and near surface, also trona, burkeite, and halite as surface salts and in sub-surface saline muds up to 3 m thick	Brine composition variable Na_2CO_3 8.5% NaCl 17.0 Na_2SO_4 4.5 $Na_2B_4O_7$ 0.5	Present drying up of lake resulted from 1917 diversion of Owen River water; intermittent soda ash producer; contains several tens of millions of tons of Na_2CO_3; last plant shut 1968; Dub, 1947
Searles Lake, north-west San Bernardino County	83 km^2 saline playa in mountain valley	Complex salt body with interstitial brine, underlies 83 km^2 of surface	Concentrated high-quality brine Na_2CO_3 4–5.0% NaCl 16.0 Na_2SO_4 7.0 KCL 3–5.0 $Na_2B_4O_7$ 1.5 Spec. Grav. 1.3	Soda ash together with sodium sulfate, borate potash, lithium, and bromine extracted from brines; contains several hundred million tons of sodium carbonate; Smith, 1966, 1979

(*cont.*)

Table 3. Occurrences of Sodium Carbonate in Various Parts of the World (cont.)

Name and Location	Physical Character	Geologic Character	Chemical Composition	Remarks and References
UNITED STATES (*cont.*)				
Colorado				
Piceance Basin, northwest Colorado	Saline part of Green River formation in center part of basin	Beds of nahcolite and disseminated nahcolite and dawsonite in oil shale	Several thousand meters of thickness contain about 10% dawsonite and nahcolite	Possible future extraction as byproduct from shale oil production; Smith and Milton, 1966, Dyni, 1980
Nebraska				
Jesse Lake and other alkaline lakes in northwest Nebraska, chiefly in Cherry and Sheridan Counties	Many lakes among sand hills, more than 1 km^2, rolling terrain	Natural drainage through sandy soils accumulates in hollow with impervious mud bottom	Jesse Lake (near Antioch) TDS 190.0 g/L K_2SO_4 26.5 K_2CO_3 21.5 Na_2HCO_3 36.2 $NaCO_3$ 9.2 NaCl 5.8 SiO_2 0.2	Potash produced briefly; Barbour, 1916
Nevada				
Soda Lakes (Ragtown Lakes)	One lake about 1.2 km^2; smaller lake about 0.8 km^2	Lakes in cinder cone of sandy pyroclastic debris surrounded by sandy lake beds; trona precipitated	Dissolved solids include: Na_2CO_3 20.0% NaCl 80.0 Na_2SO_4 20.0	Had some production between 1893 and 1969; est. 0.5 Mt Na_2CO_3 present; Chartard, 1890, Morrison, 1964
Railroad Valley, northeast Nye County	Buried salts in lake beds at 200 to 300 m below playa lake in broad valley	Beds of both solid gaylussite and of gaylussite crystals in clay of Plio-Pleistocene lake; brine and solids, trona gaylussite, and natron	Gaylussite $Na_2CO_3 \cdot CaCO_3 \cdot 5H_2O$	Large reserves of a material with no known commercial utilization; Free, 1913
Oregon				
Lake Abert, Lake County, southeast Oregon	Area 130 to 150 km^2; 1.5 to 1.8 m deep, dries up	Alkaline semi-permanent lake in downfaulted block in volcanic rocks; in-flow mostly the Chewaucan River	Brine Na_2CO_3 10–20.0 g/L $NaHCO_3$ 4–13.0 NaCl 17–38.0 Na_2SO_4 1–1.5 KCl 1–1.5 pH 9.0+	Up to 141 TDS in playa interstitial water; Allison, 1947, Phillips and Van Denburgh, 1971
Alkali Lake 31 km north-northeast of Albert Lake	Area 13 to 15 km^2, water in very shallow pond	Playa containing pot holes containing mostly natron and brine; recharged by ground water; local trona crust	Brine HCO_3+CO_3 10–20.0 g/L NaCl 76.0 Na_2SO_4 65.0 K_2SO_4 29.0	Allison, 1947, Phillips and Val Denburgh, 1971
Harney Lake, central Harney County	120 km^2 lake, maximum depth 1.4 m; dries up	Surface water dilute substurface inter-stitial brine up to 50 g/L dissolved solids; no solid salts	Brine Na_2CO_3 3.6 g/L Na_2SO_4 1.2 NaCl 4.8 KCl 0.4 $Na_2B_4O_7$ 1.9	TDS recorded 1912; 22.0 g/L; Phillips and Van Denburgh, 1971
Summer Lake, Lake County, southeast Oregon	Area 260 to 470 km^2, 0.50 to 1 m deep; dries up	Alkaline semi-perennial lake in downfaulted block; inflow maintained by ground water recharge and rain	Brine Na_2CO_3 8–14.0 g/L $NaHCO_3$ 3–7.0 NaCl 5–9.0 Na_2SO_4 1–2.0 KCl 0.5–1	Up to 130 g/L total solids in playa mud water, pH 9.0; Allison, 1947, Phillips and Van Denburgh, 1971
Utah				
Uinta Basin, Duchesne, County	Area of 223 km^2 at 732 to 1 280 m deep	Halite and sodium carbonate minerals about 5.5 m thick	Bedded sodium salts	Possible extraction using solution mining; Dyni et al., 1985
Deep Creek, Tooele County	Alkali deposits		Salts Na_2CO_3 25.1% $NaHCO_3$ 14.8 NaCl 38.0 Na_2SO_4 17.4 K_2SO_4 4.7	Chartard, 1890

(*cont.*)

Table 3. Occurrences of Sodium Carbonate in Various Parts of the World (cont.)

Name and Location	Physical Character	Geologic Character	Chemical Composition		Remarks and References
UNITED STATES (*cont.*)					
Washington					
Carbonate Lake southeast Grant County	0.6 km^2, very shallow in small valley rimmed by basalt flow	Brine in pools and under ground crystalline salts occurred in pans or pot holes	Concentrated brine		Destroyed by construction of dam; former efforts to produce soda ash; Bennett, 1962
			Na_2CO_3	128.0 g/L	
			$NaHCO_3$	—	
			NaCl	55.0	
			Na_2SO_3	38.0	
			KCl	21.0	
			Na	91.0	
			Sp. Gr.	1.3	
Omak Lake, south-central Okahagan County	15 km^2 lake, 30 to 90 m deep in steep-sided valley in crystalline rocks	Stream fed in abandoned valley, vertically stratified	Very dilute water		3.6 Mt of Na_2CO_3 present in water; Bennett, 1962
			Na_2CO_3	4.0 g/L	
			NaCl	0.3	
			Na_2SO_4	1.9	
			K	10	
Soap Lake, north-central Grant County	3.5 km^2 lake, avg. depth 7.5 m at Grand Covlee in abandoned valley of Columbia River, floored by basalt	The last of a chain of four lakes fed by subsurface flow; density stratified	Rather dilute and variable		570 kt of Na_2CO_3 present in brine; contains water-soluble heavy oil; Bennett, 1962
			Na_2CO_3	21.0 g/L	
			NaCl	8	
			Na_2SO_4	10	
Wyoming					
Berthaton and Yale Lakes (claims), 50 km southwest of Casper	Alkaline lakes 0.3 to 216 km^2 in area	Lakes in area of Tertiary rocks; brine and crystalline salts present	Old Analyses		Schultz, 1910
			Brine		
			Na_2CO_3	66.0%	
			NaCl	14	
			Na_2SO_4	20	
			Crystal Deposit		
			Na_2CO_3	50	
			$NaHCO_3$	6	
			NaCl	6	
			Na_2SO_4	37	
Green River trona, Sweetwater County	Buried trona beds of 120 to 900 m underlie 2 800 km^2 in Green River Basin	42 trona beds up to 10 m thick in organic marlstone and oil shale deposited in Eocene saline lake	Mined trona		Five producers, 10.7 Mtpy soda ash capacity; Bradley and Eugster, 1969, Deardorff and Mannion, 1971
			Na_2CO_3		
			$NaHCO_3$	90.26%	
			NaCl	0.040	
			Na_2SO_4	0.024	
			SiO_2	0.003	
			Organic	Trace	
			Water insol	10.0	
VENEZUELA					
Lago de Laguinillus State of Merida	Small pond which may dry up	In an area of alkalic rocks	Trona and gaylussite form from evaporation		Not an important source; Liddle, 1946; Indians mined about 50 tpy of cured trona (Urao), Chartard, 1890.
			Na_2CO_3	47.0%	
			$NaHCO_3$	37.2	
			H_2O	14.8	
			Insol	1.0	

to 1918 and refined soda ash from 1919–1922. The majority of the calcined trona was used to neutralize the acidity of the mine water at the Witwatersrand mines (Wagner, 1922). A total of about 19 kt of trona were produced from the Pretoria Salt Pan during its brief operation.

At Bir El Natron in Sudan, a large deposit of trona is found mixed with sodium chloride and burkeite. Trona comprises about 35% of the ore body. Some production has been reported but quantitative information is not available. Chatard (1890) reported the occurrence of trona near Fezzan in the Sahara. In Tanzania, Lake Natron is a large sodium carbonate deposit containing an estimated 150 Mt of soda ash resource covering an area of 1 039 km^2. It is located about 30 km south of Lake Magadi in Kenya. Total reserves may be about 1 Gt when including the additional sodium carbonate brines and crystals intermixed with mud layers. Lake Natron lies in the eastern rift valley of Africa that contains other soda ash deposits in other countries. The brine concentration and composition of Lake Natron is nearly identical to that of Lake Magadi in Kenya. The source of the sodium carbonate in Lake Natron originates from three carbonatite volcanoes, including the active Oldoinyo Lengai volcano that erupts with sodium carbonate lava and ash. Several studies have been made to develop the deposit, the most recent being in 1990. Lake Eyasi in Tanzania is also an alkaline lake that has produced some crude soda ash on a limited scale.

Lake Katwe in Uganda contains trona, burkeite, and hanksite that were formed by about 50 small springs reacting with carbonatite and other volcanic strata in the area. Approximately 5 kt of crude trona are recovered from the lake annually.

Antarctica: Although there are no reported deposits of trona on this continent, occurrences of efflorescences of various sodium salts have been documented in the Vestfold Hills, Bunger's Oasis, and McMurdo Sound regions (Dort, Jr. and Dort, 1970). Some of evaporite minerals found include halite, mirabilite, thenardite, bloedite, which are commonly associated with sodium carbonate deposits elsewhere in the world.

Asia: China, India, Mongolia, Tibet, and the former USSR are known to have several sodium carbonate occurrences. Some of the geologic and mining information regarding sodium carbonate deposits in Asia are old, nonexistent, or unavailable for many of the occurrences. Some new information have been published within the past decade and are included.

Most of the information pertaining to Chinese soda ash deposits are obtained from older references, and the names or spellings of many of the deposits have changed making correlations between the new and former names difficult. China has several alkali lakes and at least two major buried trona deposits, which are located in the 260 km^2 Wucheng Basin and at Anpeng in the 13 km^2 Biyang Depression, both in Henan.

The origin of the late-Eocene Wucheng deposit is unknown but is believed to be similar to that of the Eocene Green River deposit in Wyoming. The ore body contains about 77% trona and 23% nahcolite, and lies 650 to 970 m deep. The beds range from 0.5 to 3 m in thickness with the ore grade averaging 50% soda ash equivalent. There are 36 ore beds, each with brown oil shale below and silt-bearing argillaceous dolomite above each trona layer. The upper 21 beds contain halite and trona and generally are 1 to 3 m thick, with a maximum of 4.56 m. The lower 15 trona beds typically are 0.5 to 1.5 m thick, with a maximum of 2.38 m. Reserves are estimated to be about 130 Mt of soda ash equivalent.

The Anpeng Deposit in Tongbai Province was discovered in 1976 and consists of nahcolite and trona with significant sodium carbonate brines. Two phases of trona deposition are reported with a total of 10 evaporation cycles containing trona beds ranging from several to 10 m thick. The shallow phase varies from 1 288 to 1 686 m in depth and the deeper phase ranging from 2 055 to 2 408 m in depth. One exploratory oil drill hole located 11 trona beds having a cumulative thickness of 30 m of trona (Garrett, 1992). This deposit contains an estimated 120 Mt of soda ash equivalent. At least one soda ash plant has been built in Henan to process the trona that reportedly has an annual production capacity of about 10 kt.

Several soda lakes (called Nors) are found in Inner Mongolia, such as in Ih Ju League and Xilin Gol League, and other areas of China, such as in Hami in Xinjiang Province. Construction began in 1971 in Xilin Gol (in Sonid Right Banner in northeast Inner Mongolia) on a soda ash plant (known as the Chagannuoer Alkali mine in one reference and the Ximengchaganor mine in another) that was to produce 150 kt of natural soda ash. Transportation, communication, water supply, and other infrastructure were started; however, construction stopped in 1979 because funding was diverted for other State construction projects. The project resumed in 1986 and was to be completed by 1989. In addition to soda ash, the plan was to produce 50 kt of caustic soda, 50 kt sodium bicarbonate, and 30 kt sodium sulfate. A second natural soda ash facility with an annual capacity of 200 kt was planned for the Wulan Buh desert area of Inner Mongolia. Startup was planned for 1991, but there was no confirmation by 1992 that the plant was in operation. Total soda ash production for Inner Mongolia in 1990 was reported to be 50 kt, which indicates that one or both of the natural operations is working at only partial capacity.

Sodium carbonate is found with mixed salts in India in alkaline soils, lakes, and brines. Sambhar Lake and Lonar Lake, which is the world's only known impact crater in basalt, contain about 22 Mt of soda ash reserves (Fredrikson et al., 1973). Lonar Lake contained about 85 m^3 of brine in 1910 and produced about 600 t of soda ash in 1903. Up to 20 kt of crude trona has been mined annually from various sources at different times during the 20th century.

Several sodium carbonate-bearing lakes, such as Zabuye Caka, Bangkog Cuo, and Guogaling Cuo, are reported in the Qinghai-Xizang Plateau of Tibet. Trona, natron, thermonatrite, halite, northupite, mirabilite, and hydromagnesite are commonly associated with the brine. Potassium, lithium, rubidium, and cesium alkali metals are associated with sodium, chloride, sulfate, and carbonate ions—similar to the chemical composition of Searles Lake, CA, and certain lakes in South America.

The Kazakhstan and the Kulundinsk depression of Altay near Slovgorod and Barnaul in Siberia have produced some natural soda ash from alkali lakes containing natron and brine. Alkali lakes have also been reported near Chita in eastern Siberia, the Araxes Plain of Armenia, in Kazakhstan, Krasnaya Polyana, Kazakh-Tauz region of Azerbaydzhan, and in the Carpathians. Most of these lakes are too small and remote for large scale commercial development. Plants to produce soda ash from natural soda sources have been built at Lake Tanatar and the Petukhov Lakes, both near Mikhaylovskiy (Batalin et al., 1974). Natural soda ash production in the USSR reached its peak in the 1950s with only about 60 kt of output (Shabad and Sagers, 1987). Soda ash has also been produced as a byproduct of alumina refining from nepheline syenite (a sodium-potassium-aluminum silicate) at several locations, including the Pikalevo plant in Leningrad Oblast. The Kola Peninsula is the primary location where the nepheline syenite is a coproduct with apatite. The nepheline is sintered with limestone in rotary kilns to produce alumina, soda ash, potash, and cement. The alumina content in the nephelite was only about half that of bauxite, the process was considered to be economical because it produced coproducts.

Australia: Although salt pans containing sodium chloride and sodium sulfate are common in Australia, there are no significant surface occurrences of sodium carbonate. Alkaline springs have been discovered near Herbertson, Queensland. In the Bowen and Sydney Basins of Queensland, dawsonite has been found, some of which reportedly occurs as alterations of other minerals from sodium carbonate mineralization formed by the leaching of alkaline or ultrabasic rocks.

A sodium bicarbonate-bearing brine (1.6 to 2.5% $NaHCO_3$) was discovered at a depth of 600 to 1 100 m while exploring for petroleum. Reserves have been estimated to be between 500 Mt to 1 Gt of soda ash equivalent. Additional exploration and drilling may locate a trona or nahcolite source of the brine.

Europe: Saline and alkaline lakes are not common in Europe because the present climate is not conducive for development. Sodium carbonate efflorescences have been reported near Szegedin in the Danube Valley in Hungary. Processing of the crude soda (known as *szekso*) into refined soda ash was reported by Chatard (1890). Alkaline brines have been found in Fulophaza, Hungary, at depths of 30 m. Trona and searlesite have been reported in Miocene lacustrine marls and clays in northern Bosnia, Yugoslavia.

Turkey is the third country in the world that has a buried trona deposit. It is located approximately 15 km northwest of Beypazari, about 30 km west of Ankara. The deposit has an area of about 8 km^2, and is of Middle to Upper Miocene age. It is very similar to the Wucheng deposit in China and has some resemblance to the Green River deposit. Seventeen trona beds located 160 to 260 m from the upper part of the Hirka Formation, which shows similar lithological and mineral paragenesis as those of the Wilkins Peak Section of the Wyoming Green River Formation. The trona beds range in thickness from 2 cm to 11.5 m, providing a cumulative thickness of 38 m. Total reserves are reported to be 235 Mt (Dundar, 1988). Feasibility studies and joint venture proposals to develop the Beypazari resource have been conducted since the mid-1980s. As of mid-1992, FMC Corp. and Solvay SA appear to have the competitive advantage over other candidates to co-develop the deposit with Etibank, the Turkish government-owned mining agency.

North America: Soda ash occurs in natural deposits found in Canada, Mexico, and the United States. The largest trona resource discovered to date is found near Green River, WY.

Canada. There are several small soda ash deposits in British Columbia, in western Canada. They tend to be located in the same zone that extends into the saline lakes of Washington. There was reportedly some commercial production of soda ash from some of the deposits between 1928–45. Most of deposits are located in shallow basins in glacial drift overlying basaltic lava flows. The sodium carbonate is primarily found in brines with some surface

Fig. 2. Spiral Solar Concentrator, Lake Texaco, Mexico.

crystal layers and crusts. Cummings (1940) estimated that the soda lakes in British Columbia contained about 103 kt of sodium carbonate reserves.

Mexico. Several alkaline thermal springs, subterranean brines, thin crusts, and surface efflorescences contain sodium carbonate in various locations in Mexico. The only other natural soda ash operation in the Western Hemisphere outside of the United States is at Lake Texcoco 20 km northeast of Mexico City at San Cristobal Ecatepec. A large part of the region was covered in about 1900 by three lakes: Lakes Texcoco, Xochimilco, and Chalco. The latter two contained potable water while Texcoco consisted of brackish water. Fearing that Mexico City may become flooded, Texcoco was allowed to dry up by diverting feeder streams to canals. The hot climate accelerated the evaporation rate, and concentrated the brines which percolated down through the lake floor clay sediments. The brine tends to concentrate on top of two impermeable zones about 53 m deep. The upper zone is richer in sodium carbonate and contains about 120 Mt of soda ash resource while the lower zone contains about 40 Mt. Additional sodium carbonate probably was introduced by local volcanic activity from two volcanoes located 120 km to the south of the deposit—Popocatepetl (garden) and Ixtaccihuatl (sleeping woman). Andesite mountains also occur east and west of Mexico City. The resource covers an area of about 52 km^2.

A small pilot plant was constructed in 1940; however, production of soda ash first began in 1948 with a capacity of about 100 tpd. More than 512 wells located over the deposit collect the underground brines which are sent to a spiral solar evaporator (El Caracol) that has an area of 841 ha and is 50 to 60 cm deep. The spiral concentrator is shown in Fig. 2. The incoming brine is concentrated from 4% (sodium carbonate content) to 6 to 7% in the rainy season and 9 to 10% in the dry season. Weaker brine stays in the outside perimeter of the concentrator for about one year and in the inside perimeter, where the maximum concentration occurs, for about 6 months. The concentrated brine is pumped from the center of the spiral concentrator to the nearby soda ash refinery where it is carbonated into sodium bicarbonate and then calcined into soda ash. Until September 1988, the facility was operated by a subsidiary of a government-controlled holding company. The 200 ktpy operation is now owned by a private group.

Another sodium carbonate resource in Mexico is located at Bahia de Adair on the northern coast of the Gulf of California which was first described in 1898. The area consists of four small playa lakes. The highest concentration of carbonate occurs at a depth from 6 to 19 m from the surface. Crusts found on the surface average about 31% sodium carbonate. The total identified resource in the area is estimated to contain about 250 to 400 kt of soda ash. Subsequent exploration of the area did not detect any significant buried trona deposit. Other occurrences in Mexico include deposits in Chihuahua, the Durango, Sayula, and Zacatecas basins, Lake Cuitzeo, Valle de Santiago, and Laguna de Xaltocan and Laguna de Zumpango, both located in the Valley of Mexico (Garrett, 1992).

United States. Sodium carbonate occurrences are found in Alaska, California, Colorado, Nebraska, Nevada, Oregon, Utah, Washington, and Wyoming.

OWENS LAKE, CA—Oscar Loew of the Lt. George M. Wheeler Surveys West of the 100th Meridian estimated in 1876 that Owens Lake contained enough sodium carbonate to supply the country for about 100 years at that time period's rate of consumption. In 1886, the average depth of the lake was about 5.2 m; however, with the diverting of the Owens River into the Los Angeles aqueduct beginning in 1917, Owens Lake has become a dry evaporite lakebed 4.9 km long and 3 km wide with an area of 87 km^2. The deposit contains about 53 Mt consisting of approximately 35% brine and 65% solid salts (by volume) with an average thickness of about 1 m and a maximum of 3 m (Alderman, Jr., 1983).

The first natural soda ash operation in California began in 1887 at Keeler, on the northeast shore of Owens Lake (Ver Planck, 1959). The Inyo Development Co. constructed solar evaporation ponds that were filled with lake water at the beginning of warm weather. The brine was allowed to evaporate until it reached a density of 30° Be, after which trona began to crystallize. If unwanted salts began to form, the operator drew off some of the mother liquor and replace it with fresh lake water. At the beginning of October, the ponds were drained and the harvesting of trona, which was about 1.9 cm thick, began. The trona was converted to a 95% pure soda ash product by heating it in beehive furnaces to remove the water of crystallization and excess carbon dioxide contained in the sodium bicarbonate. By 1900, Inyo had an annual capacity of 10 kt. Other natural soda ash companies set up operations on the perimeter of Owens Lake in the early 20th century and prospered for several decades until the late-1960s, when competition from Wyoming natural soda ash increased. Today, Lake Minerals Corp., a subsidiary of Cominco American which has been harvesting crude trona from the surface, is contemplating constructing a 600 kt soda ash plant with Vulcan Chemical Co. as it partner.

SEARLES LAKE, CA—Searles Lake is a large playa lake soda deposit located in the lowest part of the Searles Valley, which is a closed structural basin filled with alluvium and non-marine evaporites. The lowermost muds were deposited about 3 m.y. ago with the last lake filling approximately 3.5 k years ago (Smith, 1979). It is the last of a series of chain lakes (which include Mono and Owens Lakes to the north) that was fed by glacial melt waters originating from the east side of the Sierra Nevada Range. Changes in the inflow of upstream water and in regional precipitation caused the lake waters to evaporate to a salt playa. The deposit is 16 km long by 10 km wide with an area of about 104 km^2.

The Searles Lake deposit varies in thickness from zero near the periphery to more than 900 m in the deeper sections. It consists of alternating mud beds and brine-saturated salt beds, both of which are thicker toward the lake center. Halite is the primary mineral found, with trona, hanksite, borax, thenardite, and burkeite.

The main production intervals are called the "Upper and Lower Salt structures" that occur between 15 to 107 m below the surface. From the surface to the bottom, they are: 1) Overburden Mud, which is from zero to 6 m thick; 2) Upper Salt, which ranges from zero to 23 m thick at the lake center, and consists of two members—a lower unit composed of trona and halite with minor quantities of burkeite and thenardite, and an upper unit containing halite and hanksite with some beds of trona and minor amounts of borax and sulfohalite; 3) Parting Mud, 3 m thick in the center to 8 m thick toward the outer areas; 4) Lower Salt, which is 12 m thick at the center to zero meters at the edge of the lake, consists of seven saline beds are mainly trona (with some borax, burkeite, nahcolite, and northupite) interbedded with mud layers; 5) Bottom Mud, which is a 30 m-thick bowl-shaped structure that has clay layers containing gaylussite with minor quantities of dolomite, calcite, halite, thenardite, mirabilite, trona, and borax; and 6) Mixed Layer, which is a series of mud and saline layers 67 to 95 m below the surface, that consists of two sections—an Upper Mixed Layer comprising three units (A,B, and C zones—A being high in carbonate and low in chloride with trona and nahcolite predominating, and B and C rich in trona, halite, and thenardite) based on electrical conductivity, and a Lower Mixed Layer (D,E, and F zones) based on mineral composition. The vertical and lateral extent of the Mixed Layer has not fully been explored (Moulton, Jr., 1981).

The economic potential of Searles Lake was first recognized in 1862 by John Wemple Searles, who discovered borax crystals on the surface. Although he and his brother Dennis formed the San Bernardino Borax Mining Co. in 1873, it was not until 1908 when the California Trona Co. was formed to produce soda ash using the equipment from the Searles' old borax plant. The operation never succeeded, and the plant went into receivership in 1909. The American Trona Corp. was organized on June 12, 1913, and acquired the properties of the California Trona Co. An embargo on German potash because of the war prompted the company to produce potash rather than soda ash. On June 4, 1926, the American Potash and Chemical Co. (which acquired American Trona Corp.) was formed, but it was not until 1931 when the first soda ash was produced at the Trona Plant. The company merged with Kerr-McGee Chemical Corp. on Dec. 29, 1967. Soda ash production at the Trona Plant ended in July 1979 because of equipment obsolescence, rising operating costs, and because the company brought onstream its Argus soda ash facility, located adjacent to the Trona Plant that continued to produce boric acid, potash, and sodium sulfate coproducts.

The Argus Plant was designed to produce only natural soda ash. It was commissioned in July 1978, and presently has an annual production capacity of 1 Mt.

Kerr-McGee also operated a second soda ash plant known as the Westend Plant that it acquired from Stauffer Chemical Co. in October 1974. The plant originally was built in 1918 to produce borax but experienced operating problems. Borax was successfully produced in 1926 and soda ash in 1927 by the Westend Chemical Co, which ultimately merged with the Stauffer Chemical Co. in 1956. The operation continued to produce soda ash until January 1988. All the Searles Lake facilities were sold by Kerr-McGee to North American Chemical Co. on Dec. 1, 1990.

GREEN RIVER BASIN, WY—This deposit is the largest buried trona resource in the world, covering more than 3 100 km^2. It is the result of the evaporation of Lake Gosiute, a freshwater lake that formed in the Eocene more than 50 m.y. ago, and named by Clarence King in 1878. The sodium and the carbonate constituents of trona owe their respective origins to the leaching of the extensive pyroxene-andesite volcanic ash layers and the seasonal influxes of carbonate-rich sediments into the local depositional basin. Subterranean thermal springs also contributed a significant quantity of dissolved alkaline carbonates as well as sodium, calcium, silica, and bicarbonate ions. Trona can only precipitate in the presence of abundant carbon dioxide, which most likely was supplied from the atmosphere and/or the biogenic decay of Eocene plant and aquatic remains (Bradley and Eugster, 1969).

The Wilkins Peak Member of the Eocene Green River Formation in southwest Wyoming contains 42 trona beds, 25 of which have a thickness of 0.91 m or more (which is the lower limit of economic significance) and each covering an area of about 175 km^2. Eleven of these beds exceed 1.83 m in thickness. The beds dip slightly to the southwest, and are numbered from bottom to top. Although many investigators have studied, described and mapped the Green River trona deposit, the most quantitative resource evaluation was done by Culbertson (1966) and updated by Burnside and Culbertson (1979). Their resource estimates were derived using planimetering isopach maps and using a conversion factor of 2 631 t per acre-foot of trona. The total resource is estimated to contain 122 Gt of trona and mixed trona-halite. Approximately 103 Gt of trona are contained in trona and trona-halite beds greater than 1.22 m thick, which is the lower limit of economic extraction. Assuming that 1.8 t of trona yields about 1.0 t of soda ash, about 47 Gt of identified soda ash resources could be obtained from the 56 Gt of bedded trona and the 47 Gt of interbedded or intermixed trona and halite that are in beds greater than 1.22 m thick. Approximately 34 Gt of reserve base soda ash could be extracted from the 36 Gt of halite-free trona and the 25 Gt of interbedded or intermixed trona and halite in beds greater than 1.83 m thick. These reserve base estimates were derived when the state-of-the-art technology was room and pillar extraction, which assumed an extraction rate of about 50% (leaving the remaining 50% behind as pillars); however, if solution mining technology improves in the future, new reserve base and reserve data must be derived for this deposit as well as others.

Because this deposit perhaps is the most important geologic sequence of trona deposition in the world and will continue to supply trona for world consumption for centuries to come, Table 4 lists the more important information about the economic trona beds. Fig. 3 shows a map of the Green River basin with the approximate outer limits of the major economic trona beds, and the location of the five soda ash producers.

OTHER—The Yukon Flats Basin of Alaska contains a series of shallow saline and freshwater lakes. Potash salts in the area have been known since 1929. In 1978, trona samples were identified in efflorescent crusts about 1 cm thick in certain lake margins and dry lake beds. The lakes trend in a northeasterly direction and occupy an area of about 240 by 48 km.

The Piceance Creek Basin in northwest Colorado contains two potential sources of soda ash—nahcolite, natural sodium bicarbonate. The word is formed from the elements contained: *na*, sodium, *h*, hydrogen, *c*, carbon, and *o*, oxygen, and dawson*ite* (sodium-aluminum carbonate). This deposit represents the second-largest known sodium carbonate resource in the world. Although there has been and will continue to be limited sodium bicarbonate production, undoubtedly this nahcolite deposit will be commercially developed for its sodium carbonate content long after the Green River, Wyoming trona deposits are depleted a few centuries from now.

Identified resources of 29 Gt of nahcolite and 17 Gt of dawsonite, equivalent to 18 and 6 Gt, respectively, of soda ash resources are available as single mineral extraction or as a byproduct of oil shale processing. These evaporite minerals belong to the Green River Formation that were deposited in the Eocene time. Nahcolite occurs as aggregates (62%), disseminated crystals (24%), and impure beds (14%) (Dyni, 1980). Commercial development to mine nahcolite in conjunction with oil shale was proposed in the early 1980s; however, leasing obstacles and economic problems prevented successful commercialization. Single mineral extraction using solution mining on sodium leases began as a pilot program in the mid-1980's; however, large scale development did not begin until the late 1980s but was subsequently stalled because of declining market conditions for nahcolite used for flue gas desulfurization. In 1989, NaTec Ltd., a partnership between Industrial Resources Inc., and CRS Sirrine, was formed to solution mine nahcolite for sulfur dioxide scrubbing applications. The company experience difficulties because of depressed markets, and entered a joint venture relationship with North American Chemical Co. in 1991.

Table 4. Estimated Identified Resources of Trona and Intermixed Trona-Halite, by Bed, in the Green River Basin, WY (areas are in square kilometers and tons are in billions)

	Total Resources				Trona Resources				Mixed Trona-Halite Resources						Active Producing Company
			Trona and mixed trona-halite		<1.22 m thick		≥1.22 m thick		<1.22 m thick		≥1.22 m thick				
Bed No.	Depth to Bed From Surface[1]	Bed Thick-ness[2]	area	tons	area	tons	area	tons	area	tons	area	tons	% Halite	Area of Halite	Name
25	228.60	5.18	271.95	1.09	129.50	.18	142.45	.91	—	—	—	—	0	—	Rhone Poulenc of Wyoming
24	243.84	5.18	396.27	1.27	238.28	.27	157.99	1.00	—	—	—	—	0	—	Rhone Poulenc of Wyoming
23	316.99	1.52	758.87	1.09	694.12	.91	62.16	.18	—	—	—	—	0	—	—
22	411.48	1.83	691.53	1.09	606.06	.82	85.47	.27	—	—	—	—	0	—	—
21	414.53	2.44	802.90	1.18	751.10	1.00	51.80	.18	—	—	—	—	0	—	—
20	417.58	2.59	562.03	1.72	266.77	.36	295.26	1.36	—	—	—	—	0	—	Tg Soda Ash, Inc.
19	449.58	4.57	538.72	1.36	305.62	.36	233.10	1.00	—	—	—	—	0	—	Tg Soda Ash, Inc.
18	792.48	1.98	714.84	1.27	269.36	.36	15.54	.09	303.03	.36	126.91	.45	50–100	small	—
17	600.46	6.10	2253.30	11.16	515.41	.64	1111.11	6.35	7.77	<.09	603.47	4.17	10–60	medium	FMC, General Chemical, Solvay
16	731.52	4.27	2227.40	6.53	1357.16	1.72	303.03	1.18	—	—	567.21	3.63	10–60	medium	—
15	786.38	5.79	2196.32	10.70	600.88	.82	701.89	3.36	5.18	<.09	888.37	6.53	60–100	200–400	—
14	838.20	6.10	2004.66	10.89	391.09	.54	963.48	4.90	—	—	650.09	5.44	10–60	medium	—
13	576.07	2.13	958.30	1.63	789.95	1.00	168.35	.64	—	—	—	—	0–small	<130	—
12	847.34	4.57	2100.49	9.71	520.59	.64	890.96	4.08	—	—	688.94	4.99	10–60	medium	—
11	795.53	4.88	1592.85	6.26	608.65	.82	82.88	.36	5.18	<.09	896.14	5.08	60–100	200–400	—
10	826.01	3.66	1188.81	4.35	145.04	.18	67.34	.27	308.21	.36	668.22	3.54	60–100	200–400	—
9	914.40	7.92	1898.47	8.71	699.30	.91	181.30	.73	75.11	.09	942.76	6.99	60–100	200–400	—
8	533.40	1.83	1116.29	1.72	966.07	1.27	150.22	.45	—	—	—	—	0–small	<130	—
7	929.64	1.77	1598.03	3.27	986.79	1.27	572.39	1.91	23.31	<.09	15.54	.09	0–small	<130	
6	928.12	3.72	1520.33	5.44	567.21	.73	202.02	.82	—	—	751.10	3.90	10–60	medium	—
5	826.01	6.10	1844.08	8.62	572.39	.73	1038.59	5.99	12.95	<.09	220.15	1.91	0–small	<130	—
4	624.84	4.27	1367.52	5.72	492.10	.64	875.42	5.08	—	—	—	—	0	—	—
3	560.83	2.59	1144.78	3.18	626.78	.82	502.46	2.27	—	—	15.54	.09	minor	—	—
2	576.38	10.67	1214.71	9.07	259.00	.36	942.76	8.53	—	—	12.95	.18	minor	—	FMC Wyoming Corp.
1	674.22	11.28	839.16	4.90	269.36	.36	569.80	4.54	—	—	—	—	0–small	<130	FMC Wyoming Corp.
Total	—	—	—	121.93	—	17.69	—	56.43	—	1.27	—	46.99	—	—	—

[1]Depth from surface to top of bed at maximum thickness, in meters. [2]Maximum observed thickness, in meters.

Note: Of the 42 trona beds, 25 are economic with the beds averaging 85 to 95% trona, and the remaining 17 minor beds range from 0.3 to 1.2 m thick with areas ranging from 26 to 1 000 square kilometers. To calculate soda ash resources, divide trona data by 1.8. Table modified after Burnside and Culbertson (1979), Culbertson (1971), and Leigh (personal communication, 1992).

Although the sand hills area in northwest Nebraska has numerous small lakes that contain primarily potassium carbonate, sodium carbonate occurs in some of the alkaline lakes. In Jess Lake, the soda ash content is about 36% (Barbour, 1916). Water draining through the sandy soils collects at the bowl of impervious mud bottoms. Soda ash also occurs in several lakes in south-central Oregon. Abert, Alkali, and Summer Lakes contain the most sodium carbonate of all the late Pleistocene lakes in the region. Alkali Lake apparently was mined for a period of time beginning in 1955 by A.M. Matlock (Anon., 1955). Deposits of mixed salts formed in potholes, which were small brine pools that concentrated and evaporated the salts from solution. The potholes resembled those formed in some Canadian lakes, such as those in Last Chance Lake. Natron, sodium chloride, sodium sulfate, and some potassium salts were obtained from the deposit. Approximately 100 t had been mined by Sep. 1955.

Trona, nahcolite, and other authigenic sodium minerals have been found while core drilling in the southwestern part of the Uinta Basin near Duchesne, UT. The bedded deposit, with a maximum thickness of about 34 m, has an area of about 223 km^2 and lies about 73 to 128 m below the surface. Because the deposits are deep, solution mining would probably have to be used if this deposit were developed. Although current data indicates that this location contains less trona than in the Green River Basin in Wyoming and less nahcolite than in the Piceance Creek Basin in Colorado, the presence of sodium-carbonate bearing springs indicates the possibility of larger commercial deposits in the region (Dyni, 1985).

Washington has several soda ash deposits found in lakes in the north-central portion of the state. The lakes with the highest sodium carbonate content are Mitchell Lake (5 ha area), Omak Lake (1 550 ha area), Soap Lake (350 ha area), and Salt Lake. These lakes contain varying concentrations of sodium salts, including natron, gaylussite, sodium chloride, and Glauber's salt. Carbonate Lake and Mitchell Lake produced commercial soda ash prior to World War II. Carbonate Lake was permanently flooded in 1950 when a dam was constructed as part of the Columbia irrigation project. The deposit no longer exists as a sodium carbonate resource (Bennett, 1962).

Other sodium carbonate occurrences have been reported in Arizona, Montana, and North Dakota; however, detailed information regarding their character is scarce.

South America: Although South America has some of the world's largest salt pans (salars), it apparently lacks significant resources of natural sodium carbonate. Salt pans occur at high and low elevations but the majority of the known surface salts are sulfate or sulfate-chloride varieties. Occurrences have been reported in the Los Andes Province of northwestern Argentina, in Bolivia, western Minas Gerais, in Brazil, Chile, southern Peru, and Venezuela. The

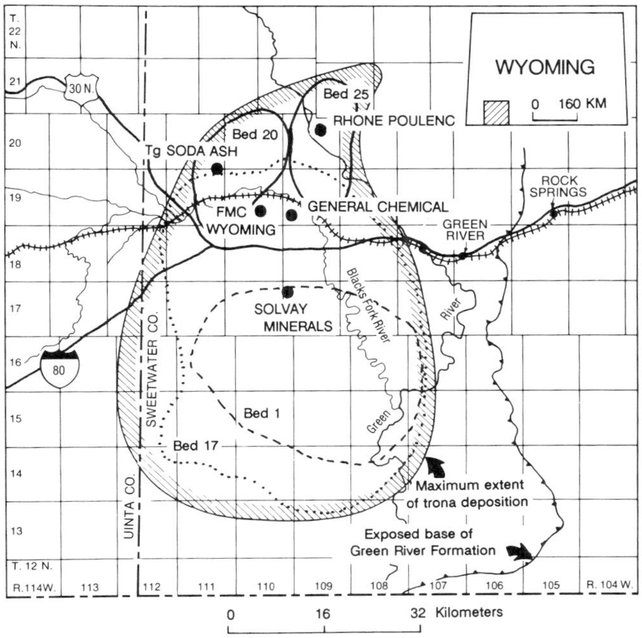

Fig. 3. Map of Green River Basin, WY.

sodium carbonate deposits that occur in the area where Argentina, Brazil, and Chile meet are small and are considered economically insignificant (Catalano, 1926, Cordini, 1967).

Approximately eight sodium carbonate occurrences are found in the Santa Cruz Province of Argentina. Perhaps one of the best described deposits in the area is the Salina Juan Domingo playa lake, which averages 56.3% sodium carbonate and 8.3% sodium bicarbonate (Bernasconi and Cangioli, 1982). Natron occurs on the surface with trona and other carbonates found in the bottom muds. The dried lake may contain about 1.5 Mt of soda ash equivalent. La Ernestina Lake in the Ultracan area and other nearby lakes, such as Laguna el Carancho in la Pampa, are reported to contain sodium carbonate (Cordini, 1967).

Two alkaline lakes in Bolivia contain sodium carbonate. They are Cochi Laguna, located 50 km south of Chiquana that contains a trona and thermonatrite crust 6 to 10 cm thick, and Collpe Laguna, located east of Salar de Chaivira. The 800 m lake contains an alkaline crust and subsurface sodium carbonate-bearing brine in the bottom muds.

Several salt lakes occur in the Upper Sao Francisco area northeast of Brasilia, Brazil. Some of the lakes contain soda ash crusts and shallow brine with the average content ranging between 23 to 41 wt % soda ash. Sodium carbonate occurrences also are found in Urucuia, Carinhanha, Rio Correntes, and Rio Grande.

Trona and gaylussite occur in the surface crusts of the Lago de Laguinillus valley of Venezuela. The salts average about 47 wt % soda ash. Between 50 to 80 tpy of crude trona (urao) has been produced by local Indians. Another deposit is located at Urao Lagoon.

TECHNOLOGY

Mining

Commercial mining of Wyoming trona began in 1948 by the Intermountain Chemical Co., a subsidiary of the Food Machinery Corp. (now FMC Corp.). Because trona is similar to coal, except that trona is somewhat harder, early trona mining used some of the conventional coal-mining equipment. As mining technology improved in the next few decades, newer equipment was utilized and adapted for trona mining such as continuous miners in the early 1960s, with longwall mining, shortwall mining, and solution mining being introduced in the early 1980s. Dredging of sodium carbonate minerals from tailing ponds is another type of mining that began about 1985.

All the Wyoming trona mines use room and pillar mining with multiple entry systems. Most use a combination of one or more type of mining (conventional, continuous, longwall, shortwall, or solution). In addition, each company normally has several crews operating at different places in the mine during an 8 hour shift. Normally, room and pillar mining has an ore extraction efficiency rate of about 45% (the remaining 55% of ore remains as pillars for structural integrity. It is part of the reserve base used for economic evaluation of ore reserves). Some rooms are enlarged to panels of varying dimensions, depending on the type of mining technology used. The heights and widths of the rooms and pillars depends on several factors, including the local conditions and the depth of the particular mine. The deeper the mine, the greater the overburden pressure on the pillars which may require larger sized pillars to support the overlying strata. Usually, pillar dimensions are greater in areas of long-term support, such as near main production and ventilation shafts and main entries. The removal of the ore does create about 2 m of subsidence on the surface, although it is virtually unnoticeable. An average pillar measures about 21 by 24 m. The maze of interconnecting underground tunnels and passageways in each of the five Wyoming mines is extensive. FMC, for example, has more than 3 200 km of underground streets in its 39 km^2 area; more than the total length of the ''streets'' in the city of San Francisco.

Conventional mining had been the primary method of mining trona ore until continuous mining gradually has been replacing it in order to reduce labor costs. In the early 1950s, coal mining equipment and technology was used to cut, drill, blast, and load the trona. A conventional mining crew today normally consists of nine men with productivity averaging about 680 t per shift. A five step mining sequence consists of roof bolting, top cutting, drilling, blasting, and loading and hauling (Steele, 1971). Dynamite had been the traditional explosive used to blast the ore from the work face until ANFO (ammonium nitrate and fuel oil) was introduced. Efforts were made to increase the extraction efficiency by decreasing the size of the pillars and experimenting with arched roofs; however, roof failure and floor heave proved to be problems (Post, 1985) and other mining methods were tested.

In 1960, two foreign methods were introduced to Wyoming: the French, or pseudo longwall, and the English, or pseudo shortwall methods. Both used a combination of borer miners and conventional equipment. Mining problems were reduced once extraction was limited to about 50%. Basically, there are two types of continuous miners—the ripper-type, fixed drum miner and the borer type. Fig. 4 shows a drum miner cutting a section of trona. The drum type is used for short- to medium-life mine developments (10 to 18 years). Each mine operates on the advance and retreat methods; therefore, it is imperative that parts of the development remain in good condition for worker safety, ore haulage, and air ventilation requirements. A typical drum miner with a five man crew can cut a rectangular opening about 2.1 m high and 2.6 m wide (for a double-pass width of 5.2 m wide). Each drum miner weighs approximately 82 t, and cuts through the trona at about 20 to 25 cm/min., or about 20 to 30 t of trona mined per minute. This equals approximately 12.2 m per shift, or about 327 tpd, which is less that obtained from conventional mining. There is however, an 18% operating cost savings with continuous mining due to lower labor and blasting material costs (Chadwick, 1991). A miner is equipped with me-

Fig. 4. Underground mining of trona ore using a drum type continuous miner.

chanical arms that gathers the free trona at a rate of about 11 t/min, and feeds the ore to a conveyor in the middle of the continuous miner where it is unloaded from the discharge tail to a shuttle car or a feeder-breaker unit.

The borer continuous miner is used for long term mainline entries (30 years and more). Instead of the rectangular-shaped cut made by a drum miner, the borer miner with its six man crew makes an ovaloid-shaped cut 2.6 m high by 4.3 m wide. The arched roof and floor design provides greater structural support and reduces rib slabbing in the pillars. This is especially important in mines that operate at deeper depths where the overburden pressure is greater. A typical borer miner weighs 73 t, is 9.4 m long and 2.8 m wide. It has a set of two 2.6 m-long arms that cover an area of 11 m^2 that rotate perpendicular to the advance of the machine, producing at a rate of 450 t/shift for mainline development. Each twin-rotor miner usually is accompanied by two shuttle cars and one single boom roof bolter except in those mines that use mobile conveyors and feeder-breaker units located behind the miners. In this situation, the borer miner with continuous haulage is used for mine production, not mine development, with output exceeding 1 100 t/shift (Refsdal, 1991). The setup has the potential of producing about 1 800 t/shift.

The FMC Wyoming Corp. is the only natural soda ash producer that uses longwall mining to augment its mining technology. It is reportedly the only non-coal longwall operation in the United States. Allied Chemical Co. (now known as General Chemical Corp.) was the first company to use longwall mining beginning in the 1973 but abandoned the method after several years of use. FMC first installed longwall equipment in March 1981 but was hindered by the slow cutting rate of the equipment as well as slabbing and spalling problems with the trona. From July 1987 until May 1989, longwall mining temporarily ended as improvements in the machinery took place, including slab retaining plates on the canopies of the shield supports that hold the trona wall in place and automatically retract as the shearer advances. Compared with continuous miners, longwall equipment initially is more expensive; however, the higher production rate, elimination of roof bolting requirements, and greater safety for the miners more than offsets the expense. The longwall unit consists of a 550 kW, 92-t electric shearer with a lead and trail drum (each about 1.5 m in diameter by 0.6 m wide), an armored face conveyor, and face shields. At a maximum travelling speed of 15 m/min., the lead drum cuts 1.5 m of the top seam face while the trail drum takes the remaining 1.2 to 1.6 m to the floor of the face. On the return pass across the face, the lead drum becomes the trail drum and vice versa. It takes about one year to completely mine a 122 by 1 067 m panel of trona (approximately 725 kt). The areas between panels are used for roof support. The roof of the mined out areas ultimately collapses to relieve the stresses, creating massive piles of rubble called gob. A longwall unit consists of a seven man crew, plus a mechanic and a shift foreman that produce between 900 to 1 100 t per shift. The method also has an ore recovery rate of about 75% compared with only 45% for normal room and pillar mining with conventional or continuous mining equipment.

Shortwall mining is used exclusively by Tg Soda Ash, Inc., and incorporates some of the concepts used in longwall mining and continuous mining. The shortwall method uses a 60 t continuous mining machine in conjunction with a large hydraulic roof support unit equipped with 34 shields that each weigh about 18 t. Each shield exerts about 560 tons of resistance between the floor and the roof, temporarily providing the needed support to prevent the weight of the overlying rock from collapsing in the working area of the mine. The continuous miner makes a cut about 4.9 m across by about 3 m high across the length of the work face, which is approximately 55 m long. Ultimately, the roof collapses to relieve the stresses after the shortwall mining unit advances to the next section to be mined. Similar to longwall mining, shortwall mining has about a 75% extraction rate.

The concept of roof bolting was first used in 1949 to replace conventional timbering methods of roof support. After about 6 m of trona have been mined in the main tunnels and the long term developed areas (such as near shafts and maintenance shops), the overlying strata is roofbolted using independent roof-bolting drill rigs or equipment mounted on continuous miners. Because trona is much stronger than the overlying oil shale, about 0.6 m of trona is left on the roof to provide a safer roof. The water-cooled rotary drill operates at about 1 200 rpm, and drills a 1.5 m long hole at approximately 5.6 m/sec, or about 15 seconds. Expansion shell roof bolts with epoxy resin are inserted and installed with a fitted metal plate and nut on the roof side on 1.6 m centers. The entire process, from drilling to roof bolting, takes about 40 to 50 seconds. About 160 bolts are installed during each 8-hr shift, or approximately one bolt every two minutes (Chadwick, 1991). Roof bolting reportedly is the single-most highest expense in underground mining costs. One company uses about 60 000 bolts each month. Rib bolts on about 4 m centers are also used to reinforce some of the main pillars to minimize spalling or pillar failure. Several roof bolt plates with the roof bolts secured in place are shown in Fig. 4.

After the ore has been blasted or cut, it must be conveyed to central underground areas where the trona is crushed before it is transported on conveyor belts at a rate of about 8 t/min to the production shaft, where the ore is hoisted to the surface for storage and conversion to soda ash. Traditionally, a fleet of diesel or electric shuttle cars with 10 t capacities (with some up to 20 t capacities) were used to haul the ore from the discharge tail of the continuous miner at the mine face to the primary crushers. A miner can load about 13 t in a shuttle car in about 1 minute. Since the late 1980s, most Wyoming soda ash companies have installed continuous haulage systems that replaced the shuttle cars. Typically, a mobile system is placed behind a feeder-breaker unit (used to crush and feed the ore at a rate of 1 080 tph) that is in back of an advancing continuous miner that feeds the cut ore into the system. As the miner moves forward or backward, the length of the belt automatically advances or retracts from a belt storage magazine that can store up to about 365 to 1 220 m of belt, depending on the manufacturer. Belt benders are employed to permit the system to turn corners around pillars or other obstructions. Another type of mobile system uses flexible steel conveyors that has pneumatically-operated legs that retract and allow the conveyor to crawl on the floor to the next location.

The practice of solution mining reportedly was used more than

2,200 years ago in China to recover salt from deep underground formations. FMC conducted a single-well solution mining test in Wyoming in 1945 but encountered poor results and decided to pursue underground mining instead. Carl A. Bays endorsed the use of hydraulic fracturing used by the oil and gas industry to interconnect well pairs in the mid-1950s. Between 1955 and 1958, solution mining was conducted using this technique with moderate success. After years of further work and testing, FMC announced on Jan. 14, 1981, its intention to construct a demonstration plant by mid-1981 using solution mining to recover trona from the deeper beds on State lease Section 36, T17N, R110W, which was about 26 km south of its existing refinery.

A well pair consists of an injection well and a production well separated a couple of hundred meters apart. After the wells were drilled to about 655 m (bottom of Bed 1), steel casing was set into place and cemented to the well wall from the surface to below the bottom of the bed. The casing and cement were breached at the desired location of the fracture by using one of several techniques—cutting tools, jet notching, or explosive perforations. Several monitor wells were also drilled along the perimeter of the section to check if there were any excursions of the injection fluid. Water was slowly introduced through the casing to initiate hydraulic fracturing. Each injection well was pressurized to exceed the overburden lifting pressure plus the tensile strength of the trona formation. Once this occurred, the formation ruptured and the solution entered the bed and followed the path of least resistance. Ideally, the fracture should move horizontally toward the production well. Once communication was made with the production well, a solvent (caustic soda) was added to the water to completely dissolve the trona. If only water were used, the sodium bicarbonate contained in the trona would precipitate, deposit at the bottom of the solution-mined cavity, and cover the surface of the soluble trona, thereby reducing the quantity of sodium carbonate in solution and minimizing the utilization of the resource.

An extraction rate of 30% for trona solution mining was derived years ago based on information obtained from engineers engaged in solution mining of salt, potash, and nahcolite. Although there are no published reports to confirm or refute this estimate, it is conceivable that the extraction rate could be higher as more well pairs are established within an area and horizontal drilling is used to establish better communication and contact with the solvent within the trona beds.

FMC has been discharging tailings and spent solutions to its tailing ponds since about 1953. For about 40 years, sodium carbonate decahydrate crystals have been collecting on the bottom of these ponds at the rate of about 20 to 30 cm per year, and represent a large, alternate resource available for recovery. Beginning about 1985, FMC used a bucket wheel dredge to extract about 160 kt of crystals annually from the 485-hectare pond. The crystals are slurried, dewatered, melted, and processed into soda ash in the refinery. Other producers have similar ponds where sodium carbonate decahydrate has collected since the plants were built. At some time in the future, they may collect and process these crystals as well. Because of environmental considerations and the cost of constructing and maintaining tailing ponds, all the producers are involved in injecting waste tailings underground. This will reduce the future buildup of sodium decahydrate in the tailing ponds. In addition, most of the Wyoming soda ash producers are selling their spent purge liquors that contain several percent of dissolved sodium carbonate in solution. Local power utility companies purchase the liquors for Ph control in process water.

Processing

The processing of natural soda ash is significantly different than the chemical processes used to make synthetic soda ash. The processing of trona ore and sodium carbonate-bearing brines involves different mineral separation technologies whereas the manufacture of synthetic soda ash can be accomplished using one of several techniques.

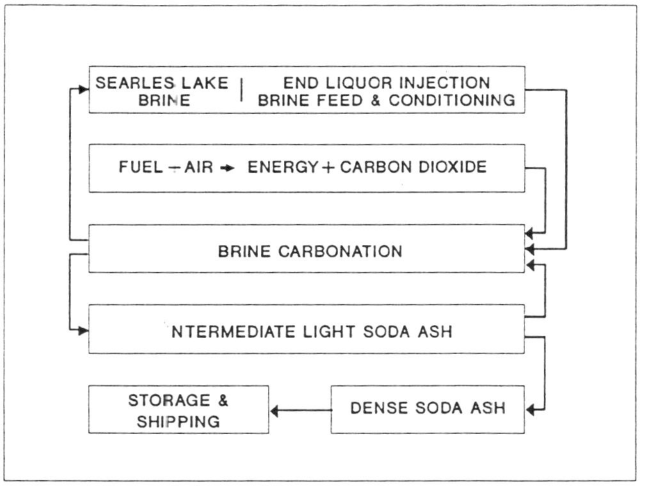

Fig. 5. Argus total system schematic.

Natural Soda Ash: Although trona ore averages approximately 90% pure sodium sesquicarbonate (with 10% accessory minerals), further refining of the ore must be done to produce a $99^{+}\%$ soda ash product. Sodium carbonate-bearing brines contain a variety of other soluble minerals, including sodium sulfate, sodium chloride, potassium chloride, and borax, which must be physically and chemically removed from the brine in order to process the sodium carbonate content into refined soda ash.

Searles Lake Brine Process. A vast array of production wells covers the surface of Searles Lake to supply the Argus soda ash refinery with sodium carbonate-bearing brine feedstock obtained from the Mixed Layer brine and the Lower Salt brine. These brines are pumped to the plant and are carbonated in carbonation towers with carbon dioxide gas obtained from calciner gas (from the conversion of soda ash from sodium bicarbonate) and from the monoethanolamine (MEA) towers that processes the power plant's flue gases into purified CO_2 gas. The sodium carbonate in the brine is converted to sodium bicarbonate, which precipitates, is collected, filtered, and calcined to light soda ash. This intermediate product is transferred to surge bins, which feed into one of three rotary bleaching units that operate at 427°C to remove residual organics. The material is concentrated and sent to sodium monohydrate crystallizers operating at 88°C that produces a larger-sized crystal. The product is screened, dewatered, washed in centrifuges, and dried and dehydrated in steam tube dryers at 260°C. The crystals are finally screened at 16 mesh and sent to storage. The final product is 99.7% sodium carbonate. The barren, or spent, solutions from the Mixed Layer brine and burkeite liquors are returned to the lake. The Lower Salt brine is treated separately at the Westend plant to recover borax and sodium sulfate. Fig. 5 shows a flow diagram of the Argus soda ash process.

Wyoming Trona Process. The sesquicarbonate process and the monohydrate process are the two techniques to refine soda ash from trona ore. Both use the same procedures but only in different sequences. It takes approximately 1.8 t of trona ore to make 1.0 t of finished soda ash. The monohydrate method is the primary process used today to make soda ash. Fig. 6 shows a flow diagram of both technologies.

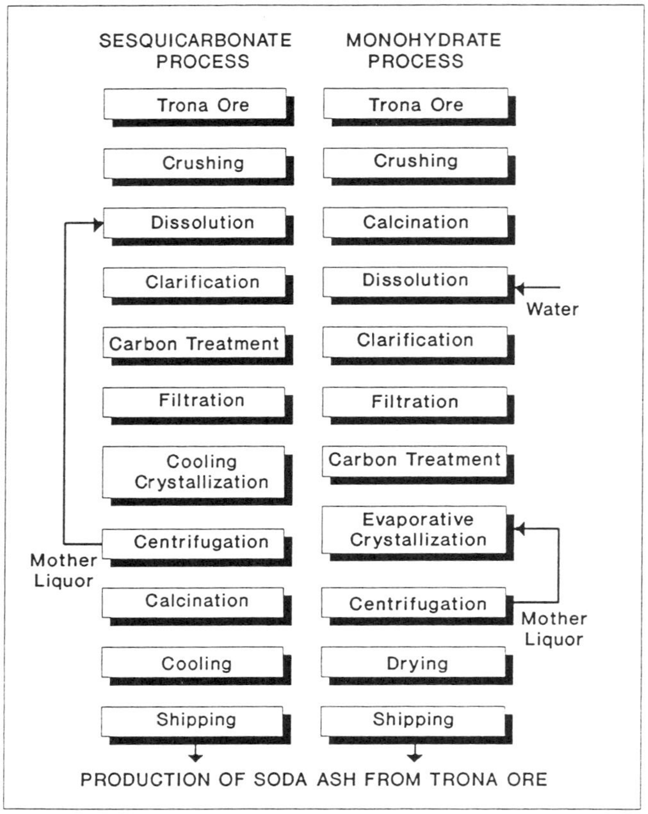

Fig. 6. Production of soda ash from trona ore.

The monohydrate process was developed by FMC Corp. in 1960. Although it patented and licensed the technology to other producers, FMC did not use the technique until 1972. In the monohydrate process, trona is crushed and calcined in rotary gas-fired calciners at 150 to 300°C to remove water and carbon dioxide from the ore, which reduces the weight of the ore by 27% leaving an impure product containing 85% soda ash and 15% insolubles. The thermal decomposition of trona involves the following reaction:

$$2Na_2CO_3 \cdot NaHCO_3 \cdot 2H_2O \text{ (trona)+heat} \rightarrow 3Na_2CO_3+CO_2+5H_2O$$

The calcinate is dissolved with hot water and the clear, hot solution containing about 30% sodium carbonate by weight is sent to evaporative multiple-effect crystallizers or mechanical vapor recompression (MVR) crystallizers where sodium carbonate monohydrate crystals ($Na_2CO_3 \cdot H_2O$) precipitate at 40 to 100°C, which is below the transition temperature of monohydrate to anhydrous soda ash. The insoluble portion of the ore containing shale and shortite are collected by clarifiers, filtered, and washed to recover any additional alkali before they are piped in a slurry to tailing ponds or injected underground in abandoned sections of the mine. Some companies pass the liquor through activated carbon beds prior to crystallization to remove trace organics solubilized from the oil shale so that the organics do not interfere with crystal growth rate and crystal habit in the crystallizers. The crystals are sent from the crystallizers to hydroclones and dewatered in centrifuges to between 2 and 6% free moisture. The centrifuge cake is conveyed to steam tube dryers where the crystals are dehydrated into anhydrous soda ash at 150°C, screened, and sent to storage or shipment. The final product made by the monohydrate process is dense soda ash with a bulk density of 0.99 to 1.04 g/mL, and an average particle size of about 250μm.

The sesquicarbonate process produces light to intermediate grades of soda ash crystals with a bulk density of about 0.89 g/mL. The trona is crushed, dissolved in hot (95°C) mother liquor, clarified, filtered, and passed to cooling crystallizers where crystals of sodium sesquicarbonate ($Na_2CO_3 \cdot NaHCO_3 \cdot 2H_2O$) precipitate. Activated carbon is added to filters to control the organics that interfere with crystal growth. The sodium sesquicarbonate is hydrocloned, centrifuged, and calcined at 110 to 175°C using gas or indirect steam heat. Dense soda ash can be made by calcining the sodium sesquicarbonate at 350°C.

Synthetic Soda Ash: Synthetic soda ash can be made by one of several chemical reactions that use common raw materials for feedstock, such as limestone, salt, and coal. Other technologies produce soda ash as a byproduct of a process.

Leblanc Process. The process was developed in 1791 by Nicolas Leblanc in the first effort to commercially synthesize soda ash. Common salt was combined with sulfuric acid to produce sodium sulfate and hydrochloric acid. The sodium sulfate was heated in a rotary furnace with coal and limestone (or charcoal and chalk) to produce a black ash, which contained sodium carbonate, calcium sulfide, and some unreacted coal. The mixture was leached with cold water, which was evaporated or dried to obtain crystalline sodium carbonate. The chemical reaction was:

$$Na_2SO_4 + 2C \rightarrow Na_2S + 2CO_2$$

$$Na_2S + CaCo_3 \rightarrow Na_2CO_3 + CaS$$

The Leblanc process was used primarily in France and England. It was displaced by the Solvay ammonia-soda process by about 1915; however, one plant in China reportedly was producing about 250 t of soda ash per year as of 1950. The Leblanc technology was not used successfully in the United States except during a short period from July 1884 to January 1885 at Laramie, WY.

Solvay Process. Also known as the ammonia-soda process, the Solvay process was developed by Alfred and Ernest Solvay in 1861 as an inexpensive way to manufacture soda ash on a large scale. To manufacture 1.0 t of synthetic soda ash requires about 2.8 t of steam, 1.7 t of salt, 1.4 t of limestone, 0.6 t of coal for the boilers and about 0.2 t of coal for the dryers. The process discharges approximately 1.7 t of waste products that include sodium chloride and calcium chloride. In the process, a sodium chloride brine is treated with milk of lime to remove magnesium ions and with soda ash to remove calcium ions. The brine is then saturated with ammonia, which acts as a carrier, and carbon dioxide gas in an absorption tower to produce ammonium bicarbonate, which reacts with the salt in solution to form sodium bicarbonate and ammonium chloride. The sodium bicarbonate precipitates, after which it is filtered and calcined at 175 to 225°C to light soda ash (based on its bulk density of 0.51 to 0.62 g/cm^3) and the gases produced are recycled back to the liquid phase. The solution containing ammonium chloride is reacted with milk of lime to recover the ammonia for reuse and to produce byproduct calcium chloride for either sale or disposal. Limestone and coke are required to make the milk of lime. Dense soda ash (bulk density of 0.96 to 1.06 g/cm^3) is made by hydrating light soda ash to change the bulk density. The chemical reactions in the Solvay process are:

$$CaCO_3 \rightarrow CaO + CO_2$$

$$C_{(amorph.)} + O_2 \rightarrow CO_2$$

$$CaO + H_2O \rightarrow Ca(OH)_2$$

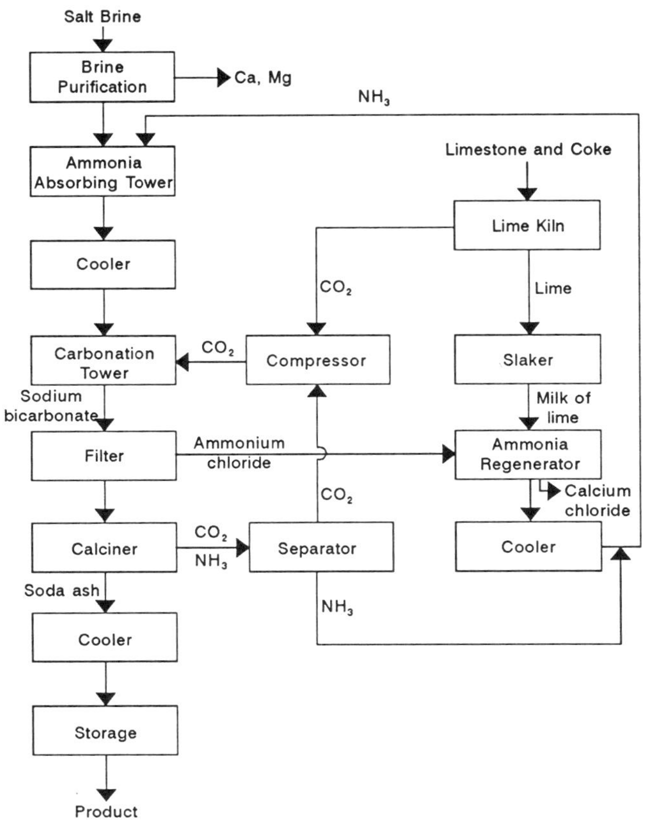

Fig. 7. Flow diagram of the Solvay ammonia-soda process.

$$NH_3 + H_2O \rightarrow NH_4OH$$

$$2NH_4OH + CO_2 \rightarrow (NH_4)_2CO_3 + H_2O$$

$$(NH_4)_2CO_3 + CO_2 + H_2O \rightarrow 2NH_4HCO_3$$

$$NH_4HCO_3 + NaCl \rightarrow NH_4Cl + NaHCO_3$$

$$2NaHCO_3 \rightarrow Na_2CO_3 + CO_2 + H_2O$$

$$2NH_4Cl + Ca(OH)_2 \rightarrow 2NH_3 + CaCl_2 + 2H_2O$$

The Solvay process is the dominant method used in the world today to manufacture synthetic soda ash. Approximately 70% of the total world output of soda ash is synthetic, the majority of which is from the Solvay process and the remainder from the ammonium chloride, New Asahi, and caustic carbonation processes. Fig. 7 shows a flow diagram of the Solvay ammonia-soda process.

Electrolytic Process. Electrolytic soda ash is produced by introducing an electric current through a cell containing a saturated solution of sodium chloride. Sodium hydroxide forms at the cathode, and as carbon dioxide gas is passed through the solution around the cathode, sodium bicarbonate precipitates. The material is collected and dried to produce soda ash. In 1939, three electrolytic plants owned by the West Virginia Pulp and Paper Co. were in operation in Tyrone, PA, Covington, VA, and Luke, MD. The plants also utilized the coproduct chlorine from the cells for bleaching applications.

Caustic Carbonation Process. Dow Chemical Co. operated the only caustic carbonation plant in the United States from 1955 to 1970 at Freeport, TX. This plant, and a few others in the world, converted sodium hydroxide obtained as a coproduct with chlorine manufacture into soda ash. The sodium hydroxide in the electrolytic cell is treated with flue gas. The basic reaction is:

$$2NaOH + CO_2 \rightarrow Na_2CO_3 + H_2O$$

Higher energy costs associated with the electrolytic process that produces both chlorine and caustic soda normally results in higher market prices for caustic soda compared with soda ash; therefore, this technique is not used extensively in the world.

Caprolactum Pyrolysis. A patent was issued to Proyectos S de Mexico to pyrolize caprolactum residue to obtain a 95% sodium carbonate-5% carbon product. A 40 tpd plant began production at Salamca Quanaquato in 1989. The soda ash-carbon material was being sold to a pulp and paper operation that converted the soda ash to caustic soda. The small quantity of carbon in the soda ash reportedly would not be a problem (Garrett, 1992).

Ammonium Chloride (AC) Process. The process was developed by Asahi Glass Co. of Japan in 1949 at the Kitakyushu Factory. It is essentially a modification of the conventional Solvay process. Because salt for synthetic soda ash manufacture must be imported, and therefore is expensive, a new technique was developed to improve the utilization of salt. In the Solvay process, about 70% of the sodium content in the salt is converted to soda ash with the remaining 30% and all of the chlorine discharged as waste. In terms of salt usage, only 28% (by weight) of imported salt was used to produce soda ash; the remaining 72% was lost as waste. The AC process utilizes more than 90% of the sodium from the salt. In the Solvay process, the ammonium chloride that is produced is converted to ammonia and recycled; however in the AC process, it is crystallized by cooling. Ammonium chloride is an excellent nitrogenous fertilizer used in Asia mainly for paddy rice; however, experimental applications indicate it has beneficial uses in wheat, sugar cane, cotton plant, coconut, palm oil trees, bananas, and jute (Tsunashima and Nakaya, 1982). Another difference between the Solvay and AC methods is that the AC method uses solid salt rather than salt brine as feedstock. In addition, the mother liquor is recycled to dissolve additional salt.

New Asahi (NA) Process. The NA process was developed in the early 1970s to improve on the AC and Solvay processes by reducing energy consumption, labor requirements, equipment maintenance. The process has two variations that can be used depending on the ammonium chloride demand situation; the NA mono-production process and the NA coproduction process. In the former process, solid ammonium chloride reacts with a lime slurry to release the ammonia which is recycled with a 25% calcium chloride solution (compared with 10% solution in the Solvay process) discharged as waste or recovered for sale. In the latter variation of the process, byproduct solid ammonium chloride is produced by direct contact cooling crystallization provided the facility is equipped with an ammonia plant. The NA Co-production process has lower production costs compared with the NA mono-production technique.

Akzo Process. Akzo Zout Chemie of The Netherlands developed a process to make vinyl chloride and soda ash from salt using an amine-solvent system catalyzed by a copper-iodide mixture. The process holds US Patent No. 4 256 719 (3/17/81). Van Andel (1983) indicated that the method was an alternative to salt electrolysis, direct ethylene chlorination, and oxychlorination techniques. It produces soda ash rather than caustic soda, and uses steam instead of electricity resulting in total energy consumption being reduced by about one half that of the conventional process.

Ormiston Mining Process. Although this process mainly was

developed by Ormiston Mining and Smelting of Canada to produce sodium hydroxide, it can be altered to produce soda ash. The method involves the electrolysis of a sodium sulfate solution in an electrolytic cell that generates sulfuric acid. Ammonia is added to neutralize the acid and the process ultimately produces sodium hydroxide and ammonium sulfate. The caustic soda can be reacted with lime and converted to soda ash. Although the operating economics appear to be cost-prohibitive, some nations without soda ash resources but with abundant sodium sulfate deposits may find this process promising (Patent issued to J.S. Thompson and D. Genders, No. 5098532, 3/24/92).

Hüls Process. This process produces soda ash and hydrochloric acid from salt using an amine-solvent system (Rauh, 1991). The technique was developed by Chemische Werke Hüls AG and was awarded US Patent No. 4 320 106 (3/16/82). The byproduct hydrochloric acid may be easier sold than other byproduct chemicals recovered from other processes, depending on market conditions.

END USES

Soda ash obtained from either natural or synthetic production processes serve a variety of end use markets. The markets in developed nations are mature and are not expected to grow appreciably but will generally parallel the population growth rates of those nations. In developing countries, where the glass, chemical, and industrial sectors are emerging, consumption is expected to grow more significantly as the demand for consumer products increases.

For more than 5,500 years, since the days of the Egyptians, soda ash has been used in glass and soap manufacture and continues to be an important end use today. Older uses for soda ash, such as in embalming the dead, treatment of colic pains, and for making bread, have given way to newer uses in the twentieth century, such as for pH control in swimming pools, for flue gas desulfurization for cleaner air, and in glass fiber insulation for warmer homes. Although we are surrounded everyday by many consumer products that contain soda ash, most people are unfamiliar with what soda ash really is. Based on reported industry sales in 1992, the manufacture of glass products represented about 49% of domestic soda ash consumption, with the glass container and beverage sector composing 59% of this end use; flat glass, 27%; fiber glass, 7%; and specialty (i.e., glass tumblers, medicine bottles, incandescent bulbs, etc.), 7%. The other end uses included chemicals (mainly sodium bicarbonate, sodium chromates, sodium phosphates, and sodium silicates), 24%; soap and detergents, 12%; distributors (which buy soda ash directly from the producers and resell it to various downstream consumers), 6%; flue gas desulfurization, 3%; pulp and paper, water treatment, and miscellaneous, 2%, each. Table 5 lists the historical trend in US soda ash consumption since 1970.

Glass: The glass industry is the largest consumer of soda ash, especially the glass container sector. Soda ash is a source of sodium oxide that is used as a fluxing agent in container, flat, fiber, and specialty glass manufacture to reduce the temperature at which the raw materials melt. Soda ash decomposes into sodium oxide and carbon dioxide, which rises through the glass melt and helps mixes the ingredients.

Soda-rich glass is softer than other more refractory types of glass; therefore, forming is easier. The quantity of soda ash added to glass batches varies with the type of glass being manufactured and the percentage of recycled glass (also known as cullet) being used. Glass containers, for example, may require more than 180 kg of dense soda ash for every 450 kg of sand. A typical batch for clear glass containers may consist of 155 kg soda ash, 172 kg limestone, 145 kg feldspar, and 5 kg sodium sulfate. Flat glass batches require about 115 kg of soda ash per 450 kg of sand used. The growing nationwide effort to recycle glass has benefited many of the 76 domestic glass-container manufacturers because cullet substitutes for part of the raw material requirements in a glass batch, and cullet melts at lower temperatures (about 20 to 25% less), thereby reducing glass production costs. The increased use of cullet has conversely affected soda ash consumption. In 1988, the national recycling rate for all glass containers was 22%; however, by 1992, the rate rose to 33%.

Table 5. Consumption of US Soda Ash by End Use

	1970	1975	1980	1985	1990
Glass	3 048	3 257	3 494	3 084	3 177
Container	N/A	N/A	2 516	1 996	1 909
Flat	N/A	N/A	559	635	851
Fiber	N/A	N/A	175	249	228
Specialty	N/A	N/A	244	204	189
Chemicals	1 796	1 479	1 294	1 361	1 573
Soaps and Detergents	172	299	453	544	835
Pulp and Paper	272	218	227	318	132
Water Treatment	181	181	227	272	91
Flue Gas Desulfurization	N/A	N/A	N/A	159	207
Distributors	N/A	N/A	N/A	N/A	376
Other	647	721	777	387	136
Total US consumption	6 116	6 155	6 472	6 125	6 527
Undistributed[1]	21	(181)	—	—	317
Total apparent consumption	6 137	5 974	6 472	6 125	6 844

N/A Not Available, [1]Discrepancies between US consumption and apparent consumption data were because of the fluctuating balance of inventory in transit from plants to domestic or export destinations. Reported consumption data available after 1987; data before 1987 were estimated.

The US soda ash consumption in the glass container industry also declined because of increased use of polyethylene terephthalate (PET) plastic containers.

Chemicals: For many years, soda ash was primarily the only chemical product produced by the US soda ash industry. As an intermediate product, it was sold to various chemical manufacturers for many downstream uses. As domestic soda ash prices and consumption stagnated in the early to mid-1980s, several soda ash suppliers diversified their product lines and began producing different soda ash-based products that had higher added value, such as sodium bicarbonate, sodium phosphates, and sodium hydroxide. Domestic sales of major sodium chemicals in 1990 amounted to $7.3 billion. The percent distribution by sector output was sodium hydroxide, 44%; sodium tripolyphosphate, 14%; soda ash, 11%; sodium bicarbonate, 7%; sodium chlorate, 5%; sodium bichromate and sodium silicates, 3% each; and miscellaneous sodium chemicals, 13%.

The inorganic chemical industry is the second largest use of soda ash comprising 24% of total domestic demand. Although there are many sodium-based compounds that are indirectly produced from soda ash, there are only six major chemicals that are directly manufactured from soda ash feedstock.

Sodium bicarbonate. Sodium bicarbonate, or baking soda, is manufactured by percolating carbon dioxide gas through a carbonation tower containing a saturated soda ash solution. The sodium bicarbonate precipitate is collected, filtered, centrifuged, dried, screened, and packaged. Approximately 0.68 t of soda ash is required to make one ton of sodium bicarbonate. The major uses of sodium bicarbonate include household consumer products, 26%; animal feed, 21%; food, baking, and beverage, 16%; plastics and chemicals, 14%; exports, 9%; pharmaceuticals and personal care products, 6%; fire extinguishers, 2%; soaps and detergents, 25%; flue gas scrubbing, 2%; and other, 2%. Five producers operating six plants have a combined annual production capacity of about 470 kt. Three of the companies are also soda ash producers.

Sodium hydroxide (caustic soda). Caustic soda is mainly produced as a coproduct with chlorine gas in the electrolysis of salt brine in either mercury, diaphragm, or membrane cells. However, non-electrolytic caustic soda, also known as chemical caustic soda, is also presently being made from lime and soda ash by three US soda ash companies. The lime-soda process was the primary method of caustic soda manufacture at the beginning of the twentieth century before the electrolytic process gained favor. Chemical caustic soda is produced by the following reaction:

$$Ca(OH)_2 + Na_2CO_3 \rightarrow CaCO_3 + 2NaOH$$

The lime is slaked (CaO converted to $Ca(OH)_2$) and added to 1.33 t of dissolved soda ash to produce one ton of chemical caustic soda. Calcium carbonate is precipitated from the reaction, calcined back to lime, and recycled. Two Wyoming soda ash producers, FMC Wyoming Corp. and Solvay Minerals Inc., currently have a combined production capacity of about 127 kt of chemical caustic soda with Tg Soda Ash, Inc. came onstream in late 1992 with 90 kt of capacity. FMC uses about half of its caustic soda to produce captive sodium cyanide for precious metal recovery at its gold operation. The balance of the chemical caustic is sold to various regional customers for different applications, including pulp and paper, water treatment, and inorganic chemical preparation.

Sodium sesquicarbonate. Sodium sesquicarbonate is a hydrated compound containing soda ash and sodium bicarbonate. Trona ore is first crushed and dissolved to separate the insoluble impurities. The sodium compounds in solution are then clarified, filtered, crystallized, centrifuged, calcined, and recovered as long needle-shaped monoclinic crystals of sodium sesquicarbonate. FMC Wyoming Corp. in Wyoming has a 50 ktpy sodium sesquicarbonate facility that sells mainly to the detergent industry.

Sodium chromates. Chrome ore is roasted with soda ash in a rotary kiln to produce sodium chromate or sodium bichromate that are used to produce other chromium compounds, such as chromic acid and chromic sulfate. The chemical reaction is:

$$2Cr_2O_3 + 4Na_2CO_3 + 3O_2 \rightarrow 4Na_2CrO_4 + 4CO_2$$

$$2Na_2CrO_4 + H_2SO_4 \rightarrow Na_2Cr_2O_7 + Na_2SO_4 + H_2O$$

For every ton of sodium bichromate produced, 0.79 t of soda ash is required. Two producers with a combined annual production capacity of 149 kt operated in 1990. The end use distribution of sodium bichromate is chromic acid, 55%; chromium oxide, 10%; leather tanning, 8%; pigments, 7%; wood preservatives, 2%; drilling mud additives, 2%; other (metal finishing, water treatment, textiles, catalysts), 3%; and exports, 13%.

Sodium phosphates. Sodium phosphates are the largest soda ash consuming sector of the industrial inorganic chemicals. The phosphate chemicals, and the corresponding quantity of soda ash required to make one ton of the chemical, are as follows: sodium tripolyphosphate ($Na_5P_3O_{10}$), 0.76; monosodium phosphate (NaH_2PO_4), 0.45; disodium phosphate (Na_2HPO_4), 0.747; trisodium phosphate (Na_3PO_4), 0.68, sodium metaphosphate ($NaPO_3$) 0.557, sodium acid pyrophosphate ($Na_2H_2P_2O_7$) 0.478, and tetrasodium pyrophosphate ($Na_4P_2O_7$), 0.84. The sodium phosphates have exceptional cleaning properties, especially sodium tripolyphosphate, and are also used in baking powder, carbonated beverages, milk products, dental materials, fire-resistant products, metal surface treatment, and water treatment.

Sodium silicates. Sodium silicate is produced from silica sand and soda ash in a process similar to glass manufacture except the sodium silicates are water soluble and glass is not. The two raw materials are melted at 2 450°C in an open hearth furnace. The melt is withdrawn, cooled, crushed, and dissolved under pressure with steam. The three types of sodium silicates, and the corresponding quantity of soda ash needed to produce one ton of the chemical are: sodium orthosilicate (0.8), sodium metasilicate pentahydrate (0.33), and water glass (0.37). Sodium silicates are used in soaps and detergents, 29%; silica-type catalysts and gels, 28%; pigments, 22%; paper adhesives, 8%; water, paper, and ore treatment, 6%; and other, 7%. Twelve companies operating 34 plants have an annual production capacity of 1.133 Mt of sodium silicates.

Other sodium chemicals. There are many other sodium-based chemicals that are derived from soda ash. Sodium aluminate, sodium cyanide, sodium fluoride, and sodium sulfite are just a few examples.

Soaps and Detergents: The third largest end use of soda ash is for soap and detergent manufacture, accounting for 12% of total domestic consumption. Soda ash is used directly and indirectly in this sector with different functions. First, soda ash is added directly to different detergent formulations to serve as an agglomerating aid, surfactant carrier, and as a source of alkalinity. Secondly, soda ash is the feedstock to make other sodium-based chemicals used as builders. Builders are agents that enhance the cleaning action in the liquid washload by tieing up the calcium and magnesium ions to improve surfactant activity. The builders include sodium tripolyphosphate (which is an excellent builder but contributes to the environmental problem of eutrophication), sodium silicates, and synthetic zeolites. A combination of one or all of these are used in various detergent formulations. Lastly, soda ash is used to soften laundry water.

The end use data presented pertain to the direct use of soda ash in soap and detergent manufacture; the quantity of soda ash used to produce other sodium chemicals used in formulations are included in the chemicals sector, which is the primary, first-tier consumer. Soda ash is used in powdered home and industrial detergents but not in the newer liquid detergent formulations. Many states have legislation to restrict or ban the use of phosphatic detergents because of environmental considerations; therefore, detergent companies have modified product formulations to accommodate regional market restrictions. As a result, national sodium tripolyphosphate consumption has been decreasing, thereby reducing soda ash usage. To compensate for the decline, additional soda ash has been added to detergents as alternate medium-strength builders, such as in synthetic zeolites and in sodium silicates. Despite the growth of liquid detergents, which represent about 40% of all home laundry detergents, the net effect is that soda ash consumption in the soap and detergent sector remains strong.

Flue Gas Desulfurization: Flue gas desulfurization is a method to reduce sulfur dioxide emissions in stack gases from the burning of fuel. Soda ash, and other soda ash-based compounds such as sodium bicarbonate and sodium sesquicarbonate, are very effective dry sorbents of sulfur and nitrogen compounds. The reaction is as follows:

$$Na_2CO_3 + SO_2 + \frac{1}{2}O_2 \rightarrow Na_2SO_4 + CO_2$$

Compared to calcium-based compounds such as lime and limestone, soda ash-based scrubbing reagents have been proven to be more effective scrubbers because of their greater surface area that enhances the reaction with sulfur and nitrogen. However, soda ash remains more expensive and geographically restrictive than calcium materials. Most of the power plants that burn high-sulfur coal are located east of the Mississippi River where there are plentiful limestone quarries that can be mined more economically and transported shorter distances than soda ash produced in the West. Therefore, the quantity of soda ash sold for this end use is restrictive because of the economic advantages of lime and limestone. Some soda ash, however, is blended with lime for use in the dual-alkali process. The flue gas desulfurization market is 3% of total domestic soda ash sales.

Distributors: Distributors offer a variety of services to many soda ash consumers that may not be available from a soda ash producer directly, such as buying in bags, small volumes, or on an infrequent basis. Soda ash sold to distributors for resale comprised 6% of domestic reported consumption in 1991.

Pulp and Paper: Soda ash is used in pulp digestion and bleaching operations and in water treatment and paper coating formulations. Its main use is in making the sodium sulfite buffer solution used in the neutral sulfite pulp manufacturing process, in which the soda ash neutralizes the corrosive effects of the recovered black liquor. Although many pulp mills had used caustic soda because of its high sodium oxide content and its strong alkalinity, high prices and supply shortages of caustic soda prompted several mills to switch to soda ash that was less expensive and more available. The total market represented about 2% of US soda ash sales in 1992.

Water Treatment: Soda ash is used to chemically alter the pH of municipal and industrial water supplies and as a precipitant to remove impurities in brine and industrial process water. In the basic water treatment process, soda ash is added to adjust the acidity or alkalinity of water. Generally, it is added to acidic water to raise the pH and reduce the corrosivity of the water and the accumulation of mineral scale, thereby extending the life of metal pipes and equipment. One example of this application is in swimming pool chemicals where soda ash is added to adjust to pH of the water.

As a precipitant, it removes calcium and magnesium ions and other contaminants from water, which lessens the hardness of water. Most chlorine and caustic soda manufacturers pre-treat the sodium chloride-rich brine feedstock with soda ash to remove any calcium and magnesium impurities, thereby improving feedstock composition and product quality.

Since 1987, when the US Bureau of Mines began collecting reported soda ash consumption data, a certain quantity of soda ash equivalent contained in purge liquors, soda liquors, and mine waters have been sold to coal-fired power plants for process water treatment. These data are included under the water treatment category rather than under flue gas desulfurization.

Another unique soda ash application in water treatment involves the treatment of surface waters (e.g., lakes and rivers) contaminated from acid rain or natural causes. Although soda ash is a more expensive alkali compared with less expensive calcium-based reagents such as limestone, soda ash reportedly is more effective than lime in neutralizing the effects of acidification. The concept has been conducted in lakes in Sweden and New York with positive results; however, the practice has been criticized for merely treating the result of acid rain rather than attempting to prevent the spread of pollution from the source, which would involve flue gas desulfurization. The total quantity of soda ash used for all water treatment applications is only about 2% of total soda ash demand.

Miscellaneous Uses: Although soda ash is used in several minor applications, specific end use data is unavailable; however, the total category is 2% of domestic reported consumption. The uses include agriculture, brine treatment, corn syrup manufacture, drilling mud additives, dyes and pigments, enamels, food processing, leather tanning, metal refining, perfume, pharmaceuticals, and textiles.

Table 6. Time-Price-Value Relationships for Soda Ash (dollars per ton)

	Wyoming list price		Actual value[1]	Constant 1992 dollars[2]	
Year	short ton[3]	metric ton	metric ton	List price	Actual value
1970	49.00	54.01	23.18	186.03	79.84
1971	49.00	54.01	23.38	176.48	76.40
1972	35.00	38.58	24.56	120.21	76.53
1973	38.00	41.89	27.95	122.63	81.82
1974	42.00	46.30	37.34	124.67	100.54
1975	49.00	54.01	46.52	132.72	114.31
1976	51.00	56.22	54.78	129.96	126.63
1977	55.00	60.63	59.73	131.13	129.18
1978	61.00	67.24	60.09	134.81	120.48
1979	76.00	83.78	71.15	154.64	131.33
1980	86.00	94.80	99.04	159.85	167.00
1981	92.00	101.41	100.52	155.39	154.03
1982	84.00	92.59	97.39	133.58	140.51
1983	84.00[4]	92.59	84.82	128.37	117.60
1984	90.00[5]	99.21	73.85	131.81	98.12
1985	90.00[6]	99.21	74.76	127.06	95.75
1986	83.00[7]	91.49	71.97	114.15	89.80
1987	83.00	91.49	73.61	110.61	88.99
1988	93.00	102.51	73.81	119.28	85.89
1989	93.00	102.51	84.96	114.33	94.76
1990	98.00	108.03	91.33	115.68	97.80
1991	98.00	108.03	92.79	111.63	95.88
1992	98.00	108.03	89.21	108.03	89.21

[1]Values are the combined total revenue of California and Wyoming natural soda ash sold at list prices, spot prices, discount, long-term contracts, and for export, divided by the quantity of soda ash sold on a bulk, f.o.b. basis as reported to the US Bureau of Mines. The average annual value may not necessarily correspond to posted list prices for soda ash.
[2]Annual price and values, per metric ton, compared with final 1992 implicit price deflators for gross domestic product based on 1987=100.
[3]Wyoming soda ash always quoted in short tons.
[4]A $15 Temporary Voluntary Allowance (TVA) initiated lowers list price to $69 market price.
[5]TVA reduced to $11; list price increase brings price from 73 to $79.
[6]TVA reduced to $7; brings price to $83.
[7]TVA now included in price of $83.

ECONOMIC FACTORS

Prices

For three-quarters of the twentieth century, the synthetic soda ash industry with several nationwide plants, established the lead for setting the soda ash selling price leaving the smaller natural soda ash industry to follow by paralleling any price movement made by the former. Rising fuel costs after the 1973 energy crisis coupled with implementing pollution-abatement equipment required by environmental legislation, adversely affected the synthetic industry more than the natural operators, thereby creating a two-tier sales price; one for natural soda ash and another for synthetic soda ash.

Table 6 lists the historical time-price-value relationships of soda ash compared with 1992 constant dollars for gross domestic product. The posted list price of soda ash differs from the value that is reported to the US Bureau of Mines annually. The value may or may not correspond to the published list price for the commodity. The values which tend to be synonymous with the average market price are the combined total revenue of California and Wyoming natural bulk, dense soda ash sold on an FOB basis at list prices, spot prices, discount, long term contracts, and for export, divided by the

Table 7. US Soda Ash Statistics (thousand tons)

Year	Production: Natural	Production: Synthetic	Production: Total	Imports	Exports	World Production
1970	2 430	4 005	6 434	—	305	19 133
1971	2 599	3 878	6 477	—	396	19 815
1972	2 919	3 910	6 829	—	435	20 463
1973	3 376	3 482	6 858	9	386	21 368
1974	3 683	3 181	6 864	29	512	22 435
1975	3 927	2 542	6 468	2	480	23 365
1976	4 709	2 034	6 744	(1/2)	585	26 082
1977	5 650	1 642	7 292	1	689	26 443
1978	6 152	1 361[e]	7 513[e]	7	707	27 210
1979	6 699	787	7 487	37	905	28 188
1980	6 743	764	7 507	17	993	28 346
1981	6 795	718	7 513	11	953	28 014
1982	6 446	649	7 095	16	1 006	26 807
1983	7 007	674	7 681	18	1 484	27 892
1984	7 053	669	7 721	15	1 495	29 234
1985	7 092	669	7 721	51	1 584	29 142
1986	7 614	41	7 655	96	1 859	29 360
1987	8 066	0	8 066	136	2 017	30 203
1988	8 738	0	8 738	121	2 238	31 306
1989	8 995	0	8 995	129	2 648	32 032
1990	9 156	0	9 156	146	2 392	31 938
1991	9 005	0	9 005	134	2 734	31 053
1992	9 379	0	9 379	72	2 955	31 067

[e]estimated because two of three synthetic plants closed by yearend 1978; therefore data was considered proprietary.

aggregate quantity of soda ash sold. Bagging, handling, and insurance costs are not included.

Foreign Trade

Prior to the startup of synthetic soda ash manufacturing in the United States in 1884 and apart from limited natural soda ash production in the West, imported soda ash supplied the majority of domestic supply requirements for the glass and detergent industries. As domestic soda ash production capacity expanded, the United States rose from being import dependent to a net exporting nation. Table 7 lists the salient import and export data for the United States, including world soda ash production.

Imports: England had been a major source of soda ash imports during the late nineteenth century and into the twentieth century. Synthetic soda ash plants were built in Canada and Mexico to primarily satisfy those countries' requirements; however, because of their close geographic location, these nations began trading with the United States to solve supply shortages and/or to provide alternate soda ash grades. The Solvay plant in Amherstburg, ON, is a subsidiary of a US soda ash company, and provides light soda ash for certain applications in exchange for US dense material.

Exports: US soda ash exports have grown significantly in the last two decades because of world demand increasing, favorable currency exchange rates, lifting or easement in foreign tariffs and embargoes, and synthetic plant closures. Another important factor has been the creation of the domestic export association known as the American Natural Soda Ash Corp. (ANSAC), which was reorganized in 1983 from the Soda Ash Export Trading Association that was assembled in 1982. ANSAC has been very successful in raising export sales by consolidating the efforts of six individual US soda ash producers in the world market. In 1981, prior to ANSAC, exports were only 13% of US production, but by 1991 had risen to more than 30% of total output.

BIBLIOGRAPHY AND REFERENCES

Anon., 1921, "New Precedent Set As U.S. Wins Soda Lake Water Suit," *The Bulletin*, Nov. 21, 1921, San Francisco.

Anon., 1955, "Mining of Alkali Lake Sodium Deposits Begun," *The Ore-Bin*, Vol. 17, No. 9, Sep., p. 68.

Anon., 1972, "An Economic Analysis of a White Nahcolite Installation in Colorado," Option I, Open File Report 31–72, July, National Technical Information Service, PB-212 046, US Bureau of Mines, 69 pp.

Anon., 1986, "Chagannvoer Alkali Mine of Inner Mongolia," *China Economic Weekly*, p. 15.

Ahlfeld, F., and Schneider-Scherbena, A., 1964, "Los Yacimientos Minerales y de Hidercarbures de Bolivia," Bottetin No. 5, Minister de Mineras y Petroleo, La Paz.

Aitala, R., and Aitala, M., 1992, "Development of New Products in Wyoming's Trona Industry," *Proceedings*, 10th Industrial Minerals International Congress, San Francisco, pp. 24–26.

Aitala, R., and Henrickson, F., 1982, "The Natural-Synthetic Soda Ash Rivalry Intensifies—Fundamental Cost Structures for Natural and Synthetic Soda Ash Operations," *Proceedings*, 5th Industrial Minerals International Congress, Madrid, pp. J1–14.

Aitala, R., 1984, "Exports of Natural Soda Ash From the USA Explode in 1984," *Proceedings*, 6th Industrial Minerals International Congress, Toronto, 11 pp.

Alderman, S.S., Jr., 1983, "Geology of the Owens Lake Evaporite Deposit," 6th Symposium on Salt, B.C. Schreiber and H.L. Harner, eds., Vol. 1, The Salt Institute, Alexandria, VA, pp. 75–83.

Allison, I.S., 1947, "Sodium Salts of Lake County, Oregon," GMI Short Paper No. 17, Oregon Dept. of Geology and Mineral Industries, 12 pp.

Atia, A.K.M., Hilmy, E.E., and Bolous, S.M., 1970, "Mineralogy of the Salt Deposits of Wadi El Natrun," Arab Republic of Egypt, Desert Institute Bulletin, Vol. 20, pp. 49–74.

Averill, C.V., 1947, "Mines and Mineral Resources of Lake County, Cal," *California Journal of Mines and Geology*, Vol. 43, No. 1, pp. 16–40.

Baker, B.H., 1958, "Geology of the Magadi Area," Report 42, Kenya Geological Survey, 81 pp.

Batalin, Y.V., Stankevich, Y.F., and Kasimov, B.S., 1974, "Distribution Patterns and Possible Discoveries of Deposits of Natural Soda in the USSR," *International Geology Review*, Vol. 16, No. 4, pp. 446–458.

Barbour, E.H., 1916, "A Preliminary Report on the Alkali Resources of Nebraska," Miscellaneous Papers, Vol. 4, Pt. 28, Nebraska Geogical Survey, pp. 405–438.

Barnes, J.W., ed., 1961, "The Mineral Resources of Uganda," Bulletin No. 4, Geological Survey of Uganda, 46 pp.

Baumann, A., Förstner, U., and Rohde, R., 1975, "Lake Shala: Water Chemistry, Mineralogy, and Geochemistry of Sediments in an Ethiopian Rift Lake," *Geol. Rdsch.*, Vol. 64, pp. 593–609.

Bennett, W.A.G., 1962 "Saline Lake Deposits of Washington," Bulletin No. 49, Washington Division of Mines and Geology, pp. 19–49.

Bernasconi, O.H., and Cangioli, 1982, "Geoeconomic Possibilities of Saline Manifestations of Western Santa Cruz, Argentina," Quinto Congreso Latin America de Geologea, Vol. 2, pp. 597–610; Vol. 5, pp. 369–380.

Bond, G.W., 1946, "A Geochemical Survey of the Underground Water Supplies of the Union of South Africa," Memoir No. 41, Union of South Africa Geological Survey, pp. 44–47.

Botha, J.C., 1977, "The World Soda Ash Situation Excluding South Africa," Internal Report No. 11, Republic of South Africa Department of Mines, 71 pp.

Bradley, W.H., 1964, "Geology of the Green River Formation and Associated Eocene Rocks in Southwestern Wyoming and Adjacent Part of Colorado and Utah," Professional Paper 496-A, US Geological Survey, 86 pp.

Bradley, W.G., and Eugster, H.P., 1969, "Geochemistry and Paleolimnology of the Trona Deposit and Associated Authogenic Minerals of the Green River Formation of Wyoming," US Geological Survey, Professional Paper 496-B, 71 pp.

Bui, E.N., and Wilding, L.P., 1988, "Pedogenesis and Mineralogy of a Halaquept in Niger," *Geoderma*, Vol. 43, No. 1, pp. 49–64.

Bukhshtein V.M., and Zakina, A.E., 1949, "The Solubility Polytherm of Sodium Carbonate in the Brine of the Mikhailov Lakes," Transactions, All-Union Institute for the Study of Salts, No. 21, pp. 223–242, Chemical Abstracts, Vol. 46, 1952, p. 8911.

Burnside, M.J., and Culbertson, W.C., 1979, "Trona Deposits in the Green River Basin, Sweetwater, Uinta, and Lincoln Counties, Wyoming," Open File Report 79–737, US Geological Survey, 10 pp.

Carstens, G.L., 1982, ''Sodium Chemicals—Owens Lake, California,'' No. 42, Northwest Mining Convention, 21 pp.

Catalano, L.R., 1926, ''Geologia Economica de Los Yacimientes de Borate y Materiales de las Cuencas—Solar Cauchari,'' Publication No. 23, Direccion General de Minas Geologia y Hidrologica, Buenos Aires, 110 pp.

Chadwick, J., 1991, ''Big Island Trona Mine,'' *Mining Magazine,* Vol. 165, No. 1, July, pp. 6–10.

Chatard, T.M., 1890, ''Natural Soda: Its Occurrence and Utilization,'' Bulletin 60, US Geological Survey, pp. 162–163, 237–243.

Christie, W.A.K., 1912, ''Note on the Lonar Soda Deposit,'' Records, Geological Survey of India, Pt. 4, Vol. 41, pp. 276–285.

Clarke, F.W., 1924, ''The Data of Geochemistry,'' Bulletin 770, US Geological Survey, 841 pp.

Clautice, K.H., and Mowatt, T.C., 1981, ''Trona Occurrences Within the Yukon Flats Basin, Alaska,'' Open File Report 69–81, US Bureau of Mines, 34 pp.

Cordini, I.R., 1967, ' Reservas Salinas de Argentina,'' Anales No. 13, Instituto Nacional de Geologia y Mineria, Buenos Aires, 108 pp.

Cotter, G.D., 1923, ''The Alkaline Lakes and the Soda Industry of Sind,'' Memoir 47, Pt., 2. Geological Survey of India, pp. 202–311.

Culbertson, W.C., 1966, ''Trona in the Wilkins Peak Member of the green River Formation, Southwestern Wyoming,'' Professional Paper 550-B, US Geological Survey, pp. B159–164.

Culbertson, W.C., 1971, ''Stratigraphy of the Trona Deposits in the Green River Formation, Southwest, Wyoming,'' *Contributions to Geology,* Vol. 10, No. 1, R.B. Parker, ed., University of Wyoming, Laramie, pp. 25–27.

Cummings, J.M., 1949, ''Saline and Hydromagnesite Deposits of British Columbia,'' Bulletin No. 4, 1940, British Columbia Department of Mines, 160 pp.

Deardorff, D.L., and Mannion, L.E., 1971, ''Wyoming Trona Deposits,'' Contributions to Geology, Vol. 10, No. 1, R.B. Parker, ed., University of Wyoming, Laramie, pp. 25–27.

Dewey, J.R., and Aman, J.P., 1976, ''Longwall Mining of Trona,'' *Mining Congress Journal,* Vol. 62, No. 12, pp. 52–75.

Dort, W., Jr., and Dort, D.S., 1970, ''Low Temperature Origin of Sodium Sulfate Deposits, Particularly in Antarctica,'' 3rd Symposium on Salt, Vol. 1, J.L. Rau and L.F. Dellwig, eds., Northern Ohio Geological Society, Cleveland, pp. 181–203.

Dub, G.D., 1947, ''Owens Lake—Source of Sodium Minerals,'' Technical Publication No. 2235, AIME, 13 pp.

Dundar, A., and Tenekeci, O., 1987, ''The Geology and Some Mineralogical Characteristics of Ankara- Beypazari (Natural Soda) Bed,'' Industrial Minerals of Turkey Conference, Istanbul, Sep., 17 pp.

Dundar, A., 1988, ''Ankara-Beypazari Trona Deposit—Geology and Mineralogical Characteristics,'' *Industrial Minerals,* No. 246, Mar., pp. 6–9.

Durand-Chastel, M.H., 1968, ''L'exploitation de Saumures de Carbonate de Soude au Mexique,'' *Science et Techniques,* No. 12, pp. 23–30.

Dutoit, A.L., 1948, *The Geology of South Africa,* Hafner, New York, p. 428.

Dyni, J.R., 1980, ''Colorado Nahcolite Deposits: Geology and Outlook For Development,'' Preprint 80–83, SME-AIME Annual Meeting, Las Vegas, Feb., 11 pp.

Dyni, J.R., Milton, C., and Cashion, W.B., Jr., 1985, ''The Saline Facies of the Upper Part of the Green River Formation Near Duchesne, Utah,'' *Geology and Energy Resources, Uinta Basin of Utah,* M.D. Picard, ed., Utah Geological Association, Salt Lake City, pp. 51–60.

Eugster, H.P., 1970, ''Chemistry and Origin of the Brines of Lake Magadi, Kenya,'' Special Paper, Vol. 3, Mineralogical Society of America, pp. 213–235.

Eugster, H.P., 1971, ''Origin and Deposition of Trona,'' *Contributions to Geology,* Vol. 10, No. 1, R.B. Parker, ed., University of Wyoming, Laramie, pp. 15–23.

Eugster, H.P., 1980a, ''Geochemistry of Evaporitic Lacustrine Deposits,'' *Annual Review, Earth and Planetary Sciences,* Vol. 8, pp. 35–63.

Eugster, H.P., 1980b, ''Lake Magadi, Kenya, and Its Precursors,'' *Hypersaline Brines and Evaporitic Environments,* Elsevier, Amsterdam, pp. 195–232.

Eugster, H.P., and Jones, B.F., 1979, ''Behavior of Major Solutes During Closed-Basin Brine Evolution,'' *American Journal of Science,* Vol. 279, pp. 609–631.

Eugster, H.P., and Maglione, G., 1979, ''Brines and Evaporites of the Lake Chad Basin, Africa,'' *Geochemica et Cosmochimica Acta,* Vol. 43, pp. 973–981.

Eugster, H.P., and Smith, G.I., 1965, ''Mineral Equilibria in the Searles Lake Evaporites,'' *Journal of Geology,* Vol. 6, pp. 473–522.

Eugster, H.P., and Surdam, R.C., 1973, ''Depositional Environment of the Green River Formation of Wyoming: A Preliminary Report,'' *Geological Society of America Bulletin,* Vol. 84, pp. 1115–1120.

Fahey, J.J., 1962, ''Saline Minerals of the Green River Formation,'' Professional Paper 405, US Geological Survey, 50 pp.

Farag, A.A., Mulla, T.H., and Skander, J.L., 1975, ''Concentration of Sodium Carbonate From Trona by Flotation,'' Bulletin, Faculty of Engineering, University of Alexandria, Egypt, *Chemical Engineering,* Vol. 14, No. 1, pp. 31–41.

Fawley, A.P., 1958, ''Diamond Drilling in the Bahi Swamp,'' *Records,* Tanganyika Geological Survey, Vol. 6 (1956), pp. 47–51.

Fife, W.E., 1974, ''Continuous Mining of Trona Ore,'' *4th Symposium on Salt,* Vol. 1, A.H. Coogan, ed., Northern Ohio Geological Society, Cleveland, pp. 483–489.

Fredrikson, K., et al., 1973, ''Lonar Lake, India: An Impact Crater in Basalt,'' *Science,* Vol. 180, pp. 862–864.

Free, E.E., 1913, ''Progress in Potash Prospecting in Railroad Valley, Nevada,'' *Mining and Scientific Press,* Vol. 107, No. 5, Aug., pp. 176–178.

Freise, F.W., 1932, ''Beobachtungen au Brasilianischen Salzsumpfen,'' *Chemie der Erde,* Band 7, Heft 1, pp. 24–34.

Frint, W.R., 1971, ''Processing of Wyoming Trona,'' *Contributions to Geology,* Vol. 10, No. 1, R.B. Parker, ed., University of Wyoming, Laramie, pp. 25–27.

Frint, W.R., 1985, ''FMC's Newest Goal: Commercial Solution Mining of Trona,'' *Engineering and Mining Journal,* Vol. 186, No. 9, pp. 26–35.

Gale, H.S., 1914, ''Salines in the Owens, Searles and Panamint Basins, Southeastern California,'' Bulletin 580L, US Geological Survey, pp. 251–323.

Garrels, R.M., and Mackenzie, F.T., 1967, ''Origin of the Chemical Composition of Some Springs and Lakes,'' *Equilibrium Concepts in Natural Water Systems,* Advances in Chemistry Series 57, American Chemical Society, Washington, pp. 222–242.

Garrett, D.E., 1992, *Natural Soda Ash—Occurrences, Processing, and Use,* Van Nostrand Reinhold, New York, 636 pp.

Garrett, D.E., and Phillips, J.F., 1960, ''Sodium Carbonate from Natural Sources in the United States,'' *Industrial Minerals and Rocks,* 3rd ed., J.L. Gillson, ed., AIME, New York, pp. 799–809.

Gonzalez-Reyna, J., 1956,''Riqueza Minera y Yacimiento,'' *Minerales de Mexico,* 3rd ed., 20th Congreso Geologico Internacional, Banco de Mexico S.A. Dept. de Investigaciones Industriales, Mexico, pp. 431–439.

Guest, N.J., 1956, *Records,* Geological Survey of Tanganyika, Vol. 4, p. 7.

Harben, P.W., and Bates, R.L., 1984, ''Sodium Carbonate,'' *Geology of the Nonmetallics,* Metal Bulletin, New York, pp. 225–230.

Hardie, L.A., and Eugster, H.P., 1970, ''The Evolution of Closed-Basin Brines,'' Special Paper 3, Mineralogical Society of America, pp. 273–290.

Harness, C.L., and Coons, A.T., 1942, ''Sodium Carbonate,'' Information Circular 7212, June, US Bureau of Mines, 33 pp.

Harris, R.E., 1984, ''An Overview of the Geology and Production of Wyoming Trona,'' Preprint 84–326, SME-AIME Fall Meeting, Denver, Oct., 11 pp.

Horino, F.G., and Hooker, V.E., 1978, ''Mechanical Properties of Cores Obtained from the Unleached Saline Zone, Piceance Creek Basin Rio Blanco County, CO,'' Report of Investigations 8297, US Bureau of Mines, 21 pp.

Hou, T.P., 1942, *Manufacture of Soda,* Monograph Series No. 65, American Chemical Society, Reinhold, New York, 590 pp.

Jackson, D., 1981, ''Trona Solution Mining—FMC Goes After Deep Wyoming Deposits,'' *Engineering and Mining Journal,* Vol. 182, No. 12, pp. 68–69.

Jakucs, L., 1977, *Morphogenetics of Karst Regional (Variants of Karst Evolution),* trans. from Hungarian B. Balhay, Halsted Press, New York, pp. 29–33.

Jones, B.F., 1965, ''The Hydrology and Mineralogy of Deep Springs Lake, Inyo County, California,'' Professional Paper 502-A, US Geological Survey, 55 pp.

Karajian, H.A., 1920, Mineral Resources of Armenia and Anatolia, Armen Technical Book Co., New York, pp. 121–125.
King, C., 1948, "Soda Ash and Saltcake in California," California Journal of Mines and Geology, Vol. 44, pp. 189–200.
Klingman, C.L., 1972, "Sodium and Sodium Compounds," *Minerals Yearbook 1972,* US Bureau of Mines, pp. 1147–1151.
Kostick, D.S., 1992, "Soda Ash," *Mining Engineering,* Vol. 44, No. 6, p. 573.
Kostick, D.S., 1992, "Soda Ash," *Mineral Commodity Summaries—1992,* US Bureau of Mines, pp. 162–163.
Kostick, D.S., 1991, "Soda Ash," *Annual Report—1990,* Sep. 1991, Preprint, US Bureau of Mines, 21 pp.
Kostick, D.S., 1985, "U.S. Soda Ash Industry—The Next Decade," *Mining Engineering,* Vol. 37, No. 10, pp. 1209–1212.
Kostick, D.S., 1985, "Soda Ash and Sodium Sulfate," *Mineral Facts and Problems,* Bulletin 675, US Bureau of Mines, pp. 741–755.
Kostick, D.S., 1983, "U.S. Short-Term Demand Outlook for Soda Ash and Factors Affecting Availability in the World Market," *Chemistry and Industry,* No. 4, Feb. 21, pp. 161–165.
Kostick, D.S., 1983, "The U.S. Soda Ash Industry—Past, Present, and Future," 5th Industrial Minerals International Congress, Madrid, pp. 161–165.
Kostick, D.S., 1983, "The Economic Significance of the Forms of Trona Occurrences," The Wyoming Geological Association Earth Science Bulletin, Vol. 16, pp. 1–8.
Kostick, D.S., 1981, "The Influence of Solution Mining on the World Soda Ash Market," Preprint 81–368, SME-AIME Fall Meeting, Denver, Nov., 8 pp.
Kostick, D.S., and Foster, R.J., 1979, "Soda Ash (Sodium Carbonate), Sodium Sulfate, and Sodium," Mineral Commodity Profiles, US Bureau of Mines, 14 pp.
Kuchin, M.I., 1928, "Salt Lakes of the Slavgorod Region in Siberia," Journal of Chemical Industry (Moscow), Vol. 5, pp. 292–297; Chemical Abstracts, Vol. 22, p. 3963.
Lee, C.H., and Clark, W.O., 1916, "Report on Soda Lakes Investigation: Truckee-Carson Project Near Fallon, Nev.," US Geological Survey, pp. 657–706.
Lefond, S.J., 1981, "Botswana—An Opportunity," Preprint 81–73, SME-AIME Annual Meeting, Chicago, IL, Feb., 3 pp.
Leigh, R.T., 1989, "Trona: An Overview," Preprint 89–115, SME-AIME Annual Meeting, Las Vegas, Feb.-Mar., 15 pp.
Liddle, R.A., 1946, *The Geology of Venezuela and Trinidad,* 2nd ed., Paleontological Research Institution, Cayuga Press, Ithaca, NY, 589 pp.
Lindeman, H.B., 1954, "Sodium Carbonate Brine and Trona Deposits in Sweetwater County, Wyoming," Circular 235, US Geological Survey, 10 pp.
Lindsay, J.D., 1938, "Producing Natural Soda Ash in Washington," Chemical and Metallurgical Engineering, Vol. 45, No. 3, pp. 144–145.
Livingston, D.A., 1963, "Chemical Composition of Lakes and Rivers," Chap. G, *Data of Geochemistry,* 6th ed., M. Fleischer, ed., Professional Paper 440-G, US Geological Survey, pp. 617, 633.
Lucas, A., 1912, "Natural Soda Deposits in Egypt," Paper No. 22, Ministry of Finance, Survey Department, Government Press, Cairo, 29 pp.
McCall, G.J., 1967, "Geology of the Nakuru-Thompson Falls-Lake Hannington Area," Report No. 78, Kenya Geological Survey, pp. 1–122.
McLeod, J.E., 1975, "Trona Mine—Soda Ash Production," *The Glass Industry,* Vol. 50, No. 8, pp. 12–16.
Maglione, G., 1972, "New Data on the Processes at Present Giving Rise to Saline Concentrations on the North-East Periphery of Lake Chad," *Geology of Saline Deposits,* G. Richter-Bernberg, ed., UNESCO, Paris, pp. 211–213.
Makanjuola, A.A., and Beetlestone, J.G., 1973, "Some Chemical and Mineralogical Notes on Kaum," *Journal of Mineralogy and Geology,* Vol. 10, No. 1–2, pp. 31–41.
Mannion, L.E., 1975, "Sodium Carbonate Deposits," *Industrial Minerals and Rocks,* 4th ed., S.J. Lefond, ed., AIME, New York, pp. 1061–1079.
Mannion, L.E., 1975, "Wyoming Trona," Preprint 75-H-309, SME-AIME Fall Meeting, Salt Lake City, Sep., 15 pp.
Mannion, L.E., 1983, "Sodium Carbonate Deposits," *Industrial Minerals and Rocks,* 5th ed., S.J. Lefond, ed., AIME, New York, pp. 1187–1206.
Massey, N.W.D., 1973, "Resources Inventory of Botswana: Industrial Minerals and Rocks," Mineral Report No. 3, Geological Survey Dept., Lobatse, Botswana, 30 pp.
Milani, J.L., Jr., 1983, "The Development of Natural Soda Ash in the United States," *Chemistry and Industry,* No. 4, Feb. 21, pp. 1–7.
Milton, C., 1971, "Authigenic Minerals in the Green River Formation," *Contributions to Geology,* Vol. 10., No. 1, R.B. Parker, ed., University of Wyoming, Laramie, pp. 57–63.
Morrison, R.B., 1964, "Lake Lahontan: Geology of Southern Carson Desert, Nevada," Professional Paper 401, US Geological Survey, p. 116.
Moulton, G.F., Jr., 1981, "Compendium of Searles Lake Operations," *Trans. AIME,* Vol. 270, pp. 1918–1922.
Murdock, T.G., 1949, "Industrial Minerals of Ethiopia," *Trans. AIME,* Vol. 181, pp. 376–384.
Niinomy, K., 1930, "A New Locality of Gaylussite in Eastern Mongolia with Associated Natural Soda," *Economic Geology,* Vol. 25, No. 7, pp. 758–763.
Orr, D., and Grantham, D.R., 1931, "Some Salt Lakes of the North Rift Zone," Short Paper No. 8, Geological Survey Dept., Tanganyika Territory, 21 pp.
Papke, K.G., 1976, "Evaporites and Brines in Nevada Playas," Bulletin 87, Nevada Bureau of Mines and Geology, pp. 25–27.
Pelletier, R.A., 1964, *Mineral Resources of South-Central Africa,* Oxford University Press, Oxford, p. 131.
Perelman, A.I., 1967, *Geochemistry of Epigenesis,* trans. from Russian N.N. Kohanowski, Plenum Press, New York, pp. 176–181.
Phelps, R.W., 1990, "Trona—A Tale of Two Companies," *Mining Engineering,* Vol. 191, No. 11, pp. 20- 25.
Phillips, K.N., and Van Denburgh, A.S., 1971, "Hydrology and Geochemistry of Abert, Summer, and Goose Lakes and Other Closed-Basin Lakes in South-Central Oregon," Professional Paper 502-B, US Geological Survey, 86 pp.
Post, L.N., 1985, "FMC's Westvaco Soda Ash Operation Uses A Variety of Mining Techniques," *Mining Engineering,* Vol. 37, No. 10, pp. 1200–1203.
Purcell, R.W., 1982, "Soda Ash—The View From Europe," 5th Industrial Minerals International Congress, Madrid, Apr., 21 pp.
Raleigh, C.W., 1985, "Sodium Carbonate," Ceramic Bulletin, Vol. 64, No. 5, pp. 665–667.
Rau, E., 1969, "Sodium Compounds," *Encyclopedia of Chemical Technology,*" Vol. 18, 2nd ed., Kirk-Othmer, eds., Interscience, New York, pp. 458–468.
Rauh, F., 1991, "Alkali and Chlorine Products—Sodium Carbonate," *Encyclopedia of Chemical Technology,* Vol. 1, 4th ed., Kirk-Othmer, eds., Wiley, New York, pp. 1025–1039.
Refsdal, L.C., 1991, "Trona Mining at Tenneco Soda Ash Operation," 97th Annual Northwest Mining Association Convention, Spokane, WA, Dec. 4–6, 1991, 15 pp.
Rettig, S.L., and Jones, B.F., 1965, "Determination of Carbonate, Bicarbonate, and Total CO_2 in Carbonate Brines," Professional Paper 501-D (1964), US Geological Survey, pp. D134–137.
Risacher, F., Echenique, A., and Ballivian, O., 1978, "Informe Geologico—Economico de la Laguna Collpa y de la Laguna Hedronda," unpublished paper 19 pp.
Roy, B.C., 1959, "The Economic Geology and Mineral Resources of Rajasthan and Ajmer," Memoirs, Geological Survey of India, Vol. 86, pp. 271–296.
Sahni, M.R., 1951, "Alkaline Soils and Lakes of India," Records, Geological Survey of India, Vol. 81, Pt. 2, pp. 241–296.
Said, R., 1962, The Geology of Egypt, Elsevier, Amsterdam, p. 272.
Sayre, J.A., 1964, "A Bonanza in Borax: The Story of John Wemple Searles and Searles Lake," *Journal of the West,* Vol. 3, No. 3, July, pp. 277–290.
Schneider, G.I.C., and Genis, G., 1991, "Soda Ash and Thenardite," Open File Report MRS 24, Namibia Ministry of Mines and Energy, 18 pp.
Schultz, A.R., 1910, "Deposits of Sodium Salts in Wyoming," Bulletin 430, US Geological Survey, pp. 570–589.
Schultze, G., 1953, "Los Conglomerados Terciario de la Region Lagunera de Durango y Coahuila, y Sus Relaciones Igneas. Geomorfológicas y Climatologicas," Boletin 30, Cons. de Rec. Nat. No Renovables.
Shabad, T., and Sagers, M.J., 1987, "The Chlor-Alkali Industries in the USSR," *Soviet Geography,* Vol. 28, June 1987, p. 437.
Shamel, R.E., 1992, "The Alkali Imbalance: Implications for Soda Ash and

Calcined Trona,'' *10th Industrial Minerals International Congress,* San Francisco, pp. 27–30.

Shamel, R.E., 1986, ''The Global Soda Ash Business: Review and Forecast,'' *7th Industrial Minerals International Congress,* Vol. 1, Monte Carlo, pp. 191–198.

Singewald, J.T., 1949, ''Mineral Resources of Colombia,'' Bulletin 964-B, US Geological Survey, pp. 186–192.

Smith, G.I., 1964, ''Character and Distribution of Non-clastic Minerals in the Searles Lake Evaporite Deposit, California,'' Bulletin 1181-P, US Geological Survey, 58 pp.

Smith, G.I., 1966, ''Geology of Searles Lake—A Guide to Prospecting for Buried Continental Salines,'' *2nd Symposium on Salt,* Vol. 1, J.L. Rau, ed., Northern Ohio Geological Society, Cleveland, pp. 137–140.

Smith, G.I., 1979, ''Subsurface Stratigraphy and Geochemistry of Late Quaternary Evaporites, Searles Lake, California,'' Professional Paper 1043, US Geological Survey, 130 pp.

Smith, G.I., and Friedman, I., 1975, ''Chemical Sedimentation and Diagenesis of Pleistocene Evaporites in Searles Lake, California,'' *9th International Sedimentation Congress,* Nice, pp. 137–140.

Smith, J.W., and Milton, C., 1966, ''Dawsonite in the Green River Formation of Colorado,'' *Economic Geology,* Vol. 61, No. 6, Sep.-Oct., pp. 1029–1042.

Stanley, D.J., and Sheng, H., 1979, ''Trona in Nile Cone Late Quaternary Sediments: Probable Redepositional Origin,'' *Marine Geology,* Elsevier, Amsterdam, pp. M21–28.

Steele, R.J., 1971, ''The Mining of Wyoming Trona,'' *Contributions to Geology,* Vol. 10, No. 1, R.B. Parker, ed., University of Wyoming, Laramie, pp. 39–41.

Stoffers, P., and Holdship, S., 1975, ''Diagenesis of Sediments in an Alkaline Lake: Lake Manyara, Tanzania,'' *9th International Congress of Sedimentology,* Nice, pp. 211–216.

Surdam, R.C., and Wolfbauer, C.A., 1975, ''Green River Formation, Wyoming, A Playa lake Complex,'' Geological Society of America, Bulletin, Vol. 86, pp. 335–345.

Sutherland, N., Ward, V.C., and Carman, E.H., 1987, ''The Separation of Sodium Salts,'' South African Patent ZA 879599, Dec. 22, 14 pp.

Takeuchi, M., 1990, ''The Current Aspect on Supply/Demand of Soda Ash in Japan,'' *1st East Asia Chlor-Alkali Industry Conference,* Tokyo, Apr., pp. 39–42.

Talling, J.F., and Talling, I.B., 1965, ''The Chemical Composition of African Lake Waters,'' *Int. Rev. Ges. Hydrobiol.,* Vol. 50, pp. 421–463.

Taylor, S.L., 1987–1988, ''Nevada's Underwater Mystery in the Desert,'' *In Focus,* Annual Journal of the Churchill County Museum Association, Fallon, NV, Vol. 1, No. 1, pp. 14–18.

Torgasheff, B.P., 1929, ''Soda in China, Manchuria, and Neighboring Countries,'' *Chinese Economic Journal,* Bureau of Industrial and Commercial Information, National Government of the Republic of China, Vol. 5, No. 2, Aug., pp. 662–681.

Tsunashima, K., and Nakaya, K., 1982, ''The New Asahi (NA) Process for Synthetic Soda Ash Production,'' *5th Industrial Minerals International Congress,* Madrid, pp. H1–10.

Tulus, R., 1944, ''Die Analyse des Wassers vom Van-see (Ostanatolium),'' *Review Faculte Science,* University of Istanbul, Vol. 9, pp. 61–63.

Van Andel, E., 1983, ''An Electrolysis-Free Route to Vinyl Chloride Monomer and Soda Ash,'' *Chemistry and Industry,* Feb. 21, No. 4, pp. 139–141.

Vanderburg, W.O., 1940, ''Reconnaissance of Mining Districts in Churchill County, Nev.,'' Information Circular 7093, US Bureau of Mines, pp. 43–45.

Ver Planck, W.E., 1959, ''Soda Ash Industry of Owens Lake—1887–1959,'' *Mineral Information Service,* Vol. 12, No. 10, Oct., 6 pp.

Wagner, P.A., 1922, ''The Pretoria Salt-Pan—A Soda Caldera,'' Union of South Africa Geological Survey, Memoir No. 20, 136 pp.

Wang, J., 1987, ''A Preliminary Study of the Characteristics and Conditions for Forming the Anpeng Trona Deposit,'' *Petroleum Exploration and Development,* Vol. 5, pp. 93–99.

Watson, I., 1981, ''US Soda Ash—Green River's Solution—Europe's Problem,'' *Industrial Minerals,* No. 162, Mar., pp. 37–38.

Williams, L.A., 1972, ''Geology of the Amboseli Area,'' Kenya Geological Survey Report No. 90, 86 pp.

Youxun, Z., 1985, ''Geology of the Wucheng Trona Deposit in Henan, China,'' *6th Symposium on Salt,* Vol. 1, B.C. Schreiber and H.L. Harner, eds., The Salt Institute, Alexandria, VA, pp. 67–73.

Yuretich, R.F., 1979, ''Modern Sediments and Sedimentary Processes in Lake Rudolf, Eastern Rift Valley, Kenya,'' *Sedimentology,* Vol. 26, pp. 313–331.

Zeller, A.F., 1991, ''Trona—Soda Ash Capacity Increasing,'' *Engineering and Mining Journal,* Vol. 192, No. 3, pp. 51–53.

Zhang, M., et al. 1989, *Saline Lakes on the Qinghai-Xizang (Tibet) Plateau,* Beijing Science and Technique Press, Beijing, pp. 36–43.

SODIUM SULFATE RESOURCES

SID McILVEEN, JR. AND ROBERT L. CHEEK, JR.

Sodium sulfate is an important industrial chemical. As recently as ten years ago it was produced and consumed in the United States in quantities exceeding 1 Mtpy. Since then, both its production and use have declined; however, approximately half the production still comes from natural sources. Fig. 1 illustrates the history of production of natural sodium sulfate in the United States. Production of natural sodium sulfate from various types of deposits is the main source of this chemical in Canada and Mexico, and probably in Argentina, Chile, Iran, Spain, and the Russian Republics.

MINERALOGY AND PHYSICAL PROPERTIES

Sodium sulfate is widespread in occurrence and is a common constituent of many mineral waters, as well as seawater. Atmospheric precipitation contains sulfate; it is one of the major dissolved constituents of rain and snow (Davis and DeWiest, 1966). Many of the saline lakes throughout the world contain varying amounts of sodium sulfate. Because sodium is usually the dominant cation, some workers make an anionic distinction, referring to lakes containing predominantly sulfate as bitter lakes and those containing predominantly carbonate as alkali or soda lakes (Bateman, 1950). Sodium sulfate in its natural form is found in two principal minerals, mirabilite and thenardite. Mirabilite, the hydrous form, is commonly called Glauber's salt. It was discovered by the German chemist, J.R. Glauber (1603–1668), who derived its name from the Latin, *sal mirabile*, meaning wonderful salt. Thenardite, the anhydrous variety, was named for the French chemist, Louis Jacques Thenard (1777–1857) of the University of Paris (Mitchell, 1979). The largest quantities occur in the form of mirabilite. Sodium sulfate is found in varying degrees of purity, from theoretically pure efflorescent crystals of mirabilite to combinations and admixtures of other salts and impurities. It is a common constituent of some brines; from this source much is extracted commercially.

Sodium sulfate also is found in compounds, such as the minerals glauberite, the double salt of anhydrous sodium and calcium sulfate, bloedite, the hydrous double-salt of sodium sulfate and magnesium sulfate, and burkeite, the anhydrous double-salt of sodium carbonate and sodium sulfate.

Over 40 minerals contain sodium sulfate in varying proportions; many are of special interest because of their frequent occurrence.

Table 1 lists some sodium-sulfate bearing minerals. The reader is referred to other publications (Cole, 1926, Dana, 1932, Grabau, 1920, and Dietrich, 1969) for descriptions of these minerals. Only mirabilite and thenardite will be described herein. Mirabilite $Na_2SO_4 \cdot 10H_2O$, contains 55.9% water of crystallization. It is noted for its efflorescence or spontaneous loss of water. On dehydration it changes to the anhydrous form, Na_2SO_4. Mirabilite is an opaque to colorless, water-soluble mineral that tastes first cool, then slightly bitter. It has a specific gravity of 1.48. It frequently forms as efflorescent, needlelike monoclinic crystals, but generally is found in the massive form.

Thenardite, the anhydrous mineral, Na_2SO_4, contains 43.68% Na_2O and 56.32% SO_3. It ranges from colorless to white and may be tinted shades of gray or brown. It is a water soluble mineral with a slightly salty taste. Its specific gravity (2.67) and hardness (2.5 to 3) exceed those of mirabilite. It commonly occurs in the massive form without visible crystals. Its crystals are frequently tabular pyramids of the orthorhombic system.

Sodium sulfate also occurs as a heptahydrate, containing seven molecules of water, but this is unstable and has not been found in the natural environment.

The solubility of sodium sulfate has an important effect on the crystallization of the salt in nature, as well as in its production. Its solubility in water generally increases as a nonlinear function of temperature. Below 1.2°C, ice and mirabilite form. As the temperature is increased above 0°C, increasing amounts of sodium sulfate become soluble. At 32.4°C, a transition point on the solubility curve is reached, as the decahydrate melts in its own water of crystallization and the anhydrous form crystallizes. With increasing temperatures, solubility decreases somewhat. The presence of other dissolved salts changes the transition temperature and solubility characteristics of sodium sulfate.

CLASSIFICATION OF DEPOSITS

Essentially all commercial deposits of sodium sulfate have resulted from accumulation and evaporation of surface and ground water in basins with interior drainage called playas. They are found in arid to semiarid regions. Thus these deposits fall within the broad classification of evaporites. They can be further classified in two categories: those occurring as crystalline beds of mirabilite or glauberite and/or brines within or underlying playa lakes; and those found as buried beds of thenardite, glauberite, and associated minerals. The first category includes the principal deposits of the world. Notable among these are the lake deposits of western Canada, Great Salt Lake, UT, the vast mirabilite-glauberite deposits of the Kara-Bogaz-Gol Gulf on the eastern shore of the Caspian Sea, and the brine deposits of the western United States and Mexico. These require pumping, dredging, or reservoiring for exploitation.

Commercial deposits of the second category are relatively rare. Deposits of thenardite and glauberite are presently being mined in Spain, and thenardite has been mined in the past at Camp Verde, AZ, and at Rhodes Marsh near Mina, NV. Conventional mining techniques are used for extraction.

ORIGIN OF SODIUM SULFATE DEPOSITS

Evaporation of surface waters, with consequent concentration of salts, is the principal agent in the formation of alkali and bitter

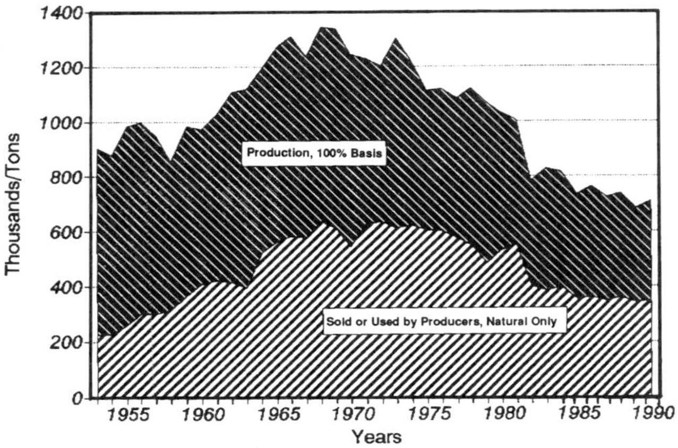

Fig. 1. Graph of sodium sulfate production in the United States, manufactured and natural.

Table 1. A List of Minerals That Contain Sodium Sulfate

Mineral	Composition	$\%Na_2SO_4$
Thenardite (anhydrous)*	Na_2SO_4	100
Hanksite	$9Na_2SO_4 \cdot 2Na_2CO_3 \cdot KCl$	81.7
D'ansite	$9Na_2SO_4 \cdot MgSO_4 \cdot 3NaCl$	81.2
Lecontite	$(Na,NH_4,K)_2\ SO_4 \cdot 2H_2O$	<79.8
Vanthoffite	$3Na_2SO_4 \cdot MgSO_4$	78
Hectorfloresite**	$4Na_2SO_4 \cdot NaClO_3$	74.1
Sulphohalite	$2Na_2SO_4 \cdot NaCl \cdot NaF$	73.9
Burkeite	$Na_6(SO_4)_2(CO_3)$	72.8
Eugsterite (Fritzsche's salt)	$2Na_2SO_4 \cdot CaSO_4 \cdot 2H_2O$	62.3
Darapskite	$NaNO_3 \cdot Na_2SO_4 \cdot H_2O$	58.0
Hydro-glauberite	$5Na_2SO_4 \cdot 3CaSO_4 \cdot 6H_2O$	57.9
Glauberite	$Na_2SO_4 \cdot CaSO_4$	51.1
Loeweite	$MgSO_4 \cdot Na_2SO_4 \cdot 2\frac{1}{2}H_2O$	46.2
Ferrinatrite	$3Na_2SO_4Fe_2(SO_4)_3 \cdot 6H_2O$	45.6
Mirabilite (Glauber's salt)	$Na_2SO_4 \cdot 10H_2O$	44.1
Bleodite (astrakanite)	$MgSO_4 \cdot Na_2SO_4 \cdot H_2O$	42.5
Kroehnkite	$CuSO_4 \cdot Na_2SO_4 \cdot 2H_2O$	42.1
Nickelbloedite	$Na_2Ni(SO_4)_2 \cdot 4H_2O$	40.3
Sideronatrite	$Na_2Fe(SO_4)_2(OH) \cdot 3H_2O$	38.9
Caracolite	$Pb(OH)Cl \cdot Na_2SO_4$	35.4
Palmierite	$(K,Na)_2Pb(SO_4)_2$	<31.9
Tychite	$2MgCO_3 \cdot 2Na_2CO_3 \cdot Na_2SO_4$	27.2
Aphthitalite (glaserite)	$(K,Na)_3Na(SO_4)_2$	21–38
Tamarugite	$Na_2SO_4 \cdot Al_2(SO_4)_3 \cdot 12H_2O$	20.3
Natrochalcite	$Cu_4(OH)_2(SO_4)_3 \cdot Na_2SO_4 \cdot 2H_2O$	18.8
Almeriite	$Na_2SO_4 \cdot Al_2(SO_4)_3 \cdot 5Al(OH)_3 \cdot H_2O$	15.9
Mendozite (soda alum)	$Na_2SO_4 \cdot Al_2(SO_4)_3 \cdot 24H_2O$	15.5
Natrojarosite	$Na_2Fe_6(OH)_{12}(SO_4)_4$	14.7
Noselite	$3Na_2Al_2Si_2O_8 \cdot Na_2SO_4$	14.3
Slavikite	$(Na,K)_2SO_4 \cdot Fe_{10}(OH)_6(SO_4)_{12} \cdot 63H_2O$	<4.6

*Salt cake is anhydrous sodium sulfate with a purity lower than 98%
**Erickson, et al., 1989.

lakes from which sodium sulfate and other salts are extracted. Because of the great difference in composition between seawater salts and those of playa lake deposits, it is generally accepted that these deposits are recent accumulations from nonmarine sources.

The brines of playa lake deposits are more diverse in composition than the oceans because each deposit contains soluble salts from discrete sources within its drainage area. The salinity of the brines is much greater because of the high rates of evaporation characteristic in arid interior basins where the deposits occur.

The evolution of brine and subsequent precipitation of minerals involve complex geochemical factors. These include the weathering of solutes from the watershed, their transport to the playa basin, and their evaporative concentration. Jones (1966) points out the importance of lithology and weathering processes to the dominant anion in playa lakes of the western United States. Krauskopf (1967) discusses the fate of various elements and radicals during transport and the eventual arrival of sodium as the dominant cation and either sulfate or chloride as the chief anion. Hardie and Eugster (1970) discuss the evolution of dilute waters in closed basins into four groups: Group A (Na-CO_3-SO_4-Cl brines), Group B (Na_2SO_4-$NaCl$ brines), Group C (Na-Mg-Ca-Cl brines) and Group D (Na-Mg-SO_4-Cl brines). The precipitation of calcite, sepiolite, and gypsum are important points of bifurcation during the evolutionary process.

In nature two processes, evaporation and cooling, cause the deposition of sodium sulfate. Evaporation decreases the volume of water, and cooling decreases the solubility of the salt. Below the freezing point of water, ice decreases the available water for solution and the solubility of sodium sulfate minerals, causing mirabilite to form. This natural phenomenon accounts for the accumulation of intermittent beds of crystalline sodium sulfate in the bitter lakes of Canada, Russia, and the United States from which sodium sulfate may be harvested during the winter. Crystallization from cooling also results from day and night temperature changes, particularly at high elevations where these changes can be extreme.

In warmer climates, evaporation is the principal agent of precipitation of sodium sulfate. In those climates natural cooling is not employed in the production of sodium sulfate, but concentrated brines are pumped to the surface and cooled by refrigeration to yield mirabilite crystals.

Most sodium sulfate deposits of commercial significance contain permanent crystalline beds and brines at or near the surface. The crystalline beds can range in thickness from a few millimeters to several meters. They usually are interstratified with other salts, clay, silt, and organic matter. The impervious clays and silts help preserve a portion of the intermittent beds from redissolving. Brines generally occur within crystal beds, but in some deposits crystal beds are lacking.

In most deposits sodium sulfate predominates, with sodium and magnesium chlorides, magnesium sulfates, and carbonates also present. In more complex deposits, concentrations of potassium chloride and borates can be appreciable, depending on the location of the deposit in relation to the source of soluble salts.

The origin of sodium sulfate deposits has prompted the development of many theories. Variations in these theories mainly concern the source of alkalic salts that make up the deposits. Other factors considered are the origin of the basin and the agent of transportation and deposition of the sodium sulfate (Ruffel, 1970).

Early theories generally attributed the source of the salts to decomposition of granitic rocks. It was assumed that the feldspars in granites yielded soluble salts that were transported by surface waters, and that pyrite and other sulfides in these rocks were oxidized to sulfuric acid, which immediately dissolved some of the basic oxides to produce soluble sulfates (Wells, 1923).

Following an extensive study of the deposits of western Canada, L.H. Cole (1926) considered four possible sources of the sodium sulfate. These are: 1) underlying sands and shales, which contain small amounts of alkalies, and interbedded volcanic-ash beds, which contain an appreciable percentage of these salts; 2) connate waters in underlying strata; 3) springs in or near the deposits; and 4) surface deposits of bentonitic glacial drift. Cole concluded that the probable source was the salts contained in the bentonites, which are widely distributed in the glacial drift. Surface waters carrying calcium salts in solution react by base exchange with the alkali silicates, releasing the salts contained in the bentonite in the form of soluble sulfates, etc. These in turn are concentrated and deposited in undrained lake basins.

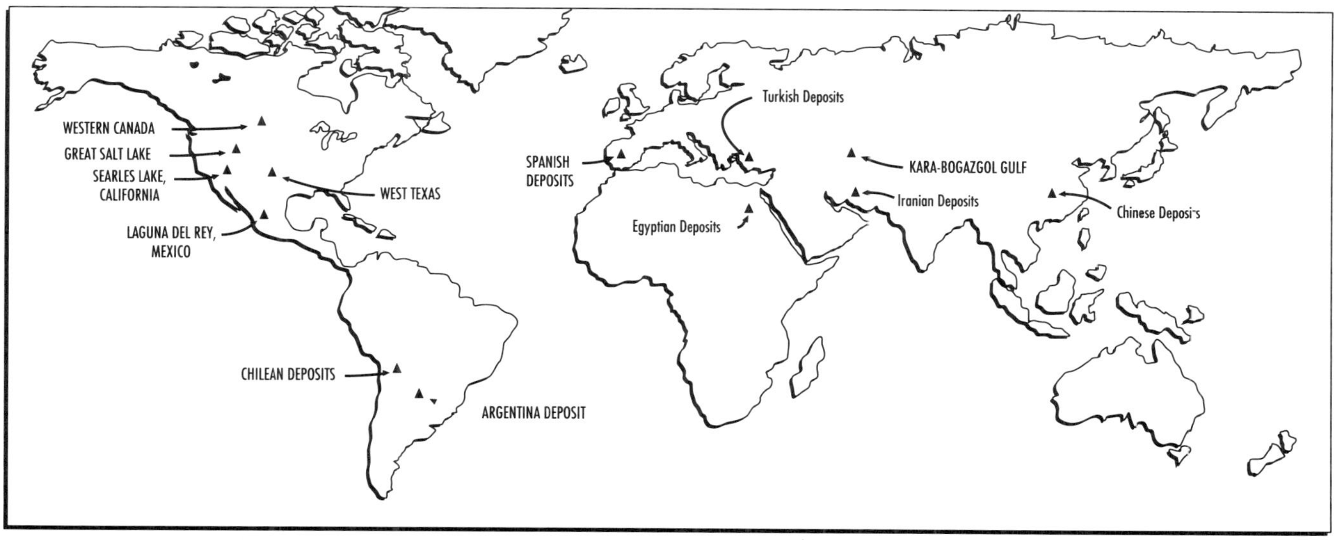

Fig. 2. Principal producing sodium sulfate deposits of the world.

More recent investigators have attributed the source of some sodium sulfate to the decomposition of rocks and sediments, such as bentonite, that also contain sulfates, and from the leaching of buried beds of sulfates. Organic rich clay and shale also would produce sulfate through the oxidation of pyrite and marcasite (Davis, 1966). A subaerial means of transport is provided because many playa lakes are hydrologically coupled with the surrounding water table. Sulfides are found in the organic rich muds (containing stromatolites, etc.) and vegetative debris of some playa lake sediments. The fermentation of organics under anaerobic conditions would produce hydrogen sulfide which could oxidize to sulfate (Reeves, Jr., 1968).

Grossman (1968) studied the sodium sulfate deposits of the Northern Great Plains of Canada and the United States and attributed the origin of those deposits to the solution of deeply buried marine evaporites. In his model, ground water from the Rocky Mountains moved downward and eastward into the Saskatchewan Sub-basin during the Quaternary, leaching the buried evaporities of the Prairie Formation (Devonian). Ascending mineralized ground water during the late Pleistocene discharged into stratified drift where freezing segregated pure crystals of sodium sulfate in meromictic (chemically stratified and stagnant) lakes.

In a later study, Last and Slezak (1986) question the dissolution of deeply buried Paleozoic evaporites as the ultimate source of the salts based upon the incompatible brine chemistry. Their paper describes the origin of the basins, in addition to the hydrochemistry, mineralogy, sedimentology, and diagenesis of the deposits in western Canada.

The US Geological Survey has studied the saline lakes in West Texas in order to investigate the hypothesis that the source of salts in the basins originated from connate waters or brines migrating from deeply buried Paleozoic evaporites or from Cretaceous rocks that crop out in the lake basins. They concluded that the origin of the deposits was, and is, by evaporative concentration of the runoff of potable ground water discharging from the shallow High Plains Aquifer. Their study dealt with an analysis of hydraulic heads and pore volume flushing in addition to evaluation of solute ratios and isotopic chemistry (Wood and Jones, 1990).

A subsequent article (Wood and Sanford, 1990) dealt with the ratio of groundwater outflow to inflow (flux ratio) in hydrologically open basins and the resultant mineralogy and volume of precipitated salts. In addition, possible mechanisms by which brine could escape from the system are discussed. The authors developed a groundwater model, PHRQPITZ, to define the effects of various flux ratios on the mineralogy and thickness of deposits. Another paper (Sanford and Wood, 1991) investigates two typical waters, seawater and Na-HCO_3 ground water, in order to demonstrate the importance of basin leakage on brine evolution.

OCCURRENCE, RESERVES, AND PRODUCTION

Sodium sulfate deposits of sufficient size and purity for commercial mining are found in several countries (Fig. 2). The principal producing countries are Canada, Mexico, the United States, the former USSR, and Spain. Less important are Argentina, Chile, Iran, and Turkey.

Mexico

Mexico produced 650 kt of natural sodium sulfate during 1990, (Kostick, 1990) making it the largest producing country in the world.

A large evaporite deposit containing a high concentration of sodium sulfate is located at Laguna del Ray in northern Coahuila, approximately 160 km north of Torreon. The evaporite lens consists of glauberite and bloedite, this last one as thin beds and irregular aggregates.

The crystal bed is 20 km long and 5 km wide, with 30 m thickness at the center. The main concentration of bloedite (a double sulfate salt of sodium and magnesium) is found as a layer 1 to 2 m thick, located about 8 m below the top of the evaporite mass. Mirabilite and epsomite occur in minor proportions. The formation underlying the crystal bed is impervious clay and silt containing some sodium chloride. Glauberite occurs as thin tabular crystals, while bloedite is found in massive concentrations. The intercrystalline spaces, amounting to as much as 25% of the evaporite mass, are partially filled with clay, and the rest with a brine in equilibrium with the solid phase.

A composite average analysis of brine from a depth of 20 m from several wells in the lake bed showed 18.46% sodium sulfate, 5.91% sodium chloride, and 5.28% magnesium sulfate.

Analysis of several cores of the crystal bed showed 35 to 36% sodium sulfate and only 1.5 to 1.8% sodium chloride. Magnesium sulfate ranged from 4.7 to 12.7%.

Estimated reserves are approximately 160 Mt of anhydrous sodium sulfate.

Sodium sulfate is produced at Laguna del Rey by Quimica del Rey, S.A. de C.V., a subsidiary of Industrias Penoles, S.A. de C.V. The plant was built in 1963 with a capacity of 30 ktpy. In 1988 the plant was expanded to its present capacity of 510 ktpy. It is reportedly the largest sodium sulfate plant in the world (personal communication).

Other deposits are known to occur in northern Coahuila and in the state of Chihuahua, but in size and sodium sulfate content they are smaller than the deposit at Laguna del Rey.

Sulfate de Viesca recovers 22 ktpy of sodium sulfate as a byproduct from production of sodium chloride from natural brine. The plant is near Viesca in the southwestern part of Coahuila.

Deposits of thenardite occur in the state of Sonora near the municipality of Bacadehuachi some 190 km northeast of Hermosillo. Outcroppings of thenardite and efflorescences in the soil occur discontinuously along the east side of the Bacadehuachi Valley in an area approximately 6 km in length and 4 km in width. The deposits occur within the Baucrit Group (Oligocene). The outcroppings are covered by unconsolidated surface material, which varies in thickness between 10 and 30 m.

Four workings have been developed within this area. In the largest pit, the Oro Blanco, the thenardite is reported to attain a thickness of 20 m. Analyses of selected samples from the Oro Blanco ran between 95 to 99% Na_2SO_4 (Anon., 1981).

During 1982, Para Mex, the Mexicanized affiliate of the Spanish company Tolsa, had conducted some exploration of the Oro Blanco and was evaluating the commercial possibilities of producing sodium sulfate from this area (personal communication).

Quimica del Rey, the country's main producer, states it produced 483 kt in 1990. They estimate their 1991 production to be 500 kt. Approximately 27% of 1990 production was destined for South and Central American markets (Quimica del Rey, personal communication).

Spain

Sodium sulfate from deposits of solid thenardite is produced in Spain. The El Castillar deposit is located in Villarrubia de Santiago, Toledo Province, where beds of thenardite 5 to 6 m thick are mined by room and pillar methods. The deposit extends over an area of 30 km^2 and contains probable reserves of 200 Mt. The composition of the ore is reported as 70 to 75% sodium sulfate, 11 to 17% calcium sulfate, 0.4 to 0.6% sodium chloride, 3 to 9% clay, and 1 to 5% moisture (Rovira, 1960). The mine, and a plant producing anhydrous sodium sulfate, is operated by Union Salinera de Espana S.A.

Ciraderos Minerales y Derivados S.A. (Crimidesa) mines glauberite at Cerezo del Rio Tiron in Burgos Province.

Another company, S.A. Sulquisa, produces sodium sulfate by solution mining of glauberite. It is a joint venture of Industrias Penoles S.A. de C.V. of Mexico and Spanish mining interests. S.A. Sulquisa increased its capacity in 1987. Their production has increased from 83 ktpy in 1987 to 116 ktpy in 1990 (Ross, personal communication, 1991).

Total production of natural sodium sulfate from Spain has increased. For 1990 it was estimated to be 460 kt, ranking it second in world production (Kostick, 1991).

Canada

Numerous saline lakes containing sodium sulfate are known on the western plains of Canada. Seventeen deposits contain in excess of 1 Mt each. Only one of the large deposits is in Alberta; the remainder are in Saskatchewan. Their commercial exploitation began in 1918 at Mukiki Lake, near Saskatoon, Saskatchewan (Last, 1986). At present, only four lakes and three companies account for essentially all sodium sulfate production (personal communication). Two companies use dredges to mine the permanent crystal beds. One additional lake in Saskatchewan is worked only intermittently.

An interesting occurrence of sodium sulfate has been reported at Castleguard Cave, which drains the Columbian Icefields of the Canadian Rocky Mountains. The mirabilite occurs in association with gypsum and epsomite (Younge and Krouse, 1987). Though of geological interest, this occurrence is apparently noncommercial.

Commercial deposits of sodium sulfate are located in the Prairie Provinces of Canada where they occur in basins and depressions in glacial drift that lack apparent outlets. In general, the deposits are crystalline beds that underlie saturated brine lakes containing a few millimeters to approximately a meter of brine in the spring and summer. The deposits contain two recognizable phases of crystalline beds, intermittent and permanent.

The intermittent beds form on the bottom and along the shores of the lakes. They crystallize and redissolve with fluctuations in temperature. This can occur in a matter of hours, but as winter sets in, less solution of the crystals occurs, causing the beds to become thicker and more compact.

The permanent beds underlying the lake usually are interstratified with thin layers of mud, clay, and organic matter. These beds range in thickness from less than one to several meters. At Ingebrigt Lake, permanent beds are reported (Cole, L.H., 1926) to cover an area of 2.7 km^2 with an average thickness of 6.7 m. In one part of the deposit a thickness of 42.4 m was measured. Cole estimated the tonnage of hydrous salt in Ingebrigt Lake to be 22.7 Mt. A more recent evaluation is 8.2 Mt of anhydrous sodium sulfate (Broughton, 1977).

Most of the deposits presently in production contain several million tons of sodium sulfate. The bulk of the Canadian reserve occurs in Saskatchewan. One deposit in Alberta contains 2.7 Mt (Watson, 1980). However, this deposit is currently idle (personal communication). Estimates of total reserves in Saskatchewan range between 54.4 and 181 Mt. Another evaluation places the total anhydrous sodium sulfate reserve at 54.4 to 63.5 Mt and evaluated commercial reserves at 27.2 to 36.2 Mt (Broughton, 1977). An updated estimate of reserves would account for the tonnage mined since then.

Production of natural sodium sulfate in 1990 was 347 kt. Production in 1991 is expected to be less than in 1990 due to the idling of plants at Metiskow, Cabri and Hardene by Agassiz Resources Ltd. Approximately 151 kt of the 327 kt produced in 1989 were exported.

Current capacity is reported at approximately 613 ktpy though production has never been close to reported capacity. The termination of Agassiz's production reduced this figure to an effective capacity of approximately 453 ktpy. This could presumably increase by 160 ktpy should a new owner reopen the Agassiz operations. About 253 ktpy of capacity belongs to Saskatchewan Minerals which is owned by Dickenson Mines, Limited (Huissier, personal communication, 1991).

Commonwealth of Independent States (CIS formerly USSR)

Many deposits containing sodium sulfate occur in the Russian Republics in a large area from the Black Sea to western China and farther north into west-central and eastern Siberia. The economically important deposits are in the Aral and Caspian depressions and in the Kalunda Steppe in Kazakhstan. While literature on sodium sulfate deposits in the Russian Republics is reportedly fairly extensive, it is not readily available and requires translation. The reader is referred to a paper (Dort, Jr. and Dort, 1970) that contains a summary of several publications on Russian deposits.

The largest known occurrence of sodium sulfate in the world is in the Kara-Bogaz-Gol Gulf, an embayment on the eastern side of the Caspian Sea. It is separated from the Caspian by a bar with a narrow strait through which water flows from the sea into the embayment. Sixty-five years ago the gulf covered 18 000 km^2 and had a depth of 12 m. Owing to evaporation, the Gulf has shrunk in area to 10 000 km^2 and in depth to about 3 m and had a much higher saline content (Garbell, 1963).

Kolosov et al. (1974) reported that Kara-Bogaz-Gol experienced a final phase of intense desiccation between 1939 and 1948. Seawater then moved across the bar from the Caspian Sea at a reduced rate of 9 to 10 km^3 per year. For a time, the movement of seawater into the bay apparently reached equilibrium with evaporation. After 1957 or 1958 no decline in water level or constriction of the bay's area or appreciable change in brine composition was observed. Gypsum and glauberite were being precipitated out of the saturated brine throughout the bay. Bloedite (astrakanite) and halite were being deposited locally. Some carbonates were being deposited in the zone of mixing adjacent to the strait.

As a result of concern over the declining water levels in the Caspian Sea, a dam was constructed in 1980 across the strait that had once replenished the Kara-Bogaz-Gol with Caspian seawater. This accelerated the severe desiccation of the Gulf and, by 1983, no surface brine remained. Winds carrying sulfate dust traveled hundreds of kilometers causing severe ecological problems (Levine, 1988). Seawater was released to the Gulf in 1984 at a rate of about 2 km^3/yr (Kurilenko et al., 1988). However, this was apparently inadequate to solve the problem. Soviet scientists predicted that 80% of the minerals in the Gulf would be irretrievably lost without adequate recharge from the Caspian Sea (Levine, 1988).

The first commercial utilization of salts at Kara-Bogaz-Gol, which began in 1929, used surface waters as a source of raw material. However, this source was abandoned in 1953 due to the fluctuating hydrological and hydrochemical regimes of the Gulf and the operators began pumping the interstitial brines of the buried salt deposits as a feed source. Four recognizable salt horizons occur to a depth of 18 m, each with distinct hydrological and mineralogical characteristics. Recent hydrological proposals by scientists from Leningrad State University and the All-Union Research Institute of Halurgy have recommended the following changes to the local brine industry: locating and developing production wells based upon optimum hydrogeology and stratigraphy, lowering the brine output of each well from 500 to 250 m^3/hr, and utilizing the end liquors from the plants to recharge the pay horizons (Kurilenko et al., 1988).

In 1924, Kara-Bogaz-Gol was estimated to contain 1 Gt of sodium sulfate (Grabau, 1920). Later estimates of 2 Gt have been reported (Goudge and Tomkins, 1960, Hemy, 1969). More recently, Russian sources indicate that reserves at Kara-Bogaz-Gol may be lower than was previously thought (Kurilenko and Frolovskii, 1982).

The Aral Sea is considered to be another rich source of sodium sulfate but the diversion of water from its two major tributaries, the Amu Darya and the Syr Rivers for agricultural purposes has resulted in a severe constriction of the Aral. Unless replenishment can be reestablished in order to counteract the evaporation loss, it has been predicted that the Aral Sea could be gone within another 30 years (Ellis, 1990). Some sulfate is produced in the central Asian republics that adjoin it. Lakes in western Siberia and in Kazakhstan are good sources, but are not known to be presently producing except on a local scale. The steppe region around Kalunda, in Alta Kray, has a number of lakes rich in sodium sulfate. A sulfate combine has been established on Lake Kuchuk, the largest of these (Hemy, 1969).

Sodium sulfate and magnesium salts are produced from Kara-Bogaz-Gol Gulf by Karabogazgolsulphate Combine. The CIS ranks fourth in world production, producing an estimated 365 kt of natural sodium sulfate in 1990 (Kostick, 1990), mostly from the Kara-Bogaz-Gol deposit. A 37.5 ktpy sodium sulfate unit was commissioned there in 1988 (Levine, 1988).

United States

Deposits of sodium sulfate are found in Arizona, California, Colorado, Idaho, Montana, Nevada, New Mexico, North Dakota, Oregon, Texas, Utah, Washington, and Wyoming. At the present time, production is chiefly from brine deposits at Searles Lake, CA, at Great Salt Lake, UT, and in West Texas. Small quantities have been produced from saline lake deposits in Wyoming. Most of the southwest brines are contained in dry lakes or playas underlain by beds of mirabilite, thenardite, and other evaporite salts. In the northwestern United States, the deposits associated with saline or intermittent dry lakes are similar in origin and composition to the Canadian deposits. Except for the brine deposits in West Texas, the deposits of the southwest are chemically more complex than the northern deposits.

California: California has several sodium sulfate deposits. Dale Lake, Durmid Hills, Soda Lake, and Danby Lake have a record of past production but are not now producing (Smith, 1966). Owens Lake has reserves of 10.9 Mt but has not produced any sodium sulfate (Kostick, 1980).

At Searles Lake, minerals have been extracted from brines since 1873. Sodium sulfate was first produced in 1914. Searles Lake currently is the leading source of natural sodium sulfate in the United States (Moulton, 1980). It probably has produced more sodium sulfate than any other deposit in the world. The annual value of the minerals production to date exceeds $30 million and total production to date exceeds $1 billion (Smith, 1979). The deposit was operated by the Kerr-McGee Chemical Corporation for many years but in November 1990 the operation at Trona, CA was purchased by the North America Chemical Company (Kostick, 1990b).

Sodium sulfate and other usable salts are produced at Searles Lake from brine underlying a large playa. The surface of the playa covers an estimated 155 km^2 but the central part, which forms the main deposit, is about 30 km^2 in area (Graubau, 1920). A hole drilled by Kerr-McGee Corp. in 1968 encountered 930 m of basin fill and bottomed in quartz-monzonite (Moulton, 1980). The salt deposits are shown in cross section on Fig. 3.

Brine occurs in the interstices of the salt deposits. Ions present in the brine are sodium, potassium, carbonate, bicarbonate, sulfate, chloride, and borate. The upper salt, lower salt, and mixed layer are the productive zones, each contributing a distinctive brine to meet the individual requirements of the plants. In some instances, solar concentration of the produced brine is used to improve plant efficiency (Moulton, 1980).

A plant expansion underway in 1977 was to bring capacity to 590 ktpy (Parkinson, 1977). The plant never operated as planned and, in fact, in 1982 Kerr-McGee shut down about half of their sodium sulfate capacity, so output dropped to about 200 ktpy.

Utah: Sodium sulfate is produced from natural brine at Great Salt Lake, UT. Thick beds of mirabilite, intercalated with thin clay layers, have been found beneath the lake at several localities. Mirabilite crystallizes and sinks to the bottom of the lake during the winter. A layer of mirabilite 0.15 to 0.30 m thick occurs seasonally

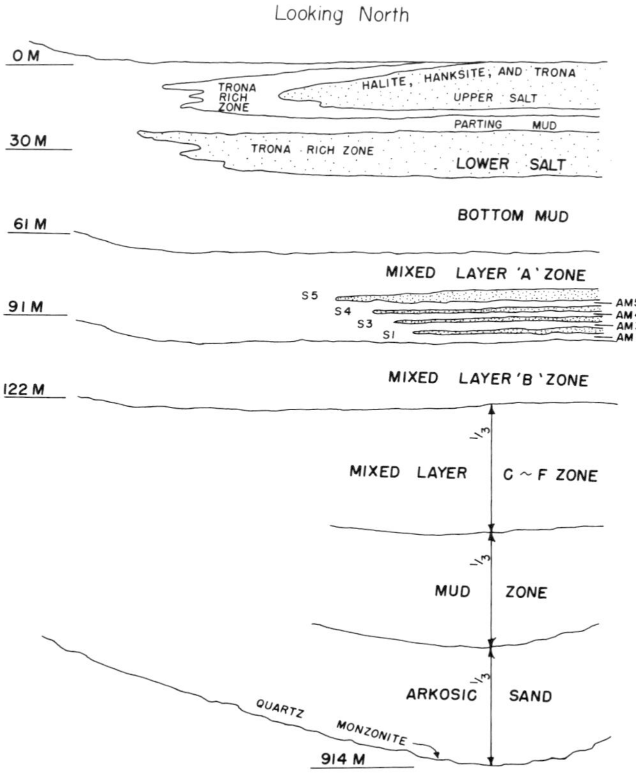

Fig. 3. Cross-section of Searles Lake (*courtesy:* North American Chemicals, Inc.).

Fig. 4. Ozark-Mahoning Company's Cedar Lake plant in West Texas.

over large areas of the lake. Under favorable conditions, it is carried by winds and currents to the shore where it accumulates to a depth of 0.3 m or more. Strand deposits as thick as 3 m have been observed.

The Great Salt Lake is estimated to contain 3.8 Gt of salt, 342 Mt of which are sodium sulfate (Cohenour, 1966). Great Salt Lake Minerals and Chemical Corp. estimates the playa contains 408 Mt of sodium sulfate (Kostick, 1980).

On the north end of the lake, near Ogden, Great Salt Lake Minerals and Chemicals Corp., an affiliate of North America Chemical Co., operates 69 km^2 of evaporation ponds and a plant complex for producing potassium sulfate and magnesium chloride. Sodium sulfate and common salt are byproducts. First production was reported in 1970. The operation has a sodium sulfate capacity between 22.7 and 32.9 ktpy.

Texas: Four sodium sulfate deposits were operated in West Texas by the Ozark-Mahoning Company until 1987 when three of them, namely, Rich, Mound, and Brownfield Lakes, were shut down and mothballed due to economic and climatologic conditions. The company has produced sodium sulfate from brine in West Texas continuously since 1933 (Fig. 4). The Texas deposits underlie playas and occur as brine and crystalline mirabilite. The brine contains mainly sodium chloride and sodium sulfate, with lesser amounts of magnesium and potassium salts, and is produced from 30.5 m deep wells. A fifth deposit, Soda Lake, near Monahans in Ward County, was abandoned in 1970 after nearly 40 years of continuous operations, during which time over 1 Mt of anhydrous sodium sulfate were produced.

Cedar Lake deposit, near Seagraves, covers a surface area of about 21 km^2. It is the only deposit currently in production. Beds of crystalline salts are minimal, but strata containing saturated brines are found from near the surface to a depth of 30.5 m. The shallower strata contain brines of low salt content and are less productive. Richer brines come from beds 12 to 30.5 m below the surface. Brines from the Cedar Lake deposit contain about 10.5% sodium sulfate, 14% sodium chloride, and 1% each of magnesium and potassium.

Brownfield Lake deposit, 16 km east of Brownfield, covers an area of about 4 km^2. The main deposit lies outside the boundary of the present dry lake surface. Saturated brine was produced from crystal beds at depths from 9 to 30.5 m. These beds, which contain thin layers of clay and silt, range from a meter to as much as 4 m in thickness. The brine contains 15% sodium sulfate and 10% sodium chloride. A solution mining project to extract the crystalline reserve was developed.

Rich Lake deposit, 19 km north of Brownfield Lake, produced sodium sulfate bearing brine from a crystal bed averaging 2 m in thickness. The brine was transported by pipeline to the Brownfield Lake plant for processing. Ozark-Mahoning operated plants at Brownfield Lake and Cedar Lake. The combined capacity of the two plants was 159 ktpy.

A fourth deposit, Mound Lake, is located 13 km north of the Brownfield Lake plant. The lake has an area of 5 km^2 and lacustrine sediments occur to a depth of 24 m. The major portion of the reserve is brine similar in composition to that of Cedar Lake. A significant tonnage of mirabilite also exists. Mound Lake began producing in April 1981, and the brine also was transported to the Brownfield plant for processing.

The Brownfield Plant, in addition to its satellite deposits at Rich and Mound Lakes, is being kept on a care and maintenance basis and could be reopened pending an improvement in the sodium sulfate market.

Arizona: At the Camp Verde deposit in Yavapai County, solid beds of thenardite of varying thickness were mined for many years by Western Chemical Co. The deposit is unusual, as pure anhydrous sodium sulfate occurring in solid crystal form is rare. The deposit consists of layers of thenardite and other sodium salts interbedded with clay, which form a part of the Verde Formation. The total thickness of the beds is at least 46 m. Mining was done in a large quarry and by underground room and pillar methods. The last mining was done by Arizona Chemical Co., a subsidiary of American Cyanamid Co., in 1934 (Ralston, 1949).

Nevada: Nevada's principal producing deposit in years past was Rhodes Marsh, near Mina in Mineral County. The deposit covers an area of 28 km^2. Production was confined mainly to the central portion of the deposit, where a 1 to 2 m thick bed of solid thenardite was mined (Ralston, 1949).

New Mexico: A sodium sulfate deposit is known near Alamogordo in Dona Ana County. In Eddy County, near the potash mines, brine wells containing appreciable amounts of sodium sulfate and magnesium sulfate have been reported (Lang, 1941). These wells were produced by Ozark-Mahoning from 1951 through 1957; the brines were trucked to Ozark's plant near Monahans, TX for processing. The wells depleted rapidly and were abandoned.

North Dakota: The sodium sulfate deposits of northwestern North Dakota are located in Divide, Williams, Mountrail, and Ward Counties. They are a southeasterly extension of the deposits in Saskatchewan. Sodium sulfate in North Dakota was first noted by the Lewis and Clark Expedition. Several lakes are known. Most of them were investigated and prospected by the Federal Emergency Relief Administration (FERA) in 1934. An important reevaluation, based on drilling, was made by the US Bureau of Mines in 1948 (Binyon, 1952).

The sodium sulfate content is generally high; sodium chloride, magnesium sulfate, and calcium sulfate are the main impurities. The US Bureau of Mines reported tonnages occurring in 18 lakes. Nine of the largest were estimated to each contain over 1 Mt of Glauber's salt. The major portion of the reserve occurs in the permanent and intermittent crystal beds and only a minor portion occurs in the form of brine. Total reserves are reported to be 25 Mt of Glauber's salt or 11 Mt of anhydrous sodium sulfate (Binyon, 1952, Crosby, 1973).

Grenora Lake No. 2, the largest deposit, covers 2 km^2. Drilling done by FERA showed the maximum thickness of the permanent crystal bed to be 24 m and the average thickness to be 3.66 m. The deposit was estimated by the US Bureau of Mines to contain 5 Mt of Glauber's salt or 2.2 Mt of anhydrous sodium sulfate (Binyon, 1952). Drilling conducted by Ozark-Mahoning in 1977 generally corroborated the Bureau's estimate. Grenora Lake No. 2 is owned by Ozark-Mahoning, but is not favorably located for economical exploitation.

Other States: A number of deposits in Washington, Wyoming, and Montana are not considered economically important because of their location and size, even though their composition is similar to the Canadian deposits.

Iran

Iran produced 185 ktpy of sodium sulfate in the year ending March 1990; down somewhat from their peak production of 265 ktpy in the year ending March 1988. This ranks Iran sixth in world production. In 1990, production was from 13 mines located in the provinces of Markazi, Semnan, and Teheran (Antonides, personal communication, 1992).

Turkey

Natural sodium sulfate is produced in Turkey, although output is not large. First production recorded was in 1963, when 1 264 t were produced. Output has increased steadily, and production in 1990 was about 80 kt, making it seventh in world production (Kostick, 1991).

Several saline lakes in southwestern Turkey contain appreciable amounts of sodium sulfate. Tersakan Lake is located about 160 km south of Ankara, near the town of Cihanbeyli. The lake, 13 km long and 4 km wide, covers an area of 52 km^2. Its sodium sulfate content is in excess of 8%. Sodium chloride and magnesium sulfate are present in amounts less than 5 and 2%, respectively, and as the calcium lake water has been impounded to induce evaporation, the sodium sulfate content of the brine increases. In the bottom of the ponds crystalline layers over 95% pure have formed.

At Aci Gol, a large lake 56 km east of Denizli, the saline content is similar to that of Tersakan Lake, although the sodium content is slightly lower and the content of sodium chloride slightly higher. Aci Gol is over 26 km long, nearly 10 km wide, and has a surface area of about 200 km^2.

Three companies currently produce sodium sulfate from Aci Gol: Alkim Alkali Kimya (in addition to two small lakes), Sodas Sodyum Sanayi, and Otuzbir Kimya Ltd. (Karayazici, 1988).

Egypt (Africa)

A series of salt pans, some of which contain sodium sulfate, occur in the deserts and velds of the South African uplands. To date, none have been commercially exploited. In 1968, however, Associated Sulfate (Pty) Ltd. was established to build a plant to produce sodium sulfate using raw materials from solar salt pans and effluent from the Rustenburg platinum plant.

About 110 Mt of sodium sulfate reserves exist in the Makgadikgadi Basin of Botswana (Kostick, 1980, Massey, 1973). For over 40 years, various parties have considered, and/or attempted, to develop this large deposit, which covers more than 900 km^2.

In March 1991, a plant was completed at Sua Pan by Soda Ash Botswana (Pty.) Ltd., owned 52% by a South African consortium and 48% by the Government of Botswana.

Production of 300 ktpy of soda ash and 650 ktpy of common salt is predicted to occur some two years after the initial startup of the plant. Sodium sulfate and potassium chloride are mentioned as possible byproducts (Antonides, 1988). Currently, however, sodium sulfate is reportably being returned to the lake as spent bitterns (Antonides, personal communication, 1992).

Production of natural sodium sulfate in South Africa was insignificant in 1990 (USBM data).

Sodium sulfate deposits are found in the northern part of the Western Desert of Egypt, some 80 km west of the Nile Delta. Twelve principal lakes occur for 25 km along the axis of Wadi El Natron, a minor tectonic depression. Some of the larger lakes remain permanently flooded, whereas the smaller lakes dry up seasonally. Upon exploration, one of the latter, Beida Lake, was discovered to contain probable reserves of the B catagory, exceeding 1.1 Mt of thenardite and about 100 kt of halite (Nakhla et al., 1986). Egypt produced 45 kt of natural sodium sulfate in 1990, making it the eighth largest producer in the world (Kostick, 1991).

China

China produced 27 kt of natural sodium sulfate in 1990, placing it ninth in the ranking of world producers (USBM data).

Information on sodium sulfate in China, though available, is not readily attainable; therefore, not much is known about the geology of deposits. One known occurrence is a salt lake at Yuncheng, north of the Zhongtiao Mountains in Shanxi, which has been a source of salt for centuries. Chemical plants around the lake use brine containing significant amounts of boron, bromine, calcium, and iodine in addition to sodium sulfate and other sulfur compounds. Production from Yuncheng reportably accounts for 60% of China's output of mirabilite (Chin, 1986).

Mirabilite has been reported at the 200 000 km^2 Qaidam Basin, in Qinghai, which contains 20 major salt lakes. Mineralization in the basin includes mirabilite and borate in addition to sylvite and salt (Chin, 1983).

Argentina

Many salt lakes and playas, some of which contain rich deposits of sodium sulfate, are located in Argentina, on the arid plateaus and in the intermontane valleys of the Andes (Wells, 1923).

A large playa deposit of mirabilite is located in northwestern Argentina in Jujuy Province. The deposit is in the Andes at an elevation of 3 353 m. A permanent bed of mirabilite covering 53 km^2 occurs in the middle of a dry lakebed. The permanent bed, covered by a thin layer of sodium chloride, calcium sulfate, and sand, is reported to be 4.6 m thick at the center of the deposit and gradually thinner toward the edges. Subsurface waters from melting snow in the surrounding mountains feed into the lake and thus furnish a supply of brine year round.

Production of sodium sulfate in Argentina decreased from 32 ktpy in 1986 to 11 ktpy in 1990. Sufarentina SAMIC, a subsidiary of CADIPSA, produces sodium sulfate at LaLoberia, Santa Cruz Province, and Salar del Rio Grande, Salta Province.

Chile

Small amounts of natural sodium sulfate are produced from playa deposits in northern Chile. The deposits are in the Atacama Desert, which lies along the west side of the Andes between the mountains and the Pacific Ocean, at an elevation of about 1 067 m.

Chile produced 10 kt of natural sodium sulfate in 1990. Some additional output is byproduct from the nitrate industry.

Antarctica

Since mirabilite is the first mineral, after ice, to appear in freezing seawater, it is not surprising that several occurrences have been noted in Antarctica. Mirabilite efflorescences commonly occur on the surfaces of rocks, glaciers, and sea ice; some occurrences are also reported farther inland.

Sodium sulfate has been described in coastal lakes which have been elevated and separated from the sea due to isostatic rebound. Other occurrences are reported in impoundments behind glaciers. The inland deposits are probably a result of seawater-derived salts transported with snow and accumulated in local depressions.

Though these occurrences are not economic, they are of geological interest. The reader will find them described by Dort, Jr. and Dort (1970) along with a discussion of the low temperature origin of sodium sulfate.

SYNTHETIC SODIUM SULFATE

Synthetic production was formerly grouped together under the byproduct listings, which is still the predominant segment. It includes one of the faster growing segments of sodium sulfate production—that of reclaiming sodium sulfate from waste streams. For many years, synthetic sodium sulfate has been a byproduct or coproduct from the production of other chemicals. Now due to its importance, synthetic sodium sulfate from reclamation processes has been added to this list. Important sources are:

1) Manufacturing of viscose rayon. Sodium sulfate is generated in an acid spin bath by the neutralization of sodium hydroxide used to prepare the viscose solution. In the past this source has been of almost equal importance with natural sodium sulfate on a worldwide basis. However, growing popularity of other synthetic fibers coupled with environmental concerns has resulted in reduced production. Of particular significance has been the replacement of rayon tire cord by other synthetic fibers, fiberglass, and steel. Nevertheless, rayon manufacturing continues to be an important source of supply. Avtex Fibers, Inc., a major manufacturer with nameplate capacity of 77 ktpy ceased operation in November of 1989 after encountering difficulty meeting environmental standards (Kostick, 1990). On the other hand, Formosa Plastics Corporation is constructing a rayon plant in Louisiana that is going to have a 68 kt annual production capacity due onstream in 1993.

2) Manufacture of hydrochloric acid from sodium chloride and sulfuric acid in the Mannheim furnace, and similar processes. In the Hargreaves process, sulfur, steam, and air are substituted for sulfuric acid. Many years ago the Mannheim furnace was the main source of sodium sulfate, and this chemical was truly a coproduct rather than a byproduct. Because the salt discharge from the furnace tended to form cakes, it was named salt cake. This name remains as a trade designation for low grade sodium sulfate. Large amounts of hydrochloric acid are produced today as a byproduct in the manufacture of chlorinated hydrocarbons. As a result, the Mannheim furnace process is generally considered to be obsolete; however, in certain geographical areas and in particular raw material situations it continues to be an economically significant source of sodium sulfate. Only the Climax operation in Hobbs, NM is listed as a current United States byproduct manufacturer.

3) As a byproduct in numerous other processes wherein certain sodium salts are converted to acids by reaction with sulfuric acid, or sodium alkalies are neutralized with sulfuric acid as part of the process. Current operating (or under construction) examples are the processing of chromium ores to produce sodium dichromate, and the production of lithium salts, vanadium pentoxide, boric acid, ascorbic acid, phenol, formic acid, silica pigments, and various catalysts. One aspect of several of these producers, both in this category and the preceeding, needs to be noted: their ability to change the byproduct. In the Mannheim furnace for instance, if potassium chloride is substituted for the sodium chloride then the byproduct would be potassium sulfate instead of sodium sulfate. The Climax producer in Hobbs has shown this capability. It is an economic decision and might have to include some changes in some of the processing equipment. But, it is not just limited to this one process, and should the market become extremely depressed, or perhaps if alternative opportunities arise, there could be further switching by the byproduct and coproduct manufacturers. Announcement of the construction of a Mannheim process in Japan was recently noted and is designed to produce potassium sulfate (Anon., 1991d).

4) A newly added category is that from reclamation. This includes reclamation of sulfate values with final form of sodium sulfate from liquid waste streams; from flue gas desulfurization units and reclamation of lead from discarded batteries. In the former, there have been a number of articles about the AquaTech process by Allied-Signal, Inc., which treats sulfate bearing streams from pulp mills, thus generating sodium sulfate for both internal process use or sale to other pulp mills with excess available for sale to the open market (Anon., 1991a). ICI recently introduced a unit which appeared to be quite similar (Anon., 1991c). Both of these processes are probable sources of synthetic production in the future (Fleming, 1991).

In 1985 the US Bureau of Mines published its pilot work with the citric process for the flue gas desulfurization (FGD) that has been installed in several coal-fired electric power generating plants. There have been a number of inquiries by engineering firms about the processing from the Glauber's salt phase, the final product shown in the report, to a final anhydrous material for sale on the open market. If left with Glauber's salt as a final form, most areas at present would permit it to go to a landfill. The latest annual report by the US Bureau of Mines shows two small producers with flue gas desulfurization as sources, which perhaps incorporate the citric or a similar process. It may be this process is more attractive than FGD units which generate gypsum.

Natural sodium sulfate remains the source for about half the market in the United States and has remained important worldwide during the past decades. Although vast resources are available, producers of the natural material currently are aware of the threat

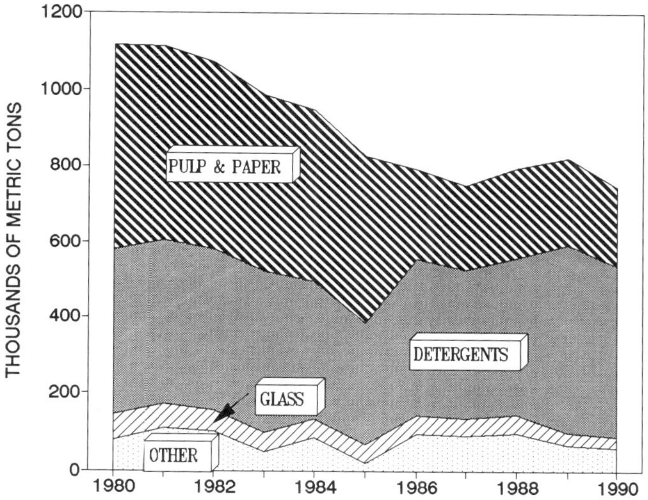

Fig. 5. Sodium sulfate consumption in the United States.

from byproduct sources. The outlook for the natural sodium sulfate business is certainly not the most promising it has ever been.

End Uses

Detergents: In the United States, the detergent industry has displaced the pulp and paper industry as the leading market for anhydrous sodium sulfate during the past decade. It did so by simply remaining at the same level. It rebounded from a large drop in the middle of the decade to end up at just about the same level in 1990 as it was in 1980 while the pulp and paper industry experienced a steady decline over the same period. We may witness the same phenomena with the detergent industry in the coming decade as the major formulators have made the decision to drastically reduce the amount of sodium sulfate used as a filler in their concentrate detergents (Anon., 1991b). Though it is very much an economic issue—supermarket shelf space being an extremely expensive item—there are environmental benefits which have appeal to the household consumer.

As used by the detergent industry, sodium sulfate is an inexpensive filler and diluent, since it is white, non-corrosive, and neutral. In addition, it is claimed to have very mild detergent properties. Formerly, about one-third of dry home laundry detergent was sodium sulfate. Now it is much less than that, perhaps a sixteenth or even as low as a sixty-fourth.

Kraft Paper: At the beginning of the 1980s, the manufacture of kraft paper pulp was the principal use of sodium sulfate. It has now dropped to a distant second place, being almost the sole segment of decline for the consumption of sodium sulfate. In the early part of the decade, the substitution of emulsified sulfur and caustic soda, which was much cheaper than salt cake, first made the impact on salt cake usage. At this time synthetic suppliers who had formerly produced truly a salt cake—notably the rayon industry and similar byproduct producers—had to upgrade their facilities to produce higher purity material. This trend began in 1982 and finally leveled out in 1986 at about the current level of just over the 200 kt mark.

The potential for this industry to be a net producer of sodium sulfate was mentioned earlier in the Synthetic Sodium Sulfate section.

Glass: Another important consumer of sodium sulfate is the glass industry. Sodium sulfate is known to have two favorable effects in the glass melter. It helps early reaction with unmelted batch, and its influence in the early stages of glass formation. It was noted by Peter Harben (Harben, 1991) that there is about twice as much sodium sulfate in a batch of container glass as there is in flat glass, running about 1 and .5%, respectively. This is in line with the concept that the more complicated the shape of the glass mold, the more sodium sulfate would be used, it being an aid to the pliability of the glass. It does appear that flat glass is much more demanding in certain aspects, such as particle size and limits of certain chemical contaminants. Both commodities demand the least variation possible for all of their raw materials—sodium sulfate is very definitely included.

Other: Perhaps the only bright spot in today's sodium sulfate marketplace is in the textile dyeing industry; although earlier noted as used in Europe and Japan, it was formerly a minor market in the United States. The leading manufacturer of dyeing equipment used in the world, Gaston County Dyeing Machine Company, reports that the use of sodium sulfate reduces corrosion of the stainless steel equipment used in the dyeing industry. A big impetus for the switchover from common salt to salt cake has come from the recent formulation of dyes so that sodium sulfate can be as efficient as common salt in penetrating the dye to the center of the thread.

A number of other uses for sodium sulfate still are important though the consumption is spread out to a large number of small users rather than the few large users of the previous industries. It is used as an ingredient of stock foods and remedies, in the manufacture of inks tanning of leather, manufacturing synthetic sponges, as well as many other uses.

The literature abounds with articles and patents for the use of Glauber's salt as one of the main ingredients in passive solar energy devices. The advantage it offers is, with appropriate other salts, it has its phase change at room temperature. It absorbs significant amounts of energy in the process of crystallizing, then as it melts it gives off energy at the same temperature. Thus, with the appropriate phase change temperature, by selection of the appropriate salts in the mixture, it can act to moderate the heat load in warming an environment or, conversely, moderate the cooling load in keeping an enclosure cool. No commercial applications have made a lasting impact though. The problems (no single source is cited) stem from the mixture being able to survive only a finite number of phase changes, then the mixture has to be either recharged or replaced. New and better mixtures are claimed, but as yet they have not made their way to the marketplace with any success.

It is difficult to estimate worldwide consumption patterns because of great variations from one country to the next. Canada, Sweden, and Finland are large producers of kraft paper pulp, and perhaps 95% of the sodium sulfate consumed in those countries is used for this purpose. We believe environmental concerns are having impacts overseas similar to those being experienced in the United States. In the United Kingdom, France, Germany, Japan, and many other countries, paper production is small, so the major consumption of sodium sulfate is in detergents, glassmaking, production of other chemicals, dyeing processes, and miscellaneous uses. Worldwide consumption patterns probably are not far from the estimate shown for the United States.

Product Specifications

There being no industry standards for designation of grades, types, or specifications, there is a proliferation of designations, most of which are not fully applicable throughout the industry they are supposed to describe. Detergent grade, for instance, has in common only purity—and even that is not completely uniform. A product that is 99.7% pure, yet has a trace of coloring, would not be acceptable to most detergent manufacturers. Those in the detergent industry are further differentiated by their tolerance for varying amounts of common salt, iron, and also have individual particle size considerations, sometimes completely at different ends of the spectrum from each other.

When perusing the literature, or perhaps when discussing the uses of sodium sulfate with those familiar with the industry and members of the fraternity of producers, it is not rare to hear or see the term salt cake applied loosely to mean all production of sodium sulfate. This term technically applies to sodium sulfate that is discolored, lumpy, and/or has contaminants which preclude its use except by the pulp and paper industry or other industries that can tolerate such attributes and where it is sold at reduced prices. Salt cake in this latter, and more accurately narrow connotation, can range in purity from 90 to 99% Na_2SO_4. As mentioned in the earlier discussion of the Mannheim Process, the sodium sulfate produced was relatively impure, discolored, and came out in the form of cakes—thus the term salt cake. It is synonomous with low purity sodium sulfate. Only a few natural resource producers have selected this grade for production, most of the time as captive sources for the pulp paper industry. They are able to accomplish this production with very little refining required in the beneficiation process.

High purity, also referred to as technical grade, refers to anhydrous sodium sulfate with a purity of 99% or greater. Different consumers for the same end-use may have quite different specifications. Particle size and/or strict limits on trace contaminants are important to almost all consumers of this grade of material. The glass industry, in particular, can be very strict about iron, copper, or any of the noble metals with limits in the parts per million. Most production from the natural resource producers can meet these stringent limitations, but only with some earnest effort.

Glauber's salt (mineralogically, mirabilite), the decahydrate of sodium sulfate, was once an item of commerce. Because the melting point of Glauber's salt is only slightly above room temperature, along with efflorescencing when exposed to ambient conditions, anhydrous is preferred for most volume usage. However, small quantities of Glauber's salt are still sold for some special uses. .

Processing

No two operations are exactly alike for any of the natural producers. It is only after the point where Glauber's salt is delivered to the evaporative process that strong similarities are apparent. Even within the same companies, or for different companies operating in the same vicinity, the stages up to and sometimes through the crystallization of Glauber's salt are different. It ranges from minor, though notable, differences, to extreme differences. After the Glauber's salt is fed to the evaporative system, however, the systems could be interchangable for the same final product being desired. That is, paper grade systems could be interchanged with each other or technical grade (high purity) systems might be interchanged and have little or no effect on the end product. It is the mining and extraction procedures up to this point that differentiate the producers. This is true worldwide. Each deposit differs somewhat in composition. Some of the deposits, such as those in Saskatchewan and Texas, are simple salts with only one product extracted. Those of the Great Salt Lake and Searles Lake, host a complex set of minerals, and a number of them are extracted in sophisticated processes in various forms and varieties. Yet, once the Glauber's salt is extracted, not too much difference remains in the final steps.

Canada: A number of structural changes have been made in the Canadian industry since the last edition of this book. Saskatchewan Minerals, formerly a Province Corporation, is now owned by Dickenson Mines, Limited. The Francana Operations now include Alberta Sulfate's Metiskow facility, which along with the Capri and Alsask operations, were acquired by Agassiz Resources, Limited. All of the Francana operations were reported producing in 1989, but in early 1991 went on standby status. At present, only four of the natural producers' locations are in production. These are: Orminston Mining and Smelting at Ormiston; two of Saskatchewan Minerals plants: Chaplin and Fox Valley; and Midwest Chemicals at Palo, all located in the province of Saskatchewan. Many Canadian producers utilize seasonal temperature differences to assist in the processing of the sodium sulfate deposits. Brine accumulates on the lakebed and becomes nearly saturated in the late summer months. Then it is pumped to large reservoirs which may be over 3 m deep. As cold weather sets in, Glauber's salt crystallizes from the saturated brine. The residual mother liquor, containing most of the magnesium and sodium impurities, is drained off. After draining, the Glauber's salt remains in the reservoirs until the coldest part of the season, usually January or February, when it is harvested and accumulated in stockpiles at the plant site.

A dredging system, pioneered by Orminston, is used at two of the lakes to supplement the stockpiled supply during the warmer season. Besides Orminston, Saskatchewan Minerals at Chaplin uses a dredge in the warm weather.

Obviously the Canadian plants are somewhat dependent on nature to produce the crystals of Glauber's salt. Drought or other unusual conditions can affect the production cycle. The plants try to keep more than one year's supply of Glauber's salt in their stockpiles to keep production from being interrupted by such conditions. This method has proved very successful in Canada, producing a salt cake of high quality.

To prepare a salable anhydrous product the Glauber's salt must be dehydrated. Because of the inverse solubility of sodium sulfate (i.e., less soluble at higher temperatures), this is not a simple, straightforward operation. In the early days dehydration was done in large rotary kilns, some of which were fired by lignite. None of the operating producers are using this method.

When Glauber's salt is melted, the solubility of anhydrous sodium sulfate in the water of crystallization is exceeded, causing precipitation. The balance of the water must be evaporated to produce an anhydrous product. Ozark-Mahoning's submerged combustion evaporators are used for this purpose by Orminston Mining and Smelting, Midwest Chemicals, and Saskatchewan Minerals at Fox Valley. The Saskatchewan Minerals plant at Chaplin uses a triple effect vacuum evaporator.

The choice of evaporative system is an economic issue, hinging mainly on two issues: Initial Capital Cost and Energy Efficiency. The choice is either low capital cost, such as Submerged Combustion Evaporators and other direct fired units with high operating cost for energy and maintenance, or high energy efficiency evaporators, such as the multiple-effect or mechanical vapor recompression evaporators, which have higher initial capital cost but enjoy low energy and operating costs. If high energy efficiency is chosen, a second issue is choosing between Multiple-Effect and Mechanical Vapor Recompression Evaporators. Then the critera of selection is the relative cost of electricity and gas or other heating fuel. The Multiple-Effect Evaporators use more heating fuel and less electrical energy than the Vapor Recompression Evaporators, but can be more cost effective depending on the relative cost. It was the relative cost of fuel that prompted Saskatchewan Mineral's Bishopric plant to choose the triple-effect over the vapor recompression evaporators when they replaced their Holland evaporator in 1983.

At all plants, the anhydrous salts precipitated in the evaporators may be separated from the saturated solution by mechanical means, including centrifuges and hydroclones, and the wet salt finally dried in direct-fired rotary kilns. Parallel-flow dryers are now preferred for ease of control and fuel economy, whereas in the older processes countercurrent flow was used in the dryers.

United States: The United States has also seen changes in corporate ownership. Kerr-McGee sold its Searles Lake plant to North American Chemicals, Inc. in 1990. North American Chemicals is an affiliate of Great Salt Lake Minerals and Chemicals Corp., another US natural sodium sulfate producer. The largest production of natural sodium sulfate in the United States still comes from Searles Lake, CA, even though output is much less than in the past. Prior to 1982, the Kerr-McGee Chemical Corp. plants made about 500 kt of sodium sulfate annually. The Trona plant, representing about half of the total sodium sulfate capacity, was shut down in 1982. The West End plant now is North American's only source of sodium sulfate.

The Searles Lake brine is complex and a number of salable products are extracted. At the West End plant, sodium sulfate is recovered along with soda ash and borax. Mixed brines are first carbonated with carbon dioxide to precipitate sodium bicarbonate, which is removed by filtration. The decarbonated brine is cooled to crystallize borax. This borax is separated from the brine, which is then cooled further to produce a crop of Glauber's salt. After it is removed, a final cooling recovers a second crop of borax. By heating, the sodium bicarbonate is converted to soda ash and the borax is either crystallized as a hydrate or dehydrated to anhydrous form. The Glauber's salt is washed, melted, and recrystallized as anhydrous sodium sulfate. The following description of the Trona plant is included because it is of interest, even though the sodium sulfate part is shut down.

In the Trona main plant cycle there were three major process plants. The brine from the lake was cooled to produce a crop of Glauber's salt which was removed and dried to sodium sulfate. The resultant mother liquor from this step was processed further to recover potash and borax. A different brine was treated in a liquid-liquid solvent extraction plant to recover boron values as boric acid. After it was removed, mixed sodium and potassium sulfates were precipitated by evaporation and these mixed sulfates were fed to the potassium sulfate plants. The Trona carbonation plant was somewhat similar to the West End plant in that carbon dioxide precipitated sodium bicarbonate which was filtered and removed. Borax was then precipitated by cooling. Finally, burkeite, a double salt of sodium sulfate and sodium carbonate, was recovered and fed to the Argus process.

In the Argus plant, which is the newest addition, brines from the lake are mixed with burkeite solutions and streams from other plants to recover additional soda ash. Nothing has been published since Moulton (1980), so the details of present plant practice at Trona have not been revealed.

Ozark-Mahoning Company, the nation's second largest natural producer, has seen a change of ownership occur recently. Ozark-Mahoning had been a wholly owned subsidiary of Pennwalt since 1974. Pennwalt was purchased by Elf Aquitaine, a French Company, in 1989. Pennwalt, along with Atochem, Inc., M & T Chemicals, was merged into the present Elf Atochem North America, Inc., in 1990. One of Ozark's two sodium sulfate operations was closed in September 1987 because of poor market conditions and increasing cost associated with the water table rising in the plant area. It has been placed on standby status and could resume production if market conditions improve. The Brownfield plant, built in 1956 and expanded to its present capacity of 64 kt in 1959, was the smaller of the two plants operated by Ozark-Mahoning from brine deposits in West Texas. The other, near Seagraves, was built in 1965 and expanded to its present capacity of 141 kt in 1985 while incorporating the technically more advanced evaporation system that had been installed at Brownfield in 1982. Both of the plants use a refrigerization process to crystallize the Glauber's salts that had been developed in the early 1930s for a plant near Monahans, TX. The Monahans plant closed in 1970 because the deposits were essentially exhausted. The Ozark submerged combustion burner used in the evaporative section of the old Monahans plant probably represented the first commercial application of submerged combustion in the United States.

The Texas brine contains both sodium chloride and sodium sulfate, as well as smaller amounts of magnesium and potassium salts. At the Brownfield plant a portion of the brine was pumped through the Permian salt bed to introduce sodium chloride. The high sodium chloride content depressed the solubility of sodium sulfate, improving it extraction. At the Seagraves plant sufficient sodium chloride is present in the brine; additional salt is not needed. The Texas climate does not lend itself to the production of Glauber's salt by atmospheric chilling, so mechanical refrigeration and heat exchange are required. The Glauber's salt formed in this manner is filtered and washed, yielding a product of high purity.

Melting and dehydration of Glauber's salt is accomplished by a mechanical vapor recompression evaporation system, which has much improved fuel efficiency over the former submerged combustion units, even after the Seagraves plant had been updated with a heat recovery system in conjunction with the submerged combustion units. Hydroclones and centrifuges separate the anhydrous crystals from the saturated solution, which is returned to the evaporators. The crystals are dried in parallel-flow rotary kilns.

The other producer of natural sodium sulfate in the United States is Great Salt Lake Minerals & Chemicals Corp. The plant began production late in 1970. The initial rated capacity announced was 15 kt of sodium sulfate, plus 220 kt of potassium sulfate and 90 kt of anhydrous magnesium chloride. Solar evaporation is an important step in processing, with several complex salts harvested from evaporation ponds for further refining in the plant by proprietary processes. Early operations were beset with difficulties, but these problems were overcome. Very little has been published on production processes used by the company. A flood on May 5, 1984, took this operation out of production for a prolonged period (Searls, 1984).

Great Salt Lake contains tremendous reserves of sodium sulfate and no doubt will remain a factor of some importance in the future.

Mexico: The brine processed at Quimica del Rey, S.A. de C.V., is very strong and probably saturated. It contains more than 26% sodium sulfate and appreciable amounts of chlorides and magnesium. About 65 wells produce the brine from a depth of approximately 12 m. The brine temperature is 25°C before precooling with the spent brine. The cooled brine is fed to a vacuum crystallizer where it is cooled to 9°C by evaporative cooling, causing Glauber's salt to crystallize. The slurry is thickened, filtered, and melted. Evaporation is accomplished in triple-effect evaporators. Final drying is in co-current-fired rotary kilns. Part of the mother liquor is processed further for recovery of magnesia products while the balance goes to waste.

The most recent expansion of this plant occurred in 1988. Seven separate production lines are capable of producing 510 ktpy at which time the dryers were converted from counter-current to co-current firing.

Other Countries: Other than for Spain, little is published about processing in other countries. The little that has been seen is consistent with the earlier statement that the mining portion differs, extracting being by incorporating some form for crystallization of Glauber's salt, while the evaporative portion is in line with the techniques applicable for the particular end use of the material. It can be reasonably assumed that methods used are similar to those used in the United States, Mexico, and Canada, wherein chilling by natural or mechanical means precipitates Glauber's salt, allowing separation from other salt. The Glauber's salt would then be dried to anhydrous sodium sulfate.

In Argentina saturated brine from a lakebed is pumped to crystallization tanks, where great night and day temperature variation results in precipitation of Glauber's salt. This is possible because the plant site is at an elevation of 3 353 m. The Glauber's salt is melted by heating; the precipitated anhydrous sodium sulfate is then separated and dried. There is no evaporation of water so the saturated solutions probably are recycled through the system.

Deposits in Spain, one of the few countries where thenardite is the starting material, are mined underground and present special problems because the minerals contain an appreciable amount of calcium sulfate in the form of glauberite ($Na_2SO_4 \cdot CaSO_4$). Countercurrent leaching at temperatures between 35 and 40°C dissolves most of the sodium sulfate, leaving calcium sulfate undissolved. The solids, which probably contain other insoluble impurities, are settled in large Dorr thickners. A second leaching step with water at 29°C results in maximum recovery of sodium sulfate. Evaporation is carried out in a single-effect vacuum evaporator with forced circulation. Final drying is done in a fluid bed dryer with hot air heated by steam (Rovira, 1960).

Prices

Published price information for sodium sulfate usually only approximates the true market price. In the United States prices usually are based on quotations from byproduct sources located near consuming points. Because producers of the natural salt are located at great distances from most customers, the FOB plant value for natural material may be much less than the published price. On the other hand, in periods of shortages and rising prices, published information often lags behind actual selling levels.

The Dec. 6, 1991, *Chemical Marketing Reporter* quoted sodium sulfate in bulk carload lots, works, freight equalized, in the East, at 126 to $128 per ton. It also quoted sodium sulfate, in the West, bulk, carload lots, freight equalized, at 136 to $138 per ton. Salt cake was quoted at $72 per .9 ton. United States tariff rates for sodium sulfate are as follows:

	Most Favored Nation	Non-most Favored Nation
Crude (salt cake)	Free	Free
Anhydrous	36.4¢/t	$3.30/t

For tax purposes a depletion allowance of 14% for both domestic and foreign deposits is permitted by the Internal Revenue Service.

ACKNOWLEDGMENTS

The authors acknowledge the valuable assistance and suggestions received from Dennis Kostick, Sodium Compounds Commodity Specialist, US Bureau of Mines, as well as his equivalent in Energy, Mines and Resources Canada, Patrick Morel-a-l'Huissier. We wish to thank S.W. Fletcher of the North American Chemical Company for review and permission to print the graphics on Searles Lake. We appreciate the review of the sections on Mexico and Spain by Alberto Ross of Servicios Industriales Penoles, S.A. de C.V. We thank Tariq Mahmood, Allen Karpman of Elf Atochem and Ruby Felan of Ozark-Mahoning for their chemical review, Russian translations, and manuscript preparation, respectively. We thank Warren Wood of the US Geological Survey for his help. In addition, we thank the following specialists with the US Bureau of Mines: Lloyd Antonides (Iran and Botswana), Hendrik Van Oss (Turkey), and Richard M. Levine (Russian Republics).

BIBLIOGRAPHY AND REFERENCES

Anon., 1967, "The Sodium Sulfate Industry in Saskatchewan," unpublished report, Industrial Minerals Division, Saskatchewan Department Mineral Resources, 8 pp.

Anon., 1970a, "Great Salt Lake Starts $30 Million Pond-Plant Complex," *Chemical & Engineering News*, Vol. 48, No. 51, Dec. 7, p. 17.

Anon., 1970b, "Utah Company Gets Set to Tap Mineral Wealth of Great Salt Lake," *Engineering & Mining Journal*, Vol. 171, No. 4, pp. 67–70.

Anon., 1971a, "Chemical Profile, Sodium Sulfate," 2nd rev., *Chemical Marketing Reporter*, Oct. 1.

Anon., 1971b, "1971 World Review," *Pulp and Paper*, World Review ed., pp. 150–154.

Anon., 1972a, "Business Newsletter," *Chemical Week*, Vol. 110, No. 20, May 17, p. 17.

Anon., 1972b, "Harvest of Sulfate of Potashes to be 3–4 Times Greater in 1972," *The Mining Record*, Oct. 11, p. 3.

Anon., 1972c, "Petrochemicals Get the Emphasis in Mexico Buildup," *Chemical Week*, Vol. 110, No. 20, May 17, p. 33.

Anon., 1972d, "Sodium Sulphate: World Survey of Production, Consumption and Prices with Special References to Future Trends," Roskill Information Services, Ltd., London.

Anon., 1980, *Industrial Research and Development*, Vol. 22, No. 10, pp. 67–69.

Anon., 1981, "Evaluacion Preliminar del Yacimiento de Sulfato de Sodio en Bacadehuachi, Son.," Informe de Avance, Direction de Mineria, Geologica y Energeticos, State of Sonora, Mexico, Oct.

Anon., 1983, "Mineral Notes," *Industrial Minerals*, No. 187, Apr., p. 68.

Anon., 1991a, "Allied Sulfate Recycler To Be Marketed by Ahlstrom," *Chemical Marketing Reporter*, Vol. 237, No. 24, June 11, p. 9.

Anon., 1991b, "Detergent Compact Shift Makes Sulfate Outlook Grim," *Chemical Marketing Reporter,* Vol. 239, No. 17, Apr. 29, p. 5.

Anon., 1991c, "Company News," *Industrial Minerals*, No. 286, July, p. 53.

Anon., 1991d, "Mineral Notes," *Industrial Minerals*, No. 286, July, p. 54.

Anon., 1991e, "World of Minerals," *Industrial Minerals*, No. 286, July, p. 9.

Antonides, L., 1988, "Botswana," *Minerals Yearbook*, Vol. III, US Bureau of Mines, 167 pp.

Ashley, B.E., 1972, "Mineral Industry of Mexico," Preprint, *Minerals Yearbook*, US Bureau of Mines, 13 pp.

Bateman, A.M., 1950, *Economic Mineral Deposits*, 2nd ed., John Wiley and Sons, New York, pp. 194–196.

Binyon, E.O., 1952, "North Dakota Sodium Sulfate Deposits," Report of Investigations 4880, US Bureau of Mines, 41 pp.

Broughton, P.L., 1977, "Natural Salt Cake in Canada," *Industrial Minerals*, No. 121, Oct., pp. 51–59.

Chin, E., 1983, "Mineral Industry of China," *Minerals Yearbook*, 193 pp.

Chin, E., 1986, "Mineral Industry of China," *Minerals Yearbook*, 218 pp.

Cohenour, R.E., 1966, "Salt Inventory of Great Salt Lake," *2nd Symposium on Salt*, Vol. 1, Northern Ohio Geological Society, Cleveland, p. 212.

Cole, J.W., 1970, "Mineral Industry of Chile," *Minerals Yearbook*, Vol. 3, US Bureau of Mines, pp. 197–206.

Cole, L.H., 1926, "Sodium Sulphate of Western Canada," No. 646, Canada Department of Mines, pp. 26–27, 74–78, 113.

Corrick, J.D., 1972, "Mineral Industry of Spain," Preprint, *Minerals Yearbook*, US Bureau of Mines, 21 pp.

Crosby, E.J., 1973, "Sodium Sulfate," Mineral and Water Resources of North Dakota, Bulletin 63, North Dakota Geological Survey, pp. 151–154.

Dana, E.S., 1932, *A Textbook of Mineralogy*, 4th ed., John Wiley and Sons, New York, pp. 809–810.

Davis, S.N., and DeWiest, R.J.M., 1966, *Hydrogeology*, John Wiley and Sons, New York, pp. 108–109.

Dewey, L.S., 1972, "Mineral Industry of Argentina," Preprint, *Minerals Yearbook*, Vol. 3, US Bureau of Mines, 11 pp.

Dietrich, R.V., 1969, *Mineral Tables*, McGraw-Hill, New York.

Dort, W., Jr., and Dort, D.S., 1970, "Low Temperature Origin of Sodium Sulfate Deposits, Particularly in Antarctica," *3rd Symposium on Salt*, Vol. 1, Northern Ohio Geological Society, Cleveland, pp. 181–203.

Eissenberg, D., and Wyman, C., 1980, "What's in Store for Phase Change?" *Solar Age*, Vol. 5, No. 5, pp. 12–16.

Ellis, W.S., 1990, "A Soviet Sea Lies Dying," *National Geographic*, Vol. 177, No. 2, 73 pp.

Erickson, G.E., et al., 1989, "Mineralogical Studies of the Nitrate Deposits of Chile; VI, Hectorfloresite, $Na_9(IO_3)(So_4)_4$, a new saline mineral," *American Mineralogist*, Vol. 74, Nos. 9–10, pp. 1207–1214.

Fleming, B.I., 1991, "Chlorine & Caustic in Pulp Bleaching: Future

Trends,'' Paper, 3rd Annual Canadian Conference On Markets For Industrial Minerals, Industrial Minerals '91, Oct. 22–23, Toronto, p. 17.

Fletcher, S.W., Jr., 1979, Talk Presented to the Fall 1979 Seminar, Raw Material Section, American Paper Institute, Oct.

Garbell, M., 1963, ''Sea that Spills into Desert,'' *Scientific American*, Vol. 209, Aug., pp. 94–100.

Goudge, M.F., and Tomkins, R.V., 1960, ''Sodium Sulfate from Natural Sources,'' *Industrial Minerals and Rocks*, 3rd ed., J.L. Gillson, ed., AIME, New York, pp. 809–818.

Grabau, A.W., 1920, ''Geology of the Non-Metallic Mineral Deposits,'' *Principles of Salt Deposition*, 1st ed., Vol. 1, McGraw-Hill, New York, pp. 29–40, 277–279.

Grossman, I.G., 1949, ''Sodium Sulfate Deposits of Western North Dakota,'' Report of Investigation No. 1, North Dakota Geological Survey, 66 pp.

Grossman, I.G., 1968, ''Origin of the Sodium Sulfate Deposits of the Northern Great Plains of Canada and the United States,'' Professional Paper 600-B, US Geological Survey, pp. B104-B105.

Harben, P., 1991, ''Glass Raw Materials, Aspects of Quality, Quantity, and Prices,'' *International Minerals*, No. 286, pp. 31–43.

Hardie, L.A., and Eugster, H.P., 1970, ''The Evolution of Closed-Basin Brines,'' Special Paper 3, Mineralogical Society of America, pp. 273–290.

Havighorst, C.R., 1963, ''AP&CC's New Process Separates Borates From Ore by Extraction,'' *Chemical Engineering*, Vol. 70, No. 23, Nov. 11, pp. 228–232.

Hemy, G.W., 1969, ''Chemical Minerals in the U.S.S.R,'' Part I, *Chemical and Process Engineering*, Vol. 50, No. 8, pp. 63–68.

Holland, E., 1980, ''The Mattapoisett House,'' *Solar Age*, Vol. 5, No. 5, pp. 24–28.

Jolly, J.L.W., 1976, ''The Mineral Industry of Iran,'' *Minerals Yearbook*, US Bureau of Mines, pp. 12–13.

Jones, B.F., 1966, ''Geochemical Evolution of Closed Basin Water in the Western Great Basin,'' *2nd Symposium on Salt*, Vol. 1, Northern Ohio Geological Society, Cleveland, pp. 181–200.

Karayazici, F., 1988, ''Industry Structure: Past, Present and Future,'' *Industrial Minerals*, Supplement, Mar., 19 pp.

Klingman, C.L., 1972, ''Sodium and Sodium Compounds,'' *Minerals Yearbook* 1972, US Bureau of Mines, pp. 1147–1151.

Koelling, G.W., 1971 ''Mineral Industry of Argentina,'' *Minerals Yearbook*, Vol. 3, US Bureau of Mines, pp. 87–98.

Kolosov, A.S., Pustyl'nikov, A.M., and Fedlin, V.P., 1974, ''Precipitation of Glauberite and Conditions of Deposition of Glauberite-Bearing Sediments in the Kara Bogaz Gol,'' *Doklady Akademii Nauk USSR*, Vol. 219, No. 6, pp. 1457–1460.

Kostick, D.S., 1978–1979, ''Sodium and Sodium Compounds,'' Preprint, *Minerals Yearbook, 1978–1979*, US Bureau of Mines 8 pp.

Kostick, D.S., 1980, ''Sodium and Sodium Compounds,'' *Minerals Yearbook, 1980*, US Bureau of Mines.

Kostick, D.S., 1982, ''Sodium Sulfate,'' *Mineral Commodity Summaries*, 1982, US Bureau of Mines, pp. 144–145.

Kostick, D.S., 1990, ''Sodium Sulfate,'' *Annual Report*, US Bureau of Mines, p. 12.

Kostick, D.S., 1991, ''Sodium Sulfate,'' *Annual Report*, US Bureau of Mines, July, 12 pp.

Krauskopf, K.B., 1967, *Introduction to Geochemistry*, McGraw-Hill, Inc., New York, p. 320.

Kurilenko, V.V., Ruday, I.G., and Shvarts, A.A., 1988, ''Formation Conditions and Suitability for Industrial Use of Subsurface Brines from the Northern Part of the Kara-Bogaz-Gol Gulf [USSR],'' *International Geology Review*, Vol. 30, No. 11, p. 1238.

Kurilenko, V.V., and Frolovskii, E.E., 1982, ''Hydrogeochemical and Hydrodynamic Regime of Intercrystalline Subsurface Brines of the Kurguzul Inlet,'' Nov. Dannye Geol., Goekhim; Podzemn. Vodam Ploezn. Iskop. Solenosnykh Basseinov, [Vses. Solevoe Soveshch. ''Solenosnye Form. Prakt. Znach. Ikh Izuch.''], 2nd Meeting Date 1979, 67 pp., A.L. Yanshin and M.A. Zharkov, eds., Izd. Nauka, Sib. Otd., Novosibirsk, USSR.

Lang, W.B., 1941, ''New Sources of Sodium Sulfate in New Mexico,'' *Bulletin*, Vol. 25, Jan., American Association of Petroleum Geologists, pp. 152–160.

Last, W.M., and Slezak, L.A., 1986, ''Sodium Sulfate Deposits of Western Canada: Geology, Mineralogy and Origin,'' *Economic Minerals of Saskatchewan*, Special Publication Number 8, C.F. Gilboy and L.W. Vigrass, eds., Saskatchewan Geological Society, pp. 197–205.

Levine, R.M., 1988, ''Mineral Industry of U.S.S.R.,'' *Minerals Yearbook*, Vol. 3, US Bureau of Mines, pp. 789–790.

Massey, N.W.D., 1973, ''Resources Inventory of Botswana: Industrial Rocks and Minerals,'' Mineral Resources Report No. 3, Geological Survey Department, 39 pp.

McCoy, M., 1991, 'Detergent Compact Shift Makes Sulfate Outlook Grim,'' *Chemical Marketing Reporter*, Vol. 239, No. 17, Apr. 29, p. 5.

Mitchell, R.S., 1979, *Mineral Names: What Do They Mean?* Van Nostrand Reinhold Company, New York, pp. 155, 190.

Moore, L., 1972, ''Mineral Industry of Chile,'' Preprint, *Minerals Yearbook*, Vol. 3, US Bureau of Mines, 12 pp.

Moulton, G.F., 1980. ''Compendium of Searles Lake Operations,'' Preprint 80–133, SME-AIME Annual Meeting, Las Vegas, Feb., 6 pp.

Nakhla, F.M., Saleh, S.A., and Gad, N.L., 1986, ''Geology and Mineralogy of Thenardite (Na_2SO_4) Deposit of Beida Lake, Wadi El-Natrun, Egypt,'' *Crystal Chemistry of Minerals*, papers and proceedings of the General Meeting-International Mineralogical Association. Vol. 13, pp. 891–898, Sep. 19–25, 1982, Varna, Bulgaria.

Nissen, W.I., et al., 1985, ''The Citrate Process for Flue Gas Desulfurization,'' Bulletin 686, US Bureau of Mines.

Parkinson, G., 1977, ''Kerr-McGee Expands Soda Ash Output Nine-fold from Searles Lake Brines,'' *Engineering and Mining Journal*, Vol. 178, No. 10, pp. 71–75.

Rathjen, J.A., 1972, ''Minerals Industry of Canada,'' Preprint *Minerals Yearbook 1972*, Vol. 3, US Bureau of Mines, 21 pp.

Ralston, O.C., 1949, ''Natural Sodium Carbonate and Sodium Sulphate,'' *Industrial Minerals and Rocks*, 2nd ed., S.H. Dolbear and O. Bowles, eds., AIME, New York, pp. 945–962.

Reeves, C.C., Jr., 1968, *Introduction to Paleolimnology*, Elsevier Publishing Co., pp. 68–69.

Rovira, J.M.R., 1960, ''Sodium Sulfate: Its Recovery from the Minerals Containing It,'' *ION*, No. 233, Dec.

Ruffel, P.G., 1970, ''Natural Sodium Sulfate in North America,'' *3rd Symposium on Salt*, Vol. 1, Northern Ohio Geological Society, pp. 429–451.

Sanford, W.E., and Wood, W.W., 1991, ''Brine Evolution and Mineral Deposition in Hydrologically Open Evaporite Basins,'' *American Journal of Science*, Vol. 291, Sep., pp. 687–710.

Searls, J.P., 1984. ''Sodium Compounds,'' Preprint, *1984 Minerals Yearbook*, US Bureau of Mines, p. 2.

Shekarchi, E., 1972, ''Mineral Industry of Turkey,'' Preprint, *Minerals Yearbook*, Vol. 3, US Bureau of Mines, 11 pp.

Smith, G.I., 1966, ''Sodium Sulfate,'' Bulletin 191, Mineral Resources of California, California Division of Mines and Geology, pp. 389–392.

Smith, G.I., 1979, ''Subsurface Stratigraphy and Geochemistry of the Late Quaternary Evaporites, Searles Lake, Calif.,'' Professional Paper 1043, US Geological Survey.

Taylor, R., 1990, ''Minor ingredients and their effects on glass melting and properties,'' *Glass Production Technology International*, p. 33.

Watson, I., 1980, ''Sodium Sulphate: Present Shortage on Short-term Basis,'' *Industrial Minerals*, No. 152, May, pp. 61–77.

Weisman, W.I., and Anderson, R.C., 1953, ''The Production of Sodium Sulfate from Natural Brines at Monahans, Texas,'' *Mining Engineering*, Vol. 5, No. 7.

Weisman, W.I., and Tandy, C.W., 1975, ''Sodium Sulfate Deposits,'' *Industrial Minerals and Rocks*, 4th ed., S.J. Lefond, ed., AIME, New York, pp. 1081–1092.

Wells, R.G., 1923, ''Sodium Sulphate: Its Sources and Uses,'' Bulletin 717, US Geological Survey, pp. 17–18.

Wood, W.W., and Blair, B.F., 1990, ''Origin of Solutes in Saline Lakes and Springs on the Southern High Plains of Texas and New Mexico,'' *Geologic Framework and Regional Hydrology: Upper Cenozoic Blackwater Draw and Ogallala Formations, Great Plains*, T.C. Gustavson, ed., Texas Bureau of Economic Geology, pp. 193–207.

Wood, W.W., and Sanford, W.E., 1990, ''Ground-Water Control of Evaporite Deposition,'' *Economic Geology*, Vol. 85, pp. 1226–1235.

Younge, C.J., and Krouse, H.R., 1987, ''The Origin of Sulfates in Castleguard Cave, Columbia Icefields, Canada,'' *Chemical Geology*, Isotope Geoscience Sec., Vol. 65, No. 3–4, Aug. 20, pp. 427–433.

STAURÓLITE

ROBERT B. FULTON, III

Staurolite, an iron aluminum silicate mineral, is used industrially as a high value-in-use sandblasting agent, as a premium-grade foundry sand, and as the source of aluminum in portland cement manufacture in areas where the aluminum constituent is not economically available from shale or argillaceous limestone.

Staurolite is produced by E.I. duPont de Nemours and Company as a coproduct of titanium mineral production in Florida, where it was first recovered on a commercial scale in 1952 from the Pleistocene sand deposit known as Trail Ridge. At this deposit it comprises about one-fifth of the heavy mineral suite, the principal other heavy minerals being ilmenite, leucoxene, rutile, zircon, kyanite, sillimanite, and tourmaline. Staurolite and other nonconductive silicate heavy minerals are separated from the conductive titanium minerals on high-tension electrostatic separators. The silicates are then separated from one another magnetically, the staurolite being relatively magnetic. In the total process, 200 t of raw sand yield 7 t of heavy mineral concentrate from which about 1 t of staurolite is recovered. The staurolite concentrates have the characteristics given in Table 1.

Staurolite concentrates are sold in three grades having the size distribution of subrounded grains shown in Table 2. All three grades are used in sandblasting for cleaning metal, for paint removal, and for cleaning masonry buildings.

Staurolite compares favorably with lower priced abrasives, such as boiler slags and silica sand, in applications where the abrasive is used only once and is not recovered and recycled. Because it is hard, fine, clean, and dense it can reduce blasting time, and less material is required to clean a specific area. Compared to higher priced abrasives such as aluminum oxide and garnet, which are recycled, staurolite costs significantly less per kilogram and cleans at equal or faster rates. Its breakdown factor, which is a measure of recycling performance, is intermediate between those for aluminum oxide and garnet.

Table 1. Characteristics of Staurolite Concentrates

Typical Mineral Composition, %	
Staurolite	85
Titanium minerals	10
Quartz	3
Typical Chemical Composition, %	
Al_2O_3	45 (min.)
Fe_2O_3	18 (max.)
ZrO_2	3 (max.)
TiO_2	5 (max.)
Free silica	4 (max.)
Physical Characteristics	
Bulk density	2 176 kg/m^3
Specific gravity	3.60–3.65
Hardness (Mohs)	7.0–7.5
Melting point	1 537°C
Coefficient of expansion	7.8×10^{-6}

Tables 3 and 4 show the blast cleaning performance of the three staurolite grades compared to other abrasives. Table 5 compares the longevity of premium priced abrasives during recycling. Table 6 shows a cost comparison between silica sand and staurolite.

Fringe benefits claimed for staurolite's use as an abrasive are:

1. Good flow properties and freedom from the lumps or slivers found in silica sand or boiler slag abrasives.
2. Low dusting affords good visibility, permits faster working, and reduces cleanup problems.
3. Heated storage is not required during freezing conditions, because the mineral is nonhygroscopic.

Table 2. Particle Size Distribution of Three Staurolite Products

		% Retained on Sieve		
US Standard Sieve No.	mm	Coarse	Biasill®	Starblast®
20	0.84	1	—	—
30	0.59	3	—	—
40	0.42	26	1	3
50	0.30	58	5	13
70	0.21	9	19	25
100	0.149	—	52	36
140	0.105	—	19	18
200	0.074	—	3	3
270	0.053	—	1	1
Pan	—	3	Trace	None

Table 3. Cleaning Rate and Profile for Blast Cleaning Abrasives

Abrasive	Cleaning Rate m^2/hr	Abrasive Flow Rate kg/hr	Profile mm
Starblast®	22	272	0.028
Biasill ®	22	263	0.023
Coarse Staurolite	19	243	0.038
Rotoblast® G80A	14	318	0.030
Rotoblast® G40A	11	340	0.053
Garnet RT-80	21	290	0.025
Garnet RT-60	18	238	0.025
Aluminum Oxide			
30 FDT	16	302	0.038
54/70	21	261	0.033
80/100	23	245	0.028
Black Beauty® 50	15	340	0.038
Flintbrasive® #3	11	236	0.043
Flintshot®	19	318	0.043
Flint Silica	20	331	0.036

Tests conducted at DuPont Engineering Test Center in 1970. 100% mill scale on new steel plate. All surfaces to Class 1 white metal. Conditions: 0.95 cm nozzle; 634-655 × 10^3 Pa nozzle pressure.

Table 4. Blast Cleaning Field Trials

Abrasive	Degree of Cleaning	Cleaning Rate m^2/hr	Abrasive Flow Rate kg/hr	Profile mm
	New Pipe—Intact Mill Scale			
Polygrit®	Commercial	16	617	0.044
Flintshot®	Near white	29	383	0.033
Biasill®	White	27	258	0.033
Coarse Staurolite	Commercial/ near white	29	293	0.029
Starblast®	White	25	287	0.030
	New Pipe—Mill Scale and Rust			
Polygrit®	Commercial	21	748	0.049
Flintshot®	Commercial	22	399	0.037
Biasill®	Near white/ white	20	247	0.035
Coarse Staurolite	Commercial	21	394	0.038
Starblast®	Near white	22	290	0.037

Tests conducted by William W. Pearson, Inc., Chester, PA, 1971.
Conditions: 1.27 cm nozzle; 517–522 × 10^3 Pa nozzle pressure.

Table 5. Breakdown Factors for Abrasives

Abrasive	Number of Uses	Breakdown Factor*
Aluminum Oxide 30 FDT	1	0.80
	3	0.61
Garnet RT 60	1	0.61
	3	0.38
Starblast®	1	0.77
	3	0.45
Flintshot®	1	0.68
	3	0.36
Flintbrasive®	1	0.58
	3	0.32

*Breakdown factors can range from 1 to 0. A material showing no change after blasting has a factor of 1, while one that breaks completely into dust has a factor of 0.
Tests conducted at DuPont Engineering Test Center, 1970.
Conditions: 0.64 cm nozzle; 634–655 × 10^3 Pa nozzle pressure; 0.64 cm steel plate target.

Table 6. Cost Comparison Using Silica Sand and Staurolite

	Silica Sand	Staurolite
Labor Cost to Blast 100 m^2 ($)	247	113
Abrasive Cost ($/ton)	27	90
Total Cost to Blast 100 m^2 ($)	334	205
Abrasive Cost to Blast 100 m^2 ($)	86	91
%Total Cost for Labor	74	55

Source: Modified from Seavey, 1985.

The ownership of trademarks for the abrasives shown in Tables 2, 3, and 4 is as follows:

- Starblast® and Biasill®—staurolite from E.I. duPont de Nemours and Co., Inc.
- Rotoblast®—manufactured metallic grit from Pangborn Corporation.
- Black Beauty®—slag grit from H. B. Reed Corporation.
- Flintbrasive®—abrasive grit from The Independent Gravel Company.
- Flintshot®—silica sand from Ottawa Silica Company (now US Silica Company).
- Flint Silica—silica sand from Ottawa Silica Company.
- Polygrit®—slag grit from MDC Industries.

BIBLIOGRAPHY

Anon., 1972, "Starblast Abrasive," Technical Bulletin, E.I. du Pont de Nemours & Co.

Huskonen, W.D., 1970, "Spotlight on Specialty Sands," *Foundry Magazine,* Oct., pp. 40–45.

Laux, J.P., and Lipton, J.M., 1972, "An Introduction to Staurolite Sand," *Foundry Magazine,* May, pp. 158, 160.

Seavey, M., 1985, "Abrasive Blasting Above 100 PSI," *Journal of Protective Coatings,* Vol. 2, No. 7, pp. 26–37.

Wagner, A.J., 1972, "Biasil Staurolite Sand," *Modern Casting,* Feb., pp. 64–65.

STONE, CRUSHED

DEAN H. HERRICK

The production of crushed stone in the United States has exceeded 1.0 Gtpy since 1987. Eighty percent of the crushed stone produced is used in the aggregate industry, accounting for over half of the quantity of natural aggregate consumed in the United States. Approximately 71% of the crushed stone produced in 1989 was limestone and dolomite followed by granite, traprock, sandstone and quartzite, miscellaneous stone, marble, shell, calcareous marl, volcanic cinder and scoria, and slate in order of descending volume (Table 1).

Of the remaining 20% crushed stone produced in the United States and not used as aggregate, 12% is used in the manufacture of cement which is an important construction material. Only 8% of crushed stone is used in specialty products not associated with the construction industry. Therefore, this chapter will concentrate on crushed stone as used in aggregates, with occasional reference to other uses or to chapters which discuss specific nonconstruction uses of crushed stone.

Construction aggregates are hard, essentially inert materials suitable for being formed into a stable mass by the addition of cementing or binding materials that produce a concrete by compaction, or by natural weight to produce a road or foundation fill. Aggregates are usually from 80 to 100% of the material volume of the construction masses in which they are used.

Conditions necessary for a rock deposit to be developed for construction crushed stone include:

- Quality—should pass rigid specifications for strength and durability;
- Quantity—adequate volume of rock is needed to have a production life of at least 10 years, preferably 20 or more years;
- Market—must have adequate market to support a new crushed stone operation;
- Transportation—transportation cost must be competitive for the intended market;
- Environmental—the impacts of mining, processing and transportation of crushed stone on the environment at or surrounding the mine site must be within acceptable limits;
- Permitability—mining, processing and transportation of crushed stone must usually be permitted by governing agency.

The following discussion will touch upon all facets of crushed stone operations from the mineral deposit and its geology to technology, uses, economic factors and end with governmental regulations of mining crushed stone. The sections are not designed to answer all questions, but rather to be an introduction to the crushed stone industry. Where more information is required, *The Aggregate Handbook* by Barksdale (1991) is recommended as the best general source.

GEOLOGY

Mineralogy

Because rocks which are suitable for crushed stone are among the most common sedimentary, igneous, and metamorphic rocks, the major minerals which are found in crushed stone products are the most common in the crust of the earth. Calcite, dolomite, feldspars, quartz (and other forms of SiO_2), pyroxenes, amphiboles, and micas are the major minerals found in most rocks from which crushed stone can be made.

Excessive amounts of certain minerals may make rocks weak, excessively sensitive to changes in moisture, or chemically reactive and may be objectionable for some crushed stone uses. These minerals will be discussed under chemical and physical properties in the sections following.

Chemical Properties

Chemical properties of rocks are important in determining the quality of a deposit for crushed stone use. Ideally, aggregate is an inert filler and should not change chemically in use. However, some rocks contain minerals that are chemically reactive in portland cement, bituminous concrete, or in other environments. In portland cement concrete, two types of chemical reactions between the aggregate and cement can occur. Alkali-silica reactivity occurs when silica reacts with alkali in the cement to form a gel around the aggregate. The gel has an increased volume and causes deterioration of the concrete. Rocks containing silica in the form of natural glass, chalcedony, opal, chert, finely divided quartz or strained quartz are potentially reactive with high alkali cement. Alkali-carbonate reactivity takes place when dolomitic limestones of moderate to high clay content interact with high alkali cements (Rogers, 1979). Expansion appears to be due to microfracturing throughout the cement and rock particles as well as in reaction rims surrounding the carbonate rock particles. Little or no alkali silica gel occurs and does not appear to be the cause of expansion.

Sulfides, primarily pyrite, marcasite, and pyrrhotite, react with oxygen and water to form iron hydroxides and sulfates. When excessive iron sulfides are present in crushed stone pavement, discoloration may be objectionable and some expansion or weakening may occur.

Aggregates may also present a problem in bituminous mixes when the bituminous film strips or separates from the aggregate. The stripping can be related to the surface electrical charge of the mineral aggregate. Materials with a negative surface charge may attract moisture and thus promote stripping. Rocks with a high silica

Table 1. Crushed Stone Sold or Used in the United States, By Kind[1]

Kind	1987 Number of quarries	1987 Quantity (thousand tons)	1987 Value (thousands)	1987 Unit value	1989 Number of quarries	1989 Quantity (thousand tons)	1989 Value (thousands)	1989 Unit value
Limestone[2]	2 496	718 897	$3,249,713	$ 4.52	2 533	731 775	$3,234,317	$ 4.42
Dolomite	119	44 140	206,904	4.69	136	44 728	216,957	4.85
Marble	51	5 058	62,335	12.32	63	11 088	137,481	12.40
Calcareous marl	16	3 768	7,650	2.03	15	4 643	11,614	2.50
Shell	19	7 622	51,028	6.69	18	6 280	34,794	5.54
Granite	735	163 268	900,682	5.52	376	152 800	864,962	5.66
Traprock	813	93 815	505,187	5.38	446	89 957	524,271	5.82
Sandstone and quartzite	523	29 479	157,934	5.36	241	28 887	148,696	5.15
Slate	10	2 114	14,258	6.74	7	1 894	12,926	6.82
Volcanic cinder and scoria	359	3 318	14,952	4.51	118	3 293	20,206	6.14
Miscellaneous stone	304	17 261	77,954	4.52	110	25 460	119,538	4.70
TOTAL[3]:	XX	1 088 700	5,248,600	4.82	XX	1 039 700	5,325,800	5.12

XX Not applicable.
[1]Table adapted from Tepordei, 1991.
[2]Includes "Limestone/dolomite," reported with no distinction between the two.
[3]Data may not add to totals shown because of independent rounding.

content can have negative surface charges; thus, some quartzites, gneisses, granites and schists with high quartz contents may have stripping problems. Surfactant chemicals can be added to the asphalt mix to reduce stripping.

The end use of carbonate rocks for crushed stone is often determined by the chemistry of the rock. Chemically pure forms of limestone and dolomite can be used for fillers and extenders, chemical stone, scrubber stone, and cement. Each use of the stone must meet a variety of specifications. Cement manufacture commonly requires a limestone with low MgO and low total Na_2O and K_2O contents. Other oxides, such as SiO_2, Al_2O_3, and Fe_2O_3 which are required in portland cement, may be present in the limestone or may be introduced from an outside source. Fillers and extenders typically require a high $CaCO_3$ content and a low silica and iron content. High quality fillers are used in paints, rubber, plastic, and other filler uses. For a more detailed discussion on chemical properties and the effects on usage, please see other chapters in this book, including Limestone and Dolomite and Fillers and Coatings.

Physical Properties

The major physical properties which are critical for crushed stone are strength or durability, porosity and pore size, and volume integrity, i.e., maintaining relatively constant volume when subjected to variable moisture or freezing and thawing conditions. Also important is a tendency to break into equant or relatively cubic particles on crushing.

Igneous Texture: Aggregate quality is determined by the strength of the interlocking bonds between the grains. Fine to medium grained igneous rocks such as traprocks (usually diabase, gabbro, or basalt) are more resistant to abrasion due to the strength of the interlocking grains. However, glassy, highly porous, and coarsely crystalline igneous rocks are easily shattered under impact, and are usually not acceptable for most aggregate uses.

Igneous rocks with excessive amounts of olivine tend to be weak, because the olivine crystals tend to be somewhat rounded and do not interlock with other minerals. Quartz, however, tends to be tightly interlocked with other minerals and is a major source of strength for many rocks.

Metamorphic Rock Variations: Physical properties of metamorphic rocks generally are related to the degree of metamorphism and the mineralogical composition. The higher the degree of metamorphism, usually the higher the quality of the rock. Slates tend to be platy and generally are poor quality aggregates due to the generation of flat particles upon crushing. Although schists are generally better than slates, excessive flat particles may result from crushing because of the mica layers. Gneisses with a low concentration of mica (less than 10%) are usually of suitable quality for aggregate use.

Quartzites and marbles are often excellent aggregate, although the quality will vary with the crystal size, the coarser sizes tending to be excessively brittle. Amphiboles, which may be found in metamorphosed carbonate rocks and in metamorphosed silica-poor igneous rocks, may be highly oriented and create a weak, brittle rock.

Sedimentary Rock Variations: Abrasion resistance is usually determined by the degree of cementation of sedimentary rocks, with well cemented rocks usually dense and of high quality. Typically, older rocks are more compacted and better cemented than younger rocks.

Clay-rich or argillaceous rocks such as shales tend to form flats when crushed and disintegrate upon wetting and drying, a phenomenon called slaking. Shales may be in the form of paper-thin layers, very thick layers continuous over wide areas, or as lenses which are laterally discontinuous. Rocks such as limestone and sandstone which contain shale partings may be unsound as an aggregate due to the slaking along the shale layers.

Dolomitic limestones may be unsound if they contain clay. Growing dolomite crystals tend to push clay minerals to the outside of the crystals, the so-called rejection texture (Dunn and Hudec, 1972). This clay envelope can create a continuous network of clay particles which can cause expansion and contraction on wetting and drying, weakening the rock.

Weathering and Diagenetic Effects: Most rocks deteriorate to varying degrees in the weathering environment. Feldspars tend to weather to argillaceous minerals; mafics tend to alter to iron oxides or other secondary minerals; calcite and dolomite tend to dissolve or otherwise alter; and sulfides oxidize. Virtually all of these changes weaken rocks and tend to make them less satisfactory for crushed stone use.

Certain rocks like sandstone and conglomerates tend to recement at their outcrops (case harden) and may be of higher quality near the surface than at depth.

Table 2. Classification of Rocks Commonly Used for Crushed Stone

Group	General Classification	Rock	Approx. Specific Gravity	Los Angeles Abrasion Test* Avg.	Los Angeles Abrasion Test* Mid-Range†
		Gabbro	2.9	18	
		Granite	2.6	38	27–49
	Intrusive	Syenite	2.7	24	15–27
		Diorite	2.8		
Igneous		Peridotite	2.9		
		Felsite‡	2.6	18	
	Extrusive	Traprocks:			
		Basalt	2.8	14	10–17
		Diabase	2.9	18	13–21
		Dolomite	2.7	25	18–31
		Limestone	2.6	26	19–30
	Calcareous	Coquina			
		Coral	<2.6		
		Shell			
Sedimentary					
		Chert	2.5	26	
	Siliceous	Sandstone	2.6	38	24–48
		Graywacke	2.6	17	14–20§
		Amphibolite	3.0	35	22–46
	Foliated	Gneiss	2.7	45	33–57
		Schist	2.8	38	26–49
Metamorphic					
		Marble	2.7	47	26–64
	Nonfoliated	Quartzite	2.7	28	20–35
		Serpentine	2.3–2.6		

**Source:* Woolf, 1953, modified by Schenck and Torries, 1983.
†Highest and lowest value after excluding top one-fifth and lowest one-fifth of tests reported by D.O. Woolf.
‡Including andesite, dacite, rhyolite, and trachyte.
§Private correspondence.

Distribution of Major Deposits

Many rock types meet the basic physical requirements established for aggregates (Table 2). Limestone and dolomite account for 71% and granite for 15% of crushed stone aggregates as shown in Table 1.

Rocks suitable for crushed stone uses are well distributed across the United States. However, certain regions, such as portions of the Gulf Coast, the Atlantic Coastal Plain, and North Dakota are deficient in readily accessible deposits that meet typical national quality specifications. In these areas, rocks of lower quality may be utilized for some purposes or crushed stone may have to be imported for more rigorous uses. Where suitable alluvial deposits are available, their lower production costs will favor them over stone.

The predominant rock used varies from region to region. Georgia, North and South Carolina, and Virginia account for 70% of the US production of crushed granite. Illinois, Indiana, Ohio, New York, and Pennsylvania account for about 27% of the limestone and dolomite produced, while Missouri, Florida, Texas, and Tennessee account for 30%. Shell is used in declining quantities in the Gulf states and over half is produced in Louisiana. Traprock production is scattered in pockets across the northern, western and southwestern, and eastern portions of the country. In decreasing order, production in Oregon, New Jersey, Washington, Connecticut, and Virginia accounts for nearly 60% of the traprock (Tepordei, 1991).

TECHNOLOGY

Exploration

Exploration for new crushed stone sources in the United States is largely a matter of understanding the distribution of rock types. Nearly all of the country has been mapped in some detail by geologists from the US Geological Survey, state geology surveys, colleges and universities, private industry, and consultants. An exploration program should begin with a regional evaluation of previous geologic work in the area of interest. Once the regional data is gathered, specific areas can be targeted for detailed evaluation. Geologic interpretation is necessary for relevant variations in lithology, structure, and rock quality. In areas where the geology is very complex, such as in the Adirondacks of New York, geologic mapping is essential in comprehending rock quality variations and setting up a sampling program.

The purpose of a sampling and testing program is to determine quality, quantity, and configuration of the reserves along with determining those factors which influence mineability. When a layer is sampled for testing, the whole width should be sampled and multiple samples of each zone are needed over the mineable area to evaluate any lateral changes in quality. Duplicate samples need to be collected to check on laboratory results. The engineering tests used must conform with the requirements of the specifying agencies in whose jurisdiction the producer markets his products.

The types of samples commonly taken for testing of crushed stone quality include grab samples, shot samples, and core samples. Grab samples are preliminary only and are rarely typical of a deposit. Grab samples can be taken from outcrops, stockpiles, core, or quarry faces. Care must be taken that samples are reasonably representative of the zones in question. In some cases it may be possible to take a representative shot from an outcrop or quarry face. However, if only part of the shot is sampled, it must be considered a grab sample. A shot sample, although large in volume, is only representative of a limited area.

The preferred method of sampling a deposit is through core drilling. The sample is normally continuous, which allows for proper detailed description of the rock (logging) and allows composite sampling of entire zones. Taking half of split cores is usually the best way to sample, thus making it possible to retain a duplicate sample for future reference. Correlation can later be made between test results and rock units by looking at the unsampled core. In addition, if more tests are required on the rock units, the remaining core is available without performing additional drilling. The size of the core which is selected is determined by cost, handleability, recoverability, quantity and testing needs. The best size is approximately 5 cm in diameter (NX), but cores can range from 1.6 to 91 cm in diameter.

The spacing and number of core holes are dependent on the complexity of the geology and on the purpose of the drilling. A wide spacing, such as 300 m centers, may be appropriate when drilling flat lying very homogeneous rock units. However, for evaluating complex geologic units, or units where the chemistry is very critical, closer spacing of core holes is usually required. When possible, drilling should be perpendicular to the bedding, to the schistosity or to other known variations in quality. Drilling and other testing are sufficient when all critical characteristics of quality, configuration and mineability are known within a block of stone which is large enough to produce enough stone to amortize the total investment.

A quality control program will assure that the sampling technique was properly performed. Sample splits, duplicate samples and blind samples are the most commonly used methods to monitor sampling procedures.

Table 3. Some Common Tests and Specifications for Crushed Stone

Physical Test	ASTM Test No.	AASHTO Test No.	AASHTO Spec. M79 Aggregate for Bituminous Concrete Surface Course	AASHTO Spec. M80 Course Aggregate for Portland Cement Concrete
Standard Specification	C33			
Abrasion				
Los Angeles	C131	T96	<40% loss	<40% loss
Machine	C535			
Soundness				
Sodium sulfate loss (5 cycles)	C88	T104	<12% loss (5 cycles)	<12% (5 cycles)
Freeze and thaw	C682	T103		
Gradation				
Sieve analysis	C136	T27	No. 7 and No. 57	No. 357, 467 (blended from sizes 3 or 4 and 57 or 67), or size No. 57
Alkali reactivity	C227 C289 C586			
Deleterious material				
Organic materials	C40	T21		
Clay lumps	C142	T112		<0.25%
Soft fragments	C235	T189		<2.0%
Coal and lignite	C123	T113		<0.25%
Chert	See Alkali reactivity			
Excess fine material				
-200 mesh	C117	T11		<0.5%
Specific gravity and absorption				
Coarse aggregate	C127	T85		
Slag			1 121 kg/m^3	1 121 kg/m^3
Fine aggregate	C128	T84		
Crushed count				
Retained on No. 4 Sieve, 1 frac. face			>50%	
Particle shape				
Thin or elongated	D4791			Commonly <15%
Petrographic analysis	C295			
Sampling method	D75	T2		

Source: Modified from Schenck and Torries, 1983.

Reserve calculations of mineable stone must take property setbacks, bench heights and widths, and any regulatory factors which may impede availability into consideration.

When exploring for a new source, a deposit's permitability must not be overlooked. A property with high quality and quantity of stone reserves located in an ideal market setting may appear to be viable. However, most areas of the United States require some type of mining permit, at the state or local level or both. Without proper permitting, mining the deposit may be impossible. Therefore, knowledge of the local mining laws is an essential part of exploration which cannot be overlooked.

Specifications and Testing

Specifications for crushed stone to be used for construction are modeled after those recommended by organizations such as the American Society for Testing and Materials (ASTM), American Association of State Highway and Transportation Officials (AASHTO), US Department of Transportation (Federal Highway Administration), the US Army Corps of Engineers, and the various state departments of transportation. The specifications set by the individual departments of transportation are often adopted as the standard specification for all private construction within each state.

Depending upon climatic conditions and upon the availability of materials, different areas in the United States have different specifications for crushed stone. In addition, specifications are continually being changed by the various agencies in an attempt to guarantee satisfactory materials for particular applications without being overly restrictive or incurring unnecessary costs.

Standardized tests have been designed to evaluate the physical properties of an aggregate. A summary of the different types of tests commonly used, the ASTM and AASHTO code numbers for the tests, and typical AASHTO specification limits are given in Table 3. For the details on how each test is conducted, see the appropriate ASTM or AASHTO test identified by the code number.

Abrasion Tests: The most common test that measures the resistance of an aggregate to abrasion is called the Los Angeles Abrasion Test (ASTM method C131 or C535). Rocks in which the individual grains are weak or are loosely cemented are not desirable as aggregate because these materials break down under pressure from use or during the handling and mixing procedures. The breakdown of the coarse aggregate may cause a deviation from the desired gradation because an excess of fine particles may be generated. Examples of rock types that commonly fail the Los Angeles test are weakly cemented sandstones, some argillaceous limestones, and coarse grained granitic rocks or marble. It should be noted that some types of rock may have relatively high abrasion losses but still may perform satisfactorily.

Soundness Tests: The two types of soundness tests commonly used are the sodium sulfate (or magnesium sulfate) soundness test and freeze-thaw tests. Repeated cycles of freezing and thawing or immersing the stone in a sodium or magnesium sulfate solution followed by drying cause some rocks to disintegrate. The apparent cause is the expansion of ice or salt crystals in the pores of the rock, although for some slightly argillaceous rocks, expansion may be caused by wetting alone (Dunn and Hudec, 1972). Deterioration in these tests is a function of mineralogy, porosity, permeability, and tensile strength of the rock. Rock types that commonly fail the soundness test include porous sandstones and argillaceous sedimentary rocks (refer to ASTM method C88). Individual specifying agencies may have variations of freeze-thaw or other soundness tests to suit their particular needs.

Gradation Tests: Crushed stone is manufactured in a variety of sizes and gradations. In general, aggregate to be used in concrete should be graded to give as dense a mixture as possible and still allow the concrete to be workable. The coarser sizes are used in more massive structures and the finer sizes where spaces between steel reinforcement bars and forms are small (Severinghaus, 1960). Conversely, aggregate used for drainage purposes should be uniformly sized to have as great a volume of voids among the pieces as possible and should be relatively free of fine particles. Several sizes of stone are blended with manufactured or natural sand to maintain uniformity in gradation of the aggregate used in blacktop and concrete mixes.

Deleterious Materials: Deleterious materials include organic matter, chert, clay, coal and coke, shale, and soft particles. The amount of deleterious material allowed depends upon the type of deleterious material and its construction use.

Color: Crushed stone used for decorative purposes must be of specific colors and hues. The colors must be permanent and not change with exposure to the weather or to atmospheric conditions.

Fines: A wash test is generally used to determine the amount of minus 74 mm (-200 mesh) material in the aggregate. An excess of fine material may prevent the aggregate from draining freely, may form an unwanted coating in the coarse particles, and may require excessive amounts of binder to coat each particle.

Specific Gravity and Water Absorption Tests: Specific gravity of the crushed stone is used to determine the solid volume occupied by the aggregate so that the portland cement or bitumen concrete mixture can be designed correctly. Absorption is a measure of the water absorbed by a rock and is used to determine the proper amount of water to be added to portland cement concrete mixes. If high absorption of water occurs and is not taken into account, the resulting concrete mixes would be too dry. The ability of a rock to absorb liquid is also an important consideration in the manufacture of bituminous concrete mixes. All aggregate used in bituminous concrete must be dry, and water in the interior of more porous stone particles is difficult and expensive to remove. In addition, porous rock requires more bitumen than nonporous rock.

Particle Shape: Crushed stone particles should be as cubical as possible with a minimum of flat and elongated pieces so the rock will pack uniformly. Some metamorphosed or sheared rocks have pronounced platy cleavage and are difficult to crush without producing an unacceptable quantity of flat and elongated particles.

Concrete Tests: In addition to meeting all other requirements for concrete aggregate, specimens of portland cement concrete made with the aggregate in question may be subjected to strength tests. In this manner, the concrete mix design, the cement, and the concrete aggregate are all tested simultaneously. If an aggregate meets all other specifications for concrete aggregate and the strengths of the resulting concrete are too low, the trouble probably is due to an improper mix design or improper curing of the concrete specimen rather than to the aggregate.

Bituminous concretes are similarly tested largely to determine the bonding between the aggregate and the bitumen as well as the absorption of bitumen.

Skid Resistance: Much research has been done concerning the type of aggregate to be used in the surface courses of highways to minimize accidents caused by skidding vehicles (Anon., 1972). In theory, a soft, homogeneous rock that polishes easily and quickly should form a slick surface which would promote skidding, especially when the surface is wet. On the other hand, the skid-resisting properties of a highway surface are improved if the aggregate used retains a rough, hard surface in use. The subject is complex, and it is not yet possible to predict accurately the skid-resisting properties of a particular aggregate. Skid performance tests on existing highway surfaces permit evaluation of material in the surface course of these highways.

Byproducts and Coproducts

Most byproducts and coproducts associated with crushed stone production are due to unique chemical or physical characteristics of the rock or overburden being mined. The unique qualities allow for uses other than coarse and fine aggregate. When the rock is a limestone or dolomite, the chemical content plays a part in the economic feasibility of producing certain products. Byproducts and coproducts produced for their unique chemistry are used in agricultural lime, mineral food, cement stone, lime, dead-burned dolomite, flux stone, chemical stone, glass manufacture, sulfur oxide removal, acid water treatment, fillers and extenders, whitings, roofing granules, and other miscellaneous uses.

In addition to the chemically pure rocks used in the products listed above, overburden above the crushed stone deposit often has value as a coproduct. The overburden can have a range of physical characteristics which determine its end use. The most common products derived from the overburden are sand and gravel, clay, fill, and topsoil. The amount of overburden which can be sold can influence the available reserves of the crushed stone.

USES

Almost 100% of crushed stone is used in the construction sector of the national economy. The major construction uses of crushed stone can be divided into four end-use categories: highway, residential, nonresidential (commercial), and government projects such as dams. Highways account for the largest share of total use. The crushed stone used for various purposes is estimated below largely by using engineering estimates based upon a set of assumptions.

Highway Construction

Historical usage of stone per million dollars in the construction of highways is given in Table 4. The percentage of stone costs relative to the overall cost of highway construction has been decreasing because the cost of stone is relatively stable compared to the rising trend of total highway construction costs.

The following uses were derived based on engineering estimates and the assumptions given for each type of highway.

Interstate Highway Construction: Assuming that highways are constructed with 25 cm of concrete, 36 cm of subbase, have paved shoulders and that 8.1% of total aggregate used is natural sand (which is excluded from the use factors):

Table 4. Use of Aggregate in Highway Construction (Three-Year Moving Averages)

A. Thousands of tons of aggregate used per million $ of highway construction cost	1962–1964	1968–1970	1974–1976	1976–1978	1978–1980
Federal aid-interstate					
Rural	89	75	45	37	34
Urban	42	41	24	21	21
Rural and urban	70	63	35	30	29
Federal aid-noninterstate					
Rural	115	69	49	40	35
Urban	57	50	32	26	21
Rural and urban	90	62	41	34	29
All federal aid	77	62	38	32	29
B. Cost of purchased aggregate as a % of federal total highway aid					
Interstate					
Rural		12	9	10	12
Urban		10	8	8	10
Total		12	9	9	11
Non-interstate primary					
Rural		13	13	12	12
Urban		14	12	10	9
Total		14	12	11	11

Source: US Federal Highway Administration, "Highway Construction Use Factors for Aggregates," Dept. of Transportation, Washington, DC, annual; Schenck and Torries, 1983. †All costs exclude right-of-way and engineering costs.

1) Six lane interstate highways use 152 kt of coarse aggregate per kilometer.

2) Four lane interstate highways with bridges and interchanges as would be typical of an urban setting use 123 kt of coarse aggregate per kilometer.

3) Four lane rural interstate highways with no bridges or interchanges use 94 kt of coarse aggregate per kilometer.

Primary Two Lane Highway Construction: Assuming that highways are paved with 10 cm of bituminous concrete and 20 cm of subbase, and that 94% of the total aggregate used must be coarse or graded aggregate including subbase:

1) Two lane highways with bridges and interchanges use 35 kt of coarse aggregate per kilometer.

2) Two lane highways without bridges and interchanges use 18 kt of coarse aggregate per kilometer.

Secondary Two Lane Highway Construction: Assuming that secondary roads are paved with a maximum of 10 cm of bituminous concrete and a maximum of 20 cm of subbase:

1) Municipal roads use 18 kt of aggregate per kilometer.

2) County roads use 14 kt of aggregate per kilometer.

Highway Maintenance: Based on operating statistics of the Pennsylvania Department of Highways in eastern Pennsylvania in 1969 (Torries, 1972), the per capita use of aggregate used in highway maintenance in this particular state was 0.079 tons per person. Approximately 65% of this total is classified as coarse aggregate. Bituminous concrete accounts for almost all repaving.

Residential and Nonresidential Construction: Regression analysis on data from eastern Pennsylvania (Torries, 1972) indicated that a per capita average of $80 of residential and nonresidential construction was spent in the region each year from 1964 to 1969. On a 1991 basis this would amount to approximately $271. Use factors of 0.014 and 0.013 tons of aggregate per dollar of residential and nonresidential construction, respectively, have been reported (McCarl, 1969). About 72% of the total quantity of aggregate used in residential and nonresidential construction is coarse aggregate.

Mining and Processing

The activities that contribute to the total cost of processed aggregate include acquiring the deposit, stripping, drilling, blasting, crushing, screening, stockpiling, transporting the finished product, and meeting environmental and safety regulations. The following discussion concerns the location of the quarry and the methods by which the stone is to be quarried and processed.

Quarry Location: There must be an identifiable market for crushed stone and a workable deposit of suitable material at a location where zoning and other public-related constraints permit the quarrying and crushing of stone and related activities. The choice among several possible locations is based upon estimates of production and transportation costs, price, and risks over time. The possible addition of related activities at the production site, such as bituminous or portland cement concrete production, may be included in the analysis. The residual value of the land after the depletion of the deposit should also be considered when a production life of generally 15 years or less is anticipated.

Production Capacity of the Operation: The expected size of the market over time, recoverable reserves of stone, and economies of scale in production need to be considered in determining optimal plant capacity. The average capacity of crushed stone operations in the United States is approximately 350 ktpy with the range varying from 100 kt to over 5 Mtpy.

General Site Considerations: The site of initial production, future expansion of the quarry, temporary and permanent site screening, and pollution control should be planned at the time a new quarry is started. The nature and type of overburden (for quarry operations), land-use controls such as zoning, setback and depth restrictions, the thickness and attitude of the deposit, planned use of the land after the deposit is depleted, and the nature of ground water encountered influence decisions related to quarrying and reclamation procedures. It must be decided if the operation is to be worked by open pit, underground, or by a combination of methods. Quarrying below water level may result in increased reserves of clean stone with no additional overburden removal, but pumping cost, costs of disposing of the water, and the final end use of the depleted quarry (or cost of reclaiming the land) must be considered.

General Plant Arrangement: The relative locations of the primary crusher, crushing and screening plant, stockpiles, and ancillary equipment determine how efficiently materials can move through the plant. Items to be considered in designing a site layout include the quarry plan, topography, zoning and setback requirements, the types and locations of possible portland cement or bituminous concrete mixing plants, flooding problems, and intraplant material handling rates.

It is often desirable to locate the primary crusher close to the quarry face to minimize the pit haul. This scheme may not be possible when more than one face is being worked or where there are serious water problems and the danger of flooding exists. It is usually desirable to screen the plant from the sight of neighbors and motorists with landscaping such as planted embankments. For examples of typical layouts see Bowers et al., 1991.

Stripping Plan: Primary factors influencing decisions pertaining to stripping include pollution, zoning, and reclamation considerations; the amount and type of material to be stripped; the distance the overburden is to be moved; equipment available; and time constraints. In some political jurisdictions, the amount of land that can be stripped and left bare at any one time is regulated by law. If large quantities of materials are to be stripped, it may be more profitable for the quarry operator to contract the stripping to a company that specializes in moving large quantities of earth materials.

Trucks and shovels or front end loaders are commonly used where the overburden must be moved long distances (over 600 m) or over public roads. Self-loading units such as scrapers are used for shorter hauls off public roads. Draglines and hoe-type excavators are commonly used where the overburden is sticky or is under water or where the rock surface is irregular. Under certain conditions draglines and bulldozers can be used to strip and place the dirt into previously quarried portions of the quarry so that stripping, quarrying, and reclaiming are accomplished simultaneously.

Drilling and Blasting Procedure: The nature and geometry of the deposits determine the type of drill to use, hole size, type of explosive, drill hole configuration, and the size and frequency of the blasts. Generally, large diameter holes are less expensive per ton of stone produced than small diameter holes. However, a quarry operation of less than 350 ktpy may not require the full time services of a large drill. A contract driller who services two or three quarries with a large drill can often do so at a lower unit cost than if each of the quarries did its own drilling.

Safety, zoning, and other public-related constraints; thickness of the deposit and bench heights; the geologic attitude of the deposit (inclination and thickness of the beds and the amount and type of fractures in the deposit); desired degree of rock fragmentation; and the number and configurations of the working faces must be considered when drilling and blasting procedures are selected.

Prevention of damage and public apprehension caused by blasting and its attendant noise is a major concern to the quarry operator. Ground vibrations are commonly measured by a recording seismograph to insure that the velocity and amplitude of the seismic waves are within acceptable levels. The generally accepted level at which ground vibrations cause no structural damage is 51 mm/s (Nicholls et al., 1971). However, people near a quarry (less than 900 m from the blast) may notice the ground vibrations and the airblast even though no property damage is done. Weather conditions may have a profound effect on airblast which can be minimized by use of proper techniques. For more details on blasting see Archibald (1991).

Types of Processing Required: Crushed stone plants can be generally classified as wet or dry plants. The preferred type of plant depends upon the classification of stone in the deposit, the types of contaminants in the deposit, the availability of pond space and water, and zoning and pollution problems. Washing generally is used to remove undesirable amounts of clay from stone. Once the clays are removed, the wash water and the contained solids are sent to settling ponds. Wet plants do not have the dust problems that are often associated with dry process plants. Regulations of dry process plants usually require dust collectors which may present problems of waste disposal similar to those of wet plants.

Permanent vs. Portable Plants. A quarry operator may use a permanent, a portable, or a combination of the two types of plants to process quarried stone. Factors to be considered in making the choice include plant capacity, the life of the deposit, the types of products required, labor rates, equipment available, impact of unionization and time constraints in obtaining equipment, setting up the plant, and obtaining production.

The equipment of a portable plant is often mounted on wheels and is of such a size and weight that each piece of machinery can be transported over public roads. Since portable plants are designed to be used under a variety of conditions, a portable plant may not be as efficient as a comparable permanent plant for any one location. Making a wide range of products at one time can sometimes be awkward in a portable plant.

Portable plants are often used to supply stone for a specific project such as the construction of a section of a highway or a dam. Since these plants are operated as an adjunct to the primary construction project, the labor rates paid in the quarry usually follow the same schedule as the labor rates used on the main construction project.

The delivery and set up time for a new permanent 270 to 160 t/h crushing and screening plant may be as much as six to nine months. If a stone producer has a portable plant available, the portable plant can often be set up and be in production within a few days.

Portable plants can be used in conjunction with permanent plants to meet surge demands and to process materials that cannot be processed efficiently in the permanent plant.

Types of Crushers. The selection of the proper size, type, and number of crushers depends upon the compressive strength and abrasiveness of the rock, the products desired, the quantity and type of deleterious materials in the deposit, the screening capacity, and the capital cost of the various types of crushers vs. the operating and maintenance costs associated with each type.

In general, crushers fragment rock by rapid impact, by slow compression, or by a combination of the two actions. Impact type crushers have greater capacity-to-cost ratios than do compression-type crushers, but abrasive stone (usually greater than 8% free SiO_2) may cause accelerated wear on impact-type crushers. Crusher designers and manufacturers or engineers should be consulted to determine the proper types and numbers of crushers required to process a particular stone (see also Bowers et al., 1991).

Screening, Classifying, and Washing Plant. Gradation specifications of crushed stone are met by screening, classifying, and washing the crushed stone. Factors to be considered in designing the screening, classifying, or washing plant include the number and types of products required, the nature of the material to be processed, amount of wet or sticky material, particle shape, amount of oversize and fine particles, and the densities of the various materials, and capital cost vs. the operating and maintenance costs of alternate equipment.

Stockpile Methods: Often stockpiles and reclaim systems are automated. Automated systems require less manpower and fewer trucks and loaders but are generally not economical in small plants. Additional factors to be considered include the storage space available, the method used to transport the product, and the desired sustained and peak capacity of the plant. It is especially important to avoid segregation of stone sizes during stockpiling and loading (see Bowers et al., 1991).

Pollution Controls: Air pollution control systems are classified as dry, wet, or combination systems. Dry systems include cyclones and baghouses. Cyclones operate by imparting centrifugal forces on particulate matter which forces the matter against a cylindrical wall. The particles are therefore removed from the air stream and fall to the bottom of the cyclone where they are recovered. Baghouses pass air through electrostatically charged filters. The particles precipitate on the filters and clean air is removed from the system. The charged filters are periodically turned off whereby the particulate matter falls to the bottom of the baghouse and is removed. The recovered dry fines may either be disposed of or, in some cases, may be salable.

Wet systems can utilize sprays for control of air pollution or can be an integral part of a conventional washing plant. Spray systems require relatively small amounts of water, with the use of chemical wetting agents further reducing water consumption. In a some spray systems, the water that is used is combined with the dust and both are fed into the crushed stone product. Caution must be exercised to keep the dust content below permissible amounts of 74 µm (−200 mesh) particles in the finished product.

Table 5. Representative Open Pit Production Cost,* 272.2 t/h Plant Producing 453.6 ktpy, All Day Operation

	Sandstone, Limestone-Dolomite, $	Traprock, Gabbro, Granite, $	Diabase, Basalt, $
Stripping	0.04	0.04	0.04
Drilling	0.09	0.11	0.18
Shooting	0.13	0.13	0.15
Loading	0.65	0.65	0.65
Hauling	0.08	0.08	0.08
Processing	0.18	0.26	0.31
Stockpiling	0.18	0.18	0.18
Maintenance	0.73	0.90	1.17
Dust control	0.05	0.05	0.05
Royalty	0.13	0.13	0.13
Selling	0.08	0.08	0.08
Total average variable cost	1.75	2.02	2.42
Depreciation	0.51	0.78	0.78
Office overhead	0.13	0.13	0.13
Total average fixed cost	0.65	0.90	0.90
Total average cost	2.40	2.94	3.33
Total plant equipment cost, x 1000	2593	3886	3886

*Based on 1972 costs escalated at a rate of price increases from 1972 through 1991. Totals may not add because of rounding. *Source:* Schenck and Torries, 1983.

Table 6. Labor Required for a 300 tph Crushed Stone Plant†

	No. Required		
Item	Pit	Plant	Total
Shovel operator	1		1
50-ton pit vehicle operator	3		3
Driller	1		1‡
Driller helper	1		1‡
Blaster	1		1‡
Blaster helper	1		1‡
Mechanic	0.5	0.5	1
Oiler	0.5	0.5	1
Welder	0.5	0.5	1
Crusher operator		1	1
Conveyor and bin operator		1	1
Loader operator		1	1
Stock truck driver		3	3
Superintendent		1	1
Lab man		1	1
Scale man		1	1
Laborer		1	1
Drop ball operator	1		
Total:	10.5	11.5	22
Productivity per man @ 300 tons per man-hour	28.5	26.0	13.6

†Assumes stripping is done by contract work, limestone crushing is done by conventional open pit operations, and there are no unusual problems. ‡May be contracted. *Source:* Schenck and Torries, 1983.

Under standard water pollution regulations, wash water and runoff from a quarry requires impoundment in a series of settling ponds. It is generally desirable to recycle the water from the settling ponds within the operation, so that turbid wash water does not enter local streams.

An integrated approach to pollution control is usually required for the entire operation from quarry face to transport of the product to consumers. The type of control program selected depends upon the amount and type of dust generated, the availability of pond space and water, zoning, air and water pollution regulations, the type of processing plant utilized, amount of dust acceptable in the product, and capital cost vs. operating and maintenance costs. Special regulations are imposed on siliceous dusts because of the danger of silicosis and the possibility that silica dust may cause cancer.

Pollution regulations and limits vary, depending upon location and the enforcing agency. Solutions to pollution problems may require substantial capital investment.

ECONOMIC FACTORS

Costs

Production costs for the quarrying and processing of three different types of materials are shown in Table 5. The costs are based on the labor requirements shown in Table 6 and on the following assumptions:

1) Only normal products are manufactured and these products meet all specifications for use in heavy construction.

2) There are sufficient reserves of materials available.

3) Open pit quarrying is possible with a level quarry floor and bench heights of 12 m.

4) Ground water, overburden, and environmental problems do not require any unusual expenditures.

For underground mining, it can be assumed that total average mining costs in 1990 range from about \$4.96/t to \$7.16/t for room and pillar mining.

The size and gradation of the stone produced influence production costs. In general, finer sizes are more expensive to produce per ton; however, very large stone such as jetty stone, may be very expensive because each stone may have to be separately produced and individually handled.

A given crushed stone plant and its associated stone deposit will have a limited combination of gradations that are feasible to produce. In addition, there will be excessive fine sizes that cannot be avoided, and the operator must then pay special attention to their disposal by finding a market, such as manufactured sand, by stockpiling for some future sale, or by disposing of them in an environmentally acceptable manner.

Markets

Most crushed stone is sold to construction contractors who buy their stone in large quantities at standard discount prices or at prices set by competitive bidding.

Demand and Competition: Because demand for aggregates is derived from the demand for other goods such as highways, buildings, and public works, the markets virtually always coincide with urban population concentrations. Transportation costs relative to costs of competitor's transportation costs generally define the market area that can profitably be serviced by each stone quarry.

Commonly, only a few producers are conveniently located to serve a particular market. However, pricing is often competitive because of the relative ease with which additional firms can enter the stone business (Table 7). As population centers increase in size, those quarries already in existence may be able to charge higher prices due to locational advantages, because new quarries can be opened only at larger distances from market areas.

Industry Structure: There were about 4 063 crushed stone plants in the United States in 1989 (Tepordei, 1991). In the United

Table 7. Factors That Affect Competition in Crushed Stone Markets

Factors Tending to Promote Competition	Factors Tending to Inhibit Competition
1. Widely understood technology and management.	1. Lack of available quarry sites. a. Reserves geologically absent b. Reserves depleted c. Buildings covering reserves d. Legal restrictions (zoning and pollution control)
2. Equity requirement in medium six figures, balance of financing in loans from equipment manufacturers.	2. Restrictive material specifications.
3. Many alternative underdeveloped quarry sites.	3. Quarry ownership by major portion of consumers.
4. Roving demand.	4. Dominant producers.
5. Costs and profits easily estimated.	5. Restrictive practices (favoritism) by labor, management, or government officials.
6. Consuming firms often with many times the financial strength of producers.	6. Inadequate transport capacity.
7. Availability of developed quarries due to completion of heavy construction projects.	7. Stable market.
8. Cheaper substitute (gravel) often available.	
9. Surplus transport capacity.	
10. Easily portable plants with operating costs competitive with those of stationary plants.	
11. Expanding market.	
12. Competitive bidding for large contracts.	
13. Substitutes available as byproducts from other sources.	

Source: Schenck and Torries, 1983.

States, 275 quarries produced 900 kt of stone per year or more, for a total output of about 480 Mtpy or 44% of the US production (Tepordei, 1991). Conversely, 43% of the quarries operating during 1989 produced less than 91 ktpy and collectively accounted for only 4.3% of total production.

The structure of the crushed stone industry generally is competitive, localized, free of significant barriers to entry (with important local exceptions), and is not dominated by firms of national scope. Continuation of this industry structure is favored primarily by: 1) economies of scale reached at a size far below 1% of national demand, 2) a high ratio of transportation cost to production cost, 3) widespread availability of a common substitute (gravel), and 4) easily acquired managerial and financial requirements. The present trends to increased governmental involvement with the industry and increased limitations on availability of sites in urban areas in the future may inhibit entry of new firms.

Vertical integration exists to some degree with stone users, such as construction firms, portland cement, and steel producers entering the industry. A few large nationwide stone companies have been organized by conglomerate industries. Forward integration exists for a few firms that market crushed stone aggregates as a joint product of other mineral production such as mining of chemical limestone or sale of waste rock from mining.

Transportation

Of the total crushed stone produced in the United States in 1989, 88% was transported by truck, 7% by rail, and 5% by waterway (1989 US Bureau of Mines adjusted figures). The method of transportation used is determined by cost and availability. Prices of hauling crushed stone are usually figured in terms of cost per ton-km (historically cost per ton-mile), ranging from less than $0.01 per ton-km for Great Lakes and ocean-going ships to over $0.20 per ton-km for trucks in metropolitan areas. Average rail rates lie between these extremes. Actual costs of transportation are quite variable with considerable overlap between the different methods. Factors determining cost include variables in supply, demand, haul distances, traffic, quantities involved, size of carriers, potential for back haul, and other competitive factors.

Hauling by truck is the most common mode of transportation due to its versatility. The truck routes may be limited by weight restrictions or capacities, but typically, county, state, and interstate highways are all capable of handling haul truck traffic. The main disadvantages of using truck transportation is the cost. Commonly, 50 to 75 km is the maximum distance crushed stone is hauled by truck. If rail is available, then transportation costs may be cut from 50 to 70%, thereby allowing the crushed stone to be hauled for much longer distances. The disadvantages to rail transport are limited rail sites, inflexible schedules, and up front capital costs in developing rail sidings and handling facilities.

By far the cheapest method of transporting crushed stone is by water, either by barge or by ship. Ocean-going vessel transport is less expensive than barge, mainly due to the carrying capacity. Barge capacity ranges from 1 to 10 kt. Ships with capacities of up to 64 kt or more are used on the Great Lakes and ocean routes.

Alternative Materials

Although gravel is the most frequently substituted material, other alternatives, such as sand, steel, wood, lightweight aggregate, rubber tires, bricks, oyster shells, shale, recycled concrete, glass, aluminum, dimension stone, or slag are possible. Less aggregate can also be used in highway construction when using portland cement concrete rather than bituminous concrete.

It is sometimes possible to blend materials in various proportions to make an acceptable product. Crushed gravel that does not have enough broken faces for a particular use can sometimes be blended with a small quantity of crushed stone to make a blended aggregate that meets all requirements. The soundness and Los Angeles test results can be similarly upgraded by blending higher quality materials. When two or more types of materials are blended to make aggregate, care must be taken to insure that the individual types are compatible. For example, great differences in specific gravity may cause segregation.

GOVERNMENTAL CONSIDERATIONS

Government agencies have been increasingly involved in the activities of crushed stone producers through specifications, regulations, and taxation. Environmental controls, land use regulations, and worker health and safety standards have had a significant impact on capital and operating costs of plants. Numerous permits are generally required. Examples of government programs include USBM safety regulations, state pollution control permits, and reclamation regulations as well as the more common commercial requirements such as highway use permits and business licenses.

The public needs to be aware of the consequences of inhibiting the production of crushed stone because production of alternative materials could incur greater environmental damage. Most quarry sites can be used after all reserves have been depleted, including uses that directly benefit the public, such as parks or landfills. The public benefits from the present competitive organization of the crushed stone aggregates industry through lower cost public and private construction than would be possible if a philosophy inhibiting mineral production were to be imposed.

Some governments are recognizing that planning for future reserves near consuming areas may minimize transportation costs for aggregates. Several states and cities have enacted statutes designed to preserve aggregate resources. Fairfax County, Virginia, perhaps the first to recognize the necessity of resource conservation, enacted appropriate zoning ordinances in 1961. Similar actions have been taken in California, Colorado, and Vermont.

Specifications

Most quality specifications for crushed stone products are set by government agencies. The specifications are usually selected from among those written by ASTM as appropriate considering such factors as availability of materials, costs, and public safety. Specifications change as new technical knowledge is added and as governmental priorities change.

Taxes, Subsidies, and Depletion Provisions

In contrast to some other minerals, federal income tax accounting for aggregates is straightforward. Percentage depletion for construction aggregate and chemical limestone is 5 and 14% of gross sales, respectively.

State and local severance taxes and local property taxes on minerals should be checked before opening a quarry in an unfamiliar area. A few localities impose ad valorem or severance taxes of a set amount per ton on mineral production. For example, in 1972 Ohio instituted a severance tax on all minerals which for crushed stone aggregates amounts to $0.01 per ton. Such taxes are easy for the taxing authority to administer, do not require the services of a highly skilled assessor, and are difficult to challenge once their legality is established under the state constitution. Sales taxes based on FOB prices exist in many areas and taxes on quarry land may change when an operation is started.

Land Use and Zoning

A quarry requires at least 4 to 8 ha for processing and stockpiling and at least another 8 ha for stone reserves. A quarry in which 270 to 450 ktpy are produced commonly needs about 81 ha of land. Such a parcel of suitably zoned land on the fringes of metropolitan areas will usually be valuated at 4,000 to well over $8,000 per hectare. High land costs in urban areas generally force new operations to more rural areas. Suitable acreage in such locations would cost, in the early 1990s, from 1,200 to $2,400 per hectare.

Much of the semirural land near metropolitan areas is zoned as is almost all urban and suburban land. Many presently active quarries were established prior to the imposition of zoning and either have been specifically zoned industrial or exist as nonconforming prior uses within areas where mining would otherwise be excluded. When stone is exhausted within a parcel, the operator will either try to expand the area of nonconforming use or move to a site more distant from the market.

There are, in general, five types of zoning regulations that pertain to quarrying:

1) Specific exclusion of quarrying or processing of stone

2) Quarrying and processing of stone permitted in specified areas, not by right, but subject to approval by the zoning board

3) Quarrying and processing of stone permitted by right in specified areas

4) No mention of quarrying or processing of stone in the zoning ordinance

5) Quarrying and processing of stone, either by right or subject to approval, but with stringent regulations on the quarry development and reclamation that effectively prohibit the operation of a stone quarry.

It can be seen that although zoning regulations may consider production of crushed stone, the decision as to whether a quarry is allowed for a specific area can still be open to question.

Environmental Controls

With the increased environmental awareness of the 1980s and early 1990s, the growth of new housing developments near quarries which were formally in very rural areas has often created environmental conflicts. Truck traffic, dust, and blasting noise, which may not have been problems in a remote area, often cause difficulties when the population surrounds an operation.

Particulate air emissions are associated with processing plants and truck traffic. Wet processing using spray bars on screen decks is commonly used to suppress dust during processing, but arid climates do not always have adequate supplies of water for suppression. When dust becomes a major problem in arid climates, enclosing processing areas and collecting dust in bag houses may be employed. When drying and packaging fine particle products such as fillers and extenders, dust is a major problem and enclosure of the plant in buildings is the standard practice to reduce dust.

The mineralogy of crushed stone may result in regulatory controls or may create a requirement that the stone be designed as potentially carcinogenic or as potentially causing pulmonary problems. Minerals such as free silica (quartz, tridymite, or cristobolite), actinolite-tremolite-anthophyllite (ATA) or asbestos may be regulated to varying degrees.

Turbid water from an operation is not generally allowed to enter streams. Process water is typically cleared by a series of settling ponds that allow fine particles to settle out. Standard practice is to recirculate the processed water once the fine particles have settled out of suspension. This not only reduces the suspended solids in the water but also reduces the water needs. With the growth of housing developments comes the increase in need of water. Water drawdown by a stone operation may be caused by either process water needs or by pumping of water from the quarry. Mining hydrology studies may be performed to determine the effects and extent of impact of mining on the local water table.

Blasting, front end loaders, haul trucks, and the processing plant are sources of noise and may be subject to local regulations. Proper planning of quarry operations which include location of the plant, mining sequence, and retention of natural noise barriers such as trees and topography will aid in noise reduction. Significant progress has been made by blasting companies to reduce noise and vibration while maintaining production levels. Backup alarms are the main noise problem for front end loaders. New practices to reduce this noise include using backup alarms equipped with radar devices to limit the alarm to times when objects are behind the equipment. Haul trucks generally are equipped with rubber tires and noise reducing mufflers to control noise.

Mine sites with the lowest number of accidents and fatalities are typically the ones with the best health and safety training programs. Knowledge and awareness of potential hazards and how to deal with them on the job are the best prevention of accidents. Mandatory safety requirements such as hard hat and safety eyeglass areas, maximum speed limits, traffic control signs, and seat belt laws go a long way in reducing potential accidents.

Truck Traffic

Truck traffic in builtup areas around quarries may be the cause of complaints. Frequently local governments may limit times or routes that trucks may travel.

Public concerns regarding final land use can be diminished by performing reclamation concurrent with mining. Current reclamation practices can not only enhance the beauty of exhausted mining sites but create additional value, not only to the operator, but to the community as well. Innovative reclamation ideas can help foster positive community relations when the local town benefits from the final land use.

PROBLEMS AND FUTURE TRENDS

The greatest uncertainty for producers of crushed stone will be the level of government regulation and restrictions associated with production and consumption. These uncertainties will be especially great for proposed new stone producing facilities that must compete with older, more favorably situated facilities.

Demand

Demand for crushed stone should continue to parallel construction activity. Market areas for the individual producer will continue to be delineated by the interrelationship of the cost of truck transportation, production costs, and the location of competitive producing plants. Depletion of sand and gravel deposits adjacent to certain metropolitan markets could result in a continuation of the shift to increased use of crushed stone where quarry sites are available. Manufactured sand should increasingly penetrate the fine aggregate market.

Specifications for aggregates can be expected to change slowly as researchers continue to develop improved material performance tests such as for frost susceptibility (ASTM C682) and to develop modified requirements for special aggregate applications such as skid-resistant surfaces for highways.

Governmental Impacts

The structure of the industry may gradually change if external forces such as governmental regulations tend to inhibit competition. The short term effect of federal, state, and local regulations in quarries and crushed stone operations is to decrease the number of operations. This will increase the price of stone to the consumer because fewer quarries mean longer average truck hauls and decreased competition. The costs of government regulations will be passed on to the consumers of crushed stone.

If the quarry industry continues to reduce the undesirable features of quarries, quarry operations may become more acceptable to the public in the long run. Such a change in attitude could result in more permissive zoning that would tend to increase the number of quarries operating in the future. This could promote competition and produce a relative decline in the delivered cost of stone to the consumer.

Supply

There is no significant physical limit to reserves of stone suitable for supplying aggregate that meets quality specifications as they are presently established throughout the country. Locally, however, land use regulations may continue to reduce availability of stone. In response to regulations on quarrying, the number of underground operations should increase.

Productivity

Despite anticipated productivity gains, total average cost of producing stone should increase at a rate higher than warranted by inflation. The primary causes will be the need to install expensive pollution control systems, the reduction of reserves by expanded buffer zones surrounding quarries, increases in taxation, and increased transportation costs. An increase in the number of states and localities imposing severance taxes can be anticipated as a result of the growing public interest in the operation of natural resource industries. Productivity gains due to technical improvements in equipment should offset most increases in labor costs.

Size of Companies

Firms that are large enough to develop the specialized managerial talent required to respond to governmental activities affecting the crushed stone aggregates industry will tend to expand their share of the industry. The size of such large firms will be increased and their profit performance as a percentage of financial assets should exceed that of smaller firms that are without the resources to respond to regulatory changes.

BIBLIOGRAPHY AND REFERENCES

Anon., 1966, *Standard Specifications for Highway Materials and Methods of Sampling and Testing*, Pt. I, "Specifications," 463 pp; Pt. II, "Methods of Sampling and Testing," 693 pp., 9th ed., American Association of State Highway Officials, Washington, DC.

Anon., *Highway Construction Usage Factors for Aggregates*, US Federal Highway Administration, US Government Printing Office, Washington, DC, annual issues.

Anon., 1972, "Skid Resistance," *Synthesis of Highway Practice No. 14*, National Cooperative Highway Research Program, National Research Board, Washington, DC, 66 pp.

Anon., 1974, "Quality Concrete with Crushed Stone Aggregate," National Crushed Stone Association, Washington, DC, 31 pp.

Anon., 1992, *1992 Annual Book of ASTM Standards*, Vol. 04.02, "Concrete and Mineral Aggregates," American Society for Testing and Materials, Philadelphia, PA, 840 pp.

Anon., 1982, "The Seven Best Crushed Stone Plants," *Rock Products*, series of articles available as a reprint, 32 pp.

Archibald, R.D., 1991, "Extraction Principles," *The Aggregate Handbook*, R.D. Barksdale, ed., National Stone Association, Washington, DC.

Barksdale, R.D., 1991, *The Aggregate Handbook*, National Stone Association, Washington, DC., various pages.

Bowers, R.L., II, et al., 1991, "Processing Plant Principles," *The Aggregate Handbook*, R.D. Barksdale, ed., National Stone Association, Washington, DC, pp. 8-1–8-84.

Dunn, J.R., and Hudec, P.P., 1972, "Frost and Sorption Effect in Argillaceous Rocks," *Frost Action in Soils*, Highway Research Record, No. 393, pp. 65–78.

French, R.R., 1968, "Transportation of Mineral Aggregates in Indiana," *Indiana Academy of Science Proceedings for 1968*, Vol. 78, pp. 348–354.

McCarl, H.N., 1969, "The Mineral Aggregate Industry in the Vicinity of Baltimore, Maryland," unpublished PhD thesis, The Pennsylvania State University, University Park, PA, Dec., 157 pp.

Nicholls, H.R., et al. 1971, "Blasting Vibrations and Their Effects on Structures," Bulletin 656, US Bureau of Mines, 105 pp.

Ozol, M.A., and Locke, S.R., 1974, "Study to Determine the Feasibility of an Experiment to Transfer Technology to the Crushed Stone Industry," National Science Foundation, Contract NSF-C826, 94 pp.

Robertson, J.L., 1978, 'Iowa Firm's Portable Plants Blanket Market Area," *Rock Products*, Vol. 81, No. 7, pp. 62–65.

Rogers, C.A., 1979, "Alkali Aggregate Reactions, Concrete Aggregate Testing and Problem Aggregates," *Ontario, A Review*, Report EM-31, Ontario, 27 pp.

Schenck, G.K., 1971, "Who Needs a Private Weather Service," *Mining Engineering*, Vol. 23, No. 4, pp. 54–57.

Schenck, G.K., and Torries, T.F., 1983, "Aggregates-Crushed Stone," *Industrial Minerals and Rocks*, 5th ed., Vol. 1, S.J. Lefond, ed., AIME, New York, pp. 60–80.

Schultz, G., 1981, "Aggregate Plant Design: The Planned Approach," *Rock Products*, series of articles available as a reprint, 32 pp.

Schwochow, S.D., 1980, "The Tortuous Trail to Permit Approval," *Rock Products*, Aug., pp. 50–53.

Severinghaus, N., 1960, "Crushed Stone," *Industrial Minerals and Rocks*, 3rd ed., J.L. Gillson, ed., AIME, New York, pp. 285–302.

Singleton, R.H., 1980, "Directory of Principal Stone Producers in the United States in 1980," *Mineral Industry Surveys*, US Bureau of Mines, 10 pp.

Singleton, R.H., 1980, "Stone," *Minerals Yearbook*, US Bureau of Mines, pp. 765–793.

Singleton, R.H., 1980, "Stone," *Mineral Facts and Problems*, US Bureau of Mines, pp. 853–868.

Singleton, R.H., 1982, "Stone," *Mineral Commodity Summaries*, US Bureau of Mines, pp. 146–147.

Soles, J.A., 1979, Petrography in the Evaluation of Aggregates and Concrete, CANMET, Report 80–3E, Energy Mines and Resources Department, Ottawa, Dec., 14 pp.

Stearn, E.W., 1978, "A Look at Two Challenges to Quarrying," *Rock Products*, Vol. 81, No. 9, pp. 71–76.

Tepordei, V.V., 1991, "Crushed Stone," *Minerals Yearbook*, US Bureau of Mines, pp. 1–28.

Torries, T.F., 1972, "Economic Descriptions and Simulations of the Construction Aggregate Industry in the Allentown-Bethlehem-Easton Area, Pennsylvania," unpublished PhD thesis, The Pennsylvania State University, University Park, PA, Sep., 188 pp.

Woolf, D.O., 1953, "Results of Physical Tests of Road Building Aggregates," Physical Research Branch, Bureau of Public Roads, US Department of Commerce, Washington, DC, 225 pp.

STONE, DIMENSION

W. ROBERT POWER

Stone is considered by many to be the premium material for beauty and durability in all kinds of construction. Its use dates to the dawn of civilization and only buildings made of stone have survived from ancient times. The use of stone relative to other materials declined from the beginning of the present century reaching a low point in the third quarter, but there has been a resurgence in the use of stone in construction during the past decade, due probably, to new techniques for producing stone, a greater recognition of its beauty and durability, and greater affluence among consumers. The aesthetic appeal of stone is attested to by the many attempts to simulate it in synthetic materials. Today, natural stone is considered by many to be the ultimate in aesthetic appeal, durability, and ease of maintenance.

Consumption of dimension stone has increased dramatically during the past decade. The US Bureau of Mines reported that the total value of US consumption in 1978 was $113 million and that total imports were valued at $65 million in 1979 (Anon., 1980). In 1989 consumption of domestic dimension stone was valued at $208.3 million and imports at $525 million. This dramatic increase occurred after demand had decreased at an average rate of 2.5% in the two decades prior to 1978 (Anon., 1980). More than half of the imports come from Italy, but production of dimension stone has increased worldwide spectacularly.

Meaningful statistics on dimension stone are hard to acquire. The finished piece of stone is a highly value-added product, but stone may be sold as rough blocks, sawed slabs, or finished product. The same piece of stone may therefore be sold and reported three or more times. Furthermore, although the quantity of stone is most commonly reported in tons, stone cladding is more often sold by the surface area and at the same price for different thicknesses. By any measure, however, the production and consumption of dimension stone has increased dramatically during the past decade.

A significant indication of the resurgence of stone was the introduction of two new trade publications during the 1980s. *Stone World* (Tradelink Publishing Co., Inc., Oradell, NJ), introduced in 1984 is published monthly and *Dimensional Stone* (Dimensional Stone Institute, Inc., Woodland Hills, CA), introduced in 1985 is published monthly. Both magazines provide useful information on new technology, market trends, producing areas and companies, and other information not previously available.

DEFINITIONS

In this paper the term *dimension stone* will refer to stone that is finished to specific dimensions and shapes. Most commonly it is quarried in large rectangular blocks which are then sawed into slabs for further finishing. It is used in buildings, monuments, furniture, industrial applications, and other uses. Other stone is sold in either natural or broken sizes and shapes which are sorted into size ranges, but not finished or dressed to specific dimensions. This stone may be used for building, paving, decorative, or other purposes. It is sold as field stone, flagging, rubble, and under similar names.

The American Society for Testing and Materials (ASTM) and the US Bureau of Mines include all of these groups as Dimension Stone. Nonetheless the distinction seems important. The dimension stone industry as defined here is highly sophisticated, using complex, often computerized equipment, and large quantities of industrial diamonds and other abrasives. It is capital intensive and becoming more so. The production of stone that is not cut or finished to specific dimensions remains a less sophisticated, more labor intensive industry. The equipment used is simpler—drills, front end loaders, and trucks—and the products are hand picked and sorted. One thing that is common to both, however, is sophisticated marketing techniques. While some non-dimension building stones are produced by mom-and-pop type operations serving local markets only, others are marketed nationally and/or internationally.

Three of the most widely distributed trade magazines reflect the distinction. *Dimensional Stone* covers dimension stone as here defined with articles on new machinery, production techniques, market trends, and similar information. An annual directory and buyers guide is published every December. *Stone World* also is devoted to dimension stone, publishing similar articles, and a buyers guide every November. *Building Stone Magazine* (Building Stone Institute, Purdys, NY) covers the full range of building stones including rubble, flagging, and field stone. It contains more articles describing the end product (finished buildings, homes, etc.) and fewer on technique. Much of the advertising in this magazine is for stone that is not quarried in blocks and dressed to specific size. All three magazines have extensive advertiser indices and offer reader services to their advertisers.

There is no simple way to logically organize *Dimension and Building Stone* into a neat outline. Building stone includes both dressed and undressed stone. Therefore dimension stone can be considered a subset of building stone. Dimension stone (as here defined) can be used for building, furniture, rolling mills, and other industrial uses. Therefore building stone can be considered a subset of dimension stone. The problem is probably of little interest to anyone except those faced with the task of writing a chapter to include both, and a few academics, but since this chapter is expected to cover both dimension and building stone, here is the solution offered.

In this chapter dimension and building stone will be subdivided into two classes: 1) stone that is cut and finished to specified sizes and shapes and 2) stone that is sold in natural or broken sizes and shapes. The former may be used for buildings, monuments, paving, furniture, and decorative objects; the latter may be used for building, paving, rough construction, landscaping, and erosion control.

Stone that is cut and finished to specific sizes and shapes is typically quarried in rectangular blocks, then sawed and finished to specification. It is sometimes called cut stone. Stone sold in natural sizes and shapes may be collected from the surface, broken out of a quarry, or be a byproduct of a block quarry. It is typically hand sorted into ranges of size and shapes.

Both may be used in building. When so used, broken stone will produce an irregular, or natural appearance. Horizontal lines will be wavy. Cut stone used in building will produce a regular or geometric appearance. Horizontal lines will be straight. Cut stone is considered by many to be the only true dimension stone.

Dimension and building stone are also classified lithologically. The lithologic classification can create great confusion among geologists and engineers. The dimension stone industry is one of the oldest in the world and many terms used by it pre-date the advent of modern geology. For example the *Oxford English Dictionary* lists five columns of definitions for marble, many of which have no relationship to a modern geologic definition. In Italy, marble includes all hard stone that is capable of taking a polish, including granite and other rock types (Anon., 1987). The ASTM committee on dimension stone has gradually modified its definitions to bring them more into line with modern geology, but large disparities remain. This can become very important in case law where original intent governs. Most commercial people dealing in stone are not well versed in modern scientific definitions, nor is there any compelling reason why they should be. It is important for the geologist or engineer who becomes involved with the commercial stone business to learn the trade language.

Stone Sold and Used in Natural or Broken Sizes

This includes all stone that is gathered from the surface, broken out of the quarry, or is an undressed byproduct from a block quarry. In building construction, it produces a natural look and in today's market is especially popular in home and small commercial building. Some is produced and sold locally by mom-and-pop operations, but much is distributed nationally.

Field Stone: Field stone is stone that is gathered from the surface and used to make stone walls, fireplaces, and similar structures. Its use dates to antiquity and was an important building material in early America. There is no connotation as to size, shape, or rock type except that it can be set by hand. Rock shops may sell field stone, but there is no generally accepted commercial connotation as to the nature of the stone.

Rough Construction and Jetty Stone: Large blocks of rough-hewn granite or other stone have long been used in such structures as retaining walls, seawalls, and bridge work. Requirements for such stone are that it resist weathering and erosion and that it break naturally into rectangular shapes that are easily set. Granite fits these requirements admirably and is the most widely used stone for this purpose. Other good sources are waste or rejected blocks from dimension stone quarries.

Rubble: Smaller sizes of broken stone are very popular as wall facing material in building construction, both for homes and small commercial buildings. The primary purpose is aesthetic—it replaces brick or other veneer—but ease of setting, weather resistance, lightweight, and the ability to bond well to mortar are also important. The stone can be produced as a byproduct from other stone operations or it can be produced from small pits with a minimum of capital investment by small operators. The only requirements are that it find a market and that it can be transported easily and cheaply to that market.

Because of the ease of production there are many small producers supplying local markets. In 1981, some 260 companies produced dimension stone in 425 quarries located in 39 states (Singleton, 1982). Some of the sellers quarry their own stone, but some also contract for the exclusive right to sell stone produced by a small operator in return for supplying advertising and distribution services.

The rock types advertised by the Building Stone Institute include all the standard dimension stone types as well as obsidian (which is actually pumice) and specialty stones. Some of these have national distribution.

Rubble is set in random patterns. The size of pieces ranges from about 0.05 to 0.5 m^2 of exposed rock. A few years ago slabs of thin-bedded natural flagstone were commonly used with the bed set parallel to the wall, but recently the three dimension effect has become more popular and natural flagstones are more commonly set in courses with the bed perpendicular to the wall (see next section). Nonetheless slab-shaped rocks have an advantage in that they cover more surface with less volume and weight. Low density rocks such as pumice and other porous rock have several advantages. Shipping costs are lower, setting is easier for the stone mason, fewer or no anchors are required to tie the stone veneer to the wall, and the wall need not be so strong.

Flagstone: Flagstone consists of thin slabs of stone used as paving. Slate and fine-grained sandstone are the most used rock types, but any naturally fissile stone can be used provided it has sufficient abrasion resistance.

Flagstone slabs are generally 25 to 75 mm thick but thinner slabs may be used where they are placed on an exceptionally firm base and subjected to only light traffic. Flagstone may be sawed into rectangular shapes or may be produced in naturally irregular shapes. It may be sold in either cut or broken sizes. The largest production of cut flagstone is produced from bluestone and slate (see next section). Much slate flagstone is produced from quarry blocks that are not suitable for mill stock or other higher priced products. Random-shaped flagging is produced by small producers for local markets. A standard for abrasion resistance of stone subjected to foot traffic is given in ASTM designation C241–90.

Paving Blocks: Small rectangular blocks of granite were formerly used to pave roads, docks, freight yards, and other areas subjected to heavy traffic. This market has all but disappeared owing in part to the substitution of rubber tired for steel rim wheels. A small amount of granite paving is used for ornamental paving and in a similar application much granite is used to cover slopes near bridge abutments and other areas where slope stability and erosion is a problem.

Stone Cut to Size

This group includes all stone that is cut to specific dimensions on at least two sides. Most of it is quarried as large, rectangular blocks which are sawed into slabs and pieces of the desired dimension.

Dimension Stone: Dimension stone includes all building stone that is cut to specific dimensions on all sides. The surface may be textured, smoothed, or polished, but is always controlled and finished to specifications. Dimension stone may be quarried and finished by a single producer or it may be finished by manufacturers who purchase quarry blocks or sawed slabs from a quarrier.

Accurate data on production of cut dimension stone are difficult to obtain. The US Bureau of Mines publishes annual reports on the production and value of various lithologies, but these data are deceptive because they are reported in mass units whereas most cut stone used in construction is sold by volume. The production and finishing of cut stone is essentially a value-added business where actual cost of raw material is a small portion of total cost.

Specifications for dimension stone are stricter than for other categories of building stone. The most important physical tests as specified by the ASTM are listed in Table 1. In addition to the tests listed in Table 1, the coefficient of expansion and thermal conductivity are important in some applications.

Most dimension stone is used as non-load bearing, curtain walls

Table 1. ASTM Specifications for Building Stone

Specifications	
C 568–89	Limestone Dimension Stone
C 503–89	Marble Dimension Stone (Exterior)
C 406–89	Roofing Slate
C 615–85	Granite Dimension Stone
C 616–89	Quartz-based Dimension Stone
C 629–89	Slate Dimension Stone
Test Methods	
C 241–90	Abrasion Resistance of Stone Subjected to Foot Traffic
C 97–90	Absorption and Bulk Specific Gravity of Dimension Stone
C 170–90	Compressive Strength of Dimension Stone
C 889–89	Flexural Strength of Dimension Stone
C 120–90	Flexural Testing of Slate (Modulus of Rupture, Modulus of Elasticity)
C 99–87	Modulus of Rupture of Dimension Stone
C 121–90	Water Absorption of Slate
C 217–85	Weather Resistance of Dimension Stone
Definitions of Terms	
C 119–90	Dimension Stone

or veneer. The use of natural stone as structural members in buildings is almost non-existent in modern architecture, but columns, lintels, and other structural members may be faced with natural stone. The stone is hung using one of many systems on the market. The 1990 Directory and Buyers Guide issue of *Dimensional Stone* (December 1990) lists fourteen suppliers of anchor systems. The magazine also publishes regular updates on the systems which are constantly being improved to meet the demands of modern architecture.

Stone panels ranging in thickness from about 20 to 100 mm are set as a veneer with horizontal support at each floor and appropriate anchors to the structural wall behind. The load-bearing requirements of the stone are therefore limited to self support for a floor height despite the rather high requirements for compressive strength called for in the ASTM designations.

Thin panels offer two advantages: savings in the cost of the stone, and savings in weight resulting in the use of less structural steel to support the stone. Because of the increased cost of structural steel, there is constant pressure to reduce the weight and hence the thickness of panels. When carried to excess this can result in failure of the panels. The problem is especially acute when very large panels are used—a practice commonly favored by architects.

Slate generally is used in natural thicknesses, commonly ranging from 6 to 50 mm and with natural cleft or face showing. Honing the surface destroys the natural beauty of the stone and increases the chance of deleterious weathering. Slabs too thick for a particular application may be ground on the reverse side to specification. Most slate is used for roofing and flooring. Other uses are for billiard tables, electric panels, blackboards, mantels, window sills, and similar uses.

Natural stone used in conjunction with other structural materials is a significant development in the use of dimension stone. This approach developed naturally in conjunction with the use of very thin slabs in the marble industry; now limestone, granite, and slate are being individually suspended on the face of the building so as to allow movement at the interface. Elastic jointing compounds have made possible weathertight seals. Stone is commonly preassembled into units such as column covers and window frames using epoxy glue and metal supports. The units are furnished with brackets for easy attachment to hangers built into structural members of the building.

A trend in the application of all kinds of dimension stone is toward use of larger slabs. It is not uncommon for limestone and granite panels to be used in single floor to floor spans. Marble slabs are more limited in size (2 m^2), but marble-faced panels backed with precast concrete may be produced in single units for floor to floor installation. Because many of the large slabs are supported at top and bottom only, producers and producer associations have adopted wind load and other design specifications relating size of the panel to the thickness of the stone in order to give adequate strength. Furthermore, these requirements have put additional burdens on the quarrier because no physical flaws in the stone along which separation might occur can be tolerated. Whereas smaller sized pieces could be cut between natural flaws in a block, the quarrier is now called on to produce large, virtually flaw-free blocks of stone.

Color and pattern are inherent qualities of stone, but they can be influenced by the finish. Polished finishes tend to darken color and bring vivid contrasts. Rough textured surfaces tend to lighten color and subdue patterns but tend to catch dirt. Polished marble is used extensively on interior walls where the natural color and veining of many exotic marbles are used to advantage. In exterior work where uniform color and subdued pattern are currently in favor, rough-textured finishes may be used except where air-borne soot is a problem. These surfaces are achieved by abrading the surface, or in the case of granite, flame treating.

Granite and serpentine marble (verde antique) are favored stone for use in the base course and entrance features of large buildings. These are the only natural stones that will hold a polish in exterior use and polished surfaces are easily maintained as well as attractive.

Although seldom actually specified, the long term availability of a particular stone may be an important factor in its selection for some uses. This is because of the possibility of additions or repair work which must match the original structure. A related factor is that many architects seem to prefer established building stones with which they are familiar. These considerations give long established quarries an advantage over new or untried stone despite any other considerations.

Tile: Tile is thin (2 to 3 cm) stone cut into square or rectangular shapes almost always with a polished surface. In the United States, tile is still sold in 0.3 m "one-foot" square pieces or sizes that are integrally divisible into one foot. The principal use is flooring, but it may also be used on walls and countertops. Consumption of tile in the United States has greatly expanded in recent years, mostly because of the advent of automated tile cutting and polishing machines. Most tile currently produced is marble because of the ease of cutting and polishing, but granite tile is increasingly produced and consumed. Blocks not suitable for making large panels can be used for tile, an advantage to the producer.

Monumental Stone: Monumental stone encompasses a variety of products ranging from simple grave markers, to statues, mausoleums, and elaborate structures such as the Lincoln and Jefferson Memorials. In construction and fabrication the larger memorials and mausoleums differ little from building construction except that they may contain more elaborate carving.

Granite and marble, particularly white marble, are the preferred stones for monumental use. According to US Bureau of Mines statistics, more than half the granite quarried is used for monumental purposes. Georgia and Vermont are the largest producers of monumental marble, but the producing companies produce more building than monumental stone.

Standard specifications for building stones do not apply to monumental stone unless the purchaser chooses. The principal requirements are for uniformity of texture and color, freedom from flaws, suitability for polishing and carving, and resistance to weathering. These qualities either are aesthetic, and hence not amenable to standard specifications, or are qualities best determined by experience. So far no one has developed a satisfactory test to predetermine the resistance of a stone to weathering.

Ashlar: Ashlar consists of rectangular pieces of stone of nonuniform size that are set randomly in a wall. The exposed surface of each piece is generally less that 0.4 m^2. The surface may be sawed (smooth face) or rough-hewn (rock face). Ashlar blocks are prepared to specified thickness, usually about 75 mm. Ashlar may be sawed from quarry blocks that are not suitable for finishing to cut dimension stone or it may be made from rocks that break naturally into rectangular pieces. Although smooth face ashlar was very popular in the past the rough-hewn look of rock face is more popular today.

Split-Faced Ashlar: Split-faced ashlar consists of brick-shaped blocks of stone that are laid in courses. Blocks of marble or limestone are sawed into modular thicknesses and then split with a power-driven wedge called a guillotine. The blocks are trimmed and laid in courses with the broken face showing. Natural flagstones (sandstone, slate, and limestone) are similarly broken into brick-shaped pieces differing only in that the thickness is not precisely controlled. In the limestone and marble industry, split-faced ashlar consumes much stone that is considered waste in the production of dimension stone.

Roofing Slate: The annual dollar volume of roofing slate declined from $5,186,000 in 1908 to $2,302,000 in 1978. It is now back up to $9,132,000 (Taylor, 1991) or more. It is used in new construction and for repairs and restoration of older buildings, and is considered by many to be the most permanent type of roof covering. Standard specifications for roofing slate are given in ASTM designation C406–89.

Mill Stock Slate: Smooth-finished slabs of slate are used for electric switchboard panels, billiard tables, blackboards, counter tops, and similar purposes. Electric slate must have high electrical resistance and be free of magnetite and other conducting minerals.

Slate Dimension Stone: Either cleaved or smooth-finished slate may be used as hearthstones, mantels, steps, sills, or other structural purposes. Standard specifications for structural slate are given in ASTM designation C629–89.

Curbing: Granite curbing is used in large quantities because of its superior resistance to abrasion and weathering. In areas of abundant freeze-thaw cycling where deicing salts are used, granite has proved an especially resistant material. Granite curbing is broken out along the rift and grain of the stone, but it is also possible to break out curved pieces by close drilling. Mt. Airy Granite Co. of North Carolina specifies a sawed upper surface, but the face and back are smooth quarry split. Curbing has exceeded all other forms of dressed granite produced for the past several years (Taylor, 1988).

Miscellaneous Uses: Other uses of dimension stone include laboratory furniture and sinks in which soapstone is considered the premium material. Sandstone, quartzite, and granite are used to line tube mills for grinding ore or other material. Precision surface plates for mounting optical instruments and fine machine work are made from granite. Sandstone and soapstone have been used as refractory brick. In the past large numbers of grindstones and millstones were made from sandstone or granite. Arkansas novaculite, a fine quartzite, is fashioned into hones. Granite, because of its resistance to chemical corrosion and its favorable physical properties, is used to make cylinders up to 9.1 m long and 1.5 m or more in diameter for grinding pulp in paper mills. It is also used as skid caps and tank liners in steel pickling mills (Richter, 1988).

LITHOLOGIC CLASSIFICATION

Almost every variety of rock can or has been used as dimension stone. The suitability of a particular stone for use as dimension stone is governed primarily by physical properties and aesthetic appeal. Mineralogy and chemistry are, for the most part, important only indirectly as they bear on the physical properties. The exception to this statement is the resistance of a rock to weathering which is profoundly affected by mineralogy and chemistry as well as by physical properties. Standard geologic nomenclature and classification of rocks is needlessly cumbersome for an industry which is dominated by nonprofessional, sales-oriented people and architects.

The ASTM has adopted standard definitions for granite, limestone, marble, greenstone, sandstone, slate, and certain subdivisions and varieties of each (ASTM designation C 119–90). The Building Stone Institute formerly gave a glossary of words including rock types in its annual *Stone Catalog* (Anon., 1972). However, this catalog has been discontinued. Commercial definitions recognize and respect history and use within the industry, and at the same time attempt to keep the commercial definitions compatible with scientific terminology. This sometimes creates confusion. The following sections discuss both scientific and commercial definitions.

Granite

Commercial Definition: Commercial granite includes all feldspathic rocks of visibly granular or gneissic texture. It thus includes granite family rocks as defined below plus syenite, gabbro, anorthosite, and other plutonic igneous rocks. In commercial usage fine-grained igneous rocks such as diabase or basalt may be called black granite.

Commercial granites are commonly named for the quarry or location where they are found, with names modified by adjectives giving color, texture, or some exotic description.

Geology: To the petrologist granite is a visibly crystalline rock with interlocking texture and composed essentially of alkali feldspar and quartz. It is a member of the granite family of rocks which include all rocks with granitic texture and composed of essential quartz and feldspar. The granite family is subdivided into specific members such as granite (proper) and granodiorite on the basis of the kind and proportion of the three major feldspar varieties in the rock. All members of the family contain quartz and most contain minor accessory minerals such as biotite, muscovite, hornblende, and pyroxene.

Granitoid rocks containing essential feldspar, but with no quartz, are called syenite if alkali feldspar predominates, diorite if calcium feldspar predominates, or monzonite if the feldspar are about equal in abundance. These rocks typically contain considerable amounts of ferromagnesian minerals such as biotite, hornblende, and pyroxene as accessories. Rocks with approximately equal proportions of calcium feldspar and pyroxene are classed as gabbro or norite. These are included in the typical black granites of commercial usage. Rocks with mineral composition of the granite family, but with a texture that shows distinct planar or linear properties owing to the parallel alignment of mineral grains, are called gneiss.

Most of the rocks discussed in this section are classed as plutonic by the geologist, meaning that they formed deep within the earth's crust and characteristically have a visibly granular texture. Exceptions are basalt family rocks, generally known as traprock and sold as black granite. All but gneiss are typically of igneous origin, meaning that they formed by consolidation from molten material—magma. Most gneisses and some granite family members, however, are thought to have been formed by metamorphism; that is, by transformation from preexisting sedimentary or other rocks.

Granite family rocks may occur as dikes, sills, or large subjacent masses of irregular or globular shape. They may be considered bottomless for practical economic purposes. The mode of emplacement may be by forceful injection where the molten magma shoulders aside the country rock, or more passively by stoping where the rising magma engulfs fragments of the country rock which either sink into or are resorbed by the magma. Most granites are thought to have been emplaced in the roots of mountains that form at convergent plate margins. They are exposed only after profound erosion has removed the covering rocks. Granite emplacement may be contemporaneous with orogenic folding (synorogenic) or may follow the principal orogenic movements (postorogenic). Postorogenic granites, passively emplaced by block stoping, are more apt to be free of internal strain and usually make better dimension stone.

Sandstone

Commercial Definition: The ASTM defines a commercial sandstone as a ''sedimentary rock composed mostly of mineral and rock fragments within the sand size range (2 to 0.06 mm) and having a minimum of 60% free silica, cemented or bonded to a greater or lesser degree by various materials including silica, iron oxides, carbonates, or clay, and which fractures around (not through) the constituent grains'' (ASTM designation C119—90). This definition unnecessarily excludes many arkoses which have less than 60% free silica, including some brownstones from the Connecticut Valley (see Pettijohn, 1975).

ASTM defines quartzitic sandstone as having 90% free silica and quartzite as ''highly indurated, typically metamorphosed sandstone containing at least 95% free silica, which fractures conchoidally through the grains'' (ASTM designation C119–90). This definition would exclude many metamorphosed sandstones mapped as quartzites by geologists.

Common commercial varieties of sandstone quarried in the United States include bluestone—''a dense, hard, fine-grained, commonly feldspathic sandstone of medium to dark greenish-gray or bluish-gray color that may split readily along original bedding planes to form thin slabs'' and brownstone—''a dense medium-grained stone, locally grading to conglomerate, with a distinctive dark brown to red color'' (ASTM C119–90). Typical brownstones are arkosic sandstones quarried from Triassic basins in eastern United States. Freestone is sandstone or limestone that splits with equal ease in any direction.

Geology: Sandstone is a clastic sedimentary rock composed of indurated sand grains, most of which fall within the size range $\frac{1}{16}$ to 2 mm. The interstices typically contain cementing material, but enough voids may remain to give the rock considerable porosity and permeability. Rocks composed of clastic fragments larger than sand-size are called conglomerate if the pieces are rounded, breccia if angular. Rocks composed of clastic particles smaller than sand size are siltstone or shale.

Sandstones are deposited by waning currents, mostly in water, rarely by wind. They are the end product of a weathering and erosion cycle that starts with the disintegration and decomposition of source rock, followed by erosion and transportation of the resulting material to a final resting place where the sediment is deposited in horizontal layers. Quartz is the only abundant sand-size material in the source rocks that strongly resists alteration or decomposition by weathering. Typically feldspar and other labile materials are depleted during the cycle, leaving a concentrated residue of quartz. The amount of quartz concentration is a measure of the maturity of the sand. Quartz-rich sandstones, sometimes called orthoquartzites, tend to be well sorted and to be cemented with secondarily precipitated calcite or silica.

Immature sandstones containing abundant feldspar are called feldspathic sandstones (5 to 25% feldspar) or arkoses (more than 25% feldspar). Alternatively, immature sandstone containing abundant rock fragments is called lithic sandstone, subgraywacke, or graywacke. Immature sandstones tend to be poorly sorted with the interstices commonly filled with clay, iron oxide, and other detritus.

Perhaps the most conspicuous feature of sandstone is bedding or stratification which has formed by successive influxes of sediment. Bedding planes commonly are parting planes. The direction of bedding is called rift in the industry. Those bedding planes along which the rock splits with greatest ease are called reeds. Some sandstones have a second direction of easy splitting that is nearly perpendicular to bedding. The second direction, called run, can be of great aid to the quarrier.

Sandstones with thin, even, and regular bedding along which the rock easily splits are natural flagstones. Sandstones required for sawing into cut dimension stone must have massive beds as the saw cuts are generally made perpendicular to bedding. Some sandstone layers have internal crossbedding that is not parallel to the principal stratification. If the stone parts along the crossbedding, the stone is unsuitable for dimension stone.

Limestone

Commercial Definition: Limestone is ''a rock of sedimentary origin composed principally of calcium carbonate . . . or the double carbonate of calcium and magnesium . . . or some combination of these'' (ASTM designation C119–90). In its most recent designation the ASTM (C119–90) recognizes as special varieties: calcarenite, coquina, dolomite, microcrystalline limestone, oolitic limestone, and recrystallized limestone. Of these terms, only calcarenite, coquina, and oolitic limestone are commonly used in commerce.

Limestones are commonly designated by location and texture, for example, Indiana Oolitic Limestone. The color may be included in a name such as Cream Kasota Vein, or such other descriptive or exotic terminology may be used as salesmanship dictates.

The Building Stone Institute, following usage, has defined an oolitic limestone as a stone formed of shells and shell fragments. The fragments are of more or less uniform size and generally 2 mm or less in greatest dimension. The original fossil texture is well preserved. Such a rock should be called calcarenite by the present ASTM definition. Many commercial oolitic limestones are composed in whole or part of oolites so would also fit the scientific definition of oolitic limestone; others do not.

Shell limestones generally contain larger unbroken fossil shells enclosed in a fine-grained matrix. Coquinas are composed of loosely cemented shells and shell fragments.

Travertine and recrystallized limestone may be considered either limestone or marble, based on the ability to take a polish.

Geology: Limestone is a sedimentary rock composed of one or both of the two polymorphs of calcium carbonate—calcite and aragonite. Aragonite is common in limestones of Recent age, but as it is unstable and inverts to calcite, it is rare or absent in ancient limestones. Sedimentary rock composed of the mineral dolomite (calcium-magnesium carbonate) is called dolomite or dolostone. Dolomitic limestone contains calcite and dolomite; magnesian limestone contains magnesium carbonate but not necessarily as dolomite. Many limestones and dolostones contain abundant sand or clay in which case they may be called sandy or argillaceous limestone (or dolostone).

The vast majority of limestones are composed of shells or shell fragments. When deposited in turbulent water the fragments are broken, sorted, and reworked. Resulting features such as size sorting and bedding are the same as for sandstones and the rocks have been called calcarenites, where the word ''arenite'' designates sand-sized particles. In quieter water, the sturdier shells may remain intact but other shells readily disintegrate into ultrafine particles which may wash out leaving a residue of larger shells sorted as to size. In very quiet water the ultrafine material plus the shells of one-celled microorganisms may form a matrix in which larger fossils may or may not be imbedded. Rocks composed of finely commuted calcite are called calcilutites. Limestones composed predominantly or entirely of the shells of microorganisms are called chalk.

The only limestones known to form by direct chemical precipitation in open seas are in the form of oolites. Oolites are tiny (1 mm or less) spheres of calcite or aragonite, commonly formed with a concentric structure around a nucleus consisting of a tiny fossil or other fragment. Modern studies show that oolites form in shallow, turbulent, warm seas where loss of carbon dioxide to the atmosphere causes supersaturation of calcium carbonate. Oolitic limestones may be composed entirely of oolites or of a mixture of oolites and shell fragments.

Modern petrologic studies show that much of the finely commuted material in limestones is reworked by burrowing organisms and pelletized by filter feeders. Modern carbonate classifications are based on the texture of particles, be they fossil fragments, pellets, oolites, or other fragments, and the degree of sorting of the particles. For further information, the reader is referred to various texts on sedimentary petrology.

The origin of dolostone is less clear. Some appear to form by in situ replacement of calcium by magnesium and the resulting stone commonly retains the texture and structure of the original limestone. Other dolomites are thought to form by direct precipitation in hypersaline waters where the climate is hot and dry.

Marble

Commercial Definition: Commercial marble is any crystalline rock composed predominantly of calcite, dolomite, or serpentine that is capable of taking a polish (ASTM designation C119–90). Marble is probably the oldest term used for dimension stone. In ancient Rome, the root word for marble—*marmore*—was used for all hard stones that could be polished. The same practice is followed in Italy today where the term marble is used for all hard stone that will take a polish, including granite. The practice is not followed in the United States today where commercial marble encompasses true marble in the geologic sense as well as many crystalline limestones, travertines, and serpentine, but not other lithologies.

Travertine is a cellular limestone deposited from hot springs. Onyx marble is a dense, crystalline form of limestone, deposited from cold water solutions and commonly found in caves. Verde antique is massive serpentine, commonly crisscrossed by veinlets of calcite, and capable of taking a high polish.

Commercial trade names commonly give the quarry location, color, and a descriptive adjective describing the pattern or texture of the stone. Where a pattern is referred to as veined, the stone is generally cut across the bedding or layering. Fleuri patterns are achieved by cutting parallel to layering.

Geology: Marble in scientific usage is a metamorphosed limestone or dolostone which is so thoroughly recrystallized that much or all of the sedimentary and biologic textures are obliterated. An interlocking, mosaic texture results. Bedding may be partially preserved in the form of compositional layering or banding. Where deformation is intense the layering may be folded, stretched, and pulled apart producing a swirly marbleized appearance that gives marble cake its name.

Impurities present in the original carbonate sediment form a characteristic suite of magnesium and lime silicate accessory minerals. Most limestones are slightly magnesian to dolomitic and the chief mineral impurities are quartz and clay minerals. Common accessory minerals are talc, chlorite, amphibole, and pyroxene. Accessory minerals are colorless to golden-brown magnesian varieties if iron is absent, but green if iron is present. Organic material entrapped within the rock forms graphite that is commonly accompanied by finely disseminated pyrite. Pure calcite marble is white, but tiny amounts of impurities will color it significantly. Graphite and pyrite commonly color marble gray. Finely disseminated hematite will color marble pink.

Most true marbles are deformed rocks in which original bedding planes have been folded, crumpled, or otherwise distorted. The crumpling is generally most intense along the axes of folds. The layering may remain straight and planar on the limbs. Such is the case in West Rutland, VT, where the character of the marble is very consistent parallel to the layering, but changes markedly across the layering. On the limbs of folds the layers may be stretched rather than crumpled. In some of the Alabama deposits brittle interbeds of dolomite marble have broken and pulled apart during deformation and the more plastic calcite marble squeezed into the spaces around the broken pieces. At Tate, GA, a combination of intense crumpling and stretching has practically obliterated all sense of original bedding and the entire deposit is a nearly homogeneous mass of swirly veined marble.

Crystalline limestones that take a polish are classed as marble commercially. Such limestones are not considered marble by petrologists, particularly if sedimentary and biologic textures are well preserved. The rocks may show recrystallization of the matrix and be thoroughly cemented with secondary calcite. They are dense, nonporous rocks. Many of the more decorative marbles are of this class.

Serpentine, classed as marble commercially, is not so classed by petrologists. Most serpentine results from metamorphism of peridotite and related rock. Peridotites, originally composed of pyroxene and olivine, are recrystallized at relatively low metamorphic temperatures in the presence of water to form serpentine minerals (hydrous magnesian silicates) and amphibole. Iron released in the reaction commonly forms magnetite; and lime released from pyroxene, or introduced from outside, commonly forms calcite veinlets. Serpentine commonly forms in tabular or lens-shaped bodies associated with other metamorphic rocks.

Travertine is a dense, commonly banded and cellular limestone deposited by chemical precipitation from springs or rivers. It is a hard, dense form of tufa. Much of the cellular structure is thought to have formed from the weathering out of plants or other debris around which the travertine was precipitated.

Slate

Commercial slate is defined as a microgranular metamorphic rock derived from argillaceous sediments, and is characterized by closely spaced parting planes called rock cleavage. This is entirely independent of original bedding. The cleavage permits the rock to split easily into thin slabs (ASTM designation C119–90). This definition applies for both commercial and scientific slate. Slates are formed by dynamic metamorphism. Beds of differing composition that form color stripes on cleaved surfaces are called ribbons. Ribbons are considered deleterious in high-grade slate. The best slate therefore comes from rocks having thick, massive, original bedding.

Quartz is usually the dominant mineral, but extremely fine-grained mica and other platy minerals are always present. The platy minerals tend to be aligned with the slaty cleavage and impart a sheen to the surface. Most slates contain pyrite or iron oxide and many contain graphite. The color of slate is determined principally by the oxidation state of the iron and the amount of graphite or other coloring material present. Graphite and pyrite impart a dark gray or black color; ferric iron oxides color the slate red. Green slates owe their color to the presence of chlorite, other iron silicates, or ferrous iron oxides.

Other Stone

Other rock types sold as dimension stone include greenstone, soapstone, and traprock or basalt.

Greenstone: Greenstone is a metamorphic rock containing "one or more of the following minerals: chlorite, epidote, or actinolite" (ASTM designation C119–90). It is derived by metamorphism of basalt, diabase, or other mafic igneous rocks.

Basalt: Basalt is a microcrystalline volcanic rock composed of pyroxene and calcic plagioclase. Traprock is a common term for dike rock of the same composition. The correct petrologic term for traprock is diabase or dolerite. Diabase may have columnar jointing, but lacks flow structures and commonly is coarser grained than basalt. When sold for dimension stone it is commonly called black granite or traprock.

Soapstone: Soapstone is a metamorphic rock derived from peridotite or other ultrabasic rocks and composed principally of talc. It has great heat-retaining properties and chemical inertness. For this reason it is used for laboratory furniture, stoves, and fireplaces.

QUARRY METHODS

Quarry operations typically involve isolating a mass of stone by cutting it free on all sides but one from the parent mass. The isolated mass is then lifted or separated from the parent mass by breaking it free or by undercutting with a wire or chainsaw. The freed stone may be a block of size suitable for sale or further processing—a mill block—or it may be much larger—a quarry block. Quarry blocks are subdivided by breaking into smaller blocks that can be removed from the quarry and sent to the mill. Typical mill blocks range in size from 12 to 30 t, but may be as large as 65 t. Quarry operations thus involve cutting the stone, breaking the stone, and removing the stone from the quarry.

Quarry Plan

The quarry plan is determined by the geometry of the deposit, the amount of overburden, and the necessity of handling rectangular blocks weighing up to 65 t. Most quarries are the open pit type. They have relatively small surface area and they are worked downward in a pit below the general surface level. Open pit quarries have several advantages: they can be located on outcrops or where overburden is thin; they are within reach of fixed derricks with movable booms; electric service, air compressors, loading docks, and other service facilities can be more or less permanently fixed near the working area. Traditionally the quarry size is limited by the reach of the derrick boom. Larger open pit quarries generally require several derricks, both for lifting out blocks and for emplacing and moving quarry machinery. In recent years many operators have used large tractors, forklifts, and other machinery to move equipment and quarry blocks across the quarry floor thus extending the reach of the derrick boom. Some have eliminated derricks altogether and built ramps to give access to fork lifts and other mobile equipment for moving blocks and equipment.

Open pit quarries are particularly well suited to massive deposits such as granite and some marble formations (Fig. 1). Steeply dipping layered deposits are also quarried by this method and the hanging wall is commonly undercut. Wall support is generally not required where a single opening by a single derrick is made. However, if successive quarries are opened along the strike, a web between quarries may be left for wall support.

The depth of open pit quarries is limited by the thickness or depth of good stone, by spalling or so called pressure breaks, or by excessive time loss in entering and leaving the quarry and in lifting out blocks. Open pits seldom reach depths of more than 75 m.

Flat-lying, layered deposits may be worked as open shelf quarries in a hillside. In this case the quarry floor is level with the ground

Fig. 1. Deep marble quarry in north Georgia.

on one side. If the deposit is thick, the quarry may be changed into an open pit as deeper layers are removed. The quarry may also be extended under the hill as an underground quarry if the overburden is too thick for removal on the high side.

Blocks are nearly always cut parallel to layering if the layering is pronounced and regular. Therefore, where layering is tilted, the quarry floors are not flat. This increases the difficulty of moving equipment and removing blocks. Movable equipment, such as tractors and forklifts, cannot be used, and enough derricks must be set to reach the entire working area of the quarry.

Underground quarries are room and pillar mines. Generally, no more than 25% and commonly as little as 15% of the stone is left for roof support. Underground derricks are set for loading blocks. These may be of fixed-boom type as blocks are generally skidded under them for loading onto trucks or other conveyances. In recent years, forklifts strong enough to lift quarry blocks have become available, eliminating the need for underground derricks in flat-floored quarries.

The initial opening of a tunnel quarry may be made by cutting out blocks in a manner similar to removing key blocks in an open quarry. The cost of removing these key or tunnel blocks is higher than for removing floor blocks. Alternatively, some underground marble quarries in Vermont have been opened by removing an overlying layer of rock with conventional mining methods (i.e., blasting and mucking). More efficient quarry equipment has made this unnecessary. After the initial opening is made quarry methods are generally the same as for open quarries except for block removal.

Fig. 2. Slot drill used for horizontal and vertical line drilling (*photo courtesy:* Tamrock Co.).

Cutting the Stone

Channeling Machines: One of the oldest methods for cutting stone is with channeling machines. These cut with the chopping action of several chisel-edged steel bars clamped together. The machine travels back and forth on tracks and a channel 50 to 64 mm wide is cut. As the channel is deepened longer bars are substituted in the machine. The method is slow and the channel bars require frequent sharpening so that the quarrier is practically obliged to have a blacksmith shop. Channeling has generally been replaced by other quarry methods.

Drilling and Broaching: Drilling and broaching is another long established method for cutting stone. Drill and broaching tools are mounted on a quarry bar built into a rigid frame. The tools slide along the bar which serves to align a series of closely spaced holes. The web between these is cut out by the chiseling and chopping action of the broaching tool. The method is suitable for all kinds of stone, but is slow, and like channeling steel, the broaching steel requires frequent sharpening. The method has largely been replaced by line drilling in modern quarries.

Line Drilling: Line drilling, also called slot drilling, consists of drilling a line of overlapping holes. The drill is mounted on a quarry bar or frame which serves to align the holes and hold the drill rigidly in position. A key to successful line drilling is a guide attached to the drill that is inserted into the last complete hole in the line, and serves to insure that successive holes overlap. The Tamrock Company has developed a machine (Slot Drill 600) that mechanically spaces 64 mm holes, then drills the web between them using a special guide (Fig. 2). Tamrock claims a maximum depth of 6 m for the system. This machine illustrates a trend toward mechanization which cuts labor costs.

Jet Channeling: Jet channeling is the most common method of cutting granite. The method, also known as flame-cutting, consists of disintegrating the rock with intense heat from the combustion of fuel oil and oxygen. The thermal shock causes the rock to spall and the fragments are blown from the channel. The torch, which is passed back and forth over the rock, is commonly attached to a frame in order to keep the cut true. In some granite quarries, large blocks are freed using this method and then are subdivided by using cheaper methods of breaking the rock. Jet channeling works well in granite and other quartz-bearing igneous rocks that spall. It does not work in quartz-free rocks, or carbonate rocks that calcine or fuse, rather than spall.

Jet channeling has two significant disadvantages. The noise level is so high as to constitute a serious health hazard; and the kerf is wide and irregular, thus wasting stone. For these reasons, jet channeling is being replaced by other methods.

Wire Saws Using Abrasives: Wire saws using quartz sand or silicon carbide have been used for many years to cut marble and more rarely, granite and limestone. The method, once dominant in some marble districts, has been and is being replaced by diamond wire saws (see following section).

The wire saw consists of a single, double, or triple strand of wire that runs over sheaves and is fed under tension into the stone. Sand, aluminum oxide, silicon carbide, or other abrasive is fed to the wire and wears a groove or channel in the stone. The sheaves are mounted on a tower or track so that they can be moved into the cut. In order to press the wire into the cut, there must be two free and parallel sides to the block of stone being cut or else a large hole to admit the sheaves must be drilled. In some cases the hole need only be large enough to admit the axle, hub, and supporting arm of the sheave. The sheave is designed to cut its own groove as it is pressed into the stone. In some operations, other quarry cutting methods are used to cut parallel channels between which the stone may be wire sawed.

The wire itself abrades and must last for the entire cut. If it breaks before the cut is finished the kerf may be too small to insert a new wire and the cut is lost. Therefore, the wire is commonly several thousand meters long.

A single wire may be used to make many parallel or other cuts by guiding it with appropriate sheaves. At least one movable sheave must be set up with appropriate counterweights in order to keep tension on the wire. Wire-sawing produces a smooth cut that minimizes subsequent finishing in the shop and saves stone in the mill.

Diamond Wire Saws: Diamond wire saws were introduced into the stone industry during the 1970s, first in mills and shops, then into the quarries. It has and continues to replace the older wire saw techniques. Detailed descriptions of the diamond wire are given in Hawkins et al. (1990) and Cai and Mancini (1989). The wire is a steel cable on which small, diamond-impregnated or diamond-bonded, cylindrical beads are mounted. The beads are separated by either wire springs for soft, non-abrasive rock, or plastic for hard, abrasive rock. The wire is looped through drill holes and is driven by a motor-driven drive wheel mounted on tracks. Tension is provided by pulling the drive wheel assembly away from the cut on the tracks. Cuts may be either vertical or horizontal.

Most cuts require the drilling of two intersecting holes. The wire must, therefore, make a sharp angle when the cut is started, the cut becomes rounded as cutting proceeds. This requires care in providing correct tension during various phases of the cut. Cooling and lubrication is provided by water.

The technique requires precise drilling of the intersecting holes. In order to loop the wire, a nylon line is attached to a ball or wad of paper or other material that is forced through the holes hydraulically or with compressed air. The diamond wire can then be pulled through the intersecting holes with the nylon line. Further details can be provided by the Diamond Bort Company, Columbia, SC.

Chain and Belt Saws: Chain saws using carbide teeth were used more than 30 years ago to cut soapstone in Virginia. More sophisticated saws made in Europe and also using carbide teeth were used to cut limestone and soft marble, but were unsuccessful for cutting harder stone. Finally, with the incorporation of diamond set teeth, chain saws became practical for use in marble and slate. The most recent development is a belt saw in which a neoprene belt carrying diamonds imbedded in metal segments replaces the chain (Figs. 3 and 4). Cooling and lubricating water flows through the blade, which is hollow. Blades up to 5 m long are used in softer limestone, shorter blades in harder stone such as marble. Belt saws are also used in sandstone and slate, but apparently not yet in granite. It seems probable that ultimately belt or chain saws will be used successfully in granite and other hard stone.

Fig. 3. W.F. Myers diamond belt saw.

Fig. 4. W.F. Myers diamond belt saw.

Water Jet: A promising new development in cutting granite is the water jet. A pulsating jet of water at pressures ranging from 96.5 to 275.8 MPa is directed at the granite and causes it to disintegrate. It is less noisy, produces less dust, and promises to be more economical than jet piercing. Rocky Mountain Rose Red, Inc., of Lyons, CO, is currently using one (Anon., 1991a, Zielinski, 1992). The Elberton Granite Association in Georgia has tested two experimental machines (Anon., 1979), but they have apparently been less successful than diamond wire saws.

Breaking and Lifting the Stone

In open pit quarries, vertical cuts are made dividing the floor of the quarry into appropriately sized blocks. An initial key block may be broken free by drilling a hole in the center to the depth of the vertical cuts. An explosive charge in this hole frees the block with or without damage to the block. Alternatively, the block may be wedged free or forced free with an air bladder. In underground quarries, essentially the same technique is used except the initial cuts are horizontal and the key block is taken out of the tunnel face.

Once the key block is removed, a series of closely spaced holes

Fig. 5. Trimmer 100 used for trimming and squaring blocks (*photo courtesy:* Tamrock Co.).

is drilled underneath or behind the remaining blocks and they are broken free with the use of feathers and wedges, with explosives, or in some cases, wire saws. Feathers are pieces of half-round steel that are inserted into the hole. Steel wedges are driven between the feathers to produce the break.

Dimension stone quarriers have traditionally shied away from use of explosives, but recent experience indicates that if used properly, explosives can be used successfully and economically without damaging the stone. Most modern quarriers use explosives to break the block free. Prima cord or light plastic-enclosed charges inserted into the holes generally provide enough force and do not damage the stone. When a block of stone is broken free, it may be further subdivided by the same technique.

Granite typically has two directions of easy breakage called rift and grain or rift and run. In some quarries large blocks containing many thousand cubic meters of stone are cut free by jet channeling or other means. These blocks are then broken into mill-size blocks along the rift and run directions. Some granite ledge quarries rely entirely on breakage to produce dimension blocks. Holes drilled for feathers and wedges in granite commonly need penetrate only a few centimeters of stone. If the rift is strong, the break produced runs true for the entire length of the block. If the rift is not strong, however, drill holes may penetrate the entire width of the block as in the so called Finnish Method (see following). Often, odd-shaped blocks are trimmed and squared by line drilling and breaking (Fig. 5).

Marble and limestone are often cut into pieces of appropriate height and width for mill blocks, but of much greater length. These are then lifted (broken free from the quarry floor), using feathers and wedges. They are also broken into appropriate lengths by the same technique. Drill holes for the feathers and wedges generally penetrate the entire width of the block.

The stone may also be broken free with air pressure. A quarry block is cut free on all sides but the bottom. The bottom is drilled as though for using feathers and wedges. A large reinforced bladder is inserted in the vertical kerf that separates the block from the quarry. Air pressure applied to the bladder breaks the block free and sets it out from the quarry face. Usually 23 kg of pressure or less is sufficient. In some cases the air bladder can also be used to break loose key blocks.

A serious problem in some quarries is the failure of stone to break along predetermined directions. This may happen in stone that does not have a natural grain or in which the quarry direction does not parallel the natural grain. In some deep marble quarries in Georgia, stone that is cut free on the sides has a tendency to break

Fig. 6. Large flat quarry using diamond belt saw and forklifts.

loose from the floor along arch-shaped fractures rather than parallel to the quarry floor. This not only ruins a mill block, but presents a costly cleanup problem in order to keep the quarry floor flat. These breaks, called rainbow breaks by the quarrymen, are most common in large, deep quarries and where a long time has elapsed between cutting the block free on the sides and lifting it from the quarry floor. Some deep quarries have been abandoned because of rainbow breaks even though the stone was good otherwise.

Some quarriers use a wire saw to cut quarry blocks with very large surface area. These slabs are broken free, turned over on their side, and broken into appropriate sized mill blocks. This procedure not only eliminates much expensive cutting, but allows the quarrier more flexibility in choosing the size and position of the mill blocks to be cut from the slab. The slab can often be subdivided in such a manner that a minimum number of blocks contain natural flaws, whereas with a set floor pattern the quarrier has less control.

Removing Blocks from the Quarry

Mill blocks are traditionally removed from the quarry with fixed derricks. The procedure is slow and all other operations under the derrick boom must be halted while blocks are removed. The length of the boom also limits the size of some quarries. Blocks may be skidded from greater distance with tractors and crawler-type derricks, which can also add flexibility. A more efficient method the use of forklifts that are now available in sizes large enough to handle mill blocks. Because fork lifts require a flat or nearly flat floor and access ramps, they are not appropriate for all quarries. However, the economics are such that for large quarries, an investment in ramps is probably justified (Fig. 6).

Mill Operations

Most stone is sold as blocks, slabs, or finished product. Mill operations involve cutting blocks into slabs and then cutting the slabs into appropriate sizes and shapes for the final product. Finally the stone surface is given a final finish.

Sawing Slabs: Primary saw cuts may be made with a gang saw, wire saw, or circular saw.

Gang saws consist of a group of flat steel blades set in a frame that moves with a reciprocating motion. The frame is automatically fed into the block at a rate that depends on the type of stone and the abrasive or cutting edge used. A sand-water mixture may be fed into the saw for cutting soft stone. For harder stone such as granite or quartzite, heavier notched blades have been used and the cutting media is steel shot. For faster cutting, the blade may be fed with aluminum oxide, carborundum, or other abrasive. Nearly all modern mills use set teeth of diamond or carbide. Carbide may be used for softer limestones, but harder rocks such as marble and granite require diamonds. Diamonds are set in segments which are brazed to the gang saw blade. Set teeth generally cut faster and smoother than loose abrasives. As many as 40 blades may be set in a single gang saw when thin stock is being cut.

In most gang saws, the blade is horizontal and the frame is pressed downward through the block. The size of the block that can be sawed is thus limited by the size of the frame and blades. A newer development is vertically set gangs through which the block moves horizontally producing longer slabs.

One of the problems in gang saws is keeping equal tension on the blades. If a blade loosens it may wander or buckle and jam the saw or ruin the cut. The traditional method of holding a blade in position is with wedges driven through a hole in the blade outside the frame. Considerable skill is required to maintain in a single frame a large number of blades under equal tension. A more modern method involves the use of hydraulic tensioners.

Wire saws are commonly used to make primary cuts in granite blocks. These may be grouped into multiple or ganged wire saws. The wire saws may be either twisted wire using abrasive or diamond wire saws.

Wire saws are also used to make irregular or curved cuts. The saw is set vertically and the stone is fed into it as wood is fed into a band saw. Wire saws have also been used to turn large columns. Here the saw is usually set horizontal and the stone is slowly turned into it.

A recent development in the granite industry is the use of large-diameter circular saws to make primary cuts. The saw makes repeated passes across the block cutting 5 to 10 mm at each pass. The resulting cut is remarkably smooth so that increased sawing costs are more than compensated for. The largest circular saw in operation is reported to be 4.2 m in diameter and capable of 1.6 m total depth of cut (Anon., 1989). Another recent development is the use of the W.F. Meyers' narrow-belt, block saw for primary cuts in quarries and mills in the Indiana Limestone District.

Secondary Cuts: Circular saws are used for secondary and later cuts. The saws, sometimes called bridge saws, are set in a frame which guides the saw and in most modern versions are automatically controlled, often by computer.

Shaping, Grinding, and Polishing: A wide variety of methods is used for finishing the stone. Shaping may be done by hand-used pneumatic tools, by wire saws that cut as a band saw cuts wood, or by carborundum or diamond wheels that grind the stone to a reciprocal shape just as a wood-router bit shapes a board. For softer stones large milling machines using carbide teeth may be used to gouge shapes into the stone. Diamonds have almost universally replaced other abrasives in modern mills and automated machines are common.

Cylindrical shapes are made on a lathe that turns the stone into a carbide or diamond bit or into a wire saw. Boring machines may be set with diamonds or with carbide.

In older mills, flat surfaces may be ground on a large horizontal cast iron wheel called a rubbing bed. A water-sand mixture is used as abrasive. More modern plants have replaced rubbing beds with automatic surfacing machines using impregnated diamonds. The stone is belt fed under a series of successively finer horizontal wheels. Some machines fine grind the edge of the stone as well as the surface. Much polishing is still done by a conventional grinding-polishing machine that moves a spindle over the stone surface. Wheels using successively finer abrasives are set on the spindle.

A significant development of the last several years is the introduction of automated tile-making machines that trim and polish the tile at lower cost and with more accuracy than heretofore possible. These machines are largely responsible for the much greater production and consumption of stone tile.

Another significant development is of automated shaping machines that can duplicate intricate, carved pieces.

Traditionally much of the finishing is hand work by skilled stoneworkers. These skills are being lost and much effort has gone into automation of the finishing process.

DISTRIBUTION OF DEPOSITS AND RESERVES

The ultimate reserves of dimension stone must be considered inexhaustable. There can only be shortages of particular types or colors that can be discovered and produced at a competitive price. The distribution of various lithologies is a result of geologic factors. The location of producing quarries is governed by economics. Dimension stone, particularly marble and granite, is traded internationally in blocks, slabs, and finished products. Therefore every continent and nearly every country is a producer or potential producer of dimension stone.

Following is a brief summary of the major producing areas, described by lithologies, first in North America, then in the rest of the world.

The United States and Canada

The distribution of the major lithologic types is governed by geology. Currier (1960) divided the United States into five geologic provinces for the purpose of describing this distribution. All but one extend into Canada.

Atlantic and Gulf Coastal Plain: The Atlantic and Gulf Coastal Plain lies east and south of the Appalachian Mountains extending from New York to Alabama, and south of the interior sedimentary province from Alabama through Texas. It is underlain by poorly consolidated sedimentary rocks of Cretaceous and younger age that lie in nearly horizontal strata. Little dimension stone is produced in the province, but soft coquina that hardens on exposure to air has been used as building stone in Florida since the 17th century and Cretaceous limestone is quarried near Liberty Hill, TX.

Appalachian Crystalline Province: This province extends from Alabama to Maine and beyond through eastern Canada to Newfoundland. It is sometimes called the "older Appalachians" because it is the exposed root zone of an ancient mountain chain that formed during mid-Paleozoic time. Granite, gneiss, marble, serpentine, slate, and other crystalline rocks are produced from this region.

Granite. The principal granite-producing areas are in Georgia and New England. North Carolina is also a significant producer.

GEORGIA—Georgia produces more granite than any other state or province in North America. The industry is centered about the town of Elberton in northeast Georgia. Unlike many districts, Elberton is not dominated by one company, but rather has many producers. In 1990, the Elberton Granite Association had more than 90 members. Some members specialize in quarrying blocks, some in sawing slabs, and some in making finished products. Still other companies are vertically integrated performing all operations from quarry to finished product. According to T.A. Robinson, Executive Vice President of the Elberton Granite Association (written communication, 1991), the production is principally aimed at the memorial market, but structural stone such as building facing and floor tiles, are a growing market. More than 40 granite quarries were in operation in the Elberton batholith within 40 km of Elberton at the beginning of 1991 (Robinson, written communication, 1991), but Elberton firms also operate quarries or buy stone in several other states or countries.

The prevailing method of cutting stone is jet burner channeling, but some operators are switching to high speed drilling using Tamrock's slot drill, or to diamond wire saws. This trend will probably continue because of higher efficiency and the noise and waste associated with jet burning. A summary of new quarry techniques is given in the Elberton *Graniteer* (Anon., 1990d).

NEW ENGLAND—The three New England states of Vermont, New Hampshire, and Massachusetts produced more than 68 km^3 of granite in 1989. Connecticut was also a significant producer, but data was withheld by the US Bureau of Mines. The largest granite producing district is Barre, VT, where the Rock of Ages Co. is the sole producer.

Descriptions of the Barre district are given by Richter (1987) and Ratte' and Ogden (1989). Granite has been quarried from a mid-Paleozic pluton since the mid-nineteenth century. Most of the production is for monuments, but building facing panels, surface plates, and paper rolls up to 10 m long and 2 m in diameter are also produced. Quarriers use natural rift and grain to remove blocks averaging 5 to 6 m^3 (Richter, 1987).

OTHER—North Carolina is also a major porducer of granite, principally from the Mt. Airy District. South Carolina and Maine are minor producers.

Sandstone. Arkosic sandstone has been produced from down-faulted basins of Triassic age for more than 100 years. These basins form a sub-province within the Appalachian Crystalline Province (Currier, 1960). The stone, known as brownstone, was formerly used extensively for buildings and residences in New England and the mid-Atlantic states, but production is much reduced today.

Marble. True marble (i.e., marble formed by metamorphism of limestone or dolostone) occurs throughout the province and most states within the province have produced marble at one time or another. Today, however, nearly all production comes from Georgia and Vermont.

GEORGIA—The Georgia Marble Company operates several large dimension stone quarries in Pickens County, GA. The stone is medium-to-coarse grained, white or light gray, and most commonly veined with darker material. The stone is cut with diamond belt saws and quarry bars; quarry blocks are broken free and subdivided using mild explosives in spaced drill holes. Mill blocks are removed with large forklifts. It is a large-volume, low-recovery operation. The stone is used for monuments and buildings.

VERMONT—The Vermont Marble Company quarries white marble from a large underground quarry in the town of Danby, VT. A 10- to 25-m-thick layer of Ordovician marble lies in a broad, shallow syncline. Tunnel cuts are made by making a horizontal cut with a diamond-toothed chainsaw and cutting the sides with a diamond wire saw. Vertical floor cuts are then made with a diamond chain or belt saw and bottom cuts with a diamond wire saw. Average block size is about 20 t (Ratte' and Ogden, 1989).

The company also operates a black marble quarry at Isle la Motte, VT, and a verde antique quarry at Rochester, VT.

OTHER—Alabama was once a major producer of white marble from the Sylacauga marble belt, but the principal producers have switched to industrial fillers. Recent attempts to open new quarries in the district have not yet put blocks on the market. Marble has been produced from the Murphy Marble Belt in western North Carolina. White marble from the Cockeysville Formation in Maryland was used on the Washington Monument. White marble from Lee, MA, was used in the nation's Capitol in the last century, and recently blocks from the quarry were extracted by a Canadian firm. None of these districts have significant production at the present time.

Slate. Vermont and Pennsylvania account for 70% of US production of slate. In Vermont, lower Cambrian slates are quarried into blocks by drilling and blasting followed by manual splitting. The high angle of dip of the slaty cleavage makes conventional quarry techniques difficult. Blocks are taken to the mill where they are cut by diamond saws and trimmed to size for manual splitting into roofing slates. Floor tile and flagging are gauged by planers to specified thickness. Structural slate (mantels, electric panels, and similar products) are finished in the mill using standard cutting, honing, and polishing equipment (Ratte' and Ogden, 1989).

Pennsylvania slates of upper Ordovician age in the Lehigh-Northampton district have a nearly horizontal axial plane cleavage. The slate is quarried in benches 4 m thick using wire saws with silica sand abrasive to isolate quarry blocks. These are then lifted by drilling and wedging. Channeling machines are still sometimes used in freezing weather. Quarry blocks are reduced to manageable size (mill blocks) by splitting along "watersplits" and a natural grain or sculp which is perpendicular to the slaty cleavage. In the mill, blocks are finished by diamond saws, splitters, and various other tools into the finished product (Berkheiser, Jr., 1984, Colley, 1990).

Slate is also produced in Virginia (Sweet, 1990), in New York, and intermittently in Nova Scotia and Newfoundland (Vagt and Bergeron, 1990).

Soapstone. Soapstone is produced in Nelson County, VA. Blocks are cut out by a large German-made chain saw and finished into fireplaces, cookstoves, and other products benefitting from the heat retaining properties of the stone (Sweet, 1990). This district was formerly a major producer of laboratory furniture.

Interior Sedimentary Province: This province extends from the Appalachian Mountains to the Rockies, and from Mexico north into the prairie provinces of Canada. It is underlain by nearly flat-lying, mostly marine sedimentary rocks ranging in age from Cambrian to Cretaceous. These sedimentary rocks cover an older, Precambrian basement which is an extension of the Canadian Shield, and is exposed in several isolated uplifts. Limestone, sandstone, and marble (crystalline limestone), are produced from the sedimentary rocks and granite from basement rocks exposed in the isolated uplifts.

Limestone. The amount of dimension limestone produced in Indiana exceeds the amount of all dimension stone produced in any other state. Indiana limestone has probably been the dominant building stone in the United States for more than a century. The industry is centered around the towns of Bloomington and Bedford in south-central Indiana where quarries have been operated since 1827. The district and its history have been described by Patton and Carr (1982).

The stone has been called Bedford Rock, Bedford Oolitic Limestone, Indiana Oolitic Limestone, and Indiana Limestone. Today Indiana Limestone is a generic term for building stone quarried from the Salem Limestone of Mississippian age in the Bedford-Bloomington district (Patton and Carr, 1982). The stone is a "light gray to bluish-gray, massive, even-grained, granular, porous, and cross-bedded calcarenite" (Patton and Carr, 1982). The industry markets stone buff, gray, and variegated, a combination of the two colors.

Because of the district's long history many kinds of quarry methods have been used to extract stone. Today most quarriers in the district combine W.F. Meyers diamond-belt, quarry saws with wire saws using sand as an abrasive (McDonald, written communication, 1991).

In northwestern Alabama similar stone of the same age is quarried in an underground mine. Here tunnel cuts are made with three quarry bars mounted on a massive frame and so aligned that the top block tapers to the rear making extraction easy.

Limestone is also produced from the Oneota Formation of Ordovician age in the upper mid-westerm states. In central Texas limestone in produced from lower Paleozoic rocks.

Marble. Some crystalline limestones within the interior province take a polish and, therefore, are classed as marble. The most notable is the Holston formation in east Tennessee from which vast quantities of Tennessee Marble were produced during the first half of the century. The district lies within the Valley and Ridge province of the Appalachian Mountains where the rocks are steeply folded and faulted. Blocks must be quarried parallel to bedding; and because of the steep dip, it is impossible to use forklift trucks, and belt or chainsaws. Wire saws are used however (Zielinski, 1989). Most of the best deposits have been worked as far as possible without underground tunneling. For these reasons the district is in deep decline although Tennessee Marble remains one of the attractive and durable dimension stones in use. One quarry near Friendsville, TN, remains in operation as does the Imperial Black Marble quarry in Grainer County (Upham, written communication, 1991).

Sandstone. Most sandstone production in the United States probably comes from the interior province. However, because most sandstone is probably sold in broken and irregular sizes it does not qualify as dimension stone in this chapter. Bluestone from New York and Pennsylvania is sawed and dressed and a small amount of sandstone from Tennessee is cut and polished for furniture.

Granite. The sedimentary rocks of the interior province are underlain by crystalline rocks that are an extension of the Canadian Shield. These basement rocks are exposed in localized uplifts in southeastern Missouri, central Texas, the Arbuckle and Wichita Mountains of Oklahoma, and the Black Hills in South Dakota. All of these areas have produced or are producing dimension granite.

Canadian Shield: The Canadian Shield includes the Lake Superior crystalline province of Currier (1960) and underlies most of Canada west of the Maritime Provinces and east of the Rocky Mountains. It is underlain by Precambrian rocks, mostly granite and gneiss, but extensive marble occurs in the Grenville subprovince of southern Ontario and Quebec.

Granite. In the United States, the principal producer from this province is the Cold Springs Granite Company of Cold Spring, MN, which operates a number of quarries in Minnesota and elsewhere, including Texas and Manitoba.

In Canada, granite is produced in Manitoba, Ontario, and Quebec (Anon., 1990c). Quebec is the largest producer with about 25 active quarries (Nantel, 1983, Anon., 1990c).

Although there are few active quarries in the other provinces, the potential is very large. Several of the provinces have active programs to find and promote dimension stone deposits. Ontario and Saskatchewan publish catalogs depicting potential sources.

Other Stone. Only small amounts of other dimension stone including marble and quartzite are produced from the Canadian Shield.

Western Province: The Western province includes a wide variety of subprovinces that encompass nearly every conceivable geologic environment. Nearly every type of rock can be found and at one time or another has been quarried. It is the only province in which which recent volcanic rocks are found and is the only region from which travertine is produced.

Travertine. Travertine deposits are typically associated with recent geologic activity such as volcanism and faulting. For this reason, the Western Province is the only province in the United States and Canada from which travertine is produced. Currently, the only production is from New Mexico, Idaho, and Montana, but potential deposits are known to occur in Oregon (Geitgey, written communication, 1991), and possibly could be found in other states.

Two travertine deposits in southeast Idaho are operated by Idaho Travertine, Inc. of Idaho Falls, ID. Production since 1968 is said to have averaged 2 000 m^3 per year (Neumann, 1991). The principal deposit is just west of Swann Valley, ID. It is a boulder deposit meaning that it consists of large masses or boulders of travertine imbedded in a highly fractured matrix. The boulders are excavated by dozers and loaders. A D-9 dozer fitted with a ripping bar on the front end is used to pry loose the boulders. Those that are too large to transport to the mill are trimmed by line drilling at 15 to 20 cm spacing and blasting with prima chord. Recovery is said to range from 5 to 10% of volume removed.

The Montana deposit is near Gardiner, MT, north of Yellowstone Park. This is also a boulder deposit and quarrying methods are similar to that of Idaho Travertine. Five to ten ton blocks are removed. Slabs, tile, and ashlar are produced at the shop in Livingston, MT.

About 50 discrete deposits of travertine have been identified in New Mexico (Barker, 1986). New Mexico Travertine Co. quarries one of these from a deposit approximately 40 km west of Belen, NM, producing five color-varieties of travertine and one limestone (Austin and Barker, 1990). The stone is quarried using a concrete saw to remove cap rock, and diamond belt and wire saws to cut the travertine. The mill just west of Belen finishes the stone to polished slabs and split-face ashlar.

Other Stone. Nearly every type of stone can be found and at one time or another has been quarried in this province, particularly in the early part of the present century. Most of the quarries were abandoned long ago and no permanent centers of dimension stone production have developed within the province similar to those in Georgia, Indiana, and Vermont. Information on quarry sites and past activity can be obtained from most state or provincial surveys. For example, see Parks (1914), Goldman (1957), Moen (1967), Berg (1974), White (1987), or Harris (1991).

The Colorado Yule Marble Quarry, Gunnison County, Colorado, which supplied white marble for the Lincoln Memorial and the Tomb of the Unknown Soldier, has recently been reopened.

Outside the United States and Canada

International trade in dimension stone is in the form of finished stone ready for installation, in dressed slabs, and in rough blocks or slabs. Granite and marble (including travertine) are the principal rock types. The great bulk of international trade is still within the countries of western Europe and North America, but countries outside this trading block are becoming increasingly important.

In 1989, total United States imports of marble exceeded $260 million, of granite, $194 million; all other stone, $70 million. More than half of the imports were from Italy (Taylor, 1991).

Italy dominates international trade, being the leading importer of rough blocks and the leading exporter of finished material. It is also probably the leading manufacturer of stone-working machinery.

Marble: Most of the world's commercial marble formed from carbonate rocks that were deposited in the great Tethyan seaway that stretched from the Iberian Peninsula to the Himalayas. These rocks were metamorphosed into marble during the Alpine orogenies and they have been quarried since classic times.

The Carrara District in northern Italy produces what is probably the world's most famous marble. Travertine has been produced from Tivoli since classic times. Marble in many patterns and colors have been quarried since Roman times and the classic names for these have come into generic use in much of the rest of the world.

Marble has been quarried in Greece from at least the 6th century BC. The Greek marble industry has undergone a significant revival recently and Greece has become the second largest exporter of marble to the United States. *Dimensional Stone* (Maynard and Zielinsky, 1991) and *Stone World* (Anon., 1991b) have recently published articles describing the Grecian marble industry.

Taiwan is the second largest exporter of marble to the United States. Other countries that exported significant amounts of marble to the United States in 1989 are Spain, Portugal, China, Mexico, France, Germany, and the Philippines (Taylor, 1991).

Granite: Granite is the most abundant rock in the continents. It is found on all continents and most countries. Italy, the major exporter to the United States, gets its stone from Sardinia or by importing blocks and shipping finished stone. Other countries that export significant granite to the United States are Canada, Brazil, India, and the Republic of South Africa. Notably missing from the list are the Scandinavian countries, Finland, Sweden, and Norway, which produce large quantities of granite for the European market. Like eastern Canada, most of Scandinavia is underlain by an exposed Precambrian Shield of crystalline rocks and so have enormous reserves of exposed granite. Finnish quarriers are also significant innovators of quarry machinery and quarry techniques (Hyyasalo, 1990).

EVALUATION OF DEPOSITS

In order to properly evaluate a potential dimension stone deposit, one must determine the marketability of the stone, its soundness, its size and shape, the overburden, the distance to market, the kinds of transportation available, the availability of labor, the availability of power and other utilities, and similar concerns. Most of these must be determined in the evaluation of any mineral deposit, but several are unique to dimension stone deposits.

Marketability

The marketability of a stone is governed largely by aesthetic factors not subject to quantitative measurement. Aesthetic factors include color, pattern, and texture. Unless the deposit is an extension of a proved quarry or an exact duplicate of a popular stone, the judgment should be left to an experienced architect or marketing person.

It is important that a reliable sample be obtained. The best sample is a block from which slabs 0.3 m^2 or larger can be sawed. These slabs are used to evaluate color and pattern and also can be used for making required physical tests. The more varied the color and pattern the larger the sample should be. Drill cores are commonly deceptive, but may be suitable for fine-grained uniform textured stone or for extensions of known deposits.

It is the responsibility of the examining geologist or engineer to determine the size, extent, and uniformity of the deposit and to obtain samples that are truly representative. In the case of layered deposits, samples from each separate stratum should be obtained and compared for uniformity.

Purchasers of stone for large buildings generally make their choice on the basis of the history and reputation of the quarry, samples submitted by the quarrier, or more commonly, both. Swenson (1991) observed that the buyer is often disappointed because the color and texture of the final product does not meet the expectations of the buyer. He pointed out that "stone continually changes during the life of the quarry" (Swenson, 1991) and samples do not always represent what the quarry is currently producing. The fault may be over optimism on the part of the salesperson, lack of sufficient quarry development, or lack of knowledge of the deposit. Swenson offered a number of suggestions to architects, but said "Geologic studies are effective in determining the long term resource . . . , but may indicate nothing about the short term design needs of a project'' (Swenson, 1991). This, it seems, indicates a lack of communication or understanding of the capabilities of geological investigations. Few quarriers have geologists on their staff or even hire consulting geologists. Fewer still, map their quarries. Some, like Vermont Marble Co., which remained under the same management for years and had competent geological staffs, know their deposits intimately and can predict quality and grade with great accuracy. Good, detailed geologic mapping, sometimes supplemented with core drilling, should allow a quarrier to predict changes in the nature of the stone.

One of the problems is that few in the quarry business are knowledgeable enough about modern geology to recognize that if a consultant is hired, that consultant must be either well versed in dimension stone problems or be a specialist in the lithologic type under consideration or both. A limestone or crystalline limestone requires a carbonate petrologist, granite requires an igneous petrologist-structural geologist. Whether it is cost effective to hire the specialist is for the quarriers and architect to decide, but even geologic specialists are generally less expensive than lawyers.

Soundness

Soundness or freedom from flaws is one of the most critical factors in evaluating a dimension stone deposit and one of the most difficult to determine. Every joint and fracture should be mapped. This should be supplemented by core drilling parallel to the direction in which the stone is to be sawed. If at all possible, the deposit should be drilled in two intersecting directions in order to pick up all possible weakness directions that will limit the size of blocks and slabs. If mapping shows a pronounced joint direction, the drilling should be perpendicular to it. The drilling must be done with extreme care and nothing smaller than NX size should be used. If the core breaks on anything but a natural weakness, it will have failed in its purpose of proving sound stone.

Any directions of natural splitting that would aid in quarrying should be determined. Granite and many limestones and sandstones have such directions that are known as rift, run, grain, and similar names. These can be beneficial, whereas close-spaced open fractures are deleterious. The investigator should remember that the ultimate object is to free large blocks that are without internal fractures that will appear only after the block is sawed.

Some rocks that have been subjected to great stress are still strained. In many cases such stone will develop new fractures and split after blocks are quarried. Some hidden fractures show up only after the block has been sawed. This is costly as the flaw appears only after a large investment has been made in the stone. This type of fracture is probably less common in horizontally layered sedimentary rocks and in postorogenic granites that have intruded passively than in rocks that have been subjected to tectonic stress.

Other Factors

The investigator should remember that quarrying blocks for cut and sized dimension stone is still a high labor input industry. He or she should also keep in mind that each salable unit—a mill block—will weigh 9 t or more and will require special handling methods to extract and transport to market. The total output of a quarry is small as compared to other industrial rocks and minerals. It takes a large operation to extract more than a few blocks a day and few if any deposits would justify building new roads, railroads or expensive utility lines just to service a quarry.

The physical tests required by ASTM specifications (Anon, 1991c) require a sample from which blocks adequate for the required testing can be cut (see Table 1).

BIBLIOGRAPHY AND REFERENCES

Anon., 1972, Stone Catalog, Building Stone Institute, Purdys, NY.

Anon., 1979, ''Water Jet Machines Tested in Elberton,'' *Elberton Graniteer*, Vol. 23, pp. 4–5.

Anon., 1980, ''Dimension Stone,'' *Mineral Facts and Problems*, Bulletin 671, US Bureau of Mines, pp. 861–868.

Anon., 1987, *Marble, Italian Culture, Technology and Design*, Italian Institute for Foreign Trade, Italian Trade Commission, New York, 173 pp.

Anon., 1989, ''World's Largest Diamond Saw Blade Now in Operation in the US,'' *Dimensional Stone*, Vol. 5, No. 5, p. 11.

Anon., 1990a, ''Elberton Quarries Embrace New Technology,'' *Elberton Graniteer*, Elberton Granite Association, Vol. 34, No. 2, pp. 6–7.

Anon., 1990b, ''Carl Meyer Celebrates 100th Anniversary with First Diamond Gang Saw for Granite,'' *Dimensional Stone*, Vol. 6, No. 8, pp. 56–58.

Anon., 1990c, ''Stone,'' *1990 Canadian Minerals Yearbook*, Department of Energy, Mines, and Resources, pp. 59.1- 59.23.

Anon., 1990d, ''Elberton Granite, the Quarrying and Manufacturing Process,'' Elberton Granite Association, Elberton, GA.

Anon., 1991a, ''Water Also Quarries Stone,'' *Stone World*, Vol. 8, No. 5, p. 47.

Anon., 1991b, ''Greece: Modern Stoneworking with an Ancient Tradition,'' *Stone World*, Vol. 8, No. 4, pp. G1-G22.

Anon., 1991c, *Book of ASTM Standards*, Vol. 04.08, American Society for Testing & Materials, Philadelphia, pp. 1–33.

Austin, G.S., and Barker, J.M., 1990, ''Commercial Travetine in New Mexico,'' *New Mexico Geology*, Vol. 12, pp. 50–58.

Barker, J.M., 1986, ''Travertine,'' Open-File Report 229, New Mexico Bureau of Mines and Mineral Resources, pp. 94–199.

Barton, W.R., 1968, ''Dimension Stone,'' Information Circular 8391, US Bureau of Mines, 147 pp.

Berg, R.B., 1974, ''Building Stone in Montana,'' Bulletin 94, Montana Bureau of Mines and Geology, 41 pp.

Berkheiser, S.W., Jr., 1984, ''Pennsylvania's Slate Industry, Alive and Well,'' *Proceedings*, Special Paper No. 2, 20th Forum on the Geology of Industrial Minerals, Maryland Geological Survey, pp. 23–33.

Bowles, O., 1939, *The Stone Industries*, 2nd ed., McGraw-Hill, New York, 519 pp.

Bowles, O., 1955a, ''Memorial Stone,'' Information Circular 7720, US Bureau of Mines, 6 pp.

Bowles, O., 1955b, ''Slate,'' Information Circular 7719, US Bureau of Mines, 12 pp.

Bowles, O., 1956a, ''Granite as Dimension Stone,'' Information Circular 7753, US Bureau of Mines, 18 pp.

Bowles, O., 1956b, ''Limestone and Dolomite,'' Information Circular 7738, US Bureau of Mines, 29 pp.

Bowles, O., 1958, ''Marble,'' Information Circular 7829, US Bureau of Mines, 31 pp.

Bowles, O., and Barton, W.R., 1963, ''Sandstone as Dimension Stone,'' Information Circular 8182, US Bureau of Mines, 30 pp.

Cia, O., and Mancini, R., 1989, ''Diamond Wire for Cutting Hard Rock,'' *Dimensional Stone*, Vol. 5, No. 2, pp. 24–27.

Colley, D., 1990, ''New Interest in Ribbon Slate,'' *Stone World*, Vol. 7, No. 6, pp. 43–48.

Cooper, J.D., 1970, ''Stone,'' *Mineral Facts and Problems*, Bulletin 650, US Bureau of Mines, pp. 1219–1235.

Currier, L.W., 1960, ''Geologic Appraisal of Dimension Stone Deposits,'' Bulletin 1109, US Geological Survey, pp. 7-14.

Dickie, G.B., ''Building Stone in Nova Scotia,'' Information Circular 12, Nova Scotia Department of Mines and Energy, 20 pp.

Goldman, H.B., 1957, ''Dimension Stone,'' Bulletin 176, California Division of Mines and Geology, pp. 591–606.

Harris, R.E., 1991, ''Decorative Stones of Wyoming,'' Public Information Circular No. 31, Geological Survey of Wyoming, 27 pp.

Hawkins, A.C., et al., 1990, ''The Diamond Wire Saw,'' *Dimensional Stone*, Vol. 6, No. 8, pp. 44–50, 64.

Hockman, A., 1953, ''Physical Properties of Currently Produced Marbles,'' Circular LC1010, National Bureau of Standards, 13 pp.

Hyyasalo, H., 1990, ''The Finnish Method,'' *Dimensional Stone*, Vol. 6, No. 8, pp. 60–64.

Ladabouche, J.J., 1973, ''Dimension Stone,'' Mining Engineering, Vol. 25, No. 1, p. 32.

Maynard, J., and Zielinsky, B., 1991, ''The Greek Marble Industry Today,'' *Dimensional Stone*, Vol. 7, No. 4, pp. 67–84.

Moen, W.S., 1967, ''Building Stone of Washington,'' Bulletin 55, Washington Department of Conservation, Division of Mines and Geology, 85 pp.

Nantel, S., *Architectural and Ornamental Granite Quarries in Quebec*, PRO 88–03, Ministere de l'Energie et des Resources, Quebec, 4 pp.

Nantel, S., 1983, ''Dimension Stone of Quebec: Geological Aspects of Commercial Granite Deposits,'' *Proceedings*, Misc. Paper 114, 19th Forum on the Geology of Industrial Minerals, Ontario Geological Survey, pp. 96–108.

Neumann, T.R., 1991, ''Dimension Stone,'' *Principal Deposits of Industrial Minerals of Idaho*, Bureau of Mines Special Report, US Bureau of Mines, pp. 56–60.

Parks, H.M., 1914, ''Preliminary Report on Building Stone in Oregon,'' *The Mineral Resources of Oregon*, Oregon Bureau of Mines and Geology, pp. 10–46.

Patton, J.R., and Carr, D.D., 1982, *The Salem Limestone in the Indiana Building-Stone Industry*, Occasional Paper 38, Indiana Geological Survey, 31 pp.

Pettijohn, F.J., 1975, *Sedimentary Rocks*, 3rd ed., Harper and Row, New York, 628 pp (see p. 215, Table 7).

Power, W.R., 1972, ''An Evaluation of Building Dimension Stone Deposits,'' *Mining Engineering*, Vol. 24, No. 5, pp. 42–44.

Ratte', C.A., and Ogden, D.G., 1989, ''Marble, Granite and Slate Industries of Vermont,'' Field Trip Guidebook T362, American Geophysical Union, 18 pp.

Richter, D.A., 1987, ''Barre Granite Quarries, Barre, Vermont,'' Centennial Field Guide, Northeast Section, Geological Society of America, pp. 239–242.

Richter, D.A., 1988, ''Granite as Chemically Resistant Masonry,'' *Corrosion and Chemical Resistant Masonry Materials Handbook*, W.L. Sheppard, ed., Noyes Publications, Park Ridge, NJ, pp. 192–196.

Shadmon, A., 1972, *Stone in Israel*, Ministry of Development, Natural Resources Research Organization, Stone Technology Center, Technion, Haifa, 64 pp.

Shelton, J.E., and Drake, H.J., 1975, ''Stone,'' *Mineral Facts and Problems*, Bulletin 667, US Bureau of Mines, pp. 145–147.

Singleton, R., 1982, ''Stone,'' *Mineral Commodity Summaries 1982*, US Bureau of Mines, pp. 146–147.

Sweet, P.C., 1990, ''Present and Future Dimension Stone Industry in Virginia,'' *Proceedings*, 24th Forum on the Geology of Industrial Minerals, South Carolina Geological Survey, pp. 129–135.

Swenson, J.M., 1991, ''Quality from the Quarry,'' *Dimensional Stone*, Vol. 7, No. 9, pp. 56–59, 79.

Taylor, H.A., 1985, ''Dimension Stone,'' *Mineral Facts and Problems*, Bulletin 675, US Bureau of Mines, pp. 769–776.

Taylor, H.A., 1988, *Dimension Stone, 1988*, US Bureau of Mines, 10 pp.

Taylor, H.A., 1991, *Dimension Stone, 1989*, US Bureau of Mines, 9 pp.

Vagt, O., and Bergeron, M., 1990, ''Stone,'' *Canadian Minerals Yearbook for 1990*, Ministry of Energy, Mines, and Resources Canada, pp. 59.1–59.23.

White, G.V., 1987, ''Dimension Stone Quarries in British Columbia,'' Paper 1987–1, British Columbia Ministry of Energy, Mines and Petroleum Resources, pp. 309–342.

Wood, S.O., Jr., and Carrales, M., Jr., 1969, ''Stone,'' *Minerals Yearbook*, Vol. I-II, US Bureau of Mines, pp. 1019–1040.

Zielinski, B., 1989, ''Marble Quarrying in America, a Cautious Revolution,'' *Dimensional Stone*, Vol. 5, No. 8, pp. 27–29.

Zielinski, B., 1992, ''Waterjet Technology: The Shape of Things to Come in the Quarry,'' *Dimensional Stone*, Vol. 8, No. 5, pp. 30–31, 75.

STRONTIUM MINERALS

JOYCE A. OBER

Strontium occurs commonly in nature, averaging 0.034% of all igneous rock; however, only two minerals, celestite and strontianite, contain strontium in sufficient quantities to make its recovery practical, and these minerals are found in sedimentary rocks. Of the two, only celestite has been found to occur in deposits of sufficient size to make development of mining facilities currently attractive. Celestite, the most common strontium mineral, consists primarily of strontium sulfate ($SrSO_4$). Strontianite, the second most common mineral, consists primarily of strontium carbonate ($Sr\ CO_3$). Strontianite would be the more useful of the two common minerals because strontium is used most commonly in the carbonate form, but few deposits have been discovered that are suitable for development.

The physical properties of strontium ores are listed in Table 1.

PROCESSING

Strontium commonly occurs with barium and calcium, two elements with properties very similar to strontium, thus making separation difficult. Because removing many impurities from celestite is difficult and energy intensive, current strontium chemical producers require material to contain at least 90% strontium sulfate. Most of the currently operating celestite facilities can produce sufficient supplies with only minimal processing necessary to achieve acceptable specifications. Hand sorting and some washing are all that are necessary at many strontium mines; only a few operations use froth flotation or gravity separation to beneficiate their ore.

The black ash method and the soda ash method are the two most common recovery techniques. The black ash method, known alternatively as the calcining method, produces chemical-grade strontium carbonate, which contains at least 98% strontium carbonate. The soda ash, or direct-conversion method, produces technical-grade strontium carbonate, containing at least 97% strontium carbonate.

The black ash method received its name because in the first step in the procedure the crushed and screened celestite are mixed with powdered coal, making a black mixture. The mixture is then heated to about 1 100°C, expelling oxygen in the form of carbon dioxide from the insoluble strontium sulfate to form water-soluble strontium sulfide. Strontium sulfide is dissolved in water, and the resulting solution filtered. Carbon dioxide then passes through the solution or soda ash is added. Either compound provides the necessary carbon and oxygen for strontium carbonate to form and precipitate from the solution. The precipitated strontium carbonate is then removed from the solution by filtering and is dried, ground, and packaged. The sulfur released in the process is either recovered as elemental sulfur or in other byproduct sulfur compounds.

In the soda ash method, ground celestite is washed and most of the water removed. The thickened mixture is then mixed with soda ash and treated with steam for 1 to 3 hours. During this time, celestite and soda ash react to form strontium carbonate and sodium sulfate. Sodium sulfate is water soluble, making it possible to separate the insoluble strontium carbonate by centrifuging. Although the soda ash method is a simpler process, the lower grade of the product causes it to be the less preferred method of recovery. The black ash method is the most common method of strontium carbonate production, and most new production facilities use black ash technology.

MAJOR DEPOSITS AND PRODUCTION

United States

Although no mines were active between World War I and World War II, celestite deposits have been identified nationwide. All demand was met from foreign sources. During World War II, celestite was mined in California and Texas.

Only one company, Chemical Products Corporation (CPC) produces strontium carbonate, the most important strontium compound by far. Most of the celestite CPC uses is imported from Mexico.

CPC uses the black ash method of strontium carbonate production at its facility.

Other strontium compounds are produced but on a minor scale. CPC purchased the strontium nitrate production facilities from FMC Corp. in Modesto, CA, when that company discontinued production of carbonate and nitrate in 1984. The company moved the equipment to Cartersville and now is also the sole US producer of strontium nitrate.

Several US companies produce strontium compounds from strontium carbonate. Mallinkrodt Inc. of St. Louis, MO, produces strontium chloride, and Mineral Pigments Corp. of Beltsville, MD, produces strontium chromate. A few other companies produce downstream strontium compounds, but on a very small scale.

In 1944, domestic strontium deposits were operated at the following sites: near Blanket, in Brown County, TX; in Nolan County, TX; in the Fish Mountains in Imperial County, CA; and near Ludlow, CA (Gwinn, 1944). At that time, the major use for strontium chemicals was for pyrotechnic applications, such as signal flares and tracer bullets required for military applications. Immediately following World War II, this demand disappeared, causing domestic production of celestite to decrease quickly and eventually disappear.

Resources in the United States have been estimated at 3.2 Mt, with an identified reserve base of 1.4 Mt. The reserve figure includes material containing no more than 60% strontium sulfate, which is much too low grade to meet today's rigid specifications. In addition to deposits mined in the early 1940s, celestite has been discovered in Arizona, Arkansas, Kentucky, Michigan, Missouri,

Table 1. Physical Properties of Strontium Minerals

Property	Celestite	Strontianite
Cleavage	Perfect basal and prismatic	Imperfect prismatic
Color	Primarily white to pale grayish blue Sometimes shades of yellow and red	Primarily white or grayish white Sometimes pale green or yellowish
Fracture	Uneven	Conchoidal
Hardness	3–3.5	3.5–4.0
Index of refraction	1.622–1.631	1.520–1.667
Luster	Vitreous, pearly on cleavage	Vitreous
Specific gravity	3.9–4.0	3.6–3.8
Streak	White	White
Transparency	Translucent to opaque Rarely, nearly transparent	Translucent to transparent

New York, Ohio, Pennsylvania, Tennessee, Utah, and Washington (Schreck and Arundale, 1959).

Other Countries

Huge deposits of high grade celestite have been discovered throughout the world. In almost all instances, celestite deposits occur in remote, undeveloped locations far from population centers where inexpensive labor is available for mining.

Detailed information on most world resources is not readily available. Because many of the large deposits are in remote, sparsely inhabited areas, very little formal exploration has been done. Other deposits may be well identified but are in countries from which specific mineral information is not easily obtained.

The data in Table 2 are annual rated capacity for chemical processing plants as of Dec. 31, 1990. Rated capacity is defined as the maximum quantity of product that can be produced in a period of time on a normally sustainable long term operating rate, based on the physical equipment of the plant, and given acceptable routine operating procedures involving labor, energy, materials, and maintenance. Capacity includes both operating plants and temporarily closed plants that, in the judgment of the author, can be brought into production within a short period of time with minimum capital expenditure.

Two types of strontium capacity exist, ore production capacity and compound production capacity. Strontium compound capacity was deemed the more important for several reasons. Essentially, all strontium is consumed in compound form, primarily as the carbonate and to a lesser extent as the nitrate. Strontium carbonate is also the intermediate product in production of downstream compounds. At least 12 countries mined celestite in 1989. Virtually all of the celestite mined annually is consumed in the production of strontium carbonate and other strontium compounds. Very little strontium was consumed directly in the mineral form.

Listing both ore and compound production capacity could be misleading, possibly resulting in double counting of strontium capacity. Because most ore is produced for market by means of very simple sorting and beneficiation techniques, capacity can easily be raised to meet increased demand. Strontium carbonate production capacity is the limiting factor in strontium production, requiring complicated and sophisticated processing facilities. Therefore, strontium carbonate production capacity, reported in terms of contained strontium, was selected for tabulation. Construction to expand strontium carbonate capacity has increased because of the continued growth in international sales of color television sets. Capacity information was compiled from published reports.

Table 2. World Strontium Carbonate Annual Production Capacity, Dec. 31, 1990 (tons)

Country	Rated capacity[1]
North America:	
Mexico	65 000
United States[2,3]	22 000
Total	87 000
Europe:	
Germany, Federal Republic of[2]	15 000
Spain	8 000
USSR	(4)
United Kingdom	(4)
Total	23 000
Asia:	
China	1 800
Japan[2]	31 000
Korea, Republic of	40 000
Total	72 800
World total	182 000

[1] Included capacity at operating plants as well as plants on a standby basis. [2]Production was entirely from imported celestite. [3]Includes strontium nitrate production capacity. [4]Strontium carbonate production capacity existed in these countries, but specific data are not available.

Algeria: Celestite is mined in Algeria from a surface deposit near Beni Mansour. Estimated reserves are more than 1 Mt. The deposit is mined by Enterprise des Produits Non-ferreux et des Substances Utiles (ENOF) and is controlled by L'Enterprise Nationale de Developpement Minière (Edemines), a division of Société Nationale de Recherches et d'Exploitations Minières (SONAREM). No strontium has been exported in recent years, but, in the past, as much as 3.6 ktpy has been exported to nations in Eastern Europe and the former Federal Republic of Germany. Although the celestite is relatively low grade, only about 70% strontium sulfate, it contains low concentrations of barite, silicate, gypsum, and iron, making it possible to concentrate the ore to acceptable levels using only gravity separation. The deposit has not been fully exploited, but there is interest in further development (Anon., 1988b).

Argentina: Celestite mining is controlled by the Argentine Mining Union at the San Juan Mine in the Province of Mendoza. Celestite has been identified at several other small mines, the Maria Del Carmen and Don Luis Mines, also in Mendoza Province, and the Rayoso, Julio, Cerro Partido, and Llao Mines in the Province of Neuquen. It is unlikely that any production is being carried out on a regular basis at these mines. Most Argentine production is exported to Brazil (Anon., 1988c).

Canada: Canada has produced celestite ore in the past from the McRae deposit, also known as the old Kaiser Celestite Mining Ltd. mine, at Enon, Cape Breton County, Nova Scotia. Mineral rights were acquired by Timminco Metals, a division of Timminco Ltd. of Toronto. Exploration identified reserves of more than 1 Mt of celestite, with grades ranging from 60 to 65% strontium sulfate. Reopening the mine has been considered, but the low grade of the ore makes economic feasibility questionable. Other strontium deposits have been identified in British Columbia, Newfoundland, and Ontario (Dawson, 1985), but none as significant as the deposit in Nova Scotia. Timminco Metals is the largest producer of strontium metal in the world and the only producer of strontium metal in North America. The company operates a processing facility in Westmeath, ON. The strontium production facilities in Westmeath and facilities in Haley, ON, produce strontium metal by the aluminothermic reduction of strontium oxide.

China: Celestite is mined in China almost exclusively for domestic consumption. An ore-dressing facility at the Nanjing mine in Jiang Su Province has a capacity of 10 ktpy. The Nanjing deposit is reported to be high grade and close enough to the surface for open pit mining (Anon., 1988d). Past import data indicated that strontium chemical plants must exist, but no specific details are available. Reports indicate that Japan imports strontium carbonate from China and that strontium carbonate from China has also been imported into the United States.

Cyprus: Celestite mining began in Cyprus in 1985 at Vassliko, near Limassol, in southern Cyprus. The mine is operated by Hellenic Mining Co. Ltd. The ore averaged about 54% strontium sulfate and, therefore, requires concentration to make it marketable. The ore is beneficiated through a flotation process to reach 94% strontium sulfate. The reserves have been determined to be 200 kt of celestite (Griffiths, 1984). No mining has occurred since 1987.

Germany, Federal Republic of: Kali-Chemie AG of Hannover produces strontium carbonate from imported celestite at its plant at Bad Hoenningen. The major western European producer uses the black ash method in the production facilities. Plant capacity is estimated to be 15 ktpy. Kali-Chemie imports most of its celestite from Spain and Turkey; it exports 80 to 90% of the production, most of which goes to the United States and the Republic of Korea. Kali-Chemie also produces strontium hydroxide and strontium nitrate. The nitrate is produced by its Italian subsidiary, Societa Bario e Derivati SpA, in Massa (Griffiths, 1985a).

Iran: One of the world's largest celestite deposits is in the northwestern part of the Dasht-e-Kavir salt desert. The deposit is mined by Iran Strontium Co., a subsidiary of Cherkate Sahami Sanati Va Maadani Irani (Simiran). Average strontium sulfate values for the deposit have been reported at more than 91%, but the levels of barium sulfate and calcium sulfate are unfavorably high. Proven reserves total 1.8 Mt of celestite, 430 kt of which has no overburden (Shiebel, 1978). Most of the celestite produced from this deposit is believed to go to the former USSR.

Italy: Mining has been done by Minera Chimica Farnesiana SpA in Tarquinia, near Rome. The product was 75% strontium sulfate and 5% barium sulfate. Most production was sold domestically. Annual production capacity at the mine was 7.3 ktpy, but production ceased in 1988. Some of the processed ore was exported to the USSR. Societa Bario e Derivati, a subsidiary of Kali-Chemie AG of the Federal Republic of Germany, produces strontium nitrate at Massa. Production and capacity figures are not available, but it is known that most of the production is exported to the United States (Anon., 1986). Deposits in the interior of Sicily are not presently being mined.

Japan: No celestite is mined in Japan, but that country is the largest consumer of strontium carbonate in the world, mostly due to its large television and electronics industry. Four companies produce strontium carbonate from imported celestite. Honjo Chemical Corp. operates a production facility for 18 ktpy of carbonate at Neyagawa in Osaka using celestite from Spain and China. Sakai Chemical Industries Co. also produces strontium carbonate, by the black ash method in Osaka at a 12-ktpy plant using Chinese, Mexican, and Spanish celestite. Japan Special Chemicals (Nihon Tokushu Kasei) and Dowa Chemicals also produce strontium carbonate, but on a much smaller scale. The total capacity for both companies is only about 1.1 ktpy. These two companies used celestite from Spain (Anon., 1988e).

Korea, Republic of: Although celestite deposits have not been identified in Korea, its growing electronics and television industry prompted a European company to form a joint venture with a Korean firm. Kali-Chemie and Samsung Corning Ltd., Korea, formed Daehan Specialty Chemicals Co. Ltd. to build a strontium carbonate and barium carbonate plant. The plant began operation in 1990 with a production capacity of 40 ktpy of strontium carbonate and barium carbonate, combined. The production will be targeted to the growing television industry in that country. The black ash method will be used (Anon., 1988a).

Mexico: Mexico exports more strontium compounds and minerals to the United States than any other country (Table 3). Two companies announced plans to build production facilities in the north. CPC planned to build a strontium carbonate plant in Reinosa, Tamaulipas, close to the US border. Capacity should be about 22 ktpy (Anon., 1990c). Quimica Dinámica SA de CV, a Monterrey company that currently produces 8 kt of strontium carbonate, formed a joint venture with an unnamed Japanese company to create Cistron Mexicana SA de CV. The new company is expected to build a strontium carbonate plant initially designed to produce 30 ktpy, with a possible doubling of capacity in a second phase of construction (Anon., 1990b).

Although strontium carbonate production has been a recent development in Mexico, the huge celestite resources in the country make it a very attractive location for additional production facilities. After a major US strontium carbonate and strontium nitrate producer (FMC Corp.) closed its plant in 1984, Cia. Minera La Valenciana SA (CMV) bought the strontium carbonate processing equipment from the FMC plant in California. The plant, which had used the soda ash method of carbonate production and was converted to the black ash method, was relocated to Torreón in Coahuila State, near a company-operated mine. The plant capacity is 50 ktpy. Shipments from the plant, targeted for the television and electronics industry in the Far East, began in May 1987. Sales y Oxidos (SYOSA), 49% owned by Kali-Chemie, built a new strontium carbonate plant near Monterrey with a capacity of 15 ktpy. The company's old strontium carbonate operation will be converted to produce barium carbonate. SYOSA, which produced carbonate by the black ash method from celestite it mines nearby, marketed its production in the United States (Anon., 1990a).

Mexico is one of the world's largest producers of celestite. Reports indicate that 11 companies are currently mining celestite. CMV mines celestite from the San Agustín deposit near Torreón, and Compañía Minera Ocampo SA mines a deposit near Saltillo in Nuevo Leon State. There also is some small production in Hidalgo, Aguascalientes, and Chihuahua states. Reserves of the San Agustín mine alone have been estimated at more than 725 kt. SYOSA mines west of Monterrey. Other small operations are in Coahuila, Aguascalientes, Chihuahua, and Nuevo Leon. Mexican celestite is primarily high grade, and only hand sorting is required to achieve at least 92% strontium sulfate with low barium content. Abandoned mines and easily identified deposits that have not yet been developed are common throughout a large area in northern Mexico. Reserves are believed to be vast, but these deposits have not been extensively explored (Anon., 1988f).

Pakistan: All mines in Pakistan are the property of Provincial governments and are operated by private companies for the gov-

Table 3. US Imports for Consumption of Strontium Ore and Compounds, by Country

	1989		1990	
	kg	Value[1]	kg	Value[1]
Celestite:				
Mexico	25 640 000	1 956 000	46 484 000	3 727 000
Spain	—	—	2 234 000	425 000
Other	—	—	6 000	37 000
Total	25 640 000	1 956 000	48 724 000	4 189 000
Strontium carbonate:				
Canada	18 000	11 256	17 781	2 978
China	—	—	75 000	45 558
France	216 000	131 602	—	—
Germany, Federal Republic of	4 427 276	2 992 291	2 806 028	2 018 791
Italy	3 648	3 520	—	—
Japan	934 883	635 860	6 300	33 899
Netherlands	—	—	15 506 833	9 210 192
Mexico	12 872 502	6 669 611	36 000	24 760
United Kingdom	6 137	11 012	16	12 670
Total	18 478 446	10 455 152	18 447 958	11 348 848
Strontium nitrate:				
Germany, Federal Republic of	600	8 536	—	—
Italy	43 003	40 725	17 000	16 016
Mexico	571 849	546 291	1 661 385	1 589 517
Spain	18 000	17 100	36 000	30 618
Total	633 452	612 652	1 714 385	1 633 151

[1]Customs value. *Source:* US Bureau of the Census.

ernments. Recent celestite production figures have been reported for two active mines, one near Dawood Khail and one near Karachi (Anon., 1988g). Tawakkal Mineral Exports Corp. recovered celestite from these mines in the Dadu District of Sind Province. No beneficiation was necessary to produce 94% strontium sulfate. Reserves of celestite have not been quantified at this location (Griffiths, 1987). Other deposits, with 500 kt of reported reserves, have been identified in the Punjab Province (Anon., 1986).

Spain: Spain is a large celestite producer. Celestite is produced from the Montevive deposit, which is operated by Herederos de Aurelio Farjardo Vilches, with sales and marketing handled by Bruno SA. Selective mining and hand sorting are all that is necessary to produce ore grades of more than 92% strontium sulfate, although a concentration plant has been constructed. Reserves are believed to be at least 2 Mt. Most Spanish production is exported to Japan, some remains in Spain, and most of the rest is exported to the Federal Republic of Germany (Anon., 1986). Promotora de Industria del Sur (Proinsur SA) produces strontium carbonate and strontium nitrate at a combined facility near Granada. The plant has a theoretical design capacity of 8 ktpy of carbonate, but has never reached this capacity. Strontium carbonate is produced by the soda ash method. The plant also has a production capacity of about 3 ktpy of nitrate, which has not yet been fully used (Griffiths, 1985b).

Turkey: Turkey competes with Mexico, Spain, and Iran in claiming the world's largest strontium reserves. Barit Maden Turk AS produces celestite from a mine near Sivas. Run-of-mine ore is gravity separated to produce a concentrate with a minimum of 95% strontium sulfate. Because of the harsh climate in the region, the mine can be operated only on a seasonal basis, from May to October. Another mine near Sivas was formerly operated by Bilfer Madencilik AS, which reestablished concessions for future celestite mining. Identified reserves have been placed at 550 kt, and further reserve potential is estimated to be greater than 2 Mt. Turkish celestite was primarily exported to the Federal Republic of Germany (Karayazici, 1987).

USSR: Very little is known about Soviet production of strontium minerals. The following deposits are known: the Karakum Desert, Zaunguz Plateau; near Bashkir; caprocks of the Romy and Isachkov salt domes; in the Pinega area, Archangel Province; near Yakutsk, eastern Siberia; and in Turkestan from eastern Fergana to the shores of the Caspian Sea, and from southeastern Bucharia to the shores of the Sea of Aral. Reserves are believed to be very large, but the ore grade is probably not high (Kishimoto, 1982). No details are available concerning the production of strontium carbonate in the USSR except that there is production, probably from domestic celestite, as well as from imports from Iran and Turkey.

United Kingdom: Celestite deposits in the Bristol area are mined by Bristol Minerals Co. Ltd. The ore is crushed, washed, and graded to achieve a product with 95% strontium sulfate. Reserve estimates vary from 100 to 500 kt. The British deposits continue to be encroached upon by large scale housing developments, which limit possible exploitation of the minerals to the near future. This is one of the few sites where celestite deposits are not in remote locations (Griffiths, 1984).

ECONOMIC FACTORS

Uses

Approximately 80% of all strontium is consumed in ceramics and glass manufacture, primarily in television faceplate glass and ceramic ferrite magnets (Table 4), and in smaller amounts in other ceramic and glass applications. In the past 20 years, color television production has become the major consumer of strontium. Because of unique properties that strontium compounds possess, technologies have developed that take advantage of those properties for specific applications. All color televisions and other devices containing color cathode-ray tubes sold in the United States are required

Table 4. US Estimated Distribution of Primary Strontium Compounds, by End Use (%)

End use	1988	1989	1990
Electrolytic production of zinc	4	4	3
Ferrite ceramic magnets	11	10	9
Pigments and fillers	4	3	4
Pyrotechnics and signals	10	10	11
Television picture tubes	68	70	70
Other	3	3	3
Total	100	100	100

by law to contain strontium in the faceplate glass of the picture tube. Strontium blocks X-rays better than barium, which was previously used, and, although lead is a better X-ray barrier than strontium, it causes a browning of the glass. Major manufacturers of television picture tube glass incorporate about 8% strontium oxide, by weight, in their glass faceplate material. The strontium is added to the glass melt in the form of strontium carbonate, and during processing it is converted to strontium oxide. In addition to blocking X-rays, the strontium improves the appearance of the glass, increasing the brilliance and improving the quality of the picture.

Trends in television production show a shift to larger, flatter tubes that require thicker glass, and therefore more strontium. Although the television industry in the United States is considered mature, the demand for replacement televisions as well as additional sets in large numbers of households continues. The trend to personal computers and sophisticated, computerized instrumentation increased the demand for strontium in color monitors for these devices.

Permanent ceramic magnets are another large end use for strontium compounds, in the form of strontium ferrite. When these magnets were first developed, they were used primarily as magnetic closures for refrigerator doors. Applications have expanded to include extensive use in small direct current (dc) motors, especially for automotive applications such as windshield wiper motors, as well as loudspeakers, other electronic equipment, toys, and magnetically attached decorative items.

Strontium ferrites are used in permanent ceramic magnets because they have high coercive force, high thermal and electrical resistivity, and are chemically inert. They retain their magnetism well, are not adversely affected by electrical currents or high temperatures, and do not react with most chemical solvents. Other properties that make the strontium magnets more attractive for specific applications are their resistance to demagnetization and lower density, making them more desirable in applications where weight is a factor.

Barium or lead can replace the strontium in ferrite magnets, but strontium ferrites have been found to possess the best combination of properties necessary for superior magnets.

One of the most consistent and continuing applications for strontium has been in pyrotechnic devices. Strontium burns with a brilliant red flame, and no other material has been found to be better in this application.

The strontium compound used most frequently in pyrotechnic devices is strontium nitrate. Some strontium compounds are slightly hygroscopic, but strontium nitrate takes on very little water and imparts the desired brilliant red. Strontium carbonate, strontium oxalate, strontium sulfate, and strontium chlorate can be used in pyrotechnic applications, but strontium nitrate is used in significantly larger quantities than any of these.

Pyrotechnic devices are used in military and nonmilitary applications. Military pyrotechnic applications that contain strontium include tracer ammunition, military flares, and marine distress signals. Nonmilitary applications include warning devices and fireworks.

Strontium is used to remove lead impurities during the electrolytic production of zinc. Zinc used in diecasting alloys is required to contain less than 0.003% lead. Addition of strontium carbonate in sulfuric acid to the electrolyte reduces the lead content of the electrolyte and of the zinc that is deposited on the cathode.

The addition of strontium chromate to paint creates a coating that is resistant to corrosion. It is an effective coating for aluminum, most notably on aircraft fuselages and ships. These paints are used to some degree on aluminum packaging to prevent package corrosion. The nitrate and chloride contents of strontium chromate paint pigment are very strictly controlled to prevent corrosion.

Consumption of metallic strontium is still a very limited factor in total strontium consumption. Small amounts of strontium are added to molten aluminum to improve the castability of the metal, making it much more suitable for casting items that have been traditionally made from steel. The addition of strontium to the melt improves the machinability of the casting. The use of cast aluminum parts is currently gaining popularity in the automotive industry. Aluminum parts were impractical before modification techniques were developed. The reduction in the weight of the automobile by using cast aluminum parts instead of steel helps to improve energy efficiency of the cars incorporating these parts.

At the present time, other strontium end uses consume only small amounts of strontium and strontium compounds. As mentioned above, the presence of strontium in glass applications improves the brilliance of the glass. It also improves the quality of certain ceramic glazes as well as eliminates the toxicity that may be present in glazes containing lead or barium. One high-tech strontium ceramic is strontium titanate, which is sometimes used as substrate material for semiconductors and also in some optical and piezoelectric applications.

Strontium chloride is used in toothpaste for sensitive teeth. For this application, the content of impurities must be strictly controlled, with limits for some of them in the parts per million range.

Strontium phosphate is used in the manufacture of fluorescent lights, and the entire range of strontium chemicals is used in analytical chemistry laboratories.

Prices

The average customs value for celestite imported from Mexico in 1990 was about \$80 per ton. Mexico was the source of 89% of all celestite imports in 1990. Another 10%, valued at \$190 per ton, was imported from Spain. Average value for total strontium imports was about \$85 per ton. This value was slightly higher than the average customs value reported for 1989. Values of imported strontium compounds varied according to the type of compound and the country of origin. Table 5 lists salient statistics.

Foreign Trade

According to the Port Import/Export Reporting Service of *The Journal of Commerce*, US exports of strontium compounds were about 2.2 kt in 1990, a 58% decrease from the exports reported from the same source in 1989. Of these exports, 98% was strontium carbonate that went to Taiwan. Other reported strontium exports were chloride, chromate, ferrite, nitrate, and peroxide, all in relatively small quantities.

The Bureau of the Census began using the Harmonized Tariff Schedule in 1989 to identify material passing through US customs. This system makes classification codes for imports and exports consistent internationally. With the adoption of the new codes, imports and export codes for strontium changed, and some formerly

Table 5. Salient Strontium Statistics
(metric tons of contained strontium unless otherwise noted[1])

	1986	1987	1988	1989	1990
United States:					
Production, strontium minerals	—	—	—	—	—
Imports for consumption:					
Strontium minerals	13 200	16 900	18 100	12 000	21 500
Strontium compounds	4 300	5 000	7 800	11 300	11 800
Exports[2] (compounds)	680	1 600	3 000	3 200	1 300
Shipments from Government stockpile excesses	—	—	—	—	—
Price, average value of mineral imports at port of exportation, dollars per ton	$107	$95	$85	$76	$86
World production[3] (celestite)	150 000	182 000	231 000[p]	244 000[e]	235 000[e]

[e]estimated, [p]preliminary, [r]revised. [1]The strontium content of celestite is 43.88%, which was used to convert units to celestite. [2]Export data are obtained from *The Journal of Commerce* Port Import/Export Reporting Service. [3]Excludes China and the USSR.

listed details are no longer available. Trade information for strontium metal was grouped with barium under a classification of other alkaline earth metals, making specific information concerning strontium unavailable. Strontium chromate data were eliminated. Other strontium compounds experienced changes in classification numbers, and the grouping of compounds differed.

According to reports from the Bureau of the Census, imports of celestite increased 90% from the levels reported in 1989. Nearly all was from Mexico and a small amount from Spain.

Mexico continued to grow in importance as a source for imported strontium compounds; the Federal Republic of Germany was second. Imports of strontium carbonate were virtually the same as in 1989, with imports from Mexico becoming a larger percentage of total imports. Strontium carbonate from The Republic of Germany continued to decrease as Mexico took over more of that market. Imports of strontium nitrate nearly tripled for the second consecutive year.

PROBLEMS AND FUTURE TRENDS

Problems

In 1990, the Government of Mexico requested a waiver of the competitive need limit for strontium carbonate. Under the General System of Preferences, Mexican strontium carbonate entered the United States duty free. A tariff was assessed against imports of strontium carbonate from Mexico when those imports represented more than 50% of total strontium carbonate imports. The duty free status was revoked in July 1990. The waiver was expected to be approved because there was no opposition from the domestic strontium carbonate producer, and the consumers favored the proposal.

Strontium-90, a radioactive isotope of strontium, is a component of nuclear waste from reactors and nuclear fallout. It presents special problems for waste disposal and storage because it generates a significant amount of heat as it decays. When released into the atmosphere through nuclear testing or accidents, it easily enters the food supply in milk and vegetables because it is so similar chemically to calcium. Scientists at Argonne National Laboratory have developed a process for removing strontium-90 from liquid nuclear waste. The method known as SREX dissolves more than 99.999% of the strontium-90 from the waste stream in a special solvent. Pure strontium-90 is then removed from the solvent with water or dilute acid.

The recovered strontium-90 could be used in radio-isotopic thermal generators that are used in devices such as harbor markers, which require long-lasting, unattended power supplies (Worthy, 1990).

Future Trends

The future continues to look bright for the strontium industry for the near term, although, if both planned operations in Mexico are actually accomplished, a serious oversupply could occur. Continued growth is expected in the color television industry, and larger screens are expected to increase in popularity. Demand for strontium carbonate for television application should expand accordingly. Other markets will probably continue to increase at their current, slower rate. Development of a technology to produce an affordable flat television display could severely reduce the demand for strontium carbonate. Although a high-quality, large, flat screen is not yet available, small models are, and research continues to perfect the technology. The question remains whether a new display system can be developed that will be economically attractive to the general public. Initial devices are expected to find application in military hardware and other sophisticated medical and scientific instrumentation. Flat screen display systems may eventually replace cathode-ray tubes, and, at that point, strontium producers may experience a serious setback.

ACKNOWLEDGMENTS

This chapter was adapted from the US Bureau of Mines *Strontium 1990* Annual Report. Domestic survey data were prepared by Maureen Nash, mineral data assistant, and the international production table was prepared by Virginia A. Woodson, international data coordinator.

The sole US strontium carbonate producer voluntarily provided domestic production data to the US Bureau of Mines. Production and stock data, however, were withheld from publication to avoid disclosing company proprietary data.

The Bureau also estimated the distribution of strontium compounds by end use. Of the 10 operations to which a survey request was sent, 8 responded. The information collected from this survey represents nearly 100% of the end-use data shown in Table 4. Consumption for the nonrespondents was estimated using reported prior year consumption levels.

REFERENCES

Anon., 1986, *Economics of Strontium*, 4th ed., Roskill Information Services Ltd., London, pp. 13–15, 35.

Anon., 1988a, "Solvay Invests in Strontium Carbonate Facility, as CMV Doubles Capacity," *Industrial Minerals*, No. 262, p. 8.
Anon., 1988b, "Bureau of Mines High Tech Defense Material Supply Study," State Department Telegram 01397, Mar. 9, US Embassy, Algiers, Algeria, 1 p.
Anon., 1988c, "USBM High Tech Defense Material Supply Study," State Department Telegram 03429, Mar. 31, US Embassy, Buenos Aires, Argentina, 1 p.
Anon., 1988d, "PRC Strontium Ore," State Department Telegram 16410, June 20, US Embassy, Beijing, China, 1 p.
Anon., 1988e, "Japanese Data on Gallium, Indium, Silicon, and Strontium," State Department Telegram 10820, June 9, US Embassy, Tokyo, Japan, p. 2.
Anon., 1988f, "Strontium Ore Production in Mexico," State Department Telegram 08752, Apr. 14, US Embassy, Mexico City, Mexico, 2 pp.
Anon., 1988g, "USBM Study: High Tech Defense Material Supply," State Department Telegram 09360, Apr. 28, US Embassy, Islamabad, Pakistan, p. 1.
Anon., 1990a, "New Strontium Carbonate Plant Commissioned and CPC to Build Another," *Industrial Minerals*, No. 274, p. 12.
Anon., 1990b, "Strontium Entry Poised for Launch," *Chemical Marketing Reporter*, Vol. 237, No. 19, p. 4.
Anon., 1990c, "Strontium Unit Planned by CPC," *Chemical Marketing Reporter*, Vol. 237, No. 19, p. 4.
Dawson, K.R., 1985 "Geology of Barium, Strontium, and Fluorine Deposits in Canada," Economic Geology Report 34, Geological Survey of Canada, 136 pp.
Griffiths, J., 1984, "The Industrial Minerals of Cyprus," *Industrial Minerals*, No. 207, p. 35.
Griffiths, J., 1985a, "Celestite: New Production and Processing Developments," *Industrial Minerals*, No. 218, p. 25.
Griffiths, J., 1985b, "Spain's Industrial Minerals," *Industrial Minerals*, No. 217, p. 40.
Griffiths, J., 1987, "Pakistan's Mineral Potential," *Industrial Minerals*, No. 238, pp. 20, 31.
Gwinn, G.R., 1944, "Minor Nonmetals," *Minerals Yearbook 1944*, US Bureau of Mines, pp. 1544–1545.
Karayazici, F., 1987, "Industry Structure Past, Present, and Future," *Proceedings*, Turkey's Industrial Minerals Conference, Metal Bulletin PLC, London, p. 14.
Kishimoto, F., 1982, *New Metal Resources Within USSR*, Geological Survey Bureau of Japan, 25 pp.
Schreck, A., and Arundale, J., 1959, "Strontium-Materials Survey," Information Circular 7933, US Bureau of Mines, 45 pp.
Shiebel, W., 1978, "New Strontium Deposits in Iran," *Industrial Minerals*, No. 132, pp. 54–59.
Worthy, W., 1990, "Process Removes ^{90}Sr from Nuclear Wastes," *Chemical and Engineering News*, Vol. 68, No. 37, p. 6.

OTHER SOURCES OF INFORMATION
US Bureau of Mines Publications

"Strontium," *Mineral Commodity Summary*, annual.
"Strontium," *Supply, Demand, and Technology*, Information Circular 9213, 1989.

Other Sources

American Ceramic Society Bulletin, monthly
Chemical Marketing Reporter, weekly
Engineering and Mining Journal, monthly
Industrial Minerals, monthly
Mining Annual Review
Mining Engineering, monthly
Mining Journal, weekly
Mining Magazine, monthly
World Mining, monthly.

SULFUR RESOURCES

GREGORY R. WESSEL

Sulfur is a nonmetallic element widely distributed in nature and of great physical and economic importance. It is the 14th most abundant element in the Earth's crust (0.06%) and an important constituent of animal and plant life.

Sulfur has been known and used since ancient times for a number of medicinal and industrial uses. At present, most sulfur is used to generate sulfuric acid that is used in a wide variety of industrial processes, particularly the production of fertilizer. Because of this, sulfuric acid (and hence sulfur) consumption is often regarded as a good index of a nation's industrial development.

In the past, sulfur was mined from surface occurrences in several geologic environments, and was used in relatively small amounts. With time, the uses of sulfur and sulfuric acid expanded, as has the need for larger quantities of these commodities. Sulfur is now mined from both surface and underground deposits, and is recovered as a byproduct from a number of industrial processes.

Despite valiant efforts and years of work by sulfur explorationists and others, many aspects of sulfur mineralization remain controversial. Almost as controversial is the spelling of the word sulfur. The English spelling *sulphur* commonly is used outside America and in the American sulfur mining industry, but sulfur is the correct American spelling as approved by the American Chemical Society, the American Geological Institute, and many others. Those new to the American sulfur industry often find it puzzling to be reprimanded for using the correct American spelling.

Sulfur resources are abundant and widespread, but the extent to which they can be classified as reserves is constrained by prevailing prices and extraction technologies. At present, sulfur can be economically mined from very few deposits.

The sulfur industry is roughly divisible into two sectors: voluntary (or discretionary) and involuntary (or nondiscretionary). In voluntary production, the mining of sulfur or pyrites is the sole objective, and the recovery of the resource is as complete as economic conditions will allow. During involuntary production, sulfur or sulfuric acid (termed recovered sulfur) are produced as byproducts, and the quantity of the output is dictated by the demand for the primary product. Voluntary sulfur now accounts for only about 35% of the elemental sulfur produced worldwide, and most investigators believe that voluntary sulfur will be less important in the future.

Sulfur sources and products are described as follows (after Barker, 1983):

Sulfur Sources:

Combined sulfur—sulfur that occurs in nature combined with other elements, commonly referring to sulfides and sulfates.

Cupriferous pyrites—pyrite containing minor amounts of copper sulfides.

Hydrogen sulfide—a toxic gas that occurs in petroleum and natural gas.

Involuntary sulfur—sulfur produced as a byproduct in response to legislative or process mandates.

Native sulfur—naturally occurring elemental sulfur.

Nonferrous metal sulfides—copper, lead, zinc, nickel, and molybdenum sulfides that are processed for their metal content.

Organic sulfur—complex organic sulfur compounds that occur in petroleum, coal, oil shale, and tar sands.

Pyrites—iron sulfide minerals that include pyrite, marcasite, and pyrrhotite.

Sulfate sulfur—sulfur contained in anhydrite and gypsum.

Voluntary sulfur—sulfur produced in response to market demand.

Basic Sulfur Products:

Acid sludge—contaminated sulfuric acid usually returned to acid plants for reconstitution.

Brimstone—synonymous with crude sulfur.

Bright sulfur—crude sulfur free of discoloring impurities and bright yellow in color.

Broken sulfur—solid crude sulfur crushed to -8 in.

Byproduct sulfuric acid—sulfuric acid produced as a byproduct of a metallurgical or industrial process, generally relating to combined sulfur sources.

Crude sulfur—commercial nomenclature for elemental sulfur.

Dark sulfur—crude sulfur discolored by minor quantities of hydrocarbons, ranging up to 0.3% carbon content.

Elemental sulfur—processed sulfur in the elemental form produced from native sulfur or combined sulfur sources, generally with a minimum sulfur content of 99.5%.

Formed sulfur—elemental sulfur cast or pressed into particular shapes to enhance handling and to suppress dust generation and moisture retention.

Frasch sulfur—elemental sulfur produced from native sulfur sources by the Frasch mining process.

Liquid sulfur—synonymous with molten sulfur.

Liquid sulfur dioxide—purified sulfur dioxide compressed to the liquid phase.

Molten sulfur—crude sulfur in the molten state.

Prilled sulfur—solid crude sulfur in the form of pellets produced by cooling molten sulfur in air or water.

Recovered sulfur—elemental sulfur produced from combined sulfur sources (including byproduct hydrogen sulfide, but sometimes referring only to sulfur from fossil fuels) by any method.

Slated sulfur—solid crude sulfur in the form of slate-like lumps, produced by allowing molten sulfur to solidify on a moving belt.

Specialty sulfur—prepared or refined grades of elemental sulfur that include amorphous, colloidal, flowers, precipitated, wettable, flour, and paste sulfur.

Sulfur ore—unprocessed ore containing native sulfur.

Sulfuric acid—sulfuric acid of commerce as produced from all sources of sulfur, generally reported in terms of 100% H_2SO_4 with a 32.69% sulfur content.

Nearly 58 Mt of sulfur is estimated to have been mined or recovered as a byproduct during 1990 (Table 1). Production was obtained in nearly 70 countries, of which relatively few accounted for most of the production. The United States and the former USSR are the leading producers, but both are also net importers of sulfur.

Sulfur occurs naturally as a native element in subsurface deposits within salt domes and bedded evaporites (gypsum and anhydrite), many of which are mined in situ using the Frasch mining method. Such deposits are thought to form through the actions of bacteria in the presence of hydrocarbons and calcium sulfate, and are termed bioepigenetic deposits. Bioepigenetic sulfur sources currently supply 25 to 30% (about 11 Mt) of the total sulfur demand (Tables 1 and 2), but have accounted for larger percentages of the total production in the recent past. Bioepigenetic sulfur sources constitute most of the voluntary sulfur production. Large bioepigenetic deposits are exploited in the United States, Mexico, Poland, Iraq, Russia, the Ukraine, and other former Soviet republics.

Native sulfur also occurs in solfatara-type deposits associated with volcanism and recent hydrothermal activity. These deposits are common in volcanic regions, such as the western United States, Central America, throughout the Andes, and the Philippines. Volcanic deposits are often contaminated with toxic elements, and so are rarely utilized, generally for non-agricultural purposes.

Sulfur also occurs in combination with base metals as ferrous sulfides (pyrites) and nonferrous sulfides, from which sulfur or sulfuric acid is produced as a primary product or byproduct. This source of sulfur currently supplies about 15 to 17% (about 1 Mt) of the total demand for sulfur and sulfur compounds, and is an important source of sulfur in China, Finland, Italy, Japan, Norway, the Philippines, the Republic of South Africa, Spain, Sweden, the former USSR, and Yugoslavia (Table 1).

Large quantities of sulfur are present in natural gas and petroleum as hydrogen sulfide and organic compounds from which elemental sulfur is commonly recovered. Sulfur recovered from natural gas and petroleum refining currently supplies about 40% (roughly 23 Mtpy) of worldwide demand. In the United States, fuel processing plants accounted for 57% of the total output of sulfur in all forms in 1990. Regions that are prolific producers of sour gas and byproduct sulfur include Canada, France, Germany, the United States, the former USSR, Saudi Arabia, Iran and Kuwait.

Sulfur also occurs in organic and combined forms in tar sands, oil shale, and coal, although little sulfur is recovered from these sources. Enormous amounts of sulfur are present as sulfate (anhydrite and gypsum) in evaporite basins throughout the world, but economic methods have not been developed to recover sulfur from sulfate.

Commercial grades of elemental sulfur are classified as crude (run-of-mine, bright or dark) and recovered. Both types of elemental sulfur are shipped in liquid or solid form, and are guaranteed to be 99.55% pure. The only significant impurity is usually carbon, present as hydrocarbons dispersed throughout the sulfur (dark sulfur may contain up to 0.3% carbon). Most end uses require relatively pure sulfur free of contaminating elements such as arsenic, selenium, and tellurium.

Most elemental sulfur in the United States is marketed in the liquid state. In international trade, sulfur is generally shipped as broken or formed sulfur. Formed sulfur (prills, slates, etc.) is preferred to broken sulfur because it causes fewer dust problems. Speciality grades (such as refined sulfur of various shapes and sublimed sulfur) are prepared from commercial grades and distilled sulfur.

Sulfuric acid may be sold at various strengths, but it is generally marketed on a 100% basis. Acid is generally shipped at 66° Baume (93% H_2SO_4), as 98% acid, or as 20 to 22% fuming oleum, which is sulfur trioxide dissolved in sulfuric acid (Morse, 1991).

HISTORICAL BACKGROUND

Sulfur has been known and used by man since at least 4000 BC as a pigment, fumigant, and for medicinal and religious purposes (Haynes, 1959). Earlier peoples throughout the world probably used sulfur as well, but the rapid oxidation of sulfur in surface environments tends to remove it from the archaeological record.

As early as 2000 BC, the Egyptians used sulfur in the bleaching of linen textiles (Morse, 1990, 1991). During the Peloponnesian War, 5th century BC, the Greeks may have employed the first chemical weapon when they burned a sulfur-pitch mixture to generate suffocating gases. The Romans used sulfur to create incendiary weapons, and they mined much of the sulfur they used from open cuts and shallow underground mines in the evaporite-hosted stratiform deposits in Sicily. The Romans also mined sulfur at other sites in their empire, such as in present day Syria. Muslim alchemists probably produced the first sulfuric acid about 700 AD.

The early uses of sulfur were relatively limited and only small quantities were mined. The invention of gunpowder by the Chinese in the 10th century led to its widespread use in warfare and made sulfur an important mineral commodity. Gunpowder was used effectively at the Battle of Crécy (1346), and its introduction led to profound social and political changes in Europe. Throughout this time, the Sicilian deposits continued to satisfy European demand.

Early in the 17th century, Johann Glauber, a German alchemist, accidentally discovered a process of making sulfuric acid by burning sulfur and niter under a glass bell. Glauber's process was destined to become the cornerstone of the sulfuric acid industry, but at that time, sulfuric acid was generally used only for medicinal purposes. As others reproduced Glauber's efforts, and sulfuric acid production increased, industrial uses for sulfur developed and expanded. An English physician, John Roebuck, began manufacturing large quantities of sulfuric acid using a less expensive lead-lined container, and soon similar acid manufactories appeared. Because it is the cheapest and most versatile of the mineral acids, sulfuric acid became the workhorse of numerous chemical industries. By 1820, there were 23 acid plants in England alone (Haynes, 1959). The first American acid production was in 1793.

The Sicilian sulfur deposits continued to supply much of the world's sulfur, but an attempt by French controlling interests in 1838 to stabilize the price of Sicilian sulfur at a high level led to the use of other sources of sulfur for the production of sulfuric acid. In 1833, a French chemist, Michel Perrett, developed a reasonably economical method of obtaining sulfur dioxide by roasting pyrites. His invention led to the development of a rotary furnace in 1870 that made pyrites a profitable raw material for acid. By 1880, sulfuric acid throughout the world was produced from pyrites, except in the United States where import tariffs on iron and copper prevented the shift from native sulfur. These import tariffs were removed in 1890 and 1894 respectively, and by 1900 pyrites had become the principal acid-making material in the United States.

Native sulfur deposits were known in America, and several attempts were made to mine sulfur from salt dome cap rock in Louisiana between 1870 and 1890 using conventional underground mining techniques. These efforts were all frustrated by the nature of the sulfur ore and the overlying strata. In 1894, Herman Frasch developed a method of mining sulfur in situ using hot water and compressed air. Sulfur was first commercially obtained using the Frasch method in 1895 by the Union Sulphur Company at Sulphur Mines dome, Calcasieu Parish, LA. The Frasch method was also

Table 1. World Production of Sulfur in All Forms, by Country and Source,[1] in Units of Thousand Tons (modified from Morse, 1991)

Country[2] and source[3]	1986	1987	1988	1989	1990[e]
Algeria					
Byproduct: Natural gas and petroleum[e]	20	20	20	20	20
Australia					
Byproduct: Metallurgy	174	194	200[e]	200[e]	200
Petroleum	10	9	9[e]	10[e]	10
Austria					
Byproduct: Metallurgy	11	10	11	12	12
Natural gas and petroleum	29	25	36	37[r]	37
Gypsum	24	13	—	—	—
Bahrain					
Byproduct: Petroleum	46	48	48	48[e]	48
Belgium					
Byproduct: All sources[e]	300	300	310	320	320
Bolivia					
Native	5	9	7	8	1
Brazil					
Frasch	6	6	6	6	6
Pyrites	92	77	103	72[r]	90
Byproduct: Metallurgy	100	153	152	164[r]	160
Petroleum	74	77	61	60[r]	60
Bulgaria[e]					
Pyrites	80	80	70	70	70
Byproduct: All sources	62	65	60	60	60
Canada					
Byproduct: Metallurgy	758[r]	851[r]	944[r]	896[r]	1 029
Natural gas	5 161	5 268	5 415[r]	5 161[r]	5 194[5]
Petroleum[e]	129	142	152	149[5,r]	223
Tar sands	424	426	485	491	501[5]
Chile					
Native: Refined	13	15	17	16	28[5]
From caliche	44	22	21	20[e]	21
Byproduct: Metallurgy	41[r]	335[r]	416[r]	400[e,r]	400
China[e]					
Native	300	300	300	300	300
Pyrites	2 500	3 700	3 900	4 000	4 000
Byproduct: All sources	300	500	550	600	600
Colombia					
Native	36	41	43	46[r]	45
Byproduct: Petroleum[e]	10	10	8[5]	8	8
Cuba					
Byproduct: Petroleum[e]	5	5	5	5	5
Cyprus[6]					
Pyrites	25	41	51	26[r]	26
Czechoslovakia[e]					
Native	6	6	6	6	6
Pyrites	60	60	60	50[r]	50
Byproduct: All sources	42[5]	40[r]	40[r]	40[r]	40
Denmark					
Byproduct: Petroleum	13	13[r]	14[r]	14[e,r]	15
Ecuador[e]					
Native	4	5	5	4	4
Byproduct: Natural gas	5	5	5	5	5
Petroleum	5	5	5	5	5
Egypt					
Byproduct: Natural gas and petroleum[e]	7	8	8	8	8
Finland					
Pyrites	275	313[r]	271[r]	365[r]	360
Byproduct: Metallurgy[e]	260	230[r]	240	230[r]	230
Petroleum	42	40[e]	47	41[r]	42
France					
Byproduct: Natural gas	957[r]	883[r]	775[r]	647[r]	650
Petroleum[e]	180	200	226	239[r]	235
Unspecified[e]	180	180	180	150[r]	160
Germany					
Eastern states					
Byproduct: All sources[e]	315	315	315	290	265
Western states					
Byproduct: Metallurgy[e,7]	300	300	310	315	320
Natural gas	998	1 029[r]	952[r]	1 050[e]	1 000
Petroleum[e]	190	210	205	210	215
Unspecified[e]	285	285	280	310	300

(cont.)

Table 1. World Production of Sulfur in All Forms, by Country and Source,[1] in Units of Thousand Tons (modified from Morse, 1991) (cont.)

Country[2] and source[3]	1986	1987	1988	1989	1990[e]
Greece[e]					
Pyrites	66	70	70[r]	70	70
Byproduct: Natural gas	135[r]	135[r]	135[r]	135[r]	135
Petroleum	5	5	5	5	5
Hungary[e]					
Pyrites	1	1	1	1	1
Byproduct: All sources	10	10	10	10	10
India					
Pyrites	8	14	12	16[r]	16
Byproduct: Metallurgy[e]	120	120	125	125	125
Petroleum[e]	1	—	1	5	5
Indonesia[6]					
Native	5	4	4[r]	4	4
Iran[e]					
Native	30	30	30	30	30
Byproduct: Natural gas and petroleum	250	300	300	300	300
Iraq[e]					
Frasch	600	707	700	900	800
Byproduct: Natural gas and petroleum	200	250	350	370	250
Israel					
Byproduct: Natural gas and petroleum	29	40	68[r]	68[r]	64
Italy[e]					
Pyrites	309[5]	314[5,r]	310[5,r]	320[r]	300
Byproduct: All sources[8]	185[r]	190[r]	190[r]	190[r]	190
Japan					
Pyrites	158	79	71[r]	62[r]	53[5]
Byproduct: Metallurgy	1 255[r]	1 250[r]	1 268[r]	1 384[r]	1 450
Petroleum	985	1 020[r]	1 093[r]	1 176[r]	1 200
Korea, North[e]					
Pyrites	200	200	200	200	200
Byproduct: Metallurgy	30	30	30	30	30
Korea, South[e]					
Byproduct: Metallurgy	55	55	55	55	55
Petroleum	35	35	35	36	36
Kuwait					
Byproduct: Natural gas and petroleum[e]	260	310	360	575	400
Libya					
Byproduct: Natural gas and petroleum[e]	14	14	14	15[r]	15
Mexico					
Frasch	1 588	1 806	1 628	1 528[r]	1 450
Byproduct: Metallurgy[e]	134	154	240	286[r]	290
Natural gas and petroleum	456[r]	501[r]	510	555	600
Namibia					
Pyrites	134	74	135	130[e]	125
Netherlands					
Byproduct: Petroleum[e]	250	211[r]	215[r]	260[r]	280
Netherlands Antilles					
Byproduct: Petroleum[e]	40	60	60	60	60
New Zealand					
Byproduct: All sources	1[e]	1[e]	4	1[r]	4
Norway					
Pyrites	181	179	152[r]	122[r]	125
Byproduct: Metallurgy	67	80[e]	80[e]	75[r]	75
Petroleum[e]	13	15	10	13[r]	15
Oman					
Byproduct: Unspecified[e]	16	17	17	17	17
Pakistan					
Native	1	1	1	—[r]	
Byproduct: All sources[9]	26	26	25	25	25
Peru[e]					
Native	[5,9]	[9]	[9]	[9]	[9]
Byproduct: All sources	66	66	66	66	66
Philippines					
Pyrites	113	158	160[e]	195[e]	185
Byproduct: Metallurgy[e]	120	140	150	185	185
Poland[10]					
Frasch	4 437	4 410	4 411[r]	4 300[r]	4 200
Native	550	694	679[r]	660[r]	650
Byproduct: Metallurgy[e]	170	170	150	150	140
Petroleum[e]	30	30	20	20	20
Gypsum[e]	20	20	20	20	20

(cont.)

Table 1. World Production of Sulfur in All Forms, by Country and Source,[1] in Units of Thousand Tons (modified from Morse, 1991) (cont.)

Country[2] and source[3]	1986	1987	1988	1989	1990[e]
Portugal					
Pyrites	141[r]	123[r]	110[r]	91[r]	95
Byproduct: All sources[e]	5	5	5	3[r]	3
Qatar					
Byproduct: Natural gas	49	53	54[e]	52[e,r]	52
Romania[e]					
Pyrites	150	150	150	150	130
Byproduct: All sources	140	130	120	120	100
Saudi Arabia					
Byproduct: Natural gas and petroleum	1 446	1 432	1 378[r]	1 400[e]	1 500
Singapore					
Byproduct: Petroleum[e]	50	50	50	55	55
South Africa, Republic of					
Pyrites	499	468	505[r]	461[r]	452[5]
Byproduct: Metallurgy[e]	108	105	113[r]	104[r]	111
Petroleum[e,11]	110	110	120	120	125
Spain					
Pyrites	1 195	960	1 057[r]	938[r]	950
Byproduct: Coal (lignite) gasification[e]	2	2	2	2	2
Metallurgy[e]	210	225	110[r]	110[r]	125
Petroleum	8	8	8	8	8
Sweden					
Pyrites	227	215[r]	286[r]	233[e,r]	235
Byproduct: Metallurgy[e]	125[5]	130[r]	125	125	125
Petroleum[e]	49[5]	50	45	40[r]	40
Switzerland					
Byproduct: Petroleum	3	4	4[e]	4[e]	4
Syria					
Byproduct: Natural gas and petroleum[e]	35	40	40	40	40
Taiwan					
Byproduct: All sources	63	89	86	76	80
Trinidad and Tobago					
Byproduct: Petroleum[e]	5	5	5	5	5
Turkey					
Native	40[r]	39	30[r]	23[r]	20
Pyrites	4	22	75[r]	83[e,r]	85
Byproduct: Petroleum	7	9	16	13	16
Unspecified[e]	22[r]	11[r]	5[r]	49[r]	55
USSR[e]					
Frasch	1 100	1 100	1 100	1 100	1 000
Native	2 400[r]	2 400[r]	2 400[r]	2 350[r]	2 000
Pyrites	2 090	2 150	2 150	2 150	1 900
Byproduct: Metallurgy	1 050	1 250	1 375	1 350	1 200
Natural gas	1 625	2 400	3 290	2 500	2 500
Petroleum	400	450	450	450	425
United Arab Emirates, Abu Dhabi[e]					
Byproduct: Natural gas	104	97	100[r]	70[r]	100
Petroleum	11	8	10[r]	10[r]	10
United Kingdom					
Byproduct: Metallurgy	58	51	55	55[e]	50
Petroleum	105	119	129	130[e]	135
United States					
Frasch	4 043	3 202	3 174	3 888	3 726[5]
Pyrites	W	W	W	W	W
Byproduct: Metallurgy	919	1 003	1 125	1 190	1 294[5]
Natural gas	2 246	2 536	2 501	2 537	2 336[5]
Petroleum	3 570	3 624	3 943	3 973	4 200[5]
Unspecified	309	173	3	4	4[5]
Uruguay					
Byproduct: Petroleum[e]	2	2	2	2	2
Venezuela					
Byproduct: Natural gas and petroleum	99	125[e]	125	125[e]	125
Yugoslavia					
Pyrites and pyrrhotite	320	258	252	255[e]	250
Byproduct: Metallurgy[e]	175	175	170	170	170
Petroleum[e]	3	3	3	3	3
Zaire					
Byproduct: Metallurgy[e]	39	35	35	35	34
Zambia					
Pyrites	8	19	27	29[e]	29
Byproduct: Metallurgy[e]	74	74	74	74	74

(cont.)

Table 1. World Production of Sulfur in All Forms, By Country and Source,[1] in Units of Thousand Tons (modified from Morse, 1991) (cont.)

Country[2] and source[3]	1986	1987	1988	1989	1990[e]
Zimbabwe[e]					
Pyrites	25	21	18	25[r]	25
Byproduct: All sources	5	5	5	5	5
GRAND TOTAL[4]	53 736	56 582	58 528	58 717	57 668
Of which: Frasch	11 774	11 231	11 019	11 722	11 182
Native	3 434	3 566	3 543	3 467	3 109
Pyrites	8 861	9 746	10 196	10 114	9 822
Byproduct: Coal (lignite) gasification	2	2	2	2	2
Metallurgy	6 353	7 120	7 553	7 720	7 884
Natural gas	11 280	12 406	13 227	12 157	11 972
Natl. gas/petr. undif.	2 845	3 065	3 209	3 513	3 359
Petroleum	6 386	6 577	7 004	7 177	7 515
Tar sands	424	426	485	491	501
Unspecified sources	2 332	2 408	2 271	2 336	2 304
Gypsum	44	33	20	20	20
% Contribution of Sulfur Sources					
Frasch	22	20	19	20	19
Native	6	6	6	6	5
Pyrites	16	17	17	17	17
Byproduct: Coal (lignite) gasification	−1	−1	−1	−1	−1
Metallurgy	12	13	13	13	14
Natural gas	21	22	23	21	21
Natl. gas/petr. undif.	5	5	5	6	6
Petroleum	12	12	12	12	13
Tar sands	1	1	1	1	1
Unspecified sources	4	4	4	4	4
Gypsum	−1	−1	−1	−1	−1

[e]estimated, [r]revised, W-withheld to avoid disclosing company proprietary data, included with "Byproduct: Unspecified sources."
[1]table includes data available through May 17, 1991.
[2]in addition to the countries listed, a number of other nations may produce sulfur or sulfur compounds as byproducts, but output, if any, is not reported and cannot be estimated. Countries in this category that may recover sulfur from petroleum include Albania, Bangladesh, Brunei, Burma, Costa Rica, Guatemala, Honduras, Jamaica, Malaysia, Nicaragua, Paraguay, and the People's Democratic Republic of Yemen. Albania and Burma may also produce sulfur from oil and gas extraction.
[3]"source" refers to both means of collecting sulfur and type of raw material. Sources listed include: 1. Frasch recovery, 2. Native, comprising all production of elemental sulfur by traditional (non-Frasch) mining methods, 3. Pyrites (regardless in what form sulfur is recovered), 4. Byproduct recovery from listed processes, 5. Recovery from mined gypsum. Recovery of sulfur in the form of sulfuric acid from artificial gypsum produced as a byproduct of phosphatic fertilizer production is excluded to prevent double counting. Production of Frasch sulfur, other native sulfur, pyrites-derived sulfur, mined gypsum-derived sulfur, byproduct sulfur from extraction of oil and natural gas, and recovery from tar sands are all credited to the country of origin of the extracted raw material; in contrast, byproduct recovery from metallurgical operations, petroleum refining, and spent oxides is credited to the nation where recovery takes place, which is not always the original source of the crude product.
[4]data may not add because of independent rounding.
[5]reported figure.
[6]in addition, may produce limited quantities of byproduct sulfur from oil refining.
[7]includes only the elemental sulfur equivalent of sulfuric acid produced as a byproduct from metallurgical furnaces; additional output may be included under "Byproduct: Unspecified sources."
[8]includes recovery from gypsum, if any.
[9]less than 1/2 unit.
[10]official Polish sources report total Frasch and native mined elemental sulfur output annually, undifferentiated; this figure has been divided between Frasch and other native sulfur on the basis of supplementary data.
[11]estimates for 1986–90 include byproduct production from synthetic fuels.

employed in 1912 at Bryan Mound, Brazoria County, TX, and later at many other domes (Ellison, Jr., 1971, Halbouty, 1979).

Shipping problems during World War I, and a growing preference for brimstone over pyrites to use in the contact process of acid production (first developed in the 1880s to generate larger quantities of high strength acid), led to expanded use of Frasch sulfur. By 1919, 48% of American acid production utilized brimstone; by 1929, 69% came from this source (Haynes, 1959).

World War II disrupted sulfur production and transportation outside North America, and recovery after the war was slow. Foreign consumers became more dependent upon US Frasch producers as they attempted to rehabilitate and expand their acid industries.

In 1950, Frasch sulfur from the US Gulf Coast and sulfuric acid from pyrite-roasting plants in Europe were the major sources of the western world's sulfur. Competition from low cost US mines caused a rapid decline in sulfur production from Italy.

A supply shortfall during the Korean war led to short-lived controls imposed on the industry by the US government. The controls were lifted after the cessation of hostilities, and concurrently new mines were opened at several sites. Producers opened new mines at Garden Island Bay and Damon Mound, and Mexico entered the Frasch market with production from San Cristobal in 1954 (Salas, 1991). Also of great significance was the expansion and accelerated growth of recovered sulfur production in Canada beginning at about this same time. Recovered sulfur production

Table 2. Sulfur Supply (in All Forms) From Natural Resources (after data from Nieć, 1986a)

Source	Major methods of recovery	World resources, Gt	% of supply
Native deposits:			
Bioepigenetic	Frasch method, surface mining	approx. 2.5	30
Volcanic	Surface mining, smelting/flotation	approx. 0.5	−1
Pyrites and base-metal sulfides	Sulfuric acid and sulfur dioxide from roasting, acid from stack gases	approx. 1.5	15
Natural gas	Hydrogen sulfide recovered using Claus process	0.7	45
Petroleum	Desulfurization during refining	0.5	10
Tar sands	Desulfurization during refining	approx. 2.0	−1
Coal	Byproduct pyrite and chemical processing	+1.0	−1
Gypsum and anhydrite	Surface mining, sulfate reduction	very great	−1

began in France in 1957, and Texasgulf opened the world's first liquid sulfur terminal in 1958 (Barker, 1983, Giusti, 1984).

The first Frasch sulfur from the huge deposits in Poland reached the market in 1960. Canadian and French recovered sulfur production continued to expand and the US share of world exports fell 19% in four years to 31%. Mexico surpassed the United States as the dominant exporter in 1964. Additional Frasch mines were opened in Poland in 1966 and 1967, and the large deposit at the Culberson mine in West Texas was discovered by Duval in 1967. As time progressed, recovered sulfur output exceeded Frasch sulfur output, even with the opening of the Frasch mine at Mishraq, Iraq, in 1972.

Price offensives by the oil and phosphate cartels in 1974 and 1975 led to recession and the drawdown of existing sulfur stocks. This resulted in a major cutback in the United States and Mexico, the beginning of a gradual decline in North American Frasch production that may continue into the future. Despite intense exploration, particularly during the late 1960s in West Texas, only one large sulfur deposit has been discovered in the United States since the discovery of the Culberson deposit, the new offshore Main Pass mine in the US Gulf Coast.

The three largest producers of mined sulfur in the United States are Pennzoil Sulphur Co., Freeport Sulphur Co. (part of Freeport-McMoRan Resource Partners Inc.), and Texasgulf Inc. These three companies accounted for 32% of US production of sulfur in all forms in 1990 (Loughbrough, 1991).

Pennzoil produces sulfur from its Culberson mine in West Texas and is currently engaged in exploration activities in West Texas, the Gulf Coast, and elsewhere. Texasgulf currently operates one mine in Texas, Boling dome, extracting about 140 000 tpy of sulfur, although this mine is scheduled to close in the mid-1990s. Freeport currently operates two mines in the Gulf Coast, Caminada and the large (68 Mt) Main Pass mine, and two other Freeport mines recently closed (Garden Island Bay and Grand Isle). The Main Pass mine is currently the largest known sulfur reserve in the Western Hemisphere.

The United States is still the world's foremost producer and consumer of sulfur and sulfuric acid, with production from Frasch, recovered, and byproduct sources. However, the United States is a net importer of sulfur, with most of the imports coming from Canada and Mexico. The former USSR is currently the second largest producer (followed by Canada and Poland), but recent developments in the former Soviet Union's recovered production and independence movements within the various republics will certainly alter this status.

Sulfur in various forms is produced worldwide, and no single nation dominates as a supplier to world markets. Poland is currently the largest producer of native sulfur, producing about five million tons per year from three mines. The main producer of sulfur in Poland is Siarkopol, and the product is marketed by the Ciech foreign trade enterprise.

Frasch sulfur is also produced in large quantities in the former USSR and from the Mishraq deposit in Iraq, located near Mosul in the Kurdish north of the country. Iraq's status as a major exporter was severely affected during the United Nation's trade embargo and the Gulf War of late 1990 and early 1991.

Recovered elemental sulfur is the major sulfur source in Canada, France, Germany, Kuwait (devastated as a result of the Iraqi invasion), Saudi Arabia, and the United States, and Canada is the world's largest sulfur exporter. China is currently the fifth largest producer and the largest pyrites producer, and is the third largest consumer. Future developments and environmental considerations may lead to an increase in involuntary sulfur production, possibly at the expense of Frasch producers.

GEOLOGY OF NATIVE SULFUR DEPOSITS

Mineralogy and Classification of Deposits

Elemental sulfur occurs naturally by replacement reactions with gypsum and anhydrite in salt dome cap rocks and bedded evaporites. It is also common in volcanic environments, and associated with thermal springs.

Native sulfur is readily oxidized, particularly in the presence of sulfur-oxidizing bacteria, and so is somewhat ephemeral in nature. Surface exposures are relatively rare, and are generally restricted to regions of ongoing or very recent mineralization. Most known sulfur deposits are thought to be Tertiary in age or younger, but some fossil deposits (from which the sulfur has been removed) are known that may be older.

The element sulfur occurs as sulfate (gypsum, anhydrite) throughout the world, and in sulfides, both ferrous and nonferrous. All of these occurrences of sulfur are locally utilized as a source of elemental sulfur and sulfuric acid. Sulfur is present in organic compounds and as hydrogen sulfide in petroleum and natural gas, and in coal, tar sands, and oil shale.

The atomic weight of sulfur is 32.064, and it has an atomic number of 16. Sulfur has four oxidation states (+6, +4, 0, and -2) of which the S^{-2} sulfide ion and the S^{+6} sulfate cation are the most important in nature.

There are four stable isotopes of sulfur—^{32}S, ^{33}S, ^{34}S, and ^{36}S—that respectively comprise 95.02, 0.75, 4.21, and 0.02 percent of the total terrestrial sulfur. The isotopic composition of sulfur is expressed in terms of $\delta^{34}S$ (Faure, 1977). The most important cause for variation in the isotopic composition of sulfur in nature is fractionation by sulfate-reducing bacteria. These bacteria split oxygen from sulfate ions and excrete hydrogen sulfide that is enriched in ^{32}S relative to the sulfate. The extent of the fractionation is variable, and is dependent upon the rate of reduction, the temperature, the nature and availability of the food supply, and the size of the sulfate reservoir. Native sulfur from salt dome cap rock and bedded evaporites (as described below) is enriched in

^{32}S, and the residual anhydrite/gypsum and associated sulfate minerals such as barite are enriched in ^{34}S, lending evidence to the origin of sulfur in these environments by sulfate-reducing bacteria (Feely and Kulp, 1957, Thode, 1972).

Native sulfur is polymorphous; although it commonly occurs in the orthorhombic or alpha (α) sulfur form, it is also known in two monoclinic forms, beta (β) sulfur and gamma (γ) sulfur. The monoclinic forms are unstable at normal surface temperatures and readily convert to the orthorhombic form. Orthorhombic sulfur, when heated to 95.5°C, changes to monoclinic crystals. Natural monoclinic sulfur is rare, but it is known from several sites including Moravia where it has been given the name of rosickyite (Ford, 1949). Native sulfur also occurs in an amorphous state.

Native sulfur crystals are commonly dipyramidal and sometimes thick tabular on c(001) or sphenoidal. Sulfur also occurs in irregular masses, in reniform shapes, as incrustations and earthy masses, and in stalactitic/stalagmitic forms. Cleavage is imperfect, and fracture is conchoidal to uneven. Native sulfur is brittle to imperfectly sectile, and has a hardness of 1.5 to 2.5, and a resinous luster. Sulfur has a specific gravity of 2.05 to 2.09. The average weight of orthorhombic sulfur is 2.07 g/ml. The average weight of monoclinic sulfur 1.96 g/ml, and amorphous sulfur is 1.92 g/ml (Ellison, Jr., 1971). The color of sulfur varies from bright yellow to honey yellow, yellowish brown, greenish, or reddish to yellowish gray. The color is dependent upon the temperature and discoloring impurities such as hydrocarbons and clay.

Sulfur is a nonconductor of electricity, and a poor conductor of heat. Orthorhombic sulfur melts at 112.8°C, monoclinic sulfur at 119.2°C, and amorphous sulfur at 120.0°C. Molten sulfur is fairly mobile near the melting point, but its viscosity increases with increasing temperature to 188°C, then decreases in viscosity to its boiling point at 444.6°C. The specific gravity of molten sulfur is 1.79 at 132°C, and the flash point (modified Cleveland open cup) is 168.3 to 187.3°C.

Sulfur is insoluble in water and is nonhydroscopic, but is soluble in benzene, ether, toluene, alcohol, and carbon disulfide. Sulfur is unaffected by most acids. Sulfur is readily ignited in air, burning at 270°C with a blue flame and evolution of sulfur dioxide. Corrosivity is nil for dry sulfur with slow generation of sulfuric acid when wet. Vapor pressure is less than 0.1 μm Hg at 20°C.

Native sulfur deposits are classified according to genesis by Ruckmick et al. (1979) in the following fashion:

- I. Biogenic deposits
  - A. Bioepigenetic
    - 1. Cap rock
    - 2. Stratabound
  - B. Biosyngenetic
- II. Volcanic deposits
- III. Oxidative deposits
- IV. Thermogenic accumulations

The biogenic deposits listed include all types of mineralization thought to originate through the actions of sulfate-reducing bacteria. The bioepigenetic deposits are the most economically significant, and include deposits in salt dome cap rocks ("cap rock" above) and replacement deposits in bedded gypsum and anhydrite ("stratabound" above; "stratiform" as discussed below). The biosyngenetic deposits include small occurrences of native sulfur in modern marine or lacustrine sediments generated by bacterial reduction of sulfate. None of the significant sulfur occurrences in the world are known to be biosyngenetic.

Native sulfur is common in volcanic terrains, generally as vein and fracture fillings, and associated with thermal springs. Volcanic occurrences are generally small, but a few have been mined and are locally important.

The oxidative classification of Ruckmick et al. refers to widespread, but generally small, occurrences of sulfur formed at the surface or at depth wherever hydrogen sulfide (regardless of origin) is oxidized. This class of deposit is discussed in more detail in the section on Stratiform Deposits where two oxidative deposits are described.

Native sulfur is also known to originate from high temperature/high pressure reactions in deep strata, here termed thermogenic accumulations. This type of deposit occurs in association with sour natural gas and is discussed in the section on recovered elemental sulfur.

Salt Domes

Salt domes, pillows, and anticlines are common in evaporite basins throughout the world. Salt domes host sulfur mineralization at numerous widespread sites, including Tunisia, the Canadian Arctic, and the Gulf Coast region of the United States and Mexico (Fig. 1), but only those salt domes in the Gulf Coast are presently known to host economic sulfur deposits. Salt domes are also important traps for oil and gas accumulations, and the presence of hydrocarbons is essential for the development of native sulfur.

Numerous references discuss the origins of salt domes and cap rock, including previous editions of this work. Additional overviews of salt domes and pertinent Gulf Coast geology can be found in Ellison, Jr. (1971), Halbouty (1979), Kyle and Posey (1991), Lerche and O'Brien (1987), Salas (1991), Seni and Kyle (1986), and Warren (1989) among others. Many recent papers present data collected as part of efforts to evaluate salt domes for the storage of radioactive waste, and data pertaining to fluid migration, cap rock development, and metallic sulfide occurrences.

Salt domes are generally described as diapirs of halite that have intruded younger strata in response to early differential loading and subsequent density inversion. However, it is misleading to believe that salt is capable of intruding many kilometers of strata. Diapirs begin forming early in the sequence of burial, probably as a response to differential loading, and pass to a stage of diapirism as overlying strata are loaded on the salt bed in thicknesses greatest where the salt is thinning and migrating toward the diapir (Jackson and Talbot, 1986, Posey and Kyle, 1988, Trusheim, 1960). The top of the salt plug maintains a more or less constant elevation relative to the sediment-water interface, although the buoyancy of the salt and continued upward movement may cause minor faulting and deformation in strata above, and the salt may breach the surface.

In the Gulf Coast, the salt that forms the domes has its source in the Jurassic Louann Formation, a thick deposit of halite thousands of meters below the surface. The Louann halite is relatively pure, but contains about 5% insoluble material, mostly anhydrite crystals with small amounts of dolomite, calcite, quartz, and traces of limonite, barite, and other minerals.

Subsequent sedimentation on the top of the halite deformed the salt and initiated movement, eventually directed to distinct salt plugs that grew in response to continued loading. Late in the history of each salt diapir, the top of the salt plug is dissolved through the actions of migrating fresh waters. As salt movement continues upward, the halite at the top is removed and the anhydrite residue accumulates, eventually becoming compacted and cemented. Further dissolution and residue underplating result in the formation of a banded anhydrite cap rock that grows younger downward. The salt dissolution at the upper surface of the salt stock is relatively flat. The location of active dissolution probably migrates across the salt dome with time, and may be only locally apparent. Only those salt domes that penetrate to within about 2 km of the surface exhibit anhydrite cap rock (Halbouty, 1979).

The upper (or outermost) portions of the anhydrite cap rock are

Fig. 1. Past and present sulfur-producing domes, Gulf Coast region, United States and Mexico (modified after Anon., 1985a).

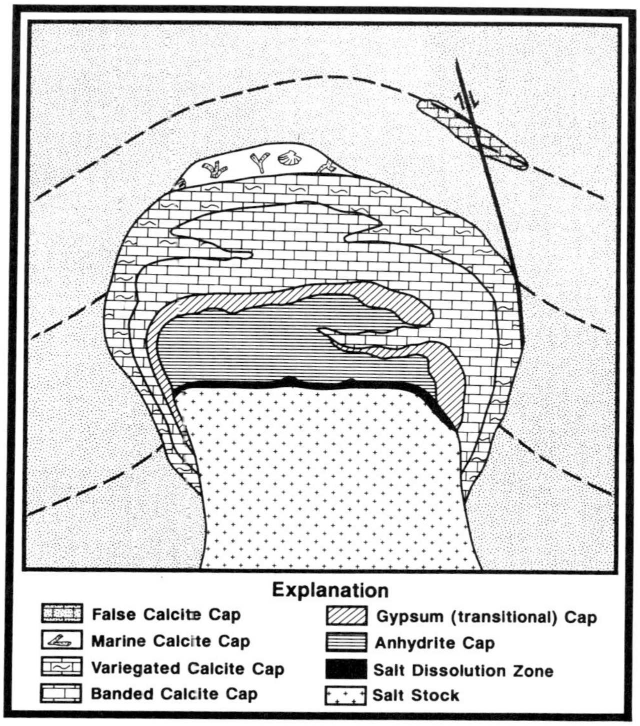

Fig. 2. Schematic cross section of a salt dome cap rock showing an idealized sequence of caprock lithologies (modified after Kyle and Posey, 1991).

commonly replaced by calcite cap rock. A schematic cross section of a typical salt dome is illustrated in Fig. 2. The calcite cap rock is roughly divisible into two parts: an upper variegated calcite cap, and a lower banded calcite cap (Kyle and Posey, 1991) both of which may be intensely brecciated and veined with crystalline calcite. The calcite cap is separated from the anhydrite by a transitional zone of gypsum. The underlying anhydrite may be locally absent, particularly at the edges of the dome, if calcite replacement has been complete.

Native sulfur may be found throughout the cap rock and adjacent super-cap strata in minor amounts, but it is generally concentrated in the lower portion of the calcite cap. The upper portion of the calcite cap is often barren of sulfur, possibly because of nondeposition and/or subsequent oxidation. The thickest accumulations of calcite cap, and also of native sulfur, are commonly on the basinward sides of the salt domes, which in the US Gulf Coast is predominately to the south.

Celestite and barite locally accompany the sulfur as minor accessory minerals. Several recent studies have also documented the occurrence of Mississippi Valley-type metallic sulfides in both calcite and anhydrite cap rock. Documented sulfide minerals include pyrite, marcasite, pyrrhotite, hauerite, alabandite, sphalerite, galena, chalcopyrite, and acanthite (Price et al., 1983, Kyle and Price, 1986). Uranium minerals are also found in super-cap strata adjacent to salt domes, and realgar has been noted associated with silver values at one dome (Wessel, unpublished data, 1991).

The origin of sulfur deposits in salt dome cap rock (and in replacements in bedded evaporites, described below) is now widely accepted to be associated with the actions of sulfate-reducing bacteria in the presence of a sulfate source (anhydrite/gypsum) and hydrocarbons. At relatively shallow depths (generally less than 750 m, but locally deeper) and at temperatures below about 60°C, sulfate-reducing bacteria can thrive in the subsurface given an adequate energy source. Sulfate-reducing bacteria are anaerobic, but they are ubiquitous in nature and exist (and probably thrive) with aerobic bacteria in microenvironments.

The origin of the sulfur deposits is complex and dependent upon several parameters. The host anhydrite must be relatively near the surface and exposed to migrating meteoric waters, so that the anhydrite is hydrated to gypsum and put into solution. Into this system, petroleum and/or natural gas is introduced from depth, generally along faults and joints or permeable strata. Sulfate-reducing and other bacteria introduced with the meteoric water oxidize the hydrocarbons and reduce the sulfate to hydrogen sulfide, represented by the following reaction (greatly simplified):

$$CaSO_4 + CH_4 \text{ (hydrocarbons)} + \text{bacteria} \rightarrow H_2S + CaCO_3 + H_2O$$

The hydrogen sulfide is oxidized to elemental sulfur with molecular oxygen, presumably present in the groundwater, represented by:

$$2H_2S + O_2 \rightarrow 2S^\circ + 2H_2O$$

Hydrogen sulfide may also be oxidized to sulfur by ferric iron, polysulfide disproportionation, or microbially by colorless and colored sulfur-oxidizing bacteria (Machel, 1992). The oxidation of hydrogen sulfide by excess sulfate ions has also been suggested as a mechanism for producing native sulfur (Davis and Kirkland, 1979, Ruckmick et al., 1979). In more oxidizing environments, hydrogen sulfide may be oxidized to valences higher than zero, and no elemental sulfur may be formed.

The reactions above basically represent only the end products of sulfate reduction and sulfur deposition. Numerous intermediate reactions have been omitted for clarity, but are important to understanding the nature of sulfur generation. Regardless of the mechanism of oxidation of the hydrogen sulfide, it appears that critical to the generation of a significant sulfur deposit is the confinement of the hydrogen sulfide to a restricted area where the native sulfur may be deposited and preserved. Dissipation of hydrogen sulfide to the atmosphere or elsewhere will prevent the deposition of economic quantities of sulfur. Similarly, subsequent oxidation of sulfur mineralization, especially due to the actions of sulfur-oxidizing bacteria such as *Thiobacillus* sp., may leave calcite replacement bodies barren of native sulfur. Overlying impermeable strata, such as shales, are generally credited with providing a barrier to hydrogen sulfide migration and later oxidation.

One particular species of sulfate-reducing bacteria, *desulfovibrio desulfuricans*, is commonly credited with being the most significant contributor to sulfur mineralization, but several species of *desulfovibrio* are probably involved, as well as species of *clostridium*, *desulfotomaculum*, and possibly others. These sulfate-reducing bacteria are strict anaerobes, but they can exist and thrive in microenvironments distributed throughout oxidized strata containing aerobic bacteria. The sulfate-reducing bacteria do not degrade hydrocarbons, but utilize only simple organic substrates such as acetate, lactate, pyruvate, fumarate, and formate (Ivanov, 1968, Kuznetsov et al., 1963, Postgate, 1979, Skyring, 1987). Some are thought to oxidize methane, but for hydrocarbons to be considered an energy source, aerobic bacteria must be present to generate appropriate substrates. The process of native sulfur deposition from sulfate-reduction likely involves numerous types of bacteria in a complex and evolving ecosystem.

The amount of hydrocarbons necessary to produce a significant sulfur deposit is fairly large. It is thought that 2 to 4 bbl of oil must be consumed for each ton of sulfur deposited (Ruckmick et al.,

1979). The actual amount of hydrocarbons required probably varies depending upon local geologic conditions.

Concurrent with the generation of hydrogen sulfide and the deposition of sulfur, calcite in multiple generations (sometimes referred to as biocalcite) is deposited as a replacement of the anhydrite and forms the calcite cap rock. Early generations of calcite may be massive and fine-grained; later calcite filling veins and vugs is often finely to coarsely crystalline, white to tan, and commonly petroliferous. Much of the calcite cap rock is a breccia containing clasts of wall rock and earlier generations of calcite and associated minerals. The calcite breccia may be barren of sulfur because of non-deposition, later oxidation and removal, or perhaps because of remobilization and nearby reprecipitation as described by Nieć (1992).

The role of sulfate-reducing bacteria in the formation of sulfur deposits is supported by isotope studies of both sulfur and carbon. Sulfate-reducing bacteria fractionate the sulfur isotopes and selectively concentrate the light isotope. The sulfur is enriched in ^{32}S, and the remaining sulfate is enriched in ^{34}S. The calcite cap rock is isotopically similar to the carbon in associated crude oil, and significantly enriched in ^{12}C with respect to sedimentary limestone (Faure, 1977, Feely and Kulp, 1957).

There are 167 onshore salt domes and 180 offshore domes presently known in the Gulf Coast province of Texas (excluding south Texas) and Louisiana (Eager and Salisbury, unpublished data, 1991). Of the 167 onshore domes, 90 are shallow enough to be prospective for sulfur. Twenty-four of these domes have already produced sulfur (included in Table 3); 12 have produced more than 5 Mt, 5 have produced more than 10 Mt, and 3 have produced more than 20 Mt. Mexico is also a significant producer of salt dome-hosted sulfur, but published data on Mexican production and geology are not extensive (Salas, 1991). Two of the most important salt domes in the Gulf Coast are described in the following sections.

Boling Dome: Boling dome is located in Wharton County, Texas, about 100 km southwest of Houston. Discovered in 1922, Boling dome has been in production since 1928. Currently operated by Texasgulf Inc., Boling has produced over 82 Mt of sulfur from more than 8 000 wells (Samuelson, 1986, 1992, Seni, 1986, 1987). A cogeneration plant, opened in 1985, generates electricity that is marketed to a local utility. Sulfur production is currently at a rate of about 361 tpd, much less than the peak production in 1948 of 8 510 tpd. Boling dome is nearly exhausted of sulfur, and would have been closed before now were it not for the cogeneration plant. Sulfur production is only planned for another couple of years, and the property is currently on the market.

Boling dome is the largest salt dome in Texas with an areal extent of 38 km^2 at the $-1\ 219$ m elevation contour. The minimum depth to cap rock is 117 m, and the margins of the cap rock plunge steeply below a depth of 1 524 m.

The cap rock is divisible into three zones: an upper barren calcite zone, a middle calcite/sulfur zone, and a lower anhydrite zone. The transitional gypsum zone, present at other domes, is poorly developed at Boling. The calcite cap rock is composed of light- to dark-gray, fine- to medium-grained calcite that has been brecciated and veined with crystalline calcite. Rhythmically banded calcite, common at other domes, is rare at Boling except near the base of the calcite cap rock. The anhydrite cap rock is dense, dark gray, and composed of interlocking euhedral anhydrite grains cemented by anhydrite. Uncemented anhydrite sands are locally present at the contact with the underlying salt. The top of the salt diapir is relatively flat (Fig. 3).

Relatively large amounts of detrital material are incorporated within the cap rock. Lenses of calcitic sandstone and shale, transported to the surface with the salt, are present in both the calcite and anhydrite cap rock, and locally hamper mining operations.

The main sulfur deposit is within the calcite/sulfur zone in the southeastern third of the dome (Fig. 4). The richest areas of sulfur mineralization roughly correspond to the thickest calcite cap rock, which ranges to over 122 m thick. Some wells in the thickest calcite intersected sulfur mineralization averaging 20% sulfur by volume for the entire 122 m interval.

Accessory minerals include barite, pyrite, hauerite, sphalerite, and galena. The barite occurs filling cavities and as massive bodies in the upper part of the calcite cap. The sulfide minerals are concentrated along the east flank of the dome, but are not present in quantities of economic interest.

Main Pass: The Main Pass deposit, located in Main Pass Block 299 in US federal waters about 27 km east of Venice, LA, was discovered in 1988 by a joint-venture effort of Freeport-McMoRan Inc., IMC Fertilizer Inc., and Felmont Oil Corporation (now Homestake Sulphur Co.). The deposit contains proven sulfur reserves of 68 Mt. The Main Pass deposit is situated in about 63 m of water (Ackerman, 1992, Anon., 1991a, Edwards et al., 1992).

Prior to exploration for sulfur, Chevron discovered oil and gas in strata flanking the salt dome. Freeport's sulfur drilling in 1988 intersected oil and gas reserves within the cap rock and overlying strata. Freeport purchased Chevron's hydrocarbon reserves and consolidated hydrocarbon and sulfur interests under the Main Pass Joint Venture. Both sulfur and hydrocarbons are produced. Recoverable oil and gas reserves are 39.4 million barrels and 0.2 Gm^3 respectively. The stacked sulfur/oil/gas deposits present challenging engineering problems for the development staff.

At the time of this writing, available geologic data for the deposit are few. The Main Pass 299 salt dome is roughly circular in plan (Fig. 5 and Fig. 6) with an approximate diameter of 3 000 m at the 900 m sub-sea contour (Ackerman, 1992, Edwards et al., 1992). The top of the dome is relatively flat about 400 m below sea level. Cap rock extends over about 680 hectares and consists of calcite and anhydrite. Total cap rock thickness varies from about 40 to 160 m. Sulfur-bearing cap rocks up to 70 m thick were intercepted in the exploration program.

The mine complex will consist of 15 offshore sulfur mining platforms and related facilities. Production began in 1992, and the mine will reach full production in 1993. Production is expected to reach 2 Mtpy over a mine life of 30 years.

Stratiform Deposits

Stratiform deposits of native sulfur associated with bedded gypsum and anhydrite are relatively common in oil-producing evaporite basins throughout the world, but large deposits suitable for mining are few in number. Significant stratiform sulfur deposits have been or are currently mined in Sicily, Poland, Iraq, the former USSR, and West Texas. Similar occurrences are known in many countries including Angola, Argentina, Canada, China, Egypt, England, Greece, Iran, Israel, northern Mexico, Romania, Spain, Syria, Turkey, and numerous others. In addition, recent exploration efforts have delineated potentially valuable stratiform deposits on the Mediterranean coast of Egypt (Anon., 1988a). Several deposits that are currently important are described herein.

The origin of the stratiform deposits is very similar to that of the salt dome-hosted deposits described above. Shallow gypsum or anhydrite beds are replaced by bioepigenetic calcite and sulfur through the actions of sulfate-reducing bacteria in the presence of oil and migrating meteoric water.

Some authors, including Dessau et al. (1962) for the deposits in Sicily, have proposed that the deposits are syngenetic, but this theory is no longer widely accepted. Known syngenetic deposits are limited to minor occurrences in recent sediments.

Many of the stratiform deposits are in arid regions where shallow evaporites are more likely to be preserved, and are com-

Table 3. Production of Frasch Sulfur in the United States and Mexico by Mine and Company, 1894 Through 1990 (revised from Barker, 1983)

Mine	Company	Years operated	Total output, 1 000 t
UNITED STATES:			
Sulphur Mine, LA	Union	1894–1924	9 563
Bryan Mound, TX	Freeport	1912–1935	5 081
Gulf (Big Hill), TX	Texasgulf	1919–1936	12 548
Hoskins Mound, TX	Freeport	1923–1955	11 069
Big Creek, TX	Union	1925–1926	2
Palangana, TX	Duval	1928–1935	241
Boling, TX	Union	1928–1929	8
Boling, TX	Texasgulf	1929–present	81 833
Long Point, TX	Texasgulf	1930–1938	408
Jefferson Island, TX	Jefferson Lake	1932–1936	438
Grande Ecaille, LA	Freeport	1933–1978	41 044
Boling, TX	Duval	1935–1940	580
Boling, TX	Baker-Williams	1935–1935	1
Clemens, TX	Jefferson Lake	1937–1960	3 024
Orchard, TX	Duval	1938–1970	5 583
Long Point, TX	Jefferson Lake	1946–1982	8 710
Moss Bluff, TX	Texasgulf	1948–1982	9 456
Starks, LA	Jefferson Lake	1951–1960	853
Spindletop, TX	Texasgulf	1952–1976	9 994
Bay Ste. Elaine, LA	Freeport	1952–1959	1 149
Damon Mound, TX	Standard Sulphur	1953–1957	142
Garden Island Bay, LA	Freeport	1953–1991	28 065
Nash, TX	Freeport	1954–1956	153
Chacahoula, LA	Freeport	1955–1962	1 218
Fannett, TX	Texasgulf	1958–1977	3 533
High Island, TX	U.S. Sulfur	1960–1962	38
Grand Isle, offshore LA	Freeport	1960–1991	26 813
Lake Pelto, LA	Freeport	1960–1975	5 712
Gulf (Big Hill), TX	Texasgulf	1965–1970	510
Sulphur Mine, LA	Allied Chemical	1966–1970	69
Nash, TX	Phelan Sulphur	1966–1969	174
Heiner, W. Texas	Duval	1967–1970	588
Bryan Mound, TX	Hooker Chemical	1967–1968	5
Chacahoula, LA	John W. Mecom	1967–1970	156
Ft. Stockton, W. Texas	Farmland/Atl. Rich.	1968–1985	1 864
Lake Hermitage, LA	Jefferson Lake	1968–1972	448
Caminada, offshore LA	Freeport	1968–1969	330
Bully Camp, LA	Texasgulf	1968–1978	1 753
High Island, TX	Pan Am Petroleum	1968–1971	111
Culberson, W. Texas	Duval/Pennzoil	1969–present	38 620
Comanche Creek, W. Texas	Texasgulf	1975–1983	1 727
Phillips Ranch, W. Texas	Pennzoil	1980–1982	86
Caillou Island, LA	Freeport	1980–1984	803
Comanche Creek, W. Texas	Texasgulf	1988–1989	228
Caminada, offshore LA	Freeport	1988–present	1 273
Main Pass 299, LA	Freeport	1992?–	n.a.
MEXICO:			
San Cristobal	Azufre Golfo Mex.	1953–1957	165
Jaltipan	Azufrera Panamer.	1954–present	33 965
Salinas	Azufre Veracruz	1956–1969	3 931
Nopalapa	Explor. del Istmo	1957–1960	327
Nopalapa	Explor. del Istmo	1967–1971	354
Texistepec	Central Minero	1959–1961	9
Texistepec	Explor. del Istmo	1971–present	12 468
Coachapa (Salinas mine)	Azufrera Panamer.	1981–present	1 990
Petapa	Azufrera Panamer.	1984–?	?
Otapan	Azufrera Panamer.	1987–present	89

monly proximal to a major river or other recognized source of meteoric water. The deposits are usually at the edges of the evaporite basins, and within or adjacent to structures (such as drape folds, anticlines, and faults) that have focused hydrocarbon migration and/or surface water channeling and evaporite dissolution. As described here, these deposits are difficult exploration targets and are often found by accident during the drilling of oil wells.

West Texas: Stratiform sulfur deposits are widespread in the Permian basin of West Texas. Sulfur is currently produced from one Frasch mine, Pennzoil's Culberson mine, and was produced in the past from several smaller deposits. Exploration efforts are currently limited, but the Permian basin has been the site of intense exploration, most notably in the late 1960s. There are two known sulfur districts in West Texas, the Ft. Stockton district, and the Rustler

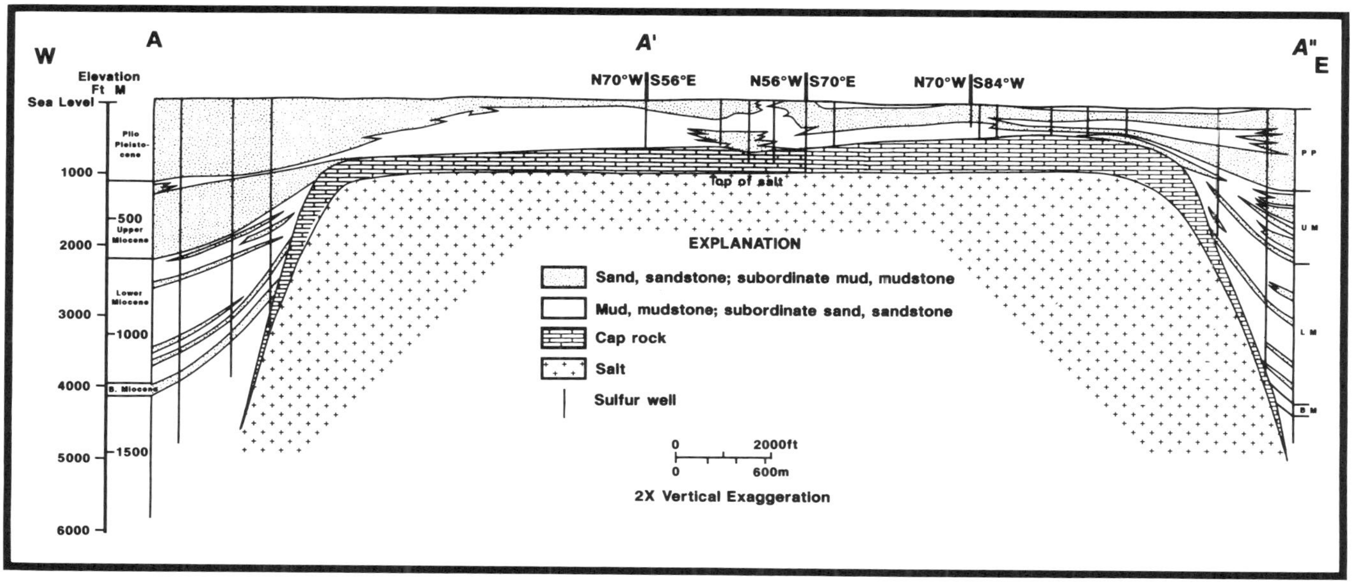

Fig. 3. West to east cross section of cap rock at Boling dome, Wharton Co., Texas (after Seni, 1986). Location of cross section is shown in Fig. 4.

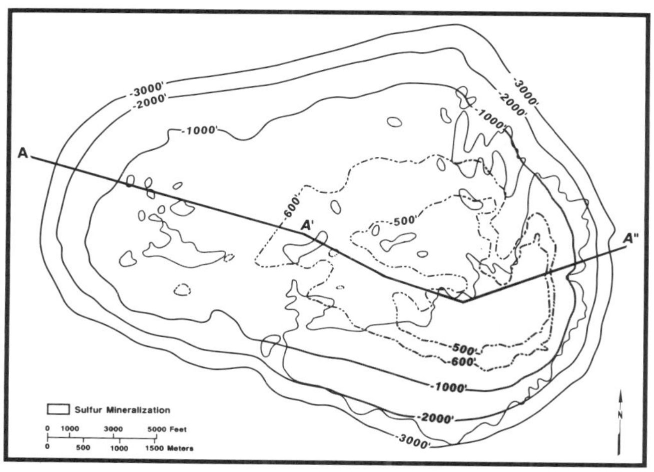

Fig. 4. Depth to top of cap rock (in feet) and location of sulfur mineralization at Boling dome, Wharton Co., Texas. Cross section A-A″ is shown in Fig. 3 (after Samuelson, 1992).

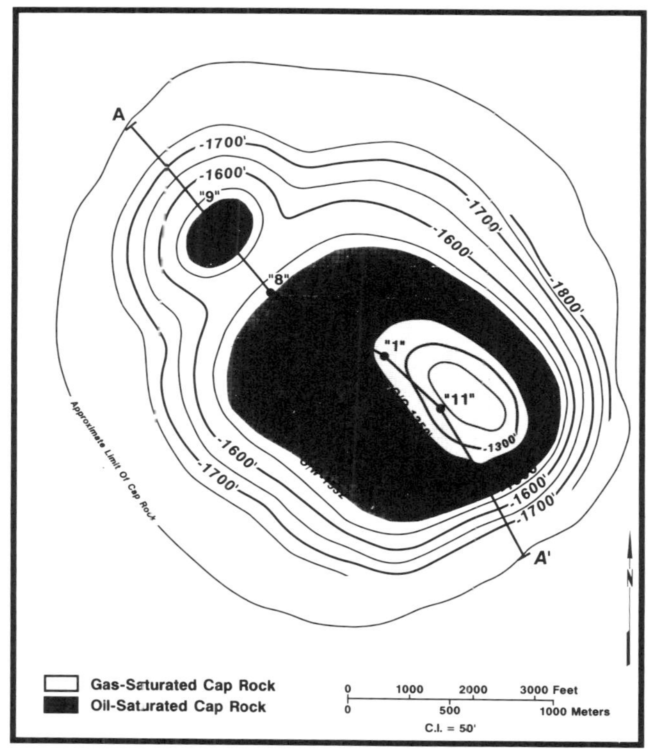

Fig. 5. Depth to top of calcite cap rock (in feet, datum sea level), Main Pass mine, offshore Louisiana, with location of gas and oil zones and cross section shown in Fig. 6, contour interval = 50 feet (after Edwards et al., 1992).

Springs district. The Ft. Stockton district is located on the Central Basin platform that separates the Midland and Delaware basins (Fig. 7). The Rustler Springs district and the Culberson mine are located in the Delaware basin portion of the Permian basin.

Sulfur in the Ft. Stockton district occurs with calcite and replaces anhydrite of the Permian Salado, Tansill, and Yates Formations. The host formations, as well as formations above and below, drape over the Pennsylvanian-Permian Central Basin platform and define a northwest-trending anticline locally termed the Heiner anticline (Fig. 8). Structural highs along the axis of the Heiner anticline served as foci for hydrocarbon entrapment and subsequent sites of sulfate reduction and sulfur mineralization (Ellison, Jr., 1971, Hentz et al., 1989, McNeal and Hemenway, 1972, Zimmerman and Thomas, 1969).

Native sulfur was first noted in the Ft. Stockton district in an oil well drilled in about 1900 (Adkins, 1927). Subsequent oil drilling noted other occurrences of sulfur, but it was not until the exploration surge of the mid-1960s that the Ft. Stockton district was discovered. The majority of the district is covered by Quaternary alluvium, and the sulfur mineralization is not exposed at the surface.

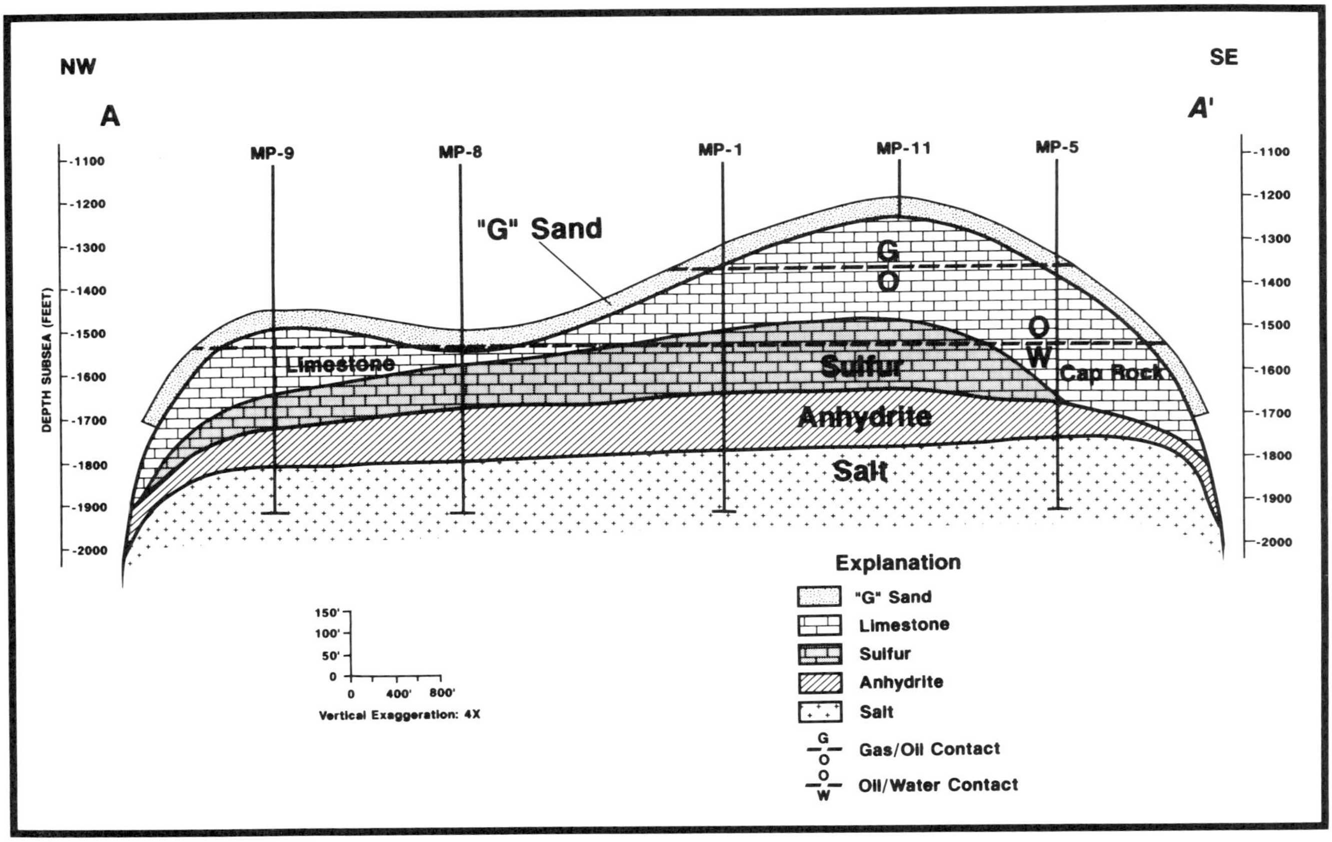

Fig. 6. Cross section A-A', Main Pass mine, offshore Louisiana, depths shown in feet from sea level (after Edwards et al., 1992).

Production from the Ft. Stockton district has been limited to three small Frasch mines: Duval's (Pennzoil's) Heiner mine, Sinclair's (Atlantic Richfield's) Ft. Stockton mine, and Texasgulf's Comanche Creek mine. All three mines are now closed, although the Comanche Creek mine operated briefly in 1988 and 1989. The Comanche Creek deposit is thought to have been the largest with 7.6 Mt of original in-place sulfur, but none of the deposits produced more than 2.0 Mt. The discontinuous mineralization in the Ft. Stockton district combined with high production costs make these deposits uneconomic at present.

The Rustler Springs district contains numerous small sulfur deposits, and one very large deposit that is currently producing Frasch sulfur, Pennzoil's Culberson mine. The Culberson mine has an operating capacity of 2.5 Mtpy, but currently produces at a rate of about 1.9 Mtpy (Crawford, 1990). About 40 Mt of sulfur have been produced from the mine, and reserves totaled 30 Mt at the end of 1990. Other production from the Rustler Springs district is limited to the Phillips Ranch deposit, from which 86 360 t were produced in 1980 to 1982. Currently, an 813 tpd Frasch plant is being constructed at the Phillips Ranch site by Addwest Minerals, but production has been postponed until economic conditions are more favorable (Guilinger and Nestlerode, 1992).

Mineralization at the Culberson mine is hosted by bedded anhydrite/gypsum in the Ochoan Castile, Salado, and Rustler Formations, but most of the ore is within the Salado Formation (Crawford, 1990, Hentz et al., 1989, Mussey and Tyree, 1985, Smith, 1980, Wallace and Crawford, 1992). The sulfur occurs with calcite and accessory barite and celestite in three general modes (Crawford, 1990): 1) as microcrystalline disseminations in gray diagenetic (replacement) limestone or cave-fill breccias, 2) as orthorhombic crystals lining cavernous voids in the replacement limestone, and 3) as coarsely crystalline crusts or agglomerates that nearly fill all voids in the limestone. The diagenetic limestone (calcite) is a product of biogenic replacement of gypsum/anhydrite solution-collapse breccias, and original breccia textures are commonly preserved. Similarly, calcite and sulfur locally replace and preserve varves in the Castile evaporite. The ores commonly contain up to 30 wt% sulfur, and average about 25% porosity.

The Culberson ore body is elliptical in shape and extends from 100 to 720 m below the surface (Fig. 9). The ores are intimately associated with dissolution collapse and cave-fill breccias (Fig. 10) that preferentially occur along selected horizons in the Salado Formation (Wallace and Crawford, 1992, Miller, 1992). The collapse and cave breccias presumably channeled both hydrocarbons and oxygenated meteoric water, and led to the reduction and replacement of evaporite-bearing breccia clasts and adjacent evaporitic strata. The dissolution collapse and cave breccias appear to be localized along northeast-trending Tertiary normal faults and joints (Fig. 11). Rarely visible at the surface, such fractures can be inferred from the linear distribution of collapse sinks, minor offsets indicated by oil well data, and sulfur-related alteration features. The presence of a residual silty clay cap over the Culberson deposit appears to have been critical to the preservation of hydrogen sulfide and to the generation and preservation of native sulfur (Fig. 10).

The presence of sulfur in the Rustler Springs area has been known since 1854. Numerous surface occurrences of sulfur and erosional remnants of bioepigenetic calcite (castiles, Kirkland and Evans, 1976) are known, and there were several attempts beginning in the early 1900s to mine surface ores (Evans, 1946, Porch, Jr.,

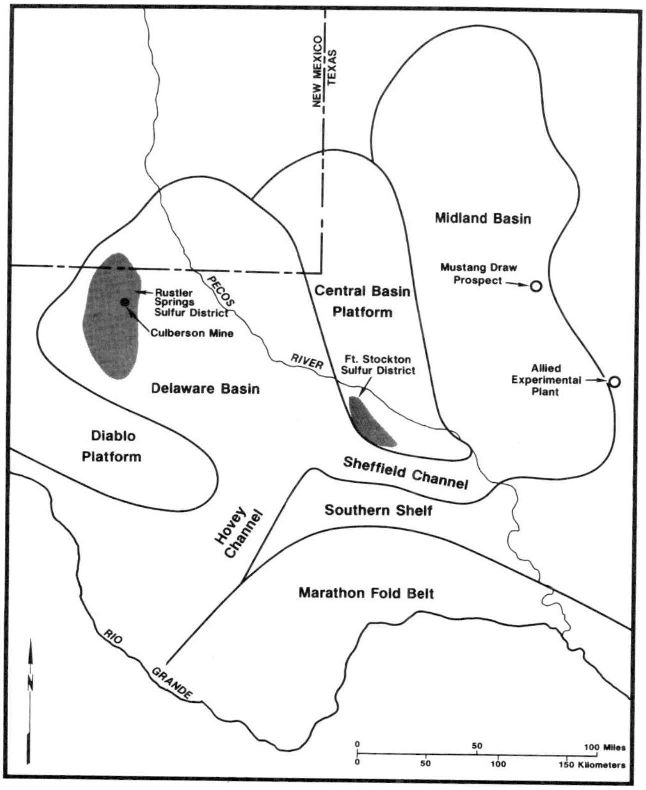

Fig. 7. Location of sulfur districts and deposits in Permian Basin, Texas, with respect to structural and stratigraphic features of Pennsylvanian and Permian age.

1917). The Culberson mine was discovered in 1967 by Duval Corporation near the site of some old surface works.

Poland: Native sulfur deposits in Poland occur in Upper Miocene strata on the northern (platform) border of the Carpathian foredeep (Fig. 12). Sulfur occurs in bioepigenetic calcite replacements of a gypsum unit of middle Badenian age, and is accompanied by barite and celestite. The sulfur deposits are located on uplifted blocks near the northern margin of the Carpathian foredeep at depths from a few to about 360 m (Anon., 1988b, Anon., 1990a, Nieć, 1992, Pawłowski et al., 1979, Preisner, 1991).

The sulfur-bearing rocks vary in thickness from a few meters to over 40 m, and they are subhorizontal. Where mined today, the sulfur deposits are not exposed at the surface. The host strata in nearly the entire region are covered by Quaternary deposits of variable thickness.

The sulfur content of the ore is quite variable, and ranges from a few percent up to more than 50% in selected samples. Rocks containing less than 5% sulfur are considered barren. The average sulfur content of individual deposits varies from 22 to 33%. Ore textures commonly mimic textures in the original gypsum.

Small surface deposits of native sulfur have been exploited on a limited scale for hundreds of years. The earliest documented sulfur mining in Poland occurred in 1415, but it was not until 1953 that exploration efforts resulted in a discovery of sulfur-bearing limestone in the subsurface near Tarnobrzeg, now the center of sulfur mining.

There are two main sulfur-producing areas in Poland: the Staszów-Tarnobrzeg region containing four deposits, and the Lubaczów area where one deposit is dissected by the Polish-Ukrainian border. The regional trend of sulfur mineralization follows the northern margin of the Carpathian foredeep into Ukraine, where sulfur is mined at several deposits using both Frasch and open pit methods. Westward from Staszów-Tarnobrzeg, only small sulfur deposits have been found, which were mined prior to 1921.

The known Polish sulfur deposits together contain about 600 Mt of demonstrated sulfur reserves and about 500 Mt of inferred and hypothetical resources (Nieć, 1992). Opencast mining was begun in 1956 at the Piaseczno mine, and Frasch mining modified to fit the unconfined ores of the Polish deposits (the thermo-hydrodynamic or hydrodynamic method) has been successfully applied since 1966 (Anon., 1986a). Currently, four mines are active (one opencast and three hydrodynamic) and one is under construction. Since the first production in 1956, the total annual sulfur production has increased to over 5 Mtpy, and it now averages between 4 and 5 Mtpy. The ultimate recovery from the sulfur deposits varies between 40 and 70%.

Iraq: Sulfur deposits in Iraq are found in the folded portion of the Mesopotamian depression that separates the nappes and the Zagros Mountains to the east from the unfolded Arabian platform to the west (Barker et al., 1979). Within this region is the largest stratiform sulfur deposit in the world, the Mishraq deposit. Located about 40 km southeast of Mosul and 315 km north of Baghdad, the Mishraq deposit consists of at least three ore bodies thought to contain at least 100 Mt of native sulfur. Sketchy reports of recent discoveries suggest that recoverable reserves may total as much as 500 to 600 Mt (Babbitt, 1991).

Native sulfur and bioepigenetic calcite at Mishraq are present in three main zones in the basal Lower Fars Formation, a sequence of middle Miocene interbedded bituminous carbonates and anhydrite/gypsum with minor shale and sandstone (Figs. 13 and 14). Sulfur mineralization is thickest in areas where intense folding and faulting enhanced solution, fluid migration, and mixing of hydrocarbons and meteoric waters. The combined thickness of the three sulfur-bearing zones varies from 2 to 123.9 m, with a maximum ore section of 107.9 m and an average grade of 23.14 wt% sulfur. Seven ore types have been differentiated; the predominate ore type is coarsely crystalline sulfur interbedded with bands of secondary limestone.

Sulfur mineralization and hydrogen sulfide springs were first noted in the Mishraq area during oil exploration in the 1930s. Russian geologists conducted sulfur exploration in the 1960s, and they discovered an ore body while drilling several prospective structures.

The Mishraq mine, currently operated by the Mishraq State Sulphur Enterprise (MSSE), has been producing since 1972 using the Polish hydrodynamic (modified Frasch) method. The plant was built with a nominal capacity of one million tons per year. Numerous problems have prevented reaching design capacity, and as a result only 11 Mt were produced between 1972 and 1991.

Production wells at Mishraq are located on a bluff several hundred feet above the Tigris River. In the recent past, excessive quantities of mining fluids were being injected into the ore and large quantities of fluid and molten sulfur were observed to be pouring out of the bluff and into the Tigris River, causing severe contamination (Babbitt, 1991). Fluids were also leaking into adjacent barren structures as far as 3 km away. This is indicative of excessive costs due to fluid loss and overheating of the host rock.

The inclusion of carbon, hydrocarbons, and ash in Mishraq sulfur has long been a problem in marketing Iraqi Frasch sulfur. Early filtering systems were inadequate and routinely resulted in losses of up to 50%. Recent efforts by Freeport under contract to MSSE to construct a filtering plant were halted by Iraq's invasion of Kuwait. The resulting war with the allied powers led to the destruction of some recovered sulfur plants, but there were no reports of significant damage to the Mishraq facility.

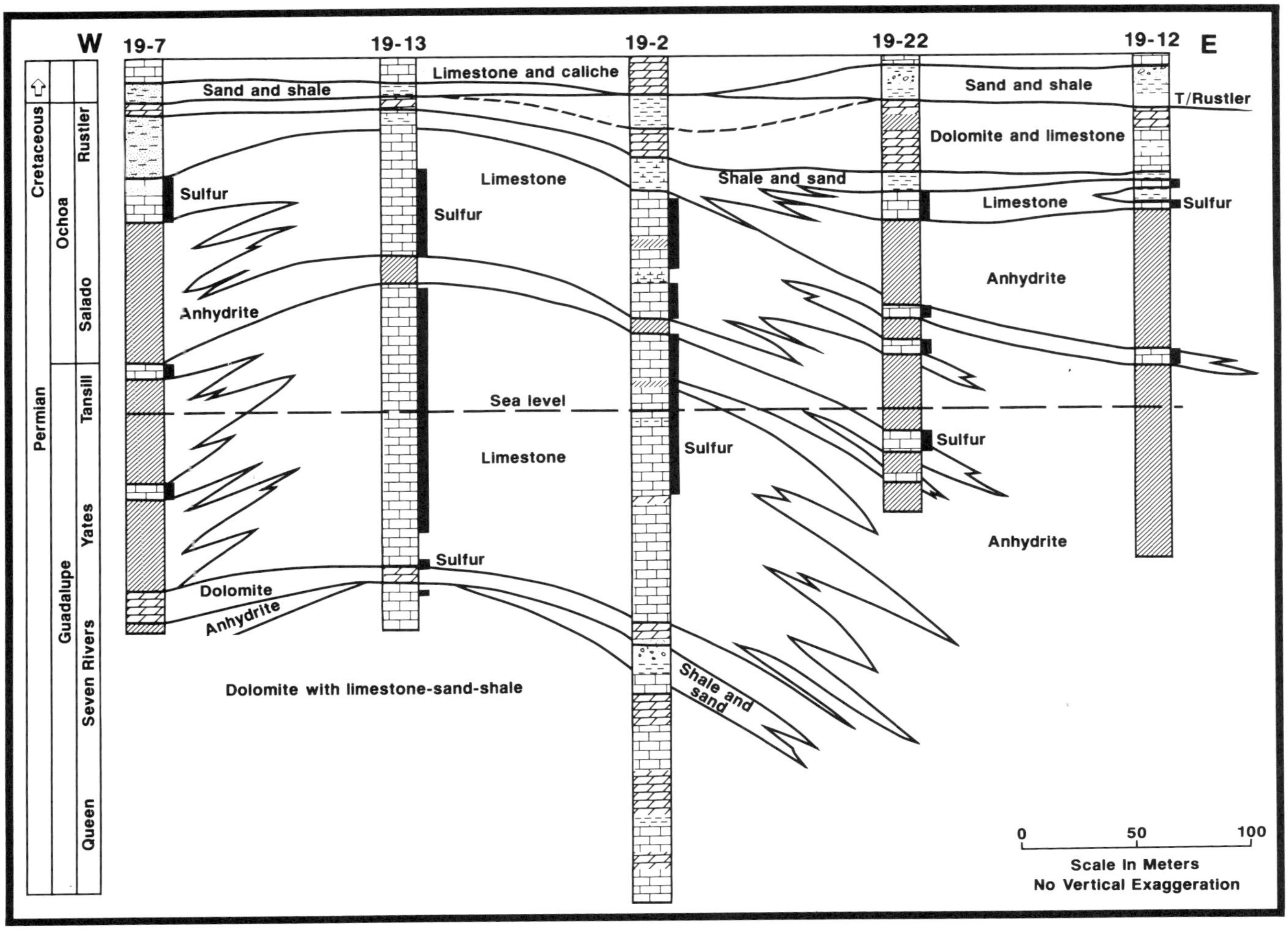

Fig. 8. Cross section showing stratigraphic distribution of sulfur, Ft. Stockton mine, Pecos Co., Texas (after Zimmerman and Thomas, 1969).

During Iraq's war with Iran in the 1980s, production at Mishraq varied widely. Total Iraqi sulfur production (of which about 80% comes from Mishraq) in 1981 was 145 kt; in 1987, it was 850 kt (Anon., 1989a). Prior to the invasion of Kuwait, Iraq was producing about 1.8 Mtpy, with Mishraq producing at a rate of about 1.4 Mtpy. MSSE had previously announced that production at Mishraq would expand to 2.0 Mtpy by 1991, but it is probable that continuing political and economic instability will adversely affect production for some time. When the country achieves stability, a large increase in sulfur production is likely.

Oxidative Deposits: Ruckmick et al. (1979) also include a category of deposits that they term oxidative. In the strictest sense, almost all native sulfur deposits are oxidative in that they are formed from the oxidation of hydrogen sulfide, but this category properly includes only those deposits that are not associated with volcanic or biogenic sources of sulfur. They probably are formed as a result of the oxidation of hydrogen sulfide from a remote source, resulting in native sulfur deposition in existing voids in porous host rock. The hydrogen sulfide may originate in high temperature diagenetic settings, such as those associated with thermal maturation of crude oil and thermochemical sulfate reduction (Machel, 1992), but the source of the hydrogen sulfide is commonly difficult to identify. Oxidative deposits are widespread, but they are generally small and are not currently utilized as a source of sulfur.

Oxidative deposits of native sulfur have recently been described from Lechuguilla Cave, located in the Capitan reef at the northern margin of the Delaware basin, West Texas and New Mexico. At Lechuguilla Cave, hydrogen sulfide dissolved in basinal brines was oxidized to native sulfur, and is intimately associated with speleogenesis there and at Carlsbad Caverns (Spirakis and Cunningham, 1992).

Numerous occurrences of native sulfur in porous carbonate strata of the San Andres Formation, Permian basin, West Texas, are also probably oxidative deposits of this type. One such deposit, the Mustang Draw occurrence in Glasscock Co., Texas, was a site of exploration activity from 1987 to 1990. A resource of 7.4 Mt of sulfur was identified in rock that locally contains 6.5% sulfur by volume (Ruckmick et al., 1992). A similar occurrence in the Permian Clearfork Formation south of Mustang Draw was the site of an unsuccessful pilot Frasch test in 1968 (Ellison, Jr., 1971, Zimmerman and Thomas, 1969).

Volcanic Deposits

Sulfur of volcanic origin is probably the most widely distributed of native sulfur deposits. Deposits of this type are known in nearly all volcanic regions of the world, particularly in the Pacific Rim

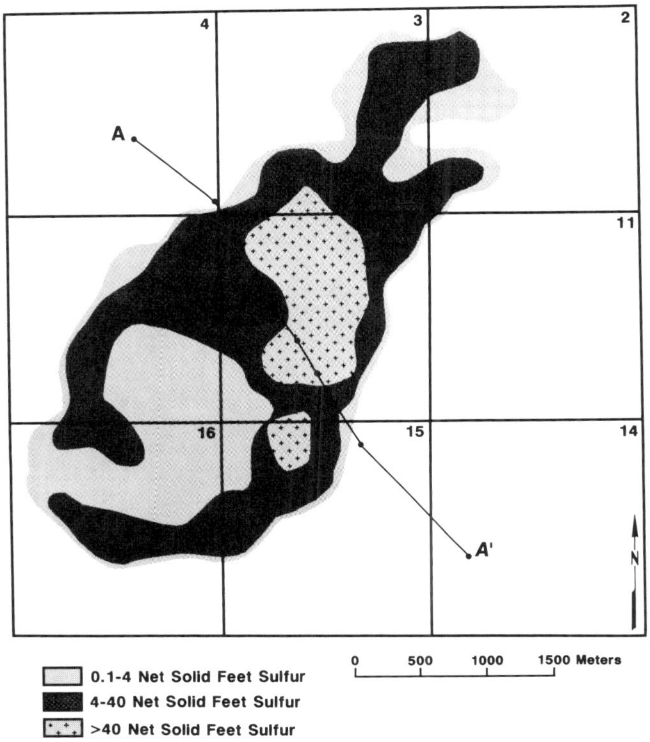

Fig. 9. Total net solid feet of sulfur, Culberson mine, Culberson Co., Texas, and location of cross section shown in Fig. 10 (after Wallace and Crawford, 1992).

region. Volcanic deposits are present in western North and South America, Central America, the Kamchatka Peninsula, the Kurile Islands, Japan, the Philippines, Indonesia, Taiwan, and New Zealand, as well as in volcanic regions in western Turkey, Greece, Italy, and elsewhere.

Volcanic sulfur deposits can be subdivided into four types by mode of emplacement (Barker, 1983, Nieć, 1986a): 1) impregnation and replacement deposits, usually formed by the actions of hot acid springs that are rich in sulfurous gases, often in tuffs and volcanic breccias, with sulfur in irregular or lens-shaped ore bodies aligned along fracture zones; accompanied by sulfides and alteration (alunitization, argillization, silicification) that locally completely replaces the host rock; economically the most important of the volcanic deposits; 2) sublimation deposits, formed by gaseous or fumarolic activity near volcanic craters, where sulfur coats fractures and cavities near the surface in small deposits that are locally mined; 3) sedimentary deposits, formed by precipitation of sulfur in hot crater lakes; and 4) sulfur flows, small flow deposits that probably originated from older deposits that were remelted, of local importance only. Volcanic sulfur deposits are usually found in tuffs, lava flows, and other volcanic rocks, but are also found in sedimentary, intrusive and other rock types in volcanic regions.

Volcanic deposits have been mined in the past in many areas, but significant production of volcanic sulfur has been limited to Japan, Turkey, Mexico, Italy, and the Andean countries of South America. Mining methods used include standard underground and surface mining methods as described below.

Mined volcanic sulfur dominated Japanese production until the mid-1960s, when recovered sulfur challenged its importance (Anon., 1985a). The mining of volcanic ores in Japan ceased completely in the mid-1970s. Sulfur deposits in a highly altered rhyolite dike near Keciborlu in southwest Turkey have in the past supplied much of Turkey's domestic demand. In the United States, several deposits have been mined to supply local needs, including the Leviathan Deposit in Alpine County, California, which was mined from 1953 to 1962 to supply sulfuric acid for Anaconda's Yerington, NV, copper operation. In recent years, numerous projects have been initiated to recover sulfur from volcanic ores, including sites in South America (Anon., 1987a), Romania (Fodor et al., 1989), the Philippines (Anon., 1986b), Utah (Slim, 1986), and California (Ward, 1992).

In comparison to bioepigenetic deposits, volcanic sulfur deposits are generally small. Few contain more than 10 Mt of reserves, but some (such as the Matsou Deposit in Japan) are significantly larger. Volcanic deposits are also at a disadvantage because they are sometimes located far from markets and economical transportation. In addition, volcanic sulfur is commonly impure and contains elements (such as arsenic, selenium, and tellurium) that make it unsuitable for some applications. Some ores cannot be economically processed using existing technology. Commercial interest in volcanic deposits is high only when premium or protected markets exist.

EXPLORATION TECHNIQUES

Exploration for native sulfur deposits is a difficult task, especially for Frasch-minable deposits. There is often little indication at the surface of subsurface sulfur deposits, and surface exposures of sulfur are rare because of the instability of elemental sulfur (it is readily oxidized to soluble sulfate). Indirect indicators may be all the explorationist has to suggest the presence of sulfur, and subsequent drilling may be necessary to fully test such blind targets.

For bioepigenetic deposits, the genetic relationship between hydrocarbons and sulfur dictates that exploration be directed to appropriate host rocks in areas that contain oil or gas. For this reason, sulfur exploration commonly makes extensive use of oil and gas well data, so much so that sulfur exploration is sometimes almost indistinguishable from efforts to locate shallow oil and gas reserves. The sulfur explorationist is essentially looking for a potential oil trap adjacent to or within evaporitic strata that has been biodegraded and is now replaced by calcite and sulfur. Exploration drilling for sulfur sometimes intersects oil and gas accumulations, just as oil wells commonly intersect sulfur shows.

Barker et al. (1979) and Nieć (1986a) summarized the general conditions necessary for the generation of stratiform sulfur deposits: 1) thick gypsum or anhydrite beds, generally in the border regions of evaporite basins, 2) proximity to hydrocarbon accumulations in correlative or underlying strata, 3) hydrodynamic communication between sulfates and hydrocarbons, often along joints, faults, solution-derived porosity, or other structural or stratigraphic pathways, 4) a stratigraphic or structural trap (anticlines, monoclines, horsts, etc.) to contain the precursors of the sulfur-forming reaction, 5) serial, coincident, or coexisting reducing and oxidizing environments where hydrocarbons are biodegraded, and sulfate is reduced and then oxidized to sulfur, and 6) an overlying barrier to fluid migration capable of containing reaction products and preventing later oxidation of sulfur.

Tools useful in locating stratiform deposits are few and locally unreliable. Standard field and photogeologic techniques are commonly used to identify prospective structures and surface alteration. Detailed mapping such as that described by Wessel (1992) is valuable where surface exposures are adequate. Surface occurrences of sulfur or bioepigenetic calcite (verified by isotope studies) are most encouraging, as are occurrences of sulfur and bioepigenetic calcite in oil and gas well cuttings. Anomalous occurrences

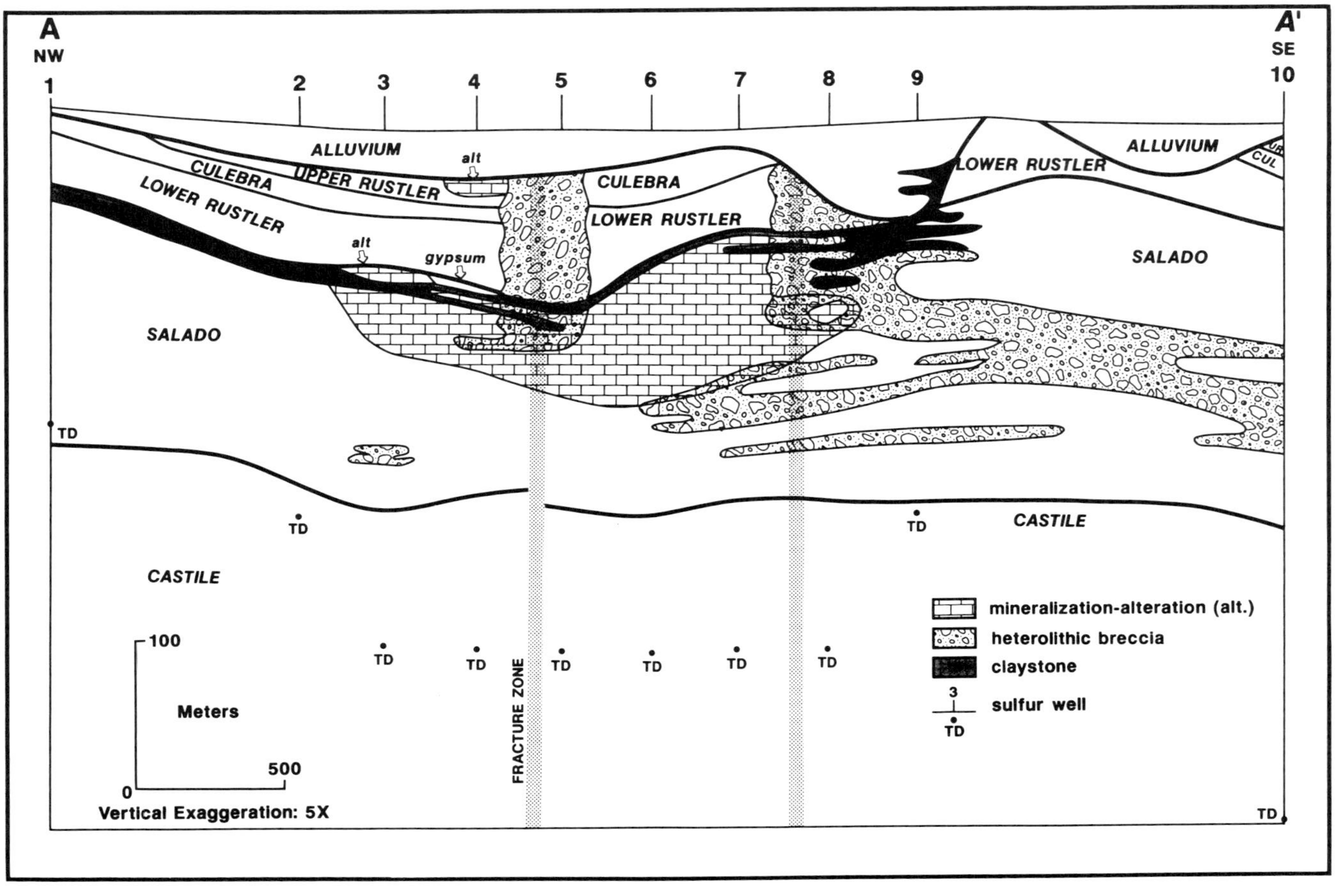

Fig. 10. Cross section A-A′, Culberson mine, Culberson Co., Texas. Location of cross section shown in Fig. 9 (after Wallace and Crawford, 1992).

of silica and iron staining or bleaching are sometimes considered evidence of leakage of reactants from below. A number of surface geophysical surveys (including gamma ray spectrometer, spontaneous potential, resistivity, induced polarization, and ground and airborne magnetic) are utilized to locate hydrogen sulfide plumes and alteration haloes (Salisbury, 1992). Thermal infrared, remote sensing, and seismic data have been locally useful, and some surface geochemical techniques (soil gas, airborne gas spectrometry, etc.) have been tested.

Gravity surveys are routinely used to locate stratiform sulfur deposits because of the void development and brecciation associated with sulfur mineralization. Unaltered anhydrite has a density of about 2.96, and sulfur ore bodies vary in density from 2.00 to 2.40, yielding a potentially measurable negative gravity anomaly. The amount of variation depends upon the nature of the sulfur mineralization, void development, depth of mineralization, and numerous unknown stratigraphic and structural variables. Gravity data are best used in combination with other data, but interpretation is always difficult (Salisbury, 1992, Schmoker, 1979). Actual use has resulted in few exploration successes.

Downhole logs of oil and gas wells are commonly used to establish formation control and identify low density zones that may be indicative of sulfur mineralization. The use of these logs is limited where well density is low or shallow strata are not logged.

Salt domes present perhaps the most straightforward exploration problem. The locations of most salt domes are known in areas such as the Gulf Coast of the United States and Mexico as a result of past oil exploration efforts. In the United States, the onshore domes have been intensely explored, and little potential remains for exploration success. However, many offshore domes remain untested for sulfur, and the likelihood of another discovery like Main Pass remains high. Given that the location of a specific salt dome is known, the prudent sulfur explorationist will evaluate its potential based upon the depth and thickness of cap rock, the presence (if known) of calcite cap rock, and the presence (if any) of shows of mineralization or alteration in oil wells. Gravity and seismic data may be used to define the shape of the cap rock. Gravity data may also indicate potential ore bodies in the cap rock, but as described above the interpretation of gravity data is often difficult. Oil well shows of sulfur and bioepigenetic calcite are the most commonly used guides to mineralization. Hydrogen sulfide springs may be present at the surface, but they are not reliable indicators of mineralization. Other surface indications and techniques as described above are often not applicable, particularly in the Gulf Coast.

Exploration for volcanic sulfur follows the more traditional style associated with exploration for hydrothermal deposits. Similar to bioepigenetic deposits, some volcanic deposits are discovered by accident during metals exploration (Ward, 1992). Exploration for other sources of sulfur, such as pyrites and sour gas, is uncommon because they generally supply sulfur only as a byproduct.

For many sulfur deposits, the most successful exploration tool is the drilling rig. Nevertheless, great care must be taken to correctly describe and interpret cuttings, core, and other drilling data. There is no substitute for an observant and skilled field geologist.

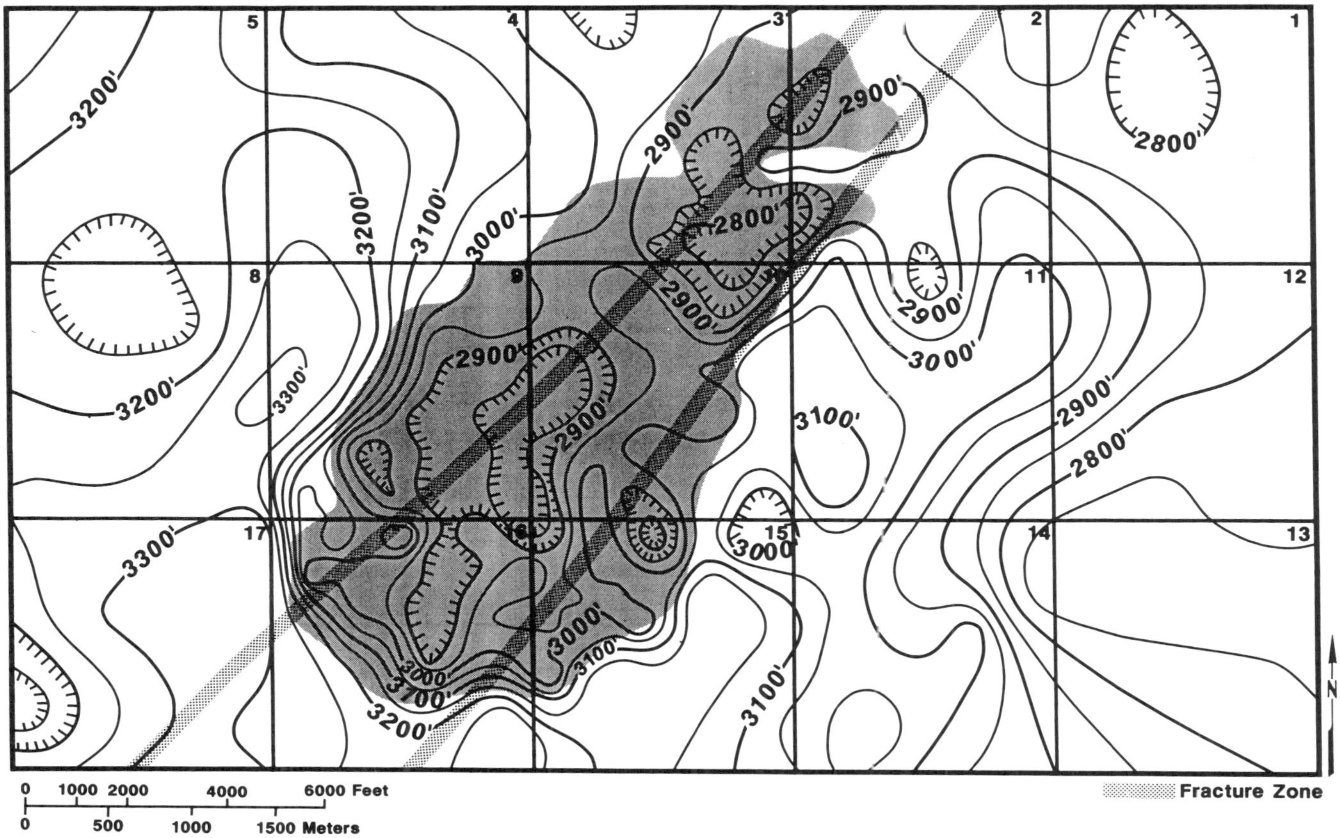

Fig. 11. Elevation of the top of the Salado Formation (in feet), Culberson mine, Culberson Co., Texas. Shading shows area of orebody (after Wallace and Crawford, 1992).

MINING AND PROCESSING OF NATIVE SULFUR ORES

In sulfur mining, three different mining techniques are applied: conventional underground methods, conventional open pit methods, and the Frasch mining (or underground melting) method. The last method is specific for sulfur owing to the low temperature at which sulfur melts. About 90% of all sulfur mined is obtained through Frasch mining; most of the remaining is recovered using surface mining techniques, and very little is obtained by underground mining methods (Nieć, 1986b).

For surface and near-surface sulfur ores, surface mining methods predominate, but conventional underground methods are also used, particularly in steeply dipping volcanic deposits. Techniques utilized include open pit, tunneling, room and pillar, cut-and-fill systems, and various forms of stoping. These methods are generally confined to shallow high grade ores with no overlying impermeable barrier that might allow utilization of the cheaper Frasch mining technique. Underground mining techniques are seldom used because of relatively high costs, and because of technical problems related to the common abundance of subsurface water and hydrogen sulfide. Ore recovered from surface and underground mines is processed to separate the sulfur from the host rocks and associated contaminants by various combinations of melting, distillation, agglomeration, flotation, and solvent extraction (Dale, 1981; Fodor et al., 1989).

The Frasch mining process was devised by Herman Frasch, the Research and Development Director for Standard Oil Co. and the founder of the Union Sulphur Co. First used in 1894 at Sulphur Mine dome, Louisiana, the Frasch technique is now used to extract native sulfur from salt domes in the United States and Mexico, and (with local modifications) from stratiform deposits in West Texas, Poland, the former USSR, and Iraq.

The Frasch process capitalizes upon several characteristics of sulfur: the low melting point (about 115°C), insolubility and immiscibility in water, and specific gravity of 1.8. Basically, the process involves heating water to a temperature of about 165°C and pumping the heated water through the deposit. The heat from the water melts the sulfur, which sinks to accumulate in a pool below the hot water, and is air-lifted out through nested pipes in the same hole used to inject the water.

Requisites for the economical production of sulfur using the Frasch process are: 1) a large sulfur deposit (preferably in excess of several million tons) hosted in a porous and permeable formation, generally cavernous limestone breccia, 2) impermeable or partly permeable stratigraphic seals bounding the ore body so that it is hydrologically isolated, 3) a large, dependable supply of water, and 4) an inexpensive fuel supply. An attractive deposit would contain more than 10 Mt of recoverable sulfur, average 20% or more of sulfur over an interval of at least 30 m, lie between 65 and 800 m deep, have a uniform porosity of 10 to 15%, and be overlain by an impervious bed (Donner and Wornat, 1973). Deposits less attractive than this may still be very profitable, but deposits exhibiting poor characteristics may be mined only in special cases (Nieć, 1986b).

A typical Frasch well is shown in Fig. 15. A single well consists of four concentric pipes varying in diameter. The outer pipe acts as

Fig. 12. Sulfur and hydrocarbon deposits in the Carpathian foredeep, Poland and Ukraine (after Nieć, 1992).

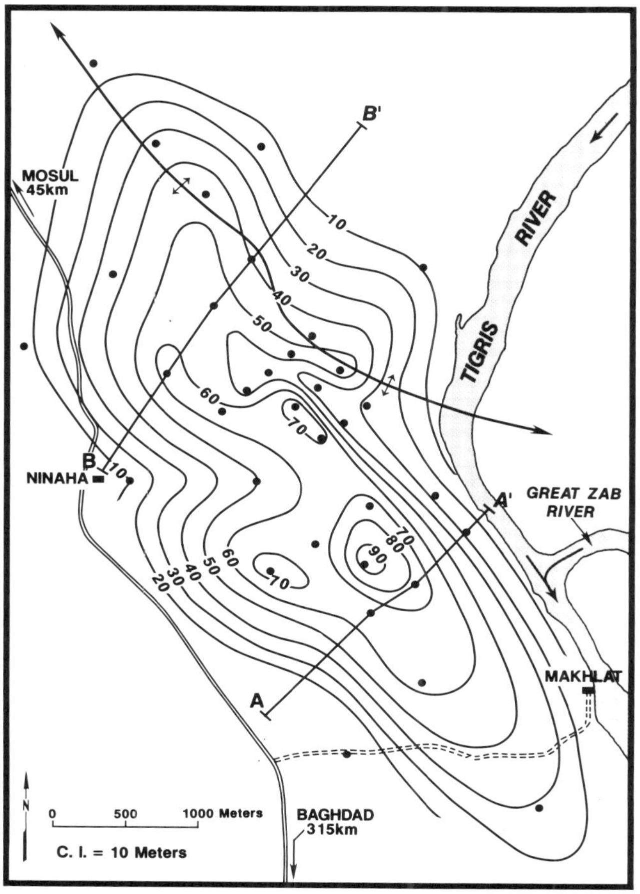

Fig. 13. Map of total thickness of sulfur-bearing rock, Mishraq deposit, northern Iraq (modified from Barker et al., 1979), showing locations of cross sections illustrated in Fig. 14.

a protective casing and extends to the top of the barren cap rock. The next pipe extends to the bottom of the ore body and carries the flow of hot water, which is forced into the sulfur-bearing formation through perforations at the bottom of the pipe. As sulfur melts, it percolates down through the porous host rock to form a pool of liquid sulfur at the bottom of the well. Of the two inner pipes, the smaller one in the center carries compressed air to foam and lift the sulfur through the larger pipe to the surface, where the air is removed in a separator prior to shipment via heated pipeline, barge, or rail car to storage, or to a sulfur-forming device to be solidified in various forms. The volume of the compressed air, and hence the rate of production, is closely monitored so that the production rate equals the sulfur melting rate in order not to deplete the sulfur pool and cause the well to produce water. If the pool is depleted, production is stopped and additional water is injected until the sulfur pool is reestablished (Anon., 1990b, Donner and Wornat, 1973). Bleed wells that extract cooled water from the deposit are generally located at the edges of the mining area. Individual sulfur wells usually recover sulfur from an area less than 2 000 m^2, and may be operated only 3 to 4 months before exhaustion. In a large operation, there is a continual program of drilling to replace wells that become exhausted. On average, about 70% of the sulfur is ultimately recovered.

The philosophy of Frasch mining is to keep the temperature inside the deposit as close as possible to 150°C, at which the viscosity of sulfur is at its lowest. At temperatures over 160°C, the viscosity dramatically increases, and sulfur may become unrecoverable. Overheating of the host rock and excessive fluid loss have caused severe problems at the Mishraq, Iraq, deposit.

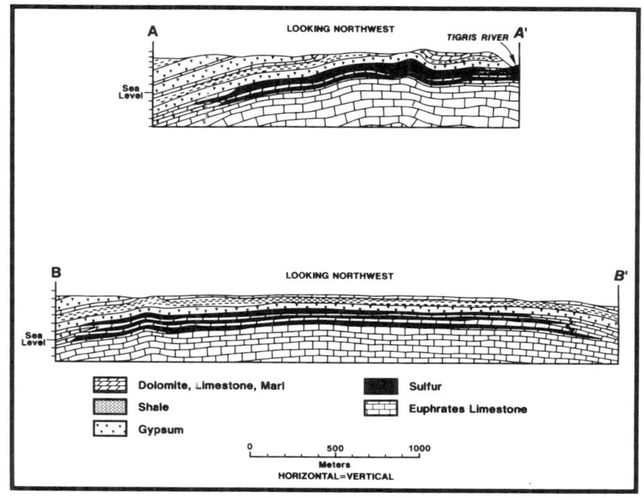

Fig. 14. Cross sections A-A' and B-B', Mishraq deposit, northern Iraq (after Barker et al., 1979). Locations of cross sections shown on Fig. 13.

The costs and availability of water and fuel are often critical to the economic operation of a Frasch mine. A large quantity of water is injected to mine the sulfur, up to 57 m^3 or more per ton of sulfur mined, and a dependable supply of good quality water is very desirable. Fresh water is generally preferred, but sea water has been used successfully at several sites in the Gulf of Mexico (Barker, 1983), and is currently being used at the Main Pass mine (Ackerman, 1992, Edwards et al., 1992).

In areas like West Texas where water is scarce, excessive fluid loss cannot be tolerated. Water from the bleed wells, a mixture of spent injection water and connate water that is contaminated with corrosive and scale-forming solutes, is locally treated, reheated, and reinjected. This also eliminates environmental problems associated with the disposal of contaminated bleed water.

Fuel cost increases during the energy crises of the 1970s also led to increased reinjection of spent waters and the development of several new recycling methods (Anon., 1990b) in order to minimize heat losses. One system developed by Freeport recycles mine water underground by mixing it in place with "new" superheated water. This system was stated to reduce the energy consumption of a well producing over 2 tph to a third of the original amount. Rising energy costs have also been circumvented through fuel substitution and cogeneration, as at Texasgulf's Boling dome mine.

In stratiform deposits where the sulfur-bearing rocks are not hydrologically isolated from adjacent strata, fluid loss through underlying or overlying strata can prevent economic sulfur recovery. Such a situation exists in Poland, where in contrast to the salt domes of the Gulf Coast that are at moderate depth and spatially isolated, the sulfur deposits are at shallow depths, are only 10 to 15 m thick, and overlie permeable strata. They are also permeated by meteoric water For these reasons, the hydrodynamic method (a variation of the Frasch process) was developed and has been utilized since 1966. The hydrodynamic method is also used in Iraq, and a similar technique is used to exploit the Culberson deposit in West Texas (Anon., 1990b).

The hydrodynamic method uses the same concentric well pip-

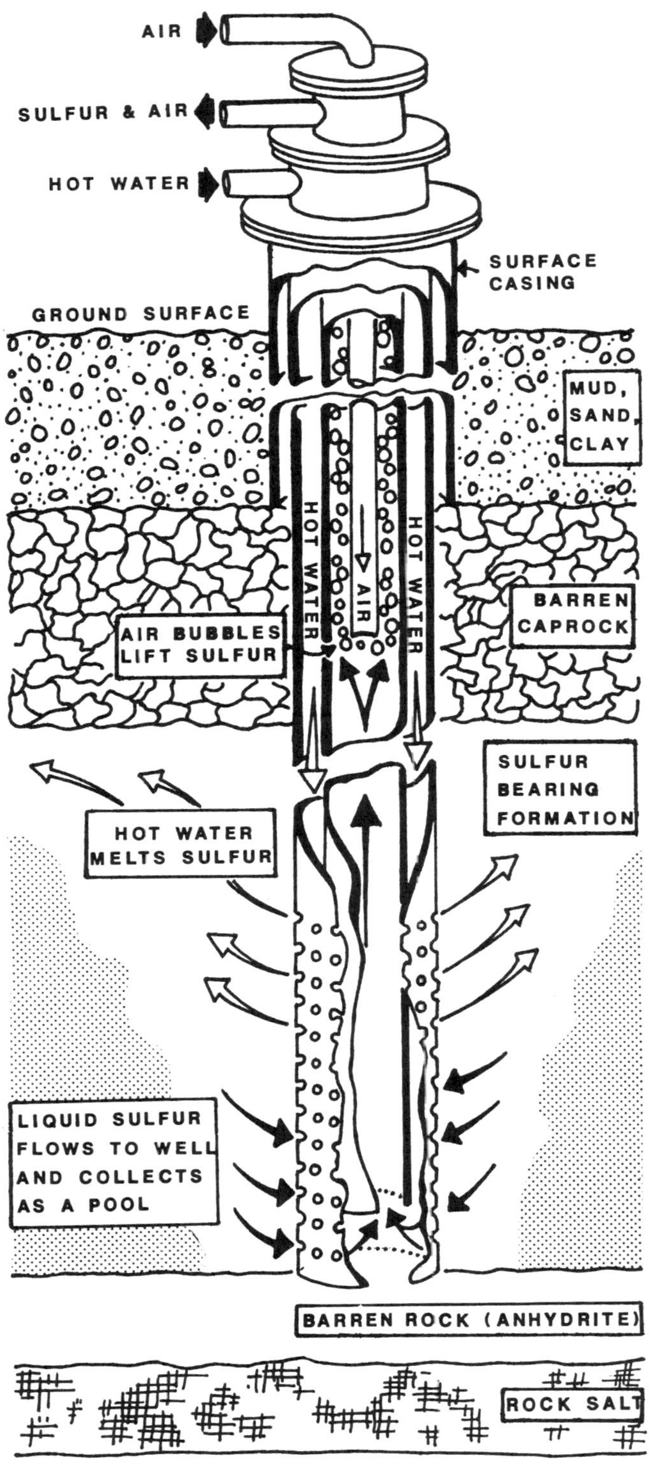

Fig. 15. Schematic cutaway view of a typical Frasch well (after Donner and Wornat, 1973).

ing as in the Frasch process, but the wells are arranged in rows along an exploitation front. Because the sulfur-bearing rock cannot retain the heated water under pressure, special measures are used to direct the flow of water underground, including mud injection to seal off areas, judicious blasting to create permeability, and water withdrawal through depression wells placed in advance of the exploitation front. The injected water is thus directed to the depression wells rather than being lost in all directions.

Molten sulfur tends to wet the host limestone, leaving about 5% residual sulfur in the ore. For this reason, and because islands of unmelted sulfur remain between production holes, few Frasch operations recover more than about 70% of the sulfur in place, and many sulfur deposits that have been abandoned yielded considerably less than that (Anon., 1990b). As a result, several secondary and tertiary recovery techniques have been developed, including the Thermofluid technique described by Zakiewicz (1986). Secondary recovery techniques are most popular during times of high sulfur prices.

RECOVERED ELEMENTAL SULFUR

Recovered elemental sulfur as discussed here is largely a nondiscretionary byproduct of petroleum refining, natural gas processing, and coking plants. Recovered sulfur is produced primarily to comply with environmental regulations applicable directly to processing facilities or indirectly by restricting the sulfur content of fuels sold or used.

In 1990, involuntary sources of sulfur contributed nearly 65% of the elemental sulfur produced worldwide. The combination of recovered elemental sulfur and byproduct sulfuric acid comprised more than 58% of the total world production of sulfur in all forms.

The principal sources of recovered sulfur are hydrogen sulfide in sour natural gas and organic sulfur compounds in crude oil. Recovery is mainly in the elemental form, although some is converted directly to sulfuric acid. Much smaller quantities of sulfur are recovered from tar sand, oil shale, and coal.

Recovered sulfur is the predominant sulfur source in Canada, France, Germany, the United States, Saudi Arabia, and Kuwait. Recovered elemental sulfur is also an important source in Iran, Iraq, Japan, and the former USSR.

Sour Natural Gas and Crude Oil

The growth in the recovery of sulfur from natural gas and crude oil since the mid-1950s had been one of the most significant trends in the sulfur industry. Generally expanding at the expense of Frasch production, recovered sulfur production has steadily increased for reasons independent of sulfur demand. The introduction of stricter environmental controls, combined with increased market demand for lighter petroleum products, a trend towards refining heavier crudes, and increased utilization of sour gases, have led to the availability of greater quantities of recovered sulfur, commonly at a substantially lower price than Frasch sulfur. Additional transportation costs are usually incurred by recovered sulfur producers, because many of the production facilities are far from present users, but their product is always price competitive. Because it is an involuntary byproduct, the cost of production attributed to recovered sulfur is more a question of accounting practices.

In times of low oil and gas prices, Frasch production is favored because of the impact of energy prices on Frasch production and also because of a coincident decline in hydrocarbon production and sulfur recovery. In times of high oil and gas prices, sulfur recovery increases, but exceptionally high prices will adversely affect hydrocarbon demand and sulfur recovery will suffer. For the foreseeable future, recovered sulfur production will likely increase at the continued expense of the Frasch producers.

Sulfur in oil occurs as organic sulfur compounds such as mercaptans (thiols), alkyl and aryl sulfides and disulfides, thiophenes,

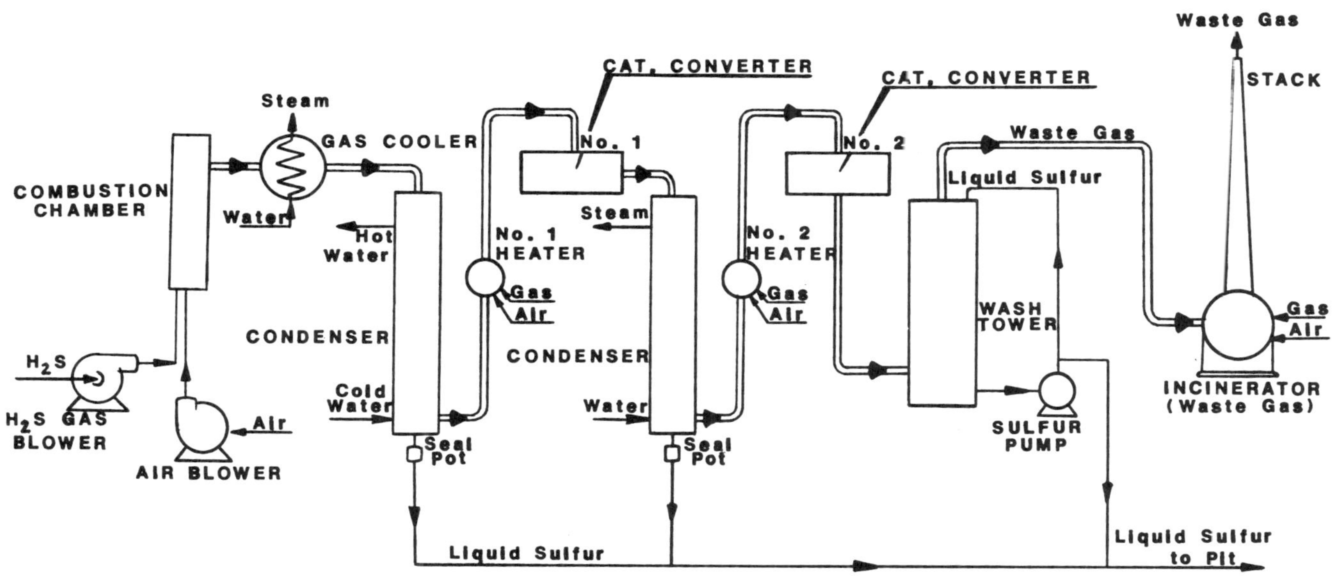

Fig. 16. Generalized flowsheet of a Claus plant for recovery of sulfur from hydrogen sulfide (after Barker, 1983).

and complex condensed aromatic heterocyclic sulfur compounds, which must be converted to hydrogen sulfide prior to the recovery of sulfur. Sulfur in natural gas is mainly in the form of hydrogen sulfide, which must be removed from the gas at the wellhead because of its toxicity, corrosiveness, and chemical activity. Hydrogen sulfide forms in high-temperature diagenetic settings (such as in deep oil and gas reservoirs) through the thermal maturation of crude oil and through thermochemical sulfate reduction. Thermochemical reduction of sulfate by hydrocarbons is probably responsible for the bulk of the hydrogen sulfide found in this setting (Machel, 1992).

Elemental sulfur is also present in high temperature hydrocarbon reservoirs as a result of oxidation of contained hydrogen sulfide by sulfate and other processes. The elemental sulfur generally occurs as disseminated droplets of liquid sulfur. Large quantities of sulfur also can be present as polysulfides and/or dissolved in the hydrocarbon fluids. The lowering of reservoir temperature and pressure (as in a well string) can result in exsolution of elemental sulfur and plugging of rock formations and production equipment. Such occurrences are termed thermogenic by Ruckmick et al. (1979). Hydrogen sulfide originating from thermochemical sulfate reduction may also migrate to near-surface environments and become oxidized to native sulfur. This mechanism may explain some of the enigmatic oxidative occurrences discussed previously.

Sulfur in crude oil is recovered during the refining process, and the amount of sulfur recovered depends upon the sulfur content of the crude, the amount of crude processed, the refining process, and the products produced. Refineries in general are equipped to remove the sulfur only from the lighter refinery products, and much of the sulfur in the crude oil remains in the refinery emission gases or the residual fuel oils and is ultimately emitted to the atmosphere when these oils are burned (although increasing amounts of sulfur are recovered from combustion gases, as discussed below). Approximately 15% of the sulfur in crude oil is recovered worldwide, but recovery is higher in the United States where residual oil is a smaller part of refinery output, and this amount is increasing due to the increasing utilization of heavier and sulfur-rich crudes.

The organic sulfur compounds in crude oil must be removed from the refinery feed and converted to hydrogen sulfide by a hydrogenation process before the sulfur can be recovered. The sulfur in natural gas is already present in the form of hydrogen sulfide, separable from the hydrocarbon gases by a variety of processes (Anon., 1987b, 1987c, Hyne, 1982). Hydrogen sulfide from both sources is converted to elemental sulfur by a process developed by C.F. Claus in Germany about 1880. Over the years, the Claus process has been improved and modified to fit a variety of situations, so that it has become known as the modified Claus process. The basic process is now used in hundreds of sulfur recovery units of various configuration throughout the world, collectively referred to simply as Claus plants. An excellent overview of the Claus process in its various forms can be found in Anon. (1986c).

In a Claus plant, concentrated hydrogen sulfide is fired in a combustion chamber connected to a waste heat boiler as shown in the generalized flowsheet in Fig. 16. Air is regulated to the combustion chamber so that part of the hydrogen sulfide is burned to produce sulfur dioxide, water vapor, and sulfur vapor. The high temperature gases are cooled in a waste heat boiler and sulfur is removed in a condenser. The efficiency of the process is raised by adding one, two, or three further stages in which the gases leaving the sulfur condenser are reheated and passed through catalytic converters and additional condensers. Finally, the total gas stream is incinerated to convert all remaining sulfur-bearing gases to sulfur dioxide before emission to the atmosphere. The sulfur is collected in the liquid form; individual plants can produce to up 2 000 tpd or more. Two converter plants may attain recoveries approaching 96%, although many plants operate at higher efficiencies because of air pollution regulations (Anon., 1986c, 1991b).

During 1990 in the United States, recovered elemental sulfur was produced by 58 companies at 151 plants in 26 states, one plant in Puerto Rico, and one plant in the US Virgin Islands (Morse, 1991; see also Anon., 1983a, 1985b). Most of these plants are of small size, with only 15 reporting annual production exceeding 100 kt. By source, 64% of the recovered elemental sulfur was produced at 84 refineries or satellite plants treating refinery gases and three coking plants. The remainder was produced by 27 companies at 64 natural gas treatment plants. The five largest American recovered-sulfur producers in 1990 were Chevron USA Inc., Exxon Co. USA, Shell Oil Co., Standard Oil Co. (Indiana), and Star Enterprise. These companies collectively accounted for 57% of recovered elemental sulfur production during 1990.

Canada is one of the largest involuntary producers of sulfur in the world, with a 1990 production of 5.9 Mt, all from oil and gas (Loughbrough, 1991). Canada is also the world's largest sulfur exporter. During the past 20 years, world sulfur price leadership has shifted from dominance by the US Frasch producers to the Canadian exporters. Canada's ability to export most of its production from Vancouver and its willingness to ship large volumes of sulfur from vatted stocks have contributed to its dominance of the sulfur market. A large portion of Canada's sulfur is sold through Prism Sulphur Corp. (successor to the now disbanded Cansulex), an export organization that represents Canadian producers in markets outside North America, aided by Sultran Ltd., a transportation arm of the Canadian sulfur industry.

The majority of Canadian sulfur production comes from sour natural gas processing plants in the Western Canadian sedimentary basin in Alberta, although production in British Columbia is increasing (Anon., 1991c). Occurrences of native sulfur are also known in Alberta, British Columbia, Nova Scotia and elsewhere, some of which are oxidative deposits (Hollister, 1984, Hora and Hamilton, 1992).

Recovered sulfur production in Alberta began in 1952 (Anon., 1984a, 1985a), but sales did not match production until 1979 when adequate transportation and marketing facilities were established. Prior to 1979, sulfur was stockpiled, reaching a level of over 21 Mt in 1977. Sales since then have only recently reduced the Canadian sulfur inventory to low levels (Schwalm and Doyle, 1987). The largest single sulfur producer in Canada is Shell Canada Ltd., which recovers 1.3 Mtpy from various gas and oil operations.

The production of sour gas continues at a high rate in Canada, but two new developments have the potential to boost Canadian production to even higher levels. Shell Canada Ltd. completed a demonstration plant to process ultra-sour gas (90% hydrogen sulfide) at Bearberry in Alberta. At present, production from the Bearberry plant is only about 75 ktpy, but under full scale development, the field could be producing 3 500 tpd within five years. The sulfur resource at Bearberry is estimated at between 70 and 100 Mt.

Shell Canada is also leading a consortium of companies in the development of the Caroline gas field in Alberta, discovered by Shell in 1986 and the largest gas reservoir found in Alberta in the past 20 years. Recovered sulfur production from gas containing 30 to 35% hydrogen sulfide at Caroline is expected to begin during 1993, and will eventually reach 1.5 Mtpy. Sulfur reserves at Caroline are estimated to be nearly 25 Mt.

The former USSR is also one of the world's largest producers and consumers of sulfur. During 1990, the former Soviet Union produced between 8 and 9 Mt of sulfur, of which about one third was recovered from oil and gas production. Despite the level of production, the former Soviet Union is also a net importer of sulfur, but this may change soon with the commissioning of new recovery plants at Astrakhan in the Volga Delta region of the Caspian Sea, and at Tengiz in Kazakhstan.

At Astrakhan, sulfur is recovered from natural gas that contains 24% hydrogen sulfide and 16% carbon dioxide (Anon., 1985c, 1988c, 1989b). The Astrakhan project consists of a three phase program, of which only Phase I has been completed. Phase I has operated intermittently since 1987, and production rates of 1.6 Mtpy have been achieved, but the plant is plagued with production and environmental problems and equipment to complete the project is currently lacking. If the technical problems at Astrakhan I can be remedied, both Astrakhan I and II should be on-stream by the end of 1992 producing 4 Mtpy, almost a six-fold increase over 1989.

The Tengiz project is also a three phase project of which only Phase I has been completed. The completion of Tengiz II is scheduled for 1992, and Tengiz III is scheduled for 1993, but problems have also been encountered with this venture, and one of the four major companies involved in the project withdrew in 1990. If the final two phases at Tengiz are completed on schedule, production by 1993 should be about 1.2 Mtpy.

Sulfur production in the former Soviet Union may increase dramatically in the next few years, adding to a worldwide surplus of sulfur, but accurate predictions are difficult given the rapidly changing political and economic situation. With the recent division of the USSR into independent republics, it is difficult to foresee the future. Numerous companies, including Freeport, have begun investing in various mining and oil projects, but the investment climate is not yet settled.

In western Europe, roughly 70% of the total sulfur production is supplied by oil and gas processing plants in France and Germany. In France, production is dominated by the Lacq natural gas field in the southwest part of the country. Discovered and operated by Société Nationale Elf Aquitaine, the Lacq field produces gas containing 15% hydrogen sulfide and 9.6% carbon dioxide (Anon., 1984b). Current reserves at Lacq are in decline, but production is expected into the next century. Production in Germany has been relatively stable in recent years, but sulfur production from power plants is becoming increasingly significant as environmental restrictions tighten in both the eastern and western states.

The rate of growth of recovered sulfur production in the Middle East has been phenomenal, with production from gasfields and oil refineries nearly tripling during the 1980s. The largest producer has been Saudi Arabia, but significant quantities are also recovered in Kuwait, Iran, and Iraq. The recent invasion of Kuwait by Iraq and subsequent war with the United States and others severely affected Kuwaiti and Iraqi production, which has not returned to pre-war levels.

Tar Sands

Sulfur is presently recovered in relatively small amounts from synthetic crude oil derived from tar sands in Canada. Known for over 200 years, the Athabasca tar sands, located in the Ft. McMurray area of Alberta, contain about 626 000 billion bbl of bitumen in quartz sands averaging 11% bitumen by weight, and represent one of the world's largest resources of hydrocarbons. The bitumen contains about 5% sulfur as organic sulfur compounds similar to those found in crude oil. Other tar sands deposits are found in surrounding regions, but none are currently exploited. The bitumen occurs in the Lower Cretaceous Wabiskaw-McMurray sandstones. The tar sands have an average thickness of 18 m (Anon., 1985a, 1985d, 1985e, Hyndman et al., 1982).

The Athabasca deposits locally outcrop and are visible at the surface, but most of the tar sand is buried up to depths of 50 m. Mining is by conventional surface techniques, and synthetic crude is generated on site. Sulfur is recovered during the generation of synthetic crude.

In the past, several tar sand projects have been proposed or tested, but at present only two commercial plants are operating in Canada. The Suncor Inc. plant, located at Mildred Lake, currently produces about 100 ktpy of sulfur. The Syncrude Canada Ltd. plant at Ft. McMurray produces approximately 400 ktpy.

Other large deposits of tar sands are known in Venezuela and Colombia. The United States has deposits in Utah, Kentucky, California, New Mexico, and elsewhere, but they are comparatively small. Interest in recovering oil from tar sands, and consequently sulfur, is highly dependent upon the price of crude oil. Tar sands presently supply about 1% of the sulfur produced worldwide, and production is not likely to increase substantially.

Oil Shale

More important than tar sands as a potential source of synthetic crude oil (and byproduct sulfur) for the United States are its oil shale deposits, particularly those in Colorado, Utah, and Wyoming (Gray et al., 1982). In past years, several companies have attempted to recover oil from shales, primarily in western Colorado, but the relatively high costs of extraction coupled with the problems of disposal of the spent shale have combined to prevent large scale utilization of oil shale. As with tar sands, interest in processing oil shale tends to wax and wane with the economic fortunes of the oil industry, although local conditions, such as political isolation or foreign currency restrictions, may spur the development of oil shale resources.

The sulfur in oil shale is present as organic sulfur compounds, as contained sulfate, and as pyrite. The total sulfur content of oil shale ranges between about 0.3 and 6.2% worldwide. Oil shales in the United States contain less than 0.75% sulfur, of which 80% is present as pyrite. Only a portion of the sulfur is recovered in the processing of synthetic crude; the sulfur as pyrite is not recovered because the pyrite remains with the spent shale.

Outside the United States, oil shale deposits are known in Brazil, the former USSR, and Africa. Oil has reportedly been recovered from shale in China, the former USSR, Israel, and Brazil, but no significant amount of sulfur has been recovered from these operations. In Brazil, Petrobrás completed a synthetic crude plant in 1972 in the southern part of the country, and has operated it partly on a semi-commercial basis. The Petrobrás plant was constructed to utilize shale from the Irati Formation that contains about 4% sulfur by weight, of which about 25% is recovered from the oil. For at least the past several years, sulfur production from oil shale in Brazil has not been great enough to be noted in compilations of production statistics (Anon., 1985a, 1985e).

Coal

Sulfur is present in coal as organic sulfur compounds and as contained pyrite, marcasite, and other metal sulfides. Some 40 to 80% of the sulfur in coal occurs as pyrite, which is physically mixed with the other constituents.

In the United States, electric utilities burn roughly 600 Mt of coal annually with an average sulfur content of 1.5% (Harben and Bates, 1990). Essentially none of the sulfur in coal is recovered prior to its combustion, although coal-cleaning technologies are capable of removing a substantial portion of the contained sulfides. Pyrite recovered from coal-bearing strata was an important source of sulfur in China, but its importance is declining (Anon., 1985a). Much of the sulfur contained in coal is recovered from coke oven and combustion gases as described under Sulfur Dioxide Emission Controls.

The organic sulfur compounds in coal can be removed only by processes of hydrogenation, liquefaction, or gasification, which change the form of the coal (Gray et al., 1982). Sulfur is produced from coal gas liquefaction in Europe (principally Germany; 100 kt in 1987), and in the Republic of South Africa (130 kt in 1987). Lignite gasification in Spain presently supplies about 2 000 tons of sulfur each year, less than 1% of the world's total sulfur production.

COMBINED SULFUR

Sulfur is also recovered from sources where it is chemically combined with other elements, principally as a sulfide or sulfate. Such compounds are processed directly to recover the sulfur as sulfuric acid or other forms, and sulfur is recovered from gaseous pollutants arising from the processing of sulfides in metal ores, coal, and other sources.

Pyrites and Other Metal Sulfides

Metal sulfides, widely distributed in nature, have long been an important source of sulfur. Sulfide ores of commercial interest for their sulfur value alone belong to the ferrous group collectively known as pyrites. Most important of these are the minerals pyrite, marcasite, and pyrrhotite. Pyrites commonly contain traces of non-ferrous metals. Pyrites are mined in several countries specifically for their sulfur content, most notably Spain and Portugal (Russell, 1989).

Sulfur is also recovered (ultimately as sulfuric acid) from non-ferrous metal sulfides that are mined and processed for their non-ferrous metal content. The most important nonferrous metal sulfides include those of copper (chalcopyrite, chalcocite), lead (galena), zinc (sphalerite), molybdenum (molybdenite), nickel (pentlandite), mercury (cinnabar), arsenic (arsenopyrite, realgar), antimony (stibnite), and silver (argentite and various sulfosalts). Associated iron sulfides are also recovered as a byproduct flotation concentrate during the processing of nonferrous ores. This source of pyrite has displaced some voluntary pyrite sources, but it is also dependent upon the market for non-ferrous metals.

Ferrous and nonferrous sulfides are an important source of sulfur in China, the former Soviet Union, the Republic of South Africa, Spain, Italy, Portugal, Norway, Sweden, Finland, the Philippines, and Yugoslavia (Russell, 1989). Sulfide ore deposits occur in a variety of geological environments. Mineralization occurs in massive lenses, tabular masses, in veins, and disseminated throughout a variety of rock types.

Pure pyrite contains 53.4% sulfur and 46.6% iron. Commercial grade pyrite ores are usually between 40 and 50% sulfur and frequently contain small amounts of arsenic, lead, copper, cobalt, gold, zinc, thallium, and other trace elements. These associated nonferrous metals can be recovered as a byproduct, but they can also be a source of environmental problems.

Pure pyrrhotite has varying chemical compositions, but it is sulfur-poor relative to pyrite. When pure, it contains about 38% sulfur and 62% iron. Pyrrhotite is softer than pyrite, and it is weakly to strongly magnetic. Cobalt, nickel, and copper are frequently found in association with pyrrhotite. Silver, gold, platinum and other trace elements are less common.

Sulfur is recovered from pyrites primarily by roasting to produce sulfur dioxide gas as a feed for the manufacture of sulfuric acid or for liquid sulfur dioxide, although sulfur may also be recovered from pyrites in the elemental form. The iron-bearing cinder from the roasting operation may be leached for the recovery of byproduct nonferrous metals and the final residue sold as iron ore. However, the iron-bearing cinder is frequently not utilized, and upon disposal may become a source of environmental contaminants.

Several processes have been developed over the years to recover sulfur from pyrites. The Noranda process, developed by Noranda Mines Ltd., controls temperature to recover about half of the sulfur in the elemental form and the remainder as sulfur dioxide for acid production.

The Orkla process, developed in the early 1930s in Norway, uses a shaft furnace for direct smelting of pyritic copper ore in the presence of a carbonaceous reducing agent (Barker, 1983). Heat in the upper zone is sufficient to drive off the labile atom of sulfur. The FeS formed is oxidized in the bottom zone to convert the remaining sulfur to sulfur dioxide, which is reduced to sulfur by the coke in the middle zone.

The Outokumpu process, commonly used today, was developed in Finland for processing ores at Pyhasalmi. Dried pyrite concentrate is blown into the top of a flash smelter where oxygen-free combustion gases decompose the pyrite to elemental sulfur and an FeS matte that is tapped from the bottom of the furnace. Sulfur vapor in the sulfur-laden gases from the furnace is condensed to liquid sulfur. The FeS matte is roasted in fluidized-bed roasters to generate sulfur dioxide for acid manufacture.

Pyrites were the second major source of sulfur in the mid- to late-19th century following the rise in the price of Sicilian sulfur. Pyrites lost predominance to Frasch sulfur in the early 20th century, and pyrite production continues to decline. Ferrous and nonferrous sulfides together supplied about 17% of sulfur in all forms produced worldwide in 1990, down from 21% in 1978.

One of the major reasons for this decline is the capital investment required for a pyrite-based sulfuric acid plant, which is two to three times that required for a comparable plant based on elemental sulfur (Loughbrough, 1991). Environmental problems have also plagued pyrite producers, because the disposal of cinder residues from pyrite processing is a great obstacle. These factors, coupled with the increasing availability of elemental sulfur, may restrict the future use of pyrites. Several acid plants in Europe have recently been converted from pyrite-based to elemental sulfur-based, and this trend is likely to continue. The pyrite mines in Spain and Portugal, which produce from Paleozoic sedimentary exhalative ores, have suffered as a result (Anon., 1985a, Harben and Bates, 1990, Loughbrough, 1991, Russell, 1989).

Sulfur is also recovered during the smelting of nonferrous sulfides. Sulfur dioxide in the smelter gases is converted to sulfuric acid, liquid sulfur dioxide, or oleum. The manner in which these concentrates are processed is similar to the way that pyrite is treated, but sulfur recovery is more difficult because of widely varying sulfur contents in fluctuating gas streams. This source of sulfur will probably increase in significance in the future as a result of progressively tighter anti-pollution legislation, although marketing of the byproduct acid is sometimes difficult because the acid plants are generally far from major markets.

In the United States, byproduct sulfuric acid from the nation's nonferrous smelters and roasters supplied about 11% of the total domestic production of sulfur in all forms in 1990. Seven acid plants operated in conjunction with copper smelters, and eight were accessories to lead, molybdenum, and zinc smelting and roasting operations. The five largest producers of byproduct sulfuric acid were ASARCO Inc., Cyprus Miami Mining Corp., Kennecott, Magma Copper Co., and Phelps Dodge Corp. Their eight plants produced 87% of the 1990 total (Morse, 1991).

Sulfur Dioxide Emission Control

In addition to being a commercial source of sulfur, sulfur dioxide is also one of the most common manmade pollutants. About 220 Mt of sulfur is discharged into the world's atmosphere annually, of which about a third comes from manmade sources. The bulk of this pollution comes from the burning of coal and petroleum products. Although these fuels often contain less that 3% sulfur by weight, large quantities of fuel are burned by utilities and industrial users, leading to the discharge of large quantities of sulfur in dilute gases.

Much of the remaining manmade emissions come from the smelting of nonferrous ores. Higher concentrations of sulfur dioxide are emitted from pyrometallurgical operations than from power plants, but there are many fewer installations worldwide.

The environmental problems associated with sulfur emissions have been known and studied for some time. The downwind deposition of sulfur-bearing compounds, principally sulfuric acid, leads to acidification of lakes and streams, loss of soil nutrients, and damage to plant life. Widespread deforestation has occurred locally in Europe (where it is estimated that pollution-related disease has spread to over half of the forested land in western Germany) and North America (Anon., 1985f, Anon., 1991d).

The effects of acid rain have been documented at many sites in the northeastern United States and eastern Canada, as well as some sites in the Rocky Mountains. Much of the pollution in Canada has its source in the United States, but 60% of the Canadian emissions come from only six metals operations, whereas in the United States, numerous utilities are responsible for over 70% of the total sulfurous emissions. For this reason, pollution control in Canada is much simpler.

Sulfur compounds are but one of numerous air pollutants, but they are also one of the easiest to control, and several useful compounds can be recovered. A comprehensive review of flue-gas desulfurization (FGD) methods is beyond the scope of this chapter, but the reader is referred to a previous chapter on that subject. FGD methods may be classified broadly as regenerative or throwaway processes depending upon whether the sulfur values are recovered in a useful form or disposed as solid waste. Both recovery methods may be carried out in wet or dry systems. Existing processes utilize a variety of compounds (such as limestone, sodium carbonate, magnesium oxide, and others) to neutralize or collect the sulfur dioxide. End products include gypsum, sulfuric acid, liquid sulfur dioxide, and elemental sulfur, all of which can be utilized if local markets exist. If local markets do not exist, large quantities of gypsum or sulfuric acid may have to be neutralized or otherwise disposed of, which may create other environmental problems.

The world's largest FGD project is to be implemented at the Drax power station in Yorkshire, England. The Drax station will use a limestone/gypsum process to produce about 100 kt of gypsum annually. The gypsum will be sold to British Gypsum (Anon., 1991e).

In the United States, the Clean Air Act of 1990 will provide a 50% reduction in sulfur dioxide emissions by 2000. This is to be achieved by a combination of alternative clean fuels, such as natural gas and low-sulfur coal, and innovative FGD technology.

Pollution control legislation, by requiring the neutralization and recovery of environmentally damaging compounds, effectively provides additional quantities of sulfur to the market regardless of the economic viability of these activities. As pollution legislation continues to tighten, increasing quantities of sulfur will be recovered to the detriment of voluntary sulfur producers. However, despite the predictions of increased tonnages of sulfur recovered from sulfur dioxide emissions, the impact on total sulfur production will not be dramatic. A high proportion of the recovered sulfur will be in the form of gypsum, and useful sulfur values are likely to be less than 10% of total production (Anon., 1985f).

Sulfates

Sulfate minerals, particularly gypsum and anhydrite, represent the world's largest sulfur resource. Thick deposits of naturally occurring anhydrite and gypsum are found in evaporite basins worldwide, and large amounts of gypsum are produced as byproduct waste materials from a number of industrial and pollution control processes. Nevertheless, sulfates are rarely utilized as a source of sulfur owing to the high cost of the processing.

Sulfur values (in the form of sulfuric acid and ammonium sulfate) are or have recently been recovered from sulfate in only a few countries, including Austria, Germany, India, Pakistan, Poland, Turkey, and the United Kingdom (Anon., 1985a). Sulfuric acid is produced from gypsum and anhydrite using the Muller-Kuhne process, developed in Germany during World War I. Ammonium sulfate is manufactured from natural and byproduct sulfate using the Merseberg process, in which calcium sulfate is reacted with ammonium carbonate (Barker, 1983).

Elemental sulfur can also be recovered from sulfate, but the process has never been economic. Although there have been several attempts to design processes for the recovery of elemental sulfur from sulfate, only one commercial attempt has been made, that of the Elcor Chemical Corp. in the late 1960s. Located in West Texas, the Elcor plant utilized gypsum from Permian evaporites in a two step process in which calcium sulfide produced in a kiln is hydrolyzed to hydrogen sulfide, which then yields sulfur by the Claus process. The process is energy-intensive, and high fuel costs combined with technical problems caused the demise of the project (Anon., 1968, 1983b, Ellison, Jr., 1971). Despite occasional efforts by researchers to produce sulfur economically from sulfates, it is unlikely that sulfates will be extensively utilized because of the increasing availability of recovered sulfur and the relatively low price of the commodity.

About one billion tons of phosphogypsum, a byproduct of the product of phosphoric acid, has been generated worldwide, of which about 685 Mt are stockpiled in Florida and Louisiana (Jolly and Chamberlain, 1992). Phosphogypsum utilization studies are underway at several organizations, including such groups as the Florida Institute of Phosphate Research and the Institute for Recyclable Materials at Louisiana State University, but economically viable alternatives to phosphogypsum impoundment or disposal have not been forthcoming. Recent studies indicate that recovery of sulfur from phosphogypsum will not be economically feasible until the sulfur price is double what it is today, a situation not likely to occur for many years.

FORMED SULFUR

Prior to the early 1950s, much of the world's tonnage of export sulfur was handled as crushed bulk. Because of safety, environmental, and economic concerns arising from breakage of the solid sulfur and generation of sulfur dust, sulfur forming techniques were developed to overcome difficulties inherent in solid sulfur operations.

Basically, liquid sulfur is formed, upon leaving the source, into solid moderately-sized particles of selected shapes. These particles are designed to resist attrition, be unwettable, and handle easily during stockpiling, loading, transport, and unloading.

Early sulfur forming was limited to the production of flakes and slates, but as sulfur output from sour gas increased in the 1960s, so did the development of new processes for sulfur forming. A variety of forms are now produced, many of which are spherical (Raymont and Hyne, 1983).

Production of Formed Sulfur

The main purpose in sulfur forming is to prevent breakage of the solid sulfur particles during handling. Two factors emerge as the most important in determining resistance to breakage: particle size and sphericity (Hyne, 1990). The larger the particle size, the greater the tendency to break apart during handling (friability). The optimum size for spherically formed sulfur is between about 1.5 and 3.5 mm. Sphericity is also important in that the more spherical the form is, the less likely it is to have easily ruptured protrusions. These factors are especially important in the so-called premium products, where fines (less than 0.3 mm) or dust production is warranted to be less than 2%. However, highly spherical products will have a low angle of repose and angle of climb (the maximum inclination that can be used to lift a product on a conveyer belt without it running back), which may generate other problems during transportation and storage.

With reduced friability, the environmental impact of solid sulfur handling is also reduced. Upon exposure to water and air, elemental sulfur reacts to form sulfurous and sulfuric acid, which can create localized environmental problems and lead to excessive corrosion of plant facilities. The reaction is bacterially mediated, and the rate of reaction is inversely proportional to the sulfur particle size. The production of acid can be minimized if dust levels are controlled or bactericides used.

There are several primary forms into which sulfur is made: flakes, slates, prills (or pellets), nuggets, granules, pastilles, and briquettes (Table 4). Flakes are the easiest, but least popular, of the forms to produce. Sulfur is cooled and solidified on the outside of large rotating drums, from which it peels off into small flakes (Anon., 1979a).

Slating, the least expensive process, is widely used but also produces particles having relatively high friability. Molten sulfur is cast onto a continuous conveyer belt, and is cooled with air or water so that it solidifies into a thin sheet. As it reaches the end of the belt, the sheet is broken, or breaks naturally, into smaller pieces.

Prilling sulfur is a more complicated forming method, but it is also widely used. There are two fundamental types of sulfur prilling: air and water. In air prilling, molten sulfur is sprayed from the top of a tower against an upward flow of air. As it falls, the sulfur breaks into small droplets and cools into prills. In water prilling, the sulfur is sprayed into tanks containing water, from which the prills are collected and dried. Prilling is also referred to as pelleting, a legacy from the production of lead shot in towers. Minor modifications to prilling techniques are used to produce nuggets.

Granulation involves applying successive coats of sulfur to solid particles of sulfur in a granulator until the particle size reaches the required value. In the case of the Procor GX granulation process, liquid sulfur is sprayed into a rotating drum in which small seed particles of sulfur are recycled from the end of the process (Hyne, 1990). The product granules are spherical, dry, hard, and free from the voids and structural weaknesses often associated with prills. The end product size can also be varied to meet customer requirements.

Pastilles are individual droplets of molten sulfur that have been dropped on a steel belt and cooled by conduction. The Sandvik Rotoform process uses a patented Rotoformer (essentially a perforated drum) to distribute the sulfur on the belt. At the end of the belt, the pastilles are scraped off and fall onto a collecting conveyer. The resulting pastilles are spherical and flattened on one side.

Characteristics of Formed Sulfur

The characteristics of the sulfur form have a direct bearing on the suitability of the form for transport. Important characteristics include friability, moisture performance, and the handling characteristics, of which friability is probably the most important (Anon., 1979a, Raymont and Hyne, 1983). The friability of a sulfur form is influenced by the particle size, surface texture, internal voids, and crystallinity.

Moisture performance is defined by the moisture content of the newly formed product and the moisture uptake and retention when it is stockpiled. Those forms produced by water quenching have a higher initial moisture content than forms produced by dry processes, but the moisture content can be reduced by mechanical dewatering and drying. The moisture uptake and retention of a sulfur form are particularly important because sulfur is commonly stored where it is exposed to the atmosphere. Formed sulfur usually has reasonable water rejection properties because sulfur itself is hydrophobic. The spherical forms exhibit superior water rejection properties; water will often penetrate only the first few centimeters into a stockpile and then drain away.

The handling characteristics of a sulfur form depend on particle size and shape, uniformity, surface texture, density, and hardness.

Table 4. Forming Processes (modified from Barker, 1983). List Is Incomplete, and Not All Processes Are Currently In Use.

Process type/name	Developer	Country of origin	Prime mover	Heat sink	Shape	Density (kg/m^3)	Angle of repose	Friability
SLATING:								
V&E Process	Procor Sul. Services	Canada	rubber belt	air or water	slates		32.4	high
Sandvik	Sandvik Steel	Sweden	steel belt	water	slates			high
Candybar					bevelled slates			high
WATER PELLETING/NUGGETING:								
Sulpel	Elliott Assoc. Dev.	UK	gravity/ pressure	water	sphere		27.0	low
Capsul	Cambrian Eng.	Canada	gravity/ pressure	water	sphere	1 152–1 169 (dry)	28.7	low
Chemsource	Chemsource	USA	gravity/ pressure	water	sphere	1 121–1 281	31.9	low-mod.
Elliott-Litwin	Elliott		gravity	water	sphere			low
Fletcher	R.B. Fletcher & Assoc.	Canada	gravity	water	semi-flat, irreg.	1 121	25–30	moderate
P.V. Com. System	P.V. Container Systems	Canada	gravity	water	nuggets	1 121–1 217	32	moderate
	Liq. Terminals Inc.	USA	pressure	water	nuggets	1 025	41	moderate
Petrosul								
AIR PRILLING								
Kaltenbach	Kaltenbach and Cie.	France	gravity	air	sphere			low
Outokumpu	Outokumpu Oy	Finland	gravity/ pressure	water	sphere	1 200–1 300	33–35	
Ciech	Ciech	Poland	gravity/ pressure	air	sphere	1 040–1 200	24.2	low
Stearns-Roger	Stearns-Roger Canada Ltd.	Canada	gravity	air	sphere	1 120–1 280	20	low
Windgate			gravity	air	sphere			low
Air Prill	Cominco	Canada	gravity	air	sphere			low
MISC. PRILLING								
Hydroprill	Ross Thermal Systems	Canada	gravity/ pressure	air & water	sphere			low
Popcorn	Unocal	USA	high pressure	air & water	"popcorn"	1 000	37	mod.-high
Dix Nugget			pressure	water	irreg. pellets			moderate
GRANULATION								
Perlomatic	APC & PEC Engineering	France	pressure	air	granules	1 000	20–32	low
Procor GX	Procor Sul. Services	Canada	pressure	air	granules	1 150	30	low
Stamicarbon			pressure	air	granules			low
PASTILLES								
Rotoform	Sandvik	Sweden	pressure	air/water	pastilles			low
Rollomat	Gebr. Kaiser	Germany			pastilles			low
MISC. SHAPES								
Briquettes	Komarek	USA		air	"egg"			high
Pyramids					pyramidal			mod.-high
BULK SULFUR								
	various			air	mixed and irreg.			very high

The properties of bulk density and the angle of repose are two of the most important handling characteristics. Normal dry bulk sulfur has a bulk density of between 1 345 and 1 440 kg/m^3 and an angle of repose of about 35°. Formed sulfur should have a density approaching bulk sulfur and as high an angle of repose as possible, but these values vary considerably, even for forms produced by the same technique. The density and angle of repose are also affected by the moisture content.

REFINED SULFUR

The category of refined sulfur includes a wide range of manufactured products of which ground sulfur is the most prominent. Sometimes referred to as speciality sulfur, this category also includes wettable, insoluble, precipitated (also referred to as sublimed or flowers of sulfur), colloidal, micronized, and double-refined sulfur, as well as special grades used in manufacturing pharmaceutical and photochemical products (Anon., 1979b, 1992).

Although it is the oldest end use of brimstone, refined sulfur having been used since antiquity particularly in pharmacology and alchemy, demand for refined sulfur continues to grow. Nevertheless, of all the brimstone consumed in the western world, only a very small percentage (about 2%) is used as refined sulfur or speciality grades. Even so, it is a business important in its own right by supplying essential and highly specialized products for agriculture and industry.

In agriculture, refined sulfur in powdered (dry or wettable), micronized, sublimed, and other forms often combined with metal compounds, is used in direct application on crops, in viticulture, and in horticulture as, first and foremost, a protection against mildew, powdery mildew, and other fungi. However, in these end uses, crude recovered sulfur increasingly is used without being refined, because recovered sulfur with a purity of 99.9% meets the specifications.

Refined sulfur, mainly ground, is also used in wine making and sugar refining, as a pesticide (since ancient times), in fertilizer as a soil conditioner, and as an additive to compound animal feeds. However, owing to cost, fertilizer sulfur now tends to be crude sulfur formed into granules or pellets prior to grinding, or is the fines separated from bulk sulfur that is to be used in single superphosphate mixtures.

In industry, the rubber and tire sector is the prime user of special sulfur, making use of ventilated, micronized, and insoluble forms, all of which are necessary in the vulcanization process. Much smaller amounts of special grades are used in the manufacture of such diverse items as pharmaceuticals, matches, fireworks, and textiles.

Although ground and refined sulfur are used to some extent almost everywhere in the world, the manufacture of ground and refined sulfur is confined to a relatively small number of enterprises, most of which are concentrated in areas where the level of economic development is high such as North America, Europe, and Japan. Other producers are found where there is significant domestic demand, such as Australia, the Republic of South Africa, India, Mexico, Brazil, and Chile, or where large quantities of recovered sulfur are available for export, such as Kuwait and Saudi Arabia (Anon., 1984c).

The manufacturing of ground sulfur must be done in highly controlled conditions to prevent explosions. Because ground sulfur will autoignite at 190°C, sulfur grinding is done in an inert atmosphere with special equipment to minimize the buildup of heat and static electricity.

A healthy international trade in refined sulfur exists because almost all economies need some, but not necessarily enough to justify establishing a domestic industry. In general, the pattern of trade is from North to South America in the Western Hemisphere, and from Europe and Japan to nonproducing countries in the Eastern Hemisphere.

END USES

Sulfur is consumed mainly in the form of acid or other intermediate chemicals used to manufacture a final product that often does not contain much, if any, sulfur. In 1990, the largest sulfur end use, sulfuric acid, represented 86% of shipments for domestic consumption (Morse, 1991). The balance is used for a variety of nonacid industrial purposes, detailed for the United States in Fig. 17 and Tables 5 and 6.

The largest single use of sulfuric acid is in the production of phosphatic fertilizers, followed by copper ore leaching, production of other inorganic chemicals, synthetic rubber and plastic materials, industrial organic chemicals, and petroleum and coal refining. Combined agricultural uses of sulfur in all forms account for over 60% of domestic use.

In the past, a variety of commonly used fertilizers contained large amounts of sulfur (such as superphosphate and ammonium sulfate), which although relatively low in primary nutrients, supplied significant amounts of sulfur. The increase in usage of high-analysis fertilizers (such as urea), combined with the progressive decrease in atmospheric sulfur, the decrease in the use of sulfur as a pesticide and fungicide, and increasing demands for higher crop yields, has led to widespread soil sulfur deficiencies in agricultural regions throughout the world. The manufacture of sulfur-bearing fertilizers (referred to as plant nutrient sulfur or PNS) will expand in the future to remedy this problem (Anon., 1983c, 1986d, 1991f, Messick, 1991).

Sulfuric Acid

Sulfuric acid, because of its desirable properties, is the most universally used mineral acid and the largest volume inorganic chemical in terms of the quantity produced and consumed (Morse, 1991). World sulfuric acid production totaled about 159 Mt in 1990 (100% H_2SO_4), of which 39 Mt was produced in the United States (Table 7).

The reasons for sulfuric acid's dominance as the most desirable mineral acid are as follows (Barker, 1983, Shelton, 1980):

1) Raw materials and acid plant operating costs are low.
2) Shipping, handling, and storage problems are fewer than for other acids.
3) Wide adaptation to numerous acidulation and neutralization processes.
4) Excellent dehydrating properties.
5) Effective catalysis in many hydrocarbon and organic chemical syntheses.
6) Readily forms organic sulfates with many hydrocarbons, which are easily hydrolyzed to form other basic products.
7) High boiling point, which limits volatilization losses due to high temperatures required in many processes.
8) Can be partially used in one process, reused in a second or even a third process, and/or be readily regenerated, fortified, and concentrated for still further uses.

Sulfuric acid is produced near consumption sites by either of two processes: the older chamber process, or the newer and more important contact process. To produce sulfuric acid, elemental sulfur or any number of sulfur-bearing materials are burned to produce sulfur dioxide. The chamber process produces a rather weak acid, about 70% H_2SO_4; the contact process produces 98% acid and stronger grades containing excess sulfur trioxide, called oleums. The chamber process is characterized by large, lead reaction chambers and the use of gaseous nitrogen oxides as the catalyst in a complex reaction to form sulfuric acid. The contact process uses a solid contact substance as a catalyst to convert sulfur dioxide and excess oxygen to sulfur trioxide.

Up to 99.5% or better conversion of sulfur to sulfur trioxide can be obtained in acid plants using a modification introduced by Farbenfabriken Bayer of Germany in 1964. The Bayer process

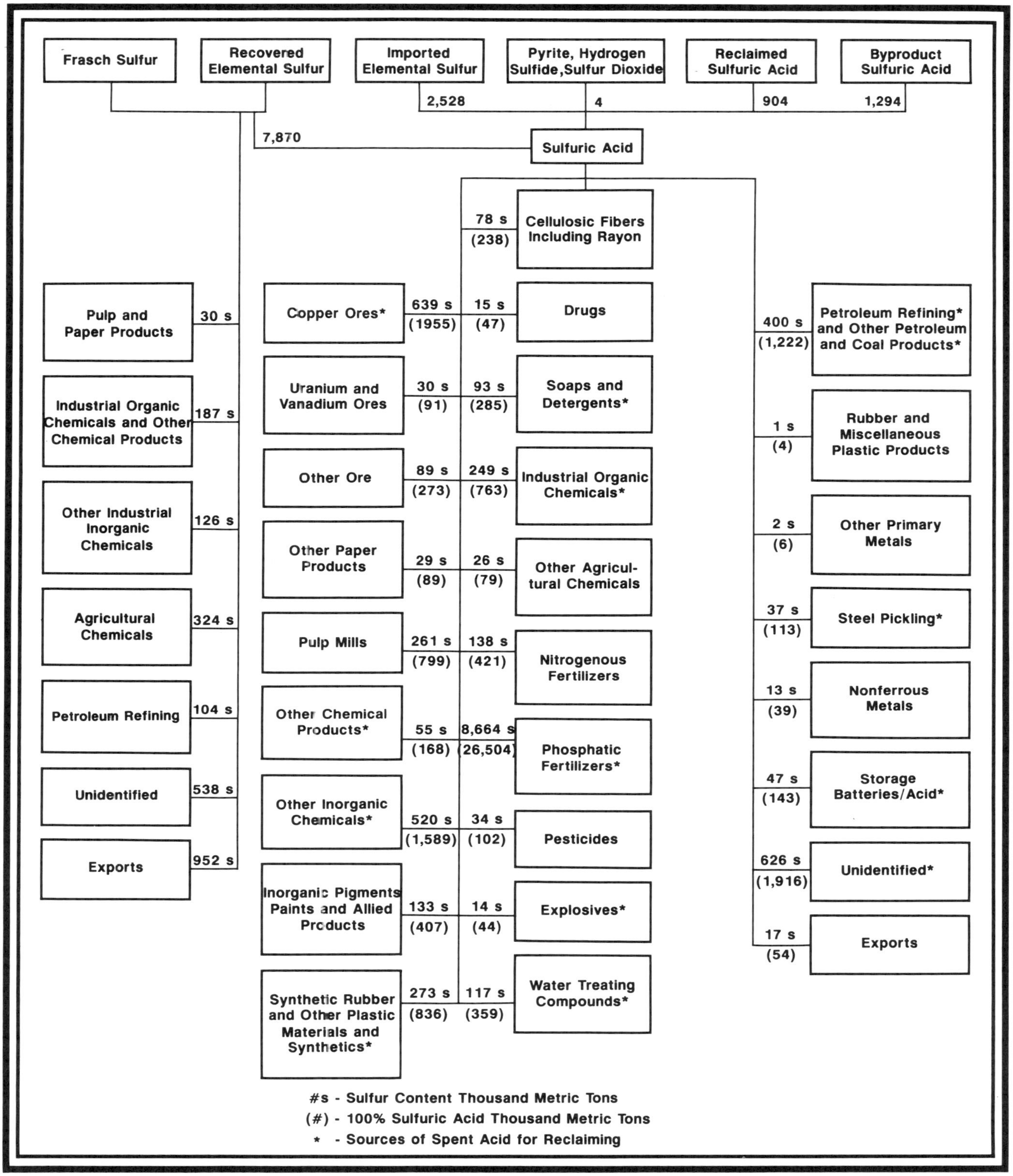

Fig. 17. Sulfur-sulfuric acid supply and end-use relationship in 1990 (after Morse, 1991).

Table 5. US Sulfur Supply-Demand Relationships, 1986–1990, Sulfur In All Forms, In Thousand Tons (Morse, 1991)

	1986	1987	1988	1989	1990
	WORLD PRODUCTION				
United States	11 087	10 539	10 746	11 592	11 560
Rest of world	42 649	46 043	47 782	47 125	46 108
Total	53 736	56 582	58 528	58 717	57 668
COMPONENTS AND DISTRIBUTION OF US SUPPLY					
Production:					
Frasch mines	4 043	3 202	3 174	3 888	3 726
Recovered elemental:					
Natural gas	2 246	2 536	2 501	2 537	2 336
Petroleum	3 570	3 624	3 943	3 973	4 200
Other forms	1 228	1 176	1 128	1 194	1 298
Total all forms	11 087	10 539	10 746	11 592	11 560
Industry stocks, Jan 1	2 799	2 748	2 316	1 112	1 301
Imports	1 347	1 599	1 996	2 260	2 571
Total US supply	15 233	14 886	15 058	14 964	15 432
Distribution of US supply:					
Industry stocks, Dec. 31	2 748	2 316	1 112	1 301	1 423
Exports	1 895	1 242	1 223	1 024	972
Demand	10 586	11 323	12 712	12 685	13 056
Apparent surplus (+), deficit (−) of supply	+4	+5	+11	−46	−19
	US DEMAND PATTERN				
Agriculture:					
Phosphoric acid	6 973	7 556	8 404	8 642	8 664
Other	683	642	598	431	522
Total agriculture	7 656	8 198	9 002	9 073	9 186
Drugs and food products	6 532	27	28	18	
Soaps, detergent, water treatment	249	204	221	200	215
Plastic and synthetic products	338	352	458	501	421
Paper products	266	294	288	328	320
Paints	119	118	122	138	133
Metal mining and processing	376	426	567	723	773
Explosives	30	49	44	17	14
Petroleum refining	947	973	958	825	504
Iron and steel production	69	67	74	59	37
Storage batteries (acid)	50	41	51	46	47
Other	421	569	900	747	1 388
Total non-agricultural demand	2 930	3 125	3 710	3 612	3 870
Total US primary demand	10 586	11 323	12 712	12 685	13 056

utilizes multiple conversion stages, and the process is pushed to thermodynamic equilibrium by removing the reaction products and recycling unconverted gas (Fig. 18). Bayer process plants are also referred to as double-contact, double-absorption, or double-catalysis plants.

For years, the double-absorption process was used with few changes, but recently substantial and important improvements have been made in the thermal efficiency of acid plants, as well as the capability to process dilute sulfur dioxide streams economically. Other developments have increased the capability to produce sulfuric acid from waste products (Anon., 1990c).

New Uses for Sulfur

Various industrial, academic, and governmental groups have been researching new uses for sulfur for some time. The growing supply of sulfur in both elemental and combined forms has led to the development of a number of new products, several of which are presently in commercial use. The practicality of some of these applications is at least partially a function of the price of sulfur in relation to competing materials.

Among the new products that have been researched or are currently in use are: 1) sulfur-asphalt compositions for paving applications to give increased durability and performance over a wider range of temperature, 2) Sulphex, a nonasphalt paving with a sulfur binding, 3) an extender or recycling agent for used asphalt, 4) foamed sulfur with high compressive strength and insulating properties, to be used for insulation, pavement subbases, and possibly light weight structural applications, 5) an impregnator of a variety of compounds to increase strength, durability, and water resistance, 6) a coating, mixed with fibers or modifiers, on block or brick buildings, linings for ponds, and other applications where a durable coating is useful, 7) sulfur concretes, mortars, and aggregates with enhanced corrosion and salt resistance, 8) sewage and waste-water treatment, 9) electrodes in alkali metal batteries, and 10) molded sulfur blocks for tile and brick (Barker, 1983, Muir, 1982, Terrel. 1982).

Paving materials containing sulfur have been tested at numer-

Table 6. Sulfur and Sulfuric Acid Sold or Used in the US, By End Use; Thousand Tons, Sulfur Content (Morse, 1991)

End Use	Elemental sulfur[1]		Sulfuric acid (sulfur equivalent)		Total	
	1989	1990	1989	1990	1989	1990
Copper ores	—	—	588	639	588	639
Uranium and vanadium ores	—	—	44	30	44	30
Other ores	—	—	40	89	40	89
Food and kindred products	W	W	—	—	W	W
Pulpmills and paper products	8	30	320	290	328	320
Inorganic pigments, paints and allied products, industrial organic chemicals, other chemical products[2]	122	187	424	382	546	569
Other inorganic chemicals	188	126	285	520	473	646
Synthetic rubber, and other plastic materials and synthetics	W	W	383	273	383	273
Cellulosic fibers, including rayon	—	—	52	78	52	78
Drugs	—	—	25	15	25	15
Soaps and detergents	W	W	85	93	85	93
Nitrogenous fertilizers	—	—	93	138	93	138
Phosphatic fertilizers	—	—	8 642	8 664	8 642	8 664
Pesticides	—	—	32	34	32	34
Other agricultural chemicals	277	324	29	26	306	350
Exposives	—	—	17	14	17	14
Water-treating compounds	—	—	115	117	115	117
Other chemical products	—	—	78	55	78	55
Petr. refining, other petr. and coal products	142	104	683	400	825	504
Rubber and misc. plastic products	—	—	2	1	2	1
Steel pickling	—	—	59	37	59	37
Nonferrous metals	—	—	49	13	49	13
Other primary metals	—	—	2	2	2	2
Storage batteries (acid)	—	—	46	47	46	47
Exported sulfuric acid	—	—	11	17	11	17
TOTAL IDENTIFIED	737	771	12 104	11 974	12 841	12 745
Unidentified	372	538	459	626	831	1 164
GRAND TOTAL	1 109	1 309	12 563	12 600	13 672	13 909

W, withheld to avoid disclosing proprietary data, included with "Unidentified;" [1]does not include elemental sulfur used for production of sulfuric acid; [2]no elemental sulfur was used in inorganic pigments and paints and allied products.

ous sites, and have been used in commercial applications (Saylak and Conger, 1982, Love and Harrigan, 1982). The use of sulfur concretes is also widespread, and has particular application to high-corrosion environments such as industrial flooring, acid storage tanks, electrolytic cells, sewer pipes, and bridge decking (Funke and McBee, 1982, Shrive et al., 1982).

ECONOMIC FACTORS

Costs

Economic factors and costs of production in the sulfur industry must be considered separately for the voluntary and involuntary producers. As with most industrial minerals, the sulfur industry is market driven. The degree of utilization of sulfur and sulfuric acid is dependent upon worldwide economic activity and the health of industries that use sulfur-bearing compounds for the production of other goods. Within this market, the voluntary and involuntary producers constantly strive to maximize their market share, at the expense of competing producers.

The voluntary sector of the sulfur industry is based on the orderly mining of discrete deposits of native sulfur, with the objective of obtaining as nearly a complete recovery of the resource as economic conditions will allow (Morse, 1985). This requires a relatively assured market and a price structure attractive to both producers and consumers. The assured market is commonly based upon the relatively close proximity of sulfur users, frequently fertilizer production centers, and the availability of cheap transportation to end users. To a significant extent, stability of sulfur supply and price is accomplished by the maintenance of large stocks of sulfur by the producers.

For the Frasch producers, the main operating expenses are fuel for heating the process water, chemicals for treating the water, and the cost of drilling production wells. Work force requirements are relatively low. Additional costs include plant maintenance, royalty payments to land owners (frequently a significant cost), taxes, and transportation.

High fuel costs or excessive subsurface water loss can increase the cost to sulfur recovery to a point where it is no longer profitable to operate the plant. Water loss is a serious problem at a number of Frasch mines, and has contributed to the closing of several, including Texasgulf's Comanche Creek mine in Pecos County, Texas.

In contrast, involuntary producers are often forced to recover sulfur from their primary product because of environmental restraints or other mandates. The economic desirability of producing sulfur is often subordinate to the necessity of producing it, and the sulfur revenues may represent only a small portion of the revenues (sometimes not even enough to pay for recovery) from the primary product. In such cases, there may be no ready market for the sulfur

Table 7. Sulfuric Acid Production and Consumption, 1988–1990; Thousand Tons 100% Sulfuric Acid

	Production			Consumption		
	1988	1989	1990	1988	1989	1990
WORLD TOTAL	166 776	161 427	158 708	168 047	162 509	160 917
WESTERN WORLD	115 890	111 765	112 395	117 412	113 091	114 940
WEST EUROPE	26 816	25 786	22 991	27 166	26 213	24 112
Belgium/Luxembourg	2 168	1 956	1 858	2 823	2 701	2 753
Finland	1 327	1 380	1 361	1 322	1 390	1 350
France	4 084	4 157	3 768	4 000	3 980	3 700
Germany	4 486	4 482	3 724	4 610	4 600	3 890
Greece	984	1 025	946	999	1 028	952
Italy	2 679	2 201	1 910	2 480	2 000	1 807
Netherlands	1 440	1 450	1 433	1 967	1 873	1 750
Spain	3 475	3 303	2 908	3 130	3 057	2 574
Sweden	942	885	870	915	890	880
United Kingdom	2 270	2 148	1 996	2 439	2 318	2 155
AFRICA	17 273	13 534	15 696	17 297	13 484	15 646
Morocco	7 815	3 770	6 670	7 816	3 842	6 670
South Africa	3 760	3 860	3 071	3 765	3 770	3 071
Tunisia	3 550	3 650	3 409	3 550	3 650	3 409
NORTH AMERICA	42 548	42 813	44 183	42 369	43 119	44 583
Canada	3 805	3 658	4 050	3 084	2 856	2 799
United States	38 743	39 155	40 133	39 285	40 263	41 784
LATIN AMERICA	9 175	9 110	8 884	9 874	9 287	9 175
Brazil	3 610	3 425	3 098	4 145	3 770	3 352
Chile	650	775	820	720	953	955
Mexico	3 960	4 025	3 974	4 030	3 830	3 876
ASIA	17 693	18 512	18 871	18 271	18 863	19 604
India	3 160	3 645	4 226	3 400	3 580	4 220
Indonesia	965	975	985	1 000	1 000	1 064
Iraq	1 010	1 100	790	1 010	1 100	790
Japan	6 763	6 882	6 880	6 716	6 873	6 905
South Korea	2 060	2 000	1 980	2 110	2 055	2 134
Philippines	685	795	896	855	940	1 054
Taiwan	905	945	725	1 010	1 150	1 025
OCEANIA	2 385	2 010	1 770	2 435	2 125	1 820
Australia	2 060	1 670	1 400	2 110	1 785	1 450
CENTRALLY PLANNED ECONOMIES	50 886	49 662	46 313	50 635	49 418	45 977
EASTERN EUROPE	9 409	8 782	6 459	9 230	8 648	6 399
Bulgaria	781	846	522	800	846	522
Czechoslovakia	1 249	1 142	1 033	1 250	1 150	1 033
Poland	3 166	3 100	2 010	3 030	2 980	1 950
Romania	1 900	1 700	1 200	1 900	1 700	1 200
Yugoslavia	1 731	1 442	1 350	1 730	1 442	1 350
USSR	29 430	28 300	27 300	29 300	28 300	27 024
OTHER CPEs	12 047	12 580	12 554	12 105	12 470	12 554
China	10 980	11 523	11 969	11 040	11 415	11 969

Source: British Sulphur Corp., 1990 statistical supplement in *Sulphur*, No. 216, pp. 49–52.

or sulfuric acid, and the producer may be forced to dump the product on the market regardless of the cost. The sulfur may also be transported to a market far from the source, where the transportation charge constitutes a large part of the sulfur cost. Involuntary sulfur production is dependent upon demand for the primary product, and as such is always independent of the need for sulfur.

In the United States, the quoted liquid Frasch sulfur price, FOB terminal Tampa, FL, is the bench mark price. Recovered sulfur producers commonly undercut this price by 5 to $15 per ton to ensure prompt sale, reduce storage costs, and guarantee long term contracts (Morse, 1991).

During the past 20 years, world sulfur price leadership has shifted from the US Frasch producers to the Canadian recovered sulfur exporters. This shift was aided by the the ability of Canada to export most of its production from the Port of Vancouver and to ship large volumes from vatted stocks. In addition, the Canadian sulfur producers have a well organized and skilled transportation arm (Sultran Ltd.).

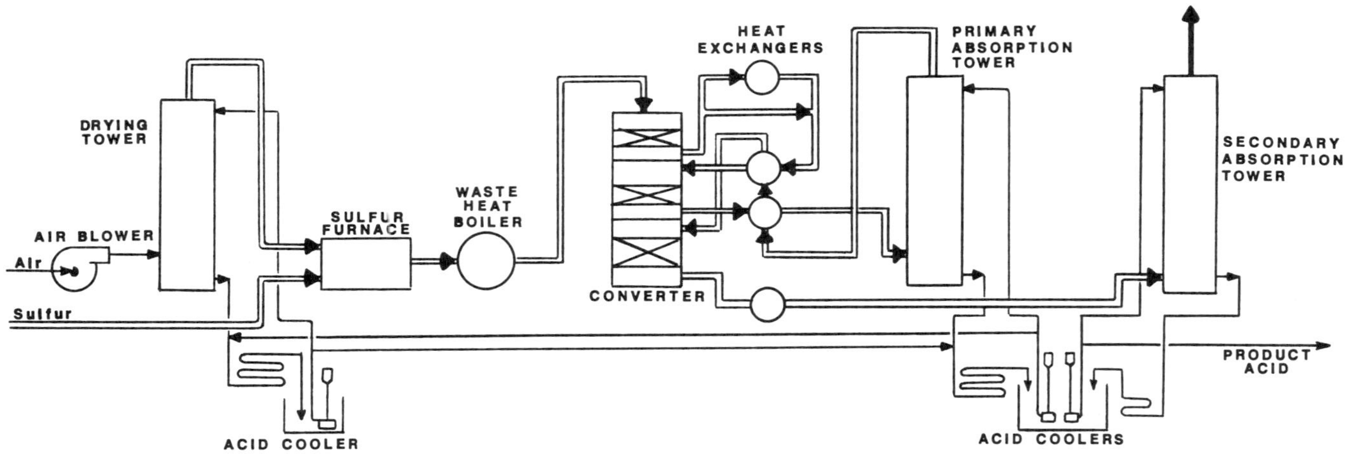

Fig. 18. Simplified flowsheet, double contact sulfuric acid plant (after Barker, 1983).

Markets

Crude sulfur is traded extensively by several relatively large exporting nations. At present, roughly 40% of worldwide brimstone production enters the international marketplace (Balazs, 1990). In 1990, exports from Canada, Poland, Saudi Arabia, Mexico, the United States, Germany, Iraq, and France (in descending order of the quantity shipped) accounted for 90% of the sulfur trade (Morse, 1991). Major sulfur importers were the United States, Morocco, India, the former USSR, Tunisia, Brazil, the United Kingdom, France, and Jordan.

With the collapse of the Soviet Union and the opening of markets in eastern Europe, traditional trade patterns in those countries are undergoing rapid change. As an example, a significant portion of Polish sulfur exports was previously restricted to the Soviet Union and other Communist Bloc nations, but consumption in eastern Europe is now much reduced and Polish production is down.

Sulfuric acid is also traded internationally, but to a much smaller degree than crude sulfur. Sulfuric acid trade is extensive in Europe, and among the United States, Canada, and Mexico. Japan exports sulfuric acid to the western Pacific region (Morse, 1985).

Transportation

Sulfur can be transported either in liquid or in solid states. Conversion to sulfuric acid, the principal end use, is easiest using liquid sulfur so this has become the preferred form, especially when environmental effects are considered. When sulfur cannot be transported to a consumer in the liquid form, it is converted to a variety of solid forms for transportation or is stockpiled. Stockpiling of sulfur is cheapest by far in large solid blocks (vats) that are broken or melted prior to shipment. Because sulfur block re-melting is expensive, as is liquid sulfur-handling equipment, transport and handling of formed solid sulfur is often more economical except where liquid-handling facilities are in place.

Transporting sulfur in the liquid phase offers certain advantages, including ease of handling, absence of fugitive dust, and reduced losses. However, unprotected personnel may be exposed to hazardous concentrations of hydrogen sulfide and sulfur dioxide during handling. Problems associated with solid sulfur include environmental degradation from fugitive dust, possible ignition of sulfur or explosion of sulfur dust, and corrosion of unprotected metal (Morse, 1985).

Sulfur transportation costs can be the critical factor in determining the economic viability of a sulfur production facility. For example, sulfur in West Texas must be shipped to the east by rail to compete with Frasch production in the Gulf Coast that is located adjacent to inexpensive water transportation. Similarly, sulfur in western Canada must be shipped to Vancouver by rail, adding a significant charge to the cost of the sulfur. During late 1991, ocean-going freight rates between Vancouver and markets in South America, Asia and North Africa varied between 15 and $40/ton (Anon., 1991g).

Alternative Materials

For many uses, there are no alternative materials that can replace sulfur or sulfuric acid. However, there are a number of diverse areas where sulfuric acid can be substituted by hydrochloric or nitric acid. The prices of these acids generally dictate which one will be used, and sulfuric acid has been the most widely used because of its relatively low cost. Sulfuric acid has been replaced by hydrochloric acid in the production of titanium dioxide, and recent developments suggest that hydrochloric acid (which now can be manufactured at a very low cost) will replace sulfuric acid in other applications as well. A significant switch to hydrochloric acid could seriously impact the merchant sulfuric acid business based on sulfur burning (Anon., 1990d).

Environmental Considerations

Sulfur-bearing compounds, particularly sulfuric acid, are of prime concern in legislation restricting releases of contaminants to the atmosphere and bodies of water. The broad topic of sulfur as a pollutant is well beyond the scope of this discussion, but it should be remembered that sources of sulfur pollutants are now utilized as sources of recovered elemental sulfur, as discussed above.

As legislation affecting atmospheric discharge of sulfur oxides or sulfuric acid is tightened, larger quantities of sulfur (generally as acid or calcium sulfate) will be available for use in other industries. Similarly, discharges of sulfuric acid to bodies of water is increasingly prohibited. Thin acid dumping at sea is now restricted by the European Economic Community, and this measure, coupled with similar efforts elsewhere, has prompted an increase in the recycling of waste acid. Formerly neutralized or otherwise disposed of, spent sulfuric acid is now regenerated at numerous sites, where it is reburned with additional sulfur to produce regenerated acid and additional virgin acid.

Environmental concerns also affect sulfuric acid consumption indirectly. Because of environmental pressures, manufacturers of titanium dioxide, a pigment commonly used in paint, have largely converted their manufacturing process from one using sulfuric acid to a chloride-based process. Sulfuric acid is also used to manufacture hydrofluoric acid for use in chlorofluorocarbons (CFCs). The decline in CFC manufacturing may result in a decline in sulfuric acid use, but alternative materials may consume additional hydrofluoric acid, and so the actual impact on the sulfur industry is unknown.

The environmental impacts of native sulfur mining and transportation are generally minimal and restricted to small areas. Frasch mining operations commonly experience varying degrees of subsidence due to sulfur removal, but at many mines, particularly in West Texas and offshore Gulf of Mexico, subsidence is mainly an engineering challenge. Bleedwater recovered during Frasch mining may have to be treated or cooled before it is disposed of, but in some cases water is scarce enough that the bleedwater is reused. Native sulfur spilled at the surface quickly reacts to form sulfurous and sulfuric acid, which may lead to acid drainage and corrosion damage, but these effects are generally restricted and of short duration.

The sulfur mines of Poland appear to be one exception to this generalization. The pollution of surface and subsurface waters, air, and soil is a growing concern in Poland, particularly in the Tarnobrzeg region where serious contamination has been documented (Preisner, 1991). Of the 5 000 hectares of land in the Tarnobrzeg area disturbed by sulfur mining (most of which is farmland), only 10% has been reclaimed.

FUTURE TRENDS

There have been a few major surprises in the last few years in respect to the outlook for world elemental sulfur production. The discovery of the Main Pass deposit offshore Louisiana, the invasion of Kuwait by Iraq and subsequent destruction of Kuwaiti production facilities and embargo of Iraqi sulfur, and the rapid collapse of the Soviet Union, have all necessitated revisions of supply and demand forecasts. The long term effects on production and consumption in Iraq and the former Soviet republics are indeterminate at present, and so predictions offered here (in part formulated prior to these events) are probably partly in error. Further developments in environmental legislation and product substitution will also affect the supply/demand relationship.

Outlook for Demand

Demand for elemental sulfur and sulfur-bearing products has continued to grow steadily as the world population and economy have grown, but temporary slumps in the market for sulfur have occurred concurrent with economic recessions. Morse (1991) predicts that world sulfur demand, from 1991 to the year 2000, will increase at an annual rate of 1.4 to 1.8% per year. Morse projects that world demand will attain 63.5 Mt in 1995 and increase to more that 70 Mt by 2000. However, citing recent declines in consumption, Inoue (1992) forecasts the total demand for elemental sulfur to grow by 1 to 1.5% per year.

Outlook for Supply

Most forecasters predict that the supply of sulfur will rise at a rate exceeding the demand, but these forecasts were made before the collapse of the Soviet Union. Inoue (1992) states that between 1980 and 1990, world production of elemental sulfur grew more slowly than demand, at a rate of 0.7% per annum. The increase came solely from involuntary sources, and an opposite trend was observed for voluntary producers. Overall, these developments resulted in an increased market share for the involuntary producers, up to 66% in 1990 from 54% in 1980.

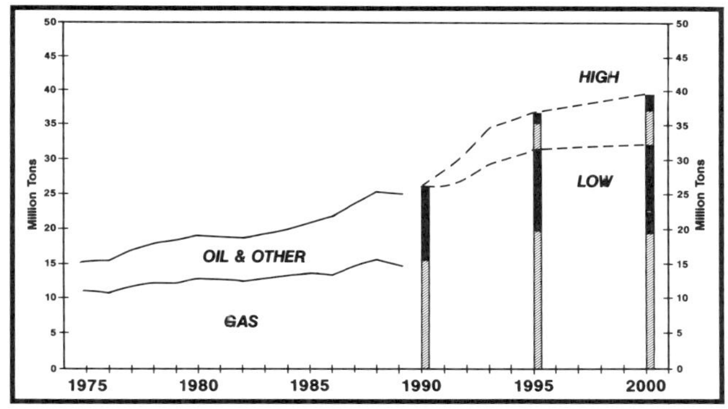

Fig. 19. World recovered sulfur production from 1975 to 1989, with projections to the year 2000 in tons (after Paulson, 1991).

Involuntary production is expected to rise further during the 1990s. In Canada, sulfur production is expected to rise (mainly because of the Caroline field) to 6.5 Mtpy by 1995, and to stay at this level to the year 2000. The Astrakhan and Tengiz projects in the former Soviet Union were expected to substantially impact sulfur availability, but continuing political and technical problems have hampered their development.

Morse (1991) states that the long term prospect is that 85% or more of the world sulfur supply will come from environmentally regulated sources (including oil and natural gas processing), and that after the turn of the century, no new operation that produces sulfur as its primary product will be developed except where it may be deemed necessary for political or social reasons. However, this pessimistic (from a voluntary producer's standpoint) view is not shared by all industry analysts. Paulson (1991) predicts that recovered sulfur should account for about 70% of world brimstone production by the turn of the century, amounting to between 33 and 39 Mtpy (Fig. 19). In addition, Loughbrough (1991) reminds us that recovered sulfur production is not driven by the requirements of the sulfur market, and that as a result there should always be a market for mined elemental sulfur. This is partly because recovered sulfur will always be insufficient to meet the total demand for sulfur, and that a decrease in the amount of recovered sulfur production can only be offset by increased voluntary production. Unfortunately, this process works the other way, as recovered sulfur producers stockpile excess production and sell it later to drive down prices. Resulting market volatility makes it that much more difficult for voluntary producers to justify expansion or development of new deposits. Loughbrough states that the production deficit between recovered sulfur output and sulfur demand will grow at about 3% per year, a gap that will be filled by the voluntary producers.

Projections by Freeport Sulphur Company (Foster, 1991) show that Frasch sulfur production in the United States will rise during the early and mid-1990s as a result of the development of the Main Pass mine (Fig. 20), but that Main Pass production will only replace production lost from other closures. World Frasch and other native sulfur production is expected to increase (Fig. 21) as a result of production from Main Pass and Iraq, but after 1995 a decline is expected as a result of reduced production in Mexico and at other mines in the United States.

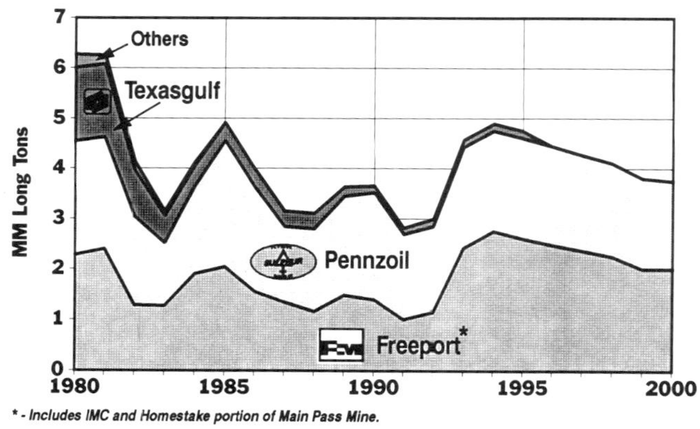

Fig. 20. Historic and projected US Frasch sulfur production, by company, 1980–2000 (after Foster, 1991).

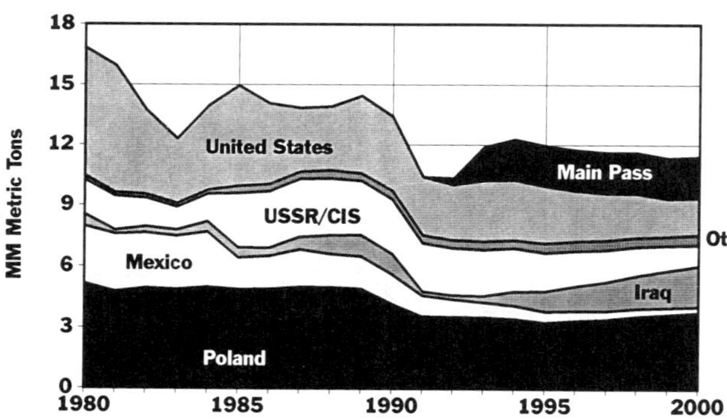

Fig. 21. Historic and projected world Frasch and other native sulfur production, 1980–2000 (after Foster, 1991).

ACKNOWLEDGMENTS

The author would like to thank Freeport Sulphur Co., Pennzoil Sulphur Co., and Texasgulf Inc. for supplying unpublished data presented in this chapter and for assisting with the drafting of the figures. Many of the figures were drafted by Kelly Larson of Pennzoil Sulphur Co., whose artistic talents are readily apparent. The text was critically read by James Barker (New Mexico Bureau of Mines and Mineral Resources), John Empsall (Pennzoil Sulphur Co.), Stephen Schwochow (Colorado School of Mines), and Bryan Wimberly (Texasgulf Inc.). Their advice and comments are greatly appreciated.

BIBLIOGRAPHY AND REFERENCES

Anon., 1968, "Elcor Chemical Corp.'s Rock House Project," *Sulphur*, No. 79, pp. 29–30.

Anon., 1979a, "Sulphur Forming Processes," *Sulphur*, No. 142, pp. 30–39.

Anon., 1979b, "Refined Sulphur—A Small Industry with Growing Prospects," *Sulphur*, No. 141, pp. 19–22.

Anon., 1983a, "Natural Gas-Based Recovered Sulphur Production in the United States," *Sulphur*, No. 166, pp. 18–22.

Anon., 1983b, "Recovering the Sulphur Values From Gypsum," *Sulphur*, No. 167, pp. 38–39.

Anon., 1983c, "Sulphur in Agriculture," *Sulphur*, No. 164, pp. 22–25.

Anon., 1984a, "Outlook for Alberta's Sulphur Production," *Sulphur*, No. 171, pp. 26–30.

Anon., 1984b, "Sulphur Activities of Elf Aquitaine," *Sulphur*, No. 171, pp. 21–24.

Anon., 1984c, "Saudi Arabia—A Source of Refined Sulphur," *Sulphur*, No. 171, pp. 33–35.

Anon., 1985a, *World Survey of Sulphur Resources*, 3rd ed., British Sulphur Corporation, London, 215 pp.

Anon., 1985b, "US Recovered Sulphur—Future Assured," *Sulphur*, No. 179, pp. 17–21.

Anon., 1985c, "Update on USSR Developments," *Sulphur*, No. 179, p. 13.

Anon., 1985d, "Canadian Oil Sands—A Revival of Interest," *Sulphur*, No. 176, pp. 17–18.

Anon., 1985e, "Oil Shales and Tar Sands: Recovering Oil and Sulphur," *Sulphur*, No. 181, pp. 37–42.

Anon., 1985f, "Pollution Sulphur," *Sulphur*, No. 181, pp. 29–33.

Anon., 1986a, "Poland, 25 Years as a Sulphur Exporter," *Sulphur*, No. 184, pp. 21–23.

Anon., 1986b, "New Sulphur Venture Launched in the Philippines," *Sulphur*, No. 185, pp. 18–19.

Anon., 1986c, "The Status of Claus Sulphur Recovery Processes," supplement to *Sulphur*, No. 187, 20 pp.

Anon., 1986d, "Sulphur: The Fourth Major Nutrient?" *Sulphur*, No. 184, pp. 17–20.

Anon., 1987a, "Recovering Sulphur From Marginal Sources," *Sulphur*, No. 193, pp. 33–36.

Anon., 1987b, "Sweetening Natural Gas: I," *Sulphur*, No. 192, pp. 30–40.

Anon., 1987c, "Sweetening Natural Gas: II," *Sulphur*, No. 193, pp. 26–31.

Anon., 1988a, "Freeport Makes Sulphur Discovery," *Industrial Minerals*, No. 244, p. 9.

Anon., 1988b, "Poland, Supply Shortfall?," *Sulphur*, No. 199, pp. 19–21

Anon., 1988c, "Astrakhan—A Testing Challenge," *Sulphur*, No. 199, p. 29.

Anon., 1989a, "Middle East, A Review of Sulphur Production, On the Up and Up," *Sulphur*, No. 201, pp. 13–17.

Anon., 1989b, "The Third World Superpower," *Sulphur*, No. 205, pp. 14–16.

Anon., 1990a, "Geared to Change, Ciech, Poland's Chemical Trading Company Celebrates 45 Years," *Sulphur*, No. 211, pp. 13–14.

Anon., 1990b, "Improved Techniques in Underground Sulphur Mining," *Sulphur*, No. 209, pp. 29–38.

Anon., 1990c, "Recent Advances in Sulphuric Acid Production Processes," *Sulphur*, No. 210, pp. 24–33.

Anon., 1990d, "Sulphuric Acid: A New Threat?" *Sulphur*, No. 211, p. 17,

Anon., 1991a, "Main Pass on Schedule," *Sulphur*, No. 217, pp. 21–23.

Anon., 1991b, "What Happens to Hydrogen in a Claus Plant?" *Sulphur*, No. 214, pp. 53–60.

Anon., 1991c, "A Bright Future?" *Sulphur*, No. 213, pp. 27–29.

Anon., 1991d, "FGD: Still Much To Achieve," *Sulphur*, No. 217, pp. 39–41.

Anon., 1991e, "Clearing the Air," *Sulphur*, No. 217, pp. 28–32.

Anon., 1991f, "Sources of Plant Nutrient Sulphur," *Sulphur*, No. 215, pp. 39–41.

Anon., 1991g, "On Course for Firmer Rates?" *Sulphur*, No. 217, p. 18.

Anon., 1992, "A Small But Important Market," *Sulphur*, No. 219, pp. 19–21.

Ackerman, J.M., 1992, "Main Pass—Frasch Sulphur Mine and Oil and Gas Development," Proceedings, *Native Sulfur—Developments in Geology and Exploration*, G.R. Wessel and B.H. Wimberly, eds., Society for Mining, Metallurgy, and Exploration, Littleton, CO, pp. 73–80.

Adkins, W.S., 1927, "The Geology and Mineral Resources of the Fort Stockton Quadrangle," Bulletin No. 2738, Bureau of Economic Geology, University of Texas-Austin, 166 pp.

Babbitt, J.F., 1991, "Impact on World Sulphur Supply and Trade," *Sulphur*, No. 213, pp. 18–22.

Balazs, C., 1990, "International Trade Patterns for Brimstone," *Sulphur*, No. 209, pp. 22–27.

Barker, J.M., 1983, "Sulfur," *Industrial Minerals and Rocks*, 5th ed., Vol. 2, S.J. Lefond, ed., AIME, New York, pp. 1235–1274.

Barker, J.M., Cochran, D.E., and Semrad, R., 1979, "Economic Geology of the Mishraq Native Sulfur Deposit, Northern Iraq," *Economic Geology*, Vol. 74, pp. 484–495.

Crawford, J.E., 1990, "Geology and Frasch-Mining Operations of the Culberson Sulfur Mine, Culberson County, West Texas," *Industrial*

Mineral Resources of the Delaware Basin, Texas and New Mexico, Guidebook Series, Vol. 8, J.R. Kyle, ed., Society of Economic Geologists, pp. 141–162.

Dale, J.M., 1981, "Sulfur From Surface Ores," *Mining Engineering*, Vol. 33, No. 9, pp. 1340–1342.

Davis, J.B., and Kirkland, D.W., 1979, "Bioepigenetic Sulfur Deposits," *Economic Geology*, Vol. 74, pp. 462–468.

Dessau, G., Jensen, M.L., and Nakai, N., 1962, "Geology and Isotopic Studies of Sicilian Sulfur Deposits," *Economic Geology*, Vol. 57, pp. 410–438.

Donner, W.S., and Wornat, R.O., 1973, "Mining Through Boreholes—Frasch Sulfur Mining System," *Mining Engineering Handbook*, Vol. 2, A.B. Cummins and I.A. Given, eds., SME, New York, pp. 21/60–21/67.

Edwards, A.F., et al., 1992, "The Geology of the Main Pass Sulphur and Oil Deposit," Preprint No. 92–254, Annual Meeting, Phoenix, AZ, SME, Littleton, CO, 14 pp.

Ellison, S.P., Jr., 1971, "Sulfur in Texas," Handbook No. 2, Bureau of Economic Geology, University of Texas-Austin, 48 pp.

Evans, G.L., 1946, "The Rustler Springs Sulphur Deposits as a Source of Fertilizer," Report of Investigations No. 1, Bureau of Economic Geology, University of Texas-Austin, 13 pp.

Faure, G., 1977, *Principles of Isotope Geology*, Wiley, New York, 464 pp.

Feely, H.W., and Kulp, J.L., 1957, "Origin of Gulf Coast Salt Dome Sulfur Deposits," *American Association of Petroleum Geologists Bulletin*, Vol. 41, pp. 1802–1853.

Fodor, I., et al., 1989, "A New Process for Extracting Sulphur From Low-Grade Ore," *Sulphur*, No. 200, pp. 36–40.

Ford, W.E., 1949, *Dana's Textbook of Mineralogy*, 4th ed., Wiley, New York, 851 pp.

Foster, R.B., 1991, "The Outlook for Sulphur Production From Frasch and Native Sources," paper presented to 13th Annual Phosphate-Sulphur Symposium, The Sulphur Institute, Washington, DC, 12 pp. with figs.

Funke, R.H., Jr., and McBee, W.C., 1982, "An Industrial Application of Sulfur Concrete," *Sulfur: New Sources and Uses*, M.E.D. Raymont, ed., Symposium Series 183, American Chemical Society, Washington, DC, pp. 195–208.

Giusti, G.P., 1984, "Texasgulf, Seventy-Five Years," *Sulphur*, No. 174, pp. 16–19.

Gray, D., et al., 1982, "The Potential Impact of Synthetic Fuels on the Sulfur Industry," *Sulfur: New Sources and Uses*, M.E.D. Raymont, ed., Symposium Series 183, American Chemical Society, Washington, DC, pp. 83–100.

Guilinger, J.R., and Nestlerode, E., 1992, "The Geology and Development of the Phillips Ranch Sulfur Deposit," Proceedings, *Native Sulfur—Developments in Geology and Exploration*, G.R. Wessel and B.H. Wimberly, eds., Society for Mining, Metallurgy, and Exploration, Littleton, CO, pp. 125–134.

Halbouty, M.T., 1979, *Salt Domes, Gulf Region, United States and Mexico*, 2nd ed., Gulf Publishing Co., Houston, TX, 561 pp.

Harben, P.W., and Bates, R.L., 1990, *Industrial Minerals, Geology and World Deposits*, Industrial Minerals Division, Metal Bulletin Plc, London, pp. 265–273.

Haynes, W., 1959, *Brimstone, The Stone That Burns*, Van Nostrand, Princeton, NJ, 308 pp.

Hentz, T.F., Price, J.G., and Gutierrez, G.N., 1989, "Geologic Occurrence and Regional Assessment of Evaporite-Hosted Native Sulfur, Trans-Pecos Texas," Report of Investigations No. 184, Bureau of Economic Geology, University of Texas-Austin, 70 pp.

Hollister, V.F., 1984, "Sulphur Deposits of Nova Scotia," *The Geology of Industrial Minerals in Canada*, G.R. Guillet and W. Martin, eds., Canadian Institute of Mining and Metallurgy, Special Volume 29, pp. 31–32.

Hora, Z.D., and Hamilton, W.N., 1992, "Native Sulfur Resource Potential of Western Canada," Proceedings, *Native Sulfur—Developments in Geology and Exploration*, G.R. Wessel and B.H. Wimberly, eds., SME, Littleton, CO, pp. 51–57.

Hyndman, A.W., Liu, J.K., and Denney, D.W., 1982, "Sulfur Recovery from Oil Sands," *Sulfur: New Sources and Uses*, M.E.D. Raymont, ed., Symposium Series 183, American Chemical Society, Washington, DC, pp. 69–82.

Hyne, J.B., 1982, "Recent Developments in Sulfur Production from Hydrogen Sulfide-Containing Gases," *Sulfur: New Sources and Uses*, M.E.D. Raymont, ed., Symposium Series 183, American Chemical Society, Washington, DC, pp. 37–56.

Hyne, J.B., 1990, "Sulphur Forming, Handling, and Transport: Still Not a Completely Routine Matter," *Sulphur*, No. 208, pp. 35–47.

Inoue, A., 1992, "Outlook for World Elemental Sulphur," *Sulphur*, No. 218, pp. 12–15.

Ivanov, M.V., 1968, "Microbiological Processes in the Formation of Sulfur Deposits," Academy of Sciences of the USSR, Institute of Microbiology, published by Israel Program for Scientific Translations, Jerusalem, 298 pp.

Jackson, M.P.A., and Talbot, C.J., 1986, "External Shapes, Strain Rates, and Dynamics of Salt Structures," *Geological Society of America Bulletin*, Vol. 97, pp. 305–323.

Jolly, J., and Chamberlain, D.L., 1992, "Progress in Programs to Utilize Phosphogypsum, Freeport-McMoRan Experience," paper presented to the symposium Sulphur Markets—Today and Tomorrow, The Sulphur Institute, Washington, DC, 18 pp. with figs.

Kirkland, D.W., and Evans, R., 1976, "Origin of Limestone Buttes, Gypsum Plain, Culberson County, Texas," *American Association of Petroleum Geologists Bulletin*, Vol. 60, pp. 2005–2018.

Kuznetsov, S.I., Ivanov, M.V., and Lyalikova, N.N., 1963, "The Role of Microorganisms in the Formation and Destruction of Sulfur Deposits," *Introduction to Geological Microbiology*, McGraw-Hill, New York, pp. 100–123.

Kyle, J.R., and Posey, H.H., 1991, "Halokinesis, Cap Rock Development, and Salt Dome Mineral Resources," *Evaporites, Petroleum, and Mineral Resources*, J.L. Melvin, ed., Elsevier, New York, pp. 413–474.

Kyle, J.R., and Price, P.E., 1986, "Metallic Sulfide Mineralization in Salt-Dome Cap Rocks, Gulf Coast, U.S.A.," *Transactions*, (Sect. B: Appl. Earth Sci.), Institute of Mining and Metallurgy, Vol. 95, pp. B6-B16.

Lerche, I., and O'Brien, J.J., eds., 1987, *Dynamical Geology of Salt and Related Structures*, Academic Press, Orlando, FL, 832 pp.

Loughbrough, R., 1991, "Natural Versus Refined Sulphur, A Breath of Frasch Air," *Industrial Minerals*, No. 286, pp. 19–29.

Love, G.D., and Harrigan, E.T., 1982, "Road Construction Using Sulfur," *Sulfur: New Sources and Uses*, M.E.D. Raymont, ed., Symposium Series 183, American Chemical Society, Washington, DC, pp. 209–224.

Machel, H.G., 1992, "Low-Temperature and High-Temperature Origin of Native Sulfur in Diagenetic Environments," Proceedings, *Native Sulfur—Developments in Geology and Exploration*, G.R. Wessel and B.H. Wimberly, eds., SME, Littleton, CO, pp. 3–22.

McNeal, R.P., and Hemenway, G.A., 1972, "Geology of the Fort Stockton Sulfur Mine Pecos County, Texas," *American Association of Petroleum Geologists Bulletin*, Vol. 56, pp. 26- 37.

Messick, D.L., 1991, "Growth Potential," *Sulphur*, No. 215, pp. 19–23.

Miller, L.J., 1992, "Sulphur Ore Controls Within the Salado and Castile Formations of West Texas," Proceedings, *Native Sulfur—Developments in Geology and Exploration*, G.R. Wessel and B.H. Wimberly, eds., SME, Littleton, CO, pp. 165–192.

Morse, D.E., 1985, "Sulfur," *Mineral Facts and Problems*, Bulletin 675, US Bureau of Mines, pp. 783–797.

Morse, D.E., 1990, "Sulfur," *Minerals Yearbook 1989*, Vol. 1, Metals and Minerals, US Department of Interior, Bureau of Mines, pp. 1027–1052.

Morse, D.E., 1991, "Sulfur," Preprint, *Minerals Yearbook 1990*, US Bureau of Mines, 26 pp.

Muir, D.R., 1982, "New Product Opportunities for Sulfur," *Sulfur: New Sources and Uses*, M.E.D. Raymont, ed., Symposium Series 183, American Chemical Society, Washington, DC, pp. 127–136.

Mussey, J.W., and Tyree, P.O., 1985, "Geology and Production of West Texas-Type Sulphur Deposits," Proceedings, *Applied Mining Geology: Problems of Sampling and Grade Control*, R.A. Metz, ed., SME-AIME, New York, pp. 113–120.

Nieć, M., 1986a, "Geology of Sulphur Deposits," *Fertilizer Minerals in Asia and the Pacific*, Vol. 1, Mineral Concentrations and Hydrocarbon Accumulations in the ESCAP Region, United Nations, Economic and Social Commission for Asia and the Pacific, pp. 84–92.

Nieć, M., 1986b, "Mining of Sulphur Deposits," *Fertilizer Minerals in Asia and the Pacific*, Vol. 1, Mineral Concentrations and Hydrocarbon Accumulations in the ESCAP Region, United Nations, Economic and Social Commission for Asia and the Pacific, pp. 143–146.

Nieć, M., 1992, "Native Sulfur Deposits in Poland," Proceedings, *Native*

Sulfur—Developments in Geology and Exploration, G.R. Wessel and B.H. Wimberly, eds., SME, Littleton, CO, pp. 23–50.

Paulson, C.O., 1991, ''Recovered Sulphur Production to the Year 2000,'' paper presented to 13th Annual Phosphate-Sulphur Symposium, The Sulphur Institute, Washington, DC, 8 pp. with figs.

Pawłowski, S., Pawłowska, K., and Kubica, B., 1979, ''Geology and Genesis of the Polish Sulfur Deposits,'' *Economic Geology*, Vol. 74, pp. 475–483.

Porch, E.L., Jr., 1917, ''The Rustler Springs Sulphur Deposits,'' Bulletin No. 1722, Bureau of Economic Geology, University of Texas-Austin, 71 pp.

Posey, H.H., and Kyle, J.R., 1988, ''Fluid-Rock Interactions in the Salt Dome Environment: An Introduction and Review,'' *Chemical Geology*, Vol. 74, pp. 1–24.

Postgate, J.R., 1979, *The Sulphate-Reducing Bacteria*, Cambridge University Press, Cambridge, UK, 125 pp.

Preisner, L., 1991, ''Sulphur Mining in Poland,'' Working Paper No. 47, Centre for Resource Studies, Queen's University, Kingston, ON, 26 pp.

Price, P.E., Kyle, K.R., and Wessel, G.R., 1983, ''Salt Dome Related Zinc-Lead Deposits,'' *Proceedings*, International Conference on Mississippi Valley Type Lead-Zinc Deposits, G. Kisvarsanyi, et al., eds., University of Missouri-Rolla, pp. 558–571.

Raymont, M.E.D., and Hyne, J.B., 1983, ''Ten Years Experience in Sulphur Forming,'' *ASR Quarterly Bulletin*, Vol. 20, No. 2.

Ruckmick, J.C., Wimberly, B.H., and Edwards, A.F., 1979, ''Classification and Genesis of Biogenic Sulfur Deposits,'' *Economic Geology*, Vol. 74, pp. 469–474.

Ruckmick, J.C., et al., 1992, ''The Mustang Draw Sulfur Deposit in Glasscock County, Texas,'' Proceedings, *Native Sulfur—Developments in Geology and Exploration*, G.R. Wessel and B.H. Wimberly, eds., SME, Littleton, CO, pp. 135–138.

Russell, A., 1989, ''Pyrites As A Sulphur Source, Down But Not Out,'' *Industrial Minerals*, No. 261, pp. 41–52.

Salas, G.P., 1991, ''History of Exploration for Sulfur in Southeast Mexico,'' *Economic Geology, Mexico*, Volume P-3, The Geology of North America, G.P. Salas, ed., Geological Society of America, pp. 415–424.

Salisbury, B.K., 1992, ''Geophysical and Geochemical Surveys in Delaware Basin Sulfur Exploration,'' Proceedings, *Native Sulfur—Developments in Geology and Exploration*, G.R. Wessel and B.H. Wimberly, eds., SME, Littleton, CO, pp. 81–90.

Samuelson, S.F., 1986, ''Sulfur Mining at Boling Dome,'' *Comparison of Cap Rocks, Mineral Resources, and Surface Features of Salt Domes in the Houston Diapir Province*, S.J. Seni and J.R. Kyle, eds., Field Trip Guidebook, Geological Society of America Annual Meeting, San Antonio, TX, pp. 147–154.

Samuelson, S.F., 1992, ''Anatomy of an Elephant: Boling Dome,'' Proceedings, *Native Sulfur—Developments in Geology and Exploration*, G.R. Wessel and B.H. Wimberly, eds., SME, Littleton, CO, pp. 59–71.

Saylak, D., and Conger, W.E., 1982, ''A Review of the State of the Art of Sulfur Asphalt Paving Technology,'' *Sulfur: New Sources and Uses*, M.E.D. Raymont, ed., Symposium Series 183, American Chemical Society, Washington, DC, pp. 155–194.

Schmoker, J.W., 1979, ''Interpretation of Borehole Gravity Surveys in a Native Sulfur Deposit, Culberson County, Texas,'' *Economic Geology*, Vol. 74, pp. 1462–1470.

Schwalm, W.J., and Doyle, K.B., 1987, ''Alberta's Sulphur Inventory: The Present Situation and Future Recoverability,'' *Sulphur*, No. 190, pp. 20–23.

Seni, S.J., 1986, ''Evolution and Lithofacies of Boling Dome Cap Rock,'' *Comparison of Cap Rocks, Mineral Resources, and Surface Features of Salt Domes in the Houston Diapir Province*, S.J. Seni and J.R. Kyle, eds., Field Trip Guidebook, Geological Society of America Annual Meeting, San Antonio, TX, pp. 155–168.

Seni, S.J., 1987, ''Evolution of Boling Dome Cap Rock With Emphasis on Included Terrigenous Clastics, Fort Bend and Wharton Counties, Texas,'' *Dynamical Geology of Salt and Related Structures*, I. Lerche and J.J. O'Brien, eds., Academic Press, Orlando, FL, pp. 543–591.

Seni, S.J., and Kyle, J.R., eds., 1986, *Comparison of Cap Rocks, Mineral Resources, and Surface Features of Salt Domes in the Houston Diapir Province*, Field Trip Guidebook, Geological Society of America Annual Meeting, San Antonio, TX, 260 pp.

Shelton, J.E., 1980, ''Sulfur,'' Preprint, Bulletin 671, US Bureau of Mines, 22 pp.

Shrive, N.G., et al., 1982, ''The Potential and Properties of Sulfur Concretes,'' *Sulfur: New Sources and Uses*, M.E.D. Raymont, ed., Symposium Series 183, American Chemical Society, Washington, DC, pp. 137–154.

Skyring, G.W., 1987, ''Sulfate Reduction in Coastal Ecosystems,'' *Geomicrobiology Journal*, Vol. 5, No. 3/4, pp. 295–374.

Slim, B.A., 1986, ''Sulphurdale, Utah, A Feasibility Study for an Open Pit Mine,'' *Sulphur*, No. 182, pp. 16–20.

Smith, A.R., 1980, ''Sulfur Deposits in Ochoan Rocks of the Gypsum Plain, Southeast New Mexico and West Texas,'' *Trans-Pecos Region, Southeastern New Mexico and West Texas*, 31st Field Conference Guidebook, P.W. Dickerson and J.M. Hoffer, eds., New Mexico Geological Society, pp. 277–282.

Spirakis, C.S., and Cunningham, K.I., 1992, ''Genesis of Sulfur Deposits in Lechuguilla Cave, Carlsbad Caverns National Park, New Mexico,'' Proceedings, *Native Sulfur—Developments in Geology and Exploration*, G.R. Wessel and B.H. Wimberly, eds., SME, Littleton, CO, pp. 139–145.

Terrel, R.L., 1982, ''The Potential for New Sulfur Products in the Middle East,'' *Sulfur: New Sources and Uses*, M.E.D. Raymont, ed., Symposium Series 183, American Chemical Society, Washington, DC, pp. 225–250.

Thode, H.G., 1972, ''Sulfur Isotope Fractionation—In Biological Processes,'' *Encyclopedia of Geochemistry and Environmental Sciences*, Encyclopedia of Earth Sciences Series, Vol. IVA, R.W. Fairbridge, Jr., ed., Van Nostrand Reinhold, New York, pp. 1152–1157.

Trusheim, F., 1960, ''Mechanisms of Salt Migration in Northern Germany,'' *American Association of Petroleum Geologists Bulletin*, Vol. 44, pp. 1519–1540.

Wallace, C.S.A, and Crawford, J.E., 1992, ''Geology of the Culberson Ore Body,'' Proceedings, *Native Sulfur—Developments in Geology and Exploration*, G.R. Wessel and B.H. Wimberly, eds., SME, Littleton, CO, pp. 91–105.

Ward, J.M., 1992, ''The Cinnabar Canyon Sulfur Deposit,'' Proceedings, *Native Sulfur—Developments in Geology and Exploration*, G.R. Wessel and B.H. Wimberly, eds., SME, Littleton, CO, pp. 159–164.

Warren, J.K., 1989, *Evaporite Sedimentology*, Prentice Hall, Englewood Cliffs, NJ, 285 pp.

Wessel, G.R., 1992, ''Surface Mapping and Subsurface Structure: Application to Permian Basin, West Texas,'' Abstract, *West Texas Geological Society Bulletin*, Vol. 31, No. 7, p. 6.

Zakiewicz, B., 1986, ''Recent Developments in Sulphur Mining by Underground Melting,'' *Sulphur*, No. 184, pp. 26–32.

Zimmerman, J.B., and Thomas, E., 1969, ''Sulfur in West Texas: Its Geology and Economics,'' Geological Circular 69–2, Bureau of Economic Geology, University of Texas-Austin, 35 pp.

TALC

ROBERT J. PINIAZKIEWICZ, EDWARD F. McCARTHY, AND NOEL A. GENCO

Talc, when it can be isolated as a pure mineral, has a composition of 63.36% SiO_2, 31.89% MgO, and 4.75% H_2O. However, as a commercial commodity less than half of all talc sold has a purity exceeding 95%. Nevertheless, impure talc products find a multitude of uses and few substitutes in many industrial applications.

Minerals commonly associated with and sold together in talcose mixtures are chlorite, dolomite, mica, magnesite, and tremolite.

Steatite was originally a mineralogical term applied to pure talc. Today steatite usually denotes a high-talc ceramic body used as an electrical insulator, and the talc used in such applications is known as steatitic talc. Steatitic talc generally is found in massive, high-purity form and in the past was machined directly into electrical parts, which rarely is done today. Impure varieties of massive or block talc are still commonly termed soapstone. Soft massive talc, suitable for crayon manufacture, has been referred to as French chalk.

It has been common practice to discuss talc, soapstone, and pyrophyllite under the same general heading. In the case of soapstone and talc this is natural, since many different types of platy, soft minerals exhibiting a high degree of lubricity (commonly referred to as slip) have been grouped together and called soapstone or talc.

Further, there are mineralogical relationships between pyrophyllite and montmorillonite and between talc and hectorite. Thus, talc and pyrophyllite are sometimes referred to as clay minerals. When finely divided talc or pyrophyllite are combined with water in proportions to make a slurry, the end product does have the appearance of a clay-water mixture.

Soapstone for utensils and ornaments was mined by prehistoric Native Americans on Santa Catalina Island, CA. In the mid-1800s soapstone from deposits along the western foothills of the Sierra Nevada Mountains was used by white settlers as building and ornamental stone and for the linings and foundations of furnaces (Anon., 1956). Previous to 1916 the annual recorded production in California did not exceed 1.8 kt. From 1912 to 1918, however, output rose sharply when the Talc City, Western, and Silver Lake mines were put into operation. From 1916 to 1935, the state's annual output of talc was in the range of 8.2 to 18.2 kt. There was another increase in the mid-1930s, when the use of talc in wall tile grew rapidly. In California the production of talc grew to 57.2 kt in 1943. The post-war building boom helped California's production to grow to 108.9 kt in 1951. Production stood at 149.8 kt in 1968, 131.6 kt in 1969, 167.9 kt in 1970, and 140.7 kt in 1972. In the mid-seventies and early eighties, however, California's talc production declined to approximately 20 kt in 1992 due to reserves exhaustion and the material's tremolite content, which raised significant product liability concerns at the time.

Talc mining in New York State dates from about 1878, when a Colonel Palmer and associates opened the first commercial talc mine on the Nelson Freeman farm near Talcville. In 1893 this operation was sold to International Pulp Co., which changed its name to International Talc Co. in 1944. The company was acquired by R.T. Vanderbilt Co., Inc., in 1974.

A new talc operation, Gouverneur Talc Co., owned by R.T. Vanderbilt Co., began operations near Balmat, NY, in 1948. The initial capacity of the processing plant was 180 tpd (Gillingham, 1950). Subsequent expansions have increased capacity to more than 725 tpd.

Talc was discovered in the area of Johnson, VT, in 1902, marking the birth of the Vermont talc industry. American Minerals Co. initiated plant operations at Johnson in 1904. Magnesia Talc Co. opened a plant at Waterbury, VT, in 1913 (Burmeister, 1963, Trauffer, 1964) and acquired American Minerals Co. in 1923. The Waterbury mine had a long history as a producer of talc crayons. Eastern Magnesia Talc Co. was formed in 1924 by the merger of Eastern Talc Co. and Magnesia Talc Co. In 1956 Vermont Mineral Co., with a talc mine at Hammondsville, VT, was also acquired. A new plant was then built at Gassetts to produce roofing products.

In order to produce high grade products for several markets, Eastern Magnesia Talc Co. began operating a modern froth flotation plant at West Windsor, VT, in 1964. The ore was provided by the Hammondsville mine. This operation, known as Windsor Minerals, Inc., continued from 1972 to 1989 as a subsidiary of Johnson and Johnson Co. In 1989 Cyprus Minerals Co. acquired Vermont Talc, Inc. and Windsor Minerals, Inc., thus consolidating the talc mining and processing operations at Hamm, Windham, Johnson, Chester, Ludlow, West Windsor, Troy, and Reading, VT, under one ownership. As the result of Cyprus' sale of its worldwide talc assets to the RTZ Group in 1992, its US talc operations are now part of Luzenac America, Inc.

The Texas talc deposits were first mined commercially in 1952 and by 1970 ranked fourth in US production behind New York, California, and Vermont. By 1990, however, talc production increased to about 300 ktpy, making Texas the third-ranking producer behind Montana and Vermont. Principal use of this talc is in the ceramics industry. Lundquist (1970) reports on a process for acid treating, calcining, and grinding Texas talc and talc-kaolin mixtures. Using this process, Lundquist increased the brightness value from 65 to more than 94.

Talc production in Montana is centered in the mountains around Ennis, Alder, and Dillon. Currently Montana is the largest talc-producing state in the United States. Prior to 1970 Montana produced significantly less tonnage annually than either New York or California. Since 1970 production has increased due to an increasing use of Montana talc for controlling pitch during paper manufacture, the use of very fine (e.g., 6 Hegman) talc for flatting and sheen control in coatings, and a shift in talc's usage patterns due to US government regulations. As fibrous minerals have come under increasing environmental scrutiny, consumers in paint, plastics, and other markets that used New York or California talcs have switched to Montana talcs to avoid dealing with fibrous mineral regulations.

Table 1. Typical Chemical Analyses of Talc Ores and Products, %

	Pure Talc (Theoretical)	1	2	3	4	5
SiO_2	63.36	35.98	59.15	62.65	59.80	54.92
MgO	31.89	32.95	31.34	30.23	27.45	27.20
Fe_2O_3	—	0.65	3.36	1.51	0.05	0.46
TiO_2	—	0.02	—	—	—	—
Al_2O_3	—	0.43	0.26	0.31	0.57	—
CaO	—	0.00	0.15	trace	6.80	5.76
K_2O	—	0.00	—	0.05	—	—
Na_2O	—	0.00	—	0.15	—	—
MnO	—	0.41	—	—	0.39	—
S	—	0.06	—	—	(SO_3) 0.07	—
NiO	—	0.21	—	—	—	
Cr_2O_3	—	0.18	—	—	—	
CoO	—	0.01	—	—	—	—
FeO	—	5.96	—	—	0.15	—
Less O for S	—	0.05	—	—	—	—
P_2O_3	—	0.01	—	—	—	—
LOI	4.75	23.18	6.06	5.14	5.93	10.76
	100.00	100.10	100.32	100.04	101.21	99.10

Source: 1–4, Chidester et al., 1964; 5, Pence, 1955. 1. Typical Vermont talc-magnesite ore, 2. Flotation product, Johnson, Vermont mill, 3. Steatite, Yellowstone mine, Montana, 4. Average talc ore, Talcville, Gouverneur District, New York, 5. Texas Talc.

END USES

Talc is an extremely versatile mineral. Despite the relatively low purity of most ores mined, talc has found a steadily increasing number of uses. With the exception of pure steatite grades, hand-picked platy cosmetic talcs, and a few products from wet processing plants, industrial products are really mixtures of many minerals. High purity talc is used in cosmetics, steatite and cordierite ceramics, for pitch control in the paper industry, and as a reinforcing filler in rubber. Products of intermediate purity (e.g., 75 to 95% talc) are used as paper fillers, for plastics reinforcing, and in wall tile, paint, and dusting compounds for rubber. Lower purity products (e.g., <75% talc) are used in roofing materials, polyester body patch compounds, flooring, and fertilizer formulations. Accessory minerals can have an important effect in some applications. Chlorite, being platy like talc, is acceptable in most end uses except steatite and cosmetics. Tremolite of acicular particle shape is beneficial in wall tile bodies, where it provides porosity, and in coatings, where it lowers viscosity; however, it is a negative in plastics due to abrasion and in cosmetics due to health concerns.

Table 1 lists chemical analyses of typical talc ores and products.

Talc, along with many other filler-type industrial minerals, is subject to increasing demands for higher purity and better quality control. A good example is talc sold to the cosmetic and toiletry goods industries. Prior to the end of 1970 a wide variety of talc and talcose products was accepted for cosmetic use if the products met rather loose and uncomplicated specifications. A good cosmetic talc had to be reasonably light colored, have good slip, be free of gritty or hard mineral particles, and measure up to a certain fragrance-retention standard. Now the cosmetic talc market is a demanding one. Not only has the industry itself set higher and new standards through the Cosmetics, Toiletries, and Fragrance Association (CTFA), but new rulings from federal agencies have made heavy demands on talc product quality.

Table 2 lists the various properties of talc that are important in specific markets or industries.

Table 2. Talc Properties Important in Specific Markets

- Paint Industry
  - Whiteness
  - Platy particle shape
  - Low oil absorption
  - Opacifying power
  - Fine particle size (Hegman gage)
- Paper Industry
  - Paper Filler:
  - Whiteness
  - Brightness (GE 78 minimum)
  - Controlled top size (50 μm maximum)
  - Platy particle shape
  - Opacifying power
  - Low abrasion
  - Particle size (8 to 12 μm median)
  - Pitch Control:
  - Surface area (12 m^2/g minimum)
  - Whiteness
  - Brightness (GE 78 minimum)
  - Low abrasion
  - Particle size (2 to 5 μm median)
  - Good dispersion without surfactants
- Ceramics Industry
  - Uniform chemical composition
  - Constant amount of shrinkage on firing
  - Fired color
  - Particle size distribution
- Cosmetics Industry
  - Quartz content (1.0% maximum; may be lowered to 0.1% maximum)
  - Tremolite content (0.1% maximum)
  - Loss on ignition (6.0% maximum at 1 000°C)
  - Neutral pH
  - Acid-soluble substances (2.0% maximum)
  - Water-soluble substances (0.1% maximum)
  - Arsenic content (3 ppm maximum)
  - Lead content (20 ppm maximum)
  - Odor, slip (lubricity), particle size, fragrance retention, and whiteness according to customer specification
- Plastics Industry
  - Platy particle shape (reinforcing ability)
  - Whiteness
  - Chemical inertness
  - Low iron content
  - Electrical resistivity
  - Superfine particle size
  - Low abrasion
  - Powder bulk density
- Roofing Industry—Asphalt Backing and Surfacing
  - Platy particle shape
  - Particle size distribution and its consistency
  - Low oil absorption
  - Minimal dust content
- Automotive Industry—Lubricants, Body Putty, and Asphaltic Undercoating
  - Free from grit (pure platy talc for lubricants)
  - Chemically inert
  - Nonwicking (undercoating)
- Rubber Industry
  - Particle size (<2 μm median; controlled top size)
  - Platy particle shape
  - Chemical inertness
  - Electrical resistivity
  - Good lubricity

Paint

In 1990 an estimated 154 kt of talc valued at approximately $24 million were used as extender and filler pigments in the US paint industry. This represented 19% of total US talc production of about 816 kt for that year. As a paint pigment extender talc's consumption is exceeded only by calcium carbonate and kaolin.

Some of the important properties of talc used in paint are color, fineness of grind, oil absorption, chemical inertness, and resulting viscosity. Measurement of oil absorption is important to the paint industry, because it helps predict the final viscosity of a coating film. This property is measured by the American Society for Testing and Materials (ASTM) Test Method D 281–84, which measures the number of parts of acid-refined linseed oil required to produce a coherent paste with 100 parts of talc pigment.

Talc helps to impart integrity, reinforce paint films, control viscosity, and prevent sagging of paint films. It improves suspension and prevents settling of solids to the bottom of the paint container. Because talc is very soft (viz., 1 on the Mohs hardness scale), it causes only minimal abrasion to process and application equipment. It disperses readily in both aqueous and solvent-based paints. Talc having a surface area greater than 12 m^2/g is used for gloss control in paints when an exact degree of sheen or luster is required. Gloss is usually determined with a reflectance meter at 60° inclination. Sheen values are determined with the same meter and with the film at 85° inclination.

A useful, uncomplicated method for testing the fineness and degree of dispersion of paint and other wet-ground products involves the Hegman gage. This gage can also be called a "fineness of grind gage" (Anon., 1970). It is a precision-machined, smooth, flat, hardened steel block with one or two grooves or paths about 133 mm long and 38 mm wide. The groove depth ranges from 100 μm (about 4 mils) at one end to zero at the other end. A scraper blade is the second component of the gage. In use on talc products, an aliquot of a mixture of talc and linseed oil (about one part talc to four parts of oil) is placed in the tapered part of the groove. Then the mixture is wiped down toward the zero depth end (i.e., 8 Hegman) of the gage with the stainless steel wiper blade. The fineness is determined by holding the gage to a familiar light source and observing the area where the coarse particles begin to appear through the resulting film. The angle between the face of the gage and the line of vision should be not more than 30° or less than 20°. The fineness readings can be reported in microns or mils. A reading of 6 on the Hegman gage indicates a particle size of about 25 μm.

Paper

One of the major consumers of talc, especially in Europe and Japan, is the paper industry. As a paper filler talc has a minimal negative effect on sheet strength but improves ink receptivity. It also reduces gloss but enhances opacity and brightness. Its retention in the paper fibrils usually approaches 100%. The main reason for the growth of talc consumption in paper is the availability of high quality micronized talc manufactured to exacting specifications. Talc fillers should not contain more than 2 to 5% calcium carbonate or similar minerals that react with the alum used in acid papermaking systems. Median particle sizes range from 8 to 12 μm, and Einlehner Abrasion values should be lower than 50 g/m^2 so as to minimize wear of papermaking and printing equipment.

When talc is used as a titanium dioxide extender in paper it has a distinct weight advantage, a specific gravity of 2.8 vs. 4.2 for TiO_2. This advantage becomes more important as postal rates increase.

High-purity talc products are used in paper in competition with titanium dioxide pigments. Considerable history already has been made in the paper industry in Finland, where talc is widely used as a titanium dioxide extender. In 1990 titanium dioxide (anatase) sold for less than $2/kg. By comparison, the best grades of talc sold for only $0.25/kg. This price difference indicates that talc, when it can be substituted, even partially, will continue to be a formidable competitor in titanium dioxide markets.

The unique property of preferentially wetting oily substances in the presence of water makes talc extremely effective in pitch control. Pitch and other oleoresinous components of paper pulps cause serious manufacturing problems if they are not controlled. Talc, micronized to a surface area of 12 m^2/g or more, adsorbs the pitch and helps prevent accumulations that would otherwise deposit on the rolls, wire, and other parts of the paper machine. Competitive materials such as diatomaceous earth and clay usually are either too abrasive or much less effective in the presence of water.

Ultrafine talc products (e.g., 5 μm maximum particle size) now finding use in paper coating colors have similar ranges of particle size distribution as coating clays and calcium carbonate. The talc products can be used to pigment latex, starch, or alpha protein coating colors. Ultrafine talc products can be used with clay, calcium carbonate, and titanium dioxide pigments to control rheological properties, calendering, gloss, ink holdout, opacity, and brightness.

Ceramics

Ceramic applications are the largest single end use for talc in North America, accounting for 28% of the total tonnage consumed in 1991. Ceramic markets for talc in Europe and Japan are much smaller because of their use of different technology.

The wall tile segment of the semivitreous whitewares market is very large but declining in the US, where it consumed nearly 190 kt of Texas and New York talcs in 1991. A typical wall tile body consists of 50 to 70% talc and 25 to 35% ball clay, with wollastonite making up the balance. This mixture forms a solid solution between 800 and 1 050°C during a 16 to 30 hr firing cycle. Such wall tile bodies have high porosity and controlled, predictable thermal expansion. After being cooled they are coated with a glaze and then refired.

Newer, European tilemaking technology uses a triaxial body consisting of kaolin, flint, and feldspar. The tiles are pressed under high pressure, precoated with a glaze, and fired once to about 1 200°C over a very short cycle. Total process time amounts to between 3 and 5 hr. The glaze melts at between 900 and 950°C. Because talc does not lose its lattice water at temperatures below about 960°C, it is almost unusable in wall tiles made with this technology.

As long ago as 1930 reports began to appear praising the improved craze resistance of semivitreous bodies in which New York talc was used. New York and Texas talcs are considered to be ideal raw materials for wall tile production for the reasons listed below.

1. The formation of enstatite as talc dissociates during firing produces high thermal expansion bodies, which results in glazes being put into compression and thereby minimizes crazing (i.e., hairline cracks in the glaze).
2. Low moisture expansion bodies are produced, which results in good resistance to delayed crazing.
3. The massive nature of the talc aids in dry pressing and in the preparation of good casting slips.
4. Low firing temperatures are possible.
5. Fast firing schedules are possible, as fast as 1 hr cold to cold.
6. The shrinkage and absorption properties of bodies are fairly constant over a wide temperature range.
7. Good white-fired bodies are possible.
8. Glazes of unusual brilliance and attractiveness can be fitted readily to high-talc bodies.

Pence (1955) gives an early report on white-firing talc from West Texas for ceramic uses. The talc described was from claims located in Hudspeth County about 37 km west of Van Horn, TX. Its chemical analysis revealed 54.92% SiO_2, 27.20% MgO, 5.76% CaO, 1.36% Al_2O_3, 0.46% Fe_2O_3, and 10.76% ignition loss. By firing standard 108 × 108 × 9.5 mm hand-pressed tiles, Pence obtained satisfactory results in terms of color, density, glaze fit, and other properties required of ceramic wall tiles.

Hobbyware bodies, which are very similar to wall tile bodies and are used by individual, home-based artisans, are still a major consumer of white-firing New York and Texas talcs. In pottery, talc may constitute 40% of the body mix. The thickness of the object being made, the texture and porosity required, and the firing conditions used all influence the amount of talc used in the mix. Tableware bodies often contain 15% talc.

Cordierite bodies have the chemical formula $2MgO{\cdot}2Al_2O_3{\cdot}5SiO_2$ and are made from Montana talc mixed with clay and aluminum oxide. Pure cordierite has an extremely low coefficient of thermal expansion ($<2.5 \times 10^{-7}$mm/°C) and has long been used in manufacturing kiln furniture. A major new application is the molded, honeycomb-like catalyst supports in auto emission reduction devices, for which cordierite is uniquely suited because of its ability to withstand repeated thermal shock.

Talc also is used in steatite bodies, which are manufactured into high-frequency electrical insulators, a segment of the ceramic whitewares market. These bodies typically consist of 85% Montana talc, 10% plastic kaolin, and 5% barium carbonate and are fired to cone 12, or 1 326°C. Steatite-grade talc must be very pure. It should contain less than 1.5% CaO, less than 4.0% Al_2O_3, and less than 1.5% Fe_2O_3, and it also must be low in sodium and potassium. It should contain no more than 5 to 10% nontalc minerals, preferably only 1 or 2%.

A minor use of talc is as an additive in dinnerware bodies composed of clay, flint, and feldspar. Comprising as much as 6% of such bodies, talc provides greatly improved craze resistance.

Plastics

Talc is an important functional filler for compounded thermoplastic and thermoset polymers. In 1990 the plastics industry in the US consumed 70 kt of talc at an average selling price of $147/t. Talc is used mostly in polypropylene, with lesser amounts going into vinyl, polyethylene, nylon, polyester, and engineering polymers.

Because of its platy particle shape talc is used in plastic primarily as a reinforcing filler. It increases the stiffness of polypropylene by 80% at a 20% loading and by 150% at a 40% loading. It also increases the heat resistance of polypropylene, reduces mold shrinkage, improves melt rheology, and reduces molding cycle time. Compared to other reinforcing fillers, such as fiberglass and mica, talc is much less abrasive. It gives polypropylene superior low-temperature shock resistance.

In polyethylene film talc is used as an antiblock agent, an additive that roughens the surface of tacky film to prevent it from adhering to itself. Loading levels range from 0.5 to 1.0%. In vinyl talc is used to improve the hot strength of profile extrusions and to reduce the shrinkage of floor tile. Fine particle size grades are used as nucleating agents for semicrystalline polyolefins, nylon, polyesters, and chlorinated polyethylene. Talc can be used in polyolefins at loading levels ranging from 1 to 50% by weight (Poppe et al., 1972).

Most talc is not used in plastics simply as a diluent or resin extender. Talc has a specific gravity of 2.8, compared to 0.93 for polypropylene. With commodity polypropylene priced at $0.65/kg or less, talc would have to be delivered to a compounder for a price of less than $0.21/kg in order to reduce the cost per unit weight of the compounded material on a one-for-one replacement basis. Few talc products that are suitable for use in plastics can be delivered for that price.

Processes for the chemical treatment of talc fillers are becoming increasingly sophisticated. Silane coupling agents are used to improve performance of talc fillers in plastic materials. Ranney et al. (1972) report that the tensile strength of 50% (by weight) loaded polyethylene can be increased by 25 to 30% over that of a non-silane-containing standard.

Silane coupling agents are used with clays, talc, and other mineral fillers. These agents are a family of organosilicon monomers that characteristically possess two different kinds of chemical functionality. The general formula for the molecule is $R\text{-}Si(OX)_3$, where R is the organofunctional group attached to the silicon atom in a thermally and hydrolytically stable manner. In this formula OX designates hydrolyzable groups connected to silicon. These groups are generally believed to be the means whereby silane coupling agents interact with siliceous inorganic surfaces (e.g., talc) to form a more durable bond with the filler mineral. With this enhanced bonding it is often possible to increase the filler loading without sacrificing the properties required in the polymer system.

Roofing Products

Talc is used both as a filler and as a surfacing material in the manufacture of roofing. When used as a filler, talc acts as a stabilizer for the melted asphaltic constituents. It helps increase stability and weather resistance. When talc is sprinkled on the surface of either asphalt shingles or roll roofing, it prevents the roofing from sticking during manufacture and storage before use.

Typical particle size specifications for talc used as a surfacing agent for roofing products are:

Screen Size	Range, Wt%
500 μm (+35 mesh)	None
212 μm (+65 mesh)	10 to 20
−212+150 μm (−65+100 mesh)	15 to 30
−150+75 μm (−100+200 mesh)	25 to 35
−75 μm (−200 mesh)	20 to 25

The use of low-grade talc as a surfacing agent, filler, and backing for smooth-surface roofing materials and as a filler and backing for asphaltic roofing shingles is a fairly stable market. When used as a surface coating, the product's color and staining properties are important. When talc is used as backing, the primary qualifications are particle size range, particle shape, absorbency, and above all, low cost.

Miscellaneous Uses

Talc is used in a wide variety of textile materials. Textile sizing compositions include binders and mineral fillers. Some textiles use sizings for temporary stiffness and improved appearance for sales appeal. Because the degree of whiteness is often very important, talc or other extremely white minerals are favored. Where extreme whiteness of color is important, talc often competes with titanium dioxide at a considerable cost advantage.

Talc and pyrophyllite are used as fillers in the integral foamed latex rubber backing for carpets, rugs, and parquet hardwood floor panels. Impurities such as calcium, magnesium, iron, and manganese can chelate with latex rubber and are therefore undesirable in such fillers. Talc is used in the felt backing of sheet flooring.

The unique lubricity of talc and its ease of dispersion are important for its use in lubricants and greases, especially those that must function over a wide temperature range without freezing or running off the bearing surface. A typical grease capable of use over an extreme temperature range might contain a silicone oil, a colloidal silica thickener, molybdenum disulfide, talc, and several chemical additives to prevent corrosion of the bearing surface or decomposition of the lubricant (Elstrom, 1961).

Table 3. Approximate Composition of Minerals Present in Talc Ores, %

	SiO_2	MgO	CaO	CO_2	Fe	Al_2O_3	K_2O	Na_2O	H_2O	TiO_2
Talc	63	32							3–7	
Serpentine (antigorite)	44	43							8–13	
Chlorite	33	36				18			5–14	
Anthophyllite	58	30	2						2.2–15	
Tremolite	57	28	13						2.3–15	
Actinolite	52	5	9		34				3	
Diopside	56	18	26							
Feldspar	65					18	17			
Magnesite		48		52						
Dolomite		22	30	48						
Calcite			56	44						
Quartz	100									
Muscovite	46	0.1				39	10	0.6	4.3	
Magnetite (Fe_3O_4)					72					
Ilmenite ($FeTiO_4$)					33					48.0
Pyrite (FeS_2)					47					
Tourmaline[a]	36	11	0.5			33	0.6	2.3	3.8	0.2
Graphite (100% C)										

[a] 10% B_2O_3.

In the cosmetics industry talc is used in baby and body powders, facial products, antiperspirant sticks, and various types of creams and lotions. It is used primarily because of its softness and smoothness, although its whiteness, chemical inertness, and hydrophobicity are also important. Highly purified talcs are used in pharmaceutical tablets as excipients (inert carriers) and in chewing gum to detackify the resins.

Talc is widely used as a dusting and mold release agent in the manufacture of all types of molded and extruded thermoset elastomers. Fine particle size grades serve as a process aid and reinforcing agent in mechanical rubber goods. Lower purity talc is used as the principal ingredient in polyester body patch compounds for automotive repair work, where the mineral's softness gives cured compounds excellent sanding properties. For the same reason it is used in gypsumboard tape joint compounds

In fertilizer manufacture talc is used as a coating for ammonium nitrate prills. It serves as an inert carrier for sensitive herbicides and as a process aid and flow aid for animal feed.

Both platy and nodular talc powders are covered by US patents for use in corrosion-proofing compositions such as for automotive undercoating (Miller and Holtzapfel, 1970). The talc is about 50% by weight finer than 10 μm and no more than 3% coarser than 44 μm. Talc plays an important role in preventing "wicking" of the composition. Undercoating products containing fillers such as asbestos are often subject to failure due to wicking, which causes moisture to travel through a corrosion-proofing coating.

The use of talc in dry fire-extinguishing powders has been patented (Cawood, 1962). Talc is used in quantities ranging from 10 to 15 parts per 100 parts of product. It is used primarily to maintain the free-flowing condition that is essential for dry powders used in fire extinguishers.

Talc is used in the textile industry both for loading and bleaching materials such as cotton sacks, cordage, rope, and string. For such purposes a good grade of low-abrasion white talc is required. Talc used for backfilling textiles can be of lower quality.

Block talc is still carved by native peoples in northern Quebec, and talc carving is a highly developed art form in China and Korea. Block talc also is cut into panels for insulating wood stoves in New England and into crayons for marking steel.

Other uses for talc include:

- Cereal polishing (e.g., rice, corn, barley)
- Bleaching agents
- Odor absorption from foods
- Floor wax
- Water filtration
- Leather treatment (for oil adsorption)
- Joint fillers and grouts
- Insecticides
- Shoe polish
- Welding rod coatings
- Printing inks (for rheology control)
- Coating for iron ore pellets in direct reduction processes
- Source of magnesium in plant foods
- Mild abrasives.

GEOLOGY

Mineralogy

Talc is a hydrous magnesium silicate with the theoretical chemical formula $Mg_3Si_4O_{10}(OH)_2$. The chemical composition of talc ores found in nature, however, is quite variable. In industry the term talc covers a broad variety of rocks and rock products. Talc is an extremely soft mineral and is rated No. 1 on the Mohs scale of hardness. Color variations range from snow-white to black, including greenish gray and various shades of green, pink, and even red. Talc's specific gravity ranges from 2.58 to 2.83. Talc usually forms as a secondary mineral by alteration of parent rocks that contain silica and magnesium, such as serpentinite, pyroxenite, and siliceous dolomite.

A sheet silicate mineral, talc is structurally similar to pyrophyllite, except that octahedral sites in the layers or sheets are occupied by magnesium instead of aluminum. Also, none of the octahedral sites are vacant, which is not the case with pyrophyllite. Chemically pure talc probably has never been found in nature in commercial quantities. Practically all of the talc products offered are impure. Even hand-sorted, selectively mined cosmetic-grade talcs contain extraneous mineral components such as dolomite and chlorite and elemental components such as iron.

When the major talc products in world trade are examined, it becomes obvious that only a minuscule percentage of world reserves are nearly pure talc. When impure talc is processed by froth flotation or similar selective separation methods, very high-purity talc products can be produced. It would seem proper to call a talc product high purity if it contains upwards of 95% $Mg_3Si_4O_{10}(OH)_2$.

Table 3 lists the approximate composition of minerals present in talc ores.

Tremolite, which has the theoretical chemical formula $2CaO{\cdot}5MgO{\cdot}8SiO_2{\cdot}H_2O$, is common in talc ores, often at the expense of the mineral talc itself. This relationship is found throughout large talc mining districts, such as in New York and California-Nevada. The main isomorphous components of the tremolite group are ferric iron, manganese, and aluminum. Tremolite can contain 2 to 3% sodium, and the group includes three varieties—iron tremolite, manganese tremolite (hexagonite), and fluorine tremolite. Other talclike minerals that constitute various proportions of some talc ores are anthophyllite, antigorite, chlorite, pyrophyllite, and serpentine.

Povarennykh (1972) includes pyrophyllite, talc, minnesotaite, and willemseite in the pyrophyllite group of minerals. In his descriptions of this group he includes Mg-pyrophyllite, Fe-pyrophyllite, Fe^{2+}-talc, $Fe^{2+}Fe^{3+}$-talc, Ni-talc, and Mg-willemseite. Talc can contain as much as 1% nickel, pyrophyllite as much as 9% magnesium. These chemical variations within the pyrophyllite group suggest a wide variety and possibility of occurrence of talc-like pyrophyllite.

At the Jatoba talc mine near Ponta Grossa, Brazil, a claylike antigorite has been produced for use as a diluent in insecticides and as talc in pharmaceutical applications (Brindley and Santos, 1971). This is another example of a talc-like mineral finding commercial acceptance as talc.

Faust and Fahey (1962) provide a good review of extensive studies by others relating to the distribution of minor elements in magnesium-rich minerals and rocks. There is agreement by scientists that "rocks of the ultrabasic suite and their hydrothermally altered forms rich in serpentine contain an unusually high concentration of nickel, chromium, cobalt, and scandium. Moreover, some of the investigators have observed that the carbonate-rich rocks, in contrast, contain very small to practically negligible concentrations of these elements." This is in agreement with the high nickel content of some Vermont talcs and its complete absence in some western carbonate-talc ore bodies. It should be noted, however, that some western talc-carbonate ore bodies do exhibit nickel anomalies in comparison to unmineralized carbonates in the same area.

Differential thermal analysis-thermogravimetric analysis (DTA-TGA) and X-ray diffraction constrained by loss on ignition (LOI) both provide good analytical methods for the identification of talc and many of the impurities associated with talc ores. A large endothermic peak and weight loss at 965°C is typical of relatively pure talc analyzed by DTA-TGA methods.

The fluorine content of talc is a matter of concern to ceramics manufacturers because of government regulations regarding stack emissions. On the basis of fluorine analyses of 22 talcs from deposits worldwide, it has been noted that talc from metamorphosed sedimentary rocks contains 0.11 to 0.48% fluorine. Talcs derived from ultramafic rocks contain less fluorine (Ross and Smith, 1968).

Later work by Troll and Gilbert (1972) on fluorine-bearing tremolites puts the theoretical maximum fluorine content of pure fluor-tremolite at 4.65%. One sample of tremolite from Richville, NY, contained 1.51% fluorine. Analysis of a sample of lavender-colored tremolite showed 0.30% fluorine. The US Bureau of Mines (USBM) and the National Institute of Environmental Health Sciences (NIEHS) have reported data that characterize other samples, including nonfibrous tremolite (Campbell et al., 1980).

Origin and Mode of Occurrence

Talc deposits occur on every continent, and their geological features are generally similar to those in the United States. The most common hosts for talc ore bodies are dolomite and ultramafic rocks. Chidester et al. (1964) consider that there are three major types of talc deposits in the United States: deposits associated with ultramafic igneous rocks; deposits associated with mafic igneous rocks; and deposits associated with sedimentary rocks. The sedimentary rock category can be subdivided into metasomatic (hydrothermal) deposits and metamorphic deposits, so talc deposits can generally be classified into one of four types.

Types of Deposits: The formation of talc deposits is usually accomplished by one of the four processes described here. The quality and economics of each deposit are controlled by its origin.

Type No. 1: Ultramafic Origin
- Parent: Ultramafic rocks (e.g., peridotite)
- Step 1: Serpentinization of parent (hydration with no loss or partial loss of Mg and Si)
- Product: Serpentinite
- Step 2: Carbonatization of serpentinite (influx of fluid containing more than 5% CO_2)
- Product: Talc-carbonate rock
- Step 3: Steatization of talc-carbonate (influx of fluid containing Si)
- Product: Talc
- Localities: Vermont, Massachusetts, Pennsylvania, Georgia, California, Arkansas, Texas, Canada, southern Spain, Finland, Egypt

Type No. 2: Mafic Origin
- Parent: Mafic rocks (e.g., gabbro)
- Step 1: Serpentinization of parent (hydration with no loss or partial loss of Mg and Si)
- Product: Serpentinite
- Step 2: Carbonatization of serpentinite (influx of fluid containing more than 5% CO_2)
- Product: Talc-carbonate rock
- Step 3: Steatization of talc-carbonate (influx of fluid containing Si)
- Product: Talc
- Localities: Virginia (Schuyler), Georgia, Egypt

Type No. 3: Metasedimentary Origin
- Parent: Dolostone
- Step 1: Metasomatic (hydrothermal) alteration of parent (influx of fluid containing Si and Mg)
- Product: Talc
- Localities: Alabama, Montana, Australia (Mount Seabrook and Three Springs), California (Talc City), Washington (Metaline Falls), Korea, China, West Texas, northern Spain

Type No. 4: Metamorphic Origin
- Parent: Siliceous to sandy dolostone or siliceous talc-carbonate rock
- Step 1: Regional or contact metamorphism of parent
- Product: Tremolite- or actinolite-containing dolomitic marble
- Step 2: Steatization of tremolite/actinolite dolomitic marble (influx of fluid containing Si)
- Product: Talc, tremolite, or actinolite dolomitic marble
- Step 3: Steatization of talc/tremolite/actinolite dolomitic marble (influx of fluid containing Si)
- Product: Talc
- Localities: North Carolina (Murphy Marble Belt), California (southern Death Valley-Kingston Range), New York (Balmat), Georgia (Chatsworth), Italy, Brazil (Brumado)

All deposits may contain amphibole as a minor accessory mineral, but Type No. 4 contains amphibole as a major component of the talc ore.

The most abundant talc resource worldwide is Type No. 1. This type of ore is used as is in low grade filler applications or is beneficiated to produce high grade concentrates. The quantity of talc produced commercially from Type No. 1 deposits is second to that from Type No. 3 deposits.

Resource Type No. 2 yields predominantly an impure soapstone that is rarely of economic importance. Talc deposits associated with mafic igneous rocks are found in the central Appalachian Mountains and are the source of most of the soapstone quarried in the United States. These rocks, which contain relatively minor amounts of the mineral talc, were probably formed by the hydrothermal alteration of gabbroic rocks.

Resource Type No. 3 is typically found in the form of talc veins in dolostone, which are readily separated by sorting. Production from these deposits accounts for most of the high-quality talc used by industrial consumers worldwide.

Resource Type No. 4 has been a larger source of talc in the past, but the presence of amphibole minerals in finished products has negatively impacted their acceptance by some consumers.

Talc Formation: The easiest method of forming pure or relatively pure talc deposits is shown in the reaction presented by Winkler (1974):

$$3 \text{ dolomite} + 4 \text{ quartz} + H_2O = \text{talc} + 3 \text{ calcite} + 3\ CO_2.$$

The geological settings of US talc deposits are as varied and diverse as would be expected for any naturally occurring mineral commodity of secondary origin, but there are nevertheless some broad similarities. First, talc is invariably a secondary mineral formed in preexisting rocks either directly from those rocks or through the introduction of new material. Second, all economically important deposits formed under conditions of low grade regional metamorphism, although local hydrothermal processes may have been the principal agents of formation. Third, commercial ore bodies are generally tabular and concordant, having been molded after the shape and fabric of the parent material, which is most commonly a sedimentary or metasedimentary rock unit. Fourth, most talc ore bodies are found as Proterozoic mineralization within Precambrian rock units. The few that are not found within Precambrian rocks are no younger than Early Paleozoic.

The reason for this almost total restriction of talc to Precambrian rocks is unclear. It may be easier to comprehend in the case of ultramafic host rocks, which tend to be much more abundant in Precambrian than in younger geological systems. The reason is somewhat more difficult to understand with regard to dolomitic rocks, however, which are certainly much more common in Cambrian and younger geological systems than in Precambrian terranes. The key may lie in understanding the geochemical conditions that were present and yielded hydrothermal fluids rich in both silica and magnesium.

Distribution of Major Deposits

Talc deposits are found in many countries. Good quality talc ores are found in Africa, North America, Australia, Brazil, People's Republic of China, Finland, France, Greece, India, Italy, North and South Korea, Spain, Manchuria, Norway, Romania, and the Commonwealth of Independent States (former USSR).

Table 4 summarizes geologic and commercial data for the principal talc districts in the United States (Chidester et al., 1964).

Current and pending government regulations concerning asbestiform minerals and quartz (crystalline silica) are likely to increase consumer demand for purer talc products.

United States: Thirteen US companies produce talc from 22 mines in 10 states (Virta, 1991). Talc or soapstone is currently mined in Alabama, Arkansas, California, Georgia, Montana, New York, Oregon, Texas, Vermont, and Virginia. Production from Montana, Vermont, Texas, and New York constitutes 96% of US output. The USBM reports specific production figures only for Montana (431.8 kt in 1990) and Texas (227 kt in 1990). Other states collectively produced 494 kt in 1990, for a total of nearly 1.153 Mt. The five largest US producers of talc in 1990 were Cyprus Industrial Minerals Co. (now Luzenac America), Gouverneur Talc Co. (R.T. Vanderbilt Co.), Montana Talc Co., Pfizer Inc. (now Barretts Minerals, Inc.), and Dal Minerals (Virta, 1991).

Deposits of steatite-grade talc in Georgia, Alabama, and Pennsylvania were mined by Native Americans as a source of material for making soapstone bowls and other implements before European colonization.

Alabama. The three talc occurrences near Winterboro in Talladega County are hydrothermal replacement deposits (Blount and Helbig, 1987) that contain high-purity white talc. Recent work has shown that this white to light gray talc was formed by the alteration of material that filled fault zones in white dolomite as well as of adjacent Cambrian dolostones and to a lesser degree Cambrian quartzites. Talc pseudomorphs after dolomite crystals are relatively common. Massive white dolomite and milky quartz fault fillings are common within the Sylacauga marble district, but hydrothermal alteration to talc appears to be restricted to these three unique occurrences. The Holmes/Ward open pit mine was opened in 1987 and is now the only producing talc mine in the southern Appalachian region.

Occurrences of talc and anthophyllite near Dadeville in Tallapoosa and Chambers Counties are associated with mafic and ultramafic rocks (Neathery, 1968). These deposits are no longer considered a viable commercial source of talc due to their intimate association with fibrous anthophyllite.

Arkansas. The Arkansas talc district in Saline County contains three small soapstone-serpentinite bodies oriented along a nearly east-west axis about 2.4 km long. The soapstone-serpentinite bodies are derived from ultramafic rocks (Wicklein, 1967) that were intruded into the Womble Shale or Bigfork Chert. The intrusions altered these country rocks into white talc phyllite having a thickness ranging from several centimeters to more than 5 m. Each of the deposits has been open pit mined, but only one now remains in operation producing minor amounts of low grade soapstone.

California. TALC CITY DISTRICT—The original California talc district is located in the Talc City Hills south of the Inyo Mountains in Inyo County. This region is a tightly folded and faulted complex of dolostones and limestones intruded by numerous felsite and diorite dikes. The talc deposits were developed within Ordovician dolostones (Hall and MacKeveh, 1958) by the hydrothermal alteration of felsite dikes and the surrounding dolostone. Alteration of diorite dikes produced veins of chlorite.

The talc ranges from white to black, depending on its origin. White to light-gray talcs are derived from felsite and light-colored dolostones, whereas the dark gray to black varieties are derived from dark gray dolostones or limestones. This district was extensively mined during World War II as the only US source of steatite for use in manufacturing ceramic bodies critical to the production of radio insulators (Page, 1951). A number of similar talc occurrences exist along the eastern slope of the Inyo Mountains and are accessible from the Saline Valley. All talc mining in this district was conducted underground with stoping techniques.

SOUTHERN DEATH VALLEY–KINGSTON DISTRICT—This district extends from the southern part of Death Valley eastward to the Kingston Range in Inyo and San Bernardino Counties. The talc is confined to the Precambrian Crystal Springs Formation, the oldest formation in the Pahrump Series. This series has been intruded by diabase sills. The lowest sill appears to be continuous and responsible for the contact metamorphism that produced white,

Table 4. Summary of Geological and Commercial Data for the Principal Talc Districts of the United States

State and District	Form and Size of Talc Body	Country Rock and Parent Rock	Paragenetic Sequence in Alteration of Parent Rock to Talc	Mineral Composition and Impurities	Physical and Chemical Characteristics of the Commercial Talc	Comercial Use
California: Silver Lake-Yucca Grove	Lenses, mainly of massive tremolitic rocks, as much as 2.45 m long, 5 m thick, and 60–90 m deep; schistose/talc rock forms layers and lenses 30 cm or more thick, mostly along the footwall.	Feldspar-quartz-diopside-carbonate hornfels, commonly separated from the talc body by phologopite-feldspar-tremolite schist a few centimeters to several meters thick. Tremolite and talc apparently derived from beds of sedimentary dolomite 1.5–5 m thick that form parts of pendants in bodies of granitic rock.	Dolomite → tremolite → talc.	Chiefly tremolite and talc with minor forsterite, serpentine, and carbonate. Impurities include phlogopite, chlorite, quartz, carbonate, and mafic dike rocks.	Tremolitic rock mostly massive; talc rock commonly schistose.	Wall tile, pottery, rubber, insecticides.
Southern Death Valley, Kingston Range	Tabular to lenticular bodies as much as 1 525 m long, 25 m thick, and 120 m downdip; most are of schistose to massive talc rock, but some contain interlayered tremolitic rock in various proportions, and some are entirely of tremolitic rock.	Cherty and noncherty dolomite, diabase, and argillite, commonly silicified at the contact. In most deposits one kind of rock forms the hanging wall, another the footwall. Talc and tremolite derived from generally massive and commonly cherty or siliceous beds of dolomite.	Carbonate+chert → tremolite → serpentine → talc. Intermediate steps in the process did not take place in many deposits.	Principally talc; tremolite mostly subordinate but in places predominant. Impurities chiefly carbonate, iron oxide stains, and wallrock inclusions.	Talc rock blocky and massive to schistose or laminated. Particle shape acicular to platy.	Paint, ceramics, paper, textiles.
Inyo Range-North Panamint Range	Small pods, lenses and tabular bodies as much as 150 m long, 15 m wide, and 120 m downdip. Some bodies show large-scale variations related to differences in parent rock, permitting selective mining.	Principally domomite and quartzite, in a few places granitic rock and limestone. Talc derived principally from massive beds of dolomite and quartzite, locally from granite or diorite.	Dolomite and quartzite → talc; also granite and diorite → talc.	Chiefly talc; tremolite rare. Impurities generally very minor; chiefly chlorite, carbonate, quartz, pyrite, and iron oxide.	Talc rock mostly massive and blocky; ranging in color from dark gray to white or pale green; shows large-scale variations in color and texture.	Steatite, ceramics, cosmetics, pharmaceuticals, paints.
Georgia: Chatsworth	Lenticular bodies as much as 1 525 m long, 45 m thick, and 90 m downdip. Lenticular and spindle-shaped bodies of massive talc rock (crayon talc) are dispersed through talc-carbonate rock that shows large-scale variations in color, texture, and mineral composition.	Schist, slate, granite, and granite gneiss. Mode of origin doubtful; parent rock either sedimentary carbonate rock, mafic igneous or volcanic rock or ultramafic igneous rock.	Probably: dolomite → serpentine → (chlorite) → talc. Possibly: ultramafic or mafic igneous rock → serpentine → (Chlorite) → talc+carbonate.	Talc rock principally talc, minor magnetite; talc-carbonate rock chiefly talc and carbonate with minor magnetite, pyrite, chlorite, quartz, feldspar, actinolite, and sericite.	Talc rock varies from massive to schistose, translucent medium green. Talc-carbonate rock varies from medium to dark gray, and varies widely in texture and composition.	Crayons, cosmetics, paint, rubber, roofing, insecticides, lubricants, foundry facings, dusting agents.
Maryland	Irregular tabular masses adjacent to pegmatite dikes in serpentinite; irregular envelopes surrounding a serpentinite core; and lenticular masses.	Schist, gneiss, pegmatite, and serpentinite. Talc derived by replacement of parts or the whole of serpentinite bodies that vary widely in size.	Dunite and peridotite → serpentine → talc+carbonate.	Talc rock chiefly talc; impurities, chiefly chlorite, variable. Talc-carbonate rock essentially talc and carbonate with chlorite, magnetite, serpentine, and other impurities.	Talc rock associated with pegmatites massive, fairly high in iron. Other talc rock and talc-carbonate rock commonly schistose.	Nonsteatite block talc, low-grade ground talcs.
Montana: Dillon-Ennis	Small veins and pods to lenses as much as 215 m long, 30 m thick, and more than 60 m in vertical extent.	Dolomite and quartz-mica schist. Talc derived from beds of dolomitic marble that range widely in size.	Dolomite →talc; also dolomite → tremolite → talc.	Essentially talc with minor amounts of impurities such as chlorite, graphite, dolomite, manganese and iron oxides, and inclusions of country rock.	Massive and blocky to schistose, nearly pure talc; ranges in color from white to pale green and buff; particles range from subequant to flaky.	Steatite, possible block steatite, cosmetics, paints, paper, ceramics.
Nevada: Palmetto-Oasis	Elongate zones of discontinuous pods and lenses to tabular masses 1.5–150 m long; generally 0.3–5 m, but as much as 15 m thick; and 45 m downdip.	Chiefly dolomitic marble, hornfels, and diabase; locally granitic rock, phyllite, and schist, commonly one kind of rock forms the hanging wall, another kind the footwall. Talc derived chiefly from extensive beds of dolomitic marble and hornfels, but locally from phyllite, schist, and granitic rock; commonly occurs at fault contacts.	Chiefly: dolomite →talc. Locally: hornfels, granite, schist, and phyllite →talc.	Chiefly talc with minor chlorite and carbonate, locally some tremolite. Much of the material is nearly pure talc.	Massive to schistose talc rock, ranging in color from dark gray to pale green and white. Talc particles commonly finely platy.	Cosmetics, pharmaceuticals, paints, paper, insulators. Much probably suitable for steatite.

New Mexico: Hembrillo Canyon-San Andres Mountains	Lenses as much as 90 m long, 7.5 m thick, and of undetermined depth.	Chiefly argillite; locally carbonate rock and silica carbonate rock. Talc probably derived from extensive sedimentary units of argillite.	Not determined	Chiefly talc with minor chlorite; locally stained with iron oxide.	White to gray talc rock with platy particles.	Cosmetics. Some hand-sorted material probably suitable for steatite.
New York	Sheetlike to lenticular zones as much as several kilometers long and 90 m wide.	Dolomitic and calcitic marble, diopsidic marble, and thin quartz-diopside layers. Talc derived chiefly from beds of dolomite and siliceous dolomite that form units as much as 1 525 m thick and several kilometers long; some talc derived from granite, syenite, quartzite, and amphibolite.	Dolomite+quartz→ diopside → tremolite → anthophyllite → serpentine → talc. One or more minerals may be omitted in the sequence.	The proportion of tremolite, anthophyllite, talc, and serpentine varies widely; any one may predominate, but tremolite rock is most common. Quartz, carbonate, diopside, manganese and iron oxides, and gypsum are minor impurities.	Varieties of commercial talc range from schistose to massive. Particles range from nearly equant to bladed and fibrous. Very white.	Largely paint and ceramics; also rubber, insecticides, foundry facings, plastics, and other uses.
North Carolina: Murphy	Small pods to lenses as much as 215 m long, 15 m thick, and 60 m downdip. Many deposits exhibit large-scale variations in color, texture, and distribution of impurities, permitting selective mining.	Calcite marble, locally highly siliceous and commonly tremolitic at the borders of the talc deposits.	Dolomite+quartz →silicated rock →talc.	Essentially talc. Impurities are chiefly minor amounts of quartz, carbonate, iron oxide stain, and rarely tremolite.	Massive to schistose talc rock, ranging in color from light gray to white. Particles subequant to flaky.	Crayons, cosmetics, paint, paper, textiles, ceramics; some suitable for steatite.
Texas: Allamoore	Largely unknown, but some tabular masses are several thousand meters long and several hundred meters wide, with tabular and irregular inclusions of carbonate rock, phyllite, and masses of chert that range widely in size.	Cherty dolomite; some conglomerate, phyllite, and diabase. Talc probably derived from extensive beds of sedimentary or pyroclastic rock such as dolomitic marl or magnesium-rich tuff.	Not determined	Essentially talc. Impurities consist of varying amounts of chert, carbonate, wall rock inclusions, surface weathering products, and organic material.	No information	Chiefly wall tile; also insecticides.
Llano	Mostly small lenses and pods in schist or along margins of serpentinite masses as much as 6.5 km long and 1.5 km wide.	Schist, gneiss, and serpentitite. Talc derived from lenticular bodies of serpentinite and from tabular units of schist, gneiss, mafic volcanic rock, and carbonate rock.	Details unknown. Serpentinite and metamorphosed sedimentary and volcanic rock altered variously to one or more of talc, carbonate, tremolite, and anthophyllite.	Talc; talc and tremolite; talc and anthophyllite; anthophyllite. Commonly contains impurities of quartz, magnetite, chlorite, and other materials.	Variable in composition and physical properties. Small amounts of massive talc rock.	Chiefly low-grade uses. Some suitable for crayons and for carvings.
Vermont	Shell of talc rock 0.3–1.0 m thick surrounds a zone of talc-carbonate rock that ranges in size from a shell 0.3–5 m thick around a serpentinite core to the entire mass of the original serpentinite body.	Schist, greenstone, and serpentinite. Talc rock and talc-carbonate rock derived almost wholly from serpentinite bodies that range in size from small pods to lenses several kilometers long and 1–2 km wide; a small proportion of the talc rock is derived from schist.	Dunite and peridotite →serpentinite. Serpentinite+CO_2 →talc+carbonate. Serpentinite+SiO_2 →talc. Schist or greenstone →talc+chlorite.	Talc rock is essentially talc with minor chlorite. Talc-carbonate rock essentially talc and magnesite with minor serpentine, magnetite, chlorite and sulfides.	Talc rock generally schistose, talc-carbonate rock schistose to massive; both are medium to light gray. Talc ranges from fine to coarsely flaky.	Paper, rubber, textiles, paint, roofing, cosmetics, filler, foundry facings, crayons.
Virginia: Schuyler	Lenticular bodies of the same extent as the original igneous rock body.	Gneiss, schist, and amphibolite. The talc (soapstone) was derived from lenticular bodies, probably originally hypersthene gabbro, as much 55 m thick and 455 m long.	Hypersthene gabbro →amphibole+chlorite →talc+carbonate.	Talc, chlorite, carbonate, amphibole, and magnetite.	Massive, felted texture.	Sawed and shaped slabs, insecticides.

fine-grained talc and white tremolite rock from cherty, dolomitic marble.

The ore bodies in this district are extensive. The largest is thought to be the Western-Acme mine complex near Tecopa, CA. The ore from this district has been used mostly in the ceramics and paint industries. The only currently active operation is the Omega mine in the Kingston Range of San Bernardino County.

SILVER LAKE-YUCCA GROVE DISTRICT—This talc district extends from the Silurian Hills northeast of Baker eastward to Yucca Grove in northeastern San Bernardino County. The lens-shaped deposits of talc were formed by metamorphic alteration of certain dolomitic strata to tremolitic marble, which was then partially altered to talc by steatization. Most of the mining has been conducted underground using stoping techniques.

SOAPSTONE—Soapstone deposits along the western foothills of the Sierra Nevada have been mined for decorative stone or low grade fillers. The Red Hill open pit mine in Calaveras County produces crude ore for filler-grade products. The soapstone is associated with serpentinite and is a typical low-brightness talc-carbonate. Similar spotty occurrences are known to the north to a point south of Medford, OR.

Georgia. Talc occurrences are associated with the Burks Mountain Serpentinite of the Kiokee Belt in Columbia County, with verde antique deposits near Holly Springs in Cherokee County, in fault zones near the margins of metagabbro bodies in Paulding County, and in other minor exposures extending from Paulding County north to Rabun County. In addition, portions of the Murphy Marble Belt contain small talc occurrences as far south as Whitestone, GA. The majority of these occurrences contain tremolite and anthophyllite.

The major producing district in Georgia has been in Murray County, where production continued from about 1890 to 1990. The ore in Murray County is a mixture of talc and chlorite with actinolite. The deposits were thought initially to have been derived from altered dolomitic portions of the Cohutta Schist (Furcron et al., 1947), but Needham (1972) considers the talc to have been formed by the serpentinization of peridotite followed by metasomatism and metamorphism. Production from Georgia in 1989 was reported as 24 kt (Virta, 1991). The bulk of this production was obtained from underground mines and was used in roofing products and as low-end fillers.

Michigan. Two talc-carbonate areas are located in Michigan near Iron Mountain and Ishpeming. The talc-carbonate near Iron Mountain is a coarse-grained, light green to light gray, poorly banded, foliated ore composed of 45% talc and 45% dolomite, plus accessory chlorite, magnesite, magnetite, magheinatite, chromite, millerite, and pyrite. The talc-carbonate near Ishpeming is a fine-grained, light gray to gray talc-magnesite rock with chlorite, dolomite, magnesite, chromite, and millerite. Both of these large talc-carbonate deposits are associated with serpentinite derived from Precambrian ultramafic rocks. No talc has been produced commercially from either of them.

Montana. The talc deposits in Montana range from tabular veins to lenticular or podiform masses of steatite. The veins are oriented generally parallel to the foliation of the carbonate host rocks. The talc was formed by the metasomatic replacement of magnesium- rich marble. Relict host rock structures are commonly preserved. Cross-cutting fault zones and joints that served as ore-forming fluid conduits often contain minor blooms or subsidiary veins of talc. A small amount of talc at some deposits is found as open-space fillings. The open-space fillings commonly have a botryoidal form, especially at the Yellowstone and Johnny Gulch mines in the Gravelly Range. Although talc in Montana can be found in virtually any color, it is predominately white to light green or dark green to gray and is weakly to moderately translucent. Its crystal size varies from deposit to deposit depending on host rock and paragenetic factors. Vein thicknesses range from microscopic to more than 200 m at the Yellowstone mine. Other minerals associated with talc in Montana include magnesite, siderite, ankerite, calcite, pyrite, graphite, chlorite, serpentine, quartz, iron oxides, and at some locations tremolite (commonly replaced by talc). Tremolite is found spatially associated with some talc deposits in Montana but is not within commercially mined ore bodies.

Winkler's (1974) generally accepted reaction for talc formation involves the introduction of silica and water to a dolomitic host and the removal of calcite and carbon dioxide. No significant amount of calcite has ever been noted near the talc deposits in Montana, however. The calcite could have been formed in the near-surface and then eroded. Anderson, et al. (1990) note that, for six deposits studied in the Ruby Range, talc formed under constant volume conditions. Such conditions require the addition of magnesium in order to offset an otherwise 32% volume reduction. Piniazkiewicz (1984) notes the lack of field evidence for volume reduction, magnesium deficiencies in host marbles, and the theoretical equation's failure to account for open-space filling by talc. Piniazkiewicz proposes the following reaction:

$$2\text{ dolomite} + (X)Mg + H_2O + 4SiO_2 = \text{talc} + 2HCO_3 + 2CO_2 + 2Ca$$

Anderson's (1987) petrographic work provides further evidence for the addition of magnesium by observing dolomite replacing calcite in the marbles adjacent to the talc bodies. Piniazkiewicz's model proposes that the additional magnesium came from pore fluid in the dolomitic marble, however. More recently, magnesium-rich seawater, possibly from the Proterozoic Belt Seaway, has been proposed as the source (Anderson, et al., 1990, Brady et al., 1991). Anderson, et al. (1990) calculate that 480 000 L of water per cubic meter of dolomite were required, for a water:rock ratio of 600:1, and suggest a near-surface hot springs environment for talc formation (e.g., $P < 2 \times 10^3$kPa, $T < 450°C$). Brady et al. (1991) determined with oxygen isotopes that dolomite, talc, and associated chlorite in the Ruby Range formed at 250°C.

All talc produced in Montana, with one negligible exception, has been mined from deposits formed by the replacement of dolomitic marbles in the Archean Wyoming Province (James, 1990) exposed in southwestern Montana (Berg, 1991). Archean basement rocks are exposed in the cores of mountain ranges in southwestern Montana as Laramide-type foreland block uplifts that have undergone subsequent Basin-and-Range extension (Schmidt and Garihan, 1983). The Archean cores of these ranges are overlain by Phanerozoic sedimentary rocks. Archean metamorphic rocks in the area consist of quartzofeldspathic gneisses, amphibolites, and intercalated schists, quartzites, and marbles with megascopic and mesoscopic, tight to isoclinal folds.

In most of southwestern Montana the Archean rocks reached metamorphic grades of amphibolite to granulite (e.g., 700 to 750°C and 6 to 8×10^3kPa; Dahl, 1979); however, in the southern Madison Range and parts of the Gravelly Range, only greenschist facies rocks are exposed. The date of this dynamothermal metamorphism (M1) was found by James and Hedge (1980), using the Rb-Sr whole-rock method, to be approximately 2750 m.y. ago. A regional Proterozoic retrograde thermal event (M2) overprinted the M1 amphibolite-granulite assemblages with greenschist assemblages. Brady et al. (1991) detected a thermal event of 1700 m.y. ago in the Ruby Range by using 40Ar/39Ar age spectra on phlogopite, biotite, and hornblende. This is consistent with Giletti's (1966) determination of 1500 to 1700 m.y. ago for the M2 regional event by using Rb-Sr whole-rock dates.

In the Ruby Range the Archean rocks are cut by northwest-trending faults along which recurrent movement has occurred since Proterozoic time. Several past workers have suggested that spatial relationships to talc deposits indicate these structures were ore-forming fluid conduits during the M2 thermal event (Okuma, 1971, Garihan, 1973, Berg, 1979a). Berg (1991) notes that recent mapping (Pearson et al., 1990) has not detected these spatial relationships in the Ruby Range, in the Tobacco Root Mountains, or as noted by Berg (1979a) himself in the Greenhorns. Fluid movement along local, layer-parallel structures is presently suggested for the Ruby Range (Anderson, et al., 1990). Deposits in the Gravelly Range appear to be controlled by foliation parallel to north-northeast structures and cross-cutting, east-northeast structures.

The time of talc formation is confined to the Proterozoic. As noted by almost all previous workers, Paleozoic dolomitic sediments lack talc occurrences within the southwestern Montana talc province. Anderson (1987) notes from petrographic evidence that dolomite and serpentine from the M2 event were replaced by talc, and he proposes that talc formation occurred either as a late stage of the M2 event or as a later, talc-forming M3 event. Brady et al. (1991) also propose a maximum age of 1700 m.y. Many workers believe that dolomitization (M2) and subsequent talc formation followed at least the incipient stages of Proterozoic Belt Basin rifting (Berg, 1991).

Talc mined in Montana represents about 37% of the tonnage and 61% of the value of US production (Virta, 1991). All of the talc in Montana is produced by three companies from four mines in Madison County. Three of the mines are open pits. The Beaverhead Mine in the Ruby Range is an underground operation that produces 11 ktpy of cosmetic-grade talc by underhand cut-and-fill methods. The open pit mines use hydraulic shovels (1- to 10-m^3) and trucks (32- to 60-ton) and supply the paper, paint, plastics, rubber, and cosmetics industries.

Nevada. The primary talc-producing region is part of a 40 km belt in southern Esmeralda County between Oasis, CA, and Lida, NV, and is commonly known as the Palmetto-Oasis District. The total historical production from this district is estimated at 138 kt of talc (Papke, 1975), but all mines are inactive at this time.

The talc is present as small, elongate pods or lenses along the contact between a porphyritic quartz monzonite and dolomite. This altered zone has the appearance of having been formed by contact metamorphism, but Papke (1975) considers it to have formed by the low temperature replacement of dolomite. The talc is white and the chlorite dark gray, with color gradations common. Production from these mines was used in markets ranging from cosmetics to insecticides, depending on the color and purity of the talc.

New York. The second most significant source of talc in eastern North America is the Precambrian Grenville Province of northern New York and southern Ontario. The largest deposit being worked is in the Gouverneur (Balmat) area of New York. The talc is associated with a siliceous dolomite that was converted into tremolite by dynamic metamorphism. Subsequent retrograde metamorphism with declining temperatures formed the talc. A smaller deposit in the Grenville Province is located near Madoc in Ontario. The talc at this location formed by the hydrothermal replacement of a marble unit, and subsequent folding and faulting have offset and thickened the ore shoots.

All of the known commercial talc in this area is found within the Balmat-Edwards, or Gouverneur, mining district. A comprehensive study of talc in the United States by the US Geological Survey (Chidester et al., 1964) reports that New York in the early 1960s had almost 75% of the measured and indicated talc ore reserves in the United States and about 40% of its inferred talc ore reserves. With relatively minor exceptions, all known talc in this district lies within a narrow belt that probably averages only 150 m wide and is only about 13 km long (Roe and Olson, 1983).

The New York talc district was long the most important in the United States in terms of total annual production, being rivaled or surpassed only by Vermont. Over the past couple of decades, however, both Montana and Texas have also pulled ahead of New York. It is noteworthy that the New York district, although dominant in production for so much of the past century and even more dominant in terms of proven and potential ore reserves, is the smallest in area of all six talc mining districts in the United States (Roe and Olson, 1983).

The bedrock complex of the Gouverneur mining district consists exclusively of metamorphic rocks of the Precambrian Grenville Series, but two specific lithologic types, gneiss and marble, predominate. The gneiss is thought to be the older of the two units and is about 900 m thick; the marble, considered to be the younger unit, has a maximum known thickness of about 600 m. Thin sheets and lenses of amphibolite are present within both the marble and the gneiss but are relatively minor in total volume. Brown and Engel (1956) subdivide the marble sequence into 15 units, of which only Unit 13, the "talc unit," is known to contain commercial deposits of talc, tremolite, or zinc minerals. Unpublished geological investigations indicate that the composition of most of the marble within this district is relatively simple and probably does not and did not originally vary appreciably along strike. The talc unit, a dolomitic marble, shows evidence of being only about 75 m thick where least contorted. Greater thicknesses are probably due to structural flowage (Roe and Olson, 1983).

The zinc and talc ores are found within the same stratigraphic unit but not together (except at a few localities). The siliceous dolomite of Unit 13 was converted to tremolite locally under conditions of dynamic metamorphism or metasomatism, or both, and talc, anthophyllite, and serpentine subsequently formed locally from the tremolite. In the northeasternmost part of the district anthophyllite is the major ore constituent at some localities (Roe and Olson, 1983).

New York talc ores commonly consist of more than 50% tremolite. Tremolitic ore has been found in underground workings to downdip depths of at least 335 m and by core drilling (generally in search of commercial zinc deposits) to downdip depths of more than 900 m. Bodies of almost tremolite-free, platy or scaly talc have been discovered in surface and underground mines near Talcville. These relatively small deposits are situated at or near the hanging wall contact and are difficult to mine by themselves. Field mapping by the US Geological Survey has indicated that tremolite, talc, and associated secondary minerals were formed prior to 10 billion years ago (Roe and Olson, 1983).

The talc ore zones in New York locally pinch and swell, but they are generally conformable with the layers of dolomitic marble. Departures from the regional average dip of about 45°NW are only of local significance. Tight and contorted cross-folds superimposed upon the main talc belt have resulted in the local thickening of the talc sequence by at least 50%. All of the rocks are complexly folded, and metamorphism to the amphibolite facies is common if not ubiquitous (Roe and Olson, 1983).

Both open pit and underground mining methods are employed. A large crushing facility is installed within one underground mine. The deposits of the New York talc mining district may well constitute the largest reserve of tremolitic talc ore in the world, for nothing described in the literature at the present time is of similar magnitude. New York talc ores are milled as mined, without any prior beneficiation. Milling is accomplished in ball mills and fluid energy mills, and the product is then air classified (Roe and Olson, 1983).

North Carolina. According to Van Horn (1948), talc of low brightness is located in Macon, Jackson, and Madison Counties in association with basic and ultrabasic igneous rocks. The talc or soapstone consists of a mixture of talc and chlorite, plus amphibole and minor amounts of magnesite, dolomite, mica, and quartz. The primary talc district is the Murphy Marble Belt, which began producing in 1859 (Van Horn, 1948) and closed in 1986. The bulk of production was from underground mines. The talc was used in cosmetics, as crayons, and fillers in other products.

The deposits in this district range from very small to moderate in size (e.g., 227 kt) and from irregular to lenslike in shape. They are found consistently in a dolomitic zone of the Murphy Marble. This zone consists of siliceous, dolomitic marble that was altered to tremolitic marble by regional metamorphism. Subsequent hydrothermal alteration produced a dolomitic marble containing white talc or white talc-tremolite. Deposits lie in a belt extending northward from Whitestone, GA, to about 11 km north of Topton, NC.

Oregon. Steatite in southern Oregon yields carving and decorative stone from small, north-south trending, nearly vertical and zoned lenses of talc, talc-carbonate, and serpentinite. The lenses are associated with the Condrey Mountain Schist, which crops out southwest of Medford on Elliot Creek Ridge. The talc is fine grained, very light silver-gray, and of fair to moderate brightness.

Other deposits of talc are located in northeastern Oregon in Baker County north of Huntington and near Sumpter. The Huntington deposit consists of light green to light brown talc-carbonate, and the small Sumpter deposit is light blue-green talc associated with serpentinite that contains chrysotile.

Texas. LLANO DISTRICT—This district, the oldest soapstone producing area in the state, consists of Precambrian metamorphic rocks in south-central Llano County, extreme northeastern Gillespie County, and extreme northwestern Blanco County. The gray to grayish brown tremolitic soapstone appears to be a product of the hydrothermal alteration of tremolitic rocks that were altered by contact metamorphism from serpentinite originally derived from harzburgite (Garrison, Jr. and Mohr, 1984). Production from this region was restricted to minor building materials and ground carriers and diluents for insecticides.

ALLAMORE DISTRICT—The Allamore District extends east-west about 32 km and is as much as 8 km wide. Its ore deposits are in the Allamore Formation of Precambrian age, which consists of thousands of meters of carbonate rocks, volcanic rocks, and phyllite (King and Flawn, 1953). The talc appears to be associated with dolomite but also to interfinger with phyllite in certain parts of the district. Edwards (1980) feels that phyllite is invariably the host rock, but Bourbon (1981) believes the talc formed in Precambrian time by the replacement of primary sedimentary magnesite. Field relationships indicate that noncalcitic carbonate rock is the host for these talc deposits.

The ore-bearing strata, which make up part of the lower plate beneath a large overthrust fault (the upper plate having been almost entirely eroded), dip either vertically or steeply to the south. The bottom of no ore body reportedly has ever been found by drilling or mining. Large, iron oxide-stained ''horses'' of either carbonate or volcanic rock are common within the talc zones but usually are easily eliminated by the use of selective mining methods. Talc commonly is found from the grass roots down, and the soil cover is generally less than 30 cm thick. Caliche is abundant in the uppermost 1.5 m of these dark-colored ore bodies but seldom causes problems at greater depths.

Initial, but far from complete, petrographic and X-ray studies of the ores indicate that tremolite is not present to any significant extent. Dolomite, however, is so finely admixed with the talc that presently known beneficiation techniques are unlikely to produce pure talc in significant quantities from known ore bodies. There is little indication of a hydrothermal origin for these ores.

The Allamore District talc ore is generally strongly foliated, with individual foliae ranging from paper-thin to as thick as 6 mm. This dark Texas crude bears little resemblance to other talc ores, even to the trained observer. One's first tendency is to identify it as black shale, perlite, or some other nontalc-bearing rock. However, despite its dark or even black color and its strikingly untalclike appearance, this ore is a superior ceramic crude with superb pressing qualities and excellent firing characteristics.

Flawn (1958) describes the Allamore District in its stage of infancy, and King and Flawn (1953) describe in detail the overall geology of the district and surrounding areas. All production has been from large, low cost open pits with the exception of minor amounts of light-colored (generally pink to white) talc, which have been mined from small underground workings and then finely ground into filler and extender products. Almost all of the talc ore produced is shipped out of the area, even into Mexico, for use in the ceramics industry and particularly for the manufacture of wall tile. This is the youngest major talc district in the United States. Production began in 1952 and amounted to 220 kt in 1989 (Virta, 1991).

Vermont-Appalachian Orogenic Belt. The largest tonnages of talc produced in the eastern United States come from altered ultramafic rock (peridotite) units of Precambrian age. The major producing districts are associated with the Appalachian Orogenic Belt as it extends through Vermont into Quebec and the Superior Province of eastern Ontario. Similar deposits are found in southern Spain, Norway, Finland, and the State of Minas Gerais in Brazil.

In the Appalachian Belt talc-carbonate rocks are associated with ophiolite complexes that were tectonically emplaced into Cambro-Ordovician metasediments. Talc-carbonate mineralization generally is found as an alteration zone rimming a serpentinite core. Subsequent shearing has complicated the structures and has created zones of high grade talc.

The talc and talc-carbonate ores are typically medium-grained, light gray to gray-green, and contain chlorite, dolomite, and calcite as accessory minerals. Minor amounts of sulfides, quartz, and amphibole are also present.

In Vermont a talc-carbonate serpentinite belt extends into the state from Quebec near North Troy and trends south through Johnson, Stowe, Moretown, Stockbridge, Ludlow, Windham, and Newfane, and into Massachusetts near Rowe. The North Troy deposit is a large, metasomatically altered serpentinite body. Alteration was complete, and the talc-carbonate ore body contains 48 to 51% talc with a narrow, 7-m thick rim of high talc-content ore. South of Johnson the talc bodies typically represent the altered rims of serpentinite cores. In this area serpentinite was emplaced along the tectonic contact between graphitic schists of the Cram Hill Formation and garnetiferous gneiss of the Moretown Formation.

The altered serpentinite bodies in Vermont are typically zoned with a relict serpentinite core grading outward through talc-carbonate, talc, and finally a thin blackwall (biotite-chlorite schist in contact with the surrounding country rock). These altered bodies range from several hundred meters long and 1 m wide to about 1.3 km long and 150 m wide.

The talc ore bodies in the Superior Province of eastern Ontario are talc-carbonate alteration zones associated with peridotite komatiite lavas and sills. The sills were emplaced into the Archean Greenstone Belt of metavolcanics. The talc zones are also associated with transcurrent shear zones. The only active mine in this district, in Penhorwood Township, is reported to be producing ore with a talc content of 45%.

The more massive variety of steatite found in New Hampshire, Vermont, Massachusetts, and Connecticut was used from the mid-1800s to the early 1900s for stoves and cooking and heating utensils. This business was gradually replaced by the production of dry-ground talc for industrial products. The need for higher grade talc products led in the early 1930s to the adoption of flotation for further beneficiation of these talc ores. The two most recent ore deposit developments in the northeastern United States and eastern Canada (namely, North Troy, VT, and Penhorwood Township, ON) beneficiate lean ores to upgraded products.

Virginia. Another type of talc deposit present in the eastern

United States was formed by the alteration of mafic rock (e.g., gabbro). This type is found in the Schuyler, VA, area. The end product is dimension stone, and the ore has no potential for ground talc production.

Washington. There are 25 reported deposits of talc and soapstone in Washington (Valentine, 1960). Talc has been produced in three counties—Skagit, Stevens, and Lincoln—but there is currently no known production in the state. Other talc discoveries have been reported in Chelan, Ferry, and Pend Oreille Counties. The deposits of talc are generally small, although those of soapstone are reported to be quite large. Glover (1936) classifies these deposits into two main types. Most of those in the western and central portions of the state were formed by alteration along shear zones within mafic and ultramafic rocks, whereas those in the eastern portion of the state are associated with altered calcareous schists and carbonate rocks.

Skagit County contains the largest number of reported talc deposits. Valentine (1960) reports seven past producers of talc or soapstone or both in this county. The deposits, most of which are located near Marblemount, are reported to be associated with serpentinite and to have resulted from alteration of basic rocks in the Northern Cascades (Joseph, 1990).

Deposits apparently containing cosmetic-grade talc have been recently explored in northern Pend Oreille County east of Metaline Falls. These deposits are found in silicified dolomitic units of the Precambrian Z Monk Formation.

Other Countries:

Australia. Talc mining takes place primarily in Western Australia and South Australia. Australian production is reported to have been 177 kt in 1989 (Virta, 1991). About 16 ktpy comes from South Australia, primarily from the Mount Fitton deposit. Most of the remaining tonnage is mined in Western Australia at the Three Springs and Mount Seabrook mines. Western Mining Corp. reported 147 kt of production from its Three Springs mine in 1989 (Anon., 1990).

The Mount Fitton deposit was formed by the hydrothermal alteration of dolomite and marble. Keeling, et al., (1990) report that the vein is 10 to 15 m thick, extends 20 m downdip, and is several hundred meters long. The vein is mined by open pit methods. The principal consumers are domestic pharmaceutical and paint manufacturers. Most of Australia's domestic needs are served from this deposit.

The Three Springs deposit is a shallowly dipping metasedimentary ore body. Dolerite (diabase) dikes truncate the deposit to the north and west. The ore body exceeds 20 m in thickness and contains very few inclusions of clay or dolomite. Mining is by open pit methods with shovels and trucks. Three Springs produces a cryptocrystalline, high-purity talc similar to that obtained from the Yellowstone and Johnny Gulch Mines in Montana and León Province in Spain. Because Three Springs is close to the port of Geraldton, significant quantities of talc are exported.

The Mount Seabrook talc mine is situated in an area of lower Proterozoic quartzite, quartz-muscovite-chlorite schist, dolomite, and talc. Steeply dipping talc veins were formed by the metasomatic alteration of refolded dolomitic horizons. Talc veins inherited the structural complexity of their dolomite host. A weathered, weakly foliated granite crops out 0.5 km east of the mine. Silicified Tertiary conglomeratic channel-fill deposits as much as 15 m thick are found along fault zones, schistose units, and lithologic contacts. Brick-red Tertiary silcrete forms a 1- to 2-m thick capping, and a subhorizontal zone of silicification extends 2 to 3 m below this capping.

The talc veins at Mount Seabrook are as much as 20 m thick and 140 m long. They have at least three and as many as five different foliation planes. Their fabric and vein boundaries dip steeply. The talc is classified into three types based on the degree of weathering: hard, glassy talc that will break only with a firm hammer blow; medium talc that can be gouged with a fingertip; and soft talc that can be crushed in the hand. The veins usually contain in excess of 90% talc. The talc is white, pale green, green, or pale brown. Premium grades are derived from white or pale green, hard and medium ores.

Austria. Annual production of 150 kt of talc comes from three open pit mines located at Rabenwald, Lassing, and Wiesskirchen. The Rabenwald deposit contains high-chlorite, low-iron material that is used in ceramics and refractories. Ore from the Lassing deposit contains as much as 70% dolomite and yields a very white product that is used in paint, paper, and plastics.

Brazil. Two talc mining areas are located about 81 km west of Casa Nova in Bahía Both localities are mined by open pit methods. The Brumado talc deposits are hosted by a thick sequence (250 to 300 m) of white to gray magnesite with tourmaline, tremolite, quartz, and dolomite as accessory constituents. The talc ore is white to very light gray, dense, and comprises layers 1 to 30 cm in thickness. Zones of deformation have contorted the layers into irregular pods and laminae. The talc appears to be the result of hydrothermal alteration of siliceous marble or calc-silicates.

The Casa Nova deposit of light green to gray talc is associated with mica schist, biotite gneiss, and amphibolite. This metamorphic complex was deformed into a series of small, irregular talc schist pods and stringers that may have been derived from the amphibolite.

Canada. BRITISH COLUMBIA—A comprehensive study of talc in British Columbia has been conducted by the Mineral Resources Division of the Ministry of Energy, Mines, and Petroleum Resources (MacLean, 1987). That study outlines most of the talc-carbonate ore deposits associated with serpentinite altered from ultramafic rocks. Several dolomite-hosted talc localities with amphibole as an accessory mineral have been reported in southeastern British Columbia.

The only currently active mining project was initiated by Pacific Talc near North Bend. The project is based on a deposit of pale green to light gray talc-carbonate associated with small, discontinuous zones of serpentinite.

NEWFOUNDLAND—Unexploited but extensive deposits of talc-carbonate rock in an ultramafic sequence are present near Ming's Bight just northeast of Baie Verte. The talc is associated with serpentinites of the Dunnage Zone. The talc-carbonate or soapstone deposits consist of very coarse-grained, gray rock with occasional thin, pale green talc veins. The talc-carbonate rocks are very similar to those in Vermont and Quebec and probably have the same origin.

China. Officially, the production of talc in China is estimated at 900 ktpy (Virta, 1989), which is probably conservative. China rivals the United States as the world leader in crude talc production. Although China has 120 talc mines distributed throughout the country, most talc is produced in three provinces—Liaoning, Shandong, and the Guangxi Zhuang Autonomous Region (Guangxi Province). Nearly half of China's production is exported to Japan for use in the paper industry. China is the world's leading exporter of crude talc (Virta, 1989).

LIAONING—Liaoning Province is the largest producer of talc in China. The deposits formed as metasomatic replacements of magnesitic and dolomitic carbonates of the Late Proterozoic Liaohe-Dashiqiao Group. The carbonates are dominantly magnesite, but some tremolitic deposits lie within sequences of dolomite and calc-silicate rocks. The principal rock fabric orientation is east-northeast. The parent rocks have undergone metamorphism ranging from medium-pressure greenschist facies to medium-pressure high amphibolite facies. The higher ranking metamorphic rocks are associated with calc-silicate assemblages.

High-purity talc veins are generally podiform and sheared. Mined widths range from 2 to 15 m, and strike length ranges from tens to hundreds of meters.

Most deposits are mined underground. Drilling is done with air-jacklegs. The face is shot with explosives and then timbered as square sets with 15-cm-diameter oak posts. Most operators mine the talc veins along heavily timbered 2- by 2-m drifts. Portions of the veins that exceed 2 m in width are mined using a slash round and retreat system. Development headings in magnesite or dolomite are 3 m wide by 2 m high and are usually unsupported. Underground mines in calc-silicate rocks normally require less support.

Rock shot from the mine face is manually loaded into ore cars and is hoisted or trammed to the surface. Additional hand sorting is performed at the surface after the 3-cm fraction is removed. The talc is sorted according to brightness and color, which ranges from light green to gray-green and pale pink.

SHANDONG—The talc deposits of Shandong Province belong to the Late Proterozoic Jutuen-Zhanggezhuang Group. The rock types within this group include schists, magnesium carbonates, and gneisses. The talc formed as a metasomatic replacement of magnesium carbonates. The talc in Shandong is generally of lower brightness than the talcs in Liaoning and Guangxi. High grade veins have minable widths ranging from 7 to 30 m. The veins commonly extend for hundreds of meters downdip and along strike. Magnesite is the most abundant waste rock in these deposits.

Intense folding and faulting preceded talc formation. Within the folds the talc veins exhibit thickening along the axial planes and thinning on the limbs.

Most of the talc produced in Shandong comes from underground mines, of which there are more than 100 in operation. The mining and sorting methods used are similar to those used in Liaoning. The talc is most commonly green to dark green.

GUANGXI—The talc deposits in Guangxi Province are the largest, high-purity talc veins in China. They belong to the Middle Proterozoic Banxi-Hetong Group. The principal mines in the district are located along a N30°E fracture zone within a belt 2 km wide by 8 km long. The host rock is dolomitic marble that is off-white to gray in outcrop. Fresh surfaces are off-white and exhibit a fine-grained texture. The marble is contorted into high-amplitude, tight to isoclinal folds and commonly contains three foliation planes. At most locations a residual, reddish clay blanket overlies the talc.

The talc in Guangxi is of high purity and high brightness. The veins are concordant, lenticular bodies that average 5 m in thickness. Vein thicknesses approach 80 m locally, and strike lengths reach hundreds of meters. The veins have sharp but irregular contacts with the marble. They are light green but contain widely spaced (10 cm), thin gray bands. The talc appears to be fine grained under a hand lens. Dolomite, magnesite, chlorite, and quartz are the main impurities in the talc veins, which contain about 2% hand-sortable internal waste.

All mining in Guangxi is in open pits. Overburden and waste are removed with dozers; scrapers are used for clay removal and shovels and trucks for rock removal. Talc-dolomite sequences are drilled, shot, and then mined manually. The talc is graded visually and hand sorted according to purity and brightness. Some mines are beginning to mechanize their operations. At such mines the talc-dolomite sequences are drilled, shot, selectively mined with shovels, and then hauled to a sorting complex for hand grading.

Egypt. The talc deposits in Egypt are confined to the southeastern deserts and are associated with Precambrian metamorphic rocks of the Nubian-Arabian Shield. The complex of rocks consists predominantly of metavolcanics and granodiorite with gneiss, schist, and serpentinites. The talc deposits are typically small veins and lens-shaped bodies associated with serpentinite and meta-andesite and located along shear zones. The quality of the serpentinite-hosted ore is low to moderate, whereas the quality of the metavolcanic-hosted ore varies from poor to excellent. Accessory minerals typically include tremolite or actinolite, and malachite staining is common.

Finland. Extensive research and development have promoted the use of talc as a coating material and filler in the local paper industry. Two principal mines, one in central Finland near Kajaani and the other in eastern Finland, supply most of the ore for the 380 kt of finished talc produced annually. The talc deposits formed either from metasomatically altered serpentinites or hydrothermally altered schists and gneisses.

France. Talc production in France amounted to 300 kt in 1989. Most of the production came from the Trimouns Mine in the Pyrenees 120 km south of Toulouse. The Trimouns talc vein is 10 to 80 m thick, more than 1 500 m long, and dips at 20 to 60°. The talc was formed when magnesium-bearing fluids altered footwall silica- and alumina-rich mica schists and gneisses (Longeaux, et al., undated). Chlorite formed in the micaceous and feldspathic footwall rocks during magnesium enrichment, and the silica that was released during chloritization transformed the overlying dolomites into talc. Unmineralized dolomite forms the hanging wall of the deposit. The current open pit mine has a stripping ratio of 12:1. Twelve different types of ore are recovered. Pharmaceutical and cosmetic-grade talc is hand mined, whereas industrial grades are extracted and sorted with small hydraulic shovels. Mining is seasonal due to harsh winter weather. The Trimouns Mine uses innovative technology such as an overhead cableway for mine-to-plant transport, laser-operated optical sorting, and video-operated optical sorting.

India. There is large (e.g., 400 ktpy in 1990) production capacity for a wide range of talc products from four states in India. The most significant capacity is in Rajasthan, where high-purity talc is mined from the contact between basic dikes and dolomitic marble.

Italy. High-purity talc for cosmetic and pharmaceutical products is produced from veins in gneiss, mica schist, and dolomite. Additional lower grade talc/magnesite is produced for the paper, paint, and plastics industries. The annual production of 200 kt supplies the domestic market and is exported to North and South America and elsewhere in Europe.

Norway. Talc processed from serpentinite-type ore bodies supplies the paint, paper, and plastics markets. Lower grade talcs are used in roofing products, animal feeds, and fertilizers.

South Korea. Processed talc from serpentinite-type ore bodies, produced in excess of 200 ktpy, is consumed domestically and is exported.

Spain. Spain has four principal talc-producing areas in the provinces of Almería, Gerona, León, and Málaga.

ALMERÍA—This province in southeastern Spain yields talc from the Sierra de las Estáncias District, which produces small quantities of very white, cosmetic-grade talc as well as dark brown talc. The talc is found associated with fault zones in gray dolomitic marble and mica schist. Mining of these small veins is carried out underground.

GERONA—This district is located in the eastern Pyrenees near the French border. The talc-chlorite ore is of moderate to low brightness and ranges from green to banded black and white, black, and gray. It is used predominantly in the paper industry, as industrial fillers, and in ceramics.

The ore was formed by the hydrothermal alteration of dolomitic marble along major fault zones, which produced vein deposits 3 to 4 m wide. The host rock is part of a metamorphic complex of mica schist and marble with layers and zones of quartzite and amphibolite. Mining is conducted selectively underground, and the ore is blended to ensure consistent mill feed.

LEÓN—This talc district is located about 53 km north of the city of León in northwestern Spain. The talc was formed by the hydrothermal replacement of carboniferous limestones (Galan-

Huertos and Rodas, 1973) along a major fault zone. It comprises irregular pods and ranges from white to light green, gray, and black. Commonly the pods and veins exhibit botryoidal structures that indicate low temperature replacement. The deposits are mined in open pits. The white talc is used in top quality products, and the gray to black grades, which fire white, are used in ceramics.

MÁLAGA—The talc-carbonate deposits in this province are located just northwest of Fuengirola. They comprise an ENE-trending zone that extends for 4 km along a major thrust fault. The talc-carbonate was formed by the alteration of serpentinite derived from ultramafic rocks. It is typically light gray but contains occasional 1- to 3-m wide bands of white talc that resulted from steatization along shear zones.

These talc-carbonate deposits pinch and swell along strike, dip steeply toward the southeast, and range up to 100 m thick. Mining originally began underground along narrow steatized zones to obtain cosmetic-grade talc. Open pit techniques were adopted later, when industrial uses were developed for the talc-carbonate ores.

Mexico, Germany, the Republic of South Africa, Chile, Sweden, and the United Kingdom (Shetland Islands) produce less than 20 kt of talc and talc-related products each year. Additionally, Venezuela and Ireland have reported potential resources of talc that have been investigated for development.

TECHNOLOGY

Exploration Techniques

Many geophysical exploration techniques have been field tested as means of locating unexposed talc deposits. Such methods include very low frequency electromagnetometry (VLF-EM), magnetometry, gravimetry, seismic refraction, self-potential (SP), and induced polarization (IP). None of these methods have been demonstrated to be definitive indicators of talc deposits when tested where talc is present as a metasomatic replacement of dolomitic marble. Some methods can, however, identify features that are commonly present in talc-bearing regions. VLF-EM assists in identifying fracture trends, and seismic refraction can locate dolomitic marble units within gneisses, schists, and quartzites.

Geochemical exploration methods have demonstrated potential in locating talc deposits. Detailed mapping and sampling at some active mines have revealed that certain minerals and elements are either depleted or enriched in proximity to talc ore bodies compared to unmineralized country rocks. These indicator minerals and elements are specific to each talc district, so different types of talc occurrences can have different geochemical signatures.

The most successful and practical exploration techniques involve mapping and sampling. Talc formation requires geological conditions that produce long lasting features on both mesoscopic and macroscopic scales. Such features include alteration assemblages, fault zones, and other structural conditions that created permeability or sites of low pressure conducive to the migration of and precipitation from mineralizing fluids.

Of course the most diagnostic tool in exploration is the actual presence of talc either in outcrop or as a component of soil. Talc is physically weak but chemically stable; therefore, the presence of talc deposits may be indicated by a disproportionate amount of talc in the overlying soil.

Mining Methods

Of the total talc produced in the United States in 1985, 93% came from open pit mines (Virta, 1989). Underground mine closures in Vermont and California since 1985 have increased this percentage. The average stripping ratio at US open pit mines in 1985 was 6.8:1.0. Generally, mines in the East have stripping ratios at or below this average, whereas mines in the West exceed it.

Open Pit Mining: Open pit mining methods include selective hand mining into baskets or ore cars (e.g., China, Brazil), selective mechanical mining with small backhoes or shovels, and selective mechanical mining with larger (e.g., 6- to 10-m^3) shovels. In all cases it is the operator's responsibility to monitor ore grade and quality, largely by visual means. Operators must be sensitive to one or more quality parameters critical to the consumer, such as iron for plastics compounders, loss on ignition (LOI) for ceramics manufacturers, arsenic for cosmetics producers, and brightness for paper companies. The critical quality parameter will usually be one of many that are tested during development drilling and mine planning. Prior to actual mining, talc ore is often further tested by means of blasthole or face samples.

Mine operators have developed specific techniques to maximize ore body recovery. Talc ore is sensitive to dilution during mining. Dilution not only lowers the talc content but may also introduce contaminants that are volumetrically insignificant but qualitatively disastrous. Miners in Montana prefer to approach high-purity talc veins from the hanging wall, which allows the operator to clean waste rock from the face of an ore body. Mechanically it is much easier to peel rock layers from a dip slope than to approach a vein parallel to its strike.

Blasting practice depends on the type of sorting or beneficiation methods to be used. Minimal blasting to yield large fragments is desired where high-purity talc is to be sorted by hand or mechanically or where sculpture-grade talc is produced. More intense fragmentation is desired where the talc is to be beneficiated by flotation or where unsorted, unbeneficiated talc is to be ground.

The largest open pit mines in the world are drill-and-blast, shovel-and-truck operations. Most mines produce more than one grade of talc, so careful mine planning and production scheduling are essential to ensure the availability of all grades.

Underground Mining: Underground mining is more common outside the United States because of generally lower labor costs. China has hundreds of single-shaft underground talc mines that produce 5 to 10 ktpy each.

Since the closure in 1990 of the Hammondsville underground mine at Reading, VT, only three active underground talc mines remain in eastern North America: R.T. Vanderbilt's mine in the Gouverneur (Balmat) District of New York, Canada Talc's Henderson mine near Madoc in Ontario, and Bakertalc's Van Reet mine near South Bolton in Quebec. Of these three the first two augment their underground ore with less costly ore from adjacent surface mining operations. The underground operations yield more selectively mined and segregated ores for specific products and end uses. The surface mines produce a greater tonnage, but more emphasis has to be placed on ore control, in-pit blending, and in some cases ore segregation in order to provide uniform mill feed for consistent quality products. Dry-processed industrial products must meet specifications that address accessory mineral content, particle sizing, particle shape, and brightness in order to satisfy customer requirements. Wet-processed products require consistent mill feed in order for flotation circuits to operate efficiently, but less pure plant feed can be tolerated if haulage distances, tailings disposal, and power factors permit an economic operation.

The level of difficulty in underground mining increases in direct proportion to ore body purity. High-purity talcs tend to be sheared or brecciated or to be found as thin veins within competent wall rock. Hard talcs that contain magnesite, dolomite, or tremolite are more stable and can be mined in room and pillar operations where dips are shallow (e.g., Chatsworth, GA, and Hammondsville, VT—both closed) or by shrinkage stoping where dips are steeper (see Fig.1).

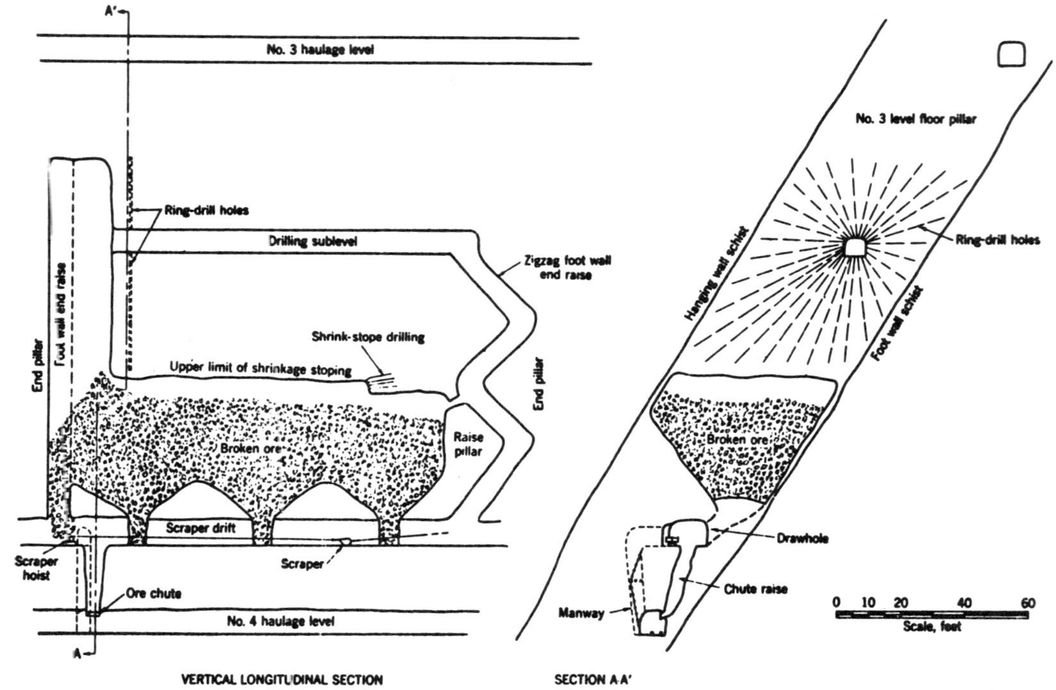

Fig. 1. Drifts, raises, and combined shrinkage and sublevel stopes (Burmeister, 1963).

High-purity talc veins with moderate to steep dips are most commonly mined in overhand or underhand cut-and-fill operations. Underhand cut-and-fill with cemented backfill is used at the Beaverhead mine in Montana and the Val Chisone mine in Italy. Overhand cuts in high-purity talc require large amounts of support, usually in the form of square-set timber. Overhand cuts in pure talc are dangerous even when timbered, because the rapid deterioration of exposed talc generates locally large loads for the timber.

Thin (e.g., 2- to 5-m wide) veins of high-purity talc usually exhibit intense shearing or brecciation. Branching, splaying, and irregular solution or soaker contacts with wall rocks complicate underground mining. Vein chasing is costly and often dangerous at overhand mining operations. Underhand methods offer the advantage of working under a solid back. Water in any amount can create very unstable conditions for high-purity talc veins.

Underground mines compete with difficulty in the United States. Low grade talc veins are more stable, but consumer concerns about health issues and the development of lower cost open pits have caused numerous low grade mine closures during the 1980s. High-purity vein mines in the United States face competition, therefore, from lower cost crudes produced in other countries.

Processing

Beneficiation: Talc processing is a function of two factors—the nature of the deposit, especially its purity, and the end use intended for the finished product. Massive talc deposits, such as those in Montana and China, are usually sorted by color and composition at the mine. This is typically done manually, where larger (+5 cm) rock is sorted on slowly moving belts. Cyprus Industrial Minerals (Luzenac America) recently installed proprietary mechanical sorting units at its Yellowstone, MT, mine.

Talc-carbonate deposits, typical of Vermont and eastern Canada, can also be sorted by color, but the talc content of the finished product will be only 60 to 80%, which is far too low for many applications. In these situations froth flotation is typically used to improve product purity.

At flotation plants the ore is first milled to liberation size, which can vary from −0.4 to −0.08 mm. For this purpose roller mills are typically used, but ball mills, hammer mills, rod mills, or any other type of mill that reduces the oversize without generating excess fines also can be used. Magnetic separators will remove magnetite, tramp iron, and other colored magnetic material. Talc is a natural floater, so only small amounts (≤1 000 ppm) of methyl-isobutyl-carbinol (MIBC) usually are needed to generate enough froth for economic flotation. Dolomite, calcite, quartz, chlorite, magnesite, and tremolite are efficiently removed from talc by this process. However, if the concentration of amphibole minerals in the feed is greater than 3%, flotation will not economically remove it down to the Cosmetic, Toiletry and Fragrance Association's standard of less than 1 000 ppm amphibole-in-talc. The flotation concentrate, typically purer than 95% talc, is thickened by gravity settling and then filtered and dried.

The third major type of economic talc deposit is typified by the talc-chlorite ore bodies of Talc de Luzenac in the Pyrenees of southern France. Although the talc-chlorite mixture has value as is, because chlorite is also a platy mineral. Some of the ore is also hand and mechanically sorted in lump form to upgrade color and talc content for specific applications.

Milling: The final processing of talc for marketing is totally dependent on the intended application, with color and particle size usually the major factors. For the roofing materials market, where talc is used as a parting agent, the ore is roller milled to −0.6 mm or hammer milled and screened to a −0.6+0.15 mm fraction. For the ceramics market, where talc is used primarily in tile and hobbyware bodies, the ore is roller milled or ball milled to a −0.15 mm grind.

The finer standard grades of talc, 74 and 44 µm grinds, are produced primarily in ring-type roller mills. These products go into such markets as flat latex paints, body patch compounds, baby and body powders, reinforced thermoplastics, and wall joint compounds. The roller mill is preferred because it preserves the platy structure of talc during milling, whereas other types of equipment, such as ball mills and hammer mills, tend to destroy it.

Micronized grades of talc are produced in either hammer mills, fluid energy mills, or jet mills. Hammer mills of either the Raymond Vertical Mill type or the Hosokawa Victory Mill type used to be most common. These mills have high speed hammers for grinding and a whizzer-type bladed air classification section. The major drawback to these units is that the classifier blades and grinding hammers are attached to the same shaft, so that if a change in particle size distribution is required the mill must be shut down to change blades or hammers. More modern hammer mills typified by Pulverizing Machinery's Air Classifying Mill (ACM) have separate drives for the milling and classification sections as well as a more efficient cage-type classifier. Talc can be milled to a 2 μm median and 25 μm topsize by using these mills, or by using air- or steam-powered fluid energy mills, often in closed circuit with high-efficiency air classifiers. In Japan air-powered jet mills with integral, high-efficiency air classifiers are used to produce products as fine as 1 μm with a topsize as small as 10 μm.

Packaging: Talc is generally sold in powder form in 22.68 kg bags, supersacks, or bulk. Packaging and conveying is generally done with impeller-type packers and dilute-phase air conveying systems. Micronized talc has a very low bulk density, less than 160 kg/m^3, and is expensive to transport in this form. Large volumes of talc are sold to the paper and rubber industries in compacted form. The talc is wetted with water in a pug-type mixer using about 20 wt% water and is then pelletized in a California Pellet Mill. The pellets are dried in a fluid-bed dryer to less than 2% moisture. The compacted product has a bulk density of 600 to 1 000 kg/m^3, a fivefold to tenfold increase over the feed powder.

Talc companies in Finland have pioneered the production of talc slurries for the Scandinavian paper industry. Slurries are supplied at 40 to 55% solids for paper filling and pitch control applications where no surfactants are used and at up to 70% solids for coating applications where a proprietary dispersant is used to lower viscosity.

In Japan, Europe, and North America, a small but growing tonnage of micronized talc is being sold as a 50 to 80% loading in plastic resin. These products are prepared in Banbury-type batch rubber mixers or continuous plastic/rubber mixers. The user then lets down the product to its final application level by mixing it in the same resin.

ECONOMIC FACTORS

Production and Markets

Statistics for production and consumption of talc worldwide in 1988, as estimated by Roskill Information Services (Anon., 1990), are as follows:

Area	Production (ktpy)	Consumption (ktpy)
North America	1 270	915
Europe	1 131	1 299
Russia	530	530
China	998	454
Japan	54	682
South Korea	146	121
Remainder of Asia	570	549
Central & South America	497	852
Australia	205	40
Total	5 403	5 442

The United States is the leading world producer with 1988 output of 1.15 Mt, followed by China, the former USSR, Brazil, India, Finland, France, and Italy. The United States is also the leading consumer, using 850 kt in 1988, followed by Japan, the USSR, Brazil, China, India, Finland, France, Mexico, and western Germany.

Talc consumption in specific markets is well defined only in North America and Europe. Roskill Information Services (Anon., 1990) breaks down the North American market as follows:

Market	1988 Sales (kt)	% of Total
Ceramics	311	34.0
Paint	130	14.2
Paper	122	13.3
Roofing	96	10.5
Plastics	91	9.9
Cosmetics	45	4.9
Rubber	18	2.0
Other[a]	101	11.0
Total	914	99.8

[a]Includes asphalt filler, joint compounds, floor tiles, and rice polishing.

The European market is broken down by Roskill (Anon., 1990) as follows:

Market	1988 Sales (kt)	% of Total
Paper	650	50.0
Paint	150	11.5
Plastics	90	6.9
Roofing	90	6.9
Ceramics	80	6.1
Animal Feed	65	5.0
Fertilizers	40	3.1
Cosmetics	30	2.3
Other Uses	105	8.2
Total	1 300	100.0

The major difference between the United States and Europe in terms of talc consumption lies in the relative position of and differences between their paper and ceramics markets. In North America the major portion of the ceramics market is wall tile, where talc comprises about 60% of a porous slow fired body. Electrical whitewares and hobbyware are minor markets. In Europe, however, electrical whitewares are the major ceramic end use.

Talc's principal role in the European paper market is as a filler. This is especially the case in Finland, Scandinavia, and France, where talc is less expensive than imported kaolin. Pitch control is only 20% of the European paper market. In North America more than 90% of the paper market is for micronized talc in pitch control. Because of its natural oleophilicity and surfactancy, talc has the ability to adsorb natural pitch resins and to keep them dispersed in the pulping slurry. In the processing of recycled paper talc performs similar functions in terms of ink and resins (so-called stickies) in the pulping process.

The world markets for talc are growing overall, but there are significant changes taking place in some segments. The ceramics market is stable, but demand in the paper market is declining. Talc as a paper filler is being replaced by calcium carbonate due to the worldwide switch to alkaline papermaking, but talc in pitch control is growing strongly due to its cost effectiveness and to the increasing use of secondary fiber.

The use of talc in plastics and roofing materials is growing rapidly due to growth in the plastics and synthetic roofing industries. The paint market for talc is stagnant, whereas the cosmetics

Table 5. End Uses for Ground Talc

Use	Thousand tons		
	1988	1989	1990
Ceramics	311	307	235
Cosmetics[a]	45	50	36
Insecticides	—	—	6
Paint	130	142	155
Paper	122	138	142
Plastics[b]	91	97	48
Refractories	1	1	4
Roofing	96	97	93
Rubber	18	20	20
Other uses[c]	100	100	109
Total[d]	914	952	848

Source: Virta 1990, 1989, 1988. [a]Incomplete data. Some cosmetic talc known to be included in "Other uses." [b]Incomplete data. Some plastics talc known to be included in "Other uses." [c]Includes art sculpture, asphalt filler, crayons, floor tile, foundry facings, rice polishing, stucco, and other uses not specified. [d]Data may not add to totals shown because of independent rounding.

market is growing steadily. Other markets such as adhesives and sealants, joint compounds, and body patch compounds are also showing steady growth, especially outside the United States.

US mine production of talc has remained at a little more than 1 Mtpy since at least 1970. World production averaged 1.7 Mtpy for the period 1945 to 1960, but by 1969 it rose to 5 Mtpy. By 1990 world talc production was estimated to have reached 7 to 8 Mtpy.

Many of the important large-tonnage uses for talc are closely related to the degree of industrialization of a country. Over the 15 years between 1945 and 1960 the United States accounted for about 38% of total world production of talc, soapstone, and pyrophyllite; by 1962 this dropped to 26%. Over the ten years preceding 1973 US production stabilized at about 22% of worldwide production. Due largely to increased production of talc in China, US talc output dropped to 14% of world talc production in 1990.

In 1990 US talc was produced by 13 companies from mines in 10 states. Vermont, Montana, New York, and Texas were the top producers, accounting for 96% of the output. Georgia, North Carolina, Arkansas, Nevada, Oregon, Virginia, and California produced the remaining tonnage. US production of talc in 1990 amounted to 1.185 Mt.

Table 5 shows the tonnages of ground talc sold or used by US producers in 1988, 1989, and 1990.

FUTURE TRENDS

The uses for talc in putties, caulks, and many types of sealants and adhesives are increasing. This is especially true of those products requiring close control of color (white and tinted) and strict control of other talc specifications to ensure high quality.

Talc competes with several other minerals. Feldspar, wollastonite, and nepheline syenite compete with it in ceramics. Mica competes with it in roofing markets. Titanium dioxide is losing ground as a competitive filler in some markets due to its high cost. Specially purified, calcined, and otherwise treated clays compete with talc in paint, paper, and rubber.

One important factor in competition for talc fillers in paint, paper, and plastics is the rapid penetration of these markets by limestone and dolomite filler products. Calcium carbonate is often considered the most versatile commercially produced mineral filler. The US market for calcium carbonate fillers and extenders exceeds the total domestic market for talc and pyrophyllite. The world market for carbonate fillers is growing at the fastest rate of any mineral filler. The widespread availability of large carbonate deposits strategically located for low cost transportation will encourage continuing competition in the white mineral filler market. A detailed comparison of calcium carbonate and talc fillers is presented by Roe (1980). The reader is also referred to the chapter on Fillers and Coatings in this book.

As more industrial mineral operations change over to wet beneficiation methods, especially flotation, there may be considerable amounts of byproduct talc produced. The very successful introduction of wet high intensity magnetic separation in the beneficiation of kaolin clays and other industrial minerals has added another dimension to the options in talc ore processing.

ACKNOWLEDGMENTS

The co-authors of this 6th edition chapter wish to acknowledge their debt to Lawrence A. Roe and Richard H. Olson, co-authors of this chapter in the 5th edition. Significant contributions to sections on the geology and mining of talc deposits worldwide were made by Ernest H. Reade and Lance P. Meade, both of Cyprus Industrial Minerals Co. (now Luzenac America), and by Erik L. Hoovestol of Cyprus Talc Corp. (now Luzenac America) We also extend special appreciation to Nikhil C. Trivedi of Specialty Minerals Inc., who provided invaluable assistance in combining and editing all material provided by the co-authors and other contributors.

BIBLIOGRAPHY

Anon., 1956, "Talc," California Mineral Information Service, Vol. 9, No. 11, Nov. 1, 8 pp.

Anon., 1966, "An Alternate Route to Size Reduction," *Chemical Processing*, Aug., p. 82.

Anon., 1969, "Process Break-Through Revitalizes 100 Year Old Mine," *Mining in Canada*, Oct., pp. 27–30.

Anon., 1970, "Standard Method of Test for Fineness of Dispersion of Pigment-Vehicle Systems," ASTM Designation D 1210–64 (reapproved 1970).

Anon., 1971a, "The Operations of A/S Norwegian Talc," *Industrial Minerals*, Jan., pp. 27–28.

Anon., 1971b, "Pyrenean Producer: S.A. des Talcs de Luzenac," *Industrial Minerals*, Jan., p. 25.

Anon., 1971c, "Talc: Micronized Grades Lead the Way," *Industrial Minerals*, Jan., pp. 9–18.

Anon., 1971d, "Talc Operations of Engelhard Minerals and Chemicals," *Industrial Minerals*, Feb., p. 46.

Anon., 1971e, "Talc: UK Consumption Shows Gradual Rise," *Industrial Minerals*, Jan., pp. 19–24.

Anon., 1971f, "Windham Plant Grinds Talc for Specialized Applications," *Rock Products*, Vol. 74, No. 6, p. 59.

Anon., 1972, "Chemical Profile—Titanium Dioxide," *Chemical Marketing Reporter*, Vol. 202, No. 24, Dec. 11, p. 9.

Anon., 1982, "Talc, Stability in a Soft Market," *Industrial Minerals*, Dec., pp. 59–73.

Anon., 1990, *The Economics of Talc and Pyrophyllite*, 6th ed., Roskill Information Services, London, 350 pp.

Anon., undated, "1989 Annual Report to Shareholders," Western Mining Corporation, Melbourne, Vic., Australia, 72 pp.

Anderson, D.L., 1987, "Timing and Mechanism of Formation of Selected Talc Deposits in the Ruby Range, Southwestern Montana," MS thesis, Montana State University, Bozeman, 90 pp.

Anderson, D.L., Mogk, D.W., and Childs, J.F., 1990, "Petrogenesis and Timing of Talc Formation in the Ruby Range, Southwestern Montana," *Economic Geology and the Bulletin of the Society of Economic Geologists*, Vol. 85, pp. 585–600.

Anderson, H.N., 1942, "Helena and Ennis Talc Deposits," BS thesis, Montana School of Mines (Montana Tech), Butte, 30 pp.

Ashton, W.H., and Russell, R.S., 1972, "Talc Beneficiation," US Patent 3 684 197, Aug. 15.

Barnes, D.F., 1958, "Infrared Luminescence of Minerals," Bulletin 1052-C, US Geological Survey, p. 113.

Bates, R.L., and Harben, P.W., 1989, "Talc," *Industrial Minerals Geology and World Deposits*, Metal Bulletin, London, pp. 274–281.

Bennett, F.E., 1972, "Tile Council of America Research in the Seventies," American Ceramic Society, Fall Meeting, Sep. 27–29, Bedford, PA.

Berg, R.B., 1979a, "Talc and Chlorite Deposits in Montana," Memoir 45, Montana Bureau of Mines and Geology, 66 pp.

Berg, R.B., 1979b, "Precambrian Geology of the West Part of the Greenhorn Range, Madison County, Montana," Geologic Map 6, Scale 1:28 570, Montana Bureau of Mines and Geology.

Berg, R.B., 1991, "Geology of Talc and Chlorite Deposits in Montana," *Proceedings*, 27th Forum on the Geology of Industrial Minerals, Information Series 115, Alberta Geological Survey, 26 pp.

Blount, A.M., and Vassiliou, A.H., 1980, "The Mineralogy and Origin of the Talc Deposits Near Winterboro, Alabama," *Economic Geology*, Vol. 75, No. 1, pp. 107–116.

Blount, A.M., and Helbig, S.R., 1987, "Talc and Chlorite in the Winterboro Area, Talladega County, Alabama," Circular 129, Geological Survey of Alabama, pp. 33–36.

Bourbon, W.B., 1981, "The Origin and Occurrences of Talc in the Allamoore District, Texas," Program with Abstracts, 17th Annual Forum on the Geology of Industrial Minerals, Albuquerque, NM.

Brady, J.B., et al., 1991, "Metasomatic Talc Deposits in Southwestern Montana: Geochemical Evidence for Deep Circulation of Water-rich Fluids," *Geological Society of America Abstracts with Programs*, Vol. 23, p. A263.

Brindley, G.W., and Santos, P. de S., 1971, "Antigorite—Its Occurrence as a Clay Mineral," *Clays and Clay Minerals*, Vol. 19, pp. 187–191.

Brown, J.S., and Engel, A.E.J., 1956, "Revision of Grenville Stratigraphy and Structure in the Balmat-Edwards District, Northwest Adirondacks, New York," *Bulletin*, Geological Society of America, Vol. 67, No. 12, pp. 1599–1622.

Brown, W.E., and MacDonald, R.D., 1963, "Talc Beneficiation," US Patent 3 102 855, Sep. 3.

Burger, J., 1991, "Talc—USA & Europe Competitive Markets," *Industrial Minerals*, Jan., pp. 17–27.

Burmeister, H.L., 1963, "Mining Methods and Costs, Eastern Magnesia Talc Co., Johnson Mine, Johnson, Vermont," Information Circular 8142, US Bureau of Mines, 44 pp.

Campbell, M.E., 1972, "Solid Lubricants, A Survey," NASA SP-5059(01), publication prepared under contract by Midwest Research Institute, Kansas City, MO, 126 pp.

Campbell, W.J., et al., 1980, "Chemical and Physical Characterization of Ampite, Chrysotile, Crocidolite, and Nonfibrous Tremolite for Oral Ingestion Studies by the National Institute of Environmental Health Sciences," Report of Investigations 8452, US Bureau of Mines, 63 pp.

Cawood, E.E.C., 1962, "Fire Extinguishing Composition," US Patent 3 063 940, Nov. 13.

Chase, W.E., 1963, "Platy Talc Beneficiation," US Patent 3 102 856, Sep. 3.

Chidester, A.H., Billings, M.P., and Cady, W.M., 1951, "Talc Investigations in Vermont," Preliminary Report, Circular 95, US Geological Survey, 33 pp.

Chidester, A.H., et al., 1964, "Talc Resources of the United States," Bulletin 1167, US Geological Survey, pp. 1–61.

Christman, R.A., 1959, "Geology of the Mount Mansfield Quadrangle, Vermont," Bulletin 12, Vermont Geological Survey, 70 pp.

Christman, R.A., and Secor, D.T., Jr., 1961, "Geology of the Camels Hump Quadrangle, Vermont," Bulletin 15, Vermont Geological Survey, 70 pp.

Clifton, R.A., 1980, "Talc and Pyrophyllite," Preprint 1978–1979, US Bureau of Mines, 7 pp.

Clifton, R.A., 1981, "Talc and Pyrophyllite," *Minerals Yearbook 1980*, US Bureau of Mines, pp. 815–820.

Cooper, J.D., and Hartwell, J.W., 1970, "Talc, Soapstone and Pyrophyllite," *Mineral Facts and Problems*, Bulletin 650, US Bureau of Mines, pp. 1267–1281.

Dahl, P., 1979, "Comparative Geothermometry Based on Major Element and Oxygen Isotope Distributions in Precambrian Metamorphic Rocks from Southwestern Montana," *American Mineralogist*, Vol. 64, pp. 1280–1293.

Dickson, T., 1982, "North American Talc," *Industrial Minerals*, Dec., pp. 75–78.

Dillender, R.D., Jr., and Gower, I.W., 1953, "Preliminary Report on the Froth Flotation of Wake County Talc," Bulletin No. 7, Dept. of Engineering Research, North Carolina State College, Raleigh, Nov., 16 pp.

Drake, M.J., and Brett, B.A., 1962, "Talc, Soapstone and Pyrophyllite," *Minerals Yearbook 1962*, US Bureau of Mines, pp. 1189–1198.

Edwards, G., 1980, "Tumbledown Mountain Talc Deposit, Allamoore District, Culberson County, Texas," *Guidebook*, 31st Field Conference, New Mexico Geological Society, pp. 245–250.

Elstron, L.W., 1961, "Slipping and Sliding," *Research Engineer*, Georgia Institute of Technology, Feb., pp. 15–18.

Engel, A.E.J., 1949, "New York Talcs, Their Geological Features, Mining, Milling and Uses," *Trans. AIME*, Vol. 184, pp. 345–348.

Engel, A.E.J., and Wright, L.A., 1960, "Talc and Soapstone," *Industrial Minerals and Rocks*, 3rd ed., J.L. Gillson, ed., AIME, New York, pp. 835–850.

Erdman, G.E., 1984, "Talc Milling in the 80's," SME Preprint 84–307, SME-AIME Fall Meeting, Denver, Oct., 8 pp.

Erdman, G.R., 1973, "Dust Control at Gouverneur Talc Company," *Trans. SME-AIME*, Vol. 254, No. 2, June, pp. 161–165.

Faust, G.T., and Fahey, J.J., 1962, "The Serpentine-Group Minerals," Professional Paper 384-A, US Geological Survey, pp. 80–82.

Flawn, P.T., 1958, "Texas Miners Boost Talc Output," *Engineering and Mining Journal*, Vol. 159, No. 1, pp. 104–105.

Frommer, P.W., and Fine, M.M., 1956, "Laboratory Flotation of Talc from Arkansas and Texas Sources," Report of Investigations 5241, US Bureau of Mines, June, 5 pp.

Furcron, A.S., Teague, K.F., and Calver, J.L., 1947, "Talc Deposits of Murray County, Georgia," Bulletin 53, Georgia Geological Survey, 75 pp.

Galan-Huertos, E., and Rodas, M., 1973, "Contribución al estudio mineralógico de los depósitos de talco de pueblo de Lillo (León, España)," *Boletín Geológico y Mínero 84–95*, pp. 347–365.

Garihan, J.M., 1973, "Geology and Talc Deposits of the Central Ruby Range, Madison County, Montana," PhD dissertation, Pennsylvania State University, University Park, 209 pp.

Garrison, J.R., Jr., and Mohr, D., 1984, "Geology of the Precambrian Rocks of the Llano Uplift, Central Texas," Field Trip Notes Guidebook 5, Austin Geological Society, Reprinted from: Geological Society of America, South-Central Section, 1983 Field Trip—Central Mineral Region Crystalline Rocks Llano Uplift, Central Texas, p. 38.

Giletti, B., 1966, "Isotopic Ages from Southwestern Montana," *Journal of Geophysical Research*, Vol. 71, pp. 4029–4036.

Gillingham, W.P., 1950, "Grinding Talc to Superfine Size," *Compressed Air Magazine*, Feb., pp. 32–37.

Glover, S.L., 1936, "Nonmetallic Mineral Resources of Washington," Bulletin 33, Washington Division of Geology, 135 pp.

Griffiths, J., 1989, "South Africa's Minerals—Diversity in Adversity," *Industrial Minerals*, Aug., pp. 18–53.

Hall, W.E., and MacKeveh, E.M., 1958, "Economic Geology of the Darwin Quadrangle, Inyo Co., California," Special Report 51, California Division of Mines, 7 pp.

Harrah, H.W., 1956, "Eastern Magnesia Talc Company, Inc.," *Deco Trefoil*, Denver Equipment Co., May-June, pp. 7–14.

Harrell, G.O., and Harrell, D.E., 1971, "Survey of Mineral Fillers in Selected Industries, Phase I—Plastics and Textiles," The Industrial Extension Service, School of Engineering, North Carolina State University, Raleigh, Apr., 105 pp.

Herod, B.C., 1954, "Pioneer Talc Leads Way in West Texas Field," *Pit and Quarry*, Vol. 54, Apr., pp. 134–135, 156.

Heystek, H., and Planz, E., 1964, "Mineralogy and Ceramic Properties of Some California Talcs," *Bulletin*, American Ceramic Society, Vol. 43, No. 8, pp. 555–561.

Hunt, R.E., 1969, "How to Select Fillers and Reinforcements for Thermoplastics," *Plastics Technology*, Nov., pp. 38–43.

James, H.L., 1990, "Precambrian Geology and Bedded Iron Deposits of the Southwestern Ruby Range, Montana," Professional Paper 1495, US Geological Survey, 39 pp.

James, H.L., and Hedge, C.E., 1980, "Age of Basement Rocks of Southwest Montana," *Geological Society of America Bulletin*, Vol. 91, pp. 11–15.

Joseph, N.L., 1990, "Industrial Minerals in Washington—Production and Potential," *Washington Geologic Newsletter*, Vol. 18, No. 4, Washington State Department of Natural Resources, 16 pp.

Karmazin, V.I., et al., 1972, "Grinding and Concentrating Capacity of Talc-Magnesite Rocks of the Pravda Deposit of the Ukranian SSR," Obogasheh, Polez, Iskop, 1971, No. 9, p. 28–29 (Russian), *Chemical Abstracts*, Vol. 77, 92190c, p. 262.

Katz, H.S., and Mileski, J.V., 1978, *Handbook of Fillers and Reinforcements for Plastics*, Van Nostrand Reinhold, New York, 652 pp.

Keeling, J., Pain, A., and Valentine, J., 1990, "Industrial Minerals in South Australia—Current Operations and Future Trends," *Proceedings*, 9th Industrial Minerals International Congress, Sydney, NSW, Australia, Mar. 25–28, pp. 30–46.

King, P.B., and Flawn, P.T., 1953, "Geology and Mineral Deposits of Precambrian Rocks of the Van Horn Area, Texas," No. 5301, Texas University Publications, 218 pp.

Klassen, V.I., and Mokrousov, V.A., 1963, *An Introduction to the Theory of Flotation*, Butterworth's, London, 493 pp.

Kretschmar, U., and Kretschmar, D., 1986, "Talc Magnesite and Asbestos Deposits in the Timmins-Kirkland Lake Area, Districts of Timiskaming and Cochrane," Study 28, Ontario Geological Survey, 100 pp.

Lamb, F.D., and Ruppert, J., 1950, "Flotation of a North Carolina Pyrophyllite Ore," Report of Investigations 4674, US Bureau of Mines, 7 pp.

Lepetic V.M., 1972, "Flotation of Copper and Molybdenum Sulfides Without Collector," Preprint 72B308, SME-AIME Fall Meeting, Birmingham, AL, Oct. 18–20, pp. 1–23.

Longeaux, J., Grange, J.P., and Atschit, L., undated, "A Rigorous Procedure to Achieve and Maintain Product Quality," Luzenac Group brochure, Toulouse, France.

Lundquist, J.D., 1970, "Talc Treatment and Talc-Containing Pigments," US Patent 3 533 821, Oct. 13.

MacLean, M., 1987, "Talc and Pyrophyllite in British Columbia," Mineral Resource Division, Geological Survey Branch, 108 pp.

McMurray, L., and Bowles, E., 1941, "The Talc Deposits of Talladega County Alabama," Circular 16, Geological Survey of Alabama, 31 pp.

Meade, L.P., 1992, "Troy Talc—An Old Deposit—A New Development," Preprint 92–286, SME Annual Meeting, Phoenix, AZ, 5 pp.

Miller, C.R., and Holtzapfel, P.J., 1970, "Corrosion Proofing Composition and Method," US Patent 3 549 391, Dec. 22.

Mulryan, H.T., 1971, "World-Wide Operations of United Sierra," *Industrial Minerals*, Feb., pp. 43–45.

Neathery, T.L., 1968, "Talc and Anthophyllite Asbestos in Tallapoosa and Chambers Counties, Alabama," Bulletin 90, Geological Survey of Alabama, 98 pp.

Neathery, T.L., 1970, "Geology and Mining of Low Grade Talc Deposits, Tallapoosa County, Alabama," Preprint 70H311, SME-AIME Fall Meeting, St. Louis, MO, 15 pp.

Neathery, T.L., et al., 1967, "Talc and Asbestos at Dadeville, Alabama," Report of Investigations 7045, US Bureau of Mines, 57 pp.

Needham, R.E., 1972, "The Geology of the Murray County, Georgia Talc District," MS thesis, Pennsylvania State University, University Park, 107 pp.

Norman, J.E., et al., 1939, "Froth Flotation of Talc Ores from Gouverneur, New York," *Bulletin*, American Ceramic Society, Vol 18, No. 8, pp. 292–297.

Okuda, S., et al., 1969, "Negative Surface Charges of Pyrophyllite and Talc," *Proceedings of the International Clay Conference*, Tokyo, Vol. 1, pp. 31–44, Vol. 2, pp. 14–16.

Okuma, A.F., 1971, "Structure of the Southwestern Ruby Range near Dillon, Montana," PhD dissertation, Pennsylvania State University, University Park, 122 pp.

Olson, R.H., 1976, "The Geology of Montana Talc Deposits," *Proceedings*, 11th Industrial Minerals Forum, L.F. Rooney and R.B. Berg, eds., Special Publication 74, Montana Bureau of Mines and Geology, pp. 99–143.

Owens, M.H., 1968, "Petrologic Study of Talc Mineralization in the Murphy Marble in Southwestern North Carolina," MS thesis, University of Tennessee, Knoxville, 64 pp.

Page, B.E., 1951, "Talc Deposits of Steatite Grade in Inyo County, California," Special Report 8, California Division of Mines, 5 pp.

Papke, K.G., 1975, "Talcose Minerals in Nevada—Talc, Chlorite and Pyrophyllite," Bulletin 84, Nevada Bureau of Mines and Geology, 62 pp.

Parsons, A.B., 1948, *Seventy-Five Years of Progress in the Mineral Industry, 1871–1946*, AIME, New York, p. 355.

Pearson, R.C., et al., 1990, "Map Showing Mineral Resource Assessment for Silver, Cobalt, and Base Metals in Proterozoic Sedimentary Rocks and for Iron, Chromium, Nickel, Talc, Chlorite, Gold, and Graphite in Archean Crystalline Rocks," Dillon 1 x 2° Quadrangle, Idaho and Montana, Map 1–1803-D, Scale 1:250 000, US Geological Survey.

Pence, F.K., 1955, "Commercially Proven White-Firing Talc Occurring in West Texas," *Bulletin*, American Ceramic Society, Vol. 34, No. 4, pp. 122–123.

Pereira, F.W., 1989, "Introducing A New Industrial Mineral Company—The Montana Talc Company," Preprint 89–120, SME Annual Meeting, Las Vegas, Feb. 27–Mar. 2, 2 pp.

Perry, A., 1972, "Supply and Demand for Ceramic Tile in the Seventies," American Ceramic Society, Fall Meeting, Sep. 27–29, Bedford, PA.

Perry, E.S., 1948, "Talc, Graphite, Vermiculite and Asbestos in Montana," Memoir 27, Montana Bureau of Mines and Geology, 44 pp.

Piniazkiewicz, R.J., 1984, "Geology of and Exploration Techniques for Pre-Beltian Talc Deposits of the Malesich Ranch, Ruby Range, Madison County, Montana," MS thesis, University of Arizona, Tucson, 101 pp.

Poppe, W., et al., 1972, "Talc Filled Metallizable Polyolefins," US Patent 3 663 260, May 16.

Povarennykh, A.S., 1972, *Crystal Chemical Classification of Minerals*, Vol. 1, trans. from Russian by J.E.S. Bradley, Plenum Press, New York, pp. 415–416.

Ranney, M.W., et al., 1972, "Silane Coupling Agents in Particulate Mineral-Filled Composites," 27th Reinforced Plastics Technical and Management Conference, SPI, Washington, DC, Feb. 8–11, pp. 1–29.

Rayner, J.H., and Brown, G., 1966, "Structure of Pyrophyllite," *Proceedings*, 13th National Conference on Clays and Clay Minerals, Pergamon Press, New York, pp. 73–84.

Roe, L.A., 1972, "High-Purity Talc from Western Ores," Preprint 72H312, SME-AIME Fall Meeting, Birmingham, AL, Oct. 18–20, 16 pp.; US Patent 3 806 043, Apr. 23, 1974.

Roe, L.A., 1980, "A Comparison of Calcium Carbonate and Talc Fillers," Preprint 80–301, SME-AIME Fall Meeting, Minneapolis, MN, Oct.

Roe, L.A., and Olson, R.H., 1983, "Talc," *Industrial Minerals and Rocks*, 5th ed., Vol. 2, S.J. Lefond, ed., AIME, New York, pp. 1275–1301.

Ross, C.P., Andrews, D.A., and Witkind, I.J., 1955, "Geologic Map of Montana," US Geological Survey.

Ross, M., and Smith, W.L., 1968, "Triclinic Talc and Associated Amphiboles from Gouverneur Mining District, New York," *American Mineralogist*, Vol. 53, May-June, pp. 751–769.

Russell, A., 1991, "India's Industrial Minerals—Completing the Picture," *Industrial Minerals*, Sep., pp. 45–54.

Sanford, R.F., 1982, "Growth of Ultramafic Reaction Zones in Greenschist to Amphibolite Facies Metamorphism," *American Journal of Science*, May, pp. 543–616.

Schmidt, C.J., and Garihan, J.M., 1983, "Laramide Tectonic Development of the Rocky Mountain Foreland of Southwestern Montana," *Rocky Mountain Foreland Basins and Uplifts*, J.D. Lowell, ed., Rocky Mountain Association Geologist, pp. 271–294.

Seymour, R.B., et al., 1972, "Fibrous Reinforcements for Polymers," *1972–73 Modern Plastics Encyclopedia*, pp. 365–394.

Sloan, E., 1908, "Catalogue of the Mineral Localities of South Carolina," South Carolina Geological Survey, 505 pp.

Trauffer, W.E., 1964, "New Vermont Talc Plant Makes High-Grade Flotation Product for Special Uses," *Pit and Quarry*, Vol. 57, No. 6, Dec., pp. 72–76, 101.

Troll, G., and Gilbert, M.C., 1972, "Fluorine-Hydroxyl Substitution in Tremolite," *American Mineralogist*, Vol. 57, pp. 1386–1403.

Valentine, G.M., 1960, "Inventory of Washington Minerals, Pt.1, Nonmetallic Minerals," 2nd ed. rev., M.T. Huntting, ed., Bulletin 37, Washington Division of Mines and Geology.

Van Horn, E.C., 1948, "Talc Deposits of the Murphy Marble Belt," Bulletin 56, North Carolina Department of Conservation and Development, 54 pp.

Virta, R.L., 1989, "The Talc Industry—An Overview," Information Circular 9220, US Bureau of Mines, 11 pp.

Virta, R.L., 1991, ''Talc and Pyrophyllite,'' Annual Report, US Bureau of Mines, 7 pp.

Wells, J.R., 1973, ''Talc, Soapstone, and Pyrophyllite in 1973,'' Annual Advance Summary, US Bureau of Mines, 2 pp.

Wicklein, P.C., 1967, ''Geology of Nickeliferous Soapstone Deposits of Saline County, Arkansas,'' MS thesis, University of Missouri, Columbia, 68 pp.

Winkler, H.G.F., 1974, *Petrogenesis of Metamorphic Rocks*, Springer-Verlag, New York, 320 pp.

Wright, L.A., 1950, ''California Talcs,'' *Trans. AIME*, Vol. 187, pp. 122–128.

Wright, L.A., 1968, ''Talc Deposits of the Southern Death Valley-Kingston Range Region, California,'' Special Report 95, California Division of Mines and Geology, 79 pp.

Zen, E-an, 1961, ''Mineralogy and Petrology of the System AL_2O_3-SiO_2-H_2O in Some Pyrophyllite Deposits of North Carolina,'' *American Mineralogist*, Vol. 46, Jan.-Feb., pp. 52–66.

TITANIUM MINERALS

THOMAS E. GARNAR AND KERRY J. STANAWAY

Titanium was discovered in 1791 and, four years later, was named after the Titans of Greek mythology. Titanium metal was first made in 1910, but was not used commercially until the 1950s when it was nick-named The Wonder Metal of the Age. The first titanium dioxide (TiO_2) white pigments were produced commercially in 1915; more than 90% of the titanium minerals produced in the world today are used in the manufacture of white pigments. The remaining 10% are used to manufacture titanium metal and other uses such as welding rod fluxes.

Titanium mineral mining began in the United States between 1880 and 1900 in Chester County, Pennyslvania. Rutile from a deposit near Roseland, VA, was mined in 1901. The use for the rutile was limited to production of a few titanium chemicals and for coloring ceramics. Larger scale mining for ilmenite, rutile, and zircon began in Florida to provide chemical materials for use in World War I in 1918. This mining ended in 1928. Large scale mining of titanium minerals in Australia began when Berrick, Anderson, and Porter (later called Tamco) started mining the beach sands at Yamba in New South Wales in 1935. Ilmenite production from beach sands in the State of Kerala in southwestern India also began in the 1930s.

The 1940s saw many changes in titanium mineral mining. Extensive deposits of ilmenite and ilmenite-bearing iron ores were discovered in eastern Quebec. The Pigments Department of E.I. Du Pont de Nemours & Co. lost their high grade Indian ilmenite supply after World War II began, and they were forced to use lower grade ilmenite from National Lead's (now NL Industries) Sanford Lake Deposit, near Tahawus, NY, in the Adirondack Mountains. At the end of the war, Du Pont undertook an intense search for a domestic ilmenite source. With the cooperation of the US Bureau of Mines, Du Pont geologists found a very large deposit east of Starke, FL, in 1946. In 1948 Du Pont entered into a contract with Humphreys Gold Corp., who operated a heavy minerals mine near Jacksonville, FL, to construct a plant and operate it for a period of ten years. The plant started up in 1949 and operated under Humphreys' management until the contract expired in 1958. Du Pont has operated the plant since that time. In 1956, a second plant was designed and built by Du Pont a few kilometers northeast of the town of Lawtey. The 1960s saw titanium mineral production expand in the southeastern US with Humphreys' contract mining Du Pont's ore body at Folkston, GA.

In the early 1970s, American Cyanamid and Union Camp Paper Co. formed a joint venture company, Titanium Enterprises (TE), to mine and separate titanium minerals from Union Camp's ore deposit southwest of Green Cove Springs, FL. In 1975 the TE plant was shutdown and later sold to Associated Minerals (USA), Inc. (AMC), a subsidiary of Renison Goldfields Consolidated Ltd., an Australian company. It now operates under the name RGC (USA). Mineral production grew in the 1970s with the development of extensive heavy mineral deposits at Eneabba in Western Australia and Richards Bay in the Republic of South Africa.

The 1980s saw the construction of a very large ilmenite mine and synthetic rutile processing plant in the state of Orissa southwest of Calcutta, India. At the same time, concern for titanium mineral shortages expected by the end of this century, gave rise to a flurry of exploration activity during the mid 1980s. As a result, large deposits were found in Virginia and North Carolina. New deposits were also discovered in Australia. Fig. 1 shows the distribution of world titanium deposits and mines. The world TiO_2 production capacities are given in Table 1.

MINERALOGY

Titanium is found in more than 45 different minerals but most commonly occurs as titanomagentite, ilmenite, and rutile. Titanomagnetite normally contains less than 10% TiO_2 making it of little or no value as an ore of titanium.

Ilmenite: Ilmenite ($FeTiO_3$) forms an incomplete series with geikielite ($MgTiO_2$) and pyrophanite ($MnTiO_2$). Ilmenite and its alteration products are the most abundant TiO_2 sources on Earth. There are a number of types of ilmenite mined for TiO_2. The first are hard rock deposits associated with gabbros, diorites, and anorthosites. They are used primarily in producing TiO_2 slag. Ilmenite is very resistant to chemical attack and is concentrated by weathering and detrital action. Chemical analyses of several different ilmenites are given in Table 2. Hard rock ilmenites may be strongly magnetic (in association with magnetite) but may also be weakly magnetic or paramagnetic. Fresh, unaltered ilmenite is black with a metallic to submetallic luster.

Ilmenite alters very slowly from crystalline $FeTiO_3$ to an amorphous mixture of FeO, Fe_2O_3, and TiO_2. During this process some of the iron oxides are leached out of the grains thereby upgrading the TiO_2 content. The end product of this alteration is either rutile or anatase. Fresh unaltered ilmenite is soluble in H_2SO_4 and HCL, but as alteration progresses, its solubility decreases proportionately. Its end products are insoluble in acids.

Leucoxene (Pseudorutile): As the ilmenite alteration process continues, the grains lighten in color and signs of recrystalization become detectable by X-ray diffraction. At this point the color of the grains are tan or gray and have a waxy luster. The colors can be enhanced for accurate identification by heating to 1 000°C for ten minutes in an oxidizing atmosphere. At this stage of alteration, the TiO_2 content of the altered grains has increased to 80 to 90%; they are only partially soluble in acids, and are only slightly paramagnetic. For many years these grains were called leucoxene. Temple (1966) proposed the name pseudorutile for these minerals. Fleischer (1987) gives the chemical formula of pseudorutile as $Fe_2Ti_3O_9$. The name leucoxene continues to be used in referring to highly altered ilmenite.

Anatase: Although anatase occurs in nature as one of the three polymorphs of TiO_2, it also commonly occurs as the end product of the ilmenite alteration process in beach sands. Major deposits of anatase are found in carbonatites at Tapira and Salitre southeast of

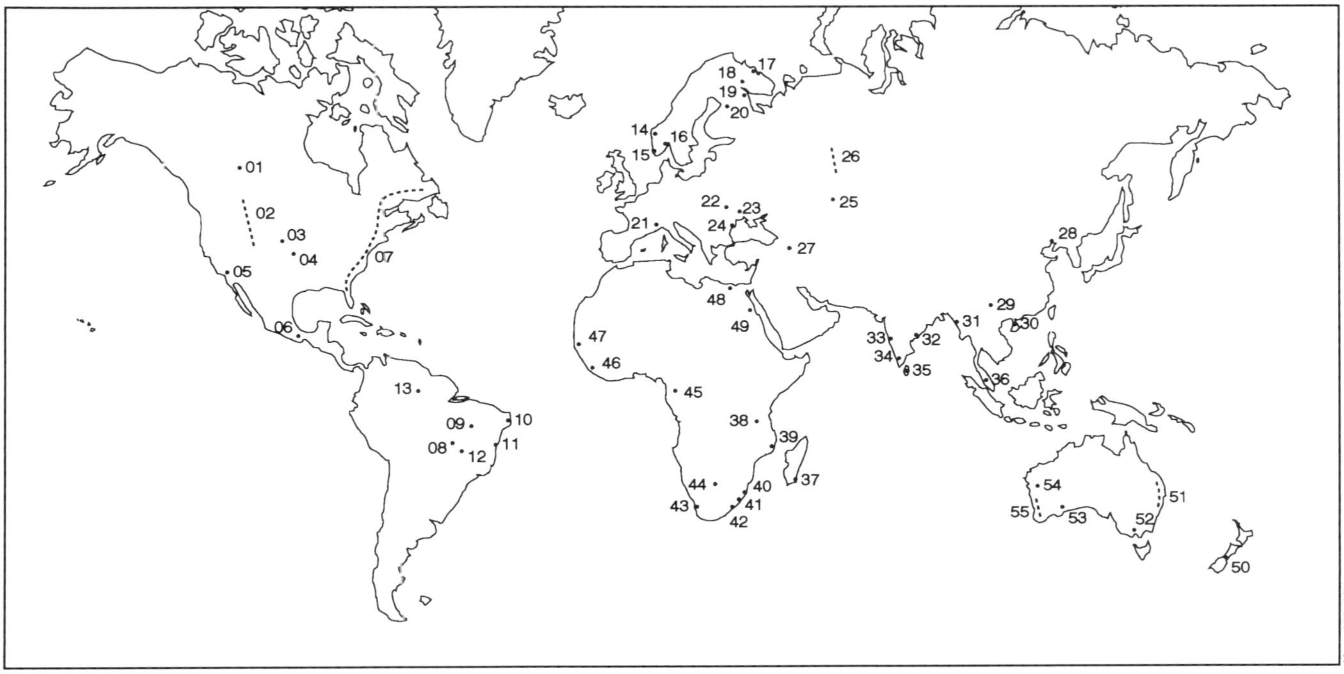

01 Athabasca
02 West Shore Cretaceous Interior Seaway
03 Powderhorn
04 Wichita Mountains
05 San Gabriel Mountains
06 Pluma Hidalgo
07 East North America Province
08 Brazilian Carbonatites
09 Campo Alegro de Lourdes
10 Natal Coast District
11 Vitoria Coast District
12 Sapucai
13 Maicuru
14 Rodsand
15 Tellnes
16 Kragero, Kodel
17 Tsangirsk
18 Afrikanda
19 Yelet Ozero
20 Otanmaki
21 Piampaludo
22 Kiev District
23 Dnepropetrovsk District
24 Chituk, Glogova, Tigveni
25 Shubino Village
26 Kopansk
27 Kamkorskoye
28 Heishan
29 Panzhihua
30 Hainan District
31 Cox's Bazaar
32 Chatrapur
33 Rathnagiri
34 Chavara
35 Pulmoddai
36 Malaysia Tin Province
37 Fort Dauphin
38 Lingana
39 Pebane Congolone
40 Richards Bay
41 Umbgaba
42 Transkei
43 Western Cape
44 Bothaville
45 Akonolinga
46 Gbangbama
47 Senegal
48 Nile Delta
49 Abu Ghalaka
50 Westport - Barrytown
51 East Australia Province
52 Murray Basin (WIM)
53 Eucla Basin
54 Murchison District
55 West Australia Province

Fig. 1. Map showing the titanium resources of the world.

Araxa in Minas Gerais, Brazil. Here they occur as pseudomorphs, often octahedral, with intergrowths of fine quartz. In beach sand deposits grains of secondary anatase are slightly translucent and are yellowish to gray in color with indices of refraction over 2.5. At first glance under the microscope, these grains appear to be zircons, but under reflected light they show slight shades of yellow or gray.

Rutile: Rutile used in titanium commerce today is in two forms: primary rutile, that makes up the major portion, and secondary rutile. Most of the primary rutile comes from beach sand deposits in Australia and the Republic of South Africa and from continental deposits in Sierra Leone. Secondary rutile is found in beach sand deposits and is the end product of the ilmenite alteration process.

Pseudobrookite: Pseudobrookite is not a commercial ore of titanium because of its low TiO_2. It commonly occurs as a pneumatolytic or fumarolic product in volcanic igneous rocks. It also forms when ilmenite or altered ilmenite are heated to about 1 400°C.

Perovskite: Perovskite deposits have been of interest as titanium ores for a number of years. Private sector and government (US Bureau of Mines) efforts to find an economic process to utilize perovskite have not been successful to date. Work by the USBM in 1973 resulted in the production of a chlorinatable product from titanium carbide made on a laboratory scale by smelting perovskite with carbon (and cereal binder). More recently the USBM announced that they had successfully produced TiO_2 pigment from perovskite using an acid sulfation method.

Other titanium minerals: Two common but lesser known titanium minerals are ulvospinel ($TiFe_2O_4$) and sphene, also called titanite ($CaTiSiO_5$). Ulvospinel occurs as exsolution patches or lamellae in magnetite. Sphene occurs as an accessory mineral in hornblende granites, syenites, and diorites. It also is found in metamorphic rocks (especially schists and gneisses) that are rich in magnesium and iron.

ORIGIN OF TITANIUM MINERAL DEPOSITS

Force (1991) states that two processes are the most directly responsible for economic titanium deposits: the first is immiscible, high specific gravity molten droplets in cooling magmas and the second weathering-resistant, high specific gravity mineral grains sorted during clastic sedimentation. In addition, it is possible that two other mineralizing processes could receive economic attention. They include contact metasomatic titanium from alkalic igneous intrusives and relict titanium dioxide in titanium-rich source rocks.

Magmatic Segregation Deposits: Magmatic segregation of TiO_2 takes place in tholeiitic basaltic magmas that pond under the crust and slowly crystallize, de-gas, and fractionate to yield molten residuals increasingly enriched in titanium, phosphorous and iron (Emslie, 1978). The increasing titanium and iron finally result in the formation of immiscible liquid oxide droplets that fall to the base of the magma chamber onto a mush of earlier formed and precip-

Table 1. World Titanium Annual Production Capacity, Dec. 31, 1990 [Thousand Tons of TiO_2 Content of Concentrates, Gross Weight of Sponge (More Than 99% Ti) and Gross Weight of Pigment (Average about 95% TiO_2)]

	Mineral		Synthetic	Sponge	TiO_2 Pigment		
	Ilmenite	Rutile	Rutile	Metal	Sulfate	Chloride	Total
North America							
Canada	880[a]				74	38	112
Mexico						79	79
United States	217	30	100	30	117	969	1 086
TOTAL	1 097	30	100	30	191	1 277	1 277
South America							
Brazil	83	15			55		55
Europe							
Belgium					70		70
Czechoslovakia					22		22
Finland					80		80
France					225		225
Germany, Rep. of					240	80	320
Italy					75		75
Netherlands						45	45
Norway	450				27		27
Poland					36		36
Spain					65		65
USSR	250	10		52	100	100	
United Kingdom				5	136	158	294
Yugoslovia					28		28
Total	700	10		57	1 104	283	1 387
Africa							
Sierra Leone	35	120			35		35
South Africa, Rep. of	640[b]	94			35		35
Total	675	214			70		70
Asia							
China	90			3	25		25
India	200	19	138		12	22	34
Japan			46	27	274	54	328
Korea, Rep. of					18		18
Malaysia	275		96				
Singapore						36	36
Sri Lanka	80	13					
Taiwan					10		10
Thailand	16						
Total	661	32	280	30	339	112	451
Australia	1 100	260	250		35	70	105
World Totals	4 316	561	630	117	1 759	1 742	3 310

[a]Contained in 80% TiO_2 slag made from ilmenite, [b]Contained in 85% TiO_2 slag made from ilmenite. *Source:* US Bureau of Mines, 1990.

itated crystals. Here the droplets can: 1) adopt positions interstitial to the older cumulate crystals; 2) coalesce to form layers parallel to the layering of the earlier crystal sediment; and 3) fill cracks and fractures that narrow downward (Force, 1991). Good examples of droplet and coalescing droplet textures can clearly be seen in the San Gabriel Mountain ilmenite deposits of California. Many titanium-iron oxide deposits accumulate as lenses in depressions on the magma chamber floor.

Titanium segregation deposits occur in a ferrodiorite or iron, titanium, and phosphorous enriched magma fractionated product separated from one magma that has intruded the fractionated product of another; the new host being the hot crystalline-mush precursor to a rock that eventually becomes an anorthosite. Large bodies of anorthosite, called massifs, appear in terrains undergoing high grade metamorphism. Here they associate with granulite and with large masses of marginal hybrid igneous and metamorphic rocks (e.g., charnokite, mangerite and jotunite). Anorthosites composed predominantly of andesine are more highly fractionated (Emslie, 1978) and tend to produce ilmenite free of intergrowths with magnetite. The less fractionated labradorite anorthosites also have iron and titanium oxides, magnetite, ulvospinel and ilmenite, fully intergrown for effective separation (Herz, 1976, Force, 1991).

The nature of the iron-titanium oxide species that form in magmatic segregation deposits is determined from oxygen fugacity, cooling history, fractionation history or magma composition, and iron loss into any aqueous phase (Elsdon, 1975). Titanium mineral concentration values derive not only from the factors just listed but also from resorption where at later stages of magma crystallization, deuteric biotite and amphibole can replace the oxides. Very late stage intrusives, highly enriched in titanium and phosphorus (nelsonites), can adopt three forms: 1) layered, concordant, discontinuous units at the base of ferrodiorites, 2) injections into the rock beneath, and 3) anastomising intrusive veinlets in granulite and anorthosite. It is these nelsonites which are made up of ilmenite ±apatite, ±rutile, and ±zircon, that constitute the titanium resources

Table 2. Chemical Analyses of Commercial Ilmenites

Constituent	AMC Capel	WSL Yoganup	AMC Eneabba	Tiwest	CRL*	Quilon (Chevera) India	Ceylon	Tellnes	Brazil	Florida	Beenup
TiO_2	55.3	55–58	60.8	62.1	50.7	59.2	52.4	44.0	54.3	65.0	51.0
Fe total	29.4	25–33	31.0	32.0	33.6	24.0	28.5	34.4	25.3	16.9	35.4
Fe_2O_3	21.6	10–16	—	—	29.2	—	—	13.1	24.6	24.1	13.8
FeO	18.4	23.2–28.0	—	—	17.0	—	—	32.4	10.4	33.0	
SiO_2	0.8	0.6	0.77	1.10	0.39	1.7	1.6	3.2	2.28	0.60	0.18
Al_2O_3	1.35	0.6	0.98	1.20	0.47	1.6	1.2	0.76	2.45	1.2	0.12
ZrO_2	0.17	0.20	0.23	0.20	0.13	0.3	1.1	n.a.	1.13	0.2	0.04
MnO	1.31	1.40	1.07	1.10	1.20	0.4	0.9	0.30	1.53	1.1	1.50
P_2O_5	0.03	0.05	0.08	0.11	0.06	0.2	0.2	<0.023	0.1	0.15	0.05
MgO	0.18	0.20	0.27	0.25	0.91	0.8	0.7	4.48	n.a.	0.21	0.16
Cr_2O_3	0.04	0.04	0.20	0.10	0.28	0.15	0.12	0.09	0.09	0.07	0.02
V_2O_5	0.11	0.20	0.22	0.20	0.18	0.21	0.23	0.23	0.09	0.18	0.10
Nb_2O_5	0.19	0.16	0.12	0.18	0.06	0.2	<0.1	<0.014	0.15	0.17	0.17
Th (ppm)	35	120	400	140	<20	218	224	n.a.	n.a.	33	8
U (ppm)	20	0	15	<10	<10	13	13	n.a.	n.a.	10	6

*upgraded. AMC—Associated Minerals Consolidated, CRL—Consolidated Rutile Limited, WSL—Westralian Sands Limited.

of the Roseland District of Virginia and of the San Gabriels in California.

The magmatic segregation deposits just described as occurring in anorthosite massifs can also be found in stratiform or layered mafic intrusions. The essential difference between the two is that the anorthosites associated with the stratiform intrusives remain in physical contact with their more basic differentiates and do not form separate bodies as do the anorthosites of the massif type. Like the titanium resources tied up in anothorsite massifs, those in stratiform bodies are not always economically extractable due to the iron and titanium oxides being either intergrown (forming interlocking crystals) or forming exsolution lamella within parent magnetites.

Alkalic Intrusive Deposits: Alkalic intrusives are another igneous rock-type rich in titanium, iron, and phosphorus. Herz (1976) ascribes the high concentrations of these elements again to the formation of immiscible molten liquids within the alkalic melts. The principal titanium minerals found in the rocks include ilmenite, titanomagnetite, sphene, perovskite, rutile, anatase, and brookite. Most titanium minerals form late in the crystallization history. In general, the alkalic rocks are not considered titanium resources because perovskite or sphene form rather than rutile or ilmenite due to the abundance of calcium.

To date, chemical processes to convert perovskite to form TiO_2 have not been proven commercially viable in Western countries. However, there are indications that perovskite has been mined for this purpose from the Afrikanda Deposit in the Kola Peninsula of Russia. Under conditions of tropical weathering, however, calcium titanium minerals such as perovskite and sphene can leach to form anatase. Carbonatite intrusives in Brazil (Tapira, Salitre, and Catalao) have weathered to depths averaging 30 m and can contain up to 24% TiO_2 as anatase weathered from perovskite (Harben, 1984). The resource is huge, but mineral beneficiation problems, together with high radioactivity and contaminating elements in the concentrate products preclude commercial potential

High trace element contents such as niobium in the rutile as well as small deposit size have prevented large scale economic utilization of titanium minerals from the Magnet Cove District deposits in Arkansas. Trace element contents in alkalic intrusive deposits probably derive from the highly differentiated volatile-rich nature of alkalic magmas at the time the titanium minerals crystallized.

Contact-Metamorphic Deposits: Apparently during the final crystallization stages of alkalic magmas, titanium contents in the original melt and compositions of the coexisting aqueous solutions can be appropriate for partitioning of titanium into final residual aqueous fluids. These fluids can leave veins rich in the titanium oxides, rutile and brookite. This aqueous phase mobility of titanium during the final crystallization of alkalic intrusives together with the potential for rutile and brookite (TiO_2) formation instead of the usual sphene or perovskite constitute a potentially interesting economic target for future exploration. Swarms of feldspar-carbonate veins containing rutile in phonolite were mined on a small scale at Magnet Cove between 1932 and 1944. In addition, in these Arkansas deposits, brookite occurs in feldspar-sulfide veins in the jacupirangite, and finally as quartz-limonite veinlets and disseminations in novaculite, a cherty country rock.

Relict Titanium Dioxide Deposits: Outside the special case just mentioned, titanium is immobile via either solution in aqueous phases or via solid state diffusion under normal crustal conditions. This suggests the possibility of economic titanium dioxide deposits by another mechanism. Relict titanium dioxide deposits (giving minerals such as pseudorutile, rutile, brookite, and anatase) will be created when under normal crustal conditions titanium remains immobile even as other elements are leached or diffused away. This can take place during rock weathering when calcium and silica leach from perovskite and sphene to form anatase (TiO_2) as has already been described in Brazil. It can also occur in oxidizing acidic sediments where pseudorutile replaces ilmenite to form altered ilmenite and leucoxene in placer deposits such as those of Florida (Temple, 1966, Mucke and Chaudhuri, 1991).

Under conditions of hydrothermal alteration, relict TiO_2 occurs as rutile (and perhaps anatase) where aqueous fluids strip other elements from the site of a previous titanium-bearing species such as ilmenite, sphene, biotite, or hornblende. This feature, apparently first recorded by Williams and Cesbron (1977) from porphyry coppers, develops best in the biotite, potassium feldspar alteration zone where rutile appears contemporaneous with the sulfides. Apparently ilmenite formation is prevented by removal of the iron as newly forming sulfides while sphene and perovskite fail to form because the aqueous fluids remove calcium to the outer, lower temperature, alteration zones.

A related process leaves relict TiO_2 in the form of rutile during regional metamorphism. Blake and Morgan (1976) report rutile in high pressure metamorphic blue-schist and ecologite facies rocks. The rutile forms when the titanium-containing mineral species of lower grade metamorphic facies are replaced by titanium-intolerant and iron- and calcium-favoring species of higher grades (Force, 1976).

Another example of a relict TiO_2 mineralizing mechanism is reported by Force for the areas of Kragero, Norway, and Beaver Creek, NY. Here titanium-poor albitite magmas intruded titanium-bearing gneisses undergoing metamorphism. Contact areas and xenoliths of the country rock became extensively resorbed into the magma. All relicts containing rutile were left when the other elements diffused into the melt leaving the TiO_2 in place.

Without the exceptional case of metamorphism of high titanium content rocks, it appears that the relict TiO_2 mechanism will only rarely yield economic deposits in metamorphic rocks. Grains of rutile and ilmenite formed by the mechanism in various rocks may, however, constitute a major source for some placer deposits.

Placer Deposits: Just as moving water can sort clastic grains according to size as in gravels, sands, or silts, and to shape as in micaceous and non-micaceous silts, so also can grains be sorted according to specific gravity to give placer deposits. The requirements for a placer are appropriate source rock, appropriate weathering, and appropriate flow regimes to sort clastic grains. Placer deposits can form in nearly all clastic sedimentary environments.

Exploration for titanium placers, also known as heavy mineral (HM) sands, has revealed that there are, in fact, two placer deposit types whose distinction it is important for the titanium prospector to realize. Stanaway (1992) characterizes the two types as: 1) trap placers rarely more than one meter thick, and usually a lot less, which develop on scour, erosion, and unconformity surfaces as a result of the reworking of sediment and the relative non-removal of the placer component, and 2) bed placers which can reach meters to tens of meters in thickness and result from higher specific gravity minerals accumulating in direct response to the fluid and grain flow regime operating at, or close to, the time of their deposition.

Trap placers are important for high value minerals such as gold, but only bed placers can supply the volumes necessary for bulk commodities such as ilmenite or zircon. The two types often occur together. Sometimes bed placers have high grade zones of reworking where trap placer features are apparent, while some economic trap placers may arise from rework of nearby non-economic low grade bed placers.

Shoreline Deposits: Force (1991), Whitworth (1956), and others have all described in detail the mechanical processes by which the titanium bearing heavy minerals such as ilmenite and rutile accumulate on beaches while Collins et al. (1986) and Roehler (1989) have discussed the structures seen, the facies revealed, and the internal and external stratigraphy of strandline heavy mineral deposits.

It appears that significant shoreline heavy mineral placer formation requires marine regression with the accumulation migrating seaward. Mineralization favors the foreshore (sub-horizonal laminae and tabular bedded units), but can also occur in the upper shoreface or surf-zone (current bedded units) where hydrodynamics and sediment supply are favorable. Care should be taken to distinguish foreshore laminae from low amplitude, hummocky, cross stratification which indicates an offshore, below wave base, environment.

Under conditions of marine transgression, resulting sediment rework favors high heavy mineral concentrations, but accumulations rarely reach economic size. Generally formation of significant bed placers during transgression requires either the removal of the upgraded products of winnowing to eolian dunes where they are stored, or multiple transgressions with each succeeding not destroying all the previous deposits. An example of the first is Trail Ridge, FL, and of the second is Eneabba, Western Australia.

At Trail Ridge favorable winds allowed a large dunal accumulation to pile up. A dune paralleling the shore formed because a landward, swampy lagoonal environment evidenced by in-site cypress stumps (Force and Garnar, 1985) prevented dune migration inland to establish dunefields paralleling the wind direction. Inland migration and grain dispersive dune field formation characterizes areas of coastal eolian sand availability lacking water or topographic sand migration barriers. The results where an area lacks such barriers can be seen today in the Wanganui-Foxton area of North Island, New Zealand. Dispersal has also occurred in the Carolinas where small dunes are scattered over a wide area, probably dissipating the potential for an economic deposit on the oldest most inland fossil Pleistocene shorelines. At Eneabba, a succession of transgressions each culminating at a lower elevation is suggested by the geological section of the deposit given by Lissiman and Oxenford (1975) and Shepherd (1990), although the first authors postulate an equally plausible still-stand during overall marine regression. An origin incorporating both ideas is likely, i.e., temporary transgressions during an overall regressive trend.

Alluvial: Despite the production prominence of shoreline beach and dune placers, placers from river deposits are also important. Production is derived from those at Gbangbama in Sierra Leone (Lang, 1970 Raufuss, 1973, Stanaway, 1992); those in the Kiev District of the Ukraine (Borisenko, 1977); an alluvial fan near Soledad Canyon adjacent to the San Gabriel Mountains in California and from Xun Jiang in Guangxi, China (Towner et al., 1988). In addition there is good evidence from drill holes for believing heavy mineral deposits in New Jersey are in fact alluvial (Stanaway 1992).

The theoretical basis for the existence of unidirectional flow heavy mineral placers comes from the work of Slingerland and Smith (1986). The ideas they quantified of grain settling, entrainment, differential transport, and shear, all contribute to the hydrodynamic equivalence that allows placers to form, if appropriate hydrodynamic and mineral supply conditions are met. In any deposit one factor can have greater effect than the others. The experimental flume work of Brady and Jobson (1973) indicates that higher velocity flows have heavy minerals concentrated in their first precipitated deposits (i.e., more heavy minerals get entrained in the faster flows, but they also fall out first). What this means is that a continuous supply (entrainment) of heavy minerals, under conditions of transition and upper flow regime, can cause a continual deposition of these same minerals in sands accumulating at the first point of energy diminution.

Offshore: Deposits found offshore can arise not only by the drowning of shoreline and alluvial deposits, but also as submarine deposits from the flow regimes outlined for alluvial deposits. The WIM 150 of the Murray Basin of Victoria, Australia, may have arisen this way since it possesses hummocky cross-stratification which is characteristic of deposits formed below fairweather wave base during storms (Williams, 1990). During storms, wave energy penetrates to greater depths than normal which can set in motion huge masses of sand by liquefaction. Such storm surges also set in motion offshore, along-shore relaxation, and geostrophic flows (Walker, 1984). Under such conditions velocities giving transition and upper flow regime conditions favoring heavy mineral concentration could occur. Barrie et al. (1988) have studied the flow conditions under which heavy mineral deposits have formed on the sea floor along the British Columbia Coast. They established the bottom shear velocities required for the removal of all grains and for the removal of quartz grains only, then compared the frequency of occurrence of these placer constructive and destructive currents during storms. Their predicted stable heavy mineral surface deposit depth at between 100 and 140 m depth was close to the actual stable accumulations depth between 80 and 120 m.

DISTRIBUTION OF NORTH AMERICAN DEPOSITS

Eastern North America Province: The annual production capacity of TiO_2 minerals in North America is 2.5 Mt. The major sources of titanium production in North America include Trail

Ridge which began production in 1949, Green Cove Springs which commenced in 1972, and Lac Tio (Allard Lake) which began in 1951 (Towner et al., 1988). The Florida deposits, which yield valuable tonnages of zircon and kyanite-sillimanite coproducts, together can produce annually about 260 kt of ilmenite, averaging around 65% TiO_2 and about 30 kt of rutile (Crowson, 1990). The titanium products, because of their high TiO_2 and high Fe_2O_3/FeO ratio, all find their outlet with chloride route pigment producers. When the new 100 ktpy plant at Maxville (an extension of Trail Ridge) comes on-stream, mine life in Florida can be expected to last until 2010. Closures of at least two of the three present Florida operations are expected by the end of this century.

The Canadian Lac Tio ilmenite concentrate averages 37% TiO_2 and is used in the production of Sorel slag with about 80% TiO_2 that is used in sulfate route TiO_2 pigment manufacture. Mine life is 50 to 100 years. Current ores contain 25 to 30% TiO_2 and consist of 78% combined ilmenite and hematite oxides (Walker, 1990).

Fig. 2 shows the distribution of deposits in North America and how the shoreline and river deposits relate to their sources. Source rocks appear to be the Grenville Event and other high grade metamorphics (Force, 1976, Puffer and Cousminer, 1982). The source potential of these rocks is well shown with the elevated titanium of stream sediment samples taken during the National Uranium Resource Evaluation (NURE) Program of the late 1970s. Grenville rocks in the United States host anorthosites in Virginia, Pennsylvania, and New York (Bartholomé et al., 1985). Between 1900 about 1949, 180 kt of ilmenite were produced from the Roseland (Piney River) District in Virginia. From 1942 until the mine shutdown in the 1970s, the anorthosites in the Adirondacks of New York (Sanford Lake, Tahawus) yielded about 10 Mt of ilmenite.

Fig. 2 shows a huge area of outcropping Grenville Event rocks in Canada with many anorthosites containing titanium mineralization (Rose, 1969). The current Lac Tio production comes from an area with exceptional grade and ease of ilmenite separation. Additionally, placer deposits have formed along the Gulf of St. Lawrence coast where sediment has accumulated as river deltas, (e.g., Natasquan).

North American placer deposits clearly lie along well established shoreline trends. Winker and Howard (1977) correlated three major shoreline sequences on the Atlantic Coastal Plain and found that the highest and innermost sequence is the point of maximum transgression in the Pliocene. This highest, oldest sequence hosts the Trail Ridge and Highland (Trail Ridge extension) deposits described by a number of authors. The newly discovered Bailey, Aurelian, Brink, Ringwood, and Old Hickory deposits of North Carolina and Virginia (Fig. 2) very likely also lie on this same shoreline. The shoreline is not everywhere at the same elevation. This is because of subsequent deformation with relative tectonic uplift in northern Florida and an even greater uplift on the Cape Fear Arch around the border area between North and South Carolina. The Folkston, Boulogne, and Green Cove Springs deposits of Georgia and Florida sit on Winker and Howard's Effingham Shoreline Sequence of Pleistocene age and the deposits have been well described by Pirkle et al. (1974). The lowest and youngest regressional beach ridge sand accumulations, the so called Chatham Sequence of Winker and Howard seem to be separable into two units in both Georgia and the Carolinas. Deposits found with the inland set of beach ridges of the Chatham Sequence in Florida include the Jacksonville and Yulee deposits; in Georgia, the Cabin Bluff and Altama; and in North Carolina the Aurora and Chowan. Yulee, Cabin Bluff, and Altama have been described again by Pirkle et al. (1984, 1989, 1991). The Jacksonville deposit was mined after World War II from 1944 to 1964 (Garnar, Jr., 1972). None of the others have been mined. The deposits on or near the present day coast of the outer edge of the Chatham Sequence include Mineral City (now Ponte Vedra) near Jacksonville which was mined during World War I.

The general absence of deposits in the central Atlantic Coastal Plain between the Carolinas could partly be due to the fact that the rivers draining there reach only to low grade metamorphics, a poor source-rock terrain known as the Carolina Slate Belt. In New Jersey heavy mineral deposits have been mined from Miocene-Pliocene boundary-aged sands of the Cohansey Formation. During their operation, the two mines at Lakehurst and Manchester yielded between 2 to 2.5 Mt of ilmenite averaging about 60% TiO_2. The Lakehurst mine closed in 1978 and Manchester in 1982. These deposits first described by Markiewicz (1969) as alluvial were later thought to be shoreline (Puffer and Cousminer, 1982). Stanaway (1992), on the basis of several stacked fining upward sequences and the erosional Kirkwood-Cohansey contact correlatable in several tens of holes, concluded the deposits within the Cohansey to be alluvial. He ascribed the lagoonal deposit of peats found at the top of some fining upward sequences to marine incursions into valleys such as can be seen today in Tidewater, Virginia, and North Carolina. Deposits in the underlying Kirkwood could be either shoreline or marine in origin.

Other alluvial deposits derived from titanium-rich source rocks of the Appalachians have been described in Virginia by Minard et al. (1976), Herz et al. (1970), and in New York by Force and Stone (1990). Fantel et al. (1986) report a large resource of fine grained ilmenite and zircon in the Cretaceous of Tennessee with about 8 Mt of TiO_2. Analysis of stratigraphy suggests a submarine origin some distance offshore. These may resemble the WIM 150 deposit of the Murray Basin in Victoria, Australia.

Following the creation of Offshore Exclusive Economic Zones there has been an upsurge of government interest in locating heavy minerals offshore. Industry follow-up has been minimal due to the poor knowledge of deposit thickness, the need to develop new technology for mineral separation on an unsteady plant, and fears of the public's perception of the ecological consequences. Berquist (1990) has reported on offshore Virginia. Grosz (1987) summarizes the available information from surficial sea floor surface sampling and his high grade zones containing over 6% total heavy mineral (THM) are shown in Fig. 2. He writes that the heavy mineral assemblages, however, contain only about 30% of the economic heavy minerals (EHM) rutile, zircon, ilmenite, kyanite, sillimanite, and monazite.

The low economic mineral assemblages just described are part of a trend. The thickest sands with the most stable mineral, high proportion EHM assemblages occur on the oldest deposits (i.e., those furthest inland called the Trail Ridge Sequence). In addition, the most highly altered ilmenites with the highest TiO_2 content occur both inland in older accumulations and toward the south where climates are warmer. Vegetation abundance may also contribute to ilmenite upgrading in weathering environments. Diagrams illustrating the variation of altered or high TiO_2 ilmenite and its decrease with depth are shown for the Trail Ridge deposit by Force and Rich (1989).

An interesting new class of deposit has been revealed offshore from Maine (Valentine and Commeau, 1989). Here fine rutile (2 to 12 μ) occurs in muds with grades averaging 0.7 to 1.0% TiO_2. These sediments, the authors believe, contain about 600 Mt of rutile, most of it derived from the breakdown by weathering of titanium minerals such as ilmenite in the source Carboniferous and Triassic aged sediments on land. Such titanium enriched muds may represent the final end result of intense weathering of ilmenite to leucoxene in sediments together with subsequent mechanical breakdown of friable leucoxene grains (Dimanche and Bartholomé, 1976).

Other North American Deposits: Magmatic segregation deposits are known to exist in the Duluth Gabbro of Minnesota, the Laramie Range of Wyoming (Pinnell and Marsh, 1954), the San

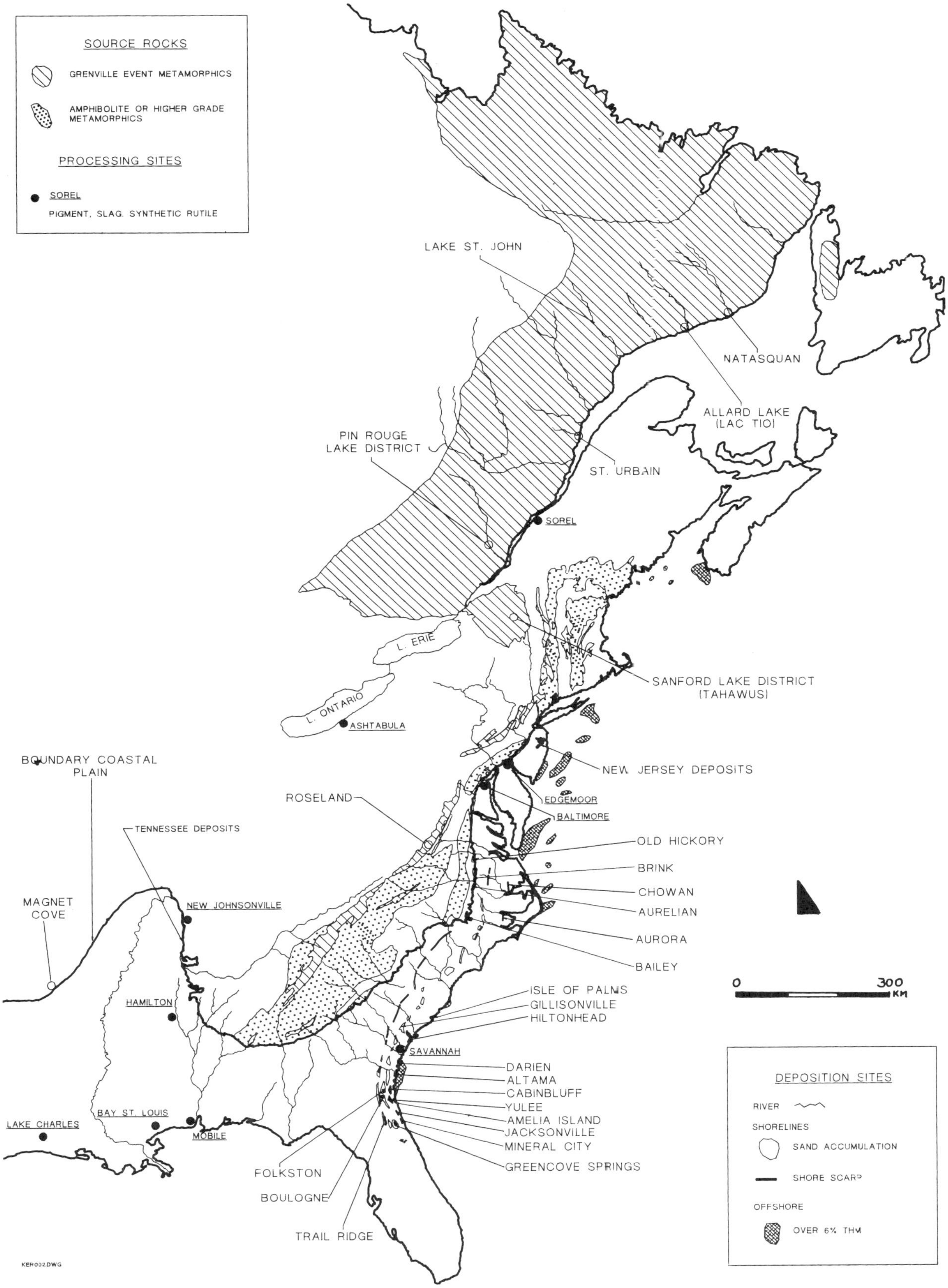

Fig. 2. Eastern North American titanium province.

Gabriels of California (Force, 1991), and, possibly, the Wichita Mountains of Oklahoma. All have various problems preventing their economic development. Good ilmenite concentrates, however, can be made from the rock deposits of California but, the commercial product derived so far, about 20 kt, comes from derived placers in Soledad Canyon. As for alkalic intrusives, it is reported that a company recently drilled the perovskite of the Powderhorn Deposit in Colorado. Reserves are huge (Fantel et al., 1986). The deposits at Magnetic Cove, however, appear moribund. They were mined from 1932 to 1944 and produced 5 kt of rutile.

DISTRIBUTION OF WORLD DEPOSITS

Australia: Beach sand mining began on the East Coast at Byron Bay in 1933 and moved to the West Coast with the mines at Capel in 1956. Australia produces about half of the world's rutile and over 40% of its synthetic rutile. The East Australian Titanium Province is shown in Fig. 3. On the East Coast large heavy mineral reserves with up to 1.3 Mt on Fraser and Moreton Islands have been closed to dredging by environmental lobbying (Wallis and Oakes, 1990).

Elsewhere on the East Coast, national parks have been created on restored mine areas and even on potential heavy mineral reserves. According to Towner et al. (1988), North Stradbroke Island, where production began in 1966, annually produces 80 kt of rutile and 300 kt of ilmenite from three dredging operations where 50 Mt of high dune sand is processed, but only about 150 kt of the ilmenite is sold. Reserves appear sufficient for another 10 to 15 years (Wallis and Oakes, 1990). Five other east coast producers together mine about 140 kt of rutile and maybe 20 kt of ilmenite. The newest of these began in 1990 at Newrybar near Ballina in New South Wales, but has recently closed.

A unique feature of the coastal East Australian Province, especially south of Stradbroke Island, is that rutile and zircon, together make up 60% of the heavy mineral assemblages (McKellar, 1975). A negative feature of all coastal deposits however, has been the high chrome content of the ilmenite. Only recently has some of this ilmenite become marketable.

Heavy mineral sands containing rutile are also known in the Pillaga area of New South Wales and on King Island, but by far the most interesting new development in the East Australian Province has been the discovery of huge reserves in the Murray Basin of Victoria. The WIM 150 deposit, one of several found recently, is a resource of 4.9 Gt of sand grading 2.8% THM (Williams, 1990). The heavy minerals contain 9% rutile, 12% leucoxene, 31% ilmenite, together with 13% zircon. Unsuitable chemical composition of the ilmenite, together with very fine mineral grain sizes, which average 45 μ, and high radioactivity zircon mean no mine developments are likely in the foreseeable future. Ilmenite cannot be used in the chloride pigment process without expensive leaching.

The Western Australian Titanium Province is shown in Fig. 4. The largest producer of the Western Australian Province, AMC had two mines at Eneabba. This area went into production in 1974 and current yields vary up to 85 kt of rutile and 450 kt of ilmenite annually, together with 250 kt of zircon. AMC added a third production unit at Eneabba West in 1990 (since, temporarily closed down) with capacity scheduled at 40 kt of rutile, 180 kt of ilmenite and 70 kt of zircon. About 230 kt of the ilmenite is to be converted to about 180 kt of synthetic rutile annually and synthetic rutile capacity is scheduled to increase to 260 kt. The new Eneabba West dredge mine has reserves for 13 years. Two other Eneabba mines will shutdown in the next ten years after the ore body is exhausted. Another major project, the Cooljarloo Joint Venture of TiWest has probable reserves for a 20 year operation which, with other deposits around Jurien Bay, could extend the mine life to 30 years. Annual production is slated at around 20 kt rutile and 350 kt of ilmenite. At Muchea, about 220 kt of the ilmenite will be upgraded to 130 kt of synthetic rutile (i.e., from 62 to 92% TiO_2).

The Eneabba, Jurien, and Cooljarloo deposits all appear to be part of the Yoganup Formation, a complex of dunes and beach ridges relating to a prograding shoreline now buried under other younger, mostly dune sands that lie against the Gin-Gin Scarp (Baxter, 1988). To the south the Wichener Scarp defines a similar sequence with deposits mined at Yoganup. At Eneabba deposits occur on eroded benches at 75, 82, 85, 91, 97, 103, 109, 115, and 128 m above sea level. At Yoganup benches occur at 26, 28, 36, 43, 46 and 66 m above sea level. The elevation differences could derive from different shorelines or result from subsequent deformation of the same shoreline since the areas are 400 km apart.

The Capel deposits occur on the Munbinea-Capel shoreline and Bassendean Dune system. These have developed at 5 m above sea level in the southern part of the Coastal Plain, but reach 36 m above sea level 50 km to the north. The youngest and outermost coastal belt of deposits consists of parallel and transverse, mobile and fixed dunes. The base of the deposits varies from sea level to plus 25 m (Baxter, 1988). The deposits here include Wonnerup, Minninup, and Koombana.

The Capel deposits are mined by AMC, Ravensthorpe, and Westralian Sands. Westralian Sands' production from the deposits at Capel and in the Yoganup area is around 370 kt of ilmenite, with another 100 kt of synthetic rutile produced. The latter is scheduled to double. The Ravensthorpe capacity is 90 kt of ilmenite per annum. The coastal belt of deposits are mined by Cable Sands at Minninup and Wonnerup with about 250 kt of ilmenite annual capacity.

Like the Eastern United States deposits, Australia's older more inland deposits seem to have the thickest sand volumes together with better heavy mineral assemblages and higher TiO_2 ilmenites. Lissiman and Oxenford (1975) give heavy mineral assemblages for Eneabba as variable, particularly in zircon content, with rutile generally around 8%, ilmenite from 30 to 60%, and zircon from 7 to 60% on the various shorelines. Variations in TiO_2 content of the ilmenite parallel those in the southeastern USA. At Yoganup, TiO_2 content of the ilmenite decreases with depth from 61% at the surface to about 54% at depths of 5 to 6 m where sands are also more clayey and thus possibly also less permeable (Welch et al., 1975).

At the younger Capel deposits, TiO_2 in the ilmenite is only 54%, while for the youngest deposits near the present coast Baxter (1977) gives chemical analyses for deposits along the Minninup line (Koombana, Wonnerup, and Minninup) ranging from 52 to 54% TiO_2. Another feature of these deposits is that ilmenite plus leucoxene contents of the heavy mineral assemblage increase to the south with values of 30 to 60% around Eneabba rising to between 70 and 90% at Minninup, Yoganup, and Capel (Baxter, 1977).

A feature these West Australian Province shoreline placers share with the new discoveries in North Carolina and Virginia is their high slimes content. For Western Australia, slimes are defined as material finer than 10 μ, essentially clay and very fine silt. In other world mining areas slimes can include coarser silt as well as all material below 40 μ in size. At Eneabba slimes average 12 to 25% and are thought to be derived from the breakdown of feldspar. Another source for such material in these deposits could be eolian dust blown from deserts or semi-arid regions.

A new discovery in the very south of the West Australian Province is the huge Beenup Deposit. Of presumed Cretaceous age, and possible alluvial origin, this deposit averages 4% THM with 60% of it ilmenite in a deposit of over 1 Gt of sandy sediments up to 60 m deep. A 500 ktpy ilmenite production is under study. There is very little rutile or zircon in the heavy mineral. Another new project in this area is the Jangardup deposit of Cable Sands which is expected to produce 200 ktpy of ilmenite with about 58% TiO_2.

New Zealand: On the west coast of the South Island of New

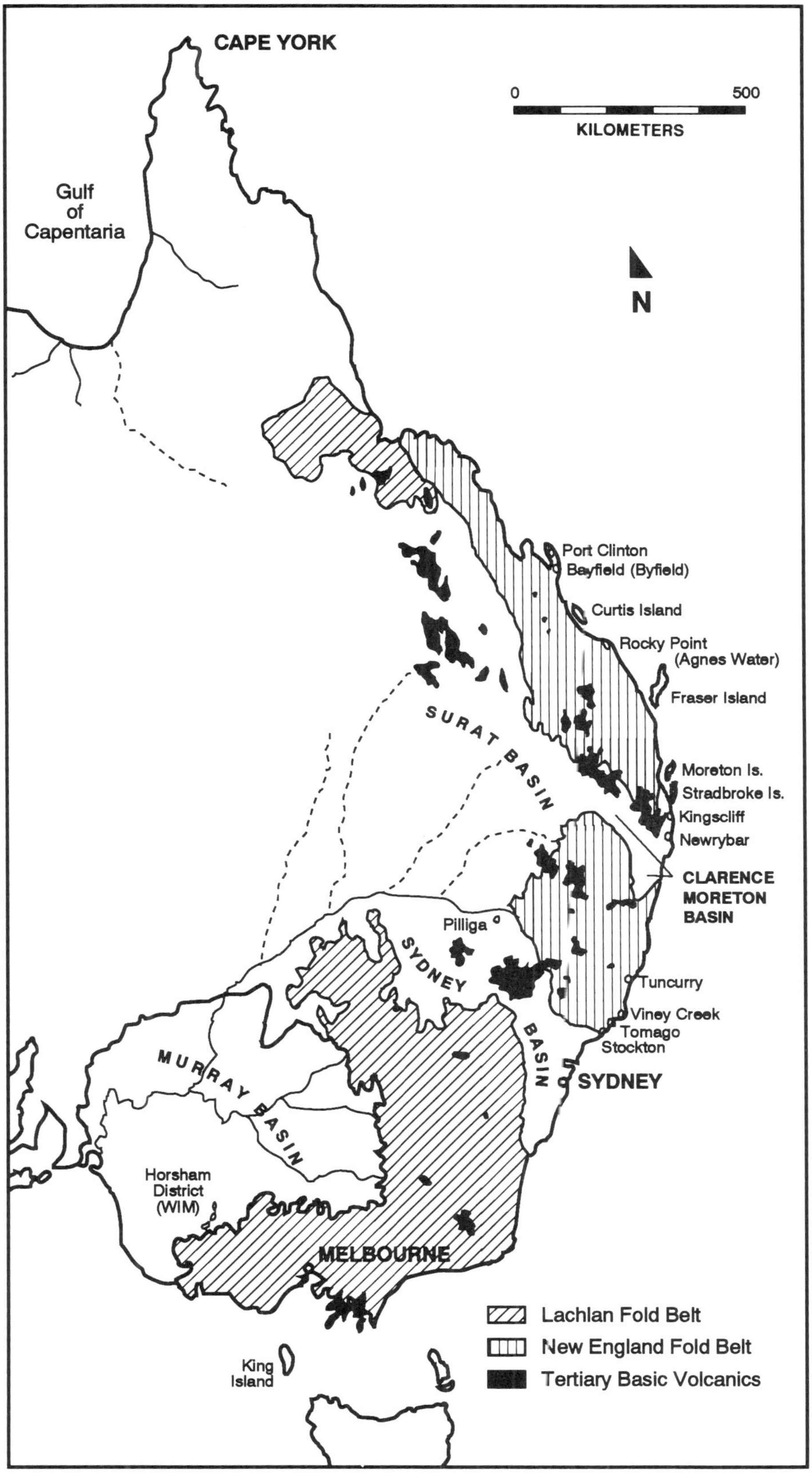

Fig. 3. Eastern Australian titanium province.

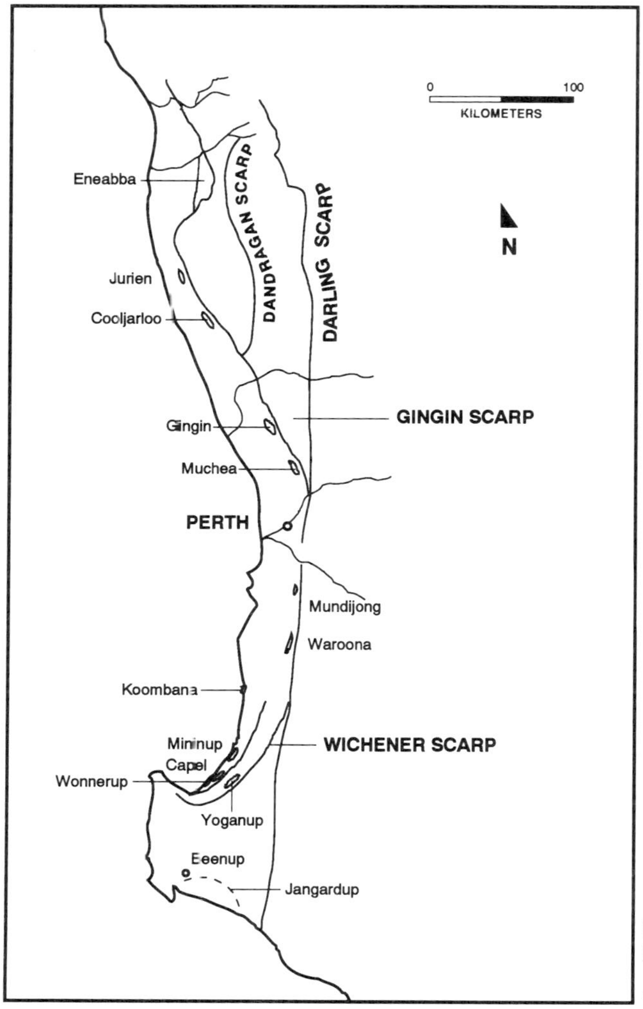

Fig. 4. Western Australian titanium province.

Zealand, ilmenite mineralization is well known at Barrytown and in the Westport area. From Barrytown 10 Mt of ilmenite grading 13% of the shoreline sands have been reported. Between 17 and 31 Mt of ilmenite have been drilled near Westport, with grades well over 5% in-ground. The principal problem with these resources is the presence of silica inclusions in the ilmenite grains. This results in the ilmenite analyzing a low 47% TiO_2 with high silica (6%), alumina (2%), and CaO (1%). Other problems are the high garnet associated with the ilmenite which increases the cost of mineral separation. Small, almost economic grades of zircon, rutile and gold are also present in the sands.

Africa: The African continent is well known for its abundant natural resources so it is not surprising that it is a major producer of titanium raw materials. Current production comes from two areas: Richards Bay on the east coast of the Republic of South Africa and Sierra Leone In addition, there are numerous potential titanium resources in Madagascar, Mozambique, Tanzania, Senegal, and Egypt.

At Richards Bay, mining occurs on present day coastal dunes at three sites. The dunefield sits on Cretaceous and Pliocene aged sediments and reaches up to 1.5 km wide and generally 60 m high with dunes subparallel to the coast in sympathy with prevailing wind directions. The inland boundary is often marked by streams, lakes, and lagoons (e.g., St. Lucia). Initial reserves were reported by Fockema (1986) to be 770 Mt sand grading 13.8% THM or 5.9% economic heavy minerals (EHM): ilmenite 35%, zircon 4%, rutile 2%. Other minerals present in the assemblage are hornblende, garnet, pyroxene, magnetite, epidote, staurolite, kyanite, sphene, and pyrite. An expansion has been completed so that it now has the annual capacity to produce 110 kt of rutile, 300 kt of zircon and around 2 Mt of ilmenite. Production began in 1977 and reserves are believed sufficient for mining until 2007 to 2012. The ilmenite, averaging 46% TiO_2, is slagged on-site to produce an 85% TiO_2 slag.

Further north along the east coast of Africa prospects are being actively promoted from the Congolone area of Mozambique. Source rock is again Proterozoic metamorphics and intrusives with schists, gneisses, migmatites, granites, and charnokites. Reserves are reported as 167 Mt at 3.25% THM (ilmenite 77%, zircon 7%, rutile 2%). More deposits have been prospected in the Pebane area, some 150 km to the southwest along the same coast. All of these Mozambique deposits occur in the coastal dunes similar to those in South Africa. Ilmenite with 54 to 57% TiO_2 can be recovered, but high chrome from 0.3 to 1.0% has been shown to concentrate with the more enriched ilmenites.

The second major producer on the African continent is Sierra Leone where 140 kt of rutile and 55 kt of ilmenite were produced in 1990. The ore contains around 1.5% rutile in-ground from a heavy mineral assemblage running 60 to 65% rutile and about 30% ilmenite. Deposits have been remobilized short distances into drainage depressions from weathered charnokites and gneiss of the Kasila Gneiss (Stanaway, 1992). In addition to these reserves in the Gbangbama area, similar mineralization can be found to the north around Sembehun and even further north at Rotifunk where low grade (0.6% rutile) resources have been found.

A significant prospect for rutile has been known since before World War II in the Cameroon around Akonolinga and to the north. Stream valley sediments over a very large area have rutile (some quite coarse) derived from adjacent Precambrian metamorphics.

A large deposit has been delineated by QIT near Fort Dauphin in SE Madagascar. High grade metamorphics in the mountains to the west of the deposit supplied the minerals to the shoreline dunes. The deposit grades are not known, but the heavy mineral tonnages are reportedly quite large of which 80% is ilmenite and 5% is rutile/anatase. The ilmenite has been leached and now contains 60% TiO_2 with low impurities. Monazite is also present.

Other coastal African prospects occur near Olifants on the west coast of South Africa, north of Dakar in Senegal, and on the Nile delta.

In Senegal, long known coastal sand/dune deposits were prospected by Du Pont. Ilmenite reportedly ranges between 53 to 56% TiO_2 with low chrome.

A persistent player on the prospecting scene has been the deposits on the Nile delta, particularly those near Rosetta. Assemblages run 55% ilmenite, 8% zircon, 1.5% rutile, 1% monazite, with the rest basically of no value. The ilmenite grades about 46% TiO_2 which is too low for use as anything but a titanium slag candidate. Concentrations are very high at the surface, but decline with depth on the present day shoreline with drilling showing average grades of around 5% EHM (ilmenite, rutile, zircon, monazite) to 15 m depths. A number of major mining companies have evaluated the deposit but none have shown active interest in developing it.

Hard rock deposits, probably of the magmatic segregation type, in layered basic intrusives, also occur in Africa. One at Abu Ghalaga near the Red Sea Coast of Egypt is estimated to have over 41 Mt of ilmenite. This ilmenite is readily separable from the associated gangue but with concentrate grades of only 34 to 41% TiO_2. Another deposit of interest is near Linganga in Tanzania and appears to contain about 6 Mt of TiO_2. Whether the TiO_2 can be separated remains unknown. Biersdorf, Kudrass, and Stackelberg (1980) report the existence of large concentrations of heavy minerals off the Zambezi delta of Mozambique. More than 270 Mt of heavy minerals at an average grade of 4.8% has been reported with much of it in less than 50 m of water. The assemblage averages 45% ilmenite, 4% zircon and 1% rutile with the rest is of little value. Ilmenite TiO_2 grade is reported to be 49%.

Europe: The principal producers in Europe are Norway, Russia, and the Ukraine. Production in Norway continued from 1900 to 1980 from the Rodsand deposit which produced 40% TiO_2 ilmenite, probably from magmatic segregations in stratiform intrusives. Current production comes from the Tellnes Deposit opened after the nearby similar Storgangen Deposit was depleted in 1960. The present ilmenite capacity is 1 Mt with 45% TiO_2 in the finished product.

The Norwegian titanium deposits are discussed by Force (1991) in detail. Norway appears to have magmatic segregation deposits (Tellnes) and alkalic intrusive (Kodal), as well as the relict TiO_2 type (Kragero). The Tellnes ore averages about 40% ilmenite in about 350 Mt of rock which gives 140 Mt ilmenite.

In the former USSR, production is reported from Ukraine placer deposits in the Kiev and Dnepropetrovsk districts and from the huge Kopansk stratiform intrusive. In the second case there are several iron mining ventures with ilmenite available as a minor slag byproduct. Placers in the Ukraine appear to be both marine, at Dnepropetrovsk, and alluvial/residual around Kiev. They have been derived from weathering of the elevated titanium mineral contents near the Volyn anorthosite labradorite massif. The massif itself, however, is not reported to have grades sufficient to mine from the unweatherd rock. Much of the Ukraine ilmenite appears to have high (60%+) grades of TiO_2. Total USSR production of ilmenite is believed to have been about 300 kt and of rutile about 20 kt.

Undeveloped prospects of interest in Europe are the Tertiary and recent placer sands of Romania and the rutile-bearing ecologite rock in Italy.

South America: Brazil is the only titanium mineral producer in South America. Large 500 million ton reserves containing 15 and 20% TiO_2 are slated for production in the mid-1990s from the Tapira and Salitre carbonatite deposits of Brazil but high calcium, phosphorus and radioactive element concentrations in mineral concentrates have prevented development. The geology of these deposits has already been discussed.

Current production in Brazil comes from mines along the east coast around Victoria (60 ktpy of ilmenite) and around Natal (80 ktpy of ilmenite). Only minor rutile is produced. The ilmenite ranges from 48 to 55% TiO_2. Deposits along the Brazil coast appear limited in size because of Pleistocene drowning of the coastline with sediment only in estuaries and in narrow strips along the coast (Gillson, 1950). Inland prospects occur in alluvials at Sapucai and Maicuru. Hard rock mineralization of apparent magmatic segregation origin occurs at Campo Alegre de Lourdes with 60 Mt of ore containing possibly 40 Mt of ilmenite (20 Mt TiO_2). Whether the ilmenite can be separated, and what its grade is, remains unknown.

Asia: Malaysia, India, Sri Lanka, and China are the Asian producers of titanium minerals. Little is known from China other than that reported by Towner et al. (1988) and Fantel et al. (1986). Total production may be around 70 kt. Placer reserves and resources around Hainan Island and the nearby Guangdong province appear to be most important. Marubeni of Japan and the Malaysian Mining Corporation have set up a joint venture to produce 50 ktpy beginning in late 1992. Of this, 35 ktpy would be exported to Japan and the remainder used to produce 10 ktpy of TiO_2 pigment on Hainan Island. Another south China province, Guangxi, is the location for an alluvial placer operation at Xun Jiang and shoreline placers around Behai. These are shown in Fig. 1 as the Hainan District. Magmatic segregation deposits of iron with titanium occur at Heishan in Hebei Province and Panzihua in Szechuan Province.

Coastal deposits in the Coxes Bazaar area of Bangladesh are very small. Sixteen known coastal dune deposits contain 18 Mt of sand grading 20% THM. Heavy mineral assemblages average 27% ilmenite, 2% rutile, 2% leucoxene and 4% zircon.

In Sri Lanka deposits occur on the northeastern coast where production is around 50 ktpy of ilmenite. The best of these is at Pulmoddai where an assemblage of 70% ilmenite, 8% rutile and 9% zircon occurs. Resources have been assessed at 4 Mt at nearly 80% EHM. The ilmenite contains 54 to 59% TiO_2.

In Malaysia, ilmenite is recovered as a byproduct of the alluvial tin mining operations. Presumably it came from the same granites that provided the tin. Production on the order of 200 kt is current but is likely to be reduced in future because of contamination from monazite making it radioactive. Most of this product was sold to Japan. Ilmenite from here assays at 54% TiO_2.

India is the major and oldest producer of titanium minerals in Asia with shoreline deposits deriving their mineral from the abundant high grade Precambrian metamorphics (as in Sri Lanka) together with the Deccan Trap plateau basalts. About 280 kt of ilmenite are mined with 90 kt used to make over 40 kt of synthetic rutile. Less than 10 kt of natural rutile is produced. The Chavara mining location is also the famous Quilon of Kerala State where ilmenite has been harvested from present beaches since early this century. Ilmenite grades at Chavara range from 55 to 60% TiO_2. The largest production comes from the Orissa Mineral Complex just south of Calcutta. Production from the Orissa beach sands began in 1983. The deposit (Chatrapur) contains 20 Mt of ilmenite (51% TiO_2).

ORE RESERVES

Total US ore reserves in 1990 were 8.1 Mt of TiO_2 with an additional 34.7 Mt as a reserve base. The total world reserves, shown in Table 3, are 289 Mt of TiO_2 with 594 million additional tons as a reserve base. As newly discovered deposits in the US and other countries are developed, it appears that there will be sufficient TiO_2 reserves to last 20 to 30 years at current world production levels.

TECHNOLOGY

Exploration Techniques: Exploration methods used for titanium minerals exploration vary with the type of mineral being sought. Hard rock ilmenite exploration uses aero and ground magnetic geophysical surveys. In almost every case, hard rock ilmenite responds with positive or negative magnetic intensities because of inherent magnetic, especially remnant magnetic properties. After a magnetic anomaly is found, verification of a deposit is determined by field examination or by using diamond drilling. If there is sufficient ilmenite, the rock can be metallurgically tested on a laboratory or pilot plant scale to determine its amenability to various ilmenite separation schemes. The products are analyzed chemically and submitted to potential consumers for their evaluation.

In general, beach sand ilmenite deposits do not respond well to magnetic surveys or other geophysical methods. A variety of other exploration methods can be used after a target area is selected. In the US, targets are usually where large volumes of sand are present and where beach ridges and other features associated with shorelines are known to occur. Early American explorationists used US Geological Survey topographic maps, physiographic references, aerial photographs, and often word of mouth to find target areas. Modern methods include Earth Resource Technology Satellite

Table 3. World Titanium Reserves and Reserve Base in 1990 (thousand tons of TiO_2 content)

	Reserves			Reserve base		
	Ilmenite	Rutile and anatase[1]	Total	Ilmenite and perovskite[2]	Rutile and anatase[1]	Total
North America:						
Canada	27 000	—	27 000	73 000	—	73 000
United States	7 800	300	8 100	33 300	1 400	34 700
Total	35 000	300	35 000	106 000	1 400	108 000
South America:						
Brazil	1 600	66 000	68 000	1 600	86 000	88 000
Europe:						
Finland	1 400	1 400	1 400	1 400		
Italy	—	—	—	9 000	21 000	30 000
Norway	32 000	32 000	90 000	90 000		
USSR	5 900	2 500	8 4400	13 000	2 500	15 500
Total	39 000	2 500	42 000	113 000	24 000	137 000
Africa:						
Egypt	—	—	—	1 400	—	1 400
Madagascar	—	—	—	15 000	—	15 000
Mozambique	2 300	100	2 400	2 300	100	2 400
Sierra Leone	1 000	2 000	3 000	1 000	2 000	3 000
South Africa, Republic of	36 000	3 600	39 600	45 000	4 500	49 500
Total	39 000	5 700	45 000	65 000	6 600	71 000
Asia:						
China	30 000	—	30 000	41 000	—	41 000
India	31 000	4 400	35 400	31 000	4 400	35 400
Malaysia	—	—	—	1 000	—	1 000
Sri Lanka	3 600	800	4 400	3 800	800	4 600
Total	65 000	5 200	70 000	77 000	5 200	82 000
Oceania:						
Australia	24 000	5 300	29 000	66 000	42 000	108 000
World total	204 000	85 000	289 000	429 000	165 000	594 000

[1]Mainly anatase in Brazil; rutile elsewhere, [2]Ilmenite except for 9.8 Mt in Colorado of perovskite. *Source:* US Bureau of Mines, 1990.

(ERTS) imagery, manned space flight photography, and new developments in computer science. A valuable tool in American titanium exploration has been the data collected by the National Uranium Resource Evaluation (NURE). Fig. 5 (Koch, 1986) shows how effective this work has been. The recently discovered deposits near Wilson and Kenly, NC, were found using NURE data published in map form by the US Department of Energy (Carpenter and Carpenter, 1991). The NURE program sampled stream sediments in the southeastern US and analyzed the fine fraction for several elements including Ti, Fe, and Zr. These elements and others typical for placer deposits, were found to be lined up in a north-south trend in the Wilson-Kenly area suggesting the presence of a feature containing concentrations of heavy minerals. The heavy-mineral deposits to the north in southeastern Virginia were found by a Virginia Geological Survey geologist who was looking for black sands to test a spiral concentrator that was to be used in evaluating offshore occurrences of heavy minerals. In some cases, such as Eneabba, Western Australia, deposits have been discovered by farmers who owned the land.

If there are sufficient quantities of radioactive minerals (i.e., monazite and zircon) present in surficial deposits, they may be detected using air radiometrics. However, rainfall or other water covering such a deposit will mask the radiation. Unless the instruments used discrimate uranium and thorium radiation, anomalous results will be shown by the presence of potassium in fertilized fields or potassium-rich clays. The use of radiometrics is discussed in detail by Force et al. (1982) and Grosz et al., (1989). Other airborne techniques are Side Aperature Radar (SAR) and Ground Penetrating Radar (GPR).

A Geiger counter or scintillometer can also be useful on the ground to detect heavy mineral deposits. Again, care must be taken not to confuse radiation from other elements such as potassium and also to be aware that some black-top roadways are sufficiently radioactive to interfere with a ground survey.

Geophysical methods (magnetometer survey, induced polarization, etc.) have been tested, but are not widely used in heavy-mineral exploration.

A variety of drilling systems are available for beach sand exploration. These include hand auger, vehicle mounted power augers, vibracore, and reverse-circulation drills. The reverse circulation drills use either water or air to force the minerals to the surface. Relatively inexpensive hand drilling of surface or near-surface deposits is generally the most cost effective method. Thick deposits extending below the water table usually require machine drilling. There are numerous papers in the literature covering the various drilling methods. One of the more recent is by MacDonald (1983).

Hole spacing depends on the stage of development of a deposit. Holes may be drilled along roadways or tracks across a target area during early stages of exploration. Grid drilling is required for development to better estimate tonnage and grade. After an ore body has been defined and production forecast, drill holes may be as close as 30 m centers. A geologist should log exploration drill holes with comments on lithology particle size, textures, color, silt or clay and other materials finer than 50 µm (slimes). If time allows, an estimate of the valuable minerals will be helpful in directing future drilling. Samples should be taken on meter or two meter intervals, being careful to preserve their integrity, and sent to a laboratory for more accurate analysis.

Field and Laboratory Separation: Gold pans of various sizes

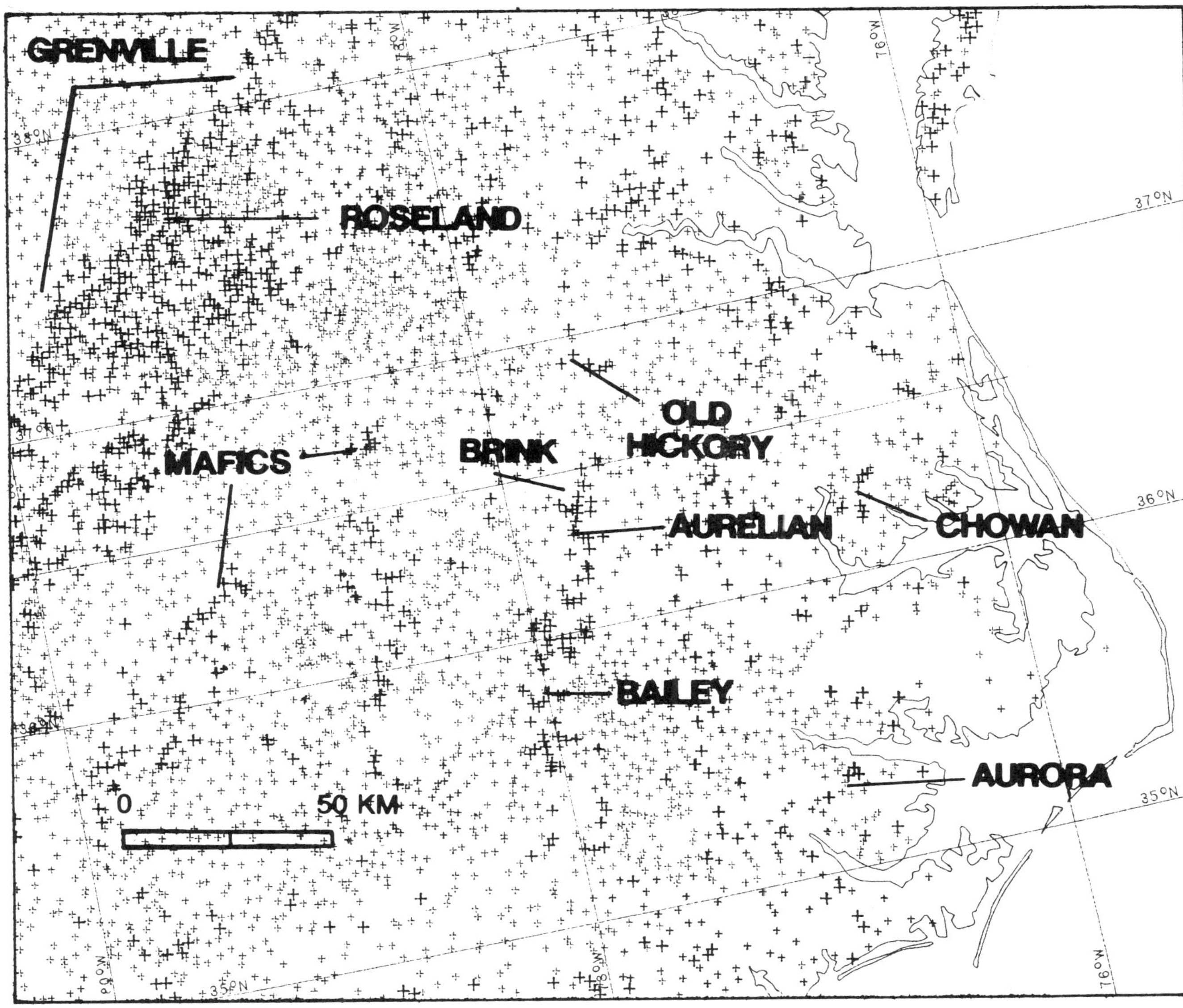

+ ⩾2% Ti in Stream Sediment (Minus 180 Micron Fraction)
+ <2% Ti in Stream Sediment (Minus 180 Micron Fraction)

Fig. 5. Map showing National Uranium Resource Evaluation (NURE) data for the Bailey-Old Rickory area, North Carolina.

and materials of construction, as well as vanning plaques, can be used in the field to recognize minerals such as ilmenite, leucoxene, rutile, zircon, pyroxene, amphiboles, and garnet with the help of a hand-lens. Reportedly, the rutile placers of Sierra Leone were discovered in the 1920s by British geologists panning stream sediments. The traditional methods to determine THM use heavy liquids (tetrabromoethane, bromoform, methylene iodide, or Clerichi solutions). These heavy liquids are considered toxic, so there is risk involved in their use. As a result, some of them have been banned in some countries (e.g., Canada, England, etc.). Newly invented non-toxic tungstate solutions that use water instead of organic solvents are considered much safer than the traditional heavy liquids. Another new development in heavy mineral separation equipment is the Magstream separator, which separates heavy-minerals by specific gravity (up to 21) using a non-toxic ferro-solution rotating in a magnetic field. The specific gravity is set by adjusting the rotational speed of separation tube. In addition to gravity separations, the Magstream can also make magnetic separations. It is a very useful tool in heavy-mineral evaluations because it can be used to concentrate the titanium minerals, zircon and the other heavy minerals for chemical analysis as well as physical testing. If there are sufficient valuable minerals present in a beach sand prospect, the ore is metallurgically tested on a pilot plant scale to determine its amenability to various mineral separation schemes. The products are analyzed chemically and submitted to potential consumers for their evaluation.

Analytical Methods

For many years, wet analytical chemistry methods were used in the analysis of titanium minerals. This type of analysis is very difficult because some of the titanium minerals (leucoxene and rutile) are insoluble in acids and can be taken into solution only by fusion methods. In the last 20 years, X-ray fluorescence (XRF) has replaced chemical methods of analysis almost entirely. Standards for use with XRF are available from the US Bureau of Standards and commercial supply houses.

In the past, grain counting (point counting) was a popular and useful means of identifying minerals present in heavy mineral concentrates and determining the amounts of each. The method is fairly accurate for the major minerals present, but accuracy is much less for the minor minerals (Garnar, Jr., 1984). Development and refinements of the X-ray microprobe, scanning electron microscope (SEM), associated XRF techniques, and modern computer science have been combined to produce the most accurate and sophisticated means for analyzing heavy-mineral concentrates. It is now possible to accurately identify each mineral present in a sample, measure each particle's size, analyze it chemically, and then give an accurate percentage for each individual mineral present in the sample, its average chemical composition, and its particle size distribution. This powerful tool has replaced grain counting in some laboratories but microscopy is still widely practiced in other labs and in the field where it can be done with a small microscope and does not require electricity.

Mining: Hard rock ilmenites are mined by open pit methods in Quebec where the ilmenite is removed selectively by power shovels and front end loaders. Since the ilmenite occurs in a relatively pure state, there is no backfill and the pit is left open. The ilmenite is shipped by ore carriers up the St. Lawrence River to Sorel for smelting. The Norwegian hard rock deposits are also mined by open pit methods. The ore is blasted with slurry explosives that produce about 150 kt of ore per blast. The ore is hauled out of the pit by means of trucks to a primary crusher where it is crushed to 15 to 20 cm and then transported by an underground conveyor belt (through a mountain) to a crushing and milling plant.

Beach sand minerals are, for the most part, mined using high capacity dredges. The dredges are cutter-head suction types and can mine up to 2 500 tph. A few bucket wheel dredges are also used. The Serria rutile dredge is a bucket ladder type designed after the Malaysian tin dredges. Each ore is different and the dredge is chosen that best fits the individual ore. Beach sand deposits contain between 90 to 98% gangue that is separated and used to backfill the dredge pond or mined area. Practically all beach sand tailings are reclaimed by returning the land to about the same contour and elevation as it was prior to mining. Topsoil, removed before mining, is then placed over the tailings, fertilized, and trees or grass planted to complete the reclamation process.

Processing: The Canadian hard rock ilmenite is further upgraded for smelting using screens, spirals, and heavy media to remove lighter gangue minerals. The Norwegian ore is crushed and milled to particles smaller than 0.2 mm. The milled ore is fed to a processing plant and concentrated by means of magnetic separation. The ilmenite is further concentrated by means of gravity separation and flotation. Standard practice in beach sand mining is to remove oversize materials on a conditioner unit ahead of a multi-stage gravity separation plant. The undersize fraction is fed to one of a variety of spiral separators that have been introduced since the Humphrey's spiral patents expired twenty years ago. The original Humphrey's spirals were available in only two types: sand spirals and coal spirals. Spirals are now available that either use wash water or do not, collect concentrate at each turn or at the discharge of the spiral, and also have a number of different diameters and pitches. The Reichert cone is sometimes used as the first stage of gravity

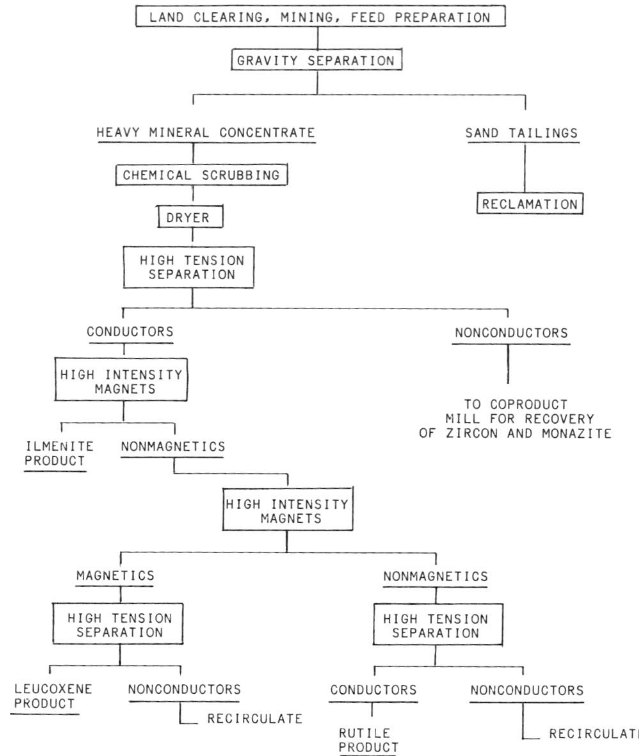

Fig. 6. Simplified flow diagram for the mining and processing of beach sand titanium mineral ores.

unit. Where the water table is near the surface, the feed conditioner unit and gravity separation plant float on the dredge pond. The alternative to floating gravity plants is to pump the ore from the mine to a land-based wet mill.

The gravity concentrates are further treated by high tension separators to concentrate the titanium minerals. In many cases, grain coatings make it necessary to wash and chemically treat the concentrates to clean the grain surfaces. The US producers use NaOH scrubbers to remove organic (humate) grain coatings ahead of their dry mills. The wet concentrates are dried and the titanium minerals separated from the silicate minerals by multiple stage high tension treatment. Rutile is separated from the ilmenite and leucoxene by high intensity magnetic separation. A typical flow diagram is shown in Fig. 6. If other valuable minerals (i.e., monazite, zircon, etc.) are present, they are further treated to recover them in byproduct or coproduct mills (see Zirconium Mineral chapter).

USES

Most of the titanium minerals produced today are used to make TiO_2 white pigments. Annual US TiO_2 pigment production capacity is shown in Table 4. A small portion go to produce titanium metal or for use as a welding rod flux.

Sulfate Pigment Manufacturing: For many years TiO_2 pigments were manufactured by reacting ilmenite with H_2SO_4. Leucoxene and rutile cannot be used in this process because they are either insoluble, or only partially soluble in H_2SO_4. In this exothermic process, ilmenite and H_2SO_4 are reacted to form titanyl sulfate and iron sulfate. The titanyl sulfate is chemically clarified and then calcined to form TiO_2 (anatase), ground, and then further

Table 4. Capacities of US Titanium Dioxide Pigment Producers (Dec. 31, 1990) (pigment capacity ktpy)

Company and plant location	Sulfate Proc.	Chloride Proc.
E. I. Du Pont de Nemours & Co.		
Antioch, CA	—	36
DeLise, MS	—	160
Edge Moor, DE	—	129
New Johnsonville TN	—	297
Kemira Inc., Savannah, GA	54	91
Kerr-McGee Chemical Corp., Hamilton, MS	—	96
SCM Chemicals Inc., Hanson Indust. USA		
Ashtabula, OH	—	111
Baltimore, MD	63	49
TOTALS	117	969

Source: US Bureau of Mines, 1990.

treated by precipitating coatings of various compounds on the particle surfaces to improve the product. The finished pigments are then used in paints, paper, plastic, synthetic fibers, etc. Lynd and Lefond (1983) state that the sulfate process produces large quantities of spent acid and iron sulfate (copperas)—up to 2 kg of waste per 0.5 kg of TiO_2 pigment. The spent acid and copperas are very noxious materials which make their disposal difficult in these environmentally regulated times.

Specifications: Specifications for titanium minerals vary with the buyer and the application. Sulfate pigments producers look for ilmenite products that are high in ferrous iron with less than 2% insoluble titanium minerals. Ores with low chromium are also required in order to eliminate yellowing caused by this element. Titanium slags in Canada and Norway contain 70 to 75% TiO_2 with 4 to 8% MgO and 4 to 5% SiO_2. Their high MgO and SiO_2 make them unacceptable to chloride pigments producers.

Chloride TiO_2 Pigment Manufacture: Du Pont pioneered chloride pigment manufacture at their New Johnsonville, TN, plant in 1959. This process revolutionized the pigments industry with superior pigments at a lower cost. Pigment is manufactured by reacting TiO_2 raw materials with hot chlorine gas producing $TiCl_4$ and $FeCl_4$. The $TiCl_4$ is oxidized forming TiO_2 (rutile). The chlorine from this step is recycled back to the reactor and is used over and over again. The TiO_2 is ground and the particles coated with compounds designed to improve its pigment properties. The waste streams from the chloride process vary with the feedstock used but are significantly lower in volume than the waste materials from the sulfate process Lynd and Lefond (1983) indicate four times more waste from sulfate than chloride). Stringent environmental regulations are making high TiO_2 feedstocks necessary for compliance, thus most producers are using slag, synthetic and natural rutile instead of ilmenite feedstocks.

Chloride Specifications: Certain elements (e.g, Mg, Mn, Ca, P, etc.) form chlorides that have boiling points higher than the temperature in the titanium mineral/chlorine reactor. These cause "the bed to stick" and eventually the reactor has to be shutdown in order to remove them. The more of these impurities present in chloride feedstocks, the more frequent the shutdowns are. The presence of radioactive elements (U and Th) are also unacceptable in titanium ores. Each producer has their own specified amounts of these elements that can be tolerated. In general slags and synthetic rutiles produced from altered (beach sand) ilmenite and natural rutile are relatively free of objectionable impurities. Specifications for chloride feedstocks and chemical analyses of commercial ores are given in Table 5.

Ilmenite Beneficiation: There are large tonnages of ilmenite throughout the world that cannot be used directly in pigment manufacture because the TiO_2 is too low or the impurity levels are too high. Quebec Iron Titanium (QIT) mines low grade ilmenite at their Allard Lake deposit. At their Sorel smelter, upgraded ore, containing about 36.8% TiO_2, is mixed with 0 to 10% lime and 8 to 14% low ash coal. The mixture is then heated to 1 500 to 1 700°C in electric arc smelting furnaces. During the smelting process the lighter titanium slag (80% TiO_2) floats to the top and the iron sinks to the bottom. The iron is drawn off from the bottom as a form of pig iron and sold as Sorelmetal. The titanium slag is cooled, crushed and screened prior to shipment. Because the Sorel slag contains impurities that are detrimental to the chloride process, only sulfate process pigments producers can use it.

Table 5. Typical Titanium Ore Specifications for Chloride Pigments Feedstocks and Chemical Analyses of Commercial Ores

Compound	Typical Specification	RBM Slag RSA	Synthetic Rutile: Kerr-McGee USA	Synthetic Rutile: Ishahara Japan	Natural Rutile
Total Ti as TiO_2	85% Min.	86.3%	90.5–93	96.1	95.2
Ti3+	?	—	—	—	—
CaO	0.2% Max.	.15	<0.35	0.01	0.02
MgO(a)	1–2% Max.	1.20	0.05	0.07	0.07
MnO(a)	0.1–1% Max.	1.78	0.10	0.03	0.01
Al_2O_3	1.0% Max.	1.13	<1.0	0.46	0.20
SiO_2	Varies	1.79	1.5–2.5	0.50	1.0
Th and U	100 ppm Max.	22 ppm	[b]	[b]	[b]
Particle Size	−30+200 mesh	[b]	[b]	[b]	[b]

[a] Some companies specify that the combination of these two compounds should not exceed 1%, [b]not available.

Richards Bay Minerals operates a similar slagging operation in South Africa using QIT technology. The major difference is the smelter feedstock which runs about 48% TiO_2 and is very low in impurities. The resulting slag is 80 to 85% TiO_2 and can be used in either chloride or sulfate pigments manufacture. The Norwegians pelletize the upgraded TiO_2 ore with bentonite. The pellets are reduced with coal. The coal ash and fines from this step are removed by magnetic separation and the reduced pellets are then smelted in an electric furnace. The TiO_2 slag is crushed and loaded for shipment by means of a belt conveyor having a capacity of 800 tph. Chemical analyses for titanium slag are shown in Table 2.

Synthetic Rutile: Federal and state regulation of toxic chemical waste has made it almost mandatory that chloride pigments producers use high TiO_2 feedstocks in order to minimize the amount of waste material to be disposed. Several processes, using oxidation, reduction, leaching, and/or chlorination to remove the iron, have been developed to produce high TiO_2 materials from low grade ilmenite. At present, all of the producers use beach sand ilmenite as starting material. The end products contain 90 to 97% TiO_2 with very low impurity levels. Kerr-McGee Chemical Corporation is the only producer of synthetic rutile in the US. Since 1977, they have produced synthetic rutile at their plant near Mobile, AL, using the Benelite Process which uses HCl to leach iron from reduced ilmenite. They recycle their waste streams so there are no liquid wastes discharged from the plant. Lynd and Lefond (1983) give details of the various synthetic rutile processes.

Titanium Metal Manufacture: Only 5 to 6% of world titanium mineral production is used to make titanium metal sponge. This is produced using the Kroll process in which titanium minerals are reacted with elemental chlorine. The resulting titanium tetrachloride is then reduced using molten magnesium forming a sponge.

Table 6. US Titanium Metal Production Capacity in 1990

Company	Ownership	Plant location	Capacity	
			Sponge	Ingot
Howmet Corp. Ingot Div.	Pechiney France	Whitehall, MI	6.3	
Int. Light Metals Corp.	Martin Marietta Nippon 60% Kokan KK 40%	Torrance, CA	5.0	
A. Johnson Metals Corp.	Ax. Johnson Grp. Sweden	Lionville, PA		1.8
Lawrence Aviation Industries Inc.	Self	Port Jefferson, NY		1.4
Oregon Metallurgical Corp.	Oremet employees 67% Public, 33%	Albany, OR	6.8	10.0
RMI Co.	USX Corp. 50% Public 50%	Ashtabula OH	10.9	—
RMI	USX Corp. 50% Public 50%	Niles, OH	—	16.3
Teledyne Allvac	Teledyne Inc.	Monroe, NC	—	5.7
Teledyne Wah Chang Albany	Teledyne Inc.	Albany, OR		.9
Titanium Metals Corp. of Am.	Baroid Corp.	Henderson, NV	12.7	15.9
Viking Metallurgical Corp.	Quanex Corp.	Verdi, NV	—	3.2
Wyman-Gordon Co.	Self	Worcester, MA	—	2.3
Totals			30.4	68.8

Source: US Bureau of Mines, 1990.

One kilogram of metal is produced from 4.37 kg of tetrachloride by reducing with 1.41 kg of magnesium. Finished products are made from the sponge. At present there are only a few sponge manufacturers in the world. Sponge and ingot manufacturers in the US are shown in Table 6.

Miscellaneous Uses: Small amounts of titanium minerals go into other uses. Ilmenite has been used in filter beds, sand blast abrasives, foundry sands and as high density fillers. Small amounts of rutile are used as a welding rod coating where it aids stabilization of the arc and frees the metal of slag. It is also used as an opacifier in ceramic glazes and to produce tan-colored glass.

ECONOMIC FACTORS

Costs: Production costs for mining and processing titanium minerals vary widely with the complexity of the mineral suite and ore body, the equipment requirements, availability and cost of energy, manpower costs, taxes, and environmental costs. In general, the mining step is energy intensive, whereas the gravity and dry milling costs are both energy and labor intensive. Labor costs may be cheap in developing third world countries, but this is offset by high energy costs and lack of infrastructure. The free world has abundant energy and good infrastructure, but labor costs are often significantly higher. Titanium mineral mining is not a high profit venture unless energy and manpower costs are balanced and effectively minimized. During periods when demand for titanium minerals peak, producers may operate at very large profit margins.

MARKETS

TiO_2 Pigments: Finished pigments are available in both the rutile and anatase forms. TiO_2 pigments' most important properties are: high refractive index giving them superior hiding power (opacity); high degree of reflectivity giving them brightness; chemical inertness to solvents except HF and hot H_2SO_4; thermal stability; and non-toxicity. The rutile form tends to be more abrasive and is used in the paints, plastics, printing inks, and cosmetics industries. The anatase form is better suited for use in the manufacture of paper, textiles, latex rubber, soap and pharmaceuticals. Distribution of US domestic titanium pigments is shown in Table 7. Over 90% of US pigment production goes into paints, paper, and plastics. The remainder is used in printing inks, rubber, floor coverings, etc. The rutile form is a semiconductor. It is an insulator at lower temperatures (20°C), but becomes a conductor at temperatures over 420°C. Because of this unique property, demand for rutile pigments in the electronics and electro-ceramics industry is growing.

Titanium Metal: The metal has many good properties including exceptional strength-to-weight ratio, corrosion resistance, high melting point, low thermal conductivity, low coefficient of expansion, and high electrical resistivity. These properties make titanium valuable in casting jet engines, aerospace applications, aircraft structure, marine hardware, chemical equipment, etc. Current use of titanium in large commercial aircraft represents about 6% of the aircraft's empty weight.

Transportation: Most titanium minerals are shipped in large quantities. The type of conveyance used depends on the distance between mineral supplier and consumer. Australian, South African, Indian, and Sri Lankan minerals are shipped via ocean freight. In the US, Du Pont and AMC ship their products to domestic consumers via railroad. Pigments and titanium metal are often shipped in small quantities by truck and railroad.

Alternative Materials: There are no potential alternatives to ilmenite, altered ilmenite, or rutile for making slag, synthetic rutile, or TiO_2 pigment. As technology develops, perovskite and Brazilian anatase may be used in place of the traditional titanium ores. As for TiO_2 pigment, it is extremely doubtful that anything will ever be found that will produce a pigment with TiO_2's properties at a competitive price.

GOVERNMENTAL CONSIDERATIONS

Federal and state governments in all of the free world's mineral and pigments producing countries have stepped in to regulate these industries. Because the world mining community, as a whole, has been regarded as hazardous to workers' health and lives, environmentally destructive, polluting, and a generally undesirable neighbor, governments have imposed stringent laws and rules governing all mining. In spite of the fact that beach sand mining is safe, non-polluting, and restores mine-out lands to be equal to or better

Table 7. US Distribution of Domestic Titanium Pigment Shipments, % Titanium Dioxide Content, by Industry

Industry	(%) 1986	1987	1988	1989	1990
Ceramics	W	W	0.4	0.4	W
Coated fabrics & textiles	0.4	0.3	W	.4	0.2
Floor coverings	.9	1.2	1.2	1.1	1.0
Paint, varnish, lacquer	52.6	49.5	48.1	50.0	48.9
Paper	20.7	24.3	24.2	25.8	26.5
Plastics	15.8	17.0	17.0	16.3	16.6
Printing ink	1.4	1.2	1.7	.8	1.6
Roofing granules	.4	.4	.4	.3	.4
Rubber	2.0	1.8	1.7	1.4	1.5
Other	5.8	4.3	5.3	3.5	3.3
Total	100.0	100.0	100.0	100.0	100.0

W withheld to avoid disclosing company proprietary data; included in Other. *Source:* US Bureau of Mines, 1990.

Table 8. US Import Duties

Tariff item	HTS No.	Most Favored Nation (MFN) Jan. 1, 1990	Non-MFN Jan. 1, 1990
Ilmenite	2614.00.6020	Free	Free
Titanium slag	2620.90.500	do.	Do.
Rutile concentrate	2614.00.6040	do.	Do.
Synthetic rutile	2614.00.300	5.0% ad val	30.0% ad val
Waste and scrap metal	8108.10.1000	Free	Free
Unwrought metal, including sponge	8108.10.50	10.5% ad val	25.0% ad val
Wrought metal:			
Articles of titanium	8108.90.30	5.5% ad val	45.0% ad val
Other	8108.90.60	150% ad val	45.0% ad val
Titanium dioxide pigments	3206.10.00	6.0% ad val	30.0% ad val
Titanium oxides	2823.00.000	6.0% ad val	30.0% ad val

Source: US Bureau of Mines, 1990.

than they were in their natural state, the strict governmental regulations are also applied to it.

Taxes: In the 1970s, the State of Florida imposed a severance tax on minerals produced in that state. The purpose of the tax was to ensure that mining companies responsibly restored lands that they had mined. The giant phosphate mining industry in central Florida was the target of the tax, because some of the mining companies had been less than responsible in reclaiming mined-out land. Now, if a company does not restore mined lands to the state's expectation, money from the severance tax is used to complete restoration. If a company restores its mined lands, part of the severance tax they have paid is refunded.

Depletion Provisions: The depletion allowance for ilmenite and rutile is 22% when mined from domestic deposits and 14% from foreign deposits.

Tariffs: Import duties as of 1990 are shown in Table 8.

Environmental: Beach sand mining presents few serious environmental hazards. The major environmental problem in the titanium industry is the disposal of wastes from pigment-producing processes that use ilmenite. Solutions to this problem make it necessary for sulfate pigment producers to neutralize and control their effluents. Sulfate producers in the United States are neutralizing their effluents by reacting them with limestone and lime to produce gypsum and iron oxides as byproducts. Neutralization is reported to add $100 a ton to sulfate pigment production costs. Synthetic rutile processes regenerate the original leaching agent and recycle it. Chloride pigment wastes are also a problem. At least one manufacturer eliminates some ferric chloride waste by selling it as a water treatment chemical.

Safety: Open pit and underground mining operations are generally regarded as high risk and require careful safety monitoring by the government. Beach sand mining is a safe industry at mines where adequate safety rules are established and enforced. Du Pont's Florida mine has received the Sentinels of Safety Award for four years (three of which were in a row) for having more than 400 000 exposure hours in a year without a lost time injury. The award is co-sponsored by the US Department of Labor, the Mine Safety and Health Administration, and the American Mining Congress. The Du Pont plants also have operated for over 10 years without a major lost time injury and won numerous company safety awards.

FUTURE TRENDS

White TiO_2 pigments have established a well defined growth pattern over the years of about 2% per year (except during brief periods of US and world economic recessions). As the pigments industry grows, so must titanium mineral production. Pigment demand grew through the 1980s to the point that additional world capacity was required. Pigment producers responded by adding 541 kt of pigment capacity, bringing world capacity to 3.8 Mtpy. At the same time the world ore suppliers expanded to meet the new demand. One of the major concerns confronting the expanding pigment industry is the effect of new environmental regulations. In some cases old sulfate plants may have to be retired and replaced by new, less polluting plants, while others will have to build additional facilities to treat hazardous and toxic wastes. As the pigment industry grows, present day mines will be mined out and new ones developed to take their places, and as more environmental regulations are passed, more high-grade pigment feedstocks will be required. To achieve this, high TiO_2 slag and synthetic rutile production capacity will have to be increased. Overall, the titanium industry is economically sound and will continue to grow in the future while working to meet future challenges and needs of the changing world.

BIBLIOGRAPHY AND REFERENCES

Barrie, J.V., et al., 1988, "Origin of Modern Heavy Mineral Deposits, Northern British Columbia Continental Shelf," Vol. 84, *Marine Geology,* pp. 43–51.

Bartholomé, M.J., et al., eds., 1985, "The Grenville Event in the Appalachians and related topics," Special Paper 194, Geological Society of America, 287 pp.

Baxter, J.L., 1977, "Heavy Mineral Sand Deposits of Western Australia," Mineral Resources Bulletin 10, Geological Survey of Western Australia.

Baxter, J.L., 1988, "History of Mineral Sands Mining in Western Australia," Exploitation of Mineral Sands, course sponsored by Australian Development Bureau, organized by The Western Australian School of Mines, and WAIT-Aid Ltd., Kalgoolie, WA, pp. 61–79.

Belperio, A.P., and Bluck, R.G., 1990, "Coastal Paleogeography and Heavy Mineral Sand Exploration Targets in the Western Murray Basin, South Australia," *Proceedings*, Australasian Institute of Mining and Metallurgy, Parkville, Vic., Vol. 295, No. 1, pp. 5–10.

Berquist, C.R., Jr., ed., 1990 "Heavy Mineral Studies—Virginia Inner Continental Shelf," Publication 103, Virginia Division of Mineral Resources, 3 pp.

Berquist, C.R., Jr., 1987, "Minerals in High-Level Gravel Deposits Along the Fall Zone of Virginia," *Virginia Minerals*, Vol. 33, No. 4, Charlottesville, VA, pp. 37–40.

Biersdorf, H., et al., 1980, "Placer Deposits of Ilmenite and Zircon on the Zambezi Shelf," *Geologisches Jahrbuch*, Heft 36, Feihe D, pp. 5–85.

Blake, M.C., and Morgan, B.A., 1976, "Rutile and Sphene in Blueschist and Related High-Pressure Facies Rocks," Professional Paper 959C, US Geological Survey, 6 pp.

Borisenko, L.F., 1977, "Deposits of Titanium," *Ore Deposits of the USSR*, Vol. 1, V.I. Smirnov, ed., Pitman Publishing, London, pp. 237–255.

Brady, L.L., and Jobson, H.E., 1973, "An Experimental Study of Heavy Mineral Segregation Under Alluvial-flow Conditions," Professional Paper 562K, US Geological Survey, 13 pp.

Brimhall, G.H., et al., 1988, "Metal Enrichment in Bauxites By Deposition of Chemically Mature Aeolian Dust," *Nature,* Vol. 333, pp. 819–824.

Carpenter, R.H., and Carpenter, S.F., 1991, "Heavy Mineral Deposits in the Upper Coastal Plain of North Carolina and Virginia," *Economic Geology*, Vol. 86, pp. 1657–1671.

Clarke, G., 1988, "Titanium Dioxide Pigment, Expansions Worldwide," *Industrial Minerals*, Aug., pp. 79–91.

Collins, L.B., Hockwimmer, B., and Baxter, J.L., 1986, "Depositional Facies and Heavy Mineral Deposits of the Yoganup Shoreline, southern Perth Basin in Australia: a World Source of Ilmenite, Rutile, Monozite and Zircon," Conference, Australasian Institute of Mining and Metallurgy, Parkville, Vic.

Crowson, P., 1990, *Minerals Handbook 1990–91: Statistics and Analysis of the Worlds Mineral Industry*, Stockton Press, New York.

Dimanche, F., and Bartholomé, P., 1976, "The Alteration of Ilmenite in Sediments," *Minerals Science and Engineering*, Vol. 8, No. 3, pp. 187–201.

Elsdon, R., 1975, "Iron-titanium oxide minerals in igneous and metamorphic rocks," *Minerals Science and Engineering*, Vol. 17, pp. 48–69.

Emslie, R.F., 1978, "Anorthosite Massifs, rapakivi granites and Late Proterozoic rifting of North America," *Precambrian Research*, Vol. 7, pp. 48–70.

Fantel, R.J., Buckingham, D.A., and Sullivan, D.E., 1986, "Titanium minerals availability—Market Economy Countries," Information Circular 9061, US Bureau of Mines, 28 pp.

Fleischer, M., 1987, *Glossary of Mineral Species*, 5th ed., Mineralogical Record Inc., Tucson, AZ, 227 pp.

Fockema, 1986, "The heavy mineral deposits north of Richards Bay," C.R. Arnhaensser, ed., *Mineral Deposits of South Africa*, Johannesburg Geological Society of South Africa, pp. 2301–07.

Force, E.R., 1976, "Titanium contents and titanium partioning in rocks in Geology and Resources of Titanium," Professional Paper 959A, US Geological Survey, 10 pp.

Force, E.R., 1980, "Is the United States Geologically Dependent On Imported Rutile?" *Proceedings of the 4th Industrial Minerals International Congress*, Atlanta, GA, pp. 43–47.

Force, E.R., 1991, "Geology of Titanium—Mineral Deposits," Special Paper 259, Geological Society of America, 112 pp.

Force, E.R., et al., 1982, "Aeroradioactivity Maps in Heavy-mineral Exploration—Charleston, South Carolina Area," Professional Paper 1218, US Geological Survey, 19 pp.

Force, E.R., and Garnar, T.E., 1985, "High Angle Aeolian Cross-Bedding at Trail Ridge, Florida," *Industrial Minerals*, Aug., pp. 55–59.

Force, E.R., and Rich, F.J., 1989, "Geologic Evolution of Trail Ridge Aeolian Heavy Mineral Sand and Underlying Peat, Northern Florida," Bulletin 1874, US Geological Survey, 16 pp.

Force, E.R., and Stone, B.D., 1990, "Heavy Mineral Dispersal and Deposition in Sandy Deltas of Glacial Lake Quinebaug, Connecticut," US Geological Survey, 21 pp.

Garnar, T.E., Jr., 1972, "Economic Geology of Florida Heavy Mineral Deposits," *Proceedings from the 7th Forum on the Geology of Industrial Minerals,* H. Puri, ed., Florida Bureau of Geology, pp. 17–21.

Garnar, T.E., Jr., 1984, "Mineralogical Methods Used in the Heavy Mineral Industry," *Applied Mineralogy*, Proceedings of the 2nd International Congress on Applied Mineralogy in the Minerals Industry, Los Angeles, CA, pp. 297–312.

Gillson, J.L., 1950, "Deposits of Heavy Minerals on the Brazilian Coast," *Trans.-AIME*, Vol. 187, New York, pp. 685–93.

Grosz, A.E., 1987, "Nature and Distribution of Potential Heavy Mineral Resources Offshore of the Atlantic Coast of the United States," *Marine Mining*, Vol. 6, pp. 339–358.

Grosz, A.E., et al., 1989, "Geologic Interpretation of the Gamma Ray Aeroradiometric Map of Central and Northern Florida," Professional Paper 1416, US Geological Survey, 48 pp.

Hahn, A.D., and Fine, A.D., 1960, "Examination of Ilmenite-bearing Sands in Otter Creek Valley, Kiowa and Tillman Counties, Oklahoma," Report of Investigations 5577, US Bureau of Mines, 77 pp.

Hanks, C.L., 1978, "Petrology and Occurrence of Magnetite-Ilmenite-Olivine Rock in the Wichita Mountains," Geological Society of America Abstracts with Programs, Vol. 10, p. 7.

Harben, P., 1984, "Titanium Minerals in Brazil—Progress and Potential," *Industrial Minerals*, Jan., pp. 45–49.

Hendrick, J.F., and Waked, L., 1989, "Heavy Minerals Including Monazite in Egypt's Black Sand Deposits," *Journal of the Less Common Metals*, Vol. 184, pp. 79–84.

Herz, N., et al., 1970, "Rutile and Ilmenite Placer Deposits, Roseland District, Nelson and Amherst Counties," Bulletin 1312F, US Geological Survey, 19 pp.

Herz, N., 1976, "Titanium Deposits in Anorthosite Massifs," Geology and Resources of Titanium, Professional Paper 959D, US Geological Survey, 6 pp.

Herz, N., and Force, E.R., 1987, "Geology and Mineral Deposits of the Roseland District of Central Virginia," Professional Paper 1371, US Geological Survey, 56 pp.

Houston, R.S., and Murphy, J.F., 1962, "Titaniferous Black Sandstone Deposits of Wyoming," Wyoming Geological Survey, 120 pp.

Koch, G.S., 1986, "A Geochemical Atlas of Georgia," *Geological Atlas No. 3*, Georgia Geologic Survey, 44 pp.

Lang, H.P., 1970, "Sekundaere Rutillagerstataetten in Sierra Leone (Westafrika)," *Erzmetall*, Vol. 23, pp. 179–183.

Lissiman, J.C., and Oxenford, R.J., 1975, "Enneabba Rutile-Zircon-Ilmenite Deposit, W.A," *Economic Geology of Australia and Papua New Guinea*, Metals, Vol. 1, C.L. Knight, ed., Monograph 5, Australasian Institute of Mining and Metallurgy, Parkville, Vic., pp. 1062–1070.

Lynd, L.E., and Lefond, S.J., 1983, "Titanium Minerals," *Industrial Minerals and Rocks*, 5th ed., Vol. 2, S.J. Lefond, ed., Society of Mining Engineers, AIME, New York, pp. 1303–1362.

Mann, G., and James, D., 1989, "The Evaluation and Development of the Barrytown Ilmenite Deposit as a Source of Titanium Dioxide Pigment in Australia: a World Source of Ilmenite, Rutile, Monazite, and Zircon," Conference, Sep.-Oct., Australasian Institute of Mining and Metallurgy, pp. 139–142.

Marsh, S.P., and Sheridan, D.M., 1976, "Rutile in Precambrian Sillimanite Quartz Gneiss and Related Rocks, East Central Front Range, Colorado," Professional Paper 959G, US Geological Survey, 17 pp.

Markiewicz, F.J., 1969, "Ilmenite Deposits of the New Jersey Coastal Plain," *Geology of selected areas in New Jersey and eastern Pennsylvania*, S. Subritzky, ed., Rutgers University Press, New Brunswick, NJ, pp. 363–382.

MacDonald, E.H., 1983, "Alluvial Mining," *The Geology, Technology and Economics of Placers*, Chapman and Hall, London, New York, 508 pp.

McGowan, J.H., and Groat C.G., 1971, "Van Horn Sandstone, West Texas, an alluvial fan model for mineral Exploration," Report of Investigations 72, Bureau of Economic Geology, Austin, TX.

McKellar, J.B., 1975, "The East Australian Rutile Province," *Economic Geology of Australia and Papua New Guinea*, Vol. 1, Metals, C.L. Knight, ed., Monograph 5, Australasian Institute of Mining and Metallurgy, Parkville, Vic., pp. 1055–1062.

McPherson, R.O., 1978, "Geology of Quaternary Ilmenite Bearing Coastal Deposits at Westport," Bulletin 87, New Zealand Geological Survey, 94 pp.

Minard, J.P., et al., 1976, "Alluvial Ilmenite Placer Deposits, Central Virginia," Professional Paper 595H, US Geological Survey, 15 pp.

Mucke, A., and Chaudhuri, J.N.B., 1991, "The Continuous Alteration of Ilmenite through Pseudorutile to Leucoxene," *Ore Geology Reviews*, Vol. 6, pp. 25–44.

Pinnell, D.B., and Marsh, J.A., 1954, "Summary Geological Report on the Titaniferous Iron Deposits of the Laramie Range, Albany County, Wyoming," *Wyoming Mines Magazine*, Vol. 44, pp. 31–33, 53, 54.

Pirkle, E.C., and Yoho, W.H., 1970, "The Heavy Mineral Orebody of Trail Ridge, Florida," *Economic Geology*, Vol. 65, pp. 17–30 .

Pirkle, E.C., et al., 1974, "The Green Cove Springs and Boulogne Heavy Mineral Sand Deposits of Florida," *Economic Geology*, Vol. 67, pp. 1127–1137.

Pirkle, E.C., 1977, "The Highland Heavy Mineral Sand Deposit on Trail Ridge in Northern Peninsula, Florida," Report of Investigations 84, Florida Bureau of Geology, 50 pp.

Pirkle, E.C., et al., 1984, "The Yulee Heavy Mineral Sand Deposits of Northeastern Florida," *Economic Geology*, Vol. 79, pp. 725–737.

Pirkle, E.C., et al., 1989, "Altama Heavy Mineral Deposits in Southeastern Georgia," *Economic Geology*, Vol. 84, pp. 425–433.

Pirkle, F.L., et al., 1991, "Cabin Bluff Heavy Mineral Deposit of Southeastern Georgia," *Economic Geology*, Vol. 86, pp. 436–44

Puffer, J.H., and Cousminer, H.L., 1982, "Factors Controlling the Accumulation of Titanium-Iron Oxide Rich Sands in the Cohansey Formation, Lakehurst Area, New Jersey," *Economic Geology*, Vol. 77, pp. 379–391.

Raufuss, W., 1973, "Structur Schwermineralfuhrung Genese Und Bergban der Sedimentaren Rutil Lagerstatten in Sierra Leone (Westafrika)," *Geologische Jahrbuch*, Reihe D, pp. 3–52.

Reid, A.F., et al., 1984, "QEM-SEM Image Analysis of Ore Minerals, Volume Fraction, Liberation, and Observational Variances," *Applied Mineralogy*, Proceedings of 2nd International Congress on Applied Mineralogy in the Minerals Industry, TMS-AIME Publications, Warrendale, PA, pp. 191–204.

Rose, E.R., 1969, "Geology of Titanium and Titaniferous Deposits of Canada," Economic Geology Report 25, Canada Geological Survey, 177 pp.

Roehler, H.W., 1989, "Origin and Distribution of Six Heavy Mineral Placer Deposits in Coastal Marine Sandstones in the Upper Cretaceous McCourt Sandstone Tongue of the Rock Springs Formation, Southwest Wyoming," Bulletin 1867, US Geological Survey, 34 pp.

Scholl, D.W., et al., 1989, "Geology and Resource Potential of the Continental Margin of Western North America and Adjacent Ocean Basins, Beaufort Sea to Baja California," Circum Pacific Council for Energy and Mineral Resources, *Earth Science Series*, Vol. 14, pp. 113–147.

Shepherd, M.S., 1990, "Eneabba Heavy Mineral Sand Placers Western Australia," *Geology of the Mineral Deposits of Australia and Papua New Guinea*, F.E. Hughes, ed., Australasian Institute of Mining and Metallurgy, Parkville, Vic., pp. 1591–1594.

Slingerland, R., and Smith, N.D., 1986, "Occurrence and formation of waterlaid placers," *Earth and Planetary Science*, Annual Review, Vol. 14, pp. 113–147.

Stanaway, K.J., 1992, "Heavy Mineral Placers," *Mining Engineering*, Vol. 44, No. 4, pp. 352–358.

Temple, A.K., 1966, "Alteration of Ilmenite," *Economic Geology*, Vol. 61, pp. 695–710.

Towner, R.R., et al., 1988, "International Strategic Minerals Inventory Summary Report—Titanium," Circular 930G, US Geological Survey, 58 pp.

Trevoy, L.W., et al., 1979, "Development of the Heavy Minerals Potential of the Athabasca Tar Sands," *Canadian Mining and Metallurgical Bulletin*, Vol. 71, pp. 175–180.

Valentine, P.C., and Commeau, J.A., 1989, "The Gulf of Maine Rutile Province Accumulation of Fine-grained Authegenic Titanium Dioxide From Sandstone and Shale Source-rocks," Open File Report 89–532, US Geological Survey, 26 pp.

Walker, R.G., 1984, "Shelf and Shallow Marine Sands: in Facies Models," R.G. Walker, ed., Geoscience Canada Reprint Series 1, pp. 141–170.

Walker, S., 1990, "The Mine at the End of the Line," *International Mining*, Nov., pp. 25–26.

Wallis, D.S., and Oakes, G.M., 1990, "Heavy Mineral Sand in Eastern Australia," *Geology of the Mineral Deposits of Australia and Papua New Guinea*, F.E. Hughes, ed., Australasian Institute of Mining and Metallurgy, Parkville, Vic., pp. 1599–1608.

Welch, B.K., Sofoulis, J., and Fitzgerald, A.C.F., 1975, "Mineral Sands of the Capel Area, W.A.," *Economic Geology of Australia and Papua New Guinea*, Monograph 5, C.L. Knight, ed., Monograph 5, Australasian Institute of Mining and Metallurgy, Parkville, Vic., pp. 1071–1088.

Whitworth, H.F., 1956, "The Zircon-rutile Deposits on the Beaches of the East Coast of Australia With Special Reference to Their Mode of Occurrence and the Origin of the Minerals," New South Wales Department of Mines Technical Reports, Vol. 4, 59 pp.

Winker, C.D., and Howard, J.D., 1977, "Correlation of Tectonically Deformed Shorelines on the Southern Atlantic Coastal Plain," *Geology*, Vol. 5, pp. 123–292.

Williams, S.A., and Cesbron, F.P., 1977, "Rutile and Apatite; Useful Prospecting Guides for Porphry Copper Deposits," *Mineralogical Magazine*, Vol. 41, pp. 288–292.

Williams, V.A., 1990, "WIM150 Detrital Heavy Mineral Deposit," *Geology of the mineral deposits of Australia and Papua, New Guinea*, F.E. Hughes, ed., Australasian Institute of Mining and Metallurgy, Parkville, Vic., pp. 1609–1614.

TRIPOLI

RICHARD B. BERG AND CHARLES T. STEUART

Tripoli and the related mineral commodities such as microcrystalline silica have been mined for more than 100 years for their abrasive properties. Although abrasive and buffing compound markets are still very important, within the last 15 years the filler and extender markets in paint, plastics, rubber, adhesives, and sealants have increased substantially. Tripoli or microcrystalline silica consists almost entirely of very small quartz crystals, many less than one micrometer in length. Material mined from different districts differs in crystal shape, grain size, and texture of the rock, all of which influence markets. Most US deposits are now mined by surface methods, and both air floated and micronized products are marketed. Deposits of tripoli now mined in the United States occur in chert-bearing Paleozoic limestone in the central part of the country with producing districts in southern Illinois, central Arkansas, and eastern Oklahoma. Although deposits within each district are confined to specific formations and extend laterally within those formations, individual bodies form minable deposits that are typically several hectares in areal extent. Tripoli is white to cream to rose and characterized by high porosity and ease of disaggregation.

DEFINITIONS

Tripoli

In the United States, tripoli was first used to describe the fine-grained, easily disaggregated material from Seneca, MO, because of its similarity to a rock from the Tripoli region of North Africa (Hovey, 1894). The North African rock is actually diatomaceous earth, a material that is similar in appearance to the rock from Seneca, but is of entirely different origin having formed by the accumulation of siliceous remains of microscopic marine or freshwater animals. Tripoli is best defined as a very fine-grained, generally porous rock that consists of microcrystalline quartz, typically formed by the alteration of a chert-bearing limestone.

Tripolite

A term used to describe a rock from the vicinity of Tripoli in North Africa which is diatomaceous earth (Quirk and Bates, 1978).

Microcrystalline Silica

Microcrystalline silica is the same material as tripoli, but the distinction between the use of these two names is dictated largely by convention and markets. Material produced from southern Illinois deposits and used in white pigment and filler applications is generally referred to as microcrystalline silica, whereas that used in abrasive applications, both from the Illinois district and from other states, is commonly called tripoli.

Amorphous Silica

Amorphous silica, a term formerly used to describe the material produced from the deposits in southern Illinois, is now replaced by the term microcrystalline silica. Amorphous silica came into use when even optical methods for the identification of very fine-grained quartz were not widely available and the Illinois product, composed of quartz grains too small to be seen with the unaided eye, was thought to consist of amorphous silica. The Illinois material is clearly crystalline quartz, as shown by X-ray diffraction analysis and scanning electron microscopy (Fig. 1).

Novaculite

Although originally used to describe a rock suitable for the manufacture of whetstones, novaculite is now defined more generally "—as a homogeneous, mostly white or light colored rock, translucent on thin edges, with a waxy or dull luster, and almost entirely composed of microcrystalline quartz" (Steuart et al., 1983). The more compact rock mined in central Arkansas from the Arkansas Novaculite is referred to as novaculite, whereas the more porous rock is referred to as tripoli.

Rottenstone

The commodity rottenstone is sometimes included within the general mineral commodity category of tripoli. Rottenstone is mined in Northumberland County in eastern Pennsylvania and formed by the weathering of a siliceous shale of Devonian age (Faill, 1979, Berkheiser, private communication, 1991). This material is used as a filler and extender, but is apparently unlike tripoli both in origin and physical properties.

Spiculite

Spiculite is a rock consisting of siliceous sponge spicules having formed by the removal by solution of the carbonate matrix of a spicule-bearing limestone. Spiculite has been mined in Texas and, because it resembles tripoli in several aspects, is included in the discussion of deposits.

MINERALOGY AND CHEMISTRY

Tripoli by definition is essentially a monomineralic rock composed of microcrystalline quartz. Its unique features are that it is an aggregate of individual microcrystalline quartz crystals loosely held

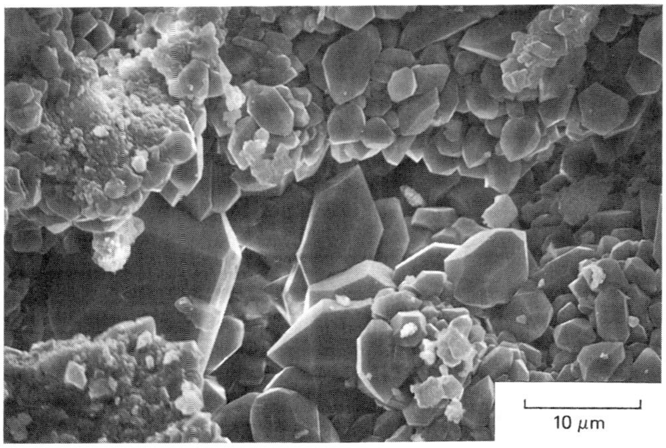

Fig. 1. Scanning electron micrograph of microcrystalline silica from the Elco district, southern Illinois.

Fig. 2. Scanning electron micrograph of tripoli from the Missouri-Oklahoma district.

together and that it is unusually porous. This texture makes it possible to easily disaggregate tripoli during processing to produce an ultra-fine material. Size of individual quartz crystals ranges down to less than 0.1 μm in southern Illinois material (Fig. 1). Tripoli from northwest Alabama is similar in texture to the tripoli or microcrystalline silica mined in southern Illinois (Rheams and Richter, 1988). Textural differences between material from various deposits may indicate differences in the relative importance of processes involved in its formation. The Missouri-Oklahoma tripoli consists of groups of quartz crystals which at high magnification show a lack of sharp crystal terminations (Fig. 2). In contrast, tripoli from the Arkansas district, formed by weathering of the Arkansas Novaculite, has a much more compact texture (Fig. 3) in which some individual quartz crystals have triple point junctions. This feature was interpreted to indicate that this rock has been metamorphosed (Keller et al., 1977).

Hematite and limonite are the major impurities in much tripoli and are responsible for pink to rose and, in some deposits, reddish brown colors. The darker shades are unsuitable for filler and pigment markets. Material which doesn't contain hematite or contains hematite in very low concentration is almost pure white and is thus desirable for filler and pigment applications where high brightness is important. Color is not as important for material used in abrasive and buffing applications and thus rose and pink tripoli can find application in these markets.

Tripoli from commercial deposits generally contains more than 98 wt% SiO_2 and some even exceeds 99 wt% (Table 1). Rose tripoli from the Missouri-Oklahoma district contains 0.83 wt% Fe_20_3, whereas the cream material from the same district contains only 0.20 wt%. The Al_2O_3 reported in some analyses can be attributed to a low concentration of clay.

GEOLOGY

All of the known domestic deposits of tripoli and related siliceous materials occur in the central and southern United States in the states of Texas, Missouri, Oklahoma, Arkansas, Illinois, Tennessee, Mississippi, Alabama, and Georgia (Fig. 4). There are substantial textural differences between these deposits, but they all appear to have similarities in their geologic setting. Except for the occurrence of tripoli in the Precambrian Murphy Marble in Georgia (Crickmay, 1937), all of the known deposits occur in Paleozoic limestones that contain chert layers (Table 2).

Fig. 3. Scanning electron micrograph of tripoli from the central Arkansas district.

Origin is generally ascribed to the thorough and extended weathering that resulted in removal of the carbonate from cherty limestone leaving a fine-grained siliceous residue. Authors differ on the extent to which actual leaching of the chert was responsible for the fine-grained tripoli. For instance, in the Missouri-Oklahoma district, Perry (1917) suggested that extensive weathering of a thick, homogeneous chert bed was responsible for the formation of those deposits, whereas Quirk and Bates (1978) suggested that beds originally produced by the deposition of colloidal silica and alkaline salts were later leached to leave the fine-grained silica. Siebenthal and Mesler (1908) suggested a slightly different precursor for the deposits in the Missouri-Oklahoma district, a siliceous limestone from which the carbonate was leached.

Other authors have interpreted the occurrence of euhedral quartz crystals as indication that secondary silica was deposited, perhaps after or accompanying solution of carbonate (Rheams and Richter, 1992, Berg and Masters, in press). It has also been suggested that the tripoli deposits in southern Illinois are largely hydrothermal (Berg and Masters, in press). Crickmay (1937), in discussing the Georgia deposits, summed up the uncertainty concerning the origin of the Georgia and other tripoli deposits by saying that there are two plausible explanations. The silica (quartz) was either deposited at the same time as the enclosing rock or it was

Table 1. Representative Analyses of Tripoli and Microcrystalline Silica (wt %)

	1 wt %	2 wt %	3 wt %	4 wt %	5 wt %	6 wt %	7 wt %
SiO_2	98.792	98.95	94.03	98.33	99.49	98.12	95.6
TiO_2	0.028	0.02	0.08	0.04	0.015		
Al_2O_3	0.532	0.40	2.34	0.71	0.102	0.62	3.0
Fe_2O_3	0.074	0.05	0.83	0.20	0.039	0.21	1.4
FeO			0.03	0.02			
MnO		0.003	0.01	0.03		0.01	
MgO	0.024	0.07	0.23	0.06	0.021	0.04	
CaO	0.027	0.10	0.89	0.24	0.014	0.19	
Na_2O	0.010		0.03	0.04			
K_2O	0.059	0.07	0.16	0.09			
P_2O_5		0.02	0.01	0.02			
CO_2			0.03	0.01			
Other			1.33	0.19			
LOI	0.454	0.61	1.85	0.55		0.83	
Total	100.00	100.3	101.85	100.53	99.68	100.02	100.00

1—Typical analysis of processed microcrystalline silica from Illinois (Unimin Specialty Minerals Inc.).
2—Average of three analyses of field samples of microcrystalline silica in the Clear Creek Chert, southern Illinois (Berg and Masters, in press).
3—Typical analysis of processed rose tripoli from the Missouri-Oklahoma district (American Tripoli Inc.).
4—Typical analysis of processed cream tripoli from the Missouri-Oklahoma district (American Tripoli Inc.).
5—Typical analysis of processed Arkansas Novaculite (Malvern Minerals Company).
6—Average of eleven field samples of tripoli in the Fort Payne Chert from Lauderdale and Colbert Counties, northwest Alabama (Rheams and Richter, 1988).
7—Average of five samples of tripoli from Georgia. Analyses estimated from ternary plot (Crickmay, 1937).

introduced into the rock, possibly by hydrothermal fluids. More than forty years later this problem is still largely unresolved.

Some general observations can be made about the occurrence of tripoli deposits. An understanding of the significance of these features would lead to a better comprehension of the geologic processes responsible for their formation.

1) Known US deposits are confined to a modest number of relatively small areas, mainly in the midcontinent.

2) They are generally confined to cherty limestones of Paleozoic age.

3) Within an individual district, deposits do not extend laterally for large distances at one stratigraphic horizon but, in most instances, consist of discrete bodies.

SOUTHERN ILLINOIS

Illinois has been a major producer of microcrystalline silica (tripoli) for most of this century. First production from deposits in southern Illinois was earlier than 1906 at which time three mills were in operation (Bain, 1907). At least nine companies have operated mills for the production of microcrystalline silica from deposits in Alexander and Union Counties. Through closure and consolidation, there were two mills in operation in 1991. Unimin Specialty Minerals Inc., a wholly owned subsidiary of UNIMIN Corporation, operates plants at Tamms and Elco, IL. In 1991, there were two active surface mines and one underground mine. Unimin Specialty Minerals, Inc. operates an open pit mine and also the Birk #2, an underground room and pillar mine. Lone Star Industries, Inc. operates an open pit mine that produces a mixture of microcrystalline silica and chert for use in the manufacture of cement at their Cape Girardeau, MO, plant. Iron staining of the microcrystalline silica at the Lone Star pit makes it unsuitable for the filler and extender markets. All of these active mines, as well as over 100 inactive mines and prospect adits, are situated within an area of approximately 150 km^2, located west of Elco in Alexander and Union Counties. There are also several inactive mines 17 km northwest of Elco in the Wolf Lake area.

Various names have been used to describe the siliceous material produced from southern Illinois deposits. It has been referred to as amorphous silica, decomposed silica, and soft silica, all in reference to its very small grain size and ease of disaggregation (Lamar, 1953). In recent years it is more generally called microcrystalline silica or tripoli. This material is characterized by the small size of euhedral quartz crystals; many are less than 2 μm long (Fig. 1). With the electron microscope it can be seen that even the smallest crystals have recognizable crystal faces.

Microcrystalline silica deposits occur within Devonian sedimentary rocks exposed along a north-south trending belt in Union and Alexander Counties on the southwestern flank of the Illinois Basin. Deposits now mined, and most of those previously mined, occur in the Lower Devonian Clear Creek Chert that is estimated to be at least 90 m thick in southwestern Illinois (Weller and Ekblaw, 1940). Outside of the area of microcrystalline silica deposits, the Clear Creek Chert consists of dolomitic limestone with interbedded chert, but within these districts Clear Creek Chert lacks carbonates. The Backbone Limestone, which underlies the Clear Creek Chert in southwestern Illinois, has not been recognized in the vicinity of the silica deposits, presumably having been removed by solution. The Grassy Knob Chert of Lower Devonian age directly underlies the Clear Creek Chert in the Elco area. Outside of the Elco area, the Grassy Knob Chert is a very cherty limestone, but as with the Clear Creek Chert, limestone does not occur in this formation in the Elco district. The Bailey Limestone, also of Lower Devonian age, underlies the Grassy Knob Chert and consists of clayey limestone with chert pods and layers. Microcrystalline silica deposits in a transitional zone between the Grassy Knob Chert and Bailey Limestone have been mined in two open pit mines and one underground mine in the vicinity of Olive Branch 15 km southwest of Elco (Berg and Masters, in press). The Clear Creek Chert is unconformably overlain by the Dutch Creek Sandstone Member of the Grand Tower Limestone. The Dutch Creek Sandstone is estimated to be only about 1.5 m thick in the Elco district.

Fig. 4. Tripoli and microcrystalline silica deposits in the United States (modified slightly from Bradbury and Ehrlinger, 1983).

In the Elco district the uppermost 20 to 30 m of the Clear Creek Chert consists of massive chert beds that cap flat-topped ridges. Beneath these chert beds the Clear Creek Chert is composed of prominent hard chert beds <30 cm thick, interbedded with beds ranging from softer chert to friable microcrystalline silica. All gradations exist between hard massive chert with small irregular cavities, to friable microcrystalline silica. Thin beds of gray kaolinitic clay, usually only about 2 cm thick, occur throughout the Clear Creek Chert. Commercial beds of microcrystalline silica typically range from 2 to 12 m in thickness (Levine, 1973), and bedding is generally close to horizontal. Individual microcrystalline silica beds are not laterally traceable throughout the district. Because the major uses of microcrystalline silica from southern Illinois require a material of high brightness, beds stained by hematite and limonite must be avoided in mining. Staining by iron minerals is more pronounced within a few meters of the surface where beds are typically colored red. Below this depth, staining is more selective with chert beds more heavily stained than the microcrystalline silica. Alternating white and red beds are common.

The absence of carbonates suggests that these deposits formed during a period of deep weathering when carbonates were removed in solution (Weller and Ekblaw, 1940). If so, the very small euhedral quartz crystals forming the microcrystalline silica could simply be the residual material left after this extended weathering. However other observations indicate that, although these deposits may have been subjected to weathering, the dominant process in their formation was hydrothermal (Berg and Masters, in press). The depth to which all carbonate has been removed in the Elco district defines a broad saucer-shaped depression that is approximately coincidental with a prominent positive magnetic anomaly attributed to a mafic pluton in the Precambrian basement (Heigold, 1976). A similar, but less pronounced, positive magnetic anomaly is centered in the Wolf Lake area. The spatial association of microcrystalline silica deposits with these magnetic anomalies suggests a causal relationship. Perhaps ground water, containing dissolved silica and heated by these cooling plutons, moved upward convectively, dissolving carbonate minerals and precipitating fine-grained quartz during cooling.

MISSOURI-OKLAHOMA

Tripoli deposits in Newton County in southwest Missouri and Ottawa County in northeast Oklahoma have a long history of production. Tripoli was first discovered near Seneca, MO, in 1869 and mining began in 1871 (Fellows, 1967). The first grinding mill was installed in 1885 and the American Tripoli Company was organized in 1892. A hundred years later this company, now American Tripoli Inc., a subsidiary of NYCO Minerals, Inc., continues to mine and process tripoli. Deposits occur in three areas—southeast of Peoria, OK, north of Seneca, MO, along the Missouri-Oklahoma border, and approximately 11 km northeast of Seneca. Tripoli is now mined from deposits near Peoria, OK and processed at Seneca.

Most of the tripoli in the Missouri-Oklahoma district is in strata assigned to the Warsaw Formation with some found in what is

Table 2. Tripoli and Microcrystalline Silica Districts in the United States

District	Producer	Host Formation and Age	Most Recent References
Southern Illinois	Unimin Specialty Minerals, Inc.	Clear Creek Chert—Early Devonian Bailey Limestone-Grassy Knob transition zone of minor importance—Early Devonian	Berg and Masters, in press
Missouri-Oklahoma	American Tripoli, Inc.	Warsaw Formation and Burlington-Keokuk Limestone—Middle Mississippian	Fellows, 1967
Arkansas	Malvern Minerals Company	Arkansas Novaculite (Upper Division) Early Mississippian	Steuart, et al., 1983
Alabama and adjacent Mississippi and Tennessee	No Current Production	Fort Payne Chert—Middle Mississippian Chepultepec Dolomite—Cambro-ordovician Copper Ridge Dolomite—Cambro-ordovician	Rheams and Richter, 1992
Georgia	No Current Production	Knox Dolomite—Cambro-Ordivician Shady Limestone—Cambrian Bangor Limestone—Late Mississippian Murphy Marble—Precambrian	Crickmay, 1937
Texas	No Current Production	Marble Falls Formation—Pennsylvanian	Damon, 1946

probably the underlying Burlington-Keokuk Limestone, both of Mississippian age (Fellows, 1967, Thompson, 1986). In prior nomenclature this stratigraphic interval was referred to as the Boone Formation of Mississippian age. Although confined to these two formations, deposits do not seem to be limited to specific stratigraphic intervals within them. The host rock in all instances is a cherty limestone. Deposits range from 1.3 to 6.6 m thick and are overlain by relatively thin overburden, generally less than 3.3 m thick (Fellows, 1967). The overburden generally consists of red clay, irregular fragments of tripoli, and chert; some deposits are overlain by a flint-like cap rock. Lenses of tripoli range from a few tens of meters across to a deposit south of Peoria, OK, that covers an area of 3.6 hectares (Quirk and Bates, 1978). Deposits occur on broad, flat-topped hills, but not in valleys. Color ranges from cream to rose depending on the iron content. In addition to a higher iron content, the rose tripoli contains more alumina and titania than the cream (Table 1). Some chert nodules occur within the tripoli deposits and if in too great a concentration make the deposit uneconomical to mine (Dobbs, 1960). Secondary silica has been deposited along some steeply dipping fractures in the deposits.

Several explanations have been offered over the past 80 years for the formation of these deposits. All of the explanations require the removal by solution of some constituent of the original sedimentary rock to form the tripoli. Quirk and Bates (1978) suggested that beds originally formed by deposition of colloidal silica and alkaline salts, and that removal of the soluble alkaline salts left the fine-grained silica or tripoli. A somewhat similar explanation was offered earlier by Siebenthal and Mesler (1908) who suggested that the precursor was a siliceous limestone from which all of the carbonate was dissolved forming tripoli. They produced a poor grade of tripoli by using acid to dissolve calcite from a specimen of siliceous limestone. However, it has not been possible to trace a bed of tripoli laterally into the unaltered rock and, as observed by Siebenthal and Mesler (1908), if leaching of siliceous limestone was solely responsible for the formation of these deposits, one would expect to find relict bedding and marine fossils in the tripoli. The deposits of tripoli are massive with no evidence of bedding and no recognizable fossils. Siebenthal and Mesler (1908) mentioned the occurrence of cherty slickensides within some of the tripoli deposits, suggesting that secondary silica was introduced into fractures along which there was later movement.

A third explanation, presented by Perry (1917), suggests that extensive weathering of a thick homogeneous chert bed was responsible for the formation of these deposits. A persistent chert bed is exposed in approximately the same stratigraphic position about 3 km north of a deposit which was being mined at the time of his writing. He noted that wherever tripoli was found, chert was observed to grade into it. SEM micrographs of chert from these deposits show a tripolitic rim on hard chert cores, indicating that at least some of the tripoli formed by the alteration of chert (Keller, 1978).

ARKANSAS

The term novaculite, when first applied to the rock found in the eastern Ouachita Mountain Range near Hot Springs, AR, referred to a white siliceous rock whose textural characteristics and hardness made it useful as a sharpening stone or whetstone. Griswold (1892) described the fine-grained whetstone known as Arkansas Stone as a "true novaculite satisfying all the necessary conditions regarding homogeneity, grittiness, finely granular structure and siliceous composition; it is translucent on the edges and has a marked conchoidal fracture." Keller et al. (1977) proposed that the word novaculite be restricted to use as a petrologic name for a thermally metamorphosed chert that shows polygonal triple-point texture. Steuart et al. (1983) define novaculite as "a homogeneous, mostly white or light-colored rock, translucent on thin edges, with a waxy to dull luster, and almost entirely comprised of microcrystalline quartz." Tripoli mined from the Arkansas Novaculite might better be defined as a homogeneous, white to tan, mostly poor to well cemented, friable rock, comprised almost entirely of microcrystalline quartz.

All of the tripolitic novaculite occurs in the Upper Division of the Arkansas Novaculite in the Ouachita Mountain Range (Fig. 5) and was determined to be early Mississippian in age by Hass (1951) on the basis of conodont studies at Caddo Gap, AR. The Upper Division of the formation is thickest in the Hot Springs area and thins abruptly to the north and more gradually to the south and west. Individual beds vary in thickness from several centimeters to 1.3 m.

Not all of the Upper Novaculite Division is tripolitic; much is calcareous and has weathered to a light brown or buff-colored, punky rock with a gritty texture giving it the appearance of a porous siltstone (Holbrook and Stone, 1978). Weathered novaculite varies from a firm, slightly porous rock to tripoli depending on the amount of carbonate present in the original rock. Originally the rock contained as much as 30% carbonate; rhodochrosite, calcite, manganiferous calcite, or ankerite. These tripolitic deposits range from only several centimeters thick to over 33 m in true thickness. The particle sizes range from 0.5 to 10 μm and particles are essentially equidimensional (Fig. 3). Color varies from white through lighter shades of tan to brown, depending on amounts of clays, iron, and manganese oxides formed during the weathering process. It is

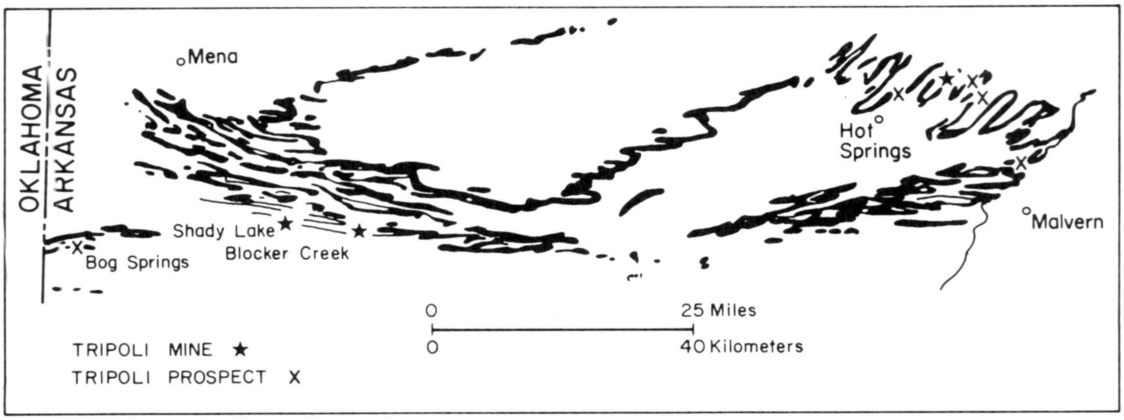

Fig. 5. Outcrop map of novaculite showing tripoli mines and prospects (Holbrook and Stone, 1978).

interesting to note that all of the good tripolitic material mined in this district is in overturned beds.

Park and Croneis (1969) grouped all of the existing explanations of the origin into three categories: direct precipitation of silica by chemical and/or organic means; replacement of non-siliceous deposits; and recrystallization and silicification of volcanic-rich material. As a result of their studies, they proposed that novaculite was formed from organically precipitated silica particles in the 1 to 10 µm range and that the ultimate texture was the result of diagenetic processes. Lowe (1975) agreed that siliceous organisms were the local source of the silica in the novaculite, but proposed further that volcanism distant from the immediate sites of accumulation may have provided much of the primary silica. Most investigators, including Sholes and McBride (1975), considered the Arkansas Novaculite to be a deepwater marine deposit, whereas Folk (1973) and a few others gave evidence that some, if not all of the formation, is a shallow marine (mostly peritidal) deposit. The evidence for the deepwater marine deposit is more convincing.

ALABAMA

Northwest Alabama

Large undeveloped tripoli deposits represent a significant resource in Lauderdale and Colbert counties of extreme northwestern Alabama. These deposits are limited stratigraphically to the upper part of the Fort Payne Chert of Early Mississippian age (Rheams and Richter, 1988). This formation, which is 46 to 61 m thick, consists of bedded chert and siliceous limestone. Tripoli generally occurs sporadically in horizontal beds 0.3 to 1.2 m thick but some occurrences range up to approximately 24 m in thickness. The Fort Payne Chert is overlain by the Tuscumbia Limestone which is overlain by the Pride Mountain Formation, all of Mississippian age. These formations are overlain by Cretaceous sediments. The tripoli is white to pinkish gray to very pale orange and consists of euhedral quartz crystals, mainly between 1 and 4 µm in length (Rheams and Richter, 1992). Texturally this material resembles the microcrystalline silica or tripoli mined in southern Illinois (Fig. 1) and purity is similar to the Illinois material (Table 1).

Scanning electron micrographs of specimens from the Fort Payne Chert outside of the area of tripoli deposits indicate a lack of euhedral quartz crystals indicating that the euhedral quartz crystals have grown in the restricted area of the tripoli deposits, an area of approximately 500 km^2. This observation indicates that the tripoli could not have formed simply by solution and removal of carbonates from the Fort Payne Chert leaving microcrystalline silica as a residual product (Rheams and Richter, 1988). Those authors suggest that ground water, which had become alkaline when it percolated through limestone, dissolved chert in the Fort Payne Chert, and that silica was reprecipitated in euhedral quartz crystals. Evidence of solution of limestone from a host rock in this area is suggested by what appears to be a paleokarst surface between the Fort Payne Chert and Cretaceous sediments (Rheams and Richter, 1992).

Northeast Alabama

Although not as extensive as deposits in northwest Alabama, tripoli and tripolitic material occur in the Mississippian Fort Payne Chert and in the Cambrian-ordovician Chepultepec and Copper Ridge Dolomites. From the early 1900s to the 1920s tripoli was produced near Tredegar, in Calhoun County (Weigel, 1927). Industrial Fillers, Inc. attempted to develop a deposit near Piedmont in northeastern Calhoun County in the mid 1960s (Bradbury and Ehrlinger, 1983). However there was no significant production from this deposit. Tripoli and tripolitic material from this area ranges from white to very pale orange to pinkish gray and ranges in size from very fine-grained material to coarser material mixed with numerous chert fragments (Rheams and Richter, 1988). Clay is abundant in some of the tripoli zones, particularly in the material from the Chepultepec and Copper Ridge Dolomites. Tripoli zones range from individual layers approximately a meter thick, to an exposure of tripoli with interbedded chert covering a stratigraphic interval of 30 m (Rheams and Richter, 1988).

Unlike the tripoli from northwest Alabama that consists of euhedral quartz crystals, that from northeast Alabama, both from the Fort Payne Chert and from the Chepultepec and Copper Ridge Dolomites, consists predominately of frosted, subangular to rounded quartz grains. On the basis of the lack of euhedral quartz crystals, these deposits are interpreted to have formed by localized subaerial weathering that caused the solution of dolomite and calcite, as well as quartz, to produce a rock consisting of fine-grained quartz and chert fragments (Rheams and Richter, 1988). Deposits in this area generally do not show as much potential for development as those in northwestern Alabama.

TENNESSEE—MISSISSIPPI

Tripoli deposits in south-central Tennessee and extreme northeast Mississippi are of similar geologic setting to those in northwestern Alabama. In the area of Tennessee along the Tennessee River Valley, tripoli has been mined near Collinwood, in Wayne

County and near Parsons, in Decatur County (Floyd, 1965). Tripoli has also been mined in southeastern Tennessee near Cleveland, in Bradley County and near Butler, in Johnson County in the northeast corner of the state. The deposits in Wayne County and adjacent Hardin County to the southwest are the most thoroughly described and perhaps represent the greatest potential (Spain, Jr., 1938, Born, 1936). These deposits occur in the upper part of the Fort Payne Chert of Mississippian age (Rheams and Richter, 1992). Tripoli beds range from a few meters to 20 m in thickness; larger deposits average about 8 m (Spain, Jr., 1938). The color is usually white with some brown streaks caused by secondary iron minerals, and the texture ranges from very compact to more friable material. Beds are horizontal with some folds developed in response to solution of carbonate from the Fort Payne Chert. Deposits are overlain by coarse, residual chert that remained after solution of the other constituents of this formation. There is general agreement that the tripoli deposits in south-central Tennessee and northeast Mississippi formed by solution of carbonate. Spain, Jr. (1938) suggested that this material was produced mainly by solution of a siliceous limestone leaving quartz, and that a lesser amount formed by the leaching of chert and reprecipitation of silica. Rheams and Richter (1992) relate the development of tripoli deposits in the Fort Payne Chert in adjacent parts of Tennessee, Mississippi and Alabama to lineament intersections. They suggest that these lineaments, seen on aerial imagery, are faults that controlled the movement of ground water, enhancing and localizing solution of carbonates and silica. Where there was greater flow of ground water, greater solutional effects caused the formation of tripoli deposits by solution of carbonates and silica, followed by the redeposition of quartz.

GEORGIA

Tripoli was discovered near Dalton in northern Georgia in the 1850s and a small amount of the material has been mined in Chatooga, Murray, Walker and Whitfield Counties, probably all before 1937 (Crickmay, 1937). All of the known deposits are in the northwestern part of the state where most occur in the Knox Dolomite (Cambro-Ordovician). Other occurrences are in the Bangor Limestone (Mississippian), Shady Limestone (Cambrian), and Murphy Marble (Precambrian). Tripoli occurs interbedded with limestone in horizontal beds attaining a thickness of 5 m at one deposit. Color of most Georgia tripoli is cream to pink but there are also shades from ivory to red. It is reported to be similar in texture to that from the Oklahoma-Missouri deposits, but a little coarser in grain size (Crickmay, 1937). Crickmay observed that many of the tripoli deposits appear to occur at one stratigraphic horizon in the Knox Dolomite suggesting that these siliceous beds were a primary feature of this formation. However, at two localities he noted quartz veinlets along bedding planes and along joints indicating the introduction of silica.

TEXAS

Deposits of spiculite occur in southern Lampasas and northern Burnet Counties in central Texas (Dietrich and Lonsdale, 1958). Spiculite has formed by the leaching of calcite from limestone containing siliceous sponge spicules of the Lower Pennsylvanian Marble Falls Formation, leaving a rock composed of siliceous sponge spicules. Spiculite was mined for two years prior to 1958 from a deposit in Burnet County for use in special drilling muds (Dietrich and Lonsdale, 1958). Deposits range up to 10 m in thickness and consist of well-stratified, friable spiculite that is white when fresh but weathers to a slightly yellowish material. Beds of spiculite alternate with chert beds in some of the deposits. Spiculite is almost pure silica except for some residual clay. A small quantity of this rock has been quarried for building stone from a quarry about 15 km south of Richland Springs in San Saba County, just east of Lampasas County (Barnes et al., 1947). The removal of carbonate from the spicule-bearing limestone of the Marble Falls Formation to form spiculite was not the result of surface weathering. Solution of carbonate is thought to have been caused by ground water moving along bedding surfaces because, in some instances, spiculite is found underlying unaltered spicule-bearing limestone of the Marble Falls Formation (Damon, 1946).

OTHER US DEPOSITS REPORTED AS TRIPOLI

Investigation by the Wyoming Geological Survey of four reported tripoli occurrences in Wyoming showed that none could be considered commercial tripoli (Harris, 1989). Of these four reported deposits, only one was verified to be tripoli and that occurrence was limited to weathering rinds developed on small agate nodules in gravel deposits. Tripoli production was reported from California in the 1930s (Bradbury and Ehrlinger, 1983), but this was probably material ranging from siliceous marl to altered volcanic rock, not what would generally be considered tripoli or microcrystalline silica.

FOREIGN DEPOSITS

Chandler (1960) reported that in the 1950s there was production of tripoli from deposits in Spain, Portugal, and South Australia. The Provinces of Albacete and Jaen in southeastern Spain were sources of tripoli production at that time. A small amount of material incorrectly referred to as tripoli has been produced from deposits in the vicinity of Adelaide, South Australia (Kalix and Barrie, 1957). This material from the locality in the Hundred of Kondoparinga, South Australia, is probably derived from the attrition of quartz and quartzite and thus not a true tripoli. Tripoli is also reported to occur in Germany and France (Chandler, 1960).

EXPLORATION AND DEVELOPMENT

The occurrence of tripoli in all of the districts of the United States has been known for many decades and, in the case of the Missouri deposits, since at least the late 1800s (Quirk and Bates, 1978). In the early days, exploration for new deposits consisted of digging small prospect pits and driving short exploratory adits. Pickering et al. (1986) estimated that the success rate of finding minable deposits in southern Illinois by exploratory drifting was less than 10%. In more recent exploration, diamond core drilling and reverse circulation drilling have been employed in delineating deposits. Initial drilling consists of widely spaced holes and is followed up with more closely spaced holes when several "tens of feet" of white tripoli are encountered. Pre-mining development holes are spaced 15 to 30.5 m apart in the Illinois district (Pickering et al., 1986).

Arkansas deposits, like those in southern Illinois, are stratigraphically limited, being totally confined to the Upper Division of the Arkansas Novaculite. These beds have a characteristic topographic expression that is seen as a series of low knobs or ridges that parallel the higher Lower Division novaculite ridge formers. This is because the softer tripolitic beds were eroded, and the more resistant beds left on either side. Float blocks of punky, vuggy, usually tan to red tripolitic material are found on the surface and help to locate the underlying deposits. Once a probable area is located, a strip, perpendicular to the strike, is cleared of vegetation and topsoil and trenches are cut across the bed to test for thickness and quality. If it is justified, the area is drilled to ascertain the limits of the deposit.

The most sought after material for those supplying the filler and extender markets is white microcrystalline silica or tripoli. Because processing does not involve decolorization procedures, mill feed must be of high brightness. Another important parameter is the ease with which the ore can be pulverized, an important consideration in energy costs. The ease of disaggregation can be judged in the raw material on the basis of porosity. The more porous the microcrystalline silica or tripoli, the more readily it can be pulverized.

MINING

With the exception of one underground mine in the southern Illinois district, all tripoli and microcrystalline silica now mined in the United States is mined by surface methods. Early mining of tripoli and microcrystalline silica in the United States was largely from underground mines. Until 1983, when Illinois Minerals began mining by open pit, all microcrystalline silica from the southern Illinois deposits was obtained from underground mines (Pickering et al., 1986). Beds in this area are flat-lying and underground mining is by the room and pillar method with horizontal entries. In the early part of this century many of these mines were small wagon mines where small amounts of microcrystalline silica were intermittently mined. More recent underground mines in Illinois have had large passageways (4.6 to 6.1 m high and 6.1 m wide) enabling front end loaders to load the shot rock directly into haul trucks. Selective mining of high quality microcrystalline silica can be more efficient in surface mines than in underground mines because it is easier to identify the most desirable white material in natural light than it is with artificial illumination underground. In both underground mines and surface pits, microcrystalline silica and interbedded chert in these Illinois deposits are very stable. Ground support is rarely required in underground workings and in one open pit mine the near-vertical wall stands more than 30 m high with no evidence of instability. Because of the high permeability of the formations in this district, water drains rapidly from mine workings. Microcrystalline silica need only be ripped, but overlying beds of massive chert at the top of the Clear Creek Chert must be drilled and shot.

Similar open pit mining methods are used in the deposits near Peoria, OK. After stripping overburden, the surface of the tripoli bed is cleaned to remove clay and the tripoli is blasted into large blocks (Quirk and Bates, 1978). Overburden to ore ratio in this district is usually no more than 1:1. Tripoli from these deposits is graded into the rose and cream products at the pit.

Mining of the Arkansas deposits consists of stripping away topsoil, drilling and blasting the Lower and Middle Division novaculite and stockpiling all three materials for future reclamation. The tripoli is then excavated with a track-hoe, loaded into trucks and hauled to the plant in Hot Springs, AR, where it is processed.

A number of factors have contributed to increased mining costs throughout the industry. One producer reported that costs have risen as much as 350% in the last 15 years. Fuel costs have increased, haul distances have become greater as deposits closest to the plant are depleted, and overburden to tripoli ratio has become greater. More stringent government regulations pertaining to mining and reclamation have also contributed to these rising mining costs.

PROCESSING

Processing of tripoli and microcrystalline silica consists essentially of drying the crude material and pulverizing it. For example, Illinois material is first crushed to less than 10 mm in a hammer mill and then dried in a fluid bed drier (Pickering et al., 1986). Coarser grades are ground in a ceramic rod mill, then classified by air flotation. The finer micronized grades are produced in self attrition mills where steam or compressed air is the energy medium.

Processing of tripoli from the Missouri-Oklahoma district begins with air drying the crude ore in drying sheds for several months as the mined material contains 25 to 35% water (Quirk and Bates, 1978). The air-dried material is then crushed, dried in a rotary drier, and ground in a hammermill and tube mill. The product from these mills is classified by air separators.

Similar processing methods are employed in Arkansas where the tripoli is hauled from the mine to storage sheds. The material then goes through a jaw crusher, a hammermill, through an external gas-fired furnace, and into the ball mills. After air classification, the material is ready for packaging according to the preference of the customer, whether it be paper bags, supersacks, boxes, or bulk.

Table 3. Typical Properties of Novacite® Tripolitic, Microcrystalline Silica Produced by Malvern Minerals Company from Arkansas Deposits

Specific Gravity	2.65
Index of Refraction	1.550
Color (Dry)	White
Color (Wet)	Gray-Tan
Oil Adsorption (Spatula) Rub Out Method	17–20%
Particle Shape (1–7 microns)	Euhedral, Platy
Particle Shape (Over 7 microns)	Clusters
pH	6.0–7.8
Acid Number	0 to (-0–18)
Thermal Stability	Absolute
(In its phase)	0 to 573°C.
Hardness	7 Mohs Scale
Loss on Ignition (LOI) (1 000°C 30 minutes)	0.20%
Loose Packed	800 kg/m^3
Dense Packed	1 281 kg/m^3

PRODUCT

Tripoli and microcrystalline silica markets demand a high-quality product with uniform characteristics. Color, or more correctly lack of color, is a very important consideration for the filler and pigment markets. Purity and uniformity of size are important in all markets. In order to compete effectively in these markets, producers must maintain high quality standards. Competition is strong and comes not only from other producers of tripoli and microcrystalline silica but also from other products used in abrasives and in the filler and extender markets. Material as close to white as possible is required by the filler and extender markets. Microcrystalline silica destined for markets range from 85 to 90% on the G.E. brightness scale (Harben, 1983). The major impurity affecting brightness adversely is iron, usually in the form of hematite. Brightness is not a primary consideration in the abrasive markets.

Purity of product is important not only because it affects brightness but also because impurities affect other properties that make microcrystalline silica desirable in specific applications. Such properties include hardness (7 on Mohs scale for quartz), high dielectric constant, and low absorption of oils (Table 3). Clay layers are avoided in mining as are chert layers. Chert is a great deal harder

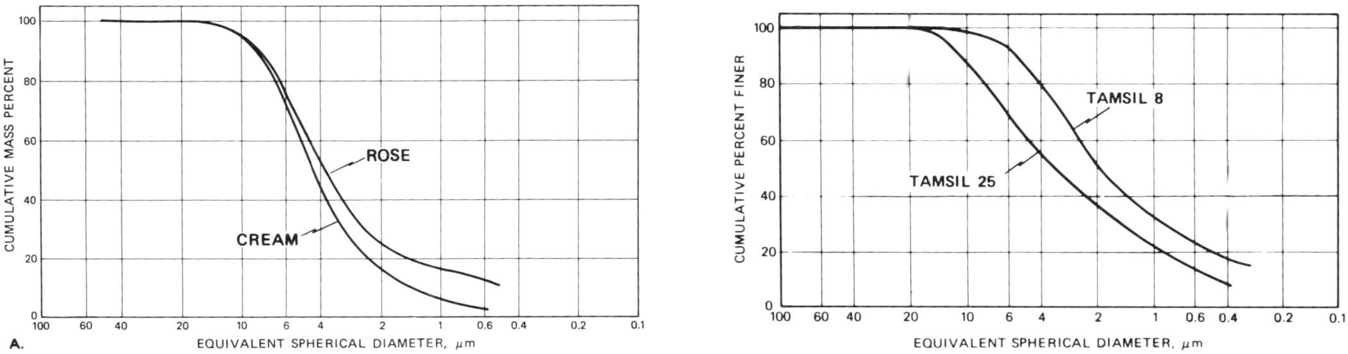

Fig. 6. Size distribution of processed tripoli. A. Seneca Rose Premium and Seneca Cream Premium (American Tripoli, Inc., Seneca, Missouri). B. TAMSIL™ 8 and TAMSIL™ 25 (Unimin Specialty Minerals, Inc., Tamms, IL).

to pulverize than microcrystalline silica or tripoli and the material produced by pulverizing chert has a different particle shape.

Particle shape, although important in all markets, is particularly important in the abrasive market. Pulverized or micronized tripoli and microcrystalline silica, unlike silica flour, do not contain quartz grains with very sharp edges. The product milled from tripoli or microcrystalline silica consists of small clusters of very small quartz crystals in which projecting crystals present a milder abrasive action than fractured quartz with knife-sharp edges.

Particle size is very important in all markets. Pulverizing and micronizing are the major steps in the production of the various particle size distributions required by different markets. Material to be used in buffing compounds must be carefully sized as the presence of coarser grains will cause scratches on an otherwise well-polished surface. The coarser grades are classified by air flotation after pulverizing. Fig. 6A shows the particle size distribution of tripoli destined for the buffing compound and abrasive markets. Finer grades are micronized in attrition mills which can produce material with a median particle size of 2 μm equivalent spherical diameter (Fig. 6B). Producers provide a range of products sized to suit particular applications.

Microcrystalline silica is receptive to all organofunctional coupling agents. These include silanes, titanates, and zircoaluminates. Attaching a coupling agent to microcrystalline silica filler allows better dispersion, higher loading, and improved adhesion which imparts increased impact and flexural strength to the composite. A coupling agent can make silica hydrophobic, thus making the product water repellant. Surface-modified products are currently used in all fields of plastics and coatings, as well as in adhesives, sealants, and rubber (Kuhn, private communication, 1991). Novakup®, produced by Malvern Minerals Company, is an example of surface-modified microcrystalline silica.

Producers must market material of uniform quality, i.e., uniform color, size distribution, particle shape and purity. To accomplish this, continual quality control through mining and milling is required. Testing procedures have advanced significantly in the last couple of decades. Size analysis, previously done by wet screening, is now carried out using instruments such as the Coulter Counter and Sedigraph. Reflectometers are used to determine the brightness or reflectance of the product at specific wavelengths of light under carefully controlled conditions.

MARKETS

Abrasives

The abrasive and buffing compound market has long standing as an important market for tripoli. Because this material is composed of quartz, with a hardness of 7 on Mohs scale, it can easily abrade many substances such as glass, painted surfaces and most metals. Also because tripoli can be easily pulverized into fine-grained material, this was an early recognized use for tripoli. Crickmay (1937) reported that before the Civil War, trainmen would stop at an exposure of tripoli along railroad tracks in Georgia to collect a supply of tripoli for polishing brass. Finer grades of tripoli are used in automotive buffing compounds. Coarser grades are used in cleansers and metal polishes. Tripoli is also used in finishing optical lenses. Color is not an important consideration in the abrasive and buffing compound markets, but particle shape and size is very important. Tripoli from the Missouri-Oklahoma district is particularly well suited for use in buffing compounds and a major portion of the production from this district is supplied to the automobile industry for this purpose. Production from the Arkansas and Illinois districts is also sold in the abrasive markets.

Paint

Microcrystalline silica is used as a pigment in paint because it can replace some of the much more expensive TiO_2 pigment and imparts desirable properties to the paint. Several properties make this commodity a desirable pigment in some paint formulations. Its greater hardness as compared to other materials such as talc, kaolin, or calcium carbonate adds durability to the paint. This is particularly important for paints used in applications where the surface is subject to abrasion such as floor paints, marine paints, and highway paints. In addition to its physical durability, quartz is chemically resistant to most chemicals including acids, with the exception of HF. Because of the chemical durability of quartz, paint formulated with microcrystalline silica can be used in those environments where the surface coating is subjected to various chemicals. Also because of the durability of quartz, paint incorporating microcrystalline silica has high resistance to weathering and chalking. Because quartz has low oil absorption properties, it is possible to load the paint with additional pigment without significantly increasing the viscosity, a consideration that is important in some paints. Low oil absorption also makes it easier to disperse microcrystalline silica than some other pigments used in paints. Just as the hardness of quartz contributes to the durability of paints, it also can cause increased wear in equipment used in the application of paints.

Plastics

Tripoli or microcrystalline silica is important in plastics where it finds application as a functional filler because it imparts the following desirable properties to the plastic.

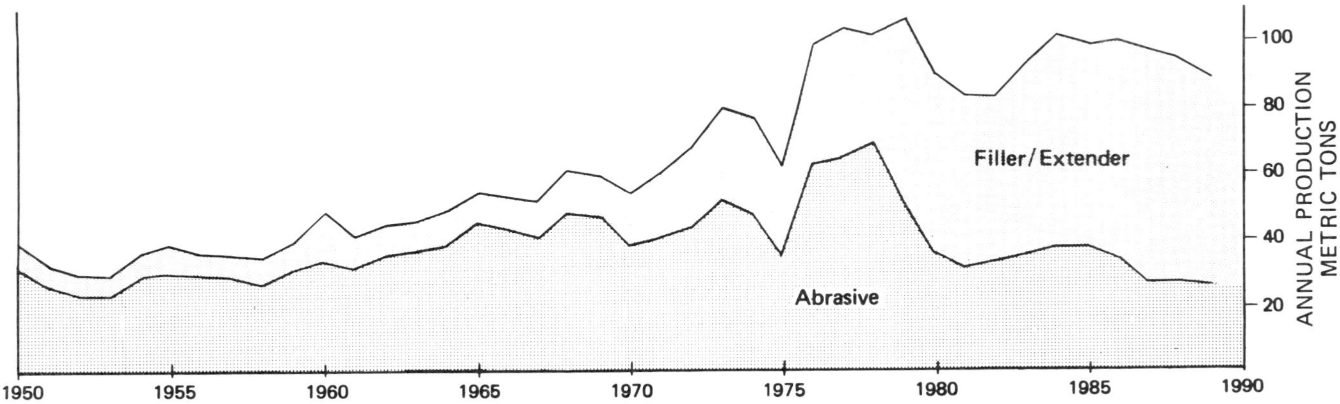

Fig. 7. Processed tripoli sold or used by producers in the United States including microcrystalline silica and Pennsylvania rottenstone. Abrasive and filler/extender uses only shown. In addition to these uses much smaller quantities are used in other applications including foundry facings. The sources of information are: 1950, Chandler and Tucker (1953); 1951–1955, Chandler and Tucker (1958); 1956–1960, Chandler and Tucker (1961); 1961–1965, Ambrose (1966); 1966–1970, Clarke (1972); 1971–1975, Clarke (1977); 1976–1980, Anon. (1981); 1981–1984, Smoak (1985); 1985–1989, Austin (1992).

1) Provides a white base for the plastic, adding to its opacity. This enables the plastic to accept a color pigment more effectively and to provide a good level of color saturation.

2) Increases the strength of the plastic, adding rigidity to the object.

3) The low oil absorption of tripoli allows a high level of loading of this functional filler without undue increase in viscosity that would inhibit molding of the plastic. Loading levels as high as 65 to 75% can be achieved with some resins.

4) In electrical cable coverings and other applications where insulating properties are important, the dielectric properties of quartz make tripoli a desirable filler.

5) Tripoli is much cheaper than the organic resins used in plastics.

Rubber

Tripoli is used as the non-reinforcing filler in rubber. Use of tripoli filler in rubber used in the fabrication of printing rollers improves wear-resistance.

Adhesives and Sealants

Tripoli is used in a variety of grouts, sealants, and high-performance epoxies. Some epoxies can be loaded with 60% of this filler. Tripoli competes in the filler market with a variety of minerals including talc, wollastonite, mica, calcium carbonate, and kaolin.

SALES AND EXPORTS

Sales of all varieties of microcrystalline silica, including tripoli and rottenstone from Pennsylvania, have increased in the United States from an average of 35 864 t for the years 1951–1955 (Chandler and Tucker, 1961) to 105 220 t in 1979, the year of maximum sales (Anon., 1981). In 1979, consumption of tripoli for filler (and extender) use exceeded that used in abrasives for the first time (Fig. 7). Sales in 1989 amounted to 95 490 t (Austin, 1992). During the last forty years, consumption of tripoli for abrasives increased through 1978, then decreased abruptly to levels close to those of the 1950s. In 1989, almost 2.5 times as much of this commodity was used in filler and extender applications than in abrasives (Austin, 1992). Foreign competition has not been a significant factor in these markets because significant production is largely confined to the United States (Harben, 1983).

The United States is an exporter of tripoli with export from the Missouri-Oklahoma district reported as early as 1894 (Fellows, 1967). Foreign markets for tripoli continue to be important. For instance, Arkansas tripoli, in addition to supplying domestic customers, is exported to Canada, Taiwan, most of Europe, Israel, Turkey, Australia, New Zealand, the Republic of South Africa, Saudi Arabia, and South America.

TRANSPORTATION

Tripoli and microcrystalline silica are shipped bagged, in supersacks and in bulk, using pneumatic carriers. Transportation is one of the major costs and is a particularly important factor for West Coast markets as all producing districts are situated in the central and eastern United States. Truck transportation is important for domestic markets with overseas shipments transported to the nearest shipping port by container carrier.

GOVERNMENTAL REGULATIONS

Within recent years there has been increased concern about health risks associated with silica. In 1969 the International Agency for Research on Cancer (IARC), established by the World Health Organization, initiated a program to evaluate the carcinogenic risk of chemicals to humans. IARC has since provided information on the carcinogenic risk of a variety of commercially produced mineral commodities and specific minerals in some of these commodities, all based on review of published research results. An IARC monograph devoted to silica and some silicates states that "There is limited evidence for the carcinogenicity of crystalline silica to humans" (Anon., 1987a). Also in 1987, Supplement 7 to the IARC monographs classified crystalline silica in group "2-A," indicating that, in their opinion, "The agent is probably carcinogenic to humans" (Anon., 1987b). The Occupational Safety and Health Administration (OSHA) cited IARC Monographs as a source to be used in establishing that a chemical is a carcinogen or a potential carcinogen (Anon., 1987c). OSHA goes on to explain in this Hazard Communication Standard that, if a product contains greater than 0.1% of a material designated as a carcinogen or potential carcinogen, it must be labelled with a warning regarding this substance. In response to concern about this Hazard Communication Standard and the possible carcinogenic risk attributed to crystalline silica, the Chemical Manufacturers Association (Chemstar Division) formed a panel composed of producers and consumers of materials containing crystalline silica (Shapiro, 1990). The Crystalline Silica Panel was formed to initiate a careful review by experts of all evidence that would have a bearing on the relationship, if there is one, between exposure to crystalline silica and cancer in humans. Other tasks of this panel deal with the regulatory and legal aspects of crystalline silica regulation and analytical methods employed in the detection of crystalline silica at low levels of concentration.

At this time it appears that the issue of the possible health risks of exposure to crystalline silica contains many unknowns and is far from settled. Producers of tripoli, at additional expense, have taken the required measures to protect their workers and customers from any possible health risk associated with this commodity.

FUTURE TRENDS

The domestic resources of microcrystalline silica and tripoli in the known districts, both those now producing and those with past production, appear to be large. Deposits in northwestern Alabama represent a large potential resource very similar in particle shape to microcrystalline silica in southern Illinois. Competition from foreign deposits is not a major consideration at this time. Government regulations always present an unknown factor. Regulations recently enacted pertaining to crystalline silica have had their impact on the industry, but the effects of these regulations have been largely accommodated.

Transportation costs will always be an important consideration for producers and those deposits situated nearest to major markets will continue to have a competitive advantage. Deposits with relatively inexpensive access to deep water ports will have an advantage in selling to foreign markets. Expansion of markets in developing countries is likely. Mining techniques will probably not change significantly in the near future as they are what might be considered commonly employed techniques for a variety of commodities. The introduction of open pit mining for the Illinois deposits in 1983 represented a significant change in one district. It seems likely that the most significant changes in this industry in the next decade will be new developments in processing of these commodities and perhaps the development of new markets. Improvements on the sophisticated processing techniques now employed may allow production of even brighter and finer-sized products. Processes such as wet grinding, chemical leaching and high-intensity magnetic separation could be employed if the high end of the market justified these additional costs. Foreign competition does not seem to be a major consideration at this time, but competing minerals such as kaolin, talc, calcium carbonate, and mica are an ever present competitive challenge in the filler and extender markets. Although each of these mineral commodities has certain desirable properties suited to particular markets, there is some market overlap. In the long-important abrasive market, tripoli and microcrystalline silica compete with manufactured abrasives.

ACKNOWLEDGMENTS

We sincerely appreciate the assistance of numerous individuals who helped substantially in the preparation of this chapter by providing information and making suggestions and reviewing portions of the manuscript. They are Gordon T. Austin, William A. Bratney, Bruce J. O'Connor, Sam Pickering, Karen F. Rheams, Ardel W. Rueff, Michael T. Cerino, Robert L. Virta, Dwight Vowels, and Doss White.

REFERENCES

Anon., 1981, "Abrasive Materials," *Minerals Yearbook 1980*, US Bureau of Mines, pp. 51–51.

Anon., 1987a, "Silica and Some Silicates," *The Evaluation of the Carcinogenic Risk of Chemicals to Humans*, Monographs Vol. 42, International Agency for Research on Cancer, Lyon, France, 289 pp.

Anon., 1987b, "Overall Evaluations of Carcinogenicity: An Updating of IARC Monographs Volumes 1–42," *The Evaluation of the Carcinogenic Risk of Chemicals to Humans*, Monographs Supplement 7, International Agency for Research on Cancer, Lyon, France, 440 pp.

Anon., 1987c, "Hazard Communication," Vol. 52, No. 163, Federal Register, Aug. 24, pp. 31852–31886.

Ambrose, P.M., 1966, "Abrasive Materials," *Minerals Yearbook 1965*, US Bureau of Mines, pp. 147–164.

Austin, G.T., 1992, "Abrasive Materials," *Minerals Yearbook 1989*, US Bureau of Mines, pp. 77–96.

Bain, H.F., 1907, "Analysis of Certain Silica Deposits," *Year Book for 1906*, Bulletin 4, Illinois State Geological Survey, pp. 185, 186.

Barnes, V.E., Dawson, R.F., and Parkinson, G.A., 1947, "Building Stones of Central Texas," Publication 4246, The University of Texas, Austin, 198 pp.

Berg, R.B., and Masters, J.M., in press, "Geology of Microcrystalline Silica (Tripoli) Deposits, Southernmost Illinois," Circular 555, Illinois State Geological Survey.

Born, K.E., 1936, "Summary of the Mineral Resources of Tennessee," *Resources of Tennessee (2nd Series)*, Tennessee Department of Education, Division of Geology, 102 pp.

Bradbury, J.C., and Ehrlinger, H.P., III, 1983, "Tripoli," *Industrial Minerals and Rocks*, 5th ed., Vol. 2, S.J. Lefond, ed., AIME, New York, pp. 1363–1374.

Chandler, H.P., 1960, "Tripoli," *Industrial Minerals and Rocks*, 3rd ed., J.L. Gillson, ed., AIME, New York, pp. 881–887.

Chandler, H.P., and Tucker, G.E., 1953, "Abrasive Materials," *Minerals Yearbook 1950*, US Bureau of Mines, pp. 90–107.

Chandler, H.P., and Tucker, G.E., 1958, "Abrasive Materials," *Minerals Yearbook 1955*, US Bureau of Mines, pp. 121–140.

Chandler, H.P., and Tucker, G.E., 1961, "Abrasive Materials," *Minerals Yearbook 1960*, US Bureau of Mines, pp. 145–163.

Clarke, R.G., 1972, "Abrasive Materials," *Minerals Yearbook 1970*, US Bureau of Mines, pp. 153–164.

Clarke, R.G., 1977, "Abrasive Materials," *Minerals Yearbook 1975*, US Bureau of Mines, pp. 165–177.

Crickmay, G.W., 1937, "Tripoli Deposits of Georgia," Information Circular No. 9, Division of Geology, Georgia Department of Forestry and Geological Development, 8 pp.

Damon, H.G., 1946, "The Origin and Distribution of Spiculite Near Lampasas, Lampasas County, Texas," *Texas Mineral Resources*, Publication 4301, University of Texas, Austin, pp. 271–282.

Dietrich, J.W., and Lonsdale, J.T., 1958, "Mineral Resources of the Colorado River Industrial Development Association Area," Report of Investigations No. 37, Texas Bureau of Economic Geology, 84 pp.

Dobbs, E.H., 1960, "Missouri-Oklahoma Type Tripoli," Preprint 60H41, SME Annual Meeting, Feb., 5 pp.

Faill, R.T., 1979, "Geology and Mineral Resources of the Montoursville South and Muncy Quadrangles and Part of the Hughesville Quadrangle, Lycoming, Northumberland, and Montour Counties, Pennsylvania," Atlas 144ab, Pennsylvania Bureau of Topographic and Geologic Survey, 114 pp.

Fellows, L.D., 1967, "Tripoli," *Mineral and Water Resources of Missouri*, Vol. 43, 2nd Series, Missouri Division of Geological Survey and Water Resources, pp. 220–223.

Floyd, R.J., 1965, "Tennessee Rock and Mineral Resources," Bulletin 66, Tennessee Department of Conservation, Division of Geology, 118 pp.

Folk, R.L., 1973, "Evidence for Peritidal Deposition of Devonian Caballos Novaculite, Marathon Basin, Texas," *American Association of Petroleum Geologists Bulletin*, Vol. 57, pp. 702–705.

Griswold, L.S., 1892, "Whetstones and the Novaculites of Arkansas," *Annual Report of the Geological Survey of Arkansas for 1890*, Vol. 3, 443 pp.

Harben, P., 1983, "Tripoli and Novaculite—The Little Known Relations," *Industrial Minerals*, No. 184, pp. 28–32.

Harris, R.E., 1989, "Tripoli (Tripolite) in Wyoming," Open-File Report 89–4, The Geological Survey of Wyoming, 10 pp.

Hass, W.H., 1951, "Age of Arkansas Novaculite," *American Association of Petroleum Geologists Bulletin*, Vol. 35, pp. 2526–2541.

Heigold, P.C., 1976, "An Aeromagnetic Survey of Southwestern Illinois," Circular 495, Illinois State Geological Survey, 28 pp.

Holbrook, D.F., and Stone, C.G., 1978, "Arkansas Novaculite—A Silica Resource," *Proceedings*, 13th Annual Forum on the Geology of Industrial Minerals, K.S. Johnson and J.A. Russell, eds., Circular 79, Oklahoma Geological Survey, pp. 51–58.

Hovey, E.O., 1894, "American Tripoli," *Scientific American Supplement No. 969*, p. 15487.

Kalix, Z., and Barrie, J., 1957, "Natural Abrasives," *Mineral Resources of Australia*, Summary Report No. 41, Bureau of Mineral Resources, Geology and Geophysics, Department of National Development, 48 pp.

Keller, W.D., 1978, "Textures of Tripoli Illustrated by Scanning Electron Micrographs," *Economic Geology*, Vol. 73, pp. 442–446.

Keller, W.D., Viele, G.W., and Johnson, C.H., 1977, "Texture of Arkansas Novaculite Indicates Thermally Induced Metamorphism," *Journal of Sedimentary Petrology*, Vol. 47, pp. 834–843.

Lamar, J.E., 1953, "Siliceous Materials of Extreme Southern Illinois," Report of Investigations 166, Illinois State Geological Survey, 39 pp.

Levine, C.R., 1973, "Geology of the Clear Creek Tripoli Deposits, Alexander County, Illinois," MS thesis, Southern Illinois University, Carbondale, 70 pp.

Lowe, D.R., 1975, "Regional Controls on Silica Sedimentation in the Ouachita System," *Geological Society of America Bulletin*, Vol. 86, pp. 1123–1127.

Park, D.E., Jr., and Croneis, C., 1969, "Origin of Caballos and Arkansas Novaculite Formations," *American Association of Petroleum Geologists Bulletin*, Vol. 53, pp. 94–111.

Perry, E.S., 1917, "Tripoli Deposits of Oklahoma," Bulletin No. 28, Oklahoma Geological Survey, 32 pp.

Pickering, S.M., et al., 1986, "Southern Illinois White Tripoli as Filler, Extender, and Abrasive," *Mining Engineering*, Vol. 38, No. 12, pp. 1125–1127.

Quirk, W.F., and Bates, A.K., 1978, "Tripoli Deposits of Southwest Missouri and Northeast Oklahoma," *Proceedings*, 13th Annual Forum on the Geology of Industrial Minerals, K.S. Johnson and J.A. Russell, eds., Circular 79, Oklahoma Geological Survey, pp. 47–50.

Rheams, K.F., and Richter, K.E., 1988, "Tripoli Deposits in Northern Alabama—A Preliminary Investigation," Circular 135, Geological Survey of Alabama, 54 pp.

Rheams, K.F., and Richter, K.E., 1992, "Occurrence, Genesis, and Industrial Potential of Tripoli Deposits in North Alabama and South-Central Tennessee," *Industrial Minerals of the Southeastern United States*, K.F. Rheams and G.H. McClellan, eds., Circular 161, Alabama Geological Survey, pp. 27–59.

Shapiro, J.C., 1990, "Chemical Manufacturers Association Chemstar Crystalline Silica Panel—Update," Preprint, Regulation of Crystalline Silica Meeting, Washington, DC, Apr. 9, SME, Littleton, CO, 3 pp.

Sholes, M.A., and McBride, E.F., 1975, "Arkansas Novaculite, Sedimentology of Paleozoic Flysch and Associated Deposits, Ouachita Mountains-Arkoma Basin Oklahoma," G. Briggs et al., eds., Guidebook, Dallas Geological Society, Apr., pp. 69–87.

Siebenthal, C.E., and Mesler, R.D., 1908, "Tripoli Deposits Near Seneca, Mo.," *Contributions to Economic Geology, 1907*, Bulletin 340, US Geological Survey, pp. 429–437.

Smoak, J.F., 1985, "Abrasive Materials," *Minerals Yearbook 1984*, US Bureau of Mines, pp. 71–89.

Spain, E.L., Jr., 1938, "Tripoli Deposits of the Western Tennessee Valley," *Trans. AIME*, Vol. 129, pp. 501–515.

Steuart, C.T., Holbrook, D.F., and Stone, C.G., 1983, "Arkansas Novaculite: Indians, Whetstones, Plastics and Beyond," *Proceedings*, 19th Forum on the Geology of Industrial Minerals, S.E. Yundt, ed., Miscellaneous Paper 114, Ontario Geological Survey, pp. 156–167.

Thompson, T.L., 1986, "Paleozoic Succession in Missouri, Part 4—Mississippian System," Report of Investigations 70, Missouri Department of Natural Resources, Division of Geology and Land Survey, 189 pp.

Weigel, W.M., 1927, "Technology and Uses of Silica Sand," Bulletin 266, US Bureau of Mines, 204 pp.

Weller, J.M., and Ekblaw, G.E., 1940, "Preliminary Geologic Map of Parts of the Alto Pass, Jonesboro, and Thebes Quadrangles with Explanation and Stratigraphic Summary by Weller, J.M.," Report of Investigations 70, Illinois State Geological Survey, 26 pp.

VERMICULITE

JAMES R. HINDMAN

Vermiculite is the mineralogical name generally applied to a group of hydrated, trioctahedral ferromagnesian aluminum platty or micaceous silicate minerals. In microscopic particles the mineral as mineral is differentiated from montmorillonite by a higher Si:Al ratio of approximately 3:1 and a higher layer charge and higher cation exchange capacities. Commercial vermiculite, the vermiculite which is mined, beneficiated, and discussed in this chapter, is an important industrial mineral which is formed by the alteration of macroscopic particles of biotite and iron-bearing phlogopite. In natural form it has the size and shape of mica, while possessing the ion exchangeable properties of zeolites and some clays.

Within the internal structure of vermiculite lie water molecules, which when rapidly heated to high temperature, transform into steam causing the vermiculite particles to increase in volume. This process of thermal exfoliation produces a lightweight product which finds use in various construction products, agriculture, horticulture, and other industrial applications. In exfoliated form it has the low density and biological inertness of expanded perlite, while maintaining a chemically active surface. Most of the uses of vermiculite are for the thermally exfoliated material.

In addition, the cation exchange potential of vermiculite and its layer-silicate structure are properties which can be used as the basis for numerous products including intumescent coatings and gaskets, inorganic films, and the treatment of nuclear and other toxic waste. When chemically and physically modified, it possesses some properties common to organoclays and synthetic compounds. In high purity concentrates and manufactured products, these properties can be altered and optimized in applications to produce superior performance.

GEOLOGY

Mineralogy

Vermiculite was first described as a mineral by Thomas H. Webb in 1824. Webb believed vermiculite to be a variety of talc and described its diagnostic property of thermal exfoliation thusly: ''If subjected to the flame of a blowpipe, or that of a common lamp, it expands and shoots out into a variety of fanciful forms, resembling most generally small *worms* having the *vermicular motion exact*.'' Observing that with one exception, all commercial vermiculite has formed by the alteration of iron-bearing phlogopite, a simple definition of vermiculite might be the alteration product of trioctahedral micas, macroscopic particles of which display marked exfoliation when heated rapidly to temperatures above 374°C (the critical temperature of water).

The first vermiculite described was found near Worcester, MA, but it was later pointed out by Brush (1866) that vermiculite had been known for some time in Japan as a children's novelty. Because the techniques for mineral analysis were somewhat primitive compared to those routinely available today, and because the structure and crystal chemistry of the micas and related minerals were not widely understood, the early literature generated at least 20 ''new'' minerals which were nothing more than variations in the color and chemistry displayed by vermiculite. Trade names such as Dugarlite, Mandolite™, Strong-Lite™, and Zonolite™ are sometimes used to denote commercial vermiculite, thermally exfoliated vermiculite and their products.

Vermiculite particles resemble mica and vary in color from colorless through tan and green to black, depending on the chemical composition. Most physical and optical properties are close to those of biotite, with perfect basal cleavage, hardness between 1.5 and 2.5 Mohs, and specific gravity between 2.2 to 2.8. Fine-sized particles will feel talcose, especially when wet. However, for commercial vermiculite the diagnostic test is that of thermal exfoliation.

Vermiculite can display a wide range of compositions, depending on the composition of the original mica and the change of chemistry during weathering and ion exchange. Chemical data for some commercial vermiculites are given in Table 1. There are numerous analyses for vermiculite given in the literature, and the analyses of vermiculites from commercial deposits will fall within the range indicated by the average analysis in the table. Many analyses separate the adsorbed water (H_2O^-) from crystalline water and water as hydroxyl (H_2O^+). The basis of separation of adsorbed water is generally considered to be 100°C. However, hydrate water within the vermiculite interlayer begins to be lost at 75°C and lower. Therefore, measured water content and loss on ignition are combined in the table as H_2O[total].

The average chemical component values calculated from 65 vermiculite analyses suggest a typical formula of:

$$(Na_{0.21},K_{0.39},Mg_{0.19},Ca_{0.13},6H_2O)$$
$$(Mg_5,Fe^{+2}_{0.2},Fe^{+3}_{0.8})[Si_{5.5},Al_{2.5},O_{20}](OH)_4$$

where the components in the first set of parenthesis represent the ion-exchangeable layer, the components in the second set of parenthesis compose the cations of the octahedral layer, and the components within the brackets comprise the tetrahedral layer.

The crystal structure of vermiculite has been investigated by many authors. The basic structure of the mineral is identical to talc and the micas: a 2:1 silicate sheet composed of two flat layers of silica and alumina tetrahedra (the tetrahedral layers) which are joined together in a layer composed of apical oxygen atoms, and magnesium, iron, and hydroxyl molecules (the octahedral layer). Between the 2:1 sheets lies the ion exchangeable layer. This layer will change in thickness depending on the interlayer cation present and the arrangement of waters of hydration associated with it.

Interlayer water is an essential component of vermiculite. When a particle of vermiculite is rapidly heated, the interlayer water transforms into steam. The pressure of the steam forces the silicate layers apart from one another and the particle is deformed from the normally flat plate into an elongated, concertina-like particle, which

Table 1. Chemical Analyses Of Vermiculite As Reported For Commercial Concentrates (Wt %)

Analysis	A-65	V-1	V-2	V-3	V-4	V-5	V-6	V-7	V-8
Li_2O	0.03					0.03			0.01
Na_2O	0.74	2.00	0.10	1.61	1.95	0.80	0.03		1.72
K_2O	2.11	0.50	0.50	5.97	1.36	2.46	2.56	4.24	6.63
MgO	20.34	19.20	23.60	24.22	17.16	23.37	21.39	20.04	16.38
CaO	0.79	0.21	3.80	0.93	2.18	1.46	0.20	0.75	1.12
BaO	0.17		0.20				0.03	0.12	
MnO	0.23	0.18		0.05	0.18	0.30		0.07	0.14
FeO	1.58			1.54	0.48	1.17	3.56		
Al_2O_3	13.86	14.50	10.20	12.68	14.23	12.08	10.01	17.36	12.85
Cr_2O_3	0.33						0.20	0.50	0.23
Fe_2O_3	7.59	10.50	5.80	4.60	8.07	5.45	1.90	8.45	8.80
SiO_2	36.94	36.50	45.10	41.20	37.67	39.37	43.05	38.66	38.34
TiO_2	1.12	0.75	0.70	1.38	0.20	1.25	1.00		1.66
H_2O[total]	15.41	15.30	10.20	5.82	15.71	11.20	15.70	8.71	10.66
Total	99.27	99.64	100.00	99.99	99.19	98.97	99.4	98.61	99.34

A-65 Average of 65 analyses statistically selected from 116 published analyses of vermiculites from various international localities.
V-1 Munglinup, Western Australia; IMDEX Minerals, Jandakot, Australia.
V-2 Massape Mine, Piaui, Brazil; Eucatex Mineral Ltda, São Paulo, Brazil.
V-3 Qieganbulake, China; Ying and Hu (1988) page 5.
V-4 Hafafit, Egypt; El Shazly et al (1975), page 95.
V-5 Phalaborwa, Republic of South Africa; American Vermiculite Corp., Marietta, GA.
V-6 Kovdor, Russia; Justo et al. (1989), page 512.
V-7 Enoree, SC; Vermiculite Reference Manual, Zonolite Division, W.R. Grace & Company, Chicago, IL, page E7.
V-8 Louisa, VA; Virginia Vermiculite Ltd., Louisa, VA.

forms the basis of most commercial applications of this mineral. The increase in volume caused by the thermal exfoliation is normally in the range of 10 to 20 times, with typical exfoliated densities of 90 to 110 kg/m^3.

The basal d-spacing of vermiculite obtained by X-ray diffraction will vary according to the composition of the ion exchangeable layer. The arrangement of hydrated magnesium ions in 14 angstrom (Å) vermiculite was detailed by Mathieson and Walker (1954). Other common basal d-spacings occur at 12 Å (i.e., barium, sodium, and strontium vermiculites), and 10 Å (i.e., potassium and ammonium vermiculites). More detailed discussions of the crystal structures of vermiculite can be found in Brindley and Brown (1980) and de la Calle and Suquet (1988).

The hydrated interlayer cations are easily subject to ion exchange. Typical ion exchange values for vermiculite concentrates are 50 to 150 milliequivalents/100 gram. If properly exfoliated, thermally exfoliated vermiculite will also have high ion exchange capacities which are diminished only 5 to 10% from the untreated material. The kinetics of ion exchange are extremely slow with potassium vermiculite, owing to the lack of interlayer separation.

Two common misconceptions are found in much of the literature dealing with the crystal structure of vermiculite as revealed by powder X-ray diffraction. Gruner (1934) used the term hydrobiotite to indicate regularly alternating layers of biotite and vermiculite. This mixed structure would produce a 12 Å basal reflection. A 12 Å reflection is common in commercial vermiculites, which can be shifted or eliminated by ion-exchange without producing an average reflection. This indicates that this 12 Å phase likely represents a discrete hydrated layer in some cases and not the alternating hydrobiotite structure.

The other common assumption in X-ray diffraction studies is the assignment of the 10 Å reflection to the presence of biotite. Potassium vermiculite is a common component in commercial vermiculites, and is indistinguishable from biotite using routine X-ray diffraction techniques. Not all researchers are aware that potassium vermiculite does contain interlayer water. This water is fills structural vacancies caused by leaching of excess potassium during vermiculitization of biotite. Much commercially desireable vermiculites will contain the 10 Å reflection their X-ray diffraction spectra, one may conclude that these reflections often represent the hydrated potassium vermiculite. This misinterpretation of the X-ray diffraction data may be a source of confusion in studies on the formation and ion exchange charactertistics of vermiculite.

Origin

There have been a number of studies which indicate that vermiculite can be formed from a number of silicate minerals. However, since the vermiculite which is mined and processed for industrial applications must produce salable concentrates of relatively large particles, the parent mineral is invariably an iron-bearing phlogopite or biotite. The well known exception to this is the unique deposit at Llano, TX, where iron-free phlogopite occurs in magnesite, and the vermiculite which is formed upon weathering is a pure magnesium vermiculite variety widely used in crystallographic and geochemical studies.

The alteration of biotite into vermiculite takes place at near-surface conditions so that all known deposits are supergene in origin. Studies by Roy and Romo (1957) indicate that vermiculite will not form as a primary mineral and that it is not stable under hydrothermal conditions. The alteration of biotite into vermiculite appears to take place in two steps which will be illustrated using a hypothetical biotite and ferric iron in solution. First, the ferrous iron is oxidized within the octahedral layer:

$$K_2(Mg_4,Fe^{+2}{}_2)[Si_6Al_2O_{20}](OH)_4$$
$$+Fe^{++} \rightarrow K_2(Mg_4Fe^{+2}Fe^{+3})\ [Si_6Al_2O_{20}]\ (OH)_4 + Fe^{+++}$$

This process, unless accompanied by the loss of interlayer potassium, results in an imbalance of charge and an increased separation between the 2:1 silicate layers. The repeat distance of the crystal structure increases from the typical 10 Å value to approximately 11 Å.

The second step in vermiculite formation begins when the silicate layers of the mica are separated sufficiently to allow for ion

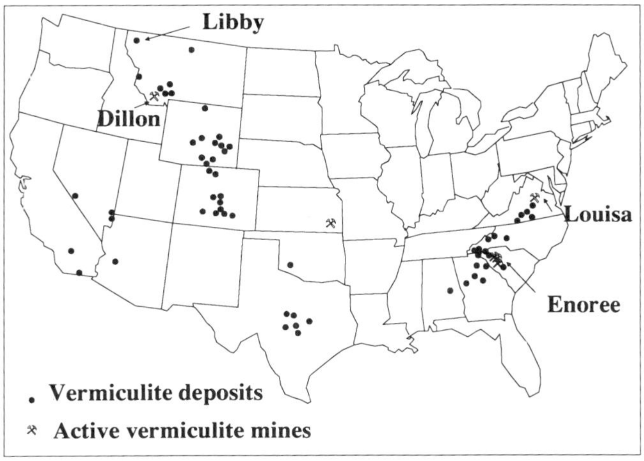

Fig. 1. United States vermiculite mines and deposits.

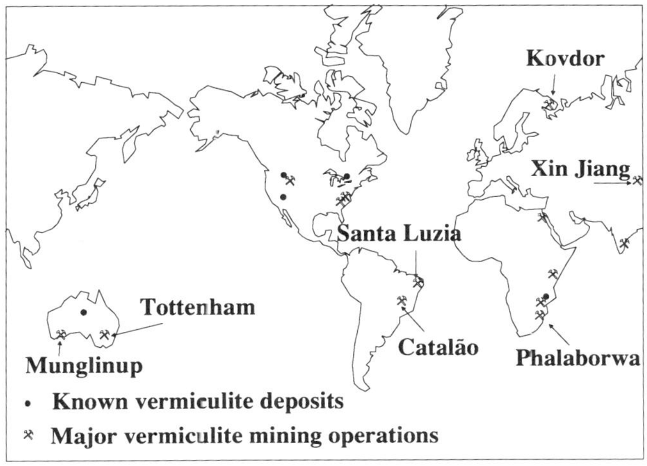

Fig. 2. Principal producing vermiculite deposits and major deposits under development (1991).

exchange. The excess potassium content of the interlayer is taken into the ground water according to:

$$K_2(Mg_4,Fe^{+2},Fe^{+3})\ [Si_6Al_2O_{20}]\ (OH)_4 \rightarrow K(Mg_4,Fe^{+2},Fe^{+3})[Si_6Al_2O_{20}](OH)_4 + K^+$$

and the content of the remaining exchangeable-ion interlayer is altered to reflect the ionic content of the ground water within the deposit by:

$$K(Mg_4,Fe^{+2},Fe^{+3})[Si_6Al_2O_{20}](OH)_4 + 0.25\ Ca^{++} + 0.25\ Mg^{++} + 3\ H_2O \rightarrow (Ca^{++}{}_{0.25},Mg^{++}{}_{0.25},\cdot 3\ H_2O)\ K(Mg_4,Fe^{+2}Fe^{+3})[Si_6Al_2O_{20}]\ (OH)_4$$

The transformation of biotite into vermiculite requires an increase in particle volume of 10 to 40%. Although not immediately obvious, this volume increase requirement places restrictions on the size and placement of a vermiculite deposit. In order to form vermiculite, the alteration products of biotite must literally lift all of the overlying rock body. Unless the lithostatic pressure can be exceeded by the forces caused by the oxidation of iron within biotite and the subsequent ion exchange, no vermiculite formation can take place. This pressure has not been measured directly, although some related data is available (i.e., Norrish and Rausell-Colom, 1963, Viani et al., 1985). Empirical geological evidence suggests that the pressure will coincide with the 250 to 300 m maximum ore depth found in many deposits.

Distribution of Major Deposits

Vermiculite is currently being mined in at least nine countries. Those current sources, as well as known potential sources of commercial vermiculite are shown in Figs. 1 and 2.

The first production of commercial vermiculite began in the United States in 1915 with the unsuccessful marketing of Tung Ash, the name given to vermiculite mined near Hecla, CO. Actually, vermiculite was inadvertently mined in the 1800s as a major accessory mineral in the Jenks Mine, NC, corundum deposit (Cooke, 1874).

The vermiculite mine started by the Zonolite Company at Libby, MT, in 1921 was the first successful venture in the vermiculite industry. Until recently, the Libby deposit was the oldest producing vermiculite mine and had been developed and enlarged for almost 70 years. This property was closed in 1990 and that closure is expected to have a major impact on the vermiculite industry.

Mining of vermiculite has taken place in other locations in the United States. Excellent summaries of statewide vermiculite deposits and mining have been prepared for Colorado (Bush, 1951); North Carolina (Murdock and Hunter, 1946), Texas (Claybaugh and Barnes, 1959) and Wyoming (Hagner, 1944). Other discussions of vermiculite deposits in the United States include Bush and Sweeney (1968) on the Appalachian Region, and Bush (1976) for the entire country. With regard to individual deposits doctoral dissertations have dealt with Libby (Boettcher, 1966) and Enoree (Libby, 1975).

Current production in the United States is restricted to mines in South Carolina, Virginia, and Montana. Several mining operations are located in northwestern South Carolina. The two major producers, W.R. Grace & Company at Enoree, and Carolina Vermiculite at Woodruff, account for most of the production in that state. A smaller operation, Patterson Vermiculite, is in the same area and serves a smaller regional market.

The other major domestic vermiculite mine is near Louisa, VA, and is operated by Virginia Vermiculite. Virginia Vermiculite purchased Carolina Vermiculite in 1990 and the combined operation is approaching the production levels of Grace as the second major producer of vermiculite in the United States.

Vermiculite is also mined in Kansas, although the product is not comparable to other commercial vermiculite concentrates. Micro-Lite, Inc., operates a mine in Woodson County from which an altered lamproite with an appreciable vermiculite content provides a material useful in animal feed supplements.

The closure of the Libby operation left the western United States with no major domestic source of vermiculite. Several known deposits have been considered for development or reopening as mining operations. One small deposit near Dillon, MT, produced small tonnages in 1990 and increased production started in 1991 with the construction of a small mill (Hindman, 1990). Deposits near Hamilton, MT, Encampment, WY, and Clark County, NV, are also possible new or renewed sources of vermiculite.

With the closure of the Libby mine and mill, the Republic of South Africa appears to now be the leader in worldwide vermiculite production, with 1990 and 1991 production of 217 738 and 197 768 tons, respectively. The single mining operation at Phalaborwa is operated by the Palabora Mining Company. The milling operation was redesigned and new areas put into production in 1991 and 1992. The Phalaborwa deposit is also known as Loolekop, and been described in a number of papers such as Gevers (1949).

Other current vermiculite producing countries include Australia, Brazil, China, Egypt, India, Japan, Kenya, Russia, and Zimbabwe. Recent data (Potter, 1991) suggests that Russia may have accounted for 15% of the known world production in 1989, as compared with estimates of 40% for the United States and 36% for South Africa. According to US Bureau of Mines (USBM) estimates, Brazil is the fourth largest producer of vermiculite. However USBM data does not include production from Australia and China, and production data for Russia is known for only a few scattered years.

Although vermiculite production is known to be small, new deposits are currently being developed. China is known to have large vermiculite deposits which are now being exploited.

TECHNOLOGY

Exploration

The exploration for vermiculite deposits is straightfoward, since they are the products of surface and near surface weathering. Since one criteria of a commercial vermiculite is large grain size, exploration is directed towards geologic structures which might contain large amounts of large biotite crystals. The most common structures are ultramafic intrusives, such as coarse-grained pyroxenites and metamorphic bodies containing biotite schists and gneisses.

Good vermiculite ore is friable and easily broken with a standard prospectors hammer. A small propane torch is useful for field identification purposes. The basic criteria for vermiculite ore is that the individual particles can be seen with the unaided eye, and that they will exfoliate when subjected to high heat. An average content of from 20 to 35% vermiculite in the ore will produce mill feed which is easily processed with current technologies. Higher grade feeds may allow for a smaller mill design and operating costs, and lower grade ores can be processed if the quality of the concentrates obtained is acceptable. The amount of vermiculite in particle sizes less than 0.25 mm or 65 mesh Tyler are not generally included in the valuation of an ore deposit because there is no significant market for this material and it is intentionally rejected during beneficiation.

Bush (1976) classifies vermiculite into three classes: Type-1 deposits which are formed within large ultramafic intrusive masses, such as pyroxenite plutons, that are cut by syenite or alkalic granite, and by carbonatitic rocks and pegmatites. The deposits at Libby, MT, and Phalaborwa, Republic of South Africa, are Type-1. The rock masses in Type-1 deposits are often zoned. Type-2 deposits are formed in association with smaller ultramatic intrusive masses such as dunite and peridotite and unzoned pyroxenite. The deposits in North Carolina are of this type. The Type-3 deposits are formed from metamorphic rocks such as biotite schists and amphibole schists. The deposits at Enoree, SC, and Dillon, MT, are Type-3, as are most of the deposits in Colorado, Nevada, Texas, and Wyoming.

Other classification schemes for vermiculite have been proposed. The classification of Borovikov (1962) is comprised of: Group 1—Deposits in ultrabasic and alkaline rock complexes; Group 2—Deposits in altered carbonate rock complexes; Group 3—Vermiculite occurrences in reaction zones of pegmatites, talc, corundum, asbestos, and other deposits, and also in metasomatic veins in serpentines; and Group 4—Deposits and occurrences of vermiculite in micaceous gneisses and other metamorphic rocks. The major deposits at Kovdor and Buldym are Group 1 deposits.

A detailed discussion of vermiculite deposits on a worldwide basis has been prepared by L'vova (1974). Varley (1952) has described most of the known vermiculite deposits and provided a discussion of the technical and economic aspects of commercial vermiculite which is still applicable after forty years.

Mining and Beneficiation

The mining and milling of vermiculite is dependent on a number of considerations not found in other mineral commodities. Unlike most other ores, the value of a vermiculite ore is dependent on the distribution and efficient recovery of relatively large particles. A vermiculite ore will normally contain vermiculite, partially altered mica, unaltered mica, and associated gangue minerals.

In the very early years of the industry, the ore bodies were selectively mined (high graded) and the sized ore was the final product of beneficiation. As these areas of relatively pure vermiculite were depleted, and the demand for coarse-sized products increased, mills were designed to beneficiate the ore. Although high grade areas may still be encountered in mining, modern milling techniques requires feeds more consistent with the overall character of the ore body.

The mining of vermiculite is normally performed using open pit methods. There have been reported instances of underground mining, such as at the Quaintance Deposit in Jackson County, CO (Goldstein, Jr., 1946), but these operations were on a small scale would not be economically feasible within the current industry situation.

Modern vermiculite mines are invariably developed by excavating the deposit with conventional equipment including graders, shovels, loaders, and trucks. Drilling and blasting may be necessary in areas where dikes, sills, and other hard bodies of rock occur within the ore body. The ore is transferred to the mill using large off-highway trucks when the mill is at the minesite. Some operations, such as at Enoree, SC, and Louisa, VA, use smaller capacity trucks operating on highways and improved roads to bring ore from deposits which are some distance from the millsite. In either case, the ore is taken to some area near the mill where the larger pieces of waste rock are removed, and the ore is blended and stockpiled. Blending can be accomplished by a number of methods ranging from spreading the ore with a dozer, to using a water monitor to wash the ore from the stockpile into the mill.

Beneficiation techniques used with vermiculite take advantage of one or more properties of the mineral. These properties include shape, density, chemistry, and resistance to comminution. Some of these properties may be applicable to both wet and dry processes, but most mills do not beneficiate with both wet and dry processes.

A stylized flow diagram for a dry beneficiation mill is shown in Fig. 3. Typically, ore processed in a dry mill will require some drying before sizing. Air classification is the normal beneficiation process in a dry mill and separation is effected by winnowing and other techniques such as air tabling. Electrostatic and high intensity magnetic techniques may be effective in some instances, as are morphology separations using roll and slot screening.

The selectivity of air classification is assisted by separating the feed into narrow size ranges, which correspond to the finished concentrates. Comminution is primarily directed at reducing the size of gangue minerals for removal during recycling through the air separators. In many ores, major gangue minerals are friable or possess distinct cleavage. In these cases the vermiculite particles may pass through impact or roll crushers relatively unaffected, while the bulk of the gangue particles are reduced to sizes which are easily removed by air separators or screening during the recycle process.

Wet beneficiation of vermiculite generally falls into two categories: froth flotation for fine particle sizes below 1 mm, and beneficiation techniques based on morphology for larger particles. Since many vermiculite ores contain major amounts of dense

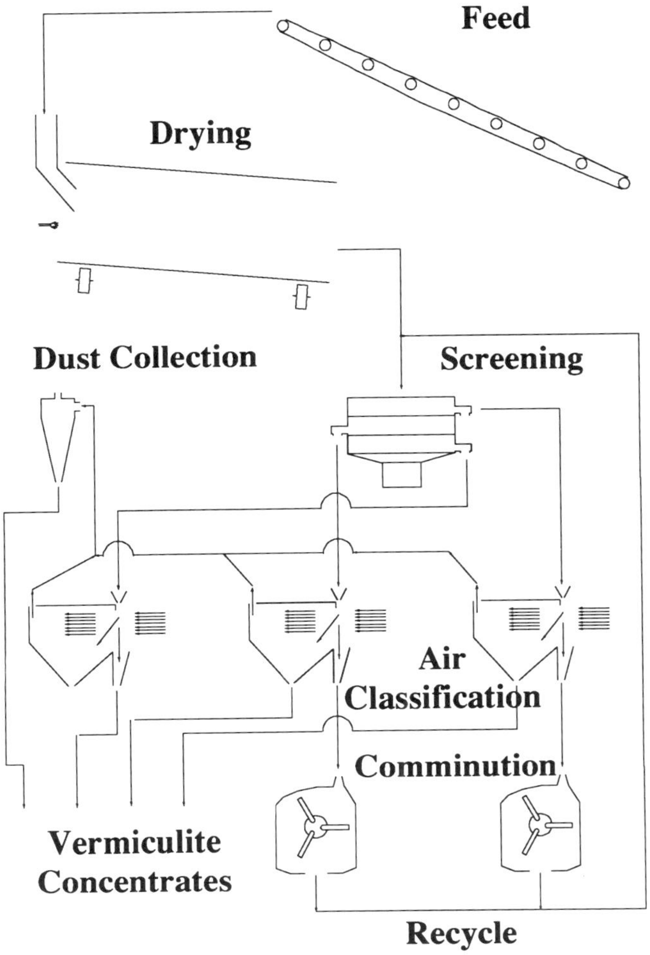

Fig. 3. Generalized flow diagram for the dry beneficiation of vermiculite.

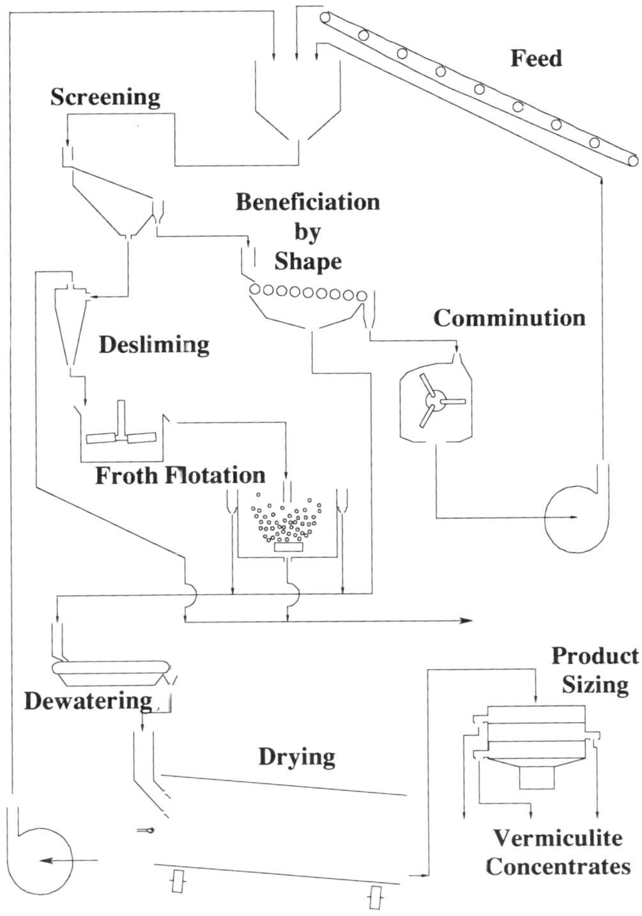

Fig. 4. Generalized wet beneficiation of vermiculite.

gangue minerals, processing steps using heavy media separation, jigs, spirals, and shaking tables can be effectively used. Simple hydroclassification may also be effective in treating fine-sized ores and flotation circuits can be improved and possibly replaced by hydrocyclones, hydraulic classifiers, and teeter chambers in some instances.

In Fig. 4, a hypothetical wet processing circuit is shown. Processing of fine-sized material is straightforward, with most common flotation cells and tables being effective. Processing of larger particle sizes is more difficult and most processes take advantage of the tendency of the vermiculite plates to adhere to, or travel along a smooth surface. Moving water films on plates and other flat surfaces can sometimes be used with excellent selectivity. Vaplon, Erickson, and Young (1978) claim wet roll screening to be effective in the processing of large particles of vermiculite. This is the separation device shown in Fig. 4 as an example of beneficiation by shape.

Vermiculite concentrates are marketed with specified particle size distributions. The design of the beneficiation circuitry in dry processing plants may produce the finished product sizes without further treatment. In wet processing plants, the mill product is produced by drying the combined mill concentrates. In this case the dried concentrate is processed into the sized products in an additional screening facility.

Processing

Most vermiculite is consumed in thermally exfoliated form. Typically, unexfoliated vermiculite concentrate, weighing 640 to 1 120 kg/m^3, is shipped in 100 ton bulk railcars from the mills to exfoliating plants located throughout the United States and Canada. In the United States, most exfoliation is performed using vertical furnaces where the sized vermiculite concentrate is allowed to fall through a hot zone produced by gas or oil burners. Exfoliation begins during the brief residence time in this zone, at temperatures of 900°C or hotter, and continues during accumulation in a hopper at the base of the furnace. The density and appearance of the exfoliated vermiculite is determined by the temperature and residence time in the hot zone, as well as the oxidative or reductive nature of the burner flame.

Other furnace types have been used in the exfoliation of vermiculite and the Torbed design seems to be effective with some concentrates that require higher temperatures and/or longer residence times. The Torbed furnace treats the vermiculite as a rotating fluid bed. Rotary furnaces, such as those used to expand perlite, generally do not produce acceptable results with vermiculite.

After the vermiculite has been exfoliated, the nonexfoliating gangue material is removed using an air separator. Purified product is then packaged in 0.1 m^3 paper bags for shipment. In larger plants the exfoliated vermiculite may be ground into powder or mixed with other materials to produce horticultural blends and industrial pre-

Table 2. United States Vermiculite Exfoliation Plants

Company	City	State
A-Tops Corporation (Austin Coating)	84 PA	PA
A-Tops Corporation (Austin Coating)	Beaver Falls	PA
Anitox Corporation	Buford	GA
Fafard, Inc.	Anderson	SC
Grace-Sierra Horticultural Products Company	Hope	AR
Grace-Sierra Horticultural Products Company	Travelers Rest	SC
A.H. Hoffman Plant Food	Landisville	PA
J.J. Brouk Company & Company	St Louis	MO
James River Limestone Company	Buchanan	VA
Koos, Inc.	Kenosha	WI
Mandoval Vermiculite Products Inc.	Houston	TX
O.M. Scott & Sons Company	Marysville	OH
P.V.P. Industries, Inc.	North Bloomfield	OH
Patterson Vermiculite Company	Enoree	SC
Pro-Grow Products, Inc.	Elizabeth City	NC
Pyrok, Inc.	Dallas	TX
The Schundler Company	Metuchen	NJ
Southwest Vermiculite, Inc. (SOLICO)	Albuquerque	NM
Strong-Lite Company	Pine Bluff	AR
Strong-Lite Products (Mica Pellets)	DeKalb	IL
Therm-O-Rock Industries, Inc.	Chandler	AZ
Therm-O-Rock, Inc.	New Eagle	PA
Thermic Refractories Inc.	Girard	IL
Verlite Company (2 Plants)	Tampa	FL
Vermiculite Industrial Corp.	East Palestine	OH
Vermiculite Ltd.	Clifton	NJ
Vermiculite Products, Inc.	Houston	TX
W.R. Grace & Company	Beltsville	MD
W.R. Grace & Company	Dallas	TX
W.R. Grace & Company	Easthampton	MA
W.R. Grace & Company	Enoree	SC
W.R. Grace & Company	Irondale	AL
W.R. Grace & Company	New Castle	PA
W.R. Grace & Company	Newark	CA
W.R. Grace & Company	Phoenix	AZ
W.R. Grace & Company	Pompano Beach	FL
W.R. Grace & Company	Portland	OR
W.R. Grace & Company	Santa Ana	CA
W.R. Grace & Company	Trenton	NJ
W.R. Grace & Company	West Chicago	IL
W.R. Grace & Company	Wilder	KY

mixes. Some exfoliation plants may also expand perlite. A list of vermiculite exfoliation plants in operation at the beginning of 1992 is given in Table 2.

Specifications and Testing

Vermiculite concentrates are invariably screened into various size ranges before shipment from the mill and further processing. In the United States screening is performed and results in finished products with a mean particle diameter either one-half or double that of coarser and finer grades. Approximate top and bottom screen sizes used for various grades are 4.75 to 2.36 mm (4 and 8 mesh Tyler) for Grade Number 1; 2.35 to 1.18 mm (8 and 14 mesh Tyler) for Grade Number 2; 1.18 to 0.60 mm (14 and 28 mesh Tyler) for Grade Number 3, etc. Imported concentrates are similarly sized. Approximate top and bottom sizes for these concentrates are 8.0 to 4.0 mm (2 1/2 mesh to 5 mesh Tyler) for Grade Large; 4.0 to 2.0 mm (5 to 9 mesh Tyler) for Grade Medium; 2.0 to 1.0 mm (9 to 16 mesh Tyler) for Grade Fine, etc. Typical weight distributions by size for domestic concentrates and their relationships to imported concentrates are shown in Fig. 5.

Ores and Concentrates are normally analysed for both grade and particle size distributions. The dried sample is screened into various

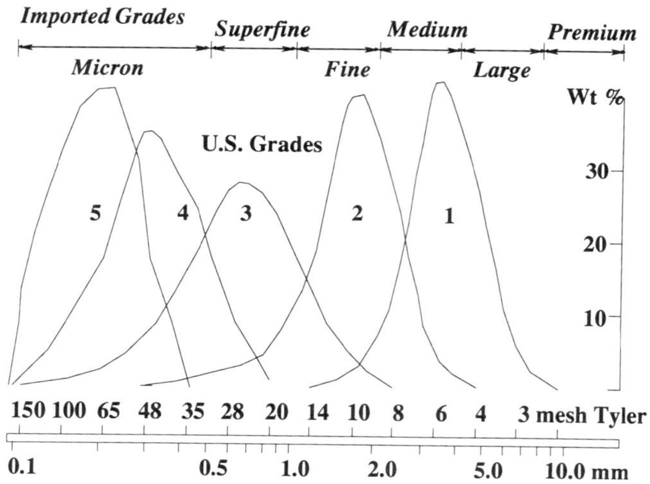

Fig. 5. Nomenclature and particle size ranges for US and metric vermiculite concentrate grades.

size fractions using sieves and a mechanical shaking device. In the United States this is typically a "Ro-Tap" and a stack of US or Tyler screens. Each screen fraction is weighed and then exfoliated in a furnace at 800 to 950°C. After cooling, the samples are again weighed and the exfoliated vermiculite is separated from the gangue material by water flotation. The remaining gangue is dried and weighed. The vermiculite concentrate is then calculated by simple difference or, if the vermiculite has been well characterized, can be calculated using a formula which accounts for excess moisture or overdrying of the sample.

Because of the cost of mining and beneficiation, as well as the cost of transportation of the finished concentrate, it is desirable to mine and produce high grade ores and products. Although lower grades can be tolerated in some cases, ore feed grades are normally in the range of 20 to 35% vermiculite. Very high grade ores are sometimes encountered which contain 80% vermiculite.

An average grade of 90% is common for finished concentrates, with some concentrates occasionally approaching 99% vermiculite. The concentrates also have specifications for the relative performance of the material during thermal exfoliation. This can be measured from samples of concentrate by laboratory exfoliation in specially designed furnaces which simulate the performance of commercial exfoliating plants.

Quality control of the exfoliated vermiculite is routinely determined by either weighing the packaged product as 0.1 m^3 bags or by counting the number of bags produced from known amounts of concentrate.

USES

Vermiculite finds use in a number of diverse applications of which a representative few are discussed here. It is probably recognized by the largest number of consumers as a common component in potting soils and other horticultural products. The exfoliated product is also used as a carrier in fertilizers, herbicides, and insecticides. Turf Builder™ (O.M. Scott) and similar lawn care products use a finer-sized concentrate than is used in growth media.

Cementitious mixtures of thermally exfoliated vermiculite and binding agents, such as gypsum plaster, have been important products. Monokote™, a product marketed by W.R. Grace & Com-

Table 3. United States Vermiculite Production and Trade Data

Year	Tons[a]	Exports[b]	Imports[c]	Consumption[d]	Average FOB Price $US/ton	
					Concentrate[e]	Exfoliated[f]
1980	306 000	27 000	24 000	303 000	$ 76.86	$213.79
1981	209 000	28 000	24 000	286 000	$ 90.26	$235.75
1982	287 000	20 000	19 000	286 000	$ 99.42	$260.33
1983	256 000	17 000	22 000	261 000	$106.32	$256.88
1984	286 000	20 000	29 000	295 000	$110.23	$235.92
1985	285 000	17 830	32 000	299 170	$113.74	$204.66
1986	288 000	20 858	32 000	299 142	$119.62	$231.79
1987	275 000	21 599	29 000	286 878	$120.42	$238.84
1988	276 000	30 647	32 000	278 923	$122.92	$243.93
1989	249 000	41 961	45 893	255 875	$122.26	$250.20
1990	209 000	44 068	43 531	208 463	$137.48	$276.56
1991	200 000	45 840	52 163	206 323	$165.41	$283.00

[a]Production data restated as metric tons from USBM published values.
[b]Export data 1980–1984, USBM; 1985–1989, Roskill and quoted sources; 1990–1991 US Customs and other sources. Value for 1991 estimated from January-October totals.
[c]Import data 1980–1988, USBM; 1989–1991 US Customs and other sources. Value for 1991 estimated from January-October totals.
[d]Consumption calculated as Production + Imports—Exports.
[e]Values for 1980–1989, USBM average values; 1990–1991, average US Customs values including vermiculite exported to Canada. Includes domestic and South African sources.
[f]Values for 1980–1990 from USBM data. Value for 1991estimated from 1985–1990 values.

pany, is applied to structural steel members in commercial buildings and has consumed major amounts of domestic vermiculite production.

Vermiculite is used in building boards of various types. Fine-sized, untreated vermiculite concentrates are included in the preparation of fire resistant plaster board. Thermally exfoliated vermiculite forms the basis of some light weight wall boards. Ground exfoliated vermiculite is used in various refractory board products.

The high absorbancy and chemical inertness of thermally exfoliated vermiculite has been used in industrial applications for many years. A recent product consists of long fabric tubes or socks filled with exfoliated vermiculite. These tubes are marketed as "pigs" by the New Pig Corporation and are distributed in the United States and exported worldwide. This product is used primarily for the containment of oil and similar liquids.

Ground thermally exfoliated vermiculite is produced by two domestic vermiculite companies and is also imported from the Republic of South Africa. The major use of this material is in friction materials, such as a replacement for asbestos in brake linings. Ground vermiculite is also slurried with clay minerals for refractory mold release and other coatings.

Ground chemically exfoliated vermiculite is marketed as aqueous dispersions for various coating and film forming applications. This technology was first developed at the Commonwealth Scientific and Industrial Research Organization (CSIRO) in Australia by George F. Walker. Research was later carried on by Imperial Chemical Industries (ICI) with product applications being developed for increasing the thermal stability of glass cloth, flexible films, and plastic fillers. Similar research has also been performed by Armstrong World Industries, Corning Glass Works, and Hercules Chemicals. W.R. Grace & Company is currently the major supplier of vermiculite dispersions in the United States, having acquired much of the patented technologies from ICI.

The temperature at which vermiculite begins to undergo thermal exfoliation varies according to the composition of the ion exchangeable layer. Langer and Marlor (1981) took advantage of this phenomenon to prepare intumescent gasketing material using low temperature exfoliating NH_4-vermiculite. This gasketing material is used extensively as wrap around mats to cushion catalytic converter assemblies in automotive applications.

Untreated vermiculite concentrates also find use in some intumescent gasket applications. In addition, untreated vermiculite concentrates are used to reduce slag build up on boiler tubes, primarily in coal fired electric utilities (Engstrom and Bain, 1983). Vermiculite used in this application is marketed as Firemate® by the Dearborn Chemical Company.

Although the immediate market effects of new domestic and imported sources of vermiculite are not clear, the entry of new companies into the industry should produce new opportunities for growth. Vermiculite dispersions constitute only one new area. One should expect that vermiculite, in all of its forms, will find application in detoxification of water and soil, nuclear waste containment and removal, and industrial spill containment and cleanup. There are a number of other high technology uses for vermiculite which may result in new products during the next few years.

ECONOMIC FACTORS

Many of the end uses of vermiculite are associated with construction and industrial uses. The consumption of vermiculite tracks the national economy to some extent, and the amount of vermiculite produced and consumed in the United States has continued to increase during the last 70 years. It is difficult to make a detailed analysis of the vermiculite market because most vermiculite companies treat all business data as proprietary. The recent closure of the Libby operation has significantly affected the production data and requires a cautious approach to their interpretation. It is to be expected that the stockpile of concentrates at Libby will continue to be the major source of vermiculite in the western United States through the second quarter of 1992.

The recent vermiculite trade data for the United States given in Table 3 is gathered from several sources. It has been a common practice to gauge the outlook for the domestic vermiculite industry based on reported production, consumption, and pricing. Although

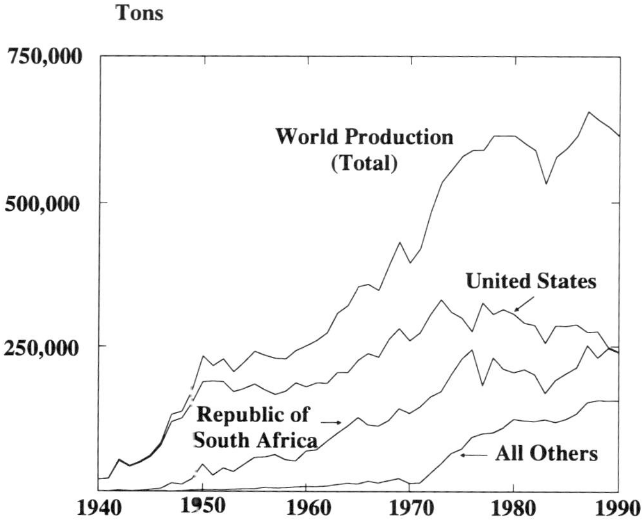

Fig. 6. Worldwide vermiculite production, 1940–1990.

this may still be a valid approach, the variables that affect the data should be considered. First, data reported by the USBM is obtained primarily through voluntary response to an annual survey. Response is not legally required, nor is accuracy enforceable. Second, the value-added aspect of vermiculite products may go well beyond the listed valuations of thermally exfoliated vermiculite. Third, other factors, such as asbestos contamination, may adversely affect sales of vermiculite from certain sources in spite of continuing demand for vermiculite concentrates.

The worldwide production of vermiculite over the last half century is shown in Fig. 6. Current data suggests that annual vermiculite production approximately 700 kt. However, better access to data from China and Russia may raise this value. The trend in domestic production is clearly in conflict with the general increase in worldwide production. An excellent review of the vermiculite industry is available from Roskill (Anon., 1991), and a thorough analysis of the western US vermiculite industry has been prepared by Marshall (1991).

The market price of vermiculite concentrates varies with particle size, with the larger-sized products commanding higher prices relative to the more abundant finer sizes. The Libby closure has eliminated the only major source of coarse-sized concentrates in the United States, and by the end of 1991 essentially all +2 mm vermiculite was obtained from foreign sources. Typical market prices for vermiculite concentrates in 1990–1991 were Number 1 (Coarse), 200–$225/t; Number 2 (Medium), $160/t; Number 3 (Fine), $125/t; and Number 4 (Superfine), 70–$80/t.

Vermiculite became an important industrial mineral because of uses developed for its thermally exfoliated products. Although there are significant current and developing markets for nonexfoliated and ion-exchanged vermiculite concentrates, at least 90% of commercial vermiculite is marketed after thermal exfoliation. It is not surprising, therefore, that the basic unit of vermiculite trade after leaving the mine and mill is bags, the major description of quality is bag yield, and the profitability of a processing plant is keyed to the cost per bag.

The cost per bag of exfoliated vermiculite is a combination of the price of vermiculite concentrate at the mine, plus the total cost of transportation to the exfoliation plant, plus the cost of exfoliation, packaging, and distribution. The value is calculated based on the number of 0.1 m^3 bags of exfoliated vermiculite obtained from total amount of concentrate processed, including the gangue minerals (also known as grit or rock). It is not uncommon for the cost of transportation to approach 50% of the cost per bag.

The bag yield (i.e., quality) of vermiculite concentrates is dependent on several variables. These include the grade of the concentrate and the size distribution of the vermiculite particles. Although the actual bag yield is dependent on the type of furnace used and the operating conditions, some measure of the thermal exfoliability of a vermiculite concentrate can be obtained by various laboratory measurements. The average value for all exfoliated vermiculite produced in the United States is between 225 and $250 per .9 ton according to recent USBM estimates. These values indicate the actual cost per 0.1 m^3 bag of exfoliated vermiculite to be in the approximate range of 2.50 to $3.75, or 22 to $33/$m^3$.

PROBLEMS AND FUTURE TRENDS

Vermiculite, as with other industrial minerals, is adversely affected by any contained asbestos. Although not given as a reason for its closure by Grace, the Libby ore deposit contains asbestiform tremolite, which has been shown to have deleterious effects. No other current source of vermiculite in the United States, or major source of imported vermiculite appears to have a problem with contained asbestos. Quartz and other varieties of SiO_2 are found only in minor amounts in vermiculite deposits, thus, free silica is not a problem in vermiculite. Vermiculite, by itself, is not considered to pose a health problem.

Vermiculite may continue to face substitution by other materials in specific applications, especially if traditional vermiculite consumers are unwilling to accept increased concentrate prices or unable to secure appropriate concentrate sizes. Perlite, with a basic price several times lower than vermiculite, will continue to be substituted for vermiculite, although this trend may be tempered by the concern of possible free-silica exposure.

The use of vermiculite in cementitious compounds, such as Monokote™, may continue to decline even with increased activity in building construction. This is because new formulations are being marketed which replace vermiculite with polystyrene. Since the major vendor of these cementitious compounds also manufactures polystyrene, and since the basic patents dealing with vermiculite-based material MK-5™ have recently expired, a decision to preferentially market the polystyrene MK-6™ product may reflect corporate strategies and not market demand.

The change of formulations away from vermiculite compounds may be offset by new vermiculite sources and marketing efforts by new companies in the United States. For many years the American Vermiculite Company (a subsidiary of the Palabora Mining Company) has been the sole source of South African concentrates in the United States, while Mandoval (another subsidiary of Palabora) has distributed the material in Europe. Mandoval has recently opened a vermiculite exfoliation facility in Texas to supply cementitious compounds (Mandokote™ and related products to the American market. New products dealing with waste treatment and disposal may create significant new markets for vermiculite. Major tests at an electric power plant in Ohio began in 1992 to evaluate thermally exfoliated vermiculite coated with magnesium compounds to treat flue gases. Initial test work indicates significant reductions of sulfur and nitrogen gas emissions (Nelson, 1988) and success of this technology on a large scale could foreshadow a significant increase in demand for the mineral.

Vermiculite in both untreated and thermally exfoliated forms has catalytic properties. It is expected that continued research will find uses for vermiculite of both forms, and vermiculite modified by organic and inorganic chemical treatment, nuclear storage, water purification, and toxic waste treatment.

SOURCES OF CURRENT INFORMATION

Yearly updates on the vermiculite industry are published in *Mining Engineering* (INDUSTRIAL MINERALS Issue, June) and by the US Bureau of Mines (Mineral Industry Surveys). Other sources of current information include the following:

British Vermiculite Association, Nottingham, United Kingdom.

European Vermiculite Association, Lightwater, United Kingdom.

Roskill Information Services Ltd., London United Kingdom.

United States Bureau of Mines, Vermiculite Specialist, Washington, DC.

The Vermiculite Association, Chicago, IL (North American Office) and Rabotstratt Ghent, Belgium (European Office).

Vermiculite Technology Newsletter, Salt Lake City, UT.

BIBLIOGRAPHY

Anon., 1991, *The Economics of Vermiculite*, 6th ed., Roskill Information Services Ltd., London, 152 pp.

Boettcher, A.L., 1966, ''Vermiculite, Hydrobiotite, and Biotite in the Rainy Creek Igneous Complex Near Libby, Montana,'' *Clay Minerals*, Vol. 6, pp. 283–296.

Borovikov, P.P., 1962, ''Genetic Types, Conditions of Formation and Economic Evaluation of Vermiculite Deposits,'' *Perlite and Vermiculite (Geology, Exploration and Production Technology)*, V.Kh. Daragan, ed., translated from Russian and Published by the Indian National Scientific Documentation Centre, New Delhi, pp. 139–176.

Brindley, G.W., and Brown, G., eds., 1980, *Crystal Structures of Clay Minerals and Their X-Ray Identification*, Mineralogical Society Monograph No. 5, Mineralogical Society, London, 495 pp.

Brush, G.J., 1866, ''Jefferisite, a New Mineral Species,'' American Journal of Science, 2nd Series, Vol. 41, pp. 248.

Bush, A.L., 1951, ''Sources of Lightweight Aggregates in Colorado,'' *Proceedings*, Colorado Scientific Society, Vol. 15, pp. 301–368.

Bush, A.L., 1976, ''Vermiculite in the United States,'' *11th Industrial Minerals Forum*, Montana Bureau Mines Special Publication 74, pp. 146–55.

Bush, A.L., and Sweeney, J.W., 1968, ''Mineral Resources of the Appalachian Region,'' *Mineral Resources of the Appalachian Region*, Professional Paper 580, US Geological Survey, pp. 210–224.

Claybaugh, S.E., and Barnes, V.E., 1959, ''Vermiculite in Central Texas,'' Report of Investigations No. 40, Bureau of Economic Geology, Austin, TX, 32 pp.

Cooke, J.P., 1874, ''The Vermiculites; Their Crystallographic and Chemical Relation to the Micas; Together with a Discussion of the Cause of the Variation of the Optical Angle in These Minerals,'' *Philosophical Magazine*, Vol. 47, pp. 241–272.

Curtis, C.D., Brown, P.E., and Somogyi, V.A., 1969, ''A Naturally Occurring Sodium Vermiculite from Unst, Shetland,'' *Clay Minerals*, Vol. 8, pp. 15–19.

de la Calle, C., and Suquet, H., 1988, ''Vermiculite,'' *Hydrous Phyllosilicates*, Reviews in Mineralogy, Vol. 19, pp. 455–496.

El Shazly, E.M., et al., 1975, ''Geology, ''Petrogenesis and Mode of Formation of Asbestos-Vermiculite Deposits at Hafafit, Egypt,'' *Egyptian Journal of Geology*, Vol. 19, pp. 87–104.

Engstrom, G.G., and Bain, D.I., 1983, ''Vermiculite as a Deposit Modifier in Coal Fired Boilers,'' US Patent 4,369,719, Jan. 25.

Gevers, T.W., 1949, ''Vermiculite at Loolekop, Palabora, North East Transvaal,'' *Transactions of the Geological Society of South Africa*, Vol. 51, pp. 133–178.

Goldstein, A.G., Jr., 1946, ''The Vermiculites and Their Utilization,'' Quarterly, Colorado School of Mines, Golden, Vol. 41, No. 4, 64 pp.

Gruner, J.W., 1934, ''The structures of vermiculite and their collapse by dehydration,'' *American Mineralogist*, Vol. 19, pp. 557–575.

Hagner, A.F., 1944, ''Wyoming Vermiculite Deposits,'' Bulletin 34, Geological Survey of Wyoming, 47 pp.

Hindman, J.R., 1990, ''Vermiculite Mining at Elk Gulch [Montana],'' *Vermiculite Technology Newsletter*, Vol. 1, pp. 18–22.

Justo, A., et al., 1989, ''Expansibility of Some Vermiculites,'' *Applied Clay Science*, Vol. 4, pp. 509–519.

Langer, R.L., and Marlor, A.J., 1981, ''Intumescent Sheet Material,'' US Patent 4,305,992, Dec. 15.

Libby, S.C., 1975, ''The Origin of Potassic Ultramafic Rocks in the Enoree Vermiculite District, South Carolina,'' PhD thesis, Pennsylvania State University, University Park, 136 pp.

L'vova, I.A., 1974, *Vermiculite Deposits of the U.S.S.R. (Formational Types and Regularities of Distribution)*, Leningrad, Vaesoiuznyi Nauchno-Issledovatel'skii Geologicheskii Institut (VSEGEI), tom 216, 231 pp., translated for the US Geological Survey by D.B. Vitaliano, Apr., 1977.

Marshall, R.C., 1991, *The U.S. Vermiculite Industry*, USDA-Forest Service, Northern Region, Missoula, MT, 18 pp.

Mathieson, A. McL., and Walker, G.F., 1954, ''Crystal Structure of Magnesium-Vermiculite,'' *American Mineralogist*, Vol. 39, pp. 231–255.

Murdock, T.G., and Hunter, C.E., 1946, ''The Vermiculite Deposits of North Carolina,'' Bulletin 50, North Carolina Department of Conservation and Development, 44 pp.

Nelson, S.G., 1988, ''Toxic Gas Absorbent and Processes for Making Same,'' US Patent 4,721,582, Jan. 26.

Norrish, K., and Rausell-Colom, J.A., 1963, ''Low-Angle X-Ray Diffraction Studies on the Swelling of Montmorillonite and Vermiculite,'' *Clays and Clay Minerals*, Vol. 10, pp. 123–49.

Potter, M.J., 1991, ''Vermiculite in 1990,'' USBM Mineral Industry Survey, 7 pp.

Roy, R., and Romo, L.A., 1957, ''Weathering Studies. 1. New Data on Vermiculite,'' *Journal of Geology*, Vol. 65, pp. 603–610.

Varley, E.R., 1952, *Vermiculite*, Mineral Resources Division, Colonial Geological Surveys, Imperial Institute Building, London, 70 pp.

Vaplon, G.G., Erickson, E., and Young, J.L., 1978, ''Concentration of plate-shaped minerals,'' US Patent 4,102,502, Jul 25.

Viani, B.E., Roth, C.B., and Low, P.F., 1985, ''Direct Measurement of the Relation Between Swelling Pressure and Interlayer Distance in Li-Vermiculite,'' *Clays and Clay Miner*, Vol. 33, pp. 244–250.

Webb, T.H., 1824, ''New Localities of Tourmalines and Talc,'' *American Journal of Science*, 1st Series, Vol. 7, pp. 55.

Ying, W., and Hu, Z., 1988, ''Qieganbulake Vermiculite and Its Applications,'' *Feijinshukuang*, Vol. 3, pp. 5–8.

WELL DRILLING MATERIALS

AUGUST V. CASTELLI

Drilling fluids have changed over the years from water and/or water frequently mixed with local clays to scientifically compounded materials (mud). This change has occurred due to changes in drilling methods from percussion to rotary: depth maximum from 21 m in the first oil well to more than 9 000 m; and variety in types of well or hole to gas, oil, water, and mineral exploration. Environmental concerns about the composition of the fluid as to heavy metals and oil on the cuttings has recently caused changes in the materials used in the mud.

The drilling fluid should perform the following functions (Darley et al., 1980):

1) Carry cuttings from beneath the bit, transport them up the annulus, and permit their separation at the surface.
2) Cool and clean the bit.
3) Reduce friction between the drilling string and the sides of the hole.
4) Maintain the stability of uncased sections of the borehole.
5) Prevent the inflow of fluids—oil, gas, or water—from permeable rocks penetrated.
6) Form a thin, low-permeability filter cake which seals pores and other openings in formations penetrated by the bit.

In addition to these, the fluid should:

1) Not injure drilling personnel nor be damaging or offensive to the environment.
2) Not require unusual or expensive methods of completion of the drilled hole.
3) Not interfere with the normal productivity of the fluid-bearing formation.
4) Not corrode or cause excessive wear of drilling equipment.

A number of industrial minerals, chemicals, and other materials are used as additives in the drilling fluid to accomplish these functions. Industrial minerals make up the highest percentage of the additives in a drilling fluid.

Drilling fluids are classified according to their principal components: water, oil, and gas. If the major component is either water or oil, the component along with the suspension of solids, cuttings and additives is referred to in the trade as a mud. The largest use of some industrial minerals, such as barite and bentonite, is in drilling muds. For other materials, such as gypsum and limestone, drilling usage is only a small percentage of total demand.

RAW MATERIALS

Table 1 lists the industrial minerals used in drilling fluids, their form, specifications, and uses.

Each raw material will be discussed as to its mineralogy, physical and chemical properties, and the location of major deposits with respect to drilling fluids.

Attapulgite

Attapulgite is a crystalline hydrated magnesium silicate clay mineral, prized for its viscosity in saline waters. The particles are needle-like rather than the platelets common to most clay minerals. The specific gravity is 2.5 and the Mohs hardness is 1 to 1.5. Attapulgite is one of the materials sometimes referred to as fuller's earth.

The major producing areas in the United States are southern Georgia and northwestern Florida. It also is mined in Spain, Senegal, the Republic of South Africa, Turkey, India, and Australia.

Barite

Barite is crystalline barium sulfate. It has a high specific gravity for a non-metallic mineral, 4.5 when pure, and is used as a weighting agent. Generally, the specific gravity ranges from 4.20 to 4.35. It is relatively soft with a Mohs hardness of 3 to 3.5 and is not very abrasive. Barite is inert, non-toxic, and clean to handle. Common contaminants are quartz, iron minerals, gypsum, and calcium carbonate. If there is considerable gypsum in the barite, it will thicken the mud.

The major deposits in the United States are in north central Nevada, near Battle Mountain. The major producing countries, other than the United States, are China, India, Mexico, the former USSR, Turkey, and Morocco. The major deposits in India and Nevada, and some in China, are stratiform, commonly quite pure bodies in bedded sedimentary rocks.

Bentonite

Bentonite is a hydrous aluminum silicate primarily composed of the clay mineral smectite (montmorillonite). It is the principal material used to control viscosity of drilling muds. Bentonites are two types, depending on the ratio of exchangeable calcium and sodium ions. Sodium bentonite is the preferred type for drilling fluids as it swells more and provides better viscosity than calcium. Both have plate-like particles, specific gravities of 2.5, and Mohs hardness of 1 to 1.5. Common contaminants are gypsum, quartz, mica, and illite clays. About a third of the bentonite sold in the United States goes into drilling muds.

In the United States the major deposits of sodium bentonite (Wyoming or western type) are in South Dakota, Wyoming, and Montana. True sodium bentonite is not mined outside the United States.

The major deposits of calcium bentonites are in Alabama, Mississippi, and Texas. The major deposits outside the United States are in Argentina, Brazil, Cyprus, Greece, India, Italy, Mex-

Table 1. Industrial Minerals Used In Well Drilling (modified from Davis and Lefond, 1983)

Additives	Form	Specifications	Use
Attapulgite*	Powder −75 μm	30 min. reading on a motor-driven, direct indicating viscometer at 600 RPM with a mixture or 20.0 g material in 350 cm^3 saturated salt solution. Residue greater than 75 μm: 8% max. Moisture: 16% max.	Viscosifier in saltwater
Barite*	Powder −75 μm	Density: 4.20 min. Water soluble alkaline earth metals as calcium: 250 mg/kg max. Residue greater than 75 μm: 3% max. Particles less than 6 μm: 30% max.	Weighting agent
Bentonite* (sodium)	Powder −75 μm	30 min. reading on a motor-driven, direct indicating viscometer at 600 RPM with a mixture of 22.5 g material in 350 cm^3 distilled water. Yield point/plastic viscosity ratio: 3 max. Filtrate: 15 cm^3 max. Residue greater than 75 μm: 4% max. Moisture: 10% max.	Viscosifier in freshwater
Nontreated* bentonite	Powder −75 μm	30 min. reading on a motor-driven, direct indicating viscometer at 600 RPM with a mixture of 25.0 g material in 350 cm^3 distilled water. Yield point/plastic viscosity ratio: 1.5 max. Dispersed plastic. Viscosity: 10 cp min. Filtrate: 12.5 cm^3 max.	Viscosifier in freshwater
OCMA Grade* Bentonite (calcium)	Powder −75 μm	30 min. reading on a motor-driven, direct indicating viscometer at (calcium) 600 RPM with a mixture of 22.5 g material in 350 cm^3 distilled water. Yield Point/plastic viscosity ratio: 6 max. Filtrate: 16 cm^3 max. Residue greater than 75 μm: 2.5% max. Moisture: 13% max.	
Diatomaceous Earth	Powder	None	Lost circulation material
Galena	Powder	Density: 6.5 min. 70% −325 mesh	Weighting agent
Gilsonite	Powder	None	Stabilizer
Graphite	Flake	None	Lubricant
Gypsum	Powder	−100 mesh	Stabilizer
Hematite*	Powder −75 μm	Density: 5.05 min. water soluble alkaline earths metal as calcium: 100 mg/kg max. Residue greater than 75 μm: 1.5% max. Residue greater than 45 μm: 15% max. Particles less than 6 μm: 15% max.	Weighting agent
Leonardite	Powder	Soluble humic acid: 75% min. −140 mesh	Thinner
Limestone	Powder	None	Low density weighting agent
Mica	Flake	Muscovite: 90% min. Coarse: thickness 100 μ max. and screen size −6 mesh + 16 mesh Fine: thickness 50 μm thick max. and screen size −20 mesh + 60 mesh	Lost circulation material
Perlite	Powder	None	Lost circulation material
Quartz Sand	Sand	Rounded grains commonly 20–40 mesh size	Propping agent
Salt	Granular	None	Stabilizer
Sepiolite*	Powder −75 μm	30 min. on a motor-driven, direct indicating viscometer salt water at 600 RPM reading 20 g material in 350 cm^3 saturated salt solution. Residue greater than 75 μm: 8% max. Moisture: 16% max.	Viscosifier in saltwater
Siderite	Powder	None	Low density weighting agent
Soda Ash	Powder	None	Remove calcium salts
Vermiculite	Flake	None	Lost circulation material

*American Petroleum Institute Specifications (Anon., 1991).

ico, and the former USSR. Some calcium bentonite is treated with soda ash and/or polyacrylates and used in drilling muds.

Diatomite (Diatomaceous Earth)

Diatomite is hydrous silica formed of skeletons of diatoms. It has limited use in the prevention of drilling fluid loss in a permeable formation. Diatomite has a Mohs hardness of 4 to 4.5 and a specific gravity of 1.95 to 2.3. The major deposits in the United States are in California and Nevada. Outside the United States the largest deposits are in France, Spain, Romania, and the USSR.

Galena

Galena, natural lead sulfide, has a Mohs hardness of 2.5 and a specific gravity of 7.4 to 7.6. Because of its high specific gravity, it is sometimes used as a weighting agent. The largest deposits in the United States are in Missouri. Other large deposits are in Australia, Canada, China, and the USSR.

Gilsonite

Gilsonite is a naturally occurring solid bitumen; it is dark colored, brittle, and has a low specific gravity. It is used as a stabilizer. The largest deposits are in northeastern Utah.

Graphite

Graphite, a naturally occurring carbon, may be used in drilling fluids as a lubricant. It has a Mohs hardness of 1 to 2 and a specific gravity of 2.3. Graphite is not now mined in the United States. The major deposits are in Brazil, China, and Madagascar.

Gypsum

A dihydrate form of calcium sulfate, gypsum has a Mohs hardness of 2 and a specific gravity of 2.32. It is used as a stabilizer in drilling muds, but non-drilling uses account for most of the consumption. Major deposits are present in many states; the largest producers are Oklahoma, Iowa, Michigan, Texas, California, Nevada, and Indiana. Outside the United States, Canada, China, Japan, and France are the largest producers of gypsum.

Iron Minerals

Hematite, siderite, and magnetite have been used to give weight to drilling fluids, but only the first two are presently being used. The hematite is the specularite variety which is micaceous and less abrasive than other varieties. It is iron oxide with a Mohs hardness of 5.5 to 6 and a specific gravity of 5 to 5.26. Hematite is produced in the United States, but not for drilling fluids. Brazilian specularite is the preferred source.

Siderite, an iron carbonate, has a Mohs hardness of 3.5 to 4 and specific gravity of 3.8. The only siderite presently being used is from deposits in East Texas (Darley et al., 1980).

Magnetite is no longer being used due to a higher abrasiveness.

Leonardite (lignite)

This is weathered and oxidized lignite, high in soluble humic acids. It is used as a thinner in the drilling fluid. The major deposits being mined solely for use in drilling fluids are in North Dakota and Wyoming.

Limestone

Limestone is a rock composed mostly of calcite, calcium carbonate, which has a Mohs hardness of 3 to 3.5 and a specific gravity of 2.7. It is mined throughout the United States and the rest of the world and has many industrial applications. It is used as a weight material in workover fluids.

Mica

Muscovite is the preferred mica mineral for use in drilling fluids, where it is used to reduce lost circulation. It has a Mohs hardness of 2 to 2.5, a specific gravity of 2.76 to 3.1, is flexible, and is insoluble in acids. The major US deposits are in North and South Carolina, South Dakota, and Connecticut. Outside the United States, the largest deposits are in Brazil, Canada, France, India, and the USSR. Of the several types, scrap and/or mica produced by flotation are used in drilling fluids.

Perlite

Perlite is a hydrated volcanic glass which expands greatly when heated. Expanded material is used to stop loss of drilling fluid into the adjacent formation. The largest deposits are in northern New Mexico. Outside the United States, the major deposit is in Greece.

Quartz Sand

The sand is silicon dioxide with a Mohs hardness of 7 and a specific gravity of 2.7. Only high quality, clean, carefully sized sands are used in oil or gas wells, where they are injected into the oil or gas producing horizons to increase permeability. The largest deposits of this material are near Ottawa, IL.

Salt

Chemically, salt is sodium chloride; in drilling fluids it functions as a stabilizer. Salt is produced either from brines or rock salt (halite). Halite has a Mohs hardness of 2.5 and a specific gravity of 2.16. The major US deposits are in Louisiana, Michigan, New York, Ohio, and Utah. Outside the United States the major deposits are found in Canada, China, Germany, India, and the USSR.

Sepiolite (Meerschaum)

Sepiolite is a hydrated magnesium silicate clay mineral. It has a specific gravity of 1 to 2 and a Mohs hardness of 2 to 2.5. The material is used as a viscosifier in saltwater and in some geothermal wells. Commonly the material has lath-like particles. Some sepiolite is produced in southern Nevada, but Spain is the largest producer. Small tonnages are mined in Turkey and China.

Soda Ash

Soda ash is sodium carbonate; it is produced either from brines or trona deposits. The largest production is from trona, which has a Mohs hardness of 3 and specific gravity of 3.23. The largest trona deposits in the world are in southwestern Wyoming. Kenya and Mexico are the largest producers outside of the United States.

Vermiculite

Vermiculite is a micaceous material that expands when heated. The expanded form is used to prevent drilling fluid loss in permeable formations. The principal source in the United States is in South Carolina since major production in Montana ceased several years ago. Elsewhere, the largest deposits are in the Republic of South Africa.

Asbestos

Asbestos is no longer used as a lost-circulation material in drilling fluids because it can be hazardous in handling.

INDUSTRY STRUCTURE

The industrial minerals used in drilling fluids in major amounts are barite, sodium bentonite, hematite, attapulgite, leonardite, and mica. Except for silica sand, which is used as a propping agent, the other industrial minerals are used in only minor amounts.

The materials used in drilling fluids generally are supplied directly to the well site by drilling fluid companies. These companies maintain warehouses and stockpoints in the major drilling areas throughout the world. If requested they also provide drilling fluid technologists (mud engineers) who supervise the mixing and use of the drilling mud.

There are three major worldwide drilling fluids companies: Baroid Drilling Fluids, Inc., a Baroid Corporation company; Milpark Drilling Fluids, a Baker Hughes company; and M-I Drilling Fluids Co., a Dresser Haliburton company. Additionally, there are numerous smaller, regional companies throughout the United States and world. The drilling fluid companies either mine and/or grind the major products, or purchase them from other producers.

In the United States at the present time, nine plants along the Louisiana-Texas Gulf Coast grind barite for drilling fluids. In addition, two other grinding plants in the same region custom grind barite for drilling fluid companies or provide their own ground barite.

Eight of the eleven plants are owned by three major mud companies. The plants along the Gulf Coast supply 85% of the drilling mud barite in the United States.

In the west, two plants operate near Battle Mountain, NV, the largest barite mining area in the United States, and one operates near Salt Lake City, UT. The ground product from these areas supplies the western United States and Alaska. Additionally, there are two other plants, one in Oklahoma and another in West Texas, that grind barite for the drilling mud market.

There are other plants throughout the United States that grind barite for the drilling mud market, but their primary business is in other minerals or for the filler-grade barite market. In addition to the operating plants, several plants have been mothballed since the downturn in the market in 1982.

Sodium bentonite is the second largest tonnage industrial mineral used in drilling fluids. Five companies with twelve operating plants produce drilling-grade products: American Colloid, Bentonite Corporation (formerly part of Baroid), Black Hills Bentonite, M-I Drilling Fluids, and WYO-BEN Corporation. Eleven of the plants are in Wyoming, five are in the Big Horn Basin in northwestern Wyoming, two in the Casper area, and four in the northeastern part. The twelfth plant is in the northwestern corner of South Dakota. The plants produce ground and granular bentonite for both the domestic and foreign markets as well as crushed and dried bentonite for export to grinding plants in the other major drilling mud markets in the world.

Specularite hematite is imported from Brazil by M-I Drilling Fluids and Densitech for the drilling-fluid market. Both companies have two plants on the Gulf Coast where they screen, dry, and if necessary, grind the filter cake. These companies supply the other drilling fluid companies with the product.

Attapulgite (fuller's earth) is produced in the Georgia-Florida area by three companies that provide a ground product for the drilling mud market as well as other markets. The three are Engelhard Mineral and Chemical Company, Floridin Company, and the Milwhite Company. The products are packaged in bags bearing the names of the drilling fluid companies. Attapulgite/seipiolite is also ground and packed in Spain and Senegal.

Leonardite is produced in southwestern North Dakota and near Glenrock, WY, for the drilling fluids market. The three major suppliers are Black Hills Bentonite in Wyoming and American Colloid and Geo Resources in North Dakota. The material is supplied in powder form in multiwalled paperbags in the producer's bags or customer's bags. The producers also sell leonardite in crushed and dried lumps for further processing.

Mica is supplied to the drilling mud industry in ground form. The two major producers for the drilling mud market are KMG Minerals in North Carolina and Mineral Mining Corporation in South Carolina. Most of the material is a flotation product or from scrap. The producers package the ground or screened product in paper bags with the customer's trade name.

Canada and India supply mica to the drilling mud industry in some of the international market areas.

Other minerals used in minor amounts in drilling are obtained from suppliers either in generic bags or with the customers' name. These minerals are only a very small portion of the producers' total output.

TECHNOLOGY

The processing of attapulgite, barite, bentonite, and leonardite are basically the same. All the raw materials used to produce the finished products are crushed, dried, ground, and packed in multiwallpaper bags. Prior to this processing, the clays and leonardite are field dried either at the mine or on the stockpiles at the plant. This decreases the amount of water that must be removed in drying and makes the material easier to crush. Drying of the clays is done in rotary driers at relatively low temperatures. If too high temperatures are used, molecular and spatial water is driven off and the swelling ability is decreased. It is not uncommon in producing drilling mud bentonite from calcium bentonite to pug or extrude the raw clay prior to drying.

The same process is sometimes used with attapulgite and sepiolite. The American Petroleum Institute (API) permits the use of additives to peptize all types of bentonite except the untreated bentonite. Both sodium and calcium bentonite can be peptized with soda ash and/or long chain sodium polyacrylates to increase the viscosity up to the specifications required. The calcium base bentonites require considerably more additives than the sodium bentonites to meet specifications. When bentonites are treated it can decrease the shelf life of the product and increase the fluid loss beyond specifications. This is especially true for calcium bentonites. It is not uncommon to treat attapulgite and sepiolite with magnesium oxide and/or soda ash. The clays, barite, and leonardite can be ground in ring roller mills, but leonardite is sometimes ground in hammermills. Leonardite must also be dried at low temperature to avoid igniting the organic material.

Specularite hematite processed in the United States is a filter cake produced during the separation process in Brazil. A small amount of agglomerated material may have formed, and some tramp material may have been picked up during transportation and storage. Due to these possibilities and the 8 to 10% moisture content, the material is dried either in a rotary drier or a flash drier (on the mill), and screened prior to packing. The agglomerates are broken up in either a hammer mill or ring roller mill.

Mica for drilling-mud products is dry-ground flake mica. It is ground further and screened to the sizes required for drilling mud use.

All these products are packaged in multi-walled paper bags and/or plastic one-ton superbags. Barite and hematite are also delivered in bulk.

Specifications

Specifications for industrial minerals used in well drilling are given in Table 1. API publishes specifications for barite, bentonite (sodium), hematite, non-treated bentonite (sodium), Oil Companies Material Association (OCMA) bentonite (calcium), attapulgite, and sepiolite (Anon., 1991). These specifications are revised periodically. A few users set additional specifications for some products. For barite, the amount of caustic soluble material is usually limited to 3 000 mg/l for carbonates and 50 mg/l for sulfides. The amount an API bentonite can be extended with polyacrylates and still have an acceptable yield point is limited to 1.81 kg/t of polyacrylate and a minimum yield point of 60. The specifications for the other minerals other than these mentioned are usually set by the drilling fluid companies. Additionally, the Environmental Protection Agency (EPA) and some states have set specifications on the amount of heavy metals contained in barite, muds, and cuttings to be used and/or disposed.

Testing

The API has set standard test procedures along with their specifications of the previously listed minerals. The EPA has standard test procedures for measuring mercury and cadmium. Most drilling fluid companies have their own test methods for materials other than those prior named, and some users may also have test methods for certain additional specifications. Some states require certification by the National Sanitation Foundation (NSF) of processing facilities for products used in water-well drilling.

Specialty Products

Attapulgite, sodium bentonite, calcium bentonite, and leonardite are used to make organophilic clays and caustic and potassium thinners. These organophilic clays are used in oil base muds as viscosifiers and/or filtrate reducers. These minerals make up over 50% of the products. The products are proprietary. Other industrial minerals are used in these specialty products in very small amounts as fillers.

USES

Industrial minerals in drilling fluids have six basic uses: weighting agents, viscosifiers, thinners, lost-circulation materials, stabilizers, and lubricants.

Weighting Agents

Barite, hematite, and galena are weighting agents with high specific gravities; they help to offset pressure, lower liquid flow into the hole from the formation, and keep the hole open. Galena is very seldom used anymore due to hazards in the handling and availability of non-floated galena. Barite is the most common weight material followed by hematite for heavy muds. Siderite and limestone are used for low weight muds and for workover and completion fluids because they are acid soluble.

Viscosifiers

Sodium bentonite and peptized calcium bentonites are the most common viscosifiers for freshwater muds, whereas attapulgite and sepiolite are the most common for saltwater muds. The organophilic clays are used as viscosifiers in oil muds and special applications, but they can also be used for fresh or saltwater applications. A viscosifier carries the cuttings to the surface, builds a cake on the wall of the hole to seal small openings, and lubricates the outer drill surfaces.

Thinners

Leonardite, caustic, and potassium specialty products are used as thinners to develop better flow properties in muds.

Lost Circulation Materials

Gilsonite, mica, diatomaceous earth, and expanded perlite and vermiculite are used to plug or seal holes and cracks in the formation being drilled. Mica is the most common industrial mineral used to reduce lost circulation. The bentonites and organophilic clays also help with lost circulation problems.

Stabilizers

Gilsonite and salt stabilize a hole when drilling through shale or salt formations. Gypsum is used as a stabilizer when drilling anhydrite or gypsum. The addition of salt or gypsum prevents the drilling fluid from dissolving the salt, gypsum, or anhydrite contained in the formation being drilled, and gilsonite coats shales preventing the shale from absorbing water and expanding and then caving into the hole.

Lubricants

Graphite and bentonites are used to lubricate the bit and pipe during drilling. They also help to prevent the pipe joints from sticking.

Proppants

Sized silica sand is used as a propping agent for sand fracturing, a process that holds the fractures open and enables the oil to flow out of the formation into the well.

Plugs

Aggregate and granular-size bentonite is now used for plugging shallow exploration and seismic holes. The completed holes are filled with this material to prevent inflow of water and prevent contamination of the formations drilled.

ECONOMIC FACTORS

Costs

The largest industrial use of barite and bentonite is in well drilling. The other industrial minerals used in well drilling make up only a small portion of the total tonnages produced each year. The major cost items to produce barite and bentonite and most of the other industrial minerals are mining, grinding, and packaging. Mining must be economical; therefore, surface mining is usually practiced. Generally, it is prohibitive to concentrate barite for the drilling mud market by any means other than screening, washing, and/or jigging. Dry grinding with ring roller mills is more economical than grinding with ball mills.

Packaging is becoming more automated with the use of palletizers to reduce labor costs. Other costs that continue to increase are those related to quality control, environmental considerations, and reclamation.

Markets

Most of the well-drilling outlay in the United States and worldwide is for oil and gas. The API sets the specifications for the major additives used in the drilling industry. If a company wishes to use the API logo on its products, it must have an API-approved Q-1 program (quality control).

The largest single market area for drilling muds in the United States is the Louisiana-Texas Gulf Coast. Nine barite grinding plants on the Gulf Coast supply this market. The other major market areas for drilling muds are California and Alaska, the Rocky Mountains, and mid-continent (Oklahoma and Kansas). Warehouses are maintained by the drilling fluid companies to supply their customers in these areas.

Transportation

Most of the industrial minerals used in well drilling have unique physical properties and are found only in a few areas. Materials of restricted occurrence that are shipped long distances, such as sodium bentonite, leonardite, and attapulgite, usually are shipped by rail in carload lots.

Lump barite and specularite hematite are mostly imported. Water freight of barite from the foreign producing countries in boatload quantities is less than the rail freight from Nevada to the US Gulf Coast, the largest consuming area.

GOVERNMENTAL CONSIDERATIONS

Taxes and Depletion

Most states have severance and/or ad valorem taxes on the industrial minerals used in well drilling. The trend is for increases in these taxes. Depletion allowances apply to all the industrial minerals used in well drilling. Import taxes are collected on most of the industrial minerals used in well drilling. The rates are different and depend on the source country, whether a most favored nation or not; for a developing country, the imports are duty free.

Land Use

All mining operations in the United States must now reclaim the mined areas no matter whether on private or public lands. Usually a like area must be reclaimed for the area mined each year. Each state has slightly different regulations.

Environmental Regulations

The EPA has implemented limits on cadmium and mercury contained in barite used in Alaska: 1 ppm mercury and 2 ppm cadmium. A similar regulation is being proposed for drilling in the western Gulf of Mexico, offshore Louisiana, and Texas. Louisiana has some limits on heavy metals contained in the mud stored in mud pits.

In the United States there are regulations on rainwater runoff from mines and processing plants, acid mine drainage, and dust emissions.

Safety Regulations

Mines, concentrating plants, and grinding plants in the United States are regulated under the Mine Safety Health Act.

PROBLEMS AND FUTURE TRENDS

The trend in the United States is to import oil rather than produce it due to the economics of drilling in the United States vs. drilling elsewhere. This trend will reduce the amount of industrial minerals used in domestic well drilling. Increased governmental regulations on the environment, land use, and land withdrawals, and increases in severance and ad valorem taxes all increase the cost of production, thus reducing the competitiveness of the United States.

BIBLIOGRAPHY

Anon., 1991, *Specifications for Drilling-Fluid Material*, 13A, 14th ed., American Petroleum Institute, Washington, DC, pp. 8–30.

Ampian, S.G., 1991, "Barite," *Minerals Yearbook, 1989*, US Bureau of Mines, pp. 135–147.

Ampian, S.G., 1991a, "Clays," *Minerals Yearbook, 1989*, US Bureau of Mines, pp. 271–304.

Darley, H.C.H., Gray, G.R., and Rogers, W.F., 1980, *Composition and Properties of Well Drilling Fluids*, 4th ed., Gulf Publishing Company, Houston, TX, pp. 4–5, 535.

Davis, L.L., 1991, "Mica," *Mineral Yearbook, 1989*, US Bureau of Mines, pp. 709–717.

Davis, N., and Lefond, S.J., 1983, "Well Drilling Materials," *Industrial Minerals and Rocks*, 5th ed., Vol. 1, S.J. Lefond, ed., AIME, New York, pp. 387–390.

Lujan, M., 1991, *Mineral Commodity Summaries, 1991*, US Bureau of Mines, 196 pp.

WOLLASTONITE

RONALD R. BAUER, JOSEPH R. COPELAND, AND KEN SANTINI

Wollastonite, named after W.H. Wollaston, an English chemist and mineralogist, is a calcium metasilicate having the chemical formula $CaSiO_3$. Although it is a unique mineral, wollastonite has a short history of industrial use. Wollastonite is the only naturally occurring, nonmetallic, white acicular mineral. Its acicularity is the main reason for its ascendancy in the late 1970s and 1980s as a reinforcing filler and as a replacement for short-fiber asbestos and short-fiber milled fiberglass. Wollastonite is used in applications ranging from bowling balls to automobile fenders, from ceramic tile to automobile brakes, and from thermal insulation board to protective industrial coatings.

The United States is the largest producer of wollastonite, with China, India, and Finland as the other three major sources. The United States, Europe, and the Far East are where most of the production is consumed.

GEOLOGY

Mineralogy and Physical and Chemical Properties

Wollastonite, a member of the pyroxenoid mineral group, has the theoretical composition of 48.3% CaO and 51.7% SiO_2. Iron, manganese, magnesium, or strontium may substitute for some of the calcium. Commonly associated minerals include garnet, diopside, calcite, quartz, epidote, apatite, and sphene.

Some of the physical properties of wollastonite are listed in Table 1.

Wollastonite is largely chemically inert; however, it can be decomposed in concentrated hydrochloric acid. Some wollastonite will fluoresce under shortwave or longwave ultraviolet light, or both, with colors ranging from pink-orange to yellow-orange and, more rarely, bluish green. In addition, wollastonite may show phosphorescence.

Because of its unique cleavage properties wollastonite breaks down during crushing and grinding into lath-like or needle-shaped particles of varying acicularity. This particle morphology imparts high strength and is therefore of considerable importance in many of the markets for wollastonite. The acicularity of particles is defined by their length:width, or length:diameter, ratio (i.e., aspect ratio). Individual particles, even the smallest ones, commonly exhibit an aspect ratio of 7:1 or 8:1 and have an average diameter of 3.5 µm. Low aspect ratio products (powder wollastonite, or milled grades) with 3:1 to 5:1 aspect ratios are used as general fillers, in ceramics, and in metallurgical fluxing. High aspect ratio products with ratios of 15:1 to 20:1 are used as functional fillers for reinforcing thermoplastic and thermoset polymer compounds and as a replacement for asbestos.

The dry brightness and whiteness of wollastonite are also important in determining its suitability for certain filler and ceramic applications. When pure, the mineral is brilliantly white, but impurities may color it cream or gray. Brightness is determined by measuring the reflectance of finely ground powder against a standard that is assigned a brightness of 100. Magnesium oxide and barium sulfate are the two standards used. G.E. brightness, a term used in North America, refers to brightness measured with a General Electric reflectometer. Commercial wollastonite products usually have a G.E. brightness ranging from 85 to 95. The Hunter method is also used to measure the brightness of wollastonite.

Loss on ignition (LOI), another important property in determining applications for wollastonite, is the amount of volatile matter driven off when the mineral is heated to 1 000°C. Commercial wollastonite products have an ignition loss ranging from 0.5 to 2.0%.

Wollastonite's naturally high pH of 9.9 (10% water slurry) is of major importance to its use in the coatings industry. Coatings formulators use milled grades as a pH stabilizer in interior and exterior PVA (polyvinyl acetate) and acrylic latex paints.

There are two polymorphs of calcium silicate: wollastonite, a low-temperature form, and pseudowollastonite, a high-temperature form. Conversion of wollastonite to pseudowollastonite occurs at about 1 120°C and results in an increase in the coefficient of expansion and a color change. Pure white wollastonite may change to a cream tint or to various shades of red or brown. This color change is thought to be due to the presence of iron and/or strontium.

Mode of Origin and Occurrence

Wollastonite is formed by both metamorphic and magmatic processes that usually involve carbonate rocks and intrusive igneous rocks. Two types of deposits are known: skarn deposits (thermal metamorphic and metasomatic) and carbonatites (magmatic).

Wollastonite-rich skarn deposits are formed by the thermal metamorphism of siliceous limestone during regional deformation or as a result of local intrusive igneous activity. Wollastonite can also form by the metasomatism of limestone by siliceous hydrothermal solutions along contacts with igneous intrusions or along fractures and faults.

Wollastonite is formed by the following reaction:

$$SiO_2 + CaCO_3 = CaSiO_3 + CO_2$$

At low temperatures quartz and calcite will form a stable assemblage, as in certain types of marble. However, if the temperature rises to approximately 400 to 450°C the reaction begins and proceeds to the right until the supply of calcite or quartz is exhausted, at a pressure of about 1×10^5 Pa. The carbon dioxide produced causes pressure to increase unless this gas escapes; and as the pressure rises, the temperature required for the reaction to proceed increases, possibly to as much as 950°C.

The mineral composition of each wollastonite deposit is dependent upon geologic conditions, so the assemblage of associated

Table 1. Physical Properties of Wollastonite

Crystal system:	triclinic
Form and habit:	bladed crystal masses; acicular
Color:	white, sometimes cream, gray, or very pale green
Hardness:	4.5 to 5 on the Mohs scale
Streak:	white
Luster:	vitreous to pearly
Cleavage:	2 perfect at 84° and 1 good
Specific gravity:	2.87 to 3.09
Melting point:	1 540°C
Inversion temperature:	pseudowollastonite (high-temperature polymorph) at 1 120°C ± 10°C

gangue minerals differs from one deposit to another and even within a single deposit. Nevertheless, there are three basic assemblages generally associated with wollastonite. They are: wollastonite-garnet (grossularite/andradite)–diopside (e.g., Lewis, NY); wollastonite-calcite-graphite (e.g., Gouverneur, NY); and wollastonite-diopside-quartz (e.g., Lac Ste Jean, PQ). Each assemblage reflects the original composition of the host rock and the nature of subsequent metamorphic and intrusive igneous events.

The occurrence and origin of wollastonite deposits held by NYCO Minerals, Inc. in the Adirondack Mountains, Essex County, New York, have been studied by a number of workers. de Rudder (1962) states that the ore bodies are typified by a simple equilibrium assemblage of wollastonite-grossularite-diopside. Accessory minerals are apatite and sphene. Wollastonite is intercalated with bands and lenses of grossularite-andradite and diopside-hedenbergite. The formation of wollastonite depended upon the initial composition of the Precambrian Grenville carbonate rocks and was independent of silica metasomatism. Thermal metamorphism produced wollastonite in siliceous limestones, diopside in siliceous dolomites, and garnet in argillaceous limestones. Minor alumina metasomatism from adjacent argillaceous sedimentary rocks caused the development of thick grossularitic skarns bordering the orebodies.

The wollastonite in Essex County is associated with mafic metasedimentary rocks (amphibolites and pyroxene gneisses) having mineral assemblages characteristic of both hornblende-granulite and pyroxene-granulite subfacies of regional metamorphism. de Rudder (1962) concludes that the gneissic structure of these metasedimentary rocks, their granoblastic texture, the lack of recognizable metamorphic halos, and the high grade mineral assemblages indicate that relatively high temperatures and pressures prevailed for long periods of time. The mafic metasedimentary rocks enclose a wollastonite-grossularite-diopside assemblage that is considered typical of either the hornblende-hornfels or the pyroxene-hornfels facies of contact metamorphism. Therefore, de Rudder proposes that the conditions of metamorphism can be best termed "thermal-load metamorphism," and he defines this term in the context of thermal metamorphism due to the proximity of magma at great depths and very high, uniform pressure.

The environment of metamorphism at NYCO's deposits must be inferred from field relations, mineral assemblages, thermodynamics, and experimental data. Based on these criteria de Rudder (1962) believes that the range of temperature and pressure most consistent with known data is 750 to 950° C and 6 to 12 x 10^8 Pa. This would correspond to a depth of burial of approximately 30 ± 10 km.

O'Hara (1976) disputes deRudder's statement that the origin of NYCO's wollastonite was a contact metamorphic event triggered by the intrusion of anorthosite into granulite facies rocks at a depth of 30 km and a temperature of 750 to 950° C. In contrast, O'Hara suggests that uplift occurred after regional metamorphism and before the anorthosite intrusion. Thus, the anorthosite intruded the metasediments in a low-pressure environment (e.g., about 2 x 10^8 Pa and 650° C). The heat from the intrusion caused thermal metamorphism of metasedimentary rocks and the crystallization of wollastonite.

Wollastonite is also found in carbonatites, where it crystallized from carbonatite magma. Carbonatite deposits generally are found as near-surface intrusions associated with nepheline syenite or nephelinite in complexes of elliptical or circular outline. Wollastonite deposits in carbonatites are of lower grade than skarn deposits but are potentially economic. It is improbable that a carbonatite would be mined for wollastonite alone, but the mineral might be obtained as a byproduct.

Worldwide Distribution of Deposits

United States: Numerous wollastonite deposits are known within the United States, and the best-documented ones are shown in Fig. 1. Commercial production is currently limited to the State of New York. Two companies produce wollastonite in New York: NYCO Minerals, Inc. and R.T. Vanderbilt Company, Inc. The NYCO operation is near Lake Champlain, and the Vanderbilt operation is about 160 km to the west in the Gouveneur District.

NYCO is the largest wollastonite producer in the world, with 1990 production estimated at 70 to 100 kt. Ore is currently mined at the Lewis Deposit in Essex County. From 1943–1982, NYCO mined wollastonite from the Fox Knoll Deposit, initially by open pit and later by underground methods. That deposit is about 4 km southwest of NYCO's processing plant at Willsboro. The Lewis Deposit is approximately 23 km southwest of the plant. Wollastonite comprises as much as 60% of the Lewis ore, with 30% garnet and 10% diopside accounting for the balance. An estimated 3.2 Mt of reserves grading 49% wollastonite remained at the Lewis mine in 1989. Additional underground reserves are available nearby at NYCO's Oak Hill Deposit, where an estimated 5.7 Mt containing 45% wollastonite have been identified. A third deposit, Deerhead, is about 8 km northeast of the Lewis mine. It contains approximately 500 kt grading 25% wollastonite. All of NYCO's deposits were formed by contact metamorphism and metasomatism of Precambrian siliceous carbonates by an anorthosite intrusion.

NYCO's plant at Willsboro produces a full product line ranging from a -45 µm (-325 mesh) powder grade to surface-modified, -10 µm (-1250 mesh), high aspect ratio (i.e., 15:1 to 20:1) products. Magnetic separation, using permanent roll magnets, removes accessory garnet and diopside from the wollastonite. Jet mills are used for producing high aspect ratio products. NYCO's products are used in asbestos replacement, and in plastic, ceramic, metallurgical, and paint applications. The company is owned by Canadian Pacific Ltd.

The remaining wollastonite production in the United States is by the R.T. Vanderbilt Company, Inc. near Harrisville, NY. Vanderbilt's operations began in late 1977, with both underground and open pit methods being used. The deposit was formed by contact metamorphism and metasomatism of Precambrian siliceous carbonates by a syenite intrusion. It differs from NYCO's deposits in that garnet is absent. The ore does contain some prehnite, graphite, magnetite, and diopside. The ore body is differentiated, with large pods of wollastonite being separate from gangue zones. This allows Vanderbilt to selectively mine the ore zones and waste other areas of the deposit. Reserves are estimated at 900 kt. Wollastonite ore is crushed at the mine site and then trucked approximately 19 km to a plant near Balmat. This plant serves as the processing facility for the company's wollastonite and talc operations. Vanderbilt produces about 40 ktpy of finished wollastonite products, which consist of powder grades ranging from +75 to -45 µm (+200 to -325 mesh) in particle size. The products are utilized in paint, ceramic, and metallurgical applications.

Fig. 1. Location of selected US wollastonite deposits.

In 1986 Vanderbilt, in a joint venture with Ankersmit Maalbedrijven BV, established a new processing plant in The Netherlands. Crushed ore from New York is used as feed, and a new product having a higher aspect ratio than the company's traditional grades is produced.

Until 1987 wollastonite was produced in the western United States by Pfizer, Inc. Pfizer selectively mined material from the Little and Big Maria Mountains about 32 km northwest of Blythe in southern California. These deposits consist of high grade pods of wollastonite interbedded with diopside and thin horizons of limestone and extend for as much as 600 m in strike length. The ore was processed at Pfizer's multi-commodity plant at Victorville, CA, about 330 km northwest of the deposits. Several hundred tons of powder-grade products were sold annually to the ceramics industry.

There are two large deposits on the northwest and southwest ends of Death Valley National Monument in California. The Hunter Mountain Deposit on the northwest end was formed by contact metamorphism of Paleozoic limestones. Wollastonite is found in an altered skarn along the contact zone between the Lost Burro–Tin Mountain Limestones and the Cretaceous Hunter Mountain Quartz Monzonite. Wollastonite is associated with subordinate amounts of garnet, epidote, and amphiboles. It comprises a north-dipping horizon more than 8 km long and as much as 100 m wide. Reserves are estimated to exceed 22.5 Mt grading 65 to 85% wollastonite.

South of Hunter Mountain is the Warm Spring Canyon Deposit. This deposit is a lens of almost pure wollastonite about 250 m long and approximately 12 m thick within moderately dipping carbonate rocks.

Numerous other wollastonite prospects have been reported in the western United States. Recent activity has been focused on the Gilbert property of White Plains Resources Corp., 40 km west of Tonopah, NV. On the surface wollastonite is found over an area about 2 070 m long and 260 m wide, grading more than 50% wollastonite in some locations. Accessory minerals include calcite, quartz, and diopside.

The locations of selected deposits worldwide are depicted in Fig. 2.

China: Wollastonite has been produced from a number of provinces in China. An estimated 70 to 100 kt were produced in 1990, ranking China second behind the United States.

In 1990 China had three operating wollastonite mines. Lishu Wollastonite Mining Corp. operated the Changweizi and Dadingshan mines near Panshi in Jilin Province. The Dadingshan Deposit is comprised of several ore zones in marble that contain 60 to 90% wollastonite associated with calcite, siliceous aggregates, and minor garnet and diopside. It is mined by open pit methods. The operation produces three lump grades (10–30, 30–50, and 50–150 mm), two powder grades (10–20 and 20–150 μm), and a number of chemically modified powder grades. In addition, three micronized grades with average particle sizes of 45, 20, and 10 μm are produced. They are all reported to have brightnesses of more than 90.

A third operating mine, Xiaoshuanpu, is in Hubei Province near Daye. It produces powder-grade wollastonite. In 1989 a new deposit was discovered at Yengchong in Yunnan province. It is estimated to contain 100 Mt of reserves. Past production has come from other deposits in Jilin Province and Anhui Province.

India: In 1969 the Belkappahar wollastonite deposit was discovered near the village of Khila in the Jodhpur Division of Raj-

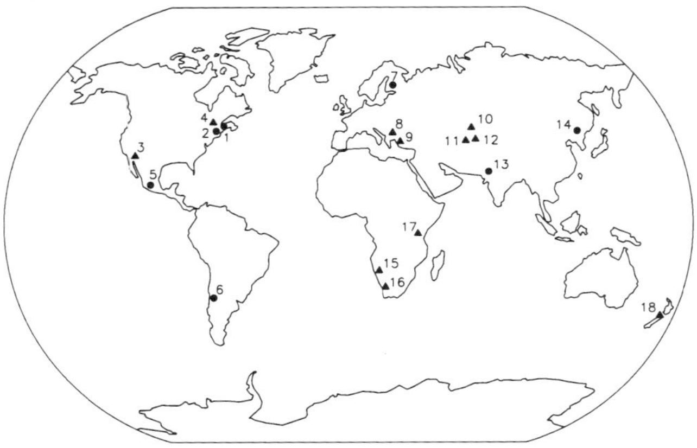

Fig. 2. Location of selected world wollastonite deposits.

Table 2. Index of Selected World Wollastonite Deposit Locations (see Fig. 2)

1	NYCO—New York State, USA
2	R.T. Vanderbilt—New York State, USA
3	Blythe, California, USA
4	Deloro, Ontario, Canada
5	General de Minerales S.A.—Zacatecas State, Mexico
6	District V, Chile
7	Oy Partek—Lappeenranta, Finland
8	Kimmeria, Greece
9	Bursa, Turkey
10	Kazakhstan, CIS.
11	Uzbekistan, CIS.
12	Tadzhikistan, CIS.
13	Wolkem Pvt. Ltd.—Rajasthan State, India
14	Jilin Province, China
15	Damaraland, Namibia
16	Modderfontein, Republic of South Africa
17	Lolkidongai, Kenya
18	Mintech NZ Ltd.—South Island, New Zealand

asthan State. After a period of exploration Wolkem Pvt. Ltd. was formed in 1972 to develop the deposit and construct a grinding plant.

The Belkappahar wollastonite comprises three zones interbedded with pyroxenites and garnet-bearing gneisses that have replaced limestone. These zones range in length from 250 to 600 m, in width from 20 to 300 m, and extend to known depths of 100 to 315 m. In 1975 proven reserves were approximately 50 Mt, with probable reserves estimated at 200 Mt.

The run-of-mine ore contains about 96% wollastonite, with calcite, garnet, diopside, and quartz making up the remaining four percent. Wolkem processes the ore into powder and high aspect ratio grades. Four powder grades, sold under the trade name Kemolit, are A-60 (250 μm), S-1 (150 μm), S-2 (75 μm) and S-3 (50 μm). The grinding mill is located near the Sirohi Road railway station. Wolkem's 95 μm high aspect ratio product is reportedly showing considerable promise in markets for asbestos replacement and plastics reinforcement. Estimated production in 1990 was 35 kt. Present plant capacity is about 60 ktpy.

Finland: In southeastern Finland, approximately 3 km south of Lappeenranta, Oy Partek controls the fourth largest wollastonite operation in the world. Production in 1990 was 33 to 40 kt. The Lappeenranta wollastonite deposit makes up the central portion of an elliptical body of Archean limestone, which strikes north-south and dips eastward at 60 to 70°. It is approximately 3 km long, 1 km wide, and completely surrounded by granite. Thin beds of quartz-feldspar rock (leptite) are interbedded with the limestone, which has been intruded by two sets of dikes. The wollastonite zone covers an area about 650 m long by 50 m wide. The thickness of the deposit varies but averages about 50 m. Within the wollastonite zone are stringers of calcite, dolomite, quartz, diopside, and leptite. In contrast to NYCO's Lewis Deposit, garnet is completely absent. The average wollastonite content is about 20%, although it reaches 60% in certain areas.

Partek's current reserves are estimated at 4 Mt. In addition, the company holds leases on a recently discovered deposit approximately 30 km from Lappeenranta at Savitapale. Although this deposit will be further explored, the initial estimate of reserves is about 20 Mt grading 25 to 30% wollastonite.

Limestone in the Lappeenranta deposit was mined for many years as a raw material for cement manufacture. It was not until 1957, however, that wollastonite was mined. Initially it was used as the main component in porous wall tile bodies by Paraisten Kalkki Oy's (Partek's former name) Turum Kaakeli ceramics plant in Turku. During that time the material was mined by open pit methods, first being blasted, then crushed, and finally hand sorted into wollastonite, limestone, and waste piles. Hand cobbing was used when necessary.

In 1967 beneficiation of wollastonite ore by froth flotation was introduced to provide growing markets with a better product. This was a two stage process. First a bulk silicate fraction containing the wollastonite was floated using an amine collector, leaving calcite behind. The second stage involved separating wollastonite from the rest of the silicate fraction using an amine collector and an anionic collector. The complexity of this system and use of costly chemical additives caused many problems. In addition, the market potential for high-purity, high-brightness calcite in paper coating led Partek to devise an alternative flotation technique in which calcite is recovered as float concentrate and wollastonite is recovered as tailings. In this new system, after wet grinding in a rod mill, hydrocycloning, and removal of the calcite, the wollastonite tailings are subjected to magnetic separation, thickening, vacuum filtration, and final grinding. A typical analysis of the finished product is 90% wollastonite, 2% quartz, 3% calcite, and 5% other silicates.

Another development took place in 1979, when an optical ore sorter was installed, to be followed by additional units later. These sorters enabled Partek to eliminate manual ore sorting and improve product quality at the flotation plant. The plant had a 20 ktpy capacity in 1979. In 1986 the company's flotation technique, in conjunction with improved high-intensity magnetic separation and the installation of new automation, resulted in wollastonite production capacity being doubled to approximately 40 ktpy. This doubling of capacity was not driven as much by growth in wollastonite markets as by the prospects for the calcite co-product in paper coating applications. Between 200 and 250 kg of wollastonite are recovered from each ton of ore processed. The calcite yield is effectively twice this amount at 500 to 600 kg/t.

Partek's powder (milled) wollastonite is produced in five main grades: FW 70 (212 μm), FW 200 (75 μm), FW 325 (45 μm), FW 425 (35 μm), and FW 635 (20 μm). However, with aspect ratios of only 3:1 to 4:1, these products are restricted to markets where reinforcement is not particularly important. In an effort to increase their aspect ratio, Partek has installed a jet mill grinding facility, which should help exploit the wollastonite's cleavage properties and produce aspect ratios of perhaps as much as 20:1. However, this could reduce the range of particle sizes that can be produced with high ratios. Micronized material is available in four sizes from 63 to 10 μm. This material has potential as a substitute for asbestos and in the reinforced plastics market. Partek also provides chemical modification according to customers' specifications.

Mexico: Wollastonite has been mined for many years in Mexico as an ornamental stone. An estimated 15 kt were produced in 1990. In the early 1960s, in response to increased world demand, Wollastonita de Mexico S.A. began large scale production in the La Blanca District of Zacatecas State. The mine was closed in 1989, however, after being sold to Arcillas Industriales de Durango S.A. de C.V. The deposit is comprised of tabular wollastonite horizons averaging 2.5 m thick within folded limestones adjacent to granitic intrusions. It lies between the villages of El Saucito and Noria del Cerro. Mining consisted of open pit and underground methods. Hand-sorted material was trucked to the railhead at Berriozabal for transport to consuming centers, mainly at tile factories in Zacatecas City and Mexico City.

General de Minerales S.A., owned by the Lamosa Group, produces wollastonite from two open pit operations at Santa Fe and Panfilo Natura, both approximately 50 to 60 km from Guadalupe in Zacatecas State. The wollastonite is found in thin, steeply dipping horizons grading as much as 50% and is associated with calcite, quartz, garnet, and diopside. Ore is processed at the company's plant at Guadalupe, where it is hand sorted, crushed, and milled to produce +75 µm, 6.3 mm, 12.5 mm, and 76.2 mm products for use in ceramics.

Minerales y Maquilas del Norte S.A. operates an open pit mine in the La Blanca District, approximately 20 km northeast of Ojacaliente in Zacatecas State. About 15 kt are mined annually to produce approximately 9 kt of product in a plant at Santa Catarina. Four grades of high and low aspect ratio wollastonite are made: 38.1 mm, 12.5 mm, 3% + 75 µm, and 1% +45 µm. Most of the product is sold to domestic and US ceramics manufacturers for use in floor and wall tiles.

The Santa Fe mine, approximately 40 km south of Pichucalco in northern Chiapas State, is owned by Compania Minera de Cerralvo and has produced some wollastonite as a byproduct at the copper operation. In the 1970s about 1 000 t of wollastonite were produced for the domestic paint industry. The deposit is an elliptical, dome-shaped mass of wollastonite skarn about 400 m long, 120 m wide, and 100 m thick.

Two large wollastonite deposits northwest of Hermosillo in the state of Sonora have attracted considerable attention from a number of companies. They are the Pilares and San Hector deposits. Most recently Cia Minera Constelacion S.A. de C.V., a subsidiary of Cominco Resources International Ltd., has conducted extensive work in the area. The deposits are situated in a north-dipping sequence of Paleozoic calcitic marbles and quartzites in close proximity to granitic rocks of Cretaceous age. Wollastonite grades exceeding 70% in beds tens of meters thick can be traced thousands of meters along strike, interbedded with calcitic marble horizons containing 40% or less wollastonite. Quartz, calcite, diopside, and garnet are the commonly associated minerals. Reserves grading more than 50% wollastonite and amounting to more than 30 Mt are represented by the two deposits combined. Beneficiation based on hand sorting has produced material averaging 90% wollastonite.

Chile: Wollastonite is mined from a deposit about 60 km northwest of Santiago. In 1990 production was about 500 t for domestic filler, paint, and ceramics markets.

Kenya: Until a few years ago Kenya was a supplier to the European wollastonite market from Lolkidongai in the center of Masailand south of Nairobi. The deposits, first discovered in the mid-1960s, were selectively mined and hand cobbed, yielding a wollastonite-calcite material that was further beneficiated by froth flotation at Nairobi. The final product consisted of about 87% wollastonite with the remainder being predominantly quartz. Kenyan wollastonite was available commercially in a 250 µm grade.

Namibia: A wollastonite milling operation at Gravellote in the Republic of South Africa was used to process ore obtained from an extensive belt of wollastonite at Damaraland in Namibia. Martin and Robson of Johannesburg operated a mine at Usakos in 1984 and was producing approximately 4 800 tpy of very pure, white material. The ore was transported to the mill at Gravellote, where it was reduced to 75 µm and bagged. It is understood that this operation is idle.

New Zealand: Mintech NZ Ltd., part of the Newmans Group, was producing approximately 500 tpy from an open pit at Takaka Hill in northwest Nelson in 1988. The ore was hand sorted, crushed, and ground at a processing plant at Mapua, where a 50 µm product (trade name WOL 18) was produced. About 60% of production was consumed by the domestic ceramic tile industry, with the remainder being sold for asbestos replacement and plastics reinforcement.

Japan: In the past two producers have supplied 10 to 15 ktpy of wollastonite to the domestic ceramics and welding rod industries.

Turkey: Turkey has produced some wollastonite for the domestic ceramics industry. Production is believed to have ceased, but a small operation was being planned in Bursa, south of Istanbul, during 1990.

Commonwealth of Independent States (CIS): Many deposits of wollastonite are known in the CIS, some of which are potentially economic while others are of mineralogical interest only. They are distributed over widely separated areas.

Central Asia More than 50 deposits of wollastonite are known, the most important of which are described here:

UZBEKISTAN—The Koytash Deposit is in the Nuratau Mountains about 72 km north of Samarkand. Wollastonite is found in a calc-silicate hornfels at the contact between argillaceous limestones and a granodiorite intrusion. The hornfels crops out for nearly 1 km and contains 30 to 40% wollastonite, 30% calcite, and smaller quantities of feldspar, garnet, diopside, quartz, idocrase, and pyroxene. Run-of-mine material was used by the Central Asiatic Scientific Investigation Institute for Geology and Mineral Deposits in the development of a successful flotation process. Wollastonite produced by this process is suitable for certain types of ceramics and has been used domestically.

The Nakpai Deposit between the towns of Almalyk and Tashkent has been traced over a distance of about 400 m and is almost 61 m thick. Reserves of approximately 1.5 Mt are estimated.

About 112 km south of Samarkand is the Lyangar Deposit, where wollastonite is found in a calc-silicate hornfels. Although no details of thickness or aerial extent are available, the deposit is very large. Wollastonite from it has been used in the manufacture of electrical ceramics.

Other deposits in Uzbekistan are associated with skarns at Ingichke in the Zirabulau Mountains, at Mynbulak in the Karzhantau Mountains northeast of Tashkent, and at Taskazgan near Bukharsk.

TADZHIKISTAN—Wollastonite is common in northern Tadzhikistan, and there has been some production from the Kansay Deposit about 29 km north of Leninabad. This deposit is similar in origin and mineralogy to the Lyangar Deposit described previously.

KAZAKHSTAN—Very little information is available on the 30 deposits known in this region, with the exception of the Bosaginski Deposit. They are all large and are associated with calcite or quartzite. The Bosaginski prospect is about 270 km south of Karaganda. Based on geologic mapping it has a strike length of 450 m and an outcrop width of as much as 70 m. It extends to a known depth of about 270 m. Wet beneficiation tests have produced commercial grade powder and high aspect ratio material.

In the general area of the Ural Mountains, wollastonite has been found only in calc-silicate zones within the Bashmakov schists.

Eastern Russia. Wollastonite is associated with skarns near Emel'dzhak. The thickness of the skarns is only 1 to 3 cm, however, and the wollastonite content ranges from 70 to 97%.

Northwest Russia. More than a dozen wollastonite occurrences are known, but most are of mineralogical interest only.

ST. PETERSBURG OBLAST'—At Kalkkitekhda near St. Petersburg a deposit of limestone contains as much as 50% wollastonite. It is of low quality and has not been worked. A suggested use is as ornamental stone. An area that has been recommended for investigation is Pilkom-selga, where contact rocks of the Karel complex and peripheral rocks of ultrabasic-alkalic intrusions might have potential for wollastonite.

Southwest Russia. Nineteen occurrences of wollastonite have been identified in Lower Doy and North Caucasus areas, but most of them are of mineralogical interest only.

KABARDIN-BALKARSKAYA—The wollastonite-bearing skarn body is located at Tyryauz, a town in the northern Caucasus on the Baksan River, about 38 km southwest of Nal'chik. The wollastonite is of poor quality and warrants no further exploration. Other deposits in this part of Russia are considered likely to contain potentially economic deposits of wollastonite and have been suggested for exploration.

Greece: Metallic & Industrial Minerals Mevior S.A. has conducted exploration and beneficiation tests on a wollastonite deposit near Kimmeria close to the Bulgarian border. The deposit consists of a series of horizons as much as 8 m thick and grading 50 to 80% wollastonite. Reserves are estimated at 300 kt. Dry magnetic separation tests produced a wollastonite product that was unacceptable in the ceramic industry because of its high calcite content. Froth flotation and wet magnetic separation have produced material that is suitable for ceramics applications. During 1991 the company was planning to construct a pilot plant at Assiros with a wollastonite product capacity of 0.5 tph.

Republic of South Africa: Pella Refractory Ores planned a 1991 startup for an operation at Modderfontein, about 18 km northeast of Garies in Namaqualand. The deposit contains more than 2 Mt grading 50 to 60% wollastonite. Beneficiation using electrostatic, high intensity magnetic separation, and flotation is expected to yield 13 600 tpy of +90% wollastonite product.

Serbia: A deposit in the vicinity of Mount Kopaonik, approximately 200 km south of Belgrade, contains 60 to 70% wollastonite with associated quartz and calcite. Geological investigations have been performed by the Government Research Organization in Zagreb, and laboratory-scale beneficiation has produced concentrates grading as much as 90% wollastonite.

Australia: Australia has no production of wollastonite, although a number of deposits are known in Queensland, New South Wales, Tasmania, and South Australia. Adelaide Wallaroo Fertilizer Pty. Ltd. has been investigating a deposit at Ethiudna in northeastern South Australia. A resource of 1 Mt containing 40% wollastonite has been identified.

Canada: The Deloro Deposit, 190 km east of Toronto in the province of Ontario, has been explored and evaluated by a joint venture of Cominco Ltd. and Platinova Resources Ltd. Skarn mineralization is developed near a granitic intrusion in silicified marble within a sequence of Late Precambrian carbonate metasediments, amphibolites, and metawackes. Reserves were estimated initially at 2 Mt grading 40% wollastonite. Additional geological work has identified four smaller horizons, ranging from 28 to 38% wollastonite and totaling almost 2 Mt. Flotation and magnetic separation tests have produced concentrates containing more than 90% wollastonite with aspect ratios of as much as 8:1.

A second deposit located at Mt. Grove, Olden Township in eastern Ontario, has been evaluated by Ram Petroleums Ltd. Drilling during 1990 defined a zone 35 m wide and 75 m deep that contains approximately 2.8 Mt grading 35% wollastonite. Minor calcite and diopside are also present.

The St. Lawrence Deposit, near Seeley's Bay in southeastern Ontario, has been recently explored and evaluated by U.S. Borax, Inc. More than 9 Mt grading 40% wollastonite have been identified.

Near Prairie Lake, 48 km northeast of Terrace Bay in Ontario, is another recently explored deposit. It is a carbonatite complex with a nepheline-rich core. Drilling has indicated a large deposit containing tens of millions of tons averaging 25% wollastonite.

In British Columbia nine wollastonite deposits are known. The Wormy Lake Deposit has been the site of extensive work by Tri-Sil Minerals, Inc. It is about 45 km northwest of Vancouver. The skarn deposit occurs in a roof pendant and crops out intermittently over a distance of 600 m. Estimated reserves are 50 Mt. Another deposit is in the vicinity of Sechelt, 2.5 km southeast of Wormy Lake. It appears to be part of the same skarn belt.

Archer, Cathro and Associates have discovered a deposit in the Yukon Territory about 450 km from the port town of Haines, AK. Surface samples contain as much as 85% wollastonite.

Other Areas: In addition to the countries discussed here, wollastonite deposits are known in Poland, Spain, France, Sweden, Italy, Switzerland, Morocco, Lithuania, and Argentina.

TECHNOLOGY

Exploration and Evaluation of Deposits

Exploration for commercial wollastonite deposits begins with defining the target parameters. These parameters include the desired location, mineralogy, and minimum reserve. An example could be Ontario, Canada, with 5 Mt averaging +50% wollastonite with no calcite. The next step is to develop an exploration plan, which may include researching known deposits, mining districts, and/or other areas that have a geological environment suitable for the formation of wollastonite. Favorable geological features may include igneous intrusions, such as anorthosites and syenites within a Precambrian metamorphic terrane containing calc-silicate rocks, such as in New York, Canada, and Finland, or Jurassic/Cretaceous granites within a Paleozoic metamorphic terrane containing calc-silicate rocks, such as in the western United States and Mexico.

Once exploration targets have been established, an initial reconnaissance geologic examination of the areas selected is conducted. Various exploration techniques used include field checking for host lithologies based on existing geologic maps, geologic mapping of host lithologies using aerial photos, stream sediment sampling (e.g., for associated garnet and diopside), grab and/or channel sampling, and mineralogical, physical, and chemical testing of field samples. Geophysical techniques seldom have been used during regional reconnaissance, because wollastonite-rich rocks exhibit no pronounced gravimetric, electrical, or magnetic properties.

In some areas, such as NYCO's Fox Knoll, Lewis, Oak Hill, and Deerhead deposits, wollastonite-rich rock does not usually crop out. This is partially due to its relatively low resistance to weathering in humid climates and can serve as an impediment during early stages of exploration. In such areas there may be geomorphological features that can be used in tracing wollastonite horizons. Wollastonite that is very coarse grained and/or contaminated with quartz can remain exposed after long periods of chemical weathering. Wollastonite is typically more resistant to weathering in arid climates, such as in certain parts of Mexico and the western United States.

After prospects have been identified, a more detailed geologic examination is conducted. This includes geologic mapping, core drilling, and trenching/sampling. Wollastonite samples generated during this phase of exploration are subjected to X-ray diffraction for mineralogical analysis and X-ray fluorescence for chemical analysis. In addition, loss on ignition, brightness, and aspect ratio are determined. Petrographic examination of thin-sections from selected samples can be useful.

If this phase of exploration produces favorable results, it is then necessary to extract a bulk sample of representative material from the deposit for additional testing to determine what marketable products can be produced. The bulk sample is subjected to the same tests as the drill core samples. Based on analytical results, the bulk sample may be beneficiated by dry or wet laboratory-scale methods to upgrade the material to standard commercial specifications by removal of gangue minerals. Laboratory-scale test work may include combinations of the following steps depending on whether a dry or wet process is required: hand sorting, crushing, screening, pebble or ball mill grinding, flotation, high intensity magnetic separation, thickening, filtration, drying, and pebble and jet mill grinding. Products resulting from beneficiation tests are again subjected to the same analyses as were the core samples. In addition, products are tested for surface area, bulk density, oil absorption, and particle size distribution. Scanning electron microscope (SEM) photographs are also taken of each product.

Processing

The mineralogy of a wollastonite ore is a major determinant of one's ability to produce a high-purity concentrate from it, and it is therefore a key issue addressed during the evaluation of a prospect. Ideally, an ore processing system should be capable of recovering a concentrate containing at least 97% if not 99% wollastonite. Important mineralogical factors include the types of gangue minerals present, the grain size, and the extent to which grains of constituent minerals are intergrown.

Deposits of wollastonite commonly contain garnet and diopside, which are nonwhite and must be beneficiated from the ore. Both of these minerals are weakly magnetic and therefore can be removed from wollastonite with high-intensity magnetic separators. Wollastonite is also found associated with large amounts of calcium carbonate. For some applications, such as plastics, calcium carbonate must be removed or essentially removed. This is commonly achieved by froth flotation.

In order for a deposit to supply all wollastonite markets it must be able to yield high aspect ratio grades as well as milled (powder) grades of product. The particle size at which high aspect ratio grades can be produced is equally important. For example, wollastonite particles having a 10:1 aspect ratio and measuring 20×2 µm are not used in the same applications as particles measuring 200×20 µm. Fines must be removed from high aspect ratio products to ensure their optimum performance as a reinforcing filler.

Processing improvements keyed to new product development are being focused now on the following targets:

- High aspect ratio, fine particle size products for use as reinforcements in competition with fine particle size (e.g., 4.5 µm) milled fiberglass and other synthetic fibers and whiskers.
- Fine particle size products having a top size of 3 to 5 µm for use in electronic applications.
- Flexible chemical modifiers that give fine particle size wollastonite a competitive advantage over short fibers in terms of impact strength, other mechanical properties, and surface appearance in polymer composites.

Specifications

Specifications for milled grades include particle size. For example, 45 µm wollastonite is 99% −45 µm in particle size. Fine particle size grades are governed by top size and median particle size (e.g., 10 µm) specifications. Depending on the end use, other specifications for milled and fine particle size grades may include color/whiteness, oil absorption, LOI, particle size distribution, and loose and tapped bulk density. High aspect ratio grades are specified by sedimentation value and loose and tapped bulk density. Specifications for chemically modified wollastonite include the percentage of chemical modifier used, percent moisture, and, depending on which of the grades is treated, one or more of the other properties already mentioned.

Wollastonite Byproducts/Coproducts

An often critical aspect in the development of an industrial mineral deposit is the sale of byproducts or coproducts. This is somewhat true of wollastonite but is perhaps not as important as for some lower priced minerals. At one operation wollastonite itself is a coproduct. At Partek's operation in Lappeenranta, Finland, wollastonite is found in the center of a deposit of limestone, which is the principal material produced. In this case the wollastonite would have to be mined and wasted if it were not recovered as a coproduct. Partek has converted it from waste to one of the highest quality wollastonite products in the world.

In another example NYCO Minerals at Willsboro, NY, has produced byproduct garnet and diopside for years. Only recently, however, has a combination of technology and health issues raised this former waste material to the status of a valuable byproduct. On the technological side, it was formerly impossible to recover a pure magnetic concentrate of garnet-diopside from the Essex County ores. The recent installation of permanent magnetic separators having high separation efficiency has permitted the production of garnet-diopside concentrates containing more than 99% magnetics. This high quality concentrate is increasingly in demand because of the toxic or otherwise hazardous nature of alternative products. In the area of sandblasting, for example, silica sand is losing favor because of the risk of silicosis associated with its use. Also, processed coal and smelter slags are known to contain heavy metals that can cause serious neurological disorders if exposures are high. Garnet is a nontoxic, nonhazardous, and cost effective substitute in this application.

One of the largest applications for NYCO's garnet-diopside byproduct is as a filter medium in the mixed media and multi-media filtration of water. In this use the relatively high specific gravity of garnet contributes to excellent stability of the filter bed when it is backflushed. Additionally, closely sized concentrates having a high coefficient of uniformity facilitate the design and engineering of filter beds for optimum fluid flow.

At other wollastonite deposits around the world the production of byproducts or coproducts is less important. Wollastonite is frequently found associated with calcite, which some producers recover as a byproduct and market as a source of lime for agricultural uses.

Synthetics

Synthetic calcium metasilicates have been produced commercially for a number of years in the United States, Denmark, Italy, Germany, and the CIS. Most of these synthetics are hydrous and, unlike natural wollastonite, have high absorptivity. None of the hydrous compounds produced have the acicular crystal morphology of natural, anhydrous wollastonite.

Synthetic calcium metasilicates are made in Denmark from silica sand and chalk, which are abundant there, and are marketed under the trade name Synopal. Chalk and sand are slurried with small amounts of dolomite, which acts as a flux. The slurry is heated in a kiln at about 1 560°C, which forms a viscous paste that is then water cooled. When solidified, the material is crushed and screened and then reheated in a second kiln at 1 250°C to permit crystallization. The finished product is subangular, white, opaque, and consists of about 50% wollastonite and 50% gehlenite ($Ca_2AL_2SiO_7$) and akermanite ($MgCa_2Si_2O_7$). Uses of Synopal are in road surfacing, mosaic tiles, flooring, and roofing felts.

A synthetic product known as Wollanita is made in Italy by heating a mixture of silica sand, chalk, and dolomite. Production methods are similar to those used in Denmark. Wollanita is used as road material, as an abrasive, and in ceramics.

Both hydrous and anhydrous calcium metasilicates are manufactured in the United States, as Silene by PPG Industries and as Micro-Cel by Manville Products Corp. Many grades of both products are produced by autoclaving lime, silica sand, and dolomite slurries at high pressure and temperature.

Actual and potential uses for synthetic calcium silicates are about the same as for natural wollastonite. In countries where wollastonite is abundant and can be produced cheaply, it is preferred over synthetic varieties. Common uses of both wollastonite and its synthetic equivalents include ceramic tilemaking and as functional extenders in the paint industry. Other applications for the synthetic material are as a flow-aid agent in powdered materials, in foam, and in mineral wool.

The tonnage of synthetic wollastonite produced annually is unpublished, but it is believed to be minor in comparison to natural wollastonite production. Prices of the synthetic equivalents are about twice those of natural wollastonite.

ECONOMIC FACTORS

Demand and Supply

Estimated world demand for wollastonite was between 270 and 323 kt in 1990. Commercial products can be classified into four categories:

1) milled grades—powders ground to 45 or 75 μm, for example;
2) fine particle size grades—products milled and air classified to a controlled top size of 10 or 15 μm, for example;
3) high aspect ratio grades—products specially ground to a typical 15:1 or 20:1 aspect ratio; and
4) chemically treated grades—products from the other three categories which are surface modified with various chemicals to improve their performance in resin-based systems.

In 1991 high aspect ratio products accounted for more than 35% of the wollastonite consumed, and chemically modified grades accounted for approximately 10%. Both of these product groups should continue to grow in demand, with high aspect ratio grades accounting for more than 50% of world consumption and chemically modified products growing to a 15% share by 1993.

The United States is the largest producer of wollastonite, accounting for between 41 and 43% of total world supply in 1990. The United States and India are the major suppliers of high aspect ratio grades together accounting for more than 90% of total sales. The United States and Finland are the principal producers of chemically treated wollastonite and account for more than 80 and 15% of total sales, respectively.

Construction materials and polymer-based products such as plastics, gaskets, and adhesives consume more than 60% of world wollastonite production. Other markets include coatings, refractories, metal fluxes, soil enrichment, and a myriad of smaller uses. The United States consumes more than 35% of world production, Europe more than 25%, and Asia about 30%.

Industry Structure

Two large producers in the United States together account for 41 to 43% of world wollastonite supply, followed by China with 26 to 31%. India and Finland each have a major wollastonite producer and together account for 21 to 28% of world supply. Several companies in Europe import wollastonite from India and China and grind it for local markets. In Asia there are also companies that grind Chinese wollastonite for local markets, such as in Japan and Korea. Smaller producers are located in Africa, Mexico, and Greece.

Producers in the United States, India, and Finland market their products worldwide through sales agent/distributor systems. Producers in the United States supplement this distribution method with their own direct sales and marketing personnel.

Pricing

Wollastonite is a unique mineral due to its acicular particle shape, as mentioned above. Its commercial value is determined by its physical properties and in-use performance. Wollastonite is not a widely occurring mineral, so freight costs to users are high. In US dollars, prices in 1991 were:

1) milled grades, 0.11 to $0.22/kg;
2) fine particle size grades, 0.40 to $0.55/kg;
3) high aspect ratio grades, 0.24 to $0.33/kg; and
4) chemically modified grades, 0.55 to $1.10/kg. Pricing is FOB the producing plant, and price lists are published by producing companies.

Health and Safety

Because wollastonite is a bladed, acicular, or needle-like mineral, some concern has been expressed regarding its health effects. This concern has been fueled by the controversy surrounding asbestos and all other mineral fibers in general. In order to determine whether wollastonite constitutes a health hazard, three different but complementary types of studies have been performed on wollastonite products. These are:

1) in vitro (in the petri dish) tests using cells, tissues, or organs from humans or other animals to evaluate biological and potential pathological activity;
2) in vivo (animal) studies to examine disease mechanisms; and
3) epidemiological (human) investigations, which study the distribution of disease and its determinants within specific populations.

Each of these types of studies produces information that often complements and correlates with that generated by the others. In none of the studies carried out to date has evidence been found that suggests wollastonite represents a health hazard. The long term health effects of inhaling wollastonite dust therefore appear to be negligible, and the mineral has been given a Group 3 classification by the International Agency for Research on Cancer (IARC).

The Fraunhofer Institute for Toxicology in Hannover, Germany, is currently conducting an investigation of the durability of wollastonite in the lung. This study will show wollastonite to be a nondurable fiber in the environment of the lung, which should result in wollastonite being excluded in future fiber legislation. A complete Health and Safety Manual is available from NYCO Minerals, Inc., the producer at Willsboro, NY.

MARKETS

Wollastonite became a growth mineral in the late 1970s and early 1980s as a replacement for short-fiber asbestos. Since that time consumption of wollastonite worldwide has doubled. High aspect ratio wollastonite has general application as a cost effective reinforcement substituting for short-fiber asbestos and short-fiber milled fiberglass. For some applications this type of wollastonite is chemically modified to enhance its performance. The major uses for wollastonite are in fire-resistant board, plastics, coatings, ceramics, and the continuous casting of steel.

Fire-Resistant Products

These products include interior and exterior construction boards, roofing tiles and slates, shaped insulation, and siding. Most are either calcium silicate based, fiber-cement based, or asbestos-fiber-cement based. They are produced by slurry, slurry-net, extrusion, or spray processes. Wollastonite is used in these products for its properties of thermal stability, reinforcement, light weight, and sound deadening. It also increases the hardness of board surfaces. Worldwide, this is the largest single market for high aspect ratio grades. For calcium silicate boards a special high aspect ratio product is produced in the United States.

Ceramics

The advantages of using wollastonite in ceramics are twofold. It improves the mechanical properties of ceramic ware, and it greatly reduces the warping and cracking of ceramic materials that are rapidly fired. Wollastonite fluxes readily with silica and alumina at low temperatures, thereby reducing thermal expansion and minimizing cracking, crazing, and shrinking. Tiles have higher green strength and better pressing qualities because of the needle-like shape of wollastonite particles. Drying rates are improved, because the interstices between wollastonite particles facilitate the rapid release of moisture from the body. Wollastonite contains no chemically bound water or carbonates, so its use in ceramic materials minimizes the liberation of gases that can cause cracks and laminations. In addition to its other merits, wollastonite contains almost 50% combined silica, so its use in partial substitution for crystalline silica such as sand, potter's flint, and china stone can reduce the incidence of silicosis among workers in the ceramics industry.

Metallurgy

Because wollastonite is a natural, low temperature flux, it has found acceptance in fluxing formulations used in the continuous casting of steel. When a stream of molten steel is poured continuously from the bottom of a refining ladle into a refractory tundish, a flux composite powder is applied to it to maintain the surface of the metal in a molten state. This practice minimizes surface defects such as scratching, prevents reoxidation of the steel surface, lubricates the mold wall, and absorbs nonmetallic inclusions. Wollastonite is also used in welding powder formulations, where it provides improved burn characteristics and inhibits sparking during welding.

Paints and Coatings

Wollastonite has been used extensively in paints and coatings for more than 35 years. It finds widespread application in trade sales paints and architectural, powder, and industrial coatings.

Milled grades of wollastonite are used in trade sales paints, where they provide reinforcement, film hardness, scrubability, gloss reduction, good color retention, low oil absorption, and low water demand. In textured coatings, stucco and block fillers, and roof coatings high aspect ratio grades are used. They offer mud crack resistance, reinforcement, high brightness, and low sag.

For powder coatings, including decorative, functional, and pipe coatings, fine particle size grades of wollastonite are used. They provide controlled particle size distribution, cohesive strength and reinforcement, edge coating, hot water resistance, color retention, and controlled melt flow. In industrial coatings, depending on the quality of the finished product, fine particle size and chemically modified wollastonite provides a wide processing window for PVC (polyvinyl chloride), good performance with all types of vehicles, ease of dispersion, low oil absorption, blister and corrosion resistance, and synergism with corrosion-inhibiting pigments.

The worldwide coatings industry represents a major growth market for wollastonite, as coatings producers formulate to high-solids, waterborne systems, and eliminate heavy metal pigments.

Plastics

All grades of wollastonite are used in the plastics industry. The three that have the best growth prospect are high aspect ratio, fine particle size, and chemically modified wollastonite.

Wollastonite has many properties that are important to varying degrees in different polymer systems. In general, it is used because of its low water absorption, good thermal stability, low resin demand, good thermal conductivity, chemical purity, and reinforcing properties. The benefits of using wollastonite in specific resin systems are listed in Table 3. The effects of wollastonite on the processing properties of plastics are summarized below for both treated and untreated grades.

Property	Untreated	Chemically Modified
Viscosity	Increase	No effect or moderate increase
Melt Flow	Decrease	No effect or moderate decrease
Processing Temp.	Increase needed	Little or no effect
Injection Pressure	No effect or slight increase	No effect or slight decrease
Flow in Mold	Moderate reduction	No effect or slight improvement
Mold Shrinkage	Decrease	Decrease
Mold Cycle Time	Decrease	Decrease

In most plastic compounds either wollastonite that has been chemically modified by the producer is used, or the compounder provides the chemical treatment in situ. The benefits of chemically modified wollastonite (and other chemically modified minerals) vs. unmodified grades include:

- Improved mineral dispersion and wetting;
- Deagglomeration/reagglomeration;
- Improved processing;
- Improved mineral-polymer bonding;
- Improved electrical properties;
- Upgraded mechanical properties; and
- Reduced water sensitivity of the finished compound.

There are also advantages to using premodified minerals vs. in situ treatment, or integral addition, such as:

- Consistent compound properties and handling characteristics from one batch to the next;
- Development of critical wet electrical and wet strength properties;
- Lower melt viscosities and therefore better flow characteristics;
- Eliminated need for chemical level adjustments from one formulation to another;
- Reduced sensitivity to process equipment design;
- Reduced sensitivity to the sequence of adding formula components;

Table 3. General Properties of Wollastonite—Reinforced Plastics

NYLONS
- Low moisture absorption
- Industry standard for platable grades
- Cost effective milled glass replacement
- Improved mold flow
- Uniform heat distribution
- Better dimensional stability
- Improved surface characteristics
- Higher heat distortion temperature
- Upgraded mechanical properties
- Reduced wall thickness

PHENOLIC MOLDING COMPOUNDS
- Low resin demand; better wet-out
- High loadings; good flow at high loadings
- Faster mold cycle time
- High reinforcement
- Cost effective milled glass replacement
- Improved machinability of molded parts
- Low moisture absorption
- Enhanced compressive strength
- Excellent dimensional stability
- Improved physical and electrical properties at elevated temperatures

POLYESTERS
- Viscosity reduction
- Low moisture absorption
- Improved surface appearance
- Better platability
- Excellent heat stability
- Improved flexural strength
- Better electrical insulating properties
- Cost effective partial glass replacement

POLYURETHANES/POLYUREAS
- Cost effective milled glass replacement in RRIM
- Improved surface appearance (distinctness of image)
- Reproducible, low-viscosity increases in slurry
- More anisotropy than with milled glass
- Nonsettling; no hardpanning in casting compounds
- Low moisture absorption
- Lower coefficient of thermal expansion
- Reduced heat sag
- Cost reduction

- Broader processing window; and
- Improved economics.

The worldwide plastics industry is the key market for growth in demand for chemically modified wollastonite.

FUTURE OUTLOOK

Secondary suppliers are entering the marketplace, such as independent grinders in Europe and Asia. China will undoubtedly become an even more significant producer during the 1990s. Production capacity is being expanded in the United States and India. Demand for wollastonite products is increasing worldwide. Wollastonite should therefore enjoy substantial growth during the 1990s. Worldwide consumption in the year 2000 is being projected as 500 kt.

In order to satisfy this growth in demand secondary suppliers will need to improve the quality of their products, such as by removing impurities and closer conformity to customers' specifications. Prompt service and efficient distribution systems are additional key requirements. Primary suppliers will need to invest in reserve acquisition and new mine development, plant capacity expansion, the implementation of statistical process controls, health and safety studies, and global distribution and service systems.

Wollastonite is certainly no longer an unknown mineral.

ACKNOWLEDGMENTS

The authors wish to thank Linda Marcotte of NYCO Minerals, Inc. and Jeanine Gibson and Sandy Prebynsky of Pincock, Allen & Holt, Inc. (PAH) for their efforts in typing the manuscript. In addition, thanks are given to Lee Aga of PAH for producing the figures in the manuscript.

BIBLIOGRAPHY AND REFERENCES

Andrews, R.W., 1970, "Wollastonite," Monograph, Institute of Geological Sciences, Her Majesty's Stationery Office, London, 114 pp.

Clark, D.W., 1980, "Hunter Mountain Wollastonite, Northern Death Valley Area, California," *Geology and Mineral Wealth of the California Desert*, D.L. Fife and A.R. Brown, eds., South Coast Geological Society, Santa Ana, CA, pp. 294–298.

Copeland, J., and Rush, O.W., 1978, "Wollastonite, Short Fiber Filler Reinforcement," *Plastics Compounding*, Nov.—Dec., pp. 26–41.

Copeland, J., and Rush, O.W., 1979, "Reinforcing with Wollastonite Filler Makes for a Tougher Polypropylene," Reprint, *Modern Plastics*, Mar., 2 pp.

de Rudder, R.D., 1962, "Mineralogy, Petrology and Genesis of the Willsboro Wollastonite Deposit, Willsboro Quadrangle, New York," PhD thesis, Indiana University, Bloomington, 136 pp.

Elevatorski, E.A., and Roe, L.A., 1983, "Wollastonite," *Industrial Minerals and Rocks*, 5th ed., S.J. Lefond, ed., AIME, New York, pp. 1383–1390.

Engelhardt, C.L., 1979, "Calcium Metasilicate—An Extender Pigment," Reprint, *American Paint and Coatings Journal*, Sep. 10, 7 pp.

Fischl, P., 1991, "Wollastonite and Tremolite Occurrences in British Columbia," Open File 1991–17, Ministry of Energy, Mines and Petroleum Resources, British Columbia Geological Survey Branch, 37 pp.

Ladoo, R.B., 1960, "Wollastonite," *Industrial Minerals and Rocks*, 3rd ed., J.L. Gillson, ed., AIME, New York, pp. 897–899.

O'Driscoll, M., 1990, "Wollastonite Production—Tempo Rises as Markets Grow," *Industrial Minerals*, No. 279, Dec., pp. 15–23.

O'Hara, P., 1976, "The Economic Geology, Petrology and Origin of the Willsboro Wollastonite Deposit, New York," MS thesis, Queens College, City University of New York, New York, 126 pp.

Power, T., 1986, "Wollastonite-Performance Filler Potential," *Industrial Minerals*, No. 220, Jan., pp. 19–34.

Santini, K., 1989, "Preliminary Feasibility Study, Bosaginski Wollastonite Deposit, Kazakhstan U.S.S.R.," unpublished report, 80 pp.

Smith, M., 1981, "Wollastonite—Production and Consumption to Climb," *Industrial Minerals*, No. 167, Aug., pp. 25–33.

Troxel, B.W., 1957, "Wollastonite, Mineral Commodities of California," California Division of Mines and Geology Bulletin, Vol. 176, pp. 693–697.

Troxel, B.W., and Morton, P.K., 1962, "Mines and Mineral Deposits of Kern County, California," County Report 1, California Division of Mines and Geology, pp. 344–345.

Vukovich, M., et al., 1957, "The Use of Wollastonite in Ceramic Bodies," Engineering Experiment Station Bulletin No. 164, Vol. 26, No. 2, Ohio State University, Columbus, 76 pp.

ZEOLITES

DAVID A. HOLMES

Zeolites are crystalline, hydrated aluminosilicates of the alkali and alkaline-earth metals. Their crystalline framework is based on a three dimensional network of SiO_4 tetrahedra with all four oxygens shared by adjacent tetrahedra. The alkali and alkaline-earth cations are loosely bound within this framework structure and may be exchanged, to varying degrees, with other cations or molecular water. Most zeolites can be reversibly dehydrated. At present, close to 40 natural zeolite minerals have been identified and over 100 zeolites have been synthesized.

Zeolites are a family of hydrous aluminosilicate minerals whose physical properties provide unique features for a wide variety of practical applications. Zeolites have been known for more than 200 years, but it was not until the middle of this century that their economic potential and widespread occurrence in sedimentary rocks were known. Since 1950, literally thousands of articles and patents have appeared in technical literature, and commercial production of natural zeolites may now exceed 1 Mt worldwide annually. In the past 40 years, the zeolite mineral group has changed from that of a museum curiosity to one of an established industrial mineral commodity.

According to Breck (1974), zeolites are characterized by the following properties:

1) High degree of hydration,
2) Low density and large void volume when dehydrated,
3) Stability of the crystal structure of many zeolites when dehydrated,
4) Cation exchange properties,
5) Uniform molecular-sized channels in the dehydrated crystals,
6) Ability to adsorb gases and vapors,
7) Catalytic properties.

All zeolites are molecular sieves, but not all molecular sieves are zeolites. Activated carbon, activated clays, alumina powder, and silica gels are also molecular sieves. Molecular sieves are materials that can selectively adsorb molecules on the basis of their size, shape, or electrical charge (Clifton, 1987). Commercial application of zeolites is based on the following physical properties: molecular sieving, ion exchange, adsorption, and catalysis.

Potential commercial utilization includes hundreds of possible applications. The principal uses at this time, however, are as follows:

- Ammonium-ion removal in waste stream treatment, sewage treatment, pet litter, and aquaculture,
- Odor control,
- Heavy metal ion removal from nuclear, mine and industrial wastes,
- Agricultural use, such as soil conditioner and animal feed supplement.

An estimated 800 kt to 1 Mt of natural zeolites were produced and consumed worldwide in 1991. Eight small operators reported collective production and consumption of an estimated 18 kt of natural zeolites in the United States during 1991, mostly for feed additives and for odor control, aquacultural, agricultural, catalytic, and ion exchange applications (Eyde, 1992).

HISTORY

The zeolites were first recognized as a new group of minerals by Baron A.F. Cronstedt, a Swedish mineralogist, in 1756 in samples from the Svappavari Copper mine in Sweden. He named the mineral family zeolite, based on the Greek words *zein* and *lithos*, meaning boiling stones, because of the bubbling and frothing achieved by heating the crystals with a blowpipe (Mumpton, 1978). The mineral group received little interest before the 1920s, when the first analysis of a zeolite crystal structure was achieved. Weigel and Steinhoff (1925) were the first to determine that chabazite can selectively absorb smaller organic molecules and reject larger ones. McBain (1932) later described this as molecular sieving. In the late 1940s research on zeolite minerals increased dramatically in the London laboratories of Professor R.M. Barrer at the Imperial College, in the laboratories of J. Sameshima in Japan, and in the research laboratories of Union Carbide Corporation. Barrer's work included the identification of molecular sieve types, quantitative studies of molecular sieving, and the use of ion exchange to modify the exchange properties of molecular sieves (Clifton, 1987).

Synthesis of zeolites was reported as early as 1862, but Barrer is credited with the first synthesis of analcime-type zeolites substantiated by X-ray diffraction (Clifton, 1987). Barrer (1938) tried unsuccessfully to synthesize chabazite and mordenite, but instead produced a synthetic zeolite now designated as Linde type A zeolite, still one of the most commercially important synthetic molecular-sieve sorbents (Milton, 1968). Union Carbide workers subsequently developed processes to manufacture other synthetic zeolites using heat and pressure in very carefully controlled reactions in closed vessels. This procedure was adapted to large scale production.

For many years after their discovery, natural zeolites were considered mineralogical curiosities that mostly filled amygdules or fractures in igneous rocks. In 1933, however, Bramlette and Posniak (1933) published a landmark paper reporting the occurrence of zeolite minerals as alteration products of vitric ash and similar materials, particularly in fine-grained sedimentary rocks. Natural zeolite research increased greatly during the 1950s with several outstanding international studies. Sudo (1950) conducted extensive studies on the Green Tuff zeolites in Japan, and Russian articles described mordenite-like zeolites in sedimentary rocks of the former USSR. Coombs (1954) reported several species of zeolites in low grade metamorphic rocks in New Zealand. Sersale (1958) identified a variety of zeolite minerals in the Neapolitan Yellow Tuff in Italy. Subsequently, Ames et al., (1958), Deffeyes (1959b),

Table 1. Typical Formulae and Physical Properties of Some Economic Zeolites (after Mumpton, 1977, Breck, 1974)

Zeolite	Typical Unit-Cell Formula[1]	Crystal System	Void Volume[1] (%)	Specific Gravity	Channel Dimensions (Å)	Thermal Stability	Ion-Exchange Capacity[2] (meq/g)
Analcime	$Na_{16}(Al_{16}Si_{32}O_{96})\cdot 16H_2O$	Cubic	18	2.24–2.29	2.6	High	4.54
Chabazite	$(Na_2Ca)_6(Al_{12}Si_{24}O_{72})\cdot 40H_2O$	Hexagonal	47	2.05–2.10	3.7 × 4.2	High	3.81
Clinoptilolite	$(Na_4K_4)(Al_8Si_{40}O_{96})\cdot 24H_2O$	Monoclinic	39?	2.16	3.9 × 5.4	High	2.54
Erionite	$(Na_2Ca_6K)_9(Al_9Si_{27}O_{27})\cdot 27H_2$	Hexagonal	35	2.02–2.08	3.6 × 5.2	High	3.12
Ferrierite	$(Na_2Mg_2)(Al_6Si_{30}O_{70})\cdot 18H_2O$	Orthorhombic	—	2.14–2.21	4.3 × 5.5 3.4 × 4.8	High	2.33
Heulandite	$Ca_4(Al_8Si_{28}O_{72})\cdot 24H_2O$	Monoclinic	39	2.10–2.20	4.0 × 5.5 4.4 × 7.2 4.1 × 4.7	Low	2.91
Laumontite	$Ca_4(Al_8Si_{16}O_{48})\cdot 16H_2O$	Monoclinic	34	2.20–2.30	4.6 × 6.3	Low	4.25
Mordenite	$Na_8(Al_8Si_{40}O_{96})\cdot 24H_2O$	Orthorhombic	28	2.12–2.15	2.9 × 5.7	High	2.29
Phillipsite	$(Na_2K)_{10}(Al_{10}Si_{22}O_{62})\cdot 20H_2O$	Orthorhombic	31	2.15–2.20	4.2 × 4.4 2.8 × 4.8 3.3	Low	3.87
Linde A[3]	$Na_{12}(Al_{12}Si_{12}O_{48})\cdot 27H_2O$	Cubic	47	1.99	4.2	High	5.48
Linde X[3]	$Na_{66}(Al_{66}Si_{106}O_{364})\cdot 264H_2O$	Cubic	50	1.93	7.4	High	4.73

[1]Taken from Mumpton (1977), adapted from Breck (1974), Meier and Olson, 1971, [2]calculated from unit-cell formula, [3]Linde A and Linde X are synthetic phases.

Mumpton (1960), and others discovered large resources of natural zeolite minerals in sedimentary deposits in the western United States. Later exploration identified other very large zeolite deposits, particularly of clinoptilolite, at other sites in the western United States and in more than 40 other countries, both in sedimentary rocks and hydrothermal deposits. Because synthetic zeolites had captured many of the markets, however, natural zeolite products are in the unusual position of challenger to their synthetic counterparts (Harben and Bates, 1984).

Research and production of synthetic zeolites has advanced greatly since the 1950s. Synthetic zeolites can be chemically tailored to meet industrial specifications, whereas the natural zeolites have a fixed range of properties and are more practically used where they are more economical or, for the few uses, where they excel.

Mumpton (1983) described the early days of zeolite exploration in North America, based principally on the search for natural zeolites in the United States by the Linde Division of Union Carbide Corporation during the period of 1957–1962.

Efforts to develop strong natural zeolite supply and marketing positions in North America led The Anaconda Minerals Company and Occidental Minerals Corporation into intense exploration and product development programs between 1974 and 1983. General withdrawal of such oil companies from the mineral industry caused both companies to end their programs, but the remnants of their reserve positions, product development, and marketing are still supporting smaller entrepreneurial efforts to gain a market status.

An Updated 6th Edition

The 5th edition of *Industrial Minerals and Rocks* included a zeolite section of four outstanding individual papers by industry leaders, each discussing one aspect of zeolites as an industrial mineral commodity. This 41-page section was not in the standard format for the volume, however, but it possessed an excellent if unwieldy bibliography. The zeolite chapter in this 6th edition is presented in the standard format to improve clarity. It presents a somewhat updated and more concise bibliography and is less than half the length of the 5th edition chapter. Readers may wish to consult the 5th edition chapter for more detailed maps showing domestic zeolite occurrences, as well as for more detailed information on synthetic zeolite manufacture.

GEOLOGY

Mineralogy

The following discussion on Mineralogy, Chemical Properties, and Physical Properties draws heavily on the works of Breck (1974), Mumpton (1977, 1978), and Flanigen (1977), leaders in the study of these complex silicate minerals.

Zeolites occur in a variety of geologic settings, mostly as alteration or authigenic minerals, low temperature-low pressure minerals in metamorphic systems, secondary minerals in weathered zones, or in veins. Commercial zeolites are presently limited to authigenic and alteration settings in finely crystalline sedimentary rocks. Nine zeolites commonly occur in sedimentary rocks: analcime, chabazite, clinoptilolite, erionite, ferrierite, heulandite, laumontite, mordenite, and phillipsite. Analcime and clinoptilolite are by far the most abundant. All nine of these zeolites show a considerable range in Si:Al ratios and cation contents. Except for heulandite and laumontite, they generally are alkalic and more siliceous than their counterparts in mafic igneous rocks (Mumpton, 1978).

Zeolite minerals with commercial potential are limited to five at present: chabazite, clinoptilolite, erionite, mordenite, and phillipsite. These are the only ones that are both abundant in nature and have favorable ion exchange, absorbency, or molecular-sieving capabilities. Ferrierite and faujasite may also have economic potential, but they are uncommon and are known at only a few significant sites in the world. With time and changing technology, other zeolite minerals will probably have economic applications.

Table 1 lists the chemical composition and selected properties of commercially significant zeolite species, as adapted from Mumpton (1977) and based on the Breck classification.

Breck (1974) originally listed 34 species of zeolite minerals in nature and additional natural species may be considered. Nearly 100 zeolites have also been synthesized. Zeolites are crystalline, hydrated aluminosilicates of alkali and alkaline-earth cations having infinite, three dimensional structures. They are further characterized by an ability to lose and gain water reversibly and to exchange constituents without major change of structure. The potential commercial application of both natural and synthetic zeolites stems from their fundamental chemical and physical properties, which are di-

Table 2. Breck's Classification of Zeolites (after Breck, 1974)

Group 1 (S4R)	Group 5 (T_5O_{10})
Analcime	Natrolite
Harmotome	Scolecite
Phillipsite	Mesolite
Gismondine	Thomsonite
Paulingite	Gonnardite
Laumontite	Edingtonite
Yugawaralite	
	Group 6 (T_8O_{16})
Group 2 (S6R)	Mordenite
Erionite	Dachiardite
Offretite	Ferrierite
Levynite	Epistibite
Sodalite	Bikitaite
Group 3 (D4R)	Group 7 ($T_{10}O_{20}$)
A-Type Zeolites	Heulandite
	Clinoptilolite
Group 4 (D6R)	Stilbite
Faujasite	Brewsterite
X	
Y	
Chabazite	
Gmelinite	

(Figures in parentheses refer to types of zeolite frameworks, e.g., D4R = double 4-ring, T_5O_{10} = a unit of 5 tetrahedons.)

rectly related to their chemical compositions and crystalline structure (Mumpton, 1977).

The basic structure of zeolites consists of $(AlSi)O_4$ tetrahedra, wherein each oxygen is shared by two tetrahedra: thus, the atomic ratio O:(Si+Al) is 2. The net negative charge of the structure is balanced by exchangeable cations, which are loosely held within the central cavities and surrounded by water molecules. Although their chemical compositions are similar, each species has its own unique crystal structure and thereby its own physical and chemical characteristics.

A classification of natural and synthetic zeolites proposed by Breck (1974) is shown as Table 2. Breck's classification is based on the crystal structure of the zeolites in seven groups. In each group, the zeolites have a common sub-unit of structure which is a specific array of $(AlSi)O_4$ tetrahedra. Well crystallized, zeolite minerals may vary considerably in chemical composition. Several different zeolite minerals may have nearly identical chemical compositions.

The nature of void spaces and the interconnected channels in dehydrated zeolites are important in determining the physical and chemical properties. Three types of channel systems were identified: a one dimensional system, two dimensional systems, and two varieties of three dimensional, intersecting channel systems (Breck, 1974).

The character of water in hydrated zeolite crystals varies, as it may include molecular clustering or direct bonding between the cations and the framework oxygen molecules. The intercrystalline volume which may be occupied by water may constitute up to 50% of the volume of the crystal. The capacity of the zeolite for adsorption is generally related to the free space or pore volume as determined by the quantity of contained water when fully hydrated (Breck, 1974).

Adsorption and ion-exchange functions in both hydrated and dehydrated zeolite crystal structures are related to channel opening characteristics. The apertures are bounded by oxygen atoms of

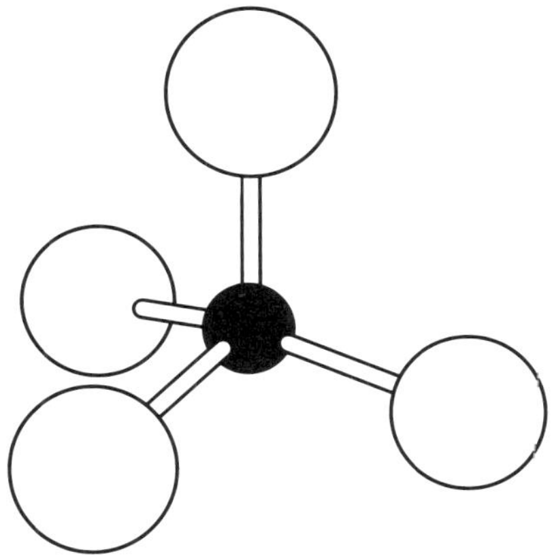

Fig. 1. Basic tetrahedron of zeolite framework, showing relationship of one silica ion to four oxygen ions.

connected tetrahedra. The limiting size of the aperture is governed by the size of the rings, which include 6, 8, 10 or 12 oxygen atoms (Breck, 1974).

Other general studies of zeolite mineralogy are Gottardi (1978), Meier (1968), and Meier and Olson (1971). Gottardi (1978) proposed a classification with twenty-three types of zeolite framework, each corresponding to a particular species of zeolite, and a more specific classification of five framework groups for the natural zeolites.

Most crystal structures can be visualized as SiO_4 and AlO_4 tetrahedra linked together in simple geometrical form. Fig. 1 illustrates a single zeolite tetrahedron showing the relationship of oxygen ions (white) with silicon and aluminum (black). Zeolite polyhedra may be connected in a number of ways, establishing the crystalline structure of the mineral.

Chemical Properties

Applications of natural zeolites make use of one or more of several chemical properties, generally including ion exchange, adsorption, or dehydration and rehydration. These properties are functions of the specific crystal structure of each species, and its framework and cationic composition. The descriptions here closely follow Mumpton (1977) in defining these properties.

Adsorption Properties: Crystalline zeolites are unique adsorbent materials. The large central cavities and entry channels of zeolites are filled with water molecules forming hydration spheres around the exchangeable cations. If the water is removed, molecules having cross-sectional diameters small enough for them to pass through the entry channels are readily adsorbed in the dehydrated channels and central cavities. Molecules too large to pass through the entry channels are excluded, giving rise to the molecular sieving property of most zeolites.

Cation Exchange Properties: The exchangeable cations of a zeolite are only loosely bonded to the tetrahedral framework and can be exchanged and removed easily by washing with a strong solution of another ion. Crystalline zeolites are very effective ion exchangers. The ion-exchange capacity is basically a function of the degree of substitution of aluminum for silicon in the framework structure. The greater the substitution, the greater the charge deficiency, and the greater the number of alkali or alkaline-earth cations required for electrical neutrality. Cation exchange behavior depends on other factors also, including: 1) the concentration of the specific cations in the solution, 2) temperature, 3) the nature of the cation species (size, charge, etc.), and 4) the structural characteristics of the particular zeolite.

Cations may be trapped in structural positions that are relatively inaccessible, thereby reducing the effective ion exchange charge capacity. Cation sieving may also take place if the cation in solution is too large to pass through the entry ports into the central cavities. Unlike most non-crystalline ion exchangers, the framework of a zeolite dictates its selectivity toward competing ions, and different structures offer different sites for the same cation. The hydration strength of some ions prevent their close approach to the seat of charge in the framework: in many zeolites, ions with low field strength are more tightly held and are more selectively taken up from solutions than the other ions.

Dehydration/Rehydration. Based on dehydration behavior, zeolites may be classified as: a) those that show no major structural changes during dehydration and exhibit continuous weight loss as a function of temperature, and b) those that undergo major structural changes, including collapse, during dehydration and exhibit discontinuities in their weight loss. Typical of the first type are such natural zeolites as clinoptilolite, mordenite, erionite, and chabazite and synthetic zeolites such as Zeolites A and X, which are thermally stable to 700 or 800°C. The dehydration behavior of zeolites in the second type is such that these exhibit reversible water loss at low temperatures, but a major structural change occurs at an elevated temperature, and the material loses its zeolitic character.

Physical Properties

The physical properties of a zeolite must be considered in two ways:

First, as a mineralogical description of the zeolite in terms of its natural properties, including morphology, crystal habit, specific gravity, density, color, grain or crystal size, degree of crystallinity, presence of corrosion or etching, presence of contaminants, and any other descriptive physical features.

Second, in terms of its physical performance as a product for any specific application, entailing such characteristics as brightness, color, Brookfield viscosity, Hercules viscosity, surface area, particle size, hardness, abrasion resistance, wet abrasion durability, thermal expansion, and resistance to weathering.

Proper characterization of any zeolite always includes determination of its basic descriptive mineralogical characteristics, and an assessment of whatever performance characteristics are believed relevant in considering it for specific commercial applications.

Origin

Zeolite minerals occur in a variety of geologic settings and can form from a variety of starting materials under a wide range of physio-chemical conditions. They occur in rocks deposited in diverse geologic settings and ages.

Volcanic glasses of acid to intermediate composition are the most common starting materials for the formation of zeolite minerals. Other common starting materials include montmorillonitic clays, poorly crystalline clays and amorphous materials, finely crystalline quartz, feldspars, and precursor zeolite minerals. Nearly all the minable zeolite deposits in the world occur as altered vitric volcaniclastic rocks.

The formation of a particular zeolite mineral is dependant on interrelated physical and chemical factors. Pressure, temperature, and time are the three physical considerations most strongly affecting zeolitic alteration. The composition of the starting materials and the composition of the reacting pore water, particularly in regard to the nature of dissolved species such as H^+, Na^+, Ca^{++}, K^+, and H_4SiO_4, are important factors.

Most zeolites in sedimentary rocks formed from volcanic ash or other pyroclastic material by reaction of the amorphous, aluminosilicate glass with pervading pore waters. Others originated by the alteration of pre-existing feldspars, feldspathoids, biogenic silica, or poorly crystalline clay minerals. Zeolites in sedimentary rocks probably formed by means of dissolution-reprecipitation reactions, with or without an intermediate gel stage.

Earlier formed zeolite minerals may alter to form different zeolites, both in near-surface environments and in deep buried settings. This is generally a result of changing physical and/or chemical parameters within the host rock.

Natural zeolites have long been known to occur in basaltic rocks as veins and fillings of vugs and cavities and were later recognized as one of the most common groups of authigenic silicates in sedimentary rocks of volcanic origin.

Based on their geologic setting, mineralogy, and genesis, zeolite deposits have been classified by Sheppard (1973), Mumpton (1973), and Munson and Sheppard (1974) into the following types:

1) *Closed System:* Deposits formed from volcanic materials in hydrologically closed, saline/alkaline lake systems;

2) *Open System:* Deposits formed in hydrologically open, freshwater lake or ground water systems;

3) *Burial Metamorphic:* Deposits formed by low grade, burial metamorphism;

4) *Hydrothermal or Hot Spring:* Deposits formed by hydrothermal or hot spring activity;

5) *Deep Marine:* Deposits formed in deep marine environments; and

6) *Weathered Zone:* Deposits formed in soils, most commonly from volcanic materials.

Commercial interest is mainly in deposits of the open and closed systems.

Closed System: These are zeolite deposits found in closed or restricted hydrologic systems in which water is unable to move freely in and out. Saline or alkaline lake deposits in hydrologically closed basins, particularly those containing vitric ash-fall tuff beds, are common economic sources. Sheppard and Gude (1968, 1969a, 1973), starting with the study of a playa lake near Tecopa, CA, found that tuffs deposited in such hydrologically closed basins are characterized by a later zonation: fresh ash near the margins of a basin, succeeded inward by a zone of almost complete zeolitization, a zone of analcime, and finally a central core of authigenic potassium feldspar. Similar relationships are typically found in other closed systems, such as the Green River Formation of Wyoming, Utah, and Colorado (Surdam and Parker, 1972); Lake Magadi, Kenya (Hay, 1968, Surdam and Eugster, 1976); and Lake Natron, Tanzania (Hay, 1966). The zeolite mineral occurrence depended on high alkalinity and the original lake brines must have been rich in HCO_3. Ash falls of glassy tuff into an alkaline-saline playa or rift lake provided the ideal setting for the formation of high grade, closed-system deposits. Erionite, chabazite, and phillipsite characterize the saline, alkaline-lake deposits. Zeolitic tuffs in saline-lake deposits are generally from a few centimeters to a few meters thick. They commonly contain monomineralic or mixed zones of the larger pore zeolites erionite and chabazite, which are relatively uncommon in other types of deposits.

Open System: Open hydrologic systems are those in which percolating ground water moved freely through the host rock. As the ground water circulated, it was chemically modified by dissolution of reactive materials or hydrolysis.

Clinoptilolite and mordenite are the principal zeolites of open-system deposits. Two examples of thick, open-system zeolite deposits are the John Day Formation of north-central Oregon and the Campianian Tuff and Neapolitan Yellow Tuff near Naples, Italy; these were formed by alteration of volcanic ash by descending ground waters. The overlying fresh glass-montmorillonite facies of the John Day Formation contrasts sharply with the clinoptilolite facies, both formed by descending ground water with increasingly higher pH and ionic strength as it reacted with volcanic glass. Where the water was relatively fresh, the glass altered to montmorillonite; as it became more and more alkalic, clinoptilolite formed.

Burial Metamorphic: Zeolite may originate in thick volcanoclastic sequences affected by metamorphism. Burial metamorphic deposits, also known as burial diagenetic deposits, are widespread in the Green Tuff region of Japan. Iijima and Utada (1972) and Iijima (1974) correlated descending zones of: 1) fresh ash, 2) alkali clinoptilolite, 3) clinoptilolite-mordenite, 4) analcime, and 5) albite with the depth of burial and related increases in temperature, pressure, and degree of metamorphism.

Coombs (1954) first recognized zeolites of the burial metamorphic type in Triassic rocks in the Southland syncline, New Zealand. Coombs et al (1959) later established a zonation of mineral assemblages, characterized by a downward succession of clinoptilolite-heulandite-analcime, laumontite-albite, and prehnite-pumpellyite-albite.

The zeolites and associated minerals of the burial metamorphic type commonly occur in thick sequences of marine volcanogenic strata. The vertical succession of mineral assemblages is one of decreasing hydration and increasing temperature of formation with depth. Zeolite rocks of this type have also been recognized in Australia (Packham and Crook, 1960), Puerto Rico (Otalora, 1964), the USSR (Kossovskaya and Shutov, 1963), Japan (Seki, 1969, Utada, 1970), BC, Canada (Surdam, 1968), and the United States. Laumontite-bearing rocks of the burial metamorphic type occur in Central Oregon (Dickinson, 1962, Brown and Thayer, 1963), in Mount Rainier National Park, WA (Fiske et al., 1963), and near Cache Creek, CA (Dickinson et al., 1969).

Hydrothermal or Hot Spring: This type includes zeolites associated with hydrothermal alteration, such as those at the East Tintic district, Utah (Lovering and Sheppard, 1960), and especially those associated with hot springs. Deposits of the geothermal type occur at Yellowstone National Park, WY (Honda and Muffler, 1970); Wairakei, New Zealand (Steiner, 1953, 1955); and Onikobe, Japan (Seki et al., 1969). Zeolites in geothermal areas commonly show a vertical zonation, and the downward succession of mineral assemblages appears to correlate with an increase in temperature. At Wairakei and Onikobe, for example, the downward succession is characterized by mordenite, laumontite, and finally wairaikeite.

Deep Marine: The most common marine zeolites are clinoptilolite and phillipsite, formed in siliceous tephra and in low-silica basaltic tephra respectively. Marine zeolites form at low temperatures and may replace as much as 80% of the marine sediments. Deepsea drilling has found abundant clinoptilolite and phillipsite in cores from many locations in the Atlantic, Pacific, and Indian Oceans. Neither zeolite appears to be in equilibrium with the sea water, but probably is a transient phase that ultimately alters to feldspar and quartz (Iijima, 1978). From studies of cores from deep sediments in the Indian Ocean, Kolla and Biscaye (1973) found that phillipsite is more common in Miocene to Recent sediments, and clinoptilolite is more common in Lower Tertiary and Cretaceous age sediments.

Weathered Zone: Zeolite occurrences within weathered zones are scattered and of little economic significance. Analcime was reported in alkaline, saline soils in the eastern San Joaquin Valley, CA (Baldar and Whittig, 1968). Hay (1963a) recognized analcime, chabazite, natrolite, and phillipsite in alkaline saline soil profiles at Olduvai Gorge, Tanzania.

DISTRIBUTION OF MAJOR DEPOSITS

United States

The western United States has several hundred known occurrences of natural zeolites, including all zeolite mineral species with known commercial potential. Resources of clinoptilolite, mordenite, and erionite are very large, while potential sources of chabazite and phillipsite are more limited. Ferrierite sources are restricted to two sites in Nevada, and there is currently no market for this mineral. Fig. 2 is a map of the western United States, showing the location and mineral species of principal domestic economic zeolite deposits.

Alabama-Mississippi: Low grade clinoptilolite occurs in the Tallahatta Formation, the Porter Creek Beds, Nanafalia Formation, Clayton Formation, Lisbon Formation, and Meridian Sand in Alabama and eastern Mississippi (Reynolds, 1970).

Clinoptilolite occurs as a diagenetic mineral in several horizons of the Tallahatta Formation as sequences of thin beds comprising minable members and grading laterally into cristobalite and argillaceous sedimentary rocks (Ky Lieu et al., 1988, Roquemore, 1984, and Spencer, 1983). Most of the potentially economic material typically contains 40 to 60% very fine-grained, finely crystalline zeolite with low cation-exchange capacity. The poor quality of the Tallahatta Formation clinoptilolite may be offset by its proximity to eastern markets, particularly for regional agricultural use and poultry feed. Genesis of the Tallahatta clinoptilolite remains unresolved; hypotheses are that the clinoptilolite is a product of 1) open-system zeolitization of Tertiary ash-fall tuffs, possibly related to volcanism in the Yucatan Peninsula, or 2) diagenesis in back- bay sedimentary environments in the Gulf Coast sediments. Reynolds and Surdam (1980) described the occurrence of phillipsite-rich strata in the Tuscahoma Formation near McKenzie, AL.

Occurrences in the Porter Creek Beds and Meridian Sand typically have less than 50% clinoptilolite, and their commercial potential is limited to pet litter and other low value uses.

Alaska: Mordenite, heulandite, and laumontite deposits occur in the vicinity of Sheep Mountain and Horn Mountain in the Upper Matanuska Valley. Zeolitization is the result of burial diagenesis and regional metamorphism of volcanic rocks and related sediments of the Lower Jurassic Talkeetna Formation and the lower units of the Upper Cretaceous Matanuska Formation (Hawkins, 1973, Madonna, 1973).

Laumontite is the predominant zeolite in the Sheep Mountain area, occurring with prehnite and possibly wairakite in vugs and veins in massive, fractured volcanic rocks and as an alteration product of vitric tuffs. In the Horn Mountain area, mordenite occurs in fine-grained vitric tuffs at the head of Albert Creek and as an alteration product of volcanic sandstones of the Matanuska Formation. Heulandite is present as tabular crystals in veins and vugs in the massive volcanic rocks of Horn Mountain and, to a lesser extent, in altered tuffs. A single laumontite occurrence in the Horn Mountain area is in altered tuffs.

Arizona: Forty-six natural zeolite occurrences have been identified in Arizona (Anon., 1977). These include deposits in three geologic settings: open system, closed system, and hydrothermal/igneous modified by open system alteration.

Most of the open system zeolites are in Cretaceous to late Tertiary rocks. Generally, the open system mordenite and clinoptilolite zeolite deposits are in thick sections of ash-flow tuffs or tuff breccias or clastic tuff units, are highly variable in quality, and include much unzeolitized fragmentary material. Large open system occurrences in the Black Mountains, Ajo Mountains, Coffee Pot Mountains, and at the Whitlock Mountain clinoptilolite deposit offer potential for low quality clinoptilolite or mordenite-clinoptilolite.

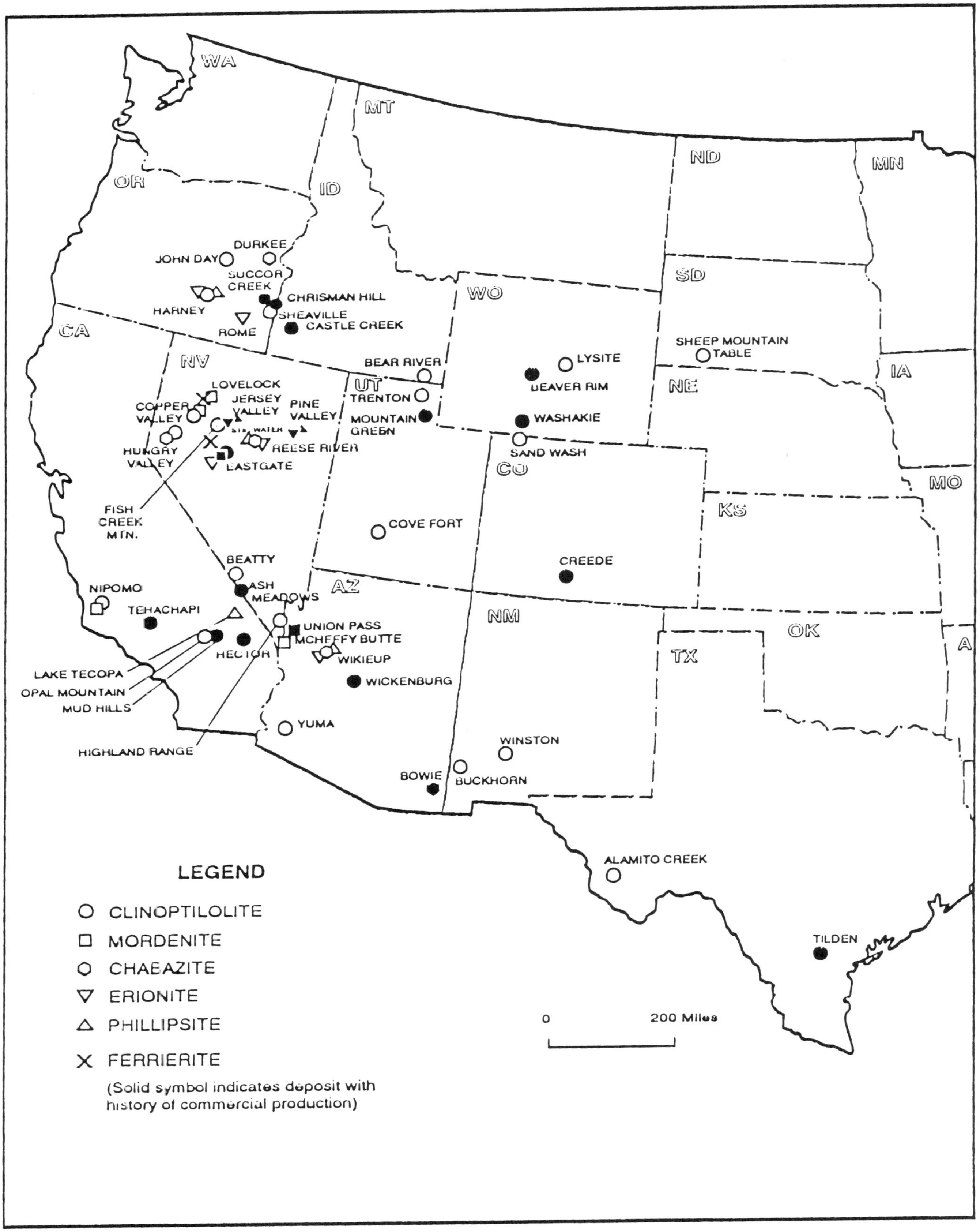

Fig. 2. Some major zeolite deposits of the United States (Holmes, unpublished, 1992).

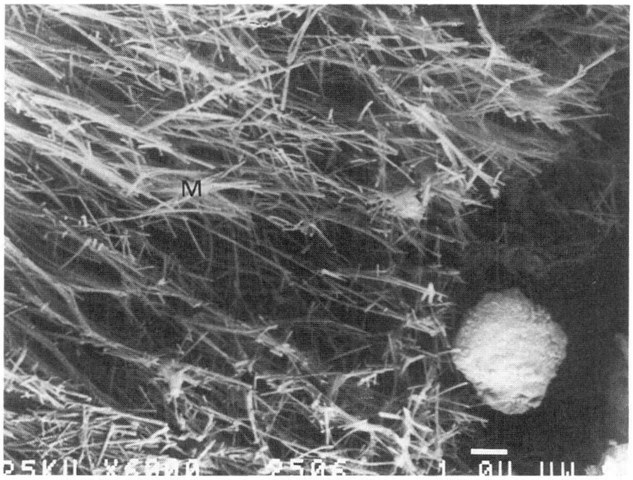

Fig. 3. Scanning electron micrograph showing fibrous, wiry crystals of mordenite (M) in zeolitized ash-flow tuffs at McHeffy Butte, AZ (magnified 6000X).

Water-laid deposits of zeolitized ash-fall tuffs more commonly consist of higher quality clinoptilolite, with lesser variability, and little contaminating fragmental material. The Wickenburg deposit is a rhyolitic ash-fall tuff partially altered to clinoptilolite located about 8 km east of Wickenburg, AZ. Crushed and screened material from this deposit has been sold to regional absorbent markets. A 5 m thick, massive, hard, white sedimentary clinoptilolite bed occurs in the Muggins Mountains northwest of Wellton and east of Yuma in the Yuma Army Testing Station. The Yuma clinoptilolite deposit is situated in a thick, altered pumiceous tuff bed on the east side of US Highway 95, about 80 km north of Yuma.

Hydrothermal/igneous occurrences are principally those of mordenite and mordenite-clinoptilolite in Tertiary volcanic rocks in the Oatman and Union Pass areas and the Black Mountains in Mohave County. Mordenite and clinoptilolite occur as alteration minerals in a halo that surrounds the Oatman gold/silver mineralized area, suggesting that the zeolitic alteration was produced by hydrothermal solutions. The Union Pass mordenite deposit is located about 32 km west of Kingman, on State Highway 68 at the summit of Union Pass. Union Carbide Corporation shipped several carloads of high purity mordenite from this deposit (Anon., 1977). Mordenite and lesser clinoptilolite occur in ash-flow tuffs and in the matrix of ash-flow breccias in the McHeffy Butte area, about 14 km south of Oatman and east of old US Highway 66 (Anon., 1977). Fig. 3 is a scanning electron micrograph showing fibrous, wiry mordenite crystals from ash-flow tuff at the McHeffy Butte locality.

Closed system zeolites are present in some Tertiary to Pleistocene intermontane basins located from Wikieup in Mohave County southeasterly to Bowie in the southeast part of the state. The Bowie deposit is the largest and most important of these.

Bowie Deposit. The Bowie deposit is the largest known high-grade deposit of chabazite in the United States. It is located in the San Simon Valley in Graham and Cochise Counties, about 19 km north of the town of Bowie (Eyde and Eyde, 1987, Sheppard et al., 1978).

The chabazite-erionite ore bed occurs as the basal layer of a zeolitized ash-fall tuff bed in Plio-Pleistocene lake beds composed of interbedded, poorly consolidated to unconsolidated clays, sandstones, and conglomerates.

The zeolitized ash-fall tuff, informally termed the marker bed, ranges in thickness from 22 to 125 cm where observed. It consists of two layers. The basal layer, informally termed the ore bed, is 10 to 20 cm thick and contains 60 to 85% chabazite or mixed chabazite-erionite with lesser clinoptilolite. Only the lower unit is mined for its zeolite content, and is remarkably uniform in thickness throughout most of the deposit.

The overlying layer consists of thin-bedded grayish-white or brownish-gray zeolitic tuff containing 30 to 50% chabazite and is not regarded as a chabazite source with current mining practices.

Several companies have mined chabazite from the Bowie deposit. The ore bed, generally less than 20 cm thick, is extracted very carefully. The mined rock is trucked to a railroad at Bowie, thence shipped to distant processing facilities where it is ground, pelletized, and activated (Mondale et al., 1978). The Linde Division of Union Carbide Corporation has operated this mining venture since 1963, extracting as much as 1 ktpy. The product is used for natural gas purification, acid-gas drying, hydrogen drying, chlorinated and fluorinated hydrocarbon purification, HCl removal from byproduct hydrogen streams, and nuclear waste stream treatment. Other companies, including W.R. Grace and Company, NRG Incorporated, Norton Company, Filtrol Corporation, and several small absorbent product producers, have also mined the deposit on an intermittent basis.

Other Occurrences. Arizona has other notable chabazite occurrences, although none has the size or quality to rival the Bowie Deposit (Eyde, 1982). There are five chabazite occurrences in the Lake Roosevelt area. The Tonto Basin deposit is the largest of these, with a bed ranging from 1.3 to 2.5 m in thickness and consisting of thin, white to pinkish-white ashy zeolitized tuffs containing chabazite and trace amounts of clinoptilolite and phillipsite.

Three chabazite-bearing zeolitic tuffs crop out on the southwest side of the Gila River Valley near Bear Springs in the Safford District. These are in the Plio-Pleistocene Solomonsville Beds, equivalent in age to the Bowie Deposit. They contain chabazite and small amounts of erionite and phillipsite.

Phillipsite and phillipsite-erionite of closed-system origin occur in thin ash-fall tuff beds in Tertiary valley sediments near Kirkland, Kirkland Junction, Verde Valley, and Cottonwood Basin, all in Yavapai County.

Vitric tuff beds, altered to phillipsite, chabazite, erionite, analcite, and at one locality, clinoptilolite, crop out in the Pliocene Big Sandy Formation near Wikieup, in southeastern Mohave County (Sheppard and Gude, 1973).

California: More than 100 zeolite occurrences are known in California and are well documented by Stinson (1988). Most are in late Tertiary rocks in open or closed systems in an east-west belt in the Mojave Desert, extending from Mojave-Rosamond easterly about 190 km to Hector and Midway. A second belt of zeolite occurrences, east of Death Valley, extends about 60 km from Tecopa and Shoshone north to the Ash Meadows District, northeast of Death Valley Junction.

Zeolite deposits of burial metamorphic or sedimentary origin are present in San Luis Obispo County, where mordenite with clinoptilolite replaced volcanic glass in tuffs of the marine Obispo Formation. Alteration by hydrothermal processes is represented by zeolitization of rhyolitic tuffs and pumice by hot-spring activity near Casa Diablo Hot Springs in Mono County.

Inyo County. Two major zeolite deposits are located in Inyo County: the Ash Meadows clinoptilolite mine and the Lake Tecopa zeolite deposit.

ASH MEADOWS CLINOPTILOLITE MINE—Very large clinoptilolite resources exist in the Ash Meadows Deposit, northeast of Death Valley Junction on the California-Nevada border. The zeolitized tuff belt occurs in an 8 km long belt in unnamed late Tertiary sedimentary rocks and extends north into Nye County, NV. White to yellowish-white zeolitized ash-flow tuffs and lapilli tuffs typically contain 80% or more clinoptilolite and exhibit high ammonia cation exchange capacity. The exposed, altered tuff beds strike north-south, typically dip 25° east, and range in thickness from 100 to 130 m. Zeolitization is irregular, leaving some zones unaltered or weakly altered, and requiring careful mining control (Denny and Drewes, 1965).

Anaconda Minerals Company drilled and developed this deposit with a small open pit, removing and stockpiling about 18 kt of material for bulk-sample testing and to initiate production of clinoptilolite products (Santini and Knostman, 1979). American Resource Corporation controls this deposit, removing material from the old Anaconda stockpile and the open pit at the California site and processing it in a crushing/screening facility at Ash Meadows Rancho in Nevada (Shelton, personal communication, 1992).

LAKE TECOPA ZEOLITE DEPOSIT—The Lake Tecopa phillipsite deposit is in the Pleistocene Lake Tecopa basin, between the towns of Shoshone and Tecopa. The Pleistocene sediments consist mostly of interbedded mudstones and altered to unaltered rhyolitic ash-fall tuffs; these interfinger with coarser clastic sediments toward the margin of the basin (Sheppard and Gude, 1968, Chesterman, 1973). Sheppard and Gude (1968) established that a fresh-glass facies occurs along the lake margin, succeeded basinward by a zeolite facies, and then by a central zone of potassium feldspar with sparse amounts of searlesite ($NaBSi_2O_6.H_2O$), reflecting the changes in the chemical environment of the lake. Phillipsite is the predominant mineral in the zeolite facies, with lesser clinoptilolite and erionite and minor analcime and chabazite. Anaconda Minerals Company conducted an extensive drilling and sampling program on this deposit.

Other Occurrences. Other zeolite occurrences include phillipsite and minor clinoptilolite in altered rhyolitic tuffs in the Waucoba Lake beds east of Big Pine (Hay, 1964); clinoptilolite in thin tuff beds in the Furnace Creek Formation in Death Valley National Monument (Sheppard, 1971); and phillipsite and clinoptilolite in the Owens Lake sediments.

KERN COUNTY—Zeolite-bearing tuffs have been reported in several Tertiary formations in Kern County. These include the Ricardo Formation in the El Paso Mountains, the Gem Hills Formation south of Mojave, the Kinnick Formation east of Tehachapi, and the Tropico Group northwest of Boron (Stinson, 1988).

Clinoptilolite and lesser phillipsite, mordenite, and analcime occur in several tuff beds in the Pliocene Ricardo Formation in Last Chance Canyon and on the northwestern flank of the El Paso Mountains. The zeolitized zones range from 1 to 7 m in thickness (Dibble, Jr., 1952, Sheppard and Gude, 1965).

Clinoptilolite and lesser mordenite occur at many sites in the Gem Hills Formation in the western Mojave Desert. This Miocene formation consists of rhyolitic lithic tuffs, tuffaceous sediments, and tuff breccias. At one site, the zeolitized tuff is more than 16 m thick and contains more than 50% clinoptilolite and mordenite.

Clinoptilolite and mordenite of open-system origin occur in altered tuffs and tuffaceous sediments of the Miocene Kinnick Formation east of Tehachapi (Dibble, Jr., and Louke, 1970). The clinoptilolite-rich tuffs were mined and used as pozzolan in 1912 for concrete in the Los Angeles aqueduct (Mumpton, 1973).

The Tropico Group is a sequence of non-marine sedimentary, pyroclastic, and volcanic rocks of Miocene and Pliocene age exposed in the vicinities of Rosamond, Mojave, Castle Butte, the Kramer borate district, and the Kramer Hills (Dibble, Jr., 1958). Zeolitized white to tannish or greenish-gray tuffs and tuff breccias are common in the lower part of the Tropico Group in the Castle Butte area. Clinoptilolite typically comprises more than 50%. Zeolitization is spotty elsewhere with lower content in the tuffs and tuff breccias in the Kramer Hills and Kramer borate district.

MONO COUNTY—Clinoptilolite, minor phillipsite, and authigenic feldspar occur in the altered tuff, lapilli tuff, and pumice about 8 km east of Casa Diablo Hot Spring (Stinson, 1988).

SAN BERNARDINO COUNTY—This county has abundant resources of clinoptilolite in altered tuffs and tuff breccias in numerous Tertiary formations. The Mud Hills clinoptilolite mine is located about 13 km north of Barstow and the Hector clinoptilolite-bentonite mine about 65 km east of Barstow. Both mines have produced high quality clinoptilolite. Large resources of clinoptilolite also occur in lithic tuffs and related sediments in the Pickhandle Formation in the Opal Mountain-Black Canyon area and in the Spanish Canyon Formation near Clewes Ridge and the Alvord Mountains (Stinson, 1988, Holmes, unpublished data). Clinoptilolite occurs in several prominent ash-fall tuff beds within the Miocene Barstow Formation. The formation is widespread throughout the Barstow area, but most of the zeolite potential is in the Mud Hills area. Sheppard and Gude (1969a) described extensive clinoptilolite mineralization in the Skyline Tuff unit, where the Mud Hills mine was operated by Occidental Minerals Corporation in 1982 to supply British Nuclear Fuels Ltd. (BNFL) with high quality ion exchange materials. Two thousand tons of carefully mined and processed clinoptilolite product were shipped to Sellafield, UK, for use in the BNFL nuclear fuel reprocessing facility (Holmes, unpublished data). The Skyline Tuff is 3 to 5 m thick at the Mud Hills mine site and typically contains 70 to 90% clinoptilolite with high cation exchange capacity values. The mine has operated intermittently since 1982 supplying clinoptilolite for use as absorbents for several successive owners.

The Hector deposit is a small but high quality clinoptilolite source about 5 km west of Hector siding (Ames, Jr. et al., 1958). It lies several kilometers north of the unique hectorite (hydrous magnesium silicate) deposit mined by Rheox, Inc. and has been mined intermittently for high-quality clinoptilolite.

Abundant clinoptilolite is found in the lithic tuffs, tuff breccias, and sandy ash-fall tuffs of the lower to middle Miocene Pickhandle Formation of the Barstow area (Stinson, 1988). Clinoptilolite sources have been developed in the Opal Mountain and Black Canyon areas for low grade absorbent uses (Holmes, unpublished data).

Clinoptilolite occurs in the Miocene Spanish Canyon Formation in the Alvord Mountains east of Barstow (Sheppard and Gude, 1964). A tuffaceous unit in the lower part of the formation has been zeolitized on both flanks of the Spanish Canyon anticline defined by Byers, Jr. (1960). Zeolitized tuffs on Clewes Ridge include two beds of clinoptilolite, each 1 m thick, separated by 1 m of bentonite. Zeolitized tuffs near the northern end of Spanish Canyon on the western limb of the anticline include a 3 m bed with a clinoptilolite content of 50 to 75% (Holmes, unpublished data).

SAN LUIS OBISPO COUNTY—Stinson (1988) reported 12 occurrences of mordenite and lesser clinoptilolite in altered tuff, lapilli tuff, and tuff breccia in the Obispo Formation and the Rincon Shale in San Luis Obispo County. The Obispo Formation (or Obispo Tuff Member of the Monterey Formation) is thickest between the towns of San Luis Obispo and Nipomo with maximum thickness of 1 050 m. Some beds are more than 30 m thick and consist of more than 75% mordenite. Surdam et al (1970) studied the distribution and genesis of authigenic silicate minerals in this formation and found that the tuff member has been extensively converted to mordenite, clinoptilolite, phillipsite, analcime, and montmorillonite, apparently by alteration of rhyolitic glass in a marine environment.

An area 6.4 to 8 km long northeast of Morro Bay is underlain by a zeolitized tuff unit of the lower Miocene and Oligocene Rincon Shale (Hall and Prior, 1975). Zeolitized tuff is also reported in the Rincon Shale east and northeast of Cambria. This tuff is hard, extremely fine grained, and contains mordenite and clinoptilolite.

OTHER COUNTIES—Murata and Whiteley (1973) reported fourteen occurrences of zeolites in the Briones Sandstone, Cierbo Sandstone, and Oursan Sandstone in a long belt extending through Santa Clara, Alameda, and Contra Costa Counties between Martinez and San Jose. Clinoptilolite is the predominant mineral, with lesser heulandite, mordenite, laumontite, stilbite, and analcime. The zeolite minerals replaced glass shards and occur as fracture fillings.

Colorado: The Creede clinoptilolite deposit is in the Oligocene Creede Formation, about 3 km south of Creede in Mineral County. Altered ash-fall tuff units up to 7 m thick, consist of white to light-gray, dense, massive clinoptilolite. The tuffs were deposited in a structural moat around the periphery of the Creede Caldera. Intermittent mining operations have tapped this deposit as a source for regional absorbent markets (Holmes, unpublished data).

Dense, massive, blue-green altered ash-fall tuffs comprise the Sand Wash clinoptilolite deposit, about 67 km west-northwest of Craig in Moffatt County. This zeolitized tuff is in Eocene sedimentary rocks of the Sand Wash Basin and is the stratigraphic equivalent of the Robin's Egg Blue clinoptilolite tuff member of the Washakie Formation, located due north in the Washakie Basin in southwestern Wyoming. The Sand Wash clinoptilolite unit varies in thickness from less than 3 m to at least 33 m, but actually is a sequence of tuffs and tuffaceous siltstones with thin interbeds of siltstone and sandstone (Roehler, 1973b).

Idaho: Southern Idaho has five large clinoptilolite deposits with commercial potential.

Bear River Clinoptilolite. Thick beds of green zeolitized tuff in the Eocene Salt Lake Formation are widespread in Franklin County in southeastern Idaho. Major occurrences are known in Bear River Canyon and the vicinity of the town of Malad. These beds are equivalent in age and physical character to the Mountain Green clinoptilolite deposit in the Salt Lake Formation east of Provo, UT (Holmes, unpublished data).

Castle Creek District. A soft, fine-grained ash-fall tuff unit in the Miocene Chalk Hills Formation is altered to fine-grained, white clinoptilolite in the Castle Creek drainage in Owyhee County. Double Eagle Petroleum and Mining Company discovered the Castle Creek deposit several kilometers south of the town of Oreana and developed economic resources of clinoptilolite.

Sheppard (1991, 1992) and Teague Mineral Products subsequently identified additional sources of high quality clinoptilolite in extensions of the Castle Creek bed to the east and southeast. Teague Mineral Products is actively mining and processing the material for sale to industrial and agricultural markets (Teague, personal communication, 1991).

Chrisman Hill Deposit. A thick, massive to thin-bedded zeolitized tuff bed occurs in the Miocene Sucker Creek Formation, about 30 km north of Jordan Valley, OR. The deposit consists of high quality, white to light-gray clinoptilolite very similar in appearance to the Sheaville, OR, deposit, 11 km to the south. Thickness of the zeolitized zone is 20 m, with millions of tons of potential resources. The deposit partially straddles the Idaho-Oregon border and is mined by Teague Mineral Products for supply to its processing plant in Adrian, OR. The Chrisman Hill deposit is on a weak hydrothermal-alteration trend, and the zeolitized zones show evidence of hydrothermal alteration as well as obliteration of sedimentary structures (Teague, personal communication, 1991).

Upper Sucker Creek Clinoptilolite. Thick, high quality clinoptilolite beds are present in the Sucker Creek Formation that extends into Idaho along the upper Succor Creek drainage from adjacent Malheur County, OR. The Sheaville, OR, deposit along Succor Creek has been drilled and developed and has large proven reserves of commercial-quality clinoptilolite. This deposit is described in the Oregon section (Holmes, unpublished data).

Challis Range Clinoptilolite. Clinoptilolite occurs in tuffs and tuffaceous sediments in the Germer Tuffaceous Member of the Challis Volcanics of Eocene or Oligocene age in the Challis Range in eastern Idaho.

Montana: Clinoptilolite is reported from eleven localities in southwestern Montana, all in tuffaceous beds thought to be of Tertiary age and deposited in intermontane basins (Berg, 1969, 1993, Sheppard, 1976). Clinoptilolite and mordenite occur in Cretaceous tuffs in Beaverhead County (Pearson, 1987).

Nevada: Nevada has huge resources of zeolites, primarily in two geologic settings: late Tertiary erionite-clinoptilolite-phillipsite-chabazite ash-fall tuff occurrences of closed system origin in northern Nevada; and thick Tertiary zeolitized ash-flow tuff occurrences of clinoptilolite and mordenite.

Closed System Deposits. Papke (1972) described four late Tertiary closed system deposits in north-central Nevada: Eastgate, Jersey Valley, Pine Valley, and Reese River. All the deposits occur in silicic ash-fall tuffs as thick as 5 m in sections of lacustrine sedimentary rocks. Common associations of zeolites are erionite-clinoptilolite, erionite-clinoptilolite-phillipsite, chabazite-clinoptilolite, and erionite alone. Mordenite and analcime occur rarely. The zeolites were formed by post-depositional diagenesis of vitric volcanic ash that fell into saline, alkaline lakes (Papke, 1972).

PINE VALLEY DEPOSIT—Erionite and erionite-phillipsite occur in altered ash-fall tuffs in a thick sequence of Pliocene lacustrine rocks of the Hay Ranch Formation about 55 km south of Carlin in Eureka County (Regnier, 1960). The principal erionite bed strikes north-south, dips 2 to 3° east and has been developed intermittently since 1957 by the Union Carbide Corporation (Papke, 1972). The principal zeolitized tuff bed is as much as 2 m thick and typically analyses 80 to 90% erionite (Holmes, unpublished data). Three to five thin beds of high purity phillipsite occur in bentonitic sediments at the south end of the prospect.

EASTGATE DEPOSIT—This deposit is 88 km east of Fallon and 5 km east of the junction of US Highway 50 and Nevada Highway 2 in Churchill County. The beds occur in the Monarch Mill Formation, an early to middle Pliocene sedimentary sequence (Papke, 1972).

The zeolite beds are of closed-system origin in air-fall tuffs. One or more zeolite beds may be present in a zone ranging in thickness from a few centimeters to as much as 5 m. The beds strike north to northwest, are exposed intermittently, and dip easterly at low angles. The most common zeolite associations are erionite and erionite-clinoptilolite with varying amounts of unaltered glass. Locally, chabazite, and phillipsite, erionite-phillipsite, and mordenite are also present.

Anaconda Minerals Company staked mining claims over this deposit in 1970 and conducted detailed exploration. East-West Minerals, Ltd. later acquired the deposit and built a processing plant nearby. They attempted to mine the southern part of the deposit, a short distance south of State Highway 2, and process the material for pet litter and absorbent uses. The venture was unsuccessful and the plant was scrapped and removed.

JERSEY VALLEY DEPOSIT—Ash-fall tuff beds within late Tertiary lacustrine sediments are altered to erionite, erionite-clinoptilolite, and erionite-phillipsite-clinoptilolite in Jersey Valley about 67 km southeast of Battle Mountain. Eighteen zeolite-rich tuff beds are described by Papke (1972), of which two have minable thicknesses of 2 to 3.5 m. Mobil Oil Corporation mined these zeolites by underground and open pit methods from 1965 to 1970 to produce raw material for catalysts in oil refining and natural gas cleanup.

REESE RIVER DEPOSIT—Deffeyes (1959b) identified erionite-clinoptilolite-chabazite and erionite-clinoptilolite-phillipsite in ash-fall tuffs within Miocene or Pliocene lacustrine sedimentary rocks about 56 km north of Austin in Lander County where Nevada Highway 8 passes through Reese River Canyon. Eight zeolite beds thicker than 33 cm are present; one is as much as 2.5 m thick.

Fig. 4. Scanning electron micrograph showing mixed chabazite (CHY) and clinoptilolite (CL) with trace erionite (E) from zeolitized ash-fall bed at Hungry Valley, NV (magnified 1500X).

Analcime and mordenite are present in trace amounts (Papke, 1972).

COPPER VALLEY DEPOSIT—This deposit is about 43 km southwest of Lovelock in the Trinity Range in Churchill County. The zeolitized vitric ash-fall tuffs are late Miocene to early Pliocene in age and occur in a sequence of interfingering basalt and andesite flows and tuffaceous sediments exposed over a strike length of approximately 1 000 m. A 33-m-thick measured section of ash-fall tuff and tuffaceous sediment includes 11 m of massive to thin-bedded, white zeolitized tuff. Clinoptilolite is the dominant zeolite in this deposit, but clinoptilolite-erionite and clinoptilolite-erionite-phillipsite are common. Mordenite occurs in the upper few meters at the surface but was not penetrated at depth by drilling (Holmes, unpublished data).

HUNGRY VALLEY DEPOSIT—This deposit is located about 38 km northeast of Reno in Washoe County. The principal zeolitized ash-fall tuff is 1 to 3 m thick, crops out over 1 400 m along strike, and consists of clinoptilolite, clinoptilolite-chabazite, and clinoptilolite-erionite (see Fig. 4). The bed generally strikes north-south, dips 15 to 40° west, and thins to the north. It occurs in a thick sequence of tuffaceous siltstone, bentonite, ash-fall tuff, silty sandstone, and minor conglomerate of the Pliocene High Rock Sequence (Papke, 1972, Holmes, unpublished data).

STILLWATER RANGE DEPOSIT—This deposit is the second largest known ferrierite concentration in the United States and is located on the east side of the Stillwater Range in Churchill County, about 8 km north of the junction of US Highway 50 and State Route 839. The zeolites occur in a sequence of lacustrine sediments and subordinate rhyolitic tuffs of probable Pliocene age. Zeolites occur principally in two units separated by about 40 m. The lower unit, about 5 m thick, consists of alternating mordenite-rich and ferrierite- and mordenite-rich beds. The upper unit is about 1 m thick and contains mordenite, ferrierite, and clinoptilolite (Papke, personal communication, 1992).

Open-System Deposits. Many large deposits of clinoptilolite and mordenite occur in ash-flow tuffs, tuff breccias, and lithic tuffs in Nevada. Most of these occurrences are open system in origin, although some appear to be related to hydrothermal alteration as well. Typically, a large ash-flow zeolite deposit contains 30 to 80% zeolite, with abundant lithic fragments and unzeolitized crystalline and vitric material. Ammonia cation exchange capacities are significantly lower than values in closed-system ash-fall tuffs.

LOVELOCK DEPOSIT—Ferrierite-mordenite-clinoptilolite and ferrierite-mordenite occur in altered Miocene or Pliocene ash-fall tuffs about 17 km northwest of Lovelock in Pershing County. Clinoptilolite is relatively minor and occurs mostly in the upper part of the stratigraphic section.

A remarkable concentration of ferrierite is present in the central part. Here, a unit with an exposed thickness of about 17 m contains 75 to 95% ferrierite and minor amounts of mordenite and clinoptilolite. Apparently, this is the largest known concentration of ferrierite in the world (Papke, personal communication, 1992).

ASH MEADOWS DEPOSIT—An ash-flow zeolite deposit with large resources of high quality clinoptilolite is present at Ash Meadows in Nye County. The Ash Meadows deposit straddles the Nevada-California border, northeast of Death Valley Junction, CA, and is described more fully in the section on California zeolites. The Nevada portion includes large resources of green clinoptilolite-rich Pliocene ash-flow and lapilli tuffs located close to Ash Meadows Rancho.

FISH CREEK MOUNTAINS DEPOSIT—This deposit, located about 80 km southwest of Battle Mountain (Mumpton, 1978), consists of thick zeolitic ash-flow tuffs exposed on a 2 500-m-long ridge. The tuff contains abundant highly altered lapilli but relatively few lithic fragments, and typically contains 75% clinoptilolite. Due to repetition by extensive folding and faulting, mining thicknesses as much as 80 m might be developed.

BEATTY AREA—The Beatty area in Nye County is rich in low grade zeolite resources. A white Pliocene ash-flow tuff over 60 m thick in Beatty Wash, about 10 km east-northeast of Beatty, typically contains 75% clinoptilolite. Altered ash-flow tuffs tens of meters thick containing 50 to 70% erionite and mixed erionite-clinoptilolite are exposed along the Silicon Mine road, about 9 km northeast of Beatty (Holmes, unpublished data). Ash-flow tuffs altered to clinoptilolite abound in the Bullfrog, Transvaal, and Indian Springs areas west and northwest of Beatty. The Bullfrog district has clinoptilolite and mordenite values of as much as 65%, whereas the Indian Springs area includes clinoptilolite and lesser mordenite in a range of 50 to 85%. Beatty Mountain, immediately east of the town, has Pliocene ash-flow tuffs with mordenite, mordenite-clinoptilolite, and clinoptilolite beds analyzing as much as 70 to 85% zeolite (Olson, personal communication, 1977).

OTHER DEPOSITS—Eight large undeveloped ash-flow tuff occurrences offer major low grade clinoptilolite and clinoptilolite-mordenite resource potential (Olson, personal communication, 1977), as follows:

1) *Nivloc Road*, 5 km west of Silverpeak, Esmeralda County
2) *Montezuma Range*, west of Goldfield, Esmeralda County
3) *Rhyolite Ridge*, 21 km southwest of Silverpeak, Esmeralda County
4) *Slate Range*, northwest of Scotty's Junction in Esmeralda County
5) *Reveille Range*, on Nevada Highway 25 near Twin Springs Ranch, Nye County
6) *Tonopah area*, several km south of Tonopah, Esmeralda County
7) *Nevada Test Site/Lathrop Wells*, numerous sites in Nye County (Gibbons et al., 1960, Hoover, 1968)
8) *Highland Range*, 5 km west of Searchlight, Clark County.

New Mexico: The Buckhorn clinoptilolite deposit consists of two flat-lying clinoptilolite-bearing ash-fall tuff and reworked tuff beds which outcrop along the west side of the Duck Creek valley about 1.5 km south of Buckhorn, NM. It occurs in the Cactus Flat beds of Pliocene-Pleistocene age. The lower bed is a zeolitized ash-fall tuff ranging in thickness from 1 to 1.6 m and consisting of 70 to 90% clinoptilolite, with chabazite, heulandite and analcime locally. The upper bed is a zeolitized ash-fall tuff about 30 cm thick, consisting of over 60% clinoptilolite and a trace of erionite (Eyde, 1982).

Hills of white to light-gray altered Tertiary volcanic rocks south of the town of Winston in Sierra County comprise the Winston clinoptilolite deposit. The host rocks are ash-flow tuffs and related tuffaceous breccias; the zeolite mineralization is predominantly clinoptilolite-heulandite. This deposit is of open system origin, but may also overlap zones of weak hydrothermal alteration (Olson, unpublished data, 1977). St. Cloud Mining Company currently mines and processes Winston deposit zeolites for sale into consumer, industrial, and agricultural markets throughout North America.

Oregon: Eastern Oregon has an abundance of natural zeolite deposits with economic potential. Minor zeolite occurrences in marine sedimentary rocks in western Oregon, however, have no apparent commercial potential.

Durkee Chabazite. This deposit is in a basin near the village of Durkee in Baker County. The basin covers 120 km^2, of which 18 are underlain by zeolites and authigenic feldspar. Possibly economic zeolite minerals include chabazite, erionite, and clinoptilolite with minor analcime, mordenite, and phillipsite. The zeolites were derived from siliceous tuffs deposited in a late Miocene saline/alkaline lake and formed diagenetically at low temperature and pressure. The zeolitized ash-fall and reworked tuffs and one ash-flow tuff unit are interbedded with diatomite, mudstone, sandstone, siltstone, and conglomerate. Detailed studies have been carried out on this deposit by the US Geological Survey (Sheppard and Gude, 1969b, 1980, Gude and Sheppard, 1978, 1986).

Exploration by Anaconda Minerals Company, Occidental Minerals Corporation, and Filtrol Corporation defined promising target areas with possible large tonnage of chabazite, but their exploration results remain unknown.

Harney Basin Zeolites. Zeolitized tuffaceous rocks in near-surface Miocene beds of the Harney-Malheur Basin comprise huge resources of clinoptilolite, with lesser erionite, chabazite, phillipsite, and mordenite on the basin margins. The principal mineralized zones underlie more than 20 km^2 about 40 km south of Burns in Harney County (Sheppard, personal communication, 1977, 1992, Holmes, unpublished data).

Zeolitic coarse- to fine-grained ash-fall tuffs and reworked zeolitic tuffaceous sediments of lacustrine and fluvial origin occur in the partly eroded Tertiary basin fill. Greene et al. (1972) mapped the host rocks as tuffaceous sedimentary rocks. Two members have been identified in this sedimentary sequence in the central basin facies. The lower member, the more favorable host rock for zeolitization, is mostly coarse to fine, altered ash-fall tuff beds of interfingering lacustrine and fluvial origin. The upper member is chiefly a mixture of pumice, scoria, other lithic fragments, plagioclase, and glass shards in a clay matrix.

A 20-m-thick massive clinoptilolite bed is present over much of the basin. Toward the southern margin of the basin, the clinoptilolite-rich tuff interfingers with thinner beds of erionite, chabazite, and phillipsite. The zeolite deposits of the Harney Basin formed as a result of diagenetic alteration of ash-fall tuffs and reworked tuffaceous fluvial sediments within a closed hydrologic basin.

Potential clinoptilolite resources in the Harney Basin deposit are huge, although the overall quality is lower than the major deposits of the Owyhee Region. Erionite resources are scattered in thinner beds in lacustrine and fluvial sediments near the southern margin of the basin. The chabazite resource potential is largely unknown, due to limited drilling and trenching. Phillipsite resources are poorly defined and are scattered through marginal areas.

John Day Zeolites. Thick beds of zeolitic tuffs occur in the Oligocene-Miocene-age John Day Formation in central Oregon (Hay, 1963b). Clinoptilolite is the dominant zeolite in this open system setting, but the overall quality is lower than the major deposits in the Owyhee region to the southeast. Early studies of this clinoptilolite as a possible source for nuclear waste treatment at the Hanford atomic works in Washington State and other sites have spurred continuing interest in this district. Calcic siliceous chabazite has also been reported in the John Day Formation (Sheppard and Gude, 1970).

Fig. 5. Scanning electron micrograph showing acicular erionite (E) from zeolitized ash-fall tuffs beds at Rome Cliffs, near Rome, OR (magnified 660X).

Rome Zeolites. The Rome zeolite district has large potential resources of high-quality erionite, mordenite, and mixed mordenite-clinoptilolite. This area is about 60 km west of the town of Jordan Valley in Malheur County near the Owyhee River and Crooked Creek in a desert upland region dissected by steep-walled canyons (Sheppard and Gude, 1983, Wolf and Ellison, 1971).

The zeolites occur in several tuffaceous members of the Rome beds, a Miocene fluvio-lacustrine sequence. The formation dips several degrees northeasterly in the Rome Cliffs area several kilometers north of the village of Rome and is about 130 m thick. Two zeolitized beds in the lower member have significant commercial promise. Each bed is about 3 m thick and contains 50 to 70% erionite. The beds are slightly harder than the surrounding unconsolidated sediments, forming resistant ridge tops and mesas. Fig. 5 is a scanning electron micrograph of acicular erionite from the prominant zeolitized ash-fall tuffs at the Billboard Cliffs.

The upper Crooked Creek zeolitic member ranges from 5 to 8 m in thickness and consists of two light gray, fine-grained ash-fall tuff units separated by a prominent thin orange or yellow tuff unit known as the Orange Layer. The member crops out in the Crooked Creek drainage in the southwest part of the basin. Both the lower and upper gray units are predominantly mordenite with some clinoptilolite, especially in the lower one.

Sandy erionite-rich tuff beds also occur in small dimension-stone quarries where US Highway 95 crosses Crooked Creek. These beds are more than 10 m thick locally and contain 40 to 70% erionite. Blocks of the soft but weather-resistant erionite were sawed in these quarries for use in local buildings.

Sheaville District. This district covers at least 10 km^2 where the westward-flowing Succor Creek crosses the Oregon-Idaho state border and the axis of a north-south trending anticline. The area is 19 km north of Jordan Valley, Malheur County, between US Highway 95 and the state boundary. The Sheaville District has large resources of high quality clinoptilolite in a favorable mining setting on lands controlled largely by Steelhead Resources Corporation.

The Sheaville clinoptilolite sequence is in the Miocene Sucker Creek Formation, a lacustrine formation of interbedded tuff, zeolitized tuff, bentonite, and carbonaceous siltstone (Almner and Grim, 1990, Sheppard and Gude, 1983).

As many as five thick, massive units of white to light gray, zeolitized ash-fall tuff crop out in the area. Clinoptilolite is the predominant zeolite, with trace amounts of mordenite. The zeolite beds are interbedded with bentonite and carbonaceous siltstone. Zeolitization is most intense near the axis of the anticline. The more intensely altered zeolitic rocks appear to have hydrothermal alteration superimposed on open-system clinoptilolite. The altered rock is massive; bedding and sedimentary structures were largely obliterated (Holmes, 1990).

Past zeolite production by Norton Chemicals, Inc. has been confined to the uppermost clinoptilolite bed on the western flank of the anticline. The bed is well exposed several hundred meters east of US Highway 95, near the western edge of the zeolitized area.

The lowest clinoptilolite bed is the thickest in the district and crops out on the south side of Succor Creek. It is as much as 22 m thick and consists of white to light-gray massive clinoptilolite. Occidental Minerals Corporation drilled this bed between 1977 and 1981 and opened a large test pit near the anticline axis in 1982 (Holmes, unpublished data).

Succor Creek Deposits. Scattered bodies of white to light gray, massive and sandy zeolitic tuff occur in the Miocene Sucker Creek Formation along the lower part of Succor Creek, both within and downstream from Succor Creek State Park (Almner and Grim, 1990). Clinoptilolite is the predominant zeolite, with content ranging from 30 to 70%. Zeolite thicknesses are as much as 26 m. These occurrences have a hydrothermal alteration overprint (Almner and Grim, 1990) and generally are more variable and less altered than the Sheaville and Chrisman Hill deposits of comparable age. Teague Mineral Products operates an open pit in the lower Succor Creek area to extract clinoptilolite for pet litter and odor control uses.

South Dakota: Economic occurrence of natural zeolites in this state is limited to the Sheep Mountain Table clinoptilolite deposit.

Sheep Mountain Table Clinoptilolite. Very large resources of clinoptilolite occur in zeolitic tuff in the Rockyford Member of the Miocene Sharps Formation which extensively caps hills and mesas in the badlands on the Pine Ridge Sioux Indian Reservation and adjacent Badlands National Monument to the north (Raymond, 1986). The zeolite-rich tuffs are typically as much as 33 m thick and are of open system genesis. The clinoptilolite content is commonly in the 50 to 60% range, with ammonia cation exchange capacity values of 0.6 to 1.40 meq/g (Desborough, 1989). This deposit is the farthest east of the large clinoptilolite sources in the United States with the exception of the lower quality Alabama-Mississippi interbedded clinoptilolite and cristobalite.

Texas: Texas has large resources of clinoptilolite in two regions (Walton et al., 1972). In south Texas a belt of zeolitized ash-fall tuff in the Wellborn and Yegua Formations of the Eocene Jackson Group of Gulf Coast sediments extends from Zapata County on the Rio Grande River northeasterly, cropping out west of Tilden in McMullen County, to the vicinity of the town of Chestahova in Karnes County. The zeolite mineralization is spotty and not continuous. The zeolitic tuffs dip gently to the east toward the Gulf of Mexico. The clinoptilolite-rich ash fall tuffs are white, massive to thin bedded and fine grained in beds typically 3 to 5 m thick but as much as 10 m. Overall the resource is very large.

A cluster of clinoptilolite prospects is located 8 to 13 km west to northwest of Tilden. The Buck Martin pit has a 3 m thickness of flat-lying, white, flaggy, zeolitized ash-fall tuff with a clinoptilolite content as much as 80%. The nearby Kuykendall pit has a 4 m bed of similar material with clinoptilolite content as much as 90%. The Tilden area zeolite resources are large, but are in scattered pods and small bodies. Several producers mine and process clinoptilolite locally to make pet litter and other absorbent products; this material contains as much as 20% bentonite (Holmes, unpublished data).

Large resources of clinoptilolite also occur in zeolitized ash-fall tuff beds and tuffaceous sediments in the Oligocene Tascotal Formation in the Trans-Pecos region between Presidio and Marfa (Bostros, 1976, Walton and Henry, 1979). The ash-rich beds are hundreds of meters thick and cover many square kilometers in the broad Alamito Creek Valley. In the vicinity of the village of Casa Piedra, clinoptilolite occurs in numerous beds of altered ash-fall tuffs, but commercial interest is concentrated on three well-exposed white, massive clinoptilolite beds just east of the community. Thickness of these is typically 1.5 to 3 m. The clinoptilolite content of the beds commonly is 50 to 70% (Walton, 1975).

Extensive clinoptilolite resources are present in the Tascotal Formation in the Wire Gap area, east of Casa Piedra. The clinoptilolite content of the richer beds generally is 50 to 70% with abundant detrital impurities. Much of the material is reworked ash-fall and ash-flow tuffs. Individual bed thicknesses range from a few centimeters to more than 10 m, but their limited lateral extent suggests that deposition occurred in shallow, local basins (Holmes, unpublished data).

Utah: Zeolite occurrences have been identified at 56 localities in Utah. Of these, 29 are clinoptilolite or analcime in altered tuffs or tuffaceous sediments and are principally of open system origin. Six of these clinoptilolite occurrences appear to have economic potential. The sedimentary deposits range from Triassic to Miocene in age, but mostly are Tertiary. The other 27 occurrences are of hydrothermal/igneous origin and include the following zeolite minerals: analcime, chabazite, heulandite, laumontite, mordenite, natrolite, phillipsite, stellerite, stilbite, and thomsonite. Ages of the rocks hosting hydrothermal/igneous occurrences are Pennsylvanian to Miocene. None of the hydrothermal/igneous occurrences appear to have economic potential (Mayes and Tripp, 1991).

The sedimentary deposits are present in ten counties as replacement of tuffs or tuffaceous sediments. Clinoptilolite occurs in all ten and analcime in six. Most of the sedimentary analcime occurrences are in the Eocene Green River Formation in the Uinta Basin. The sedimentary zeolites are of open-system origin, except for two in closed-basin settings: ancient Lake Uinta of the Green River Formation and Lake T'oo'dichi' in the Jurassic Morrison Formation on the Colorado Plateau. The Mountain Green and Trenton occurrences are paleo-lake deposits, and their origin has not been clearly established (Mayes and Tripp, 1991, Lindsey, 1975a,b).

Six sedimentary clinoptilolite deposits have economic potential. The Mountain Green area in Morgan County contains estimated zeolite resources greater than 240 Mt. The clinoptilolite content commonly is 50 to 80% and ammonia cation exchange capacity is moderate to high. A large clinoptilolite deposit in the Salt Lake Group crops out near Trenton in Cache County, extends northward into Idaho, and is probably the northern equivalent of the Mountain Green occurrence (Mayes and Tripp, 1991). Major clinoptilolite deposits in the Thomas Range and at Keg Mountain in Juab county contain 50 to 90% clinoptilolite (Lindsey, 1975a,b). Large clinoptilolite occurrences at Cove Fort and Three Creeks Reservoir in the Tushar Mountains in Beaver and Milliard Counties contain as much as 85% clinoptilolite. Based on current knowledge, large sedimentary clinoptilolite deposits at Telluride Power Plant No. 2 in Beaver County and near Panguitch in Garfield and Iron Counties appear to be too low grade to be of economic interest (Mayes and Tripp, 1991).

The 27 hydrothermal/igneous zeolites are in 11 counties,

Fig. 6. Scanning electron micrograph showing large clinoptilolite crystals (CL) from Robin's Egg Blue ash-fall tuff bed in Washakie Basin, WY (magnified 4800X).

mostly as veins, amygdule fillings, cavity linings, replacement minerals, and cement, or as alteration products surrounding igneous intrusions and hydrothermal ore bodies. At the Indian Hollow occurrence in Pinto County, hydrothermally altered rhyolitic flows contain as much as 75% mordenite (Mayes and Tripp, 1991).

The Trenton clinoptilolite mentioned above is a zeolitized air-fall tuff in the Tertiary Salt Lake Formation. It is located along the west side of Cache County, about 35 km north-northwest of Logan and extends northward into Idaho. The zeolitized green, fine-grained tuffs constitute a very large potential resource, containing as much as 70% clinoptilolite. It is a high-potassium clinoptilolite deposit, very similar to the Mountain Green occurrence to the south (Mayes and Tripp, 1991, Southard and Kolesar, 1978). Danzl (1985) reported clinoptilolite in the Salt Lake Formation at Oneida Narrows, 50 km northeast of the Trenton occurrence.

Wyoming: Natural zeolites have been described at 88 locations in Wyoming. Potentially minable zeolites, in order of importance, include clinoptilolite deposits in the Washakie Basin, along Beaver Rim, and on Lysite Mountain (Harris and King, 1990). Other promising zeolite occurrences are reported in Tertiary tuffaceous rocks in the state by Surdam (1972), Boles and Surdam (1971), and Van Houten (1964). None of the known Wyoming occurrences of clinoptilolite, heulandite, and analcime in sedimentary rock of Cretaceous and Triassic age exhibit the quality and grade adequate for economic development.

Washakie Basin. The Washakie Basin clinoptilolite was discovered and described by Roehler (1972, 1973b), who identified clinoptilolite and mordenite in tuffs of the Laney Shale Member of the Green River Formation and in tuffs in the Robin's Egg Blue (REB) tuff bed of the Adobe Town Member of the Washakie Formation. The REB bed is a light-blue to greenish-blue zeolitized ash-fall tuff altered to 50 to 90% clinoptilolite (see Fig. 6). The bed ranges from 1 to 8 m in thickness and varies in sand content throughout the Washakie Basin. The bed also locally splits into a series of beds that are about 30 m thick in aggregate (Roehler, 1973b). Extensive drilling and testing have established that the REB bed is locally very rich in clinoptilolite and suitable for a wide range of applications, including nuclear waste clean-up.

Rocky Mountain Energy Corporation (now Union Pacific Resources-Minerals) attempted to develop reserves of REB clinoptilolite on its checker-board sections on the north rim of the Washakie Basin between 1978 and 1983. The venture also produced limited tonnages of the blue-green rock for cat litter and other deodorant use (Hulbert, 1987). The US Zeolite Corporation subsequently gained lease control of the Rocky Mountain Energy fee lands and claims and obtained State mineral leases covering other promising REB occurrences along the rim. The company is now doing testing and marketing studies and has produced small tonnages of clinoptilolite for absorbent and industrial uses (Van Fleet, personal communication, 1991).

Beaver Rim Zeolites. Several beds of clinoptilolite, mixed clinoptilolite-chabazite, and clinoptilolite-erionite are present in ash-fall tuffs in the Eocene Wagon Bed Formation in the Beaver Rim District, Fremont county (Deffeyes, 1959a, Boles, 1968, Boles and Surdam, 1971, Van Houten, 1964). Some phillipsite and analcime also occur. The US Energy Corporation produced and sold small tonnages of clinoptilolite during the 1980s to western aquacultural markets. Significant resources of clinoptilolite along with promising occurrences of chabazite offer economic potential (Harris and King, 1990).

Lysite Mountain Area. Zeolites have been identified in several zones in the Eocene-age Wagon Bed Formation in Hot Springs, Washakie, and Fremont Counties (Bay, 1969, Harris and King, 1990). Clinoptilolite is the predominant zeolite and clinoptilolite-rich tuffs range in thickness from a few centimeters to 30 m. These tuffs also contain minor analcime, erionite, and mordenite.

Foreign Zeolites

International attention to natural zeolites has greatly increased over the past decade. Many new deposits have been identified, research flourishes in industrial and particularly agricultural applications, and local zeolite mining and processing operations have opened up in many countries, e.g., Argentina, Australia, Cuba, and Indonesia. Much of this activity has not reached the international trade journals or scientific publications, and the industry remains hampered by a lack of translated foreign articles and papers for research and commercial study.

Antigua: Clinoptilolite and mordenite are abundant in a thick Oligocene tuffaceous series in the Central Plains of the island of Antigua in the Leeward Islands of the eastern Caribbean Sea. The Antiguan zeolites were identified by a United Nations study team and appear similar to some clinoptilolite-mordenite deposits in the Green Belt district of Japan (Nowakowski, private communication, 1983).

Argentina: Cristamine SA processes and sells clinoptilolite from local sources for use in animal feeds.

Australia: Natural zeolites occur in Paleozoic rocks in the Drummond Basin, Queensland, and in rocks from Late Carboniferous to Tertiary age in New South Wales. Zeolite Australia Ltd. has produced and sold clinoptilolite from sources in New South Wales since 1988 (Stephen and Gout, 1993).

The occurrence of natural zeolites in New South Wales is discussed in numerous state survey reports by P.W. Fredrickson, G.G. Holmes, S.R. Pecover, and others (Holmes and Pecover, 1987).

Abundant zeolites are associated with acid to intermediate pyroclastic and epiclastic rocks in several Late Carboniferous continental to shallow marine formations in the Tamworth Belt in New South Wales. These zeolites formed primarily by alteration of vitric materials in non-welded ignimbrite ash-flow and ash-fall tuffs. Typical zeolite mineralogy includes clinoptilolite, clinoptilolite-heulandite, mordenite, and laumontite (Holmes and Pecover, 1987, Flood and Stephen, 1993).

Diagenetic alteration of silicic ash-fall tuffs and associated tuffaceous sediments yielded more than 60% heulandite-mordenite in the 16 m thick Escott Zeolite Unit of the Late Carboniferous

Currabubula Formation, Werris Syncline, New South Wales (Flood, 1991).

Zeolite minerals including chabazite, natrolite, stilbite, heulandite, and analcime occur in veins and as cavity filling materials in fractures, vugs, and amygdules in some Tertiary basaltic rocks in New South Wales. Most of these occurrences are low grade and in hard rocks presenting difficult separation problems. Several localities with highly weathered host rock and zeolite content approaching 20% may offer opportunities to recover chabazite and other zeolite minerals.

Minor amounts of erionite, chabazite, and offretite have been identified in non-marine Tertiary sedimentary interbeds in volcanic series that include diatomite and tuffs in New South Wales.

Bulgaria: Commercial deposits of zeolitic rock are abundant and widespread in Bulgaria and they represent an important mineral raw material for the country.

The two principal areas of occurrence are the Srednogorian zone south of the Balkan Mountains and the Eastern Rhodopes (Gottardi and Obradovic, 1978).

The Srednogorian zone extends in a longitudinal belt between the Balkan Mountains and the Rhodopes and has zeolites in three geologic settings:

1) Volcanic sedimentary rocks (andesitic tuffs and breccias) containing mordenite-heulandite-analcime, laumontite with stilbite and prehnite, and stilbite with minor laumontite in the Senonian complex (Kostov, et al., 1971);

2) Green to grayish-green Obrochishche tuffs containing as much as 75 to 80% phillipsite and rare clinoptilolite;

3) Oligocene clays, bentonites, and tuffs in which clinoptilolite is the dominant diagenetic mineral derived from volcanic glass.

Industrially significant zeolitic deposits are widespread in the northeastern Rhodopes. The zeolitic rocks are commonly rich in clinoptilolite and associated with Oligocene rhyolitic pyroclastic rocks (Aleksiev and Djourova, 1975, Djourova and Aleksiev, 1988, Djourova and Milakovska, 1993). Sedimentary clinoptilolite and mordenite are reported in the Kurdjali area and a bed of white clinoptilolite as much as 35 m thick is derived from tuffs near the village of Perperek. Oligocene tuffaceous sediments between the villages of Beli Plast and Mest contain as much as 110 m of clinoptilolite. Oligocene acid tuffs of the Madzharevo area host clinoptilolite, mordenite, and analcime. Silica-rich zeolites such as laumontite and stilbite are common in and around the margins of the East Rhodopes depression (Kostov, 1970).

Canada: The Bay of Fundy in Nova Scotia is a classic locality of zeolite mineral occurrence, as fillings in fractures and vugs, and of interest only as mineralogical curiosities. Surdam (1973) identified laumontite and wairakite of burial metamorphic origin in Triassic marine volcanic rocks of the Triassic Karmutsen Group on Vancouver island in British Columbia.

Sedimentary zeolites discovered in Tertiary basins in interior British Columbia during the 1980s offer commercial potential. Clinoptilolite-heulandite occur in 5 distinct tephra lenses in the Princeton Basin (Hora and Kwong, 1984, Read, 1987). These tuffs range in thickness from 7 to 22 m and also include some layers of mixed zeolite and bentonite. The zeolite replaces original glass shards in waterlain tuff as well as glass in volcanic breccia lenses of the Allenby Formation. Weak clinoptilolite-heulandite mineralization is also present in silicic waterlain tuffs in the Tertiary Lower Volcanic Formation and Allenby Formation in the Tulameen Basin (Read, 1987). Heulandite-clinoptilolite occur in Tertiary sedimentary rocks of the Cache Creek, McAbee, and Battle Creek areas. These rocks are mostly bedded rhyolitic tuffs of Eocene age and range from less than a meter up to 6 m thick (Read, 1987). Weak clinoptilolite-heulandite mineralization was also discovered in altered Eocene silicic tuffs in Hat Creek Valley (Read, 1987).

Chile: Zeolitized rhyolitic ash-fall tuffs and tuffaceous sediments in Chile host open-system clinoptilolite mineralization.

China: Clinoptilolite and mordenite deposits occur at a number of sites throughout China, most notably in altered volcanic rocks of open system origin. Most of the 3 Mt of zeolite produced annually in China go into pozzolanic applications, and the balance is applied principally to agricultural and industrial uses. Chabazite, phillipsite and erionite are reported in closed system settings in interior desert playas (personal files).

Cuba: Exploration for natural zeolites began in the 1970s has developed resources of more than 300 Mt, principally clinoptilolite-heulandite and mordenite (Gonzales and Costa de los Reyes, 1993, Griffiths, 1987, Fuentes and Gonzales, 1991).

Clinoptilolite-heulandite resources include the Tasajeras-Piojillo, San Antonio, Najasa, and La Pita deposits. The Palmarito and San Cayataro deposits are nearly pure mordenite. Most deposits consists of altered volcanic tuffs within volcanogenic sedimentary rocks of Lower Cretaceous and Middle Eocene age. More than 20 zeolite deposits have been identified. Four deposits are currently being exploited by five processing plants, with an overall annual production capacity of 500 kt (Anon., 1988, Gonzales and Costa de los Reyes, 1993).

Czechoslovakia: Natural zeolites were discovered at Nizny Hrabovec in eastern Slovakia in 1976. Reserves of more than 7 Mt of clinoptilolite ranging in grade from 40 to 60% clinoptilolite have a mean cation-exchanges capacity of 0.78 meq/g (Samajova and Kraus, 1977).

France: Exploration in France for economic zeolite deposits began in 1989, principally in the Massif Central region. Three areas of promising discoveries include clinoptilolite and heulandite in ash-fall sediments of the Limagne Bourbonnaise basin and phillipsite in ash-pumice tuffs originating from stratovolcanoes of the Cantal and Mont-Dore regions (Rocher, et al., 1991).

Clinoptilolite occurs in Upper Cretaceous marls and chalks in the Anjou district on the west side of the Paris Basin. It has also been detected in Paleogene marine sediments of the Paris Basin. Analcime and clinoptilolite are present in sediments containing volcanic debris in the Limagne de Clermont-Ferrand (Gottardi and Obradovic, 1978).

Germany: Zeolitic rocks are mined at several sites in Germany as pozzolans and construction materials. Clinoptilolite in Upper Cretaceous marls and marly limestone is mined as a raw material for cement near Hannover and surrounding areas in northwest Germany.

Quaternary volcanic rocks of the Bonn and Laacher See area of the Rhine region are mined as pozzolans from basaltic lavas, pumiceous tuffs, and highly zeolitized ash-flow tuffs. Zeolites include chabazite, phillipsite, and analcime, derived from altered lapilli and other volcanic glass.

Tertiary sediments around the Nordlinger Ries impact crater in Bavaria contain clinoptilolite and analcime, as well as trace erionite, phillipsite, and harmotone (Fruchtbauer et al., 1977, Stoffler et al., 1977).

Franke & Schulte GmbH and H.J. Schmidt Mineraltechnik GmbH & Co. custom process and sell zeolite products as trading companies.

Greece: Zeolite-containing tuffaceous formations have been identified in northern Greece, Samos Island and the Cyclades Islands (Griffiths, 1987). Deposits of clinoptilolite occur with diagenetic smectite and cristobalite in Eocene volcaniclastic sedimentary rocks in the Metaxades area, Thrace (Tsolis-Katagas, and Katagas 1990, Tsirambides et al., 1993).

Hungary: Major deposits of clinoptilolite are present in altered Sarmatian rhyolitic tuffs and ignimbrites in the Tokay Mountains, where clinoptilolite or volcanic glass typically make up 65 to 95% of the rock (Gottardi and Obradovic, 1978). Sedimentary rocks of the Zemplen Mountains contain clinoptilolite and mordenite. The volcanic tuff at Bodrogkeresztur contains mordenite. At Ratka, the sedimentary rocks contain clinoptilolite and at Ujhuta the rhyolitic tuffs contain mixed mordenite and clinoptilolite (Griffiths, 1987).

Clinoptilolite and mordenite are mined at Mad-Suba and Mad-Harcsa respectively. Hungarian production of natural zeolites was of the order of 40 to 50 kt in 1987 (Griffiths, 1987).

Clinoptilolite occurs in two horizons within Miocene tuffs and tuffaceous sediments in the Mecsek Mountains, southwest Hungary (Polgari et al., 1993).

Iceland: Zeolites formed in basaltic flows and related volcanoclastic sedimentary rocks in low temperature geothermal areas. Four distinct zeolite zones in order of increasing rock temperature are: chabazite, mesolite-scolecite, stilbite, and laumontite. Analcime occurs in all zones. Mordenite, heulandite, laumontite, and analcime are the zeolites most commonly found in the high temperature geothermal areas (Kristmanndottir and Tomasson, 1978).

Indonesia: Altered volcanic rocks hosting natural zeolites have been identified on several islands in Indonesia. Newly developed sources, primarily of clinoptilolite, are providing zeolite products to regional agricultural and aquacultural industries.

High quality clinoptilolite in prospects at Mt. Ratai and Tarahan in Lampung Province, Sumatra, is under development by a Canadian-Australian joint venture with local interests for use as a cattle feed additive, animal waste absorbent, and slow release fertilizer media. Another zeolite deposit 20 km south of Panjung, the major port in Lampung Province, is currently producing crushed and screened clinoptilolite product for use in local aquacultural markets. Clinoptilolite and mordenite are produced from a mine on Java and made into products for sale throughout the Far East.

Italy: Italy has a great number of zeolite occurrences, ranging from the huge Neapolitan and Campanian zeolitic tuff systems in central Italy near Naples and Rome to zeolites of marine origin in Sicily to burial metamorphic and hydrothermal-origin tuffs in the Apennines in Northern Italy (Sersale, 1978, de Gennaro and Adabbo, 1993).

The Neapolitan Tuff and Campanian Tuff and equivalent volcanic formations crop out over extensive areas around Rome and the Gulf of Napoli and comprise the most promising zeolite sources in Italy. Neapolitan Yellow Tuff has been used extensively since Roman days as dimension stone (Aiello, 1993) and consists of phillipsite and chabazite within Quaternary age tuffaceous rocks of 20 to 120 m thickness, probably related to a buried caldera (de Gennaro et al., 1983). Campanian Tuff occurs over a wider area than the Neapolitan Tuff, consists of altered tuffs and ash-flow tuffs. It is as much as 60 m thick, and has also been a major source of dimension stone in this region since Roman times (Aiello and Porcelli, 1975, de Gennaro et al., 1984). Both tuff systems are sources of raw materials for a wide range of industrial and agricultural applications.

Coarse pyroclastic rocks from Roccamonfina Volcano, located about 50 km northwest of Naples, contain significant amounts of chabazite and analcime and are quarried as building stone.

Tuff layers in a thick Quaternary pyroclastic sequence from the calderas of the Vulcano Laziale, about 20 km east of Rome, contain chabazite and phillipsite. The tufo lionato unit is the most widespread tuffaceous rock in the area.

Chabazite is present in altered zones in tuffs and ignimbrites within Quaternary calderas in the Latium region near Rome. Lenzi and Passaglia (1975) described widespread chabazite occurrences Bracciano Lake, commonly making up more than 50% of the rock.

Hydrothermal occurrences commonly contain zeolite mineral assemblages with no economic potential, such as the heulandite, stilbite, chabazite, analcime, natrolite, scolecite, and thomsonite at Trentino and Alto Adiger or other complex mineralogies as in Sicily or on the island of Elba (Alietti et al., 1967a).

The most promising marine occurrence is the Mount Epomeo Green Tuff on the island of Ischia in the Gulf of Naples. Phillipsite and minor analcime and chabazite in the Mount Epomeo Green Tuff represent altered Quaternary volcanic sediments within the dominantly volcanic lithology of the island (Gottardi and Obradovic, 1978, Ylagan et al., 1993). Up to 50% phillipsite occurs in Tertiary marine volcanic rocks at Garbagna, and in lesser contents at Montegibbio and Contignaco in the northern Apennines (Sersale et al., 1963, Gianello and Gottardi, 1969).

Clinoptilolite interbedded with diatomite and gypsum occurs in closed-system Tertiary lagoonal sedimentary rocks near Zovencedo, Vicenza, in northern Italy. The clinoptilolite content is 10 to 20% of the host rock (Alietti and Ferrarese, 1967b).

Analcime constitutes as much as 35% of the pietra verde, a Ladinian formation widespread in the Dolomites. The pietra verde is probably an acid tuff which was metamorphosed in a burial metamorphic setting to greenschist facies rock (Callegari, 1964, Callegari and Jobstraibizer, 1964).

Japan: Japan is a leading producer of natural zeolites. It utilizes zeolites for a wide range of agricultural, industrial, and consumer uses and is a world leader in development and application of zeolite technology. Annual production has been 60 to 100 kt or more since the mid-1960s. Japanese zeolite occurrences have been extensively studied by many workers and reviewed by Utada (1970, 1971) and Iijima and Utada (1972). Torii (1978) surveyed Japan's utilization of natural zeolites.

Zeolites occur in rocks ranging from Triassic to Quaternary in age, but the majority are in Miocene pyroclastic and volcanogenic sedimentary rocks. Most zeolite occurrences in Japan are related to low grade metamorphism. Many occurrences show zonation typical of burial metamorphism; others show zoning caused by hydrothermal effects of igneous intrusion. Boles (1977) describes the typical metamorphic zonation of Japanese occurrences and the mineralogy of the five zones Clinoptilolite and mordenite predominate in zones I through III; laumontite and analcime in zones IV and V.

Clinoptilolite and mordenite are the only zeolites currently mined in Japan, where the minerals have been used for a wide range of industrial and agricultural applications since 1949. Both minerals occur extensively in the green tuff belt of metavolcanic rocks. Clinoptilolite is the most common zeolite mineral in Japan.

At least 14 companies produced natural zeolites in Japan in 1987, but of these only half are producing significant quantities and only two had an output above 10 ktpy. Clinoptilolite is quarried at the Itaya mine in Yamagata Prefecture, Honshu, and processed by the Zeaklite Chemical & Mining Company, Ltd. for use as paper filler as well as agricultural and other applications. The company produces at least 60 ktpy and is Japan's largest zeolite producer. Sun Zeolite Ind. Co. Ltd, produces predominantly clinoptilolite from its deposit in Akita Prefecture for agricultural use (Griffiths, 1987). Nippon Kasseihakudo Co. Ltd. produces byproduct clinoptilolite from its calcium bentonite operations. They also produce synthetic zeolite A using clinoptilolite as a raw material source.

Smaller Japanese producers include Shinshin Development Co., Ltd. (mordenite + montmorillonite), Hamamichi Kohsan Co., Ltd. (zeolite + montmorillonite), and Asahi Kasei Kogyo Co., Ltd. (mordenite) (Griffiths, 1987). Other markets for Japanese zeolites are mainly in adsorption and water treatment processes.

Korea: Clinoptilolite-heulandite and mordenite have been found in both countries forming the Korean peninsula.

Within the Democratic Peoples' Republic of Korea, zeolites are found within Pliocene acid volcanic rocks in the Kilcho-Mjong Chong Province in the northeastern part of the Korean Peninsula. Clinoptilolite-rich and mordenite-rich tuffs are found in the northeastern part of the province, near the villages of Han Zin and Pho

Ha. Two clinoptilolite deposits in the southwestern part of the province contain clinoptilolite-heulandite, phillipsite, and chabazite (Han et al., 1985, Filizova et al., 1993).

Clinoptilolite and mordenite occur within altered Miocene tuffs and tuffaceous sediments in the Guryongpo area in the Republic of Korea (Kim and Choi, 1993, Noh and Kim, 1982).

Mexico: Natural zeolites occur in altered Tertiary volcanic rocks throughout Mexico, notably at Agua Prieta, El Cajon, Tetuachi, Arizpe, and San Pedro, all in Sonora Province, and at Etla Oaxaca, San Antonio, Oaxaca, San Luis Potosi, Puebla and Guanajuato. Large buildings, including a cathedral at Oaxaca, have been built of tuffaceous rocks altered to clinoptilolite and clinoptilolite-mordenite over four centuries in southern and central Mexico because of its strength, ease of working, and durability. Erionite has also been tested for the sorption of fission products from Mexican sources (personal files). Boles (personal communication, 1992) reported that associates have identified clinoptilolite, analcime, laumontite, mesolite, and natrolite occurring as alteration products of volcanic glass in Mesozoic sedimentary rocks in Baja California.

New Zealand: Natural zeolites are abundant in the Waitemata Group, an early Miocene sedimentary sequence in Auckland Province. The main part of the Group consists of flysch sediments where zeolites are present in interbedded volcanoclastic rocks. Alteration products of the volcanic glass in these beds include clinoptilolite, chabazite, mordenite, erionite, phillipsite, and analcime. The zeolitization appears related to hot-spring activity (Sameshima, 1977). The zeolites in the Waitemata Group are generally beds that are too thin to be commercial sources.

Laumontite, heulandite, and analcime occur within burial metamorphosed Triassic volcanic rocks in the Southland Syncline on South Island. Mineral distribution in the Taringatura Hills was studied extensively by Coombs (1954, 1958) and in the Hokonui Hills on the Southland Syncline by Boles (1971, 1974) and Boles and Coombs (1975, 1977).

Poland: Low grade clinoptilolite occurs at several sites within the country. First, clinoptilolite makes up as much as 40% of the volume of the upper part of the Upper Eocene radiolaria shale in the Miedzybrodzia quarry near Sanok in the Carpathian flysch. The zeolite is in a bentonitic sequence and may be an alteration product of montmorillonite.

Wieser also reported clinoptilolite as disseminated crystals and lenticular bands in Lower Eocene variegated shales of the External Flysch Carpathians. The discovery is from a drill core which intersected a 5 m thickness of zeolite in claystones and shales. The clinoptilolite forms as much as 10 to 30% of the volume and appears to be of deep sea origin.

Republic of South Africa: Clinoptilolite is produced from altered Jurassic volcanic rocks at the Pratley Perlite Mining Co. Pty. Ltd. mine at Hluhluwe in Northern Zululand in the Province of Natal. Both clinoptilolite and mordenite are associated separately with perlite and white calcium bentonite materials, which are also mined and processed (Griffiths, 1987).

Romania: Many natural zeolite occurrences are reported in Romania, but most are related to hydrothermal conditions during magmatism. Zeolites are less common in open hydrologic system type deposits and burial diagenitic-low grade metamorphic occurrences (Istrate, 1980).

More than 150 hydrothermal zeolite occurrences are known, but none have any apparent economic potential. Most of these are related to hot spring and shallow thermal activity related to shallow intrusions, volcances, or lava flows of three Alpine lithologic zones in northwest Romania. Hydrothermal zeolite minerals include natrolite, mesolite, scolecite, mordenite, heulandite, clinoptilolite, stilbite, stellerite, chabazite, and analcime. Burial metamorphic zeolites are less common and include principally stilbite and laumontite in metamorphic and pre-tertiary sedimentary rocks (Istrate, 1980).

Large deposits of clinoptilolite and mordenite occur in late Tertiary tuffs in the Transylvanian Basin and in the Precarpathian zone. These tuffs are commonly tens of meters thick and have been used for hundreds of years as building stone. They are of open system origin and offer significant economic potential (Istrate, 1980).

Altered volcanic tuff from the Mirsid deposit in the Salaj district contains 65% clinoptilolite and is a major economic zeolite source (Axente et al., 1983, Bedeleanu and Stoici, 1984).

Spain: Lower Miocene tuffs that contain analcime and clinoptilolite crop out in the La Vinuela Formation about 30 km northeast of Malaga. The green tuff appears to be of marine origin and contains 10 to 15% clinoptilolite and 10 to 20% analcime, as well as smectites and some unaltered glass.

Mordenite occurs in an Upper Miocene bentonite sequence near Cabo de Gata, about 14 km from Almeria. Montmorillonite is the major component of the bentonites, with mordenite the next most common mineral. Minas Volcan SA has capacity to mine 10 ktpy zeolites from quarries in Almeria province. Bentonitas Especiales SA mines and processes zeolites in Madrid Province in conjunction with its bentonite mining operations.

United Kingdom: Gottardi and Obradovic (1978) describe nine zeolite occurrences in the United Kingdom, mostly clinoptilolite and some laumontite and analcime. None of these sites has major commercial potential, except possibly the clinoptilolite presence in fuller's earth deposits in the Lower Greensand Formation (Lower Cretaceous) of the Baulking and Fernham areas, Berkshire, and of the Nutfield area, Surrey. The fuller's earth beds have been a principal source of absorbent products for many years.

USSR: The former Soviet Union had developed a zeolite industry second only to Japan in technological capability and range of applications. Soviet production and use of zeolites is more oriented to agricultural applications, but encompasses industrial uses such as nuclear waste treatment material as well.

Gottardi and Obradovic (1978) present a cursory overview of Soviet natural zeolite occurrences and geology. Senderov and Khitarov (1970) and Tsitsishvili (1977, 1988, et al., 1984) summarize earlier information on Soviet zeolite research and commercial application, including zeolite synthesis. Mikhavlov et al., (1971), Fersman (1962), Gvakharia (1952) and Kossovskaya (1973) discuss zeolite genesis and describe many deposits and occurrences in the various states. Clinoptilolite is the most predominantly found zeolite, and zeolite production is principally in the western republics. Zeolitic materials are widely used as cement additives in the cement industry.

Several large clinoptilolite deposits occur in the Ukraine. The Sokirnitsa clinoptilolite source in tuffs in the Transcapathian Ukraine contains 60% clinoptilolite and has a cation-exchange capacity of 1.44 meq/g.

Lower Carboniferous beds of North Onega in the North Russian Basin contain minor analcime and phillipsite.

The South Russian Basin includes the southern part of the Russian Platform, an area near the Black Sea, and the region of Povolzje as well as the area of Jaroslav and Kirov. Zeolites occur here in Jurassic, Cretaceous, and Early Tertiary sedimentary rocks. Clinoptilolite is the predominant zeolite; mordenite and analcime are less common.

Clinoptilolite and analcime are present in the Ural Basin in rocks ranging in age from Late Permian to Tertiary.

Yugoslavia: Zeolites are widespread in Yugoslavia, including minerals of open system, hydrothermal, and burial metamorphic origin. Clinoptilolite and mordenite are the principal economic zeolites and occur in altered pyroclastic rocks of Triassic to Late Tertiary age. Clinoptilolite is mined for local use, mostly agricultural, from several deposits.

Late Tertiary zeolitized pyroclastic rock sources in South Serbia include clinoptilolite in Mio-pliocene tuffs and tuffaceous sediments at Zlatokop, Meckovac, Katalenac, and Jovacka Reka, and clinoptilolite, mordenite, and erionite in tuffs at Toponica. Mordenite and analcime are present in pyroclastic and clastic rocks in the Jarandol coal basin of Serbia (Stojanovic, 1972).

Clinoptilolite occurs in the Miocene tuffs around Beograd and in the area of Fruska Gora Mountain in Serbia (Nikolic et al., 1975). Deep coring penetrated thick Tertiary tuffs partly altered to clinoptilolite, mordenite, and analcime (Obradovic and Kemenci, 1975).

Early Tertiary zeolite occurrences include clinoptilolite from two localities. Oligocene tuffs as much as 60 m thick at Donje Jesenje, near Zagreb, host clinoptilolite. Oligocene tuffs and bentonitic tuffs between Mozirje and Celje, Slovenia contain clinoptilolite and analcime (Grimsicar, 1967).

The Triassic sedimentary series contains a prominent sequence of green tuffs (locally called pietra verde) with burial metamorphic origin zeolites. Clinoptilolite, analcime and laumontite occur in green tuffs of the Budva-Bar area, Montenegrin littoral (Obradovic and Stojanovic, 1972), and the area of Knin and Sinu, Croatia, and Vrsic Mountain, Slovenia.

Obradovic later reported the presence of laumontite and clinoptilolite in green tuffs of the Tjentiste region, Bosnia.

Zeolites are mined and processed at Zaloska Gorica, Slovenija. The operation produces ground and micronized zeolites too cement additive and agricultural use, and is expanding production capability to reach environmental and additional agricultural markets.

Other: Natural zeolite occurrences have also been reported in the following countries, but the occurrences are either insignificant and uneconomic or nothing has been published on the sources or it is generally unavailable in North America:

Algeria	Namibia
Austria	Netherlands
Chad	New Guinea
Denmark	Pakistan
Egypt	Philippines
Iran	Puerto Rico
Israel	Senegal
Jamaica	Spain
Kenya	Switzerland
Malawi	Taiwan
Mali	Tunisia
Mauritania	Turkey
Mongolia	Uganda

TECHNOLOGY

Exploration Techniques

Most economic natural zeolite deposits occur in altered ash-fall and ash-flow tuffs or tuffaceous sediments of Tertiary or, to a much lesser extent, Mesozoic age. Although zeolites in other geologic settings may be pursued in the future for special uses, it is expected that bedded zeolites will remain the predominant natural source for several decades at least. Prospecting for such deposits is difficult as they are finely crystalline, difficult to identify, and commonly appear similar to altered or unaltered tuffaceous rocks, bentonite, diatomite, and other fine-grained sediments.

Prospecting is commonly done by examining and sampling known zeolite occurrences and scouting adjacent areas for bodies of similar alteration and appearance. Many known zeolite deposits in the world, and even in North America and Europe, have not been sampled and characterized adequately, so there are many opportunities to develop new zeolite sources without entering costly grassroots exploration.

As the natural zeolites develop as a commodity, however, and known deposits and occurrences have been studied and sampled, innovative exploration will be necessary to find new sources. Exploration will then require development of target models based on the geology of specific regions, known zeolite occurrences, and desired parameters for a deposit which might yield specific zeolite products.

Field exploration for zeolites must rely on careful geologic examination and sampling. Zeolitized tuffs usually appear as obviously altered rocks, but locally may be indistinguishable from unaltered rocks. Zeolitized rocks are commonly white to gray in color, massive and blocky in outcrop character, resistant to weathering, and possess identifiable surface textures. Natural zeolites may attract iron oxides or other contaminants and be stained varying shades of red, brown, yellow, or green.

X-ray diffraction analysis of outcrop and drill samples is the technique most effective in identifying zeolite minerals. Specific zeolite minerals or mixtures of minerals may be identified, and semi-quantitative estimates made of their abundance. Ammonia cation exchange capacity (CEC) analysis of samples is also a very effective early screening technique to estimate the ion exchange capacity of zeolitized materials, particularly for clinoptilolite, chabazite, erionite, and phillipsite. This also screens out heulandite and analcime, which appear similar, but have lesser if any commercial potential. Mordenite samples require X-ray diffraction and gas adsorption or separation tests to gauge their value, as they are less responsive to ammonia adsorption.

Once a zeolite ore body has been identified, mapped, and drilled sufficiently to establish its physical parameters and likely tonnage, it should be thoroughly characterized by bulk sampling and testing to define its suitability to supply raw materials for specific applications. Sufficient drilling should be done to define adequate proven and probable reserves to establish the reliability of the source to produce specific products for sustained periods.

Mining Methods

Because of their low unit value, bedded sedimentary zeolites are mined by open pit methods. Excavation is generally done using conventional earth moving equipment; in thicker bodies, it may include mine benching methods. Such mining minimizes costs, as it employs low cost explosives, rubber-tired or tracked earth moving equipment, and direct truck loading in the pit for transport to a processing plant. Variations in ore quality may be handled by selective mining and blending during mining or stockpiling. Quality control is commonly done by sampling through drilling, periodic face sampling, visual assessment of the material in place, and systematic sampling of truck loads or the process stream.

Zeolites for special high value uses may be recovered by very selective open pit mining. Chabazite-erionite raw material mined from a 15 cm-thick bed at Bowie, AZ, is used by Union Carbide Corporation to make high value molecular sieve and catalytic products. Mining at the Mud Hills clinoptilolite deposit in southern California recovered ore with strict specifications in 30 cm slices from a 4.6 m-thick deposit to yield a very high value product for ion exchange material used by British Nuclear Fuels, Ltd.

Processing

Natural zeolites are sold as crushed and screened, finely ground, or micronized to ultra-fine ground products. Crushed and screened materials may be used for low value, coarser fraction applications, such as soil conditioner or pet litter, that can tolerate a fairly wide range of particle sizes and require low product cost. Many zeolite products are crushed, ground, and air classified to a size range of −60 to +325 mesh, using Raymond mills and comparable grinding equipment. Micronized products as fine as 5 to 10 μm and ultrafine products down to 1 μm are prepared for special uses, e.g., paper filler.

Natural zeolites may be enhanced in performance by washing them with acid or NaCl solutions to raise the H^+ or Na^+ content respectively. Clinoptilolite products are particularly enhanced in ion-exchange capacity by washing to replace the slower reacting K^+ ions with Na^+ ions. Bowie, AZ, chabazite products are commonly pelletized and lightly calcined to reduce their overall friability.

Specifications

Specifications depend on the uses of the zeolite products and vary widely because of the broad range of natural zeolite products, serving many markets. The American Society for Testing and Materials (ASTM) Committee No. D-32 has been established to set testing methods and standard specifications for zeolites in the United States.

Specifications and standards in Europe and Japan are commonly set by the producing companies in a market-driven setting. Zeolite producers deal with specifications in two ways: sell materials on a custom basis to specifications negotiated with the buyer; or sell materials on a product-line basis, where each zeolite product has a name or number designation and specific physical and/or chemical characteristics. In the United States, zeolite products are commonly sold under a trade name rather than as a mineral variety, e.g., clinoptilolite.

Testing for Zeolite Mineral Identification

Natural zeolite minerals are identified primarily by their crystal habit. Chemical analyses alone are not an effective method of identification, as many zeolites have a similar chemical composition. Macroscopic zeolites, particularly those occurring in vesicles and fractures in basaltic rocks, may be identified by careful visual examination. Virtually all economic natural zeolite occurrences, however, are of microscopic grain size. Positive identification and semi-quantitative determination of such fine-grained materials can be done only by laboratory study. The principal methods of identification are by X-ray diffraction and scanning electron microscopy, with lesser use of optical microscopy and differential thermal analysis.

X-ray Diffraction (XRD): X-ray diffraction is the most reliable and widely used method for identification of the zeolites and semi-quantitative estimation (Papke, 1972). This method has a high degree of accuracy and can readily be used to identify individual zeolite minerals in mixtures of zeolite and non-zeolite minerals. It requires only a technician to operate the X-ray diffraction equipment, which may be automated and able to run multiple determinations unattended and to interpret the results. XRD analysis is a non-destructive method and the powder used for the analysis can be further tested. The method is least reliable in mixtures of minerals; the lower detection limit for an individual zeolite is about 5%. The basic equipment for analysis has moderate cost, but an experienced diffractionist may be needed to interpret the results.

A representative portion of a sample is pulverized and a powder pack prepared. A full-scan X-ray diffraction pattern is run at a rate of 1°2θ per minute A scan of 2 to 16° usually is adequate to identify all common zeolite and most clay minerals, but a scan of 2 to 35° is necessary to identify all zeolites and impurities. A scan of only a few degrees may be run when the examiner intends to check only for the presence of a specific zeolite. Computerized XRD systems may be programmed to read and interpret semi-quantitative zeolite mineralogy. Special templates showing the diagnostic peaks of clinoptilolite, mordenite, chabazite, erionite, and other common zeolites assist in interpretation of the X-ray diffraction trace. More detailed descriptions of the XRD method are given by Papke (1972) and Slansky (1987).

Scanning Electron Microscopy (SEM): The scanning electron microscope is effective in identifying zeolite minerals and establishing their size, shape, genesis, and mineralogic setting. Magnifications of several thousand times or more allow visual examination of the zeolite minerals in small samples mounted on a specimen holder. Representative samples must be used to avoid misleading results. SEM analysis is relatively slow and costly, requiring an expensive electron microscope and an experienced operator to be effective.

Optical Microscopy: Optical microscopic methods may be used for the identification of zeolite minerals in thin sections or individual crystals. This can determine the size, distribution, and relationship of the zeolite and other minerals present. The method is somewhat limited by the very fine grain size of the zeolite minerals, their low birefringence, and variation of their indices of refraction with changes in chemical composition. This method is slow and requires an experienced mineralogist to be effective.

Differential Thermal Analysis (DTA): Differential thermal analysis may be used to identify zeolite minerals, but the method is less effective when several zeolites or certain contaminants are present or for small amounts of a mineral. Variations in the exchangeable-cation content may also cause differences in the differential thermal curve. The method is very useful in distinguishing clinoptilolite from the closely related mineral heulandite (Mumpton, 1960). Gottardi and Galli (1985) have published a comprehensive collection of DTA curves for natural zeolites.

Other Methods: Other analytical methods may be used in special circumstances, but are generally complex, very costly, and not currently practical for exploration or quality control. Such methods include infrared absorption spectrography, Moessbauer spectroscopy, electron spin resonance spectroscopy, electron spin echo spectrometry, solid state nuclear magnetic resonance, neutron diffraction, and synchrotron X-ray diffraction (Slansky, 1987).

Helfferich (1964) devised a field test based on ion exchange and molecular sieving properties to distinguish zeolites from clay minerals, feldspars, and volcanic glass. It may have limited field application, but it is not regarded as a reliable testing method; it is far less practical than screening with X-ray diffraction and cation-exchange capacity methods.

Several other practical field tests are used; while not reliable, they still have some value. A geologic pick driven into a hand-sized block of zeolite-rich rock may hold the rock firmly, allowing it to be picked up. This would not be likely with an unzeolitized tuff. A second and surprisingly useful test is to place a small fragment on one's tongue. A highly zeolitized rock will stick to the tongue due to the water absorbency of the material. Diatomite and bentonite may also exhibit this ability, although to a lesser degree. A variety of this test is to pour fluid on a rock. If it is a zeolite, the water may be absorbed very quickly, in as little as a few seconds; if it is unaltered rock or bentonite, it will run off the surface or cause swelling.

Testing for Characterization

In characterizing zeolitic materials for commercial uses, it is important to quantify other physical and chemical properties, generally tailored toward the desired application. The following are physical and chemical properties and tests that may typically be required: wet chemical analysis; cation exchange capacity (CEC);

specific gravity and bulk density; brightness, whiteness and color; hydration/dehydration testing; gas adsorption; attrition in water; surface area (internal and external).

Byproducts and Coproducts

Deposits containing two or more zeolites may yield several individual varietal products or mixtures of the zeolite minerals. For example, zeolite products from the Itaya mine in Japan, a clinoptilolite and mordenite source, include products of mordenite, clinoptilolite, and mixed mordenite-clinoptilolite, depending on the selectivity of mining and processing.

Zeolite mining may yield bentonite as a byproduct or coproduct. Bentonite commonly occurs in the same geologic setting as natural zeolites. At the Teague Mineral Products operation in Oregon, both minerals are mined from nearby pits and processed in the same grinding plant.

USES

Agricultural Products

Zeolites are widely used as soil conditioner and fertilizer extender, particularly in Japan and Eastern Europe. They buffer acidic soils, improve moisture content, and increase soil cation exchange capacity. Zeolites act as a slow-release media for ammonia, and to a lesser extent potassium. The zeolite crystal structure shields the ammonium ion from rapid consumption by nitrifying bacteria (Ferguson, 1984), allowing decreased use of nitrogen fertilizers. Natural zeolites, particularly clinoptilolite, are effective carriers of herbicides, pesticides, and fungicides because of their high ion-exchange and adsorption capacities.

Zeolites also act as scavengers for undesirable heavy-metal ions (e.g., Cd, Cu, Pb, Zn) in soils, particularly those contaminated by sludge or industrial waste (Semmens and Sayfarth, 1977). Clinoptilolite added to soils contaminated with strontium-90 and cesium-137 decreases the uptake of these radioisotopes by plants, aiding considerably in the treatment of a radioactive fallout area (Nishita and Haug, 1972).

Animal Nutrition

The addition of certain natural zeolites, particularly clinoptilolite, to animal feeds in controlled amounts increases the overall feed efficiency and decreases the ammonia stress in animal digestive systems. Controlled feeding to chickens, turkeys, and pigs has generally shown faster growth rates, improved feed efficiency, less diarrhea and related health problems, and drier and less odoriferous excrement. Also, excrement containing zeolites may be a more valuable fertilizer with slow-release fertilizer characteristics, as well as a drier, less odoriferous form (Mumpton, 1984, Tsitsishvili et al., 1984).

Feeding of clinoptilolite, and to a lesser extent mordenite, to animals is an accepted practice in Japan, Cuba, Bulgaria, and Hungary (Iijima et al., 1986, Griffiths, 1987). The use is growing slowly and steadily in the United States and the former Soviet Union, but it is still not widely accepted (Pond and Mumpton, 1984). Synthetic zeolites may also find acceptance for animal feed supplement (Griffiths, 1987, Shane, 1987).

Clinoptilolite fed to broiler chickens in amounts of 1 to 3% of the total feed increases feed efficiency and lowers the time required to reach a marketable size. Laying chickens fed supplemental zeolite experienced a lesser number of days to the first egg, an increased number of eggs lain per chicken, and a general improvement in health through a reduction in fecal bacteria (Griffiths, 1987).

Clinoptilolite also reduces the incidence of post-weaning diarrhea when added to the starter diet of young pigs. A zeolite used as a feed supplement may also reduce the levels of antibiotics needed in feeds, particularly for swine.

Aquaculture

Zeolites are valuable in controlling the environment of fish farms and fisheries, particularly for ammonia removal to avoid buildup of toxic levels. As more fish hatcheries are turning to self-contained systems with reuse of water, a sizable market for natural zeolites may develop (Mumpton, 1977). Konikoff (1973) and Johnson and Sieburth (1974) showed clinoptilolite to be effective in the removal of ammonia ions from recirculating fish-culture systems. In oxygen-poor environments, concentrations exceeding several ppm ammonia can lead to damage to gill tissue, various gill diseases, reduction in the growth rate, and even death (Mumpton, 1977).

Methods for the removal of ammonia from catfish pond water using zeolites, particularly clinoptilolite, have been investigated in Alabama and Mississippi (Reynolds and Williford, 1988, Williford et al., 1991).

Ammonia-removal kits containing granular clinoptilolite as the ion-exchange media are available in many pet stores for home aquarium use. The US Fish and Wildlife Service has also successfully use clinoptilolite-charged systems for recirculating water in tank trucks transporting live fish.

Zeolite adsorption units utilizing natural mordenite are used in Japan for oxygen enhancement in aerating waters for fish culture and the transportation of live fish (Minato, 1974).

Catalysis and Petroleum Refining

Zeolites with large channel sizes have large internal surface areas in their molecular structures and are capable of catalyzing many types of reactions (Dwyer, 1984, Rees, 1984). They are particularly useful in purification applications in the petroleum and petrochemical industries, removing water and carbon dioxide from gaseous hydrocarbons, removing HCl from gas streams, assisting in hydrogen or chlorine drying, and in chlorinated and fluorinated hydrocarbon purification. Synthetic zeolites are generally used because of their larger channel sizes, greater adsorption capacities, and uniformity.

Recent catalytic developments include the use of high-silica synthetic zeolites for the production of high octane liquid fuels from a wide range of simple organic molecules, such as methanol from natural gas. High quality clinoptilolite, chabazite, and mordenite show particular promise for catalysis and natural gas separation, but synthetic zeolites currently dominate the market for these applications (Flanigen, 1984, Iijima et al., 1986).

Coal Gasification

More than 25% of the coal reserves in the United States are in deep-lying, unminable seams that will require underground gasification if they are to be utilized. Natural zeolites have several beneficial uses in coal gasification to produce commercial natural gas (Mumpton, 1978). Oxygen-enriched air can be produced using certain zeolites to selectively adsorb nitrogen from the exhaust fumes derived by the combustion of underground coal. Exhaust streams may also be treated through separate zeolite adsorption systems to remove SO_2 and other contaminants from the resulting gas stream.

Dessicants

Natural zeolites, particularly clinoptilolite, have been shown suitable for use as dessicants in some industrial applications. Clinoptilolite from the Itaya Deposit in Japan is used in a pressure-swing unit to treat humid air. European uses include drying natural gases, carbon dioxide streams, refrigerator freon, and organic chemical streams such as transformer oil and xylene (Mayes and Tripp, 1991).

Partially dehydrated zeolites are useful as anti-caking or preservative agents in bulk agricultural and chemical products. Small amounts of zeolite dessicants are added to animal feed, fertilizers, or chemicals before sealing them into bags (Iijima et al., 1986).

Dimension Stone

Zeolitic tuffs have been used for thousands of years as lightweight, durable dimension stone. The physical properties of such stone are the important factors; zeolitic tuffs have low bulk density, weather well, and can be easily cut into attractive dimension stone.

Many buildings have been constructed in ancient and modern Rome and Naples from local zeolite sources, particularly the Neapolitan yellow tuff. Mumpton (1978) described many buildings and cathedrals in central Europe constructed with zeolitic blocks from regional sources. Building blocks from the Laacher See area in Germany, Kurdzali in Bulgaria, and Tokaj in Hungary are notable examples.

Dimension stone has been produced for centuries from several sources in Japan, particularly from pumiceous marine tuffs in the Green Tuff region, Tochigi Prefecture. A principal source is a quarry near Otsunomiya City, where a 108-m-thick green tuff has been mined for hundreds of years (Mumpton, 1978).

Lightweight tuffaceous rock consisting mostly of clinoptilolite was used for many years in Antigua to construct buildings, fortress walls, and several churches (Holmes, unpublished data, 1992). Zeolites have been widely used since Mayan times as a dimension stone in Mexico. Mumpton (1973) reported that many buildings at the Mayan centers of Mitla and Monte Alban were constructed with blocks of green massive zeolitized tuff. Tuffs from this and other sources are still being quarried at Etla, Tecoatlan, and Tejupan in Oaxaca for regional construction materials.

Zeolite use for dimension stone in the United States has been quite limited, mostly because the rocks occur in the West and are not close to urban centers. Ranch houses in the West, particularly in Nevada and Oregon, have been built of erionite quarried from local sources. A small erionite-rich tuff quarry several miles southwest of Rome, OR produced several tons per year of cut stone for local construction needs (Mumpton, 1978).

Gas Separation

Zeolites have proven effective in separating various gases, based on the channel size in various zeolite crystals. Separation of nitrogen and oxygen to produce oxygen-rich air is the predominant application (Mumpton, 1978). Based on studies by Barrer (1938), Dominue and Hay (1968) used several natural zeolites to selectively adsorb nitrogen from air and produce gas streams with as much as 95% oxygen. Mordenite appears to be the best natural zeolite for oxygen generation, but clinoptilolite and chabazite also have desirable properties (Tamura, 1971). Minato and Tamura (1978) discussed the oxygen-nitrogen separation process and the range of applications in Japan.

Oxygen-generating units using zeolites, designed in Japan and Europe, provide relatively pure oxygen for steel mills, smelters, and other industrial consumers. They are in service in municipal waste water treatment plants, metal smelters, the pulp and paper industry for reoxygenation of downstream waters, secondary sewage treatment plants, and oxygenation of fish tanks during the transportation of live fish.

Ion Exchange Processes

The ability to exchange cations is one of the most important and early recognized properties of natural and synthetic zeolites. Zeolites have higher cation exchange capacities than most other minerals. Cation-exchange capacity is a measure of the number of cations per unit weight available for exchange, usually expressed as milliequivalents per 100 g of material. The ammonia ion is most commonly used for CEC testing, and the result therefrom is referred to as the ammonia ion exchange capacity.

Several applications are particularly promising: chabazite and clinoptilolite use in radioactive waste treatment, and clinoptilolite use for water pollution control. A third and greatly increasing application has been the use of synthetic zeolites as a phosphate substitute in powdered detergents. Sodium ions from the zeolite structure replace calcium and magnesium in the water, essentially softening the water and reducing undesirable precipitation onto fabrics and other surfaces. The reduction in the use of phosphates relieves stream pollution and entrophization of streams and lakes and also reduces product cost.

Natural zeolites are less commonly used in detergents and similar applications due to their lower CEC, smaller channel size, and possible presence of impurities. Natural zeolites are favored over synthetic zeolites, however, for uses such as in radioactive waste cleanup or for solutions with low pH.

Fine grinding of zeolitic materials tends to improve their cation exchange capacity, especially where the zeolite-bearing rocks are dense and have low permeability or where the zeolite crystals are large. Natural zeolites with high CEC potential include clinoptilolite, phillipsite, chabazite, and mordenite. Minerals with generally low CEC potential include heulandite, laumontite, and analcime.

Lightweight Aggregate

Expanded, lightweight aggregate similar to perlite or expanded shale has been prepared by calcining or ''popping'' zeolitic tuffs at temperatures between 1 200 and 1 400°C (Stojanovic, 1972). Moisture control is very important as the expansion is due to water absorbed in the zeolite.

Medical Applications

Natural phillipsite has been used experimentally to separate ammoniacal nitrogen from hemodialysis liquids in recycle-dialysis systems (Andersson et al., 1975). Clinoptilolite has been used successfully as a polishing agent in fluoride-containing toothpaste, allowing a higher level of fluoride to remain in anionic form (Kato et al., 1969, 1970). With its unusual ability to absorb NH_4^+ ions from both air and water, clinoptilolite has shown promise in products such as baby powder, filler in plastic baby diapers, and an ammonia control agent in diaper pails. Mordenite is used in gas separation units to provide oxygen-enriched air for hospitalized or bedridden patients.

Natural Gas Purification

Natural zeolites have been used successfully since 1968 to remove CO_2, SO_2, and H_2S from contaminated and sour natural gas

(Mumpton, 1978). Natural gas sources also include those that generate methane through decomposition of organic wastes such as sanitary landfills, municipal sewage systems, and animal waste treatment facilities. Although it is technically feasible to separate methane from CO_2, SO_2, H_2S, and other gases, the low price of natural gas from conventional sources has delayed development of methane production from such alternative sources.

The NRG Corporation developed a pressure-swing adsorption process in 1975 to extract the methane gas generated in the Palo Verde sanitary landfill in the Los Angeles, CA, area, using mixed zeolites (principally chabazite-erionite) from Bowie, AZ. Raw gas containing about 50% CH_4 and 40% CO_2 is treated to produce about 30 000 m^3 of methane meeting commercial specifications each day (Anon., 1975).

Methane gas can also be produced by treatment of solid waste materials in sewage systems using bacterial digestion, much like the anaerobic treatment of sewage sludge using clinoptilolite. Natural gases generated by digestion of animal feedlot manures offers another good potential source by separating the methane gas from CO_2 and lesser H_2S. Hydrogen sulfide removal would help greatly in odor control, another positive contribution. Goeppner and Hasselmann (1974) estimated that over 30 Mm^3 of 700 BTU gas could be produced daily by treatment of the wastes from the beef cattle in US feedlots. Jewell (1974) estimated that methane produced from treatment of animal wastes on a typical New York dairy farm would generate sufficient methane gas to satisfy the farm's total fossil fuel requirements.

Nuclear Waste Treatment and Handling

Zeolites offer considerable promise in the treatment and handling of nuclear wastes and in engineering structures for storage of such materials (Grant et al., 1987, Saha, 1985, Nelson and Mercer, 1963, Mercer et al., 1970). Several zeolites have been shown to have an extraordinary ability to selectively remove Sr^{90}, Cs^{137}, and other dangerous isotopes from solutions with high concentrations of competing ions. Natural zeolites are less expensive than organic ion-exchange resins and much more resistant to nuclear degradation. As silicates in glass and cement, they effectively entrain radioactive cations in the final concrete product.

The Hanford Laboratories of Battelle Northwest in Richland, WA, initially demonstrated the high efficiency of clinoptilolite from Hector, CA, in removing radioactive Cs^{137} and Sr^{90} from low-level waste streams at nuclear facilities (Ames, 1959, 1960). Radioactive solution treatment continued at Hanford to handle millions of gallons of solutions, and comparable Cs^{137}/Sr^{90} waste treatment followed at the National Reactor Testing Station at Arco, ID, West Valley, NY, and other nuclear facilities (Knoll, 1963, Hawkins and Short, 1965, Grant et al., 1987, 1988, 1989). Chabazite-erionite products from the Bowie deposit were used in the clean-up of the Three Mile Island nuclear power plant site (Eyde, personal communication, 1991, Griffiths, 1987).

Ion-exchange treatment utilizing clinoptilolite has also been used for nuclear-waste treatment in other countries, including the USSR, Japan, Hungary, Bulgaria, Canada, Italy, the United Kingdom, France, and Germany. Clinoptilolite from a Georgian deposit in the USSR was reported by Nikashina et al. (1974) to sorb strontium and by Vdovina et al. (1976) to effectively sorb Cs^{137} from nuclear waste streams. Large tonnages of clinoptilolite were used in direct application at the devastated Chernobyl nuclear power plant to sorb radioactive contaminants spilled during the accident (Griffiths, 1987, Filizova, 1993).

Mixed clinoptilolite and mordenite have been used for Cs^{137} removal in Japan (Kato, 1974). Clinoptilolite from the Tokaj district in Hungary has been used to encapsulate Sr^{90} for solid waste disposal (Kakasy et al., 1973) and removal of Sr^{90} and Cs^{137} from effluent waste streams (Adam et al., 1971). In Canada, clinoptilolite was used to recover Sr^{90} and Cs^{137} from waste solutions and placed in underground disposal sites at Chalk Run, ON (Mathers and Watson, 1962).

Chabazite and phillipsite in the Neapolitan Yellow Tuff at Naples, Italy, has been used to remove Cs^{137} and Sr^{90} from effluent waste streams and placed in underground disposal sites at Cassaccig, Italy (Anon., 1972).

British Nuclear Fuels Ltd. uses Mud Hills, CA, clinoptilolite to selectively sorb CS^{137}, Sr^{90}, and other radioisotopes from effluent waste streams at its nuclear fuel reprocessing plant at Sellafield, UK (Anon., 1990). Levels of Cs^{137} and Sr^{90} in the plant effluent, which is discharged into the Irish Sea, have dropped to very low levels.

Battelle Pacific Northwest Laboratory determined that clinoptilolite is also an effective media for engineered sorbent barrier structures in low-level nuclear waste storage (Freeman and Jones, 1989, Freeman et al., 1986).

Odor Control

Natural zeolites, particularly clinoptilolite, absorb ammonia ions, enabling control of animal waste and other odors. Hydrated zeolites may also take up some ammonia vapor; the gas enters the zeolite pores, where it bonds to or replaces crystalline water (Frederickson, 1987).

In Europe and the United States, clinoptilolite is used as pet litter and to suppress odors in horse stalls (Eyde, 1987). In Japan, clinoptilolite and mordenite are used to control odors in poultry houses and piggeries. Crushed and sized zeolites, principally clinoptilolite, have proven effective over the past three decades in controlling odoriferous ammonia fumes and the drying of animal excrement (Mumpton, 1977, Iijima et al., 1986, Griffiths, 1987). Zeolites are generally applied directly to the floors in pens and stables and in litter trays for cats and small pets, but they may also be added to animal feeds to gain benefits from nutritional effects and odor control in the wastes.

Natural zeolites, notably clinoptilolite, chabazite, and phillipsite, have also been used for odor control in personal products, e.g., in sport shoes. Room air cleaners with zeolite filters have also been used in living quarters and to remove high-ammonia urine odors in rest homes and similar facilities. Ground clinoptilolite is effective as a dry carpet cleaner, particularly to remove pet odors. Clinoptilolite products may also be used in baby diaper pails and as a filler in diaper plastics.

Oil Spill Cleanup

Mumpton (1978) reported that natural zeolites may be used as a sorbent to cleanup oil spills. Mixed with expanded perlite, sodium carbonate, and organic binders, activated zeolites are able to float on the water surface and sorb oil.

Paper Filler

Natural zeolites are used as fillers in some paper products, substituting for kaolin or calcium carbonate when they impart physical properties that benefit the final product (Mumpton, 1978). White clinoptilolite and mordenite have proven effective as high-bulking fillers in paper, as a desirable ink absorbent in carbonless paper, and possibly as a pitch-control agent. Due to their lower brightness, they cannot compete with kaolin and calcium carbonate for high-brightness pigment use, particularly in coating applications.

Clinoptilolite and mordenite are used in Japan as a filler in specialty and lower quality paper (Minato, 1968, Kobayashi, 1970). A relatively soft material is produced by fine grinding followed by wet- or dry-cyclone classification to yield a product with a blue-filter brightness of 80 (Takasaka, 1975). Other physical properties, such as viscosity and abrasion resistance, may limit zeolite use in paper making. Clinoptilolite is mined for use as paper filler in the Tokaj District of eastern Hungary (Nemecz and Varju, 1962). Teague Mineral Products in Oregon has supplied paper manufacturers in the Pacific Northwest with numerous samples of clinoptilolite products for test use.

Pozzolan

Pozzolan is a naturally occurring cementaceous material that can replace as much as one quarter of the portland cement in concrete to improve the quality and reduce the cost of the product. Zeolitic tuffs are used as pozzolan in many places in the world. The ancient Romans employed pozzolanic materials in building aqueducts, public buildings, and highways using the Neapolitan yellow tuff near Pozzuoli, Italy (Sersale, 1958).

In Europe zeolitic rocks are currently mined in Serbia, Germany, Bulgaria, and other countries for use as pozzolan. The minerals used include clinoptilolite, analcime, and chabazite.

Zeolitic ash-flow tuffs were mined near Tehachapi, CA, by Monolith Portland Cement Company as pozzolan in cement. According to Mumpton (1978), the pozzolanic cements were used in the construction of the 400 km long Los Angeles aqueduct in 1912, replacing up to 25% of the portland cement and saving several million dollars. Zeolitic tuffs occurring at many other sites in the West have similar potential as future pozzolan sources.

Sewage Treatment

Some natural zeolites are effective in removal of ammoniacal nitrogen and, to a lesser extent, some heavy metals from sewage and other effluent streams. Acting through ion exchange, clinoptilolite has been established as a very effective material (Ames, 1967, Mercer et al., 1970). The ammonia removed can be harmlessly discharged into the atmosphere, recovered as a gas byproduct, or converted to an ammonium sulfate fertilizer by adding sulfuric acid. Clinoptilolite also removes traces in the waste water of some heavy metals which would be detrimental in some end uses (Chelischev et al., 1974, Hashimoto, 1974, Yoshida et al., 1976, Mumpton, 1978). Plants at Truckee, CA, Rosemont, MN, and Manassas, VA, have operated successfully using clinoptilolite in tertiary sewage treatment processes to remove ammonia.

Environmental considerations are the driving force behind the probable increased use of zeolites in tertiary sewage waste treatment. Ammonia ions are toxic to fish and other aquatic life, and they contribute to the rapid growth of algae and the eutrophication of streams and lakes.

Nitrification in activated-sludge processes is enhanced by the addition of clinoptilolite (Simms, 1972). The zeolite selectively exchanges ammonia from the waste water and provides an ideal surface for the attachment of nitrifying bacteria which then oxidize the ammonia ions to nitrate (Mumpton, 1978).

Solar Energy and Heat Exchange

Chabazite and clinoptilolite are used to absorb and release heat from solar radiation for air conditioning, refrigeration, and water heating. The dehydration of the zeolite by day and its rehydration by night result in the exchange of several hundred BTU's per kilogram of zeolite, sufficient to operate refrigerators or even to cool small buildings (Tchernev, 1978). Tchernev estimated that one ton of zeolite in solar panels spread over 19 m^2 of roof surface will produce one ton of air conditioning. Recreational use, such as beer coolers on yachts or at playing fields, and use in less developed countries where power is not available to refrigerate medical supplies and food, have proven to be particularly suitable applications. Heat exchange systems have also been designed using Tchernev's principle with heat sources such as automobile exhaust systems.

Stack-Gas Cleanup

Natural zeolites offer considerable promise in the treatment of stack gases to remove SO_2 and other pollutants, particularly in coal burning power plants. Although federal and local air pollution regulations now limit the permissible amount of SO_2 in stack-gas emissions to about 100 ppm, few plants are able to meet these standards using current technology. The relatively low concentrations of SO_2 in such emissions (about 3000 ppm) elude scrubbing operations that principally use limestone. Limestone and/or sodium bicarbonates scrubbing systems followed by zeolite adsorption may remove most of the SO_2.

Clinoptilolite and mordenite are capable of adsorbing as much as 200 mg of SO_2 per gram of zeolite under static conditions and as much as 40 mg under dynamic conditions, even in the presence of copious amounts of CO_2. These zeolites are especially suited to the low pH and high temperature conditions of exhaust gas systems.

ECONOMIC FACTORS

Markets

Zeolites are a relatively new mineral commodity, with little commercial interest shown before the 1960s except for building stone and pozzolan uses. No formal commodity markets exist and sales for lower value products are confined mostly to the countries with zeolite resources. Higher value products may be sold internationally, but their total value worldwide is still small. Synthetic zeolites have established major, growing markets through silicate chemical manufacturers, mostly in North America, Western Europe, and Japan.

Because of their very wide range of applications, natural zeolites sell into diverse markets. Japan has a particularly wide range of zeolite markets for agricultural, industrial, and consumer uses. North American sales are principally for agricultural and pet litter applications. Markets in Western Europe, the former Eastern bloc countries, and Cuba are primarily for agricultural products, with a growing industrial market sector.

Steady growth is anticipated for the rest of the 1990s for agricultural, industrial, and consumer applications. The strongest areas of market growth in North America are expected to be in sewage treatment, deodorants, pet litter, soil treatment, and nuclear waste treatment and containment.

Values and Costs

Cost of zeolite products depends mostly on the type and degree of processing that must be done to satisfy specific market specifications. Mining costs are generally fairly low, typically 3 to \$6/t, unless very selective mining is done.

Most natural zeolites are sold into low value industrial or agricultural markets, commonly selling for 30 to \$70/t for granular products down to 40 mesh, and 50 to \$120/t for ground materials, in a range of −40 to −325 mesh. Consumer products such as pet litter, fish-tank media or deodorant materials commonly sell for 0.50 to $4.50/kg at the retail level.

Products for very special industrial applications, such as rad-waste filter media or catalysts in petroleum refining, may show values ranging up to thousands of dollars per ton, although their market demand may be very limited.

Transportation

Zeolite products are generally transported by highway or rail carriers in bulk, in one-ton super-sacks or in multi-wall paper bags, usually palletized. They present no special problems, subject to proper labeling and handling as a respirable silica-containing material. Natural zeolites, particularly clinoptilolite, that have been dried may absorb several percent by weight of water en route to their destination.

Zeolites prepared as dimension stone are shipped as any other stone products, generally palletized and by truck transport.

Transportation costs strongly affect the distance a lower value zeolite can travel to profitably reach markets. Agricultural grade clinoptilolite from the western United States can reach markets in the eastern half of the country, but future imports from Cuba, Antigua, and other tidewater sources could lower prices to the point where this is impossible. High value zeolites can generally be shipped nationwide and even to international points from their processing sources.

Alternate Materials

When natural zeolites enter the marketplace as new products, they may have to compete with other mineral products that are already well established. There are established alternate materials for most zeolite applications. Market entry of natural zeolites requires demonstration of technical equivalency or superiority and/or some cost advantage in each application in order to gain some of the market.

Synthetic zeolites (molecular sieves) are the major alternate materials to natural zeolites. Synthetic zeolites can be tailored in physical and chemical characteristics to serve many applications more closely and they are more uniform in quality than their natural equivalents. Natural zeolites hold advantage over synthetic materials in some applications (e.g., cesium and strontium adsorption in radioactive waste cleanup) and are also able to function at lower pH levels. They are also generally much lower in cost than synthetic zeolite products.

Activated carbon, silica gel, and similar materials are more effective than zeolites for many ion-exchange applications and are not disproportionately more expensive. Bentonite, attapulgite, and other minerals show selective high adsorbency and are also available in a competitive price range.

GOVERNMENTAL CONSIDERATIONS

Taxes, Subsidies, and Depletion Allowance

There are no specific taxes related to zeolites and the material is not subject to any government subsidies in the United States and Canada. The depletion allowance is 14% for natural zeolites mined in the United States.

Zoning and Land Use

Natural zeolites face zoning and land use restrictions similar to most other fine grained silicate minerals, e.g., kaolin or diatomite. Most active zeolite sources are located in country or rural settings, so their development rarely raises issues competitive with urban development. The very large zeolite deposits around Naples, Italy, are a conspicuous exception, but they are so extensive that many other sites are available in the region without disturbing residential or industrial lands.

Environmental Regulations

Environmental regulations vary between states and countries, and they can be a source of conflict between government regulators and mine operators. Natural zeolites are relatively innocuous and present no particular environmental problems, with three exceptions:

1) Several zeolite minerals (e.g., erionite and some mordenite) are fibrous in form and may be classified as asbestiform materials.

2) Finely crystalline silica commonly occurs in zeolite deposits and the finely ground products may contain more than 0.1% respirable silica.

3) Zeolite mines and dry processing plants tend to generate dust, causing air quality problems.

Dust from mining and plant facilities may be considered a local environmental pollutant. Most zeolites contain silica in the form of amorphous silica (e.g., opal or opal-CT) and/or crystalline silica (e.g., quartz). Processing plants may therefore be required to use efficient air pollution control systems ranging from standard baghouse dust collectors to electrolytic precipitators to minimize worker exposure to fugitive dust at bagging stations or grinding mills and to comply with local air quality control requirements. Most zeolite production uses dry processing methods. Zeolite processing will inevitably move into water washing and wet classification methods, requiring proper handling of plant effluent and tailings pond discharges.

Zeolite minerals are generally considered to be chemically inert, and most are non-fibrous. Erionite is established as a fibrous, acicular mineral and has been labeled a possible carcinogen on the basis of medical studies of high mesothelioma incidence in the Capadoccia District of Turkey. Mordenite is also a fibrous mineral, but is presently not regarded as a potential carcinogen.

Safety Regulations

Natural zeolites are dusty materials when crushed and ground, and air quality standards should be maintained carefully during mining and processing. As several of the natural zeolites are fibrous in nature (erionite and mordenite), appropriate care should be taken in handling these materials. The natural zeolites do not offer any toxic threat to plants or animals, and there are no established procedures for their remedial handling.

FUTURE TRENDS AND PROBLEMS

Future Trends

Overall, the zeolites present a healthy, growing industry with continued expansion into new applications and steady demand in industrial markets where they have achieved acceptance. Most of this activity and growth, however, has been in the synthetic zeolite field. Natural zeolites have only a small portion of the markets, commonly those with lower costs or very specific uses, and growth in North America has generally languished. Because the technology has been firmly established, an excellent range of high quality natural zeolite deposits identified and characterized, and a record of successful sales for products with consistent specifications achieved, growth of the natural zeolite field appears likely to continue through the 1990s on a slow and steady basis.

Natural zeolites have established a strong domestic market pattern for several uses: catalysis and petroleum refining, nuclear waste treatment, and odor control. These marketing areas can be expected to continue and expand. Natural zeolite use is particularly tied to 1) pollution control and 2) energy cost and efficiency issues; greater emphasis on these areas through the decade would markedly increase the probable adaption of zeolite technology. Higher energy costs and greater environmental demands will spur zeolite production and sales significantly.

Japan, the former Soviet Union, and a few other countries have evolved strong natural zeolite industries based on the availability, low cost, suitability (agricultural and industrial), and consumer applications of these minerals. Through the decade, natural zeolites should emerge as a better-defined mineral commodity, and North America will become a leading producer.

Problems

The synthetic zeolites are materials with few if any problems, as they are readily made from abundant raw materials and present no toxic or environmental problems. The natural zeolites must penetrate markets where other materials are already used and accepted, and they are also faced with a stigma of being presented as a panacea for too many material supply problems in the 1970s. Suppliers of competing materials have capitalized on this past over-selling of zeolites.

Several natural zeolites are fibrous minerals, and their use or presence even in trace amounts may suppress the use or consideration of zeolites for some new applications. Erionite has been classified as a hazardous material, which precludes its use for some applications, particularly in consumer products. Mordenite is also a fibrous mineral, but apparently has no record of carcinogenic problems. Natural zeolites commonly contain some crystalline silica, which may require stringent hazardous materials labeling for finished products

The natural zeolites lack any designation of industrial standards on a national or international basis, which inhibits their sale and use, principally in industrial and consumer markets. Creation of standards through ASTM or trade associations would alleviate this problem.

BIBLIOGRAPHY

A number of recent references are made to the Program and Abstracts from "Zeolite '93," the 4th International Conference on the Occurrence, Properties, and Utilization of Natural Zeolites, Boise, ID, June 1993. These are primarily extended abstracts and most will be reprinted in full in the Proceedings volume.

Introduction

Barrer, R.M., 1978, *Zeolite and Clay Minerals as Sorbents and Molecular Sieves*, Academic Press, New York, 499 pp.

Breck, D.W., 1974, *Zeolite Molecular Sieves*, Wiley-Interscience, New York, 771 pp.

Clifton, R.A., 1987, "Natural and Synthetic Zeolites," Information Circular 9140, US Bureau of Mines, 21 pp.

Eyde, D.T., and Eyde, E.H., 1984, "Natural Zeolites—What's The Score?" *Proceedings*, 6th Industrial Minerals International Congress, Toronto, ON.

Eyde, T.H., 1987, "Zeolites," Industrial Minerals 1986, *Mining Engineering*, Vol. 39, No. 6, pp. 422–423.

Eyde, T.H., 1992, "Zeolites," Industrial Minerals 1991, *Mining Engineering*, Vol. 44, No. 6, pp. 578–579.

Flanigen, E.M., and Mumpton, F.A., 1977, "Commercial Properties of Natural Zeolites," *Mineralogy and Geology of Natural Zeolites*, F.A. Mumpton, ed., Mineralogical Society of America Short Course Notes, Vol. 4, pp. 165–175.

Gottardi, G., and Galli, E., 1985, *Natural Zeolites*, Springer Verlag, 409 pp.

Harben, P.W., and Bates, R.L., 1984, *Geology of the Nonmetallics*, Metal Bulletin Plc., London, 392 pp.

Harben, P.W., and Bates, R.L., 1990, *Industrial Minerals: Geology and World Deposits*, Metals Bulletin Plc, London, 312 pp.

Kallo, D., and Sherry, H.S., 1988, *Occurrence, Properties and Utilization of Natural Zeolites*, Akademia Kiado, Budapest, Hungary 857 pp.

Mumpton, F.A., 1973, "Worldwide Deposits and Utilization of Natural Zeolites," *Industrial Minerals*, No. 73, Oct., pp. 30–45.

Mumpton, F.A., 1977, "Natural Zeolites," *Mineralogy and Geology of Natural Zeolites*, F.A. Mumpton, ed., Mineralogical Society of America Short Course Notes, Vol. 4, pp. 1–17.

Mumpton, F.A., 1978, "Natural Zeolites: A New Industrial Mineral Commodity," *Natural Zeolites: Occurrences, Properties, Use*, L.B. Sand and F.A. Mumpton, eds., Pergamon Press, New York, pp. 3–27.

Mumpton, F.A., 1983, "Zeolite Exploration: The Early Days," *Proceedings*, 6th International Zeolite Conference, Reno, NV, D. Olson and A. Bisio, eds., Butterworth of Col., Ltd., pp. 68–86.

Munson, R.A., and Sheppard, R.A., 1974, "Natural Zeolites: Their Properties, Occurrence, and Uses," *Mineral Sci. Engineering*, Vol. 6, Jan., pp. 19–34.

Olson, R.H., 1983, "Zeolites: Introduction," *Industrial Minerals and Rocks*, 5th ed., Vol. 2, S.J. Lefond, ed., SME-AIME, New York, pp. 1391–1399.

Rees, L.V.C., ed., 1980, *Proceedings*, 5th International Conference on Zeolites, Naples, Italy, Heyden & Son, Ltd., 902 pp.

Sand, L.B., and Mumpton, F.A., eds., 1978, *Natural Zeolites: Occurrence, Properties, Use*, Pergamon Press, New York, 546 pp.

Sheppard, R.A., 1971, "Zeolites in Sedimentary Deposits in the United States: A Review," *Molecular Sieve Zeolites—I*, R.F. Gould, ed., Advances in Chemistry Series 101, American Chemical Society, pp. 279–310.

Tschernich, R.W., 1992, *Zeolites of the World*, Geoscience Press, Phoenix, AZ, 563 pp.

Vaughan, D.E.W., 1978, "Properties of Natural Zeolites," *Natural Zeolites: Occurrence, Properties, Use*, L.B. Sand and F.A. Mumpton, eds., Pergamon Press, New York, pp. 353–371.

Geology, Mineralogy, and Origin

Alberti, A., 1993, "Crystal Structure and Chemistry of Newly Discovered Zeolites: A Review," Program and Abstracts, "Zeolite '93," (extended abstract), 4th International Conference on the Occurrence, Properties, and Utilization of Natural Zeolites, Boise, ID, June 1993, pp. 28–30.

Aleksiev, B., and Djourova, E.G., 1974, "Zeolite Rocks: Classification and Nomenclature," *Bulgarian Akademy Nauk Doklady*, Vol. 27, No. 3, pp. 373–374.

Baldar, N.D., and Whittig, L.D., 1968, "Occurrence and Synthesis of Soil Zeolites," *Proceedings*, Soil Science Society of America, Vol. 32, pp. 235–238.

Boles, J.R., 1977, "Zeolites in Low-Grade Metaphorphic Grades," *Mineralogy and Geology of Natural Zeolites*, F.A. Mumpton, ed., Mineralogical Society of America Short Course Notes, Vol. 4, pp. 103–135.

Boles, J.R., 1977a, "Zeolites in Deep-Sea Sediments," *Mineralogy and Geology of Natural Zeolites*, F.A. Mumpton, ed., Mineralogical Society of America Short Course Notes, Vol. 4, pp. 137–163.

Boles, J.R., and Wise, W.S., 1978, "Nature and Origin of Deepsea Clinoptilolite," *Natural Zeolites: Occurrence, Properties, Use*, L.B. Sand and F.A. Mumpton, eds., Pergamon Press, New York, pp. 175–198.

Bramlette, M.N., and Posniak, E., 1933, "Zeolitic Alteration of Pyroclastics," *American Mineralogist*, Vol. 18, No. 4, Apr., pp. 161–171.

Breck, D.W., 1974, *Zeolite Molecular Sieves: Structure, Chemistry and Use*, John Wiley, New York, 771 pp.
Coombs, D.S., et al., 1959, "The Zeolite Facies, with Comments on the Interpretation of Hydrothermal Syntheses," *Geochimica Et Cosmochimica Acta*, Vol. 17, pp. 53–107.
Deffeyes, K.S., 1959a, "Zeolites in Sedimentary Rocks," *Sedimentary Petrology*, Vol. 29, No. 4, pp. 602–609.
Flanigen, E.M., 1977, "Crystal Structure and Chemistry of Natural Zeolites," *Mineralogy and Geology of Natural Zeolites*, F.A. Mumpton, ed., Mineralogical Society of America Short Course Notes, Vol. 4, pp. 19–52.
Gibbons, A.B., Hinrichs, E.N., and Botinelly, T., 1960, "The Role of Impermeable Rocks in Controlling Zeolitic Alteration of Tuff," Professional Paper 400B, US Geological Survey, pp. B473–B475.
Gottardi, G., 1978, "Mineralogy and Crystal Chemistry of Zeolites," *Natural Zeolites: Occurrence, Properties, Use*, L.B. Sand and F.A. Mumpton, eds., Pergamon Press, New York, pp. 31–44.
Hay, R.L., 1963a, "Zeolitic Weathering in Olduvai Gorge, Tanganyika," Geological Society of America Bulletin, Vol. 74, pp. 1281–1286.
Hay, R.L., 1964, "Phillipsite of saline lake and soils," *American Mineralogist*, Vol. 49, No. 9–10, pp. 1366–1387.
Hay, R.L., 1966, "Zeolites and Zeolitic Reactions in Sedimentary Rocks," Special Paper 85, Geological Society of America, 130 pp.
Hay, R.L., 1977, "Geology of Zeolites in Sedimentary Rocks," *Mineralogy and Geology of Natural Zeolites*, F.A. Mumpton, ed., Mineralogical Society of America Short Course Notes, Vol. 4, pp. 53–64.
Hay, R.L., 1978, "Geologic Occurrence of Zeolites," *Natural Zeolites: Occurrence, Properties, Use*, L.B. Sand and F.A. Mumpton, eds., Pergamon Press, New York, pp. 135–143.
Hay, R.L., 1993, "New Developments in the Geology of Natural Zeolites," Program and Abstracts, "Zeolite '93," (extended abstract), 4th International Conference on the Occurrence, Properties, and Utilization of Natural Zeolites, Boise, ID, June 1993, pp. 120–123.
Hay, R.L., and Sheppard, R.A., 1977, "Zeolites in Open Hydrologic Systems," *Mineralogy and Geology of Natural Zeolites*, F.A. Mumpton, ed., Mineralogical Society of America Short Course Notes, Vol. 4, pp. 93–102.
Helfferich, F., 1964, "A Simple Identification Reaction for Zeolites (Molecular Sieves)," *American Mineralogist*, Vol. 49, pp. 1752–1754.
Iijima, A., 1974, "Genetical Connection of Sedimentary Zeolites Between Burial Diagenesis and Alkali-Saline Lake Deposits," Seminar on Occurrence, Origin, and Utilization of Sedimentary Zeolites in the Circum-Pacific Region: US-Japan Conservative Science Program, unpublished abstract, Menlo Park, CA, July.
Iijima, A., 1978, "Geologic Occurrences of Zeolites in Marine Environments," *Natural Zeolites: Occurrence, Properties, Use*, L.B. Sand and F.A. Mumpton, eds., Pergamon Press, New York, pp. 175–198.
Iijima, A., 1980, "Geology of Natural Zeolites and Zeolitic Rocks," *Proceedings*, 5th International Conference on Zeolites, L.V.C. Rees, ed., Heyden & Sons, Ltd., London, pp. 103–118.
Iijima, A., and Utada, M., 1966, "Zeolites in Sedimentary Rocks, with Reference to the Depositional Environments and Zonal Distribution," *Sedimentology*, Vol. 7, pp. 327–357.
Kolla, V., and Biscaye, P.E., 1973, "Deep-sea Zeolites: Variations in Space and Time in Sediments of the Indian Ocean," *Marine Geol.*, Vol. 15, pp. 11–17.
Meier, W.M., 1968, "Zeolite Structures," *Molecular Sieves*, Society of Chemical Industry, London, pp. 10–25.
Meier, W.M., and Olson, D.H., 1971, "Zeolite Frameworks," Advances in Chemistry, Series 101, American Chemical Society, pp. 155–170.
Mondale, K.D., 1978, "The Characteristics, Beneficiation and Ion Exchange Properties of Natural Sedimentary Zeolites," MS thesis, Pennsylvania State University, University Park, 203 pp.
Mumpton, F.A., 1960, "Clinoptilolite Redefined," *American Mineralogist*, Vol. 45, pp. 351–369.
Mumpton, F.A., 1973, "Scanning Electron Microscopy and the Origin of Sedimentary Zeolites," *Proceedings*, 3rd International Molecular Sieve Conference, pp. 159–161.
Packham, G.H., and Crook, K.A.W., 1960, "The Principle of Diagenetic Facies and Some of Its Implications," *Journal of Geology*, Vol. 68, pp. 392–407.
Rinaldi, R., 1984, "Mineralogy of Natural Zeolites: Present Status," *Proceedings*, 6th International Zeolites Conference, pp. 570–583.
Seki, Y., 1969, "Facies Series in Low-Grade Metamorphism," *Geological Journal*, Geological Society of Japan, Vol. 75, pp. 255–266.
Sheppard, R.A., 1973, "Zeolites in Sedimentary Rocks," Professional Paper 820, US Geological Survey, pp. 689–695.
Slansky, E., 1987, "Mineralogical Identification of Natural Zeolites," *Natural Zeolites of New South Wales*, G.G. Holmes and S.R. Pecover, eds. New South Wales Geological Survey, pp. 39–50.
Steiner, A., 1955, "Hydrothermal Rock Alteration," Industrial Research Bulletin 117, New Zealand Department of Science, pp. 21–26.
Surdam, R.A., 1977, "Zeolites in Closed Hydrologic Systems," *Mineralogy and Geology of Natural Zeolites*, F.A. Mumpton, ed., Mineralogical Society of America Short Course Notes, Vol. 4, pp. 65–91.
Surdam, R.C., and Sheppard, R.A., 1978, "Zeolites in Saline, Alkaline Lake Deposits," *Natural Zeolites: Occurrence, Properties, Use*, L.B. Sand and F.A. Mumpton, eds., Pergamon Press, New York, pp. 145–174.
Tschernich, R.W., 1992, *Zeolites of the World*, Geoscience Press, Inc., 563 pp.

US Deposits

Anon., 1977, "Arizona Zeolites," Arizona Department of Mineral Resources, Mineral Report No. 1, 40 pp.
Almner, S.P., and Grim, R.E., 1990, "Mineralogy, chemistry, and diagensis of tuffs in the Sucker Creek Formation (Miocene), eastern Oregon," *Clays and Clay Mineralogy*, Vol. 38, pp. 561–572.
Ames, L.L., Jr., Sands, L.B., and Goldich, S.S., 1958, "A Contribution on the Hector, California, Bentonite Deposit," *Economic Geology*, Vol. 53, pp. 22–37.
Bay, K.W., 1969, "Stratigraphy of Eocene Rocks in the Lysite Mountains area, Hot Springs, Fremont, and Washakie Counties, Wyoming," PhD thesis, University of Wyoming, Laramie, 181 pp.
Berg, R.B., 1969, "Bentonite in Montana," Bulletin 74, Montana Bureau of Mines and Geology, 34 pp.
Berg, R.B., and Lonn, J., 1993, "Reconnaissance Survey of Bedded Zeolite Occurrences in Montana," Open File Report MBMG-267, Montana Bureau of Mines and Geology, July, 21 pp.
Boles, J.R., 1968, "Zeolites and Authigenic Feldspar along a part of the Beaver Rim, Freemont County, Wyoming," MS thesis, University of Wyoming, Laramie, 64 pp.
Boles, J.R., and Surdam, R.C., 1971, "Authigenises of the Wagon Bed Formation, Central Wyoming," Contributions to Geology, University of Wyoming, Laramie, Vol. 10, pp. 141–144.
Bostros, E., 1976, "Diagenetic Alteration in the Tuffaceous Sediments of the Duff Formation, Trans-Pecos, Texas," unpublished MA thesis, University of Texas, Arlington.
Bramlette, M.N., 1946, "The Monterey Formation of California and the Origin of Its Siliceous Rocks," Professional Paper 212, US Geological Survey, pp. 22–29.
Brown, C.E., and Thayer, T.P., 1963, "Low-Grade Mineral Facies in Upper Triassic and Lower Jurassic Rocks of the Aldrich Mountains," *Journal of Sedimentary Petrology*, Vol. 33, pp. 411–425.
Byers, F.M., Jr., 1960, "Geology of the Alvord Mountain Quadrangle, San Bernardino County, California," Bulletin 11089-A, US Geological Survey, pp. 22–26.
Chesterman, C.W., 1973, "Geology of the Northeast Quarter of the Shoshone Quadrangle, Inyo County, California," California Divisions of Mines and Geology Map Sheet 18.
Danzl, R.B., 1985, "Depositional and Diagenetic Environment of the Salt Lake Group near Oneida Narrows, Southeastern Idaho," unpublished MS thesis, Idaho State University, Pocatello, 177 pp.
Denny, C.S., and Drewes, H., 1965, "Geology of the Ash Meadows Quadrangle, California-Nevada," Bulletin 1181-L, US Geological Survey, pp. L18-L19.
Deffeyes, K.S., 1959b, "Erionite from Cenozoic Tuffaceous Sediments, Central Nevada," *American Mineralogist*, Vol. 44, pp. 501–509.
Desborough, G.A., 1989, "Preliminary Study of Certain Cation-Exchange Properties of Clinoptilolite-bearing Rocky Ford Ash Member of the Miocene Sharps Formation at Craven Creek Near Wanblee, South Dakota," Open-file Report 89–0289, US Geological Survey, 22 pp.
Dibble, T.W., Jr., 1952, "Geology of the Saltdale Quadrangle, California," Bulletin 160, California Division of Mines and Geology, pp. 25–30.
Dibble, T.W., Jr., 1958, "Tertiary Stratigraphy of the Western Mojave

Desert, California,'' American Association of Petroleum Geologist Bulletin, Vol. 42, No. 1, pp. 135–140.

Dibble, T.W., Jr., and Louke, G.P., 1970, ''Geologic Map of the Tehachapi Quadrangle, Kern County, California,'' Miscellaneous Geologic Investigations Map I-461, US Geological Survey.

Dickinson, W.R., 1962, ''Petrology and Diagenesis of Jurassic Andesitic Strata in Central Oregon,'' *American Journal of Science*, Vol. 260, pp. 481–500.

Dickinson, W.R., et al., 1969, ''Burial Metamorphism of the Late Mesozoic Great Valley Sequence, Cache Creek, California,'' Geological Society of America Bulletin, Vol. 80, pp. 519–526.

Eyde, T.H., 1974, ''Exploration for Deposits of Natural Zeolite Minerals in the Western U.S.,'' SME Preprint 74–11–336.

Eyde, T.H., 1982, ''Zeolite Deposits in the Gila and San Simon Valleys of Arizona and New Mexico,'' *Proceedings*, 17th Forum on the Geology of Industrial Minerals, G.S. Austin, ed., Circular 182, New Mexico Bureau of Mines and Min. Resources, pp. 65–71.

Eyde, T.H., and Eyde, D T., 1987, ''The Bowie Chabazite Deposit,'' *Proceedings*, 21st Forum on the Geology of Industrial Minerals, H.W. Peirce, ed., Special Paper 4, Arizona Bureau of Geology and Mineral Technology.

Fiske, R.S., et al., 1963, ''Geology of Mount Rainier National Park, Washington,'' Professional Paper 444, US Geological Survey, 93 pp.

Greene, R.C., Walker, G.W., and Corcoran, F.E., 1972, ''Geologic Map of the Burns Quadrangle,'' Miscellaneous Geological Investigation Map I-680, US Geological Survey.

Gude, A.J., III, and Sheppard, R.A., 1978, ''Chabazite in Siliceous Tuffs of a Pliocene Lacustrine Deposit near Durkee, Baker County, Oregon,'' Journal Research, US Geological Survey, Vol. 6, pp. 467–472.

Gude, A.J., III, and Sheppard, R.A., 1986, ''Zeolitic Diagenesis of Tuffs in an Upper Miocene Lacustrine Deposit near Durkee, Baker County, Oregon,'' Bulletin 1578, US Geological Survey, pp. 301–333.

Hall, C.A., and Prior, S.W., 1975, ''Geologic Map of the Cayucos-San Luis Obispo Region, San Luis Obispo County, California,'' Miscellaneous Field Studies Map MF-686, US Geological Survey.

Harris, R.E., and King, J.K., 1990, ''Natural Zeolites in Wyoming,'' Open-File Report 90–4, Wyoming Geological Survey, 51 pp.

Hawkins, D.B., 1973, ''Sedimentary Zeolite Deposits of the Upper Matanuska Valley, Alaska,'' Special Report No. 6., State of Alaska, Department of Natural Resources, Division of Geological and Geophysical Surveys, College, AK.

Hay, R.L., 1963a, ''Stratigraphy and Zeolitic Diagenesis of the John Day Formation of Oregon,'' *Geological Science*, California University Publications, Vol. 42, No. 5, pp. 199–262.

Hay, R.L., 1966, ''Zeolites and Zeolitic Reactions in Sedimentary Rocks,'' Special Paper 85, Geological Society of America, 130 pp.

Holmes, D.A., 1990, ''Pacific Northwest Zeolite Update,'' *Proceedings*, 25th Forum on the Geology of Industrial Minerals and Rocks of the Pacific Northwest, pp. 79–88.

Honda, S., and Muffler, L.P.J., 1970, ''Hydrothermal Alteration in Core from Research Drill Hole Y-1, Upper Geyser Basin, Yellowstone National Park, Wyoming,'' *American Mineralogist*, Vol. 55, pp. 1714–1737.

Hoover, K.L., 1968, ''Genesis of Zeolites, Nevada Test Site,'' Nevada Test Site, E.B. Eckel, ed., Memoir 110, Geological Society of America, pp. 275–284.

Hulbert, M.H., 1987, ''Sodium, Calcium, and Ammonium Exchange on Clinoptilolite from the Fort LaClede Deposit, Sweetwater County, Wyoming,'' *Clays and Clay Minerals*, Vol. 35, pp. 458–462.

Ky, L., Williford, C.W., and Reynolds, W.R., 1988, ''Cation Exchange Characteristics of Gulf Coast Clinoptilolite,'' *Occurrence, Properties and Utilization of Natural Zeolites*, D. Kallo and H.S. Sherry, eds., H. Stillman Publishers, Boca Raton, FL, pp. 451–550.

Lindsey, D.A., 1975a, ''Mineralization Halos Diagenesis in Water-Laid Tuff on the Thomas Range, Utah,'' Professional Paper 818-B, US Geological Survey, 59 pp.

Lindsey, D.A., 1975b, ''The Effect of Sedimentation and Diagenesis on Trace Element Composition of Water-Laid Tuff in the Key Mountain Area, Utah,'' Miscellaneous Investigations Series Map I-1176, US Geological Survey.

Lovering, T.S., and Sheppard, A.O., 1960, ''Hydrothermal Alteration Zones Caused by Halogen Acid Solutions, East Tintic District, Utah,'' *American Journal of Science*, Vol. 258-A, pp. 215–229.

Madonna, J.A., 1973, ''Zeolite Occurrences in Alaska,'' unpublished MS thesis, Geology Department, University of Alaska, Fairbanks, AK.

Mayes, B.H., and Tripp, B.T., 1991, ''Zeolite Minerals in Utah,'' Open File Report 210, Utah Geological Survey.

Mondale, K.D., Mumpton, F.A., and Aplan, F.A., 1978, ''Beneficiation of Natural Zeolites from Bowie, Arizona: A Preliminary Report,'' *Natural Zeolites: Occurrence, Properties, Use*, L.B. Sands and F.A. Mumpton, eds., Permagon Press, New York, pp. 527–537.

Murata, K.J., and Whiteley, K.R., 1973, ''Zeolites in the Miocene Briones Sandstone and Related Formations of the Central Coast Ranges, California,'' *Journal of Research,* US Geological Survey, Vol. 1, pp. 255–265.

Otalora, G., 1964, ''Zeolites and Related Minerals in Cretaceous Rocks of East-Central Puerto Rico,'' *American Journal of Science*, Vol. 262, pp. 726–734.

Papke, K.G., 1970, ''Montmorillonite, Bentonite and Fuller's Earth Deposits in Nevada,'' Bulletin 76, Nevada Bureau of Mines, 47 pp.

Papke, K.G., 1972, ''Erionite and Other Associated Zeolites in Nevada,'' Bulletin 79, Nevada Bureau of Mines and Geology, 32 pp.

Pearson, R., 1987, ''Zeolites in Cretaceous Rocks in Beaverhead County,'' Open-file Report 89–297, US Geological Survey, 10 pp.

Raymond, W.H., 1986, ''Clinoptilolite Deposits in the Pine Ridge Indian Reservation, South Dakota, U.S.A.,'' *Proceedings*, 7th International Zeolite Conference, New Developments in Zeolite Technology, Y. Murakami, A. Iijima, and J.W. Ward, eds., pp. 79–85.

Regnier, J., 1958, ''Cenozoic Geology in the Vicinity of Carlin, Nevada,'' unpublished PhD thesis, Columbia University, New York.

Regnier, J., 1960, ''Cenozoic Geology in the Vicinity of Carlin, Nevada,'' Geological Society of America Bulletin, Vol. 71, pp. 1189–1210.

Reynolds, W.R., 1970, ''Mineralogy and Stratigraphy of Lower Tertiary Clays and Claystones of Alabama,'' *Journal of Sedimentary Petrology*, Vol. 40, No. 3, pp. 829–838.

Reynolds, W.R., and Surdam, R.C., 1980, ''What's So Great about the McKenzie Exposure?'' Abstract, 17th Annual Meeting of the Clay Minerals Society, Waco, TX.

Rice, S.B., Papke, K.G., and Vaughan, D.E.W., 1992, ''Chemical Controls on Ferrierite Crystallization during Diagenesis of Silicic Pyroclastic Rocks near Lovelock, Nevada,'' *American Mineralogist*, Vol. 77, pp. 314–328.

Roehler, H.W., 1972, ''Zonal Distribution of Montmorillonites and Zeolites in the Laney Shale Member of the Green River Formation in the Washakie Basin, Wyoming,'' Bulletin 800-B, US Geological Survey, pp. 121–124.

Roehler, H.W., 1973a, ''Stratigraphy of the Washakie Formation in the Washakie Basin, Wyoming,'' Bulletin 1369, US Geological Survey, 40 pp.

Roehler, H.W., 1973b, ''Mineral Resources in the Washakie Basin, Wyoming and Sand Wash Basin, Colorado,'' Wyoming Geological Association 25th Annual Field Conference Guidebook, pp. 47–56.

Roehler, H.W., 1973c, ''Stratigraphic Divisions and Geologic History of the Laney Member of the Green River Formation in the Washakie Basin in Southwestern Wyoming,'' Bulletin 1372-D, US Geological Survey, 28 pp.

Roquemore, S.K., 1984, ''Clinoptilolite Occurrence in the Tallahatta Formation (Middle Eocene) of Southeast Mississippi,'' MS thesis, University of Mississippi, University, MS, 106 pp.

Santini, K.N., and Knostman, R.W., 1979, ''Geology of the Ash Meadows Clinoptilolite Deposit, Inyo County, California and Nye County, Nevada,'' Program and Abstracts, SME Fall Meeting and Exhibit, Oct. 17–19, 1979, Tucson, AZ.

Sheppard, R.A., 1971, ''Zeolites in Sedimentary Deposits in the United States: A Review,'' *Molecular Sieve Zeolites—I*, R.F. Gould, ed., Advances in Chemistry Series 101, American Chemical Society, pp. 279–310.

Sheppard, R.A., 1973, ''Zeolites in Sedimentary Rocks,'' Professional Paper 820, US Geological Survey, pp. 689–695.

Sheppard, R.A., 1976, ''Zeolites in Sedimentary Deposits of the Northwestern United States—Potential Industrial Minerals,'' Special Publication 74, Montana Bureau of Mines and Geology, 186 pp.

Sheppard, R.A., 1991, ''Zeolitic Diagenesis of Tuffs in the Miocene Chalk Hills Formation, Western Snake River Plant, Idaho,'' Bulletin 1963, US Geological Survey, 27 pp.

Sheppard, R.A., 1993, ''Geology and Diagenetic Mineralogy of the Castle

Creek Zeolite Deposit and the Be-Jel Bentonite Deposit, Chalk Hills Formation, Oreana, Idaho,'' Program and Abstracts, ''Zeolite '93,'' 4th International Conference on the Occurrence, Properties, and Utilization of Natural Zeolites, Boise, ID, June, pp. 14–24.

Sheppard, R.A., and Gude, A.J., III, 1964, ''Reconnaissance of Zeolite Deposits in Tuffaceous Rocks of the Western Mojave Desert and Vicinity, California,'' Professional Paper 501C, US Geological Survey, pp. C114-C116.

Sheppard, R.A., and Gude, A.J., III, 1965, ''Zeolitic Authigenesis of Tuffs in the Ricardo Formation, Kern County, Southern California,'' Geological Survey Research 1965, Professional Paper 525-D, US Geological Survey, pp. D44-D47.

Sheppard, R.A., and Gude, A.J., III, 1968, ''Distribution and Genesis of Authigenic Silicate Minerals in Tuffs of Pleistocene Lake Tecopa, Inyo County, California,'' Professional Paper 597, US Geological Survey, 38 pp.

Sheppard, R.A., and Gude, A.J., III, 1969a, ''Diagenesis of Tuffs in the Barstow Formation, Mud Hills, San Bernardino County, California,'' Professional Paper 634, US Geological Survey, 35 pp.

Sheppard, R.A., and Gude, A.J., III, 1969b, ''Chemical Composition and Physical Properties of the Related Zeolites Offretite and Erionite,'' *American Mineralogist*, Vol. 54, pp. 875–886.

Sheppard, R.A., and Gude, A.J., III, 1970, ''Calcic Siliceous Chabazite from the John Day Formation, Grant County, Oregon,'' Professional Paper 700-D, US Geological Survey, pp. D176-D180.

Sheppard, R.A., and Gude, A.J., III, 1973, ''Zeolites and Associated Authigenic Silicate Minerals in Tuffaceous Rocks of the Big Sandy Formation, Mohave County, Arizona,'' Professional Paper 830, US Geological Survey, 36 pp.

Sheppard, R.A., and Gude, A.J., III, 1980, ''Chemistry of Zeolites and Zeolitic Tuffs from a Pliocene Lacustrine Deposit near Durkee, Backer County, Oregon,'' Open-file Report 80–506, US Geological Survey, 8 pp.

Sheppard, R.A., and Gude, A.J., III, 1983, *Guide to Selected Zeolite Deposits in Eastern Oregon, Southwestern Idaho, and Northwestern Nevada*, F.A. Mumpton, ed., International Committee on Natural Zeolites, Allen Press, Lawrence, KS, 72 pp.

Sheppard, R.A., Gude, A.J., III, and Elson, G.M., 1978, ''Bowie Zeolite Deposit, Cochise and Graham Counties, Arizona,'' *Natural Zeolites: Occurrences, Properties, Use*, L.B. Sand and F.A. Mumpton, eds., Pergamon Press, New York, pp. 319–328.

Southard, A.R., and Kolesar, P.T., 1978, ''An Exotic Source of Extractable Potassium in some Soils of Northern Utah,'' *Soil Science Society of America Journal,* Vol. 42, No. 3, pp. 528–530.

Spencer, J.R., 1983, ''Distribution and Petrology of Clinoptilolite in the Tallahatta Formation (Middle Eocene) of Clarke County, Mississippi,'' MS thesis, University of Mississippi, University, 139 pp.

Stinson, M.C., 1988, ''Zeolites in California,'' Bulletin 208, California Division of Mines and Geology, pp. 1–74.

Surdam, R.C., 1972, ''Economic Potential of Zeolite-rich Sedimentary Rocks in Wyoming,'' Wyoming Geological Association Earth Science Bulletin, Vol. 5, No. 1, pp. 5–8.

Surdam, R.C., and Hall, C.A., 1968, ''Zeolitization of the Obispo Formation, Coast Ranges of California (Abstracts) for 1966,'' Special Paper 101, Geological Society of America, p. 338.

Surdam, R.C., Turner, D.L., and Hall, C.A., 1970, ''Distribution and Genesis of Authigenic Silicates in the Obispo Formation,'' Geological Society of America Abstracts, No. 2, pp. 151–152.

Surdam, R.C., and Parker, R.B., 1972, ''Authigenic Alumino-Silicate Minerals in the Tuffaceous Rocks of the Green River Formation, Wyoming,'' Geological Society of America Bulletin, Vol. 83, pp. 689–700.

Van Houten, F.B., 1964, ''Tertiary Geology of the Beaver Rim Area, Fremont and Natrona Counties, Wyoming,'' Bulletin 1164, US Geological Survey, 99 pp.

Walton, A.M., 1975, ''Zeolite Diagenesis in Oligocene Volcanic Sediments, Trans-Pecos, Texas,'' Geological Society of America Bulletin, Vol 86, pp. 615–624.

Walton, A., et al., 1972, ''Zeolites of Texas,'' Open-file Report No. OF-BD-1972–1, Texas Bureau of Economic Geology.

Walton, A.W., and Henry, C.D., eds., 1979, ''Cenozoic Geology of the Trans-Pecos Volcanic Field of Texas,'' Texas Bureau of Econ. Geology Guidebook 19.

Wolf, K.H., and Ellison, B., 1971, ''Sedimentary Geology of the Zeolitic Volcanic Lacustrine Pliocene Rome Beds, Oregon,'' *Sedimentary Geology*, Vol. 6, pp. 271–302.

Foreign Deposits

Anon., 1988, *Mining Journal*, London, Oct. 14.

Aiello, R., 1993 ''Zeolitic Tuffs as Building Materials in Italy: A Review,'' Program and Abstracts, Zeolite '93, (extended abstract), 4th International Conference on the Occurrence, Properties, and Utilization of Natural Zeolites, Boise, ID, June 1991, pp. 27–28.

Alexiev, B., 1968, ''Clinoptilolite des Rhodopes du Nord-Est,'' *C.R. Acad. Bulgare Sci.*, Vol. 21, pp. 1093–1096.

Alietti, A., et al., 1967a, *Le Zeoliti in Italia*, lst Mineralogical University of Modena, pp. 245–314.

Alietti, A,, and Ferrarese, G., 1967b, ''Clinoptilolite, Na-Montmorillonite e Ossidi di Manganese in una Formazione Sedimentaria a Zovencedo (Vicenza),'' *Miner. Petr. Acta*, Vol. 13, pp. 139–146.

Askings, P.W., 1980, ''Report on Investigations: Zeolites, Gad's Hill Area,'' Open File Report, Camalco Ltd., Report (unpublished), Department of Mines, Tasmania.

Axente, D., Abrudean, M., and Baldea, A., 1983, ''Adsorption on Romanian Natural Clinoptilolite,'' *Zeolites*, Vol. 3, pp. 259- 260.

Bedeleanu, I., and Stoici, S.D., eds., 1984, *Zeoliti,* Technina, Bucharest, 227 p.

Boles, J.R., 1971 ''Stratigraphy, Petrology, Mineralogy, and Metamorphism of Mainly Triassic Rocks, Hokonai Hills, Southland, New Zealand,'' unpublished PhD thesis, University of Otago, Dunedin, New Zealand, 406 pp.

Boles, J.R., 1974, ''Structure, Stratigraphy, and Petrology of Mainly Triassic Rocks, Hokonai Hills, New Zealand,'' *New Zealand Journal of Geology and Geophysics*, Vol. 17, pp. 337–374.

Boles, J.R., and Coombs, D.S., 1975, ''Mineral Reactions in Zeolitic Triassic Tuff, Hokonai Hills, New Zealand,'' Geological Society of America Bulletin, Vol. 86, pp. 163–173.

Callegari, E., 1964. ''Rapporti fra Distribuzione di Analcime e Fenomeni di Albitizzazione Nella 'Pictra Verde' Degli di Livinallongo Nella Regione Dolomitica,'' *Studi Trentini Sci. Nat.*, Vol. 61, pp. 25–43.

Callegari, E., and Jobstraibizer, P.G., 1964, ''Sulla Presenza di Analcime Nella 'Pictra Verde' Degli Strati di Livinallongo Della Localita Tipo (Alta Valle del Corderuole Dolomiti),'' *Rend. Soc. Miner. Italia*, Vol. 20, pp. 83–99.

Coombs, D.S., 1954, ''The Nature and Alteration of Some Triassic Sediments from Southland, New Zealand,'' *Transactions*, Royal Society of New Zealand, Vol. 82, pp. 65–109.

Coombs, D.S., 1958, ''Zeolitized Tuffs from the Kuttung Glacial Beds near Seaham, New South Wales,'' *Australian Journal of Science*, Vol. 21, No. 1, pp. 18–19.

de Gennaro, M. and Adabbo, M., 1993, ''Hypotheses on the Genesis of Zeolites in Some European Volcaniclastic Deposits,'' Program and Abstracts, ''Zeolite '93,'' (extended abstract), 4th International Conference on the Occurrence, Properties, and Utilization of Natural Zeolites, Boise, ID June 1991, pp. 77–79.

de Gennaro, M., Colella, C., Franco, E., and Aiello, R., 1983, ''Italian Zeolites 1: Mineralogical and Technical Features of Neapolitan Yellow Tuff,'' *Industrial Minerals*, No. 186, pp. 47–53.

de Gennaro, M., et al., 1984, ''Italian Zeolites 2: Mineralogical and Technical Features of Campanian Tuff,'' *Industrial Minerals*, No. 204, pp. 97–109.

Djourova, E., and Aleksiev, B., 1988, ''Zeolitic Rocks,'' *Mineral Deposits in Bulgaria,* Part 1, in Bulgarian, Exogenetic Industrial Minerals and Rocks, Technika, Sofia, pp. 233–247.

Djourova, E.G. and Milakovska, Z.I., 1993, ''Detrital Zeolitic Rocks in the Northeastern Rhodopes, Bulgaria,'' Program and Abstracts, ''Zeolite '93,'' (extended abstract), 4th International Conference on the Occurrence, Properties, and Utilization of Natural Zeolites, Boise, ID, June 1993, pp. 82–83.

Eugster, H.P., 1970, ''Chemistry and Origin of the Brines of Lake Magadi, Kenya,'' Special Paper 3, Mineralogical Society of America, pp. 215–235.

Fersman, A.E., 1962, *Zeolites of Russia and their Mineralogy*, Vol. 1, Akad. Nauk SSSR, Moscow.

Filizova, L.D., Petrov, O.E., and Han, B.Z., 1993, ''Heulandite-Clinoptilolite-Group Minerals from the Northeastern Part of the Korean Peninsula,'' Program and Abstracts, ''Zeolite '93,'' (extended abstract), 4th International Conference on the Occurrence, Properties, and Utilization of Natural Zeolites, Boise, ID, pp. 91–93.

Flood, P.G., 1991, ''Prospecting for Natural Zeolites: An Exploration Model based on an Occurrence of 300 Ma Old Rocks near Werris Creek, New South Wales, Australia,'' *Transactions*, Sec. B: Appl. Earth Science, Institution of Mining and Metallurgy, Vol. 100, pp. B-9 to B-13, Jan.-Apr.

Flood, P.G., and Stephen, G.J., 1993, ''Occurrence and Origin of Australia Natural Zeolites,'' Program and Abstracts, ''Zeolite '93,'' (extended abstract), 4th International Conference on the Occurrence, Properties, and Utilization of Natural Zeolites, Boise, ID, pp. 93–94.

Frederickson, P.W., 1985, ''Characterization of NSW Zeolite Deposits,'' New South Wales Department of Mineral Resources, Mineral Resources Development Laboratory Report 85/29, unpublished.

Fruchtbauer, H., et al., 1977, ''Tertiary Lake Sediments of the Ries, Research Bore Hole Nordlingen 1973—A Summary,'' *Geol. Bau.*, Vol. 75, No. 18, pp. 18–22.

Fuentes, G.R., and Gonzales, J.A., eds., 1991, ''Zeolite '91,'' *Proceedings*, 3rd International Conference on the Occurrence, Properties, and Utilization of Natural Zeolites, Havana, Cuba.

Gianello, A., and Gottardi, G., 1969, ''Sulla Zeolitizzazione del Livello Cineritico detto 'Tripoli di Contignaco','' *Min. Petr. Acta*, Vol. 15, pp. 1–4.

Gonzales, J.L., and Costa de los Reyes, A., 1993, ''Cuban Experiences in the Production and Marketing of Natural Zeolites,'' Program and Abstracts, ''Zeolite '93,'' (extended abstract), 4th International Conference on the Occurrence, Properties, and Utilization of Natural Zeolites, Boise, ID, June, 1993, pp. 101–103.

Gottardi, G., and Obradovic, J., 1978, ''Sedimentary Zeolites in Europe,'' *Fortschr. Miner.*, Vol. 56, No. 2, pp. 316–366.

Grimsicar, A., 1967, ''Zeoliti v Oligocenskih Tufih med Mozirjem in Celje,'' *Razprava Procila*, Vol. 10, pp. 239–245.

Han, B., et al., 1985, ''Natural Zeolites,'' Science and Encyclopedia, Pyongyong, Democratic People's Republic of Korea, 232 pp.

Hay, R.L., 1968, ''Chert and its Sodium-Silicate Precursors in Sodium Carbonate Lakes of East Africa,'' *Countr. Mineral. Petrol.*, Vol. 17, pp. 255–274.

Holmes, G.G., Lishmund, S.R., and Oakes, G.M., 1982, ''A Review of Industrial Minerals and Rocks in New South Wales,'' Bulletin 30, New South Wales Geological Survey, 249 pp.

Holmes, G.G., and Pecover, S.R., 1987, ''Natural Zeolites of New South Wales,'' 65 1987/145, New South Wales Geological Society, 135 pp.

Hora, Z.D., and Kwong, Y.T.J., 1984, ''Industrial Zeolites and Ratile,'' Paper 1984–1, Geological Field Work, 1983, British Columbia Ministry of Energy, Mines, and Petroleum Resources, pp. 211–212.

Iijima, A., 1980, ''Geology of Natural Zeolites and Zeolitic Rocks,'' *Proceedings*, 5th International Conference on Zeolites, L.V.C. Rees, ed., Heyden Press, London, pp. 103–118.

Iijima, A., and Utada, M., 1972, ''A Critical Review on the Occurrence of Zeolites in Sedimentary Rocks in Japan,'' *Japanese Journal of Geology*, Vol. 42, pp. 61–83.

Iijima, A., et al., 1986, ''Zeo-Trip '86 in Japan. An excursion to selected zeolite deposits in northern Kanto and southern Tohoku, Northeast Japan, August 23–25,'' Japan Association of Zeolites—Excursion Guide, 67 pp.

Istrate, G., 1980, ''Zeolites of Romania: Their Nature and Occurrences,'' *Proceedings*, 15th International Conference on Zeolites, L.V.C. Rees, ed., Heyden Press, London, pp. 149–154.

Kim, S.J., and Choi, Y-S., 1993, ''Clinoptilolite and Mordenite from the Tertiary Volcanic Sediments in the Guryongpo Area, Korea,'' Program and Abstracts, ''Zeolite '93,'' (extended abstracts), 4th International Conference on the Occurrence, Properties, and Utilization of Natural Zeolites, Boise, ID, June 1993, pp. 133–135.

Kossovskaya, A.G., 1973, ''Genetic Association of Sedimentary Zeolites in the Soviet Union,'' *Advances in Chemistry Series*, Vol. 121, pp. 200–208.

Kossovskaya, A.G., and Shutov, V.D., 1963, ''The Correlation of Zones of Regional Epigenesis and Metagenesis in Terrigenous and Volcanic Rocks,'' *Doklady, Earth Sciences Secs.*, Akademii Nauk SSSR, Vol. 139, pp. 732–736.

Kostov, I., 1970, ''Tectonomagmatic Significance of Zeolites in the Srednogorian Zone and the Rhodopes,'' Bulletin of Geological Institute, Bulgarian Academy of Science, Ser. Geochem. Minera. Petr., Vol. 19, pp. 235–241.

Kostov, I., Radonova, T., and Tsonev, D., 1971, ''Zeolitizacija v Senonskija Kompleks na Panagjurskija Rajon,'' Bulletin Geological Institute, Bulgarian Academy of Science, Ser. Geochem. Minera. Petr., Vol. 20, pp. 123–151.

Kristmanndottir, H., and Tomasson, J., 1978, ''Zeolite Zones in Geothermal Areas in Iceland,'' *Natural Zeolites: Occurrences, Properties, Use*, L.B. Sand and F.A. Mumpton, eds., Permagon Press, New York, pp. 277–284.

Lenzi, G., and Passaglia, E., 1975, ''Fenomeni di Zeolitizzazione Nello Formazioni Vulcaniche della Regione Sabatina,'' Vol. 93, Bulletin Geological Society of Italy, pp. 623–645.

Ludwig, U., and Schwiete, H.E., 1962, ''Beitrag zur Konstitution Einiger Rheinischer Trasse,'' *Zement-Kalk-Gips*, Vol. 15, No. 4, pp. 160–165.

Martini, J., and Vuagnat, M., 1970, ''Metamorphose Niedrigst Temperierten Grades in Den Westalpen,'' *Fortschrittre Mineralogie*, Vol. 47, pp. 52–64.

Mikhaylov, A.S., Krinari, A.I. and Maslov, S.E., 1971, ''Zeolitic Rocks of South Turkmeniia,'' *Doklady Akademie*, Nauk SSSR, Vol. 199, pp. 696–700.

Minato, H., 1968, ''Characteristics and Uses of Natural Zeolites,'' *Koatsugasu*, Vol. 5, pp. 536–547.

Minato, H., and Utaja, M., 1969, ''Zeolite,'' *The Clays of Japan*, International Clay Conference, Japan Geological Survey, pp. 121–134.

Mumpton, F.A., 1973b, ''First Reported Occurrence of Zeolite in Sedimentary Rocks in Mexico,'' *American Mineralogist*, Vol. 58, pp. 287–290.

Nemecz, E., and Varju, G., 1962, ''Sodium Bentonization, Clinoptilolitization, and Adularization in the Rhyolitic Tuffs of the Scerencz Piedmont Area,'' *Acta Geol. Sci. Hung.*, Vol. 6, pp. 389–426.

Nikolic, D., Poharc, V., and Markovic, D., 1975, ''Mineralosko-Hemijako Proucnvanje Tufa Okoline Visnjico i Mogucnosti Njegove Primene,'' *Proceedings*, Zbornik Radova, Simpoz. Cementna Ind. Novi. Snd. Bcocin, pp. 33–39.

Noh, J.H., and Kim, S.J., 1982, ''Mineralogy and Genesis of Zeolites from the Tertiary Tuffaceous Rocks in the Guryongpo Area,'' *Journal*, Geological Society of Korea, Vol. 18, pp. 1- 10.

Obradovic, J., and Kemenci, R., 1975, ''Dijagenetske Promene Miocenskih Piroklastita iz Dubokih Busotina Severne Backe,'' *Geol. An. Balke. Pol.*, Vol. 39, pp. 333–355.

Obradovic, J., and Stojanovic, D., 1972, ''Occurrence of 'Pietra Verde', Similar to that Occurring in Italian Dolomites in Volcanic Sedimentary Series in the Budva-Bar Aea (Montenegro, Yugoslavia),'' *Studi Trent. Sci. Nat. Acad.*, Vol. 49, pp. 161–172.

Polgari, M., et al., 1993, ''Mineralogy, Chemistry, and Genesis of Miocene Zeolite-bearing Rhyolite Tuff, Southwest Hungary,'' Program and Abstracts, ''Zeolite '93,'' (extended abstract), 4th International Conference on the Occurrence, Properties, and Utilization of Natural Zeolites, Boise, ID, June 1993, pp. 164–165.

Read, P.B., 1987, ''Industrial Minerals in Some Tertiary Basins, Southern British Columbia,'' Geological Fieldwork, 1986, British Columbia Ministry of Energy, Mining and Petroleum Resources, pp. 247–254.

Rocher, P., et al., 1991, Program and Abstracts, ''Zeolites '91,'' 3rd International Conference on the Occurrence, Properties and Utilization of Natural Zeolites, Havana, Cuba, Apr.

Samajova, E., and Kraus, I., 1977, ''Zeolitic Volcanoclasts—A New Perspective Ore in SSR,'' *Geologicky Pruzkum*, Vol. 8, pp. 230–233.

Sameshima, T., 1977, ''Zeolites in Tuff Beds of the Waitemata Group, Auckland Province, New Zealand,'' *Natural Zeolites: Occurrence, Properties, Use*, L.B. Sand and F.A. Mumpton, eds., Pergamon Press, New York, pp. 309–317.

Seki, Y., et al., 1969, ''Metamorphism in the Tansawa Mountains, Central Japan,'' *Journal*, Japanese Association of Mineralogists, Petrologists, and Economic Geologists, Vol. 61, pp. 1–75.

Sersale, R., 1958, ''Genesi e Costituzione del Tufo Giallo Napoletano,'' *Rend. Accad. Sci. Fis. Mat.*, Vol. 25, Naples, pp. 181–207.

Sersale, R., 1978, ''Occurrence and Uses of Zeolites in Italy,'' *Natural Zeolites: Occurrences, Properties, Use*, L.B. Sand and F.A. Mumpton, eds., Pergamon Press, New York, pp. 285–302.

Sersale, R., Ajello, R., and Frigone, G., 1963, ''Sulla Presenza di Orizzonti

Zeolitici Nella Serie Oligo-Miocenica di Garbagna,'' *Atti Accad. Sci.*, Torino, Vol. 97, pp. 1–15.

Steiner, A., 1953, ''Hydrothermal Rock Alteration at Wairakei, New Zealand,'' *Economic Geology*, Vol. 48, pp. 1–13.

Stephen, G.J., and Gout, M.J., 1993, ''Production and Marketing of Australian Natural Zeolites,'' Program and Abstracts, ''Zeolite '93,'' (extended abstract), 4th International Conference on the Occurrence, Properties, and Utilization of Natural Zeolites, Boise, ID, June 1993, pp. 193–195.

Stoffler, D.U., et al., 1977, ''Research Drilling Nordlingen 1973 (Ries): Composition and Texture of Polymict Impact Breccias,'' *Geologica Bavarica*, Vol. 75, pp. 119–132.

Stojanovic, D., 1972, ''Vulkanski Tufovi i Sedimentne Steno u Srbiji an Andrzajem Zeolita,'' *Zapisnici Srpskog Geol.*, Beograd, Drustva za 1968, 1969, i 1970, pp. 9–12.

Sudo, T., 1950, ''Mineralogical Studies on the Zeolite-Bearing Pumice Tuffs Near Yokotemachi, Akita Prefecture,'' *Journal*, Geological Society of Japan, Vol. 56, pp. 13–16.

Surdam, R.C., 1968, ''Low-Grade Metamorphism of the Karmutsen Group, Central Vanouver Island, British Columbia, Canada,'' Abstracts for 1967, Special Paper 115, Geological Society of America, pp. 217–218.

Surdam, R.C., 1973, ''Low-Grade Metamorphism of Tuffaceous Rocks in the Karmutsen Group, Vancouver Island, British Columbia,'' Bulletin, Geological Society of America, Vol. 84, pp. 1911–1922.

Surdam, R.C., and Eugster, H.P., 1976, ''Mineral Reactions in the Sedimentary Deposits of the Lake Magadi Region, Kenya,'' Bulletin, Geological Society of America, Vol. 87, pp. 1739–1752.

Tsirambides, A., Filippidis, A., and Kassoli-Fournaraki, A., 1993, ''Zeolitic Alteration of Eocene Volcaniclastic Sediments at Metaxades, Thrace, Greece,'' *Applied Clay Science*, Vol. 7, pp. 509–526.

Tsitsishvili, G.V., 1978, ''Adsorption and Catalytic Properties of some Soviet Natural Zeolites,'' *Natural Zeolites: Occurrence, Properties, Use*, L.B. Sand and F.A. Mumpton, eds., Pergamon Press, New York, pp. 397–402.

Tsolis-Katagas, P., and Katagas, C., 1990, ''Zeolitic Diagenesis of Oligocene Pyroclastic Rocks of the Metaxades Area, Thrace, Greece,'' *Mineral. Magazine*, Vol. 54, pp. 95–103.

Utada, M., 1970, ''Occurrence and Distribution of Authigenic Zeolites in the Neogene Pyroclastic Rocks in Japan,'' Science Papers Collection, General Education, Tokyo University, Vol. 20, pp. 191–262.

Utada, M, 1971, ''Zeolitic Zoning of the Neogene Pyroclastic Rocks in Japan,'' Science Papers Collection, College of General Education, University of Tokyo, Vol. 21, pp. 189–221.

Varju, G., ''Ratkai Trasz (Klinoptilolitos Riollittufa) Elofodulas Foldtani Viszonyai,'' *Foldtani Kutastas*, Vol. 9, No. 3, pp. 21–30.

Ylagan, R.F., et al., 1993, ''Zeolite and Clay Mineral Diagenesis of the Mount Epomeo Green Tuff, Ischia Island, Italy,'' Program and Abstracts, ''Zeolite '93,'' (extended abstract), 4th International Conference on the Occurrence, Properties, and Utilization of Natural Zeolites, Boise, ID, June 1993, pp. 218–220.

Commercial Utilization of Natural Zeolites

Anon., 1972, ''Use of Local Minerals in the Treatment of Radioactive Wastes,'' Technical Report 136, 97-B, International Atomic Energy Agency, Vienna.

Anon., 1975, ''NuFuel, A New Source of Energy,'' Brochure of NRG NuFuel Company, 6 pp.

Anon., 1990, unpublished material, British Nuclear Fulls Ltd.

Adam, L., Kakasy, Gy., and Pallos, I., 1971, ''Removal of Radioactive Pollutants from Water Using Mineral Substances of Natural Domestic Origin,'' *Hungarian Mining Research Institute*, Vol. 14, pp. 209–212.

Ames, L.L., 1959, ''Zeolitic Extraction of Cesium from Aqueous Solutions,'' Unclassified Report HW-62607, US Atomic Energy Commission, 23 pp.

Ames, L.L., 1960, ''The Cation Sieve Properties of Clinoptilolite,'' *American Mineralogist*, Vol. 45, pp. 689–700.

Ames, L.L., 1967, ''Zeolitic Removal of Ammonium Ions from Agriculture Wastewaters,'' *Proceedings*, 13th Pacific Northwest Industrial Waste Conference, Washington State University, Pullman, pp. 135–152.

Andersson, S., et al., 1975, ''Separation of Poisons from the Dialysis Liquid of a Recycle Dialysis System,'' Ger. Offen. 2,512,212, Sep. 25, 1975, 30 pp.

Barrer, R.M., 1938, ''The Sorption of Polar and Non-Polar Gases on Zeolites,'' *Proceedings*, Royal Society of London, Vol. 167A, pp. 392–419.

Chelischev, N.F., et al., 1974, ''Ion Exchange of Heavy Metals on Clinoptilolite,'' *Doklady Akademie*, Nauk SSSR, Vol. 217, pp. 1140–1141.

Domine, D., and Hay, L., 1968, ''Process for Separating Mixtures of Gases by Isothermal Adsorption: Possibilities and Applications,'' *Molecular Sieves*, Society Chem. Ind., London, pp. 204–216.

Dwyer, J., 1984, ''Zeolite Structure, Composition and Catalysis,'' *Chemistry and Industry*, Vol. 7, pp. 258–268.

Ferguson, G.A., 1984, ''Clinoptilolite Zeolite: Effect on Ammonium Exchange Reactions, Nitrification, and Plant N-use Efficiency,'' unpublished PhD thesis, University of Arizona, Tucson.

Filizova, L., 1993, ''The Use of Zeolites to Counteract Chernobyl Fallout,'' Program and Abstracts, ''Zeolite '93,'' (extended abstract), 4th International Conference on the Occurrence, Properties and Utilization of Natural Zeolites, Boise, ID, June 1993, pp. 88–90.

Flanigen, E.M., 1984, ''Adsorption Properties of Molecular Sieve Zeolites,'' *Zeo-Agriculture: Use of Natural Zeolites in Agriculture and Aquaculture*, W.G. Pond and F.A. Mumpton, eds., Westview Press, Boulder, CO, pp. 55–68.

Frederickson, P.W , 1987, ''Properties and Uses of Natural Zeolites,'' *Natural Zeolites of New South Wales*, G.C. Holmes and S.R. Pecover, eds., New South Wales Geological Survey.

Freeman, H.D., and Jones, E.O., 1989, ''Application of Engineered Sorbent Barriers: Summary of Laboratory Data for FY 1988,'' Pacific Northwest Laboratory, Battelle Memorial Institute, PNL-6997, UC-512, unpublished report.

Freeman, H.D., Mitchell, S.J., and Buelt, J.L., 1986, ''Engineered Sorbent Barriers for Low-level Waste Disposal,'' Pacific Northwest Laboratory, Battelle Memorial Instutute, PNL-6051, UC-70B, unpublished report, 39 pp.

Goeppner, J. and Hasselmann, D.E., 1974, ''Digestion Byproduct May Give Answer to Energy Problem,'' *Water and Waste Engineer*, Vol. 11, No. 4, pp. 30–35.

Grant, D.C., et al., 1986, ''Effect of Process Variables on the Removal of Contaminants from Radioactive Waste Streams using Zeolites,'' Westinghouse R & D Center/West Valley Nuclear Services Co., Inc., presented at Waste Management '87 Symposium, Tucson, AZ, Mar.

Grant, D.C., Skriba, M.C., and Saha, A.K., 1987, ''Removal of Radioactive Contaminants from West Valley Waste Streams using Natural Zeolites,'' *Environmental Progress*, Vol. 6, No. 2, pp. 104–109.

Grant, D.C., et al., 1988, ''Design and Optimization of a Zeolite Ion Exchange System for Treatment of Radioactive Wastes,'' American Institute of Chemical Engineers, Symposium Series, Vol. 264, pp. 13–22.

Grant, D.C., et al., 1989, ''Full-scale Zeolite Ion Exchange Systems for Treating Radioactive Wastes,'' Westinghouse Science and Technology Center/West Valley Nuclear Services, Co., Inc., unpublished report, pp. 1–13.

Griffiths, J., 1987, ''Zeolites Cleaning Up—From the Laundry to Three Mile Island,'' *Industrial Minerals*, No. 232, Jan., pp. 19–33.

Hashimoto, A., 1974, ''Adsorption of Heavy Metals and Application to Watersupply Systems, II: Removal of Ca, Cd, and N as Ammonia,'' *Mizu Shori Gijutsu*, Vol. 15, pp. 1085–1091.

Hawkins, D.B. and Short, H.L., 1965, ''Equations for the Sorption of Cesium and Strontium on Soil and Clinoptilolite,'' IDO-12046, US Atomic Energy Commission, 33 pp.

Jewell, W.J., 1974, ''Methane . . . The Energy-Sufficient Farm,'' *The News*, State University of New York, Vol. 3, No. 2, p. 4.

Johnson, P.W., and Sieburth, J.M., 1974, ''Ammonia Removal by Selective Ion Exchange, A Backup System for Microbiological Filters in Closed-System Aquaculture,'' *Aquaculture*, Vol. 4, pp. 61–68.

Kakasy, I., Pallos, I., and Adam, L., 1973, ''Etude des Roches Clinoptilolitiques de la Hongrie au Point de Vue de la Fixation des Dechets Radioactifs Liquides,'' Hungarian Mining Research Institute, No. 16, pp. 169–175.

Kato, K., 1974, ''Removal of Cs^{137} from Aqueous Solutions by Zeolite,'' *Hoken Butsuri*, Vol. 9, pp. 11–16.

Kato, K., et al., 1969, ''Zeolite as Polishing Agent for Dentrifice, I: (Chemical) Properties of Zeolite,'' *Iwate Daigaku Nagakubu Hokoku*, Vol. 3, pp. 22–35.

Kato, K., et al., 1970, ''Zeolites as Polishing Agent for Dentrifice, II: Abrasive Action and Active Fluoride Ion in Zeolite-Containing Fluoride,'' *Iyokizikai Kenkyusho Hokoku*, Vol. 4, pp. 115–128.

Kobayashi, Y., 1970, ''Natural Zeolite-Fillers for Paper,'' Japan Kokai 70,041,044, 2 pp.

Konikoff, M.A., 1973, ''Comparison of Clinoptilolite and Biofillers for Nitrogen Removal in Recirculating Fish Culture Systems,'' PhD dissertation, Southern Illinois University, Carbondale, 98 pp.

Mathers, W.G., and Watson, L.C., 1962, ''A Waste Disposal Experiment Using Mineral Exchange on Clinoptilolite,'' AECL-1521, Atomic Energy of Canada, Ltd.

Mercer, B.W., et al., 1970, ''Ammonia Removal from Secondary Effluents by Selective Ion Exchange,'' *Water Pollution Control Federation Journal*, Vol. 42, No. 2, Part 2, pp. R95-R107.

Milton, R.M., 1968, ''Commercial Development of Molecular Sieve Technology,'' *Molecular Sieves*, Soc. Chem. Ind., London, pp. 199–203.

Minato, H., 1974, ''Properties of Sedimentary Zeolites for Industrial Application,'' Seminar on the Occurrence, Origin, and Utilization of Sedimentary Zeolites in the Circum-Pacific Region, US-Japan Cooperative Science Program, Menlo Park, California, unpublished abstract, July.

Minato, H., and Tamura, T., 1978, ''Production of Oxygen and Nitrogen with Natural Zeolites,'' *Natural Zeolites: Occurrence, Properties, Use*, L.B. Sand and F.A. Mumpton, eds., Pergamon Press, New York, pp. 509–516.

Mumpton, F.A., 1984, 'Natural Zeolites,'' *Zeo-Agriculture: Use of Natural Zeolites in Agriculture and Aquaculture*, W.G. Pond, and F.A. Mumpton, eds., Westview Press, Boulder, CO, pp. 33–44.

Nelson, J.L., and Mercer, B.W., 1963, ''Ion Exchange Separation of Cesium from Alkaline Waste Supernatant Solutions,'' HY-76449, US Atomic Energy Commission.

Nikashina, V.A., et al., 1974, ''Selective Sorption of Strontium by Natural Clinoptilolite from Aqueous Solutions,'' *Radiokhim*, Vol. 16, pp. 753–756.

Nishita, H., and Haug, R.M., 1972, ''Influences of Clinoptilolite on Sr-90 and Cs-137 Uptakes by Plants,'' *Soil Science*, Vol. 114, pp. 149–157.

Pond, W.G., and Mumpton, F.A., eds., 1984, *Zeo-Agriculture: Use of Natural Zeolites in Agriculture and Aquaculture*, Westview Press, Boulder, CO, 296 pp.

Rees, L.V.C., 1984, ''Adsorption and Diffusion of Gases in Zeolites,'' *Chemistry and Industry*, pp. 252–257.

Reynolds, W.R., and Williford, C.W., 1988, ''Zeolite Ammonia Removal from Catfish Pond Water,'' Project G1431–05, Technical Completion Report, Water Resources Research Institute, Mississippi State University, State College, 16 pp.

Saha, A.K., 1985, ''Natural Zeolites and Clays: The Promising Media for Selective Radioactive Partition/Immobilization,'' *Waste Management' 85*, Vol. 3, pp. 197–201.

Semmons, M.J., and Sayfarth, M., 1977, ''The Selectivity of Clinoptilolite for Certain Heavy Metals,'' *Natural Zeolites: Occurrence, Properties, Use*, L.B. Sand and F.A. Mumpton, eds., Pergamon Press, New York, pp. 517–526.

Shane, S.M., 1987, ''Synthetic Zeolite Fee Component for Laying Hens,'' *Poultry Tribune*, Apr., p. 23.

Simms, R.C., 1972, ''Enhancement of Nitrification in Activated Sludge by Addition of Hector Clinoptilolite,'' *Environmental Science Engineering Notes*, Vol. 9, pp. 2–4.

Takasaka, A., 1975, ''Properties and Use of the Itaya Zeolite from Japan,'' *Eunsai*, Vol. 20, pp. 127–134, 142.

Tamura, T., 1971, ''Oxygen Concentration Process,'' US Patent 3,533,221, Oct. 13, 1970, 10 pp.

Tchernev, D.I., 1978, ''Solar Energy Applications of Natural Zeolites,'' *Natural Zeolites: Occurrence, Properties, Uses*, L.B. Sand and F.A. Mumpton, eds., Pergamon Press, New York, pp. 479–485.

Torii, K., 1978, ''Utilization of Natural Zeolites in Japan,'' *Natural Zeolites: Occurrence, Properties, Use*, L.B. Sand and F.A. Mumpton, eds., Pergamon Press, New York, pp. 127–142.

Tsitsishvili, G.V., 1977, ''Clinoptilolite and Other Natural Zeolites,'' *Proceedings*, Symposium Problems of Clinoptilolite Studies and Application, Nov. 1974, Tbilisi, G.S.S.R., pp. 72–21.

Tsitsishvili, G.V., 1988, ''Perspectives of Natural Zeolite Applications,'' *Occurrence, Properties, and Utilization of Natural Zeolites*, D. Kallo and H.S. Sherry, eds., Akademini Kiado, Budapest, pp. 367–393.

Tsitsishvili, G.V., et al., 1984, ''Agricultural Applications of Natural Zeolites in the Soviet Union,'' *Zeo-Agriculture: the Use of Natural Zeolites in Agriculture and Acquaculture*, W.G. Pond and F.A. Mumpton, eds., Westview Press, Boulder, CO, pp. 189–221.

Vdovina, E.D., Radyuk, R.I., and Sultanov, A.S., 1976, ''Use of Uzbekistan Natural Zeolites for the Purification of Low-Level Wastewaters, I. Sorption of Radioactive Cesium,'' *Radiokhimiya*, Vol. 18, pp. 422–423.

Weigel, D., and Steinhoff, E., 1925, ''Adsorption of Organic Liquid Vapors by Chabazite,'' *Z. Kristallogr.*, Vol. 61, pp. 125–154.

Williford, C.W., Reynolds, W.R., and Quiros, M., 1991, ''Clinoptilolite Removal of Ammonia from Simulated and Natural Catfish Pond Waters,'' *Applied Clay Science*, Vol. 6, pp. 277–291.

Yoshida, H., Kurata, A., and Sanga, S., 1976, ''Removal of Heavy Metal Ions from Wastewater Using Zeolites,'' *Mizu Shori Gijutsu*, Vol. 17, pp. 219–226.

ZIRCONIUM AND HAFNIUM MINERALS

THOMAS E. GARNAR

Zirconium and hafnium are the twin elements of the periodic table because they are always found together. They resemble each other so closely that hafnium was not discovered until 135 years after the discovery of zirconium. Zirconium was discovered by Klaproth in 1789 and isolated 35 years later by Berzelius. Because hafnium's chemical properties are so close to those of zirconium, it was not discovered until 1923 when Coster and von Hevesy detected it using X-ray spectrographic analysis. These elements occur most commonly in nature as the mineral zircon ($ZrSiO_4$) and less commonly as baddeleyite (ZrO_2). They also are found as a variety of other silicates. Zircon was identified as a component in alluvial and beach sands in 1895, but it was not produced in any quantity until 20 years later. During World War I, it was produced as a coproduct of beach sand mining for titanium minerals just south of Jacksonville Beach, FL, and was patented as a refractory. It was not until the 1930s, when Zircon Rutile, Ltd., began mining at Byron Bay on the east coast of Australia, that zircon was first used as foundry sand. Later in the 1940s, NL Industries and Humphreys Gold Corp. began production of zircon sands from fossil beaches in northeast Florida. Baddeleyite first became available as a commercial product in 1916, but never in the quantity of zircon.

Zircon holds a unique position as an industrial mineral because it is used for both its physical and chemical properties and as an ore of zirconium and hafnium metals. Australia is the major zircon producer followed by the Republic of South Africa and the United States.

Zircon is also produced in India, Sierra Leone, Sri Lanka, Malaysia, the former USSR, China, Thailand, and Brazil.

The location of world zircon mines, baddeleyite mines, and undeveloped zircon ores are shown in Fig. 1. World production of zirconium mineral concentrates is shown in Table 1.

MINERALOGY AND GEOLOGY

In addition to zircon and baddelyite, zirconium and hafnium also occur as silicates in combination with iron, calcium, sodium, manganese, and other elements. More rarely, they combine with titanium, thorium, calcium, and iron to form oxides.

Zircon is extremely resistant to weathering; therefore, it is present in many ancient beach sands and placer deposits. Although resistant to alteration from external sources, it does undergo internal alteration (metamictization) as the result of the substitution of thorium and uranium (usually < 500 ppm) for zirconium either in the zircon lattice or in solid solution.

Alteration to the metamict state takes place as radioactive emanations from these elements disorder the crystal lattice. This process is accompanied by hydration, reduction in specific gravity, and changes in color. The mineralogy of zircon and baddeleyite and variation of physical and chemical properties are shown in Table 2.

The effects of zircon's internal alteration and darkened color can be corrected by heating to 1 000°C for 30 min or so. A time-temperature relationship makes it possible for decolorization and reordering of the crystal lattice to take place when zircon is heated at higher temperatures for shorter periods or at lower temperatures for longer periods. After the mineral is heated, X-ray diffraction patterns are sharper, and, except for particles of malacon, the grains will be white or colorless in the absence of surface coatings. The color change is only temporary and the particles slowly darken again with time.

Zircon fluoresces light yellow in shortwave ultraviolet light. If exposed to ultraviolet light for extended periods of time, white (calcined) zircon turns purple. With time, the induced color begins to fade and eventually disappears.

Zircon originates as an accessory mineral in a variety of igneous and metamorphic rocks, especially those containing sodic feldspars such as granite, syenite, and diorite. It is one of the earliest minerals to crystallize from a cooling magma and sometimes incorporates inclusions. Much of the zircon in these rocks crystallizes in tetragonal prisms with pyramidal terminations; however, it also commonly occurs as rounded grains in igneous rocks. It is sometimes found in metamorphic rocks such as gneiss and schists. One occurrence has been reported in a meteorite.

Before the zircon particles can become part of a beach sand deposit, their host rocks must undergo a series of events that will ultimately liberate the zircon for transportation to a seacoast. After subaerial weathering, the rocks are broken down into smaller fragments and after further weathering decompose to the point at which the smaller zircon grains are liberated. The mineral particles are transported by streams and ultimately are deposited along a marine shoreline. Here, the action of the waves, tidal currents, and wind may remove lighter quartz and form a heavy mineral deposit rich in zircon. Some of the older beach sand deposits have been consolidated and the grains cemented together to form sandstones containing concentrations of titanium minerals, zircon, monazite, and other heavy minerals. At present, all commercial zircon products are mined and separated from relatively young (Pleistocene or younger) beach sand deposits on or near active coastlines.

OCCURRENCE OF ZIRCON AND BADDELEYITE

Australian Zircon Deposits

Australia is the major world producer and exporter of zircon sand. Deposits are found on both the east and west coasts of Australia. In eastern Australia, major heavy mineral sand deposits lie between Broken Bay, New South Wales, to the south of Cape Clinton, north to Rockhampton, Queensland. The richest occur-

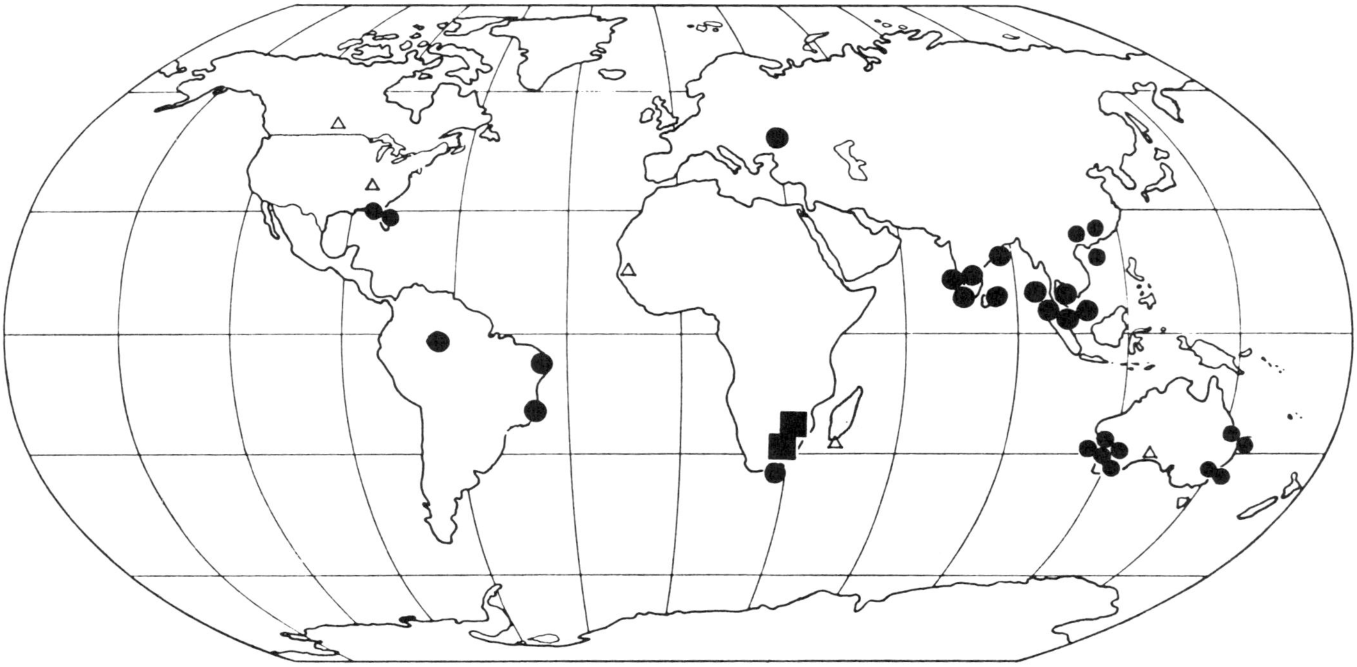

Fig. 1. World zircon and baddeleyite mines and undeveloped ores. Legend: active zircon mine ●, active baddeleyite mine ■, undeveloped zircon deposit △.

Table 1. Zirconium Mineral Concentrates: World Production, by Country (tons)

Country	1986	1987	1988	1989	1990
Australia	451 824	456 590	480 049	511 000	442 000
Brazil	15 116	18 131	28 029	32 970	33 000
China	15 000	15 000	15 000	15 000	15 000
India	10 000	15 000	17 000	18 000	18 000
Malaysia	12 633	17 828	25 671	18 704	4 279
South Africa, Republic of	140 000	140 000	150 000	180 000	180 000
Sri Lanka	4 000	4 000	3 000	3 000	3 000
Thailand	1 705	1 532	5 098	1 496	2 000
Former USSR	85 000	90 000	90 000	90 000	90 000
United States	W	W	117 606	118 388	102 073
TOTAL	741 278	753 081	929 453	987 758	889 352

W—withheld, *Source: Minerals Yearbooks*, US Bureau of Mines.

rences are concentrated in the area between Yamba and North Stradbroke Island. Although zircon had been produced from a mine at Ponte Vedra, FL, between 1918 and 1929, the eastern Australian mines, which opened in the 1930s, were the first major producers of commercial zircon sands.

The eastern Australian heavy mineral beach sand deposits are of both Pleistocene and Quaternary age. Zircon and its coproduct rutile have been derived from the breakdown and erosion of well sorted silica sandstone in the late Paleozoic and Mesozoic basins of eastern Australia. Much zircon-bearing sand as mined contains less than 0.5% zircon. Zircon from the east coast of Australia ranges in color from dark tan to pale lavender. Because of iron-bearing grain surface coatings, these zircon sands become brownish orange to orange when heated to 1 000°C. Grain size and shape vary from deposit to deposit. The eastern Australia zircon is characterized by subangular to rounded particles with many elongated prismatic grains. Particle size ranges from 90 to 100 μm depending upon which mine produced the zircon. Four companies currently produce zircon from beach sands along the east coast of Australia.

Heavy mineral sand deposits in Western Australia are found along ancient beaches that lie at elevations ranging from 10 to 200 m above present sea level. The strand-line deposits are part of a series of Pleistocene and Recent sediments that form as a thin veneer over much older marine and continental deposits throughout the Perth Basin.

Zircon has become a major export of Western Australia since the discovery of the deposits near the town of Eneabba. The Eneabba heavy minerals occur at strand lines ranging from 82 to 120 m above sea level. The Eneabba mineral grains are extensively coated with ferruginous aluminum silicates as a result of laterization. The grain surface coatings cause the zircon to take on a light orange color when heated to 1 000°C. Four companies currently produce zircon in Western Australia.

South African Deposits

Heavy minerals are produced from a high dune about 160 km long and 2 km wide north of Durban paralleling the Indian Ocean and reported to contain 700 Mt of ore. The life of the present ore body is estimated at 30 years.

The zircon grains are rounded to irregular or prismatic and are slightly finer than Du Pont Florida zircon, but coarser than Associated Minerals, US, zircon. Like Australian zircon, the grains have surface coatings and turn orange on heating. Similar concentrations of heavy minerals in dunes to the north of Richards Bay are reported.

Baddeleyite is recovered from mill tailings at the Palabora copper mine in the Transvaal. The content of ZrO_2 plus HfO_2 ranges from 97 to 99%. Phosphate Development Corp. (FOSKOR) is also a producer.

Baddeleyite may occur either as a primary or secondary mineral. It is sometimes found in alluvial deposits associated with

Table 2. Mineralogy of Zircon and Baddeleyite

	Normal Zircon	Altered Zircon (Hyacinth)	Much Altered Zircon (Malacon)	Baddeleyite
Crystal	Ditetragonal Dipyramidal	Ditetragonal Dipyramidal	Amorphous	Monoclinic Prismatic
Specific gravity	4.6–4.7	4.2–4.6	3.9–4.2	5.4–5.7
Mohs hardness	7.5	7	6–7	6.5
Color	White	Purplish	Dark	Variable
Index of refraction	1.92–1.96	1.90–1.92	1.76–1.90	2.19
Cleavage	110 poor	110 poor	None	001 Perfect
Loss on ignition	0.1%	0.5–1%	10%	

ilmenite, zirkelite, apatite, and perovskite. It is also found in Brazil, Sri Lanka, Italy, and the United States.

Deposits in the United States

A small amount of zircon was produced from beach sand near Ponte Vedra, FL, between 1918 and 1929. No companies produced heavy minerals in Florida from 1929–39. In 1940 a small quantity of heavy minerals, including zircon, was produced from surface concentrates found in the beach near Melbourne, FL. That operation was later moved to a dune area near Vero Beach and operated until about 1963. Between 1943 and 1963, zircon was produced from a deposit at Arlington, a suburb of Jacksonville, FL.

In 1963 production began from a heavy mineral deposit near Folkston, GA. When this deposit was mined out in 1974, the mining and concentrating equipment were moved a few miles south to a deposit near Boulogne, FL, where mining continued until the ore was mined out in 1979. Zircon production began from an old shoreline deposit near Green Cove Springs, FL, in 1973.

The largest domestic supply of zircon sand comes from the Trail Ridge deposit in north-central Florida. This large deposit was discovered by Du Pont in 1946 and has been in continuous production of zircon since 1950. The deposit was formed at the maximum of Florida's submergence during Pleistocene times. Reworked sands transported from older sand features were redeposited forming the ridge. Waves, currents, and wind action removed part of the lighter silica sand, leaving an enrichment of heavy minerals in sand dunes on the western flank of this prominent geographic feature.

Zircon from the Trail Ridge Deposit is extremely uniform in particle size, grain morphology, and chemical purity. The grains are free of surface coatings and have been slightly etched. Because the grains have no surface coatings, the zircon turns white on heating.

EXPLORATION AND EVALUATION

Because sands that contain zircon also contain titanium minerals, exploration techniques are the same for both mineral types. Most commonly, target areas are selected where large volumes of sand are present and where beach ridges and other features associated with shorelines are known to occur. Early explorationists used US Geological Survey quadrangle maps, physiographic references, aerial photographs, and often word of mouth to find target areas. Modern methods have become much more refined since the advent of Earth Resource Technology Satellite (ERTS) imagery, manned space flight photography, and new developments in computer science. The recently discovered deposits near Wilson and Kenly, NC, were found using National Uranium Resource Evaluation (NURE) data published in map form by the US Department of Energy (Carpenter and Carpenter, 1991). The NURE program analyzed the fine fraction of stream sediments collected in the southeastern United States for several elements including titanium, iron, and zircon The north-trending alignment of anomalous concentrations of these elements in the Wilson-Kenly area suggested the presence of a feature containing concentrations of heavy minerals (see Fig. 5 in Titanium Minerals). The heavy mineral deposits to the north in southeastern Virginia were found by a Virginia Geological Survey geologist who was looking for black sands to test a spiral concentrator that was to be used in evaluating off-shore occurrences of heavy minerals (Berquist, 1987). In some cases, such as at Eneabba, deposits have been discovered by farmers who owned the land.

A Geiger counter or scintillometer can be useful in detecting heavy mineral deposits where radioactive monazite is present. Although geophysical methods (magnetometer survey, induced polarization, etc.) have been tested, they are not widely used in heavy mineral prospecting.

A variety of drilling systems are available for beach sand exploration. These include hand auger, vehicle-mounted power auger, and reverse-circulation drills using both water and air to force the mineral to the surface. Each has its advantage, and one is chosen because it is thought to be the best for a particular type of deposit.

Physical Separation

Gold pans of various sizes and materials of construction as well as vanning plaques can be used in the field to estimate the percentage of heavy minerals present in a sand. Evaluation of prospect samples and ores in the laboratory usually requires separation of the heavy minerals as a first step. The traditional methods use heavy liquids (tetrabromoethane, bromoform, methylene iodide, or Clerichi solutions). The heavy liquids are toxic, so some risk is involved in their use, a fact that has caused them to be banned in some countries (e.g., England). Newly invented non-toxic tungstate solutions that use water instead of organic solvents are considered much safer than the traditional heavy liquids. Another new development in heavy mineral separation equipment is the Magstream separator, which separates heavy minerals by specific gravity (up to 21) using a non-toxic ferro-solution rotating in a magnetic field. The specific gravity is set by adjusting the rotational speed of the separation tube. In addition to gravity separations, the Magstream can also make magnetic separations. It is a very useful tool in heavy mineral evaluations because it can be used to concentrate zircon and the other heavy minerals for chemical analysis, as well as physical testing.

Analytical Methods

For many years, wet analytical chemistry methods were used in the analysis of zirconium-bearing materials. This type of analysis is very difficult because zircon is inert and can be taken into solution only with fusion methods with a strong flux. Because zirconium and hafnium have almost identical chemical properties, wet chemical methods cannot be used routinely to determine composition by individual element. Instead, wet analyses report combined ZrO_2 and HfO_2 as 67.2% ZrO_2 for pure zircon. In the past 20 years, X-ray fluorescence (XRF) has replaced chemical methods of analysis almost entirely.

For many years, grain counting (point counting) was a popular and useful means of identifying minerals present in heavy mineral concentrates and determining the amounts of each. The method is fairly accurate for the major minerals present, but accuracy is much less for the minor minerals (Garnar, 1984). Development and refinements of the X-ray microprobe, scanning electron microscope (SEM), associated XRF techniques, and modern computer science have been combined to produce the most accurate and sophisticated means for analyzing heavy mineral concentrates. It is now possible to identify accurately each mineral present in a sample, measure each particle's size, analyze it chemically, and then give an accurate percentage for each individual mineral present in the sample, its average chemical composition, and its particle-size distribution.

MINING AND PROCESSING

Most of the heavy mineral operations in the world are similar. The sand ores are mined with dredges, both cutterhead-suction or, in some cases, bucketline. In areas where there is insufficient water for dredging, ores are mined using elevating scrapers and bulldozers.

The heavy minerals are concentrated by removing the quartz and other light minerals in gravity concentrators. The gravity concentrates are scrubbed, if necessary, to remove any surface coatings that might interfere with dry processing. The concentrates are then dried and separated into ilmenite, leucoxene, and rutile by magnetic and high tension techniques. The remaining materials are further treated by some combination of spirals, gravity tables, high tension separators, magnets, and air tables to produce zircon. A flow diagram is shown in Fig. 2.

In the absence of surface coatings, the zircon concentrates may be calcined to produce a white color. The white zircon has long served as a trademark for sands produced in the southeastern United States. Because of grain coatings, most zircon from other parts of the world turns a yellowish-orange color when heated.

ZIRCON QUALITY STANDARDS

When zircon was originally marketed in the 1950s, only one grade was available from each producer. As competition increased and the consuming markets became more sophisticated, different grades evolved with names selected by each producer to describe the product. Sometimes, however, the same name did not mean the same thing to different producers. For example, premium zircon may mean one or more of the following: low TiO_2, low Fe_2O_3, low Al_2O_3, high ZrO_2, calcined, classified, etc. In addition, such terms as foundry grade, standard grade, ceramic grade, etc., are also applied to zircon products.

Typical zircon sand specifications are as follows:

$ZrO_2 + HfO_2$	65–66% (min)
Fe_2O_3	0.02–0.10% (max)
TiO_2	0.10–0.35% (max)
Al_2O_3	0.20–2.00% (max)

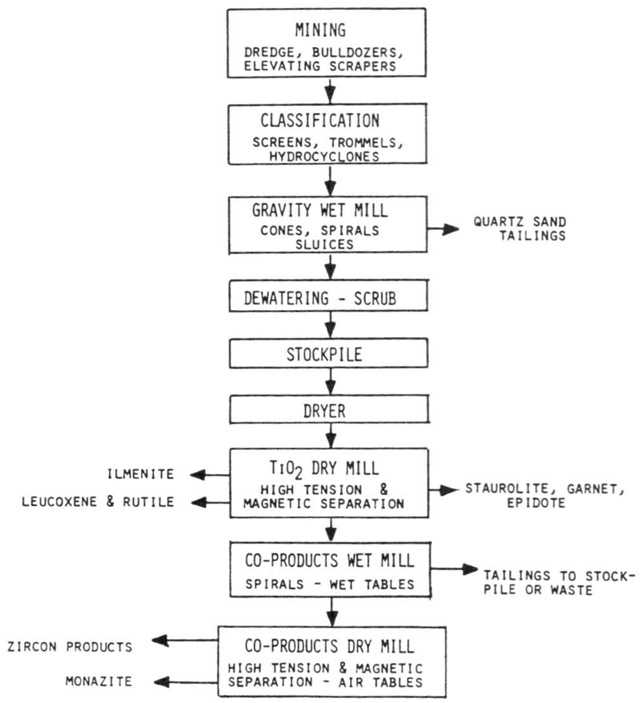

Fig. 2. Simplified flow diagram of a heavy mineral processing plant.

Du Pont also markets two special grades of zircon called Zircon-T®, a foundry grade containing as much as 2% rutile, and Zirclean® a mixture of zircon and magnetic heavy minerals used to clean turbine blades.

Zircon particle-size distribution and shape are important in most foundry applications. Zircon sands from each deposit are different in particle size (Table 3). The Eneabba ores produce the coarsest zircon particles presently available. The finest particle size comes from the Green Cove Springs Deposit in Florida. Zircon from a new deposit under development in Victoria, Australia, is less than 50 μm in size.

MARKETS AND USES

Zircon sands are used in a variety of markets ranging from foundry sands to zirconium-metal manufacture. The major uses in the United States are:

Foundry uses	35%
Zircon refractories	20%
Aluminum-zircon-silica refractories	10%
Zirconia and alumina-zirconia abrasives	11%
Alloys	4%
Other*	20%

*Includes chemicals, welding rods, $ZrCl_4$, sandblasting, and miscellaneous uses.

Foundry Use

Widespread use of zircon as a steel foundry sand began about 1950 when large quantities became available after Du Pont opened

Table 3. Typical Particle-size Distribution of Commercial Zircon Sands

Wt % Retained, US Standard Series	Opening in μm	Eneabba Deposit, Western Australia	Capel, Western Australia	Queensland, Australia	Starke, Florida	Green Cove Springs, Florida	Richards Bay, South Africa
50	300	Trace	1	—	1	—	—
70	212	14	3	1	1	Trace	Trace
100	150	46	30	6	13	1	8
140	106	30	52	47	48	25	42
200	75	9	13	44	36	63	46
270	53	1	1	2	2	10	4
Pan	53	—	—	1	Trace	1	Trace
Average Particle size in μm		162	141	111	120	98	111

its mines near Starke, FL. Low thermal expansion, clean rounded grains that use less binder, and neutral or slightly acid pH make zircon ideally suited as a foundry sand.

Zircon sand is used in the foundry for shell cores and mold linings. It is also dry ground to flour, then used to make mold washes, which are painted on mold facings to give castings smoother surface finish. The investment casting industry uses zircon sand as a stucco grain and the flour for smoother finishes.

Basic Steel Production

Zircon sand is used in basic steel production in ladle brick because it reduces erosion and extends ladle-lining life. It is used in coatings, mortars, and as ladle-nozzle fill.

Refractory Use

Zircon sands are fused to make refractory brick, which are used to line kilns, glass-melting furnaces, etc. High temperature boats, dishes, and other laboratory ware are also made from fused zircon. ZrO_2 refractories can be used at temperatures to 1 960°C. ZrO_2 melts completely at 2 760°C.

Abrasives

Zircon sand is converted into ZrO_2 and used to manufacture zirconia and alumina-zirconia abrasive products. Zircon sand is also used to clean turbines in electrical generating plants.

Zircon-Flour Opacifier

Zircon sand is wet ground, spray dried, and marketed as an opacifier for porcelain glazes. Because of its high index of refraction, it acts much like TiO_2 pigment in paint, in which material of a much lower index of refraction gives it an opaque white appearance.

Zirconium Metal

Zirconium metal is produced from zircon sand by reacting it with chlorine in the presence of carbon and then converting the tetrachloride to sponge metal, which is further worked to produce a variety of shapes ranging from zirconium plate to zirconium bolts. Major metal producers are Teledyne Wah Chang, Albany, OR, Western Zirconium, Ogden, UT, and Pechniney Ugine Kuhlman in France. The sponge contains about 2.5% Hf, which is very difficult to separate. Most commercial metal contains hafnium, but reactor-grade zirconium for use in atomic work must be hafnium-free.

Commercially pure zirconium is not a high strength metal but has properties similar to those of iron. It is valued for construction of nuclear reactors because of its low neutron-capture cross section, thermal stability, and corrosion resistance. The metal is 99.99% pure and is used to make rods, plates, sheets, foil, and wire. The unalloyed metal is difficult to roll and is usually worked at temperatures to 480°C. The metal is not attacked by nitric, sulfuric, or hydrochloric acids, but is dissolved by hydrofluoric acid.

Small amounts of zirconium are used in steel. It is a powerful deoxidizer, removes the nitrogen, and combines with the sulfur, reducing hot-shortness and giving ductility. Steels with small amounts of zirconium are fine grained and resist shock and fatigue.

Zirconium Compounds

Zirconium compounds are prepared and marketed for the following uses:

Zirconium oxychloride	Textile coatings
Zirconium carbide	Abrasive
Zirconium tetrachloride	Refine Al and Mg
Sodium zirconium sulfate	Precipitation of proteins, pigment stabilizer, and paper opacifer
Zirconium carbonate	Poison ivy ointment
Zirconium hydride	Neutron moderator

Other Zircon Products and Uses

Some zircon goes into special welding rod coats where, with other ingredients, it forms slag to shield the weld pool.

During periods of zircon shortages in the late 1960s, Du Pont extended its Florida supplies by offering Zircon-T®, which had all of the foundry properties of regular zircon, but contained more than normal TiO_2 (2% max). Du Pont also offers a specialty foundry sand made up of a mixture of zircon and aluminum silicates (kyanite and sillimanite), which is marketed under the trade name Zircore®. In some applications, this highly refractory sand has some advantages over pure zircon.

Hafnium Metal

Hafnium, though relatively abundant in nature, is sparsely disseminated and very costly to extract. Zircon sand is the only commercial source of this metal. Hafnium's melting point is 2 200°C, which is higher than the melting point of zirconium. Heat-resisting parts for special purposes have been made by compacting hafnium powder to a density of 98%. The metal has a close-packed hexagonal structure. It has excellent resistance to a wide range of

D

E

H

Q

R

S